PRIN
OF E

28 SEP 11
15 NOV 2012
14 DEC 2012
13
BAS
10/19
2020 ok
WITHDRAWN
CL.69(02/06.)

Get **more** out of libraries

Please return or renew this item by the last date shown.
You can renew online at www.hants.gov.uk/library
by phoning 0845 603 5631

Hampshire
County Council

D1512677

C014034785

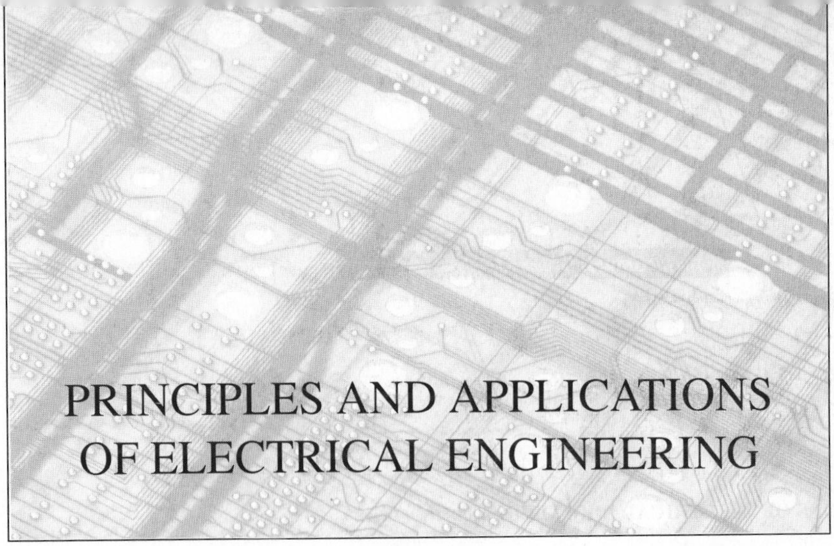

PRINCIPLES AND APPLICATIONS OF ELECTRICAL ENGINEERING

Fifth Edition

Giorgio Rizzoni
The Ohio State University

McGraw Hill

Boston Burr Ridge, IL Dubuque, IA Madison, WI New York San Francisco St. Louis
Bangkok Bogotá Caracas Kuala Lumpur Lisbon London Madrid Mexico City
Milan Montreal New Delhi Santiago Seoul Singapore Sydney Taipei Toronto

The McGraw·Hill Companies

PRINCIPLES AND APPLICATIONS OF ELECTRICAL ENGINEERING
Fifth Edition
International Edition 2007

Exclusive rights by McGraw-Hill Education (Asia), for manufacture and export. This book cannot be re-exported from the country to which it is sold by McGraw-Hill. The International Edition is not available in North America.

Published by McGraw-Hill, a business unit of The McGraw-Hill Companies, Inc., 1221 Avenue of the Americas, New York, NY 10020. Copyright © 2007 by The McGraw-Hill Companies, Inc. All rights reserved. No part of this publication may be reproduced or distributed in any form or by any means, or stored in a database or retrieval system, without the prior written consent of The McGraw-Hill Companies, Inc., including, but not limited to, in any network or other electronic storage or transmission, or broadcast for distance learning.

Some ancillaries, including electronic and print components, may not be available to customers outside the United States.

Part Opening Photos: 1: © PhotoDisc RF/Getty; 2: Photo by Giorgio Rizzoni; 3: © Sylvain Grandadam/Getty; 4: Courtesy Ford Motor Company.

10 09 08 07 06 05 04 03 02 01
20 09 08 07 06
CTF BJE

Library of Congress Control Number: 2005052214

When ordering this title, use ISBN 007-125444-7

Printed in Singapore

www.mhhe.com

HAMPSHIRE COUNTY LIBRARIES	
C 014034785	
Macaulay	09/08/07
621.3	£ 42.99
9780071254441	

In memoria di mamma

About the Author

Giorgio Rizzoni, The Ford Motor Company Chair of ElectroMechanical Systems, received the B.S., M.S., and Ph.D. degrees, all in electrical engineering, from the University of Michigan. He is currently a professor of mechanical and electrical engineering at The Ohio State University, where he teaches undergraduate courses in system dynamics, measurements, and mechatronics and graduate courses in automotive power train modeling and control, hybrid vehicle modeling and control, and system fault diagnosis.

Dr. Rizzoni has been involved in the development of innovative curricula and educational programs throughout his career. At the University of Michigan, he developed a new laboratory and curriculum for the circuits and electronics engineering service course for non-electrical engineering majors. At Ohio State, he has been involved in the development of undergraduate and graduate curricula in mechatronic systems with funding provided, in part, by the National Science Foundation through an interdisciplinary curriculum development grant. The present book has been profoundly influenced by this curriculum development.

Dr. Rizzoni and his colleagues have also developed a unique year-long graduate course sequence titled "Powertrain Modeling and Control" in collaboration with General Motors. This course sequence has been offered to General Motors employees as a series of distance-learning courses and on campus to Ohio State electrical and mechanical engineering students since 1995. In 1998, Dr. Rizzoni and colleagues were awarded funding from the U.S. Department of Energy to establish a *Graduate Automotive Technology Education Center* of excellence on Hybrid Vehicle Drivetrains and Control Systems. This activity has resulted in the development of a graduate curriculum and of research laboratories devoted to the study of hybrid-electric and fuel cell vehicle propulsion technologies.

Since 1999, Dr. Rizzoni has served as director of the Ohio State University Center for Automotive Research, an interdisciplinary research center serving the U.S. government and the automotive industry worldwide. The center conducts research in areas related to vehicle safety, energy efficiency, environmental impact, and passenger comfort. Dr. Rizzoni has published more than 200 papers in peer-reviewed journals and conference proceedings, and he has received a number of recognitions, including a 1991 NSF Presidential Young Investigator Award.

Dr. Rizzoni is a Fellow of IEEE, a Fellow of SAE, and a member of ASME and ASEE; he has served as an Associate Editor of the *ASME Journal of Dynamic Systems, Measurements, and Control* (1993 to 1998) and of the *IEEE Transactions on Vehicular Technology* (1988 to 1998). He has also served as Guest Editor of Special Issues of the *IEEE Transactions on Control System Technology,* of the *IEEE Control Systems Magazine,* and of *Control Engineering Practice;* Dr. Rizzoni is a past Chair of the ASME Dynamic Systems and Control Division, and has served as Chair of the Technical Committee on Automotive Control for the International Federation of Automatic Control (IFAC).

FIND IT ON THE WEB

Giorgio Rizzoni is the Ohio State University SAE student branch faculty adviser, and has led teams of electrical and mechanical engineering students through the development of an electric vechicle that established various land speed records in 2003 and 2004. He is also coadviser of the Ohio State University FutureTruck and Challenge-X hybrid-electric vehicle competition teams sponsored by the U.S. Department of Energy, and by General Motors and Ford.
http://car.eng.ohio-state.edu

Brief Contents

Contents

Preface

The pervasive presence of electronic devices and instrumentation in all aspects of engineering design and analysis is one of the manifestations of the electronic revolution that has characterized the second half of the 20th century. Every aspect of engineering practice, and even of everyday life, has been affected in some way or another by electrical and electronic devices and instruments. Computers are perhaps the most obvious manifestations of this presence. However, many other areas of electrical engineering are also important to the practicing engineer, from mechanical and industrial engineering, to chemical, nuclear, and materials engineering, to the aerospace and astronautical disciplines, to civil and the emerging field of biomedical engineering. Engineers today must be able to communicate effectively within the interdisciplinary teams in which they work.

OBJECTIVES

Engineering education and engineering professional practice have seen some rather profound changes in the past decade. The integration of electronics and computer technologies in all engineering academic disciplines and the emergence of digital electronics and microcomputers as a central element of many engineering products and processes have become a common theme over the nearly 20 years since the conception of this book.

The principal objective of the book is to present the *principles* of electrical, electronic, and electromechanical engineering to an audience composed of non–electrical engineering majors, and ranging from sophomore students in their first required introductory electrical engineering course, to seniors, to first-year graduate students enrolled in more specialized courses in electronics, electromechanics, and mechatronics.

A second objective is to present these principles by focusing on the important results and applications and presenting the students with the most appropriate *analytical and computational tools* to solve a variety of practical problems.

Finally, a third objective of the book is to illustrate, by way of concrete, fully worked examples, a number of relevant *applications* of electrical engineering principles. These examples are drawn from the author's industrial research experience and from ideas contributed by practicing engineers and industrial partners.

The three objectives listed above are met through the use of a number of pedagogical features. The next two sections of this preface describe the organization of the book and the major changes that have been implemented in this fourth edition.

ORGANIZATION AND CONTENT

The book is divided into three parts, devoted to *circuits, electronics, and electromechanics*. Changes in the contents are described next.

Part I: Circuits

The first part of the book remains essentially unchanged, after the significant revisions brought by the fourth edition. The only major change is the addition of approximately 110 new homework problems.

Part: II Electronics

Part II, on electronics, presents some new features in the treatment of transistors. Chapter 10, on bipolar transistors, and Chapter 11, on field-effect transistors, have been significantly reorganized to focus on the use of these devices in simple but useful circuits. Modeling emphasis is limited to large-signal models, which are sufficient for the intended purpose. New examples include the design of simple electric motor drivers and of battery chargers. These two chapters now present a new, uncomplicated, and practical treatment of the analysis and design of simple amplifiers and switching circuits using large-signal models. The revisions were based on a conscious decision to completely eliminate all of the material related to small-signal models of amplifiers. Chapter 12, on power electronics, includes two new examples describing power stage amplifier characteristics. The remainder of the electronics section, Chapters 8 and 9, and 13, 14, and 15, are mostly unchanged, except for the addition of a handful of new application-oriented examples. Nearly 100 new homework problems have been added to Part II.

Part III: Communication Systems

New in the fifth edition is the inclusion of two chapters on communications. These chapters have been added at the request of numerous schools, where it is felt that a modern engineer needs to have exposure to basic principles of communication systems. Chapter 16 is a revised edition of the an analog communications chapter that has been available on the book website since the third edition. The intent of the chapter is to present the basic principles of analog communications systems, leading to a basic understanding of analog AM and FM systems. Chapter 17, courtesy of Dr. Michael Carr, of the Ohio State University ElectroScience Laboratory, introduces the basic principles of digital communications systems. Both chapters focus on applications.

Part IV: Electromechanics

Part IV on Electromechanics has been revised for accuracy and pedagogy, but its contents are largely unchanged. This part has been used for many years by the author as a supplement in a junior-year "System Dynamics" course for mechanical engineers. The chapters include some New examples and approximately 20 new problems.

FIND IT ON THE WEB

Instructors will find additional suggestions on the organization of course materials at the book's website **http://www.mhhe.com/rizzoni.** Suggestions and sample curricula from users of the book are welcome!

FEATURES OF THE FIFTH EDITION

Pedagogy

The fifth edition continues to offer all of the time-tested pedagogical features available in the earlier editions.

- **Learning Objectives** offer an overview of key chapter ideas. Each chapter opens with a list of major objectives, and throughout the chapter the learning objective icon indicates targeted references to each objective.

- **Focus on Methodology** sections summarize important methods and procedures for the solution of common problems and assist the student in developing a methodical approach to problem solving.
- **Clearly Illustrated Examples** illustrate relevant applications of electrical engineering principles. The examples are fully integrated with the "Focus on Methodology" material, and each one is organized according to a prescribed set logical steps.
- **Check Your Understanding** exercises follow each example in the text and allow students to confirm their mastery of concepts.
- **Make the Connection** sidebars present analogies to students to help them see the connection of electrical engineering concepts to other engineering disciplines.
- **Focus on Measurements** boxes emphasize the great relevance of electrical engineering to the science and practice of measurements.
- **Find It on the Web** links included throughout the book give students the opportunity to further explore practical engineering applications of the devices and systems that are described in the text.

Supplements

The book includes a wealth of supplements, many available in electronic form. These include

- A **CD-ROM** containing computer-aided example solutions, a list of Web references for further research, device data sheets, Hewlett-Packard Instrumentation examples, and a motor control tutorial.
- A **website** (Online Learning Center) will be updated to provide students and instructors with additional resources for teaching and learning. You can find this site at **http://www.mhhe.com/rizzoni**

Online Learning Center

(http://www.mhhe.com/rizzoni)
Resources on this site include:

For Students:

- **Algorithmic Problems** that allow step-by-step problem-solving using a recursive computational procedure to create an infinite number of problems.
- **Device Data Sheets**
- **Hewlett-Packard Instrumentation Examples**
- A **Motor Control Tutorial,** and more…

For Instructors:

- **PowerPoint presentation slides** of important figures from the text
- **Instructor's Solutions Manual** with complete solutions (for instructors only)
- ***COSMOS*** (Complete Online Solutions Manual Organizing System)
- **MATLAB** Solution files for selected problems

For Instructors and Students:

- ***Find It on the Web links***, which give students the opportunity to explore, in greater depth, practical engineering applications of the devices and systems that are described in the text. In addition, several links to tutorial sites extend the boundaries of the text recnt research developments, late-breaking science and technology news, learning resources, and study guides to help you in your studies and research
- **News feeds** provide current daily news from *The New York Times* and other reliable online news resources related to the topics in the text. While most students and instructors have access to currents news online, these feeds are selected based on the topics presented in each chapter of Rizzon's text.

ACKNOWLEDGMENTS

This edition of the book requires a special acknowledgment for the effort put forth by my friend Tom Hartley of the University of Akron, who has become a mentor, coach, and inspiration for me throughout this project. Professor Hartley, who is an extraordinary teacher and a devoted user of this book, has been closely involved in the development of the fifth edition by suggesting topics for new examples and exercises, creating new homework problems, providing advice and coaching through all of the revisions, and sometimes just by lifting my spirits. I look forward to many more years of such collaborations.

I would also like to recognize the focused effort of Dr. Michael Carr, of the Ohio State University ElectroScience Laboratory, who is responsible for creating Chapter 17 on Digital Communications. His efforts and the assistance provided by graduate students Adam Margetts and Aditi Kothiyal are very much appreciated.

This book has been critically reviewed by the following people.

- Ravel Ammerman, Colorado School of Mines
- Glen Archer, Michigan Technological University
- Ray Bellem, Embry-Riddle Aeronautical University
- Keith Burgers, University of Arkansas

- Doroteo Chaverria, University of Texas—San Antonio
- Randy Collins, Clemson University
- Marcelo J. Dapino, The Ohio State University
- Alexandros Eleftheriadis, Columbia University
- Otto M. Friedrich, University of Texas—Austin
- Takis Kasparis, University of Central Florida
- Rasool Kenarangui, University of Texas—Arlington
- Andy Mayers, Penn State University
- Nathan Shenck, U.S. Naval Academy
- B. J. Shrestha, University of Missouri—Rolla
- Miklos N. Szilagyi, University of Arizona
- Albert H. Titus, SUNY Buffalo
- Trac D. Tran, Johns Hopkins University

In addition, I would like to thank the many colleagues who have pointed out errors and inconsistencies in the fourth edition. We have gladly accepted all of their suggestions. The following is a a partial list of those who have contributed to improving the accuracy of this book.

- Suresh Kumar R., Amrita School of Engineering, India
- Thomas Schubert, University of San Diego
- T. S. Liu, Professor, Chiao Tung University, Taiwan
- Mohan Krishnan, University of Detroit Mercy
- Stephen Deese
- Rony Shahidain
- Jing Sun, University of Michigan

The author is also grateful to Mr. David Mikesell and Mr. Qi Ma, graduate students at Ohio State, for help and advice.

Book prefaces have a way of marking the passage of time. When the first edition of this book was published, the birth of our first child, Alex, was nearing. Each of the following two editions was similarly accompanied by the births of Maria and Michael. Now that we have successfully reached the fifth edition (but only the third child) I am observing that Alex is beginning to understand some of the principles exposed in this book through his passion for the FIRST Lego League and the Lego Mindstorms robots. Through the years, our family continues to be the center of my life, and I am grateful to Kathryn, Alessandro, Maria, and Michael for all their love.

GUIDED TOUR

Learning Objectives offer an overview of key chapter ideas. Each chapter opens with a list of major objectives and throughout the chapter. The learning objective icon indicates targeted references to each objective.

Learning Objectives

1. Compute the solution of circuits containing linear resistors and independent and dependent sources by using *node analysis*. *Sections 3.2 and 3.4.*
2. Compute the solution of circuits containing linear resistors and independent and dependent sources by using *mesh analysis*. *Sections 3.3 and 3.4.*
3. Apply the *principle of superposition* to linear circuits containing independent sources. *Section 3.5.*
4. Compute *Thévenin and Norton equivalent circuits* for networks containing linear resistors and independent and dependent sources. *Section 3.6.*
5. Use equivalent-circuit ideas to compute the *maximum power transfer* between a source and a load. *Section 3.7.*
6. Use the concept of equivalent circuit to determine voltage, current, and power for nonlinear loads by using *load-line analysis* and analytical methods. *Section 3.8.*

3.1 Network Analysis

The analysis of an electric network consists of determining each of the unknown branch currents and node voltages. It is therefore important to define all the relevant variables as clearly as possible, and in systematic fashion. Once the known and unknown variables have been identified, a set of equations relating these variables is constructed, and these equations are solved by means of suitable techniques. The analysis of electric circuits consists of writing the smallest set of equations sufficient to solve for all the unknown variables. The procedures required to write these equations are the subject of Chapter 3 and are very well documented and codified in the form of simple rules. The analysis of electric circuits is greatly simplified if some standard conventions are followed.

Example 3.1 defines all the voltages and currents that are associated with a specific circuit

Focus on Methodology section summarize important methods and procedures for the solution of common problems and assist the student in developing a methodical approach to problem solving.

FOCUS ON METHODOLOGY

COMPUTING THE THÉVENIN VOLTAGE

1. Remove the load, leaving the load terminals open-circuited.
2. Define the open-circuit voltage v_{OC} across the open load terminals.
3. Apply any preferred method (e.g., node analysis) to solve for v_{OC}.
4. The Thévenin voltage is $v_T = v_{OC}$.

The actual computation of the open-circuit voltage is best illustrated by e ples; there is no substitute for practice in becoming familiar with these computa To summarize the main points in the computation of open-circuit voltages, cor the circuit of Figure 3.36, shown again in Figure 3.44 for convenience. Recall th equivalent resistance of this circuit was given by $R_T = R_3 + R_1 \parallel R_2$. To con v_{OC}, we disconnect the load, as shown in Figure 3.45, and immediately observ no current flows through R_3, since there is no closed-circuit connection at that br Therefore, v_{OC} must be equal to the voltage across R_2, as illustrated in Figure Since the only closed circuit is the mesh consisting of v_S, R_1, and R_2, the answ are seeking may be obtained by means of a simple voltage divider:

$$v_{OC} = v_{R2} = v_S \frac{R_2}{R_1 + R_2}$$

It is instructive to review the basic concepts outlined in the example by sidering the original circuit and its Thévenin equivalent side by side, as shov Figure 3.47. The two circuits of Figure 3.47 are equivalent in the sense that th

EXAMPLE 3.8 Mesh Analysis

LO2

Problem

Write the mesh current equations for the circuit of Figure 3.19.

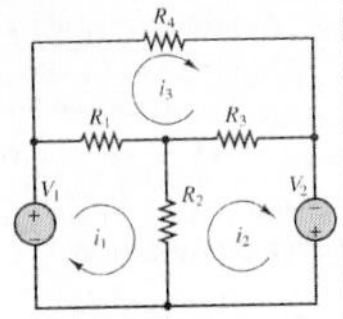

Figure 3.19

Solution

Known Quantities: Source voltages; resistor values.

Find: Mesh current equations.

Schematics, Diagrams, Circuits, and Given Data: $V_1 = 12$ V; $V_2 = 6$ V; $R_1 = 3\ \Omega$; $R_2 = 8\ \Omega$; $R_3 = 6\ \Omega$; $R_4 = 4\ \Omega$.

Analysis: We follow the Focus on Methodology steps.

1. Assume clockwise mesh currents i_1, i_2, and i_3.
2. We recognize three independent variables, since there are no current sources. Starting from mesh 1, we apply KVL to obtain

$$V_1 - R_1(i_1 - i_3) - R_2(i_1 - i_2) = 0$$

KVL applied to mesh 2 yields

$$-R_2(i_2 - i_1) - R_3(i_2 - i_3) + V_2 = 0$$

while in mesh 3 we find

$$-R_1(i_3 - i_1) - R_4 i_3 - R_3(i_3 - i_2) = 0$$

Clearly Illustrated Examples illustrate relevant applications of electrical engineering principles. The examples are fully integrated with the "Focus of Methodology" material, and each one is organized according to a prescribed set of common sense steps.

CHECK YOUR UNDERSTANDING

Find the current i_L in the circuit shown on the left, using the node voltage method.

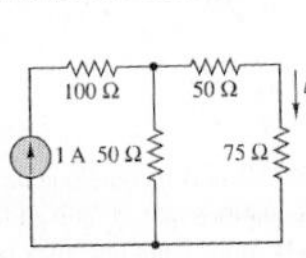

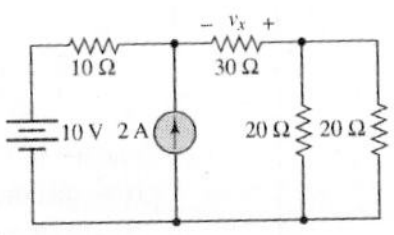

Find the voltage v_x by the node voltage method for the circuit shown on the right.

Show that the answer to Example 3.3 is correct by applying KCL at one or more nodes.

Answers: 0.2857 A; −18 V

Check Your Understanding exercises follow each example in the text and allow students to confirm their mastery of concepts.

EXAMPLE 3.5

LO1

Problem

Use the node voltage analysis to determine the voltage v in the circuit of Figure 3.9. Assume that $R_1 = 2\ \Omega$, $R_2 = 1\ \Omega$, $R_3 = 4\ \Omega$, $R_4 = 3\ \Omega$, $I_1 = 2$ A, and $I_2 = 3$ A.

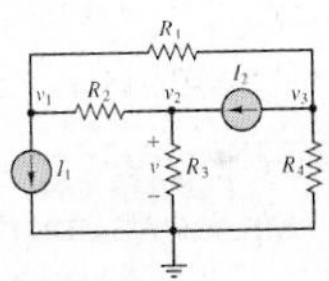

Figure 3.9 Circuit for Example 3.5

Solution

Known Quantities: Values of the resistors and the current sources.

Find: Voltage across R_3.

Analysis: Once again, we follow the steps outlined in the Focus on Methodology box.

1. The reference node is denoted in Figure 3.9.
2. Next, we define the three node voltages v_1, v_2, v_3, as shown in Figure 3.9.
3. Apply KCL at each of the $n - 1$ nodes, expressing each current in terms of the adjacent node voltages.

$$\frac{v_3 - v_1}{R_1} + \frac{v_2 - v_1}{R_2} - I_1 = 0 \qquad \text{node 1}$$

$$\frac{v_1 - v_2}{R_2} - \frac{v_2}{R_3} + I_2 = 0 \qquad \text{node 2}$$

$$\frac{v_1 - v_3}{R} - \frac{v_3}{R} - I_2 = 0 \qquad \text{node 3}$$

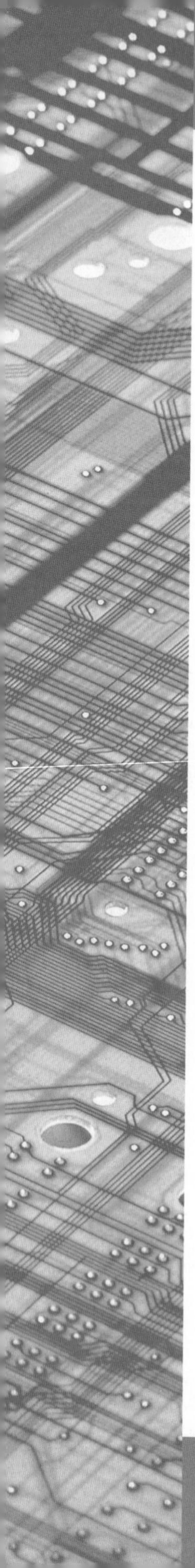

Make the Connection sidebars present analogies to students to help them see the connection of electrical engineering concepts to other engineering disciplines.

equations obtained at nodes a and b (verify this, as an exercise). This observation confirms the statement made earlier:

LO1

> In a circuit containing n nodes, we can write at most $n - 1$ independent equations.

Now, in applying the node voltage method, the currents i_1, i_2, and i_3 are expressed as functions of v_a, v_b, and v_c, the independent variables. Ohm's law requires that i_1, for example, be given by

$$i_1 = \frac{v_a - v_c}{R_1} \tag{3.5}$$

since it is the potential difference $v_a - v_c$ across R_1 that causes current i_1 to flow from node a to node c. Similarly,

$$i_2 = \frac{v_a - v_b}{R_2}$$
$$i_3 = \frac{v_b - v_c}{R_3} \tag{3.6}$$

Substituting the expression for the three currents in the nodal equations (equations 3.2 and 3.3), we obtain the following relationships:

$$i_S - \frac{v_a}{R_1} - \frac{v_a - v_b}{R_2} = 0 \tag{3.7}$$

$$\frac{v_a - v_b}{R_2} - \frac{v_b}{R_3} = 0 \tag{3.8}$$

Equations 3.7 and 3.8 may be obtained directly from the circuit, with a little practice. Note that these equations may be solved for v_a and v_b, assuming that i_S, R_1, R_2, and R_3 are known. The same equations may be reformulated as follows:

$$\left(\frac{1}{R_1} + \frac{1}{R_2}\right) v_a + \left(-\frac{1}{R_2}\right) v_b = i_S$$
$$\left(-\frac{1}{R_2}\right) v_a + \left(\frac{1}{R_2} + \frac{1}{R_3}\right) v_b = 0 \tag{3.9}$$

Examples 3.2 through 3.4 further illustrate the application of the method.

LO1

EXAMPLE 3.2 Node Analysis

Problem

Solve for all unknown currents and voltages in the circuit of Figure 3.5.

Thermal Circuit Model

The *conduction resistance* of the shaft is described by the following equation:

$$q = \frac{kA_1}{L}\Delta T$$
$$R_{\text{cond}} = \frac{\Delta T}{q} = \frac{L}{kA_1}$$

where A_1 is a cross sectional area and L is the distance from the inner core to the surface. The convection resistance is described by a similar equation, in which convective heat flow is described by the film coefficient of heat transfer, h:

$$q = hA_2\Delta T$$
$$R_{\text{conv}} = \frac{\Delta T}{q} = \frac{1}{hA_2}$$

where A_2 is the surface area of the shaft in contact with the water. The equivalent thermal resistance and the overall circuit model of the crankshaft quenching process are shown in the figures below.

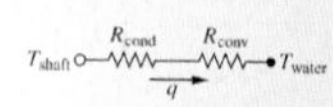

Thermal resistance representation of quenching process

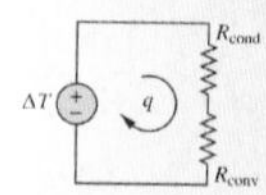

Electrical circuit representing the quenching process

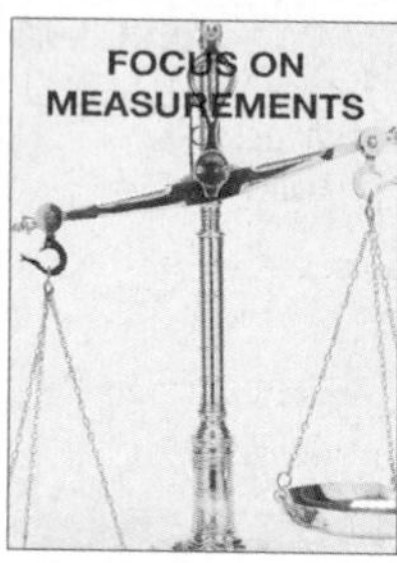

Experimental Determination of Thévenin Equivalent Circuit

Problem:

Determine the Thévenin equivalent of an unknown circuit from measurements of open-circuit voltage and short-circuit current.

Solution:

Known Quantities—Measurement of short-circuit current and open-circuit voltage. Internal resistance of measuring instrument.

Find—Equivalent resistance R_T; Thévenin voltage $v_T = v_{OC}$.

Schematics, Diagrams, Circuits, and Given Data—Measured $v_{OC} = 6.5$ V; measured $i_{SC} = 3.75$ mA; $r_m = 15\ \Omega$.

(Continued)

Focus on Measurements boxes emphasize the great relevance of electrical engineering to the science and practice of measurements.

Network connected for measurement of open-circuit voltage (ideal voltmeter)

Figure 3.68

Comments—Note how easy the experimental method is, provided we are careful to account for the internal resistance of the **measuring instruments.**

Find it on the web links included throughout the book give students the opportunity to further explore practical engineering applications of the devices and systems that are described in the text.

One last comment is in order concerning the practical measurement of the internal resistance of a network. In most cases, it is not advisable to actually short-circuit a network by inserting a series ammeter as shown in Figure 3.67; permanent

CHAPTER 1

INTRODUCTION TO ELECTRICAL ENGINEERING

The aim of this chapter is to introduce electrical engineering. The chapter is organized to provide the newcomer with a view of the different specialties making up electrical engineering and to place the intent and organization of the book into perspective. Perhaps the first question that surfaces in the mind of the student approaching the subject is, Why electrical engineering? Since this book is directed at a readership having a mix of engineering backgrounds (including electrical engineering), the question is well justified and deserves some discussion. The chapter begins by defining the various branches of electrical engineering, showing some of the interactions among them, and illustrating by means of a practical example how electrical engineering is intimately connected to many other engineering disciplines. In Section 1.2 *mechatronic systems engineering* is introduced, with an explanation of how this book can lay the foundation for interdisciplinary mechatronic product design. This design approach is illustrated by two examples. A brief historical perspective is also provided, to outline the growth and development of this relatively young engineering specialty. Section 1.3 introduces the Engineer-in-Training (EIT) national examination. In Section 1.5 the fundamental physical quantities and the system of units are defined, to set the stage for the chapters that follow. Finally, in Section 1.6 the organization of the book is discussed, to give the student, as well as the teacher, a sense of continuity in the development of the different subjects covered in Chapters 2 through 20.

Table 1.1 Electrical engineering disciplines

Circuit analysis
Electromagnetics
Solid-state electronics
Electric machines
Electric power systems
Digital logic circuits
Computer systems
Communication systems
Electro-optics
Instrumentation systems
Control systems

1.1 ELECTRICAL ENGINEERING

The typical curriculum of an undergraduate electrical engineering student includes the subjects listed in Table 1.1. Although the distinction between some of these subjects is not always clear-cut, the table is sufficiently representative to serve our purposes. Figure 1.1 illustrates a possible interconnection between the disciplines of Table 1.1. The aim of this book is to introduce the non-electrical engineering student to those aspects of electrical engineering that are likely to be most relevant to his or her professional career. Virtually all the topics of Table 1.1 will be touched on in the book, with varying degrees of emphasis. Example 1.1 illustrates the pervasive presence of electrical, electronic, and electromechanical devices and systems in a very common application: the automobile. As you read through the examples, it will be instructive to refer to Figure 1.1 and Table 1.1.

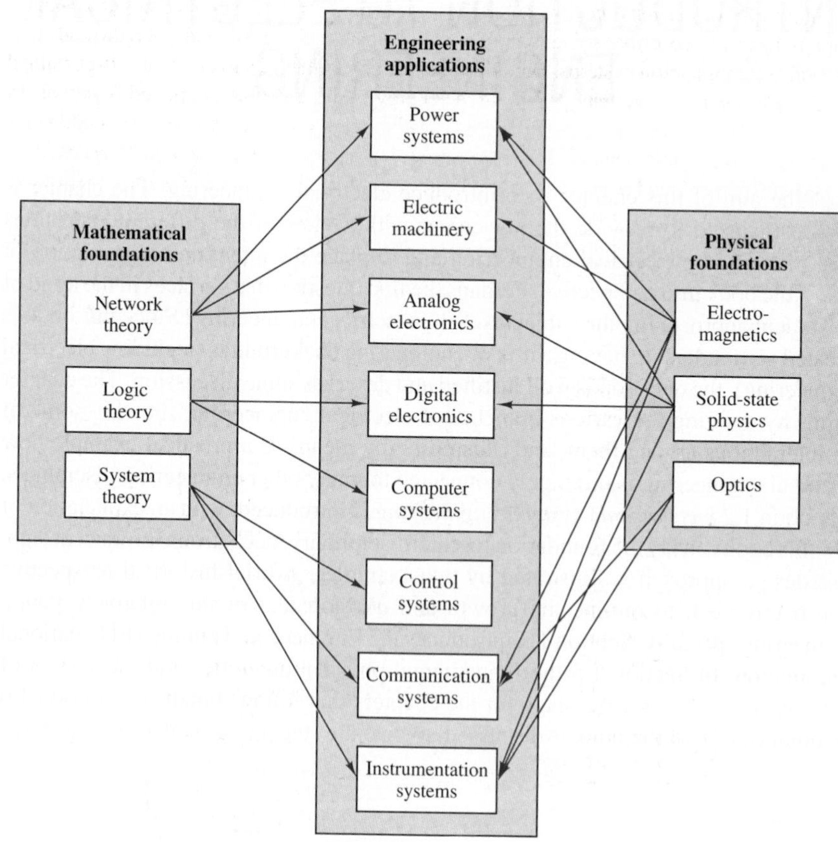

Figure 1.1 Electrical engineering disciplines

EXAMPLE 1.1 Electrical Systems in a Passenger Automobile

A familiar example illustrates how the seemingly disparate specialties of electrical engineering actually interact to permit the operation of a very familiar engineering system: the automobile. Figure 1.2 presents a view of electrical engineering systems in a modern automobile. Even in older vehicles, the electrical system—in effect, an *electric circuit*—plays a very important part in the overall operation. (Chapters 2 and 3 describe the basics of electric circuits.) An inductor coil generates a sufficiently high voltage to allow a spark to form across the spark plug gap, and to ignite the air-fuel mixture; the coil is supplied by a DC voltage provided by a lead-acid battery. Ignition circuits are studied in some detail in Chapter 5. In addition to providing the energy for the ignition circuits, the battery supplies power to many other electrical components, the most obvious of which are the lights, the windshield wipers, and the radio. Electric power (Chapter 7) is carried from the battery to all these components by means of a wire harness, which constitutes a rather elaborate electric circuit (see Figure 2.12 for a closer look). In recent years, the conventional electric ignition system has been supplanted by *electronic* ignition; that is, solid-state electronic devices called *transistors* have replaced the traditional breaker points. The advantage of transistorized ignition systems over the conventional mechanical ones is their greater reliability, ease of control, and life span (mechanical breaker points are subject to wear). You will study transistors and other electronic devices in Chapters 8, 9, and 10.

Other electrical engineering disciplines are fairly obvious in the automobile. The on-board radio receives electromagnetic waves by means of the antenna, and decodes the communication signals to reproduce sounds and speech of remote origin; other common *communication systems* that exploit *electromagnetics* are CB radios and the ever more common cellular phones. Chapters 16 and 17 describes some of the technology that is behind AM and FM radio and other common communication systems. But this is not all! The battery is, in effect, a self-contained 12-VDC *electric power system,* providing the energy for all the aforementioned functions. In order for the battery to have a useful lifetime, a charging system, composed of an alternator and of power electronic devices, is present in every automobile. Electric power systems are covered in Chapter 7 and power electronic devices in Chapter 10. The alternator is an *electric machine*, as are the motors that drive the power mirrors, power windows, power seats, and other convenience features found in luxury cars. Incidentally, the loudspeakers are also electric machines! All these devices are described in Chapters 18, 19, and 20.

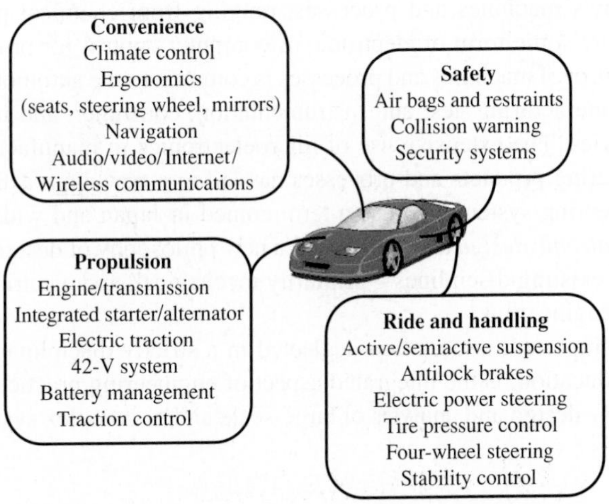

Figure 1.2 Electrical engineering systems in the automobile

The list does not end here, though. In fact, some of the more interesting applications of electrical engineering to the automobile have not been discussed yet. Consider *computer systems*. You are certainly aware that in the last two decades, environmental concerns related to exhaust emissions from automobiles have led to the introduction of sophisticated engine emission *control systems*. The heart of such control systems is a type of computer called a *microprocessor.* The microprocessor receives signals from devices (called *sensors*) that measure relevant variables—such as the engine speed, the concentration of oxygen in the exhaust gases, the position of the throttle valve (i.e., the driver's demand for engine power), and the amount of air aspirated by the engine—and subsequently computes the optimal amount of fuel and the correct timing of the spark to result in the cleanest combustion possible under the circumstances. We present a brief overview of computer systems in Chapter 14. The measurement of the aforementioned variables falls under the heading of *instrumentation*, and the interconnection between the sensors and the microprocessor is usually made up of *digital circuits*. Chapter 15 is devoted to the subject of measurements and instrumentation, although you will find a feature titled "Focus on Measurements" in most chapters. Digital circuits are covered in Chapters 13 and 14. As the presence of computers on board becomes more pervasive—in areas such as antilock braking, electronically controlled suspensions, four-wheel steering systems, and electronic cruise control—communications among the various on-board computers will have to occur at faster and faster rates. Someday in the not-so-distant future, these communications may occur over a fiber-optic network, and *electro-optics* will replace the conventional wire harness. Note that electro-optics is already present in some of the more advanced displays that are part of an automotive instrumentation system.

Finally, today's vehicles also benefit from the significant advances made in *communication systems*. Vehicle navigation systems can include *Global Positioning System*, or GPS, technology, as well as a variety of communications and networking technologies, such as wireless interfaces (e.g., based on the "Bluetooth" standard) and satellite radio and driver assistance systems, such as the GM "OnStar" system.

1.2 ELECTRICAL ENGINEERING AS A FOUNDATION FOR THE DESIGN OF MECHATRONIC SYSTEMS

Many of today's machines and processes, ranging from chemical plants to automobiles, require some form of electronic or computer control for proper operation. Computer control of machines and processes is common to the automotive, chemical, aerospace, manufacturing, test and instrumentation, consumer, and industrial electronics industries. The extensive use of microelectronics in manufacturing systems and in engineering products and processes has led to a new approach to the design of such engineering systems. To use a term coined in Japan and widely adopted in Europe, *mechatronic design* has surfaced as a new philosophy of design, based on the integration of existing disciplines—primarily mechanical, and electrical, electronic, and software engineering.[1]

A very important issue, often neglected in a strictly disciplinary approach to engineering education, is the integrated aspect of engineering practice, which is unavoidable in the design and analysis of large-scale and/or complex systems. One aim

[1] D. A. Bradley, D. Dawson, N. C. Burd, and A. J. Loader, 1991, *Mechatronics, Electronics in Products and Processes,* Chapman and Hall, London. See also ASME/IEEE *Transactions on Mechatronics*, vol 1, no. 1, 1996.

of this book is to give engineering students of different backgrounds exposure to the integration of electrical, electronic, and software engineering into their domain. This is accomplished by making use of modern computer-aided tools and by providing relevant examples and references. Section 1.6 describes how some of these goals are accomplished.

Examples 1.2 and 1.3 illustrate some of the thinking behind the mechatronic system design philosophy through two examples: the recently introduced Ford Escape hybrid-electric SUV and a land speed record electric vehicle.

EXAMPLE 1.2 Mechatronic Systems—The Ford Escape Hybrid-Electric SUV

An example of a mechatronic system that is becoming increasingly familiar is found in hybrid-electric vehicles. A hybrid-electric vehicle (HEV) employs two different energy sources to provide motive power. **Today's HEVs** employ a combination of advanced internal combustion engine technology together with electric machine and battery technology, to provide a propulsion system that exploits the best features of thermal engines and of electric machines to provide a greater energy conversion efficiency, and therefore lower fuel consumption. You will find **additional information on HEVs** online.

This example focuses on the Ford Escape HEV, shown in Figure 1.3, recently introduced in the North American market. The Escape HEV is the first commercially produced hybrid-electric sport utility vehicle. It incorporates a number of features aimed at increasing fuel economy while maintaining the comfort, performance, and utility that customers have come to expect from vehicles of this class.

The Ford Escape HEV delivers V-6–like performance and feel with a 4-cylinder engine and a hybrid drivetrain. Figure 1.4 depicts the appearance of the overall layout of the hybrid

Figure 1.3 Ford Escape HEV (*Courtesy: Ford Motor Company*)

(a) Layout of Escape HEV hybrid drivetrain

(b) Escape HEV engine compartment

(c) Escape HEV transaxle

Figure 1.4 Ford Escape HEV hybrid drivetrain (*Courtesy of Ford Motor Company*) drivetrain. The vehicle features a 4-cylinder gasoline engine operated according to the Atkinson cycle (shown in Figure 1.4(b)), an asymmetrical four-stroke cycle (with expansion ratio greater than the compression ratio) that is more fuel efficient than the traditional Otto cycle in use in the majority of gasoline fueled vehicles today. The combination of a more efficient (but slightly less performing) Atkinson cycle engine with the two electric machines incorporated in the power-split transaxle [Figure 1.4(c)] is capable of torque and acceleration comparable to that of a V-6 engine, while providing the customer with fuel economy of 35–40 mpg in the city, and 29–31 mpg on the highway.

Figure 1.5 Details of Escape HEV transaxle (*Courtesy: Ford Motor Company*)

The transaxle, shown in a cut-away view in Figure 1.5, consists of two electric machines, both permanent magnet AC machines,[2] one rated at 45 kW maximum power, used primarily as a generator, the other rated at 70 kW and used primarily as a motor. The two machines are mechanically connected through a planetary gear set, similar to the gear set used in conventional

---

[2]These types of electric machines are described in Chapters 19 and 20 of this book.

automatic transmissions. The transaxle also incorporates a power converter,[3] or inverter, that converts the DC power supplied by the battery pack [Figures 1.6(a) and (b)] to the variable frequency AC power[4] required to operate the AC electric machines. In addition to the power converter, the electronics incorporated into the transaxle include a control module, which includes various analog and digital electronic circuits,[5] including a microcontroller that receives information from various sensors to determine the best speed and power of operation of the engine and of each of the two electric machines in the transaxle.

In addition to the energy provided by the fuel stored in a conventional tank, a hybrid vehicle also stores a limited amount of energy in a battery pack. Figure 1.6(a) depicts the location of the battery pack; note that the battery pack has been purposely designed to have a flat profile [see Figure 1.6(b)], so as not to reduce useful cargo space. The battery pack consists of a 330-V NiMH (Nickel-Metal Hydride) system, made up of numerous battery modules connected in series and parallel.[6]

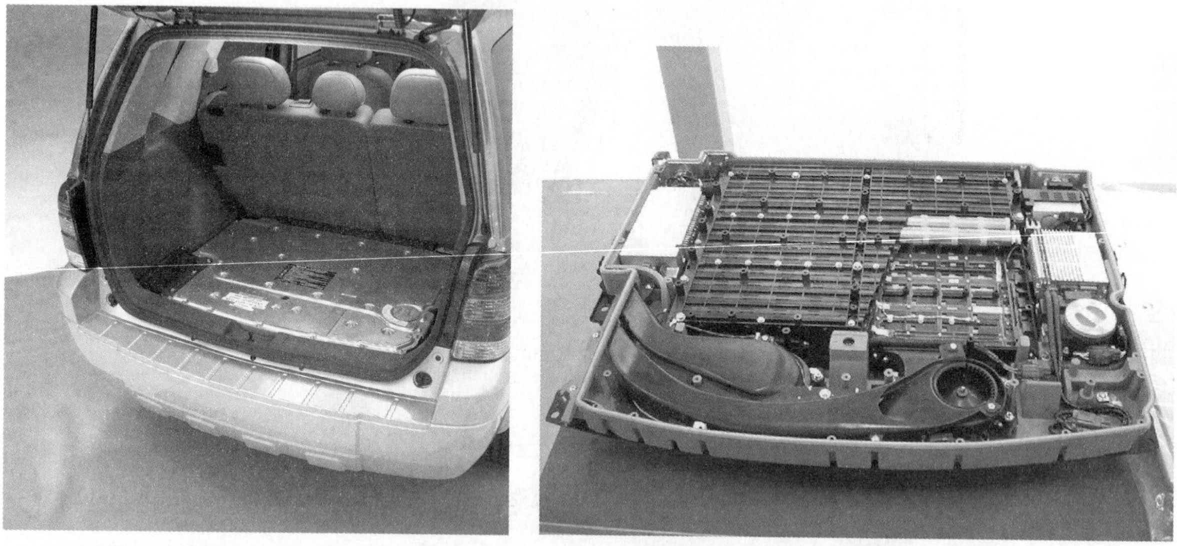

(a) Escape HEV battery compartment (b) Escape HEV battery pack

Figure 1.6 Ford Escape HEV battery pack (*Courtesy of Ford Motor Company*)

Finally, Figure 1.7 depicts the general system operation, showing how the engine and the generator motor are connected to two of the planetary gear set inputs, while the traction motor is connected to the output of the gear set, and therefore to the front axle. This picture illustrates the various possible modes of operation of the vehicle, showing that the vehicle can be powered by the engine alone, through the mechanical transmission elements, or by the traction motor alone, or by a combination of the engine and two electric machines. Further, when the vehicle is decelerating, the mechanical energy stored in the vehicle motion can be in part converted into electrical energy by the generator motor, and stored in the battery pack for later use. Another interesting feature of the Escape HEV is that the traction motor is capable of launching the

[3]Power electronics is the subject of Chapter 12 in this book.

[4]AC power is the subject of Chapter 7 in this book.

[5]Analog and digital electronics are covered in Chapters 8–15.

[6]Series and parallel circuits are explained in Chapters 2 and 3.

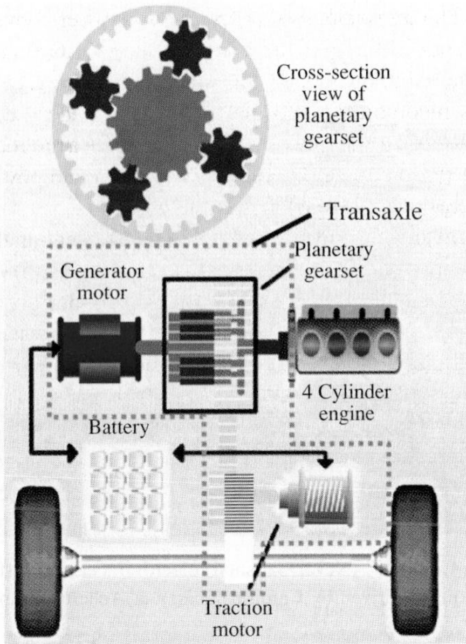

Figure 1.7 Ford Escape HEV strategy
(*Courtesy of Ford Motor Company*)

vehicle from zero speed without the need for any engine power. Thus, the engine can be safely shut off whenever the vehicle stops at a traffic light or in traffic, with significant fuel savings resulting from the complete elimination of engine idle.

EXAMPLE 1.3 Mechatronic Systems—The Buckeye Bullet Electric Land Speed Record Vehicle

Land Speed Record Racing:

In the early years of modern automobile engineering, it was electric traction that provided most of the excitement (a surprise, perhaps, to modern internal combustion engine racing enthusiasts). The vehicle shown in Figure 1.8 was the first vehicle to exceed the 100 km/hr mark—a record that lasted several years, until the internal combustion engine became the dominant propulsion system for most of the last one hundred years. In recent years, interest in **electric land speed record racing** has once again increased, as both power electronics, electric motor, and battery technology have seen significant advances, mostly motivated by the emergence of a small but growing market for electric and especially hybrid-electric vehicles.

Modern land speed record (LSR) racing consists of vehicles traveling on a designated track to obtain the fastest speed for that specific vehicle's class. Since the early 1950s, amateurs and professionals have raced toward land speed records in myriad categories on **the Bonneville Salt Flats,** an ancient lake bed spread across 30,000 acres in Utah, 90 miles west of Salt Lake City. The Salt Flats surface consists of packed salt, making it an ideal place to stage a speed race due to the perfectly flat surface and to the high-altitude location with lower air density.

The Southern California Timing Association (SCTA) governs LSR racing at Bonneville, and imposes precise rules that must be respected in order to attempt the record. In addition to this, the SCTA establishes the standard for the race and the car depending on vehicle class.

Figure 1.8 "La Jamais Contente" (Never Satisfied) electric racer, the first vehicle to surpass the 100 km/hr mark (105.880 km/hr), driven by Camillo Jenatzy at Achères, France, on April 29, 1899

Figure 1.9 The Buckeye Bullet at the Bonneville Salt Flats/*Photo by Giorgio Rizzoni*

FIND IT ON THE WEB

Electric vehicles (E class) have no body configuration restriction but there are three different classes based on the vehicle's weight, less that of the driver. The vehicle described in this example, the **Buckeye Bullet,** shown in Figure 1.9, belongs to class E.III, the unlimited weight electric class. The race course configuration, shown in Figure 1.10, consists of three different parts: the first two miles are used for the acceleration of the vehicle, while the average speed is recorded over the third, fourth, and fifth miles. After the fifth mile, the vehicle has two miles

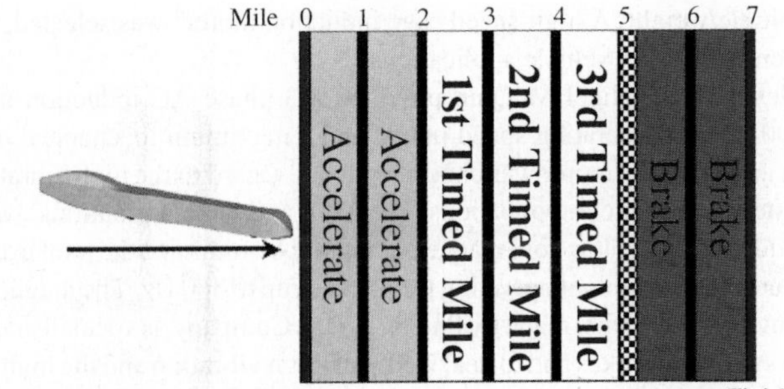

Figure 1.10 Bonneville course layout/*Source: Giorgio Rizzoni*

to decelerate using two drag parachutes. If an international record is attempted, the vehicle has only one hour to make a return run, while for a United States record, four hours are allowed between runs.

Design of an Electric Streamliner:

Vehicle Design

The Buckeye Bullet design layout is shown in Figure 1.11. The type of vehicle shown in the drawing is called a *streamliner,* a term that indicates a vehicle with enclosed wheels and a long, aerodynamic shape. The solid model of Figure 1.11 shows that the battery pack is located at the front of the vehicle, with the power converter, motor, transmission, and driveline following it. The driver is located at the rear of the vehicle in a specially designed roll cage that provides significant protection. The tail end of the vehicle contains the two parachutes. The design of the vehicle involved structural design, packaging optimization, the design of front and rear independent suspension and brake systems, the design of the body shape using computational fluid dynamics and wind tunnel testing, and dynamic and aerodynamic stability analysis. In this brief example, we describe only the propulsion system, composed of a motor, power converter (inverter), and battery pack.

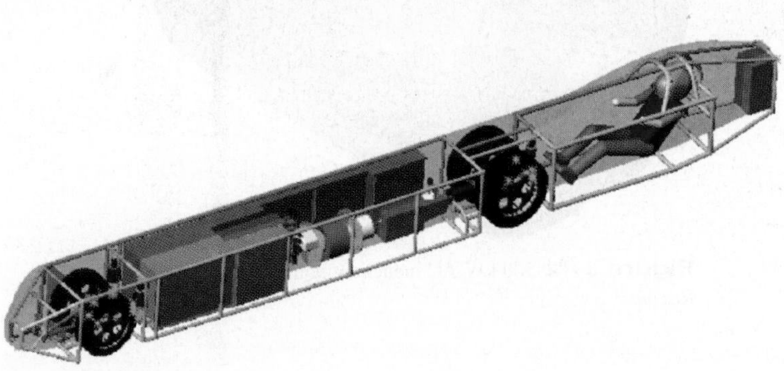

Figure 1.11 Layout of the Buckeye Bullet/*Photo by Giorgio Rizzoni*

Motor

The required power of the drive motor was determined through simulations of the vehicle; since motors with higher operating speeds have higher power densities (thus minimizing the weight for a particular power rating) the motor was designed for a maximum speed of 12,000 rev/min, and to achieve maximum power in a band between 8,500 and 10,500 rev/min. Also, since the motor had a specified duty cycle of just 2 minutes at full power in any 45-minute period, the motor selection was based on a design that allowed the highest specific loading conditions for the minimum volume/weight. A high-speed cage induction motor[7] was selected, using design procedures proven from other vehicle applications.

The final design from Nigel McQuin provided a 3-phase AC induction machine with 4 poles, 0–10,500 rev/min operating speed range, and a maximum mechanical over-speed of 12,000 rpm. The air gap of the motor was chosen with due regard to the high vibration and rotor dynamics, consistent with vehicle applications. The stator and rotor laminations[8] were manufactured of high-frequency grade, low-loss magnetic material to minimize the total iron losses from both hysteresis and eddy currents at operating frequencies up to 360 Hz. The magnet wire for the stator winding, inverter duty type, made by Phelps Dodge Company, is specially designed to resist high-voltage switching spikes and to deal well with high vibration and the high-temperature environment. The rotor cage construction is of cast copper, which minimizes the rotor heating losses and provides a stiff motor characteristic well suited to inverter drive applications.

To minimize the total weight of the drive motor, an aluminum housing was used. Based on previous vehicle experience, bearings, shaft seals, and terminal box configurations were selected for the necessary power rating and operating speeds. In view of the harsh salt environment, and the limited duty cycle of operation, a detailed thermal study determined that the design could be made total-enclosed with no cooling system. The thermal inertia of the motor housing and stator/rotor iron cores was more than adequate to store the heat losses during the race trial, with sufficient surface cooling available in the rest period to allow the return race trial to be completed.

Figure 1.12 500-kW AC induction motor/*Photo by Giorgio Rizzoni*

The drive motor and housing assembly was manufactured by Shoemaker Industrial Solutions, which coordinated the supply of the custom-cut laminations, coil winding, machining services, final assembly, and qualification testing. Figure 1.12 depicts a solid model of the motor.

Inverter

To operate an AC induction motor from a battery pack, it is necessary to convert the DC current and voltage of the battery pack into AC currents and voltages of variable amplitude

[7]Induction motors are discussed in Chapter 19 of this book.

[8]You will find a discussion of laminations and magnetic losses in Chapter 18.

and frequency. This task is performed by a power electronic system called a DC-AC power converter, or more simply *inverter*. The inverter used in the Buckeye Bullet (one view is shown in Figure 1.13) is a modified version of a system produced by Saminco, Inc. for use in electric buses. The inverter, rated at 900 VDC, uses insulated gate bipolar transistors[9] (IGBT) and operates at a switching frequency greater than 8 kHz. The inverter is provided with a liquid cooling system to limit the maximum device temperatures. The continuous power rating of the inverter is 250 kW, but thanks to the liquid cooling it is possible to operate it at significantly higher power for short periods of time. This power converter includes a microcontroller that makes the DC-AC power conversion tunable to any motor to optimize the efficiency of the system. By adjusting the switching frequency of the inverter, it is possible to control motor torque so as to limit the acceleration and jerk. These actions result in reduced tire slip and impact loading on the drivetrain.

Figure 1.13 Details of the inverter/*Photo by Giorgio Rizzoni*

Another important feature of the inverter is *dynamic braking*. During gear shifting the drive motor is slowed from approximately 10,000 rev/min to 7,000 rev/min allowing the the rotational energy to be temporarily dissipated through a special resistor to permit synchronization of the rotating parts for the closure of the clutch.

Batteries

The battery pack used in the Buckeye Bullet is made up of over 400 NiMH battery modules arranged in 20 packs to create a 900 VDC bus, and 20 kW/hr capacity. Figure 1.14 depicts the battery pack in the vehicle during charging. To maintain a safe system, the pack was designed such that every part was limited to a maximum voltage of 250 VDC. The packs were designed to be quickly assembled and disassembled to fulfill the one-hour turn-around requirement for international speed records.

Temperature plays a key role in the efficiency and power ability of the batteries. For this reason, they were closely monitored before, during, and after each run. The batteries

[9]IGBTs and inverters are described in Chapter 12.

Figure 1.14 Battery pack (with cooling fans, used during charging and after a run)/*Photo by Giorgio Rizzoni*

were precooled before the vehicle left the starting line to approximately 15°C; the maximum allowable temperature is 60°C. The cooling fans shown in Figure 1.14 are used to cool the pack during charging.

Current Records and Future Plans

The Buckeye Bullet became the first electric powered vehicle to exceed 300 mph on October 13, 2004, when it recorded an average speed over two runs of 314.958 mph during the 56th Annual World Finals at Bonneville Speedway, claiming the new U.S. record and reaching a top speed of 321.834 mph, the highest ever recorded for an electrically powered vehicle. Figure 1.15 shows the official timing slip for one of the two runs leading to the U.S. record. In addition,

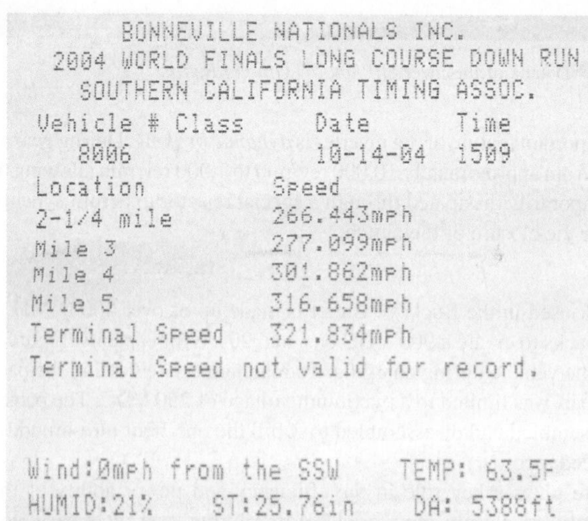

BONNEVILLE NATIONALS INC.
2004 WORLD FINALS LONG COURSE DOWN RUN
SOUTHERN CALIFORNIA TIMING ASSOC.

Vehicle # Class	Date	Time
8006	10-14-04	1509

Location	Speed
2-1/4 mile	266.443mph
Mile 3	277.099mph
Mile 4	301.862mph
Mile 5	316.658mph
Terminal Speed	321.834mph

Terminal Speed not valid for record.

Wind:0mph from the SSW TEMP: 63.5F
HUMID:21% ST:25.76in DA: 5388ft

Figure 1.15 Timing slip from a land speed record run/*Source: Giorgio Rizzoni*

in separate attempts, the Bullet also established a new international record of 271. 737 mph. Figure 1.16 shows data corresponding to the timing slip of Figure 1.15.

Currently, a new vehicle, named Buckeye Bullet II, is being planned. The greatest novelty in the new car will be a hydrogen fuel cell electric power supply system.

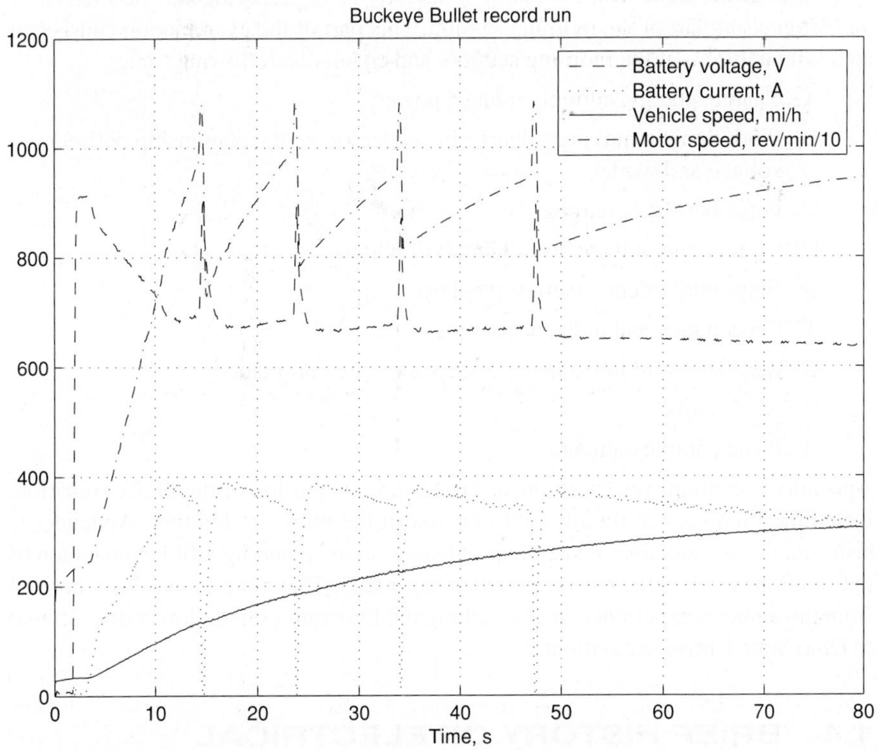

Figure 1.16 Data from the record run of Figure 1.15/*Source: Giorgio Rizzoni*

1.3 FUNDAMENTALS OF ENGINEERING EXAM REVIEW

To become a professional engineer it is necessary to satisfy four requirements. The first is the completion of a B.S. degree in engineering from an accredited college or university (although it is theoretically possible to be registered without having completed a degree). The second is the successful completion of the *Fundamentals of Engineering* (FE) *Examination.* This is an eight-hour exam that covers general undergraduate engineering education. The third requirement is two to four years of engineering experience after passing the FE exam. Finally, the fourth requirement is successful completion of the *Principles and Practice of Engineering or Professional Engineer* (PE) *Examination.*

The FE exam is a two-part national examination, administered by the **National Council of Examiners for Engineers and Surveyors** (NCEES) and given twice a year (in April and October). The exam is divided into two four-hour sessions, consisting of 120 questions in the four-hour morning session, and 60 questions in the four-hour afternoon session. The morning session covers general background in twelve different areas, one of which is *Electricity and Magnetism*. The afternoon session requires the examinee to choose among seven modules – Chemical, Civil, Electrical, Environmental, Industrial, Mechanical and Other/General engineering.

One of the aims of this book is to assist you in preparing for the *Electricity and Magnetism* part of the morning session. This part of the examination consists of approximately 9% of the morning session, and covers the following topics:

A. Charge, energy, current, voltage, power
B. Work done in moving a charge in an electric field (relationship between voltage and work)
C. Force between charges
D. Current and voltage laws (Kirchhoff, Ohm)
E. Equivalent circuits (series, parallel)
F. Capacitance and inductance
G. Reactance and impedance, susceptance and admittance
H. AC circuits
I. Basic complex algebra

Appendix C contains review of the electrical circuits portion of the FE examination, including references to the relevant material in the book. In addition, Appendix C also contains a collection of sample problems – some including a full explanation of the solution, some with answers supplied separately. This material has been derived from the author's experience in co-teaching the FE exam preparation course offered to Ohio State University seniors.

1.4 BRIEF HISTORY OF ELECTRICAL ENGINEERING

The historical evolution of electrical engineering can be attributed, in part, to the work and discoveries of the people in the following list. You will find these scientists, mathematicians, and physicists referenced throughout the text.

William Gilbert (1540–1603), English physician, founder of magnetic science, published *De Magnete*, a treatise on magnetism, in 1600.

Charles A. Coulomb (1736–1806), French engineer and physicist, published the laws of electrostatics in seven memoirs to the French Academy of Science between 1785 and 1791. His name is associated with the unit of charge.

James Watt (1736–1819), English inventor, developed the steam engine. His name is used to represent the unit of power.

Alessandro Volta (1745–1827), Italian physicist, discovered the electric pile. The unit of electric potential and the alternate name of this quantity (voltage) are named after him.

Hans Christian Oersted (1777–1851), Danish physicist, discovered the connection between electricity and magnetism in 1820. The unit of magnetic field strength is named after him.

André Marie Ampère (1775–1836), French mathematician, chemist, and physicist, experimentally quantified the relationship between electric current and the magnetic field. His works were summarized in a treatise published in 1827. The unit of electric current is named after him.

Georg Simon Ohm (1789–1854), German mathematician, investigated the relationship between voltage and current and quantified the phenomenon of resistance. His first results were published in 1827. His name is used to represent the unit of resistance.

Michael Faraday (1791–1867), English experimenter, demonstrated electromagnetic induction in 1831. His electric transformer and electromagnetic generator marked the beginning of the age of electric power. His name is associated with the unit of capacitance.

Joseph Henry (1797–1878), U.S. physicist, discovered self-induction around 1831, and his name has been designated to represent the unit of inductance. He had also recognized the essential structure of the telegraph, which was later perfected by Samuel F. B. Morse.

Carl Friedrich Gauss (1777–1855), German mathematician, and **Wilhelm Eduard Weber** (1804–1891), German physicist, published a treatise in 1833 describing the measurement of the earth's magnetic field. The gauss is a unit of magnetic field strength, while the weber is a unit of magnetic flux.

James Clerk Maxwell (1831–1879), Scottish physicist, discovered the electromagnetic theory of light and the laws of electrodynamics. The modern theory of electromagnetics is entirely founded upon Maxwell's equations.

Ernst Werner Siemens (1816–1892) and **Wilhelm Siemens** (1823–1883), German inventors and engineers, contributed to the invention and development of electric machines, as well as to perfecting electrical science. The modern unit of conductance is named after them.

Heinrich Rudolph Hertz (1857–1894), German scientist and experimenter, discovered the nature of electromagnetic waves and published his findings in 1888. His name is associated with the unit of frequency.

Nikola Tesla (1856–1943), Croatian inventor, emigrated to the United States in 1884. He invented polyphase electric power systems and the induction motor and pioneered modern AC electric power systems. His name is used to represent the unit of magnetic flux density.

1.5 SYSTEM OF UNITS

This book employs the International System of Units (also called SI, from the French *S*ystème *I*nternational des Unités). SI units are commonly adhered to by virtually all engineering professional societies. This section summarizes SI units and will serve as a useful reference in reading the book.

SI units are based on six fundamental quantities, listed in Table 1.2. All other units may be derived in terms of the fundamental units of Table 1.2. Since, in practice, one often needs to describe quantities that occur in large multiples or small fractions of a unit, standard prefixes are used to denote powers of 10 of SI (and derived) units. These prefixes are listed in Table 1.3. Note that, in general, engineering units are expressed in powers of 10 that are multiples of 3.

For example, 10^{-4} s would be referred to as 100×10^{-6} s, or 100 μs (or, less frequently, 0.1 ms).

Table 1.2 **SI units**

Quantity	Unit	Symbol
Length	Meter	m
Mass	Kilogram	kg
Time	Second	s
Electric current	Ampere	A
Temperature	Kelvin	K
Luminous intensity	Candela	cd

Table 1.3 **Standard prefixes**

Prefix	Symbol	Power
atto	a	10^{-18}
femto	f	10^{-15}
pico	p	10^{-12}
nano	n	10^{-9}
micro	μ	10^{-6}
milli	m	10^{-3}
centi	c	10^{-2}
deci	d	10^{-1}
deka	da	10
kilo	k	10^{3}
mega	M	10^{6}
giga	G	10^{9}
tera	T	10^{12}

1.6 SPECIAL FEATURES OF THIS BOOK

This book includes a number of special features designed to make learning easier and to allow students to explore the subject matter of the book in greater depth, if so desired, through the use of computer-aided tools and the Internet. The principal features of the book are described below.

➲ Learning Objectives

1. The principal *learning objectives* are clearly identified at the beginning of each chapter.
2. The symbol ➲ is used to identify definitions and derivations critical to the accomplishment of a specific learning objective.
3. Each example is similarly marked.

EXAMPLES

The examples in the book have also been set aside from the main text, so that they can be easily identified. All examples are solved by following the same basic methodology: A clear and simple problem statement is given, followed by a solution. The solution consists of several parts: All known quantities in the problem are summarized, and the problem statement is translated into a specific objective (e.g., "Find the equivalent resistance R").

Next, the given data and assumptions are listed, and finally the analysis is presented. The analysis method is based on the following principle: All problems are solved symbolically first, to obtain more general solutions that may guide the student in solving homework problems; the numerical solution is provided at the very end of the analysis. Each problem closes with comments summarizing the findings and tying the example to other sections of the book.

The solution methodology used in this book can be used as a general guide to problem-solving techniques well beyond the material taught in the introductory electrical engineering courses. The examples in this book are intended to help you develop sound problem-solving habits for the remainder of your engineering career.

CHECK YOUR UNDERSTANDING

Each example is accompanied by at least one drill exercise.

Answer: The answer is provided right below the exercise.

This feature is devoted to helping the student *make the connection* between electrical engineering and other engineering disciplines. Analogies to other fields of engineering will be found in nearly every chapter.

FOCUS ON METHODOLOGY

Each chapter, especially the early ones, includes "boxes" titled "Focus on Methodology." The content of these boxes (which are set aside from the main text) summarizes important methods and procedures for the solution of common problems. They usually consist of step-by-step instructions, and are designed to assist you in methodically solving problems.

As stated many times in this book, the need for measurements is a common thread to all engineering and scientific disciplines. To emphasize the great relevance of electrical engineering to the science and practice of measurements, a special set of examples focuses on measurement problems. These examples very often relate to disciplines outside electrical engineering (e.g., biomedical, mechanical, thermal, fluid system measurements). The "Focus on Measurements" sections are intended to stimulate your thinking about the many possible applications of electrical engineering to measurements in your chosen field of study. Many of these examples are a direct result of the author's work as a teacher and researcher in both mechanical and electrical engineering.

Find It on the Web!

The use of the Internet as a resource for knowledge and information is becoming increasingly common. In recognition of this fact, website references have been included in this book to give you a starting point in the exploration of the world of electrical engineering. Typical web references give you information on electrical engineering companies, products, and methods. Some of the sites contain tutorial material that may supplement the book's contents.

Website

The list of features would not be complete without a reference to the book's website: **http://www.mhhe.com/engcs/electrical/rizzoni**. Create a bookmark for this site now! The site is designed to provide up-to-date additions, examples, errata, and other important information.

HOMEWORK PROBLEMS

1.1 List five applications of electric motors in the common household.

1.2 By analogy with the discussion of electrical systems in the automobile, list examples of applications of the electrical engineering disciplines of Table 1.1 for each of the following engineering systems:

a. A ship.

b. A commercial passenger aircraft.

c. Your household.

d. A chemical process control plant.

1.3 Electric power systems provide energy in a variety of commercial and industrial settings. Make a list of systems and devices that receive electric power in

a. A large office building.

b. A factory floor.

c. A construction site.

PART I

CIRCUITS

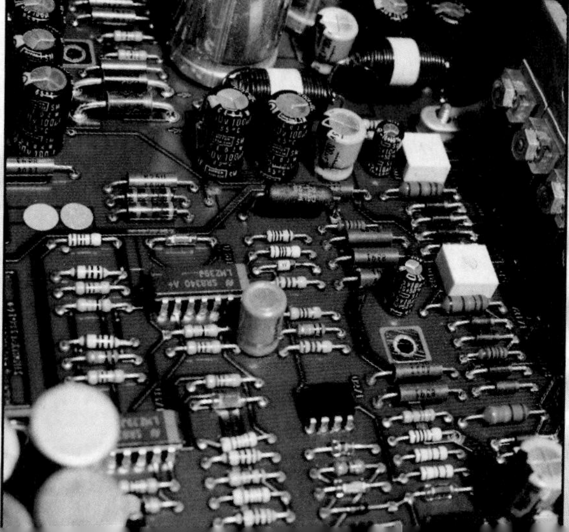

CHAPTER 2

FUNDAMENTALS OF ELECTRIC CIRCUITS

Chapter 2 presents the fundamental laws that govern the behavior of electric circuits, and it serves as the foundation to the remainder of this book. The chapter begins with a series of definitions to acquaint the reader with electric circuits; next, the two fundamental laws of circuit analysis are introduced: Kirchhoff's current and voltage laws. With the aid of these tools, the concepts of electric power and the sign convention and methods for describing circuit elements—resistors in particular—are presented. Following these preliminary topics, the emphasis moves to basic analysis techniques—voltage and current dividers, and to some application examples related to the engineering use of these concepts. Examples include a description of strain gauges, circuits for the measurements of force and other related mechanical variables, and of the study of an automotive throttle position sensor. The chapter closes with a brief discussion of electric measuring instruments. The following box outlines the principal learning objectives of the chapter.

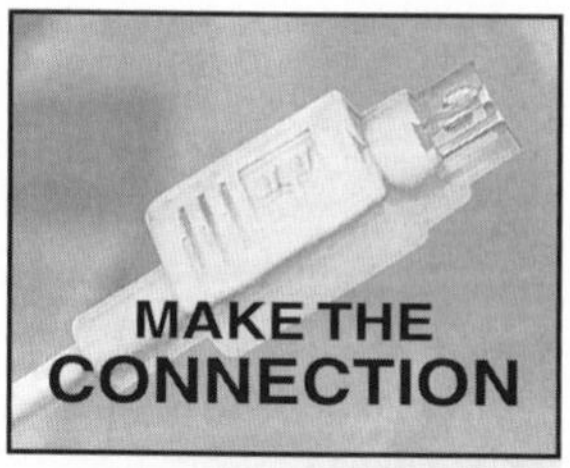

Mechanical (Gravitational) Analog of Voltage Sources

The role played by a voltage source in an electric circuit is equivalent to that played by the force of gravity. Raising a mass with respect to a reference surface increases its potential energy. This potential energy can be converted to kinetic energy when the object moves to a lower position relative to the reference surface. The voltage, or potential difference across a voltage source plays an analogous role, raising the electrical potential of the circuit, so that current can flow, converting the potential energy within the voltage source to electric power.

Learning Objectives

1. Identify the principal *elements of electric circuits*: nodes, loops, meshes, branches, and voltage and current sources. *Section 2.1*.
2. Apply *Kirchhoff's laws* to simple electric circuits and derive the basic circuit equations. *Sections 2.2 and 2.3*.
3. Apply the *passive sign convention* and compute the power dissipated by circuit elements. Calculate the power dissipated by a resistor. *Section 2.4*.
4. Apply the *voltage and current divider laws* to calculate unknown variables in simple series, parallel, and series-parallel circuits. *Sections 2.5 and 2.6*.
5. Understand the rules for connecting *electric measuring instruments* to electric circuits for the measurement of voltage, current, and power. *Sections 2.7 and 2.8*.

2.1 DEFINITIONS

In this section, we formally define some variables and concepts that are used in the remainder of the chapter. First, we define voltage and current sources; next, we define the concepts of *branch, node, loop,* and *mesh,* which form the basis of circuit analysis.

Intuitively, an ideal source is a source that can provide an arbitrary amount of energy. **Ideal sources** are divided into two types: voltage sources and current sources. Of these, you are probably more familiar with the first, since dry-cell, alkaline, and lead-acid batteries are all voltage sources (they are not ideal, of course). You might have to think harder to come up with a physical example that approximates the behavior of an ideal current source; however, reasonably good approximations of ideal current sources also exist. For instance, a voltage source connected in series with a circuit element that has a large resistance to the flow of current from the source provides a nearly constant—though small—current and therefore acts very nearly as an ideal current source. A battery charger is another example of a device that can operate as a current source.

Ideal Voltage Sources

An **ideal voltage source** is an electric device that generates a prescribed voltage at its terminals. The ability of an ideal voltage source to generate its output voltage is not affected by the current it must supply to the other circuit elements. Another way to phrase the same idea is as follows:

LO1

> An ideal voltage source provides a prescribed voltage across its terminals irrespective of the current flowing through it. The amount of current supplied by the source is determined by the circuit connected to it.

Figure 2.1 depicts various symbols for voltage sources that are employed throughout this book. Note that the output voltage of an ideal source can be a function of time. In general, the following notation is employed in this book, unless otherwise noted. A generic voltage source is denoted by a lowercase v. If it is necessary to emphasize that the source produces a time-varying voltage, then the notation $v(t)$ is

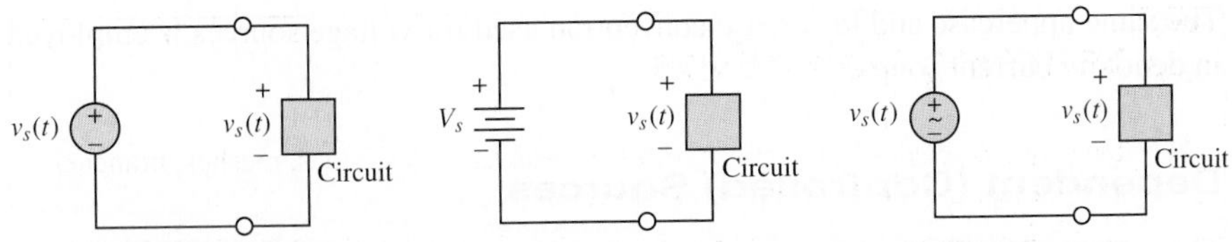

General symbol for ideal voltage source. $v_S(t)$ may be constant (DC source).

A special case: DC voltage source (ideal battery)

A special case: sinusoidal voltage source, $v_S(t) = V \cos \omega t$

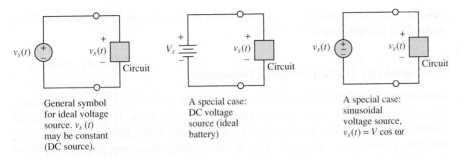

Figure 2.1 Ideal voltage sources

employed. Finally, a constant, or *direct-current,* or *DC,* voltage source is denoted by the uppercase character V. Note that by convention the direction of positive current flow out of a voltage source is *out of the positive terminal.*

The notion of an ideal voltage source is best appreciated within the context of the source-load representation of electric circuits. Figure 2.2 depicts the connection of an energy source with a passive circuit (i.e., a circuit that can absorb and dissipate energy). Three different representations are shown to illustrate the conceptual, symbolic, and physical significance of this source-load idea.

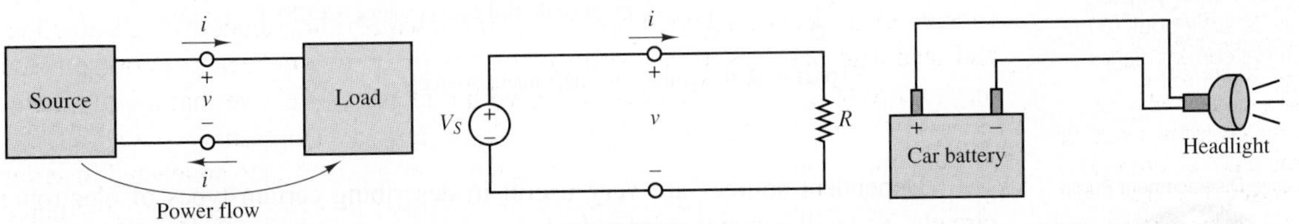

(a) Conceptual representation

(b) Symbolic (circuit) representation

(c) Physical representation

Figure 2.2 Various representations of an electrical system

In the analysis of electric circuits, we choose to represent the physical reality of Figure 2.2(c) by means of the approximation provided by ideal circuit elements, as depicted in Figure 2.2(b).

Ideal Current Sources

An **ideal current source** is a device that can generate a prescribed current independent of the circuit to which it is connected. To do so, it must be able to generate an arbitrary voltage across its terminals. Figure 2.3 depicts the symbol used to represent ideal current sources. By analogy with the definition of the ideal voltage source just stated, we write that

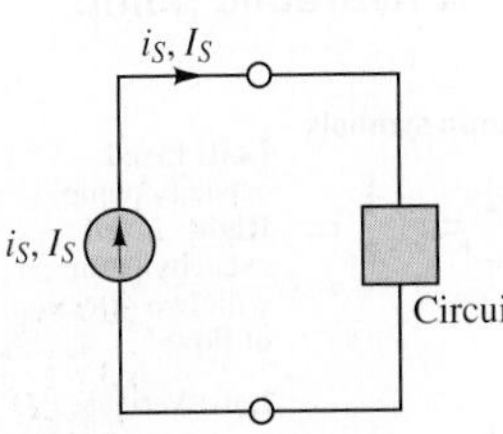

Figure 2.3 Symbol for ideal current source

An ideal current source provides a prescribed current to any circuit connected to it. The voltage generated by the source is determined by the circuit connected to it.

LO1

The same uppercase and lowercase convention used for voltage sources is employed in denoting current sources.

Dependent (Controlled) Sources

The sources described so far have the capability of generating a prescribed voltage or current independent of any other element within the circuit. Thus, they are termed *independent sources*. There exists another category of sources, however, whose output (current or voltage) is a function of some other voltage or current in a circuit. These are called **dependent** (or **controlled**) **sources.** A different symbol, in the shape of a diamond, is used to represent dependent sources and to distinguish them from independent sources. The symbols typically used to represent dependent sources are depicted in Figure 2.4; the table illustrates the relationship between the source voltage or current and the voltage or current it depends on—v_x or i_x, respectively—which can be any voltage or current in the circuit.

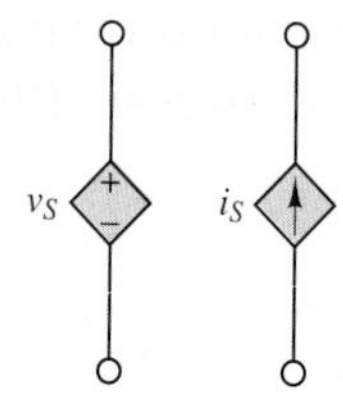

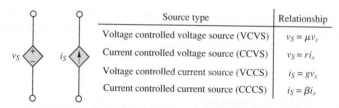

Source type	Relationship
Voltage controlled voltage source (VCVS)	$v_S = \mu v_x$
Current controlled voltage source (CCVS)	$v_S = r i_x$
Voltage controlled current source (VCCS)	$i_S = g v_x$
Current controlled current source (CCCS)	$i_S = \beta i_x$

Figure 2.4 Symbols for dependent sources

Dependent sources are very useful in describing certain types of electronic circuits. You will encounter dependent sources again in Chapters 8, 10, and 11, when electronic amplifiers are discussed.

An **electrical network** is a collection of elements through which current flows. The following definitions introduce some important elements of a network.

Branch

A **branch** is any portion of a circuit with two terminals connected to it. A branch may consist of one or more circuit elements (Figure 2.5). In practice, any circuit element with two terminals connected to it is a branch.

Node

A **node** is the junction of two or more branches (one often refers to the junction of only two branches as a *trivial node*). Figure 2.6 illustrates the concept. In effect, any connection that can be accomplished by soldering various terminals together is a node. It is very important to identify nodes properly in the analysis of electrical networks.

It is sometimes convenient to use the concept of a **supernode.** A supernode is obtained by defining a region that encloses more than one node, as shown in the rightmost circuit of Figure 2.6. Supernodes can be treated in exactly the same way as nodes.

MAKE THE CONNECTION

Hydraulic Analog of Current Sources

The role played by a current source in an electric circuit is very similar to that of a pump in a hydraulic circuit. In a pump, an internal mechanism (pistons, vanes, or impellers) forces fluid to be pumped from a reservoir to a hydraulic circuit. The volume flow rate of the fluid q, in cubic meters per second, in the hydraulic circuit, is analogous to the electrical current in the circuit.

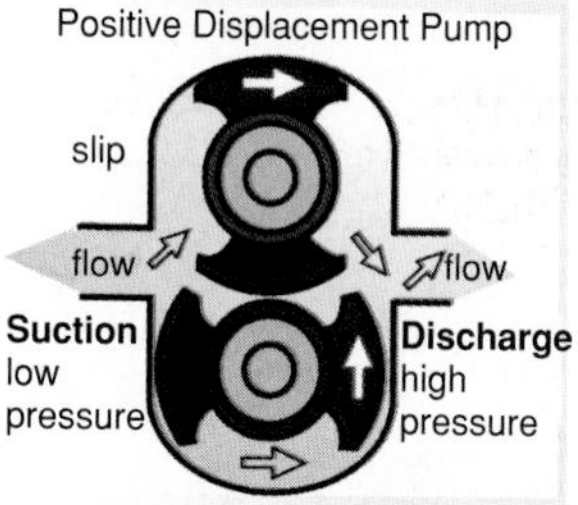

A hydraulic pump

Pump symbols

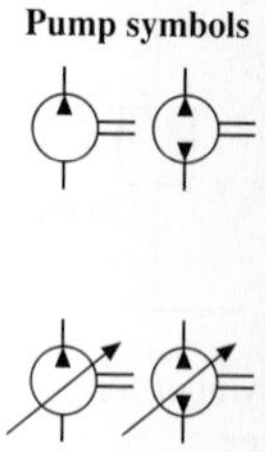

Left: Fixed capacity pump. **Right:** Fixed capacity pump with two directions of flow.

Left: Variable capacity pump. **Right:** Variable capacity pump with two directions of flow.

Courtesy: Department of Energy

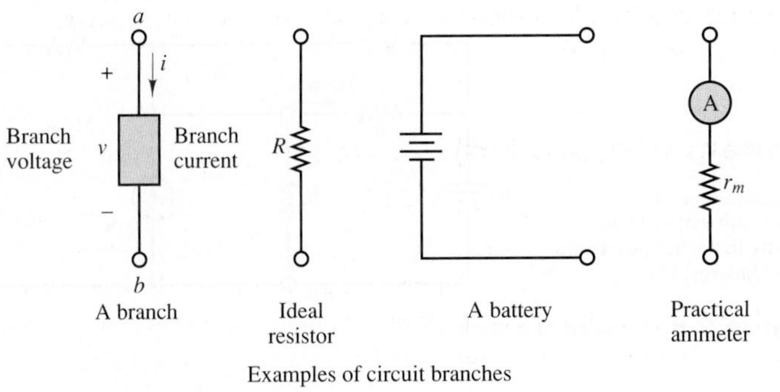

Figure 2.5 Definition of a branch

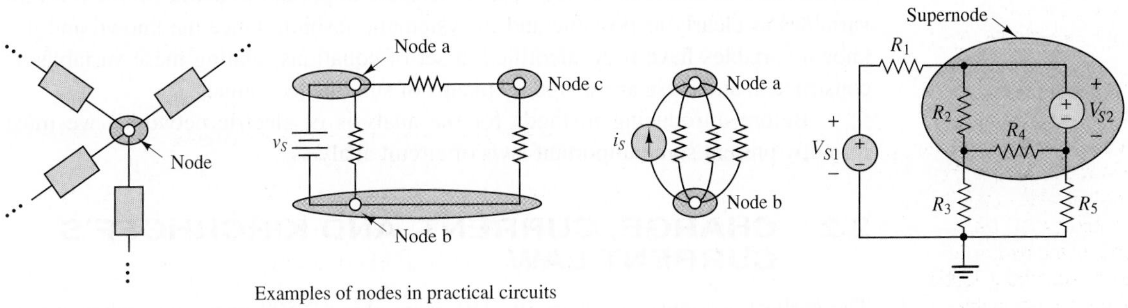

Figure 2.6 Definitions of node and supernode

Loop

A **loop** is any closed connection of branches. Various loop configurations are illustrated in Figure 2.7.

LO1

Note how two different loops in the same circuit may include some of the same elements or branches.

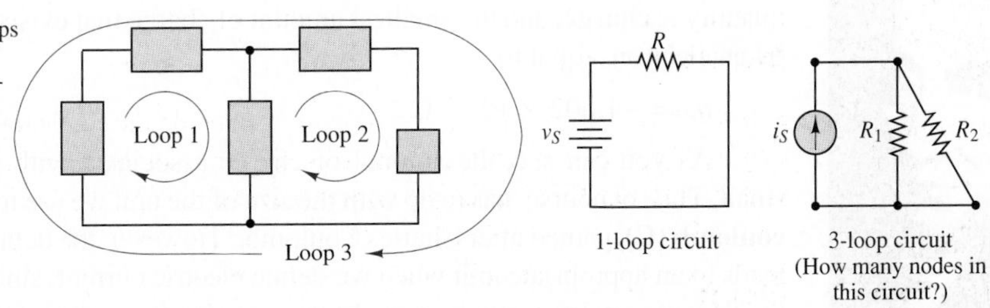

Figure 2.7 Definition of a loop

Mesh

LO1

A **mesh** is a loop that does not contain other loops. Meshes are an important aid to certain analysis methods. In Figure 2.7, the circuit with loops 1, 2, and 3 consists of two meshes: Loops 1 and 2 are meshes, but loop 3 is not a mesh, because it encircles both loops 1 and 2. The one-loop circuit of Figure 2.7 is also a one-mesh circuit. Figure 2.8 illustrates how meshes are simpler to visualize in complex networks than loops are.

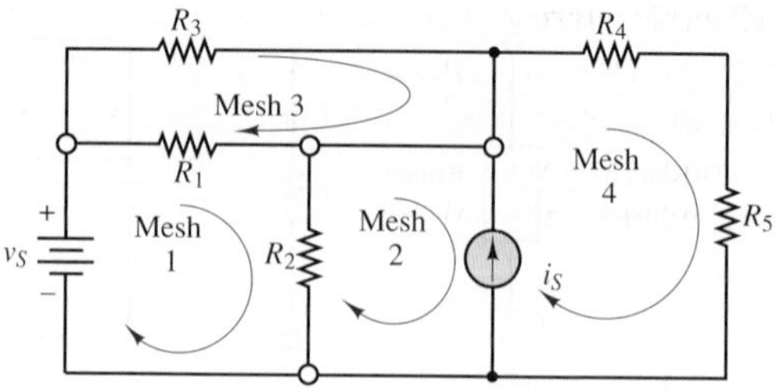

How many loops can you identify in this four-mesh circuit? (Answer: 15)

Figure 2.8 Definition of a mesh

Network Analysis

The analysis of an electric network consists of determining each of the unknown branch currents and node voltages. It is therefore important to define all the relevant variables as clearly as possible and in systematic fashion. Once the known and unknown variables have been identified, a set of equations relating these variables is constructed, and these are solved by means of suitable techniques.

Before introducing methods for the analysis of electric networks, we must formally present some important laws of circuit analysis.

2.2 CHARGE, CURRENT, AND KIRCHHOFF'S CURRENT LAW

The earliest accounts of electricity date from about 2,500 years ago, when it was discovered that static charge on a piece of amber was capable of attracting very light objects, such as feathers. The word *electricity* originated about 600 B.C.; it comes from *elektron,* which was the ancient Greek word for amber. The true nature of electricity was not understood until much later, however. Following the work of Alessandro Volta[1] and his invention of the copper-zinc battery, it was determined that static electricity and the current that flows in metal wires connected to a battery are due to the same fundamental mechanism: the atomic structure of matter, consisting of a nucleus—neutrons and protons—surrounded by electrons. The fundamental electric quantity is **charge,** and the smallest amount of charge that exists is the charge carried by an electron, equal to

$$q_e = -1.602 \times 10^{-19} \text{ C} \qquad \textbf{(2.1)}$$

Charles Coulomb (1736–1806). *Photograph courtesy of French Embassy, Washington, District of Columbia*

As you can see, the amount of charge associated with an electron is rather small. This, of course, has to do with the size of the unit we use to measure charge, the **coulomb (C),** named after Charles Coulomb.[2] However, the definition of the coulomb leads to an appropriate unit when we define electric current, since current consists of the flow of very large numbers of charge particles. The other charge-carrying particle in an atom, the proton, is assigned a plus sign and the same magnitude. The charge of a proton is

$$q_p = +1.602 \times 10^{-19} \text{ C} \qquad \textbf{(2.2)}$$

Electrons and protons are often referred to as **elementary charges.**

[1]See brief biography on page 16.

[2]See brief biography on page 16.

Electric current is defined as the time rate of change of charge passing through a predetermined area. Typically, this area is the cross-sectional area of a metal wire; however, we explore later a number of cases in which the current-carrying material is not a conducting wire. Figure 2.9 depicts a macroscopic view of the flow of charge in a wire, where we imagine Δq units of charge flowing through the cross-sectional area A in Δt units of time. The resulting current i is then given by

$$i = \frac{\Delta q}{\Delta t} \qquad \frac{\text{C}}{\text{s}} \tag{2.3}$$

If we consider the effect of the enormous number of elementary charges actually flowing, we can write this relationship in differential form:

$$i = \frac{dq}{dt} \qquad \frac{\text{C}}{\text{s}} \tag{2.4}$$

The units of current are called **amperes,** where 1 ampere (A) = 1 coulomb/second (C/s). The name of the unit is a tribute to the French scientist André-Marie Ampère.[3] The electrical engineering convention states that the positive direction of current flow is that of positive charges. In metallic conductors, however, current is carried by negative charges; these charges are the free electrons in the conduction band, which are only weakly attracted to the atomic structure in metallic elements and are therefore easily displaced in the presence of electric fields.

Current $i = dq/dt$ is generated by the flow of charge through the cross-sectional area A in a conductor.

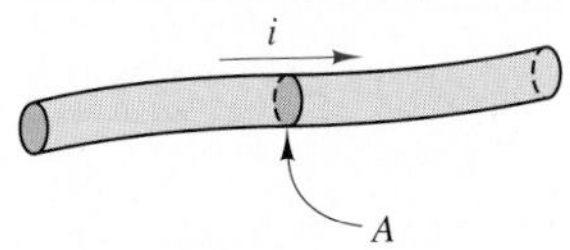

Figure 2.9 Current flow in an electric conductor

EXAMPLE 2.1 Charge and Current in a Conductor

Problem

Find the total charge in a cylindrical conductor (solid wire) and compute the current flowing in the wire.

Solution

Known Quantities: Conductor geometry, charge density, charge carrier velocity.

Find: Total charge of carriers Q; current in the wire I.

Schematics, Diagrams, Circuits, and Given Data:

Conductor length: $L = 1$ m.
Conductor diameter: $2r = 2 \times 10^{-3}$ m.
Charge density: $n = 10^{29}$ carriers/m^3.
Charge of one electron: $q_e = -1.602 \times 10^{-19}$.
Charge carrier velocity: $u = 19.9 \times 10^{-6}$ m/s.

Assumptions: None.

Analysis: To compute the total charge in the conductor, we first determine the volume of the conductor:

Volume = length × cross-sectional area

$$V = L \times \pi r^2 = (1\text{ m})\left[\pi\left(\frac{2 \times 10^{-3}}{2}\right)^2 \text{m}^2\right] = \pi \times 10^{-6}\text{ m}^3$$

[3]See brief biography on page 16.

Next, we compute the number of carriers (electrons) in the conductor and the total charge:

Number of carriers = volume × carrier density

$$N = V \times n = (\pi \times 10^{-6}\,\text{m}^3)\left(10^{29}\frac{\text{carriers}}{\text{m}^3}\right) = \pi \times 10^{23}\ \text{carriers}$$

Charge = number of carriers × charge/carrier

$$Q = N \times q_e = (\pi \times 10^{23}\ \text{carriers}) \times \left(-1.602 \times 10^{-19}\frac{\text{C}}{\text{carrier}}\right) = -50.33 \times 10^3\ \text{C}$$

To compute the current, we consider the velocity of the charge carriers and the charge density per unit length of the conductor:

Current = carrier charge density per unit length × carrier velocity

$$I = \left(\frac{Q}{L}\ \frac{\text{C}}{\text{m}}\right) \times \left(u\ \frac{\text{m}}{s}\right) = \left(-50.33 \times 10^3\ \frac{\text{C}}{\text{m}}\right)\left(19.9 \times 10^{-6}\ \frac{\text{m}}{\text{s}}\right) = -1\ \text{A}$$

Comments: Charge carrier density is a function of material properties. Carrier velocity is a function of the applied electric field.

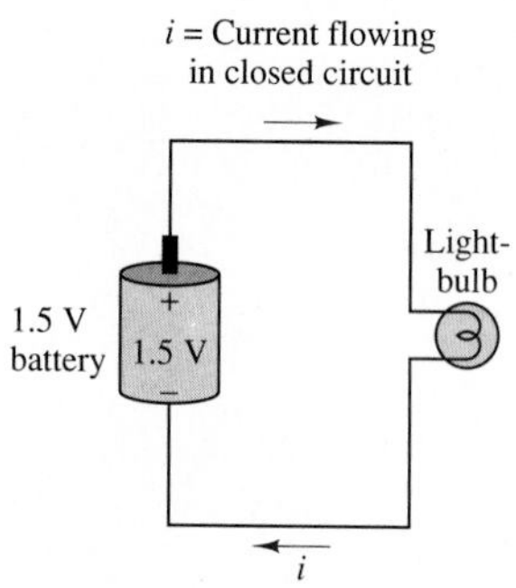

Figure 2.10 A simple electric circuit

In order for current to flow, there must exist a closed circuit.

Figure 2.10 depicts a simple circuit, composed of a battery (e.g., a dry-cell or alkaline 1.5-V battery) and a lightbulb.

Note that in the circuit of Figure 2.10, the current i flowing from the battery to the lightbulb is equal to the current flowing from the lightbulb to the battery. In other words, no current (and therefore no charge) is "lost" around the closed circuit. This principle was observed by the German scientist G. R. Kirchhoff[4] and is now known as **Kirchhoff's current law (KCL).** Kirchhoff's current law states that because charge cannot be created but must be conserved, *the sum of the currents at a node must equal zero*. Formally,

$$\sum_{n=1}^{N} i_n = 0 \qquad \text{Kirchhoff's current law} \tag{2.5}$$

LO2

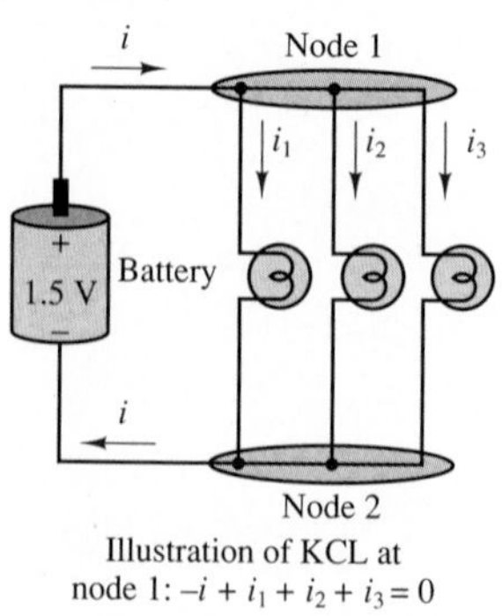

Figure 2.11 Illustration of Kirchhoff's current law

The significance of Kirchhoff's current law is illustrated in Figure 2.11, where the simple circuit of Figure 2.10 has been augmented by the addition of two lightbulbs (note how the two nodes that exist in this circuit have been emphasized by the shaded areas). In this illustration, we define currents entering a node as being negative and

[4]Gustav Robert Kirchhoff (1824–1887), a German scientist, published the first systematic description of the laws of circuit analysis. His contribution—though not original in terms of its scientific content—forms the basis of all circuit analysis.

currents exiting the node as being positive. Thus, the resulting expression for node 1 of the circuit of Figure 2.11 is

$$-i + i_1 + i_2 + i_3 = 0$$

Note that if we had assumed that currents entering the node were positive, the result would not have changed.

Kirchhoff's current law is one of the fundamental laws of circuit analysis, making it possible to express currents in a circuit in terms of one another. KCL is explored further in Examples 2.2 through 2.4.

EXAMPLE 2.2 Kirchhoff's Current Law Applied to an Automotive Electrical Harness

LO2

Problem

FIND IT ON THE WEB

Figure 2.12 shows an **automotive battery** connected to a variety of circuits in an automobile. The circuits include headlights, taillights, starter motor, fan, power locks, and dashboard panel. The battery must supply enough current to independently satisfy the requirements of each of the "load" circuits. Apply KCL to the automotive circuits.

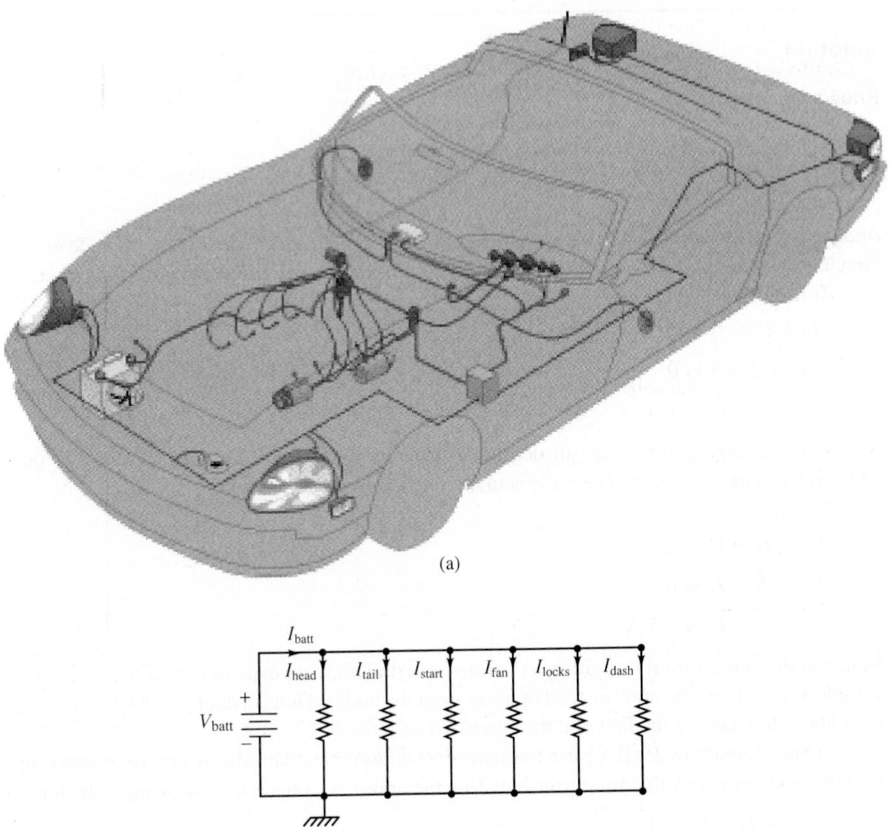

Figure 2.12 (a) Automotive circuits; (b) equivalent electric circuit

Solution

Known Quantities: Components of electrical harness: headlights, taillights, starter motor, fan, power locks, and dashboard panel.

Find: Expression relating battery current to load currents.

Schematics, Diagrams, Circuits, and Given Data: Figure 2.12.

Assumptions: None.

Analysis: Figure 2.12(b) depicts the equivalent electric circuit, illustrating how the current supplied by the battery must divide among the various circuits. The application of KCL to the equivalent circuit of Figure 2.12 requires that

$$I_{\text{batt}} - I_{\text{head}} - I_{\text{tail}} - I_{\text{start}} - I_{\text{fan}} - I_{\text{locks}} - I_{\text{dash}} = 0$$

LO2

EXAMPLE 2.3 Application of KCL

Problem

Determine the unknown currents in the circuit of Figure 2.13.

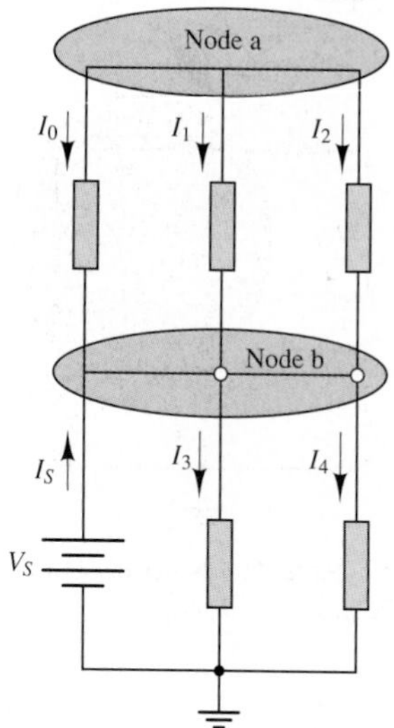

Figure 2.13 Demonstration of KCL

Solution

Known Quantities:

$$I_S = 5\text{ A} \qquad I_1 = 2\text{ A} \qquad I_2 = -3\text{ A} \qquad I_3 = 1.5\text{ A}$$

Find: I_0 and I_4.

Analysis: Two nodes are clearly shown in Figure 2.13 as node a and node b; the third node in the circuit is the reference (ground) node. In this example we apply KCL at each of the three nodes.

At node a:

$$I_0 + I_1 + I_2 = 0$$

$$I_0 + 2 - 3 = 0$$

$$\therefore \qquad I_0 = 1\text{ A}$$

Note that the three currents are all defined as flowing away from the node, but one of the currents has a negative value (i.e., it is actually flowing toward the node).

At node b:

$$I_S - I_3 - I_4 = 0$$

$$5 - 1.5 - I_4 = 0$$

$$\therefore \qquad I_4 = 3.5\text{ A}$$

Note that the current from the battery is defined in a direction opposite to that of the other two currents (i.e., toward the node instead of away from the node). Thus, in applying KCL, we have used opposite signs for the first and the latter two currents.

At the reference node: If we use the same convention (positive value for currents entering the node and negative value for currents exiting the node), we obtain the following equations:

$$-I_S + I_3 + I_4 = 0$$

$$-5 + 1.5 + I_4 = 0$$

$$\therefore \qquad I_4 = 3.5\text{ A}$$

Comments: The result obtained at the reference node is exactly the same as we already calculated at node b. This fact suggests that some redundancy may result when we apply KCL at each node in a circuit. In Chapter 3 we develop a method called *node analysis* that ensures the derivation of the smallest possible set of independent equations.

CHECK YOUR UNDERSTANDING

Repeat the exercise of Example 2.3 when $I_0 = 0.5$ A, $I_2 = 2$ A, $I_3 = 7$ A, and $I_4 = -1$ A. Find I_1 and I_S.

Answer: $I_1 = -2.5$ A and $I_S = 6$ A

EXAMPLE 2.4 Application of KCL

LO2

Problem

Apply KCL to the circuit of Figure 2.14, using the concept of supernode to determine the source current I_{S1}.

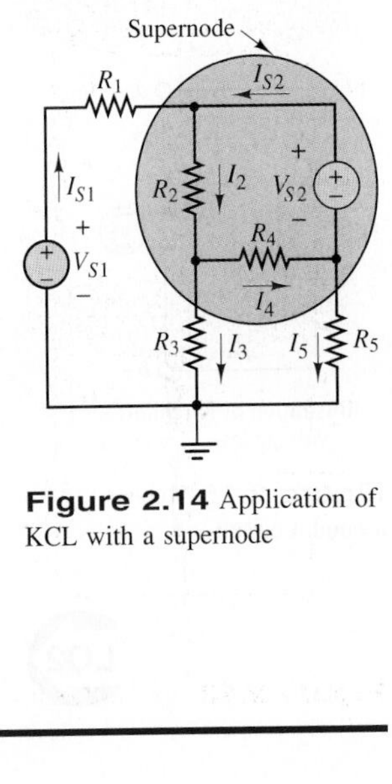

Figure 2.14 Application of KCL with a supernode

Solution

Known Quantities:

$$I_3 = 2 \text{ A} \qquad I_5 = 0 \text{ A}$$

Find: I_{S1}.

Analysis: Treating the supernode as a simple node, we apply KCL at the supernode to obtain

$$I_{S1} - I_3 - I_5 = 0$$

$$I_{S1} = I_3 + I_5 = 2 \text{ A}$$

Comments: The value of this analysis will become clear when you complete the drill exercise below.

CHECK YOUR UNDERSTANDING

Use the result of Example 2.4 and the following data to compute the current I_{S2} in the circuit of Figure 2.14.

$$I_2 = 3 \text{ A} \qquad I_4 = 1 \text{ A}$$

Answer: $I_{S2} = 1$ A

2.3 VOLTAGE AND KIRCHHOFF'S VOLTAGE LAW

Gustav Robert Kirchhoff (1824–1887). *Photograph courtesy of Deutsches Museum, Munich.*

Charge moving in an electric circuit gives rise to a current, as stated in Section 2.2. Naturally, it must take some work, or energy, for the charge to move between two points in a circuit, say, from point a to point b. The total *work per unit charge* associated with the motion of charge between two points is called **voltage.** Thus, the units of voltage are those of energy per unit charge; they have been called **volts** in honor of Alessandro Volta:

$$1 \text{ volt (V)} = 1 \frac{\text{joule (J)}}{\text{coulomb (C)}} \tag{2.6}$$

The voltage, or **potential difference,** between two points in a circuit indicates the energy required to move charge from one point to the other. The role played by a voltage source in an electric circuit is equivalent to that played by the force of gravity. Raising a mass with respect to a reference surface increases its potential energy. This potential energy can be converted to kinetic energy when the object moves to a lower position relative to the reference surface. The voltage, or potential difference, across a voltage source plays an analogous role, raising the electrical potential of the circuit, so that charge can move in the circuit, converting the potential energy within the voltage source to electric power. As will be presently shown, the direction, or polarity, of the voltage is closely tied to whether energy is being dissipated or generated in the process. The seemingly abstract concept of work being done in moving charges can be directly applied to the analysis of electric circuits; consider again the simple circuit consisting of a battery and a lightbulb. The circuit is drawn again for convenience in Figure 2.15, with nodes defined by the letters a and b. Experimental observations led Kirchhoff to the formulation of the second of his laws, **Kirchhoff's voltage law,** or **KVL.** The principle underlying KVL is that no energy is lost or created in an electric circuit; in circuit terms, the sum of all voltages associated with sources must equal the sum of the load voltages, so that *the net voltage around a closed circuit is zero.* If this were not the case, we would need to find a physical explanation for the excess (or missing) energy not accounted for in the voltages around a circuit. Kirchhoff's voltage law may be stated in a form similar to that used for KCL

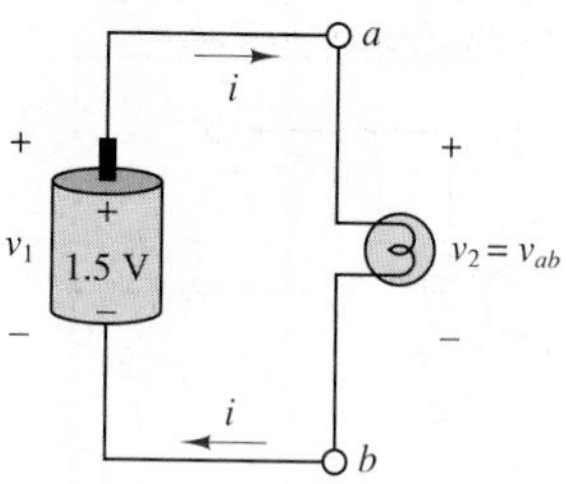

Illustration of Kirchhoff's voltage law: $v_1 = v_2$

Figure 2.15 Voltages around a circuit

LO2

$$\sum_{n=1}^{N} v_n = 0 \qquad \text{Kirchhoff's voltage law} \tag{2.7}$$

where the v_n are the individual voltages around the closed circuit. To understand this concept, we must introduce the concept of **reference voltage.**

In Figure 2.15, the *voltage across the lightbulb* is the difference between two node voltages, v_a and v_b. In a circuit, any one node may be chosen as the reference node, such that all node voltages may be referenced to this **reference voltage.** In Figure 2.15, we could select the voltage at node b as the reference voltage and observe that the battery's positive terminal is 1.5 V *above the reference voltage.* It is convenient to assign a value of zero to reference voltages, since this simplifies the voltage assignments around the circuit. With reference to Figure 2.15, and assuming that $v_b = 0$, we can write

$$v_1 = 1.5 \text{ V}$$
$$v_2 = v_{ab} = v_a - v_b = v_a - 0 = v_a$$

but v_a and v_1 are the same voltage, that is, the voltage at node a (referenced to node b). Thus

$$v_1 = v_2$$

One may think of the work done in moving a charge from point a to point b and the work done moving it back from b to a as corresponding directly to the *voltages across individual circuit elements*. Let Q be the total charge that moves around the circuit per unit time, giving rise to current i. Then the work W done in moving Q from b to a (i.e., across the battery) is

$$W_{ba} = Q \times 1.5 \text{ V} \tag{2.8}$$

Similarly, work is done in moving Q from a to b, that is, across the lightbulb. Note that the word *potential* is quite appropriate as a synonym of voltage, in that voltage represents the potential energy between two points in a circuit: If we remove the lightbulb from its connections to the battery, there still exists a voltage across the (now disconnected) terminals b and a. This is illustrated in Figure 2.16.

A moment's reflection upon the significance of voltage should suggest that it must be necessary to specify a sign for this quantity. Consider, again, the same dry-cell or alkaline battery where, by virtue of an electrochemically induced separation of charge, a 1.5-V potential difference is generated. The potential generated by the battery may be used to move charge in a circuit. The rate at which charge is moved once a closed circuit is established (i.e., the current drawn by the circuit connected to the battery) depends now on the circuit element we choose to connect to the battery. Thus, while the voltage across the battery represents the potential for *providing energy* to a circuit, the voltage across the lightbulb indicates the amount of work done in *dissipating energy*. In the first case, energy is generated; in the second, it is consumed (note that energy may also be stored, by suitable circuit elements yet to be introduced). This fundamental distinction requires attention in defining the sign (or polarity) of voltages.

We shall, in general, refer to elements that provide energy as **sources** and to elements that dissipate energy as **loads.** Standard symbols for a generalized source-and-load circuit are shown in Figure 2.17. Formal definitions are given later.

The presence of a voltage, v_2, across the open terminals a and b indicates the potential energy that can enable the motion of charge, once a closed circuit is established to allow current to flow.

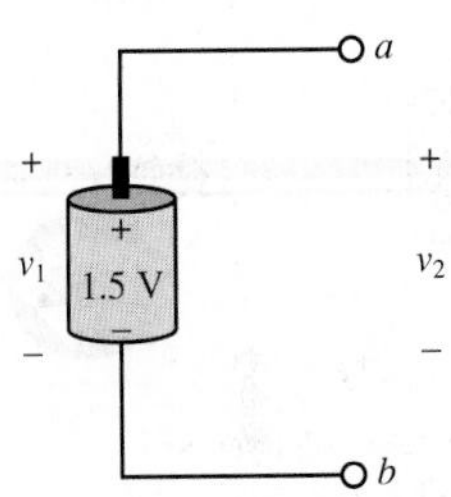

Figure 2.16 Concept of voltage as potential difference

A symbolic representation of the battery–lightbulb circuit of Figure 2.15.

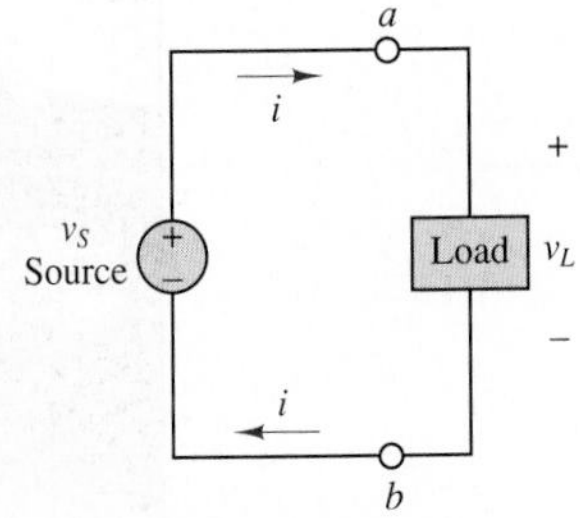

Figure 2.17 Sources and loads in an electric circuit

Ground

The concept of reference voltage finds a practical use in the *ground voltage* of a circuit. Ground represents a specific reference voltage that is usually a clearly identified point in a circuit. For example, the ground reference voltage can be identified with the case or enclosure of an instrument, or with the earth itself. In residential electric circuits, the ground reference is a large conductor that is physically connected to the earth. It is convenient to assign a potential of 0 V to the ground voltage reference.

The choice of the word *ground* is not arbitrary. This point can be illustrated by a simple analogy with the physics of fluid motion. Consider a tank of water, as shown in Figure 2.18, located at a certain height above the ground. The potential energy due to gravity will cause water to flow out of the pipe at a certain flow rate. The pressure that forces water out of the pipe is directly related to the head $h_1 - h_2$ in such a way that this pressure is zero when $h_2 = h_1$. Now the point h_3, corresponding to the ground level, is defined as having zero potential energy. It should be apparent that the pressure acting on the fluid in the pipe is really caused by the difference in potential energy $(h_1 - h_3) - (h_2 - h_3)$. It can be seen, then, that it is not necessary to assign a precise energy level to the height h_3; in fact, it would be extremely cumbersome to do so, since the equations describing the flow of water would then be different, say, in Denver, Colorado (h_3 = 1,600 m above sea level), from those that would apply in Miami, Florida (h_3 = 0 m above sea level). You see, then, that it is the relative difference in potential energy that matters in the water tank problem.

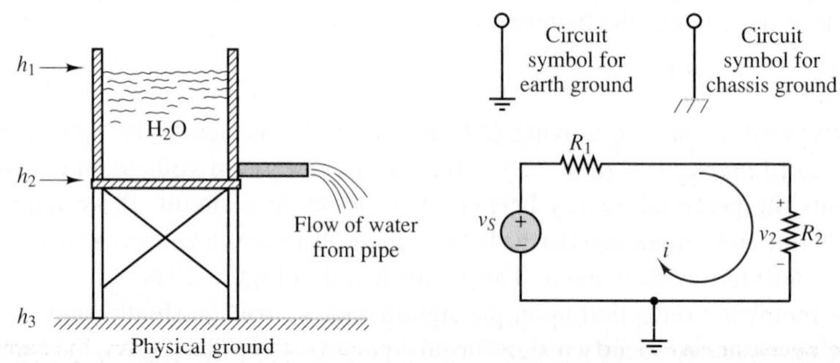

Figure 2.18 Analogy between electrical and earth ground

LO2

EXAMPLE 2.5 Kirchhoff's Voltage Law—Electric Vehicle Battery Pack

Problem

Figure 2.19(a) depicts the battery pack in the Smokin' Buckeye electric race car. In this example we apply KVL to the series connection of 31 twelve-volt batteries that make up the battery supply for the electric vehicle.

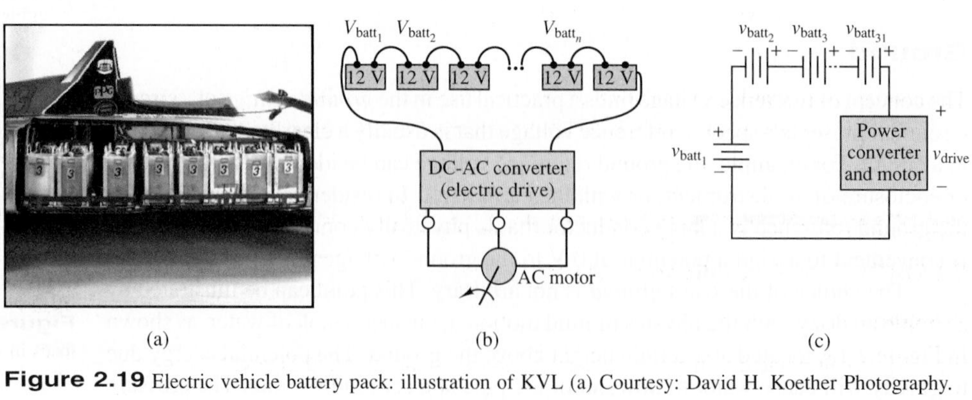

Figure 2.19 Electric vehicle battery pack: illustration of KVL (a) Courtesy: David H. Koether Photography.

Solution

FIND IT ON THE WEB

Known Quantities: Nominal characteristics of **Optima[TM] lead-acid batteries.**

Find: Expression relating battery and electric motor drive voltages.

Schematics, Diagrams, Circuits, and Given Data: $V_{batt} = 12$ V; Figure 2.19(a), (b) and (c).

Assumptions: None.

Analysis: Figure 2.19(b) depicts the equivalent electric circuit, illustrating how the voltages supplied by the battery are applied across the electric drive that powers the vehicle's 150-kW three-phase induction motor. The application of KVL to the equivalent circuit of Figure 2.19(b) requires that:

$$\sum_{n=1}^{31} V_{batt_n} - V_{drive} = 0$$

Thus, the electric drive is nominally supplied by a $31 \times 12 = 372$-V battery pack. In reality, the voltage supplied by lead-acid batteries varies depending on the state of charge of the battery. When fully charged, the battery pack of Figure 2.19(a) is closer to supplying around 400 V (i.e., around 13 V per battery).

EXAMPLE 2.6 Application of KVL

LO2

Problem

Determine the unknown voltage V_2 by applying KVL to the circuit of Figure 2.20.

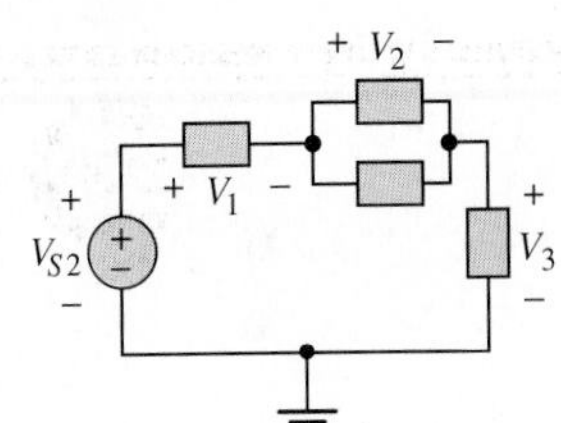

Figure 2.20 Circuit for Example 2.6

Solution

Known Quantities:

$V_{S2} = 12$ V $\quad V_1 = 6$ V $\quad V_3 = 1$ V

Find: V_2.

Analysis: Applying KVL around the simple loop, we write

$$V_{S2} - V_1 - V_2 - V_3 = 0$$

$$V_2 = V_{S2} - V_1 - V_3 = 12 - 6 - 1 = 5 \text{ V}$$

Comments: Note that V_2 is the voltage across two branches in parallel, and it must be equal for each of the two elements, since the two elements share the same nodes.

EXAMPLE 2.7 Application of KVL

LO2

Problem

Use KVL to determine the unknown voltages V_1 and V_4 in the circuit of Figure 2.21.

By KCL: $i_1 - i_2 - i_3 = 0$. In the node voltage method, we express KCL by

$$\frac{v_a - v_b}{R_1} - \frac{v_b - v_c}{R_2} - \frac{v_b - v_d}{R_3} = 0$$

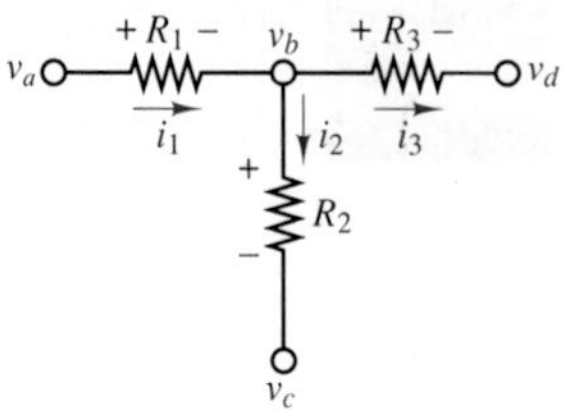

Figure 2.21 Circuit for Example 2.7

Solution

Known Quantities:

$$V_{S1} = 12 \text{ V} \qquad V_{S2} = -4 \text{ V} \qquad V_2 = 2 \text{ V} \qquad V_3 = 6 \text{ V} \qquad V_5 = 12 \text{ V}$$

Find: V_1, V_4.

Analysis: To determine the unknown voltages, we apply KVL clockwise around each of the three meshes:

$$V_{S1} - V_1 - V_2 - V_3 = 0$$
$$V_2 - V_{S2} + V_4 = 0$$
$$V_3 - V_4 - V_5 = 0$$

Next, we substitute numerical values:

$$12 - V_1 - 2 - 6 = 0$$
$$V_1 = 4 \text{ V}$$
$$2 - (-4) + V_4 = 0$$
$$V_4 = -6 \text{ V}$$
$$6 - (-6) - V_5 = 0$$
$$V_5 = 12 \text{ V}$$

Comments: In Chapter 3 we develop a systematic procedure called *mesh analysis* to solve this kind of problem.

CHECK YOUR UNDERSTANDING

Use the outer loop (around the outside perimeter of the circuit) to solve for V_1.

Answer: same as above

2.4 ELECTRIC POWER AND SIGN CONVENTION

The definition of voltage as work per unit charge lends itself very conveniently to the introduction of power. Recall that power is defined as the work done per unit time. Thus, the power P either generated or dissipated by a circuit element can be represented by the following relationship:

$$\text{Power} = \frac{\text{work}}{\text{time}} = \frac{\text{work}}{\text{charge}}\frac{\text{charge}}{\text{time}} = \text{voltage} \times \text{current} \qquad \textbf{(2.9)}$$

Thus,

The electric power generated by an active element, or that dissipated or stored by a passive element, is equal to the product of the voltage across the element and the current flowing through it.

LO3

$$P = VI \tag{2.10}$$

It is easy to verify that the units of voltage (joules per coulomb) times current (coulombs per second) are indeed those of power (joules per second, or watts).

It is important to realize that, just like voltage, power is a signed quantity, and it is necessary to make a distinction between *positive* and *negative power*. This distinction can be understood with reference to Figure 2.22, in which two circuits are shown side by side. The polarity of the voltage across circuit A and the direction of the current through it indicate that the circuit *is doing work in moving charge from a lower potential to a higher potential*. On the other hand, circuit B is dissipating energy, because the direction of the current indicates that *charge is being displaced from a higher potential to a lower potential*. To avoid confusion with regard to the sign of power, the electrical engineering community uniformly adopts the **passive sign convention,** which simply states that *the power dissipated by a load is a positive quantity* (or, conversely, that the power generated by a source is a positive quantity). Another way of phrasing the same concept is to state that if current flows from a higher to a lower voltage (plus to minus), the power is dissipated and will be a positive quantity.

It is important to note also that the actual numerical values of voltages and currents do not matter: Once the proper reference directions have been established and the passive sign convention has been applied consistently, the answer will be correct regardless of the reference direction chosen. Examples 2.8 and 2.9 illustrate this point.

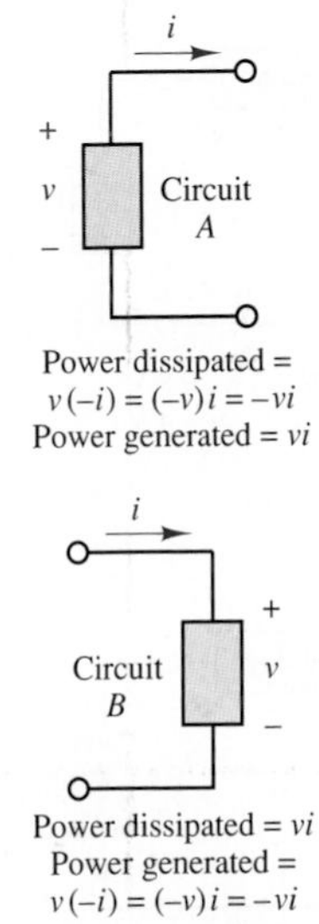

Figure 2.22 The passive sign convention

FOCUS ON METHODOLOGY

LO3

THE PASSIVE SIGN CONVENTION

1. Choose an arbitrary direction of current flow.
2. Label polarities of all active elements (voltage and current sources).
3. Assign polarities to all passive elements (resistors and other loads); for passive elements, current always flows into the positive terminal.
4. Compute the power dissipated by each element according to the following rule: If positive current flows into the positive terminal of an element, then the power dissipated is positive (i.e., the element absorbs power); if the current leaves the positive terminal of an element, then the power dissipated is negative (i.e., the element delivers power).

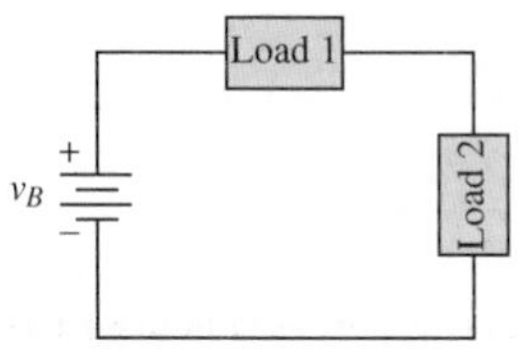

Figure 2.23

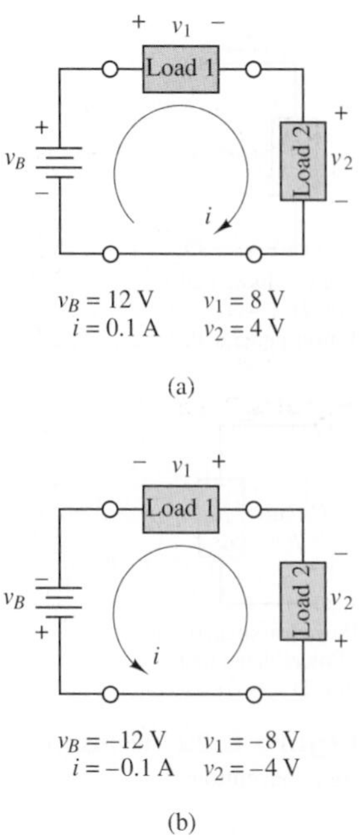

Figure 2.24

EXAMPLE 2.8 Use of the Passive Sign Convention

Problem

Apply the passive sign convention to the circuit of Figure 2.23.

Solution

Known Quantities: Voltages across each circuit element; current in circuit.

Find: Power dissipated or generated by each element.

Schematics, Diagrams, Circuits, and Given Data: Figure 2.24(a) and (b). The voltage drop across load 1 is 8 V, that across load 2 is 4 V; the current in the circuit is 0.1 A.

Assumptions: None.

Analysis: Note that the sign convention is independent of the current direction we choose. We now apply the method twice to the same circuit. Following the passive sign convention, we first select an arbitrary direction for the current in the circuit; the example will be repeated for both possible directions of current flow to demonstrate that the methodology is sound.

1. Assume clockwise direction of current flow, as shown in Figure 2.24(a).
2. Label polarity of voltage source, as shown in Figure 2.24(a); since the arbitrarily chosen direction of the current is consistent with the true polarity of the voltage source, the source voltage will be a positive quantity.
3. Assign polarity to each passive element, as shown in Figure 2.24(a).
4. Compute the power dissipated by each element: Since current flows from − to + through the battery, the power dissipated by this element will be a negative quantity:

$$P_B = -v_B \times i = -12 \text{ V} \times 0.1 \text{ A} = -1.2 \text{ W}$$

that is, the battery *generates* 1.2 watts (W). The power dissipated by the two loads will be a positive quantity in both cases, since current flows from plus to minus:

$$P_1 = v_1 \times i = 8 \text{ V} \times 0.1 \text{ A} = 0.8 \text{ W}$$

$$P_2 = v_2 \times i = 4 \text{ V} \times 0.1 \text{ A} = 0.4 \text{ W}$$

Next, we repeat the analysis, assuming counterclockwise current direction.

1. Assume counterclockwise direction of current flow, as shown in Figure 2.24(b).
2. Label polarity of voltage source, as shown in Figure 2.24(b); since the arbitrarily chosen direction of the current is not consistent with the true polarity of the voltage source, the source voltage will be a negative quantity.
3. Assign polarity to each passive element, as shown in Figure 2.24(b).
4. Compute the power dissipated by each element: Since current flows from plus to minus through the battery, the power dissipated by this element will be a positive quantity; however, the source voltage is a negative quantity:

$$P_B = v_B \times i = (-12 \text{ V})(0.1 \text{ A}) = -1.2 \text{ W}$$

that is, the battery *generates* 1.2 W, as in the previous case. The power dissipated by the two loads will be a positive quantity in both cases, since current flows from plus to minus:

$$P_1 = v_1 \times i = 8\text{ V} \times 0.1\text{ A} = 0.8\text{ W}$$

$$P_2 = v_2 \times i = 4\text{ V} \times 0.1\text{ A} = 0.4\text{ W}$$

Comments: It should be apparent that the most important step in the example is the correct assignment of source voltage; passive elements will always result in positive power dissipation. Note also that energy is conserved, as the sum of the power dissipated by source and loads is zero. In other words: *Power supplied always equals power dissipated.*

EXAMPLE 2.9

LO3

Problem

For the circuit shown in Figure 2.25, determine which components are absorbing power and which are delivering power. Is conservation of power satisfied? Explain your answer.

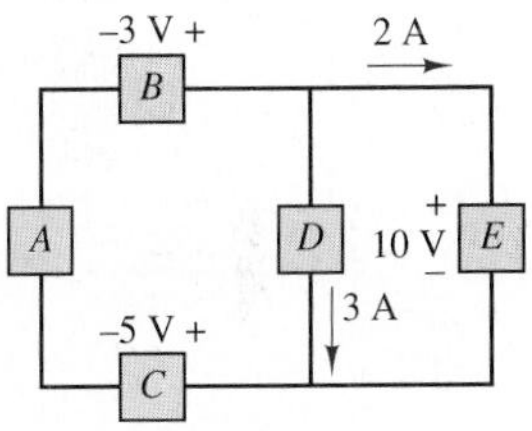

Figure 2.25

Solution

Known Quantities: Current through elements D and E; voltage across elements B, C, E.

Find: Which components are absorbing power, which are supplying power; verify the conservation of power.

Analysis: By KCL, the current through element B is 5 A, to the right. By KVL,

$$-v_a - 3 + 10 + 5 = 0$$

Therefore, the voltage across element A is

$$v_a = 12\text{ V} \qquad \text{(positive at the top)}$$

A supplies (12 V)(5 A) = 60 W
B supplies (3 V)(5 A) = 15 W
C absorbs (5 V)(5 A) = 25 W
D absorbs (10 V)(3 A) = 30 W
E absorbs (10 V)(2 A) = 20 W
Total power supplied = 60W + 15W = 75W
Total power absorbed = 25W + 30W + 20W = 75W
Total power supplied = Total power absorbed, so conservation of power is satisfied

Comments: The procedure described in this example can be easily conducted experimentally, by performing simple voltage and current measurements. Measuring devices are introduced in Section 2.8.

CHECK YOUR UNDERSTANDING

Compute the current flowing through each of the headlights of Example 2.2 if each headlight has a power rating of 50 W. How much power is the battery providing?

Determine which circuit element in the following illustration (left) is supplying power and which is dissipating power. Also determine the amount of power dissipated and supplied.

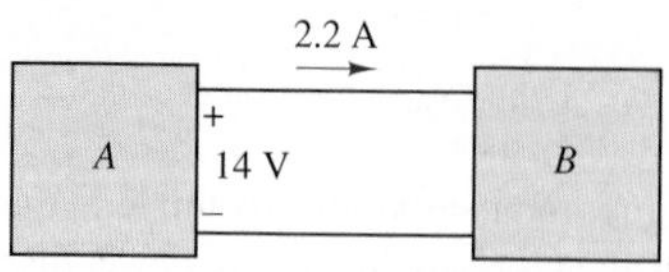

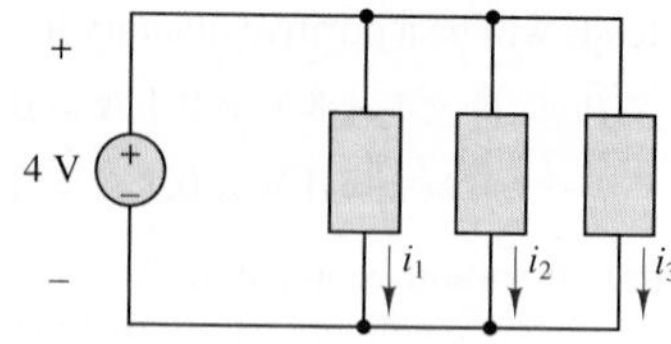

If the battery in the accompanying diagram (above, right) supplies a total of 10 mW to the three elements shown and $i_1 = 2$ mA and $i_2 = 1.5$ mA, what is the current i_3? If $i_1 = 1$ mA and $i_3 = 1.5$ mA, what is i_2?

Answers: $I_P = I_B = 4.17$ A; 100 W. A supplies 30.8 W; B dissipates 30.8 W. $i_3 = -1$ mA; $i_2 = 0$ mA.

2.5 CIRCUIT ELEMENTS AND THEIR *i-v* CHARACTERISTICS

The relationship between current and voltage at the terminals of a circuit element defines the behavior of that element within the circuit. In this section we introduce a graphical means of representing the terminal characteristics of circuit elements. Figure 2.26 depicts the representation that is employed throughout the chapter to denote a generalized circuit element: The variable i represents the current flowing through the element, while v is the potential difference, or voltage, across the element.

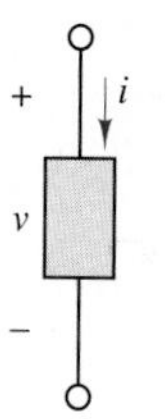

Figure 2.26 Generalized representation of circuit elements

Suppose now that a known voltage were imposed across a circuit element. The current that would flow, as a consequence of this voltage, and the voltage itself form a unique pair of values. If the voltage applied to the element were varied and the resulting current measured, it would be possible to construct a functional relationship between voltage and current known as the ***i-v* characteristic** (or **volt-ampere characteristic**). Such a relationship defines the circuit element, in the sense that if we impose any prescribed voltage (or current), the resulting current (or voltage) is directly obtainable from the i-v characteristic. A direct consequence is that the power dissipated (or generated) by the element may also be determined from the i-v curve.

Figure 2.27 depicts an experiment for empirically determining the i-v characteristic of a tungsten filament lightbulb. A variable voltage source is used to apply various voltages, and the current flowing through the element is measured for each applied voltage.

We could certainly express the i-v characteristic of a circuit element in functional form:

$$i = f(v) \qquad v = g(i) \tag{2.11}$$

In some circumstances, however, the graphical representation is more desirable, especially if there is no simple functional form relating voltage to current. The simplest form of the i-v characteristic for a circuit element is a straight line, that is,

$$i = kv \tag{2.12}$$

with k being a constant.

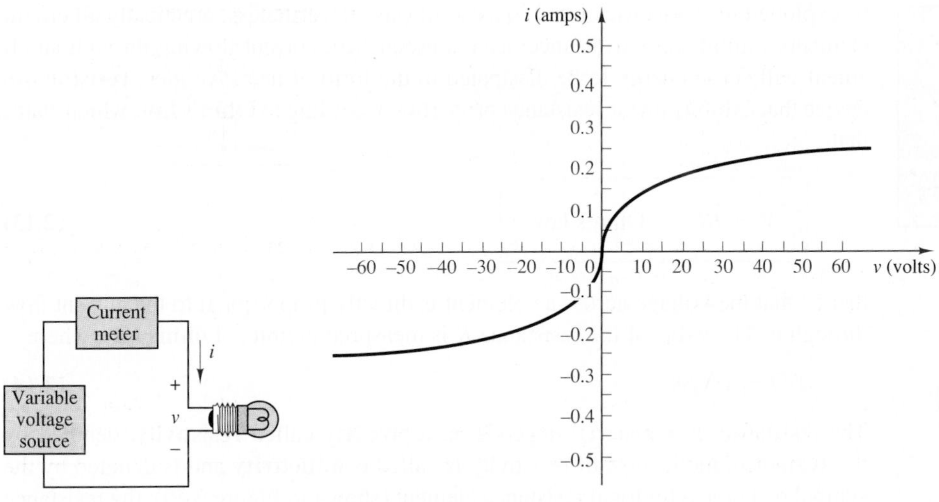

Figure 2.27 Volt-ampere characteristic of a tungsten lightbulb

We can also relate the graphical i-v representation of circuit elements to the power dissipated or generated by a circuit element. For example, the graphical representation of the lightbulb i-v characteristic of Figure 2.27 illustrates that when a positive current flows through the bulb, the voltage is positive, and conversely, a negative current flow corresponds to a negative voltage. In both cases the power dissipated by the device is a positive quantity, as it should be, on the basis of the discussion of Section 2.4, since the lightbulb is a passive device. Note that the i-v characteristic appears in only two of the four possible quadrants in the i-v plane. In the other two quadrants, the product of voltage and current (i.e., power) is negative, and an i-v curve with a portion in either of these quadrants therefore corresponds to power generated. This is not possible for a passive load such as a lightbulb; however, there are electronic devices that can operate, for example, in three of the four quadrants of the i-v characteristic and can therefore act as sources of energy for specific combinations of voltages and currents. An example of this dual behavior is introduced in Chapter 9, where it is shown that the photodiode can act either in a passive mode (as a light sensor) or in an active mode (as a solar cell).

The i-v characteristics of ideal current and voltage sources can also be useful in visually representing their behavior. An ideal voltage source generates a prescribed voltage independent of the current drawn from the load; thus, its i-v characteristic is a straight vertical line with a voltage axis intercept corresponding to the source voltage. Similarly, the i-v characteristic of an ideal current source is a horizontal line with a current axis intercept corresponding to the source current. Figure 2.28 depicts these behaviors.

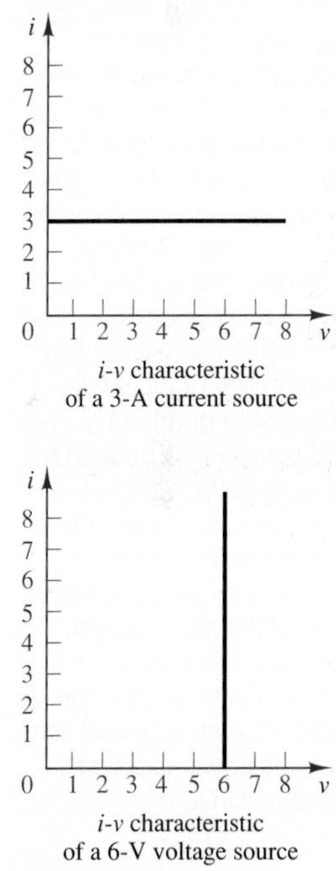

Figure 2.28 i-v characteristics of ideal sources

2.6 RESISTANCE AND OHM'S LAW

When electric current flows through a metal wire or through other circuit elements, it encounters a certain amount of **resistance,** the magnitude of which depends on the electrical properties of the material. Resistance to the flow of current may be undesired—for example, in the case of lead wires and connection cable—or it may

Electric Circuit Analogs of Hydraulic Systems—Fluid Resistance

A useful analogy can be made between the flow of electric current through electric components and the flow of incompressible fluids (e.g., water, oil) through hydraulic components. The analogy starts with the observation that the volume flow rate of a fluid in a pipe is analogous to current flow in a conductor. Similarly, pressure drop across the pipe is analogous to voltage drop across a resistor. The figure below depicts this relationship graphically. The fluid resistance opposed by the pipe to the fluid flow is analogous to an electrical resistance: The pressure difference between the two ends of the pipe causes fluid flow, much as a potential difference across a resistor forces a current flow. This analogy is explored further in Chapter 4.

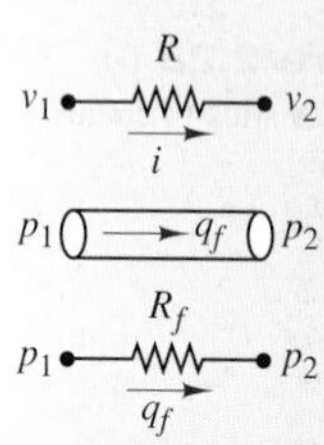

Analogy between electrical and fluid resistance

be exploited in an electric circuit in a useful way. Nevertheless, practically all circuit elements exhibit some resistance; as a consequence, current flowing through an element will cause energy to be dissipated in the form of heat. An ideal **resistor** is a device that exhibits linear resistance properties according to **Ohm's law,** which states that

$$V = IR \qquad \text{Ohm's law} \tag{2.13}$$

that is, that the voltage across an element is directly proportional to the current flow through it. The value of the resistance R is measured in units of **ohms (Ω)**, where

$$1\ \Omega = 1\ \text{V/A} \tag{2.14}$$

The resistance of a material depends on a property called **resistivity,** denoted by the symbol ρ; the inverse of resistivity is called **conductivity** and is denoted by the symbol σ. For a cylindrical resistance element (shown in Figure 2.29), the resistance is proportional to the length of the sample l and inversely proportional to its cross-sectional area A and conductivity σ.

$$R = \frac{l}{\sigma A} \tag{2.15}$$

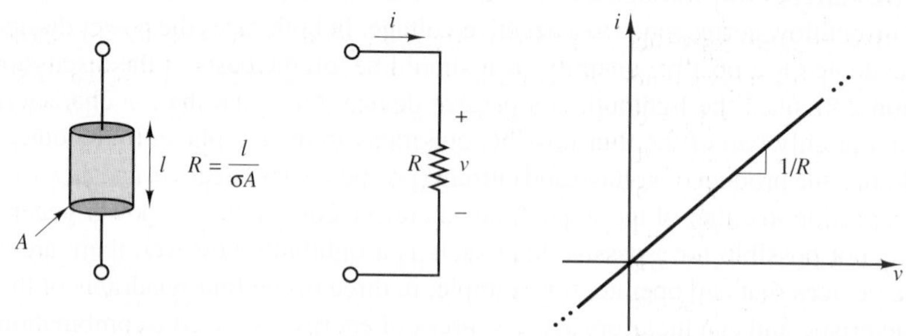

Physical resistors with resistance R. Typical materials are carbon, metal film.

Circuit symbol

i-v characteristic

Figure 2.29 The resistance element

It is often convenient to define the **conductance** of a circuit element as the inverse of its resistance. The symbol used to denote the conductance of an element is G, where

$$G = \frac{1}{R} \qquad \text{siemens (S)} \qquad \text{where} \qquad 1\ \text{S} = 1\ \text{A/V} \tag{2.16}$$

Thus, Ohm's law can be restated in terms of conductance as

$$I = GV \tag{2.17}$$

Ohm's law is an empirical relationship that finds widespread application in electrical engineering because of its simplicity. It is, however, only an approximation of the physics of electrically conducting materials. Typically, the linear relationship between voltage and current in electrical conductors does not apply at very high voltages and currents. Further, not all electrically conducting materials exhibit linear behavior even for small voltages and currents. It is usually true, however, that for some range of voltages and currents, most elements display a linear i-v characteristic. Figure 2.30 illustrates how the linear resistance concept may apply to elements with nonlinear i-v characteristics, by graphically defining the linear portion of the i-v characteristic of two common electrical devices: the lightbulb, which we have already encountered, and the semiconductor diode, which we study in greater detail in Chapter 9. Table 2.1 lists the conductivity of many common materials.

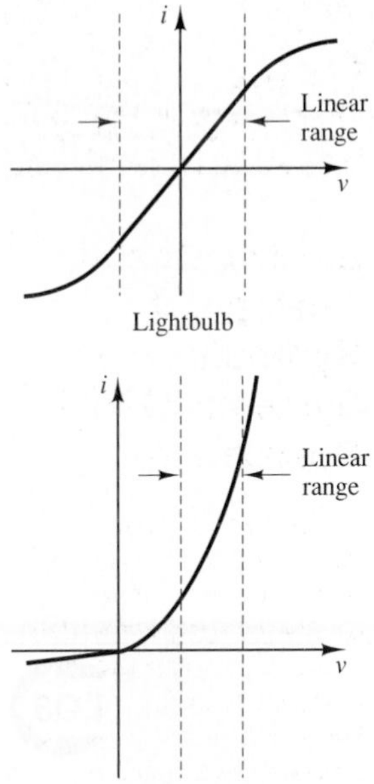

Figure 2.30

Table 2.1 Resistivity of common materials at room temperature

Material	Resistivity (Ω-m)
Aluminum	2.733×10^{-8}
Copper	1.725×10^{-8}
Gold	2.271×10^{-8}
Iron	9.98×10^{-8}
Nickel	7.20×10^{-8}
Platinum	10.8×10^{-8}
Silver	1.629×10^{-8}
Carbon	3.5×10^{-5}

The typical construction and the circuit symbol of the resistor are shown in Figure 2.29. Resistors made of cylindrical sections of carbon (with resistivity $\rho = 3.5 \times 10^{-5}$ Ω-m) are very common and are commercially available in a wide range of values for several power ratings (as explained shortly). Another common construction technique for resistors employs metal film. A common power rating for resistors used in electronic circuits (e.g., in most consumer electronic appliances such as radios and television sets) is $\frac{1}{4}$ W. Table 2.2 lists the standard values for commonly used resistors and the color code associated with these values (i.e., the common combinations of the digits $b_1 b_2 b_3$ as defined in Figure 2.31). For example, if the first three color bands on a resistor show the colors red ($b_1 = 2$), violet ($b_2 = 7$), and yellow ($b_3 = 4$), the resistance value can be interpreted as follows:

$$R = 27 \times 10^4 = 270{,}000 \ \Omega = 270 \ \text{k}\Omega$$

Table 2.2 Common resistor values ($\frac{1}{8}$-, $\frac{1}{4}$-, $\frac{1}{2}$-, 1-, 2-W rating)

Ω	Code	Ω	Multiplier	kΩ	Multiplier	kΩ	Multiplier	kΩ	Multiplier
10	Brn-blk-blk	100	Brown	1.0	Red	10	Orange	100	Yellow
12	Brn-red-blk	120	Brown	1.2	Red	12	Orange	120	Yellow
15	Brn-grn-blk	150	Brown	1.5	Red	15	Orange	150	Yellow
18	Brn-gry-blk	180	Brown	1.8	Red	18	Orange	180	Yellow
22	Red-red-blk	220	Brown	2.2	Red	22	Orange	220	Yellow
27	Red-vlt-blk	270	Brown	2.7	Red	27	Orange	270	Yellow
33	Org-org-blk	330	Brown	3.3	Red	33	Orange	330	Yellow
39	Org-wht-blk	390	Brown	3.9	Red	39	Orange	390	Yellow
47	Ylw-vlt-blk	470	Brown	4.7	Red	47	Orange	470	Yellow
56	Grn-blu-blk	560	Brown	5.6	Red	56	Orange	560	Yellow
68	Blu-gry-blk	680	Brown	6.8	Red	68	Orange	680	Yellow
82	Gry-red-blk	820	Brown	8.2	Red	82	Orange	820	Yellow

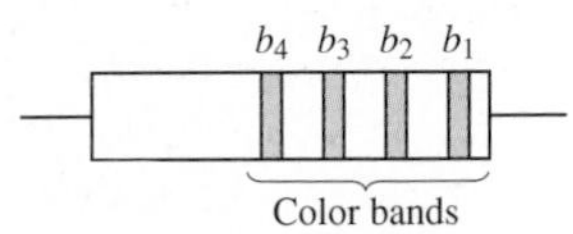

black	0	blue	6
brown	1	violet	7
red	2	gray	8
orange	3	white	9
yellow	4	silver	10%
green	5	gold	5%

Resistor value = $(b_1 b_2) \times 10^{b_3}$;
b_4 = % tolerance in actual value

Figure 2.31 Resistor color code

In Table 2.2, the leftmost column represents the complete color code; columns to the right of it only show the third color, since this is the only one that changes. For example, a 10-Ω resistor has the code brown-black-*black*, while a 100-Ω resistor has the code of brown-black-*brown*.

In addition to the resistance in ohms, the maximum allowable power dissipation (or **power rating**) is typically specified for commercial resistors. Exceeding this power rating leads to overheating and can cause the resistor to literally burn up. For a resistor R, the power dissipated can be expressed, with Ohm's law substituted into equation 2.10, by

$$P = VI = I^2R = \frac{V^2}{R} \tag{2.18}$$

That is, the power dissipated by a resistor is proportional to the square of the current flowing through it, as well as the square of the voltage across it. Example 2.10 illustrates how you can make use of the power rating to determine whether a given resistor will be suitable for a certain application.

LO3

EXAMPLE 2.10 Using Resistor Power Ratings

Problem

FIND IT ON THE WEB

Determine the minimum **resistor size** that can be connected to a given battery without exceeding the resistor's $\frac{1}{4}$-W power rating.

Solution

Known Quantities: Resistor power rating = 0.25 W. Battery voltages: 1.5 and 3 V.

Find: The smallest size $\frac{1}{4}$-W resistor that can be connected to each battery.

Schematics, Diagrams, Circuits, and Given Data: Figure 2.32, Figure 2.33.

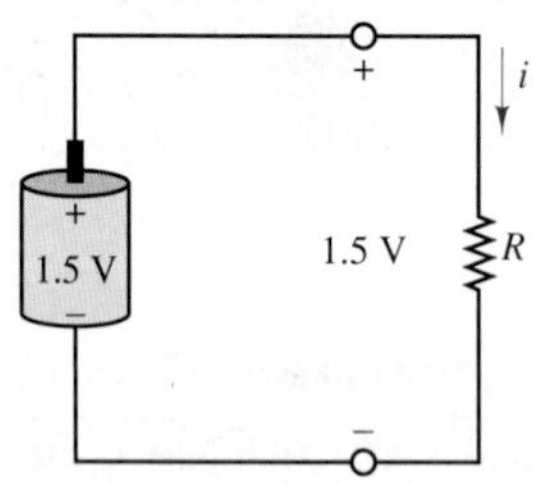

Figure 2.32

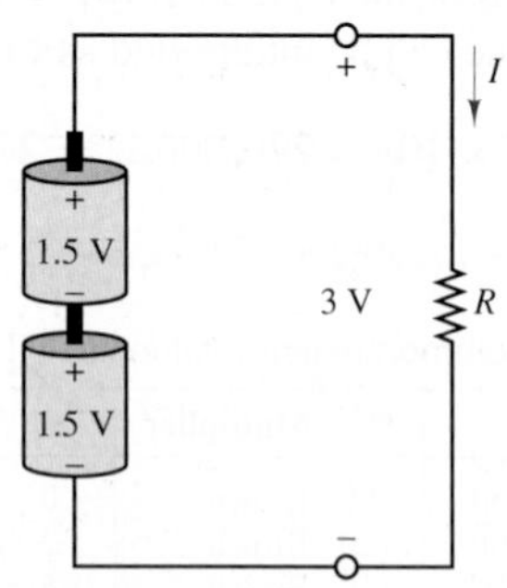

Figure 2.33

Analysis: We first need to obtain an expression for resistor power dissipation as a function of its resistance. We know that $P = VI$ and that $V = IR$. Thus, the power dissipated by any resistor is

$$P_R = V \times I = V \times \frac{V}{R} = \frac{V^2}{R}$$

Since the maximum allowable power dissipation is 0.25 W, we can write $V^2/R \le 0.25$, or $R \ge V^2/0.25$. Thus, for a 1.5-V battery, the minimum size resistor will be $R = 1.5^2/0.25 = 9\ \Omega$. For a 3-V battery the minimum size resistor will be $R = 3^2/0.25 = 36\ \Omega$.

Comments: Sizing resistors on the basis of power rating is very important in practice. Note how the minimum resistor size quadrupled as we doubled the voltage across it. This is because power increases as the square of the voltage. Remember that exceeding power ratings will inevitably lead to resistor failure!

CHECK YOUR UNDERSTANDING

A typical electronic power supply provides ±12 V. What is the size of the smallest $\frac{1}{4}$-W resistor that could be placed across (in parallel with) the power supply? (*Hint:* You may think of the supply as a 24-V supply.)

The circuit in the accompanying illustration contains a battery, a resistor, and an unknown circuit element.

1. If the voltage $V_{battery}$ is 1.45 V and $i = 5$ mA, find power supplied to or by the battery.
2. Repeat part 1 if $i = -2$ mA.

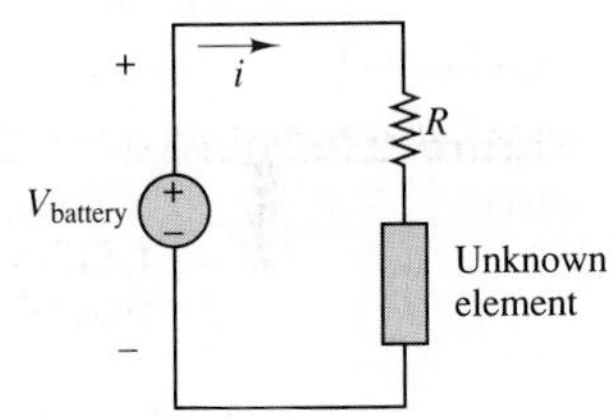

The battery in the accompanying circuit supplies power to resistors R_1, R_2, and R_3. Use KCL to determine the current i_B, and find the power supplied by the battery if $V_{battery} = 3$ V.

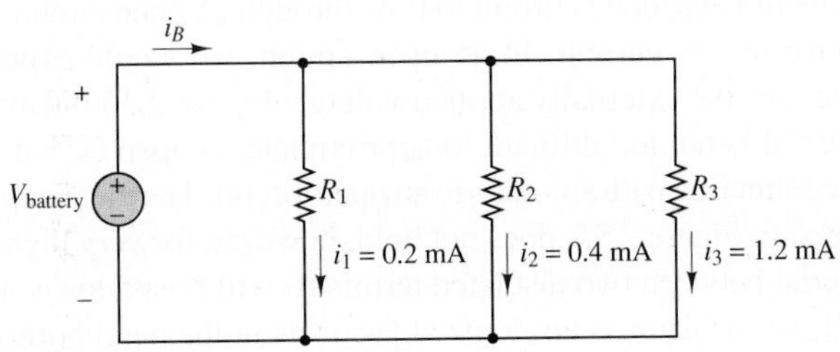

Answers: 2,304 Ω; $P_1 = 7.25 \times 10^{-3}$ W (supplied by); $P_2 = 2.9 \times 10^{-3}$ W (supplied to); $i_B = 1.8$ mA; $P_B = 5.4$ mW

Open and Short Circuits

Two convenient idealizations of the resistance element are provided by the limiting cases of Ohm's law as the resistance of a circuit element approaches zero or infinity. A circuit element with resistance approaching zero is called a **short circuit.** Intuitively, we would expect a short circuit to allow for unimpeded flow of current. In fact, metallic conductors (e.g., short wires of large diameter) approximate the behavior of a short circuit. Formally, a *short circuit* is defined as a circuit element across which the voltage is zero, regardless of the current flowing through it. Figure 2.34 depicts the circuit symbol for an ideal short circuit.

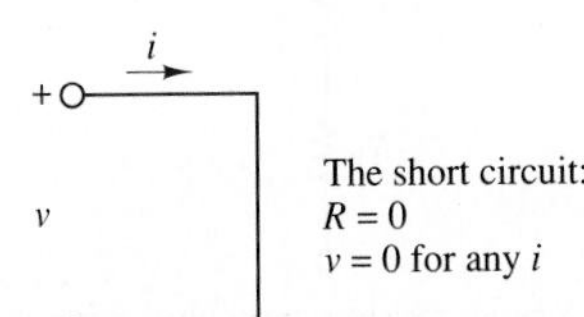

Figure 2.34 The short circuit

Physically, any wire or other metallic conductor will exhibit some resistance, though small. For practical purposes, however, many elements approximate a short circuit quite accurately under certain conditions. For example, a large-diameter copper pipe is effectively a short circuit in the context of a residential electric power supply, while in a low-power microelectronic circuit (e.g., an FM radio) a short length of 24-gauge wire (refer to Table 2.3 for the resistance of 24-gauge wire) is a more than adequate short circuit. Table 2.3 summarizes the resistance for a given length of some commonly used gauges of electrical wire. Additional information on **American Wire Gauge Standards** may be found on the internet.

Table 2.3 Resistance of copper wire

AWG size	Number of strands	Diameter per strand (in)	Resistance per 1,000 ft (Ω)
24	Solid	0.0201	28.4
24	7	0.0080	28.4
22	Solid	0.0254	18.0
22	7	0.0100	19.0
20	Solid	0.0320	11.3
20	7	0.0126	11.9
18	Solid	0.0403	7.2
18	7	0.0159	7.5
16	Solid	0.0508	4.5
16	19	0.0113	4.7
14	Solid	0.0641	2.52
12	Solid	0.0808	1.62
10	Solid	0.1019	1.02
8	Solid	0.1285	0.64
6	Solid	0.1620	0.4
4	Solid	0.2043	0.25
2	Solid	0.2576	0.16

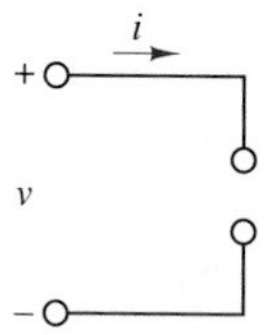

The open circuit: $R \to \infty$, $i = 0$ for any v

Figure 2.35 The open circuit

A circuit element whose resistance approaches infinity is called an **open circuit.** Intuitively, we would expect no current to flow through an open circuit, since it offers infinite resistance to any current. In an open circuit, we would expect to see zero current regardless of the externally applied voltage. Figure 2.35 illustrates this idea.

In practice, it is not too difficult to approximate an open circuit: Any break in continuity in a conducting path amounts to an open circuit. The idealization of the open circuit, as defined in Figure 2.35, does not hold, however, for very high voltages. The insulating material between two insulated terminals will break down at a sufficiently high voltage. If the insulator is air, ionized particles in the neighborhood of the two conducting elements may lead to the phenomenon of arcing; in other words, a pulse of current may be generated that momentarily jumps a gap between conductors (thanks to this principle, we are able to ignite the air-fuel mixture in a spark-ignition internal combustion engine by means of spark plugs). The ideal open and short circuits are useful concepts and find extensive use in circuit analysis.

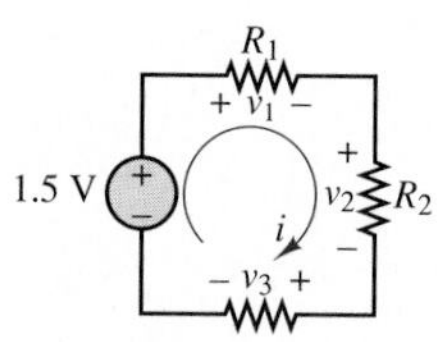

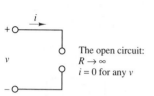

The current i flows through each of the four series elements. Thus, by KVL,

$$1.5 = v_1 + v_2 + v_3$$

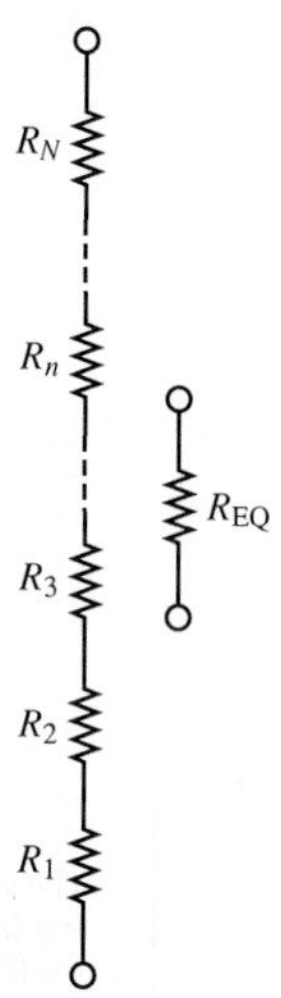

N series resistors are equivalent to a single resistor equal to the sum of the individual resistances.

Figure 2.36

Series Resistors and the Voltage Divider Rule

Although electric circuits can take rather complicated forms, even the most involved circuits can be reduced to combinations of circuit elements *in parallel* and *in series.* Thus, it is important that you become acquainted with parallel and series circuits as early as possible, even before formally approaching the topic of network analysis. Parallel and series circuits have a direct relationship with Kirchhoff's laws. The objective of this section and the next is to illustrate two common circuits based on series and parallel combinations of resistors: the voltage and current dividers. These circuits form the basis of all network analysis; it is therefore important to master these topics as early as possible.

For an example of a series circuit, refer to the circuit of Figure 2.36, where a battery has been connected to resistors R_1, R_2, and R_3. The following definition applies:

Definition

Two or more circuit elements are said to be **in series** if the current from one element exclusively flows into the next element. From KCL, it then follows that all series elements have the same current.

LO4

By applying KVL, you can verify that the sum of the voltages across the three resistors equals the voltage externally provided by the battery

$$1.5 \text{ V} = v_1 + v_2 + v_3$$

And since, according to Ohm's law, the separate voltages can be expressed by the relations

$$v_1 = iR_1 \qquad v_2 = iR_2 \qquad v_3 = iR_3$$

we can therefore write

$$1.5 \text{ V} = i(R_1 + R_2 + R_3)$$

This simple result illustrates a very important principle: To the battery, the three series resistors appear as a single equivalent resistance of value R_{EQ}, where

$$R_{EQ} = R_1 + R_2 + R_3$$

The three resistors could thus be replaced by a single resistor of value R_{EQ} without changing the amount of current required of the battery. From this result we may extrapolate to the more general relationship defining the equivalent resistance of N series resistors

$$R_{EQ} = \sum_{n=1}^{N} R_n \qquad \text{Equivalent series resistance} \tag{2.19}$$

which is also illustrated in Figure 2.36. A concept very closely tied to series resistors is that of the **voltage divider.** This terminology originates from the observation that the source voltage in the circuit of Figure 2.36 divides among the three resistors according to KVL. If we now observe that the series current i is given by

$$i = \frac{1.5 \text{ V}}{R_{EQ}} = \frac{1.5 \text{ V}}{R_1 + R_2 + R_3}$$

we can write each of the voltages across the resistors as:

$$v_1 = iR_1 = \frac{R_1}{R_{EQ}}(1.5 \text{ V})$$

$$v_2 = iR_2 = \frac{R_2}{R_{EQ}}(1.5 \text{ V})$$

$$v_3 = iR_3 = \frac{R_3}{R_{EQ}}(1.5 \text{ V})$$

That is,

LO4

> The voltage across each resistor in a series circuit is directly proportional to the ratio of its resistance to the total series resistance of the circuit.

An instructive exercise consists of verifying that KVL is still satisfied, by adding the voltage drops around the circuit and equating their sum to the source voltage:

$$v_1 + v_2 + v_3 = \frac{R_1}{R_{EQ}}(1.5\text{ V}) + \frac{R_2}{R_{EQ}}(1.5\text{ V}) + \frac{R_3}{R_{EQ}}(1.5\text{ V}) = 1.5\text{ V}$$

since $\quad R_{EQ} = R_1 + R_2 + R_3$

Therefore, since KVL is satisfied, we are certain that the voltage divider rule is consistent with Kirchhoff's laws. By virtue of the voltage divider rule, then, we can always determine the proportion in which voltage drops are distributed around a circuit. This result is useful in reducing complicated circuits to simpler forms. The general form of the voltage divider rule for a circuit with N series resistors and a voltage source is

LO4

$$v_n = \frac{R_n}{R_1 + R_2 + \cdots + R_n + \cdots + R_N} v_S \qquad \text{Voltage divider} \qquad \textbf{(2.20)}$$

LO4

EXAMPLE 2.11 Voltage Divider

Problem

Determine the voltage v_3 in the circuit of Figure 2.37.

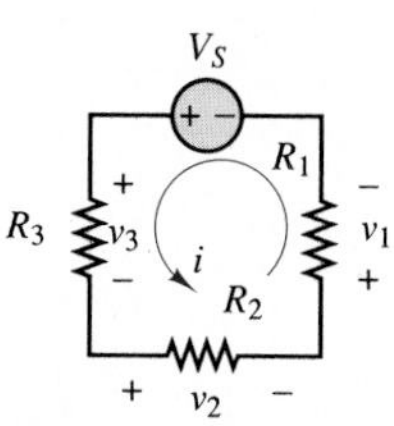

Figure 2.37

Solution

Known Quantities: Source voltage, resistance values.

Find: Unknown voltage v_3.

Schematics, Diagrams, Circuits, and Given Data: $R_1 = 10\ \Omega$; $R_2 = 6\ \Omega$; $R_3 = 8\ \Omega$; $V_S = 3$ V. Figure 2.37.

Analysis: Figure 2.37 indicates a reference direction for the current (dictated by the polarity of the voltage source). Following the passive sign convention, we label the polarities of the three resistors, and apply KVL to determine that

$$V_S - v_1 - v_2 - v_3 = 0$$

The voltage divider rule tells us that

$$v_3 = V_S \times \frac{R_3}{R_1 + R_2 + R_3} = 3 \times \frac{8}{10 + 6 + 8} = 1\text{ V}$$

Comments: Application of the voltage divider rule to a series circuit is very straightforward. The difficulty usually arises in determining whether a circuit is in fact a series circuit. This point is explored later in this section, and in Example 2.13.

CHECK YOUR UNDERSTANDING

Repeat Example 2.11 by reversing the reference direction of the current, to show that the same result is obtained.

Parallel Resistors and the Current Divider Rule

A concept analogous to that of the voltage divider may be developed by applying Kirchhoff's current law to a circuit containing only parallel resistances.

Definition

Two or more circuit elements are said to be **in parallel** if the elements share the same terminals. From KVL, it follows that the elements will have the same voltage.

LO4

Figure 2.38 illustrates the notion of parallel resistors connected to an ideal current source. Kirchhoff's current law requires that the sum of the currents into, say, the top node of the circuit be zero:

$$i_S = i_1 + i_2 + i_3$$

But by virtue of Ohm's law we may express each current as follows:

$$i_1 = \frac{v}{R_1} \qquad i_2 = \frac{v}{R_2} \qquad i_3 = \frac{v}{R_3}$$

since, by definition, the same voltage v appears across each element. Kirchhoff's current law may then be restated as follows:

$$i_S = v\left(\frac{1}{R_1} + \frac{1}{R_2} + \frac{1}{R_3}\right)$$

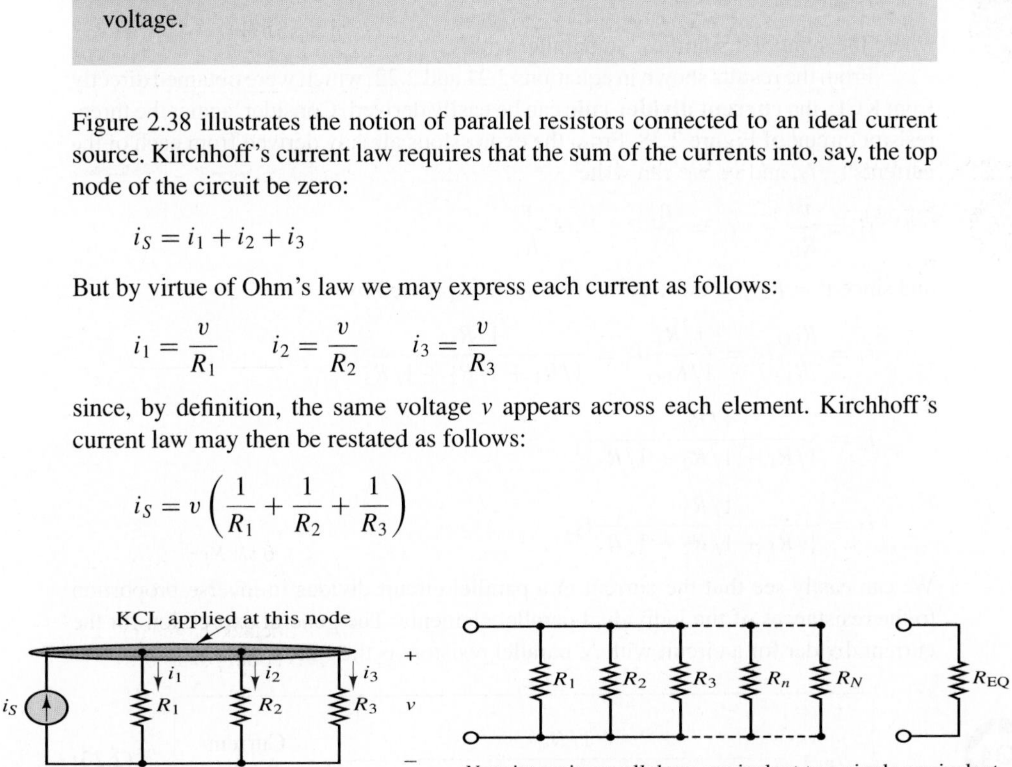

The voltage v appears across each parallel element; by KCL, $i_S = i_1 + i_2 + i_3$

N resistors in parallel are equivalent to a single equivalent resistor with resistance equal to the inverse of the sum of the inverse resistances.

Figure 2.38 Parallel circuits

Note that this equation can be also written in terms of a single equivalent resistance

$$i_S = v\frac{1}{R_{\text{EQ}}}$$

where $$\frac{1}{R_{\text{EQ}}} = \frac{1}{R_1} + \frac{1}{R_2} + \frac{1}{R_3}$$

As illustrated in Figure 2.38, we can generalize this result to an arbitrary number of resistors connected in parallel by stating that N resistors in parallel act as a single equivalent resistance R_{EQ} given by the expression

$$\frac{1}{R_{\text{EQ}}} = \frac{1}{R_1} + \frac{1}{R_2} + \cdots + \frac{1}{R_N} \tag{2.21}$$

or

$$R_{\text{EQ}} = \frac{1}{1/R_1 + 1/R_2 + \cdots + 1/R_N} \qquad \text{Equivalent parallel resistance} \tag{2.22}$$

Very often in the remainder of this book we refer to the parallel combination of two or more resistors with the notation

$$R_1 \parallel R_2 \parallel \cdots$$

where the symbol $\parallel$ signifies "in parallel with."

From the results shown in equations 2.21 and 2.22, which were obtained directly from KCL, the **current divider rule** can be easily derived. Consider, again, the three-resistor circuit of Figure 2.38. From the expressions already derived from each of the currents i_1, i_2, and i_3, we can write

$$i_1 = \frac{v}{R_1} \qquad i_2 = \frac{v}{R_2} \qquad i_3 = \frac{v}{R_3}$$

and since $v = R_{\text{EQ}} i_S$, these currents may be expressed by

$$i_1 = \frac{R_{\text{EQ}}}{R_1} i_S = \frac{1/R_1}{1/R_{\text{EQ}}} i_S = \frac{1/R_1}{1/R_1 + 1/R_2 + 1/R_3} i_S$$

$$i_2 = \frac{1/R_2}{1/R_1 + 1/R_2 + 1/R_3} i_S$$

$$i_3 = \frac{1/R_3}{1/R_1 + 1/R_2 + 1/R_3} i_S$$

We can easily see that the current in a parallel circuit divides in inverse proportion to the resistances of the individual parallel elements. The general expression for the current divider for a circuit with N parallel resistors is the following:

LO4

$$i_n = \frac{1/R_n}{1/R_1 + 1/R_2 + \cdots + 1/R_n + \cdots + 1/R_N} i_S \qquad \text{Current divider} \tag{2.23}$$

Example 2.12 illustrates the application of the current divider rule.

EXAMPLE 2.12 Current Divider

LO4

Problem

Determine the current i_1 in the circuit of Figure 2.39.

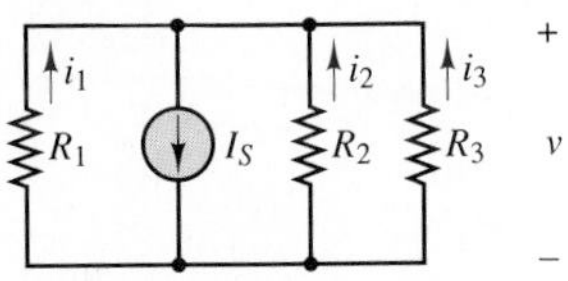

Figure 2.39

Solution

Known Quantities: Source current, resistance values.

Find: Unknown current i_1.

Schematics, Diagrams, Circuits, and Given Data: $R_1 = 10\ \Omega$; $R_2 = 2\ \Omega$; $R_3 = 20\ \Omega$; $I_S = 4$ A. Figure 2.39.

Analysis: Application of the current divider rule yields

$$i_1 = I_S \times \frac{1/R_1}{1/R_1 + 1/R_2 + 1/R_3} = 4 \times \frac{\frac{1}{10}}{\frac{1}{10} + \frac{1}{2} + \frac{1}{20}} = 0.6154 \text{ A}$$

Comments: While application of the current divider rule to a parallel circuit is very straightforward, it is sometimes not so obvious whether two or more resistors are actually in parallel. A method for ensuring that circuit elements are connected in parallel is explored later in this section, and in Example 2.13.

CHECK YOUR UNDERSTANDING

Verify that KCL is satisfied by the current divider rule and that the source current i_S divides in inverse proportion to the parallel resistors R_1, R_2, and R_3 in the circuit of Figure 2.39. (This should not be a surprise, since we would expect to see more current flow through the smaller resistance.)

Much of the resistive network analysis that is presented in Chapter 3 is based on the simple principles of voltage and current dividers introduced in this section. Unfortunately, practical circuits are rarely composed of only parallel or only series elements. The following examples and Check Your Understanding exercises illustrate some simple and slightly more advanced circuits that combine parallel and series elements.

EXAMPLE 2.13 Series-Parallel Circuit

LO4

Problem

Determine the voltage v in the circuit of Figure 2.40.

Solution

Known Quantities: Source voltage, resistance values.

Find: Unknown voltage v.

Schematics, Diagrams, Circuits, and Given Data: See Figures 2.40, 2.41.

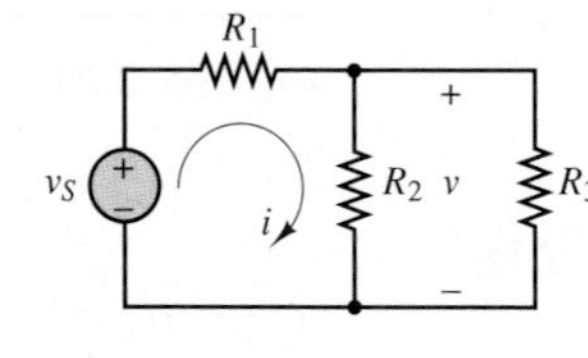

Figure 2.40

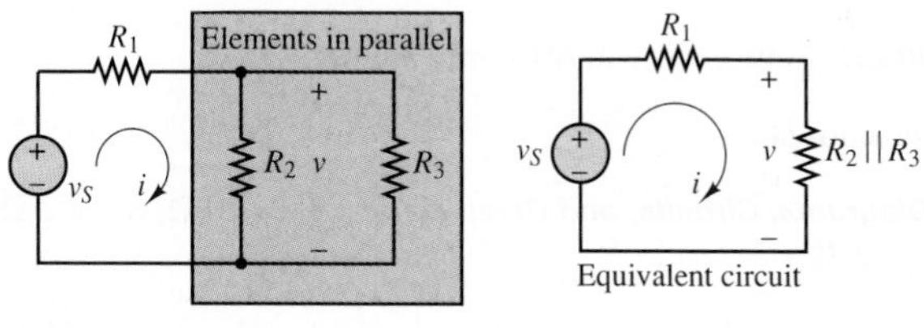

Figure 2.41

Analysis: The circuit of Figure 2.40 is neither a series nor a parallel circuit because the following two conditions do not apply:

1. The current through all resistors is the same (series circuit condition).
2. The voltage across all resistors is the same (parallel circuit condition).

The circuit takes a much simpler appearance once it becomes evident that the same voltage appears across both R_2 and R_3 and, therefore, that these elements are in parallel. If these two resistors are replaced by a single equivalent resistor according to the procedures described in this section, the circuit of Figure 2.41 is obtained. Note that now the equivalent circuit is a simple series circuit, and the voltage divider rule can be applied to determine that

$$v = \frac{R_2 \| R_3}{R_1 + R_2 \| R_3} v_S$$

while the current is found to be

$$i = \frac{v_S}{R_1 + R_2 \| R_3}$$

Comments: Systematic methods for analyzing arbitrary circuit configurations are explored in Chapter 3.

CHECK YOUR UNDERSTANDING

Consider the circuit of Figure 2.40, without resistor R_3. Calculate the value of the voltage v if the source voltage is $v_S = 5$ V and $R_1 = R_2 = 1$ kΩ.
Repeat when resistor R_3 is in the circuit and its value is $R_3 = 1$ kΩ.
Repeat when resistor R_3 is in the circuit and its value is $R_3 = 0.1$ kΩ.

Answers: $v = 2.50$ V; $v = 1.67$ V; $v = 0.4167$ V

EXAMPLE 2.14 The Wheatstone Bridge

Problem

The **Wheatstone bridge** is a resistive circuit that is frequently encountered in a variety of measurement circuits. The general form of the bridge circuit is shown in Figure 2.42(a), where R_1, R_2, and R_3 are known while R_x is an unknown resistance, to be determined. The circuit may also be redrawn as shown in Figure 2.42(b). The latter circuit is used to demonstrate the voltage divider rule in a mixed series-parallel circuit. The objective is to determine the unknown resistance R_x.

1. Find the value of the voltage $v_{ab} = v_{ad} - v_{bd}$ in terms of the four resistances and the source voltage v_S. Note that since the reference point d is the same for both voltages, we can also write $v_{ab} = v_a - v_b$.
2. If $R_1 = R_2 = R_3 = 1$ kΩ, $v_S = 12$ V, and $v_{ab} = 12$ mV, what is the value of R_x?

FIND IT ON THE WEB

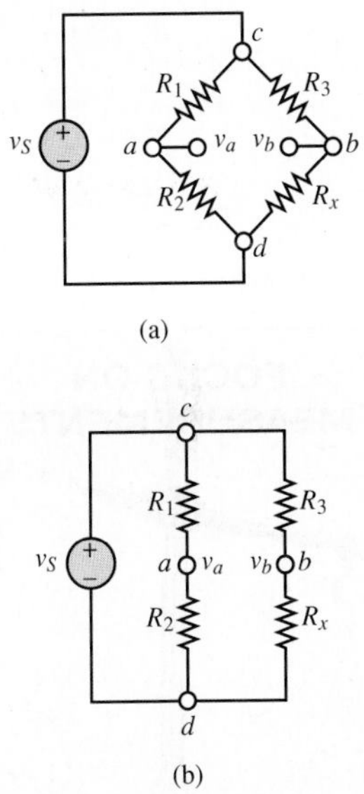

Figure 2.42 Wheatstone bridge circuits

Solution

Known Quantities: Source voltage, resistance values, bridge voltage.

Find: Unknown resistance R_x.

Schematics, Diagrams, Circuits, and Given Data: See Figure 2.42.
$R_1 = R_2 = R_3 = 1$ kΩ; $v_S = 12$ V; $v_{ab} = 12$ mV.

Analysis:

1. First we observe that the circuit consists of the parallel combination of three subcircuits: the voltage source, the series combination of R_1 and R_2, and the series combination of R_3 and R_x. Since these three subcircuits are in parallel, the same voltage will appear across each of them, namely, the source voltage v_S.

   Thus, the source voltage divides between each resistor pair $R_1 - R_2$ and $R_3 - R_x$ according to the voltage divider rule: v_{ad} is the fraction of the source voltage appearing across R_2, while v_{bd} is the voltage appearing across R_x:

$$v_{ad} = v_S \frac{R_2}{R_1 + R_2} \qquad \text{and} \qquad v_{bd} = v_S \frac{R_x}{R_3 + R_x}$$

   Finally, the voltage difference between points a and b is given by

$$v_{ab} = v_{ad} - v_{bd} = v_S \left(\frac{R_2}{R_1 + R_2} - \frac{R_x}{R_3 + R_x} \right)$$

   This result is very useful and quite general.

2. To solve for the unknown resistance, we substitute the numerical values in the preceding equation to obtain

$$0.012 = 12 \left(\frac{1{,}000}{2{,}000} - \frac{R_x}{1{,}000 + R_x} \right)$$

   which may be solved for R_x to yield

$$R_x = 996\ \Omega$$

Comments: The Wheatstone bridge finds application in many measurement circuits and instruments.

CHECK YOUR UNDERSTANDING

Use the results of part 1 of Example 2.14 to find the condition for which the voltage $v_{ab} = v_a - v_b$ is equal to zero (this is called the balanced condition for the bridge). Does this result necessarily require that all four resistors be identical? Why?

Answer: $R_1 R_x = R_2 R_3$

FOCUS ON MEASUREMENTS

Resistive Throttle Position Sensor

Problem:

The aim of this example is to determine the calibration of an **automotive resistive throttle position sensor,** shown in Figure 2.43(a). Figure 2.43(b) and (c) depicts the geometry of the throttle plate and the equivalent circuit of the throttle sensor. The throttle plate in a typical throttle body has a useful measurement range of just under 90°, from closed throttle to wide-open throttle. The possible mechanical range of rotation of the sensor is usually somewhat greater.

Figure 2.43 (a) Picture of throttle postion sensor. (*Courtesy: CTS Corporation.*)

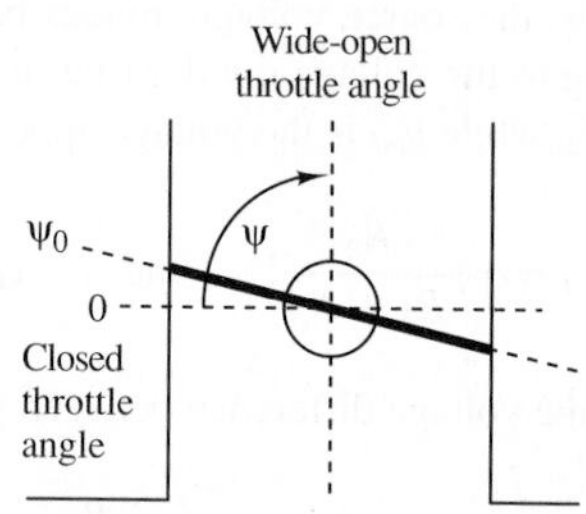

Figure 2.43 (b) Throttle blade geometry

Solution:

Known Quantities**—**Functional specifications of throttle position sensor.

(*Continued*)

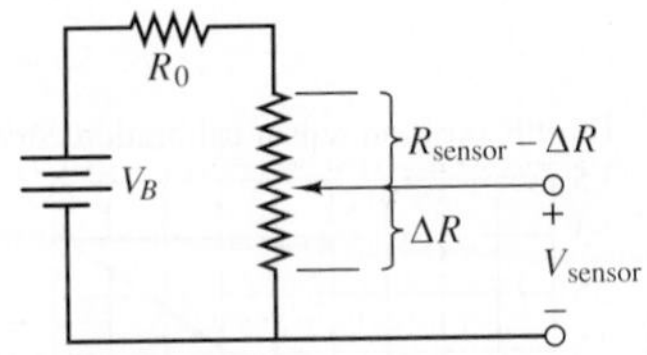

Figure 2.43 (c) Throttle position sensor equivalent circuit

Find— Calibration of sensor in volts per degree of throttle plate opening.

Schematics, Diagrams, Circuits, and Given Data—

Functional specifications of throttle position sensor

Total resistance $= R_{sensor} + R_0$	12 kΩ
R_0	3 kΩ
Input V_B	5 V ± 4% regulated
Output V_{sensor}	5% to 95% V_B
Current draw I_S	≤ 20 mA
Recommended load R_L	≤ 220 kΩ
Electrical travel,[1] maximum	112°

[1] Note that in actual operation the sensor will only be actuated between 2° and 90°.

Assumptions— Assume a nominal supply voltage of 5 V and total throttle plate travel of 88°, with a closed-throttle angle of 2° and a wide-open throttle angle of 90°.

Analysis— The equivalent circuit describing the resistive potentiometer that makes up the sensor is shown in Figure 2.43(c). The *wiper arm,* that is, the moving part of the potentiometer, defines a voltage proportional to position. The actual construction of the potentiometer is in the shape of a circle—the figure depicts the potentiometer resistor as a straight line for simplicity. The range of the potentiometer (see specifications above) is 2° to 112° for a resistance of 3 to 12 kΩ; thus, the *calibration constant* of the potentiometer is

$$k_{pot} = \frac{112 - 2}{12 - 3} = 12.22°/\text{k}\Omega, \text{ such that } \theta = k_{pot}\Delta R$$

The sensor voltage is proportional to the ratio $\Delta R / R_{sensor}$, such that

$$V_{sensor} = V_B \left(\frac{\Delta R}{R_0 + R_{sensor}}\right) = (5\text{ V})\left(\frac{\Delta R}{12}\right)$$

$$= 0.417\ \Delta R \quad \text{V} \quad (\Delta R \text{ in k}\Omega)$$

The calibration of the throttle position sensor is

$$V_{sensor} = 0.417\ \Delta R = 0.417\frac{\theta}{k_{pot}}$$

The *calibration curve* for the sensor is shown in Figure 2.43(d).

(Continued)

(Concluded)

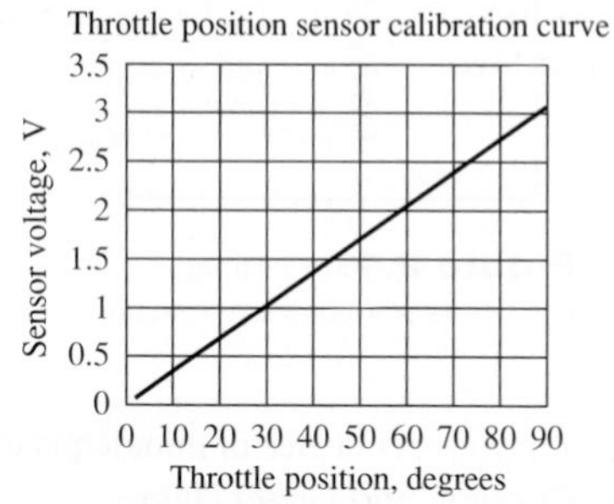

Figure 2.43 (d) Calibration curve for throttle position sensor

So if the throttle is closed, the sensor voltage will be

$$V_{\text{sensor}} = 0$$

and if the throttle is wide open

$$V_{\text{sensor}} = 0.417\ \Delta R = 0.417\frac{\theta}{k_{\text{pot}}} = 0.417\ \frac{\text{V}}{\text{k}\Omega}\ \frac{90^\circ}{12.22^\circ/\text{k}\Omega} = 3.07\ \text{V}$$

Comments—The fixed resistor R_0 prevents the wiper arm from shorting to ground. Note that the throttle position measurement does not use the entire range of rotation of the sensor.

Find It on the Web!—If you are interested in learning more about the throttle position sensors described in this example, and about **potentiometers** and **resistive position sensors**, their web links are included in the CD-ROM.

FIND IT ON THE WEB

Resistance Strain Gauges

FIND IT ON THE WEB

Another common application of the resistance concept to engineering measurements is the resistance **strain gauge.** Strain gauges are devices that are bonded to the surface of an object, and whose resistance varies as a function of the surface strain experienced by the object. Strain gauges may be used to perform measurements of strain, stress, force, torque, and pressure. Recall that the resistance of a cylindrical conductor of cross-sectional area A, length L, and conductivity σ is given by the expression

$$R = \frac{L}{\sigma A}$$

If the conductor is compressed or elongated as a consequence of an external force, its dimensions will change, and with them its resistance. In particular, if the conductor is stretched, its cross-sectional area decreases and the resistance increases. If the conductor

(Continued)

(Concluded)

is compressed, its resistance decreases, since the length L decreases. The relationship between change in resistance and change in length is given by the gauge factor GF, defined by

$$GF = \frac{\Delta R/R}{\Delta L/L}$$

And since the *strain* ϵ is defined as the fractional change in length of an object by the formula

$$\epsilon = \frac{\Delta L}{L}$$

the change in resistance due to an applied strain ϵ is given by

$$\Delta R = R_0 GF \epsilon$$

where R_0 is the resistance of the strain gauge under no strain and is called the *zero strain resistance*. The value of GF for resistance strain gauges made of metal foil is usually about 2.

Figure 2.44 depicts a typical foil strain gauge. The maximum strain that can be measured by a foil gauge is about 0.4 to 0.5 percent; that is, $\Delta L/L = 0.004 - 0.005$. For a 120-Ω gauge, this corresponds to a change in resistance on the order of 0.96 to 1.2 Ω. Although this change in resistance is very small, it can be detected by means of suitable circuitry. Resistance strain gauges are usually connected in a circuit called the Wheatstone bridge, which we analyze later in this chapter.

R_G
Circuit symbol for the strain gauge
The foil is formed by a photo-etching process and is less than 0.00002 in thick. Typical resistance values are 120, 350, and 1,000 Ω. The wide areas are bonding pads for electrical connections.

Figure 2.44 Metal-foil resistance strain gauge.

Comments—Resistance strain gauges find application in many measurement circuits and instruments. The measurement of force is one such application, shown next.

The Wheatstone Bridge and Force Measurements

Strain gauges are frequently employed in the measurement of force. One of the simplest applications of strain gauges is in the measurement of the force applied to a cantilever beam, as illustrated in Figure 2.45. Four strain gauges are employed in this case, of which two are bonded to the upper surface of the beam at a distance L from the point where the external force F is applied and two are bonded on the lower surface, also at a distance L. Under the influence of the external force, the beam deforms and causes the upper gauges to extend and the lower gauges to compress. Thus, the resistance of the upper gauges

(Continued)

(Concluded)

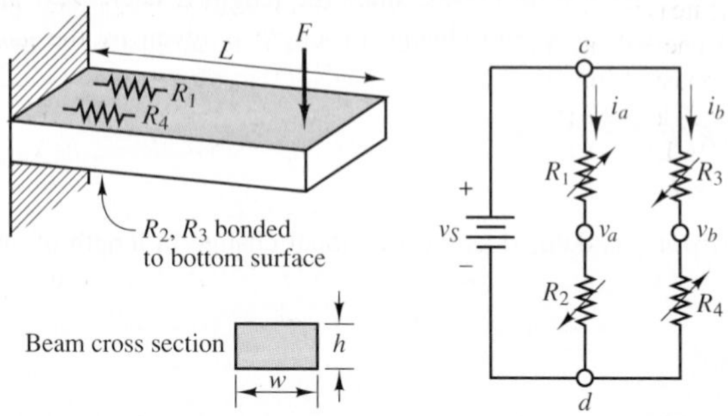

Figure 2.45 A force-measuring instrument

will increase by an amount ΔR, and that of the lower gauges will decrease by an equal amount, assuming that the gauges are symmetrically placed. Let R_1 and R_4 be the upper gauges and R_2 and R_3 the lower gauges. Thus, under the influence of the external force, we have

$$R_1 = R_4 = R_0 + \Delta R$$
$$R_2 = R_3 = R_0 - \Delta R$$

where R_0 is the zero strain resistance of the gauges. It can be shown from elementary statics that the relationship between the strain ϵ and a force F applied at a distance L for a cantilever beam is

$$\epsilon = \frac{6LF}{wh^2Y}$$

where h and w are as defined in Figure 2.45 and Y is the beam's modulus of elasticity.

In the circuit of Figure 2.45, the currents i_a and i_b are given by

$$i_a = \frac{v_S}{R_1 + R_2} \qquad \text{and} \qquad i_b = \frac{v_S}{R_3 + R_4}$$

The bridge output voltage is defined by $v_o = v_b - v_a$ and may be found from the following expression:

$$\begin{aligned} v_o = i_b R_4 - i_a R_2 &= \frac{v_S R_4}{R_3 + R_4} - \frac{v_S R_2}{R_1 + R_2} \\ &= v_S \frac{R_0 + \Delta R}{R_0 + \Delta R + R_0 - \Delta R} - v_S \frac{R_0 - \Delta R}{R_0 + \Delta R + R_0 - \Delta R} \\ &= v_S \frac{\Delta R}{R_0} = v_S GF \epsilon \end{aligned}$$

where the expression for $\Delta R / R_0$ was obtained in "Focus on Measurements: Resistance Strain Gauges." Thus, it is possible to obtain a relationship between the output voltage of the bridge circuit and the force F as follows:

$$v_o = v_S GF \epsilon = v_S GF \frac{6LF}{wh^2Y} = \frac{6 v_S GFL}{wh^2Y} F = kF$$

where k is the calibration constant for this force transducer.

***Comments*—Strain gauge bridges** are commonly used in mechanical, chemical, aerospace, biomedical, and civil engineering applications (and wherever measurements of force, pressure, torque, stress, or strain are sought).

FIND IT ON THE WEB

CHECK YOUR UNDERSTANDING

Compute the full-scale (i.e., largest) output voltage for the force-measuring apparatus of "Focus on Measurements: The Wheatstone Bridge and Force Measurements." Assume that the strain gauge bridge is to measure forces ranging from 0 to 500 newtons (N), $L = 0.3$ m, $w = 0.05$ m, $h = 0.01$ m, $GF = 2$, and the modulus of elasticity for the beam is 69×10^9 N/m^2 (aluminum). The source voltage is 12 V. What is the calibration constant of this force transducer?

Answer: v_o (full scale) = 62.6 mV; $k = 0.125$ mV/N

2.7 PRACTICAL VOLTAGE AND CURRENT SOURCES

The idealized models of voltage and current sources we discussed in Section 2.1 fail to consider the internal resistance of practical voltage and current sources. The objective of this section is to extend the ideal models to models that are capable of describing the physical limitations of the voltage and current sources used in practice. Consider, for example, the model of an ideal voltage source shown in Figure 2.1. As the load resistance R decreases, the source is required to provide increasing amounts of current to maintain the voltage $v_S(t)$ across its terminals:

$$i(t) = \frac{v_S(t)}{R} \tag{2.24}$$

This circuit suggests that the ideal voltage source is required to provide an infinite amount of current to the load, in the limit as the load resistance approaches zero. Naturally, you can see that this is impossible; for example, think about the ratings of a conventional car battery: 12 V, 450 ampere-hours (A-h). This implies that there is a limit (albeit a large one) to the amount of current a practical source can deliver to a load. Fortunately, it is not necessary to delve too deeply into the physical nature of each type of source to describe the behavior of a practical voltage source: The limitations of practical sources can be approximated quite simply by exploiting the notion of the internal resistance of a source. Although the models described in this section are only approximations of the actual behavior of energy sources, they will provide good insight into the limitations of practical voltage and current sources. Figure 2.46 depicts a model for a practical voltage source, composed of an ideal voltage source v_S in series with a resistance r_S. The resistance r_S in effect poses a limit to the maximum current the voltage source can provide:

$$i_{S\max} = \frac{v_S}{r_S} \tag{2.25}$$

Typically, r_S is small. Note, however, that its presence affects the voltage across the load resistance: Now this voltage is no longer equal to the source voltage. Since the current provided by the source is

$$i_S = \frac{v_S}{r_S + R_L} \tag{2.26}$$

the load voltage can be determined to be

$$v_L = i_S R_L = v_S \frac{R_L}{r_S + R_L} \tag{2.27}$$

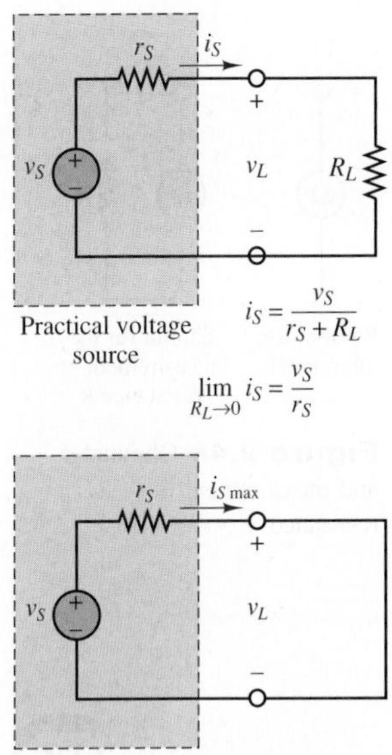

Practical voltage source

$$i_S = \frac{v_S}{r_S + R_L}$$

$$\lim_{R_L \to 0} i_S = \frac{v_S}{r_S}$$

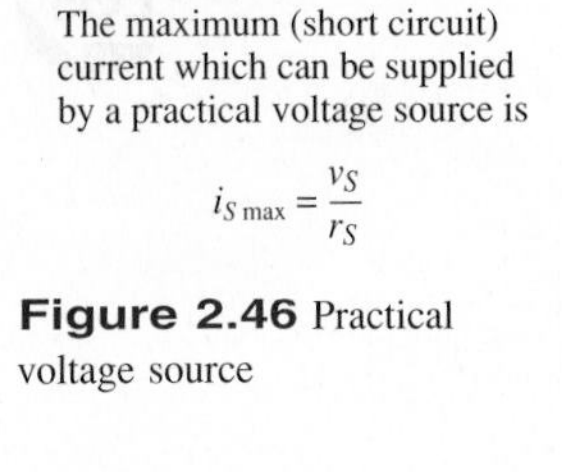

The maximum (short circuit) current which can be supplied by a practical voltage source is

$$i_{S\max} = \frac{v_S}{r_S}$$

Figure 2.46 Practical voltage source

Thus, in the limit as the source internal resistance r_S approaches zero, the load voltage v_L becomes exactly equal to the source voltage. It should be apparent that a desirable feature of an ideal voltage source is a very small internal resistance, so that the current requirements of an arbitrary load may be satisfied. Often, the effective internal resistance of a voltage source is quoted in the technical specifications for the source, so that the user may take this parameter into account.

A similar modification of the ideal current source model is useful to describe the behavior of a practical current source. The circuit illustrated in Figure 2.47 depicts a simple representation of a practical current source, consisting of an ideal source in parallel with a resistor. Note that as the load resistance approaches infinity (i.e., an open circuit), the output voltage of the current source approaches its limit

$$v_{S\max} = i_S r_S \tag{2.28}$$

A good current source should be able to approximate the behavior of an ideal current source. Therefore, a desirable characteristic for the internal resistance of a current source is that it be as large as possible.

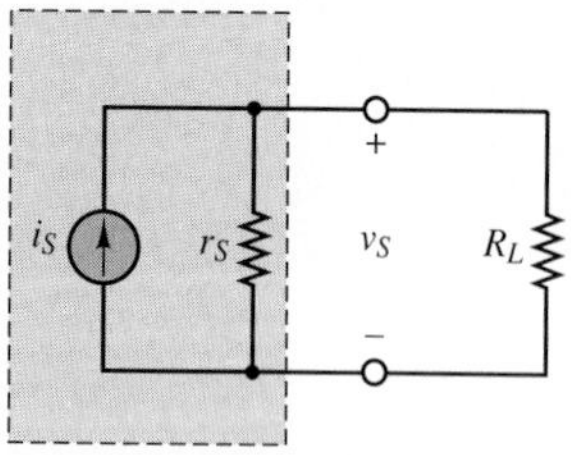

A model for practical current sources consists of an ideal source in *parallel* with an internal resistance.

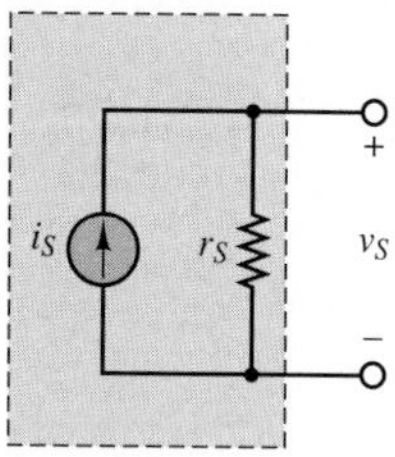

Maximum output voltage for practical current source with open-circuit load:

$v_{S\max} = i_S r_S$

Figure 2.47 Practical current source

2.8 MEASURING DEVICES

In this section, you should gain a basic understanding of the desirable properties of practical devices for the measurement of electrical parameters. The measurements most often of interest are those of current, voltage, power, and resistance. In analogy with the models we have just developed to describe the nonideal behavior of voltage and current sources, we similarly present circuit models for practical measuring instruments suitable for describing the nonideal properties of these devices.

The Ohmmeter

The **ohmmeter** is a device that when connected across a circuit element, can measure the resistance of the element. Figure 2.48 depicts the circuit connection of an ohmmeter to a resistor. One important rule needs to be remembered:

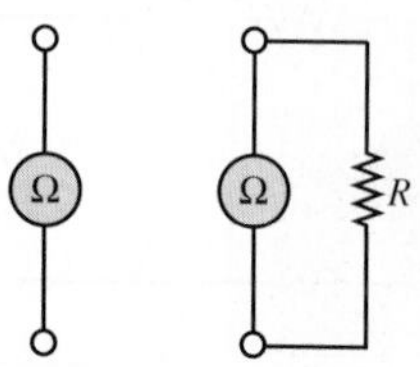

Symbol for ohmmeter

Circuit for the measurement of resistance R

Figure 2.48 Ohmmeter and measurement of resistance

> The resistance of an element can be measured only when the element is disconnected from any other circuit.

LO5

The Ammeter

The **ammeter** is a device that when connected in series with a circuit element, can measure the current flowing through the element. Figure 2.49 illustrates this idea. From Figure 2.49, two requirements are evident for obtaining a correct measurement of current:

LO5

1. The ammeter must be placed in series with the element whose current is to be measured (e.g., resistor R_2).
2. The ammeter should not restrict the flow of current (i.e., cause a voltage drop), or else it will not be measuring the true current flowing in the circuit. *An ideal ammeter has zero internal resistance.*

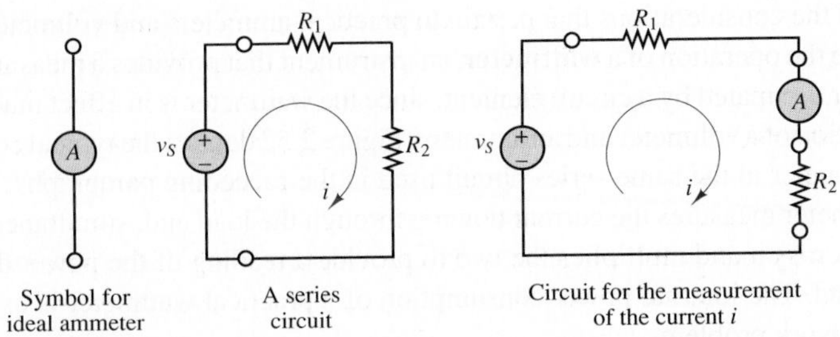

Figure 2.49 Measurement of current

The Voltmeter

The **voltmeter** is a device that can measure the voltage across a circuit element. Since voltage is the difference in potential between two points in a circuit, the voltmeter needs to be connected across the element whose voltage we wish to measure. A voltmeter must also fulfill two requirements:

LO5

1. The voltmeter must be placed in parallel with the element whose voltage it is measuring.
2. The voltmeter should draw no current away from the element whose voltage it is measuring, or else it will not be measuring the true voltage across that element. Thus, *an ideal voltmeter has infinite internal resistance.*

Figure 2.50 illustrates these two points.

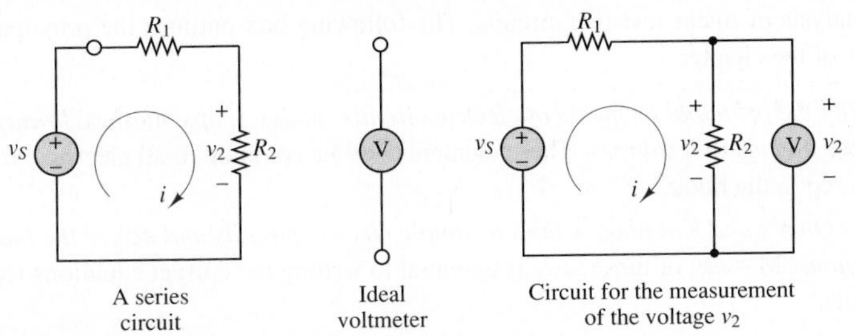

Figure 2.50 Measurement of voltage

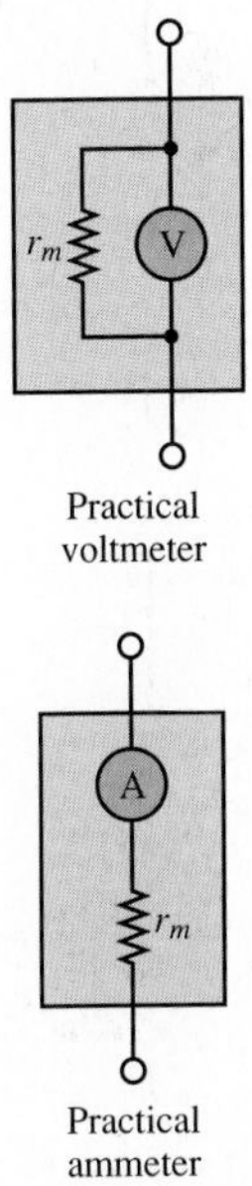

Figure 2.51 Models for practical ammeter and voltmeter

Once again, the definitions just stated for the ideal voltmeter and ammeter need to be augmented by considering the practical limitations of the devices. A practical ammeter will contribute some series resistance to the circuit in which it is measuring current; a practical voltmeter will not act as an ideal open circuit but will always draw some current from the measured circuit. The homework problems verify that these practical restrictions do not necessarily pose a limit to the accuracy of the measurements obtainable with practical measuring devices, as long as the internal resistance of the measuring devices is known. Figure 2.51 depicts the circuit models for the practical ammeter and voltmeter.

All the considerations that pertain to practical ammeters and voltmeters can be applied to the operation of a **wattmeter,** an instrument that provides a measurement of the power dissipated by a circuit element, since the wattmeter is in effect made up of a combination of a voltmeter and an ammeter. Figure 2.52 depicts the typical connection of a wattmeter in the same series circuit used in the preceding paragraphs. In effect, the wattmeter measures the current flowing through the load and, simultaneously, the voltage across it and multiplies the two to provide a reading of the power dissipated by the load. The internal power consumption of a practical wattmeter is explored in the homework problems.

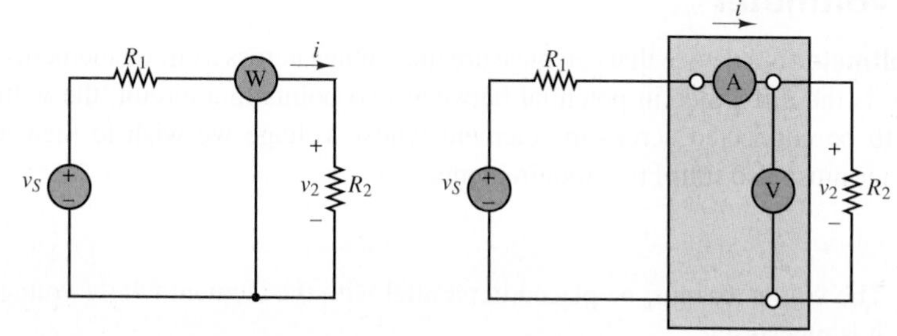

Measurement of the power dissipated in the resistor R_2: $P_2 = v_2 i$

Internal wattmeter connections

Figure 2.52 Measurement of power

Conclusion

The objective of this chapter was to introduce the background needed in the following chapters for the analysis of linear resistive circuits. The following box outlines the principal learning objectives of the chapter.

1. *Identify the principal elements of electric circuits: nodes, loops, meshes, branches, and voltage and current sources.* These elements will be common to all electric circuits analyzed in the book.
2. *Apply Ohm's and Kirchhoff's laws to simple electric circuits and derive the basic circuit equations.* Mastery of these laws is essential to writing the correct equations for electric circuits.
3. *Apply the passive sign convention and compute the power dissipated by circuit elements.* The passive sign convention is a fundamental skill needed to derive the correct equations for an electric circuit.
4. *Apply the voltage and current divider laws to calculate unknown variables in simple series, parallel, and series-parallel circuits.* The chapter includes examples of practical circuits to demonstrate the application of these principles.
5. *Understand the rules for connecting electric measuring instruments to electric circuits for the measurement of voltage, current, and power.* Practical engineering measurement systems are introduced in these sections.

HOMEWORK PROBLEMS

Section 2.1: Definitions

2.1 An isolated free electron is traveling through an electric field from some initial point where its coulombic potential energy per unit charge (*voltage*) is 17 kJ/C and velocity = 93 Mm/s to some final point where its coulombic potential energy per unit charge is 6 kJ/C. Determine the change in velocity of the electron. Neglect gravitational forces.

2.2 The unit used for voltage is the volt, for current the ampere, and for resistance the ohm. Using the definitions of voltage, current, and resistance, express each quantity in fundamental MKS units.

2.3 The capacity of a car battery is usually specified in ampere-hours. A battery rated at, say, 100 A-h should be able to supply 100 A for 1 h, 50 A for 2 h, 25 A for 4 h, 1 A for 100 h, or any other combination yielding a product of 100 A-h.

a. How many coulombs of charge should we be able to draw from a fully charged 100 A-h battery?

b. How many electrons does your answer to part a require?

2.4 The charge cycle shown in Figure P2.4 is an example of a *two-rate charge*. The current is held constant at 50 mA for 5 h. Then it is switched to 20 mA for the next 5 h. Find

a. The total charge transferred to the battery.

b. The energy transferred to the battery.

Hint: Recall that energy w is the integral of power, or $P = dw/dt$.

2.5 Batteries (e.g., lead-acid batteries) store chemical energy and convert it to electric energy on demand. Batteries do not store electric charge or charge carriers. Charge carriers (electrons) enter one terminal of the battery, acquire electrical potential energy, and exit from the other terminal at a lower voltage. Remember the electron has a negative charge! It is convenient to think of positive carriers flowing in the opposite direction, that is, conventional current, and exiting at a higher voltage. All currents in this course, unless otherwise stated, are conventional current. (Benjamin Franklin caused this mess!) For a battery with a rated voltage = 12 V and a rated capacity = 350 A-h, determine

a. The rated chemical energy stored in the battery.

b. The total charge that can be supplied at the rated voltage.

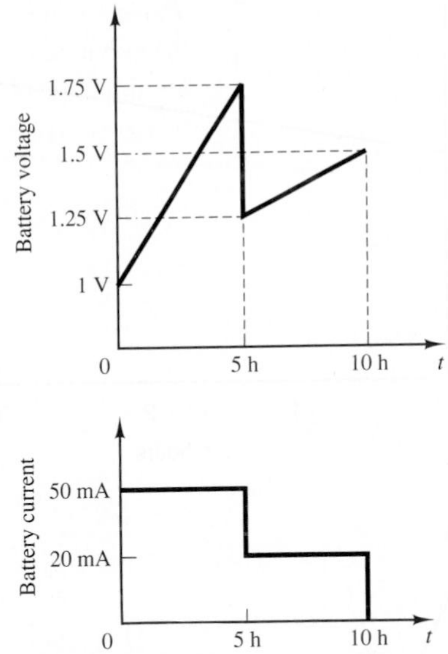

Figure P2.4

2.6 What determines the following?

a. How much current is supplied (at a constant voltage) by an ideal voltage source.

b. How much voltage is supplied (at a constant current) by an ideal current source.

2.7 An automotive battery is rated at 120 A-h. This means that under certain test conditions it can output 1 A at 12 V for 120 h (under other test conditions, the battery may have other ratings).

a. How much total energy is stored in the battery?

b. If the headlights are left on overnight (8 h), how much energy will still be stored in the battery in the morning? (Assume a 150-W total power rating for both headlights together.)

2.8 A car battery kept in storage in the basement needs recharging. If the voltage and the current provided by the charger during a charge cycle are shown in Figure P2.8,

a. Find the total charge transferred to the battery.

b. Find the total energy transferred to the battery.

2.9 Suppose the current flowing through a wire is given by the curve shown in Figure P2.9.

a. Find the amount of charge, q, that flows through the wire between $t_1 = 0$ and $t_2 = 1$ s.

b. Repeat part a for $t_2 = 2, 3, 4, 5, 6, 7, 8, 9$, and 10 s.

c. Sketch $q(t)$ for $0 \le t \le 10$ s.

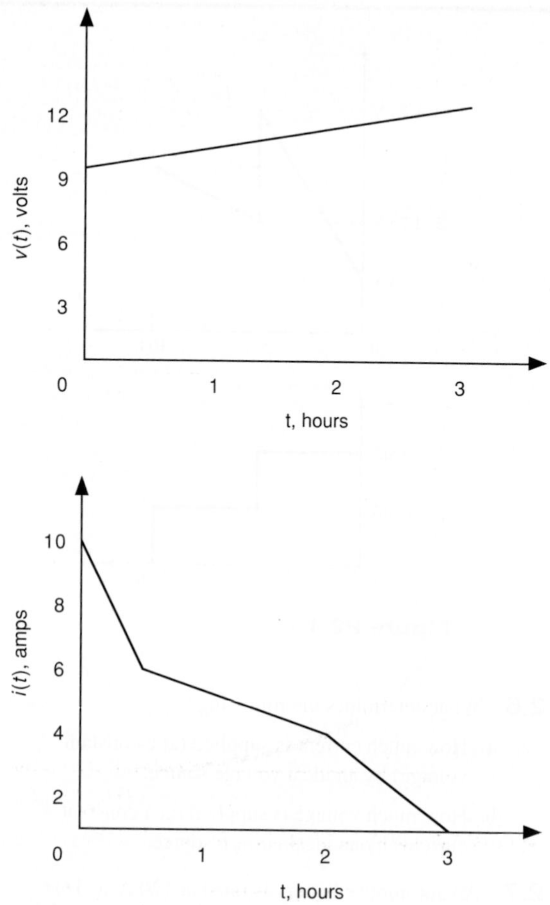

Figure P2.8

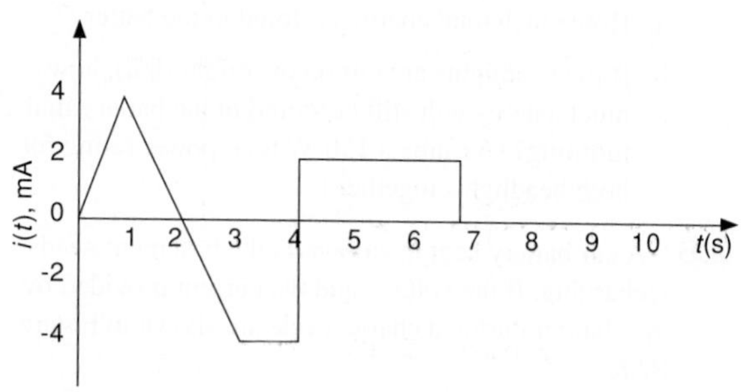

Figure P2.9

2.10 The charging scheme used in Figure P2.10 is an example of a constant-voltage charge with current limit. The charger voltage is such that the current into the battery does not exceed 100 mA, as shown in Figure P2.10. The charger's voltage increases to the maximum of 9 V, as shown in Figure P2.10. The battery is charged for 6 h. Find:

a. The total charge delivered to the battery.

b. The energy transferred to the battery during the charging cycle.

Hint: Recall that the energy, w, is the integral of power, or $P = dw/dt$.

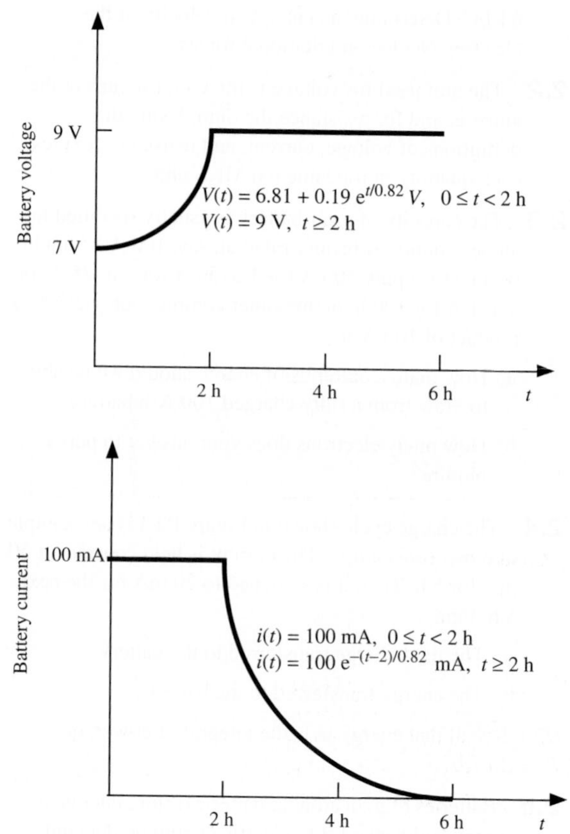

Figure P2.10

2.11 The charging scheme used in Figure P2.11 is an example of a constant-current charge cycle. The charger voltage is controlled such that the current into the battery is held constant at 40 mA, as shown in Figure P2.11. The battery is charged for 6 h. Find:

a. The total charge delivered to the battery.

b. The energy transferred to the battery during the charging cycle.

Hint: Recall that the energy, w, is the integral of power, or $P = dw/dt$.

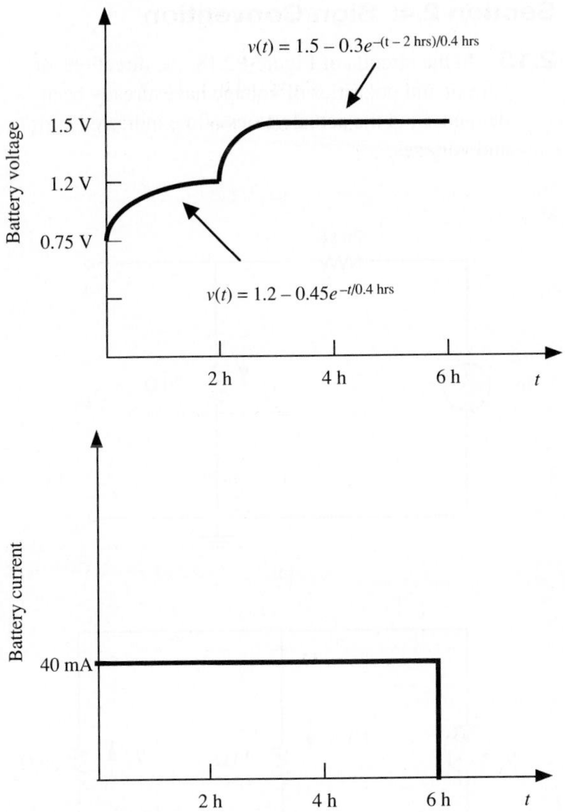

Figure P2.11

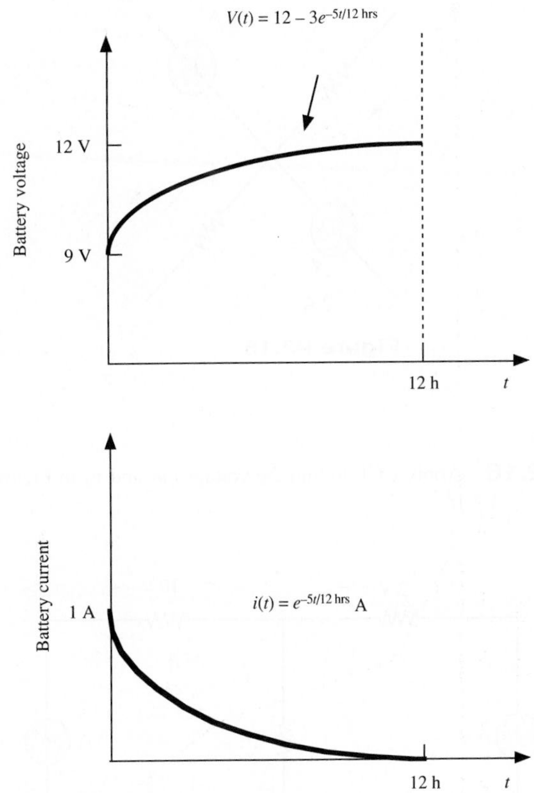

Figure P2.12

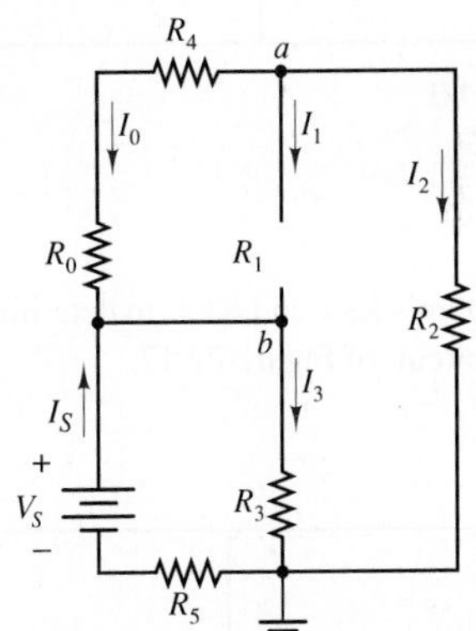

Figure P2.13

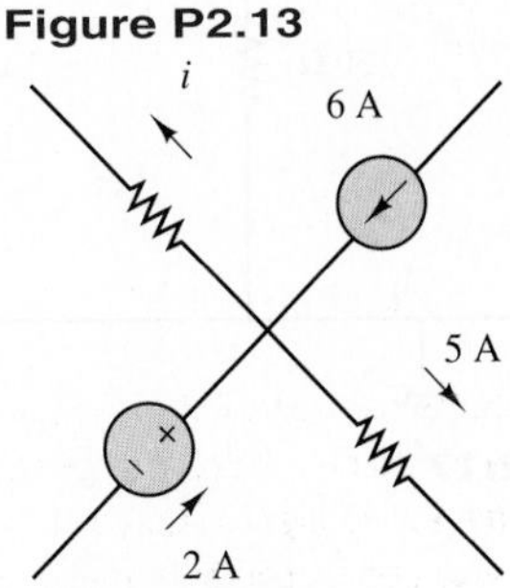

Figure P2.14

2.12 The charging scheme used in Figure P2.12 is called a *tapered-current charge cycle*. The current starts at the highest level and then decreases with time for the entire charge cycle, as shown. The battery is charged for 12 h. Find:

a. The total charge delivered to the battery.

b. The energy transferred to the battery during the charging cycle.

Hint: Recall that the energy, w, is the integral of power, or $P = dw/dt$.

Sections 2.2, 2.3: KCL, KVL

2.13 Use Kirchhoff's current law to determine the unknown currents in the circuit of Figure P2.13. Assume that $I_0 = -2$ A, $I_1 = -4$ A, $I_S = 8$ A, and $V_S = 12$ V.

2.14 Apply KCL to find the current i in the circuit of Figure P2.14.

2.15 Apply KCL to find the current I in the circuit of Figure P2.15.

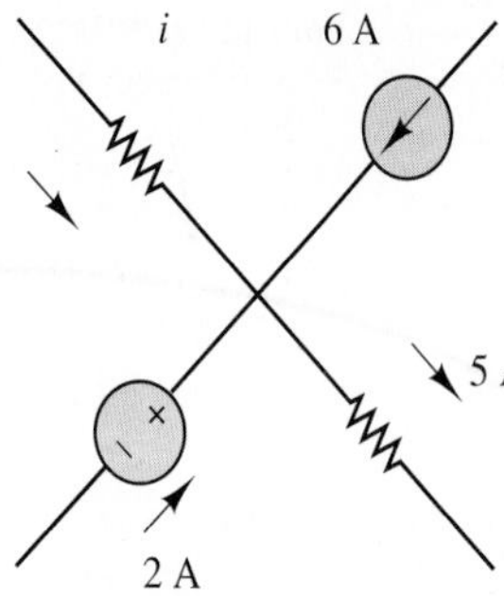

Figure P2.15

2.16 Apply KCL to find the voltages v_1 and v_2 in Figure P2.16.

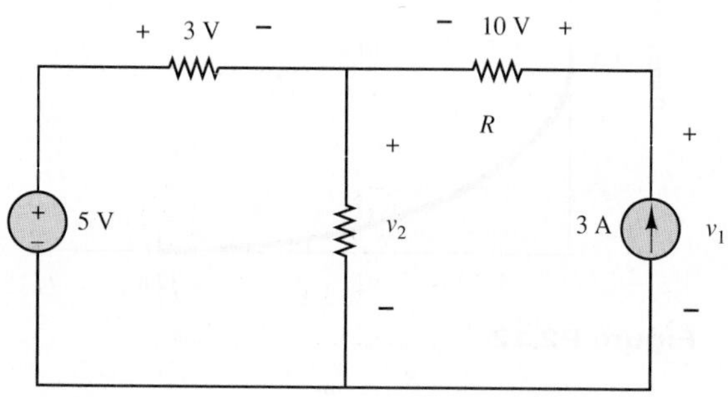

Figure P2.16

2.17 Use Ohm's Law and KCL to determine the current I_1 in the circuit of Figure P2.17.

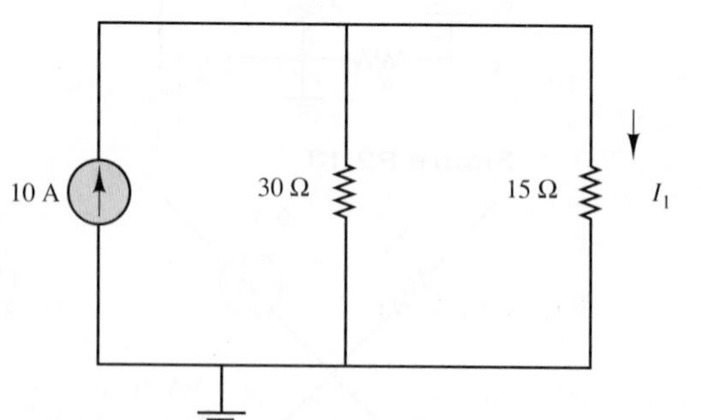

Figure P2.17

Section 2.4: Sign Convention

2.18 In the circuits of Figure P2.18, the directions of current and polarities of voltage have already been defined. Find the actual values of the indicated currents and voltages.

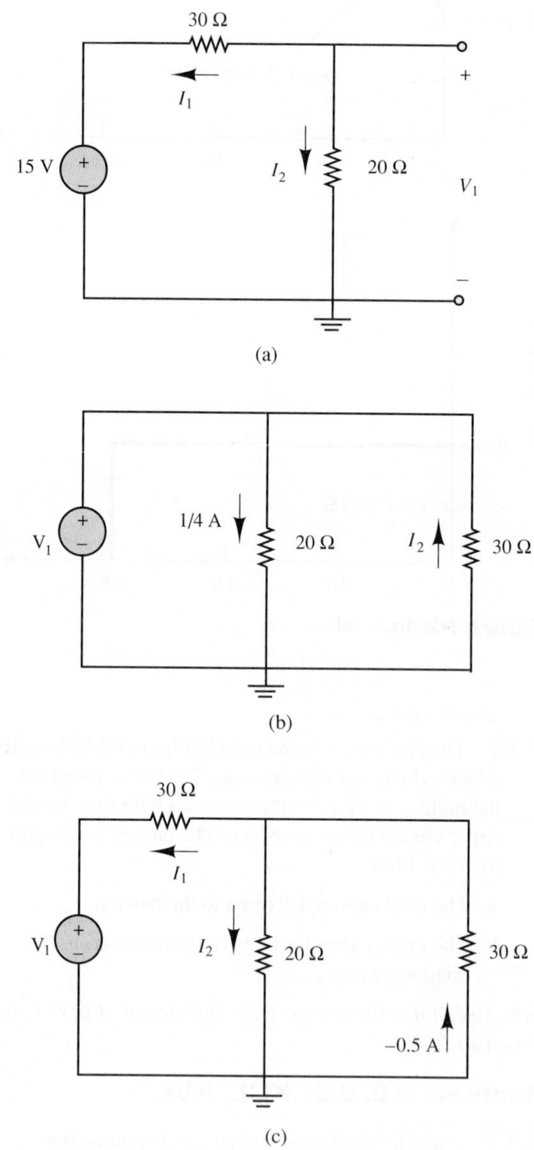

Figure P2.18

2.19 Find the power delivered by each source in the circuits of Figure P2.19.

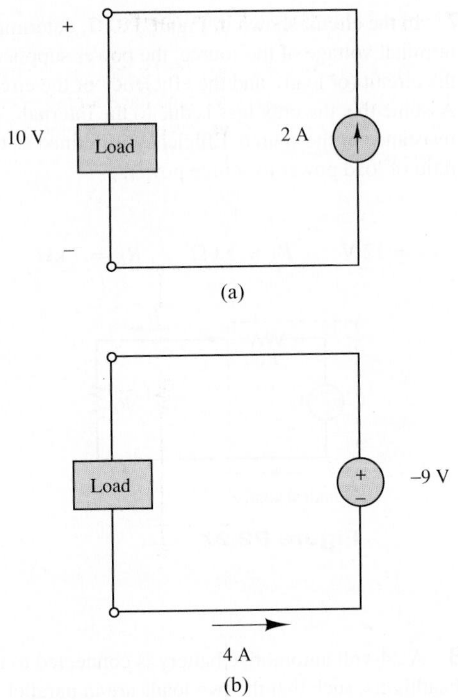

Figure P2.19

2.20 Determine which elements in the circuit of Figure P2.20 are supplying power and which are dissipating power. Also determine the amount of power dissipated and supplied.

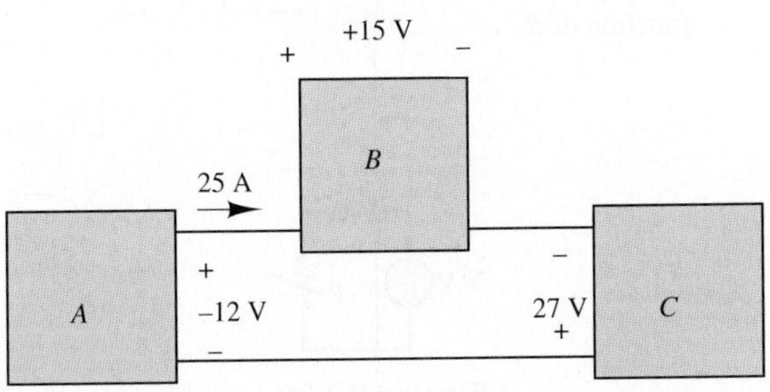

Figure P2.20

2.21 In the circuit of Figure P2.21, determine the power absorbed by the resistor R and the power delivered by the current source.

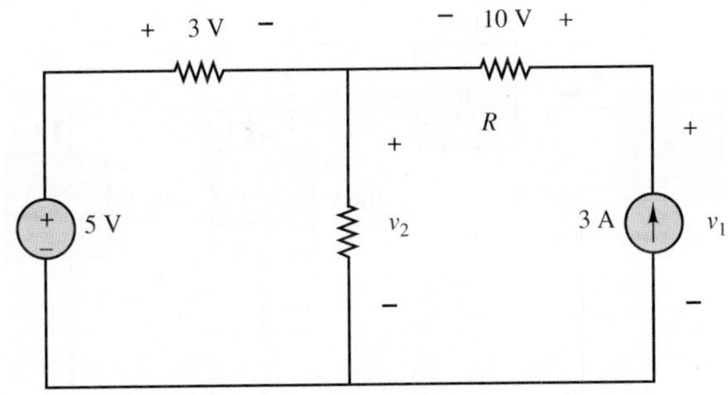

Figure P2.21

2.22 For the circuit shown in Figure P2.22:

a. Determine which components are absorbing power and which are delivering power.

b. Is conservation of power satisfied? Explain your answer.

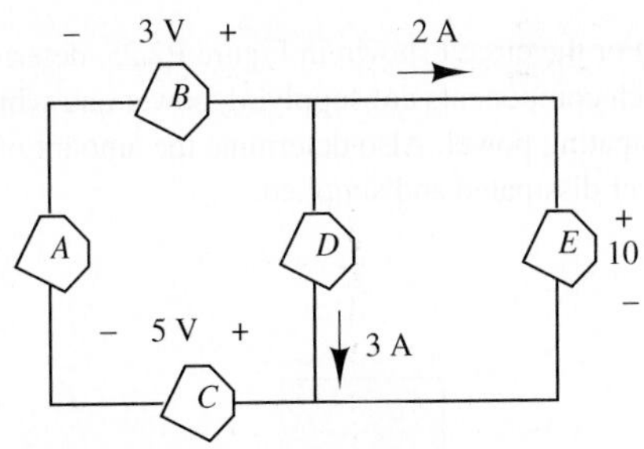

Figure P2.22

2.23 For the circuit shown in Figure P2.23, determine the power absorbed by the 5 Ω resistor.

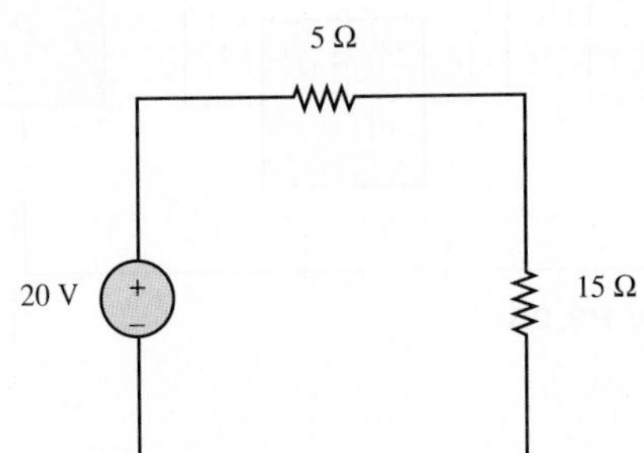

Figure P2.23

2.24 For the circuit shown in Figure P2.24, determine which components are supplying power and which are dissipating power. Also determine the amount of power dissipated and supplied.

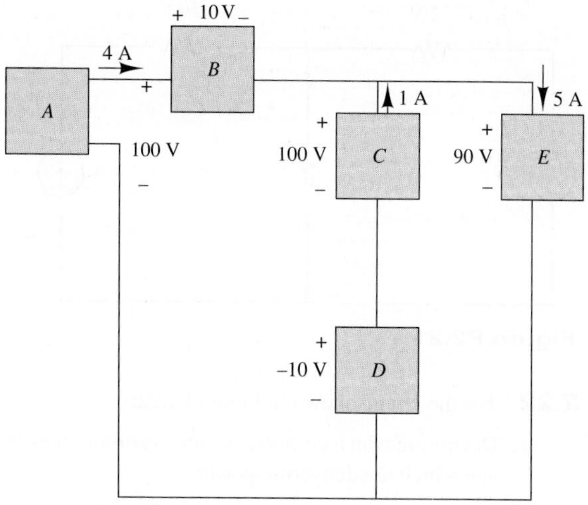

Figure P2.24

2.25 For the circuit shown in Figure P2.25. determine which components are supplying power and which are dissipating power. Also determine the amount of power dissipated and supplied.

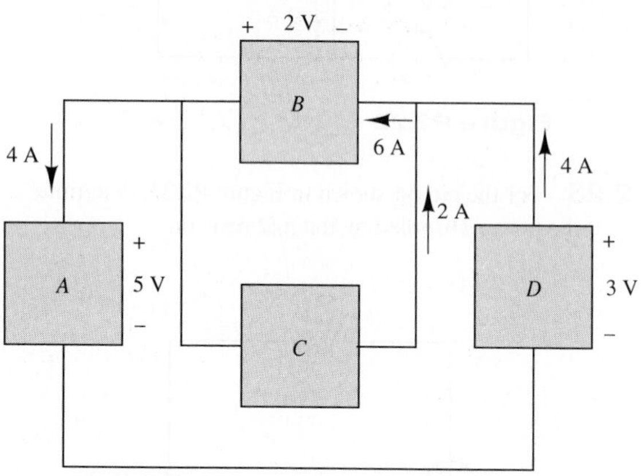

Figure P2.25

2.26 If an electric heater requires 23 A at 110 V, determine

a. The power it dissipates as heat or other losses.

b. The energy dissipated by the heater in a 24-h period.

c. The cost of the energy if the power company charges at the rate 6 cents/kWh.

2.27 In the circuit shown in Figure P2.27, determine the terminal voltage of the source, the power supplied to the circuit (or load), and the efficiency of the circuit. Assume that the only loss is due to the internal resistance of the source. Efficiency is defined as the ratio of load power to source power.

$V_S = 12$ V $\quad R_S = 5$ kΩ $\quad R_L = 7$ kΩ

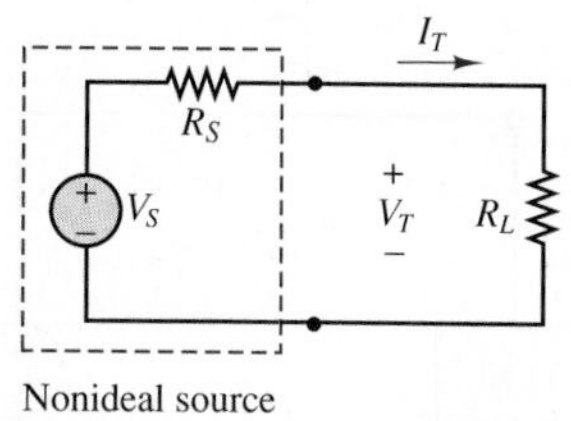

Figure P2.27

2.28 A 24-volt automotive battery is connected to two headlights, such that the two loads are in parallel; each of the headlights is intended to be a 75-W load, however, a 100-W headlight is mistakenly installed. What is the resistance of each headlight, and what is the total resistance seen by the battery?

2.29 What is the equivalent resistance seen by the battery of Problem 2.28 if two 15-W taillights are added (in parallel) to the two 75-W (each) headlights?

2.30 For the circuit shown in Figure P2.30, determine the power absorbed by the variable resistor R, ranging from 0 to 20 Ω. Plot the power absorption as a function of R.

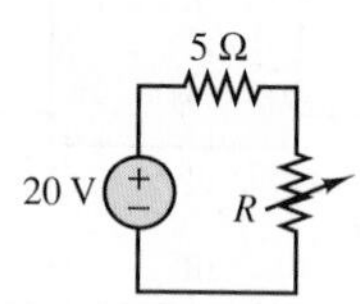

Figure P2.30

2.31 Refer to Figure P2.31.

a. Find the total power supplied by the ideal source.

b. Find the power dissipated and lost within the nonideal source.

c. What is the power supplied by the source to the circuit as modeled by the load resistance?

d. Plot the terminal voltage and power supplied to the circuit as a function of current.

Repeat $I_T = 0, 5, 10, 20, 30$ A.

$V_S = 12$ V $\quad R_S = 0.3\ \Omega$

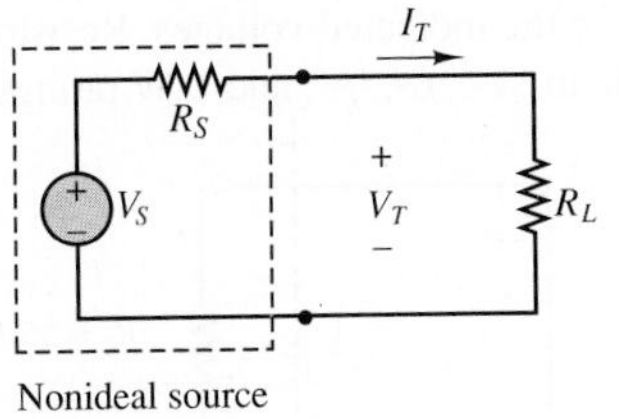

Figure P2.31

2.32 In the circuit of Figure P2.32, if $v_1 = v/4$ and the power delivered by the source is 40 mW, find R, v, v_1, and i. Given: $R_1 = 8\ \text{k}\Omega$, $R_2 = 10\ \text{k}\Omega$, $R_3 = 12\ \text{k}\Omega$.

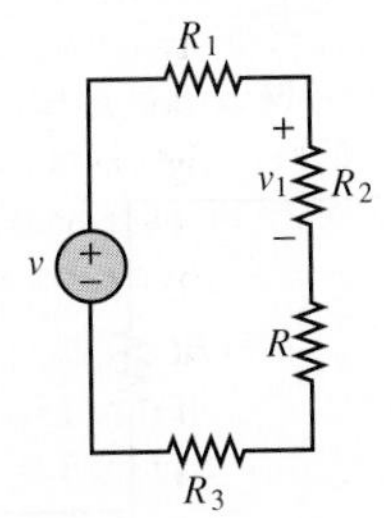

Figure P2.32

2.33 A GE SoftWhite Longlife lightbulb is rated as follows:

$P_R =$ rated power $= 60$ W

$P_{OR} =$ rated optical power $= 820$ lumens (lm) (average)

1 lumen $= \frac{1}{680}$ W

Operating life $= 1{,}500$ h (average)

$V_R =$ rated operating voltage $= 115$ V

The resistance of the filament of the bulb, measured with a standard multimeter, is 16.7 Ω. When the bulb is connected into a circuit and is operating at the rated values given above, determine

a. The resistance of the filament.

b. The efficiency of the bulb.

2.34 An incandescent lightbulb rated at 100 W will dissipate 100 W as heat and light when connected across a 110-V ideal voltage source. If three of these bulbs are connected in series across the same source, determine the power each bulb will dissipate.

2.35 An incandescent lightbulb rated at 60 W will dissipate 60 W as heat and light when connected across a 100-V ideal voltage source. A 100-W bulb will dissipate 100 W when connected across the same source. If the bulbs are connected in series across the same source, determine the power that either one of the two bulbs will dissipate.

2.36 For the circuit shown in Figure P2.36, find

a. The equivalent resistance seen by the source.

b. The current i.

c. The power delivered by the source.

d. The voltages v_1 and v_2.

e. The minimum power rating required for R_1.

Given: $v = 24$ V, $R_0 = 8\ \Omega$, $R_1 = 10\ \Omega$, $R_2 = 2\ \Omega$.

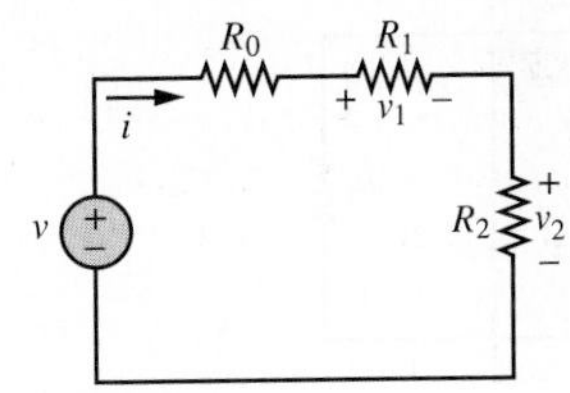

Figure P2.36

2.37 For the circuit shown in Figure P2.37, find

a. The currents i_1 and i_2.

b. The power delivered by the 3-A current source and by the 12-V voltage source.

c. The total power dissipated by the circuit.

Let $R_1 = 25\ \Omega$, $R_2 = 10\ \Omega$, $R_3 = 5\ \Omega$, $R_4 = 7\ \Omega$, and express i_1 and i_2 as functions of v. (*Hint:* Apply KCL at the node between R_1 and R_3.)

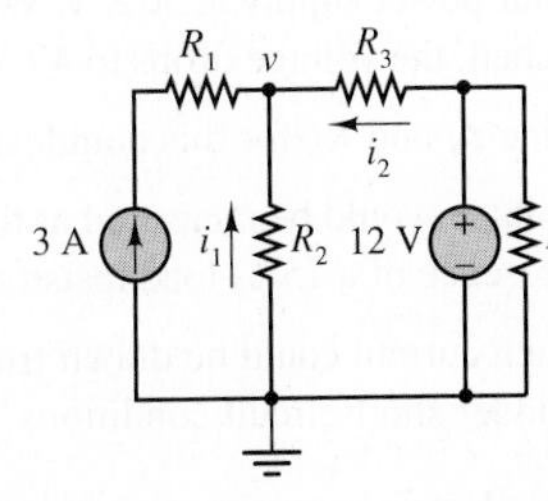

Figure P2.37

2.38 Determine the power delivered by the dependent source in the circuit of Figure P2.38.

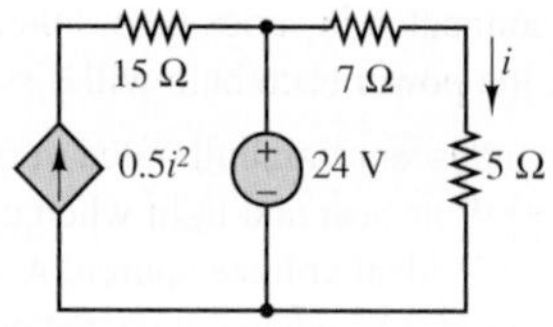

Figure P2.38

2.39 Consider NiMH hobbyist batteries shown in the circuit of Figure P2.39.

a. If $V_1 = 12.0$ V, $R_1 = 0.15\ \Omega$ and $R_L = 2.55\ \Omega$, find the load current I_L and the power dissipated by the load.

b. If we connect a second battery in parallel with battery 1 that has voltage $V_2 = 12$ V and $R_2 = 0.28\ \Omega$, will the load current I_L increase or decrease? Will the power dissipated by the load increase or decrease? By how much?

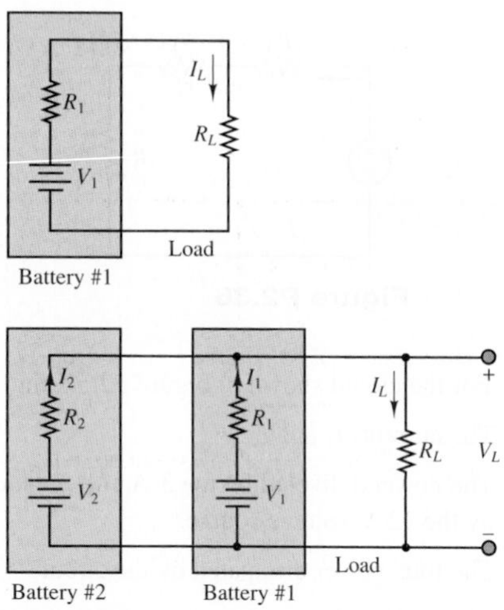

Figure P2.39

2.40 With no load attached, the voltage at the terminals of a particular power supply is 50.8 V. When a 10-W load is attached, the voltage drops to 49 V.

a. Determine v_S and R_S for this nonideal source.

b. What voltage would be measured at the terminals in the presence of a 15-Ω load resistor?

c. How much current could be drawn from this power supply under short-circuit conditions?

2.41 A 220-V electric heater has two heating coils which can be switched such that either coil can be used independently or the two can be connected in series or parallel, yielding a total of four possible configurations. If the warmest setting corresponds to 2,000-W power dissipation and the coolest corresponds to 300 W, find

a. The resistance of each of the two coils.

b. The power dissipation for each of the other two possible arrangements.

Sections 2.5, 2.6: Resistance and Ohm's Law

2.42 For the circuits of Figure P2.42, determine the resistor values (including the power rating) necessary to achieve the indicated voltages. Resistors are available in $^1/_8$-, $^1/_4$-, $^1/_2$-, and 1-W ratings.

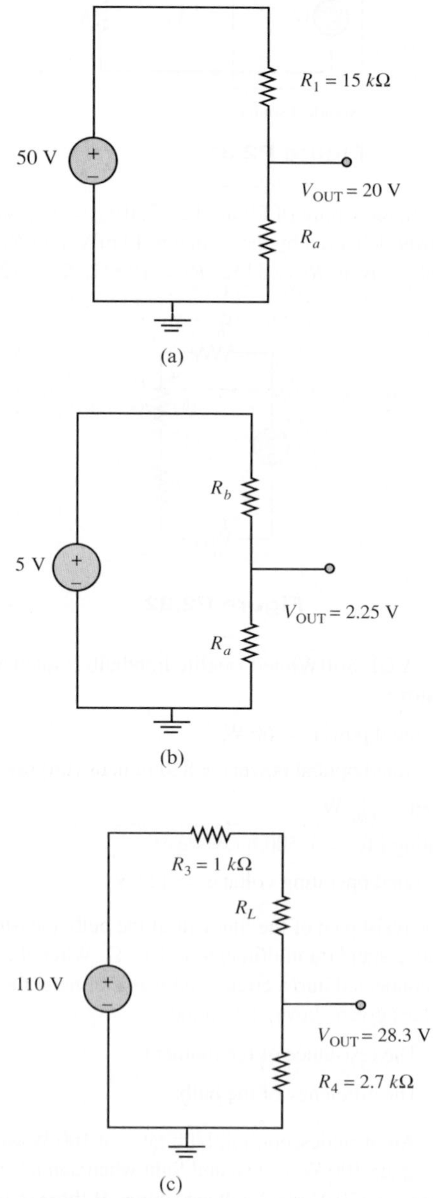

Figure P2.42

2.43 For the circuit shown in Figure P2.43, find

a. The equivalent resistance seen by the source.

b. The current i.

c. The power delivered by the source.

d. The voltages v_1, v_2.

e. The minimum power rating required for R_1.

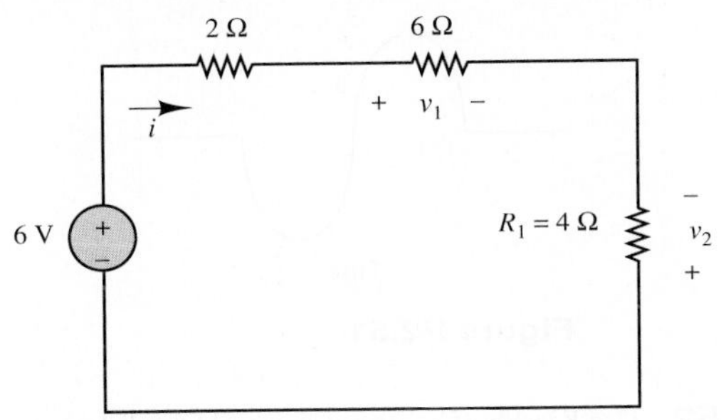

Figure P2.43

2.44 Find the equivalent resistance seen by the source in Figure P2.44, and use result to find i, i_1, and v.

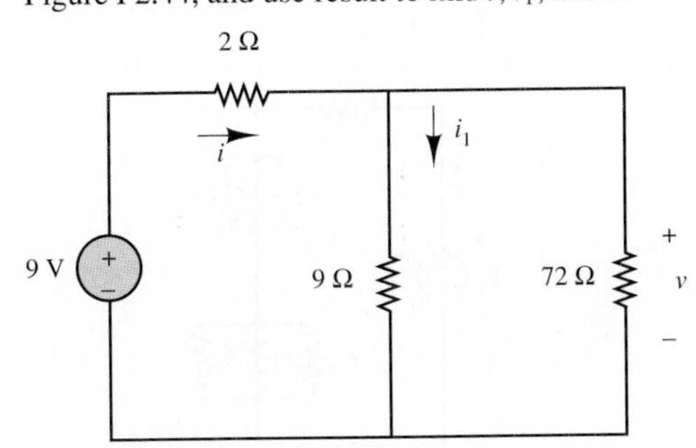

Figure P2.44

2.45 Find the equivalent resistance seen by the source and the current i in the circuit of Figure P2.45.

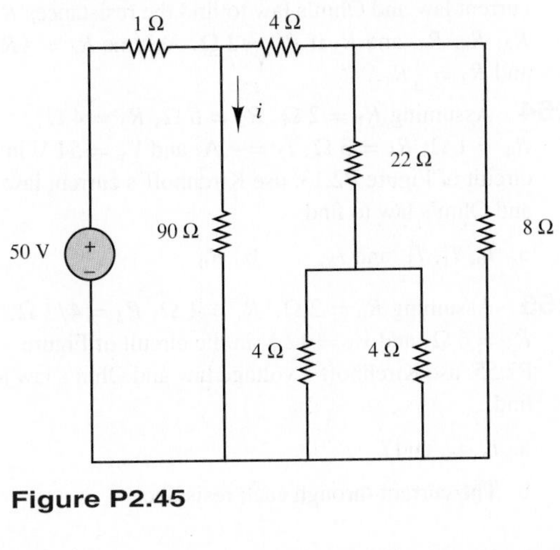

Figure P2.45

2.46 In the circuit of Figure P2.46, the power absorbed by the 15-Ω resistor is 15 W. Find R.

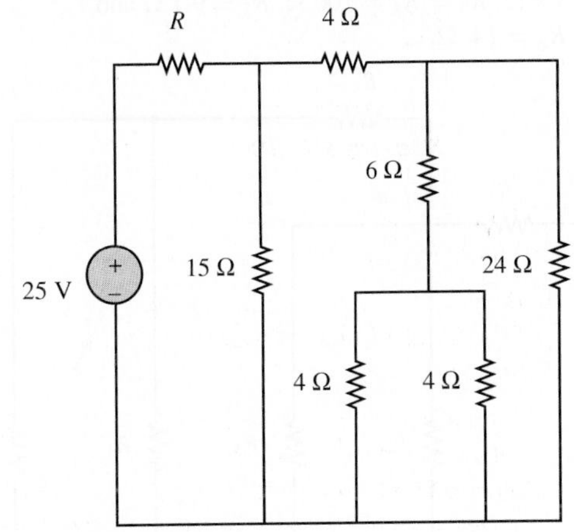

Figure P2.46

2.47 Find the equivalent resistance between terminals a and b in the circuit of Figure P2.47.

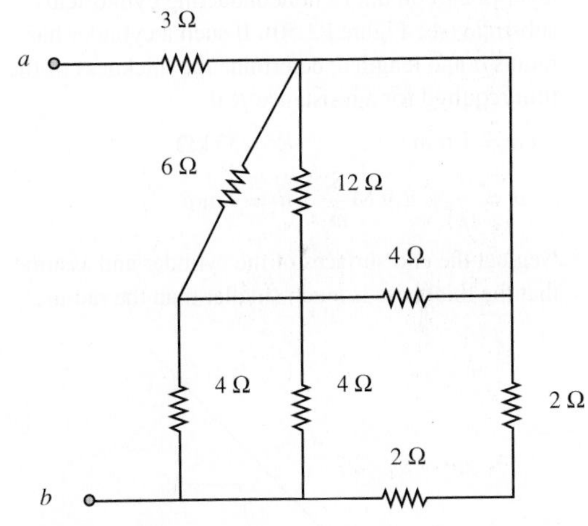

Figure P2.47

2.48 For the circuit shown in Figure P2.48, find the equivalent resistance seen by the source. How much power is delivered by the source?

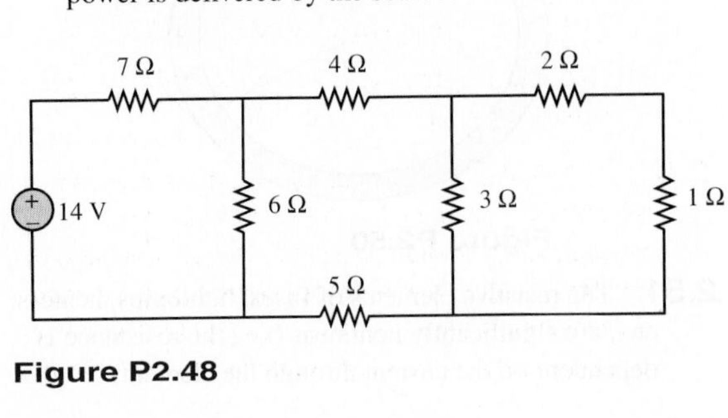

Figure P2.48

2.49 For the circuit shown in Figure P2.49, find the equivalent resistance, where $R_1 = 5\ \Omega$, $R_2 = 1\ \text{k}\ \Omega$, $R_3 = R_4 = 100\ \Omega$, $R_5 = 9.1\ \Omega$ and $R_6 = 1\ \text{k}\ \Omega$.

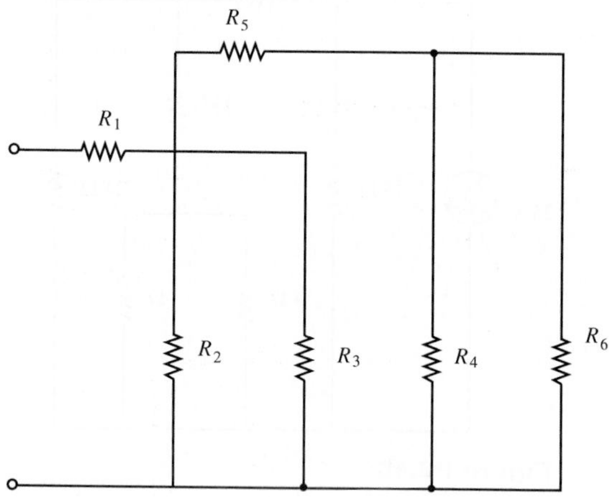

Figure P2.49

2.50 Cheap resistors are fabricated by depositing a thin layer of carbon onto a nonconducting cylindrical substrate (see Figure P2.50). If such a cylinder has radius a and length d, determine the thickness of the film required for a resistance R if

$$a = 1\text{ mm} \qquad R = 33\text{ k}\Omega$$

$$\sigma = \frac{1}{\rho} = 2.9\text{ M}\frac{\text{S}}{\text{m}} \qquad d = 9\text{ mm}$$

Neglect the end surfaces of the cylinder and assume that the thickness is much smaller than the radius.

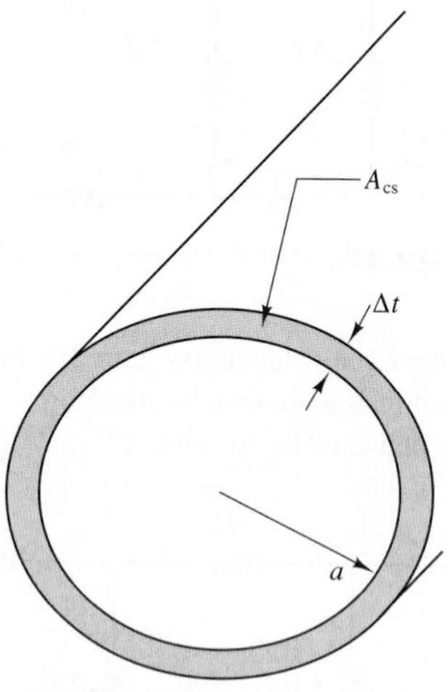

Figure P2.50

2.51 The resistive elements of fuses, lightbulbs, heaters, etc., are significantly nonlinear (i.e., the resistance is dependent on the current through the element). Assume the resistance of a fuse (Figure P2.51) is given by the expression $R = R_0[1 + A(T - T_0)]$ with $T - T_0 = kP$; $T_0 = 25°\text{C}$; $A = 0.7[°\text{C}]^{-1}$; $k = 0.35°\text{C/W}$; $R_0 = 0.11\ \Omega$; and P is the power dissipated in the resistive element of the fuse. Determine the rated current at which the circuit will melt and open, that is, "blow" (*Hint:* The fuse blows when R becomes infinite.)

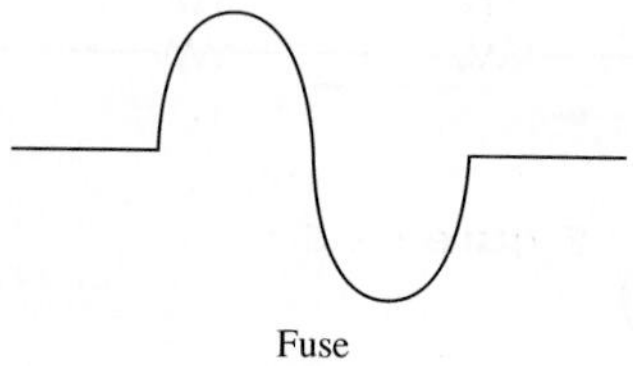

Figure P2.51

2.52 Use Kirchhoff's current law and Ohm's law to determine the current in each of the resistors R_4, R_5, and R_6 in the circuit of Figure P2.52. $V_S = 10$ V, $R_1 = 20\ \Omega$, $R_2 = 40\ \Omega$, $R_3 = 10\ \Omega$, $R_4 = R_5 = R_6 = 15\ \Omega$.

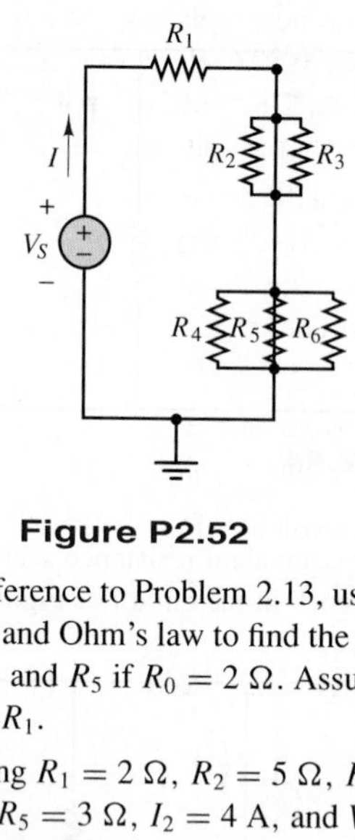

Figure P2.52

2.53 With reference to Problem 2.13, use Kirchhoff's current law and Ohm's law to find the resistances R_1, R_2, R_3, R_4, and R_5 if $R_0 = 2\ \Omega$. Assume $R_4 = \frac{2}{3}R_1$ and $R_2 = \frac{1}{3}R_1$.

2.54 Assuming $R_1 = 2\ \Omega$, $R_2 = 5\ \Omega$, $R_3 = 4\ \Omega$, $R_4 = 1\ \Omega$, $R_5 = 3\ \Omega$, $I_2 = 4$ A, and $V_S = 54$ V in the circuit of Figure P2.13, use Kirchhoff's current law and Ohm's law to find

a. I_0, I_1, I_3, and I_S. b. R_0.

2.55 Assuming $R_0 = 2\ \Omega$, $R_1 = 1\ \Omega$, $R_2 = 4/3\ \Omega$, $R_3 = 6\ \Omega$, and $V_S = 12$ V in the circuit of Figure P2.55, use Kirchhoff's voltage law and Ohm's law to find

a. i_a, i_b, and i_c.

b. The current through each resistance.

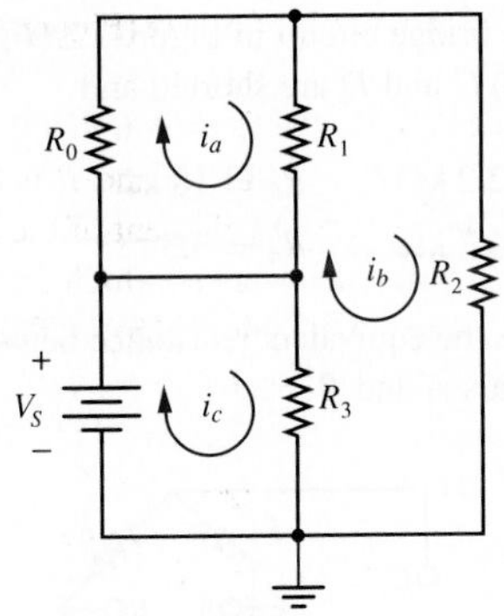

Figure P2.55

2.56 Assuming $R_0 = 2\ \Omega$, $R_1 = 2\ \Omega$, $R_2 = 5\ \Omega$, $R_3 = 4$ A, and $V_S = 24$ V in the circuit of Figure P2.55, use Kirchhoff's voltage law and Ohm's law to find

a. i_a, i_b, and i_c.

b. The voltage across each resistance.

2.57 Assume that the voltage source in the circuit of Figure P2.55 is now replaced by a current source, and $R_0 = 1\ \Omega$, $R_1 = 3\ \Omega$, $R_2 = 2\ \Omega$, $R_3 = 4$ A, and $I_S = 12$ A. Use Kirchhoff's voltage law and Ohm's law to determine the voltage across each resistance.

2.58 The voltage divider network of Figure P2.58 is expected to provide 5 V at the output. The resistors, however, may not be exactly the same; that is, their tolerances are such that the resistances may not be exactly 5 kΩ.

a. If the resistors have ± 10 percent tolerance, find the worst-case output voltages.

b. Find these voltages for tolerances of ± 5 percent.

Given: $v = 10$ V, $R_1 = 5$ kΩ, $R_2 = 5$ kΩ.

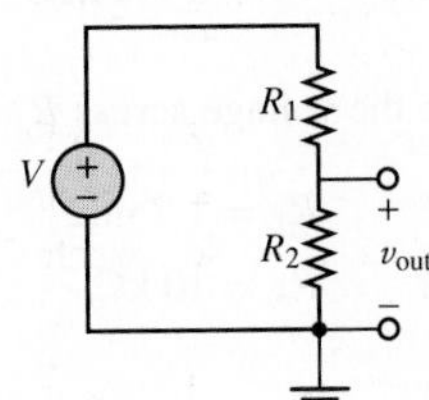

Figure P2.58

2.59 Find the equivalent resistance of the circuit of Figure P2.59 by combining resistors in series and in parallel. $R_0 = 4\ \Omega$, $R_1 = 12\ \Omega$, $R_2 = 8\ \Omega$, $R_3 = 2\ \Omega$, $R_4 = 16\ \Omega$, $R_5 = 5\ \Omega$

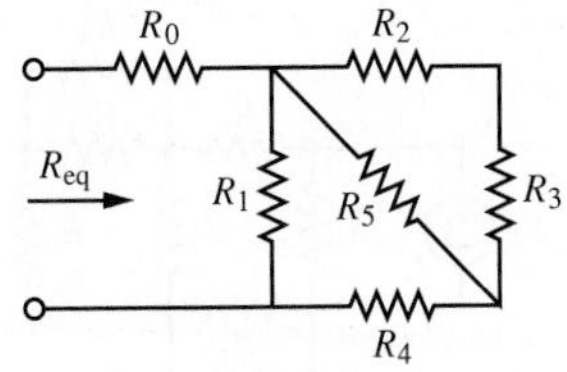

Figure P2.59

2.60 Find the equivalent resistance seen by the source and the current i in the circuit of Figure P2.60. Given: $V_s = 12$ V, $R_0 = 4\ \Omega$, $R_1 = 2\ \Omega$, $R_2 = 50\ \Omega$, $R_3 = 8\ \Omega$, $R_4 = 10\ \Omega$, $R_5 = 12\ \Omega$, $R_6 = 6\ \Omega$.

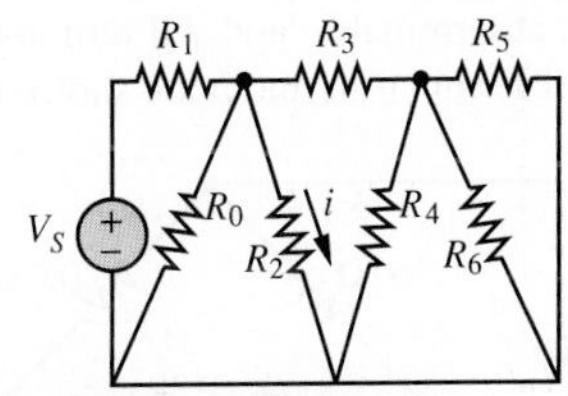

Figure P2.60

2.61 In the circuit of Figure P2.61, the power absorbed by the 20-Ω resistor is 20 W. Find R. Given: $V_s = 50$ V, $R_1 = 20\ \Omega$, $R_2 = 5\ \Omega$, $R_3 = 2\ \Omega$, $R_4 = 8\ \Omega$, $R_5 = 8\ \Omega$, $R_6 = 30\ \Omega$.

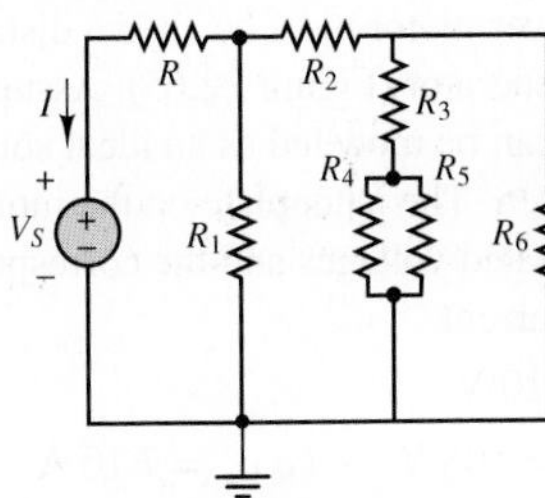

Figure P2.61

2.62 Determine the equivalent resistance of the infinite network of resistors in the circuit of Figure P2.62.

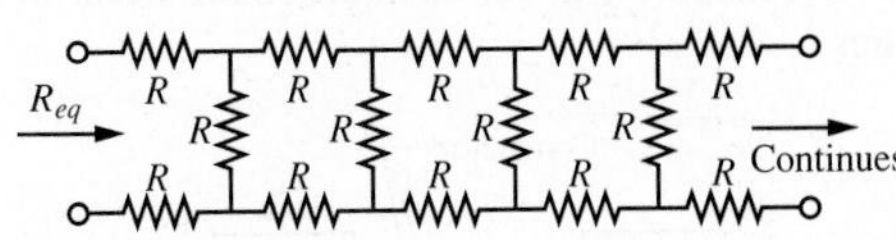

Figure P2.62

2.63 For the circuit shown in Figure P2.63 find

a. The equivalent resistance seen by the source.

b. The current through and the power absorbed by the 90-Ω resistance. Given: $V_S = 110$ V, $R_1 = 90\ \Omega$, $R_2 = 50\ \Omega$, $R_3 = 40\ \Omega$, $R_4 = 20\ \Omega$, $R_5 = 30\ \Omega$, $R_6 = 10\ \Omega$, $R_7 = 60\ \Omega$, $R_8 = 80\ \Omega$.

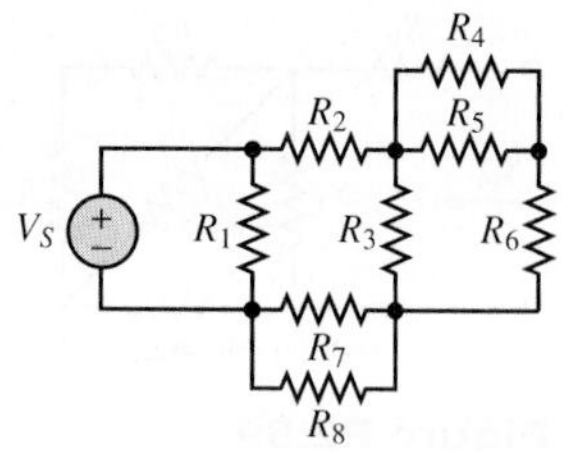

Figure P2.63

2.64 In the circuit of Figure P2.64, find the equivalent resistance looking in at terminals *a* and *b* if terminals *c* and *d* are open and again if terminals *c* and *d* are shorted together. Also, find the equivalent resistance looking in at terminals *c* and *d* if terminals *a* and *b* are open and if terminals *a* and *b* are shorted together.

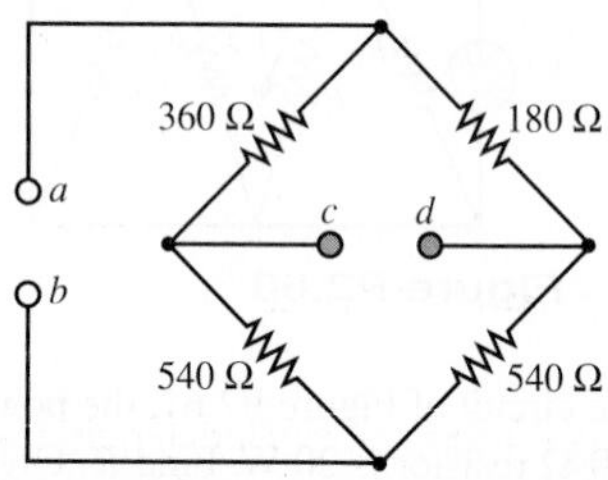

Figure P2.64

2.65 At an engineering site which you are supervising, a 1-horsepower motor must be sited a distance *d* from a portable generator (Figure P2.65). Assume the generator can be modeled as an ideal source with the voltage given. The nameplate on the motor gives the following rated voltages and the corresponding full-load current:

$V_G = 110$ V

$V_{M\,\text{min}} = 105\text{ V} \rightarrow I_{M\text{ FL}} = 7.10\text{ A}$

$V_{M\,\text{max}} = 117\text{ V} \rightarrow I_{M\text{ FL}} = 6.37\text{ A}$

If $d = 150$ m and the motor must deliver its full-rated power, determine the minimum AWG conductors which must be used in a rubber-insulated cable. Assume that the only losses in the circuit occur in the wires.

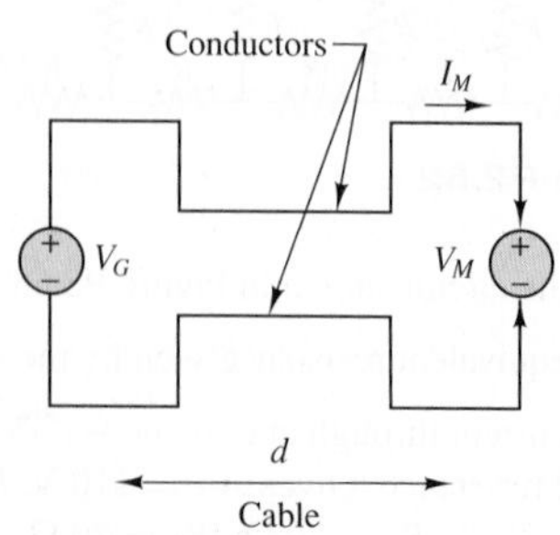

Figure P2.65

2.66 In the bridge circuit in Figure P2.66, if nodes (or terminals) *C* and *D* are shorted and

$R_1 = 2.2\text{ k}\Omega \qquad R_2 = 18\text{ k}\Omega$

$R_3 = 4.7\text{ k}\Omega \qquad R_4 = 3.3\text{ k}\Omega$

determine the equivalent resistance between the nodes or terminals *A* and *B*.

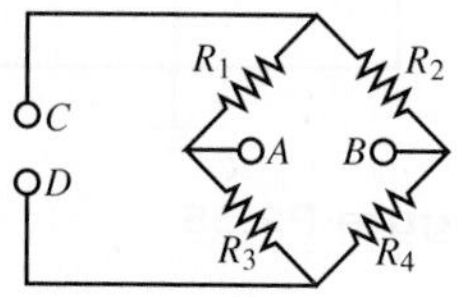

Figure P2.66

2.67 Determine the voltage between nodes *A* and *B* in the circuit shown in Figure P2.67.

$V_S = 12$ V

$R_1 = 11\text{ k}\Omega \qquad R_3 = 6.8\text{ k}\Omega$

$R_2 = 220\text{ k}\Omega \qquad R_4 = 0.22\text{ m}\Omega$

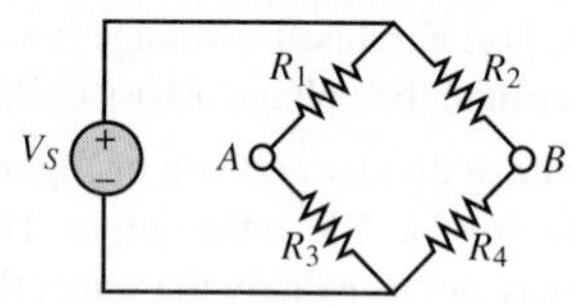

Figure P2.67

2.68 Determine the voltage between nodes *A* and *B* in the circuit shown in Figure P2.67.

$V_S = 5$ V

$R_1 = 2.2\text{ k}\Omega \qquad R_2 = 18\text{ k}\Omega$

$R_3 = 4.7\text{ k}\Omega \qquad R_4 = 3.3\text{ k}\Omega$

2.69 Determine the voltage across R_3 in Figure P2.69.

$V_S = 12\text{ V} \qquad R_1 = 1.7\text{ m}\Omega$

$R_2 = 3\text{ k}\Omega \qquad R_3 = 10\text{ k}\Omega$

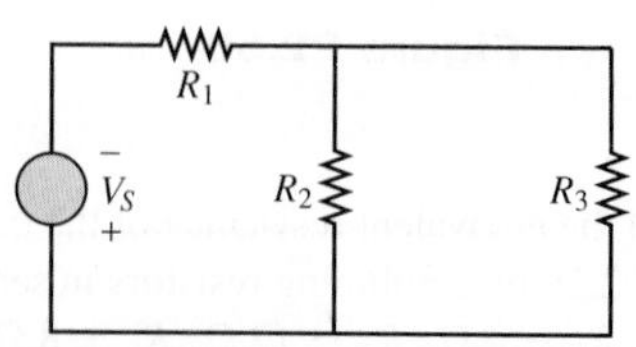

Figure P2.69

Sections 2.7, 2.8: Practical Sources and Measuring Devices

2.70 A *thermistor* is a nonlinear device which changes its terminal resistance value as its surrounding temperature changes. The resistance and temperature generally have a relation in the form of

$$R_{th}(T) = R_0 e^{-\beta(T-T_0)}$$

where R_{th} = resistance at temperature T, Ω

R_0 = resistance at temperature $T_0 = 298$ K, Ω

β = material constant, K^{-1}

T, T_0 = absolute temperature, K

a. If $R_0 = 300\ \Omega$ and $\beta = -0.01\ K^{-1}$, plot $R_{th}(T)$ as a function of the surrounding temperature T for $350 \le T \le 750$.

b. If the thermistor is in parallel with a 250-Ω resistor, find the expression for the equivalent resistance and plot $R_{th}(T)$ on the same graph for part a.

2.71 A moving-coil meter movement has a meter resistance $r_M = 200\ \Omega$, and full-scale deflection is caused by a meter current $I_m = 10\ \mu A$. The movement must be used to indicate pressure measured by the sensor up to a maximum of 100 kPa. See Figure P2.71.

a. Draw a circuit required to do this, showing all appropriate connections between the terminals of the sensor and meter movement.

b. Determine the value of each component in the circuit.

c. What is the linear range, that is, the minimum and maximum pressure that can accurately be measured?

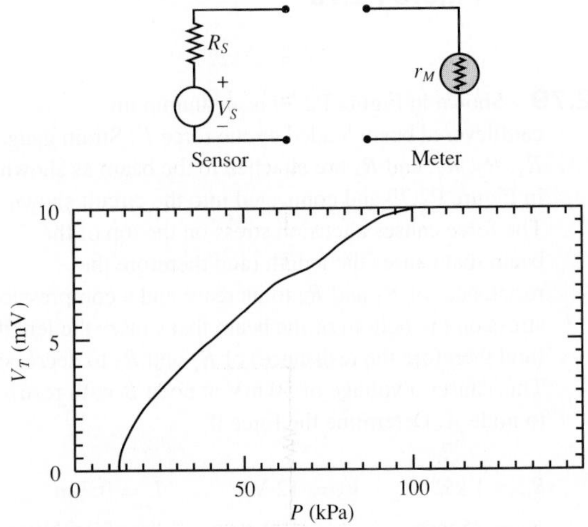

Figure P2.71

2.72 The circuit of Figure P2.72 is used to measure the internal impedance of a battery. The battery being tested is a NiMH battery cell.

a. A fresh battery is being tested, and it is found that the voltage V_{out}, is 2.28 V with the switch open and 2.27 V with the switch closed. Find the internal resistance of the battery.

b. The same battery is tested one year later, and V_{out} is found to be 2.2 V with the switch open but 0.31 V with the switch closed. Find the internal resistance of the battery.

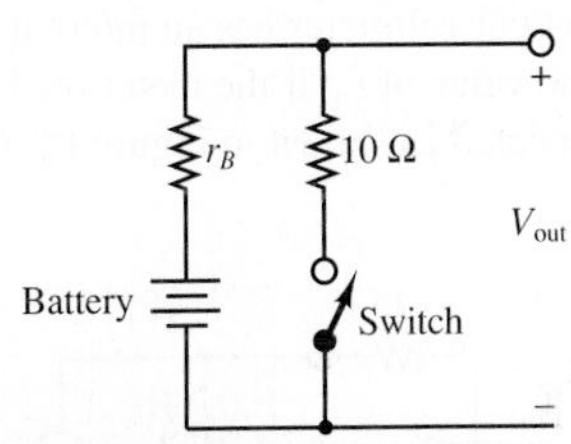

Figure P2.72

2.73 Consider the practical ammeter, described in Figure P2.73, consisting of an ideal ammeter in series with a 1-kΩ resistor. The meter sees a full-scale deflection when the current through it is 30 μA. If we desire to construct a multirange ammeter reading full-scale values of 10 mA, 100 mA, and 1 A, depending on the setting of a rotary switch, determine appropriate values of R_1, R_2, and R_3.

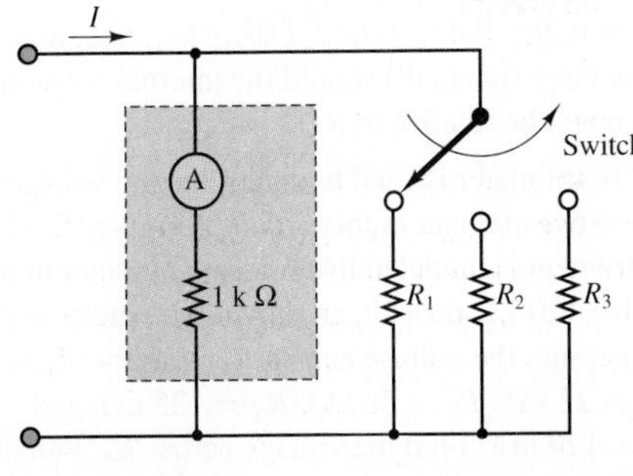

Figure P2.73

2.74 A circuit that measures the internal resistance of a practical ammeter is shown in Figure P2.74, where $R_S = 50{,}000\ \Omega$, $V_S = 12$ V, and R_p is a variable resistor that can be adjusted at will.

a. Assume that $r_a \ll 50{,}000\ \Omega$. Estimate the current i.

b. If the meter displays a current of 150 μA when $R_p = 15\ \Omega$, find the internal resistance of the meter r_a.

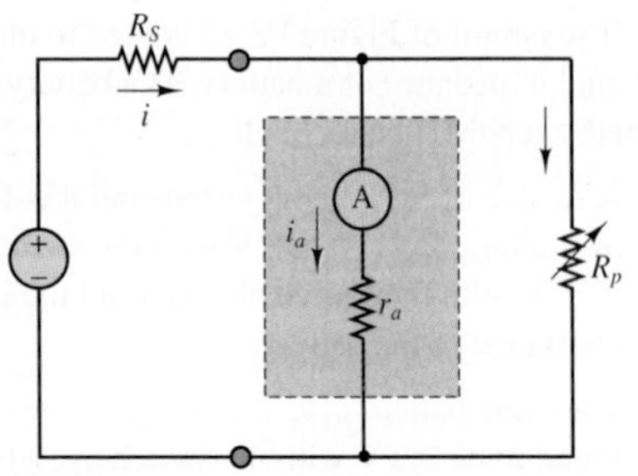

Figure P2.74

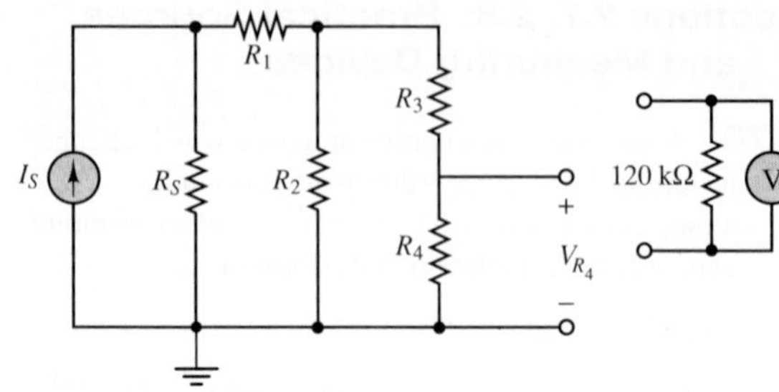

Figure P2.77

2.75 A practical voltmeter has an internal resistance r_m. What is the value of r_m if the meter reads 11.81 V when connected as shown in Figure P2.75.

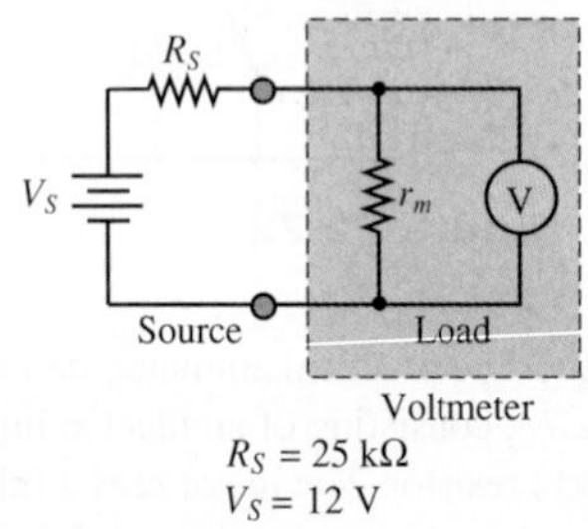

Figure P2.75

2.76 Using the circuit of Figure P2.75, find the voltage that the meter reads if $V_S = 24$ V and R_S has the following values:
$R_S = 0.2r_m, 0.4r_m, 0.6r_m, 1.2r_m, 4r_m, 6r_m$, and $10r_m$. How large (or small) should the internal resistance of the meter be relative to R_S?

2.77 A voltmeter is used to determine the voltage across a resistive element in the circuit of Figure P2.77. The instrument is modeled by an ideal voltmeter in parallel with a 120-kΩ resistor, as shown. The meter is placed to measure the voltage across R_4. Assume $R_1 = 8$ kΩ, $R_2 = 22$ kΩ, $R_3 = 50$ kΩ, $R_S = 125$ kΩ, and $I_S = 120$ mA. Find the voltage across R_4 with and without the voltmeter in the circuit for the following values:

a. $R_4 = 100\ \Omega$

b. $R_4 = 1$ kΩ

c. $R_4 = 10$ kΩ

d. $R_4 = 100$ kΩ

2.78 An ammeter is used as shown in Figure P2.78. The ammeter model consists of an ideal ammeter in series with a resistance. The ammeter model is placed in the branch as shown in the figure. Find the current through R_5 both with and without the ammeter in the circuit for the following values, assuming that $R_S = 20\ \Omega$, $R_1 = 800\ \Omega$, $R_2 = 600\ \Omega$, $R_3 = 1.2$ kΩ, $R_4 = 150\ \Omega$, and $V_S = 24$ V.

a. $R_5 = 1\ k\Omega$

b. $R_5 = 100\ \Omega$

c. $R_5 = 10\ \Omega$

d. $R_5 = 1\ \Omega$

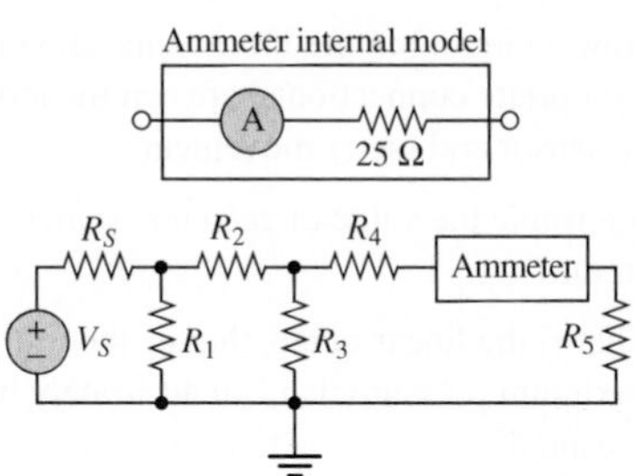

Figure P2.78

2.79 Shown in Figure P2.79 is an aluminum cantilevered beam loaded by the force F. Strain gauges R_1, R_2, R_3, and R_4 are attached to the beam as shown in Figure P2.79 and connected into the circuit shown. The force causes a tension stress on the top of the beam that causes the length (and therefore the resistance) of R_1 and R_4 to increase and a compression stress on the bottom of the beam that causes the length (and therefore the resistance) of R_2 and R_3 to decrease. This causes a voltage of 50 mV at node B with respect to node A. Determine the force if

$R_o = 1$ kΩ	$V_S = 12$ V	$L = 0.3$ m
$w = 25$ mm	$h = 100$ mm	$Y = 69$ GN/m^2

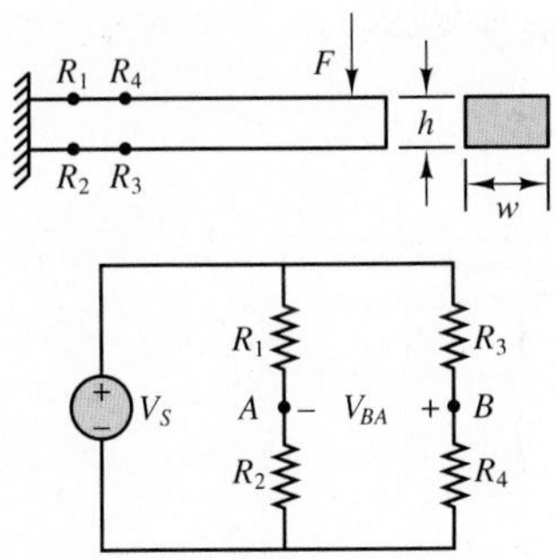

Figure P2.79

2.80 Shown in Figure P2.80 is a structural steel cantilevered beam loaded by a force F. Strain gauges R_1, R_2, R_3, and R_4 are attached to the beam as shown and connected into the circuit shown. The force causes a tension stress on the top of the beam that causes the length (and therefore the resistance) of R_1 and R_4 to increase and a compression stress on the bottom of the beam that causes the length (and therefore the resistance) of R_2 and R_3 to decrease. This generates a voltage between nodes B and A. Determine this voltage if $F = 1.3$ MN and

$$R_o = 1\ \text{k}\Omega \qquad V_S = 24\ \text{V} \qquad L = 1.7\ \text{m}$$

$$w = 3\ \text{cm} \qquad h = 7\ \text{cm} \qquad Y = 200\ \text{GN/m}^2$$

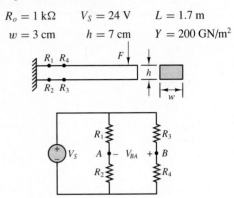

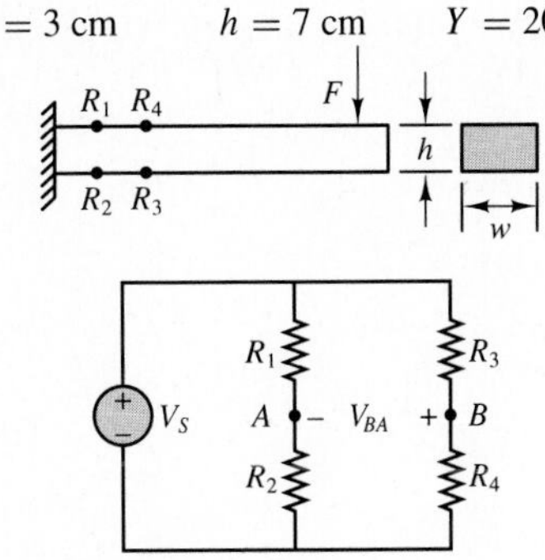

Figure P2.80

C H A P T E R

3

RESISTIVE NETWORK ANALYSIS

Chapter 3 illustrates the fundamental techniques for the analysis of resistive circuits. The chapter begins with the definition of network variables and of network analysis problems. Next, the two most widely applied methods—*node analysis* and *mesh analysis*—are introduced. These are the most generally applicable circuit solution techniques used to derive the equations of all electric circuits; their application to resistive circuits in this chapter is intended to acquaint you with these methods, which are used throughout the book. The second solution method presented is based on the *principle of superposition,* which is applicable only to linear circuits. Next, the concept of *Thévenin and Norton equivalent circuits* is explored, which leads to a discussion of *maximum power transfer* in electric circuits and facilitates the ensuing discussion of nonlinear loads and *load-line analysis.* At the conclusion of the chapter, you should have developed confidence in your ability to compute numerical solutions for a wide range of resistive circuits. The following box outlines the principal learning objectives of the chapter.

Learning Objectives

1. Compute the solution of circuits containing linear resistors and independent and dependent sources by using *node analysis*. *Sections 3.2 and 3.4.*
2. Compute the solution of circuits containing linear resistors and independent and dependent sources by using *mesh analysis*. *Sections 3.3 and 3.4.*
3. Apply the *principle of superposition* to linear circuits containing independent sources. *Section 3.5.*
4. Compute *Thévenin and Norton equivalent circuits* for networks containing linear resistors and independent and dependent sources. *Section 3.6.*
5. Use equivalent-circuit ideas to compute the *maximum power transfer* between a source and a load. *Section 3.7.*
6. Use the concept of equivalent circuit to determine voltage, current, and power for nonlinear loads by using *load-line analysis* and analytical methods. *Section 3.8.*

3.1 Network Analysis

The analysis of an electric network consists of determining each of the unknown branch currents and node voltages. It is therefore important to define all the relevant variables as clearly as possible, and in systematic fashion. Once the known and unknown variables have been identified, a set of equations relating these variables is constructed, and these equations are solved by means of suitable techniques. The analysis of electric circuits consists of writing the smallest set of equations sufficient to solve for all the unknown variables. The procedures required to write these equations are the subject of Chapter 3 and are very well documented and codified in the form of simple rules. The analysis of electric circuits is greatly simplified if some standard conventions are followed.

Example 3.1 defines all the voltages and currents that are associated with a specific circuit.

EXAMPLE 3.1

Problem

Identify the branch and node voltages and the loop and mesh currents in the circuit of Figure 3.1.

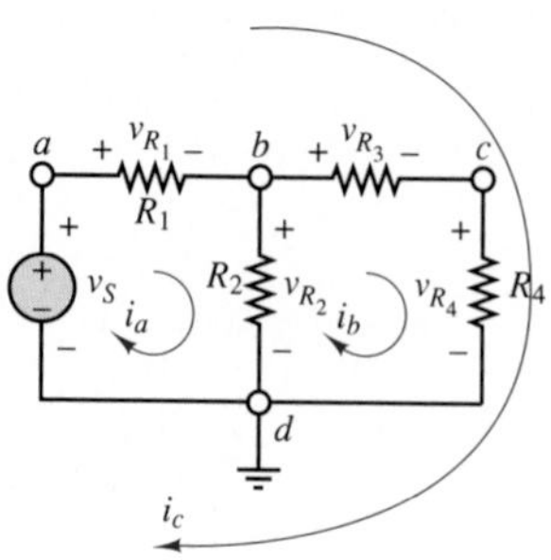

Figure 3.1

Solution

The following node voltages may be identified:

Node voltages	Branch voltages
$v_a = v_S$ (source voltage)	$v_S = v_a - v_d = v_a$
$v_b = v_{R_2}$	$v_{R_1} = v_a - v_b$
$v_c = v_{R_4}$	$v_{R_2} = v_b - v_d = v_b$
$v_d = 0$ (ground)	$v_{R_3} = v_b - v_c$
	$v_{R_4} = v_c - v_d = v_c$

Comments: Currents i_a, i_b, and i_c are loop currents, but only i_a and i_b are mesh currents.

In the example, we have identified a total of 9 variables! It should be clear that some method is needed to organize the wealth of information that can be generated simply by applying Ohm's law at each branch in a circuit. What would be desirable at this point is a means of reducing the number of equations needed to solve a circuit to the minimum necessary, that is, a method for obtaining N equations in N unknowns. The remainder of the chapter is devoted to the development of systematic circuit analysis methods that will greatly simplify the solution of electrical network problems.

3.2 THE NODE VOLTAGE METHOD

Node voltage analysis is the most general method for the analysis of electric circuits. In this section, its application to linear resistive circuits is illustrated. The **node voltage method** is based on defining the voltage at each node as an independent variable. One of the nodes is selected as a **reference node** (usually—but not necessarily—ground), and each of the other node voltages is referenced to this node. Once each node voltage is defined, Ohm's law may be applied between any two adjacent nodes to determine the current flowing in each branch. In the node voltage method, *each branch current is expressed in terms of one or more node voltages;* thus, currents do not explicitly enter into the equations. Figure 3.2 illustrates how to define branch currents in this method. You may recall a similar description given in Chapter 2.

Once each branch current is defined in terms of the node voltages, Kirchhoff's current law is applied at each node:

$$\sum i = 0 \tag{3.1}$$

Figure 3.3 illustrates this procedure.

In the node voltage method, we assign the node voltages v_a and v_b; the branch current flowing from a to b is then expressed in terms of these node voltages.

$$i = \frac{v_a - v_b}{R}$$

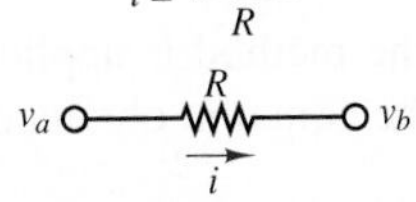

Figure 3.2 Branch current formulation in node analysis

By KCL: $i_1 - i_2 - i_3 = 0$. In the node voltage method, we express KCL by

$$\frac{v_a - v_b}{R_1} - \frac{v_b - v_c}{R_2} - \frac{v_b - v_d}{R_3} = 0$$

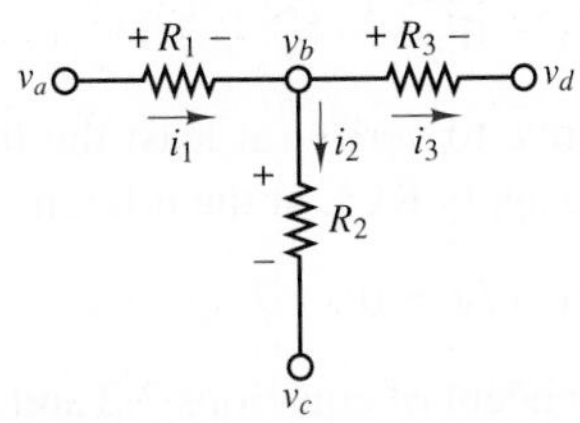

Figure 3.3 Use of KCL in node analysis

The systematic application of this method to a circuit with n nodes leads to writing n linear equations. However, one of the node voltages is the reference voltage and is therefore already known, since it is usually assumed to be zero (recall that the choice of reference voltage is dictated mostly by convenience, as explained in Chapter 2). Thus, we can write $n - 1$ *independent linear equations* in the $n - 1$ independent variables (the node voltages). Node analysis provides the minimum number of equations required to solve the circuit, since any branch voltage or current may be determined from knowledge of node voltages.

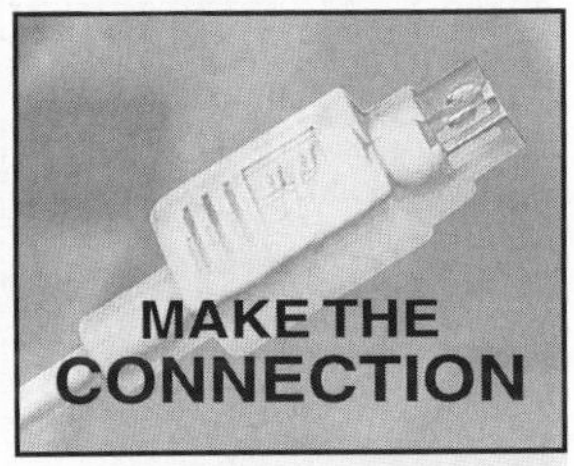

Thermal Systems

A useful analogy can be found between electric circuits and thermal systems. The table below illustrates the correspondence between electric circuit variables and thermal system variables, showing that the difference in electrical potential is analogous to the temperature difference between two bodies. Whenever there is a temperature difference between two bodies, Newton's law of cooling requires that heat flow from the warmer body to the cooler one. The flow of heat is therefore analogous to the flow of current. Heat flow can take place based on one of three mechanisms: (1) conduction, (2) convection, and (3) radiation. In this sidebar we only consider the first two, for simplicity.

Electrical variable	Thermal variable
Voltage difference v, [V]	Temperature difference ΔT, [°C]
Current i, A	Heat flux q, [W]
Resistance R, [Ω/m]	Thermal resistance R_t [°C/W]
Resistivity ρ, [Ω/m]	Conduction heat-transfer coefficient $k, \left[\frac{W}{m - °C}\right]$
(No exact electrical analogy)	Convection heat-transfer coefficient, or film coefficient of heat-transfer $h, \left[\frac{W}{m^2 - °C}\right]$

Thermal Resistance

To explain thermal resistance, consider a heat treated engine crankshaft that has just completed some thermal treatment. Assume that the shaft is to be quenched in a water bath at ambient temperature (see the figure below). Heat flows from within the shaft to the surface of the shaft, and then from the shaft surface to the water. This process continues until the temperature of the shaft is equal to that of the water.

The first mode of heat transfer in the above description is called *conduction*, and it occurs because the thermal conductivity of steel causes heat to flow from the higher temperature inner core to the lower-temperature surface. The heat transfer conduction coefficient k is analogous to the resistivity ρ of an electric conductor.

The second mode of heat transfer, *convection*, takes place at the boundary of two dissimilar materials (steel and water here). Heat transfer between the shaft and water is dependent on the surface area of the shaft in contact with the water A and is determined by the heat transfer convection coefficient h.

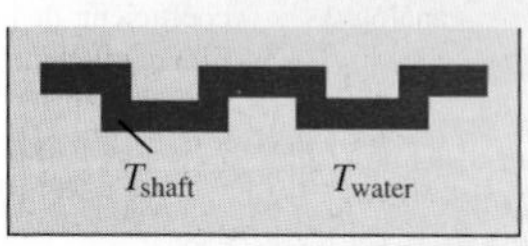

Engine crankshaft quenched in water bath.

The node analysis method may also be defined as a sequence of steps, as outlined in the following box:

LO1

FOCUS ON METHODOLOGY

NODE VOLTAGE ANALYSIS METHOD

1. Select a reference node (usually ground). This node usually has most elements tied to it. All other nodes are referenced to this node.
2. Define the remaining $n - 1$ node voltages as the independent or dependent variables. Each of the m voltage sources in the circuit is associated with a dependent variable. If a node is not connected to a voltage source, then its voltage is treated as an independent variable.
3. Apply KCL at each node labeled as an independent variable, expressing each current in terms of the adjacent node voltages.
4. Solve the linear system of $n - 1 - m$ unknowns.

Following the procedure outlined in the box guarantees that the correct solution to a given circuit will be found, provided that the nodes are properly identified and KCL is applied consistently. As an illustration of the method, consider the circuit shown in Figure 3.4. The circuit is shown in two different forms to illustrate equivalent graphical representations of the same circuit. The circuit on the right leaves no question where the nodes are. The direction of current flow is selected arbitrarily (assuming that i_S is a positive current). Application of KCL at node a yields

$$i_S - i_1 - i_2 = 0 \tag{3.2}$$

whereas at node b

$$i_2 - i_3 = 0 \tag{3.3}$$

It is instructive to verify (at least the first time the method is applied) that it is not necessary to apply KCL at the reference node. The equation obtained at node c,

$$i_1 + i_3 - i_S = 0 \tag{3.4}$$

is not independent of equations 3.2 and 3.3; in fact, it may be obtained by adding the

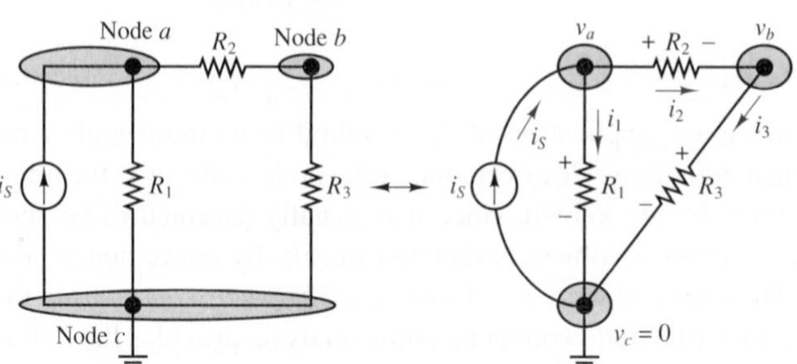

Figure 3.4 Illustration of node analysis

equations obtained at nodes a and b (verify this, as an exercise). This observation confirms the statement made earlier:

> In a circuit containing n nodes, we can write at most $n - 1$ independent equations.

LO1

Now, in applying the node voltage method, the currents i_1, i_2, and i_3 are expressed as functions of v_a, v_b, and v_c, the independent variables. Ohm's law requires that i_1, for example, be given by

$$i_1 = \frac{v_a - v_c}{R_1} \tag{3.5}$$

since it is the potential difference $v_a - v_c$ across R_1 that causes current i_1 to flow from node a to node c. Similarly,

$$i_2 = \frac{v_a - v_b}{R_2} \qquad i_3 = \frac{v_b - v_c}{R_3} \tag{3.6}$$

Substituting the expression for the three currents in the nodal equations (equations 3.2 and 3.3), we obtain the following relationships:

$$i_S - \frac{v_a}{R_1} - \frac{v_a - v_b}{R_2} = 0 \tag{3.7}$$

$$\frac{v_a - v_b}{R_2} - \frac{v_b}{R_3} = 0 \tag{3.8}$$

Equations 3.7 and 3.8 may be obtained directly from the circuit, with a little practice. Note that these equations may be solved for v_a and v_b, assuming that i_S, R_1, R_2, and R_3 are known. The same equations may be reformulated as follows:

$$\left(\frac{1}{R_1} + \frac{1}{R_2}\right) v_a + \left(-\frac{1}{R_2}\right) v_b = i_S$$
$$\left(-\frac{1}{R_2}\right) v_a + \left(\frac{1}{R_2} + \frac{1}{R_3}\right) v_b = 0 \tag{3.9}$$

Examples 3.2 through 3.4 further illustrate the application of the method.

Thermal Circuit Model

The *conduction resistance* of the shaft is described by the following equation:

$$q = \frac{kA_1}{L}\Delta T$$

$$R_{\text{cond}} = \frac{\Delta T}{q} = \frac{L}{kA_1}$$

where A_1 is a cross sectional area and L is the distance from the inner core to the surface. The convection resistance is described by a similar equation, in which convective heat flow is described by the film coefficient of heat transfer, h:

$$q = hA_2\Delta T$$

$$R_{\text{conv}} = \frac{\Delta T}{q} = \frac{1}{hA_2}$$

where A_2 is the surface area of the shaft in contact with the water. The equivalent thermal resistance and the overall circuit model of the crankshaft quenching process are shown in the figures below.

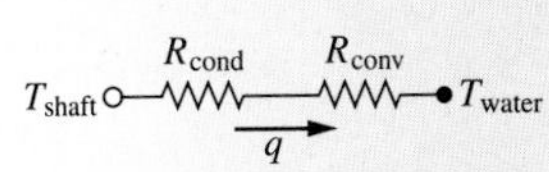

Thermal resistance representation of quenching process

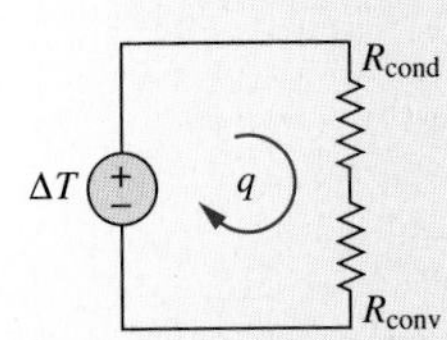

Electrical circuit representing the quenching process

LO1

EXAMPLE 3.2 Node Analysis

Problem

Solve for all unknown currents and voltages in the circuit of Figure 3.5.

Solution

Known Quantities: Source currents, resistor values.

Find: All node voltages and branch currents.

Schematics, Diagrams, Circuits, and Given Data: $I_1 = 10$ mA; $I_2 = 50$ mA; $R_1 = 1$ kΩ; $R_2 = 2$ kΩ; $R_3 = 10$ kΩ; $R_4 = 2$ kΩ.

Analysis: We follow the steps outlined in the Focus on Methodology box:

1. The reference (ground) node is chosen to be the node at the bottom of the circuit.
2. The circuit of Figure 3.5 is shown again in Figure 3.6, and two nodes are also shown in the figure. Thus, there are two independent variables in this circuit: v_1, v_2.

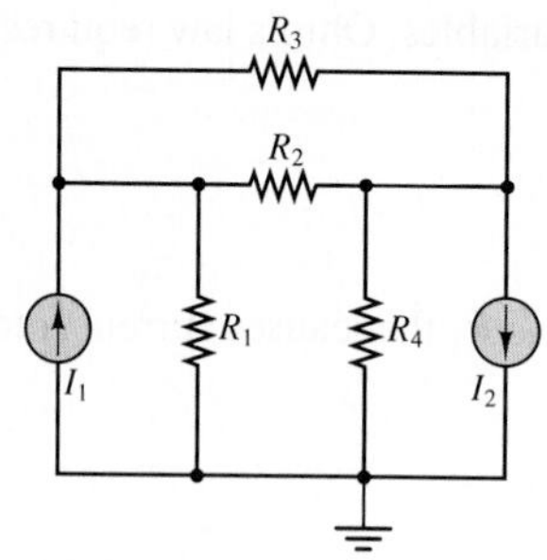

Figure 3.5

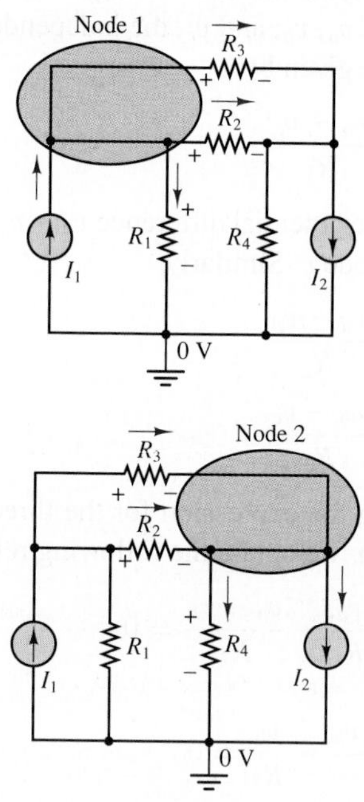

Figure 3.6

3. Applying KCL at nodes 1 and 2, we obtain

$$I_1 - \frac{v_1 - 0}{R_1} - \frac{v_1 - v_2}{R_2} - \frac{v_1 - v_2}{R_3} = 0 \qquad \text{node 1}$$

$$\frac{v_1 - v_2}{R_2} + \frac{v_1 - v_2}{R_3} - \frac{v_2 - 0}{R_4} - I_2 = 0 \qquad \text{node 2}$$

Now we can write the same equations more systematically as a function of the unknown node voltages, as was done in equation 3.9.

$$\left(\frac{1}{R_1} + \frac{1}{R_2} + \frac{1}{R_3}\right) v_1 + \left(-\frac{1}{R_2} - \frac{1}{R_3}\right) v_2 = I_1 \qquad \text{node 1}$$

$$\left(-\frac{1}{R_2} - \frac{1}{R_3}\right) v_1 + \left(\frac{1}{R_2} + \frac{1}{R_3} + \frac{1}{R_4}\right) v_2 = -I_2 \qquad \text{node 2}$$

4. We finally solve the system of equations. With some manipulation, the equations finally lead to the following form:

$$1.6v_1 - 0.6v_2 = 10$$
$$-0.6v_1 + 1.1v_2 = -50$$

These equations may be solved simultaneously to obtain

$$v_1 = -13.57 \text{ V}$$
$$v_2 = -52.86 \text{ V}$$

Knowing the node voltages, we can determine each of the branch currents and voltages in the circuit. For example, the current through the 10-kΩ resistor is given by

$$i_{10\text{ k}\Omega} = \frac{v_1 - v_2}{10{,}000} = 3.93 \text{ mA}$$

indicating that the initial (arbitrary) choice of direction for this current was the same as the actual direction of current flow. As another example, consider the current through the 1-kΩ resistor:

$$i_{1\text{ k}\Omega} = \frac{v_1}{1{,}000} = -13.57 \text{ mA}$$

In this case, the current is negative, indicating that current actually flows from ground to node 1, as it should, since the voltage at node 1 is negative with respect to ground. You may continue the branch-by-branch analysis started in this example to verify that the solution obtained in the example is indeed correct.

Comments: Note that we have chosen to assign a plus sign to currents entering a node and a minus sign to currents exiting a node; this choice is arbitrary (we could use the opposite convention), but we shall use it consistently in this book.

EXAMPLE 3.3 Node Analysis

LO1

Problem

Write the nodal equations and solve for the node voltages in the circuit of Figure 3.7.

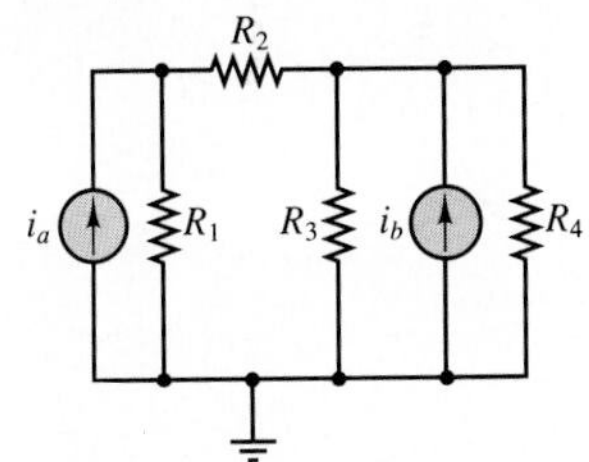

Figure 3.7

Solution

Known Quantities: Source currents, resistor values.

Find: All node voltages and branch currents.

Schematics, Diagrams, Circuits, and Given Data: $i_a = 1$ mA; $i_b = 2$ mA; $R_1 = 1$ kΩ; $R_2 = 500$ Ω; $R_3 = 2.2$ kΩ; $R_4 = 4.7$ kΩ.

Analysis: We follow the steps of the Focus on Methodology box.

1. The reference (ground) node is chosen to be the node at the bottom of the circuit.
2. See Figure 3.8. Two nodes remain after the selection of the reference node. Let us label these a and b and define voltages v_a and v_b. Both nodes are associated with independent variables.
3. We apply KCL at each of nodes a and b:

$$i_a - \frac{v_a}{R_1} - \frac{v_a - v_b}{R_2} = 0 \qquad \text{node } a$$

$$\frac{v_a - v_b}{R_2} + i_b - \frac{v_b}{R_3} - \frac{v_b}{R_4} = 0 \qquad \text{node } b$$

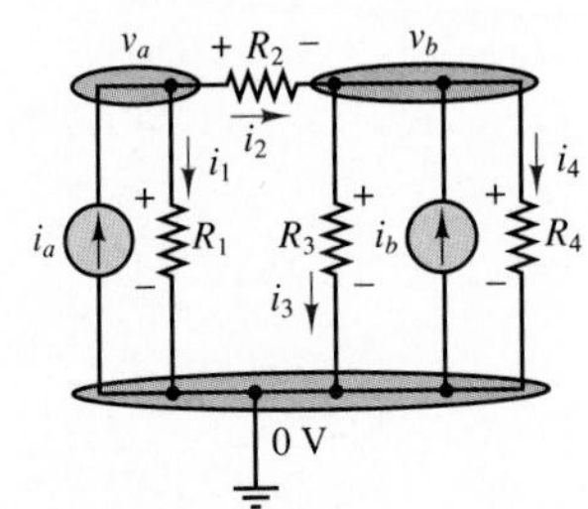

Figure 3.8

and rewrite the equations to obtain a linear system:

$$\left(\frac{1}{R_1}+\frac{1}{R_2}\right)v_a+\left(-\frac{1}{R_2}\right)v_b=i_a$$

$$\left(-\frac{1}{R_2}\right)v_a+\left(\frac{1}{R_2}+\frac{1}{R_3}+\frac{1}{R_4}\right)v_b=i_b$$

4. Substituting the numerical values in these equations, we get

$$3\times 10^{-3}v_a-2\times 10^{-3}v_b=1\times 10^{-3}$$
$$-2\times 10^{-3}v_a+2.67\times 10^{-3}v_b=2\times 10^{-3}$$

or

$$3v_a-2v_b=1$$
$$-2v_a+2.67v_b=2$$

The solution $v_a = 1.667$ V, $v_b = 2$ V may then be obtained by solving the system of equations.

LO1

EXAMPLE 3.4 Solution of Linear System of Equations Using Cramer's Rule

Problem

Solve the circuit equations obtained in Example 3.3, using Cramer's rule (see Appendix A).

Solution

Known Quantities: Linear system of equations.

Find: Node voltages.

Analysis: The system of equations generated in Example 3.3 may also be solved by using linear algebra methods, by recognizing that the system of equations can be written as

$$\begin{bmatrix} 3 & -2 \\ -2 & 2.67 \end{bmatrix}\begin{bmatrix} v_a \\ v_b \end{bmatrix}=\begin{bmatrix} 1 \\ 2 \end{bmatrix}$$

By using Cramer's rule (see Appendix A), the solution for the two unknown variables v_a and v_b can be written as follows:

$$v_a=\frac{\begin{vmatrix} 1 & -2 \\ 2 & 2.67 \end{vmatrix}}{\begin{vmatrix} 3 & -2 \\ -2 & 2.67 \end{vmatrix}}=\frac{(1)(2.67)-(-2)(2)}{(3)(2.67)-(-2)(-2)}=\frac{6.67}{4}=1.667\text{ V}$$

$$v_b=\frac{\begin{vmatrix} 3 & 1 \\ -2 & 2 \end{vmatrix}}{\begin{vmatrix} 3 & -2 \\ -2 & 2.67 \end{vmatrix}}=\frac{(3)(2)-(-2)(1)}{(3)(2.67)-(-2)(-2)}=\frac{8}{4}=2\text{ V}$$

The result is the same as in Example 3.3.

Comments: While Cramer's rule is an efficient solution method for simple circuits (e.g., two nodes), it is customary to use computer-aided methods for larger circuits. Once the nodal equations have been set in the general form presented in equation 3.9, a variety of computer aids may be employed to compute the solution. You will find the solution to the same example computed using MathCad™ in the electronic files that accompany this book.

CHECK YOUR UNDERSTANDING

Find the current i_L in the circuit shown on the left, using the node voltage method.

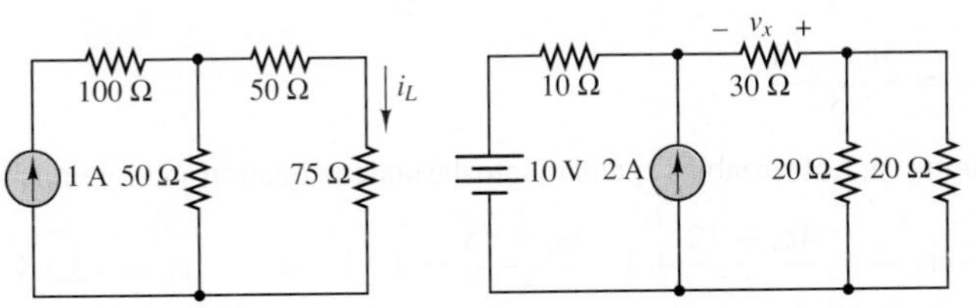

Find the voltage v_x by the node voltage method for the circuit shown on the right.
Show that the answer to Example 3.3 is correct by applying KCL at one or more nodes.

Answers: 0.2857 A; −18 V

EXAMPLE 3.5

LO1

Problem

Use the node voltage analysis to determine the voltage v in the circuit of Figure 3.9. Assume that $R_1 = 2\ \Omega$, $R_2 = 1\ \Omega$, $R_3 = 4\ \Omega$, $R_4 = 3\ \Omega$, $I_1 = 2$ A, and $I_2 = 3$ A.

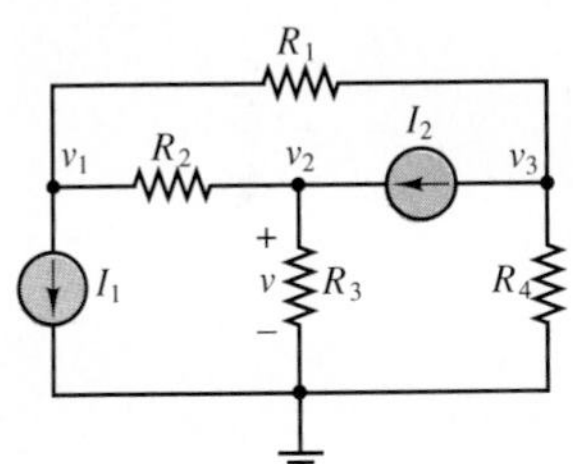

Figure 3.9 Circuit for Example 3.5

Solution

Known Quantities: Values of the resistors and the current sources.

Find: Voltage across R_3.

Analysis: Once again, we follow the steps outlined in the Focus on Methodology box.

1. The reference node is denoted in Figure 3.9.
2. Next, we define the three node voltages v_1, v_2, v_3, as shown in Figure 3.9.
3. Apply KCL at each of the $n - 1$ nodes, expressing each current in terms of the adjacent node voltages.

$$\frac{v_3 - v_1}{R_1} + \frac{v_2 - v_1}{R_2} - I_1 = 0 \qquad \text{node 1}$$

$$\frac{v_1 - v_2}{R_2} - \frac{v_2}{R_3} + I_2 = 0 \qquad \text{node 2}$$

$$\frac{v_1 - v_3}{R_1} - \frac{v_3}{R_4} - I_2 = 0 \qquad \text{node 3}$$

4. Solve the linear system of $n - 1 - m$ unknowns. Finally, we write the system of equations resulting from the application of KCL at the three nodes associated with independent

variables:

$$(-1-2)v_1 + 2v_2 + 1v_3 = 4 \qquad \text{node 1}$$
$$4v_1 + (-1-4)v_2 + 0v_3 = -12 \qquad \text{node 2}$$
$$3v_1 + 0v_2 + (-2-3)v_3 = 18 \qquad \text{node 3}$$

The resulting system of three equations in three unknowns can now be solved. Starting with the node 2 and node 3 equations, we write

$$v_2 = \frac{4v_1 + 12}{5}$$

$$v_3 = \frac{3v_1 - 18}{5}$$

Substituting each of variables v_2 and v_3 into the node 1 equation and solving for v_1 provides

$$-3v_1 + 2 \cdot \frac{4v_1 + 12}{5} + 1 \cdot \frac{3v_1 - 18}{5} = 4 \qquad \Rightarrow \qquad v_1 = -3.5 \text{ V}$$

After substituting v_1 into the node 2 and node 3 equations, we obtain

$$v_2 = -0.4 \text{ V} \qquad \text{and} \qquad v_3 = -5.7 \text{ V}$$

Therefore, we find

$$v = v_2 = -0.4 \text{ V}$$

Comments: Note that we have chosen to assign a plus sign to currents entering a node and a minus sign to currents exiting a node; this choice is arbitrary (the opposite sign convention could be used), but we shall use it consistently in this book.

CHECK YOUR UNDERSTANDING

Repeat the exercise of Example 3.5 when the direction of the current sources becomes the opposite. Find v.

Answer: $v = 0.4$ V

LO1

Node Analysis with Voltage Sources

In the preceding examples, we considered exclusively circuits containing current sources. It is natural that one will also encounter circuits containing voltage sources, in practice. The circuit of Figure 3.10 is used to illustrate how node analysis is applied to a circuit containing voltage sources. Once again, we follow the steps outlined in the Focus on Methodology box.

Step 1: *Select a reference node (usually ground). This node usually has most elements tied to it. All other nodes will be referenced to this node.*

The reference node is denoted by the ground symbol in Figure 3.10.

Step 2: *Define the remaining $n - 1$ node voltages as the independent or dependent variables. Each of the m voltage sources in the circuit will be associated with a* *dependent variable. If a node is not connected to a voltage source, then its voltage is treated as an independent variable.*

Next, we define the three node voltages v_a, v_b, v_c, as shown in Figure 3.10. We note that v_a is a dependent voltage. We write a simple equation for this dependent voltage, noting that v_a is equal to the source voltage v_S: $v_a = v_S$.

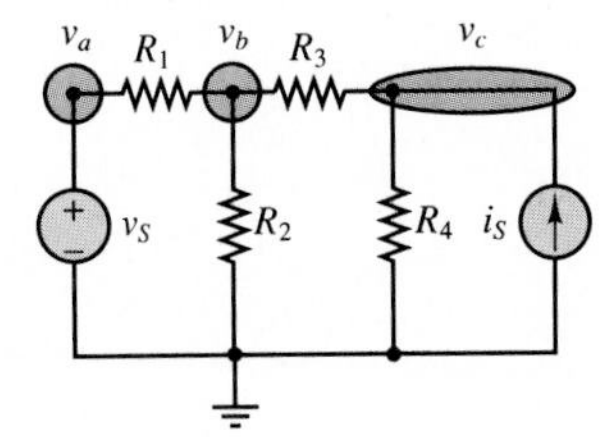

Figure 3.10 Node analysis with voltage sources

Step 3: *Apply KCL at each node labeled as an independent variable, expressing each current in terms of the adjacent node voltages.*

We apply KCL at the two nodes associated with the independent variables v_b and v_c:

At node *b*:

$$\frac{v_a - v_b}{R_1} - \frac{v_b - 0}{R_2} - \frac{v_b - v_c}{R_3} = 0$$

or

$$\frac{v_S - v_b}{R_1} - \frac{v_b}{R_2} - \frac{v_b - v_c}{R_3} = 0 \tag{3.10a}$$

At node *c*:

$$\frac{v_b - v_c}{R_3} - \frac{v_c}{R_4} + i_S = 0 \tag{3.10b}$$

Step 4: *Solve the linear system of $n - 1 - m$ unknowns.*

Finally, we write the system of equations resulting from the application of KCL at the two nodes associated with independent variables:

$$\begin{aligned}\left(\frac{1}{R_1} + \frac{1}{R_2} + \frac{1}{R_3}\right) v_b + \left(-\frac{1}{R_3}\right) v_c &= \frac{1}{R_1} v_s \\ \left(-\frac{1}{R_3}\right) v_b + \left(\frac{1}{R_3} + \frac{1}{R_4}\right) v_c &= i_S\end{aligned} \tag{3.11}$$

The resulting system of two equations in two unknowns can now be solved.

EXAMPLE 3.6

LO1

Problem

Use node analysis to determine the current *i* flowing through the voltage source in the circuit of Figure 3.11. Assume that $R_1 = 2\ \Omega$, $R_2 = 2\ \Omega$, $R_3 = 4\ \Omega$, $R_4 = 3\ \Omega$, $I = 2$ A, and $V = 3$ V.

Solution

Known Quantities: Resistance values; current and voltage source values.

Find: The current *i* through the voltage source.

Analysis: Once again, we follow the steps outlined in the Focus on Methodology box.

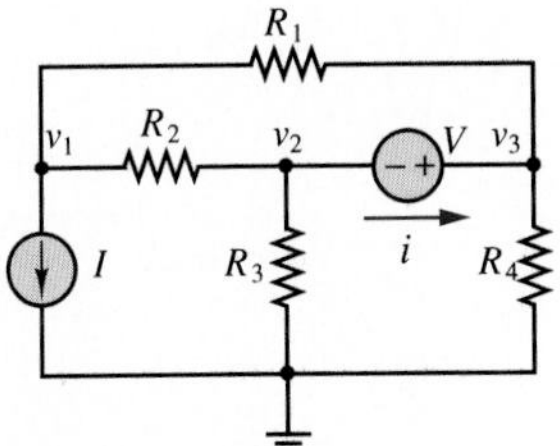

Figure 3.11 Circuit for Example 3.6

1. The reference node is denoted in Figure 3.11.
2. We define the three node voltages v_1, v_2, and v_3, as shown in Figure 3.11. We note that v_2 and v_3 are dependent on each other. One way to represent this dependency is to treat v_2

as an independent voltage and to observe that $v_3 = v_2 + 3$ V, since the potential at node 3 must be 3 V higher than at node 2 by virtue of the presence of the voltage source. Note that since we have an expression for the voltage at node 3 in terms of v_2, we will only need to write two nodal equations to solve this three-node circuit.

3. We apply KCL at the two nodes associated with the independent variables v_1 and v_2:

$$\frac{v_3 - v_1}{R_1} + \frac{v_2 - v_1}{R_2} - I = 0 \qquad \text{node 1}$$

$$\frac{v_1 - v_2}{R_2} - \frac{v_2}{R_3} - i = 0 \qquad \text{node 2}$$

where $$i = \frac{v_3 - v_1}{R_1} + \frac{v_3}{R_4}$$

Rearranging the node 2 equation by substituting the value of i yields

$$\frac{v_1 - v_2}{R_2} - \frac{v_2}{R_3} - \frac{v_3 - v_1}{R_1} - \frac{v_3}{R_4} = 0 \qquad \text{node 2}$$

4. Finally, we write the system of equations resulting from the application of KCL at the two nodes associated with independent variables:

$$-2v_1 + 1v_2 + 1v_3 = 4 \qquad \text{node 1}$$

$$12v_1 + (-9)v_2 + (-10)v_3 = 0 \qquad \text{node 2}$$

Considering that $v_3 = v_2 + 3$ V, we write

$$-2v_1 + 2v_2 = 1$$

$$12v_1 + (-19)v_2 = 30$$

The resulting system of the two equations in two unknowns can now be solved. Solving the two equations for v_1 and v_2 gives

$$v_1 = -5.64 \text{ V} \qquad \text{and} \qquad v_2 = -5.14 \text{ V}$$

This provides

$$v_3 = v_2 + 3 \text{ V} = -2.14 \text{ V}$$

Therefore, the current through the voltage source i is

$$i = \frac{v_3 - v_1}{R_1} + \frac{v_3}{R_4} = \frac{-2.14 + 5.64}{2} + \frac{-2.14}{3} = 1.04 \text{ A}$$

Comments: Knowing all the three node voltages, we now can compute the current flowing through each of the resistances as follows: $i_1 = |v_3 - v_1|/R_1$ (to left), $i_2 = |v_2 - v_1|/R_2$ (to left), $i_3 = |v_2|/R_3$ (upward), and $i_4 = |v_3|/R_4$ (upward).

CHECK YOUR UNDERSTANDING

Repeat the exercise of Example 3.6 when the direction of the current source becomes the opposite. Find the node voltages and i.

Answer: $v_1 = 5.21$ V, $v_2 = 1.71$ V, $v_3 = 4.71$ V, and $i = 1.32$ A

3.3 THE MESH CURRENT METHOD

The second method of circuit analysis discussed in this chapter employs **mesh currents** as the independent variables. The idea is to write the appropriate number of independent equations, using mesh currents as the independent variables. Subsequent application of Kirchhoff's voltage law around each mesh provides the desired system of equations.

In the mesh current method, we observe that a current flowing through a resistor in a specified direction defines the polarity of the voltage across the resistor, as illustrated in Figure 3.12, and that the sum of the voltages around a closed circuit must equal zero, by KVL. Once a convention is established regarding the direction of current flow around a mesh, simple application of KVL provides the desired equation. Figure 3.13 illustrates this point.

The current i, defined as flowing from left to right, establishes the polarity of the voltage across R.

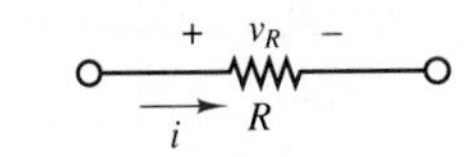

Figure 3.12 Basic principle of mesh analysis

The number of equations one obtains by this technique is equal to the number of meshes in the circuit. All branch currents and voltages may subsequently be obtained from the mesh currents, as will presently be shown. Since meshes are easily identified in a circuit, this method provides a very efficient and systematic procedure for the analysis of electric circuits. The following box outlines the procedure used in applying the mesh current method to a linear circuit.

Once the direction of current flow has been selected, KVL requires that $v_1 - v_2 - v_3 = 0$.

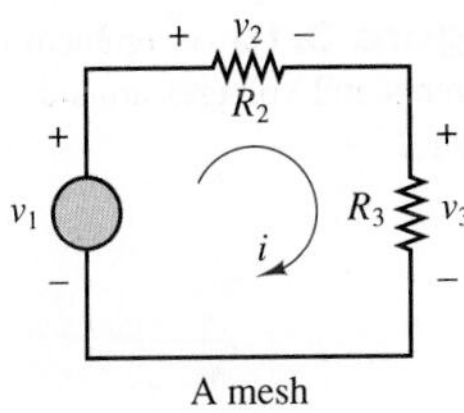

Figure 3.13 Use of KVL in mesh analysis

In mesh analysis, it is important to be consistent in choosing the direction of current flow. To avoid confusion in writing the circuit equations, unknown mesh currents are defined exclusively clockwise when we are using this method. To illustrate the mesh current method, consider the simple two-mesh circuit shown in Figure 3.14. This circuit is used to generate two equations in the two unknowns, the mesh currents i_1 and i_2. It is instructive to first consider each mesh by itself. Beginning with mesh 1, note that the voltages around the mesh have been assigned in Figure 3.15 according to the direction of the mesh current i_1. Recall that as long as signs are assigned consistently, an arbitrary direction may be assumed for any current in a circuit; if the resulting numerical answer for the current is negative, then the chosen reference direction is opposite to the direction of actual current flow. Thus, one need not be concerned about the actual direction of current flow in mesh analysis, once the directions of the mesh currents have been assigned. The correct solution will result, eventually.

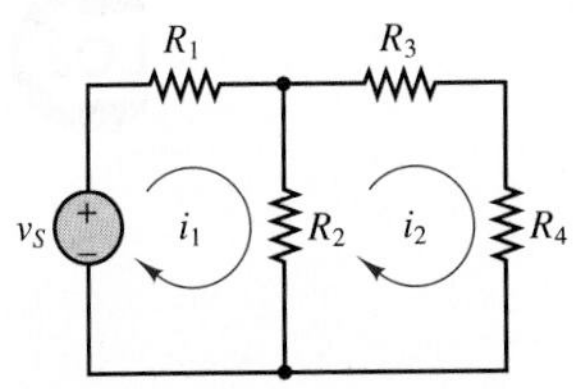

Figure 3.14 A two-mesh circuit

According to the sign convention, then, the voltages v_1 and v_2 are defined as shown in Figure 3.15. Now, it is important to observe that while mesh current i_1 is equal to the current flowing through resistor R_1 (and is therefore also the branch current through R_1), it is not equal to the current through R_2. The branch current through R_2 is the difference between the two mesh currents $i_1 - i_2$. Thus, since the polarity of voltage v_2 has already been assigned, according to the convention discussed in the previous paragraph, it follows that the voltage v_2 is given by

Mesh 1: KVL requires that $v_S - v_1 - v_2 = 0$, where $v_1 = i_1R_1$, $v_2 = (i_1 - i_2)R_1$.

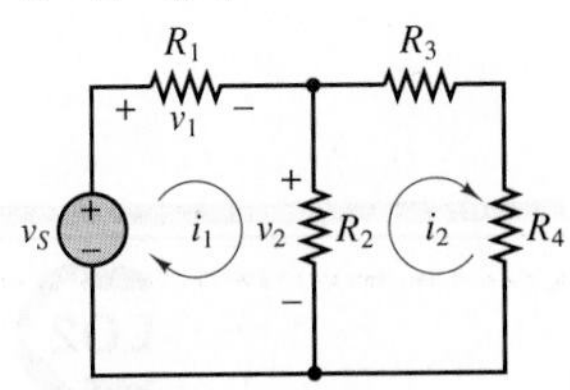

Figure 3.15 Assignment of currents and voltages around mesh 1

$$v_2 = (i_1 - i_2)R_2 \tag{3.12}$$

Finally, the complete expression for mesh 1 is

$$v_S - i_1R_1 - (i_1 - i_2)R_2 = 0 \tag{3.13}$$

The same line of reasoning applies to the second mesh. Figure 3.16 depicts the voltage assignment around the second mesh, following the clockwise direction of mesh current i_2. The mesh current i_2 is also the branch current through resistors R_3 and R_4; however, the current through the resistor that is shared by the two meshes, denoted by R_2, is now equal to $i_2 - i_1$; the voltage across this resistor is

$$v_2 = (i_2 - i_1)R_2 \tag{3.14}$$

Mesh 2: KVL requires that

$$v_2 + v_3 + v_4 = 0$$

where

$$v_2 = (i_2 - i_1)R_2$$

$$v_3 = i_2 R_3$$

$$v_4 = i_2 R_4$$

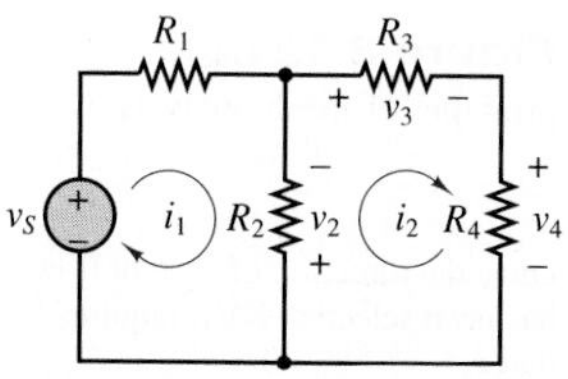

Figure 3.16 Assignment of currents and voltages around mesh 2

and the complete expression for mesh 2 is

$$(i_2 - i_1)R_2 + i_2 R_3 + i_2 R_4 = 0 \qquad \textbf{(3.15)}$$

Why is the expression for v_2 obtained in equation 3.14 different from equation 3.12? The reason for this apparent discrepancy is that the voltage assignment for each mesh was dictated by the (clockwise) mesh current. Thus, since the mesh currents flow through R_2 in opposing directions, the voltage assignments for v_2 in the two meshes are also opposite. This is perhaps a potential source of confusion in applying the mesh current method; you should be very careful to carry out the assignment of the voltages around each mesh separately.

Combining the equations for the two meshes, we obtain the following system of equations:

$$\begin{aligned}(R_1 + R_2)i_1 - R_2 i_2 &= v_S \\ -R_2 i_1 + (R_2 + R_3 + R_4)i_2 &= 0\end{aligned} \qquad \textbf{(3.16)}$$

These equations may be solved simultaneously to obtain the desired solution, namely, the mesh currents i_1 and i_2. You should verify that knowledge of the mesh currents permits determination of all the other voltages and currents in the circuit. Examples 3.7, 3.8 and 3.9 further illustrate some of the details of this method.

LO2

FOCUS ON METHODOLOGY

MESH CURRENT ANALYSIS METHOD

1. Define each mesh current consistently. Unknown mesh currents will be always defined in the clockwise direction; known mesh currents (i.e., when a current source is present) will always be defined in the direction of the current source.
2. In a circuit with n meshes and m current sources, $n - m$ independent equations will result. The unknown mesh currents are the $n - m$ independent variables.
3. Apply KVL to each mesh containing an unknown mesh current, expressing each voltage in terms of one or more mesh currents.
4. Solve the linear system of $n - m$ unknowns.

LO2

EXAMPLE 3.7 Mesh Analysis

Problem

Find the mesh currents in the circuit of Figure 3.17.

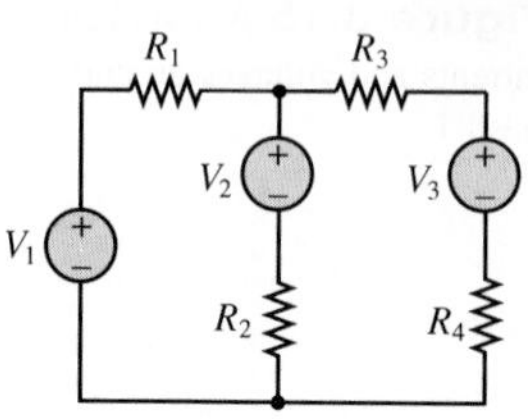

Figure 3.17

Solution

Known Quantities: Source voltages; resistor values.

Find: Mesh currents.

Schematics, Diagrams, Circuits, and Given Data: $V_1 = 10$ V; $V_2 = 9$ V; $V_3 = 1$ V; $R_1 = 5\ \Omega$; $R_2 = 10\ \Omega$; $R_3 = 5\ \Omega$; $R_4 = 5\ \Omega$.

Analysis: We follow the steps outlined in the Focus on Methodology box.

1. Assume clockwise mesh currents i_1 and i_2.
2. The circuit of Figure 3.17 will yield two equations in the two unknowns i_1 and i_2.
3. It is instructive to consider each mesh separately in writing the mesh equations; to this end, Figure 3.18 depicts the appropriate voltage assignments around the two meshes, based on the assumed directions of the mesh currents. From Figure 3.18, we write the mesh equations:

$$V_1 - R_1 i_1 - V_2 - R_2(i_1 - i_2) = 0$$
$$R_2(i_1 - i_2) + V_2 - R_3 i_2 - V_3 - R_4 i_2 = 0$$

Rearranging the linear system of the equation, we obtain

$$15i_1 - 10i_2 = 1$$
$$-10i_1 + 20i_2 = 8$$

4. The equations above can be solved to obtain i_1 and i_2:

$$i_1 = 0.5 \text{ A} \qquad \text{and} \qquad i_2 = 0.65 \text{ A}$$

Comments: Note how the voltage v_2 across resistor R_2 has different polarity in Figure 3.18, depending on whether we are working in mesh 1 or mesh 2.

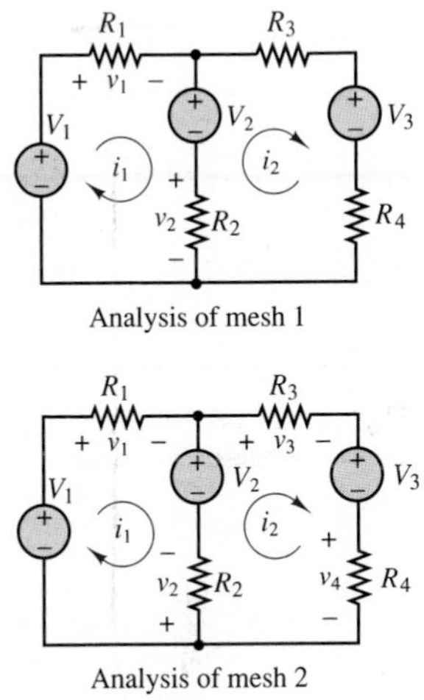

Figure 3.18

EXAMPLE 3.8 Mesh Analysis

Problem

Write the mesh current equations for the circuit of Figure 3.19.

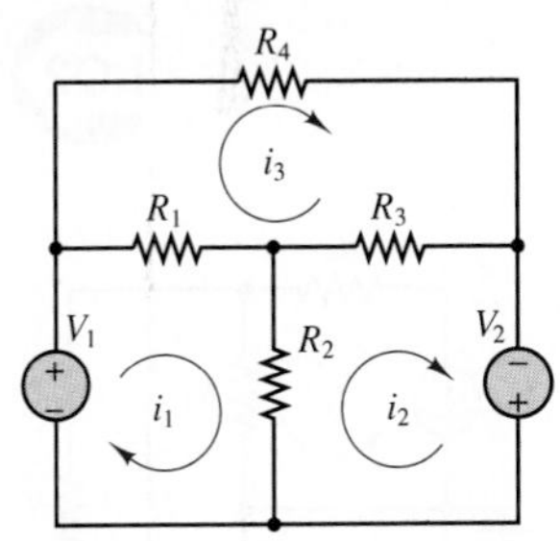

Figure 3.19

Solution

Known Quantities: Source voltages; resistor values.

Find: Mesh current equations.

Schematics, Diagrams, Circuits, and Given Data: $V_1 = 12$ V; $V_2 = 6$ V; $R_1 = 3\ \Omega$; $R_2 = 8\ \Omega$; $R_3 = 6\ \Omega$; $R_4 = 4\ \Omega$.

Analysis: We follow the Focus on Methodology steps.

1. Assume clockwise mesh currents i_1, i_2, and i_3.
2. We recognize three independent variables, since there are no current sources. Starting from mesh 1, we apply KVL to obtain

$$V_1 - R_1(i_1 - i_3) - R_2(i_1 - i_2) = 0$$

KVL applied to mesh 2 yields

$$-R_2(i_2 - i_1) - R_3(i_2 - i_3) + V_2 = 0$$

while in mesh 3 we find

$$-R_1(i_3 - i_1) - R_4 i_3 - R_3(i_3 - i_2) = 0$$

These equations can be rearranged in standard form to obtain

$$(3+8)i_1 - 8i_2 - 3i_3 = 12$$
$$-8i_1 + (6+8)i_2 - 6i_3 = 6$$
$$-3i_1 - 6i_2 + (3+6+4)i_3 = 0$$

You may verify that KVL holds around any one of the meshes, as a test to check that the answer is indeed correct.

CHECK YOUR UNDERSTANDING

Find the unknown voltage v_x by mesh current analysis in the circuit on the left.

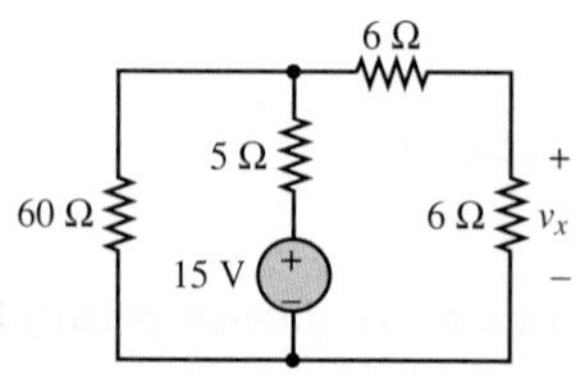

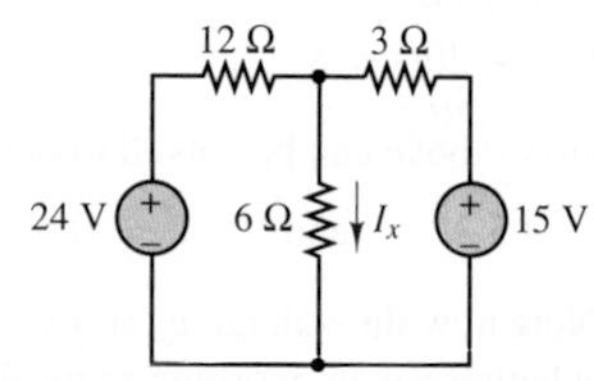

Find the unknown current I_x, using the mesh current method in the circuit on the right.

Answers: 5 V; 2 A

LO2

EXAMPLE 3.9 Mesh Analysis

Problem

The circuit of Figure 3.20 is a simplified DC circuit model of a three-wire electrical distribution service to residential and commercial buildings. The two ideal sources and the resistances R_4 and R_5 represent the equivalent circuit of the distribution system; R_1 and R_2 represent 110-V lighting and utility loads of 800 and 300 W, respectively. Resistance R_3 represents a 220-V heating load of about 3 kW. Determine the voltages across the three loads.

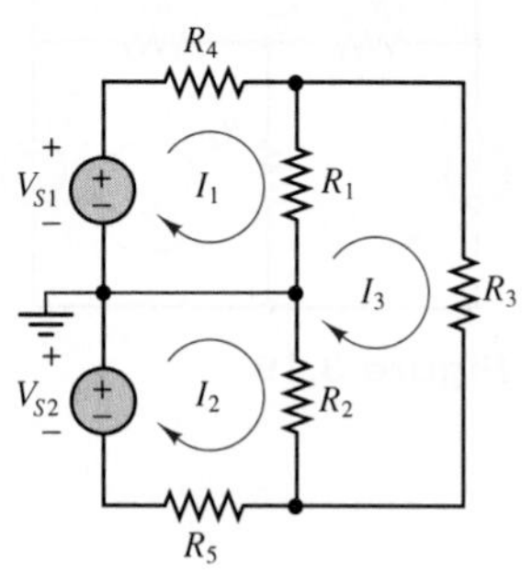

Figure 3.20

Solution

Known Quantities: The values of the voltage sources and of the resistors in the circuit of Figure 3.20 are $V_{S1} = V_{S2} = 110$ V; $R_4 = R_5 = 1.3\ \Omega$; $R_1 = 15\ \Omega$; $R_2 = 40\ \Omega$; $R_3 = 16\ \Omega$.

Find: v_1, v_2, and v_3.

Analysis: We follow the mesh current analysis method.

1. The (three) clockwise unknown mesh currents are shown in Figure 3.20. Next, we write the mesh equations.
2. No current sources are present; thus we have three independent variables. Applying KVL to each mesh containing an unknown mesh current and expressing each voltage in terms

of one or more mesh currents, we get the following:

Mesh 1:

$$V_{S1} - R_4 I_1 - R_1(I_1 - I_3) = 0$$

Mesh 2:

$$V_{S2} - R_2(I_2 - I_3) - R_5 I_2 = 0$$

Mesh 3:

$$-R_1(I_3 - I_1) - R_3 I_3 - R_2(I_3 - I_2) = 0$$

With some rearrangements, we obtain the following system of three equations in three unknown mesh currents.

$$\begin{aligned} -(R_1 + R_4)I_1 + R_1 I_3 &= -V_{S1} \\ -(R_2 + R_5)I_2 + R_2 I_3 &= -V_{S2} \\ R_1 I_1 + R_2 I_2 - (R_1 + R_2 + R_3)I_3 &= 0 \end{aligned}$$

Next, we substitute numerical values for the elements and express the equations in a matrix form as shown.

$$\begin{bmatrix} -16.3 & 0 & 15 \\ 0 & -41.3 & 40 \\ 15 & 40 & -71 \end{bmatrix} \begin{bmatrix} I_1 \\ I_2 \\ I_3 \end{bmatrix} = \begin{bmatrix} -110 \\ -110 \\ 0 \end{bmatrix}$$

which can be expressed as

$$[R][I] = [V]$$

with a solution of

$$[I] = [R]^{-1}[V]$$

The solution to the matrix problem can then be carried out using manual or numerical techniques. In this case, we have used Matlab™ to compute the inverse of the 3×3 matrix. Using Matlab™ to compute the inverse matrix, we obtain

$$[R]^{-1} = \begin{bmatrix} -0.1072 & -0.0483 & -0.0499 \\ -0.0483 & -0.0750 & -0.0525 \\ -0.0499 & -0.0525 & -0.0542 \end{bmatrix}$$

The value of current in each mesh can now be determined:

$$[I] = [R]^{-1}[V] = \begin{bmatrix} -0.1072 & -0.0483 & -0.0499 \\ -0.0483 & -0.0750 & -0.0525 \\ -0.0499 & -0.0525 & -0.0542 \end{bmatrix} \begin{bmatrix} -110 \\ -110 \\ 0 \end{bmatrix} = \begin{bmatrix} 17.11 \\ 13.57 \\ 11.26 \end{bmatrix}$$

Therefore, we find

$$I_1 = 17.11 \text{ A} \qquad I_2 = 13.57 \text{ A} \qquad I_3 = 11.26 \text{ A}$$

We can now obtain the voltages across the three loads, keeping in mind the ground location:

$$\begin{aligned} V_{R_1} &= R_1(I_1 - I_3) = 87.75 \text{ V} \\ V_{R_2} &= -R_2(I_2 - I_3) = -92.40 \text{ V} \\ V_{R_3} &= R_3 I_3 = 180.16 \text{ V} \end{aligned}$$

CHECK YOUR UNDERSTANDING

Repeat the exercise of Example 3.9, using node voltage analysis instead of the mesh current analysis.

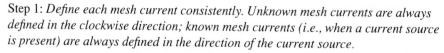

Answer: $V_{R1} = 87.75$ V, $V_{R2} = -92.40$ V, $V_{R3} = 180.16$ V

LO2 Mesh Analysis with Current Sources

In the preceding examples, we considered exclusively circuits containing voltage sources. It is natural to also encounter circuits containing current sources, in practice. The circuit of Figure 3.21 illustrates how mesh analysis is applied to a circuit containing current sources. Once again, we follow the steps outlined in the Focus on Methodology box.

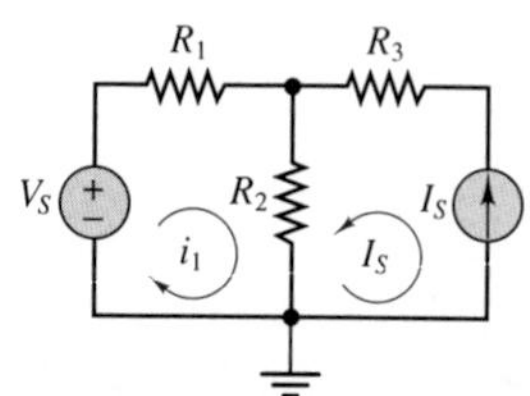

Figure 3.21 Circuit used to demonstrate mesh analysis with current sources

Step 1: *Define each mesh current consistently. Unknown mesh currents are always defined in the clockwise direction; known mesh currents (i.e., when a current source is present) are always defined in the direction of the current source.*

The mesh currents are shown in Figure 3.21. Note that since a current source defines the current in mesh 2, this (known) mesh current is in the counterclockwise direction.

Step 2: *In a circuit with n meshes and m current sources, n − m independent equations will result. The unknown mesh currents are the n − m independent variables.*

In this illustration, the presence of the current source has significantly simplified the problem: There is only one unknown mesh current, and it is i_1.

Step 3: *Apply KVL to each mesh containing an unknown mesh current, expressing each voltage in terms of one or more mesh currents.*

We apply KVL around the mesh containing the unknown mesh current:

$$V_S - R_1 i_1 - R_2(i_1 + I_S) = 0$$

or

$$(R_1 + R_2)i_1 = V_S - R_2 I_S \qquad (3.17)$$

Step 4: *Solve the linear system of n − m unknowns.*

$$i_1 = \frac{V_S - R_2 I_S}{R_1 + R_2} \qquad (3.18)$$

LO2 EXAMPLE 3.10 Mesh Analysis with Current Sources

Problem

Find the mesh currents in the circuit of Figure 3.22.

Solution

Known Quantities: Source current and voltage; resistor values.

Find: Mesh currents.

Schematics, Diagrams, Circuits, and Given Data: $I = 0.5$ A; $V = 6$ V; $R_1 = 3\ \Omega$; $R_2 = 8\ \Omega$; $R_3 = 6\ \Omega$; $R_4 = 4\ \Omega$.

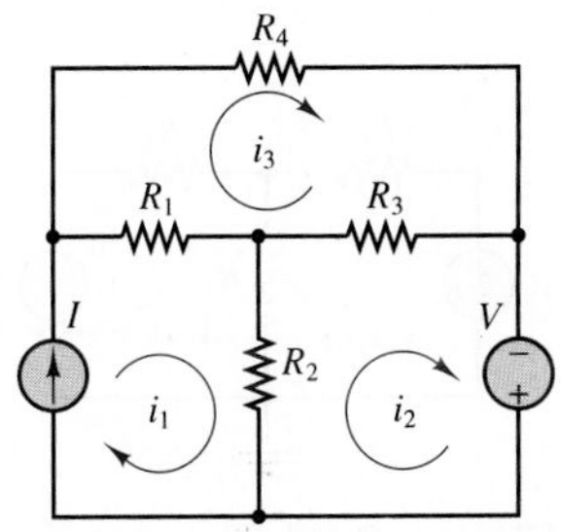

Figure 3.22

Analysis: We follow the Focus on Measurements steps.

1. Assume clockwise mesh currents i_1, i_2, and i_3.
2. Starting from mesh 1, we see immediately that the current source forces the mesh current to be equal to I:

$$i_1 = I$$

3. There is no need to write any further equations around mesh 1, since we already know the value of the mesh current. Now we turn to meshes 2 and 3 to obtain

$$-R_2(i_2 - i_1) - R_3(i_2 - i_3) + V = 0 \qquad \text{mesh 2}$$
$$-R_1(i_3 - i_1) - R_4 i_3 - R_3(i_3 - i_2) = 0 \qquad \text{mesh 3}$$

Rearranging the equations and substituting the known value of i_1, we obtain a system of two equations in two unknowns:

$$14i_2 - 6i_3 = 10$$
$$-6i_2 + 13i_3 = 1.5$$

4. These can be solved to obtain

$$i_2 = 0.95 \text{ A} \qquad i_3 = 0.55 \text{ A}$$

As usual, you should verify that the solution is correct by applying KVL.

Comments: Note that the current source has actually simplified the problem by constraining a mesh current to a fixed value.

CHECK YOUR UNDERSTANDING

Show that the equations given in Example 3.10 are correct, by applying KCL at each node.

EXAMPLE 3.11 Mesh Analysis with Current Sources

LO2

Problem

Find the unknown voltage v_x in the circuit of Figure 3.23.

Solution

Known Quantities: The values of the voltage sources and of the resistors in the circuit of Figure 3.23: $V_S = 10$ V; $I_S = 2$ A; $R_1 = 5\ \Omega$; $R_2 = 2\ \Omega$; and $R_3 = 4\ \Omega$.

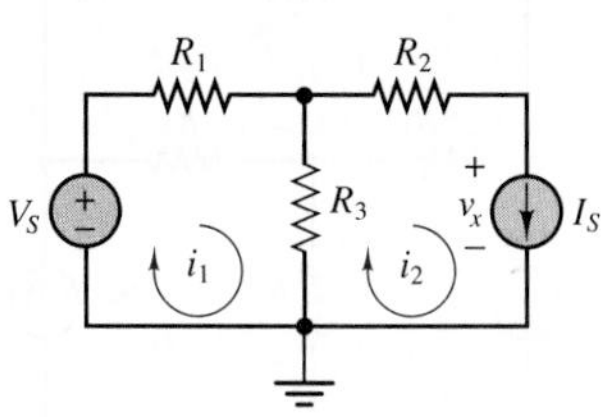

Figure 3.23 Illustration of mesh analysis in the presence of current sources

Find: v_x.

Analysis: We observe that the second mesh current must be equal to the current source:

$$i_2 = I_S$$

Thus, the unknown voltage, v_x, can be obtained applying KVL to mesh 2:

$$(i_2 - i_2)R_3 - i_2 R_2 - v_x = 0$$

$$v_x = I_S\,(R_2 + R_3)$$

To find the current i_1 we apply KVL to mesh 1:

$$V_S - i_1 R_1 - (i_1 - i_2)\,R_2 = 0$$

$$V_S + i_2 R_2 = i_1\,(R_1 + R_2)$$

but, since $i_2 = I_S$,

$$i_1 = \frac{V_S + I_S R_2}{(R_1 + R_2)} = \frac{10 + 2 \times 2}{5 + 2} = 2\text{ A}$$

Comments: Note that the presence of the current source reduces the number of unknown mesh currents by one. Thus, we were able to find v_x without the need to solve simultaneous equations.

CHECK YOUR UNDERSTANDING

Find the value of the current i_1 if the value of the current source is changed to 1 A.

Answer: 1.71 A

3.4 NODE AND MESH ANALYSIS WITH CONTROLLED SOURCES

LO1, LO2

The methods just described also apply, with relatively minor modifications, in the presence of dependent (controlled) sources. Solution methods that allow for the presence of controlled sources are particularly useful in the study of *transistor amplifiers* in Chapters 8 and 9. Recall from the discussion in Section 2.1 that a dependent source generates a voltage or current that depends on the value of another voltage or current in the circuit. When a dependent source is present in a circuit to be analyzed by node or mesh analysis, we can initially treat it as an ideal source and write the node or mesh equations accordingly. In addition to the equation obtained in this fashion, there is an equation relating the dependent source to one of the circuit voltages or currents. This **constraint equation** can then be substituted in the set of equations obtained by the techniques of node and mesh analysis, and the equations can subsequently be solved for the unknowns.

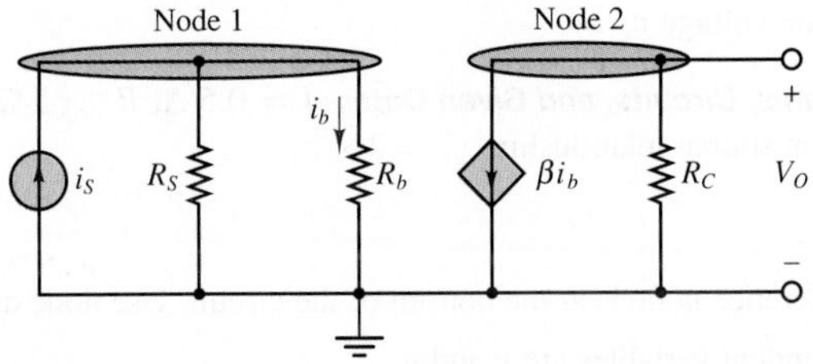

Figure 3.24 Circuit with dependent source

It is important to remark that once the constraint equation has been substituted in the initial system of equations, the number of unknowns remains unchanged. Consider, for example, the circuit of Figure 3.24, which is a simplified model of a bipolar transistor amplifier (transistors are introduced in Chapter 9). In the circuit of Figure 3.24, two nodes are easily recognized, and therefore node analysis is chosen as the preferred method. Applying KCL at node 1, we obtain the following equation:

$$i_S = v_1 \left(\frac{1}{R_S} + \frac{1}{R_b} \right) \tag{3.19}$$

KCL applied at the second node yields

$$\beta i_b + \frac{v_2}{R_C} = 0 \tag{3.20}$$

Next, observe that current i_b can be determined by means of a simple current divider:

$$i_b = i_S \frac{1/R_b}{1/R_b + 1/R_S} = i_S \frac{R_S}{R_b + R_S} \tag{3.21}$$

This is the *constraint equation,* which when inserted in equation 3.20, yields a system of two equations:

$$\begin{aligned} i_S &= v_1 \left(\frac{1}{R_S} + \frac{1}{R_b} \right) \\ -\beta i_S \frac{R_S}{R_b + R_S} &= \frac{v_2}{R_C} \end{aligned} \tag{3.22}$$

which can be used to solve for v_1 and v_2. Note that, in this particular case, the two equations are independent of each other. Example 3.12 illustrates a case in which the resulting equations are not independent.

EXAMPLE 3.12 Analysis with Dependent Sources

LO1, LO2

Problem

Find the node voltages in the circuit of Figure 3.25.

Solution

Known Quantities: Source current; resistor values; dependent voltage source relationship.

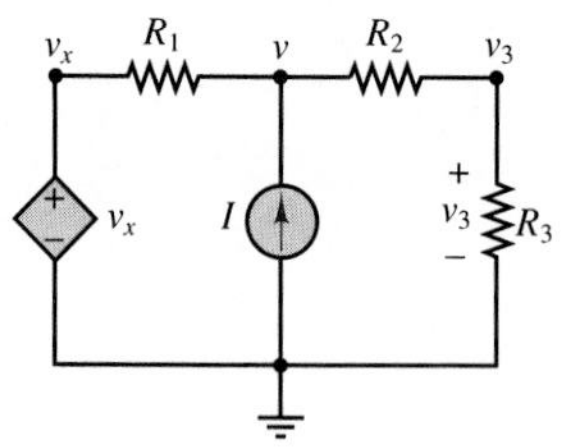

Figure 3.25

Find: Unknown node voltage v.

Schematics, Diagrams, Circuits, and Given Data: $I = 0.5$ A; $R_1 = 5\ \Omega$; $R_2 = 2\ \Omega$; $R_3 = 4\ \Omega$. Dependent source relationship: $v_x = 2 \times v_3$.

Analysis:

1. Assume the reference node is at the bottom of the circuit. Use node analysis.
2. The two independent variables are v and v_3.
3. Applying KCL to node v, we find that

$$\frac{v_x - v}{R_1} + I - \frac{v - v_3}{R_2} = 0$$

Applying KCL to node v_3, we find

$$\frac{v - v_3}{R_2} - \frac{v_3}{R_3} = 0$$

If we substitute the dependent source relationship into the first equation, we obtain a system of equations in the two unknowns v and v_3:

$$\left(\frac{1}{R_1} + \frac{1}{R_2}\right) v + \left(-\frac{2}{R_1} - \frac{1}{R_2}\right) v_3 = I$$

$$\left(-\frac{1}{R_2}\right) v + \left(\frac{1}{R_2} + \frac{1}{R_3}\right) v_3 = 0$$

4. Substituting numerical values, we obtain

$$0.7v - 0.9v_3 = 0.5$$
$$-0.5v + 0.75v_3 = 0$$

Solution of the above equations yields $v = 5$ V; $v_3 = 3.33$ V.

CHECK YOUR UNDERSTANDING

Solve the same circuit if $v_x = 2I$.

Answer: $v = \frac{21}{11}$ V; $v = \frac{14}{11}$ V

LO1, LO2

EXAMPLE 3.13 Mesh Analysis with Dependent Sources

Problem

Determine the voltage "gain" $A_v = v_2/v_1$ in the circuit of Figure 3.26.

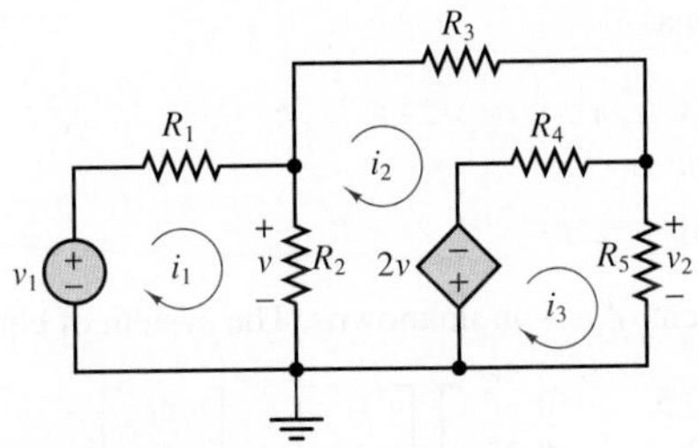

Figure 3.26 Circuit containing dependent source

Solution

Known Quantities: The values of the voltage sources and of the resistors in the circuit of Figure 3.26 are $R_1 = 1\ \Omega$; $R_2 = 0.5\ \Omega$; $R_3 = 0.25\ \Omega$; $R_4 = 0.25\ \Omega$; $R_5 = 0.25\ \Omega$.

Find: $A_v = v_2/v_1$.

Analysis: We note first that the two voltages we seek can be expressed as follows: $v = R_2(i_1 - i_2)$, and $v_2 = R_5 i_3$. Next, we follow the mesh current analysis method.

1. The mesh currents are defined in Figure 3.26.
2. No current sources are present; thus we have three independent variables, the currents i_1, i_2, and i_3.
3. Apply KVL at each mesh.

For mesh 1:

$$v_1 - R_1 i_1 - R_2(i_1 - i_2) = 0$$

or rearranging the equation gives

$$(R_1 + R_2)i_1 + (-R_2)i_2 + (0)i_3 = v_1$$

For mesh 2:

$$v - R_3 i_2 - R_4(i_2 - i_3) + 2v = 0$$

Rearranging the equation and substituing the expression $v = -R_2(i_2 - i_1)$, we obtain

$$-R_2(i_2 - i_1) - R_3 i_2 - R_4(i_2 - i_3) - 2R_2(i_2 - i_1) = 0$$
$$(-3R_2)i_1 + (3R_2 + R_3 + R_4)i_2 - (R_4)i_3 = 0$$

For mesh 3:

$$-2v - R_4(i_3 - i_2) - R_5 i_3 = 0$$

substituting the expression for $v = R_2(i_1 - i_2)$ and rearranging, we obtain

$$-2R_2(i_1 - i_2) - R_4(i_3 - i_2) - R_5 i_3 = 0$$
$$2R_2 i_1 - (2R_2 + R_4)i_2 + (R_4 + R_5)i_3 = 0$$

Finally, we can write the system of equations

$$\begin{bmatrix} (R_1 + R_2) & (-R_2) & 0 \\ (-3R_2) & (3R_2 + R_3 + R_4) & (-R_4) \\ (2R_2) & -(2R_2 + R_4) & (R_4 + R_5) \end{bmatrix} \begin{bmatrix} i_1 \\ i_2 \\ i_3 \end{bmatrix} = \begin{bmatrix} v_1 \\ 0 \\ 0 \end{bmatrix}$$

which can be written as

$$[R][i] = [v]$$

with solution

$$[i] = [R]^{-1}[v]$$

4. Solve the linear system of $n - m$ unknowns. The system of equations is

$$\begin{bmatrix} 1.5 & -0.5 & 0 \\ -1.5 & 2 & -0.25 \\ 1 & -1.25 & 0.5 \end{bmatrix} \begin{bmatrix} i_1 \\ i_2 \\ i_3 \end{bmatrix} = \begin{bmatrix} v_1 \\ 0 \\ 0 \end{bmatrix}$$

Thus, to solve for the unknown mesh currents, we must compute the inverse of the matrix of resistances R. Using Matlab™ to compute the inverse, we obtain

$$[R]^{-1} = \begin{bmatrix} 0.88 & 0.32 & 0.16 \\ 0.64 & 0.96 & 0.48 \\ -0.16 & 1.76 & 2.88 \end{bmatrix}$$

$$\begin{bmatrix} i_1 \\ i_2 \\ i_3 \end{bmatrix} = [R]^{-1} \begin{bmatrix} v_1 \\ 0 \\ 0 \end{bmatrix} = \begin{bmatrix} 0.88 & 0.32 & 0.16 \\ 0.64 & 0.96 & 0.48 \\ -0.16 & 1.76 & 2.88 \end{bmatrix} \begin{bmatrix} v_1 \\ 0 \\ 0 \end{bmatrix}$$

and therefore

$$i_1 = 0.88v_1$$

$$i_2 = 0.32v_1$$

$$i_3 = 0.16v_1$$

Observing that $v_2 = R_5 i_3$, we can compute the desired answer:

$$v_2 = R_5 i_3 = R_5(0.16v_1) = 0.25(0.16v_1)$$

$$A_v = \frac{v_2}{v_1} = \frac{0.04v_1}{v_1} = 0.04$$

Comments: The Matlab™ commands required to obtain the inverse of matrix R are listed below.

```
R=[1.5 -0.5 0; -1.5 2 -0.25; 1 -1.25 0.5];
Rinv=inv(R);
```

The presence of a dependent source did not really affect the solution method. Systematic application of mesh analysis provided the desired answer. Is mesh analysis the most efficient solution method? *Hint:* See the exercise below.

CHECK YOUR UNDERSTANDING

Determine the number of independent equations required to solve the circuit of Example 3.13 using node analysis. Which method would you use?

The current source i_x is related to the voltage v_x in the figure on the left by the relation

$$i_x = \frac{v_x}{3}$$

Find the voltage across the 8-Ω resistor by node analysis.

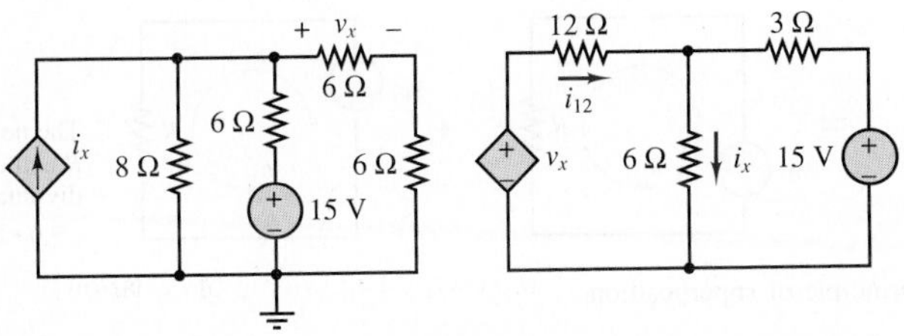

Find the unknown current i_x in the figure on the right, using the mesh current method. The dependent voltage source is related to current i_{12} through the 12-Ω resistor by $v_x = 2i_{12}$.

Answers: Two; 12 V; 1.39 A

Remarks on Node Voltage and Mesh Current Methods

The techniques presented in this section and the two preceding sections find use more generally than just in the analysis of resistive circuits. These methods should be viewed as general techniques for the analysis of any linear circuit; they provide systematic and effective means of obtaining the minimum number of equations necessary to solve a network problem. Since these methods are based on the fundamental laws of circuit analysis, KVL and KCL, they also apply to electric circuits containing nonlinear circuit elements, such as those to be introduced later in this chapter.

You should master both methods as early as possible. Proficiency in these circuit analysis techniques will greatly simplify the learning process for more advanced concepts.

3.5 THE PRINCIPLE OF SUPERPOSITION

This brief section discusses a concept that is frequently called upon in the analysis of linear circuits. Rather than a precise analysis technique, like the mesh current and node voltage methods, the principle of superposition is a conceptual aid that can be very useful in visualizing the behavior of a circuit containing multiple sources. The *principle of superposition* applies to any linear system and for a linear circuit may be stated as follows:

In a linear circuit containing N sources, each branch voltage and current is the sum of N voltages and currents, each of which may be computed by setting all but one source equal to zero and solving the circuit containing that single source.

LO3

An elementary illustration of the concept may easily be obtained by simply considering a circuit with two sources connected in series, as shown in Figure 3.27.

The circuit of Figure 3.27 is more formally analyzed as follows. The current i flowing in the circuit on the left-hand side of Figure 3.27 may be expressed as

$$i = \frac{v_{B1} + v_{B2}}{R} = \frac{v_{B1}}{R} + \frac{v_{B2}}{R} = i_{B1} + i_{B2} \quad \textbf{(3.23)}$$

Figure 3.27 also depicts the circuit as being equivalent to the combined effects of

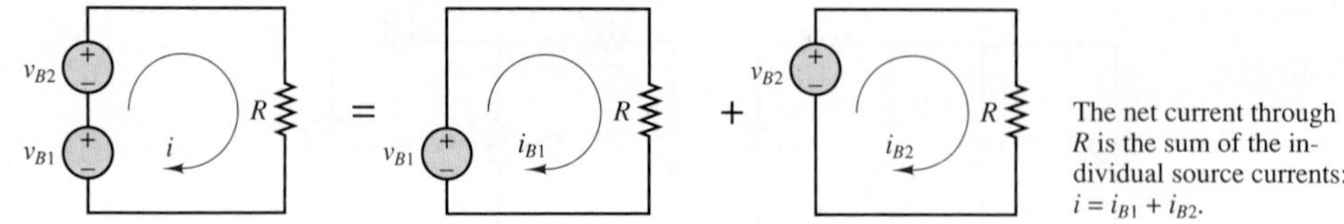

Figure 3.27 The principle of superposition

two circuits, each containing a single source. In each of the two subcircuits, a short circuit has been substituted for the missing battery. This should appear as a sensible procedure, since a short circuit, by definition, will always "see" zero voltage across itself, and therefore this procedure is equivalent to "zeroing" the output of one of the voltage sources.

If, on the other hand, we wished to cancel the effects of a current source, it would stand to reason that an open circuit could be substituted for the current source, since an open circuit is, by definition, a circuit element through which no current can flow (and which therefore generates zero current). These basic principles are used frequently in the analysis of circuits and are summarized in Figure 3.28.

LO3

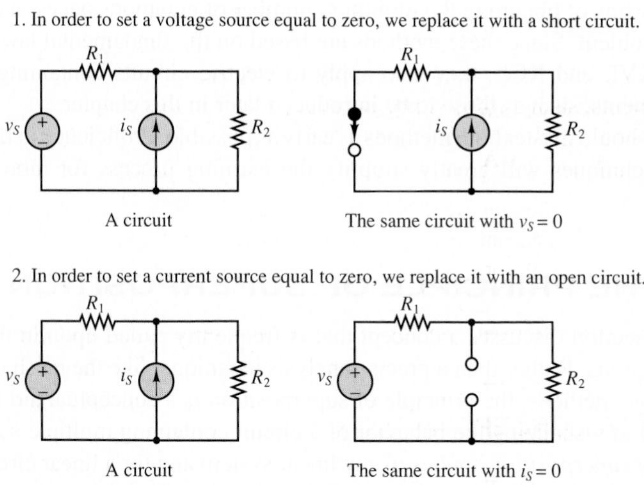

1. In order to set a voltage source equal to zero, we replace it with a short circuit.
R_1
v_S
i_S
R_2
A circuit
The same circuit with $v_S = 0$
2. In order to set a current source equal to zero, we replace it with an open circuit.
A circuit
The same circuit with $i_S = 0$

Figure 3.28 Zeroing voltage and current sources

The principle of superposition can easily be applied to circuits containing multiple sources and is sometimes an effective solution technique. More often, however, other methods result in a more efficient solution. Example 3.14 further illustrates the use of superposition to analyze a simple network. The Check Your Understanding exercises at the end of the section illustrate the fact that superposition is often a cumbersome solution method.

LO3

EXAMPLE 3.14 Principle of Superposition

Problem

Determine the current i_2 in the circuit of Figure 3.29(a), using the principle of superposition.

Solution

Known Quantities: Source voltage and current values; resistor values.

Find: Unknown current i_2.

Given Data: $V_S = 10$ V; $I_S = 2$ A; $R_1 = 5\ \Omega$; $R_2 = 2\ \Omega$; $R_3 = 4\ \Omega$.

Assumptions: Assume the reference node is at the bottom of the circuit.

Analysis: *Part 1:* Zero the current source. Once the current source has been set to zero (replaced by an open circuit), the resulting circuit is a simple series circuit shown in Figure 3.29(b); the current flowing in this circuit i_{2-V} is the current we seek. Since the total series resistance is $5 + 2 + 4 = 11\ \Omega$, we find that $i_{2-V} = 10/11 = 0.909$ A.

Part 2: Zero the voltage source. After we zero the voltage source by replacing it with a short circuit, the resulting circuit consists of three parallel branches shown in Figure 3.29(c): On the left we have a single 5-Ω resistor; in the center we have a −2-A current source (negative because the source current is shown to flow into the ground node); on the right we have a total resistance of $2 + 4 = 6\ \Omega$. Using the current divider rule, we find that the current flowing in the right branch i_{2-I} is given by

$$i_{2-I} = \frac{\frac{1}{6}}{\frac{1}{5} + \frac{1}{6}}(-2) = -0.909 \text{ A}$$

And, finally, the unknown current i_2 is found to be

$$i_2 = i_{2-V} + i_{2-I} = 0 \text{ A}$$

Comments: Superposition is not always a very efficient tool. Beginners may find it preferable to rely on more systematic methods, such as node analysis, to solve circuits. Eventually, experience will suggest the preferred method for any given circuit.

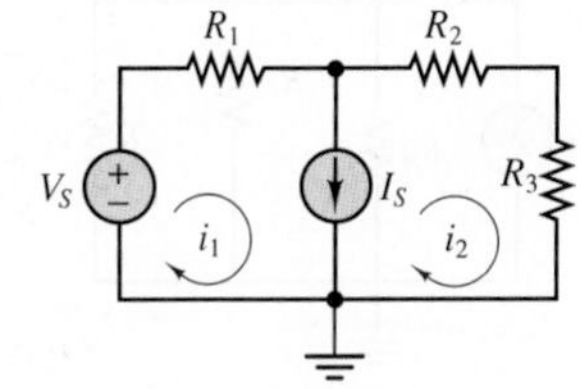

(a)

Figure 3.29 (a) Circuit for the illustration of the principle of superposition

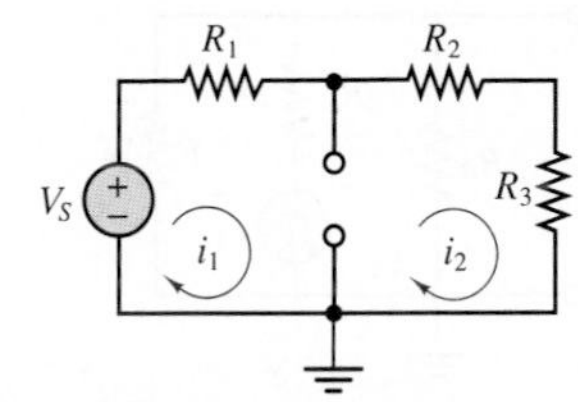

(b)

Figure 3.29 (b) Circuit with current source set to zero

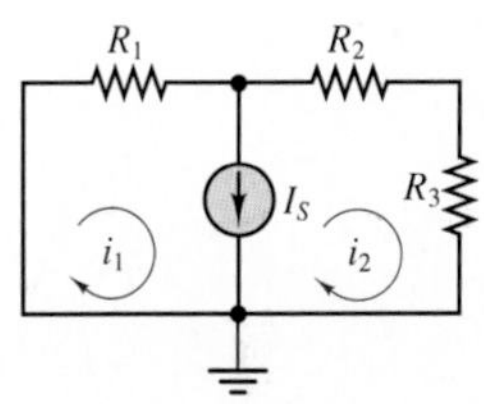

(c)

Figure 3.29 (c) Circuit with voltage source set to zero

CHECK YOUR UNDERSTANDING

In Example 3.15, verify that the same answer is obtained by mesh or node analysis.

EXAMPLE 3.15 Principle of Superposition

LO3

Problem

Determine the voltage across resistor R in the circuit of Figure 3.30.

Solution

Known Quantities: The values of the voltage sources and of the resistors in the circuit of Figure 3.30 are $I_B = 12$ A; $V_G = 12$ V; $R_B = 1\ \Omega$; $R_G = 0.3\ \Omega$; $R = 0.23\ \Omega$.

Find: The voltage across R.

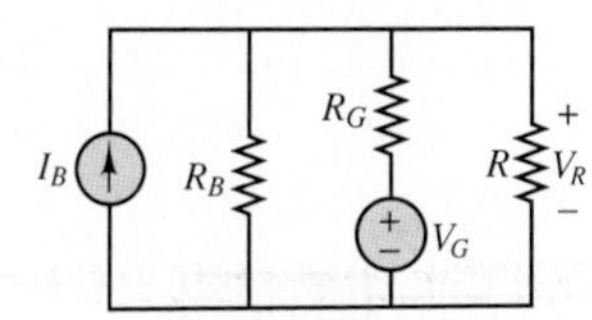

(a)

Figure 3.30 (a) Circuit used to demonstrate the principle of superposition

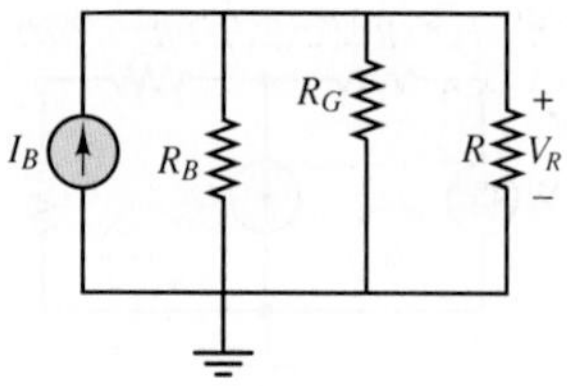

(b)

Figure 3.30 (b) Circuit obtained by suppressing the voltage source

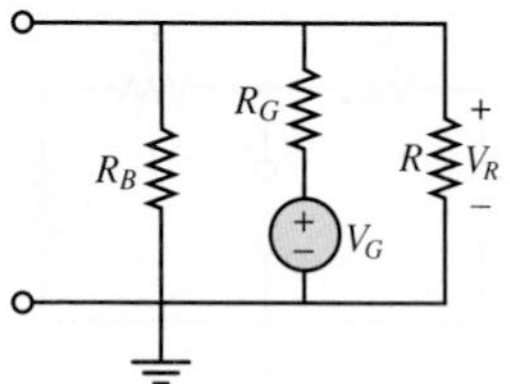

(c)

Figure 3.30 (c) Circuit obtained by suppressing the current source

Analysis: Specify a ground node and the polarity of the voltage across R. Suppress the voltage source by replacing it with a short circuit. Redraw the circuit, as shown in Figure 3.30(b), and apply KCL:

$$-I_B + \frac{V_{R-I}}{R_B} + \frac{V_{R-I}}{R_G} + \frac{V_{R-I}}{R} = 0$$

$$V_{R-I} = \frac{I_B}{1/R_B + 1/R_G + 1/R} = \frac{12}{1/1 + 1/0.3 + 1/0.23} = 1.38 \text{ V}$$

Suppress the current source by replacing it with an open circuit, draw the resulting circuit, as shown in Figure 3.30(c), and apply KCL:

$$\frac{V_{R-V}}{R_B} + \frac{V_{R-V} - V_G}{R_G} + \frac{V_{R-V}}{R} = 0$$

$$V_{R-V} = \frac{V_G/R_G}{1/R_B + 1/R_G + 1/R} = \frac{12/0.3}{1/1 + 1/0.3 + 1/0.23} = 4.61 \text{ V}$$

Finally, we compute the voltage across R as the sum of its two components:

$$V_R = V_{R-I} + V_{R-V} = 5.99 \text{ V}$$

Comments: Superposition essentially doubles the work required to solve this problem. The voltage across R can easily be determined by using a single KCL.

CHECK YOUR UNDERSTANDING

In Example 3.15, verify that the same answer can be obtained by a single application of KCL.

Find the voltages v_a and v_b for the circuits of Example 3.7 by superposition.

Solve Example 3.7, using superposition.

Solve Example 3.10, using superposition.

3.6 ONE-PORT NETWORKS AND EQUIVALENT CIRCUITS

You may recall that, in the discussion of ideal sources in Chapter 2, the flow of energy from a source to a load was described in a very general form, by showing the connection of two "black boxes" labeled *source* and *load* (see Figure 2.2). In the same figure, two other descriptions were shown: a symbolic one, depicting an ideal voltage source and an ideal resistor; and a physical representation, in which the load was represented by a headlight and the source by an automotive battery. Whatever the form chosen for source-load representation, each block—source or load—may be viewed as a two-terminal device, described by an *i-v* characteristic. This general circuit representation is shown in Figure 3.31. This configuration is called a **one-port network** and is particularly useful for introducing the notion of equivalent circuits. Note that the network of Figure 3.31 is completely described by its *i-v* characteristic; this point is best illustrated by Example 3.16.

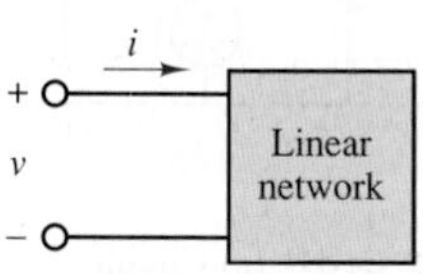

Figure 3.31 One-port network

EXAMPLE 3.16 Equivalent Resistance Calculation

LO4

Problem

Determine the source (load) current i in the circuit of Figure 3.32, using equivalent resistance ideas.

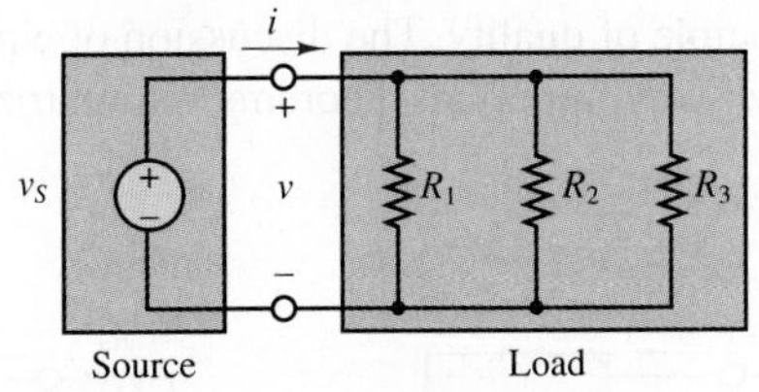

Figure 3.32 Illustration of equivalent-circuit concept

Solution

Known Quantities: Source voltage, resistor values.

Find: Source current.

Given Data: Figures 3.32 and 3.33.

Assumptions: Assume the reference node is at the bottom of the circuit.

Analysis: *Insofar as the source is concerned,* the three parallel resistors appear identical to a single equivalent resistance of value

$$R_{\mathrm{EQ}} = \frac{1}{1/R_1 + 1/R_2 + 1/R_3}$$

Thus, we can replace the three load resistors with the single equivalent resistor R_{EQ}, as shown in Figure 3.33, and calculate

$$i = \frac{v_S}{R_{\mathrm{EQ}}}$$

Comments: Similarly, *insofar as the load is concerned,* it would not matter whether the source consisted, say, of a single 6-V battery or of four 1.5-V batteries connected in series.

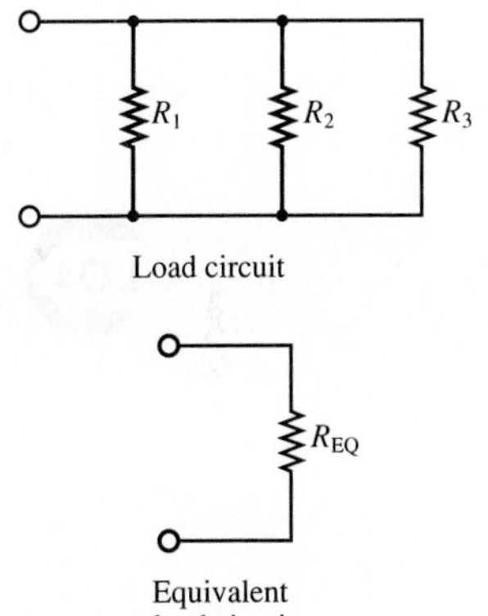

Figure 3.33 Equivalent load resistance concept

For the remainder of this section, we focus on developing techniques for computing equivalent representations of linear networks. Such representations are useful in deriving some simple—yet general—results for linear circuits, as well as analyzing simple nonlinear circuits.

Thévenin and Norton Equivalent Circuits

This section discusses one of the most important topics in the analysis of electric circuits: the concept of an **equivalent circuit.** We show that it is always possible to view even a very complicated circuit in terms of much simpler *equivalent* source and load circuits, and that the transformations leading to equivalent circuits are easily managed, with a little practice. In studying node voltage and mesh current analysis, you may have observed that there is a certain correspondence (called **duality**) between current sources and voltage sources, on one hand, and parallel and series circuits, on the other. This duality appears again very clearly in the analysis of equivalent circuits: It will shortly be shown that equivalent circuits fall into one of two classes, involving either voltage or current sources and (respectively) either series or parallel resistors, reflecting this same principle of duality. The discussion of equivalent circuits begins with the statement of two very important theorems, summarized in Figures 3.34 and 3.35.

LO4

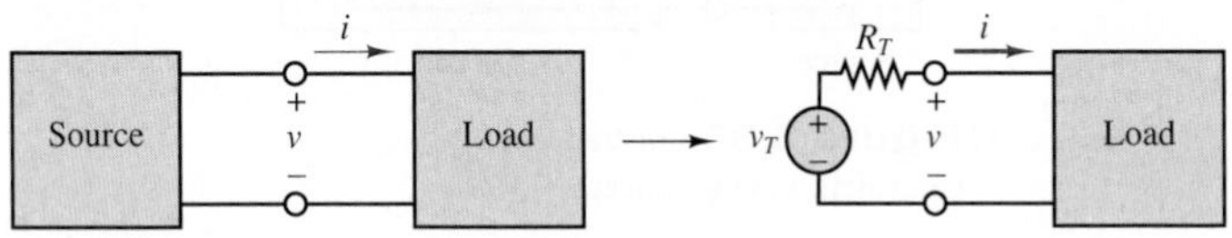

Figure 3.34 Illustration of Thévenin theorem

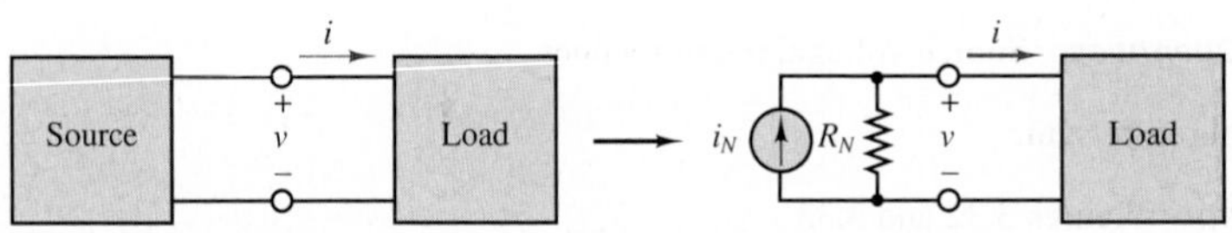

Figure 3.35 Illustration of Norton theorem

LO4

The Thévenin Theorem

When viewed from the load, any network composed of ideal voltage and current sources, and of linear resistors, may be represented by an equivalent circuit consisting of an ideal voltage source v_T in series with an equivalent resistance R_T.

LO4

The Norton Theorem

When viewed from the load, any network composed of ideal voltage and current sources, and of linear resistors, may be represented by an equivalent circuit consisting of an ideal current source i_N in parallel with an equivalent resistance R_N.

The first obvious question to arise is, How are these equivalent source voltages, currents, and resistances computed? The next few sections illustrate the computation of these equivalent circuit parameters, mostly through examples. A substantial number of Check Your Understanding exercises are also provided, with the following caution: The only way to master the computation of Thévenin and Norton equivalent circuits is by patient repetition.

Determination of Norton or Thévenin Equivalent Resistance

In this subsection, we illustrate the calculation of the equivalent resistance of a network containing only linear resistors and independent sources. The first step in computing a Thévenin or Norton equivalent circuit consists of finding the equivalent resistance presented by the circuit at its terminals. This is done by setting all sources in the circuit equal to zero and computing the effective resistance between terminals. The voltage and current sources present in the circuit are set to zero by the same technique used with the principle of superposition: Voltage sources are replaced by short circuits; current sources, by open circuits. To illustrate the procedure, consider the simple circuit of Figure 3.36; the objective is to compute the equivalent resistance the load R_L "sees" at port *a*-*b*.

To compute the equivalent resistance, we remove the load resistance from the circuit and replace the voltage source v_S by a short circuit. At this point—seen from the load terminals—the circuit appears as shown in Figure 3.37. You can see that R_1 and R_2 are in parallel, since they are connected between the same two nodes. If the total resistance between terminals a and b is denoted by R_T, its value can be determined as follows:

$$R_T = R_3 + R_1 \parallel R_2 \tag{3.24}$$

An alternative way of viewing R_T is depicted in Figure 3.38, where a hypothetical 1-A current source has been connected to terminals a and b. The voltage v_x appearing across the *a*-*b* pair is then numerically equal to R_T (only because $i_S =$ 1 A!). With the 1-A source current flowing in the circuit, it should be apparent that the source current encounters R_3 as a resistor in series with the parallel combination of R_1 and R_2, prior to completing the loop.

Summarizing the procedure, we can produce a set of simple rules as an aid in the computation of the Thévenin (or Norton) equivalent resistance for a linear resistive circuit that does not contain dependent sources. The case of circuits containing dependent sources is outlined later in this section.

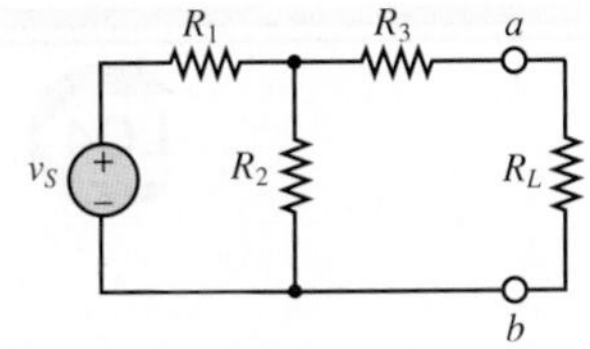

Complete circuit

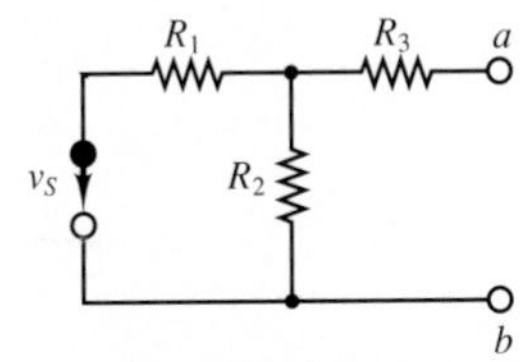

Circuit with load removed for computation of R_T. The voltage source is replaced by a short circuit.

Figure 3.36 Computation of Thévenin resistance

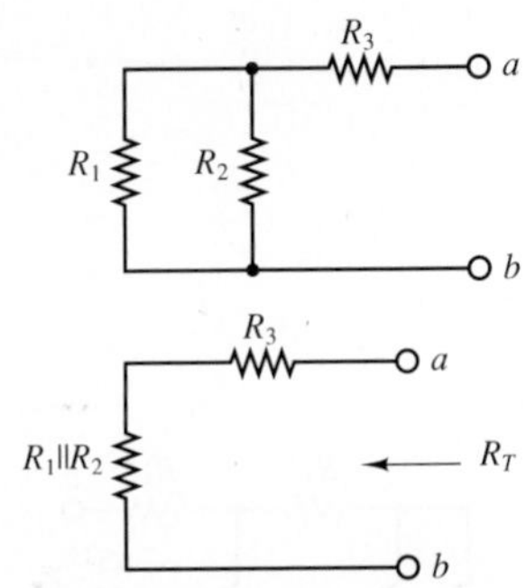

Figure 3.37 Equivalent resistance seen by the load

What is the total resistance the current i_S will encounter in flowing around the circuit?

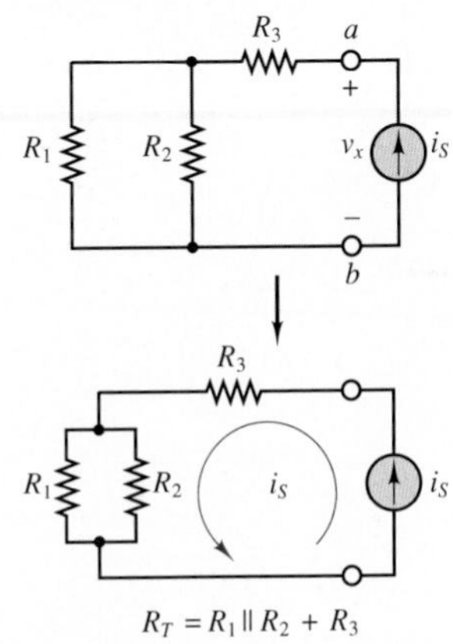

$R_T = R_1 \parallel R_2 + R_3$

Figure 3.38 An alternative method of determining the Thévenin resistance

LO4

FOCUS ON METHODOLOGY

COMPUTATION OF EQUIVALENT RESISTANCE OF A ONE-PORT NETWORK THAT DOES NOT CONTAIN DEPENDENT SOURCES

1. Remove the load.
2. Zero all independent voltage and current sources.
3. Compute the total resistance between load terminals, *with the load removed*. This resistance is equivalent to that which would be encountered by a current source connected to the circuit in place of the load.

We note immediately that this procedure yields a result that is independent of the load. This is a very desirable feature, since once the equivalent resistance has been identified for a source circuit, the equivalent circuit remains unchanged if we connect a different load. The following examples further illustrate the procedure.

EXAMPLE 3.17 Thévenin Equivalent Resistance

LO4

Problem

Find the Thévenin equivalent resistance seen by the load R_L in the circuit of Figure 3.39.

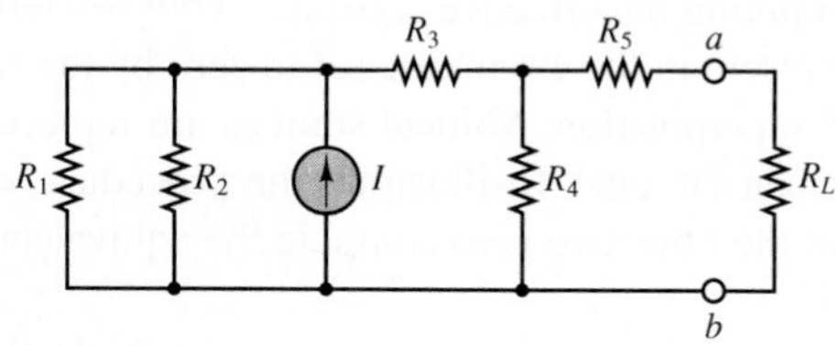

Figure 3.39

Solution

Known Quantities: Resistor and current source values.

Find: Thévenin equivalent resistance R_T.

Schematics, Diagrams, Circuits, and Given Data: $R_1 = 20\ \Omega$; $R_2 = 20\ \Omega$; $I = 5$ A; $R_3 = 10\ \Omega$; $R_4 = 20\ \Omega$; $R_5 = 10\ \Omega$.

Assumptions: Assume the reference node is at the bottom of the circuit.

Analysis: Following the Focus on Methodology box introduced in this section, we first set the current source equal to zero, by replacing it with an open circuit. The resulting circuit is depicted in Figure 3.40. Looking into terminal *a-b*, we recognize that, starting from the left (away from the load) and moving to the right (toward the load), the equivalent resistance is given by the expression

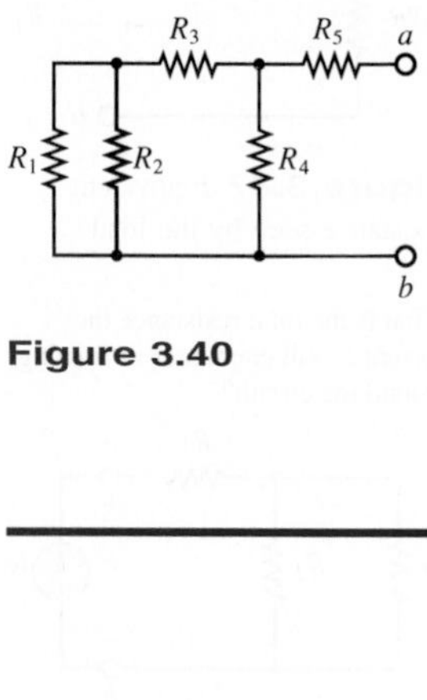

Figure 3.40

$$R_T = [((R_1||R_2) + R_3)\,||R_4] + R_5$$
$$= [((20||20) + 10)\,||20] + 10 = 20\ \Omega$$

Comments: Note that the reduction of the circuit started at the farthest point away from the load.

CHECK YOUR UNDERSTANDING

Find the Thévenin equivalent resistance of the circuit below, as seen by the load resistor R_L.

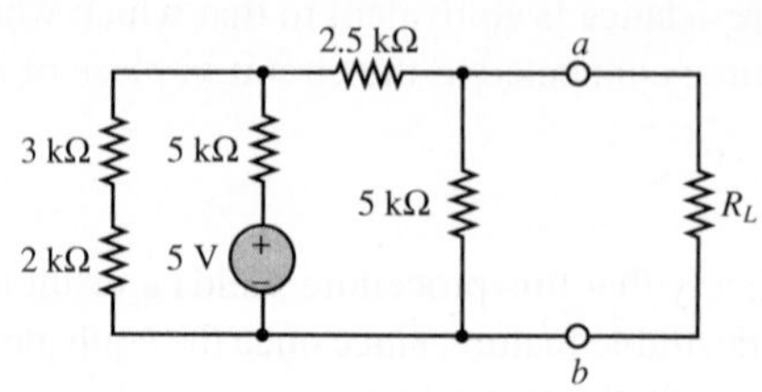

Find the Thévenin equivalent resistance seen by the load resistor R_L in the following circuit.

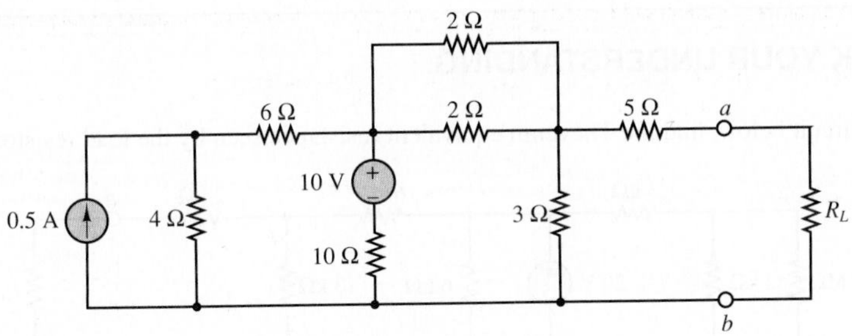

Answers: $R_T = 2.5\ \text{k}\Omega$; $R_T = 7\ \Omega$

EXAMPLE 3.18 Thévenin Equivalent Resistance

LO4

Problem

Compute the Thévenin equivalent resistance seen by the load in the circuit of Figure 3.41.

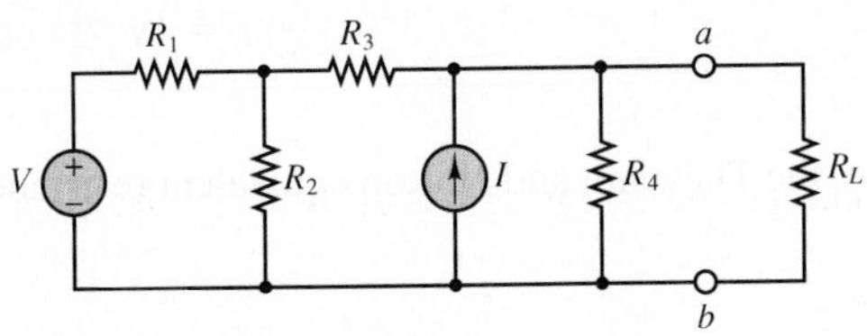

Figure 3.41

Solution

Known Quantities: Resistor values.

Find: Thévenin equivalent resistance R_T.

Schematics, Diagrams, Circuits, and Given Data: $V = 5$ V; $R_1 = 2\ \Omega$; $R_2 = 2\ \Omega$; $R_3 = 1\ \Omega$; $I = 1$ A, $R_4 = 2\ \Omega$.

Assumptions: Assume the reference node is at the bottom of the circuit.

Analysis: Following the Thévenin equivalent resistance Focus on Methodology box, we first set the current source equal to zero, by replacing it with an open circuit, then set the voltage source equal to zero by replacing it with a short circuit. The resulting circuit is depicted in Figure 3.42. Looking into terminal *a-b*, we recognize that, starting from the left (away from the load) and moving to the right (toward the load), the equivalent resistance is given by the expression

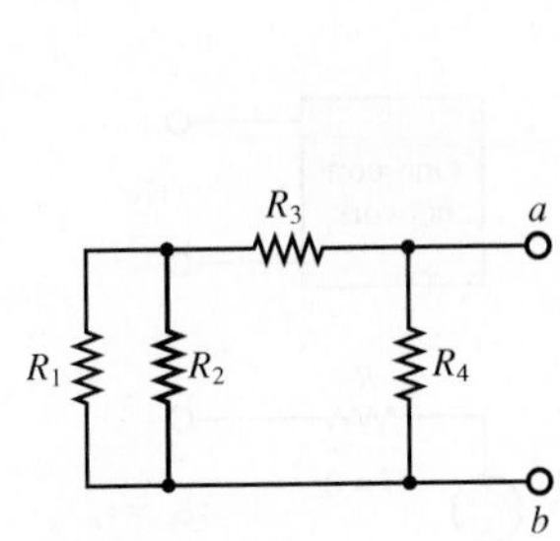

Figure 3.42

$$\begin{aligned} R_T &= ((R_1||R_2) + R_3)\,||R_4 \\ &= ((2||2) + 1)\,||2 = 1\ \Omega \end{aligned}$$

Comments: Note that the reduction of the circuit started at the farthest point away from the load.

CHECK YOUR UNDERSTANDING

For the circuit below, find the Thévenin equivalent resistance seen by the load resistor R_L.

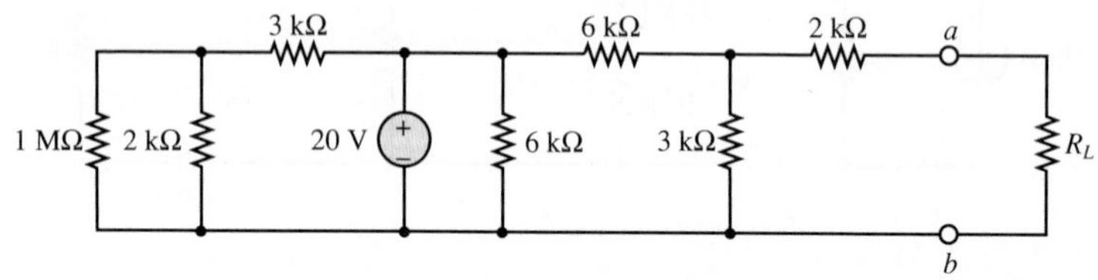

For the circuit below, find the Thévenin equivalent resistance seen by the load resistor R_L.

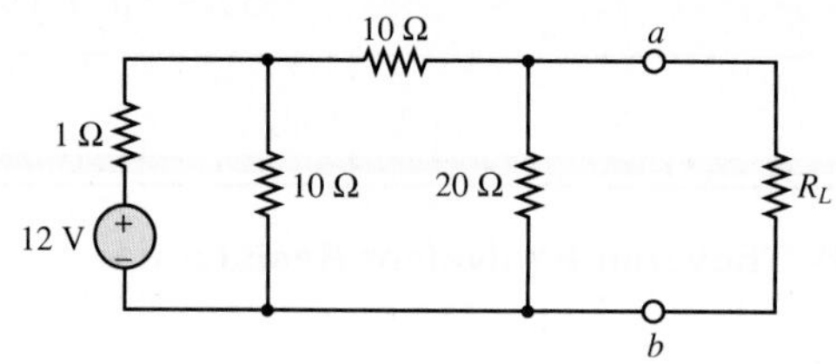

Answers: $R_T = 4.23$ kΩ; $R_T = 7.06$ Ω

As a final note, the Thévenin and Norton equivalent resistances are one and the same quantity:

$$R_T = R_N \tag{3.25}$$

Therefore, the preceding discussion holds whether we wish to compute a Norton or a Thévenin equivalent circuit. From here on, we use the notation R_T exclusively, for both Thévenin and Norton equivalents.

Computing the Thévenin Voltage

This section describes the computation of the Thévenin equivalent voltage v_T for an arbitrary linear resistive circuit containing independent voltage and current sources and linear resistors. The *Thévenin equivalent voltage* is defined as follows:

> The equivalent (Thévenin) source voltage is equal to the **open-circuit voltage** present at the load terminals (with the load removed).

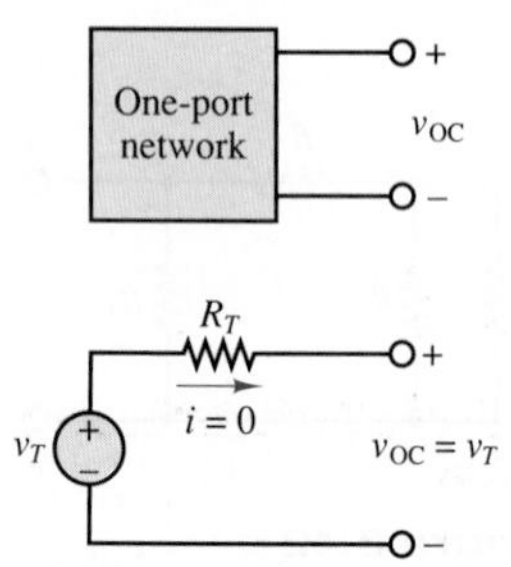

Figure 3.43 Equivalence of open-circuit and Thévenin voltage

This states that to compute v_T, it is sufficient to remove the load and to compute the open-circuit voltage at the one-port terminals. Figure 3.43 illustrates that the open-circuit voltage v_{OC} and the Thévenin voltage v_T must be the same if the Thévenin theorem is to hold. This is true because in the circuit consisting of v_T and R_T, the voltage v_{OC} must equal v_T, since no current flows through R_T and therefore the voltage across R_T is zero. Kirchhoff's voltage law confirms that

$$v_T = R_T(0) + v_{OC} = v_{OC} \tag{3.26}$$

FOCUS ON METHODOLOGY

COMPUTING THE THÉVENIN VOLTAGE

1. Remove the load, leaving the load terminals open-circuited.
2. Define the open-circuit voltage v_{OC} across the open load terminals.
3. Apply any preferred method (e.g., node analysis) to solve for v_{OC}.
4. The Thévenin voltage is $v_T = v_{OC}$.

LO2

The actual computation of the open-circuit voltage is best illustrated by examples; there is no substitute for practice in becoming familiar with these computations. To summarize the main points in the computation of open-circuit voltages, consider the circuit of Figure 3.36, shown again in Figure 3.44 for convenience. Recall that the equivalent resistance of this circuit was given by $R_T = R_3 + R_1 \parallel R_2$. To compute v_{OC}, we disconnect the load, as shown in Figure 3.45, and immediately observe that no current flows through R_3, since there is no closed-circuit connection at that branch. Therefore, v_{OC} must be equal to the voltage across R_2, as illustrated in Figure 3.46. Since the only closed circuit is the mesh consisting of v_S, R_1, and R_2, the answer we are seeking may be obtained by means of a simple voltage divider:

$$v_{OC} = v_{R2} = v_S \frac{R_2}{R_1 + R_2}$$

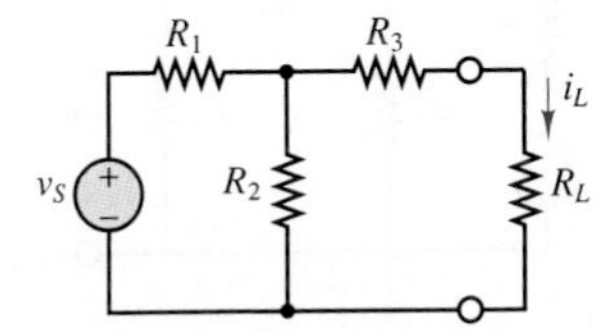

Figure 3.44

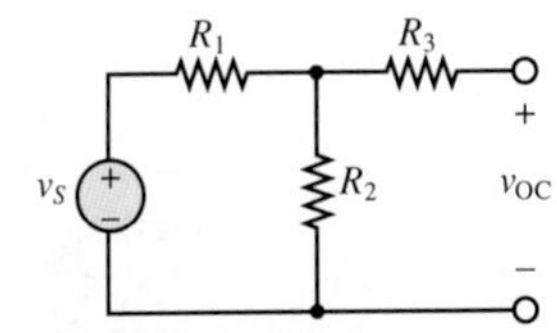

Figure 3.45

It is instructive to review the basic concepts outlined in the example by considering the original circuit and its Thévenin equivalent side by side, as shown in Figure 3.47. The two circuits of Figure 3.47 are equivalent in the sense that the current drawn by the load i_L is the same in both circuits, that current being given by

$$i_L = v_S \cdot \frac{R_2}{R_1 + R_2} \cdot \frac{1}{(R_3 + R_1 \parallel R_2) + R_L} = \frac{v_T}{R_T + R_L} \quad (3.27)$$

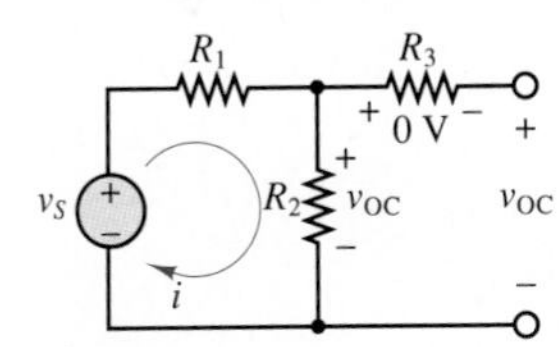

Figure 3.46

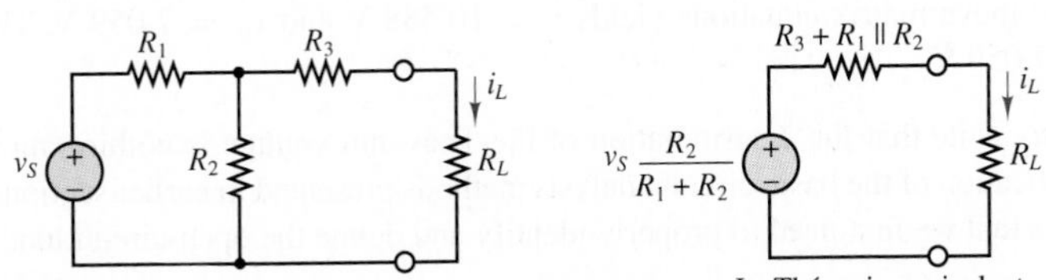

A circuit — Its Thévenin equivalent

Figure 3.47 A circuit and its Thévenin equivalent

The computation of Thévenin equivalent circuits is further illustrated in Examples 3.19 and 3.20.

EXAMPLE 3.19 Thévenin Equivalent Voltage (Open-Circuit Voltage)

LO4

Problem

Compute the open-circuit voltage v_{OC} in the circuit of Figure 3.48.

Solution

Known Quantities: Source voltage, resistor values.

Find: Open-circuit voltage v_{OC}.

Schematics, Diagrams, Circuits, and Given Data: $V = 12$ V; $R_1 = 1\ \Omega$; $R_2 = 10\ \Omega$; $R_3 = 10\ \Omega$; $R_4 = 20\ \Omega$.

Assumptions: Assume the reference node is at the bottom of the circuit.

Analysis: Following the Thévenin voltage Focus on Methodology box, first we remove the load and label the open-circuit voltage v_{OC}. Next, we observe that since v_b is equal to the reference voltage (i.e., zero), the node voltage v_a will be equal, numerically, to the open-circuit voltage. If we define the other node voltage to be v, node analysis is the natural technique for arriving at the solution. Figure 3.48 depicts the original circuit ready for node analysis. Applying KCL at the two nodes, we obtain the following two equations:

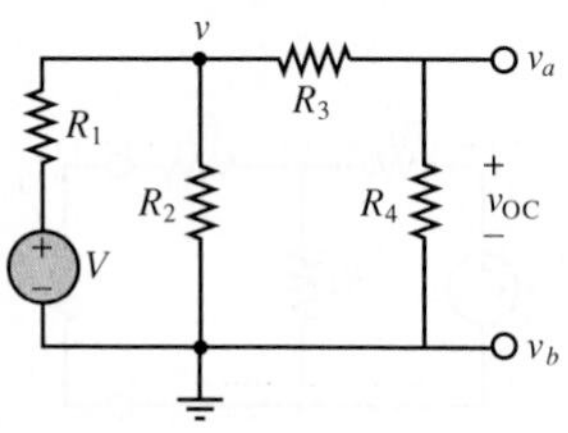

Figure 3.48

$$\frac{V - v}{R_1} - \frac{v}{R_2} - \frac{v - v_a}{R_3} = 0$$

$$\frac{v - v_a}{R_3} - \frac{v_a}{R_4} = 0$$

Substituting numerical values gives

$$\frac{12 - v}{1} - \frac{v}{10} - \frac{v - v_a}{10} = 0$$

$$\frac{v - v_a}{10} - \frac{v_a}{20} = 0$$

In matrix form we can write

$$\begin{bmatrix} 1.2 & -0.1 \\ -0.1 & 0.15 \end{bmatrix} \begin{bmatrix} v \\ v_a \end{bmatrix} = \begin{bmatrix} 12 \\ 0 \end{bmatrix}$$

Solving the above matrix equations yields $v = 10.588$ V and $v_a = 7.059$ V. Thus, $v_{OC} = v_a - v_b = 7.059$ V.

Comments: Note that the determination of the Thévenin voltage is nothing more than the careful application of the basic circuit analysis methods presented in earlier sections. The only difference is that we first need to properly identify and define the open-circuit load voltage.

CHECK YOUR UNDERSTANDING

Find the open-circuit voltage v_{OC} for the circuit of Figure 3.48 if $R_1 = 5\ \Omega$.

Answer: 4.8 V

EXAMPLE 3.20 Load Current Calculation by Thévenin Equivalent Method

LO4

Problem

Compute the load current i by the Thévenin equivalent method in the circuit of Figure 3.49.

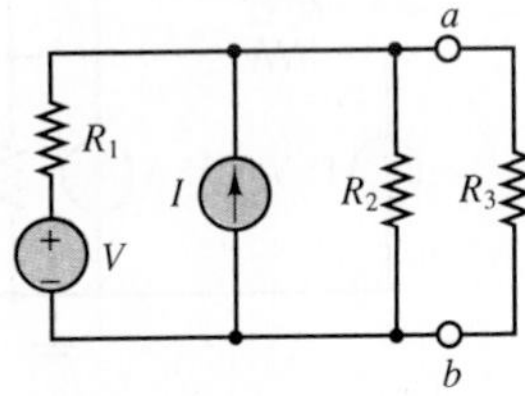

Figure 3.49

Solution

Known Quantities: Source voltage, resistor values.

Find: Load current i.

Schematics, Diagrams, Circuits, and Given Data: $V = 24$ V; $I = 3$ A; $R_1 = 4\ \Omega$; $R_2 = 12\ \Omega$; $R_3 = 6\ \Omega$.

Assumptions: Assume the reference node is at the bottom of the circuit.

Analysis: We first compute the Thévenin equivalent resistance. According to the method proposed earlier, we zero the two sources by shorting the voltage source and opening the current source. The resulting circuit is shown in Figure 3.50. We can clearly see that $R_T = R_1 \| R_2 = 4\|12 = 3\ \Omega$.

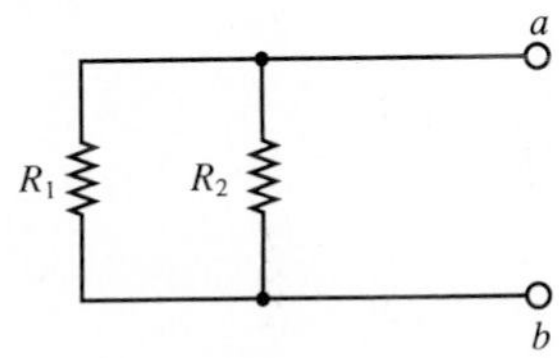

Figure 3.50

Following the Thévenin voltage Focus on Methodology box, first we remove the load and label the open-circuit voltage v_{OC}. The circuit is shown in Figure 3.51. Next, we observe that since v_b is equal to the reference voltage (i.e., zero), the node voltage v_a will be equal, numerically, to the open-circuit voltage. In this circuit, a single nodal equation is required to arrive at the solution:

$$\frac{V - v_a}{R_1} + I - \frac{v_a}{R_2} = 0$$

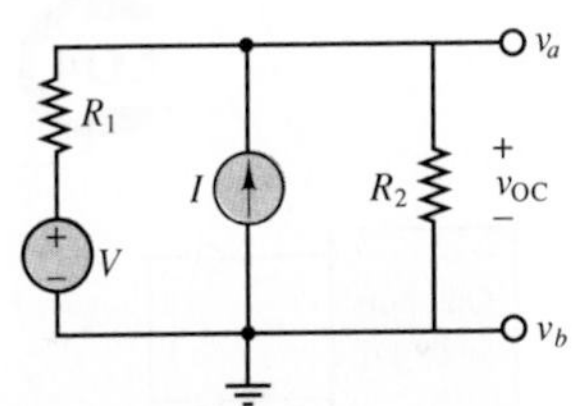

Figure 3.51

Substituting numerical values, we find that $v_a = v_{OC} = v_T = 27$ V.

Finally, we assemble the Thévenin equivalent circuit, shown in Figure 3.52, and reconnect the load resistor. Now the load current can be easily computed to be

$$i = \frac{v_T}{R_T + R_L} = \frac{27}{3+6} = 3 \text{ A}$$

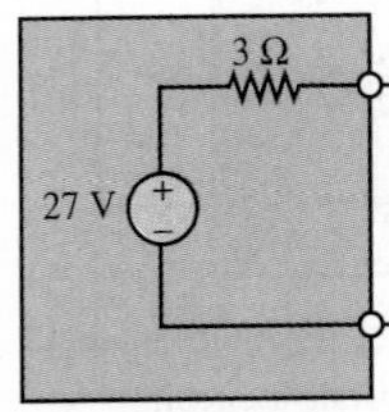

Figure 3.52 Thévenin equivalent

Comments: It may appear that the calculation of load current by the Thévenin equivalent method leads to more complex calculations than, say, node voltage analysis (you might wish to try solving the same circuit by node analysis to verify this). However, there is one major advantage to equivalent circuit analysis: Should the load change (as is often the case in many practical engineering situations), the equivalent circuit calculations still hold, and only the (trivial) last step in the above example needs to be repeated. Thus, knowing the Thévenin equivalent of a particular circuit can be very useful whenever we need to perform computations pertaining to any load quantity.

CHECK YOUR UNDERSTANDING

With reference to Figure 3.44, find the load current i_L by mesh analysis if $v_S = 10$ V, $R_1 = R_3 = 50\ \Omega$, $R_2 = 100\ \Omega$, and $R_L = 150\ \Omega$.

Find the Thévenin equivalent circuit seen by the load resistor R_L for the circuit in the figure on the left.

Find the Thévenin equivalent circuit for the circuit in the figure on the right.

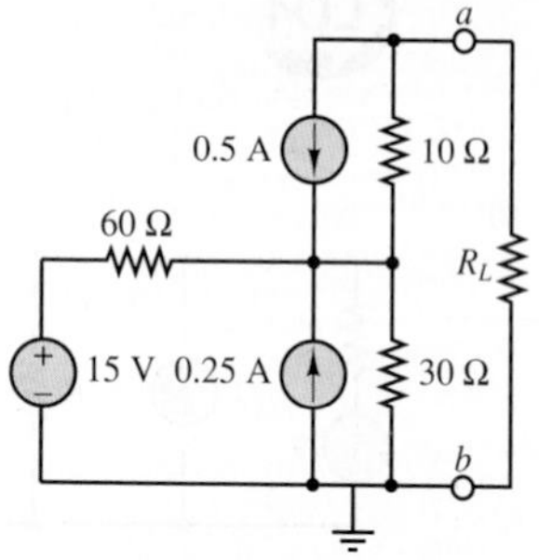

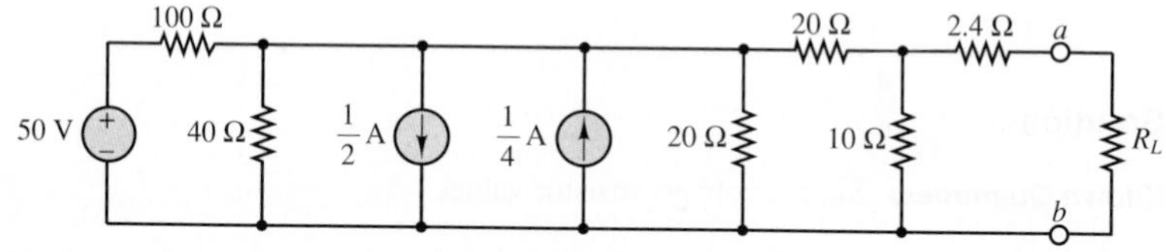

Answers: $i_L = 0.02857$ A; $R_T = 30\ \Omega$; $v_{OC} = v_T = 5$ V; $R_T = 10\ \Omega$; $v_{OC} = v_T = 0.704$ V.

Computing the Norton Current

The computation of the Norton equivalent current is very similar in concept to that of the Thévenin voltage. The following definition serves as a starting point:

LO4

Definition

The Norton equivalent current is equal to the **short-circuit current** that would flow if the load were replaced by a short circuit.

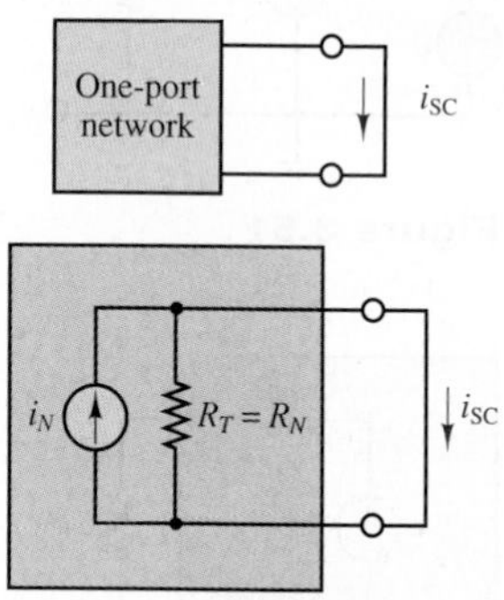

Figure 3.53 Illustration of Norton equivalent circuit

An explanation for the definition of the Norton current is easily found by considering, again, an arbitrary one-port network, as shown in Figure 3.53, where the one-port network is shown together with its Norton equivalent circuit.

It should be clear that the current i_{SC} flowing through the short circuit replacing the load is exactly the Norton current i_N, since all the source current in the circuit of Figure 3.53 must flow through the short circuit. Consider the circuit of Figure 3.54, shown with a short circuit in place of the load resistance. Any of the techniques presented in this chapter could be employed to determine the current i_{SC}. In this particular case, mesh analysis is a convenient tool, once it is recognized that the short-circuit current is a mesh current. Let i_1 and $i_2 = i_{SC}$ be the mesh currents in the circuit of Figure 3.54. Then the following mesh equations can be derived and solved for the short-circuit current:

$$(R_1 + R_2)i_1 - R_2 i_{SC} = v_S$$
$$-R_2 i_1 + (R_2 + R_3)i_{SC} = 0$$

An alternative formulation would employ node analysis to derive the equation

$$\frac{v_S - v}{R_1} = \frac{v}{R_2} + \frac{v}{R_3}$$

leading to

$$v = v_S \frac{R_2 R_3}{R_1 R_3 + R_2 R_3 + R_1 R_2}$$

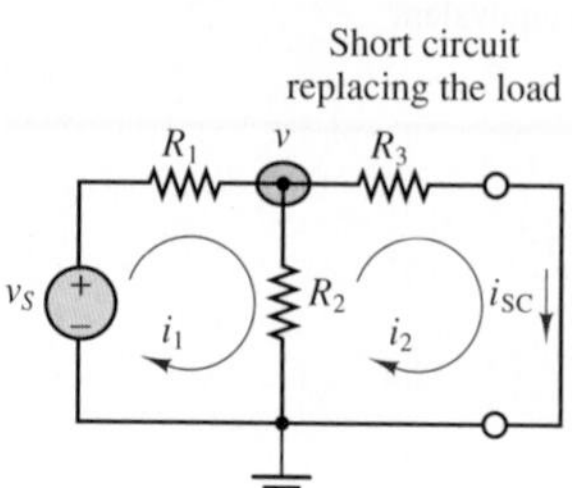

Figure 3.54 Computation of Norton current

FOCUS ON METHODOLOGY

COMPUTING THE NORTON CURRENT

LO4

1. Replace the load with a short circuit.
2. Define the short-circuit current i_{SC} to be the Norton equivalent current.
3. Apply any preferred method (e.g., node analysis) to solve for i_{SC}.
4. The Norton current is $i_N = i_{SC}$.

Recognizing that $i_{SC} = v/R_3$, we can determine the Norton current to be

$$i_N = \frac{v}{R_3} = \frac{v_S R_2}{R_1 R_3 + R_2 R_3 + R_1 R_2}$$

Thus, conceptually, the computation of the Norton current simply requires identifying the appropriate short-circuit current. Example 3.21 further illustrates this idea.

EXAMPLE 3.21 Norton Equivalent Circuit

LO4

Problem

Determine the Norton current and the Norton equivalent for the circuit of Figure 3.55.

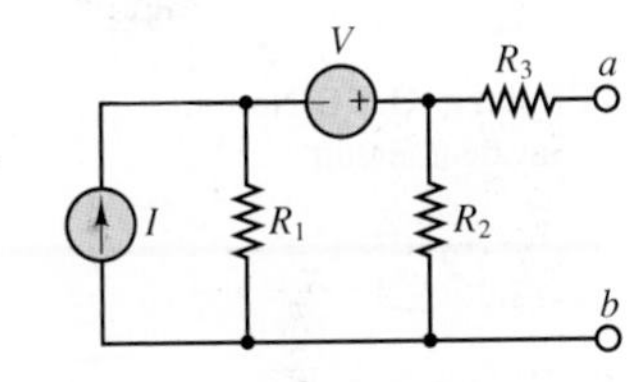

Figure 3.55

Solution

Known Quantities: Source voltage and current; resistor values.

Find: Equivalent resistance R_T; Norton current $i_N = i_{SC}$.

Schematics, Diagrams, Circuits, and Given Data: $V = 6$ V; $I = 2$ A; $R_1 = 6\ \Omega$; $R_2 = 3\ \Omega$; $R_3 = 2\ \Omega$.

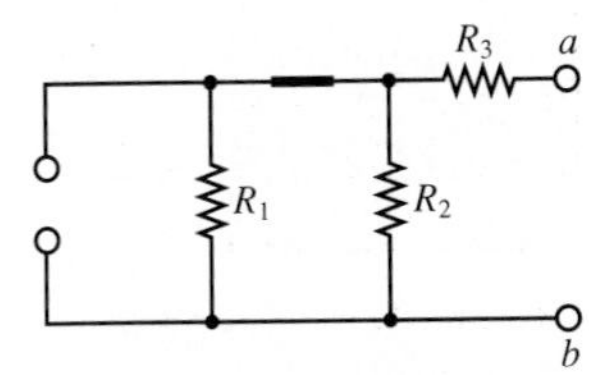

Figure 3.56

Assumptions: Assume the reference node is at the bottom of the circuit.

Analysis: We first compute the Thévenin equivalent resistance. We zero the two sources by shorting the voltage source and opening the current source. The resulting circuit is shown in Figure 3.56. We can clearly see that $R_T = R_1 \| R_2 + R_3 = 6\|3 + 2 = 4\ \Omega$.

Next we compute the Norton current. Following the Norton current Focus on Methodology box, first we replace the load with a short circuit and label the short-circuit current i_{SC}. The circuit is shown in Figure 3.57 ready for node voltage analysis. Note that we have identified two node voltages v_1 and v_2, and that the voltage source requires that $v_2 - v_1 = V$. The unknown current flowing through the voltage source is labeled i.

Now we are ready to apply the node analysis method.

1. The reference node is the ground node in Figure 3.57.
2. The two nodes v_1 and v_2 are also identified in the figure; note that the voltage source imposes the constraint $v_2 = v_1 + V$. Thus only one of the two nodes leads to an independent equation. The unknown current i provides the second independent variable, as you will see in the next step.

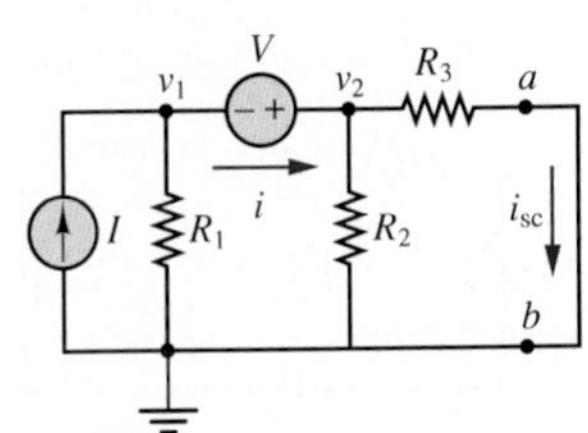

Figure 3.57 Circuit of Example 3.21 ready for node analysis

3. Applying KCL at nodes 1 and 2, we obtain the following set of equations:

$$I - \frac{v_1}{R_1} - i = 0 \qquad \text{node 1}$$

$$i - \frac{v_2}{R_2} - \frac{v_2}{R_3} = 0 \qquad \text{node 2}$$

Next, we eliminate v_1 by substituting $v_1 = v_2 - V$ in the first equation:

$$I - \frac{v_2 - V}{R_1} - i = 0 \qquad \text{node 1}$$

and we rewrite the equations in matrix form, recognizing that the unknowns are i and v_2. Note that the short-circuit current is $i_{SC} = v_2/R_3$; thus we will seek to solve for v_2.

$$\begin{bmatrix} 1 & \frac{1}{R_1} \\ -1 & \frac{1}{R_2} + \frac{1}{R_3} \end{bmatrix} \begin{bmatrix} i \\ v_2 \end{bmatrix} = \begin{bmatrix} I + \frac{V}{R_1} \\ 0 \end{bmatrix}$$

Substituting numerical values, we obtain

$$\begin{bmatrix} 1 & 0.1667 \\ -1 & 0.8333 \end{bmatrix} \begin{bmatrix} i \\ v_2 \end{bmatrix} = \begin{bmatrix} 3 \\ 0 \end{bmatrix}$$

and we can numerically solve for the two unknowns to find that $i = 2.5$ A and $v_2 = 3$ V. Finally, the Norton or short-circuit current is $i_N = i_{SC} = v_2/R_3 = 1.5$ A.

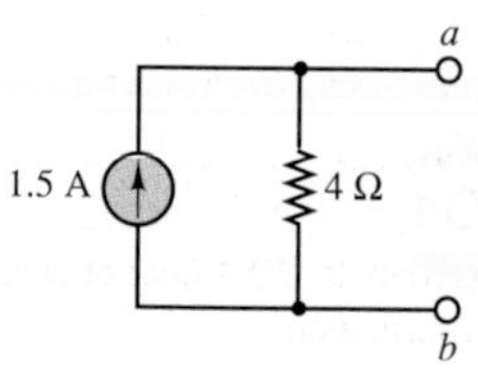

Figure 3.58 Norton equivalent circuit

Comments: In this example it was not obvious whether node analysis, mesh analysis, or superposition might be the quickest method to arrive at the answer. It would be a very good exercise to try the other two methods and compare the complexity of the three solutions. The complete Norton equivalent circuit is shown in Figure 3.58.

CHECK YOUR UNDERSTANDING

Repeat Example 3.21, using mesh analysis. Note that in this case one of the three mesh currents is known, and therefore the complexity of the solution will be unchanged.

Source Transformations

This section illustrates **source transformations,** a procedure that may be very useful in the computation of equivalent circuits, permitting, in some circumstances, replacement of current sources with voltage sources and vice versa. The Norton and Thévenin theorems state that any one-port network can be represented by a voltage source in series with a resistance, or by a current source in parallel with a resistance, and that either of these representations is equivalent to the original circuit, as illustrated in Figure 3.59.

An extension of this result is that any circuit in Thévenin equivalent form may be replaced by a circuit in Norton equivalent form, provided that we use the following relationship:

LO4

$$v_T = R_T i_N \qquad \textbf{(3.28)}$$

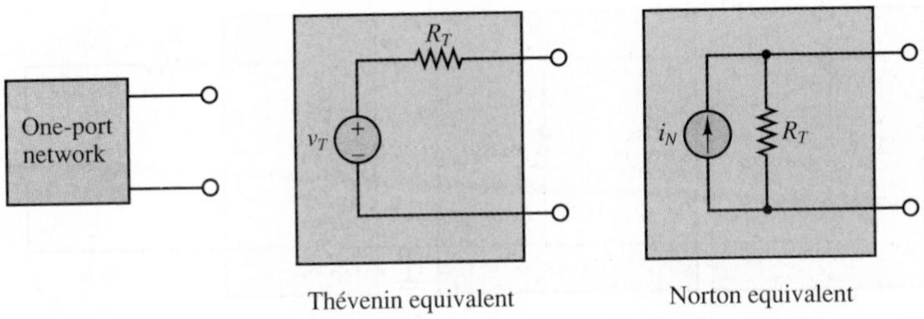

Figure 3.59 Equivalence of Thévenin and Norton representations

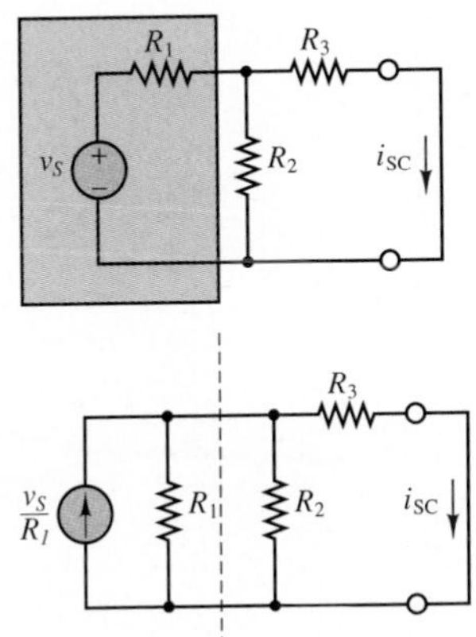

Figure 3.60 Effect of source transformation

Thus, the subcircuit to the left of the dashed line in Figure 3.60 may be replaced by its Norton equivalent, as shown in the figure. Then the computation of i_{SC} becomes very straightforward, since the three resistors are in parallel with the current source and therefore a simple current divider may be used to compute the short-circuit current. Observe that the short-circuit current is the current flowing through R_3; therefore,

$$i_{SC} = i_N = \frac{1/R_3}{1/R_1 + 1/R_2 + 1/R_3} \frac{v_S}{R_1} = \frac{v_S R_2}{R_1 R_3 + R_2 R_3 + R_1 R_2} \qquad \textbf{(3.29)}$$

which is the identical result obtained for the same circuit in the preceding section, as you may easily verify. This source transformation method can be very useful, if employed correctly. Figure 3.61 shows how to recognize subcircuits amenable to such source transformations. Example 3.22 is a numerical example illustrating the procedure.

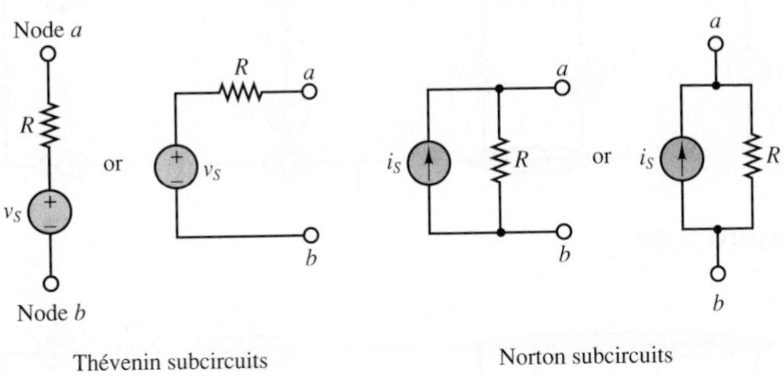

Figure 3.61 Subcircuits amenable to source transformation

EXAMPLE 3.22 Source Transformations

LO4

Problem

Compute the Norton equivalent of the circuit of Figure 3.62 using source transformations.

Solution

Known Quantities: Source voltages and current; resistor values.

Find: Equivalent resistance R_T; Norton current $i_N = i_{SC}$.

Schematics, Diagrams, Circuits, and Given Data: $V_1 = 50$ V; $I = 0.5$ A; $V_2 = 5$ V; $R_1 = 100\ \Omega$; $R_2 = 100\ \Omega$; $R_3 = 200\ \Omega$; $R_4 = 160\ \Omega$.

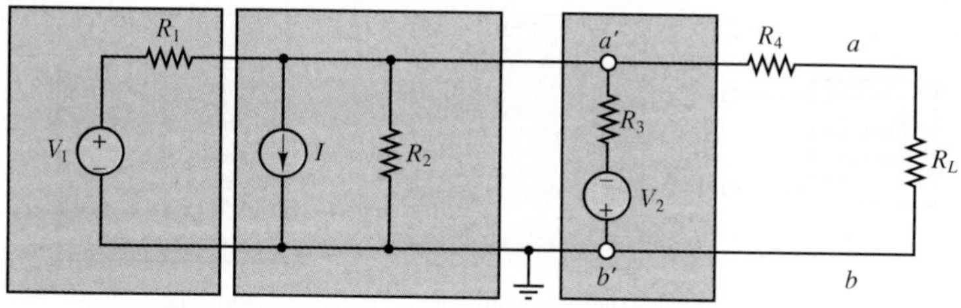

Figure 3.62

Assumptions: Assume the reference node is at the bottom of the circuit.

Analysis: First, we sketch the circuit again, to take advantage of the source transformation technique; we emphasize the location of the nodes for this purpose, as shown in Figure 3.63. Nodes a' and b' have been purposely separated from nodes a'' and b'' even though these are the same pairs of nodes. We can now replace the branch consisting of V_1 and R_1, which appears between nodes a'' and b'', with an equivalent Norton circuit with Norton current source V_1/R_1 and equivalent resistance R_1. Similarly, the series branch between nodes a' and b' is replaced by an equivalent Norton circuit with Norton current source V_2/R_3 and equivalent resistance R_3. The result of these manipulations is shown in Figure 3.64. The same circuit is now depicted in Figure 3.65 with numerical values substituted for each component. Note how easy it is to visualize the equivalent resistance: If each current source is replaced by an open circuit, we find

$$R_T = R_1||R_2||R_3|| + R_4 = 200||100||100 + 160 = 200\ \Omega$$

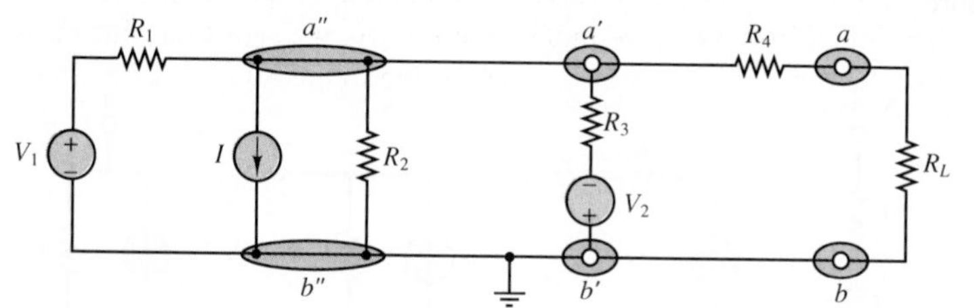

Figure 3.63

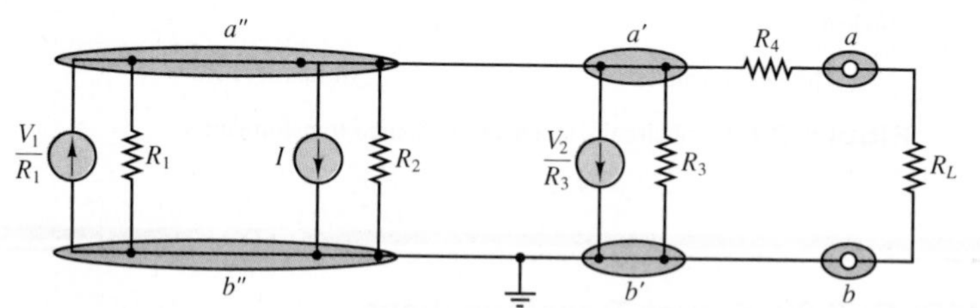

Figure 3.64

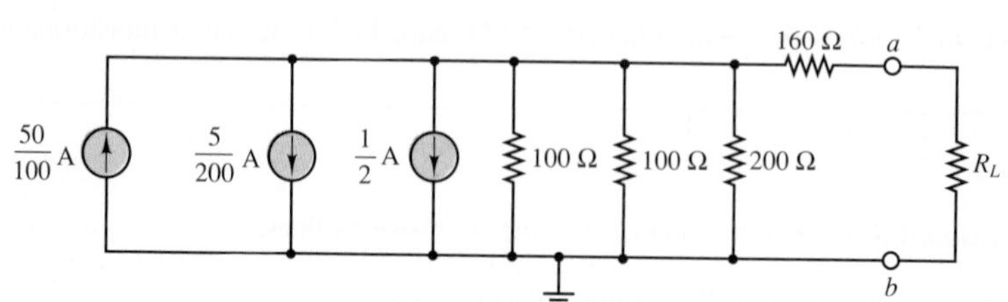

Figure 3.65

The calculation of the Norton current is similarly straightforward, since it simply involves summing the currents:

$$i_N = 0.5 - 0.025 - 0.5 = -0.025 \text{ A}$$

Figure 3.66 depicts the complete Norton equivalent circuit connected to the load.

Comments: It is not always possible to reduce a circuit as easily as was shown in this example by means of source transformations. However, it may be advantageous to use source transformation as a means of converting parts of a circuit to a different form, perhaps more naturally suited to a particular solution method (e.g., node analysis).

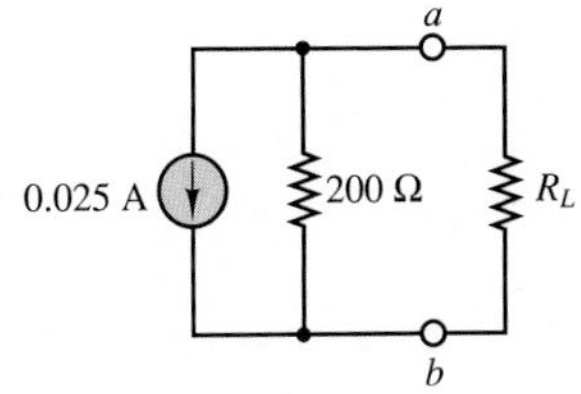

Figure 3.66

Experimental Determination of Thévenin and Norton Equivalents

The idea of equivalent circuits as a means of representing complex and sometimes unknown networks is useful not only analytically, but in practical engineering applications as well. It is very useful to have a measure, for example, of the equivalent internal resistance of an instrument, so as to have an idea of its power requirements and limitations. Fortunately, Thévenin and Norton equivalent circuits can also be evaluated experimentally by means of very simple techniques. The basic idea is that the Thévenin voltage is an open-circuit voltage and the Norton current is a short-circuit current. It should therefore be possible to conduct appropriate measurements to determine these quantities. Once v_T and i_N are known, we can determine the Thévenin resistance of the circuit being analyzed according to the relationship

$$R_T = \frac{v_T}{i_N} \tag{3.30}$$

How are v_T and i_N measured, then?

Figure 3.67 illustrates the measurement of the open-circuit voltage and short-circuit current for an arbitrary network connected to any load and also illustrates that the procedure requires some special attention, because of the nonideal nature of any practical measuring instrument. The figure clearly illustrates that in the presence of finite meter resistance r_m, one must take this quantity into account in the computation of the short-circuit current and open-circuit voltage; v_{OC} and i_{SC} appear between quotation marks in the figure specifically to illustrate that the measured "open-circuit voltage" and "short-circuit current" are in fact affected by the internal resistance of the measuring instrument and are not the true quantities.

You should verify that the following expressions for the true short-circuit current and open-circuit voltage apply (see the material on nonideal measuring instruments in Section 2.8):

$$\begin{aligned} i_N &= \text{“}i_{SC}\text{”}\left(1 + \frac{r_m}{R_T}\right) \\ v_T &= \text{“}v_{OC}\text{”}\left(1 + \frac{R_T}{r_m}\right) \end{aligned} \tag{3.31}$$

where i_N is the ideal Norton current, v_T is the Thévenin voltage, and R_T is the true Thévenin resistance. If you recall the earlier discussion of the properties of ideal ammeters and voltmeters, you will recall that for an ideal ammeter, r_m should approach zero, while in an ideal voltmeter, the internal resistance should approach an

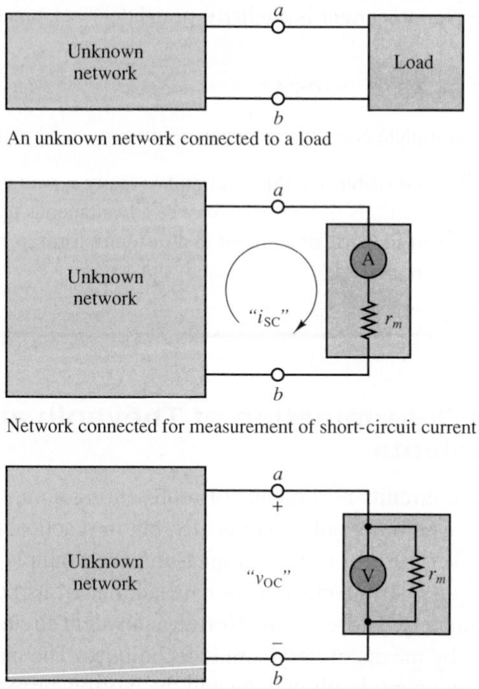

Figure 3.67 Measurement of open-circuit voltage and short-circuit current

open circuit (infinity); thus, the two expressions just given permit the determination of the true Thévenin and Norton equivalent sources from an (imperfect) measurement of the open-circuit voltage and short-circuit current, provided that the internal meter resistance r_m is known. Note also that, in practice, the internal resistance of voltmeters is sufficiently high to be considered infinite relative to the equivalent resistance of most practical circuits; on the other hand, it is impossible to construct an ammeter that has zero internal resistance. If the internal ammeter resistance is known, however, a reasonably accurate measurement of short-circuit current may be obtained. The following Focus on Measurements box illustrates the point.

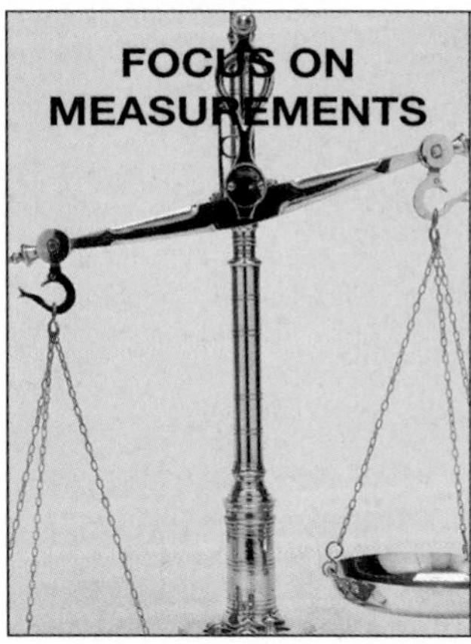

Experimental Determination of Thévenin Equivalent Circuit

Problem:

Determine the Thévenin equivalent of an unknown circuit from measurements of open-circuit voltage and short-circuit current.

Solution:

***Known Quantities*—**Measurement of short-circuit current and open-circuit voltage. Internal resistance of measuring instrument.

***Find*—**Equivalent resistance R_T; Thévenin voltage $v_T = v_{OC}$.

***Schematics, Diagrams, Circuits, and Given Data*—**Measured $v_{OC} = 6.5$ V; measured $i_{SC} = 3.75$ mA; $r_m = 15\ \Omega$.

(*Continued*)

(*Concluded*)

Assumptions—The unknown circuit is a linear circuit containing ideal sources and resistors only.

Analysis—The unknown circuit, shown on the top left in Figure 3.68, is replaced by its Thévenin equivalent and is connected to an ammeter for a measurement of the short-circuit current (Figure 3.68, top right), and then to a voltmeter for the measurement of the open-circuit voltage (Figure 3.68, bottom). The open-circuit voltage measurement yields the Thévenin voltage:

$$v_{OC} = v_T = 6.5 \text{ V}$$

To determine the equivalent resistance, we observe in the figure depicting the voltage measurement that, according to the circuit diagram,

$$\frac{v_{OC}}{i_{SC}} = R_T + r_m$$

Thus,

$$R_T = \frac{v_{OC}}{i_{SC}} - r_m = 1{,}733 - 15 = 1{,}718 \ \Omega$$

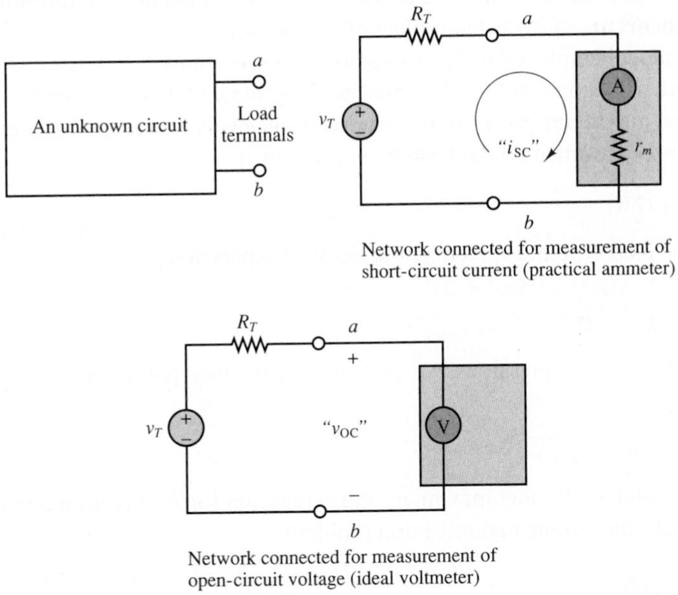

Figure 3.68

Comments—Note how easy the experimental method is, provided we are careful to account for the internal resistance of the **measuring instruments.**

FIND IT ON THE WEB

One last comment is in order concerning the practical measurement of the internal resistance of a network. In most cases, it is not advisable to actually short-circuit a network by inserting a series ammeter as shown in Figure 3.67; permanent damage to the circuit or to the ammeter may be a consequence. For example, imagine that you wanted to estimate the internal resistance of an automotive battery; connecting a laboratory ammeter between the battery terminals would surely result in immediate loss of the instrument. Most ammeters are not designed to withstand currents of such magnitude. Thus, the experimenter should pay attention to the capabilities of the ammeters and voltmeters used in measurements of this type, as well as to the (approximate) power ratings of any sources present. However, there are established techniques especially designed to measure large currents.

3.7 MAXIMUM POWER TRANSFER

The reduction of any linear resistive circuit to its Thévenin or Norton equivalent form is a very convenient conceptualization, as far as the computation of load-related quantities is concerned. One such computation is that of the power absorbed by the load. The Thévenin and Norton models imply that some of the power generated by the source will necessarily be dissipated by the internal circuits within the source. Given this unavoidable power loss, a logical question to ask is, How much power can be transferred to the load from the source under the most ideal conditions? Or, alternatively, what is the value of the load resistance that will absorb maximum power from the source? The answer to these questions is contained in the **maximum power transfer theorem,** which is the subject of this section.

LO5

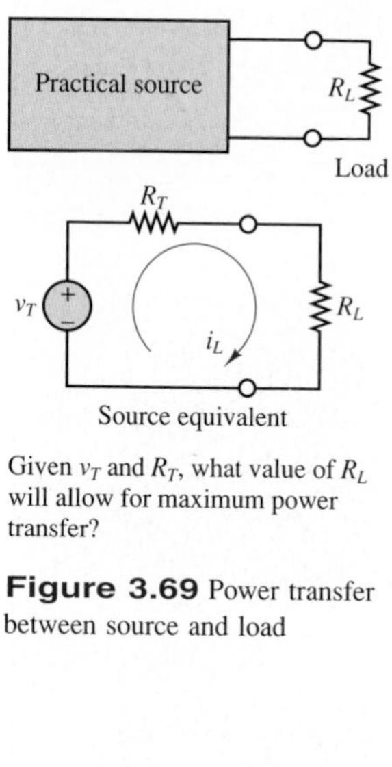

Given v_T and R_T, what value of R_L will allow for maximum power transfer?

Figure 3.69 Power transfer between source and load

The model employed in the discussion of power transfer is illustrated in Figure 3.69, where a practical source is represented by means of its Thévenin equivalent circuit. The maximum power transfer problem is easily formulated if we consider that the power absorbed by the load P_L is given by

$$P_L = i_L^2 R_L \tag{3.32}$$

and that the load current is given by the familiar expression

$$i_L = \frac{v_T}{R_L + R_T} \tag{3.33}$$

Combining the two expressions, we can compute the load power as

$$P_L = \frac{v_T^2}{(R_L + R_T)^2} R_L \tag{3.34}$$

To find the value of R_L that maximizes the expression for P_L (assuming that V_T and R_T are fixed), the simple maximization problem

$$\frac{dP_L}{dR_L} = 0 \tag{3.35}$$

must be solved. Computing the derivative, we obtain the following expression:

$$\frac{dP_L}{dR_L} = \frac{v_T^2 (R_L + R_T)^2 - 2v_T^2 R_L (R_L + R_T)}{(R_L + R_T)^4} \tag{3.36}$$

which leads to the expression

$$(R_L + R_T)^2 - 2R_L (R_L + R_T) = 0 \tag{3.37}$$

It is easy to verify that the solution of this equation is

$$R_L = R_T \tag{3.38}$$

Thus, to transfer maximum power to a load, the equivalent source and load resistances must be **matched,** that is, equal to each other. Figure 3.70 depicts a plot of the load power divided by v_T^2 versus the ratio of R_L to R_T. Note that this value is maximum when $R_L = R_T$.

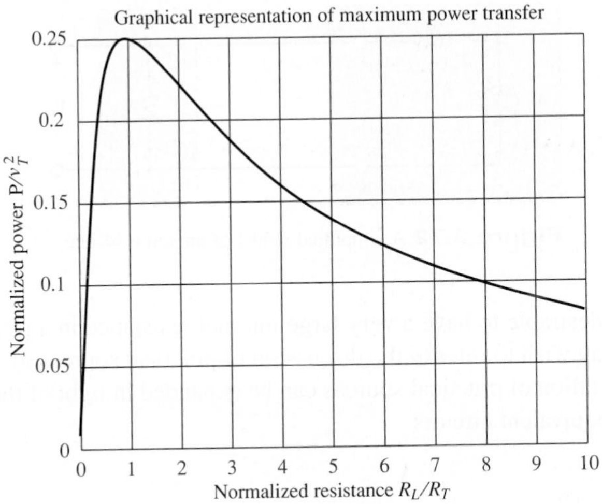

Figure 3.70 Graphical representation of maximum power transfer

This analysis shows that to transfer maximum power to a load, given a fixed equivalent source resistance, the load resistance must match the equivalent source resistance. What if we reversed the problem statement and required that the load resistance be fixed? What would then be the value of source resistance that maximizes the power transfer in this case? The answer to this question can be easily obtained by solving the Check Your Understanding exercises at the end of the section.

A problem related to power transfer is that of **source loading.** This phenomenon, which is illustrated in Figure 3.71, may be explained as follows: When a practical voltage source is connected to a load, the current that flows from the source to the load will cause a voltage drop across the internal source resistance v_{int}; as a consequence, the voltage actually seen by the load will be somewhat lower than the *open-circuit voltage* of the source. As stated earlier, the open-circuit voltage is equal to the Thévenin voltage. The extent of the internal voltage drop within the source depends on the amount of current drawn by the load. With reference to Figure 3.72, this internal drop is equal to iR_T, and therefore the load voltage will be

$$v_L = v_T - iR_T \tag{3.39}$$

It should be apparent that it is desirable to have as small an internal resistance as possible in a practical voltage source.

In the case of a current source, the internal resistance will draw some current away from the load because of the presence of the internal source resistance; this current is denoted by i_{int} in Figure 3.71. Thus the load will receive only part of the *short-circuit current* available from the source (the Norton current):

$$i_L = i_N - \frac{v}{R_T} \tag{3.40}$$

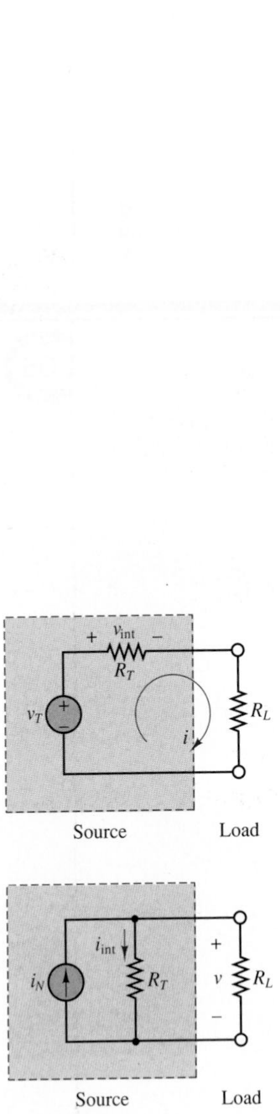

Figure 3.71 Source loading effects

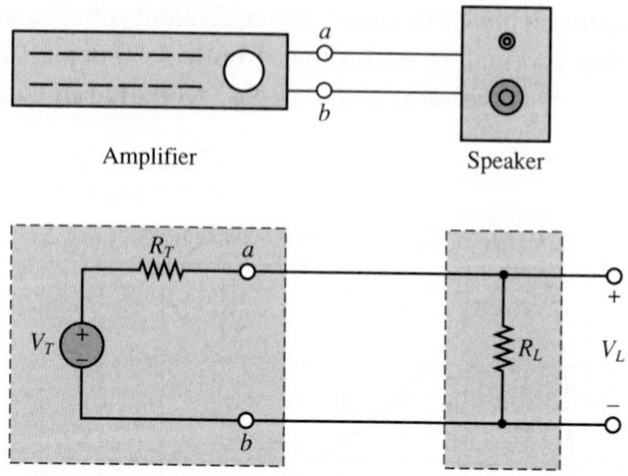

Figure 3.72 A simplified model of an audio system

It is therefore desirable to have a very large internal resistance in a practical current source. You may wish to refer to the discussion of practical sources to verify that the earlier interpretation of practical sources can be expanded in light of the more recent discussion of equivalent circuits.

LO5

EXAMPLE 3.23 Maximum Power Transfer

Problem

Use the maximum power transfer theorem to determine the increase in power delivered to a loudspeaker resulting from matching the speaker load resistance to the amplifier equivalent source resistance.

Solution

Known Quantities: Source equivalent resistance R_T; unmatched speaker load resistance R_{LU}; matched loudspeaker load resistance R_{LM}.

Find: Difference between power delivered to loudspeaker with unmatched and matched loads, and corresponding percentage increase.

Schematics, Diagrams, Circuits, and Given Data: $R_T = 8\ \Omega$; $R_{LU} = 16\ \Omega$; $R_{LM} = 8\ \Omega$.

Assumptions: The amplifier can be modeled as a linear resistive circuit, for the purposes of this analysis.

Analysis: Imagine that we have unknowingly connected an 8-Ω amplifier to a 16-Ω speaker. We can compute the power delivered to the speaker as follows. The load voltage is found by using the voltage divider rule:

$$v_{LU} = \frac{R_{LU}}{R_{LU} + R_T} v_T = \frac{2}{3} v_T$$

and the load power is then computed to be

$$P_{LU} = \frac{v_{LU}^2}{R_{LU}} = \frac{4}{9} \frac{v_T^2}{R_{LU}} = 0.0278 v_T^2$$

Let us now repeat the calculation for the case of a matched 8-Ω speaker resistance R_{LM}. Let the new load voltage be v_{LM} and the corresponding load power be P_{LM}. Then

$$v_{\mathrm{LM}} = \frac{1}{2} v_T$$

and

$$P_{\mathrm{LM}} = \frac{v_{\mathrm{LM}}^2}{R_{\mathrm{LM}}} = \frac{1}{4} \frac{v_T^2}{R_{\mathrm{LM}}} = 0.03125 v_T^2$$

The increase in load power is therefore

$$\Delta P = \frac{0.03125 - 0.0278}{0.0278} \times 100 = 12.5\%$$

Comments: In practice, an audio amplifier and a speaker are not well represented by the simple resistive Thévenin equivalent models used in the present example. Circuits that are appropriate to model amplifiers and loudspeakers are presented in later chapters. The audiophile can find further information concerning hi-fi circuits in Chapters 7 and 16.

FIND IT ON THE WEB

Focus on Computer-Aided Tools: A very nice illustration of the **maximum power transfer theorem** based on MathCad™ may be found in the Web references.

CHECK YOUR UNDERSTANDING

A practical voltage source has an internal resistance of 1.2 Ω and generates a 30-V output under open-circuit conditions. What is the smallest load resistance we can connect to the source if we do not wish the load voltage to drop by more than 2 percent with respect to the source open-circuit voltage?

A practical current source has an internal resistance of 12 kΩ and generates a 200-mA output under short-circuit conditions. What percentage drop in load current will be experienced (with respect to the short-circuit condition) if a 200-Ω load is connected to the current source?

Repeat the derivation leading to equation 3.38 for the case where the load resistance is fixed and the source resistance is variable. That is, differentiate the expression for the load power, P_L with respect to R_S instead of R_L. What is the value of R_S that results in maximum power transfer to the load?

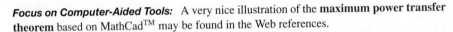

Answers: 58.8 Ω; 1.64%; $R_s = 0$

3.8 NONLINEAR CIRCUIT ELEMENTS

Until now the focus of this chapter has been on linear circuits, containing ideal voltage and current sources, and linear resistors. In effect, one reason for the simplicity of some of the techniques illustrated earlier is the ability to utilize Ohm's law as a simple, linear description of the i-v characteristic of an ideal resistor. In many practical instances, however, the engineer is faced with elements exhibiting a nonlinear i-v characteristic. This section explores two methods for analyzing nonlinear circuit elements.

Description of Nonlinear Elements

There are a number of useful cases in which a simple functional relationship exists between voltage and current in a nonlinear circuit element. For example, Figure 3.73 depicts an element with an exponential i-v characteristic, described by the following equations:

$$
\begin{aligned}
i &= I_0 e^{\alpha v} \qquad v > 0 \\
i &= -I_0 \qquad v \le 0
\end{aligned}
\tag{3.41}
$$

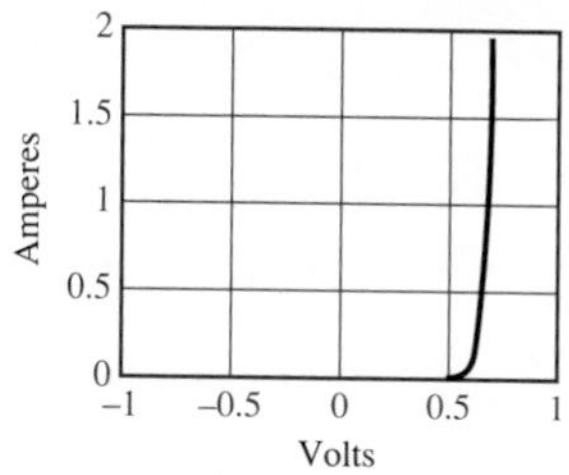

Figure 3.73 The i-v characteristic of exponential resistor

There exists, in fact, a circuit element (the semiconductor diode) that very nearly satisfies this simple relationship. The difficulty in the i-v relationship of equation 3.41 is that it is not possible, in general, to obtain a closed-form analytical solution, even for a very simple circuit.

With the knowledge of equivalent circuits you have just acquired, one approach to analyzing a circuit containing a nonlinear element might be to treat the nonlinear element as a load and to compute the Thévenin equivalent of the remaining circuit, as shown in Figure 3.74. Applying KVL, the following equation may then be obtained:

Nonlinear element as a load. We wish to solve for v_x and i_x.

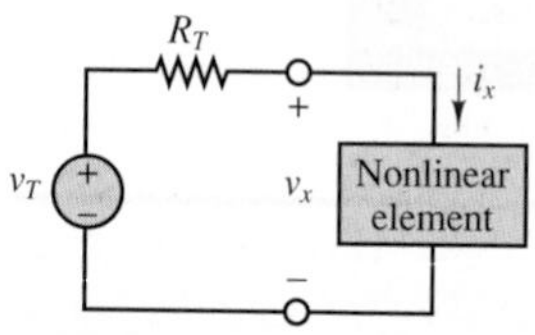

Figure 3.74 Representation of nonlinear element in a linear circuit

$$
v_T = R_T i_x + v_x \tag{3.42}
$$

To obtain the second equation needed to solve for both the unknown voltage v_x and the unknown current i_x, it is necessary to resort to the i-v description of the nonlinear element, namely, equation 3.41. If, for the moment, only positive voltages are considered, the circuit is completely described by the following system:

$$
\begin{aligned}
i_x &= I_0 e^{\alpha v_x} \qquad v_x > 0 \\
v_T &= R_T i_x + v_x
\end{aligned}
\tag{3.43}
$$

The two parts of equation 3.43 represent a system of two equations in two unknowns; however, one of these equations is nonlinear. If we solve for the load voltage and current, for example, by substituting the expression for i_x in the linear equation, we obtain the following expression:

$$
v_T = R_T I_0 e^{\alpha v_x} + v_x \tag{3.44}
$$

or

$$
v_x = v_T - R_T I_0 e^{\alpha v_x} \tag{3.45}
$$

Equations 3.44 and 3.45 do not have a closed-form solution; that is, they are *transcendental equations*. How can v_x be found? One possibility is to generate a solution numerically, by guessing an initial value (for example, $v_x = 0$) and iterating until a sufficiently precise solution is found. This solution is explored further in the homework problems. Another method is based on a graphical analysis of the circuit and is described in the following section.

Graphical (Load-Line) Analysis of Nonlinear Circuits

The nonlinear system of equations of the previous section may be analyzed in a different light, by considering the graphical representation of equation 3.42, which may also be written as

$$
i_x = -\frac{1}{R_T} v_x + \frac{v_T}{R_T} \tag{3.46}
$$

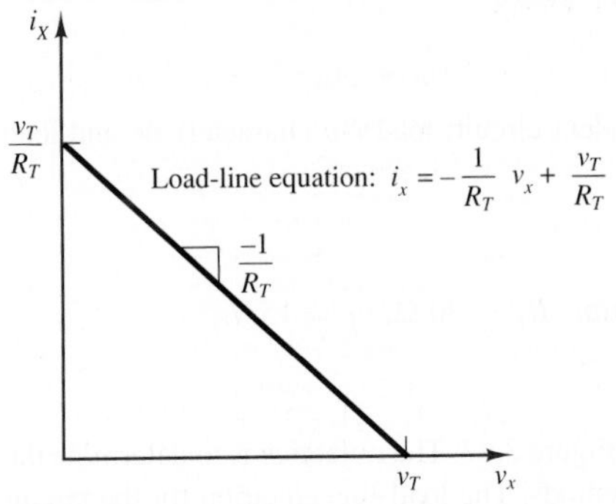

Figure 3.75 Load line

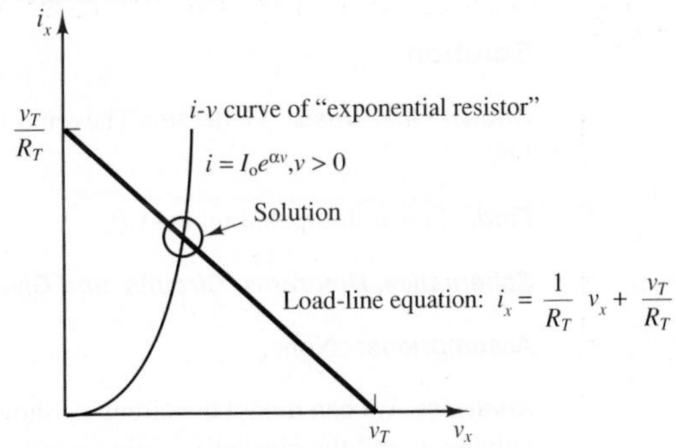

Figure 3.76 Graphical solution of equations 3.44 and 3.45

We notice, first, that equation 3.46 describes the behavior of any load, linear or nonlinear, since we have made no assumptions regarding the nature of the load voltage and current. Second, it is the equation of a line in the $i_x v_x$ plane, with slope $-1/R_T$ and i_x intercept V_T/R_T. This equation is referred to as the **load-line equation;** its graphical interpretation is very useful and is shown in Figure 3.75.

The load-line equation is but one of two i-v characteristics we have available, the other being the nonlinear-device characteristic of equation 3.41. The intersection of the two curves yields the solution of our nonlinear system of equations. This result is depicted in Figure 3.76.

Finally, another important point should be emphasized: The linear network reduction methods introduced in the preceding sections can always be employed to reduce any circuit containing a single nonlinear element to the Thévenin equivalent form, as illustrated in Figure 3.77. The key is to identify the nonlinear element and to treat it as a load. Thus, the equivalent-circuit solution methods developed earlier can be very useful in simplifying problems in which a nonlinear load is present. Examples 3.24 and 3.25 further illustrate the load-line analysis method.

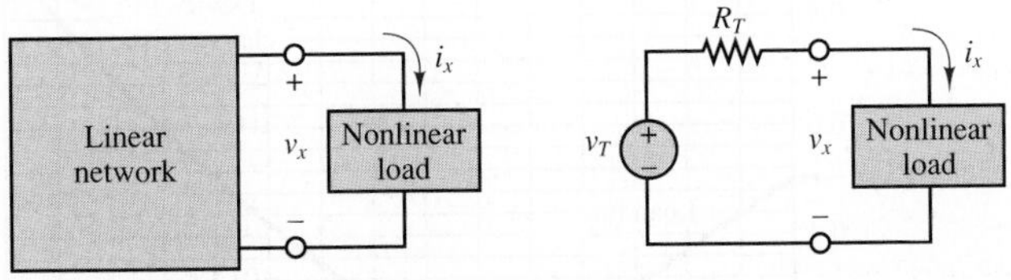

Figure 3.77 Transformation of nonlinear circuit of Thévenin equivalent

EXAMPLE 3.24 Nonlinear Load Power Dissipation

Problem

A linear generator is connected to a nonlinear load in the configuration of Figure 3.77. Determine the power dissipated by the load.

Solution

Known Quantities: Generator Thévenin equivalent circuit; load i-v characteristic and load line.

Find: Power dissipated by load P_x.

Schematics, Diagrams, Circuits, and Given Data: $R_T = 30\ \Omega$; $v_T = 15$ V.

Assumptions: None.

Analysis: We can model the circuit as shown in Figure 3.77. The objective is to determine the voltage v_x and the current i_x, using graphical methods. The load-line equation for the circuit is given by the expression

$$i_x = -\frac{1}{R_T}v_x + \frac{v_T}{R_T}$$

or

$$i_x = -\frac{1}{30}v_x + \frac{15}{30}$$

This equation represents a line in the $i_x v_x$ plane, with i_x intercept at 0.5 A and v_x intercept at 15 V. To determine the operating point of the circuit, we superimpose the load line on the device i-v characteristic, as shown in Figure 3.78, and determine the solution by finding the intersection of the load line with the device curve. Inspection of the graph reveals that the intersection point is given approximately by

$$i_x = 0.14 \text{ A} \qquad v_x = 11 \text{ V}$$

and therefore the power dissipated by the nonlinear load is

$$P_x = 0.14 \times 11 = 1.54 \text{ W}$$

It is important to observe that the result obtained in this example is, in essence, a description of experimental procedures, indicating that the analytical concepts developed in this chapter also apply to practical measurements.

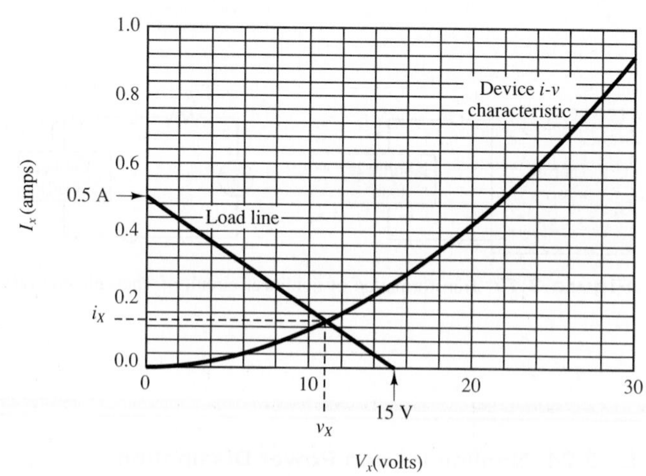

Figure 3.78

CHECK YOUR UNDERSTANDING

Example 3.24 demonstrates a graphical solution method. Sometimes it is possible to determine the solution for a nonlinear load by analytical methods. Imagine that the same generator of Example 3.24 is now connected to a "square law" load, that is, one for which $v_x = \beta i_x^2$, with $\beta = 0.1$. Determine the load current i_x. [*Hint:* Assume that only positive solutions are possible, given the polarity of the generator.]

Answer: $i_x = 0.5$ A

EXAMPLE 3.25 Load Line Analysis

LO6

Problem

A temperature sensor has a nonlinear i-v characteristic, shown in the figure on the left. The load is connected to a circuit represented by its Thévenin equivalent circuit. Determine the current flowing through the temperature sensor. The circuit connection is identical to that of Figure 3.77.

Solution

Known Quantities: $R_T = 6.67\ \Omega$; $V_T = 1.67$ V. $i_x = 0.14 - 0.03v_x^2$.

Find: i_x.

Analysis: The figure on the left depicts the device i-v characteristic. The figure on the right depicts a plot of both the device i-v characteristic and the load line obtained from

$$i_x = -\frac{1}{R_T}v_x + \frac{v_T}{R_T} = -0.15v_x + 0.25$$

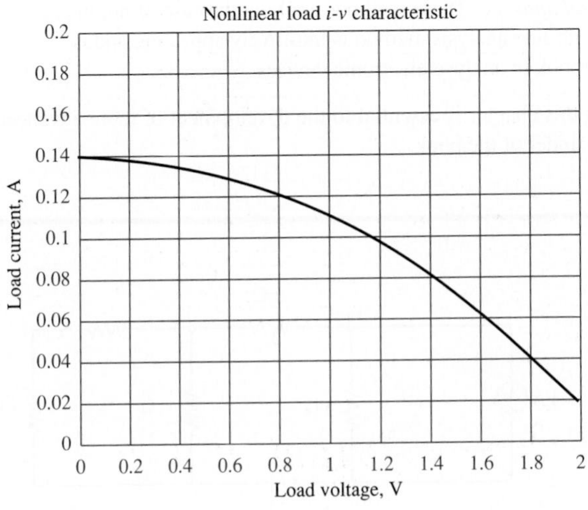

(a)

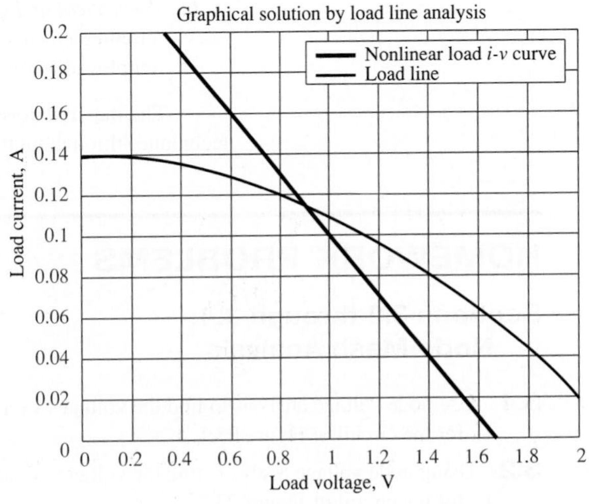

(b)

The solution for v_x and i_x occurs at the intersection of the device and load-line characteristics: $i_x \approx 0.12$ A, $v_x \approx 0.9$ V.

CHECK YOUR UNDERSTANDING

Knowing that the load i-v characteristic is given exactly by the expression $i_x = 0.14 - 0.03v_x^2$, determine the load current i_x. [*Hint:* Assume that only positive solutions are possible, given the polarity of the generator.]

Answer: $i_x = 0.116$ A

Conclusion

The objective of this chapter is to provide a practical introduction to the analysis of linear resistive circuits. The emphasis on examples is important at this stage, since we believe that familiarity with the basic circuit analysis techniques will greatly ease the task of learning more advanced ideas in circuits and electronics. In particular, your goal at this point should be to have mastered six analysis methods, summarized as follows:

1., 2. *Node voltage and mesh current analysis.* These methods are analogous in concept; the choice of a preferred method depends on the specific circuit. They are generally applicable to the circuits we analyze in this book and are amenable to solution by matrix methods.

3. *The principle of superposition.* This is primarily a conceptual aid that may simplify the solution of circuits containing multiple sources. It is usually not an efficient method.
4. *Thévenin and Norton equivalents.* The notion of equivalent circuits is at the heart of circuit analysis. Complete mastery of the reduction of linear resistive circuits to either equivalent form is a must.
5. *Maximum lower transfer.* Equivalent circuits provide a very clear explanation of how power is transferred from a source to a load.
6. *Numerical and graphical analysis.* These methods apply in the case of nonlinear circuit elements. The load-line analysis method is intuitively appealing and is employed again in this book to analyze electronic devices.

The material covered in this chapter is essential to the development of more advanced techniques throughout the remainder of the book.

HOMEWORK PROBLEMS

Sections 3.2 through 3.4: Node Mesh Analysis

3.1 Use node voltage analysis to find the voltages V_1 and V_2 for the circuit of Figure P3.1.

3.2 Using node voltage analysis, find the voltages V_1 and V_2 for the circuit of Figure P3.2.

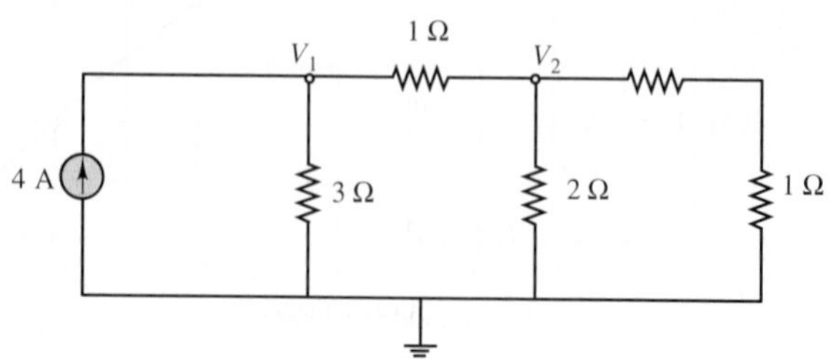

Figure P3.1

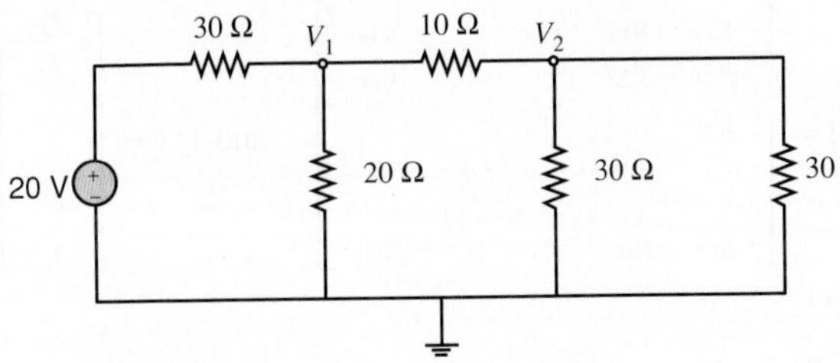

Figure P3.2

3.3 Using node voltage analysis in the circuit of Figure P3.3, find the voltage v across the 0.25-ohm resistance.

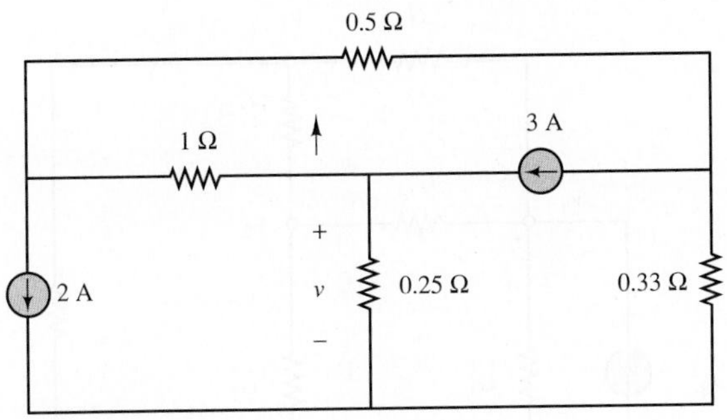

Figure P3.3

3.4 Using node voltage analysis in the circuit of Figure P3.4, find the current i through the voltage source.

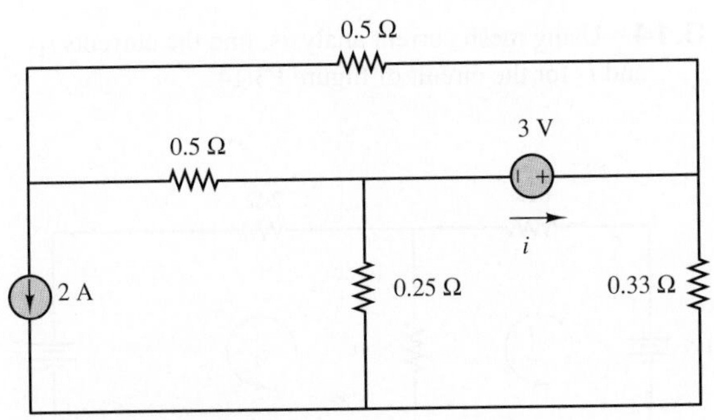

Figure P3.4

3.5 In the circuit shown in Figure P3.5, the mesh currents are

$$I_1 = 5\text{ A} \quad I_2 = 3\text{ A} \quad I_3 = 7\text{ A}$$

Determine the branch currents through:

a. R_1. b. R_2. c. R_3.

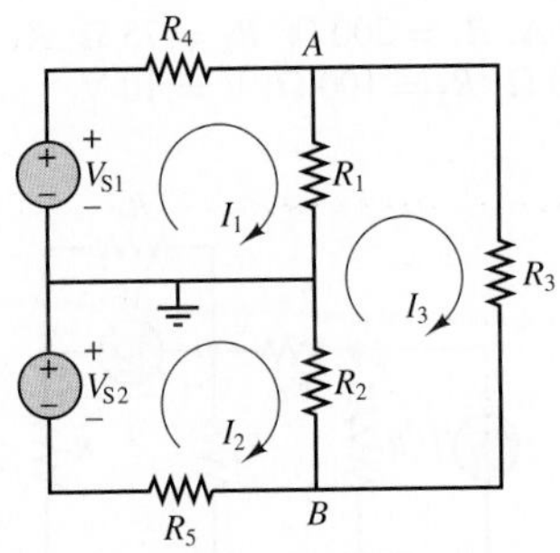

Figure P3.5

3.6 In the circuit shown in Figure P3.5, the source and node voltages are

$$V_{S1} = V_{S2} = 110\text{ V}$$
$$V_A = 103\text{ V} \quad V_B = -107\text{ V}$$

Determine the voltage across each of the five resistors.

3.7 Using node voltage analysis in the circuit of Figure P3.7, find the currents i_1 and i_2. $R_1 = 3\ \Omega$; $R_2 = 1\ \Omega$; $R_3 = 6\ \Omega$.

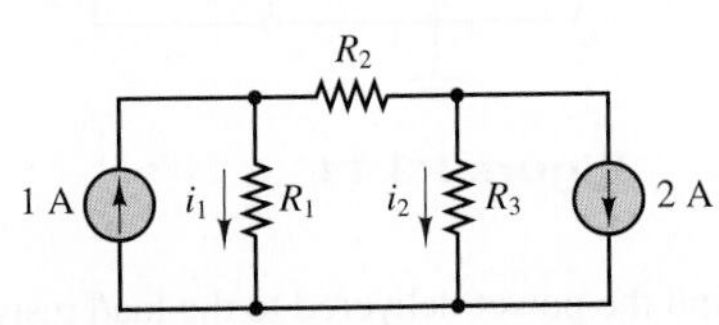

Figure P3.7

3.8 Use the mesh analysis to determine the currents i_1 and i_2 in the circuit of Figure P3.7.

3.9 Using node voltage analysis in the circuit of Figure P3.9, find the current i through the voltage source. Let $R_1 = 100\ \Omega$; $R_2 = 5\ \Omega$; $R_3 = 200\ \Omega$; $R_4 = 50\ \Omega$; $V = 50$ V; $I = 0.2$ A.

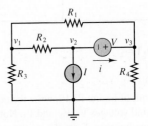

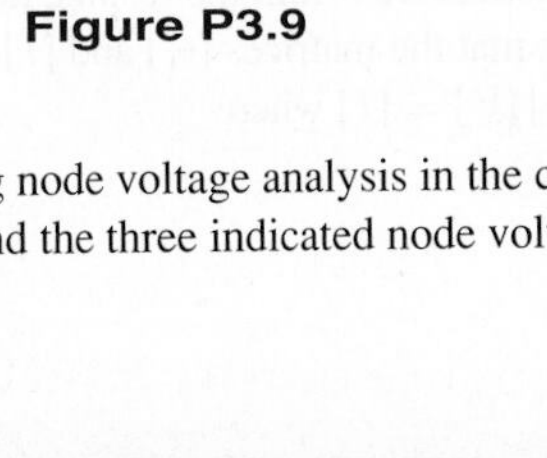

Figure P3.9

3.10 Using node voltage analysis in the circuit of Figure P3.10, find the three indicated node voltages. Let

$I = 0.2$ A; $R_1 = 200\ \Omega$; $R_2 = 75\ \Omega$; $R_3 = 25\ \Omega$; $R_4 = 50\ \Omega$; $R_5 = 100\ \Omega$; $V = 10$ V.

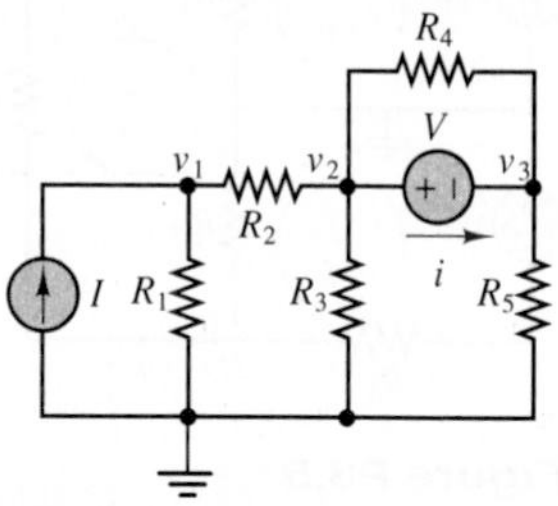

Figure P3.10

3.11 Using node voltage analysis in the circuit of Figure P3.11, find the current i drawn from the independent voltage source. Let $V = 3$ V; $R_1 = \frac{1}{2}\ \Omega$; $R_2 = \frac{1}{2}\ \Omega$; $R_3 = \frac{1}{4}\ \Omega$; $R_4 = \frac{1}{2}\ \Omega$; $R_5 = \frac{1}{4}\ \Omega$; $I = 0.5$ A.

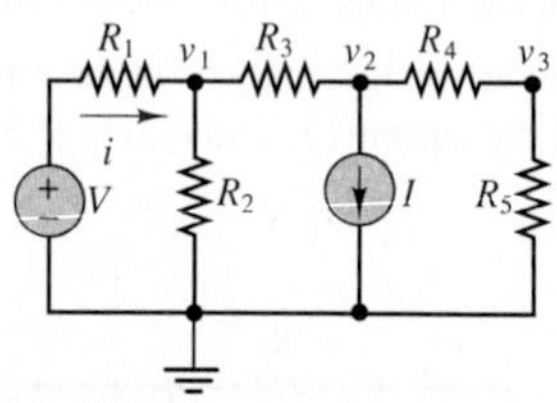

Figure P3.11

3.12 Find the power delivered to the load resistor R_L for the circuit of Figure P3.12, using node voltage analysis, given that $R_1 = 2\ \Omega$, $R_V = R_2 = R_L = 4\ \Omega$, $V_S = 4\ V$, and $I_S = 0.5$ A.

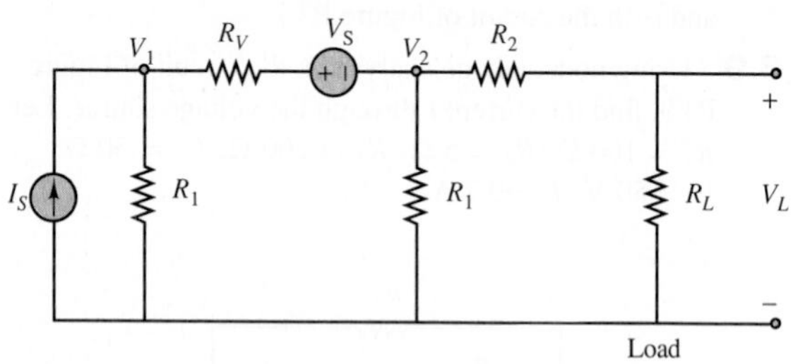

Figure P3.12

3.13

a. For the circuit of Figure P3.13, write the node equations necessary to find voltages V_1, V_2, and V_3. Note that $G = 1/R$ = conductance. From the results, note the interesting form that the matrices $[G]$ and $[I]$ have taken in the equation $[G][V] = [I]$ where

$$[G] = \begin{bmatrix} g_{11} & g_{12} & g_{13} & \cdots & g_{1n} \\ g_{21} & g_{22} & \cdots & \cdots & g_{2n} \\ g_{31} & & \ddots & & \\ \vdots & & & \ddots & \\ g_{n1} & g_{n2} & \cdots & \cdots & g_{nn} \end{bmatrix} \quad \text{and } [I] = \begin{bmatrix} I_1 \\ I_2 \\ \vdots \\ \vdots \\ I_n \end{bmatrix}$$

b. Write the matrix form of the node voltage equations again, using the following formulas:

$g_{ii} = \sum$ conductances connected to node i

$g_{ij} = -\sum$ conductances *shared* by nodes i and j

$I_i = \sum$ all *source* currents into node i

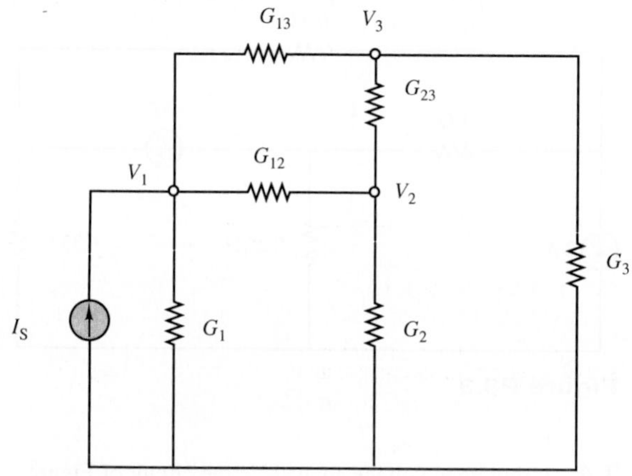

Figure P3.13

3.14 Using mesh current analysis, find the currents i_1 and i_2 for the circuit of Figure P3.14.

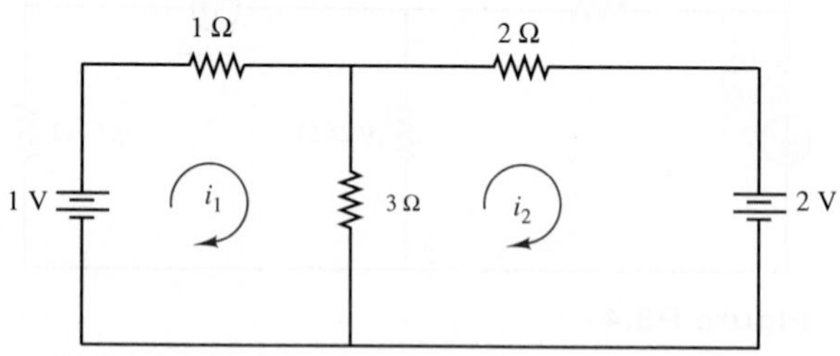

Figure P3.14

3.15 Using mesh current analysis, find the currents I_1 and I_2 and the voltage across the top 10-Ω resistor in the circuit of Figure P3.15.

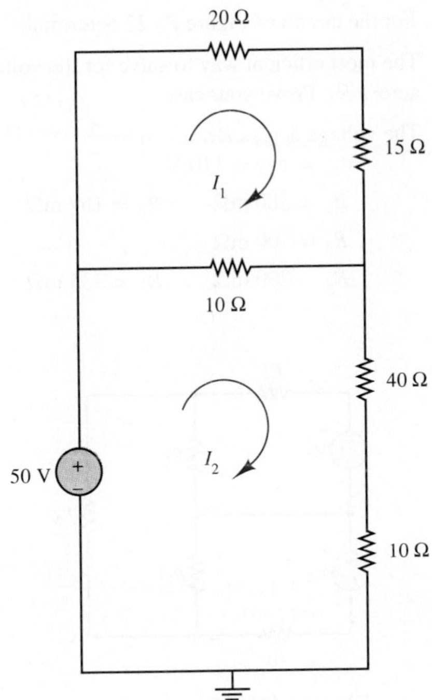

Figure P3.15

3.16 Using mesh current analysis, find the voltage, v, across the 3-Ω resistor in the circuit of Figure P3.16.

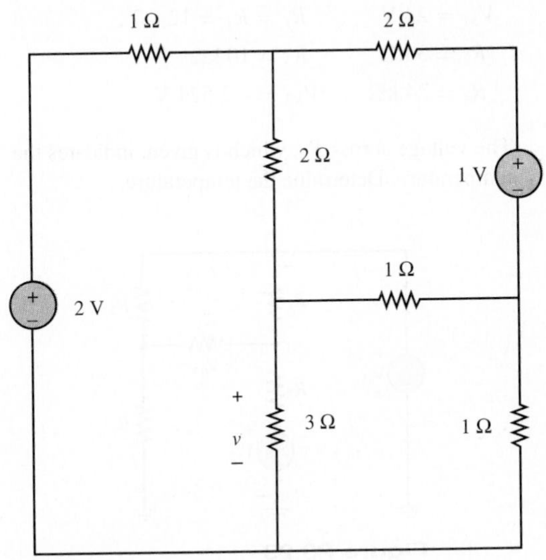

Figure P3.16

3.17 Using mesh current analysis, find the currents I_1, I_2, and I_3 and the voltage across the 40-Ω resistor in the circuit of Figure P3.17 (assume polarity according to I_2).

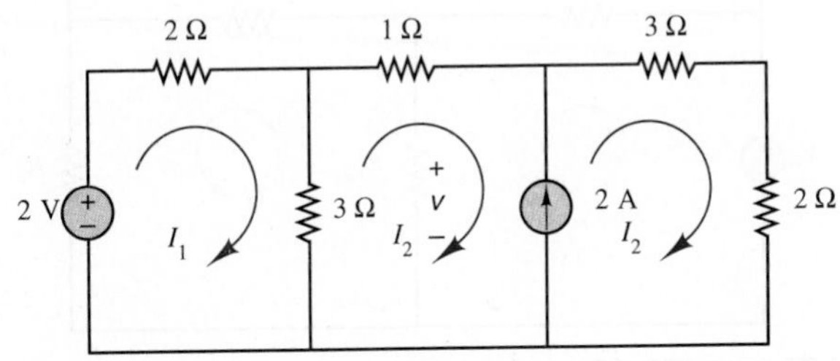

Figure P3.17

3.18 Using mesh current analysis, find the voltage, v, across the source in the circuit of Figure P3.18.

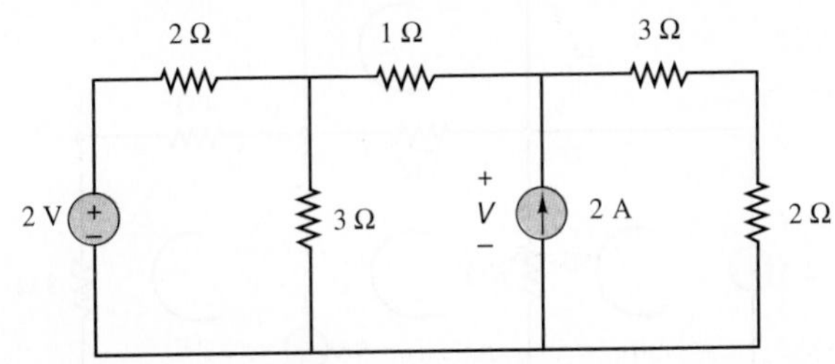

Figure P3.18

3.19 a. For the circuit of Figure P3.19, write the mesh equations in matrix form. Notice the form of the $[R]$ and $[V]$ matrices in the $[R][I] = [V]$, where

$$[R] = \begin{bmatrix} r_{11} & r_{12} & r_{13} & \cdots & r_{1n} \\ r_{21} & r_{22} & \cdots & \cdots & r_{2n} \\ r_{31} & & \ddots & & \\ \vdots & & & \ddots & \\ r_{n1} & r_{n2} & \cdots & \cdots & r_{nn} \end{bmatrix} \quad \text{and } [V] = \begin{bmatrix} V_1 \\ V_2 \\ \vdots \\ \vdots \\ V_n \end{bmatrix}$$

b. Write the matrix form of the mesh equations again by using the following formulas:

$r_{ii} = \sum$ resistances around loop i

$r_{ij} = -\sum$ resistances *shared* by loops i and j

$V_i = \sum$ source voltages around loop i

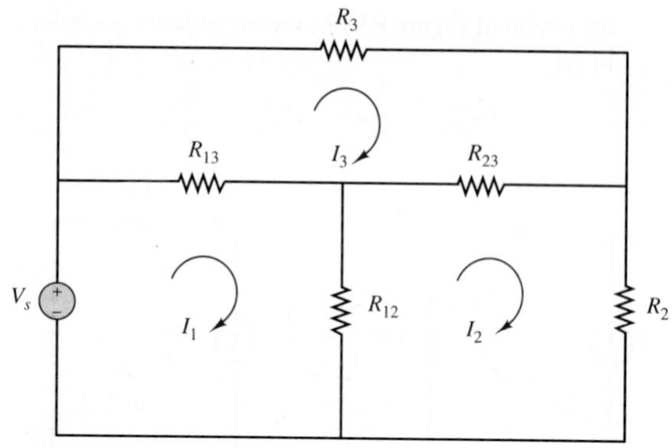

Figure P3.19

3.20 For the circuit of Figure P3.20, use mesh current analysis to find the matrices required to solve the circuit, and solve for the unknown currents. [*Hint*: you may find source transformations useful.]

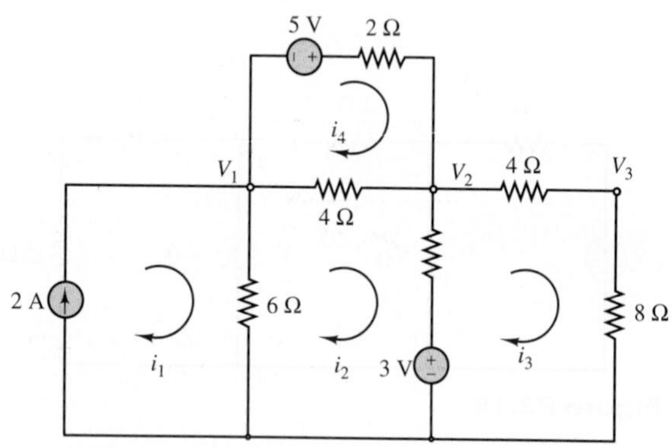

Figure P3.20

3.21 In the circuit in Figure P3.21, assume the source voltage and source current and all resistances are known.

a. Write the node equations required to determine the node voltages.

b. Write the matrix solution for each node voltage in terms of the known parameters.

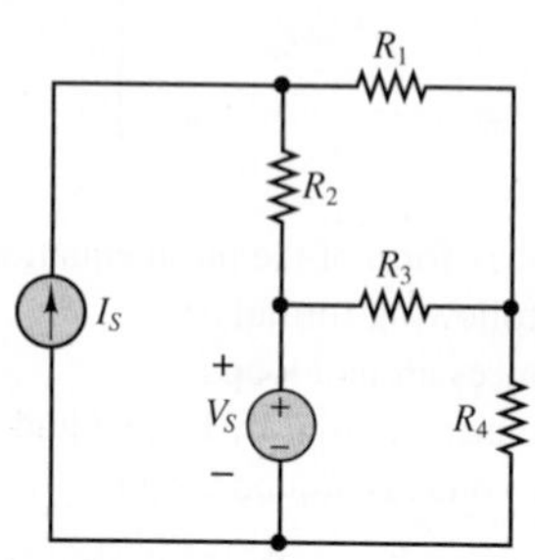

Figure P3.21

3.22 For the circuit of Figure P3.22 determine

a. The most efficient way to solve for the voltage across R_3. Prove your case.

b. The voltage across R_3.

$$V_{S1} = V_{S2} = 110 \text{ V}$$

$$R_1 = 500 \text{ m}\Omega \qquad R_2 = 167 \text{ m}\Omega$$

$$R_3 = 700 \text{ m}\Omega$$

$$R_4 = 200 \text{ m}\Omega \qquad R_5 = 333 \text{ m}\Omega$$

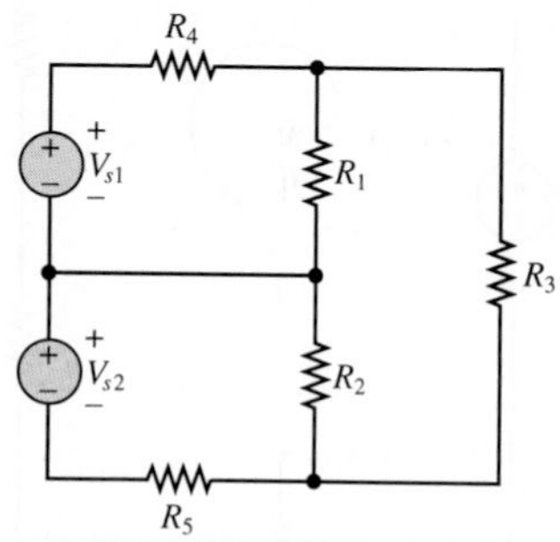

Figure P3.22

3.23 In the circuit shown in Figure P3.23, V_{S2} and R_s model a temperature sensor, i.e.,

$$V_{S2} = kT \qquad k = 10 \text{ V/°C}$$

$$V_{S1} = 24 \text{ V} \qquad R_s = R_1 = 12 \text{ k}\Omega$$

$$R_2 = 3 \text{ k}\Omega \qquad R_3 = 10 \text{ k}\Omega$$

$$R_4 = 24 \text{ k}\Omega \qquad V_{R3} = -2.524 \text{ V}$$

The voltage across R_3, which is given, indicates the temperature. Determine the temperature.

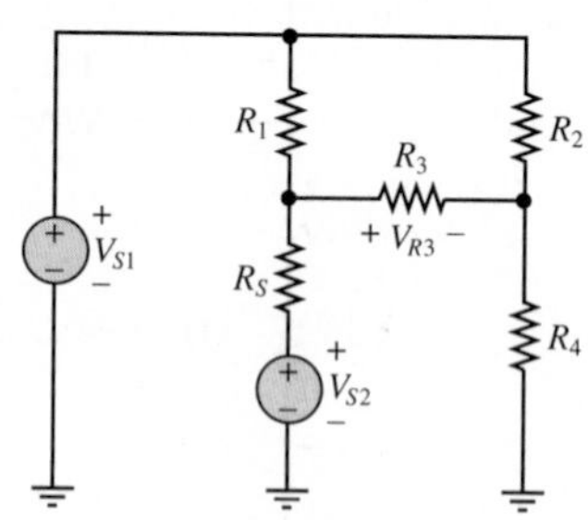

Figure P3.23

3.24 Using KCL, perform node analysis on the circuit shown in Figure P3.24, and determine the voltage across R_4. Note that one source is a controlled voltage source! Let $V_S = 5$ V; $A_V = 70$; $R_1 = 2.2$ kΩ; $R_2 = 1.8$ kΩ; $R_3 = 6.8$ kΩ; $R_4 = 220$ Ω.

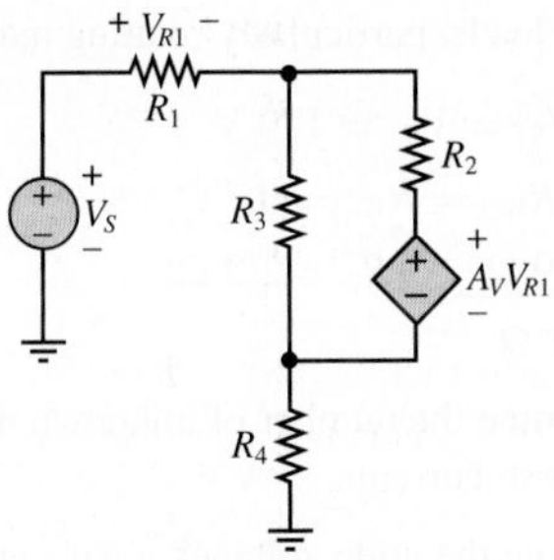

Figure P3.24

3.25 Using mesh current analysis, find the voltage v across R_4 in the circuit of Figure P3.25. Let $V_{S1} = 12$ V; $V_{S2} = 5$ V; $R_1 = 50\ \Omega$; $R_2 = R_3 = 20\ \Omega$; $R_4 = 10\ \Omega$; $R_5 = 15\ \Omega$.

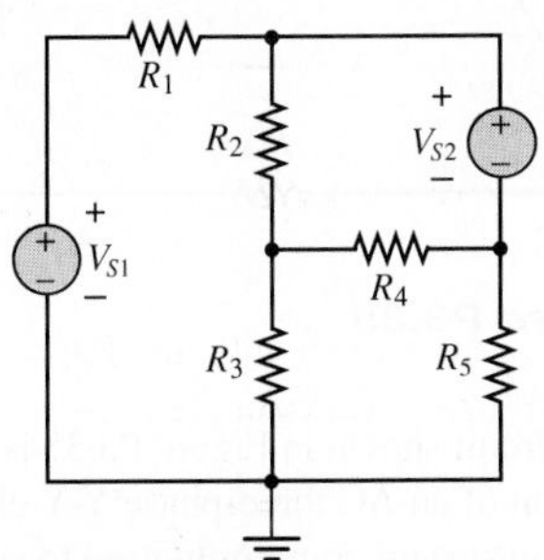

Figure P3.25

3.26 Use mesh current analysis to solve for the voltage v across the current source in the circuit of Figure P3.26. Let $V = 3$ V; $I = 0.5$ A; $R_1 = 20\ \Omega$; $R_2 = 30\ \Omega$; $R_3 = 10\ \Omega$; $R_4 = 30\ \Omega$; $R_5 = 20\ \Omega$.

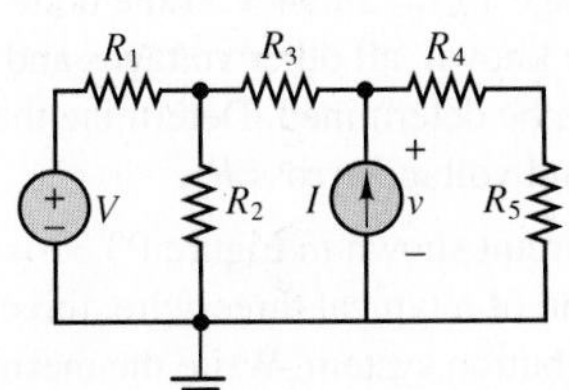

Figure P3.26

3.27 Use mesh current analysis to find the current i in the circuit of Figure P3.27. Let $V = 5.6$ V; $R_1 = 50\ \Omega$; $R_2 = 1.2\ \text{k}\Omega$; $R_3 = 330\ \Omega$; $g_m = 0.2$ S; $R_4 = 440\ \Omega$.

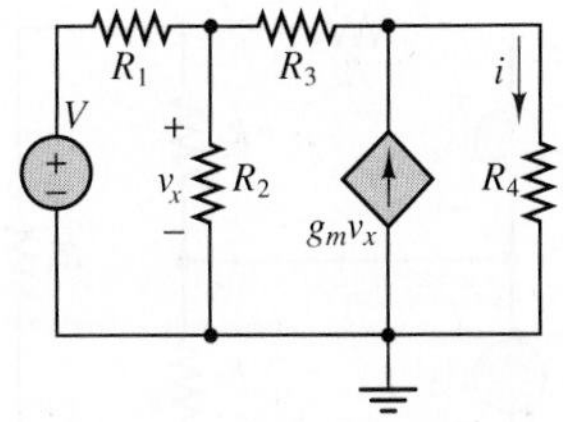

Figure P3.27

3.28 Using mesh current analysis, find the current i through the voltage source in the circuit of Figure P3.9.

3.29 Using mesh current analysis, find the current i in the circuit of Figure P3.10.

3.30 Using mesh current analysis, find the current i in the circuit of Figure P3.30.

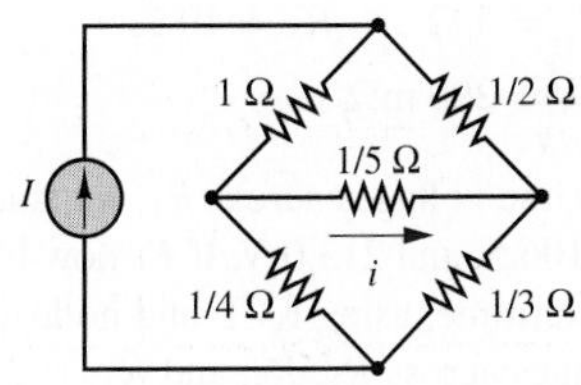

Figure P3.30

3.31 Using mesh current analysis, find the voltage gain $A_v = v_2/v_1$ in the circuit of Figure P3.31.

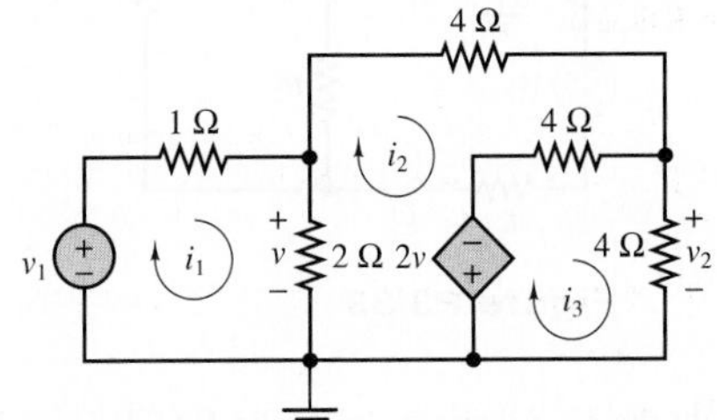

Figure P3.31

3.32 In the circuit shown in Figure P3.32:

$V_{S1} = V_{S2} = 450$ V

$R_4 = R_5 = 0.25\ \Omega$

$R_1 = 8\ \Omega \qquad R_2 = 5\ \Omega$

$R_3 = 32\ \Omega$

Determine, using KCL and node analysis, the voltage across R_1, R_2, and R_3.

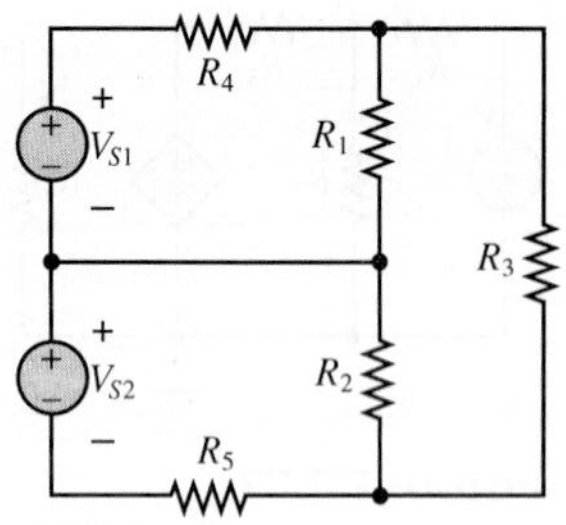

Figure P3.32

3.33 In the circuit shown in Figure P3.33, F_1 and F_2 are fuses. Under normal conditions they are modeled as a short circuit. However, if excess current flows through a fuse, its element melts and the fuse "blows" (i.e., it becomes an open circuit).

$V_{S1} = V_{S2} = 115$ V

$R_1 = R_2 = 5\ \Omega \qquad R_3 = 10\ \Omega$

$R_4 = R_5 = 200$ mΩ

Normally, the voltages across R_1, R_2, and R_3 are 106.5, −106.5, and 213.0 V. If F_1 now blows, or opens, determine, using KCL and node analysis, the new voltages across R_1, R_2, and R_3.

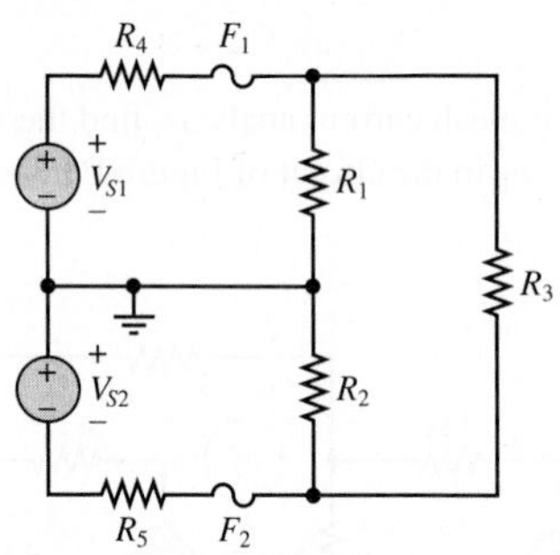

Figure P3.33

3.34 In the circuit shown in Figure P3.33, F_1 and F_2 are fuses. Under normal conditions they are modeled as a short circuit. However, if excess current flows through a fuse, it "blows" and the fuse becomes an open circuit.

$V_{S1} = V_{S2} = 120$ V

$R_1 = R_2 = 2\ \Omega \qquad R_3 = 8\ \Omega$

$R_4 = R_5 = 250$ m Ω

If F_1 blows, or opens, determine, using KCL and node analysis, the voltages across R_1, R_2, R_3, and F_1.

3.35 The circuit shown in Figure P3.35 is a simplified DC version of an AC three-phase Y-Y electrical distribution system commonly used to supply industrial loads, particularly rotating machines.

$V_{S1} = V_{S2} = V_{S3} = 170$ V

$R_{W1} = R_{W2} = R_{W_3} = 0.7\ \Omega$

$R_1 = 1.9\ \Omega \qquad R_2 = 2.3\ \Omega$

$R_3 = 11\ \Omega$

a. Determine the number of unknown node voltages and mesh currents.

b. Compute the node voltages v_1', v_2', and v_3'. With respect to v_n'.

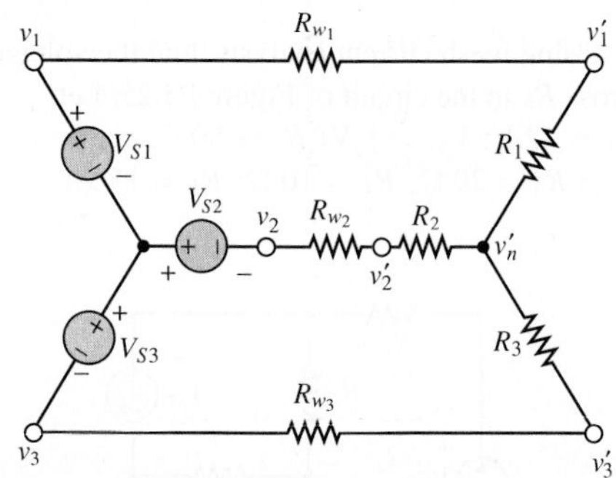

Figure P3.35

3.36 The circuit shown in Figure P3.35 is a simplified DC version of an AC three-phase Y-Y electrical distribution system commonly used to supply industrial loads, particularly rotating machines.

$V_{S1} = V_{S2} = V_{S3} = 170$ V

$R_{W1} = R_{W2} = R_{W3} = 0.7\ \Omega$

$R_1 = 1.9\ \Omega \qquad R_2 = 2.3\ \Omega$

$R_3 = 11\ \Omega$

Node analysis with KCL and a ground at the terminal common to the three sources gives the only unknown node voltage $V_N = 28.94$ V. If the node voltages in a circuit are known, all other voltages and currents in the circuit can be determined. Determine the current through and voltage across R_1.

3.37 The circuit shown in Figure P3.35 is a simplified DC version of a typical three-wire, three-phase AC Y-Y distribution system. Write the mesh (or loop) equations and any additional equations required to determine the current through R_1 in the circuit shown.

3.38 Determine the branch currents, using KVL and loop analysis in the circuit of Figure P3.35.

$V_{S2} = V_{S3} = 110\ V \qquad V_{S1} = 90$ V

$R_1 = 7.9\ \Omega \qquad R_2 = R_3 = 3.7\ \Omega$

$R_{W1} = R_{W2} = R_{W3} = 1.3\ \Omega$

3.39 In the circuit shown in Figure P3.33, F_1 and F_2 are fuses. Under normal conditions they are modeled as a short circuit. However, if excess current flows through a fuse, its element melts and the fuse blows (i.e., it becomes an open circuit).

$V_{S1} = V_{S2} = 115$ V

$R_1 = R_2 = 5\ \Omega \qquad R_3 = 10\ \Omega$

$R_4 = R_5 = 200$ mΩ

Determine, using KVL and a mesh analysis, the voltages across R_1, R_2, and R_3 under normal conditions (i.e., no blown fuses).

Section 3.5: Superposition

3.40 With reference to Figure P3.40, determine the current through R_1 due only to the source V_{S2}.

$V_{S1} = 110$ V $\qquad V_{S2} = 90$ V

$R_1 = 560\ \Omega \qquad R_2 = 3.5$ kΩ

$R_3 = 810\ \Omega$

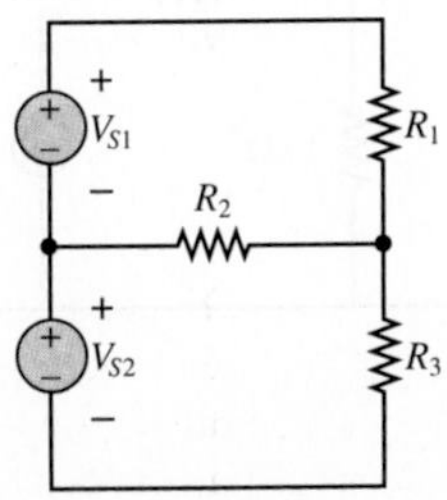

Figure P3.40

3.41 Determine, using superposition, the voltage across R in the circuit of Figure P3.41.

$I_B = 12$ A $\qquad R_B = 1\ \Omega$

$V_G = 12$ V $\qquad R_G = 0.3\ \Omega$

$R = 0.23\ \Omega$

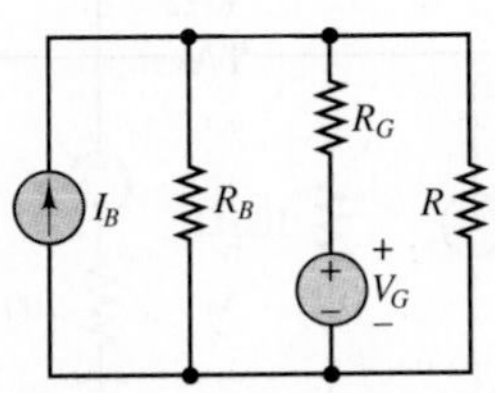

Figure P3.41

3.42 Using superposition, determine the voltage across R_2 in the circuit of Figure P3.42.

$V_{S1} = V_{S2} = 12$ V

$R_1 = R_2 = R_3 = 1$ kΩ

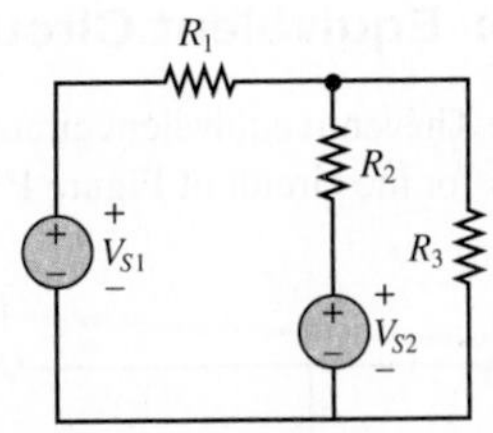

Figure P3.42

3.43 With reference to Figure P3.43, using superposition, determine the component of the current through R_3 that is due to V_{S2}.

$V_{S1} = V_{S2} = 450$ V

$R_1 = 7\ \Omega \qquad R_2 = 5\ \Omega$

$R_3 = 10\ \Omega \qquad R_4 = R_5 = 1\ \Omega$

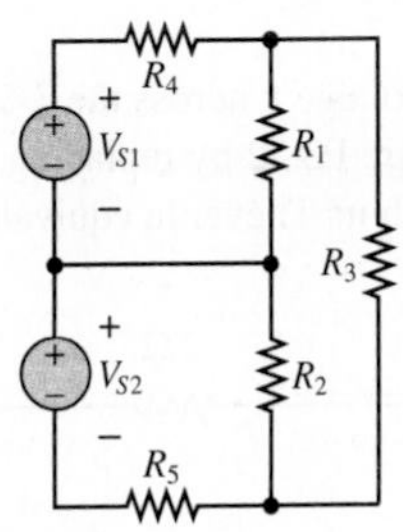

Figure P3.43

3.44 The circuit shown in Figure P3.35 is a simplified DC version of an AC three-phase electrical distribution system.

$V_{S1} = V_{S2} = V_{S3} = 170$ V

$R_{W1} = R_{W2} = R_{W3} = 0.7\ \Omega$

$R_1 = 1.9\ \Omega \qquad R_2 = 2.3\ \Omega$

$R_3 = 11\ \Omega$

To prove how cumbersome and inefficient (although sometimes necessary) the method is, determine, using superposition, the current through R_1.

3.45 Repeat Problem 3.9, using the principle of superposition.

3.46 Repeat Problem 3.10, using the principle of superposition.

3.47 Repeat Problem 3.11, using the principle of superposition.

3.48 Repeat Problem 3.23, using the principle of superposition.

3.49 Repeat Problem 3.25, using the principle of superposition.

3.50 Repeat Problem 3.26, using the principle of superposition.

Section 3.6: Equivalent Circuits

3.51 Find the Thévenin equivalent circuit as seen by the 3-Ω resistor for the circuit of Figure P3.51.

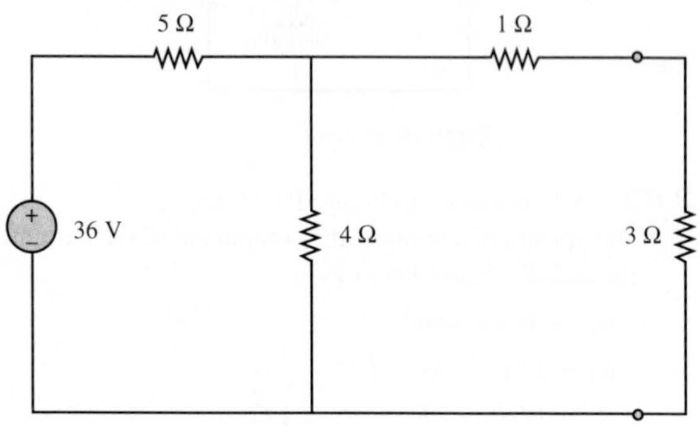

Figure P3.51

3.52 Find the voltage v across the 3-Ω resistor in the circuit of Figure P3.52 by replacing the remainder of the circuit with its Thévenin equivalent.

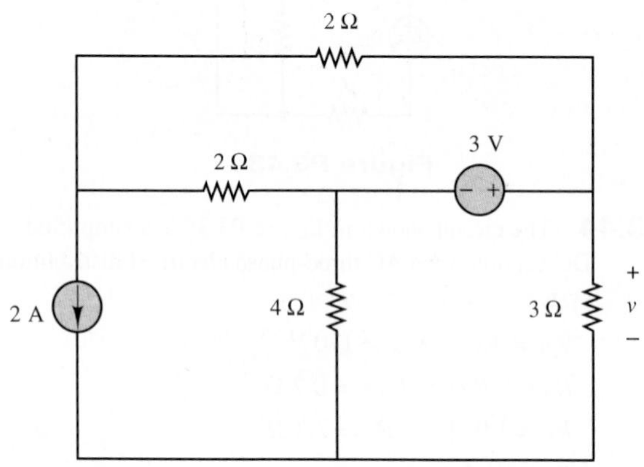

Figure P3.52

3.53 Find the Norton equivalent of the circuit to the left of the 2-Ω resistor in the Figure P3.53.

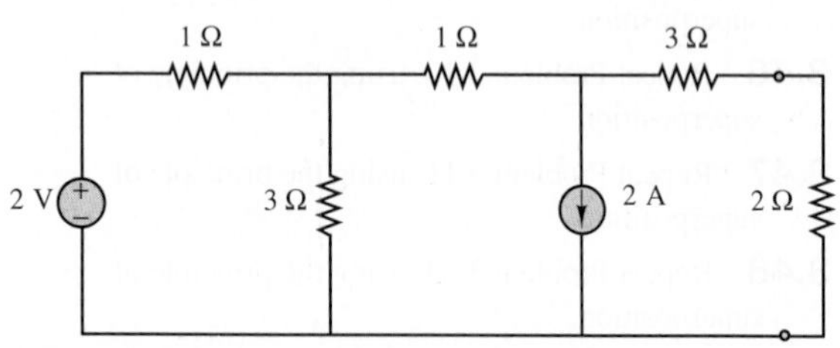

Figure P3.53

3.54 Find the Norton equivalent to the left of terminals a and b of the circuit shown in Figure P3.54.

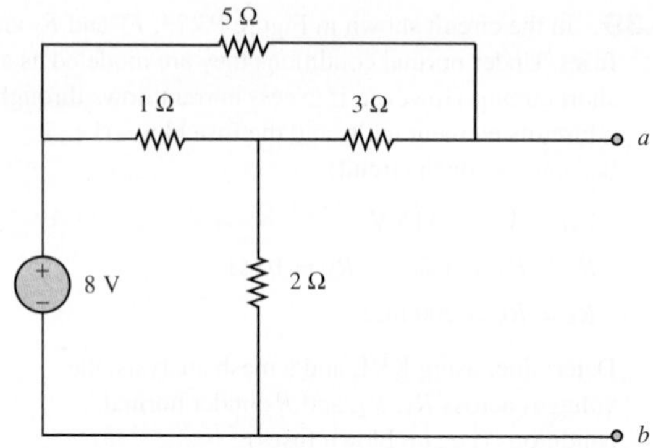

Figure P3.54

3.55 Find the Thévenin equivalent circuit that the load sees for the circuit of Figure P3.55.

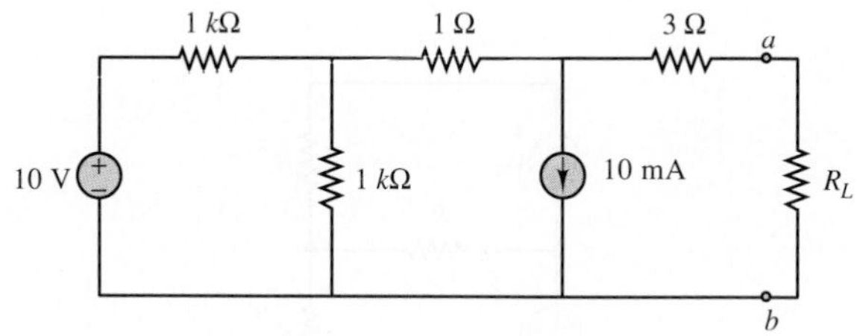

Figure P3.55

3.56 Find the Thévenin equivalent resistance seen by the load resistor R_L in the circuit of Figure P3.56.

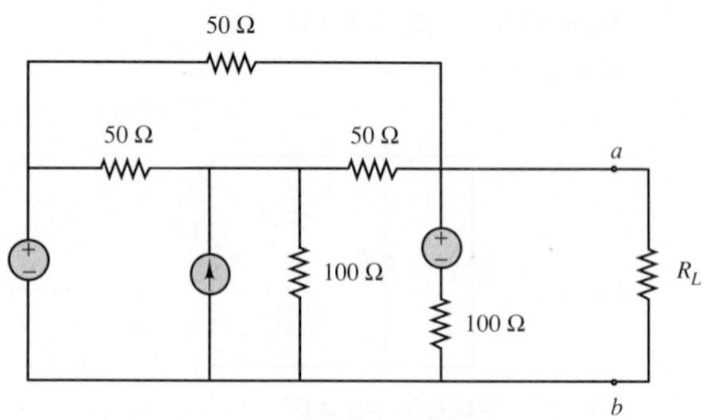

Figure P3.56

3.57 Find the Thévenin equivalent of the circuit connected to R_L in Figure P3.57.

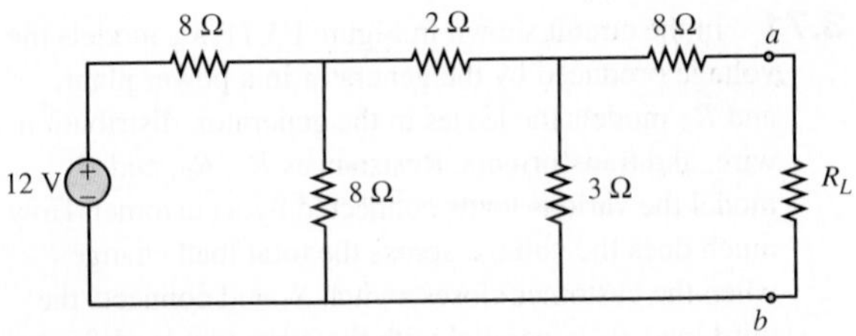

Figure P3.57

3.58 Find the Thévenin equivalent of the circuit connected to R_L in Figure P3.58, where $R_1 = 10\ \Omega$, $R_2 = 20\ \Omega$, $R_g = 0.1\ \Omega$, and $R_p = 1\ \Omega$.

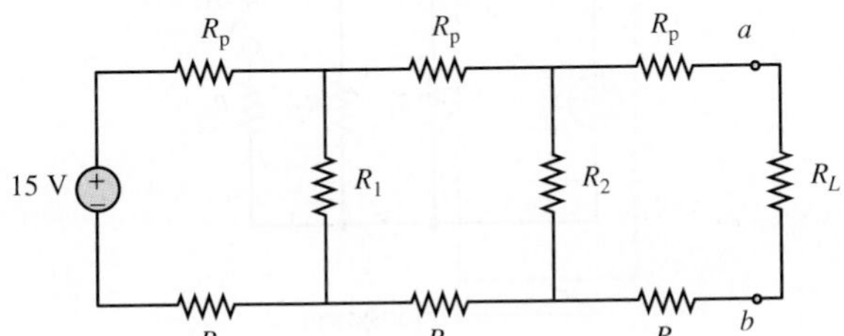

Figure P3.58

3.59 The Wheatstone bridge circuit shown in Figure P3.59 is used in a number of practical applications. One traditional use is in determining the value of an unknown resistor R_x.
Find the value of the voltage $V_{ab} = V_a - V_b$ in terms of R, R_x, and V_S.
If $R = 1\ \text{k}\Omega$, $V_S = 12$ V and $V_{ab} = 12$ mV, what is the value of R_x?

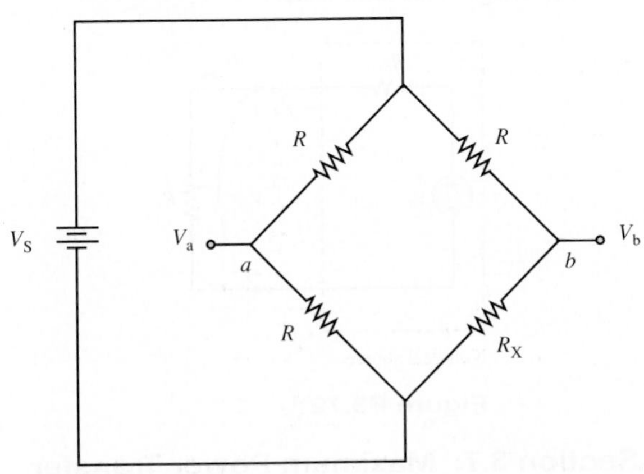

Figure P3.59

3.60 It is sometimes useful to compute a Thévenin equivalent circuit for a Wheatstone bridge. For the circuit of Figure P3.60,

a. Find the Thévenin equivalent resistance seen by the load resistor R_L.

b. If $V_S = 12$ V, $R_1 = R_2 = R_3 = 1\ \text{k}\Omega$, and R_x is the resistance found in part b of the previous problem, use the Thévenin equivalent to compute the power dissipated by R_L, if $R_L = 500\ \Omega$.

c. Find the power dissipated by the Thévenin equivalent resistance R_T with R_L included in the circuit.

d. Find the power dissipated by the bridge without the load resistor in the circuit.

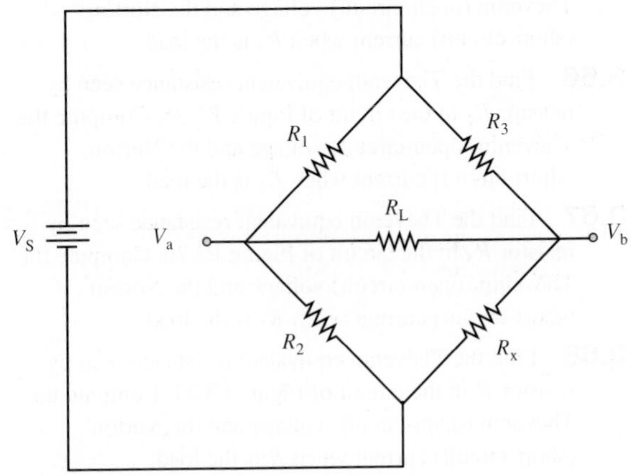

Figure P3.60

3.61 The circuit shown in Figure P3.61 is in the form of what is known as a *differential amplifier*. Find the expression for v_0 in terms of v_1 and v_2 using Thévenin's or Norton's theorem.

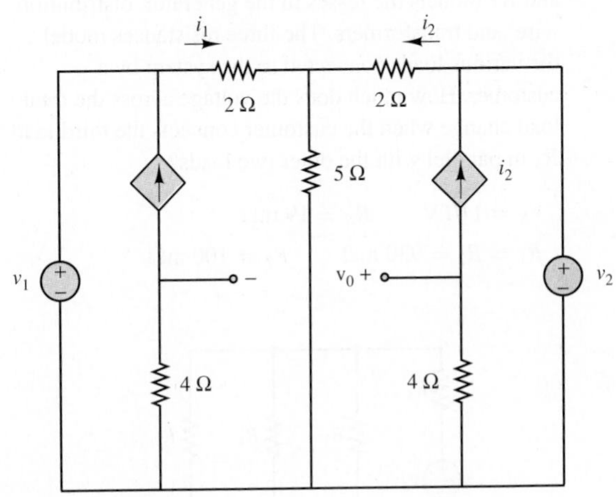

Figure P3.61

3.62 Find the Thévenin equivalent resistance seen by resistor R_3 in the circuit of Figure P3.5. Compute the

Thévenin (open-circuit) voltage and the Norton (short-circuit) current when R_3 is the load.

3.63 Find the Thévenin equivalent resistance seen by resistor R_5 in the circuit of Figure P3.10. Compute the Thévenin (open-circuit) voltage and the Norton (short-circuit) current when R_5 is the load.

3.64 Find the Thévenin equivalent resistance seen by resistor R_5 in the circuit of Figure P3.11. Compute the Thévenin (open-circuit) voltage and the Norton (short-circuit) current when R_5 is the load.

3.65 Find the Thévenin equivalent resistance seen by resistor R_3 in the circuit of Figure P3.23. Compute the Thévenin (open-circuit) voltage and the Norton (short-circuit) current when R_3 is the load.

3.66 Find the Thévenin equivalent resistance seen by resistor R_4 in the circuit of Figure P3.25. Compute the Thévenin (open-circuit) voltage and the Norton (short-circuit) current when R_4 is the load.

3.67 Find the Thévenin equivalent resistance seen by resistor R_5 in the circuit of Figure P3.26. Compute the Thévenin (open-circuit) voltage and the Norton (short-circuit) current when R_5 is the load.

3.68 Find the Thévenin equivalent resistance seen by resistor R in the circuit of Figure P3.41. Compute the Thévenin (open-circuit) voltage and the Norton (short-circuit) current when R is the load.

3.69 Find the Thévenin equivalent resistance seen by resistor R_3 in the circuit of Figure P3.43. Compute the Thévenin (open-circuit) voltage and the Norton (short-circuit) current when R_3 is the load.

3.70 In the circuit shown in Figure P3.70, V_S models the voltage produced by the generator in a power plant, and R_S models the losses in the generator, distribution wire, and transformers. The three resistances model the various loads connected to the system by a customer. How much does the voltage across the total load change when the customer connects the third load R_3 in parallel with the other two loads?

$V_S = 110$ V $\quad R_S = 19$ mΩ

$R_1 = R_2 = 930$ mΩ $\quad R_3 = 100$ mΩ

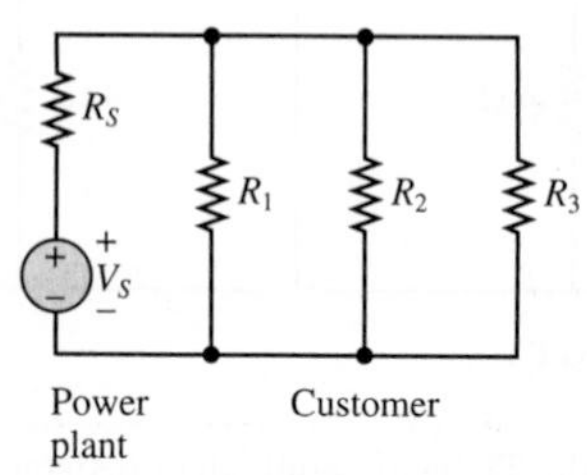

Figure P3.70

3.71 In the circuit shown in Figure P3.71, V_S models the voltage produced by the generator in a power plant, and R_S models the losses in the generator, distribution wire, and transformers. Resistances R_1, R_2, and R_3 model the various loads connected by a customer. How much does the voltage across the total load change when the customer closes switch S_3 and connects the third load R_3 in parallel with the other two loads?

$V_S = 450$ V $\quad R_S = 19$ mΩ

$R_1 = R_2 = 1.3$ Ω $\quad R_3 = 500$ mΩ

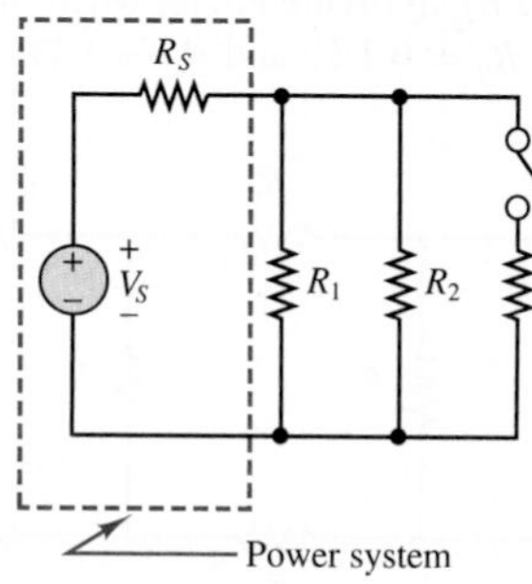

Figure P3.71

3.72 A nonideal voltage source is modeled in Figure P3.72 as an ideal source in series with a resistance that models the internal losses, that is, dissipates the same power as the internal losses. In the circuit shown in Figure P3.72, with the load resistor removed so that the current is zero (i.e., no load), the terminal voltage of the source is measured and is 20 V. Then, with $R_L = 2.7$ kΩ, the terminal voltage is again measured and is now 18 V. Determine the internal resistance and the voltage of the ideal source.

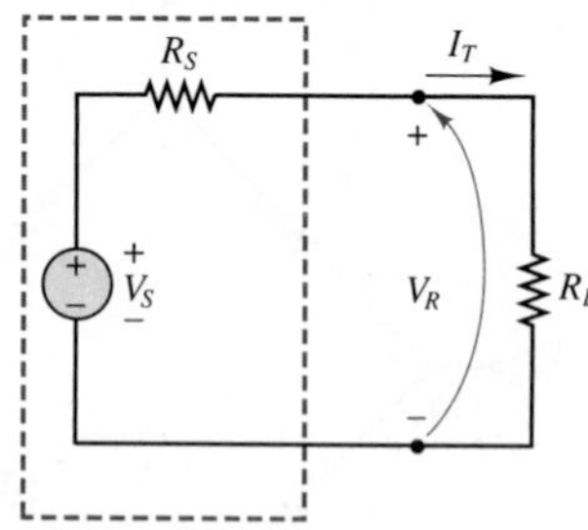

Figure P3.72

Section 3.7: Maximum Power Transfer

3.73 The equivalent circuit of Figure P3.73 has

$V_T = 12$ V $\quad R_T = 8$ Ω

If the conditions for maximum power transfer exist, determine

a. The value of R_L.
b. The power developed in R_L.
c. The efficiency of the circuit, that is, the ratio of power absorbed by the load to power supplied by the source.

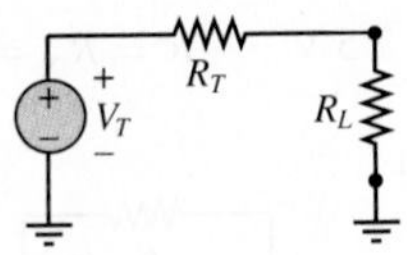

Figure P3.73

3.74 The equivalent circuit of Figure P3.73 has

$$V_T = 35 \text{ V} \qquad R_T = 600\ \Omega$$

If the conditions for maximum power transfer exist, determine

a. The value of R_L.
b. The power developed in R_L.
c. The efficiency of the circuit.

3.75 A nonideal voltage source can be modeled as an ideal voltage source in series with a resistance representing the internal losses of the source, as shown in Figure P3.75. A load is connected across the terminals of the nonideal source.

$$V_S = 12 \text{ V} \qquad R_S = 0.3\ \Omega$$

a. Plot the power dissipated in the load as a function of the load resistance. What can you conclude from your plot?
b. Prove, analytically, that your conclusion is valid in all cases.

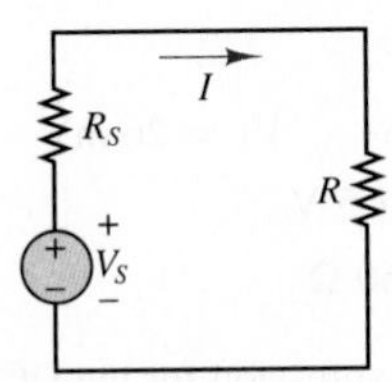

Figure P3.75

Section 3.8: Nonlinear Circuit Elements

3.76 Write the node voltage equations in terms of v_1 and v_2 for the circuit of Figure P3.76. The two nonlinear resistors are characterized by

$$i_a = 2v_a^3$$
$$i_b = v_b^3 + 10v_b$$

Do not solve the resulting equations.

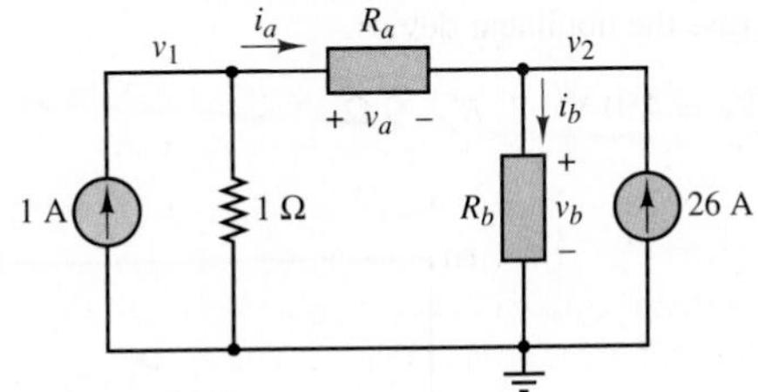

Figure P3.76

3.77 We have seen that some devices do not have a linear current–voltage characteristic for all i and v; that is, R is not constant for all values of current and voltage. For many devices, however, we can estimate the characteristics by piecewise linear approximation. For a portion of the characteristic curve around an operating point, the slope of the curve is relatively constant. The inverse of this slope at the operating point is defined as *incremental resistance* R_{inc}:

$$R_{\text{inc}} = \left.\frac{dV}{dI}\right|_{[V_0, I_0]} \approx \left.\frac{\Delta V}{\Delta I}\right|_{[V_0, I_0]}$$

where $[V_0, I_0]$ is the operating point of the circuit.

a. For the circuit of Figure P3.77, find the operating point of the element that has the characteristic curve shown.
b. Find the incremental resistance of the nonlinear element at the operating point of part a.
c. If V_T is increased to 20 V, find the new operating point and the new incremental resistance.

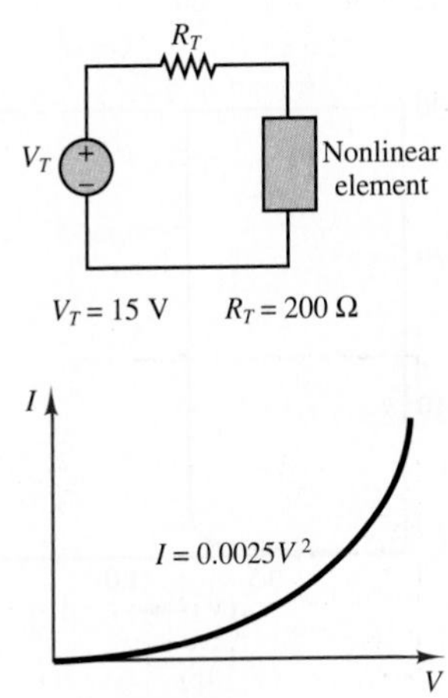

Figure P3.77

3.78 The device in the circuit in Figure P3.78 is an induction motor with the nonlinear i-v characteristic shown. Determine the current through and the voltage across the nonlinear device.

$V_S = 450$ V $\quad R = 9\ \Omega$

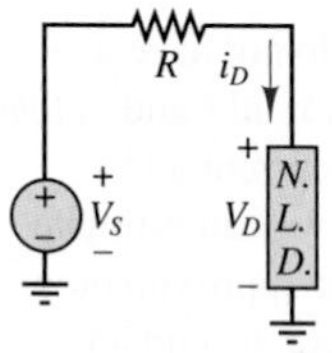

(a)

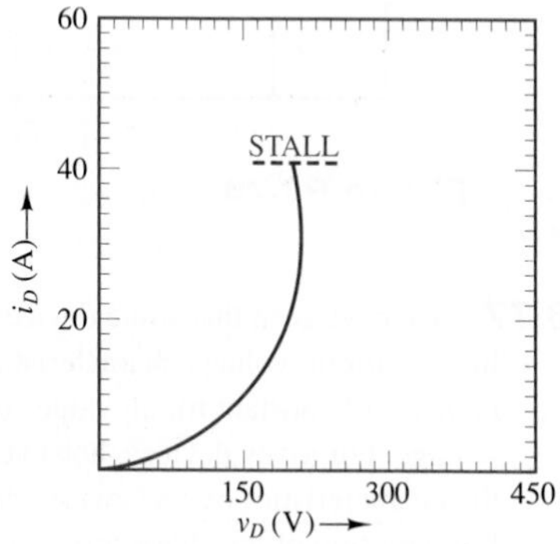

(b)

Figure P3.78

3.79 The nonlinear device in the circuit shown in Figure P3.79 has the i-v characteristic given.

$V_S = V_{TH} = 1.5$ V $\quad R = R_{eq} = 60\ \Omega$

Determine the voltage across and the current through the nonlinear device.

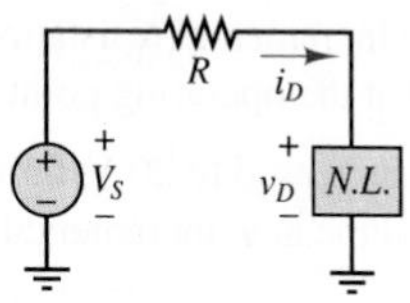

(a)

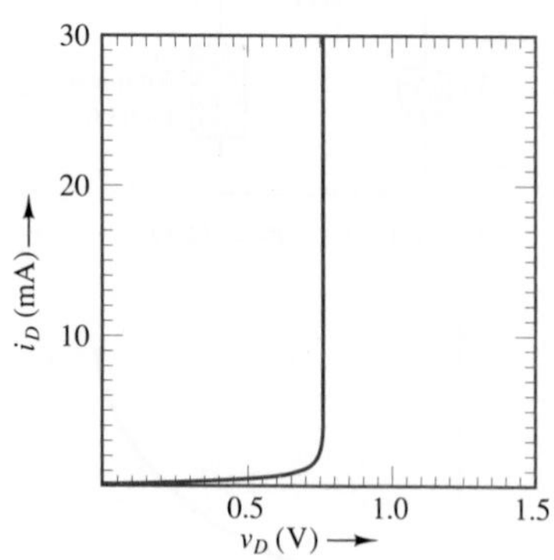

(b)

Figure P3.79

3.80 The resistance of the nonlinear device in the circuit in Figure P3.80 is a nonlinear function of pressure. The i-v characteristic of the device is shown as a family of curves for various pressures. Construct the DC load line. Plot the voltage across the device as a function of pressure. Determine the current through the device when $P = 30$ psig.

$V_S = V_{TH} = 2.5$ V $\quad R = R_{eq} = 125\ \Omega$

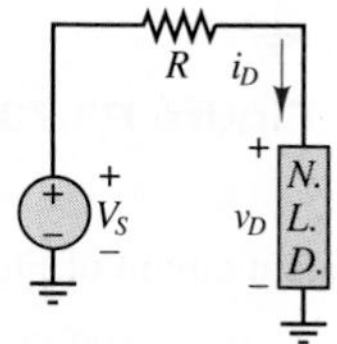

(a)

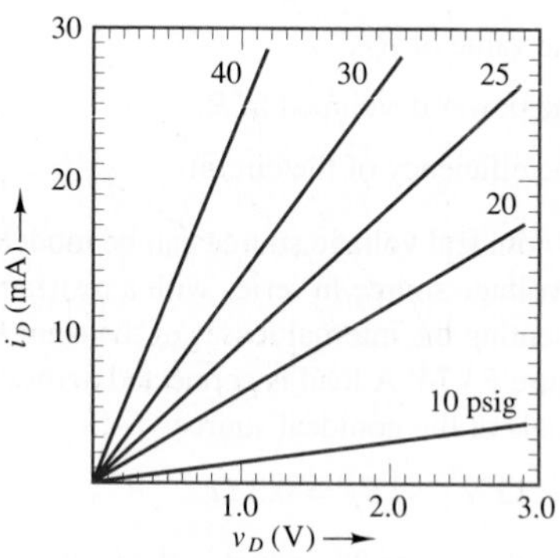

(b)

Figure P3.80

3.81 The nonlinear device in the circuit shown in Figure P3.81 has the i-v characteristic

$$i_D = I_o e^{v_D/V_T}$$

$$I_o = 10^{-15}\text{ A} \qquad V_T = 26\text{ mV}$$

$$V_S = V_{TH} = 1.5\text{ V}$$

$$R = R_{eq} = 60\ \Omega$$

Determine an expression for the DC load line. Then use an iterative technique to determine the voltage across and current through the nonlinear device.

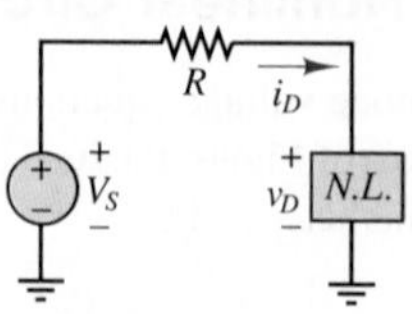

Figure P3.81

3.82 The resistance of the nonlinear device in the circuits shown in Figure P3.82 is a nonlinear function of pressure. The i-v characteristic of the device is shown as a family of curves for various pressures. Construct the DC load line and determine the current through the device when $P = 40$ psig.

$V_S = V_{TH} = 2.5$ V $R = R_{eq} = 125\ \Omega$

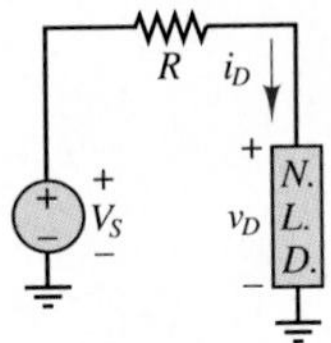

(a)

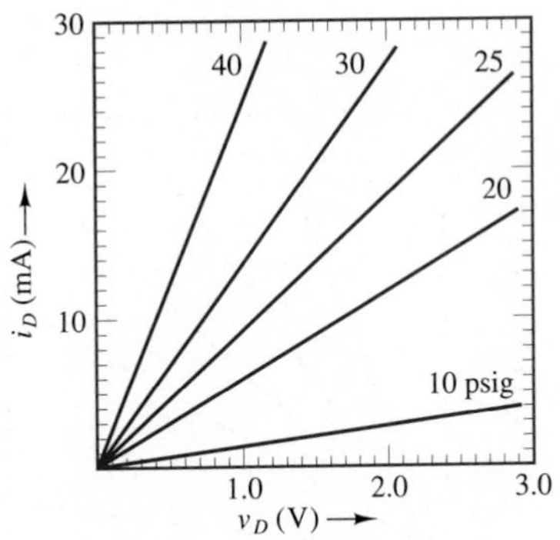

(b)

Figure P3.82

3.83 The voltage-current ($i_D - v_D$) relationship of a semiconductor diode may be approximated by the expression

$$i_D = I_{SAT}\left(\exp\left\{\frac{v_D}{kT/q}\right\} - 1\right)$$

where, at room temperature,

$$I_{SAT} = 10^{-12}\ \text{A}$$
$$\frac{kT}{q} = 0.0259\ \text{V}$$

a. Given the circuit of Figure P3.83, use graphical analysis to find the diode current and diode voltage if $R_T = 22\ \Omega$ and $V_T = 12$ V.

b. Write a computer program in Matlab™ (or in any other programming language) that will find the diode voltage and current using the flowchart shown in Figure P3.83.

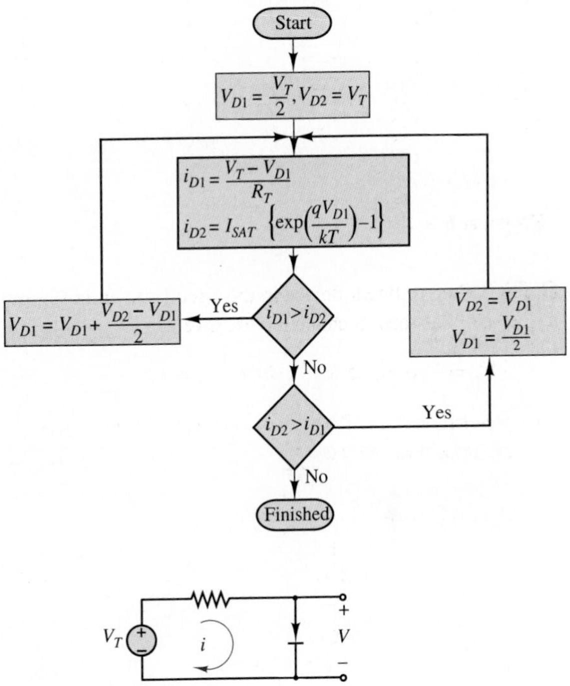

Figure P3.83

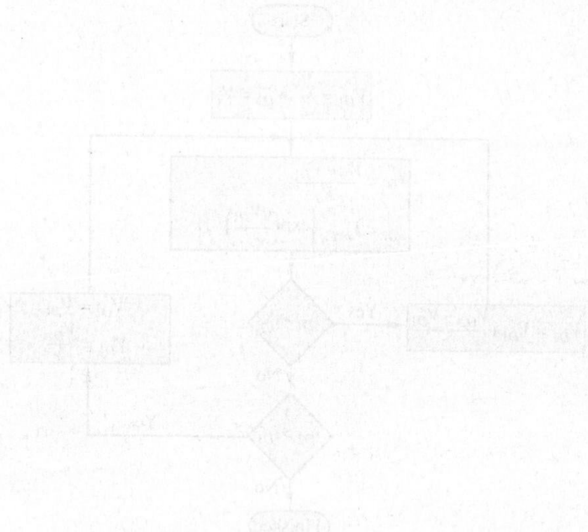

C H A P T E R

4

AC NETWORK ANALYSIS

Chapter 4 is dedicated to two main ideas: *energy storage (dynamic) circuit elements* and the analysis of *AC circuits excited by sinusoidal voltages and currents*. Dynamic circuit elements, that is, capacitors and inductors, are defined. These are circuit elements that are described by an i-v characteristic of differential or integral form. Next, *time-dependent signal sources* and the concepts of average and root-mean-square (rms) values are introduced. Special emphasis is placed on sinusoidal signals, as this class of signals is especially important in the analysis of electric circuits (think, e.g., of the fact that all electric power for residential and industrial uses comes in sinusoidal form). Once these basic elements have been presented, the focus shifts to how to write circuit equations when time-dependent sources and dynamic elements are present: The equations that result from the application of KVL and KCL take the form of differential equations. The general solution of these differential equations is covered in Chapter 5. The remainder of the chapter discusses one particular case: the solution of circuit differential equations when the excitation is a sinusoidal voltage or current; a very powerful method, *phasor analysis,* is introduced along with the related concept of *impedance*. This methodology effectively converts the circuit differential equations to algebraic equations in which complex algebra notation is used to arrive at the solution. Phasor analysis is then used to demonstrate that all the network analysis techniques of Chapter 3 are applicable to the analysis of dynamic circuits with sinusoidal excitations, and a number of examples are presented.

Learning Objectives

1. Compute currents, voltages, and energy stored in capacitors and inductors. *Section 1.*
2. Calculate the average and root-mean-square value of an arbitrary (periodic) signal. *Section 2.*
3. Write the differential equation(s) for circuits containing inductors and capacitors. *Section 3.*
4. Convert time-domain sinusoidal voltages and currents to phasor notation, and vice versa; and represent circuits using impedances. *Section 4.*
5. Apply the circuit analysis methods of Chapter 3 to AC circuits in phasor form. *Section 5.*

4.1 ENERGY STORAGE (DYNAMIC) CIRCUIT ELEMENTS

The ideal resistor was introduced through Ohm's law in Chapter 2 as a useful idealization of many practical electrical devices. However, in addition to resistance to the flow of electric current, which is purely a dissipative (i.e., an energy loss) phenomenon, electric devices may exhibit energy storage properties, much in the same way as a spring or a flywheel can store mechanical energy. Two distinct mechanisms for energy storage exist in electric circuits: **capacitance** and **inductance,** both of which lead to the storage of energy in an electromagnetic field. For the purpose of this discussion, it will not be necessary to enter into a detailed electromagnetic analysis of these devices. Rather, two ideal circuit elements will be introduced to represent the ideal properties of capacitive and inductive energy storage: the **ideal capacitor** and the **ideal inductor.** It should be stated clearly that ideal capacitors and inductors do not exist, strictly speaking; however, just like the ideal resistor, these "ideal" elements are very useful for understanding the behavior of physical circuits. In practice, any component of an electric circuit will exhibit some resistance, some inductance, and some capacitance—that is, some energy dissipation and some energy storage. The sidebar on hydraulic analogs of electric circuits illustrates that the concept of capacitance does not just apply to electric circuits.

The Ideal Capacitor

A physical capacitor is a device that can store energy in the form of a charge separation when appropriately polarized by an electric field (i.e., a voltage). The simplest capacitor configuration consists of two parallel conducting plates of cross-sectional area A, separated by air (or another **dielectric**[1] material, such as mica or Teflon). Figure 4.2 depicts a typical configuration and the circuit symbol for a capacitor.

[1] A dielectric material is a material that is not an electrical conductor but contains a large number of electric dipoles, which become polarized in the presence of an electric field.

Fluid (Hydraulic) Capacitance

We continue the analogy between electrical and hydraulic circuits. If a vessel has some elasticity, energy is stored in the expansion and contraction of the vessel walls (this should remind you of a mechanical spring). This phenomenon gives rise to **fluid capacitance** effect very similar to electrical capacitance. The energy stored in the compression and expansion of the gas is of the *potential energy* type. Figure 4.1 depicts a gas-bag accumulator: a two-chamber arrangement that permits fluid to displace a membrane separating the incompressible fluid from a compressible fluid (e.g., air). The analogy shown in Figure 4.1 assumes that the reference pressure p_0 is zero ("ground" or reference pressure), and that v_2 is ground. The analog equations are given below.

$$q_f = C_f \frac{d\Delta p}{dt} = C_f \frac{dp}{dt}$$

$$i = C \frac{d\Delta v}{dt} = C \frac{dv_1}{dt}$$

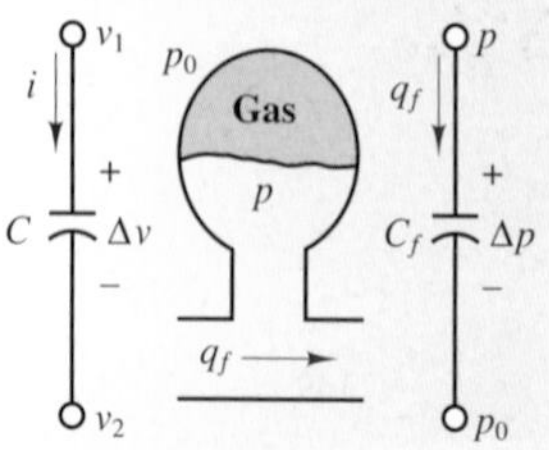

Figure 4.1 Analogy between electrical and fluid capacitance

The presence of an insulating material between the conducting plates does not allow for the flow of DC current; thus, *a capacitor acts as an open circuit in the presence of DC current*. However, if the voltage present at the capacitor terminals changes as a function of time, so will the charge that has accumulated at the two capacitor plates, since the degree of polarization is a function of the applied electric field, which is time-varying. In a capacitor, the charge separation caused by the polarization of the dielectric is proportional to the external voltage, that is, to the applied electric field

$$Q = CV \tag{4.1}$$

where the parameter C is called the *capacitance* of the element and is a measure of the ability of the device to accumulate, or store, charge. The unit of capacitance is coulomb per volt and is called the **farad (F).** The farad is an unpractically large unit for many common electronic circuit applications; therefore it is common to use microfarads (1 μF $= 10^{-6}$ F) or picofarads (1 pF $= 10^{-12}$ F). From equation 4.1 it becomes apparent that if the external voltage applied to the capacitor plates changes in time, so will the charge that is internally stored by the capacitor:

$$q(t) = Cv(t) \tag{4.2}$$

Thus, although no current can flow through a capacitor if the voltage across it is constant, a time-varying voltage will cause charge to vary in time.

The change with time in the stored charge is analogous to a current. You can easily see this by recalling the definition of current given in Chapter 2, where it was stated that

$$i(t) = \frac{dq(t)}{dt} \tag{4.3}$$

that is, electric current corresponds to the time rate of change of charge. Differentiating equation 4.2, one can obtain a relationship between the current and voltage in a capacitor:

$$i(t) = C\frac{dv(t)}{dt} \qquad i\text{-}v \text{ relation for capacitor} \tag{4.4}$$

LO1

Equation 4.4 is the defining circuit law for a capacitor. If the differential equation that defines the i-v relationship for a capacitor is integrated, one can obtain the following relationship for the voltage across a capacitor:

$$v_C(t) = \frac{1}{C}\int_{-\infty}^{t} i_C(t')dt' \tag{4.5}$$

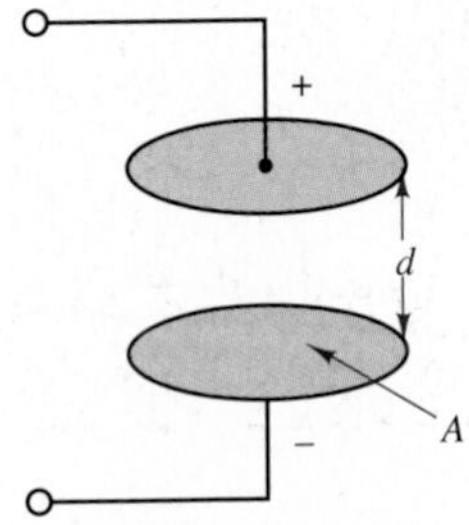

Parallel-plate capacitor with air gap d (air is the dielectric)

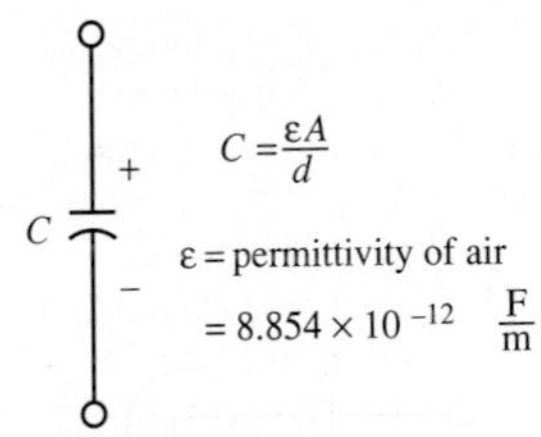

Circuit symbol

Figure 4.2 Structure of parallel-plate capacitor

Equation 4.5 indicates that the capacitor voltage depends on the past current through the capacitor, up until the present time t. Of course, one does not usually have precise information regarding the flow of capacitor current for all past time, and so it is useful to define the initial voltage (or *initial condition*) for the capacitor according to the following, where t_0 is an arbitrary initial time:

$$V_0 = v_C(t = t_0) = \frac{1}{C}\int_{-\infty}^{t_0} i_C(t')\,dt' \tag{4.6}$$

The capacitor voltage is now given by the expression

$$v_C(t) = \frac{1}{C}\int_{t_0}^{t} i_C(t')\,dt' + V_0 \qquad t \geq t_0 \tag{4.7}$$

The significance of the initial voltage V_0 is simply that at time t_0 some charge is stored in the capacitor, giving rise to a voltage $v_C(t_0)$, according to the relationship $Q = CV$. Knowledge of this initial condition is sufficient to account for the entire history of the capacitor current.

Capacitors connected in series and parallel can be combined to yield a single equivalent capacitance. The rule of thumb, which is illustrated in Figure 4.3, is the following:

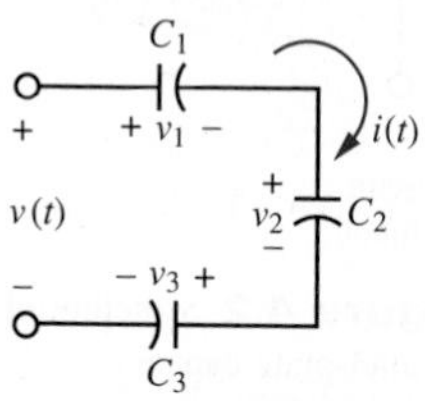

$$C_{EQ} = \frac{1}{\frac{1}{C_1} + \frac{1}{C_2} + \frac{1}{C_3}}$$

Capacitances in series combine like resistors in parallel

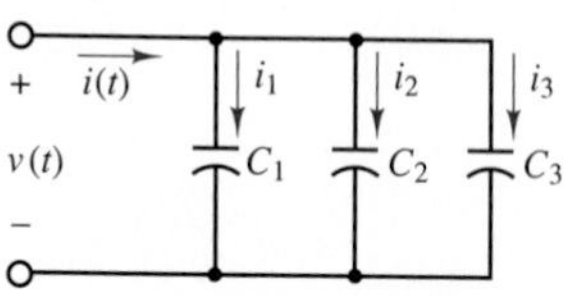

$C_{EQ} = C_1 + C_2 + C_3$

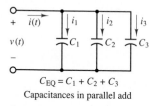

Capacitances in parallel add

Figure 4.3 Combining capacitors in a circuit

> Capacitors in parallel add. Capacitors in series combine according to the same rules used for resistors connected in parallel.

It is very easy to prove that capacitors in series combine as shown in Figure 4.3, using the definition of equation 4.5. Consider the three capacitors in series in the circuit of Figure 4.3. Using Kirchhoff's voltage law and the definition of the capacitor voltage, we can write

$$\begin{aligned} v(t) &= v_1(t) + v_1(t) + v_1(t) \\ &= \frac{1}{C_1}\int_{-\infty}^{t} i(t')\,dt' + \frac{1}{C_2}\int_{-\infty}^{t} i(t')\,dt' + \frac{1}{C_3}\int_{-\infty}^{t} i(t')\,dt' \\ &= \left(\frac{1}{C_1} + \frac{1}{C_2} + \frac{1}{C_3}\right)\int_{-\infty}^{t} i(t')\,dt' \end{aligned} \tag{4.8}$$

Thus, the voltage across the three series capacitors is the same as would be seen across a single equivalent capacitor C_{eq} with $1/C_{eq} = 1/C_1 + 1/C_2 + 1/C_3$, as illustrated in Figure 4.3. You can easily use the same method to prove that the three parallel capacitors in the bottom half of Figure 4.3 combine as do resistors in series.

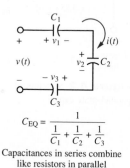

EXAMPLE 4.1 Charge Separation in Ultracapacitors

Problem

Ultracapacitors are finding application in a variety of fields, including as a replacement or supplement for batteries in hybrid-electric vehicles. In this example you will make your first acquaintance with these devices.

An ultracapacitor, or "supercapacitor," stores energy electrostatically by polarizing an electrolytic solution. Although it is an electrochemical device (also known as an electrochemical double-layer capacitor), there are no chemical reactions involved in its energy storage mechanism. This mechanism is highly reversible, allowing the ultracapacitor to be charged and discharged hundreds of thousands of times. An ultracapacitor can be viewed as two nonreactive porous plates suspended within an electrolyte, with a voltage applied across the plates. The applied potential on the positive plate attracts the negative ions in the electrolyte, while the potential on the negative plate attracts the positive ions. This effectively creates two layers of capacitive storage, one where the charges are separated at the positive plate and another at the negative plate.

Recall that capacitors store energy in the form of separated electric charge. The greater the area for storing charge and the closer the separated charges, the greater the capacitance. A conventional capacitor gets its area from plates of a flat, conductive material. To achieve high capacitance, this material can be wound in great lengths, and sometimes a texture is imprinted on it to increase its surface area. A conventional capacitor separates its charged plates with a dielectric material, sometimes a plastic or paper film, or a ceramic. These dielectrics can be made only as thin as the available films or applied materials.

An ultracapacitor gets its area from a porous carbon-based electrode material, as shown in Figure 4.4. The porous structure of this material allows its surface area to approach 2,000 square meters per gram (m^2/g), much greater than can be accomplished using flat or textured films and plates. An ultracapacitor's charge separation distance is determined by the size of the ions in the electrolyte, which are attracted to the charged electrode. This charge separation [less than 10 angstroms (Å)] is much smaller than can be achieved using conventional dielectric materials. The combination of enormous surface area and extremely small charge separation gives the ultracapacitor its outstanding capacitance relative to conventional capacitors.

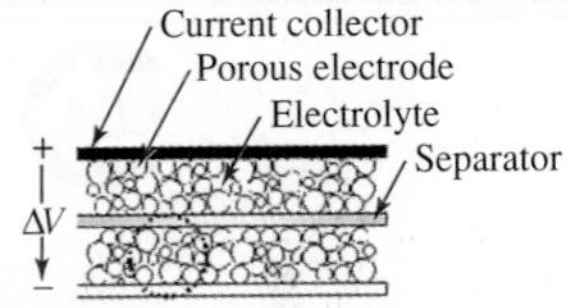

Figure 4.4 Ultracapacitor structure

Use the data provided to calculate the charge stored in an ultracapacitor, and calculate how long it will take to discharge the capacitor at the maximum current rate.

Solution

Known Quantities: Technical specifications are as follows:

Capacitance	100 F	$(-10\%/+30\%)$
Series resistance	DC	15 mΩ (±25%)
	1kHz	7 mΩ (±25%)
Voltage	Continuous	2.5 V; Peak 2.7 V
Rated current	25 A	

Find: Charge separation at nominal voltage and time to complete discharge at maximum current rate.

Analysis: Based on the definition of charge storage in a capacitor, we calculate

$$Q = CV = 100 \text{ F} \times 2.5 \text{ V} = 250 \text{ C}$$

To calculate how long it would take to discharge the ultracapacitor, we approximate the defining differential equation (4.4) as follows:

$$i = \frac{dq}{dt} \approx \frac{\Delta q}{\Delta t}$$

Since we know that the discharge current is 25 A and the available charge separation is 250 F, we can calculate the time to complete discharge, assuming a constant 25-A discharge:

$$\Delta t = \frac{\Delta q}{i} = \frac{250 \text{ C}}{25 \text{ A}} = 10 \text{ s}$$

Comments: We shall continue our exploration of ultracapacitors in Chapter 5. In particular, we shall look more closely at the charging and discharging behavior of these devices, taking into consideration their internal resistance.

CHECK YOUR UNDERSTANDING

Compare the charge separation achieved in this ultracapacitor with a (similarly sized) electrolytic capacitor used in power electronics applications, by calculating the charge separation for a 2,000-μF electrolytic capacitor rated at 400 V.

Answer: 0.8 C

LO1

EXAMPLE 4.2 Calculating Capacitor Current from Voltage

Problem

Calculate the current through a capacitor from knowledge of its terminal voltage.

Solution

Known Quantities: Capacitor terminal voltage; capacitance value.

Find: Capacitor current.

Assumptions: The initial current through the capacitor is zero.

Schematics, Diagrams, Circuits, and Given Data: $v(t) = 5(1 - e^{-t/10^{-6}})$ volts; $t \geq 0$ s; $C = 0.1\ \mu$F. The terminal voltage is plotted in Figure 4.5.

Assumptions: The capacitor is initially discharged: $v(t = 0) = 0$.

Analysis: Using the defining differential relationship for the capacitor, we may obtain the current by differentiating the voltage:

$$i_C(t) = C\frac{dv(t)}{dt} = 10^{-7}\frac{5}{10^{-6}}\left(e^{-t/10^{-6}}\right) = 0.5e^{-t/10^{-6}} \qquad \text{A} \qquad t \geq 0$$

A plot of the capacitor current is shown in Figure 4.6. Note how the current jumps to 0.5 A instantaneously as the voltage rises exponentially: The ability of a capacitor's current to change instantaneously is an important property of capacitors.

Comments: As the voltage approaches the constant value 5 V, the capacitor reaches its maximum charge storage capability for that voltage (since $Q = CV$) and no more current flows through the capacitor. The total charge stored is $Q = 0.5 \times 10^{-6}$ C. This is a fairly small amount of charge, but it can produce a substantial amount of current for a brief time. For example, the fully charged capacitor could provide 100 mA of current for a time equal to 5 μs:

$$I = \frac{\Delta Q}{\Delta t} = \frac{0.5 \times 10^{-6}}{5 \times 10^{-6}} = 0.1 \text{ A}$$

There are many useful applications of this energy storage property of capacitors in practical circuits.

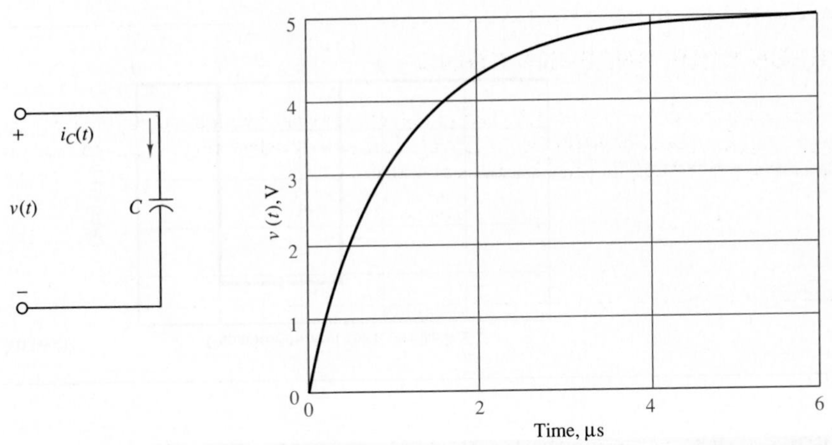

Figure 4.5

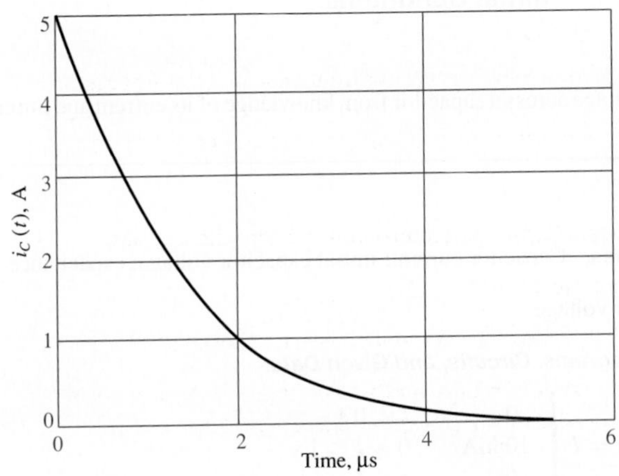

Figure 4.6

CHECK YOUR UNDERSTANDING

The voltage waveform shown below appears across a 1,000-μF capacitor. Plot the capacitor current $i_C(t)$.

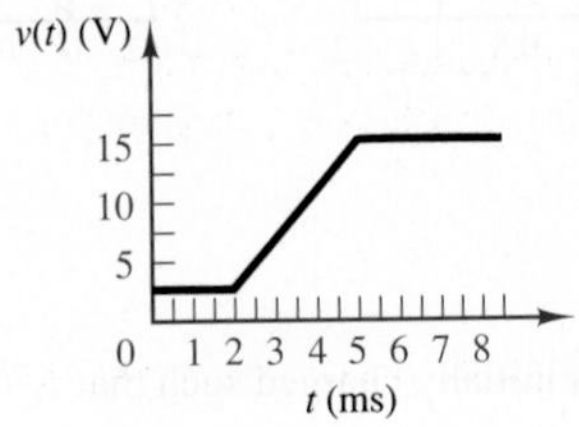

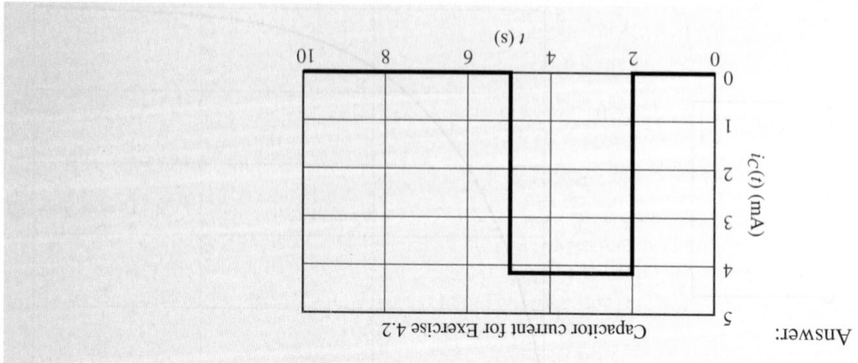

LO1

EXAMPLE 4.3 Calculating Capacitor Voltage from Current and Initial Conditions

Problem

Calculate the voltage across a capacitor from knowledge of its current and initial state of charge.

Solution

Known Quantities: Capacitor current; initial capacitor voltage; capacitance value.

Find: Capacitor voltage.

Schematics, Diagrams, Circuits, and Given Data:

$$i_C(t) = I \begin{cases} 0 & t < 0 \text{ s} \\ 10 \text{ mA} & 0 \le t \le 1 \text{ s} \\ 0 & t > 1 \text{ s} \end{cases}$$

$$v_C(t = 0) = 2 \text{ V} \qquad C = 1{,}000 \ \mu\text{F}$$

The capacitor current is plotted in Figure 4.7(a).

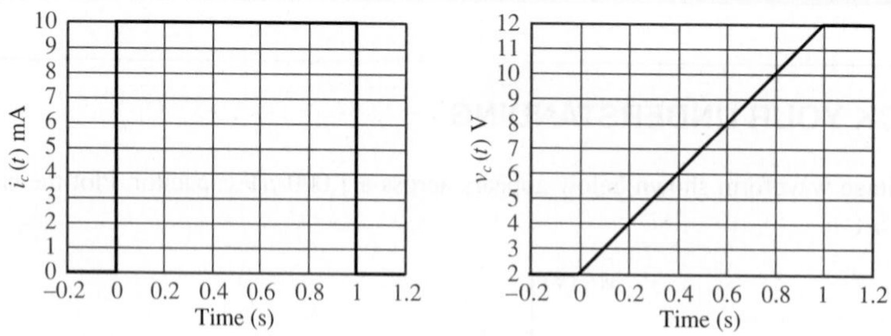

(a) (b)

Figure 4.7

Assumptions: The capacitor is initially charged such that $v_C(t = t_0 = 0) = 2$ V.

Analysis: Using the defining integral relationship for the capacitor, we may obtain the voltage by integrating the current:

$$v_C(t) = \frac{1}{C}\int_{t_0}^{t} i_C(t')\,dt' + v_C(t_0) \qquad t \geq t_0$$

$$= \begin{cases} \dfrac{1}{C}\displaystyle\int_0^1 I\,dt' + V_0 = \dfrac{I}{C}t + V_0 = 10t + 2\text{ V} & 0 \leq t \leq 1\text{ s} \\ 12\text{ V} & t > 1\text{ s} \end{cases}$$

Comments: Once the current stops, at $t = 1$ s, the capacitor voltage cannot develop any further but remains at the maximum value it reached at $t = 1$ s: $v_C(t = 1) = 12$ V. The final value of the capacitor voltage after the current source has stopped charging the capacitor depends on two factors: (1) the initial value of the capacitor voltage and (2) the history of the capacitor current. Figure 4.7(a) and (b) depicts the two waveforms.

CHECK YOUR UNDERSTANDING

Find the maximum current through the capacitor of Example 4.3 if the capacitor voltage is described by $v_C(t) = 5t + 3$ V for $0 \leq t \leq 5$ s.

Answer: 5 mA

FIND IT ON THE WEB

Physical capacitors are rarely constructed of two parallel plates separated by air, because this configuration yields very low values of capacitance, unless one is willing to tolerate very large plate areas. To increase the capacitance (i.e., the ability to store energy), physical capacitors are often made of tightly rolled sheets of metal film, with a dielectric (paper or Mylar) sandwiched in between. Table 4.1 illustrates typical values, materials, maximum voltage ratings, and useful frequency ranges for various types of capacitors. The voltage rating is particularly important, because any insulator will break down if a sufficiently high voltage is applied across it.

Table 4.1 **Capacitors**

Material	Capacitance range	Maximum voltage (V)	Frequency range (Hz)
Mica	1 pF to 0.1 μF	100–600	10^3–10^{10}
Ceramic	10 pF to 1 μF	50–1,000	10^3–10^{10}
Mylar	0.001 μF to 10 μF	50–500	10^2–10^8
Paper	1,000 pF to 50 μF	100–105	10^2–10^8
Electrolytic	0.1 μF to 0.2 F	3–600	10–10^4

Energy Storage in Capacitors

You may recall that the capacitor was described earlier in this section as an energy storage element. An expression for the energy stored in the capacitor $W_C(t)$ may be derived easily if we recall that energy is the integral of power, and that the instantaneous power in a circuit element is equal to the product of voltage and current:

$$\begin{aligned} W_C(t) &= \int P_C(t')\,dt' \\ &= \int v_C(t')i_C(t')\,dt' \\ &= \int v_C(t')C\,\frac{dv_C(t')}{dt'}\,dt' \end{aligned} \tag{4.9}$$

LO1

$$W_C(t) = \frac{1}{2}Cv_C^2(t) \qquad \text{Energy stored in a capacitor (J)}$$

Example 4.4 illustrates the calculation of the energy stored in a capacitor.

EXAMPLE 4.4 Energy Storage in Ultracapacitors

Problem

Determine the energy stored in the ultracapacitor of Example 4.1.

Solution

Known Quantities: See Example 4.1.

Find: Energy stored in capacitor.

Analysis: To calculate the energy, we use equation 4.9:

$$W_C = \frac{1}{2}Cv_C^2 = \frac{1}{2}(100\text{ F})(2.5\text{ V})^2 = 312.5\text{ J}$$

CHECK YOUR UNDERSTANDING

Compare the energy stored in this ultracapacitor with a (similarly sized) electrolytic capacitor used in power electronics applications, by calculating the charge separation for a 2,000-μF electrolytic capacitor rated at 400 V.

Answer: 160 J

Capacitive Displacement Transducer and Microphone

As shown in Figure 4.2, the capacitance of a parallel-plate capacitor is given by the expression

$$C = \frac{\varepsilon A}{d}$$

where ε is the **permittivity** of the dielectric material, A is the area of each of the plates, and d is their separation. The permittivity of air is $\varepsilon_0 = 8.854 \times 10^{-12}$ F/m, so that two parallel plates of area 1 m^2, separated by a distance of 1 mm, would give rise to a capacitance of 8.854×10^{-3} μF, a very small value for a very large plate area. This relative inefficiency makes parallel-plate capacitors impractical for use in electronic circuits. On the other hand, parallel-plate capacitors find application as *motion transducers,* that is, as devices that can measure the motion or displacement of an object. In a capacitive motion transducer, the air gap between the plates is designed to be variable, typically by fixing one plate and connecting the other to an object in motion. Using the capacitance value just derived for a parallel-plate capacitor, one can obtain the expression

$$C = \frac{8.854 \times 10^{-3} A}{x}$$

where C is the capacitance in picofarads, A is the area of the plates in square millimeters, and x is the (variable) distance in millimeters. It is important to observe that the change in capacitance caused by the displacement of one of the plates is nonlinear, since the capacitance varies as the inverse of the displacement. For small displacements, however, the capacitance varies approximately in a linear fashion.

The *sensitivity* S of this motion transducer is defined as the slope of the change in capacitance per change in displacement x, according to the relation

$$S = \frac{dC}{dx} = -\frac{8.854 \times 10^{-3} A}{2x^2} \frac{\text{pF}}{\text{mm}}$$

Thus, the sensitivity increases for small displacements. This behavior can be verified by plotting the capacitance as a function of x and noting that as x approaches zero, the slope of the nonlinear $C(x)$ curve becomes steeper (thus the greater sensitivity). Figure 4.8 depicts this behavior for a transducer with area equal to 10 mm^2.

FIND IT ON THE WEB

This simple capacitive displacement transducer actually finds use in the popular **capacitive (or condenser) microphone,** in which the sound pressure waves act to displace one of the capacitor plates. The change in capacitance can then be converted to a change in voltage or

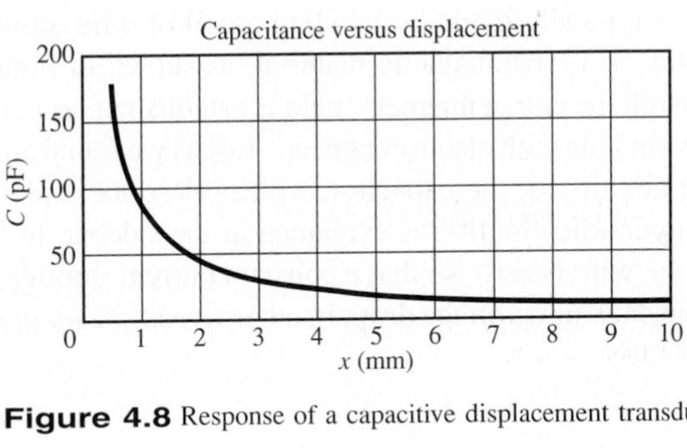

Figure 4.8 Response of a capacitive displacement transducer

(Continued)

(Concluded)

current by means of a suitable circuit. An extension of this concept that permits measurement of differential pressures is shown in simplified form in Figure 4.9. In the figure, a three-terminal variable capacitor is shown to be made up of two fixed surfaces (typically, spherical depressions ground into glass disks and coated with a conducting material) and of a deflecting plate (typically made of steel) sandwiched between the glass disks. Pressure inlet orifices are provided, so that the deflecting plate can come into contact with the fluid whose pressure it is measuring. When the pressure on both sides of the deflecting plate is the same, the capacitance between terminals b and d, denoted by C_{db}, will be equal to that between terminals b and c, denoted by C_{bc}. If any pressure differential exists, the two capacitances will change, with an increase on the side where the deflecting plate has come closer to the fixed surface and a corresponding decrease on the other side.

This behavior is ideally suited for the application of a bridge circuit, similar to the Wheatstone bridge circuit illustrated in Example 2.14, and also shown in Figure 4.9. In the bridge circuit, the output voltage v_{out} is precisely balanced when the differential pressure across the transducer is zero, but it will deviate from zero whenever the two capacitances are not identical because of a pressure differential across the transducer. We shall analyze the bridge circuit later.

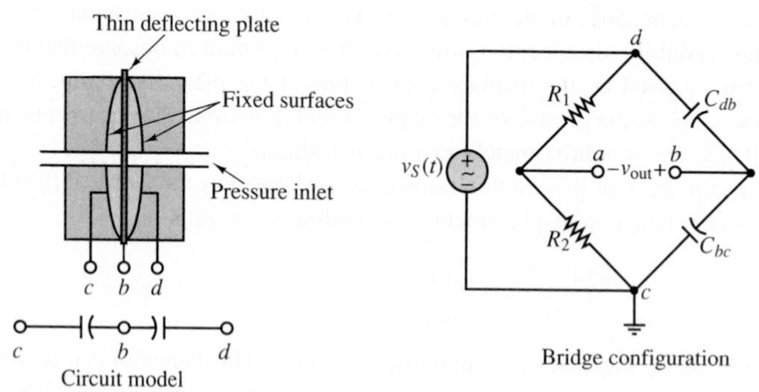

Figure 4.9 Capacitive pressure transducer and related bridge circuit

The Ideal Inductor

The ideal inductor is an element that has the ability to store energy in a magnetic field. Inductors are typically made by winding a coil of wire around a **core,** which can be an insulator or a ferromagnetic material, as shown in Figure 4.10. When a current flows through the coil, a magnetic field is established, as you may recall from early physics experiments with electromagnets.[2] Just as we found an analogy between electric and fluid circuits for the capacitor, we can describe a phenomenon similar to inductance in hydraulic circuits, as explained in the sidebar. In an ideal inductor, the resistance of the wire is zero, so that a constant current through the inductor will flow freely without causing a voltage drop. In other words, *the ideal inductor acts as a*

[2]See also Chapter 16.

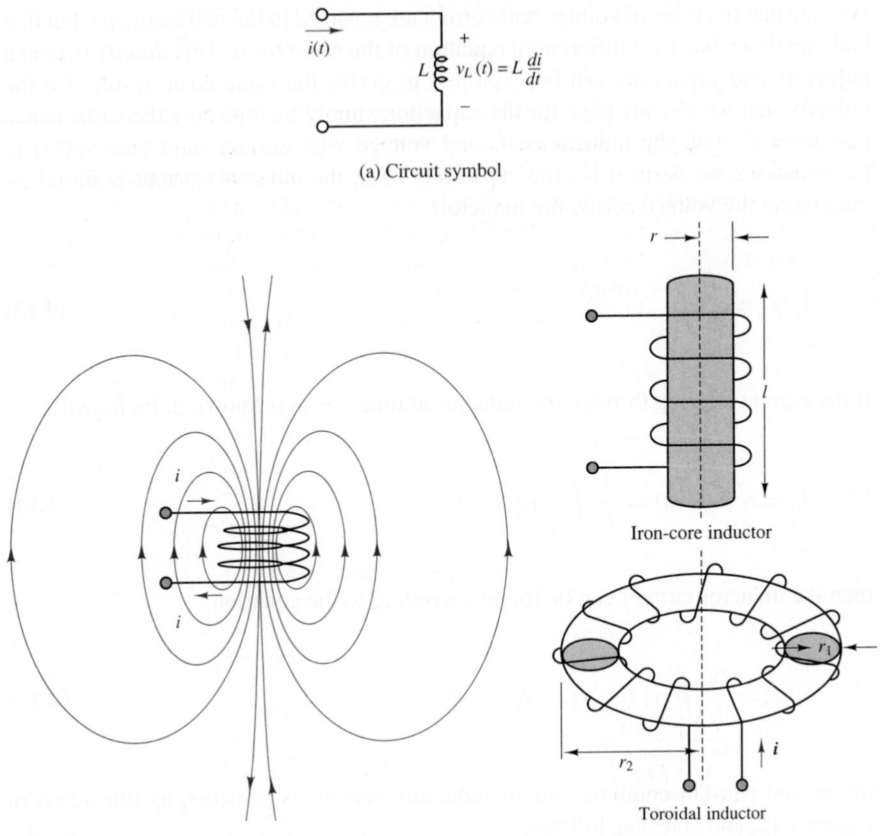

(a) Circuit symbol

(b) Magnetic flux lines in the vicinity of a current-carrying coil

(c) Practical inductors

Figure 4.10 Inductance and practical inductors

short circuit in the presence of DC. If a time-varying voltage is established across the inductor, a corresponding current will result, according to the following relationship:

$$v_L(t) = L\frac{di_L(t)}{dt} \qquad i\text{-}v \text{ relation for inductor} \tag{4.10}$$

where L is called the *inductance* of the coil and is measured in **henrys (H),** where

$$1\ \text{H} = 1\ \text{V-s/A} \tag{4.11}$$

Henrys are reasonable units for **practical inductors;** millihenrys (mH) and microhenrys (μH) are also used.

It is instructive to compare equation 4.10, which defines the behavior of an ideal inductor, with the expression relating capacitor current and voltage:

$$i_C(t) = C\frac{dv_C(t)}{dt} \tag{4.12}$$

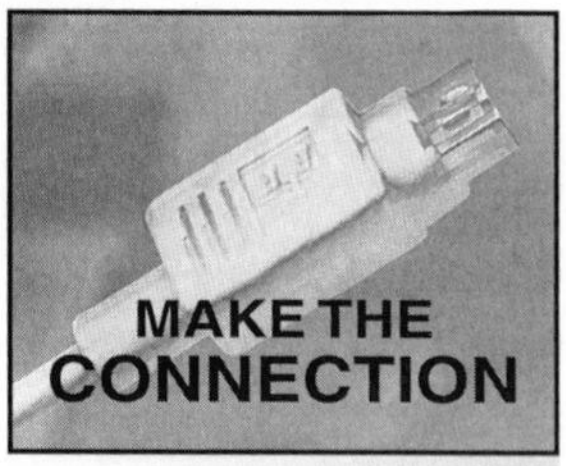

Fluid (Hydraulic) Inertance

The **fluid inertance** parameter is analogous to inductance in the electric circuit. Fluid inertance, as the name suggests, is caused by the inertial properties, i.e., the mass, of the fluid in motion. As you know from physics, a particle in motion has kinetic energy associated with it; fluid in motion consists of a collection of particles, and it also therefore must have kinetic energy storage properties. (think of water flowing out of a fire hose!). The equations that define the analogy are given below

$$\Delta p = p_1 - p_2 = I_f \frac{dq_f}{dt}$$

$$\Delta v = v_1 - v_2 = L \frac{di}{dt}$$

Figure 4.11 depicts the analogy between electrical inductance and fluid inertance. These analogies and the energy equations that apply to electrical and fluid circuit elements are summarized in Table 4.2.

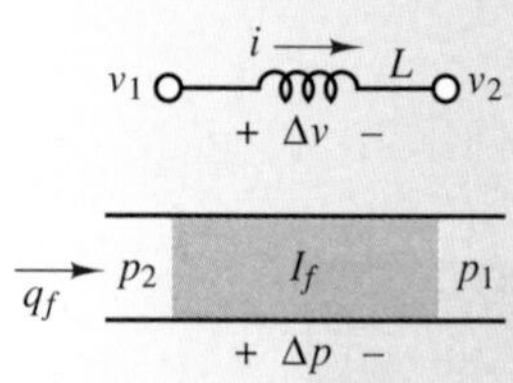

Figure 4.11 Analogy between fluid inertance and electrical inductance

We note that the roles of voltage and current are reversed in the two elements, but that both are described by a differential equation of the same form. This *duality* between inductors and capacitors can be exploited to derive the same basic results for the inductor that we already have for the capacitor, simply by replacing the capacitance parameter C with the inductance L and voltage with current (and vice versa) in the equations we derived for the capacitor. Thus, the inductor current is found by integrating the voltage across the inductor:

$$i_L(t) = \frac{1}{L} \int_{-\infty}^{t} v_L(t')\, dt' \tag{4.13}$$

If the current flowing through the inductor at time $t = t_0$ is known to be I_0, with

$$I_0 = i_L(t = t_0) = \frac{1}{L} \int_{-\infty}^{t_0} v_L(t')\, dt' \tag{4.14}$$

then the inductor current can be found according to the equation

$$i_L(t) = \frac{1}{L} \int_{t_0}^{t} v_L(t')\, dt' + I_0 \qquad t \geq t_0 \tag{4.15}$$

Series and parallel combinations of inductors behave as resistors, as illustrated in Figure 4.12, and stated as follows:

> Inductors in series add. Inductors in parallel combine according to the same rules used for resistors connected in parallel.

LO1

Table 4.2 Analogy between electric and fluid circuits

Property	Electrical element or equation	Hydraulic analogy
Potential variable	Voltage or potential difference	Pressure difference
Flow variable	Current flow	Fluid volume flow rate
Resistance	Resistor R	Fluid resistor R_f
Capacitance	Capacitor C	Fluid capacitor C_f
Inductance	Inductor L	Fluid inertor I_f
Power dissipation	$P = i^2 R$	$P_f = q_f^2 R_f$
Potential energy storage	$W_p = \frac{1}{2} C v^2$	$W_p = \frac{1}{2} C_f p^2$
Kinetic energy storage	$W_k = \frac{1}{2} L i^2$	$W_k = \frac{1}{2} I_f q_f^2$

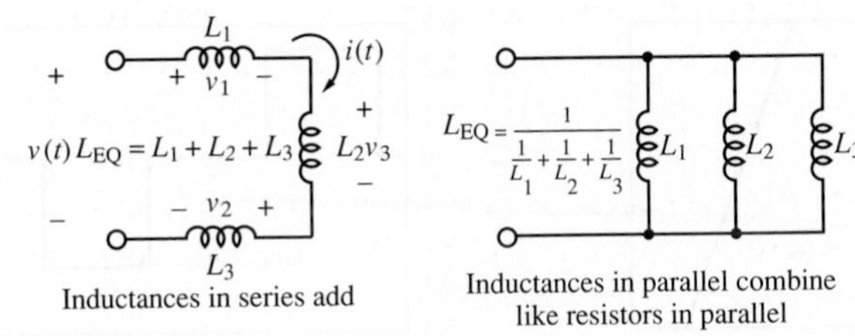

Figure 4.12 Combining inductors in a circuit

It is very easy to prove that inductors in series combine as shown in Figure 4.12, using the definition of equation 4.10. Consider the three inductors in series in the circuit on the left of Figure 4.12. Using Kirchhoff's voltage law and the definition of the capacitor voltage, we can write

$$\begin{aligned} v(t) = v_1(t) + v_1(t) + v_1(t) &= L_1 \frac{di(t)}{dt} + L_2 \frac{di(t)}{dt} + L_3 \frac{di(t)}{dt} \\ &= (L_1 + L_2 + L_3) \frac{di(t)}{dt} \end{aligned} \tag{4.16}$$

Thus, the voltage across the three series inductors is the same that would be seen across a single equivalent inductor L_{eq} with $L_{eq} = L_1 + L_2 + L_3$, as illustrated in Figure 4.12. You can easily use the same method to prove that the three parallel inductors on the right half of Figure 4.12 combine as resistors in parallel do.

EXAMPLE 4.5 Calculating Inductor Voltage from Current

LO1

Problem

Calculate the voltage across the inductor from knowledge of its current.

Solution

Known Quantities: Inductor current; inductance value.

Find: Inductor voltage.

Schematics, Diagrams, Circuits, and Given Data:

$$i_L(t) = \begin{cases} 0 \text{ mA} & t < 1 \text{ ms} \\ -\dfrac{0.1}{4} + \dfrac{0.1}{4}t \text{ mA} & 1 \le t \le 5 \text{ ms} \\ 0.1 \text{ mA} & 5 \le t \le 9 \text{ ms} \\ 13 \times \dfrac{0.1}{4} - \dfrac{0.1}{4}t \text{ mA} & 9 \le t \le 13 \text{ ms} \\ 0 \text{ mA} & t > 13 \text{ ms} \end{cases}$$

$L = 10$ H

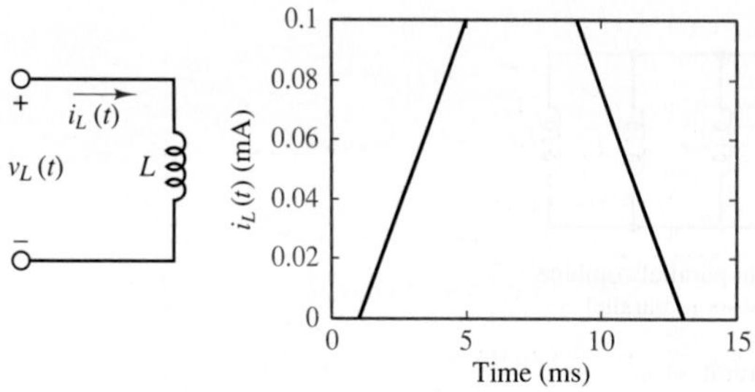

Figure 4.13

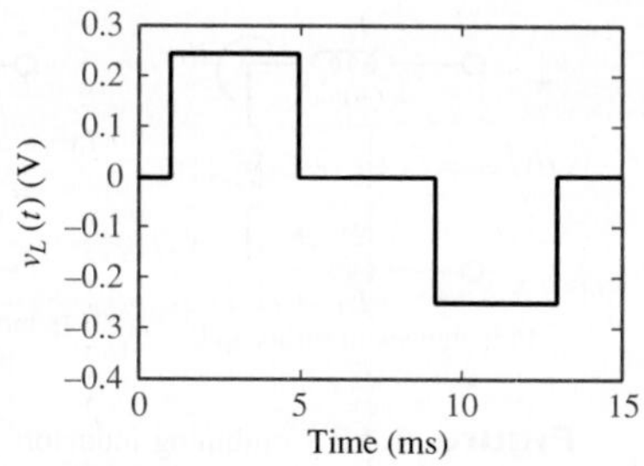

Figure 4.14

The inductor current is plotted in Figure 4.13.

Assumptions: $i_L(t = 0) \leq 0$.

Analysis: Using the defining differential relationship for the inductor, we may obtain the voltage by differentiating the current:

$$v_L(t) = L\frac{di_L(t)}{dt}$$

Piecewise differentiating the expression for the inductor current, we obtain

$$v_L(t) = \begin{cases} 0 \text{ V} & t < 1 \text{ ms} \\ 0.25 \text{ V} & 1 < t \leq 5 \text{ ms} \\ 0 \text{ V} & 5 < t \leq 9 \text{ ms} \\ -0.25 \text{ V} & 9 < t \leq 13 \text{ ms} \\ 0 \text{ V} & t > 13 \text{ ms} \end{cases}$$

The inductor voltage is plotted in Figure 4.14.

Comments: Note how the inductor voltage has the ability to change instantaneously!

CHECK YOUR UNDERSTANDING

The current waveform shown below flows through a 50-mH inductor. Plot the inductor voltage $v_L(t)$.

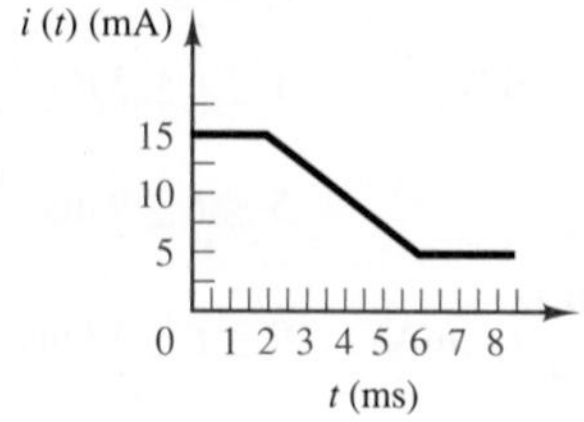

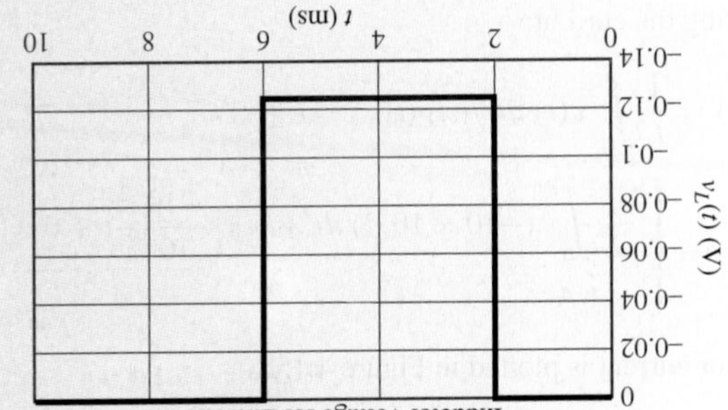

Answer:

Inductor voltage for Exercise 4.5

EXAMPLE 4.6 Calculating Inductor Current from Voltage

Problem

Calculate the current through the inductor from knowledge of the terminal voltage and of the initial current.

Solution

Known Quantities: Inductor voltage; initial condition (current at $t = 0$); inductance value.

Find: Inductor current.

Schematics, Diagrams, Circuits, and Given Data:

$$v(t) = \begin{cases} 0 \text{ V} & t < 0 \text{ s} \\ -10 \text{ mV} & 0 < t \leq 1 \text{ s} \\ 0 \text{ V} & t > 1 \text{ s} \end{cases}$$

$$L = 10 \text{ mH}; \qquad i_L(t = 0) = I_0 = 0 \text{ A}$$

The terminal voltage is plotted in Figure 4.15(a).

Assumptions: $i_L(t = 0) = I_0 = 0$.

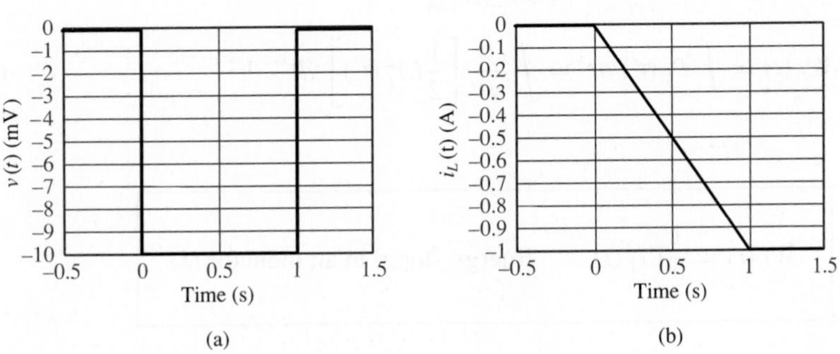

Figure 4.15

Analysis: Using the defining integral relationship for the inductor, we may obtain the voltage by integrating the current:

$$i_L(t) = \frac{1}{L}\int_{t_0}^{t} v(t)\,dt' + i_L(t_0) \qquad t \geq t_0$$

$$= \begin{cases} \dfrac{1}{L}\displaystyle\int_{0}^{t'} (-10 \times 10^{-3})\,dt' + I_0 = \dfrac{-10^{-2}}{10^{-2}}t + 0 = -t \text{ A} & 0 \leq t \leq 1 \text{ s} \\ -1 \text{ A} & t > 1 \text{ s} \end{cases}$$

The inductor current is plotted in Figure 4.15b.

Comments: Note how the inductor voltage has the ability to change instantaneously!

CHECK YOUR UNDERSTANDING

Find the maximum voltage across the inductor of Example 4.6 if the inductor current voltage is described by $i_L(t) = 2t$ amperes for $0 \leq t \leq 2$ s.

Answer: 20 mV

Energy Storage in Inductors

The magnetic energy stored in an ideal inductor may be found from a power calculation by following the same procedure employed for the ideal capacitor. The instantaneous power in the inductor is given by

$$P_L(t) = i_L(t)v_L(t) = i_L(t)L\frac{di_L(t)}{dt} = \frac{d}{dt}\left[\frac{1}{2}Li_L^2(t)\right] \qquad \textbf{(4.17)}$$

Integrating the power, we obtain the total energy stored in the inductor, as shown in the following equation:

$$W_L(t) = \int P_L(t')\,dt' = \int \frac{d}{dt'}\left[\frac{1}{2}Li_L^2(t')\right]dt' \qquad \textbf{(4.18)}$$

LO1

$$W_L(t) = \frac{1}{2}Li_L^2(t) \qquad \text{Energy stored in an inductor (J)}$$

Note, once again, the duality with the expression for the energy stored in a capacitor, in equation 4.9.

EXAMPLE 4.7 Energy Storage in an Ignition Coil

Problem

Determine the energy stored in an automotive ignition coil.

Solution

Known Quantities: Inductor current initial condition (current at $t = 0$); inductance value.

Find: Energy stored in inductor.

Schematics, Diagrams, Circuits, and Given Data: $L = 10$ mH; $i_L = I_0 = 8$ A.

Analysis:

$$W_L = \frac{1}{2}Li_L^2 = \frac{1}{2} \times 10^{-2} \times 64 = 32 \times 10^{-2} = 320 \text{ mJ}$$

Comments: A more detailed analysis of an automotive ignition coil is presented in Chapter 5 to accompany the discussion of transient voltages and currents.

CHECK YOUR UNDERSTANDING

Calculate and plot the inductor energy and power for a 50-mH inductor subject to the current waveform shown below. What is the energy stored at $t = 3$ ms? Assume $i(-\infty) = 0$.

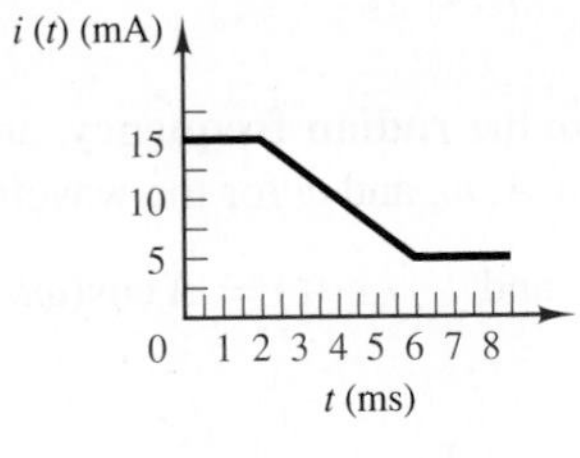

Answer:
$$w(t) = \begin{cases} 5.625 \times 10^{-6} \text{ J} & 0 \le t < 2 \text{ ms} \\ 0.156t^2 - (2.5 \times 10^{-3})t + 10^{-6} & 2 \le t < 6 \text{ ms} \\ 0.625 \times 10^{-6} & t \ge 6 \text{ ms} \end{cases}$$

$$p(t) = \begin{cases} (20 \times 10^{-3} - 2.5t)(-0.125 \text{ W}) & 2 \le t < 6 \text{ ms} \\ 0 & \text{otherwise} \end{cases}$$

$$w(t = 3 \text{ ms}) = 3.9\ \mu\text{J}$$

4.2 TIME-DEPENDENT SIGNAL SOURCES

In Chapter 2, the general concept of an ideal energy source was introduced. In this chapter, it will be useful to specifically consider sources that generate time-varying voltages and currents and, in particular, sinusoidal sources. Figure 4.16 illustrates the convention that will be employed to denote time-dependent signal sources.

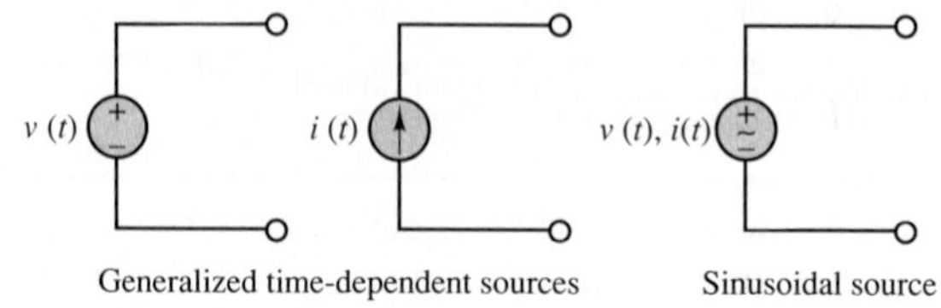

Figure 4.16 Time-dependent signal sources

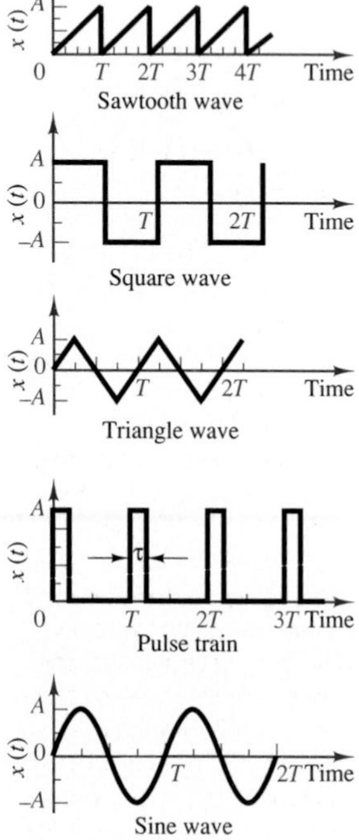

Figure 4.17 Periodic signal waveforms

One of the most important classes of time-dependent signals is that of **periodic signals.** These signals appear frequently in practical applications and are a useful approximation of many physical phenomena. A periodic signal $x(t)$ is a signal that satisfies the equation

$$x(t) = x(t + nT) \qquad n = 1, 2, 3, \ldots \tag{4.19}$$

where T is the **period** of $x(t)$. Figure 4.17 illustrates a number of periodic waveforms that are typically encountered in the study of electric circuits. Waveforms such as the sine, triangle, square, pulse, and sawtooth waves are provided in the form of voltages (or, less frequently, currents) by commercially available **signal** (or **waveform**) **generators.** Such instruments allow for selection of the waveform peak amplitude, and of its period.

As stated in the introduction, sinusoidal waveforms constitute by far the most important class of time-dependent signals. Figure 4.18 depicts the relevant parameters of a sinusoidal waveform. A generalized sinusoid is defined as

$$x(t) = A\cos(\omega t + \phi) \tag{4.20}$$

where A is the **amplitude,** ω the **radian frequency,** and ϕ the **phase.** Figure 4.18 summarizes the definitions of A, ω, and ϕ for the waveforms

$$x_1(t) = A\cos(\omega t) \qquad \text{and} \qquad x_2(t) = A\cos(\omega t + \phi)$$

where

$$\begin{aligned} f &= \text{natural frequency} = \frac{1}{T} \qquad \text{cycles/s, or Hz} \\ \omega &= \text{radian frequency} = 2\pi f \qquad \text{rad/s} \\ \phi &= 2\pi \frac{\Delta t}{T} \qquad \text{rad} \\ &= 360 \frac{\Delta t}{T} \qquad \text{deg} \end{aligned} \tag{4.21}$$

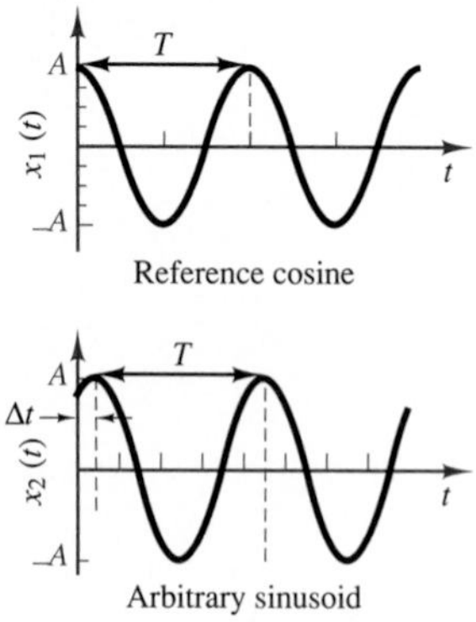

Figure 4.18 Sinusoidal waveforms

The phase shift ϕ permits the representation of an arbitrary sinusoidal signal. Thus, the choice of the reference cosine function to represent sinusoidal signals—arbitrary as it may appear at first—does not restrict the ability to represent all sinusoids. For example, one can represent a sine wave in terms of a cosine wave simply by introducing a phase shift of $\pi/2$ rad:

$$A\sin(\omega t) = A\cos\left(\omega t - \frac{\pi}{2}\right) \tag{4.22}$$

Although one usually employs the variable ω (in units of radians per second) to denote sinusoidal frequency, it is common to refer to natural frequency f in units of cycles per second, or **hertz (Hz).** The reader with some training in music theory knows that a sinusoid represents what in music is called a *pure tone;* an A-440, for example, is a tone at a frequency of 440 Hz. It is important to be aware of the factor of 2π that differentiates radian frequency (in units of radians per second) from natural frequency (in units of hertz). The distinction between the two units of frequency—which are otherwise completely equivalent—is whether one chooses to define frequency in terms of revolutions around a trigonometric circle (in which case the resulting units are radians per second) or to interpret frequency as a repetition rate (cycles per second), in which case the units are hertz. The relationship between the two is the following:

$$\omega = 2\pi f \qquad \text{Radian frequency} \tag{4.23}$$

Why Do We Use Units of Radians for the Phase Angle ϕ?

The engineer finds it frequently more intuitive to refer to the phase angle in units of degrees; however, to use consistent units in the argument (the quantity in the parentheses) of the expression $x(t) = A \sin(\omega t + \phi)$, we must express ϕ in units of radians, since the units of ωt are $[\omega]\cdot[t] = (\text{rad/s})\cdot \text{s} = \text{rad}$. Thus, we will consistently use units of radians for the phase angle ϕ in all expressions of the form $x(t) = A \sin(\omega t + \phi)$. To be consistent is especially important when one is performing numerical calculations; if one used units of degrees for ϕ in calculating the value of $x(t) = A \sin(\omega t + \phi)$ at a given t, the answer would be incorrect.

Why Sinusoids?

By now you should have developed a healthy curiosity about why so much attention is being devoted to sinusoidal signals. Perhaps the simplest explanation is that the electric power used for industrial and household applications worldwide is generated and delivered in the form of either 50- or 60-Hz sinusoidal voltages and currents. Chapter 7 will provide more details regarding the analysis of electric power circuits. The more ambitious reader may explore the box "Fourier Analysis" in Chapter 6 to obtain a more comprehensive explanation of the importance of sinusoidal signals. Note that the methods developed in this section and the subsequent sections apply to many engineering systems, not just to electric circuits, and will be encountered again in the study of dynamic-system modeling and of control systems.

Average and RMS Values

Now that a number of different signal waveforms have been defined, it is appropriate to define suitable measurements for quantifying the strength of a time-varying electric signal. The most common types of measurements are the **average** (or **DC**) **value** of a signal waveform—which corresponds to just measuring the mean voltage or current over a period of time—and the **root-mean-square** (or **rms**) **value,** which takes into account the fluctuations of the signal about its average value. Formally, the operation of computing the average value of a signal corresponds to integrating the signal waveform over some (presumably, suitably chosen) period of time. We define the time-averaged value of a signal $x(t)$ as

LO2

$$\langle x(t)\rangle = \frac{1}{T}\int_0^T x(t')\,dt' \qquad \text{Average value} \tag{4.24}$$

where T is the period of integration. Figure 4.19 illustrates how this process does, in fact, correspond to computing the average amplitude of $x(t)$ over a period of T seconds.

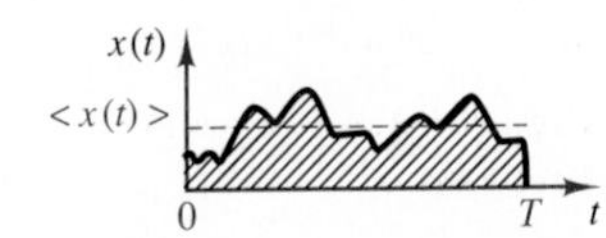

Figure 4.19 Averaging a signal waveform

LO2

EXAMPLE 4.8 Average Value of Sinusoidal Waveform

Problem

Compute the average value of the signal $x(t) = 10 \cos(100t)$.

Solution

Known Quantities: Functional form of the periodic signal $x(t)$.

Find: Average value of $x(t)$.

Analysis: The signal is periodic with period $T = 2\pi/\omega = 2\pi/100$; thus we need to integrate over only one period to compute the average value:

$$\langle x(t)\rangle = \frac{1}{T}\int_0^T x(t')\,dt' = \frac{100}{2\pi}\int_0^{2\pi/100} 10\cos(100t)\,dt$$
$$= \frac{10}{2\pi}\langle\sin(2\pi) - \sin(0)\rangle = 0$$

Comments: The average value of a sinusoidal signal is zero, independent of its amplitude and frequency.

CHECK YOUR UNDERSTANDING

Express the voltage $v(t) = 155.6\sin(377t + \pi/6)$ in cosine form. You should note that the radian frequency $\omega = 377$ will recur very often, since $377 = 2\pi(60)$; that is, 377 is the radian equivalent of the natural frequency of 60 cycles/s, which is the frequency of the electric power generated in North America.

Compute the average value of the sawtooth waveform shown in the figure below.

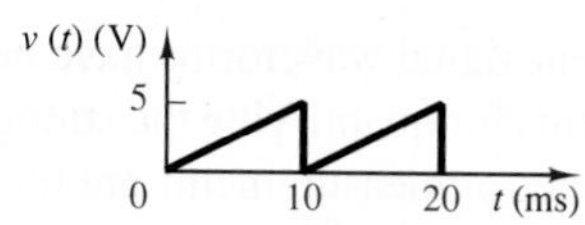

Compute the average value of the shifted triangle wave shown below.

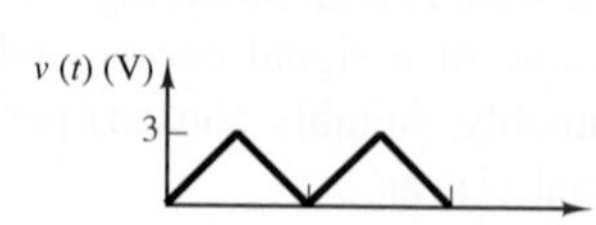

Answers: $v(t) = 155.6\cos(377t - \pi/3)$; $\langle v(t)\rangle = 2.5$ V; $\langle v(t)\rangle = 1.5$ V

The result of Example 4.8 can be generalized to state that

$$\langle A\cos(\omega t + \phi)\rangle = 0 \qquad \textbf{(4.25)}$$

a result that might be perplexing at first: If any sinusoidal voltage or current has zero average value, is its average power equal to zero? Clearly, the answer must be no. Otherwise, it would be impossible to illuminate households and streets and power industrial machinery with 60-Hz sinusoidal current! There must be another way, then, of quantifying the strength of an AC signal.

Very conveniently, a useful measure of the voltage of an AC waveform is the rms value of the signal $x(t)$, defined as follows:

$$x_{\text{rms}} = \sqrt{\frac{1}{T}\int_0^T x^2(t')\,dt'} \qquad \text{Root-mean-square value} \tag{4.26}$$

LO2

Note immediately that if $x(t)$ is a voltage, the resulting x_{rms} will also have units of volts. If you analyze equation 4.26, you can see that, in effect, the rms value consists of the square *r*oot of the average (or *m*ean) of the *s*quare of the signal. Thus, the notation *rms* indicates exactly the operations performed on $x(t)$ in order to obtain its rms value.

The definition of rms value does not help explain why one might be interested in using this quantity. The usefulness of rms values for AC signals in general, and for AC voltages and current in particular, can be explained easily with reference to Figure 4.20. In this figure, the same resistor is connected to two different voltage sources: a DC source and an AC source. We now ask, What is the *effective value* of the current from the DC source such that the *average power* dissipated by the resistor in the DC circuit is exactly the same as the *average power* dissipated by the same resistor in the AC circuit? The direct current I_{eff} is called the **effective value** of the alternating current, which is denoted by $i_{\text{ac}}(t)$. To answer this question, we assume that $v_{\text{ac}}(t)$ and therefore $i_{\text{ac}}(t)$ are periodic signals with period T. We then use the definition of average value of a signal given in equation 4.24 to compute the total energy dissipated by R during one period in the circuit of Figure 4.20(b):

$$W = TP_{\text{AV}} = T\langle p(t)\rangle = \int_0^T p(t')\,dt' = \int_0^T Ri_{\text{ac}}^2(t')\,dt' = I_{\text{eff}}^2 R \tag{4.27}$$

Thus,

$$I_{\text{eff}} = \sqrt{\int_0^T i_{\text{ac}}^2(t')\,dt'} = I_{\text{rms}} \tag{4.28}$$

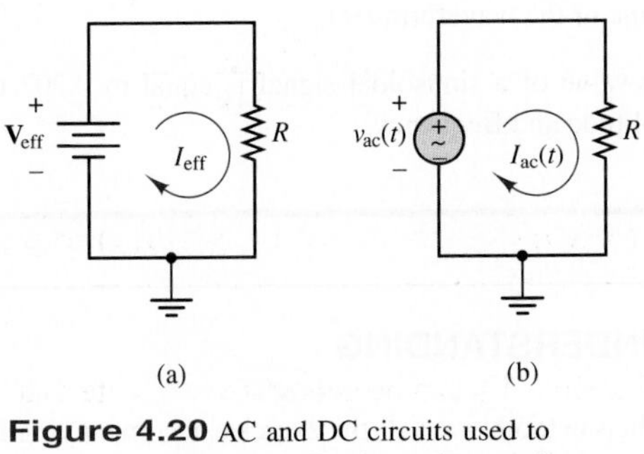

(a) (b)

Figure 4.20 AC and DC circuits used to illustrate the concept of effective and rms values

That is,

LO2

> The rms, or effective, value of the current $i_{ac}(t)$ is the DC that causes the same average power (or energy) to be dissipated by the resistor.

From here on we shall use the notation V_{rms}, or $\tilde{V}$, and I_{rms}, or $\tilde{I}$, to refer to the effective (or rms) value of a voltage or current.

LO2

EXAMPLE 4.9 RMS Value of Sinusoidal Waveform

Problem

Compute the rms value of the sinusoidal current $i(t) = I\cos(\omega t)$.

Solution

Known Quantities: Functional form of the periodic signal $i(t)$.

Find: RMS value of $i(t)$.

Analysis: Applying the definition of rms value in equation 4.26, we compute

$$i_{rms} = \sqrt{\frac{1}{T}\int_0^T i^2(t')\,dt'} = \sqrt{\frac{\omega}{2\pi}\int_0^{2\pi/\omega} I^2\cos^2(\omega t')\,dt'}$$

$$= \sqrt{\frac{\omega}{2\pi}\int_0^{2\pi/\omega} I^2\left[\frac{1}{2}+\frac{1}{2}\cos(2\omega t')\right]dt'}$$

$$= \sqrt{\frac{1}{2}I^2 + \frac{\omega}{2\pi}\int_0^{2\pi/\omega}\frac{I^2}{2}\cos(2\omega t')\,dt'}$$

At this point, we recognize that the integral under the square root sign is equal to zero (see Example 4.8), because we are integrating a sinusoidal waveform over two periods. Hence,

$$i_{rms} = \frac{I}{\sqrt{2}} = 0.707I$$

where I is the *peak value* of the waveform $i(t)$.

Comments: The rms value of a sinusoidal signal is equal to 0.707 times the peak value, independent of its amplitude and frequency.

CHECK YOUR UNDERSTANDING

Find the rms value of the sawtooth wave of the exercise accompanying Example 4.8.

Find the rms value of the half cosine wave shown in the next figure.

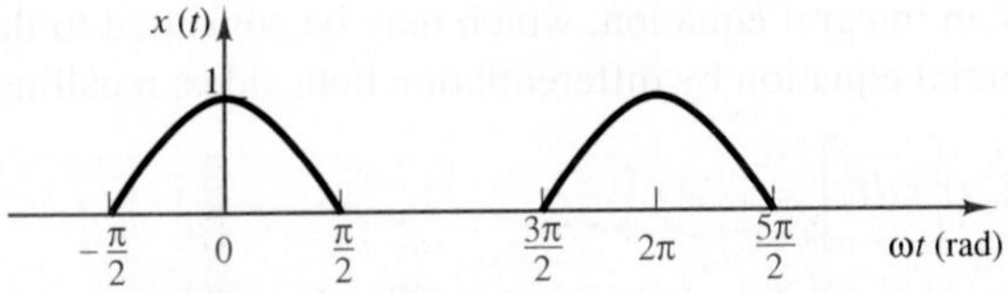

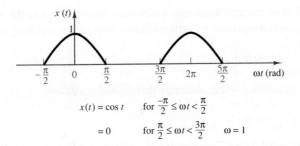

$$x(t) = \cos t \quad \text{for } \frac{-\pi}{2} \le \omega t < \frac{\pi}{2}$$
$$= 0 \quad \text{for } \frac{\pi}{2} \le \omega t < \frac{3\pi}{2} \quad \omega = 1$$

Answers: 2.89 V; 0.5 V

Example 4.9 illustrates how the rms value of a sinusoid is proportional to its peak amplitude. The factor of $0.707 = 1/\sqrt{2}$ is a useful number to remember, since it applies to any sinusoidal signal. It is not, however, generally applicable to signal waveforms other than sinusoids, as the Check Your Understanding exercises have illustrated.

4.3 SOLUTION OF CIRCUITS CONTAINING ENERGY STORAGE ELEMENTS (DYNAMIC CIRCUITS)

Sections 4.1 and 4.2 introduced energy storage elements and time-dependent signal sources. The logical next task is to analyze the behavior of circuits containing such elements. The major difference between the analysis of the resistive circuits studied in Chapters 2 and 3 and the circuits we explore in the remainder of this chapter is that now the equations that result from applying Kirchhoff's laws are differential equations, as opposed to the algebraic equations obtained in solving resistive circuits. Consider, for example, the circuit of Figure 4.21, which consists of the series connection of a voltage source, a resistor, and a capacitor. Applying KCL at the node connecting the resistor to the capacitor and using the definition of capacitor current in equation 4.4, we obtain the following equations:

A circuit containing energy-storage elements is described by a differential equation. The differential equation describing the series RC circuit shown is

$$\frac{di_C}{dt} + \frac{1}{RC} i_C = \frac{dv_S}{dt}$$

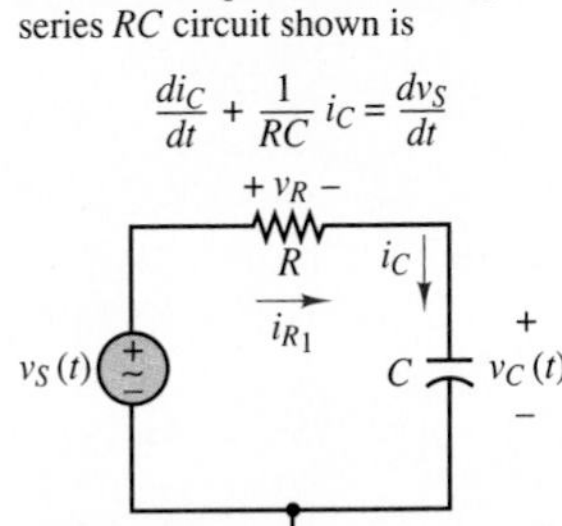

Figure 4.21 Circuit containing energy storage element

$$i_R(t) = \frac{v_S(t) - v_C(t)}{R} = i_C(t) = C\,\frac{dv_C(t)}{dt} \tag{4.29}$$

or

$$\frac{dv_C(t)}{dt} + \frac{1}{RC} v_C(t) = \frac{1}{RC} v_S(t) \tag{4.30}$$

Equation 4.30 is a first-order, linear, ordinary differential equation in the variable v_C. Alternatively, we could derive an equivalent relationship by applying KVL around the circuit of Figure 4.21:

$$-v_S(t) + v_R(t) + v_C(t) = 0 \tag{4.31}$$

Observing that $i_R(t) = i_C(t)$ and using the capacitor equation 4.5, we can write

$$-v_S(t) + Ri_C(t) + \frac{1}{C}\int_{-\infty}^{t} i_C(t')\,dt' = 0 \tag{4.32}$$

Equation 4.32 is an integral equation, which may be converted to the more familiar form of a differential equation by differentiating both sides; recalling that

$$\frac{d}{dt}\left[\int_{-\infty}^{t} i_C(t')\,dt'\right] = i_C(t) \tag{4.33}$$

we obtain the first-order, linear, ordinary differential equation

$$\frac{di_C(t)}{dt} + \frac{1}{RC} i_C(t) = \frac{1}{R}\frac{dv_S(t)}{dt} \tag{4.34}$$

Equations 4.30 and 4.34 are very similar; the principal differences are the variable in the differential equation [$v_C(t)$ versus $i_C(t)$] and the right-hand side. Solving either equation for the unknown variable permits the computation of all voltages and currents in the circuit.

> ***Note to the Instructor:*** **If so desired, the remainder of this chapter can be skipped, and the course can continue with Chapter 5 without any loss of continuity.**

Forced Response of Circuits Excited by Sinusoidal Sources

Consider again the circuit of Figure 4.21, where now the external source produces a sinusoidal voltage, described by the expression

$$v_S(t) = V \cos \omega t \tag{4.35}$$

Substituting the expression $V\cos(\omega t)$ in place of the source voltage $v_S(t)$ in the differential equation obtained earlier (equation 4.30), we obtain the following differential equation:

$$\frac{d}{dt} v_C + \frac{1}{RC} v_C = \frac{1}{RC} V \cos \omega t \tag{4.36}$$

Since the forcing function is a sinusoid, the solution may also be assumed to be of the same form. An expression for $v_C(t)$ is then

$$v_C(t) = A \sin \omega t + B \cos \omega t \tag{4.37}$$

which is equivalent to

$$v_C(t) = C \cos(\omega t + \phi) \tag{4.38}$$

Substituting equation 4.37 in the differential equation for $v_C(t)$ and solving for the coefficients A and B yield the expression

$$A\omega \cos \omega t - B\omega \sin \omega t + \frac{1}{RC}(A \sin \omega t + B \cos \omega t) = \frac{1}{RC} V \cos \omega t \tag{4.39}$$

and if the coefficients of like terms are grouped, the following equation is obtained:

$$\left(\frac{A}{RC} - B\omega\right) \sin \omega t + \left(A\omega + \frac{B}{RC} - \frac{V}{RC}\right) \cos \omega t = 0 \tag{4.40}$$

The coefficients of sin ωt and cos ωt must both be identically zero in order for equation 4.40 to hold. Thus,

$$\frac{A}{RC} - B\omega = 0$$

and **(4.41)**

$$A\omega + \frac{B}{RC} - \frac{V}{RC} = 0$$

The unknown coefficients A and B may now be determined by solving equation 4.41:

$$A = \frac{V\omega RC}{1 + \omega^2(RC)^2}$$
(4.42)
$$B = \frac{V}{1 + \omega^2(RC)^2}$$

Thus, the solution for $v_C(t)$ may be written as follows:

$$v_C(t) = \frac{V\omega RC}{1 + \omega^2(RC)^2} \sin \omega t + \frac{V}{1 + \omega^2(RC)^2} \cos \omega t \qquad \textbf{(4.43)}$$

This response is plotted in Figure 4.22.

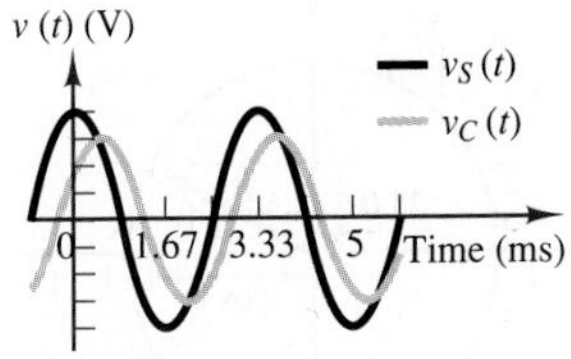

Figure 4.22 Waveforms for the AC circuit of Figure 4.21

The solution method outlined in the previous paragraphs can become quite complicated for circuits containing a large number of elements; in particular, one may need to solve higher-order differential equations if more than one energy storage element is present in the circuit. A simpler and preferred method for the solution of AC circuits is presented in Section 4.4. This brief section has provided a simple, but complete, illustration of the key elements of AC circuit analysis. These can be summarized in the following statement:

LO3

In a sinusoidally excited linear circuit, all branch voltages and currents are *sinusoids* at the *same frequency* as the excitation signal. The amplitudes of these voltages and currents are a *scaled* version of the excitation *amplitude,* and the voltages and currents may be *shifted in phase* with respect to the excitation signal.

These observations indicate that three parameters uniquely define a sinusoid: *frequency, amplitude,* and *phase*. But if this is the case, is it necessary to carry the "excess luggage," that is, the sinusoidal functions? Might it be possible to simply keep track of the three parameters just mentioned? Fortunately, the answers to these two questions are no and yes, respectively. Section 4.4 describes the use of a notation that, with the aid of complex algebra, eliminates the need for the sinusoidal functions of time, and for the formulation and solution of differential equations, permitting the use of simpler algebraic methods.

4.4 PHASOR SOLUTION OF CIRCUITS WITH SINUSOIDAL EXCITATION

In this section, we introduce an efficient notation to make it possible to represent sinusoidal signals as *complex numbers,* and to eliminate the need for solving differential equations. The student who needs a brief review of complex algebra will find a reasonably complete treatment in Appendix A, including solved examples and Check

Leonhard Euler (1707–1783). *Photograph courtesy of Deutsches Museum, Munich.*

Your Understanding exercises. For the remainder of the chapter, it will be assumed that you are familiar with both the rectangular and the polar forms of complex number coordinates; with the conversion between these two forms; and with the basic operations of addition, subtraction, multiplication, and division of complex numbers.

Euler's Identity

Named after the Swiss mathematician Leonhard Euler (the last name is pronounced "Oiler"), Euler's identity forms the basis of phasor notation. Simply stated, the identity defines the **complex exponential** $e^{j\theta}$ as a point in the complex plane, which may be represented by real and imaginary components:

$$e^{j\theta} = \cos\theta + j\sin\theta \tag{4.44}$$

Figure 4.23 illustrates how the complex exponential may be visualized as a point (or vector, if referenced to the origin) in the complex plane. Note immediately that the magnitude of $e^{j\theta}$ is equal to 1:

$$|e^{j\theta}| = 1 \tag{4.45}$$

since

$$|\cos\theta + j\sin\theta| = \sqrt{\cos^2\theta + \sin^2\theta} = 1 \tag{4.46}$$

and note also that writing Euler's identity corresponds to equating the polar form of a complex number to its rectangular form. For example, consider a vector of length A making an angle θ with the real axis. The following equation illustrates the relationship between the rectangular and polar forms:

$$Ae^{j\theta} = A\cos\theta + jA\sin\theta = A\angle\theta \tag{4.47}$$

In effect, Euler's identity is simply a trigonometric relationship in the complex plane.

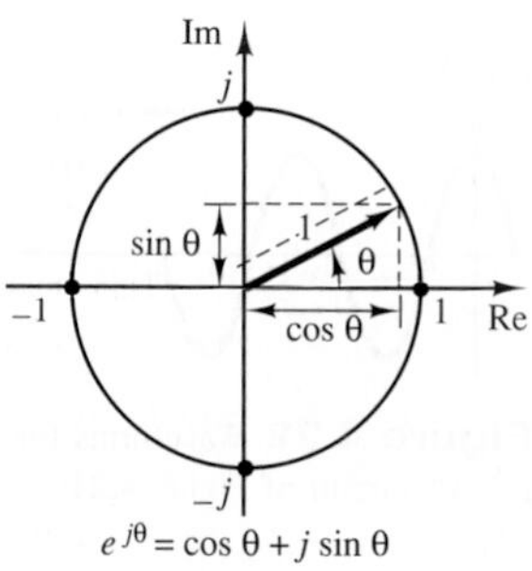

Figure 4.23 Euler's identity

Phasors

To see how complex numbers can be used to represent sinusoidal signals, rewrite the expression for a generalized sinusoid in light of Euler's equation:

$$A\cos(\omega t + \theta) = \text{Re}\,(Ae^{j(\omega t+\theta)}) \tag{4.48}$$

This equality is easily verified by expanding the right-hand side, as follows:

$$\begin{aligned}\text{Re}\,(Ae^{j(\omega t+\theta)}) &= \text{Re}\,[A\cos(\omega t+\theta) + jA\sin(\omega t+\theta)]\\ &= A\cos(\omega t+\theta)\end{aligned}$$

We see, then, that *it is possible to express a generalized sinusoid as the real part of a complex vector* whose **argument,** or **angle,** is given by $\omega t + \theta$ and whose length, or **magnitude,** is equal to the peak amplitude of the sinusoid. The **complex phasor** corresponding to the sinusoidal signal $A\cos(\omega t + \theta)$ is therefore defined to be the complex number $Ae^{j\theta}$:

$$Ae^{j\theta} = \text{complex phasor notation for } A\cos(\omega t+\theta) = A\angle\theta \tag{4.49}$$

It is important to explicitly point out that this is a *definition.* Phasor notation arises from equation 4.48; however, this expression is simplified (for convenience, as will be promptly shown) by removing the "real part of" operator (Re) and factoring out and deleting the term $e^{j\omega t}$. Equation 4.50 illustrates the simplification:

$$A\cos(\omega t+\theta) = \text{Re}\,(Ae^{j(\omega t+\theta)}) = \text{Re}\,(Ae^{j\theta}e^{j\omega t}) \tag{4.50}$$

The reason for this simplification is simply mathematical convenience, as will become apparent in the following examples; you will have to remember that the $e^{j\omega t}$ term that was removed from the complex form of the sinusoid is really still present, indicating the specific frequency of the sinusoidal signal ω. With these caveats, you should now be prepared to use the newly found phasor to analyze AC circuits. The following comments summarize the important points developed thus far in the section. Please note that the concept of phasor has no real physical significance. It is a convenient mathematical tool that simplifies the solution of AC circuits.

FOCUS ON METHODOLOGY

LO4

1. Any sinusoidal signal may be mathematically represented in one of two ways: a **time-domain form**

$$v(t) = A\cos(\omega t + \theta)$$

and a **frequency-domain** (or **phasor**) **form**

$$\mathbf{V}(j\omega) = Ae^{j\theta} = A\angle\theta$$

Note the $j\omega$ in the notation $\mathbf{V}(j\omega)$, indicating the $e^{j\omega t}$ dependence of the phasor. In the remainder of this chapter, bold uppercase quantities indicate phasor voltages or currents.

2. A phasor is a complex number, expressed in polar form, consisting of a *magnitude* equal to the peak amplitude of the sinusoidal signal and a *phase angle* equal to the phase shift of the sinusoidal signal *referenced to a cosine signal.*
3. When one is using phasor notation, it is important to note the specific frequency ω of the sinusoidal signal, since this is not explicitly apparent in the phasor expression.

EXAMPLE 4.10 Addition of Two Sinusoidal Sources in Phasor Notation

LO4

Problem

Compute the phasor voltage resulting from the series connection of two sinusoidal voltage sources (Figure 4.24).

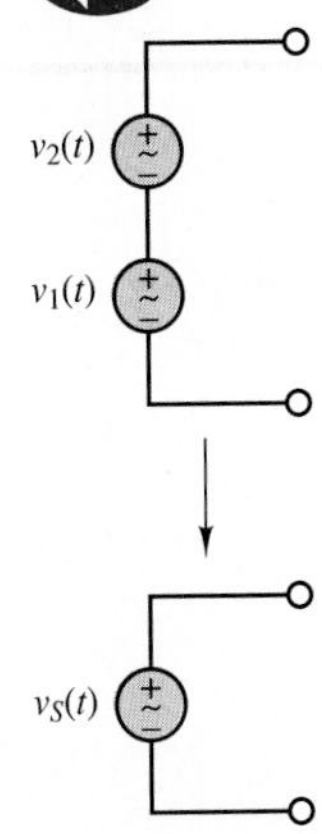

Figure 4.24

Solution

Known Quantities:

$$v_1(t) = 15\cos\left(377t + \frac{\pi}{4}\right) \quad \text{V}$$

$$v_2(t) = 15\cos\left(377t + \frac{\pi}{12}\right) \quad \text{V}$$

Find: Equivalent phasor voltage $v_S(t)$.

Analysis: Write the two voltages in phasor form:

$$\mathbf{V}_1(j\omega) = 15\angle\frac{\pi}{4} \quad \text{V}$$

$$\mathbf{V}_2(j\omega) = 15e^{j\pi/12} = 15\angle\frac{\pi}{12} \quad \text{V}$$

Convert the phasor voltages from polar to rectangular form:

$$\mathbf{V}_1(j\omega) = 10.61 + j10.61 \quad \text{V}$$

$$\mathbf{V}_2(j\omega) = 14.49 + j3.88 \quad \text{V}$$

Then

$$\mathbf{V}_S(j\omega) = \mathbf{V}_1(j\omega) + \mathbf{V}_2(j\omega) = 25.10 + j14.49 = 28.98e^{j\pi/6} = 28.98\angle\frac{\pi}{6} \quad \text{V}$$

Now we can convert $\mathbf{V}_S(j\omega)$ to its time-domain form:

$$v_S(t) = 28.98\cos\left(377t + \frac{\pi}{6}\right) \quad \text{V}$$

Comments: Note that we could have obtained the same result by adding the two sinusoids in the time domain, using trigonometric identities:

$$v_1(t) = 15\cos\left(377t + \frac{\pi}{4}\right) = 15\cos\frac{\pi}{4}\cos(377t) - 15\sin\frac{\pi}{4}\sin(377t) \quad \text{V}$$

$$v_2(t) = 15\cos\left(377t + \frac{\pi}{12}\right) = 15\cos\frac{\pi}{12}\cos(377t) - 15\sin\frac{\pi}{12}\sin(377t) \quad \text{V}$$

Combining like terms, we obtain

$$\begin{aligned} v_1(t) + v_2(t) &= 15\left(\cos\frac{\pi}{4} + \cos\frac{\pi}{12}\right)\cos(377t) - 15\left(\sin\frac{\pi}{4} + \sin\frac{\pi}{12}\right)\sin(377t) \\ &= 15[1.673\cos(377t) - 0.966\sin(377t)] \\ &= 15\sqrt{(1.673)^2 + (0.966)^2} \times \cos\left[377t + \arctan\left(\frac{0.966}{1.673}\right)\right] \\ &= 15\left[1.932\cos\left(377t + \frac{\pi}{6}\right)\right] = 28.98\cos\left(377t + \frac{\pi}{6}\right) \quad \text{V} \end{aligned}$$

The above expression is, of course, identical to the one obtained by using phasor notation, but it required a greater amount of computation. In general, phasor analysis greatly simplifies calculations related to sinusoidal voltages and currents.

CHECK YOUR UNDERSTANDING

Add the sinusoidal voltages $v_1(t) = A\cos(\omega t + \phi)$ and $v_2(t) = B\cos(\omega t + \theta)$ using phasor notation, and then convert back to time-domain form.

a. $A = 1.5$ V, $\phi = 10°$; $B = 3.2$ V, $\theta = 25°$.

b. $A = 50$ V, $\phi = -60°$; $B = 24$ V, $\theta = 15°$.

Answers: (a) $v_1 + v_2 = 4.67\cos(\omega t + 0.3526 \text{ rad})$;
(b) $v_1 + v_2 = 60.8\cos(\omega t - 0.6562 \text{ rad})$

It should be apparent by now that phasor notation can be a very efficient technique to solve AC circuit problems. The following sections continue to develop this new method to build your confidence in using it.

Superposition of AC Signals

Example 4.10 explored the combined effect of two sinusoidal sources of different phase and amplitude, but of the same frequency. It is important to realize that the simple answer obtained there does not apply to the superposition of two (or more) sinusoidal sources that *are not at the same frequency.* In this subsection, the case of two sinusoidal sources oscillating at different frequencies is used to illustrate how phasor analysis can deal with this, more general case.

The circuit shown in Figure 4.25 depicts a source excited by two current sources connected in parallel, where

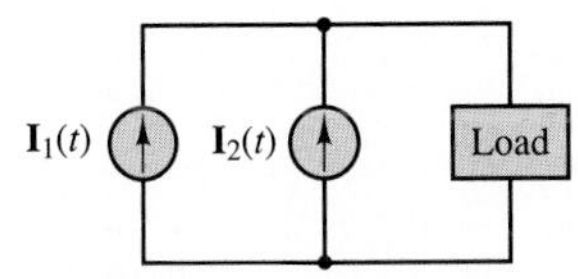

Figure 4.25 Superposition of AC

$$\begin{aligned} i_1(t) &= A_1 \cos(\omega_1 t) \\ i_2(t) &= A_2 \cos(\omega_2 t) \end{aligned} \qquad \textbf{(4.51)}$$

The load current is equal to the sum of the two source currents; that is,

$$i_L(t) = i_1(t) + i_2(t) \qquad \textbf{(4.52)}$$

or, in phasor form,

$$\mathbf{I}_L = \mathbf{I}_1 + \mathbf{I}_2 \qquad \textbf{(4.53)}$$

At this point, you might be tempted to write $\mathbf{I}_1$ and $\mathbf{I}_2$ in a more explicit phasor form as

$$\begin{aligned} \mathbf{I}_1 &= A_1 e^{j0} \\ \mathbf{I}_2 &= A_2 e^{j0} \end{aligned} \qquad \textbf{(4.54)}$$

and to add the two phasors, using the familiar techniques of complex algebra. However, this approach *would be incorrect*. Whenever a sinusoidal signal is expressed in phasor notation, the term $e^{j\omega t}$ is implicitly present, where ω is the actual radian frequency of the signal. In our example, the two frequencies are not the same, as can be verified by writing the phasor currents in the form of equation 4.50:

$$\begin{aligned} \mathbf{I}_1 &= \text{Re}\,(A_1 e^{j0} e^{j\omega_1 t}) \\ \mathbf{I}_2 &= \text{Re}\,(A_2 e^{j0} e^{j\omega_2 t}) \end{aligned} \qquad \textbf{(4.55)}$$

Since phasor notation does not *explicitly* include the $e^{j\omega t}$ factor, this can lead to serious errors if you are not careful! The two phasors of equation 4.54 cannot be added, but must be kept separate; thus, the only unambiguous expression for the load current in this case is equation 4.52. To complete the analysis of any circuit with multiple sinusoidal sources at different frequencies using phasors, it is necessary to solve the circuit separately for each signal and then add the individual answers obtained for the different excitation sources. Example 4.11 illustrates the response of a circuit with two separate AC excitations using AC superposition.

LO4

EXAMPLE 4.11 AC Superposition

Problem

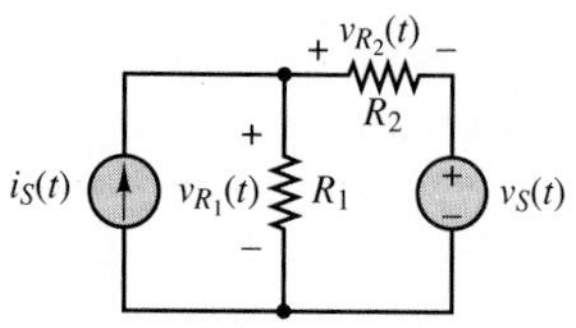

$R_1 = 150\ \Omega$, $R_2 = 50\ \Omega$

Figure 4.26

Compute the voltages $v_{R1}(t)$ and $v_{R2}(t)$ in the circuit of Figure 4.26.

Solution

Known Quantities:

$$i_S(t) = 0.5\cos[2\pi(100t)] \qquad \text{A}$$
$$v_S(t) = 20\cos[2\pi(1{,}000t)] \qquad \text{V}$$

Find: $v_{R1}(t)$ and $v_{R2}(t)$.

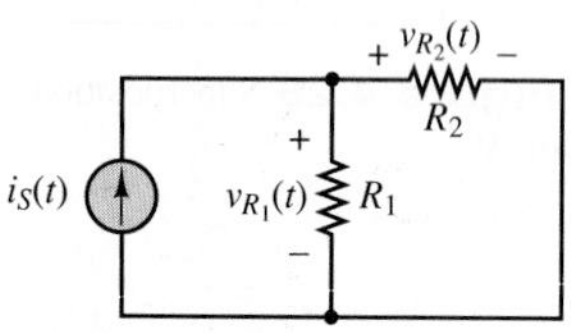

Figure 4.27

Analysis: Since the two sources are at different frequencies, we must compute a separate solution for each. Consider the current source first, with the voltage source set to zero (short circuit) as shown in Figure 4.27. The circuit thus obtained is a simple current divider. Write the source current in phasor notation:

$$\mathbf{I}_S(j\omega) = 0.5e^{j0} = 0.5\angle 0 \qquad \text{A} \qquad \omega = 2\pi 100 \text{ rad/s}$$

Then

$$\mathbf{V}_{R1}(\mathbf{I}_S) = \mathbf{I}_S \frac{R_2}{R_1 + R_2} R_1 = 0.5\angle 0 \left(\frac{50}{150+50}\right) 150 = 18.75\angle 0 \qquad \text{V}$$
$$\omega = 2\pi(100) \text{ rad/s}$$

$$\mathbf{V}_{R2}(\mathbf{I}_S) = \mathbf{I}_S \frac{R_1}{R_1 + R_2} R_2 = 0.5\angle 0 \left(\frac{150}{150+50}\right) 50 = 18.75\angle 0 \qquad \text{V}$$
$$\omega = 2\pi(100) \text{ rad/s}$$

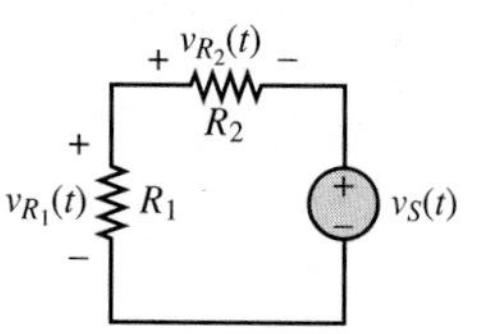

Figure 4.28

Next, we consider the voltage source, with the current source set to zero (open circuit), as shown in Figure 4.28. We first write the source voltage in phasor notation:

$$\mathbf{V}_S(j\omega) = 20e^{j0} = 20\angle 0 \qquad \text{V} \qquad \omega = 2\pi(1{,}000) \text{ rad/s}$$

Then we apply the voltage divider law, to obtain

$$\mathbf{V}_{R1}(\mathbf{V}_S) = \mathbf{V}_S \frac{R_1}{R_1 + R_2} = 20\angle 0 \left(\frac{150}{150+50}\right) = 15\angle 0 \qquad \text{V}$$
$$\omega = 2\pi(1{,}000) \text{ rad/s}$$

$$\mathbf{V}_{R2}(\mathbf{V}_S) = -\mathbf{V}_S \frac{R_2}{R_1 + R_2} = -20\angle 0 \left(\frac{50}{150+50}\right) = -5\angle 0 = 5\angle\pi \qquad \text{V}$$
$$\omega = 2\pi(1{,}000) \text{ rad/s}$$

Now we can determine the voltage across each resistor by adding the contributions from each source and converting the phasor form to time-domain representation:

$$\mathbf{V}_{R1} = \mathbf{V}_{R1}(\mathbf{I}_S) + \mathbf{V}_{R1}(\mathbf{V}_S)$$
$$v_{R1}(t) = 18.75\cos[2\pi(100t)] + 15\cos[2\pi(1{,}000t)] \qquad \text{V}$$

and

$$\mathbf{V}_{R2} = \mathbf{V}_{R2}(\mathbf{I}_S) + \mathbf{V}_{R2}(\mathbf{V}_S)$$

$$v_{R2}(t) = 18.75\cos[2\pi(100t)] + 5\cos[2\pi(1{,}000t) + \pi] \quad \text{V}$$

Comments: Note that it is impossible to simplify the final expression any further, because the two components of each voltage are at different frequencies.

CHECK YOUR UNDERSTANDING

Add the sinusoidal currents $i_1(t) = A\cos(\omega t + \phi)$ and $i_2(t) = B\cos(\omega t + \theta)$ for

a. $A = 0.09$ A, $\phi = 72°$; $B = 0.12$ A, $\theta = 20°$.

b. $A = 0.82$ A, $\phi = -30°$; $B = 0.5$ A, $\theta = -36°$.

Answers: (a) $i_1 + i_2 = 0.19\cos(\omega t + 0.733)$; (b) $i_1 + i_2 = 1.32\cos(\omega t - 0.5637)$

Impedance

We now analyze the i-v relationship of the three ideal circuit elements in light of the new phasor notation. The result will be a new formulation in which resistors, capacitors, and inductors will be described in the same notation. A direct consequence of this result will be that the circuit theorems of Chapter 3 will be extended to AC circuits. In the context of AC circuits, any one of the three ideal circuit elements defined so far will be described by a parameter called **impedance,** which may be viewed as a *complex resistance*. The impedance concept is equivalent to stating that capacitors and inductors act as *frequency-dependent resistors,* that is, as resistors whose resistance is a function of the frequency of the sinusoidal excitation. Figure 4.29 depicts the same circuit represented in conventional form (top) and in phasor-impedance form (bottom); the latter representation explicitly shows phasor voltages and currents and treats the circuit element as a generalized "impedance." It will presently be shown that each of the three ideal circuit elements may be represented by one such impedance element.

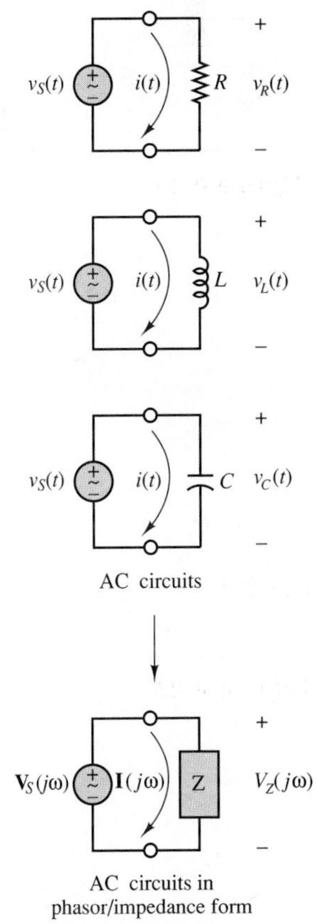

Figure 4.29 The impedance element

Let the source voltage in the circuit of Figure 4.29 be defined by

$$v_S(t) = A\cos\omega t \qquad \text{or} \qquad \mathbf{V}_S(j\omega) = Ae^{j0°} = A\angle 0 \tag{4.56}$$

without loss of generality. Then the current $i(t)$ is defined by the i-v relationship for each circuit element. Let us examine the frequency-dependent properties of the resistor, inductor, and capacitor, one at a time.

The Resistor

Ohm's law dictates the well-known relationship $v = iR$. In the case of sinusoidal sources, then, the current flowing through the resistor of Figure 4.29 may be expressed as

$$i(t) = \frac{v_S(t)}{R} = \frac{A}{R}\cos\omega t \tag{4.57}$$

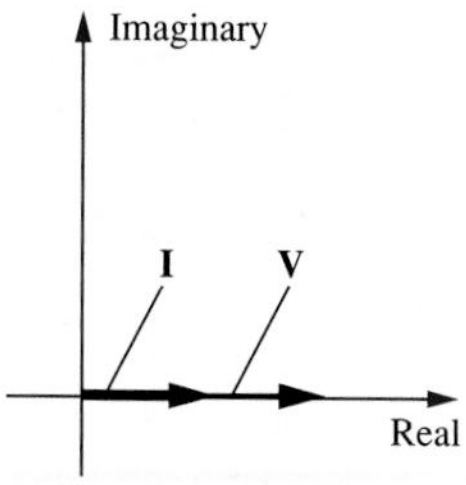

Figure 4.30 Phasor voltage and current relationships for a resistor

Converting the voltage $v_S(t)$ and the current $i(t)$ to phasor notation, we obtain the following expressions:

$$\mathbf{V}_Z(j\omega) = A\angle 0$$
$$\mathbf{I}(j\omega) = \frac{A}{R}\angle 0 \tag{4.58}$$

The relationship between $\mathbf{V}_Z$ and $\mathbf{I}$ in the complex plane is shown in Figure 4.30. Finally, the *impedance* of the resistor is defined as the ratio of the phasor voltage across the resistor to the phasor current flowing through it, and the symbol Z_R is used to denote it:

LO4

$$Z_R(j\omega) = \frac{\mathbf{V}_Z(j\omega)}{\mathbf{I}(j\omega)} = R \qquad \text{Impedance of a resistor} \tag{4.59}$$

Equation 4.59 corresponds to Ohm's law in phasor form, and the result should be intuitively appealing: Ohm's law applies to a resistor independent of the particular form of the voltages and currents (whether AC or DC, for instance). The ratio of phasor voltage to phasor current has a very simple form in the case of the resistor. In general, however, the impedance of an element is a complex function of frequency, as it must be, since it is the ratio of two phasor quantities, which are frequency-dependent. This property will become apparent when the impedances of the inductor and capacitor are defined.

The Inductor

Recall the defining relationships for the ideal inductor (equations 4.10 and 4.13), repeated here for convenience:

$$v_L(t) = L\frac{di_L(t)}{dt}$$
$$i_L(t) = \frac{1}{L}\int v_L(t') \tag{4.60}$$

Let $v_L(t) = v_S(t)$ and $i_L(t) = i(t)$ in the circuit of Figure 4.29. Then the following expression may be derived for the inductor current:

$$i_L(t) = i(t) = \frac{1}{L}\int v_S(t')\,dt'$$
$$i_L(t) = \frac{1}{L}\int A\cos\omega t'\,dt' \tag{4.61}$$
$$= \frac{A}{\omega L}\sin\omega t$$

Note how a dependence on the radian frequency of the source is clearly present in the expression for the inductor current. Further, the inductor current is shifted in phase (by 90°) with respect to the voltage. This fact can be seen by writing the inductor voltage and current in time-domain form:

$$v_S(t) = v_L(t) = A\cos\omega t$$
$$i(t) = i_L(t) = \frac{A}{\omega L}\cos\left(\omega t - \frac{\pi}{2}\right) \tag{4.62}$$

It is evident that the current is not just a scaled version of the source voltage, as it was for the resistor. Its magnitude depends on the frequency ω, and it is shifted (delayed) in phase by $\pi/2$ rad, or 90°. Using phasor notation, equation 4.62 becomes

$$\begin{aligned} \mathbf{V}_Z(j\omega) &= A\angle 0 \\ \mathbf{I}(j\omega) &= \frac{A}{\omega L}\angle -\frac{\pi}{2} \end{aligned} \tag{4.63}$$

The relationship between the phasor voltage and current is shown in Figure 4.31. Thus, the impedance of the inductor is defined as follows:

$$Z_L(j\omega) = \frac{\mathbf{V}_Z(j\omega)}{\mathbf{I}(j\omega)} = \omega L\angle\frac{\pi}{2} = j\omega L \qquad \text{Impedance of an inductor} \tag{4.64}$$

LO4

Note that the inductor now appears to behave as a *complex frequency-dependent resistor,* and that the magnitude of this complex resistor ωL is proportional to the signal frequency ω. Thus, an inductor will "impede" current flow in proportion to the sinusoidal frequency of the source signal. This means that at low signal frequencies, an inductor acts somewhat as a short circuit, while at high frequencies it tends to behave more as an open circuit.

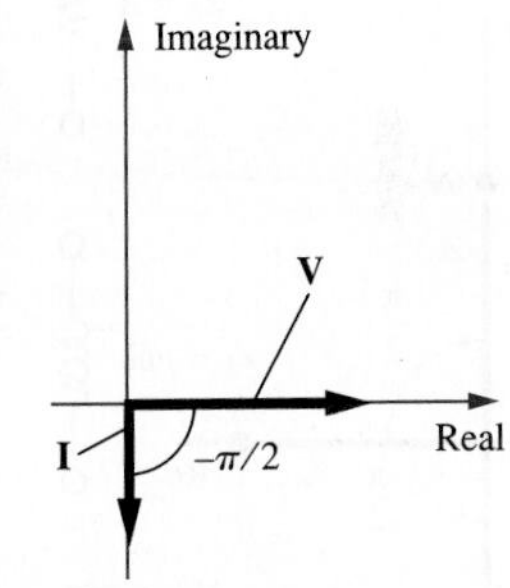

Figure 4.31 Phasor voltage and current relationships for an inductor

The Capacitor

An analogous procedure may be followed to derive the equivalent result for a capacitor. Beginning with the defining relationships for the ideal capacitor

$$\begin{aligned} i_C(t) &= C\frac{dv_C(t)}{dt} \\ v_C(t) &= \frac{1}{C}\int i_C(t')\,dt' \end{aligned} \tag{4.65}$$

with $i_C = i$ and $v_C = v_S$ in Figure 4.29, we can express the capacitor current as

$$\begin{aligned} i_C(t) &= C\frac{dv_C(t)}{dt} \\ &= C\frac{d}{dt}(A\cos\omega t) \\ &= -C(A\omega\sin\omega t) \\ &= \omega CA\cos\left(\omega t + \frac{\pi}{2}\right) \end{aligned} \tag{4.66}$$

so that, in phasor form,

$$\begin{aligned} \mathbf{V}_Z(j\omega) &= A\angle 0 \\ \mathbf{I}(j\omega) &= \omega CA\angle\frac{\pi}{2} \end{aligned} \tag{4.67}$$

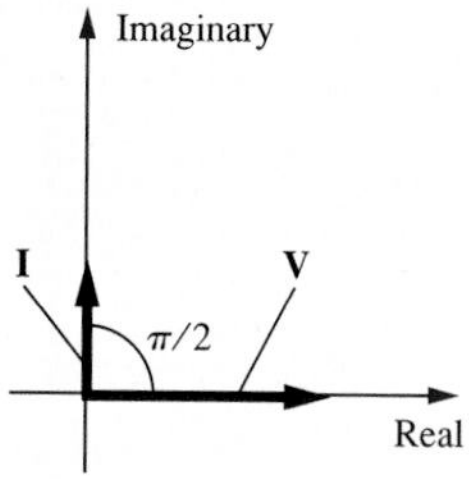

Figure 4.32 Phasor voltage and current relationships for a capacitor

The relationship between the phasor voltage and current is shown in Figure 4.32. The impedance of the ideal capacitor $Z_C(j\omega)$ is therefore defined as follows:

$$\begin{aligned} Z_C(j\omega) &= \frac{\mathbf{V}_Z(j\omega)}{\mathbf{I}(j\omega)} = \frac{1}{\omega C}\angle\frac{-\pi}{2} \\ &= \frac{-j}{\omega C} = \frac{1}{j\omega C} \end{aligned} \qquad \text{Impedance of a capacitor} \tag{4.68}$$

where we have used the fact that $1/j = e^{-j\pi/2} = -j$. Thus, the impedance of a capacitor is also a frequency-dependent complex quantity, with the impedance of the capacitor varying as an inverse function of frequency; and so a capacitor acts as a short circuit at high frequencies, whereas it behaves more as an open circuit at low frequencies. Figure 4.33 depicts $Z_C(j\omega)$ in the complex plane, alongside $Z_R(j\omega)$ and $Z_L(j\omega)$.

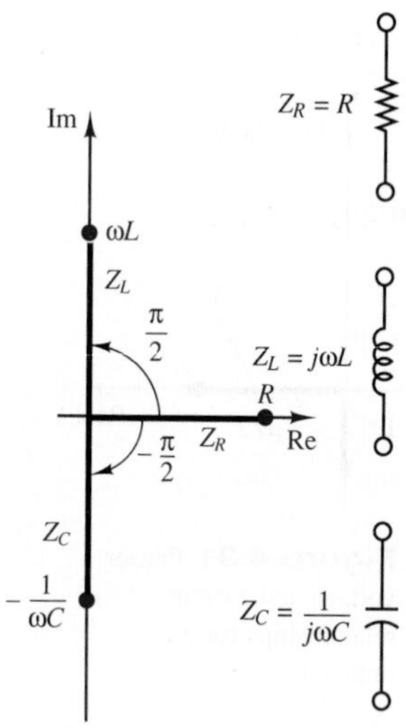

Figure 4.33 Impedances of R, L, and C in the complex plane

The impedance parameter defined in this section is extremely useful in solving AC circuit analysis problems, because it will make it possible to take advantage of most of the network theorems developed for DC circuits by replacing resistances with complex-valued impedances. Examples 4.12 to 4.14 illustrate how branches containing series and parallel elements may be reduced to a single equivalent impedance, much in the same way as resistive circuits were reduced to equivalent forms. It is important to emphasize that although the impedance of simple circuit elements is either purely real (for resistors) or purely imaginary (for capacitors and inductors), the general definition of impedance for an arbitrary circuit must allow for the possibility of having both a real and an imaginary part, since practical circuits are made up of more or less complex interconnections of different circuit elements. In its most general form, the impedance of a circuit element is defined as the sum of a real part and an imaginary part

$$Z(j\omega) = R(j\omega) + jX(j\omega) \tag{4.69}$$

where R is called the **AC resistance** and X is called the **reactance.** The frequency dependence of R and X has been indicated explicitly, since it is possible for a circuit to have a frequency-dependent resistance. Note that the reactances of equations 4.64 and 4.68 have units of ohms, and that **inductive reactance** is always positive, while **capacitive reactance** is always negative. Examples 4.12 to 4.14 illustrate how a complex impedance containing both real and imaginary parts arises in a circuit. Impedance is another useful mathematical tool that is convenient in solving AC circuits, but has no real physical significance. *Please note that the impedance $Z(j\omega)$ is not a phasor,* but just a complex number.

LO4

EXAMPLE 4.12 Impedance of a Practical Capacitor

Problem

A practical capacitor can be modeled by an ideal capacitor in parallel with a resistor. The parallel resistance represents leakage losses in the capacitor and is usually quite large. Find the impedance of a practical capacitor at the radian frequency $\omega = 377$ rad/s (60 Hz). How will the impedance change if the capacitor is used at a much higher frequency, say, 800 kHz?

Solution

Known Quantities: Figure 4.34; $C_1 = 0.001\ \mu\text{F} = 1 \times 10^{-9}$ F; $R_1 = 1\ \text{M}\Omega$.

Find: The equivalent impedance of the parallel circuit Z_1.

Analysis: To determine the equivalent impedance, we combine the two impedances in parallel.

$$Z_1 = R_1 \left\| \frac{1}{j\omega C_1} \right. = \frac{R_1(1/j\omega C_1)}{R_1 + 1/j\omega C_1} = \frac{R_1}{1 + j\omega C_1 R_1}$$

Substituting numerical values, we find

$$Z_1(\omega = 377) = \frac{10^6}{1 + j377 \times 10^6 \times 10^{-9}} = \frac{10^6}{1 + j0.377}$$

$$= 9.3571 \times 10^5 \angle(-0.3605)\ \Omega$$

The impedance of the capacitor alone at this frequency would be

$$Z_{C1}(\omega = 377) = \frac{1}{j377 \times 10^{-9}} = 2.6525 \times 10^6 \angle\left(-\frac{\pi}{2}\right)\ \Omega$$

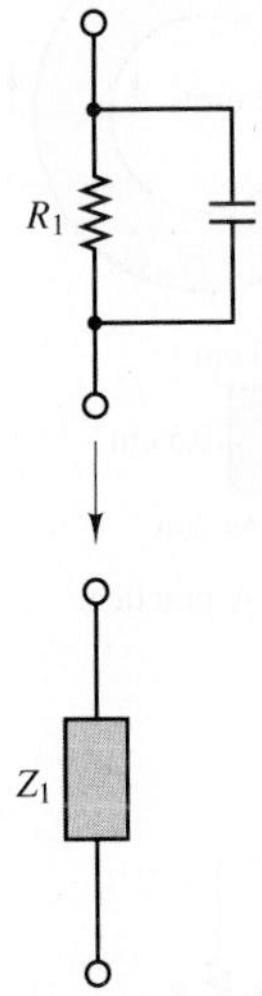

Figure 4.34

You can easily see that the parallel impedance Z_1 is quite different from the impedance of the capacitor alone, Z_{C1}.

If the frequency is increased to 800 kHz, or $1600\pi \times 10^3$ rad/s—a radio frequency in the AM range—we can recompute the impedance to be

$$Z_1(\omega = 1600\pi \times 10^3) = \frac{10^6}{1 + j1600\pi \times 10^3 \times 10^{-9} \times 10^6}$$

$$= \frac{10^6}{1 + j1600\pi} = 198.9\angle(-1.5706)\ \Omega$$

The impedance of the capacitor alone at this frequency would be

$$Z_{C1}(\omega = 1600\pi \times 10^3) = \frac{1}{j1600\pi \times 10^3 \times 10^{-9}} = 198.9\ \angle\left(-\frac{\pi}{2}\right)\ \Omega$$

Now, the impedances Z_1 and Z_{C1} are virtually identical (note that $\pi/2 = 1.5708$ rad). Thus, the effect of the parallel resistance is negligible at high frequencies.

Comments: The effect of the parallel resistance at the lower frequency (corresponding to the well-known 60-Hz AC power frequency) is significant: The effective impedance of the practical capacitor is substantially different from that of the ideal capacitor. On the other hand, at much higher frequency, the parallel resistance has an impedance so much larger than that of the capacitor that it effectively acts as an open circuit, and there is no difference between the ideal and practical capacitor impedances. This example suggests that the behavior of a circuit element depends very much on the frequency of the voltages and currents in the circuit.

EXAMPLE 4.13 Impedance of a Practical Inductor

LO4

Problem

A practical inductor can be modeled by an ideal inductor in series with a resistor. Figure 4.35 shows a *toroidal* (doughnut-shaped) inductor. The series resistance represents the resistance of the coil wire and is usually small. Find the range of frequencies over which the impedance of this practical inductor is largely *inductive* (i.e., due to the inductance in the circuit). We

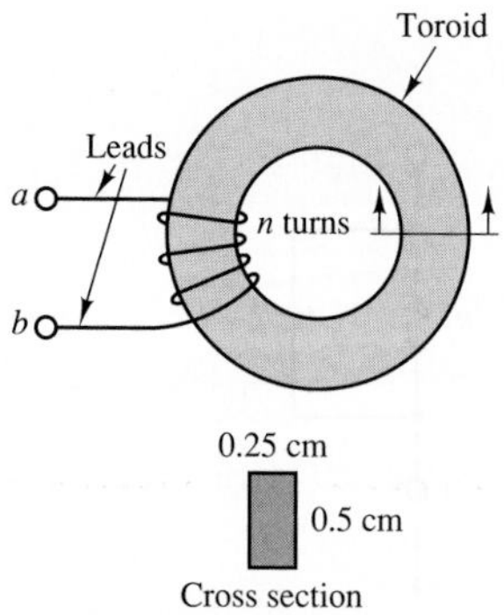

Figure 4.35 A practical inductor

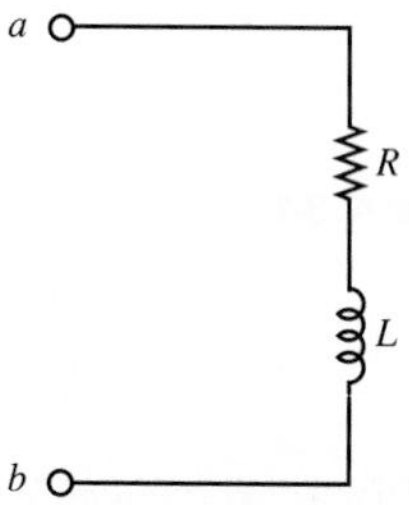

Figure 4.36

shall consider the impedance to be inductive if the impedance of the inductor in the circuit of Figure 4.36 is at least 10 times as large as that of the resistor.

Solution

Known Quantities: $L = 0.098$ H; lead length $= l_c = 2 \times 10$ cm; $n = 250$ turns; wire is 30-gauge. Resistance of 30-gauge wire $= 0.344\ \Omega$/m.

Find: The range of frequencies over which the practical inductor acts nearly as an ideal inductor.

Analysis: We first determine the equivalent resistance of the wire used in the practical inductor, using the cross section as an indication of the wire length l_w in the coil:

$$l_w = 250(2 \times 0.25 + 2 \times 0.5) = 375 \text{ cm}$$
$$l = \text{ total length} = l_w + l_c = 375 + 20 = 395 \text{ cm}$$

The total resistance is therefore

$$R = 0.344\ \Omega/\text{m} \ \times 0.395 \text{ m} = 0.136\ \Omega$$

Thus, we wish to determine the range of radian frequencies, ω, over which the magnitude of $j\omega L$ is greater than $10 \times 0.136\ \Omega$:

$$\omega L > 1.36 \qquad \text{or} \qquad \omega > \frac{1.36}{\text{L}} = \frac{1.36}{0.098} = 1.39 \text{ rad/s}$$

Alternatively, the range is $f = \omega/2\pi > 0.22$ Hz.

Comments: Note how the resistance of the coil wire is relatively insignificant. This is true because the inductor is rather large; wire resistance can become significant for very small inductance values. At high frequencies, a capacitance should be added to the model because of the effect of the insulator separating the coil wires.

LO4

EXAMPLE 4.14 Impedance of a More Complex Circuit

Problem

Find the equivalent impedance of the circuit shown in Figure 4.37.

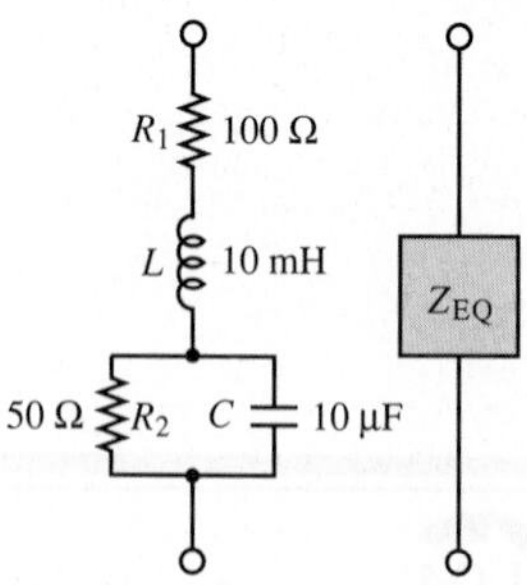

Figure 4.37

Solution

Known Quantities: $\omega = 10^4$ rad/s; $R_1 = 100\ \Omega$; $L = 10$ mH; $R_2 = 50\ \Omega$; $C = 10\ \mu$F.

Find: The equivalent impedance of the series-parallel circuit.

Analysis: We determine first the parallel impedance $Z_{||}$ of the R_2-C circuit.

$$Z_{||} = R_2 \Bigg\| \frac{1}{j\omega C} = \frac{R_2(1/j\omega C)}{R_2 + 1/j\omega C} = \frac{R_2}{1 + j\omega C R_2}$$

$$= \frac{50}{1 + j10^4 \times 10 \times 10^{-6} \times 50} = \frac{50}{1 + j5} = 1.92 - j9.62$$

$$= 9.81\angle(-1.3734)\ \Omega$$

Next, we determine the equivalent impedance Z_{eq}:

$$Z_{eq} = R_1 + j\omega L + Z_{\|} = 100 + j10^4 \times 10^{-2} + 1.92 - j9.62$$
$$= 101.92 + j90.38 = 136.2\angle 0.723\ \Omega$$

Is this impedance inductive or capacitive?

Comments: At the frequency used in this example, the circuit has an inductive impedance, since the reactance is positive (or, alternatively, the phase angle is positive).

CHECK YOUR UNDERSTANDING

Compute the equivalent impedance of the circuit of Example 4.14 for $\omega = 1{,}000$ and 100,000 rad/s.

Calculate the equivalent series capacitance of the parallel R_2C circuit of Example 4.14 at the frequency $\omega = 10$ rad/s.

Answers: $Z(1{,}000) = 140 - j10$; $Z(100{,}000) = 100 + j999$; $X_{\|} = 0.25$; $C = 0.4$ F

FOCUS ON MEASUREMENTS

Capacitive Displacement Transducer

Earlier, we introduced the idea of a capacitive displacement transducer when we considered a parallel-plate capacitor composed of a fixed plate and a movable plate. The capacitance of this variable capacitor was shown to be a *nonlinear* function of the position of the movable plate x (see Figure 4.8). In this example, we show that under certain conditions the impedance of the capacitor varies as a *linear* function of displacement—that is, the movable-plate capacitor can serve as a linear transducer.

Recall the expression derived earlier

$$C = \frac{8.854 \times 10^{-3} A}{x} \qquad \text{pF}$$

where C is the capacitance in picofarads, A is the area of the plates in millimeters square, and x is the (variable) distance in millimeters. If the capacitor is placed in an AC circuit, its impedance will be determined by the expression

$$Z_C = \frac{1}{j\omega C}$$

so that

$$Z_C = \frac{x}{j\omega 8.854 A} \qquad \Omega$$

Thus, at a fixed frequency ω, the impedance of the capacitor will vary linearly with displacement. This property may be exploited in the bridge circuit of Figure 4.9, where a differential pressure transducer was shown as being made of two movable-plate

(Continued)

capacitors, such that if the capacitance of one increased as a consequence of a pressure differential across the transducer, the capacitance of the other had to decrease by a corresponding amount (at least for small displacements). The circuit is shown again in Figure 4.38, where two resistors have been connected in the bridge along with the variable capacitors [denoted by $C(x)$]. The bridge is excited by a sinusoidal source.

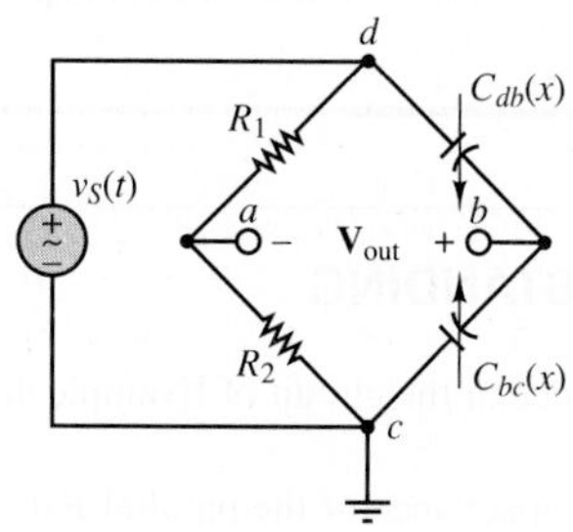

Figure 4.38 Bridge circuit for capacitive displacement transducer

Using phasor notation, we can express the output voltage as follows:

$$\mathbf{V}_{out}(j\omega) = \mathbf{V}_S(j\omega)\left(\frac{Z_{C_{bc}}(x)}{Z_{C_{db}}(x) + Z_{C_{bc}}(x)} - \frac{R_2}{R_1 + R_2}\right)$$

If the nominal capacitance of each movable-plate capacitor with the diaphragm in the center position is given by

$$C = \frac{\varepsilon A}{d}$$

where d is the nominal (undisplaced) separation between the diaphragm and the fixed surfaces of the capacitors (in millimeters), the capacitors will see a change in capacitance given by

$$C_{db} = \frac{\varepsilon A}{d - x} \quad \text{and} \quad C_{bc} = \frac{\varepsilon A}{d + x}$$

when a pressure differential exists across the transducer, so that the impedances of the variable capacitors change according to the displacement

$$Z_{C_{db}} = \frac{d - x}{j\omega 8.854A} \quad \text{and} \quad Z_{C_{bc}} = \frac{d + x}{j\omega 8.854A}$$

and we obtain the following expression for the phasor output voltage

$$\mathbf{V}_{out}(j\omega) = \mathbf{V}_S(j\omega)\left(\frac{\dfrac{d + x}{j\omega 8.854A}}{\dfrac{d - x}{j\omega 8.854A} + \dfrac{d + x}{j\omega 8.854A}} - \frac{R_2}{R_1 + R_2}\right)$$

$$= \mathbf{V}_S(j\omega)\left(\frac{1}{2} + \frac{x}{2d} - \frac{R_2}{R_1 + R_2}\right)$$

$$= \mathbf{V}_S(j\omega)\frac{x}{2d}$$

(*Continued*)

(Concluded)

if we choose $R_1 = R_2$. Thus, the output voltage will vary as a scaled version of the input voltage in proportion to the displacement. A typical $v_{out}(t)$ is displayed in Figure 4.39 for a 0.05-mm "triangular" diaphragm displacement, with $d = 0.5$ mm and $\mathbf{V}_S$ a 25-Hz sinusoid with 1-V amplitude.

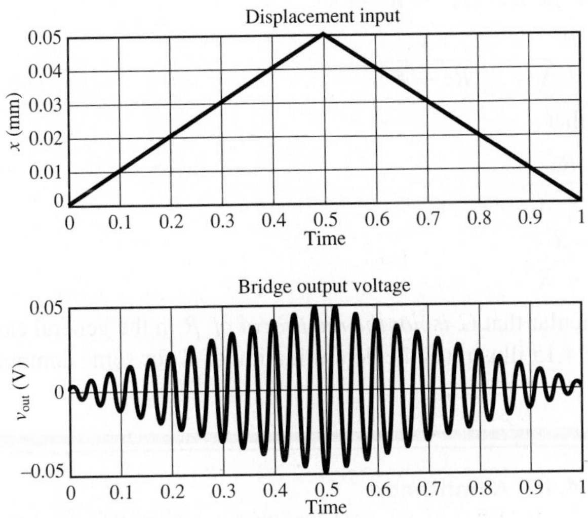

Figure 4.39 Displacement input and bridge output voltage for capacitive displacement transducer

Admittance

In Chapter 3, it was suggested that the solution of certain circuit analysis problems was handled more easily in terms of conductances than resistances. This is true, for example, when one is using node analysis, or in circuits with many parallel elements, since conductances in parallel add as resistors in series do. In AC circuit analysis, an analogous quantity may be defined—the reciprocal of complex impedance. Just as the conductance G of a resistive element was defined as the inverse of the resistance, the **admittance** of a branch is defined as follows:

$$Y = \frac{1}{Z} \quad \text{S} \tag{4.70}$$

Note immediately that whenever Z is purely real, that is, when $Z = R + j0$, the admittance Y is identical to the conductance G. In general, however, Y is the complex number

$$Y = G + jB \tag{4.71}$$

where G is called the **AC conductance** and B is called the **susceptance;** the latter plays a role analogous to that of reactance in the definition of impedance. Clearly, G and B are related to R and X. However, this relationship is not as simple as an inverse.

Let $Z = R + jX$ be an arbitrary impedance. Then the corresponding admittance is

$$Y = \frac{1}{Z} = \frac{1}{R + jX} \tag{4.72}$$

To express Y in the form $Y = G + jB$, we multiply numerator and denominator by $R - jX$:

$$\begin{aligned} Y &= \frac{1}{R + jX} \frac{R - jX}{R - jX} = \frac{R - jX}{R^2 + X^2} \\ &= \frac{R}{R^2 + X^2} - j\frac{X}{R^2 + X^2} \end{aligned} \tag{4.73}$$

and conclude that

$$\begin{aligned} G &= \frac{R}{R^2 + X^2} \\ B &= \frac{-X}{R^2 + X^2} \end{aligned} \tag{4.74}$$

Notice in particular that G *is not the reciprocal of* R in the general case!

Example 4.15 illustrates the determination of Y for some common circuits.

LO4

EXAMPLE 4.15 Admittance

Problem

Find the equivalent admittance of the two circuits shown in Figure 4.40.

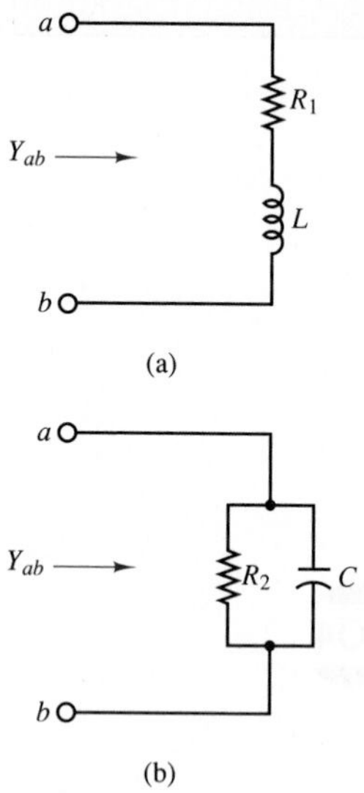

Figure 4.40

Solution

Known Quantities: $\omega = 2\pi \times 10^3$ rad/s; $R_1 = 50\ \Omega$; $L = 16$ mH; $R_2 = 100\ \Omega$; $C = 3\ \mu$F.

Find: The equivalent admittance of the two circuits.

Analysis: Circuit (a): First, determine the equivalent impedance of the circuit:

$$Z_{ab} = R_1 + j\omega L$$

Then compute the inverse of Z_{ab} to obtain the admittance:

$$Y_{ab} = \frac{1}{R_1 + j\omega L} = \frac{R_1 - j\omega L}{R_1^2 + (\omega L)^2}$$

Substituting numerical values gives

$$Y_{ab} = \frac{1}{50 + j2\pi \times 10^3 \times 0.016} = \frac{1}{50 + j100.5} = 3.968 \times 10^{-3} - j7.976 \times 10^{-3}\ \text{S}$$

Circuit (b): First, determine the equivalent impedance of the circuit:

$$Z_{ab} = R_2 \left\| \frac{1}{j\omega C} \right. = \frac{R_2}{1 + j\omega R_2 C}$$

Then compute the inverse of Z_{ab} to obtain the admittance:

$$Y_{ab} = \frac{1 + j\omega R_2 C}{R_2} = \frac{1}{R_2} + j\omega C = 0.01 + j0.019\ \text{S}$$

Comments: Note that the units of admittance are siemens (S), that is, the same as the units of conductance.

CHECK YOUR UNDERSTANDING

Compute the equivalent admittance of the circuit of Example 4.14.

Answer: $Y_{EQ} = 5.492 \times 10^{-3} - j4.871 \times 10^{-3}$

4.5 AC CIRCUIT ANALYSIS METHODS

This section illustrates how the use of phasors and impedance facilitates the solution of AC circuits by making it possible to use the same solution methods developed in Chapter 3 for DC circuits. The AC circuit analysis problem of interest in this section consists of determining the unknown voltage (or currents) in a circuit containing linear passive circuit elements (R, L, C) and excited by a sinusoidal source. Figure 4.41 depicts one such circuit, represented in both conventional time-domain form and phasor-impedance form.

The first step in the analysis of an AC circuit is to note the frequency of the sinusoidal excitation. Next, all sources are converted to phasor form, and each circuit element to impedance form. This is illustrated in the phasor circuit of Figure 4.41. At this point, if the excitation frequency ω is known numerically, it will be possible to express each impedance in terms of a known amplitude and phase, and a numerical answer to the problem will be found. It does often happen, however, that one is interested in a more general circuit solution, valid for an arbitrary excitation frequency. In this latter case, the solution becomes a function of ω. Both cases are explored in the examples.

With the problem formulated in phasor notation, the resulting solution is also in phasor form and must be converted to time-domain form. In effect, the use of phasor notation is but an intermediate step that greatly facilitates the computation of the final answer. In summary, here is the procedure that will be followed to solve an AC circuit analysis problem. Example 4.16 illustrates the various aspects of this method.

LO5

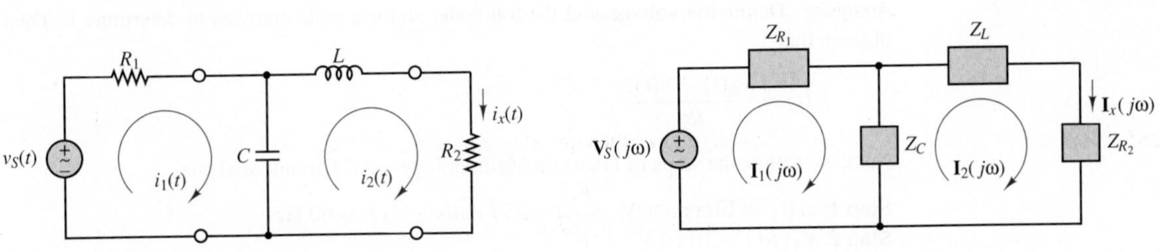

A sample circuit for AC analysis

The same circuit in phasor form

Figure 4.41 An AC circuit

FOCUS ON METHODOLOGY

LO5

AC CIRCUIT ANALYSIS

1. Identify the sinusoidal source(s) and note the excitation frequency.
2. Convert the source(s) to phasor form.
3. Represent each circuit element by its impedance.
4. Solve the resulting phasor circuit, using appropriate network analysis tools.
5. Convert the (phasor-form) answer to its time-domain equivalent, using equation 4.50.

LO5

EXAMPLE 4.16 Phasor Analysis of AC Circuit

Problem

Apply the phasor analysis method to the circuit of Figure 4.42 to determine the source current.

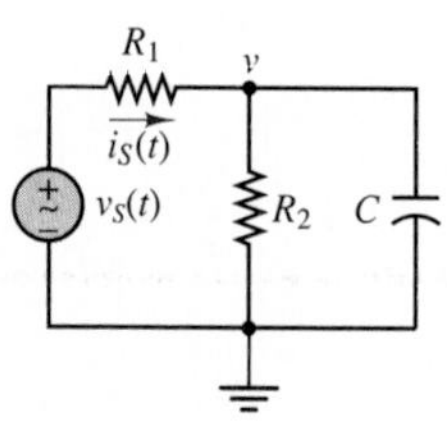

Figure 4.42

Solution

Known Quantities: Figures 4.42, 4.43, $v_S(t) = 10\cos\omega t$; $\omega = 377$ rad/s; $R_1 = 50\ \Omega$; $R_2 = 200\ \Omega$; $C = 100\ \mu$F.

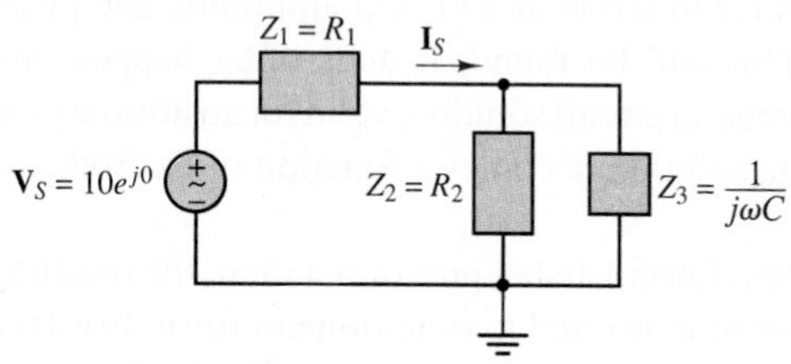

Figure 4.43

Find: The source current $i_S(t)$.

Analysis: Define the voltage v at the top node, and use node analysis to determine v. Then observe that

$$i_S(t) = \frac{v_S(t) - v(t)}{R_1}$$

Next, we follow the steps of Focus on Methodology—AC Circuit Analysis.

Step 1: $v_S(t) = 10\cos\,\omega t$ V $\omega = 377$ rad/s $(f = 60$ Hz$)$

Step 2: $\mathbf{V}_S(j\omega) = 10\angle 0$ V

Step 3: $Z_{R_1} = R_1$ $Z_{R_2} = R_2$ $Z_C = \dfrac{1}{j\omega C}$

The resulting phasor circuit is shown in Figure 4.43.

Step 4: Next, we solve for the source current, using node analysis. First we find **V**:

$$\frac{\mathbf{V}_S - \mathbf{V}}{Z_{R_1}} = \frac{\mathbf{V}}{Z_{R_2}||Z_C}$$

$$\frac{\mathbf{V}_S}{Z_{R_1}} = \mathbf{V}\left(\frac{1}{Z_{R_2}||Z_C} + \frac{1}{Z_{R_1}}\right) = \mathbf{V}\left(\frac{1}{\dfrac{R_2 \cdot (1/j\omega C)}{R_2 + (1/j\omega C)}} + \frac{1}{R_1}\right)$$

$$= \mathbf{V}\left(\frac{j\omega C R_2 + 1}{R_2} + \frac{1}{R_1}\right) = \mathbf{V}\left[\frac{(j\omega C R_2 R_1 + R_1) + R_2}{R_1 R_2}\right]$$

$$\mathbf{V} = \left[\frac{(j\omega C R_2 R_1 + R_1) + R_2}{R_1 R_2}\right]^{-1} \frac{\mathbf{V}_S}{R_1} = \left[\frac{R_1 R_2}{(j\omega C R_2 R_1 + R_1) + R_2}\right] \frac{\mathbf{V}_S}{R_1}$$

$$= \left[\frac{50 \times 200}{(j377 \times 10^{-4} \times 50 \times 200 + 50) + 200}\right] \frac{\mathbf{V}_S}{50}$$

$$= 0.4421\angle(-0.9852)\mathbf{V}_S = 4.421\angle(-0.9852)$$

Then we compute $\mathbf{I}_S$:

$$\mathbf{I}_S = \frac{\mathbf{V}_S - \mathbf{V}}{Z_{R_1}} = \frac{10\angle 0 - 4.421\angle(-0.9852)}{50} = 0.1681\angle(0.4537)$$

Step 5: Finally, we convert the phasor answer to time-domain notation:
$i_S(t) = 0.1681 \cos(377t + 0.4537)$

CHECK YOUR UNDERSTANDING

Repeat Example 4.16 by combining the parallel R_2C circuit into a single impedance $Z_{||}$ and computing the series current.

Answer: Same as above

EXAMPLE 4.17 AC Circuit Solution for Arbitrary Sinusoidal Input

LO5

Problem

Determine the general solution of Example 4.16 for any sinusoidal source, $A\cos(\omega t + \phi)$.

Solution

Known Quantities: $R_1 = 50\ \Omega$; $R_2 = 200\ \Omega$, $C = 100\ \mu$F.

Find: The phasor source current $\mathbf{I}_S(j\omega)$.

Analysis: Since the radian frequency is arbitrary, it will be impossible to determine a numerical answer. The answer will be a function of ω. The source in phasor form is represented by the expression $\mathbf{V}_S(j\omega) = A\angle\phi$. The impedances will be $Z_{R_1} = 50\ \Omega$; $Z_{R_2} = 200\ \Omega$; $Z_C = -j10^4/\omega\ \Omega$. Note that the impedance of the capacitor is a function of ω.

Taking a different approach from Example 4.16, we observe that the source current is given by the expression

$$\mathbf{I}_S = \frac{\mathbf{V}_S}{Z_{R_1} + Z_{R_2}||Z_C}$$

The parallel impedance $Z_{R_2}||Z_C$ is given by the expression

$$Z_{R_2}||Z_C = \frac{Z_{R_2} \times Z_C}{Z_{R_2} + Z_C} = \frac{200 \times 10^4/j\omega}{200 + 10^4/j\omega} = \frac{2 \times 10^6}{10^4 + j\omega 200}\ \Omega$$

Thus, the total series impedance is

$$Z_{R_1} + Z_{R_2}||Z_C = 50 + \frac{2 \times 10^6}{10^4 + j\omega 200} = \frac{2.5 \times 10^6 + j\omega 10^4}{10^4 + j\omega 200}\ \Omega$$

and the phasor source current is

$$\mathbf{I}_S = \frac{\mathbf{V}_S}{Z_{R_1} + Z_{R_2}||Z_C} = A\angle\phi \frac{10^4 + j\omega 200}{2.5 \times 10^6 + j\omega 10^4}\ \text{A}$$

Comments: The expression obtained in this example can be evaluated for an arbitrary sinusoidal excitation, by substituting numerical values for A, ϕ, and ω in the above expression. The answer can then be computed as the product of two complex numbers. As an example, you might wish to substitute the values used in Example 4.16 ($A = 10$ V, $\phi = 0$ rad, $\omega = 377$ rad/s) to verify that the same answer is obtained.

CHECK YOUR UNDERSTANDING

Compute the magnitude of the current $\mathbf{I}_S(j\omega)$ of Example 4.17 if $A = 10$ and $\phi = 0$, for $\omega = 10$, 10^2, 10^3, 10^4, and 10^5 rad/s. Can you explain these results intuitively?

Answer: $|\mathbf{I}_S| = 0.041$ A; 0.083 A; 0.194 A; 0.2 A; 0.2 A. As the frequency increases, the impedance of the capacitor goes to zero, and in the limit the capacitor acts as a short circuit. Thus, at sufficiently high frequency, $|\mathbf{I}_S| \approx |\mathbf{V}_S|/R_1 = 0.2$ A.

By now it should be apparent that the laws of network analysis introduced in Chapter 3 are also applicable to phasor voltages and currents. This fact suggests that it may be possible to extend the node and mesh analysis methods developed earlier to circuits containing phasor sources and impedances, although the resulting simultaneous complex equations are difficult to solve without the aid of a computer, even for relatively simple circuits. On the other hand, it is very useful to extend the concept of equivalent circuits to the AC case, and to define complex Thévenin (or

Norton) equivalent impedances. The fundamental difference between resistive and AC equivalent circuits is that the AC Thévenin (or Norton) equivalent circuits will be frequency-dependent and complex-valued. In general, then, one may think of the resistive circuit analysis of Chapter 3 as a special case of AC analysis in which all impedances are real.

AC Equivalent Circuits

In Chapter 3, we demonstrated that it was convenient to compute equivalent circuits, especially in solving for load-related variables. Figure 4.44 depicts the two representations analogous to those developed in Chapter 3. Figure 4.44(a) shows an *equivalent load*, as viewed by the source, while Figure 4.44(b) shows an *equivalent source* circuit, from the perspective of the load.

In the case of linear resistive circuits, the equivalent load circuit can always be expressed by a single equivalent resistor, while the equivalent source circuit may take the form of a Norton or a Thévenin equivalent. This section extends these concepts to AC circuits and demonstrates that the notion of equivalent circuits applies to phasor sources and impedances as well. The techniques described in this section are all analogous to those used for resistive circuits, with resistances replaced by impedances, and arbitrary sources replaced by phasor sources. The principal difference between resistive and AC equivalent circuits will be that the latter are frequency-dependent. Figure 4.45 summarizes the fundamental principles used in computing an AC equivalent circuit. Note the definite analogy between impedance and resistance elements, and between conductance and admittance elements.

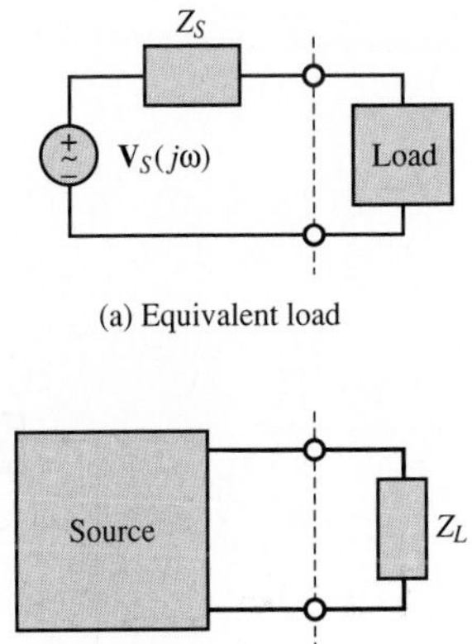

(a) Equivalent load

(b) Equivalent source

Figure 4.44 AC equivalent circuits

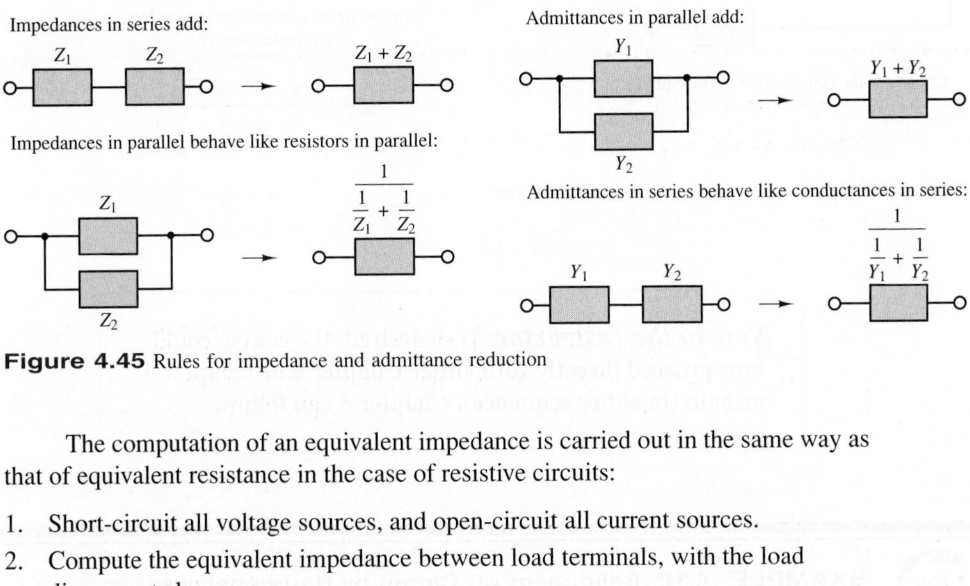

Figure 4.45 Rules for impedance and admittance reduction

LO5

The computation of an equivalent impedance is carried out in the same way as that of equivalent resistance in the case of resistive circuits:

1. Short-circuit all voltage sources, and open-circuit all current sources.
2. Compute the equivalent impedance between load terminals, with the load disconnected.

To compute the Thévenin or Norton equivalent form, we recognize that the Thévenin equivalent voltage source is the open-circuit voltage at the load terminals and the Norton equivalent current source is the short-circuit current (the current with the load replaced by a short circuit). Figure 4.46 illustrates these points by outlining the steps in the computation of an equivalent circuit. The remainder of the section will consist of examples aimed at clarifying some of the finer points in the calculation of such equivalent circuits. Note how the initial circuit reduction proceeds exactly as in the case of a resistive circuit; the details of the complex algebra required in the calculations are explored in the examples.

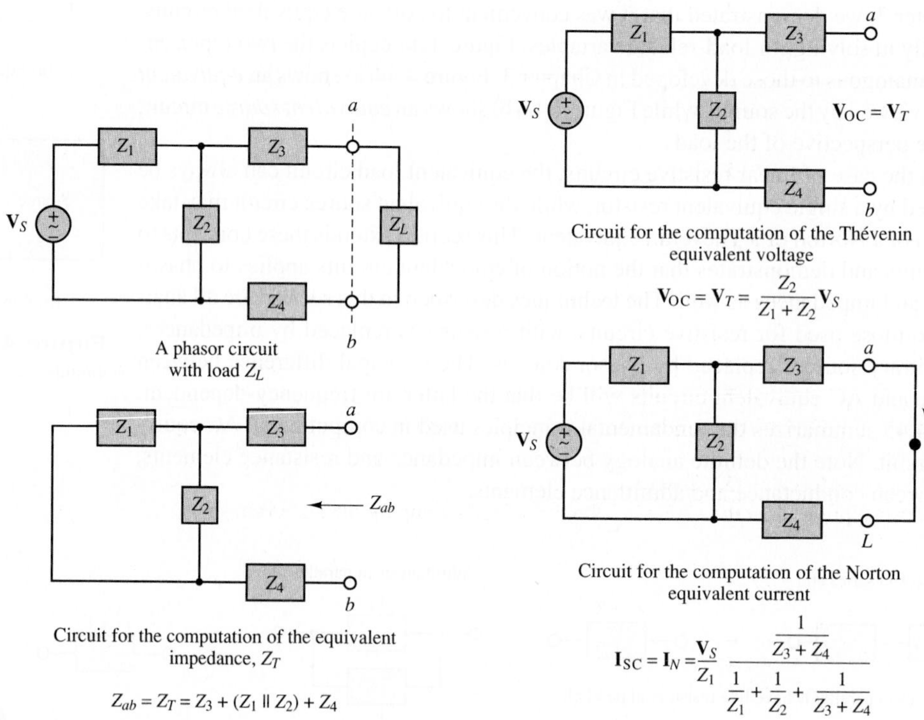

A phasor circuit with load Z_L
Circuit for the computation of the Thévenin equivalent voltage
$\mathbf{V}_{OC} = \mathbf{V}_T = \frac{Z_2}{Z_1 + Z_2}\mathbf{V}_S$
Circuit for the computation of the equivalent impedance, Z_T
$Z_{ab} = Z_T = Z_3 + (Z_1 \parallel Z_2) + Z_4$
Circuit for the computation of the Norton equivalent current
$\mathbf{I}_{SC} = \mathbf{I}_N = \frac{\mathbf{V}_S}{Z_1} \frac{\frac{1}{Z_3 + Z_4}}{\frac{1}{Z_1} + \frac{1}{Z_2} + \frac{1}{Z_3 + Z_4}}$

Figure 4.46 Reduction of AC circuit to equivalent form

> ***Note to the Instructor:*** **If so desired, the course could now proceed directly with either Chapter 6 or Chapter 7 or both (in either sequence). Chapter 5 can follow.**

LO5

EXAMPLE 4.18 Solution of AC Circuit by Node Analysis

Problem

The electrical characteristics of electric motors (which are described in greater detail in the last two chapters of this book) can be approximately represented by means of a series RL circuit. In this problem we analyze the currents drawn by two different motors connected to the same AC voltage supply (Figure 4.47).

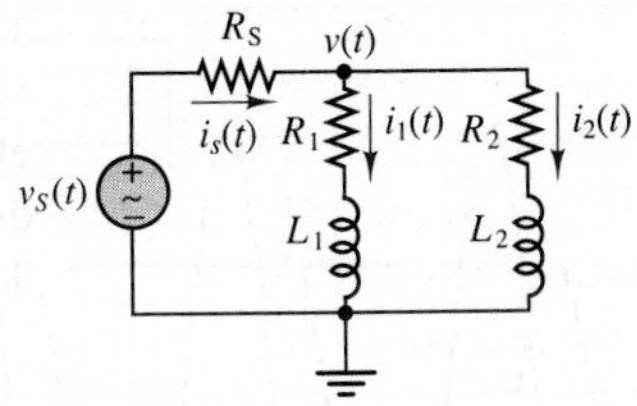

Figure 4.47 Circuit for Example 4.18

Solution

Known Quantities: $R_S = 0.5\ \Omega$; $R_1 = 2\ \Omega$; $R_2 = 0.2\ \Omega$; $L_1 = 0.1$ H; $L_2 = 20$ mH. $v_S(t) = 155\cos(377t)$ V.

Find: The motor load currents $i_1(t)$ and $i_2(t)$.

Analysis: First, we calculate the impedances of the source and of each motor:

$$Z_S = 0.5\ \Omega$$
$$Z_1 = 2 + j377 \times 0.1 = 2 + j37.7 = 37.75\angle 1.52\ \Omega$$
$$Z_2 = 0.2 + j377 \times 0.02 = 0.2 + j7.54 = 7.54\angle 1.54\ \Omega$$

The source voltage is $\mathbf{V}_S = 155\angle 0$ V.

Next, we apply KCL at the top node, with the aim of solving for the node voltage $\mathbf{V}$:

$$\frac{\mathbf{V}_S - \mathbf{V}}{Z_S} = \frac{\mathbf{V}}{Z_1} + \frac{\mathbf{V}}{Z_2}$$
$$\frac{\mathbf{V}_S}{Z_S} = \frac{\mathbf{V}}{Z_S} + \frac{\mathbf{V}}{Z_1} + \frac{\mathbf{V}}{Z_2} = \mathbf{V}\left(\frac{1}{Z_S} + \frac{1}{Z_1} + \frac{1}{Z_2}\right)$$
$$\mathbf{V} = \left(\frac{1}{Z_S} + \frac{1}{Z_1} + \frac{1}{Z_2}\right)^{-1}\frac{\mathbf{V}_S}{Z_S} = \left(\frac{1}{0.5} + \frac{1}{2 + j37.7} + \frac{1}{0.2 + j7.54}\right)^{-1}\frac{\mathbf{V}_S}{0.5}$$
$$= 154.1\angle 0.079 \text{ V}$$

Having computed the phasor node voltage $\mathbf{V}$, we can now easily determine the phasor motor currents $\mathbf{I}_1$ and $\mathbf{I}_2$:

$$\mathbf{I}_1 = \frac{\mathbf{V}}{Z_1} = \frac{154\angle 0.079}{2 + j37.7} = 4.083\angle -1.439$$
$$\mathbf{I}_2 = \frac{\mathbf{V}}{Z_2} = \frac{154\angle 0.079}{0.2 + j7.54} = 20.44\angle -1.465$$

Finally, we can write the time-domain expressions for the currents:

$$i_1(t) = 4.083\cos(377t - 1.439) \quad \text{A}$$
$$i_2(t) = 20.44\cos(377t - 1.465) \quad \text{A}$$

Figure 4.48 depicts the source voltage (scaled down by a factor of 10) and the two motor currents.

Comments: Note the phase shift between the source voltage and the two motor currents.

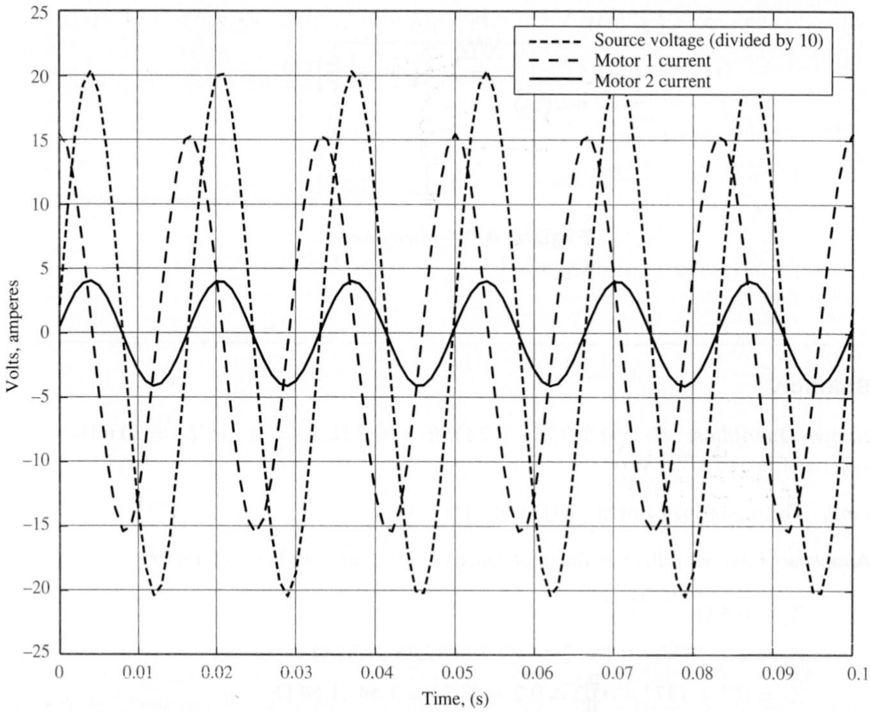

Figure 4.48 Plot of source voltage and motor currents for Example 4.18

CHECK YOUR UNDERSTANDING

Determine the Norton equivalent current in Example 4.18, assuming that the load is the $R_2 - L_2$ series circuit.

Answer: $22e^{j0}$ A

LO5

EXAMPLE 4.19 Thévenin Equivalent of AC Circuit

Problem

Compute the Thévenin equivalent of the circuit of Figure 4.49.

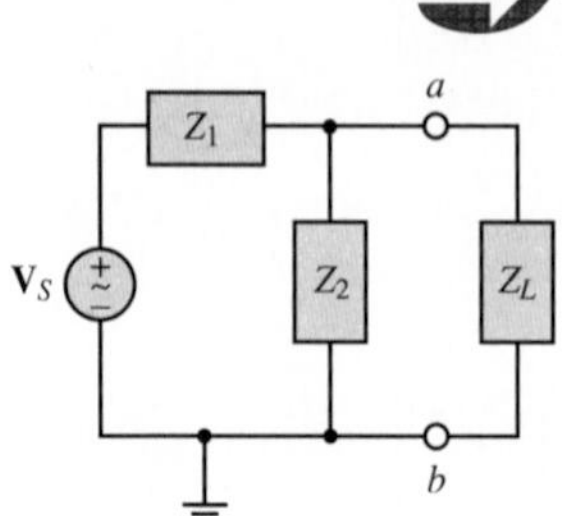

$V_S = 110\angle 0°$ $Z_1 = 5\ \Omega$ $Z_2 = j20\ \Omega$

Figure 4.49

Solution

Known Quantities: $Z_1 = 5\ \Omega$; $Z_2 = j20\ \Omega$. $v_S(t) = 110\cos(377t)$ V.

Find: Thévenin equivalent circuit.

Analysis: First compute the equivalent impedance seen by the (arbitrary) load Z_L. As illustrated in Figure 4.46, we remove the load, short-circuit the voltage source, and compute the equivalent impedance seen by the load; this calculation is illustrated in Figure 4.50.

$$Z_T = Z_1 || Z_2 = \frac{Z_1 \times Z_2}{Z_1 + Z_2} = \frac{5 \times j20}{5 + j20} = 4.71 + j1.176\ \Omega$$

Next, we compute the open-circuit voltage, between terminals a and b:

$$\mathbf{V}_T = \frac{Z_2}{Z_1 + Z_2}\mathbf{V}_S = \frac{j20}{5 + j20} 110\angle 0 = \frac{20\angle \pi/2}{20.6\angle 1.326} 110\angle 0 = 106.7\angle 0.245\ \text{V}$$

The complete Thévenin equivalent circuit is shown in Figure 4.51.

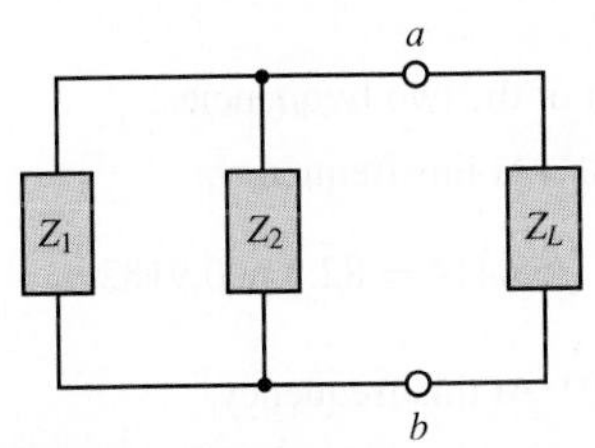

Figure 4.50

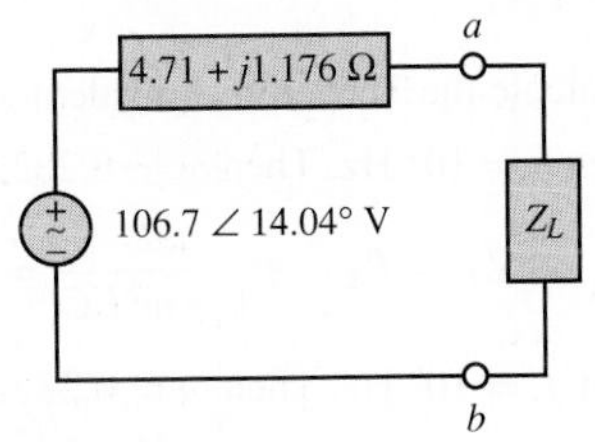

Figure 4.51

Comments: Note that the procedure followed for the computation of the equivalent circuit is completely analogous to that used in the case of resistive circuits (Section 3.6), the only difference being in the use of complex impedances in place of resistances. Thus, other than the use of complex quantities, there is no difference between the analysis leading to DC and AC equivalent circuits.

EXAMPLE 4.20 Thévenin Equivalent of AC Circuit

LO5

Problem

Determine the Thévenin equivalent circuit seen by the load in the circuit of Figure 4.52 when the input sinusoidal voltage is (a) at a frequency of 10^3 Hz and (b) at a frequency of 10^6 Hz.

Solution

Known Quantities: The values of the resistances $R_S = R_L = 50\ \Omega$, capacitance $C = 0.1\ \mu$F, inductance $L = 10$ mH.

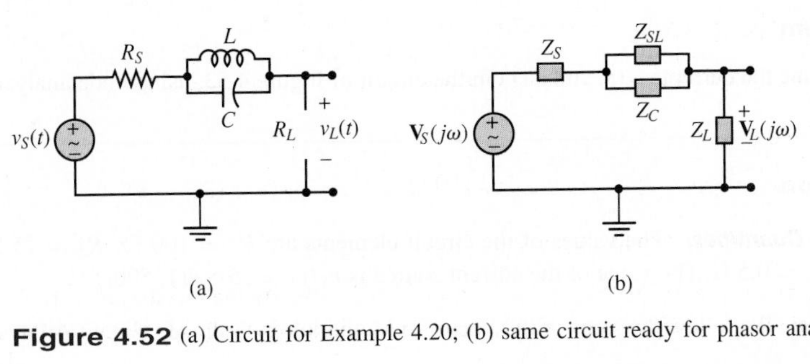

Figure 4.52 (a) Circuit for Example 4.20; (b) same circuit ready for phasor analysis

Analysis: First, we convert the circuit to phasor form, as shown in Figure 4.52(b). Next, we compute the Thévenin equivalent impedance with the load removed:

$$Z_T = Z_S + Z_L || Z_C = R_S + \frac{j\omega L \times 1/j\omega C}{j\omega L + 1/j\omega C}$$

$$= R_S + \frac{j\omega L}{j\omega L \times j\omega C + 1} = R_S + j\frac{\omega L}{1 - \omega^2 LC}$$

We observe that the Thévenin equivalent voltage is equal to the source voltage, since once the load impedance is removed, no current flows in the circuit and the voltage drop across the impedances is zero. Thus,

$$\mathbf{V}_T = \mathbf{V}_S$$

Next, we evaluate the Thévenin equivalent at each of the two frequencies.

a. Let $f = 10^3$ Hz. Then $\omega = 6.2832 \times 10^3$. At this frequency,

$$Z_T = R_S + j\frac{\omega L}{1 - \omega^2 LC} = 50 + j65.414 = 82.33\angle 0.9182$$

b. Let $f = 10^6$ Hz. Then $\omega = 6.2832 \times 10^6$. At this frequency,

$$Z_T = R_S + j\frac{\omega L}{1 - \omega^2 LC} = 50 + j1.5916 = 50\angle(-0.0318)$$

Comments: Note that at the higher frequency the equivalent impedance is very close to that of the resistor R_S. This happens because at high frequency the capacitor behaves very much as a short circuit, and the inductor as an open circuit. Thus, the two elements in parallel behave very much as a short circuit.

CHECK YOUR UNDERSTANDING

Determine the value of the capacitor and inductor impedance at the two frequencies to confirm the statement made in the "Comments" above.

Answer: At $\omega = 2\pi \times 10^3$, $Z_L = j62.832\ \Omega$, $Z_C = -j1.5915 \times 10^3\ \Omega$. At $\omega = 2\pi \times 10^6$, $Z_L = j6.2832 \times 10^4\ \Omega$, $Z_C = -j1.5915\ \Omega$.

EXAMPLE 4.21 Solution of AC Circuit by Mesh Analysis

Problem

Determine the currents $i_1(t)$ and $i_2(t)$ in the circuit of Figure 4.53, using node analysis.

Solution

Known Quantities: The values of the circuit elements are $R_1 = 100\ \Omega$, $R_2 = 75\ \Omega$, $C = 1\ \mu$F, $L = 0.5$ H. The value of the current source is $v_S(t) = 15\cos(1,500t)$ V.

Analysis: We follow the steps of the Focus on Methodology box "AC Circuit Analysis."

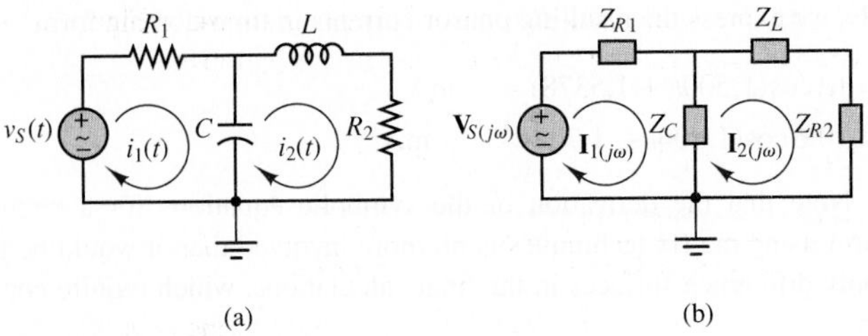

Figure 4.53 (a) Circuit for Example 4.21; (b) same circuit ready for phasor analysis

Step 1: $v_S(t) = 15\cos(1{,}500t)$ $\quad \omega = 1{,}500$ rad/s

Step 2: $V_S(j\omega) = 15\angle 0$

Step 3: $Z_{R_1} = R_1 \quad Z_{R_2} = R_2 \quad Z_C = \dfrac{1}{j\omega C} \quad Z_L = j\omega L$

The resulting phasor circuit is shown in Figure 4.53(b).

Step 4: We solve for the source current using mesh analysis. First, we write the mesh equations:

$$\mathbf{V}_S(j\omega) - Z_{R_1}\mathbf{I}_1(j\omega) - Z_C[\mathbf{I}_1(j\omega) - \mathbf{I}_2(j\omega)] = 0 \qquad \text{mesh 1}$$

$$Z_C[\mathbf{I}_2(j\omega) - \mathbf{I}_1(j\omega)] - Z_L\mathbf{I}_2(j\omega) + Z_{R_2}\mathbf{I}_2(j\omega) = 0 \qquad \text{mesh 2}$$

Next, we write the matrix form of the equations:

$$\begin{bmatrix} Z_{R_1} + Z_C & -Z_C \\ -Z_C & Z_L - Z_{R_2} \end{bmatrix} \begin{bmatrix} \mathbf{I}_1(j\omega) \\ \mathbf{I}_2(j\omega) \end{bmatrix} = \begin{bmatrix} \mathbf{V}_S(j\omega) \\ 0 \end{bmatrix}$$

and we use Cramer's rule to solve for the two currents:

$$\mathbf{I}_1(j\omega) = \frac{\begin{vmatrix} \mathbf{V}_S(j\omega) & -Z_C \\ 0 & Z_L - Z_{R_2} \end{vmatrix}}{\begin{vmatrix} Z_{R_1} + Z_C & -Z_C \\ -Z_C & Z_L - Z_{R_2} \end{vmatrix}} = \frac{Z_L - Z_{R_2}}{(Z_{R_1} + Z_C)(Z_L - Z_{R_2}) - Z_C^2}\mathbf{V}_S(j\omega)$$

$$\mathbf{I}_2(j\omega) = \frac{\begin{vmatrix} Z_{R_1} + Z_C & \mathbf{V}_S(j\omega) \\ -Z_C & 0 \end{vmatrix}}{\begin{vmatrix} Z_{R_1} + Z_C & -Z_C \\ -Z_C & Z_L - Z_{R_2} \end{vmatrix}} = \frac{Z_C}{(Z_{R_1} + Z_C)(Z_L - Z_{R_2}) - Z_C^2}\mathbf{V}_S(j\omega)$$

Now we substitute the impedance values in the above expressions:

$$\mathbf{I}_1(j\omega) = \frac{Z_L + Z_{R_2}}{(Z_{R_1} + Z_C)(Z_L - Z_{R_2}) - Z_C^2} = \frac{j\omega L - R_2}{(R_1 + 1/j\omega C)(j\omega L - R_2) - (1/j\omega C)^2}$$

$$= \frac{j\omega C(j\omega L - R_2)}{(j\omega C R_1 + 1)(j\omega L - R_2) - 1/j\omega C}$$

$$\mathbf{I}_2(j\omega) = \frac{Z_C}{(Z_{R_1} + Z_C)(Z_L - Z_{R_2}) - Z_C^2} = \frac{1}{(j\omega C R_1 + 1)(j\omega L - R_2) - 1/j\omega C}$$

and use numerical values to obtain

$$\mathbf{I}_1(j\omega) = 7.974 \times 10^{-4}\angle(1.5378)\mathbf{V}_S(j\omega) = 0.012\angle(1.5378) \text{ A}$$

$$\mathbf{I}_2(j\omega) = 7.0528 \times 10^{-4}\angle(-1.7034)\mathbf{V}_S(j\omega) = 0.0106\angle(-1.7034) \text{ A}$$

Step 5: Finally, we express the resulting phasor currents in time-domain form:

$$i_1(t) = 12\cos(1{,}500t + 1.5378) \qquad \text{mA}$$
$$i_2(t) = 10.6\cos(1{,}500t - 1.7034) \qquad \text{mA}$$

Comments: Note that the derivation of the symbolic equations for a circuit in phasor-impedance form using matrix techniques is no more involved than it would be for a resistive circuit. The only difference surfaces in the final calculations, which require complex algebra manipulations.

CHECK YOUR UNDERSTANDING

Use circuit reduction techniques (combining Z_1 and Z_{R_2} in series, then in parallel with Z_C, and again in series with Z_{R_2}) to calculate the source current, and show that it is equal to the current $i_1(t)$ computed above.

Conclusion

In this chapter we have introduced concepts and tools useful in the analysis of AC circuits. The importance of AC circuit analysis cannot be overemphasized, for a number of reasons. First, circuits made up of resistors, inductors, and capacitors constitute reasonable models for more complex devices, such as transformers, electric motors, and electronic amplifiers. Second, sinusoidal signals are ever-present in the analysis of many physical systems, not just circuits. The skills developed in Chapter 4 will be called upon in the remainder of the book. In particular, they form the basis of Chapters 5 and 6. You should have achieved the following objectives, upon completion of this chapter.

1. *Compute currents, voltages, and energy stored in capacitors and inductors.* In addition to elements that dissipate electric power, there exist electric energy storage elements, the capacitor and the inductor.
2. *Calculate the average and root-mean-square value of an arbitrary (periodic) signal.* Energy storage elements are important whenever the excitation voltages and currents in a circuit are time-dependent. Average and rms values describe two important properties of time-dependent signals.
3. *Write the differential equation(s) for circuits containing inductors and capacitors.* Circuits excited by time-dependent sources and containing energy storage (dynamic) circuit elements give rise to differential equations.
4. *Convert time-domain sinusoidal voltages and currents to phasor notation, and vice versa, and represent circuits using impedances.* For the special case of sinusoidal sources, one can use phasor representation to convert sinusoidal voltages and currents into complex phasors, and use the impedance concept to represent circuit elements.
5. *Apply the circuit analysis methods of Chapter 3 to AC circuits in phasor form.* Once a circuit is represented in phasor-impedance form, all the solution methods practiced in Chapter 3 apply.

HOMEWORK PROBLEMS

Section 4.1 Energy Storage Elements

4.1 The current through a 0.5-H inductor is given by $i_L = 2\cos(377t + \pi/6)$. Write the expression for the voltage across the inductor.

4.2 The voltage across a 100-μF capacitor takes the following values. Calculate the expression for the current through the capacitor in each case.

a. $v_C(t) = 40\cos(20t - \pi/2)$ V

b. $v_C(t) = 20\sin 100t$ V

c. $v_C(t) = -60\sin(80t + \pi/6)$ V

d. $v_C(t) = 30\cos(100t + \pi/4)$ V

4.3 The current through a 250-mH inductor takes the following values. Calculate the expression for the voltage across the inductor in each case.

a. $i_L(t) = 5\sin 25t$ A

b. $i_L(t) = -10\cos 50t$ A

c. $i_L(t) = 25\cos(100t + \pi/3)$ A

d. $i_L(t) = 20\sin(10t - \pi/12)$ A

4.4 In the circuit shown in Figure P4.4, let

$$i(t) = \begin{cases} 0 & \text{for } -\infty < t < 0 \\ t & \text{for } 0 \le t < 10 \text{ s} \\ 10 & \text{for } 10 \text{ s} \le t < \infty \end{cases}$$

Find the energy stored in the inductor for all time.

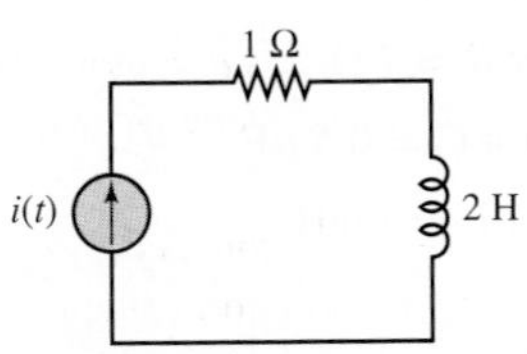

Figure P4.4

4.5 With reference to Problem 4.4, find the energy delivered by the source for all time.

4.6 In the circuit shown in Figure P4.4 let

$$i(t) = \begin{cases} 0 & \text{for } -\infty < t < 0 \\ t & \text{for } 0 \le t < 10 \text{ s} \\ 20 - t & \text{for } 10 \le t < 20 \text{ s} \\ 0 & \text{for } 20 \text{ s} \le t < \infty \end{cases}$$

Find

a. The energy stored in the inductor for all time

b. The energy delivered by the source for all time

4.7 In the circuit shown in Figure P4.7, let

$$v(t) = \begin{cases} 0 & \text{for } -\infty < t < 0 \\ t & \text{for } 0 \le t < 10 \text{ s} \\ 10 & \text{for } 10 \text{ s} \le t < \infty \end{cases}$$

Find the energy stored in the capacitor for all time.

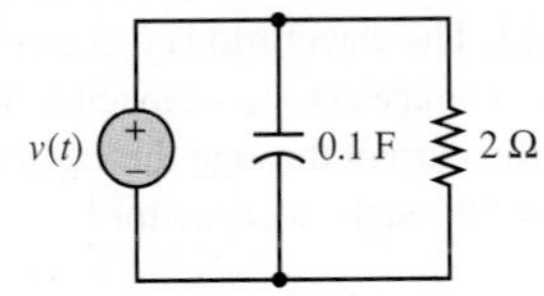

Figure P4.7

4.8 With reference to Problem 4.7, find the energy delivered by the source for all time.

4.9 In the circuit shown in Figure P4.7 let

$$i(t) = \begin{cases} 0 & \text{for } -\infty < t < 0 \\ t & \text{for } 0 \le t < 10 \text{ s} \\ 20 - t & \text{for } 10 \le t < 20 \text{ s} \\ 0 & \text{for } 20 \text{ s} \le t < \infty \end{cases}$$

Find

a. The energy stored in the capacitor for all time

b. The energy delivered by the source for all time

4.10 Find the energy stored in each capacitor and inductor, under steady-state conditions, in the circuit shown in Figure P4.10.

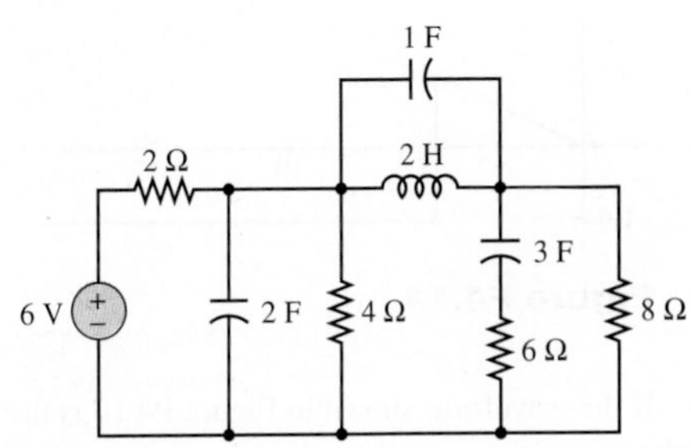

Figure P4.10

4.11 Find the energy stored in each capacitor and inductor, under steady-state conditions, in the circuit shown in Figure P4.11.

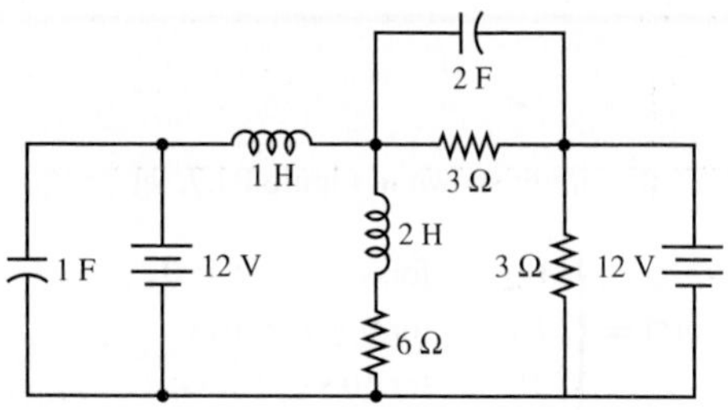

Figure P4.11

4.12 The plot of time-dependent voltage is shown in Figure P4.12. The waveform is piecewise continuous. If this is the voltage across a capacitor and $C = 80\ \mu$F, determine the current through the capacitor. How can current flow "through" a capacitor?

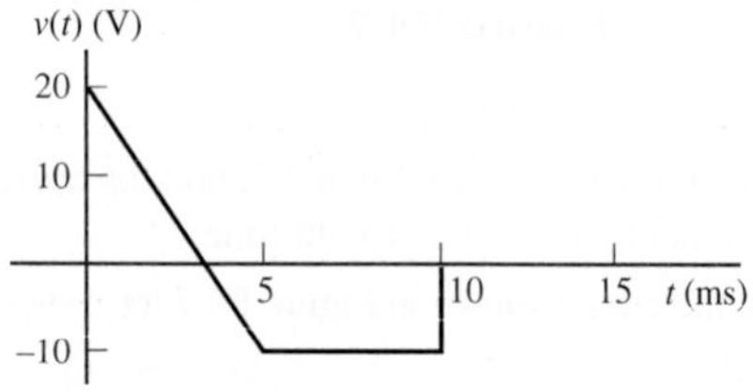

Figure P4.12

4.13 The plot of a time-dependent voltage is shown in Figure P4.12. The waveform is piecewise continuous. If this is the voltage across an inductor $L = 35$ mH, determine the current through the inductor. Assume the initial current is $i_L(0) = 0$.

4.14 The voltage across an inductor plotted as a function of time is shown in Figure P4.14. If $L = 0.75$ mH, determine the current through the inductor at $t = 15\ \mu$s.

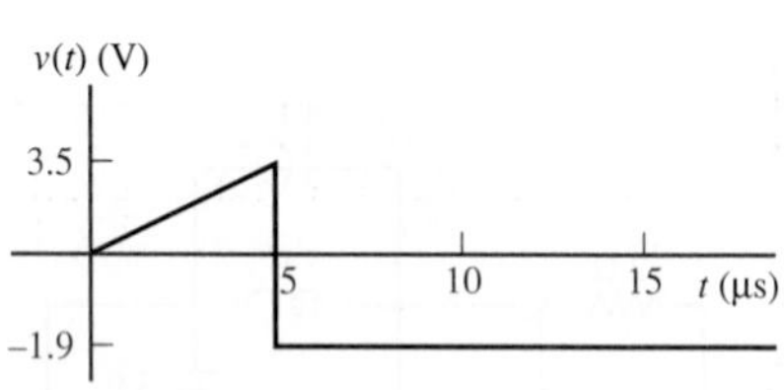

Figure P4.14

4.15 If the waveform shown in Figure P4.15 is the voltage across a capacitor plotted as a function of time with

$$v_{\text{PK}} = 20 \text{ V} \qquad T = 40\ \mu\text{s} \qquad C = 680 \text{ nF}$$

determine and plot the waveform for the current through the capacitor as a function of time.

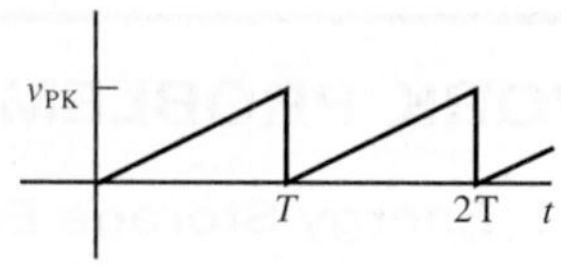

Figure P4.15

4.16 If the current through a 16-μH inductor is zero at $t = 0$ and the voltage across the inductor (shown in Figure P4.16) is

$$v_L(f) = \begin{cases} 0 & t < 0 \\ 3t^2 & 0 < t < 20\ \mu\text{s} \\ 1.2 \text{ nV} & t > 20\ \mu\text{s} \end{cases}$$

determine the current through the inductor at $t = 30\ \mu$s.

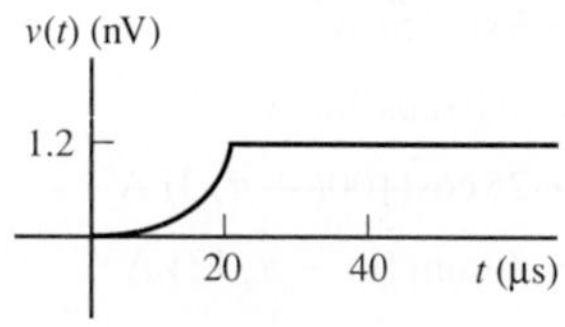

Figure P4.16

4.17 Determine and plot as a function of time the current through a component if the voltage across it has the waveform shown in Figure P4.17 and the component is a

a. Resistor $R = 7\ \Omega$

b. Capacitor $C = 0.5\ \mu$F

c. Inductor $L = 7$ mH

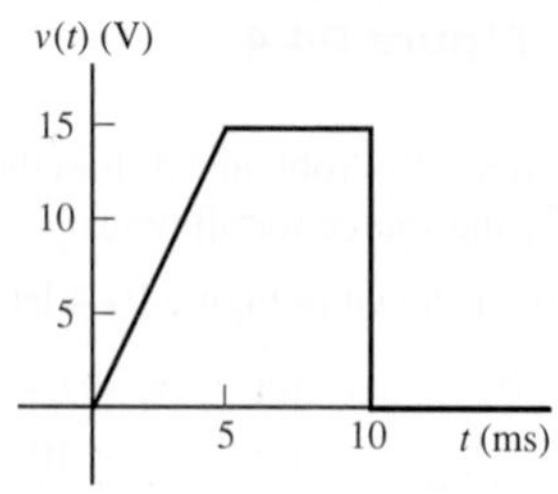

Figure P4.17

4.18 If the plots shown in Figure P4.18 are the voltage across and the current through an ideal capacitor, determine the capacitance.

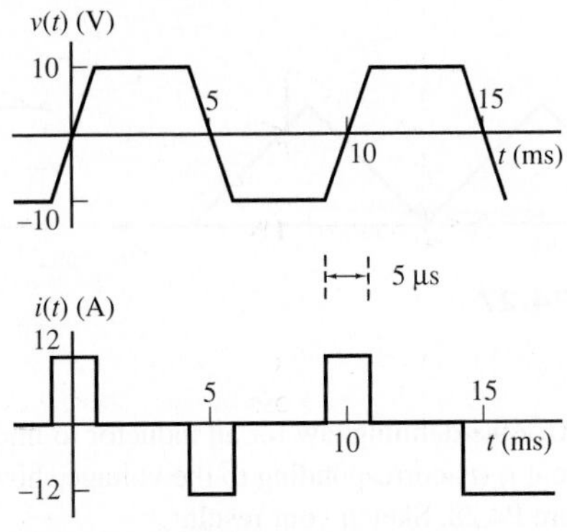

Figure P4.18

4.19 If the plots shown in Figure P4.19 are the voltage across and the current through an ideal inductor, determine the inductance.

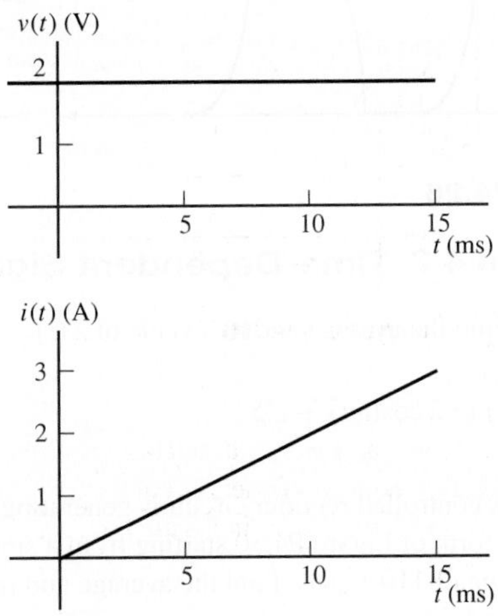

Figure P4.19

4.20 The voltage across and the current through a capacitor are shown in Figure P4.20. Determine the value of the capacitance.

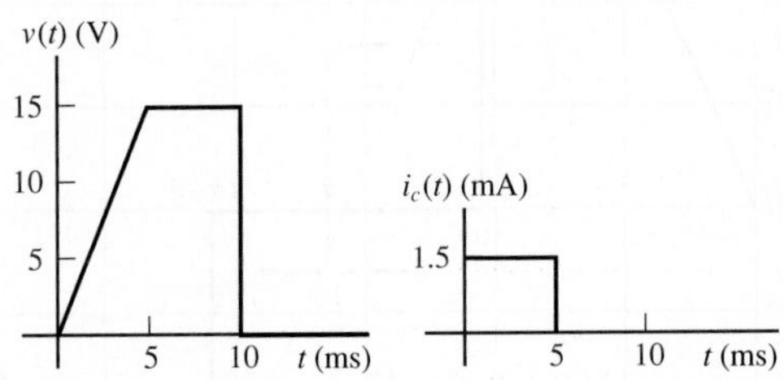

Figure P4.20

4.21 The voltage across and the current through a capacitor are shown in Figure P4.21. Determine the value of the capacitance.

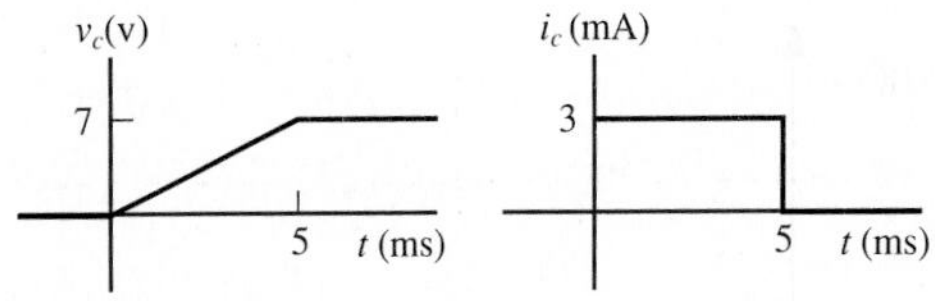

Figure P4.21

4.22 The voltage $v(t)$ shown in Figure P4.22 is applied to a 10-mH inductor. Find the current through the inductor. Assume $i_L(0) = 0$ A.

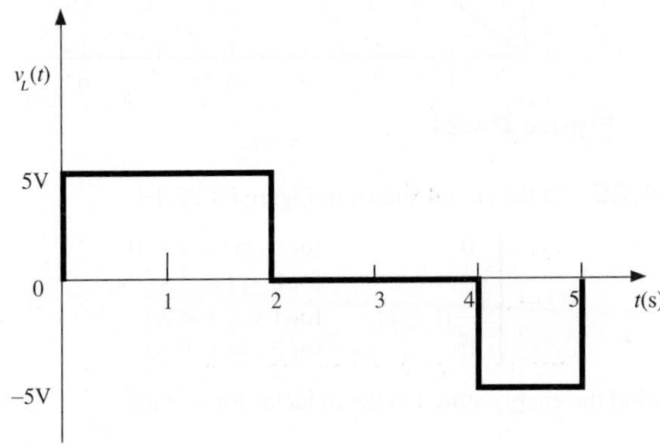

Figure P4.22

4.23 The current waveform shown in Figure P4.23 flows through a 2-H inductor. Plot the inductor voltage $v_L(t)$.

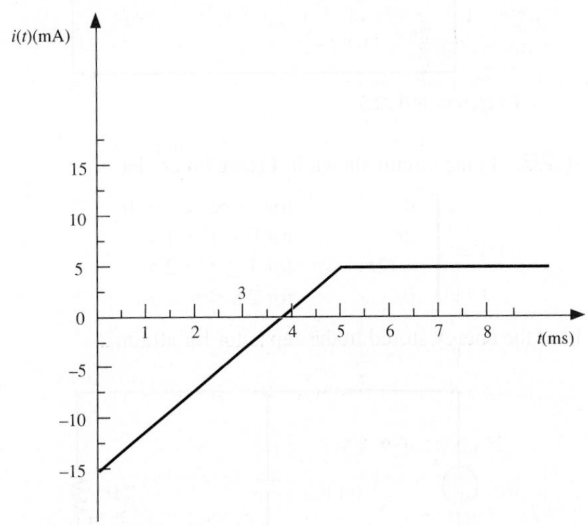

Figure P4.23

4.24 The voltage waveform shown in Figure P4.24 appears across a 100-mH inductor and a 500-μF capacitor. Plot the capacitor and inductor currents, $i_C(t)$ and $i_L(t)$, assuming $i_L(0) = 0$ A.

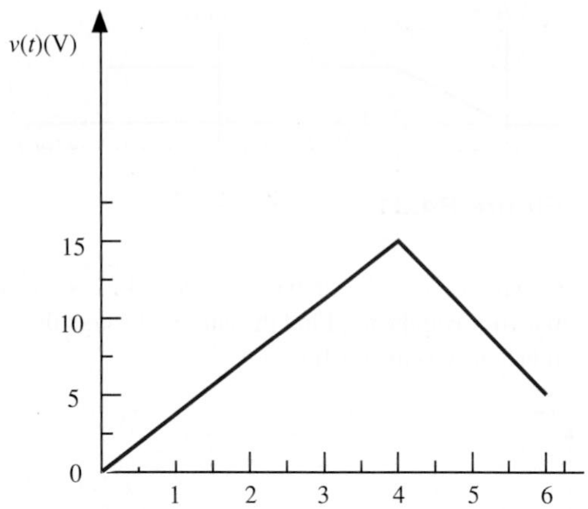

Figure P4.24

4.25 In the circuit shown in Figure P4.25, let

$$i(t) = \begin{cases} 0 & \text{for } -\infty < t < 0 \\ t & \text{for } 0 \le t < 1 \text{ s} \\ -(t-2) & \text{for } 1 \text{ s} \le t < 2 \text{ s} \\ 0 & \text{for } 2 \text{ s} \le t < \infty \end{cases}$$

Find the energy stored in the inductor for all time.

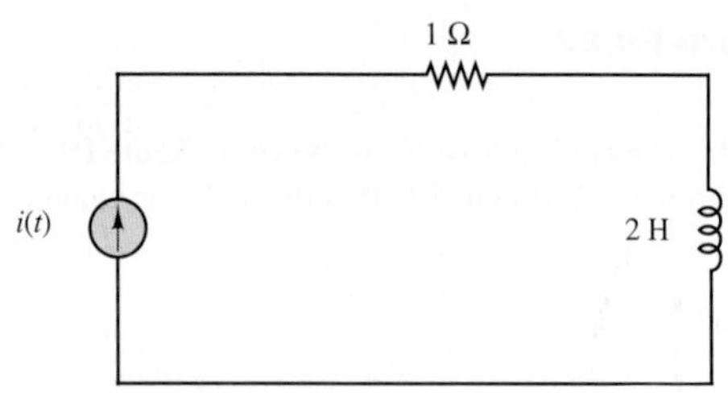

Figure P4.25

4.26 In the circuit shown in Figure P4.26, let

$$v(t) = \begin{cases} 0 & \text{for } -\infty < t < 0 \\ 2t & \text{for } 0 \le t < 1 \text{ s} \\ -(2t-4) & \text{for } 1 \le t < 2 \text{ s} \\ 0 & \text{for } 2 \text{ s} \le t < \infty \end{cases}$$

Find the energy stored in the capacitor for all time.

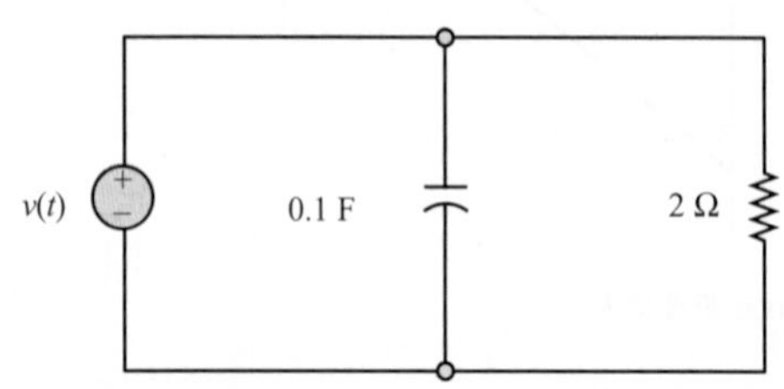

Figure P4.26

4.27 Use the defining law for a capacitor to find the current $i_C(t)$ corresponding to the voltage shown in Figure P4.27. Sketch your result.

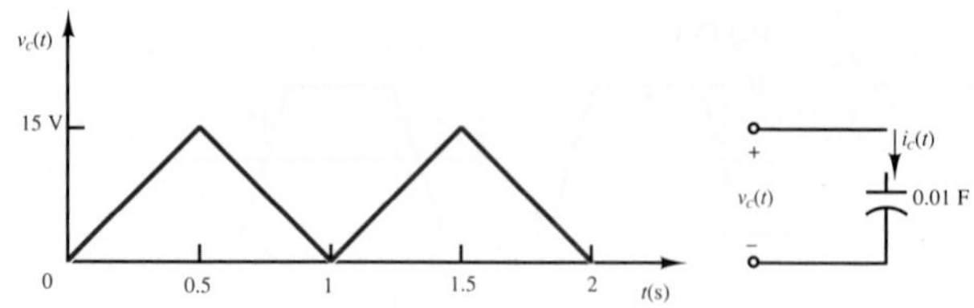

Figure P4.27

4.28 Use the defining law for an inductor to find the current $i_L(t)$ corresponding to the voltage shown in Figure P4.28. Sketch your result.

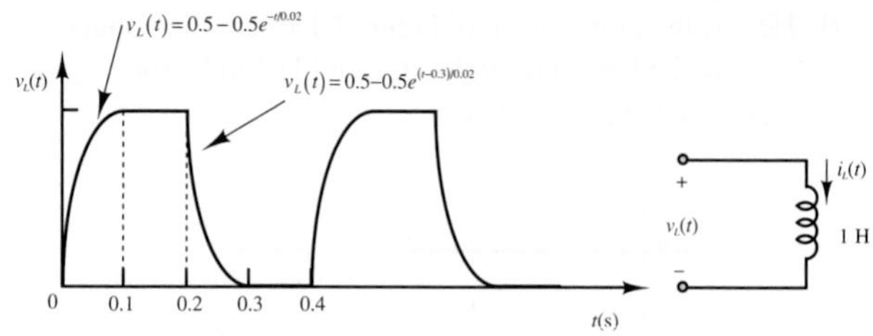

Figure P4.28

Section 4.2 Time-Dependent Signals

4.29 Find the average and rms value of $x(t)$.

$$x(t) = 2\cos(\omega t) + 2.5$$

4.30 A controlled rectifier circuit is generating the waveform of Figure P4.30 starting from a sinusoidal voltage of 110 V rms. Find the average and rms voltage.

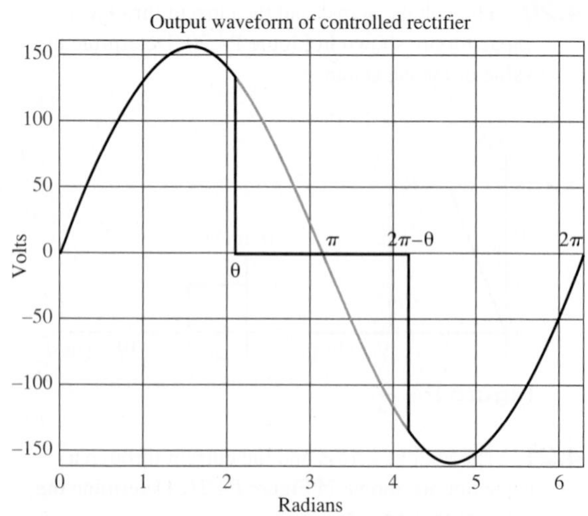

Figure P4.30

4.31 With reference to Problem 4.30, find the angle θ that corresponds to delivering exactly one-half of the total available power in the waveform to a resistive load.

4.32 Find the ratio between average and rms value of the waveform of Figure P4.32.

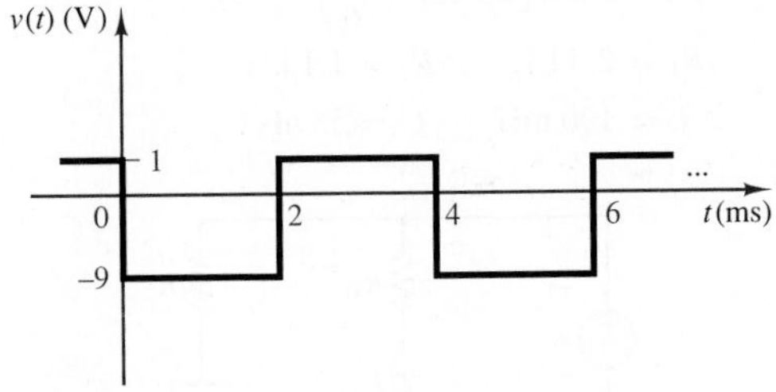

Figure P4.32

4.33 Given the current waveform shown in Figure P4.33, find the power dissipated by a 1-Ω resistor.

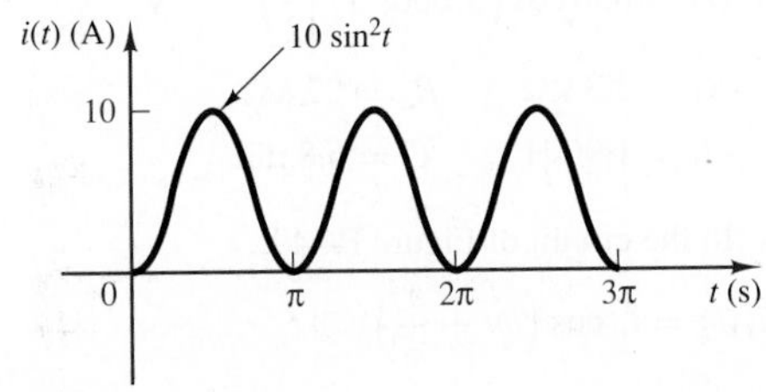

Figure P4.33

4.34 Find the ratio between average and rms value of the waveform of Figure P4.34.

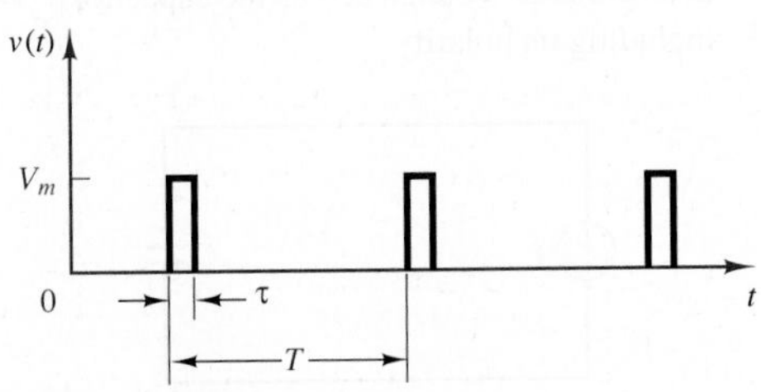

Figure P4.34

4.35 Find the rms value of the waveform shown in Figure P4.35.

4.36 Determine the rms (or effective) value of

$$v(t) = V_{DC} + v_{AC} = 50 + 70.7\cos(377t)\text{ V}$$

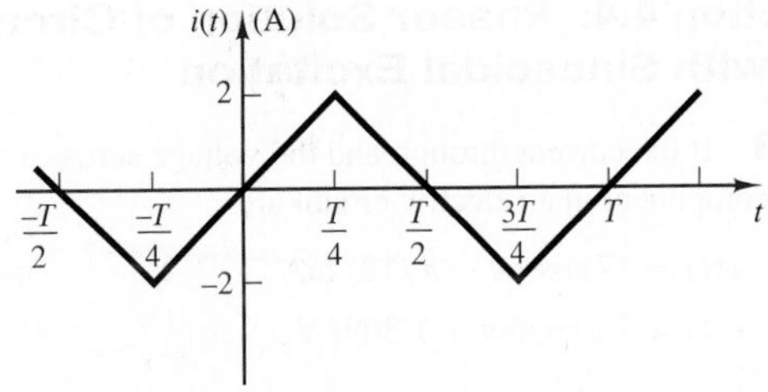

Figure P4.35

4.37 Find the phasor form of the following functions:

a. $v(t) = 155\cos(377t - 25°)$ V

b. $v(t) = 5\sin(1{,}000t - 40°)$ V

c. $i(t) = 10\cos(10t + 63°) + 15\cos(10t - 42°)$ A

d. $i(t) = 460\cos(500\pi t - 25°)$
$- 220\sin(500\pi t + 15°)$ A

4.38 Convert the following complex numbers to polar form:

a. $4 + j4$

b. $-3 + j4$

c. $j + 2 - j4 - 3$

4.39 Convert the following to polar form and compute the product. Compare the result with that obtained using rectangular form.

a. $(50 + j10)(4 + j8)$

b. $(j2 - 2)(4 + j5)(2 + j7)$

4.40 Complete the following exercises in complex arithmetic.

a. Find the complex conjugate of $(4 + j4)$, $(2 - j8)$, $(-5 + j2)$.

b. Convert the following to polar form by multiplying the numerator and denominator by the complex conjugate of the denominator and then performing the conversion to polar coordinates:

$$\frac{1 + j7}{4 + j4}, \quad \frac{j4}{2 - j8}, \quad \frac{1}{-5 + j2}.$$

c. Repeat part b but this time convert to polar coordinates before performing the division.

4.41 Convert the following expressions to real-imaginary form: j^j, $e^{j\pi}$.

4.42 Given the two voltages $v_1(t) = 10\cos(\omega t + 30°)$ and $v_2(t) = 20\cos(\omega t + 60°)$, find $v(t) = v_1(t) + v_2(t)$ using

a. Trigonometric identities.

b. Phasors.

Section 4.4: Phasor Solution of Circuits with Sinusoidal Excitation

4.43 If the current through and the voltage across a component in an electric circuit are

$$i(t) = 17\cos(\omega t - \pi/12) \text{ mA}$$
$$v(t) = 3.5\cos(\omega t + 1.309) \text{ V}$$

where $\omega = 628.3$ rad/s, determine

a. Whether the component is a resistor, capacitor, or inductor.

b. The value of the component in ohms, farads, or henrys.

4.44 Describe the sinusoidal waveform shown in Figure P4.44, using time-dependent and phasor notation.

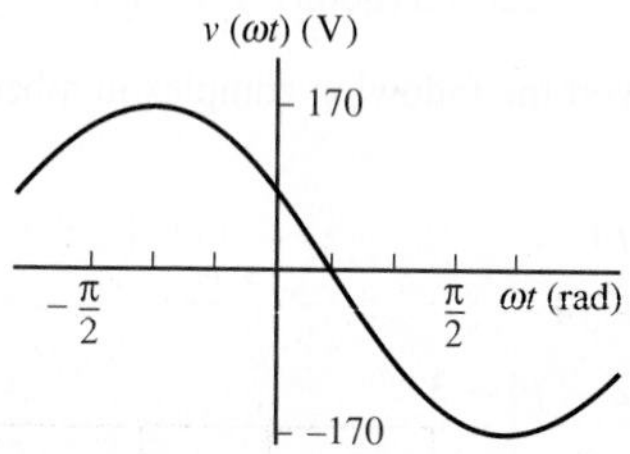

Figure P4.44

4.45 Describe the sinusoidal waveform shown in Figure P4.45, using time-dependent and phasor notation.

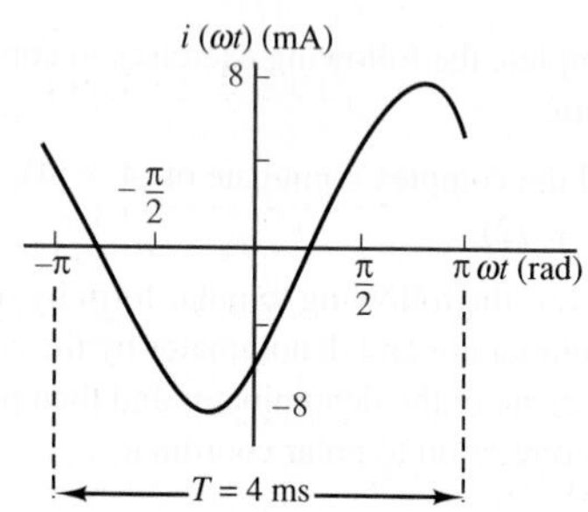

Figure P4.45

4.46 The current through and the voltage across an electrical component are

$$i(t) = I_o \cos\left(\omega t + \frac{\pi}{4}\right) \qquad v(t) = V_o \cos \omega t$$

where

$$I_o = 3 \text{ mA} \qquad V_o = 700 \text{ mV} \qquad \omega = 6.283 \text{ rad/s}$$

a. Is the component inductive or capacitive?

b. Plot the instantaneous power $p(t)$ as a function of ωt over the range $0 < \omega t < 2\pi$.

c. Determine the average power dissipated as heat in the component.

d. Repeat parts (b) and (c) if the phase angle of the current is changed to 0°.

4.47 Determine the equivalent impedance in the circuit shown in Figure P4.47:

$$v_s(t) = 7\cos\left(3{,}000t + \frac{\pi}{6}\right) \quad \text{V}$$
$$R_1 = 2.3 \text{ k}\Omega \qquad R_2 = 1.1 \text{ k}\Omega$$
$$L = 190 \text{ mH} \qquad C = 55 \text{ nF}$$

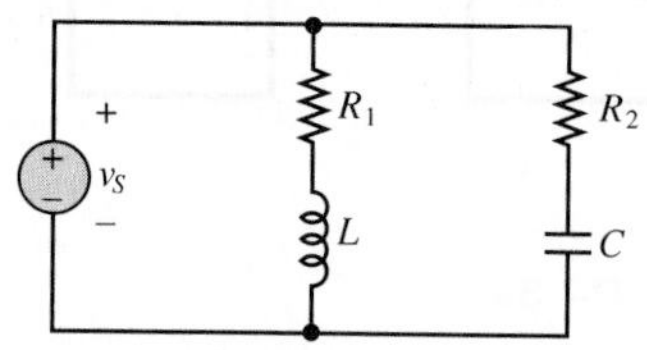

Figure P4.47

4.48 Determine the equivalent impedance in the circuit shown in Figure P4.47:

$$v_s(t) = 636\cos\left(3{,}000t + \frac{\pi}{12}\right) \quad \text{V}$$
$$R_1 = 3.3 \text{ k}\Omega \qquad R_2 = 22 \text{ k}\Omega$$
$$L = 1.90 \text{ H} \qquad C = 6.8 \text{ nF}$$

4.49 In the circuit of Figure P4.49,

$$i_s(t) = I_o \cos\left(\omega t + \frac{\pi}{6}\right)$$
$$I_o = 13 \text{ mA} \qquad \omega = 1{,}000 \text{ rad/s}$$
$$C = 0.5\ \mu\text{F}$$

a. State, using phasor notation, the source current.

b. Determine the impedance of the capacitor.

c. Using phasor notation only and showing all work, determine the voltage across the capacitor, including its polarity.

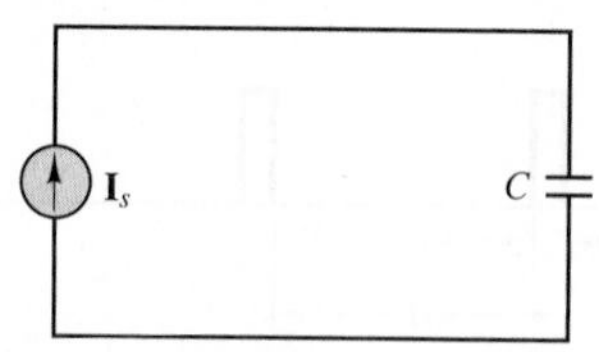

Figure P4.49

4.50 Determine $i_3(t)$ in the circuit shown in Figure P4.50 if

$$i_1(t) = 141.4\ \cos(\omega t + 2.356) \qquad \text{mA}$$
$$i_2(t) = 50\ \sin(\omega t - 0.927) \qquad \text{mA}$$
$$\omega = 377 \text{ rad/s}$$

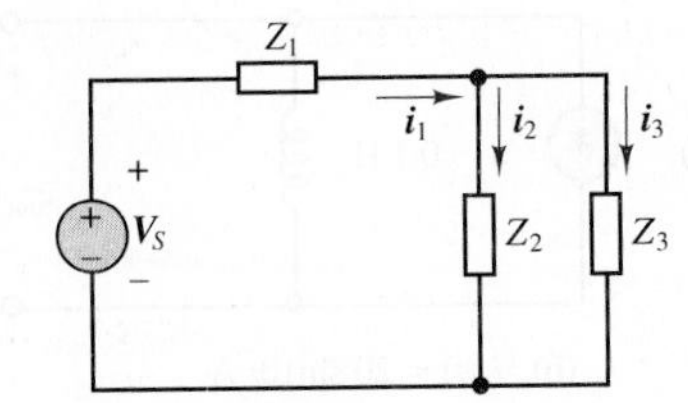

Figure P4.50

4.51 Determine the current through Z_3 in the circuit of Figure P4.51.

$v_{s1} = v_{s2} = 170\cos(377t)$ V

$Z_1 = 5.9\angle 0.122\ \Omega$

$Z_2 = 2.3\angle 0\ \Omega$

$Z_3 = 17\angle 0.192\ \Omega$

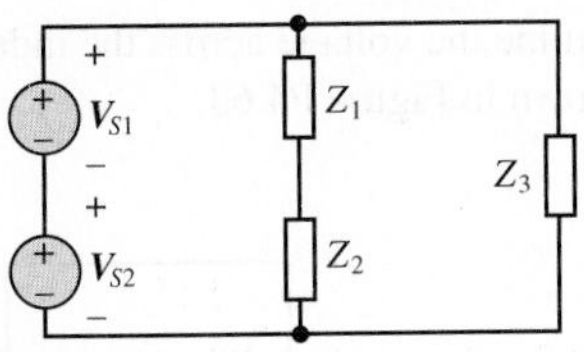

Figure P4.51

4.52 Determine the frequency so that the current I_i and the voltage V_o in the circuit of of Figure P4.52 are in phase.

$Z_s = 13{,}000 + j\omega 3\ \Omega$

$R = 120\ \Omega$

$L = 19$ mH $\quad C = 220$ pF

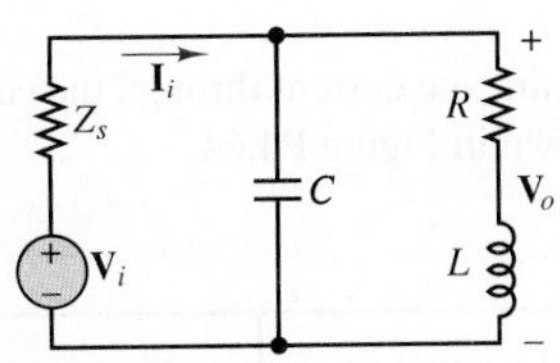

Figure P4.52

4.53 The coil resistor in series with L models the internal losses of an inductor in the circuit of Figure P4.53. Determine the current supplied by the source if

$v_s(t) = V_o\cos(\omega t + 0)$

$V_o = 10$ V $\quad \omega = 6$ Mrad/s $\quad R_s = 50\ \Omega$

$R_c = 40\ \Omega \quad L = 20\ \mu$H $\quad C = 1.25$ nF

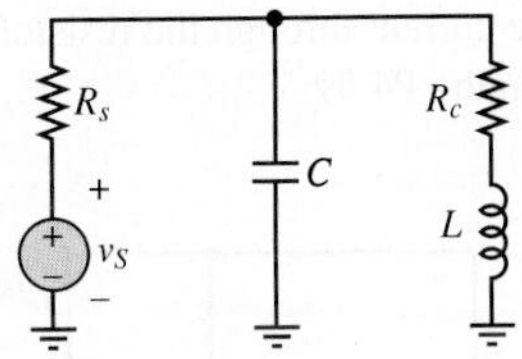

Figure P4.53

4.54 Using phasor techniques, solve for the current in the circuit shown in Figure P4.54.

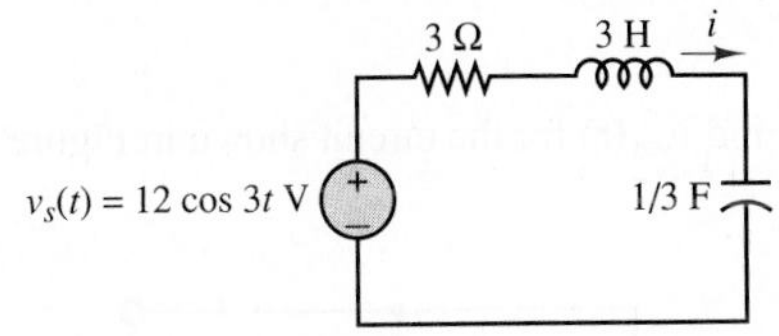

Figure P4.54

4.55 Using phasor techniques, solve for the voltage v in the circuit shown in Figure P4.55.

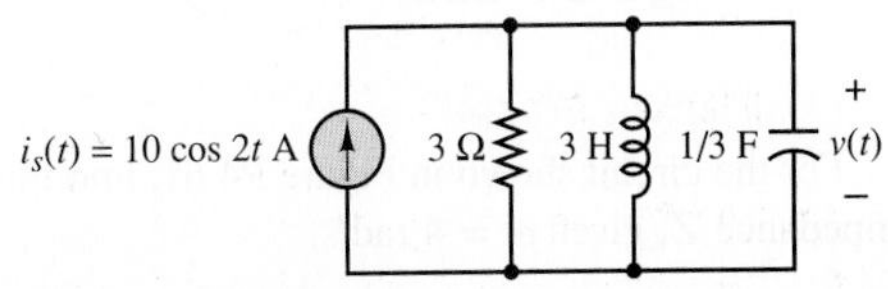

Figure P4.55

4.56 Solve for $\mathbf{I}_1$ in the circuit shown in Figure P4.56.

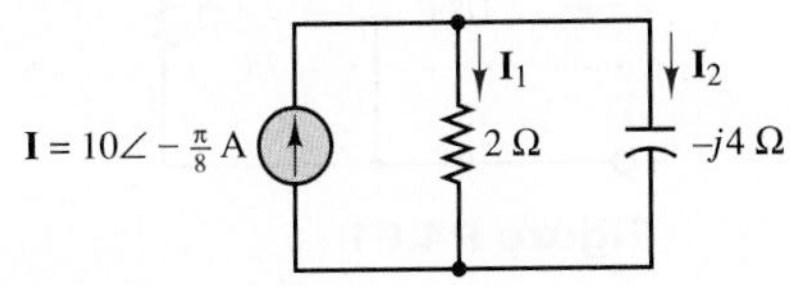

Figure P4.56

4.57 Solve for $\mathbf{V}_2$ in the circuit shown in Figure P4.57. Assume $\omega = 2$.

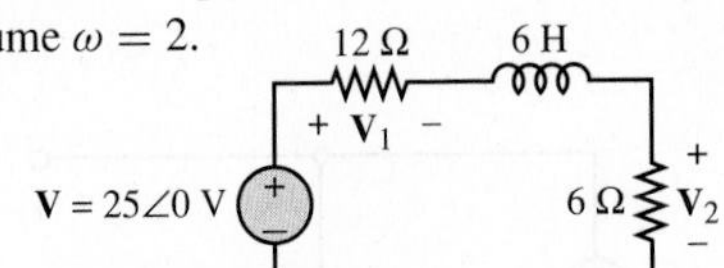

Figure P4.57

4.58 With reference to Problem 4.55, find the value of ω for which the current through the resistor is maximum.

4.59 Find the current through the resistor in the circuit shown in Figure P4.59.

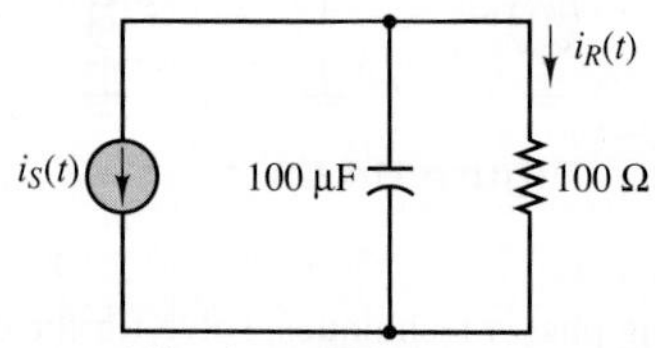

$i_S(t) = 1\cos(200\pi t)$

Figure P4.59

4.60 Find $v_{out}(t)$ for the circuit shown in Figure P4.60.

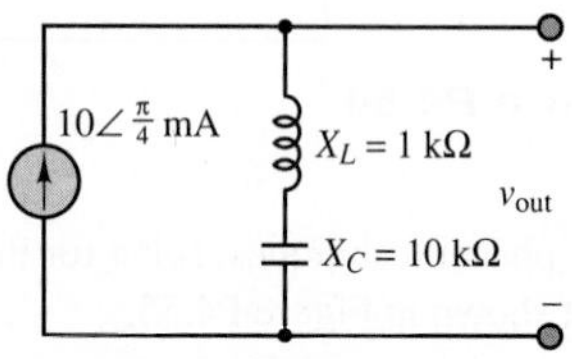

Figure P4.60

4.61 For the circuit shown in Figure P4.61, find the impedance Z, given $\omega = 4$ rad/s.

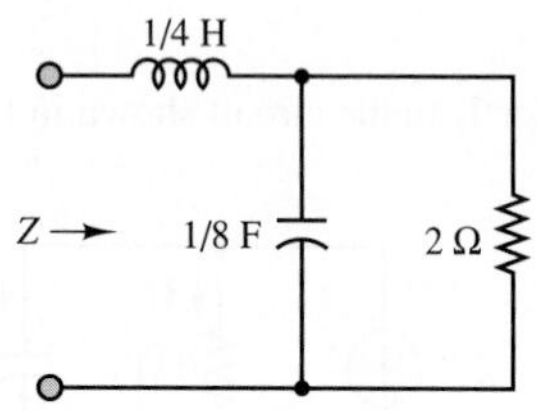

Figure P4.61

4.62 Find the sinusoidal steady-state outputs for each of the circuits shown in Figure P4.62.

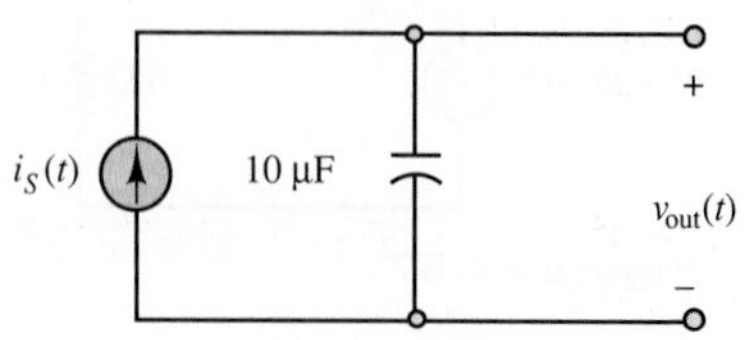

(a) $i_S(t) = 10\cos 100\pi t$ A

Figure P4.62 *(Continued)*

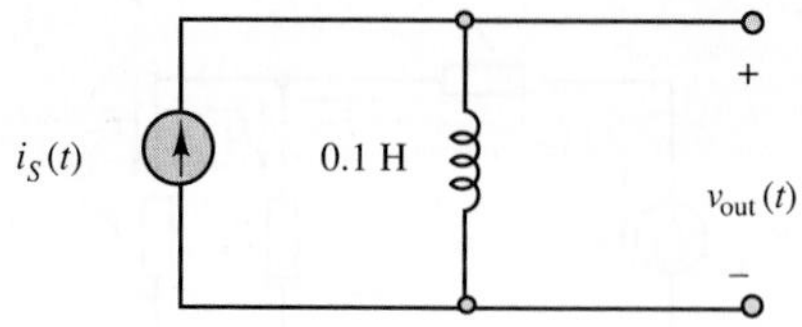

(b) $i_S(t) = 20\sin 10t$ A

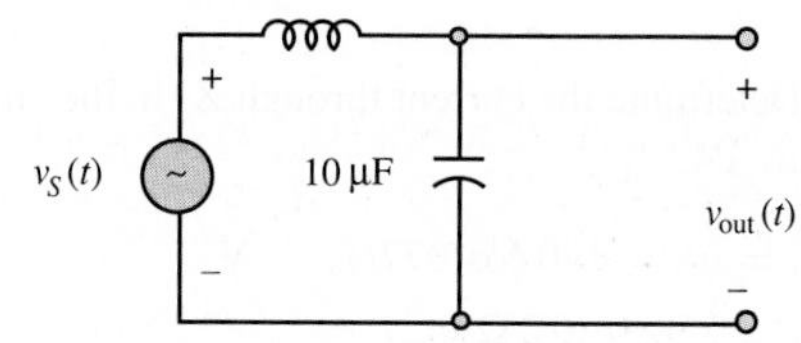

(c) $v_S(t) = 50\sin 100t$ V

Figure P4.62

4.63 Determine the voltage across the inductor in the circuit shown in Figure P4.63.

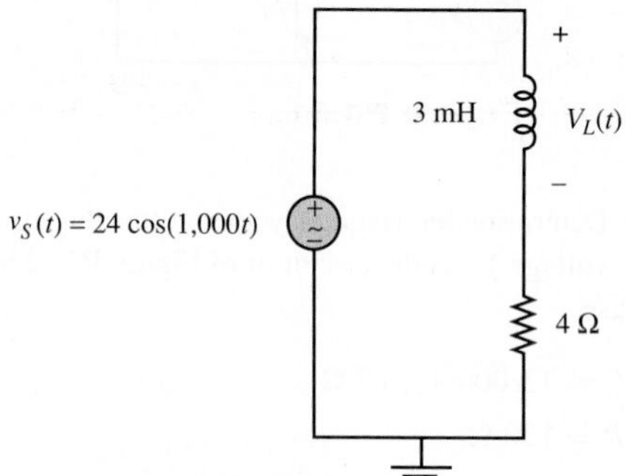

Figure P4.63

4.64 Determine the current through the capacitor in the circuit shown in Figure P4.64.

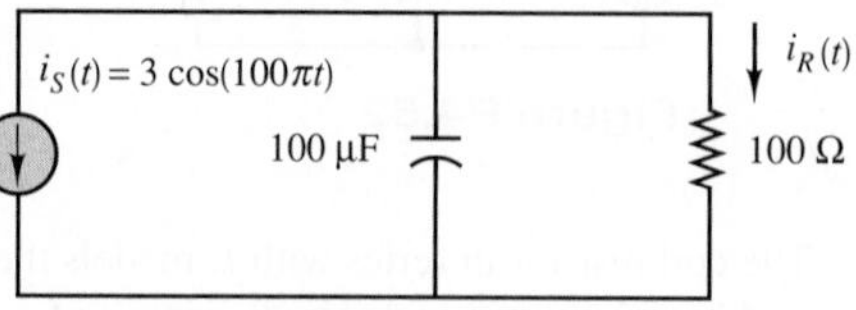

Figure P4.64

4.65 For the circuit shown in the Figure P4.65, find the frequency that causes the equivalent impedance to appear purely resistive.

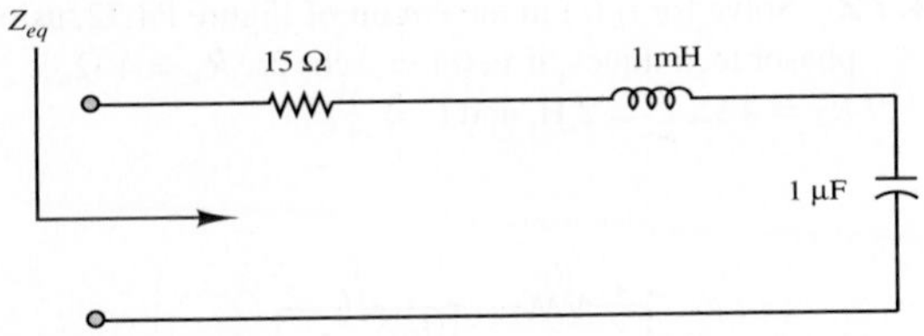

Figure P4.65

4.66

a. Find the equivalent impedance Z_L shown in Figure P4.66(a), as seen by the source, if the frequency is 377 rad/s.

b. If we wanted the source to see the load as completely resistive, what value of capacitance should we place between the terminals a and b as shown in Figure P4.66(b)? [*Hint:* Find an expression for the equivalent impedance Z_L, and then find C so that the phase angle of the impedance is zero.]

c. What is the actual impedance that the source sees with the capacitor included in the circuit?

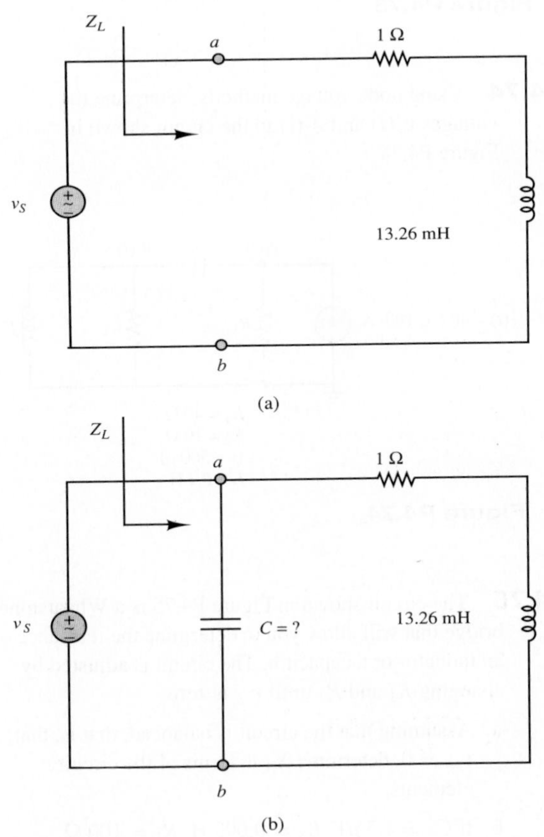

Figure P4.66

4.67 The capacitor model we have used so far has been treated as an ideal circuit element. A more accurate model for a capacitor is shown in Figure P4.67. The ideal capacitor, C, has a large "leakage" resistance, R_C, in parallel with it. R_C models the leakage current through the capacitor. R_1 and R_2 represent the lead wire resistances, and L_1 and L_2 represent the lead wire inductances.

a. If $C = 1\ \mu$F, $R_C = 100$ MΩ, $R_1 = R_2 = 1\ \mu\Omega$ and $L_1 = L_2 = 0.1\ \mu$H, find the equivalent impedance seen at the terminals a and b as a function of frequency ω.

b. Find the range of frequencies for which Z_{ab} is capacitive, i.e., $X_{ab} > 10|R_{ab}$.

[*Hint*: assume that R_C is is much greater than $1/wC$ so that you can replace R_C by an infinite resistance in part b.]

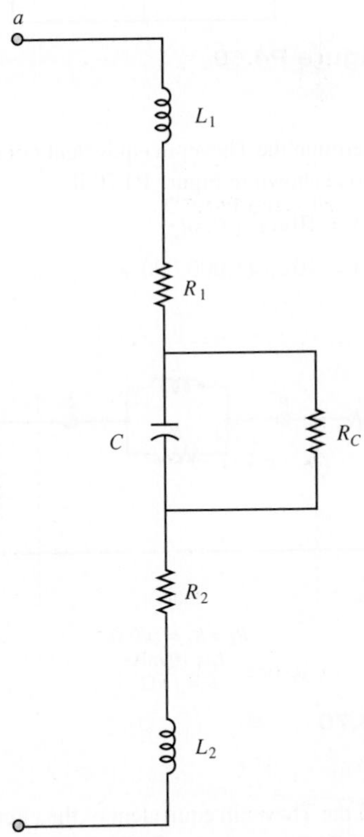

Figure P4.67

Section 4.5: AC Circuit Analysis Methods

4.68 Using phasor techniques, solve for v in the circuit shown in Figure P4.68.

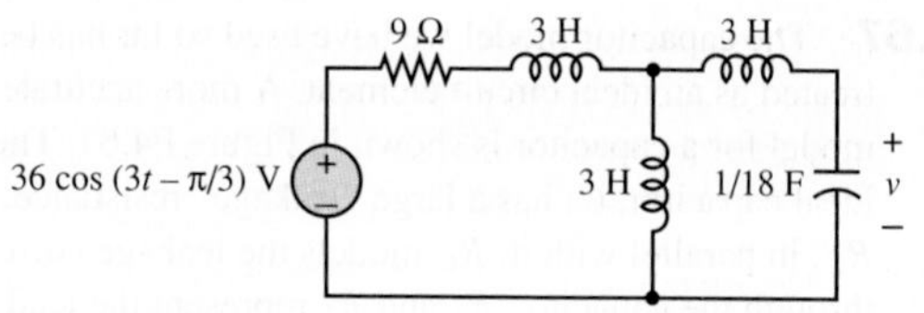

Figure P4.68

4.69 Using phasor techniques, solve for i in the circuit shown in Figure P4.69.

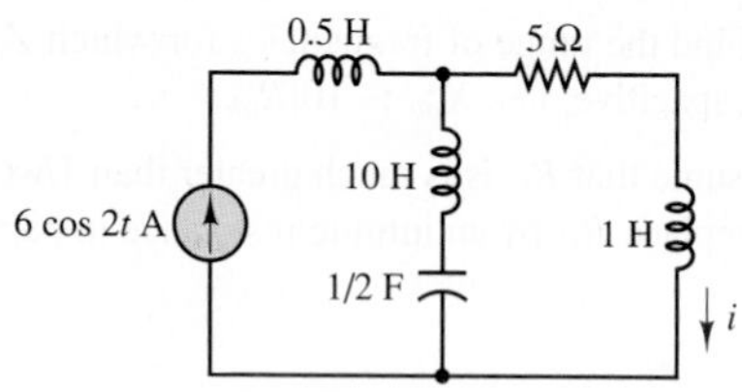

Figure P4.69

4.70 Determine the Thévenin equivalent circuit as seen by the load shown in Figure P4.70 if

a. $v_S(t) = 10\cos(1{,}000t)$

b. $v_S(t) = 10\cos(1{,}000{,}000t)$

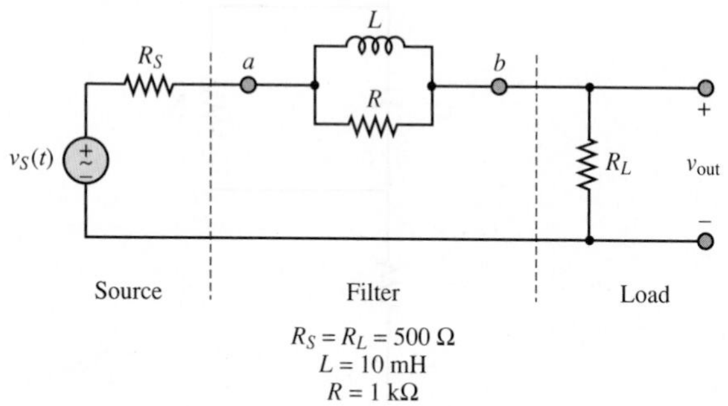

Figure P4.70

4.71 Find the Thévenin equivalent of the circuit shown in Figure P4.71 as seen by the load resistor.

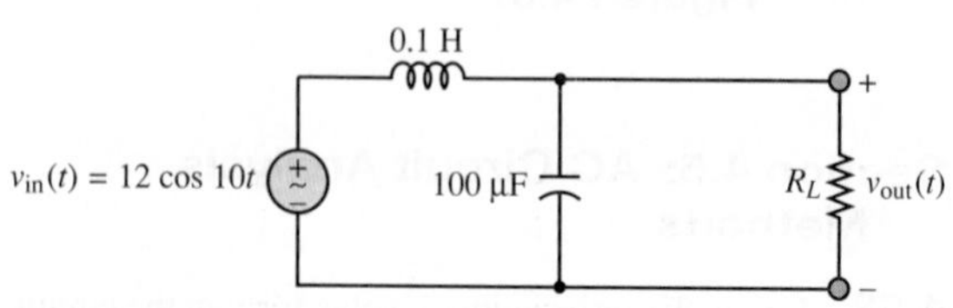

Figure P4.71

4.72 Solve for $i_L(t)$ in the circuit of Figure P4.72, using phasor techniques, if $v_S(t) = 2\cos 2t$, $R_1 = 4\ \Omega$, $R_2 = 4\ \Omega$, $L = 2$ H, and $C = \frac{1}{4}$ F.

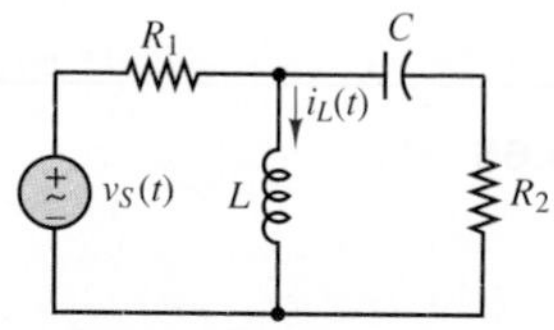

Figure P4.72

4.73 Using mesh current analysis, determine the currents $i_1(t)$ and $i_2(t)$ in the circuit shown in Figure P4.73.

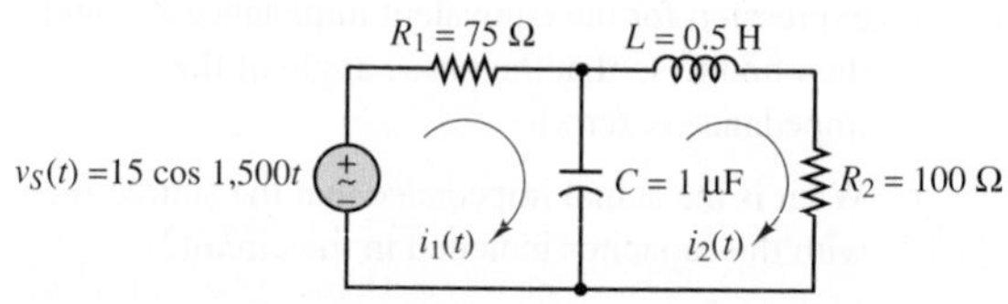

Figure P4.73

4.74 Using node voltage methods, determine the voltages $v_1(t)$ and $v_2(t)$ in the circuit shown in Figure P4.74.

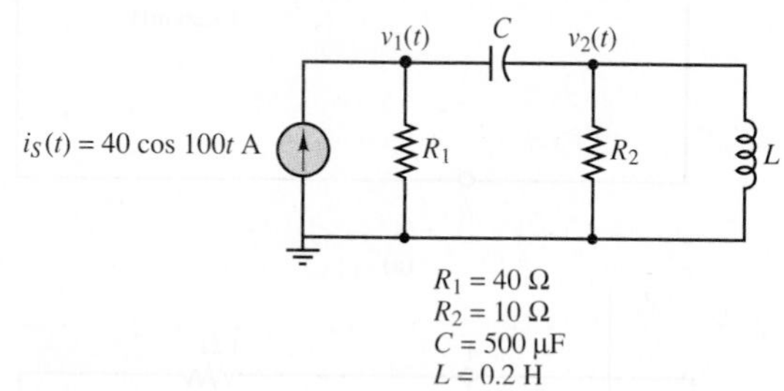

Figure P4.74

4.75 The circuit shown in Figure P4.75 is a Wheatstone bridge that will allow you to determine the reactance of an inductor or a capacitor. The circuit is adjusted by changing R_1 and R_2 until v_{ab} is zero.

a. Assuming that the circuit is balanced, that is, that $v_{ab} = 0$, determine X_4 in terms of the circuit elements.

b. If $C_3 = 4.7\ \mu$F, $L_3 = 0.098$ H, $R_1 = 100\ \Omega$, $R_2 = 1\ \Omega$, $v_S(t) = 24\sin(2{,}000t)$, and $v_{ab} = 0$, what is the reactance of the unknown circuit element? Is it a capacitor or an inductor? What is its value?

c. What frequency should be avoided by the source in this circuit, and why?

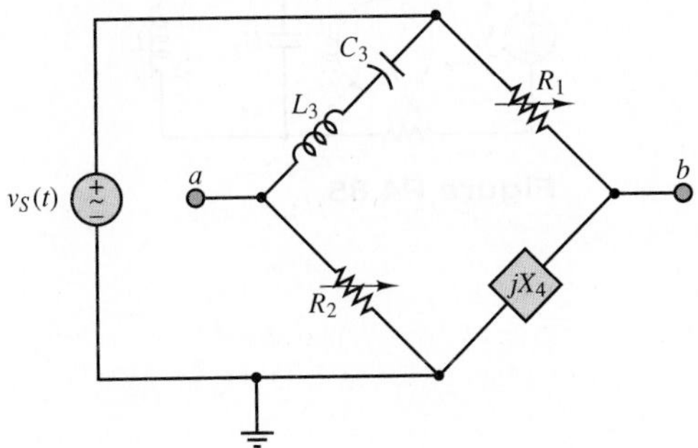

Figure P4.75

4.76 Compute the Thévenin impedance seen by resistor R_2 in Problem 4.72.

4.77 Compute the Thévenin voltage seen by the inductance L in Problem 4.74.

4.78 Find the Thévenin equivalent circuit as seen from terminals a-b for the circuit shown in Figure P4.77.

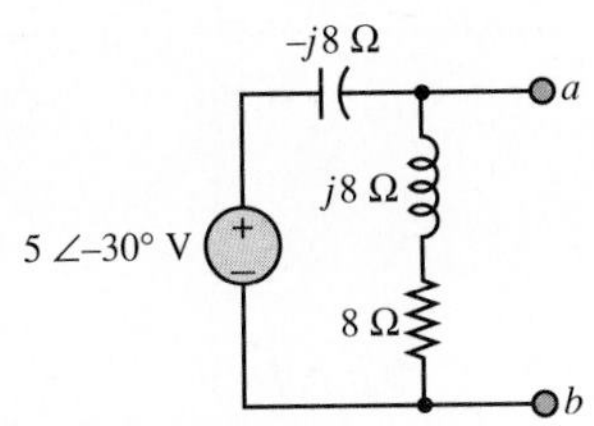

Figure P4.78

4.79 Compute the Thévenin voltage seen by resistor R_2 in Problem 4.72.

4.80 Find the Norton equivalent circuit seen by resistor R_2 in Problem 4.72.

4.81 Write the two loop equations required to solve for the loop currents in the circuit of Figure P4.81 in

a. Integral-differential form

b. Phasor form

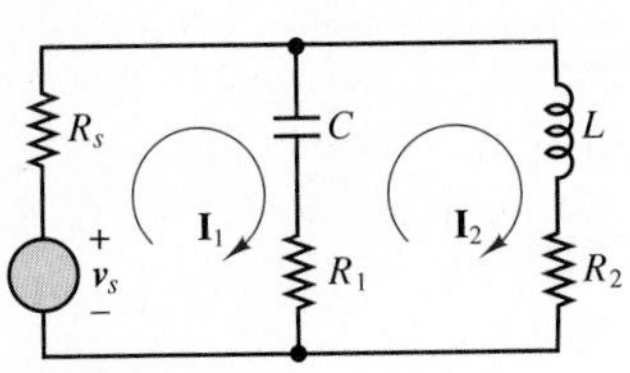

Figure P4.81

4.82 Write the node equations required to solve for all voltages and currents in the circuit of Figure P4.81. Assume all impedances and the two source voltages are known.

4.83 In the circuit shown in Figure P4.83,

$$v_{s1} = 450 \cos \omega t \quad \text{V}$$
$$v_{s2} = 450 \cos \omega t \quad \text{V}$$

A solution of the circuit with the ground at node e as shown gives

$$\mathbf{V}_a = 450\angle 0 \text{ V} \qquad \mathbf{V}_b = 440\angle \frac{\pi}{6} \text{ V}$$
$$\mathbf{V}_c = 420\angle -3.49 \text{ V}$$
$$\mathbf{V}_{bc} = 779.5\angle 0.098 \text{ V} \qquad \mathbf{V}_{cd} = 153.9\angle 1.2 \text{ V}$$
$$\mathbf{V}_{ba} = 230.6\angle 1.875 \text{ V}$$

If the ground is now moved from node e to node d, determine $\mathbf{V}_b$ and $\mathbf{V}_{bc}$.

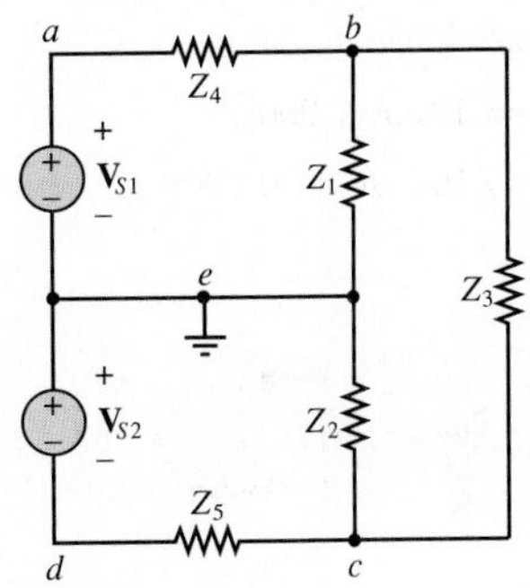

Figure P4.83

4.84 Determine V_o in the circuit of Figure P4.84 if

$$v_i = 4 \cos\left(1{,}000t + \frac{\pi}{6}\right) \quad \text{V}$$
$$L = 60 \text{ mH} \qquad C = 12.5 \ \mu\text{F}$$
$$R_L = 120 \ \Omega$$

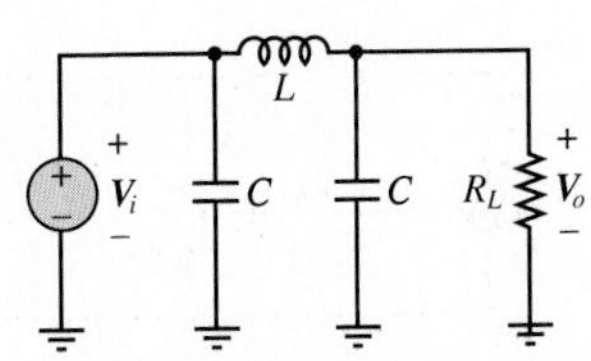

Figure P4.84

4.85 The mesh currents and source voltages in the circuit shown in Figure P4.85 are

$$i_1(t) = 3.127\cos(\omega t - 0.825) \quad \text{A}$$
$$i_2(t) = 3.914\cos(\omega t - 1.78) \quad \text{A}$$
$$i_3(t) = 1.900\cos(\omega t + 0.655) \quad \text{A}$$
$$v_{S1}(t) = 130.0\cos(\omega t + 0.176) \quad \text{V}$$
$$v_{S2}(t) = 130.0\cos(\omega t - 0.436) \quad \text{V}$$

where $\omega = 377.0$ rad/s. Determine one of the following: L_1, C_2, R_3, or L_3.

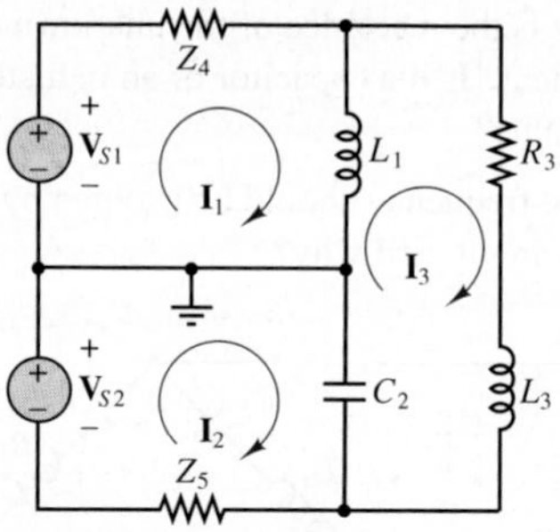

Figure P4.85

CHAPTER

5

TRANSIENT ANALYSIS

The aim of Chapter 5 is to develop a systematic methodology for the solution of first- and second-order circuits excited by switched DC sources. The chapter presents a unified approach to determining the transient response of linear *RC, RL,* and *RLC* circuits; and although the methods presented in the chapter focus only on first- and second-order circuits, the approach to the transient solution is quite general. Throughout the chapter, practical applications of first- and second-order circuits are presented, and numerous analogies are introduced to emphasize the general nature of the solution methods and their applicability to a wide range of physical systems, including hydraulics, mechanical systems, and thermal systems.

Learning Objectives

1. Understand the meaning of transients. *Section 1.*
2. Write differential equations for circuits containing inductors and capacitors. *Section 2.*
3. Determine the DC steady-state solution of circuits containing inductors and capacitors. *Section 3.*
4. Write the differential equation of first-order circuits in standard form, and determine the complete solution of first-order circuits excited by switched DC sources. *Section 4.*
5. Write the differential equation of second-order circuits in standard form, and determine the complete solution of second-order circuits excited by switched DC sources. *Section 5.*
6. Understand analogies between electric circuits and hydraulic, thermal, and mechanical systems.

5.1 TRANSIENT ANALYSIS

The graphs of Figure 5.1 illustrate the result of the sudden appearance of a voltage across a hypothetical load [a DC voltage in Figure 5.1(a), an AC voltage in Figure 5.1(b)]. In the figure, the source voltage is turned on at time $t = 0.2$ s. The voltage waveforms of Figure 5.1 can be subdivided into three regions: a *steady-state* region for $0 \le t \le 0.2$ s; a *transient* region for $0.2 \le t \le 2$ s (approximately); and a new steady-state region for $t > 2$ s, where the voltage reaches a steady DC or AC condition. The objective of **transient analysis** is to describe the behavior of a voltage or a current during the transition between two distinct steady-state conditions.

LO1

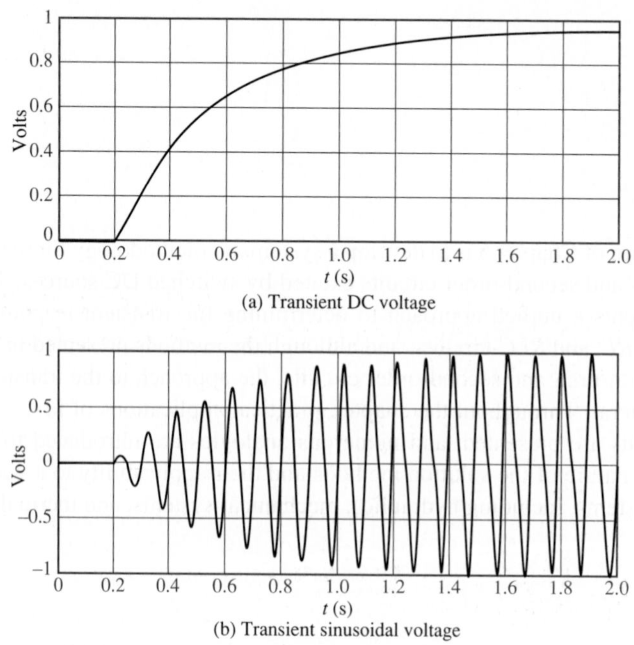

Figure 5.1 Examples of transient response

The material presented in the remainder of this chapter will provide the tools necessary to describe the *transient response* of circuits containing resistors, inductors, and capacitors. A general example of the type of circuit that is discussed in this section is shown in Figure 5.2. The switch indicates that we turn the battery power on at time $t = 0$. Transient behavior may be expected whenever a source of electrical energy is switched on or off, whether it be AC or DC. A typical example of the transient response to a switched DC voltage is what occurs when the ignition circuits in an automobile are turned on, so that a 12-V battery is suddenly connected to a large number of electric circuits. The degree of complexity in transient analysis depends on the number of energy storage elements in the circuit; the analysis can become quite involved for high-order circuits. In this chapter, we analyze only first- and second-order circuits, that is, circuits containing one or two energy storage elements, respectively. In electrical engineering practice, we typically resort to computer-aided analysis for higher-order circuits.

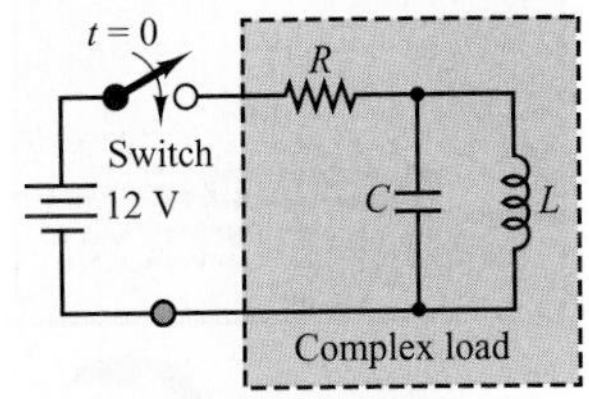

Figure 5.2 Circuit with switched DC excitation

A convenient starting point in approaching the transient response of electric circuits is to consider the general model shown in Figure 5.3, where the circuits in the box consist of a combination of resistors connected to a *single energy storage element,* either an inductor or a capacitor. Regardless of how many resistors the circuit contains, it is a **first-order circuit.** In general, the response of a first-order circuit to a switched DC source will appear in one of the two forms shown in Figure 5.4, which represent, in order, a **decaying exponential** and a **rising exponential** waveform. In the next sections, we will systematically analyze these responses by recognizing that they are exponential and can be computed very easily once we have the proper form of the differential equation describing the circuit.

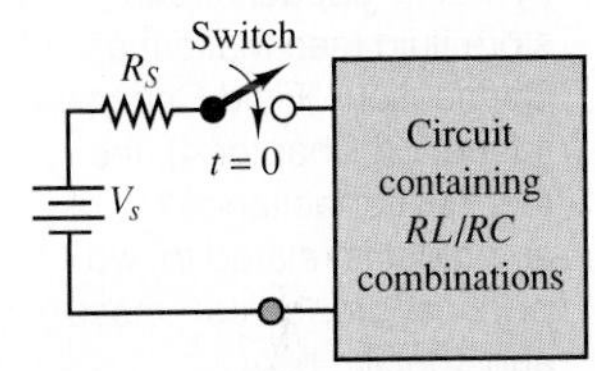

Figure 5.3 A general model of the transient analysis problem

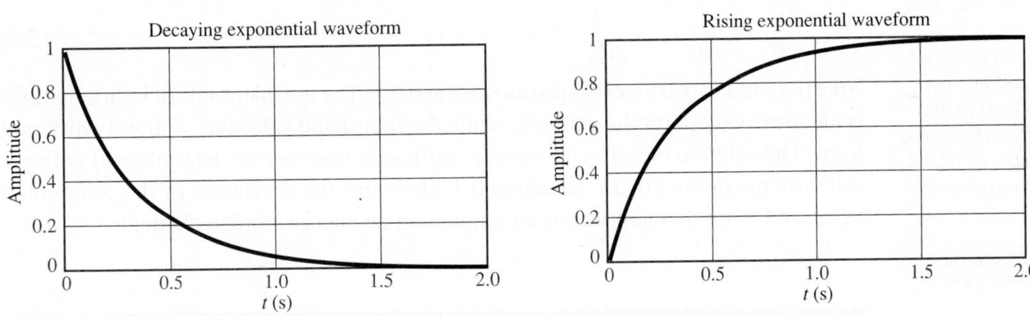

Figure 5.4 Decaying and rising exponential responses

5.2 WRITING DIFFERENTIAL EQUATIONS FOR CIRCUITS CONTAINING INDUCTORS AND CAPACITORS

A circuit containing energy-storage elements is described by a differential equation. The differential equation describing the series *RC* circuit shown is

$$\frac{di_C}{dt} + \frac{1}{RC} i_C = \frac{dv_S}{dt}$$

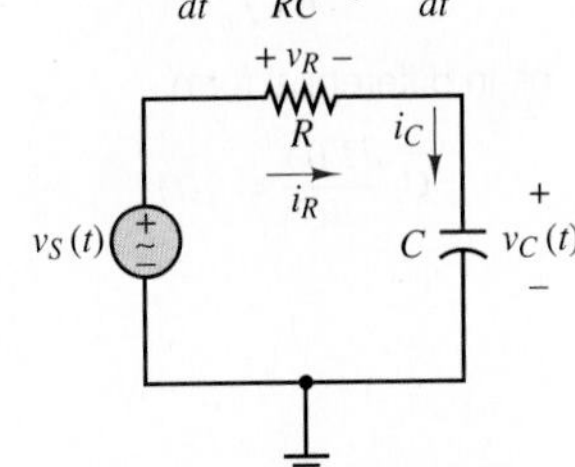

Figure 5.5 Circuit containing energy storage element

The major difference between the analysis of the resistive circuits studied in Chapters 2 and 3 and the circuits we explore in the remainder of this chapter is that now the equations that result from applying Kirchhoff's laws are differential equations, as opposed to the algebraic equations obtained in solving resistive circuits. Consider, for example, the circuit of Figure 5.5, which consists of the series connection of a voltage source, a resistor, and a capacitor. Applying KVL around the loop, we may obtain the following equation:

$$v_S(t) - v_R(t) - v_C(t) = 0 \tag{5.1}$$

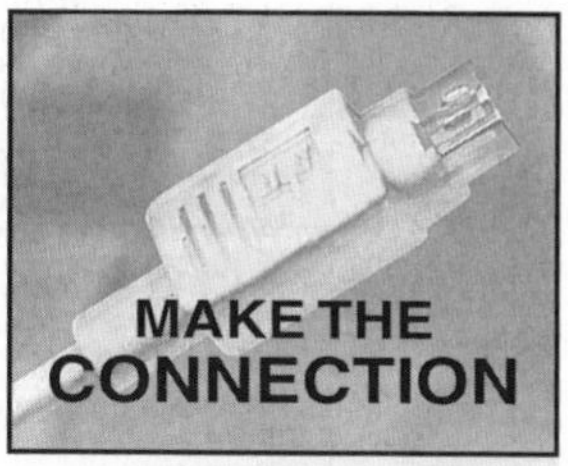

Thermal Capacitance

LO6

Just as an electrical capacitor can store energy and a hydraulic capacitor can store fluid (see Make the Connection, "Fluid Capacitance," in Chapter 4), the thermal capacitance C_t, of an object is related to two physical properties—mass and specific heat:

$$C_t = mc; \; m = \text{mass [kg]}$$

$$c = \text{specific heat [J/°C-kg]}$$

Physically, thermal capacitance is related to the ability of a mass to store heat, and describes how much the temperature of the mass will rise for a given addition of heat. If we add heat at the rate q J/s for time Δt and the resulting temperature rise is ΔT, then we can define the thermal capacitance to be

$$C_t = \frac{\text{heat added}}{\text{temperature rise}} = \frac{q\,\Delta t}{\Delta T}$$

If the temperature rises from value T_0 at time t_0 to T_1 at time t_1, then we can write

$$T_1 - T_0 = \frac{1}{C_t}\int_{t_0}^{t_1} q(t)\,dt$$

or, in differential form,

$$C_t \frac{dT(t)}{dt} = q(t)$$

Observing that $i_R = i_C$, we may combine equation 5.1 with the defining equation for the capacitor (equation 4.6) to obtain

$$v_S(t) - Ri_C(t) - \frac{1}{C}\int_{-\infty}^{t} i_C\,dt' = 0 \tag{5.2}$$

Equation 5.2 is an integral equation, which may be converted to the more familiar form of a differential equation by differentiating both sides of the equation and recalling that

$$\frac{d}{dt}\left[\int_{-\infty}^{t} i_C(t')\,dt'\right] = i_C(t) \tag{5.3}$$

to obtain the differential equation

$$\frac{di_C}{dt} + \frac{1}{RC}i_C = \frac{1}{R}\frac{dv_S}{dt} \tag{5.4}$$

where the argument t has been dropped for ease of notation.

Observe that in equation 5.4, the independent variable is the series current flowing in the circuit, and that this is not the only equation that describes the series RC circuit. If, instead of applying KVL, for example, we had applied KCL at the node connecting the resistor to the capacitor, we would have obtained the following relationship:

$$i_R = \frac{v_S - v_C}{R} = i_C = C\frac{dv_C}{dt} \tag{5.5}$$

or

$$\frac{dv_C}{dt} + \frac{1}{RC}v_C = \frac{1}{RC}v_S \tag{5.6}$$

Note the similarity between equations 5.4 and 5.6. The left-hand side of both equations is identical, except for the variable, while the right-hand side takes a slightly different form. The solution of either equation is sufficient, however, to determine all voltages and currents in the circuit. Example 5.1 illustrates the derivation of the differential equation for another simple circuit containing an energy storage element.

EXAMPLE 5.1 Writing the Differential Equation of an *RL* Circuit

LO2

Problem

Derive the differential equation of the circuit shown in Figure 5.6.

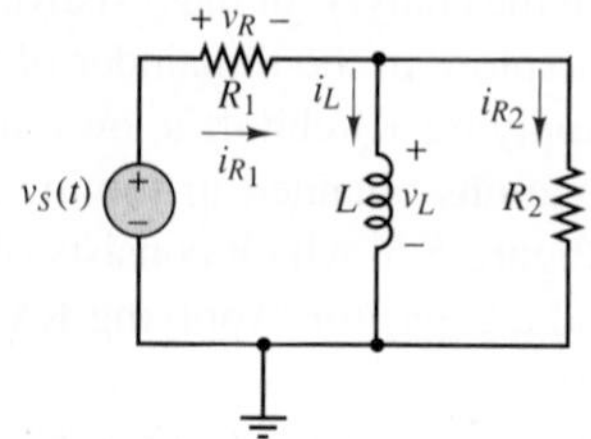

Figure 5.6

Solution

Known Quantities: $R_1 = 10\ \Omega$; $R_2 = 5\ \Omega$; $L = 0.4$ H.

Find: The differential equation in $i_L(t)$.

Assumptions: None.

Analysis: Apply KCL at the top node (node analysis) to write the circuit equation. Note that the top node voltage is the inductor voltage v_L.

$$i_{R1} - i_L - i_{R2} = 0$$

$$\frac{v_S - v_L}{R_1} - i_L - \frac{v_L}{R_2} = 0$$

Next, use the definition of inductor voltage to eliminate the variable v_L from the nodal equation.

$$\frac{v_S}{R_1} - \frac{L}{R_1}\frac{di_L}{dt} - i_L - \frac{L}{R_2}\frac{di_L}{dt} = 0$$

$$\frac{di_L}{dt} + \frac{R_1 R_2}{L(R_1 + R_2)} i_L = \frac{R_2}{L(R_1 + R_2)} v_S$$

Substituting numerical values, we obtain the following differential equation:

$$\frac{di_L}{dt} + 8.33 i_L = 0.833 v_S$$

Comments: Deriving differential equations for dynamic circuits requires the same basic circuit analysis skills that were developed in Chapter 3. The only difference is the introduction of integral or derivative terms originating from the defining relations for capacitors and inductors.

CHECK YOUR UNDERSTANDING

Write the differential equation for each of the circuits shown below.

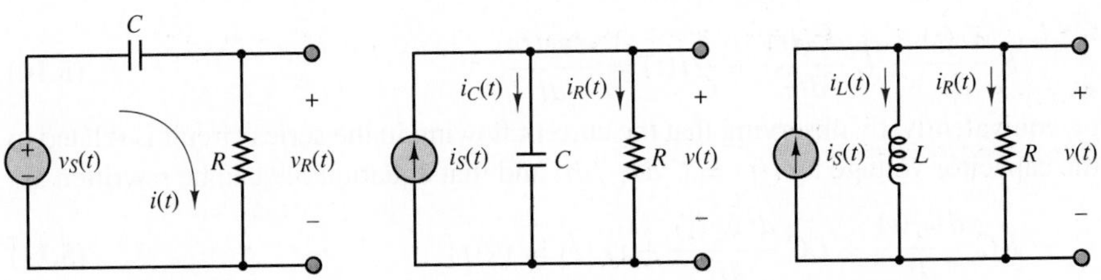

(a) (b) (c)

Answer: (a) $RC\frac{dv_C(t)}{dt} + v_C(t) = v_S(t)$; (b) $RC\frac{dv(t)}{dt} + v(t) = Ri_S(t)$;
(c) $\frac{L}{R}\frac{di_L(t)}{dt} + i_L(t) = i_S(t)$

MAKE THE CONNECTION

Thermal System Dynamics

LO6

To describe the dynamics of a thermal system, we write a differential equation based on energy balance. The difference between the heat added to the mass by an external source and the heat leaving the same mass (by convection or conduction) must be equal to the heat stored in the mass:

$$q_{in} - q_{out} = q_{stored}$$

An object is internally heated at the rate q_{in}, in ambient temperature $T = T_a$; the thermal capacitance and thermal resistance are C_t and R_t. From energy balance:

$$q_{in}(t) - \frac{T(t) - T_a}{R_t} = C_t \frac{dT(t)}{dt}$$

$$R_t C_t \frac{dT(t)}{dt} + T(t) = R_t q_{in}(t) + T_a$$

$$\tau_t = R_t C_t \qquad K_{St} = R_t$$

This first-order system is identical in its form to an electric *RC* circuit, as shown below.

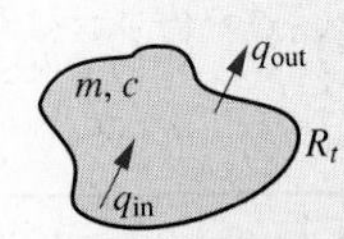

Thermal system

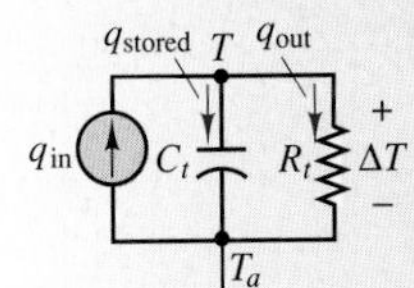

Equivalent electrical circuit

We can generalize the results presented in the preceding pages by observing that any circuit containing a single energy storage element can be described by a differential equation of the form

$$a_1 \frac{dx(t)}{dt} + a_0 x(t) = b_0 f(t) \quad \textbf{(5.7)}$$

where $x(t)$ represents the capacitor voltage in the circuit of Figure 5.5 and the inductor current in the circuit of Figure 5.6, and where the constants a_0, a_1, and b_0 consist of combinations of circuit element parameters. Equation 5.7 is a **first-order linear ordinary differential equation** with constant coefficients. The equation is said to be of first order because the highest derivative present is of first order; it is said to be ordinary because the derivative that appears in it is an ordinary derivative (in contrast to a *partial* derivative;) and the coefficients of the differential equation are constant in that they depend only on the values of resistors, capacitors, or inductors in the circuit, and not, for example, on time, voltage, or current.

Equation 5.7 can be rewritten as

$$\frac{a_1}{a_0} \frac{dx(t)}{dt} + x(t) = \frac{b_0}{a_0} f(t)$$

LO2

or **(5.8)**

$$\tau \frac{dx(t)}{dt} + x(t) = K_S f(t) \qquad \text{First-order system equation}$$

where the constants $\tau = a_1/a_0$ and $K_S = b_0/a_0$ are termed the **time constant** and the **DC gain,** respectively. We shall return to this form when we derive the complete solution to this first-order differential equation.

Consider now a circuit that contains two energy storage elements, such as that shown in Figure 5.7. Application of KVL results in the following equation:

$$-Ri(t) - L\frac{di(t)}{dt} - \frac{1}{C}\int_{-\infty}^{t} i(t')\,dt' + v_S(t) = 0 \quad \textbf{(5.9)}$$

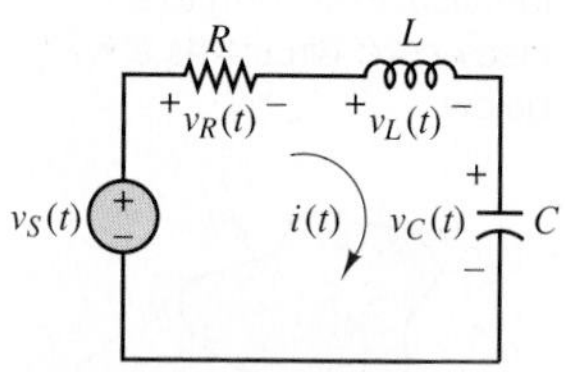

Figure 5.7 Second-order circuit

Equation 5.9 is called an *integrodifferential equation,* because it contains both an integral and a derivative. This equation can be converted to a differential equation by differentiating both sides, to obtain

$$R\frac{di(t)}{dt} + L\frac{d^2i(t)}{dt^2} + \frac{1}{C}i(t) = \frac{dv_S(t)}{dt} \quad \textbf{(5.10)}$$

or, equivalently, by observing that the current flowing in the series circuit is related to the capacitor voltage by $i(t) = C\,dv_C/dt$, and that equation 5.9 can be rewritten as

$$RC\frac{dv_C(t)}{dt} + LC\frac{d^2v_C(t)}{dt^2} + v_C(t) = v_S(t) \quad \textbf{(5.11)}$$

Note that although different variables appear in the preceding differential equations, both equations 5.10 and 5.11 can be rearranged to appear in the same general form, as follows:

$$a_2\frac{d^2x(t)}{dt^2} + a_1\frac{dx(t)}{dt} + a_0x(t) = b_0(t) \quad \textbf{(5.12)}$$

where the general variable $x(t)$ represents either the series current of the circuit of Figure 5.7 or the capacitor voltage. By analogy with equation 5.8, we call equation 5.12

a **second-order linear ordinary differential equation** with constant coefficients. Equation 5.12 can be rewritten as

$$\frac{a_2}{a_0}\frac{d^2x(t)}{dt^2} + \frac{a_1}{a_0}\frac{dx(t)}{dt} + x(t) = \frac{b_0}{a_0}f(t)$$

or **(5.13)**

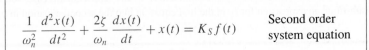

$$\frac{1}{\omega_n^2}\frac{d^2x(t)}{dt^2} + \frac{2\zeta}{\omega_n}\frac{dx(t)}{dt} + x(t) = K_S f(t) \qquad \text{Second order system equation}$$

where the constants $\omega_n = \sqrt{a_0/a_2}$, $\zeta = (a_1/2)\sqrt{1/a_0 a_2}$, and $K_S = b_0/a_0$ are termed the **natural frequency,** the **damping ratio,** and the **DC gain,** respectively. We shall return to this form when we derive the complete solution to this second-order differential equation.

As the number of energy storage elements in a circuit increases, one can therefore expect that higher-order differential equations will result. Computer aids are often employed to solve differential equations of higher order; some of these software packages are specifically targeted at the solution of the equations that result from the analysis of electric circuits (e.g., *Electronics Workbench*™).

EXAMPLE 5.2 Writing the Differential Equation of an *RLC* Circuit

Problem

Derive the differential equation of the circuit shown in Figure 5.8.

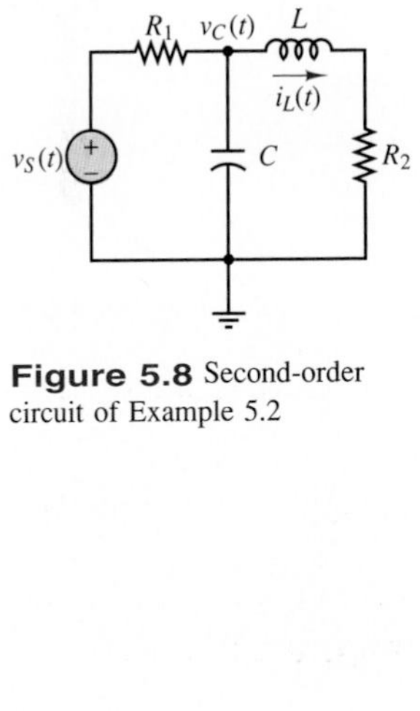

Figure 5.8 Second-order circuit of Example 5.2

Solution

Known Quantities: $R_1 = 10\ \text{k}\Omega$; $R_2 = 50\ \Omega$; $L = 10$ mH; $C = 0.1\ \mu$F.

Find: The differential equation in $i_L(t)$.

Assumptions: None.

Analysis: Apply KCL at the top node (node analysis) to write the first circuit equation. Note that the top node voltage is the capacitor voltage v_C.

$$\frac{v_S - v_C}{R_1} - C\frac{dv_C}{dt} - i_L = 0$$

Now, we need a second equation to complete the description of the circuit, since the circuit contains two energy storage elements (second-order circuit). We can obtain a second equation in the capacitor voltage v_C by applying KVL to the mesh on the right-hand side:

$$v_C - L\frac{di_L}{dt} - R_2 i_L = 0$$

$$v_C = L\frac{di_L}{dt} + R_2 i_L$$

Next, we can substitute the above expression for v_C into the first equation, to obtain a *second-order* differential equation, shown below.

$$\frac{v_s}{R_1} - \frac{L}{R_1}\frac{di_L}{dt} - \frac{R_2}{R_1}i_L - C\frac{d}{dt}\left(L\frac{di_L}{dt} + R_2 i_L\right) - i_L = 0$$

Rearranging the equation, we can obtain the standard form similar to that of equation 5.12:

$$R_1CL\frac{d^2i_L}{dt^2} + (R_1R_2C + L)\frac{di_L}{dt} + (R_1 + R_2)i_L = v_S$$

Comments: Note that we could have derived an analogous equation by using the capacitor voltage as an independent variable; either energy storage variable is an acceptable choice. You might wish to try obtaining a second-order equation in v_C as an exercise. In this case, you would want to substitute an expression for i_L in the first equation into the second equation in v_C.

CHECK YOUR UNDERSTANDING

Derive a differential equation in the variable $v_C(t)$ for the circuit of Example 5.2.

Answer: $LC\frac{d^2v_C(t)}{dt^2} + \left(\frac{L}{R_1} + R_2C\right)\frac{dv_C(t)}{dt} + \left(\frac{R_2}{R_1} + 1\right)v_C(t) = \frac{R_2}{R_1}v_S(t) + \frac{L}{R_1}\frac{dv_S(t)}{dt}$

5.3 DC STEADY-STATE SOLUTION OF CIRCUITS CONTAINING INDUCTORS AND CAPACITORS—INITIAL AND FINAL CONDITIONS

This section deals with the DC steady-state solution of the differential equations presented in Section 5.2. In particular, we illustrate simple methods for deriving the **initial** and **final conditions** of circuits that are connected to a switched DC source. These conditions will be very helpful in obtaining the complete transient solution. Further, we also show how to compute the initial conditions that are needed to solve the circuit differential equation, using the principle of continuity of inductor voltage and current.

DC Steady-State Solution

The term *DC steady state* refers to circuits that have been connected to a DC (voltage or current) source for a very long time, such that it is reasonable to assume that all voltages and currents in the circuits have become constant. If all variables are constant, the steady-state solution of the differential equation can be found very easily, since all derivatives must be equal to zero. For example, consider the differential equation derived in Example 5.1 (see Figure 5.6):

$$\frac{di_L(t)}{dt} + \frac{R_1R_2}{L(R_1 + R_2)}i_L(t) = \frac{R_2}{L(R_1 + R_2)}v_S(t) \quad \textbf{(5.14)}$$

Rewriting this equation in the general form of equation 5.8, we obtain

$$\frac{L(R_1 + R_2)}{R_1R_2}\frac{di_L(t)}{dt} + i_L(t) = \frac{1}{R_1}v_S(t)$$

or **(5.15)**

$$\tau\frac{di_L(t)}{dt} + i_L(t) = K_S v_S(t)$$

where

$$\tau = \frac{L(R_1 + R_2)}{R_1 R_2} \qquad \text{and} \qquad K_S = \frac{1}{R_1}$$

With v_S equal to a constant (DC) voltage, after a suitably long time the current in the circuit is a constant, and the derivative term goes to zero:

$$\frac{L(R_1 + R_2)}{R_1 R_2}\frac{di_L(t)}{dt} + i_L(t) = \frac{1}{R_1}v_S(t)$$

and

$$i_L = \frac{1}{R_1}v_S \qquad \text{as } t \to \infty \tag{5.16}$$

or

$$i_L = K_S v_S$$

Note that the steady-state solution is found very easily, and it is determined by the constant K_S, which we called the *DC gain* of the circuit. You can see that the general form of the first-order differential equation (equation 5.8) is very useful in finding the steady-state solution.

Let us attempt the same method for a second-order circuit, considering the solution of Example 5.2 (see Figure 5.8):

$$R_1 CL\frac{d^2 i_L(t)}{dt^2} + (R_1 R_2 C + L)\frac{di_L(t)}{dt} + (R_1 + R_2)i_L(t) = v_S(t)$$

or **(5.17)**

$$\frac{R_1 CL}{R_1 + R_2}\frac{d^2 i_L(t)}{dt^2} + \frac{R_1 R_2 C + L}{R_1 + R_2}\frac{di_L(t)}{dt} + i_L(t) = \frac{1}{R_1 + R_2}v_S(t)$$

We can express this differential equation in the general form of equation 5.13, to find that

$$\frac{1}{\omega_n^2}\frac{d^2 i_L(t)}{dt^2} + \frac{2\zeta}{\omega_n}\frac{di_L(t)}{dt} + i_L(t) = K_S v_S(t)$$

$$\frac{1}{\omega_n^2} = \frac{R_1 CL}{R_1 + R_2} \qquad \frac{2\zeta}{\omega_n} = \frac{R_1 R_2 C + L}{R_1 + R_2} \qquad K_S = \frac{1}{R_1 + R_2} \tag{5.18}$$

and that, one more time, the steady-state solution, when the derivatives are equal to zero, is

$$\frac{1}{\omega_n^2}\frac{d^2 i_L(t)}{dt^2} + \frac{2\zeta}{\omega_n}\frac{di_L(t)}{dt} + i_L(t) = K_S v_S(t)$$

and **(5.19)**

$$i_L = K_S v_S \qquad \text{as } t \to \infty$$

A different way to arrive at the same result is to start from the defining equation for the capacitor and inductor and see what happens as $t \to \infty$:

$$i_C(t) = C\frac{dv_C(t)}{dt} \qquad \text{defining equation for capacitor}$$

$$i_C(t) \to 0 \qquad \text{as } t \to \infty \qquad \text{steady-state capacitor current} \tag{5.20}$$

and

$$
\begin{aligned}
v_L(t) &= L\,\frac{di_L(t)}{dt} && \text{defining equation for inductor} \\
v_L(t) &\to 0 \quad \text{as } t \to \infty && \text{steady-state inductor voltage}
\end{aligned}
\tag{5.21}
$$

Thus, capacitor currents and the inductor voltages become zero in the DC steady state. From a circuit analysis standpoint, this means that we can very easily apply circuit analysis methods from Chapters 2 and 3 to determine the steady-state solution of any circuit containing capacitors and inductors if we observe that the circuit element for which the current is always zero is the *open circuit,* and the circuit element for which the voltage is always zero is the *short circuit*. Thus we can make the following observation:

LO3

At DC steady state, all capacitors behave as open circuits and all inductors behave as short circuits.

Prior to presenting some examples, we make one last important comment. The DC steady-state condition is usually encountered in one of two cases: *before a switch is first activated,* in which case we call the DC steady-state solution the **initial condition,** and *a long time after a switch has been activated,* in which case we call the DC steady-state solution the **final condition.** We now introduce the notation $x(\infty)$ to denote the value of the variable $x(t)$ as $t \to \infty$ (final condition) and the notation $x(0)$ to denote the value of the variable $x(t)$ at $t = 0$, that is, just before the switch is activated (initial condition). These ideas are illustrated in Examples 5.3 and 5.4.

LO3

EXAMPLE 5.3 Initial and Final Conditions

Problem

Determine the capacitor voltage in the circuit of Figure 5.9(a) a long time after the switch has been closed.

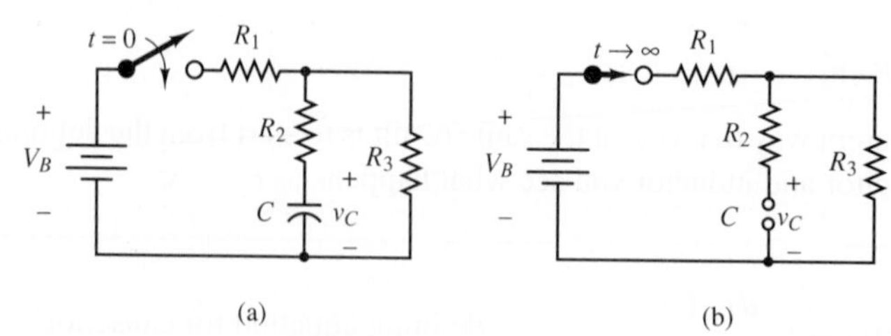

Figure 5.9 (a) Circuit for Example 5.3; (b) same circuit a long time after the switch is closed

Solution

Known Quantities: The values of the circuit elements are $R_1 = 100\ \Omega$; $R_2 = 75\ \Omega$; $R_3 = 250\ \Omega$; $C = 1\ \mu\text{F}$; $V_B = 12$ V.

Analysis: After the switch has been closed for a long time, we treat the capacitor as an open circuit, as shown in Figure 5.9(b). Now the problem is a simple DC circuit analysis problem. With the capacitor as an open circuit, no current flows through resistor R_2, and therefore we have a simple voltage divider. Let V_3 be the voltage across resistor V_3; then

$$V_3(\infty) = \frac{R_3}{R_1 + R_3} V_B = \frac{250}{350}(12) = 8.57\ \text{V}$$

To determine the capacitor voltage, we observe that the voltage across each of the two parallel branches must be the same, that is, $V_3(\infty) = v_c(\infty) + V_2(\infty)$ where $V_2(\infty)$ is the steady-state value of the voltage across resistor R_2. But $V_2(\infty)$ is zero, since no current flows through the resistor, and therefore

$$v_C(\infty) = V_3(\infty) = 8.57\ \text{V}$$

Comments: The voltage $v_C(\infty)$ is the final condition of the circuit of Figure 5.9(a).

CHECK YOUR UNDERSTANDING

Now suppose that the switch is opened again. What will be the capacitor voltage after a long time? Why?

Answer: $v_C(\infty) = 0$ V. The capacitor will discharge through R_2 and R_3.

EXAMPLE 5.4 Initial and Final Conditions

LO3

Problem

Determine the inductor current in the circuit of Figure 5.10(a) just before the switch is opened.

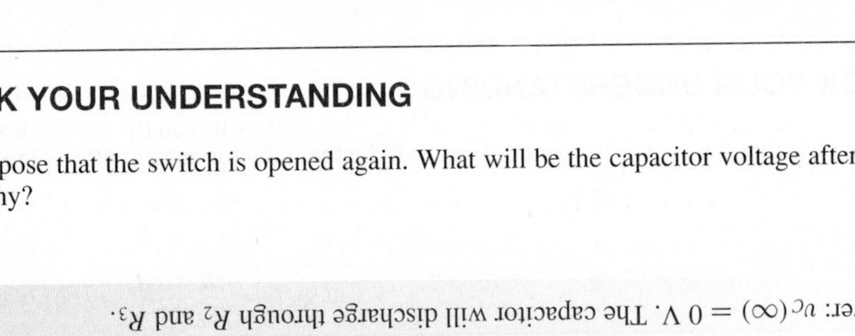

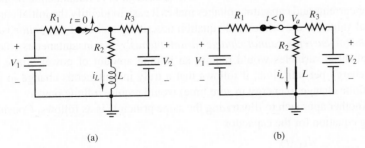

Figure 5.10 (a) Circuit for Example 5.4; (b) same circuit a long time before the switch is opened

Solution

Known Quantities: The values of the circuit elements are $R_1 = 1\ \text{k}\Omega$; $R_2 = 5\ \text{k}\Omega$; $R_3 = 3.33\ \text{k}\Omega$; $L = 0.1$ H; $V_1 = 12$ V; $V_2 = 4$ V.

Analysis: Before opening, the switch has been closed for a long time. Thus, we have a steady-state condition, and we treat the inductor as a short circuit, as shown in Figure 5.10(b). Now it is a simple DC circuit analysis problem that is best approached using node analysis:

$$\frac{V_1 - V_a}{R_1} - \frac{V_a}{R_2} + \frac{V_2 - V_a}{R_3} = 0$$

$$V_a = \left(\frac{1}{R_1} + \frac{1}{R_2} + \frac{1}{R_3}\right)^{-1} \left(\frac{V_1}{R_1} + \frac{V_2}{R_3}\right) = 8.79\ \text{V}$$

To determine the inductor current, we observe that

$$i_L(0) = \frac{V_a}{R_2} = \frac{8.79}{5{,}000} = 1.758\ \text{mA}$$

Comments: The current $i_L(0)$ is the initial condition of the circuit of Figure 5.10(a).

CHECK YOUR UNDERSTANDING

Now suppose that the switch is opened. What will be the inductor current after a long time?

Answer: $i_L(\infty) = \dfrac{V_2}{R_2 + R_3} = 0.48$ mA

Continuity of Inductor Currents and Capacitor Voltages, and Initial Conditions

As has already been stated, the primary variables employed in the analysis of circuits containing energy storage elements are *capacitor voltages* and *inductor currents*. This choice stems from the fact that the energy storage process in capacitors and inductors is closely related to these respective variables. The amount of charge stored in a capacitor is directly related to the voltage present across the capacitor, while the energy stored in an inductor is related to the current flowing through it. A fundamental property of inductor currents and capacitor voltages makes it easy to identify the initial condition and final value for the differential equation describing a circuit: *Capacitor voltages and inductor currents cannot change instantaneously*. An instantaneous change in either of these variables would require an infinite amount of power. Since power equals energy per unit time, it follows that a truly instantaneous change in energy (i.e., a finite change in energy in zero time) would require infinite power.

Another approach to illustrating the same principle is as follows. Consider the defining equation for the capacitor

$$i_C(t) = C\frac{dv_C(t)}{dt}$$

and assume that the capacitor voltage $v_C(t)$ can change instantaneously, say, from 0 to V volts, as shown in Figure 5.11. The value of dv_C/dt at $t = 0$ is simply the slope of the voltage $v_C(t)$ at $t = 0$. Since the slope is infinite at that point, because of the instantaneous transition, it would require an infinite amount of current for the voltage across a capacitor to change instantaneously. But this is equivalent to requiring an infinite amount of power, since power is the product of voltage and current. A similar argument holds if we assume a "step" change in inductor current from, say, 0 to I amperes: An infinite voltage would be required to cause an instantaneous change in inductor current. This simple fact is extremely useful in determining the response of a circuit. Its immediate consequence is that

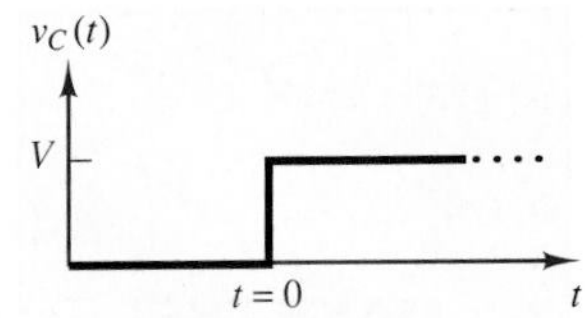

Figure 5.11 Abrupt change in capacitor voltage

The value of an inductor current or a capacitor voltage just prior to the closing (or opening) of a switch is equal to the value just after the switch has been closed (or opened). Formally,

$$v_C(0^+) = v_C(0^-) \tag{5.22}$$

$$i_L(0^+) = i_L(0^-) \tag{5.23}$$

where the notation 0^+ signifies "just after $t = 0$" and 0^- means "just before $t = 0$."

LO3

EXAMPLE 5.5 Continuity of Inductor Current

LO3

Problem

Find the initial condition and final value of the inductor current in the circuit of Figure 5.12.

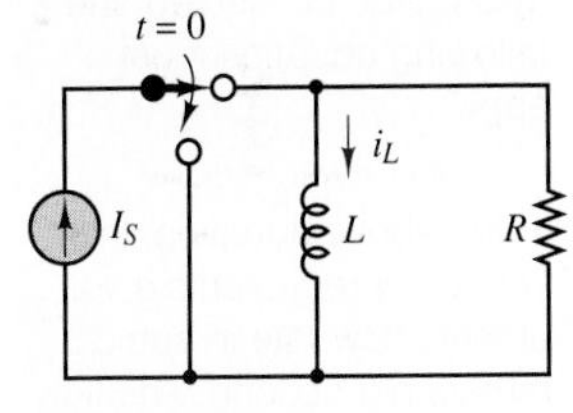

Figure 5.12

Solution

Known Quantities: Source current I_S; inductor and resistor values.

Find: Inductor current at $t = 0^+$ and as $t \to \infty$.

Schematics, Diagrams, Circuits, and Given Data: $I_S = 10$ mA.

Assumptions: The current source has been connected to the circuit for a very long time.

Analysis: At $t = 0^-$, since the current source has been connected to the circuit for a very long time, the inductor acts as a short circuit, and $i_L(0^-) = I_S$. Since all the current flows through the inductor, the voltage across the resistor must be zero. At $t = 0^+$, the switch opens and we can state that

$$i_L(0^+) = i_L(0^-) = I_S$$

because of the continuity of inductor current.

The circuit for $t \geq 0$ is shown in Figure 5.13, where the presence of the current $i_L(0^+)$ denotes the initial condition for the circuit. A qualitative sketch of the current as a function of time is also shown in Figure 5.13, indicating that the inductor current eventually becomes zero as $t \to \infty$.

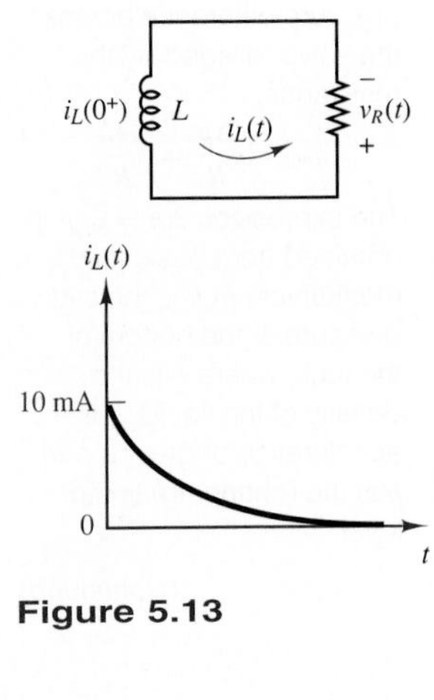

Figure 5.13

Comments: Note that the direction of the current in the circuit of Figure 5.13 is dictated by the initial condition, since the inductor current cannot change instantaneously. Thus, the current will flow counterclockwise, and the voltage across the resistor will therefore have the polarity shown in the figure.

CHECK YOUR UNDERSTANDING

The switch in the circuit of Figure 5.10 (Example 5.4) has been open for a long time, and it is closed again at $t = t_0$. Find the initial condition $i_L(t_0^+)$.

Answer: $i_L(0^+) = i_L(0^-) = 0.48$ mA

5.4 TRANSIENT RESPONSE OF FIRST-ORDER CIRCUITS

First-order systems occur very frequently in nature: Any system that has the ability to store energy in one form (potential or kinetic, but not both) and to dissipate this stored energy is a first order system. In an electric circuit, we recognize that any circuit containing a single energy storage element (an inductor or a capacitor) and a combination of voltage or current sources and resistors is a first-order circuit. We also encounter first-order systems in other domains. For example, a mechanical system characterized by mass and damping (e.g., sliding or viscous friction) but that does not display any elasticity or compliance is a first-order system. A fluid system displaying flow resistance and fluid mass storage (fluid capacitance) is also first-order; an example of a first-order hydraulic system is a liquid-filled tank with a valve (variable orifice). Thermal systems can also often be modeled as having first-order behavior: The ability to store and to dissipate heat leads to first-order differential equations. The heating and cooling of many physical objects can often be approximated in this fashion. The aim of this section is to help you develop a sound methodology for the solution of first-order circuits, and to help you make the connection with other domains and disciplines, so that someday you may apply these same ideas to other engineering systems.

Elements of the Transient Response

As explained in Section 5.1, the transient response of a circuit consists of three parts: (1) the steady-state response prior to the transient (in this chapter, we shall only consider transients caused by the switching on or off of a DC excitation); (2) the transient response, during which the circuit adjusts to the new excitation; and (3) the steady-state response following the end of the transient. The steps involved in computing the complete transient response of a first-order circuit excited by a switched DC source are outlined in the next Focus on Methodology box. You will observe that we have already explored each of the steps listed below, and that this methodology is very straightforward, provided that you correctly identify the proper segments of the response (before, during, and after the transient).

MAKE THE CONNECTION

Hydraulic Tank

LO6

The analogy between electric and hydraulic circuits illustrated in earlier chapters can be applied to the hydraulic tank shown in Figure 5.14. The tank is cylindrical with cross-sectional area A, and the liquid contained in the tank exits the tank through a valve, which is modeled by a fluid resistance R. Initially, the level, or head, of the liquid is h_0. The principle of conservation of mass can be applied to the liquid in the tank of Figure 5.14 to determine the rate at which the tank will empty. For mass to be conserved, the following equation must apply:

$$q_{in} - q_{out} = q_{stored}$$

In the above equation, the variable q represents a volumetric flow rate in cubic meters per second. The flow rate into the tank is zero in this particular case, and the flow rate out is given by the pressure difference across the valve, divided by the resistance:

$$q_{out} = \frac{\Delta p}{R} = \frac{\rho g h}{R}$$

The expression $\Delta p = \rho g h$ is obtained from basic fluid mechanics: $\rho g h$ is the static pressure at the bottom of the tank, where ρ is the density of the liquid, g is the acceleration of gravity, and h is the (changing) liquid level.

(Continued)

FOCUS ON METHODOLOGY

LO4

FIRST-ORDER TRANSIENT RESPONSE

1. Solve for the steady-state response of the circuit before the switch changes state ($t = 0^-$) and after the transient has died out ($t \to \infty$). We shall generally refer to these responses as $x(0^-)$ and $x(\infty)$.
2. Identify the initial condition for the circuit $x(0^+)$, using continuity of capacitor voltages and inductor currents [$v_C = v_C(0^-)$, $i_L(0^+) = i_L(0^-)$], as illustrated in Section 5.3.
3. Write the differential equation of the circuit for $t = 0^+$, that is, immediately after the switch has changed position. The variable $x(t)$ in the differential equation will be either a capacitor voltage $v_C(t)$ or an inductor current $i_L(t)$. It is helpful at this time to reduce the circuit to Thévenin or Norton equivalent form, with the energy storage element (capacitor or inductor) treated as the load for the Thévenin (Norton) equivalent circuit. Reduce this equation to standard form (equation 5.8).
4. Solve for the time constant of the circuit: $\tau = R_T C$ for capacitive circuits, $\tau = L/R_T$ for inductive circuits.
5. Write the complete solution for the circuit in the form

$$x(t) = x(\infty) + [x(0) - x(\infty)]e^{-t/\tau}$$

General Solution of First-Order Circuits

The methodology outlined in the preceding box is illustrated below for the general form of the first-order system equation (equation 5.8, repeated below for convenience),

$$\frac{a_1}{a_0}\frac{dx(t)}{dt} + x(t) = \frac{b_0}{a_0} f(t)$$

or **(5.24)**

$$\tau \frac{dx(t)}{dt} + x(t) = K_S f(t)$$

where the constants $\tau = a_1/a_0$ and $K_S = b_0/a_0$ are termed the **time constant** and the **DC gain,** respectively. Consider the special case where $f(t)$ is a DC forcing function, switched on at time $t = 0$. Let the initial condition of the system be $x(t = 0) = x(0)$. Then we seek to solve the differential equation

$$\tau \frac{dx(t)}{dt} + x(t) = K_S F \qquad t \geq 0 \tag{5.25}$$

As you may recall from an earlier course in differential equations, this solution consists of two parts: the **natural response** (or **homogeneous solution**), with the forcing function set equal to zero, and the **forced response** (or **particular solution**), in which we consider the response to the forcing function. The **complete response** then

MAKE THE CONNECTION

(Concluded)

The flow rate stored is related to the rate of change of the fluid volume contained in the tank (the tank stores potential energy in the mass of the fluid):

$$q_{\text{stored}} = A\frac{dh}{dt}$$

Thus, we can describe the emptying of the tank by means of the first-order linear ordinary differential equation

$$0 - q_{\text{out}} = q_{\text{stored}}$$

$$\Rightarrow \quad -\frac{\rho g h}{R} = A\frac{dh}{dt}$$

$$\frac{RA}{\rho g}\frac{dh}{dt} + h = 0$$

$$\Rightarrow \quad \tau\frac{dh}{dt} + h = 0$$

$$\tau = \frac{RA}{\rho g}$$

We know from the content of the present section that the solution of the first-order equation with zero input and initial condition h_0 is

$$h(t) = h_0 e^{-t/\tau}$$

Thus, the tank will empty exponentially, with time constant determined by the fluid properties, that is, by the resistance of the valve and by the area of the tank.

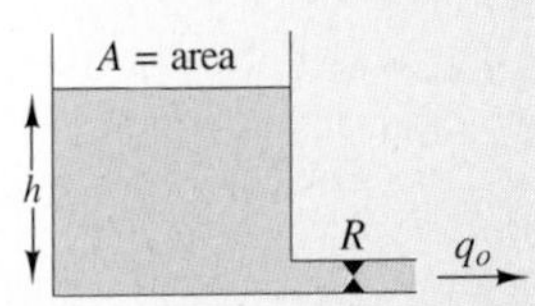

Figure 5.14 Analogy between electrical and fluid capacitance

consists of the sum of the natural and forced responses. Once the form of the complete response is known, the initial condition can be applied to obtain the final solution.

Natural Response

The natural response is found by setting the excitation equal to zero. Thus, we solve the equation

$$\tau \frac{dx_N(t)}{dt} + x_N(t) = 0$$

or **(5.26)**

$$\frac{dx_N(t)}{dt} = -\frac{x_N(t)}{\tau}$$

Where we use the notation $x_N(t)$ to denote the natural response. The solution of this equation is known to be of exponential form:

LO4

$$x_N(t) = \alpha e^{-t/\tau} \qquad \text{Natural response} \tag{5.27}$$

The constant α in equation 5.27 depends on the initial condition, and can only be evaluated once the complete response has been determined. If the system does not have an external forcing function, then the natural response is equal to the complete response, and the constant α is equal to the initial condition $\alpha = x(0)$. The Make the Connection sidebar "Hydraulic Tank" illustrates this case intuitively by considering a fluid tank emptying through an orifice. You can see that in the case of the tank, once the valve (with flow resistance R) is open, the liquid drains at an exponential rate, starting from the initial liquid level h_0, and with a time constant τ determined by the physical properties of the system. You should confirm the fact that if the resistance of the valve is decreased (i.e., liquid can drain out more easily), the time constant will become smaller, indicating that the tank will empty more quickly. The concept of time constant is further explored in Example 5.6.

LO4

EXAMPLE 5.6 First-Order Systems and Time Constants

Problem

Create a table illustrating the exponential decay of a voltage or current in a first-order circuit versus the number of time constants.

Solution

Known Quantities: Exponential decay equation.

Find: Amplitude of voltage or current $x(t)$ at $t = 0, \tau, 2\tau, 3\tau, 4\tau, 5\tau$.

Assumptions: The initial condition at $t = 0$ is $x(0) = X_0$.

Analysis: We know that the exponential decay of $x(t)$ is governed by the equation

$$x(t) = X_0 e^{-t/\tau}$$

Thus, we can create the following table for the ratio $x(t)/X_0 = e^{-n\tau/\tau}$, $n = 0, 1, 2, \ldots,$ at each value of t:

$\frac{x(t)}{X_0}$	n
1	0
0.3679	1
0.1353	2
0.0498	3
0.0183	4
0.0067	5

Figure 5.15 depicts the five points on the exponential decay curve.

Comments: Note that after three time constants, x has decayed to approximately 5 percent of the initial value, and after five time constants to less than 1 percent.

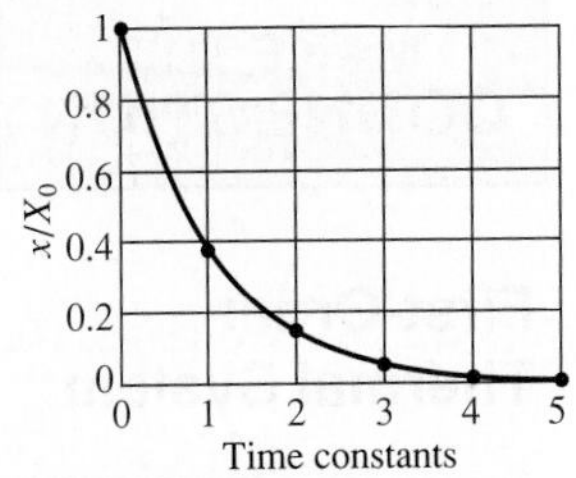

Figure 5.15 First-order exponential decay and time constants

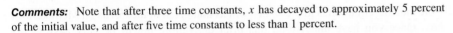

CHECK YOUR UNDERSTANDING

Find the time $t_{50\%}$ when $x(t)$ has decayed to exactly one-half of the initial value X_0.

Answer: $t_{50\%} = 0.693\tau$

Forced Response

For the case of interest to us in this chapter, the forced response of the system is the solution to the equation

$$\tau \frac{dx_F(t)}{dt} + x_F(t) = K_S F \qquad t \geq 0 \tag{5.28}$$

in which the forcing function F is equal to a constant for $t = 0$. For this special case, the solution can be found very easily, since the derivative term becomes zero in response to a constant excitation; thus, the forced response is found as follows:

$$x_F(t) = K_S F \qquad t \geq 0 \qquad \text{Forced response} \tag{5.29}$$

LO4

Note that this is exactly the **DC steady-state solution** described in Section 5.4! We already knew how to find the forced response of any *RLC* circuit when the excitation is a switched DC source. Further, we recognize that the two solutions are identical by writing

$$x_F(t) = x(\infty) = K_S F \qquad t \geq 0 \tag{5.29}$$

Complete Response

The complete response can now be calculated as the sum of the two responses:

$$x(t) = x_N(t) + x_F(t) = \alpha e^{-t/\tau} + K_S F = \alpha e^{-t/\tau} + x(\infty) \qquad t \geq 0 \tag{5.30}$$

First-Order Thermal System

An automotive transmission generates heat, when engaged, at the rate $q_{in} = 2{,}000$ J/s. The thermal capacitance of the transmission is $C_t = mc = 12$ kJ/°C. The effective convection resistance through which heat is dissipated is $R_t = 0.2$°C/W.

1. What is the steady-state temperature the transmission will reach when the initial (ambient) temperature is 5°C?

With reference to the Make the Connection sidebar on thermal capacitance, we write the differential equation based on energy balance:

$$R_t C_t \frac{dT}{dt} + T = R_t q_{in}$$

At steady state, the rate of change of temperature is zero, hence, $T(\infty) = R_t q_{in}$. Using the numbers given, $T(\infty) = 0.04 \times 2{,}000 = 80$°C.

(Continued)

To solve for the unknown constant α, we apply the initial condition $x(t = 0) = x(0)$:

$$\begin{aligned} x(t = 0) &= x(0) = \alpha + x(\infty) \\ \alpha &= x(0) - x(\infty) \end{aligned} \qquad \textbf{(5.31)}$$

so that we can finally write the complete response:

$$x(t) = [x(0) - x(\infty)]\, e^{-t/\tau} + x(\infty) \quad t \ge 0 \qquad \text{Complete response} \qquad \textbf{(5.32)}$$

LO4

Note that this equation is in the same form as that given in the Focus on Methodology box. Once you have understood this brief derivation, it will be very easy to use the simple shortcut of writing the solution to a first-order circuit by going through the simple steps of determining the initial and final conditions and the time constant of the circuit. This methodology is illustrated in the remainder of this section by a number of examples.

EXAMPLE 5.7 Complete Solution of First-Order Circuit

LO4

Problem

Determine an expression for the capacitor voltage in the circuit of Figure 5.17.

Solution

Known Quantities: Initial capacitor voltage; battery voltage, resistor and capacitor values.

Find: Capacitor voltage as a function of time $v_C(t)$ for all t.

Schematics, Diagrams, Circuits, and Given Data: $v_C(t = 0^-) = 5$ V; $R = 1$ kΩ; $C = 470$ μF; $V_B = 12$ V. Figures 5.17 and 5.18.

Assumptions: None.

Analysis:

Step 1: Steady-state response. We first observe that the capacitor had previously been charged to an initial voltage of 5 V. Thus,

$$v_C(t) = 5 \text{ V} \quad t < 0 \quad \text{and} \quad v_C(0^-) = 5 \text{ V}$$

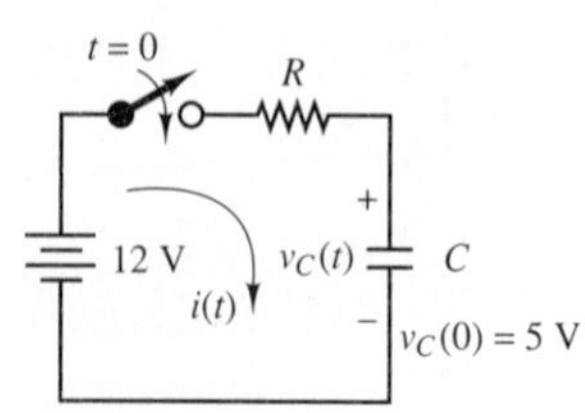

Figure 5.17

When the switch has been closed for a long time, the capacitor current becomes zero (see equation 5.20); alternatively, we can replace the capacitor by an open circuit. In either case, the fact that the current in a simple series circuit must become zero after a suitably long time tells us that

$$v_C(\infty) = V_B = 12 \text{ V}$$

Step 2: Initial condition. We can determine the initial condition for the variable $v_C(t)$ by virtue of the continuity of capacitor voltage (equation 5.22):

$$v_C(0^+) = v_C(0^-) = 5 \text{ V}$$

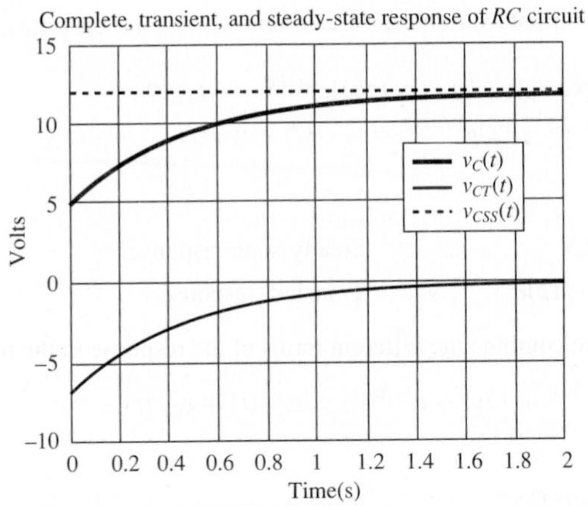

(a)

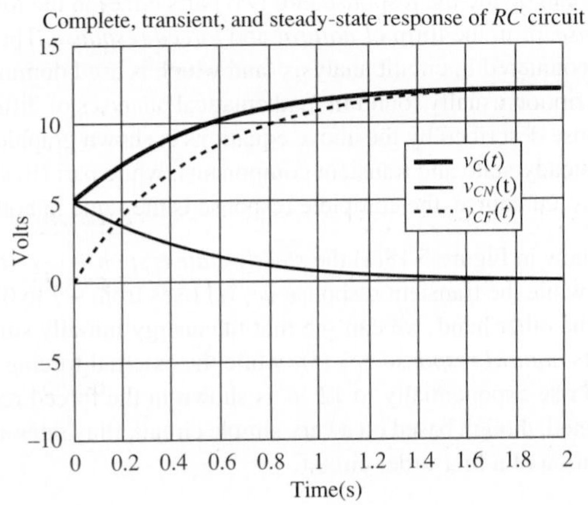

(b)

Figure 5.18 (a) Complete, transient, and steady-state responses of the circuit of Figure 5.17; (b) complete, natural, and forced responses of the circuit of Figure 5.17.

Step 3: Writing the differential equation. At $t = 0$ the switch closes, and the circuit is described by the following differential equation, obtained by application of KVL:

$$V_B - Ri_C(t) - v_C(t) = V_B - RC\frac{dv_C(t)}{dt} - v_C(t) = 0 \qquad t > 0$$

$$RC\frac{dv_C(t)}{dt} + v_C(t) = V_B \qquad t > 0$$

Step 4: Time constant. In the above equation we recognize the following variables, with reference to equation 5.22:

$$x = V_C \qquad \tau = RC \qquad K_S = 1 \qquad f(t) = V_B \qquad t > 0$$

Step 5: Complete solution. From the Focus on Methodology box, we know that the solution is of the form

$$x(t) = x(\infty) + [x(0) - x(\infty)]e^{-t/\tau}$$

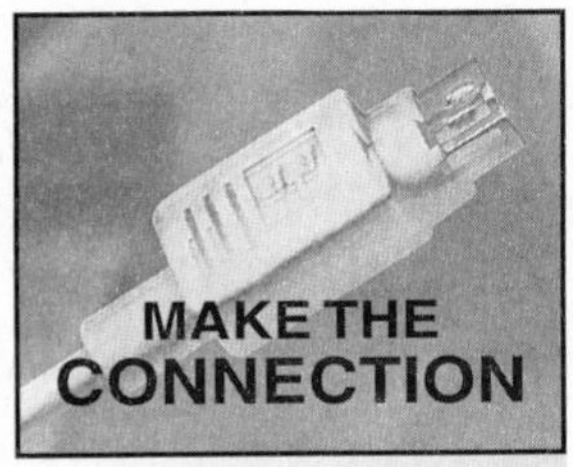

(Concluded)

2. How long will it take the transmission to reach 90 percent of the final temperature?

The general form of the solution is

$$\begin{aligned} T(t) &= [T(0) - T(\infty)]e^{-t/\tau} \\ &\quad + T(\infty) \\ &= T(0) + T(\infty) \\ &\quad \times (1 - e^{-t/\tau}) \\ &= 5 + 80(1 - e^{-t/\tau}) \end{aligned}$$

thus, the transmission temperature starts out at 5°C, and increases to its final value of 85°C, as shown in the plot of Figure 5.16.

Given the final value of 85°C, we calculate 90 percent of the final temperature to be 76.5°C. To determine the time required to reach this temperature, we solve the following equation for the argument t:

$$T(t_{90\%}) = 76.5$$
$$= 5 + 80(1 - e^{-t_{90\%}/\tau})$$
$$\frac{71.5}{80} = 1 - e^{-t_{90\%}/\tau}$$
$$0.10625 = e^{-t_{90\%}/\tau} \quad \Rightarrow$$
$$t_{90\%} = 2.24\tau = 1{,}076 \text{ s}$$
$$= 17.9 \text{ min}$$

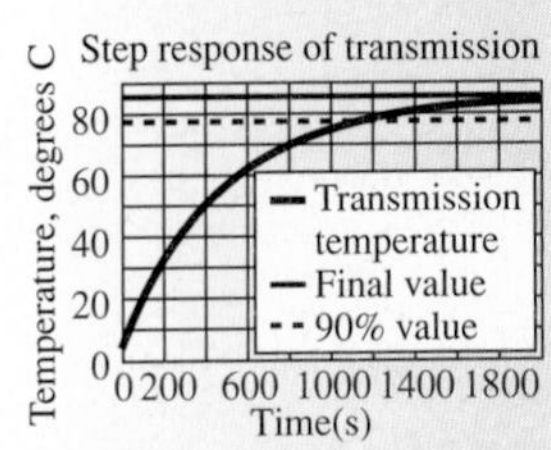

Figure 5.16 Temperature response of automotive transmission

Substituting the appropriate variables, we can write

$$v_C(t) = v_C(\infty) + [v_C(0) - v_C(\infty)]e^{-t/RC} \qquad t \geq 0$$
$$v_C(t) = 12 + (5 - 12)e^{-t/0.47} = v_{CSS}(t) + v_{CT}(t)$$

with

$$v_{CSS}(t) = 12 \text{ V} \qquad \text{Steady-state response}$$
$$v_{CT}(t) = (5 - 12)e^{-t/0.47} \text{ V} \qquad \text{Transient response}$$

Alternatively, we can combine the different terms of the response in the following form:

$$v_C(t) = 5e^{-t/0.47} + 12(1 - e^{-t/0.47}) = v_{CN}(t) + v_{CF}(t)$$

with

$$v_{CN}(t) = 5e^{-t/0.47} \text{ V} \qquad \text{Natural response}$$
$$v_{CF}(t) = 12(1 - e^{-t/0.47}) \text{ V} \qquad \text{Forced response}$$

You can see that we can divide the response into two parts either in the form of *steady-state* and *transient response*, or in the form of *natural* and *forced response*. The former is the one more commonly encountered in circuit analysis (and which is used dominantly in this book); the latter is the description usually found in mathematical analyses of differential equations. The complete response described by the above equations is shown graphically in Figure 5.18; part (a) depicts the steady-state and transient components, while part (b) shows the natural and forced responses. Of course, the complete response is the same in both cases.

Comments: Note how in Figure 5.18(a) the *steady-state response* $v_{CSS}(t)$ is simply equal to the battery voltage, while the transient response $v_{CT}(t)$ rises from −7 to 0 V exponentially. In Figure 5.18(b), on the other hand, we can see that the energy initially stored in the capacitor decays to zero via its *natural response* $v_{CN}(t)$, while the external forcing function causes the capacitor voltage to rise exponentially to 12 V, as shown in the forced response $v_{CF}(t)$. The example just completed, though based on a very simple circuit, illustrates all the steps required to complete the solution of a first-order circuit.

CHECK YOUR UNDERSTANDING

What happens if the initial condition (capacitor voltage for $t < 0$) is zero?

Answer: The complete solution is equal to the forced solution. $v_C(t) = v_{CF}(t) = 12(1 - e^{-t/0.47})$

Energy Storage in Capacitors and Inductors

It is appropriate at this time to recall that capacitors and inductors are energy storage elements. Consider, first, a capacitor, which accumulates charge according to the relationship $Q = CV$. The charge accumulated in the capacitor leads to the storage of energy according to the following equation:

$$W_C = \frac{1}{2}Cv_C^2(t) \qquad \text{Energy stored in a capacitor} \tag{5.33}$$

To understand the role of stored energy, consider, as an illustration, the simple circuit of Figure 5.19, where a capacitor is shown to have been connected to a battery V_B for a long time. The capacitor voltage is therefore equal to the battery voltage: $v_C(t) = V_B$. The charge stored in the capacitor (and the corresponding energy) can be directly determined by using equation 5.33. Suppose, next, that at $t = 0$ the capacitor is disconnected from the battery and connected to a resistor, as shown by the action of the switches in Figure 5.19. The resulting circuit would be governed by the RC differential equation described earlier, subject to the initial condition $v_C(t = 0) = V_B$. Thus, according to the results of Section 5.4, the capacitor voltage would decay exponentially according to the following equation:

$$v_C(t) = V_B e^{-t/RC} \tag{5.34}$$

Physically, this exponential decay signifies that the energy stored in the capacitor at $t = 0$ is dissipated by the resistor at a rate determined by the time constant of the circuit $\tau = RC$. Intuitively, the existence of a closed-circuit path allows for the flow of a current, thus draining the capacitor of its charge. All the energy initially stored in the capacitor is eventually dissipated by the resistor.

A very analogous reasoning process explains the behavior of an inductor. Recall that an inductor stores energy according to the expression

$$W_L = \frac{1}{2}Li_L^2(t) \qquad \text{Energy stored in an inductor} \tag{5.35}$$

Thus, in an inductor, energy storage is associated with the flow of a current (note the dual relationship between i_L and v_C). Consider the circuit of Figure 5.20, which is similar to that of Figure 5.19 except that the battery has been replaced with a current source and the capacitor with an inductor. For $t < 0$, the source current I_B flows through the inductor, and energy is thus stored; at $t = 0$, the inductor current is equal to I_B. At this point, the current source is disconnected by means of the left-hand switch, and a resistor is simultaneously connected to the inductor, to form a closed circuit.[1] The inductor current will now continue to flow through the resistor, which dissipates the energy stored in the inductor. By the reasoning in the preceding discussion, the inductor current will decay exponentially:

$$i_L(t) = I_B e^{-tR/L} \tag{5.36}$$

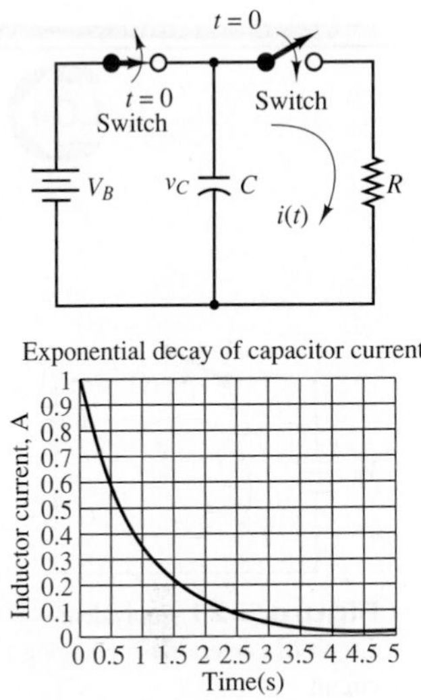

Figure 5.19 Decay through a resistor of energy stored in a capacitor

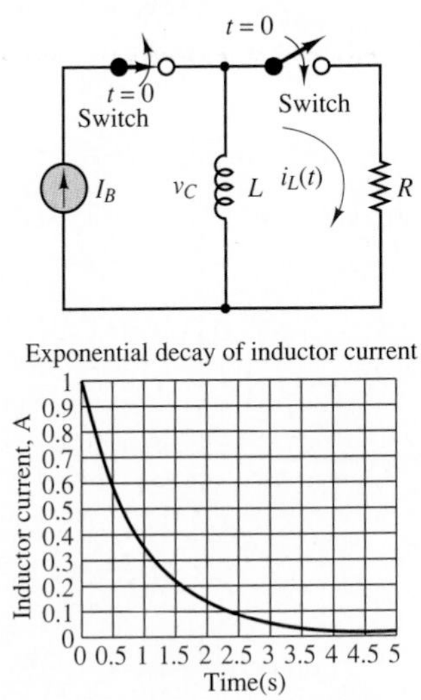

Figure 5.20 Decay through a resistor of energy stored in an inductor

[1] Note that in theory an ideal current source cannot be connected in series with a switch. For the purpose of this hypothetical illustration, imagine that upon opening the right-hand side switch, the current source is instantaneously connected to another load, not shown.

That is, the inductor current will decay exponentially from its initial condition, with a time constant $\tau = L/R$. Example 5.8 further illustrates the significance of the time constant in a first-order circuit.

LO4

EXAMPLE 5.8 Charging a Camera Flash—Time Constants

Problem

A capacitor is used to store energy in a camera flash light. The camera operates on a 6-V battery. Determine the time required for the energy stored to reach 90 percent of the maximum. Compute this time in seconds and as a multiple of the time constant. The equivalent circuit is shown in Figure 5.21.

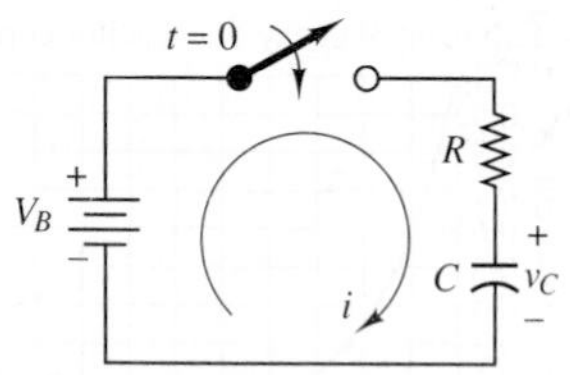

Figure 5.21 Equivalent circuit of camera flash charging circuit

Solution

Known Quantities: Battery voltage; capacitor and resistor values.

Find: Time required to reach 90 percent of the total energy storage.

Schematics, Diagrams, Circuits, and Given Data: $V_B = 6$ V; $C = 1{,}000\ \mu$F; $R = 1$ kΩ.

Assumptions: Charging starts at $t = 0$, when the flash switch is turned on. The capacitor is completely discharged at the start.

Analysis: First, we compute the total energy that can be stored in the capacitor:

$$E_{\text{total}} = \tfrac{1}{2}Cv_C^2 = \tfrac{1}{2}CV_B^2 = 18 \times 10^{-3}\ \text{J}$$

Thus, 90 percent of the total energy will be reached when $E_{\text{total}} = 0.9 \times 18 \times 10^{-3} = 16.2 \times 10^{-3}$ J. This corresponds to a voltage calculated from

$$\tfrac{1}{2}Cv_C^2 = 16.2 \times 10^{-3}$$

$$v_C = \sqrt{\frac{2 \times 16.2 \times 10^{-3}}{C}} = 5.692\ \text{V}$$

Next, we determine the time constant of the circuit: $\tau = RC = 10^{-3} \times 10^3 = 1$ s; and we observe that the capacitor will charge exponentially according to the expression

$$v_C = 6(1 - e^{-t/\tau}) = 6(1 - e^{-t})$$

Note that the expression for $v_C(t)$ is equal to the forced response, since the natural response is equal to zero (see Example 5.7) when the initial condition in $v_C(0) = 0$. To compute the time required to reach 90 percent of the energy, we must therefore solve for t in the equation

$$v_{C\text{-}90\%} = 5.692 = 6(1 - e^{-t})$$
$$0.949 = 1 - e^{-t}$$
$$0.051 = e^{-t}$$
$$t = -\ln 0.051 = 2.97\ \text{s}$$

The result corresponds to a charging time of approximately 3 time constants.

Comments: This example demonstrates the physical connection between the time constant of a first-order circuit and a practical device. If you wish to practice some of the calculations related to time constants, you might calculate the number of time constants required to reach 95 and 99 percent of the total energy stored in a capacitor.

CHECK YOUR UNDERSTANDING

If we wished to double the amount of energy stored in the capacitor, how would the charging time change?

Answer: It would also double, as C would be twice as large, thus doubling τ.

EXAMPLE 5.9 Starting Transient of DC Motor

LO4

Problem

An approximate circuit representation of a DC motor consists of a series RL circuit, shown in Figure 5.22. Apply the first-order circuit solution methodology just described to this approximate DC motor equivalent circuit to determine the transient current.

Solution

Known Quantities: Initial motor current; battery voltage; resistor and inductor values.

Find: Inductor current as a function of time $iL(t)$ for all t.

Schematics, Diagrams, Circuits, and Given Data: $R = 4\ \Omega$; $L = 0.1$ H; $V_B = 50$ V. Figure 5.22.

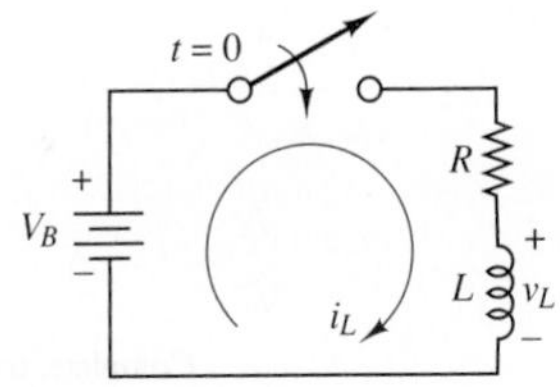

Figure 5.22 Circuit for Example 5.9

Assumptions: None.

Analysis:

Step 1: Steady-state response. The inductor current prior to the closing of the switch must be zero; thus,

$$i_L(t) = 0 \text{ A} \qquad t < 0 \qquad \text{and} \qquad i_L(0^-) = 0 \text{ A}$$

When the switch has been closed for a long time, the inductor current becomes a constant and can be calculated by replacing the inductor with a short circuit:

$$i_L(\infty) = \frac{V_B}{R} = \frac{50}{4} = 12.5 \text{ A}$$

Step 2: Initial condition. We can determine the initial condition for the variable $i_L(t)$ by virtue of the continuity of inductor current (equation 5.23):

$$i_L(0^+) = i_L(0^-) = 0$$

Step 3: Writing the differential equation. At $t = 0$ the switch closes, and the circuit is described by the following differential equation, obtained by application of KVL:

$$V_B - Ri_L(t) - L\frac{di_L(t)}{dt} = 0 \qquad t > 0$$

$$\frac{L}{R}\frac{di_L(t)}{dt} + i_L(t) = \frac{1}{R}V_B \qquad t > 0$$

Step 4: Time constant. In the equations listed in step 3, we recognize the following variables, with reference to equation 5.8:

$$x = i_L \qquad \tau = \frac{L}{R} \qquad K_S = \frac{1}{R} \qquad f(t) = V_B \qquad t > 0$$

Step 5: Complete solution. From the Focus on Methodology box, we know that the solution is of the form

$$x(t) = x(\infty) + [x(0) - x(\infty)]e^{-t/\tau}$$

Substituting the appropriate variables, we can write

$$\begin{aligned} i_L(t) &= i_L(\infty) + [i_L(0) - i_L(\infty)]e^{-Rt/L} \qquad t \ge 0 \\ &= 12.5 + (0 - 12.5)e^{-t/0.025} = i_{LSS}(t) + i_{LT}(t) \end{aligned}$$

with

$$\begin{aligned} i_{LSS}(t) &= 12.5 \text{ A} && \text{Steady-state response} \\ i_{LT}(t) &= (-12.5)e^{-t/0.025} \text{ A} && \text{Transient response} \end{aligned}$$

Alternatively, we can combine the different terms of the response in the following form:

$$i_L(t) = 0 + 12.5(1 - e^{-t/0.025}) = i_{LN}(t) + i_{LF}(t) \qquad t \ge 0$$

with

$$\begin{aligned} i_{LN}(t) &= 0 \text{ A} && \text{Natural response} \\ i_{LF}(t) &= 12.5(1 - e^{-t/0.025}) \text{ A} && \text{Forced response} \end{aligned}$$

The complete response described by the above equations is shown graphically in Figure 5.23.

Comments: Note that in practice it is not a good idea to place a switch in series with an inductor. As the switch opens, the inductor current is forced to change instantaneously, with

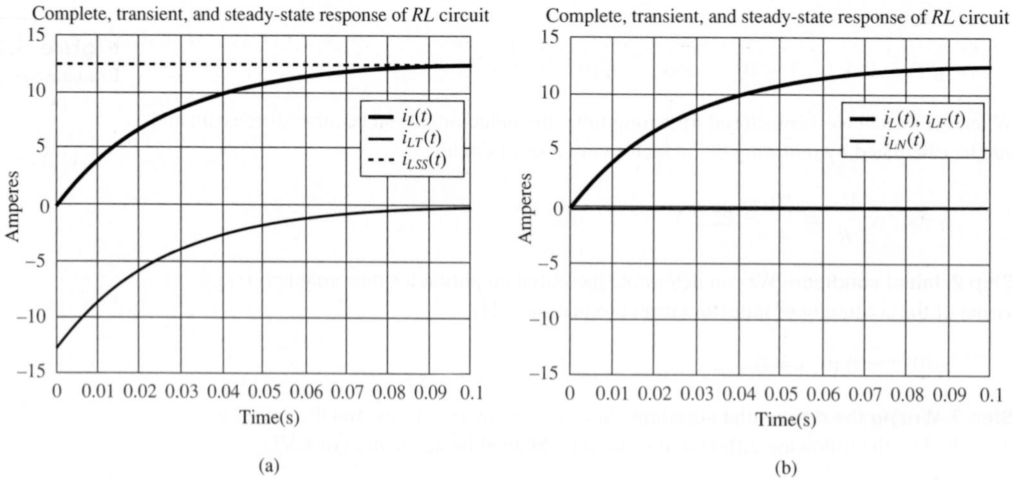

Figure 5.23 Transient response of electric motor: (a) steady-state and transient responses; (b) natural and forced responses

the result that di_L/dt, and therefore $v_L(t)$, approaches infinity. The large voltage transient resulting from this *inductive kick* can damage circuit components. A practical solution to this problem, the freewheeling diode, is presented in Section 12.5.

In the preceding examples we have seen how to systematically determine the solution of first-order circuits. The solution methodology was applied to two simple cases, but it applies in general to any first-order circuit, providing that one is careful to identify a Thévenin (or Norton) equivalent circuit, determined with respect to the energy storage element (i.e., treating the energy storage element as the load). Thus the equivalent circuit methodology for resistive circuits presented in Chapter 3 applies to transient circuits as well. Figure 5.24 depicts the general appearance of a first-order circuit once the resistive part of the circuit has been reduced to Thévenin equivalent form.

An important comment must be made before we demonstrate the equivalent circuit approach to more complex circuit topologies. Since the circuits that are the subject of the present discussion usually contain a switch, one must be careful to determine the equivalent circuits *before and after the switch changes position.* In other words, it is possible that the equivalent circuit seen by the load before activating the switch is different from the circuit seen after the switch changes position.

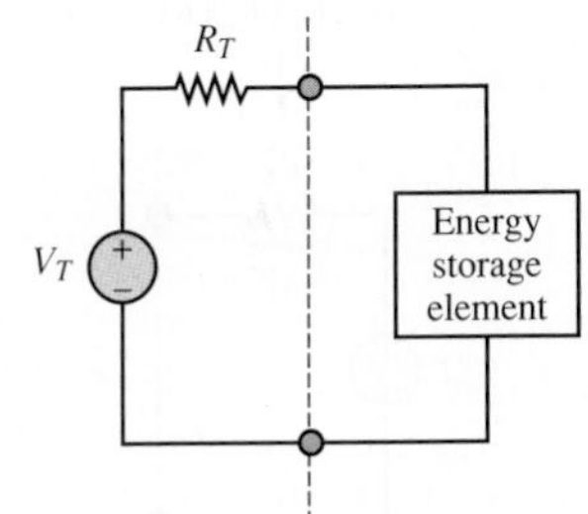

Figure 5.24 Equivalent-circuit representation of first-order circuits

To illustrate the procedure, consider the RC circuit of Figure 5.25. The objective is to determine the capacitor voltage for all time. The switch closes at $t = 0$. For $t < 0$, we recognize that the capacitor has been connected to the battery V_2 through resistor R_2. This circuit is already in Thévenin equivalent form, and we know that the capacitor must have charged to the battery voltage V_2, provided that the switch has been closed for a sufficient time (we shall assume so). Thus,

$$\begin{aligned} v_C(t) &= V_2 \qquad t \le 0 \\ V_C(0^-) &= V_C(0^+) = V_2 \end{aligned} \tag{5.37}$$

After the switch closes, the circuit on the left-hand side of Figure 5.25 must be accounted for. Figure 5.26 depicts the new arrangement, in which we have moved the capacitor to the far right-hand side, in preparation for the evaluation of the equivalent circuit. Using the Thévenin-to-Norton source transformation technique (introduced in Chapter 3), we next obtain the circuit at the top of Figure 5.27, which can be easily reduced by adding the two current sources and computing the equivalent parallel resistance of R_1, R_2, and R_3. The last step illustrated in the figure is the conversion

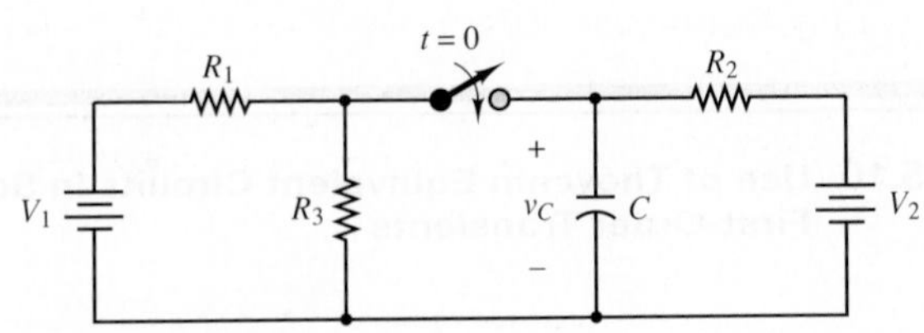

Figure 5.25 A more involved RC circuit

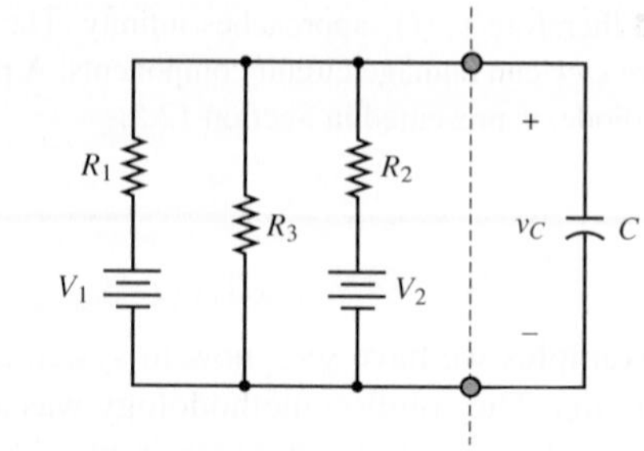

Figure 5.26 The circuit of Figure 5.22 for $t \geq 0$

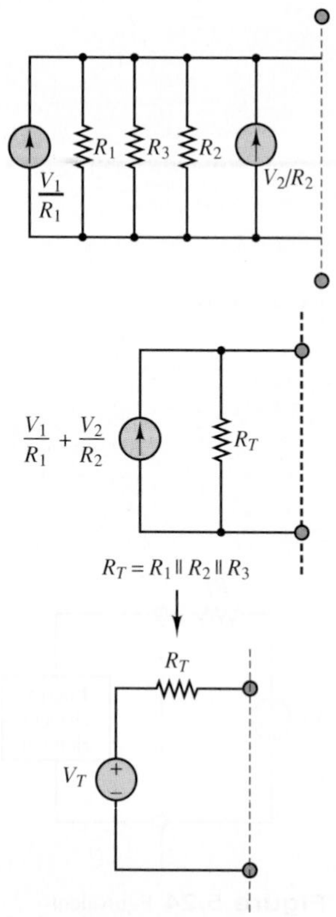

Figure 5.27 Reduction of the circuit of Figure 5.26 to Thévenin equivalent form

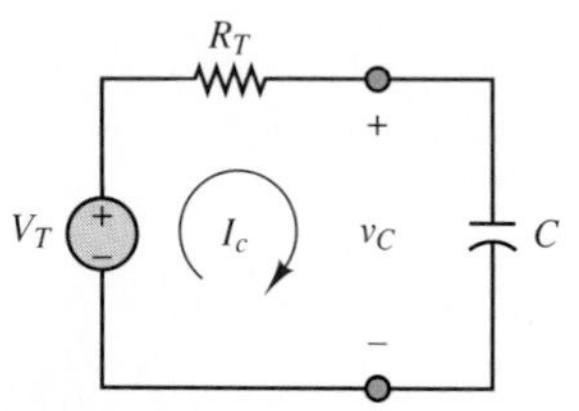

Figure 5.28 The circuit of Figure 5.25 in equivalent form for $t \geq 0$

to Thévenin form. Figure 5.28 depicts the final appearance of the equivalent circuit for $t \geq 0$.

When the switch has been closed for a long time, the capacitor sees the Thévenin equivalent circuit computed in Figures 5.27 and 5.28. Thus, when the capacitor is replaced with an open circuit, $v_C(\infty) = V_T$. Further, we can determine the initial condition for the variable $v_C(t)$ by virtue of the continuity of capacitor voltage (equation 5.22): $v_C(0^+) = v_C(0^-) = V_2$. At $t = 0$ the switch closes, and the circuit is described by the following differential equation, obtained by application of KVL for the circuit of Figure 5.28:

$$V_T - R_T i_C(t) - v_C(t) = V_T - R_T C \frac{dv_C(t)}{dt} - v_C(t) = 0 \qquad t > 0$$

$$R_T C \frac{dv_C(t)}{dt} + v_C(t) = V_T \qquad t > 0$$

with

$$R_T = R_1 || R_2 || R_3$$

$$V_T = R_T \left(\frac{V_1}{R_1} + \frac{V_2}{R_2} \right)$$

In the above equation we recognize, with reference to equation 5.22, the following variables and parameters: $x = v_C$; $\tau = R_T C$; $K_S = 1$; $f(t) = V_T$ for $t > 0$. And we can write the complete solution

$$v_C(t) = v_C(\infty) + [v_C(0) - v_C(\infty)]e^{-t/\tau} \qquad t \geq 0$$

$$v_C(t) = V_T + (V_2 - V_T)e^{-t/R_T C}$$

The use of Thévenin equivalent circuits to obtain transient responses is emphasized in the next few examples.

LO4

EXAMPLE 5.10 Use of Thévenin Equivalent Circuits in Solving First-Order Transients

Problem

The circuit of Figure 5.29 includes a switch that can be used to connect and disconnect a battery. The switch has been open for a very long time. At $t = 0$ the switch closes, and then at $t = 50$ ms the switch opens again. Determine the capacitor voltage as a function of time.

Solution

Known Quantities: Battery voltage; resistor and capacitor values.

Find: Capacitor voltage as a function of time $v_C(t)$ for all t.

Schematics, Diagrams, Circuits, and Given Data: $R_1 = R_2 = 1{,}000\ \Omega$, $R_3 = 500\ \Omega$, and $C = 25\ \mu\text{F}$. Figure 5.29.

Assumptions: None.

Analysis:

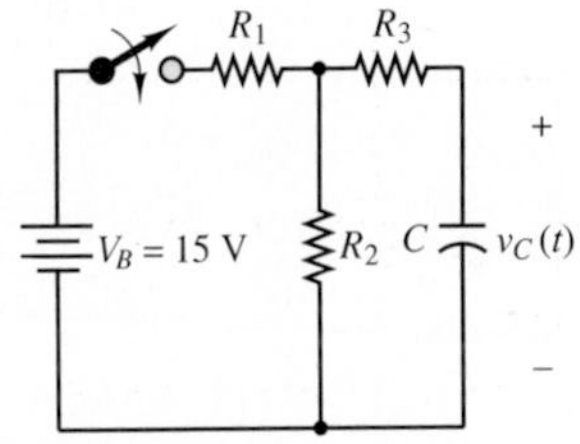

Figure 5.29 Circuit for Example 5.10

Part 1— $0 \le t < 50$ ms

Step 1: Steady-state response. We first observe that any charge stored in the capacitor has had a discharge path through resistors R_3 and R_2. Thus, the capacitor must be completely discharged. Hence,

$$v_C(t) = 0\text{ V} \qquad t < 0 \qquad \text{and} \qquad v_C(0^-) = 0\text{ V}$$

To determine the steady-state response, we look at the circuit a long time after the switch has been closed. At steady state, the capacitor behaves as an open circuit, and we can calculate the equivalent open circuit (Thévenin) voltage and equivalent resistance to be

$$v_C(\infty) = V_B \frac{R_2}{R_1 + R_2}$$
$$= 7.5\text{ V}$$
$$R_T = R_3 + R_1 \| R_2 = 1\text{ k}\Omega$$

Step 2: Initial condition. We can determine the initial condition for the variable $v_C(t)$ by virtue of the continuity of capacitor voltage (equation 5.22):

$$v_C(0^+) = v_C(0^-) = 0\text{ V}$$

Step 3: Writing the differential equation. To write the differential equation, we use the Thévenin equivalent circuit for $t \ge 0$, with $V_T = v_c(\infty)$ and we write the resulting differential equation

$$V_T - R_T i_C(t) - v_C(t) = V_T - R_T C\frac{dv_C(t)}{dt} - v_C(t) = 0 \qquad 0 \le t < 50\text{ ms}$$

$$R_T C\frac{dv_C(t)}{dt} + v_C(t) = V_T \qquad 0 \le t < 50\text{ ms}$$

Step 4: Time constant. In the above equation we recognize, with reference to equation 5.22, the following variables and parameters:

$$x = v_C; \qquad \tau = R_T C = 0.025\text{ s}; \qquad K_S = 1;$$
$$f(t) = V_T = 7.5\text{ V} \qquad 0 \le t < 50\text{ ms}$$

Step 5: Complete solution. The complete solution is

$$v_C(t) = v_C(\infty) + [v_C(0) - v_C(\infty)]e^{-t/\tau} \qquad 0 \le t < 50\text{ ms}$$
$$v_C(t) = V_T + (0 - V_T)e^{-t/R_T C} = 7.5(1 - e^{-t/0.025})\text{ V} \qquad 0 \le t < 50\text{ ms}$$

Part 2— $t \geq 50$ ms

As mentioned in the problem statement, at $t = 50$ ms the switch opens again, and the capacitor now discharges through the series combination of resistors R_3 and R_2. Since there is no external forcing function once the switch is opened, this problem only involves determining the natural response of the circuit. Recall that the natural response is of the form $x_N(t) = \alpha e^{-t/\tau}$ (see equation 5.27). We note two important facts:

1. The constant α is the initial condition *at the time when the switch is opened,* since that represents the actual value of the voltage across the capacitor from which the exponential decay will begin.
2. The time constant of the decay is now the time constant of the circuit with the switch open, that is, $\tau = (R_2 + R_3)C = 0.0375$ s.

To determine α, we must calculate the value of the capacitor voltage at the time when the switch is opened:

$$\alpha = v_C(t = 50 \text{ ms}) = 7.5(1 - e^{-0.05/0.025}) = 6.485 \text{ V}$$

Thus, we can write the capacitor voltage response for $t \geq 50$ ms as follows:

$$v_C(t) = 6.485e^{-(t-0.05)/0.0375}$$

The composite response is plotted below.

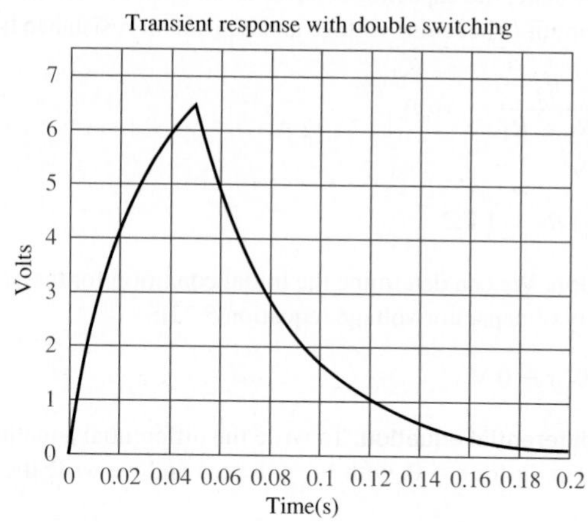

Comments: Note that the two parts of the response are based on two different time constants, and that the rising portion of the response changes faster (shorter time constant) than the decaying part.

CHECK YOUR UNDERSTANDING

What will the intial condition for the exponential decay be if the switch opens at $t = 100$ ms?

Answer: 7.363 V

EXAMPLE 5.11 Turn-Off Transient of DC Motor

LO4

Problem

Determine the motor voltage for all time in the simplified electric motor circuit model shown in Figure 5.30. The motor is represented by the series RL circuit in the shaded box.

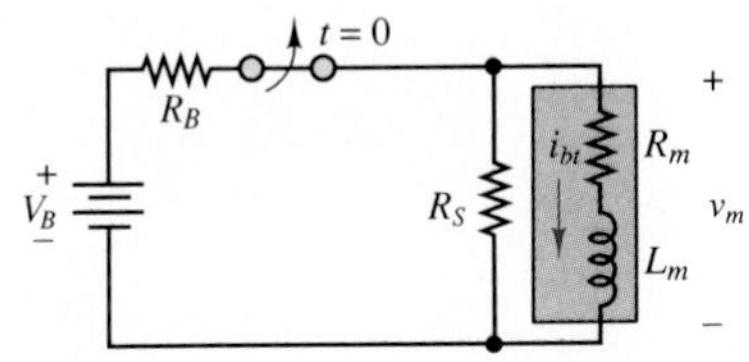

Figure 5.30

Solution

Known Quantities: Battery voltage, resistor, and inductor values.

Find: The voltage across the motor as a function of time.

Schematics, Diagrams, Circuits, and Given Data: $R_B = 2\ \Omega$; $R_S = 20\ \Omega$; $R_m = 0.8\ \Omega$; $L = 3$ H; $V_B = 100$ V.

Assumptions: The switch has been closed for a long time.

Analysis: With the switch closed for a long time, the inductor in the circuit of Figure 5.30 behaves as a short circuit. The current through the motor can then be calculated by the current divider rule in the modified circuit of Figure 5.31, where the inductor has been replaced with a short circuit and the Thévenin circuit on the left has been replaced by its Norton equivalent:

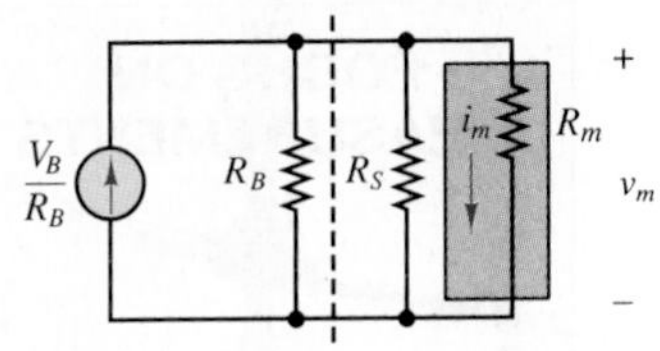

Figure 5.31

$$i_m = \frac{1/R_m}{1/R_B + 1/R_s + 1/R_m} \frac{V_B}{R_B}$$
$$= \frac{1/0.8}{1/2 + 1/20 + 1/0.8} \frac{100}{2} = 34.72 \text{ A}$$

This current is the initial condition for the inductor current: $i_L(0) = 34.72$ A. Since the motor inductance is effectively a short circuit, the motor voltage for $t < 0$ is equal to

$$v_m(t) = i_m R_m = 27.8 \text{ V} \qquad t < 0$$

When the switch opens and the motor voltage supply is turned off, the motor sees only the *shunt* (parallel) resistance R_S, as depicted in Figure 5.32. Remember now that the inductor current cannot change instantaneously; thus, the motor (inductor) current i_m must continue to flow in the same direction. Since all that is left is a series RL circuit, with resistance $R = R_S + R_m = 20.8\ \Omega$, the inductor current will decay exponentially with time constant $\tau = L/R = 0.1442$ s:

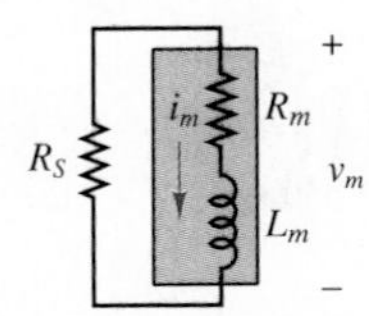

Figure 5.32

$$i_L(t) = i_m(t) = i_L(0)e^{-t/\tau} = 34.7e^{-t/0.1442} \qquad t > 0$$

The motor voltage is then computed by adding the voltage drop across the motor resistance and inductance:

$$v_m(t) = R_m i_L(t) + L\frac{di_L(t)}{dt}$$
$$= 0.8 \times 34.7e^{-t/0.1442} + 3\left(-\frac{34.7}{0.1442}\right)e^{-t/0.1442} \qquad t > 0$$
$$= -694.1e^{-t/0.1442} \qquad t > 0$$

The motor voltage is plotted in Figure 5.33.

Comments: Notice how the motor voltage rapidly changes from the steady-state value of 27.8 V for $t < 0$ to a large negative value due to the turn-off transient. This *inductive kick* is typical of *RL* circuits, and results from the fact that although the inductor current cannot change instantaneously, the inductor voltage can and does, as it is proportional to the derivative

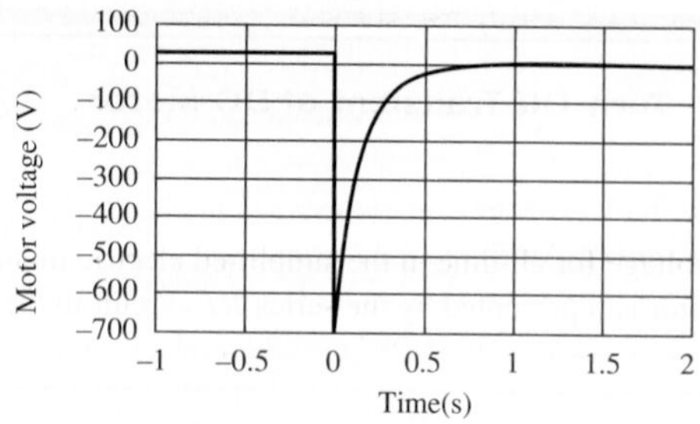

Figure 5.33 Motor voltage transient response

of i_L. This example is based on a simplified representation of an electric motor, but illustrates effectively the need for special starting and stopping circuits in electric motors. Some of these ideas are explored in Chapter 12, Power Electronics; Chapter 17, Introduction to Electric Machines; and Chapter 18, Special-Purpose Electric Machines.

FOCUS ON MEASUREMENTS

Coaxial Cable Pulse Response

LO3

FIND IT ON THE WEB

Problem:

A problem of great practical importance is the transmission of *pulses* along cables. Short voltage pulses are used to represent the two-level binary signals that are characteristic of digital computers; it is often necessary to transmit such voltage pulses over long distances through **coaxial cables,** which are characterized by a finite resistance per unit length and by a certain capacitance per unit length, usually expressed in picofarads per meter. A simplified model of a long coaxial cable is shown in Figure 5.34. If a 10-m cable has a capacitance of 1,000 pF/m and a series resistance of 0.2 Ω/m, what will the output of the pulse look like after traveling the length of the cable?

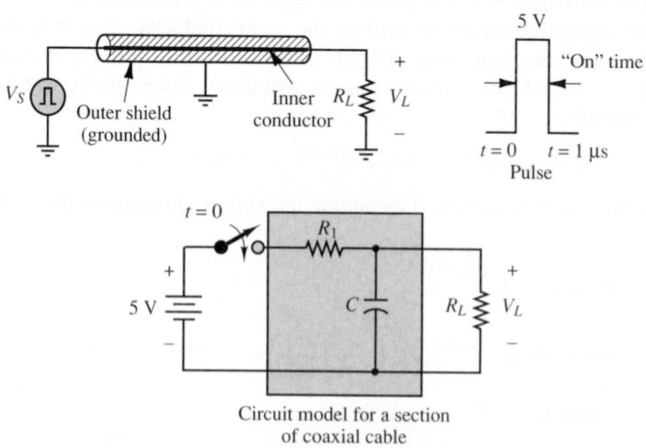

Figure 5.34 Pulse transmission in a coaxial cable

(Continued)

Solution:

Known Quantities—Cable length, resistance, and capacitance; voltage pulse amplitude and time duration.

Find—The cable voltage as a function of time.

Schematics, Diagrams, Circuits, and Given Data—$r_1 = 0.2\ \Omega/\text{m}$; $R_L = 150\ \Omega$; $c = 1{,}000$ pF/m; $l = 10$ m; pulse duration $= 1\ \mu$s.

Assumptions—The short voltage pulse is applied to the cable at $t = 0$. Assume zero initial conditions.

Analysis—The voltage pulse can be modeled by a 5-V battery connected to a switch; the switch will then close at $t = 0$ and open again at $t = 1\ \mu$s. The solution strategy will therefore proceed as follows. First, we determine the initial condition; next, we solve the transient problem for $t > 0$; finally, we compute the value of the capacitor voltage at $t = 1\ \mu$s—that is, when the switch opens again—and solve a different transient problem. Intuitively, we know the equivalent capacitor will charge for $1\ \mu$s, and the voltage will reach a certain value. This value will be the initial condition for the capacitor voltage when the switch is opened; the capacitor voltage will then decay to zero, since the voltage source has been disconnected. Note that the circuit will be characterized by two different time constants during the two transient stages of the problem. The initial condition for this problem is zero, assuming that the switch has been open for a long time.

The differential equation for $0 < t < 1\ \mu$s is obtained by computing the Thévenin equivalent circuit relative to the capacitor when the switch is closed:

$$V_T = \frac{R_L}{R_1 + R_L} V_B \qquad R_T = R_1 \| R_L \qquad \tau = R_T C \qquad 0 < t < 1\ \mu\text{s}$$

As we have already seen, the differential equation is given by the expression

$$R_T C \frac{dv_C}{dt} + v_C = V_T \qquad 0 < t < 1\ \mu\text{s}$$

and the solution is of the form

$$v_C(t) = -V_T e^{-t/R_T C} + V_T = V_T(1 - e^{-t/R_T C}) \qquad 0 < t < 1\ \mu\text{s}$$

We can assign numerical values to the solution by calculating the effective resistance and capacitance of the cable:

$$R_1 = r_1 \times l = 0.2 \times 10 = 2\ \Omega$$

$$C = c \times l = 1{,}000 \times 10 = 10{,}000\ \text{pF}$$

$$R_T = 2 \| 150 = 1.97\ \Omega \qquad V_T = \frac{150}{152} V_B = 4.93\ \text{V}$$

$$\tau_{\text{on}} = R_T C = 19.74 \times 10^{-9}\ \text{s}$$

so that

$$v_C(t) = 4.93(1 - e^{-t/19.74 \times 10^{-9}}) \qquad 0 < t < 1\ \mu\text{s}$$

At the time when the switch opens again, $t = 1\ \mu$s, the capacitor voltage can be found to be $v_C(t = 1\ \mu\text{s}) = 4.93$ V.

When the switch opens again, the capacitor will discharge through the load resistor R_L; this discharge is described by the natural response of the circuit consisting of C and R_L

(Continued)

(Concluded)

and is governed by the following values: $v_C(t = 1\ \mu s) = 4.93$ V, $\tau_{\text{off}} = R_L C = 1.5\ \mu s$. We can directly write the natural solution as follows:

$$v_C(t) = v_C(t = 1 \times 10^{-6}) \times e^{-(t-1\times10^{-6})/\tau_{\text{off}}}$$
$$= 4.93 \times e^{-(t-1\times10^{-6})/1.5\times10^{-6}} \qquad t \geq 1\ \mu s$$

Figure 5.35 shows a plot of the solution for $t > 0$, along with the voltage pulse.

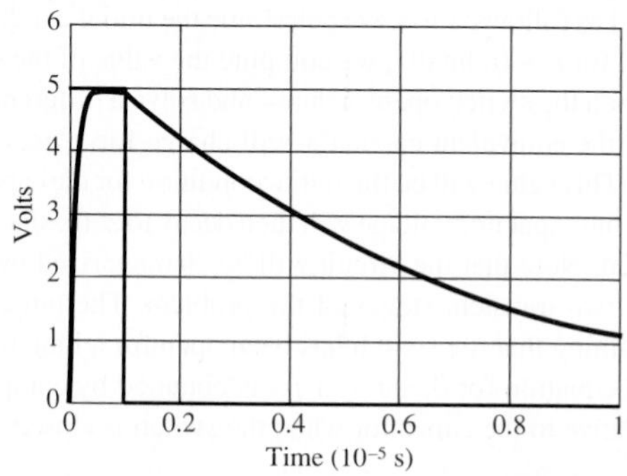

Figure 5.35 Coaxial cable pulse response

Comments—Note that the voltage response shown in Figure 5.35 rapidly reaches the desired value, near 5 V, thanks to the very short charging time constant τ_{on}. On the other hand, the discharging time constant τ_{off} is significantly slower. As the length of the cable is increased, however, τ_{on} will increase, to the point that the voltage pulse may not rise sufficiently close to the desired 5-V value in the desired time. While the numbers used in this example are somewhat unrealistic, you should remember that cable length limitations may exist in some applications because of the cable's intrinsic capacitance and resistance. In general, long cables such as electric transmission lines and very high-freqency circuits cannot be analyzed by way of the lumped-parameter methods presented here, and require *distributed circuit analysis* techniques.

EXAMPLE 5.12 Transient Response of Supercapacitors

Problem

An industrial, uninterruptible power supply (UPS) is intended to provide continuous power during unexpected power outages. Ultracapacitors can store a significant amount of energy and release it during transient power outages to ensure delicate or critical electrical/electronic systems. Assume that we wish to design a UPS that is required to make up for a temporary power glitch in a permanent power supply for 5 s. The system that is supported by this UPS operates at a nominal voltage of 50 V and has a maximum nominal voltage of 60 V, but can function with a supply voltage as low as 25 V. Design a UPS by determining the suitable number of series and parallel elements required.

Solution

Known Quantities: Maximum, nominal, and minimum voltage; power rating and time requirements; Ultracapacitor data (see Example 4.1).

Find: Number of series and parallel ultracapacitor cells needed to satisfy the specifications.

Schematics, Diagrams, Circuits, and Given Data: Figure 5.36. Capacitance of one cell: $C_{\text{cell}} = 100$ F; resistance of one cell: $R_{\text{cell}} = 15$ mΩ; nominal cell voltage $V_{\text{cell}} = 2.5$ V. (See Example 4.1.)

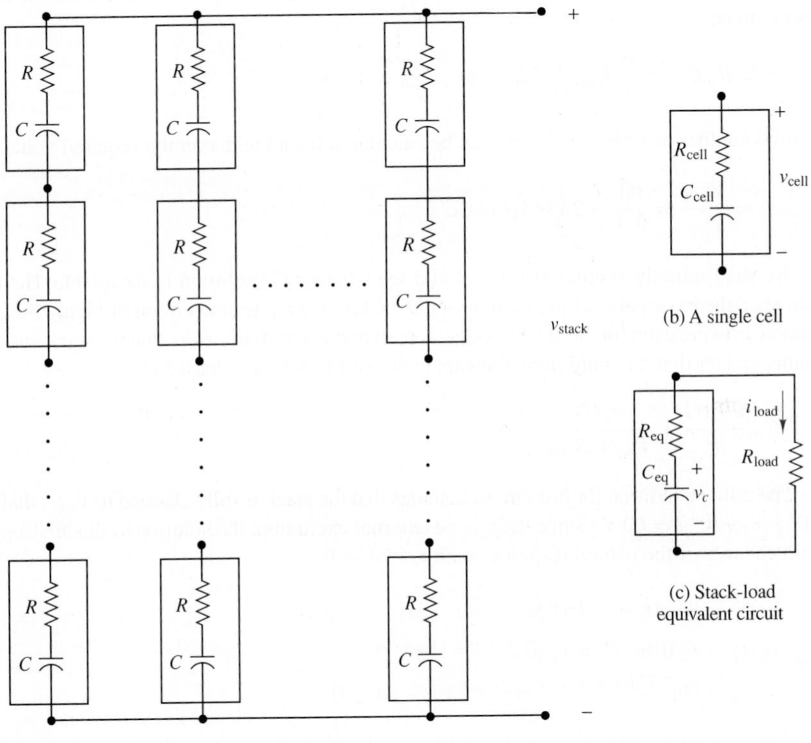

(a) An ultracapacitor stack

(b) A single cell

(c) Stack-load equivalent circuit

Figure 5.36

Assumptions: The load can be modeled as a 0.5-Ω resistance.

Analysis: The total capacitance of the "stack" required to satisfy the specifications is obtained by combining capacitors in series and parallel, as illustrated in Figure 5.36(a). Figure 5.36(b) depicts the electric circuit model of a single cell. First we define some of the important variables that are to be used in the problem.

The *allowable voltage drop* in the supply is $\Delta V = 25$ V, since the load can operate with a supply as low as 25 V and nominally operates at 50 V.

The *time interval* over which the voltage will drop (but not below the allowable minimum of 25 V) is 5 s.

The equivalent resistance of the stack consists of n parallel strings of m resistors each; thus

$$R_{eq} = mR_{cell}||mR_{cell}||\ldots||mR_{cell} \qquad n \text{ times}$$

$$R_{eq} = \frac{m}{n}R_{cell}$$

The equivalent capacitance of the stack can similarly be calculated by recalling that capacitors in series combine as do resistors in parallel (and vice versa):

$$C_{eq} = \frac{n}{m}C_{cell}$$

Finally, given the equivalent resistance and capacitance of the stack, we can calculate the time constant to be

$$\tau = R_{eq}C_{eq} = \frac{m}{n}R_{cell}\frac{n}{m}C_{cell} = R_{cell}C_{cell} = 1.5 \text{ s}$$

The total number of series capacitors can be calculated from the maximum required voltage:

$$m = \frac{V_{max}}{V_{cell}} = \frac{60}{2.5} = 24 \text{ series cells}$$

We shall initially assume that $n = 1$ and see whether the solution is acceptable. Having established the basic parameters, we now apply KCL to the equivalent circuit of Figure 5.36(c) to obtain an expression for the stack voltage. Recall that we wish to ensure that the stack voltage remains greater than the minimum allowable voltage (25 V) for at least 5 s:

$$C_{eq}\frac{dv_c(t)}{dt} + \frac{v_c(t)}{R_{eq} + R_{load}} = 0$$

The initial condition for this circuit assumes that the stack is fully charged to V_{max}, that is, $v_C(0^-) = v_C(0^+) = 60$ V. Since there is no external excitation, the solution to this first-order circuit consists of the natural response, with $v_c(\infty) = 0$

$$v_C(t) = [v_C(0) - v_C(\infty)]e^{-t/\tau} + v_C(\infty) \qquad t \ge 0$$

$$v_C(t) = v_C(0)e^{-t/\tau} = v_C(0)e^{-t/(R_{eq}+R_{load})C_{eq}}$$

$$= 60e^{-t/\{[(m/n)R_{cell}+R_{load}](n/m)C_{cell}\}} \qquad t \ge 0$$

Since we are actually interested in the load voltage, we can use a voltage divider to relate the supercapacitor voltage to the load voltage:

$$v_{load}(t) = \frac{R_{load}}{R_{eq} + R_{load}}v_c(t) = \frac{R_{load}}{(m/n)R_{cell} + R_{load}}v_c(t)$$

$$= \frac{0.5}{(m/n)(0.015) + 0.5}60e^{-t/\{[(m/n)R_{cell}+R_{load}](n/m)C_{cell}\}}$$

This relationship can be used to calculate the appropriate number of parallel strings n such that the load voltage is above 25 V (the minimum allowable load voltage) at $t = 5$ s. The solution could be obtained analytically, by substituting the known values $m = 24$, $R_{load} = 0.5$, $C_{cell} = 100$ F, $R_{cell} = 15$ mΩ, $t = 5$ s, and $v_{load}(t = 5) = 25$ V and by solving for the only unknown, n. We leave this as an exercise. Figure 5.37 plots the transient response of the stack load system for $n = 1$ to 5. You can see that for $n = 3$ the requirement that $v_{load} = 25$ V for at least 5 s is satisfied.

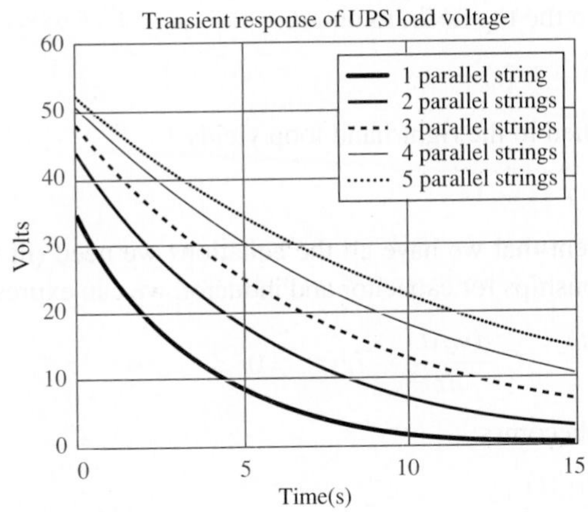

Figure 5.37 Transient response of supercapacitor circuit

CHECK YOUR UNDERSTANDING

Derive the result obtained in Example 5.12 analytically, by solving the transient response for the unknown value n.

Answer: $n = 3$

5.5 TRANSIENT RESPONSE OF SECOND-ORDER CIRCUITS

In many practical applications, understanding the behavior of first- and second-order systems is often all that is needed to describe the response of a physical system to external excitation. In this section, we discuss the solution of the second-order differential equations that characterize *RLC* circuits.

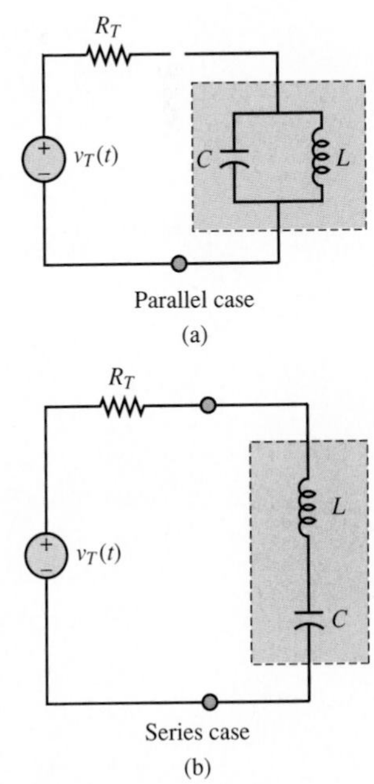

Figure 5.38 Second-order circuits

Deriving the Differential Equations for Second-Order Circuits

A simple way of introducing second-order circuits consists of replacing the box labeled "Circuit containing *RL*/*RC* combinations" in Figure 5.3 with a combination of two energy storage elements, as shown in Figure 5.38. Note that two different cases are considered, depending on whether the energy storage elements are connected in series or in parallel.

Consider the parallel case first, which has been redrawn in Figure 5.39 for clarity. Practice and experience will eventually suggest the best method for writing the circuit equations. At this point, the most sensible procedure consists of applying the basic circuit laws to the circuit of Figure 5.39. Start with KVL around the left-hand loop:

$$v_T(t) - R_T i_S(t) - v_C(t) = 0 \quad \textbf{(5.38)}$$

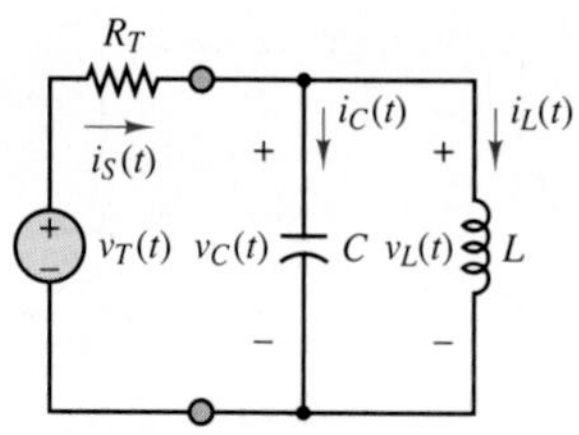

Figure 5.39 Parallel case

Then apply KCL to the top node, to obtain

$$i_S(t) - i_C(t) - i_L(t) = 0 \tag{5.39}$$

Further, KVL applied to the right-hand loop yields

$$v_C(t) = v_L(t) \tag{5.40}$$

It should be apparent that we have all the equations we need (in fact, more). Using the defining relationships for capacitor and inductor, we can express equation 5.39 as

$$\frac{v_T(t) - v_C(t)}{R_T} - C\frac{dv_C(t)}{dt} - i_L(t) = 0 \tag{5.41}$$

and equation 5.40 becomes

$$v_C(t) = L\frac{di_L(t)}{dt} \tag{5.42}$$

Substituting equation 5.42 in equation 5.41, we can obtain a differential circuit equation in terms of the variable $i_L(t)$:

$$\frac{1}{R_T}v_T(t) - \frac{L}{R_T}\frac{di_L(t)}{dt} = LC\frac{d^2i_L(t)}{dt^2} + i_L(t) \tag{5.43}$$

or

$$LC\frac{d^2i_L(t)}{dt^2} + \frac{L}{R_T}\frac{di_L(t)}{dt} + i_L(t) = \frac{1}{R_T}v_T \tag{5.44}$$

The solution to this differential equation (which depends, as in the case of first-order circuits, on the initial conditions and on the forcing function) completely determines the behavior of the circuit. By now, two questions should have appeared in your mind:

1. Why is the differential equation expressed in terms of $i_L(t)$? [Why not $v_C(t)$?]
2. Why did we not use equation 5.38 in deriving equation 5.44?

In response to the first question, it is instructive to note that, knowing $i_L(t)$, we can certainly derive any one of the voltages and currents in the circuit. For example,

$$v_C(t) = v_L(t) = L\frac{di_L(t)}{dt} \tag{5.45}$$

$$i_C(t) = C\frac{dv_C(t)}{dt} = LC\frac{d^2i_L(t)}{dt^2} \tag{5.46}$$

To answer the second question, note that equation 5.44 is not the only form the differential circuit equation can take. By using equation 5.38 in conjunction with equation 5.39, one could obtain the following equation:

$$v_T(t) = R_T[i_C(t) + i_L(t)] + v_C(t) \tag{5.47}$$

Upon differentiating both sides of the equation and appropriately substituting from equation 5.41, the following second-order differential equation in v_C would be obtained:

$$LC\frac{d^2v_C(t)}{dt^2} + \frac{L}{R_T}\frac{dv_C(t)}{dt} + v_C(t) = \frac{L}{R_T}\frac{dv_T(t)}{dt} \tag{5.48}$$

Note that the left-hand side of the equation is identical to equation 5.44, except that v_C has been substituted for i_L. The right-hand side, however, differs substantially

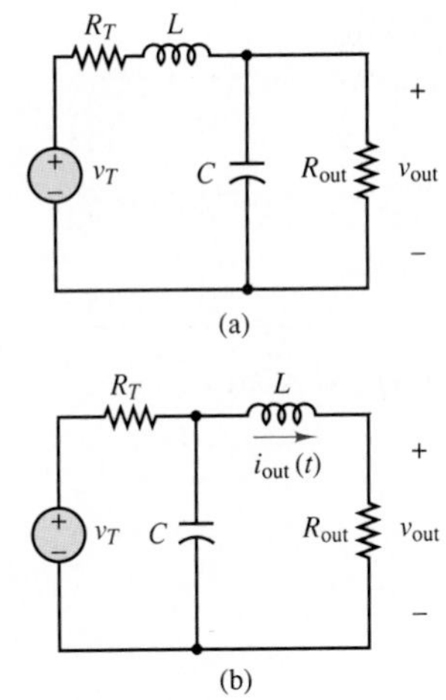

Figure 5.40 Two second-order circuits

from equation 5.44, because the forcing function is the derivative of the equivalent voltage.

Since all the desired circuit variables may be obtained either as a function of i_L or as a function of v_C, the choice of the preferred differential equation depends on the specific circuit application, and we conclude that there is no unique method to arrive at the final equation. As a case in point, consider the two circuits depicted in Figure 5.40. If the objective of the analysis were to determine the output voltage v_{out}, then for the circuit in Figure 5.40(a), one would choose to write the differential equation in v_C, since $v_C = v_{\text{out}}$. In the case of Figure 5.40(b), however, the inductor current would be a better choice, since $v_{\text{out}} = R_T i_{\text{out}}$.

Solution of Second Order Circuits

Second-order systems also occur very frequently in nature, and are characterized by the ability of a system to store energy in one of two forms—potential or kinetic—and to dissipate this stored energy; second-order systems always contain two energy storage elements. Electric circuits containing two capacitors, two inductors, or one capacitor and one inductor are usually second-order systems (unless we can combine the two capacitors or inductors into a single element by virtue of a series or parallel combination, in which case the system is of first order). Earlier in this chapter we saw that second-order differential equations can be written in the form of equations 5.12 and 5.13, repeated here for convenience.

LO5

$$a_2 \frac{d^2x(t)}{dt^2} + a_1 \frac{dx(t)}{dt} + a_0 x(t) = b_0 f(t)$$

or **(5.49)**

$$\frac{1}{\omega_n^2}\frac{d^2x(t)}{dt^2} + \frac{2\zeta}{\omega_n}\frac{dx(t)}{dt} + x(t) = K_S f(t)$$

In equation 5.48, the constants $\omega_n = \sqrt{a_0/a_2}$, $\zeta = (a_1/2)\sqrt{1/a_0 a_2}$, and $K_S = b_0/a_0$ are termed the **natural frequency,** the **damping ratio,** and the **DC gain** (or **static sensitivity**), respectively. Just as we were able to attach a very precise meaning to the constants τ and K_S in the case of first-order circuits, the choice of constants ω_n, ζ, and K_S is not arbitrary, but represents some very important characteristics of the response of second-order systems, and of second-order circuits in particular. As an illustration, let $K_S = 1$, $\omega_n = 1$, and $\zeta = 0.2$ in the differential equation 5.48, and let $f(t)$ correspond to a switched input that turns on from zero to unit amplitude[2] at $t = 0$. The response is plotted in Figure 5.41.

We immediately note three important characteristics of the response of Figure 5.41:

1. The response asymptotically tends to a *final value* of 1.
2. The response *oscillates* with a period approximately equal to 6 s.
3. The oscillations *decay* (and eventually disappear) as time progresses.

Each of these three observations can be explained by one of the three parameters defined in equation 5.49:

[2]This input is more formally called a **unit step,** and the response deriving from it is called the **unit step response.**

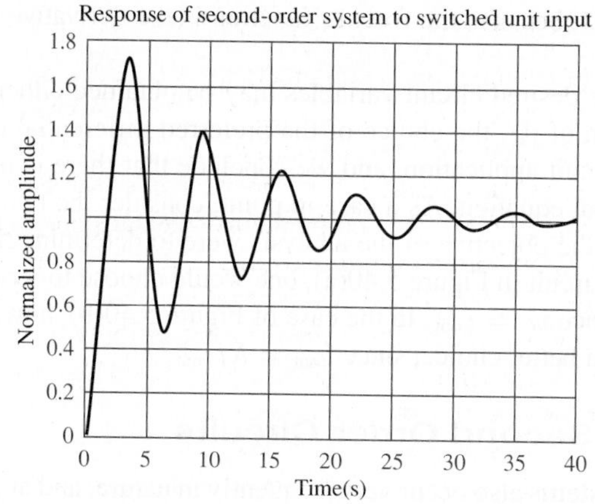

Figure 5.41 Response of switched second-order system with $K_S = 1$, $\omega_n = 1$, and $\zeta = 0.1$

1. The *final value* of 1 is predicted by the DC gain $K_S = 1$, which tells us that in the steady state (when all the derivative terms are zero) $x(t) = f(t)$.
2. The period of oscillation of the response is related to the natural frequency: $\omega_n = 1$ leads to the calculation $T = 2\pi/\omega_n = 2\pi \approx 6.28$ s. Thus, the natural frequency parameter describes the *natural frequency* of oscillation of the system.
3. Finally, the reduction in amplitude of the oscillations is governed by the damping ratio ζ. It is not as easy to visualize the effect of the damping ratio from a single plot, so we have included the plot of Figure 5.42, illustrating how the same system is affected by changes in the damping ratio. You can see that as ζ increases, the amplitude of the initial oscillation becomes increasingly smaller until, when $\zeta = 1$, the response no longer *overshoots* the final value of 1 and has a response that looks, qualitatively, like that of a first-order system.

In the remainder of this chapter we will study solution methods to determine the response of second-order circuits.

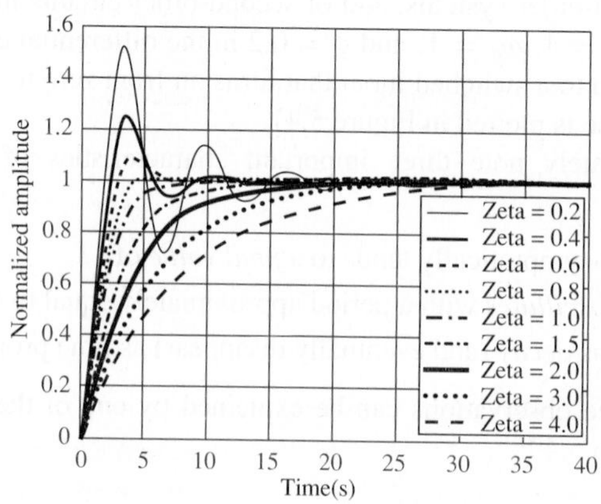

Figure 5.42 Response of switched second-order system with $K_S = 1$, $\omega_n = 1$, and ζ ranging from 0.2 to 4

Elements of the Transient Response

The steps involved in computing the complete transient response of a second-order circuit excited by a switched DC source are, in essence, the same as the steps we took in solving a first-order circuit: First we determine the initial and final conditions, using exactly the same techniques used for first-order circuits (see Section 5.4); then we compute the transient response. The computation of the transient response of a second-order circuit, however, cannot be simplified quite as much as was done for first-order circuits. With a little patience you will find that although it takes a little longer to go through the computations, the methods are the same as those used in Section 5.5.

The solution of a second-order differential equation also requires that we consider the **natural response** (or **homogeneous solution**), with the forcing function set equal to zero, and the **forced response** (or **particular solution**), in which we consider the response to the forcing function. The **complete response** then consists of the sum of the natural and forced responses. Once the form of the complete response is known, the initial condition can be applied to obtain the final solution.

Natural Response of a Second-Order System

The natural response is found by setting the excitation equal to zero. Thus, we solve the equation

$$\frac{1}{\omega_n^2}\frac{d^2x_N(t)}{dt^2}+\frac{2\zeta}{\omega_n}\frac{dx_N(t)}{dt}+x_N(t)=0 \tag{5.50}$$

where we use the notation $x_N(t)$ to denote the natural response. Just as in the case of first-order systems, the solution of this equation is known to be of exponential form:

$$x_N(t)=\alpha e^{st} \qquad \text{Natural response} \tag{5.51}$$

Substituting this expression into equation 5.50, we obtain the algebraic equation

$$\frac{1}{\omega_n^2}s^2\alpha e^{st}+\frac{2\zeta}{\omega_n}s\alpha e^{st}+\alpha e^{st}=0 \tag{5.52}$$

Equation 5.52 can be equal to zero only if

$$\frac{s^2}{\omega_n^2}+\frac{2\zeta}{\omega_n}s+1=0 \tag{5.53}$$

This polynomial in the variable s is called the **characteristic polynomial** of the differential equation, and it gives rise to two **characteristic roots** s_1 and s_2. Thus, the function $x_N(t)=\alpha e^{st}$ is a solution of the homogeneous differential equation only when $s=s_1$ and $s=s_2$. The natural response of the system is a linear combination of the response associated with each characteristic root, that is,

$$x_N(t)=\alpha_1 e^{s_1t}+\alpha_2 e^{s_2t} \tag{5.54}$$

Now, one can solve for the two characteristic roots simply by finding the roots of equation 5.53:

$$s_{1,2}=-\zeta\omega_n\pm\frac{1}{2}\sqrt{(2\zeta\omega_n)^2-4\omega_n^2}=-\zeta\omega_n\pm\omega_n\sqrt{\zeta^2-1} \tag{5.55}$$

It is immediately apparent that three possible cases exist for the roots of the natural solution of a second-order differential equation, as shown in this Focus on Methodology box.

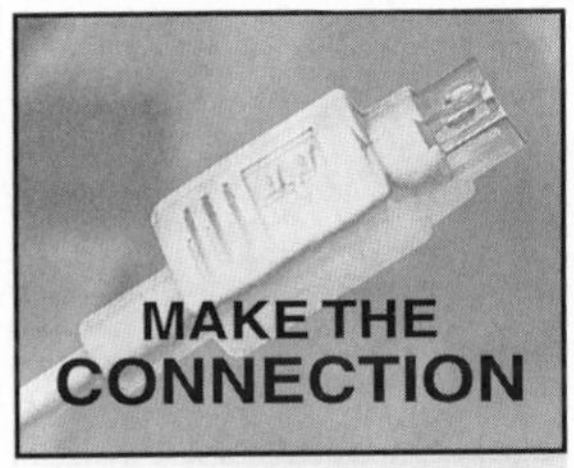

Automotive Suspension

If we analyze the mechanical system shown in Figure 5.43(a) using Newton's Second Law, $ma=\sum F$, we obtain the equation

$$m\frac{d^2x(t)}{dt^2}=F(t)-b\frac{dx(t)}{dt}-kx(t)$$

Comparing this equation with equation 5.49, we can rewrite the same equation in the standard form of a second-order system:

$$\frac{m}{k}\frac{d^2x(t)}{dt^2}+\frac{b}{k}\frac{dx(t)}{dt}+x(t)=\frac{1}{k}f(t)$$

The series electrical circuit of Figure 5.43(b) can be obtained by KVL:

$$v_s-Ri_L-v_C-L\frac{di_L}{dt}=0$$

$$i_L=i_C=C\frac{dv_C}{dt}$$

$$LC\frac{d^2v_C}{dt^2}+RC\frac{dv_C}{dt}+v_C=v_S$$

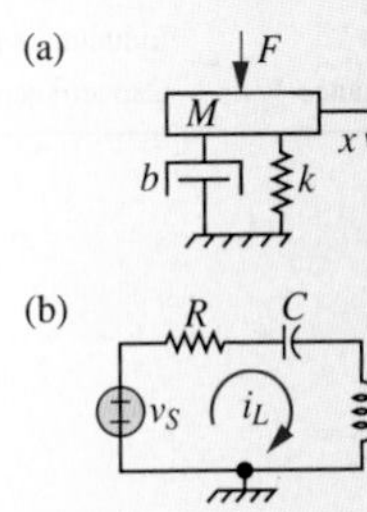

Figure 5.43 Analogy between electrical and mechanical systems

(Continued)

(Concluded)

If we now compare both second-order differential equations to the standard form of equation 5.49, we can make the following observations:

$$\frac{1}{\omega_n^2}\frac{d^2x(t)}{dt^2} + \frac{2\zeta}{\omega_n}\frac{dx(t)}{dt} + x(t) = K_S f(t)$$

$$\omega_n = \sqrt{\frac{k}{m}}$$

$$\zeta = \frac{b}{k}\frac{\omega_n}{2} = \frac{b}{2}\sqrt{\frac{1}{km}}$$

Mechanical

$$\omega_n = \sqrt{\frac{1}{LC}}$$

$$\zeta = RC\frac{\omega_n}{2} = \frac{R}{2}\sqrt{\frac{C}{L}}$$

Electrical

Comparing the expressions for the natural frequency and damping ratio in the two differential equations, we arrive at the following analogies:

Mechanical system	**Electrical system**
Damping coefficient b	Resistance R
Mass m	Inductance L
Compliance $1/k$	Capacitance C

FOCUS ON METHODOLOGY

ROOTS OF SECOND-ORDER SYSTEMS

Case 1: **Real and distinct roots.** This case occurs when $\zeta > 1$, since the term under the square root sign is positive in this case, and the roots are $s_{1,2} = -\zeta\omega_n \pm \omega_n\sqrt{\zeta^2 - 1}$. This leads to an **overdamped response.**

Case 2: **Real and repeated roots.** This case holds when $\zeta = 1$, since the term under the square root is zero in this case, and $s_{1,2} = -\zeta\omega_n = -\omega_n$. This leads to a **critically damped response.**

Case 3: **Complex conjugate roots.** This case holds when $\zeta < 1$, since the term under the square root is negative in this case, and $s_{1,2} = -\zeta\omega_n \pm j\omega_n\sqrt{1 - \zeta^2}$. This leads to an **underdamped response.**

LO5

As we shall see in the remainder of this section, identifying the roots of the second-order differential equation is the key to writing the natural solution. Example 5.13 applies these concepts to an electric circuit.

EXAMPLE 5.13 Natural Response of Second-Order Circuit

LO5

Problem

Find the natural response of the circuit shown in Figure 5.44.

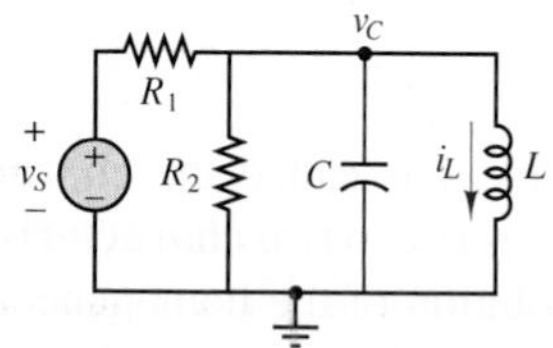

Figure 5.44

Solution

Known Quantities: Circuit elements.

Find: The natural response of the differential equation in $i_L(t)$ describing the circuit of Figure 5.44.

Schematics, Diagrams, Circuits, and Given Data: $R_1 = 8\ \text{k}\Omega$; $R_2 = 8\ \text{k}\Omega$; $C = 10\ \mu\text{F}$; $L = 1\ \text{H}$.

Assumptions: None

Analysis: To compute the natural response of the circuit, we set the source equal to zero (short circuit) and observe that the two resistors are in parallel, and can be replaced by a single resistor $R = R_1||R_2$. We apply KCL to the resulting parallel *RLC* circuit, observing that the capacitor voltage is the top node voltage in the circuit:

$$\frac{v_C}{R} + C\frac{dv_C}{dt} + i_L = 0$$

Next, we observe that

$$v_C = v_L = L\frac{di_L}{dt}$$

and we substitute the above expression for v_C into the first equation to obtain

$$LC\frac{d^2 i_L}{dt^2} + \frac{L}{R}\frac{di_L}{dt} + i_L = 0$$

This equation is in the form of equation 5.50, with $K_S = 1$, $\omega_n^2 = 1/LC$, and $2\zeta/\omega_n = L/R$. We can therefore compute the roots of the differential equation from the expression

$$s_{1,2} = -\zeta\omega_n \pm \frac{1}{2}\sqrt{(2\zeta\omega_n)^2 - 4\omega_n^2} = -\zeta\omega_n \pm \omega_n\sqrt{\zeta^2 - 1}$$

It is always instructive to calculate the values of the three parameters first. We can see by inspection that $K_S = 1$; $\omega_n = 1/\sqrt{LC} = 1/\sqrt{10^{-5}} = 316.2$ rad/s; and $\zeta = L\omega_n/2R = 0.04$. Since $\zeta < 1$, the response is *underdamped,* and the roots will be of the form $s_{1,2} = -\zeta\omega_n \pm j\omega_n\sqrt{1-\zeta^2}$. Substituting numerical values, we have $s_{1,2} = -12.5 \pm j316.2$, and we can write the natural response of the circuit as

$$\begin{aligned} x_N(t) &= \alpha_1 e^{\left(-\zeta\omega_n + j\omega_n\sqrt{1-\zeta^2}\right)t} + \alpha_2 e^{\left(-\zeta\omega_n - j\omega_n\sqrt{1-\zeta^2}\right)t} \\ &= \alpha_1 e^{(-12.5+j316.2)t} + \alpha_2 e^{(-12.5-j316.2)t} \end{aligned}$$

The constants α_1 and α_2 can be determined only once the forced response and the initial conditions have been determined. We shall see more complete examples very shortly.

Comments: Note that once the second-order differential equation is expressed in general form (equation 5.50) and the values of the three parameters are identified, the task of writing the natural solution with the aid of the Focus on Methodology box, Roots of Second-Order Systems," is actually very simple.

CHECK YOUR UNDERSTANDING

For what value of R will the circuit response become critically damped?

Answer: $R = 158.1\ \Omega$

Let us now more formally review each of the three cases of the natural solution of a second-order system.

1. OVERDAMPED SOLUTION

Real and distinct roots. This case occurs when $\zeta > 1$, since the term under the square root sign is positive, and the roots are $s_{1,2} = -\zeta\omega_n \pm \omega_n\sqrt{\zeta^2 - 1}$. In the case of an overdamped system, the general form of the solution is

LO5

$$x_N(t) = \alpha_1 e^{s_1 t} + \alpha_2 e^{s_2 t} = \alpha_1 e^{\left(-\zeta\omega_n + \omega_n\sqrt{\zeta^2-1}\right)t} + \alpha_2 e^{\left(-\zeta\omega_n - \omega_n\sqrt{\zeta^2-1}\right)t}$$
$$= \alpha_1 e^{-t/\tau_1} + \alpha_2 e^{-t/\tau_2} \tag{5.56}$$
$$\tau_2 = \frac{1}{\zeta\omega_n + \omega_n\sqrt{\zeta^2-1}} \qquad \tau_1 = \frac{1}{\zeta\omega_n - \omega_n\sqrt{\zeta^2-1}}$$

The appearance is that of the sum of two first-order systems, as shown in Figure 5.45.

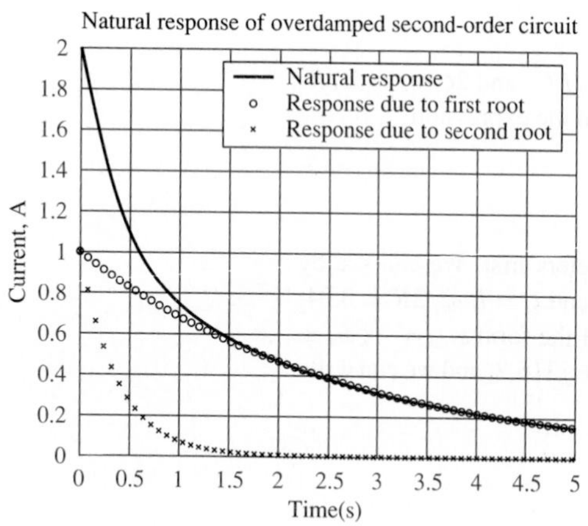

Figure 5.45 Natural response of underdamped second-order system for $\alpha_1 = \alpha_2 = 1$; $\zeta = 1.5$; $\omega_n = 1$

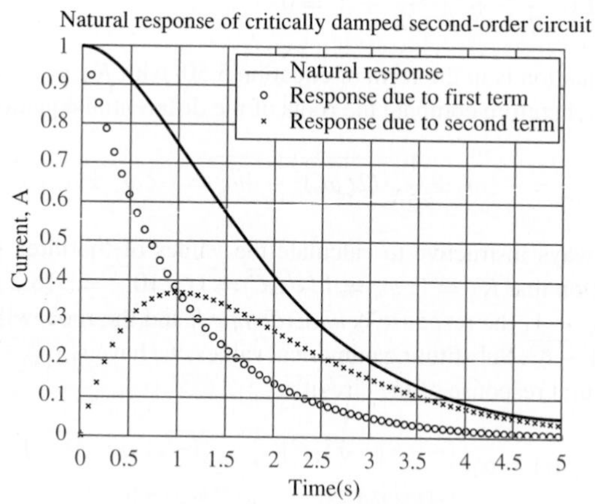

Figure 5.46 Natural response of a critically damped second-order system for $\alpha_1 = \alpha_2 = 1$; $\zeta = 1$; $\omega_n = 1$

2. CRITICALLY DAMPED SOLUTION

Real and repeated roots. This case holds when $\zeta = 1$, since the term under the square root sign is zero, and $s_{1,2} = -\zeta\omega_n = -\omega_n$. This leads to a **critically damped response.** In the case of a critically damped system, the general form of the solution is

LO5

$$x_N(t) = \alpha_1 e^{s_1 t} + \alpha_2 t e^{s_2 t} = \alpha_1 e^{-\omega_n t} + \alpha_2 t e^{-\omega_n t} = \alpha_1 e^{-t/\tau} + \alpha_2 t e^{-t/\tau}$$
$$\tau = \frac{1}{\omega_n} \tag{5.57}$$

Note that the second term is multiplied by t; thus a critically damped system consists of the sum of a first-order exponential term plus a similar term multiplied by t. The appearance is that of the sum of two first-order systems, as shown in Figure 5.46.

3. UNDERDAMPED SOLUTION

Complex conjugate roots. This case holds when $\zeta < 1$, since the term under the square root sign is negative, and $s_{1,2} = -\zeta\omega_n \pm j\omega_n\sqrt{1-\zeta^2}$. This leads to an

underdamped response. The general form of the response is

$$x_N(t) = \alpha_1 e^{\left(-\zeta\omega_n + j\omega_n\sqrt{1-\zeta^2}\right)t} + \alpha_2 e^{\left(-\zeta\omega_n - j\omega_n\sqrt{1-\zeta^2}\right)t} \tag{5.58}$$

To better understand this solution, and the significance of the complex exponential, let us assume that $\alpha_1 = \alpha_2 = \alpha$. Then we can further manipulate equation 5.58 to obtain

LO5

$$\begin{aligned} x_N(t) &= \alpha e^{-\zeta\omega_n t}\left(e^{\left(j\omega_n\sqrt{1-\zeta^2}\right)t} + e^{\left(-j\omega_n\sqrt{1-\zeta^2}\right)t}\right) \\ &= 2\alpha e^{-\zeta\omega_n t}\cos\left(\omega_n\sqrt{1-\zeta^2}\right)t \end{aligned} \tag{5.59}$$

The last step in equation 5.59 made use of Euler's equation (equation 4.44), and clearly illustrates the appearance of the underdamped response of a second-order system: The response oscillates at a frequency ω_d, called the **damped natural frequency,** where $\omega_d = \omega_n\sqrt{1-\zeta^2}$. This frequency asymptotically approaches the natural frequency as ζ tends to zero. The oscillation is *damped* by the exponential decay term $2\alpha e^{-\zeta\omega_n}$. The time constant for the exponential decay is $\tau = 1/\zeta\omega_n$, so you can see that as ζ becomes larger (more damping), τ becomes smaller and the oscillations decay more quickly. The two factors that make up the response are plotted in Figure 5.47, along with their product, which is the natural response.

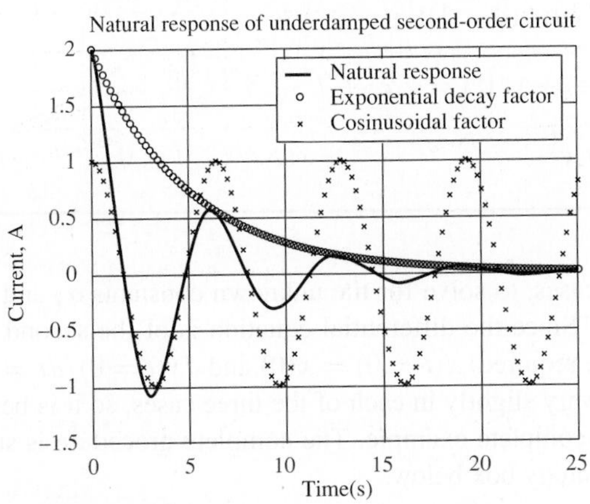

Figure 5.47 Natural response of an underdamped second-order system for $\alpha_1 = \alpha_2 = 1$; $\zeta = 0.2$; $\omega_n = 1$

Forced Response

For the case of interest to us in this chapter, that is, switched DC sources, the forced response of the system is the solution to the equation

$$\frac{1}{\omega_n^2}\frac{d^2x(t)}{dt^2} + \frac{2\zeta}{\omega_n}\frac{dx(t)}{dt} + x(t) = K_S f(t) \tag{5.60}$$

in which the forcing function $f(t)$ is equal to a constant F for $t \geq 0$. For this special case, the solution can be found very easily, since the derivative term becomes zero in response to a constant excitation; thus, the forced response is found as follows:

LO5

$$x_F(t) = K_S F \qquad t \geq 0 \qquad \text{Forced response} \tag{5.61}$$

Note that this is again exactly the **DC steady-state solution** described in Section 5.4! We already know how to find the forced response of any *RLC* circuit when the excitation is a switched DC source. Further, we recognize that the two solutions are identical by writing

$$x_F(t) = x(\infty) = K_S F \qquad t \geq 0 \tag{5.62}$$

Complete Response

The complete response can now be calculated as the sum of the two responses, and it depends on which of the three cases—overdamped, critically damped, or underdamped—applies to the specific differential equation.

LO5

Overdamped case ($\zeta > 1$):

$$x(t) = x_N(t) + x_F(t) = \alpha_1 e^{\left(-\zeta\omega_n + \omega_n\sqrt{\zeta^2-1}\right)t} + \alpha_2 e^{\left(-\zeta\omega_n - \omega_n\sqrt{\zeta^2-1}\right)t} + x(\infty) \qquad t \geq 0 \tag{5.63}$$

Critically damped case ($\zeta = 1$):

$$x(t) = x_N(t) + x_F(t) = \alpha_1 e^{-\zeta\omega_n t} + \alpha_2 t e^{-\zeta\omega_n t} + x(\infty) \qquad t \geq 0 \tag{5.64}$$

Underdamped case ($\zeta < 1$):

$$x(t) = x_N(t) + x_F(t) = \alpha_1 e^{\left(-\zeta\omega_n + j\omega_n\sqrt{1-\zeta^2}\right)t} + \alpha_2 e^{\left(-\zeta\omega_n - j\omega_n\sqrt{1-\zeta^2}\right)t} + x(\infty) \qquad t \geq 0 \tag{5.65}$$

In each of these cases, to solve for the unknown constants α_1 and α_2, we apply the initial conditions. Since the differential equation is of the second order, two initial conditions will be required: $x(t = 0) = x(0)$ and $dx(t = 0)/dt = \dot{x}(0)$. The details of the procedure vary slightly in each of the three cases, so it is best to present each case by way of a complete example. The complete procedure is summarized in the Focus on Methodology box below.

LO5

FOCUS ON METHODOLOGY

SECOND-ORDER TRANSIENT RESPONSE

1. Solve for the steady-state response of the circuit before the switch changes state ($t = 0^-$) and after the transient has died out ($t \to \infty$). We shall generally refer to these responses as $x(0^-)$ and $x(\infty)$.
2. Identify the initial conditions for the circuit $x(0^+)$, and $\dot{x}(0^+)$, using the continuity of capacitor voltages and inductor currents $[v_C(0^+) = v_C(0^-), i_L(0^+) = i + L(0^-)]$ and circuits analysis. This will be illustrated by examples.

(Continued)

FOCUS ON METHODOLOGY

(Concluded)

3. Write the differential equation of the circuit for $t = 0^+$, that is, immediately after the switch has changed position. The variable $x(t)$ in the differential equation will be either a capacitor voltage $v_C(t)$ or an inductor current $i_L(t)$. Reduce this equation to standard form (equation 5.9, or 5.49).
4. Solve for the parameters of the second-order circuit, ω_n and ζ.
5. Write the complete solution for the circuit in one of the three forms given below as appropriate:

Overdamped case ($\zeta > 1$):

$$x(t) = x_N(t) + x_F(t) = \alpha_1 e^{\left(-\zeta\omega_n + \omega_n\sqrt{\zeta^2-1}\right)t} + \alpha_2 e^{\left(-\zeta\omega_n - \omega_n\sqrt{\zeta^2-1}\right)t} + x(\infty) \qquad t \geq 0$$

Critically damped case ($\zeta = 1$):

$$x(t) = x_N(t) + x_F(t) = \alpha_1 e^{-\zeta\omega_n t} + \alpha_2 t e^{-\zeta\omega_n t} + x(\infty) \qquad t \geq 0$$

Underdamped case ($\zeta < 1$):

$$x(t) = x_N(t) + x_F(t) = \alpha_1 e^{\left(-\zeta\omega_n + j\omega_n\sqrt{1-\zeta^2}\right)t} + \alpha_2 e^{\left(-\zeta\omega_n - j\omega_n\sqrt{1-\zeta^2}\right)t} + x(\infty) \qquad t \geq 0$$

6. Apply the initial conditions to solve for the constants α_1 and α_2.

LO5

EXAMPLE 5.14 Complete response of overdamped second-order circuit

Problem

Determine the complete response of the circuit shown in Figure 5.50 by solving the differential equation for the current $i_L(t)$.

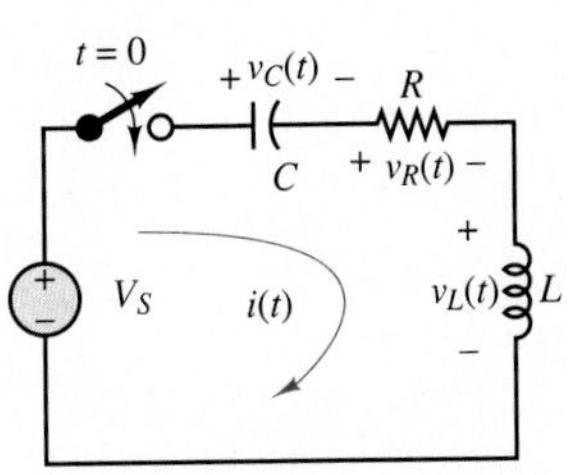

$R = 5000\ \Omega$ $L = 1$ H $C = 1\ \mu$F
$V_S = 25$ V

Figure 5.50

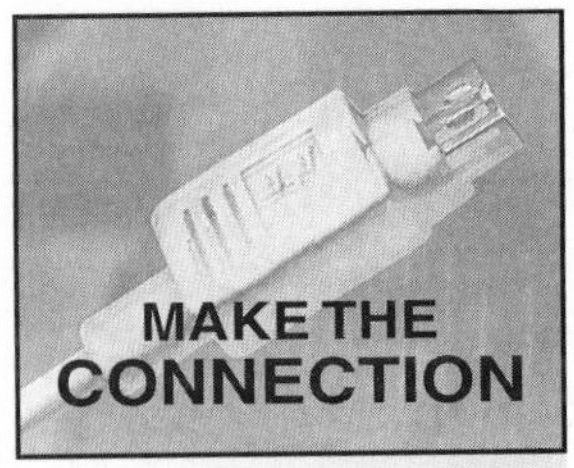

Automotive Suspension

LO6

The mechanical system model described in an earlier Make the Connection sidebar can serve as an approximate representation of an automotive suspension system. The mass m represents the vehicle mass, the spring represents the suspension strut (or coils), and the damper models the shock absorbers. The differential equation of this second-order system is given below.

$$\frac{m}{k}\frac{d^2x_{body}(t)}{dt^2} + \frac{b}{k}\frac{dx_{body}(t)}{dt} + x_{body}(t) = \frac{1}{k}x_{road}(t) + \frac{b}{k}\frac{dx_{road}(t)}{dt}$$

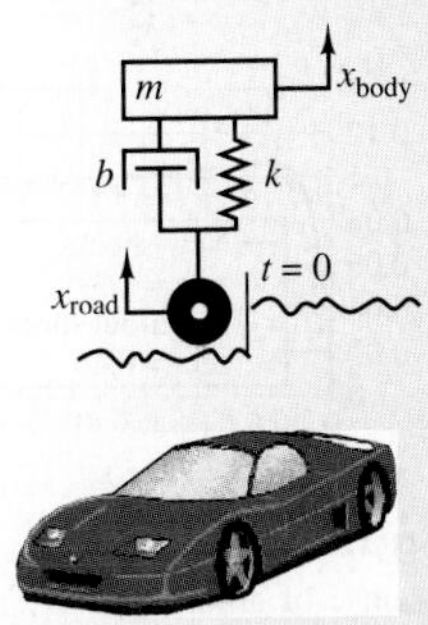

$m = 1{,}500$ kg
$k = 20{,}000$ N/m
$b_{new} = 15{,}000$ N-s/m
$b_{old} = 5{,}000$ N-s/m

Figure 5.48 Automotive suspension system

(Continued)

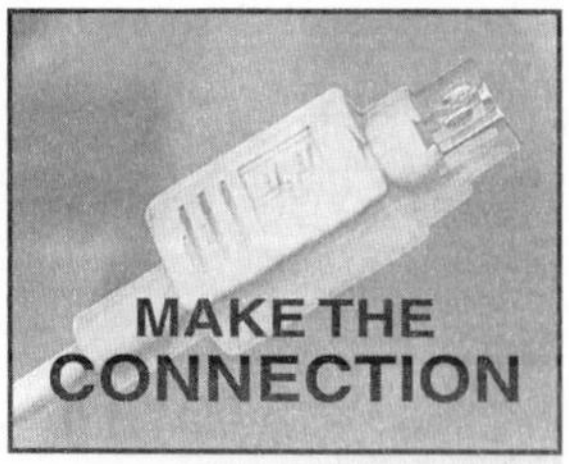

(Concluded)

The input to the suspension system is the road surface profile, which generates both displacement and velocity inputs x_{road} and $\dot{x}_{road}$. One objective of the suspension is to isolate the body of the car (i.e., the passengers) from any vibration caused by unevenness in the road surface. Automotive suspension systems are also very important in guaranteeing vehicle stability and in providing acceptable handling. In this illustration we simply consider the response of the vehicle to a sharp step of amplitude 10 cm (see Figure 5.49) for two cases, corresponding to new and worn-out shock absorbers, respectively. Which ride would you prefer?

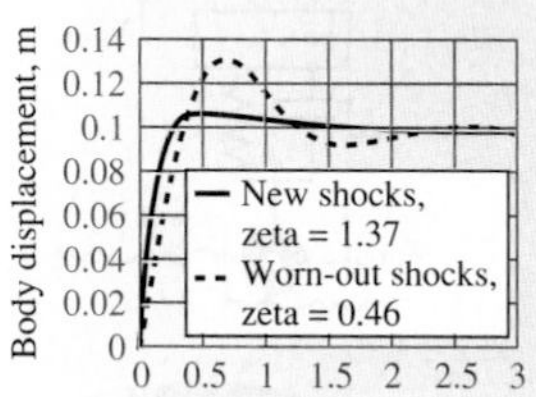

Figure 5.49 "Step" response of automotive suspension

Solution

Known Quantities: Circuit elements.

Find: The complete response of the differential equation in $i_L(t)$ describing the circuit of Figure 5.50.

Schematics, Diagrams, Circuits, and Given Data: $V_S = 25$ V; $R = 5$ kΩ; $C = 1\ \mu$F; $L = 1$ H.

Assumptions: The capacitor has been charged (through a separate circuit, not shown) prior to the switch closing, such that $v_C(0) = 5$ V.

Analysis:

Step 1: Steady-state response. Before the switch closes, the current in the circuit must be zero. We are therefore sure that the inductor current is initially zero: $i_L(0^-) = 0$. We cannot know, in general, what the state of charge of the capacitor is. The problem statement tells us that $v_C(0^-) = 5$ V. This fact will be useful later, when we determine the initial conditions.

After the switch has been closed for a long time and all the transients have died, the capacitor becomes an open circuit, and the inductor behaves as a short circuit. Since the open circuit prevents any current flow, the voltage across the resistor will be zero. Similarly, the inductor voltage is zero, and therefore the source voltage will appear across the capacitor. Hence, $i_L(\infty) = 0$ and $v_C(\infty) = 25$ V.

Step 2: Initial conditions. Recall that for a second-order circuit, we need to determine two initial conditions; and recall that, from continuity of inductor voltage and capacitor currents, we know that $i_L(0^-) = i_L(0^+) = 0$ and $v_C(0^-) = v_C(0^+) = 5$ V. Since the differential equation is in the variable i_L, the two initial conditions we need to determine are $i_L(0^+)$ and $di_L(0^+)/dt$. These can actually be found rather easily by applying KVL at $t = 0^+$:

$$V_S - v_C(0^+) - Ri_L(0^+) - v_L(0^+) = 0$$

$$V_S - v_C(0^+) - Ri_L(0^+) - L\frac{di_L(0^+)}{dt} = 0$$

$$\frac{di_L(0^+)}{dt} = \frac{V_S}{L} - \frac{v_C(0^+)}{L} = 25 - 5 = 20 \text{ V}$$

Step 3: Differential equation. The differential equation for the series circuit can be obtained by KVL:

$$V_S - v_C - Ri_L(t) - L\frac{di_L(t)}{dt} = 0$$

After substituting

$$v_C(t) = \frac{1}{C}\int_{-\infty}^{t} i_L(t')\,dt'$$

we have the equation

$$LC\frac{di_L(t)}{dt} + RCi_L(t) + \int_{-\infty}^{t} i_L(t')\,dt' = CV_S$$

which can be differentiated on both sides to obtain

$$LC\,\frac{d^2 i_L(t)}{dt^2} + RC\,\frac{di_L(t)}{dt} + i_L(t) = C\,\frac{dV_S}{dt} = 0$$

Note that the right-hand side (forcing function) of this differential equation is exactly zero.

Step 4: Solve for ω_n and ζ. If we now compare the second-order differential equations to the standard form of equation 5.50, we can make the following observations:

$$\omega_n = \sqrt{\frac{1}{LC}} = 1,000\ \text{rad/s}$$

$$\zeta = RC\frac{\omega_n}{2} = \frac{R}{2}\sqrt{\frac{C}{L}} = 2.5$$

Thus, the second-order circuit is overdamped.

Step 5: Write the complete solution. Knowing that the circuit is overdamped, we write the complete solution for this case:

$$x(t) = x_N(t) + x_F(t) = \alpha_1 e^{\left(-\zeta\omega_n + \omega_n\sqrt{\zeta^2-1}\right)t} + \alpha_2 e^{\left(-\zeta\omega_n - \omega_n\sqrt{\zeta^2-1}\right)t} + x(\infty) \qquad t \ge 0$$

and since $x_F = x(\infty) = 0$, the complete solution is identical to the homogeneous solution:

$$i_L(t) = i_{LN}(t) = \alpha_1 e^{\left(-\zeta\omega_n + \omega_n\sqrt{\zeta^2-1}\right)t} + \alpha_2 e^{\left(-\zeta\omega_n - \omega_n\sqrt{\zeta^2-1}\right)t} \qquad t \ge 0$$

Step 6: Solve for the constants α_1 and α_2. Finally, we solve for the initial conditions to evaluate the constants α_1 and α_2. The first initial condition yields

$$i_L(0^+) = \alpha_1 e^0 + \alpha_2 e^0$$

$$\alpha_1 = -\alpha_2$$

The second initial condition is evaluated as follows:

$$\frac{di_L(t)}{dt} = \left(-\zeta\omega_n + \omega_n\sqrt{\zeta^2-1}\right)\alpha_1 e^{\left(-\zeta\omega_n + \omega_n\sqrt{\zeta^2-1}\right)t} + \left(-\zeta\omega_n - \omega_n\sqrt{\zeta^2-1}\right)\alpha_2 e^{\left(-\zeta\omega_n - \omega_n\sqrt{\zeta^2-1}\right)t}$$

$$\frac{di_L(0^+)}{dt} = \left(-\zeta\omega_n + \omega_n\sqrt{\zeta^2-1}\right)\alpha_1 e^0 + \left(-\zeta\omega_n - \omega_n\sqrt{\zeta^2-1}\right)\alpha_2 e^0$$

Substituting $\alpha_1 = -\alpha_2$, we get

$$\frac{di_L(0^+)}{dt} = \left(-\zeta\omega_n + \omega_n\sqrt{\zeta^2-1}\right)\alpha_1 - \left(-\zeta\omega_n - \omega_n\sqrt{\zeta^2-1}\right)\alpha_1$$

$$= 2\left(\omega_n\sqrt{\zeta^2-1}\right)\alpha_1 = 20$$

$$\alpha_1 = \frac{20}{2\left(\omega_n\sqrt{\zeta^2-1}\right)} = 4.36 \times 10^{-3}$$

$$\alpha_2 = -\alpha_1 = -4.36 \times 10^{-3}$$

We can finally write the complete solution:

$$i_L(t) = 4.36 \times 10^{-3} e^{-208.7t} - 4.36 \times 10^{-3} e^{-4,791.3t} \qquad t \ge 0$$

A plot of the complete solution and of its components is given in Figure 5.51.

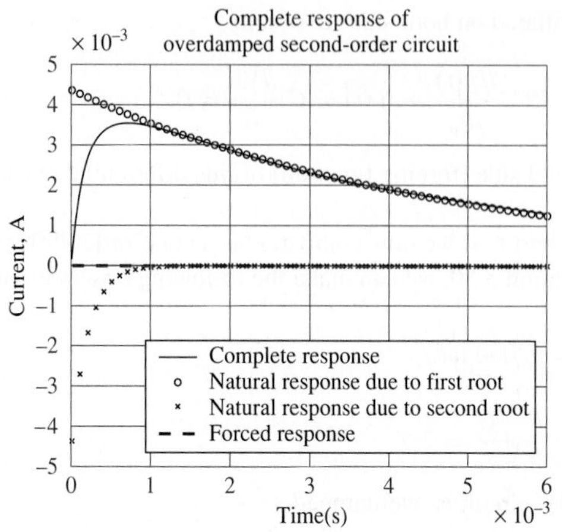

Figure 5.51 Complete response of overdamped second order circuit

CHECK YOUR UNDERSTANDING

Obtain the differential eqation of the circuit of Figure 5.50 with v_C as the independent variable.

Answer: $LC\dfrac{d^2v_C}{dt^2} + RC\dfrac{dv_C}{dt} + v_C = V_S$

LO5

EXAMPLE 5.15 Complete Response of Critically Damped Second-Order Circuit

Problem

Determine the complete response of the circuit shown in Figure 5.52 by solving the differential equation for the voltage $v(t)$.

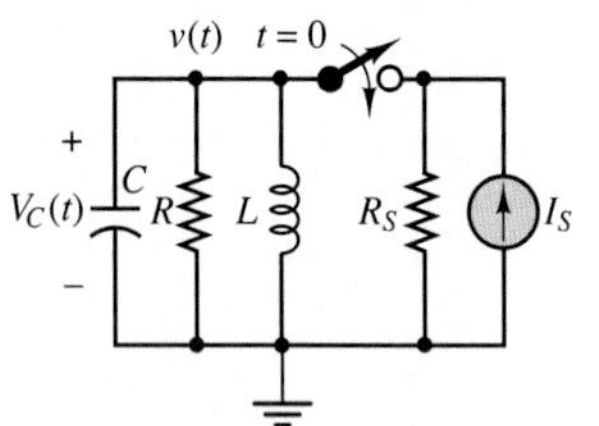

Figure 5.52 Circuit for Example 5.15

Solution

Known Quantities: Circuit elements.

Find: The complete response of the differential equation in $i_L(t)$ describing the circuit of Figure 5.52.

Schematics, Diagrams, Circuits, and Given Data: $I_S = 5$ A; $R = 500\ \Omega$; $C = 2\ \mu$F; $L = 2$ H.

Assumptions: None.

Analysis:

Step 1: Steady-state response. With the switch open for a long time, any energy stored in the capacitor and inductor has had time to be dissipated by the resistor; thus, the currents and voltages in the circuit are zero: $i_L(0^-) = 0$, $v_C(0^-) = v(0^-) = 0$.

After the switch has been closed for a long time and all the transients have died, the capacitor becomes an open circuit, and the inductor behaves as a short circuit. With the inductor behaving as a short circuit, all the source current will flow through the inductor, and $i_L(\infty) = I_S = 5$ A. On the other hand, the current through the resistor is zero, and therefore $v_C(\infty) = v(\infty) = 0$ V.

Step 2: Initial conditions. Recall that for a second-order circuit we need to determine two initial conditions; and recall that, from continuity of inductor voltage and capacitor currents, we know that $i_L(0^-) = i_L(0^+) = 0$ A and $v_C(0^-) = v_C(0^+) = 0$ V. Since the differential equation is in the variable v_C, the two initial conditions we need to determine are $v_C(0^+)$ and $dv_C(0^+)/dt$. These can actually be found rather easily by applying KCL at $t = 0^+$:

$$I_S - \frac{v_C(0^+)}{R_S} - i_L(0^+) - \frac{v_C(0^+)}{R} - C\frac{dv_C(0^+)}{dt} = 0$$

Since $v_C(0^+) = 0$ and $i_L(0^+) = 0$, we can easily determine $dv_C(0^+)/dt$:

$$\frac{dv_C(0^+)}{dt} = \frac{I_S}{C} = \frac{5}{2 \times 10^{-6}} = 2.5 \times 10^6 \ \frac{\text{V}}{\text{s}}$$

Step 3: Differential equation. The differential equation for the series circuit can be obtained by KCL:

$$I_S - \frac{v_C(t)}{R_S} - i_L(t) - \frac{v_C(t)}{R} - C\frac{dv_C(t)}{dt} = 0 \qquad t \geq 0$$

Knowing that

$$v_C(t) = v_L(t) = L\frac{di_L(t)}{dt}$$

we can obtain a relationship

$$i_L(t) = \frac{1}{L}\int_{-\infty}^{t} v_C(t')\,dt'$$

resulting in the integrodifferential equation

$$I_S - \frac{v_C(t)}{R_S} - \frac{1}{L}\int_{-\infty}^{t} v_C(t')\,dt' - \frac{v_C(t)}{R} - C\frac{dv_C(t)}{dt} = 0 \qquad t \geq 0$$

which can be differentiated on both sides to obtain

$$LC\frac{d^2v_C(t)}{dt^2} + \frac{L\,(R_S + R)}{R_S R}\frac{dv_C(t)}{dt} + v_C(t) = L\frac{dI_S}{dt} \qquad t \geq 0$$

Note that the right-hand side (forcing function) of this differential equation is exactly zero.

Step 4: Solve for ω_n and ζ. If we now compare the second-order differential equations to the standard form of equation 5.50, we can make the following observations:

$$\omega_n = \sqrt{\frac{1}{LC}} = 500 \text{ rad/s}$$

$$\zeta = \frac{L}{R_{\text{eq}}}\frac{\omega_n}{2} = \frac{1}{2R_{\text{eq}}}\sqrt{\frac{L}{C}} = 1$$

$$R_{\text{eq}} = \frac{RR_S}{R + R_S}$$

Thus, the second-order circuit is critically damped.

Step 5: Write the complete solution. Knowing that the circuit is critically damped ($\zeta = 1$), we write the complete solution for this case:

$$x(t) = x_N(t) + x_F(t) = \alpha_1 e^{-\zeta\omega_n t} + \alpha_2 t e^{-\zeta\omega_n t} + x(\infty) \qquad t \geq 0$$

and, since $v_{CF} = v_C(\infty) = 0$, the complete solution is identical to the homogeneous solution:

$$v_C(t) = V_{CN}(t) = \alpha_1 e^{-\zeta\omega_n t} + \alpha_2 t e^{-\zeta\omega_n t} \qquad t \geq 0$$

Step 6: Solve for the constants α_1 and α_2. Finally, we solve for the initial conditions to evaluate the constants α_1 and α_2. The first initial condition yields

$$v_C(0^+) = \alpha_1 e^0 + \alpha_2 \cdot 0 \cdot e^0 = 0$$
$$\alpha_1 = 0$$

The second initial condition is evaluated as follows:

$$\frac{dv_C(t)}{dt} = (-\zeta\omega_n)\alpha_1 e^{-\zeta\omega_n t} + (-\zeta\omega_n)\alpha_2 t e^{-\zeta\omega_n t} + \alpha_2 e^{-\zeta\omega_n t}$$
$$\frac{dv_C(0^+)}{dt} = (-\zeta\omega_n)\alpha_1 e^0 + \alpha_2 e^0 = \alpha_2$$
$$\alpha_2 = 2.5 \times 10^6$$

We can finally write the complete solution

$$v_C(t) = 2.5 \times 10^6 t e^{-500t} \qquad t \geq 0$$

A plot of the complete solution and of its components is given in Figure 5.53.

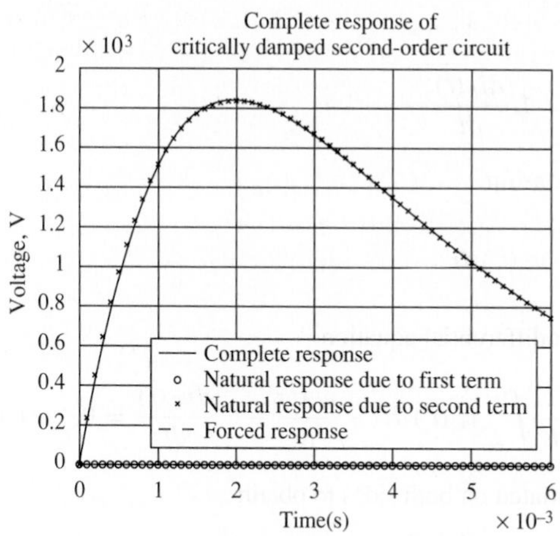

Figure 5.53 Complete response of overdamped second-order circuit

---

CHECK YOUR UNDERSTANDING

Obtain the differential equation of the circuit of Figure 5.52 with i_L as the independent variable.

Answer: $LC\frac{d^2 i_L(t)}{dt^2} + \frac{L}{R_{eq}}\frac{di_L(t)}{dt} + i_L(t) = I_S \qquad t \geq 0$

EXAMPLE 5.16 Complete Response of Underdamped Second-Order Circuit

LO5

Problem

Determine the complete response of the circuit shown in Figure 5.54 by solving the differential equation for the current $i_L(t)$.

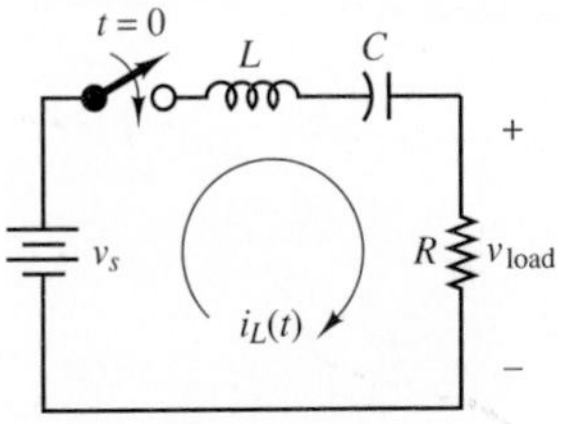

Figure 5.54 Circuit for Example 5.16

Solution

Known Quantities: Circuit elements.

Find: The complete response of the differential equation in $i_L(t)$ describing the circuit of Figure 5.54.

Schematics, Diagrams, Circuits, and Given Data: $V_S = 12$ V; $R = 200\ \Omega$; $C = 10\ \mu$F; $L = 0.5$ H.

Assumptions: The capacitor had been previously charged (through a separate circuit, not shown), such that $v_C(0^-) = v_C(0^+) = 2$ V.

Analysis:

Step 1: Steady-state response. Since the inductor current is zero when the switch is open, $i_L(0^-) = 0$; on the other hand, we have assumed that $v_C(0^-) = v(0^-) = 2$ V. After the switch has been closed for a long time and all the transients have died, the capacitor becomes an open circuit, and the inductor behaves as a short circuit. With the capacitor behaving as an open circuit, all the source voltage will appear across the capacitor, and, of course, the inductor current is zero: $i_L(\infty) = 0$ A, $v_C(\infty) = V_S = 12$ V.

Step 2: Initial conditions. Recall that for a second-order circuit we need to determine two initial conditions; and recall that, from continuity of inductor voltage and capacitor currents, we know that $i_L(0^-) = i_L(0^+) = 0$ A and $v_C(0^-) = v_C(0^+) = 2$ V. Since the differential equation is in the variable i_L, the two initial conditions we need to determine are $i_L(0^+)$ and $di_L(0^+)/dt$. The second initial condition can be found by applying KVL at $t = 0^+$:

$$V_S - v_C(0^+) - Ri_L(0^+) - v_L(0^+) = 0$$

$$V_S - v_C(0^+) - Ri_L(0^+) - L\frac{di_L(0^+)}{dt} = 0$$

$$\frac{di_L(0^+)}{dt} = \frac{V_S}{L} - \frac{v_C(0^+)}{L} = \frac{12}{0.5} - \frac{2}{0.5} = 20 \text{ A/s}$$

Step 3: Differential equation. The differential equation for the series circuit is obtained by KCL:

$$V_S - L\frac{di_L(t)}{dt} - v_C(t) - Ri_L(t) = 0 \qquad t \geq 0$$

Knowing that

$$i_L(t) = C\frac{dv_C(t)}{dt}$$

we can obtain the relationship

$$v_C(t) = \frac{1}{C}\int_{-\infty}^{t} i_L(t')\,dt'$$

resulting in the integrodifferential equation

$$V_S - L\frac{di_L(t)}{dt} - \frac{1}{C}\int_{-\infty}^{t} i_L(t')dt' - Ri_L(t) = 0 \qquad t \geq 0$$

which can be differentiated on both sides to obtain

$$LC\frac{d^2 i_L(t)}{dt^2} + RC\frac{di_L(t)}{dt} + i_L(t) = C\frac{dV_S}{dt} \qquad t \geq 0$$

Note that the right-hand side (forcing function) of this differential equation is exactly zero, since V_S is a constant.

Step 4: Solve for ω_n and ζ. If we now compare the second-order differential equations to the standard form of equation 5.50, we can make the following observations:

$$\omega_n = \sqrt{\frac{1}{LC}} = 447 \text{ rad/s}$$

$$\zeta = RC\frac{\omega_n}{2} = \frac{R}{2}\sqrt{\frac{C}{L}} = 0.447$$

Thus, the second-order circuit is underdamped.

Step 5: Write the complete solution. Knowing that the circuit is underdamped ($\zeta < 1$), we write the complete solution for this case as

$$x(t) = x_N(t) + x_F(t) = \alpha_1 e^{\left(-\zeta\omega_n + j\omega_n\sqrt{1-\zeta^2}\right)t} + \alpha_2 e^{\left(-\zeta\omega_n - j\omega_n\sqrt{1-\zeta^2}\right)t} + x(\infty) \quad t \geq 0$$

and since $x_F = i_{LF} = i_L(\infty) = 0$, the complete solution is identical to the homogeneous solution:

$$i_L(t) = i_{LN}(t) = \alpha_1 e^{\left(-\zeta\omega_n + j\omega_n\sqrt{1-\zeta^2}\right)t} + \alpha_2 e^{\left(-\zeta\omega_n - j\omega_n\sqrt{1-\zeta^2}\right)t} \qquad t \geq 0$$

Step 6: Solve for the constants α_1 and α_2. Finally, we solve for the initial conditions to evaluate the constants α_1 and α_2. The first initial condition yields

$$i_L(0^+) = \alpha_1 e^0 + \alpha_2 e^0 = 0$$

$$\alpha_1 = -\alpha_2$$

The second initial condition is evaluated as follows:

$$\frac{di_L(t)}{dt} = \left(-\zeta\omega_n + j\omega_n\sqrt{1-\zeta^2}\right)\alpha_1 e^{\left(-\zeta\omega_n + j\omega_n\sqrt{1-\zeta^2}\right)t} + \left(-\zeta\omega_n - j\omega_n\sqrt{1-\zeta^2}\right)\alpha_2 e^{\left(-\zeta\omega_n - j\omega_n\sqrt{1-\zeta^2}\right)t}$$

$$\frac{di_L(0^+)}{dt} = \left(-\zeta\omega_n + j\omega_n\sqrt{1-\zeta^2}\right)\alpha_1 e^0 + \left(-\zeta\omega_n - j\omega_n\sqrt{1-\zeta^2}\right)\alpha_2 e^0$$

Substituting $\alpha_1 = -\alpha_2$, we get

$$\frac{di_L(0^+)}{dt} = \left(-\zeta\omega_n + j\omega_n\sqrt{1-\zeta^2}\right)\alpha_1 - \left(-\zeta\omega_n - j\omega_n\sqrt{1-\zeta^2}\right)\alpha_1 = 20 \text{ V}$$

$$2\left(j\omega_n\sqrt{1-\zeta^2}\right)\alpha_1 = 20$$

$$\alpha_1 = \frac{10}{j\omega_n\sqrt{1-\zeta^2}} = -j\frac{10}{\omega_n\sqrt{1-\zeta^2}} = -j0.025$$

$$\alpha_2 = -\alpha_1 = j0.025$$

We can finally write the complete solution:

$$\begin{aligned} i_L(t) &= -j0.025e^{(-200+j400)t} + j0.025e^{(-200-j400)t} \qquad t \geq 0 \\ &= 0.025e^{-200t}(-je^{j400t} + je^{-j400t}) = 0.025e^{-200t} \times 2\sin 400t \\ &= 0.05e^{-200t}\sin 400t \qquad t \geq 0 \end{aligned}$$

In the above equation, we have used Euler's identity to obtain the final expression. A plot of the complete solution and of its components is given in Figure 5.55.

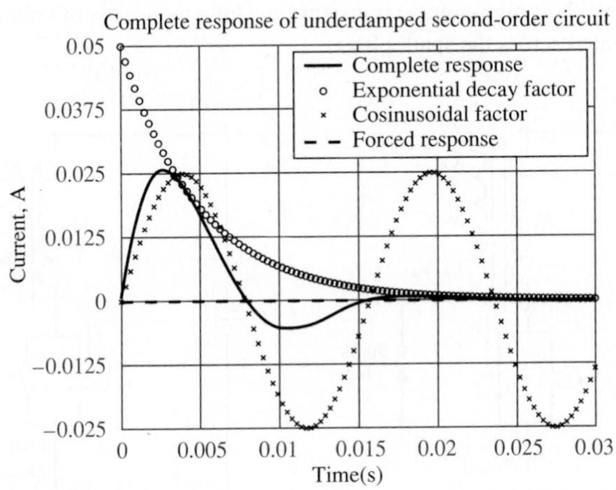

Figure 5.55 Complete response of underdamped second-order circuit

CHECK YOUR UNDERSTANDING

If the inductance is reduced to one-half of its original value (from 0.5 to 0.25 H), for what range of values of R will the circuit be underdamped?

Answer: $R \leq 316\ \Omega$

EXAMPLE 5.17 Transient Response of Automotive Ignition Circuit

LO5

Problem

FIND IT ON THE WEB

The circuit shown in Figure 5.56 is a simplified but realistic representation of an automotive ignition system. The circuit includes an **automotive battery,** a transformer[3] **(ignition coil),** a capacitor (known as *condenser* in old-fashioned automotive parlance), and a switch. The

[3]Transformers are discussed more formally in Chapters 7 and 17; the operation of the transformer in an ignition coil will be explained ad hoc in this example.

switch is usually an electronic switch (e.g., a transistor—see Chapter 10) and can be treated as an ideal switch. The circuit on the left represents the ignition circuit immediately after the electronic switch has closed, following a spark discharge. Thus, one can assume that no energy is stored in the inductor prior to the switch closing, say at $t = 0$. Furthermore, no energy is stored in the capacitor, as the short circuit (closed switch) across it would have dissipated any charge in the capacitor. The primary winding of the ignition coil (left-hand side inductor) is then given a suitable length of time to build up stored energy, and then the switch opens, say at $t = \Delta t$, leading to a rapid voltage buildup across the secondary winding of the coil (right-hand side inductor). The voltage rises to a very high value because of two effects: the *inductive voltage kick* described in Example 5.11 and the voltage multiplying effect of the transformer. The result is a very short high-voltage transient (reaching thousands of volts), which causes a spark to be generated across the spark plug.

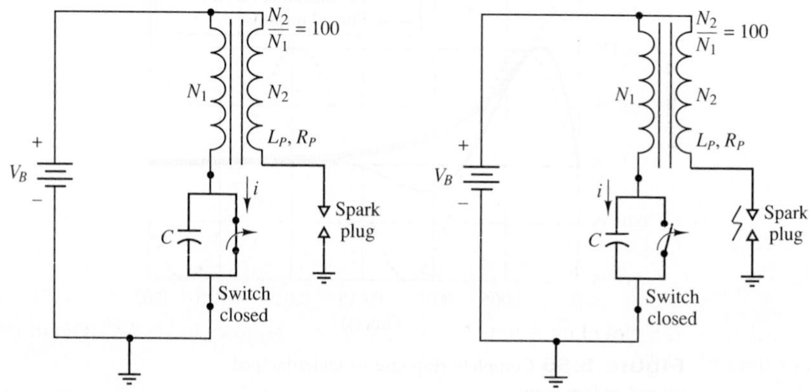

Figure 5.56

Solution

Known Quantities: Battery voltage, resistor, capacitor, inductor values.

Find: The ignition coil current $i(t)$ and the open-circuit voltage across the spark plug $v_{OC}(t)$.

Schematics, Diagrams, Circuits, and Given Data: $V_B = 12$ V; $R_p = 2\ \Omega$; $C = 10\ \mu$F; $L_p = 5$ mH.

Assumptions: The switch has been open for a long time, and it closes at $t = 0$. The switch opens again at $t = \Delta t$.

Analysis: Assume that initially no energy is stored in either the inductor or the capacitor, and that the switch is closed, as shown in Figure 5.57(a). When the switch is closed, a first-order circuit is formed by the primary coil inductance and capacitance. The solution of this circuit will now give us the initial condition that will be in effect when the switch is ready to open again. This circuit is identical to that analyzed in Example 5.9, and we can directly borrow the solution obtained from that example, after suitably replacing the final value and time constant:

$$i_L(t) = i_L(\infty) + [i_L(0) - i_L(\infty)]e^{-t/\tau} \qquad t \geq 0$$

$$i_L(t) = 6(1 - e^{-t/2.5\times 10^{-3}}) \qquad t \geq 0$$

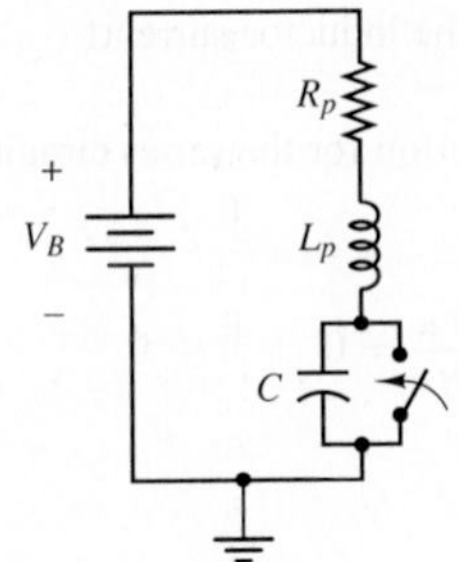

Figure 5.57 (a)
First-order transient circuit

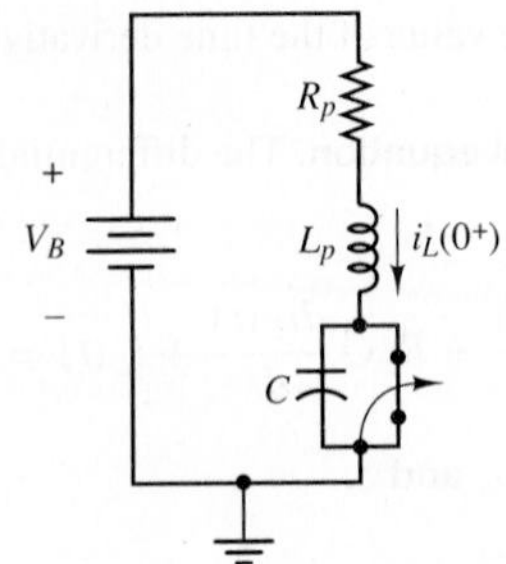

Figure 5.57 (b)
Second-order transient circuit, following opening of switch

with

$$i_{LSS}(\infty) = \frac{V_B}{R_p} = 6 \text{ A} \qquad \text{Final value}$$

$$\tau = \frac{L_p}{R_p} = 2.5 \times 10^{-3} \text{ s} \qquad \text{Time constant}$$

Let the switch remain closed until $t = \Delta t = 12.5$ ms $= 5\tau$. At time Δt, the value of the inductor current will be

$$i_L(t = \Delta t) = 6(1 - e^{-5}) = 5.96 \text{ A}$$

that is, the current reaches 99 percent of its final value in five time constants.

Now, when the switch opens at $t = \Delta t$, we are faced with a series *RLC* circuit similar to that of Example 5.16, except for the fact that the initial condition that is nonzero is the inductor current (in Example 5.16 you will recall that the capacitor had a nonzero initial condition). We can therefore borrow the solution to Example 5.16, with some slight modifications because of the difference in initial conditions, as shown below. Please note that even though the second-order circuit transient starts at $t = 12.5$ ms, we have "reset" the time to $t = 0$ for simplicity in writing the solution. You should be aware that the solution below actually starts at $t = 12.5$ ms.

Step 1: Steady-state response. At $t = \Delta t$, $i_L(0^-) = 6$ A; $v_C(0^-) = v(0^-) = 0$ V. After the switch has been closed for a long time and all the transients have died, the capacitor becomes an open circuit, and the inductor behaves as a short circuit. With the capacitor behaving as an open circuit, all the source voltage will appear across the capacitor, and, of course, the inductor current is zero: $i_L(\infty) = 0$ A, $v_C(\infty) = V_S = 12$ V.

Step 2: Initial conditions. Since the differential equation is in the variable i_L, the two initial conditions we need to determine are $i_L(0^+)$ and $di_L(0^+)/dt$. Now $i_L(0^-) = i_L(0^+) =$ 5.96 A. The second initial condition can be found by applying KVL at $t = 0^+$

$$V_B - v_C(0^+) - Ri_L(0^+) - L\frac{di_L(0^+)}{dt} = 0$$

$$\frac{di_L(0^+)}{dt} = \frac{V_B}{L} - \frac{v_C(0^+)}{L} - Ri_L(0^+) = \frac{12}{5 \times 10^{-3}} - 0 - 2 \times 5.96$$
$$= 2{,}388 \text{ A/s}$$

Note the very large value of the time derivative of the inductor current!

Step 3: Differential equation. The differential equation for the series circuit can be obtained by KCL:

$$L_pC\frac{d^2i_L(t)}{dt^2}+R_pC\frac{di_L(t)}{dt}+i_L(t)=C\frac{dV_B}{dt}=0 \qquad t\geq 0$$

Step 4: Solve for ω_n and ζ.

$$\omega_n=\sqrt{\frac{1}{L_pC}}=4{,}472\text{ rad/s}$$

$$\zeta=R_pC\frac{\omega_n}{2}=\frac{R_p}{2}\sqrt{\frac{C}{L_p}}=0.0447$$

Thus, the ignition circuit is underdamped.

Step 5: Write the complete solution.

$$i_L(t)=i_{LN}(t)=\alpha_1e^{\left(-\zeta\omega_n+j\omega_n\sqrt{1-\zeta^2}\right)t}+\alpha_2e^{\left(-\zeta\omega_n-j\omega_n\sqrt{1-\zeta^2}\right)t} \qquad t\geq 0$$

Step 6: Solve for the constants α_1 and α_2. Finally, we solve for the initial conditions to evaluate the constants α_1 and α_2. The first initial condition yields

$$i_L(0^+)=\alpha_1e^0+\alpha_2e^0=5.96\text{ A}$$
$$\alpha_1=5.96-\alpha_2$$

The second initial condition is evaluated as follows:

$$\frac{di_L(0^+)}{dt}=\left(-\zeta\omega_n+j\omega_n\sqrt{1-\zeta^2}\right)\alpha_1e^0+\left(-\zeta\omega_n-j\omega_n\sqrt{1-\zeta^2}\right)\alpha_2e^0$$

Substituting $\alpha_1=5.96-\alpha_2$, we get

$$\frac{di_L(0^+)}{dt}=\left(-\zeta\omega_n+j\omega_n\sqrt{1-\zeta^2}\right)\alpha_1$$
$$+\left(-\zeta\omega_n-j\omega_n\sqrt{1-\zeta^2}\right)(5.96-\alpha_1)=2{,}388\text{ V}$$

$$2\left(j\omega_n\sqrt{1-\zeta^2}\right)\alpha_1+5.96\left(-\zeta\omega_n-j\omega_n\sqrt{1-\zeta^2}\right)=2{,}388\text{ V}$$

$$\alpha_1=\frac{2{,}388-5.96\left(-\zeta\omega_n-j\omega_n\sqrt{1-\zeta^2}\right)}{2j\omega_n\sqrt{1-\zeta^2}}=2.98-j0.4$$

$$\alpha_2=5.96-\alpha_1=2.98+j0.4$$

and we can finally write the complete solution, as

$$i_L(t)=(2.98-j0.4)e^{\left(-\zeta\omega_n+j\omega_n\sqrt{1-\zeta^2}\right)t}$$
$$+(2.98+j0.4)e^{\left(-\zeta\omega_n-j\omega_n\sqrt{1-\zeta^2}\right)t} \qquad t\geq\Delta t$$

$$\begin{aligned} i_L(t) &= 2.98e^{-\zeta\omega_n t}\left(e^{(j\omega_n\sqrt{1-\zeta^2})t} + e^{(-j\omega_n\sqrt{1-\zeta^2})t}\right) \\ &= 2.98e^{-200t}(e^{j4,463t} + e^{-j4,463t}) = 5.96e^{-200t}\sin(-4,463t) \end{aligned}$$

A plot of the inductor current for $-10 = t = 50$ ms is shown in Figure 5.58. Notice the initial first-order transient at $t = 0$ followed by a second-order transient at $t = 12.5$ ms.

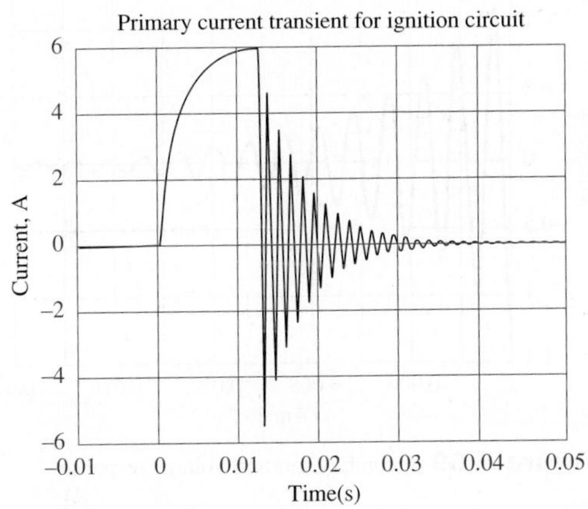

Figure 5.58 Transient current response of ignition current

To compute the primary voltage, we simply differentiate the inductor current and multiply by L; to determine the secondary voltage, which is that applied to the spark plug, we simply remark that a 1:100 transformer increases the voltage by a factor of 100, so that the secondary voltage is 100 times larger than the primary voltage.[4] Thus, the expression for the secondary voltage is

$$\begin{aligned} v_{\text{spark plug}} &= 100 \times L\frac{di_L}{dt} = 0.5 \times \frac{d}{dt}\left[5.96e^{-200t}\sin(-4,463t)\right] \\ &= 0.5 \times 5.96[-200 \times e^{-200t}\sin(-4,463t) - 4,463 \\ &\qquad \times e^{-200t}\cos(4,463t)] \\ &= 596e^{-200t}\sin(4,463t) - 13,300e^{-200t}\cos(4,463t) \end{aligned}$$

where we have "reset" time to $t = 0$ for simplicity. We are actually interested in the value of this voltage at $t = 0$, since this is what will generate the spark; evaluating the above expression at $t = 0$, we obtain

$$v_{\text{spark plug}}(t = 0) = -13,300 \text{ V}$$

One can clearly see that the result of the switching is a very large (negative) voltage spike, capable of generating a spark across the plug gap. A plot of the inductor voltage starting at the time when the switch is opened is shown in Figure 5.59, showing that

[4]The secondary current, on the other hand, will decrease by a factor of 100, so that power is conserved—see Section 7.3.

approximately 0.3 ms after the switching transient, the secondary voltage reaches approximately −12,500 V! This value is rather typical of the voltages required to generate a spark across an automotive plug.

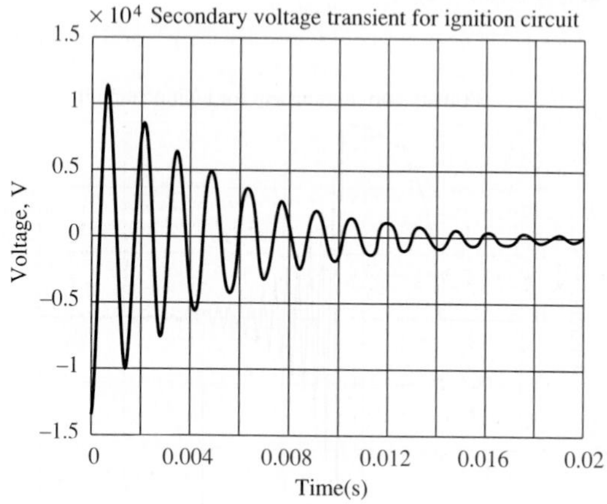

Figure 5.59 Secondary ignition voltage response

Conclusion

Chapter 5 has focused on the solution of first- and second-order differential equations for the case of DC switched transients, and it has presented a number of analogies between electric circuits and other physical systems, such as thermal, hydraulic, and mechanical.

While many other forms of excitation exist, turning a DC supply on and off is a very common occurrence in electrical, electronic, and electromechanical systems. Further, the methods discussed in this chapter can be readily extended to the solution of more general problems.

Upon completing this chapter, you should have mastered the following learning objectives:

1. *Write differential equations for circuits containing inductors and capacitors.* You have seen that writing the differential equations of dynamic circuits involves two concepts: applying Kirchhoff's laws and using the constitutive differential or integral relationships for inductors and capacitors. Often, it is convenient to isolate the purely resistive part of a circuit and reduce it to an equivalent circuit.
2. *Determine the DC steady-state solution of circuits containing inductors and capacitors.* You have learned that the DC steady-state solution of any differential equation can be easily obtained by setting the derivative terms equal to zero. An alternate method for computing the DC steady-state solution is to recognize that under DC steady-state conditions, inductors behave as short circuits and capacitors as open circuits.
3. *Write the differential equation of first-order circuits in standard form, and determine the complete solution of first-order circuits excited by switched DC sources.* First-order systems are most commonly described by way of two constants: the DC gain and the time constant. You have learned how to recognize these constants, how to compute the initial and final conditions, and how to write the complete solution of all first-order circuits almost by inspection.

4. *Write the differential equation of second-order circuits in standard form, and determine the complete solution of second-order circuits excited by switched DC sources.* Second-order circuits are described by three constants: the DC gain, the natural frequency, and the damping ratio. While the method for obtaining the complete solution for a second-order circuit is logically the same as that used for a first-order circuit, some of the details are a little more involved in the second-order case.
5. *Understand analogies between electric circuits and hydraulic, thermal, and mechanical systems.* Many physical systems in nature exhibit the same first- and second-order characteristics as the electric circuits you have studied in this chapter. We have taken a look at some thermal, hydraulic, and mechanical analogies.

HOMEWORK PROBLEMS

Section 5.2: Writing Differential Equations for Circuits Containing Inductors and Capacitors

5.1 Write the differential equation for $t > 0$ for the circuit of Figure P5.21.

5.2 Write the differential equation for $t > 0$ for the circuit of Figure P5.23.

5.3 Write the differential equation for $t > 0$ for the circuit of Figure P5.27.

5.4 Write the differential equation for $t > 0$ for the circuit of Figure P5.29.

5.5 Write the differential equation for $t > 0$ for the circuit of Figure P5.32.

5.6 Write the differential equation for $t > 0$ for the circuit of Figure P5.34.

5.7 Write the differential equation for $t > 0$ for the circuit of Figure P5.41.

5.8 Write the differential equation for $t > 0$ for the circuit of Figure P5.47. Assume $V_S = 9$ V, $R_1 = 10$ kΩ, and $R_2 = 20$ kΩ.

5.9 Write the differential equation for $t > 0$ for the circuit of Figure P5.49.

5.10 Write the differential equation for $t > 0$ for the circuit of Figure P5.52.

Section 5.3: DC Steady-State Solution of Circuits Containing Inductors and Capacitors—Initial and Final Conditions

5.11 Determine the initial and final conditions for the circuit of Figure P5.21.

5.12 Determine the initial and final conditions for the circuit of Figure P5.23.

5.13 Determine the initial and final conditions for the circuit of Figure P5.27.

5.14 Determine the initial and final conditions for the circuit of Figure P5.29.

5.15 Determine the initial and final conditions for the circuit of Figure P5.32.

5.16 Determine the initial and final conditions for the circuit of Figure P5.34.

5.17 Determine the initial and final conditions for the circuit of Figure P5.41.

5.18 Determine the initial and final conditions for the circuit of Figure P5.47. Assume $V_S = 9$ V, $R_1 = 10$ kΩ, and $R_2 = 20$ kΩ.

5.19 Determine the initial and final conditions for the circuit of Figure P5.49.

5.20 Determine the initial and final conditions for the circuit of Figure P5.52.

Section 5.4: Transient Response of First-Order Circuits

5.21 Just before the switch is opened at $t = 0$, the current through the inductor is 1.70 mA in the direction shown in Figure P5.21. Did steady-state conditions exist just before the switch was opened?

$L = 0.9$ mH $\quad V_S = 12$ V

$R_1 = 6$ kΩ $\quad R_2 = 6$ kΩ

$R_3 = 3$ kΩ

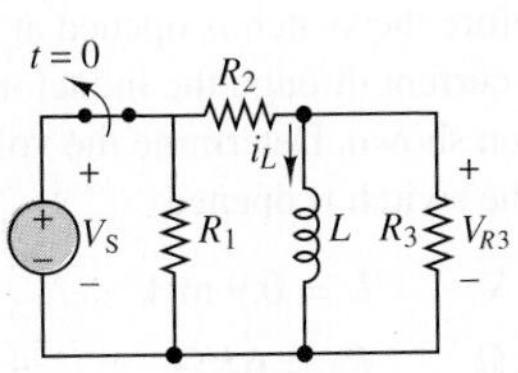

Figure P5.21

5.22 At $t < 0$, the circuit shown in Figure P5.22 is at steady state. The switch is changed as shown at $t = 0$.

$V_{S1} = 35$ V $\quad V_{S2} = 130$ V
$C = 11\ \mu$F $\quad R_1 = 17$ kΩ
$R_2 = 7$ kΩ $\quad R_3 = 23$ kΩ

Determine at $t = 0^+$ the initial current through R_3 just after the switch is changed.

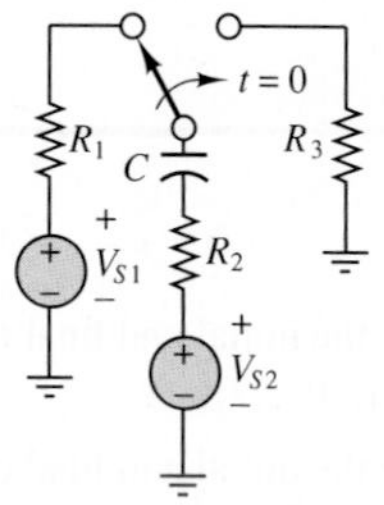

Figure P5.22

5.23 Determine the current through the capacitor just before and just after the switch is closed in Figure P5.23. Assume steady-state conditions for $t < 0$.

$V_1 = 12$ V $\quad C = 0.5\ \mu$F
$R_1 = 0.68$ kΩ $\quad R_2 = 1.8$ kΩ

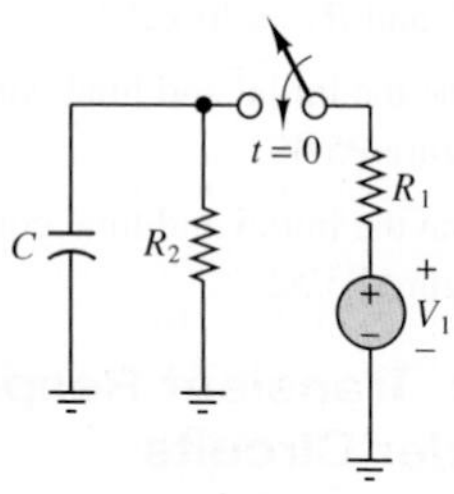

Figure P5.23

5.24 Determine the current through the capacitor just before and just after the switch is closed in Figure P5.23. Assume steady-state conditions for $t < 0$.

$V_1 = 12$ V $\quad C = 150\ \mu$F
$R_1 = 400$ mΩ $\quad R_2 = 2.2$ kΩ

5.25 Just before the switch is opened at $t = 0$ in Figure P5.21, the current through the inductor is 1.70 mA in the direction shown. Determine the voltage across R_3 just after the switch is opened.

$V_S = 12$ V $\quad L = 0.9$ mH
$R_1 = 6$ kΩ $\quad R_2 = 6$ kΩ
$R_3 = 3$ kΩ

5.26 Determine the voltage across the inductor just before and just after the switch is changed in Figure P5.26. Assume steady-state conditions exist for $t < 0$.

$V_S = 12$ V $\quad R_s = 0.7$ Ω
$R_1 = 22$ kΩ $\quad L = 100$ mH

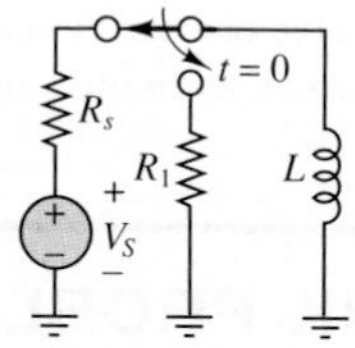

Figure P5.26

5.27 Steady-state conditions exist in the circuit shown in Figure P5.27 at $t < 0$. The switch is closed at $t = 0$.

$V_1 = 12$ V $\quad R_1 = 0.68$ kΩ
$R_2 = 2.2$ kΩ $\quad R_3 = 1.8$ kΩ
$C = 0.47\ \mu$F

Determine the current through the capacitor at $t = 0^+$, just after the switch is closed.

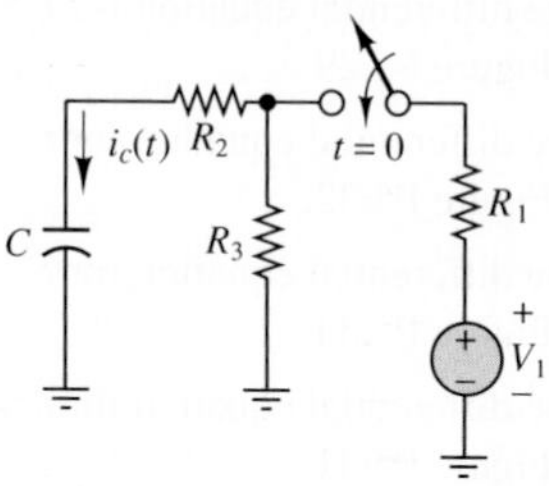

Figure P5.27

5.28 At $t > 0$, the circuit shown in Figure P5.22 is at steady state. The switch is changed as shown at $t = 0$.

$V_{S1} = 35$ V $\quad V_{S2} = 130$ V
$C = 11\ \mu$F $\quad R_1 = 17$ kΩ
$R_2 = 7$ kΩ $\quad R_3 = 23$ kΩ

Determine the time constant of the circuit for $t > 0$.

5.29 At $t < 0$, the circuit shown in Figure P5.29 is at steady state. The switch is changed as shown at $t = 0$.

$V_{S1} = 13$ V $\quad V_{S2} = 13$ V
$L = 170$ mH $\quad R_1 = 2.7$ Ω
$R_2 = 4.3$ kΩ $\quad R_3 = 29$ kΩ

Determine the time constant of the circuit for $t > 0$.

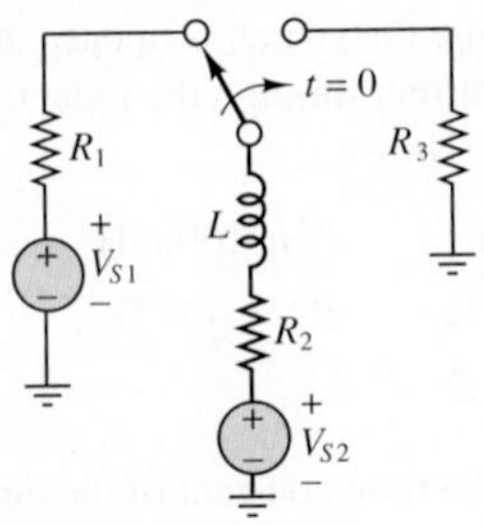

Figure P5.29

5.30 Steady-state conditions exist in the circuit shown in Figure P5.27 for $t < 0$. The switch is closed at $t = 0$.

$V_1 = 12$ V $\quad C = 0.47\ \mu$F

$R_1 = 680\ \Omega \quad R_2 = 2.2$ kΩ

$R_3 = 1.8$ kΩ

Determine the time constant of the circuit for $t > 0$.

5.31 Just before the switch is opened at $t = 0$ in Figure P5.21, the current through the inductor is 1.70 mA in the direction shown.

$V_S = 12$ V $\quad L = 0.9$ mH

$R_1 = 6$ kΩ $\quad R_2 = 6$ kΩ

$R_3 = 3$ kΩ

Determine the time constant of the circuit for $t > 0$.

5.32 Determine $v_C(t)$ for $t > 0$. The voltage across the capacitor in Figure P5.32 just before the switch is changed is given below.

$v_C(0^-) = -7$ V $\quad I_o = 17$ mA $\quad C = 0.55\ \mu$F

$R_1 = 7$ kΩ $\quad R_2 = 3.3$ kΩ

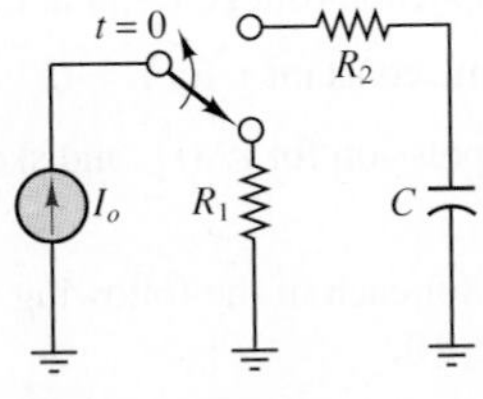

Figure P5.32

5.33 Determine $i_{R_3}(t)$ for $t > 0$ in Figure P5.29.

$V_{S1} = 23$ V $\quad V_{S2} = 20$ V

$L = 23$ mH $\quad R_1 = 0.7\ \Omega$

$R_2 = 13\ \Omega \quad R_3 = 330$ kΩ

5.34 Assume DC steady-state conditions exist in the circuit shown in Figure P5.34 for $t < 0$. The switch is changed at $t = 0$ as shown.

$V_{S1} = 17$ V $\quad V_{S2} = 11$ V

$R_1 = 14$ kΩ $\quad R_2 = 13$ kΩ

$R_3 = 14$ kΩ $\quad C = 70$ nF

Determine

a. $v(t)$ for $t > 0$

b. The time required, after the switch is operated, for $V(t)$ to change by 98 percent of its total change in voltage

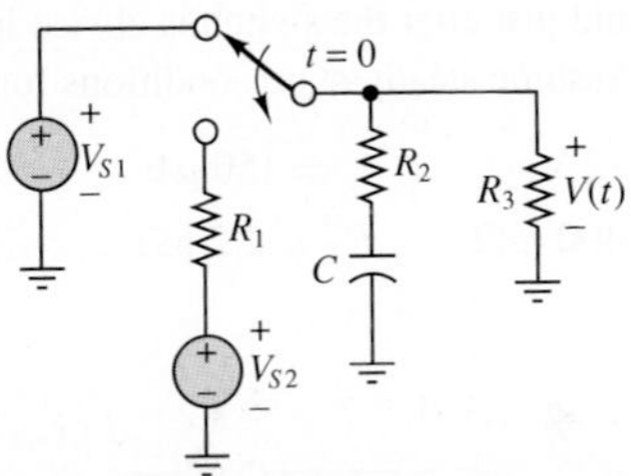

Figure P5.34

5.35 The circuit of Figure P5.35 is a simple model of an automotive ignition system. The switch models the "points" that switch electric power to the cylinder when the fuel-air mixture is compressed. And R is the resistance between the electrodes (i.e., the "gap") of the spark plug.

$V_G = 12$ V $\quad R_G = 0.37\ \Omega$

$R = 1.7$ kΩ

Determine the value of L and R_1 so that the voltage across the spark plug gap just after the switch is changed is 23 kV and so that this voltage will change exponentially with a time constant $\tau = 13$ ms.

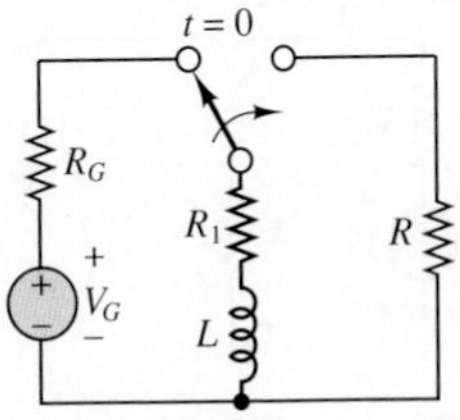

Figure P5.35

5.36 The inductor L in the circuit shown in Figure P5.36 is the coil of a relay. When the current through the coil is equal to or greater than +2 mA, the relay functions. Assume steady-state conditions at $t < 0$. If

$V_S = 12$ V

$L = 10.9$ mH $\quad R_1 = 3.1$ kΩ

determine R_2 so that the relay functions at $t = 2.3$ s.

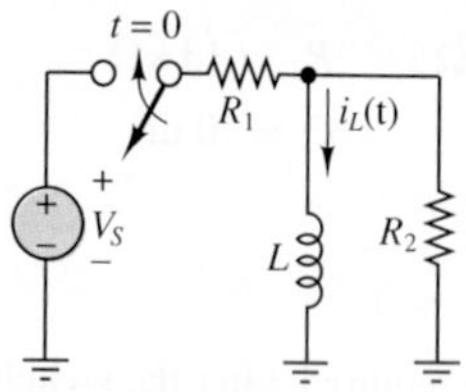

Figure P5.36

5.37 Determine the current through the capacitor just before and just after the switch is closed in Figure P5.37. Assume steady-state conditions for $t < 0$.

$V_1 = 12$ V $\quad C = 150\ \mu$F

$R_1 = 400$ mΩ $\quad R_2 = 2.2$ kΩ

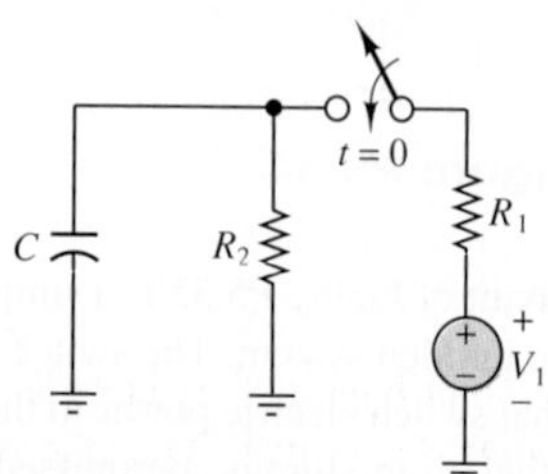

Figure P5.37

5.38 Determine the voltage across the inductor just before and just after the switch is changed in Figure P5.38. Assume steady-state conditions exist for $t < 0$.

$V_S = 12$ V $\quad R_S = 0.24\ \Omega$

$R_1 = 33$ kΩ $\quad L = 100$ mH

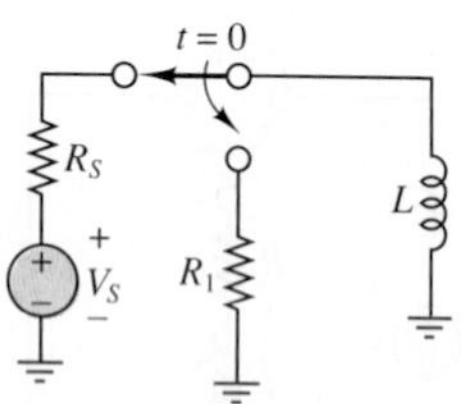

Figure P5.38

5.39 Steady-state conditions exist in the circuit shown in Figure P5.27 for $t < 0$. The switch is closed at $t = 0$.

$V_1 = 12$ V $\quad C = 150\ \mu$F

$R_1 = 4$ MΩ $\quad R_2 = 80$ MΩ

$R_3 = 6$ MΩ

Determine the time constant of the circuit for $t > 0$.

5.40 Just before the switch is opened at $t = 0$ in Figure P5.21, the current through the inductor is 1.70 mA in the direction shown.

$V_S = 12$ V $\quad L = 100$ mH

$R_1 = 400\ \Omega$ $\quad R_2 = 400\ \Omega$

$R_3 = 600\ \Omega$

Determine the time constant of the circuit for $t > 0$.

5.41 For the circuit shown in Figure P5.41, assume that switch S_1 is always open and that switch S_2 closes at $t = 0$.

a. Find the capacitor voltage $v_C(t)$ at $t = 0^+$.

b. Find the time constant τ for $t \geq 0$.

c. Find an expression for $v_C(t)$, and sketch the function.

d. Find $v_C(t)$ for each of the following values of t: $0, \tau, 2\tau, 5\tau, 10\tau$.

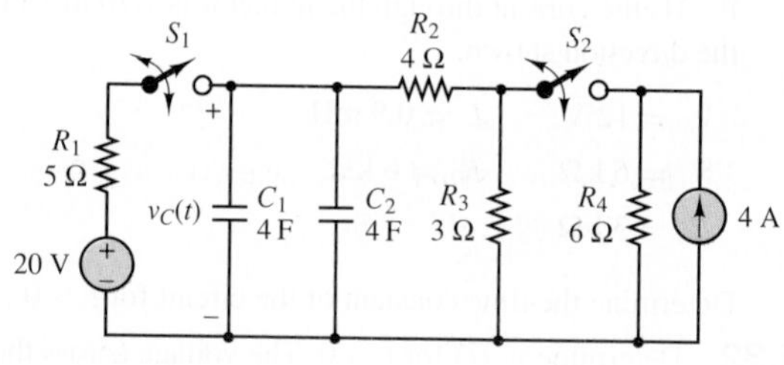

Figure P5.41

5.42 For the circuit shown in Figure P5.41, assume that switch S_1 has been open for a long time and closes at $t = 0$. Conversely, switch S_2 has been closed and opens at $t = 0$.

a. Find the capacitor voltage $v_C(t)$ at $t = 0^+$.

b. Find the time constant τ for $t \geq 0$.

c. Find an expression for $v_C(t)$, and sketch the function.

d. Find $v_C(t)$ for each of the following values of t: $0, \tau, 2\tau, 5\tau, 10\tau$.

5.43 For the circuit of Figure P5.41, assume that switch S_2 is always open, and that switch S_1 has been closed for a long time and opens at $t = 0$. At $t = t_1 = 3\tau$, switch S_1 closes again.

a. Find the capacitor voltage $v_C(t)$ at $t = 0^+$.

b. Find an expression for $v_C(t)$ for $t > 0$, and sketch the function.

5.44 Assume that S_1 and S_2 close at $t = 0$ in Figure P5.41.

a. Find the capacitor voltage $v_C(t)$ at $t = 0^+$.

b. Find the time constant τ for $t \geq 0$.

c. Find an expression for $v_C(t)$, and sketch the function.

d. Find $v_C(t)$ for each of the following values of t: $0, \tau, 2\tau, 5\tau, 10\tau$.

5.45 In the circuit of Figure P5.41, S_1 opens at $t = 0$ and S_2 opens at $t = 48$ s.

a. Find $v_C(t = 0^+)$.

b. Find τ for $0 \leq t \leq 48$ s.

c. Find an expression for $v_C(t)$ valid for $0 \leq t \leq 48$ s.

d. Find τ for $t > 48$ s.

e. Find an expression for $v_C(t)$ valid for $t > 48$ s.

f. Plot $v_C(t)$ for all time.

5.46 For the circuit shown in Figure P5.41, assume that switch S_1 opens at $t = 96$ s and switch S_2 opens at $t = 0$.

a. Find the capacitor voltage at $t = 0$.

b. Find the time constant for $0 < t < 96$ s.

c. Find an expression for $v_C(t)$ when $0 \leq t \leq 96$ s, and compute $v_C(t = 96)$.

d. Find the time constant for $t > 96$ s.

e. Find an expression for $v_C(t)$ when $t > 96$ s.

f. Plot $v_C(t)$ for all time.

5.47 For the circuit of Figure P5.47, determine the value of resistors R_1 and R_2, knowing that the time constant before the switch opens is 1.5 ms, and it is 10 ms after the switch opens. Given: $R_5 = 15\ \text{k}\Omega$, $R_3 = 30\ \text{k}\Omega$, and $C = 1\ \mu\text{F}$.

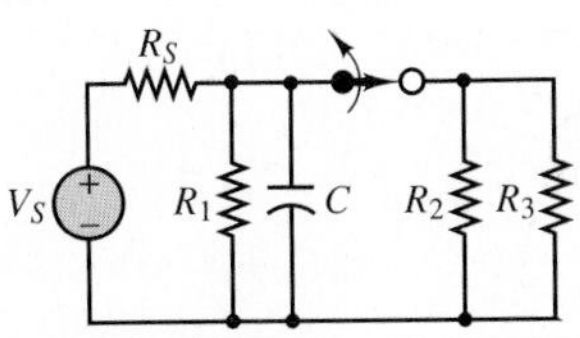

Figure P5.47

5.48 For the circuit of Figure P5.47, assume $V_S = 100$ V, $R_S = 4\ \text{k}\Omega$, $R_1 = 2\ \text{k}\Omega$, $R_2 = R_3 = 6\ \text{k}\Omega$, $C = 1\ \mu\text{F}$, and the circuit is in a steady-state condition before the switch opens. Find the value of v_C 2.666 ms after the switch opens.

5.49 In the circuit of Figure P5.49, the switch changes position at $t = 0$. At what time will the current through the inductor be 5 A? Plot $i_L(t)$.

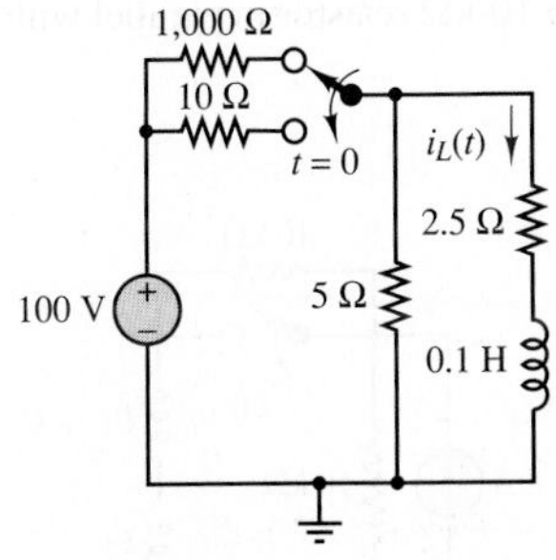

Figure P5.49

5.50 Consider the circuit of Figure P5.49, and assume that the mechanical switching action requires 5 ms. Further assume that during this time, neither switch position has electrical contact. Find

a. $i_L(t)$ for $0 < t < 5$ ms

b. The maximum voltage between the contacts during the 5-ms duration of the switching

Hint: This problem requires solving both a turn-off and a turn-on transient problem.

5.51 The circuit of Figure P5.51 includes a model of a voltage-controlled switch. When the voltage across the capacitors reaches the value v_M^c, the switch is closed. When the capacitor voltage reaches the value v_M^o, the switch opens. If $v_M^o = 1$ V and the period of the capacitor voltage waveform is 200 ms, find v_M^c.

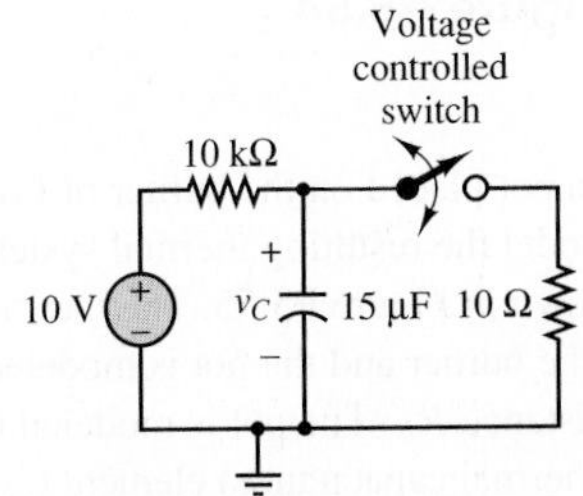

Figure P5.51

5.52 At $t = 0$, the switch in the circuit of Figure P5.52 closes. Assume that $i_L(0) = 0$ A. For $t \geq 0$, find

a. $i_L(t)$

b. $v_{L_1}(t)$

5.53 For the circuit of Figure P5.52, assume that the circuit is at steady state for $t < 0$. Find the voltage across the 10-kΩ resistor in parallel with the switch for $t \geq 0$.

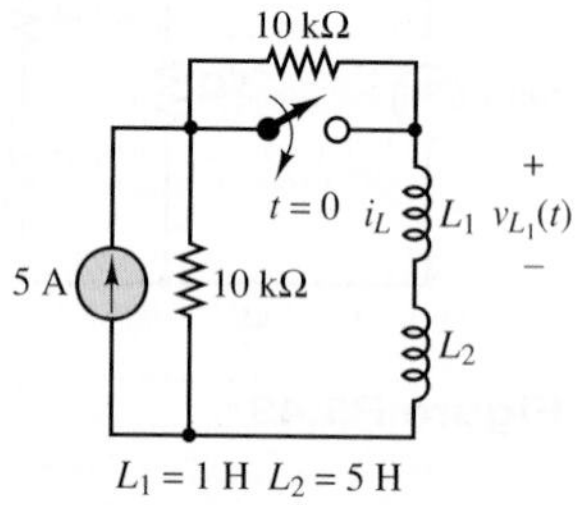

Figure P5.52

5.54 We use an analogy between electrical circuits and thermal conduction to analyze the behavior of a pot heating on an electric stove. We can model the heating element as shown in the circuit of Figure P5.54. Find the "heat capacity" of the burner, C_S, if the burner reaches 90 percent of the desired temperature in 10 seconds.

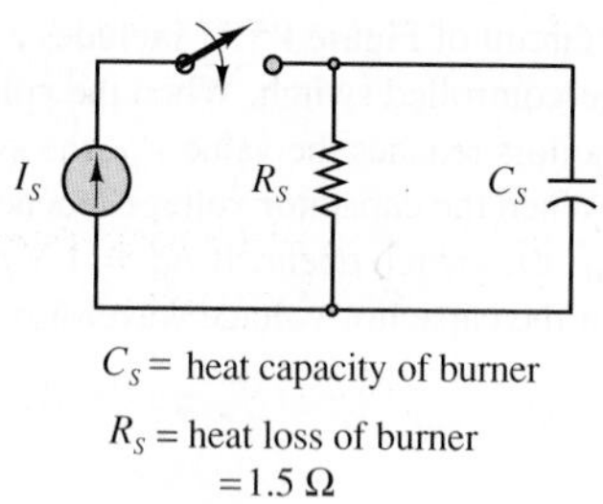

Figure P5.54

5.55 With a pot placed on the burner of Problem 5.54, we can model the resulting thermal system with the circuit shown in Figure P5.55. The thermal loss between the burner and the pot is modeled by the series resistance R_L. The pot is modeled by a heat storage (thermal capacitance) element C_P, and a loss (thermal resistance) element, R_P.

a. Find the final temperature of the water in the pot—that is, find $v(t)$ as $t \to \infty$ – if: $I_S = 75$ A; $C_P = 80$ F; $R_L = 0.8\Omega$; $R_P = 2.5\Omega$, and the burner is the same as in Problem 5.54.

b. How long will it take for the water to reach 80 percent of its final temperature?

[*Hint*: Neglect C_S since $C_S \ll C_P$.]

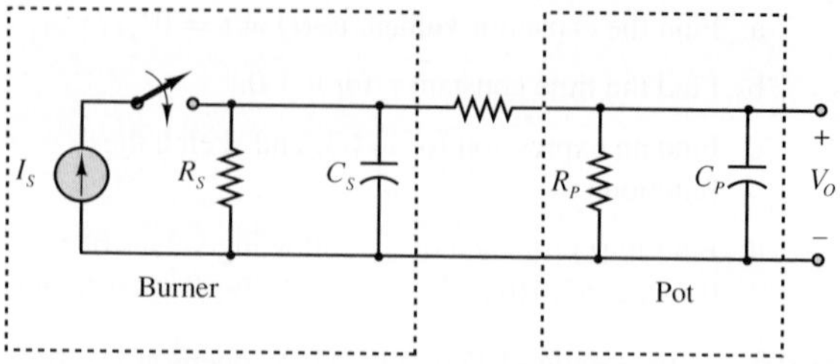

Figure P5.55

5.56 The circuit of Figure P5.56 is used as a variable delay in a burglar alarm. The alarm is a siren with internal resistance of 1 kΩ. The alarm will not sound until the current i_L exceeds 100 μA. Find the range of the variable resistor, R, for which the delay is between 1 and 2 s. Assume the capacitor is initially uncharged. This problem will require a graphical or numerical solution.

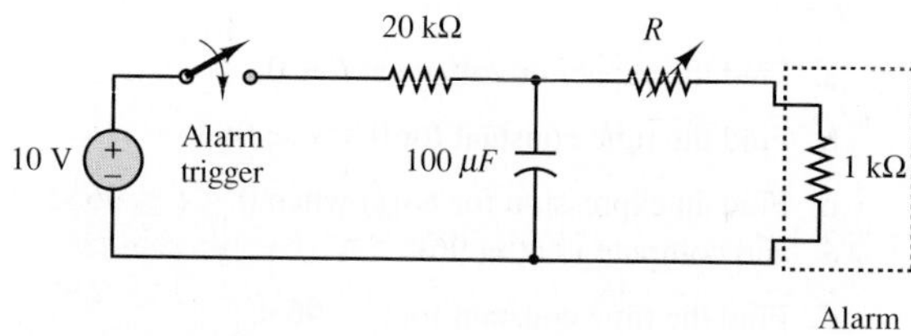

Figure P5.56

5.57 Find the voltage across C_1 in the circuit of Figure P5.57 for $t > 0$. Let $C_1 = 5\mu$F; $C_2 = 10\mu$F. Assume the capacitors are initially uncharged.

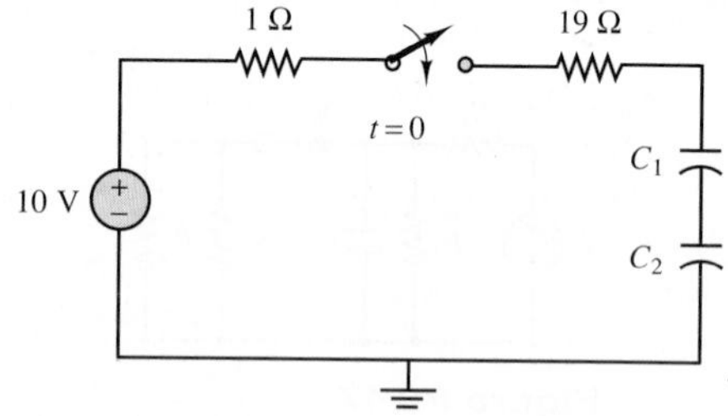

Figure P5.57

5.58 The switch in the circuit of Figure P5.58 opens at $t = 0$. It closes at $t = 10$ seconds.

a. What is the time constant for $9 < t < 10$ s?

b. What is the time constant for $t > 10$ s?

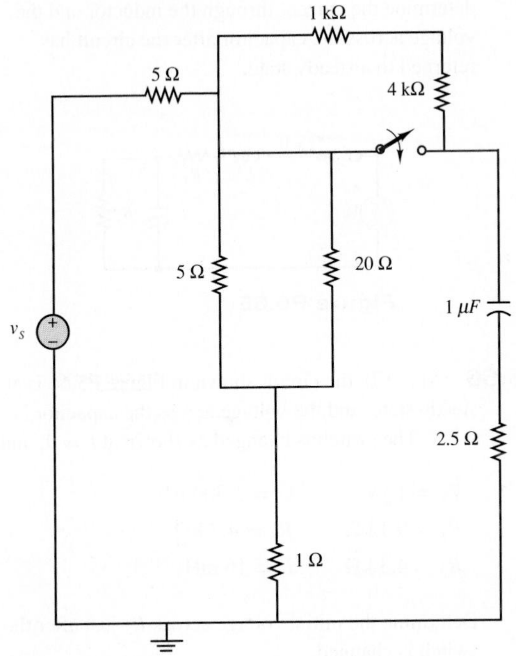

Figure P5.58

5.59 The circuit of Figure P5.59 models the charging circuit of an electronic flash for a camera. As you know, after the flash is used, it takes some time for it to recharge.

a. If the light that indicates that the flash is ready turns on when $V_C = 0.99 \times 7.5$ V, how long will you have to wait before taking another picture?

b. If the shutter button stays closed for 1/30 s, how much energy is delivered to the flash bulb, represented by R_2? Assume that the capacitor has completely charged.

c. If you do not wait till the flash is fully charged and you take a second picture 3 s after the flash is first turned on, how much energy is delivered to R_2?

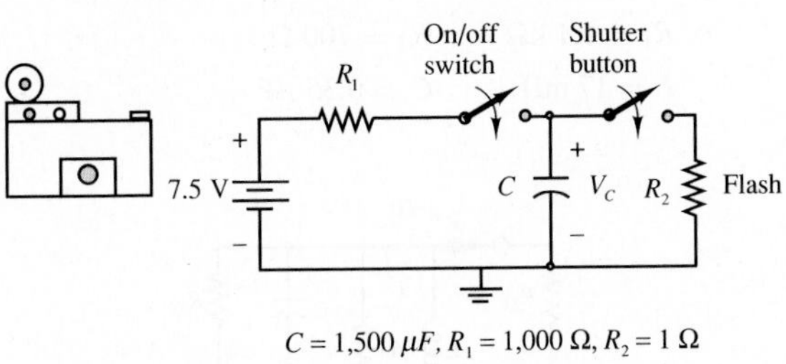

Figure P5.59

5.60 The ideal current source in the circuit of Figure P5.60 switches between various current levels, as shown in the graph. Determine and sketch the voltage across the inductor, $v_L(t)$ for t between 0 and 2 s. You may assume that the current source has been at zero for a very long time before $t = 0$.

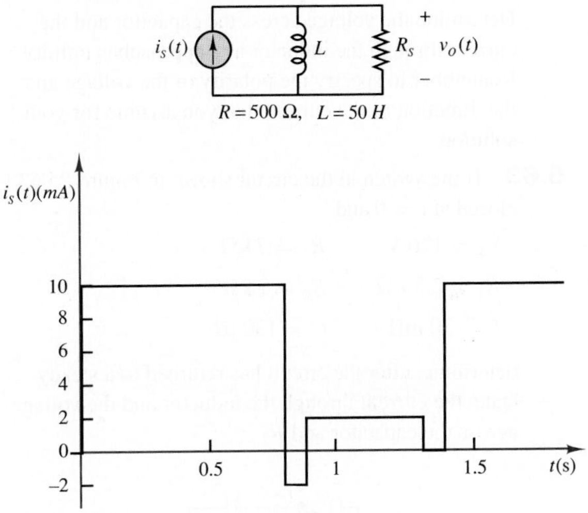

Figure P5.60

Section 5.5: Transient Response of Second-Order Circuits

5.61 In the circuit shown in Figure P5.61:

$$V_{S1} = 15\text{ V} \qquad V_{S2} = 9\text{ V}$$
$$R_{S1} = 130\ \Omega \qquad R_{S2} = 290\ \Omega$$
$$R_1 = 1.1\text{ k}\Omega \qquad R_2 = 700\ \Omega$$
$$L = 17\text{ mH} \qquad C = 0.35\ \mu\text{F}$$

Assume that DC steady-state conditions exist for $t < 0$. Determine the voltage across the capacitor and the current through the inductor and R_{S2} as t approaches infinity.

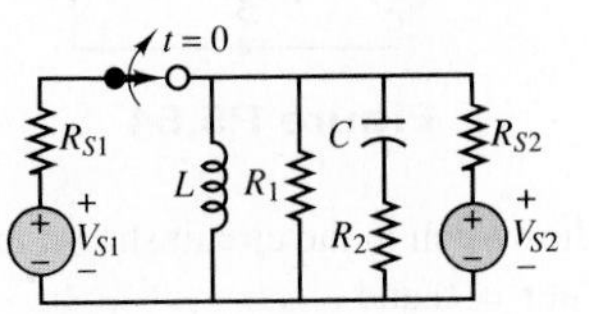

Figure P5.61

5.62 In the circuit shown in Figure P5.61

$V_{S1} = 12$ V	$V_{S2} = 12$ V
$R_{S1} = 50\ \Omega$	$R_{S2} = 50\ \Omega$
$R_1 = 2.2$ kΩ	$R_2 = 600\ \Omega$
$L = 7.8$ mH	$C = 68\ \mu$F

Assume that DC steady-state conditions exist at $t < 0$. Determine the voltage across the capacitor and the current through the inductor as t approaches infinity. Remember to specify the polarity of the voltage and the direction of the current that you assume for your solution.

5.63 If the switch in the circuit shown in Figure P5.63 is closed at $t = 0$ and

$V_S = 170$ V	$R_S = 7$ kΩ
$R_1 = 2.3$ kΩ	$R_2 = 7$ kΩ
$L = 30$ mH	$C = 130\ \mu$F

determine, after the circuit has returned to a steady state, the current through the inductor and the voltage across the capacitor and R_1.

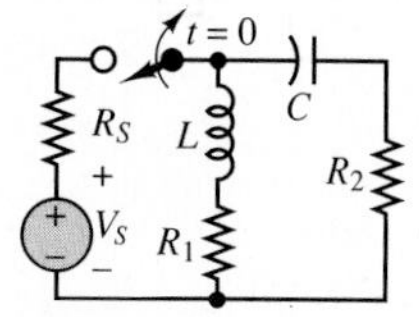

Figure P5.63

5.64 If the switch in the circuit shown in Figure P5.64 is closed at $t = 0$ and

$V_S = 12$ V	$C = 130\ \mu$F
$R_1 = 2.3$ kΩ	$R_2 = 7$ kΩ
$L = 30$ mH	

determine the current through the inductor and the voltage across the capacitor and across R_1 after the circuit has returned to a steady state.

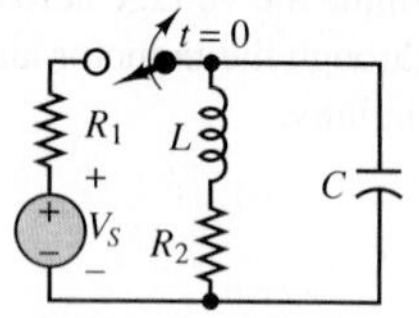

Figure P5.64

5.65 If the switch in the circuit shown in Figure P5.65 is closed at $t = 0$ and

$V_S = 12$ V	$C = 0.5\ \mu$F
$R_1 = 31$ kΩ	$R_2 = 22$ kΩ
$L = 0.9$ mH	

determine the current through the inductor and the voltage across the capacitor after the circuit has returned to a steady state.

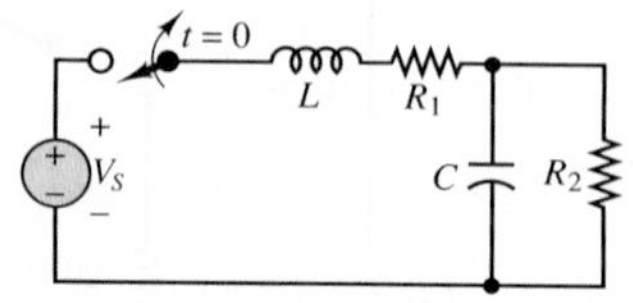

Figure P5.65

5.66 At $t < 0$, the circuit shown in Figure P5.66 is at steady state, and the voltage across the capacitor is +7 V. The switch is changed as shown at $t = 0$, and

$V_S = 12$ V	$C = 3{,}300\ \mu$F
$R_1 = 9.1$ kΩ	$R_2 = 4.3$ kΩ
$R_3 = 4.3$ kΩ	$L = 16$ mH

Determine the initial voltage across R_2 just after the switch is changed.

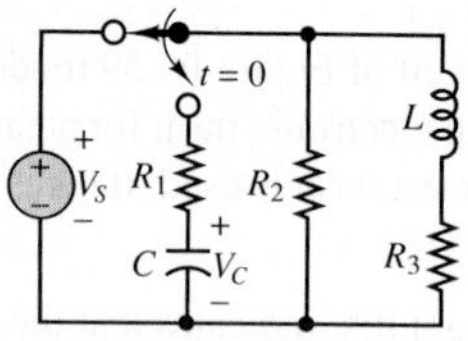

Figure P5.66

5.67 In the circuit shown in Figure P5.67, assume that DC steady-state conditions exist for $t < 0$. Determine at $t = 0^+$, just after the switch is opened, the current through and voltage across the inductor and the capacitor and the current through R_{S2}.

$V_{S1} = 15$ V	$V_{S2} = 9$ V
$R_{S1} = 130\ \Omega$	$R_{S2} = 290\ \Omega$
$R_1 = 1.1$ kΩ	$R_2 = 700\ \Omega$
$L = 17$ mH	$C = 0.35\ \mu$F

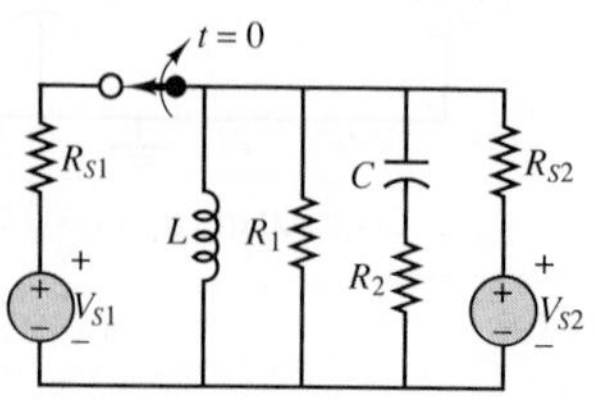

Figure P5.67

5.68 In the circuit shown in Figure P5.67,

$V_{S1} = 12$ V $\quad V_{S2} = 12$ V

$R_{S1} = 50\ \Omega \quad R_{S2} = 50\ \Omega$

$R_1 = 2.2$ k$\Omega \quad R_2 = 600\ \Omega$

$L = 7.8$ mH $\quad C = 68\ \mu$F

Assume that DC steady-state conditions exist for $t < 0$. Determine the voltage across the capacitor and the current through the inductor as t approaches infinity. Remember to specify the polarity of the voltage and the direction of the current that you assume for your solution.

5.69 Assume the switch in the circuit of Figure P5.69 has been closed for a very long time. It is suddenly opened at $t = 0$ and then reclosed at $t = 5$ s. Determine an expression for the inductor current for $t \geq 0$.

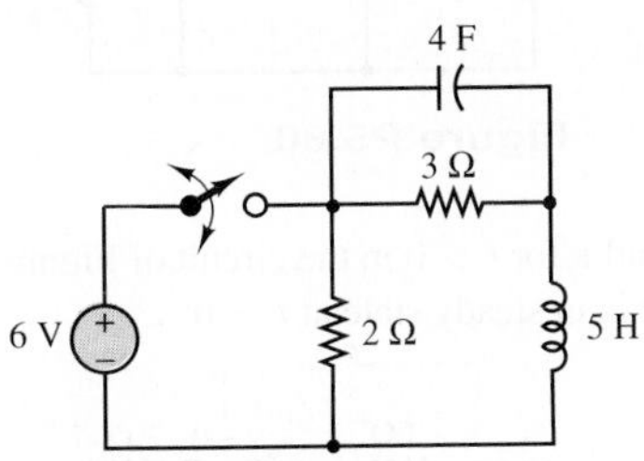

Figure P5.69

5.70 For the circuit of Figure P5.70, determine if it is underdamped or overdamped. Find also the capacitor value that results in critical damping.

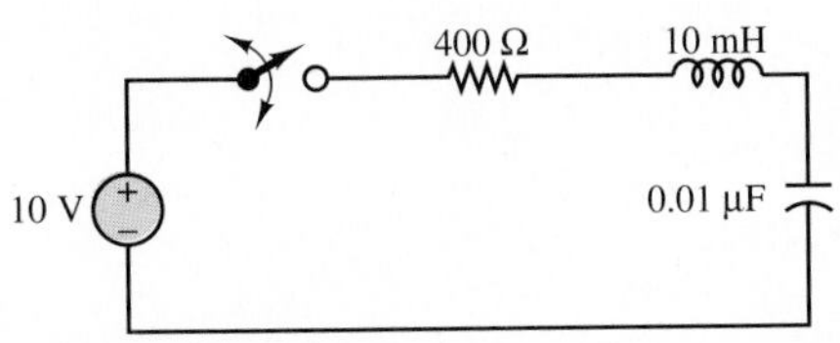

Figure P5.70

5.71 Assume the circuit of Figure P5.70 initially stores no energy. The switch is closed at $t - 0$. Find

a. Capacitor voltage as t approaches infinity

b. Capacitor voltage after 20 μs

c. Maximum capacitor voltage

5.72 Assume the circuit of Figure P5.72 initially stores no energy. Switch S_1 is open, and S_2 is closed. Switch S_1 is closed at $t = 0$, and switch S_2 is opened at $t = 5$ s. Determine an expression for the capacitor voltage for $t \geq 0$.

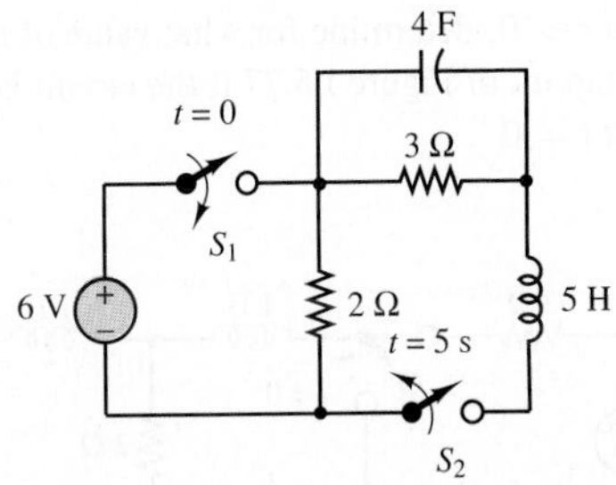

Figure P5.72

5.73 Assume that the circuit shown in Figure P5.73 is underdamped and that the circuit initially has no energy stored. It has been observed that after the switch is closed at $t = 0$, the capacitor voltage reaches an initial peak value of 70 V when $t = 5\pi/3\ \mu$s and a second peak value of 53.2 V when $t = 5\pi\ \mu$s, and it eventually approaches a steady-state value of 50 V. If $C = 1.6$ nF, what are the values of R and L?

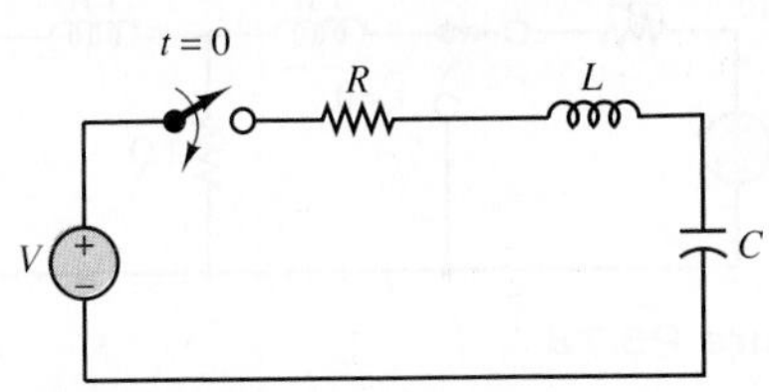

Figure P5.73

5.74 Given the information provided in Problem 5.73, explain how to modify the circuit so that the first peak occurs at $5\pi\ \mu$s. Assume that $C = 1.6\ \mu$F.

5.75 Find i for $t > 0$ in the circuit of Figure P5.75 if $i(0) = 0$ A and $v(0) = 10$ V.

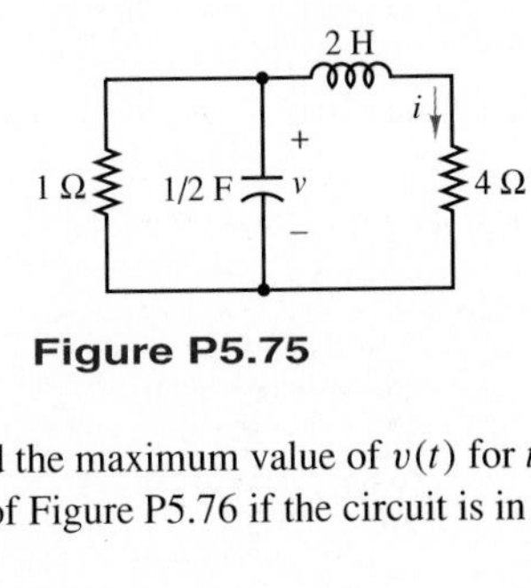

Figure P5.75

5.76 Find the maximum value of $v(t)$ for $t > 0$ in the circuit of Figure P5.76 if the circuit is in steady state at $t = 0^-$.

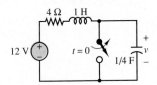

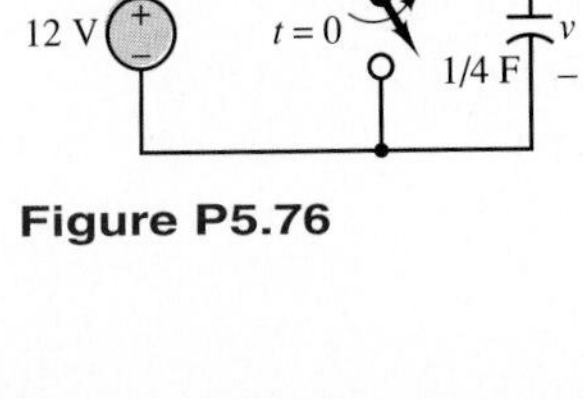

Figure P5.76

5.77 For $t > 0$, determine for what value of t $i = 2.5$ A in the circuit of Figure P5.77 if the circuit is in steady state at $t = 0^-$.

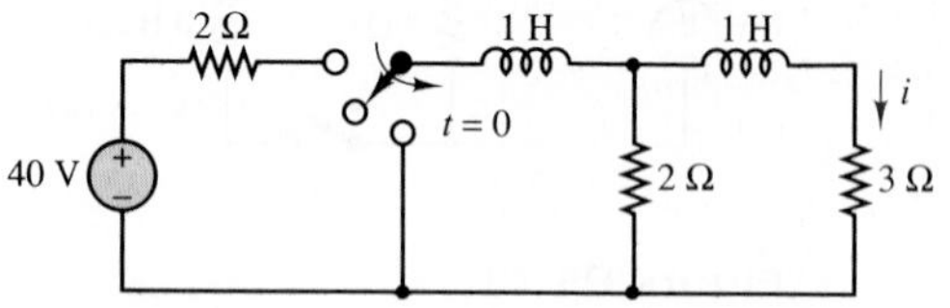

Figure P5.77

5.78 For $t > 0$, determine for what value of t $i = 6$ A in the circuit of Figure P5.78 if the circuit is in steady state at $t = 0^-$.

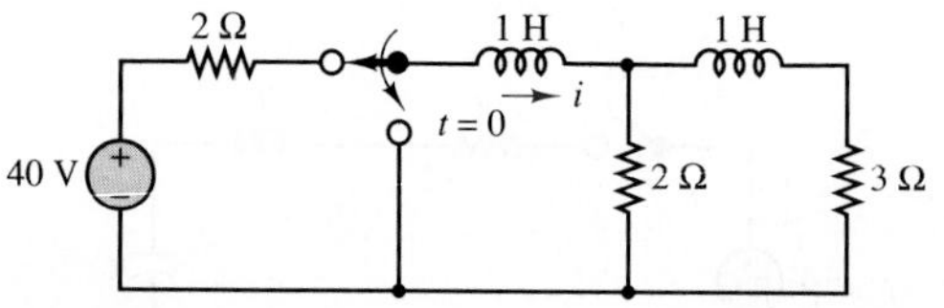

Figure P5.78

5.79 For $t > 0$, determine for what value of t $v = 7.5$ V in the circuit of Figure P5.79 if the circuit is in steady state at $t = 0^-$.

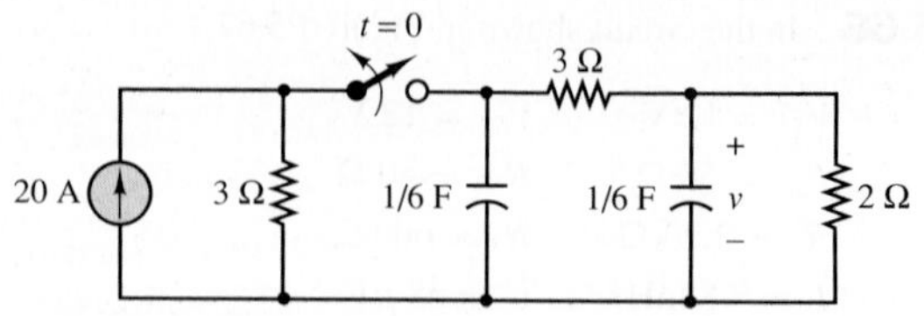

Figure P5.79

5.80 The circuit of Figure P5.80 is in steady state at $t = 0^-$. Assume $L = 3$ H; find the maximum value of v and the maximum voltage between the contacts of the switch.

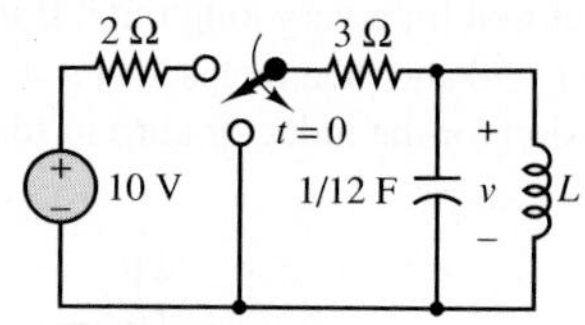

Figure P5.80

5.81 Find v for $t > 0$ in the circuit of Figure P5.81 if the circuit is in steady state at $t = 0^-$.

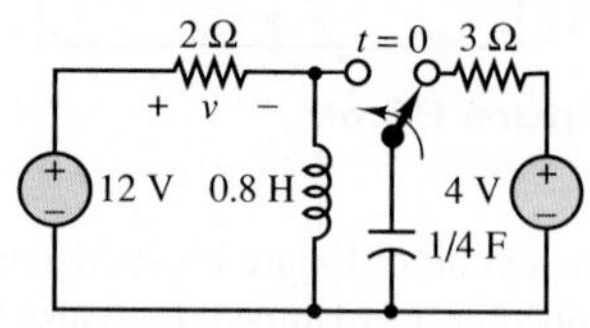

Figure P5.81

CHAPTER

6

FREQUENCY RESPONSE AND SYSTEM CONCEPTS

Chapter 4 introduced the notions of energy storage elements and dynamic circuit equations and developed appropriate tools (complex algebra and phasors) for the solution of AC circuits. In Chapter 5, we explored the solution of first- and second-order circuits subject to switching transients. The aim of this present chapter is to exploit AC circuit analysis methods to study the frequency response of electric circuits.

It is common, in engineering problems, to encounter phenomena that are frequency-dependent. For example, structures vibrate at a characteristic frequency when excited by wind forces (some high-rise buildings experience perceptible oscillation!). The propeller on a ship excites the shaft at a vibration frequency related to the engine's speed of rotation and to the number of blades on the propeller. An internal combustion engine is excited periodically by the combustion events in the individual cylinder, at a frequency determined by the firing of the cylinders. Wind blowing across a pipe excites a resonant vibration that is perceived as sound (wind instruments operate on this principle). Electric circuits are no different from other dynamic systems in this respect, and a large body of knowledge has been developed for understanding the frequency response of electric circuits, mostly based on the ideas behind phasors and impedance. These ideas, and the concept of filtering, will be explored in this chapter.

The ideas developed in this chapter will also be applied, by analogy, to the analysis of other physical systems (e.g., mechanical systems), to illustrate the generality of the concepts.

Learning Objectives

1. Understand the physical significance of frequency domain analysis, and compute the frequency response of circuits using AC circuit analysis tools. *Section 6.1.*
2. Compute the Fourier spectrum of periodic signals by using the Fourier series representation, and use this representation in connection with frequency response ideas to compute the response of circuits to periodic inputs. *Section 6.2.*
3. Analyze simple first- and second-order electrical filters, and determine their frequency response and filtering properties. *Section 6.3.*
4. Compute the frequency response of a circuit and its graphical representation in the form of a Bode plot. *Section 6.4.*

6.1 SINUSOIDAL FREQUENCY RESPONSE

The **sinusoidal frequency response** (or, simply, **frequency response**) of a circuit provides a measure of how the circuit responds to sinusoidal inputs of arbitrary frequency. In other words, given the input signal amplitude, phase, and frequency, knowledge of the frequency response of a circuit permits the computation of the output signal. Section 6.2, "Fourier Analysis," provides further explanation of the importance of sinusoidal signals. Suppose, for example, that you wanted to determine how the load voltage or current varied in response to different excitation signal frequencies in the circuit of Figure 6.1. An analogy could be made, for example, with how a speaker (the load) responds to the audio signal generated by a CD player (the source) when an amplifier (the circuit) is placed between the two.[1] In the circuit of Figure 6.1, the signal source circuitry is represented by its Thévenin equivalent. Recall that the impedance Z_S presented by the source to the remainder of the circuit is a function of the frequency of the source signal (Section 4.4). For the purpose of illustration, the amplifier circuit is represented by the idealized connection of two impedances Z_1 and Z_2, and the load is represented by an additional impedance Z_L. What, then, is the frequency response of this circuit? The following is a fairly general definition:

LO1

The frequency response of a circuit is a measure of the variation of a load-related voltage or current as a function of the frequency of the excitation signal.

[1] In reality, the circuitry in a high-fidelity stereo system is far more complex than the circuits discussed in this chapter and in the homework problems. However, from the standpoint of intuition and everyday experience, the audio analogy provides a useful example; it allows you to build a quick feeling for the idea of frequency response. Practically everyone has an intuitive idea of bass, midrange, and treble as coarsely defined frequency regions in the audio spectrum. The material presented in the next few sections should give you a more rigorous understanding of these concepts.

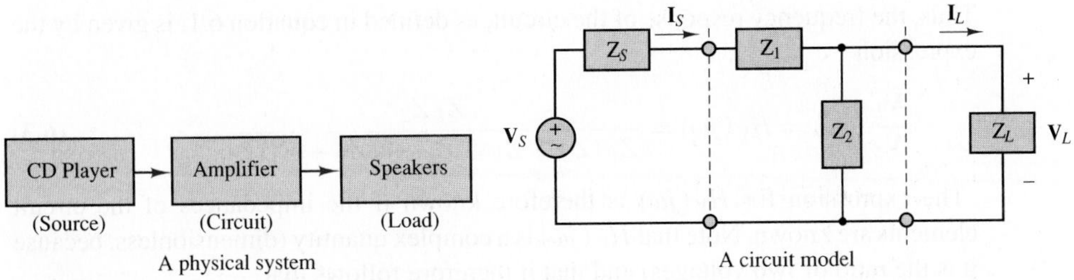

Figure 6.1 A circuit model

According to this definition, frequency response could be defined in a variety of ways. For example, we might be interested in determining how the load voltage varies as a function of the source voltage. Then analysis of the circuit of Figure 6.1 might proceed as follows.

To express the frequency response of a circuit in terms of variation in output voltage as a function of source voltage, we use the general formula

$$H_V(j\omega) = \frac{\mathbf{V}_L(j\omega)}{\mathbf{V}_S(j\omega)} \tag{6.1}$$

One method that allows for representation of the load voltage as a function of the source voltage (this is, in effect, what the frequency response of a circuit implies) is to describe the source and attached circuit by means of the Thévenin equivalent circuit. (This is not the only useful technique; the node voltage or mesh current equations for the circuit could also be employed.) Figure 6.2 depicts the original circuit of Figure 6.1 with the load removed, ready for the computation of the Thévenin equivalent.

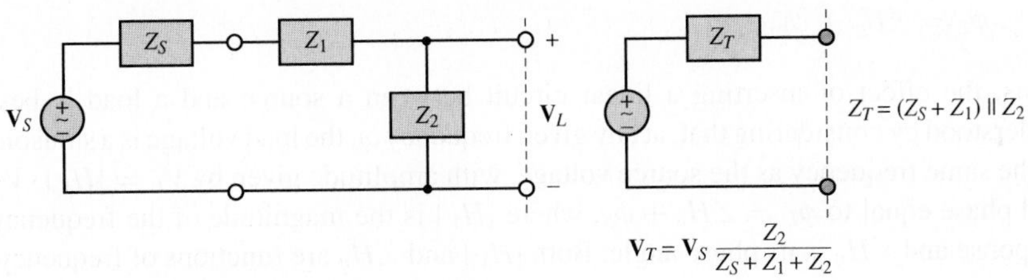

Figure 6.2 Thévenin equivalent source circuit

Next, an expression for the load voltage $\mathbf{V}_L$ may be found by connecting the load to the Thévenin equivalent source circuit and by computing the result of a simple voltage divider, as illustrated in Figure 6.3 and by the following equation:

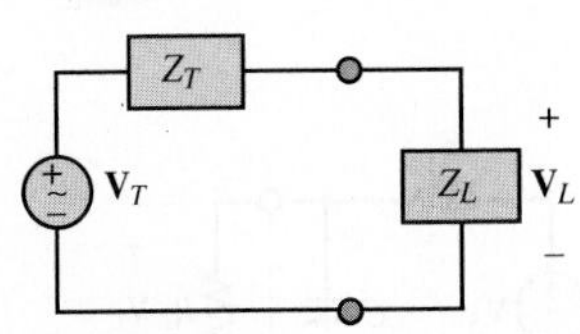

Figure 6.3 Complete equivalent circuit

$$\begin{aligned}
\mathbf{V}_L &= \frac{Z_L}{Z_L + Z_T}\mathbf{V}_T \\
&= \frac{Z_L}{Z_L + (Z_S + Z_1)Z_2/(Z_S + Z_1 + Z_2)} \cdot \frac{Z_2}{Z_S + Z_1 + Z_2}\mathbf{V}_S \\
&= \frac{Z_L Z_2}{Z_L(Z_S + Z_1 + Z_2) + (Z_S + Z_1)Z_2}\mathbf{V}_S
\end{aligned} \tag{6.2}$$

Thus, the frequency response of the circuit, as defined in equation 6.1, is given by the expression

$$\frac{\mathbf{V}_L}{\mathbf{V}_S}(j\omega) = H_V(j\omega) = \frac{Z_L Z_2}{Z_L(Z_S + Z_1 + Z_2) + (Z_S + Z_1)Z_2} \qquad \textbf{(6.3)}$$

The expression for $H_V(j\omega)$ is therefore known if the impedances of the circuit elements are known. Note that $H_V(j\omega)$ is a complex quantity (dimensionless, because it is the ratio of two voltages) and that it therefore follows that

LO1

$\mathbf{V}_L(j\omega)$ is a phase-shifted and amplitude-scaled version of $\mathbf{V}_S(j\omega)$.

If the phasor source voltage and the frequency response of the circuit are known, the phasor load voltage can be computed as follows:

$$\mathbf{V}_L(j\omega) = H_V(j\omega) \cdot \mathbf{V}_S(j\omega) \qquad \textbf{(6.4)}$$

$$V_L e^{j\phi_L} = |H_V| e^{j\angle H_v} \cdot V_S e^{j\phi_S} \qquad \textbf{(6.5)}$$

or

$$V_L e^{j\phi_L} = |H_V| V_S e^{j(\angle H_v + \phi_S)} \qquad \textbf{(6.6)}$$

where

$$V_L = |H_V| \cdot V_S$$

and **(6.7)**

$$\phi_L = \angle H_v + \phi_S$$

Thus, the effect of inserting a linear circuit between a source and a load is best understood by considering that, at any given frequency ω, the load voltage is a sinusoid at the same frequency as the source voltage, with amplitude given by $V_L = |H_V| \cdot V_S$ and phase equal to $\phi_L = \angle H_v + \phi_S$, where $|H_V|$ is the magnitude of the frequency response and $\angle H_v$ is its phase angle. Both $|H_V|$ and $\angle H_v$ are functions of frequency.

LO1

EXAMPLE 6.1 Computing the Frequency Response of a Circuit by Using Equivalent Circuit Ideas

Problem

Compute the frequency response $H_V(j\omega)$ for the circuit of Figure 6.4.

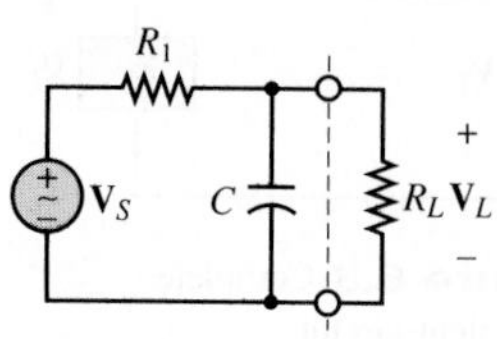

Figure 6.4

Solution

Known Quantities: $R_1 = 1\ \text{k}\Omega$; $C = 10\ \mu\text{F}$; $R_L = 10\ \text{k}\Omega$.

Find: The frequency response $H_V(j\omega) = \mathbf{V}_L(j\omega)/\mathbf{V}_S(j\omega)$.

Assumptions: None.

Analysis: To solve this problem, we use an equivalent circuit approach. Recognizing that R_L is the load resistance, we determine the equivalent circuit representation of the circuit to the left of the load, using the techniques perfected in Chapters 3 and 4. The Thévenin equivalent circuit is shown in Figure 6.5. Using the voltage divider rule and the equivalent circuit shown in the figure, we obtain the following expressions:

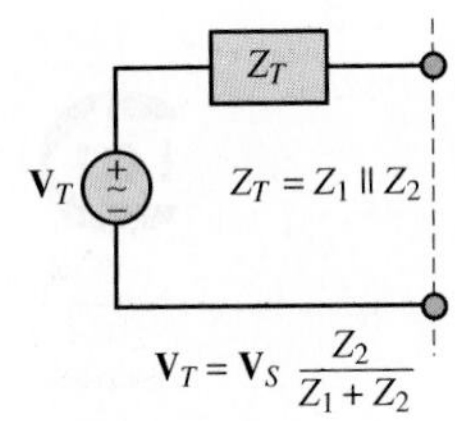

Figure 6.5

$$\begin{aligned}\mathbf{V}_L &= \frac{Z_L}{Z_T + Z_L}\mathbf{V}_T \\ &= \frac{Z_L}{Z_1 Z_2/(Z_1+Z_2) + Z_L}\frac{Z_2}{Z_1+Z_2}\mathbf{V}_S \\ &= H_V \mathbf{V}_S\end{aligned}$$

and

$$\frac{\mathbf{V}_L}{\mathbf{V}_S}(j\omega) = H_V(j\omega) = \frac{Z_L Z_2}{Z_L(Z_1+Z_2) + Z_1 Z_2}$$

The impedances of the circuit elements are $Z_1 = 10^3\ \Omega$, $Z_2 = 1/(j\omega \times 10^{-5})\ \Omega$, and $Z_L = 10^4\ \Omega$. The resulting frequency response can be calculated to be

$$H_V(j\omega) = \frac{\dfrac{10^4}{j\omega \times 10^{-5}}}{10^4\left(10^3 + \dfrac{1}{j\omega \times 10^{-5}}\right) + \dfrac{10^3}{j\omega \times 10^{-5}}} = \frac{100}{110 + j\omega}$$

$$= \frac{100}{\sqrt{110^2 + \omega^2}e^{j\arctan\left(\frac{\omega}{110}\right)}} = \frac{100}{\sqrt{110^2+\omega^2}}\angle -\arctan\left(\frac{\omega}{110}\right)$$

Comments: The use of equivalent circuit ideas is often helpful in deriving frequency response functions, because it naturally forces us to identify source and load quantities. However, it is certainly not the only method of solution. For example, node analysis would have yielded the same results just as easily, by recognizing that the top node voltage is equal to the load voltage and by solving directly for $\mathbf{V}_L$ as a function of $\mathbf{V}_S$, without going through the intermediate step of computing the Thévenin equivalent source circuit.

CHECK YOUR UNDERSTANDING

Compute the magnitude and phase of the frequency response function at the frequencies $\omega = 10, 100$, and 1,000 rad/s.

Answers: Magnitude = 0.9054, 0.6727, and 0.0994; phase (degrees) = −5.1944, −42.2737, and −83.7227

The importance and usefulness of the frequency response concept lie in its ability to summarize the response of a circuit in a single function of frequency $H(j\omega)$, which can predict the load voltage or current at any frequency, given the input. Note that the frequency response of a circuit can be defined in four different ways:

LO1

$$H_V(j\omega) = \frac{\mathbf{V}_L(j\omega)}{\mathbf{V}_S(j\omega)} \qquad H_I(j\omega) = \frac{\mathbf{I}_L(j\omega)}{\mathbf{I}_S(j\omega)}$$
$$H_Z(j\omega) = \frac{\mathbf{V}_L(j\omega)}{\mathbf{I}_S(j\omega)} \qquad H_Y(j\omega) = \frac{\mathbf{I}_L(j\omega)}{\mathbf{V}_S(j\omega)} \tag{6.8}$$

If $H_V(j\omega)$ and $H_I(j\omega)$ are known, one can directly derive the other two expressions:

$$H_Z(j\omega) = \frac{\mathbf{V}_L(j\omega)}{\mathbf{I}_S(j\omega)} = Z_L(j\omega)\frac{\mathbf{I}_L(j\omega)}{\mathbf{I}_S(j\omega)} = Z_L(j\omega)H_I(j\omega) \tag{6.9}$$

$$H_Y(j\omega) = \frac{\mathbf{I}_L(j\omega)}{\mathbf{V}_S(j\omega)} = \frac{1}{Z_L(j\omega)}\frac{\mathbf{V}_L(j\omega)}{\mathbf{V}_S(j\omega)} = \frac{1}{Z_L(j\omega)}H_V(j\omega) \tag{6.10}$$

The remainder of the chapter builds on equations (6.8) to give you all the tools needed to make use of the concept of frequency response.

LO1

EXAMPLE 6.2 Computing the Frequency Response of a Circuit

Problem

Compute the frequency response $H_Z(j\omega)$ for the circuit of Figure 6.6.

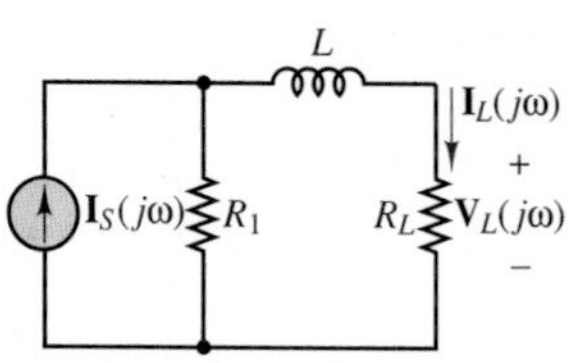

Figure 6.6

Solution

Known Quantities: $R_1 = 1\ \text{k}\Omega$; $L = 2$ mH; $R_L = 4\ \text{k}\Omega$.

Find: The frequency response $H_Z(j\omega) = \mathbf{V}_L(j\omega)/\mathbf{I}_S(j\omega)$.

Assumptions: None.

Analysis: To determine expressions for the load voltage, we recognize that the load current can be obtained simply by using a current divider between the two branches connected to the current source, and that the load voltage is simply the product of the load current and R_L.

Using the current divider rule, we obtain the following expression for $\mathbf{I}_L$:

$$\mathbf{I}_L = \frac{1/(R_L + j\omega L)}{1/(R_L + j\omega L) + 1/R_1}\mathbf{I}_S$$
$$= \frac{1}{1 + R_L/R_1 + j\omega L/R_1}\mathbf{I}_S$$

and

$$\frac{\mathbf{V}_L}{\mathbf{I}_S}(j\omega) = H_Z(j\omega) = \frac{I_L R_L}{I_S}$$

$$= \frac{R_L}{1 + R_L/R_1 + j\omega L/R_1}$$

Substituting numerical values, we obtain

$$H_Z(j\omega) = \frac{4 \times 10^3}{1 + 4 + j(2 \times 10^{-3}\omega)/10^3}$$

$$= \frac{0.8 \times 10^3}{1 + j0.4 \times 10^{-6}\omega}$$

Comments: You should verify that the units of the expression for $H_Z(j\omega)$ are indeed ohms, as they should be from the definition of H_Z.

CHECK YOUR UNDERSTANDING

Compute the magnitude and phase of the frequency response function at the frequencies $\omega = 1$, 10, and 100 rad/s.

Answer: Magnitude = 0.1857, 0.0485, and 0.0050; phase (degrees) = −21.8014, −75.9638, and −88.5679

6.2 FOURIER ANALYSIS

The aim of this section is to introduce the concept of frequency domain analysis of signals, and more specifically the **Fourier series.** In the next few pages we explain how it is possible to represent periodic signals by means of the superposition of various sinusoidal signals of different amplitude, phase, and frequency. Let the signal $x(t)$be periodic with period T, that is,

$$x(t) = x(t + T) = x(t + nT) \qquad n = 1, 2, 3, \ldots; \ T = \text{period} \qquad \textbf{(6.11)}$$

An example of a periodic signal is shown in Figure 6.7.

The signal $x(t)$ can be expressed as an infinite summation of sinusoidal components, known as a **Fourier series,** using either of the following two representations.

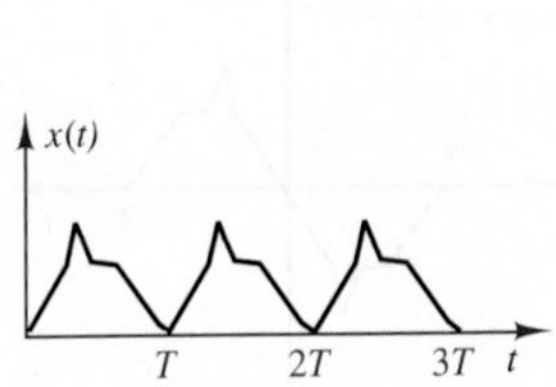

Figure 6.7 A periodic signal

LO2

Fourier Series

1. Sine-cosine (quadrature) representation

$$x(t) = a_0 + \sum_{n=1}^{\infty} a_n \cos\left(n\frac{2\pi}{T}t\right) + \sum_{n=1}^{\infty} b_n \sin\left(n\frac{2\pi}{T}t\right) \quad \textbf{(6.12)}$$

2. Magnitude and phase form

$$x(t) = c_0 + \sum_{n=1}^{\infty} c_n \sin\left(n\frac{2\pi}{T}t + \theta_n\right) \quad \textbf{(6.13)}$$

$$x(t) = c_0 + \sum_{n=1}^{\infty} c_n \cos\left(n\frac{2\pi}{T}t - \psi_n\right) \quad \textbf{(6.14)}$$

In each of these expressions, the period T is related to the **fundamental frequency** of the signal ω_0 by

$$\omega_0 = 2\pi f_0 = \frac{2\pi}{T} \quad \text{rad/s} \quad \textbf{(6.15)}$$

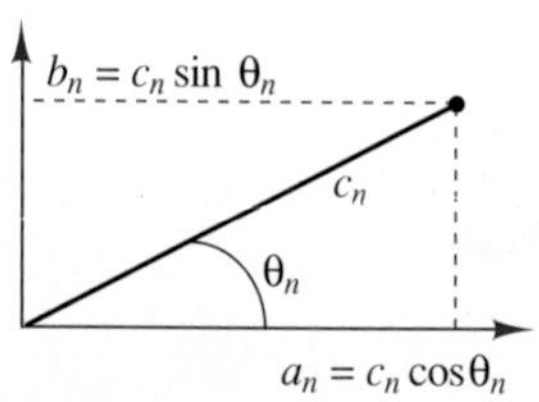

Figure 6.8 Relationship between $\{a_n, b_n\}$ and $\{c_n, \theta_n\}$ forms

It is straightforward to show that equation 6.13 is equivalent to equation 6.12, by expanding equation 6.13 by using trigonometric identities:

$$\sqrt{a_n^2 + b_n^2} = c_n \quad \text{and} \quad \frac{b_n}{a_n} = \cot(\theta_n) \quad \textbf{(6.16)}$$

Similarly, one can show that equation 6.14 is equivalent to equation 6.12 if

$$\sqrt{a_n^2 + b_n^2} = c_n \quad \text{and} \quad \frac{b_n}{a_n} = \tan(\psi_n) \quad \textbf{(6.17)}$$

Figure 6.8 is a graphical representation of the equivalence of the $\{a_n, b_n\}$ and $\{c_n, \theta_n\}$ forms of the Fourier series. Equations 6.7 and 6.8 permit easy conversion between the two forms of the Fourier series. In each of the above representations, $\omega_0 = 2\pi f_0 = 2\pi/T$ is called the **fundamental frequency** (in units of radians per second), and the frequencies $2\omega_0$, $3\omega_0$, $4\omega_0$, etc., are called its **harmonics.**

Each of the two forms of the Fourier series, equations 6.12 and 6.13 (or 6.14), has its distinct advantages. The sine-cosine representation uses odd and even functions of the independent variable. An odd function of time is one that satisfies the following relationship

$$f(-t) = -f(t) \quad \textbf{(6.18)}$$

Sine functions satisfy such a relationship and are odd. An even function of time is one that satisfies

$$f(-t) = f(t) \quad \textbf{(6.19)}$$

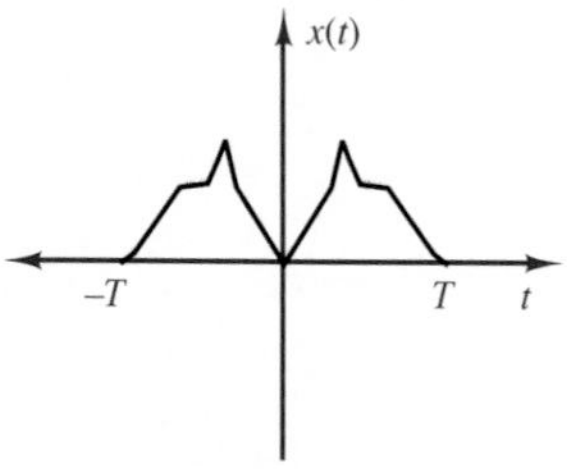

(a) Even function, $x(-t) = x(t)$

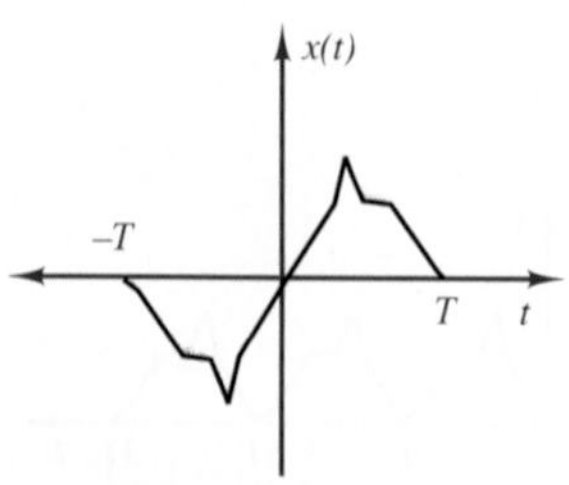

(b) Odd function, $x(-t) = -x(t)$

Figure 6.9 Definition of even and odd functions

Cosine functions are even, as is the constant value a_0. Figure 6.9 shows examples of even and odd functions.

The advantage of the representation in equation 6.12 is that if $x(t)$ is known to be odd (even), it can be represented as the sum of only odd (even) functions [i.e., using only the sine (cosine) terms], thus resulting in easier evaluation of the Fourier series coefficients.

The magnitude and phase forms, equations 6.13 and 6.14, separate out the magnitude information c_n from the phase information θ_n or ψ_n. Thus, they may be combined more readily with the magnitude and phase frequency responses of linear systems to periodic inputs, a procedure described in the following section. The magnitude and phase forms also allow a graphical representation of the Fourier series in the form of a discrete **frequency spectrum,** as shown in Figure 6.10. This plot depicts the magnitude and phase of the sinusoidal component at each of the frequencies (fundamental and its harmonics) that are contained in the signal $x(t)$.

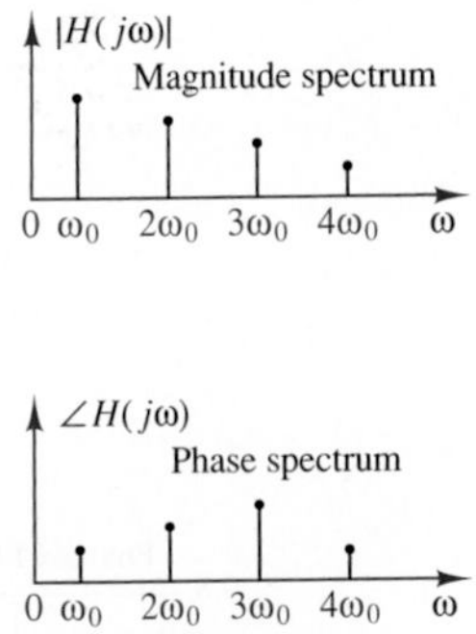

Figure 6.10 Discrete frequency spectrum

Computation of Fourier Series Coefficients

The computation of the $\{a_n, b_n\}$ or $\{c_n, \theta_n\}$ coefficients for the periodic function $x(t)$ is based on the following formulas:

$$a_0 = \frac{1}{T}\int_0^T x(t)\,dt = \frac{1}{T}\int_{-T/2}^{T/2} x(t)\,dt = \text{average value of } x(t) \quad \textbf{(6.20)}$$

$$a_n = \frac{2}{T}\int_0^T x(t)\cos\left(n\frac{2\pi}{T}t\right)dt = \frac{2}{T}\int_{-T/2}^{T/2} x(t)\cos\left(n\frac{2\pi}{T}t\right)dt \quad \textbf{(6.21)}$$

$$b_n = \frac{2}{T}\int_0^T x(t)\sin\left(n\frac{2\pi}{T}t\right)dt = \frac{2}{T}\int_{-T/2}^{T/2} x(t)\sin\left(n\frac{2\pi}{T}t\right)dt \quad \textbf{(6.22)}$$

LO2

In the above equations, the limits of the integrals have been written in two different forms, to illustrate that it does not matter where the integration starts, provided that it is carried out over one entire period. The c_n and θ_n (or ψ_n) values can be derived from the a_n and b_n coefficients by using equations 6.16 and 6.17.

To illustrate the significance of the Fourier series decomposition, we consider the square wave of Figure 6.11(a). The function depicted in the figure is an odd function, and thus we only need to compute the odd (sine) coefficients. A homework problem asks you to compute the Fourier coefficients for this square wave. The result of this calculation is shown in Figure 6.11(b), where the first six nonzero Fourier series terms are plotted. Note that the first term corresponds to a_0, that is, the average value of the function, and is a constant. The other five terms are the first five components of the Fourier series that have nonzero coefficients; note that they all correspond to n odd (1, 3, 5, 7, and 9). Note also that the coefficients for $n = 1$, 5, and 9 are positive, and those for $n = 3$ and 7 are negative (you can see this by looking at the peaks of the cosine waveforms). This alternation of positive and negative cosines is required so that each term can add or subtract from the previous terms as needed to "flatten" the waveform into a square wave. Figure 6.11(c) compares the original square wave with the Fourier series approximation. It should be evident that 10 terms are not sufficient to reproduce the sharp edges of the square wave, but it should also be clear that as we add more Fourier terms, the resulting approximation will be closer and closer to the square wave signal.

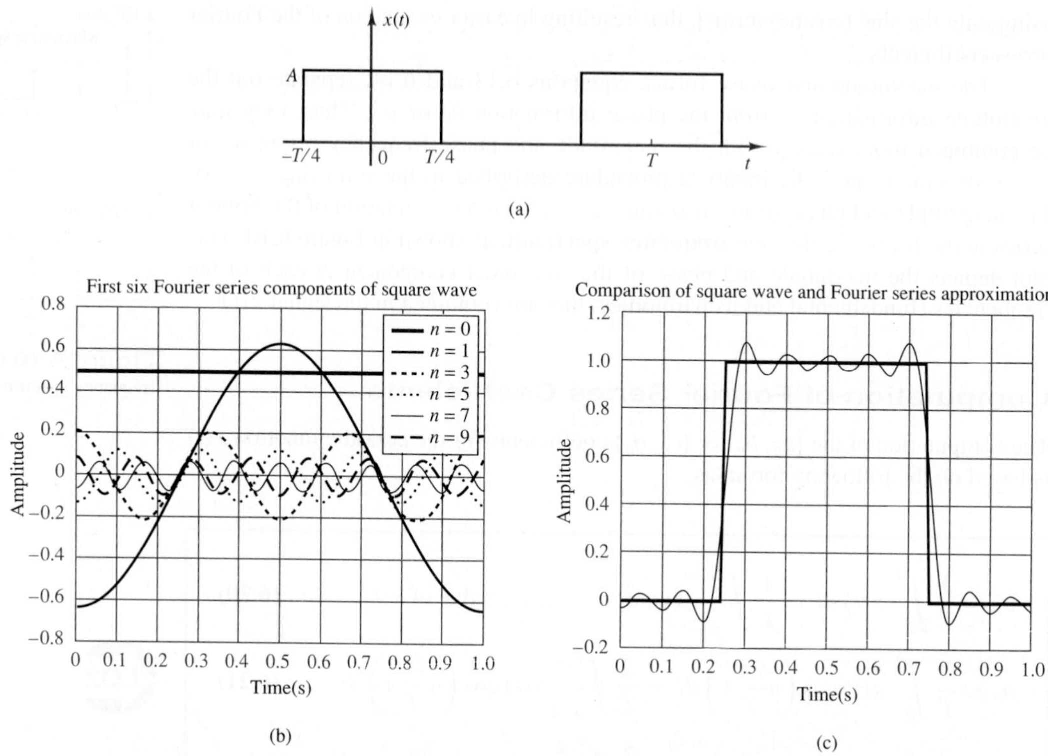

Figure 6.11 Square wave and its representation by a Fourier series. (a) Square wave (even function); (b) first three terms; (c) sum of first three terms

LO2

EXAMPLE 6.3 Computation of Fourier Series Coefficients

Problem

Compute the complete Fourier spectrum of the *sawtooth* function shown in Figure 6.12; that is, find a general expression for the coefficients a_n and b_n as a function of n, and then compute the spectrum of $x(t)$, that is, the coefficients c_n and θ_n. Plot the spectrum of the signal.

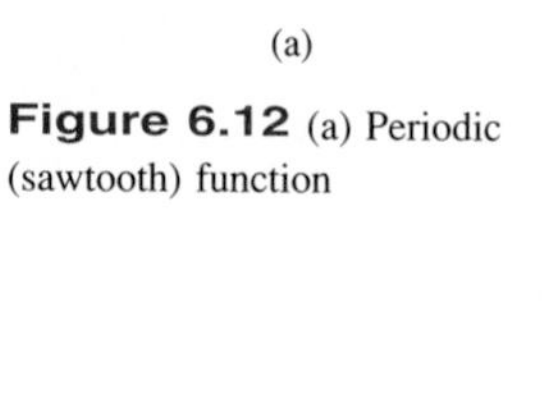

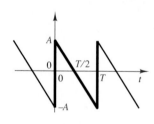

Figure 6.12 (a) Periodic (sawtooth) function

Solution

Known Quantities: Amplitude and period of sawtooth waveform.

Find: Fourier series coefficients a_n and b_n.

Schematics, Diagrams, Circuits, and Given Data: The function is periodic, with period $T = 1$ s, peak amplitude $A = 1$.

Assumptions: None.

Analysis: The function in Figure 6.11 is an odd function (convince yourself of this fact). Thus, we only need compute the b_n coefficients. First, we find an expression for $x(t)$ as an explicit function of t:

$$x(t) = A\left(1 - \frac{2t}{T}\right) \qquad 0 \le t < T$$

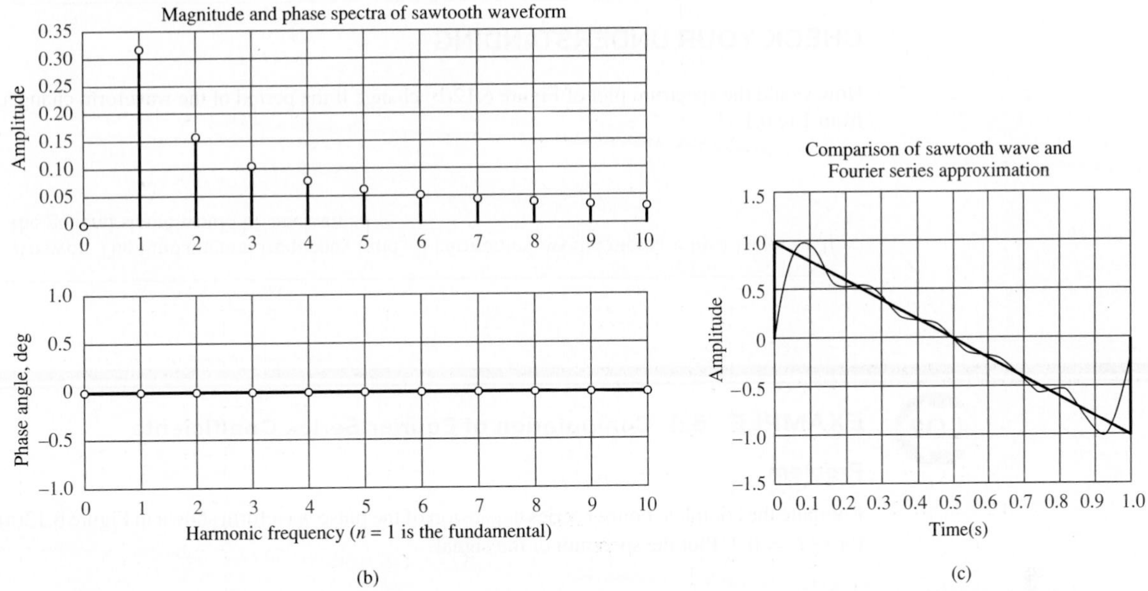

Figure 6.12 (b) spectrum of sawtooth waveform; (c) approximation of sawtooth waveform for $N = 5$

Then we evaluate the integral in equation 6.22:

$$b_n = \frac{2}{T}\int_0^T A\left(1 - \frac{2t}{T}\right)\sin\left(n\frac{2\pi}{T}t\right)dt$$

$$= \frac{2}{T}\int_0^T A\sin\left(n\frac{2\pi}{T}t\right)dt + \frac{2A}{T}\int_0^T \left(-\frac{2t}{T}\right)\sin\left(n\frac{2\pi}{T}t\right)dt$$

$$= \frac{2A}{T}\left[-\frac{T}{2n\pi}\cos\left(n\frac{2\pi}{T}t\right)\right]_0^T - \frac{4A}{T^2}\int_0^T t\cdot\sin\left(n\frac{2\pi}{T}t\right)dt$$

$$= 0 - \frac{4A}{T^2}\left[\frac{1}{n^2(2\pi/T)^2}\sin\left(n\frac{2\pi}{T}t\right) - \frac{t}{n(2\pi/T)}\cos\left(n\frac{2\pi}{T}\right)\right]_0^T$$

$$= -\frac{4A}{T^2}\left[-\frac{T^2}{2n\pi}\cos(2n\pi)\right] = \frac{2A}{n\pi} \qquad n = 1, 2, 3, \ldots$$

To compute the spectrum of the signal, we apply equation 6.16:

$$c_n = \sqrt{a_n^2 + b_n^2} = |b_n|$$

$$\theta_n = \cot^{-1}\frac{b_n}{a_n} = \cot^{-1}\frac{b_n}{0} = 0$$

The individual components of the spectrum of $x(t)$ are shown in Figure 6.11(b).

Comments: A computer program can be used to visualize the result of a Fourier series approximation when only the first N frequency components are included in the summation. Figure 6.12(c) depicts the results of this approximation.

CHECK YOUR UNDERSTANDING

How would the spectrum plot of Figure 6.12(b) change if the period of the waveform changed from 1 to 0.1 s?

Answer: The fundamental frequency (and all harmonics) would increase by a factor of 10; the general shape would be unchanged.

EXAMPLE 6.4 Computation of Fourier Series Coefficients

Problem

Compute the complete Fourier series expansion of the pulse waveform shown in Figure 6.13(a) for $\tau/T = 0.2$. Plot the spectrum of the signal.

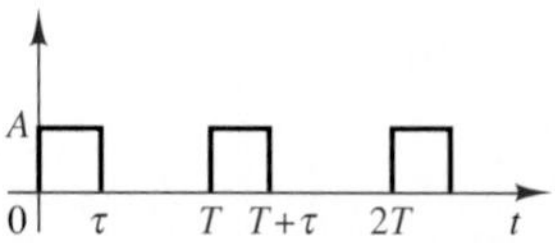

Figure 6.13 (a) Pulse train

Solution

Known Quantities: Amplitude and period of pulse train waveform.

Find: Fourier series coefficients a_n and b_n; Fourier spectrum.

Schematics, Diagrams, Circuits, and Given Data: The function is periodic, with period $T = 1$ s, peak amplitude $A = 1$.

Assumptions: None.

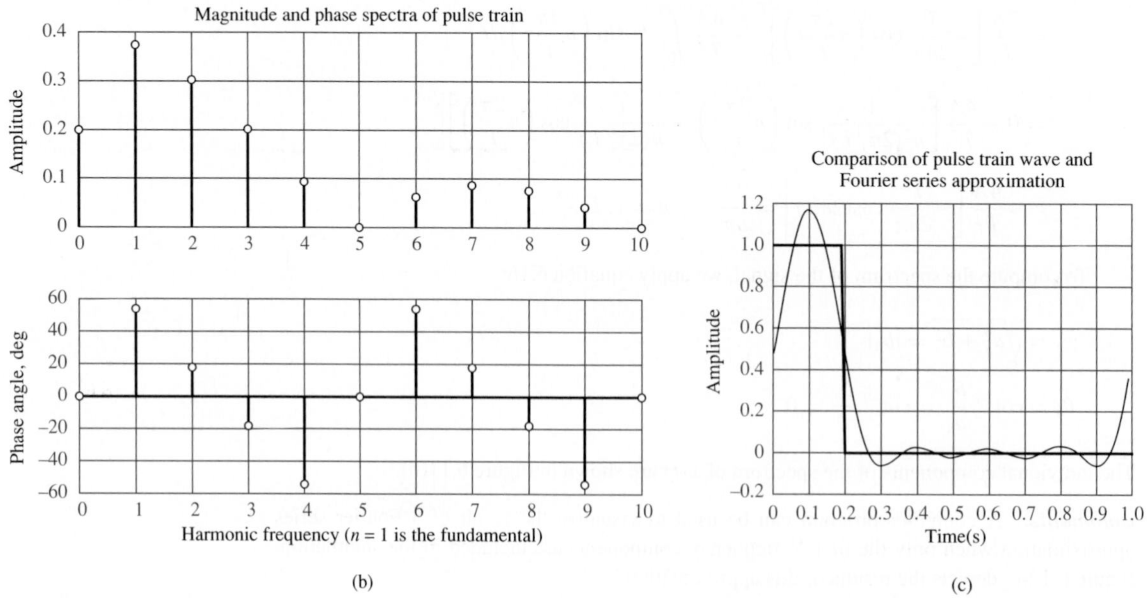

Figure 6.13 (b) signal spectrum; (c) approximation obtained using 11 Fourier coefficients

Analysis: The function in Figure 6.13 is neither odd, nor even. Thus, we need to compute both the a_n and the b_n coefficients. First, we find an expression for $x(t)$ as an explicit function of t:

$$x(t) = \begin{cases} A & 0 \le t < \tau \\ 0 & \tau \le t < T \end{cases}$$

Then we evaluate the integrals of equations 6.20 through 6.22:

$$a_0 = \frac{1}{T}\int_0^{T/5} A\,dt = \frac{A}{5}$$

$$\begin{aligned} a_n &= \frac{2}{T}\int_0^{T} A\cos\left(n\frac{2\pi}{T}t\right)dt\frac{2}{T} = \int_0^{T/5} A\cos\left(n\frac{2\pi}{T}t\right)dt + \int_{T/5}^{T} 0\cos\left(n\frac{2\pi}{T}t\right)dt \\ &= \frac{2}{T}\frac{AT}{2n\pi}\sin\left(\frac{2n\pi}{T}t\right)\Big|_0^{T/5} \\ &= \frac{2}{T}\frac{AT}{2n\pi}\left[\sin\left(\frac{2n\pi}{5}\right) - 0\right] = \frac{A}{n\pi}\sin\left(\frac{2n\pi}{5}\right) \end{aligned}$$

$$\begin{aligned} b_n &= \frac{2}{T}\int_0^{T/5} A\sin\left(n\frac{2\pi}{T}t\right)dt = \frac{2}{T}\frac{AT}{2n\pi}\left[-\cos\left(\frac{2n\pi}{T}t\right)\right]\Big|_0^{T/5} \\ &= \frac{2}{T}\frac{AT}{2n\pi}\left[-\cos\left(\frac{2n\pi}{5}\right) + \cos(0)\right] = \frac{A}{n\pi}\left[1 - \cos\left(\frac{2n\pi}{5}\right)\right] \end{aligned}$$

To compute the spectrum of the signal, we apply equation 6.16:

$$c_n = \sqrt{a_n^2 + b_n^2} = \sqrt{\left[\frac{A}{n\pi}\sin\left(\frac{2n\pi}{5}\right)\right]^2 + \left\{\frac{A}{n\pi}\left[1 - \cos\left(\frac{2n\pi}{5}\right)\right]\right\}^2}$$

$$\theta_n = \cot^{-1}\left(\frac{b_n}{a_n}\right) = \cot^{-1}\left\{\frac{(A/n\pi)[1 - \cos(2n\pi/5)]}{(A/n\pi)\sin(2n\pi/5)}\right\} = 0$$

The frequency spectrum of $x(t)$ (magnitude and phase) is shown in Figure 6.13(b). Table 6.1 lists the first seven coefficients in both forms.

Table 6.1 **Fourier coefficients of pulse train**

n	a_n	b_n	c_n	θ_n (deg)
0	0.2	0	0.2	0
1	0.3027	0.2199	0.3742	54
2	0.0935	0.2879	0.3027	18
3	−0.0624	0.1919	0.2018	−18
4	−0.0757	0.0550	0.0935	−54
5	0	0	0	0
6	0.0505	0.0367	0.0624	54
7	0.0267	0.0823	0.0865	18
8	−0.0234	0.0720	0.0757	−18
9	−0.0336	0.0244	0.0416	−54
10	0	0	0	0

Comments: A computer program can be used to visualize the result of a Fourier series approximation when only the first *10* frequency components are included in the summation. Figure 6.13(c) depicts the results of this approximation.

CHECK YOUR UNDERSTANDING

Determine which coefficients are zero when the *duty cycle* of the pulse train is $\tau/T = 0.25$.

Answer: $n = 4, 8$

LO2

EXAMPLE 6.5 Computation of Fourier Series Coefficients

Problem

Compute all the coefficients of the Fourier series expansion for the signal $x(t) = 1.5\cos(100t)$.

Solution

Known Quantities: Expression of signal waveform.

Find: Fourier series coefficients a_n and b_n.

Schematics, Diagrams, Circuits, and Given Data: The function is periodic, with period $T = 2\pi/100$ and peak amplitude of 1.5.

Assumptions: None.

Analysis: This function is already in Fourier series form, since it contains only sinusoidal terms! We recognize the following parameters: $\omega_0 = 100$; $a_0 = 0$; $a_1 = 1.5$; $b_1 = 0$; etc. (all other a_n and b_n coefficients are zero). Expressing the coefficients in magnitude-phase form, we have

$$c_1 = 1.5 \qquad \text{and} \qquad \theta_1 = \frac{\pi}{2}$$

CHECK YOUR UNDERSTANDING

Determine the a_n and b_n Fourier coefficients of the signal $y(t) = 1.5\cos(100t + \pi/4)$. (*Hint:* Use trigonometric identities to expand the cosine function.)

Answer: $a_0 = 0$, $a_1 = 1.0607$, $b_1 = 1.0607$. All other coefficients are zero.

Response of Linear Systems to Periodic Inputs

The frequency response concept is particularly useful when one deals with a system excited by a periodic input; in this case, the input may be modeled by a Fourier series consisting of a summation of sinusoids of different frequencies. Each of these sinusoids is of known amplitude and phase. Assume that the Fourier series representing a signal consists of a finite number of terms:

$$x(t) = c_0 + \sum_{n=1}^{N} c_n \sin\left(n\frac{2\pi}{T}t + \theta_n\right) \qquad \textbf{(6.23)}$$

Each of the N sinusoids is then characterized by amplitude c_n, phase θ_n, and frequency $\omega_n = n\omega_0$, where $\omega_0 = 2\pi/T$ and T is the period of the input signal. For example, the periodic input could be the periodic sawtooth waveform of Example 6.3.

Figure 6.14 illustrates the general input-output representation of a system, making use of the frequency response concept. The figure shows that if the input to a linear system $q_{in}(t)$ can be represented in phasor form, that is, by the phasor $Q_{in}(j\omega)$, then the output can be computed by multiplying the phasor form of the input by the frequency response function on the linear system. This product is, of course, a complex number consisting of a magnitude and a phase, which can be computed by multiplying the magnitude of the input phasor by the magnitude of the frequency response function, and by adding the phase angle of the input phasor to the phase angle of the frequency response function. We shall see various examples of this procedure in this section.

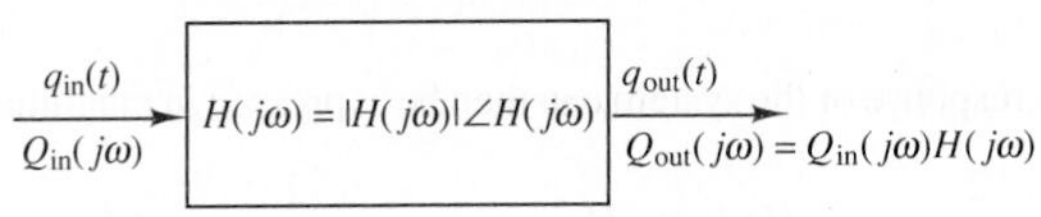

Figure 6.14 Response of a linear system to a phasor input

In the case of a periodic input expressed in terms of a (truncated) Fourier series, we must recognize that each of the input sinusoidal components propagates through the system according to the frequency response. Thus, the discrete magnitude spectrum of the periodic output signal in the steady state is equal to the discrete magnitude spectrum of the input signal multiplied by the amplitude ratio of the frequency response of the system *at the appropriate frequencies*. The phase spectrum of the output signal in the steady state is equal to the phase spectrum of the input signal added to the phase angle frequency response of the system *at the appropriate frequencies*. If $x(t)$ is the input to a linear system in the form given by equation 6.13, and if the linear system has a frequency response function $H(j\omega)$, then the output of the system $y(t)$ is given by

$$y(t) = \sum_{n=1}^{N} |H(j\omega_n)|\, c_n \sin\left[\omega_n t + \theta_n + \angle H(j\omega_n)\right] \qquad \textbf{(6.24)}$$

where $|H(j\omega_n)|$ and $\angle H(j\omega_n)$ are the magnitude and phase, respectively, of the frequency response of the system *at the frequency corresponding to the nth harmonic of the input* $n\omega_0$. Let us illustrate this idea by means of a couple of examples.

EXAMPLE 6.6 Response of Linear System to a Periodic Input

LO2

Problem

Let a linear system with $H(j\omega) = 2/(0.2j\omega + 1)$ be excited by the sawtooth waveform of Example 6.3, and assume that we are only interested in the response to the first two Fourier components of the input waveform.

Solution

Known Quantities: $T = 0.25$ s; $A = 2$.

Find: Output of system $y(t)$ in response to input $x(t)$.

Assumptions: None.

Analysis: According to the Fourier series definitions of the previous section, and using the first two terms of the Fourier series expansion of the sawtooth waveform of Example 6.3, we have

$$x(t) = \frac{2A}{\pi} \sin\left(\frac{2\pi}{0.25} t\right) + \frac{A}{\pi} \sin\left(\frac{4\pi}{0.25} t\right) = \frac{4}{\pi} \sin(8\pi t) + \frac{2}{\pi} \sin(16\pi t)$$

Thus, for this example,

$$c_1 = \sqrt{a_1^2 + b_1^2} = |b_1| = \frac{4}{\pi} \qquad \omega_1 = 1\omega_0 = 8\pi$$

and

$$c_2 = \sqrt{a_2^2 + b_2^2} = |b_2| = \frac{2}{\pi} \qquad \omega_2 = 2\omega_0 = 16\pi$$

The frequency response of the system can then be expressed in magnitude and phase form:

$$H(j\omega) = \frac{2}{0.2 j\omega + 1} = |H(j\omega)| \angle H(j\omega) = \frac{2}{\sqrt{(0.2\omega)^2 + 1}} \angle \left(-\arctan \frac{\omega}{5}\right)$$

The frequency response plots (magnitude and phase) are shown in Figure 6.14. We should observe that the system is excited only at the frequencies $\omega_1 = 8\pi = 25.1$ rad/s and $\omega_2 = 16\pi = 50.2$ rad/s. At this point, we could evaluate the frequency response of the system at these frequencies either graphically (from the frequency response plots of Figure 6.14) or analytically. We choose the latter because we can compute the answers with greater accuracy:

$$|H(j\omega_1)| = \frac{2}{\sqrt{(0.2\omega_1)^2 + 1}} = 0.3902 \Phi(j\omega_1) = -1.37 \text{ rad} = -78.75^\circ$$

$$|H(j\omega_2)| = \frac{2}{\sqrt{(0.2\omega_2)^2 + 1}} = 0.1980 \Phi(j\omega_2) = -1.47 \text{ rad} = -84.32^\circ$$

Finally, we can compute the steady-state periodic output of the system:

$$y(t) = \sum_{n=1}^{2} |H(j\omega_n)| c_n \sin\left[\omega_n t + \theta_n + \angle H(j\omega_n)\right]$$

$$= 0.3902 \times \frac{4}{\pi} \sin(8\pi t - 1.37) + 0.1980 \times \frac{2}{\pi} \sin(16\pi t - 1.47)$$

The input and output signals for the system are plotted in Figure 6.15. Note how the first two components of the Fourier series of the sawtooth waveform of Example 6.1 provide a coarse approximation of the general shape of the waveform. Given the frequency response of the system used in this example, would the accuracy of the computed response increase if higher-frequency components ($n > 2$) were included in the Fourier series expansion for the signal $x(t)$?

Comments: A computer program was used to generate the plots of Figure 6.15(a) and (b), and can be used to calculate the output of a linear system, with known frequency response function, to an arbitrary input represented by a finite summation of sinusoidal terms.

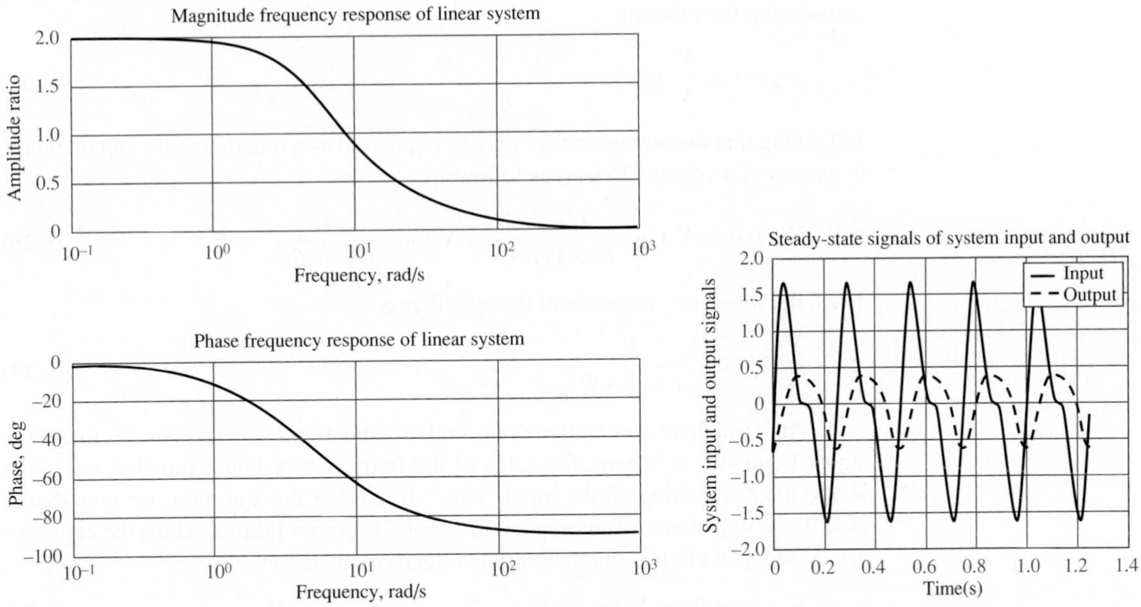

(a) (b)

Figure 6.15 (a) Frequency response of linear system; (b) input and output waveforms

---

CHECK YOUR UNDERSTANDING

Extend the result of this example of considering the third frequency component. What are the amplitude and phase of the component of $y(t)$ at the frequency $3\omega_0$?

Answer: magnitude $= 0.0562$, phase $= -1.505$ rad $= -86.2°$

6.3 FILTERS

There are many practical applications that involve filters of one kind or another. Just to mention two, filtration systems are used to eliminate impurities from drinking water, and sunglasses are used to filter out eye-damaging ultraviolet radiation and to reduce the intensity of sunlight reaching the eyes. An analogous concept applies to electric circuits: it is possible to *attenuate* (i.e., reduce in amplitude) or altogether eliminate signals of unwanted frequencies, such as those that may be caused by electrical noise or other forms of interference. This section will be devoted to the analysis of electrical filters.

RC low-pass filter. The circuit preserves lower frequencies while attenuating the frequencies above the cutoff frequency $\omega_0 = 1/RC$. The voltages $\mathbf{V}_i$ and $\mathbf{V}_o$ are the filter input and output voltages, respectively.

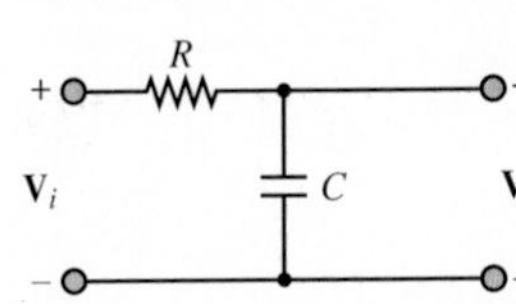

Figure 6.16 A simple *RC* filter

Low-Pass Filters

Figure 6.16 depicts a simple ***RC* filter** and denotes its input and output voltages, respectively, by $\mathbf{V}_i$ and $\mathbf{V}_o$. The frequency response for the filter may be obtained by considering the function

$$H(j\omega) = \frac{\mathbf{V}_o}{\mathbf{V}_i}(j\omega) \tag{6.25}$$

and noting that the output voltage may be expressed as a function of the input voltage by means of a voltage divider, as follows:

$$\mathbf{V}_o(j\omega) = \mathbf{V}_i(j\omega)\frac{1/j\omega C}{R + 1/j\omega C} = \mathbf{V}_i(j\omega)\frac{1}{1 + j\omega RC} \tag{6.26}$$

Thus, the frequency response of the *RC* filter is

$$\frac{\mathbf{V}_o}{\mathbf{V}_i}(j\omega) = \frac{1}{1 + j\omega CR} \tag{6.27}$$

An immediate observation upon studying this frequency response, is that if the signal frequency ω is zero, the value of the frequency response function is 1. That is, the filter is passing all the input. Why? To answer this question, we note that at $\omega = 0$, the impedance of the capacitor, $1/j\omega C$, becomes infinite. Thus, the capacitor acts as an open circuit, and the output voltage equals the input:

$$\mathbf{V}_o(j\omega = 0) = \mathbf{V}_i(j\omega = 0) \tag{6.28}$$

Since a signal at sinusoidal frequency equal to zero is a DC signal, this filter circuit does not in any way affect DC voltages and currents. As the signal frequency increases, the magnitude of the frequency response decreases, since the denominator increases with ω. More precisely, equations 6.29 to 6.32 describe the magnitude and phase of the frequency response of the *RC* filter:

$$\begin{aligned} H(j\omega) &= \frac{\mathbf{V}_o}{\mathbf{V}_i}(j\omega) = \frac{1}{1 + j\omega CR} \\ &= \frac{1}{\sqrt{1 + (\omega CR)^2}}\frac{e^{j0}}{e^{j\arctan(\omega CR/1)}} \\ &= \frac{1}{\sqrt{1 + (\omega CR)^2}} \cdot e^{-j\arctan(\omega CR)} \end{aligned} \tag{6.29}$$

or

$$H(j\omega) = |H(j\omega)|e^{j\angle H(j\omega)} \tag{6.30}$$

with

$$|H(j\omega)| = \frac{1}{\sqrt{1 + (\omega CR)^2}} = \frac{1}{\sqrt{1 + (\omega/\omega_0)^2}} \tag{6.31}$$

and

$$\angle H(j\omega) = -\arctan(\omega CR) = -\arctan\frac{\omega}{\omega_0} \tag{6.32}$$

with

$$\omega_0 = \frac{1}{RC} \tag{6.33}$$

The simplest way to envision the effect of the filter is to think of the phasor voltage $\mathbf{V}_i = V_i e^{j\phi_i}$ scaled by a factor of $|H|$ and shifted by a phase angle $\angle H$ by the filter *at each frequency*, so that the resultant output is given by the phasor $V_o e^{j\phi_o}$, with

$$V_o = |H| \cdot V_i$$
$$\phi_o = \angle H + \phi_i \tag{6.34}$$

and where $|H|$ and $\angle H$ are functions of frequency. The frequency ω_0 is called the **cutoff frequency** of the filter and, as will presently be shown, gives an indication of the filtering characteristics of the circuit.

It is customary to represent $H(j\omega)$ in two separate plots, representing $|H|$ and $\angle H$ as functions of ω. These are shown in Figure 6.17 in normalized form, that is, with $|H|$ and $\angle H$ plotted versus ω/ω_0, corresponding to a cutoff frequency $\omega_0 = 1$ rad/s. Note that, in the plot, the frequency axis has been scaled logarithmically. This is a common practice in electrical engineering, because it enables viewing a very broad range of frequencies on the same plot without excessively compressing the low-frequency end of the plot. The frequency response plots of Figure 6.17 are commonly employed to describe the frequency response of a circuit, since they can provide a clear idea at a glance of the effect of a filter on an excitation signal. For example, the *RC* filter of Figure 6.16 has the property of "passing" signals at low frequencies ($\omega \ll 1/RC$) and of filtering out signals at high frequencies ($\omega \gg 1/RC$). This type of filter is called a **low-pass filter.** The cutoff frequency $\omega = 1/RC$ has a special significance in that it represents—approximately—the point where the filter begins to filter out the higher-frequency signals. The value of $|H(j\omega)|$ at the cutoff frequency is $1/\sqrt{2} = 0.707$. Note how the cutoff frequency depends exclusively on the values of R and C. Therefore, one can adjust the filter response as desired simply by selecting appropriate values for C and R, and therefore one can choose the desired filtering characteristics.

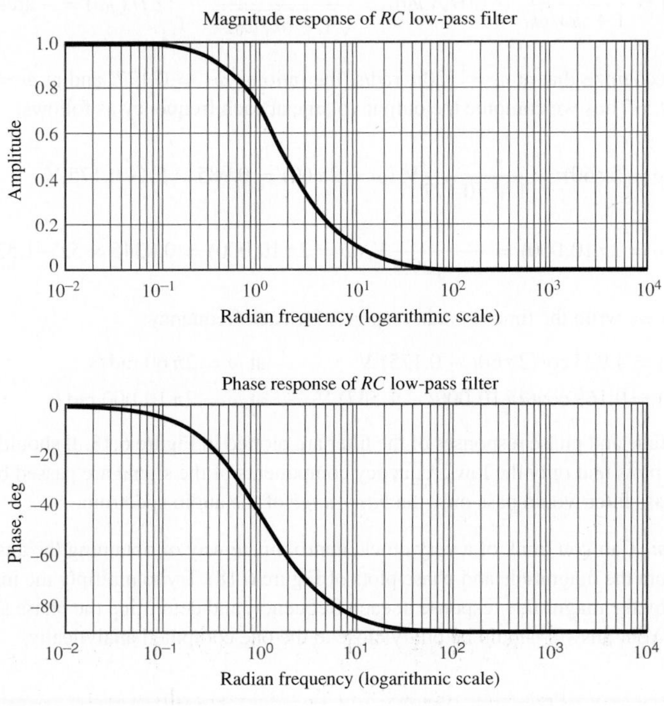

Figure 6.17 Magnitude and phase response plots for *RC* filter

LO3

EXAMPLE 6.7 Frequency Response of *RC* Filter

Problem

Compute the response of the *RC* filter of Figure 6.16 to sinusoidal inputs at the frequencies of 60 and 10,000 Hz.

Solution

Known Quantities: $R = 1\text{ k}\Omega$; $C = 0.47\ \mu\text{F}$; $v_i(t) = 5\cos(\omega t)$ V.

Find: The output voltage $v_o(t)$ at each frequency.

Assumptions: None.

Analysis: In this problem, we know the input signal voltage and the frequency response of the circuit (equation 6.29), and we need to find the output voltage at two different frequencies. If we represent the voltages in phasor form, we can use the frequency response to calculate the desired quantities:

$$\frac{\mathbf{V}_o}{\mathbf{V}_i}(j\omega) = H_V(j\omega) = \frac{1}{1 + j\omega CR}$$

$$\mathbf{V}_o(j\omega) = H_V(j\omega)\mathbf{V}_i(j\omega) = \frac{1}{1 + j\omega CR}\mathbf{V}_i(j\omega)$$

If we recognize that the cutoff frequency of the filter is $\omega_0 = 1/RC = 2{,}128$ rad/s, we can write the expression for the frequency response in the form of equations 6.31 and 6.32:

$$H_V(j\omega) = \frac{1}{1 + j\omega/\omega_0} \qquad |H_V(j\omega)| = \frac{1}{\sqrt{1 + (\omega/\omega_0)^2}} \qquad \angle H(j\omega) = -\arctan\left(\frac{\omega}{\omega_0}\right)$$

Next, we recognize that at $\omega = 120\pi$ rad/s, the ratio $\omega/\omega_0 = 0.177$, and at $\omega = 20{,}000\pi$, $\omega/\omega_0 = 29.5$. Thus we compute the output voltage at each frequency as follows:

$$\mathbf{V}_o(\omega = 2\pi 60) = \frac{1}{1 + j0.177}\mathbf{V}_i(\omega = 2\pi 60) = 0.985 \times 5\angle{-0.175}\text{ V}$$

$$\mathbf{V}_o(\omega = 2\pi 10{,}000) = \frac{1}{1 + j29.5}\mathbf{V}_i(\omega = 2\pi 10{,}000) = 0.0345 \times 5\angle{-1.537}\text{ V}$$

And finally we write the time-domain response for each frequency:

$$v_o(t) = 4.923\cos(2\pi 60t - 0.175)\text{ V} \qquad \text{at } \omega = 2\pi 60 \text{ rad/s}$$

$$v_o(t) = 0.169\cos(2\pi 10{,}000t - 1.537)\text{ V} \qquad \text{at } \omega = 2\pi 10{,}000 \text{ rad/s}$$

The magnitude and phase responses of the filter are plotted in Figure 6.18. It should be evident from these plots that only the low-frequency components of the signal are passed by the filter. This low-pass filter would pass only the *bass range* of the audio spectrum.

Comments: Can you think of a very quick, approximate way of obtaining the answer to this problem from the magnitude and phase plots of Figure 6.18? Try to multiply the input voltage amplitude by the magnitude response at each frequency, and determine the phase shift at each frequency. Your answer should be pretty close to the one computed analytically.

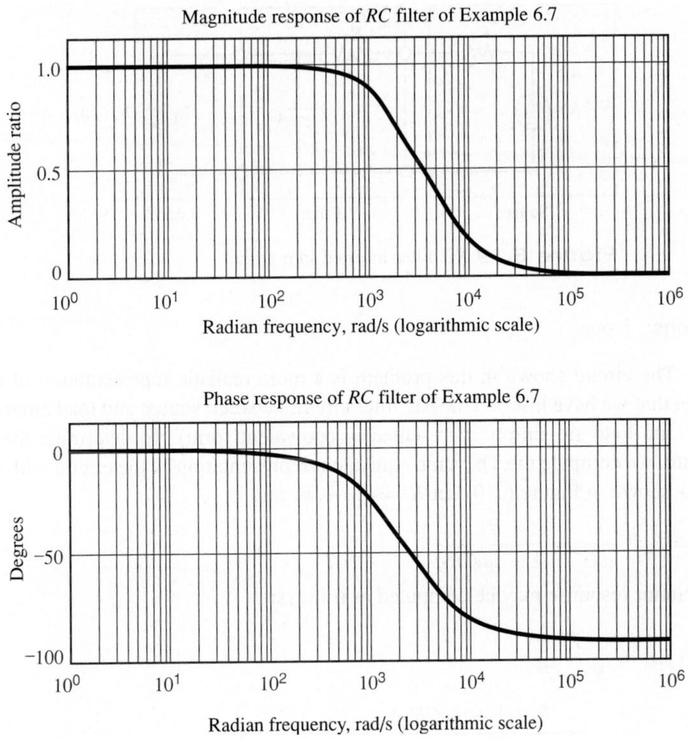

Figure 6.18 Response of *RC* filter of Example 6.7

CHECK YOUR UNDERSTANDING

A simple *RC* low-pass filter is constructed using a 10-μF capacitor and a 2.2-kΩ resistor. Over what range of frequencies will the output of the filter be within 1 percent of the input signal amplitude (i.e., when will $V_L \geq 0.99 V_S$)?

Answer: $0 \leq \omega \leq 6.48$ rad/s

EXAMPLE 6.8 Frequency Response of *RC* Low-Pass Filter in a More Realistic Circuit

LO3

Problem

Compute the response of the *RC* filter in the circuit of Figure 6.19.

Solution

Known Quantities: $R_S = 50\ \Omega$; $R_1 = 200\ \Omega$; $R_L = 500\ \Omega$; $C = 10\ \mu$F.

Find: The output voltage $v_o(t)$ at each frequency.

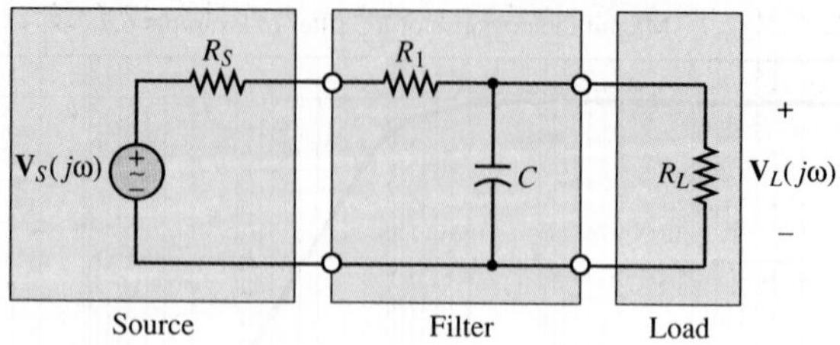

Figure 6.19 *RC* filter inserted in a circuit

Assumptions: None.

Analysis: The circuit shown in this problem is a more realistic representation of a filtering problem, in that we have inserted the *RC* filter circuit between source and load circuits (where the source and load are simply represented in equivalent form). To determine the response of the circuit, we compute the Thévenin equivalent representation of the circuit with respect to the load, as shown in Figure 6.20. Let $R' = R_S + R_1$ and

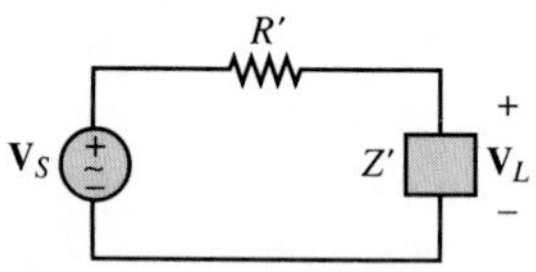

Figure 6.20 Equivalent circuit representation of Figure 6.19

$$Z' = R_L \| \frac{1}{j\omega C} = \frac{R_L}{1 + j\omega C R_L}$$

Then the circuit response may be computed as follows:

$$\begin{aligned}\frac{\mathbf{V}_L}{\mathbf{V}_S}(j\omega) &= \frac{Z'}{R' + Z'} \\ &= \frac{R_L/(1 + j\omega C R_L)}{R_S + R_1 + R_L/(1 + j\omega C R_L)} \\ &= \frac{R_L}{R_L + R_S + R_1 + j\omega C R_L (R_S + R_1)} \\ &= \frac{R_L/(R_L + R')}{1 + j\omega C R_L \| R'}\end{aligned}$$

The above expression can be written as follows:

$$H(j\omega) = \frac{R_L/(R_L + R')}{1 + j\omega C R_L \| R'} = \frac{K}{1 + j\omega C R_{\text{EQ}}} = \frac{0.667}{1 + j(\omega/600)}$$

Comments: Note the similarity and difference between the above expression and equation 6.27: The numerator is different from 1, because of the voltage divider effect resulting from the source and load resistances, and the cutoff frequency is given by the expression

$$\omega_0 = \frac{1}{C R_{\text{EQ}}}$$

CHECK YOUR UNDERSTANDING

Connect the filter of Example 6.7 to a 1-V sinusoidal source with internal resistance of 50 Ω to form a circuit similar to that of Figure 6.19. Determine the circuit cutoff frequency ω_0 if the load resistance is 470 Ω

Answer: $\omega_0 = 6{,}553.3$ rad/s

EXAMPLE 6.9 Filter Attenuation

Problem

A low-pass filter has the frequency response given by the function shown below. Determine at which frequency the output of the filter has magnitude equal to 10 percent of the magnitude of the input.

$$H(j\omega)\frac{K}{(j\omega/\omega_1 + 1)(j\omega/\omega_2 + 1)}$$

Solution

Known Quantities: Frequency response function of a filter.

Find: Frequency $\omega_{10\%}$ at which the output peak amplitude is equal to 10 percent of the input peak amplitude.

Schematics, Diagrams, Circuits, and Given Data: $K = 1$; $\omega_1 = 100$; $\omega_2 = 1{,}000$.

Assumptions: None.

Analysis: The statement of the problem is equivalent to asking for what value of ω the magnitude of the frequency response is equal to $0.1K$. Since $K = 1$, we can formulate the problem as follows.

$$|H(j\omega)| = \left|\frac{K}{(j\omega/\omega_1 + 1)(j\omega/\omega_2 + 1)}\right| = 0.1K$$

$$\frac{1}{\sqrt{(1 - \omega^2/\omega_1\omega_2)^2 + \omega^2(1/\omega_1 + 1/\omega_2)^2}} = 0.1$$

Now let $\Omega = \omega^2$, and expand the above expression:

$$\left(1 - \frac{\Omega}{\omega_1\omega_2}\right)^2 + \Omega\left(\frac{1}{\omega_1} + \frac{1}{\omega_2}\right)^2 = 100$$

$$\Omega^2 + \left[(\omega_1\omega_2)^2\left(\frac{1}{\omega_1} + \frac{1}{\omega_2}\right)^2 - 2\omega_1\omega_2\right]\Omega - 99(\omega_1\omega_2)^2 = 0$$

Substituting numerical values in the expression, we obtain a quadratic equation that can be solved to obtain the roots $\Omega = -1.6208 \times 10^6$ and $\Omega = 0.6108 \times 10^6$. Selecting the positive root as the only physically possible solution (negative frequencies do not have a physical meaning), we can then solve for $\omega = \sqrt{\Omega} = 782$ rad/s. Figure 6.21(a) depicts the magnitude response of the filter; you can see that around the frequency of 300 rad/s, the magnitude response is indeed close to 0.1. The phase response is shown in Figure 6.21(b).

Comments: This type of problem, which recurs in the homework assignments, can be solved numerically, as done above, or graphically, as illustrated in the next exercise.

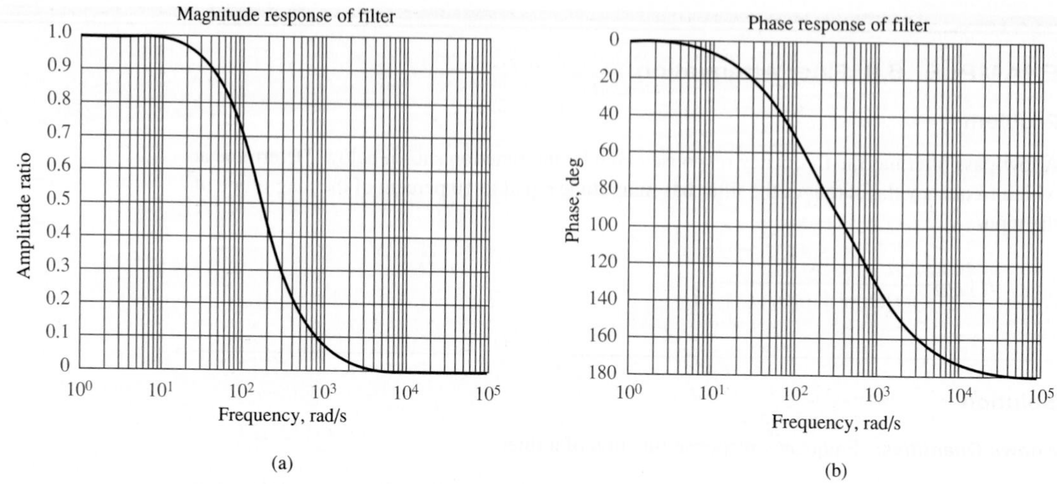

Figure 6.21 Frequency response of filter of Example 6.9. (a) Magnitude response; (b) phase response

CHECK YOUR UNDERSTANDING

Use the phase response plot of Figure 6.21(b) to determine at which frequency the phase shift introduced in the input signal by the filter is equal to $-90°$.

Answer: $\omega = 300$ rad/s (approximately)

Much more complex low-pass filters than the simple *RC* combinations shown so far can be designed by using appropriate combinations of various circuit elements. The synthesis of such advanced filter networks is beyond the scope of this book; however, we discuss the practical implementation of some commonly used filters in Chapters 8 and 15, in connection with the discussion of the operational amplifier. The next two sections extend the basic ideas introduced in the preceding pages to high-pass and bandpass filters, that is, to filters that emphasize the higher frequencies or a band of frequencies, respectively.

High-Pass Filters

RC high-pass filter. The circuit preserves higher frequencies while attenuating the frequencies below the cutoff frequency $\omega_0 = 1/RC$.

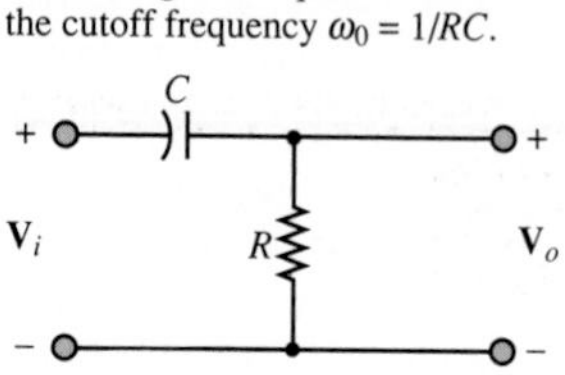

Figure 6.22 High-pass filter

Just as you can construct a simple filter that preserves low frequencies and attenuates higher frequencies, you can easily construct a **high-pass filter** that passes mainly those frequencies *above a certain cutoff frequency*. The analysis of a simple high-pass filter can be conducted by analogy with the preceding discussion of the low-pass filter. Consider the circuit shown in Figure 6.22. The frequency response for the high-pass filter

$$H(j\omega) = \frac{\mathbf{V}_o}{\mathbf{V}_i}(j\omega)$$

may be obtained by noting that

$$\mathbf{V}_o(j\omega) = \mathbf{V}_i(j\omega)\frac{R}{R + 1/j\omega C} = \mathbf{V}_i(j\omega)\frac{j\omega CR}{1 + j\omega CR} \tag{6.35}$$

Thus, the frequency response of the filter is

$$\frac{\mathbf{V}_o}{\mathbf{V}_i}(j\omega) = \frac{j\omega CR}{1 + j\omega CR} \tag{6.36}$$

which can be expressed in magnitude-and-phase form by

$$H(j\omega) = \frac{\mathbf{V}_o}{\mathbf{V}_i}(j\omega) = \frac{j\omega CR}{1 + j\omega CR} = \frac{\omega CR e^{j\pi/2}}{\sqrt{1 + (\omega CR)^2} e^{j\arctan(\omega CR/1)}}$$

$$= \frac{\omega CR}{\sqrt{1 + (\omega CR)^2}} \cdot e^{j[\pi/2 - \arctan(\omega CR)]}$$

or **(6.37)**

$$H(j\omega) = |H| e^{j\angle H}$$

with

$$|H(j\omega)| = \frac{\omega CR}{\sqrt{1 + (\omega CR)^2}} \tag{6.38}$$

$$\angle H(j\omega) = 90^\circ - \arctan(\omega CR)$$

You can verify by inspection that the amplitude response of the high-pass filter will be zero at $\omega = 0$ and will asymptotically approach 1 as ω approaches infinity, while the phase shift is $\pi/2$ at $\omega = 0$ and tends to zero for increasing ω. Amplitude-and-phase response curves for the high-pass filter are shown in Figure 6.23. These plots have been normalized to have the filter cutoff frequency $\omega_0 = 1$ rad/s. Note that, once again, it is possible to define a cutoff frequency at $\omega_0 = 1/RC$ in the same way as was done for the low-pass filter.

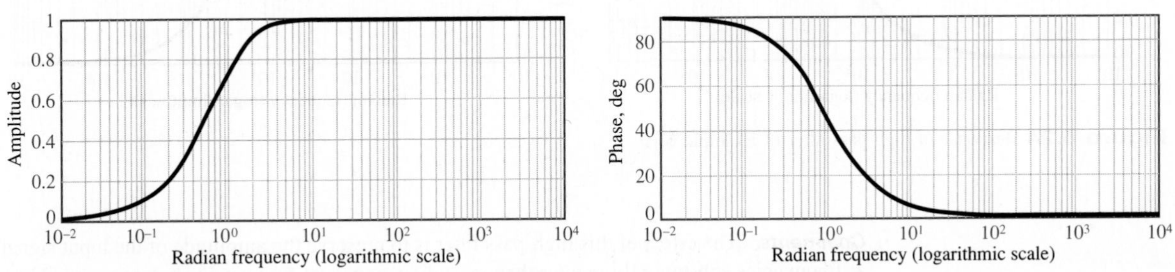

Figure 6.23 Frequency response of a high-pass filter

EXAMPLE 6.10 Frequency Response of *RC* High-Pass Filter

LO3

Problem

Compute the response of the *RC* filter in the circuit of Figure 6.22. Evaluate the response of the filter at $\omega = 2\pi \times 100$ and $2\pi \times 10{,}000$ rad/s.

Solution

Known Quantities: $R = 200\ \Omega$; $C = 0.199\ \mu\text{F}$.

Find: The frequency response $H_V(j\omega)$.

Assumptions: None.

Analysis: We first recognize that the cutoff frequency of the high-pass filter is $\omega_0 = 1/RC = 2\pi \times 4{,}000$ rad/s. Next, we write the frequency response as in equation 6.36:

$$H_V(j\omega) = \frac{\mathbf{V}_o}{\mathbf{V}_i}(j\omega) = \frac{j\omega CR}{1 + j\omega CR}$$

$$= \frac{\omega/\omega_0}{\sqrt{1 + (\omega/\omega_0)^2}} \angle\left(\frac{\pi}{2} - \arctan\left(\frac{\omega}{\omega_0}\right)\right)$$

We can now evaluate the response at the two frequencies:

$$H_V(\omega = 2\pi \times 100) = \frac{100/4{,}000}{\sqrt{1 + (100/4{,}000)^2}} \angle\left(\frac{\pi}{2} - \arctan\left(\frac{100}{4{,}000}\right)\right) = 0.025\angle 1.546$$

$$H_V(\omega = 2\pi \times 10{,}000) = \frac{10{,}000/4{,}000}{\sqrt{1 + (10{,}000/4{,}000)^2}} \angle\left(\frac{\pi}{2} - \arctan\left(\frac{10{,}000}{4{,}000}\right)\right)$$

$$= 0.929\angle 0.38$$

The frequency response plots are shown in Figure 6.24.

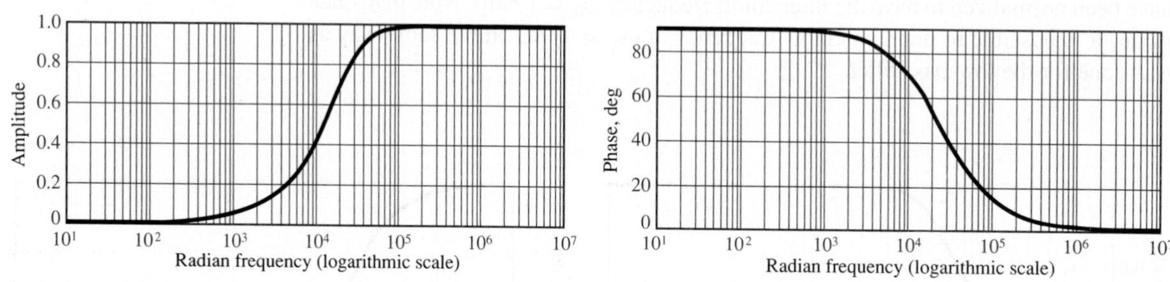

Figure 6.24 Response of high-pass filter of Example 6.10

Comments: The effect of this high-pass filter is to preserve the amplitude of the input signal at frequencies substantially greater than ω_0, while signals at frequencies below ω_0 would be strongly attenuated. With $\omega_0 = 2\pi \times 4{,}000$ (that is, 4,000 Hz), this filter would pass only the *treble range* of the audio frequency spectrum.

CHECK YOUR UNDERSTANDING

Determine the cutoff frequency for each of the four "prototype" filters shown below. Which are high-pass and which are low-pass?

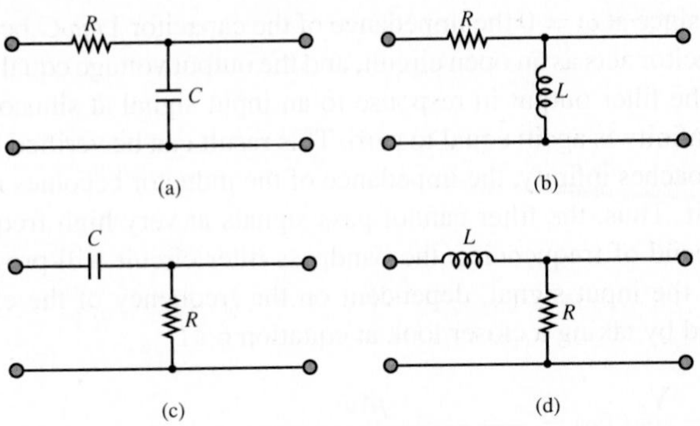

Show that it is possible to obtain a high-pass filter response simply by substituting an inductor for the capacitor in the circuit of Figure 6.16. Derive the frequency response for the circuit.

Answers: (a) $\omega_0 = \dfrac{1}{RC}$ (low); (b) $\omega_0 = \dfrac{R}{L}$ (high); (c) $\omega_0 = \dfrac{1}{RC}$ (high); (d) $\omega_0 = \dfrac{R}{L}$ (low)

$$|H(j\omega)| = \frac{\omega L/R}{\sqrt{1+(\omega L/R)^2}} \qquad \angle H(j\omega) = 90^\circ + \arctan\left(\frac{-\omega L}{R}\right)$$

Bandpass Filters, Resonance, and Quality Factor

Building on the principles developed in the preceding sections, we can also construct a circuit that acts as a **bandpass filter,** passing mainly those frequencies *within a certain frequency range*. The analysis of a simple *second-order* bandpass filter (i.e., a filter with two energy storage elements) can be conducted by analogy with the preceding discussions of the low-pass and high-pass filters. Consider the circuit shown in Figure 6.25 and the related frequency response function for the filter

RLC bandpass filter. The circuit preserves frequencies within a band.

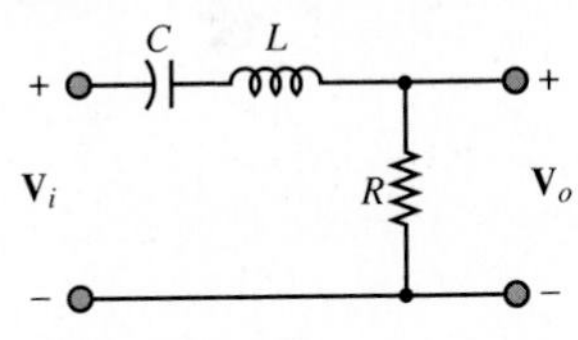

Figure 6.25 *RLC* bandpass filter

$$H(j\omega) = \frac{\mathbf{V}_o}{\mathbf{V}_i}(j\omega)$$

Noting that

$$\mathbf{V}_o(j\omega) = \mathbf{V}_i(j\omega)\frac{R}{R + 1/j\omega C + j\omega L} = \mathbf{V}_i(j\omega)\frac{j\omega CR}{1 + j\omega CR + (j\omega)^2 LC} \qquad \textbf{(6.39)}$$

we may write the frequency response of the filter as

$$\frac{\mathbf{V}_o}{\mathbf{V}_i}(j\omega) = \frac{j\omega CR}{1 + j\omega CR + (j\omega)^2 LC} \qquad \textbf{(6.40)}$$

Equation 6.40 can often be factored into the form

$$\frac{\mathbf{V}_o}{\mathbf{V}_i}(j\omega) = \frac{jA\omega}{(j\omega/\omega_1 + 1)(j\omega/\omega_2 + 1)} \qquad \textbf{(6.41)}$$

where ω_1 and ω_2 are the two frequencies that determine the **passband** (or **bandwidth**) of the filter—that is, the frequency range over which the filter "passes" the input signal—and A is a constant that results from the factoring. An immediate observation we can make is that if the signal frequency ω is zero, the response of the filter is equal to zero, since at $\omega = 0$ the impedance of the capacitor $1/j\omega C$ becomes infinite. Thus, the capacitor acts as an open circuit, and the output voltage equals zero. Further, we note that the filter output in response to an input signal at sinusoidal frequency approaching infinity is again equal to zero. This result can be verified by considering that as ω approaches infinity, the impedance of the inductor becomes infinite, that is, an open circuit. Thus, the filter cannot pass signals at very high frequencies. In an intermediate band of frequencies, the bandpass filter circuit will provide a variable attenuation of the input signal, dependent on the frequency of the excitation. This may be verified by taking a closer look at equation 6.41:

$$\begin{aligned} H(j\omega) &= \frac{\mathbf{V}_o}{\mathbf{V}_i}(j\omega) = \frac{jA\omega}{(j\omega/\omega_1 + 1)(j\omega/\omega_2 + 1)} \\ &= \frac{A\omega e^{j\pi/2}}{\sqrt{1+(\omega/\omega_1)^2}\sqrt{1+(\omega/\omega_2)^2}e^{j\arctan(\omega/\omega_1)}e^{j\arctan(\omega/\omega_2)}} \\ &= \frac{A\omega}{\sqrt{[1+(\omega/\omega_1)^2][1+(\omega/\omega_2)^2]}} e^{j[\pi/2 - \arctan(\omega/\omega_1) - \arctan(\omega/\omega_2)]} \end{aligned} \tag{6.42}$$

Equation 6.42 is of the form $H(j\omega) = |H|e^{j\angle H}$, with

$$|H(j\omega)| = \frac{A\omega}{\sqrt{[1+(\omega/\omega_1)^2][1+(\omega/\omega_2)^2]}}$$

and **(6.43)**

$$\angle H(j\omega) = \frac{\pi}{2} - \arctan\frac{\omega}{\omega_1} - \arctan\frac{\omega}{\omega_2}$$

The magnitude and phase plots for the frequency response of the bandpass filter of Figure 6.25 are shown in Figure 6.26. These plots have been normalized to have the filter passband centered at the frequency $\omega = 1$ rad/s.

The frequency response plots of Figure 6.26 suggest that, in some sense, the bandpass filter acts as a combination of a high-pass and a low-pass filter. As illustrated in the previous cases, it should be evident that one can adjust the filter response as desired simply by selecting appropriate values for L, C, and R.

Resonance and Bandwidth

The response of second-order filters can be explained more generally by rewriting the frequency response function of the second-order bandpass filter of Figure 6.25 in the following forms:

$$\begin{aligned} \frac{V_o}{V_i}(j\omega) &= \frac{j\omega CR}{LC(j\omega)^2 + j\omega CR + 1} \\ &= \frac{(2\zeta/\omega_n)j\omega}{(j\omega/\omega_n)^2 + (2\zeta/\omega_n)j\omega + 1} \\ &= \frac{(1/Q\omega_n)j\omega}{(j\omega/\omega_n)^2 + (1/Q\omega_n)j\omega + 1} \end{aligned} \tag{6.44}$$

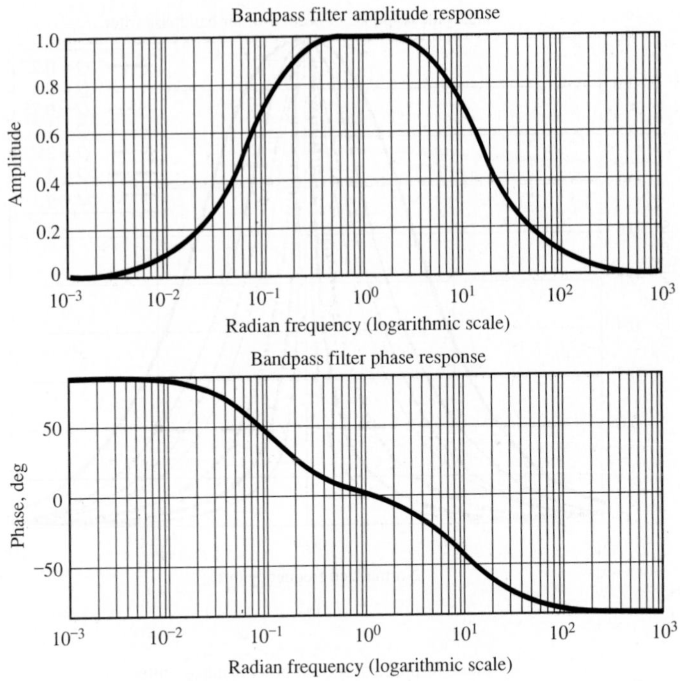

Figure 6.26 Frequency response of *RLC* bandpass filter

with the following definitions:[2]

$$\begin{aligned} \omega_n &= \sqrt{\frac{1}{LC}} = \text{natural or resonant frequency} \\ Q &= \frac{1}{2\zeta} = \frac{1}{\omega_n CR} = \frac{1}{R}\sqrt{\frac{L}{C}} = \text{quality factor} \\ \zeta &= \frac{1}{2Q} = \frac{R}{2}\sqrt{\frac{C}{L}} = \text{damping ratio} \end{aligned} \tag{6.45}$$

LO3

Figure 6.27 depicts the normalized frequency response (magnitude and phase) of the second-order bandpass filter for $\omega_n = 1$ and various values of Q (and ζ). The peak displayed in the frequency response around the frequency ω_n is called a *resonant peak*, and ω_n is the **resonant frequency.** Note that as the **quality factor** Q increases, the sharpness of the resonance increases and the filter becomes increasingly *selective* (i.e., it has the ability to filter out most frequency components of the input signals except for a narrow band around the resonant frequency). One measure of the selectivity of a bandpass filter is its **bandwidth.** The concept of bandwidth can be easily visualized in the plot of Figure 6.27(a) by drawing a horizontal line across the plot

---

[2]If you have already studied the section on second-order transient response in Chapter 5, you will recognize the parameters ζ and ω_n.

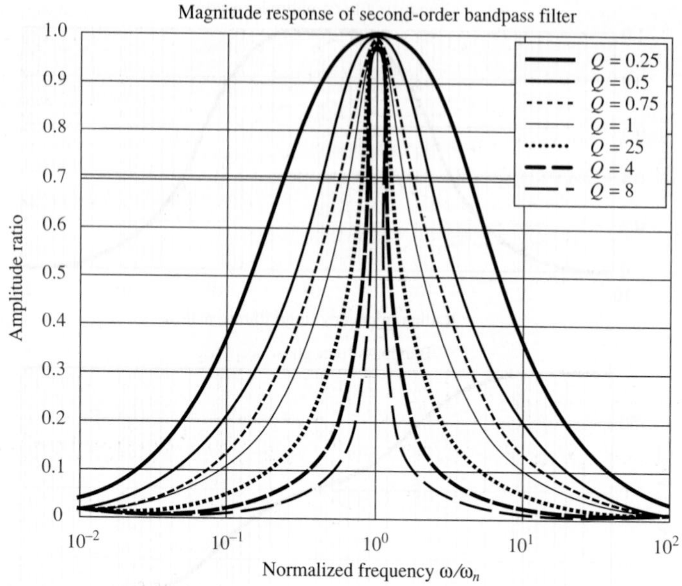

(a)

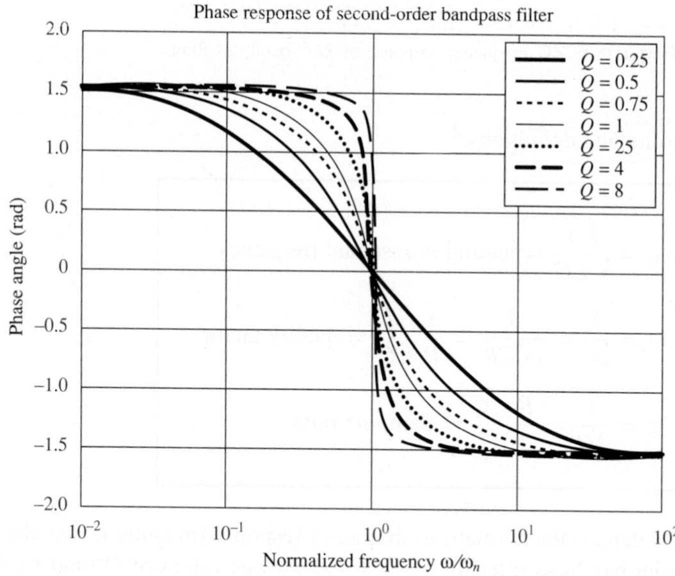

(b)

Figure 6.27 (a) Normalized magnitude response of second-order bandpass filter; (b) normalized phase response of second-order bandpass filter

(we have chosen to draw it at the amplitude ratio value of 0.707 for reasons that will be explained shortly). The frequency range between (magnitude) frequency response points intersecting this horizontal line is defined as the **half-power bandwidth** of the filter. The name *half-power* stems from the fact that when the amplitude response is equal to 0.707 (or $1/\sqrt{2}$), the voltage (or current) at the output of the filter has decreased by the same factor, relative to the maximum value (at the resonant frequency). Since power in an electric signal is proportional to the square of the voltage or current, a drop by a factor $1/\sqrt{2}$ in the output voltage or current corresponds to the power being reduced by a factor of $\frac{1}{2}$. Thus, we term the frequencies at which the intersection of the 0.707 line with the frequency response occurs the **half-power frequencies.** Another useful definition of bandwidth B is as follows. We shall make use of this definition in the following examples. Note that a high-Q filter has a narrow bandwidth, and a low-Q filter has a wide bandwidth.

$$B = \frac{\omega_n}{Q} \qquad \text{bandwidth} \tag{6.46}$$

LO3

EXAMPLE 6.11 Frequency Response of Bandpass Filter

LO3

Problem

Compute the frequency response of the bandpass filter of Figure 6.25 for two sets of component values.

Solution

Known Quantities:

(a) $R = 1\ \text{k}\Omega$; $C = 10\ \mu\text{F}$; $L = 5$ mH.

(b) $R = 10\ \Omega$; $C = 10\ \mu\text{F}$; $L = 5$ mH.

Find: The frequency response $H_V(j\omega)$.

Assumptions: None.

Analysis: We write the frequency response of the bandpass filter as in equation 6.40:

$$H_V(j\omega) = \frac{\mathbf{V}_o}{\mathbf{V}_i}(j\omega) = \frac{j\omega CR}{1 + j\omega CR + (j\omega)^2 LC}$$

$$= \frac{\omega CR}{\sqrt{\left(1-\omega^2 LC\right)^2 + (\omega CR)^2}} \angle \left[\frac{\pi}{2} - \arctan\left(\frac{\omega CR}{1-\omega^2 LC}\right)\right]$$

We can now evaluate the response for two different values of the series resistance. The frequency response plots for case a (large series resistance) are shown in Figure 6.28. Those for case b (small series resistance) are shown in Figure 6.29. Let us calculate some quantities for each case. Since L and C are the same in both cases, the *resonant frequency* of the two circuits will be the same:

$$\omega_n = \frac{1}{\sqrt{LC}} = 4.47 \times 10^3 \text{ rad/s}$$

On the other hand, the *quality factor* Q will be substantially different:

$$Q_a = \frac{1}{\omega_n CR} \approx 2.22 \qquad \text{case a}$$

$$Q_b = \frac{1}{\omega_n CR} \approx 0.022 \qquad \text{case b}$$

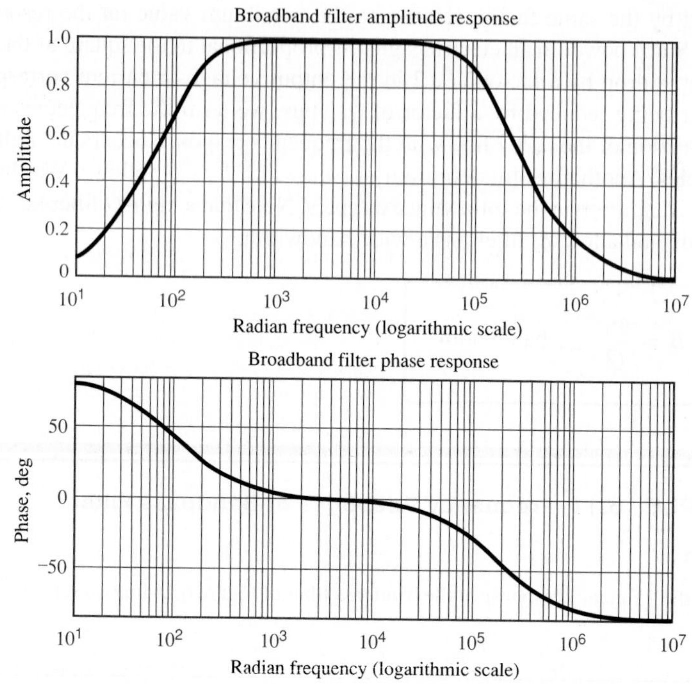

Figure 6.28 Frequency response of broadband bandpass filter of Example 6.11

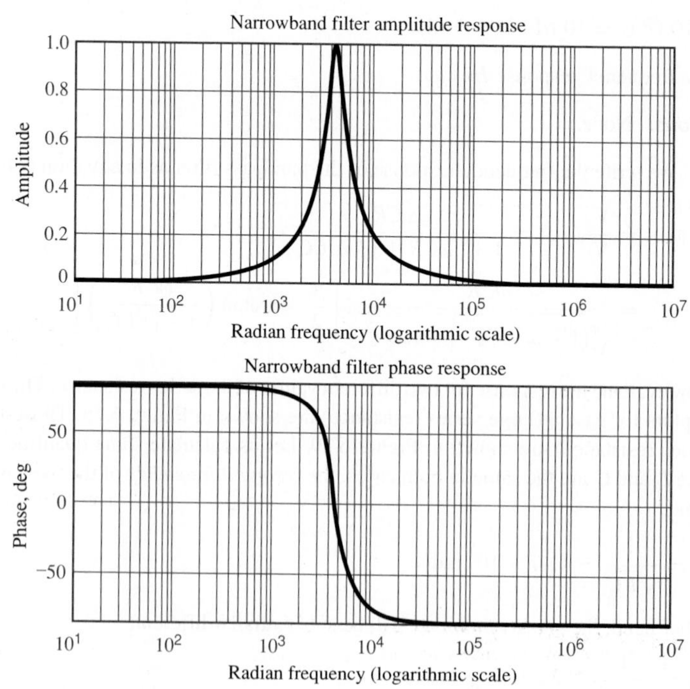

Figure 6.29 Frequency response of narrowband bandpass filter of Example 6.11

From these values of Q we can calculate the approximate bandwidth of the two filters:

$$B_a = \frac{\omega_n}{Q_a} \approx 10{,}000 \text{ rad/s} \qquad \text{case a}$$

$$B_b = \frac{\omega_n}{Q_b} \approx 100 \text{ rad/s} \qquad \text{case b}$$

The frequency response plots in Figures 6.28 and 6.29 confirm these observations.

Comments: It should be apparent that while at the higher and lower frequencies most of the amplitude of the input signal is filtered from the output, at the midband frequency (4,500 rad/s) most of the input signal amplitude passes through the filter. The first bandpass filter analyzed in this example would "pass" the *midband range* of the audio spectrum, while the second would pass only a very narrow band of frequencies around the **center frequency** of 4,500 rad/s. Such narrowband filters find application in **tuning circuits,** such as those employed in conventional AM radios (although at frequencies much higher than that of the present example). In a tuning circuit, a narrowband filter is used to tune in a frequency associated with the **carrier** of a radio station (e.g., for a station found at a setting of AM 820, the carrier wave transmitted by the radio station is at a frequency of 820 kHz). By using a variable capacitor, it is possible to tune in a range of carrier frequencies and therefore select the preferred station. Other circuits are then used to decode the actual speech or music signal modulated on the carrier wave; some of these are discussed in Chapters 9 and 19.

CHECK YOUR UNDERSTANDING

Compute the frequencies ω_1 and ω_2 for the bandpass filter of Example 6.11 (with $R = 1\text{ k}\Omega$) for equating the magnitude of the bandpass filter frequency response to $1/\sqrt{2}$ (this will result in a quadratic equation in ω, which can be solved for the two frequencies). Note that these are the **half-power frequencies.**

Answer: $\omega_1 = 99.95$ rad/s; $\omega_2 = 200.1$ krad/s

Wheatstone Bridge Filter

LO3

FIND IT ON THE WEB

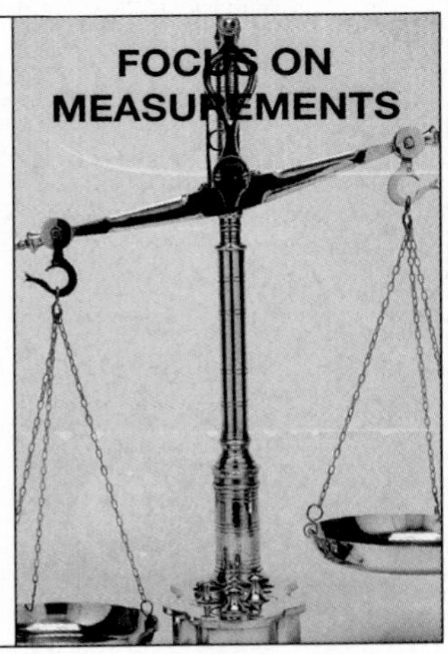

Problem:

The Wheatstone bridge circuit of Example 2.14 and Focus on Measurements, "Wheatstone Bridge and Force Measurements" in Chapter 2 is used in a number of instrumentation applications, including the **measurement of force.** Figure 6.30 depicts the appearance of the bridge circuit. When undesired noise and interference are present in a measurement, it is often appropriate to use a low-pass filter to reduce the effect of the noise. The capacitor that is connected to the output terminals of the bridge in Figure 6.30 constitutes an effective and simple low-pass filter, in conjunction with the bridge resistance. Assume that the average resistance of each leg of the bridge is 350 Ω (a standard value for strain gauges) and that we desire to measure a sinusoidal

(Continued)

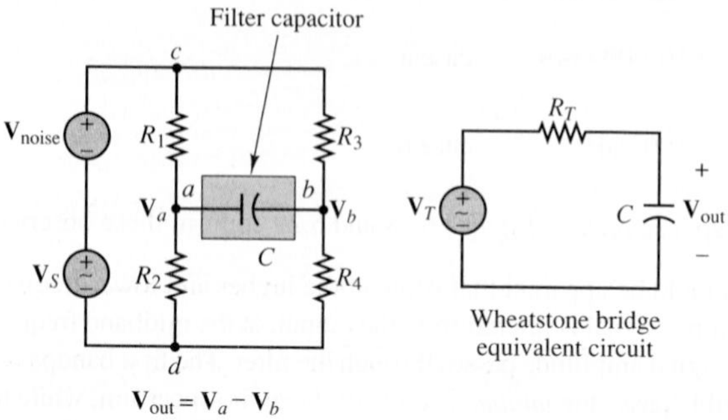

Figure 6.30 Wheatstone bridge with equivalent circuit and simple capacitive filter

force at a frequency of 30 Hz. From prior measurements, it has been determined that a filter with a cutoff frequency of 300 Hz is sufficient to reduce the effects of noise. Choose a capacitor that matches this filtering requirement.

Solution:

By evaluating the Thévenin equivalent circuit for the Wheatstone bridge, calculating the desired value for the filter capacitor becomes relatively simple, as illustrated on the right side of Figure 6.30. The Thévenin resistance for the bridge circuit may be computed by short-circuiting the two voltage sources and removing the capacitor placed across the load terminals:

$$R_T = R_1 \parallel R_2 + R_3 \parallel R_4 = 350 \parallel 350 + 350 \parallel 350 = 350\ \Omega$$

Since the required cutoff frequency is 300 Hz, the capacitor value can be computed from the expression

$$\omega_0 = \frac{1}{R_T C} = 2\pi \times 300$$

or

$$C = \frac{1}{R_T \omega_0} = \frac{1}{350 \times 2\pi \times 300} = 1.51\ \mu\text{F}$$

The frequency response of the bridge circuit is of the same form as equation 6.27:

$$\frac{\mathbf{V}_{out}}{\mathbf{V}_T}(j\omega) = \frac{1}{1 + j\omega C R_T}$$

This response can be evaluated at the frequency of 30 Hz to verify that the attenuation and phase shift at the desired signal frequency are minimal:

$$\frac{\mathbf{V}_{out}}{\mathbf{V}_T}(j\omega = j2\pi \times 30) = \frac{1}{1 + j2\pi \times 30 \times 1.51 \times 10^{-6} \times 350}$$
$$= 0.9951\angle(-5.7^\circ)$$

Figure 6.31 depicts the appearance of a 30-Hz sinusoidal signal before and after the addition of the capacitor to the circuit.

(Continued)

(Concluded)

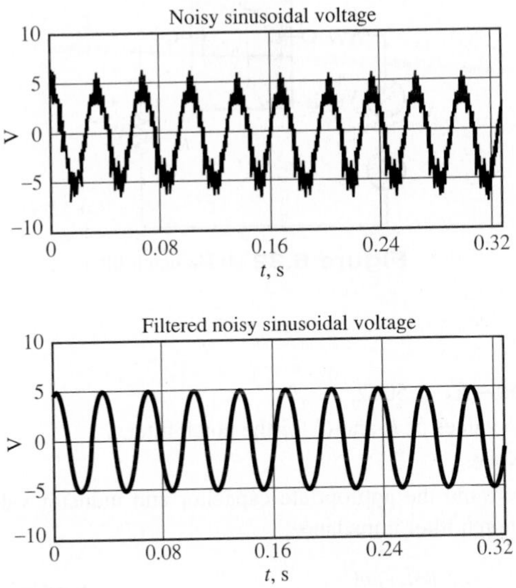

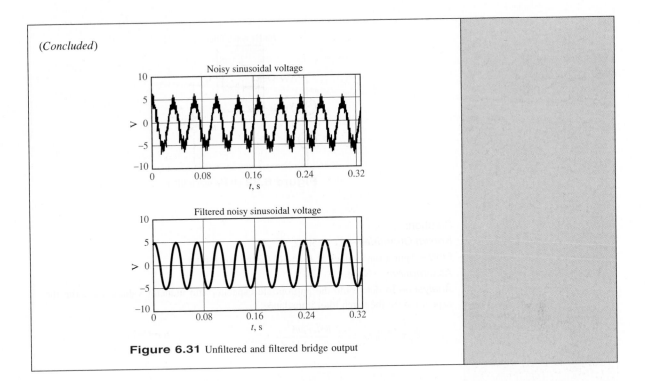

Figure 6.31 Unfiltered and filtered bridge output

AC Line Interference Filter LO3

FIND IT ON THE WEB

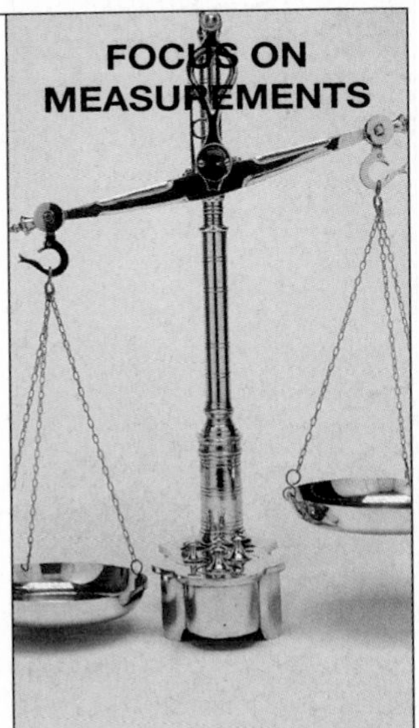

Problem:

One application of narrowband filters is seen in rejecting interference due to AC line power. Any undesired 60-Hz signal originating in the AC line power can cause serious interference in sensitive instruments. In medical instruments such as the **electrocardiograph,** 60-Hz notch filters are often provided to reduce the effect of this interference[3] on cardiac measurements. Figure 6.32 depicts a circuit in which the effect of 60-Hz noise is represented by way of a 60-Hz sinusoidal generator connected in series with a signal source ($\mathbf{V}_S$), representing the desired signal. In this example we design a 60-Hz narrowband (or *notch*) filter to remove the unwanted 60-Hz noise.

[3] See Focus on Measurements: Electrocardiogram Amplifier in Chapter 8 and Section 15.2 for further information on electrocardiograms and line noise, respectively.

(Continued)

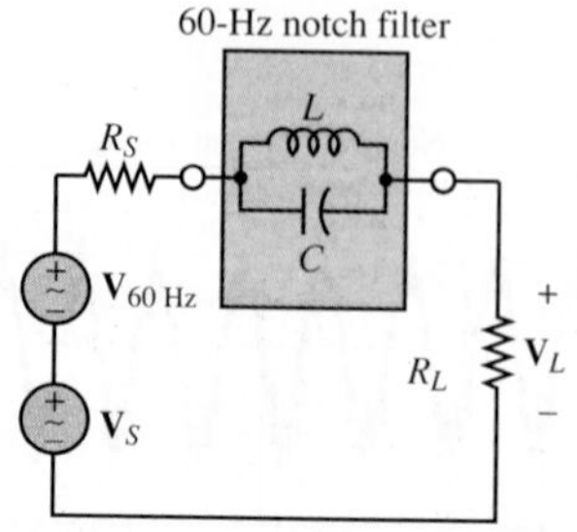

Figure 6.32 60-Hz notch filter

Solution:

Known Quantities—$R_S = 50\ \Omega$.

Find—Appropriate values of L and C for the notch filter.

Assumptions—None.

Analysis—To determine the appropriate capacitor and inductor values, we write the expression for the notch filter impedance:

$$Z_{\|} = Z_L \| Z_C = \frac{j\omega L / j\omega C}{j\omega L + 1/j\omega C}$$

$$= \frac{j\omega L}{1 - \omega^2 LC}$$

Note that when $\omega^2 LC = 1$, the impedance of the circuit is infinite! The frequency

$$\omega_0 = \frac{1}{\sqrt{LC}}$$

is the resonant frequency of the LC circuit. If this resonant frequency were selected to be equal to 60 Hz, then the series circuit would show an infinite impedance to 60-Hz currents, and would therefore block the interference signal, while passing most of the other frequency components. We thus select values of L and C that result in $\omega_0 = 2\pi \times 60$. Let $L = 100$ mH. Then

$$C = \frac{1}{\omega_0^2 L} = 70.36\ \mu\text{F}$$

The frequency response of the complete circuit is given below:

$$H_V(j\omega) = \frac{\mathbf{V}_o(j\omega)}{\mathbf{V}_i(j\omega)} = \frac{R_L}{R_S + R_L + Z_{\|}}$$

$$= \frac{R_L}{R_S + R_L + j\omega L/(1 - \omega^2 LC)}$$

and is plotted in Figure 6.33.

Comments—It would be instructive for you to calculate the response of the notch filter at frequencies in the immediate neighborhood of 60 Hz, to verify the attenuation effect of the notch filter.

(Continued)

(Concluded)

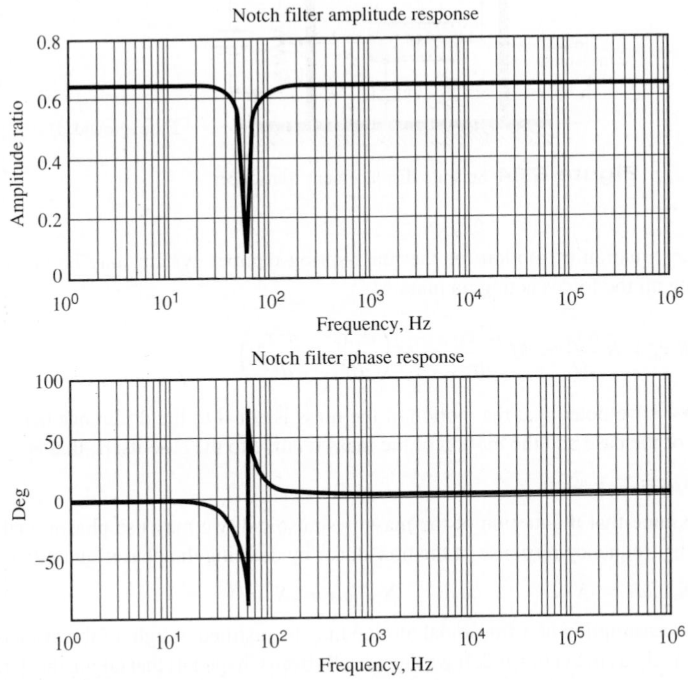

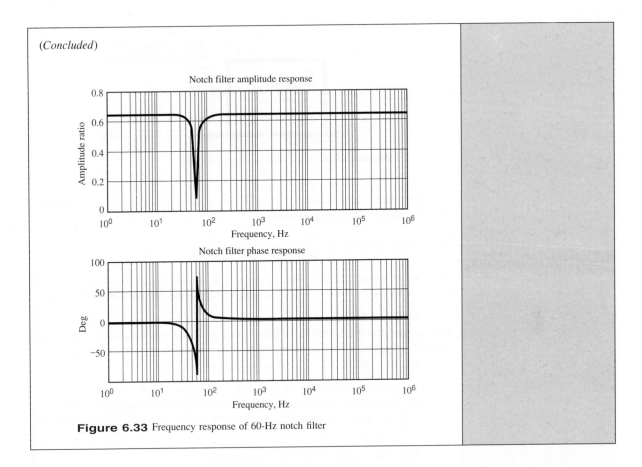

Figure 6.33 Frequency response of 60-Hz notch filter

Seismic Transducer

LO3

FIND IT ON THE WEB

This example illustrates the application of the frequency response idea to a practical displacement transducer. The frequency response of a **seismic displacement transducer** is analyzed, and it is shown that there is an analogy between the equations describing the mechanical transducer and those that describe a second-order electric circuit.

The configuration of the transducer is shown in Figure 6.34. The transducer is housed in a case rigidly affixed to the surface of a body whose motion is to be measured. Thus, the case will experience the same displacement as the body, x_i. Inside the case, a small mass M rests on a spring characterized by stiffness K, placed in parallel with a damper B. The wiper arm of a potentiometer is connected to the floating mass M; the potentiometer is attached to the transducer case, so that the voltage V_o is proportional to the *relative displacement of the mass with respect to the case* x_o.

(Continued)

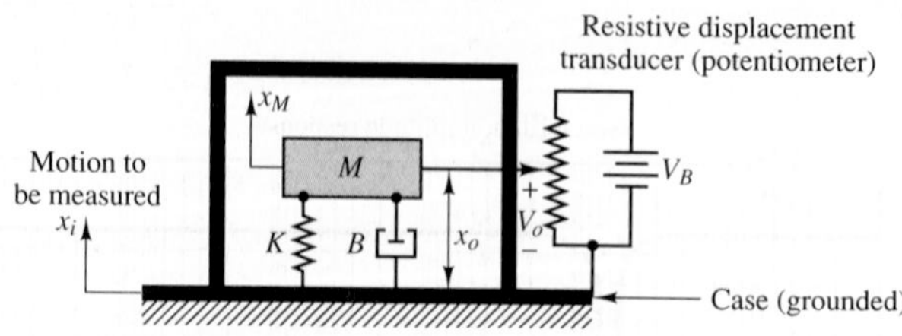

Figure 6.34 Seismic displacement transducer

The equation of motion for the mass-spring-damper system may be obtained by summing all the forces acting on mass M:

$$Kx_o + B\frac{dx_o}{dt} = M\frac{d^2x_M}{dt^2} = M\left(\frac{d^2x_i}{dt^2} - \frac{d^2x_o}{dt^2}\right)$$

where we have noted that the motion of the mass is equal to the difference between the motion of the case and the motion of the mass relative to the case itself; that is,

$$x_M = x_i - x_o$$

If we assume that the motion of the mass is sinusoidal, we may use phasor analysis to obtain the frequency response of the transducer by defining the phasor quantities

$$\mathbf{X}_i(j\omega) = |X_i|e^{j\phi_i} \quad \text{and} \quad \mathbf{X}_o(j\omega) = |X_o|e^{j\phi_o}$$

The assumption of a sinusoidal motion may be justified in light of the discussion of Fourier analysis in Section 6.2. If we then recall (from Chapter 4) that taking the derivative of a phasor corresponds to multiplying the phasor by $j\omega$, we can rewrite the second-order differential equation as follows:

$$M(j\omega)^2\mathbf{X}_o + B(j\omega)\mathbf{X}_o + K\mathbf{X}_o = M(j\omega)^2\mathbf{X}_i$$

$$(-\omega^2 M + j\omega B + K)\mathbf{X}_o = -\omega^2 M\mathbf{X}_i$$

and we can write an expression for the frequency response:

$$\frac{\mathbf{X}_o(j\omega)}{\mathbf{X}_i(j\omega)} = H(j\omega) = \frac{-\omega^2 M}{-\omega^2 M + j\omega B + K}$$

The frequency response of the transducer is plotted in Figure 6.35 for the component values $M = 0.005$ kg and $K = 1{,}000$ N/m and for three values of B:

$B = 10$ N-s/m (dotted line)

$B = 2$ N-s/m (dashed line)

and

$B = 1$ N-s/m (solid line)

The transducer clearly displays a high-pass response, indicating that for a sufficiently high input signal frequency, the measured displacement (proportional to the voltage V_o) is equal to the input displacement x_i, which is the desired quantity. Note how sensitive the frequency response of the transducer is to changes in damping: as B changes from 2 to 1, a sharp **resonant peak** appears around the frequency $\omega = 316$ rad/s (approximately 50 Hz). As B increases to a value of 10, the amplitude response curve shifts to the right.

(*Continued*)

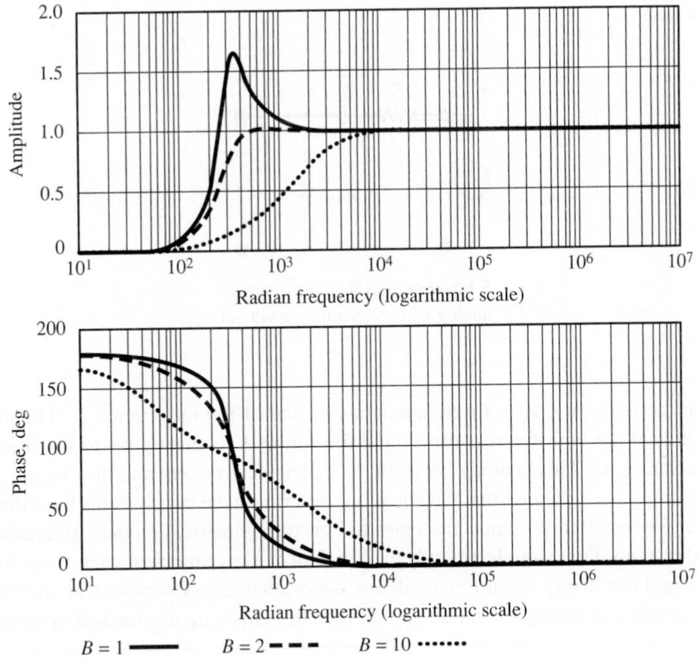

Figure 6.35 Frequency response of seismic transducer

Thus, this transducer, with the preferred damping given by $B = 2$, would be capable of correctly measuring displacements at frequencies above a minimum value, about 1,000 rad/s (or 159 Hz). The choice of $B = 2$ as the preferred design may be explained by observing that, ideally, we would like to obtain a constant amplitude response at all frequencies. The magnitude response that most closely approximates the ideal case in Figure 6.35 corresponds to $B = 2$. This concept is commonly applied to a variety of **vibration measurements.**

FIND IT ON THE WEB

We now illustrate how a second-order electric circuit will exhibit the same type of response as the seismic transducer. Consider the circuit shown in Figure 6.36. The frequency response for the circuit may be obtained by using the principles developed in the preceding sections:

$$\frac{\mathbf{V}_o}{\mathbf{V}_i}(j\omega) = \frac{j\omega L}{R + 1/j\omega C + j\omega L} = \frac{(j\omega L)(j\omega C)}{j\omega CR + 1 + (j\omega L)(j\omega C)}$$

$$= \frac{-\omega^2 L}{-\omega^2 L + j\omega R + 1/C}$$

Comparing this expression with the frequency response of the seismic transducer,

$$\frac{\mathbf{X}_o(j\omega)}{\mathbf{X}_i(j\omega)} = H(j\omega) = \frac{-\omega^2 M}{-\omega^2 M + j\omega B + K}$$

we find that there is a definite resemblance between the two. In fact, it is possible to draw an analogy between input and output motions and input and output voltages. Note

(*Continued*)

(Concluded)

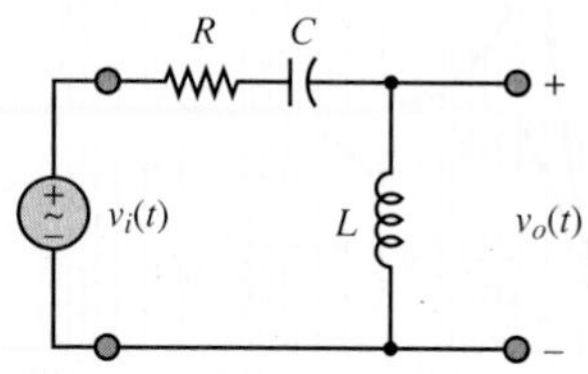

Figure 6.36 Electric circuit analog of the seismic transducer

also that the mass M plays a role analogous to that of the inductance L. The damper B acts in analogy with the resistor R; and the spring K is analogous to the inverse of the capacitance, C. This analogy between the mechanical system and the electric circuit derives simply from the fact that the equations describing the two systems have the same form. Engineers often use such analogies to construct electrical *models*, or *analogs*, of physical systems. For example, to study the behavior of a large mechanical system, it might be easier and less costly to start by modeling the mechanical system with an inexpensive electric circuit and testing the model, rather than the full-scale mechanical system.

6.4 BODE PLOTS

Frequency response plots of linear systems are often displayed in the form of logarithmic plots, called **Bode plots,** where the horizontal axis represents frequency on a logarithmic scale (base 10) and the vertical axis represents the amplitude ratio or phase of the frequency response function. In Bode plots the amplitude ratio is expressed in units of **decibels (dB),** where

$$\left|\frac{A_o}{A_i}\right|_{\text{dB}} = 20\log_{10}\frac{A_o}{A_i} \tag{6.47}$$

The phase shift is expressed in degrees or radians. Frequency is usually plotted on a logarithmic (base-10) scale as well. Note that the use of the decibel units implies that one is measuring a *ratio*. The use of logarithmic scales enables large ranges to be covered. Furthermore, as shown subsequently in this section, frequency response plots of high-order systems may be obtained easily from frequency response plots of the factors of the overall sinusoidal frequency response function, if logarithmic scales are used for amplitude ratio plots. Consider, for example, the *RC* low-pass filter of Example 6.7 (Figure 6.16). The frequency response of this filter can be written in the form

$$\frac{\mathbf{V}_o}{\mathbf{V}_i}(j\omega) = \frac{1}{j\omega/\omega_0 + 1} = \frac{1}{\sqrt{1 + (\omega/\omega_0)^2}}\angle -\tan^{-1}\left(\frac{\omega}{\omega_0}\right) \tag{6.48}$$

where $\tau = RC = 1/\omega_0$ and ω_0 is the cutoff, or half-power, frequency of the filter.

Figure 6.37 shows the frequency response plots (magnitude and phase) for this filter; such plots are termed **Bode plots,** after the mathematician Hendrik W. Bode. The normalized frequency on the horizontal axis is $\omega\tau$. One of the great advantages of

Bode plots is that they permit easy straight-line approximations, as illustrated below. If we express the magnitude frequency response of the first-order filter $(\mathbf{V}_o/\mathbf{V}_i)(j\omega) = K/(1 + j\omega/\omega_0)$ in units of decibels, we have

$$\left|\frac{\mathbf{V}_o}{\mathbf{V}_i}(j\omega)\right|_{\text{dB}} = 20\log_{10}\left|\frac{K}{1 + j\omega/\omega_0}\right| \tag{6.49}$$

$$= 20\log_{10}\frac{K}{\sqrt{1 + (\omega/\omega_0)^2}} = 20\log_{10} K - 10\log_{10}\left[1 + \left(\frac{\omega}{\omega_0}\right)^2\right]$$

If $\omega/\omega_0 \ll 1$, then

$$\left|\frac{\mathbf{V}_o}{\mathbf{V}_i}(j\omega)\right|_{\text{dB}} \approx 20\log_{10} K - 10\log_{10} 1 = 20\log_{10} K \tag{6.50}$$

Thus, the expression 6.40 is well approximated by a straight line of zero slope at very low frequencies (equation 6.50). This is the low-frequency *asymptotic approximation* of the Bode plot.

If $\omega/\omega_0 \gg 1$, we can similarly obtain a high-frequency asymptotic approximation:

$$\left|\frac{\mathbf{V}_o}{\mathbf{V}_i}(j\omega)\right|_{\text{dB}} \approx 20\log_{10} K - 20\log_{10}\frac{\omega}{\omega_0} \tag{6.51}$$

$$= 20\log_{10} K - 20\log_{10}\omega + 20\log_{10}\omega_0$$

Note that equation 6.51 represents a straight line of slope −20 dB per **decade** (factor-of-10 increase in frequency). A decade increase in ω results in an increase in log ω of unity. Note also that when ω equals the cutoff frequency ω_0, the expression for $|(\mathbf{V}_o/\mathbf{V}_i)(j\omega)|$ given by equation 6.51 equals that given by equation 6.50. In other words, the low- and high-frequency asymptotes intersect at ω_0. Thus, the magnitude response Bode plot of a first-order low-pass filter can be easily approximated by two straight lines intersecting at ω_0. Figure 6.37(a) clearly shows the approximation.

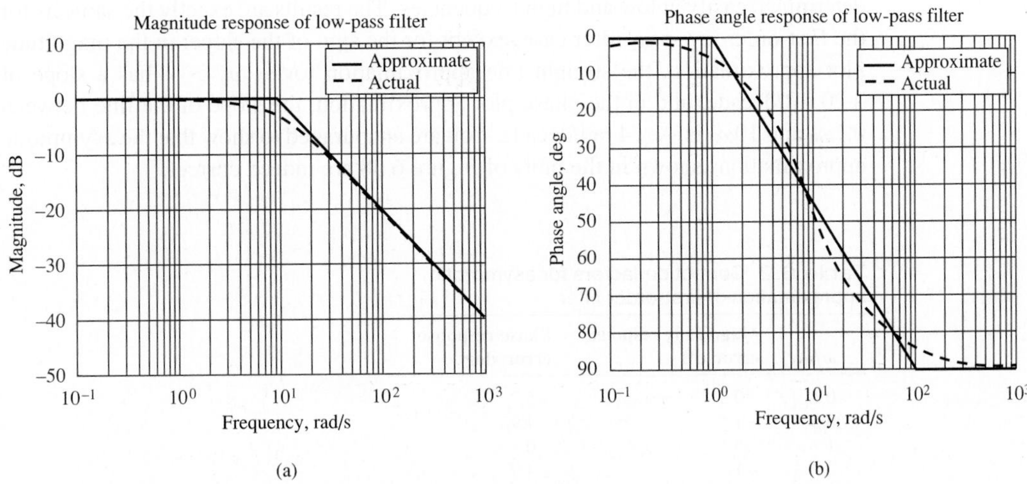

Figure 6.37 Bode plots for low-pass *RC* filter; the frequency variable is normalized to ω/ω_0. (a) Magnitude response; (b) phase response

Consider now the phase angle of the frequency response function $\angle(\mathbf{V}_o/\mathbf{V}_i)(j\omega) = -\tan^{-1}(\omega/\omega_0)$. This response can be approximated as follows:

$$-\tan^{-1}\left(\frac{\omega}{\omega_0}\right)\begin{cases} 0 & \text{when } \omega \ll \omega_0 \\ -\dfrac{\pi}{4} & \text{when } \omega = \omega_0 \\ -\dfrac{\pi}{2} & \text{when } \omega \gg \omega_0 \end{cases}$$

If we (somewhat arbitrarily) agree that $\omega < 0.1\omega_0$ is equivalent to the condition $\omega/\omega_0 \ll 1$, and that $\omega > 10\omega_0$ is equivalent to the condition $\omega/\omega_0 \gg 1$, then these approximations are summarized by three straight lines: one of zero slope (with phase equal to 0) for $\omega < 0.1\omega_0$, one with slope $-\pi/4$ rad/decade between $0.1\omega_0$ and $10\omega_0$, and one of zero slope (with phase equal to $-\pi/2$) for $\omega > 10\omega_0$. These approximations are illustrated in the plot of Figure 6.37(b). What errors are incurred in making these approximations? Table 6.2 lists the actual errors. Note that the maximum magnitude response error at the cutoff frequency is -3 dB; thus the cutoff or half-power frequency is often called **3-dB frequency.**

If we repeat the analysis done for the low-pass filter for the case of the high-pass filter (see Figure 6.22), we obtain a very similar approximation:

$$\begin{aligned}\frac{\mathbf{V}_o}{\mathbf{V}_i}(j\omega) &= \frac{j\omega CR}{1 + j\omega CR} = \frac{j(\omega/\omega_0)}{1 + j(\omega/\omega_0)} \\ &= \frac{(\omega/\omega_0)\angle(\pi/2)}{\sqrt{1 + (\omega/\omega_0)^2}\angle \arctan(\omega/\omega_0)} \\ &= \frac{\omega/\omega_0}{\sqrt{1 + (\omega/\omega_0)^2}}\angle\left(\frac{\pi}{2} - \arctan\frac{\omega}{\omega_0}\right)\end{aligned} \tag{6.52}$$

Figure 6.38 depicts the Bode plots for equation 6.52, where the horizontal axis indicates the normalized frequency ω/ω_0. Asymptotic approximations may again be determined easily at low and high frequencies. The results are exactly the same as for the first-order low-pass filter case, except for the sign of the slope: in the magnitude plot approximation, the straight-line approximation for $\omega/\omega_0 > 1$ has a slope of $+20$ dB/decade, and in the phase plot approximation, the slope of the line between $0.1\omega_0$ and $10\omega_0$ is $-\pi/4$ rad/decade. You are encouraged to show that the asymptotic approximations shown in the plots of Figure 6.38 are indeed correct.

Table 6.2 Correction factors for asymptotic approximation of first-order filter

ω/ω_0	Magnitude response error, dB	Phase response error, deg
0.1	0	−5.7
0.5	−1	4.9
1	−3	0
2	−1	−4.9
10	0	+5.7

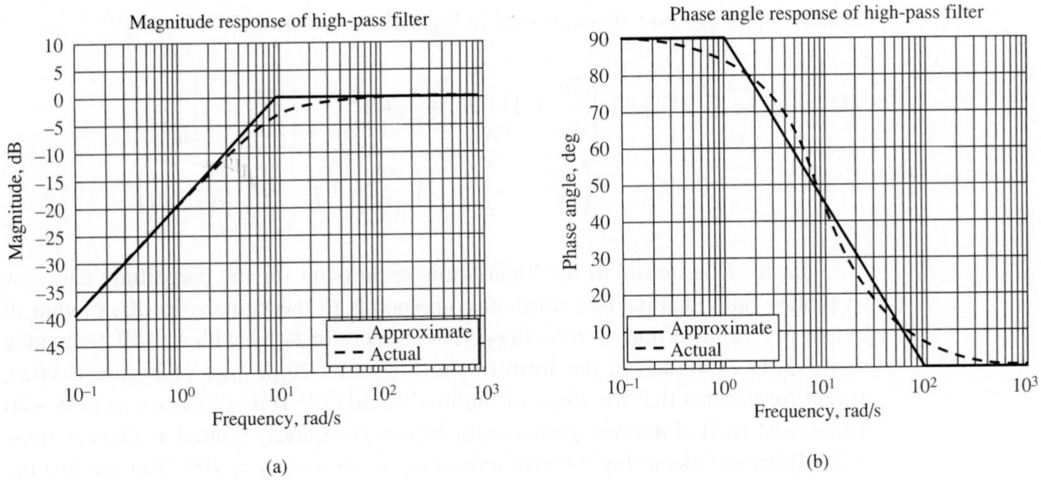

Figure 6.38 Bode plots for high-pass *RC* filter. (a) Magnitude response; (b) phase response

Bode Plots of Higher-Order Filters

Bode plots of high-order systems may be obtained by combining Bode plots of factors of the higher-order frequency response function. Let, for example,

$$H(j\omega) = H_1(j\omega)H_2(j\omega)H_3(j\omega) \quad \textbf{(6.53)}$$

which can be expressed, in logarithmic form, as

$$|H(j\omega)|_{\text{dB}} = |H_1(j\omega)|_{\text{dB}} + |H_2(j\omega)|_{\text{dB}} + |H_3(j\omega)|_{\text{dB}} \quad \textbf{(6.54)}$$

and

$$\angle H(j\omega) = \angle H_1(j\omega) + \angle H_2(j\omega) + \angle H_3(j\omega) \quad \textbf{(6.55)}$$

Consider as an example the frequency response function

$$H(j\omega) = \frac{j\omega + 5}{(j\omega + 10)(j\omega + 100)} \quad \textbf{(6.56)}$$

The first step in computing the asymptotic approximation consists of factoring each term in the expression so that it appears in the form $a_i(j\omega/\omega_i + 1)$, where the frequency ω_i corresponds to the appropriate 3-dB frequency, w_1, w_2, or w_3. For example, the function of equation 6.56 is rewritten as follows:

$$\begin{aligned} H(j\omega) &= \frac{5(j\omega/5 + 1)}{10(j\omega/10 + 1)100(j\omega/100 + 1)} \\ &= \frac{0.005(j\omega/5 + 1)}{(j\omega/10 + 1)(j\omega/100 + 1)} = \frac{K(j\omega/\omega_1 + 1)}{(j\omega/\omega_2 + 1)(j\omega/\omega_3 + 1)} \end{aligned} \quad \textbf{(6.57)}$$

Equation 6.57 can now be expressed in logarithmic form:

$$|H(j\omega)|_{\text{dB}} = |0.005|_{\text{dB}} + \left|\frac{j\omega}{5}+1\right|_{\text{dB}} - \left|\frac{j\omega}{10}+1\right|_{\text{dB}} - \left|\frac{j\omega}{100}+1\right|_{\text{dB}}$$
$$\angle H(j\omega) = \angle 0.005 + \angle\left(\frac{j\omega}{5}+1\right) - \angle\left(\frac{j\omega}{10}+1\right) - \angle\left(\frac{j\omega}{100}+1\right) \quad \textbf{(6.58)}$$

Each of the terms in the logarithmic expression for the magnitude can now be plotted individually. The constant corresponds to the value −46 dB, plotted in Figure 6.39(a) as a line of zero slope. The numerator term, with a 3-dB frequency $\omega_1 = 5$, is expressed in the form of the first-order Bode plot of Figure 6.37(a), except for the fact that the slope of the line leaving the zero axis at $\omega_1 = 5$ is +20 dB/decade; each of the two denominator factors is similarly plotted as lines of slope −20 dB/decade, departing the zero axis at $\omega_2 = 10$ and $\omega_3 = 100$. You see that the individual factors are very easy to plot *by inspection*, once the frequency response function has been normalized in the form of equation 6.57.

If we now consider the phase response portion of equation 6.58, we recognize that the first term, the phase angle of the constant, is always zero. The numerator first-order term, on the other hand, can be approximated as shown in Figure 6.37(b), that is, by drawing a straight line starting at $0.1\omega_1 = 0.5$, with slope $+\pi/4$ rad/decade (*positive because this is a numerator factor*) and ending at $10\omega_1 = 50$, where the asymptote $+\pi/2$ is reached. The two denominator terms have similar behavior, except for the fact that the slope is $-\pi/4$ and that the straight line with slope $-\pi/4$ rad/decade extends between the frequencies $0.1\omega_2$ and $10\omega_2$, and $0.1\omega_3$ and $10\omega_3$, respectively.

Figure 6.39 depicts the asymptotic approximations of the individual factors in equation 6.58, with the magnitude factors shown in part a and the phase factors in part b. When all the asymptotic approximations are combined, the complete frequency response approximation is obtained. Figure 6.40 depicts the results of the asymptotic Bode approximation when compared with the actual frequency response functions.

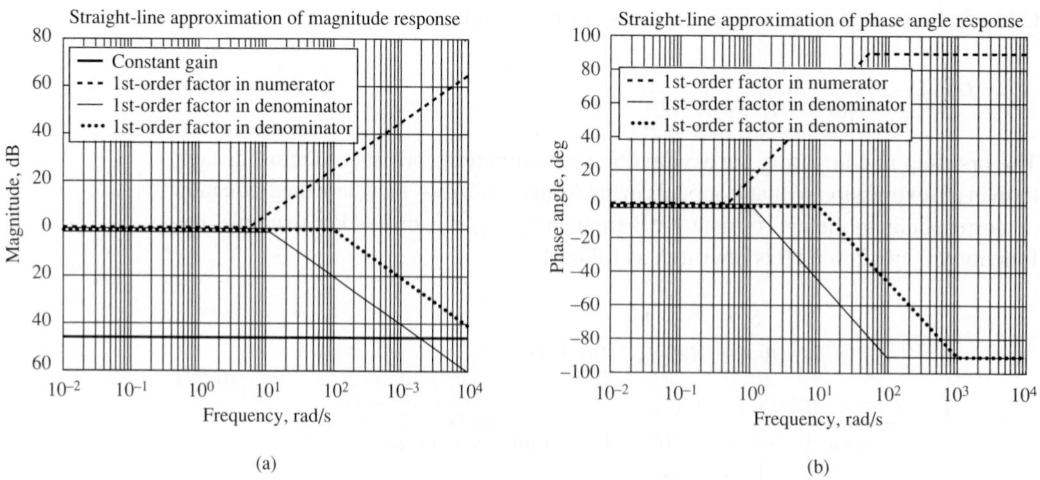

Figure 6.39 Bode plot approximation for a second-order frequency response function. (a) Magnitude; (b) phase

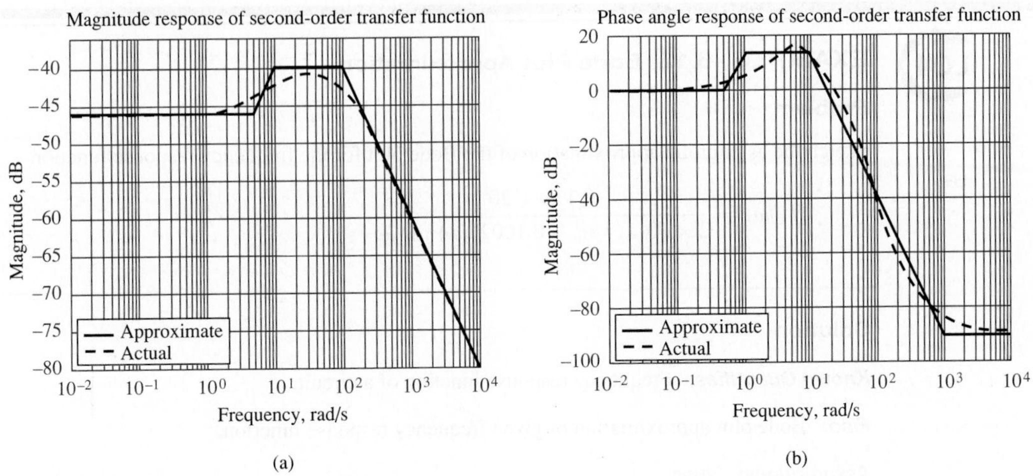

Figure 6.40 Comparison of Bode plot approximation with the actual frequency response function. (a) Magnitude; (b) phase

You can see that once a frequency response function is factored into the appropriate form, it is relatively easy to sketch a good approximation of the Bode plot, even for higher-order frequency response functions. Examples 6.12 and 6.13 illustrate some additional details. The methodology is summarized in the box below.

FOCUS ON METHODOLOGY

BODE PLOTS

LO4

This box illustrates the Bode plot asymptotic approximation construction procedure. The method assumes that there are no complex conjugate factors in the response, and that both the numerator and denominator can be factored into first-order terms with real roots.

1. Express the frequency response function in factored form, resulting in an expression similar to equation 6.57:

$$H(j\omega) = \frac{K(j\omega/\omega_1 + 1)\cdots(j\omega/\omega_m + 1)}{(j\omega/\omega_{m+1} + 1)\cdots(j\omega/\omega_n + 1)}$$

2. Select the appropriate frequency range for the semilogarithmic plot, extending at least a decade below the lowest 3-dB frequency and a decade above the highest 3-dB frequency.
3. Sketch the magnitude and phase response asymptotic approximations for each of the first-order factors, using the techniques illustrated in Figures 6.37 and 6.38.
4. Add, graphically, the individual terms to obtain a composite response.
5. If desired, apply the correction factors of Table 6.2

LO4

EXAMPLE 6.12 Bode Plot Approximation

Problem

Sketch the asymptotic approximation of the Bode plot for the frequency response function

$$H(j\omega) = \frac{0.1j\omega + 20}{2 \times 10^{-5}(j\omega)^3 + 0.1002(j\omega)^2 + j\omega}$$

Solution

Known Quantities: Frequency response function of a circuit.

Find: Bode plot approximation of given frequency response function.

Assumptions: None

Analysis: Following the Focus on Methodology box on Bode plots, we first factor the function into the standard form

$$H(j\omega) = \frac{K(j\omega/\omega_1 + 1)\cdots(j\omega/\omega_m + 1)}{(j\omega/\omega_{m+1} + 1)\cdots(j\omega/\omega_n + 1)}$$

After a little algebra, we can obtain the following frequency response function in standard form:

$$H(j\omega) = \frac{20(j\omega/200 + 1)}{j\omega(j\omega/10 + 1)(j\omega/5{,}000 + 1)}$$

We immediately notice that there is a factor of $j\omega$ in the denominator; this term needs to be treated somewhat differently. The Bode plot of the function $1/j\omega$ can be expressed in logarithmic form as follows:

$$\left|\frac{1}{j\omega}\right|_{\text{dB}} = -20\log_{10}\frac{\omega}{1}$$

$$\angle\frac{1}{j\omega} = 0 - \frac{\pi}{2} = -\frac{\pi}{2}$$

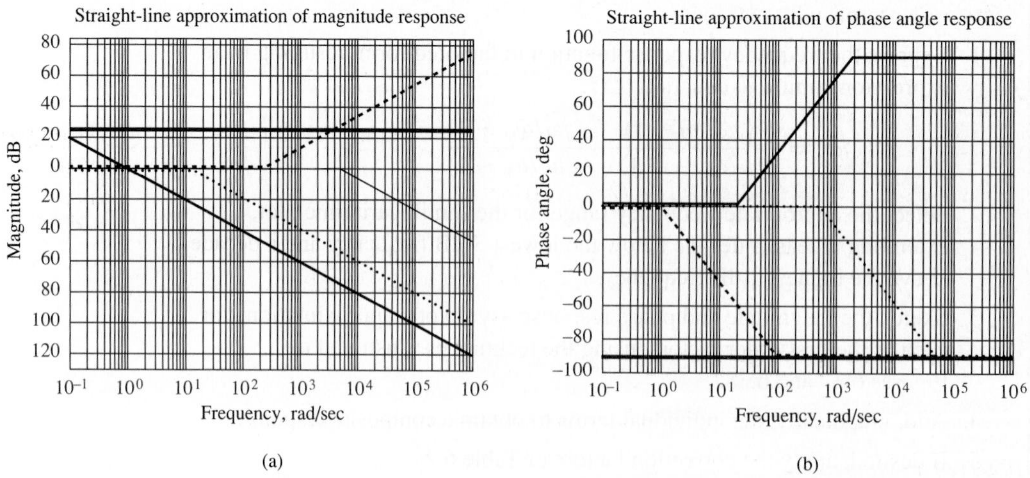

Figure 6.41 Approximate (asymptotic) frequency response of individual first-order terms. (a) Magnitude; (b) phase

That is, the magnitude of the denominator factor $j\omega$ is represented by a line with slope of -20 dB/decade intersecting the frequency (horizontal) axis at $\omega = 1$. Its phase response is a constant equal to $-\pi/2$.

Now we can sketch the magnitude and phase response of each of the individual first-order factors, as shown in Figure 6.41(a) and (b). The composite asymptotic approximations of the magnitude and phase responses are shown in Figure 6.42(a) and (b).

Comments: A computer program can be used to generate the Bode plot approximation shown in Figures 6.41 and 6.42. Note that the only real effort in generating the asymptotic approximation lies in the factoring of the frequency response function.

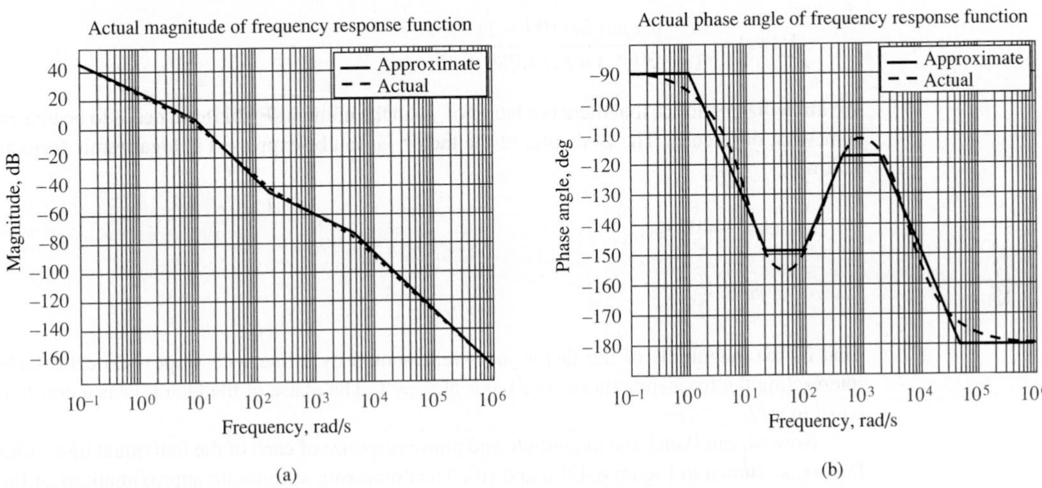

Figure 6.42 Comparison of approximate and exact frequency response. (a) Magnitude; (b) phase

CHECK YOUR UNDERSTANDING

Show that you can obtain exactly the same plot if, instead of separately sketching the factor K in the numerator and the factor $j\omega$ in the denominator, you combine these two into a single denominator factor equal to $j\omega/K$.

EXAMPLE 6.13 Bode Plot Approximation

LO4

Problem

Sketch the asymptotic approximation of the Bode plot for the frequency response function

$$H(j\omega) = \frac{10^{-3}(j\omega)^2 + 0.1\,j\omega}{[1/(9 \times 10^4)](j\omega)^2 + (3{,}030/90{,}000)\,j\omega + 1}$$

Solution

Known Quantities: Frequency response function of a circuit.

Find: Bode plot approximation of given frequency response function.

Assumptions: None

Analysis: Following the Focus on Methodology box on Bode plots, we first factor the function into the standard form

$$H(j\omega) = \frac{K(j\omega/\omega_1 + 1)\cdots(j\omega/\omega_m + 1)}{(j\omega/\omega_{m+1} + 1)\cdots(j\omega/\omega_n + 1)}$$

After a little algebra, we can obtain the following frequency response function in standard form:

$$H(j\omega) = \frac{0.1j\omega(j\omega/100 + 1)}{(j\omega/30 + 1)(j\omega/3{,}000 + 1)}$$

We immediately notice that there is a factor of $j\omega$ in the numerator; this term needs to be treated somewhat differently. The Bode plot of the factor $j\omega$ can be expressed in logarithmic form as follows:

$$|j\omega|_{\text{dB}} = 20\log_{10}\frac{\omega}{1}$$

$$\angle j\omega = \frac{\pi}{2}$$

That is, the magnitude of the factor $j\omega$ is represented by a line with slope +20 dB/decade intersecting the frequency (horizontal) axis at $\omega = 1$. The phase of the factor $j\omega$ is a constant equal to $\pi/2$.

Now we can sketch the magnitude and phase response of each of the individual first-order factors, as shown in Figure 6.43(a) and (b). The composite asymptotic approximations of the magnitude and phase responses are shown in Figure 6.44(a) and (b).

Comments: A computer program can be used to generate the Bode plot approximation shown in Figures 6.43 and 6.44. Note that the only real effort in generating the asymptotic approximation is expended in the factoring of the frequency response function.

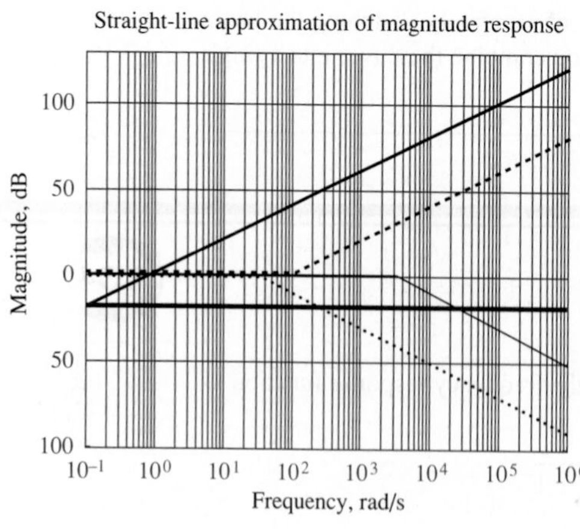

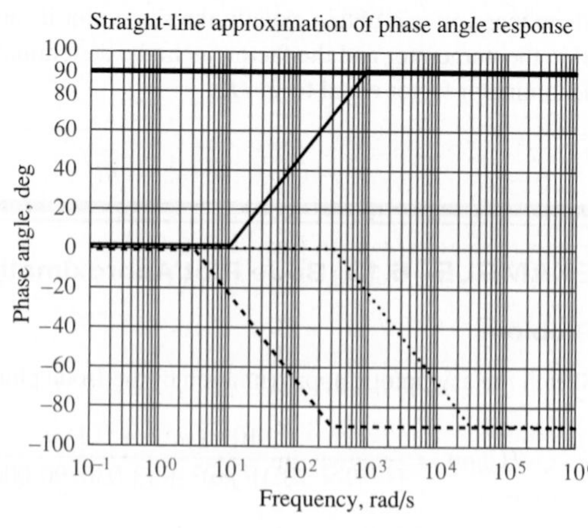

Figure 6.43 Approximate (asymptotic) frequency response of individual first-order terms. (a) Magnitude; (b) phase

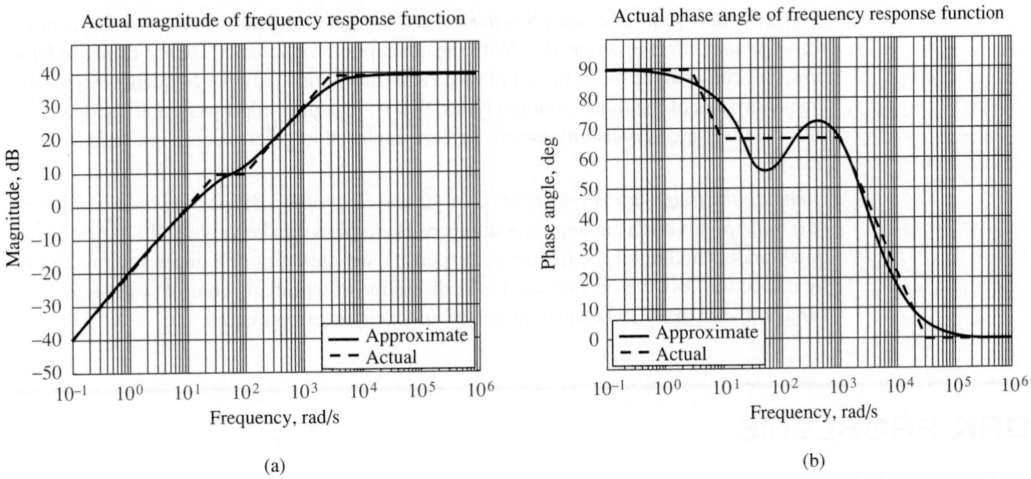

Figure 6.44 Comparison of approximate and exact frequency response. (a) Magnitude; (b) phase

CHECK YOUR UNDERSTANDING

Show that you can obtain exactly the same plot if, instead of separately sketching the factor K and the factor $j\omega$ in the numerator, you combine these two into a single denominator factor equal to $j\omega/K^{-1}$.

Conclusion

Chapter 6 focuses on the frequency response of linear circuits, and it is a natural extension of the material covered in Chapter 4. The concepts of the spectrum of a signal, obtained through the Fourier series representation for periodic signals, and of the frequency response of a filter are very useful ideas that extend well beyond electrical engineering. For example, civil, mechanical, and aeronautical engineering students who study the vibrations of structures and machinery will find that the same methods are employed in those fields.

Upon completing this chapter, you should have mastered the following learning objectives:

1. *Understand the physical significance of frequency domain analysis, and compute the frequency response of circuits by using AC circuit analysis tools.* You had already acquired the necessary tools (phasor analysis and impedance) to compute the frequency response of circuits in Chapter 4; in the material presented in Section 6.1, these tools are put to use to determine the frequency response functions of linear circuits.
2. *Compute the Fourier spectrum of periodic signals by using the Fourier series representation, and use this representation in connection with frequency response ideas to compute the response of circuits to periodic inputs.* The concept of spectrum is very important in many engineering applications; in Section 6.2 you learned to compute the Fourier spectrum of an important class of functions: those that repeat periodically. The frequency spectrum of signals makes frequency domain analysis (i.e., computing the response of circuits using the phasor domain representation of signals) very easy, even for relatively complex signals, because it allows you to decompose the signals into a summation of sinusoidal components, which can then be easily handled one at a time.

3. *Analyze simple first- and second-order electrical filters, and determine their frequency response and filtering properties.* With the concept of *frequency response* firmly in hand, now you can analyze the behavior of electrical filters and study the frequency response characteristics of the most common types, that is, low-pass, high-pass, and bandpass filters. Filters are very useful devices and are explored in greater depth in Chapters 8 and 15.
4. *Compute the frequency response of a circuit and its graphical representation in the form of a Bode plot.* Graphical approximations of Bode plots can be very useful to develop a quick understanding of the frequency response characteristics of a linear system, almost by inspection. Bode plots find use in the discipline of automatic control systems, a subject that is likely to be encountered by most engineering majors.

HOMEWORK PROBLEMS

Section 6.1: Sinusoidal Frequency Response

6.1

a. Determine the frequency response $\mathbf{V}_{out}(j\omega)/\mathbf{V}_{in}(j\omega)$ for the circuit of Figure P6.1.
b. Plot the magnitude and phase of the circuit for frequencies between 10 and 10^7 rad/s on graph paper, with a linear scale for frequency.
c. Repeat part b, using semilog paper. (Place the frequency on the logarithmic axis.)
d. Plot the magnitude response on semilog paper with magnitude in decibels.

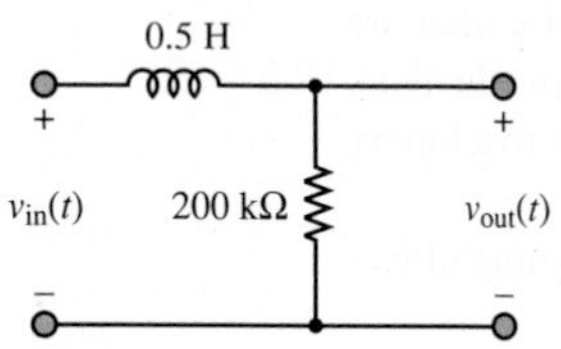

Figure P6.1

6.2 Repeat Problem 6.1 for the circuit of Figure P6.2.

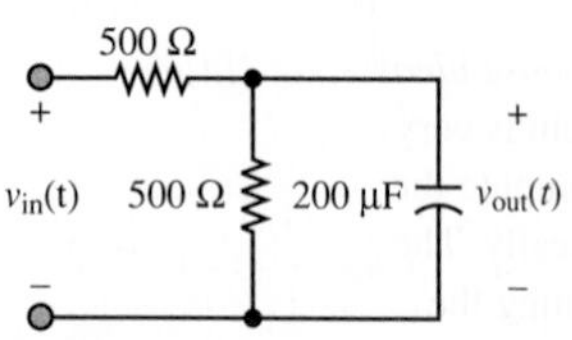

Figure P6.2

6.3 Repeat Problem 6.1 for the circuit of Figure P6.3.

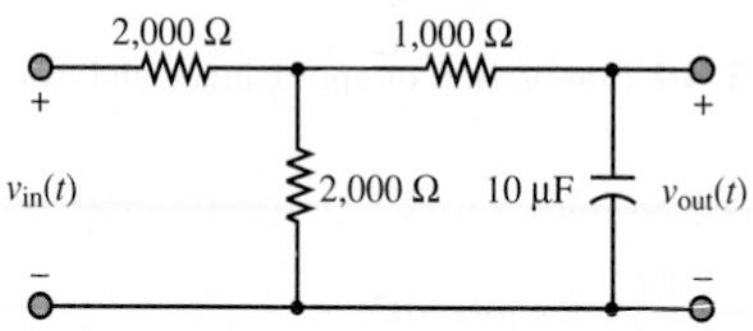

Figure P6.3

6.4 Repeat Problem 6.1 for the circuit of Figure P6.4. $R_1 = 500\ \Omega$; $R_2 = 1{,}000\ \Omega$; $L = 2$ H; $C = 100\ \mu$F.

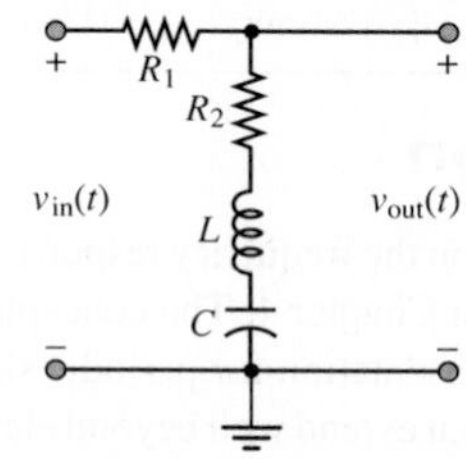

Figure P6.4

6.5 Determine the frequency response of the circuit of Figure P6.5, and generate frequency response plots. $R_1 = 20$ kΩ; $R_2 = 100$ kΩ; $C_1 = 100\ \mu$F; $C_2 = 5\ \mu$F.

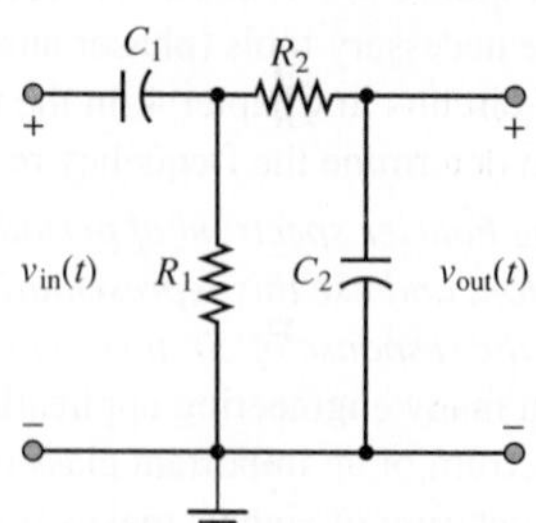

Figure P6.5

6.6 In the circuit shown in Figure P6.6, where $C = 0.5\ \mu$F and $R = 2$ kΩ,

a. Determine how the input impedance $Z(j\omega) = \frac{\mathbf{V}_i(j\omega)}{\mathbf{I}_i(j\omega)}$ behaves at extremely high and low frequencies.

b. Find an expression for the impedance.

c. Show that this expression can be manipulated into the form $Z(j\omega) = R\left[1 + j\frac{1}{\omega RC}\right]$.

d. Determine the frequency $\omega = \omega_C$ for which the imaginary part of the expression in part c is equal to 1.

e. Estimate (without computing it) the magnitude and phase angle of $Z(j\omega)$ at $\omega = 10$ rad/s and $\omega = 10^5$ rad/s.

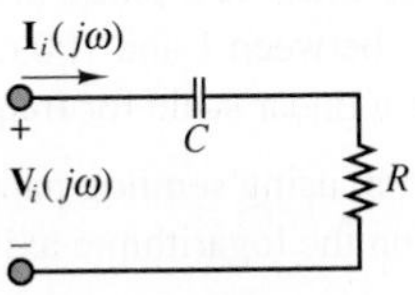

Figure P6.6

6.7 In the circuit shown in Figure P6.7, where $L = 2$ mH and $R = 2$ kΩ,

a. Determine how the input impedance $Z(j\omega) = \frac{\mathbf{V}_i(j\omega)}{\mathbf{I}_i(j\omega)}$ behaves at extremely high and low frequencies.

b. Find an expression for the impedance.

c. Show that this expression can be manipulated into the form $Z(j\omega) = R\left[1 + j\frac{\omega L}{R}\right]$.

d. Determine the frequency $\omega = \omega_C$ for which the imaginary part of the expression in part c is equal to 1.

e. Estimate (without computing it) the magnitude and phase angle of $Z(j\omega)$ at $\omega = 10^5$ rad/s, 10^6 rad/s, and 10^7 rad/s.

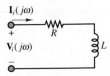

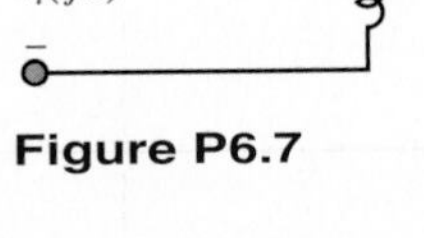

Figure P6.7

6.8 In the circuit shown in Figure P6.8, if

$L = 190$ mH $\quad R_1 = 2.3$ kΩ

$C = 55$ nF $\quad R_2 = 1.1$ kΩ

a. Determine how the input impedance behaves at extremely high or low frequencies.

b. Find an expression for the input impedance in the form

$$Z(j\omega) = Z_o\left[\frac{1 + jf_1(\omega)}{1 + jf_2(\omega)}\right]$$

$$Z_o = R_1 + \frac{L}{R_2 C}$$

$$f_1(\omega) = \frac{\omega^2 R_1 LC - R_1 - R_2}{\omega(R_1 R_2 C + L)}$$

$$f_2(\omega) = \frac{\omega^2 LC - 1}{\omega C R_2}$$

c. Determine the four frequencies at which $f_1(\omega) = +1$ or -1 and $f_2(\omega) = +1$ or -1.

d. Plot the impedance (magnitude and phase) versus frequency.

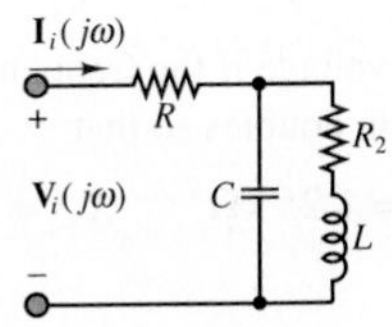

Figure P6.8

6.9 In the circuit of Figure P6.9:

$R_1 = 1.3$ kΩ $\quad R_2 = 1.9$ kΩ

$C = 0.5182$ μF

Determine:

a. How the voltage transfer function

$$H_V(j\omega) = \frac{\mathbf{V}_o(j\omega)}{\mathbf{V}_i(j\omega)}$$

behaves at extremes of high and low frequencies.

b. An expression for the voltage transfer function and show that it can be manipulated into the form

$$H_v(j\omega) = \frac{H_o}{1 + jf(\omega)}$$

where

$$H_o = \frac{R_2}{R_1 + R_2} \qquad f(\omega) = \frac{\omega R_1 R_2 C}{R_1 + R_2}$$

c. The frequency at which $f(\omega) = 1$ and the value of H_o in decibels.

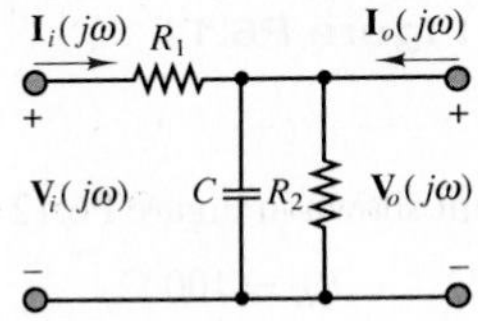

Figure P6.9

6.10 The circuit shown in Figure P6.10 is a second-order circuit because it has two reactive components (L and C). A complete solution will not be attempted. However, determine:

a. The behavior of the voltage frequency response at extremely high and low frequencies.

b. The output voltage $\mathbf{V}_o$ if the input voltage has a frequency where:

$$\mathbf{V}_i = 7.07\angle\frac{\pi}{4}\ \text{V} \qquad R_1 = 2.2\ \text{k}\Omega$$

$$R_2 = 3.8\ \text{k}\Omega \qquad X_c = 5\ \text{k}\Omega \qquad X_L = 1.25\ \text{k}\Omega$$

c. The output voltage if the frequency of the input voltage doubles so that

$$X_C = 2.5\ \text{k}\Omega \qquad X_L = 2.5\ \text{k}\Omega$$

d. The output voltage if the frequency of the input voltage again doubles so that

$$X_C = 1.25\ \text{k}\Omega \qquad X_L = 5\ \text{k}\Omega$$

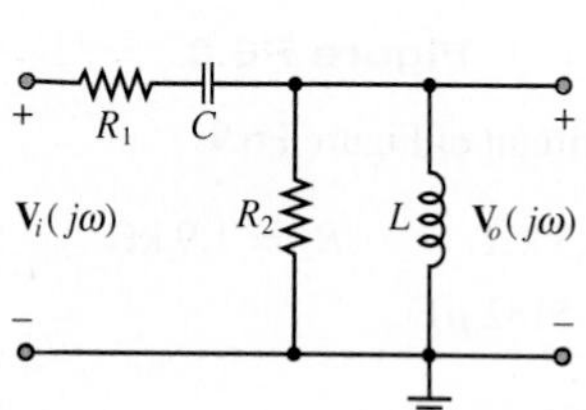

Figure P6.10

6.11 In the circuit shown in Figure P6.11, determine the frequency response function in the form

$$H_v(j\omega) = \frac{\mathbf{V}_o(j\omega)}{\mathbf{V}_i(j\omega)} = \frac{H_{vo}}{1 \pm jf(\omega)}$$

and plot $H_v(j\omega)$.

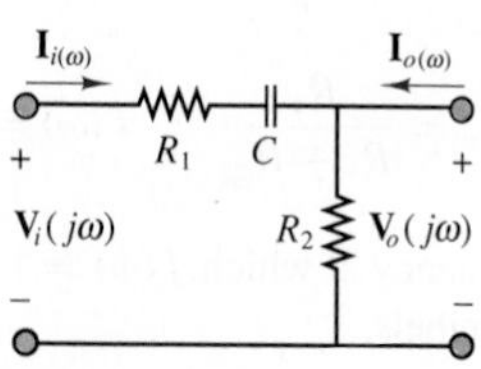

Figure P6.11

6.12 The circuit shown in Figure P6.12 has

$$R_1 = 100\ \Omega \qquad R_L = 100\ \Omega$$

$$R_2 = 50\ \Omega \qquad C = 80\ \text{nF}$$

Compute and plot the frequency response function.

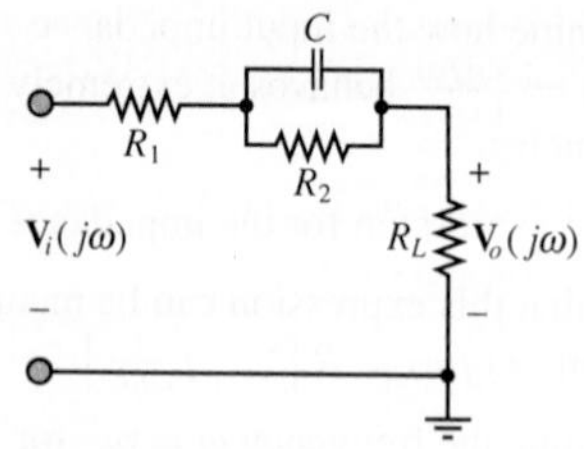

Figure P6.12

6.13

a. Determine the frequency response $\mathbf{V}_{\text{out}}(j\omega)/\mathbf{V}_{\text{in}}(j\omega)$ for the circuit of Figure P6.13.

b. Plot the magnitude and phase of the circuit for frequencies between 1 and 100 rad/s on graph paper, with a linear scale for frequency.

c. Repeat part b, using semilog paper. (Place the frequency on the logarithmic axis.)

d. Plot the magnitude response on semilog paper with magnitude in dB.

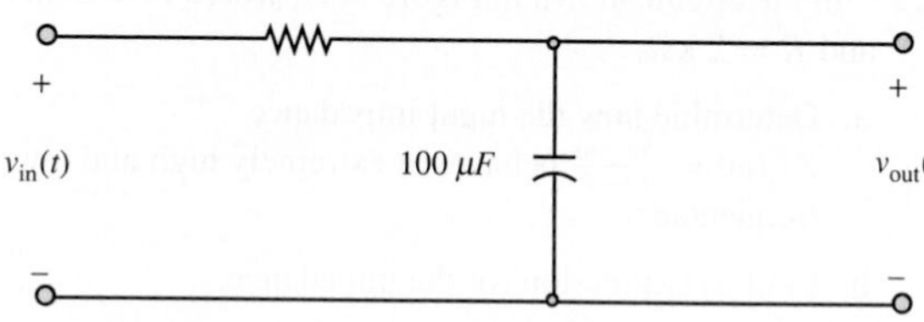

Figure P6.13

6.14 Consider the circuit shown in Figure P6.14.

a. Sketch the amplitude response of $Y = I/V_S$.

b. Sketch the amplitude response of V_1/V_S.

c. Sketch the amplitude response of V_2/V_S.

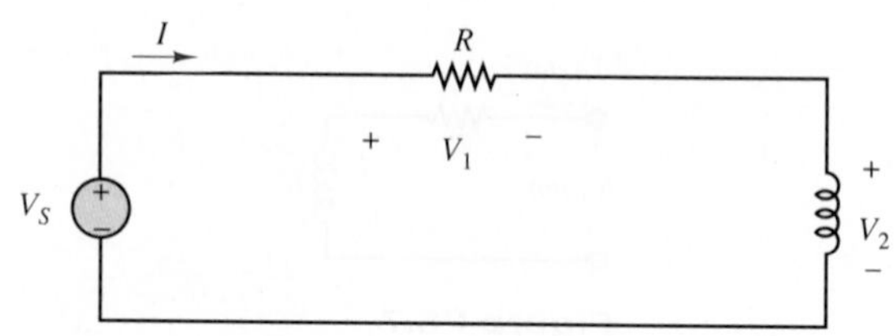

Figure P6.14

Section 6.2: Fourier Analysis

6.15 Use trigonometric identities to show that the equalities in equations 6.16 and 6.17 hold.

6.16 Derive a general expression for the Fourier series coefficients of the square wave of Figure 6.11(a) in the text.

6.17 Compute the Fourier series coefficient of the periodic function shown in Figure P6.17 and defined as

$$x(t) = \begin{cases} A & 0 \le t < \dfrac{T}{3} \\ 0 & \dfrac{T}{3} \le t < T \end{cases}$$

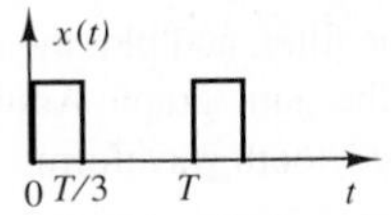

Figure P6.17

6.18 Compute the Fourier series coefficient of the periodic function shown in Figure P6.18 and defined as

$$x(t) = \begin{cases} \cos\left(\dfrac{2\pi}{T}t\right) & -\dfrac{T}{4} \le t < \dfrac{T}{4} \\ 0 & \text{else} \end{cases}$$

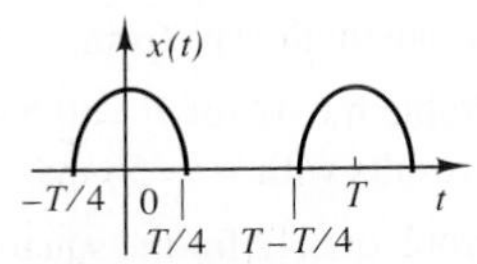

Figure P6.18

6.19 Compute the Fourier series expansion of the function shown in Figure P6.19, and express it in sine-cosine (a_n, b_n coefficients) form.

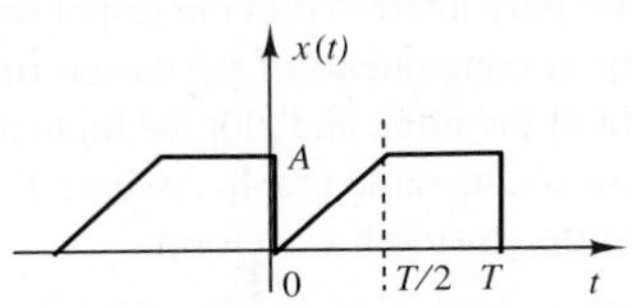

Figure P6.19

6.20 Compute the Fourier series expansion of the function shown in Figure P6.20, and express it in sine-cosine (a_n, b_n coefficients) form.

$$x(t) = \begin{cases} \sin\left(\dfrac{2\pi}{T}t\right) & 0 \le t < \dfrac{T}{2} \\ 0 & \dfrac{T}{2} \le t < T \end{cases}$$

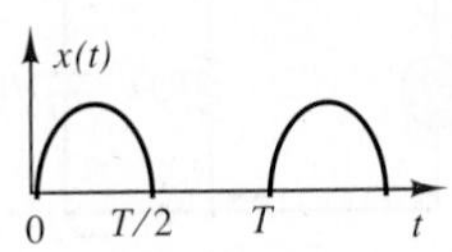

Figure P6.20

6.21 The trapezoidal function shown in Figure P6.21 is often used as a voltage excitation to brushless DC machines. Write a complete expression for the function $x(t)$ and compute the Fourier coefficients.

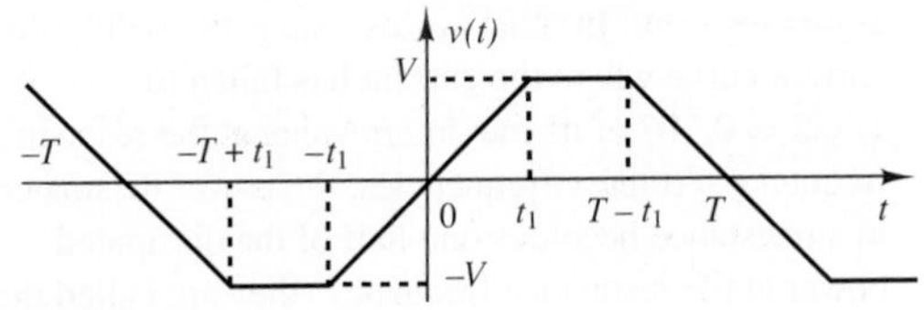

Figure P6.21

6.22 Write an expression for the signal shown in Figure P6.22, and derive a complete expression for its Fourier series.

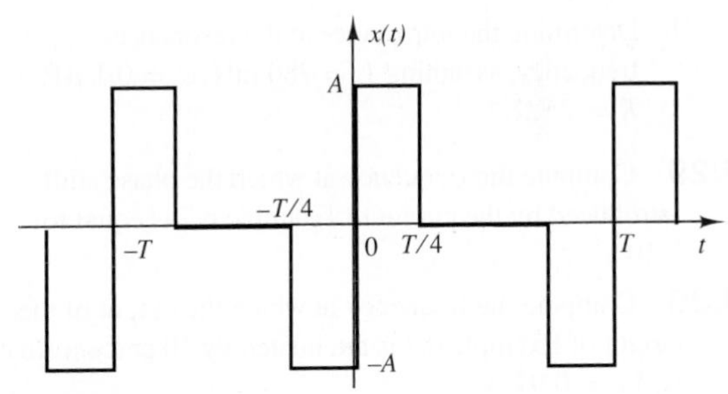

Figure P6.22

6.23 Derive all Fourier series coefficients of the function $x(t) = 10\cos(10t + \pi/6)$.

6.24 Set up but do not compute the integrals for the Fourier coefficients of the periodic function shown in Figure P6.24.

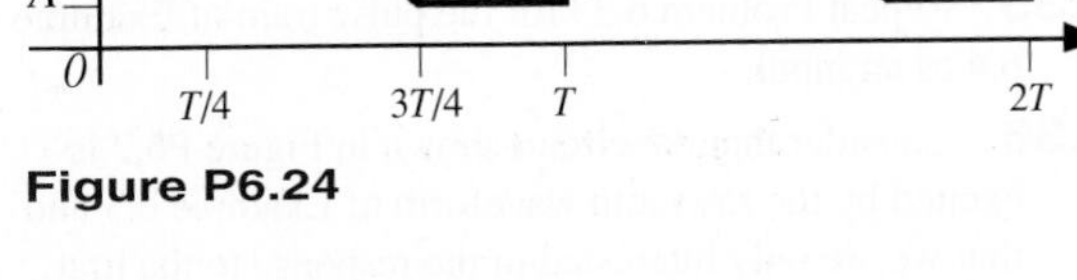

Figure P6.24

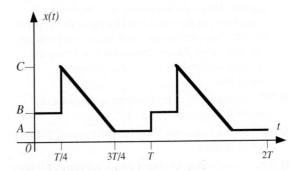

Section 6.3: Filters

6.25 Using a 15-kΩ resistance, design an *RC* high-pass filter with a breakpoint at 200 kHz.

6.26 Using a 500-Ω resistance, design an *RC* low-pass filter that would attenuate a 120-Hz sinusoidal voltage by 20 dB with respect to the DC gain.

6.27 In an *RLC* circuit, assume ω_1 and ω_2 such that $\mathbf{I}(j\omega_1) = \mathbf{I}(j\omega_2) = \mathbf{I}_{\max}/\sqrt{2}$ and $\Delta\omega$ such that $\Delta\omega = \omega_2 - \omega_1$. In other words, $\Delta\omega$ is the width of the current curve where the current has fallen to $1/\sqrt{2} = 0.707$ of its maximum value at the resonance frequency. At these frequencies, the power dissipated in a resistance becomes one-half of the dissipated power at the resonance frequency (they are called the half-power points). In an *RLC* circuit with a high quality factor, show that $Q = \omega_0/\Delta\omega$.

6.28 In an *RLC* circuit with a high quality factor:

a. Show that the impedance at the resonance frequency becomes a value of Q times the inductive resistance at the resonance frequency.

b. Determine the impedance at the resonance frequency, assuming $L = 280$ mH, $C = 0.1\ \mu$F, $R = 25\ \Omega$.

6.29 Compute the frequency at which the phase shift introduced by the circuit of Example 6.7 is equal to $-10°$.

6.30 Compute the frequency at which the output of the circuit of Example 6.7 is attenuated by 10 percent (that is, $V_L = 0.9V_S$).

6.31 Compute the frequency at which the output of the circuit of Example 6.11 is attenuated by 10 percent (that is, $V_L = 0.9V_S$).

6.32 Compute the frequency at which the phase shift introduced by the circuit of Example 6.11 is equal to 20°.

6.33 Consider that the filter shown in Figure P6.1 is excited by a sawtooth waveform and that we are only interested in the response to the first two Fourier components of the waveform. Determine the output of the filter, and plot the input and output waveforms on the same graph. Assume the period $T = 10\ \mu$s and the peak amplitude $A = 1$ for the sawtooth waveform.

6.34 Repeat Problem 6.33 with the square wave of Figure 6.11(a) as an input.

6.35 Repeat Problem 6.33 for the pulse train of Example 6.4 as an input.

6.36 Consider that the circuit shown in Figure P6.2 is excited by the sawtooth waveform of Example 6.3 and that we are only interested in the response to the first three Fourier components of the waveform. Determine the output of the filter, and plot the input and output waveforms on the same graph. Assume $T = 0.5$ s and $A = 2$ for the sawtooth waveform.

6.37 Repeat Problem 6.36 with the square wave of Figure 6.11(a) as an input.

6.38 Repeat Problem 6.36 with the pulse train of Example 6.4 as an input.

6.39 Consider that the filter shown in Figure P6.3 is excited by the sawtooth waveform of Example 6.3 and that we are only interested in the response to the first four Fourier components of the waveform. Determine the output of the filter, and plot the input and output waveforms on the same graph. Assume $T = 0.1$ s and $A = 1$ for the sawtooth waveform.

6.40 Repeat Problem 6.39 with the square wave of Figure 6.11(a) as an input.

6.41 Repeat Problem 6.39 with the pulse train of Example 6.4 as an input.

6.42 Consider that the filter shown in Figure P6.4 is excited by a sawtooth waveform of Example 6.3 and that we are only interested in the response to the first two Fourier components of the waveform. Determine the output of the filter, and plot the input and output waveforms on the same graph. Assume $T = 50$ ms and $A = 2$ for the sawtooth waveform.

6.43 Repeat Problem 6.42 for $T = 0.5$ s and 5 ms, and compare the results with $T = 50$ ms.

6.44 Repeat Problem 6.42 for the square wave of Figure 6.11(a).

6.45 Repeat Problem 6.42 with the pulse train of Example 6.4 as an input.

6.46 Consider that the filter shown in Figure P6.5 is excited by the sawtooth waveform of Example 6.3 and that we are only interested in the response to the first three Fourier components of the waveform. Determine the output of the filter, and plot the input and output waveforms on the same graph. Assume $T = 5$ s and $A = 1$ for the sawtooth waveform.

6.47 Repeat Problem 6.46 for $T = 50$ s, and compare the results with Problem 6.46.

6.48 Repeat Problem 6.46 with the square wave of Figure 6.11(a) as an input.

6.49 Repeat Problem 6.46 with the pulse train of Example 6.4 as an input.

6.50 Consider the circuit shown in Figure P6.50. Determine the resonance frequency and the bandwidth for the circuit.

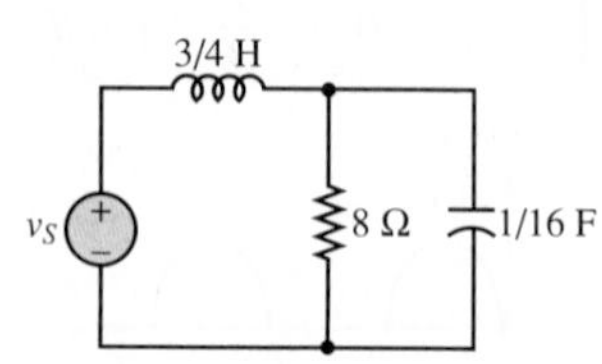

Figure P6.50

6.51 Are the filters shown in Figure P6.51 low-pass, high-pass, bandpass, or bandstop (notch) filters?

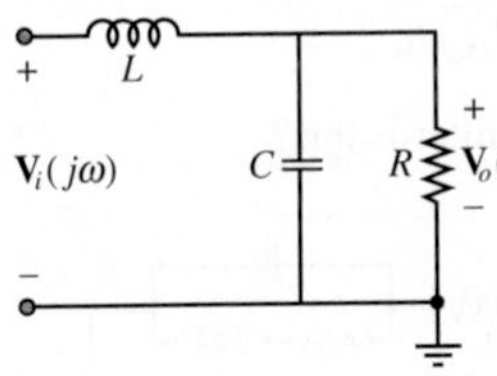

(a)

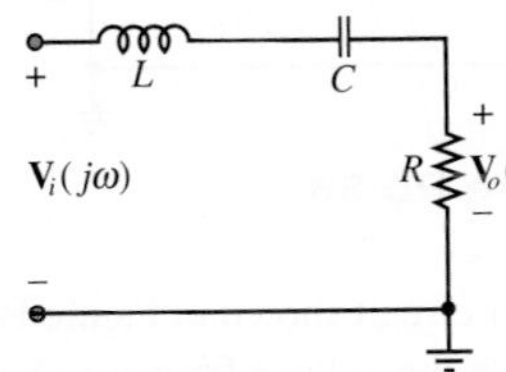

(b)

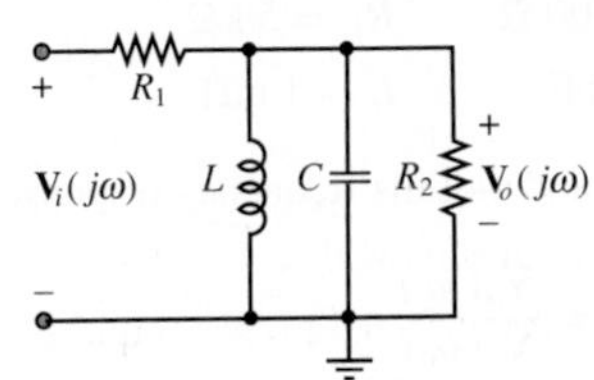

(c)

Figure P6.51

6.52 Determine if each of the circuits shown in Figure P6.52 is a low-pass, high-pass, bandpass, or bandstop (notch) filter.

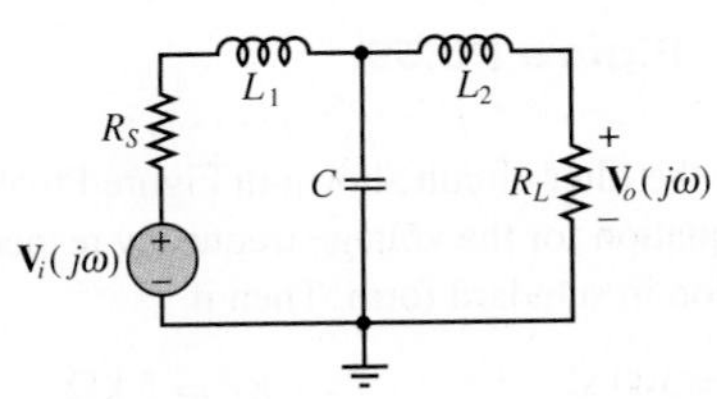

(a)

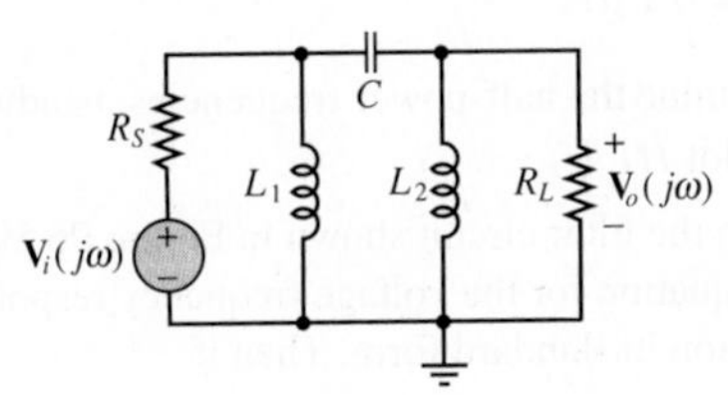

(b)

Figure P6.52 *(Continued)*

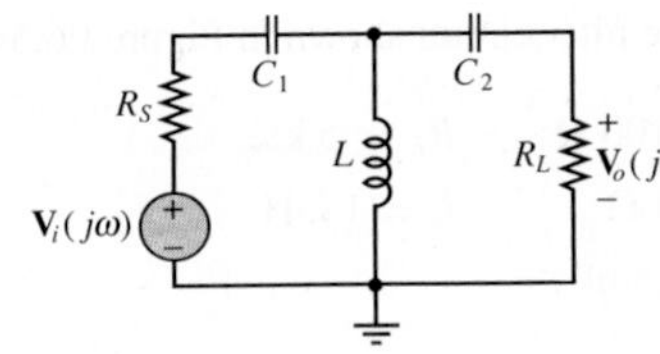

(c)

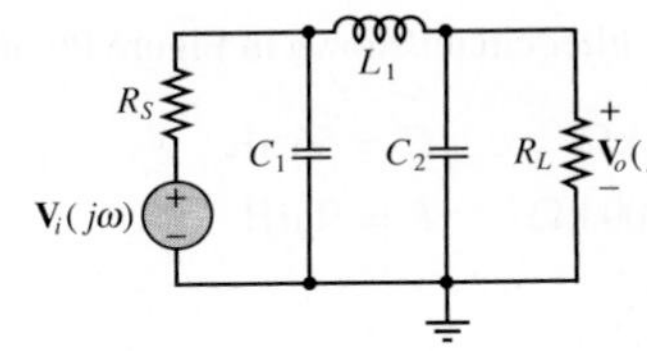

(d)

Figure P6.52

6.53 For the filter circuit shown in Figure P6.53:

a. Determine if this is a low-pass, high-pass, bandpass, or bandstop filter.

b. Compute and plot the frequency response function if

$L = 11$ mH $\quad C = 0.47$ nF

$R_1 = 2.2$ kΩ $\quad R_2 = 3.8$ kΩ

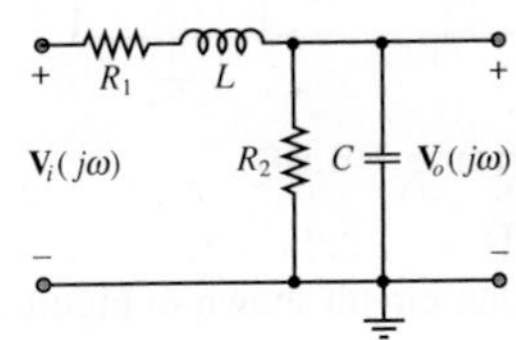

Figure P6.53

6.54 In the filter circuit shown in Figure P6.54:

$R_S = 100\ \Omega \quad R_L = 5$ kΩ

$R_c = 400\ \Omega \quad L = 1$ mH

$C = 0.5$ nF

Compute and plot the frequency response function. What type of filter is this?

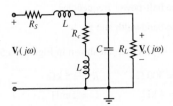

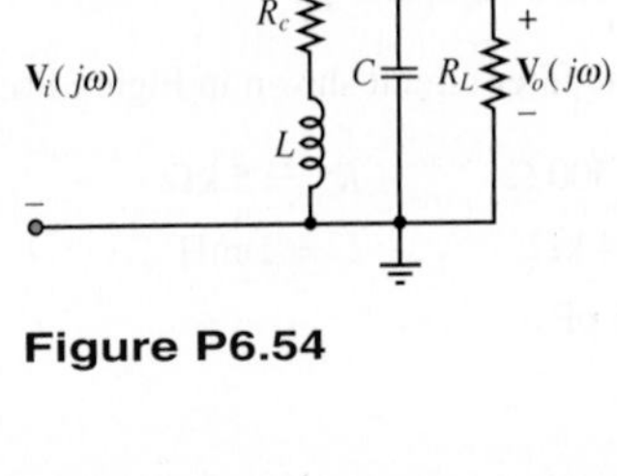

Figure P6.54

6.55 In the filter circuit shown in Figure P6.54:

$R_S = 100\ \Omega \qquad R_L = 5\ \text{k}\Omega$
$R_c = 4\ \Omega \qquad L = 1\ \text{mH}$
$C = 0.5\ \text{nF}$

Compute and plot the frequency response function. What type of filter is this?

6.56 In the filter circuit shown in Figure P6.56:

$R_S = 5\ \text{k}\Omega \qquad C = 56\ \text{nF}$
$R_L = 100\ \text{k}\Omega \qquad L = 9\ \mu\text{H}$

Determine:

a. An expression for the voltage frequency response function

$$H_v(j\omega) = \frac{\mathbf{V}_o(j\omega)}{\mathbf{V}_i(j\omega)}$$

b. The resonant frequency.

c. The half-power frequencies.

d. The bandwidth and Q.

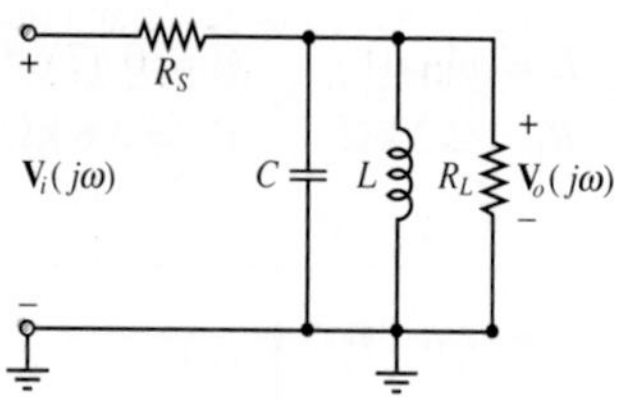

Figure P6.56

6.57 In the filter circuit shown in Figure P6.56:

$R_S = 5\ \text{k}\Omega \qquad C = 0.5\ \text{nF}$
$R_L = 100\ \text{k}\Omega \qquad L = 1\ \text{mH}$

Determine:

a. An expression for the voltage frequency response function

$$H_v(j\omega) = \frac{\mathbf{V}_o(j\omega)}{\mathbf{V}_i(j\omega)}$$

b. The resonant frequency.

c. The half-power frequencies.

d. The bandwidth and Q.

6.58 In the filter circuit shown in Figure P6.58:

$R_S = 500\ \Omega \qquad R_L = 5\ \text{k}\Omega$
$R_c = 4\ \text{k}\Omega \qquad L = 1\ \text{mH}$
$C = 5\ \text{pF}$

Compute and plot the voltage frequency response function

$$H(j\omega) = \frac{\mathbf{V}_o(j\omega)}{\mathbf{V}_i(j\omega)}$$

What type of filter is this?

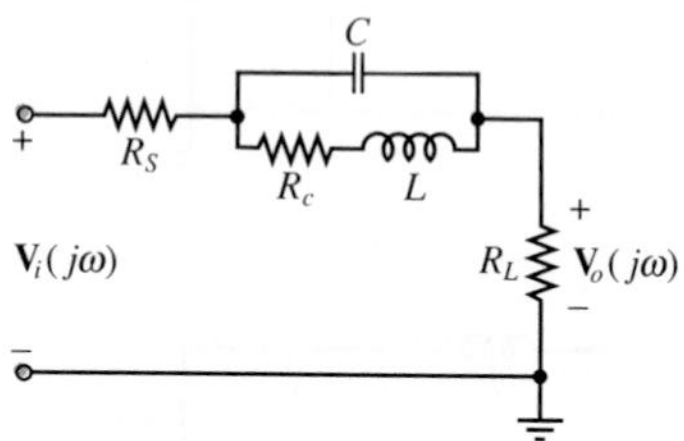

Figure P6.58

6.59 In the filter circuit shown in Figure P6.59, derive the equation for the voltage frequency response function in standard form. Then, if

$R_S = 500\ \Omega \qquad R_L = 5\ \text{k}\Omega$
$C = 5\ \text{pF} \qquad L = 1\ \text{mH}$

compute and plot the frequency response function

$$H(j\omega) = \frac{\mathbf{V}_o(j\omega)}{\mathbf{V}_i(j\omega)}$$

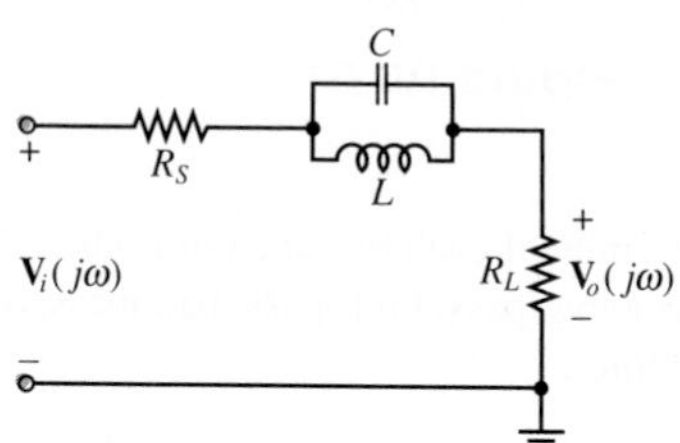

Figure P6.59

6.60 In the filter circuit shown in Figure P6.59, derive the equation for the voltage frequency response function in standard form. Then if

$R_s = 500\ \Omega \qquad R_L = 5\ \text{k}\Omega$
$\omega_r = 12.1278\ \text{Mrad/s} \qquad C = 68\ \text{nF}$
$L = 0.1\ \mu\text{H}$

determine the half-power frequencies, bandwidth, and Q. Plot $H(j\omega)$.

6.61 In the filter circuit shown in Figure P6.59, derive the equation for the voltage frequency response function in standard form. Then if

$R_s = 4.4\ \text{k}\Omega \qquad R_L = 60\ \text{k}\Omega \qquad \omega_r = 25\ \text{Mrad/s}$
$C = 0.8\ \text{nF} \qquad L = 2\ \mu\text{H}$

determine the half-power frequencies, bandwidth, and Q. Plot $H(j\omega)$.

6.62 In the bandstop (notch) filter shown in Figure P6.62:

$L = 0.4$ mH $\quad R_c = 100\ \Omega$

$C = 1$ pF $\quad R_s = R_L = 3.8\ \text{k}\Omega$

Determine:

a. An expression for the voltage frequency response function in the form

$$H_v(j\omega) = \frac{\mathbf{V}_o(j\omega)}{\mathbf{V}_i(j\omega)} = H_o \frac{1 + jf_1(\omega)}{1 + jf_2(\omega)}$$

b. The magnitude of the function at high and low frequencies and at the resonant frequency.

c. The resonant frequency.

d. The half-power frequencies.

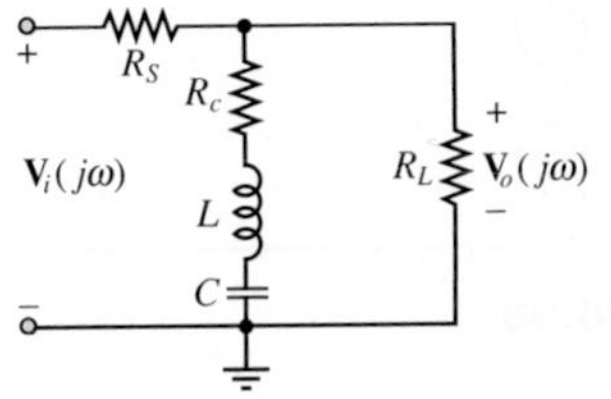

Figure P6.62

6.63 In the filter circuit shown in Figure P6.56:

$R_S = 5\ \text{k}\Omega \quad C = 5$ nF

$R_L = 50\ \text{k}\Omega \quad L = 2$ mH

Determine:

a. An expression for the voltage frequency response function

$$H_V(j\omega) = \frac{\mathbf{V}_o(j\omega)}{\mathbf{V}_i(j\omega)}$$

b. The resonant frequency.

c. The half-power frequencies.

d. The bandwidth and Q.

e. Plot $H_V(j\omega)$.

6.64 The function of a loudspeaker *crossover network* is to channel frequencies higher than a given crossover frequency, f_c, into the high-frequency speaker (tweeter) and frequencies below f_c into the low-frequency speaker (woofer). Figure P6.64 shows an approximate equivalent circuit where the amplifier is represented as a voltage source with zero internal resistance and each speaker acts as an 8 Ω resistance. If the crossover frequency is chosen to be 1,200 Hz, evaluate C and L. [*Hint:* The break frequency would be a reasonable value to set as the crossover frequency.]

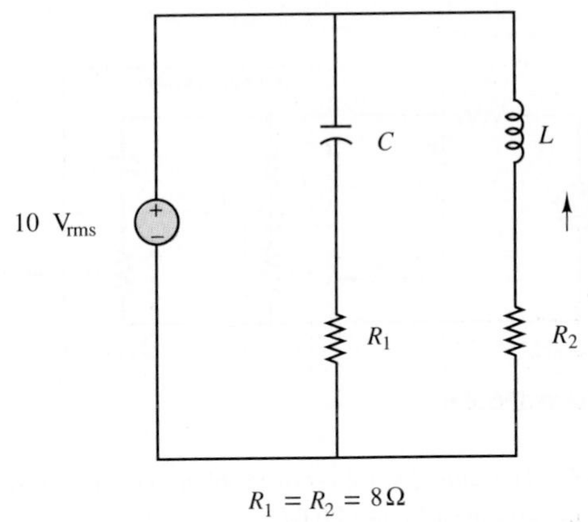

Figure P6.64

6.65 What is the frequency response, $\mathbf{V}_{out}(\omega)/\mathbf{V}_S(\omega)$, for the circuit of Figure P6.65? Sketch the frequency response of the circuit (magnitude and phase) if $R_S = R_L = 5,000\ \Omega$, $L = 10\ \mu$H, and $C = 0.1\ \mu$F.

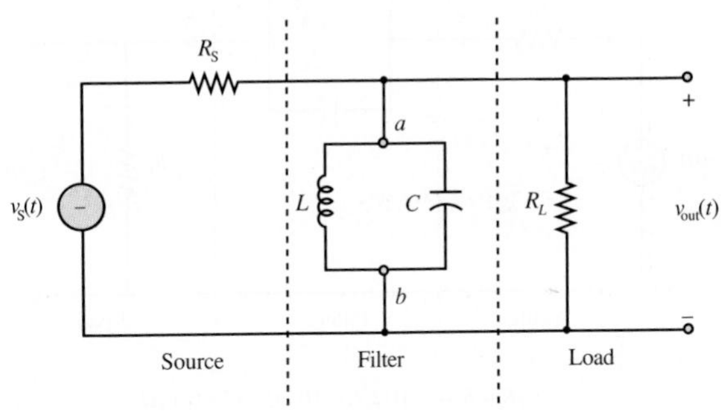

Figure P6.65

6.66 Many stereo speakers are two-way speaker systems; that is, they have a woofer for low-frequency sounds and a tweeter for high-frequency sounds. To get the proper separation of frequencies going to the woofer and to the tweeter, crossover circuitry is used. A crossover circuit is effectively a bandpass, high-pass, or low-pass filter. The system model is shown in Figure 6.66.

a. If $L = 2$ mH, $C = 125\ \mu$F, and $R_S = 4\ \Omega$, find the load impedance as a function of frequency. At what frequency is maximum power transfer obtained?

b. Plot the magnitude and phase responses of the currents through the woofer and tweeter as a function of frequency.

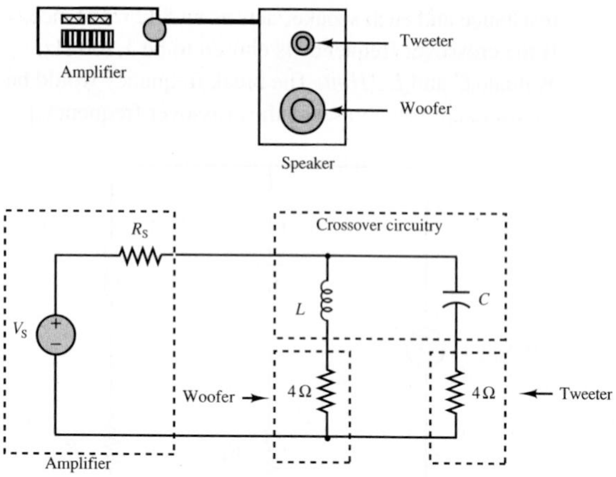

Figure P6.66

6.67 The same LC values of Problem 6.66 are used in the circuit of Figure P6.67.

a. Compute the frequency response of this circuit, $\mathbf{V}_{\text{out}}(j\omega)/\mathbf{V}_S(j\omega)$.

b. Plot the frequency response of the circuit.

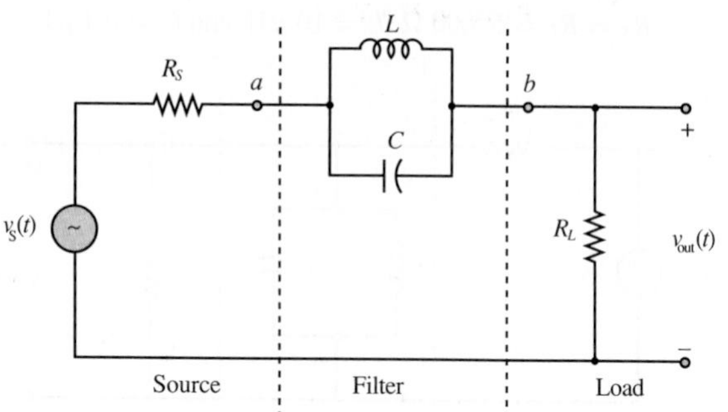

$R_S = R_L = 500\,\Omega$; $L = 10\ mH$; $C = 0.1\ \mu F$

Figure P6.67

6.68 It is very common to see interference caused by power lines, at a frequency of 60 Hz. This problem outlines the design of the notch filter, shown in Figure P6.68, to reject a band of frequencies around 60 Hz.

a. Write the impedance function for the filter of Figure P6.68 (the resistor r_L represents the internal resistance of a practical inductor).

b. For what value of C will the center frequency of the filter equal 60 Hz if $L = 100$ mH and $r_L = 5\ \Omega$?

c. Would the "sharpness," or selectivity, of the filter increase or decrease if r_L were to increase?

d. Assume that the filter is used to eliminate the 60-Hz noise from a signal generator with output frequency of 1 kHz. Evaluate the frequency response $\mathbf{V}_L/\mathbf{V}_{\text{in}}(j\omega)$ at both frequencies if:

$$v_g(t) = \sin(2\pi 1{,}000\, t)\text{V} \quad r_g = 50\ \Omega$$
$$v_n(t) = 3\sin(2\pi 60\, t) \quad R_L = 300\ \Omega$$

And if L and C are as in part b.

e. Plot the magnitude frequency response $\left|\frac{\mathbf{V}_L}{\mathbf{V}_{in}}(j\omega)\right|_{dB}$ vs. ω on a logarithmic scale and indicate the value of $\left|\frac{\mathbf{V}_L}{\mathbf{V}_{in}}(j\omega)\right|_{dB}$ at the frequencies 60 Hz and 1,000 Hz on your plot.

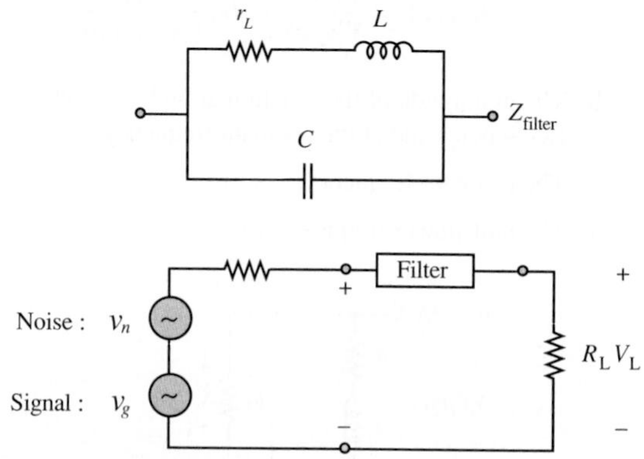

Figure P6.68

6.69 The circuit of Figure P6.69 is representative of an amplifier-speaker connection. The crossover circuit (filter) is a low-pass filter that is connected to a woofer. The filter's topography is known as a π network.

a. Find the frequency response $\mathbf{V}_o(j\omega)/\mathbf{V}_S(j\omega)$.

b. If $C_1 = C_2 = C$, $R_S = R_L = 600\ \Omega$, and $1/\sqrt{LC} = R/L = 1/RC = 2{,}000\pi$, plot $|\ \mathbf{V}_o(j\omega)/\mathbf{V}_S(j\omega)\ |$ in dB versus frequency (logarithmic scale) in the range 100 Hz $\le f \le$ 10,000 Hz.

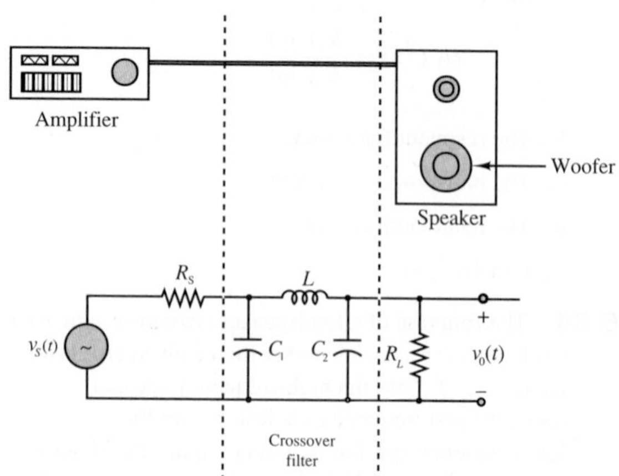

Figure P6.69

Section 6.4: Bode Plots

6.70 In the circuit shown in Figure P6.70:

a. Determine the frequency response function

$$H(j\omega) = \frac{\mathbf{V}_{out}(j\omega)}{\mathbf{V}_{in}(j\omega)}$$

b. Manually sketch a magnitude and phase Bode plot of the system, using a five-cycle semilog paper. Show the factored polynomial and all the steps in constructing the plot. Clearly show the break frequencies on the ω axis. (*Hint:* To factor the denominator polynomial, you may find it helpful to use the Matlab™ command "roots.")

c. Use Matlab™ and the *Bode* command to generate the same plot, and verify that your answer is indeed correct. Assume $R_1 = R_2 = 1\ \text{k}\Omega$, $C_1 = 1\ \mu\text{F}$, $C_2 = 1\ \text{mF}$, $L = 1\ \text{H}$.

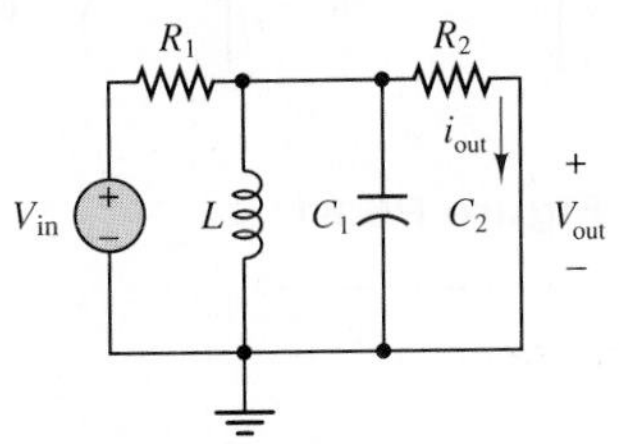

Figure P6.70

6.71 Repeat Problem 6.70 for the frequency response function

$$H(j\omega) = \frac{\mathbf{I}_{out}(j\omega)}{\mathbf{V}_{in}(j\omega)}$$

Use the same component values as in Problem 6.70.

6.72 Repeat Problem 6.70 for the circuit of Figure P6.72 and the frequency response function

$$H(j\omega) = \frac{\mathbf{V}_{out}(j\omega)}{\mathbf{I}_{in}(j\omega)}$$

Let $R_1 = R_2 = 1\ \text{k}\Omega$, $C = 1\ \mu\text{F}$, $L = 1\ \text{H}$.

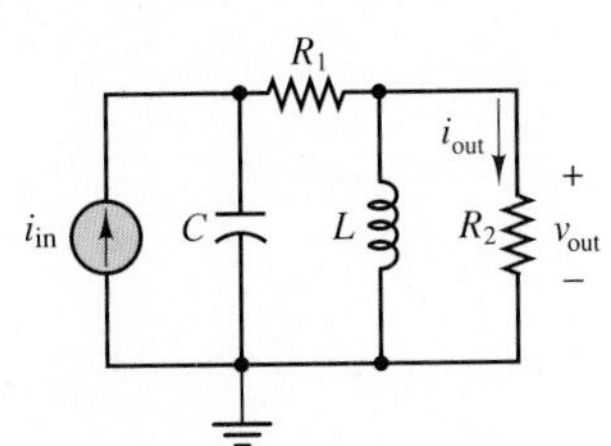

Figure P6.72

6.73 Repeat Problem 6.70 for the circuit of Figure P6.72 and the frequency response function of

$$H(j\omega) = \frac{\mathbf{I}_{out}(j\omega)}{\mathbf{I}_{in}(j\omega)}$$

Use the same values as in Problem 6.72.

6.74 Repeat Problem 6.70 for the circuit of Figure P6.74 and the frequency response function

$$H(j\omega) = \frac{\mathbf{V}_{out}(j\omega)}{\mathbf{I}_{in}(j\omega)}$$

Assume that $R_1 = R_2 = 1\ \text{k}\Omega$, $C_1 = 1\ \mu\text{F}$, $C_2 = 1\ \text{mF}$.

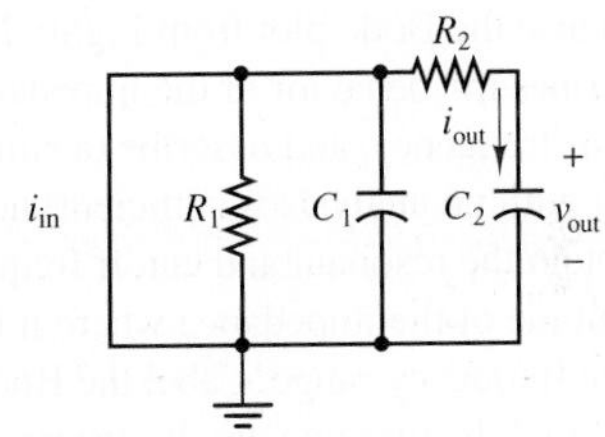

Figure P6.74

6.75 Repeat Problem 6.70 for the circuit of Figure P6.74 and the frequency response function

$$H(j\omega) = \frac{\mathbf{I}_{out}(j\omega)}{\mathbf{I}_{in}(j\omega)}$$

Use the same component values as in problem 6.74.

6.76 With reference to Figure P6.4:

a. Manually sketch a magnitude and phase Bode plot of the system using semilog paper. Show the factored polynomial and all the steps in constructing the plot. Clearly show the break frequencies on the ω axis.

b. Use Matlab™ and the *Bode* command to generate the same plot, and verify that your answer is indeed correct.

6.77 Repeat Problem 6.76 for the circuit of Figure P6.4, considering the load voltage v_L as a voltage across the capacitor.

6.78 Repeat Problem 6.76 for the circuit of Figure P6.5.

6.79 Assume in a certain frequency range that the ratio of output amplitude to input amplitude is proportional to $1/\omega^3$. What is the slope of the Bode plot in this frequency range, expressed in decibels per decade?

6.80 Assume that the output amplitude of a circuit depends on frequency according to

$$V = \frac{A\omega + B}{\sqrt{C + D\omega^2}}$$

Find:

a. The break frequency.

b. The slope of the Bode plot (in decibels per decade) above the break frequency.

c. The slope of the Bode plot below the break frequency.

d. The high-frequency limit of V.

6.81 Determine an expression for the circuit of Figure P6.81(a) for the equivalent impedance in standard form. Choose the Bode plot from Figure P6.81(b) that best describes the behavior of the impedance as a function of frequency, and describe (a simple one-line statement with no analysis is sufficient) how you would obtain the resonant and cutoff frequencies and the magnitude of the impedance where it is constant over some frequency range. Label the Bode plot to indicate which feature you are discussing.

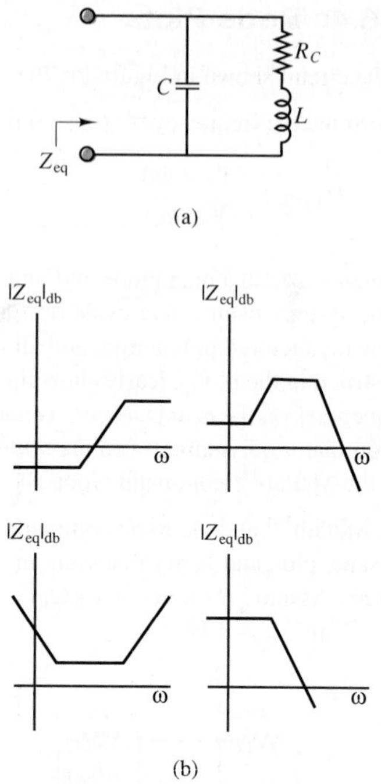

Figure P6.81

C H A P T E R

7

AC POWER

The aim of this chapter is to introduce the student to simple AC power calculations and to the generation and distribution of electric power. The chapter builds on the material developed in Chapter 4—namely, phasors and complex impedance—and paves the way for the material on electric machines in Chapters 16, 17, and 18.

The chapter starts with the definition of AC average and complex power and illustrates the computation of the power absorbed by a complex load; special attention is paid to the calculation of the power factor, and to power factor correction. The next subject is a brief discussion of ideal transformers and of maximum power transfer. This is followed by an introduction to three-phase power. The chapter ends with a discussion of electric power generation and distribution.

Learning Objectives

1. Understand the meaning of instantaneous and average power, master AC power notation, and compute average power for AC circuits. Compute the power factor of a complex load. *Section 7.1.*
2. Learn complex power notation; compute apparent, real, and reactive power for complex loads. Draw the power triangle, and compute the capacitor size required to perform power factor correction on a load. *Section 7.2.*
3. Analyze the ideal transformer; compute primary and secondary currents and voltages and turns ratios. Calculate reflected sources and impedances across ideal transformers. Understand maximum power transfer. *Section 7.3.*
4. Learn three-phase AC power notation; compute load currents and voltages for balanced wye and delta loads. *Section 7.4.*
5. Understand the basic principles of residential electrical wiring and of electrical safety. *Sections 7.5, 7.6.*

7.1 POWER IN AC CIRCUITS

The objective of this section is to introduce AC power. As already mentioned in Chapter 4, 50- or 60-Hz AC electric power constitutes the most common form of electric power distribution; in this section, the phasor notation developed in Chapter 4 will be employed to analyze the power absorbed by both resistive and complex loads.

Instantaneous and Average Power

From Chapter 4, you already know that when a linear electric circuit is excited by a sinusoidal source, all voltages and currents in the circuit are also sinusoids of the same frequency as that of the excitation source. Figure 7.1 depicts the general form of a linear AC circuit. The most general expressions for the voltage and current delivered to an arbitrary load are as follows:

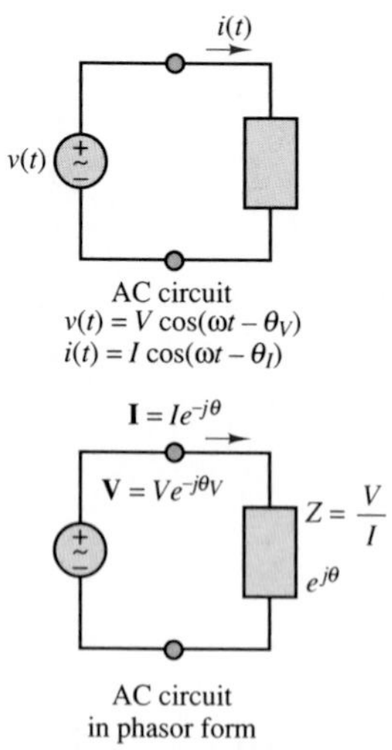

Figure 7.1 Circuit for illustration of AC power

$$\begin{aligned} v(t) &= V\cos(\omega t - \theta_V) \\ i(t) &= I\cos(\omega t - \theta_I) \end{aligned} \tag{7.1}$$

where V and I are the peak amplitudes of the sinusoidal voltage and current, respectively, and θ_V and θ_I are their phase angles. Two such waveforms are plotted in Figure 7.2, with unit amplitude and with phase angles $\theta_V = \pi/6$ and $\theta_I = \pi/3$. The phase shift between source and load is therefore $\theta = \theta_V - \theta_I$. It will be easier, for the purpose of this section, to assume that $\theta_V = 0$, without any loss of generality, since all phase angles will be referenced to the source voltage's phase. In Section 7.2, where complex power is introduced, you will see that this assumption is not necessary since phasor notation is used. In this section, some of the trigonometry-based derivations are simpler if the source voltage reference phase is assumed to be zero.

Since the **instantaneous power** dissipated by a circuit element is given by the product of the instantaneous voltage and current, it is possible to obtain a general expression for the power dissipated by an AC circuit element:

$$p(t) = v(t)i(t) = VI\cos(\omega t)\cos(\omega t - \theta) \tag{7.2}$$

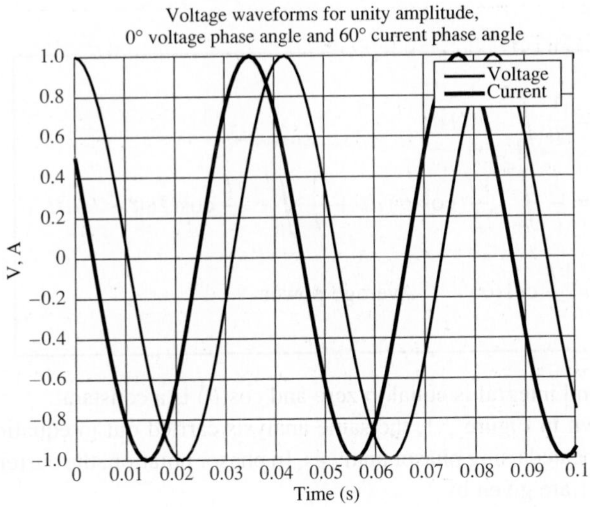

Figure 7.2 Current and voltage waveforms for illustration of AC power

Equation 7.2 can be further simplified with the aid of trigonometric identities to yield

$$p(t) = \frac{VI}{2}\cos(\theta) + \frac{VI}{2}\cos(2\omega t - \theta) \tag{7.3}$$

where θ is the difference in phase between voltage and current. Equation 7.3 illustrates how the instantaneous power dissipated by an AC circuit element is equal to the sum of an average component $\frac{1}{2}VI\cos(\theta)$ and a sinusoidal component $\frac{1}{2}VI\cos(2\omega t - \theta)$, oscillating at a frequency double that of the original source frequency.

The instantaneous and average power are plotted in Figure 7.3 for the signals of Figure 7.2. The **average power** corresponding to the voltage and current signals of equation 7.1 can be obtained by integrating the instantaneous power over one cycle of the sinusoidal signal. Let $T = 2\pi/\omega$ represent one cycle of the sinusoidal signals. Then the average power P_{av} is given by the integral of the instantaneous power $p(t)$

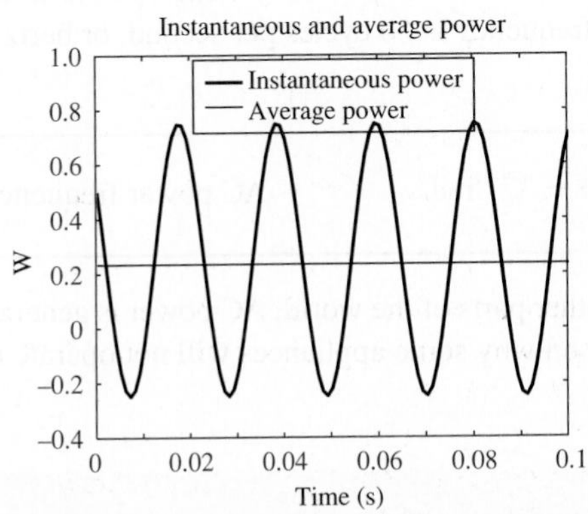

Figure 7.3 Instantaneous and average power dissipation corresponding to the signals plotted in Figure 7.2

over one cycle

LO1

$$P_{av} = \frac{1}{T}\int_0^T p(t)\,dt$$
$$= \frac{1}{T}\int_0^T \frac{VI}{2}\cos(\theta)\,dt + \frac{1}{T}\int_0^T \frac{VI}{2}\cos(2\omega t - \theta)\,dt \quad \textbf{(7.4)}$$

$$P_{av} = \frac{VI}{2}\cos(\theta) \qquad \text{Average power} \quad \textbf{(7.5)}$$

since the second integral is equal to zero and $\cos(\theta)$ is a constant.

As shown in Figure 7.1, the same analysis carried out in equations 7.1 to 7.3 can also be repeated using phasor analysis. In phasor notation, the current and voltage of equation 7.1 are given by

$$\mathbf{V}(j\omega) = Ve^{j0} \quad \textbf{(7.6)}$$

$$\mathbf{I}(j\omega) = Ie^{-j\theta} \quad \textbf{(7.7)}$$

Note further that the impedance of the circuit element shown in Figure 7.1 is defined by the phasor voltage and current of equations 7.6 and 7.7 to be

$$Z = \frac{V}{I}e^{j(\theta)} = |Z|e^{j\theta} \quad \textbf{(7.8)}$$

The expression for the average power obtained in equation 7.4 can therefore also be represented using phasor notation, as follows:

LO1

$$P_{av} = \frac{1}{2}\frac{V^2}{|Z|}\cos\theta = \frac{1}{2}I^2|Z|\cos\theta \qquad \text{Average power} \quad \textbf{(7.9)}$$

AC Power Notation

It has already been noted that AC power systems operate at a fixed frequency; in North America, this frequency is 60 cycles per second, or hertz (Hz), corresponding to a radian frequency

$$\omega = 2\pi \cdot 60 = 377 \text{ rad/s} \qquad \text{AC power frequency} \quad \textbf{(7.10)}$$

In Europe and most other parts of the world, AC power is generated at a frequency of 50 Hz (this is the reason why some appliances will not operate under one of the two systems).

LO1

Therefore, for the remainder of this chapter the radian frequency ω is fixed at 377 rad/s, unless otherwise noted.

With knowledge of the radian frequency of all voltages and currents, it will always be possible to compute the exact magnitude and phase of any impedance in a circuit.

A second point concerning notation is related to the factor $\frac{1}{2}$ in equation 7.9. It is customary in AC power analysis to employ the rms value of the AC voltages and currents in the circuit (see Section 4.2). Use of the **rms value** eliminates the factor $\frac{1}{2}$ in power expressions and leads to considerable simplification. Thus, the following expressions will be used in this chapter:

$$V_{\text{rms}} = \frac{V}{\sqrt{2}} = \tilde{V} \tag{7.11}$$

$$I_{\text{rms}} = \frac{I}{\sqrt{2}} = \tilde{I} \tag{7.12}$$

$$\begin{aligned} P_{\text{av}} &= \frac{1}{2}\frac{V^2}{|Z|}\cos\theta = \frac{\tilde{V}^2}{|Z|}\cos\theta \\ &= \frac{1}{2}I^2|Z|\cos\theta = \tilde{I}^2|Z|\cos\theta = \tilde{V}\tilde{I}\cos\theta \end{aligned} \tag{7.13}$$

Figure 7.4 illustrates the **impedance triangle,** which provides a convenient graphical interpretation of impedance as a vector in the complex plane. From the figure, it is simple to verify that

$$R = |Z|\cos\theta \tag{7.14}$$

$$X = |Z|\sin\theta \tag{7.15}$$

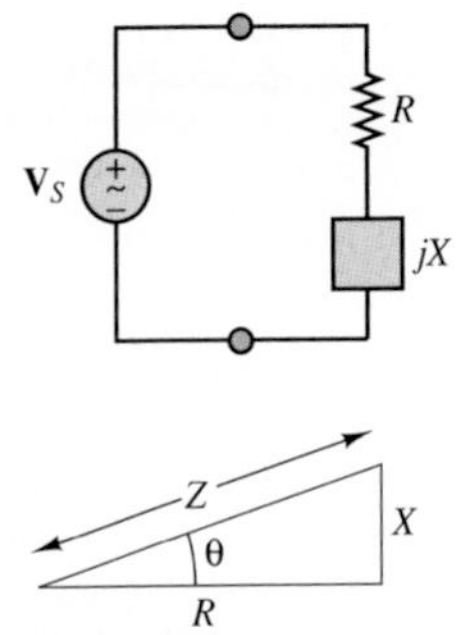

Figure 7.4 Impedance triangle

Finally, the amplitudes of phasor voltages and currents will be denoted throughout this chapter by means of the rms amplitude. We therefore introduce a slight modification in the phasor notation of Chapter 4 by defining the following **rms phasor** quantities:

$$\tilde{\mathbf{V}} = V_{\text{rms}}e^{j\theta_V} = \tilde{V}e^{j\theta_V} = \tilde{V}\angle\theta_V \tag{7.16}$$

and

$$\tilde{\mathbf{I}} = I_{\text{rms}}e^{j\theta_I} = \tilde{I}e^{j\theta_I} = \tilde{I}\angle\theta_I \tag{7.17}$$

In other words,

> Throughout the remainder of this chapter, the symbols $\tilde{V}$ and $\tilde{I}$ will denote the rms value of a voltage or a current, and the symbols $\tilde{\mathbf{V}}$ and $\tilde{\mathbf{I}}$ will denote rms phasor voltages and currents.

Also recall the use of the symbol $\angle$ to represent the complex exponential. Thus, the sinusoidal waveform corresponding to the phasor current $\tilde{\mathbf{I}} = \tilde{I}\angle\theta_I$ corresponds to the time-domain waveform

$$i(t) = \sqrt{2}\tilde{I}\cos(\omega t + \theta_I) \tag{7.18}$$

and the sinusoidal form of the phasor voltage $\mathbf{V} = \tilde{V}\angle\theta_V$ is

$$v(t) = \sqrt{2}\tilde{V}\cos(\omega t + \theta_V) \tag{7.19}$$

EXAMPLE 7.1 Computing Average and Instantaneous AC Power

Problem

Compute the average and instantaneous power dissipated by the load of Figure 7.5.

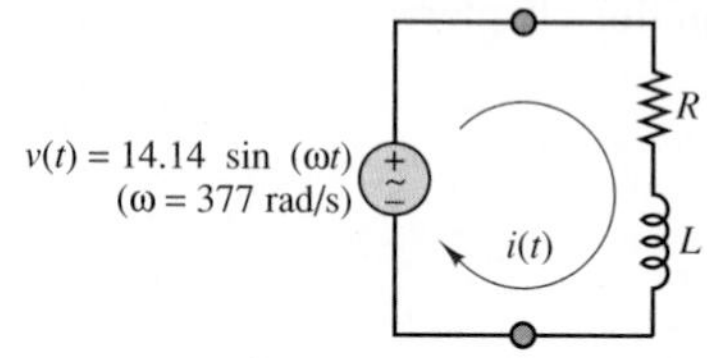

Figure 7.5

Solution

Known Quantities: Source voltage and frequency, load resistance and inductance values.

Find: P_{av} and $p(t)$ for the RL load.

Schematics, Diagrams, Circuits, and Given Data: $v(t) = 14.14\sin(377t)$ V; $R = 4\ \Omega$; $L = 8$ mH.

Assumptions: Use rms values for all phasor quantities in the problem.

Analysis: First, we define the phasors and impedances at the frequency of interest in the problem, $\omega = 377$ rad/s:

$$\tilde{\mathbf{V}} = 10\angle\left(-\frac{\pi}{2}\right) \qquad Z = R + j\omega L = 4 + j3 = 5\angle 0.644$$

$$\tilde{\mathbf{I}} = \frac{\tilde{\mathbf{V}}}{Z} = \frac{10\angle(-\pi/2)}{5\angle 0.644} = 2\angle(-2.215)$$

The average power can be computed from the phasor quantities:

$$P_{av} = \tilde{\mathbf{V}}\tilde{\mathbf{I}}\cos(\theta) = 10 \times 2 \times \cos(0.644) = 16\ \text{W}$$

The instantaneous power is given by the expression

$$p(t) = v(t) \times i(t) = \sqrt{2} \times 10\sin(377t) \times \sqrt{2} \times 2\cos(377t - 2.215) \qquad \text{W}$$

The instantaneous voltage and current waveforms and the instantaneous and average power are plotted in Figure 7.6.

Comments: Please pay attention to the use of rms values in this example: It is very important to remember that we have defined phasors to have rms amplitude in the power calculation. This is a standard procedure in electrical engineering practice.

Note that the instantaneous power can be negative for brief periods of time, even though the average power is positive.

CHECK YOUR UNDERSTANDING

Show that the equalities in equation 7.9 hold when phasor notation is used.

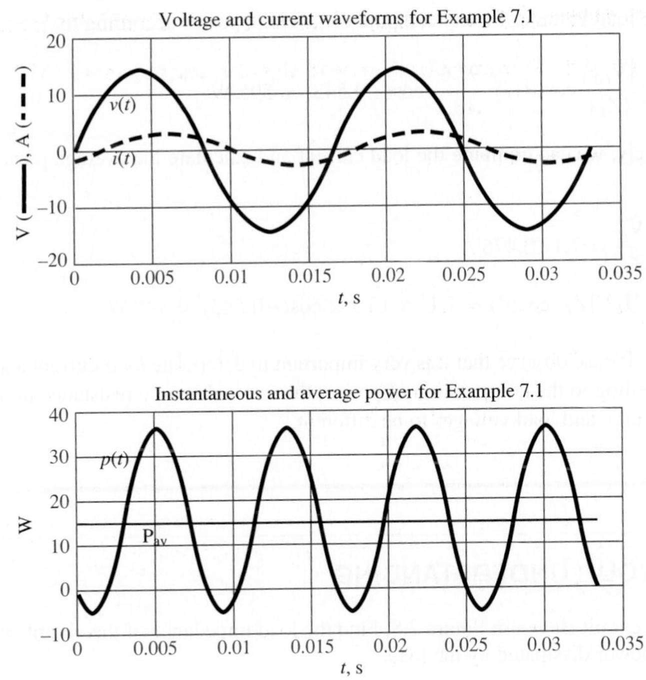

Figure 7.6

EXAMPLE 7.2 Computing Average AC Power

LO1

Problem

Compute the average power dissipated by the load of Figure 7.7.

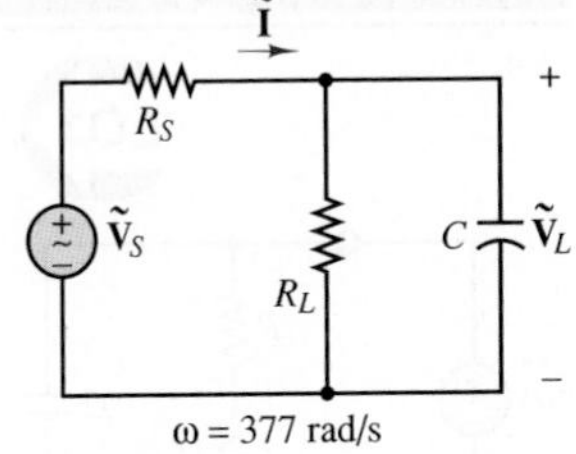

Figure 7.7

Solution

Known Quantities: Source voltage, internal resistance and frequency, load resistance and inductance values.

Find: P_{av} for the RC load.

Schematics, Diagrams, Circuits, and Given Data: $\tilde{\mathbf{V}}_s = 110\angle 0$; $R_S = 2\ \Omega$; $R_L = 16\ \Omega$; $C = 100\ \mu\text{F}$.

Assumptions: Use rms values for all phasor quantities in the problem.

Analysis: First, we compute the load impedance at the frequency of interest in the problem, $\omega = 377$ rad/s:

$$Z_L = R \| \frac{1}{j\omega C} = \frac{R_L}{1 + j\omega C R_L} = \frac{16}{1 + j0.6032} = 13.7\angle(-0.543)\ \Omega$$

Next, we compute the load voltage, using the voltage divider rule:

$$\tilde{\mathbf{V}}_L = \frac{Z_L}{R_S + Z_L}\tilde{\mathbf{V}}_S = \frac{13.7\angle(-0.543)}{2 + 13.7\angle(-0.543)} 110\angle 0 = 97.6\angle(-0.067)\ \text{V}$$

Knowing the load voltage, we can compute the average power according to

$$P_{\text{av}} = \frac{|\tilde{\mathbf{V}}_L|^2}{|Z_L|}\cos(\theta) = \frac{97.6^2}{13.7}\cos(-0.543) = 595 \text{ W}$$

or, alternatively, we can compute the load current and calculate the average power according to

$$\tilde{\mathbf{I}}_L = \frac{\tilde{\mathbf{V}}_L}{Z_L} = 7.1\angle 0.476 \text{ A}$$

$$P_{\text{av}} = |\tilde{\mathbf{I}}_L|^2 |Z_L| \cos(\theta) = 7.1^2 \times 13.7 \times \cos(-0.543) = 595 \text{ W}$$

Comments: Please observe that it is very important to determine *load* current and/or voltage before proceeding to the computation of power; the internal source resistance in this problem causes the source and load voltages to be different.

CHECK YOUR UNDERSTANDING

Consider the circuit shown in Figure 7.8. Find the load impedance of the circuit, and compute the average power dissipated by the load.

Answer: $Z = 4.8e^{-j33.5^\circ}$ Ω; $P_{\text{av}} = 2{,}103.4$ W

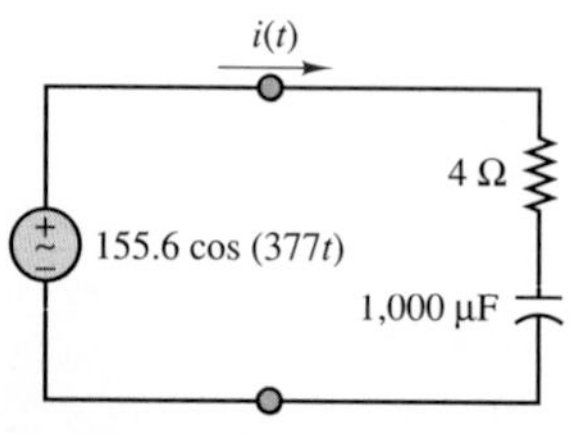

Figure 7.8

EXAMPLE 7.3 Computing Average AC Power

LO1

Problem

Compute the average power dissipated by the load of Figure 7.9.

Solution

Known Quantities: Source voltage, internal resistance and frequency, load resistance, capacitance and inductance values.

Find: P_{av} for the complex load.

Schematics, Diagrams, Circuits, and Given Data: $\tilde{\mathbf{V}}_s = 110\angle 0$ V; $R = 10$ Ω; $L = 0.05$ H; $C = 470$ μF.

Assumptions: Use rms values for all phasor quantities in the problem.

Analysis: First, we compute the load impedance at the frequency of interest in the problem, $\omega = 377$ rad/s:

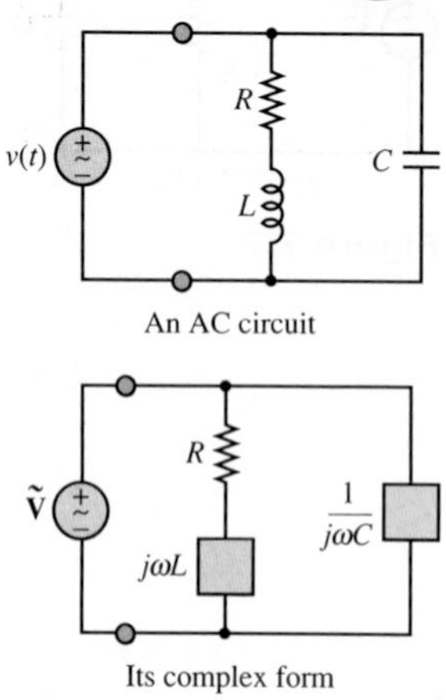

An AC circuit

Its complex form

Figure 7.9

$$Z_L = (R + j\omega L) \| \frac{1}{j\omega C} = \frac{(R + j\omega L)/j\omega C}{R + j\omega L + 1/j\omega C}$$

$$= \frac{R + j\omega L}{1 - \omega^2 LC + j\omega CR} = 1.16 - j7.18$$

$$= 7.27\angle(-1.41)\ \Omega$$

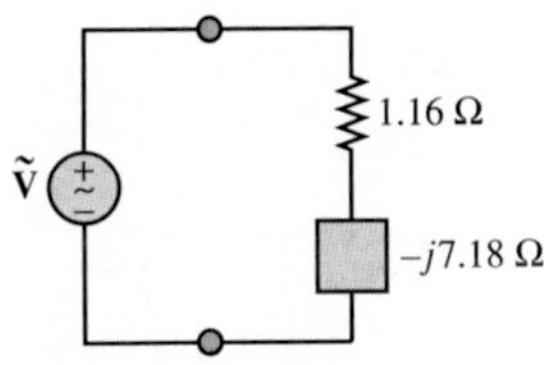

Figure 7.10

Note that the equivalent load impedance consists of a capacitive load at this frequency, as shown in Figure 7.10. Knowing that the load voltage is equal to the source voltage, we can compute the average power according to

$$P_{av} = \frac{|\tilde{\mathbf{V}}_L|^2}{|Z_L|} \cos(\theta) = \frac{110^2}{7.27} \cos(-1.41) = 266 \text{ W}$$

CHECK YOUR UNDERSTANDING

Compute the power dissipated by the internal source resistance in Example 7.2.

Use the expression $P_{av} = \tilde{I}^2 |Z| \cos(\theta)$ to compute the average power dissipated by the load of Example 7.2.

Answers: 101.46 W; See Example 7.2

Power Factor

The phase angle of the load impedance plays a very important role in the absorption of power by a load impedance. As illustrated in equation 7.13 and in the preceding examples, the average power dissipated by an AC load is dependent on the cosine of the angle of the impedance. To recognize the importance of this factor in AC power computations, the term $\cos(\theta)$ is referred to as the **power factor (pf).** Note that the power factor is equal to 0 for a purely inductive or capacitive load and equal to 1 for a purely resistive load; in every other case,

$$0 < \text{pf} < 1 \qquad \textbf{(7.20)}$$

Two equivalent expressions for the power factor are given in the following:

$$\text{pf} = \cos(\theta) = \frac{P_{av}}{\tilde{V}\tilde{I}} \qquad \text{Power factor} \qquad \textbf{(7.21)}$$

where $\tilde{V}$ and $\tilde{I}$ are the rms values of the load voltage and current, respectively.

7.2 COMPLEX POWER

The expression for the instantaneous power given in equation 7.3 may be expanded to provide further insight into AC power. Using trigonometric identities, we obtain the following expressions:

$$\begin{aligned} p(t) &= \frac{\tilde{V}^2}{|Z|}[\cos\theta + \cos\theta\cos(2\omega t) + \sin\theta\sin(2\omega t)] \\ &= \tilde{I}^2|Z|[\cos\theta + \cos\theta\cos(2\omega t) + \sin\theta\sin(2\omega t)] \\ &= \tilde{I}^2|Z|\cos\theta(1 + \cos 2\omega t) + \tilde{I}^2|Z|\sin\theta\sin(2\omega t) \end{aligned} \quad \textbf{(7.22)}$$

Recalling the geometric interpretation of the impedance Z of Figure 7.4, you may recognize that

$$|Z|\cos\theta = R$$

and **(7.23)**

$$|Z|\sin\theta = X$$

are the resistive and reactive components of the load impedance, respectively. On the basis of this fact, it becomes possible to write the instantaneous power as

$$\begin{aligned} p(t) &= \tilde{I}^2R(1 + \cos 2\omega t) + \tilde{I}^2X\sin(2\omega t) \\ &= \tilde{I}^2R + \tilde{I}^2R\cos(2\omega t) + \tilde{I}^2X\sin(2\omega t) \end{aligned} \quad \textbf{(7.24)}$$

The physical interpretation of this expression for the instantaneous power should be intuitively appealing at this point. As equation 7.24 suggests, the instantaneous power dissipated by a complex load consists of the following three components:

LO2

1. An average component, which is constant; this is called the *average power* and is denoted by the symbol P_{av}:

   $$P_{av} = \tilde{I}^2R \quad \textbf{(7.25)}$$

   where $R = \text{Re}\, Z$.
2. A time-varying (sinusoidal) component with zero average value that is contributed by the power fluctuations in the resistive component of the load and is denoted by $p_R(t)$:

   $$\begin{aligned} p_R(t) &= \tilde{I}^2R\cos 2\omega t \\ &= P_{av}\cos 2\omega t \end{aligned} \quad \textbf{(7.26)}$$

3. A time-varying (sinusoidal) component with zero average value, due to the power fluctuation in the reactive component of the load and denoted by $p_X(t)$:

   $$\begin{aligned} p_X(t) &= \tilde{I}^2X\sin 2\omega t \\ &= Q\sin 2\omega t \end{aligned} \quad \textbf{(7.27)}$$

   where $X = \text{Im}\, Z$ and Q is called the **reactive power.** *Note that since reactive elements can only store energy and not dissipate it, there is no net average power absorbed by X.*

Since P_{av} corresponds to the power absorbed by the load resistance, it is also called the **real power,** measured in units of watts (W). On the other hand, Q takes the name of *reactive power*, since it is associated with the load reactance. Table 7.1 shows the general methods of calculating P and Q.

The units of Q are **volt-amperes reactive,** or **VAR.** Note that Q represents an exchange of energy between the source and the reactive part of the load; thus, no net power is gained or lost in the process, since the average reactive power is zero. In general, it is desirable to minimize the reactive power in a load. Example 7.6 will explain the reason for this statement.

Table 7.1 Real and reactive power

Real power P_{av}	Reactive power Q
$\tilde{V}\tilde{I}\cos(\theta)$	$\tilde{V}\tilde{I}\sin(\theta)$
$\tilde{I}^2R$	$\tilde{I}^2X$

The computation of AC power is greatly simplified by defining a fictitious but very useful quantity called the **complex power** S

$$\mathbf{S} = \tilde{\mathbf{V}}\tilde{\mathbf{I}}^* \qquad \text{Complex power} \tag{7.28}$$

where the asterisk denotes the complex conjugate (see Appendix A). You may easily verify that this definition leads to the convenient expression

$$\mathbf{S} = \tilde{V}\tilde{I}\cos\theta + j\tilde{V}\tilde{I}\sin\theta = \tilde{I}^2R + j\tilde{I}^2X = \tilde{I}^2Z$$

or

$$\mathbf{S} = P_{av} + jQ \tag{7.29}$$

The complex power $\mathbf{S}$ may be interpreted graphically as a vector in the complex plane, as shown in Figure 7.11.

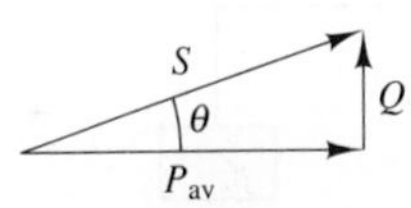

$$|S| = \sqrt{P_{av}^2 + Q^2} = \tilde{V}\cdot\tilde{I}$$

$$P_{av} = \tilde{V}\tilde{I}\cos\theta$$

$$Q = \tilde{V}\tilde{I}\sin\theta$$

Figure 7.11 The complex power triangle

The magnitude of $\mathbf{S}$, denoted by $|\mathbf{S}|$, is measured in units of **volt-amperes (VA)** and is called the **apparent power,** because this is the quantity one would compute by measuring the rms load voltage and currents without regard for the phase angle of the load. Note that the right triangle of Figure 7.11 is similar to the right triangle of Figure 7.4, since θ is the load impedance angle. The complex power may also be expressed by the product of the square of the rms current through the load and the complex load impedance:

$$\mathbf{S} = \tilde{I}^2\mathbf{Z}$$

or

$$\tilde{I}^2R + j\tilde{I}^2X = \tilde{I}^2\mathbf{Z} \tag{7.30}$$

or, equivalently, by the ratio of the square of the rms voltage across the load to the complex conjugate of the load impedance:

$$\mathbf{S} = \frac{\tilde{V}^2}{\mathbf{Z}^*} \tag{7.31}$$

The power triangle and complex power greatly simplify load power calculations, as illustrated in the following examples.

FOCUS ON METHODOLOGY

COMPLEX POWER CALCULATION FOR A SINGLE LOAD

LO2

1. Compute the load voltage and current in rms phasor form, using the AC circuit analysis methods presented in Chapter 4 and converting peak amplitude to rms values.

$$\tilde{\mathbf{V}} = \tilde{V}\angle\theta_V$$
$$\tilde{\mathbf{I}} = \tilde{I}\angle\theta_I$$

2. Compute the complex power $S = \tilde{\mathbf{V}}\tilde{\mathbf{I}}^*$ and set Re $S = P_{av}$, Im $S = Q$.
3. Draw the power triangle, as shown in Figure 7.11.
4. If Q is negative, the load is capacitive; if positive, the load is reactive.
5. Compute the apparent power $|S|$ in volt-amperes.

LO2

EXAMPLE 7.4 Complex Power Calculations

Problem

Use the definition of complex power to calculate real and reactive power for the load of Figure 7.12.

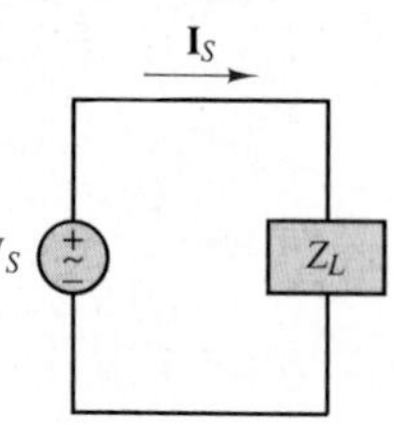

Figure 7.12

Solution

Known Quantities: Source, load voltage, and current.

Find: $S = P_{av} + jQ$ for the complex load.

Schematics, Diagrams, Circuits, and Given Data: $v(t) = 100\cos(\omega t + 0.262)$ V; $i(t) = 2\cos(\omega t - 0.262)$ A; $\omega = 377$ rad/s.

Assumptions: Use rms values for all phasor quantities in the problem.

Analysis: First, we convert the voltage and current to phasor quantities:

$$\tilde{\mathbf{V}} = \frac{100}{\sqrt{2}}\angle 0.262 \text{ V} \qquad \tilde{\mathbf{I}} = \frac{2}{\sqrt{2}}\angle(-0.262) \text{ A}$$

Next, we compute real and reactive power, using the definitions of equation 7.13:

$$P_{av} = |\tilde{\mathbf{V}}||\tilde{\mathbf{I}}|\cos(\theta) = \frac{200}{2}\cos(0.524) = 86.6 \text{ W}$$

$$Q = |\tilde{\mathbf{V}}||\tilde{\mathbf{I}}|\sin(\theta) = \frac{200}{2}\sin(0.524) = 50 \text{ VAR}$$

Now we apply the definition of complex power (equation 7.28) to repeat the same calculation:

$$S = \tilde{\mathbf{V}}\tilde{\mathbf{I}}^* = \frac{100}{\sqrt{2}}\angle 0.262 \times \frac{2}{\sqrt{2}}\angle -(-0.262) = 100\angle 0.524$$
$$= 86.6 + j50 \text{ W}$$

Therefore

$$P_{av} = 86.6 \text{ W} \qquad Q = 50 \text{ VAR}$$

Comments: Note how the definition of complex power yields both quantities at one time.

CHECK YOUR UNDERSTANDING

Use complex power notation to compute the real and reactive power for the load of Example 7.2.

Answer: $P_{av} = 593$ W; $Q = -358$ VAR

EXAMPLE 7.5 Real and Reactive Power Calculations

LO2

Problem

Use the definition of complex power to calculate real and reactive power for the load of Figure 7.13.

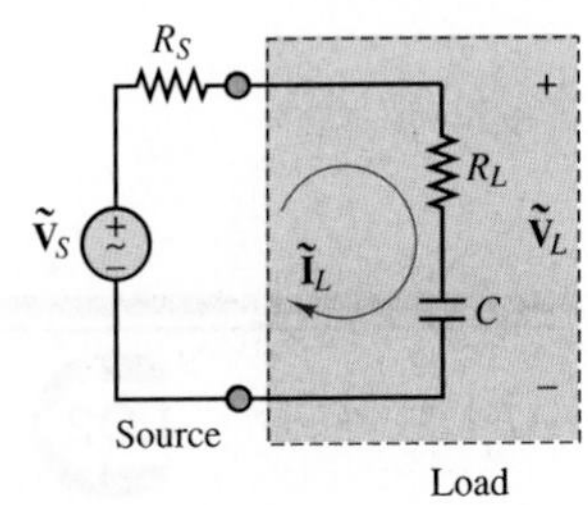

Figure 7.13

Solution

Known Quantities: Source voltage and resistance; load impedance.

Find: $S = P_{av} + jQ$ for the complex load.

Schematics, Diagrams, Circuits, and Given Data: $\tilde{\mathbf{V}}_S = 110\angle 0$ V; $R_S = 2\ \Omega$; $R_L = 5\ \Omega$; $C = 2{,}000\ \mu$F.

Assumptions: Use rms values for all phasor quantities in the problem.

Analysis: Define the load impedance

$$Z_L = R_L + \frac{1}{j\omega C} = 5 - j1.326 = 5.173\angle(-0.259)\ \Omega$$

Next, compute the load voltage and current:

$$\tilde{\mathbf{V}}_L = \frac{Z_L}{R_S + Z_L}\tilde{\mathbf{V}}_S = \frac{5 - j1.326}{7 - j1.326} \times 110 = 79.66\angle(-0.072) \text{ V}$$

$$\tilde{\mathbf{I}}_L = \frac{\tilde{\mathbf{V}}_L}{Z_L} = \frac{79.66\angle(-0.072)}{5.173\angle(-0.259)} = 15.44\angle 0.187 \text{ A}$$

Finally, we compute the complex power, as defined in equation 7.28:

$$S = \tilde{\mathbf{V}}_L\tilde{\mathbf{I}}_L^* = 79.9\angle(-0.072) \times 15.44\angle(-0.187) = 1{,}233\angle(-0.259)$$
$$= 1{,}192 - j316 \text{ W}$$

Therefore

$$P_{av} = 1{,}192 \text{ W} \qquad Q = -316 \text{ VAR}$$

Comments: Is the reactive power capacitive or inductive?

CHECK YOUR UNDERSTANDING

Use complex power notation to compute the real and reactive power for the load of Figure 7.8.

Answer: $P_{av} = 2.1$ kW; $Q = 1.39$ kVAR

Although the reactive power does not contribute to any average power dissipation in the load, it may have an adverse effect on power consumption, because it increases the overall rms current flowing in the circuit. Recall from Example 7.2 that the presence of any source resistance (typically, the resistance of the line wires in AC power circuits) will cause a loss of power; the power loss due to this line resistance is unrecoverable and constitutes a net loss for the electric company, since the user never receives this power. Example 7.6 illustrates quantitatively the effect of such **line losses** in an AC circuit.

LO2

EXAMPLE 7.6 Real Power Transfer for Complex Loads

Problem

Use the definition of complex power to calculate the real and reactive power for the load of Figure 7.14. Repeat the calculation when the inductor is removed from the load, and compare the real power transfer between source and load for the two cases.

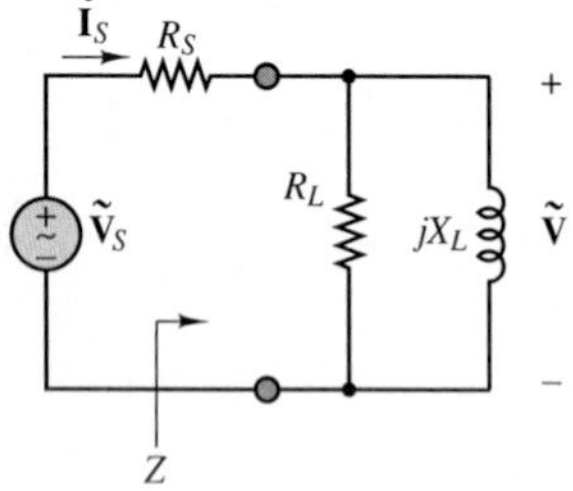

Figure 7.14

Solution

Known Quantities: Source voltage and resistance; load impedance.

Find:

1. $S_a = P_{ava} + jQ_a$ for the complex load.
2. $S_b = P_{avb} + jQ_b$ for the real load.
3. Compare P_{av}/P_S for the two cases.

Schematics, Diagrams, Circuits, and Given Data: $\tilde{\mathbf{V}}_S = 110\angle 0$ V; $R_S = 4\ \Omega$; $R_L = 10\ \Omega$; $jX_L = j6\ \Omega$.

Assumptions: Use rms values for all phasor quantities in the problem.

Analysis:

1. The inductor is part of the load. Define the load impedance.

$$Z_L = R_L \| j\omega L = \frac{10 \times j6}{10 + j6} = 5.145\angle 1.03\ \Omega$$

Next, compute the load voltage and current:

$$\tilde{\mathbf{V}}_L = \frac{Z_L}{R_S + Z_L}\tilde{\mathbf{V}}_S = \frac{5.145\angle 1.03}{4 + 5.145\angle 1.03} \times 110 = 70.9\angle 0.444\ \text{V}$$

$$\tilde{\mathbf{I}}_L = \frac{\tilde{\mathbf{V}}_L}{Z_L}\frac{70.9\angle 0.444}{5.145\angle 1.03} = 13.8\angle(-0.586)\ \text{A}$$

Finally, we compute the complex power, as defined in equation 7.28:

$$S_a = \tilde{\mathbf{V}}_L\tilde{\mathbf{I}}_L^* = 70.9\angle 0.444 \times 13.8\angle 0.586 = 978\angle 1.03$$
$$= 503 + j839\ \text{W}$$

Therefore

$$P_{\text{ava}} = 503\ \text{W} \qquad Q_a = +839\ \text{VAR}$$

2. The inductor is removed from the load (Figure 7.15). Define the load impedance:

$$Z_L = R_L = 10$$

Next, compute the load voltage and current:

$$\tilde{\mathbf{V}}_L = \frac{Z_L}{R_S + Z_L}\tilde{\mathbf{V}}_S = \frac{10}{4 + 10} \times 110 = 78.6\angle 0\ \text{V}$$

$$\tilde{\mathbf{I}}_L = \frac{\tilde{\mathbf{V}}_L}{Z_L} = \frac{78.6\angle 0}{10} = 7.86\angle 0\ \text{A}$$

Finally, we compute the complex power, as defined in equation 7.28:

$$S_b = \tilde{\mathbf{V}}_L\tilde{\mathbf{I}}_L^* = 78.6\angle 0 \times 7.86\angle 0 = 617\angle 0 = 617\ \text{W}$$

Therefore

$$P_{\text{avb}} = 617\ \text{W} \qquad Q_b = 0\ \text{VAR}$$

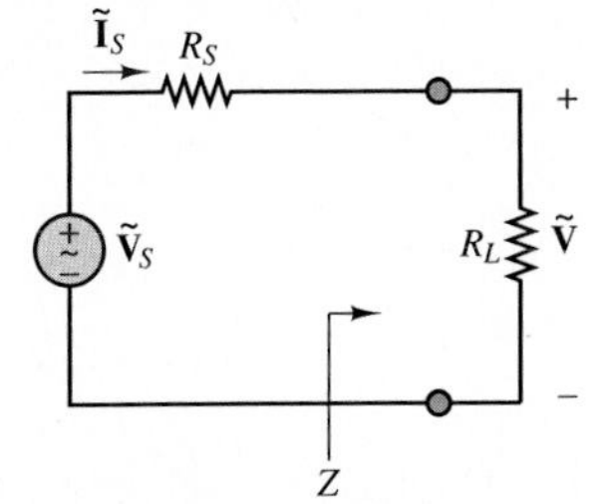

Figure 7.15

3. Compute the percent power transfer in each case. To compute the power transfer we must first compute the power delivered by the source in each case, $S_S = \tilde{\mathbf{V}}_S\tilde{\mathbf{I}}_S^*$. For Case 1:

$$\tilde{\mathbf{I}}_S = \frac{\tilde{\mathbf{V}}_S}{Z_{\text{total}}} = \frac{\tilde{\mathbf{V}}_S}{R_S + Z_L} = \frac{110}{4 + 5.145\angle 1.03} = 13.8\angle(-0.586)\ \text{A}$$

$$S_{Sa} = \tilde{\mathbf{V}}_S\tilde{\mathbf{I}}_S^* = 110 \times 13.8\angle -(-0.586) = 1{,}264 + j838\ \text{VA} = P_{Sa} + jQ_{Sa}$$

and the percent real power transfer is:

$$100 \times \frac{P_a}{P_{Sa}} = \frac{503}{1{,}264} = 39.8\%$$

For Case 2:

$$\tilde{\mathbf{I}}_S = \frac{\tilde{\mathbf{V}}_S}{Z_{\text{total}}} = \frac{\tilde{\mathbf{V}}}{R_S + R_L} = \frac{110}{4 + 10} = 7.86\angle 0\ \text{A}$$

$$S_{Sb} = \tilde{\mathbf{V}}_S\tilde{\mathbf{I}}_S^* = 110 \times 7.86 = 864 + j0\ \text{W} = P_{Sb} + jQ_{Sb}$$

and the percent real power transfer is:

$$100 \times \frac{P_b}{P_{Sb}} = \frac{617}{864} = 71.4\%$$

Comments: You can see that if it were possible to eliminate the reactive part of the impedance, the percentage of real power transferred from the source to the load would be significantly increased! A procedure that accomplishes this goal, called *power factor* correction, is discussed next.

CHECK YOUR UNDERSTANDING

Compute the change in percent of power transfer for the case where the inductance of the load is one-half of the original value.

Answer: 17.1 percent

Power Factor, Revisited

The power factor, defined earlier as the cosine of the angle of the load impedance, plays a very important role in AC power. A power factor close to unity signifies an efficient transfer of energy from the AC source to the load, while a small power factor corresponds to inefficient use of energy, as illustrated in Example 7.6. It should be apparent that if a load requires a fixed amount of real power P, the source will be providing the smallest amount of current when the power factor is the greatest, that is, when $\cos\theta = 1$. If the power factor is less than unity, some additional current will be drawn from the source, lowering the efficiency of power transfer from the source to the load. However, it will be shown shortly that it is possible to correct the power factor of a load by adding an appropriate reactive component to the load itself.

Since the reactive power Q is related to the reactive part of the load, its sign depends on whether the load reactance is inductive or capacitive. This leads to the following important statement:

LO2

If the load has an inductive reactance, then θ is positive and the current *lags* (or *follows*) the voltage. Thus, when θ and Q are positive, the corresponding power factor is termed *lagging*. Conversely, a capacitive load will have a negative Q and hence a negative θ. This corresponds to a *leading* power factor, meaning that the load current *leads* the load voltage.

Table 7.2 illustrates the concept and summarizes all the important points so far. In the table, the phasor voltage $\tilde{\mathbf{V}}$ has a zero phase angle, and the current phasor is referenced to the phase of $\tilde{\mathbf{V}}$.

Table 7.2 **Important facts related to complex power**

	Resistive load	**Capacitive load**	**Inductive load**
Ohm's law	$\tilde{\mathbf{V}}_L = Z_L\tilde{\mathbf{I}}_L$	$\tilde{\mathbf{V}}_L = Z_L\tilde{\mathbf{I}}_L$	$\tilde{\mathbf{V}}_L = Z_L\tilde{\mathbf{I}}_L$
Complex impedance	$Z_L = R_L$	$Z_L = R_L + jX_L$ $X_L < 0$	$Z_L = R_L + jX_L$ $X_L > 0$
Phase angle	$\theta = 0$	$\theta < 0$	$\theta > 0$
Complex plane sketch	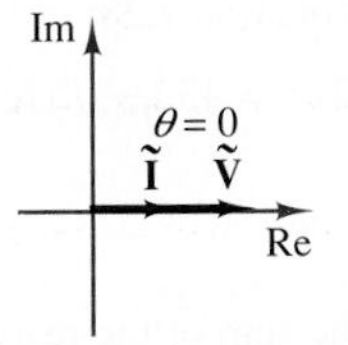	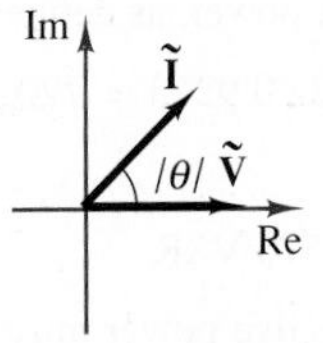	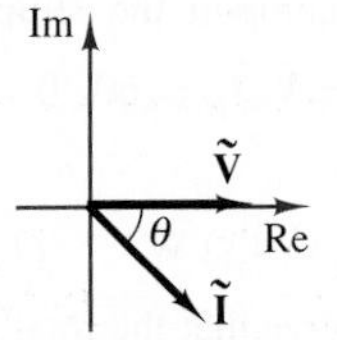
Explanation	The current is in phase with the voltage.	The current "leads" the voltage.	The current "lags" the voltage.
Power factor	Unity	Leading, < 1	Lagging, < 1
Reactive power	0	Negative	Positive

The following examples illustrate the computation of complex power for a simple circuit.

EXAMPLE 7.7 Complex Power and Power Triangle

Problem

Find the reactive and real power for the load of Figure 7.16. Draw the associated power triangle.

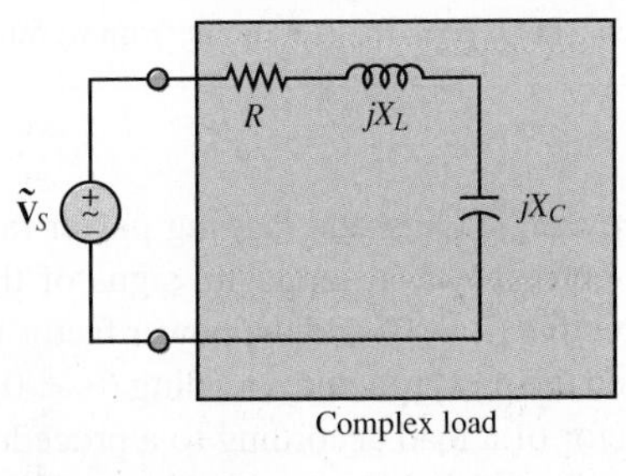

Figure 7.16

Solution

Known Quantities: Source voltage; load impedance.

Find: $S = P_{av} + jQ$ for the complex load.

Schematics, Diagrams, Circuits, and Given Data: $\tilde{\mathbf{V}}_S = 60\angle 0$ V; $R = 3\ \Omega$; $jX_L = j9\ \Omega$; $jX_C = -j5\ \Omega$.

Assumptions: Use rms values for all phasor quantities in the problem.

Analysis: First, we compute the load current:

$$\tilde{\mathbf{I}}_L = \frac{\tilde{\mathbf{V}}_L}{Z_L} = \frac{60\angle 0}{3 + j9 - j5} = \frac{60\angle 0}{5\angle 0.9273} = 12\angle(-0.9273)\ \text{A}$$

Next, we compute the complex power, as defined in equation 7.28:

$$S = \tilde{\mathbf{V}}_L\tilde{\mathbf{I}}_L^* = 60\angle 0 \times 12\angle 0.9273 = 720\angle 0.9273 = 432 + j576\ \text{VA}$$

Therefore

$$P_{\text{av}} = 432\ \text{W} \qquad Q = 576\ \text{VAR}$$

If we observe that the total reactive power must be the sum of the reactive powers in each of the elements, we can write $Q = Q_C + Q_L$ and compute each of the two quantities as follows:

$$Q_C = |\tilde{\mathbf{I}}_L|^2 \times X_C = (144)(-5) = -720\ \text{VAR}$$

$$Q_L = |\tilde{\mathbf{I}}_L|^2 \times X_L = (144)(9) = 1{,}296\ \text{VAR}$$

and

$$Q = Q_L + Q_C = 576\ \text{VAR}$$

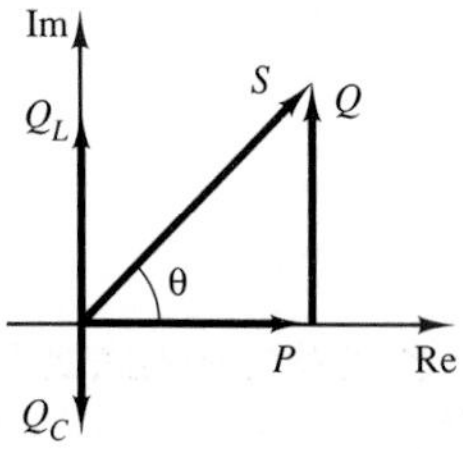

Note: $S = P_R + jQ_C + jQ_L$

Figure 7.17

Comments: The power triangle corresponding to this circuit is drawn in Figure 7.17. The vector diagram shows how the complex power S results from the vector addition of the three components P, Q_C, and Q_L.

CHECK YOUR UNDERSTANDING

Compute the power factor for the load of Example 7.7 with and without the inductor in the circuit.

Answer: $pf = 0.6$, lagging (with L in circuit); $pf = 0.5145$, leading (without L)

The distinction between leading and lagging power factors made in Table 7.2 is important, because it corresponds to opposite signs of the reactive power: Q is positive if the load is inductive ($\theta > 0$) and the power factor is lagging; Q is negative if the load is capacitive and the power factor is leading ($\theta < 0$). It is therefore possible to improve the power factor of a load according to a procedure called **power factor correction,** that is, by placing a suitable reactance in parallel with the load so that the reactive power component generated by the additional reactance is of opposite sign to the original load reactive power. Most often the need is to improve the power factor of an inductive load, because many common industrial loads consist of electric motors, which are predominantly inductive loads. This improvement may be accomplished by placing a capacitance in parallel with the load. Example 7.8 illustrates a typical power factor correction for an industrial load.

FOCUS ON METHODOLOGY

COMPLEX POWER CALCULATION FOR POWER FACTOR CORRECTION

LO2

1. Compute the load voltage and current in rms phasor form, using the AC circuit analysis methods presented in Chapter 4 and converting peak amplitude to rms values.
2. Compute the complex power $S = \tilde{\mathbf{V}}\tilde{\mathbf{I}}^*$ and set Re $S = P_{av}$, Im $S = Q$.
3. Draw the power triangle, for example, as shown in Figure 7.17.
4. Compute the power factor of the load $pf = \cos(\theta)$.
5. If the reactive power of the original load is positive (inductive load), then the power factor can be brought to unity by connecting a parallel capacitor across the load, such that $Q_C = -1/\omega C = -Q$, where Q is the reactance of the inductive load.

EXAMPLE 7.8 Power Factor Correction

LO2

Problem

Calculate the complex power for the circuit of Figure 7.18, and correct the power factor to unity by connecting a parallel reactance to the load.

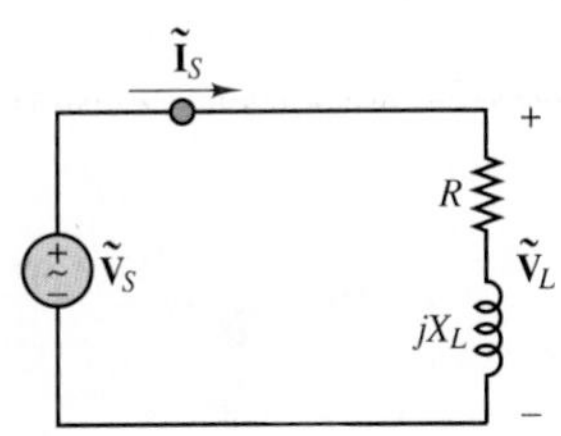

Figure 7.18

Solution

Known Quantities: Source voltage; load impedance.

Find:

1. $S = P_{av} + jQ$ for the complex load.
2. Value of parallel reactance required for power factor correction resulting in pf = 1.

Schematics, Diagrams, Circuits, and Given Data: $\tilde{\mathbf{V}}_S = 117\angle 0$ V; $R_L = 50\ \Omega$; $jX_L = j86.7\ \Omega$.

Assumptions: Use rms values for all phasor quantities in the problem.

Analysis:

1. First, we compute the load impedance:

$$Z_L = R + jX_L = 50 + j86.7 = 100\angle 1.047\ \Omega$$

Next, we compute the load current

$$\tilde{\mathbf{I}}_L = \frac{\tilde{\mathbf{V}}_L}{Z_L} = \frac{117\angle 0}{50 + j86.6} = \frac{117\angle 0}{100\angle 1.047} = 1.17\angle(-1.047)\ \text{A}$$

and the complex power, as defined in equation 7.28:

$$S = \tilde{\mathbf{V}}_L\tilde{\mathbf{I}}_L^* = 117\angle 0 \times 1.17\angle 1.047 = 137\angle 1.047 = 68.4 + j118.5\ \text{W}$$

Therefore

$$P_{\text{av}} = 68.4 \text{ W} \qquad Q = 118.5 \text{ VAR}$$

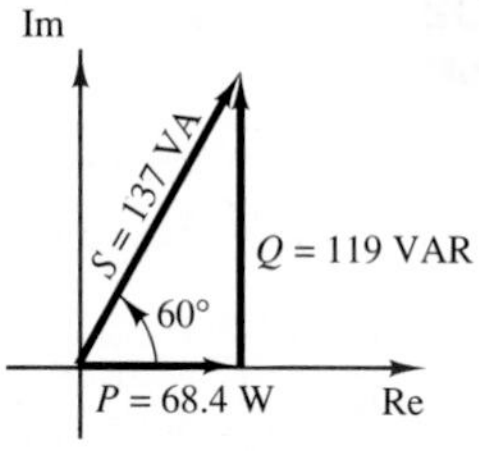

Figure 7.19

The power triangle corresponding to this circuit is drawn in Figure 7.19. The vector diagram shows how the complex power S results from the vector addition of the two components P and Q_L. To eliminate the reactive power due to the inductance, we will need to add an equal and opposite reactive power component $-Q_L$, as described below.

2. To compute the reactance needed for the power factor correction, we observe that we need to contribute a negative reactive power equal to -118.5 VAR. This requires a negative reactance and therefore a capacitor with $Q_C = -118.5$ VAR. The reactance of such a capacitor is given by

$$X_C = \frac{|\tilde{\mathbf{V}}_L|^2}{Q_C} = -\frac{(117)^2}{118.5} = -115\ \Omega$$

and since

$$C = -\frac{1}{\omega X_C}$$

we have

$$C = -\frac{1}{\omega X_C} = -\frac{1}{377(-115)} = 23.1\ \mu\text{F}$$

Comments: The power factor correction is illustrated in Figure 7.20. You can see that it is possible to eliminate the reactive part of the impedance, thus significantly increasing the percentage of real power transferred from the source to the load. Power factor correction is a very common procedure in electric power systems.

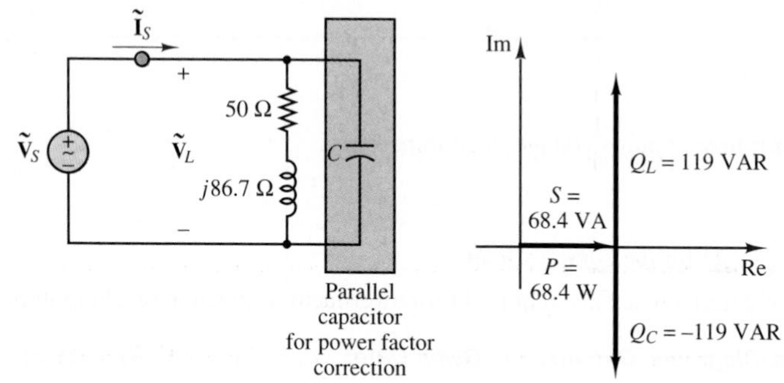

Figure 7.20 Power factor correction

CHECK YOUR UNDERSTANDING

Compute the magnitude of the current drawn by the source after the power factor correction in Example 7.8.

Answer: 0.584 A

EXAMPLE 7.9 Can a Series Capacitor Be Used for Power Factor Correction?

LO2

Problem

The circuit of Figure 7.21 proposes the use of a series capacitor to perform power factor correction. Show why this is *not* a feasible alternative to the parallel capacitor approach demonstrated in Example 7.8.

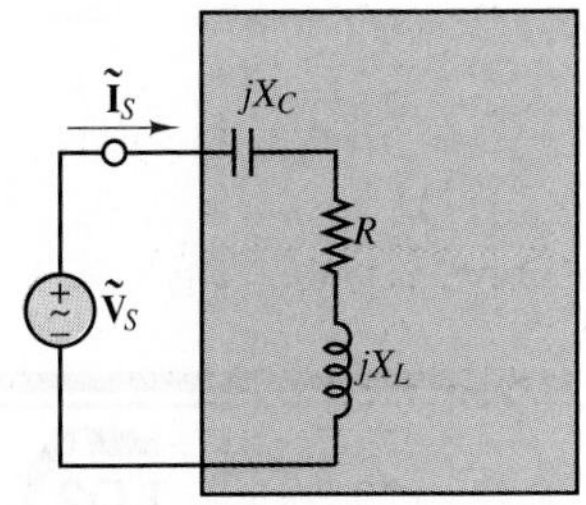

Figure 7.21

Solution

Known Quantities: Source voltage; load impedance.

Find: Load (source) current.

Schematics, Diagrams, Circuits, and Given Data: $\tilde{\mathbf{V}}_S = 117\angle 0$ V; $R_L = 50\ \Omega$; $jX_L = j86.7\ \Omega$; $jX_C = -j86.7\ \Omega$.

Assumptions: Use rms values for all phasor quantities in the problem.

Analysis: To determine the feasibility of the approach, we compute the load current and voltage, to observe any differences between the circuit of Figure 7.21 and that of Figure 7.20. First, we compute the load impedance:

$$Z_L = R + jX_L - jX_C = 50 + j86.7 - j86.7 = 50\ \Omega$$

Next, we compute the load (source) current:

$$\tilde{\mathbf{I}}_L = \tilde{\mathbf{I}}_S = \frac{\tilde{\mathbf{V}}_L}{Z_L} = \frac{117\angle 0}{50} = 2.34\ \text{A}$$

Comments: Note that a twofold increase in the series current results from the addition of the series capacitor. This would result in a doubling of the power required by the generator, with respect to the solution found in Example 7.8. Further, in practice, the parallel connection is much easier to accomplish, since a parallel element can be added externally, without the need for breaking the circuit.

CHECK YOUR UNDERSTANDING

Determine the power factor of the load for each of the following two cases, and whether it is leading or lagging.

a. $v(t) = 540\cos(\omega t + 15°)$ V, $i(t) = 2\cos(\omega t + 47°)$ A

b. $v(t) = 155\cos(\omega t - 15°)$ V, $i(t) = 2\cos(\omega t - 22°)$ A

Answer: a. 0.848, leading; b. 0.9925, lagging

The measurement and correction of the power factor for the load are an extremely important aspect of any engineering application in industry that requires the use of substantial quantities of electric power. In particular, industrial plants, construction sites, heavy machinery, and other heavy users of electric power must be aware of the power factor that their loads present to the electric utility company. As was already observed, a low power factor results in greater current draw from the electric utility and greater line losses. Thus, computations related to the power factor of complex loads are of great utility to any practicing engineer. To provide you with deeper insight into calculations related to power factor, a few more advanced examples are given in the remainder of the section.

LO2

EXAMPLE 7.10 Power Factor Correction

Problem

A capacitor is used to correct the power factor of the load of Figure 7.22. Determine the reactive power when the capacitor is not in the circuit, and compute the required value of capacitance for perfect pf correction.

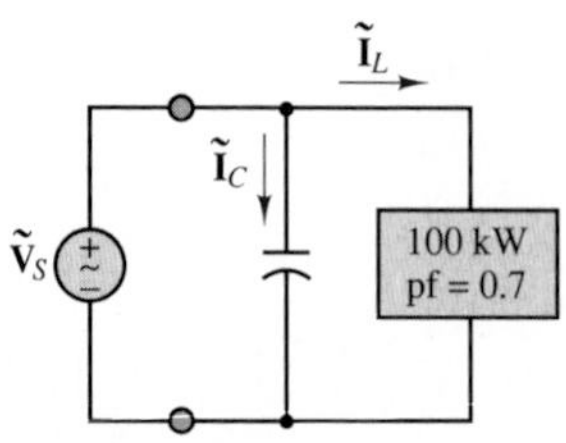

Figure 7.22

Solution

Known Quantities: Source voltage; load power and power factor.

Find:

1. Q when the capacitor is not in the circuit.
2. Value of capacitor required for power factor correction resulting in pf $= 1$.

Schematics, Diagrams, Circuits, and Given Data: $\tilde{\mathbf{V}}_S = 480\angle 0$; $P = 10^5$ W; pf $= 0.7$ lagging.

Assumptions: Use rms values for all phasor quantities in the problem.

Analysis:

1. With reference to the power triangle of Figure 7.11, we can compute the reactive power of the load from knowledge of the real power and of the power factor, as shown below:

$$|S| = \frac{P}{\cos(\theta)} = \frac{P}{\text{pf}} = \frac{10^5}{0.7} = 1.429 \times 10^5 \text{ VA}$$

Since the power factor is lagging, we know that the reactive power is positive (see Table 7.2), and we can calculate Q as shown below:

$$Q = |S| \sin(\theta) \qquad \theta = \arccos(\text{pf}) = 0.795$$

$$Q = 1.429 \times 10^5 \times \sin(0.795) = 102 \text{ kVAR}$$

2. To compute the reactance needed for the power factor correction, we observe that we need to contribute a negative reactive power equal to -102 kVAR. This requires a negative reactance and therefore a capacitor with $Q_C = -102$ kVAR. The reactance of such a capacitor is given by

$$X_C = \frac{|\tilde{\mathbf{V}}_L|^2}{Q_C} = \frac{(480)^2}{-102 \times 10^5} = -2.258$$

and since

$$C = -\frac{1}{\omega X_C}$$

we have

$$C = -\frac{1}{\omega X_C} = -\frac{1}{377 \times (-2.258)} = 1{,}175\ \mu\text{F}$$

Comments: Note that it is not necessary to know the load impedance to perform power factor correction; it is sufficient to know the *apparent* power and the power factor.

CHECK YOUR UNDERSTANDING

Determine if a load is capacitive or inductive, given the following facts:

a. pf $= 0.87$, leading

b. pf $= 0.42$, leading

c. $v(t) = 42\cos(\omega t)$ V, $i(t) = 4.2\sin(\omega t)$ A

d. $v(t) = 10.4\cos(\omega t - 22°)$ V, $i(t) = 0.4\cos(\omega t - 22°)$ A

Answer: a. Capacitive; b. capacitive; c. inductive; d. neither (resistive)

EXAMPLE 7.11 Power Factor Correction

LO2

Problem

A second load is added to the circuit of Figure 7.22, as shown in Figure 7.23. Determine the required value of capacitance for perfect pf correction after the second load is added. Draw the phasor diagram showing the relationship between the two load currents and the capacitor current.

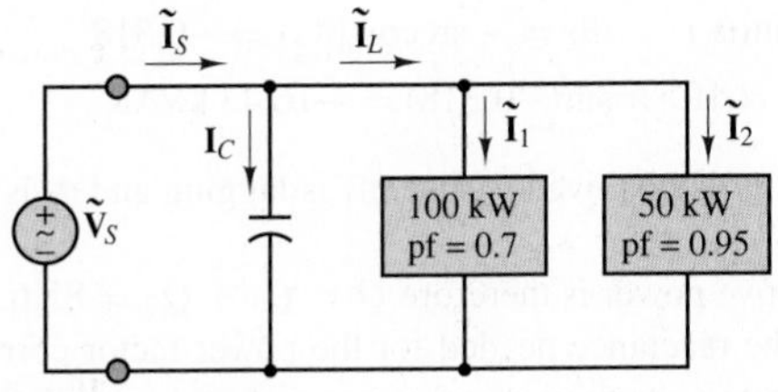

Figure 7.23

Solution

Known Quantities: Source voltage; load power and power factor.

Find:

1. Power factor correction capacitor.
2. Phasor diagram.

Schematics, Diagrams, Circuits, and Given Data: $\tilde{\mathbf{V}}_S = 480\angle 0$ V; $P_1 = 10^5$ W; $\mathrm{pf}_1 = 0.7$ lagging; $P_2 = 5 \times 10^4$ W; $\mathrm{pf}_2 = 0.95$ leading; $\omega = 377$ rad/s.

Assumptions: Use rms values for all phasor quantities in the problem.

Analysis:

1. We first compute the two load currents, using the relationships given in equations 7.28 and 7.29:

$$P = |\tilde{\mathbf{V}}_S||\tilde{\mathbf{I}}_1^*|\cos(\theta_1)$$

$$|\tilde{\mathbf{I}}_1^*| = \frac{P_1}{|\tilde{\mathbf{V}}_S|\cos(\theta_1)}$$

$$\tilde{\mathbf{I}}_1 = \frac{P_1}{|\tilde{\mathbf{V}}_S|\mathrm{pf}_1}\angle - \arccos(\mathrm{pf}_1) = \frac{10^5}{480 \times 0.7}\angle - \arccos(0.7)$$

$$= 298\angle(-0.795)\ \mathrm{A}$$

and similarly

$$\tilde{\mathbf{I}}_2 = \frac{P_2}{|\tilde{\mathbf{V}}_S|\mathrm{pf}_2}\angle - \arccos(\mathrm{pf}_2) = \frac{5 \times 10^4}{480 \times 0.95}\angle - \arccos(0.95)$$

$$= 110\angle(-0.318)\ \mathrm{A}$$

where we have selected the positive value of arccos (pf_1) because pf_1 is lagging, and the negative value of arccos (pf_2) because pf_2 is leading. Now we compute the apparent power at each load:

$$|S_1| = \frac{P_1}{\mathrm{pf}_1} = \frac{P_1}{\cos(\theta_1)} = \frac{10^5}{0.7} = 1.429 \times 10^5\ \mathrm{VA}$$

$$|S_2| = \frac{P_2}{\mathrm{pf}_2} = \frac{P_2}{\cos(\theta_2)} = \frac{5 \times 10^4}{0.95} = 5.263 \times 10^4\ \mathrm{VA}$$

and from these values we can calculate Q as shown:

$$Q_1 = |S_1|\sin(\theta_1) \qquad \theta_1 = \arccos(\mathrm{pf}_1) = 0.795$$

$$Q_1 = 1.429 \times 10^5 \times \sin(0.795) = 102\ \mathrm{kVAR}$$

$$Q_2 = |S_2|\sin(\theta_2) \qquad \theta_2 = -\arccos(\mathrm{pf}_2) = -0.318$$

$$Q_2 = 5.263 \times 10^4 \times \sin(-0.318) = -16.43\ \mathrm{kVAR}$$

where, once again, θ_1 is positive because pf_1 is lagging and θ_2 is negative because pf_2 is leading (see Table 7.2).

The total reactive power is therefore $Q = Q_1 + Q_2 = 85.6$ kVAR.

To compute the reactance needed for the power factor correction, we observe that we need to contribute a negative reactive power equal to -85.6 kVAR. This requires a negative reactance and therefore a capacitor with $Q_C = -85.6$ kVAR. The reactance of such a capacitor is given by

$$X_C = \frac{|\tilde{\mathbf{V}}_S|^2}{Q_C} = \frac{(480)^2}{-85.6 \times 10^5} = -2.694$$

and since

$$C = -\frac{1}{\omega X_C}$$

we have

$$C = \frac{1}{\omega X_C} = -\frac{1}{377(-2.692)} = 984.6\ \mu\text{F}$$

2. To draw the phasor diagram, we need only to compute the capacitor current, since we have already computed the other two:

$$Z_C = jX_C = -j2.692\ \Omega$$

$$\tilde{\mathbf{I}}_C = \frac{\tilde{\mathbf{V}}_S}{Z_C} = 178.2\angle\frac{\pi}{2}\ \text{A}$$

The total current is $\tilde{\mathbf{I}}_S = \tilde{\mathbf{I}}_1 + \tilde{\mathbf{I}}_2 + \tilde{\mathbf{I}}_C = 312.5\angle 0°$ A. The phasor diagram corresponding to these three currents is shown in Figure 7.24.

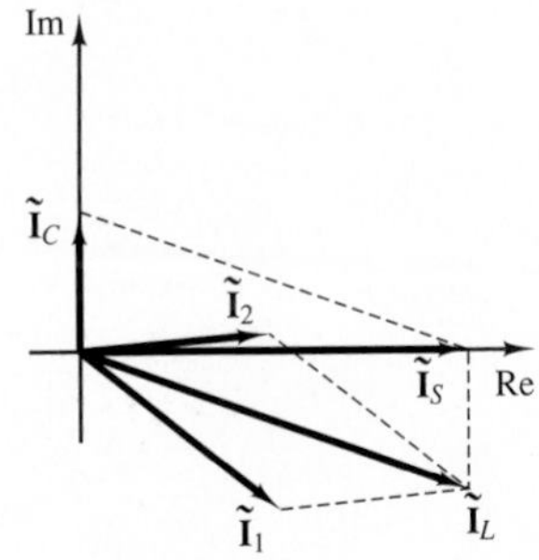

Figure 7.24

CHECK YOUR UNDERSTANDING

Compute the power factor for an inductive load with $L = 100$ mH and $R = 0.4\ \Omega$.

Answer: pf = 0.0105, lagging

The Wattmeter LO2

FIND IT ON THE WEB

The instrument used to measure power is called a **wattmeter.** The external part of a wattmeter consists of four connections and a metering mechanism that displays the amount of real power dissipated by a circuit. The external and internal appearance of a wattmeter is depicted in Figure 7.25. Inside the wattmeter are two coils: a current-sensing coil and a voltage-sensing coil. In this example, we assume for simplicity that the impedance of the current-sensing coil Z_I is zero and that the impedance of the voltage-sensing coil Z_V is infinite. In practice, this will not necessarily be true; some correction mechanism will be required to account for the impedance of the sensing coils.

A wattmeter should be connected as shown in Figure 7.26, to provide both current and voltage measurements. We see that the current-sensing coil is placed in series with the load and that the voltage-sensing coil is placed in parallel with the load. In this

(Continued)

manner, the wattmeter is seeing the current through and the voltage across the load. Remember that the power dissipated by a circuit element is related to these two quantities. The wattmeter, then, is constructed to provide a readout of the real power absorbed by the load: $P = \mathrm{Re}(S) = \mathrm{Re}(\mathbf{VI}^*)$.

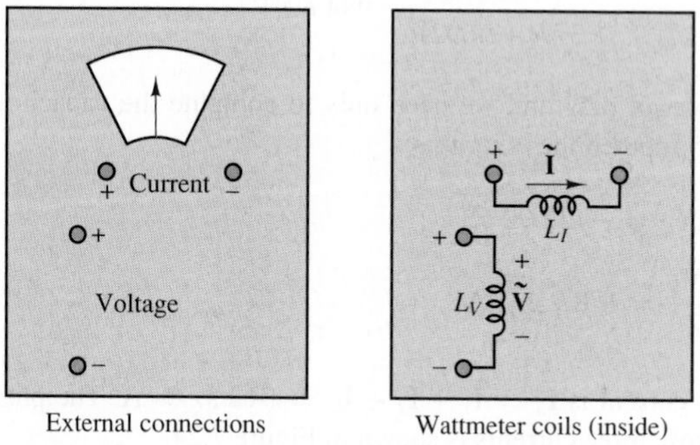

Figure 7.25

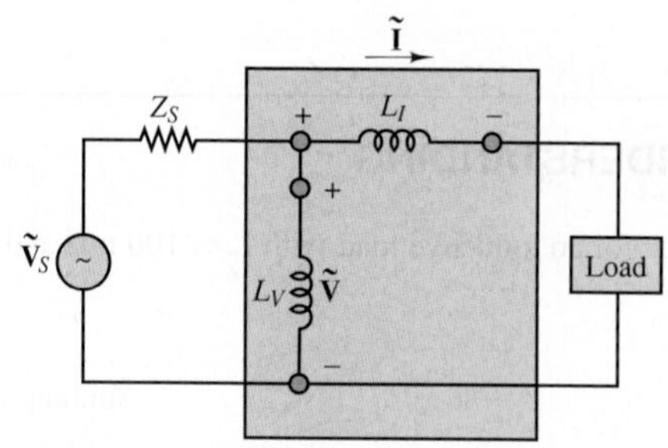

Figure 7.26

Problem:

1. For the circuit shown in Figure 7.27, show the connections of the wattmeter, and find the power dissipated by the load.
2. Show the connections that will determine the power dissipated by R_2. What should the meter read?

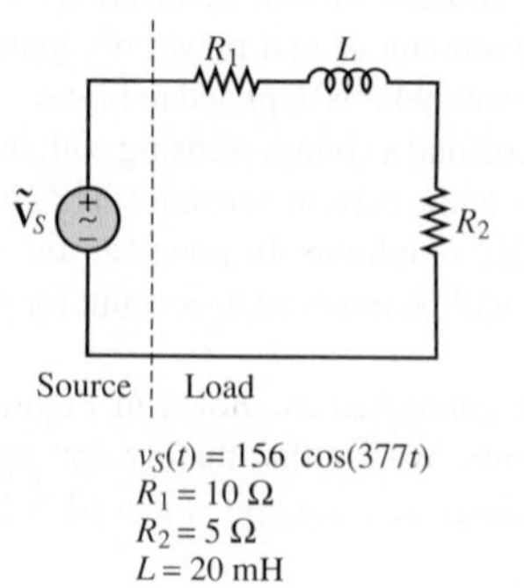

Figure 7.27

(Continued)

(*Concluded*)

Solution:

1. To measure the power dissipated by the load, we must know the current through and the voltage across the entire load circuit. This means that the wattmeter must be connected as shown in Figure 7.28. The wattmeter should read

$$P = \text{Re}(\tilde{\mathbf{V}}_S\tilde{\mathbf{I}}^*) = \text{Re}\left[\left(\frac{156}{\sqrt{2}}\angle 0\right)\left(\frac{(156/\sqrt{2})\angle 0}{R_1 + R_2 + j\omega L}\right)^*\right]$$

$$= \text{Re}\left[110\angle 0^\circ\left(\frac{110\angle 0}{15 + j7.54}\right)^*\right]$$

$$= \text{Re}\left[110\angle 0^\circ\left(\frac{110\angle 0}{16.79\angle 0.466}\right)^*\right] = \text{Re}\,\frac{110^2}{16.79\angle(-0.466)}$$

$$= \text{Re}\,(720.67\angle 0.466)$$

$$= 643.88\ \text{W}$$

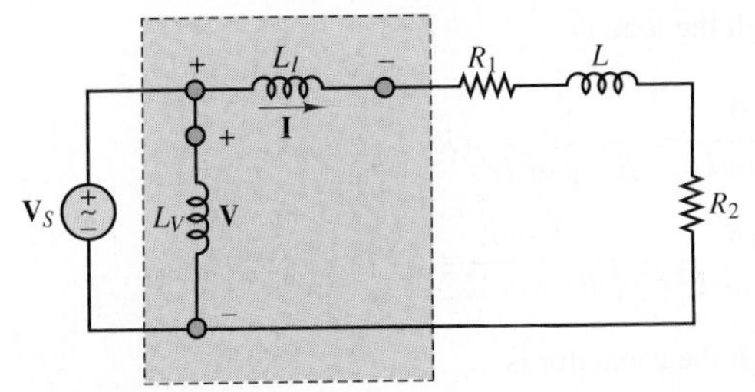

Figure 7.28

2. To measure the power dissipated by R_2 alone, we must measure the current through R_2 and the voltage across R_2 *alone*. The connection is shown in Figure 7.29. The meter will read

$$P = \tilde{I}^2 R_2 = \left[\frac{110}{(15^2 + 7.54^2)^{1/2}}\right]^2 \times 5 = \frac{110^2}{15^2 + 7.54^2} \times 5$$

$$= 215\ \text{W}$$

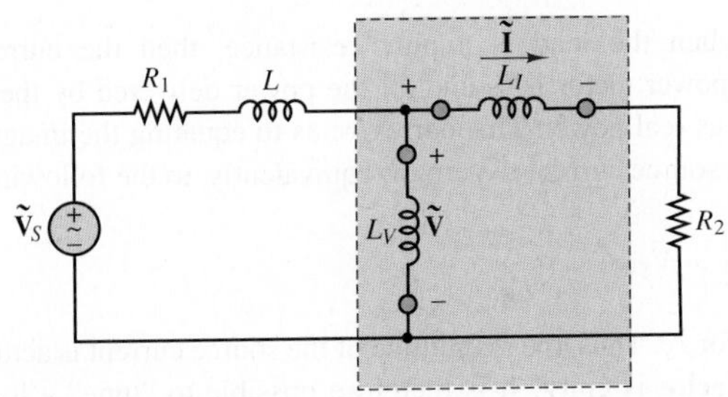

Figure 7.29

FOCUS ON MEASUREMENTS

Power Factor

Problem:

A capacitor is being used to correct the power factor to unity. The circuit is shown in Figure 7.30. The capacitor value is varied, and measurements of the total current are taken. Explain how it is possible to zero in on the capacitance value necessary to bring the power factor to unity just by monitoring the current $\tilde{\mathbf{I}}_S$.

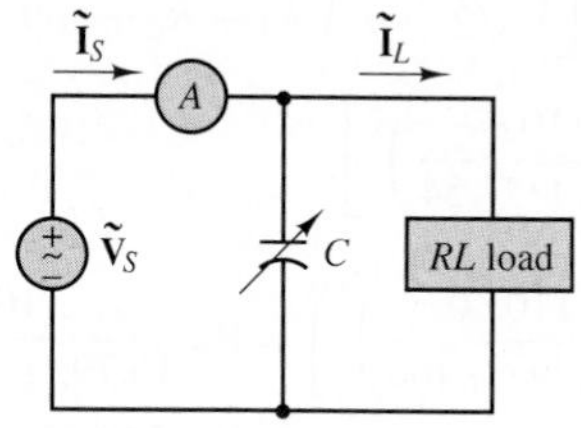

Figure 7.30

Solution:

The current through the load is

$$\tilde{\mathbf{I}}_L = \frac{\tilde{V}_S\angle 0°}{R + j\omega L} = \frac{\tilde{V}_S}{R^2 + \omega^2 L^2}(R - j\omega L)$$

$$= \frac{\tilde{V}_S R}{R^2 + \omega^2 L^2} - j\frac{\tilde{V}_S \omega L}{R^2 + \omega^2 L^2}$$

The current through the capacitor is

$$\tilde{\mathbf{I}}_C = \frac{\tilde{V}_S\angle 0°}{1/j\omega C} = j\tilde{V}_S\omega C$$

The source current to be measured is

$$\tilde{\mathbf{I}}_S = \tilde{\mathbf{I}}_L + \tilde{\mathbf{I}}_C = \frac{\tilde{V}_S R}{R^2 + \omega^2 L^2} + j\left(\tilde{V}_S\omega C - \frac{\tilde{V}_S \omega L}{R^2 + \omega^2 L^2}\right)$$

The magnitude of the source current is

$$\tilde{I}_S = \sqrt{\left(\frac{\tilde{V}_S R}{R^2 + \omega^2 L^2}\right)^2 + \left(\tilde{V}_S\omega C - \frac{\tilde{V}_S \omega L}{R^2 + \omega^2 L^2}\right)^2}$$

We know that when the load is a pure resistance, then the current and voltage are in phase, the power factor is 1, and all the power delivered by the source is dissipated by the load as real power. This corresponds to equating the imaginary part of the expression for the source current to zero or, equivalently, to the following expression:

$$\frac{\tilde{V}_S \omega L}{R^2 + \omega^2 L^2} = \tilde{V}_S\omega C$$

in the expression for $\tilde{I}_S$. Thus, the magnitude of the source current is actually a minimum when the power factor is unity! It is therefore possible to "tune" a load to a unity pf by observing the readout of the ammeter while changing the value of the capacitor and selecting the capacitor value that corresponds to the lowest source current value.

7.3 TRANSFORMERS

AC circuits are very commonly connected to each other by means of **transformers.** A transformer is a device that couples two AC circuits magnetically rather than through any direct conductive connection and permits a "transformation" of the voltage and current between one circuit and the other (e.g., by matching a high-voltage, low-current AC output to a circuit requiring a low-voltage, high-current source). Transformers play a major role in electric power engineering and are a necessary part of the electric power distribution network. The objective of this section is to introduce the ideal transformer and the concepts of impedance reflection and impedance matching. The physical operations of practical transformers, and more advanced models, is discussed in Chapter 16.

The Ideal Transformer

The ideal transformer consists of two coils that are coupled to each other by some magnetic medium. There is no electrical connection between the coils. The coil on the input side is termed the **primary,** and that on the output side the **secondary.** The primary coil is wound so that it has n_1 turns, while the secondary has n_2 turns. We define the **turns ratio** N as

$$N = \frac{n_2}{n_1} \tag{7.32}$$

Figure 7.31 illustrates the convention by which voltages and currents are usually assigned at a transformer. The dots in Figure 7.31 are related to the polarity of the coil voltage: coil terminals marked with a dot have the same polarity.

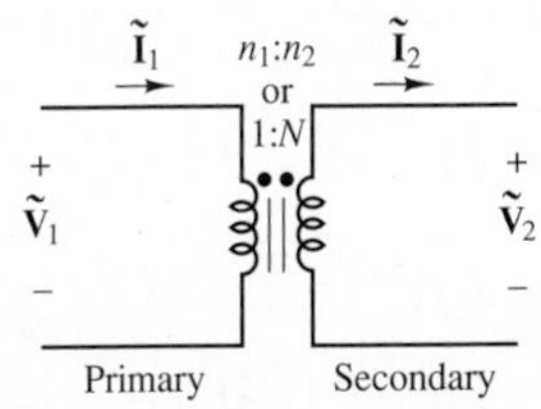

Figure 7.31 Ideal transformer

Since an ideal inductor acts as a short circuit in the presence of DC, transformers do not perform any useful function when the primary voltage is DC. However, when a time-varying current flows in the primary winding, a corresponding time-varying voltage is generated in the secondary because of the magnetic coupling between the two coils. This behavior is due to Faraday's law, as explained in Chapter 16. The relationship between primary and secondary current in an ideal transformer is very simply stated as follows:

$$\begin{aligned} \tilde{\mathbf{V}}_2 &= N\tilde{\mathbf{V}}_1 \\ \tilde{\mathbf{I}}_2 &= \frac{\tilde{\mathbf{I}}_1}{N} \end{aligned} \qquad \text{Ideal transformer} \tag{7.33}$$

> An ideal transformer multiplies a sinusoidal input voltage by a factor of N and divides a sinusoidal input current by a factor of N.

If N is greater than 1, the output voltage is greater than the input voltage and the transformer is called a **step-up transformer.** If N is less than 1, then the transformer is called a **step-down transformer,** since $\tilde{\mathbf{V}}_2$ is now smaller than $\tilde{\mathbf{V}}_1$. An ideal transformer can be used in either direction (i.e., either of its coils may be viewed as the input side, or primary). Finally, a transformer with $N = 1$ is called an **isolation transformer** and may perform a very useful function if one needs to electrically isolate two circuits from each other; note that any DC at the primary will not appear at the secondary coil. An important property of ideal transformers is the conservation of power; one can easily verify that an ideal transformer conserves power, since

$$S_1 = \tilde{\mathbf{I}}_1^* \tilde{\mathbf{V}}_1 = N \tilde{\mathbf{I}}_2^* \frac{\tilde{\mathbf{V}}_2}{N} = \tilde{\mathbf{I}}_2^* \tilde{\mathbf{V}}_2 = S_2 \tag{7.34}$$

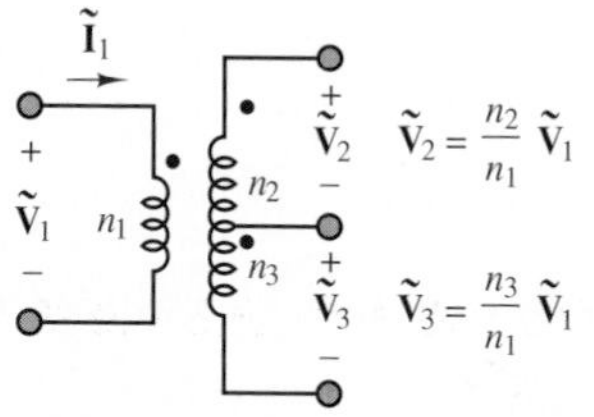

Figure 7.32 Center-tapped transformer

That is, the power on the primary side equals that on the secondary.

In many practical circuits, the secondary is tapped at two different points, giving rise to two separate output circuits, as shown in Figure 7.32. The most common configuration is the **center-tapped transformer,** which splits the secondary voltage into two equal voltages. The most common occurrence of this type of transformer is found at the entry of a power line into a household, where a high-voltage primary (see Figure 7.58) is transformed to 240 V and split into two 120-V lines. Thus, $\tilde{\mathbf{V}}_2$ and $\tilde{\mathbf{V}}_3$ in Figure 7.32 are both 120-V lines, and a 240-V line $(\tilde{\mathbf{V}}_2 + \tilde{\mathbf{V}}_3)$ is also available.

LO3

EXAMPLE 7.12 Ideal Transformer Turns Ratio

Problem

We require a transformer to deliver 500 mA at 24 V from a 120-V rms line source. How many turns are required in the secondary? What is the primary current?

Solution

Known Quantities: Primary and secondary voltages; secondary current; number of turns in the primary coil.

Find: n_2 and $\tilde{\mathbf{I}}_1$.

Schematics, Diagrams, Circuits, and Given Data: $\tilde{V}_1 = 120$ V; $\tilde{V}_2 = 24$ V; $\tilde{I}_2 = 500$ mA; $n_1 = 3{,}000$ turns.

Assumptions: Use rms values for all phasor quantities in the problem.

Analysis: Using equation 7.33, we compute the number of turns in the secondary coil as follows:

$$\frac{\tilde{V}_1}{n_1} = \frac{\tilde{V}_2}{n_2} \qquad n_2 = n_1 \frac{\tilde{V}_2}{\tilde{V}_1} = 3{,}000 \times \frac{24}{120} = 600 \text{ turns}$$

Knowing the number of turns, we can now compute the primary current, also from equation 7.33:

$$n_1 \tilde{I}_1 = n_2 \tilde{I}_2 \qquad \tilde{I}_1 = \frac{n_2}{n_1} \tilde{I}_2 = \frac{600}{3{,}000} \times 500 = 100 \text{ mA}$$

Comments: Note that since the transformer does not affect the phase of the voltages and currents, we could solve the problem by using simply the rms amplitudes.

CHECK YOUR UNDERSTANDING

With reference to Example 7.12, compute the number of primary turns required if $n_2 = 600$ but the transformer is required to deliver 1 A. What is the primary current now?

Answer: $n_1 = 3{,}000$; $\tilde{I}_1 = 200$ mA

EXAMPLE 7.13 Center-Tapped Transformer

LO3

Problem

A center-tapped power transformer has a primary voltage of 4,800 V and two 120-V secondaries (see Figure 7.32). Three loads (all resistive, i.e., with unity power factor) are connected to the transformer. The first load, R_1, is connected across the 240-V line (the two outside taps in Figure 7.32). The second and third loads, R_2 and R_3, are connected across each of the 120-V lines. Compute the current in the primary if the power absorbed by the three loads is known.

Solution

Known Quantities: Primary and secondary voltages; load power ratings.

Find: $\tilde{I}_{\text{primary}}$.

Schematics, Diagrams, Circuits, and Given Data: $\tilde{V}_1 = 4{,}800$ V; $\tilde{V}_2 = 120$ V; $\tilde{V}_3 = 120$ V; $P_1 = 5{,}000$ W; $P_2 = 1{,}000$ W; $P_3 = 1{,}500$ W.

Assumptions: Use rms values for all phasor quantities in the problem.

Analysis: Since we have no information about the number of windings or about the secondary current, we cannot solve this problem by using equation 7.33. An alternative approach is to apply conservation of power (equation 7.34). Since the loads all have unity power factor, the voltages and currents will all be in phase, and we can use the rms amplitudes in our calculations:

$$|S_{\text{primary}}| = |S_{\text{secondary}}|$$

or

$$\tilde{V}_{\text{primary}} \times \tilde{I}_{\text{primary}} = P_{\text{secondary}} = P_1 + P_2 + P_3$$

Thus,

$$4{,}800 \times \tilde{I}_{\text{primary}} = 5{,}000 + 1{,}000 + 1{,}500 = 7{,}500 \text{ W}$$

$$\tilde{I}_{\text{primary}} = \frac{7{,}500 \text{ W}}{4{,}800 \text{ A}} = 1.5625 \text{ A}$$

CHECK YOUR UNDERSTANDING

If the transformer of Example 7.13 has 300 turns in the secondary coil, how many turns will the primary require?

Answer: $n_2 = 12{,}000$

Impedance Reflection and Power Transfer

As stated in the preceding paragraphs, transformers are commonly used to couple one AC circuit to another. A very common and rather general situation is that depicted in Figure 7.33, where an AC source, represented by its Thévenin equivalent, is connected to an equivalent load impedance by means of a transformer.

It should be apparent that expressing the circuit in phasor form does not alter the basic properties of the ideal transformer, as illustrated in the following equations:

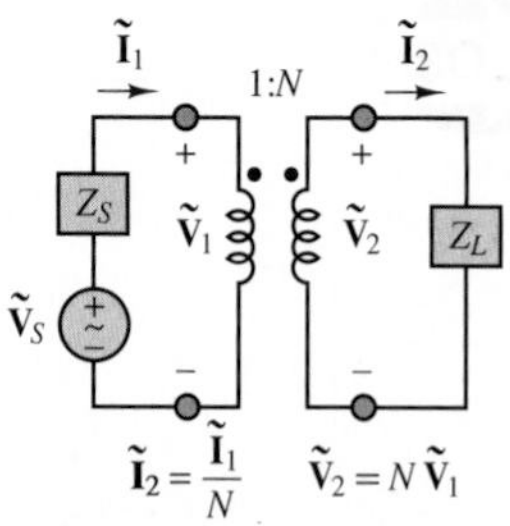

Figure 7.33 Operation of an ideal transformer

$$\tilde{\mathbf{V}}_1 = \frac{\tilde{\mathbf{V}}_2}{N} \qquad \tilde{\mathbf{I}}_1 = N\tilde{\mathbf{I}}_2$$
$$\tilde{\mathbf{V}}_2 = N\tilde{\mathbf{V}}_1 \qquad \tilde{\mathbf{I}}_2 = \frac{\tilde{\mathbf{I}}_1}{N} \tag{7.35}$$

These expressions are very useful in determining the equivalent impedance seen by the source and by the load, on opposite sides of the transformer. At the primary connection, the equivalent impedance seen by the source must equal the ratio of $\tilde{\mathbf{V}}_1$ to $\tilde{\mathbf{I}}_1$

$$Z' = \frac{\tilde{\mathbf{V}}_1}{\tilde{\mathbf{I}}_1} \tag{7.36}$$

which can be written as

$$Z' = \frac{\tilde{\mathbf{V}}_2/N}{N\tilde{\mathbf{I}}_2} = \frac{1}{N^2}\frac{\tilde{\mathbf{V}}_2}{\tilde{\mathbf{I}}_2} \tag{7.37}$$

But the ratio $\tilde{\mathbf{V}}_2/\tilde{\mathbf{I}}_2$ is, by definition, the load impedance Z_L. Thus,

$$Z' = \frac{1}{N^2}Z_L \tag{7.38}$$

That is, the AC source "sees" the load impedance reduced by a factor of $1/N^2$.

The load impedance also sees an equivalent source. The open-circuit voltage is given by

$$\tilde{\mathbf{V}}_{OC} = N\tilde{\mathbf{V}}_1 = N\tilde{\mathbf{V}}_S \tag{7.39}$$

since there is no voltage drop across the source impedance in the circuit of Figure 7.33. The short-circuit current is given by

$$\tilde{\mathbf{I}}_{SC} = \frac{\tilde{\mathbf{V}}_S}{Z_S}\frac{1}{N} \tag{7.40}$$

and the load sees a Thévenin impedance equal to

$$Z'' = \frac{\tilde{\mathbf{V}}_{OC}}{\tilde{\mathbf{I}}_{SC}} = \frac{N\tilde{\mathbf{V}}_S}{(\tilde{\mathbf{V}}_S/Z_S)(1/N)} = N^2 Z_S \tag{7.41}$$

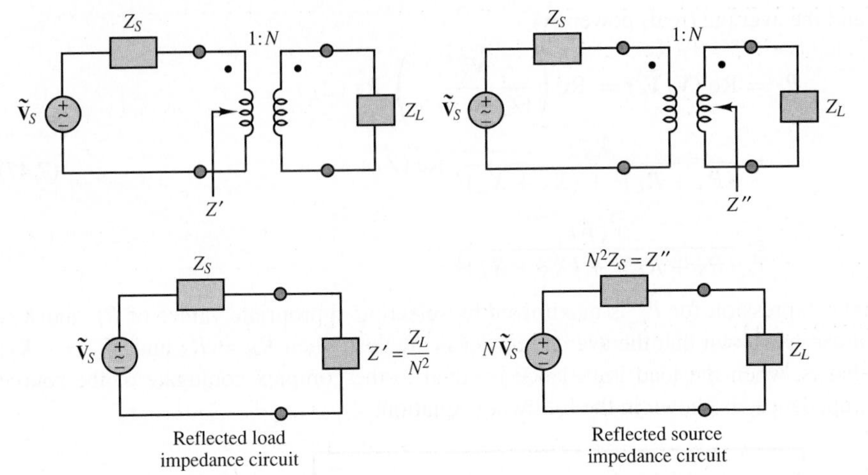

Figure 7.34 Impedance reflection across a transformer

Thus the load sees the source impedance multiplied by a factor of N^2. Figure 7.34 illustrates this **impedance reflection** across a transformer. It is very important to note that an ideal transformer changes the magnitude of the load impedance seen by the source by a factor of $1/N^2$. This property naturally leads to the discussion of power transfer, which we consider next.

Recall that in DC circuits, given a fixed equivalent source, maximum power is transferred to a resistive load when the latter is equal to the internal resistance of the source; achieving an analogous maximum power transfer condition in an AC circuit is referred to as **impedance matching.** Consider the general form of an AC circuit, shown in Figure 7.35, and assume that the source impedance Z_S is given by

$$Z_S = R_S + jX_S \tag{7.42}$$

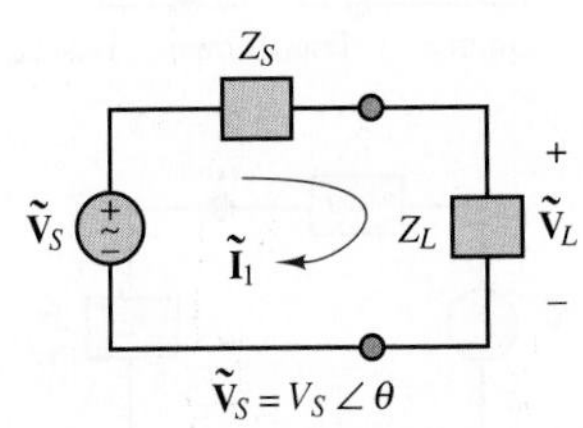

Figure 7.35 The maximum power transfer problem in AC circuits

The problem of interest is often that of selecting the load resistance and reactance that will maximize the real (average) power absorbed by the load. Note that the requirement is to maximize the real power absorbed by the load. Thus, the problem can be restated by expressing the real load power in terms of the impedance of the source and load. The real power absorbed by the load is

$$P_L = \tilde{V}_L \tilde{I}_L \cos\theta = \text{Re}\,(\tilde{\mathbf{V}}_L \tilde{\mathbf{I}}_L^*) \tag{7.43}$$

where

$$\tilde{\mathbf{V}}_L = \frac{Z_L}{Z_S + Z_L}\tilde{\mathbf{V}}_S \tag{7.44}$$

and

$$\tilde{\mathbf{I}}_L^* = \left(\frac{\tilde{\mathbf{V}}_S}{Z_S + Z_L}\right)^* = \frac{\tilde{\mathbf{V}}_S^*}{(Z_S + Z_L)^*} \tag{7.45}$$

Thus, the complex load power is given by

$$S_L = \tilde{\mathbf{V}}_L \tilde{\mathbf{I}}_L^* = \frac{Z_L \tilde{\mathbf{V}}_S}{Z_S + Z_L} \times \frac{\tilde{\mathbf{V}}_S^*}{(Z_S + Z_L)^*} = \frac{\tilde{\mathbf{V}}_S^2}{|Z_S + Z_L|^2} Z_L \tag{7.46}$$

and the average (real) power by

$$P_L = \mathrm{Re}\,(\tilde{\mathbf{V}}_L \tilde{\mathbf{I}}_L^*) = \mathrm{Re}\left(\frac{\tilde{\mathbf{V}}_S^2}{|Z_S + Z_L|^2}\right)\mathrm{Re}\,(Z_L)$$

$$= \frac{\tilde{V}_S^2}{(R_S + R_L)^2 + (X_S + X_L)^2}\mathrm{Re}\,(Z_L) \qquad \textbf{(7.47)}$$

$$= \frac{\tilde{V}_S^2 R_L}{(R_S + R_L)^2 + (X_S + X_L)^2}$$

The expression for P_L is maximized by selecting appropriate values of R_L and X_L; it can be shown that the average power is greatest when $R_L = R_S$ and $X_L = -X_S$, that is, when the load impedance is equal to the complex conjugate of the source impedance, as shown in the following equation:

$$Z_L = Z_S^* \qquad \text{Maximum power transfer}$$
that is,
$$R_L = R_S \qquad X_L = -X_S \qquad \textbf{(7.48)}$$

When the load impedance is equal to the complex conjugate of the source impedance, the load and source impedances are matched and maximum power is transferred to the load.

LO3

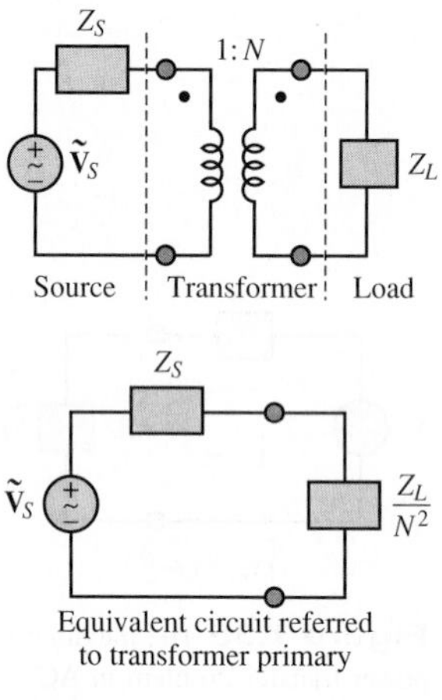

Figure 7.36 Maximum power transfer in an AC circuit with a transformer

In many cases, it may not be possible to select a matched load impedance, because of physical limitations in the selection of appropriate components. In these situations, it is possible to use the impedance reflection properties of a transformer to maximize the transfer of AC power to the load. The circuit of Figure 7.36 illustrates how the reflected load impedance, as seen by the source, is equal to Z_L/N^2, so that maximum power transfer occurs when

$$\frac{Z_L}{N^2} = Z_S^*$$

$$R_L = N^2 R_S \qquad \textbf{(7.49)}$$

$$X_L = -N^2 X_S$$

LO3

EXAMPLE 7.14 Use of Transformers to Increase Power Line Efficiency

Problem

Figure 7.37 illustrates the use of transformers in electric power transmission lines. The practice of transforming the voltage before and after transmission of electric power over long distances is very common. This example illustrates the gain in efficiency that can be achieved through the use of transformers. The example makes use of ideal transformers and assumes simple resistive circuit models for the generator, transmission line, and load. These simplifications permit a clearer understanding of the efficiency gains afforded by transformers.

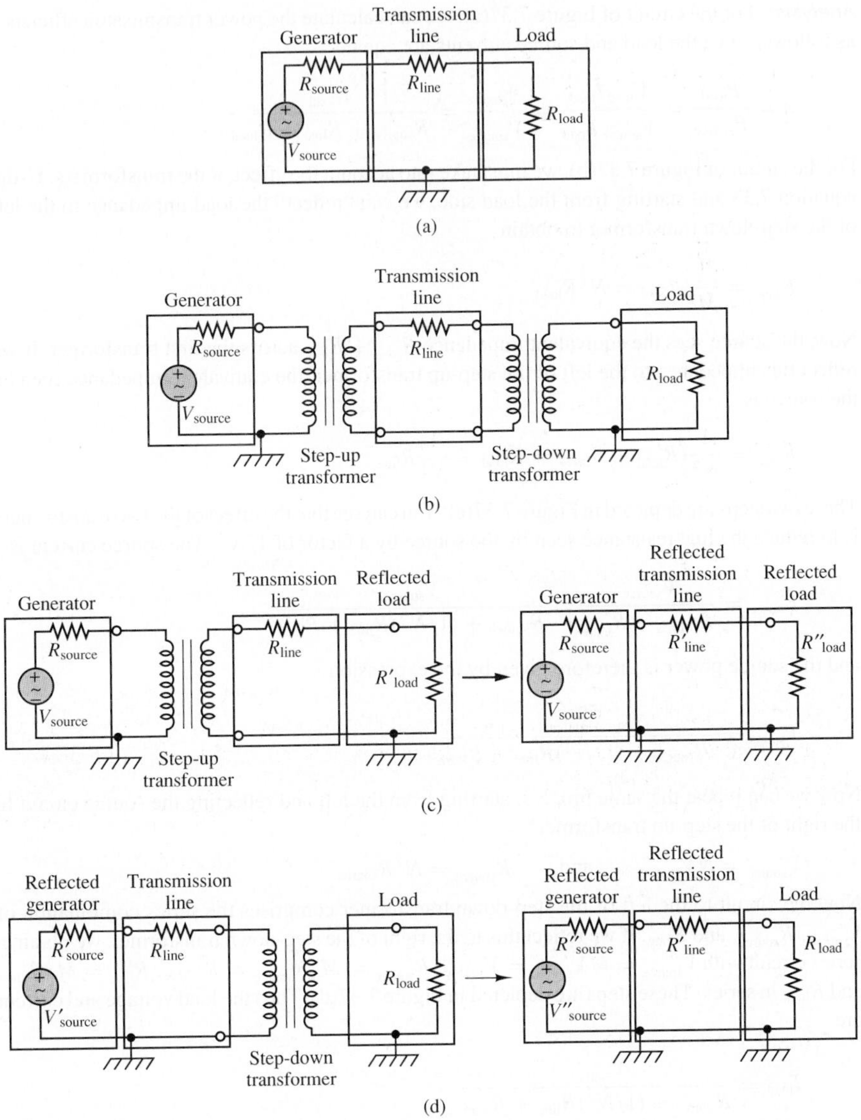

Figure 7.37 Electric power transmission: (a) direct power transmission; (b) power transmission with transformers; (c) equivalent circuit seen by generator; (d) equivalent circuit seen by load.

Solution

Known Quantities: Values of circuit elements.

Find: Calculate the power transfer efficiency for the two circuits of Figure 7.37.

Schematics, Diagrams, Circuits, and Given Data: Step-up transformer turns ratio is N, step-down transformer turns ratio is $M = 1/N$.

Assumptions: None.

Analysis: For the circuit of Figure 7.37(a), we can calculate the power transmission efficiency as follows, since the load and source currents are equal:

$$\eta = \frac{P_{\text{load}}}{P_{\text{source}}} = \frac{\tilde{V}_{\text{load}}\tilde{I}_{\text{load}}}{\tilde{V}_{\text{source}}\tilde{I}_{\text{load}}} = \frac{\tilde{V}_{\text{load}}}{\tilde{V}_{\text{source}}} = \frac{R_{\text{load}}}{R_{\text{source}} + R_{\text{line}} + R_{\text{load}}}$$

For the circuit of Figure 7.37(b), we must take into account the effect of the transformers. Using equation 7.38 and starting from the load side, we can "reflect" the load impedance to the left of the step-down transformer to obtain

$$R'_{\text{load}} = \frac{1}{M^2}R_{\text{load}} = N^2 R_{\text{load}}$$

Now, the source sees the equivalent impedance $R'_{\text{load}} + R_{\text{line}}$ across the first transformer. If we reflect this impedance to the left of the step-up transformer, the equivalent impedance seen by the source is

$$R''_{\text{load}} = \frac{1}{N^2}(R'_{\text{load}} + R_{\text{line}}) = R_{\text{load}} + \frac{1}{N^2}R_{\text{line}}$$

These two steps are depicted in Figure 7.37(c). You can see that the effect of the two transformers is to reduce the line resistance seen by the source by a factor of $1/N^2$. The source current is

$$\tilde{I}_{\text{source}} = \frac{\tilde{V}_{\text{source}}}{R_{\text{source}} + R''_{\text{load}}} = \frac{\tilde{V}_{\text{source}}}{R_{\text{source}} + (1/N^2)R_{\text{line}} + R_{\text{load}}}$$

and the source power is therefore given by the expression

$$P_{\text{source}} = \frac{\tilde{V}^2_{\text{source}}}{R_{\text{source}} + (1/N^2)R_{\text{line}} + R_{\text{load}}}$$

Now we can repeat the same process, starting from the left and reflecting the source circuit to the right of the step-up transformer:

$$\tilde{V}'_{\text{source}} = N\tilde{V}_{\text{source}} \qquad \text{and} \qquad R'_{\text{source}} = N^2 R_{\text{source}}$$

Now the circuit to the left of the step-down transformer comprises the series combination of $\tilde{V}'_{\text{source}}$, R'_{source}, and R_{line}. If we reflect this to the right of the step-down transformer, we obtain a series circuit with $\tilde{V}''_{\text{source}} = M\tilde{V}'_{\text{source}} = \tilde{V}_{\text{source}}$, $R'_{\text{source}} = M^2 R'_{\text{source}} = R_{\text{source}}$, $R'_{\text{line}} = M^2 R_{\text{line}}$, and R_{load} in series. These steps are depicted in Figure 7.37(d). Thus the load voltage and current are

$$\tilde{I}_{\text{load}} = \frac{\tilde{V}_{\text{source}}}{R_{\text{source}} + (1/N^2)R_{\text{line}} + R_{\text{load}}}$$

and

$$\tilde{V}_{\text{load}} = \tilde{V}_{\text{source}}\frac{R_{\text{load}}}{R_{\text{source}} + (1/N^2)R_{\text{line}} + R_{\text{load}}}$$

and we can calculate the load power as

$$P_{\text{load}} = \tilde{I}_{\text{load}}\tilde{V}_{\text{load}} = \frac{\tilde{V}^2_{\text{source}}R_{\text{load}}}{\left(R_{\text{source}} + (1/N^2)R_{\text{line}} + R_{\text{load}}\right)^2}$$

Finally, the power efficiency can be computed as the ratio of the load to source power:

$$\eta = \frac{P_{\text{load}}}{P_{\text{source}}} = \frac{\tilde{V}^2_{\text{source}}R_{\text{load}}}{\left(R_{\text{source}} + (1/N^2)R_{\text{line}} + R_{\text{load}}\right)^2}\,\frac{R_{\text{source}} + (1/N^2)R_{\text{line}} + R_{\text{load}}}{\tilde{V}^2_{\text{source}}}$$

$$= \frac{R_{\text{load}}}{R_{\text{source}} + (1/N^2)R_{\text{line}} + R_{\text{load}}}$$

Comparing the expression with the one obtained for the circuit of Figure 7.37(a), we can see that the power transmission efficiency can be significantly improved by reducing the effect of the line resistance by a factor of $1/N^2$.

CHECK YOUR UNDERSTANDING

Assume that the generator produces a source voltage of 480 Vrms, and that $N = 300$. Further assume that the source impedance is 2 Ω, the line impedance is also 2 Ω, and that the load impedance is 8 Ω. Calculate the efficiency improvement for the circuit of Figure 7.37(b) over the circuit of Figure 7.37(a).

Answer: 80% vs. 67%.

EXAMPLE 7.15 Maximum Power Transfer Through a Transformer

LO3

Problem

Find the transformer turns ratio and load reactance that results in maximum power transfer in the circuit of Figure 7.38.

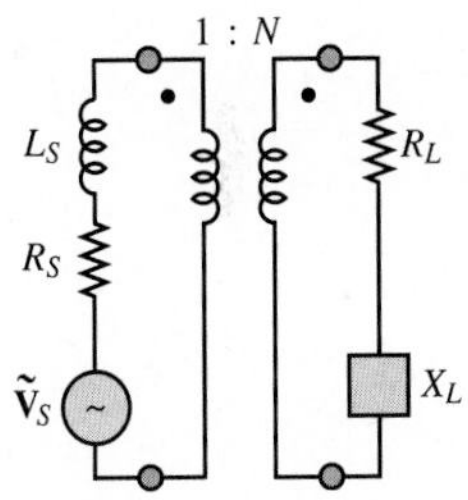

Figure 7.38

Solution

Known Quantities: Source voltage, frequency, and impedance; load resistance.

Find: Transformer turns ratio and load reactance.

Schematics, Diagrams, Circuits, and Given Data: $\tilde{\mathbf{V}}_S = 240\angle 0$ V; $R_S = 10\ \Omega$; $L_S = 0.1$ H; $R_L = 400\ \Omega$; $\omega = 377$ rad/s.

Assumptions: Use rms values for all phasor quantities in the problem.

Analysis: For maximum power transfer, we require that $R_L = N^2 R_S$ (equation 7.48). Thus,

$$N^2 = \frac{R_L}{R_S} = \frac{400}{10} = 40 \qquad N = \sqrt{40} = 6.325$$

Further, to cancel the reactive power, we require that $X_L = -N^2 X_S$, that is,

$$X_S = \omega \times 0.1 = 37.7$$

and

$$X_L = -40 \times 37.7 = -1{,}508$$

Thus, the load reactance should be a capacitor with value

$$C = -\frac{1}{X_L \omega} = -\frac{1}{(-1{,}508)(377)} = 1.76\ \mu\text{F}$$

CHECK YOUR UNDERSTANDING

The transformer shown in Figure 7.39 is ideal. Find the turns ratio N that will ensure maximum power transfer to the load. Assume that $Z_S = 1{,}800\ \Omega$ and $Z_L = 8\ \Omega$.

The transformer shown in Figure 7.39 is ideal. Find the source impedance Z_S that will ensure maximum power transfer to the load. Assume that $N = 5.4$ and $Z_L = 2 + j10\ \Omega$.

Answers: $N = 0.0667$; $Z_S = 0.0686 - j0.3429\ \Omega$

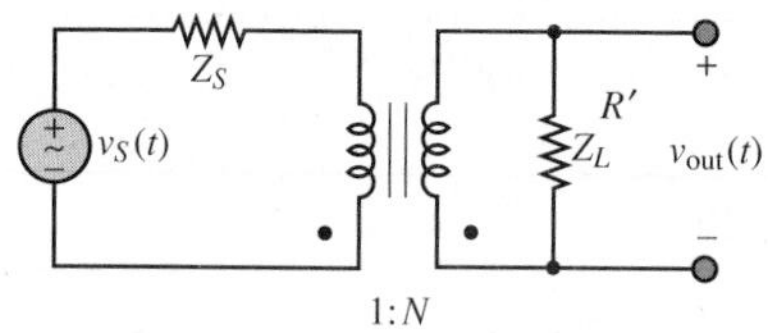

Figure 7.39

7.4 THREE-PHASE POWER

The material presented so far in this chapter has dealt exclusively with **single-phase AC power,** that is, with single sinusoidal sources. In fact, most of the AC power used today is generated and distributed as **three-phase power,** by means of an arrangement in which three sinusoidal voltages are generated out of phase with one another. The primary reason is efficiency: The weight of the conductors and other components in a three-phase system is much lower than that in a single-phase system delivering the same amount of power. Further, while the power produced by a single-phase system has a pulsating nature (recall the results of Section 7.1), a three-phase system can deliver a steady, constant supply of power. For example, later in this section it will be shown that a three-phase generator producing three **balanced voltages**—that is, voltages of equal amplitude and frequency displaced in phase by 120°—has the property of delivering constant instantaneous power.

Another important advantage of three-phase power is that, as will be explained in Chapter 17, three-phase motors have a nonzero starting torque, unlike their single-phase counterpart. The change to three-phase AC power systems from the early DC system proposed by Edison was therefore due to a number of reasons: the efficiency resulting from transforming voltages up and down to minimize transmission losses over long distances; the ability to deliver constant power (an ability not shared by single- and two-phase AC systems); a more efficient use of conductors; and the ability to provide starting torque for industrial motors.

To begin the discussion of three-phase power, consider a three-phase source connected in the **wye** (or **Y**) **configuration,** as shown in Figure 7.40. Each of the three voltages is 120° out of phase with the others, so that, using phasor notation, we may write

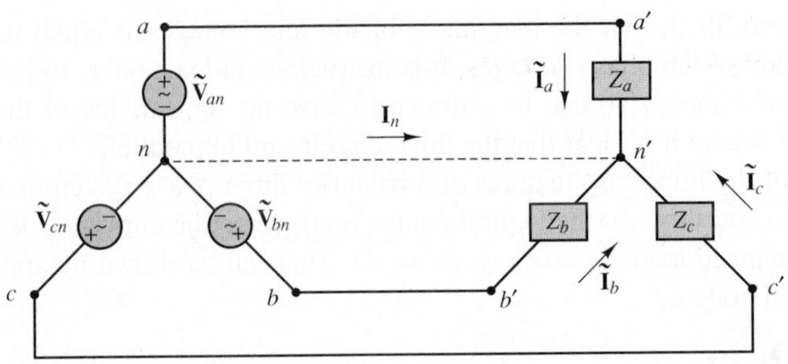

Figure 7.40 Balanced three-phase AC circuit

$$\begin{aligned}\tilde{\mathbf{V}}_{an} &= \tilde{V}_{an}\angle 0^\circ \\ \tilde{\mathbf{V}}_{bn} &= \tilde{V}_{bn}\angle -(120^\circ) \\ \tilde{\mathbf{V}}_{cn} &= \tilde{V}_{cn}\angle(-240^\circ) = \tilde{V}_{cn}\angle 120^\circ\end{aligned} \quad \text{Phase voltages} \tag{7.50}$$

LO4

where the quantities $\tilde{V}_{an}$, $\tilde{V}_{bn}$, and $\tilde{V}_{cn}$ are rms values and are equal to each other. To simplify the notation, it will be assumed from here on that

$$\tilde{V}_{an} = \tilde{V}_{bn} = \tilde{V}_{cn} = \tilde{V} \tag{7.51}$$

Chapter 17 will discuss how three-phase AC electric generators may be constructed to provide such balanced voltages. In the circuit of Figure 7.40, the resistive loads are also wye-connected and balanced (i.e., equal). The three AC sources are all connected together at a node called the *neutral node,* denoted by n. The voltages $\tilde{\mathbf{V}}_{an}$, $\tilde{\mathbf{V}}_{bn}$, and $\tilde{\mathbf{V}}_{cn}$ are called the **phase voltages** and form a balanced set in the sense that

$$\tilde{\mathbf{V}}_{an} + \tilde{\mathbf{V}}_{bn} + \tilde{\mathbf{V}}_{cn} = 0 \tag{7.52}$$

This last statement is easily verified by sketching the phasor diagram. The sequence of phasor voltages shown in Figure 7.41 is usually referred to as the **positive** (or ***abc***) **sequence.**

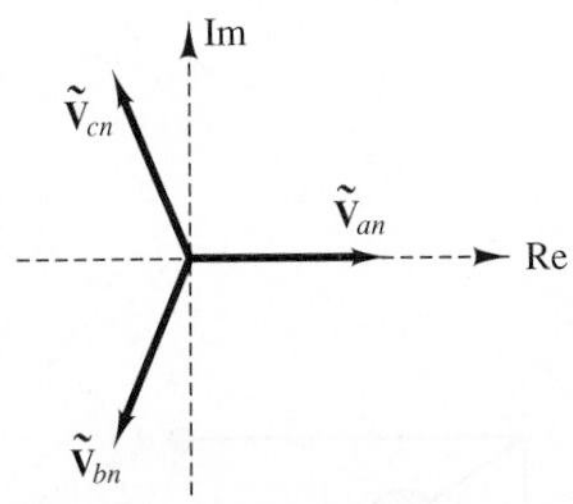

Figure 7.41 Positive, or *abc*, sequence for balanced three-phase voltages

Consider now the "lines" connecting each source to the load, and observe that it is possible to also define **line voltages** (also called *line-to-line voltages*) by considering the voltages between lines aa' and bb', lines aa' and cc', and lines bb' and cc'. Since the line voltage, say, between aa' and bb' is given by

$$\tilde{\mathbf{V}}_{ab} = \tilde{\mathbf{V}}_{an} + \tilde{\mathbf{V}}_{nb} = \tilde{\mathbf{V}}_{an} - \tilde{\mathbf{V}}_{bn} \tag{7.53}$$

the line voltages may be computed relative to the phase voltages as follows:

$$\begin{aligned}\tilde{\mathbf{V}}_{ab} &= \tilde{V}\angle 0^\circ - \tilde{V}\angle(-120^\circ) = \sqrt{3}\tilde{V}\angle 30^\circ \\ \tilde{\mathbf{V}}_{bc} &= \tilde{V}\angle(-120^\circ) - \tilde{V}\angle 120^\circ = \sqrt{3}\tilde{V}\angle(-90^\circ) \\ \tilde{\mathbf{V}}_{ca} &= \tilde{V}\angle 120^\circ - \tilde{V}\angle 0^\circ = \sqrt{3}\tilde{V}\angle 150^\circ\end{aligned} \quad \text{Line voltages} \tag{7.54}$$

LO4

It can be seen, then, that the magnitude of the line voltages is equal to $\sqrt{3}$ times the magnitude of the phase voltages. It is instructive, at least once, to point out that the circuit of Figure 7.40 can be redrawn to have the appearance of the circuit of Figure 7.42, where it is clear that the three circuits are in parallel.

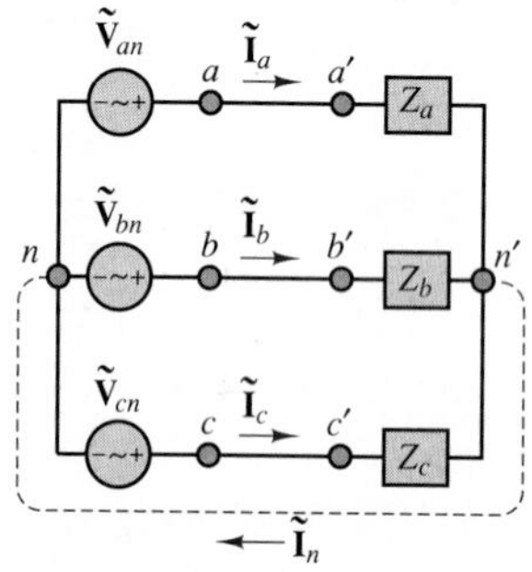

Figure 7.42 Balanced three-phase AC circuit (redrawn)

One of the important features of a balanced three-phase system is that it does not require a fourth wire (the neutral connection), since the current $\tilde{\mathbf{I}}_n$ is identically zero (for balanced load $Z_a = Z_b = Z_c = Z$). This can be shown by applying KCL at the neutral node n:

$$\begin{aligned}\tilde{\mathbf{I}}_n &= \tilde{\mathbf{I}}_a + \tilde{\mathbf{I}}_b + \tilde{\mathbf{I}}_c \\ &= \frac{1}{Z}(\tilde{\mathbf{V}}_{an} + \tilde{\mathbf{V}}_{bn} + \tilde{\mathbf{V}}_{cn}) \\ &= 0\end{aligned} \tag{7.55}$$

Another, more important characteristic of a balanced three-phase power system may be illustrated by simplifying the circuits of Figures 7.40 and 7.42 by replacing the balanced load impedances with three equal resistances R. With this simplified configuration, one can show that the total power delivered to the balanced load by the three-phase generator is constant. This is an extremely important result, for a very practical reason: Delivering power in a smooth fashion (as opposed to the pulsating nature of single-phase power) reduces the wear and stress on the generating equipment. Although we have not yet discussed the nature of the machines used to generate power, a useful analogy here is that of a single-cylinder engine versus a perfectly balanced V-8 engine. To show that the total power delivered by the three sources to a balanced resistive load is constant, consider the instantaneous power delivered by each source:

$$\begin{aligned}p_a(t) &= \frac{\tilde{V}^2}{R}(1 + \cos 2\omega t) \\ p_b(t) &= \frac{\tilde{V}^2}{R}[1 + \cos(2\omega t - 120°)] \\ p_c(t) &= \frac{\tilde{V}^2}{R}[1 + \cos(2\omega t + 120°)]\end{aligned} \tag{7.56}$$

The total instantaneous load power is then given by the sum of the three contributions:

$$\begin{aligned}p(t) &= p_a(t) + p_b(t) + p_c(t) \\ &= \frac{3\tilde{V}^2}{R} + \frac{\tilde{V}^2}{R}[\cos 2\omega t + \cos(2\omega t - 120°) \\ &\quad + \cos(2\omega t + 120°)] \\ &= \frac{3\tilde{V}^2}{R} = \text{constant!}\end{aligned} \tag{7.57}$$

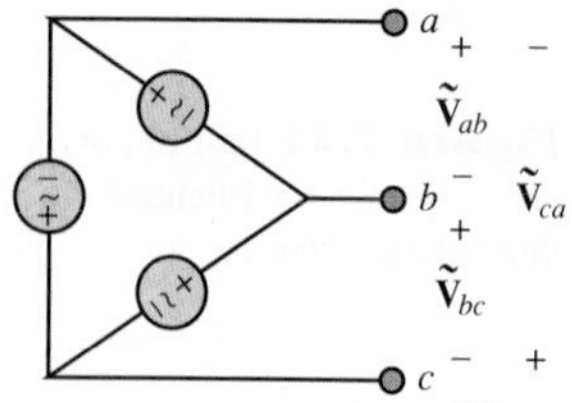

A delta-connected three-phase generator with line voltages V_{ab}, V_{bc}, V_{ca}

Figure 7.43 Delta-connected generators

You may wish to verify that the sum of the trigonometric terms inside the brackets is identically zero.

It is also possible to connect the three AC sources in a three-phase system in a **delta** (or **Δ**) **connection,** although in practice this configuration is rarely used. Figure 7.43 depicts a set of three delta-connected generators.

EXAMPLE 7.16 Per-Phase Solution of Balanced Wye-Wye Circuit

LO4

Problem

Compute the power delivered to the load by the three-phase generator in the circuit shown in Figure 7.44.

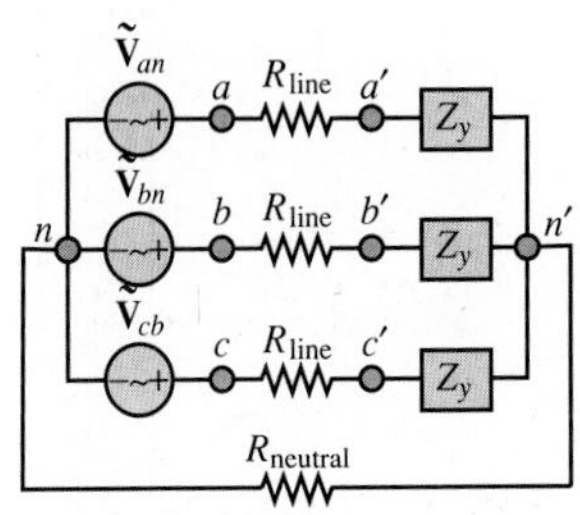

Figure 7.44

Solution

Known Quantities: Source voltage, line resistance, load impedance.

Find: Power delivered to the load P_L.

Schematics, Diagrams, Circuits, and Given Data: $\tilde{\mathbf{V}}_{an} = 480\angle 0$ V; $\tilde{\mathbf{V}}_{bn} = 480\angle(-2\pi/3)$ V; $\tilde{\mathbf{V}}_{cn} = 480\angle(2\pi/3)$ V; $Z_y = 2 + j4 = 4.47\angle 1.107\ \Omega$; $R_{\text{line}} = 2\ \Omega$; $R_{\text{neutral}} = 10\ \Omega$.

Assumptions: Use rms values for all phasor quantities in the problem.

Analysis: Since the circuit is balanced, we can use per-phase analysis, and the current through the neutral line is zero, that is, $\tilde{\mathbf{V}}_{n-n'} = 0$. The resulting per-phase circuit is shown in Figure 7.45. Using phase a for the calculations, we look for the quantity

$$P_a = |\tilde{\mathbf{I}}|^2 R_L$$

where

$$|\tilde{\mathbf{I}}| = \left|\frac{\tilde{\mathbf{V}}_a}{Z_y + R_{\text{line}}}\right| = \left|\frac{480\angle 0}{2 + j4 + 2}\right| = \left|\frac{480\angle 0}{5.66\angle(\pi/4)}\right| = 84.85 \text{ A}$$

and $P_a = (84.85)^2 \times 2 = 14.4$ kW. Since the circuit is balanced, the results for phases b and c are identical, and we have

$$P_L = 3P_a = 43.2 \text{ kW}$$

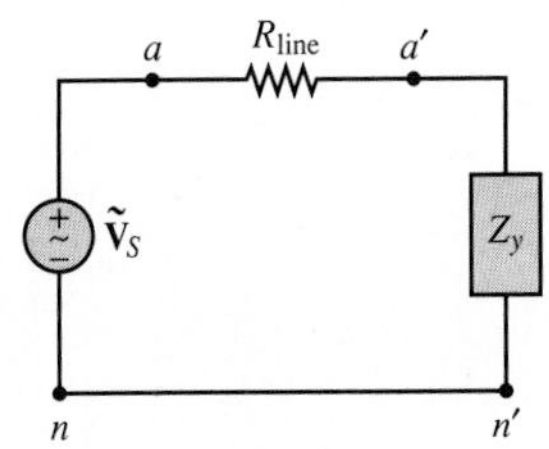

Figure 7.45 One phase of the three-phase circuit

Comments: Note that since the circuit is balanced, there is zero voltage across neutrals. This fact is shown explicitly in Figure 7.45, where n and n' are connected to each other directly. Per-phase analysis for balanced circuits turns three-phase power calculations into a very simple exercise.

CHECK YOUR UNDERSTANDING

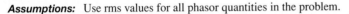

Find the power lost in the line resistance in the circuit of Example 7.16.

Compute the power delivered to the balanced load of Example 7.16 if the lines have zero resistance and $Z_L = 1 + j3\ \Omega$.

Show that the voltage across each branch of the wye load is equal to the corresponding phase voltage (e.g., the voltage across Z_a is $\tilde{\mathbf{V}}_a$).

Prove that the sum of the instantaneous powers absorbed by the three branches in a balanced wye-connected load is constant and equal to $3\tilde{\mathbf{V}}\tilde{\mathbf{I}}\cos\theta$.

Answers: $P_{\text{line}} = 43.2$ kW; $S_L = 69.12$ W $+ j207.4$ VA

Balanced Wye Loads

In the previous section we performed some power computations for a purely resistive balanced wye load. We now generalize those results for an arbitrary balanced complex load. Consider again the circuit of Figure 7.40, where now the balanced load consists of the three complex impedances

$$Z_a = Z_b = Z_c = Z_y = |Z_y|\angle\theta \tag{7.58}$$

From the diagram of Figure 7.40, it can be verified that each impedance sees the corresponding phase voltage across itself; thus, since currents $\tilde{\mathbf{I}}_a$, $\tilde{\mathbf{I}}_b$, and $\tilde{\mathbf{I}}_c$ have the same rms value $\tilde{I}$, the phase angles of the currents will differ by $\pm120°$. It is therefore possible to compute the power for each phase by considering the phase voltage (equal to the load voltage) for each impedance, and the associated line current. Let us denote the complex power for each phase by S

$$S = \tilde{\mathbf{V}}\tilde{\mathbf{I}}^* \tag{7.59}$$

so that

$$\begin{aligned} S &= P + jQ \\ &= \tilde{V}\tilde{I}\cos\theta + j\tilde{V}\tilde{I}\sin\theta \end{aligned} \tag{7.60}$$

where $\tilde{V}$ and $\tilde{I}$ denote, once again, the rms values of each phase voltage and line current, respectively. Consequently, the total real power delivered to the balanced wye load is $3P$, and the total reactive power is $3Q$. Thus, the total complex power S_T is given by

$$\begin{aligned} S_T &= P_T + jQ_T = 3P + j3Q \\ &= \sqrt{(3P)^2 + (3Q)^2}\angle\theta \end{aligned} \tag{7.61}$$

and the apparent power is

$$\begin{aligned} |S_T| &= 3\sqrt{(VI)^2\cos^2\theta + (VI)^2\sin^2\theta} \\ &= 3VI \end{aligned}$$

and the total real and reactive power may be expressed in terms of the apparent power:

$$\begin{aligned} P_T &= |S_T|\cos\theta \\ Q_T &= |S_T|\sin\theta \end{aligned} \tag{7.62}$$

Balanced Delta Loads

In addition to a wye connection, it is possible to connect a balanced load in the delta configuration. A wye-connected generator and a delta-connected load are shown in Figure 7.46.

Note immediately that now the corresponding line voltage (not phase voltage) appears across each impedance. For example, the voltage across $Z_{c'a'}$ is $\tilde{\mathbf{V}}_{ca}$. Thus, the three load currents are given by

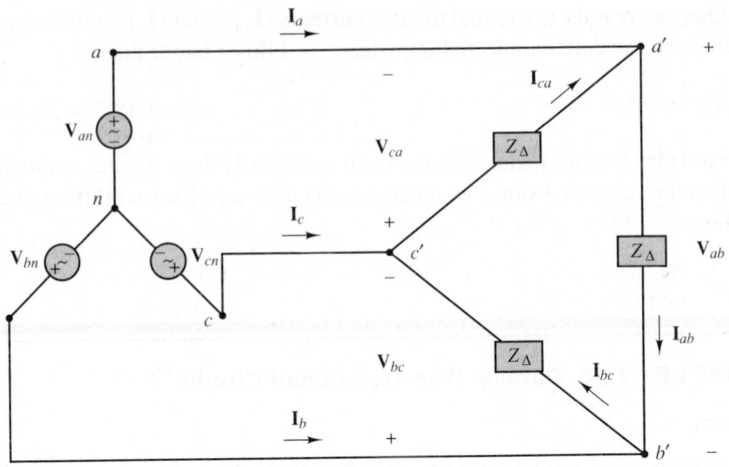

Figure 7.46 Balanced wye generators with balanced delta load

$$
\begin{aligned}
\tilde{\mathbf{I}}_{ab} &= \frac{\tilde{\mathbf{V}}_{ab}}{Z_\Delta} = \frac{\sqrt{3}V\angle(\pi/6)}{|Z_\Delta|\angle\theta} \\
\tilde{\mathbf{I}}_{bc} &= \frac{\tilde{\mathbf{V}}_{bc}}{Z_\Delta} = \frac{\sqrt{3}V\angle(-\pi/2)}{|Z_\Delta|\angle\theta} \\
\tilde{\mathbf{I}}_{ca} &= \frac{\tilde{\mathbf{V}}_{ca}}{Z_\Delta} = \frac{\sqrt{3}V\angle(5\pi/6)}{|Z_\Delta|\angle\theta}
\end{aligned}
\tag{7.63}
$$

To understand the relationship between delta-connected and wye-connected loads, it is reasonable to ask the question, For what value of Z_Δ would a delta-connected load draw the same amount of current as a wye-connected load with impedance Z_y for a given source voltage? This is equivalent to asking what value of Z_Δ would make the line currents the same in both circuits (compare Figure 7.42 with Figure 7.46).

The line current drawn, say, in phase a by a wye-connected load is

$$
(\tilde{\mathbf{I}}_{an})_y = \frac{\tilde{\mathbf{V}}_{an}}{Z} = \frac{\tilde{V}}{|Z_y|}\angle(-\theta) \tag{7.64}
$$

while that drawn by the delta-connected load is

$$
\begin{aligned}
(\tilde{\mathbf{I}}_a)_\Delta &= \tilde{\mathbf{I}}_{ab} - \tilde{\mathbf{I}}_{ca} \\
&= \frac{\tilde{\mathbf{V}}_{ab}}{Z_\Delta} - \frac{\tilde{\mathbf{V}}_{ca}}{Z_\Delta} \\
&= \frac{1}{Z_\Delta}(\tilde{\mathbf{V}}_{an} - \tilde{\mathbf{V}}_{bn} - \tilde{\mathbf{V}}_{cn} + \tilde{\mathbf{V}}_{an}) \\
&= \frac{1}{Z_\Delta}(2\tilde{\mathbf{V}}_{an} - \tilde{\mathbf{V}}_{bn} - \tilde{\mathbf{V}}_{cn}) \\
&= \frac{3\tilde{\mathbf{V}}_{an}}{Z_\Delta} = \frac{3\tilde{\mathbf{V}}}{|Z_\Delta|}\angle(-\theta)
\end{aligned}
\tag{7.65}
$$

One can readily verify that the two currents $(\tilde{\mathbf{I}}_a)_\Delta$ and $(\tilde{\mathbf{I}}_a)_y$ will be equal if the magnitude of the delta-connected impedance is 3 times larger than Z_y:

$$Z_\Delta = 3Z_y \qquad \textbf{(7.66)}$$

This result also implies that a delta load will necessarily draw 3 times as much current (and therefore absorb 3 times as much power) as a wye load with the same branch impedance.

LO4

EXAMPLE 7.17 Parallel Wye-Delta Load Circuit

Problem

Compute the power delivered to the wye-delta load by the three-phase generator in the circuit shown in Figure 7.47.

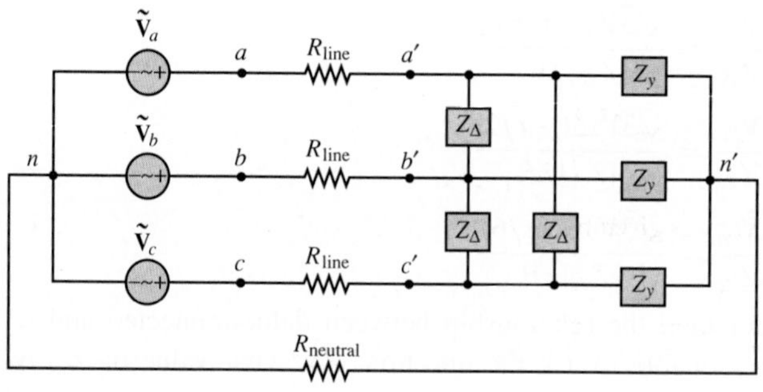

Figure 7.47 AC circuit with delta and wye loads

Solution

Known Quantities: Source voltage, line resistance, load impedance.

Find: Power delivered to the load P_L.

Schematics, Diagrams, Circuits, and Given Data: $\tilde{\mathbf{V}}_{an} = 480\angle 0$ V; $\tilde{\mathbf{V}}_{bn} = 480\angle(-2\pi/3)$ V; $\tilde{\mathbf{V}}_{cn} = 480\angle(2\pi/3)$ V; $Z_y = 2 + j4 = 4.47\angle 1.107\ \Omega$; $Z_\Delta = 5 - j2 = 5.4\angle(-0.381)\ \Omega$; $R_{\text{line}} = 2\ \Omega$; $R_{\text{neutral}} = 10\ \Omega$.

Assumptions: Use rms values for all phasor quantities in the problem.

Analysis: We first convert the balanced delta load to an equivalent wye load, according to equation 7.66. Figure 7.48 illustrates the effect of this conversion.

$$Z_{\Delta-y} = \frac{Z_\Delta}{3} = 1.667 - j0.667 = 1.8\angle(-0.381)\ \Omega.$$

Since the circuit is balanced, we can use per-phase analysis, and the current through the neutral line is zero, that is, $\tilde{\mathbf{V}}_{n-n'} = 0$. The resulting per-phase circuit is shown in Figure 7.49. Using

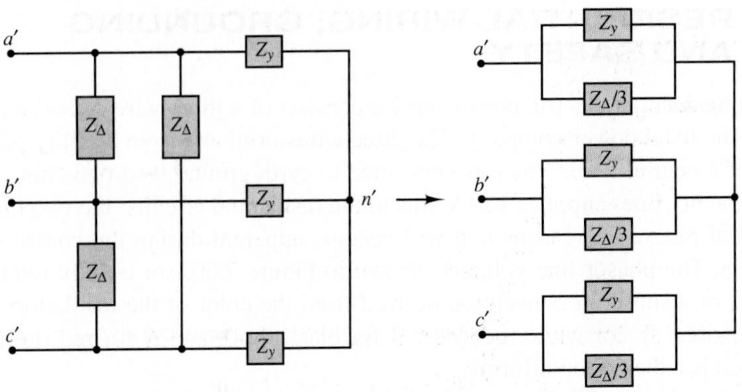

Figure 7.48 Conversion of delta load to equivalent wye load

phase a for the calculations, we look for the quantity

$$P_a = |\tilde{\mathbf{I}}|^2 R_L$$

where

$$Z_L = Z_y \| Z_{\Delta-y} = \frac{Z_y \times Z_{\Delta-y}}{Z_y + Z_{\Delta-y}} = 1.62 - j0.018 = 1.62\angle(-0.011)\ \Omega$$

and the load current is given by

$$|\tilde{\mathbf{I}}| = \left|\frac{\tilde{\mathbf{V}}_a}{Z_L + R_{\text{line}}}\right| = \left|\frac{480\angle 0}{1.62 + j0.018 + 2}\right| = 132.6 \text{ A}$$

and $P_a = (132.6)^2 \times \text{Re}(Z_L) = 28.5$ kW. Since the circuit is balanced, the results for phases b and c are identical, and we have

$$P_L = 3P_a = 85.5 \text{ kW}$$

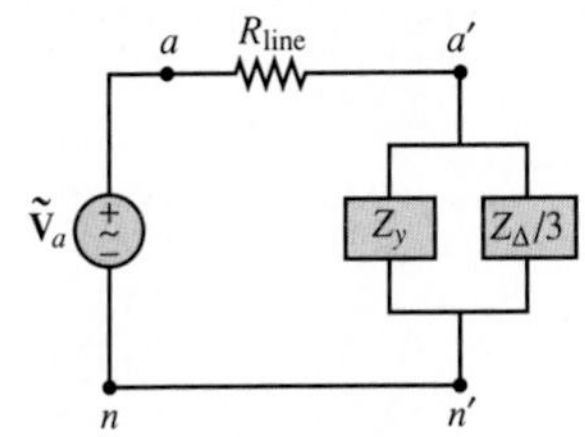

Figure 7.49 Per-phase circuit

Comments: Note that per-phase analysis for balanced circuits turns three-phase power calculations into a very simple exercise.

CHECK YOUR UNDERSTANDING

Derive an expression for the rms line current of a delta load in terms of the rms line current of a wye load with the same branch impedances (that is, $Z_Y = Z_\Delta$) and same source voltage. Assume $Z_S = 0$.

The equivalent wye load of Example 7.17 is connected in a delta configuration. Compute the line currents.

Answers: $I_\Delta = 3I_Y$; $\tilde{\mathbf{I}}_a = 189\angle 0°$ A; $\tilde{\mathbf{I}}_b = 189\angle(-120°)$ A; $\tilde{\mathbf{I}}_c = 189\angle 120°$ A

7.5 RESIDENTIAL WIRING; GROUNDING AND SAFETY

Common residential electric power service consists of a three-wire AC system supplied by the local power company. The three wires originate from a utility pole and consist of a neutral wire, which is connected to earth ground, and two "hot" wires. Each of the hot lines supplies 120 V rms to the residential circuits; the two lines are 180° out of phase, for reasons that will become apparent during the course of this discussion. The phasor line voltages, shown in Figure 7.50, are usually referred to by means of a subscript convention derived from the color of the insulation on the different wires: W for white (neutral), B for black (hot), and R for red (hot). This convention is adhered to uniformly.

The voltages across the hot lines are given by

$$\tilde{\mathbf{V}}_B - \tilde{\mathbf{V}}_R = \tilde{\mathbf{V}}_{BR} = \tilde{\mathbf{V}}_B - (-\tilde{\mathbf{V}}_B) = 2\tilde{\mathbf{V}}_B = 240\angle 0^\circ \qquad \textbf{(7.67)}$$

Thus, the voltage between the hot wires is actually 240 V rms. Appliances such as electric stoves, air conditioners, and heaters are powered by the 240-V rms arrangement. On the other hand, lighting and all the electric outlets in the house used for small appliances are powered by a single 120-V rms line.

The use of 240-V rms service for appliances that require a substantial amount of power to operate is dictated by power transfer considerations. Consider the two circuits shown in Figure 7.51. In delivering the necessary power to a load, a lower line loss will be incurred with the 240-V rms wiring, since the power loss in the lines (the I^2R **loss,** as it is commonly referred to) is directly related to the current required by the load. In an effort to minimize line losses, the size of the wires is increased for the lower-voltage case. This typically reduces the wire resistance by a factor of 2. In the top circuit, assuming $R_S/2 = 0.01\ \Omega$, the current required by the 10-kW load is approximately 83.3 A, while in the bottom circuit, with $R_S = 0.02\ \Omega$, it is approximately one-half as much (41.7 A). (You should be able to verify that the

LO5

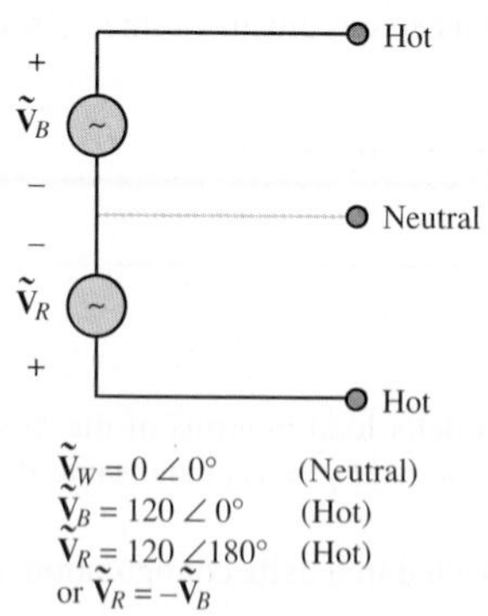

Figure 7.50 Line voltage convention for residential circuits

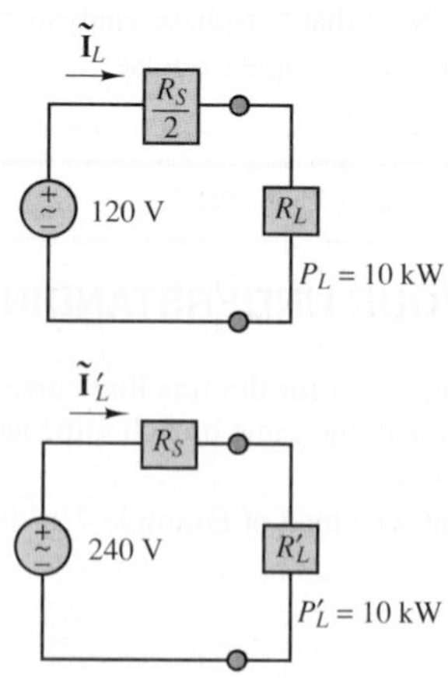

Figure 7.51 Line losses in 120- and 240-VAC circuits

approximate I^2R losses are 69.4 W in the top circuit and 34.7 W in the bottom circuit.) Limiting the I^2R losses is important from the viewpoint of efficiency, besides reducing the amount of heat generated in the wiring for safety considerations. Figure 7.52 shows some typical wiring configurations for a home. Note that several circuits are wired and fused separately.

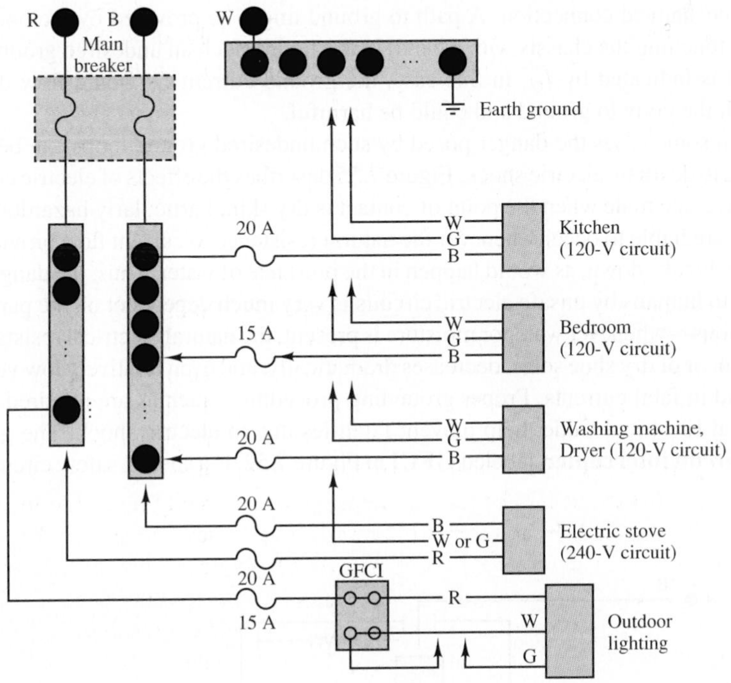

Figure 7.52 A typical residential wiring arrangement

CHECK YOUR UNDERSTANDING

Use the circuit of Figure 7.51 to show that the I^2R losses will be higher for a 120-V service appliance than a 240-V service appliance if both have the same power usage rating.

Answer: The 120-V circuit has double the losses of the 240-V circuit for the same power rating.

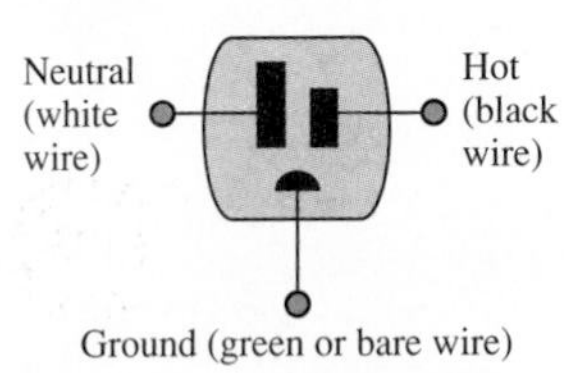

Figure 7.53 A three-wire outlet

Today, most homes have three wire connections to their outlets. The outlets appear as sketched in Figure 7.53. Then why are both the ground and neutral connections needed in an outlet? The answer to this question is *safety:* The ground connection is used to connect the chassis of the appliance to earth ground. Without this provision, the appliance chassis could be at any potential with respect to ground, possibly even at the hot wire's potential if a segment of the hot wire were to lose some insulation and come in contact with the inside of the chassis! Poorly grounded appliances can thus be a significant hazard. Figure 7.54 illustrates schematically how, even though the chassis is intended to be insulated from the electric circuit, an unintended connection (represented by the dashed line) may occur, for example, because of corrosion or a loose mechanical connection. A path to ground might be provided by the body of a person touching the chassis with a hand. In the figure, such an undesired ground loop current is indicated by I_G. In this case, the ground current I_G would flow directly through the body to ground and could be harmful.

In some cases the danger posed by such undesired ground loops can be great, leading to death by electric shock. Figure 7.55 describes the effects of electric currents on an average male when the point of contact is dry skin. Particularly hazardous conditions are liable to occur whenever the natural resistance to current flow provided by the skin breaks down, as would happen in the presence of water. Thus, the danger presented to humans by unsafe electric circuits is very much dependent on the particular conditions—whenever water or moisture is present, the natural electrical resistance of dry skin, or of dry shoe soles, decreases dramatically, and even relatively low voltages can lead to fatal currents. Proper grounding procedures, such as are required by the National Electrical Code, help prevent fatalities due to electric shock. The **ground fault circuit interrupter,** labeled **GFCI** in Figure 7.52, is a special safety circuit used

FIND IT ON THE WEB

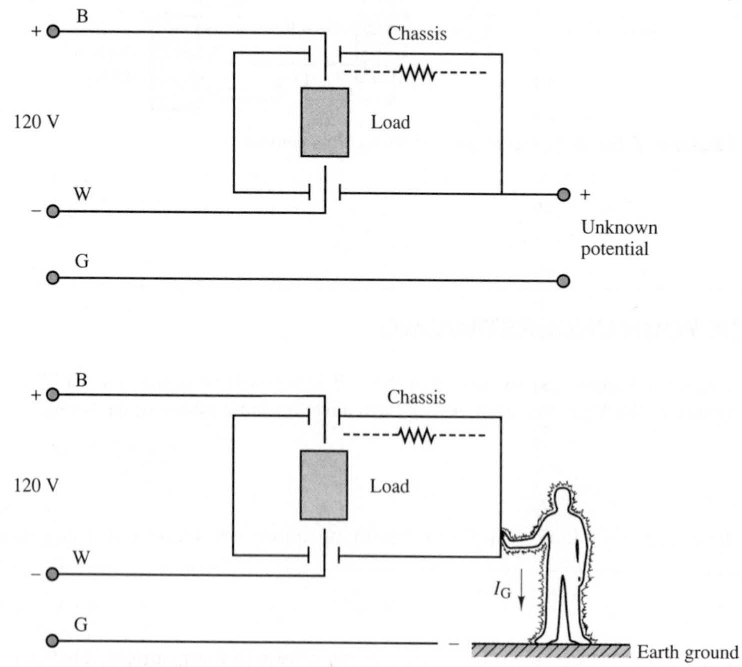

Figure 7.54 Unintended connection

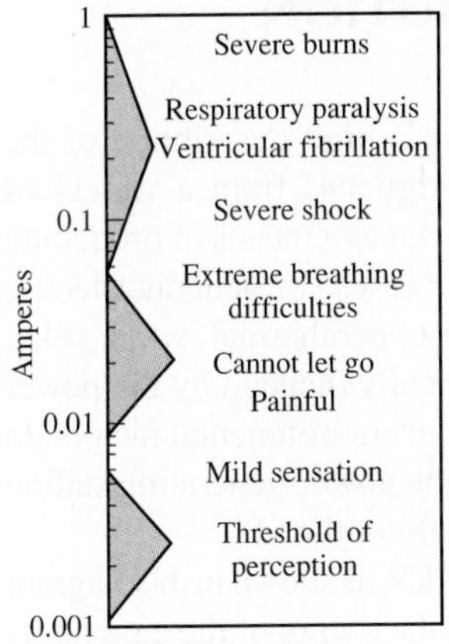

Figure 7.55 Physiological effects of electric currents

Figure 7.56 Outdoor pool

primarily with outdoor circuits and in bathrooms, where the risk of death by electric shock is greatest. Its application is best described by an example.

Consider the case of an outdoor pool surrounded by a metal fence, which uses an existing light pole for a post, as shown in Figure 7.56. The light pole and the metal fence can be considered as forming a chassis. If the fence were not properly grounded all the way around the pool and if the light fixture were poorly insulated from the pole, a path to ground could easily be created by an unaware swimmer reaching, say, for the metal gate. A GFCI provides protection from potentially lethal ground loops, such as this one, by sensing both the hot-wire (B) and the neutral (W) currents. If the difference between the hot-wire current I_B and the neutral current I_W is more than a few milliamperes, then the GFCI disconnects the circuit nearly instantaneously. Any significant difference between the hot and neutral (return-path) currents means that a second path to ground has been created (by the unfortunate swimmer, in this example) and a potentially dangerous condition has arisen. Figure 7.57 illustrates the idea. GFCIs are typically resettable circuit breakers, so that one does not need to replace a fuse every time the GFCI circuit is enabled.

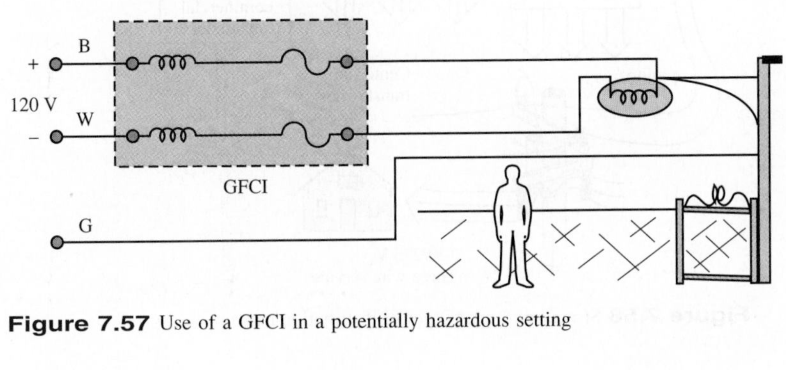

Figure 7.57 Use of a GFCI in a potentially hazardous setting

7.6 GENERATION AND DISTRIBUTION OF AC POWER

We now conclude the discussion of power systems with a brief description of the various elements of a power system. Electric power originates from a variety of sources; in Chapter 17, electric generators will be introduced as a means of producing electric power from a variety of energy conversion processes. In general, electric power may be obtained from hydroelectric, thermoelectric, geothermal, wind, solar, and nuclear sources. The choice of a given source is typically dictated by the power requirement for the given application, and by economic and environmental factors. In this section, the structure of an AC power network, from the power-generating station to the residential circuits discussed in Section 7.5, is briefly outlined.

A typical generator will produce electric power at 18 kV, as shown in the diagram of Figure 7.58. To minimize losses along the conductors, the output of the generators is processed through a step-up transformer to achieve line voltages of hundreds of kilovolts (345 kV, in Figure 7.58). Without this transformation, the majority of the power generated would be lost in the **transmission lines** that carry the electric current from the power station.

The local electric company operates a power-generating plant that is capable of supplying several hundred megavolt-amperes (MVA) on a three-phase basis. For this reason, the power company uses a three-phase step-up transformer at the generation plant to increase the line voltage to around 345 kV. One can immediately see that at the rated power of the generator (in megavolt-amperes) there will be a significant reduction of current beyond the step-up transformer.

LO5

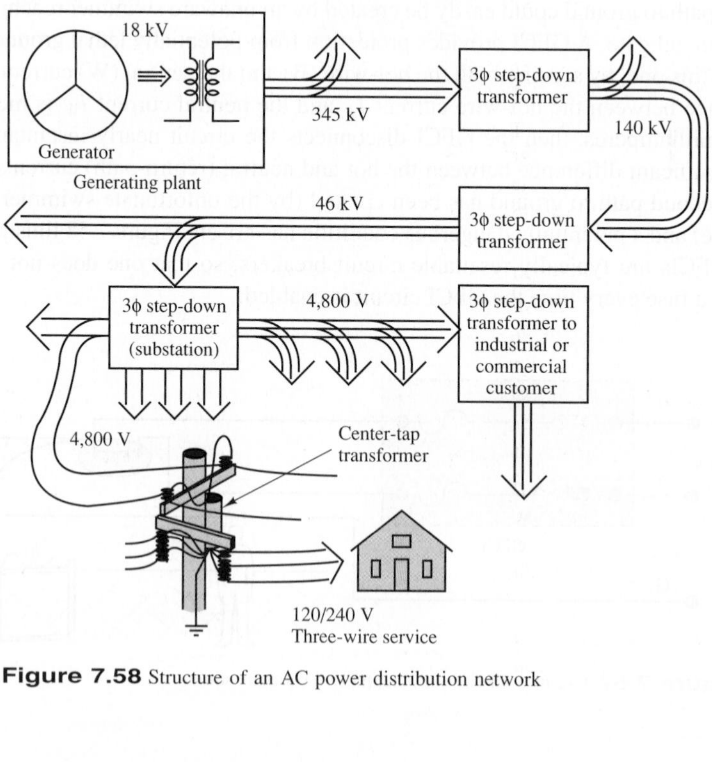

Figure 7.58 Structure of an AC power distribution network

Beyond the generation plant, an electric power network distributes energy to several **substations.** This network is usually referred to as the **power grid.** At the substations, the voltage is stepped down to a lower level (10 to 150 kV, typically). Some very large loads (e.g., an industrial plant) may be served directly from the power grid, although most loads are supplied by individual substations in the power grid. At the local substations (one of which you may have seen in your own neighborhood), the voltage is stepped down further by a three-phase step-down transformer to 4,800 V. These substations distribute the energy to residential and industrial customers. To further reduce the line voltage to levels that are safe for residential use, step-down transformers are mounted on utility poles. These drop the voltage to the 120/240-V three-wire single-phase residential service discussed in Section 7.5. Industrial and commercial customers receive 460- and/or 208-V three-phase service.

Conclusion

Chapter 7 introduces the essential elements that permit the analysis of AC power systems. AC power is essential to all industrial activities, and to the conveniences we are accustomed to in residential life. Virtually all engineers will be exposed to AC power systems in their careers, and the material presented in this chapter provides all the necessary tools to understand the analysis of AC power circuits. Upon completing this chapter, you should have mastered the following learning objectives:

1. *Understand the meaning of instantaneous and average power, master AC power notation, and compute average power for AC circuits. Compute the power factor of a complex load.* The power dissipated by a load in an AC circuit consists of the sum of an average and a fluctuating component. In practice, the average power is the quantity of interest.
2. *Learn complex power notation; compute apparent, real, and reactive power for complex loads. Draw the power triangle, and compute the capacitor size required to perform power factor correction on a load.* AC power can best be analyzed with the aid of complex notation. Complex power S is defined as the product of the phasor load voltage and the complex conjugate of the load current. The real part of S is the real power actually consumed by a load (that for which the user is charged); the imaginary part of S is called the reactive power and corresponds to energy stored in the circuit—it cannot be directly used for practical purposes. Reactive power is quantified by a quantity called the *power factor,* and it can be minimized through a procedure called *power factor correction.*
3. *Analyze the ideal transformer; compute primary and secondary currents and voltages and turns ratios. Calculate reflected sources and impedances across ideal transformers. Understand maximum power transfer.* Transformers find many applications in electrical engineering. One of the most common is in power transmission and distribution, where the electric power generated at electric power plants is stepped "up" and "down" before and after transmission, to improve the overall efficiency of electric power distribution.
4. *Learn three-phase AC power notation; compute load currents and voltages for balanced wye and delta loads.* AC power is generated and distributed in three-phase form. Residential services are typically single-phase (making use of only one branch of the three-phase lines), while industrial applications are often served directly by three-phase power.
5. *Understand the basic principles of residential electrical wiring, of electrical safety, and of the generation and distribution of AC power.*

HOMEWORK PROBLEMS

Section 7.1: Power in AC Circuits

7.1 The heating element in a soldering iron has a resistance of 30 Ω. Find the average power dissipated in the soldering iron if it is connected to a voltage source of 117 V rms.

7.2 A coffeemaker has a rated power of 1,000 W at 240 V rms. Find the resistance of the heating element.

7.3 A current source $i(t)$ is connected to a 50-Ω resistor. Find the average power delivered to the resistor, given that $i(t)$ is

a. $5 \cos 50t$ A

b. $5 \cos(50t - 45°)$ A

c. $5 \cos 50t - 2 \cos(50t - 0.873)$ A

d. $5 \cos 50t - 2$ A

7.4 Find the rms value of each of the following periodic currents:

a. $\cos 450t + 2 \cos 450t$

b. $\cos 5t + \sin 5t$

c. $\cos 450t + 2$

d. $\cos 5t + \cos(5t + \pi/3)$

e. $\cos 200t + \cos 400t$

7.5 A current of 4 A flows when a neon light advertisement is supplied by a 110-V rms power system. The current lags the voltage by 60°. Find the power dissipated by the circuit and the power factor.

7.6 A residential electric power monitoring system rated for 120-V rms, 60-Hz source registers power consumption of 1.2 kW, with a power factor of 0.8. Find

a. The rms current.

b. The phase angle.

c. The system impedance.

d. The system resistance.

7.7 A drilling machine is driven by a single-phase induction machine connected to a 110-V rms supply. Assume that the machining operation requires 1 kW, that the tool machine has 90 percent efficiency, and that the supply current is 14 A rms with a power factor of 0.8. Find the AC machine efficiency.

7.8 Given the waveform of a voltage source shown in Figure P7.8, find:

a. The steady DC voltage that would cause the same heating effect across a resistance.

b. The average current supplied to a 10-Ω resistor connected across the voltage source.

c. The average power supplied to a 1-Ω resistor connected across the voltage source.

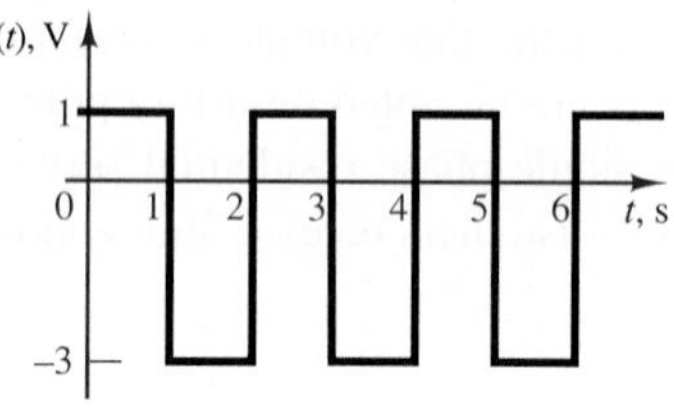

Figure P7.8

7.9 A current source $i(t)$ is connected to a 100-Ω resistor. Find the average power delivered to the resistor, given that $i(t)$ is:

a. $4 \cos 100t$ A

b. $4 \cos(100t - 50°)$ A

c. $4 \cos 100t - 3 \cos(100t - 50°)$ A

d. $4 \cos 100t - 3$ A

7.10 Find the rms value of each of the following periodic currents:

a. $\cos 377t + \cos 377t$

b. $\cos 2t + \sin 2t$

c. $\cos 377t + 1$

d. $\cos 2t + \cos(2t + 135°)$

e. $\cos 2t + \cos 33$

Section 7.2: Complex Power

7.11 A current of 10 A rms flows when a single-phase circuit is placed across a 220-V rms source. The current lags the voltage by 60°. Find the power dissipated by the circuit and the power factor.

7.12 A single-phase circuit is placed across a 120-V rms, 60-Hz source, with an ammeter, a voltmeter, and a wattmeter connected. The instruments indicate 12 A, 120 V, and 800 W, respectively. Find

a. The power factor.

b. The phase angle.

c. The impedance.

d. The resistance.

7.13 For the following numeric values, determine the average power, P, the reactive power, Q, and the complex power, S, of the circuit shown in Figure P7.13. Note: phasor quantities are rms.

a. $v_S(t) = 650 \cos(377t)$ V
$i_L(t) = 20 \cos(377t - 10°)$ A

b. $\mathbf{V}_S = 460\angle 0°$ V
$\mathbf{I}_L = 14.14\angle -45°$ A

c. $\mathbf{V}_S = 100\angle 0°$ V
$\mathbf{I}_L = 8.6\angle -86°$ A

d. $\mathbf{V}_S = 208\angle -30°$ V
$\mathbf{I}_L = 2.3\angle -63°$ A

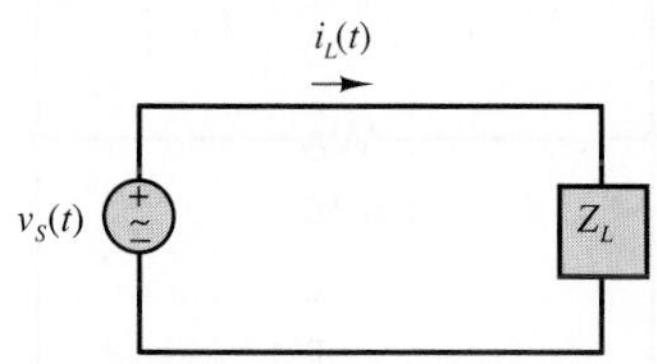

Figure P7.13

7.14 For the circuit of Figure P7.13, determine the power factor for the load and state whether it is leading or lagging for the following conditions:

a. $v_S(t) = 540 \cos(\omega t + 15°)$ V
$i_L(t) = 20 \cos(\omega t + 47°)$ A

b. $v_S(t) = 155 \cos(\omega t - 15°)$ V
$i_L(t) = 20 \cos(\omega t - 22°)$ A

c. $v_S(t) = 208 \cos(\omega t)$ V
$i_L(t) = 1.7 \sin(\omega t + 175°)$ A

d. $Z_L = (48 + j16)\ \Omega$

7.15 For the circuit of Figure P7.13, determine whether the load is capacitive or inductive for the circuit shown if

a. $pf = 0.87$ (leading)

b. $pf = 0.42$ (leading)

c. $v_S(t) = 42 \cos(\omega t)$
$i_L(t) = 4.2 \sin(\omega t)$

d. $v_S(t) = 10.4 \cos(\omega t - 12°)$
$i_L(t) = 0.4 \cos(\omega t - 12°)$

7.16 The circuit shown in Figure P7.16 is to be used on two different sources, each with the same amplitude but at different frequencies.

a. Find the instantaneous real and reactive power if $v_S(t) = 120 \cos 377t$ (i.e., the frequency is 60 Hz).

b. Find the instantaneous real and reactive power if $v_S(t) = 650 \cos 314t$ (i.e., the frequency is 50 Hz).

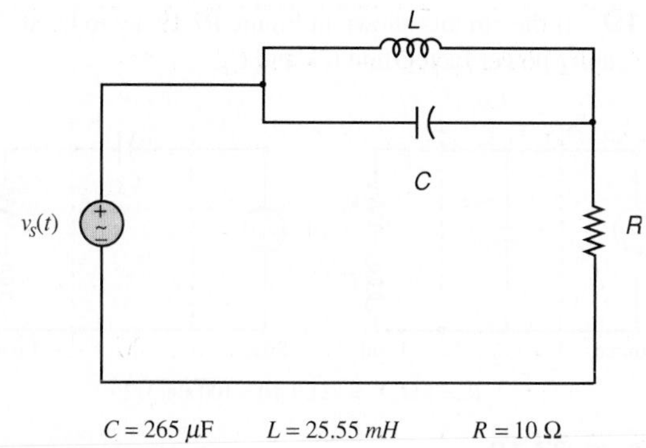

Figure P7.16

7.17 A load impedance, $Z_L = 10 + j3\ \Omega$, is connected to a source with line resistance equal to 1 Ω, as shown in Figure P7.17. Calculate the following values:

a. The average power delivered to the load.

b. The average power absorbed by the line.

c. The apparent power supplied by the generator.

d. The power factor of the load.

e. The power factor of line plus load.

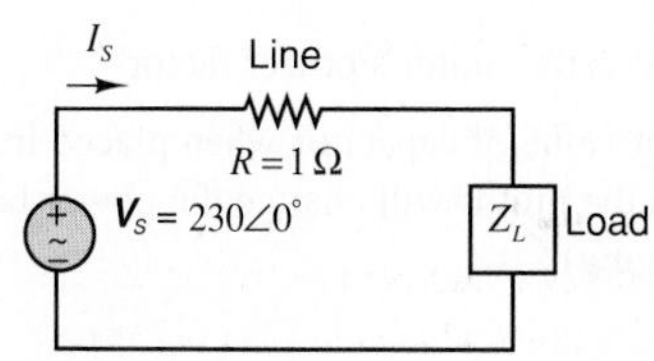

Figure P7.17

7.18 A single-phase motor draws 220 W at a power factor of 80 percent (lagging) when connected across a 200-V, 60-Hz source. A capacitor is connected in parallel with the load to give a unity power factor, as shown in Figure P7.18. Find the required capacitance.

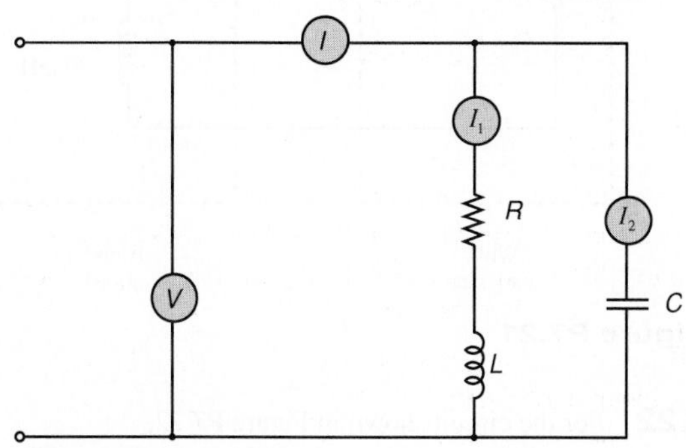

Figure P7.18

7.19 If the circuits shown in Figure P7.19 are to be at unity power factor, find C_P and C_S.

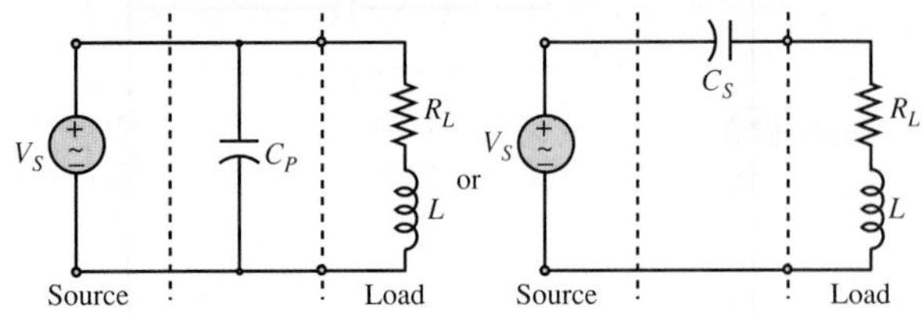

$R_L = 5\ \Omega$, $X_L = 5\ \Omega$, $v_S(t) = 100\sin(377t)$

Figure P7.19

7.20 A 1,000 W electric motor is connected to a source of 120 V_{rms}, 60 Hz, and the result is a lagging pf of 0.8. To correct the pf to 0.95 lagging, a capacitor is placed in parallel with the motor. Calculate the current drawn from the source with and without the capacitor connected. Determine the value of the capacitor required to make the correction.

7.21 The motor inside a blender can be modeled as a resistance in series with an inductance, as shown in Figure P7.21.

a. What is the average power, P_{AV}, dissipated in the load?

b. What is the motor's power factor?

c. What value of capacitor when placed in parallel with the motor will change the power factor to 0.9 (lagging)?

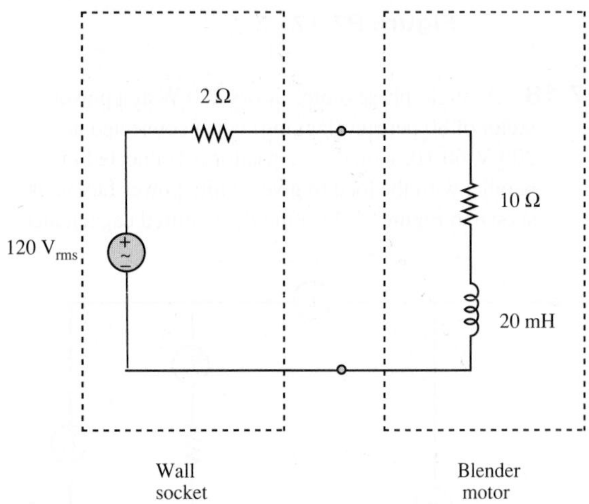

Figure P7.21

7.22 For the circuit shown in Figure P7.22,

a. Find the Thévenin equivalent circuit for the source.

b. Find the power dissipated by the load resistor.

c. What value of load impedance would permit maximum power transfer?

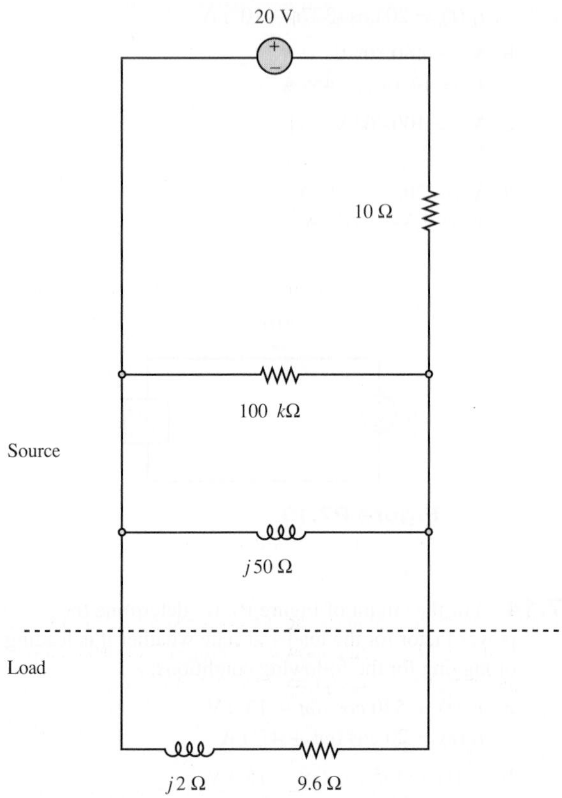

Figure P7.22

7.23 For the following numerical values, determine the average power P, the reactive power Q, and the complex power S of the circuit shown in Figure P7.23. Note: phasor quantities are rms.

a. $v_S(t) = 450\cos(377t)$ V
$i_L(t) = 50\cos(377t - 0.349)$ A

b. $\tilde{\mathbf{V}}_S = 140\angle 0$ V
$\tilde{\mathbf{I}}_L = 5.85\angle(-\pi/6)$ A

c. $\tilde{\mathbf{V}}_S = 50\angle 0$ V
$\tilde{\mathbf{I}}_L = 19.2\angle 0.8$ A

d. $\tilde{\mathbf{V}}_S = 740\angle(-\pi/4)$ V
$\tilde{\mathbf{I}}_L = 10.8\angle(-1.5)$ A

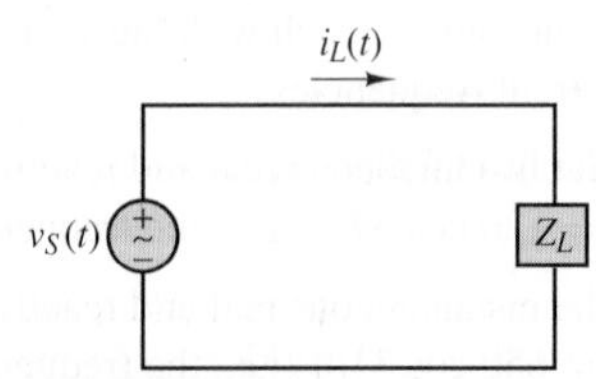

Figure P7.23

7.24 For the circuit of Figure P7.23, determine the power factor for the load and state whether it is leading or lagging for the following conditions:

a. $v_S(t) = 780\cos(\omega t + 1.2)$ V
   $i_L(t) = 90\cos(\omega t + \pi/2)$ A

b. $v_S(t) = 39\cos(\omega t + \pi/6)$ V
   $i_L(t) = 12\cos(\omega t - 0.185)$ A

c. $v_S(t) = 104\cos(\omega t)$ V
   $i_L(t) = 48.7\sin(\omega t + 2.74)$ A

d. $Z_L = (12 + j8)\ \Omega$

7.25 For the circuit of Figure P7.23, determine whether the load is capacitive or inductive for the circuit shown if

a. pf = 0.48 (leading)

b. pf = 0.17 (leading)

c. $v_S(t) = 18\cos(\omega t)$
   $i_L(t) = 1.8\sin(\omega t)$

d. $v_S(t) = 8.3\cos(\omega t - \pi/6)$
   $i_L(t) = 0.6\cos(\omega t - \pi/6)$

7.26 Find the real and reactive power supplied by the source in the circuit shown in Figure P7.26. Repeat if the frequency is increased by a factor of 3.

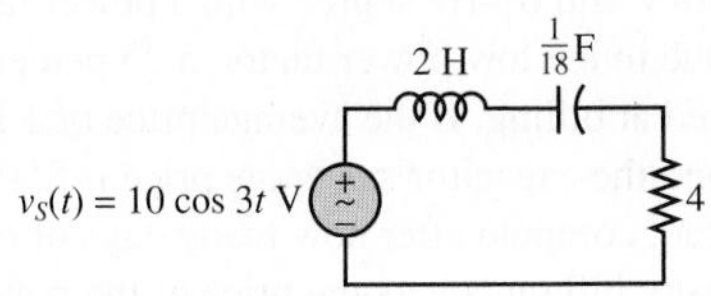

Figure P7.26

7.27 In the circuit shown in Figure P7.27, the sources are $\tilde{\mathbf{V}}_{S1} = 36\angle(-\pi/3)$ V and $\tilde{\mathbf{V}}_{S2} = 24\angle 0.644$ V. Find

a. The real and imaginary current supplied by each source.

b. The total real power supplied.

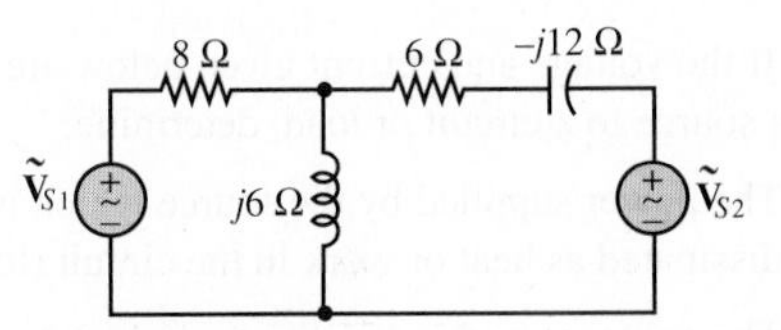

Figure P7.27

7.28 The load Z_L in the circuit of Figure P7.28 consists of a 25-Ω resistor in series with a 0.1-mF capacitor. Assuming $f = 60$ Hz, find

a. The source power factor.

b. The current $\tilde{\mathbf{I}}_S$.

c. The apparent power delivered to the load.

d. The apparent power supplied by the source.

e. The power factor of the load.

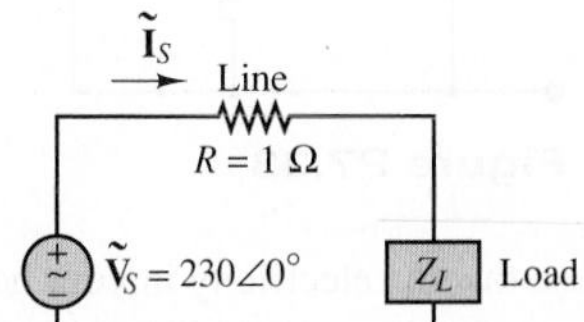

Figure P7.28

7.29 The load Z_L in the circuit of Figure P7.28 consists of a 25-Ω resistor in series with a 0.1-H inductor. Assuming $f = 60$ Hz, calculate the following.

a. The apparent power supplied by the source.

b. The apparent power delivered to the load.

c. The power factor of the load.

7.30 The load Z_L in the circuit of Figure P7.28 consists of a 25-Ω resistor in series with a 0.1-mF capacitor and a 70.35-mH inductor. Assuming $f = 60$ Hz, calculate the following.

a. The apparent power delivered to the load.

b. The real power supplied by the source.

c. The power factor of the load.

7.31 Calculate the apparent power, real power, and reactive power for the circuit shown in Figure P7.31. Draw the power triangle. Assume $f = 60$ Hz.

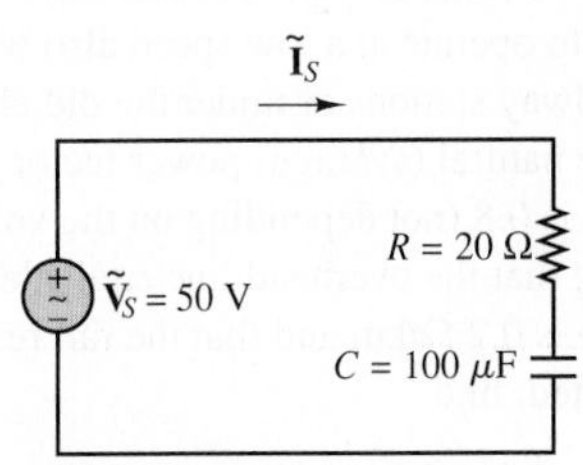

Figure P7.31

7.32 Repeat Problem 7.31 for the two cases $f = 50$ Hz and $f = 0$ Hz (DC).

7.33 A single-phase motor is connected as shown in Figure P7.33 to a 50-Hz network. The capacitor value is chosen to obtain unity power factor. If $V = 220$ V, $I = 20$ A, and $I_1 = 25$ A, find the capacitor value.

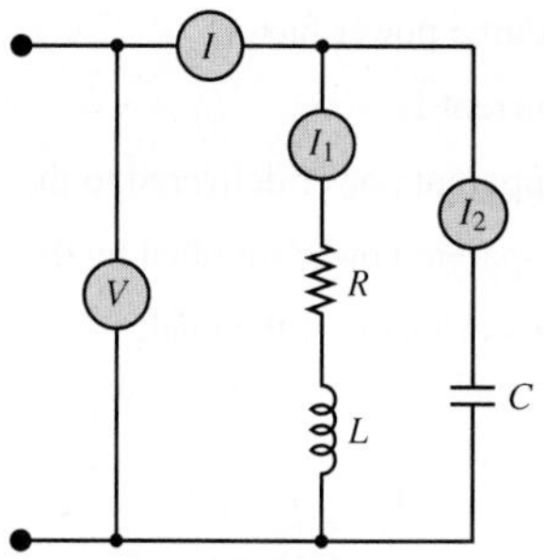

Figure P7.33

7.34 Suppose that the electricity in your home has gone out and the power company will not be able to have you hooked up again for several days. The freezer in the basement contains several hundred dollars' worth of food that you cannot afford to let spoil. You have also been experiencing very hot, humid weather and would like to keep one room air-conditioned with a window air conditioner, as well as run the refrigerator in your kitchen. When the appliances are on, they draw the following currents (all values are rms):

Air conditioner:	9.6 A @ 120 V
	pf = 0.90 (lagging)
Freezer:	4.2 A @ 120 V
	pf = 0.87 (lagging)
Refrigerator:	3.5 A @ 120 V
	pf = 0.80 (lagging)

In the worst-case scenario, how much power must an emergency generator supply?

7.35 The French TGV high-speed train absorbs 11 MW at 300 km/h (186 mi/h). The power supply module is shown in Figure P7.35. The module consists of two 25-kV single-phase power stations connected at the same overhead line, one at each end of the module. For the return circuits, the rail is used. However, the train is designed to operate at a low speed also with 1.5-kV DC in railway stations or under the old electrification lines. The natural (average) power factor in the AC operation is 0.8 (not depending on the voltage). Assuming that the overhead line equivalent specific resistance is 0.2 Ω/km and that the rail resistance could be neglected, find

a. The equivalent circuit.

b. The locomotive's current in the condition of a 10 percent voltage drop.

c. The reactive power.

d. The supplied real power, overhead line losses, and maximum distance between two power stations supplied in the condition of a 10 percent voltage drop when the train is located at the half-distance between the stations.

e. Overhead line losses in the condition of a 10 percent voltage drop when the train is located at the half-distance between the stations, assuming pf = 1. (The French TGV is designed with a state-of-the-art power compensation system.)

f. The maximum distance between the two power stations supplied in the condition of a 10 percent voltage drop when the train is located at the half-distance between the stations, assuming the DC (1.5-kV) operation at one-quarter power.

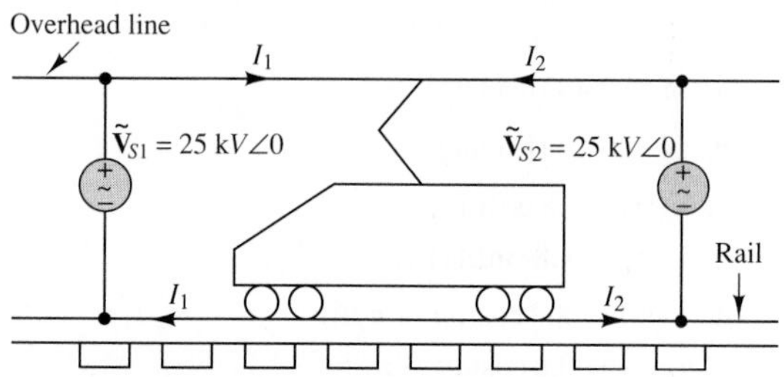

Figure P7.35

7.36 An industrial assembly hall is continuously lighted by one hundred 40-W mercury vapor lamps supplied by a 120-V and 60-Hz source with a power factor of 0.65. Due to the low power factor, a 25 percent penalty is applied at billing. If the average price of 1 kWh is \$0.01 and the capacitor's average price is \$50 per millifarad, compute after how many days of operation the penalty billing covers the price of the power factor correction capacitor. (To avoid penalty, the power factor must be greater than 0.85.)

7.37 With reference to Problem 7.36, consider that the current in the cable network is decreasing when power factor correction is applied. Find

a. The capacitor value for the unity power factor.

b. The maximum number of additional lamps that can be installed without changing the cable network if a local compensation capacitor is used.

7.38 If the voltage and current given below are supplied by a source to a circuit or load, determine

a. The power supplied by the source which is dissipated as heat or work in the circuit (load).

b. The power stored in reactive components in the circuit (load).

c. The power factor angle and the power factor.

$$\tilde{\mathbf{V}}_s = 7\angle 0.873 \text{ V} \qquad \tilde{\mathbf{I}}_s = 13\angle(-0.349) \text{ A}$$

7.39 Determine the time-averaged total power, the real power dissipated, and the reactive power stored in each of the impedances in the circuit shown in Figure P7.39 if

$$\tilde{\mathbf{V}}_{s1} = 170/\sqrt{2}\angle 0 \text{ V}$$
$$\tilde{\mathbf{V}}_{s2} = 170/\sqrt{2}\angle\frac{\pi}{2} \text{ V}$$
$$\omega = 377 \text{ rad/s}$$
$$Z_1 = 0.7\angle\frac{\pi}{6}\ \Omega$$
$$Z_2 = 1.5\angle 0.105\ \Omega$$
$$Z_3 = 0.3 + j0.4\ \Omega$$

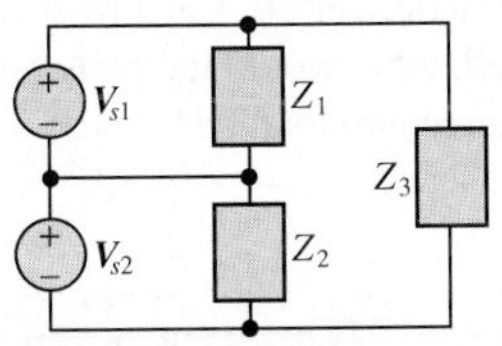

Figure P7.39

7.40 If the voltage and current supplied to a circuit or load by a source are

$$\tilde{\mathbf{V}}_s = 170\angle(-0.157) \text{ V} \qquad \tilde{\mathbf{I}}_s = 13\angle 0.28 \text{ A}$$

determine

a. The power supplied by the source which is dissipated as heat or work in the circuit (load).

b. The power stored in reactive components in the circuit (load).

c. The power factor angle and power factor.

Section 7.3: Transformers

7.41 A center-tapped transformer has the schematic representation shown in Figure P7.41. The primary-side voltage is stepped down to two secondary-side voltages. Assume that each secondary supplies a 5-kW resistive load and that the primary is connected to 120 V rms. Find

a. The primary power.

b. The primary current.

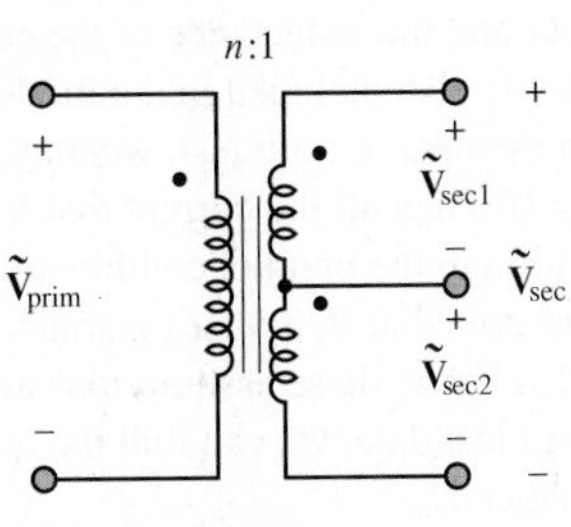

Figure P7.41

7.42 A center-tapped transformer has the schematic representation shown in Figure P7.41. The primary-side voltage is stepped down to a secondary-side voltage $\tilde{\mathbf{V}}_{\text{sec}}$ by a ratio of $n:1$. On the secondary side, $\tilde{\mathbf{V}}_{\text{sec1}} = \tilde{\mathbf{V}}_{\text{sec2}} = \frac{1}{2}\tilde{\mathbf{V}}_{\text{sec}}$.

a. If $\tilde{\mathbf{V}}_{\text{prim}} = 220\angle 0°$ V and $n = 11$, find $\tilde{\mathbf{V}}_{\text{sec}}$, $\tilde{\mathbf{V}}_{\text{sec1}}$, and $\tilde{\mathbf{V}}_{\text{sec2}}$.

b. What must n be if $\tilde{\mathbf{V}}_{\text{prim}} = 110\angle 0°$ V and we desire $|\tilde{\mathbf{V}}_{\text{sec2}}|$ to be 5 V rms?

7.43 For the circuit shown in Figure P7.43, assume that $v_g = 120$ V rms. Find

a. The total resistance seen by the voltage source.

b. The primary current.

c. The primary power.

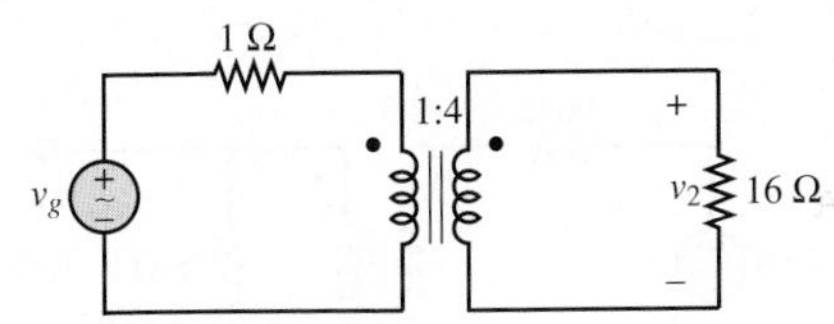

Figure P7.43

7.44 With reference to Problem 7.43 and Figure P7.43 find

a. The secondary current.

b. The installation efficiency $P_{\text{load}}/P_{\text{source}}$.

c. The value of the load resistance which can absorb the maximum power from the given source.

7.45 An ideal transformer is rated to deliver 460 kVA at 380 V to a customer, as shown in Figure P7.45.

a. How much current can the transformer supply to the customer?

b. If the customer's load is purely resistive (i.e., if pf = 1), what is the maximum power that the customer can receive?

c. If the customer's power factor is 0.8 (lagging), what is the maximum usable power the customer can receive?

d. What is the maximum power if the pf is 0.7 (lagging)?

e. If the customer requires 300 kW to operate, what is the minimum power factor with the given size transformer?

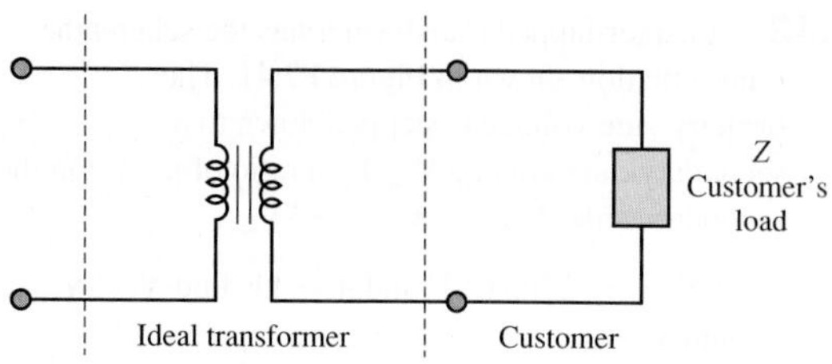

Figure P7.45

7.46 For the ideal transformer shown in Figure P7.46, consider that $v_S(t) = 294\cos(377t)$ V. Find

a. Primary current.

b. $v_o(t)$.

c. Secondary power.

d. The installation efficiency $P_{\text{load}}/P_{\text{source}}$.

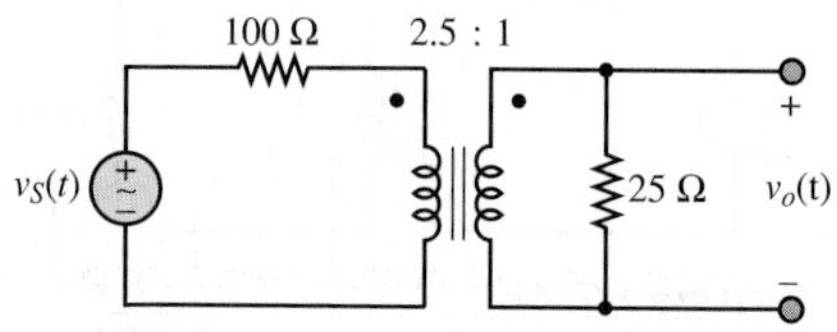

Figure P7.46

7.47 If the transformer shown in Figure P7.47 is ideal, find the turns ratio $N = 1/n$ that will provide maximum power transfer to the load.

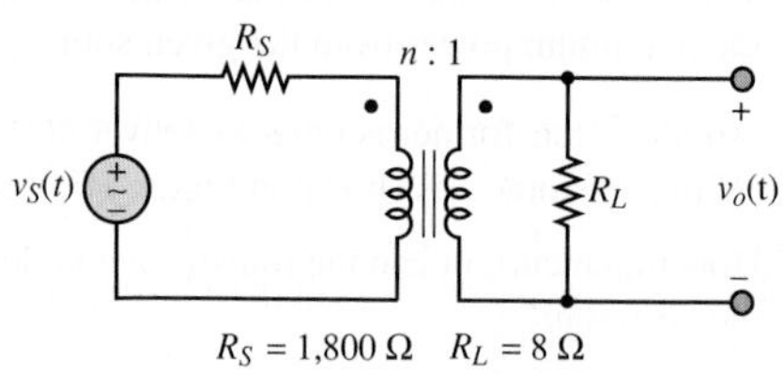

Figure P7.47

7.48 Assume the 8-Ω resistor is the load in the circuit shown in Figure P7.48. Assume $v_g = 110$ V rms and a variable turns ratio of $1 : n$. Find

a. The maximum power dissipated by the load.

b. The maximum power absorbed from the source.

c. The power transfer efficiency.

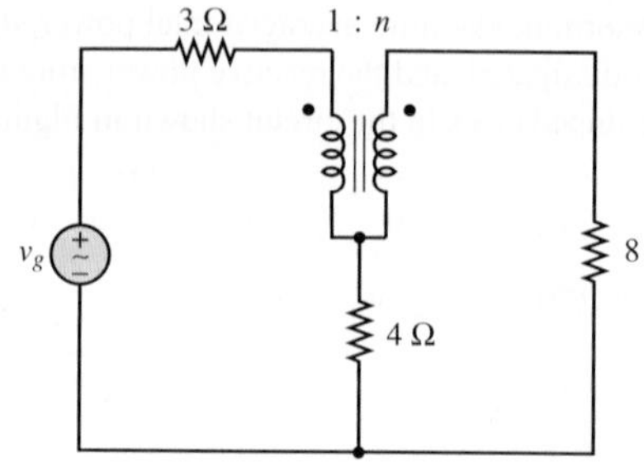

Figure P7.48

7.49 If we knew that the transformer shown in Figure P7.49 were to deliver 50 A at 110 V rms with a certain resistive load, what would the power transfer efficiency between source and load be?

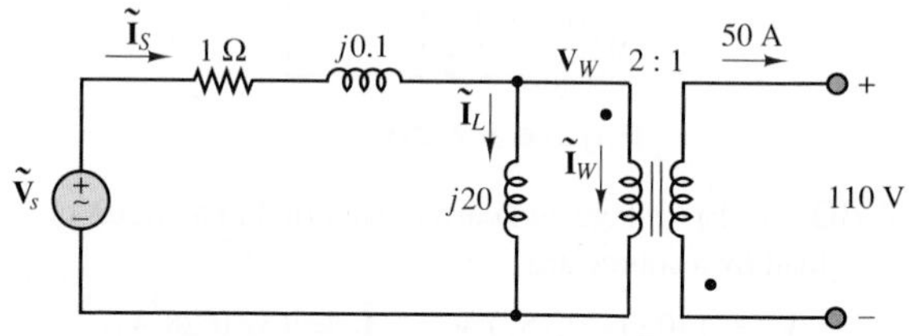

Figure P7.49

7.50 A method for determining the equivalent circuit of a transformer consists of two tests: the open-circuit test and the short-circuit test. The open-circuit test, shown in Figure P7.50(a), is usually done by applying rated voltage to the primary side of the transformer while leaving the secondary side open. The current into the primary side is measured, as is the power dissipated.

The short-circuit test, shown in Figure P7.50(b), is performed by increasing the primary voltage until rated current is going into the transformer while the secondary side is short-circuited. The current into the transformer, the applied voltage, and the power dissipated are measured.

The equivalent circuit of a transformer is shown in Figure P7.50(c), where r_w and L_w represent the winding resistance and inductance, respectively, and r_c and L_c represent the losses in the core of the transformer and the inductance of the core. The ideal transformer is also included in the model.

With the open-circuit test, we may assume that $\tilde{\mathbf{I}}_P = \tilde{\mathbf{I}}_S = 0$. Then all the current that is measured is directed through the parallel combination of r_c and L_c. We also assume that $|r_c || j\omega L_c|$ is much greater than $r_w + j\omega L_w$. Using these assumptions and the open-circuit test data, we can find the resistance r_c and the inductance L_c.

In the short-circuit test, we assume that $\tilde{\mathbf{V}}_{\text{secondary}}$ is zero, so that the voltage on the primary side of the ideal transformer is also zero, causing no current flow through the $r_c - L_c$ parallel combination. Using this assumption with the short-circuit test data, we are able to find the resistance r_w and inductance L_w.

Using the following test data, find the equivalent circuit of the transformer:

Open-circuit test: $\tilde{\mathbf{V}} = 241$ V
$\tilde{\mathbf{I}} = 0.95$ A
$P = 32$ W

Short-circuit test: $\tilde{\mathbf{V}} = 5$ V
$\tilde{\mathbf{I}} = 5.25$ A
$P = 26$ W

Both tests were made at $\omega = 377$ rad/s.

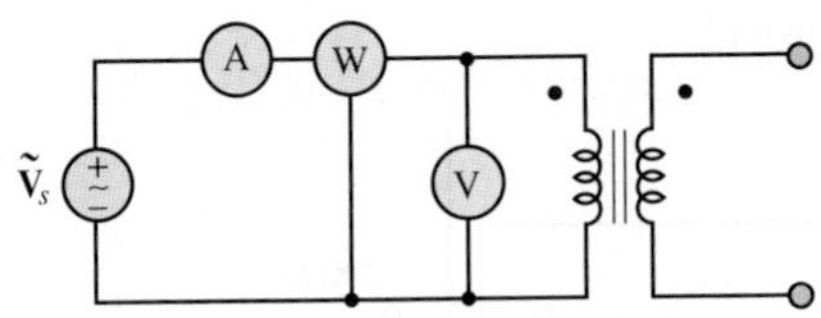

(a)

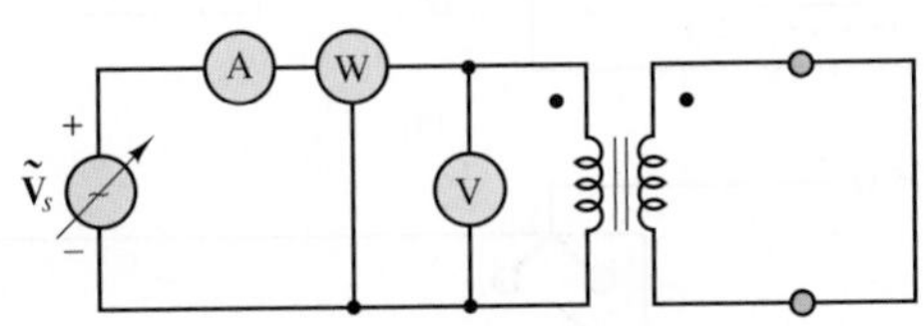

(b)

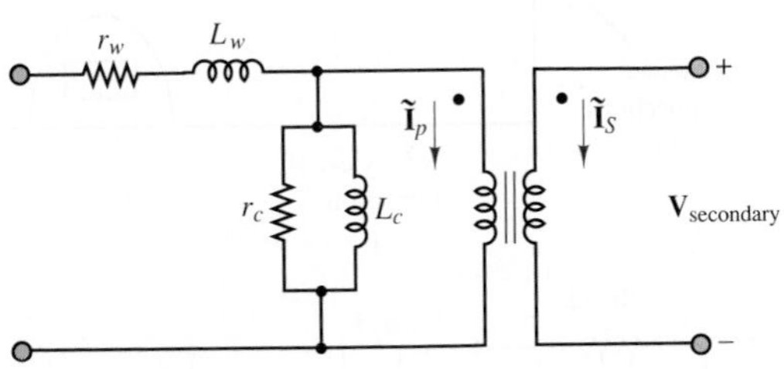

(c)

Figure P7.50

7.51 Using the methods of Problem 7.50 and the following data, find the equivalent circuit of the transformer tested:

Open-circuit test: $\tilde{\mathbf{V}}_P = 4{,}600$ V
$\tilde{\mathbf{I}}_{OC} = 0.7$ A
$P = 200$ W

Short-circuit test: $P = 50$ W
$\tilde{\mathbf{V}}_P = 5.2$ V

The transformer is a 460-kVA transformer, and the tests are performed at 60 Hz.

7.52 A method of thermal treatment for a steel pipe is to heat the pipe by the Joule effect, flowing a current directly in the pipe. In most cases, a low-voltage high-current transformer is used to deliver the current through the pipe. In this problem, we consider a single-phase transformer at 220 V rms, which delivers 1 V. Due to the pipe's resistance variation with temperature, a secondary voltage regulation is needed in the range of 10 percent, as shown in Figure P7.52. The voltage regulation is obtained with five different slots in the primary winding (high-voltage regulation). Assuming that the secondary coil has two turns, find the number of turns for each slot.

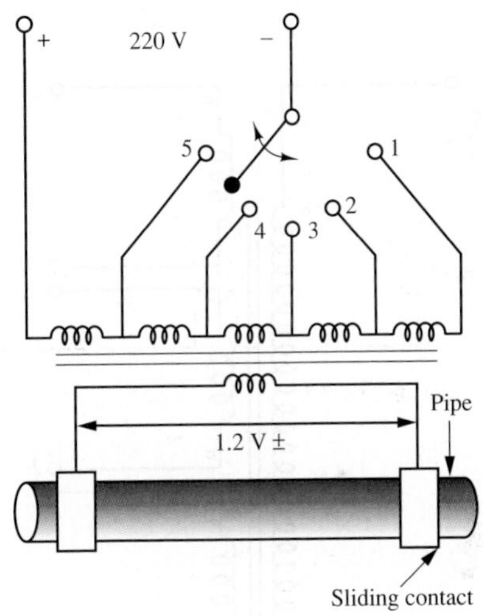

Figure P7.52

7.53 With reference to Problem 7.52, assume that the pipe's resistance is 0.0002 Ω, the secondary resistance (connections + slide contacts) is 0.00005 Ω, and the primary current is 28.8 A with pf = 0.91 Find

a. The plot number.

b. The secondary reactance.

c. The power transfer efficiency.

7.54 A single-phase transformer used for street lighting (high-pressure sodium discharge lamps) converts 6 kV to 230 V (to load) with an efficiency of 0.95. Assuming pf = 0.8 and the primary apparent power is 30 kVA, find

a. The secondary current.

b. The transformer's ratio.

7.55 The transformer shown in Figure P7.55 has several sets of windings on the secondary side. The windings have the following turns ratios:

a. $: N = 1/15$

b. $: N = 1/4$

c. $: N = 1/12$

d. $: N = 1/18$

If $\mathbf{V}_{\text{prim}} = 120$ V, find and draw the connections that will allow you to construct the following voltage sources:

a. $24.67\angle 0°$ V

b. $36.67\angle 0°$ V

c. $18\angle 0°$ V

d. $54.67\angle 180°$ V

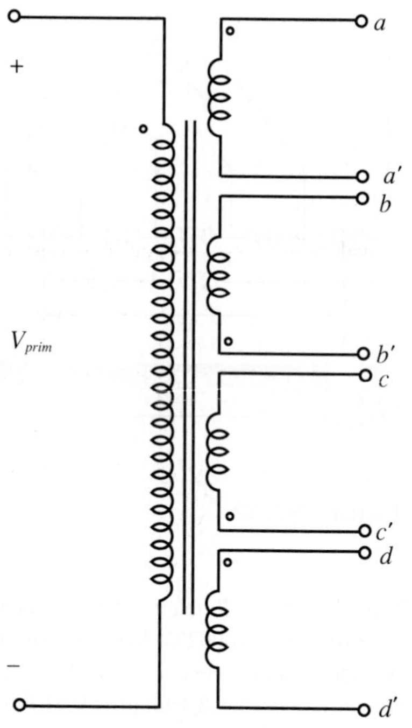

Figure P7.55

7.56 The circuit in Figure P7.56 shows the use of ideal transformers for impedance matching. You have a limited choice of turns ratios among available transformers. Suppose you can find transformers with turns ratios of 2:1, 7:2, 120:1, 3:2, and 6:1. If Z_L is $475\angle -25°\ \Omega$ and Z_{ab} must be $267\angle -25°$, find the combination of transformers that will provide this impedance. (You may assume that polarities are easily reversed on these transformers.)

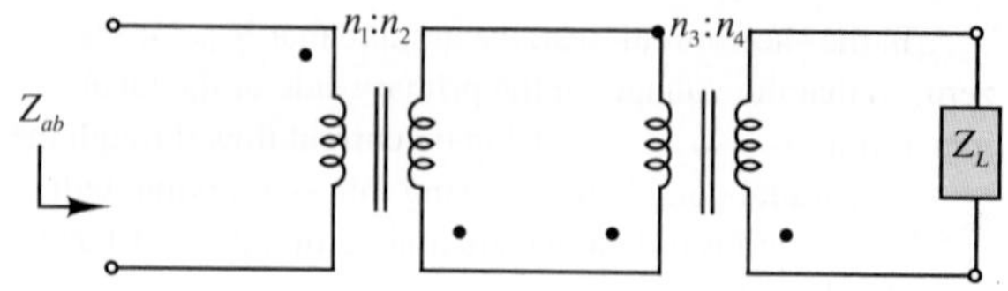

Figure P7.56

7.57 The wire that connects an antenna on your roof to the TV set in your den is a 300-Ω wire, as shown in Figure P7.57(a). This means that the impedance seen by the connections on your set is 300 Ω. Your TV, however, has a 75-Ω impedance connection, as shown in Figure P7.57(b). To achieve maximum power transfer from the antenna to the television set, you place an ideal transformer between the antenna and the TV as shown in Figure P7.57(c). What is the turns ratio, $N = 1/n$, needed to obtain maximum power transfer?

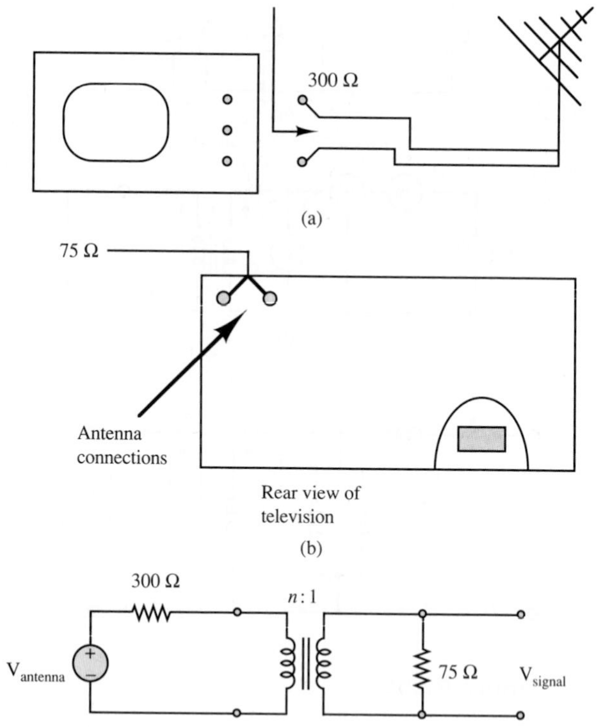

Figure P7.57

Section 7.4: Three-Phase Power

7.58 The magnitude of the phase voltage of a balanced three-phase wye system is 220 V rms. Express each phase and line voltage in both polar and rectangular coordinates.

7.59 The phase currents in a four-wire wye-connected load are as follows:

$$\tilde{\mathbf{I}}_{an} = 10\angle 0, \qquad \tilde{\mathbf{I}}_{bn} = 12\angle \frac{5\pi}{6} \qquad \tilde{\mathbf{I}}_{cn} = 8\angle 2.88$$

Determine the current in the neutral wire.

7.60 For the circuit shown in Figure P7.60, we see that each voltage source has a phase difference of $2\pi/3$ in relation to the others.

a. Find $\tilde{\mathbf{V}}_{RW}$, $\tilde{\mathbf{V}}_{WB}$, and $\tilde{\mathbf{V}}_{BR}$, where $\tilde{\mathbf{V}}_{RW} = \tilde{\mathbf{V}}_R - \tilde{\mathbf{V}}_W$, $\tilde{\mathbf{V}}_{WB} = \tilde{\mathbf{V}}_W - \tilde{\mathbf{V}}_B$, and $\tilde{\mathbf{V}}_{BR} = \tilde{\mathbf{V}}_B - \tilde{\mathbf{V}}_R$.

b. Repeat part a, using the calculations

$$\tilde{\mathbf{V}}_{RW} = \tilde{\mathbf{V}}_R \sqrt{3}\angle(-\pi/6)$$
$$\mathbf{V}_{WB} = \mathbf{V}_W \sqrt{3}\angle(-\pi/6)$$
$$\mathbf{V}_{BR} = \mathbf{V}_B \sqrt{3}\angle(-\pi/6)$$

c. Compare the results of part a with the results of part b.

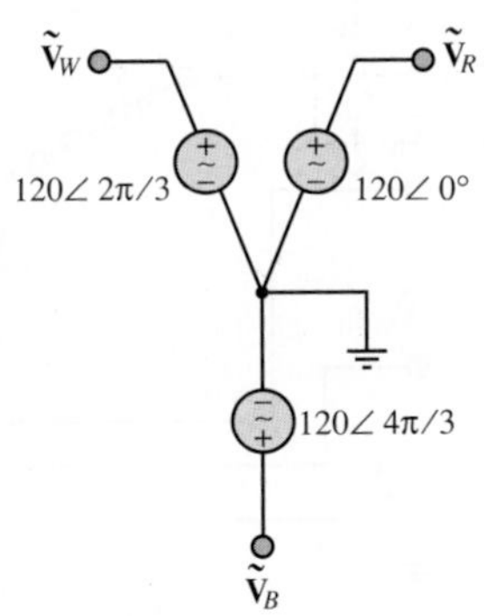

Figure P7.60

7.61 For the three-phase circuit shown in Figure P7.61, find the current in the neutral wire and the real power.

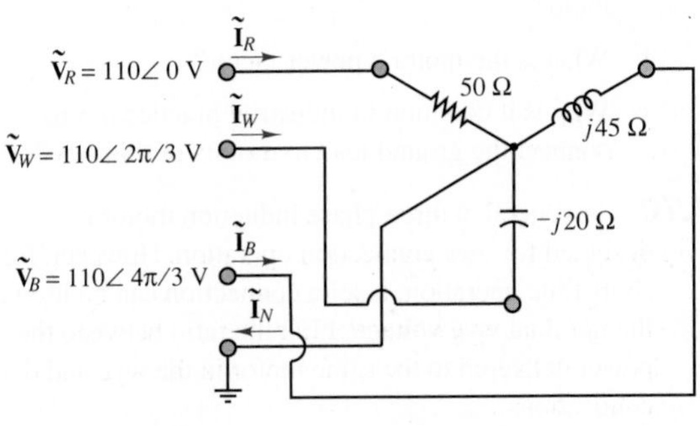

Figure P7.61

7.62 For the circuit shown in Figure P7.62, find the current in the neutral wire and the real power.

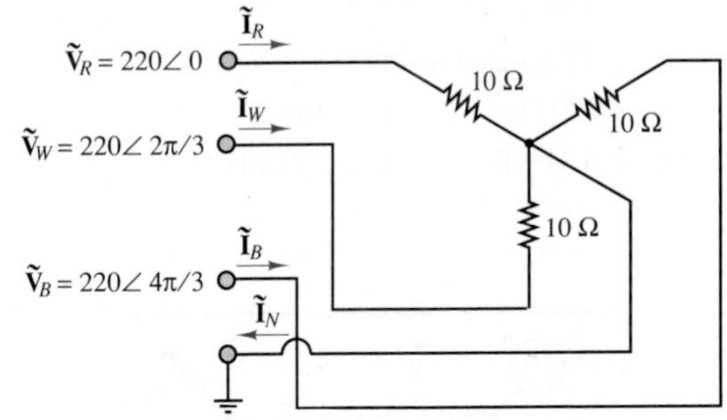

Figure P7.62

7.63 A three-phase steel-treatment electric oven has a phase resistance of 10 Ω and is connected at three-phase 380-V AC. Compute

a. The current flowing through the resistors in wye and delta connections.

b. The power of the oven in wye and delta connections.

7.64 A naval in-board synchronous generator has an apparent power of 50 kVA and supplies a three-phase network of 380 V. Compute the phase currents, the active powers, and the reactive powers if

a. The power factor is 0.85.

b. The power factor is 1.

7.65 In the circuit of Figure P7.65:

$v_{s1} = 170\cos(\omega t)$ V

$v_{s2} = 170\cos(\omega t + 2\pi/3)$ V

$v_{s3} = 170\cos(\omega t - 2\pi/3)$ V

$f = 60$ Hz $\qquad Z_1 = 0.5\angle 20°\ \Omega$

$Z_2 = 0.35\angle 0°\ \Omega \qquad Z_3 = 1.7\angle(-90°)\ \Omega$

Determine the current through Z_1, using

a. Loop/mesh analysis.

b. Node analysis.

c. Superposition.

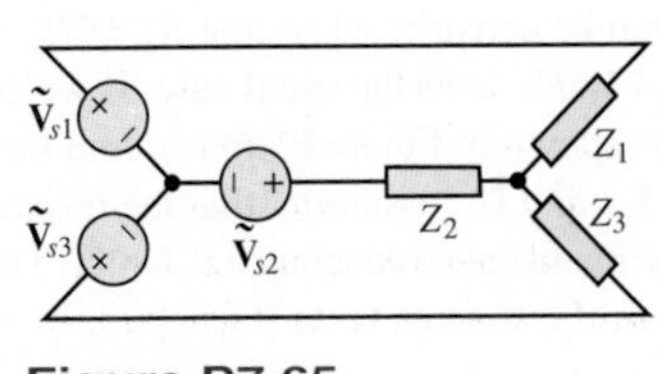

Figure P7.65

7.66 Determine the current through R in the circuit of Figure P7.66:

$$v_1 = 170\cos(\omega t) \quad \text{V}$$
$$v_2 = 170\cos(\omega t - 2\pi/3) \quad \text{V}$$
$$v_3 = 170\cos(\omega t + 2\pi/3) \quad \text{V}$$
$$f = 400 \text{ Hz} \qquad R = 100\ \Omega$$
$$C = 0.47\ \mu\text{F} \qquad L = 100 \text{ mH}$$

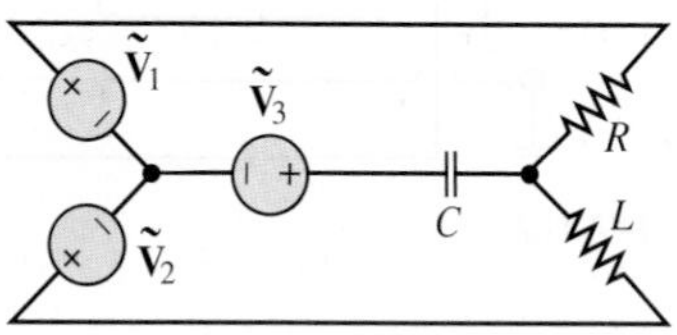

Figure P7.66

7.67 The three sources in the circuit of Figure P7.67 are connected in wye configuration and the loads in a delta configuration. Determine the current through each impedance.

$$v_{s1} = 170\cos(\omega t) \quad \text{V}$$
$$v_{s2} = 170\cos(\omega t + 2\pi/3) \quad \text{V}$$
$$v_{s3} = 170\cos(\omega t - 2\pi/3) \quad \text{V}$$
$$f = 60 \text{ Hz} \qquad Z_1 = 3\angle 0\ \Omega$$
$$Z_2 = 7\angle \pi/2\ \Omega \qquad Z_3 = 0 - j11\ \Omega$$

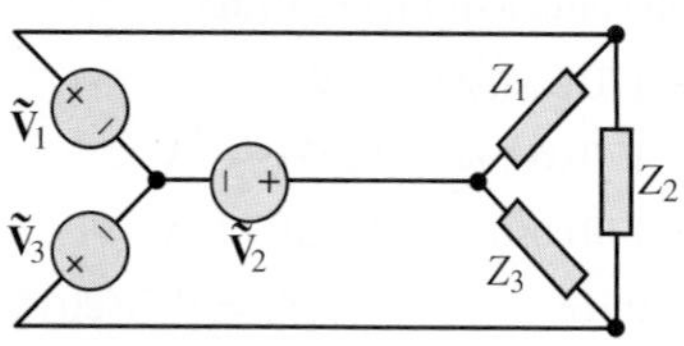

Figure P7.67

7.68 If we model each winding of a three-phase motor like the circuit shown in Figure P7.68(a) and connect the windings as shown in Figure P7.68(b), we have the three-phase circuit shown in Figure P7.68(c). The motor can be constructed so that $R_1 = R_2 = R_3$ and $L_1 = L_2 = L_3$, as is the usual case. If we connect the motor as shown in Figure P7.68(c), find the currents $\tilde{\mathbf{I}}_R$, $\tilde{\mathbf{I}}_W$, $\tilde{\mathbf{I}}_B$, and $\tilde{\mathbf{I}}_N$, assuming that the resistances are 40 Ω each and each inductance is 5 mH. The frequency of each of the sources is 60 Hz.

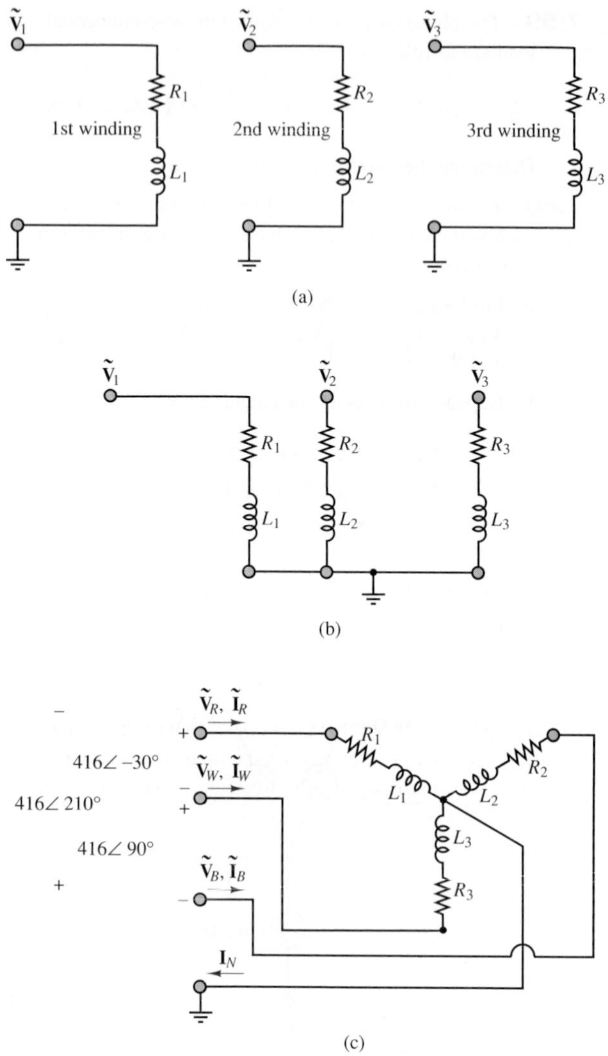

Figure P7.68

7.69 With reference to the motor of Problem 7.67,

a. How much power (in watts) is delivered to the motor?

b. What is the motor's power factor?

c. Why is it common in industrial practice *not* to connect the ground lead to motors of this type?

7.70 In general, a three-phase induction motor is designed for wye connection operation. However, for short-time operation, a delta connection can be used at the nominal wye voltage. Find the ratio between the power delivered to the same motor in the wye and delta connections.

7.71 A residential four-wire system supplies power at 220 V rms to the following single-phase appliances: On the first phase, there are ten 75-W bulbs. On the second phase, there is a 750-W vacuum cleaner with a power factor of 0.87. On the third phase, there are ten 40-W fluorescent lamps with power factor of 0.64. Find

a. The current in the netural wire.

b. The real, reactive, and apparent power for each phase.

7.72 The electric power company is concerned with the loading of its transformers. Since it is responsible for a large number of customers, it must be certain that it can supply the demands of *all* customers. The power company's transformers will deliver rated kVA to the secondary load. However, if the demand increased to a point where greater than rated current were required, the secondary voltage would have to drop below rated value. Also, the current would increase, and with it the I^2R losses (due to winding resistance), possibly causing the transformer to overheat. Unreasonable current demand could be caused, for example, by excessively low power factors at the load.

The customer, on the other hand, is not greatly concerned with an inefficient power factor, provided that sufficient power reaches the load. To make the customer more aware of power factor considerations, the power company may install a penalty on the customer's bill. A typical penalty–power factor chart is shown in Table 7.3. Power factors below 0.7 are not permitted. A 25 percent penalty will be applied to any billing after two consecutive months in which the customer's power factor has remained below 0.7.

Table 7.3

Power factor	Penalty
0.850 and higher	None
0.8 to 0.849	1%
0.75 to 0.799	2%
0.7 to 0.749	3%

Courtesy of Detroit Edison.

The wye-wye circuit shown in Figure P7.72 is representative of a three-phase motor load. Assume rms values.

a. Find the total power supplied to the motor.

b. Find the power converted to mechanical energy if the motor is 80 percent efficient.

c. Find the power factor.

d. Does the company risk facing a power factor penalty on its next bill if all the motors in the factory are similar to this one?

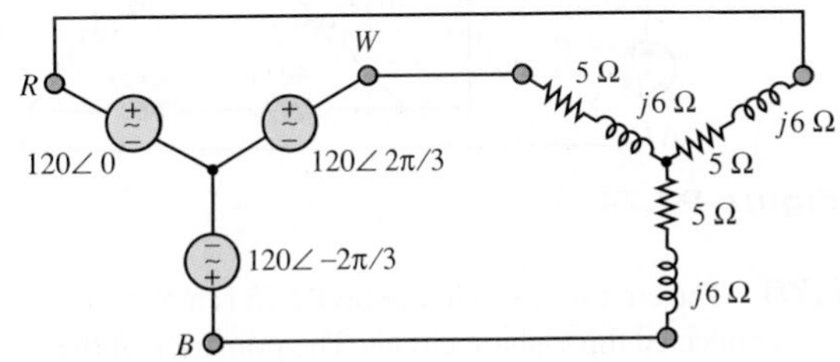

Figure P7.72

7.73 To correct the power factor problems of the motor in Problem 7.72, the company has decided to install capacitors as shown in Figure P7.73. Assume rms values.

a. What capacitance must be installed to achieve a unity power factor if the line frequency is 60 Hz?

b. Repeat part a if the power factor is to be 0.85 (lagging).

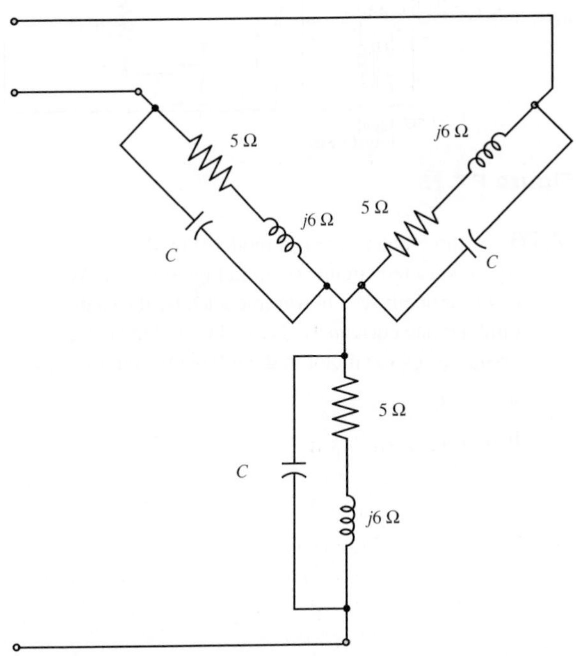

Figure P7.73

7.74 Find the apparent power and the real power delivered to the load in the Y-Δ circuit shown in Figure P7.74. What is the power factor? Assume rms values.

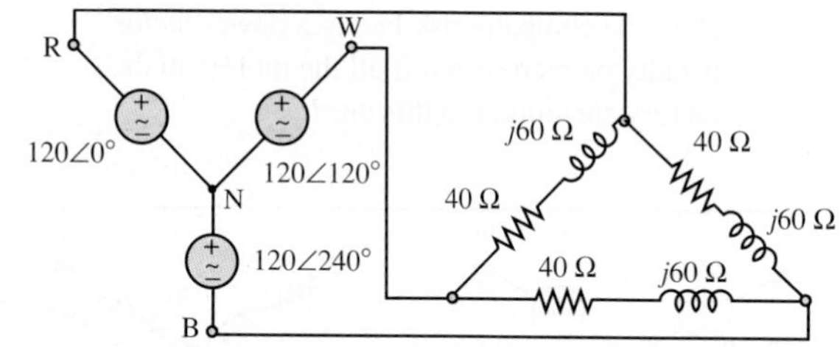

Figure P7.74

7.75 The circuit shown in Figure P7.75 is a Y-Δ-Y connected three-phase circuit. The primaries of the transformers are wye-connected, the secondaries are delta-connected, and the load is wye-connected. Find the currents $\mathbf{I}_{RP}$, $\mathbf{I}_{WP}$, $\mathbf{I}_{BP}$, $\mathbf{I}_A$, $\mathbf{I}_B$, and $\mathbf{I}_C$.

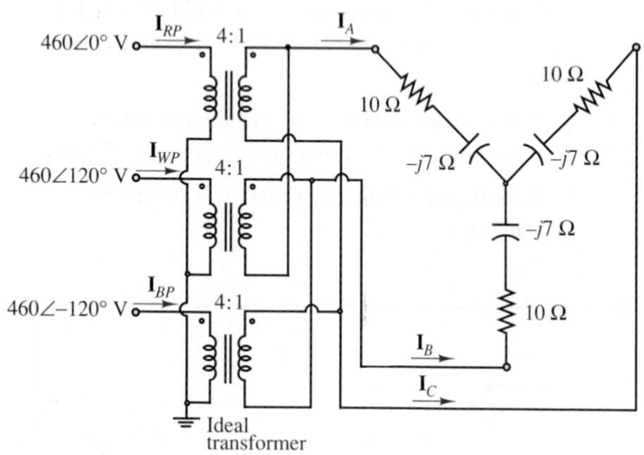

Figure P7.75

7.76 A three-phase motor is modeled by the wye-connected circuit shown in Figure P7.76. At $t = t_1$, a line fuse is blown (modeled by the switch). Find the line currents $\mathbf{I}_R$, $\mathbf{I}_W$, and $\mathbf{I}_B$ and the power dissipated by the motor in the following conditions:

a. $t \ll t_1$

b. $t \gg t_1$

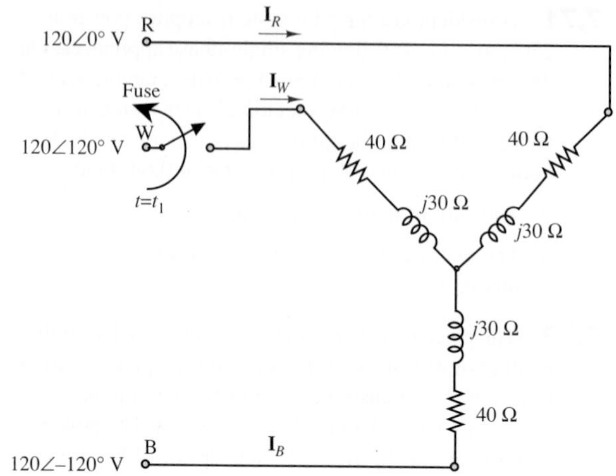

Figure P7.76

7.77 For the circuit shown in Figure P7.77, find the currents $\mathbf{I}_A$, $\mathbf{I}_B$, $\mathbf{I}_C$ and $\mathbf{I}_N$, and the real power dissipated by the load.

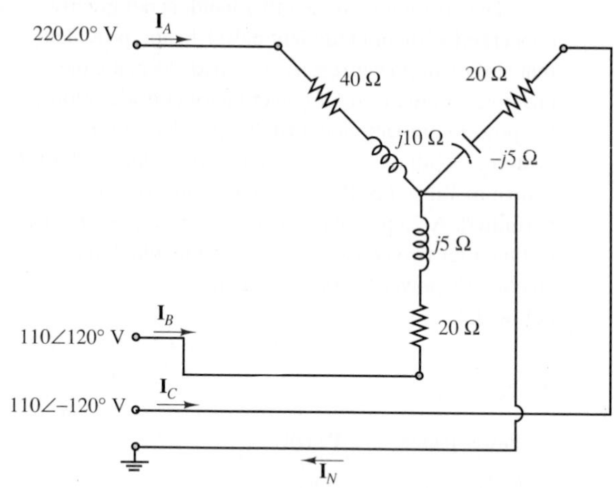

Figure P7.77

PART II

ELECTRONICS

CHAPTER

8

OPERATIONAL AMPLIFIERS

In this chapter we analyze the properties of the ideal amplifier and explore the features of a general-purpose amplifier circuit known as the *operational amplifier* (*op-amp*). Understanding the gain and frequency response properties of the operational amplifier is essential for the user of electronic instrumentation. Fortunately, the availability of operational amplifiers in integrated-circuit form has made the task of analyzing such circuits quite simple. The models presented in this chapter are based on concepts that have already been explored at length in earlier chapters, namely, Thévenin and Norton equivalent circuits and frequency response ideas.

Mastery of operational amplifier fundamentals is essential in any practical application of electronics. This chapter is aimed at developing your understanding of the fundamental properties of practical operational amplifiers. A number of useful applications are introduced in the examples and homework problems.

Learning Objectives

1. Understand the properties of ideal amplifiers and the concepts of gain, input impedance, and output impedance. *Section 8.1.*
2. Understand the difference between open-loop and closed-loop op-amp configurations; and compute the gain of (or complete the design of) simple inverting, noninverting, summing, and differential amplifiers using ideal op-amp analysis. Analyze more advanced op-amp circuits, using ideal op-amp analysis; and identify important performance parameters in op-amp data sheets. *Section 8.2.*
3. Analyze and design simple active filters. Analyze and design ideal integrator and differentiator circuits. *Sections 8.3, 8.4.*
4. Understand the structure and behavior of analog computers; design analog computer circuits to solve simple differential equations. *Section 8.5.*
5. Understand the principal physical limitations of an op-amp. *Section 8.6.*

8.1 IDEAL AMPLIFIERS

One of the most important functions in electronic instrumentation is that of amplification. The need to amplify low-level electric signals arises frequently in a number of applications. Perhaps the most familiar use of amplifiers arises in converting the low-voltage signal from a cassette tape player, a radio receiver, or a compact disk player to a level suitable for driving a pair of speakers. Figure 8.1 depicts a typical arrangement. Amplifiers have a number of applications of interest to the non–electrical engineer, such as the amplification of low-power signals from transducers (e.g., bioelectrodes, strain gauges, thermistors, and accelerometers) and other, less obvious functions that will be reviewed in this chapter, for example, filtering and impedance isolation. We turn first to the general features and characteristics of amplifiers, before delving into the analysis of the operational amplifier.

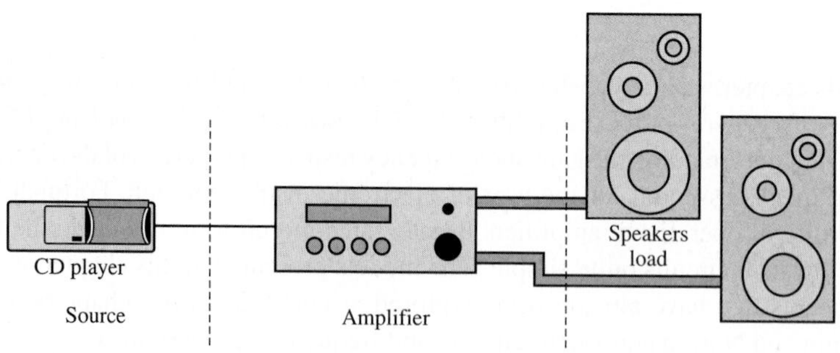

Figure 8.1 Amplifier in audio system

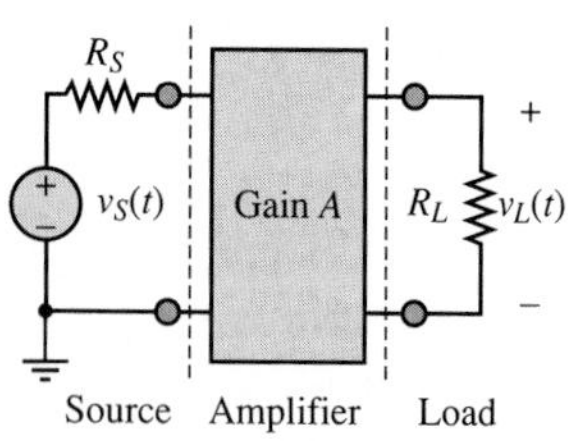

Figure 8.2 A voltage amplifier

Ideal Amplifier Characteristics

The simplest model for an amplifier is depicted in Figure 8.2, where a signal $v_S(t)$ is shown being amplified by a constant factor A, called the *gain* of the amplifier. Ideally, the load voltage should be given by the expression

$$v_L(t) = Av_S(t) \tag{8.1}$$

Note that the source has been modeled as a Thévenin equivalent, and the load as an equivalent resistance. Thévenin's theorem guarantees that this picture can be representative of more complex circuits. Hence, the equivalent source circuit is the circuit the amplifier "sees" from its input port; and R_L, the load, is the equivalent resistance seen from the output port of the amplifier.

What would happen if the roles were reversed? That is, what does the source see when it "looks" into the input port of the amplifier, and what does the load see when it "looks" into the output port of the amplifier? While it is not clear at this point how one might characterize the internal circuitry of an amplifier (which is rather complex), it can be presumed that the amplifier will act as an equivalent load with respect to the source and as an equivalent source with respect to the load. After all, this is a direct application of Thévenin's theorem. Figure 8.3 provides a pictorial representation of this simplified characterization of an amplifier. The "black box" of Figure 8.2 is now represented as an equivalent circuit with the following behavior. The input circuit has equivalent resistance R_{in}, so that the input voltage v_{in} is given by

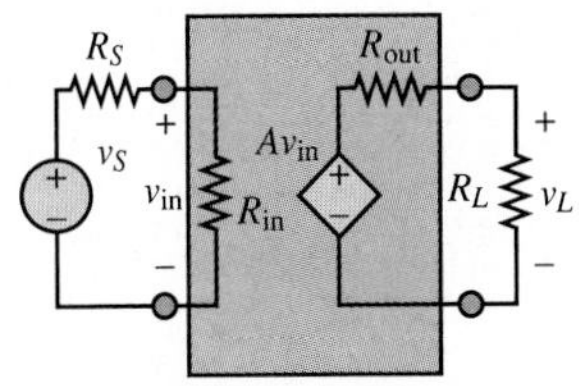

Figure 8.3 Simple voltage amplifier model

$$v_{\text{in}} = \frac{R_{\text{in}}}{R_S + R_{\text{in}}} v_S \tag{8.2}$$

The equivalent input voltage seen by the amplifier is then amplified by a constant factor A. This is represented by the controlled voltage source Av_{in}. The controlled source appears in series with an internal resistor R_{out}, denoting the internal (output) resistance of the amplifier. Thus, the voltage presented to the load is

$$v_L = Av_{\text{in}} \frac{R_L}{R_{\text{out}} + R_L} \tag{8.3}$$

or, by substituting the equation for v_{in},

$$v_L = \left(A \frac{R_{\text{in}}}{R_S + R_{\text{in}}} \frac{R_L}{R_{\text{out}} + R_L} \right) v_S \tag{8.4}$$

In other words, the load voltage is an amplified version of the source voltage.

Unfortunately, the amplification factor is now dependent on both the source and load impedances, and on the input and output resistance of the amplifier. Thus, a given amplifier would perform differently with different loads or sources. What are the desirable characteristics for a voltage amplifier that would make its performance relatively independent of source and load impedances? Consider, once again, the expression for v_{in}. If the input resistance of the amplifier R_{in} were very large, the source voltage v_S and the input voltage v_{in} would be approximately equal:

$$v_{\text{in}} \approx v_S \tag{8.5}$$

since

$$\lim_{R_{\text{in}} \to \infty} \frac{R_{\text{in}}}{R_{\text{in}} + R_S} = 1 \tag{8.6}$$

By an analogous argument, it can also be seen that the desired output resistance for the amplifier R_{out} should be very small, since for an amplifier with $R_{\text{out}} = 0$, the load voltage would be

$$v_L = Av_{\text{in}} \tag{8.7}$$

LO1 Combining these two results, we can see that as R_{in} approaches infinity and R_{out} approaches zero, the ideal amplifier magnifies the source voltage by a factor A

$$v_L = Av_S \tag{8.8}$$

just as was indicated in the "black box" amplifier of Figure 8.2.

LO1 Thus, two desirable characteristics for a general-purpose voltage amplifier are a very *large input impedance* and a very *small output impedance*. In the next sections we will show how operational amplifiers provide these desired characteristics.

8.2 THE OPERATIONAL AMPLIFIER

An **operational amplifier** is an **integrated circuit,** that is, a large collection of individual electric and electronic circuits integrated on a single silicon wafer. An operational amplifier—or op-amp—can perform a great number of operations, such as addition, filtering, and integration, which are all based on the properties of ideal amplifiers and of ideal circuit elements. The introduction of the operational amplifier in integrated-circuit (IC) form marked the beginning of a new era in modern electronics. Since the introduction of the first IC op-amp, the trend in electronic instrumentation has been to move away from the discrete (individual-component) design of electronic circuits, toward the use of integrated circuits for a large number of applications. This statement is particularly true for applications of the type the non–electrical engineer is likely to encounter: op-amps are found in most measurement and instrumentation applications, serving as extremely versatile building blocks for any application that requires the processing of electric signals.

Next, we introduce simple circuit models of the op-amp. The simplicity of the models will permit the use of the op-amp as a circuit element, or building block, without the need to describe its internal workings in detail. Integrated-circuit technology has today reached such an advanced stage of development that it can be safely stated that for the purpose of many instrumentation applications, the op-amp can be treated as an ideal device. Following the introductory material presented in this chapter, more advanced instrumentation applications are explored in Chapter 15.

The Open-Loop Model

The ideal operational amplifier behaves very much as an ideal **difference amplifier,** that is, a device that amplifies the difference between two input voltages. Operational amplifiers are characterized by near-infinite input resistance and very small output resistance. As shown in Figure 8.4, the output of the op-amp is an amplified version of the difference between the voltages present at the two inputs:[1]

$$v_{\text{out}} = A_{V(\text{OL})}(v^+ - v^-) \tag{8.9}$$

The input denoted by a plus sign is called the **noninverting input** (or terminal), while that represented with a minus sign is termed the **inverting input** (or terminal). The amplification factor, or gain, $A_{V(\text{OL})}$ is called the **open-loop voltage gain** and is quite large by design, typically on the order of 10^5 to 10^7; it will soon become apparent why a large open-loop gain is a desirable characteristic. Together with the high input resistance and low output resistance, the effect of a large amplifier open-loop voltage

[1] The amplifier of Figure 8.4 is a *voltage amplifier*; another type of operational amplifier, called a *current* or *transconductance amplifier*, is described in the homework problems.

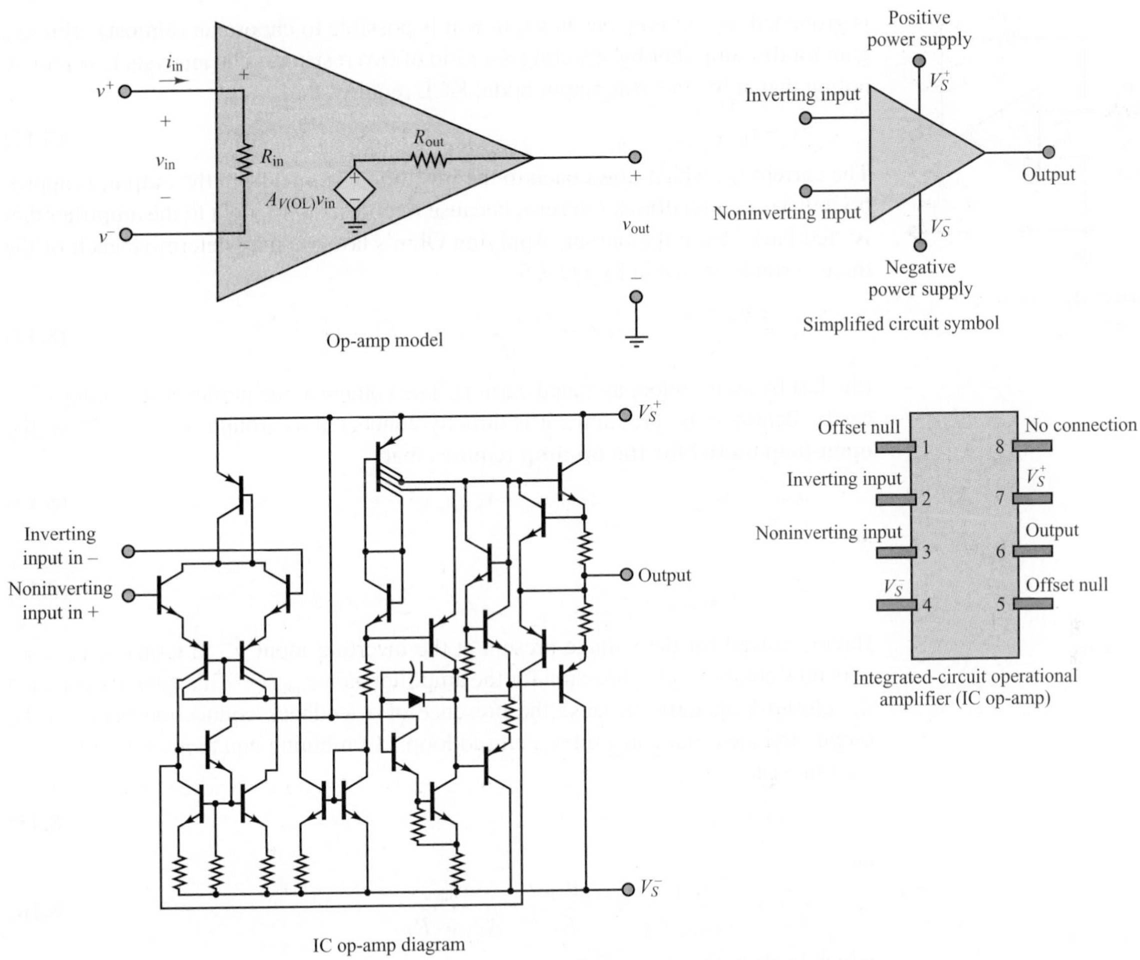

Figure 8.4 Operational amplifier model symbols, and circuit diagram

gain $A_{V(OL)}$ is such that op-amp circuits can be designed to perform very nearly as ideal voltage or current amplifiers. In effect, to analyze the performance of an op-amp circuit, only one assumption will be needed: that the current flowing into the input circuit of the amplifier is zero, or

$$i_{in} = 0 \tag{8.10}$$

FIND IT ON THE WEB

This assumption is justified by the large input resistance and large open-loop gain of the **operational amplifier.** The model just introduced will be used to analyze three amplifier circuits in the next part of this section.

The Operational Amplifier in Closed-Loop Mode

The Inverting Amplifier

One of the more popular circuit configurations of the op-amp, because of its simplicity, is the so-called inverting amplifier, shown in Figure 8.5. The input signal to be amplified is connected to the inverting terminal, while the noninverting terminal

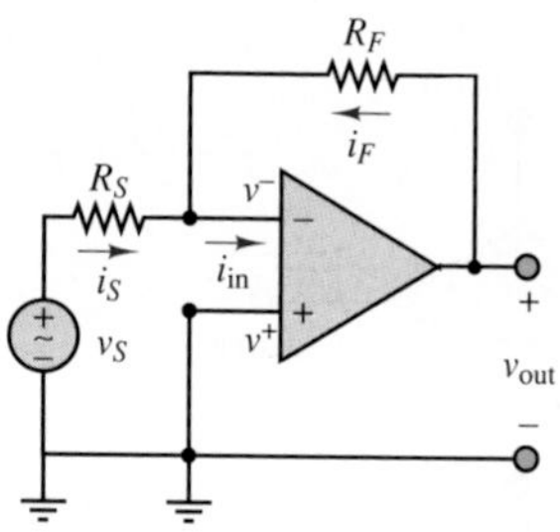

Figure 8.5 Inverting amplifier

is grounded. It will now be shown how it is possible to choose an (almost) arbitrary gain for this amplifier by selecting the ratio of two resistors. The analysis is begun by noting that at the inverting input node, KCL requires that

$$i_S + i_F = i_{in} \tag{8.11}$$

The current i_F, which flows back to the inverting terminal from the output, is appropriately termed **feedback current,** because it represents an input to the amplifier that is "fed back" from the output. Applying Ohm's law, we may determine each of the three currents shown in Figure 8.5:

$$i_S = \frac{v_S - v^-}{R_S} \qquad i_F = \frac{v_{out} - v^-}{R_F} \qquad i_{in} = 0 \tag{8.12}$$

(the last by assumption, as stated earlier). The voltage at the noninverting input v^+ is easily identified as zero, since it is directly connected to ground: $v^+ = 0$. Now, the **open-loop model for the op-amp** requires that

$$v_{out} = A_{V(OL)}(v^+ - v^-) = -A_{V(OL)}v^- \tag{8.13}$$

or

$$v^- = -\frac{v_{out}}{A_{V(OL)}} \tag{8.14}$$

Having solved for the voltage present at the inverting input v^- in terms of v_{out}, we may now compute an expression for the amplifier gain v_{out}/v_S. This quantity is called the **closed-loop gain,** because the presence of a feedback connection between the output and the input constitutes a closed loop.[2] Combining equations 8.11 and 8.12, we can write

$$i_S = -i_F \tag{8.15}$$

and

$$\frac{v_S}{R_S} + \frac{v_{out}}{A_{V(OL)}R_S} = -\frac{v_{out}}{R_F} - \frac{v_{out}}{A_{V(OL)}R_F} \tag{8.16}$$

which leads to the expression

$$\frac{v_S}{R_S} = -\frac{v_{out}}{R_F} - \frac{v_{out}}{A_{V(OL)}R_F} - \frac{v_{out}}{A_{V(OL)}R_S} \tag{8.17}$$

or

$$v_S = -v_{out}\left(\frac{1}{R_F/R_S} + \frac{1}{A_{V(OL)}R_F/R_S} + \frac{1}{A_{V(OL)}}\right) \tag{8.18}$$

If the open-loop gain of the amplifier $A_{V(OL)}$ is sufficiently large, the terms $1/(A_{V(OL)}R_F/R_S)$ and $1/A_{V(OL)}$ are essentially negligible, compared with $1/(R_F/R_S)$. As stated earlier, typical values of $A_{V(OL)}$ range from 10^5 to 10^7, and thus it is reasonable to conclude that, to a close approximation, the following expression describes the closed-loop gain of the inverting amplifier:

LO2

$$\frac{v_{out}}{v_S} = -\frac{R_F}{R_S} \qquad \text{Inverting amplifier closed-loop gain} \tag{8.19}$$

[2]This terminology is borrowed from the field of automatic controls, for which the theory of closed-loop feedback systems forms the foundation.

That is, the closed-loop gain of an inverting amplifier may be selected simply by the appropriate choice of two externally connected resistors. The price for this extremely simple result is an inversion of the output with respect to the input—that is, a minus sign.

Next, we show that by making an additional assumption it is possible to simplify the analysis considerably. Consider that, as was shown for the inverting amplifier, the inverting terminal voltage is given by

$$v^- = -\frac{v_{\text{out}}}{A_{V(\text{OL})}} \tag{8.20}$$

Clearly, as $A_{V(\text{OL})}$ approaches infinity, the inverting-terminal voltage is going to be very small (practically, on the order of microvolts). It may then be assumed that *in the inverting amplifier*, v^- is virtually zero:

$$v^- \approx 0 \tag{8.21}$$

This assumption prompts an interesting observation (which may not yet appear obvious at this point):

> The effect of the feedback connection from output to inverting input is to force the voltage at the inverting input to be equal to that at the noninverting input.

LO2

This is equivalent to stating that for an op-amp *with negative feedback*,

$$v^- \approx v^+ \tag{8.22}$$

The analysis of the operational amplifier can now be greatly simplified if the following two assumptions are made:

1.	$i_{\text{in}} = 0$	Assumptions for analysis of ideal	**(8.23)**
2.	$v^- = v^+$	op-amp with negative feedback	

LO2

This technique will be tested in the next subsection by analyzing a noninverting amplifier configuration. Example 8.1 illustrates some simple design considerations.

CHECK YOUR UNDERSTANDING

LO2

Consider an op-amp connected in the inverting configuration with a nominal closed-loop gain $-R_F/R_S = -1{,}000$ (this would be the gain if the op-amp had infinite open-loop gain). Derive an expression for the closed-loop gain that includes the value of the open-loop voltage gain as a parameter (*Hint:* start with equation 8.18, and do not assume that $A_{V(\text{OL})}$ is infinite); compute the closed-loop gain for the following values of $A_{V(\text{OL})}$: 10^7, 10^6, 10^5, 10^4. How large should the open-loop gain be if we desire to achieve the intended closed-loop gain with less than 0.1 percent error?

Answers: 999.1; 999.0; 990.1; 909.1. For 0.1 percent accuracy, $A_{V(\text{OL})}$ should equal 10^6.

Why Feedback?

Why is such emphasis placed on the notion of an amplifier with a very large open-loop gain and with negative feedback? Why not just design an amplifier with a reasonable gain, say, $\times 10$, or $\times 100$, and just use it as such, without using feedback connections? In these paragraphs, we hope to answer these and other questions, introducing the concept of **negative feedback** in an intuitive fashion.

The fundamental reason for designing an amplifier with a very large open-loop gain is the flexibility it provides in the design of amplifiers with an (almost) arbitrary gain; it has already been shown that the gain of the inverting amplifier is determined by the choice of two external resistors—undoubtedly a convenient feature! Negative feedback is the mechanism that enables us to enjoy such flexibility in the design of linear amplifiers.

To understand the role of feedback in the operational amplifier, consider the internal structure of the op-amp shown in Figure 8.4. The large open-loop gain causes any difference in voltage at the input terminals to appear greatly amplified at the output. When a negative feedback connection is provided, as shown, for example, in the inverting amplifier of Figure 8.5, the output voltage v_{out} causes a current i_F to flow through the feedback resistance so that KCL is satisfied at the inverting node. Assume, for a moment, that the differential voltage

$$\Delta v = v^+ - v^-$$

is identically zero. Then the output voltage will continue to be such that KCL is satisfied at the inverting node, that is, such that the current i_F is equal to the current i_S.

Suppose, now, that a small imbalance in voltage Δv is present at the input to the op-amp. Then the output voltage will be increased by an amount $A_{V(\text{OL})}\,\Delta v$. Thus, an incremental current approximately equal to $A_{V(\text{OL})}\,\Delta v/R_F$ will flow from output to input via the feedback resistor. The effect of this incremental current is to reduce the voltage difference Δv to zero, so as to restore the original balance in the circuit. One way of viewing negative feedback, then, is to consider it a self-balancing mechanism, which allows the amplifier to preserve zero potential difference between its input terminals.

A practical example that illustrates a common application of negative feedback is the thermostat. This simple temperature control system operates by comparing the desired ambient temperature and the temperature measured by a thermometer and turning a heat source on and off to maintain the difference between actual and desired temperature as close to zero as possible. An analogy may be made with the inverting amplifier if we consider that, in this case, negative feedback is used to keep the inverting-terminal voltage as close as possible to the noninverting-terminal voltage. The latter voltage is analogous to the desired ambient temperature in your home, while the former plays a role akin to that of the actual ambient temperature. The open-loop gain of the amplifier forces the two voltages to be close to each other, in much the same way as the furnace raises the heat in the house to match the desired ambient temperature.

It is also possible to configure operational amplifiers in a **positive feedback** configuration if the output connection is tied to the noninverting input. We do not discuss this configuration in this chapter, but present an example of it, the **voltage comparator,** in Chapter 15.

LO2

EXAMPLE 8.1 Inverting Amplifier Circuit

Problem

Determine the voltage gain and output voltage for the inverting amplifier circuit of Figure 8.5. What will the uncertainty in the gain be if 5 and 10 percent tolerance resistors are used, respectively?

Solution

Known Quantities: Feedback and source resistances, source voltage.

Find: $A_V = v_{\text{out}}/v_{\text{in}}$; maximum percent change in A_V for 5 and 10 percent tolerance resistors.

Schematics, Diagrams, Circuits, and Given Data: $R_S = 1\text{ k}\Omega$; $R_F = 10\text{ k}\Omega$; $v_S(t) = A\cos(\omega t)$; $A = 0.015$ V; $\omega = 50$ rad/s.

Assumptions: The amplifier behaves ideally; that is, the input current into the op-amp is zero, and negative feedback forces $v^+ = v^-$.

Analysis: Using equation 8.19, we calculate the output voltage:

$$v_{\text{out}}(t) = A_V \times v_S(t) = -\frac{R_F}{R_S} \times v_S(t) = -10 \times 0.015\cos(\omega t) = -0.15\cos(\omega t)$$

The input and output waveforms are sketched in Figure 8.6.

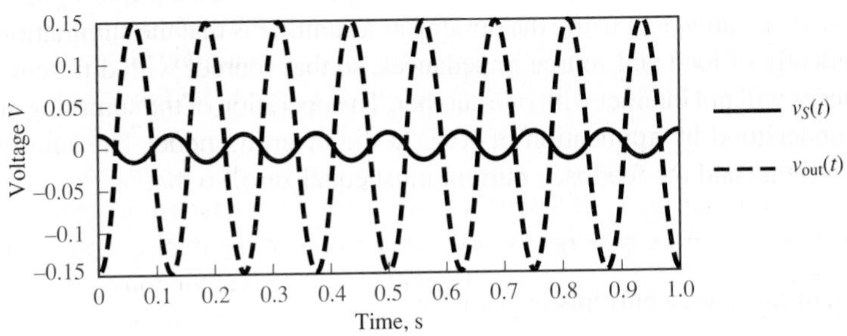

Figure 8.6

The nominal gain of the amplifier is $A_{V\text{ nom}} = -10$. If 5 percent tolerance resistors are employed, the worst-case error will occur at the extremes:

$$A_{V\text{ min}} = -\frac{R_{F\text{ min}}}{R_{S\text{ max}}} = -\frac{9{,}500}{1{,}050} = 9.05 \qquad A_{V\text{ max}} = -\frac{R_{F\text{ max}}}{R_{S\text{ min}}} = -\frac{10{,}500}{950} = 11.05$$

The percentage error is therefore computed as

$$100 \times \frac{A_{V\text{ nom}} - A_{V\text{ min}}}{A_{V\text{ nom}}} = 100 \times \frac{10 - 9.05}{10} = 9.5\%$$

$$100 \times \frac{A_{V\text{ nom}} - A_{V\text{ max}}}{A_{V\text{ nom}}} = 100 \times \frac{10 - 11.05}{10} = -10.5\%$$

Thus, the amplifier gain could vary by as much as ±10 percent (approximately) when 5 percent resistors are used. If 10 percent resistors were used, we would calculate a percent error of approximately ± 20 percent, as shown below.

$$A_{V\text{ min}} = -\frac{R_{F\text{ min}}}{R_{S\text{ max}}} = -\frac{9{,}000}{1{,}100} = 8.18 \qquad A_{V\text{ max}} = -\frac{R_{F\text{ max}}}{R_{S\text{ min}}} = -\frac{11{,}000}{900} = 12.2$$

$$100 \times \frac{A_{V\text{ nom}} - A_{V\text{ min}}}{A_{V\text{ nom}}} = 100 \times \frac{10 - 8.18}{10} = 18.2\%$$

$$100 \times \frac{A_{V\text{ nom}} - A_{V\text{ max}}}{A_{V\text{ nom}}} = 100 \times \frac{10 - 12.2}{10} = -22.2\%$$

Comments: Note that the worst-case percent error in the amplifier gain is double the resistor tolerance.

CHECK YOUR UNDERSTANDING

Calculate the uncertainty in the gain if 1 percent "precision" resistors are used.

Answer: +1.98 to −2.20 percent

The Summing Amplifier

A useful op-amp circuit that is based on the inverting amplifier is the **op-amp summer,** or **summing amplifier.** This circuit, shown in Figure 8.7, is used to add signal sources. The primary advantage of using the op-amp as a summer is that the summation occurs independently of load and source impedances, so that sources with different internal impedances will not interact with one another. The operation of the summing amplifier is best understood by application of KCL at the inverting node: The sum of the N source currents and the feedback current must equal zero, so that

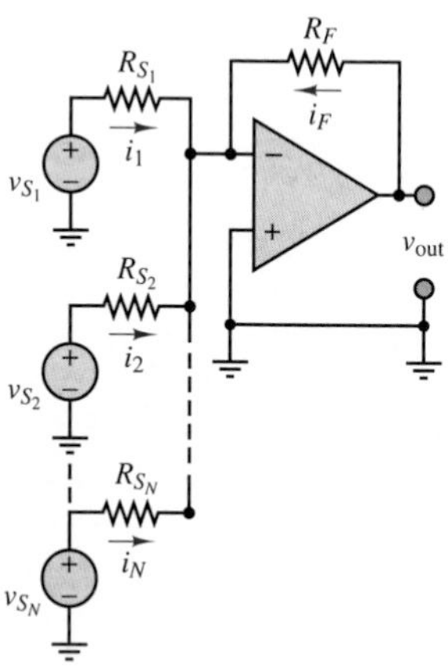

Figure 8.7 Summing amplifier

$$i_1 + i_2 + \cdots + i_N = -i_F \tag{8.24}$$

But each of the source currents is given by

$$i_n = \frac{v_{S_n}}{R_{S_n}} \qquad n = 1, 2, \ldots, N \tag{8.25}$$

while the feedback current is

$$i_F = \frac{v_{out}}{R_F} \tag{8.26}$$

Combining equations 8.25 and 8.26, and using equation 8.15, we obtain the following result:

$$\sum_{n=1}^{N} \frac{v_{S_n}}{R_{S_n}} = -\frac{v_{out}}{R_F} \tag{8.27}$$

or

LO2

$$v_{out} = -\sum_{n=1}^{N} \frac{R_F}{R_{S_n}} v_{S_n} \qquad \text{Summing amplifier} \tag{8.28}$$

That is, the output consists of the weighted sum of N input signal sources, with the weighting factor for each source equal to the ratio of the feedback resistance to the source resistance.

The Noninverting Amplifier

To avoid the negative gain (i.e., phase inversion) introduced by the inverting amplifier, a **noninverting amplifier** configuration is often employed. A typical noninverting amplifier is shown in Figure 8.8; note that the input signal is applied to the noninverting terminal this time.

The noninverting amplifier can be analyzed in much the same way as the inverting amplifier. Writing KCL at the inverting node yields

$$i_F = i_S + i_{\text{in}} \approx i_S \tag{8.29}$$

where

$$i_F = \frac{v_{\text{out}} - v^-}{R_F} \tag{8.30}$$

$$i_S = \frac{v^-}{R_S} \tag{8.31}$$

Now, since $i_{\text{in}} = 0$, the voltage drop across the source resistance R is equal to zero. Thus,

$$v^+ = v_S \tag{8.32}$$

and, using equation 8.22, we get

$$v^- = v^+ = v_S \tag{8.33}$$

Substituting this result in equations 8.29 and 8.30, we can easily show that

$$i_F = i_S \tag{8.34}$$

or

$$\frac{v_{\text{out}} - v_S}{R_F} = \frac{v_S}{R_S} \tag{8.35}$$

It is easy to manipulate equation 8.35 to obtain the result

$$\frac{v_{\text{out}}}{v_S} = 1 + \frac{R_F}{R_S} \qquad \text{Noninverting amplifier closed-loop gain} \tag{8.36}$$

LO2

which is the closed-loop gain expression for a noninverting amplifier. Note that the gain of this type of amplifier is always positive and greater than (or equal to) 1.

The same result could have been obtained without making the assumption that $v^+ = v^-$, at the expense of some additional work. The procedure one would follow in this latter case is analogous to the derivation carried out earlier for the inverting amplifier, and it is left as an exercise.

In summary, in the preceding pages it has been shown that by constructing a nonideal amplifier with very large gain and near-infinite input resistance, it is possible to design amplifiers that have near-ideal performance and provide a variable range of gains, easily controlled by the selection of external resistors. The mechanism that allows this is negative feedback. From here on, unless otherwise noted, it will

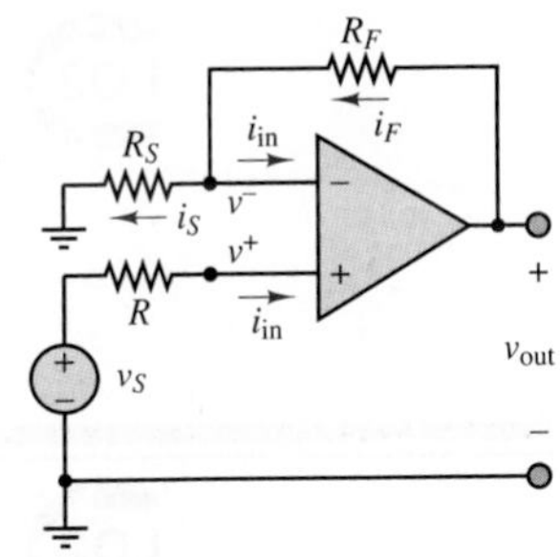

Figure 8.8 Noninverting amplifier

be reasonable and sufficient in analyzing new op-amp configurations to utilize these two assumptions:

LO2

1. $i_{in} = 0$	Approximations used for ideal
2. $v^- = v^+$	op-amps with negative feedback

(8.37)

LO2

EXAMPLE 8.2 Voltage Follower

Problem

Determine the closed-loop voltage gain and input resistance of the **voltage follower** circuit of Figure 8.9.

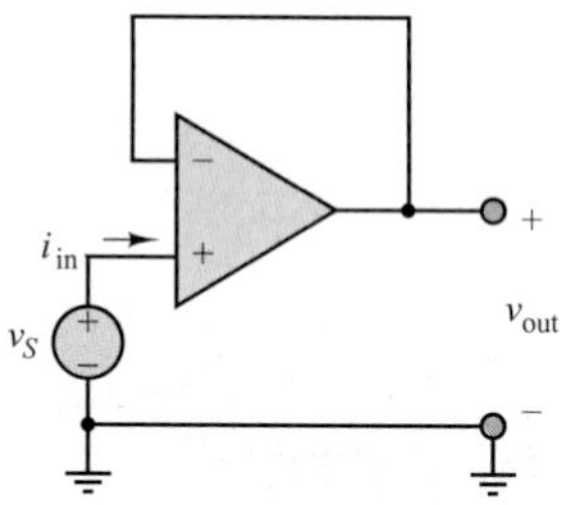

Figure 8.9 Voltage follower

Solution

Known Quantities: Feedback and source resistances, source voltage.

Find:

$$A_V = \frac{v_{out}}{v_S} \qquad r_i = \frac{v_{in}}{i_{in}}$$

Assumptions: The amplifier behaves ideally; that is, the input current into the op-amp is zero, and negative feedback forces $v^+ = v^-$.

Analysis: From the ideal op-amp assumptions, $v^+ = v^-$. But $v^+ = v_S$ and $v^- = v_{out}$, thus

$v_S = v_{out}$	Voltage follower

The name *voltage follower* derives from the ability of the output voltage to "follow" exactly the input voltage. To compute the input resistance of this amplifier, we observe that since the input current is zero,

$$r_i = \frac{v_S}{i_{in}} \to \infty$$

Comments: The input resistance of the voltage follower is the most important property of the amplifier: The extremely high input resistance of this amplifier (on the order of megohms to gigohms) permits virtually perfect isolation between source and load and eliminates *loading* effects. Voltage followers, or *impedance buffers*, are commonly packaged in groups of four or more in integrated-circuit form. The data sheets for one such IC are contained in the accompanying CD-ROM, and may also be found in the device templates for analog ICs in the Electronics Workbench™ libraries.

CHECK YOUR UNDERSTANDING

Derive an expression for the closed-loop gain of the voltage follower that includes the value of the open-loop voltage gain as a parameter. (*Hint:* follow the procedure of equations 8.11 through 8.19 with the appropriate modifications, and do not assume that $A_{V(OL)}$ is infinite.) How large should the open-loop gain be if we desire to achieve the intended closed-loop gain (unity) with less than 0.1 percent error?

Answer: The expression for the closed-loop gain is $v_{out}/v_{in} = 1 + 1/A_{V(OL)}$; thus $A_{V(OL)}$ should equal 10^4 for 0.1 percent accuracy.

The Differential Amplifier

The third closed-loop model examined in this chapter is a combination of the inverting and noninverting amplifiers; it finds frequent use in situations where the difference between two signals needs to be amplified. The basic **differential amplifier** circuit is shown in Figure 8.10, where the two sources v_1 and v_2 may be independent of each other or may originate from the same process, as they do in the Focus on Measurements box "Electrocardiogram (EKG) Amplifier."

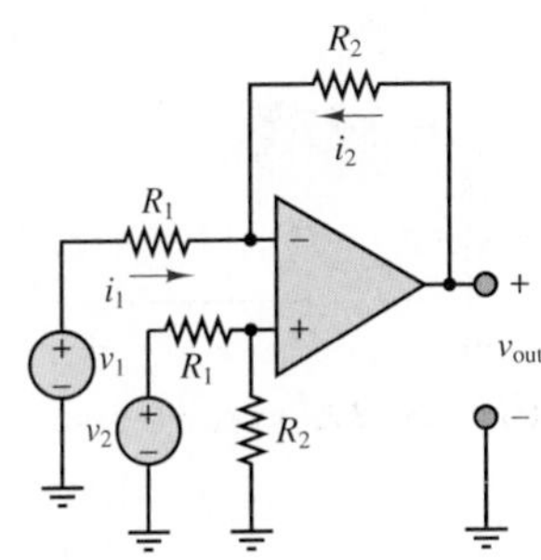

Figure 8.10 Differential amplifier

The analysis of the differential amplifier may be approached by various methods; the one we select to use at this stage consists of

1. Computing the noninverting- and inverting-terminal voltages v^+ and v^-.
2. Equating the inverting and noninverting input voltages: $v^- = v^+$.
3. Applying KCL at the inverting node, where $i_2 = -i_1$.

Since it has been assumed that no current flows into the amplifier, the noninverting-terminal voltage is given by the following voltage divider:

$$v^+ = \frac{R_2}{R_1 + R_2} v_2 \tag{8.38}$$

If the inverting-terminal voltage is assumed equal to v^+, then the currents i_1 and i_2 are found to be

$$i_1 = \frac{v_1 - v^+}{R_1} \tag{8.39}$$

and

$$i_2 = \frac{v_{out} - v^+}{R_2} \tag{8.40}$$

and since

$$i_2 = -i_1 \tag{8.41}$$

the following expression for the output voltage is obtained:

$$v_{out} = R_2 \left[\frac{-v_1}{R_1} + \frac{1}{R_1 + R_2} v_2 + \frac{R_2}{R_1(R_1 + R_2)} v_2 \right] \tag{8.42}$$

LO2

$$v_{\text{out}} = \frac{R_2}{R_1}(v_2 - v_1) \qquad \text{Differential amplifier closed-loop gain}$$

Thus, the differential amplifier magnifies the difference between the two input signals by the closed-loop gain R_2/R_1.

In practice, it is often necessary to amplify the difference between two signals that are both corrupted by noise or some other form of interference. In such cases, the differential amplifier provides an invaluable tool in amplifying the desired signal while rejecting the noise. The Focus on Measurements box "Electrocardiogram (EKG) Amplifier" provides a realistic look at a very common application of the differential amplifier.

In summary, Table 8.1 provides a quick reference to the basic op-amp circuits presented in this section.

CHECK YOUR UNDERSTANDING

Derive the result given above for the differential amplifier, using the principle of superposition. Think of the differential amplifier as the combination of an inverting amplifier with input equal to v_2 and a noninverting amplifier with input equal to v_1.

Answer: $v_{\text{out}} = \frac{R_2}{R_1}(v_2 - v_1)$

Table 8.1 Summary of basic op-amp circuits

Configuration	Circuit diagram	Closed-loop gain (under ideal assumptions of equation 8.23)
Inverting amplifier	Figure 8.5	$v_{\text{out}} = -\frac{R_F}{R_S} v_S$
Summing amplifier	Figure 8.7	$v_{\text{out}} = -\frac{R_F}{R_1} v_{S1} - \frac{R_F}{R_2} v_{S2} - \cdots - \frac{R_F}{R_n} v_{Sn}$
Noninverting amplifier	Figure 8.8	$v_{\text{out}} = \left(1 + \frac{R_F}{R_S}\right) v_S$
Voltage follower	Figure 8.9	$v_{\text{out}} = v_S$
Differential amplifier	Figure 8.10	$v_{\text{out}} = \frac{R_2}{R_1}(v_2 - v_1)$

Electrocardiogram (EKG) Amplifier

FIND IT ON THE WEB

This example illustrates the principle behind a two-lead **electrocardiogram (EKG) measurement.** The desired cardiac waveform is given by the difference between the potentials measured by two electrodes suitably

(*Continued*)

placed on the patient's chest, as shown in Figure 8.11. A healthy, noise-free EKG waveform $v_1 - v_2$ is shown in Figure 8.12.

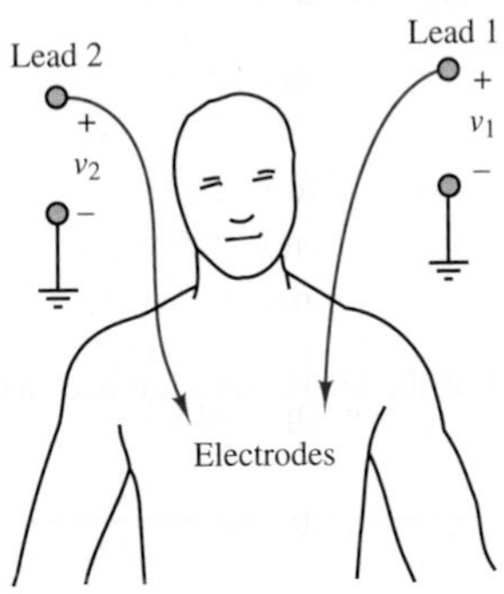

Figure 8.11 Two-lead electrocardiogram

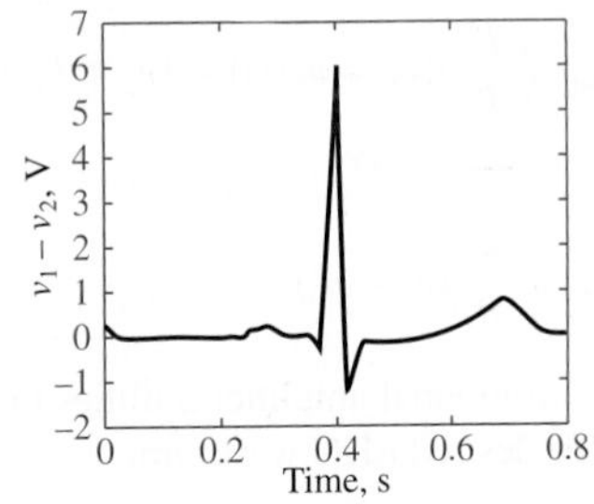

Figure 8.12 EKG waveform

Unfortunately, the presence of electrical equipment powered by the 60-Hz, 110-V AC line current causes undesired interference at the electrode leads: the lead wires act as antennas and pick up some of the 60-Hz signal in addition to the desired EKG voltage. In effect, instead of recording the desired EKG signals v_1 and v_2, the two electrodes provide the following inputs to the EKG amplifier, shown in Figure 8.13:

Lead 1:

$$v_1(t) + v_n(t) = v_1(t) + V_n \cos(377t + \phi_n)$$

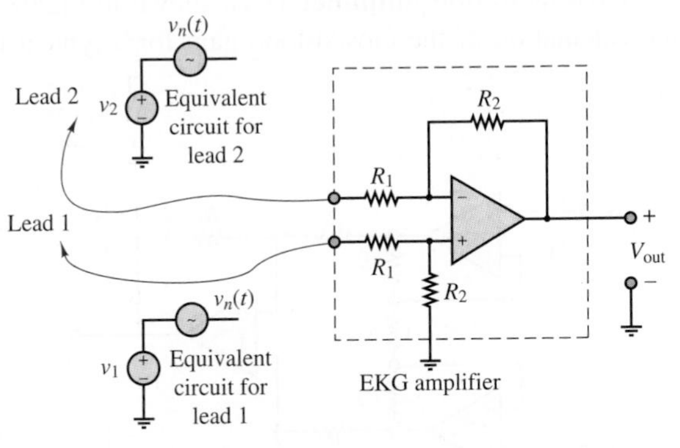

Figure 8.13 EKG amplifier

Lead 2:

$$v_2(t) + v_n(t) = v_2(t) + V_n \cos(377t + \phi_n)$$

The interference signal $V_n \cos(377t + \phi_n)$ is approximately the same at both leads, because the electrodes are chosen to be identical (e.g., they have the same lead lengths) and are in close proximity to each other. Further, the nature of the interference signal is such that

(*Continued*)

(Concluded)

it is common to both leads, since it is a property of the environment in which the EKG instrument is embedded. On the basis of the analysis presented earlier, then,

$$v_{\text{out}} = \frac{R_2}{R_1}\{[v_1 + v_n(t)] - [v_2 + v_n(t)]\}$$

or

$$v_{\text{out}} = \frac{R_2}{R_1}(v_1 - v_2)$$

Thus, the differential amplifier nullifies the effect of the 60-Hz interference, while amplifying the desired EKG waveform.

The preceding Focus on Measurements box introduces the concept of **common-mode** and **differential-mode signals.** The desired differential-mode EKG signal was amplified by the op-amp while the common-mode disturbance was canceled. Thus, the differential amplifier provides the ability to reject common-mode signal components (such as noise or undesired DC offsets) while amplifying the differential-mode components. This is a very desirable feature in instrumentation systems. In practice, rejection of the common-mode signal is not complete: some of the common-mode signal component will always appear in the output. This fact gives rise to a figure of merit called the *common-mode rejection ratio*, which is discussed in Section 8.6.

Often, to provide impedance isolation between bridge transducers and the differential amplifier stage, the signals v_1 and v_2 are amplified separately. This technique gives rise to the **instrumentation amplifier (IA),** shown in Figure 8.14. Example 8.3 illustrates the calculation of the closed-loop gain for a typical instrumentation amplifier.

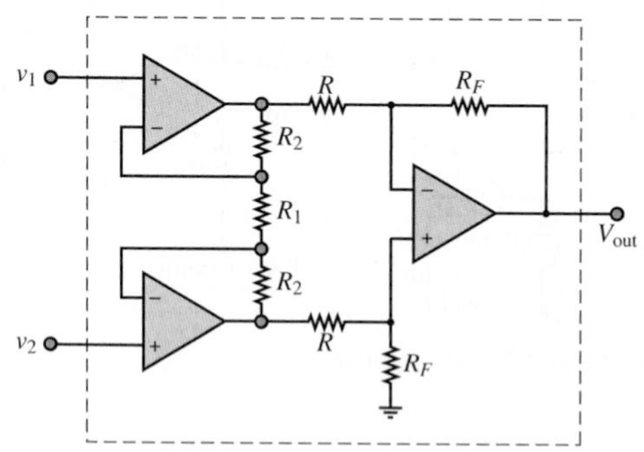

Figure 8.14 Instrumentation amplifier

LO2

EXAMPLE 8.3 Instrumentation Amplifier

Problem

Determine the closed-loop voltage gain of the instrumentation amplifier circuit of Figure 8.14.

Solution

Known Quantities: Feedback and source resistances.

Find:

$$A_V = \frac{v_{out}}{v_1 - v_2} \tag{8.43}$$

Assumptions: Assume ideal op-amps.

Analysis: We consider the input circuit first. Thanks to the symmetry of the circuit, we can represent one-half of the circuit as illustrated in Figure 8.15(a), depicting the lower half of the first *stage* of the instrumentation amplifier. We next recognize that the circuit of Figure 8.15(a) is a noninverting amplifier (see Figure 8.8), and we can directly write the expression for the closed-loop voltage gain (equation 8.36):

$$A = 1 + \frac{R_2}{R_1/2} = 1 + \frac{2R_2}{R_1}$$

Each of the two inputs v_1 and v_2 is therefore an input to the second *stage* of the instrumentation amplifier, shown in Figure 8.15(b). We recognize the second stage to be a differential amplifier (see Figure 8.10), and can therefore write the output voltage after equation 8.42:

$$v_{out} = \frac{R_F}{R}(Av_1 - Av_2) = \frac{R_F}{R}\left(1 + \frac{2R_2}{R_1}\right)(v_1 - v_2) \tag{8.44}$$

from which we can compute the closed-loop voltage gain of the instrumentation amplifier:

$$A_V = \frac{v_{out}}{v_1 - v_2} = \frac{R_F}{R}\left(1 + \frac{2R_2}{R_1}\right) \qquad \text{Instrumentation amplifier}$$

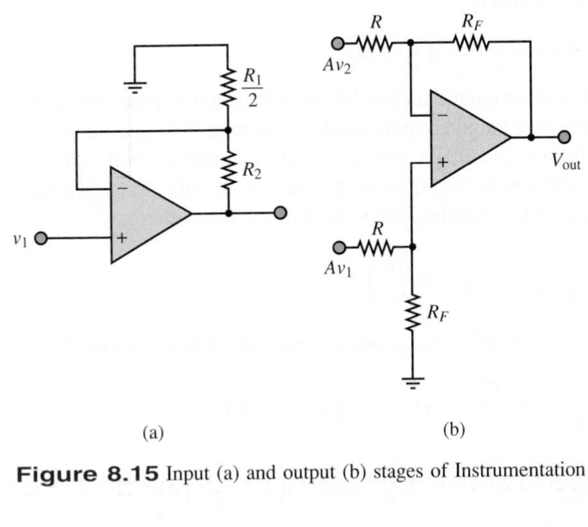

Figure 8.15 Input (a) and output (b) stages of Instrumentation amplifier

Comments: This circuit is analyzed in depth in Chapter 15.

Because the instrumentation amplifier has widespread application—and in order to ensure the best possible match between resistors—the entire circuit of Figure 8.14 is often packaged as a single integrated circuit. The advantage of this configuration is that resistors R_1 and R_2 can be matched much more precisely in an integrated circuit than would be possible by using discrete components. A typical, commercially available integrated-circuit package is the AD625. Data sheets for this device are provided in the accompanying CD-ROM.

Another simple op-amp circuit that finds widespread application in electronic instrumentation is the **level shifter.** Example 8.4 discusses its operation and its application. The following Focus on Measurements box illustrates its use in a sensor calibration circuit.

LO2

EXAMPLE 8.4 Level Shifter

Problem

The level shifter of Figure 8.16 has the ability to add or subtract a DC offset to or from a signal. Analyze the circuit and design it so that it can remove a 1.8-V DC offset from a sensor output signal.

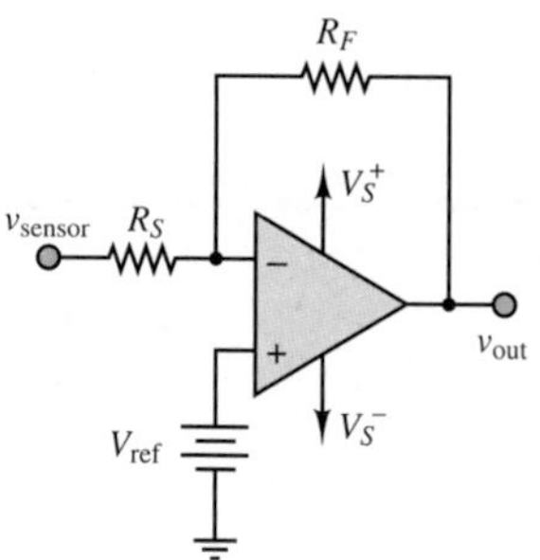

Figure 8.16 Level shifter

Solution

Known Quantities: Sensor (input) voltage; feedback and source resistors.

Find: Value of V_{ref} required to remove DC bias.

Schematics, Diagrams, Circuits, and Given Data: $v_S(t) = 1.8 + 0.1\cos(\omega t)$; $R_F = 220\text{ k}\Omega$; $R_S = 10\text{ k}\Omega$.

Assumptions: Assume an ideal op-amp.

Analysis: We first determine the closed-loop voltage gain of the circuit of Figure 8.16. The output voltage can be computed quite easily if we note that, upon applying the principle of superposition, the sensor voltage sees an inverting amplifier with gain $-R_F/R_S$, while the battery sees a noninverting amplifier with gain $1 + R_F/R_S$. Thus, we can write the output voltage as the sum of two outputs, due to each of the two sources:

$$v_{out} = -\frac{R_F}{R_S} v_{sensor} + \left(1 + \frac{R_F}{R_S}\right) V_{ref}$$

Substituting the expression for v_{sensor} into the equation above, we find that

$$\begin{aligned} v_{out} &= -\frac{R_F}{R_S}[1.8 + 0.1\cos(\omega t)] + \left(1 + \frac{R_F}{R_S}\right) V_{ref} \\ &= -\frac{R_F}{R_S}[0.1\cos(\omega t)] - \frac{R_F}{R_S}(1.8) + \left(1 + \frac{R_F}{R_S}\right) V_{ref} \end{aligned}$$

Since the intent of the design is to remove the DC offset, we require that

$$-\frac{R_F}{R_S}(1.8) + \left(1 + \frac{R_F}{R_S}\right) V_{ref} = 0$$

or

$$V_{\text{ref}} = 1.8\frac{R_F/R_S}{1 + R_F/R_S} = 1.714 \text{ V}$$

Comments: The presence of a precision voltage source in the circuit is undesirable, because it may add considerable expense to the circuit design and, in the case of a battery, it is not adjustable. The circuit of Figure 8.17 illustrates how one can generate an adjustable voltage reference by using the DC supplies already used by the op-amp, two resistors R, and a potentiometer R_p. The resistors R are included in the circuit to prevent the potentiometer from being shorted to either supply voltage when the potentiometer is at the extreme positions. Using the voltage divider rule, we can write the following expression for the reference voltage generated by the resistive divider:

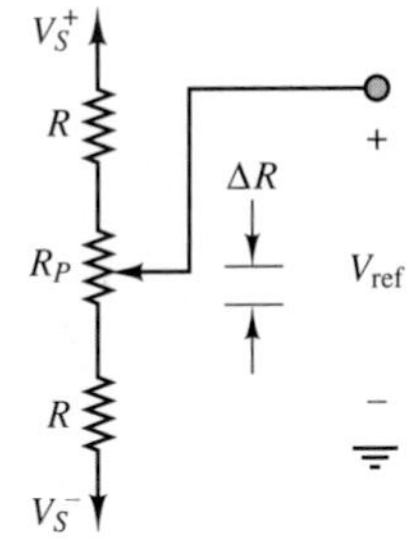

Figure 8.17

$$V_{\text{ref}} = \frac{R + \Delta R}{2R + R_p}\left(V_S^+ - V_S^-\right)$$

If the voltage supplies are symmetric, as is almost always the case, we can further simplify the expression to

$$V_{\text{ref}} = \pm\frac{R + \Delta R}{2R + R_p}V_S^+$$

Note that by adjusting the potentiometer R_p, we can obtain any value of V_{ref} between the supply voltages.

CHECK YOUR UNDERSTANDING

With reference to Example 8.4, find ΔR if the supply voltages are symmetric at ± 15 V and a 10-kΩ potentiometer is tied to the two 10-kΩ resistors.

With reference to Example 8.4, find the range of values of V_{ref} if the supply voltages are symmetric at ± 15 V and a 1-kΩ potentiometer is tied to the two 10-kΩ resistors.

Answers: $\Delta R = 6{,}714\ \Omega$; V_{ref} is between ± 0.714 V

EXAMPLE 8.5 Temperature Control Using Op-Amps

LO2

Problem

One of the most common applications of op-amps is to serve as a building block in analog control systems. The objective of this example is to illustrate the use of op-amps in a temperature control circuit. Figure 8.18(a) depicts a system for which we wish to maintain a constant temperature of 20°C in a variable temperature environment. The temperature of the system is measured via a thermocouple (see Chapter 15, *Temperature Measurements*). Heat can be added to the system by providing a current to a heater coil, represented in the figure by the resistor R_{coil}. The heat flux thus generated is given by the quantity $q_{\text{in}} = i^2 R_{\text{coil}}$, where i is the current provided by a power amplifier and R_{coil} is the resistance of the heater coil. The system is insulated on three sides, and loses heat to the ambient through convective heat transfer on the fourth side [right-hand side in Figure 18(a)]. The convective heat loss is represented by an equivalent thermal resistance, R_t. The system has mass m, specific heat c, and its thermal capacitance is $C_t = mc$ (see *Make the Connection—Thermal Capacitance*, p. 218 and *Make the Connection—Thermal System Dynamics*, p. 219 in Chapter 5).

Solution

Known Quantities: Sensor (input) voltage; feedback and source resistors, thermal system component values.

Find: Select desired value of proportional gain, K_P, to achieve automatic temperature control.

Schematics, Diagrams, Circuits, and Given Data: $R_{\text{coil}} = 5\Omega$; $R_t = 2°\text{C/W}$; $C_t = 50\ \text{J/°C}$; $\alpha = 1\ \text{V/°C}$. Figure 8.18 (a), (b), (c), (d).

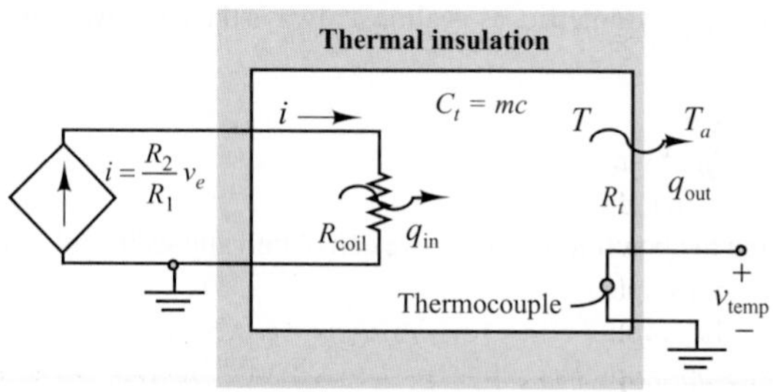

Figure 8.18 (a) Thermal system

Assumptions: Assume ideal op-amps.

Analysis: The thermal system is described by the following equation, based on conservation of energy.

$$q_{\text{in}} - q_{\text{out}} = q_{\text{stored}}$$

where q_{in} represents the heat added to the system by the electrical heater, q_{out} represents the heat lost from the system through convection to the surrounding air, and q_{stored} represents the heat stored in the system through its thermal capacitance. In the system of Figure 8.18(a), the temperature, T, of the system is measured by a thermocouple that we assume produces a voltage proportional to temperature: $v_{\text{temp}} = \alpha \text{T}$. Further, we assume that the power amplifier can be simply modeled by a *voltage-controlled current source*, as shown in the figure, such that its current is proportional to an external voltage. This *error* voltage, v_e, depends on the difference between the actual temperature of the system, T, and the reference temperature, T_{ref}, i.e., $v_e = v_{\text{ref}} - v_{\text{temp}} = \alpha(T_{\text{ref}} - T)$. With reference to the block diagram of Figure 8.18(b), we see that to maintain the temperature of the system at the desired level, we can use the *difference* voltage v_e as an input to the power amplifier [the controlled current source of Figure 8.18(a)]. You should easily convince yourself that a positive v_e corresponds to the need for heating the system, since a positive v_e corresponds to a system temperature lower than the reference temperature. Next, the power amplifier can output a positive current for a positive v_e. Thus, the block diagram shown in Figure 8.18(b) corresponds to an *automatic control system* that automatically increases or decreases the heater coil current to maintain the system temperature at the desired (reference) value. The "Amplifier" block in Figure 8.18(b) gives us the freedom to decide *how much* to increase the power amplifier output current to provide the necessary heating. The *proportional gain* of the amplifier, K_P, is a design parameter of the circuit that allows the user to optimize the response of the system for a specific design requirement. For example, a system specification could require that the automatic temperature control system be designed so as to maintain the temperature to within 1 degree of the reference temperature for external temperature disturbances as large as 10 degrees. As you shall see, we can adjust the response of the system by varying the proportional gain.

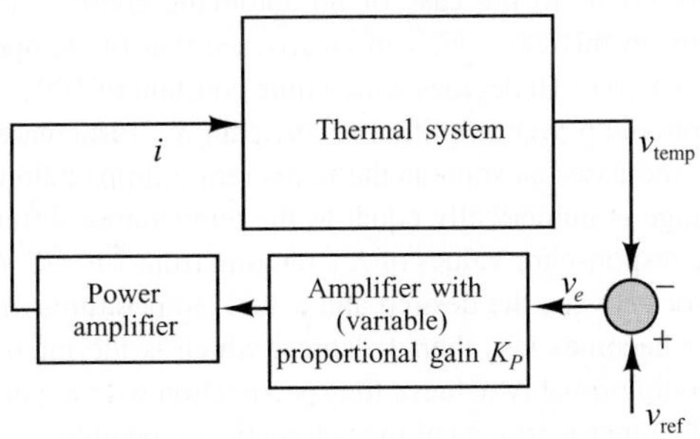

Figure 8.18 (b) Block diagram of control system

The objective of this example is to show how operational amplifiers can be used to provide two of the functions illustrated in the block diagram of Figure 8.18(b):(1) the summing amplifier computes the difference between the reference temperature and the system temperature; and (2) an inverting amplifier implements the *proportional gain* function that allows the designer to select the response of the amplifier by choosing an appropriate proportional gain. Figure 8.18(c) depicts the two-stage op-amp circuit that performs these functions. The first element is an inverting amplifier with unity gain, with the function of changing the sign of v_{ref}. The second amplifier, and inverting summing amplifier, adds v_{temp} to $-v_{\text{ref}}$ and inverts the sum of these two signals, while at the same time amplifying it by the gain R_2/R_1. Thus, the output of the circuit of Figure 8.18(c) consists of the quantity $R_2/R_1(v_{\text{ref}} - v_{\text{temp}}) = K_P(v_{\text{ref}} - v_{\text{temp}})$. In other words, *selection of the feedback resistor* R_2 *is equivalent to choosing the gain* K_P.

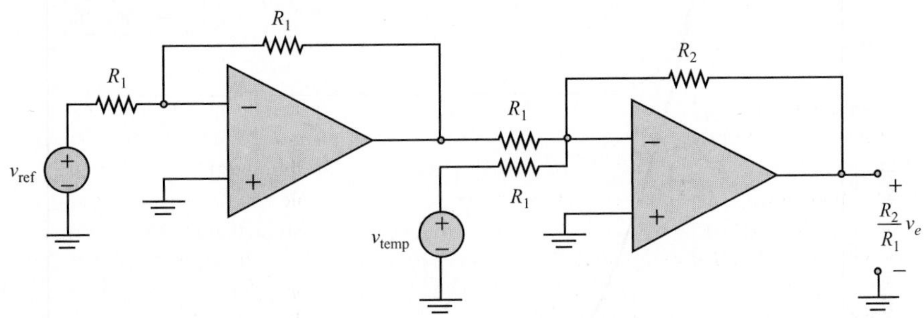

Figure 8.18 (c) Circuit for generating error voltage and proportional gain

To analyze the response of the system for various values of K_P, we must first understand how the system responds in the absence of automatic control. The differential equation describing the system is:

$$q_{\text{in}}(t) - \frac{T(t) - T_a}{R_t} = C_t \frac{dT(t)}{dt}$$

$$R_t C_t \frac{dT(t)}{dt} + T(t) = R_t q_{\text{in}}(t) + T_a$$

$$q_{\text{in}}(t) = R_{\text{coil}} i^2(t)$$

Thus, the thermal system is a first-order system, and its time constant is $\tau = R_t C_t = 2°\text{C/W} \times 50\ \text{J/°C} = 100$ s. The inputs to the system are the ambient temperature, T_a, and the heat flux, q_{in}, which is proportional to the square of the heater coil current. Imagine now that the thermal system is suddenly exposed to a 10-degree change in ambient temperature (for example, a drop from 20°C to 10°C). Figure 8.18(d) depicts the response of the system for various values of K_P. $K_P = 0$ corresponds to the case of no automatic control—that is, the *open loop* response of the system. In this case, we can clearly see that the temperature of the system drops exponentially from 20 to 10 degrees with a time constant of 100 s. This is so because no heating is provided from the power amplifier. As the gain K_P is increased to 1, the difference or "error" voltage, v_e, increases as soon as the temperature drops below the reference value. Since $\alpha = 1$, the voltage is numerically equal to the temperature difference. Figure 8.18(d) shows the temperature response for values of K_P ranging from 1 to 50. You can see that as the gain increases, the error between the desired and actual temperatures decreases very quickly. In particular, the error becomes less than 1 degree, which is the intended specification, for $K_P = 5$ (in fact, we could probably achieve the specification with a gain slightly less than 5). To better understand the inner workings of the automatic temperature control system, it is also helpful to look at the error voltage, which is amplified to provide the power amplifier output current. With reference to Figure 8.18(e), we can see that when $K_P = 1$, the current increases somewhat slowly to a final value of about 2.7 A; as the gain is increased to 5 and 10, the current response increases more rapidly, and eventually settles to values of 3 and 3.1 A, respectively. The steady state value of the current is reached in about 20 s for $K_P = 5$, and in about 10 s for $K_P = 10$.

Comments: Please note that even though the response of the system is satisfactory, the temperature error is not zero. That is, the automatic control system has a *steady-state error*. The design specifications recognized this fact by specifying that a 1°C tolerance was sufficient.

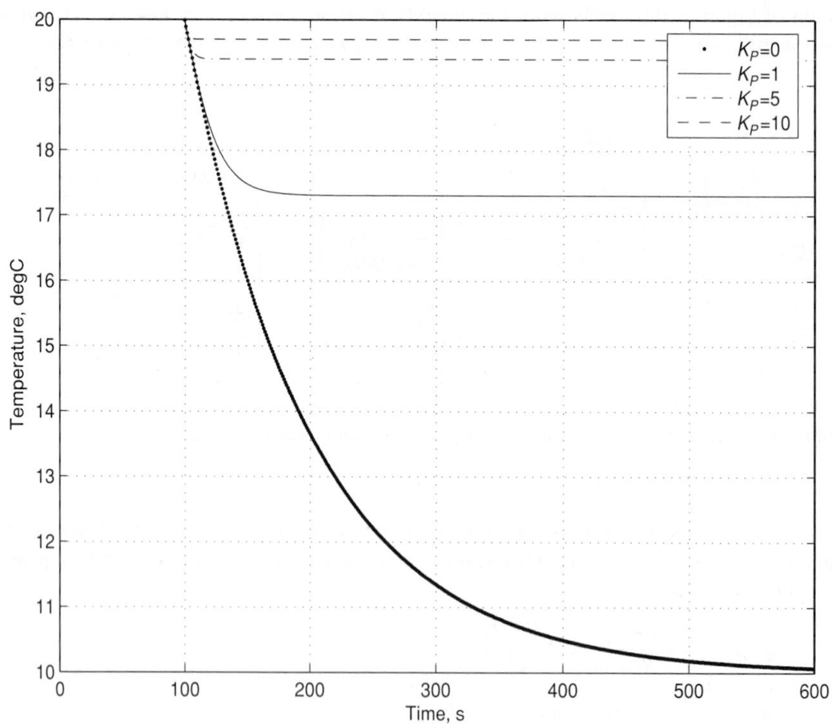

Figure 8.18 (d) Response of thermal system for various values of proportional gain, K_P

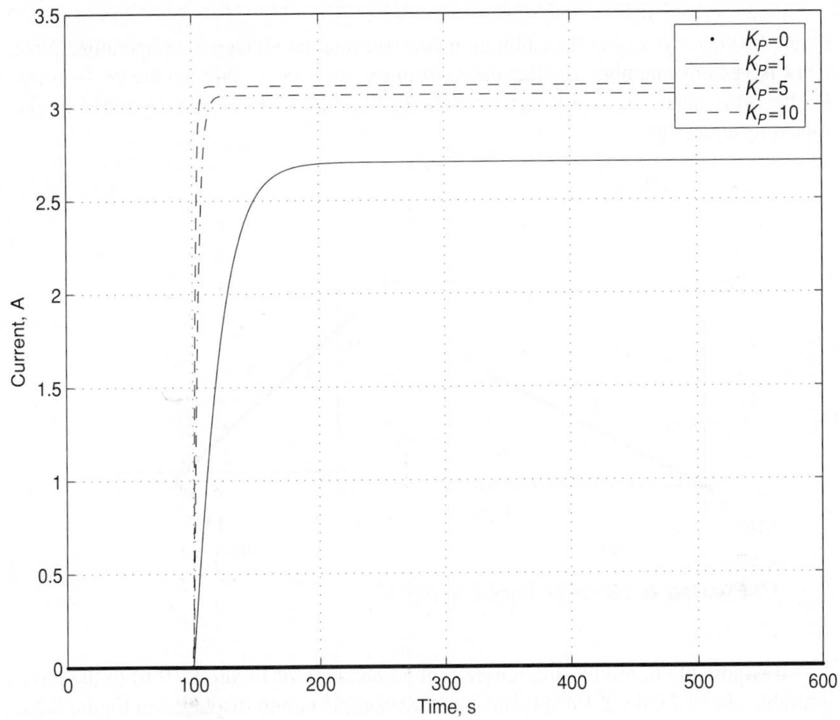

Figure 8.18 (e) Power amplifier output current for various proportional gain, K_p

CHECK YOUR UNDERSTANDING

How much steady state power, in Watts, will be input to the thermal system of Example 8.5 to maintain its temperature in the face of a 10°C ambient temperature drop for values of K_P of 1, 5, and 10?

Answers: $K_P = 1$: 36.5 W; $K_P = 5$: 45 W; $K_P = 10$: 48 W

Sensor Calibration Circuit (LO2)

In many practical instances, the output of a sensor is related to the physical variable we wish to measure in a form that requires some signal conditioning. The most desirable form of a sensor output is one in which the electrical output of the sensor (e.g., voltage) is related to the physical variable by a constant factor. Such a relationship is depicted in

(Continued)

Figure 8.19(a), where k is the calibration constant relating voltage to temperature. Note that k is a positive number, and that the *calibration curve* passes through the (0, 0) point. On the other hand, the sensor characteristic of Figure 8.19(b) is best described by the following equation:

$$v_{\text{sensor}} = -\beta T + V_0$$

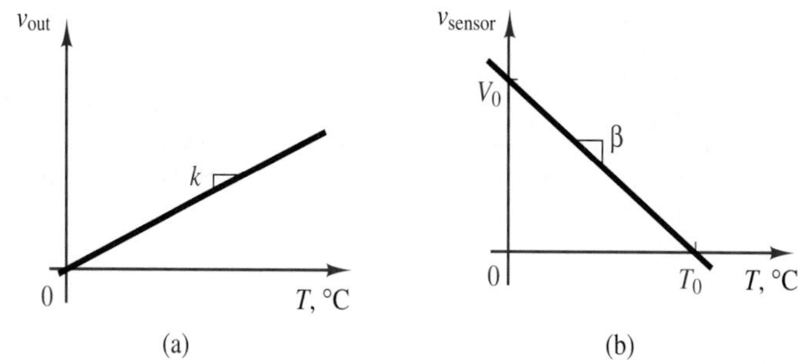

Figure 8.19 Sensor calibration curves

It is possible to modify the sensor calibration curve of Figure 8.19(b) to the more desirable one of Figure 8.19(a) by means of the simple circuit displayed in Figure 8.20. This circuit provides the desired calibration constant k by a simple gain adjustment, while the zero (or bias) offset is adjusted by means of a potentiometer connected to the voltage supplies. The detailed operation of the circuit is described in the following paragraphs.

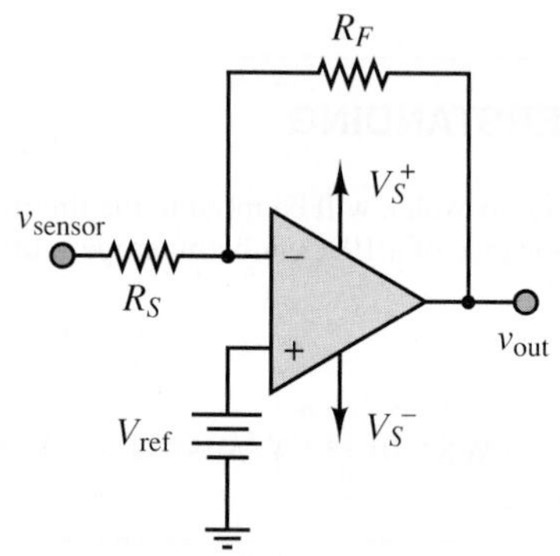

Figure 8.20 Sensor calibration circuit

As noted before, the nonideal characteristic can be described by the following equation:

$$v_{\text{sensor}} = -\beta T + V_0$$

(*Continued*)

(Concluded)

Then the output of the op-amp circuit of Figure 8.20 may be determined by using the principle of superposition:

$$v_{\text{out}} = -\frac{R_F}{F_S}v_{\text{sensor}} + \left(1 + \frac{R_F}{R_S}\right)V_{\text{ref}}$$

$$= -\frac{R_F}{R_S}(-\beta T + V_0) + \left(1 + \frac{R_F}{R_S}\right)V_{\text{ref}}$$

After substituting the expression for the transducer voltage and after some manipulation, we see that by suitable choice of resistors, and of the reference voltage source, we can compensate for the nonideal transducer characteristic. We want the following expression to hold:

$$v_{\text{out}} = \frac{R_F}{R_S}\beta T + \left(1 + \frac{R_F}{R_S}\right)V_{\text{ref}} - \frac{R_F}{R_S}V_0 = kT$$

If we choose

$$\frac{R_F}{R_S}\beta = k$$

and

$$V_{\text{ref}} = \frac{R_F/R_S}{1 + R_F/R_S}V_0$$

then $v_{\text{out}} = kT$.

Note that

$$V_{\text{ref}} \approx V_0 \qquad \text{if} \qquad \frac{R_F}{R_S} \gg 1$$

and we can directly convert the characteristic of Figure 8.19(b) to that of Figure 8.19(a). Clearly, the effect of selecting the gain resistors is to change the magnitude of the slope of the calibration curve. The fact that the sign of the slope changes is purely a consequence of the inverting configuration of the amplifier. The reference voltage source simply shifts the DC level of the characteristic, so that the curve passes through the origin.

CHECK YOUR UNDERSTANDING

With reference to the Focus on Measurements box "Sensor Calibration Circuit," find numerical values of R_F/R_S and V_{ref} if the temperature sensor has $\beta = 0.235$ and $V_0 = 0.7$ V and the desired relationship is $v_{\text{out}} = 10\,T$.

Answers: $R_F/R_S = 42.55$; $V_{\text{ref}} = 0.684$ V

Practical Op-Amp Design Considerations

LO2

The results presented in the preceding pages suggest that operational amplifiers permit the design of a rather sophisticated circuit in a few very simple steps, simply by selecting appropriate resistor values. This is certainly true, provided that the circuit component selection satisfies certain criteria. Here we summarize some important practical design criteria that the designer should keep in mind when selecting component values for op-amp circuits. Section 8.6 explores the practical limitations of op-amps in greater detail.

1. Use standard resistor values. While any arbitrary value of gain can, in principle, be achieved by selecting the appropriate combination of resistors, the designer is often constrained to the use of standard 5 percent resistor values (see Table 2.1). For example, if your design requires a gain of 25, you might be tempted to select, say, 100- and 4-kΩ resistors to achieve $R_F/R_S = 25$. However, inspection of Table 2.1 reveals that 4 kΩ is not a standard value; the closest 5 percent tolerance resistor value is 3.9 kΩ, leading to a gain of 25.64. Can you find a combination of standard 5 percent resistors whose ratio is closer to 25?
2. Ensure that the load current is reasonable (do not select very small resistor values). Consider the same example given in step 1. Suppose that the maximum output voltage is 10 V. The feedback current required by your design with $R_F = 100$ kΩ and $R_S = 4$ kΩ would be $I_F = 10/100{,}000 = 0.1$ mA. This is a very reasonable value for an op-amp, as explained in Section 8.6. If you tried to achieve the same gain by using, say, a 10-Ω feedback resistor and a 0.39-Ω source resistor, the feedback current would become as large as 1 A. This is a value that is generally beyond the capabilities of a general-purpose op-amp, so the selection of exceedingly low resistor values is not acceptable. On the other hand, the selection of 10-kΩ and 390-Ω resistors would still lead to acceptable values of current, and would be equally good. As a general rule of thumb, you should avoid resistor values lower than 100 Ω in practical designs.
3. Avoid stray capacitance (do not select excessively large resistor values). The use of exceedingly large resistor values can cause unwanted signals to couple into the circuit through a mechanism known as *capacitive coupling*. This phenomenon is discussed in Chapter 15. Large resistance values can also cause other problems. As a general rule of thumb, avoid resistor values higher than 1 MΩ in practical designs.
4. Precision designs may be warranted. If a certain design requires that the amplifier gain be set to a very accurate value, it may be appropriate to use the (more expensive) option of precision resistors: for example, 1 percent tolerance resistors are commonly available, at a premium cost. Some of the examples and homework problems explore the variability in gain due to the use of higher- and lower-tolerance resistors.

FOCUS ON METHODOLOGY

LO2

USING OP-AMP DATA SHEETS

FIND IT ON THE WEB

Here we illustrate use of **device data sheets** for two commonly used operational amplifiers. The first, the LM741, is a general-purpose (low-cost) amplifier; the second, the LMC6061, is a precision CMOS high-input-impedance single-supply amplifier. Excerpts from the data sheets are shown below, with some words of explanation. Later in this chapter we compare the electrical characteristics of these two op-amps in greater detail.

(*Continued*)

LM741 General Description and Connection Diagrams—This sheet summarizes the general characteristics of the op-amp. The connection diagrams are shown. Note that the op-amp is available in various packages: a metal can package, a dual-in-line package (DIP), and two ceramic dual-in-line options. The dual-in-line (or S.O.) package is the one you are most likely to see in a laboratory. Note that in this configuration the integrated circuit has eight connections, or *pins*: two for the voltage supplies (V^+ and V^-); two inputs (inverting and noninverting); one output; two offset null connections (to be discussed later in the chapter); and a no-connection (NC) pin.

LM741 Operational Amplifier

General Description

The LM741 series are general-purpose operational amplifiers which feature improved performance over industry standards like the LM709. They are direct, plug-in replacements for the 709C, LM201, MC1439, and 748 in most applications.

The amplifiers offer many features which make their application nearly foolproof: overload protection on the input and output, no latch-up when the common-mode range is exceeded, as well as freedom from oscillations.

The LM741C and LM741E are identical to the LM741 and LM741A except that the LM741C and LM741E have their performance guaranteed over a 0 to +70°C temperature range, instead of −55 to +125°C.

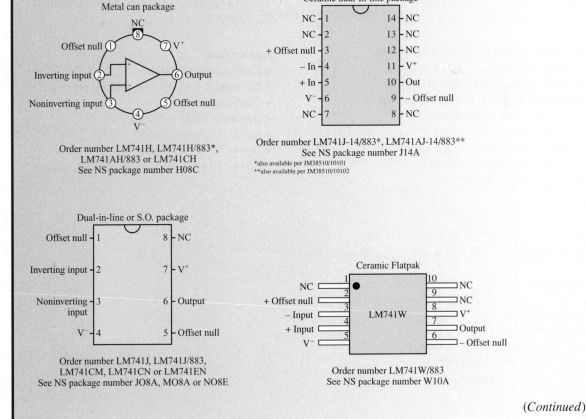

Metal can package

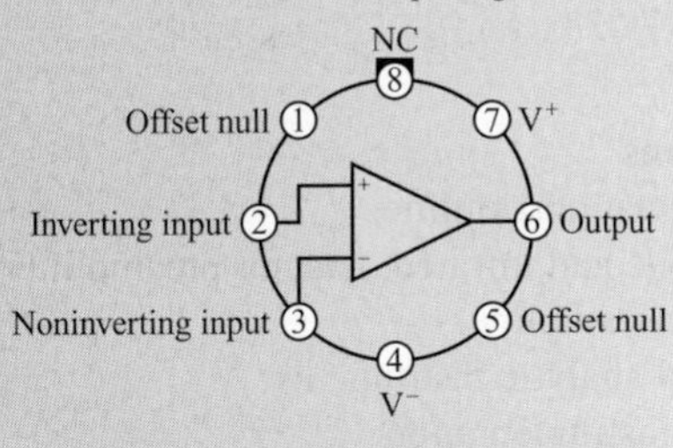

Order number LM741H, LM741H/883*, LM741AH/883 or LM741CH
See NS package number H08C

Ceramic dual-in-line package

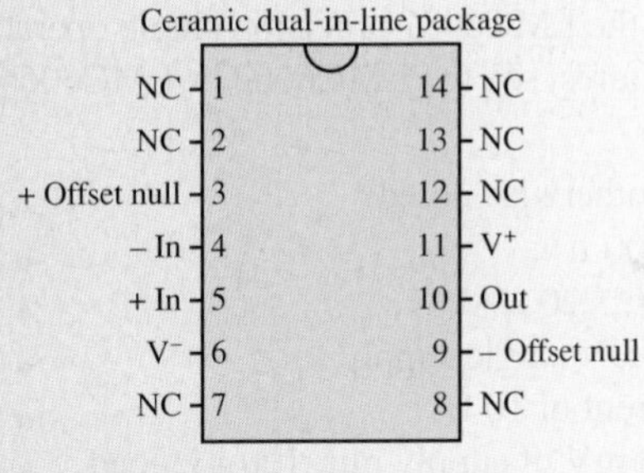

Order number LM741J-14/883*, LM741AJ-14/883**
See NS package number J14A
*also available per JM38510/10101
**also available per JM38510/10102

Dual-in-line or S.O. package

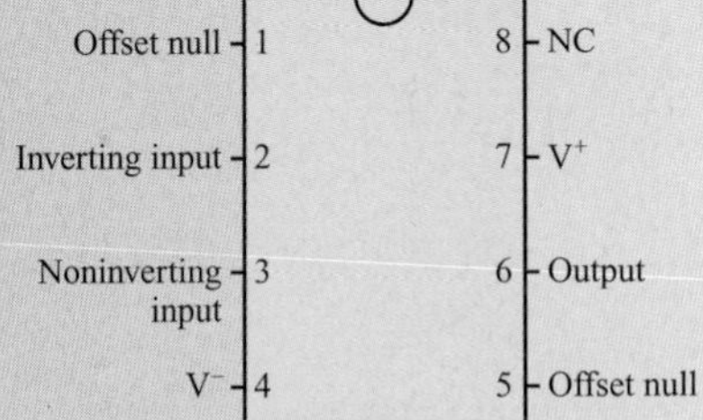

Order number LM741J, LM741J/883, LM741CM, LM741CN or LM741EN
See NS package number JO8A, MO8A or NO8E

Ceramic Flatpak

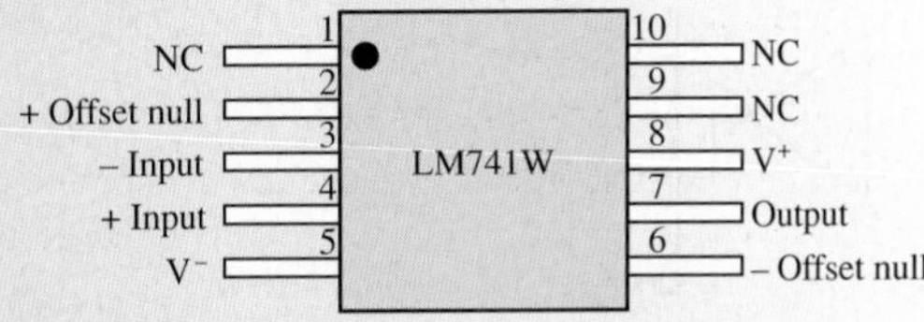

Order number LM741W/883
See NS package number W10A

(*Continued*)

(Concluded)

LMC6061 General Description and Connection Diagrams—The description and diagram below reveal several similarities between the 741 and 6061 op-amps, but also some differences. The 6061 uses more advanced technology and is characterized by some very desirable features (e.g., the very low power consumption of CMOS circuits results in typical supply currents of only 20 μA!). You can also see from the connection diagram that pins 1 and 5 (used for offset null connections in the 741) are not used in this IC. We return to this point later in the chapter. A further point of comparison between these two devices is their (1998) cost: the LM741 (in quantities of 1,000) costs \$0.32 per unit; the LMC6061 sells for \$0.79 per unit, also in quantities of 1,000 or more.

LMC6061 Precision CMOS Single Micropower Operational Amplifier

General Description

The LMC6061 is a precision single low offset voltage, micropower operational amplifier, capable of precision single-supply operation. Performance characteristics include ultralow input bias current, high voltage gain, rail-to-rail output swing, and an input common-mode voltage range that includes ground. These features, plus its low power consumption, make the LMC6061 ideally suited for battery-powered applications.

Other applications using the LMC6061 include precision full-wave rectifiers, integrators, references, sample-and-hold circuits, and true instrumentation amplifiers.

This device is built with National's advanced double-poly silicon-gate CMOS process. For designs that require higher speed, see the LMC6081 precision single operational amplifier. For a dual or quad operational amplifier with similar features, see the LMC6062 or LMC6064, respectively.

Features (typical unless otherwise noted)
- Low offset voltage: 100 μV
- Ultralow supply current: 20 μA
- Operates from 4.5- to 15-V single supply
- Ultralow input bias current of 10 fA
- Output swing within 10 mV of supply rail, 100-kΩ load
- Input common-mode range includes V^-
- High voltage gain: 140 dB
- Improved latch-up immunity

Applications
- Instrumentation amplifier
- Photodiode and infrared detector preamplifier
- Transducer amplifiers
- Handheld analytic instruments
- Medical instrumentation
- Digital-to-analog converter
- Charge amplifier to piezoelectric transducers

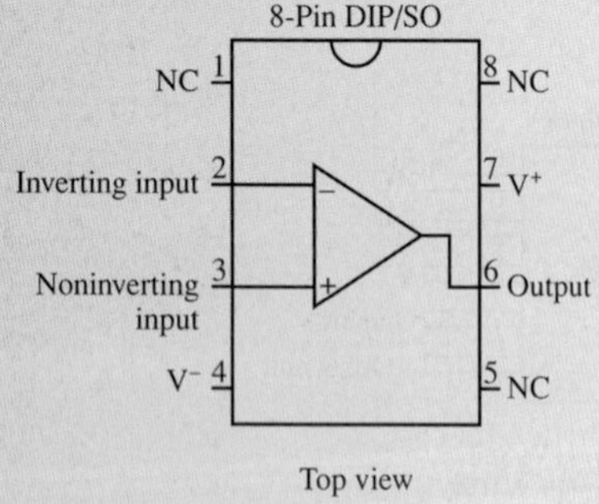

8.3 ACTIVE FILTERS

The range of useful applications of an operational amplifier is greatly expanded if energy storage elements are introduced into the design; the frequency-dependent properties of these elements, studied in Chapters 4 and 6, will prove useful in the design of various types of op-amp circuits. In particular, it will be shown that it is possible to shape the frequency response of an operational amplifier by appropriate use of complex impedances in the input and feedback circuits. The class of filters one can obtain by means of op-amp designs is called **active filters,** because op-amps can provide amplification (gain) in addition to the filtering effects already studied in Chapter 6 for passive circuits (i.e., circuits comprising exclusively resistors, capacitors, and inductors).

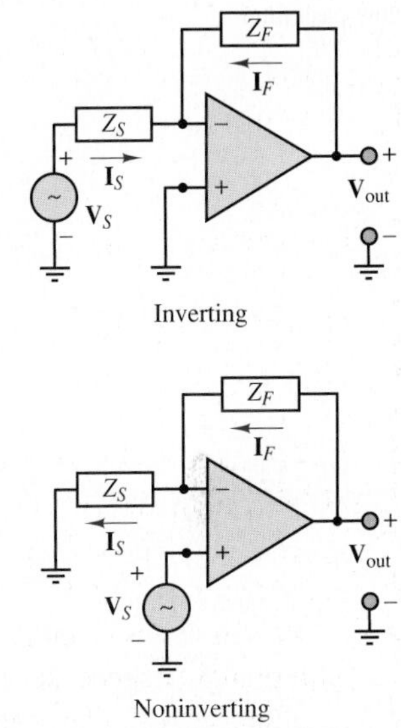

Figure 8.21 Op-amp circuits employing complex impedances

The easiest way to see how the frequency response of an op-amp can be shaped (almost) arbitrarily is to replace the resistors R_F and R_S in Figures 8.5 and 8.8 with impedances Z_F and Z_S, as shown in Figure 8.21. It is a straightforward matter to show that in the case of the inverting amplifier, the expression for the closed-loop gain is given by

$$\frac{\mathbf{V}_{\text{out}}}{\mathbf{V}_S}(j\omega) = -\frac{Z_F}{Z_S} \tag{8.45}$$

whereas for the noninverting case, the gain is

$$\frac{\mathbf{V}_{\text{out}}}{\mathbf{V}_S}(j\omega) = 1 + \frac{Z_F}{Z_S} \tag{8.46}$$

where Z_F and Z_S can be arbitrarily complex impedance functions and where $\mathbf{V}_S$, $\mathbf{V}_{\text{out}}$, $\mathbf{I}_F$, and $\mathbf{I}_S$ are all phasors. Thus, it is possible to shape the frequency response of an ideal op-amp filter simply by selecting suitable ratios of feedback impedance to source impedance. By connecting a circuit similar to the low-pass filters studied in Chapter 6 in the feedback loop of an op-amp, the same filtering effect can be achieved and, in addition, the signal can be amplified.

The simplest op-amp low-pass filter is shown in Figure 8.22. Its analysis is quite simple if we take advantage of the fact that the closed-loop gain, as a function of frequency, is given by

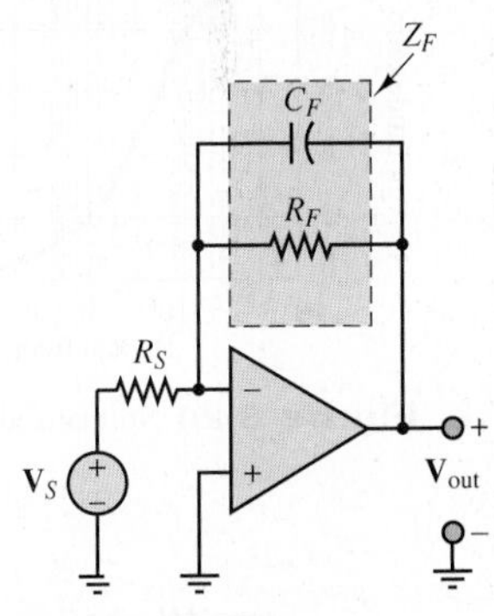

Figure 8.22 Active low-pass filter

$$A_{\text{LP}}(j\omega) = -\frac{Z_F}{Z_S} \tag{8.47}$$

where

$$Z_F = R_F \parallel \frac{1}{j\omega C_F} = \frac{R_F}{1 + j\omega C_F R_F} \tag{8.48}$$

and

$$Z_S = R_S \tag{8.49}$$

Note the similarity between Z_F and the low-pass characteristic of the passive RC circuit! The closed-loop gain $A_{\text{LP}}(j\omega)$ is then computed to be

LO3

$$A_{\text{LP}}(j\omega) = -\frac{Z_F}{Z_S} = -\frac{R_F/R_S}{1 + j\omega C_F R_F} \qquad \text{Low-pass filter} \tag{8.50}$$

This expression can be factored into two terms. The first is an amplification factor analogous to the amplification that would be obtained with a simple inverting amplifier (i.e., the same circuit as that of Figure 8.22 with the capacitor removed); the second is a low-pass filter, with a cutoff frequency dictated by the parallel combination of

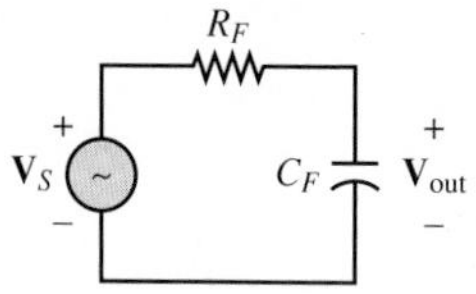

Figure 8.23 Passive low-pass filter

R_F and C_F in the feedback loop. The filtering effect is completely analogous to what would be attained by the passive circuit shown in Figure 8.23. However, the op-amp filter also provides amplification by a factor of R_F/R_S.

It should be apparent that the response of this op-amp filter is just an amplified version of that of the passive filter. Figure 8.24 depicts the amplitude response of the active low-pass filter (in the figure, $R_F/R_S = 10$ and $1/R_FC_F = 1$) in two different graphs; the first plots the amplitude ratio $\mathbf{V}_{\text{out}}(j\omega)$ versus radian frequency ω on a logarithmic scale, while the second plots the amplitude ratio $20 \log \mathbf{V}_S(j\omega)$ (in units of decibels), also versus ω on a logarithmic scale. You will recall from Chapter 6 that decibel frequency response plots are encountered very frequently. Note that in the decibel plot, the slope of the filter response for frequencies significantly higher than the cutoff frequency,

$$\omega_0 = \frac{1}{R_FC_F} \tag{8.51}$$

is -20 dB/decade, while the slope for frequencies significantly lower than this cutoff frequency is equal to zero. The value of the response at the cutoff frequency is found to be, in units of decibel,

$$|A_{\text{LP}}(j\omega_0)|_{\text{dB}} = 20 \log_{10} \frac{R_F}{R_S} - 20 \log \sqrt{2} \tag{8.52}$$

where

$$-20 \log_{10} \sqrt{2} = -3 \text{ dB} \tag{8.53}$$

Thus, ω_0 is also called the **3-dB frequency.**

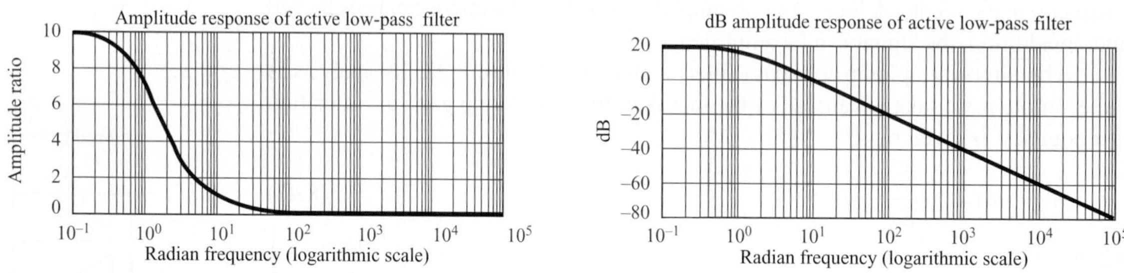

Figure 8.24 Normalized response of active low-pass filter

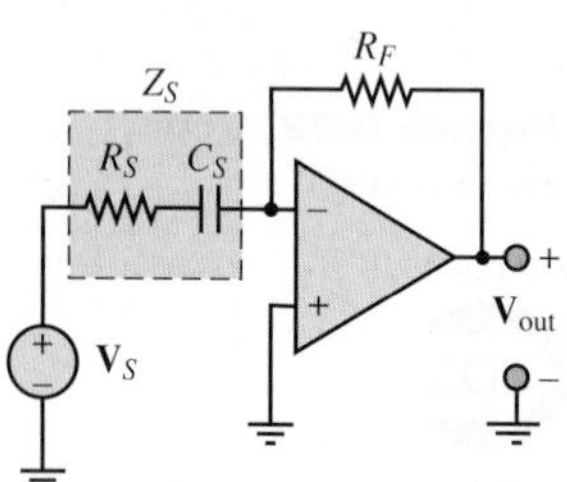

Figure 8.25 Active high-pass filter

Among the advantages of such active low-pass filters is the ease with which the gain and the bandwidth can be adjusted by controlling the ratios R_F/R_S and $1/R_FC_F$, respectively.

It is also possible to construct other types of filters by suitably connecting resistors and energy storage elements to an op-amp. For example, a high-pass active filter can easily be obtained by using the circuit shown in Figure 8.25. Observe that the impedance of the input circuit is

$$Z_S = R_S + \frac{1}{j\omega C_S} \tag{8.54}$$

and that of the feedback circuit is

$$Z_F = R_F \tag{8.55}$$

Then the following gain function for the op-amp circuit can be derived:

$$A_{\text{HP}}(j\omega) = -\frac{Z_F}{Z_S} = -\frac{j\omega C_S R_F}{1 + j\omega R_S C_S} \qquad \text{High-pass filter} \tag{8.56}$$

As ω approaches zero, so does the response of the filter, whereas as ω approaches infinity, according to the gain expression of equation 8.56, the gain of the amplifier approaches a constant:

$$\lim_{\omega\to\infty} A_{\text{HP}}(j\omega) = -\frac{R_F}{R_S} \tag{8.57}$$

That is, above a certain frequency range, the circuit acts as a linear amplifier. This is exactly the behavior one would expect of a high-pass filter. The high-pass response is depicted in Figure 8.26, in both linear and decibel plots (in the figure, $R_F/R_S = 10$ and $1/R_S C = 1$). Note that in the decibel plot, the slope of the filter response for frequencies significantly lower than $\omega = 1/R_S C_S = 1$ is +20 dB/decade, while the slope for frequencies significantly higher than this cutoff (or 3-dB) frequency is equal to zero.

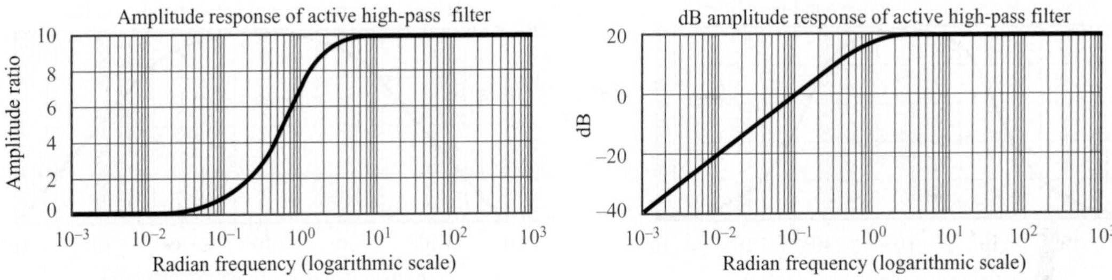

Figure 8.26 Normalized response of active high-pass filter

As a final example of active filters, let us look at a simple active band-pass filter configuration. This type of response may be realized simply by combining the high- and low-pass filters we examined earlier. The circuit is shown in Figure 8.27.

The analysis of the bandpass circuit follows the same structure used in previous examples. First we evaluate the feedback and input impedances:

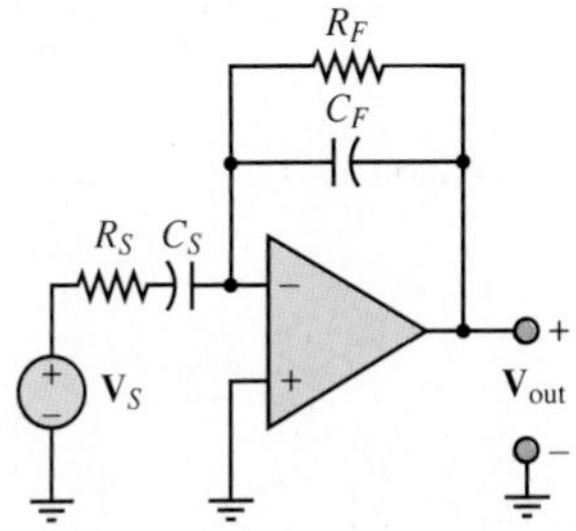

Figure 8.27 Active band-pass filter

$$Z_F = R_F \parallel \frac{1}{j\omega C_F} = \frac{R_F}{1 + j\omega C_F R_F} \tag{8.58}$$

$$Z_S = R_S + \frac{1}{j\omega C_S} = \frac{1 + j\omega C_S R_S}{j\omega C_S} \tag{8.59}$$

Next we compute the closed-loop frequency response of the op-amp, as follows:

$$A_{\text{BP}}(j\omega) = -\frac{Z_F}{Z_S} = -\frac{j\omega C_S R_F}{(1 + j\omega C_F R_F)(1 + j\omega C_S R_S)} \qquad \text{Band-pass filter} \tag{8.60}$$

The form of the op-amp response we just obtained should not be a surprise. It is very similar (although not identical) to the product of the low-pass and high-pass responses of equations 8.50 and 8.56. In particular, the denominator of $A_{BP}(j\omega)$ is exactly the product of the denominators of $A_{LP}(j\omega)$ and $A_{HP}(j\omega)$. It is particularly enlightening to rewrite $A_{LP}(j\omega)$ in a slightly different form, after making the observation that each RC product corresponds to some "critical" frequency:

$$\omega_1 = \frac{1}{R_F C_S} \qquad \omega_{LP} = \frac{1}{R_F C_F} \qquad \omega_{HP} = \frac{1}{R_S C_S} \tag{8.61}$$

It is easy to verify that for the case where

$$\omega_{HP} > \omega_{LP} \tag{8.62}$$

the response of the op-amp filter may be represented as shown in Figure 8.28 in both linear and decibel plots (in the figure, $\omega_1 = 1$, $\omega_{HP} = 1{,}000$, and $\omega_{LP} = 10$). The decibel plot is very revealing, for it shows that, in effect, the bandpass response is the graphical superposition of the low-pass and high-pass responses shown earlier. The two 3-dB (or cutoff) frequencies are the same as in $A_{LP}(j\omega)$, $1/R_F C_F$; and in $A_{HP}(j\omega)$, $1/R_S C_S$. The third frequency, $\omega_1 = 1/R_F C_S$, represents the point where the response of the filter crosses the 0-dB axis (rising slope). Since 0 dB corresponds to a gain of 1, this frequency is called the **unity gain frequency.**

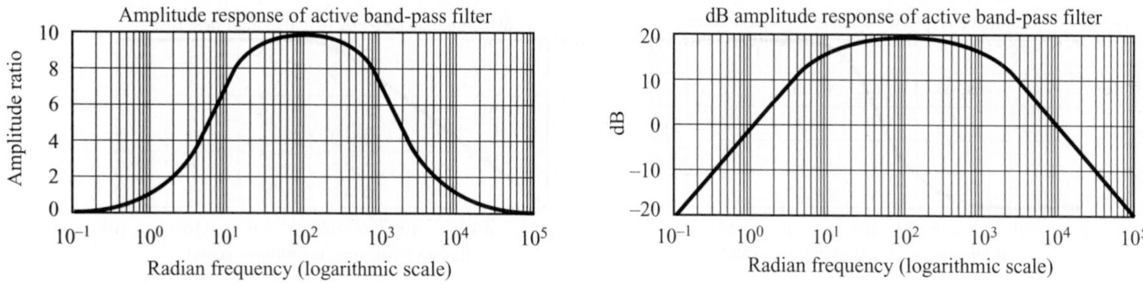

Figure 8.28 Normalized amplitude response of active band-pass filter

The ideas developed thus far can be employed to construct more complex functions of frequency. In fact, most active filters one encounters in practical applications are based on circuits involving more than one or two energy storage elements. By constructing suitable functions for Z_F and Z_S, it is possible to realize filters with greater frequency selectivity (i.e., sharpness of cutoff), as well as flatter bandpass or band-rejection functions (i.e., filters that either allow or reject signals in a limited band of frequencies). A few simple applications are investigated in the homework problems and some advanced applications in Chapter 15. One remark that should be made in passing, though, pertains to the exclusive use of capacitors in the circuits analyzed thus far. One of the advantages of op-amp filters is that it is not necessary to use both capacitors and inductors to obtain a bandpass response. Suitable connections of capacitors can accomplish that task in an op-amp. This seemingly minor fact is of great importance in practice, because inductors are expensive to mass-produce to close tolerances and exact specifications and are often bulkier than capacitors with equivalent energy storage capabilities. On the other hand, capacitors are easy to manufacture in a wide variety of tolerances and values, and in relatively compact packages, including in integrated-circuit form.

Example 8.6 illustrates how it is possible to construct active filters with greater frequency selectivity by adding energy storage elements to the design.

EXAMPLE 8.6 Second-Order Low-Pass Filter

LO3

Problem

Determine the closed-loop voltage gain as a function of frequency for the op-amp circuit of Figure 8.29.

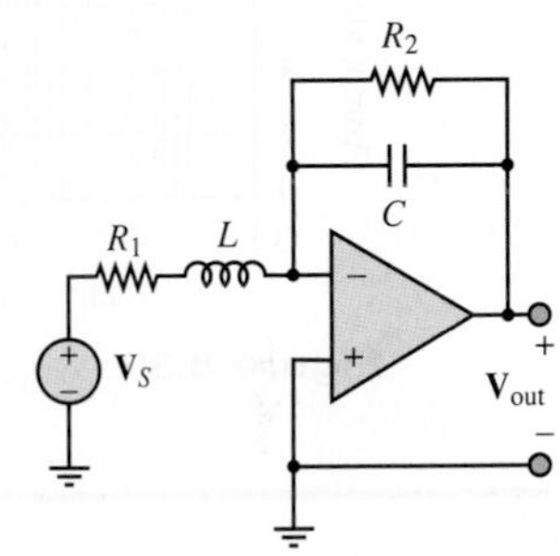

Figure 8.29

Solution

Known Quantities: Feedback and source impedances.

Find:

$$A(j\omega) = \frac{\mathbf{V}_{out}(j\omega)}{\mathbf{V}_S(j\omega)}$$

Schematics, Diagrams, Circuits, and Given Data: $R_2C = L/R_1 = \omega_0$.

Assumptions: Assume an ideal op-amp.

Analysis: The expression for the gain of the filter of Figure 8.29 can be determined by using equation 8.45:

$$A(j\omega) = \frac{\mathbf{V}_{out}(j\omega)}{\mathbf{V}_S(j\omega)} = -\frac{Z_F(j\omega)}{Z_S(j\omega)}$$

where

$$Z_F(j\omega) = R_2 || \frac{1}{j\omega C} = \frac{R_2}{1 + j\omega C R_2} = \frac{R_2}{1 + j\omega/\omega_0}$$

$$= R_1 + j\omega L = R_1\left(1 + j\omega\frac{L}{R_1}\right) = R_1\left(1 + \frac{j\omega}{\omega_0}\right)$$

Thus, the gain of the filter is

$$A(j\omega) = \frac{R_2/(1 + j\omega/\omega_0)}{R_1(1 + j\omega/\omega_0)}$$

$$= \frac{R_2/R_1}{(1 + j\omega/\omega_0)^2}$$

Comments: Note the similarity between the expression for the gain of the filter of Figure 8.29 and that given in equation 8.50 for the gain of a (first-order) low-pass filter. Clearly, the circuit analyzed in this example is also a low-pass filter, of second order (as the quadratic denominator term suggests). Figure 8.30 compares the two responses in both linear and decibel (Bode) magnitude plots. The slope of the decibel plot for the second-order filter at higher frequencies is twice that of the first-order filter (−40 versus −20 dB/decade). We should also remark that the use of an inductor in the filter design is not recommended in practice, as explained in the above section, and that we have used it in this example only because of the simplicity of the resulting gain expressions. Section 15.3 introduces design methods for practical high-order filters.

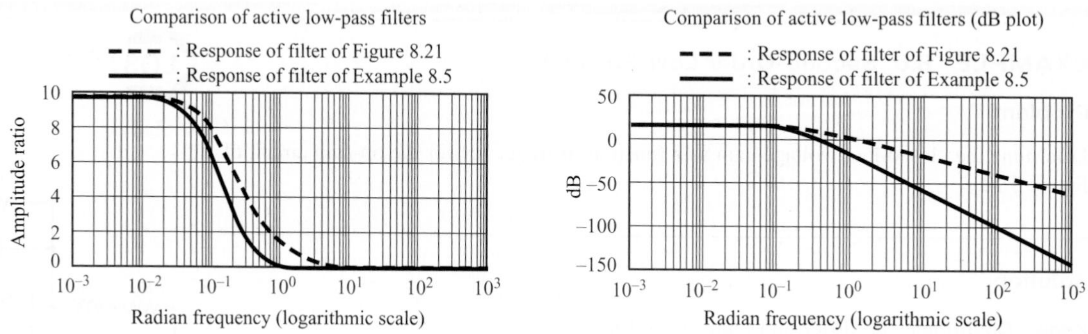

Figure 8.30 Comparison of first- and second-order active filters

CHECK YOUR UNDERSTANDING

Design a low-pass filter with closed-loop gain of 100 and cutoff (3-dB) frequency equal to 800 Hz. Assume that only 0.01-μF capacitors are available. Find R_F and R_S.

Repeat the design of the exercise above for a high-pass filter with cutoff frequency of 2,000 Hz. This time, however, assume that only standard values of resistors are available (see Table 2.1 for a table of standard values). Select the nearest component values, and calculate the percent error in gain and cutoff frequency with respect to the desired values.

Find the frequency corresponding to attenuation of 1 dB (with respect to the maximum value of the amplitude response) for the filter of the two previous exercises.

What is the decibel gain for the filter of Example 8.6 at the cutoff frequency ω_0? Find the 3-dB frequency for this filter in terms of the cutoff frequency ω_0, and note that the two are not the same.

Answers: $R_F = 19.9$ kΩ, $R_S = 199$ Ω; $R_F = 820$ kΩ, $R_S = 8.2$ kΩ; error: gain = 0 percent, $\omega_{3\text{ dB}} = 2.9$ percent; 407 Hz; −6 dB; $\omega_{3\text{ dB}} = 0.642\omega_0$

8.4 INTEGRATOR AND DIFFERENTIATOR CIRCUITS

In the preceding sections, we examined the frequency response of op-amp circuits for sinusoidal inputs. However, certain op-amp circuits containing energy storage elements reveal some of their more general properties if we analyze their response to inputs that are time-varying but not necessarily sinusoidal. Among such circuits are the commonly used integrator and differentiator; the analysis of these circuits is presented in the following paragraphs.

The Ideal Integrator

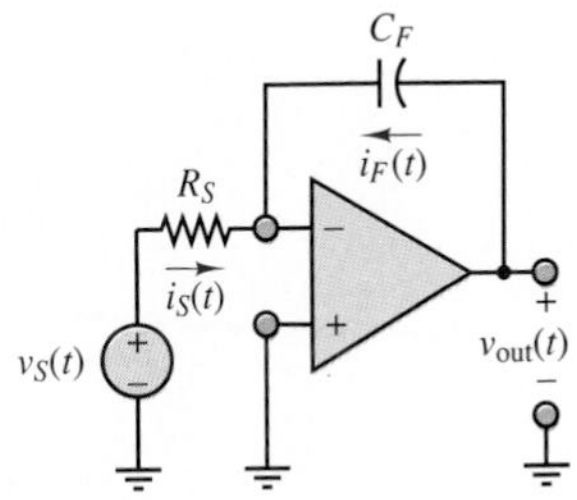

Figure 8.31 Op-amp integrator

Consider the circuit of Figure 8.31, where $v_S(t)$ is an arbitrary function of time (e.g., a pulse train, a triangular wave, or a square wave). The op-amp circuit shown provides an output that is proportional to the integral of $v_S(t)$. The analysis of the integrator circuit is, as always, based on the observation that

$$i_S(t) = -i_F(t) \tag{8.63}$$

where

$$i_S(t) = \frac{v_S(t)}{R_S} \tag{8.64}$$

It is also known that

$$i_F(t) = C_F\frac{dv_{out}(t)}{dt} \tag{8.65}$$

from the fundamental definition of the capacitor. The source voltage can then be expressed as a function of the derivative of the output voltage:

$$\frac{1}{R_SC_F}v_S(t) = -\frac{dv_{out}(t)}{dt} \tag{8.66}$$

By integrating both sides of equation 8.66, we obtain the following result:

$$v_{out}(t) = -\frac{1}{R_SC_F}\int_{-\infty}^{t} v_S(t')\,dt' \qquad \text{Op-amp integrator} \tag{8.67}$$

This equation states that the output voltage is the integral of the input voltage.

There are numerous applications of the op-amp integrator, most notably the **analog computer,** which is discussed in Section 8.5. Example 8.7 illustrates the operation of the op-amp integrator.

EXAMPLE 8.7 Integrating a Square Wave

Problem

Determine the output voltage for the integrator circuit of Figure 8.32 if the input is a square wave of amplitude $\pm A$ and period T.

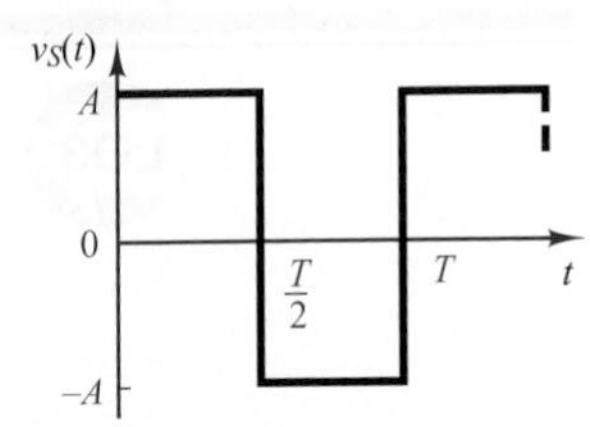

Figure 8.32

Solution

Known Quantities: Feedback and source impedances; input waveform characteristics.

Find: $v_{out}(t)$.

Schematics, Diagrams, Circuits, and Given Data: $T = 10$ ms; $C_F = 1\ \mu$F; $R_S = 10$ kΩ.

Assumptions: Assume an ideal op-amp. The square wave starts at $t = 0$, and therefore $v_{out}(0) = 0$.

Analysis: Following equation 8.67, we write the expression for the output of the integrator:

$$v_{\text{out}}(t) = -\frac{1}{R_F C_S}\int_{-\infty}^{t} v_S(t')\,dt' = -\frac{1}{R_F C_S}\left[\int_{-\infty}^{0} v_S(t')\,dt' + \int_{0}^{t} v_S(t')\,dt'\right]$$

$$= -\frac{1}{R_F C_S}\left[v_{\text{out}}(0) + \int_{0}^{t} v_S(t')\,dt'\right]$$

Next, we note that we can integrate the square wave in a piecewise fashion by observing that $v_S(t) = A$ for $0 \le t < T/2$ and $v_S(t) = -A$ for $T/2 \le t < T$. We consider the first half of the waveform:

$$v_{\text{out}}(t) = -\frac{1}{R_F C_S}\left[v_{\text{out}}(0) + \int_{0}^{t} v_S(t')\,dt'\right] = -100\left(0 + \int_{0}^{t} A\,dt'\right)$$

$$= -100At \qquad 0 \le t < \frac{T}{2}$$

$$v_{\text{out}}(t) = v_{\text{out}}\left(\frac{T}{2}\right) - \frac{1}{R_F C_S}\int_{T/2}^{t} v_S(t')\,dt' = -100A\frac{T}{2} - 100\int_{T/2}^{t} (-A)\,dt'$$

$$= -100A\frac{T}{2} + 100A\left(t - \frac{T}{2}\right) = -100A(T - t) \qquad \frac{T}{2} \le t < T$$

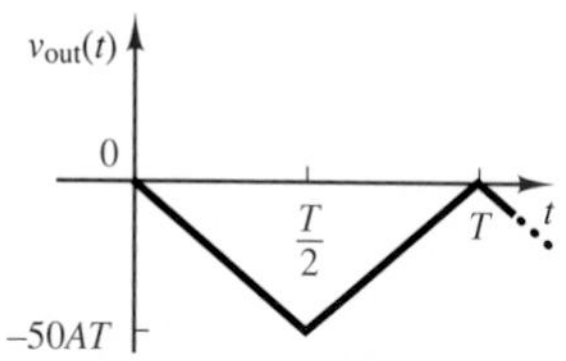

Figure 8.33

Since the waveform is periodic, the above result will repeat with period T, as shown in Figure 8.33. Note also that the average value of the output voltage is not zero.

Comments: The integral of a square wave is thus a triangular wave. This is a useful fact to remember. Note that the effect of the initial condition is very important, since it determines the starting point of the triangular wave.

CHECK YOUR UNDERSTANDING

Plot the frequency response of the ideal integrator in the form of a Bode plot. Determine the slope of the straight-line segments in decibels per decade. You may assume $R_S C_F = 10$.

Answer: −20 dB/decade

LO3

EXAMPLE 8.8 Proportional-Integral Control with Op-Amps

Problem

Consider the temperature control circuit of Example 8.5, shown again in Figure 8.34(a). The aim of this example is to illustrate the very common practice of *proportional-integral*, or *PI, control*. In Experiment 8.5, we discovered that the proportional control implemented with the gain K_P could still give rise to a steady-state error in the final temperature of the system. This error can be eliminated by using an automatic control system that feeds back a component that is proportional to the *integral of the error voltage*, in addition to the proportional term used in Example 8.5. Figure 8.34(b) depicts the block diagram of such a PI controller. Now, the design of the control system requires selecting two gains, the *proportional gain* K_P, and the *integral gain* K_I.

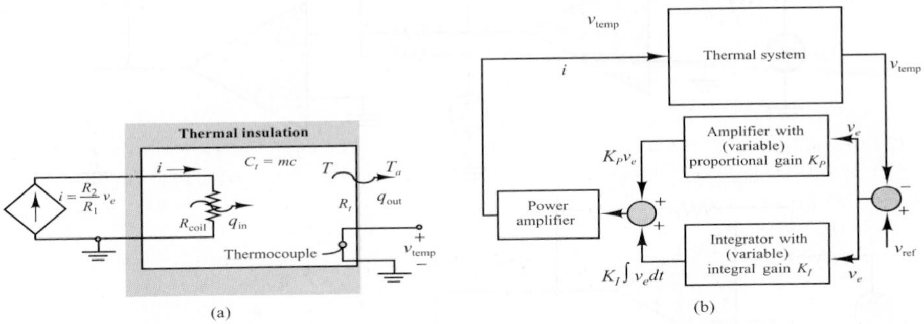

Figure 8.34 (a) Thermal system and (b) Block diagram of control system

---

Solution

Known Quantities: Sensor (input) voltage; feedback and source resistors, thermal system component values.

Find: Select desired value of proportional gain, K_P, and integral gain, K_I, to achieve automatic temperature control with zero steady-state error.

Schematics, Diagrams, Circuits and Given Data: $R_{coil} = 5\ \Omega$; $R_t = 2°\text{C/ W}$; $C_t = 50\ \text{J/°C}$; $\alpha = 1\ \text{V/°C}$.

Assumptions: Assume ideal op-amps.

Analysis: The circuit of Figure 8.34(c) shows two op-amp circuits — the top circuit generates the error voltage v_e, as was done in Example 8.5. The only difference is that in this case the circuit does not provide any gain. The bottom circuit amplifies v_e by the proportional gain, $-K_P = -R_2/R_1$ and also computes the integral of v_e times the integral gain $-K_I = -1/R_3C$. These two quantities are then summed through another inverting summer circuit, which takes care of the sign change as well.

Figure 8.34(d) depicts the temperature response of the system for $K_P = 5$ (the value we had selected in Example 8.5) and different values of K_I. Note that the steady-state error is now zero! This is a property of controllers that incorporate an integral term. Figure 8.34(e) shows the current supplied to the heater coil. Note that the response is quite fast, and that the temperature deviation is minimal.

Comments: The addition of the integral term in the controller causes the system temperature to oscillate in response to the $-10°\text{C}$ temperature disturbance described in Example 8.5 (p. 430) (for sufficiently high values of K_I). This oscillation is a characteristic of an underdamped second-order system (see Chapter 5)—but we originally started out with a first-order thermal system! The addition of the integral term has increased the order of the system, and now it is possible for the system to display oscillatory behavior, that is, to have complex conjugate roots (poles). To those familiar with thermal systems, this behavior should cause a raised eyebrow! It is well known that thermal systems cannot display underdamped behavior (that is, there is no thermal system property analogous to inductance. The introduction of the integral gain can in fact cause temperature oscillations, as if we had introduced an artificial "thermal inductor" in the system.

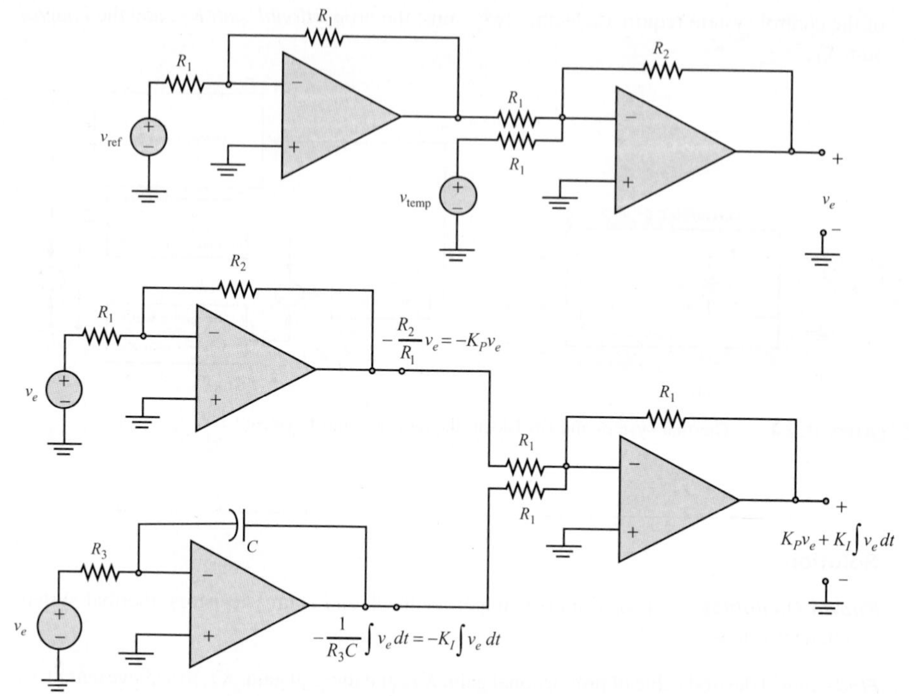

(c)

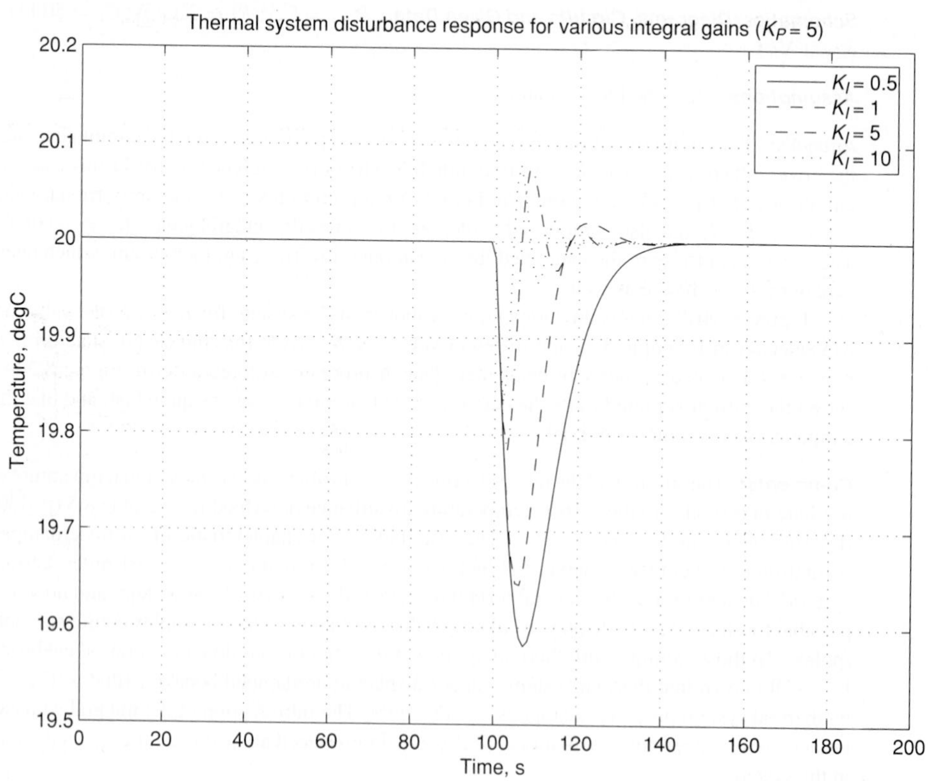

(d)

Figure 8.34 (c) Circuit for generating error voltage and proportional gain and (d) Response of thermal system for various values of integral gain, K_I ($K_P = 5$)

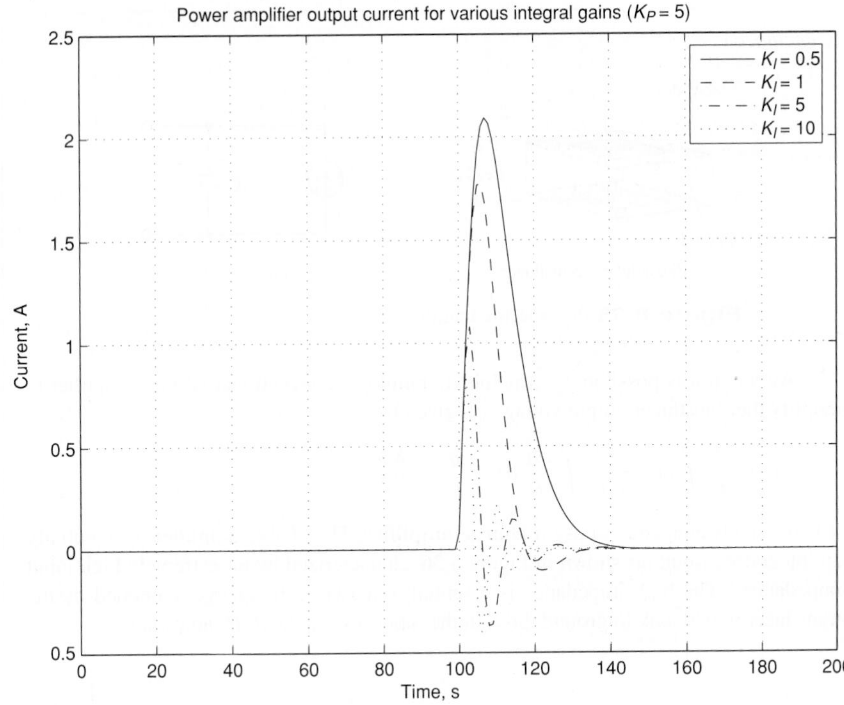

(e)

Figure 8.34 (e) Power amplifier current system for various values of integral gain, K_I ($K_P = 5$)

LO3

Charge Amplifiers

FIND IT ON THE WEB

One of the most common families of transducers for the measurement of force, pressure, and acceleration is that of **piezoelectric transducers.** These transducers contain a piezoelectric crystal that generates an electric charge in response to deformation. Thus, if a force is applied to the crystal (leading to a displacement), a charge is generated within the crystal. If the external force generates a displacement x_i, then the transducer will generate a charge q according to the expression

$$q = K_P x_i$$

Figure 8.35 depicts the basic structure of the piezoelectric transducer, and a simple circuit model. The model consists of a current source in parallel with a capacitor, where the current source represents the rate of change of the charge generated in response to an external force; and the capacitance is a consequence of the structure of the transducer, which consists of a piezoelectric crystal (e.g., quartz or Rochelle salt) sandwiched between conducting electrodes (in effect, this is a parallel-plate capacitor).

(Continued)

(Concluded)

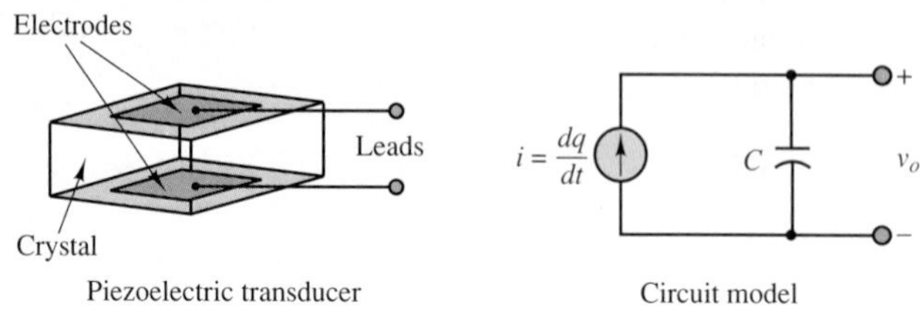

Figure 8.35 Piezoelectric transducer

Although it is possible, in principle, to employ a conventional voltage amplifier to amplify the transducer output voltage v_o, given by

$$v_o = \frac{1}{C}\int i\,dt = \frac{1}{C}\int \frac{dq}{dt}\,dt = \frac{q}{C} = \frac{K_P x_i}{C}$$

it is often advantageous to use a **charge amplifier.** The charge amplifier is essentially an integrator circuit, as shown in Figure 8.36, characterized by an extremely high input impedance.[3] The high impedance is essential; otherwise, the charge generated by the transducer would leak to ground through the input resistance of the amplifier.

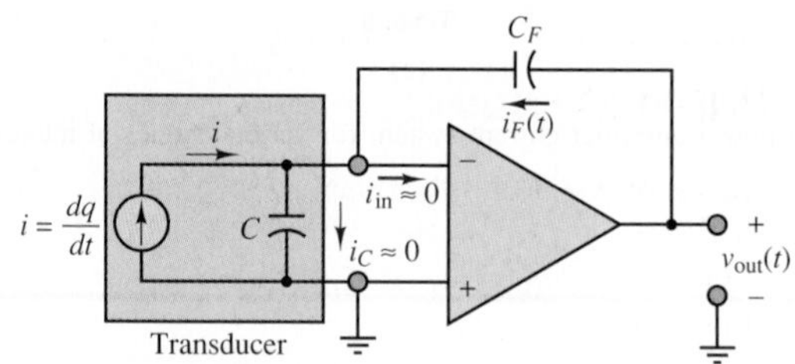

Figure 8.36 Charge amplifier

Because of the high input impedance, the input current to the amplifier is negligible; further, because of the high open-loop gain of the amplifier, the inverting-terminal voltage is essentially at ground potential. Thus, *the voltage across the transducer is effectively zero.* As a consequence, to satisfy KCL, the feedback current $i_F(t)$ must be equal and opposite to the transducer current i:

$$i_F(t) = -i$$

and since

$$v_{out}(t) = \frac{1}{C_F}\int i_F(t)\,dt$$

it follows that the output voltage is proportional to the charge generated by the transducer, and therefore to the displacement:

$$v_{out}(t) = \frac{1}{C_F}\int -i\,dt = \frac{1}{C_F}\int -\frac{dq}{dt}\,dt = -\frac{q}{C_F} = -\frac{K_P x_i}{C_F}$$

Since the displacement is caused by an external force or pressure, this sensing principle is widely adopted in the measurement of force and pressure.

[3] Special op-amps are employed to achieve extremely high input impedance, through FET input circuits.

The Ideal Differentiator

Using an argument similar to that employed for the integrator, we can derive a result for the ideal differentiator circuit of Figure 8.37. The relationship between input and output is obtained by observing that

$$i_S(t) = C_S \frac{dv_S(t)}{dt} \tag{8.68}$$

and

$$i_F(t) = \frac{v_{\text{out}}(t)}{R_F} \tag{8.69}$$

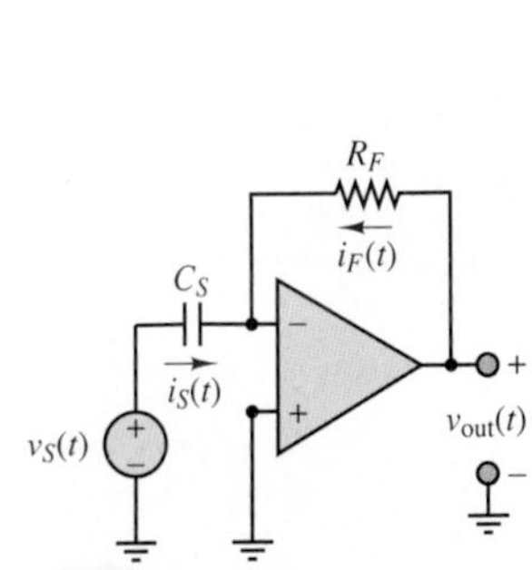

Figure 8.37 Op-amp differentiator

so that the output of the differentiator circuit is proportional to the derivative of the input:

$$v_{\text{out}}(t) = -R_F C_S \frac{dv_S(t)}{dt} \qquad \text{Op-amp differentiator} \tag{8.70}$$

LO3

Although mathematically attractive, the differentiation property of this op-amp circuit is seldom used in practice, because differentiation tends to amplify any noise that may be present in a signal.

CHECK YOUR UNDERSTANDING

Plot the frequency response of the ideal differentiator in the form of a Bode plot. Determine the slope of the straight-line segments in decibels per decade. You may assume $R_F C_S = 100$. Verify that if the triangular wave of Example 8.7 is the input to the ideal differentiator of Figure 8.37, that resulting output is a square wave.

Answer: +20 dB/decade

8.5 ANALOG COMPUTERS

FIND IT ON THE WEB

Prior to the advent of digital computers, the solution of differential equations and the simulation of complex dynamic systems were conducted exclusively by means of analog computers. **Analog computers** still find application in engineering practice in the simulation of dynamic systems. The analog computer is a device that is based on three op-amp circuits introduced earlier in this chapter: the amplifier, the summer, and the integrator. These three building blocks permit the construction of circuits that can be used to solve differential equations and to simulate dynamic systems. Figure 8.38 depicts the three symbols that are typically employed to represent the principal functions of an analog computer.

The simplest way to discuss the operation of the analog computer is to present an example. Consider the simple second-order mechanical system, shown in Figure 8.39, that represents, albeit in a greatly simplified fashion, one corner of an automobile suspension system. The mass M represents the mass of one-quarter of the vehicle, the damper B represents the shock absorber, and the spring K represents the suspension

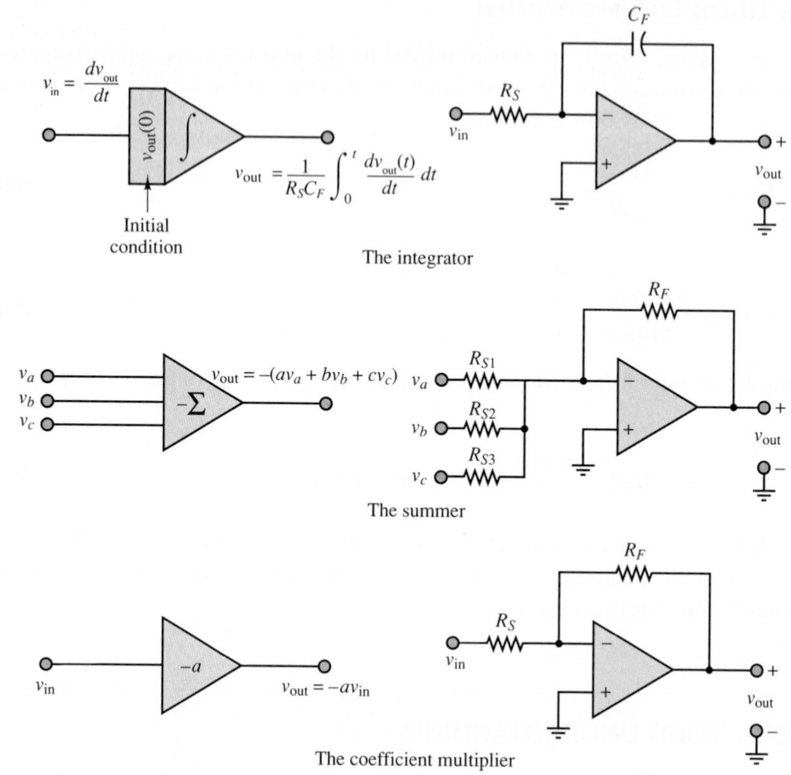

Figure 8.38 Elements of the analog computer

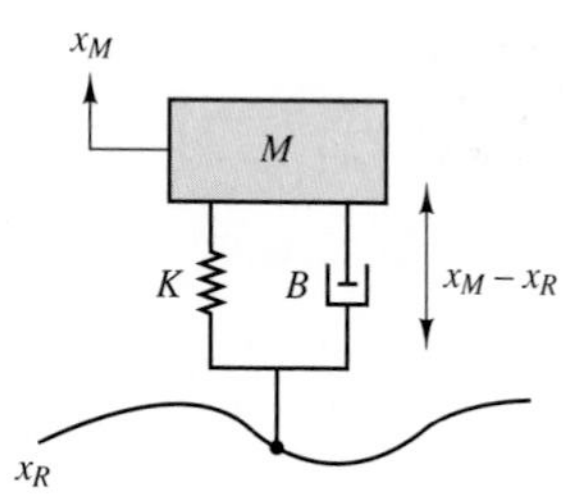

Figure 8.39 Model of automobile suspension

spring (or strut). The differential equation of the system may be derived as follows:

$$M \frac{d^2 x_M}{dt^2} + B\left(\frac{dx_M}{dt} - \frac{dx_R}{dt}\right) + K(x_M - x_R) = 0 \tag{8.71}$$

Rearranging terms, we obtain the following equation, in which the terms related to the road displacement and velocity—x_R and dx_R/dt, respectively—are the forcing functions:

$$M \frac{d^2 x_M}{dt^2} + B \frac{dx_M}{dt} + K x_M = B \frac{dx_R}{dt} + K x_R \tag{8.72}$$

Assume that the car is traveling over a "washboard" surface on an unpaved road, such that the road profile is approximately described by the expression

$$x_R(t) = X \sin(\omega t) \tag{8.73}$$

It follows, then, that the vertical velocity input to the suspension is given by the expression

$$\frac{dx_R}{dt} = \omega X \cos(\omega t) \tag{8.74}$$

and we can write the equation for the suspension system in the form

$$M \frac{d^2 x_M}{dt^2} + B \frac{dx_M}{dt} + K x_M = B\omega X \cos(\omega t) + KX \sin(\omega t) \tag{8.75}$$

It would be desirable to solve the equation for the displacement x_M, which represents the motion of the vehicle mass in response to the road excitation. The solution can be used as an aid in designing the suspension system that best absorbs the road vibration, providing a comfortable ride for passengers. Equation 8.75 may be rearranged to obtain

$$\frac{d^2x_M}{dt^2} = -\frac{B}{M}\frac{dx_M}{dt} - \frac{K}{M}x_M + \frac{B}{M}\omega X\cos(\omega t) + \frac{K}{M}X\sin(\omega t) \tag{8.76}$$

This equation is now in a form appropriate for solution by repeated integration, since we have isolated the highest derivative term; thus, it will be sufficient to integrate the right-hand side twice to obtain the solution for the displacement of the vehicle mass x_M.

Figure 8.40 depicts the three basic operations that need to be performed to integrate the differential equation describing the motion of the mass M. Note that in each of the three blocks—the summer and the two integrators—the inversion due to the inverting amplifier configuration used for the integrator is already accounted for. Finally, the basic summing and integrating blocks together with three coefficient multipliers (inverting amplifiers) are connected in the configuration that corresponds to the preceding differential equation (equation 8.76). You can easily verify that the analog computer circuit of Figure 8.41 does indeed solve the differential equation in x_M by repeated integration.

LO4

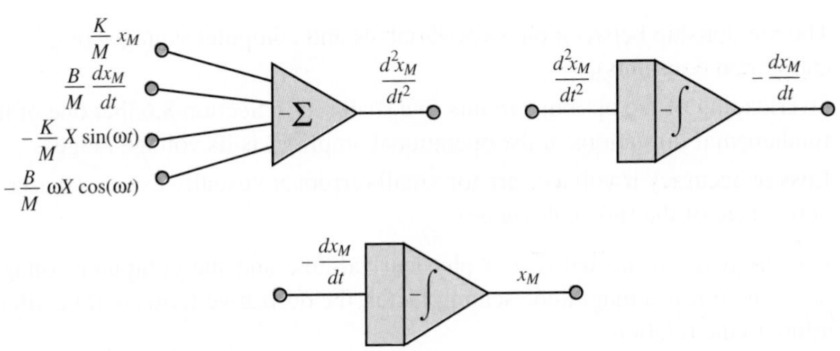

Figure 8.40 Solution by repeated integration

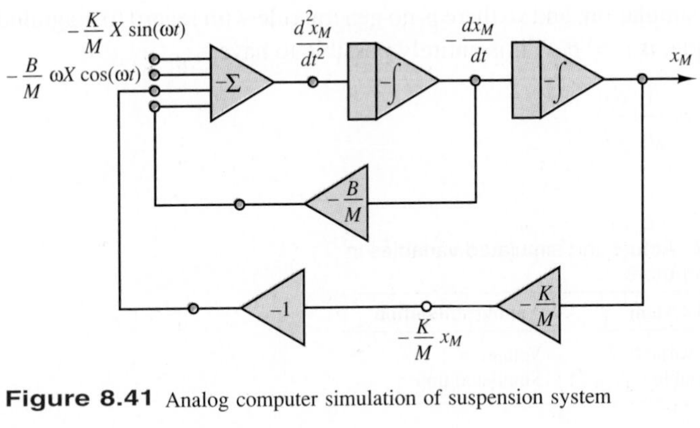

Figure 8.41 Analog computer simulation of suspension system

Scaling in Analog Computers

One of the important issues in analog computing is that of scaling. Since the analog computer implements an electrical analog of a physical system, there is no guarantee that the voltages and currents in the analog computer circuits will be of the same order of magnitude as the physical variables (e.g., velocity, displacement, temperature, or flow) that they simulate. Further, it is not necessary that the computer simulate the physical system with the same time scale; it may be desirable in practice to speed up or slow down the simulation, thus the interest in time scaling and magnitude scaling.

Let Table 8.2 represent the physical and simulation variables. Considering time scaling first, we let t denote real time and τ be the computer time variable. Then the time derivative of a physical variable can be expressed as

$$\frac{dx}{dt} = \frac{dx}{d\tau}\frac{d\tau}{dt} = \alpha\frac{dx}{d\tau} \quad \textbf{(8.77)}$$

where α is the scaling factor between real time and computer time:

$$\tau = \alpha t \quad \textbf{(8.78)}$$

For higher-order derivatives, the following relationship will hold:

$$\frac{d^n x}{dt^n} = \alpha^n \frac{d^n x}{d\tau^n} \quad \textbf{(8.79)}$$

While time scaling is likely to be prompted by a desire to speed up or slow down a computation, magnitude scaling is motivated by several different factors:

1. The relationship between physical variables and computer voltages (e.g., calibration constants).
2. Overloading of the op-amp circuits (we shall see in Section 8.6 that one of the fundamental limitations of the operational amplifier is its voltage range).
3. Loss of accuracy if voltages are too small (errors are usually expressed as a percentage of the full-scale range).

Thus, if the relationship between a physical variable and the computer voltage is $v = \beta x$, where β is a magnitude scaling factor, the derivative terms will be affected according to the relation

$$\frac{dx}{dt} = \frac{1}{\beta}\frac{dv}{dt} \quad \textbf{(8.80)}$$

Note that different scaling factors may be introduced at each point in the analog computer simulation, and so there is no general rule with regard to magnitude scaling. For example, if $v = \beta_0 x$, it is entirely possible to have

$$\frac{dx}{dt} = \frac{1}{\beta_1}\frac{dv}{dt}$$

LO4

Table 8.2 Actual and simulated variables in analog computers

Physical system	Analog simulation
Physical variable x	Voltage v
Time variable t	Simulated time τ

EXAMPLE 8.9 Analog Computer Simulation of Automotive Suspension

LO4

Problem

Implement the analog computer simulation of the automotive suspension system of Figure 8.39 using the building blocks of Figure 8.38.

Solution

Known Quantities: Mass, spring rate, and damping parameters of automotive suspension.

Find: Component values of analog computer circuit of Figure 8.41.

Schematics, Diagrams, Circuits, and Given Data: $M = 400$ kg; $K = 1.6 \times 10^5$ N/m; $B = 20 \times 10^3$ N/m-s.

Assumptions: Assume ideal op-amps. Express all resistors in megohms and all capacitors in microfarads.

Analysis: With reference to Figure 8.41, we observe that the analog computer simulation of the automotive suspension requires a four-input summer, two integrators, two coefficient multipliers, and one sign inverter.

Expressing all resistors in megohms and all capacitors in microfarads is useful because each integrator has a multiplier of $-1/RC$. Using $R = 1$ MΩ and $C = 1$ μF results in $-1/RC = -1$. Figure 8.42 depicts the integrator configuration. The four-input summing amplifier uses 1-MΩ resistors throughout, so that the gain for each input is also equal to -1. On the other hand, the two coefficient multipliers are required to have gains of $-K/M = -4{,}000$ and $-B/M = 50$, respectively. Thus we select 10-MΩ and 2.5-kΩ resistors for the first coefficient multiplier and 1-MΩ and 20-kΩ resistors for the second. Finally, the inverter can be realized with two 1-MΩ resistors. The various elements are depicted in Figure 8.42.

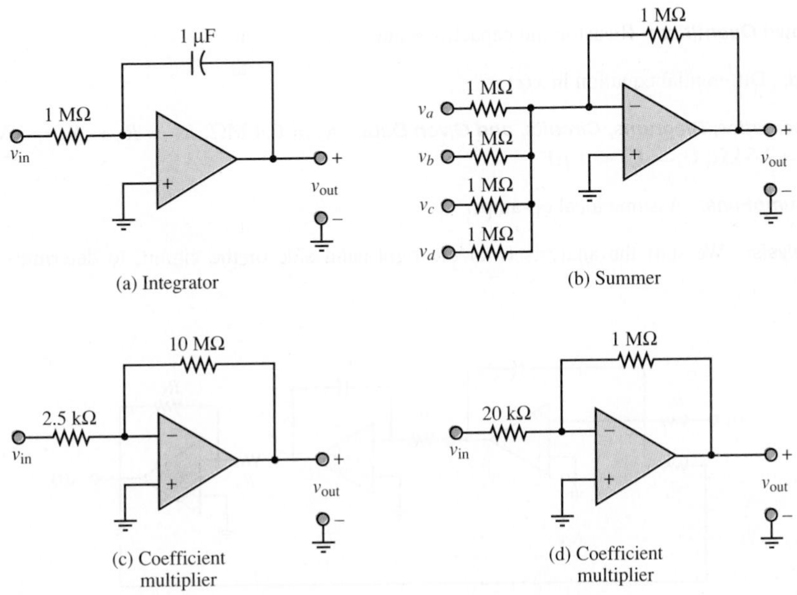

Figure 8.42 Analog computer simulation of suspension system

Comments: Note that it is not necessary to employ five op-amps in this analog simulator. The summing amplifier function and the first integrator could, for example, be combined into a single op-amp, and one of the two coefficient multipliers could be eliminated. This idea is explored further in the homework problems.

CHECK YOUR UNDERSTANDING

Modify the gains of the coefficient multipliers in Example 8.9 if you wish to slow down the simulation by a factor of 10 (that is, $\alpha = 0.1$ in equation 8.77).

For the simulation of Example 8.9, what will the largest magnitude of the voltage analog of x_M be for a road displacement $x_R = 0.01\sin(100t)$? [*Hint:* Use phasor techniques to compute the frequency response $x_M(j\omega)/F(j\omega)$, where $f(t) = B\,dx_R(t)/dt + Kx_R(t)$, and evaluate the output voltage by multiplying the input by the magnitude of the frequency response at $\omega = 100$ rad/s.]

Answers: $B/M = 5$; $K/M = 40$; $x_{M\max} = 0.0082$ m

LO4

EXAMPLE 8.10 Deriving a Differential Equation from an Analog Computer Circuit

Problem

Derive the differential equation corresponding to the analog computer simulator of Figure 8.43.

Solution

Known Quantities: Resistor and capacitor values.

Find: Differential equation in $x(t)$.

Schematics, Diagrams, Circuits, and Given Data: $R_1 = 0.4$ MΩ; $R_2 = R_3 = R_5 = 1$ MΩ; $R_4 = 2.5$ kΩ; $C_1 = C_2 = 1$ μF.

Assumptions: Assume ideal op-amps.

Analysis: We start the analysis from the right-hand side of the circuit, to determine the

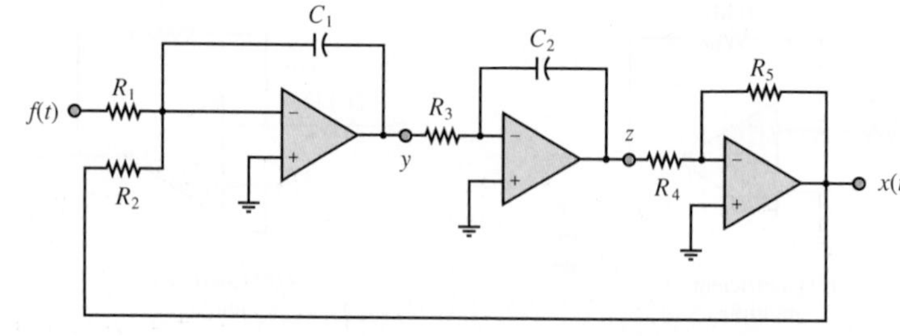

Figure 8.43 Analog computer simulation of unknown system

intermediate variable z as a function of x:

$$x = -\frac{R_5}{R_4} z = -400z$$

Next, we move to the left to determine the relationship between y and z:

$$z = -\frac{1}{R_3 C_2} \int y(t')\, dt' \qquad \text{or} \qquad y = -\frac{dz}{dt}$$

Finally, we determine y as a function of x and f:

$$y = -\frac{1}{R_2 C_1} \int x(t')\, dt' - \frac{1}{R_1 C_1} \int f(t')\, dt' = -\int \left[x(t') + 2.5 f(t')\right] dt'$$

or

$$\frac{dy}{dt} = -x - 2.5f$$

Substituting the expressions into one another and eliminating the variables y and z, we obtain the differential equation in x:

$$x = -400z$$

$$\frac{dx}{dt} = -400 \frac{dz}{dt} = 400y$$

$$\frac{d^2x}{dt^2} = 400 \frac{dy}{dt} = 400\,(x - 2.5f)$$

and

$$\frac{d^2x}{dt^2} + 400x = -1{,}000f$$

Comments: Note that the summing and integrating functions have been combined into a single block in the first amplifier.

CHECK YOUR UNDERSTANDING

Derive the differential equation corresponding to the analog computer circuit shown in the figure.

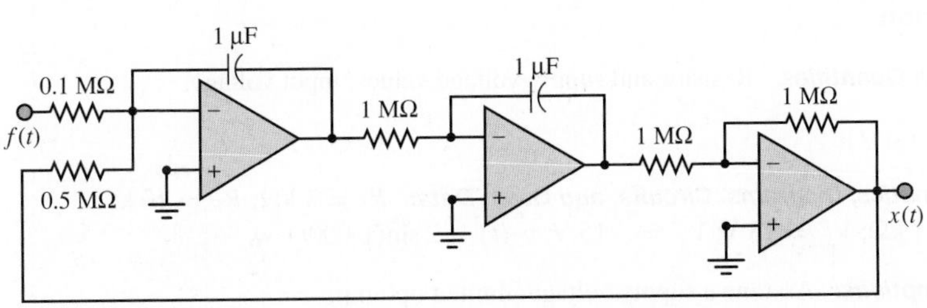

Answer: $d^2x/dt^2 + 2x = -10f(t)$

8.6 PHYSICAL LIMITATIONS OF OPERATIONAL AMPLIFIERS

Thus far, the operational amplifier has been treated as an ideal device, characterized by infinite input resistance, zero output resistance, and infinite open-loop voltage gain. Although this model is adequate to represent the behavior of the op-amp in a large number of applications, practical operational amplifiers are not ideal devices but exhibit a number of limitations that should be considered in the design of instrumentation. In particular, in dealing with relatively large voltages and currents, and in the presence of high-frequency signals, it is important to be aware of the nonideal properties of the op-amp. In this section, we examine the principal limitations of the operational amplifier.

Voltage Supply Limits

As indicated in Figure 8.4, operational amplifiers (and all amplifiers, in general) are powered by external DC voltage supplies V_S^+ and V_S^-, which are usually symmetric and on the order of ± 10 to ± 20 V. Some op-amps are especially designed to operate from a single voltage supply; but for the sake of simplicity from here on we shall consider only symmetric supplies. The effect of limiting supply voltages is that amplifiers are capable of amplifying signals *only within the range of their supply voltages;* it would be physically impossible for an amplifier to generate a voltage greater than V_S^+ or less than V_S^-. This limitation may be stated as follows:

LO5

$$V_S^- < v_{\text{out}} < V_S^+ \qquad \text{Voltage supply limitation} \tag{8.81}$$

For most op-amps, the limit is actually approximately 1.5 V less than the supply voltages. How does this practically affect the performance of an amplifier circuit? An example will best illustrate the idea.

LO5

EXAMPLE 8.11 Voltage Supply Limits in an Inverting Amplifier

Problem

Compute and sketch the output voltage of the inverting amplifier of Figure 8.44.

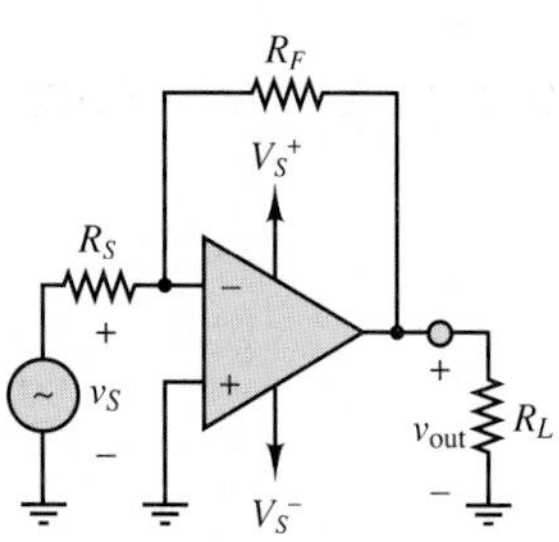

Figure 8.44

Solution

Known Quantities: Resistor and supply voltage values; input voltage.

Find: $v_{\text{out}}(t)$.

Schematics, Diagrams, Circuits, and Given Data: $R_S = 1\text{ k}\Omega$; $R_F = 10\text{ k}\Omega$; $R_L = 1\text{ k}\Omega$; $V_S^+ = 15$ V; $V_S^- = -15$ V; $v_S(t) = 2\sin(1{,}000t)$.

Assumptions: Assume a supply voltage–limited op-amp.

Analysis: For an ideal op-amp the output would be

$$v_{\text{out}}(t) = -\frac{R_F}{R_S} v_S(t) = -10 \times 2\sin(1{,}000t) = -20\sin(1{,}000t)$$

However, the supply voltage is limited to ± 15 V, and the op-amp output voltage will therefore saturate before reaching the theoretical peak output value of ± 20 V. Figure 8.45 depicts the output voltage waveform.

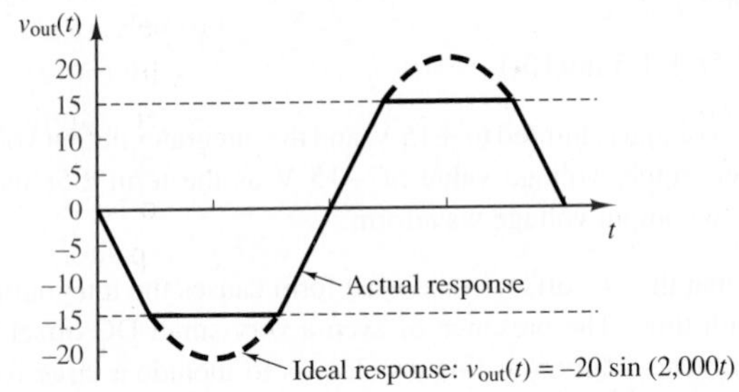

Figure 8.45 Op-amp output with voltage supply limit

Comments: In a practical op-amp, saturation would be reached at 1.5 V below the supply voltages, or at approximately ± 13.5 V.

Note how the voltage supply limit actually causes the peaks of the sine wave to be clipped in an abrupt fashion. This type of hard nonlinearity changes the characteristics of the signal quite radically, and could lead to significant errors if not taken into account. Just to give an intuitive idea of how such clipping can affect a signal, have you ever wondered why rock guitar has a characteristic sound that is very different from the sound of classical or jazz guitar? The reason is that the "rock sound" is obtained by overamplifying the signal, attempting to exceed the voltage supply limits, and causing clipping similar in quality to the distortion introduced by voltage supply limits in an op-amp. This clipping broadens the spectral content of each tone and causes the sound to be distorted.

One of the circuits most directly affected by supply voltage limitations is the op-amp integrator. Example 8.12 illustrates how saturation of an integrator circuit can lead to severe signal distortion.

EXAMPLE 8.12 Voltage Supply Limits in an Op-Amp Integrator

Problem

Compute and sketch the output voltage of the integrator of Figure 8.31.

Solution

Known Quantities: Resistor, capacitor, and supply voltage values; input voltage.

Find: $v_{out}(t)$.

Schematics, Diagrams, Circuits, and Given Data: $R_S = 10\ \text{k}\Omega$; $C_F = 20\ \mu\text{F}$; $V_S^+ = 15$ V; $V_S^- = -15$ V; $v_S(t) = 0.5 + 0.3\cos(10t)$.

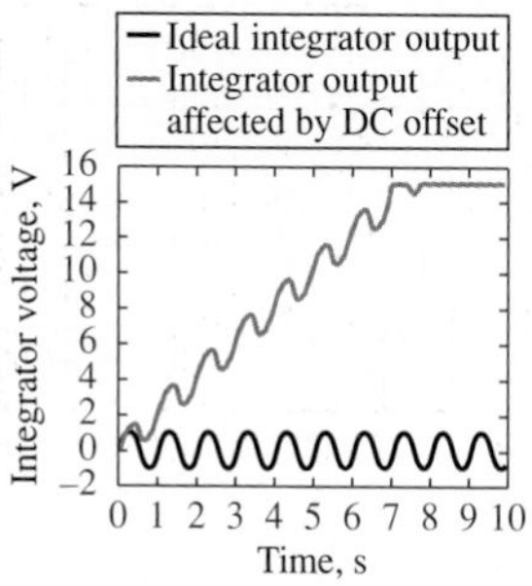

Figure 8.46 Effect of DC offset on integrator

Assumptions: Assume a supply voltage–limited op-amp. The initial condition is $v_{\text{out}}(0) = 0$.

Analysis: For an ideal op-amp integrator the output would be

$$v_{\text{out}}(t) = -\frac{1}{R_S C_F}\int_{-\infty}^{t} v_S(t')\,dt' = -\frac{1}{0.2}\int_{-\infty}^{t}\left[0.5 + 0.3\cos(10t')\right]dt'$$

$$= -2.5t + 1.5\sin(10t)$$

However, the supply voltage is limited to ± 15 V, and the integrator output voltage will therefore saturate at the lower supply voltage value of -15 V as the term $2.5t$ increases with time. Figure 8.46 depicts the output voltage waveform.

Comments: Note that the DC offset in the waveform causes the integrator output voltage to increase linearly with time. The presence of even a very small DC offset will always cause integrator saturation. One solution to this problem is to include a large feedback resistor in parallel with the capacitor; this solution is explored in the homework problems.

CHECK YOUR UNDERSTANDING

How long will it take (approximately) for the integrator of Example 8.12 to saturate if the input signal has a 0.1-V DC bias [that is, $v_S(t) = 0.1 + 0.3\cos(10t)$]?

Answer: Approximately 150 s

Frequency Response Limits

Another property of all amplifiers that may pose severe limitations to the op-amp is their finite bandwidth. We have so far assumed, in our ideal op-amp model, that the open-loop gain is a very large constant. In reality, $A_{V(\text{OL})}$ is a function of frequency and is characterized by a low-pass response. For a typical op-amp,

LO5

$$A_{V(\text{OL})}(j\omega) = \frac{A_0}{1 + j\omega/\omega_0} \qquad \text{Finite bandwidth limitation} \qquad \textbf{(8.82)}$$

The cutoff frequency of the op-amp open-loop gain ω_0 represents approximately the point where the amplifier response starts to drop off as a function of frequency, and is analogous to the cutoff frequencies of the RC and RL circuits of Chapter 6. Figure 8.47 depicts $A_{V(\text{OL})}(j\omega)$ in both linear and decibel plots for the fairly typical values $A_0 = 10^6$ and $\omega_0 = 10\pi$. It should be apparent from Figure 8.47 that the assumption of a very large open-loop gain becomes less and less accurate for increasing frequency. Recall the initial derivation of the closed-loop gain for the inverting amplifier: In obtaining the final result $\mathbf{V}_{\text{out}}/\mathbf{V}_S = -R_F/R_S$, it was assumed that $A_{V(\text{OL})} \to \infty$. This assumption is clearly inadequate at the higher frequencies.

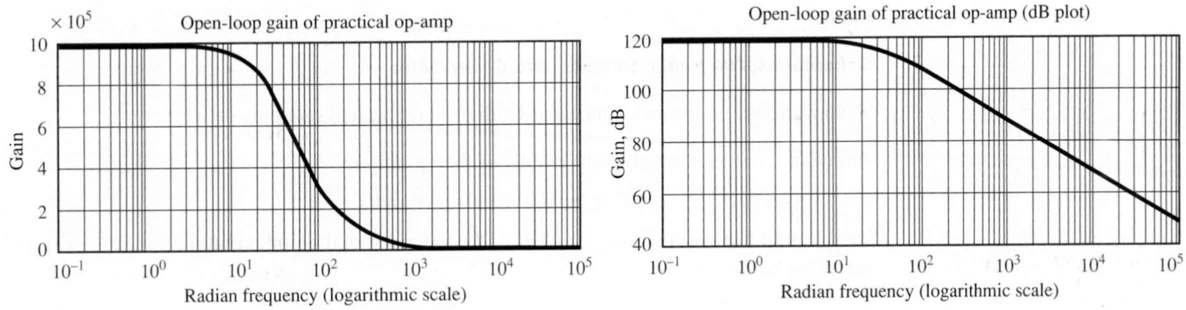

Figure 8.47 Open-loop gain of practical op-amp

The finite bandwidth of the practical op-amp results in a fixed **gain-bandwidth product** for any given amplifier. The effect of a constant gain-bandwidth product is that as the closed-loop gain of the amplifier is increased, its 3-dB bandwidth is proportionally reduced until, in the limit, if the amplifier were used in the open-loop mode, its gain would be equal to A_0 and its 3-dB bandwidth would be equal to ω_0. The constant gain-bandwidth product is therefore equal to the product of the open-loop gain and the open-loop bandwidth of the amplifier: $A_0\omega_0 = K$. When the amplifier is connected in a closed-loop configuration (e.g., as an inverting amplifier), its gain is typically much less than the open-loop gain and the 3-dB bandwidth of the amplifier is proportionally increased. To explain this further, Figure 8.48 depicts the case in which two different linear amplifiers (achieved through any two different negative feedback configurations) have been designed for the same op-amp. The first has closed-loop gain A_1, and the second has closed-loop gain A_2. The bold line in the figure indicates the open-loop frequency response, with gain A_0 and cutoff frequency ω_0. As the gain decreases from the open-loop gain A_0, to A_1, we see that the cutoff frequency increases from ω_0 to ω_1. If we further reduce the gain to A_2, we can expect the bandwidth to increase to ω_2. Thus,

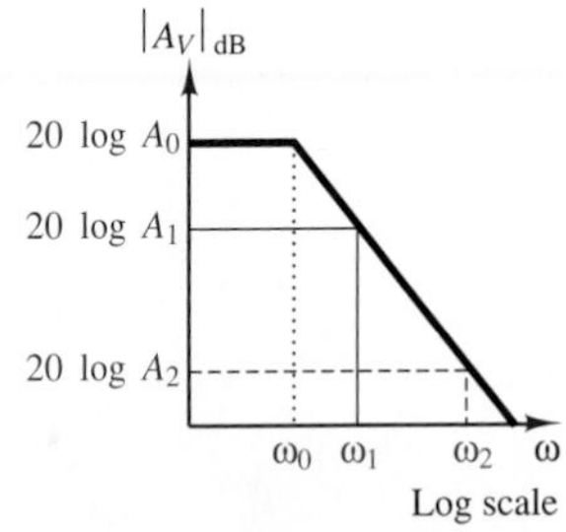

Figure 8.48

LO5

The product of gain and bandwidth in any given op-amp is constant. That is,

$$A_0 \times \omega_0 = A_1 \times \omega_1 = A_2 \times \omega_2 = K \qquad \textbf{(8.83)}$$

EXAMPLE 8.13 Gain-Bandwidth Product Limit in an Op-Amp

LO5

Problem

Determine the maximum allowable closed-loop voltage gain of an op-amp if the amplifier is required to have an audio-range bandwidth of 20 kHz.

Solution

Known Quantities: Gain-bandwidth product.

Find: $A_{V\ max}$.

Schematics, Diagrams, Circuits, and Given Data: $A_0 = 10^6$; $\omega_0 = 2\pi \times 5$ rad/s.

Assumptions: Assume a gain-bandwidth product limited op-amp.

Analysis: The gain-bandwidth product of the op-amp is

$$A_0 \times \omega_0 = K = 10^6 \times 2\pi \times 5 = \pi \times 10^7 \text{ rad/s}$$

The desired bandwidth is $\omega_{max} = 2\pi \times 20{,}000$ rad/s, and the maximum allowable gain will therefore be

$$A_{max} = \frac{K}{\omega_{max}} = \frac{\pi \times 10^7}{\pi \times 4 \times 10^4} = 250\ \frac{V}{V}$$

For any closed-loop voltage gain greater than 250, the amplifier would have reduced bandwidth.

Comments: If we desired to achieve gains greater than 250 and maintain the same bandwidth, two options would be available: (1) Use a different op-amp with greater gain-bandwidth product, or (2) connect two amplifiers in cascade, each with lower gain and greater bandwidth, such that the product of the gains would be greater than 250.

To further explore the first option, you may wish to look at the device data sheets for different op-amps and verify that op-amps can be designed (at a cost!) to have substantially greater gain-bandwidth product than the amplifier used in this example. The second option is examined in Example 8.14.

CHECK YOUR UNDERSTANDING

What is the maximum gain that could be achieved by the op-amp of Example 8.13 if the desired bandwidth is 100 kHz?

Answer: $A_{max} = 50$

LO5

EXAMPLE 8.14 Increasing the Gain-Bandwidth Product by Means of Amplifiers in Cascade

Problem

Determine the overall 3-dB bandwidth of the cascade amplifier of Figure 8.49.

Solution

Known Quantities: Gain-bandwidth product and gain of each amplifier.

Find: $\omega_{3\ dB}$ of cascade amplifier.

Schematics, Diagrams, Circuits, and Given Data: $A_0\omega_0 = K = 4\pi \times 10^6$ for each amplifier. $R_F/R_S = 100$ for each amplifier.

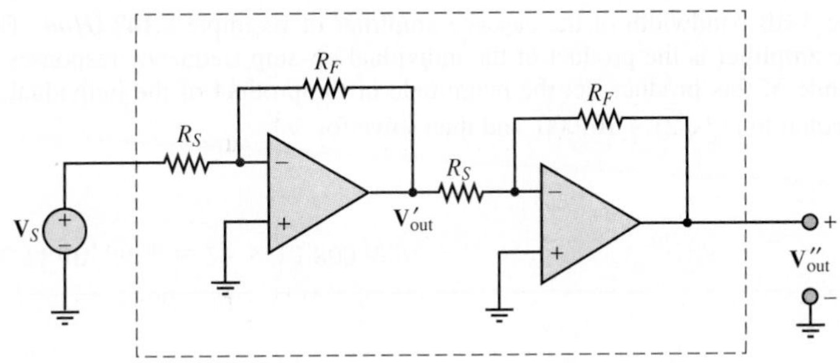

Figure 8.49 Cascade amplifier

Assumptions: Assume gain-bandwidth product limited (otherwise ideal) op-amps.

Analysis: Let A_1 and ω_1 denote the gain and the 3-dB bandwidth of the first amplifier, respectively, and A_2 and ω_2 those of the second amplifier.

The 3-dB bandwidth of the first amplifier is

$$\omega_1 = \frac{K}{A_1} = \frac{4\pi \times 10^6}{10^2} = 4\pi \times 10^4 \ \frac{\text{rad}}{\text{s}}$$

The second amplifier will also have

$$\omega_2 = \frac{K}{A_2} = \frac{4\pi \times 10^6}{10^2} = 4\pi \times 10^4 \ \frac{\text{rad}}{\text{s}}$$

Thus, the approximate bandwidth of the cascade amplifier is $4\pi \times 10^4$, and the gain of the cascade amplifier is $A_1 A_2 = 100 \times 100 = 10^4$.

Had we attempted to achieve the same gain with a single-stage amplifier having the same K, we would have achieved a bandwidth of only

$$\omega_3 = \frac{K}{A_3} = \frac{4\pi \times 10^6}{10^4} = 4\pi \times 10^2 \ \frac{\text{rad}}{\text{s}}$$

Comments: In practice, the actual 3-dB bandwidth of the cascade amplifier is not quite as large as that of each of the two stages, because the gain of each amplifier starts decreasing at frequencies somewhat lower than the nominal cutoff frequency.

CHECK YOUR UNDERSTANDING

In Example 8.14, we implicitly assumed that the gain of each amplifier was constant for frequencies up to the cutoff frequency. This is, in practice, not true, since the individual op-amp closed-loop gain starts dropping below the DC gain value according to the equation

$$A(j\omega) = \frac{A_1}{1 + j\omega/\omega_1}$$

Thus, the calculations carried out in the example are only approximate. Find an expression for the closed-loop gain of the cascade amplifier. (*Hint:* The combined gain is equal to the product of the individual closed-loop gains.) What is the actual gain in decibels at the cutoff frequency ω_0 for the cascade amplifier?

What is the 3-dB bandwidth of the cascade amplifier of Example 8.14? [*Hint:* The gain of the cascade amplifier is the product of the individual op-amp frequency responses. Compute the magnitude of this product, set the magnitude of the product of the individual frequency responses equal to $(1/\sqrt{2}) \times 10{,}000$, and then solve for ω.]

Answers: 74 dB; $\omega_{3\text{ dB}} = 2\pi \times 12{,}800$ rad/s

Input Offset Voltage

LO5

Another limitation of practical op-amps results because even in the absence of any external inputs, it is possible that an **offset voltage** will be present at the input of an op-amp. This voltage is usually denoted by $\pm V_{os}$, and it is caused by mismatches in the internal circuitry of the op-amp. The offset voltage appears as a differential input voltage between the inverting and noninverting input terminals. The presence of an additional input voltage will cause a DC bias error in the amplifier output, as illustrated in Example 8.15. Typical and maximum values of V_{os} are quoted in manufacturers' data sheets. The worst-case effects due to the presence of offset voltages can therefore be predicted for any given application.

LO5

EXAMPLE 8.15 Effect of Input Offset Voltage on an Amplifier

Problem

Determine the effect of the input offset voltage V_{os} on the output of the amplifier of Figure 8.50.

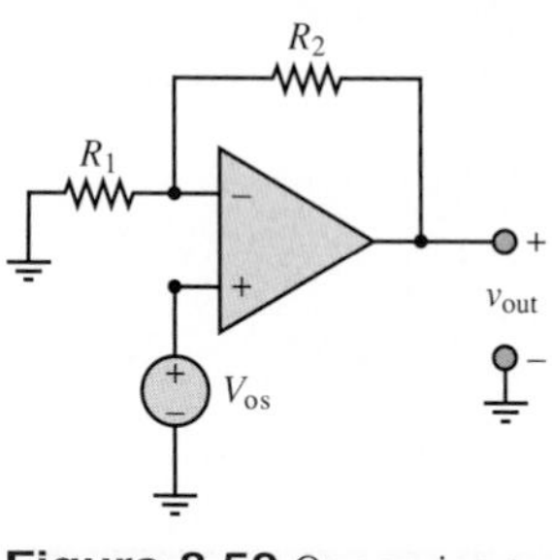

Figure 8.50 Op-amp input offset voltage

Solution

Known Quantities: Nominal closed-loop voltage gain; input offset voltage.

Find: The offset voltage component in the output voltage $v_{\text{out,os}}$.

Schematics, Diagrams, Circuits, and Given Data: $A_{\text{nom}} = 100$; $V_{os} = 1.5$ mV.

Assumptions: Assume an input offset voltage–limited (otherwise ideal) op-amp.

Analysis: The amplifier is connected in a noninverting configuration; thus its gain is

$$A_{V\text{ nom}} = 100 = 1 + \frac{R_F}{R_S}$$

The DC offset voltage, represented by an ideal voltage source, is represented as being directly applied to the noninverting input; thus

$$V_{\text{out,os}} = A_{V\text{ nom}} V_{os} = 100 V_{os} = 150 \text{ mV}$$

Thus, we should expect the output of the amplifier to be shifted upward by 150 mV.

Comments: The input offset voltage is not, of course, an external source, but it represents a voltage offset between the inputs of the op-amp. Figure 8.53 depicts how such an offset can be zeroed.

The worst-case offset voltage is usually listed in the device data sheets (see the data sheets in the accompanying CD-ROM for an illustration). Typical values are 2 mV for the 741c general-purpose op-amp and 5 mV for the FET-input TLO81.

CHECK YOUR UNDERSTANDING

What is the maximum gain that can be accepted in the op-amp circuit of Example 8.15 if the offset is not to exceed 50 mV?

Answer: $A_{V\max} = 33.3$

Input Bias Currents

Another nonideal characteristic of op-amps results from the presence of small input bias currents at the inverting and noninverting terminals. Once again, these are due to the internal construction of the input stage of an operational amplifier. Figure 8.51 illustrates the presence of nonzero input bias currents I_B going into an op-amp.

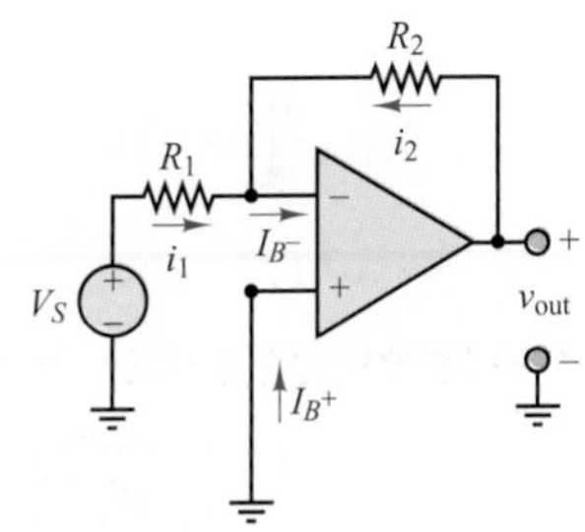

Figure 8.51

Typical values of I_B depend on the semiconductor technology employed in the construction of the op-amp. Op-amps with bipolar transistor input stages may see input bias currents as large as 1 μA, while for FET input devices, the input bias currents are less than 1 nA. Since these currents depend on the internal design of the op-amp, they are not necessarily equal.

One often designates the **input offset current** I_{os} as

$$I_{os} = I_{B+} - I_{B-} \tag{8.84}$$

LO5

The latter parameter is sometimes more convenient from the standpoint of analysis. Example 8.16 illustrates the effect of the nonzero input bias current on a practical amplifier design.

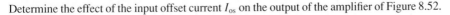

EXAMPLE 8.16 Effect of Input Offset Current on an Amplifier

LO5

Problem

Determine the effect of the input offset current I_{os} on the output of the amplifier of Figure 8.52.

Solution

Known Quantities: Resistor values; input offset current.

Find: The offset voltage component in the output voltage $v_{out,os}$.

Schematics, Diagrams, Circuits, and Given Data: $I_{os} = 1\ \mu$A; $R_2 = 10$ kΩ.

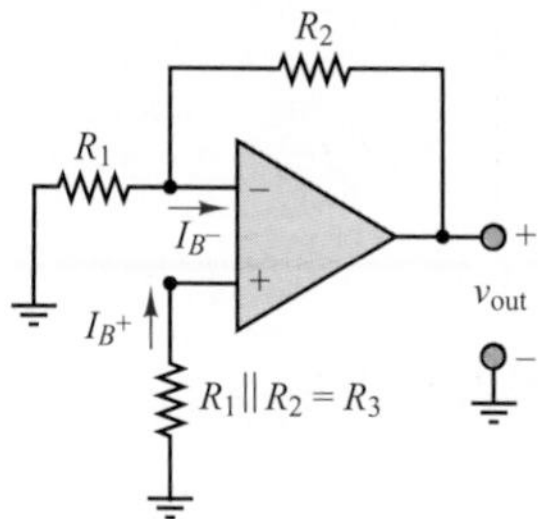

Figure 8.52

Assumptions: Assume an input offset current–limited (otherwise ideal) op-amp.

Analysis: We calculate the inverting and noninverting terminal voltages caused by the offset current in the absence of an external input:

$$v^+ = R_3 I_{B^+} \qquad v^- = v^+ = R_3 I_{B^+}$$

With these values we can apply KCL at the inverting node and write

$$\frac{v_{\text{out}} - v^-}{R_2} - \frac{v^+}{R_1} = I_{B^-}$$

$$\frac{v_{\text{out}}}{R_2} - \frac{-R_3 I_{B^+}}{R_2} - \frac{-R_3 I_{B^+}}{R_1} = I_{B^-}$$

$$v_{\text{out}} = R_2 \left[-I_{B^+} R_3 \left(\frac{1}{R_2} + \frac{1}{R_1} \right) + I_{B^-} \right] = -R_2 I_{\text{os}}$$

Thus, we should expect the output of the amplifier to be shifted downward by $R_2 I_{\text{os}}$, or $10^4 \times 10^{-6} = 10$ mV for the data given in this example.

Comments: Usually, the worst-case input offset currents (or input bias currents) are listed in the device data sheets. Values can range from 100 pA (for CMOS op-amps, for example, LMC6061) to around 200 nA for a low-cost general-purpose amplifier (for example, μA741c).

Output Offset Adjustment

Both the offset voltage and the input offset current contribute to an output offset voltage $V_{\text{out,os}}$. Some op-amps provide a means for minimizing $V_{\text{out,os}}$. For example, the μA741 op-amp provides a connection for this procedure. Figure 8.53 shows a typical pin configuration for an op-amp in an eight-pin dual-in-line package (DIP) and the circuit used for nulling the output offset voltage. The variable resistor is adjusted until v_{out} reaches a minimum (ideally, 0 V). Nulling the output voltage in this manner removes the effect of both input offset voltage and current on the output.

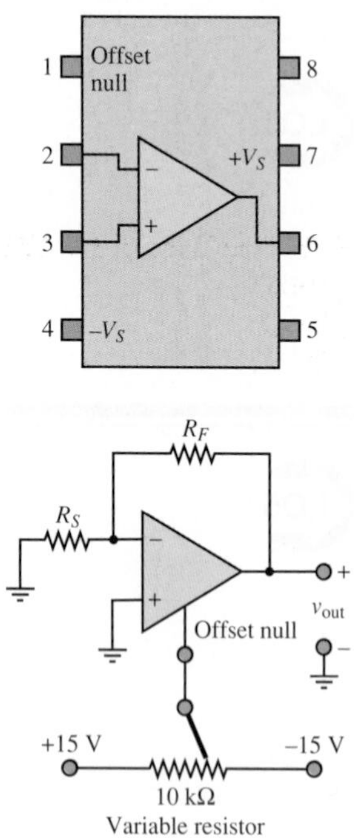

Figure 8.53 Output offset voltage adjustment

Slew Rate Limit

Another important restriction in the performance of a practical op-amp is associated with rapid changes in voltage. The op-amp can produce only a finite rate of change at its output. This limit rate is called the **slew rate.** Consider an ideal step input, where at $t = 0$ the input voltage is switched from 0 to V volts. Then we would expect the output to switch from 0 to AV volts, where A is the amplifier gain. However, $v_{\text{out}}(t)$ can change at only a finite rate; thus,

LO5

$$\left| \frac{dv_{\text{out}}(t)}{dt} \right|_{\max} = S_0 \qquad \text{Slew rate limitation} \tag{8.85}$$

Figure 8.54 shows the response of an op-amp to an ideal step change in input voltage. Here, S_0, the slope of $v_{\text{out}}(t)$, represents the slew rate.

The slew rate limitation can affect sinusoidal signals, as well as signals that display abrupt changes, as does the step voltage of Figure 8.54. This may not be obvious until we examine the sinusoidal response more closely. It should be apparent that the maximum rate of change for a sinusoid occurs at the zero crossing, as shown by Figure 8.55. To evaluate the slope of the waveform at the zero crossing, let

$$v(t) = A \sin \omega t \tag{8.86}$$

so that

$$\frac{dv(t)}{dt} = \omega A \cos \omega t \tag{8.87}$$

The maximum slope of the sinusoidal signal will therefore occur at $\omega t = 0, \pi, 2\pi, \ldots,$ so that

$$\left| \frac{dv(t)}{dt} \right|_{\max} = \omega \times A = S_0 \tag{8.88}$$

Thus, the maximum slope of a sinusoid is proportional to both the signal frequency and the amplitude. The curve shown by a dashed line in Figure 8.55 should indicate that as ω increases, so does the slope of $v(t)$ at the zero crossings. What is the direct consequence of this result, then? Example 8.17 gives an illustration of the effects of this slew rate limit.

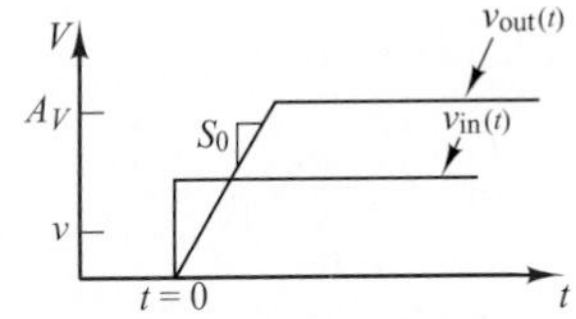

Figure 8.54 Slew rate limit in op-amps

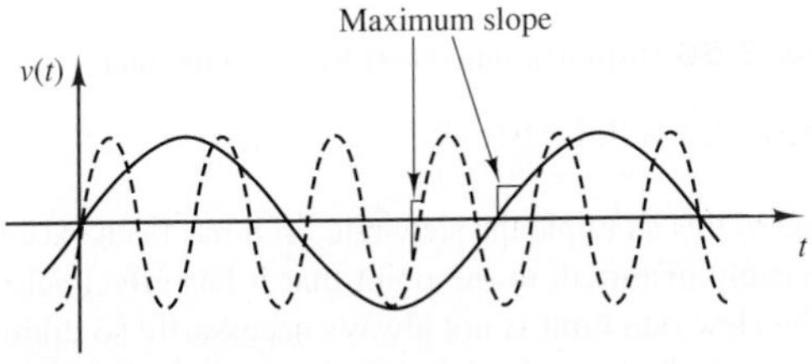

Figure 8.55 The maximum slope of a sinusoidal signal varies with the signal frequency

EXAMPLE 8.17 Effect of Slew Rate Limit on an Amplifier

LO5

Problem

Determine the effect of the slew rate limit S_0 on the output of an inverting amplifier for a sinusoidal input voltage of known amplitude and frequency.

Solution

Known Quantities: Slew rate limit S_0; amplitude and frequency of sinusoidal input voltage; amplifier closed-loop gain.

Find: Sketch the theoretically correct output and the actual output of the amplifier in the same graph.

Schematics, Diagrams, Circuits, and Given Data: $S_0 = 1$ V/μs; $v_S(t) = \sin(2\pi \times 10^5 t)$; $A_V = 10$.

Assumptions: Assume a slew rate–limited (otherwise ideal) op-amp.

Analysis: Given the closed-loop voltage gain of 10, we compute the theoretical output voltage to be

$$v_{\text{out}}(t) = -10\sin(2\pi \times 10^5 t)$$

The maximum slope of the output voltage is then computed as follows:

$$\left|\frac{dv_{\text{out}}(t)}{dt}\right|_{\max} = A\omega = 10 \times 2\pi \times 10^5 = 6.28\ \frac{\text{V}}{\mu\text{s}}$$

Clearly, the value calculated above far exceeds the slew rate limit. Figure 8.56 depicts the approximate appearance of the waveforms that one would measure in an experiment.

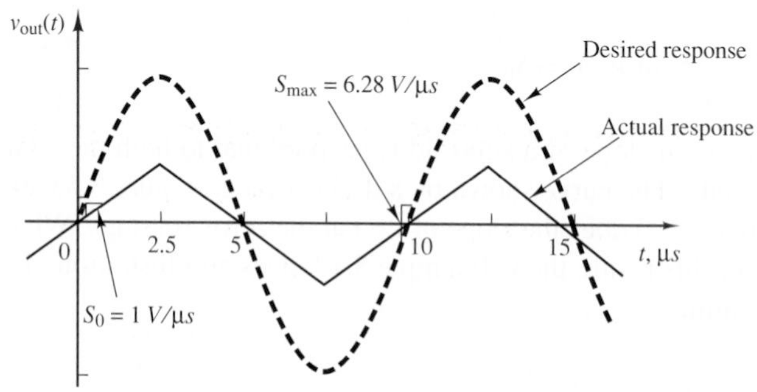

Figure 8.56 Distortion introduced by slew rate limit

Comments: Note that in this example the slew rate limit has been exceeded severely, and the output waveform is visibly distorted, to the point that it has effectively become a triangular wave. The effect of the slew rate limit is not always necessarily so dramatic and visible; thus one needs to pay attention to the specifications of a given op-amp. The slew rate limit is listed in the device data sheets. Typical values can range from 13 V/μs, for the TL081, to around 0.5 V/μs for a low-cost general-purpose amplifier (for example, μA741c).

CHECK YOUR UNDERSTANDING

Given the desired peak output amplitude (10 V), what is the maximum frequency that will not result in violating the slew rate limit for the op-amp of Example 8.17?

Answer: $f_{\max} = 15.9$ KHz

Short-Circuit Output Current

Recall the model for the op-amp introduced in Section 8.2, which represented the internal circuits of the op-amp in terms of an equivalent input resistance R_{in} and a controlled voltage source $A_V v_{\text{in}}$. In practice, the internal source is not ideal, because it cannot provide an infinite amount of current (to the load, to the feedback connection, or to both). The immediate consequence of this nonideal op-amp characteristic is that the maximum output current of the amplifier is limited by the so-called short-circuit output current I_{SC}:

LO5

$$|I_{out}| < I_{SC} \quad \text{Short-circuit output current limitation} \qquad (8.89)$$

To further explain this point, consider that the op-amp needs to provide current to the feedback path (in order to "zero" the voltage differential at the input) and to whatever load resistance, R_L, may be connected to the output. Figure 8.57 illustrates this idea for the case of an inverting amplifier, where I_{SC} is the load current that would be provided to a short-circuit load ($R_L = 0$).

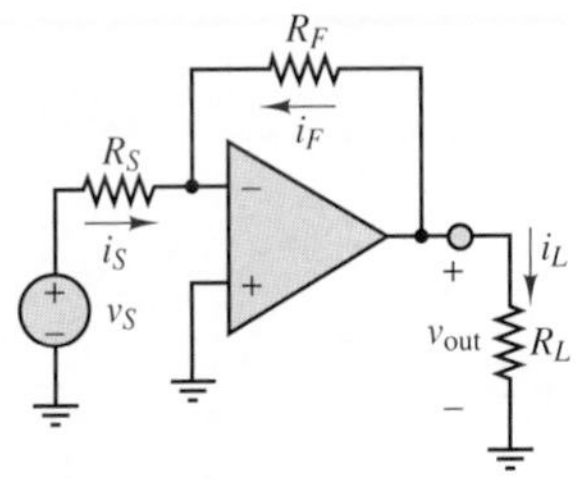

Figure 8.57

EXAMPLE 8.18 Effect of Short-Circuit Current Limit on an Amplifier

LO5

Problem

Determine the effect of the short-circuit limit I_{SC} on the output of an inverting amplifier for a sinusoidal input voltage of known amplitude.

Solution

Known Quantities: Short-circuit current limit I_{SC}; amplitude of sinusoidal input voltage; amplifier closed-loop gain.

Find: Compute the maximum allowable load resistance value $R_{L\ min}$, and sketch the theoretical and actual output voltage waveforms for resistances smaller than $R_{L\ min}$.

Schematics, Diagrams, Circuits, and Given Data: $I_{SC} = 50$ mA; $v_S(t) = 0.05\sin(\omega t)$; $A_V = 100$.

Assumptions: Assume a short-circuit current–limited (otherwise ideal) op-amp.

Analysis: Given the closed-loop voltage gain of 100, we compute the theoretical output voltage to be

$$v_{out}(t) = -A_V v_S(t) = -5\sin(\omega t)$$

To assess the effect of the short-circuit current limit, we calculate the peak value of the output voltage, since this is the condition that will require the maximum output current from the op-amp:

$$v_{out\ peak} = 5 \text{ V}$$

$$I_{SC} = 50 \text{ mA}$$

$$R_{L\ min} = \frac{v_{out\ peak}}{I_{SC}} = \frac{5 \text{ V}}{50 \text{ mA}} = 100\ \Omega$$

For any load resistance less than 100 Ω, the required load current will be greater than I_{SC}. For example, if we chose a 75-Ω load resistor, we would find that

$$v_{out\ peak} = I_{SC} \times R_L = 3.75 \text{ V}$$

That is, the output voltage cannot reach the theoretically correct 5-V peak, and would be "compressed" to reach a peak voltage of only 3.75 V. This effect is depicted in Figure 8.58.

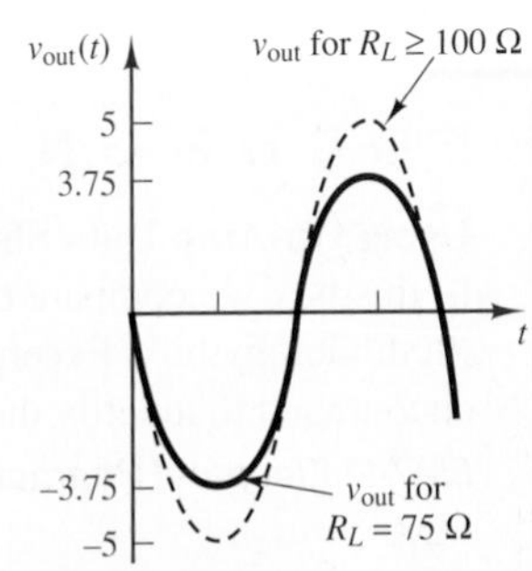

Figure 8.58 Distortion introduced by short-circuit current limit

Comments: The short-circuit current limit is listed in the device data sheets. Typical values for a low-cost general-purpose amplifier (say, the 741c) are in the tens of milliamperes.

Common-Mode Rejection Ratio (CMRR)

Early in this chapter the Focus on Measurements box "Electrocardiogram (EKG) Amplifier" introduced the notion of differential-mode and common-mode signals. If we define A_{dm} as the **differential-mode gain** and A_{cm} as the **common-mode gain** of the op-amp, the output of an op-amp can then be expressed as follows:

LO5

$$v_{out} = A_{dm}(v_2 - v_1) + A_{cm}\left(\frac{v_2 + v_1}{2}\right) \tag{8.90}$$

Under ideal conditions, A_{cm} should be exactly zero, since the differential amplifier should completely reject common-mode signals. The departure from this ideal condition is a figure of merit for a differential amplifier and is measured by defining a quantity called the **common-mode rejection ratio (CMRR).** The CMRR is defined as the ratio of the differential-mode gain to the common-mode gain and should ideally be infinite:

LO5

$$\text{CMRR} = \frac{A_{dm}}{A_{cm}} \qquad \text{Common-mode rejection ratio}$$

The CMRR is often expressed in units of decibels (dB). The common-mode rejection ratio idea is explored further in the problems at the end of the chapter and in Chapter 15.

FOCUS ON METHODOLOGY

LO5

Using Op-Amp Data Sheets—Comparison of LM741 and LMC6061

In this box we compare the LM741 and the LMC6061 op-amps that were introduced in an earlier Focus on Methodology box. Excerpts from the data sheets are shown below, with some words of explanation. You are encouraged to identify the information given below in the sheets that may be found on the web.

LM741 Electrical Characteristics—An abridged version of the electrical characteristics of the LM741 is shown next.

(Continued)

Offset Nulling Circuit

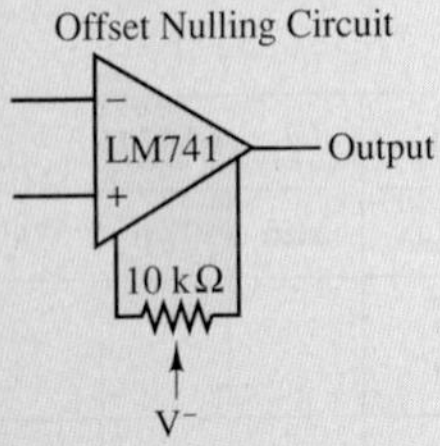

Electrical Characteristics

Parameter	Conditions	LM741A/LM741E Min	LM741A/LM741E Typ	LM741A/LM741E Max	LM741 Min	LM741 Typ	LM741 Max	LM741C Min	LM741C Typ	LM741C Max	Units
Input offset voltage	$T_A = 25°C$										
	$R_S \leq 10\ k\Omega$					1.0	5.0		2.0	6.0	mV
	$R_S \leq 50\ \Omega$		0.8	3.0							mV
	$T_{A_{MIN}} \leq T_A \leq T_{A_{MAX}}$										
	$R_S \leq 50\ \Omega$			4.0							mV
	$R_S \leq 10\ k\Omega$						6.0			7.5	mV
Average input offset voltage drift				15							$\mu V/°C$
Input offset voltage adjustment range	$T_A = 25°C$, $V_S = \pm 20\ V$	±10				±15			±15		mV
Input offset current	$T_A = 25°C$		3.0	30		20	200		20	200	nA
	$T_{A_{MIN}} \leq T_A \leq T_{A_{MAX}}$			70		85	500			300	nA
Average input offset current drift				0.5							nA/°C
Input bias current	$T_A = 25°C$		30	80		80	500		80	500	nA
	$T_{A_{MIN}} \leq T_A \leq T_{A_{MAX}}$			0.210			1.5			0.8	μA
Input resistance	$T_A = 25°C$, $V_S = \pm 20\ V$	1.0	6.0		0.3	2.0		0.3	2.0		MΩ
	$T_{A_{MIN}} \leq T_A \leq T_{A_{MAX}}$, $V_S = \pm 20\ V$	0.5									MΩ
Input voltage range	$T_A = 25°C$							±12	±13		V
	$T_{A_{MIN}} \leq T_A \leq T_{A_{MAX}}$				±12	±13					V
Large-signal voltage gain	$T_A = 25°C$, $R_L \geq 2\ k\Omega$										
	$V_S = \pm 20\ V$, $V_O = \pm 15\ V$	50									V/mV
	$V_S = \pm 15\ V$, $V_O = \pm 10\ V$				50	200		20	200		V/mV
	$T_{A_{MIN}} \leq T_A \leq T_{A_{MAX}}$, $R_L \geq 2\ k\Omega$										
	$V_S = \pm 20\ V$, $V_O = \pm 15\ V$	32									V/mV
	$V_S = \pm 15\ V$, $V_O = \pm 10\ V$				25			15			V/mV
	$V_S = \pm 5\ V$, $V_O = \pm 2\ V$	10									V/mV
Output voltage swing	$V_S = \pm 20\ V$										
	$R_L \geq 10\ k\Omega$	±16									V
	$R_L \geq 2\ k\Omega$	±15									V
	$V_S = \pm 15\ V$										
	$R_L \geq 10\ k\Omega$				±12	±14		±12	±14		V
	$R_L \geq 2\ k\Omega$				±10	±13		±10	±13		V

(*Continued*)

Electrical Characteristics

Parameter	Conditions	LM741A/LM741E			LM741			LM741C			Units
		Min	Typ	Max	Min	Typ	Max	Min	Typ	Max	
Output short-circuit current	$T_A = 25°C$	10	25	35		25			25		mA
	$T_{AMIN} \leq T_A \leq T_{AMAX}$	10		40							mA
Common-mode rejection ratio	$T_{AMIN} \leq T_A \leq T_{AMAX}$										
	$R_S \leq 10\ k\Omega$, $V_{CM} = \pm 12\ V$				70	90		70	90		dB
	$R_S \leq 50\ \Omega$, $V_{CM} = \pm 12\ V$	80	95								dB
Supply voltage rejection ratio	$T_{AMIN} \leq T_A \leq T_{AMAX}$, $V_S = \pm 20\ V$ to $V_S = \pm 5\ V$										
	$R_S \leq 50\ \Omega$	86	96								dB
	$R_S \leq 10\ k\Omega$			77	96		77	96			dB
Transient response	$T_A = 25°C$, unity gain										
rise time			0.25	0.8		0.3			0.3		μs
overshoot			6.0	20		5			5		%
Bandwidth	$T_A = 25°C$	0.437	1.5								MHz
Slew rate	$T_A = 25°C$, unity gain	0.3	0.7			0.5			0.5		$V/\mu s$
Supply current	$T_A = 25°C$					1.7	2.8		1.7	2.8	mA
Power consumption	$T_A = 25°C$										
	$V_S = \pm 20\ V$		80	150							mW
	$V_S = \pm 15\ V$					50	85		50	85	mW
LM741A	$V_S = \pm 20\ V$										
	$T_A = T_{AMIN}$			165							mW
	$T_A = T_{AMAX}$			135							mW
LM741E	$V_S = \pm 20\ V$										
	$T_A = T_{AMIN}$			150							mW
	$T_A = T_{AMAX}$			150							mW
LM741	$V_S = \pm 15\ V$										
	$T_A = T_{AMIN}$					60	100				mW
	$T_A = T_{AMAX}$					45	75				mW

LMC6061 Electrical Characteristics—An abridged version of the electrical characteristics of the LMC6061 is shown below.

DC Electrical Characteristics

Unless otherwise specified, all limits guaranteed for $T_J = 25°C$. **Boldface** limits apply at the temperature extremes. $V^+ = 5\ V$, $V^- = 0\ V$, $V_{CM} = 1.5\ V$, $V_O = 2.5\ V$, and $R_L > 1\ M\Omega$ unless otherwise specified.

Symbol	Parameter	Conditions	Typ	LMC6061AM Limit	LMC6061AI Limit	LMC6061I Limit	Units
V_{OS}	Input offset voltage		100	350 **1,200**	350 **900**	800 **1,300**	μV Max
TCV_{OS}	Input offset voltage average drift		1.0				$\mu V/°C$

(*Continued*)

DC Electrical Characteristics

Symbol	Parameter	Conditions		Typ	LMC6061AM Limit	LMC6061AI Limit	LMC6061I Limit	Units
I_B	Input bias current			0.010	**100**	**4**	**4**	pA Max
I_{OS}	Input offset current			0.005	**100**	**2**	**2**	pA Max
R_{IN}	Input resistance			>10				Tera Ω
CMRR	Common-mode rejection ratio	$0\ V \leq V_{CM} \leq 12.0\ V$ $V^+ = 15\ V$		85	75 **70**	75 **72**	66 **63**	dB Min
+PSRR	Positive power supply rejection ratio	$5\ V \leq V^+ \leq 15\ V$ $V_O = 2.5\ V$		85	75 **70**	75 **72**	66 **63**	dB Min
−PSRR	Negative power supply rejection ratio	$0\ V \leq V^- \leq -10\ V$		100	84 **70**	84 **81**	74 **71**	dB Min
V_{CM}	Input common-mode voltage range	$V^+ = 5\ V$ and 15 V for CMRR ≥ 60 dB		−0.4	−0.1 **0**	−0.1 **0**	−0.1 **0**	V Max
				$V^+ - 1.9$	$V^+ -2.3$ **V^+ −2.6**	$V^+ -2.3$ **V^+ −2.5**	$V^+ -2.3$ **V^+ −2.5**	V Min
A_V	Large-signal voltage gain	$R_L = 100\ k\Omega$	Sourcing	4,000	400 **200**	400 **300**	300 **200**	V/mV Min
			Sinking	3,000	180 **70**	180 **100**	90 **60**	V/mV Min
		$R_L = 25\ k\Omega$	Sourcing	3,000	400 **150**	400 **150**	200 **80**	V/mV Min
			Sinking	2,000	100 **35**	100 **50**	70 **35**	V/mV Min
V_O	Output swing	$V^+ = 5\ V$ $R_L = 100\ k\Omega$ to 2.5 V		4.995	4.990 **4.970**	4.990 **4.980**	4.950 **4.925**	V Min
				0.005	0.010 **0.030**	0.010 **0.020**	0.050 **0.075**	V Max
		$V^+ = 5\ V$ $R_L = 25\ k\Omega$ to 2.5 V		4.990	4.975 **4.955**	4.975 **4.965**	4.950 **4.850**	V Min
				0.010	0.020 **0.045**	0.020 **0.035**	0.050 **0.150**	V Max
		$V^+ = 15\ V$ $R_L = 100\ k\Omega$ to 7.5 V		14.990	14.975 **14.955**	14.975 **14.965**	14.950 **14.925**	V Min
				0.010	0.025 **0.050**	0.025 **0.035**	0.050 **0.075**	V Max
		$V^+ = 15\ V$ $R_L = 25\ k\Omega$ to 7.5 V		14.965	14.900 **14.800**	14.900 **14.850**	14.850 **14.800**	V Min
				0.025	0.050 **0.200**	0.050 **0.150**	0.100 **0.200**	V Max

(*Continued*)

DC Electrical Characteristics

Symbol	Parameter	Conditions	Typ	LMC6061AM Limit	LMC6061AI Limit	LMC6061I Limit	Units
I_O	Output current $V^+ = 5$ V	Sourcing, $V_O = 0$ V	22	16 **8**	16 **10**	13 **8**	mA Min
		Sinking, $V_O = 5$ V	21	16 **7**	16 **8**	16 **8**	mA Min
I_O	Output current $V^+ = 15$ V	Sourcing, $V_O = 0$ V	25	15 **9**	15 **10**	15 **10**	mA Min
		Sinking, $V_O = 13$ V	35	24 **7**	24 **8**	24 **8**	mA Min
I_S	Supply current	$V^+ = +5$ V, $V_O = 1.5$ V	20	24 **35**	24 **32**	32 **40**	mA Max
		$V^+ = +15$ V, $V_O = 7.5$ V	24	30 **40**	30 **38**	40 **48**	μA Max

AC Electrical Characteristics

Unless otherwise specified, all limits guaranteed for $T_J = 25°$C. **Boldface** limits apply at the temperature extremes. $V^+ = 5$ V, $V^- = 0$ V, $V_{CM} = 1.5$ V, $V_O = 2.5$ V, and $R_L > 1$ MΩ unless otherwise specified.

Symbol	Parameter	Conditions	Typ	LMC6061AM Limit	LMC6061AI Limit	LMC6061I Limit	Units
SR	Slew rate		35	20 **8**	20 **10**	15 **7**	V/ms Min
GBW	Gain-bandwidth product		100				kHz
θ_m	Phase margin		50				Deg
e_n	Input-referred voltage noise	$F = 1$ kHz	83				nV/$\sqrt{\text{Hz}}$
i_n	Input-referred current noise	$F = 1$ kHz	0.0002			pA/$\sqrt{\text{Hz}}$	
THD	Total harmonic distortion	$F = 1$ kHz, $A_V = -5$ $R_L = 100$ kΩ, $V_O = 2V_{PP}$ ±5-V supply	0.01				% %

Comparison:

Input offset voltage—Note that the typical input offset voltage in the 6061 is only 100 μV, versus 0.8 mV in the 741.

Input offset voltage adjustments—The recommended circuit is shown for the 741, and a range of ±15 mV is given. The 6061 does not require offset voltage adjustment.

Input offset current—The 741 sheet reports a typical value of 3 nA (3×10^{-9} A); the corresponding value for the 6061 is 0.005 pA (5×10^{-15} A)! This extremely low value is due to the MOS construction of the amplifier (see Chapter 11 for a discussion of MOS stage input impedance).

(*Continued*)

(*Concluded*)

Input resistance—The specifications related to input offset current are mirrored by the input resistance specifications. The 741 has a respectable typical input resistance of 6 MΩ; the 6061 has an input resistance greater than 10 TΩ (1 teraohm = 10^{12} Ω). Once again, this is the result of MOS construction.

Large-signal voltage gain—The 741 lists a typical value of 50 V/mV (or 5×10^4) for its open-loop voltage gain; the 6061 lists values greater than or equal to 2,000 V/mV (or 2×10^6).

CMRR—The typical common-mode rejection ratio is 95 dB for the 741 and 85 dB for the 6061.

Slew rate—0.7 V/μs for the 741 and 35 V/ms for the 6061.

Bandwidth—The bandwidth for the 741 is listed as 1.5 MHz (this would be the unity gain bandwidth), while the 6061 lists a 100-kHz gain-bandwidth product.

Output short-circuit current—25 mA for both devices.

Note that while the LMC6061 is certainly superior to the LM741 op-amp in a number of categories, there are certain features (e.g., bandwidth and slew rate) that might cause a designer to prefer the 741 for a specific application.

Manufacturers generally supply values for the parameters discussed in this section in their device data specifications. Typical data sheets for common op-amps may be found in the accompanying CD-ROM.

Conclusion

Operational amplifiers constitute the single most important building block in analog electronics. The contents of this chapter will be frequently referenced in later sections of this book. Upon completing this chapter, you should have mastered the following learning objectives:

1. *Understand the properties of ideal amplifiers and the concepts of gain, input impedance, and output impedance.* Ideal amplifiers represent fundamental building blocks of electronic instrumentation. With the concept of an ideal amplifier clearly established, one can design practical amplifiers, filters, integrators, and many other signal processing circuits. A practical op-amp closely approximates the characteristics of ideal amplifiers.
2. *Understand the difference between open-loop and closed-loop op-amp configuration; and compute the gain (or complete the design of) simple inverting, noninverting, summing, and differential amplifiers using ideal op-amp analysis. Analyze more advanced op-amp circuits, using ideal op-amp analysis, and identify important performance parameters in op-amp data sheets.* Analysis of op-amp circuits is made easy by a few simplifying assumptions, which are based on the op-amp having a very large input resistance, a very small output resistance, and a large open-loop gain. The simple inverting and noninverting amplifier configurations permit the design of very useful circuits simply by appropriately selecting and placing a few resistors.
3. *Analyze and design simple active filters. Analyze and design ideal integrator and differentiator circuits.* The use of capacitors in op-amp circuits extends the applications of this useful element to include filtering, integration, and differentiation.
4. *Understand the structure and behavior of analog computers, and design analog computer circuits to solve simple differential equations.* The properties of op-amp summing amplifiers and integrators make it possible to construct analog computers that can serve as an aid in the solution of differential equations and in the simulation of dynamic systems. While digital computer-based numerical simulations have become very popular in the last two decades, there is still a role for analog computers in some specialized applications.

5. *Understand the principal physical limitations of an op-amp.* It is important to understand that there are limitations in the performance of op-amp circuits that are not predicted by the simple op-amp models presented in the early sections of the chapter. In practical designs, issues related to voltage supply limits, bandwidth limits, offsets, slew rate limits, and output current limits are very important if one is to achieve the design performance of an op-amp circuit.

HOMEWORK PROBLEMS

Section 8.1: Ideal Amplifiers

8.1 The circuit shown in Figure P8.1 has a signal source, two stages of amplification, and a load. Determine, in decibels, the power gain $G = P_O/P_S = V_o I_o / V_S I_S$, where

$R_s = 0.6\ \text{k}\Omega$ $\quad R_L = 0.6\ \text{k}\Omega$

$R_{i1} = 3\ \text{k}\Omega$ $\quad R_{i2} = 3\ \text{k}\Omega$

$R_{o1} = 2\ \text{k}\Omega$ $\quad R_{o2} = 2\ \text{k}\Omega$

$A_{V(\text{OL})} = 100$ $\quad G_{m2} = 350\ \text{mS}$

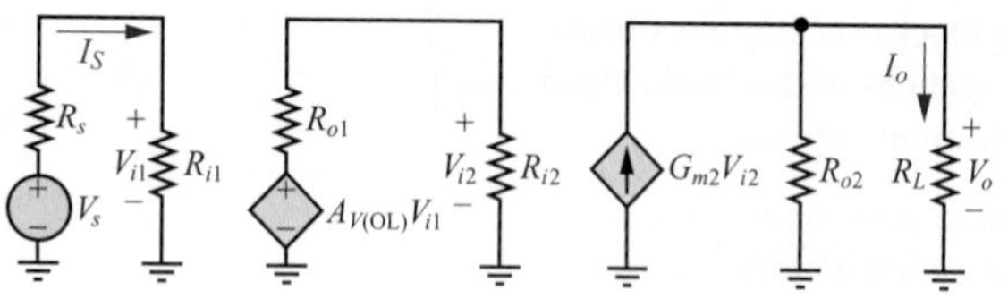

Figure P8.1

8.2 A temperature sensor in a production line under normal operating conditions produces a no-load (i.e., sensor current = 0) voltage:

$v_s = V_{so}\cos(\omega t)$ $\quad R_s = 400\ \Omega$

$V_{so} = 500\ \text{mV}$ $\quad \omega = 6.28\ \text{krad/s}$

The temperature is monitored on a display (the load) with a vertical line of light-emitting diodes. Normal conditions are indicated when a string of the bottommost diodes 2 cm in length is on. This requires that a voltage be supplied to the display input terminals where

$R_L = 12\ \text{k}\Omega$ $\quad v_o = V_o\cos(\omega t)$ $\quad V_o = 6\ \text{V}$

The signal from the sensor must be amplified. Therefore, a voltage amplifier, shown in Figure P8.2, is connected between the sensor and CRT with

$R_i = 2\ \text{k}\Omega$ $\quad R_o = 3\ \text{k}\Omega$

Determine the required no-load gain of the amplifier.

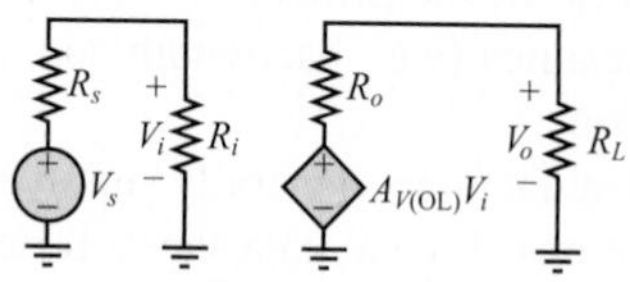

Figure P8.2

8.3 What approximations are usually made about the voltages and currents shown in Figure P8.3 for the ideal operational amplifier model?

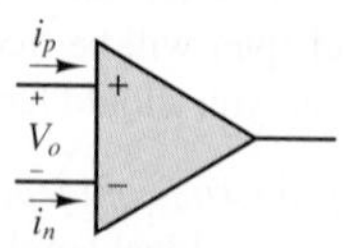

Figure P8.3

8.4 What approximations are usually made about the circuit components and parameters shown in Figure P8.4 for the ideal op-amp model?

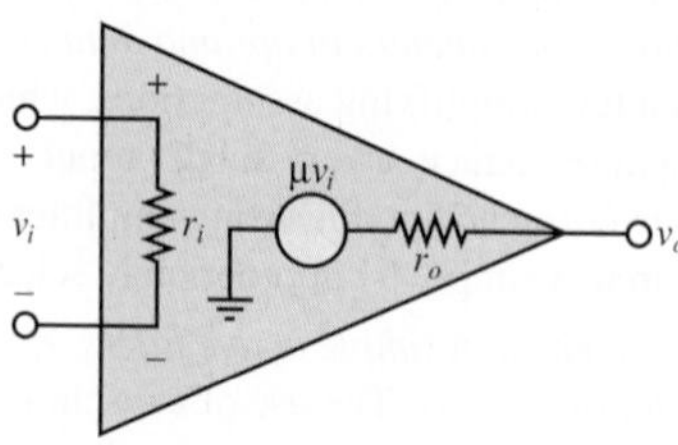

Figure P8.4

Section 8.2: The Operational Amplifier

8.5 Find v_1 in the circuits of Figure P8.5(a) and (b). Note how in the circuit of Figure P8.5(b) the op-amp voltage follower holds v_1 to the value $v_g/2$, while in the circuit of Figure P8.5(a) the 3-kΩ resistor "loads" the output.

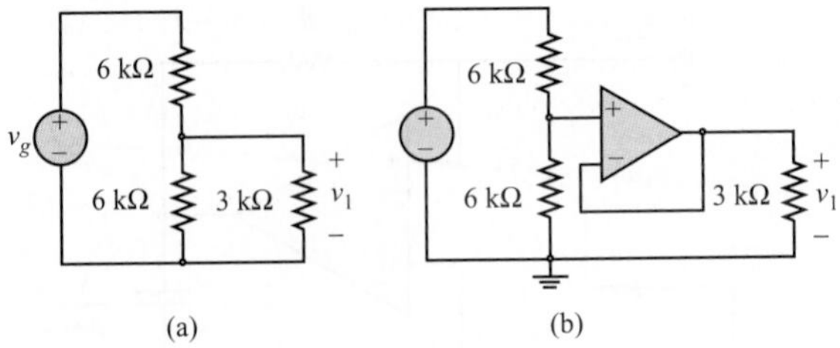

Figure P8.5

8.6 Find the current i in the circuit of Figure P8.6.

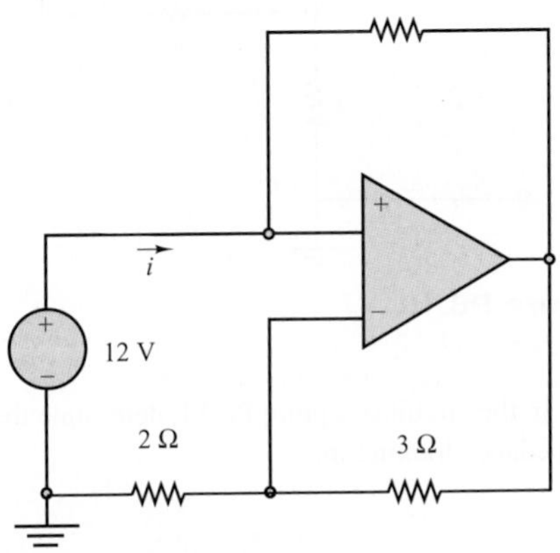

Figure P8.6

8.7 Find the voltage v_o in the circuit of Figure P8.7.

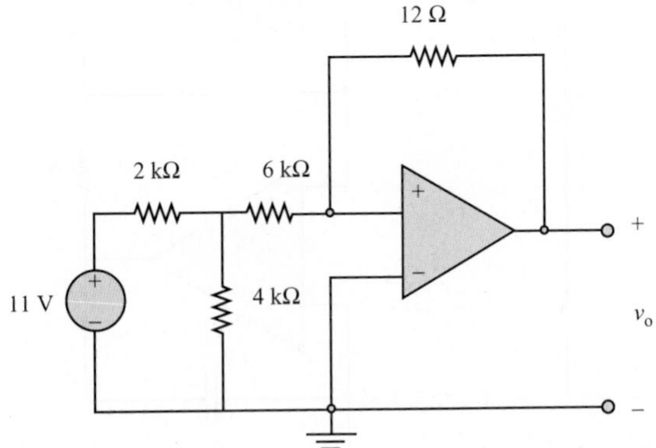

Figure P8.7

8.8 Show that the circuit of Figure P8.8 is a noninverting summing amplifier.

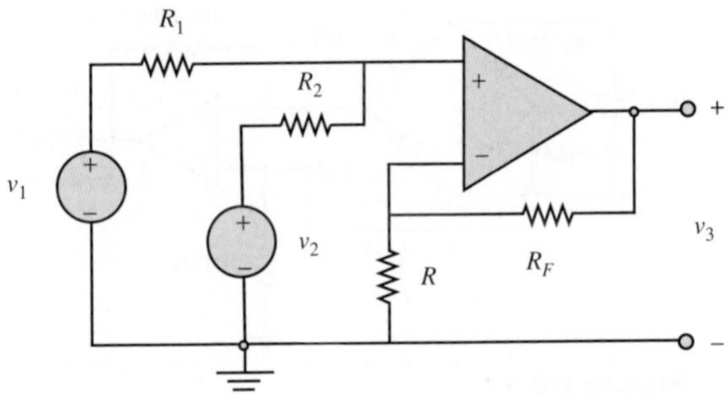

Figure P8.8

8.9 Determine an expression for the overall gain $A_v = v_o/v_i$ for the circuit of Figure P8.9. Find the input conductance, $G_{in} = i_i/v_i$ seen by the voltage source v_i. Assume that the op-amp is ideal.

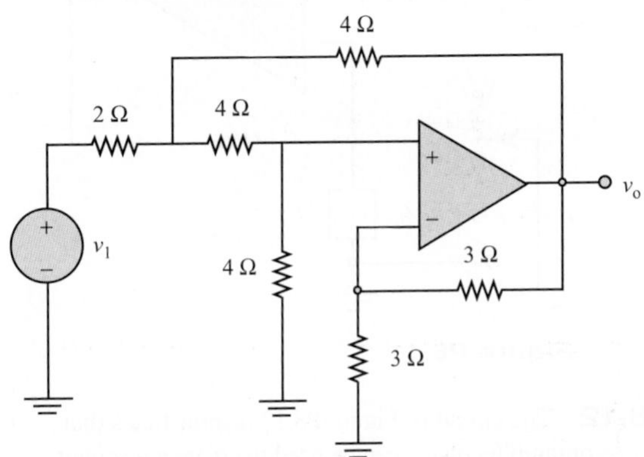

Figure P8.9

8.10 Differential amplifiers are often used in conjunction with Wheatstone bridge circuits. Consider the bridge shown in Figure P8.10, where each resistor is a temperature sensing element, and its change in resistance is directly proportion to a change in temperature—that is, $\Delta R = \alpha(\pm\Delta T)$, where the sign is determined by the positive or negative temperature coefficient of the resistive element.

a. Find the Thévenin equivalent that the amplifier sees at point a and at point b. Assume that $|\Delta R|^2 \ll R_0$.

b. If $|\Delta R| = K\Delta T$, with K a numerical constant, find an expression for v_{out} (ΔT), that is, for v_{out} as a function of the change in temperature.

8.11 The circuit shown in Figure P8.11 is called a *negative impedance converter*. Determine the impedance looking in:

$Z_{in} = \frac{V_1}{I_1}$, if

a. $Z_L = R$ and if

b. $Z_L = \frac{1}{j\omega C}$.

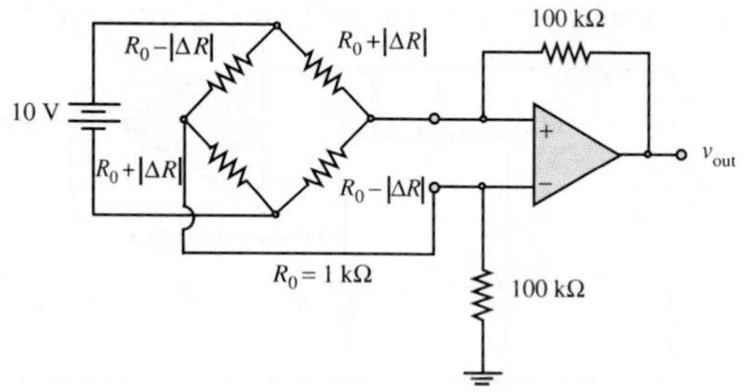

Figure P8.10

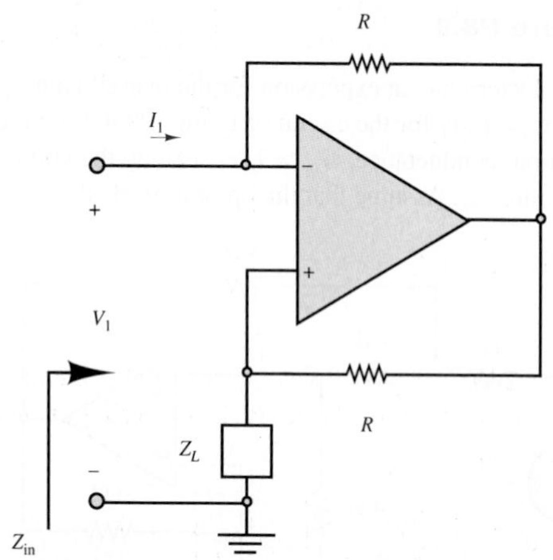

Figure P8.11

8.12 The circuit of Figure P8.12 demonstrates that op-amp feedback can be used to create a resonant circuit without the use of an inductor. Determine the gain function $\frac{V_2}{V_1}$. [*Hint*: Use node analysis.]

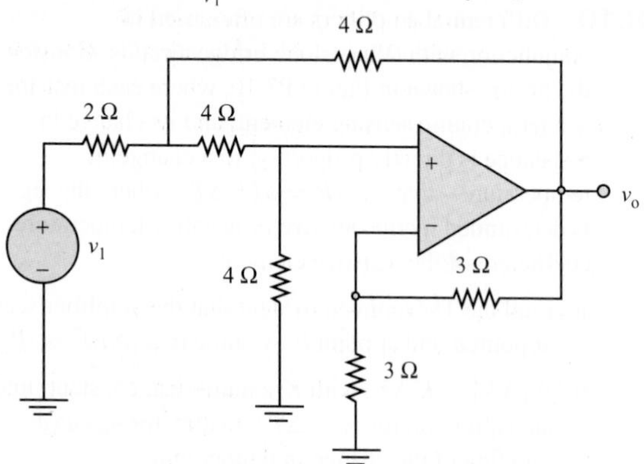

Figure P8.12

8.13 Inductors are difficult to use as components of integrated circuits due to the need for large coils of wire. As an alternative, a "solid-state inductor" can be constructed as in the circuit of Figure 8.13.

a. Determine the impedance looking in $Z_{in} = \frac{V_1}{I_1}$.

b. What is the impedance when $R = 1,000\ \Omega$ and $C = 0.02\ \mu F$?

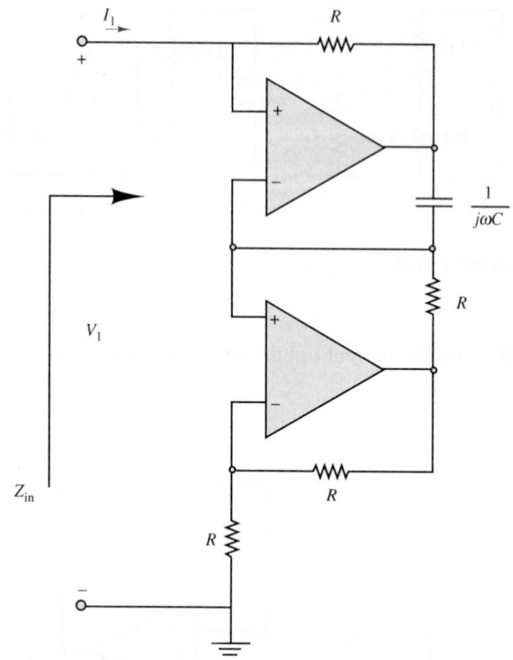

Figure P8.13

8.14 In the circuit of Figure P8.14, determine the impedance looking in.

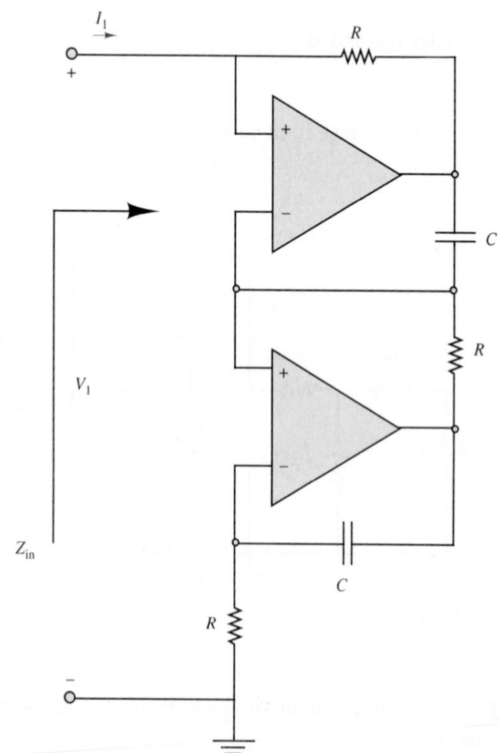

Figure P8.14

8.15 It is easy to construct a current source using an inverting amplifier configuration. Verify that the current in R_L is independent of the value of R_L, assuming that the op-amp stays in its linear operating region, and find the value of this current.

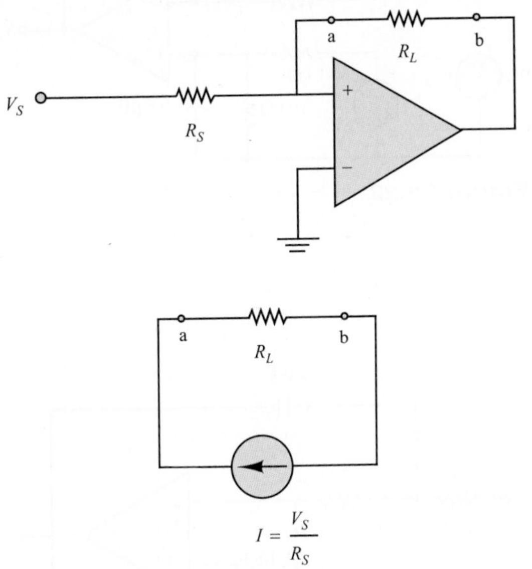

Figure P8.15

8.16 A "super diode" or "precision diode," which eliminates the diode offset voltage, is shown in Figure P8.16. Determine the output signal for the given input signal, $V_{\text{in}}(t)$.

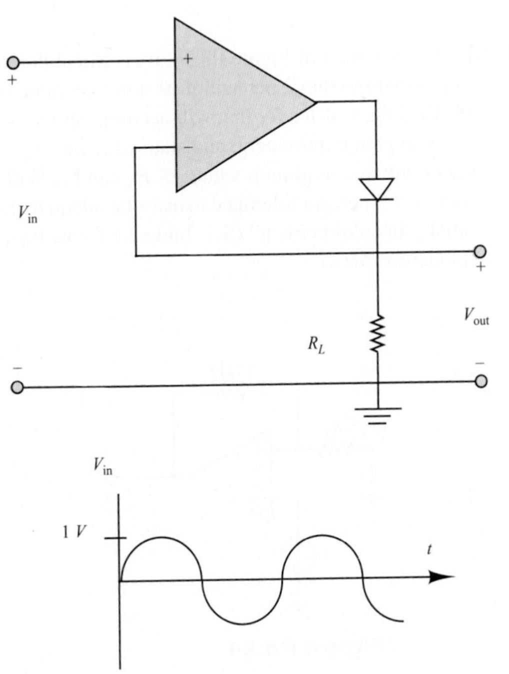

Figure P8.16

8.17 Determine the response function $\frac{V_2}{V_1}$ for the circuit of Figure P8.17.

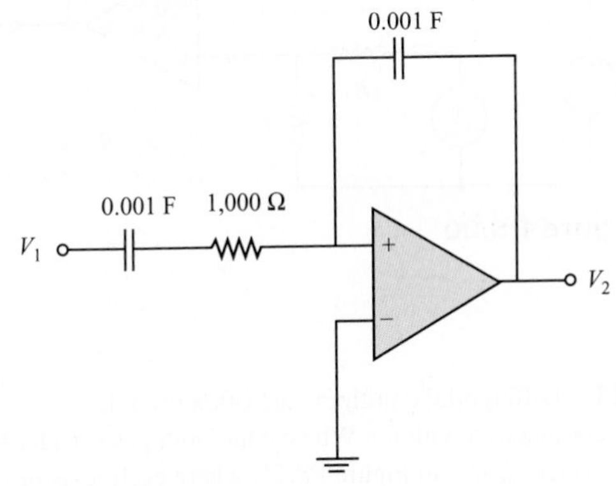

Figure P8.17

8.18 Time delays are often encountered in engineering systems. They can be approximated using Euler's definition as

$$e^{-sT} = \lim_{N \to \infty} \left[\frac{1}{\frac{sT}{N} + 1} \right]^N$$

If $T = 1$, and $N = 1$, then the approximation can be implemented by the circuit of Problem 8.17 (see Figure P8.17), with the addition of a unity gain inverting amplifier to eliminate the negative sign. Modify the circuit of Figure P8.17 as needed and use it as many times as necessary to design an approximate time delay for $T = 1$ and $N = 4$ in Euler's definition of the exponential.

8.19 Show that the circuit of Figure P8.19 is a noninverting summer.

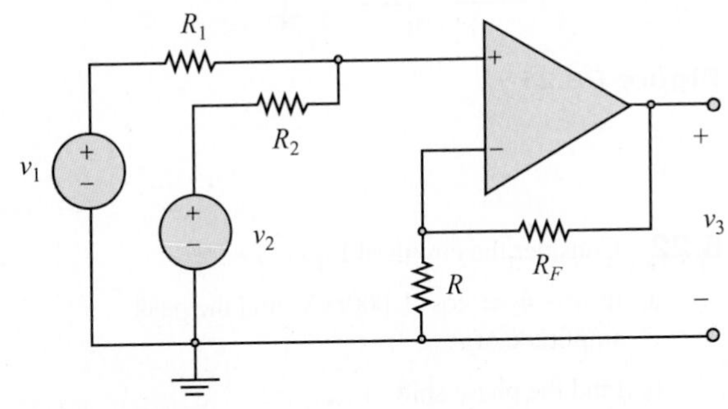

Figure P8.19

8.20 For the circuit of Figure P8.20, find the voltage v and the current i.

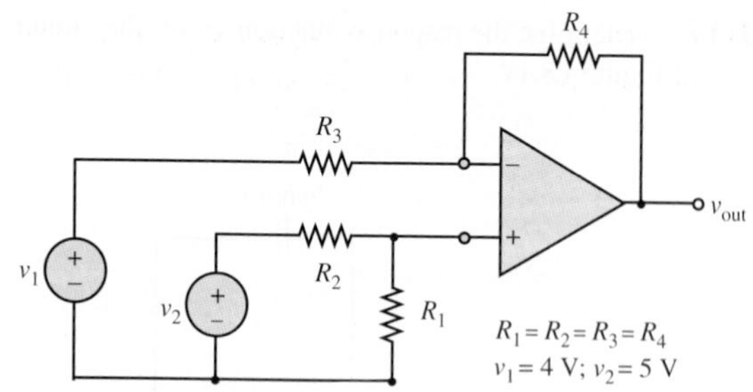

Figure P8.20

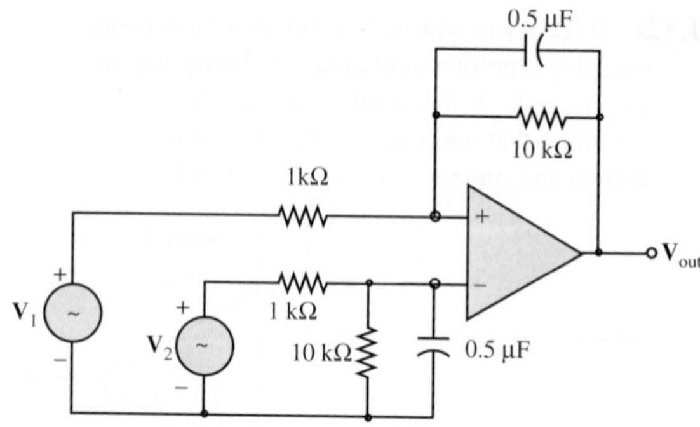

Figure P8.22

8.21 Differential amplifiers are often used in conjunction with the Wheatstone bridge. Consider the bridge shown in Figure P8.21, where each resistor is a temperature-sensing element and the change in resistance is directly proportional to a change in temperature—that is, $\Delta R = \alpha(\pm \Delta T)$, where the sign is determined by the positive or negative temperature coefficient of the resistive element.

a. Find the Thévenin equivalent that the amplifier sees at point a and point b. Assume that $|\Delta R|^2 \ll R_o$.

b. If $|\Delta R| = K\,\Delta T$, with K a numerical constant, find an expression for $v_{out}(\Delta T)$, that is v_{out} as a function of change in temperature.

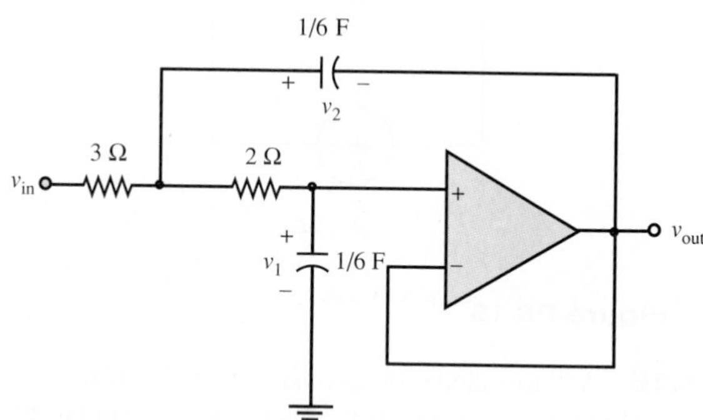

Figure P8.23

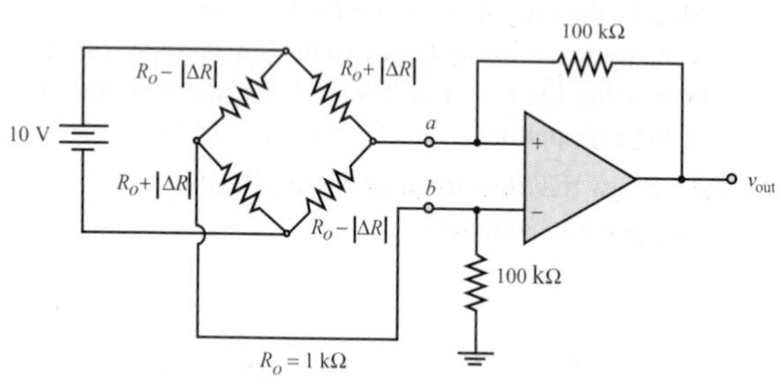

Figure P8.21

8.24 In the circuit of Figure P8.24, it is critical that the gain remain within 2 percent of its nominal value of 16. Find the resistor R_S that will accomplish the nominal gain requirement, and state what the maximum and minimum values of R_S can be. Will a *standard* 5 percent tolerance resistor be adequate to satisfy this requirement? (See Table 2.1 for resistor standard values.)

8.22 Consider the circuit of Figure P8.22.

a. If $v_1 - v_2 = \cos(1{,}000t)$ V, find the peak amplitude of v_{out}.

b. Find the phase shift of v_{out}.

[*Hint*: Use phasor analysis.]

8.23 Find an expression for the gain of the circuit of Figure P8.23.

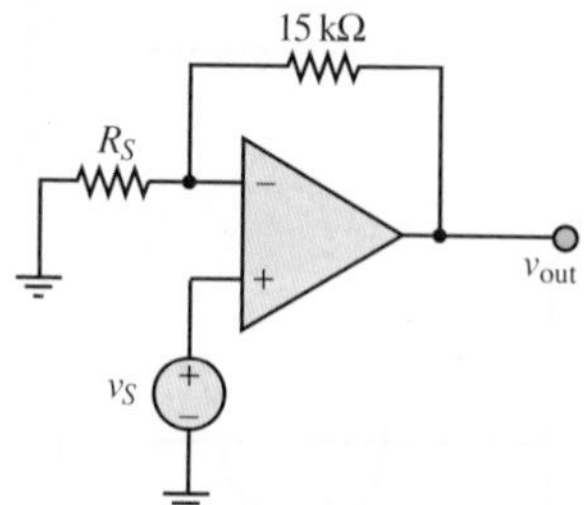

Figure P8.24

8.25 An inverting amplifier uses two 10 percent tolerance resistors: $R_F = 33$ kΩ and $R_S = 1.2$ kΩ.

a. What is the nominal gain of the amplifier?

b. What is the maximum value of $|A_V|$?

c. What is the minimum value of $|A_V|$?

8.26 The circuit of Figure P8.26 will remove the DC portion of the input voltage $v_1(t)$ while amplifying the AC portion. Let $v_1(t) = 10 + 10^{-3} \sin \omega t$ V, $R_F = 10$ kΩ, and $V_{\text{batt}} = 20$ V.

a. Find R_S such that no DC voltage appears at the output.

b. What is $v_{\text{out}}(t)$, using R_S from part a?

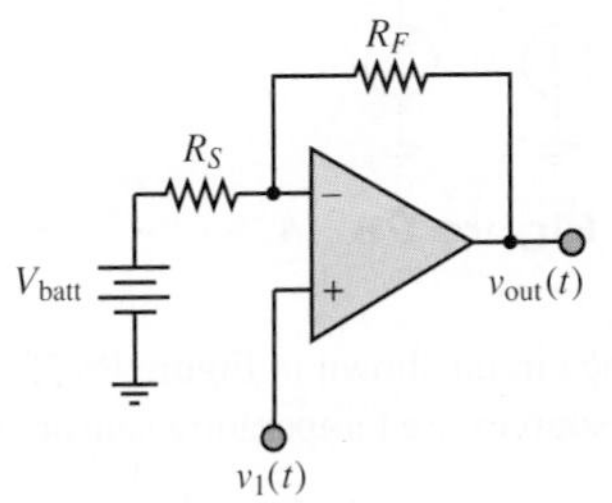

Figure P8.26

8.27 Figure P8.27 shows a simple practical amplifier that uses the 741 op-amp. Pin numbers are as indicated. Assume the input resistance is $R = 2$ MΩ, the open-loop gain $A_{V(\text{OL})} = 200{,}000$, and output resistance $R_o = 50$ Ω. Find the exact gain $A_V = v_o/v_i$.

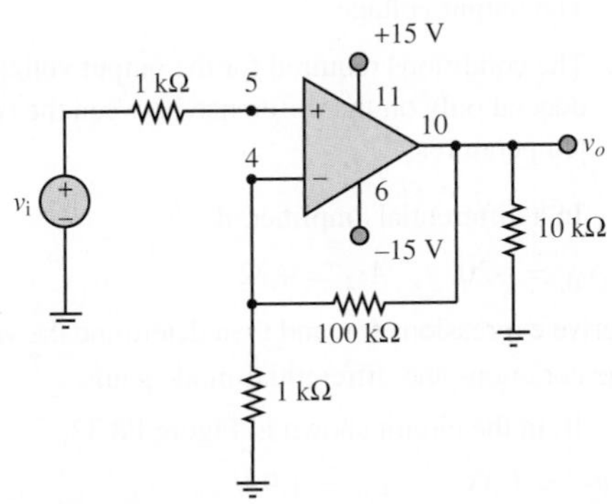

Figure P8.27

8.28 Design an inverting summing amplifier to obtain the following weighted sum of four different signal sources:

$$v_{\text{out}} = -(2 \sin \omega_1 t + 4 \sin \omega_2 t + 8 \sin \omega_3 t + 16 \sin \omega_4 t)$$

Assume that $R_F = 5$ kΩ, and determine the required source resistors.

8.29 The amplifier shown in Figure P8.29 has a signal source, a load, and one stage of amplification with

$R_s = 2.2$ kΩ $\quad R_1 = 1$ kΩ

$R_F = 8.7$ kΩ $\quad R_L = 20$ Ω

Motorola MC1741C op-amp:

$r_i = 2$ MΩ $\quad r_o = 25$ Ω

$\mu = 200{,}000$

In a first-approximation analysis, the op-amp parameters given above would be neglected and the op-amp modeled as an ideal device. In this problem, include their effects on the input resistance of the amplifier circuit.

a. Derive an expression for the input resistance v_i/i_i including the effects of the op-amp.

b. Determine the value of the input resistance including the effects of the op-amp.

c. Determine the value of the input resistance assuming the op-amp is ideal.

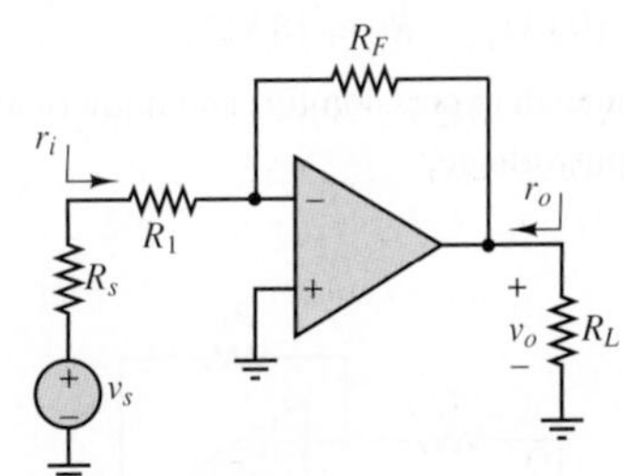

Figure P8.29

8.30 In the circuit shown in Figure P8.30, if

$R_1 = 47$ kΩ $\quad R_2 = 1.8$ kΩ

$R_F = 220$ kΩ

$v_s = 0.02 + 0.001 \cos(\omega t) \quad$ V

determine

a. An expression for the output voltage.

b. The value of the output voltage.

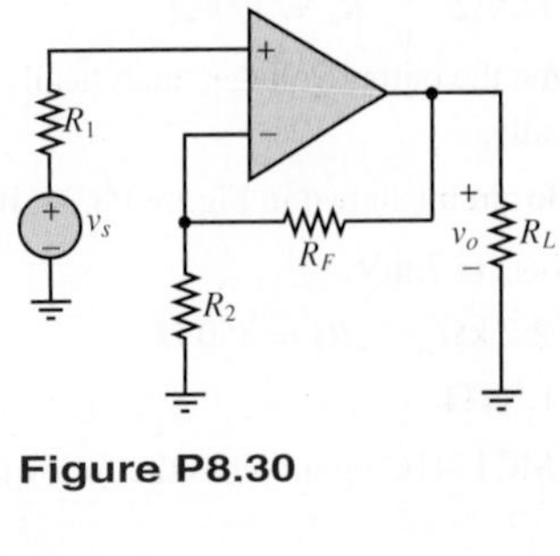

Figure P8.30

8.31 If, in the circuit shown in Figure P8.31,

$v_S = 50 \times 10^{-3} + 30 \times 10^{-3} \cos(\omega t)$

$R_s = 50\ \Omega \qquad R_L = 200\ \Omega$

determine the output voltage.

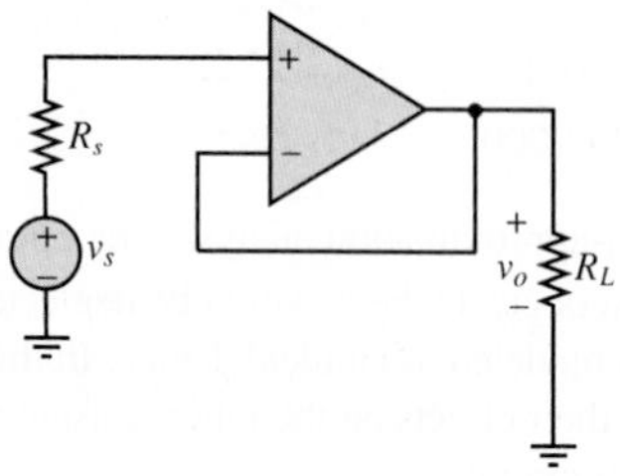

Figure P8.31

8.32 In the circuit shown in Figure P8.32,

$v_{S1} = 2.9 \times 10^{-3} \cos(\omega t) \qquad \text{V}$

$v_{S2} = 3.1 \times 10^{-3} \cos(\omega t) \qquad \text{V}$

$R_1 = 1\ \text{k}\Omega \qquad R_2 = 3.3\ \text{k}\Omega$

$R_3 = 10\ \text{k}\Omega \qquad R_4 = 18\ \text{k}\Omega$

Determine an expression for, and numerical value of, the output voltage.

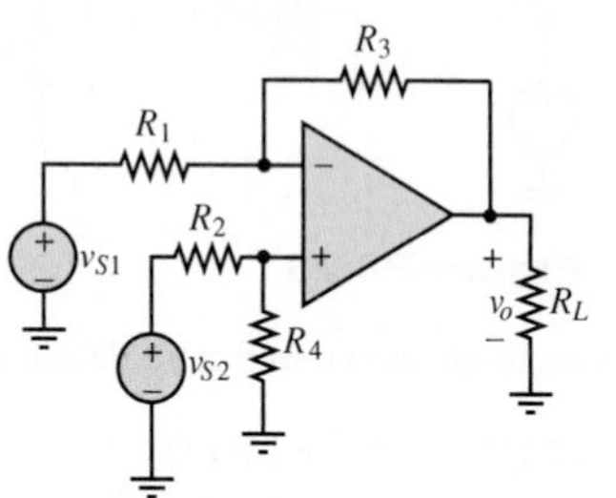

Figure P8.32

8.33 In the circuit shown in Figure P8.32,

$v_{S1} = 5\ \text{mV} \qquad v_{S2} = 7\ \text{mV}$

$R_1 = 1\ \text{k}\Omega \qquad R_2 = 15\ \text{k}\Omega$

$R_3 = 72\ \text{k}\Omega \qquad R_4 = 47\ \text{k}\Omega$

Determine the output voltage, analytically and numerically.

8.34 In the circuit shown in Figure P8.34, if

$v_{S1} = v_{S2} = 7\ \text{mV}$

$R_F = 2.2\ \text{k}\Omega \qquad R_1 = 850\ \Omega$

$R_2 = 1.5\ \text{k}\Omega$

and the MC1741C op-amp has the following parameters:

$r_i = 2\ \text{M}\Omega \qquad \mu = 200{,}000$

$r_o = 25\ \Omega$

determine

a. An expression for the output voltage.

b. The voltage gain for each of the two input signals.

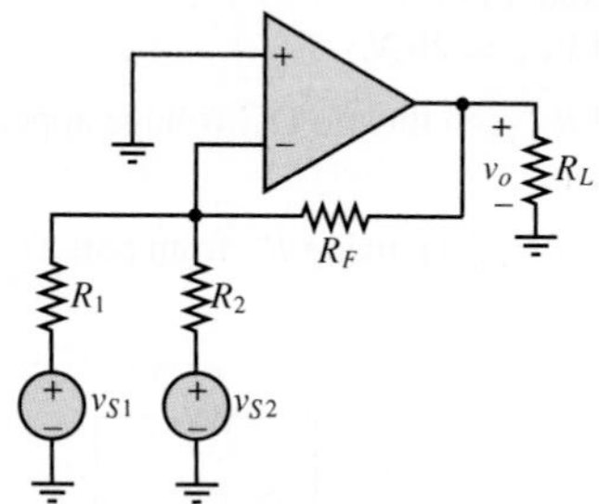

Figure P8.34

8.35 In the circuit shown in Figure P8.32, the two voltage sources are temperature sensors with a response

$v_{S1} = kT_1 \qquad v_{S2} = kT_2$

where

$k = 50\ \text{mV/°C}$

$R_1 = 11\ \text{k}\Omega \qquad R_2 = 27\ \text{k}\Omega$

$R_3 = 33\ \text{k}\Omega \qquad R_4 = 68\ \text{k}\Omega$

$T_1 = 35°\text{C}$ and $T_2 = 100°\text{C}$. Determine

a. The output voltage.

b. The conditions required for the output voltage to depend only on the difference between the two temperatures.

8.36 In a differential amplifier, if

$A_{v1} = -20 \qquad A_{v2} = +22$

derive expressions for, and then determine the value of, the common- and differential-mode gains.

8.37 If, in the circuit shown in Figure P8.32,

$v_{S1} = 1.3\text{V} \qquad v_{S2} = 1.9\ \text{V}$

$R_1 = R_2 = 4.7\ \text{k}\Omega$

$R_3 = R_4 = 10\ \text{k}\Omega \qquad R_L = 1.8\ \text{k}\Omega$

determine

a. The output voltage.

b. The common-mode component of the output voltage.

c. The differential-mode component of the output voltage.

8.38 The two voltage sources shown in Figure P8.32 are pressure sensors where, for each source and with P = pressure in kilopascals,

$$v_{S1,2} = A + BP_{1,2}$$

$$A = 0.3 \text{ V} \qquad B = 0.7 \frac{\text{V}}{\text{psi}}$$

$$R_1 = R_2 = 4.7 \text{ k}\Omega$$

$$R_3 = R_4 = 10 \text{ k}\Omega$$

$$R_L = 1.8 \text{ k}\Omega$$

If $P_1 = 6$ kPa and $P_2 = 5$ kPa, determine, using superposition, that part of the output voltage which is due to the

a. Common-mode input voltage.

b. Differential-mode input voltage.

8.39 A linear potentiometer (variable resistor) R_P is used to sense and give a signal voltage v_y proportional to the current y position of an xy plotter. A reference signal v_R is supplied by the software controlling the plotter. The difference between these voltages must be amplified and supplied to a motor. The motor turns and changes the position of the pen and the position of the "pot" until the signal voltage is equal to the reference voltage (indicating the pen is in the desired position) and the motor voltage = 0. For proper operation the motor voltage must be 10 times the difference between the signal and reference voltage. For rotation in the proper direction, the motor voltage must be negative with respect to the signal voltage for the polarities shown. An additional requirement is that $i_P = 0$ to avoid loading the pot and causing an erroneous signal voltage.

a. Design an op-amp circuit that will achieve the specifications given. Redraw the circuit shown in Figure P8.39, replacing the box (drawn with dotted lines) with your circuit. Be sure to show how the signal voltage and output voltage are connected in your circuit.

b. Determine the value of each component in your circuit. The op-amp is a μA741C.

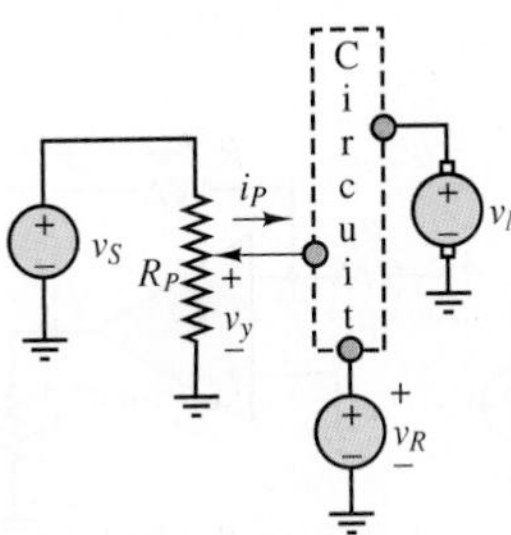

Figure P8.39

8.40 In the circuit shown in Figure P8.32,

$$v_{S1} = 13 \text{ mV} \qquad v_{S2} = 19 \text{ mV}$$

$$R_1 = 1 \text{ k}\Omega \qquad R_2 = 13 \text{ k}\Omega$$

$$R_3 = 81 \text{ k}\Omega \qquad R_4 = 56 \text{ k}\Omega$$

Determine the output voltage.

8.41 Figure P8.41 shows a simple voltage-to-current converter. Show that the current I_{out} through the light-emitting diode, and therefore its brightness, is proportional to the source voltage V_s as long as $V_s > 0$.

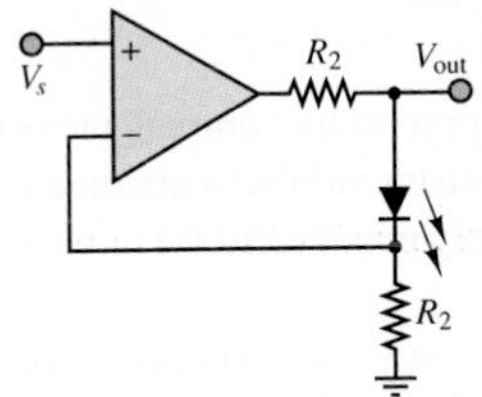

Figure P8.41

8.42 Figure P8.42 shows a simple current-to-voltage converter. Show that the voltage V_{out} is proportional to the current generated by the cadmium sulfide (CdS) solar cell. Also show that the transimpedance of the circuit V_{out}/I_s is $-R$.

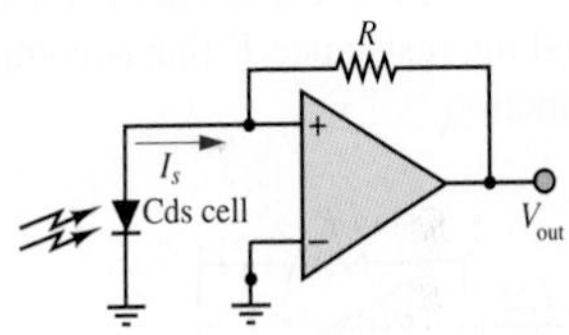

Figure P8.42

8.43 An op-amp voltmeter circuit as in Figure P8.43 is required to measure a maximum input of $E = 20$ mV. The op-amp input current is $I_B = 0.2\ \mu$A, and the meter circuit has $I_m = 100\ \mu$A full-scale deflection and $r_m = 10$ kΩ. Determine suitable values for R_3 and R_4.

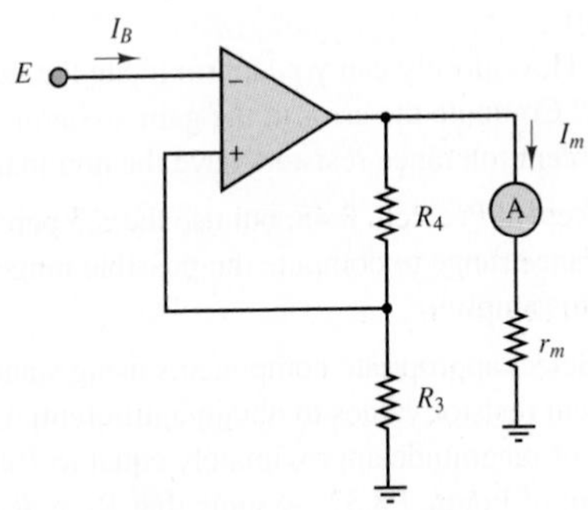

Figure P8.43

8.44 Find an expression for the output voltage in the circuit of Figure P8.44.

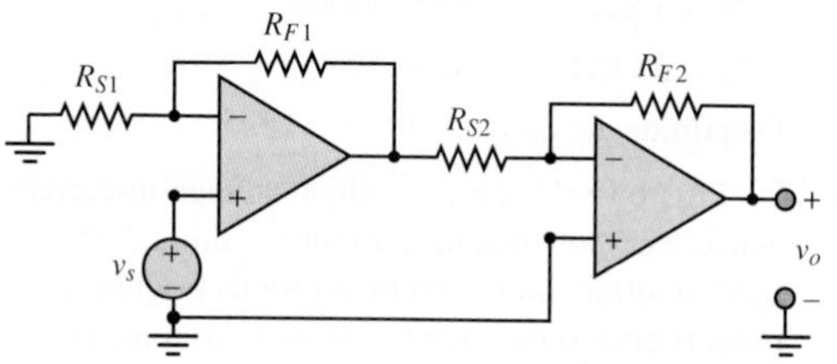

Figure P8.44

8.45 Select appropriate components using standard 5 percent resistor values to obtain a gain of magnitude approximately equal to 1,000 in the circuit of Figure P8.44.

How closely can you approximate the desired gain? Compute the error in the gain, assuming that the 5 percent tolerance resistors have the nominal value.

8.46 Repeat Problem 8.45, but use the ±5 percent tolerance range to compute the possible range of gains for this amplifier.

8.47 The circuit shown in Figure P8.47 can function as a precision ammeter. Assume that the voltmeter has a range of 0 to 10 V and a resistance of 20 kΩ. The full-scale reading of the ammeter is intended to be 1 mA. Find the resistance R that accomplishes the desired function.

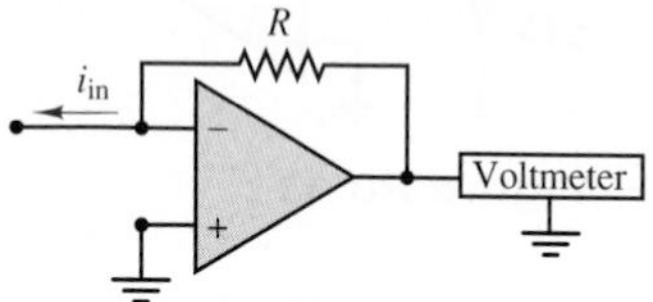

Figure P8.47

8.48 Select appropriate components using standard 5 percent resistor values to obtain a gain of magnitude approximately equal to 200 in the circuit of Figure P8.30.

How closely can you approximate the desired gain? Compute the error in the gain, assuming that the 5 percent tolerance resistors have the nominal value.

8.49 Repeat Problem 8.48, but use the ±5 percent tolerance range to compute the possible range of gains for this amplifier.

8.50 Select appropriate components using standard 1 percent resistor values to obtain a differential amplifier gain of magnitude approximately equal to 100 in the circuit of Figure P8.32. Assume that $R_3 = R_4$ and $R_1 = R_2$.

How closely can you approximate the desired gain? Compute the error in the gain, assuming that the 1 percent tolerance resistors have the nominal value.

8.51 Repeat Problem 8.50, but use the ±1 percent tolerance range to compute the possible range of gains for this amplifier. You may assume that $R_3 = R_4$ and $R_1 = R_2$.

Section 8.3: Active Filters

8.52 The circuit shown in Figure P8.52 is an active filter with

$$C = 1\ \mu\text{F} \qquad R = 10\ \text{k}\Omega \qquad R_L = 1\ \text{k}\Omega$$

Determine

a. The gain (in decibels) in the passband.

b. The cutoff frequency.

c. Whether this is a low- or high-pass filter.

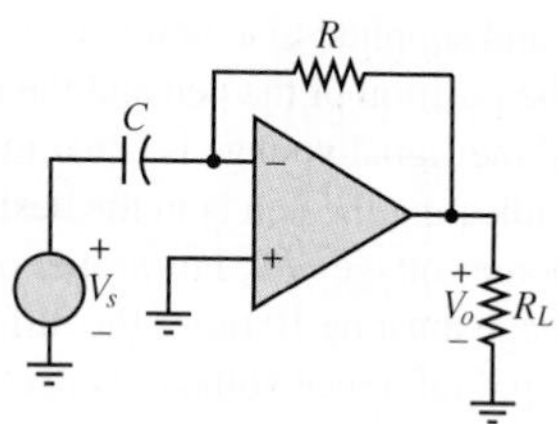

Figure P8.52

8.53 The op-amp circuit shown in Figure P8.53 is used as a filter.

$$C = 0.1\ \mu\text{F} \qquad R_L = 333\ \Omega$$

$$R_1 = 1.8\ \text{k}\Omega \qquad R_2 = 8.2\ \text{k}\Omega$$

Determine

a. Whether the circuit is a low- or high-pass filter.

b. The gain V_o/V_s in decibels in the passband, that is, at the frequencies being passed by the filter.

c. The cutoff frequency.

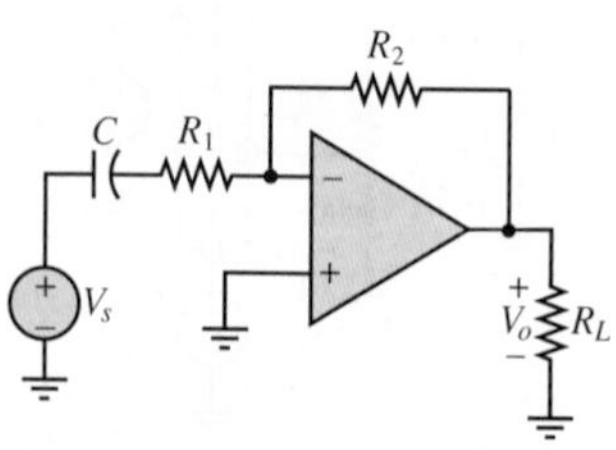

Figure P8.53

8.54 The op-amp circuit shown in Figure P8.53 is used as a filter.

$C = 200$ pF $\quad R_L = 1$ kΩ

$R_1 = 10$ kΩ $\quad R_2 = 220$ kΩ

Determine

a. Whether the circuit is a low- or high-pass filter.

b. The gain V_o/V_s in decibels in the passband, that is, at the frequencies being passed by the filter.

c. The cutoff frequency.

8.55 The circuit shown in Figure P8.55 is an active filter with

$R_1 = 4.7$ kΩ $\quad C = 100$ pF

$R_2 = 68$ kΩ $\quad R_L = 220$ kΩ

Determine the cutoff frequencies and the magnitude of the voltage frequency response function at very low and at very high frequencies.

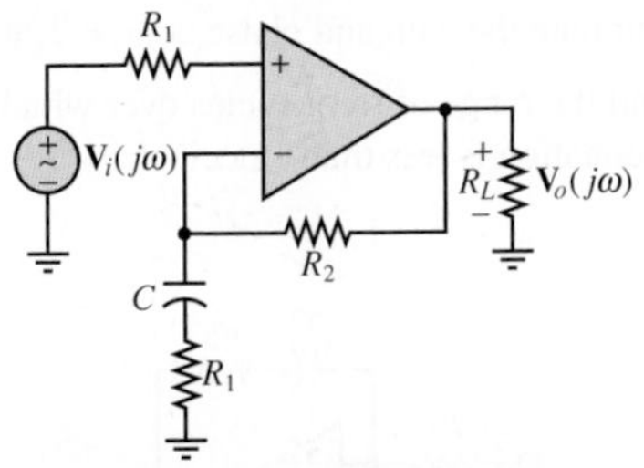

Figure P8.55

8.56 The circuit shown in Figure P8.56 is an active filter with

$R_1 = 1$ kΩ $\quad R_2 = 4.7$ kΩ

$R_3 = 80$ kΩ $\quad C = 20$ nF

Determine

a. An expression for the voltage frequency response function in the standard form:

$$H_v(j\omega) = \frac{\mathbf{V}_o(j\omega)}{\mathbf{V}_i(j\omega)}$$

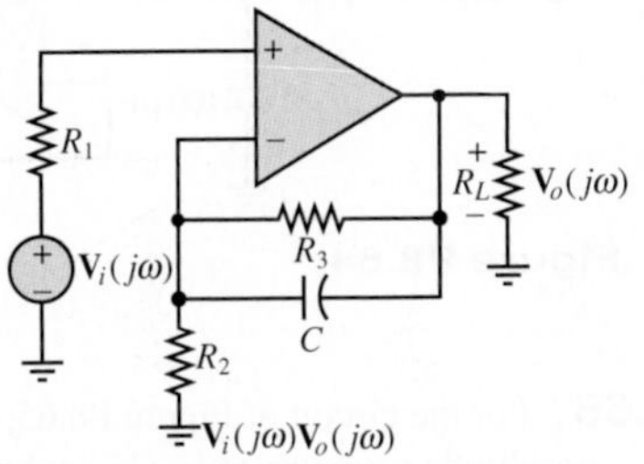

Figure P8.56

b. The cutoff frequencies.

c. The passband gain.

d. The Bode plot.

8.57 The op-amp circuit shown in Figure P8.57 is used as a filter.

$R_1 = 9.1$ kΩ $\quad R_2 = 22$ kΩ

$C = 0.47$ μF $\quad R_L = 2.2$ kΩ

Determine

a. Whether the circuit is a low- or high-pass filter.

b. An expression in standard form for the voltage transfer function.

c. The gain in decibels in the passband, that is, at the frequencies being passed by the filter, and the cutoff frequency.

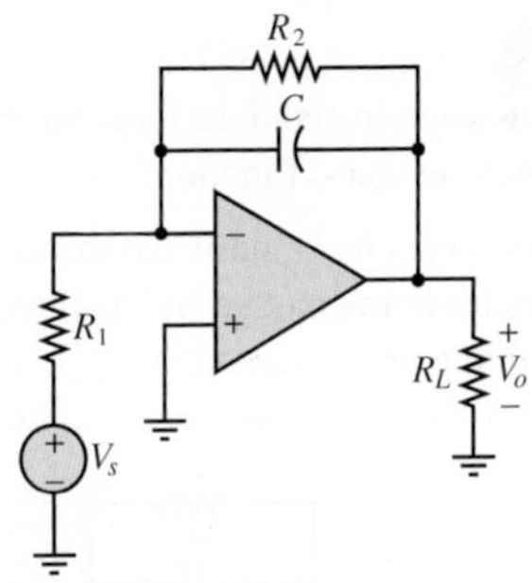

Figure P8.57

8.58 The op-amp circuit shown in Figure P8.57 is a low-pass filter with

$R_1 = 2.2$ kΩ $\quad R_2 = 68$ kΩ

$C = 0.47$ nF $\quad R_L = 1$ kΩ

Determine

a. An expression for the voltage frequency response function.

b. The gain in decibels in the passband, that is, at the frequencies being passed by the filter, and the cutoff frequency.

8.59 The circuit shown in Figure P8.59 is a bandpass filter. If

$R_1 = R_2 = 10$ kΩ

$C_1 = C_2 = 0.1$ μF

determine

a. The passband gain.

b. The resonant frequency.

c. The cutoff frequencies.

d. The circuit Q.

e. The Bode plot.

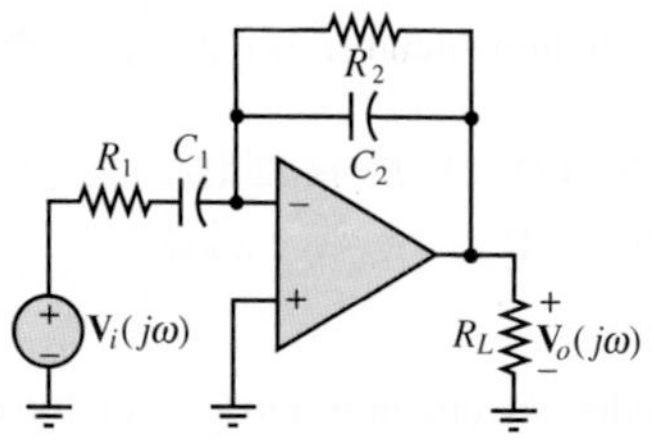

Figure P8.59

8.60 The op-amp circuit shown in Figure P8.60 is a low-pass filter with

$R_1 = 220\ \Omega \qquad R_2 = 68\ \text{k}\Omega$

$C = 0.47\ \text{nF} \qquad R_L = 1\ \text{k}\Omega$

Determine

a. An expression in standard form for the voltage frequency response function.

b. The gain in decibels in the passband, that is, at the frequencies being passed by the filter, and the cutoff frequency.

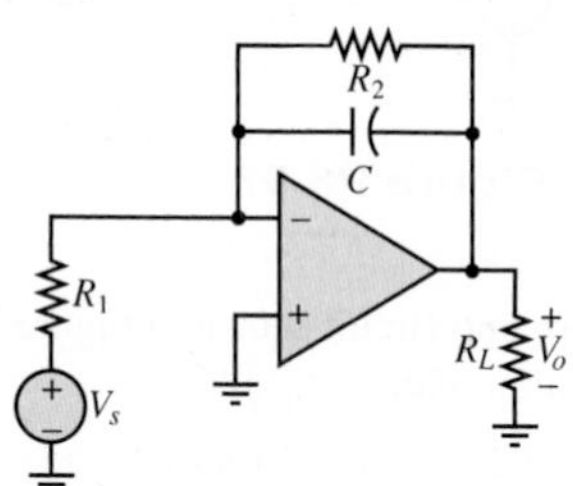

Figure P8.60

8.61 The circuit shown in Figure P8.61 is a bandpass filter. If

$R_1 = 2.2\ \text{k}\Omega \qquad R_2 = 100\ \text{k}\Omega$

$C_1 = 2.2\ \mu\text{F} \qquad C_2 = 1\ \text{nF}$

determine the passband gain.

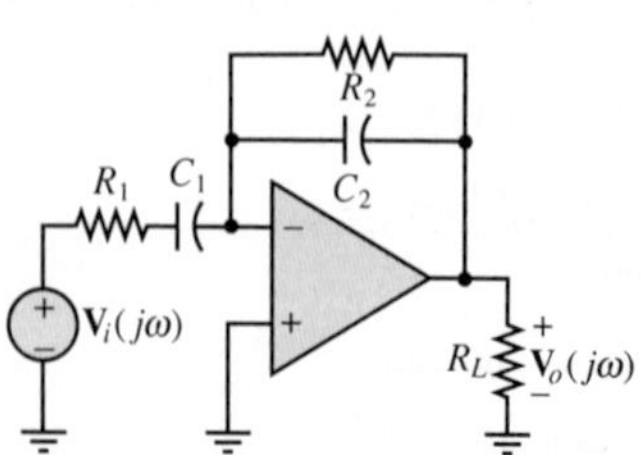

Figure P8.61

8.62 Compute the frequency response of the circuit shown in Figure P8.62.

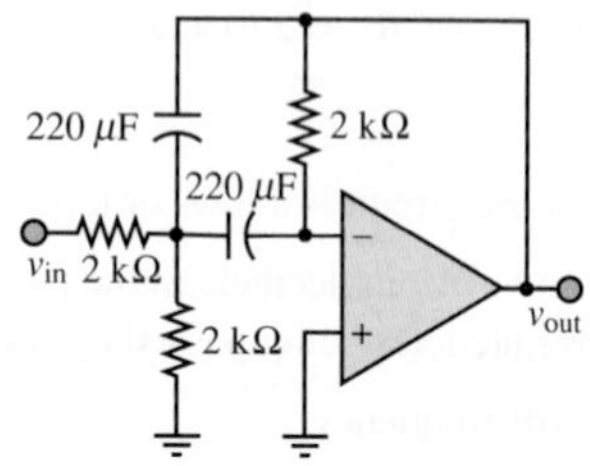

Figure P8.62

8.63 The inverting amplifier shown in Figure P8.63 can be used as a low-pass filter.

a. Derive the frequency response of the circuit.

b. If $R_1 = R_2 = 100\ \text{k}\Omega$ and $C = 0.1\ \mu\text{F}$, compute the attenuation in decibels at $\omega = 1{,}000$ rad/s.

c. Compute the gain and phase at $\omega = 2{,}500$ rad/s.

d. Find the range of frequencies over which the attenuation is less than 1 decibel.

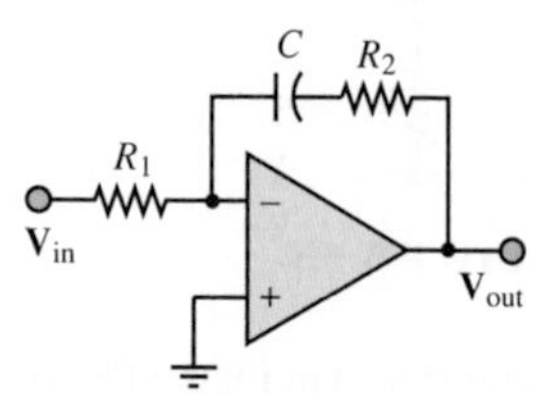

Figure P8.63

8.64 Find an expression for the gain of the circuit of Figure P8.64.

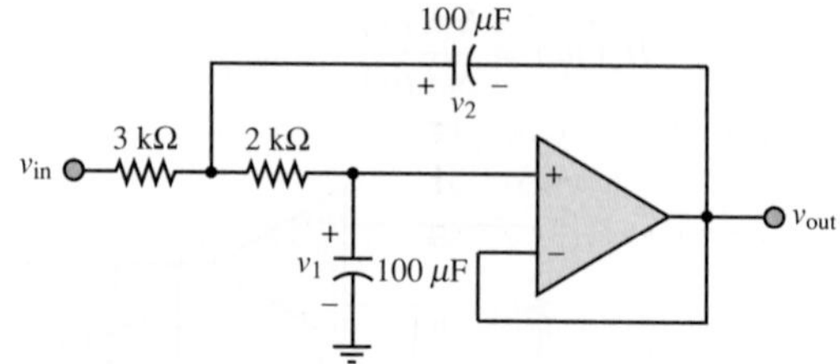

Figure P8.64

8.65 For the circuit of Figure P8.65, sketch the amplitude response of V_2/V_1, indicating the half-power frequencies. Assume the op-amp is ideal.

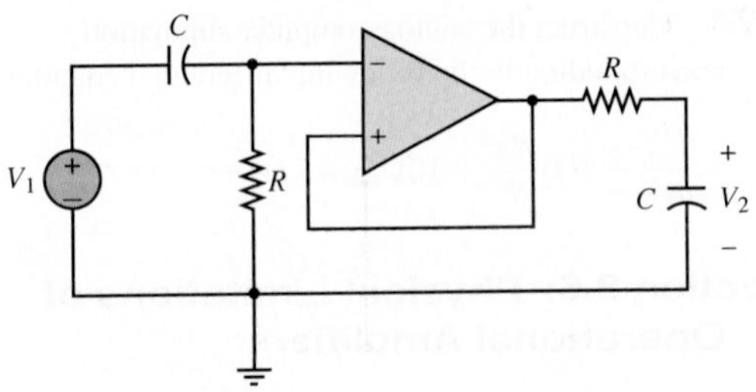

Figure P8.65

8.66 Determine an analytical expression for the circuit shown in Figure P8.66. What kind of a filter does this circuit implement?

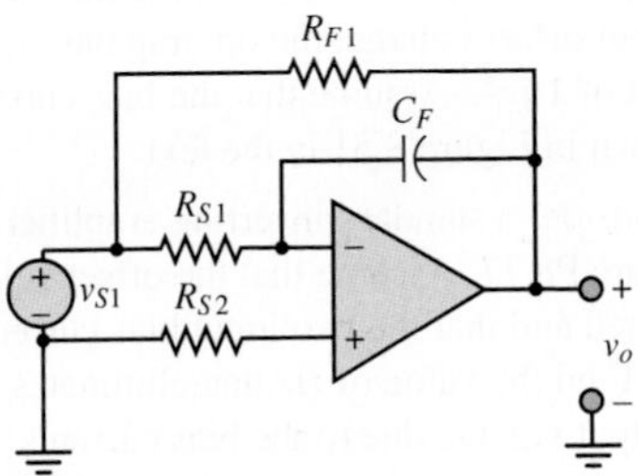

Figure P8.66

8.67 Determine an analytical expression for the circuit shown in Figure P8.67. What kind of a filter does this circuit implement?

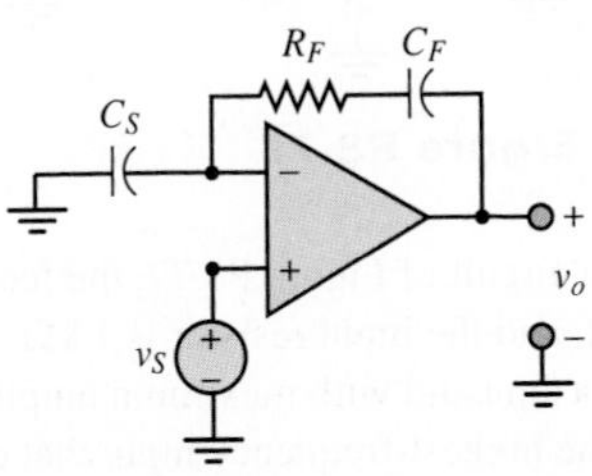

Figure P8.67

Section 8.4: Integrator and Differentiator Circuits; Section 8.5: Analog Computers

8.68 The circuit shown in Figure P8.68(a) will give an output voltage which is either the integral or the derivative of the source voltage shown in Figure P8.68(b) multiplied by some gain. If

$$C = 1\ \mu\text{F} \qquad R = 10\ \text{k}\Omega \qquad R_L = 1\ \text{k}\Omega$$

determine an expression for and plot the output voltage as a function of time.

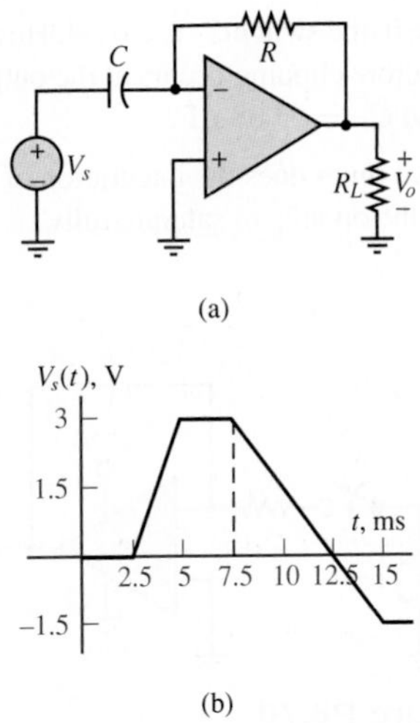

Figure P8.68

8.69 The circuit shown in Figure P8.69(a) will give an output voltage which is either the integral or the derivative of the supply voltage shown in Figure P8.69(b) multiplied by some gain. Determine

a. An expression for the output voltage.
b. The value of the output voltage at $t = 5$, 7.5, 12.5, 15, and 20 ms and a plot of the output voltage as a function of time if

$$C = 1\ \mu\text{F} \qquad R = 10\ \text{k}\Omega \qquad R_L = 1\ \text{k}\Omega$$

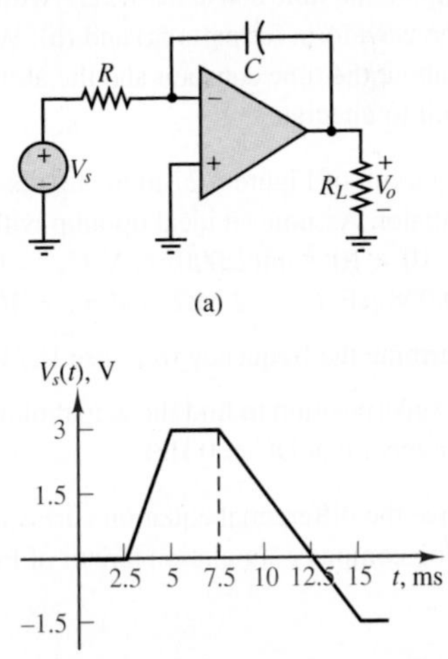

Figure P8.69

8.70 The circuit shown in Figure P8.70 is an integrator. The capacitor is initially uncharged, and the source voltage is

$$v_{\text{in}}(t) = 10 \times 10^{-3} + \sin(2{,}000\pi t) \qquad \text{V}$$

a. At $t = 0$, the switch S_1 is closed. How long does it take before clipping occurs at the output if $R_s = 10$ kΩ and $C_F = 0.008\ \mu$F?

b. At what times does the integration of the DC input cause the op-amp to saturate fully?

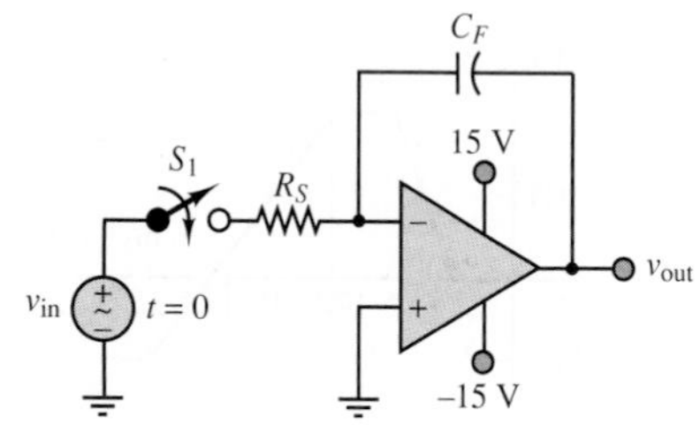

Figure P8.70

8.71 A practical integrator is shown in Figure 8.22 in the text. Note that the resistor in parallel with the feedback capacitor provides a path for the capacitor to discharge the DC voltage. Usually, the time constant $R_F C_F$ is chosen to be long enough not to interfere with the integration.

a. If $R_S = 10$ kΩ, $R_F = 2$ MΩ, $C_F = 0.008\ \mu$F, and $v_S(t) = 10$ V + $\sin(2{,}000\pi t)$ V, find $v_{out}(t)$, using phasor analysis.

b. Repeat part a if $R_F = 200$ kΩ, and if $R_F = 20$ kΩ.

c. Compare the time constants $R_F C_F$ with the period of the waveform for parts (a) and (b). What can you say about the time constant and the ability of the circuit to integrate?

8.72 The circuit of Figure 8.27 in the text is a practical differentiator. Assume an ideal op-amp with $v_S(t) = 10 \times 10^{-3} \sin(2{,}000\pi t)$ V, $C_S = 100\ \mu$F, $C_F = 0.008\ \mu$F, $R_F = 2$ MΩ, and $R_S = 10$ kΩ.

a. Determine the frequency response $V_o/V_S(\omega)$.

b. Use superposition to find the actual output voltage (remember that DC = 0 Hz).

8.73 Derive the differential equation corresponding to the analog computer simulation circuit of Figure P8.73.

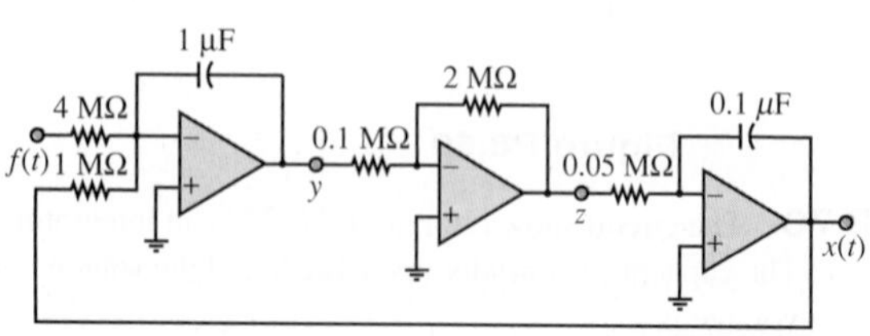

Figure P8.73

8.74 Construct the analog computer simulation corresponding to the following differential equation:

$$\frac{d^2x}{dt^2} + 100\frac{dx}{dt} + 10x = -5f(t)$$

Section 8.6: Physical Limitations of Operational Amplifiers

8.75 Consider the noninverting amplifier of Figure 8.8 in the text. Find the error introduced in the output voltage if the op-amp has an input offset voltage of 2 mV. Assume that the input bias currents are zero, and that $R_S = R_F = 2.2$ kΩ. Assume that the offset voltage appears as shown in Figure 8.50 in the text.

8.76 Repeat Problem 8.75, assuming that in addition to the input offset voltage, the op-amp has an input bias current of 1 μA. Assume that the bias current appears as shown in Figure 8.51 in the text.

8.77 Consider a standard inverting amplifier, as shown in Figure P8.77. Assume that the offset voltage can be neglected and that the two input bias currents are equal. Find the value of R_x that eliminates the error in the output voltage due to the bias currents.

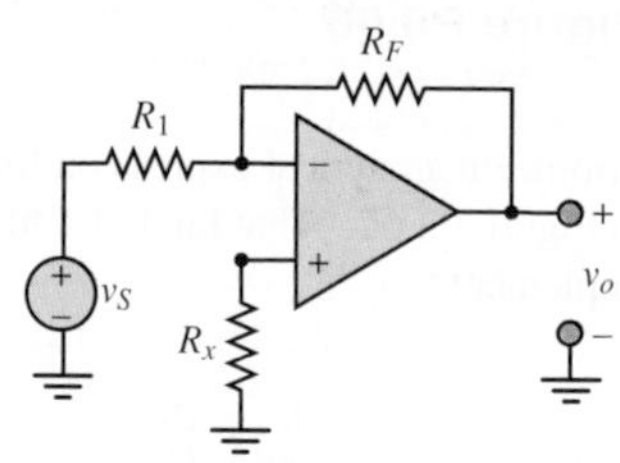

Figure P8.77

8.78 In the circuit of Figure P8.77, the feedback resistor is 3.3 kΩ, and the input resistor is 1 kΩ. If the input signal is a sinusoid with maximum amplitude of 1.5 V, what is the highest-frequency input that can be used without exceeding the slew rate limit of 1 V/μs?

8.79 An op-amp has the open-loop frequency response shown in Figure 8.47 in the text. What is the approximate bandwidth of a circuit that uses the op-amp with a closed-loop gain of 75? What is the bandwidth if the gain is 350?

8.80 The ideal charge amplifier discussed in the Focus on Measurements box "Charge Amplifiers" will saturate in the presence of any DC offsets, as discussed in Section 8.6. The circuit of Figure P8.80 represents a practical charge amplifier, in which the user is provided with a choice of three time constants—$\tau_{long} = R_L C_F$, $\tau_{medium} = R_M C_F$, $\tau_{short} = R_S C_F$—which can be selected by means of a switch. Assume that

$R_L = 10\ \text{M}\Omega$, $R_M = 1\ \text{M}\Omega$, $R_S = 0.1\ \text{M}\Omega$, and $C_F = 0.1\ \mu\text{F}$. Analyze the frequency response of the practical charge amplifier for each case, and determine the lowest input signal frequency that can be amplified without excessive distortion for each case. Can this circuit amplify a DC signal?

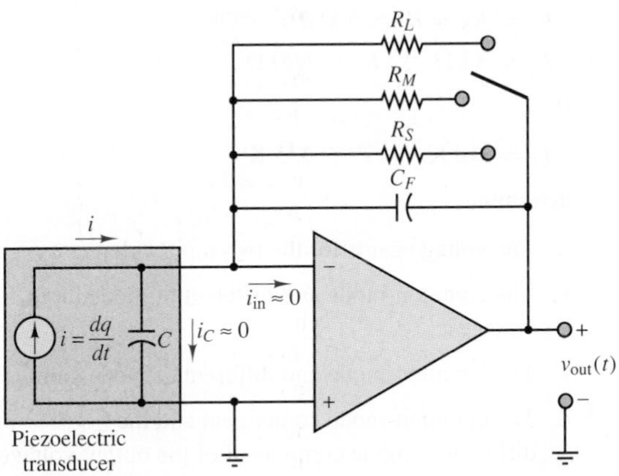

Figure P8.80

8.81 Consider a differential amplifier. We desire the common-mode output to be less than 1 percent of the differential-mode output. Find the minimum decibel common-mode rejection ratio to fulfill this requirement if the differential-mode gain $A_{dm} = 1{,}000$. Let

$$v_1 = \sin(2{,}000\pi t) + 0.1\sin(120\pi t) \quad \text{V}$$
$$v_2 = \sin(2{,}000\pi t + 180°) + 0.1\sin(120\pi t) \quad \text{V}$$
$$v_{out} = A_{dm}(v_1 - v_2) + A_{cm}\frac{v_1 + v_2}{2}$$

8.82 Square wave testing can be used with operational amplifiers to estimate the *slew rate*, which is defined as the maximum rate at which the output can change (in volts per microsecond). Input and output waveforms for a noninverting op-amp circuit are shown in Figure P8.82. As indicated, the rise time t_R of the output waveform is defined as the time it takes for that waveform to increase from 10 to 90 percent of its final value, or

$$t_R \stackrel{\triangle}{=} t_B - t_A = -\tau(\ln 0.1 - \ln 0.9) = 2.2\tau$$

where τ is the circuit time constant. Estimate the slew rate for the op-amp.

8.83 Consider an inverting amplifier with open-loop gain 10^5. With reference to equation 8.18:

a. If $R_S = 10\ \text{k}\Omega$ and $R_F = 1\ \text{M}\Omega$, find the voltage gain $A_{V(CL)}$.

b. Repeat part a if $R_S = 10\ \text{k}\Omega$ and $R_F = 10\ \text{M}\Omega$.

c. Repeat part a if $R_S = 10\ \text{k}\Omega$ and $R_F = 100\ \text{M}\Omega$.

d. Using the resistor values of part c, find $A_{V(CL)}$ if $A_{V(OL)} \to \infty$.

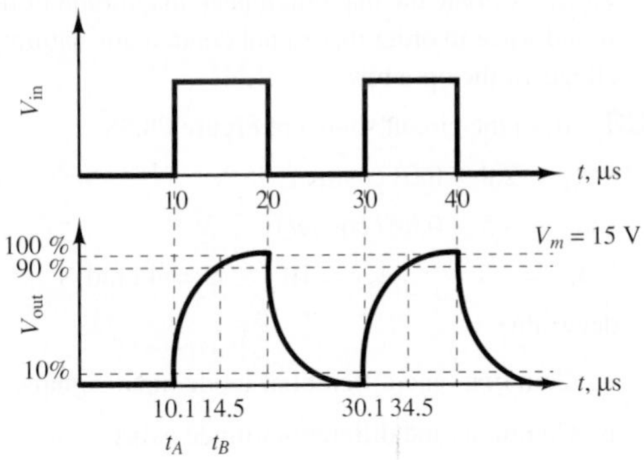

Figure P8.82

8.84

a. If the op-amp of Figure P8.84 has an open-loop gain of 45×10^5, find the closed-loop gain for $R_F = R_S = 7.5\ \text{k}\Omega$, with reference to equation 8.18.

b. Repeat part a if $R_F = 5R_S = 37{,}500\ \Omega$.

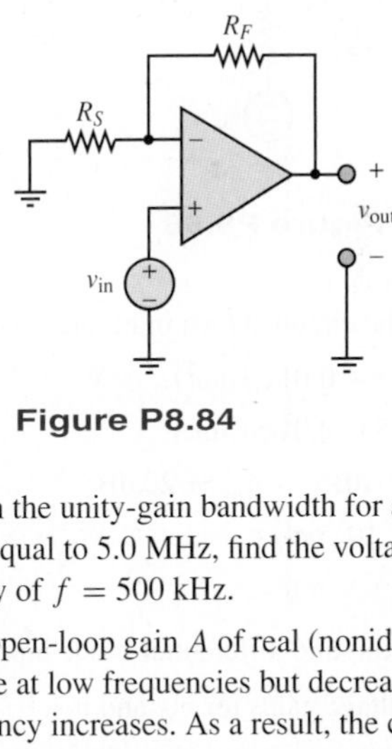

Figure P8.84

8.85 Given the unity-gain bandwidth for an ideal op-amp equal to 5.0 MHz, find the voltage gain at a frequency of $f = 500$ kHz.

8.86 The open-loop gain A of real (nonideal) op-amps is very large at low frequencies but decreases markedly as frequency increases. As a result, the closed-loop gain of op-amp circuits can be strongly dependent on frequency. Determine the relationship between a finite and frequency-dependent open-loop gain $A_{V(OL)}(\omega)$ and the closed-loop gain $A_{V(CL)}(\omega)$ of an inverting amplifier as a function of frequency. Plot $A_{V(CL)}$ versus ω. Notice that $-R_F/R_S$ is the low-frequency closed-loop gain.

8.87 A sinusoidal sound (pressure) wave $p(t)$ impinges upon a condenser microphone of sensitivity S (mV/kPa). The voltage output of the microphone v_s is amplified by two cascaded inverting amplifiers to produce an amplified signal v_O. Determine the peak amplitude of the sound wave (in decibels) if $v_O = 5$ V_{RMS}. Estimate the maximum peak magnitude of the sound wave in order that v_O not contain any saturation effects of the op-amps.

8.88 If, in the circuit shown in Figure P8.88,

$v_{S1} = 2.8 + 0.01\cos(\omega t)$ V

$v_{S2} = 3.5 - 0.007\cos(\omega t)$ V

$A_{v1} = -13 \quad A_{v2} = 10 \quad \omega = 4$ krad/s

determine

a. Common- and differential-mode input signals.
b. Common- and differential-mode gains.
c. Common- and differential-mode components of the output voltage.
d. Total output voltage.
e. Common-mode rejection ratio.

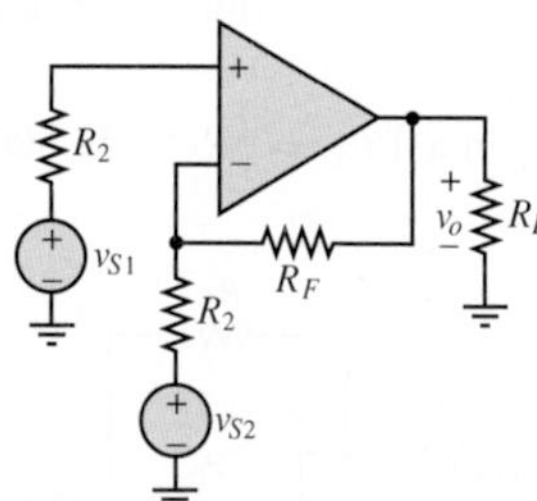

Figure P8.88

8.89 If, in the circuit shown in Figure P8.88,

$v_{S1} = 3.5 + 0.01\cos(\omega t)$ V

$v_{S2} = 3.5 - 0.01\cos(\omega t)$ V

$A_{vc} = 10$ dB $\quad A_{vd} = 20$ dB

$\omega = 4 \times 10^3$ rad/s

determine

a. Common- and differential-mode input voltages.
b. The voltage gains for v_{S1} and v_{S2}.
c. Common-mode component and differential-mode component of the output voltage.
d. The common-mode rejection ratio (CMRR) in decibels.

8.90 In the circuit shown in Figure P8.90, the two voltage sources are temperature sensors with $T =$ temperature (Kelvin) and

$v_{S1} = kT_1 \quad v_{S2} = kT_2$

where

$k = 120\ \mu$V/K

$R_1 = R_3 = R_4 = 5$ kΩ

$R_2 = 3$ kΩ $\quad R_L = 600$ Ω

If

$T_1 = 310$ K $\quad T_2 = 335$ K

determine

a. The voltage gains for the two input voltages.
b. The common-mode and differential-mode input voltages.
c. The common-mode and differential-mode gains.
d. The common-mode component and the differential-mode component of the output voltage.
e. The common-mode rejection ratio (CMRR) in decibels.

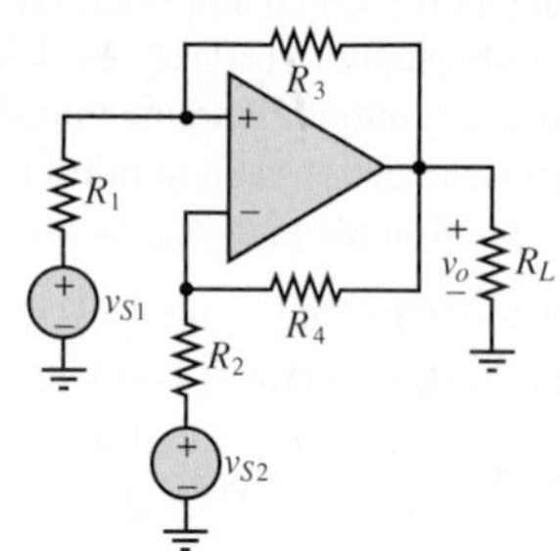

Figure P8.90

8.91 In the differential amplifier shown in Figure P8.90,

$v_{S1} = 13$ mV $\quad v_{S2} = 9$ mV

$v_o = v_{oc} + v_{od}$

$v_{oc} = 33$ mV (common-mode output voltage)

$v_{od} = 18$ V (differential-mode output voltage)

Determine

a. The common-mode gain.
b. The differential-mode gain.
c. The common-mode rejection ratio in decibels.

C H A P T E R

9

SEMICONDUCTORS AND DIODES

This chapter introduces a new topic: **solid state electronics.** You must be familiar with the marvelous progress that has taken place in this field since the invention of the *transistor.* Think of the progress that personal computing alone has made in the last 50 years! Modern electronic systems are possible because of individual discrete electronic devices that have been integrated into functioning as complex systems. Although the use of discrete electronic devices has largely been replaced by that of *integrated circuits* (e.g., the operational amplifier of Chapter 8), it is still important to understand how the individual elements function. The aim of this and of the next three chapters is to introduce the fundamental operation of *semiconductor electronic devices.* These include two principal families of elements: *diodes* and *transistors.* The focus of Chapters 9 through 12 is principally on *discrete* devices, that is, on analyzing and using individual diodes and transistors in various circuits.

This chapter explains the workings of the semiconductor diode, a device that finds use in many practical circuits used in electric power systems and in high- and low-power electronic circuits. The emphasis in the chapter is on using simple models of the semiconductor devices, and on reducing the resulting circuits to ones that we can analyze based on the circuit analysis tools introduced in earlier chapters. This is usually done in two steps: first, the i-v characteristic of the diode is analyzed, and it is shown that one can use simple linear circuit elements (ideal resistors, ideal voltage sources) to describe the operation of the diode by way of *circuit models;* second, the circuit models are inserted in a practical circuit in place of the diode, and well-known circuit analysis methods are employed to analyze the resulting *linear* circuit. Thus, once you have understood how to select an appropriate diode model, you will once again apply the basic circuit analysis principles that you mastered in the first part of this book.

Learning Objectives

1. Understand the basic principles underlying the physics of semiconductor devices in general and of the *pn* junction in particular. Become familiar with the diode equation and *i*-*v* characteristic. *Sections 9.1, 9.2*.
2. Use various circuit models of the semiconductor diode in simple circuits. These are divided into two classes: large-signal models, useful to study rectifier circuits, and small-signal models, useful in signal processing applications. *Section 9.2*.
3. Study practical full-wave rectifier circuits and learn to analyze and determine the practical specifications of a rectifier by using large-signal diode models. *Section 9.3*.
4. Understand the basic operation of Zener diodes as voltage references, and use simple circuit models to analyze elementary voltage regulators. *Section 9.4*.
5. Use the diode models presented in Section 9.2 to analyze the operation of various practical diode circuits in signal processing applications. *Section 9.5*.
6. Understand the basic principle of operation of photodiodes, including solar cells, photosensors, and light-emitting diodes. *Section 9.6*.

9.1 ELECTRICAL CONDUCTION IN SEMICONDUCTOR DEVICES

This section briefly introduces the mechanism of conduction in a class of materials called **semiconductors.** Elemental[1] or intrinsic semiconductors are materials consisting of elements from group IV of the periodic table and having electrical properties falling somewhere between those of conducting and of insulating materials. As an example, consider the conductivity of three common materials. Copper, a good conductor, has a conductivity of 0.59×10^6 S/cm; glass, a common insulator, may range between 10^{-16} and 10^{-13} S/cm; and silicon, a semiconductor, has a conductivity that varies from 10^{-8} to 10^{-1} S/cm. You see, then, that the name *semiconductor* is an appropriate one.

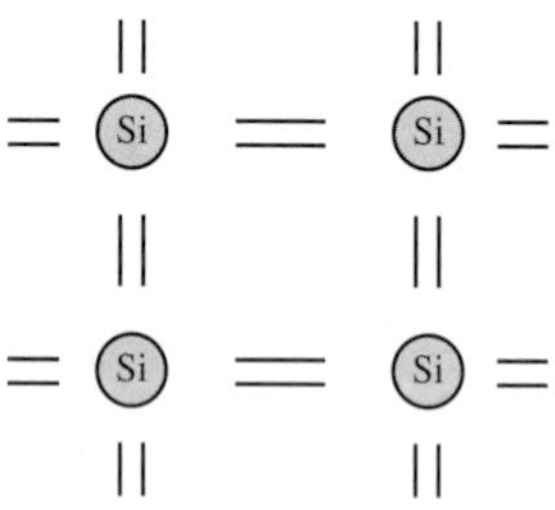

Figure 9.1 Lattice structure of silicon, with four valence electrons

A conducting material is characterized by a large number of conduction band electrons, which have a very weak bond with the basic structure of the material. Thus, an electric field easily imparts energy to the outer electrons in a conductor and enables the flow of electric current. In a semiconductor, on the other hand, one needs to consider the lattice structure of the material, which in this case is characterized by **covalent bonding.** Figure 9.1 depicts the lattice arrangement for silicon (Si), one of the more common semiconductors. At sufficiently high temperatures, thermal energy causes the atoms in the lattice to vibrate; when sufficient kinetic energy is

[1] Semiconductors can also be made of more than one element, in which case the elements are not necessarily from group IV.

present, some of the valence electrons break their bonds with the lattice structure and become available as conduction electrons. These **free electrons** enable current flow in the semiconductor. Note that in a conductor, valence electrons have a very loose bond with the nucleus and are therefore available for conduction to a much greater extent than valence electrons in a semiconductor. One important aspect of this type of conduction is that the number of charge carriers depends on the amount of thermal energy present in the structure. Thus, many semiconductor properties are a function of temperature.

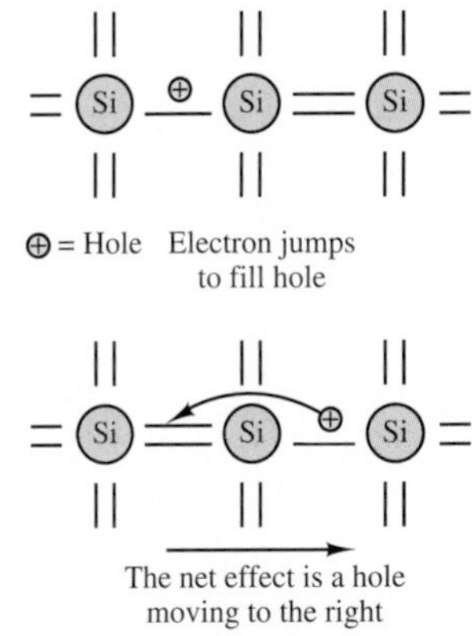

A vacancy (or hole) is created whenever a free electron leaves the structure.
This "hole" can move around the lattice if other electrons replace the free electron.

Figure 9.2 Free electrons and "holes" in the lattice structure

The free valence electrons are not the only mechanism of conduction in a semiconductor, however. Whenever a free electron leaves the lattice structure, it creates a corresponding positive charge within the lattice. Figure 9.2 depicts the situation in which a covalent bond is missing because of the departure of a free electron from the structure. The vacancy caused by the departure of a free electron is called a **hole.** Note that whenever a hole is present, we have, in effect, a positive charge. The positive charges also contribute to the conduction process, in the sense that if a valence band electron "jumps" to fill a neighboring hole, thereby neutralizing a positive charge, it correspondingly creates a new hole at a different location. Thus, the effect is equivalent to that of a positive charge moving to the right, in the sketch of Figure 9.2. This phenomenon becomes relevant when an external electric field is applied to the material. It is important to point out here that the **mobility**—that is, the ease with which charge carriers move across the lattice—differs greatly for the two types of carriers. Free electrons can move far more easily around the lattice than holes. To appreciate this, consider the fact that a free electron has already broken the covalent bond, whereas for a hole to travel through the structure, an electron must overcome the covalent bond each time the hole jumps to a new position.

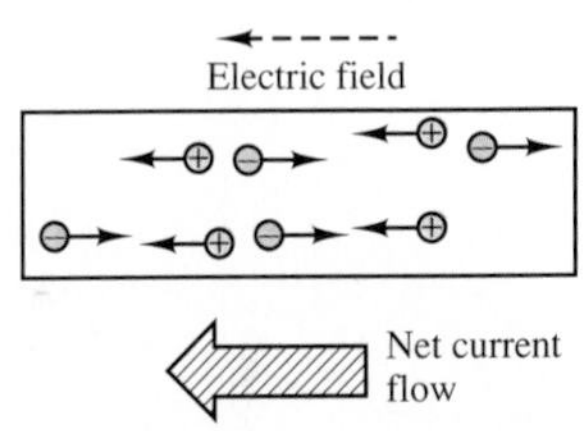

An external electric field forces holes to migrate to the left and free electrons to the right. The net current flow is to the left.

Figure 9.3 Current flow in a semiconductor

According to this relatively simplified view of semiconductor materials, we can envision a semiconductor as having two types of charge carriers—holes and free electrons—which travel in opposite directions when the semiconductor is subjected to an external electric field, giving rise to a net flow of current in the direction of the electric field. Figure 9.3 illustrates the concept.

An additional phenomenon, called **recombination,** reduces the number of charge carriers in a semiconductor. Occasionally, a free electron traveling in the immediate neighborhood of a hole will recombine with the hole, to form a covalent bond. Whenever this phenomenon takes place, two charge carriers are lost. However, in spite of recombination, the net balance is such that a number of free electrons always exist at a given temperature. These electrons are therefore available for conduction. The number of free electrons available for a given material is called the **intrinsic concentration** n_i. For example, at room temperature, silicon has

$$n_i = 1.5 \times 10^{16} \text{ electrons/m}^3 \qquad \textbf{(9.1)}$$

Note that there must be an equivalent number of holes present as well.

An additional free electron is created when Si is "doped" with a group V element.

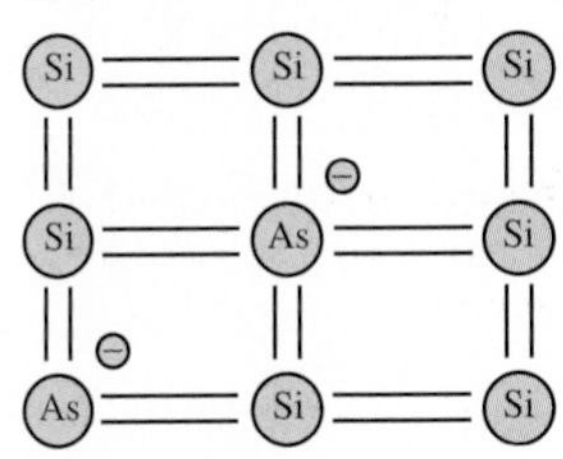

Figure 9.4 Doped semiconductor

Semiconductor technology rarely employs pure, or intrinsic, semiconductors. To control the number of charge carriers in a semiconductor, the process of **doping** is usually employed. Doping consists of adding impurities to the crystalline structure of the semiconductor. The amount of these impurities is controlled, and the impurities can be of one of two types. If the dopant is an element from the fifth column of the periodic table (e.g., arsenic), the end result is that wherever an impurity is present, an additional free electron is available for conduction. Figure 9.4 illustrates the concept. The elements providing the impurities are called **donors** in the case of group V elements, since they "donate" an additional free electron to the lattice structure. An equivalent situation arises when group III elements (e.g., indium) are used to dope silicon. In this case, however, an additional hole is created by the doping element, which is called an **acceptor,** since it accepts a free electron from the structure and generates a hole in doing so.

Semiconductors doped with donor elements conduct current predominantly by means of free electrons and are therefore called ***n*-type semiconductors.** When an acceptor element is used as the dopant, holes constitute the most common carrier, and the resulting semiconductor is said to be a ***p*-type semiconductor.** Doping usually takes place at such levels that the concentration of carriers due to the dopant is significantly greater than the intrinsic concentration of the original semiconductor. If n is the total number of free electrons and p that of holes, then in an n-type doped semiconductor, we have

$$n \gg n_i \tag{9.2}$$

and

$$p \ll p_i \tag{9.3}$$

Thus, free electrons are the **majority carriers** in an n-type material, while holes are the **minority carriers.** In a p-type material, the majority and minority carriers are reversed.

Doping is a standard practice for a number of reasons. Among these are the ability to control the concentration of charge carriers and the increase in the conductivity of the material that results from doping.

9.2 THE *pn* JUNCTION AND THE SEMICONDUCTOR DIODE

A simple section of semiconductor material does not in and of itself possess properties that make it useful for the construction of electronic circuits. However, when a section of p-type material and a section of n-type material are brought in contact to form a ***pn* junction,** a number of interesting properties arise. The *pn* junction forms the basis of the **semiconductor diode,** a widely used circuit element.

Figure 9.5 depicts an idealized *pn* junction, where on the p side we see a dominance of positive charge carriers, or holes, and on the n side, the free electrons dominate. Now, in the neighborhood of the junction, in a small section called the **depletion region,** the mobile charge carriers (holes and free electrons) come into contact with each other and recombine, thus leaving virtually no charge carriers at the junction. What is left in the depletion region, in the absence of the charge carriers, is the lattice structure of the n-type material on the right and of the p-type material on the left. But the n-type material, deprived of the free electrons, which have recombined

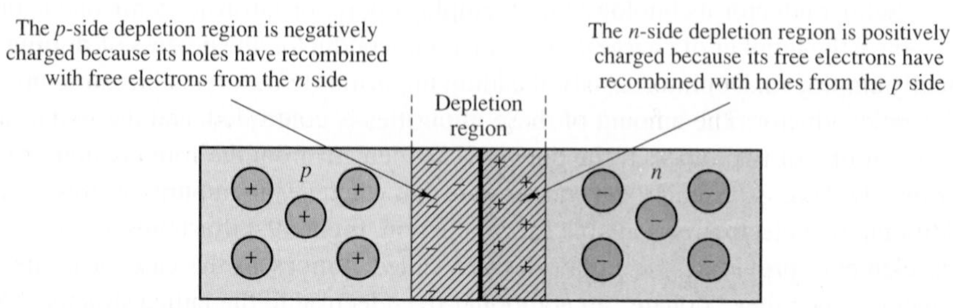

Figure 9.5 A *pn* junction

with holes in the neighborhood of the junction, is now positively charged. Similarly, the p-type material at the junction is negatively charged, because holes have been lost to recombination. The net effect is that while most of the material (p- or n-type) is charge-neutral because the lattice structure and the charge carriers neutralize each other (on average), the depletion region sees a separation of charge, giving rise to an electric field pointing from the n side to the p side. The charge separation therefore causes a **contact potential** to exist at the junction. This potential is typically on the order of a few tenths of a volt and depends on the material (about 0.6 to 0.7 V for silicon). The contact potential is also called the *offset voltage* V_γ.

Now, in the n-type materials, holes are the minority carriers; the relatively few p-type carriers (holes) are thermally generated, and recombine with free electrons. Some of these holes drift into the depletion region (to the left, in Figure 9.5), and they are pushed across the junction by the existing electric field. A similar situation exists in the p-type material, where now electrons drift across the depletion region (to the right). The net effect is that a small **reverse saturation current** I_S flows through the junction in the reverse direction (to the left) when the diode is reverse biased [see Figure 9.7(a)]. This current is largely independent of the junction voltage and is mostly determined by thermal carrier generation; that is, it is dependent on temperature. As the temperature increases, more hole-electron pairs are thermally generated, and the greater number of minority carriers produce a greater I_S [at room temperature, I_S is on the order of nanoamperes (10^{-9} A) in silicon]. This current across the junction flows opposite to the drift current and is called **diffusion current** I_d. Of course, if a hole from the p side enters the n side, it is quite likely that it will quickly recombine with one of the n-type carriers on the n side. One way to explain diffusion current is to visualize the diffusion of a gas in a room: gas molecules naturally tend to diffuse from a region of higher concentration to one of lower concentration. Similarly, the p-type material, for example, has a much greater concentration of holes than the n-type material. Thus, some holes will tend to diffuse into the n-type material across the junction, although only those that have sufficient (thermal) energy to do so will succeed. Figure 9.6 illustrates this process for a diode in equilibrium, with no bias applied.

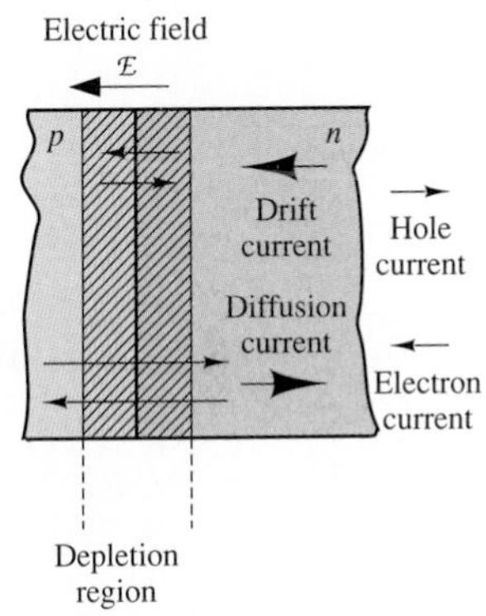

Figure 9.6 Drift and diffusion currents in a pn junction

The phenomena of drift and diffusion help explain how a pn junction behaves when it is connected to an external energy source. Consider the diagrams of Figure 9.7, where a battery has been connected to a pn junction in the **reverse-biased** direction [Figure 9.7(a)] and in the **forward-biased** direction [Figure 9.7(b)]. We assume that some suitable form of contact between the battery wires and the semiconductor material can be established (this is called an **ohmic contact**). The effect of a reverse bias is to increase the contact potential at the junction. Now, the majority carriers trying to diffuse across the junction need to overcome a greater barrier (a larger potential) and a wider depletion region. Thus, the diffusion current becomes negligible. The only current that flows under reverse bias is the very small reverse saturation current, so that the diode current i_D (defined in the figure) is

$$i_D = -I_0 = I_S \tag{9.4}$$

while the reverse saturation current above is actually a minority-carrier drift current. When the pn junction is forward-biased, the contact potential across the junction is lowered (note that V_B acts in opposition to the contact potential). Now, the diffusion of majority carriers is aided by the external voltage source; in fact, the diffusion current increases as a function of the applied voltage, according to equation 9.5

$$I_d = I_0 e^{qv_D/kT} \tag{9.5}$$

where v_D is the voltage across the pn junction, $k = 1.381 \times 10^{-23}$ J/K is Boltzmann's

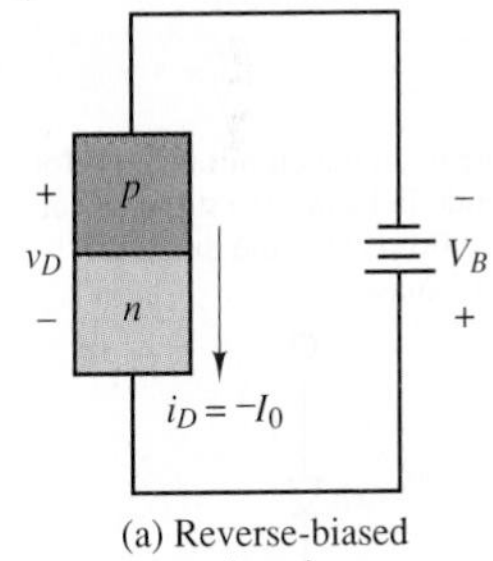

(a) Reverse-biased pn junction

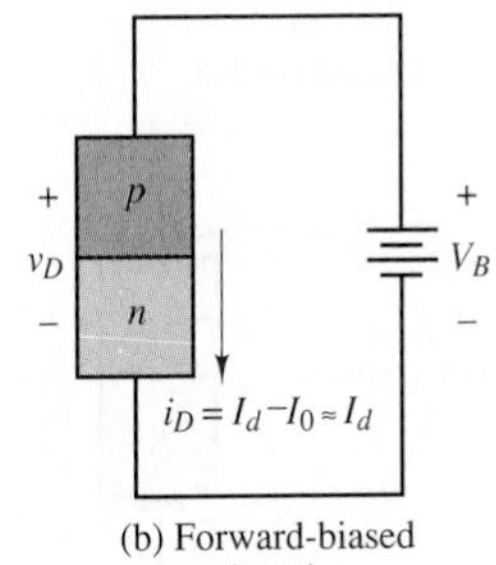

(b) Forward-biased pn junction

Figure 9.7 Forward- and reverse-biased pn junctions

constant, q is the charge of one electron, and T is the temperature of the material in kelvins (K). The quantity kT/q is constant at a given temperature and is approximately equal to 25 mV at room temperature. The net diode current under forward bias is given by equation 9.6

LO1

$$i_D = I_d - I_0 = I_0(e^{qv_D/kT} - 1) \qquad \text{Diode equation} \tag{9.6}$$

which is known as the **diode equation.** Figure 9.8 depicts the diode i-v characteristic described by the diode equation for a fairly typical silicon diode for positive diode voltages. Since the reverse saturation current I_0 is typically very small (10^{-9} to 10^{-15} A),

$$i_D = I_0 e^{qv_D/kT} \tag{9.7}$$

is a good approximation if the diode voltage v_D is greater than a few tenths of a volt.

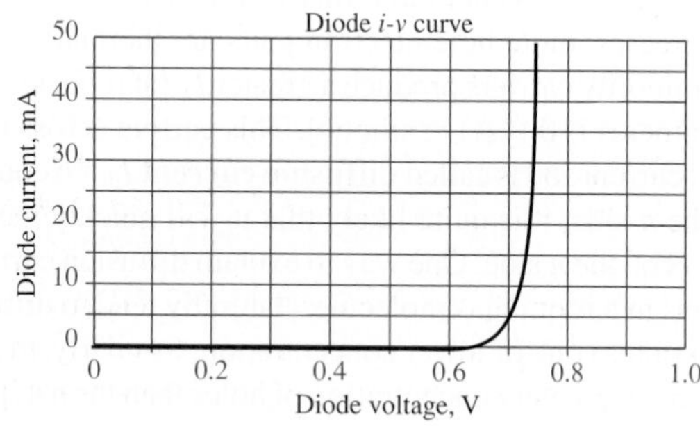

Figure 9.8 Semiconductor diode i-v characteristic

The arrow in the circuit symbol for the diode indicates the direction of current flow when the diode is forward-biased.

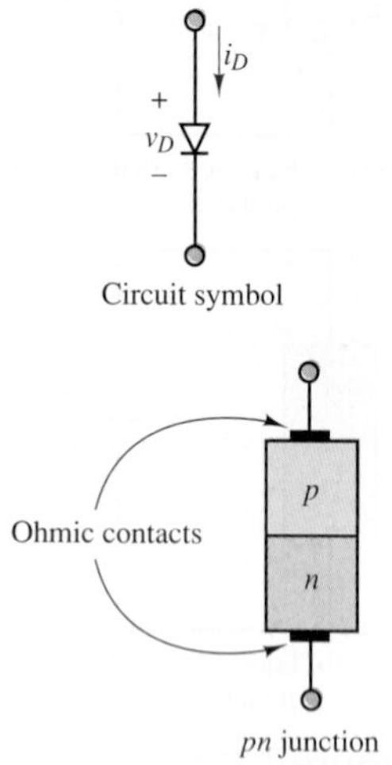

Figure 9.9 Semiconductor diode circuit symbol

The ability of the pn junction to essentially conduct current in only one direction—that is, to conduct only when the junction is forward-biased—makes it valuable in circuit applications. A device having a single pn junction and ohmic contacts at its terminals, as described in the preceding paragraphs, is called a *semiconductor diode*, or simply *diode*. As will be shown later in this chapter, it finds use in many practical circuits. The circuit symbol for the diode is shown in Figure 9.9, along with a sketch of the pn junction.

Figure 9.10 summarizes the behavior of the semiconductor diode by means of its i-v characteristic; it will become apparent later that this i-v characteristic plays an important role in constructing circuit models for the diode. Note that a third region appears in the diode i-v curve that has not been discussed yet. The **reverse breakdown** region to the far left of the curve represents the behavior of the diode when a sufficiently high reverse bias is applied. Under such a large reverse bias (greater in magnitude than the voltage V_Z, a quantity that will be explained shortly), the diode conducts current again, this time *in the reverse direction.* To explain the mechanism of reverse conduction, one needs to visualize the phenomenon of *avalanche breakdown.* When a very large negative bias is applied to the pn junction, sufficient energy is imparted to charge carriers that reverse current can flow, well beyond the normal reverse saturation current. In addition, because of the large electric field, electrons are energized to such levels that if they collide with other charge carriers at a lower energy level, some

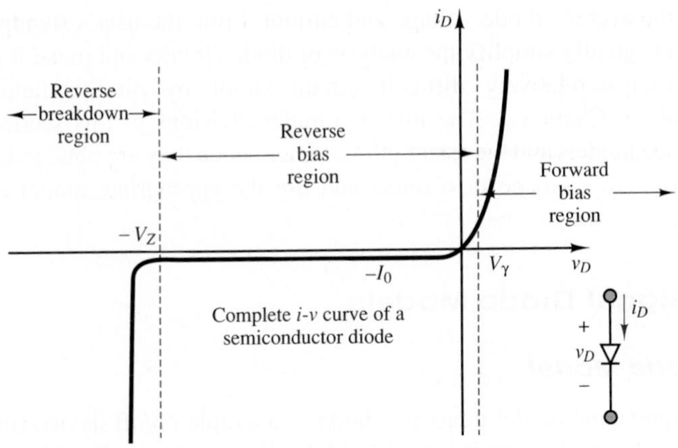

Figure 9.10 The i-v characteristic of the semiconductor diode

LO1

of their energy is transferred to the carriers with lower energy, and these can now contribute to the reverse conduction process as well. This process is called *impact ionization*. Now, these new carriers may also have enough energy to energize other low-energy electrons by impact ionization, so that once a sufficiently high reverse bias is provided, this process of conduction is very much like an avalanche: a single electron can ionize several others.

The phenomenon of **Zener breakdown** is related to avalanche breakdown. It is usually achieved by means of heavily doped regions in the neighborhood of the metal-semiconductor junction (the ohmic contact). The high density of charge carriers provides the means for a substantial reverse breakdown current to be sustained, at a nearly constant reverse bias, the **Zener voltage** V_Z. This phenomenon is very useful in applications where one would like to hold some load voltage constant, for example, in **voltage regulators,** which are discussed in a later section.

To summarize the behavior of the semiconductor diode, it is useful to refer to the sketch of Figure 9.10, observing that when the voltage across the diode v_D is greater than the offset voltage V_γ, the diode is said to be forward-biased and acts nearly as a short circuit, readily conducting current. When v_D is between V_γ and the Zener breakdown voltage $-V_Z$, the diode acts very much as an open circuit, conducting a small reverse current I_0 of the order of only nanoamperes (nA). Finally, if the voltage v_D is more negative than the Zener voltage $-V_Z$, the diode conducts again, this time in the reverse direction.

9.3 CIRCUIT MODELS FOR THE SEMICONDUCTOR DIODE

From the viewpoint of a *user* of electronic circuits (as opposed to a *designer*), it is often sufficient to characterize a device in terms of its i-v characteristic, using either load-line analysis or appropriate circuit models to determine the operating currents and voltages. This section shows how it is possible to use the i-v characteristics of the semiconductor diode to construct simple yet useful *circuit models*. Depending on the desired level of detail, it is possible to construct *large-signal models* of the diode, which describe the gross behavior of the device in the presence of relatively large voltages and currents; or *small-signal models*, which are capable of describing the behavior of the diode in finer detail and, in particular, the response of the diode to small

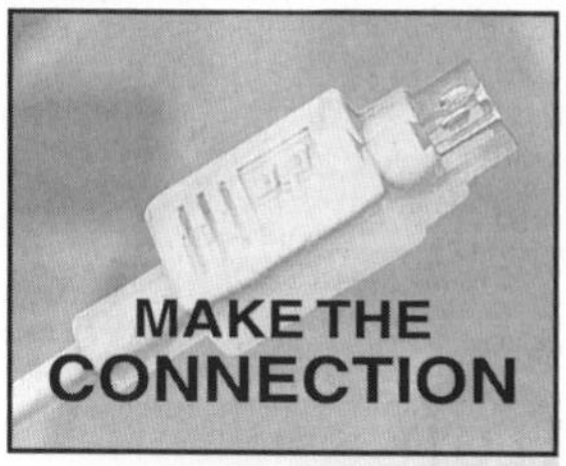

Hydraulic Check Valves

FIND IT ON THE WEB

To understand the operation of the semiconductor diode intuitively, we make reference to a very common hydraulic device that finds application whenever one wishes to restrict the flow of a fluid to a single direction and to prevent (check) reverse flow. Hydraulic **check valves** perform this task in a number of ways. We illustrate a few examples in this box.

Figure 1 depicts a *swing check valve*. In this design, flow from left to right is permitted, as the greater fluid pressure on the left side of the valve forces the swing "door" to open. If flow were to reverse, the reversal of fluid pressure (greater pressure on the right) would cause the swing door to shut.

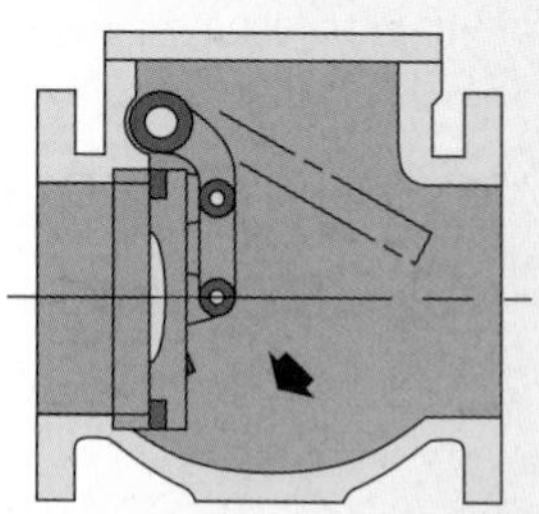

Figure 1

Figure 2 depicts a *flapper check valve*. The principle is similar to that described above for the

(*Continued*)

changes in the average diode voltage and current. From the user's standpoint, these circuit models greatly simplify the analysis of diode circuits and make it possible to effectively analyze relatively "difficult" circuits simply by using the familiar circuit analysis tools of Chapter 3. The first two major divisions of this section describe different diode models and the assumptions under which they are obtained, to provide the knowledge you will need to select and use the appropriate model for a given application.

Large-Signal Diode Models

Ideal Diode Model

Our first large-signal model treats the diode as a simple on/off device (much like a check valve in hydraulic circuits; see the "Make The Connection" sidebar "Hydraulic Check Valves").

Figure 9.11 illustrates how, on a large scale, the i-v characteristic of a typical diode may be approximated by an open circuit when $v_D < 0$ and by a short circuit when $v_D \geq 0$ (recall the i-v curves of the ideal short and open circuits presented in Chapter 2). The analysis of a circuit containing a diode may be greatly simplified by using the short-circuit–open-circuit model. From here on, this diode model will be known as the **ideal diode model.** In spite of its simplicity, the ideal diode model (indicated by the symbol shown in Figure 9.11) can be very useful in analyzing diode circuits.

> In the remainder of the chapter, ideal diodes will always be represented by the filled (black) triangle symbol shown in Figure 9.11.

Consider the circuit shown in Figure 9.12, which contains a 1.5-V battery, an ideal diode, and a 1-kΩ resistor. A technique will now be developed to determine whether the diode is conducting or not, with the aid of the ideal diode model.

Assume first that the diode is conducting (or, equivalently, that $v_D \geq 0$). This enables us to substitute a short circuit in place of the diode, as shown in Figure 9.13, since the diode is now represented by a short circuit, $v_D = 0$. This is consistent with the initial assumption (i.e., diode "on"), since the diode is assumed to conduct for

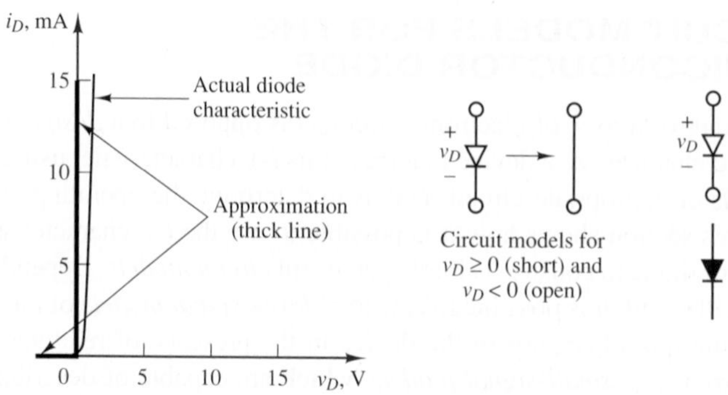

LO2

Figure 9.11 Large-signal on/off diode model

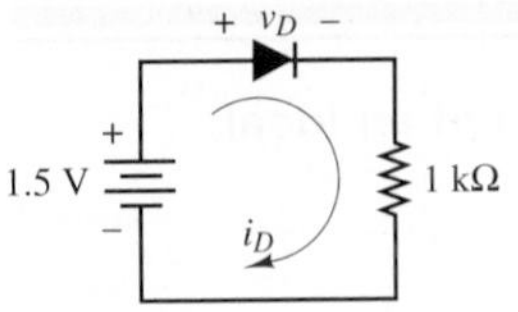

Figure 9.12 Circuit containing ideal diode

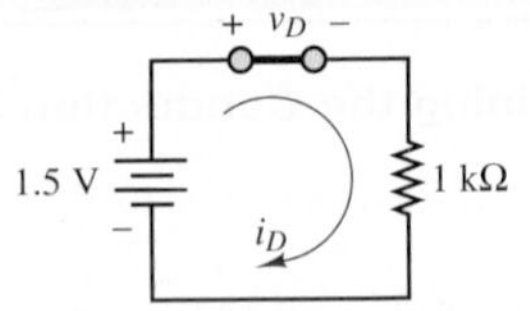

Figure 9.13 Circuit of Figure 9.12, assuming that the ideal diode conducts

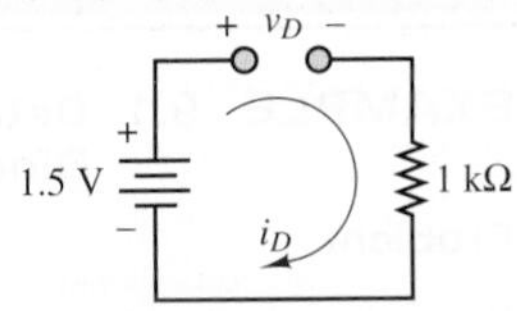

Figure 9.14 Circuit of Figure 9.12, assuming that the ideal diode does not conduct

$v_D \geq 0$ and since $v_D = 0$ does not contradict the assumption. The series current in the circuit (and through the diode) is $i_D = 1.5/1{,}000 = 1.5$ mA. To summarize, the assumption that the diode is on in the circuit of Figure 9.13 allows us to assume a positive (clockwise) current in the circuit. Since the direction of the current and the diode voltage are consistent with the assumption that the diode is on ($v_D \geq 0$, $i_D > 0$), it must be concluded that the diode is indeed conducting.

Suppose, now, that the diode had been assumed to be off. In this case, the diode would be represented by an open circuit, as shown in Figure 9.14. Applying KVL to the circuit of Figure 9.14 reveals that the voltage v_D must equal the battery voltage, or $v_D = 1.5$ V, since the diode is assumed to be an open circuit and no current flows through the circuit. Equation 9.8 must then apply.

$$1.5 = v_D + 1{,}000 i_D = v_D \tag{9.8}$$

But the result $v_D = 1.5$ V is contrary to the initial assumption (that is, $v_D < 0$). Thus, assuming that the diode is off leads to an inconsistent answer. Clearly, the assumption must be incorrect, and therefore the diode must be conducting.

This method can be very useful in more involved circuits, where it is not quite so obvious whether a diode is seeing a positive or a negative bias. The method is particularly effective in these cases, since one can make an educated guess whether the diode is on or off and can solve the resulting circuit to verify the correctness of the initial assumption. Some solved examples are perhaps the best way to illustrate the concept.

FOCUS ON METHODOLOGY

DETERMINING THE CONDUCTION STATE OF AN IDEAL DIODE

1. Assume a diode conduction state (on or off).
2. Substitute ideal circuit model into circuit (short circuit if "on," open circuit if "off").
3. Solve for diode current and voltage, using linear circuit analysis techniques.
4. If the solution is consistent with the assumption, then the initial assumption was correct; if not, the diode conduction state is opposite to that initially assumed. For example, if the diode has been assumed to be "off" but the diode voltage computed after replacing the diode with an open circuit is a forward bias, then it must be true that the actual state of the diode is "on."

(Concluded)

swing check valve. In Figure 2, fluid flow is permitted from left to right, and not in the reverse direction. The response of the valve of Figure 2 is faster (due to the shorter travel distance of the flapper) than that of Figure 1.

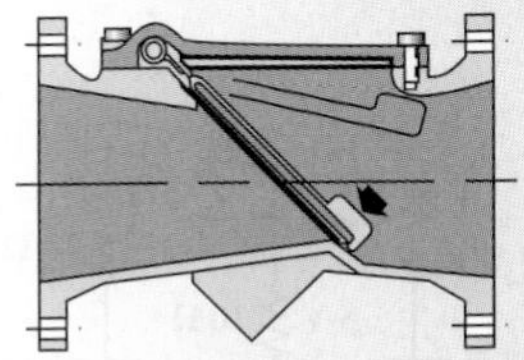

Figure 2

You will find the analysis of the diode circuits in this chapter much easier to understand intuitively if you visualize the behavior of the diode to be similar to that of the check valves shown here, with the pressure difference across the valve orifice being analogous to the voltage across the diode and the fluid flow rate being analogous to the current through the diode. Figure 3 depicts the diode circuit symbol. Current flows only from left to right whenever the voltage across the diode is positive, and no current flows when the diode voltage is reversed. The circuit element of Figure 3 is functionally analogous to the two check valves of Figures 1 and 2.

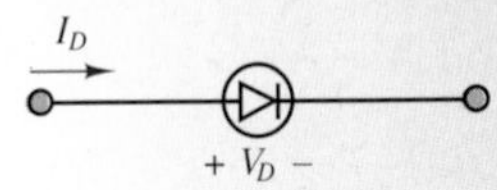

Figure 3

LO2

EXAMPLE 9.1 Determining the Conduction State of an Ideal Diode

Problem

Determine whether the ideal diode of Figure 9.15 is conducting.

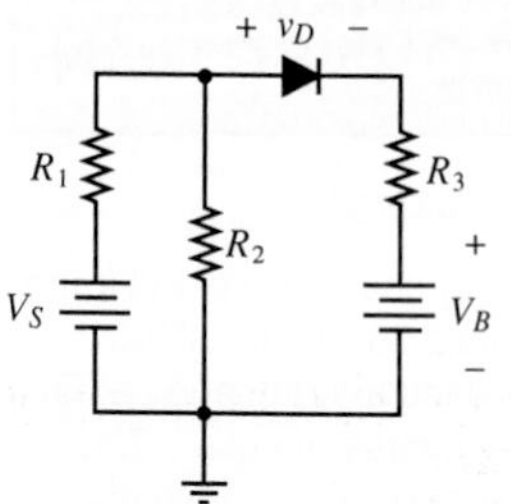

Figure 9.15

Solution

Known Quantities: $V_S = 12$ V; $V_B = 11$ V; $R_1 = 5\ \Omega$; $R_2 = 10\ \Omega$; $R_3 = 10\ \Omega$.

Find: The conduction state of the diode.

Assumptions: Use the ideal diode model.

Analysis: Assume initially that the ideal diode does not conduct, and replace it with an open circuit, as shown in Figure 9.16. The voltage across R_2 can then be computed by using the voltage divider rule:

$$v_1 = \frac{R_2}{R_1 + R_2} V_S = \frac{10}{5 + 10} 12 = 8 \text{ V}$$

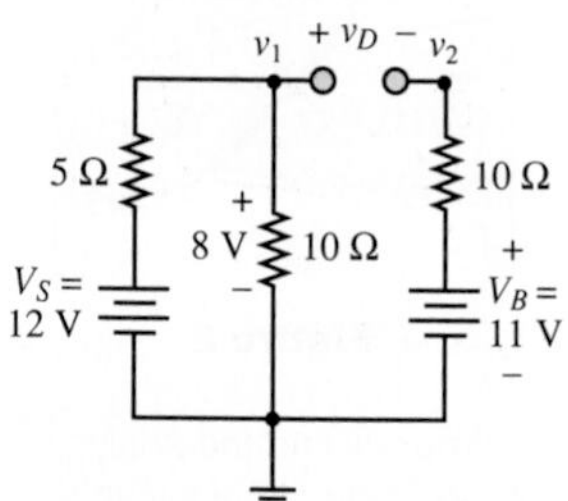

Figure 9.16

Applying KVL to the right-hand-side mesh (and observing that no current flows in the circuit since the diode is assumed off), we obtain

$$v_1 = v_D + V_B \qquad \text{or} \qquad v_D = 8 - 11 = -3 \text{ V}$$

The result indicates that the diode is reverse-biased, and confirms the initial assumption. Thus, the diode is not conducting.

As further illustration, let us make the opposite assumption and assume that the diode conducts. In this case, we should replace the diode with a short circuit, as shown in Figure 9.17. The resulting circuit is solved by node analysis, noting that $v_1 = v_2$ since the diode is assumed to act as a short circuit.

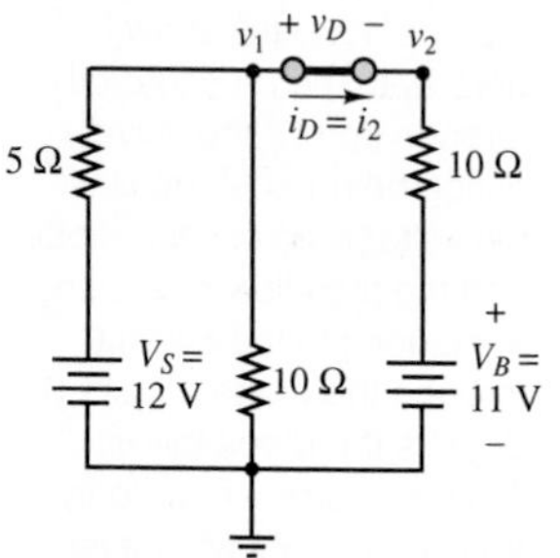

Figure 9.17

$$\frac{V_S - v_1}{R_1} = \frac{v_1}{R_2} + \frac{v_1 - V_B}{R_3}$$

$$\frac{V_S}{R_1} + \frac{V_B}{R_3} = \frac{v_1}{R_1} + \frac{v_1}{R_2} + \frac{v_1}{R_3}$$

$$\frac{12}{5} + \frac{11}{10} = \left(\frac{1}{5} + \frac{1}{10} + \frac{1}{10}\right) v_1$$

$$v_1 = 2.5(2.4 + 1.1) = 8.75 \text{ V}$$

Since $v_1 = v_2 < V_B = 11$ V, we must conclude that current is flowing in the reverse direction (from V_B to node v_2/v_1) through the diode. This observation is inconsistent with the initial assumption, since if the diode were conducting, we could see current flow only in the forward direction. Thus, the initial assumption was incorrect, and we must conclude that the diode is not conducting.

Comments: The formulation of diode problems illustrated in this example is based on making an initial assumption. The assumption results in replacing the ideal diode with either a short or an open circuit. Once this step is completed, the resulting circuit is a linear circuit and can be solved by known methods to verify the consistency of the initial assumption.

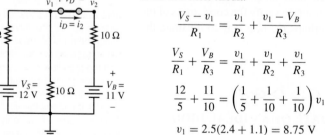

CHECK YOUR UNDERSTANDING

If the resistor R_2 is removed from the circuit of Figure 9.15, will the diode conduct?

Answer: Yes

EXAMPLE 9.2 Determining the Conduction State of an Ideal Diode

LO2

Problem

Determine whether the ideal diode of Figure 9.18 is conducting.

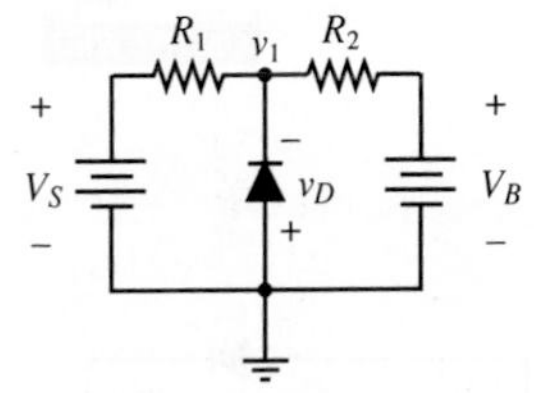

Figure 9.18

Solution

Known Quantities: $V_S = 12$ V; $V_B = 11$ V; $R_1 = 5\ \Omega$; $R_2 = 4\ \Omega$.

Find: The conduction state of the diode.

Assumptions: Use the ideal diode model.

Analysis: Assume initially that the ideal diode does not conduct, and replace it with an open circuit, as shown in Figure 9.19. The current flowing in the resulting series circuit (shown in Figure 9.19) is

$$i = \frac{V_S - V_B}{R_1 + R_2} = \frac{1}{9} \text{ A}$$

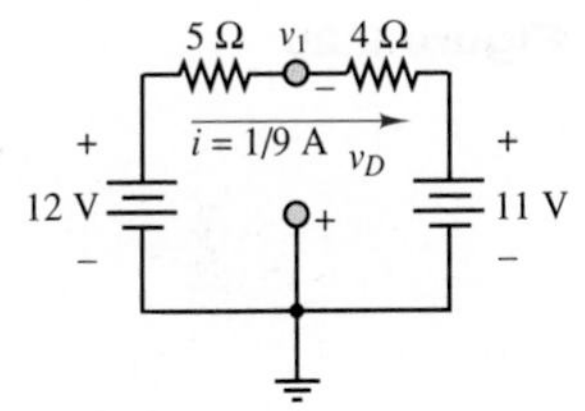

Figure 9.19

The voltage at node v_1 is

$$\frac{12 - v_1}{5} = \frac{v_1 - 11}{4}$$

$$v_1 = 11.44 \text{ V}$$

The result indicates that the diode is strongly reverse-biased, since $v_D = 0 - v_1 = -11.44$ V, and confirms the initial assumption. Thus, the diode is not conducting.

CHECK YOUR UNDERSTANDING

Repeat the analysis of Example 9.2, assuming that the diode is conducting, and show that this assumption leads to inconsistent results.

Determine which of the diodes conduct in the circuit shown in the figure, for each of the following voltages. Treat the diodes as ideal.

a. $v_1 = 0$ V; $v_2 = 0$ V
b. $v_1 = 5$ V; $v_2 = 5$ V
c. $v_1 = 0$ V; $v_2 = 5$ V
d. $v_1 = 5$ V; $v_2 = 0$ V

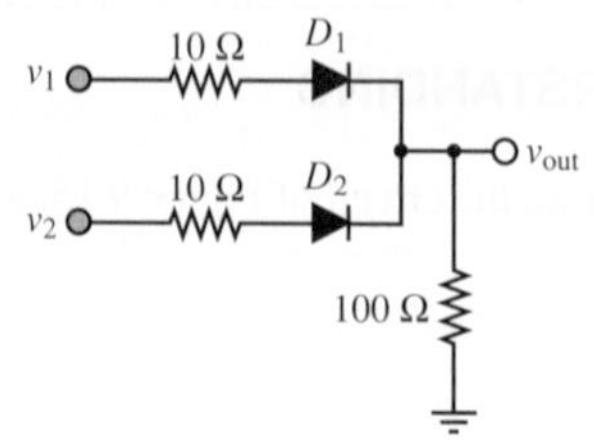

Answers: (a) Neither; (b) both; (c) D_2 only; (d) D_1 only

FIND IT ON THE WEB

One of the important applications of the semiconductor diode is in **rectification** of AC signals, that is, the ability to convert an AC signal with zero average (DC) value to a signal with a nonzero DC value. The application of the semiconductor diode as a rectifier is very useful in obtaining DC voltage supplies from the readily available AC line voltage. Here, we illustrate the basic principle of rectification, using an ideal diode—for simplicity, and because the large-signal model is appropriate when the diode is used in applications involving large AC voltage and current levels.

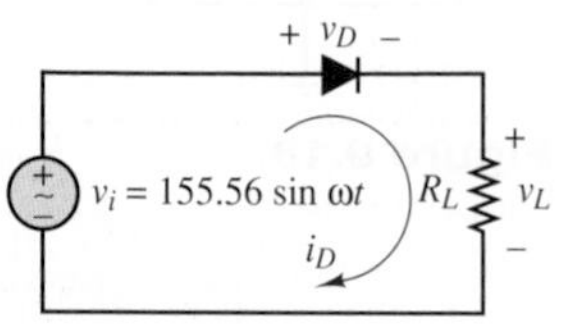

Figure 9.20

Consider the circuit of Figure 9.20, where an AC source $v_i = 155.56 \sin \omega t$ is connected to a load by means of a series ideal diode. From the analysis of Example 9.1, it should be apparent that the diode will conduct only during the positive half-cycle of the sinusoidal voltage—that is, that the condition $v_D \geq 0$ will be satisfied only when the AC source voltage is positive—and that it will act as an open circuit during the negative half-cycle of the sinusoid ($v_D < 0$). Thus, the appearance of the load voltage will be as shown in Figure 9.21, with the negative portion of the sinusoidal waveform cut off. The rectified waveform clearly has a nonzero DC (average) voltage, whereas the average input waveform voltage was zero. When the diode is conducting, or $v_D \geq 0$, the unknowns v_L and i_D can be found by using the following equations:

$$i_D = \frac{v_i}{R_L} \quad \text{when} \quad v_i > 0 \tag{9.9}$$

and

$$v_L = i_D R_L \tag{9.10}$$

The load voltage v_L and the input voltage v_i are sketched in Figure 9.21. From equation 9.10, it is obvious that the current waveform has the same shape as the load voltage. The average value of the load voltage is obtained by integrating the load voltage over one period and dividing by the period:

$$v_{\text{load, DC}} = \frac{\omega}{2\pi} \int_0^{\pi/\omega} 155.56 \sin \omega t \, dt = \frac{155.56}{\pi} = 49.52 \text{ V} \tag{9.11}$$

The circuit of Figure 9.20 is called a **half-wave rectifier,** since it preserves only one-half of the waveform. This is not usually a very efficient way of rectifying an

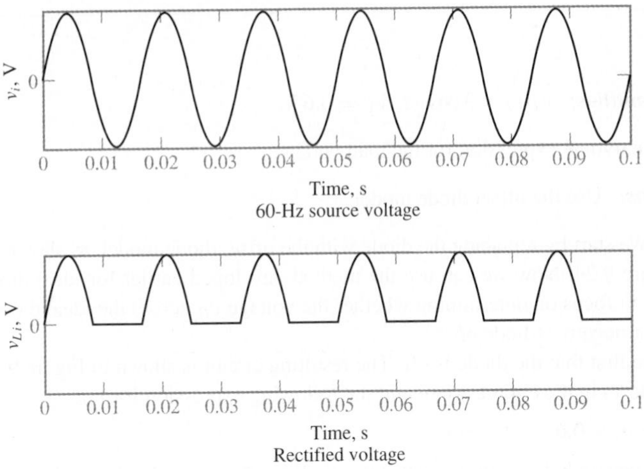

Figure 9.21 Ideal diode rectifier input and output voltages

AC signal, since one-half of the energy in the AC signal is not recovered. It will be shown in a later section that it is possible to recover also the negative half of the AC waveform by means of a *full-wave rectifier.*

Offset Diode Model

While the ideal diode model is useful in approximating the large-scale characteristics of a physical diode, it does not account for the presence of an offset voltage, which is an unavoidable component in semiconductor diodes (recall the discussion of the contact potential in Section 9.2). The **offset diode model** consists of an ideal diode in series with a battery of strength equal to the offset voltage (we shall use the value $V_\gamma = 0.6$ V for silicon diodes, unless otherwise indicated). The effect of the battery is to shift the characteristic of the ideal diode to the right on the voltage axis, as shown in Figure 9.22. This model is a better approximation of the large-signal behavior of a semiconductor diode than the ideal diode model.

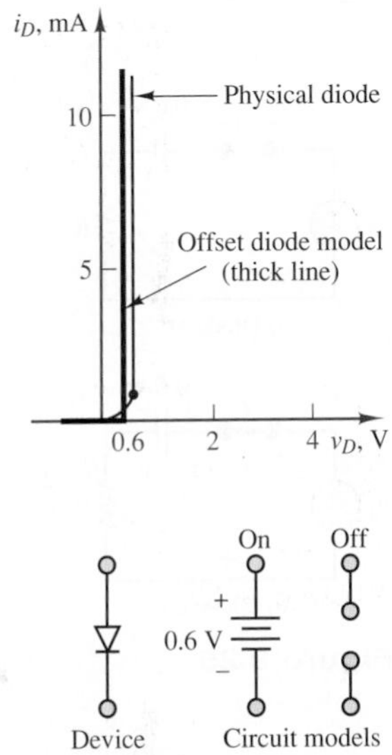

Figure 9.22

According to the offset diode model, the diode of Figure 9.22 acts as an open circuit for $v_D < 0.6$ V, and it behaves as a 0.6-V battery for $v_D \geq 0.6$ V. The equations describing the offset diode model are as follows:

$$\begin{aligned} &v_D \geq 0.6\text{ V} \qquad \text{Diode} \rightarrow \text{0.6-V battery} \\ &v_D < 0.6\text{ V} \qquad \text{Diode} \rightarrow \text{open circuit} \end{aligned} \qquad \text{offset diode model} \qquad \textbf{(9.12)}$$

The offset diode model may be represented by an ideal diode in series with a 0.6-V ideal battery, as shown in Figure 9.23. Use of the offset diode model is best described by means of examples.

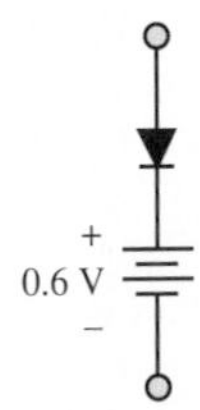

Figure 9.23 Offset diode as an extension of ideal diode model

---

EXAMPLE 9.3 Using the Offset Diode Model in a Half-Wave Rectifier

LO2

Problem

Compute and plot the rectified load voltage v_R in the circuit of Figure 9.24.

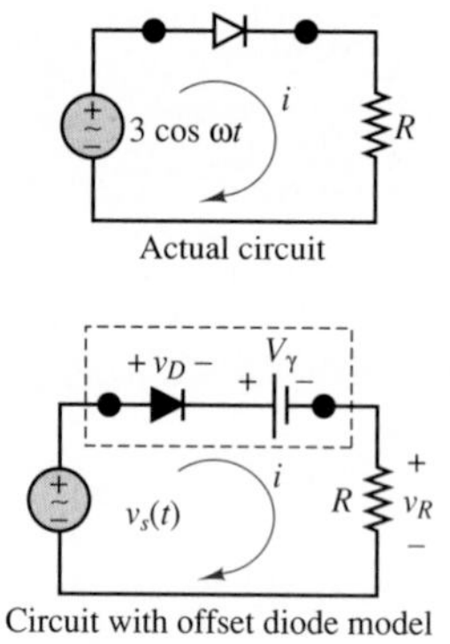

Figure 9.24

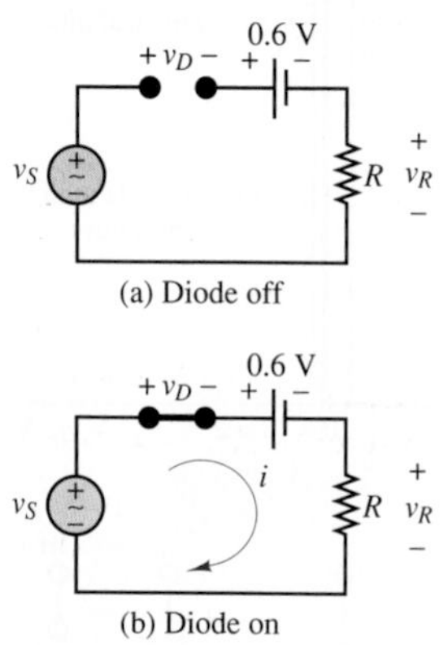

Figure 9.25

Solution

Known Quantities: $v_S(t) = 3\cos\omega t$; $V_\gamma = 0.6$ V.

Find: An analytical expression for the load voltage.

Assumptions: Use the offset diode model.

Analysis: We start by replacing the diode with the offset diode model, as shown in the lower half of Figure 9.24. Now we can use the method developed earlier for ideal diode analysis; that is, we can focus on determining whether the voltage v_D across the ideal diode is positive (diode on) or negative (diode off).

Assume first that the diode is off. The resulting circuit is shown in Figure 9.25(a). Since no current flows in the circuit, we obtain the following expression for v_D:

$$v_D = v_S - 0.6$$

To be consistent with the assumption that the diode is off, we require that v_D be negative, which in turn corresponds to

$$v_S < 0.6 \text{ V} \qquad \text{Diode off condition}$$

With the diode off, the current in the circuit is zero, and the load voltage is also zero. If the source voltage is greater than 0.6 V, the diode conducts, and the current flowing in the circuit and resulting load voltage are given by the expressions

$$i = \frac{v_S - 0.6}{R} \qquad v_R = iR = v_S - 0.6$$

We summarize these results as follows:

$$v_R = \begin{cases} 0 & \text{for } v_S < 0.6 \text{ V} \\ v_S - 0.6 & \text{for } v_S \geq 0.6 \text{ V} \end{cases}$$

The resulting waveform is plotted with v_S in Figure 9.26.

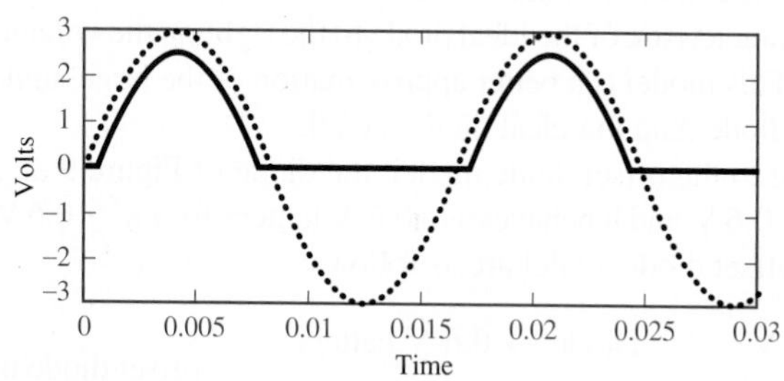

Figure 9.26 Source voltage (dotted curve) and rectified voltage (solid curve) for the circuit of Figure 9.24.

Comments: Note that use of the offset diode model leads to problems that are very similar to ideal diode problems, with the addition of a voltage source in the circuit.

Also observe that the load voltage waveform is shifted downward by an amount equal to the offset voltage V_γ. The shift is visible in the case of this example because V_γ is a substantial fraction of the source voltage. If the source voltage had peak values of tens or hundreds of volts, such a shift would be negligible, and an ideal diode model would serve just as well.

CHECK YOUR UNDERSTANDING

Compute the DC value of the rectified waveform for the circuit of Figure 9.20 for $v_i = 52 \cos \omega t$ V.

Answer: 16.55 V

EXAMPLE 9.4 Using the Offset Diode Model

LO2

Problem

Use the offset diode model to determine the value of v_1 for which diode D_1 first conducts in the circuit of Figure 9.27.

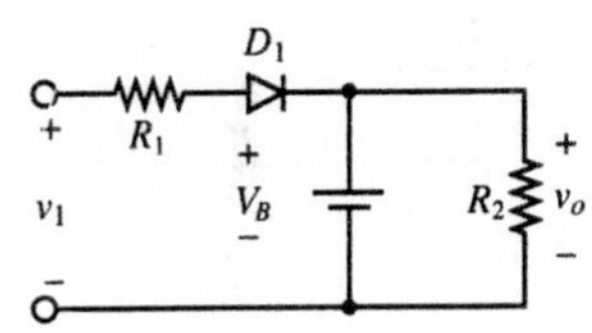

Figure 9.27

Solution

Known Quantities: $V_B = 2$ V; $R_1 = 1$ kΩ; $R_2 = 500$ Ω; $V_\gamma = 0.6$ V.

Find: The lowest value of v_1 for which diode D_1 conducts.

Assumptions: Use the offset diode model.

Analysis: We start by replacing the diode with the offset diode model, as shown in Figure 9.28. Based on our experience with previous examples, we can state immediately that if v_1 is negative, the diode will certainly be off. To determine the point at which the diode turns on as v_1 is increased, we write the circuit equation, assuming that the diode is off. If you were conducting a laboratory experiment, you might monitor v_1 and progressively increase it until the diode conducts; the equation below is an analytical version of this experiment. With the diode off, no current flows through R_1, and

$$v_1 = v_{D1} + V_\gamma + V_B$$

According to this equation,

$$v_{D1} = v_1 - 2.6$$

and the condition required for the diode to conduct is

$$v_1 > 2.6 \text{ V} \qquad \text{Diode "on" condition}$$

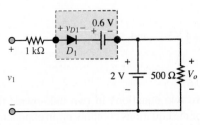

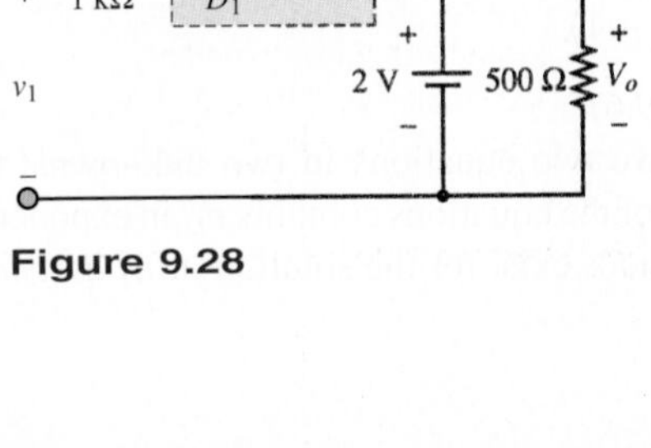

Figure 9.28

Comments: Once again, the offset diode model permits use of the same analysis method that was developed for the ideal diode model.

CHECK YOUR UNDERSTANDING

Determine which of the diodes conduct in the circuit shown below. Each diode has an offset voltage of 0.6 V.

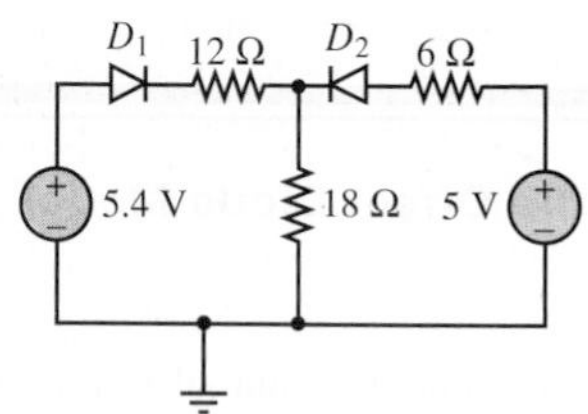

Answer: Both diodes conduct.

Small-Signal Diode Models

As one examines the diode i-v characteristic more closely, it becomes apparent that the short-circuit approximation is not adequate to represent the *small-signal behavior* of the diode. The term *small-signal behavior* usually signifies the response of the diode to small time-varying signals that may be superimposed on the average diode current and voltage. Figure 9.8 depicts a close-up view of a silicon diode i-v curve. From this figure, it should be apparent that the short-circuit approximation is not very accurate when a diode's behavior is viewed on an expanded scale. To a first-order approximation, however, the i-v characteristic resembles that of a resistor (i.e., is linear) for voltages greater than the offset voltage. Thus, it may be reasonable to model the diode as a resistor (instead of a short circuit) *once it is conducting*, to account for the slope of its i-v curve. In the following discussion, the method of load-line analysis (which was introduced in Chapter 3) is exploited to determine the **small-signal resistance** of a diode.

Consider the circuit of Figure 9.29, which represents the Thévenin equivalent circuit of an arbitrary linear resistive circuit connected to a diode. Equations 9.13 and 9.14 describe the operation of the circuit:

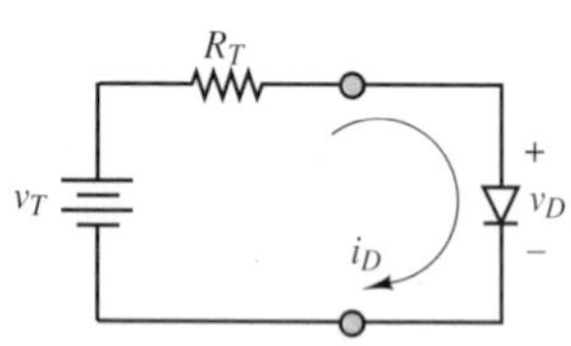

Figure 9.29 Diode circuit for illustration of load-line analysis

$$v_T = i_D R_T + v_D \qquad \textbf{(9.13)}$$

arises from application of KVL, and

$$i_D = I_0(e^{qv_D/kT} - 1) \qquad \textbf{(9.14)}$$

is the diode equation (9.6).

Although we have two equations in two unknowns, these cannot be solved analytically, since one of the equations contains v_D in exponential form. As discussed in Chapter 3, two methods exist for the solution of *transcendental equations* of this type: graphical and numerical. In the present case, only the graphical solution shall be considered. The graphical solution is best understood if we associate a curve in the i_D–v_D plane with each of the two preceding equations. The diode equation gives rise to the familiar curve of Figure 9.8. The *load-line equation*, obtained by KVL, is the equation of a line with slope $-1/R$ and ordinate intercept given by V_T/R_T.

$$i_D = -\frac{1}{R_T}v_D + \frac{1}{R_T}V_T \qquad \text{Load-line equation} \tag{9.15}$$

The superposition of these two curves gives rise to the plot of Figure 9.30, where the solution to the two equations is graphically found to be the pair of values (I_Q, V_Q). The intersection of the two curves is called the **quiescent (operating) point,** or ***Q* point.** The voltage $v_D = V_Q$ and the current $i_D = I_Q$ are the actual diode voltage and current when the diode is connected as in the circuit of Figure 9.29. Note that this method is also useful for circuits containing a larger number of elements, provided that we can represent these circuits by their Thévenin equivalents, with the diode appearing as the load.

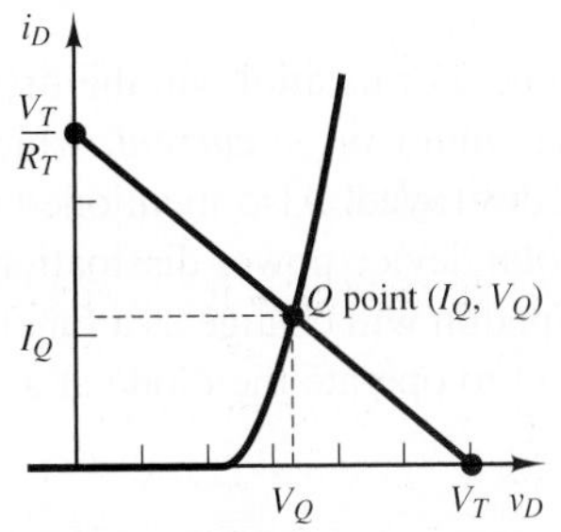

Figure 9.30 Graphical solution of equations 9.13 and 9.14

FOCUS ON METHODOLOGY

DETERMINING THE OPERATING POINT OF A DIODE

LO2

1. Reduce the circuit to a Thévenin or Norton equivalent circuit with the diode as the load.
2. Write the load-line equation (9.15).
3. Solve numerically two simultaneous equations in two unknowns (the load-line equations and the diode equation) for the diode current and voltage.

   or

4. Solve graphically by finding the intersection of the diode curve (e.g., from a data sheet) with the load-line curve. The intersection of the two curves is the diode operating point.

FOCUS ON METHODOLOGY

USING DEVICE DATA SHEETS

One of the most important design tools available to engineers is the **device data sheet.** In this box we illustrate the use of a device data sheet for the 1N400X diode. This is a *general-purpose rectifier* diode, designed to conduct average currents in the 1.0-A range. Excerpts from the data sheet are shown below, with some words of explanation.

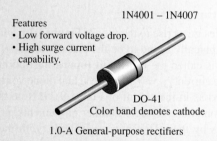

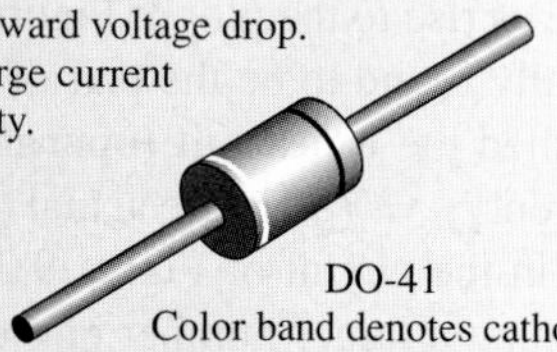
1N4001 – 1N4007
Features
• Low forward voltage drop.
• High surge current capability.
DO-41
Color band denotes cathode

1.0-A General-purpose rectifiers

ABSOLUTE MAXIMUM RATINGS

The table below summarizes the limitations of the device. For example, in the first column one can find the maximum allowable average current (1 A) and the maximum *surge current*, which is the maximum short-time burst current the diode can sustain without being destroyed. Also mentioned are the **power rating** and operating temperatures. Note that in the entry for the total device power dissipation, **derating** information is also given. Derating implies that the device power dissipation will change as a function of temperature, in this case at the rate of 20 mW/°C. For example, if we expect to operate the diode at a temperature of 100°C, we calculate a derated power of

$$P = 2.5\text{ W} - (75°\text{C} \times 0.02\text{ mW/°C}) = 1.0\text{ W}$$

Thus, the diode operated at a higher temperature can dissipate only 1 W.

Absolute Maximum Ratings* $T = 25°$C unless otherwise noted

Symbol	Parameter	Value	Units
I_0	Average rectified current 0.375-in lead length @ $T_A = 75°$C	1.0	A
$i_{f(\text{surge})}$	Peak forward surge current 8.3-ms single half-sine-wave Superimposed on rated load (JEDEC method)	30	A
P_D	Total device dissipation Derate above 25°C	2.5 20	W mW/°C
$R_{\theta JA}$	Thermal resistance, junction to ambient	50	°C/W
T_{stg}	Storage temperature range	−55 to +175	°C
T_J	Operating junction temperature	−55 to +150	°C

*These ratings are limiting values above which the serviceability of any semiconductor device may be impaired.

(*Continued*)

(*Concluded*)

ELECTRICAL CHARACTERISTICS

The section on electrical characteristics summarizes some of the important voltage and current specifications of the diode. For example, the maximum DC reverse voltage is listed for each diode in the 1N400X family. Similarly, you will find information on the maximum forward voltage, reverse current, and typical junction capacitance.

Electrical Characteristics $T = 25°C$ unless otherwise noted

	Device							
Parameter	**4001**	**4002**	**4003**	**4004**	**4005**	**4006**	**4007**	**Units**
Peak repetitive reverse voltage	50	100	200	400	600	800	1,000	V
Maximum rms voltage	35	70	140	280	420	560	700	V
DC reverse voltage (rated V_R)	50	100	200	400	600	800	1,000	V
Maximum reverse current @ rated V_R $T_A = 25°C$				5.0				μA
$T_A = 100°C$				500				μA
Maximum forward voltage @ 1.0 A				1.1				V
Maximum full-load reverse current, full cycle $T_A = 75°C$				30				μA
Typical junction capacitance $V_R = 4.0$ V, $f = 1.0$ MHz				15				pF

TYPICAL CHARACTERISTIC CURVES

Device data sheets always include characteristic curves that may be useful to a designer. In this example, we include the forward-current derating curve, in which the maximum forward current is derated as a function of temperature. To illustrate this curve, we point out that at a temperature of 100°C the maximum diode current is around 0.65 A (down from 1 A). A second curve is related to the diode forward current versus forward voltage (note that this curve was obtained for a very particular type of input, consisting of a pulse of width equal to 300 μs and 2 percent duty cycle).

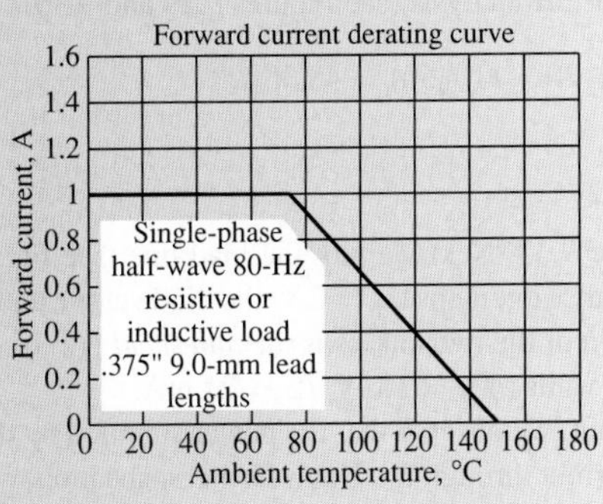

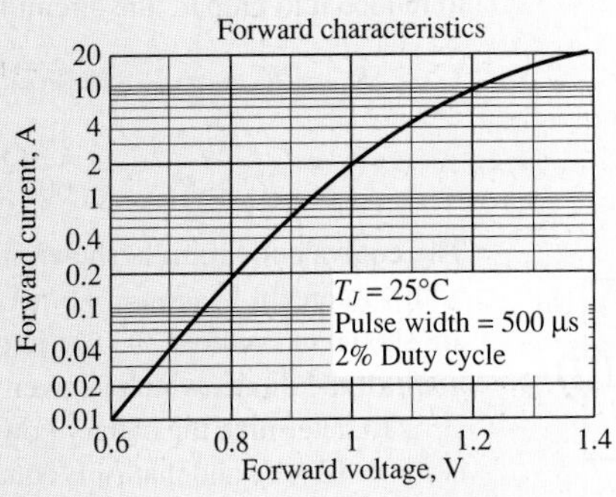

LO2

EXAMPLE 9.5 Using Load-Line Analysis and Diode Curves to Determine the Operating Point of a Diode

Problem

Determine the operating point of the 1N914 diode in the circuit of Figure 9.31, and compute the total power output of the 12-V battery.

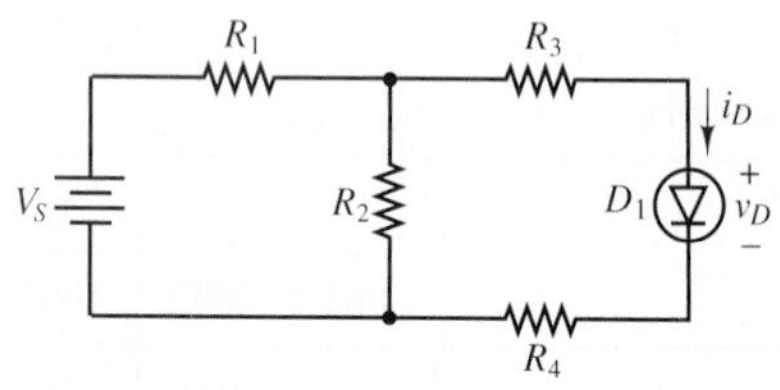

Figure 9.31

Solution

Known Quantities: $V_S = 12$ V; $R_1 = 50\ \Omega$; $R_2 = 10\ \Omega$; $R_3 = 20\ \Omega$; $R_4 = 20\ \Omega$.

Find: The diode operating voltage and current and the power supplied by the battery.

Assumptions: Use the diode nonlinear model, as described by its i-v curve (Figure 9.32).

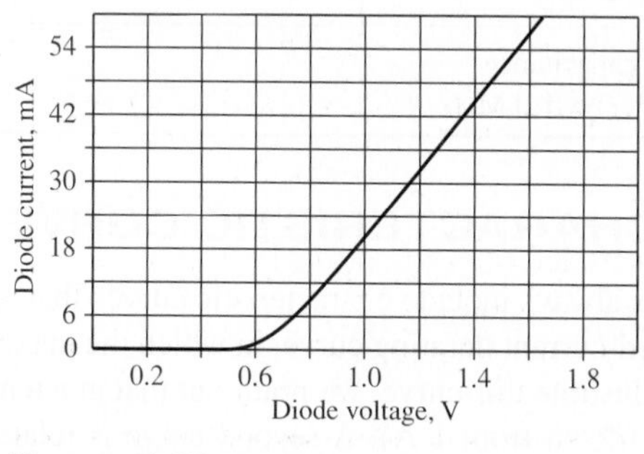

Figure 9.32 The 1N914 diode i-v curve

Analysis: We first compute the Thévenin equivalent representation of the circuit of Figure 9.31 to reduce it to prepare the circuit for load-line analysis (see Figures 9.29 and 9.30).

$$R_T = R_1 + R_2 + (R_3 \| R_4) = 20 + 20 + (10 \| 50) = 48.33\ \Omega$$

$$V_T = \frac{R_2}{R_1 + R_2} V_S = \frac{10}{60} 12 = 2\ \text{V}$$

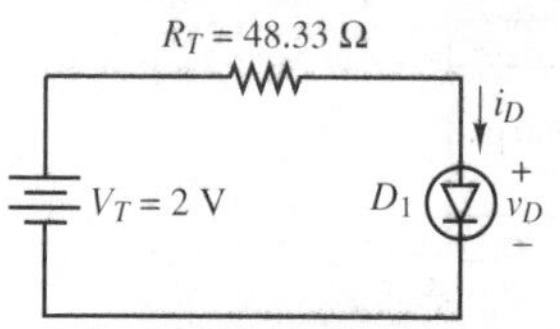

Figure 9.33

The equivalent circuit is shown in Figure 9.33. Next we plot the load line (see Figure 9.30), with y intercept $V_T/R_T = 41$ mA and with x intercept $V_T = 2$ V; the diode curve and load line are shown in Figure 9.34. The intersection of the two curves is the *quiescent* (Q) or *operating point* of the diode, which is given by the values $V_Q = 1.0$ V, $I_Q = 21$ mA.

To determine the battery power output, we observe that the power supplied by the battery is $P_B = 12 \times I_B$ and that I_B is equal to current through R_1. Upon further inspection, we see that the battery current must, by KCL, be equal to the sum of the currents through R_2 and through the diode. We already know the current through the diode I_Q. To determine the current through

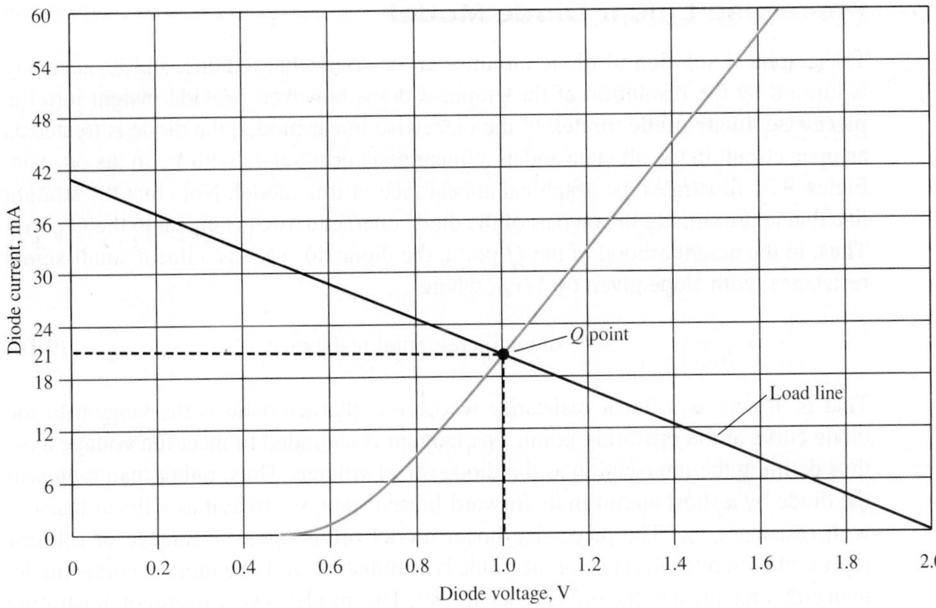

Figure 9.34 Superposition of load line and diode i-v curve

R_2, we observe that the voltage across R_2 is equal to the sum of the voltages across R_3, R_4, and D_1:

$$V_{R2} = I_Q(R_3 + R_4) + V_Q = 0.021 \times 40 + 1 = 1.84 \text{ V}$$

and therefore the current through R_2 is $I_{R2} = V_{R2}/R_2 = 0.184$ A.

Finally,

$$P_B = 12 \times I_B = 12 \times (0.021 + 0.184) = 12 \times 0.205 = 2.46 \text{ W}$$

Comments: Graphical solutions are not the only means of solving the nonlinear equations that result from using a nonlinear model for a diode. The same equations could be solved numerically by using a nonlinear equation solver.

CHECK YOUR UNDERSTANDING

Use load-line analysis to determine the operating point (Q point) of the diode in the circuit shown in figure. The diode has the characteristic curve of Figure 9.32.

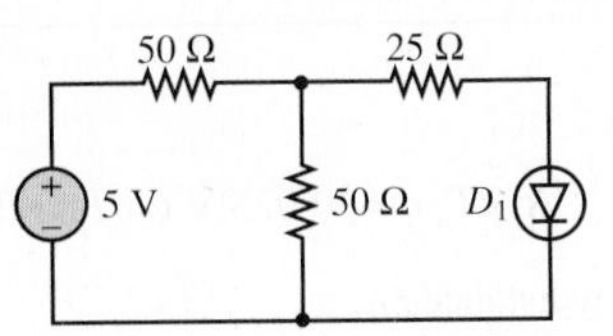

Answer: $V_Q = 1.11$ V, $I_Q = 27.7$ mA

Piecewise Linear Diode Model

The graphical solution of diode circuits can be somewhat tedious, and its accuracy is limited by the resolution of the graph; it does, however, provide insight into the **piecewise linear diode model.** In the piecewise linear model, the diode is treated as an open circuit in the off state and as a linear resistor in series with V_γ in the on state. Figure 9.35 illustrates the graphical appearance of this model. Note that the straight line that approximates the on part of the diode characteristic is tangent to the Q point. Thus, in the neighborhood of the Q point, the diode does act as a linear small-signal resistance, with slope given by $1/r_D$, where

$$\frac{1}{r_D} = \left.\frac{\partial i_D}{\partial v_D}\right|_{(I_Q, V_Q)} \qquad \text{diode incremental resistance} \qquad \textbf{(9.16)}$$

That is, it acts as a linear resistance whose i-v characteristic is the tangent to the diode curve at the operating point. The tangent is extended to meet the voltage axis, thus defining the intersection as the diode offset voltage. Thus, rather than represent the diode by a short circuit in its forward-biased state, we treat it as a linear resistor, with resistance r_D. The piecewise linear model offers the convenience of a linear representation once the state of the diode is established, and of a more accurate model than either the ideal or the offset diode model. This model is very useful in illustrating the performance of diodes in real-world applications.

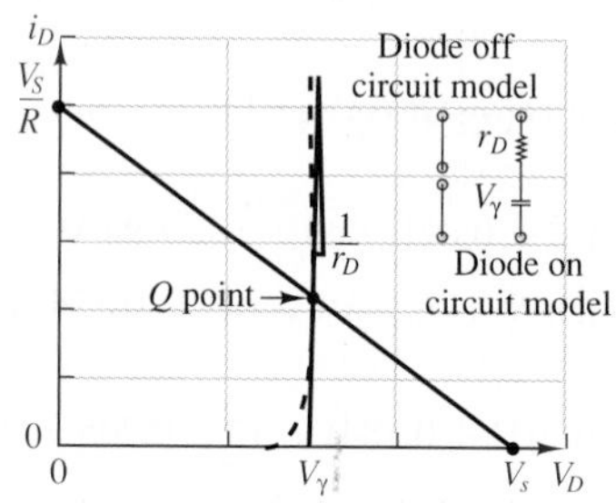

Figure 9.35 Piecewise linear diode model

LO2

EXAMPLE 9.6 Computing the Incremental (Small-Signal) Resistance of a Diode

Problem

Determine the incremental resistance of a diode, using the diode equation.

Solution

Known Quantities: $I_0 = 10^{-14}$ A; $kT/q = 0.025$ V (at $T = 300$ K); $I_Q = 50$ mA.

Find: The diode small-signal resistance r_D.

Assumptions: Use the approximate diode equation (equation 9.7).

Analysis: The approximate diode equation relates diode voltage and current according to

$$i_D = I_0 e^{qv_D/kT}$$

From the preceding expression we can compute the incremental resistance, using equation 9.16:

$$\frac{1}{r_D} = \left.\frac{\partial i_D}{\partial v_D}\right|_{(I_Q, V_Q)} = \frac{qI_0}{kT} e^{qV_Q/kT}$$

To calculate the numerical value of the above expression, we must first compute the quiescent diode voltage corresponding to the quiescent current $I_Q = 50$ mA:

$$V_Q = \frac{kT}{q} \log_e \frac{I_Q}{I_0} = 0.731 \text{ V}$$

Substituting the numerical value of V_Q in the expression for r_D, we obtain

$$\frac{1}{r_D} = \frac{10^{-14}}{0.025} e^{0.731/0.025} = 2 \text{ S} \qquad \text{or} \qquad r_D = 0.5\ \Omega$$

Comments: It is important to understand that while one can calculate the linearized incremental resistance of a diode at an operating point, this does not mean that the diode can be treated simply as a resistor. The linearized small-signal resistance of the diode is used in the piecewise linear diode model to account for the fact that there is a dependence between diode voltage and current (i.e., the diode i-v curve is not exactly a vertical line for voltages above the offset voltage—see Figure 9.35).

CHECK YOUR UNDERSTANDING

Compute the incremental resistance of the diode of Example 9.6 if the current through the diode is 250 mA.

Answer: $r_D = 0.1\ \Omega$

EXAMPLE 9.7 Using the Piecewise Linear Diode Model

LO2

Problem

Determine the load voltage in the rectifier of Figure 9.36, using a piecewise linear approximation.

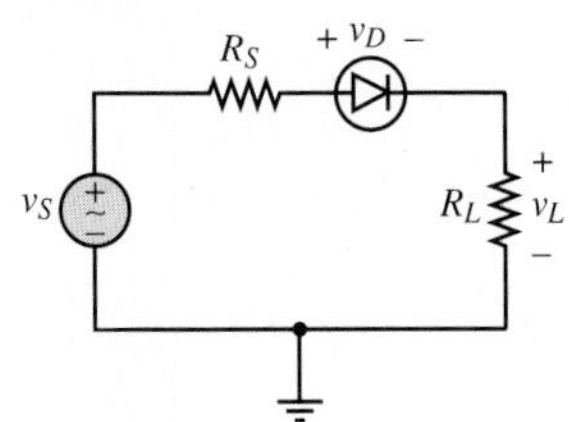

Figure 9.36

Solution

Known Quantities: $v_S(t) = 10 \cos \omega t$; $V_\gamma = 0.6$ V; $r_D = 0.5\ \Omega$; $R_S = 1\ \Omega$; $R_L = 10\ \Omega$.

Find: The load voltage v_L.

Assumptions: Use the piecewise linear diode model (Figure 9.35).

Analysis: We replace the diode in the circuit of Figure 9.36 with the piecewise linear model, as shown in Figure 9.37. Next, we determine the conduction condition for the ideal diode by

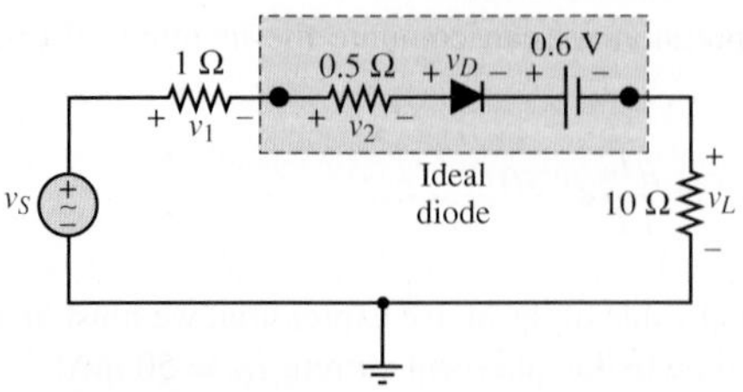

Figure 9.37

applying KVL to the circuit of Figure 9.37:

$$v_S = v_1 + v_2 + v_D + 0.6 + v_L$$

$$v_D = v_S - v_1 - v_2 - 0.6 - v_L$$

We use the above equation as was done in Example 9.4, that is, to determine the source voltage value for which the diode first conducts. Observe first that the diode will be off for negative values of v_S. With the diode off, that is, an open circuit, the voltages v_1, v_2, and v_L are zero and

$$v_D = v_S - 0.6$$

Thus, the condition for the ideal diode to conduct ($v_D > 0$) corresponds to

$$v_S \geq 0.6 \text{ V} \qquad \text{Diode on condition}$$

Once the diode conducts, we replace the ideal diode with a short circuit and compute the load voltage, using the voltage divider rule. The resulting load equations are

$$v_L = \begin{cases} 0 & v_S < 0.6 \text{ V} \\ \dfrac{R_L}{R_S + r_D + R_L}(v_S - V_\gamma) = 8.7 \cos \omega t - 0.52 & v_S \geq 0.6 \text{ V} \end{cases}$$

The source and load voltage are plotted in Figure 9.38(a).

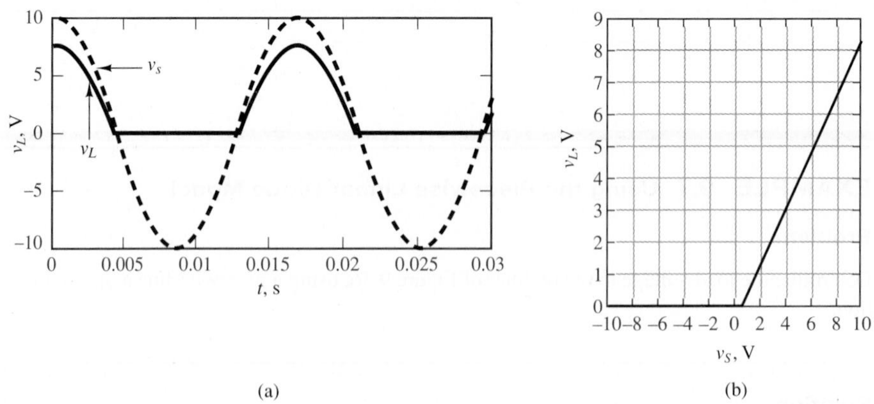

Figure 9.38 (a) Source voltage and rectified load voltage; (b) voltage transfer characteristic

It is instructive to compute the *transfer characteristic* of the diode circuit by generating a plot of v_L versus v_S. This is done with reference to the equation for v_L given above; the result is plotted in Figure 9.38(b).

Comments: The methods developed in this example will be very useful in analyzing some practical diode circuits in the next section.

CHECK YOUR UNDERSTANDING

Consider a half-wave rectifier similar to that of Figure 9.20, with $v_i = 18 \cos t$ V, and a 4-Ω load resistor. Sketch the output waveform if the piecewise linear diode model is used to represent the diode, with $V_\gamma = 0.6$ V and $r_D = 1$ Ω. What is the peak value of the rectifier output waveform?

Answer: $V_{L,\text{peak}} = 13.92$ V

9.4 RECTIFIER CIRCUITS

This section illustrates some of the applications of diodes to practical engineering circuits. The nonlinear behavior of diodes, especially the rectification property, makes these devices valuable in a number of applications. In this section, more advanced rectifier circuits (the **full-wave rectifier** and the **bridge rectifier**) will be explored, as well as **limiter** and *peak detector* circuits. These circuits will be analyzed by making use of the circuit models developed in the preceding sections; as stated earlier, these models are more than adequate to develop an understanding of the operation of diode circuits.

FIND IT ON THE WEB

The Full-Wave Rectifier

The half-wave rectifier discussed earlier is one simple method of converting AC energy to DC energy. The need for converting one form of electric energy to the other arises frequently in practice. The most readily available form of electric power is AC (the standard 110- or 220-V rms AC line power), but one frequently needs a DC power supply, for applications ranging from the control of certain types of electric motors to the operation of electronic circuits such as those discussed in Chapters 8 through 14. You will have noticed that most consumer electronic circuits, from CD players to personal computers, require AC-DC power adapters.

The half-wave rectifier, however, is not a very efficient AC-DC conversion circuit, because it fails to utilize one-half of the energy available in the AC waveform, by not conducting current during the negative half-cycle of the AC waveform. The full-wave rectifier shown in Figure 9.39 offers a substantial improvement in efficiency over the half-wave rectifier. The first section of the full-wave rectifier circuit includes an AC source and a center-tapped transformer (see Chapter 7) with 1:2N turns ratio. The purpose of the transformer is to obtain the desired voltage amplitude prior to rectification. Thus, if the peak amplitude of the AC source voltage is v_S, the amplitude of the voltage across each half of the output side of the transformer will be Nv_S; this scheme permits scaling the source voltage up or down (depending on whether N is greater or less than 1), according to the specific requirements of the application. In addition to scaling the source voltage, the transformer isolates the rectifier circuit

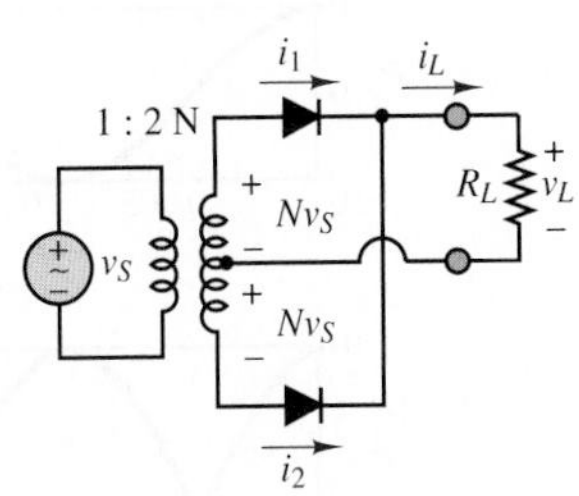

Figure 9.39 Full-wave rectifier

from the AC source voltage, since there is no direct electrical connection between the input and output of a transformer (see Chapter 16).

In the analysis of the full-wave rectifier, the diodes will be treated as ideal, since in most cases the source voltage is the AC line voltage (110 V rms, 60 Hz) and therefore the offset voltage is negligible in comparison. The key to the operation of the full-wave rectifier is to note that during the positive half-cycle of v_S, the top diode is forward-biased while the bottom diode is reverse-biased; therefore, the load current during the positive half-cycle is

$$i_L = i_1 = \frac{N v_S}{R_L} \qquad v_S \geq 0 \tag{9.17}$$

while during the negative half-cycle, the bottom diode conducts and the top diode is off, and the load current is given by

$$i_L = i_2 = \frac{-N v_S}{R_L} \qquad v_S < 0 \tag{9.18}$$

Note that the direction of i_L is always positive, because of the manner of connecting the diodes (when the top diode is off, i_2 is forced to flow from plus to minus across R_L).

The source voltage, the load voltage, and the currents i_1 and i_2 are shown in Figure 9.40 for a load resistance $R_L = 1\ \Omega$ and $N = 1$. The full-wave rectifier results in a twofold improvement in efficiency over the half-wave rectifier introduced earlier.

The Bridge Rectifier

Another rectifier circuit commonly available "off the shelf" as a single *integrated-circuit package*[2] is the *bridge rectifier*, which employs four diodes in a bridge configuration, similar to the Wheatstone bridge already explored in Chapter 2. Figure 9.41 depicts the bridge rectifier, along with the associated integrated-circuit (IC) package.

The analysis of the bridge rectifier is simple to understand by visualizing the operation of the rectifier for the two half-cycles of the AC waveform separately. The key is that, as illustrated in Figure 9.42, diodes D_1 and D_3 conduct during the positive half-cycle, while diodes D_2 and D_4 conduct during the negative half-cycle. Because of the structure of the bridge, the flow of current through the load resistor is in the same direction (from c to d) during both halves of the cycle, hence, the

[2]An integrated circuit is a collection of electronic devices interconnected on a single silicon chip.

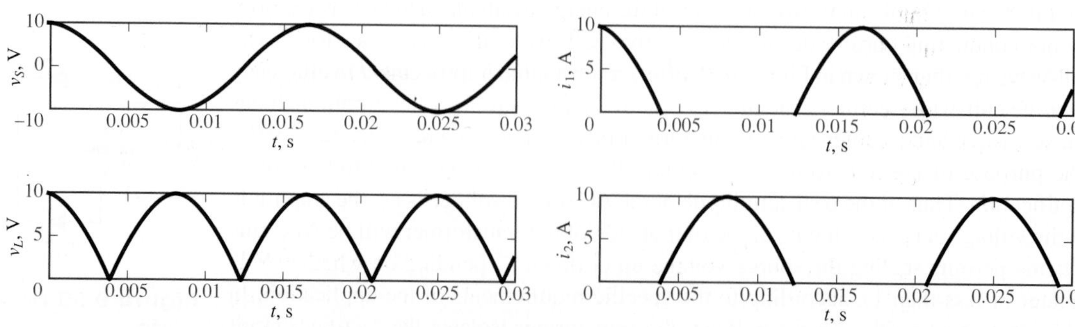

Figure 9.40 Full-wave rectifier current and voltage waveforms ($R_L = 1\ \Omega$)

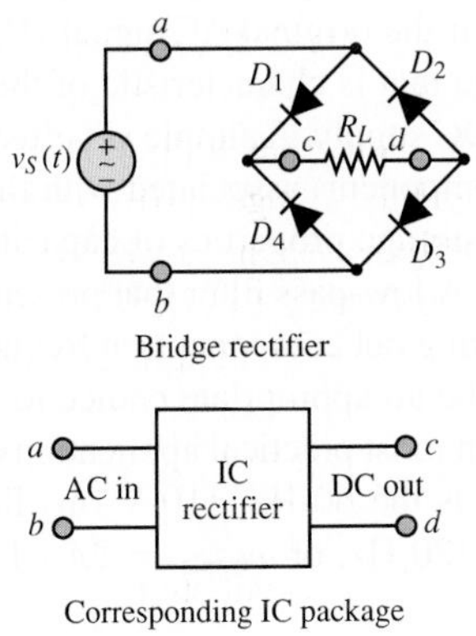

Figure 9.41 Full-wave bridge rectifier

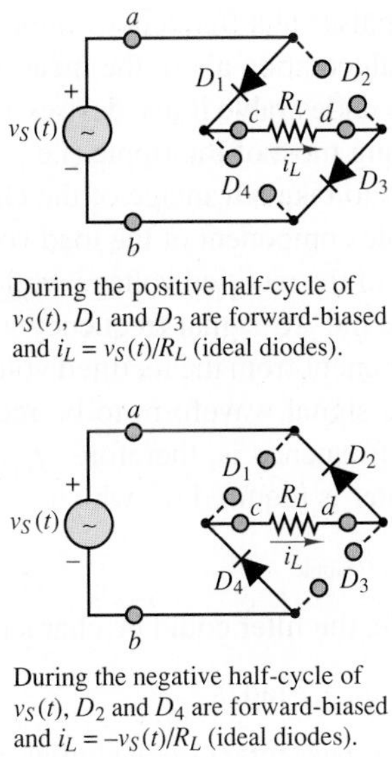

During the positive half-cycle of $v_S(t)$, D_1 and D_3 are forward-biased and $i_L = v_S(t)/R_L$ (ideal diodes).
During the negative half-cycle of $v_S(t)$, D_2 and D_4 are forward-biased and $i_L = -v_S(t)/R_L$ (ideal diodes).

Figure 9.42 Operation of bridge rectifier

full-wave rectification of the waveform. The original and rectified waveforms are shown in Figure 9.43(a) for the case of ideal diodes and a 30-V peak AC source. Figure 9.43(b) depicts the rectified waveform if we assume diodes with a 0.6-V offset voltage. Note that the waveform of Figure 9.43(b) is not a pure rectified sinusoid any longer: The effect of the offset voltage is to shift the waveform downward by twice the offset voltage. This is most easily understood by considering that the load seen by the source during either half-cycle consists of two diodes in series with the load resistor.

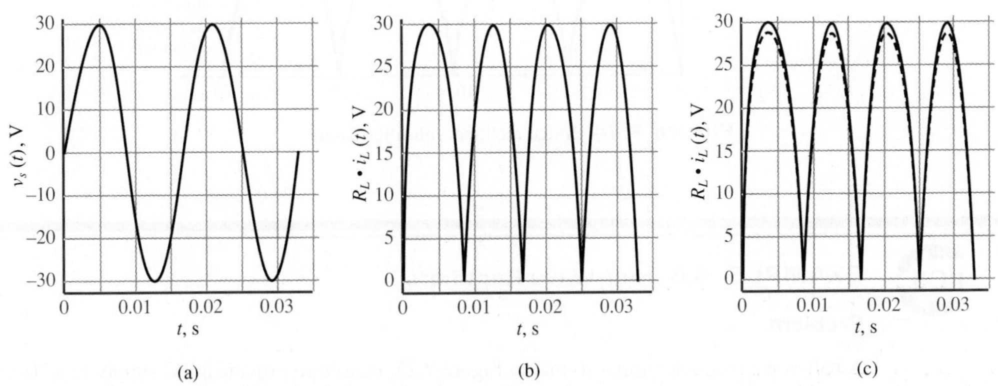

Figure 9.43 (a) Unrectified source voltage; (b) rectified load voltage (ideal diodes); (c) rectified load voltage (ideal and offset diodes)

Although the conventional and bridge full-wave rectifier circuits effectively convert AC signals that have zero average, or DC, value to a signal with a nonzero average voltage, either rectifier's output is still an oscillating waveform. Rather than provide a smooth, constant voltage, the full-wave rectifier generates a sequence of sinusoidal pulses at a frequency double that of the original AC signal. The **ripple**—that is, the fluctuation about the mean voltage that is characteristic of these rectifier circuits—is undesirable if one desires a true DC supply. A simple yet effective means of eliminating most of the ripple (i.e., AC component) associated with the output of a rectifier is to take advantage of the energy storage properties of capacitors to filter out the ripple component of the load voltage. A low-pass filter that preserves the DC component of the rectified voltage while filtering out components at frequencies at or above twice the AC signal frequency would be an appropriate choice to remove the ripple component from the rectified voltage. In most practical applications of rectifier circuits, the signal waveform to be rectified is the 60-Hz, 110-V rms line voltage. The ripple frequency is, therefore, $f_{\text{ripple}} = 120$ Hz, or $\omega_{\text{ripple}} = 2\pi \cdot 120$ rad/s. A low-pass filter is required for which

$$\omega_0 \ll \omega_{\text{ripple}} \tag{9.19}$$

For example, the filter could be characterized by

$$\omega_0 = 2\pi \cdot 2 \text{ rad/s}$$

A simple low-pass filter circuit similar to those studied in Chapter 6 that accomplishes this task is shown in Figure 9.44.

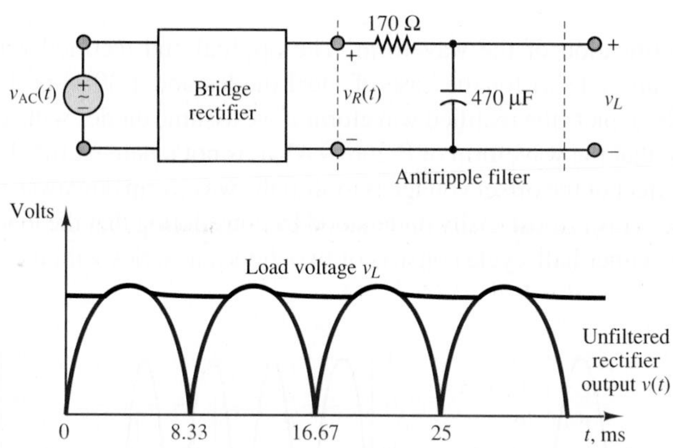

Figure 9.44 Bridge rectifier with filter circuit

LO3

EXAMPLE 9.8 Half-Wave Rectifiers

Problem

A half-wave rectifier, similar to that in Figure 9.25, is used to provide a DC supply to a 50-Ω load. If the AC source voltage is 20 V (rms), find the peak and average current in the load. Assume an ideal diode.

Solution

Known Quantities: Value of circuit elements and source voltage.

Find: Peak and average values of load current in half-wave rectifier circuit.

Schematics, Diagrams, Circuits, and Given Data: $v_S = 20$ V(rms), $R = 20\ \Omega$.

Assumptions: Ideal diode.

Analysis: According to the ideal diode model, the peak load voltage is equal to the peak sinusoidal source voltage. Thus, the peak load current is

$$i_{\text{peak}} = \frac{v_{\text{peak}}}{R_L} = \frac{\sqrt{2} v_{\text{rms}}}{R_L} = 0.567 \text{ A}$$

To compute the average current, we must integrate the half-wave rectified sinusoid:

$$\langle i \rangle = \frac{1}{T} \int_0^T i(t)\, dt = \frac{1}{T} \left[\int_0^{T/2} \frac{v_{\text{peak}}}{R_L} \sin(\omega t)\, dt + \int_{T/2}^{T} 0\, dt \right]$$

$$= \frac{v_{\text{peak}}}{\pi R_L} = \frac{\sqrt{2} v_{\text{rms}}}{\pi R_L} = 0.18 \text{ A}$$

CHECK YOUR UNDERSTANDING

What is the peak current if an offset diode model is used with offset voltage equal to 0.6 V?

Answer: 0.544 A

EXAMPLE 9.9 Bridge Rectifier

LO3

Problem

A bridge rectifier, similar to that in Figure 9.41, is used to provide a 50-V, 5-A DC supply. What is the resistance of the load that will draw exactly 5 A? What is the required rms source voltage to achieve the desired DC voltage? Assume an ideal diode.

Solution

Known Quantities: Value of circuit elements and source voltage.

Find: RMS source voltage and load resistance in bridge rectifier circuit.

Schematics, Diagrams, Circuits, and Given Data: $\langle v_L \rangle = 50$ V; $\langle i_L \rangle = 5$ A.

Assumptions: Ideal diode.

Analysis: The load resistance that will draw an average current of 5 A is easily computed to be

$$R_L = \frac{\langle v_L \rangle}{\langle i_L \rangle} = \frac{50}{5} = 10\ \Omega$$

Note that this is the lowest value of resistance for which the DC supply will be able to provide the required current. To compute the required rms voltage, we observe that the average load voltage can be found from the expression

$$\langle v_L \rangle = R_L \langle i_L \rangle = \frac{R_L}{T}\int_0^T i(t)\,dt = \frac{R_L}{T}\left[\int_0^{T/2} \frac{v_{\text{peak}}}{R_L}\sin(\omega t)\,dt\right]$$

$$= \frac{2v_{\text{peak}}}{\pi} = \frac{2\sqrt{2}v_{\text{rms}}}{\pi} = 50\text{ V}$$

Hence,

$$v_{\text{rms}} = \frac{50\pi}{2\sqrt{2}} = 55.5\text{ V}$$

CHECK YOUR UNDERSTANDING

Show that the DC output voltage of the full-wave rectifier of Figure 9.39 is $2N\,v_{S\text{peak}}/\pi$.
Compute the peak voltage output of the bridge rectifier of Figure 9.40, assuming diodes with 0.6-V offset voltage and a 110-V rms AC supply.

Answer: 154.36 V

9.5 DC POWER SUPPLIES, ZENER DIODES, AND VOLTAGE REGULATION

The principal application of rectifier circuits is in the conversion of AC to DC power. A circuit that accomplishes this conversion is usually called a **DC power supply.** In power supply applications, transformers are employed to obtain an AC voltage that is reasonably close to the desired DC supply voltage. DC power supplies are very useful in practice: Many familiar electric and electronic appliances (e.g., radios, personal computers, TVs) require DC power to operate. For most applications, it is desirable that the DC supply be as steady and ripple-free as possible. To ensure that the DC voltage generated by a DC supply is constant, DC supplies contain voltage regulators, that is, devices that can hold a DC load voltage relatively constant in spite of possible fluctuations in the DC supply. This section describes the fundamentals of voltage regulators.

A typical DC power supply is made up of the components shown in Figure 9.45. In the figure, a transformer is shown connecting the AC source to the rectifier circuit to permit scaling of the AC voltage to the desired level. For example, one might

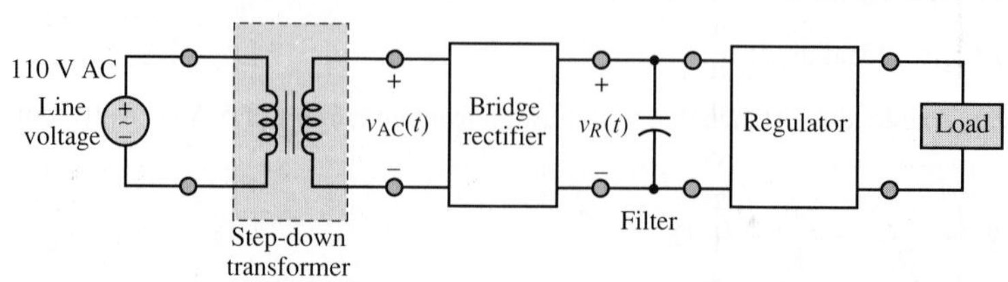

Figure 9.45 DC power supply

wish to step the 110-V rms line voltage down to a lower DC voltage by means of a transformer prior to rectification and filtering, to eventually obtain a 12-VDC regulated supply (*regulated* here means that the output voltage is a DC voltage that is constant and independent of load and supply variations). Following the step-down transformer are a bridge rectifier, a filter capacitor, a voltage regulator, and finally the load.

The most common device employed in voltage regulation schemes is the Zener diode. Zener diodes function on the basis of the reverse portion of the i-v characteristic of the diode discussed in Section 9.2. Figure 9.10 in Section 9.2 illustrates the general characteristic of a diode, with forward offset voltage V_γ and **reverse Zener voltage** V_Z. Note how steep the i-v characteristic is at the Zener breakdown voltage, indicating that in the Zener breakdown region the diode can hold a very nearly constant voltage for a large range of currents. This property makes it possible to use the Zener diode as a voltage reference.

With reference to Figure 9.10, we see that once the Zener diode is reverse-biased with a reverse voltage larger than the Zener voltage, the Zener diode behaves very nearly as an ideal voltage source, providing a voltage reference that is nearly constant. This reference is not exact because the slope of the Zener diode curve for voltages lower than $-V_Z$ is not infinite; the finite slope will cause the Zener voltage to change slightly. For the purpose of this analysis, we assume that the reverse-biased Zener voltage is actually constant. The operation of the Zener diode as a voltage regulator is simplified in the analysis shown in the rest of this section to illustrate some of the limitations of voltage regulator circuits. The circuits analyzed in the remainder of this section do not represent practical voltage regulator designs, but are useful to understand the basic principles of voltage regulation. Chapter 12, and in particular Example 12.1, describes the operation of a practical regulator circuit.

The operation of the Zener diode may be analyzed by considering three modes of operation:

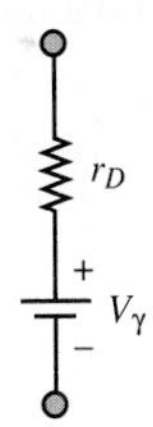

Figure 9.46 Zener diode model for forward bias

1. For $v_D \geq V_\gamma$, the device acts as a conventional forward-biased diode (Figure 9.46).

LO4

2. For $V_Z < v_D < V_\gamma$, the diode is reverse-biased but Zener breakdown has not taken place yet. Thus, it acts as an open circuit.
3. For $v_D \leq V_Z$, Zener breakdown occurs and the device holds a nearly constant voltage $-V_Z$ (Figure 9.47).

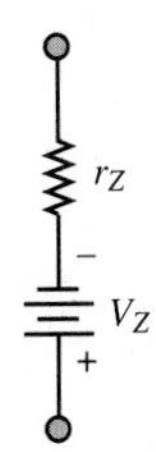

Figure 9.47 Zener diode model for reverse bias

The combined effect of forward and reverse bias may be lumped into a single model with the aid of ideal diodes, as shown in Figure 9.48.

To illustrate the operation of a Zener diode as a voltage regulator, consider the circuit of Figure 9.49(a), where the unregulated DC source V_S is regulated to the value of the Zener voltage V_Z. Note how the diode must be connected "upside down" to obtain a positive regulated voltage. Note also that if v_S is greater than V_Z, it follows that the Zener diode is in its reverse breakdown mode. Thus, one need not worry whether the diode is conducting or not in simple voltage regulator problems, provided that the unregulated supply voltage is guaranteed to stay above V_Z (a problem arises, however, if the unregulated supply can drop below the Zener voltage). Assuming that the resistance r_Z is negligible with respect to R_S and R_L, we replace the Zener diode with the simplified circuit model of Figure 9.49(b), consisting of a battery of strength V_Z (the effects of the nonzero Zener resistance are explored in the examples and homework problems).

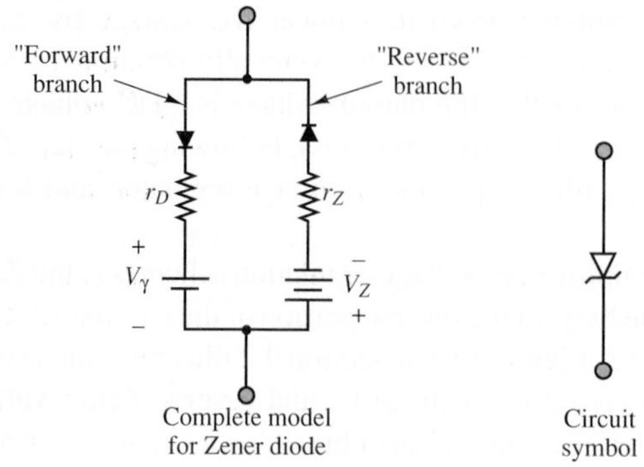

The Zener diode may be modeled as consisting of two parallel branches, representing forward and reverse operation.

Figure 9.48 Complete model for Zener diode

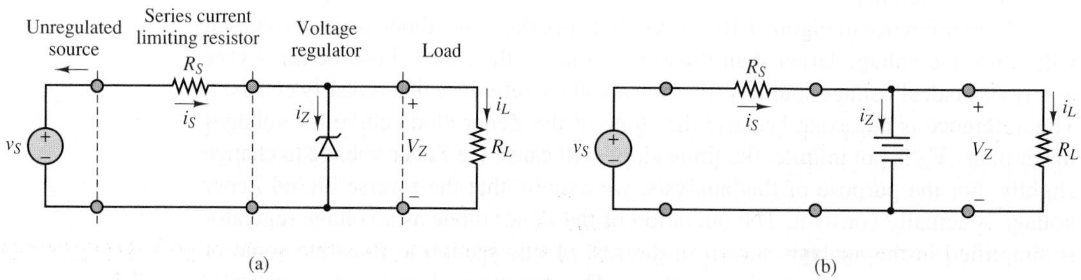

Figure 9.49 (a) A Zener diode voltage regulator; (b) simplified circuit for Zener regulator

Three simple observations are sufficient to explain the operation of this voltage regulator:

1. The load voltage must equal V_Z, as long as the Zener diode is in the reverse breakdown mode. Then

$$i_L = \frac{V_Z}{R_L} \tag{9.20}$$

2. The load current (which should be constant if the load voltage is to be regulated to sustain V_Z) is the difference between the unregulated supply current i_S and the diode current i_Z:

$$i_L = i_S - i_Z \tag{9.21}$$

This second point explains intuitively how a Zener diode operates: Any current in excess of that required to keep the load at the constant voltage V_Z is "dumped" to ground through the diode. Thus, the Zener diode acts as a sink to the undesired source current.

3. The source current is given by

$$i_S = \frac{v_S - V_Z}{R_S} \tag{9.22}$$

In the ideal case, the operation of a Zener voltage regulator can be explained very simply on the basis of this model. The examples and exercises will illustrate the effects of the practical limitations that arise in the design of a practical voltage regulator; the general principles are discussed in the following paragraphs.

The Zener diode is usually rated in terms of its maximum allowable power dissipation. The power dissipated by the diode P_Z may be computed from

$$P_Z = i_Z V_Z \tag{9.23}$$

Thus, one needs to worry about the possibility that i_Z will become too large. This may occur either if the supply current is very large (perhaps because of an unexpected upward fluctuation of the unregulated supply) or if the load is suddenly removed and all the supply current sinks through the diode. The latter case, of an open-circuit load, is an important design consideration.

Another significant limitation occurs when the load resistance is small, thus requiring large amounts of current from the unregulated supply. In this case, the Zener diode is hardly taxed at all in terms of power dissipation, but the unregulated supply may not be able to provide the current required to sustain the load voltage. In this case, regulation fails to take place. Thus, in practice, the range of load resistances for which load voltage regulation may be attained is constrained to a finite interval:

$$R_{L\,\min} \le R_L \le R_{L\,\max} \tag{9.24}$$

where $R_{L\,\max}$ is typically limited by the Zener diode power dissipation and $R_{L\,\min}$ by the maximum supply current. Examples 9.10 through 9.12 illustrate these concepts.

EXAMPLE 9.10 Determining the Power Rating of a Zener Diode

LO4

Problem

We wish to design a regulator similar to the one depicted in Figure 9.49(a). Determine the minimum acceptable power rating of the Zener diode.

Solution

Known Quantities: $v_S = 24$ V; $V_Z = 12$ V; $R_S = 50\ \Omega$; $R_L = 250\ \Omega$.

Find: The maximum power dissipated by the Zener diode under worst-case conditions.

Assumptions: Use the piecewise linear Zener diode model (Figure 9.48) with $r_Z = 0$.

Analysis: When the regulator operates according to the intended design specifications, that is, with a 250-Ω load, the source and load currents may be computed as follows:

$$i_S = \frac{v_S - V_Z}{R_S} = \frac{12}{50} = 0.24 \text{ A}$$

$$i_L = \frac{V_Z}{R_L} = \frac{12}{250} = 0.048 \text{ A}$$

Thus, the Zener current would be

$$i_Z = i_S - i_L = 0.192 \text{ A}$$

corresponding to a nominal power dissipation

$$P_Z = i_Z V_Z = 0.192 \times 12 = 2.304 \text{ W}$$

However, if the load were accidentally (or intentionally) disconnected from the circuit, all the load current would be diverted to flow through the Zener diode. Thus, the *worst-case* Zener current is actually equal to the source current, since the Zener diode would sink all the source current for an open-circuit load:

$$i_{Z\,max} = i_S = \frac{v_S - V_Z}{R_S} = \frac{12}{50} = 0.24 \text{ A}$$

Therefore the maximum power dissipation that the Zener diode must sustain is

$$P_{Z\,max} = i_{Z\,max} V_Z = 2.88 \text{ W}$$

Comments: A safe design would exceed the value of $P_{Z\,max}$ computed above. For example, one might select a 3-W Zener diode.

CHECK YOUR UNDERSTANDING

How would the power rating change if the load were reduced to 100 Ω?

Answer: The worst-case power rating would not change.

LO4

EXAMPLE 9.11 Calculation of Allowed Load Resistances for a Given Zener Regulator

Problem

Calculate the allowable range of load resistances for the Zener regulator of Figure 9.50 such that the diode power rating is not exceeded.

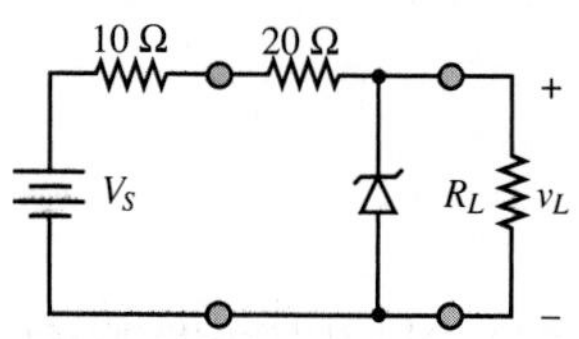

Figure 9.50

Solution

Known Quantities: $V_S = 50$ V; $V_Z = 14$ V; $P_Z = 5$ W.

Find: The smallest and largest values of R_L for which load voltage regulation to 14 V is achieved, and which do not cause the diode power rating to be exceeded.

Assumptions: Use the piecewise linear Zener diode model (Figure 9.48) with $r_Z = 0$.

Analysis:

1. *Determining the minimum acceptable load resistance.* To determine the minimum acceptable load, we observe that the regulator can at most supply the load with the amount of current that can be provided by the source. Thus, the minimum theoretical resistance can be computed by assuming that all the source current goes to the load, and that the load voltage is regulated at the nominal value:

$$R_{L\,min} = \frac{V_Z}{i_S} = \frac{V_Z}{(V_S - V_Z)/30} = \frac{14}{36/30} = 11.7 \ \Omega$$

If the load required any more current, the source would not be able to supply it. Note that for this value of the load, the Zener diode dissipates zero power, because the Zener current is zero.

2. *Determining the maximum acceptable load resistance*. The second constraint we need to invoke is the power rating of the diode. For the stated 5-W rating, the maximum Zener current is

$$i_{Z\,\max} = \frac{P_Z}{V_Z} = \frac{5}{14} = 0.357 \text{ A}$$

Since the source can generate

$$i_{S\,\max} = \frac{V_S - V_Z}{30} = \frac{50 - 14}{30} = 1.2 \text{ A}$$

the load must not require any less than $1.2 - 0.357 = 0.843$ A; if it required any less current (i.e., if the resistance were too large), the Zener diode would be forced to sink more current than its power rating permits. From this requirement we can compute the maximum allowable load resistance

$$R_{L\,\min} = \frac{V_Z}{i_{S\,\max} - i_{Z\,\max}} = \frac{14}{0.843} = 16.6\ \Omega$$

Finally, the range of allowable load resistance is $11.7\ \Omega \le R_L \le 16.6\ \Omega$.

Comments: Note that this regulator *cannot* operate with an open-circuit load! This is obviously not a very useful circuit.

CHECK YOUR UNDERSTANDING

What should the power rating of the Zener diode be to withstand operation with an open-circuit load?

Answer: $P_{Z\,\max} = 16.8$ W

EXAMPLE 9.12 Effect of Nonzero Zener Resistance in a Regulator

LO4

Problem

Calculate the amplitude of the ripple present in the output voltage of the regulator of Figure 9.51. The unregulated supply voltage is depicted in Figure 9.52.

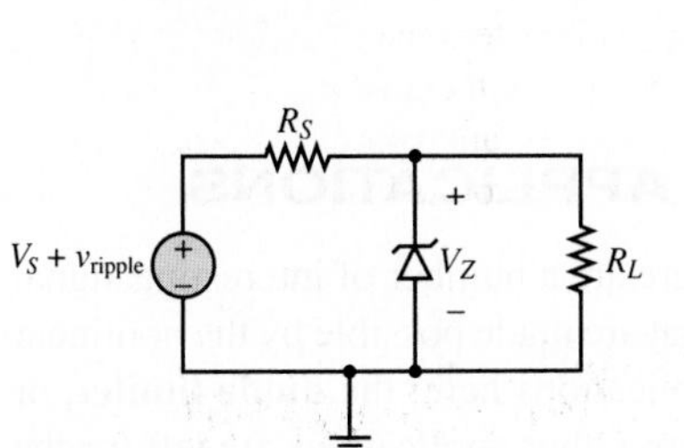

Figure 9.51

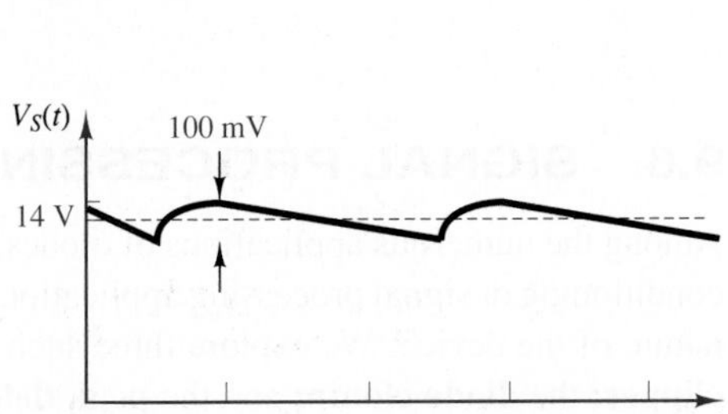

Figure 9.52

Solution

Known Quantities: $v_S = 14$ V; $v_{\text{ripple}} = 100$ mV; $V_Z = 8$ V; $r_Z = 10\ \Omega$; $R_S = 50\ \Omega$; $R_L = 150\ \Omega$.

Find: Amplitude of ripple component in load voltage.

Assumptions: Use the piecewise linear Zener diode model (Figure 9.48).

Analysis: To analyze the circuit, we consider the DC and AC equivalent circuits of Figure 9.53 separately.

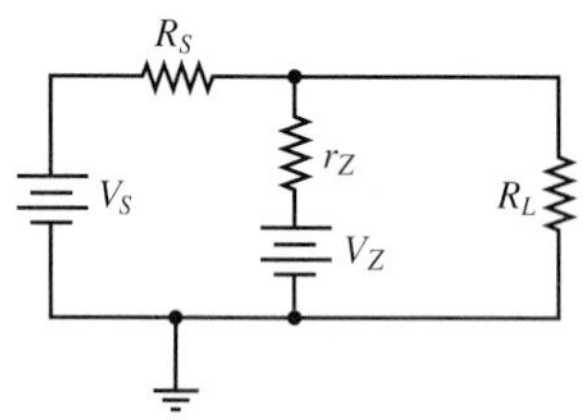

DC equivalent circuit

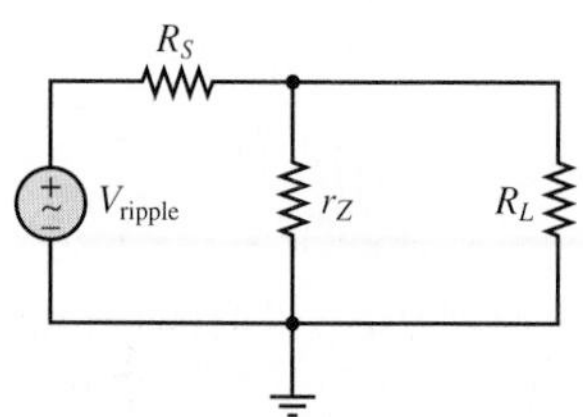

AC equivalent circuit

Figure 9.53

1. *DC equivalent circuit.* The DC equivalent circuit reveals that the load voltage consists of two contributions: that due to the unregulated DC supply and that due to the Zener diode V_Z. Applying superposition and the voltage divider rule, we obtain

$$V_L = V_S \left(\frac{r_Z \| R_L}{r_Z \| R_L + R_S} \right) + V_Z \left(\frac{R_S \| R_L}{r_Z \| R_L + R_S} \right) = 2.21 + 6.32 = 8.53 \text{ V}$$

2. *AC equivalent circuit.* The AC equivalent circuit allows us to compute the AC component of the load voltage as follows:

$$v_L = v_{\text{ripple}} \left(\frac{r_Z \| R_L}{r_Z \| R_L + R_S} \right) = 0.016 \text{ V}$$

that is, 16 mV of ripple is present in the load voltage, or approximately one-sixth the source ripple.

Comments: Note that the DC load voltage is affected by the unregulated source voltage; if the unregulated supply were to fluctuate significantly, the regulated voltage would also change. Thus, one of the effects of the Zener resistance is to cause imperfect regulation. If the Zener resistance is significantly smaller than both R_S and R_L, its effects will not be as pronounced.

CHECK YOUR UNDERSTANDING

Compute the actual DC load voltage and the percent of ripple reaching the load (relative to the initial 100-mV ripple) for the circuit of Example 9.12 if $r_Z = 1\ \Omega$.

Answer: 8.06 V, 2 percent

9.6 SIGNAL PROCESSING APPLICATIONS

Among the numerous applications of diodes, there are a number of interesting signal conditioning or signal processing applications that are made possible by the nonlinear nature of the device. We explore three such applications here: the **diode limiter,** or **clipper;** the **diode clamp;** and the **peak detector.** Other applications are left for the homework problems.

The Diode Clipper (Limiter)

The *diode clipper* is a relatively simple diode circuit that is often employed to protect loads against excessive voltages. The objective of the clipper circuit is to keep the load voltage within a range, say, $-V_{\max} \le v_L(t) \le V_{\max}$, so that the maximum allowable load voltage (or power) is never exceeded. The circuit of Figure 9.54 accomplishes this goal.

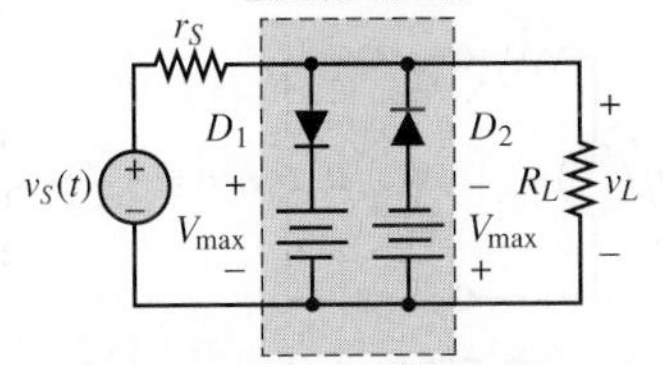

Figure 9.54 Two-sided diode clipper

The circuit of Figure 9.54 is most easily analyzed by first considering just the branch containing D_1. This corresponds to clipping only the positive peak voltages; the analysis of the negative voltage limiter is left as a drill exercise. The circuit containing the D_1 branch is sketched in Figure 9.55; note that we have exchanged the location of the D_1 branch and that of the load branch for convenience. Further, the circuit is reduced to Thévenin equivalent form. Having reduced the circuit to a simpler form, we can now analyze its operation for two distinct cases: the ideal diode and the piecewise linear diode.

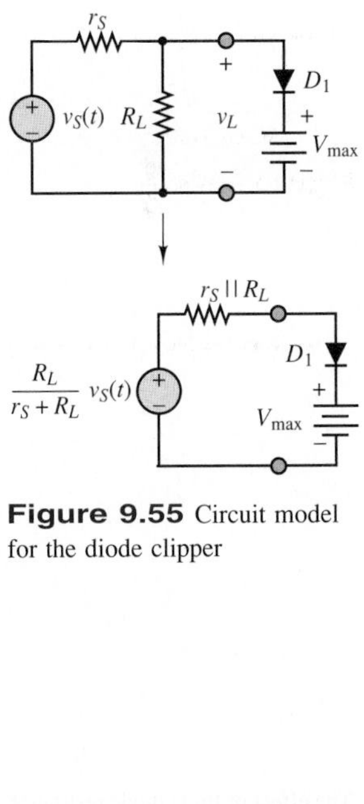

Figure 9.55 Circuit model for the diode clipper

1. Ideal diode model. For the ideal diode case, we see immediately that D_1 conducts if

$$\frac{R_L}{r_S + R_L} v_S(t) \ge V_{\max} \tag{9.25}$$

and that if this condition occurs, then (D_1 being a short circuit) the load voltage v_L becomes equal to $V_{\max}$. The equivalent circuit for the on condition is shown in Figure 9.56.

If, on the other hand, the source voltage is such that

$$\frac{R_L}{r_S + R_L} v_S(t) < V_{\max} \tag{9.26}$$

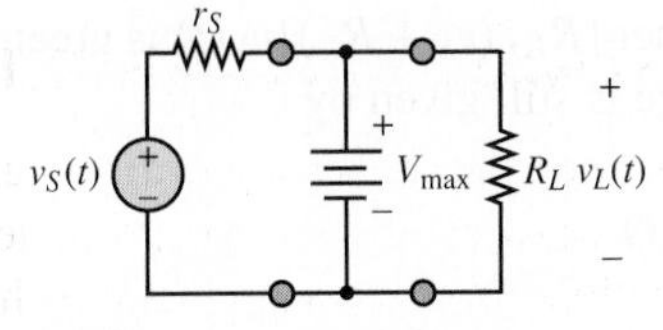

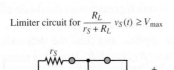

Figure 9.56 Equivalent circuit for the one-sided limiter (diode on)

then D_1 is an open circuit and the load voltage is simply

$$v_L(t) = \frac{R_L}{r_S + R_L} v_S(t) \tag{9.27}$$

The equivalent circuit for this case is depicted in Figure 9.57.

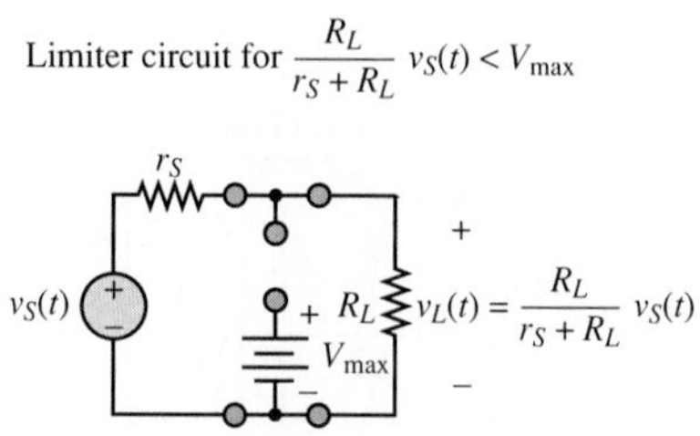

Figure 9.57 Equivalent circuit for the one-sided limiter (diode off)

The analysis for the negative branch of the circuit of Figure 9.54 can be conducted by analogy with the preceding derivation, resulting in the waveform for the two-sided clipper shown in Figure 9.58. Note how the load voltage is drastically "clipped" by the limiter in the waveform of Figure 9.58. In reality, such hard clipping does not occur, because the actual diode characteristic does not have the sharp on/off breakpoint the ideal diode model implies. One can develop a reasonable representation of the operation of a physical diode limiter by using the piecewise linear model.

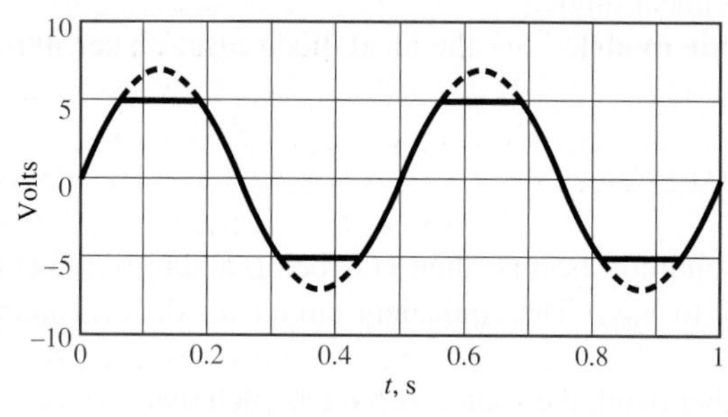

Figure 9.58 Two-sided (ideal diode) clipper input and output voltages

2. Piecewise linear diode model. To avoid unnecessary complexity in the analysis, assume that V_{max} is much greater than the diode offset voltage, and therefore assume that $V_\gamma \approx 0$. We do, however, consider the finite diode resistance r_D. The circuit of Figure 9.55 still applies, and thus the determination of the diode on/off state is still based on whether $[R_L/(r_S + R_L)]v_S(t)$ is greater or less than V_{max}. When D_1 is open, the load voltage is still given by

$$v_L(t) = \frac{R_L}{r_S + R_L} v_S(t) \tag{9.28}$$

When D_1 is conducting, however, the corresponding circuit is as shown in Figure 9.59.

The primary effect the diode resistance has on the load waveform is that some of the source voltage will reach the load even when the diode is conducting. This is

The effect of finite diode resistance on the limiter circuit.

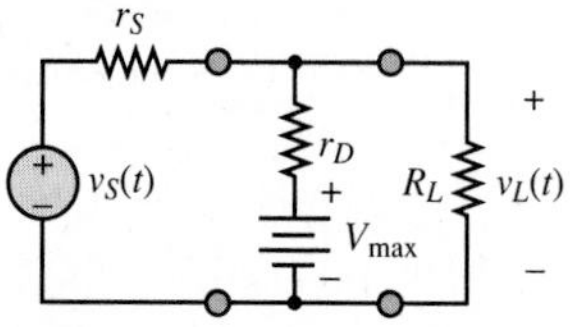

Figure 9.59 Circuit model for the diode clipper (piecewise linear diode model)

most easily verified by applying superposition; it can be readily shown that the load voltage is now composed of two parts, one due to the voltage V_{max} and the other proportional to $v_S(t)$:

$$v_L(t) = \frac{R_L \parallel r_S}{r_D + (R_L \parallel r_S)} V_{max} + \frac{r_D \parallel R_L}{r_S + (r_D \parallel R_L)} v_S(t) \tag{9.29}$$

It may easily be verified that as $r_D \to 0$, the expression for $v_L(t)$ is the same as that for the ideal diode case. The effect of the diode resistance on the limiter circuit is depicted in Figure 9.60. Note how the clipping has a softer, more rounded appearance.

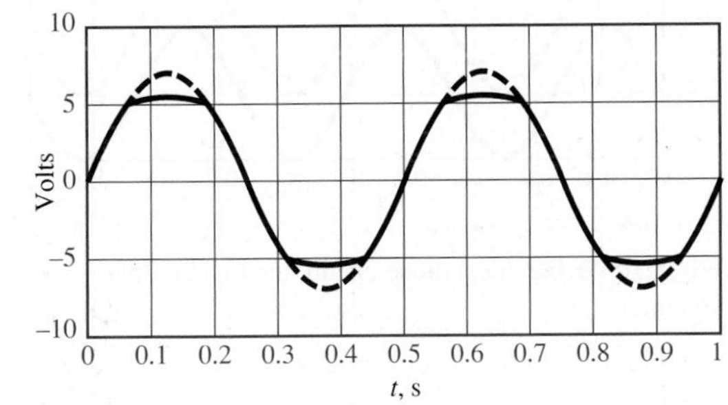

Figure 9.60 Voltages for the diode clipper (piecewise linear diode model)

CHECK YOUR UNDERSTANDING

For the one-sided diode clipper of Figure 9.55, find the percentage of the source voltage that reached the load if $R_L = 150\ \Omega$, $r_S = 50\ \Omega$, and $r_D = 5\ \Omega$. Assume that the diode is conducting, and use the circuit model of Figure 9.59.

Answer: 9.8 percent

The Diode Clamp

Another circuit that finds common application is the *diode clamp*, which permits "clamping" a waveform to a fixed DC value. Figure 9.61 depicts two different types of clamp circuits.

The operation of the simple clamp circuit is based on the notion that the diode will conduct current only in the forward direction, and that therefore the capacitor will charge during the positive half-cycle of $v_S(t)$ but will not discharge during the negative half-cycle. Thus, the capacitor will eventually charge up to the peak voltage of $v_S(t)$, denoted by V_{peak}. The DC voltage across the capacitor has the effect of shifting the source waveform down by V_{peak}, so that after the initial transient period the output voltage is

$$v_{out}(t) = v_S(t) - V_{peak} \tag{9.30}$$

and the positive peaks of $v_S(t)$ are now clamped at 0 V. For equation 9.30 to be

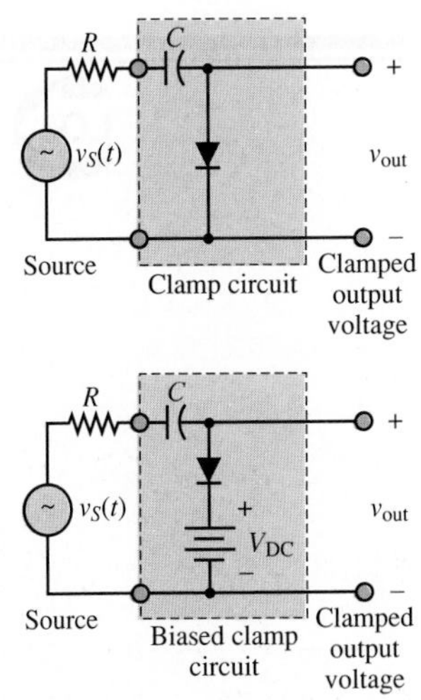

Figure 9.61 Diode clamp circuits

accurate, it is important that the RC time constant be greater than the period T of $v_S(t)$:

$$RC \gg T \tag{9.31}$$

Figure 9.62 depicts the behavior of the diode clamp for a sinusoidal input waveform, where the dashed line is the source voltage and the solid line represents the clamped voltage.

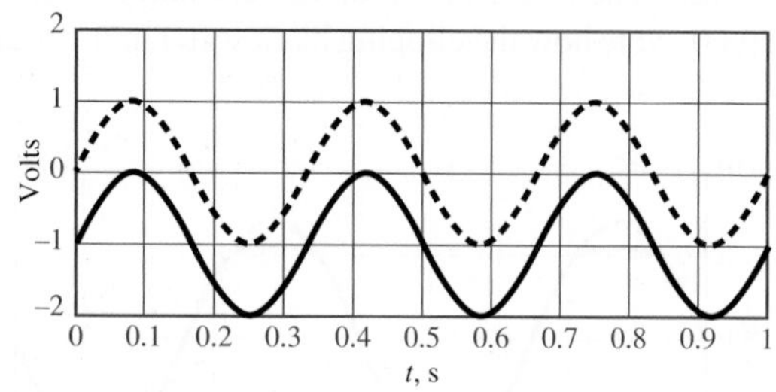

Figure 9.62 Ideal diode clamp input and output voltages

The clamp circuit can also work with the diode in the reverse direction; the capacitor will charge to $-V_{\text{peak}}$ with the output voltage given by

$$v_{\text{out}}(t) = v_S(t) + V_{\text{peak}} \tag{9.32}$$

Now the output voltage has its negative peaks clamped to zero, since the entire waveform is shifted upward by V_{peak} volts. Note that in either case, the diode clamp has the effect of introducing a DC component in a waveform that does not originally have one. It is also possible to shift the input waveform by a voltage different from V_{peak} by connecting a battery V_{DC} in series with the diode, provided that

$$V_{\text{DC}} < V_{\text{peak}} \tag{9.33}$$

The resulting circuit is called a *biased diode clamp*; it is discussed in Example 9.13.

LO5

EXAMPLE 9.13 Biased Diode Clamp

Problem

Design a biased diode clamp to shift the DC level of the signal $v_S(t)$ up by 3 V.

Solution

Known Quantities: $v_S(t) = 5 \cos \omega t$.

Find: The value of V_{DC} in the circuit in the lower half of Figure 9.61.

Assumptions: Use the ideal diode model.

Analysis: With reference to the circuit in the lower half of Figure 9.61, we observe that once the capacitor has charged to $V_{\text{peak}} - V_{\text{DC}}$, the output voltage will be given by

$$v_{\text{out}} = v_S - V_{\text{peak}} + V_{\text{DC}}$$

Since V_{DC} must be smaller than V_{peak} (otherwise the diode would never conduct!), this circuit would never permit raising the DC level of v_{out}. To solve this problem, we must invert both the diode and the battery, as shown in the circuit of Figure 9.63. Now the output voltage is given by

$$v_{out} = v_S + V_{peak} - V_{DC}$$

To have a DC level of 3 V, we choose $V_{DC} = 2$ V. The resulting waveforms are shown in Figure 9.64.

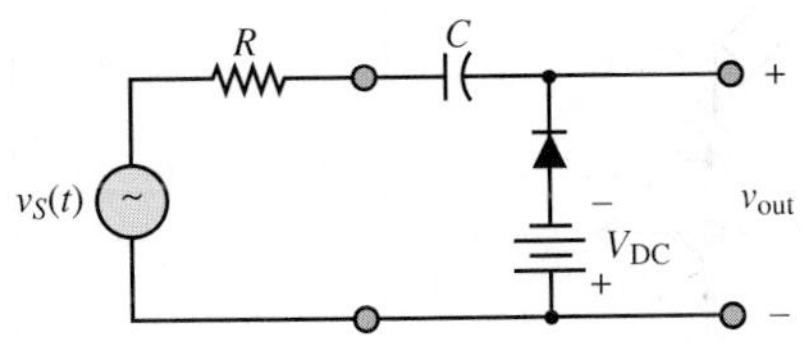

Figure 9.63

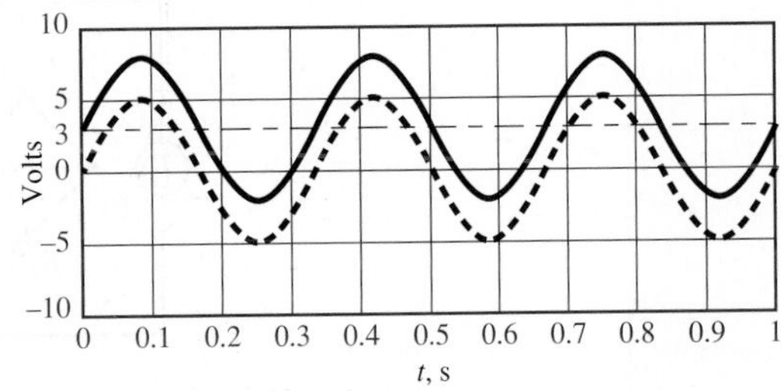

Figure 9.64

The Diode Peak Detector

Another common application of semiconductor diodes, the *peak detector*, is very similar in appearance to the half-wave rectifier with capacitive filtering of Fig. 9.44. One of its more classic applications is in the demodulation of amplitude-modulated (AM) signals. We study this circuit in the next Focus on Measurements box.

Peak Detector Circuit for Capacitive Displacement Transducer

In Chapter 4, a capacitive displacement transducer was introduced in the Focus on Measurements box "Capacitive Displacement Transducer and Microphone." It took the form of a parallel-plate capacitor composed of a fixed plate and a movable plate. The capacitance of this variable capacitor was shown to be a function of displacement; that is, it was shown that a movable-plate capacitor can serve as a linear transducer. Recall the expression derived in Chapter 4

$$C = \frac{8.854 \times 10^{-3} A}{x}$$

where C is the capacitance in picofarads, A is the area of the plates in square millimeters, and x is the (variable) distance in millimeters. If the capacitor is placed in an AC circuit, its impedance will be determined by the expression

$$Z_C = \frac{1}{j\omega C}$$

so that

$$Z_C = \frac{x}{j\omega 8.854 \times 10^{-3} A}$$

(Continued)

Thus, at a fixed frequency ω, the impedance of the capacitor will vary linearly with displacement. This property may be exploited in the bridge circuit of Figure 9.65, where a differential-pressure transducer is shown made of two movable-plate capacitors. If the capacitance of one of these capacitors increases as a consequence of a pressure difference across the transducer, the capacitance of the other must decrease by a corresponding amount, at least for small displacements (you may wish to refer to Figure 4.9 for a picture of this transducer). The bridge is excited by a sinusoidal source.

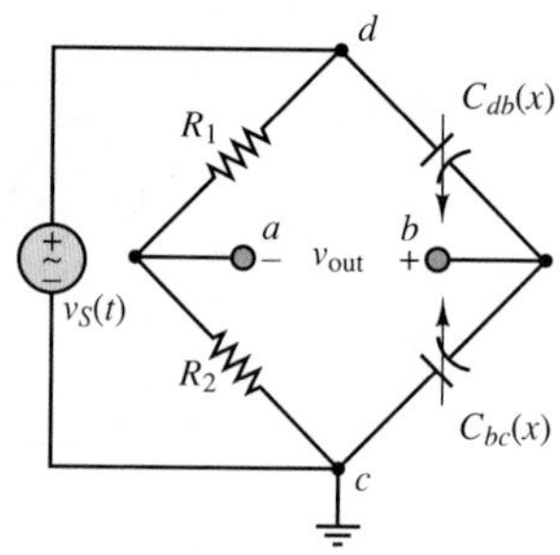

Figure 9.65 Bridge circuit for displacement transducer

Using phasor notation, in Chapter 4 we showed that the output voltage of the bridge circuit is given by

$$\mathbf{V}_{out}(j\omega) = \mathbf{V}_S(j\omega)\frac{x}{2d}$$

provided that $R_1 = R_2$. Thus, the output voltage will vary as a scaled version of the input voltage in proportion to the displacement. A typical $v_{out}(t)$ is displayed in Figure 9.66

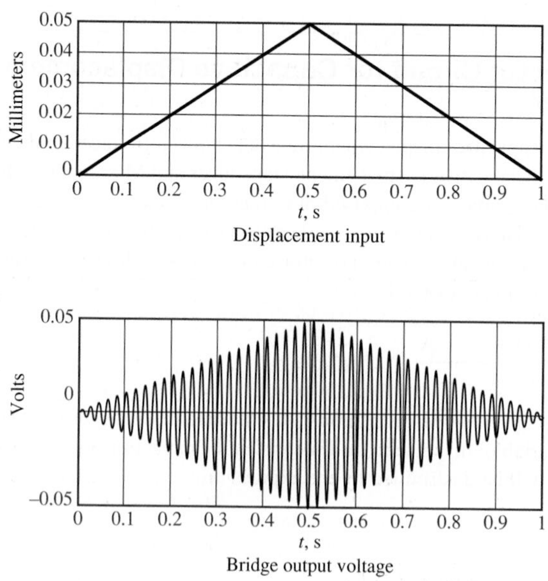

Figure 9.66 Displacement and bridge output voltage waveforms

(Continued)

(Concluded)

for a 0.05-mm "triangle" diaphragm displacement, with $d = 0.5$ mm and $\mathbf{V}_S$ a 50-Hz sinusoid with 1-V amplitude. Clearly, although the output voltage is a function of the displacement x, it is not in a convenient form, since the displacement is proportional to the amplitude of the sinusoidal peaks.

The diode peak detector is a circuit capable of tracking the sinusoidal peaks without exhibiting the oscillations of the bridge output voltage. The peak detector operates by rectifying and filtering the bridge output in a manner similar to that of the circuit of Figure 9.44. The ideal peak detector circuit is shown in Figure 9.67, and the response of a practical peak detector is shown in Figure 9.68. Its operation is based on the rectification property of the diode, coupled with the filtering effect of the shunt capacitor, which acts as a low-pass filter.

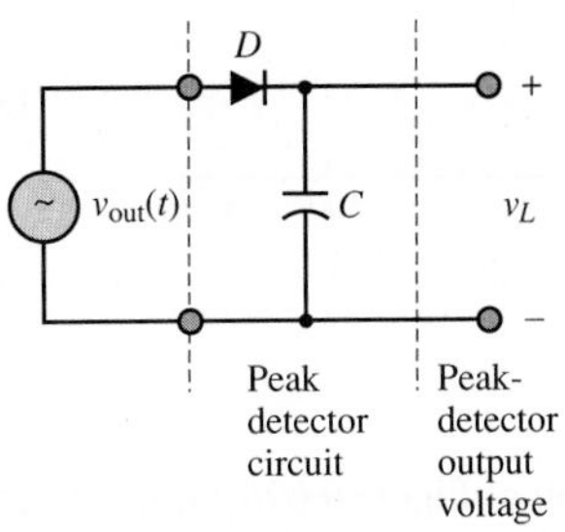

Figure 9.67 Peak detector circuit

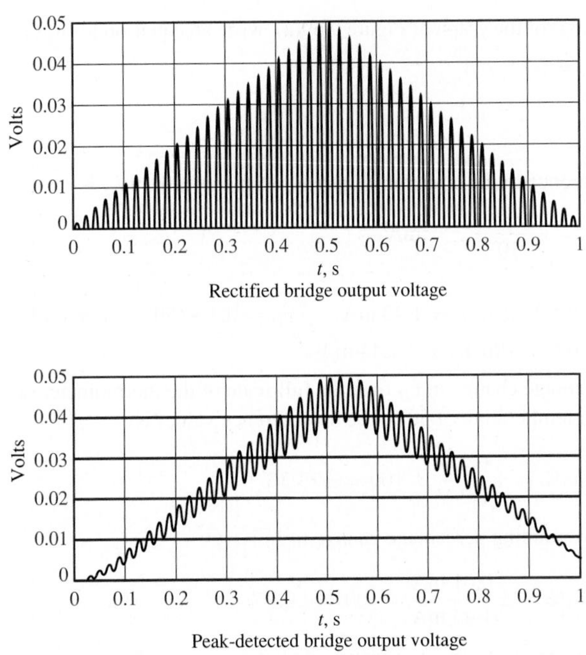

Figure 9.68 Rectified and peak-detected bridge output voltage waveforms

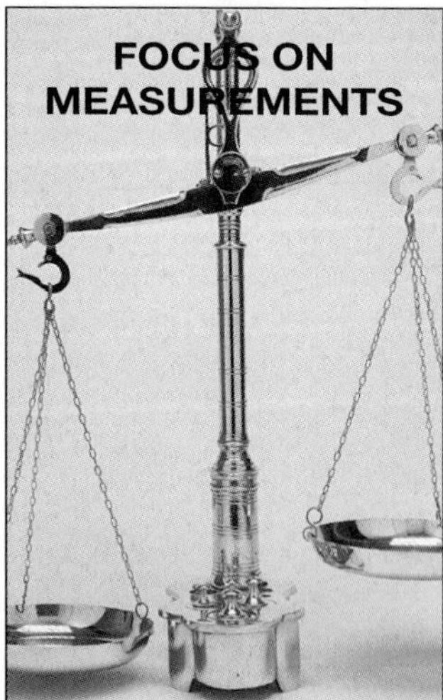

FOCUS ON MEASUREMENTS

Diode Thermometer (LO5)

Problem:

An interesting application of a diode, based on the diode equation, is an electronic thermometer. The concept is based on the empirical observation that if the current through a diode is nearly constant, the offset voltage is nearly a linear function of temperature, as shown in Figure 9.69(a).

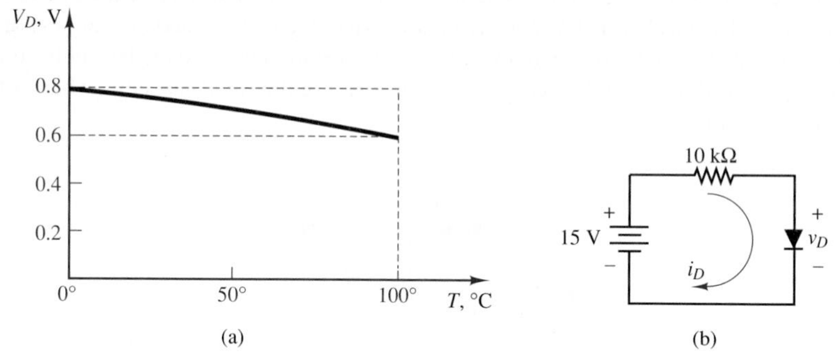

Figure 9.69

1. Show that i_D in the circuit of Figure 9.69(b) is nearly constant in the face of variations in the diode voltage, v_D. This can be done by computing the percentage change in i_D for a given percentage change in v_D. Assume that v_D changes by 10 percent, from 0.6 to 0.66 V.
2. On the basis of the graph of Figure 9.69(a), write an equation for $v_D(T^\circ)$ of the form

$$v_D = \alpha T^\circ + \beta$$

Solution:

1. With reference to the circuit of Figure 9.69(a), the current i_D is

$$i_D = \frac{15 - v_D}{10} \quad \text{mA}$$

For

$$v_D = 0.8\ \text{V}(0^\circ), i_D = 1.42\ \text{mA} \qquad v_D = 0.7\ \text{V}(50^\circ), i_D = 1.43\ \text{mA}$$
$$v_D = 0.6\ \text{V}(100^\circ), i_D = 1.44\ \text{mA}$$

The percentage change in v_D over the full scale of the thermometer (assuming the midrange temperature of 50° to be the reference value) is

$$\Delta v_D\% = \pm\frac{0.1\ \text{V}}{0.7\ \text{V}} \times 100 = \pm 14.3\%$$

The corresponding percentage change in i_D is

$$\Delta i_D\% = \pm\frac{0.01\ \text{mA}}{1.43\ \text{mA}} \times 100 = \pm 0.7\%$$

Thus, i_D is nearly constant over the range of operation of the diode thermometer.

(Continued)

(Concluded)

2. The diode voltage versus temperature equation can be extracted from the graph of Figure 9.69(a):

$$v_D(T) = \frac{(0.8 - 0.6)\text{ V}}{(0 - 100)^\circ\text{C}} T + 0.8\text{ V} = -0.02T + 0.8\text{ V}$$

***Comments*—**The graph of Figure 9.69(a) was obtained experimentally by calibrating a commercial diode in both hot water and an ice bath. The circuit of Figure 9.69(b) is rather simple, and one could fairly easily design a better constant-current source; however, this example illustrates than an inexpensive diode can serve quite well as the sensing element in an **electronic thermometer.**

FIND IT ON THE WEB

9.7 PHOTODIODES

FIND IT ON THE WEB

Another property of semiconductor materials that finds common application in measurement systems is their response to light energy. In appropriately fabricated diodes, called **photodiodes,** when light reaches the depletion region of a *pn* junction, photons cause hole-electron pairs to be generated by a process called *photoionization*. This effect can be achieved by using a surface material that is transparent to light. As a consequence, the reverse saturation current depends on the light intensity (i.e., on the number of incident photons), in addition to the other factors mentioned earlier, in Section 9.2. In a photodiode, the reverse current is given by $-(I_0 + I_p)$, where I_p is the additional current generated by photoionization. The result is depicted in the family of curves of Figure 9.70, where the diode characteristic is shifted downward by

LO6

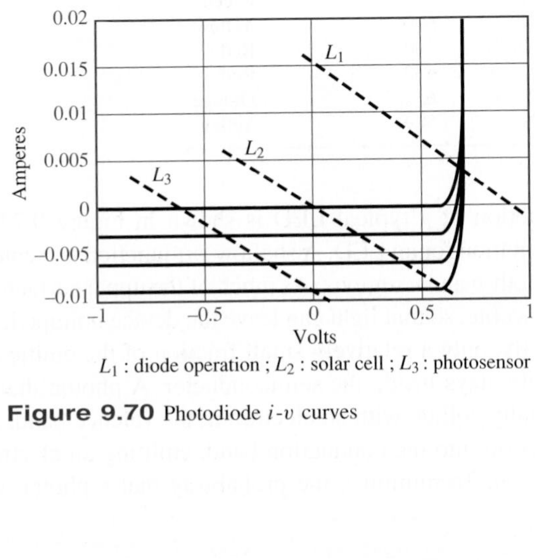

L_1 : diode operation ; L_2 : solar cell ; L_3 : photosensor

Figure 9.70 Photodiode i-v curves

an amount related to the additional current generated by photoionization. Figure 9.70 depicts the appearance of the i-v characteristic of a photodiode for various values of I_p, where the i-v curve is shifted to lower values for progressively larger values of I_p. The circuit symbol is depicted in Figure 9.71.

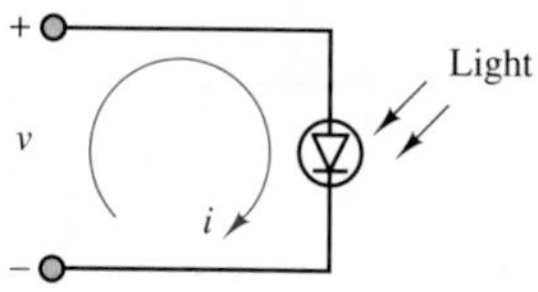

Figure 9.71 Photodiode circuit symbol

Also displayed in Figure 9.70 are three load lines, which depict the three modes of operation of a photodiode. Curve L_1 represents normal diode operation, under forward bias. Note that the operating point of the device is in the positive i, positive v (first) quadrant of the i-v plane; thus, the diode dissipates positive power in this mode and is therefore a passive device, as we already know. On the other hand, load line L_2 represents the operation of the photodiode as a **solar cell;** in this mode, the operating point is in the negative i, positive v, or fourth, quadrant, and therefore the power dissipated by the diode is *negative*. In other words, the photodiode is generating power by converting light energy to electric energy. Note further that the load line intersects the voltage axis at zero, meaning that no supply voltage is required to bias the photodiode in the solar-cell mode. Finally, load line L_3 represents the operation of the diode as a light sensor: when the diode is reverse-biased, the current flowing through the diode is determined by the light intensity; thus, the diode current changes in response to changes in the incident light intensity.

The operation of the photodiode can also be reversed, in principle, by forward-biasing the diode and causing a significant level of recombination to take place in the depletion region. Some of the energy released is converted to light energy by emission of photons. Thus, a diode operating in this mode emits light when forward-biased. Photodiodes used in this way are called **light-emitting diodes (LEDs);** they exhibit a forward (offset) voltage of 1 to 2 V. The circuit symbol for the LED is shown in Figure 9.72.

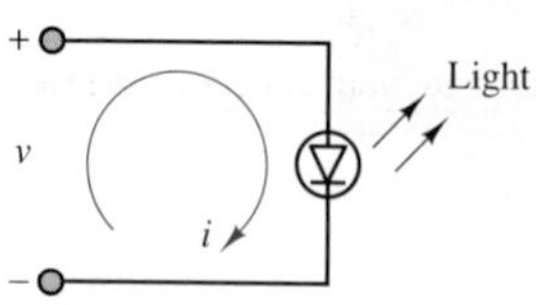

Figure 9.72 Light-emitting diode circuit symbol

Gallium arsenide (GaAs) is one of the more popular substrates for creating LEDs; gallium phosphide (GaP) and the alloy $GaAs_{1-x}P_x$ are also quite common. Table 9.1 lists combinations of materials and dopants used for common LEDs and the colors they emit. The dopants are used to create the necessary *pn* junction.

Table 9.1 LED materials and wavelengths

Material	Dopant	Wavelength, nm	Color
GaAs	Zn	900	Infrared
GaAs	Si	910–1,020	Infrared
GaP	N	570	Green
GaP	N	590	Yellow
GaP	Zn, O	700	Red
$GaAs_{0.6}P_{0.4}$		650	Red
$GaAs_{0.35}P_{0.65}$	N	632	Orange
$GaAs_{0.15}P_{0.85}$	N	589	Yellow

The construction of a typical LED is shown in Figure 9.73, along with the schematic representation for an LED. A shallow *pn* junction is created with electrical contacts made to both *p* and *n* regions. As much of the upper surface of the *p* material is uncovered as possible, so that light can leave the device unimpeded. It is important to note that, actually, only a relatively small fraction of the emitted light leaves the device; the majority stays inside the semiconductor. A photon that stays inside the device will eventually collide with an electron in the valence band, and the collision will force the electron into the conduction band, emitting an electron-hole pair and absorbing the photon. To minimize the probability that a photon will be absorbed

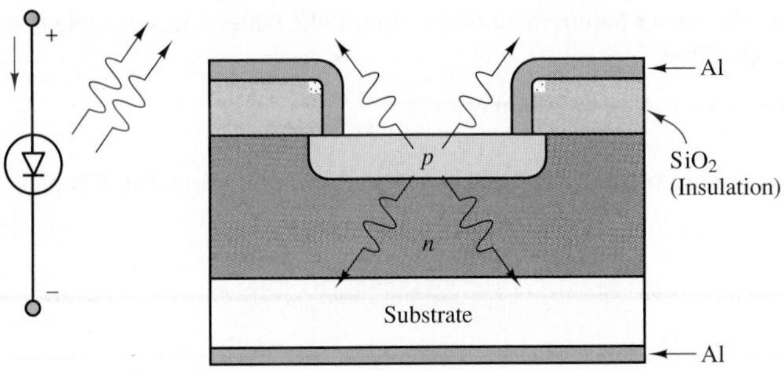

Figure 9.73 Light-emitting diode (LED)

before it has an opportunity to leave the LED, the depth of the p-doped region is left very thin. Also, it is advantageous to have most of the recombinations that emit photons occur as close to the surface of the diode as possible. This is made possible by various doping schemes, but even so, of all the carriers going through the diode, only a small fraction emit photons that are able to leave the semiconductor.

A simple LED drive circuit is shown in Figure 9.74. From the standpoint of circuit analysis, LED characteristics are very similar to those of the silicon diode, except that the offset voltage is usually quite a bit larger. Typical values of V_γ can be in the range of 1.2 to 2 V, and operating currents can range from 20 to 100 mA. Manufacturers usually specify an LED's characteristics by giving the rated operating-point current and voltage.

EXAMPLE 9.14 Analysis of Light-Emitting Diode

LO6

Problem

For the circuit of Figure 9.74, determine (1) the LED power consumption, (2) the resistance R_S, and (3) the power required by the voltage source.

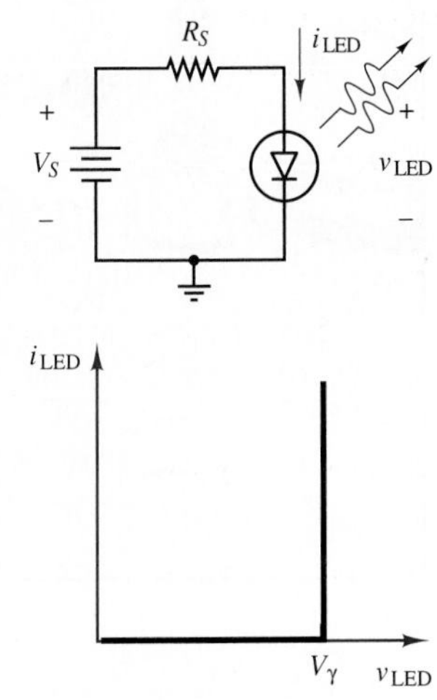

Figure 9.74 LED drive circuit and i-v characteristic based on offset model

Solution

Known Quantities: Diode operating point: $V_{LED} = 1.7$ V; $I_{LED} = 40$ mA; $V_S = 5$ V.

Find: P_{LED}; R_S; P_S.

Assumptions: Use the offset diode model.

Analysis:

1. The power consumption of the LED is determined directly from the specification of the operating point:

$$P_{LED} = V_{LED} \times I_{LED} = 68 \text{ mW}$$

2. To determine the required value of R_S to achieve the desired operating point, we apply KVL around the circuit of Figure 9.74:

$$V_S = I_{LED} R_S + V_{LED}$$

$$R_S = \frac{V_S - V_{LED}}{I_{LED}} = \frac{5 - 1.7}{40 \times 10^{-3}} = 82.5\ \Omega$$

3. To satisfy the power requirement of the circuit, the battery must be able to supply 40 mA to the diode. Thus,

$$P_S = V_S \times I_{\text{LED}} = 200 \text{ mW}$$

Comments: A more practical LED biasing circuit may be found in Chapter 10 (Example 10.7).

CHECK YOUR UNDERSTANDING

Determine the source resistance required to bias the LED of Example 9.14 if the required LED current is 24 mA.

Answer: 137.5 Ω

Opto-isolators

One of the common applications of photodiodes and LEDs is the **optocoupler,** or **opto-isolator.** This device, which is usually enclosed in a sealed package, uses the light-to-current and current-to-light conversion property of photodiodes and LEDs to provide signal connection between two circuits without any need for electrical connections. Figure 9.75 depicts the circuit symbol for the opto-isolator.

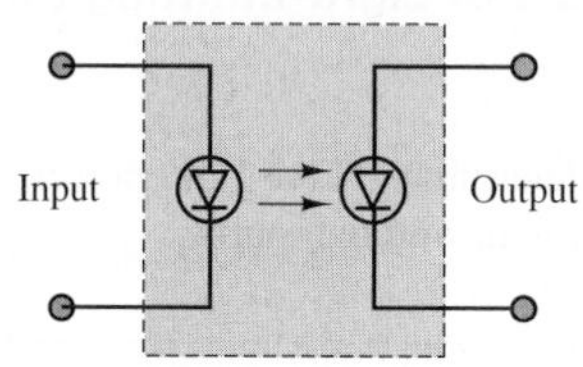

Figure 9.75 Opto-isolator

Because diodes are nonlinear devices, the opto-isolator is not used in transmitting analog signals: the signals would be distorted because of the nonlinear diode i-v characteristic. However, opto-isolators find a very important application when on/off signals need to be transmitted from high-power machinery to delicate computer control circuitry. The optical interface ensures that potentially damaging large currents cannot reach delicate instrumentation and computer circuits.

Conclusion

This chapter introduces the topic of electronic devices by presenting the semiconductor diode. Upon completing this chapter, you should have mastered the following learning objectives:

1. *Understand the basic principles underlying the physics of semiconductor devices in general and of the pn junction in particular. Become familiar with the diode equation and i-v characteristic.* Semiconductors have conductive properties that fall between

those of conductors and insulators. These properties make the materials useful in the construction of many electronic devices that exhibit nonlinear i-v characteristics. Of these devices, the diode is one of the most commonly employed.

2. *Use various circuit models of the semiconductor diode in simple circuits. These are divided into two classes: the large-signal models, useful to study rectifier circuits, and the small-signal models, useful in signal processing applications.* The semiconductor diode acts as a one-way current valve, permitting the flow of current only when it is biased in the forward direction. The behavior of the diode is described by an exponential equation, but it is possible to approximate the operation of the diode by means of simple circuit models. The simplest (ideal) model treats the diode either as a short circuit (when it is forward-biased) or as an open circuit (when it is reverse-biased). The ideal model can be extended to include an offset voltage, which represents the contact potential at the diode *pn* junction. A further model, useful for small-signal circuits, includes a resistance that models the forward resistance of the diode. With the aid of these models it is possible to analyze diode circuits by using the DC and AC circuit analysis methods of earlier chapters.
3. *Study practical full-wave rectifier circuits and learn to analyze and determine the practical specifications of a rectifier by using large-signal diode models.* One of the most important properties of the diode is its ability to rectify AC voltages and currents. Diode rectifiers can be of the half-wave and full-wave types. Full-wave rectifiers can be constructed in a two-diode configuration or in a four-diode bridge configuration. Diode rectification is an essential element of DC power supplies. Another important part of a DC power supply is the filtering, or smoothing, that is usually accomplished by using capacitors.
4. *Understand the basic operation of Zener diodes as voltage references, and use simple circuit models to analyze elementary voltage regulators.* In addition to rectification and filtering, the power supply requires output voltage regulation. Zener diodes can be used to provide a voltage reference that is useful in voltage regulators.
5. *Use the diode models presented in Section 9.2 to analyze the operation of various practical diode circuits in signal processing applications.* In addition to power supply applications, diodes find use in many signal processing and signal conditioning circuits. Of these, the diode peak detector, the diode limiter, and the diode clamp are explored in this chapter.
6. *Understand the basic principle of operation of photodiodes, including solar cells, photosensors, and light-emitting diodes.* Semiconductor material properties can also be affected by light intensity. Certain types of diodes, known as *photodiodes,* find applications in light detectors, solar cells, or light-emitting diodes.

HOMEWORK PROBLEMS

Section 9.1: Electrical Conduction in Semiconductor Devices; Section 9.2: The *pn* Junction and the Semiconductor Diode

9.1 In a semiconductor material, the net charge is zero. This requires the density of positive charges to be equal to the density of negative charges. Both charge carriers (free electrons and holes) and ionized dopant atoms have a charge equal to the magnitude of one electronic charge. Therefore the charge neutrality equation (CNE) is:

$$p_o + N_d^+ - n_o - N_a^- = 0$$

where

$n_o =$ equilibrium negative carrier density

$p_o =$ equilibrium positive carrier density

$N_a^- =$ ionized acceptor density

$N_d^+ =$ ionized donor density

The carrier product equation (CPE) states that as a

semiconductor is doped, the product of the charge carrier densities remains constant:

$$n_o p_o = \text{const}$$

For intrinsic silicon at $T = 300$ K:

$$\text{Const} = n_{io} p_{io} = n_{io}^2 = p_{io}^2$$

$$= \left(1.5 \times 10^{16} \frac{1}{\text{m}^3}\right)^2 = 2.25 \times 10^{32} \frac{1}{\text{m}^2}$$

The semiconductor material is n- or p-type depending on whether donor or acceptor doping is greater. Almost all dopant atoms are ionized at room temperature. If intrinsic silicon is doped:

$$N_A \approx N_a^- = 10^{17} \frac{1}{\text{m}^3} \qquad N_d = 0$$

Determine:

a. If this is an n- or p-type extrinsic semiconductor.

b. Which are the major and which the minority charge carriers.

c. The density of majority and minority carriers.

9.2 If intrinsic silicon is doped, then

$$N_a \approx N_a^- = 10^{17} \frac{1}{\text{m}^3} \qquad N_d \approx N_d^+ = 5 \times 10^{18} \frac{1}{\text{m}^3}$$

Determine:

a. If this is an n- or p-type extrinsic semiconductor.

b. Which are the majority and which the minority charge carriers.

c. The density of majority and minority carriers.

9.3 Describe the microscopic structure of semiconductor materials. What are the three most commonly used semiconductor materials?

9.4 Describe the thermal production of charge carriers in a semiconductor and how this process limits the operation of a semiconductor device.

9.5 Describe the properties of donor and acceptor dopant atoms and how they affect the densities of charge carriers in a semiconductor material.

9.6 Physically describe the behavior of the charge carriers and ionized dopant atoms in the vicinity of a semiconductor pn junction that causes the potential (energy) barrier that tends to prevent charge carriers from crossing the junction.

Section 9.3: Circuit Models for the Semiconductor Diode

9.7 Consider the circuit of Figure P9.7. Determine whether the diode is conducting or not. Assume that the diode is an ideal diode.

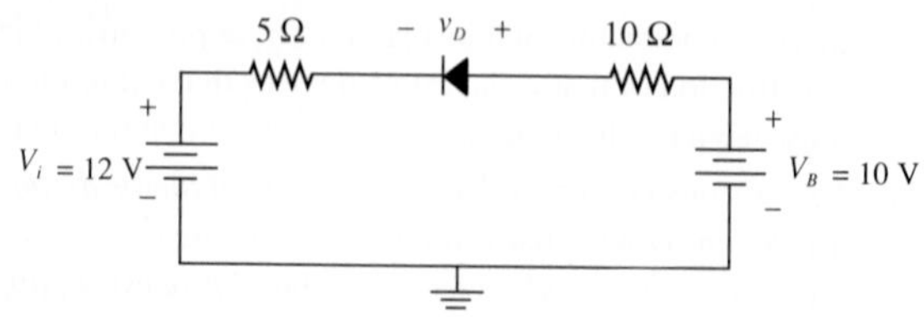

Figure P9.7

9.8 Repeat Problem 9.7 for $V_i = 12$ V and $V_B = 15$ V.

9.9 Consider the circuit of Figure P9.9. Determine whether the diode is conducting or not. Assume that the diode is an ideal diode.

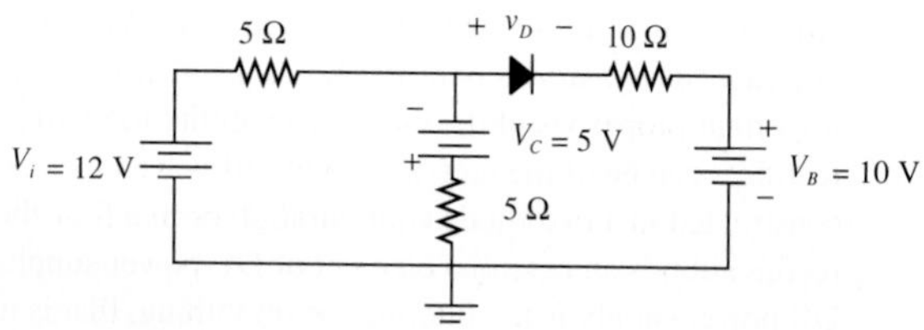

Figure P9.9

9.10 Repeat Problem 9.9 for $V_B = 15$ V.

9.11 Repeat Problem 9.9 for $V_C = 15$ V.

9.12 Repeat Problem 9.9 for $V_C = 10$ V and $V_B = 15$ V.

9.13 For the circuit of Figure P9.13, sketch $i_D(t)$ for the following conditions:

a. Use the ideal diode model.

b. Use the ideal diode model with offset ($V_\gamma = 0.6$ V).

c. Use the piecewise linear approximation with

$$r_D = 1 \text{ k}\Omega$$
$$V_\gamma = 0.6 \text{ V}$$

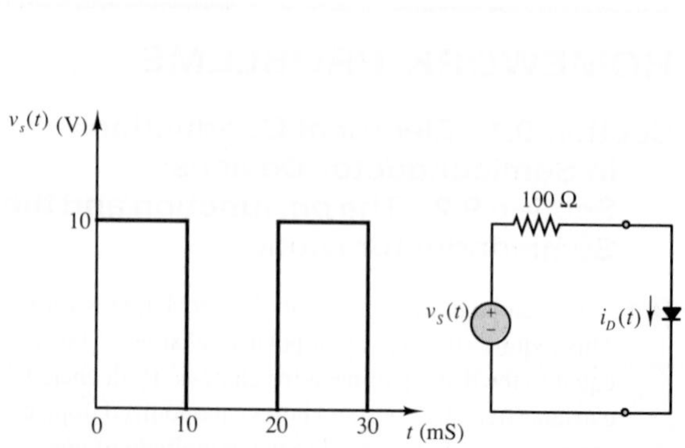

Figure P9.13

9.14 For the circuit of Figure P9.14, find the range of V_{in} for which D_1 is forward-biased. Assume an ideal diode.

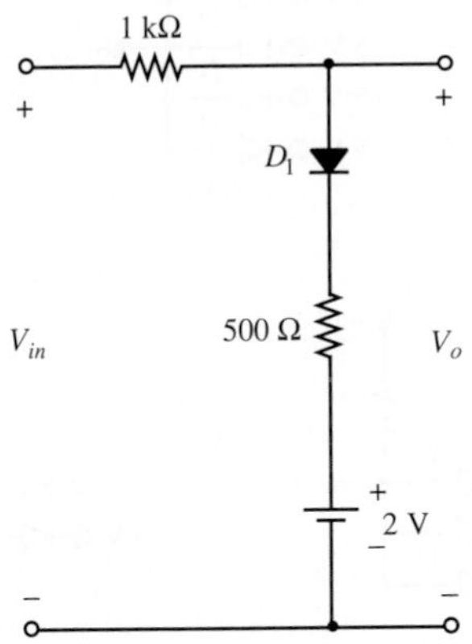

Figure P9.14

9.15 One of the more interesting applications of a diode, based on the diode equation, is an electronic thermometer. The concept is based on the empirical observation that if the current through a diode is nearly constant, the offset voltage is nearly a linear function of the temperature, as shown in Figure P9.15(a).

a. Show that i_D in the circuit of Figure P9.15(b) is nearly constant in the face of variations in the diode voltage, v_D. This can be done by computing the percent change in i_D for a given percent change in v_D. Assume that v_D changes by 10 percent, from 0.6 to 0.66 V.

b. On the basis of the graph of Figure P9.14(a), write an equation for $v_D\ (T°)$ of the form

$$v_D = \alpha T° + \beta$$

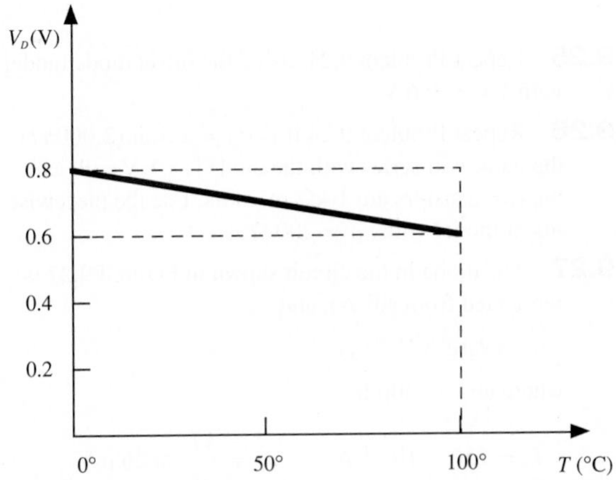

(a)

Figure P9.15 *Continued*

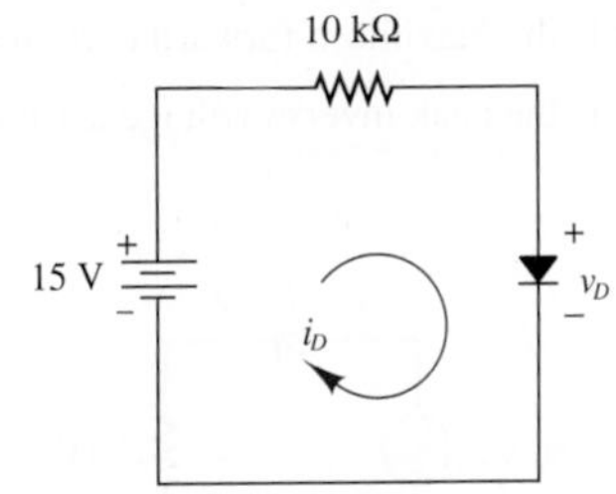

(b)

Figure P9.15

9.16 Find voltage v_L in the circuit of Figure P9.16 where D is an ideal diode, for positive and negative values of v_S. Sketch a plot of v_L versus v_S.

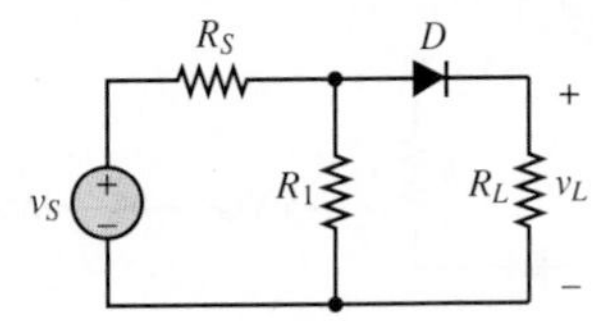

Figure P9.16

9.17 Repeat Problem 9.16, using the offset diode model.

9.18 In the circuit of Figure P9.16, $v_S = 6$ V and $R_1 = R_S = R_L = 1$ kΩ. Determine i_D and v_D graphically, using the diode characteristic of the 1N461A.

9.19 Assume that the diode in Figure P9.19 requires a minimum current of 1 mA to be above the knee of its i-v characteristic. Use $V_\gamma = 0.7$ V.

a. What should be the value of R to establish 5 mA in the circuit?

b. With the value of R determined in part a, what is the minimum value to which the voltage E could be reduced and still maintain diode current above the knee?

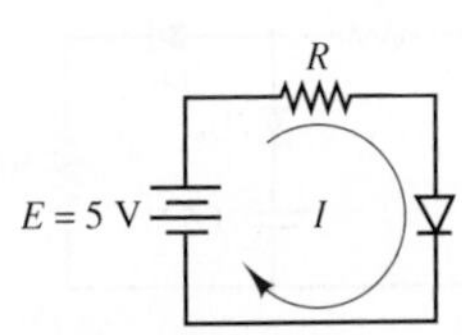

Figure P9.19

9.20 In Figure P9.20, a sinusoidal source of 50 V rms drives the circuit. Use the offset diode model for a silicon diode.

a. What is the maximum forward current?

b. What is the peak inverse voltage across the diode?

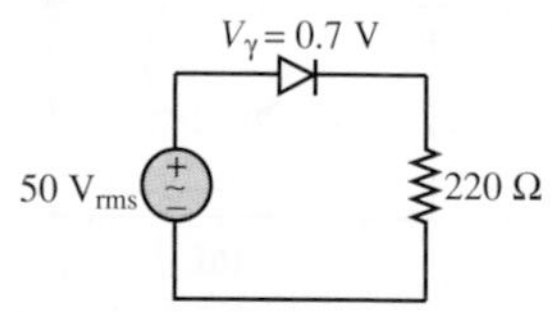

Figure P9.20

9.21 Determine which diodes are forward-biased and which are reverse-biased in each of the configurations shown in Figure P9.21.

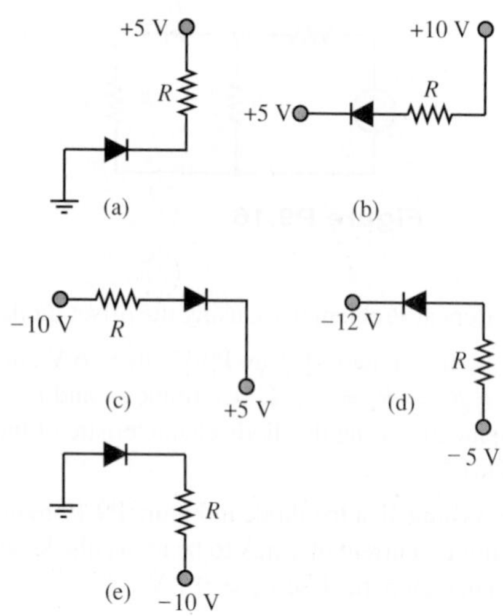

Figure P9.21

9.22 In the circuit of Figure P9.22, find the range of V_{in} for which D_1 is forward-biased. Assume ideal diodes.

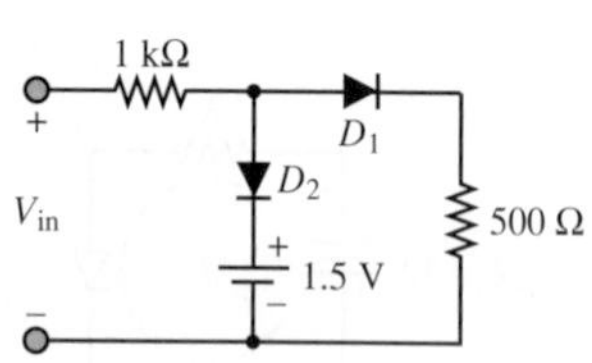

Figure P9.22

9.23 Determine which diodes are forward-biased and which are reverse-biased in the configurations shown in Figure P9.23. Assuming a 0.7-V drop across each forward-biased diode, determine the output voltage.

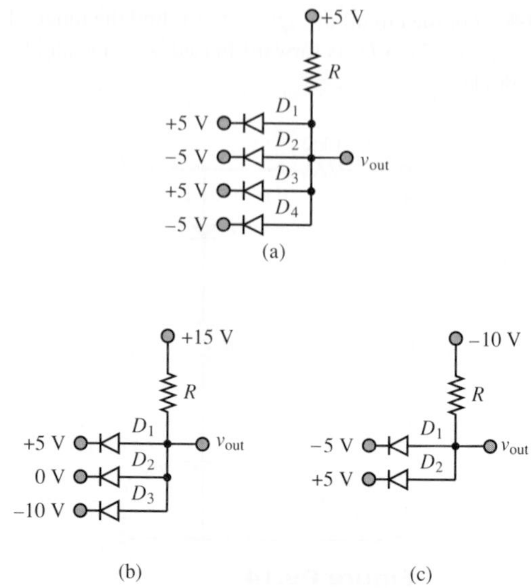

Figure P9.23

9.24 Sketch the output waveform and the voltage transfer characteristic for the circuit of Figure P9.24. Assume ideal diode characteristics, $v_S(t) = 10 \sin(2{,}000\pi t)$.

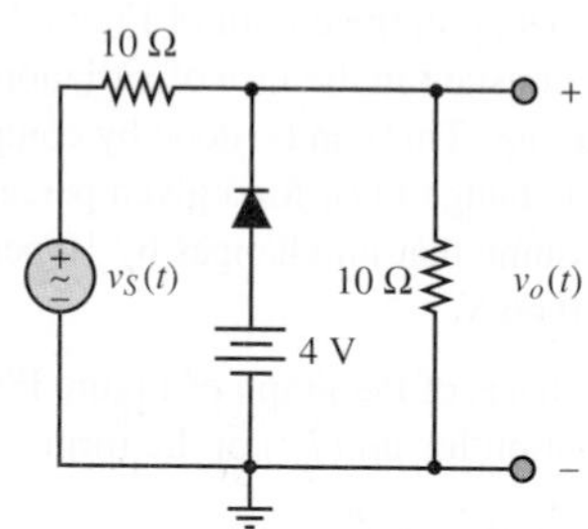

Figure P9.24

9.25 Repeat Problem 9.24, using the offset diode model with $V\gamma = 0.6$ V.

9.26 Repeat Problem 9.24 if $v_S(t) = 1.5 \sin(2{,}000\pi t)$, the battery in series with the diode is a 1-V cell, and the two resistors are 1-kΩ resistors. Use the piecewise linear model with $r_D = 200\ \Omega$.

9.27 The diode in the circuit shown in Figure P9.27 is fabricated from silicon, and

$$i_D = I_o(e^{v_D/V_T} - 1)$$

where at $T = 300$ K

$$I_o = 250 \times 10^{-12}\ \text{A} \qquad V_T = \frac{kT}{q} \approx 26\ \text{mV}$$

$$v_S = 4.2\ \text{V} + 110\cos(\omega t) \qquad \text{mV}$$

$$\omega = 377\ \text{rad/s} \qquad R = 7\ \text{k}\Omega$$

Determine, using superposition, the DC or Q point current through the diode

a. Using the DC offset model for the diode.

b. By numerically solving the circuit characteristic (i.e., the DC load-line equation) and the device characteristic (i.e., the diode equation).

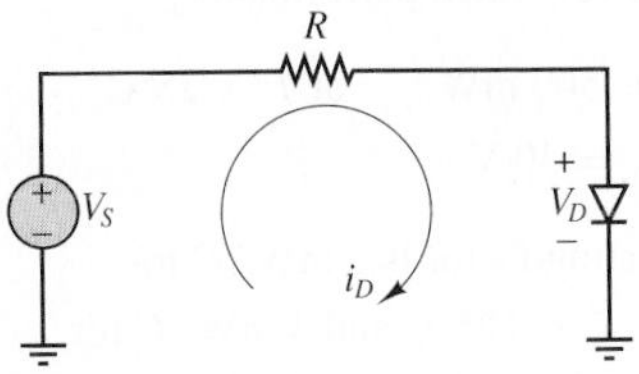

Figure P9.27

9.28 If the diode in the circuit shown in Figure P9.27 is fabricated from silicon and

$$i_D = I_o(e^{v_D/V_T} - 1)$$

where at $T = 300$ K

$$I_o = 2.030 \times 10^{-15} \text{ A} \qquad V_T = \frac{kT}{q} \approx 26 \text{ mV}$$

$$v_S = 5.3 \text{ V} + 7\cos(\omega t) \qquad \text{mV}$$

$$\omega = 377 \text{ rad/s} \qquad R = 4.6 \text{ k}\Omega$$

Determine, using superposition and the offset (or threshold) voltage model for the diode, the DC or Q point current through the diode.

9.29 A diode with the i-v characteristic shown in Figure 9.8 in the text is connected in series with a 5-V voltage source (in the forward bias direction) and a load resistance of 200 Ω. Determine

a. The load current and voltage.

b. The power dissipated by the diode.

c. The load current and voltage if the load is changed to 100 Ω and 500 Ω.

9.30 A diode with the i-v characteristic shown in Figure 9.32 in the text is connected in series with a 2-V voltage source (in the forward bias direction) and a load resistance of 200 Ω. Determine

a. The load current and voltage.

b. The power dissipated by the diode.

c. The load current and voltage if the load is changed to 100 Ω and 300 Ω.

9.31 The diode in the circuit shown in Figure P9.27 is fabricated from silicon and

$$i_D = I_o(e^{v_D/V_T} - 1)$$

where at $T = 300$ K

$$I_o = 250 \times 10^{-12} \text{ A} \qquad V_T = \frac{kT}{q} \approx 26 \text{ mV}$$

$$v_S = V_S + v_s = 4.2 \text{ V} + 110\cos(\omega t) \qquad \text{mV}$$

$$\omega = 377 \text{ rad/s} \qquad R = 7 \text{ k}\Omega$$

The DC operating point or quiescent point (Q point) and the AC small-signal equivalent resistance at this Q point are

$$I_{DQ} = 0.548 \text{ mA} \qquad V_{DQ} = 0.365 \text{ V} \qquad r_d = 47.45\ \Omega$$

Determine, using superposition, the AC voltage across the diode and the AC current through it.

9.32 The diode in the circuit shown in Figure P9.32 is fabricated from silicon and

$$R = 2.2 \text{ k}\Omega \qquad V_{S2} = 3 \text{ V}$$

Determine the minimum value of V_{S1} at and above which the diode will conduct with a significant current.

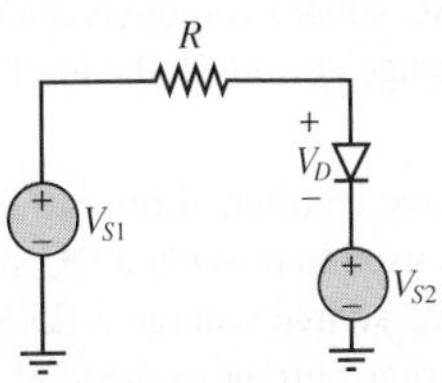

Figure P9.32

Section 9.4: Rectifier Circuits

9.33 Find the average value of the output voltage for the circuit of Figure P9.33 if the input voltage is sinusoidal with an amplitude of 5 V. Let $V_\gamma = 0.7$ V.

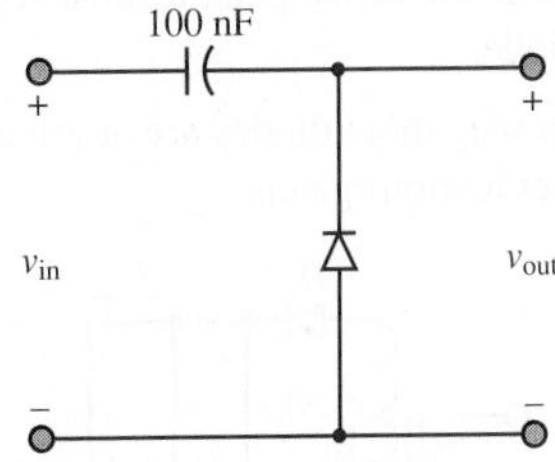

Figure P9.33

9.34 In the rectifier circuit shown in Figure P9.33, $v(t) = A\sin(2\pi 100)t$ V. Assume a forward voltage drop of 0.7 V across the diode when it is conducting. If conduction must begin during each positive half-cycle at an angle no greater than 5°, what is the minimum peak value A that the AC source must produce?

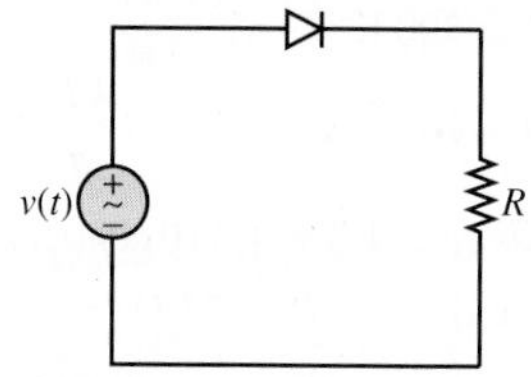

Figure P9.34

9.35 A half-wave rectifier is to provide an average voltage of 50 V at its output.

a. Draw a schematic diagram of the circuit.

b. Sketch the output voltage waveshape.

c. Determine the peak value of the output voltage.

d. Sketch the input voltage waveshape.

e. What is the rms voltage at the input?

9.36 A half-wave rectifier, similar to that of Figure 9.25 in the text, is used to provide a DC supply to a 100-Ω load. If the AC source voltage is 30 V (rms), find the peak and average current in the load. Assume an ideal diode.

9.37 A half-wave rectifier, similar to that of Figure 9.25 in the text, is used to provide a DC supply to a 220-Ω load. If the AC source voltage is 25 V (rms), find the peak and average current in the load. Assume an ideal diode.

9.38 In the full-wave power supply shown in Figure P9.38 the diodes are 1N4001 with a rated peak reverse voltage (also called peak inverse voltage) of 25 V. They are fabricated from silicon.

$$n = 0.05883$$

$$C = 80\ \mu\text{F} \qquad R_L = 1\ \text{k}\Omega$$

$$V_{\text{line}} = 170\cos(377t) \qquad \text{V}$$

a. Determine the actual peak reverse voltage across each diode.

b. Explain why these diodes are or are not suitable for the specifications given.

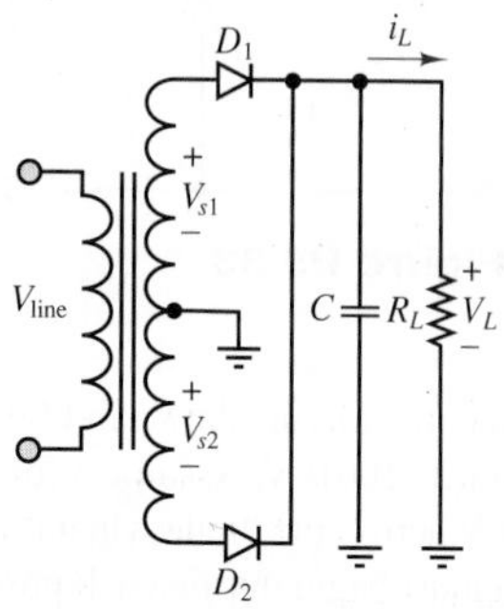

Figure P9.38

9.39 In the full-wave power supply shown in Figure P9.38,

$$n = 0.1$$

$$C = 80\ \mu\text{F} \qquad R_L = 1\ \text{k}\Omega$$

$$V_{\text{line}} = 170\cos(377t) \qquad \text{V}$$

The diodes are 1N914 switching diodes (but used here for AC-DC conversion), fabricated from silicon, with the following rated performance:

$$P_{\max} = 500\ \text{mW} \qquad \text{at } T = 25^\circ\text{C}$$

$$V_{\text{pk-rev}} = 30\ \text{V}$$

The derating factor is 3 mW/°C for $25^\circ\text{C} < T \le 125^\circ\text{C}$ and 4 mW/°C for $125^\circ\text{C} < T \le 175^\circ\text{C}$.

a. Determine the actual peak reverse voltage across each diode.

b. Explain why these diodes are or are not suitable for the specifications given.

9.40 The diodes in the full-wave DC power supply shown in Figure P9.38 are silicon. The load voltage waveform is shown in Figure P9.40. If

$$I_L = 60\ \text{mA} \qquad V_L = 5\ \text{V} \qquad V_{\text{ripple}} = 5\%$$

$$V_{\text{line}} = 170\cos(\omega t) \qquad \text{V} \qquad \omega = 377\ \text{rad/s}$$

determine the value of

a. The turns ratio n.

b. The capacitor C.

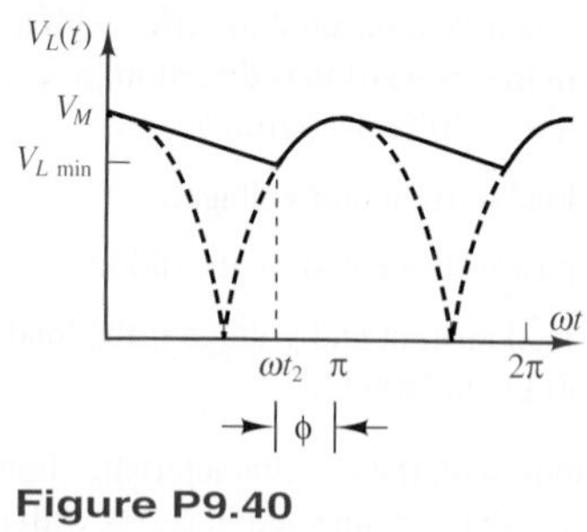

Figure P9.40

9.41 The diodes in the full-wave DC power supply shown in Figure P9.38 are silicon. If

$$I_L = 600\ \text{mA} \qquad V_L = 50\ \text{V}$$

$$V_r = 8\% = 4\ \text{V}$$

$$V_{\text{line}} = 170\cos(\omega t) \qquad \text{V} \qquad \omega = 377\ \text{rad/s}$$

determine the value of

a. The turns ratio n.

b. The capacitor C.

9.42 The diodes in the full-wave DC power supply shown in Figure P9.38 are silicon. If

$I_L = 5$ mA $\quad V_L = 10$ V

$V_r = 20\% = 2$ V

$V_{\text{line}} = 170\cos(\omega t)$ V $\quad \omega = 377$ rad/s

determine the

a. Turns ratio n.

b. The value of the capacitor C.

9.43 You have been asked to design a full-wave bridge rectifier for a power supply. A step-down transformer has already been chosen. It will supply 12 V rms to your rectifier. The full-wave rectifier is shown in the circuit of Figure P9.43.

a. If the diodes have an offset voltage of 0.6 V, sketch the input source voltage $v_S(t)$ and the output voltage $v_L(t)$, and state which diodes are on and which are off in the appropriate cycles of $v_S(t)$. The frequency of the source is 60 Hz.

b. If $R_L = 1{,}000\ \Omega$ and a capacitor, placed across R_L to provide some filtering, has a value of 8 μF, sketch the output voltage $v_L(t)$.

c. Repeat part b, with the capacitance equal to 100 μF.

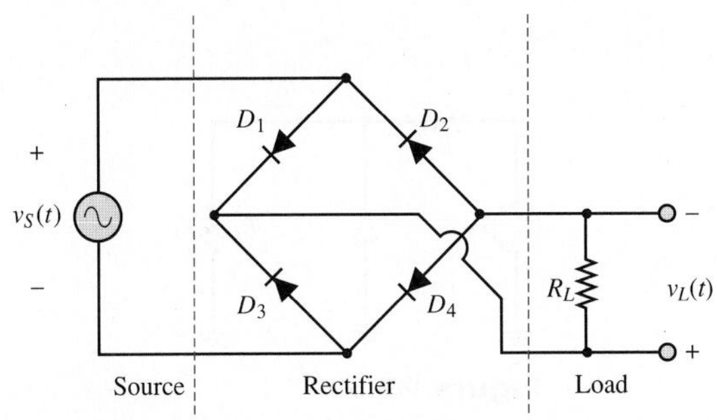

Figure P9.43

9.44 In the full-wave power supply shown in Figure P9.44 the diodes are 1N4001 with a rated peak reverse voltage (also called peak inverse voltage) of 50 V. They are fabricated from silicon.

$V_{\text{line}} = 170\cos(377t)$ V

$n = 0.2941$

$C = 700\ \mu$F $\quad R_L = 2.5$ kΩ

a. Determine the actual peak reverse voltage across each diode.

b. Explain why these diodes are or are not suitable for the specifications given.

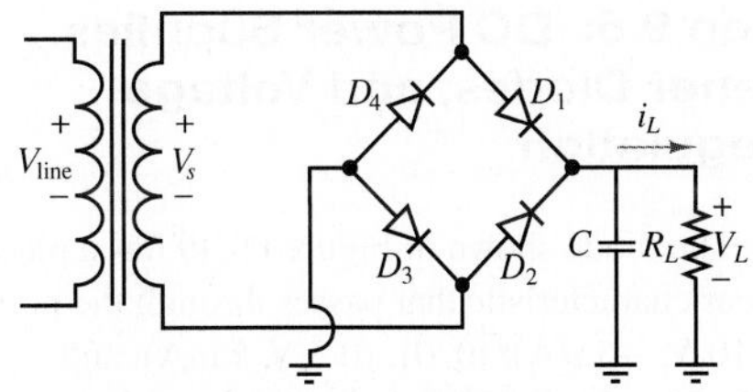

Figure P9.44

9.45 In the full-wave power supply shown in Figure P9.44 the diodes are 1N4001 general-purpose silicon diodes with a rated peak reverse voltage of 10 V and

$V_{\text{line}} = 156\cos(377t)$ V

$n = 0.04231 \quad V_r = 0.2$ V

$I_L = 2.5$ mA $\quad V_L = 5.1$ V

a. Determine the actual peak reverse voltage across the diodes.

b. Explain why these diodes are or are not suitable for the specifications given.

9.46 The diodes in the full-wave DC power supply shown in Figure P9.44 are silicon. If

$I_L = 650$ mA $\quad V_L = 10$ V

$V_r = 1$ V $\quad \omega = 377$ rad/s

$V_{\text{line}} = 170\cos(\omega t)$ V $\quad \phi = 23.66°$

determine the value of the average and peak current through each diode.

9.47 The diodes in the full-wave DC power supply shown in Figure P9.44 are silicon. If

$I_L = 85$ mA $\quad V_L = 5.3$ V

$V_r = 0.6$ V $\quad \omega = 377$ rad/s

$V_{\text{line}} = 156\cos(\omega t)$ V

determine the value of

a. The turns ratio n.

b. The capacitor C.

9.48 The diodes in the full-wave DC power supply shown in Figure P9.44 are silicon. If

$I_L = 250$ mA $\quad V_L = 10$ V

$V_r = 2.4$ V $\quad \omega = 377$ rad/s

$V_{\text{line}} = 156\cos(\omega t)$ V

determine the value of

a. The turns ratio n.

b. The capacitor C.

Section 9.5: DC Power Supplies, Zener Diodes, and Voltage Regulation

9.49 The diode shown in Figure P9.49 has a piecewise linear characteristic that passes through the points (-10 V, -5 μA), (0, 0), (0.5 V, 5 mA), and (1 V, 50 mA). Determine the piecewise linear model, and using that model, solve for i and v.

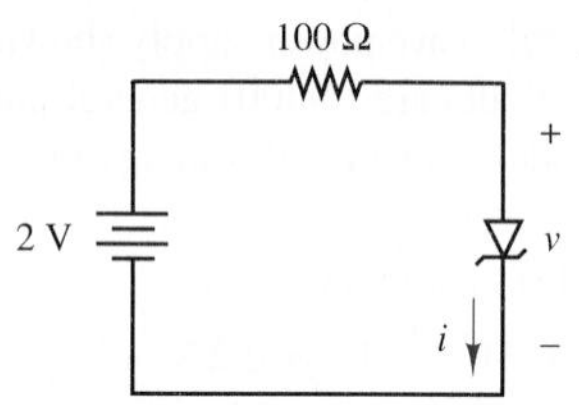

Figure P9.49

9.50 Find the minimum value of R_L in the circuit shown in Figure P9.50 for which the output voltage remains at just 5.6 V.

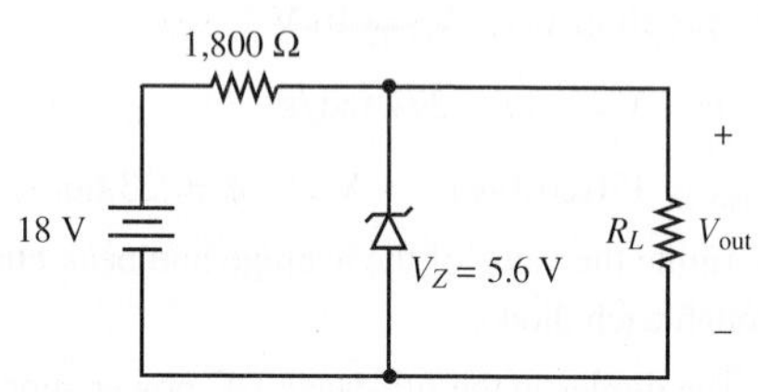

Figure P9.50

9.51 Determine the minimum value and the maximum value that the series resistor may have in a regulator circuit whose output voltage is to be 25 V, whose input voltage varies from 35 to 40 V, and whose maximum load current is 75 mA. The Zener diode used in this circuit has a maximum current rating of 250 mA.

9.52 The i-v characteristic of a semiconductor diode designed to operate in the Zener breakdown region is shown in Figure P9.52. The Zener or breakdown region extends from a minimum current at the knee of the curve, equal here to about -5 mA (from the graph), and the maximum rated current equal to -90 mA (from the specification sheet). Determine the Zener resistance and Zener voltage of the diode.

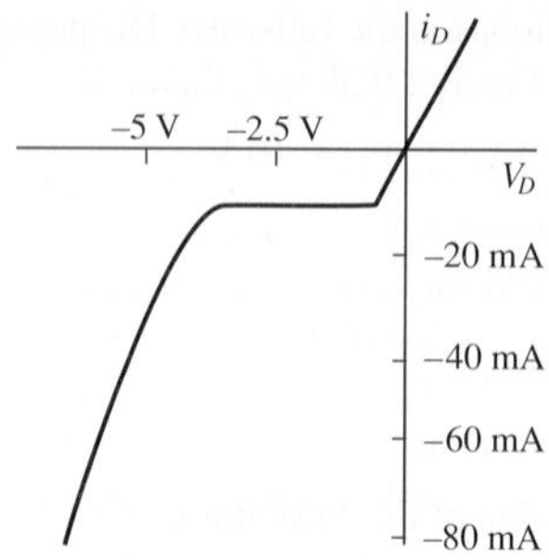

Figure P9.52

9.53 The Zener diode in the simple voltage regulator circuit shown in Figure P9.53 is a 1N5231B. The source voltage is obtained from a DC power supply. It has a DC and a ripple component

$$v_S = V_S + V_r$$

where:

$V_S = 20$ V $\quad V_r = 250$ mV

$R = 220\ \Omega \quad I_L = 65$ mA $\quad V_L = 5.1$ V

$V_z = 5.1$ V $\quad r_z = 17\ \Omega \quad P_{\text{rated}} = 0.5$ W

$i_{z\,\text{min}} = 10$ mA

Determine the maximum rated current the diode can handle without exceeding its power limitation.

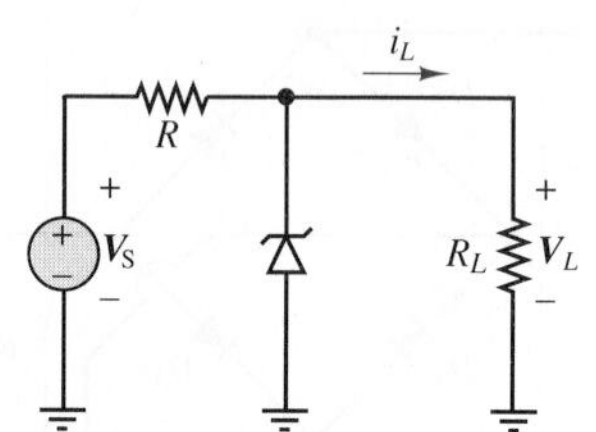

Figure P9.53

9.54 The 1N963 Zener diode in the simple voltage regulator circuit shown in Figure P9.53 has the specifications

$V_z = 12$ V $\quad r_z = 11.5\ \Omega \quad P_{\text{rated}} = 400$ mW

At the knee of curve,

$i_{zk} = 0.25$ mA $\quad r_{zk} = 700\ \Omega$

Determine the maximum rated current the diode can handle without exceeding its power limitation.

9.55 In the simple voltage regulator circuit shown in Figure P9.53, R must maintain the Zener diode current within its specified limits for all values of the source voltage, load current, and Zener diode voltage. Determine the minimum and maximum values of R

which can be used.

$V_z = 5\text{ V} \pm 10\%$ $\quad r_z = 15\ \Omega$

$i_{z\,min} = 3.5$ mA $\quad i_{z\,max} = 65$ mA

$V_S = 12 \pm 3$ V $\quad I_L = 70 \pm 20$ mA

9.56 In the simple voltage regulator circuit shown in Figure P9.53, R must maintain the Zener diode current within its specified limits for all values of the source voltage, load current, and Zener diode voltage. If

$V_z = 12\text{ V} \pm 10\%$ $\quad r_z = 9\ \Omega$

$i_{z\,min} = 3.25$ mA $\quad i_{z\,max} = 80$ mA

$V_S = 25 \pm 1.5$ V

$I_L = 31.5 \pm 21.5$ mA

determine the minimum and maximum values of R which can be used.

9.57 In the simple voltage regulator circuit shown in Figure P9.53, the Zener diode is a 1N4740A.

$V_z = 10\text{ V} \pm 5\%$ $\quad r_z = 7\ \Omega$ $\quad i_{z\,min} = 10$ mA

$P_{rated} = 1$ W $\quad i_{z\,max} = 91$ mA

$V_S = 14 \pm 2$ V $\quad R = 19.8\ \Omega$

Determine the minimum and maximum load current for which the diode current remains within its specified values.

9.58 In the simple voltage regulator circuit shown in Figure P9.53, the Zener diode is a 1N963. Determine the minimum and maximum load current for which the diode current remains within its specified values.

$V_z = 12\text{ V} \pm 10\%$ $\quad r_z = 11.5\ \Omega$

$i_{z\,min} = 2.5$ mA $\quad i_{z\,max} = 32.6$ mA

$P_R = 400$ mW

$V_S = 25 \pm 2$ V $\quad R = 470\ \Omega$

9.59 For the Zener regulator shown in the circuit of Figure P9.59, we desire to hold the load voltage to 14 V. Find the range of load resistances for which regulation can be obtained if the Zener diode is rated at 14 V, 5 W.

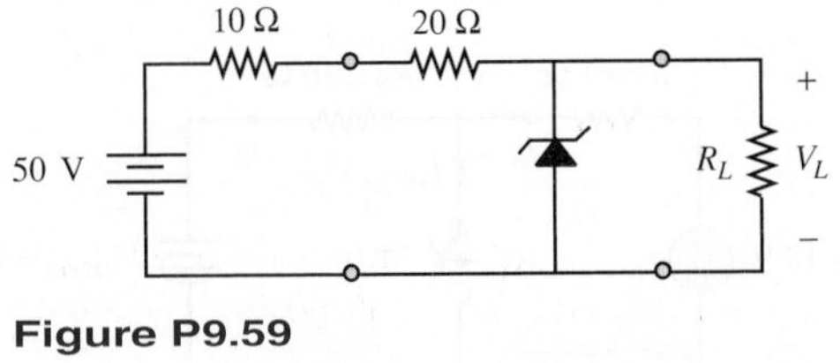

Figure P9.59

9.60 A Zener diode ideal i-v characteristic is shown in Figure P9.60(a). Given a Zener voltage, V_Z of 7.7 V, find the output voltage V_{out} for the circuit of Figure P9.60(b) if V_S is:

a. 12 V

b. 20 V

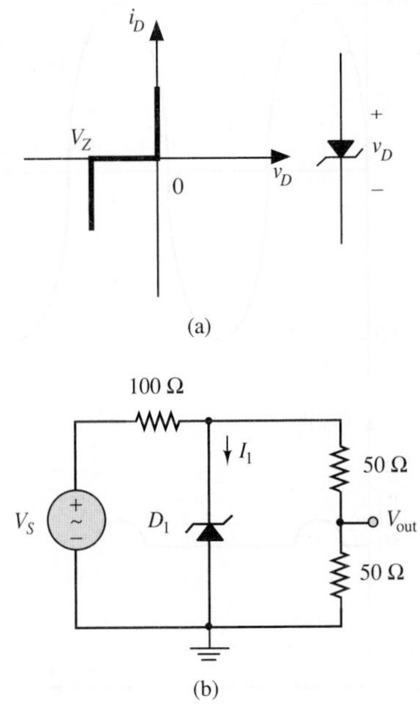

Figure P9.60

Section 9.6: Signal Processing Applications; Section 9.7: Photodiodes

9.61 For the voltage limiter circuit of Figure P9.61, plot the voltage across R_L versus v_S for $-20 < v_S < 20$ V. Assume that

$R_S = 10\ \Omega$ $\quad V_{Z1} = 10$ V $\quad R_2 = 10\ \Omega$ $\quad V_{Z4} = 5$ V

$R_1 = 1\ \Omega$ $\quad R_L = 40\ \Omega$

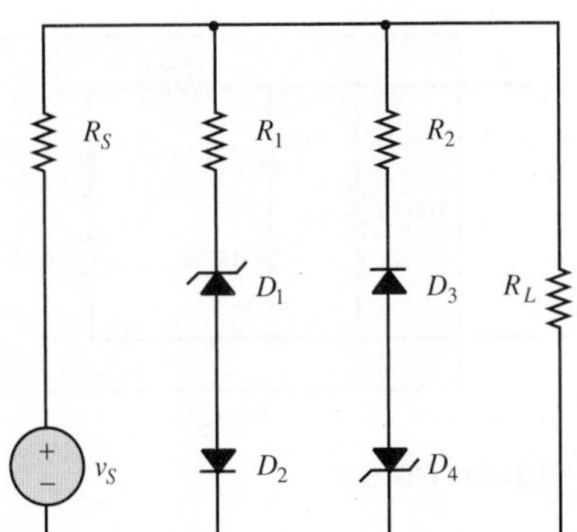

Figure P9.61

9.62 Develop a series diode clipper circuit which will separate a signal as shown in Figure P9.62.

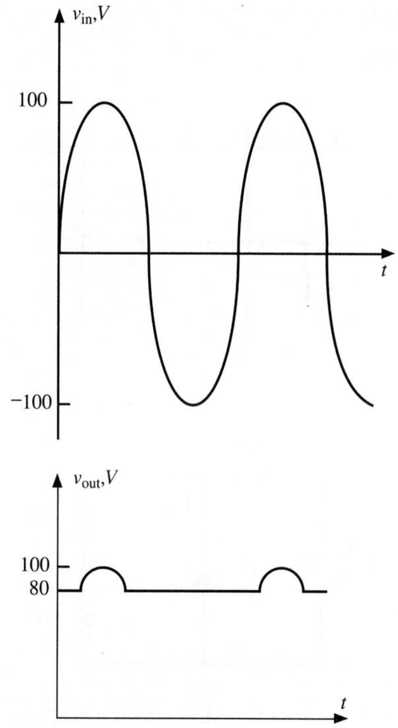

Figure P9.62

9.63 Assuming that the diodes are ideal, determine and sketch the i-v characteristics for the circuit of Figure P9.63. Consider the range $10 \geq v \geq 0$.

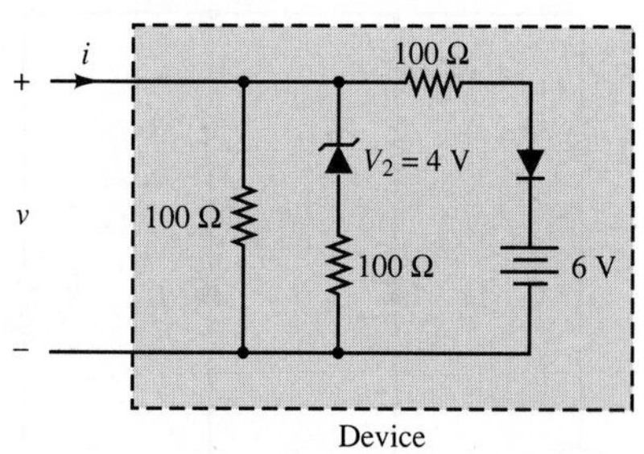

Figure P9.63

9.64 Given the input voltage waveform and the circuit shown in Figure P9.64, sketch the output voltage.

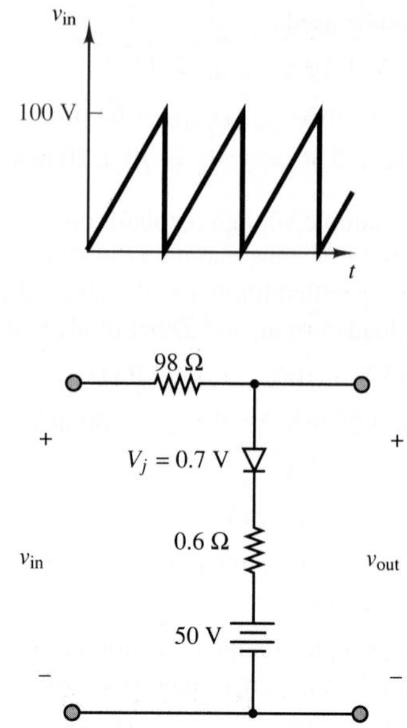

Figure P9.64

9.65 We are using a voltage source to charge an automotive battery as shown in the circuit of Figure P9.65(a). At $t = t_1$, the protective circuitry of the source causes switch S_1 to close, and the source voltage goes to zero. Find the currents I_S, I_B, and I_{SW} for the following conditions:

a. $t = t_1^-$

b. $t = t_1^+$

c. What will happen to the battery after the switch closes?

Now we are going to charge the battery, using the circuit of Figure P9.65(b). Repeat parts a and b if the diode has an offset voltage of 0.6 V.

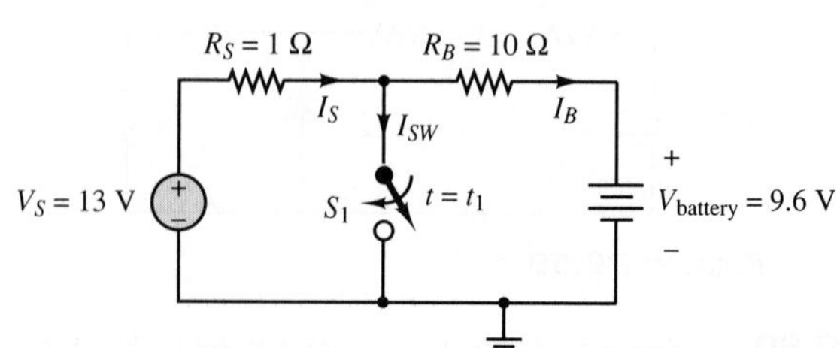

(a)

Figure P9.65 *Continued*

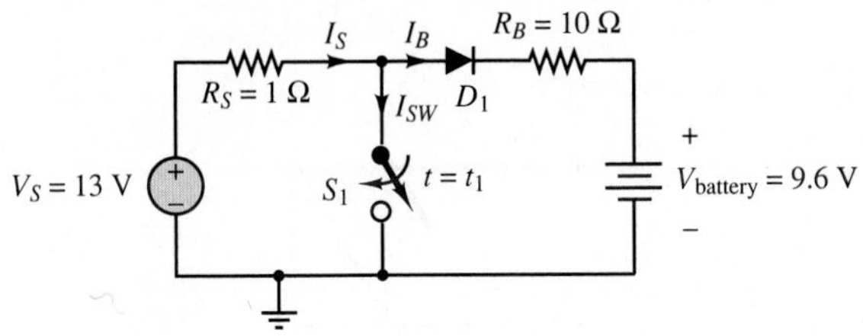

(b)

Figure P9.65

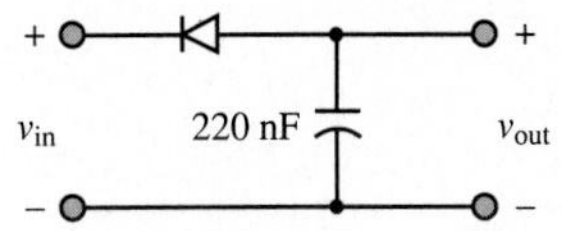

Figure P9.66

9.66 Find the output voltage of the peak detector shown in Figure P9.66. Use sinusoidal input voltages with amplitude 6, 1.5, and 0.4 V and zero average value. Let $V_\gamma = 0.7$ V.

9.67 For the LED circuit of Example 9.14, determine the LED power consumption if the LED consumes 20 mA at the same voltage. How much power is required of the source?

9.68 For the LED circuit of Example 9.14, determine the LED power consumption if the LED consumes 30 mA and the diode voltage is 1.5 V. How much power is required of the source?

CHAPTER

10

BIPOLAR JUNCTION TRANSISTORS: OPERATION, CIRCUIT MODELS, AND APPLICATIONS

Chapter 10 continues the discussion of electronic devices that began in Chapter 9 with the semiconductor diode. This chapter describes the operating characteristics of one of the two major families of electronic devices: bipolar transistors. Chapter 10 is devoted to a brief, qualitative discussion of the physics and operation of the bipolar junction transistor (BJT), which naturally follows the discussion of the *pn* junction in Chapter 9. The i-v characteristics of bipolar transistors and their operating states are presented. Large-signal circuit models for the BJT are then introduced, to illustrate how one can analyze transistor circuits by using basic circuit analysis methods. A few practical examples are discussed to illustrate the use of the circuit models.

This chapter introduces the operation of the bipolar junction transistor. Bipolar transistors represent one of two major families of electronic devices that can serve as amplifiers and switches. Chapter 10 reviews the operation of the bipolar junction transistor and presents simple models that permit the analysis and design of simple amplifier and switch circuits.

Learning Objectives

1. Understand the basic principles of amplification and switching. *Section 10.1.*
2. Understand the physical operation of bipolar transistors; determine the operating point of a bipolar transistor circuit. *Section 10.2.*
3. Understand the large-signal model of the bipolar transistor, and apply it to simple amplifier circuits. *Section 10.3.*
4. Select the operating point of a bipolar transistor circuit; understand the principle of small signal amplifiers. *Section 10.4.*
5. Understand the operation of a bipolar transistor as a switch, and analyze basic analog and digital gate circuits. *Section 10.5.*

10.1 TRANSISTORS AS AMPLIFIERS AND SWITCHES

A transistor is a three-terminal semiconductor device that can perform two functions that are fundamental to the design of electronic circuits: **amplification** and **switching.** Put simply, amplification consists of magnifying a signal by transferring energy to it from an external source, whereas a transistor switch is a device for controlling a relatively large current between or voltage across two terminals by means of a small control current or voltage applied at a third terminal. In this chapter, we provide an introduction to the two major families of transistors: *bipolar junction transistors,* or *BJTs;* and *field-effect transistors,* or *FETs.*

The operation of the transistor as a linear amplifier can be explained qualitatively by the sketch of Figure 10.1, in which the four possible modes of operation of a transistor are illustrated by means of circuit models employing controlled sources (you may wish to review the material on controlled sources in Section 2.1). In Figure 10.1, controlled voltage and current sources are shown to generate an output proportional to an input current or voltage; the proportionality constant μ is called the internal *gain* of the transistor. As will be shown, the BJT acts essentially as a current-controlled device, while the FET behaves as a voltage-controlled device.

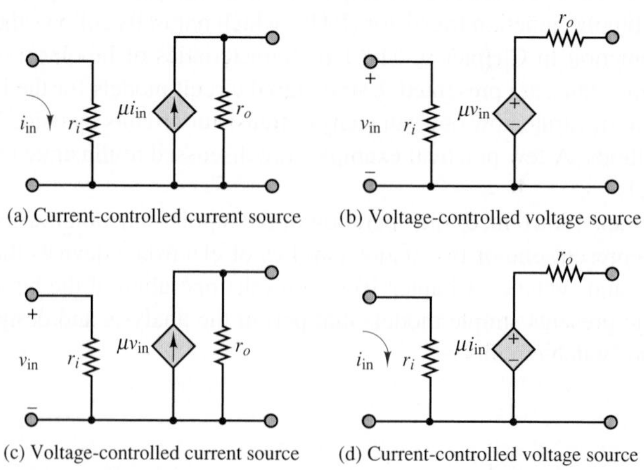

Figure 10.1 Controlled-source models of linear amplifier transistor operation

Transistors can also act in a nonlinear mode, as voltage- or current-controlled switches. When a transistor operates as a switch, a small voltage or current is used to control the flow of current between two of the transistor terminals in an on/off fashion. Figure 10.2 depicts the idealized operation of the transistor as a switch, suggesting that the switch is closed (on) whenever a control voltage or current is greater than zero and is open (off) otherwise. It will later become apparent that the conditions for the switch to be on or off need not necessarily be those depicted in Figure 10.2.

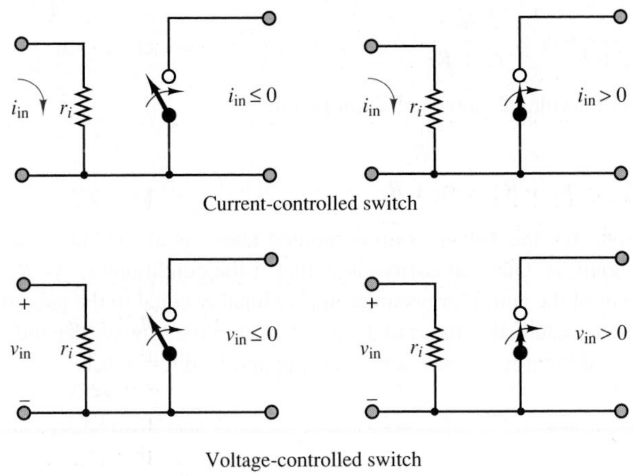

Figure 10.2 Models of ideal transistor switches

EXAMPLE 10.1 Model of Linear Amplifier

LO1

Problem

Determine the voltage gain of the amplifier circuit model shown in Figure 10.3.

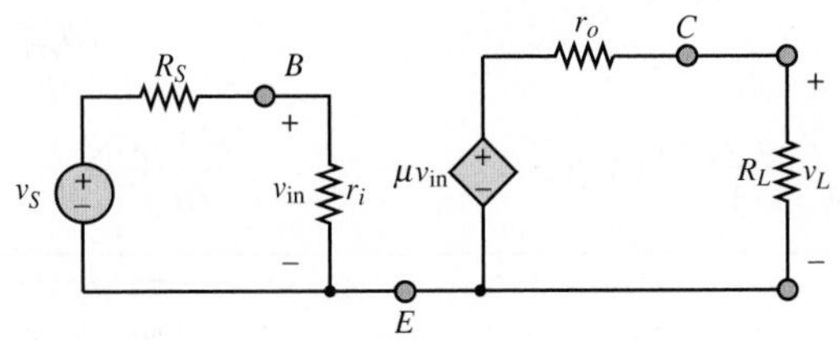

Figure 10.3

Solution

Known Quantities: Amplifier internal input and output resistances r_i and r_o; amplifier internal gain μ; source and load resistances R_S and R_L.

Find: $A_V = \dfrac{v_L}{v_S}$

Analysis: First determine the input voltage, v_{in}, using the voltage divider rule:

$$v_{\text{in}} = \frac{r_i}{r_i + R_S} v_S$$

Then, the output of the controlled voltage source is:

$$\mu v_{\text{in}} = \mu \frac{r_i}{r_i + R_S} v_S$$

and the output voltage can be found by the voltage divider rule:

$$v_L = \mu \frac{r_i}{r_i + R_S} v_S \times \frac{R_L}{r_o + R_L}$$

Finally, the amplifier voltage gain can be computed:

$$A_V = \frac{v_L}{v_S} = \mu \frac{r_i}{r_i + R_S} \times \frac{R_L}{r_o + R_L}$$

Comments: Note that the voltage gain computed above is always less than the transistor internal voltage gain, μ. One can easily show that if the conditions $r_i \gg R_S$ and $r_o \ll R_L$ hold, then the gain of the amplifier becomes approximately equal to the gain of the transistor. One can therefore conclude that the actual gain of an amplifier always depends on the relative values of source and input resistance, and of output and load resistance.

CHECK YOUR UNDERSTANDING

Repeat the analysis of Example 10.1 for the current-controlled voltage source model of Figure 10.1(d). What is the amplifier voltage gain? Under what conditions would the gain A be equal to μ/R_S?

Repeat the analysis of Example 10.1 for the current-controlled current source model of Figure 10.1(a). What is the amplifier voltage gain?

Repeat the analysis of Example 10.1 for the voltage-controlled current source model of Figure 10.1(c). What is the amplifier voltage gain?

Answers: $A = \mu \frac{1}{r_i + R_S} \frac{R_L}{r_o + R_L}$; $r_i \to 0, r_o \to 0$: $A = \frac{1}{r_i + R_S} \frac{r_o R_L}{r_o + R_L} \mu$; $A = \mu \frac{r_i}{r_i + R_S} \frac{r_o R_L}{r_o + R_L}$

10.2 OPERATION OF THE BIPOLAR JUNCTION TRANSISTOR

The *pn* junction studied in Chapter 9 forms the basis of a large number of semiconductor devices. The semiconductor diode, a two-terminal device, is the most direct application of the *pn* junction. In this section, we introduce the **bipolar junction transistor (BJT)**. As we did in analyzing the diode, we will introduce the physics of transistor devices as intuitively as possible, resorting to an analysis of their i-v characteristics to discover important properties and applications.

A BJT is formed by joining three sections of semiconductor material, each with a different doping concentration. The three sections can be either a thin n region sandwiched between p^+ and p layers, or a p region between n and n^+ layers, where the superscript plus indicates more heavily doped material. The resulting BJTs are called *pnp* and *npn* transistors, respectively; we discuss only the latter in this chapter. Figure 10.4 illustrates the approximate construction, symbols, and nomenclature for the two types of BJTs.

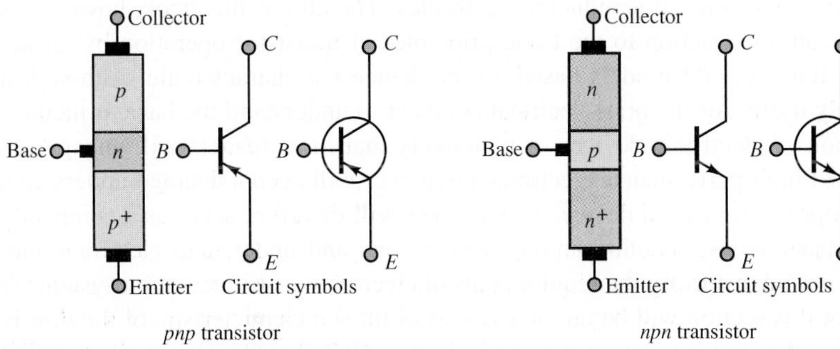

Figure 10.4 Bipolar junction transistors

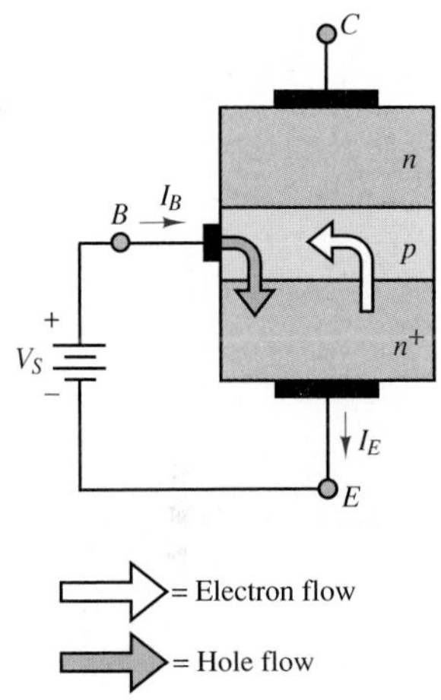

The *BE* junction acts very much as an ordinary diode when the collector is open. In this case, $I_B = I_E$.

Figure 10.5 Current flow in an *npn* BJT

The operation of the *npn* BJT may be explained by considering the transistor as consisting of two back-to-back *pn* junctions. The **base-emitter (BE) junction** acts very much as a diode when it is forward-biased; thus, one can picture the corresponding flow of hole and electron currents from base to emitter when the collector is open and the *BE* junction is forward-biased, as depicted in Figure 10.5. Note that the electron current has been shown larger than the hole current, because of the heavier doping of the n side of the junction. Some of the electron-hole pairs in the base will recombine; the remaining charge carriers will give rise to a net flow of current from base to emitter. It is also important to observe that the base is much narrower than the emitter section of the transistor.

Imagine, now, reverse-biasing the **base-collector (BC) junction**. In this case, an interesting phenomenon takes place: the electrons "emitted" by the emitter with the *BE* junction forward-biased reach the very narrow base region, and after a few are lost to recombination in the base, most of these electrons are "collected" by the collector. Figure 10.6 illustrates how the reverse bias across the *BC* junction is in such a direction as to sweep the electrons from the emitter into the collector. This phenomenon can take place because the base region is kept particularly narrow. Since the base is narrow, there is a high probability that the electrons will have gathered enough momentum from the electric field to cross the reverse-biased collector-base junction and make it into the collector. The result is that there is a net flow of current from collector to emitter (opposite in direction to the flow of electrons), in addition to the hole current from base to emitter. The electron current flowing into the collector through the base is substantially larger than that which flows into the base from the external circuit. One can see from Figure 10.6 that if KCL is to be satisfied, we must have

$$I_E = I_B + I_C \tag{10.1}$$

The most important property of the bipolar transistor is that the small base current controls the amount of the much larger collector current

$$I_C = \beta I_B \tag{10.2}$$

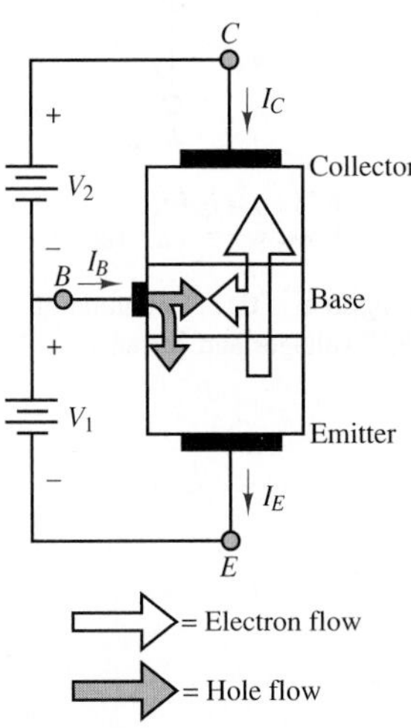

When the *BC* junction is reverse-biased, the electrons from the emitter region are swept across the base into the collector.

Figure 10.6 Flow of emitter electrons into the collector in an *npn* BJT where β is a current amplification factor dependent on the physical properties of the transistor. Typical values of β range from 20 to 200. The operation of a *pnp* transistor is completely analogous to that of the *npn* device, with the roles of the charge carriers (and therefore the signs of the currents) reversed. The symbol for a *pnp* transistor is shown in Figure 10.4.

The exact operation of bipolar transistors can be explained by resorting to a detailed physical analysis of the *npn* or *pnp* structure of these devices. The reader interested in such a discussion of transistors is referred to any one of a number of excellent books on semiconductor electronics. The aim of this book, however, is to provide an introduction to the basic principles of transistor operation by means of simple linear circuit models based on the device *i-v* characteristic. Although it is certainly useful for the non–electrical engineer to understand the basic principles of operation of electronic devices, it is unlikely that most readers will engage in the design of high-performance electronic circuits or will need a detailed understanding of the operation of each device. This chapter will therefore serve as a compendium of the basic ideas, enabling an engineer to read and understand electronic circuit diagrams and to specify the requirements of electronic instrumentation systems. The focus of this section will be on the analysis of the *i-v* characteristic of the *npn* BJT, based on the circuit notation defined in Figure 10.7. The device *i-v* characteristics will be presented qualitatively, without deriving the underlying equations, and will be utilized in constructing circuit models for the device.

The operation of the BJT is defined in terms of two currents and two voltages: i_B, i_C, v_{CE}, and v_{BE}.

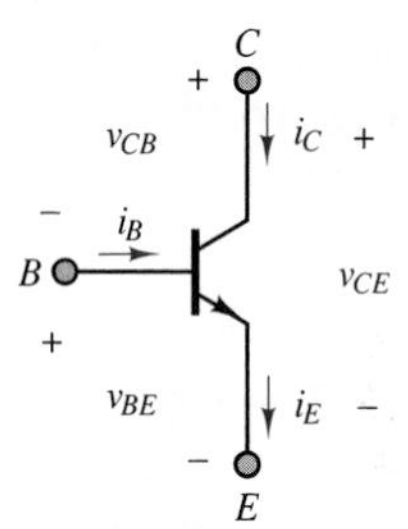

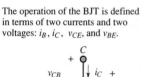

KCL: $i_E = i_B + i_C$
KVL: $v_{CE} = v_{CB} + v_{BE}$

Figure 10.7 Definition of BJT voltages and currents

The number of independent variables required to uniquely define the operation of the transistor may be determined by applying KVL and KCL to the circuit of Figure 10.7. Two voltages and two currents are sufficient to specify the operation of the device. Note that since the BJT is a three-terminal device, it will not be sufficient to deal with a single *i-v* characteristic; two such characteristics are required to explain the operation of this device. One of these characteristics relates the base current, i_B to the base-emitter voltage v_{BE}; the other relates the collector current i_C to the collector-emitter voltage v_{CE}. The latter characteristic actually consists of a *family* of curves. To determine these *i-v* characteristics, consider the *i-v* curves of Figures 10.8 and 10.9, using the circuit notation of Figure 10.7. In Figure 10.8, the collector is open and the *BE* junction is shown to be very similar to a diode. The ideal current source I_{BB} injects a base current, which causes the junction to be forward-biased. By varying I_{BB}, one can obtain the open-collector *BE* junction *i-v* curve shown in the figure.

If a voltage source were now to be connected to the collector circuit, the voltage v_{CE} and, therefore, the collector current i_C could be varied, in addition to the base

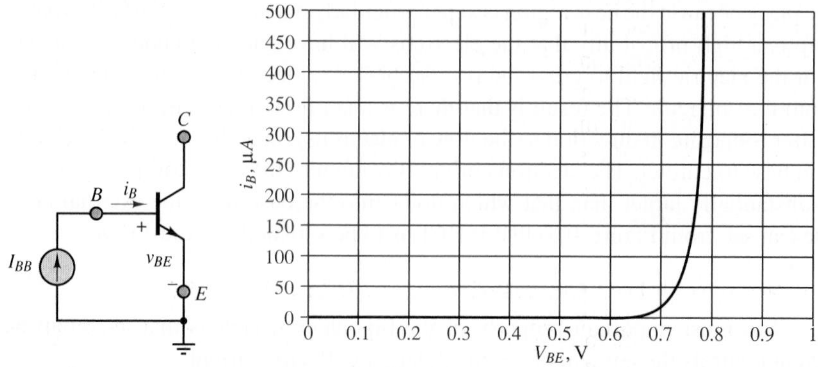

Figure 10.8 The *BE* junction open-collector curve

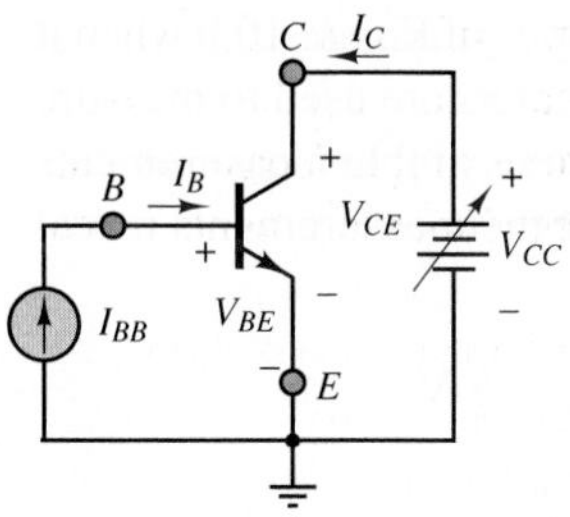

(a)

Figure 10.9(a) Ideal test circuit to determine the *i*-*v* characteristic of a BJT

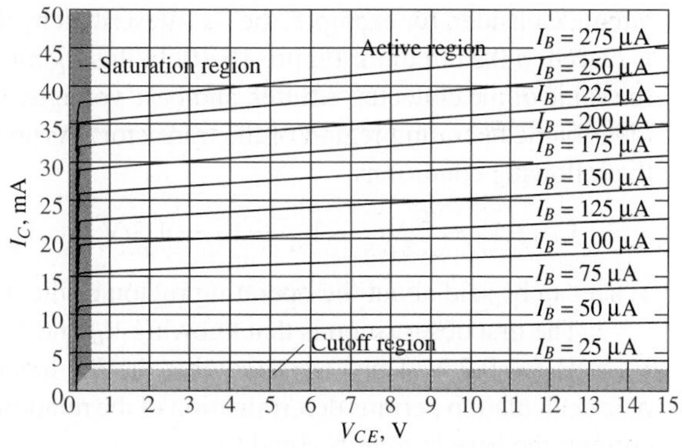

(b)

Figure 10.9(b) The collector-emitter output characteristics of a BJT

current i_B. The resulting circuit is depicted in Figure 10.9(a). By varying both the base current and the collector-emitter voltage, one could then generate a plot of the device **collector characteristic.** This is also shown in Figure 10.9(b). Note that this figure depicts not just a single i_C-v_{CE} curve, but an entire family, since for each value of the base current i_B, an i_C-v_{CE} curve can be generated. Four regions are identified in the collector characteristic:

1. The **cutoff region,** where both junctions are reverse-biased, the base current is very small, and essentially no collector current flows.
2. The **active linear region,** in which the transistor can act as a linear amplifier, where the *BE* junction is forward-biased and the *CB* junction is reverse-biased.
3. The **saturation region,** in which both junctions are forward-biased.
4. The **breakdown region,** which determines the physical limit of operation of the device.

LO2

From the curves of Figure 10.9(b), we note that as v_{CE} is increased, the collector current increases rapidly, until it reaches a nearly constant value; this condition holds until the collector junction breakdown voltage BV_{CEO} is reached (for the purposes of this book, we shall not concern ourselves with the phenomenon of breakdown, except in noting that there are maximum allowable voltages and currents in a transistor). If we were to repeat the same measurement for a set of different values of i_B, the corresponding value of i_C would change accordingly, hence, the family of collector characteristic curves.

Determining the Operating Region of a BJT

Before we discuss common circuit models for the BJT, it will be useful to consider the problem of determining the operating region of the transistor. A few simple voltage measurements permit a quick determination of the state of a transistor placed in a

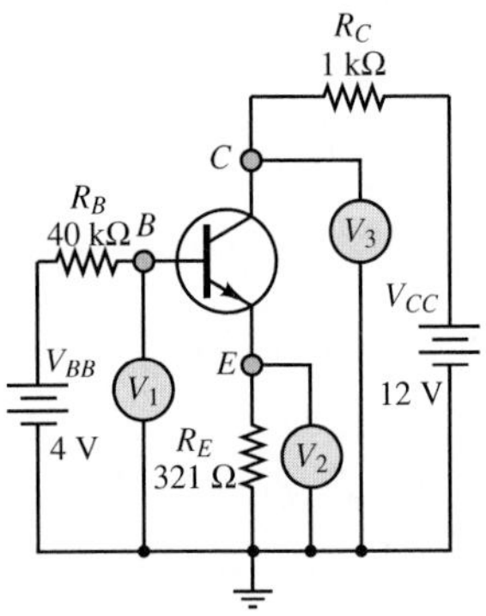

Figure 10.10 Determination of the operation region of a BJT

circuit. Consider, for example, the BJT described by the curves of Figure 10.9 when it is placed in the circuit of Figure 10.10. In this figure, voltmeters are used to measure the value of the collector, emitter, and base voltages. Can these simple measurements identify the operating region of the transistor? Assume that the measurements reveal the following conditions:

$$V_B = V_1 = 2 \text{ V} \qquad V_E = V_2 = 1.3 \text{ V} \qquad V_C = V_3 = 8 \text{ V}$$

What can be said about the operating region of the transistor?

The first observation is that knowing V_B and V_E permits determination of V_{BE}: $V_B - V_E = 0.7$ V. Thus, we know that the BE junction is forward-biased. Another quick calculation permits determination of the relationship between base and collector current: the base current is equal to

$$I_B = \frac{V_{BB} - V_B}{R_B} = \frac{4-2}{40{,}000} = 50\ \mu\text{A}$$

while the collector current is

$$I_C = \frac{V_{CC} - V_C}{R_C} = \frac{12-8}{1{,}000} = 4 \text{ mA}$$

Thus, the current amplification (or gain) factor for the transistor is

$$\frac{I_C}{I_B} = \beta = 80$$

Such a value for the current gain suggests that the transistor is in the linear active region, because substantial current amplification is taking place (typical values of current gain range from 20 to 200). Finally, the collector-to-emitter voltage V_{CE} is found to be $V_{CE} = V_C - V_E = 8 - 1.3 = 6.7$ V.

At this point, you should be able to locate the operating point of the transistor on the curves of Figures 10.8 and 10.9. The currents I_B and I_C and the voltage V_{CE} uniquely determine the state of the transistor in the I_C-V_{CE} and I_B-V_{BE} characteristic curves. What would happen if the transistor were not in the linear active region? Examples 10.2 and 10.3 answer this question and provide further insight into the operation of the bipolar transistor.

LO2

EXAMPLE 10.2 Determining the Operating Region of a BJT

Problem

Determine the operating region of the BJT in the circuit of Figure 10.10 when the base voltage source V_{BB} is short-circuited.

Solution

Known Quantities: Base and collector supply voltages; base, emitter, and collector resistance values.

Find: Operating region of the transistor.

Schematics, Diagrams, Circuits, and Given Data: $V_{BB} = 0$; $V_{CC} = 12$ V; $R_B = 40$ kΩ; $R_C = 1$ kΩ; $R_E = 500$ Ω.

Analysis: Since $V_{BB} = 0$, the base will be at 0 V, and therefore the base-emitter junction is reverse-biased and the base current is zero. Thus the emitter current will also be nearly zero. From equation 10.1 we conclude that the collector current must also be zero. Checking these observations against Figure 10.9(b) leads to the conclusion that the transistor is in the cutoff state. In these cases the three voltmeters of Figure 10.10 will read zero for V_B and V_E and $+12$ V for V_C, since there is no voltage drop across R_C.

Comments: In general, if the base supply voltage is not sufficient to forward-bias the base-emitter junction, the transistor will be in the cutoff region.

CHECK YOUR UNDERSTANDING

Describe the operation of a *pnp* transistor in the active region, by analogy with that of the *npn* transistor.

EXAMPLE 10.3 Determining the Operating Region of a BJT

LO2

Problem

Determine the operating region of the BJT in the circuit of Figure 10.11.

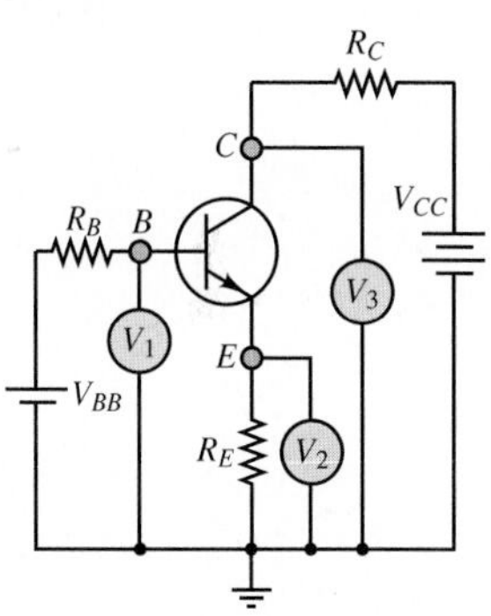

Figure 10.11

Solution

Known Quantities: Base, collector, and emitter voltages with respect to ground.

Find: Operating region of the transistor.

Schematics, Diagrams, Circuits, and Given Data: $V_1 = V_B = 2.7$ V; $V_2 = V_E = 2$ V; $V_3 = V_C = 2.3$ V.

Analysis: To determine the region of the transistor, we shall compute V_{BE} and V_{BC} to determine whether the *BE* and *BC* junctions are forward- or reverse-biased. Operation in the *saturation region* corresponds to forward bias at both junctions (and very small voltage drops); operation in the *active region* is characterized by a forward-biased *BE* junction and a reverse-biased *BC* junction.

From the available measurements, we compute:

$$V_{BE} = V_B - V_E = 0.7 \text{ V}$$

$$V_{BC} = V_B - V_C = 0.4 \text{ V}$$

Since both junctions are forward-biased, the transistor is operating in the saturation region. The value of $V_{CE} = V_C - V_E = 0.3$ V is also very small. This is usually a good indication that the BJT is operating in saturation.

Comments: Try to locate the operating point of this transistor in Figure 10.9(b), assuming that

$$I_C = \frac{V_{CC} - V_3}{R_C} = \frac{12 - 2.3}{1{,}000} = 9.7 \text{ mA}$$

CHECK YOUR UNDERSTANDING

For the circuit of Figure 10.11, the voltmeter readings are $V_1 = 3$ V, $V_2 = 2.4$ V, and $V_3 = 2.7$ V. Determine the operating region of the transistor.

Answer: Saturation

10.3 BJT LARGE-SIGNAL MODEL

The *i*-*v* characteristics and the simple circuits of the previous sections indicate that the BJT acts very much as a current-controlled current source: A small amount of current injected into the base can cause a much larger current to flow into the collector. This conceptual model, although somewhat idealized, is useful in describing a **large-signal model** for the BJT, that is, a model that describes the behavior of the BJT in the presence of relatively large base and collector currents, close to the limit of operation of the device. This model is certainly not a complete description of the properties of the BJT, nor does it accurately depict all the effects that characterize the operation of such devices (e.g., temperature effects, saturation, and cutoff); however, it is adequate for the intended objectives of this book, in that it provides a good qualitative feel for the important features of transistor amplifiers.

Large-Signal Model of the *npn* BJT

The large-signal model for the BJT recognizes three basic operating modes of the transistor. When the *BE* junction is reverse-biased, no base current (and therefore no forward collector current) flows, and the transistor acts virtually as an open circuit; the transistor is said to be in the *cutoff region*. In practice, there is always a leakage current flowing through the collector, even when $V_{BE} = 0$ and $I_B = 0$. This leakage current is denoted by I_{CEO}. When the *BE* junction becomes forward-biased, the transistor is said to be in the *active region,* and the base current is amplified by a factor of β at the collector:

$$I_C = \beta I_B \tag{10.3}$$

Since the collector current is controlled by the base current, the controlled-source symbol is used to represent the collector current. Finally, when the base current becomes sufficiently large, the collector-emitter voltage V_{CE} reaches its saturation limit, and the collector current is no longer proportional to the base current; this is called the *saturation region.* The three conditions are described in Figure 10.12 in terms of simple circuit models. The corresponding collector curves are shown in Figure 10.13.

The large-signal model of the BJT presented in this section treats the *BE* junction as an offset diode and assumes that the BJT in the linear active region acts as an ideal

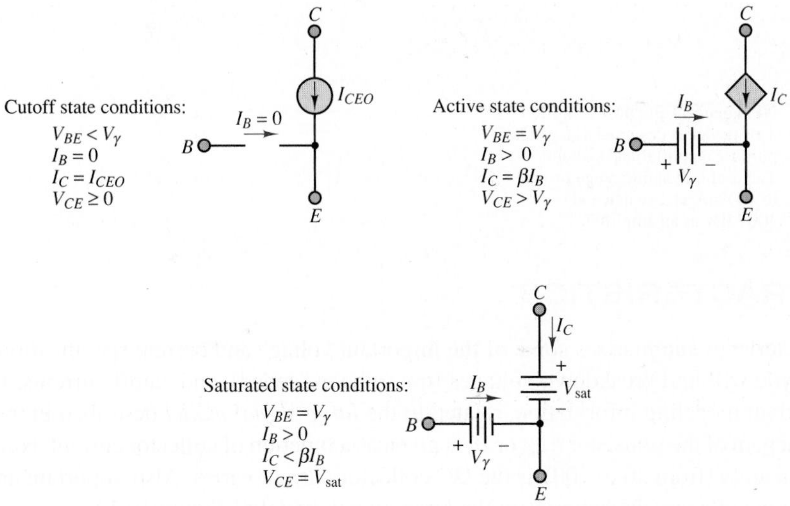

Figure 10.12 An *npn* BJT large-signal model

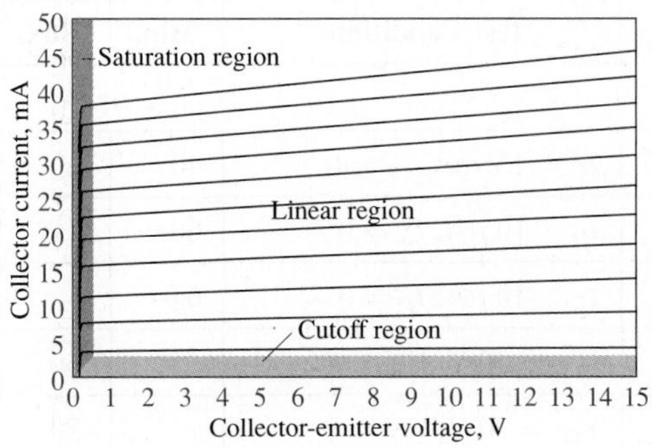

Figure 10.13 BJT collector characteristic

controlled current source. In reality, the *BE* junction is better modeled by considering the forward resistance of the *pn* junction; further, the BJT does not act quite as an ideal current-controlled current source. Nonetheless, the large-signal BJT model is a very useful tool for many applications. Example 10.4 illustrates the application of this large-signal model in a practical circuit and illustrates how to determine which of the three states is applicable, using relatively simple analysis.

FOCUS ON METHODOLOGY

USING DEVICE DATA SHEETS

One of the most important design tools available to engineers is the **device data sheet.** In this box we illustrate the use of a device data sheet for the 2N3904 bipolar transistor. This is an *npn general-purpose amplifier* transistor. Excerpts from the data sheet are shown below, with some words of explanation.

FIND IT ON THE WEB

(*Continued*)

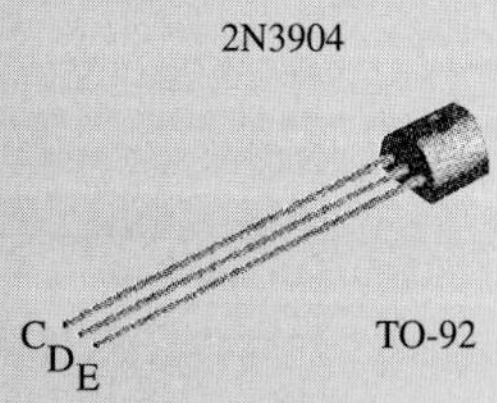

NPN general-purpose amplifier
This device is designed as a general purpose amplifier and switch. The useful dynamic range extends to 100 mA as a switch and to 100 MHz as an amplifier.

ELECTRICAL CHARACTERISTICS

The section on electrical characteristics summarizes some of the important voltage and current specifications of the transistor. For example, you will find breakdown voltages (not to be exceeded), and cutoff currents. In this section you also find important modeling information, related to the *large-signal model* described in this chapter. The large-signal current gain of the transistor h_{FE} or β, is given as a function of collector current. Note that this parameter varies significantly (from 30 to 100) as the DC collector current varies. Also important are the *CE* and *BE* junction saturation voltages (the batteries in the large-signal model of Figure 10.12).

Electrical Characteristics $T_A = 25°C$ unless otherwise noted

Symbol	Parameter	Test Conditions	Min.	Max.	Units
Off Characteristics					
$V_{(BR)CEO}$	Collector-emitter breakdown voltage	$I_C = 1.0$ mA, $I_B = 0$	40		V
$V_{(BR)CBO}$	Collector-base breakdown voltage	$I_C = 10\ \mu$A, $I_E = 0$	60		V
$V_{(BR)EBO}$	Emitter-base breakdown voltage	$I_E = 10\ \mu$A, $I_C = 0$	6.0		V
I_{BL}	Base cutoff current	$V_{CE} = 30$ V, $V_{EB} = 0$		50	nA
I_{CEX}	Collector cutoff current	$V_{CE} = 30$ V, $V_{EB} = 0$		50	nA
On Characteristics					
h_{FE}	DC gain	$I_C = 0.1$ mA, $V_{CE} = 1.0$ V $I_C = 1.0$ mA, $V_{CE} = 1.0$ V $I_C = 10$ mA, $V_{CE} = 1.0$ V $I_C = 50$ mA, $V_{CE} = 1.0$ V $I_C = 100$ mA, $V_{CE} = 1.0$ V	40 70 100 60 30	 300 	
$V_{CE(sat)}$	Collector-emitter saturation voltage	$I_C = 10$ mA, $I_B = 1.0$ mA $I_C = 50$ mA, $I_B = 5.0$ mA		0.2 0.3	V V
$V_{BE(sat)}$	Base-emitter saturation voltage	$I_C = 10$ mA, $I_B = 1.0$ mA $I_C = 50$ mA, $I_B = 5.0$ mA	0.065	0.85 0.95	V V

THERMAL CHARACTERISTICS

This table summarizes the thermal limitations of the device. For example, one can find the **power rating,** listed at 625 mW at 25°C. Note that in the entry for the total device power dissipation, **derating** information is also given. Derating implies that the device power dissipation will change as a function of temperature, in

(Continued)

(Concluded)

this case at the rate of 5 mW/°C. For example, if we expect to operate the diode at a temperature of 100°C, we calculate a derated power of

$$P = 625\text{ mW} - 75°\text{C} \times 5\text{ mW/°C} = 250\text{ mW}$$

Thus, the diode operated at a higher temperature can dissipate only 250 mW.

Thermal Characteristics $T_A = 25°\text{C}$ unless otherwise noted

		Max.		
Symbol	**Characteristic**	**2N3904**	**PZT3904**	**Units**
P_D	Total device dissipation Derate above 25°C	625 5.0	1,000 8.0	mW mW/°C
$R_{\theta JC}$	Thermal resistance, junction to case	83.3		°C/W
$R_{\theta JA}$	Thermal resistance, junction to ambient	200	125	°C/W

TYPICAL CHARACTERISTIC CURVES

Device data sheets always include characteristic curves that may be useful to a designer. In this example, we include the base-emitter "on" voltage as a function of collector current, for three device temperatures. We also show the power dissipation versus ambient temperature derating curve for three different device *packages*. The transistor's ability to dissipate power is determined by its heat transfer properties; the package shown above is the TO-92 package; the SOT-223 and SOT-23 packages have different heat transfer characteristics, leading to different power dissipation capabilities.

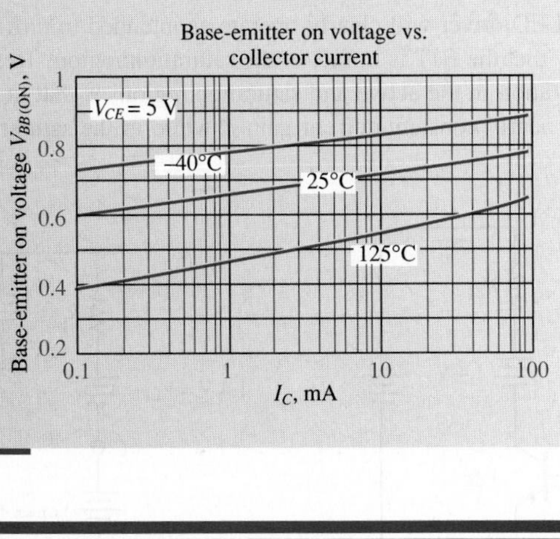

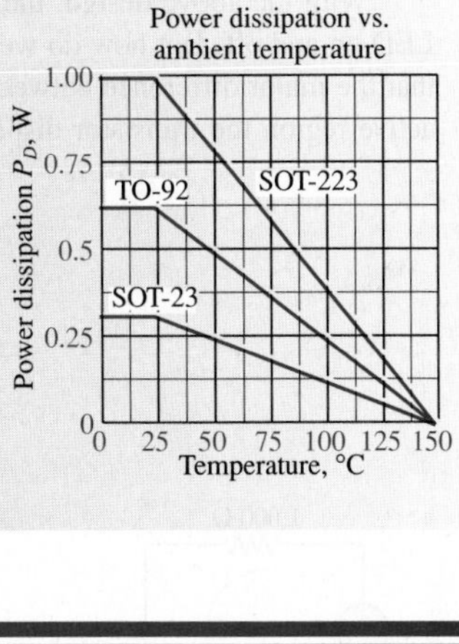

EXAMPLE 10.4 LED Driver

LO3

Problem

Design a transistor amplifier to supply a LED. The LED is required to turn on and off following the on/off signal from a digital output port of a microcomputer. The circuit is shown in Figure10.14.

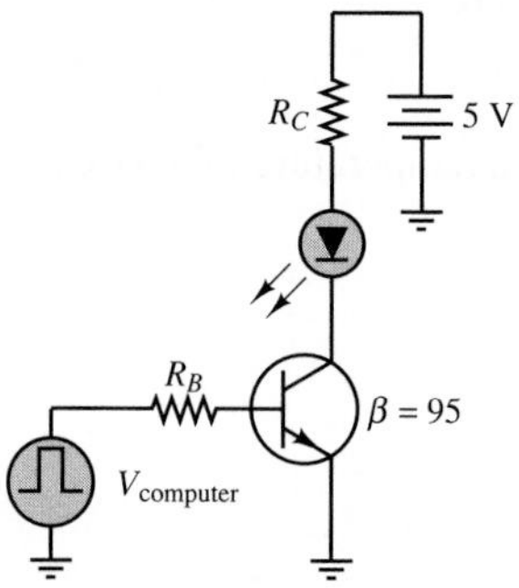

Figure 10.14 LED driver circuit

Solution

Known Quantities: Microprocessor output resistance and output signal voltage and current levels; LED offset voltage, required current, and power rating; BJT current gain and base-emitter junction offset voltage.

Find: Collector resistance R_C such that the transistor is in the saturation region when the computer outputs 5 V; power dissipated by LED.

Schematics, Diagrams, Circuits, and Given Data:
Microprocessor: output resistance $= R_B = 1\ \text{k}\Omega$; $V_{ON} = 5$ V; $V_{OFF} = 0$ V; $I = 5$ mA.
Transistor: $V_{CC} = 5$ V; $V_\gamma = 0.7$ V; $\beta = 95$; $V_{CE\text{sat}} = 0.2$ V.
LED: $V_{\gamma\text{LED}} = 1.4$ V; $I_{\text{LED}} > 15$ mA; $P_{\max} = 100$ mW.

Assumptions: Use the large-signal model of Figure 10.12.

Analysis: When the computer output voltage is zero, the BJT is clearly in the cutoff region, since no base current can flow. When the computer output voltage is $V_{ON} = 5$ V, we wish to drive the transistor into the saturation region. Recall that operation in saturation corresponds to small values of collector-emitter voltages, with typical values of V_{CE} around 0.2 V. Figure 10.15(a) depicts the equivalent base-emitter circuit when the computer output voltage is $V_{ON} = 5$ V. Figure 10.15(b) depicts the collector circuit, and Figure 10.15(c), the same collector circuit with the large-signal model for the transistor (the battery $V_{CE\text{sat}}$) in place of the BJT. From this saturation model we write

$$V_{CC} = R_C I_C + V_{\gamma\text{LED}} + V_{CE\text{sat}}$$

or

$$R_C = \frac{V_{CC} - V_{\gamma\text{LED}} - V_{CE\text{sat}}}{I_C} = \frac{3.4}{I_C}$$

We know that the LED requires at least 15 mA to be on. Let us suppose that 30 mA is a reasonable LED current to ensure good brightness. Then the value of collector resistance that would complete our design is, approximately, $R_C = 113\ \Omega$.

With the above design, the BJT LED driver will clearly operate as intended to turn the LED on and off. But how do we know that the BJT is in fact in the saturation region? Recall that the major difference between operation in the active and saturation regions is that in the active region the transistor displays a nearly constant current gain β while in the saturation

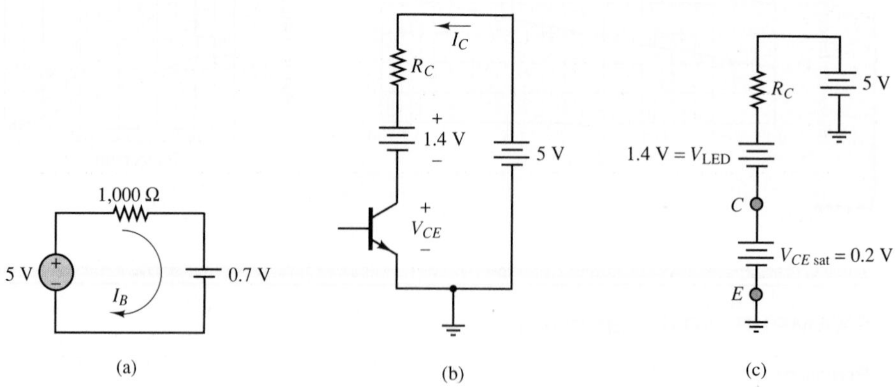

Figure 10.15 (a) *BE* circuit for LED driver; (b) equivalent collector circuit of LED driver, assuming that the BJT is in the linear active region; (c) LED driver equivalent collector circuit, assuming that the BJT is in the saturation region

region the current gain is much smaller. Since we know that the nominal β for the transistor is 95, we can calculate the base current, using the equivalent base circuit of Figure 10.15(a), and determine the ratio of base to collector current:

$$I_B = \frac{V_{ON} - V_\gamma}{R_B} = \frac{4.3}{1{,}000} = 4.3 \text{ mA}$$

The actual large-signal current gain is therefore equal to $30/4.3 = 6.7 \ll \beta$. Thus, it can be reasonably assumed that the BJT is operating in saturation.

We finally compute the LED power dissipation:

$$P_{LED} = V_{\gamma LED} I_C = 1.4 \times 0.3 = 42 \text{ mW} < 100 \text{ mW}$$

Since the power rating of the LED has not been exceeded, the design is complete.

Comments: Using the large-signal model of the BJT is quite easy, since the model simply substitutes voltage sources in place of the *BE* and *CE* junctions. To be sure that the correct model (e.g., saturation versus active region) has been employed, it is necessary to verify either the current gain or the value of the *CE* junction voltage. Current gains near the nominal β indicate active region operation, while small *CE* junction voltages denote operation in saturation.

CHECK YOUR UNDERSTANDING

Repeat the analysis of Example 10.4 for $R_S = 400\ \Omega$. In which region is the transistor operating? What is the collector current?

What is the power dissipated by the LED in Example 10.4 if $R_S = 30\ \Omega$?

Answers: Saturation; 8.5 mA; 159 mW

EXAMPLE 10.5 Simple BJT Battery Charger (BJT Current Source)

LO3

Problem

Design a constant-current battery charging circuit; that is, find the values of V_{CC}, R_1, R_2 (a potentiometer) that will cause the transistor Q_1 to act as a constant current source with selectable current range between 10 and 100 mA.

Solution

Known Quantities: Transistor large signal parameters, NiCd battery nominal voltage.

Find: V_{CC}, R_1, R_2.

Schematics, Diagrams, Circuits, and Given Data: Figure 10.16. $V_\gamma = 0.6$ V; $\beta = 100$.

Assumptions: Assume that the transistor is in the active region. Use the large-signal model with $\beta = 100$.

Analysis: According to the large-signal model, transistor Q_1 amplifies the base current by a

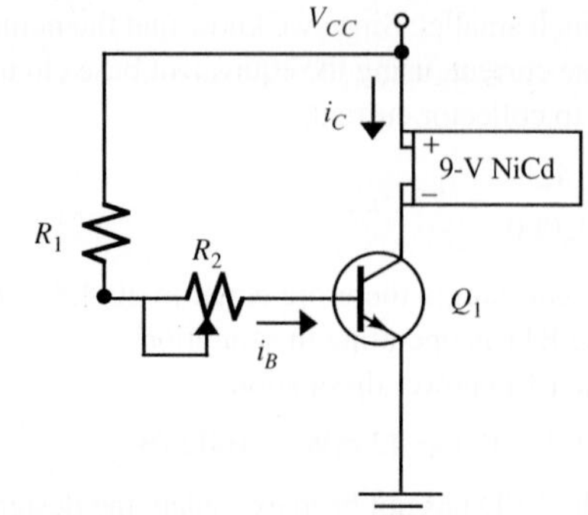

Figure 10.16 Simple battery charging circuit

factor of β. The transistor base current, i_B, is given by the expression:

$$i_B = \frac{V_{CC} - V_\gamma}{R_1 + R_2}$$

Since $i_C = \beta i_B$, the collector current, which is the battery charging current, we can solve the problem by satisfying the inequality

$$10 \text{ mA} \le i_C = \beta \left(\frac{V_{CC} - V_\gamma}{R_1 + R_2} \right) \le 100 \text{ mA}$$

The potentiometer R_2 can be set to any value ranging from zero to R_2, and the maximum current of 100 mA will be obtained when $R_2 = 0$. Thus, we can select a value of R_1 by setting

$$100 \text{ mA} = \beta \left(\frac{V_{CC} - V_\gamma}{R_1} \right) \quad \text{or } R_1 = \left(V_{CC} - V_\gamma \right) \frac{\beta}{10^{-1}} \ \Omega$$

We can select $V_{CC} = 12$ V (a value reasonably larger than the battery nominal voltage) and calculate

$$R_1 = (12 - 0.6) \frac{100}{10^{-1}} = 11,400 \ \Omega$$

Since 12 kΩ is a standard resistor value, we should select R_1= 12 kΩ, which will result in a slightly lower maximum current. The value of the potentiometer R_2 can be found as follows:

$$R_2 = \frac{\beta}{0.01} \left(V_{CC} - V_\gamma \right) - R_1 = 102,600 \ \Omega$$

Since 100-kΩ potentiometers are standard components, we can choose this value for our design, resulting in a slightly higher minimum current than the specified 10 mA.

Comments: A practical note on NiCd batteries: the standard 9-V NiCd batteries are actually made of eight 1.2-V cells. Thus the actual nominal battery voltage is 9.6 V. Further, as the battery becomes fully charged, each cell rises to approximately 1.3 V, leading to a full charge voltage of 10.4 V.

CHECK YOUR UNDERSTANDING

What will the collector-emitter voltage be when the battery is fully charged (10.4 V)? Is this consistent with the assumption that the transistor is in the active region?

Answer: $V_{CEQ} \approx 1.6$ V $> V_\gamma = 0.6$ V

EXAMPLE 10.6 Simple BJT Motor Drive Circuit

LO3

Problem

The aim of this example is to design a BJT driver for the Lego® 9V Technic motor, model 43362. Figure 10.17(a) shows the driver circuit and a picture of the motor. The motor has a maximum (stall) current of 340 mA. Minimum current to start motor rotation is 20 mA. The aim of the circuit is to control the current to the motor (and therefore the motor torque, which is proportional to the current) through potentiometer R_2.

Solution

Known Quantities: Transistor large-signal parameters, component values.

Find: Values of R_1 and R_2.

Schematics, Diagrams, Circuits, and Given Data: Figure 10.17. $V_\gamma = 0.6$ V; $\beta = 40$.

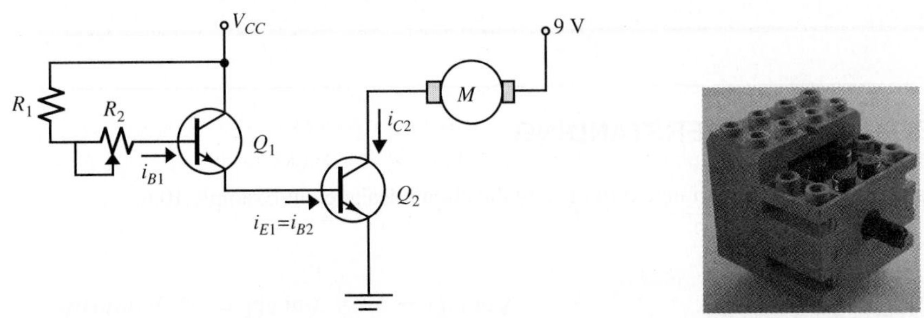

(a) BJT driver circuit

(b) Lego® 9V Technic motor, model 43362
Courtesy: Philippe "Philo" Hurbain

Figure 10.17 Motor Drive Circuit

Assumptions: Assume that the transistors are in the active region. Use the large-signal model with $\beta = 40$ for each transistor.

Analysis: The two-transistor configuration shown in Figure 10.17(a) is similar to a *Darlington pair*. This configuration is used very often and can be purchased as a single package. The emitter current from Q_1, $i_{E1} = (\beta + 1)\, i_{B1}$ becomes the base current for Q_2, and therefore,

$$i_{C2} = \beta i_{E1} = \beta(\beta + 1) i_{B1}.$$

The Q_1 base current is given by the expression

$$i_B = \frac{V_{CC} - V_\gamma}{R_1 + R_2}$$

Therefore the motor current can take the range.

$$i_{C2\,\min} \le \beta\,(\beta + 1)\left(\frac{V_{CC} - V_\gamma}{R_1 + R_2}\right) \le i_{C2\,\max}$$

The potentiometer R_2 can be set to any value ranging from zero to R_2 and the maximum (stall) current of 340 mA will be obtained when $R_2 = 0$. Thus, we can select a value of R_1 by choosing $V_{CC} = 12$ V and setting

$$i_{C2\max} = 0.34 \text{ A} = \beta(\beta + 1)\left(\frac{V_{CC} - V_\gamma}{R_1}\right) \quad \text{or } R_1 = \frac{\beta(\beta + 1)}{0.34}\left(V_{CC} - V_\gamma\right) = 54,988\ \Omega$$

Since 56 kΩ is a standard resistor value, we should select R_1= 56 kΩ, which will result in a slightly lower maximum current. The value of the potentiometer R_2 can be found from the minimum current requirement of 20 mA:

$$R_2 = \frac{\beta(\beta + 1)}{0.02}\left(V_{CC} - V_\gamma\right) - R_1 = 879,810\ \Omega$$

Since 1-MΩ potentiometers are standard components, we can choose this value for our design, resulting in a slightly lower minimum current than the specified 20 mA.

Comments: While this design is quite simple, it only permits manual control of the motor current (and torque). If we wished to, say, have the motor under computer control, we would need a circuit that could respond to an external voltage. This design is illustrated in the homework problems.

CHECK YOUR UNDERSTANDING

Compute the actual current range provided by the circuit designed in Example 10.6.

Answer: $i_{C2\max} = 334$ mA; $i_{C2\min} = 17.7$ mA

Large-Signal Amplifier for Diode Thermometer

Problem:

In Chapter 9 we explored the use of a diode as the sensing element in an electronic thermometer (see the Focus on Measurements box "Diode Thermometer"). In the present example, we illustrate the design of a transistor amplifier for such a diode thermometer. The circuit is shown in Figure 10.18.

Solution:

Known Quantities—Diode and transistor amplifier bias circuits; diode voltage versus temperature response.

Find—Collector resistance and transistor output voltage versus temperature.

Schematics, Diagrams, Circuits, and Given Data—$V_{CC} = 12$ V; large-signal $\beta =$ 188.5; $V_{BE} = 0.75$ V; $R_S = 500\ \Omega$; $R_B = 10$ kΩ.

Assumptions—Use a 1N914 diode and a 2N3904 transistor.

(Continued)

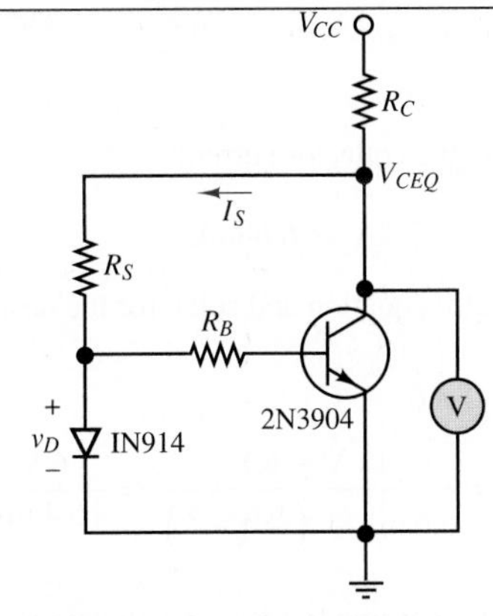

Figure 10.18 Large-signal amplifier for diode thermometer

Analysis—With reference to the circuit of Figure 10.18 and to the diode temperature response characteristic of Figure 10.19(a), we observe that the midrange diode thermometer output voltage is approximately 1.1 V. Thus, we should design the transistor amplifier so that when $v_D = 1.1$ V, the transistor output is in the center of the collector characteristic for minimum distortion. Since the collector supply is 12 V, we choose to have the Q point at $V_{CEQ} = 6$ V.

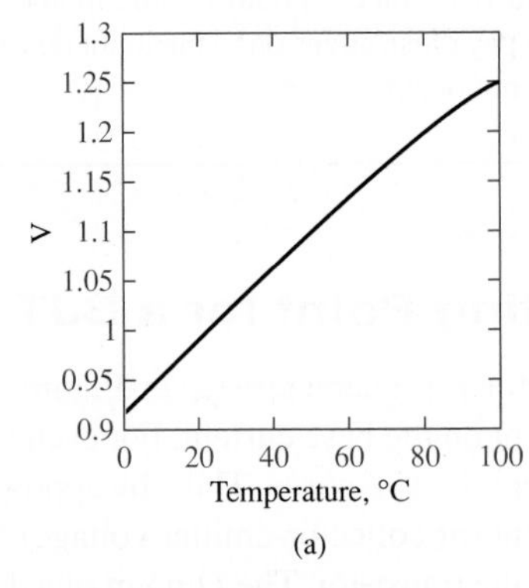

Figure 10.19(a) Diode voltage temperature dependence

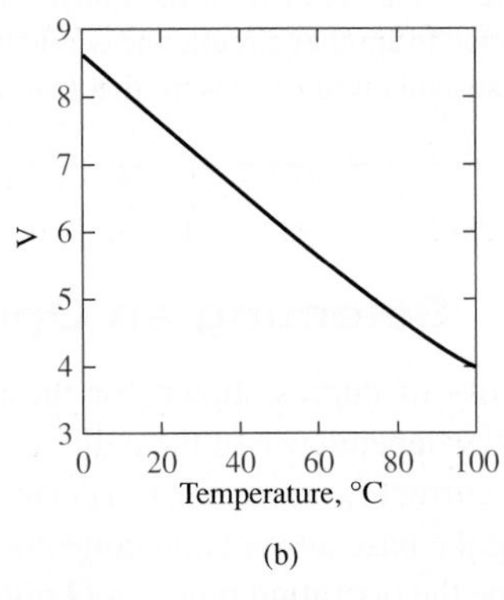

Figure 10.19(b) Amplifier output

Knowing that the diode output voltage at the quiescent point is 1.1 V, we compute the quiescent base current

$$v_D - I_{BQ} R_B - V_{BEQ} = 0$$

$$I_{BQ} = \frac{v_D - V_{BEQ}}{R_B} = \frac{1.1 - 0.75}{10{,}000} = 35\ \mu\text{A}$$

(Continued)

(Concluded)

Knowing β, we can compute the collector current:

$$I_{CQ} = \beta I_{BQ} = 188.5 \times 35\ \mu\text{A} = 6.6\ \text{mA}$$

Now we can write the collector equation and solve for the desired collector resistance:

$$V_{CC} - I_{CQ}R_C - V_{CEQ} = 0$$

$$R_C = \frac{V_{CC} - V_{CEQ}}{I_{CQ} + I_S} = \frac{12\ \text{V} - 6\ \text{V}}{6.6\ \text{mA} + \left(\frac{V_{CEQ} - v_D}{R_S}\right)} = \frac{6\ \text{V}}{16.4\ \text{mA}} = 366\ \Omega$$

Once the circuit is designed according to these specifications, the output voltage can be determined by computing the base current as a function of the diode voltage (which is a function of temperature); from the base current, we can compute the collector current and use the collector equation to determine the output voltage $v_{\text{out}} = v_{CE}$. The result is plotted in Figure 10.19(b).

Comments—Note that the transistor amplifies the slope of the temperature by a factor of approximately 6. Observe also that the common-emitter amplifier used in this example causes a sign inversion in the output (the output voltage now decreases for increasing temperatures, while the diode voltage increases). Finally, we note that the design shown in this example assumes that the impedance of the voltmeter is infinite. This is a good assumption in the circuit shown, because a practical voltmeter will have a very large input resistance relative to the transistor output resistance. Should the thermometer be connected to another circuit, one would have to pay close attention to the input resistance of the second circuit to ensure that loading did not occur.

10.4 Selecting an Operating Point for a BJT

The family of curves shown for the collector i-v characteristic in Figure 10.9(b) reflects the dependence of the collector current on the base current. For each value of the base current i_B, there exists a corresponding i_C-v_{CE} curve. Thus, by appropriately selecting the base current and collector current (or collector-emitter voltage), we can determine the operating point, or ***Q* point,** of the transistor. The Q point of a device is defined in terms of the **quiescent** (or **idle**) **currents** and **voltages** that are present at the terminals of the device when DC supplies are connected to it. The circuit of Figure 10.20 illustrates an ideal **DC bias circuit,** used to set the Q point of the BJT in the approximate center of the collector characteristic. The circuit shown in Figure 10.20 is not a practical DC bias circuit for a BJT amplifier, but it is very useful for the purpose of introducing the relevant concepts. A practical bias circuit is discussed later in this section.

By appropriate choice of I_{BB}, R_C and V_{CC}, the desired Q point may be selected.

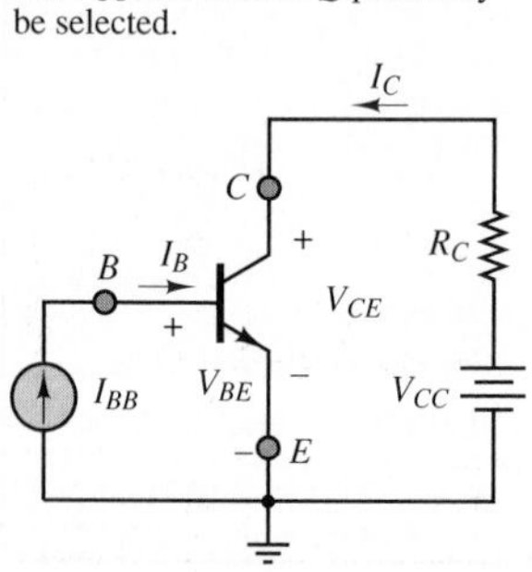

Figure 10.20 A simplified bias circuit for a BJT amplifier

Applying KVL around the base-emitter and collector circuits, we obtain the following equations:

$$I_B = I_{BB} \quad \textbf{(10.4)}$$

and

$$V_{CE} = V_{CC} - I_C R_C \quad \textbf{(10.5)}$$

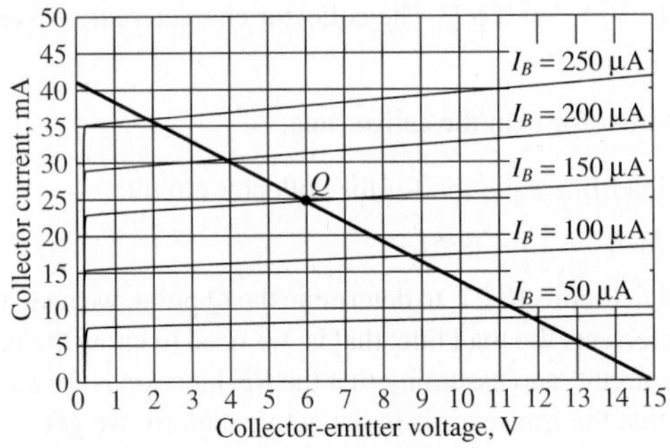

Figure 10.21 Load-line analysis of a simplified BJT amplifier

which can be rewritten as

$$I_C = \frac{V_{CC} - V_{CE}}{R_C} \tag{10.6}$$

Note the similarity of equation 10.6 to the load-line curves of Chapters 3 and 9. Equation 10.6 represents a line that intersects the I_C axis at $I_C = V_{CC}/R_C$ and the V_{CE} axis at $V_{CE} = V_{CC}$. The slope of the load line is $-1/R_C$. Since the base current I_B is equal to the source current I_{BB}, the operating point may be determined by noting that the load line intersects the entire collector family of curves. The intersection point at the curve that corresponds to the base current $I_B = I_{BB}$ constitutes the operating, or Q, point. The load line corresponding to the circuit of Figure 10.20 is shown in Figure 10.21, superimposed on the collector curves for the 2N3904 transistor. In Figure 10.21, $V_{CC} = 15$ V, $V_{CC}/R_C = 40$ mA, and $I_{BB} = 150\ \mu$A; thus, the Q point is determined by the intersection of the load line with the I_C-V_{CE} curve corresponding to a base current of 150 μA.

Once an operating point is established and direct currents I_{CQ} and I_{BQ} are flowing into the collector and base, respectively, the BJT can serve as a linear amplifier, as was explained in Section 10.2. Example 10.7 serves as an illustration of the DC biasing procedures just described.

EXAMPLE 10.7 Calculation of DC Operating Point for BJT Amplifier

LO4

Problem

Determine the DC operating point of the BJT amplifier in the circuit of Figure 10.22.

Solution

Known Quantities: Base, collector, and emitter resistances; base and collector supply voltages; collector characteristic curves; *BE* junction offset voltage.

Find: Direct (quiescent) base and collector currents I_{BQ} and I_{CQ} and collector-emitter voltage V_{CEQ}.

Schematics, Diagrams, Circuits, and Given Data: $R_B = 62.7\ \text{k}\Omega$; $R_C = 375\ \Omega$;

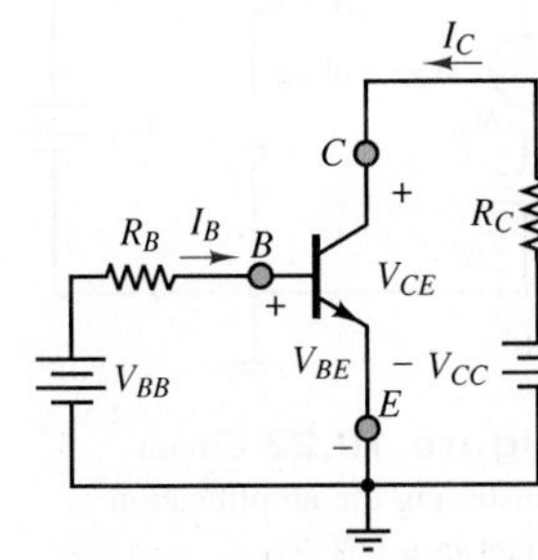

Figure 10.22

$V_{BB} = 10$ V; $V_{CC} = 15$ V; $V_\gamma = 0.6$ V. The collector characteristic curves are shown in Figure 10.21.

Assumptions: The transistor is in the active state.

Analysis: Write the load-line equation for the collector circuit:

$$V_{CE} = V_{CC} - R_C I_C = 15 - 375 I_C$$

The load line is shown in Figure 10.21; to determine the Q point, we need to determine which of the collector curves intersects the load line; that is, we need to know the base current. Applying KVL around the base circuit, and assuming that the BE junction is forward-biased (this results from the assumption that the transistor is in the active region), we get

$$I_B = \frac{V_{BB} - V_{BE}}{R_B} = \frac{V_{BB} - V_\gamma}{R_B} = \frac{10 - 0.6}{62{,}700} = 150\ \mu\text{A}$$

The intersection of the load line with the 150-μA base curve is the DC operating or quiescent point of the transistor amplifier, defined below by the three values:

$$V_{CEQ} = 7\ \text{V} \qquad I_{CQ} = 22\ \text{mA} \qquad I_{BQ} = 150\ \mu\text{A}$$

Comments: The base circuit consists of a battery in series with a resistance; we shall soon see that it is not necessary to employ two different voltage supplies for base and collector circuits, but that a single collector supply is sufficient to bias the transistor. Note that even in the absence of an external input to be amplified (AC source), the transistor dissipates power; most of the power is dissipated by the collector circuit: $P_{CQ} = V_{CEQ} \times I_{CQ} = 154$ mW.

CHECK YOUR UNDERSTANDING

How would the Q point change if the base current increased to 200 μA?

Answer: $V_{CEQ} \approx 4$ V, $I_{CQ} \approx 31$ mA

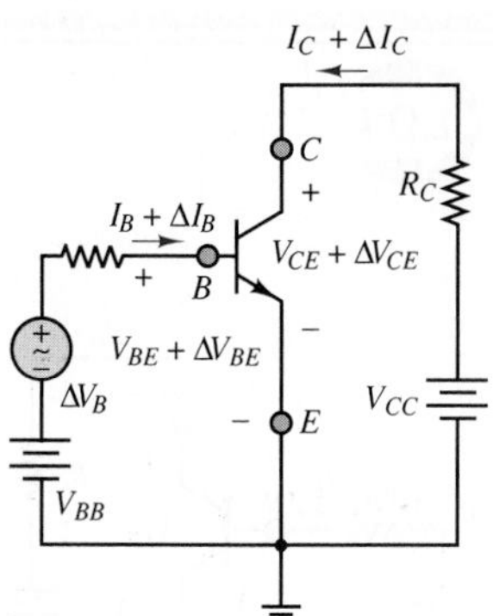

Figure 10.23 Circuit illustrating the amplification effect in a BJT

How can a transistor amplify a signal, then, given the V_{BE}-I_B and V_{CE}-I_C curves discussed in this section? The small-signal amplifier properties of the transistor are best illustrated by analyzing the effect of a small sinusoidal current superimposed on the DC flowing into the base. The circuit of Figure 10.23 illustrates the idea, by including a small-signal AC source, of strength ΔV_B, in series with the base circuit. The effect of this AC source is to cause sinusoidal oscillations ΔI_B about the Q point, that is, around I_{BQ}. A study of the collector characteristic indicates that for a sinusoidal oscillation in I_B, a corresponding, but larger oscillation will take place in the collector current. Figure 10.16 illustrates the concept. Note that the base current oscillates between 110 and 190 μA, causing the collector current to correspondingly fluctuate between 15.3 and 28.6 mA. The notation that will be used to differentiate between DC and AC (or fluctuating) components of transistor voltages and currents is as follows: DC (or quiescent) currents and voltages will be denoted by uppercase symbols, for example, I_B, I_C, V_{BE}, V_{CE}. AC components will be preceded by a Δ: $\Delta I_B(t)$, $\Delta I_C(t)$, $\Delta V_{BE}(t)$, $\Delta V_{CE}(t)$. The complete expression for one of these

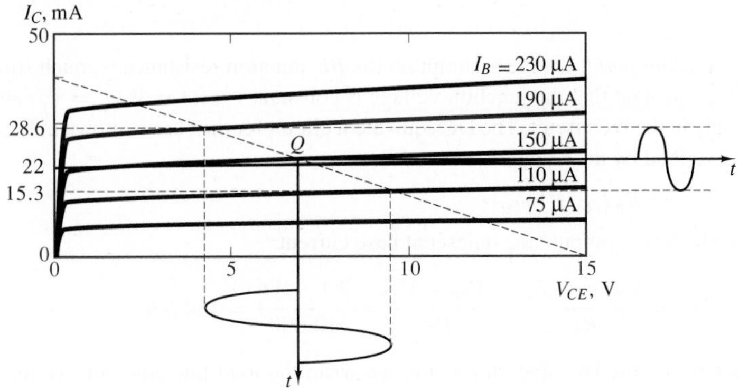

Figure 10.24 Amplification of sinusoidal oscillations in a BJT

quantities will therefore include both a DC term and a time-varying, or AC, term. For example, the collector current may be expressed by $i_C(t) = I_C + \Delta I_C(t)$.

The i-v characteristic of Figure 10.24 illustrates how an increase in collector current follows the same sinusoidal pattern of the base current but is greatly amplified. Thus, the BJT acts as a *current amplifier,* in the sense that any oscillations in the base current appear amplified in the collector current. Since the voltage across the collector resistance R_C is proportional to the collector current, one can see how the collector voltage is also affected by the amplification process. Example 10.8 illustrates numerically the effective amplification of the small AC signal that takes place in the circuit of Figure 10.23.

EXAMPLE 10.8 A BJT Small-Signal Amplifier

LO4

Problem

With reference to the BJT amplifier of Figure 10.25 and to the collector characteristic curves of Figure 10.21, determine (1) the DC operating point of the BJT, (2) the nominal current gain β at the operating point, and (3) the AC voltage gain $A_V = \Delta V_o/\Delta V_B$.

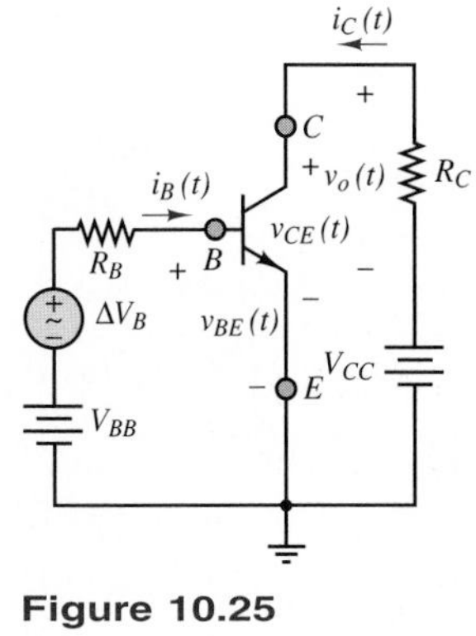

Figure 10.25

Solution

Known Quantities: Base, collector, and emitter resistances; base and collector supply voltages; collector characteristic curves; *BE* junction offset voltage.

Find: (1) DC (quiescent) base and collector currents I_{BQ} and I_{CQ} and collector-emitter voltage V_{CEQ}, (2) $\beta = \Delta I_C/\Delta I_B$, and (3) $A_V = \Delta V_o/\Delta V_B$.

Schematics, Diagrams, Circuits, and Given Data: $R_B = 10$ kΩ; $R_C = 375$ Ω; $V_{BB} = 2.1$ V; $V_{CC} = 15$ V; $V_\gamma = 0.6$ V. The collector characteristic curves are shown in Figure 10.21.

Assumptions: Assume that the *BE* junction resistance is negligible compared to the base resistance. Assume that each voltage and current can be represented by the superposition of a DC (quiescent) value and an AC component, for example, $v_O(t) = V_{OQ} + \Delta V_O(t)$.

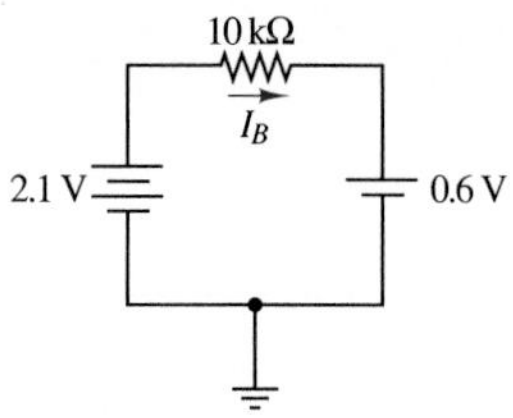

Figure 10.26

Analysis:

1. *DC operating point.* On the assumption the *BE* junction resistance is much smaller than R_B, we can state that the junction voltage is constant, $v_{BE}(t) = V_{BEQ} = V_\gamma$, and plays a role only in the DC circuit. The DC equivalent circuit for the base is shown in Figure 10.26 and described by the equation

$$V_{BB} = R_B I_{BQ} + V_{BEQ}$$

from which we compute the quiescent base current:

$$I_{BQ} = \frac{V_{BB} - V_{BEQ}}{R_B} = \frac{V_{BB} - V_\gamma}{R_B} = \frac{2.1 - 0.6}{10{,}000} = 150\ \mu\text{A}$$

To determine the DC operating point, we write the load-line equation for the collector circuit:

$$V_{CE} = V_{CC} - R_C I_C = 15 - 375 I_C$$

The load line is shown in Figure 10.27. The intersection of the load line with the 150-μA base curve is the DC operating or quiescent point of the transistor amplifier, defined below by the three values $V_{CEQ} = 7.2$ V, $I_{CQ} = 22$ mA, and $I_{BQ} = 150\ \mu$A.

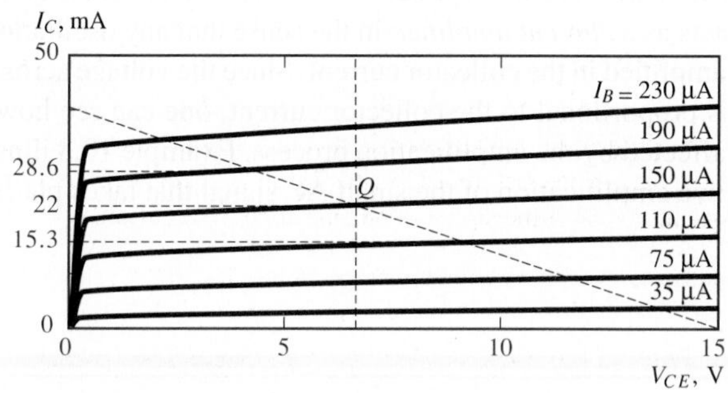

Figure 10.27 Operating point on the characteristic curve

2. *AC gain.* To determine the current gain, we resort, again, to the collector curves. Figure 10.27 indicates that if we consider the values corresponding to base currents of 190 and 110 μA, the collector will see currents of 28.6 and 15.3 mA, respectively. We can think of these collector current excursions ΔI_C from the Q point as corresponding to the effects of an oscillation ΔI_B in the base current, and we can calculate the current gain of the BJT amplifier according to

$$\beta = \frac{\Delta I_C}{\Delta I_B} = \frac{28.6 \times 10^{-3} - 15.3 \times 10^{-3}}{190 \times 10^{-6} - 110 \times 10^{-6}} = 166.25$$

Thus, the nominal current gain of the transistor is approximately $\beta = 166$.

3. *AC voltage gain.* To determine the AC voltage gain $A_V = \Delta V_o / \Delta V_B$, we need to express ΔV_o as a function of ΔV_B. Observe that $v_o(t) = R_C i_C(t) = R_C I_{CQ} + R_C\, \Delta I_C(t)$. Thus we can write:

$$\Delta V_o(t) = -R_C\, \Delta I_C(t) = -R_C \beta\, \Delta I_B(t)$$

Using the principle of superposition in considering the base circuit, we observe that $\Delta I_B(t)$ can be computed from the KVL base equation

$$\Delta V_B(t) = R_B\, \Delta I_B(t) + \Delta V_{BE}(t)$$

but we had stated in part 1 that since the BE junction resistance is negligible relative to R_B, $\Delta V_{BE}(t)$ is also negligible. Thus,

$$\Delta I_B = \frac{\Delta V_B}{R_B}$$

Substituting this result into the expression for $\Delta V_o(t)$, we can write

$$\Delta V_o(t) = -R_C \beta \, \Delta I_B(t) = -\frac{R_C \, \beta \Delta V_B(t)}{R_B}$$

or

$$\frac{\Delta V_o(t)}{\Delta V_B} = A_V = -\frac{R_C}{R_B}\beta = -6.23$$

Comments: The circuit examined in this example is not quite a practical transistor amplifier yet, but it demonstrates most of the essential features of BJT amplifiers. We summarize them as follows.

- Transistor amplifier analysis is greatly simplified by considering the DC bias circuit and the AC equivalent circuits separately. This is an application of the principle of superposition.
- Once the bias point (or DC operating or quiescent point) has been determined, the current gain of the transistor can be determined from the collector characteristic curves. This gain is somewhat dependent on the location of the operating point.
- The AC voltage gain of the amplifier is strongly dependent on the base and collector resistance values. Note that the AC voltage gain is negative! This corresponds to a 180° phase inversion if the signal to be amplified is a sinusoid.

Many issues remain to be considered before we can think of designing and analyzing a practical transistor amplifier. It is extremely important that you master this example before studying the remainder of the section.

CHECK YOUR UNDERSTANDING

Calculate the Q point of the transistor if R_C is increased to 680 Ω.

Answer: $V_{CEQ} \approx 5$ V, $I_{BQ} \approx 110$ μA, $I_{CQ} \approx 15$ mA

In discussing the DC biasing procedure for the BJT, we pointed out that the simple circuit of Figure 10.20 would not be a practical one to use in an application circuit. In fact, the more realistic circuit of Example 10.7 is also not a practical biasing circuit. The reasons for this statement are that two different supplies are required (V_{CC} and V_{BB})—a requirement that is not very practical—and that the resulting DC bias (operating) point is not very stable. This latter point may be made clearer by pointing out that the location of the operating point could vary significantly if, say, the current gain of the transistor β were to vary from device to device. A circuit that provides great improvement on both counts is shown in Figure 10.28. Observe, first, that the voltage supply V_{CC} appears across the pair of resistors R_1 and R_2, and that therefore

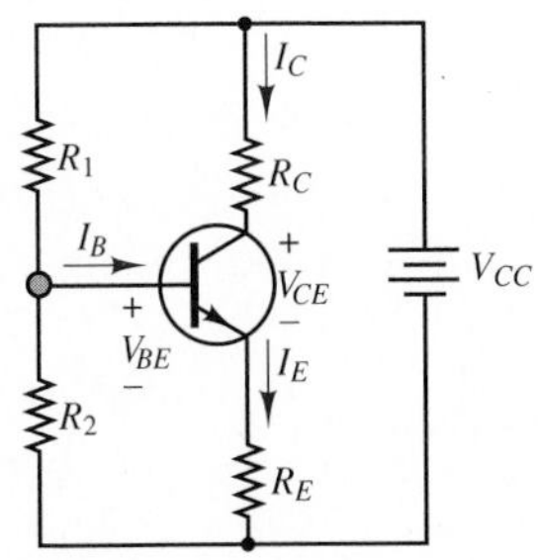

Figure 10.28 Practical BJT self-bias DC circuit

the base terminal for the transistor will see the Thévenin equivalent circuit composed of the equivalent voltage source

$$V_{BB} = \frac{R_2}{R_1 + R_2} V_{CC} \tag{10.7}$$

and of the equivalent resistance

$$R_B = R_1 \parallel R_2 \tag{10.8}$$

Figure 10.29(b) shows a redrawn DC bias circuit that makes this observation more evident. The circuit to the left of the dashed line in Figure 10.29(a) is represented in Figure 10.29(b) by the equivalent circuit composed of V_{BB} and R_B.

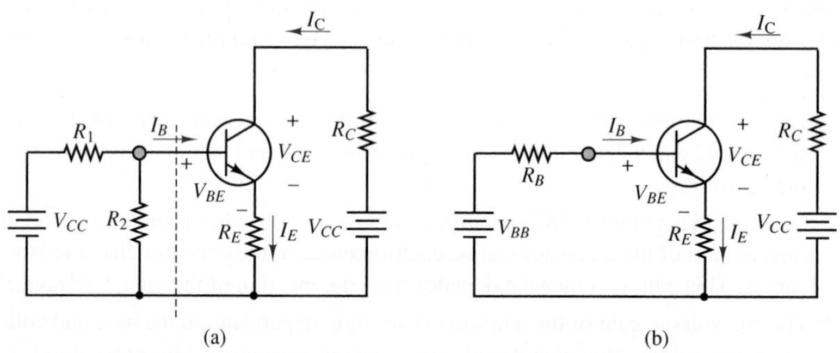

Figure 10.29 DC self-bias circuit represented in equivalent-circuit form

Recalling that the *BE* junction acts much as a diode, we note that the following equations describe the DC operating point of the self-bias circuit. Around the base-emitter circuit,

$$V_{BB} = I_B R_B + V_{BE} + I_E R_E = [R_B + (\beta + 1)R_E]I_B + V_{BE} \tag{10.9}$$

where V_{BE} is the *BE* junction voltage (diode forward voltage) and $I_E = (\beta + 1)I_B$. Around the collector circuit, on the other hand, the following equation applies:

$$V_{CC} = I_C R_C + V_{CE} + I_E R_E = I_C \left(R_C + \frac{\beta + 1}{\beta} R_E \right) + V_{CE} \tag{10.10}$$

since

$$I_E = I_B + I_C = \left(\frac{1}{\beta} + 1 \right) I_C$$

These two equations may be solved to obtain (1) an expression for the base current

$$I_B = \frac{V_{BB} - V_{BE}}{R_B + (\beta + 1)R_E} \tag{10.11}$$

from which the collector current can be determined as $I_C = \beta I_B$, and (2) an expression for the collector-emitter voltage

$$V_{CE} = V_{CC} - I_C \left(R_C + \frac{\beta + 1}{\beta} R_E \right) \tag{10.12}$$

This last equation is the load-line equation for the bias circuit. Note that the effective load resistance seen by the DC collector circuit is no longer just R_C, but is now given by

$$R_C + \frac{\beta + 1}{\beta} R_E \approx R_C + R_E$$

Example 10.9 provides a numerical illustration of the analysis of a DC self-bias circuit for a BJT.

EXAMPLE 10.9 Practical BJT Bias Circuit

LO4

Problem

Determine the DC bias point of the transistor in the circuit of Figure 10.28.

Solution

Known Quantities: Base, collector, and emitter resistances; collector supply voltage; nominal transistor current gain; *BE* junction offset voltage.

Find: DC (quiescent) base and collector currents I_{BQ} and I_{CQ} and collector-emitter voltage V_{CEQ}.

Schematics, Diagrams, Circuits, and Given Data: $R_1 = 100$ kΩ; $R_2 = 50$ kΩ; $R_C = 5$ kΩ; $R_E = 3$ kΩ; $V_{CC} = 15$ V; $V_\gamma = 0.7$ V, $\beta = 100$.

Analysis: We first determine the equivalent base voltage from equation 10.7

$$V_{BB} = \frac{R_2}{R_1 + R_2} V_{CC} = \frac{50}{100 + 50} 15 = 5 \text{ V}$$

and the equivalent base resistance from equation 10.8

$$R_B = R_1 \| R_2 = 33.3 \text{ k}\Omega$$

Now we can compute the base current from equation 10.11

$$I_B = \frac{V_{BB} - V_{BE}}{R_B + (\beta + 1) R_E} = \frac{V_{BB} - V_\gamma}{R_B + (\beta + 1) R_E} = \frac{5 - 0.7}{33{,}000 + 101 \times 3{,}000} = 12.8 \ \mu\text{A}$$

and knowing the current gain of the transistor β, we can determine the collector current:

$$I_C = \beta I_B = 1.28 \text{ mA}$$

Finally, the collector-emitter junction voltage can be computed with reference to equation 10.12:

$$V_{CE} = V_{CC} - I_C \left(R_C + \frac{\beta + 1}{\beta} R_E \right)$$

$$= 15 - 1.28 \times 10^{-3} \left(5 \times 10^3 + \frac{101}{100} \times 3 \times 10^3 \right) = 4.78 \text{ V}$$

Thus, the Q point of the transistor is given by:

$$V_{CEQ} = 4.78 \text{ V} \qquad I_{CQ} = 1.28 \text{ mA} \qquad I_{BQ} = 12.8 \ \mu\text{A}$$

CHECK YOUR UNDERSTANDING

In the circuit of Figure 10.29, find the value of V_{BB} that yields a collector current $I_C = 6.3$ mA. What is the corresponding collector-emitter voltage? Assume that $V_{BE} = 0.6$ V, $R_B = 50$ kΩ, $R_E = 200$ Ω, $R_C = 1$ kΩ, $B = 100$, and $V_{CC} = 14$ V. What percentage change in collector current would result if β were changed to 150 in Example 10.9? Why does the collector current increase less than 50 percent?

Answers: $V_{BB} = 5$ V, $V_{CE} = 6.44$ V; 3.74%. Because R_E provides negative feedback action that will keep I_C and I_E nearly constant

The material presented in this section has illustrated the basic principles that underlie the operation of a BJT and the determination of its Q point.

10.5 BJT SWITCHES AND GATES

In describing the properties of transistors, it was suggested that, in addition to serving as amplifiers, three-terminal devices can be used as electronic switches in which one terminal controls the flow of current between the other two. It had also been hinted in Chapter 9 that diodes can act as on/off devices as well. In this section, we discuss the operation of diodes and transistors as electronic switches, illustrating the use of these electronic devices as the switching circuits that are at the heart of **analog** and **digital gates.** Transistor switching circuits form the basis of digital logic circuits, which are discussed in greater detail in Chapter 13. The objective of this section is to discuss the internal operation of these circuits and to provide the reader interested in the internal workings of digital circuits with an adequate understanding of the basic principles.

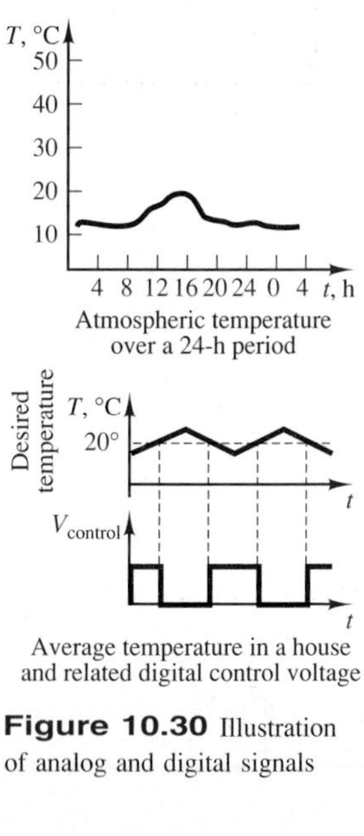

Figure 10.30 Illustration of analog and digital signals

An **electronic gate** is a device that, on the basis of one or more input signals, produces one of two or more prescribed outputs; as will be seen shortly, one can construct both digital and analog gates. A word of explanation is required, first, regarding the meaning of the words *analog* and *digital*. An analog voltage or current—or, more generally, an analog signal—is one that varies in a continuous fashion over time, in *analogy* (hence the expression *analog*) with a physical quantity. An example of an analog signal is a sensor voltage corresponding to ambient temperature on any given day, which may fluctuate between, say, 30 and 50°F. A digital signal, on the other hand, is a signal that can take only a finite number of values; in particular, a commonly encountered class of digital signals consists of **binary signals,** which can take only one of two values (for example, 1 and 0). A typical example of a binary signal would be the control signal for the furnace in a home heating system controlled by a conventional thermostat, where one can think of this signal as being "on" (or 1) if the temperature of the house has dropped below the thermostat setting (desired value), or "off" (or 0) if the house temperature is greater than or equal to the set temperature (say, 68°F). Figure 10.30 illustrates the appearance of the analog and digital signals in this furnace example.

The discussion of digital signals will be continued and expanded in Chapters 13, 14, and 15. Digital circuits are an especially important topic, because a large part of today's industrial and consumer electronics is realized in digital form.

Diode Gates

You will recall that a diode conducts current when it is forward-biased and otherwise acts very much as an open circuit. Thus, the diode can serve as a switch if properly employed. The circuit of Figure 10.31 is called an **OR gate;** it operates as follows. Let voltage levels greater than, say, 2 V correspond to a "logic 1" and voltages less than 2 V represent a "logic 0." Suppose, then, that input voltages v_A and v_B can be equal to either 0 V or 5 V. If $v_A = 5$ V, diode D_A will conduct; if $v_A = 0$ V, D_A will act as an open circuit. The same argument holds for D_B. It should be apparent, then, that the voltage across the resistor R will be 0 V, or logic 0, if both v_A and v_Bx are 0. If either v_A or v_B is equal to 5 V, though, the corresponding diode will conduct, and—assuming an offset model for the diode with $V_\gamma = 0.6$ V—we find that $v_{out} = 4.4$ V, or logic 1. Similar analysis yields an equivalent result if both v_A and v_B are equal to 5 V.

This type of gate is called an OR gate because v_{out} is equal to logic 1 (or "high") if either v_A *or* v_B is on, while it is logic 0 (or "low") if neither v_A *nor* v_B is on. Other functions can also be implemented; however, the discussion of diode gates will be limited to this simple introduction, because diode gate circuits, such as the one of Figure 10.31, are rarely, if ever, employed in practice. Most modern digital circuits employ transistors to implement switching and gate functions.

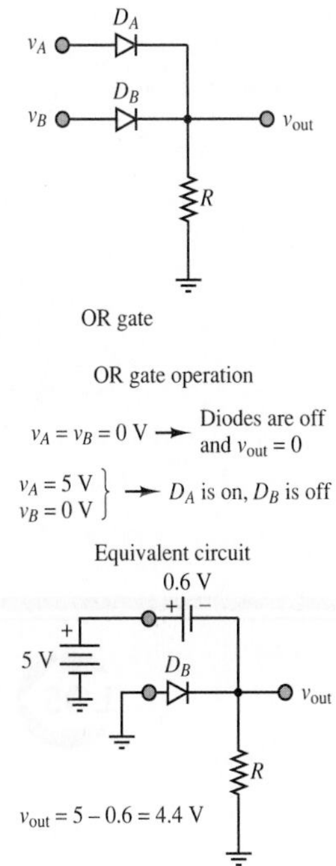

Figure 10.31 Diode OR gate

BJT Gates

In discussing large-signal models for the BJT, we observed that the i-v characteristic of this family of devices includes a *cutoff* region, where virtually no current flows through the transistor. On the other hand, when a sufficient amount of current is injected into the base of the transistor, a bipolar transistor will reach *saturation,* and a substantial amount of collector current will flow. This behavior is quite well suited to the design of electronic gates and switches and can be visualized by superimposing a load line on the collector characteristic, as shown in Figure 10.32.

The operation of the simple **BJT switch** is illustrated in Figure 10.32, by means of load-line analysis. Writing the load-line equation at the collector circuit, we have

$$v_{CE} = V_{CC} - i_C R_C \qquad \textbf{(10.13)}$$

and

$$v_{out} = v_{CE} \qquad \textbf{(10.14)}$$

Thus, when the input voltage v_{in} is low (say, 0 V), the transistor is in the cutoff region and little or no current flows, and

$$v_{out} = v_{CE} = V_{CC} \qquad \textbf{(10.15)}$$

so that the output is "logic high."

When v_{in} is large enough to drive the transistor into the saturation region, a substantial amount of collector current will flow and the collector-emitter voltage will be reduced to the small saturation value $V_{CE\,sat}$, which is typically a fraction of a volt. This corresponds to the point labeled B on the load line. For the input voltage v_{in} to drive the BJT of Figure 10.32 into saturation, a base current of approximately 50 μA will be required. Suppose, then, that the voltage v_{in} could take the values 0 or 5 V. Then if $v_{in} = 0$ V, v_{out} will be nearly equal to V_{CC}, or, again, 5 V. If, on the other hand, $v_{in} = 5$ V and R_B is, say, equal to 89 kΩ [so that the base current required for saturation flows into the base: $i_B = (v_{in} - V_\gamma)/R_B = (5 - 0.6)/89{,}000 \approx 50\ \mu$A], we have the BJT in saturation, and $v_{out} = V_{CE\,sat} \approx 0.2$ V.

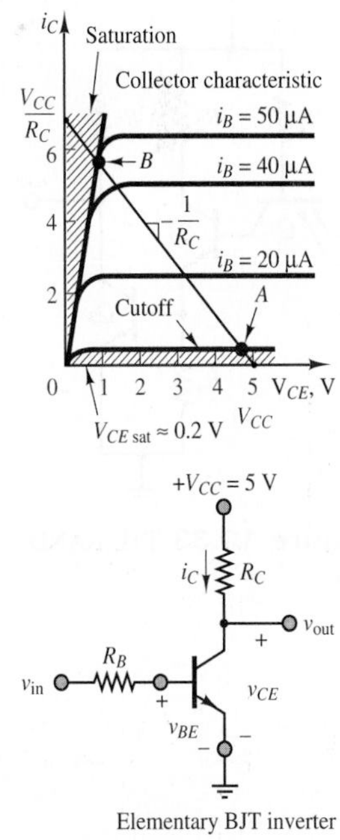

Figure 10.32 BJT switching characteristic

LO4

Thus, you see that whenever v_{in} corresponds to a logic high (or logic 1), v_{out} takes a value close to 0 V, or logic low (or 0); conversely, $v_{\text{in}} =$ "0" (logic "low") leads to $v_{\text{out}} =$ "1." The values of 5 and 0 V for the two logic levels 1 and 0 are quite common in practice and are the standard values used in a family of logic circuits denoted by the acronym **TTL,** which stands for **transistor-transistor logic.**[1] One of the more common TTL blocks is the **inverter** shown in Figure 10.32, so called because it "inverts" the input by providing a low output for a high input, and vice versa. This type of inverting, or "negative," logic behavior is quite typical of BJT gates (and of transistor gates in general).

In the following paragraphs, we introduce some elementary BJT logic gates, similar to the diode gates described previously; the theory and application of digital logic circuits are discussed in Chapter 13. Example 10.10 illustrates the operation of a **NAND gate,** that is, a logic gate that acts as an inverted AND gate (thus the prefix N in NAND, which stands for NOT).

LO5

EXAMPLE 10.10 TTL NAND Gate

Problem

Complete the table below to determine the logic gate operation of the TTL NAND gate of Figure 10.33.

v_1	v_2	State of Q_1	State of Q_2	v_{out}
0 V	0 V			
0 V	5 V			
5 V	0 V			
5 V	5 V			

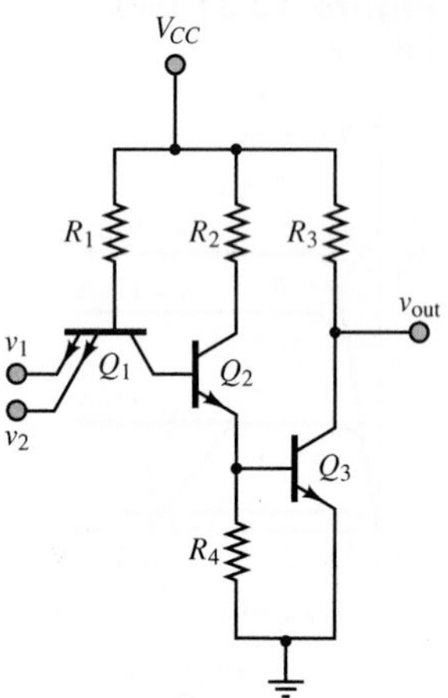

Figure 10.33 TTL NAND gate

Solution

Known Quantities: Resistor values; $V_{BE\text{on}}$ and $V_{CE\text{sat}}$ for each transistor.

Find: v_{out} for each of the four combinations of v_1 and v_2.

Schematics, Diagrams, Circuits, and Given Data: $R_1 = 5.7\ \text{k}\Omega$; $R_2 = 2.2\ \text{k}\Omega$; $R_3 = 2.2\ \text{k}\Omega$; $R_4 = 1.8\ \text{k}\Omega$; $V_{CC} = 5$ V; $V_{BE\text{on}} = V_\gamma = 0.7$ V; $V_{CE\text{sat}} = 0.2$ V.

Assumptions: Treat the *BE* and *BC* junctions of Q_1 as offset diodes. Assume that the transistors are in saturation when conducting.

Analysis: The inputs to the TTL gate, v_1 and v_2, are applied to the emitter of transistor Q_1. The transistor is designed so as to have two emitter circuits in parallel. Transistor Q_1 is modeled by the offset diode model, as shown in Figure 10.34. We now consider each of the four cases.

1. $v_1 = v_2 = 0$ V. With the emitters of Q_1 connected to ground and the base of Q_1 at 5 V, the *BE* junction will clearly be forward-biased and Q_1 is on. This result means that the base current of Q_2 (equal to the collector current of Q_1) is negative, and therefore Q_2 must be off. If Q_2 is off, its emitter current must be zero, and therefore no base current can flow

[1]TTL logic values are actually quite flexible, with v_{HIGH} as low as 2.4 V and v_{LOW} as high as 0.4 V.

into Q_3, which is in turn also off. With Q_3 off, no current flows through R_3, and therefore $v_{out} = 5 - v_{R3} = 5$ V.

2. $v_1 = 5$ V; $v_2 = 0$ V. Now, with reference to Figure 10.34, we see that diode D_1 is still forward-biased, but D_2 is now reverse-biased because of the 5-V potential at v_2. Since one of the two emitter branches is capable of conducting, base current will flow and Q_1 will be on. The remainder of the analysis is the same as in case 1, and Q_2 and Q_3 will both be off, leading to $v_{out} = 5$ V.
3. $v_1 = 0$ V; $v_2 = 5$ V. By symmetry with case 2, we conclude that, again, one emitter branch is conducting, and therefore Q_1 will be on, Q_2 and Q_3 will both be off, and $v_{out} = 5$ V.
4. $v_1 = 5$ V; $v_2 = 5$ V. When both v_1 and v_2 are at 5 V, diodes D_1 and D_2 are both strongly reverse-biased, and therefore no emitter current can flow. Thus, Q_1 must be off. Note, however, that while D_1 and D_2 are reverse-biased, D_3 is forward-biased, and therefore a current will flow into the base of Q_2; thus, Q_2 is on and since the emitter of Q_2 is connected to the base of Q_3, Q_3 will also see a positive base current and will be on. To determine the output voltage, we assume that Q_3 is operating in saturation. Then, applying KVL to the collector circuit, we have

$$V_{CC} = I_{C3}R_3 + V_{CE3}$$

or

$$I_{C3} = \frac{V_{CC} - V_{CE3}}{R_C} = \frac{V_{CC} - V_{CE\text{sat}}}{R_C} = \frac{5 - 0.2}{2{,}200} = 2.2 \text{ mA}$$

and

$$v_{out} = V_{CC} - I_C R_3 = 5 - 2.2 \times 10^{-3} \times 2.2 \times 10^{-3} = 5 - 4.84 = 0.16 \text{ V}$$

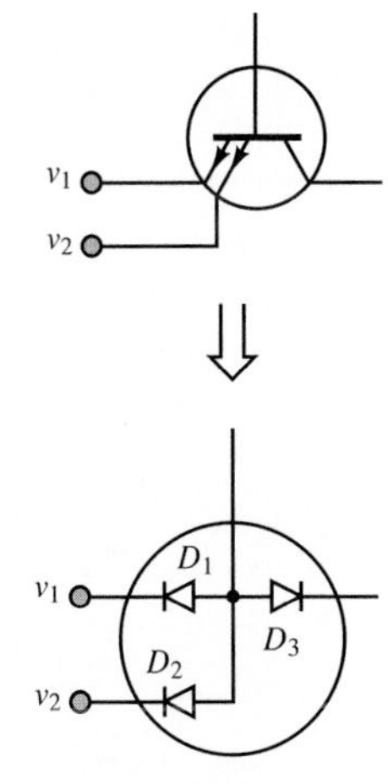

Figure 10.34

These results are summarized in the next table. The output values are consistent with TTL logic; the output voltage for case 4 is sufficiently close to zero to be considered zero for logic purposes.

v_1	v_2	State of Q_2	State of Q_3	v_{out}
0 V	0 V	Off	Off	5 V
0 V	5 V	Off	Off	5 V
5 V	0 V	Off	Off	5 V
5 V	5 V	On	On	0.16 V

Comments: While exact analysis of TTL logic gate circuits could be tedious and involved, the method demonstrated in this example—to determine whether transistors are on or off—leads to very simple analysis. Since in logic devices one is interested primarily in logic levels and not in exact values, this approximate analysis method is very appropriate.

CHECK YOUR UNDERSTANDING

Using the BJT switching characteristic of Figure 10.32, find the value of R_B required to drive the transistor to saturation, assuming that this state corresponds to a base current of 50 μA, if the minimum v_{in} for which we wish to turn the transistor on is 2.5 V.

Answer: $R_B \le 348$ kΩ

The analysis method employed in Example 10.10 can be used to analyze any TTL gate. With a little practice, the calculations of this example will become familiar. The homework problems will reinforce the concepts developed in this section.

Conclusion

This chapter introduces the bipolar junction transistor, and by way of simple circuit model demonstrates its operation as an amplifier and a switch. Upon completing this chapter, you should have mastered the following learning objectives:

1. *Understand the basic principles of amplification and switching.* Transistors are three-terminal electronic semiconductor devices that can serve as amplifiers and switches.
2. *Understand the physical operation of bipolar transistors; determine the operating point of a bipolar transistor circuit.* The bipolar junction transistor has four regions of operation. These can be readily identified by simple voltage measurements.
3. *Understand the large-signal model of the bipolar transistor, and apply it to simple amplifier circuits.* The large-signal model of the BJT is very easy to use, requiring only a basic understanding of DC circuit analysis, and can be readily applied to many practical situations.
4. *Select the operating point of a bipolar transistor circuit.* Biasing a bipolar transistor consists of selecting the appropriate values for the DC supply voltage(s) and for the resistors that comprise a transistor amplifier circuit. When biased in the forward active region, the bipolar transistor acts as a current-controlled current source, and can amplify small currents injected into the base by as much as a factor of 200.
5. *Understand the operation of a bipolar transistor as a switch and analyze basic analog and digital gate circuits.* The operation of a BJT as a switch is very straightforward, and consists of designing a transistor circuit that will go from cut-off to saturation when an input voltage changes from a high to a low value, or vice versa. Transistor switches are commonly used to design digital logic gates.

HOMEWORK PROBLEMS

Section 10.2: Operation of the Bipolar Junction Transistor

10.1 For each transistor shown in Figure P10.1, determine whether the *BE* and *BC* junctions are forward- or reverse-biased, and determine the operating region.

10.2 Determine the region of operation for the following transistors:

a. *npn*, $V_{BE} = 0.8$ V, $V_{CE} = 0.4$ V

b. *npn*, $V_{CB} = 1.4$ V, $V_{CE} = 2.1$ V

c. *pnp*, $V_{CB} = 0.9$ V, $V_{CE} = 0.4$ V

d. *npn*, $V_{BE} = -1.2$ V, $V_{CB} = 0.6$ V

10.3 Given the circuit of Figure P10.3, determine the operating point of the transistor. Assume the BJT is a silicon device with $\beta = 100$. In what region is the transistor?

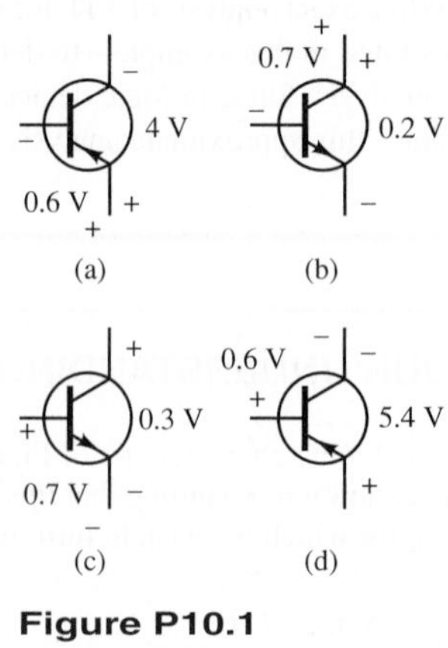

Figure P10.1

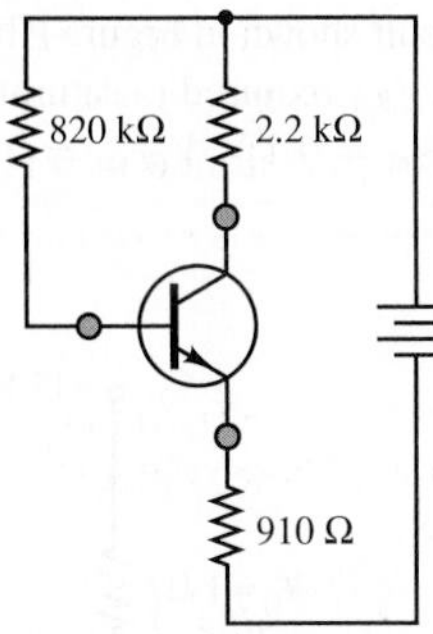

Figure P10.3

10.4 The magnitudes of a *pnp* transistor's emitter and base currents are 6 and 0.1 mA, respectively. The magnitudes of the voltages across the emitter-base and collector-base junctions are 0.65 and 7.3 V. Find

a. V_{CE}.

b. I_C.

c. The total power dissipated in the transistor, defined here as $P = V_{CE}I_C + V_{BE}I_B$.

10.5 Given the circuit of Figure P10.5, determine the emitter current and the collector-base voltage. Assume the BJT has $V_\gamma = 0.6$ V.

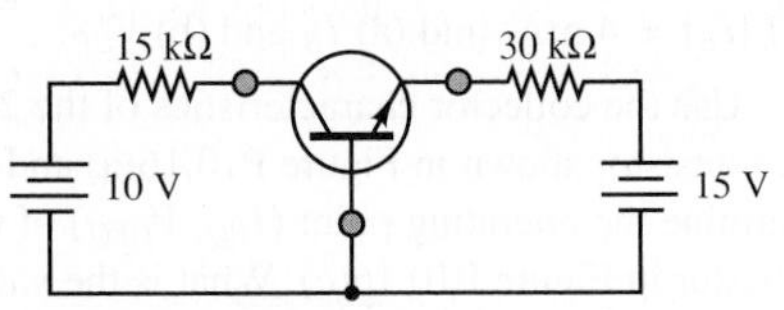

Figure P10.5

10.6 Given the circuit of Figure P10.6, determine the operating point of the transistor. Assume a 0.6-V offset voltage and $\beta = 150$. In what region is the transistor?

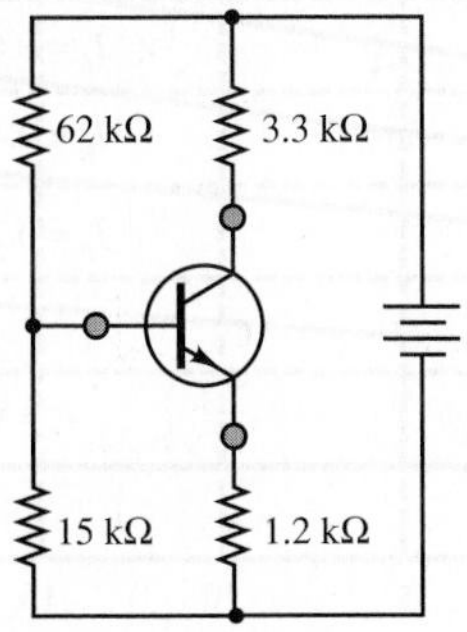

Figure P10.6

10.7 Given the circuit of Figure P10.7, determine the emitter current and the collector-base voltage. Assume the BJT has a 0.6-V offset voltage at the *BE* junction.

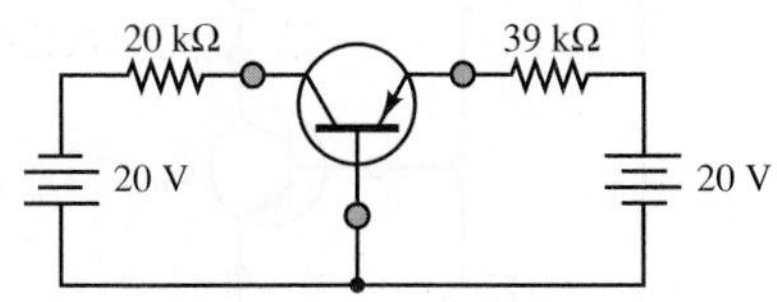

Figure P10.7

10.8 If the emitter resistor in Problem 10.7 (Figure P10.7) is changed to 22 kΩ, how does the operating point of the BJT change?

10.9 The collector characteristics for a certain transistor are shown in Figure P10.9.

a. Find the ratio I_C/I_B for $V_{CE} = 10$ V and $I_B = 100$, 200, and 600 μA.

b. The maximum allowable collector power dissipation is 0.5 W for $I_B = 500$ μA. Find V_{CE}.

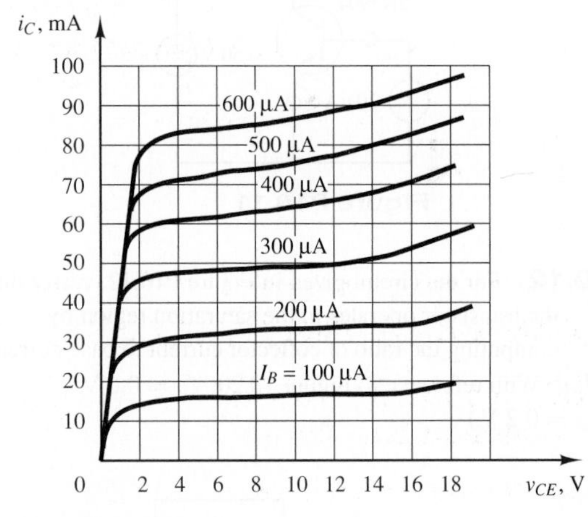

Figure P10.9

(*Hint:* A reasonable approximation for the power dissipated at the collector is the product of the collector voltage and current $P = I_C V_{CE}$, where P is the permissible power dissipation, I_C is the quiescent collector current, and V_{CE} is the operating point collector-emitter voltage.)

10.10 Given the circuit of Figure P10.10, assume both transistors are silicon-based with $\beta = 100$. Determine:

a. I_{C1}, V_{C1}, V_{CE1}

b. I_{C2}, V_{C2}, V_{CE2}

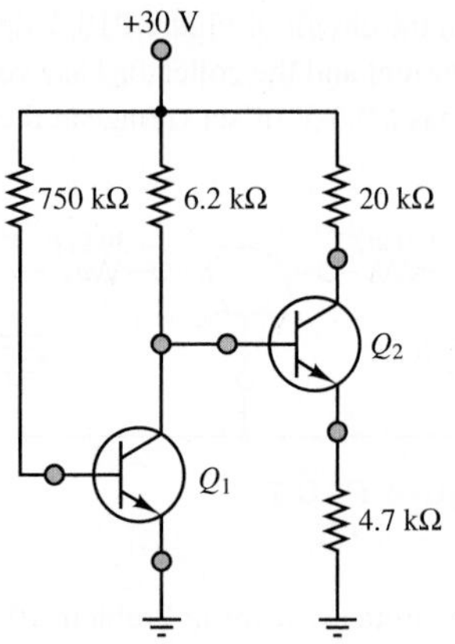

Figure P10.10

10.11 Use the collector characteristics of the 2N3904 *npn* transistor to determine the operating point (I_{CQ}, V_{CEQ}) of the transistor in Figure P10.11. What is the value of β at this point?

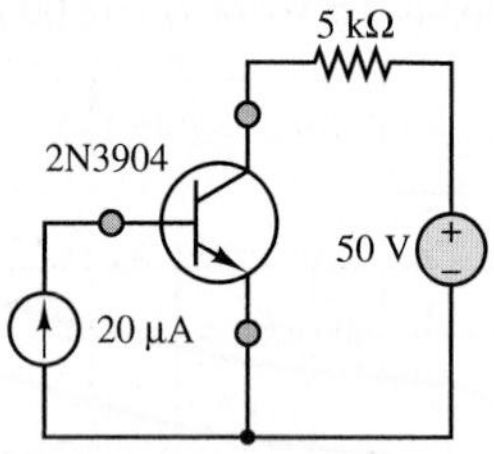

Figure P10.11

10.12 For the circuit given in Figure P10.12, verify that the transistor operates in the saturation region by computing the ratio of collector current to base current. (*Hint:* With reference to Figure 10.20, $V_\gamma = 0.6$ V, $V_{\text{sat}} = 0.2$ V.)

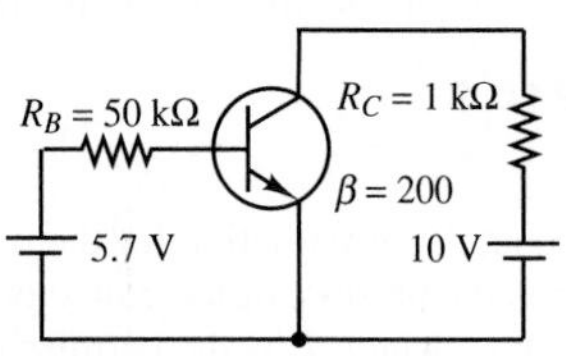

Figure P10.12

10.13 For the circuit in Figure 10.28 in the text, V_{CC} =20 V, $R_C = 5$ kΩ, and $R_E = 1$ kΩ. Determine the region of operation of the transistor if:

a. $I_C = 1$ mA, $I_B = 20\ \mu$A, $V_{BE} = 0.7$ V

b. $I_C = 3.2$ mA, $I_B = 0.3$ mA, $V_{BE} = 0.8$ V

c. $I_C = 3$ mA, $I_B = 1.5$ mA, $V_{BE} = 0.85$ V

10.14 For the circuit shown in Figure P10.14, determine the base voltage V_{BB} required to saturate the transistor. Assume that $V_{CE\text{sat}} = 0.1$ V, $V_{BE\text{sat}} = 0.6$ V, and $\beta = 50$.

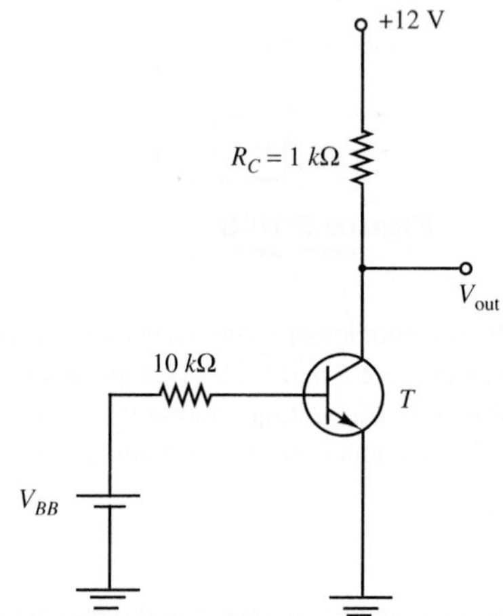

Figure P10.14

10.15 An *npn* transistor is operated in the active region with the collector current 60 times the base current and with junction voltages of $V_{BE} = 0.6$ V and $V_{CB} = 7.2$ V. If $|I_E| = 4$ mA, find (a) I_B and (b) V_{CE}.

10.16 Use the collector characteristics of the 2N3904 *npn* transistor shown in Figure P10.16(a) and (b) to determine the operating point (I_{CQ}, V_{CEQ}) of the transistor in Figure P10.16(c). What is the value of β at this point?

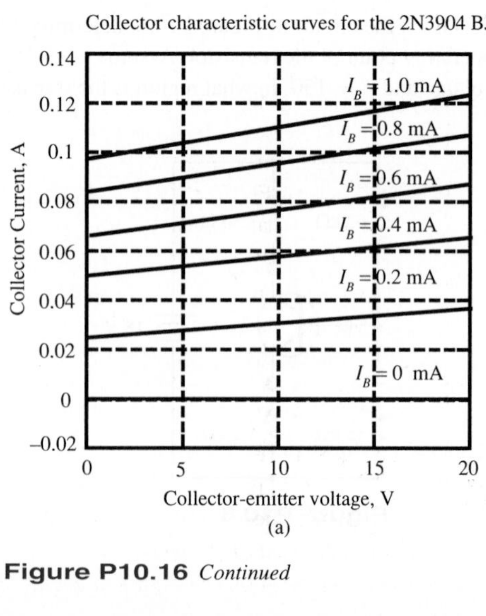

Figure P10.16 *Continued*

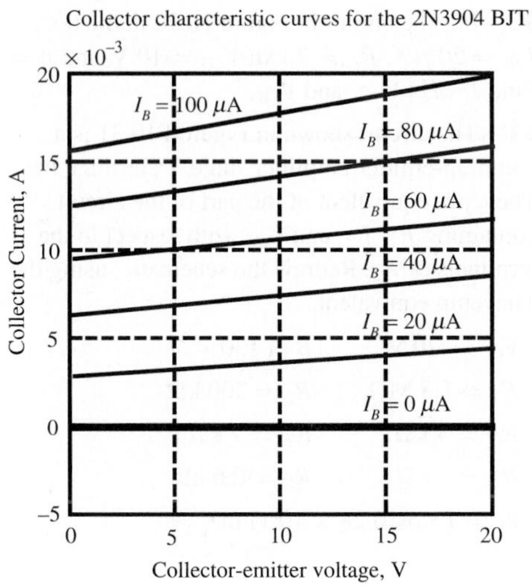

(b)

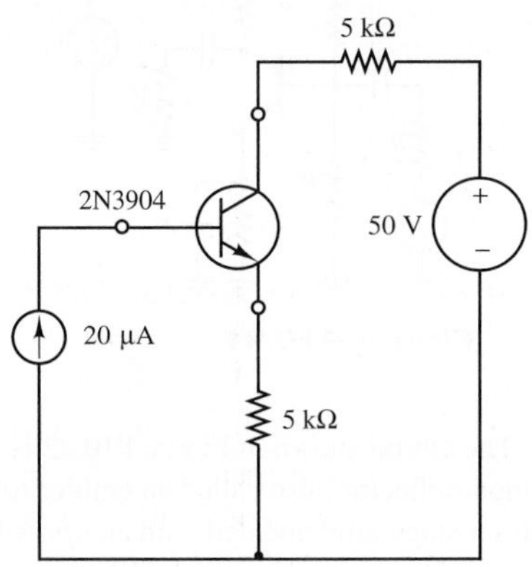

(c)

Figure P10.16

Section 10.3: BJT Large-Signal Model

10.17 With reference to the LED driver of Example 10.4, Figure 10.14 in the text, assume that we need to drive an LED that requires $I_{LED} = 10$ mA. All other values are unchanged. Find the range of collector resistance R_C values that will permit the transistor to supply the required current.

10.18 With reference to the diode thermometer of The Focus on Measurements box "Large-Signal Amplifier for Diode Thermometer," Figure 10.18 in the text, let $R_B = 33$ kΩ, $V_{CEQ} = 6$ V, v_D (voltage across diode) $= 1.1$ V, $V_{BEQ} = 0.75$ V. Find the value of R_C that is required to achieve the given Q point.

10.19 With reference to the LED driver of Example 10.4, Figure 10.14 in the text, assume that $R_C = 340\ \Omega$ that we need to drive an LED that requires $I_{LED} \geq 10$ mA, and that the maximum base current that can be supplied by the microprocessor is 5 mA. All other parameters and requirements are the same as in Example 10.4. Determine the range of values of the base resistance R_B that will satisfy this requirement.

10.20 Use the same data given in Problem 10.19, but assume that $R_B = 10$ kΩ. Find the minimum value of β that will ensure correct operation of the LED driver.

10.21 Repeat Problem 10.20 for the case of a microprocessor operating on a 3.3-V supply (that is, $V_{on} = 3.3$ V).

10.22 Consider the LED driver circuit of Figure 10.14 in the text. This circuit is now used to drive an automotive fuel injector (an electromechanical solenoid valve). The differences in the circuit are as follows: The collector resistor and the LED are replaced by the fuel injector, which can be modeled as a series *RL* circuit. The voltage supply for the fuel injector is 13 V (instead of 5 V). For the purposes of this problem, it is reasonable to assume $R = 12\ \Omega$ and $L \sim 0$. Assume that the maximum current that can be supplied by the microprocessor is 1 mA, that the current required to drive the fuel injector must be at least 1 A, and that the transistor saturation voltage is $V_{CE\text{sat}} = 1$ V. Find the minimum value of β required for the transistor.

10.23 With reference to Problem 10.22, assume $\beta = 2{,}000$. Find the allowable range of R_B.

10.24 With reference to Problem 10.22, a new generation of power-saving microcontrollers operates on 3.3-V supplies (that is, $V_{on} = 3.3$ V). Assume $\beta = 2{,}000$. Find the allowable range of R_B.

10.25 The circuit shown in Figure P10.25 is a 9-V battery charger. The purpose of the Zener diode is to provide a constant voltage across resistor R_2, such that the transistor will source a constant emitter (and therefore collector) current. Select the values of R_2, R_1, and V_{CC} such that the battery will be charged with a constant 40-mA current.

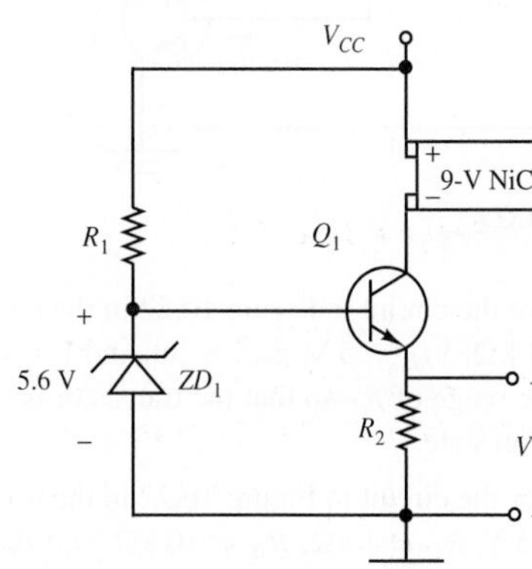

Figure P10.25

10.26 The circuit of Figure P10.26 is a variation of the battery charging circuit of Problem 10.25. Analyze the operation of the circuit and explain how this circuit will provide a decreasing charging current (taper current cycle) until the NiCd battery is fully charged (10.4 V—see note in Example 10.5). Choose appropriate values of V_{CC} and R_1 that would result in a practical design. Use standard resistor values.

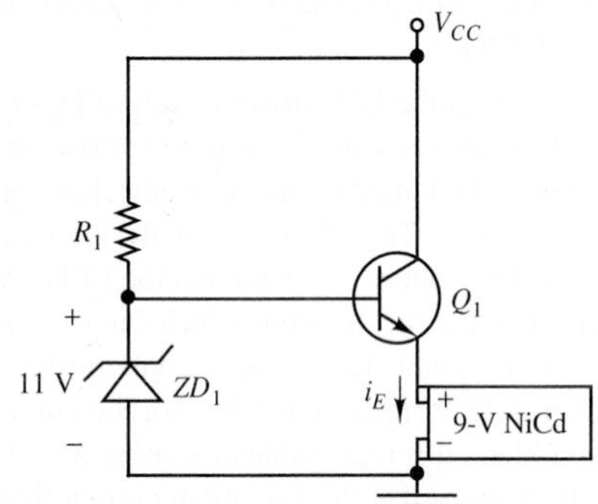

Figure P10.26

10.27 The circuit of Figure P10.27 is a variation of the motor driver circuit of Example 10.6. The external voltage v_{in} represents the analog output of a microcontroller, and ranges between zero and 5 V. Complete the design of the circuit by selecting the value of the base resistor, R_b, such that the motor will see the maximum design current when $v_{\text{in}} = 5$ V. Use the transistor β value and the design specifications for motor maximum and minimum current given in Example 10.6.

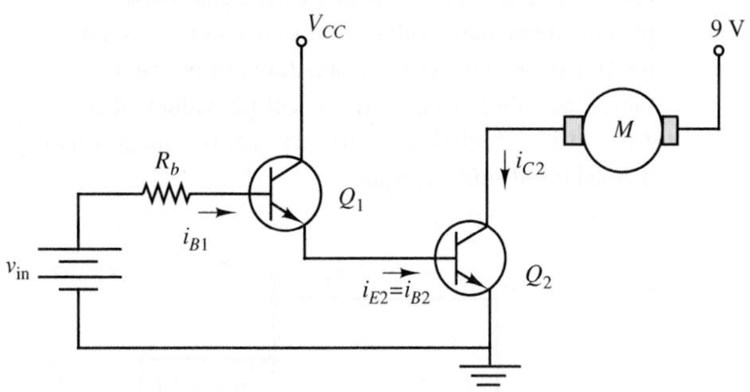

Figure P10.27

10.28 For the circuit in Figure 10.22 in the text, $R_C = 1$ kΩ, $V_{BB} = 5$ V, $\beta_{\min} = 50$, and $V_{CC} = 10$ V. Find the range of R_B so that the transistor is in the saturation state.

10.29 For the circuit in Figure 10.22 in the text, $V_{CC} = 5$ V, $R_C = 1$ kΩ, $R_B = 10$ kΩ, and $\beta_{\min} = 50$. Find the range of values of V_{BB} so that the transistor is in saturation.

10.30 For the circuit in Figure 10.20 in the text, $I_{BB} = 20\ \mu$A, $R_C = 2$ kΩ, $V_{CC} = 10$ V, and $\beta = 100$. Find I_C, I_E, V_{CE}, and V_{CB}.

10.31 The circuit shown in Figure P10.31 is a common-emitter amplifier stage. Determine the Thévenin equivalent of the part of the circuit containing R_1, R_2, and V_{CC} with respect to the terminals of R_2. Redraw the schematic, using the Thévenin equivalent.

$V_{CC} = 20$ V $\quad \beta = 130$
$R_1 = 1.8$ MΩ $\quad R_2 = 300$ kΩ
$R_C = 3$ kΩ $\quad R_E = 1$ kΩ
$R_L = 1$ kΩ $\quad R_S = 0.6$ kΩ
$v_S = 1\ \cos(6.28 \times 10^3 t)$ mV

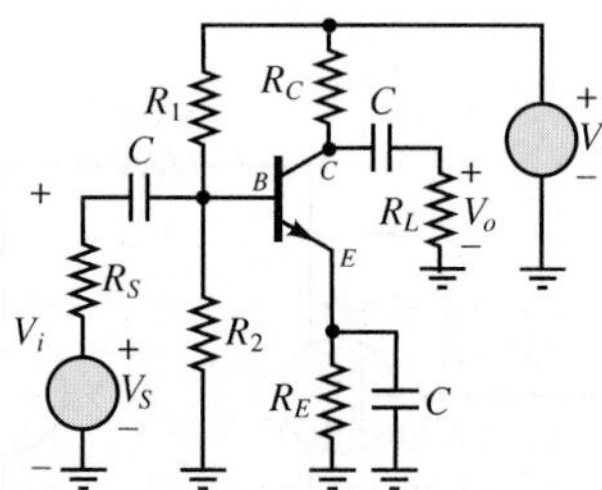

Figure P10.31

10.32 The circuit shown in Figure P10.32 is a common-collector (also called an emitter follower) amplifier stage implemented with an *npn* silicon transistor. Determine V_{CEQ} at the DC operating or Q point.

$V_{CC} = 12$ V $\quad \beta = 130$
$R_1 = 82$ kΩ $\quad R_2 = 22$ kΩ
$R_S = 0.7$ kΩ $\quad R_E = 0.5$ kΩ
$R_L = 16$ Ω

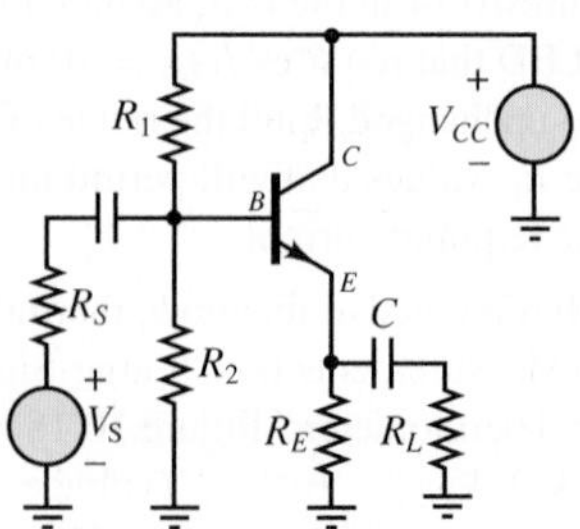

Figure P10.32

10.33 Shown in Figure P10.33 is a common-emitter amplifier stage implemented with an *npn* silicon transistor and two DC supply voltages (one positive and one negative) instead of one. The DC bias circuit connected to the base consists of a single resistor. Determine V_{CEQ} and the region of operation.

$V_{CC} = 12$ V $\quad V_{EE} = 4$ V

$\beta = 100 \quad R_B = 100$ kΩ

$R_C = 3$ kΩ $\quad R_E = 3$ kΩ

$R_L = 6$ kΩ $\quad R_S = 0.6$ kΩ

$v_S = 1\cos(6.28 \times 10^3 t)$ mV

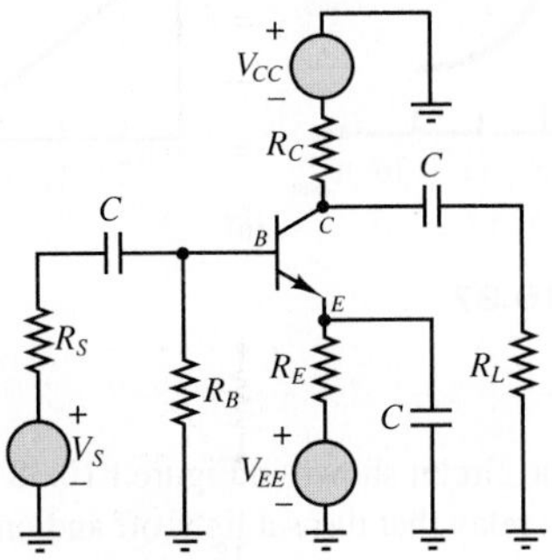

Figure P10.33

10.34 Shown in Figure P10.34 is a common-emitter amplifier stage implemented with an *npn* silicon transistor. The DC bias circuit connected to the base consists of a single resistor; however, it is connected directly between base and collector. Determine V_{CEQ} and the region of operation.

$V_{CC} = 12$ V

$\beta = 130 \quad R_B = 325$ kΩ

$R_C = 1.9$ kΩ $\quad R_E = 2.3$ kΩ

$R_L = 10$ kΩ $\quad R_S = 0.5$ kΩ

$v_S = 1\cos(6.28 \times 10^3 t)$ mV

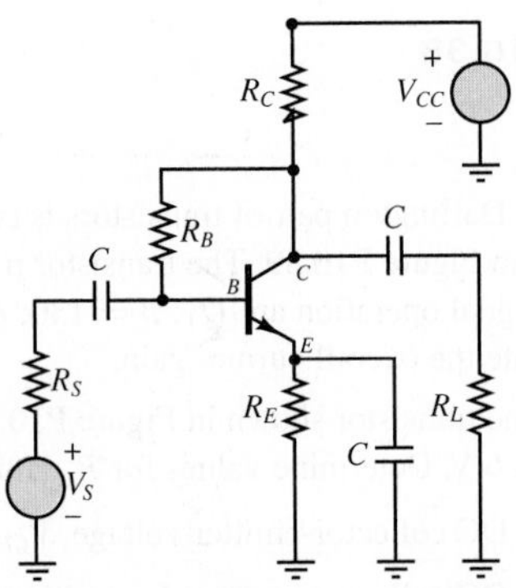

Figure P10.34

10.35 For the circuit shown in Figure P10.35 v_S is a small sine wave signal with average value of 3 V. If $\beta = 100$ and $R_B = 60$ kΩ.

a. Find the value of R_E so that I_E is 1 mA.

b. Find R_C so that V_C is 5 V.

c. For $R_L = 5$ kΩ, find the small-signal equivalent circuit of the amplifier.

d. Find the voltage gain.

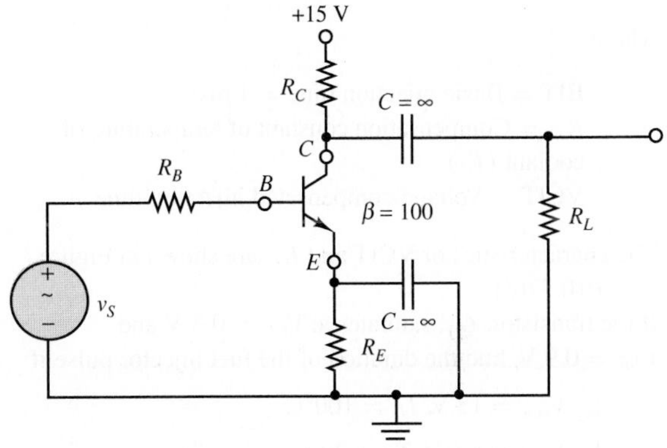

Figure P10.35

10.36 The circuit in Figure P10.36 is in the common-collector configuration. Assuming $R_C = 200\ \Omega$:

a. Find the operating point of the transistor.

b. If the voltage gain is defined as v_{out}/v_{in}, find the voltage gain. If the current gain is defined as i_{out}/i_{in}, find the current gain.

c. Find the input resistance, r_i.

d. Find the output resistance, r_o.

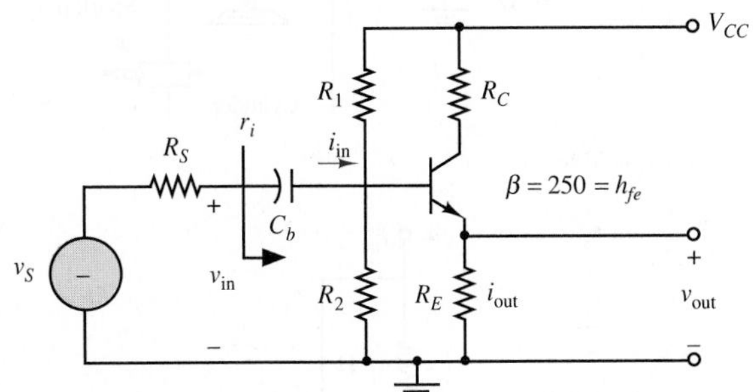

$R_E = 250\ \Omega \quad R_1 = 9{,}221\ \Omega$

$V_{CC} = 15$ V $\quad C_B = \infty$

$R_2 = 6{,}320\ \Omega$

Figure P10.36

10.37 The circuit that supplies energy to an automobile's fuel injector is shown in Figure P10.37(a). The internal circuitry of the injector can be modeled as shown in Figure P10.37(b). The injector will inject gasoline into the intake manifold when $I_{inj} \geq 0.1$ A. The voltage V_{signal} is a pulse train whose shape is as shown in Figure P10.37(c). If the engine is cold and under start-up conditions, the signal duration, τ, is determined by the equation

$$\tau = \text{BIT} \times K_C + \text{VCIT}$$

where

BIT = Basic injection time = 1 ms
K_C = Compensation constant of temperature of coolant (T_C)
VCIT = Voltage-compensated injection time

The characteristics of VCIT and K_C are shown in Figure P10.37(d).

If the transistor, Q_1, saturates at $V_{CE} = 0.3$ V and $V_{BE} = 0.9$ V, find the duration of the fuel injector pulse if

a. $V_{batt} = 13$ V, $T_C = 100°$C
b. $V_{batt} = 8.6$ V, $T_C = 20°$C

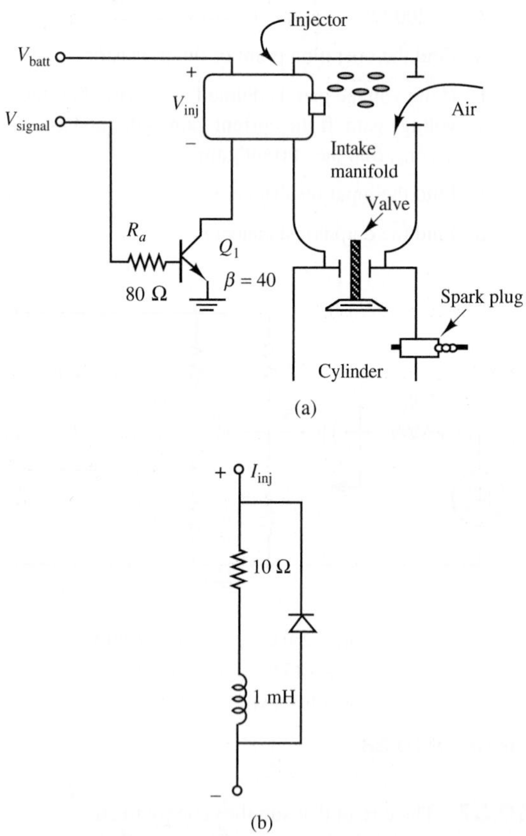

Figure P10.37 *Continued*

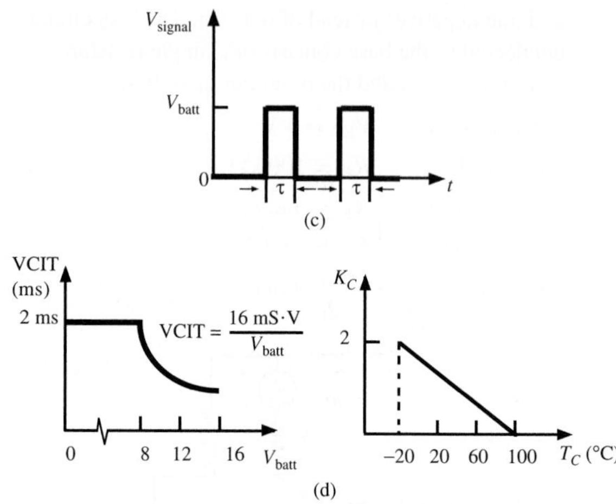

Figure P10.37

10.38 The circuit shown in Figure P10.38 is used to switch a relay that turns a light off and on under the control of a computer. The relay dissipates 0.5 W at 5 VDC. It switches on at 3 VDC and off at 1.0 VDC. What is the maximum frequency with which the light can be switched? The inductance of the relay is 5 mH, and the transistor saturates at 0.2 V, $V_\gamma = 0.8$ V.

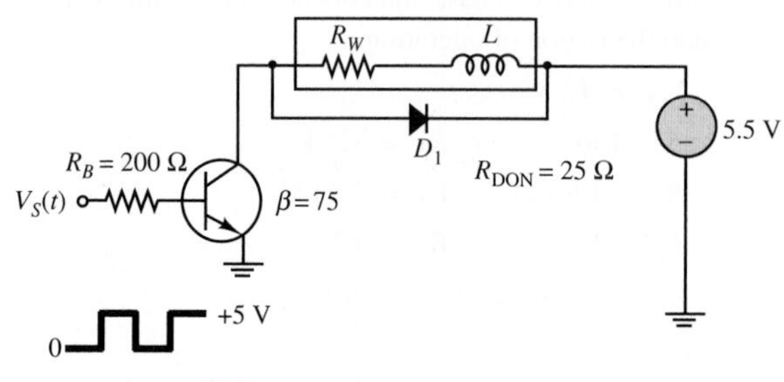

Figure P10.38

10.39 A Darlington pair of transistors is connected as shown in Figure P10.39. The transistor parameters for large-signal operation are Q_1: $\beta = 130$; Q_2: $\beta = 70$. Calculate the overall current gain.

10.40 The transistor shown in Figure P10.40 has $V_X = 0.6$ V. Determine values for R_1 and R_2 such that

a. The DC collector-emitter voltage, V_{CEQ}, is 5 V.
b. The DC collector current, I_{CQ}, will vary no more than 10% as β varies from 20 to 50.
c. Values of R_1 and R_2 which will permit maximum symmetrical swing in the collector current. Assume $\beta = 100$.

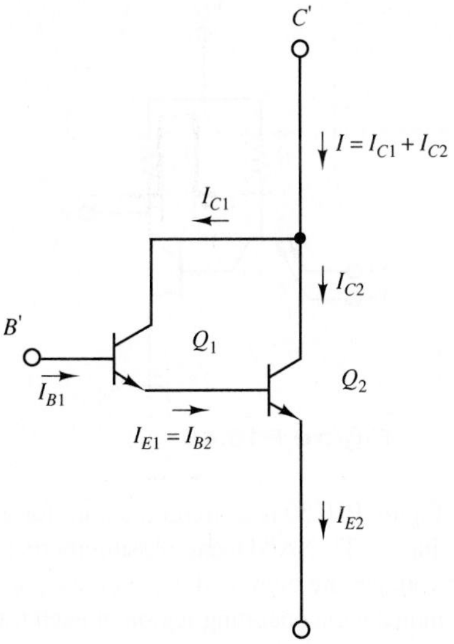

Figure P10.39

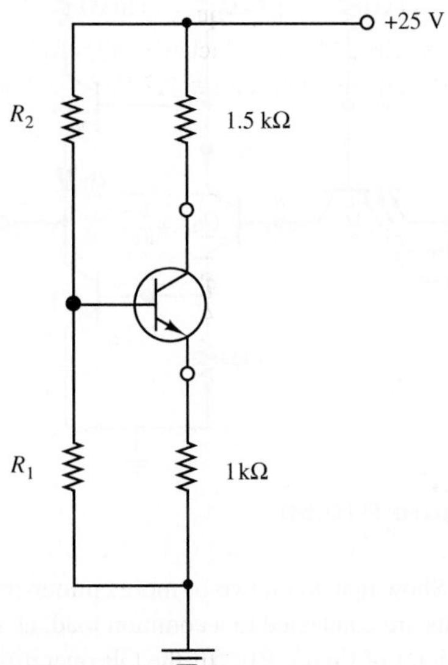

Figure P10.40

Section 10.4: BJT Switches and Gates

10.41 Show that the circuit of Figure P10.41 functions as an OR gate if the output is taken at v_{o1}.

10.42 Show that the circuit of Figure P10.41 functions as a NOR gate if the output is taken at v_{o2}.

10.43 Show that the circuit of Figure P10.43 functions as an AND gate if the output is taken at v_{o1}.

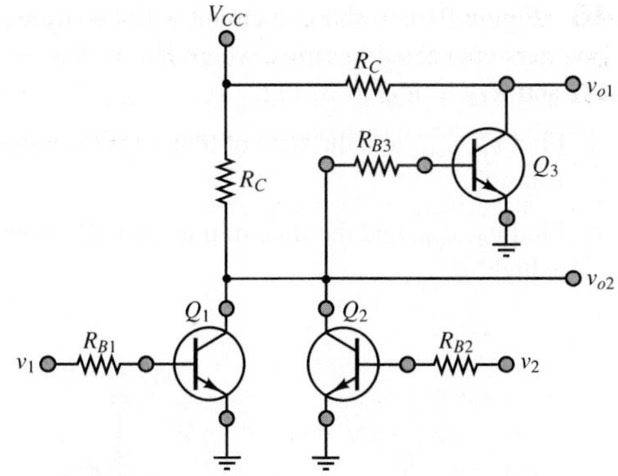

Figure P10.41

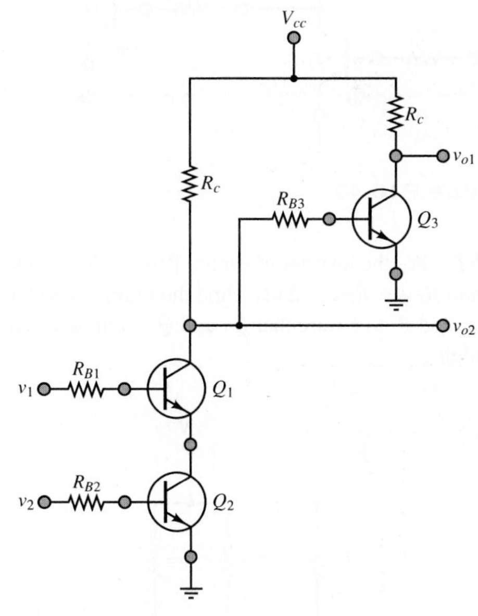

Figure P10.43

10.44 Show that the circuit of Figure P10.43 functions as a NAND gate if the output is taken at v_{o2}.

10.45 In Figure P10.45, the minimum value of v_{in} for a high input is 2.0 V. Assume that transistor Q_1 has a β of at least 10. Find the range for resistor R_B that can guarantee that the transistor Q_1 is on.

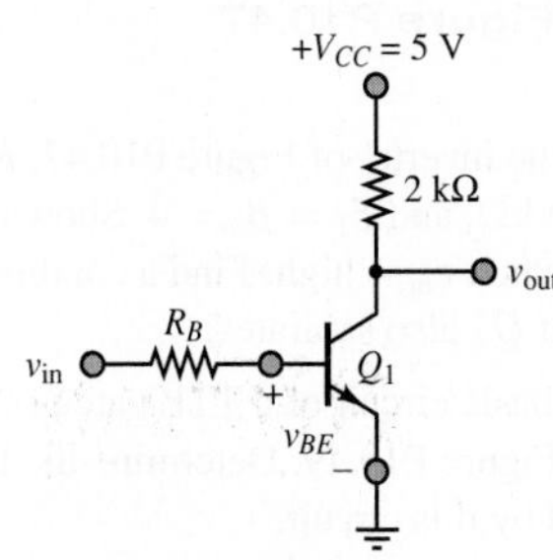

Figure P10.45

10.46 Figure P10.46 shows a circuit with two transistor inverters connected in series, where $R_{1C} = R_{2C} = 10$ kΩ and $R_{1B} = R_{2B} = 27$ kΩ.

a. Find v_B, v_{out}, and the state of transistor Q_1 when v_{in} is low.

b. Find v_B, v_{out}, and the state of transistor Q_1 when v_{in} is high.

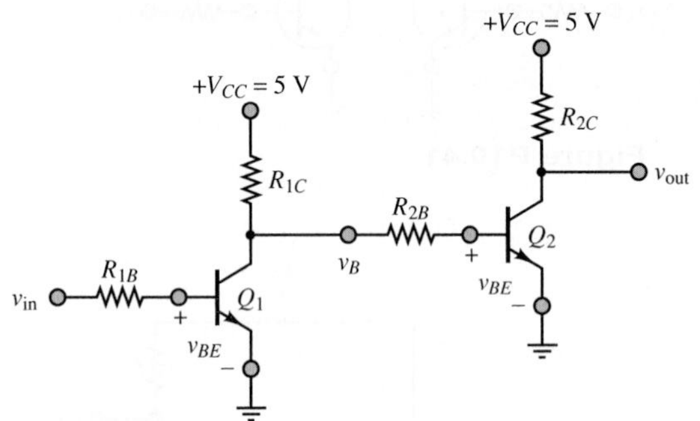

Figure P10.46

10.47 For the inverter of Figure P10.47, $R_B = 5$ kΩ and $R_{C1} = R_{C2} = 2$ kΩ. Find the minimum values of β_1 and β_2 to ensure that Q_1 and Q_2 saturate when v_{in} is high.

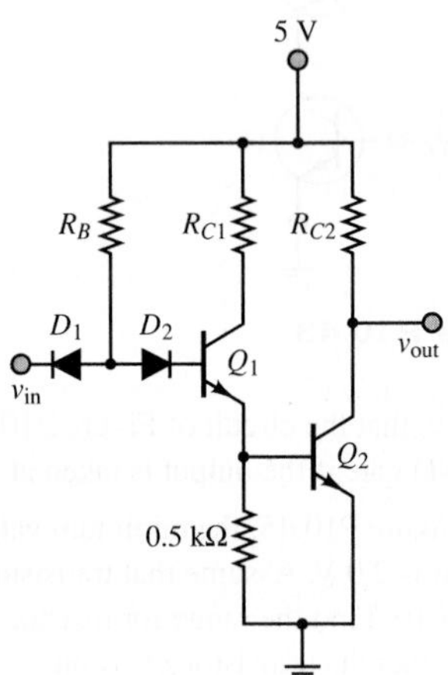

Figure P10.47

10.48 For the inverter of Figure P10.47, $R_B = 4$ kΩ, $R_{C1} = 2.5$ kΩ, and $\beta_1 = \beta_2 = 4$. Show that Q_1 saturates when v_{in} is high. Find a condition for R_{C2} to ensure that Q_2 also saturates.

10.49 The basic circuit of a TTL gate is shown in the circuit of Figure P10.49. Determine the logic function performed by this circuit.

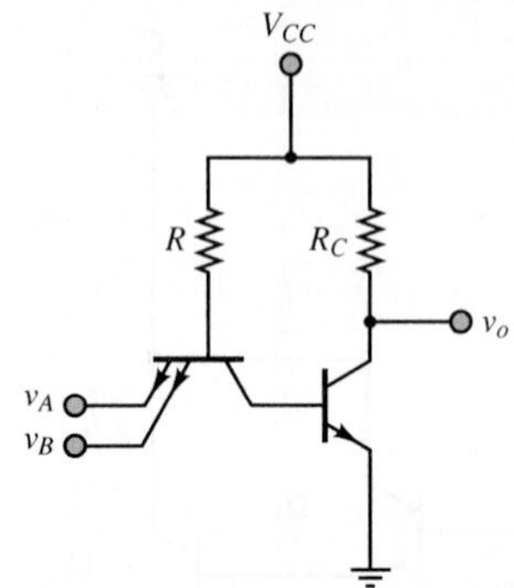

Figure P10.49

10.50 Figure P10.50 is a circuit diagram for a three-input TTL NAND gate. Assuming that all the input voltages are high, find v_{B1}, v_{B2}, v_{B3}, v_{C2}, and v_{out}. Also indicate the operating region of each transistor.

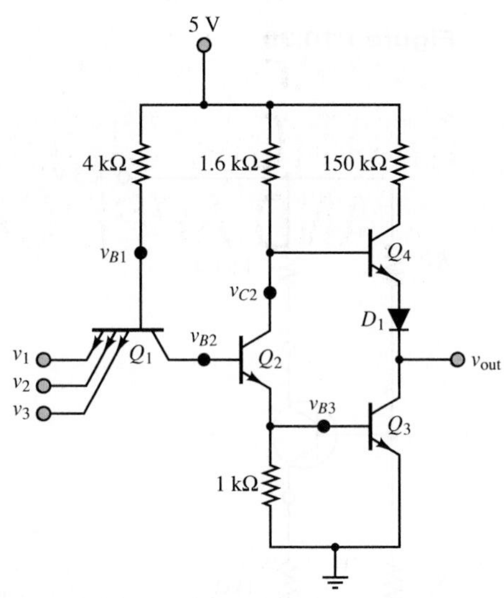

Figure P10.50

10.51 Show that when two or more emitter-follower outputs are connected to a common load, as shown in the circuit of Figure P10.51, the OR operation results; that is, $v_o = v_1$ OR v_2.

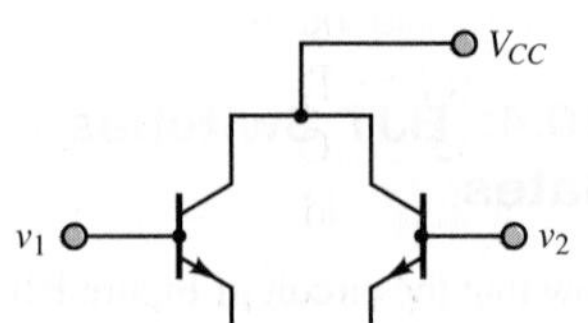

Figure P10.51

CHAPTER 11

FIELD-EFFECT TRANSISTORS: OPERATION, CIRCUIT MODELS, AND APPLICATIONS

Chapter 11 introduces the family of field-effect transistors, or FETs. The concept that forms the basis of the operation of the field-effect transistor is that an external electric field may be used to vary the conductivity of a *channel,* causing the FET to behave either as a voltage-controlled resistor or as a voltage-controlled current source.

FETs are the dominant transistor family in today's integrated electronics, and although these transistors come in several different configurations, it is possible to understand the operation of the different devices by focusing principally on one type.

In this chapter we focus on the basic operation of the enhancement-mode, metal-oxide-semiconductor FET, leading to the technologies that are commonly known as NMOS, PMOS, and CMOS. The chapter reviews the operation of these devices as large-signal amplifiers and as switches.

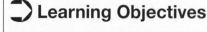

Learning Objectives

1. Understand the classification of field-effect transistors. *Section 11.1.*
2. Learn the basic operation of enhancement-mode MOSFETs by understanding their i–v curves and defining equations. *Section 11.2.*
3. Learn how enhancement-mode MOSFET circuits are biased. *Section 11.3.*
4. Understand the concept and operation of FET large-signal amplifiers. *Section 11.4.*
5. Understand the concept and operation of FET switches. *Section 11.5.*
6. Analyze FET switches and digital gates. *Section 11.5.*

LO1

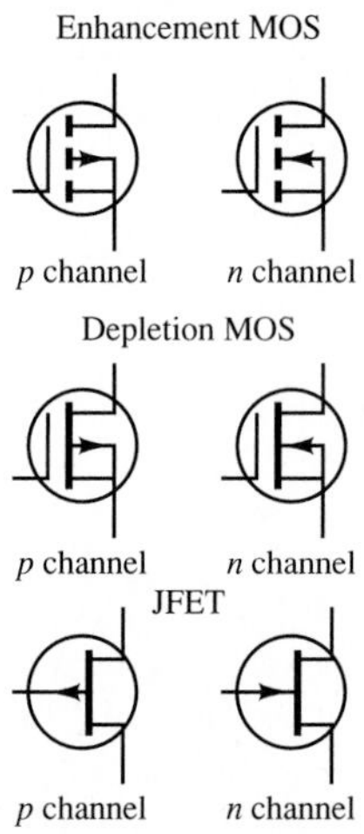

Figure 11.1 Classification of field-effect transistors

11.1 CLASSIFICATION OF FIELD-EFFECT TRANSISTORS

Figure 11.1 depicts the classification of field-effect transistors, as well as the more commonly used symbols for these devices. These devices can be grouped into three major categories. The first two categories are both types of **metal-oxide semiconductor field-effect transistors,** or **MOSFETs: enhancement-mode MOSFETs** and **depletion-mode MOSFETs.** The third category consists of **junction field-effect transistors,** or **JFETs.** In addition, each of these devices can be fabricated either as an ***n*-channel** device or as a ***p*-channel** device, where the n or p designation indicates the nature of the doping in the semiconductor channel. All these transistors behave in a very similar fashion, and we shall predominantly discuss enhancement MOSFETs in this chapter, although a brief discussion of depletion devices and JFETs is also included.

11.2 OVERVIEW OF ENHANCEMENT-MODE MOSFETS

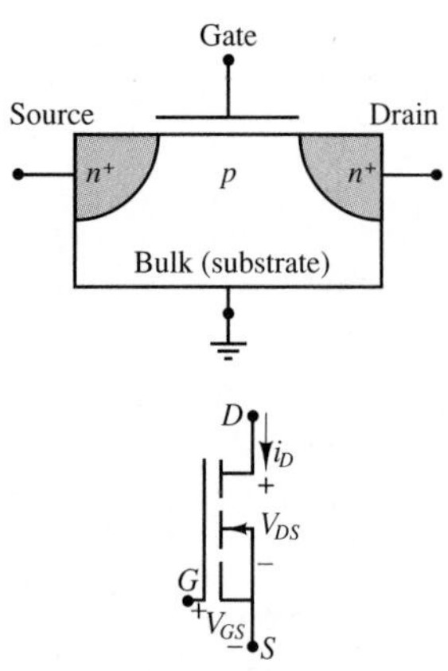

Figure 11.2 The n-channel enhancement MOSFET construction and circuit symbol

Figure 11.2 depicts the circuit symbol and the approximate construction of a typical n-channel enhancement-mode MOSFET. The device has three terminals: the **gate** (analogous to the base in a BJT), the **drain** (analogous to the collector), and the **source** (analogous to the emitter). The **bulk** or **substrate** of the device is shown to be electrically connected to the source, and therefore it does not appear in the electric circuit diagram as a separate terminal. The gate consists of a metal film layer, separated from the p-type bulk by a thin oxide layer (hence the terminology *metal-oxide semiconductor*). The drain and source are both constructed of n^+ material.

Imagine now that the drain is connected to a positive voltage supply V_{DD}, and the source is connected to ground. Since the p-type bulk is connected to the source, and hence to ground, the drain-bulk n^+p junction is strongly reverse-biased. The junction voltage for the pn^+ junction formed by the bulk and the source is zero, since both are connected to ground. Thus, the path between drain and source consists of two reverse-biased pn junctions, and no current can flow. This situation is depicted in Figure 11.3(a): in the absence of a gate voltage, the n-channel enhancement-mode MOSFET acts as an open circuit. Thus, enhancement-mode devices are *normally off.*

Suppose now that a positive voltage is applied to the gate; this voltage will create an electric field in the direction shown in Figure 11.3(b). The effect of the electric

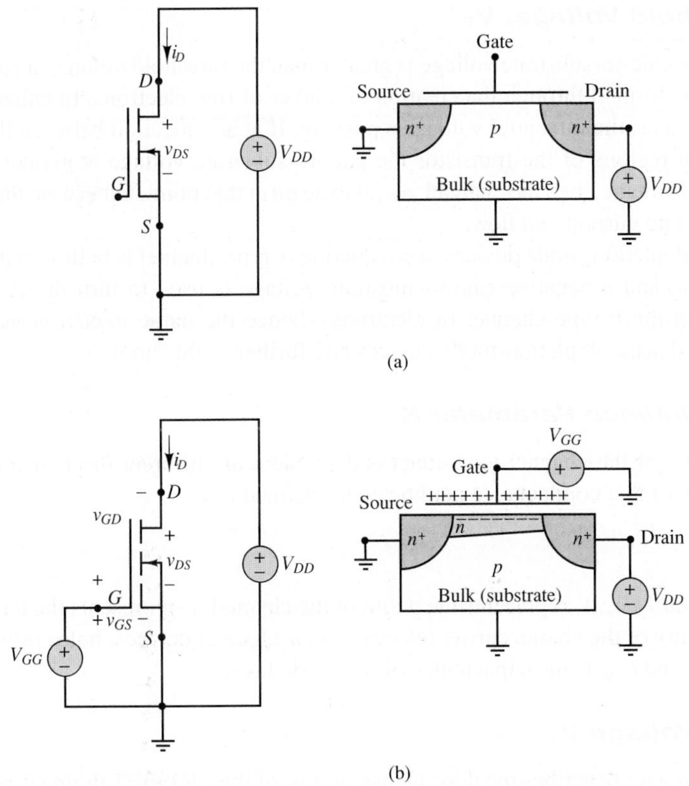

Figure 11.3 Channel formation in NMOS transistor: (a) With no external gate voltage, the source-substrate and substrate-drain junctions are both reverse-biased, and no conduction occurs; (b) when a gate voltage is applied, charge-carrying electrons are drawn between the source and drain regions to form a conducting channel.

LO2

field is to repel positive charge carriers away from the surface of the p-type bulk, and to form a narrow **channel** near the surface of the bulk in which negative charge carriers dominate and are available for conduction. For a fixed drain bias, the greater the strength of the externally applied electric field (i.e., the higher the gate voltage), the higher the concentration of carriers in the channel, and the higher its conductivity. This behavior explains the terminology *enhancement mode*, because the application of an external electric field *enhances* the conduction in the channel by creating n-type charge carriers. It should also be clear why these devices are called *field-effect*, since it is an external electric field that determines the conduction properties of the transistor.

It is also possible to create *depletion-mode* devices in which an externally applied field depletes the channel of charge carriers by reducing the effective channel width. Depletion-mode MOSFETs are normally on, and they can be turned off by application of an external electric field.

To complete this brief summary of the operation of MOS transistors, we note that, in analogy with pnp bipolar transistors, it is also possible to construct p-channel MOSFETS. In these transistors, conduction occurs in a channel formed in n-type bulk material via positive charge carriers.

We first define a few key parameters that generally apply to enhancement-mode and depletion-mode and to n-channel as well as p-channel devices.

Threshold Voltage, V_T

When the gate-to-substrate voltage is greater than the threshold voltage, a conducting channel is formed through the creation of a layer of free electrons. In enhancement-mode devices, the threshold voltage is positive. If at any location between the source and drain regions of the transistor the gate-to-substrate voltage is greater than the threshold voltage, then the channel is said to be *on* at that point. Otherwise the channel is *off*, and no current can flow.

In depletion-mode devices, a conducting n-type channel is built into the device by design, and a negative gate-to-substrate voltage is used to turn the channel off (depleting the n-type channel of electrons—hence the name *depletion mode*). We shall not discuss depletion-mode devices any further in this book.

Conductance Parameter K

The ability of the channel to conduct is dependent on different mechanisms, which are captured in a conductance parameter K, defined as

$$K = \frac{W}{L}\frac{\mu C_{ox}}{2} \tag{11.1}$$

In equation 11.1, W represents the width of the channel, L represents the length, μ is the mobility of the charge carrier (electrons in n-channel devices, holes in p-channel devices), and C_{ox} is the capacitance of the oxide layer.

Early Voltage V_A

This parameter describes the dependence on v_{DS} of the MOSFET drain current in the saturation region. It is common to assume that V_A approaches infinity, indicating that the drain current is independent of v_{DS}. The role of this parameter will become more obvious in the next section.

Operation of the *n*-Channel Enhancement-Mode MOSFET

We first focus on n-channel enhancement mode transistors, which are generally referred to as NMOS devices. The operation of these devices is most effectively explained by making reference to the four-quadrant plot of Figure 11.4. In this figure, the behavior of the NMOS device is tied to whether the channel is on or off at the source or drain end of the transistor. Recall that whenever the gate-to-substrate voltage is higher than the threshold voltage, the channel is on.

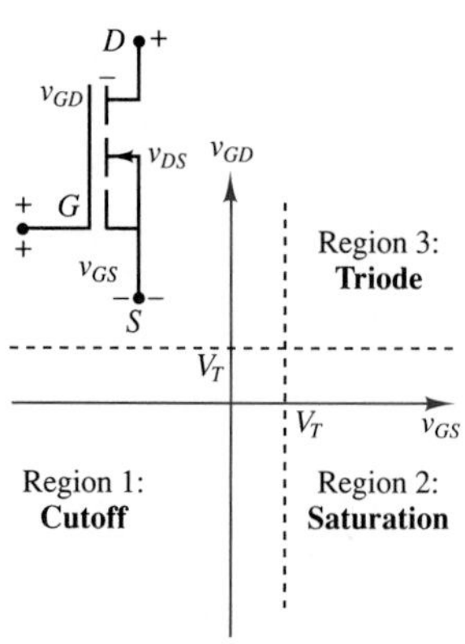

Figure 11.4 Regions of operation of NMOS transistor

Cutoff Region

When both $v_{GS} < V_T$ and $v_{GD} < V_T$, the channel is off at both the source and the drain. Thus, there is no conduction region between drain and source, and no current can be conducted. We call this the **cutoff region,** indicated in Figure 11.4 by region 1. In this region,

$$i_D = 0 \qquad \text{Cutoff region} \tag{11.2}$$

Saturation Region

When $v_{GS} > V_T$, and $v_{GD} < V_T$, the channel is on at the source end and off at the drain. In this mode, the drain current is (very nearly) independent of the drain-to-source voltage v_{DS} and depends on only the gate voltage. We call this the **saturation region,** indicated in Figure 11.4 by region 2. In this region, the MOSFET acts as a *voltage-controlled current source*. The equation for the drain current is given in equation 11.3. Note that in the more complete form of the equation, both the parameter V_A and the drain-to-source voltage v_{DS} appear. If, as is commonly done, we assume that V_A is very large, then we can use the approximate form, also shown below, which is independent of v_{DS}.

$$\begin{aligned} i_D &= K(v_{GS} - V_T)^2\left(1 + \frac{v_{DS}}{V_A}\right) \qquad \text{Saturation region} \qquad \textbf{(11.3)} \\ &\cong K(v_{GS} - V_T)^2 \end{aligned}$$

Triode or Ohmic Region

When $v_{GS} > V_T$ and $v_{GD} > V_T$, the channel is on at both ends of the device. In this mode, the drain current is strongly dependent on both the drain-to-source voltage v_{DS} and the gate-to-source voltage v_{GS}. We call this the **triode** or **ohmic region,** indicated in Figure 11.4 by region 2. The equation for the drain current is

$$i_D = K[2(v_{GS} - V_T)v_{DS} - v_{DS}^2] \qquad \text{Triode or ohmic region} \qquad \textbf{(11.4)}$$

If v_{DS} is much smaller than v_{GS}, then $v_{GD} = v_{GS} - v_{DS} \approx v_{GS}$. Thus, for small values of v_{DS} (see the drain characteristic curves in Figure 11.5), the channel is approximately equally on at both the drain and the source end. Thus, changes in the gate voltage will directly affect the conductivity of the channel. In this mode, the MOSFET behaves very much as a *voltage-controlled resistor* that is controlled by the gate voltage. This property finds much use in integrated circuits, in that it is easier to implement an integrated-circuit version of a resistor through a MOSFET than to actually build a passive resistor. There also exist other applications of the voltage-controlled resistor property of MOSFETs in tunable (variable-gain) amplifiers and in analog gates.

The three regions of operation can also be identified in the drain characteristic curves shown in Figure 11.5. In this figure, the circuit of Figure 11.3(b) is used to vary the gate and drain voltages with respect to the source and substrate (which are assumed to be electrically connected). You can see that for $v_{GS} < V_T$ and $v_{GD} < V_T$, the transistor is in the cutoff region (1) and no drain current flows. To better understand the difference between the saturation and triode (or ohmic) regions of operation, the boundary between these two regions is shown in Figure 11.5 by the curve $i_D = K v_{DS}^2$. You can see that in the saturation region (2), the transistor supplies nearly constant drain current, the value of which is dependent on the square of the gate-to-source voltage. Thus, in this region the MOSFET operates as a *voltage-controlled current source*, and it can be used in a variety of amplifier applications. On the other hand, in the triode region (3), the drain current is very strongly dependent on both the gate-to-source and the drain-to-source voltages (see equation 11.4). If, however, v_{DS} is much smaller than v_{GS}, then $v_{GD} = v_{GS} + v_{DS} \approx v_{GS}$, and the channel is on equally, or very nearly so, at both the source and drain ends. This corresponds to the region near the origin in the curves of Figure 11.5, in which the drain current curves are nearly straight lines. In this portion of the triode region, the MOSFET acts as a variable resistor, with resistance (i.e., the reciprocal of the slope in the $i_D - v_{DS}$ curves) controlled by the gate-to-source voltage. As mentioned earlier, this variable resistor characteristic of MOSFETs is widely exploited in integrated circuits. Finally, if the drain-to-source voltage exceeds the **breakdown voltage** V_{DSS}, the drain current will

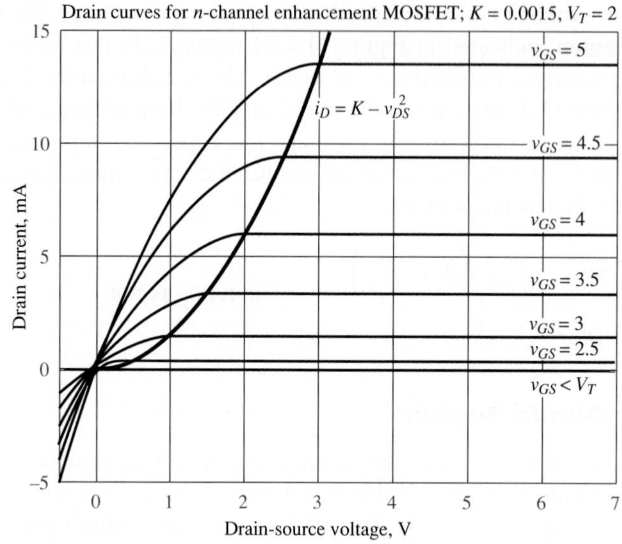

Figure 11.5 Drain characteristic curves for a typical NMOS transistor with $V_T = 2$ V and $K = 1.5$ mA/V^2

increase sharply and the result may be device failure. This **breakdown region** is not shown in Figure 11.5.

LO2

EXAMPLE 11.1 Determining the Operating State of a MOSFET

Problem

Determine the operating state of the MOSFET shown in the circuit of Figure 11.6 for the given values of V_{DD} and V_{GG} if the ammeter and voltmeter shown read the following values:

a. $V_{GG} = 1$ V; $V_{DD} = 10$ V; $v_{DS} = 10$ V; $i_D = 0$ mA; $R_D = 100\ \Omega$.

b. $V_{GG} = 4$ V; $V_{DD} = 10$ V; $v_{DS} = 2.8$ V; $i_D = 72$ mA; $R_D = 100\ \Omega$.

c. $V_{GG} = 3$ V; $V_{DD} = 10$ V; $v_{DS} = 1.5$ V; $i_D = 13.5$ mA; $R_D = 630\ \Omega$.

Solution

Known Quantities: MOSFET drain resistance; drain and gate supply voltages; MOSFET equations.

Find: MOSFET quiescent drain current, i_{DQ}, and quiescent drain-source voltage, v_{DSQ}.

Schematics, Diagrams, Circuits, and Given Data: $V_T = 2$ V; $K = 18$ mA/V^2.

Assumptions: Use the MOSFET equations 11.2–11.4 as needed.

Analysis:

a. Since the drain current is zero, the MOSFET is in the cutoff region. You should verify that both the conditions $v_{GS} < V_T$ and $v_{GD} < V_T$ are satisfied.

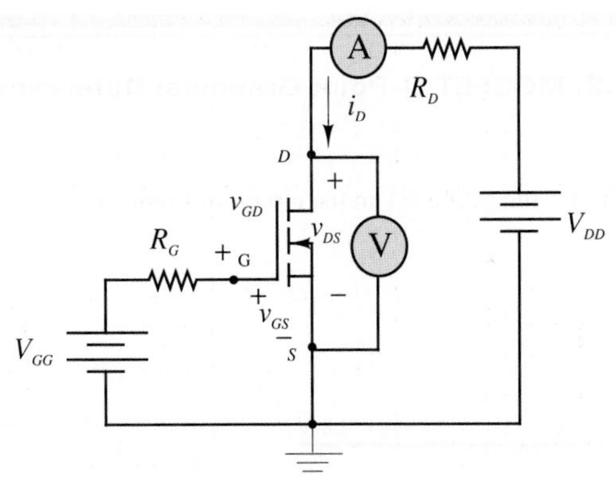

Figure 11.6 Circuit used in Example 11.1

b. In this case, $v_{GS} = V_{GG} = 4\text{ V} > V_T$. On the other hand, $v_{GD} = v_G - v_D = 4 - 2.8 = 1.2\text{ V} < V_T$. Thus, the transistor is in the saturation region. We can calculate the drain current to be: $i_D = K(v_{GS} - V_T)^2 = 18 \times (4-2)^2 = 72$ mA. Alternatively, we can also calculate the drain current as $i_D = \frac{V_{DD} - v_{DS}}{R_D} = \frac{10-2.8}{0.1\text{ k}\Omega} = 72$ mA.

c. In the third case, $v_{GS} = V_{GG} = v_G\ 3\text{ V} > V_T$. The drain voltage is measured to be $v_{DS} = v_D = 1.5$ V, and therefore $v_{GD} = 3 - 1.5 = 1.5\text{ V} < V_T$. In this case, the MOSFET is in the ohmic, or triode, region. We can now calculate the current to be $i_D = K[2(v_{GS} - V_T)v_{DS} - v_{DS}^2] = 18 \times [2 \times (3-2) \times 1.5 - 1.5^2] = 13.5$ mA. We can also calculate the drain current to be $i_D = \frac{V_{DD} - v_{DS}}{R_D} = \frac{(10-1.5)V}{0.630\text{ k}\Omega} = 13.5$ mA.

CHECK YOUR UNDERSTANDING

What is the operating state of the MOSFET of Example 11.1 for the following conditions?

$$V_{GG} = 10/3\text{ V}; V_{DD} = 10\text{ V}; v_{DS} = 3.6\text{ V}; i_D = 32\text{ mA}; R_D = 200\ \Omega.$$

Answer: Saturation

11.3 BIASING MOSFET CIRCUITS

Now that we have analyzed the basic characteristics of MOSFETs of the n-channel enhancement MOSFET and can identify its operating state, we are ready to develop systematic procedures for biasing a MOSFET circuit. This section presents two bias circuits. The first, illustrated in Examples 11.2 and 11.3, uses two distinct voltage supplies. This bias circuit is easier to understand, but not very practical—as we have already seen with BJTs, it is preferable to have a single DC voltage supply. This desire is addressed by the second bias circuit, described in Examples 11.4 and 11.5.

EXAMPLE 11.2 MOSFET Q-Point Graphical Determination

Problem

Determine the Q point for the MOSFET in the circuit of Figure 11.7.

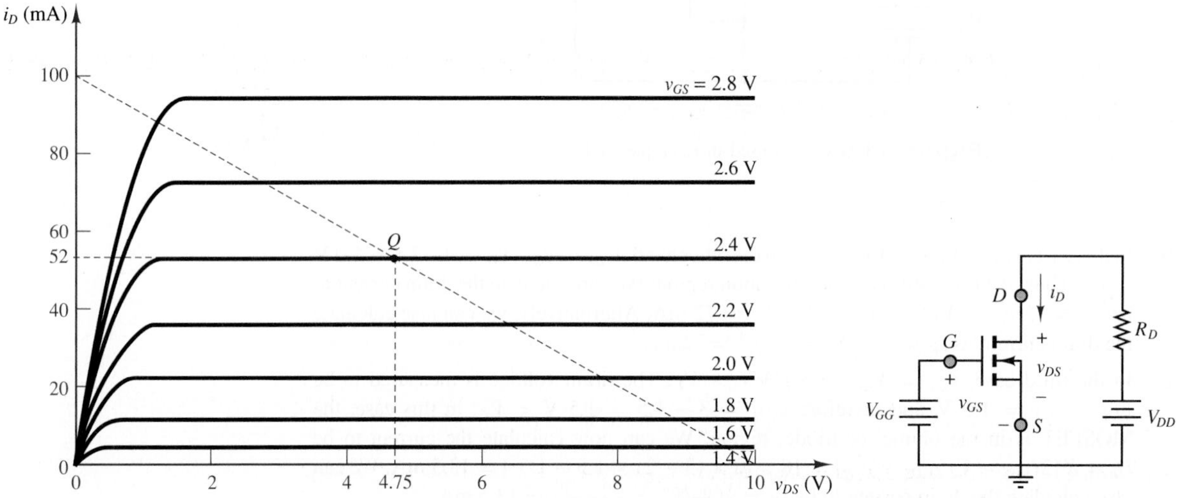

Figure 11.7 The n-channel enhancement MOSFET circuit and drain characteristic for Example 11.2

Solution

Known Quantities: MOSFET drain resistance; drain and gate supply voltages; MOSFET drain curves.

Find: MOSFET quiescent drain current i_{DQ} and quiescent drain-source voltage v_{DSQ}.

Schematics, Diagrams, Circuits, and Given Data: $V_{GG} = 2.4$ V; $V_{DD} = 10$ V; $R_D = 100\ \Omega$.

Assumptions: Use the drain curves of Figure 11.7.

Analysis: To determine the Q point, we write the drain circuit equation, applying KVL:

$$V_{DD} = R_D i_D + v_{DS}$$

$$10 = 100 i_D + v_{DS}$$

The resulting curve is plotted as a dashed line on the drain curves of Figure 11.7 by noting that the drain current axis intercept is equal to $V_{DD}/R_D = 100$ mA and that the drain-source voltage axis intercept is equal to $V_{DD} = 10$ V. The Q point is then given by the intersection of the load line with the $V_{GG} = 2.4$ V curve. Thus, $i_{DQ} = 52$ mA and $v_{DSQ} = 4.75$ V.

Comments: Note that the Q-point determination for a MOSFET is easier than for a BJT, since there is no need to consider the gate circuit, because gate current flow is essentially zero. In the case of the BJT, we also needed to consider the base circuit.

CHECK YOUR UNDERSTANDING

Determine the operating region of the MOSFET of Example 11.2 when $v_{GS} = 3.5$ V.

Answer: The MOSFET is in the ohmic region.

EXAMPLE 11.3 MOSFET Q-Point Calculation

LO3

Problem

Determine the Q point for the MOSFET in the circuit of Figure 11.7.

Solution

Known Quantities: MOSFET drain resistance; drain and gate supply voltages; MOSFET equations.

Find: MOSFET quiescent drain current i_{DQ} and quiescent drain-source voltage v_{DSQ}.

Schematics, Diagrams, Circuits, and Given Data: $V_{GG} = 2.4$ V; $V_{DD} = 10$ V; $V_T = 1.4$ V; $K = 48.5$ mA/V^2; $R_D = 100\ \Omega$.

Assumptions: Use the MOSFET equations 11.2 through 11.4 as needed.

Analysis: The gate supply V_{GG} ensures that $v_{GSQ} = V_{GG} = 2.4$ V. Thus, $v_{GSQ} > V_T$. We assume that the MOSFET is in the saturation region, and we proceed to use equation 11.3 to calculate the drain current:

$$i_{DQ} = K(v_{GS} - V_T)^2 = 48.5(2.4 - 1.4) = 48.5 \text{ mA}$$

Applying KVL to the drain loop, we can calculate the quiescent drain-to-source voltage as:

$$v_{DSQ} = V_{DD} - R_D i_{DQ} = 10 - 100 \times 48.5 \times 10^{-3} = 5.15 \text{ V}$$

Now we can verify the assumption that the MOSFET was operating in the saturation region. Recall that the conditions required for operation in region 2 (saturation) were $v_{GS} > V_T$ and $v_{GD} < V_T$. The first condition is clearly satisfied. The second can be verified by recognizing that $v_{GD} = v_{GS} + v_{SD} = v_{GS} - v_{DS} = -2.75$ V. Clearly, the condition $v_{GD} < V_T$ is also satisfied, and the MOSFET is indeed operating in the saturation region.

CHECK YOUR UNDERSTANDING

Find the lowest value of R_D for the MOSFET of Example 11.3 that will place the MOSFET in the ohmic region.

Answer: $\sim 400\ \Omega$

LO3

EXAMPLE 11.4 MOSFET Self-Bias Circuit

Problem

Figure 11.8(a) depicts a self-bias circuit for a MOSFET. Determine the Q point for the MOSFET by choosing R_S such that $v_{DSQ} = 8$ V.

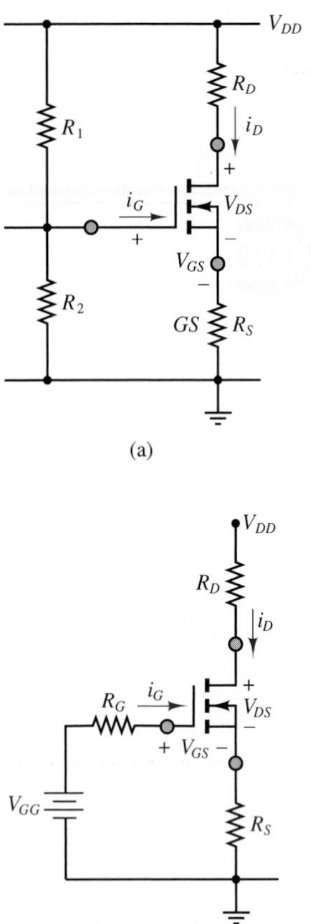

Figure 11.8 (a) Self-bias circuit for Example 11.4; (b) equivalent circuit for (a)

Solution

Known Quantities: MOSFET drain and gate resistances; drain supply voltage; MOSFET parameters V_T and K; desired drain-to-source voltage v_{DSQ}.

Find: MOSFET quiescent gate-source voltage v_{GSQ}, quiescent drain current i_{DQ}, and quiescent drain-source voltage v_{DSQ}.

Schematics, Diagrams, Circuits, and Given Data: $V_{DD} = 30$ V; $R_D = 10$ kΩ; $R_1 = R_2 = 1.2$ MΩ; $V_T = 4$ V; $K = 0.2188$ mA/V^2; $v_{DSQ} = 8$ V.

Assumptions: Assume operation in the saturation region.

Analysis: First we reduce the circuit of Figure 11.8(a) to the circuit of Figure 11.8(b), in which the voltage divider rule has been used to compute the value of the fictitious supply V_{GG}:

$$V_{GG} = \frac{R_2}{R_1 + R_2} V_{DD} = 15 \text{ V}$$

Let all currents be expressed in milliamperes and all resistances in kilohms. Applying KVL around the equivalent gate circuit of Figure 11.8(b) yields

$$v_{GSQ} + i_{GQ} R_G + i_{DQ} R_S = V_{GG} = 15 \text{ V}$$

where $R_G = R_1 \| R_2$. Since $i_{GQ} = 0$, due to the infinite input resistance of the MOSFET, the gate equation simplifies to

$$v_{GSQ} + i_{DQ} R_S = 15 \text{ V} \qquad \textbf{(a)}$$

The drain circuit equation is

$$v_{DSQ} + i_{DQ} R_D + i_{DQ} R_S = V_{DD} = 30 \text{ V} \qquad \textbf{(b)}$$

Using equation 11.3, we get

$$i_{DQ} = K(v_{GS} - V_T)^2 \qquad \textbf{(c)}$$

We can obtain the third equation needed to solve for the three unknowns v_{GSQ}, i_{DQ}, and v_{DSQ}. From equation (a) we write

$$i_{DQ} R_S = V_{GG} - v_{GSQ} = \frac{V_{DD}}{2} - v_{GSQ}$$

and we substitute the result into equation (b):

$$V_{DD} = i_{DQ} R_D + v_{DSQ} + \frac{V_{DD}}{2} - v_{GSQ}$$

or

$$i_{DQ} = \frac{1}{R_D}\left(\frac{V_{DD}}{2} - v_{DSQ} + v_{GSQ}\right)$$

Substituting the above equation for i_{DQ} into equation (c), we finally obtain a quadratic equation that can be solved for v_{GSQ} since we know the desired value of v_{DSQ}:

$$\frac{1}{R_D}\left(\frac{V_{DD}}{2} - v_{DSQ} + v_{GSQ}\right) = K\left(v_{GSQ} - V_T\right)^2$$

$$K v_{GSQ}^2 - 2K V_T v_{GSQ} + K V_T^2 - \frac{1}{R_D}\left(\frac{V_{DD}}{2} - v_{DSQ}\right) - \frac{1}{R_D} v_{GSQ} = 0$$

$$v_{GSQ}^2 - \left(2V_T + \frac{1}{KR_D}\right) v_{GSQ} + V_T^2 - \frac{1}{KR_D}\left(\frac{V_{DD}}{2} - v_{DSQ}\right) = 0$$

$$v_{GSQ}^2 - 8.457 v_{GSQ} + 12.8 = 0$$

The two solutions for the above quadratic equation are

$$v_{GSQ} = 6.48 \text{ V} \qquad \text{and} \qquad v_{GSQ} = 1.97 \text{ V}$$

Only the first of these two values is acceptable for operation in the saturation region, since the second root corresponds to a value of v_{GS} lower than the threshold voltage (recall that $V_T = 4$ V). Substituting the first value into equation (c), we can compute the quiescent drain current

$$i_{DQ} = 1.35 \text{ mA}.$$

Using this value in the gate circuit equation (a), we compute the solution for the source resistance:

$$R_S = 6.32 \text{ k}\Omega$$

Comments: Why are there two solutions to the problem posed in this example? Mathematically, we know that this should be the case because the drain universal equation is a quadratic equation. As you can see, we used the physical constraints of the problem to select the appropriate solution.

CHECK YOUR UNDERSTANDING

Determine the appropriate value of R_S if we wish to move the operating point of the MOSFET of Example 11.4 to $v_{DSQ} = 12$ V. Also find the values of v_{GSQ} and i_{DQ}. Are these values unique?

Answer: The answer is not unique. For the smaller value of $v_{GS} = 2.86$ V, $R_S = 20.7$ kΩ and $i_D = 0.586$ mA. For the larger value of v_{GS}, $R_S = 11.5$ kΩ.

EXAMPLE 11.5 Analysis of MOSFET Amplifier

LO3

Problem

Determine the gate and drain-source voltage and the drain current for the MOSFET amplifier of Figure 11.9.

Solution

Known Quantities: Drain, source, and gate resistors; drain supply voltage; MOSFET parameters.

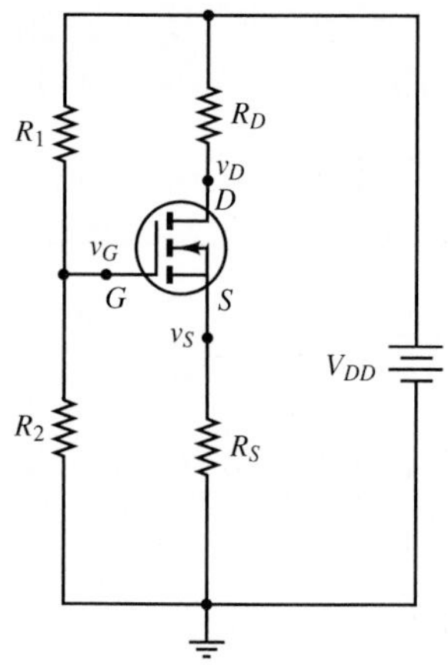

Figure 11.9

Find: v_{GS}; v_{DS}; i_D.

Schematics, Diagrams, Circuits, and Given Data: $R_1 = R_2 = 1\ \text{M}\Omega$; $R_D = 6\ \text{k}\Omega$; $R_S = 6\ \text{k}\Omega$; $V_{DD} = 10$ V. $V_T = 1$ V; $K = 0.5\ \text{mA/V}^2$.

Assumptions: The MOSFET is operating in the saturation region. All currents are expressed in milliamperes and all resistors in kilohms.

Analysis: The gate voltage is computed by applying the voltage divider rule between resistors R_1 and R_2 (remember that no current flows into the transistor):

$$v_G = \frac{R_2}{R_1 + R_2} V_{DD} = \frac{1}{2} V_{DD} = 5 \text{ V}$$

Assuming saturation region operation, we write

$$v_{GS} = v_G - v_S = v_G - R_S i_D = 5 - 6 i_D$$

The drain current can be computed from equation 11.3:

$$i_D = K\,(v_{GS} - V_T)^2 = 0.5\,(5 - 6i_D - 1)^2$$

leading to

$$36 i_D^2 - 50 i_D + 16 = 0$$

with solutions

$$i_D = 0.89 \text{ mA} \qquad \text{and} \qquad i_D = 0.5 \text{ mA}$$

To determine which of the two solutions should be chosen, we compute the gate-source voltage for each. For $i_D = 0.89$ mA, $v_{GS} = 5 - 6i_D = -0.34$ V. For $i_D = 0.5$ mA, $v_{GS} = 5 - 6i_D = 2$ V. Since v_{GS} must be greater than V_T for the MOSFET to be in the saturation region, we select the solution

$$i_D = 0.5 \text{ mA} \qquad v_{GS} = 2 \text{ V}$$

The corresponding drain voltage is therefore found to be

$$v_D = v_{DD} - R_D i_D = 10 - 6 i_D = 7 \text{ V}$$

And therefore

$$v_{DS} = v_D - v_S = v_D - i_D R_S = 7 - 3 = 4 \text{ V}$$

Comments: Now that we have computed the desired voltages and current, we can verify that the conditions for operation in the saturation region are indeed satisfied: $v_{GS} = 2 > V_T$ and $v_{GD} = v_{GS} - v_{DS} = 2 - 4 = -2 < V_T$. Since the inequalities are satisfied, the MOSFET is indeed operating in the saturation region.

Operation of the *P*-Channel Enhancement-Mode MOSFET

The operation of a p-channel enhancement-mode MOS transistor is very similar in concept to that of an n-channel device. Figure 11.10 depicts a test circuit and a sketch of the device construction. Note that the roles of n-type and p-type materials are reversed and that the charge carriers in the channel are no longer electrons, but holes. Further, the threshold voltage is now negative: $V_T < 0$. However, if we replace v_{GS} with v_{SG}, v_{GD} with v_{DG}, and v_{DS} with v_{SD}, and we use $|V_T|$ in place of V_T, then the

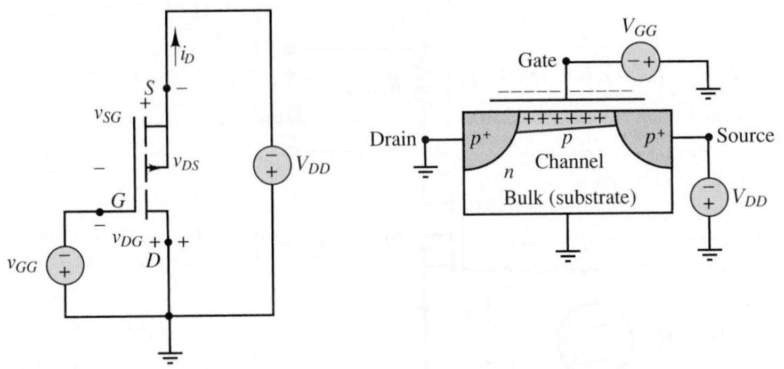

Figure 11.10 The p-channel enhancement-mode field-effect transistor (PMOS)

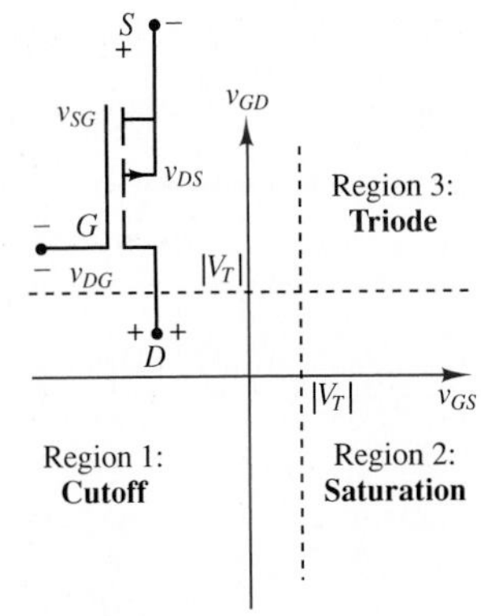

Figure 11.11 Regions of operation of PMOS transistor

analysis of the device is completely analogous to that of the n-channel MOS transistor. In particular, Figure 11.11 depicts the behavior of a PMOS transistor in terms of the gate-to-drain and gate-to-source voltages, in analogy with Figure 11.4. The resulting equations for the three modes of operation of the PMOS transistor are summarized below:

Cutoff region: $v_{SG} < |V_T|$ and $v_{DG} < |V_T|$.

$$i_D = 0 \qquad \text{Cutoff region} \tag{11.5}$$

Saturation region: when $v_{SG} > |V_T|$ and $v_{DG} < |V_T|$.

$$i_D \cong K(v_{SG} - |V_T|)^2 \qquad \text{Saturation region} \tag{11.6}$$

Triode or ohmic region: when $v_{SG} > |V_T|$ and $v_{DG} > |V_T|$.

$$i_D = K[2(v_{SG} - |V_T|)v_{SD} - v_{SD}^2] \qquad \text{Triode or ohmic region} \tag{11.7}$$

11.4 MOSFET LARGE-SIGNAL AMPLIFIERS

The objective of this section is to illustrate how a MOSFET can be used as a large signal amplifier, in applications similar to those illustrated in Chapter 10 for bipolar transistors. Equation 11.3, repeated below for convenience, describes the approximate relationship between the drain current and gate-source voltage for the MOSFET in a large-signal amplifier application. Appropriate biasing, as explained in the preceding section, is used to ensure that the MOSFET is operating in the saturation mode.

$$i_D \cong K(v_{GS} - V_T)^2 \tag{11.8}$$

MOSFET amplifiers are commonly found in one of two configurations: *common-source* and *source-follower* amplifiers. Figure 11.12 depicts a basic common-source configuration. Note that when the MOSFET is in saturation, this amplifier is essentially a voltage-controlled current source (VCCS), in which the drain current is

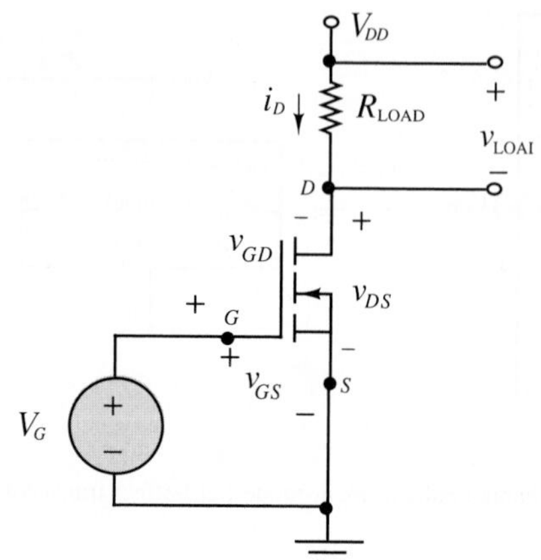

Figure 11.12 Common-source MOSFET amplifier

controlled by the gate voltage. Thus, the load voltage, across the load resistance, is given by the expression

$$v_{\text{LOAD}} = R_{\text{LOAD}} i_D \cong R_{\text{LOAD}} K(v_{GS} - V_T)^2 = R_{\text{LOAD}} K(V_G - V_T)^2 \qquad \textbf{(11.9)}$$

A source-follower amplifier is shown in Figure 11.13(a). Note that the load is now connected between the source and ground. The behavior of this circuit depends on the load current and can be analyzed for the resistive load of Figure 11.13 by observing that the load voltage is given by the expression $v_{\text{LOAD}} = R_{\text{LOAD}} i_D$, where

$$i_D \cong K(v_{GS} - V_T)^2 = K(\Delta v - v_S)^2 = K(\Delta v - v_{\text{LOAD}})^2 = K(\Delta v - R_{\text{LOAD}} i_D)^2 \qquad \textbf{(11.10)}$$

and where $\Delta v = V_G - V_T$. We can then solve for the load current from the quadratic equation:

$$\begin{aligned} i_D &= K(\Delta v - R_{\text{LOAD}} i_D)^2 = K\Delta v^2 - 2K\Delta v R_{\text{LOAD}} i_D + R_{\text{LOAD}}^2 i_D^2 \\ i_D^2 &- \frac{1}{R_{\text{LOAD}}^2}(2K\Delta v R_{\text{LOAD}} + 1) + \frac{K}{R_{\text{LOAD}}^2}\Delta v^2 = 0 \end{aligned} \qquad \textbf{(11.11)}$$

with solution

$$\begin{aligned} i_D^2 &- \frac{1}{R_{\text{LOAD}}^2}(2K\Delta v R_{\text{LOAD}} + 1) i_D + \frac{K}{R_{\text{LOAD}}^2}\Delta v^2 = 0 \\ i_D &= \frac{1}{2R_{\text{LOAD}}^2}(2K\Delta v R_{\text{LOAD}} + 1) \\ &\pm \frac{1}{2}\sqrt{\left[\frac{1}{R_{\text{LOAD}}^2}(2K\Delta v R_{\text{LOAD}} + 1)\right]^2 - \frac{4K}{R_{\text{LOAD}}^2}\Delta v^2} \end{aligned} \qquad \textbf{(11.12)}$$

Figure 11.13(b) depicts the drain current response of the source-follower MOSFET amplifier when the gate voltage varies between the threshold voltage and 5 volts for a 100-Ω load when $K = 0.018$ and $V_T = 1.2$ V. Note that the response of this amplifier is linear in the gate voltage. This behavior is due to the fact that the source voltage increases as the drain current increases, since the source voltage is proportional to i_D.

The following examples, 11.6 and 11.7, illustrate the application of simple MOSFET large-signal amplifiers as battery chargers and electric motor drivers.

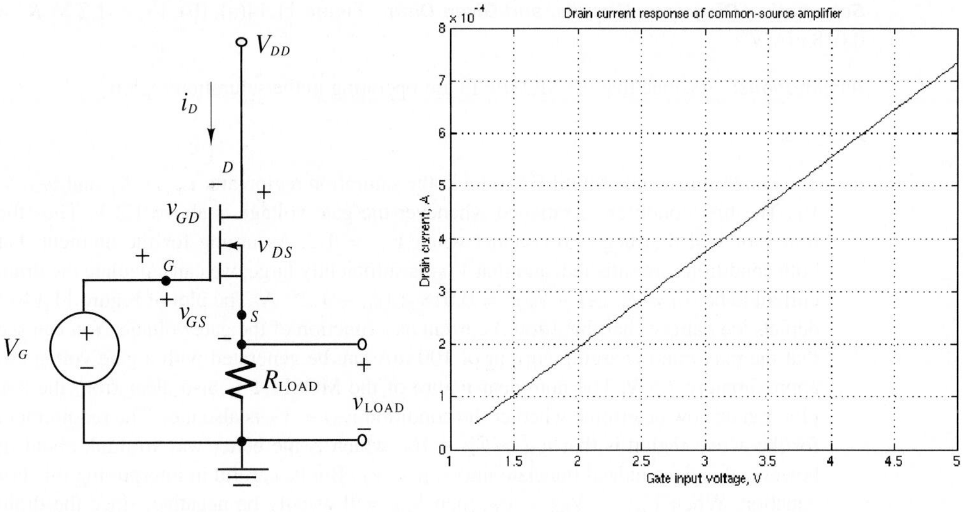

(a) (b)

Figure 11.13 (a) Source follower MOSFET amplifier. (b) Drain current response for a 100-Ω load when $K = 0.018$ and $V_T = 1.2$ V

EXAMPLE 11.6 Using a MOSFET as a Current Source for Battery Charging

Problem

Analyze the two battery charging circuits shown in Figure 11.14. Use the transistor parameters to determine the range of required gate voltages, V_G, to provide a variable charging current up to a maximum of 0.1 A. Assume that the terminal voltage of a fully discharged battery is 9 V, and of a fully charged battery 10.5 V.

Solution

Known Quantities: Transistor large-signal parameters, NiCd battery nominal voltage.

Find: V_{DD}, V_G, range of gate voltages leading to a maximum charging current of 0.1A.

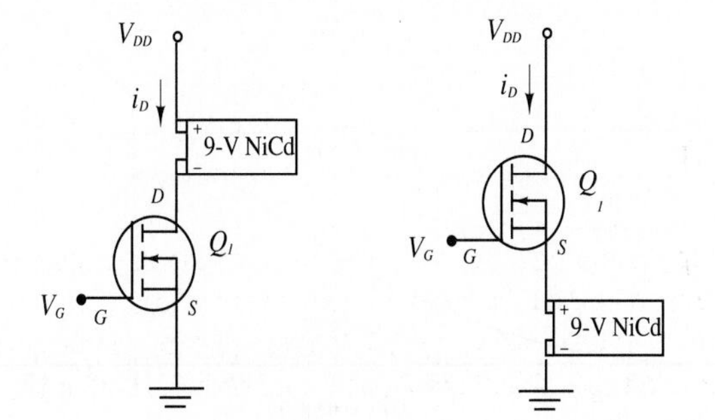

(a) Common-source current source (b) Common-drain current source

Figure 11.14 *Continued*

Schematics, Diagrams, Circuits, and Given Data. Figure 11.14(a), (b). $V_T = 1.2$ V; $K = 0.018$ mA/V^2;

Assumptions: Assume that the MOSFETs are operating in the saturation region.

Analysis:

a. The conditions for the MOSFET to be in the saturation region are: $v_{GS} > V_T$ and $v_{GD} < V_T$. The first condition is satisfied whenever the gate voltage is above 1.2 V. Thus the transistor will first begin to conduct when $V_G = 1.2$. Assuming for the moment that both conditions are satisfied, and that V_{DD} is sufficiently large, we can calculate the drain current to be: $i_D = K(v_{GS} - V_T)^2 = 0.018 \times (v_G - 1.2)^2$ A. The plot of Figure 11.14(c) depicts the battery charging (drain) current as a function of the gate voltage. You can see that the maximum charging current of 100 mA can be generated with a gate voltage of approximately 3.5 V. The nonlinear nature of the MOSFET is also clear from the i–v plot. Let us now determine whether the condition $v_{GD} < V_T$ is also met. The requirement for the active region is that $v_{DS} > v_{GS} - V_T$, which is the better way to think about it; however this does indeed translate into $v_{GD} < V_T$. But be careful in interpreting this last equation. When $V_{DS} > V_{GS} - V_T$, then V_{GD} will usually be negative, since the drain voltage should be larger than the source voltage. This can be easily done if we realize that $v_{GD} = v_G - v_D$, and that $v_D = V_{DD} - V_B$, where V_B is the battery voltage. To satisfy the condition $v_{GD} < V_T$, solve the following inequality:

$v_{GD} < V_T$

$v_G - v_D < V_T$, which is negative

$v_D > v_G - V_T$

$V_{DD} - V_B < v_G - V_T$, so actually $V_{DD} - V_B > v_G - V_T$

To ensure that the NMOS remains in the saturation region throughout the range of battery voltage, we need to make V_{DD} sufficiently large. In this case V_{DD} should be larger than 12 V.

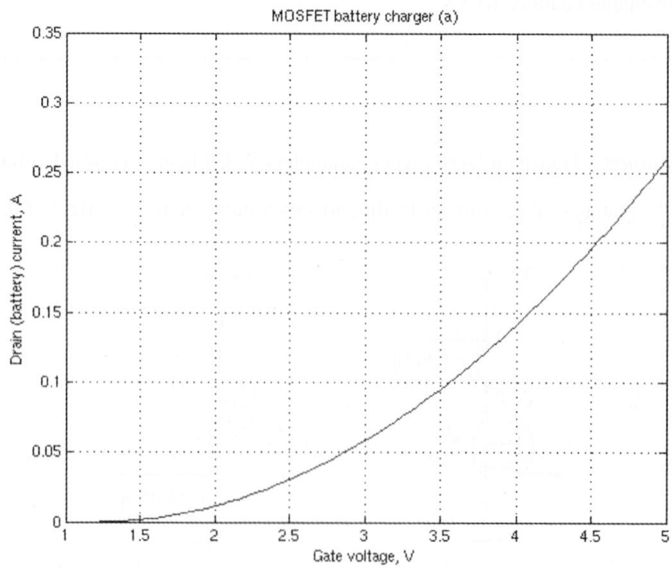

(c) Gate voltage–drain current curve for the circuit of Figure 11.14(a)

Figure 11.14 *Continued*

b. The analysis of the second circuit is based on the observation that the voltage at the source terminal of the MOSFET is equal to the gate voltage minus the threshold voltage. If we wish to charge the battery up to 10.5 V, we need to have a gate voltage of at least 11.7 V, to satisfy $v_{GS} > V_T$. If we assume that the battery is initially discharged (9 V), we can calculate the initial charging current to be

$$i_D = K(v_{GS} - V_T)^2 = K(V_G - V_B - V_T)^2 = 0.018 \times (11.7 - 9 - 1.2)^2$$
$$= 0.0405 \text{ A}.$$

If we further assume that during the charging the battery voltage increases linearly from 9 to 10.5 V over a period of 20 minutes, we can calculate the charging current as the battery voltage increases. Note that when the battery is fully charged, v_{GS} is no longer larger than the threshold voltage and the transistor is cut off. A plot of the drain (charging) current as a function of time is shown in Figure 11.14(d). Note that the charging current naturally tapers to zero as the battery voltage increases.

Comments: In the circuit of part (b), please note that the battery voltage is not likely to actually increase linearly. The voltage rise will begin to taper off as the battery begins to approach full charge. In practice, this means that the charging process will take longer than projected in Figure 11.14(d).

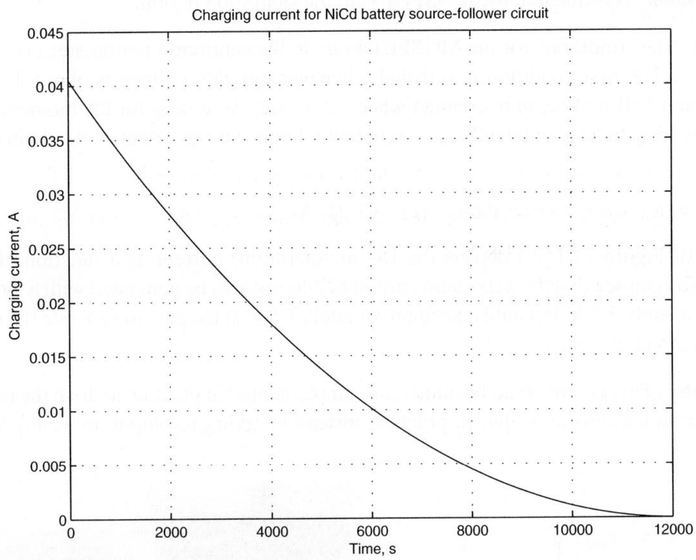

(d) Charging current profile for the circuit of Figure 11.14(b)

Figure 11.14 Simple battery charging circuits

CHECK YOUR UNDERSTANDING

What is the maximum power dissipation of the MOSFET for each of the circuits in Example 11.6?

Answers: Part (a): $P_{NMOS} = v_{DS} \times i_D = (v_D - v_S) \times i_D = 3 \times 0.1 = 300$ mW
Part (b): $P_{NMOS} = v_{DS} \times i_D = (v_D - v_S) \times i_D = 3 \times 0.0405 = 121.5$ mW

EXAMPLE 11.7 MOSFET DC Motor Drive Circuit

Problem

FIND IT ON THE WEB

The aim of this example is to design a MOSFET driver for the **Lego® 9V Technic motor, model 43362**. Figures 11.15(a) and (b) show the driver circuit and a picture of the motor, respectively. The motor has a maximum (stall) current of 340 mA. Minimum current to start motor rotation is 20 mA. The aim of the circuit is to control the current to the motor (and therefore the motor torque, which is proportional to the current) via the gate voltage.

Solution

Known Quantities: Transistor large-signal parameters, component values.

Find: Values of R_1 and R_2.

Schematics, Diagrams, Circuits, and Given Data: Figure 11.15. $V_T = 1.2$ V; $K = 0.08$ A/V^2.

Assumptions: Assume that the MOSFET is in the saturation region.

Analysis: The conditions for the MOSFET to be in the saturation region are: $v_{GS} > V_T$ and $v_{GD} < V_T$. The first condition is satisfied whenever the gate voltage is above 1.2 V. Thus the transistor will first begin to conduct when $V_G = 1.2$. Assuming for the moment that both conditions are satisfied, and that V_{DD} is sufficiently large, we can calculate the drain current to be:

$$i_D = K(v_{GS} - V_T)^2 = 0.08 \times (v_G - 1.2)^2 \text{ A}.$$

The plot of Figure 11.15(c) depicts the DC motor (drain) current as a function of the gate voltage. You can see that the maximum current of 340 mA can be generated with a gate voltage of approximately 3.3 V. It would take approximately 1.5 V at the gate to generate the minimum required current of 20 mA.

Comments: This circuit could be quite easily implemented in practice to drive the motor with a signal from a microcontroller. In practice, instead of trying to output an analog voltage, a

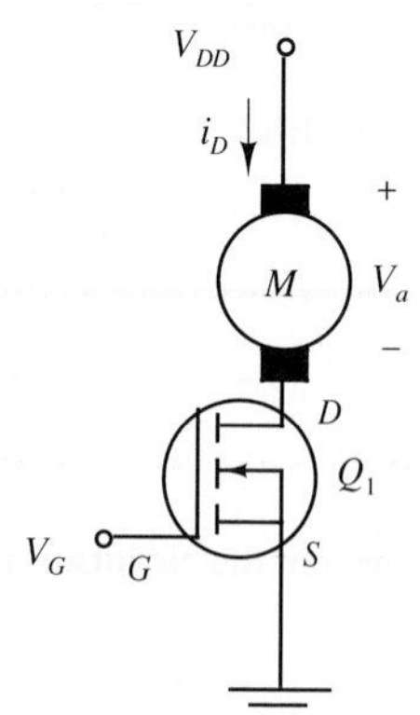

(a) MOSFET DC motor driver circuit

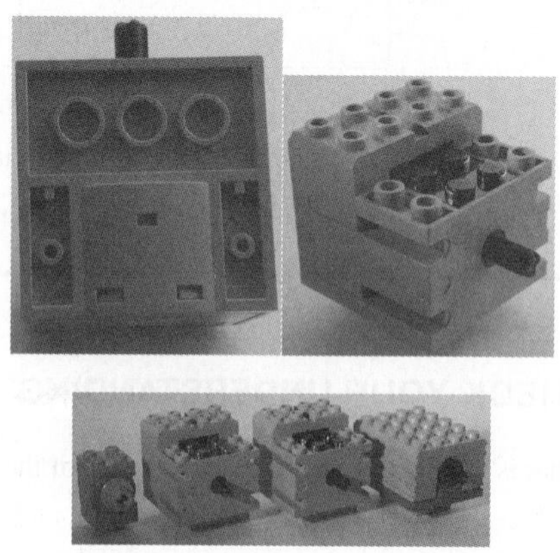

(b) Lego® 9V Technic motor, top: model 43362; bottom: family of Lego® motors
Courtesy: Philippe "philo" Hurbain

Figure 11.15 *Continued*

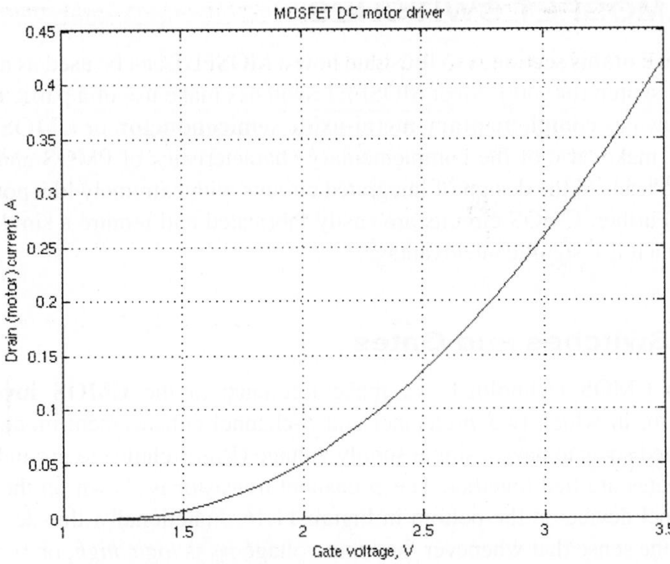

(c) Drain current–gate voltage curve for the MOSFET in saturation

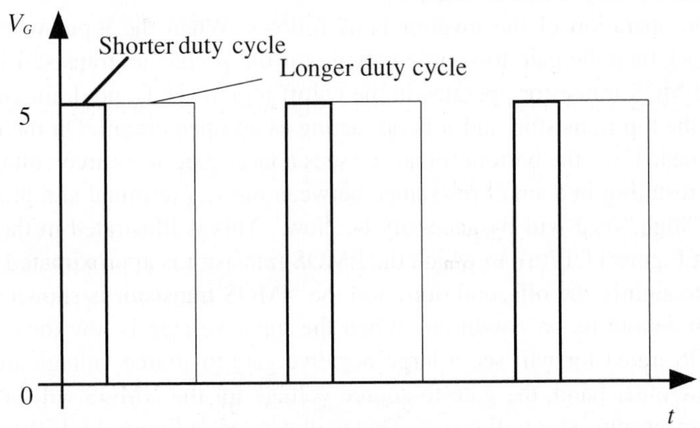

(d) Pulse-width modulation (PWM) gate voltage waveforms

Figure 11.15 DC motor drive circuit

microcontroller is better suited to the generation of a digital (On-Off) signal. For example, the gate drive signal could be a *pulse-width modulated* (PWM) 0–5 V pulse train, in which the ratio of the *On* time to the period of the waveform time is called *duty cycle*. Figure 11.15(d) depicts the possible appearance of a digital PWM gate voltage input.

CHECK YOUR UNDERSTANDING

What is the range of duty cycles needed to cover the current range of the Lego motor?

Answer: 30 to 70 percent

11.5 MOSFET SWITCHES

The objective of this section is to illustrate how a MOSFET can be used as an analog or a digital switch (or gate). Most MOSFET switches make use of a particular technology known as **complementary metal-oxide semiconductor,** or **CMOS.** CMOS technology makes use of the complementary characteristics of PMOS and NMOS devices and leads to the design of integrated circuits with extremely low power consumption. Further, CMOS circuits are easily fabricated and require a single supply voltage, which is a significant advantage.

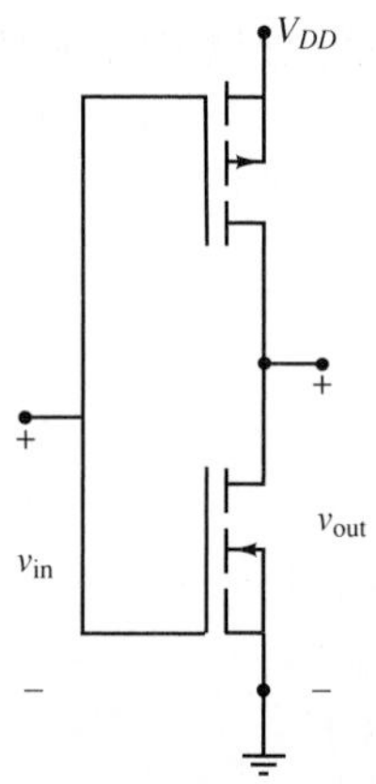

Figure 11.16 CMOS inverter

Digital Switches and Gates

To explain CMOS technology, we make reference to the **CMOS inverter** of Figure 11.16, in which two p-channel and n-channel enhancement-mode devices are connected so as to have a single supply voltage (V_{DD}, relative to ground) and so that their gates are tied together. The p-channel transistor is shown on the top and the n-channel device on the bottom in Figure 11.16. Functionally, this device is an inverter in the sense that whenever the input voltage v_{in} is *logic high*, or 1 (i.e., near V_{DD}), then the output voltage is *logic low* (or 0). If the input is logic 0, on the other hand, then the output will be logic 1.

The operation of the inverter is as follows. When the input voltage is high (near V_{DD}), then the gate-to-source voltage for the p-channel transistor is near zero and the PMOS transistor operates in the cutoff region. Thus, no drain current flows through the top transistor, and it is *off*, acting as an open circuit. On the other hand, with v_{in} near V_{DD}, the bottom transistor sees a large gate-to-source voltage and will turn *on*, resulting in a small resistance between the v_{out} terminal and ground. Thus, if v_{in} is "high," v_{out} will by necessity be "low." This is illustrated in the simplified sketch of Figure 11.17(a), in which the PMOS transistor is approximated by an open switch (to signify the off condition) and the NMOS transistor is shown as a closed switch to denote its on condition. When the input voltage is low (near 0 V), then the PMOS transistor will see a large negative gate-to-source voltage and will turn on; on the other hand, the gate-to-source voltage for the NMOS will be near zero, and the lower transistor will be off. This is illustrated in Figure 11.17(b), using ideal

LO5

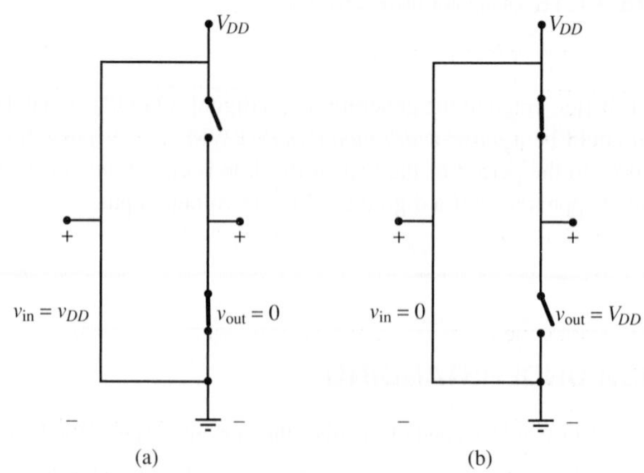

Figure 11.17 CMOS inverter approximate by ideal switches: (a) When v_{in} is "high," v_{out} is tied to ground; (b) when v_{in} is "low," v_{out} is tied to V_{DD}.

switches to approximate the individual transistors. Note that this simple logic inverter does not require the use of any resistors to bias the transistors: it is completely self-contained and very easy to fabricate. Further, it is also characterized by very low power consumption, making it ideal for many portable consumer electronic applications.

Examples 11.8 and 11.9 illustrate a number of digital switch and gate applications of MOS technology. Example 11.8 explores a NMOS switch using the drain characteristic curves; Example 11.9 analyzes a digital logic gate built using CMOS technology.

EXAMPLE 11.8 MOSFET Switch

LO5

Problem

Determine the operating points of the MOSFET switch of Figure 11.18 when the signal source output is equal to 0 and 2.5 V, respectively.

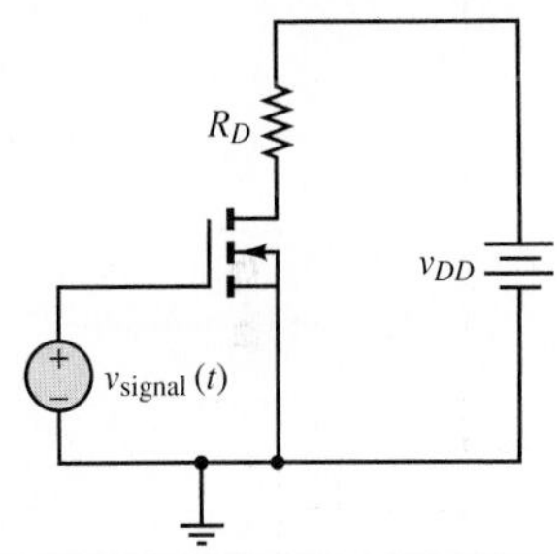

Figure 11.18

Solution

Known Quantities: Drain resistor; V_{DD}; signal source output voltage as a function of time.

Find: Q point for each value of the signal source output voltage.

Schematics, Diagrams, Circuits, and Given Data: $R_D = 125\ \Omega$; $V_{DD} = 10$ V; $v_{signal}(t) = 0$ V for $t < 0$; $v_{signal}(t) = 2.5$ V for $t = 0$.

Assumptions: Use the drain characteristic curves for the MOSFET (Figure 11.19).

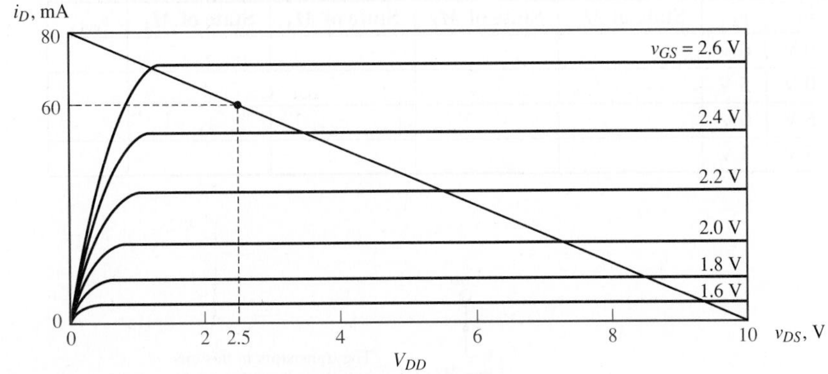

Figure 11.19 Drain curves for MOSFET of Figure 11.18

Analysis: We first draw the load line, using the drain circuit equation

$$V_{DD} = R_D i_D + v_{DS} \qquad 10 = 125 i_D + v_{DS}$$

recognizing a v_{DS} axis intercept at 10 V and an i_D axis intercept at $10/125 = 80$ mA.

1. $t < 0$ s. When the signal source output is zero, the gate voltage is zero and the MOSFET is in the cutoff region. The Q point is

$$v_{GSQ} = 0 \text{ V} \qquad i_{DQ} = 0 \text{ mA} \qquad v_{DSQ} = 10 \text{ V}$$

2. $t \geq 0$ s. When the signal source output is 2.5 V, the gate voltage is 2.5 V and the MOSFET is in the saturation region. The Q point is

$$v_{GSQ} = 0 \text{ V} \qquad i_{DQ} = 60 \text{ mA} \qquad v_{DSQ} = 2.5 \text{ V}$$

This result satisfies the drain equation, since $R_D i_D = 0.06 \times 125 = 7.5$ V.

Comments: The simple MOSFET configuration shown can quite effectively serve as a switch, conducting 60 mA when the gate voltage is switched to 2.5 V.

CHECK YOUR UNDERSTANDING

What value of R_D would ensure a drain-to-source voltage v_{DS} of 5 V in the circuit of Example 11.8?

Answer: 62.5 Ω

LO5

EXAMPLE 11.9 CMOS Gate

Problem

Determine the logic function implemented by the CMOS gate of Figure 11.20. Use the table below to summarize the behavior of the circuit.

v_1	v_2	State of M_1	State of M_2	State of M_3	State of M_4	v_{out}
0 V	0 V					
0 V	5 V					
5 V	0 V					
5 V	5 V					

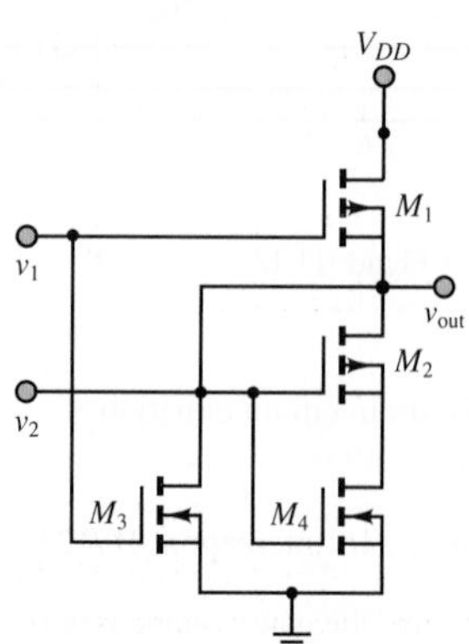

The transistors in this circuit show the substrate for each transistor connected to its respective gate. In a true CMOS IC, the substrates for the p-channel transistors are connected to 5 V and the substrates of the n-channel transistors are connected to ground.

Figure 11.20

Solution

Find: v_{out} for each of the four combinations of v_1 and v_2.

Schematics, Diagrams, Circuits, and Given Data: $V_T = 1.7$ V; $V_{DD} = 5$ V.

Assumptions: Treat the MOSFETs as open circuits when off and as linear resistors when on.

Analysis:

1. $v_1 = v_2 = 0$ V. With both input voltages equal to zero, neither M_3 nor M_4 can conduct, since the gate voltage is less than the threshold voltage for both transistors. Both M_1 and M_2 will similarly be off, and no current will flow through the drain-source circuits of M_1 and M_2. Thus, $v_{out} = V_{DD} = 5$ V. This condition is depicted in Figure 11.21.
2. $v_1 = 5$ V; $v_2 = 0$ V. Now M_2 and M_4 are off because of the zero gate voltage, while M_1 and M_3 are on. Figure 11.22(a) depicts this condition. Thus, $v_{out} = 0$.
3. $v_1 = 5$ V; $v_2 = 0$ V. By symmetry with case 2, we conclude that $v_{out} = 0$.
4. $v_1 = 5$ V; $v_2 = 5$ V. Now both M_1 and M_2 are open circuits, and therefore $v_{out} = 0$.

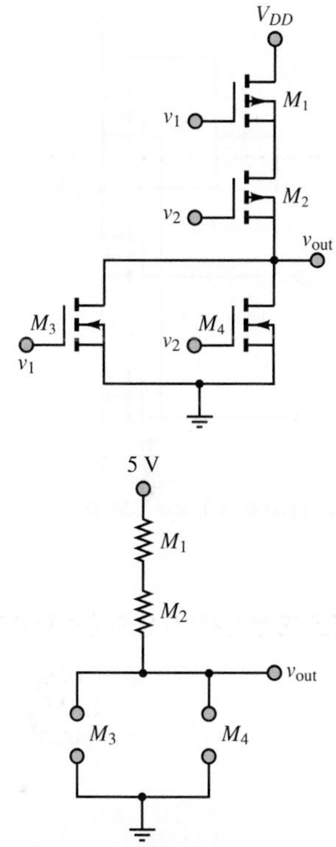

With both v_1 and v_2 at 0 V, M_3 and M_4 will be turned off (in cutoff), since v_{GS} is less than V_T (0 V < 1.7 V). Both M_1 and M_2 will be turned on, since the gate-to-source voltages will be greater than V_T.

Figure 11.21

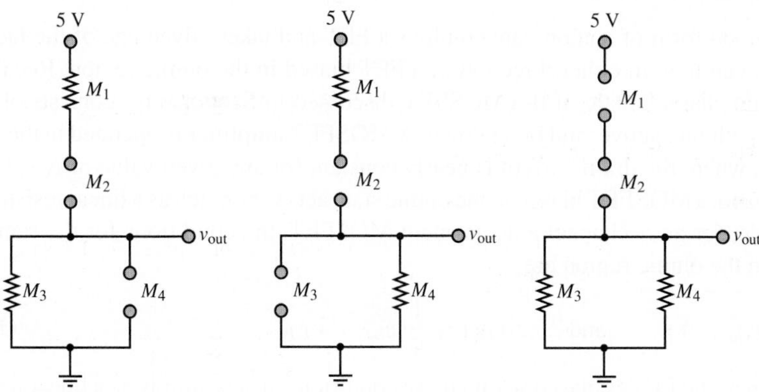

(a) (b) (c)

Figure 11.22

These results are summarized in the table below. The output voltage for case 4 is sufficiently close to zero to be considered zero for logic purposes.

v_1	v_2	M_1	M_2	M_3	M_4	v_{out}
0 V	0 V	On	On	On	Off	5 V
0 V	5 V	On	Off	Off	Off	0 V
5 V	0 V	Off	On	Off	On	0 V
5 V	5 V	Off	Off	On	On	0 V

Thus, the gate is a NOR gate.

Comments: While exact analysis of CMOS gate circuits could be tedious and involved, the method demonstrated in this example—to determine whether transistors are on or off—leads to very simple analysis. Since in logic devices one is interested primarily in logic levels and not in exact values, this approximate analysis method is very appropriate.

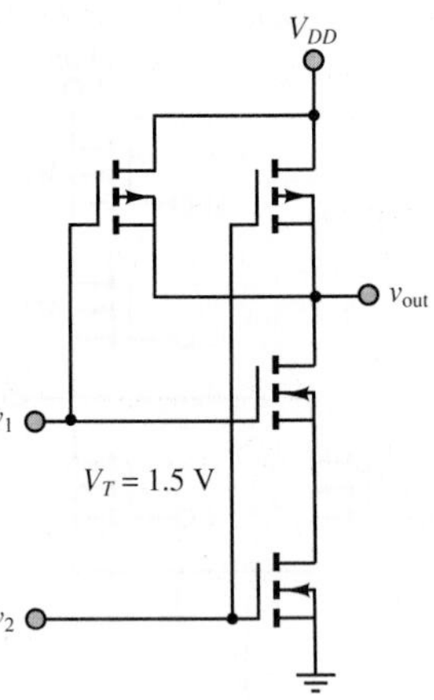

Figure 11.23 CMOS gate

CHECK YOUR UNDERSTANDING

Analyze the CMOS gate of Figure 11.23 and find the output voltages for the following conditions: (a) $v_1 = 0$, $v_2 = 0$; (b) $v_1 = 5$ V, $v_2 = 0$; (c) $v_1 = 0$, $v_2 = 5$ V; (d) $v_1 = 5$ V, $v_2 = 5$ V. Identify the logic function accomplished by the circuit.

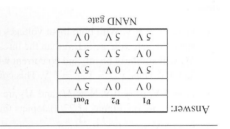

Answer:

v_1	v_2	v_{out}
0 V	0 V	5 V
5 V	0 V	5 V
0 V	5 V	5 V
5 V	5 V	0 V

NAND gate

Analog Switches

A common form of analog gate employs a FET and takes advantage of the fact that current can flow in either direction in a FET biased in the ohmic region. Recall that the drain characteristic of the MOSFET discussed in Section 11.2 consists of three regions: ohmic, active, and breakdown. A MOSFET amplifier is operated in the active region, where the drain current is nearly constant for any given value of v_{GS}. On the other hand, a MOSFET biased in the ohmic state acts very much as a linear resistor. For example, for an n-channel enhancement MOSFET, the conditions for the transistor to be in the ohmic region are

$$v_{GS} > V_T \qquad \text{and} \qquad |v_{DS}| \le \frac{1}{4}(v_{GS} - V_T) \tag{11.13}$$

As long as the FET is biased within these conditions, it acts simply as a linear resistor, and it can conduct current in either direction (provided that v_{DS} does not exceed the limits stated in equation 11.13). In particular, the resistance of the channel in the ohmic region is found to be

$$r_{DS} = \frac{1}{2K(v_{GS} - V_T)} \tag{11.14}$$

so that the drain current is equal to

$$i_D \approx \frac{v_{DS}}{r_{DS}} \qquad \text{for} \quad |v_{DS}| \le \frac{1}{4}(v_{GS} - V_T) \quad \text{and} \quad v_{GS} > V_T \tag{11.15}$$

The most important feature of the MOSFET operating in the ohmic region, then, is that it acts as a voltage-controlled resistor, with the gate-source voltage v_{GS} controlling the channel resistance R_{DS}. The use of the MOSFET as a switch in the ohmic region, then, consists of providing a gate-source voltage that can either hold the MOSFET in the cutoff region ($v_{GS} \le V_T$) or bring it into the ohmic region. In this fashion, v_{GS} acts as a control voltage for the transistor.

Consider the circuit shown in Figure 11.24, where we presume that v_C can be varied externally and that v_{in} is some analog signal source that we may wish to connect to the load R_L at some appropriate time. The operation of the switch is as follows. When $v_C \le V_T$, the FET is in the cutoff region and acts as an open circuit.

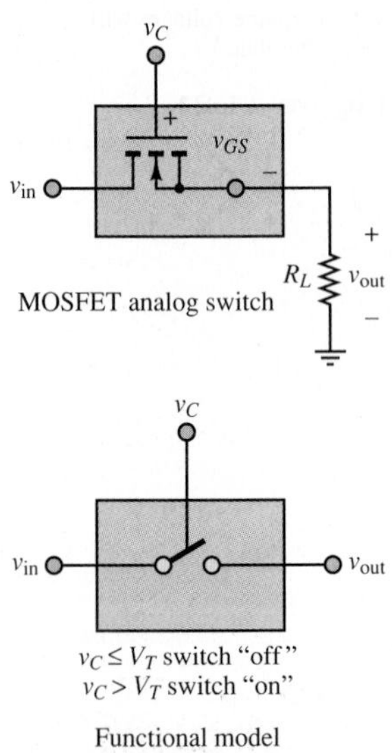

Figure 11.24 MOSFET analog switch

When $v_C > V_T$ (with a value of v_{GS} such that the MOSFET is in the ohmic region), the transistor acts as a linear resistance R_{DS}. If $R_{DS} \ll R_L$, then $v_{out} \approx v_{in}$. By using a pair of MOSFETs, it is possible to improve the dynamic range of signals one can transmit through this analog gate.

MOSFET analog switches are usually produced in integrated-circuit (IC) form and denoted by the symbol shown in Figure 11.25. A CMOS analog gate is described in the next Focus on Measurements box.

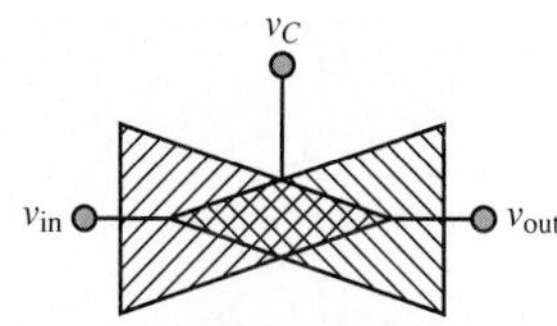

Figure 11.25 Symbol for bilateral FET analog gate

MOSFET Bidirectional Analog Gate

The variable-resistor feature of MOSFETs in the ohmic state finds application in the **analog transmission gate.** The circuit shown in Figure 11.26 depicts a circuit constructed using CMOS technology. The circuit operates on the basis of a control voltage v that can be either "low" (say, 0 V) or "high" ($v > V_T$), where V_T is the threshold voltage for the n-channel MOSFET and $-V_T$ is the threshold voltage for the p-channel MOSFET. The circuit operates in one of two modes. When the gate of Q_1 is connected to the high voltage and the gate of Q_2 is connected to the low voltage, the path between v_{in} and v_{out} is a relatively small resistance, and the transmission gate conducts. When the gate of Q_1 is connected to the low voltage and the gate of Q_2 is connected to the high voltage, the transmission gate acts as a very large resistance and is an open circuit for all practical purposes. A more precise analysis follows.

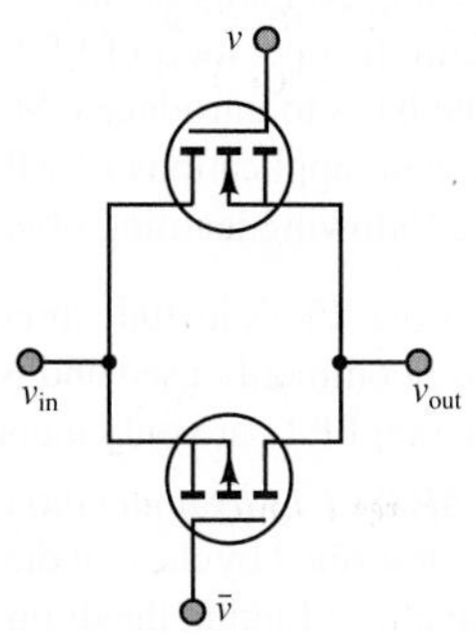

(a) CMOS transmission gate

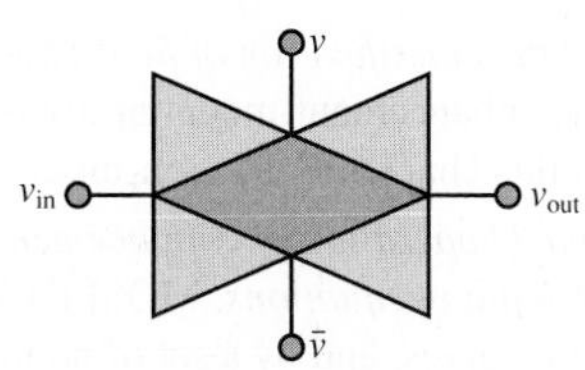

(b) CMOS transmission gate circuit symbol

Figure 11.26 Analog transmission gate

Let $v = V > V_T$ and $\bar{v} = 0$. Assume that the input voltage v_{in} is in the range $0 \le v_{in} \le V$. To determine the state of the transmission gate, we consider only the extreme cases $v_{in} = 0$ and $v_{in} = V$. When $v_{in} = 0$, $v_{GS1} = v - v_{in} = V - 0 = V > V_T$. Since V is above the threshold voltage, MOSFET Q_1 conducts (in the ohmic region). Further, $v_{GS2} = \bar{v} - v_{in} = 0 > -V_T$. Since the gate-source voltage is not more negative than the threshold voltage, Q_2 is in cutoff and does not conduct. Since one of the two possible paths between v_{in} and v_{out} is conducting, the transmission gate is on. Now consider the other extreme, where $v_{in} = V$. By reversing the previous argument, we can see that Q_1 is now off, since $v_{GS1} = 0 < V_T$. However, now Q_2 is in the ohmic state, because $v_{GS2} = \bar{v} - v_{in} = 0 - V < -V_T$. In this case, then, it is Q_2

(Continued)

(Concluded)

that provides a conducting path between the input and the output of the transmission gate, and the transmission gate is also on. We have therefore concluded that when $v = V$ and $\overline{v} = 0$, the transmission gate conducts and provides a near-zero-resistance (typically tens of ohms) connection between the input and the output of the transmission gate, for values of the input ranging from 0 to V.

Let us now reverse the control voltages and set $v = 0$ and $\overline{v} = V > V_T$. It is very straightforward to show that in this case, regardless of the value of v_{in}, both Q_1 and Q_2 are always off; therefore, the transmission gate is essentially an open circuit.

The analog transmission gate finds common application in *analog multiplexers* and *sample-and-hold* circuits, to be discussed in Chapter 15.

CHECK YOUR UNDERSTANDING

Show that the CMOS bidirectional gate described in the Focus on Measurements box "MOSFET Bidirectional Analog Gate" is off for all values of v_{in} between 0 and V whenever $v = 0$ and $\overline{v} = V > V_T$.

Conclusion

This chapter has introduced field-effect transistors, focusing primarily on metal-oxide semiconductor enhancement-mode n-channel devices to explain the operation of FETs as amplifiers. A brief introduction to p-channel devices is used as the basis to introduce CMOS technology, and to present analog and digital switches and logic gate applications of MOSFETs. Upon completing this chapter, you should have mastered the following learning objectives:

1. *Understand the classification of field-effect transistors.* FETs include three major families; the enhancement-mode family is the most commonly used and is the one explored in this chapter. Depletion-mode and junction FETs are only mentioned briefly.
2. *Learn the basic operation of enhancement-mode MOSFETs by understanding their i–v curves and defining equations.* MOSFETs can be described by the i–v drain characteristic curves, and by a set of nonlinear equations linking the drain current to the gate-to-source and drain-to-source voltages. MOSFETs can operate in one of four regions: *cutoff*, in which the transistor does not conduct current; *triode*, in which the transistor can act as a voltage-controlled resistor under certain conditions; *saturation*, in which the transistor acts as a voltage-controlled current source and can be used as an amplifier; and *breakdown* when the limits of operation are exceeded.
3. *Learn how enhancement-mode MOSFET circuits are biased.* MOSFET circuits can be biased to operate around a certain operating point, known as the Q point, by appropriately selecting supply voltages and resistors.
4. *Understand the concept and operation of FET large-signal amplifiers.* Once a MOSFET circuit is properly biased in the saturation region, it can serve as an amplifier by virtue of its voltage-controlled current source property: small changes in the gate-to-source voltages are translated to proportional changes in drain current.
5. *Understand the concept and operation of FET switches.* MOSFETs can serve as analog and digital switches: by controlling the gate voltage, a MOSFET can be turned on and off (digital switch), or its resistance can be modulated (analog switch).
6. *Analyze FET switches and digital gates.* These devices find application in CMOS circuits as digital logic gates and analog transmission gates.

HOMEWORK PROBLEMS

Section 11.2: *n*-Channel MOSFET Operation

11.1 The transistors shown in Figure P11.1 have $|V_T| = 3$ V. Determine the operating region.

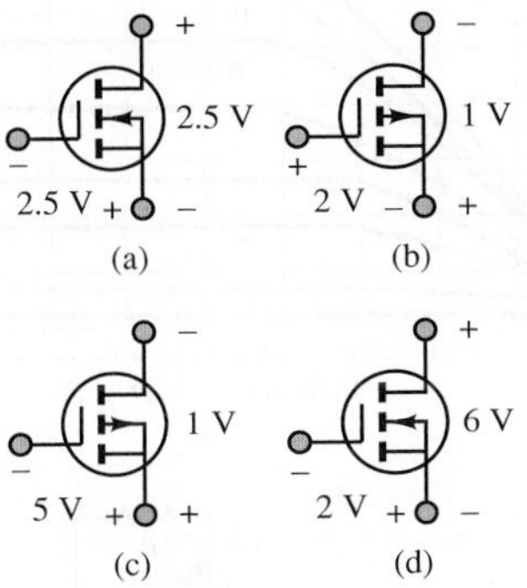

Figure P11.1

11.2 The three terminals of an n-channel enhancement-mode MOSFET are at potentials of 4, 5, and 10 V with respect to ground. Draw the circuit symbol, with the appropriate voltages at each terminal, if the device is operating

a. In the ohmic region.

b. In the active region.

11.3 An enhancement-type NMOS transistor with $V_T = 2$ V has its source grounded and a 3-V DC source connected to the gate. Determine the operating state if

a. $v_D = 0.5$ V

b. $v_D = 1$ V

c. $v_D = 5$ V

11.4 In the circuit shown in Figure P11.4, the p-channel transistor has $V_T = 2$ V and $K = 10$ mA/V^2. Find R and v_D for $i_D = 0.4$ mA.

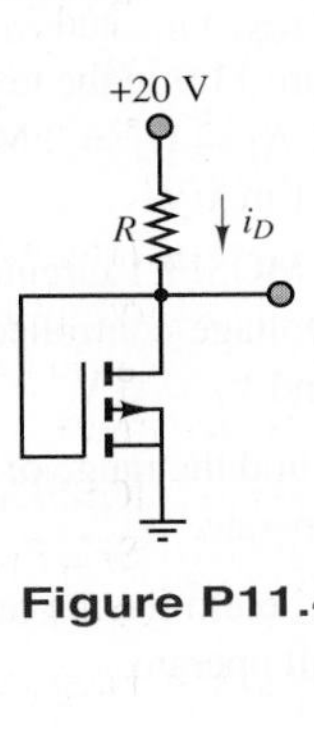

Figure P11.4

11.5 An enhancement-type NMOS transistor has $V_T = 2$ V and $i_D = 1$ mA when $v_{GS} = v_{DS} = 3$ V. Find the value of i_D for $v_{GS} = 4$ V.

11.6 An n-channel enhancement-mode MOSFET is operated in the ohmic region, with $v_{DS} = 0.4$ V and $V_T = 3.2$ V. The effective resistance of the channel is given by $R_{DS} = 500/(V_{GS} - 3.2)$ Ω. Find i_D when $v_{GS} = 5$ V, $R_{DS} = 500$ Ω, and $v_{GD} =$ 4 V.

11.7 An enhancement-type NMOS transistor with $V_T = 2.5$ V has its source grounded and a 4-V DC source connected to the gate. Find the operating region of the device if

a. $v_D = 0.5$ V

b. $v_D = 1.5$ V

11.8 An enhancement-type NMOS transistor has $V_T = 4$ V and $i_D = 1$ mA when $v_{GS} = v_{DS} = 6$ V. Neglecting the dependence of i_D on v_{DS} in saturation, find the value of i_D for $v_{GS} = 5$ V.

11.9 The NMOS transistor shown in Figure P11.9 has $V_T = 1.5$ V and $K = 0.4$ mA/V^2. If v_G is a pulse with 0 to 5 V, find the voltage levels of the pulse signal at the drain output.

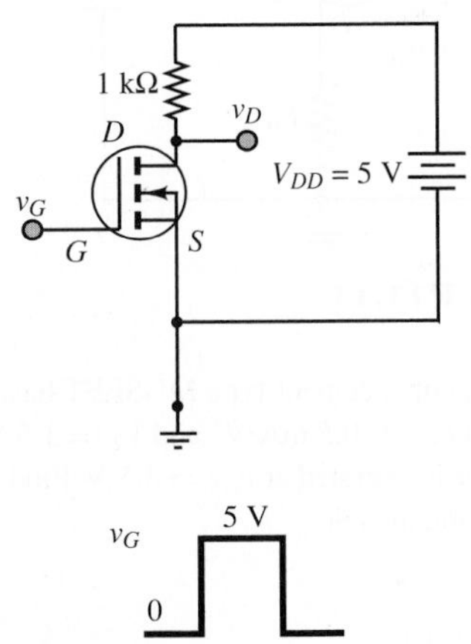

Figure P11.9

11.10 In the circuit shown in Figure P11.10, a drain voltage of 0.1 V is established. Find the current i_D for $V_T = 1$ V and $k = 0.5$ mA/V^2.

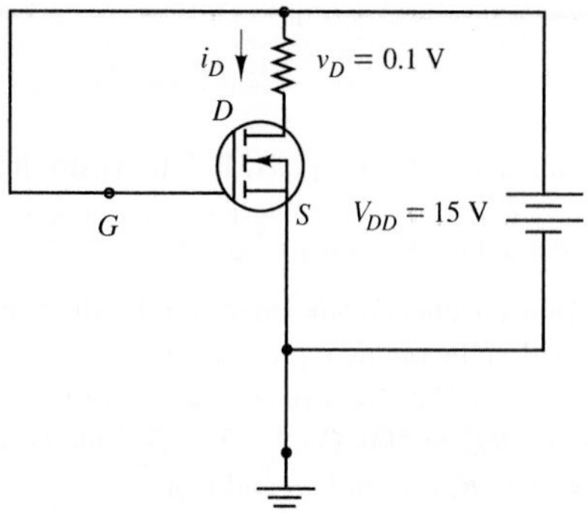

Figure P11.10

11.11 In the circuit shown in Figure P11.11, the MOSFET operates in the active region, for $i_D = 0.5$ mA and $v_D = 3$ V. This enhancement-type PMOS has $V_T = -1$ V, $k = 0.5$ mA/V^2. Find

a. R_D.

b. The largest allowable value of R_D for the MOSFET to remain in the saturation region.

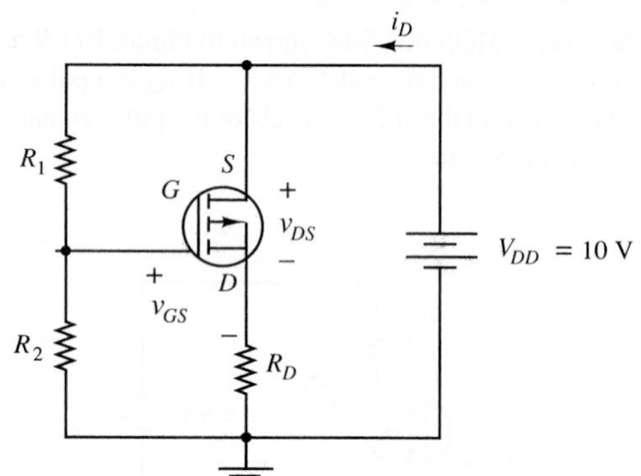

Figure P11.11

11.12 An enhancement-type MOSFET has the parameters $k = 0.5$ mA/V^2 and $V_T = 1.5$ V, and the transistor is operated at $v_{GS} = 3.5$ V. Find the drain current obtained at

a. $v_{DS} = 3$ V

b. $v_{DS} = 10$ V

Section 11.3: *n*-Channel MOSFET Amplifiers

11.13 The i–v characteristic of an n-channel enhancement MOSFET is shown in Figure P11.13(a); a standard amplifier circuit based on the n-channel MOSFET is shown in Figure P11.13(b). Determine the quiescent current i_{DQ} and drain-to-source voltage v_{DS} if $V_{GG} = 7$ V, $V_{DD} = 10$ V, and $R_D = 5\ \Omega$. In what region is the transistor operating?

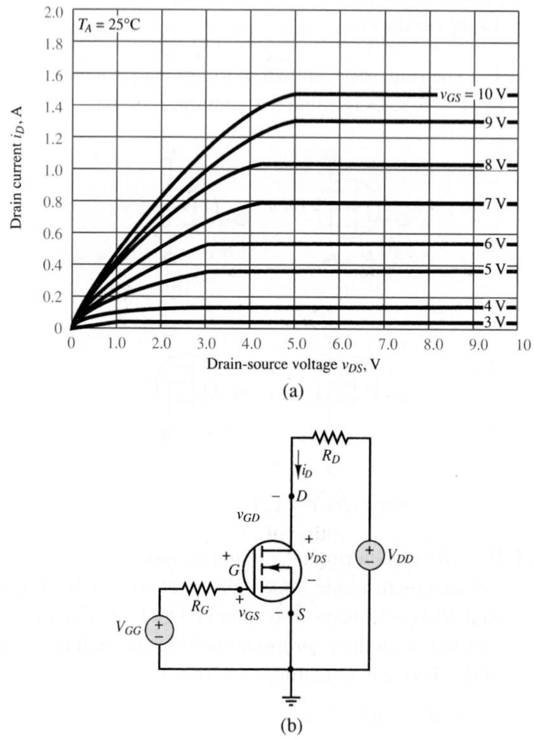

Figure P11.13

11.14 Determine the Q point and operating region for the MOSFET in the circuit of Figure P11.13(b). $V_{GG} = 7$ V; $V_{DD} = 20$ V; $V_T = 3$ V; $K = 50$ mA/V^2; $R_D = 5\ \Omega$.

11.15 Determine the Q point and operating region for the MOSFET in the circuit of Figure 11.9 in the text. Assume $V_{DD} = 36$ V, $R_D = 10$ kΩ, $R_1 = R_2 = 2$ MΩ, $V_T = 4$ V, and $K = 0.1$ mA/V^2. You will need to find R_S, v_{DSQ}, and i_{DQ}.

11.16 Determine v_{GS}, v_{DS}, and i_D for the transistor amplifier of Figure 11.9 in the text. Assume $V_{DD} = 12$ V; $R_D = 10$ kΩ; $R_1 = R_2 = 2$ MΩ; $R_S = 10$ kΩ; $V_T = 1$ V; $K = 1$ mA/V^2.

11.17 The power MOSFET circuit of Figure P11.17 is configured as a voltage-controlled current source. Let $K = 1.5$ A/V^2 and $V_T = 3$ V.

a. If $V_G = 5$ V, find the range of R_L for which the VCCS will operate.

b. If $R_L = 1\ \Omega$, determine the range of V_G for which the VCCS will operate.

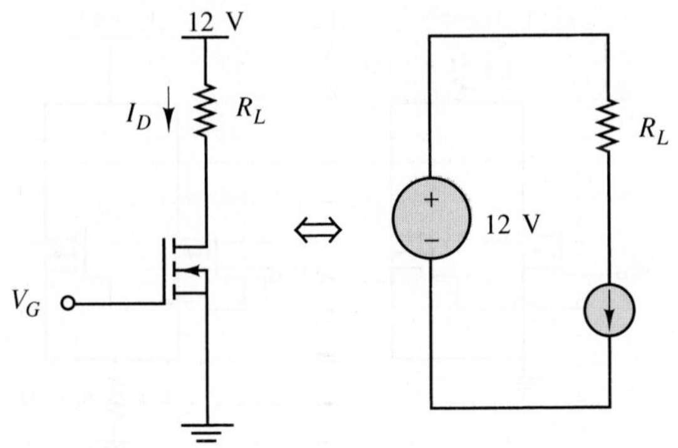

Figure P11.17

11.18 The circuit of Figure P11.18 is called a *source follower*, and acts as a voltage-controlled current source (VCCS).

a. Determine I_L if $V_G = 10$ V, $R_L = 2\ \Omega$, $K = 0.5$ A/V^2, $V_T = 4$ V.

b. If the power rating of the MOSFET is 50 W, how small can R_L be?

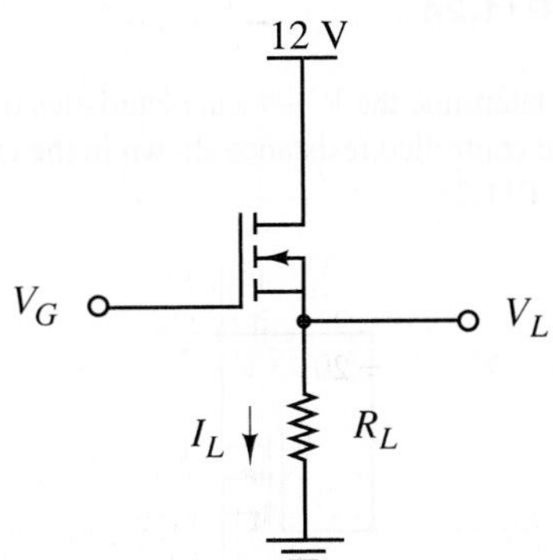

Figure P11.18

11.19 The circuit of Figure P11.19 is a Class A amplifier.

a. Determine the output current for the given biased audio tone input, $V_G = 10 + 0.1\cos(500t)$ V. Let $K = 2\text{mA/V}^2$ and $V_T = 3$ V.

b. Determine the output voltage.

c. Determine the voltage gain of the $\cos(500t)$ signal.

d. Determine the DC power consumption of the resistor and the MOSFET.

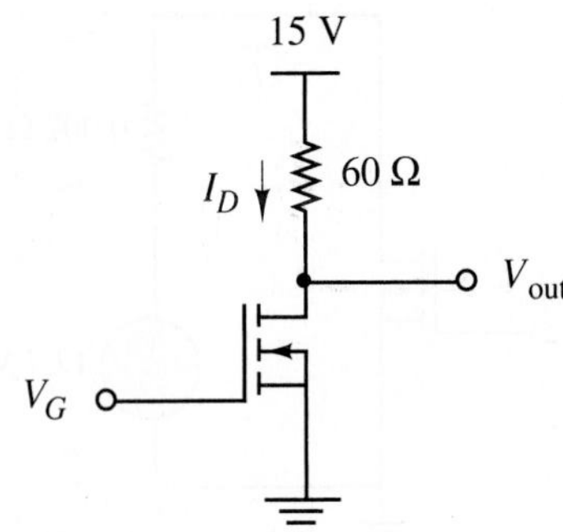

Figure P11.19

11.20 The circuit of Figure P11.20 is a source-follower amplifier. Let $K = 30\ \text{mA/V}^2$, $V_T = 4$ V, and $V_G = 9 + 0.1\cos(500t)$ V.

a. Determine the load current I_L.

b. Determine the output voltage V_{out}.

c. Determine the voltage gain for the $\cos(100t)$ signal.

d. Determine the DC power consumption of the MOSFET and R_L.

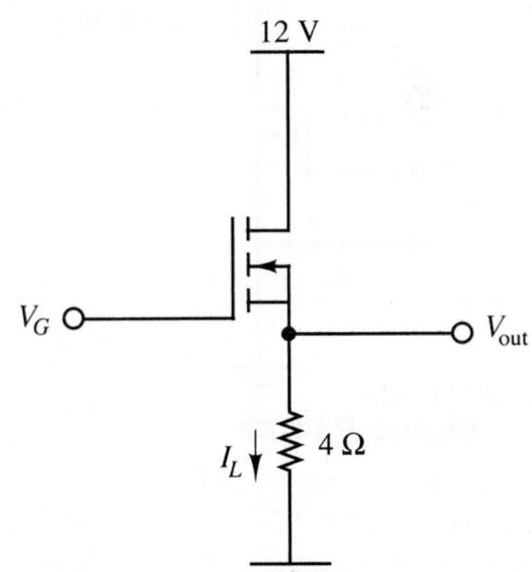

Figure P11.20

11.21 Sometimes it is necessary to discharge batteries before recharging. To do this, an electronic load can be used. A high-power electronic load is shown in Figure P11.21, for the battery discharge application. With $K = 4\ \text{A/V}^2$, $V_T = 3$ V, and $V_G = 8$ V, determine the discharging current I_D and the required MOSFET power rating.

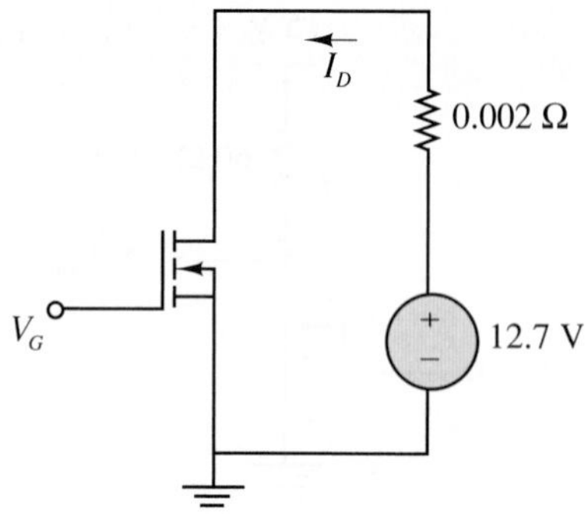

Figure P11.21

11.22 A precision voltage source can be created by driving the drain of a MOSFET. Figure P11.22 shows a circuit that will accomplish this function. With $I_{\text{Ref}} = 0.01$ A, determine the output V_G. Let $K = 0.006$ mA/V^2 and $V_T = 1.5$ V.

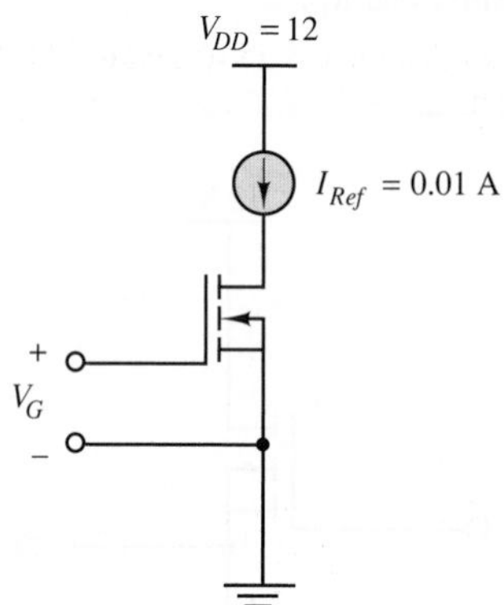

Figure P11.22

11.23 To allow more current in a MOSFET amplifier, several MOSFETs can be connected in parallel. Determine the load current in each of the circuits of Figure P11.23. Let $K = 0.2$ A/V^2 and $V_T = 3$ V.

11.24 A "push-pull amplifier" can be constructed from matched n-and-p-channel MOSFETs, as shown in Figure P11.24. Let

$K_n = K_p = 0.5$ A/V^2, $V_{Tn} = +3$ V, $V_{Tp} = -3$ V and $V_{\text{in}} = 0.8\cos(1{,}000t)$ V.

Determine V_L and I_L.

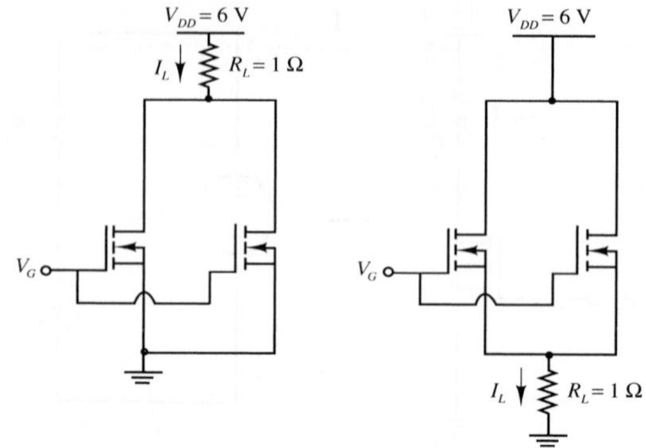

Figure P11.23

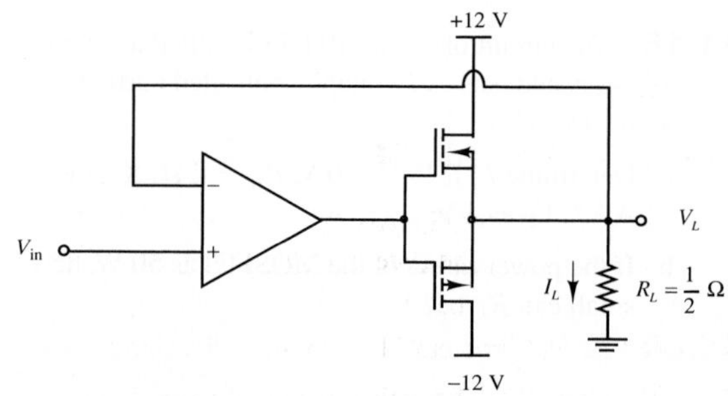

Figure P11.24

11.25 Determine the $V - I$ characteristics of the voltage controlled resistance shown in the circuit of Figure P11.25.

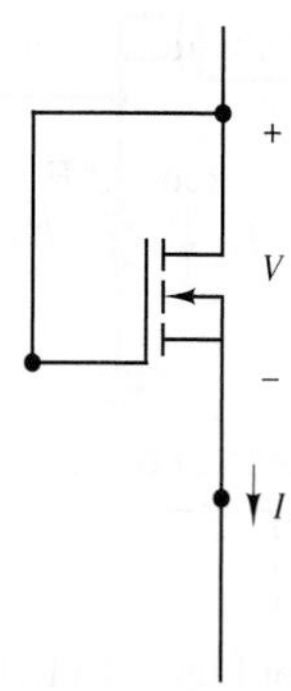

Figure P11.25

11.26 Determine V_L and I_L for the two-stage amplifier shown in the circuit of Figure P11.26, with identical MOSFETs having $K = 1$ A/V^2 and $V_T = 3$V, for

a. $V_G = 4$ V,

b. $V_G = 5$ V, and

c. $V_G = 4 + 0.1\cos(750t)$.

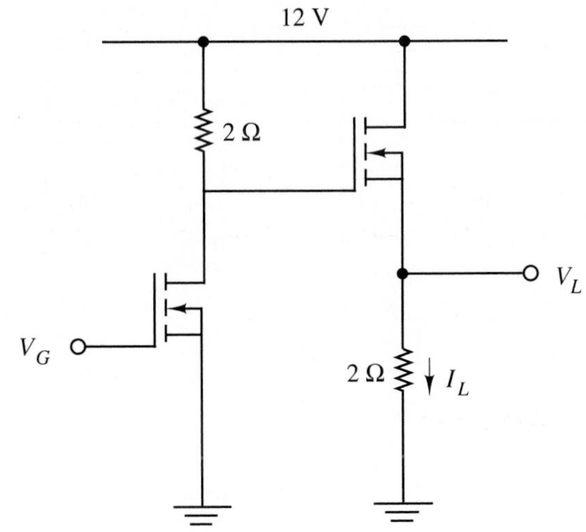

Figure P11.26

Section 11.4: MOSFET Switches

11.27 For the CMOS NAND gate of Figure 11.23 in the text identify the state of each transistor for $v_1 = v_2 = 5$ V.

11.28 Repeat Problem 11.27 for $v_1 = 5$ V and $v_2 = 0$ V.

11.29 Draw the schematic diagram of a two-input CMOS OR gate.

11.30 Draw the schematic diagram of a two-input CMOS AND gate.

11.31 Draw the schematic diagram of a two-input CMOS NOR gate.

11.32 Draw the schematic diagram of a two-input CMOS NAND gate.

11.33 Show that the circuit of Figure P11.33 functions as a logic inverter.

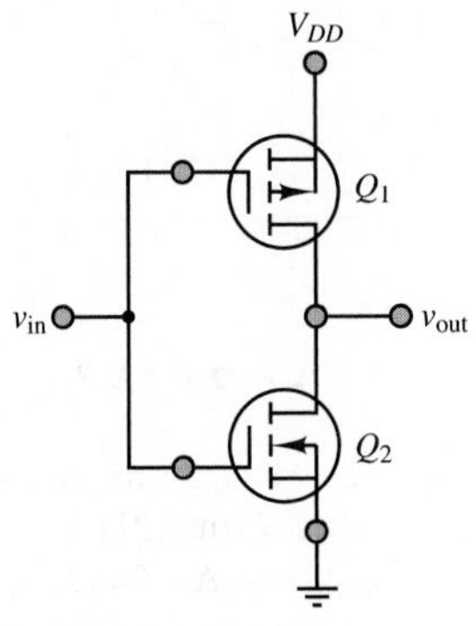

Figure P11.33

11.34 Show that the circuit of Figure P11.34 functions as a NOR gate.

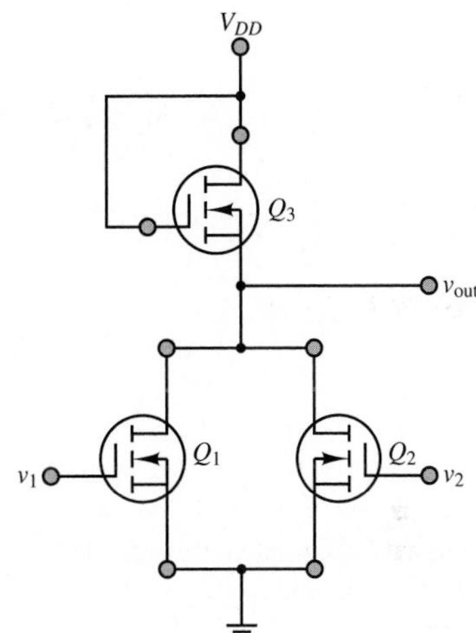

Figure P11.34

11.35 Show that the circuit of Figure P11.35 functions as a NAND gate.

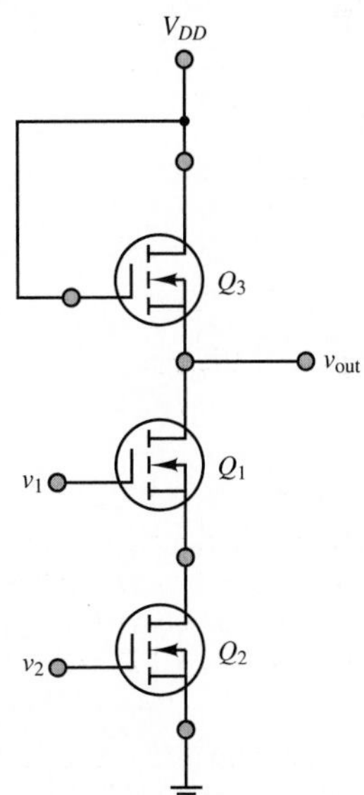

Figure P11.35

CHAPTER

12

POWER ELECTRONICS

The objective of this chapter is to present a survey of power electronic devices and systems. Power electronic devices form the "muscle" of many electromechanical systems. For example, one finds such devices in many appliances, in industrial machinery, and virtually wherever an electric motor is found, since one of the foremost applications of power electronic devices is to supply and control the currents and voltages required to power electric machines, such as those introduced in Part III of this book.

Power electronic devices are specially designed diodes and transistors that have the ability to carry large currents and sustain large voltages; thus, the basis for this chapter is the material on diodes and transistors introduced in Chapters 9 through 11.

This chapter describes the basic properties of each type of power electronic device, and it illustrates the application of a selected few, especially in electric motor power supplies. After completing the chapter, you will be able to recognize the symbols for the major power semiconductor devices and understand their principles of operation. You will also understand the operation of the principal electronic power supplies for DC and AC motors.

Learning Objectives

1. Learn the classification of power electronic devices and circuits. *Sections 12.1 and 12.2*
2. Analyze the operation of practical voltage regulators. *Section 12.3.*
3. Understand the operation and principal limitations of transistor power amplifiers. *Section 12.4.*
4. Analyze the operation of single- and three-phase controlled rectifier circuits. *Section 12.5.*
5. Understand the operation of power converters used in electric motor control, and perform simplified analysis on DC-DC converters. *Section 12.6.*

LO1

Device	Device symbol
Diode	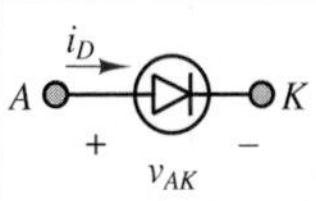
Thyristor	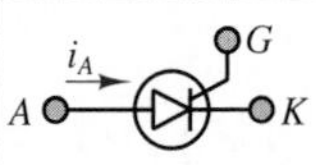
Gate turnoff thyristor (GTO)	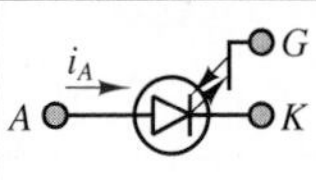
Triac	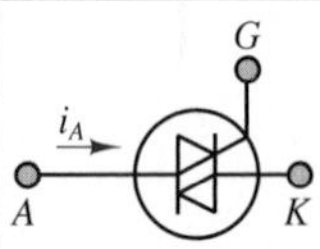
npn BJT	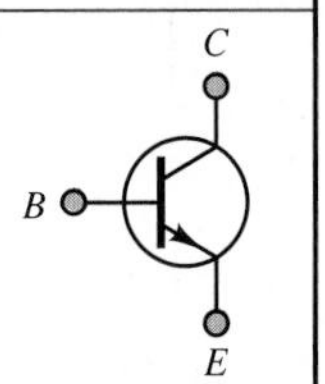
IGBT	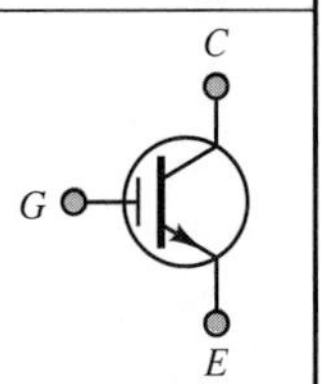
n-channel MOSFET	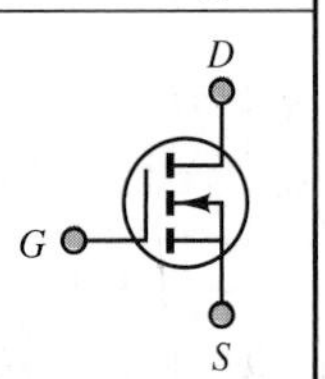

Figure 12.1 Classification of power electronic devices

12.1 CLASSIFICATION OF POWER ELECTRONIC DEVICES

Power semiconductors can be broadly subdivided into five groups: (1) power diodes, (2) thyristors, (3) power bipolar junction transistors (BJTs), (4) insulated-gate bipolar transistors (IGBTs), and (5) static induction transistors (SITs). Figure 12.1 depicts the symbols for the most common power electronic devices.

Power diodes are functionally identical to the diodes introduced in Chapter 9, except for their ability to carry much larger currents. You will recall that a diode conducts in the forward-biased mode when the anode voltage (V_A) is higher than the cathode voltage (V_K). Three types of power diodes exist: *general-purpose, high-speed (fast-recovery)*, and *Schottky*. Typical ranges of voltage and current are 3,000 V and 3,500 A for general-purpose diodes and 3,000 V and 1,000 A for high-speed devices. The latter have switching times as low as a fraction of a microsecond. Schottky diodes can switch much faster (in the nanosecond range) but are limited to around 100 V and 300 A. The forward voltage drop of power diodes is not much higher than that of low-power diodes, being between 0.5 and 1.2 V. Since power diodes are used with rather large voltages, the forward bias voltage is usually considered negligible relative to other voltages in the circuit, and the switching characteristics of power diodes may be considered near ideal. The principal consideration in choosing power diodes is their power rating.

Thyristors function as power diodes with an additional gate terminal that controls the time when the device begins conducting; a thyristor starts to conduct when a small gate current is injected into the gate terminal, provided that the anode voltage is greater than the cathode voltage (or $V_{AK} > 0$ V). The forward voltage drop of a thyristor is on the order of 0.5 to 2 V. Once conduction is initiated, the gate current has no further control. To stop conduction, the device must be reverse-biased; that is, one must ensure that $V_{AK} \leq 0$ V. Thyristors can be rated at up to 6,000 V and 3,500 A. The **turnoff time** is an important characteristic of thyristors; it represents the time required for the device current to return to zero after external switching of V_{AK}. The fastest turnoff times available are in the range of 10 μs; however, such turnoff times are achieved only in devices with slightly lower power ratings (1,200 V, 1,000 A).

Thyristors can be subclassified into the following groups: force-commutated and line-commutated thyristors, gate turnoff thyristors (GTOs), reverse-conducting thyristors (RCTs), static induction thyristors (SITs), gate-assisted turnoff thyristors (GATTs), light-activated silicon controlled rectifiers (LASCRs), and MOS controlled thyristors (MCTs). It is beyond the scope of this chapter to go into a detailed description of each of these types of devices; their operation is typically a slight modification of the basic operation of the thyristor. The reader who wishes to gain greater insight into this topic may refer to one of a number of excellent books specifically devoted to the subject of power electronics.

Two types of thyristor-based device deserve some more attention. The **triac,** as can be seen in Figure 12.1, consists of a pair of thyristors connected back to back, with a single gate; this allows for current control in either direction. Thus, a triac may be thought of as a bidirectional thyristor. The gate turnoff thyristor (GTO), on the other hand, can be turned on by applying a short positive pulse to the gate, like a thyristor, and can also be turned off by application of a short negative pulse. Thus, GTOs are very convenient in that they do not require separate commutation circuits to be turned on and off.

Power BJTs can reach ratings up to 1,200 V and 400 A, and they operate in much the same way as a conventional BJT. Power BJTs are used in power converter applications at frequencies up to around 10 kHz. **Power MOSFETs** can operate at somewhat higher frequencies (a few to several tens of kilohertz), but are limited in power (typically up to 1,000 V, 50 A). **Insulated-gate bipolar transistors (IGBTs)** are voltage-controlled (because of their insulated gate, reminiscent of insulated-gate FETs) power transistors that offer superior speed with respect to BJTs but are not quite as fast as power MOSFETs.

12.2 CLASSIFICATION OF POWER ELECTRONIC CIRCUITS

The devices that are discussed in this chapter find application in a variety of **power electronic circuits.** This section briefly summarizes the principal types of power electronic circuits and qualitatively describes their operation. The following sections will describe the devices and their operation in these circuits in greater detail.

One possible classification of power electronic circuits is given in Table 12.1. Many of the types of circuits are similar to circuits introduced in earlier chapters.

LO1

Table 12.1 Power electronic circuits

Circuit type	Essential features
Voltage regulators	Regulate a DC supply to a fixed voltage output
Power amplifiers	Large-signal amplification of voltages and currents
Switches	Electronic switches (e.g., transistor switches)
Diode rectifier	Converts fixed AC voltage (single- or multiphase) to fixed DC voltage
AC-DC converter (controlled rectifier)	Converts fixed AC voltage (single- or multiphase) to variable DC voltage
AC-AC converter (AC voltage controller)	Converts fixed AC voltage to variable AC voltage (single- or multiphase)
DC-DC converter (chopper)	Converts fixed DC voltage to variable DC voltage
DC-AC converter (inverter)	Converts fixed DC voltage to variable AC voltage (single- or multiphase)

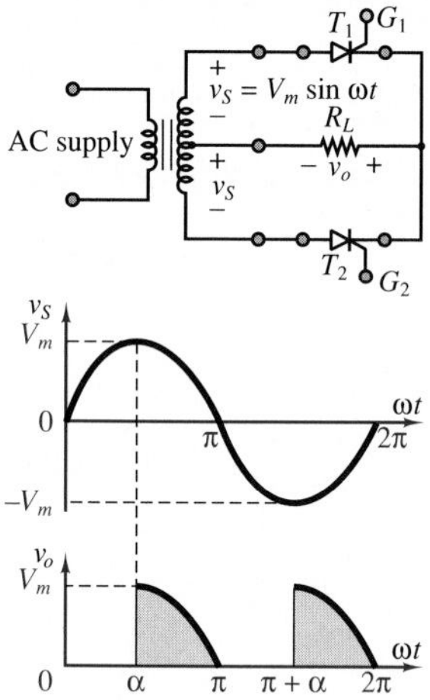

Figure 12.2 AC-DC converter circuit and waveform

Voltage regulators were introduced in Chapter 9 (see Figure 9.48); this chapter presents a more detailed discussion of practical regulators. Power electronic switches function exactly as the transistor switches described in Chapters 10 and 11 (see Figures 10.30 and 11.24); their function is to act as voltage- or current-controlled switches to turn AC or DC supplies on and off. Transistor power amplifiers are the high-power version of the BJT and MOSFET amplifiers studied in Chapter 10; it is important to consider power limitations and signal distortion very carefully in power amplifiers.

Diode rectifiers were discussed in Chapter 9 in their single-phase form (see Figures 9.39, and 9.41); similar rectifiers can also be designed to operate with three-phase sources. The operation of a single-phase full-wave rectifier was summarized in Figure 9.40. AC-DC converters are also rectifiers, but they take advantage of the controlled properties of thyristors. The thyristor gate current can be timed to "fire" conduction at variable times, resulting in a variable DC output, as illustrated in Figure 12.2, which shows the circuit and behavior of a single-phase AC-DC converter. This type of converter is very commonly used as a supply for DC electric motors. In Figure 12.2, α is the firing angle of thyristor T_1, where the device starts to conduct.

FIND IT ON THE WEB

AC-AC converters are used to obtain a variable AC voltage from a fixed AC source. Figure 12.3 shows a triac-based AC-AC converter, which takes advantage of the bidirectional capability of triacs to control the rms value of an alternating voltage. Note in particular that the resulting AC waveform is no longer a pure sinusoid even though its fundamental period (frequency) is unchanged. A **DC-DC converter,** also known as a *chopper,* or *switching regulator,* permits conversion of a fixed DC source to a variable DC supply. Figure 12.4 shows how such an effect may be obtained by controlling the base-emitter voltage of a bipolar transistor, enabling conduction at the desired time. This results in the conversion of the DC input voltage to a variable–duty-cycle output voltage, whose average value can be controlled by selecting the on time of the transistor. DC-DC converters find application as variable voltage supplies for DC electric motors used in electric vehicles.

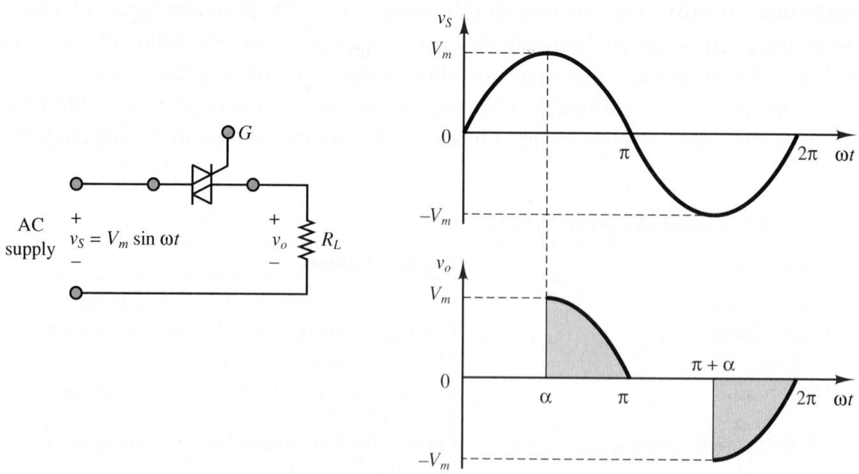

Figure 12.3 AC-AC converter circuit and waveform

FIND IT ON THE WEB

Finally, **DC-AC converters, or inverters,** are used to convert a fixed DC supply to a variable AC supply; they find application in AC motor control. The operation of these circuits is rather complex; it is illustrated conceptually in the waveforms of

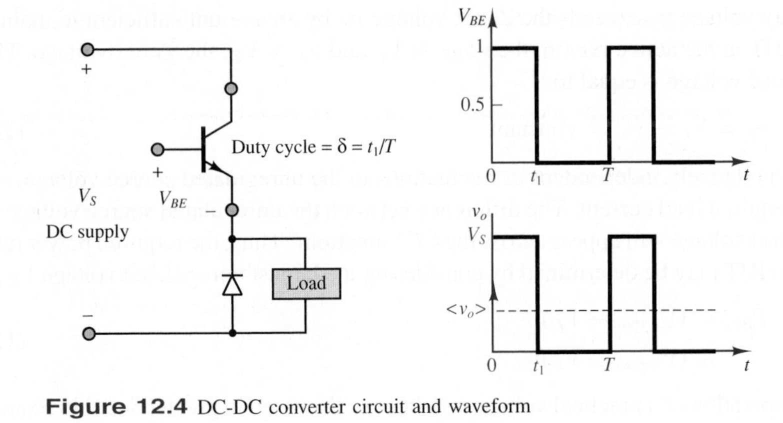

Figure 12.4 DC-DC converter circuit and waveform

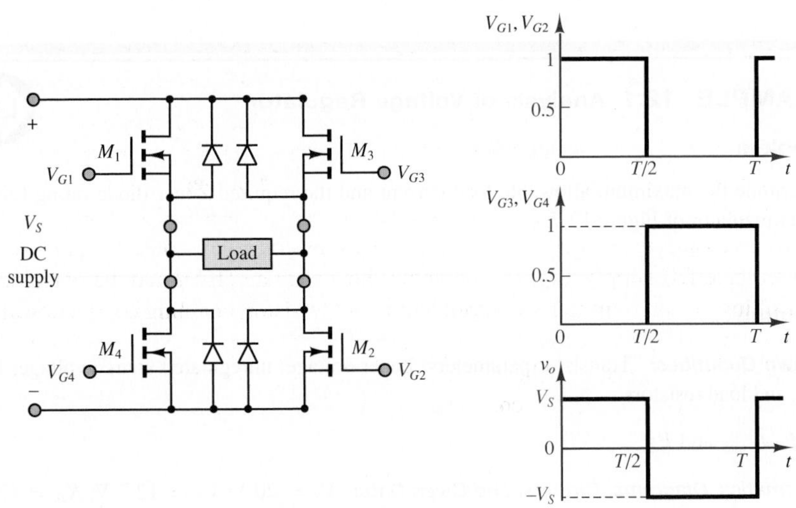

Figure 12.5 DC-AC converter circuit and waveform

Figure 12.5, where it is shown that by appropriately switching two pairs of transistors it is possible to generate an AC waveform (square wave).

Each of the circuits of Table 12.1 will be analyzed in greater detail later in this chapter.

12.3 VOLTAGE REGULATORS

You will recall the discussion of the Zener diode as a voltage regulator in Chapter 9, where we introduced a voltage regulator as a three-terminal device that acts nearly as an ideal battery. Figure 12.6 depicts the appearance of a Zener diode i–v characteristic and shows a block diagram of a three-terminal regulator.

A simple Zener diode is often inadequate for practical voltage regulation. In some cases, the Zener resistance alone might cause excessive power dissipation in the Zener diode (especially when little current is required by the load). A more practical—and often used—circuit is a regulator that includes a series pass transistor, shown in Figure 12.7. The operation of this voltage regulator is as follows. If the unregulated

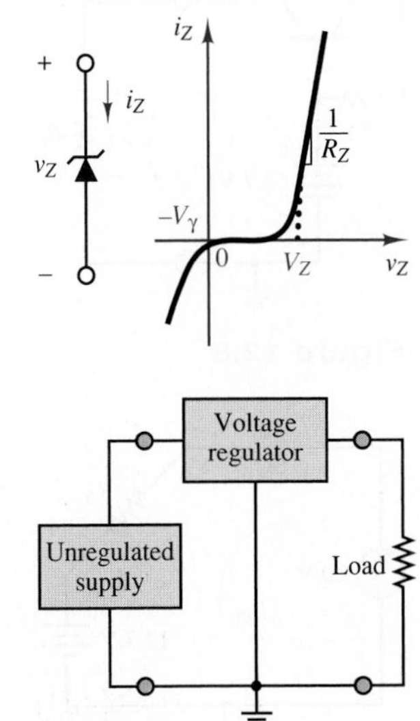

Figure 12.6 Zener diode characteristic and voltage regulator circuit

supply voltage v_S exceeds the Zener voltage v_Z by an amount sufficient to maintain the BJT in the active region, then $v_{BE} \approx V_\gamma$ and $v_Z \approx V_Z$, the Zener voltage. Thus, the load voltage is equal to

$$v_L = V_Z - V_\gamma = \text{constant} \quad \textbf{(12.1)}$$

and is relatively independent of fluctuations in the unregulated source voltage, or in the required load current. The difference between the unregulated source voltage and the load voltage will appear across the *CE* "junction." Thus, the required power rating of the BJT may be determined by considering the largest unregulated voltage $V_{S\ \max}$:

$$\begin{aligned} P_{\text{BJT}} &= (V_{S\ \max} - V_L)i_C \\ &\approx (V_{S\ \max} - V_L)i_L \end{aligned} \quad \textbf{(12.2)}$$

The operation of a practical voltage regulator is discussed in greater detail in Example 12.1.

EXAMPLE 12.1 Analysis of Voltage Regulator

LO2

Problem

Determine the maximum allowable load current and the required Zener diode rating for the Zener regulator of Figure 12.7.

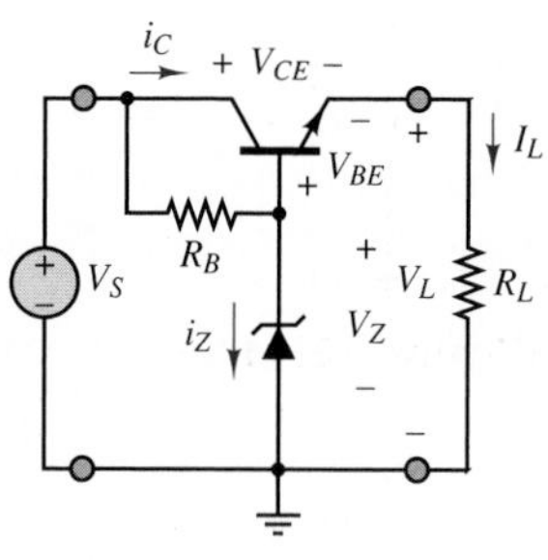

Figure 12.7 Practical voltage regulator

Solution

Known Quantities: Transistor parameters; Zener voltage; unregulated source voltage; BJT base and load resistors.

Find: $I_{L\ \max}$ and P_Z.

Schematics, Diagrams, Circuits, and Given Data: $V_S = 20$ V; $V_Z = 12.7$ V; $R_B = 47\ \Omega$; $R_L = 10\ \Omega$. Transistor data: TIP31 (see Table 12.2).

Assumptions: Use the large-signal model of the BJT. Assume that the BJT is in the active region and the Zener diode is on and therefore regulating to the nominal voltage.

Analysis: Figure 12.8 depicts the equivalent load circuit. Applying KVL, we obtain

$$V_Z = V_{BE} + R_L I$$

from which we can compute the load current:

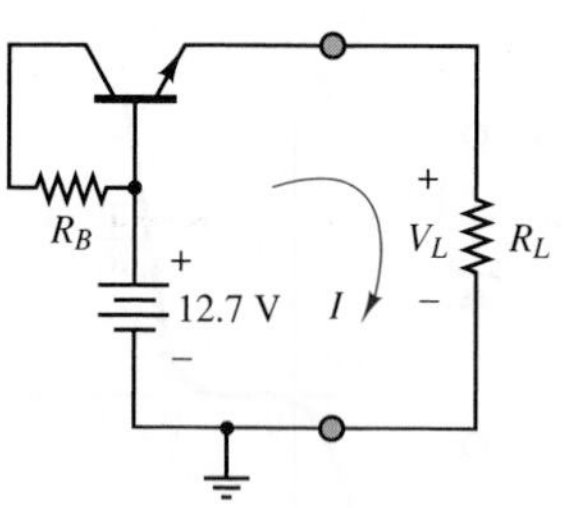

Figure 12.8

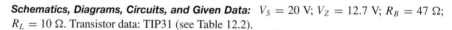

$$I = \frac{V_Z - V_\gamma}{R_L} = \frac{12.7 - 1.3}{10} = 1.14 \text{ A}$$

We then note that I is also the BJT emitter current I_E.

Applying KVL to the base circuit, shown in Figure 12.9, we compute the current through the base resistor:

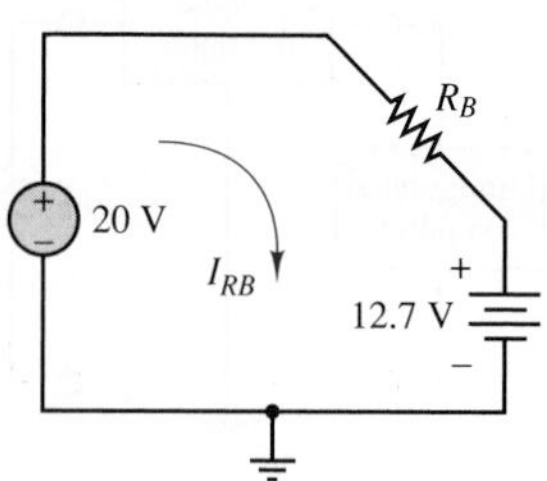

Figure 12.9

$$I_{RB} = \frac{V_S - V_Z}{R_B} = \frac{20 - 12.7}{47} = 0.155 \text{ A}$$

Knowing the base and emitter currents, we can determine whether the transistor in indeed in the active region, as assumed. With reference to Figure 12.10, we do so by computing V_{CB} and V_{BE} to determine the value of V_{CE}:

The base voltage is fixed by the presence of the Zener diode:

$$V_B = V_Z = 12.7 \text{ V}$$

$$V_{CB} = I_{RB}R_B = 0.155 \times 47 = 7.3 \text{ V}$$

$$V_E = I_L R_L = 1.14 \times 10 = 11.4 \text{ V}$$

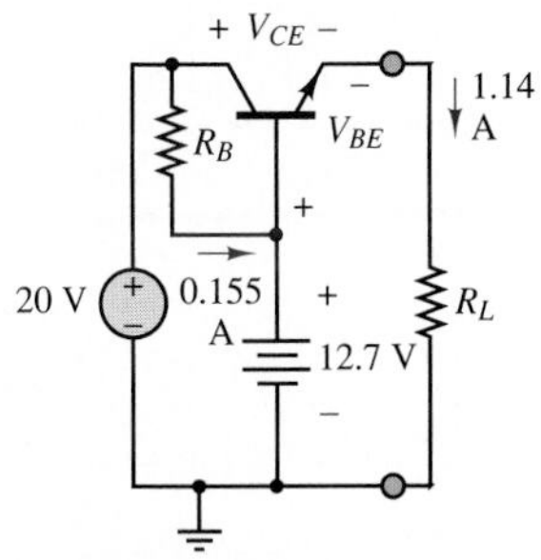

Figure 12.10

Thus,

$$V_{CE} = V_{CB} + V_{BE} = V_{CB} + (V_B - V_E) = 7.3 + (12.7 - 11.4) = 8.6 \text{ V}$$

This value of the collector-emitter voltage indicates that the BJT is in the active region. Thus, we can use the large-signal model and compute the base current and subsequently the Zener current:

$$I_B = \frac{I_E}{\beta + 1} = \frac{1.14}{10 + 1} = 103.6 \text{ mA}$$

Applying KVL at the base junction (see Figure 12.11), we find

$$I_{RB} - I_B - I_Z = 0$$

$$I_Z = I_{RB} - I_B = 0.155 - 0.1036 = 51.4 \text{ mA}$$

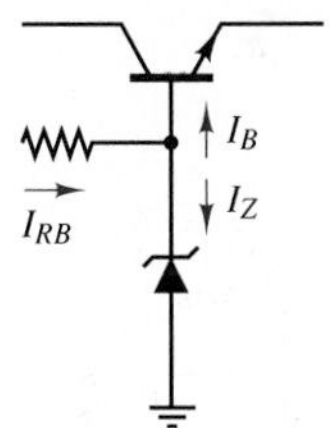

Figure 12.11

and the power dissipated by the Zener diode is

$$\boxed{P_Z = I_Z \times V_Z = 0.0514 \times 12.7 = 0.652 \text{ W}}$$

Comments: It will be instructive to compare these results with the Zener regulator examples of Chapter 9 (Examples 9.10, 9.11, and 9.12). Note that the Zener current is kept at a reasonably low level by the presence of the BJT (the load current is an amplified version of the base current).

CHECK YOUR UNDERSTANDING

Repeat Example 12.1 using the TIP31 transistor (see Table 12.2).

Three-terminal voltage regulators are available in packaged form to include all the necessary circuitry (often including protection against excess heat dissipation). Regulators are rated in terms of the regulated voltage and power dissipation. Some types provide a variable regulated voltage by means of an external adjustment. Because of their requirement for relatively large power (and therefore heat) dissipation, voltage regulators often need to be attached to a **heat sink,** a thermally conductive assembly that aids in the cooling process. Figure 12.12 depicts the appearance of typical heat sinks. Heat sinking is a common procedure with many power electronic devices.

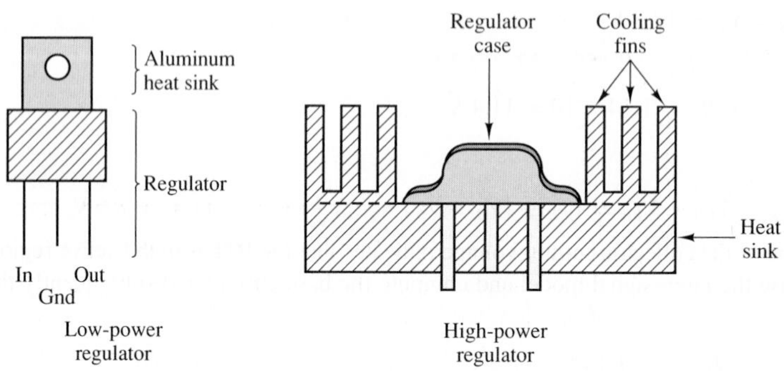

Figure 12.12 Heat sink construction for voltage regulators

12.4 POWER AMPLIFIERS AND TRANSISTOR SWITCHES

So far, we have primarily considered low-power electronic devices, either in the form of small-signal linear amplifiers or as switches and digital logic gates (the latter is discussed in greater detail in Chapter 13). There are many applications, however, in which it is desirable to provide a substantial amount of power to a load. Among the most common applications are loudspeakers (these can draw several amperes); electric motors and electromechanical actuators, which are considered in greater detail in Chapters 16 through 18, and DC power supplies, which have already been analyzed to some extent in Chapter 9. In addition to such applications, the use and control of electric power in industry require electronic devices that can carry currents as high as hundreds of amperes, and voltages up to thousands of volts. Examples are readily found in the control of large motors and heavy industrial machinery.

The aim of this section is to discuss some of the more relevant issues in the design of *power amplifiers,* such as distortion and heat dissipation, and to introduce power switching transistors.

Power Amplifiers

The brief discussion of power amplifiers in this section makes reference exclusively to the BJTs; this family of devices has traditionally dominated the field of **power amplifiers,** although in recent years semiconductor technology has made power MOSFETs competitive with the performance of bipolar devices.

You may recall the notion of breakdown from the introductory discussion of BJTs. In practice, a bipolar transistor is limited in its operation by three factors: the maximum collector current, the maximum collector-emitter voltage, and the maximum power dissipation, which is the product of I_C and V_{CE}. Figure 12.13 illustrates graphically the power limitation of a BJT by showing the regions where the maximum capabilities of the transistor are exceeded:

1. Exceeding the maximum allowable current $I_{C\ max}$ on a continuous basis will result in melting the wires that bond the device to the package terminals.
2. Maximum power dissipation is the locus of points for which $V_{CE}I_C = P_{max}$ at a case temperature of 25°C. The average power dissipation should not exceed P_{max}.
3. The instantaneous value of v_{CE} should not exceed $V_{CE\ max}$; otherwise, avalanche breakdown of the collector-base junction may occur.

LO3

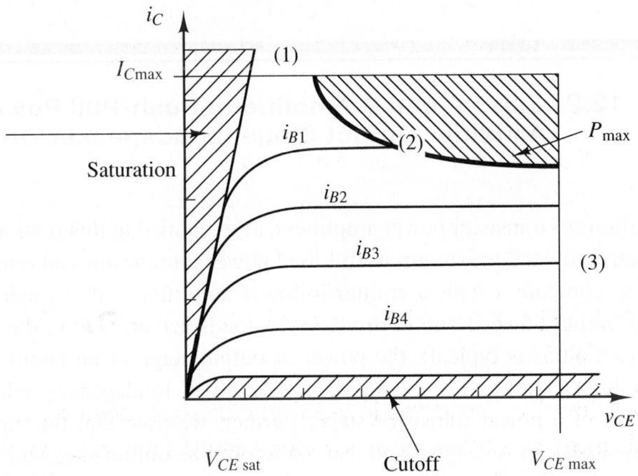

Figure 12.13 Limitations of a BJT amplifier

It is important to note that the linear operation of the transistor as an amplifier is also limited by the saturation and cutoff limits.

The operation of a BJT as a linear amplifier is rather severely limited by these factors. Consider, first, the effect of driving an amplifier beyond the limits of the linear active region, into saturation or cutoff. The result will be signal distortion. For example, a sinusoid amplified by a transistor amplifier that is forced into saturation, either by a large input or by an excessive gain, will be compressed around the peaks, because of the decreasing device gain in the extreme regions. Thus, to satisfy these limitations—and to fully take advantage of the relatively distortion-free linear active region of operation for a BJT—the Q point should be placed in the center of the device characteristic to obtain the **maximum symmetric swing.**

LO3

The maximum power dissipation of the device, of course, presents a more drastic limitation on the performance of the amplifier, in that the transistor can be irreparably damaged if its power rating is exceeded. Values of the maximum allowable collector current $I_{C\ max}$, of the maximum allowable transistor power dissipation P_{max}, and of other relevant power BJT parameters are given in Table 12.2 for a few typical devices. Because of their large geometry and high operating currents, power transistors have typical parameters quite different from those of small-signal transistors.

From Table 12.2, we can find some of these differences:

1. β is low. It can be as low as 5; the typical value is 20 to 80.
2. Typically $I_{C\ max}$ is in the ampere range; it can be as high as 100 A.
3. V_{CEO} is usually 40 to 100 V, but it can reach 500 V.

Table 12.2 Typical parameters for representative power BJTs

	MJE3055T	TIP31	MJE170
Type	*npn*	*npn*	*pnp*
Maximum I_C (continuous)	10 A	3 A	−3 A
V_{CEO}	60 V	40 V	−40 A
Power rating	75 W	40 W	12.5 W
β	20 @ $I_C = 4$ A	10 @ $I_C = 3$ A	30 @ $I_C = -0.5$ A
$V_{CE\ \text{sat}}$	1.1 V	1.3 V	−1.7 V
$V_{BE\ \text{on}}$	8 V	1.8 V	−2 V

LO3

EXAMPLE 12.2 Class A and B Amplifiers: Push-Pull Power Amplifier Output Stage (Loudspeaker Driver)

Problem

One of the limitations of transistor power amplifiers, as explained in this section, is their power dissipation, which limits the maximum useful load power a transistor can output. The aim of this example is to compare a typical emitter-follower amplifier with a **push-pull amplifier** when used as an output power stage to drive, say, a loudspeaker. Due to the low impedance of a loudspeaker (8 ohms is typical), the power or output stage of an audio amplifier needs to provide fairly high currents. For example, an 8-Ω 50-W loudspeaker will require 2.5 A, demanding the use of a power transistor stage. Further, we note that the input signal to be amplified is intrinsically an AC signal and that, therefore, the output stage current must swing above and below ground.

Solution

Known Quantities: Transistor parameters for TIP33C-34C complementary power BJT pair.

$$I_C(\text{A}) = 10$$
$$V_{CBO}(\text{V}) = 100$$
$$V_{CEO}(\text{V}) = 100$$
$$P_D(\text{W}) = 80$$
$$H_{FE}(\text{min/max}) = 20/100$$
$$I_C/V_{CE}(\text{A/V}) = 3.0/4.0$$
$$V_{CE(SAT)}(\text{V}) = 1.0$$
$$I_C/I_B(\text{A/mA}) = 3.0/300$$

Find: Analyze the two transistor amplifiers in Figures 12.14(a) and (b).

Schematics, Diagrams, Circuits, and Given Data: Figure 12.14.

Assumptions: None.

Analysis: Consider first the emitter follower of Figure 12.14(a). As explained earlier in the chapter, with reference to Figure 12.13, to ensure that the amplifier output can swing above and below ground, the biasing of the amplifier will require that a substantial (quiescent) DC current flow through the amplifier at all times, so that the Q point of the amplifier is in the middle of the characteristic curve, and the collector current can "swing" over the full range between the

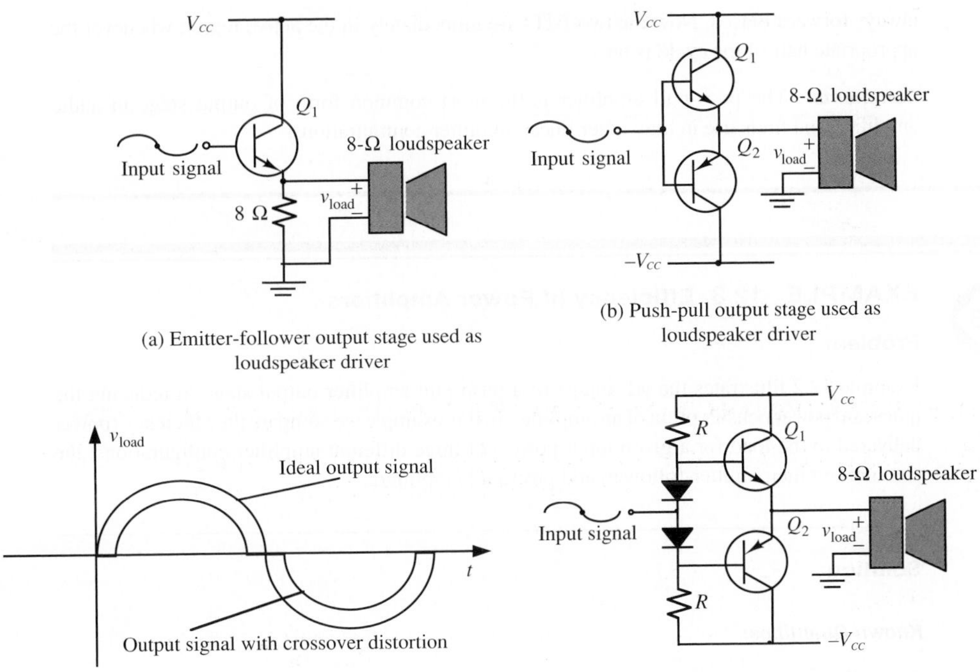

(a) Emitter-follower output stage used as loudspeaker driver

(b) Push-pull output stage used as loudspeaker driver

(c) Crossover distortion

(d) Push-pull output stage correcting for crossover distortion

Figure 12.14 Emitter-follower (class A) and push-pull (class B) amplifier

cutoff region on the low end and the maximum power dissipation (or saturation) limits shown in Figure 12.13.

This means that if we require a swing of 5 A (± 2.5A around the Q point), the amplifier will see a continuous quiescent current of 2.5 A, which will result in substantial continuous power dissipation. We can estimate this quiescent power dissipation to be the sum of the power dissipated by the 8-Ω resistor, approximately $R \times I_{CQ}^2 = 8 \times 2.5^2 = 50$ W plus the power dissipated by the transistor, $I_{CQ} \times V_{CEQ} = 2.5 \times V_{CEQ}$, which could be as much as 10 or 20 additional watts, depending on the collector-emitter bias voltage. So we are looking at a continuous power dissipation that is greater than the (50-W) power actually absorbed by the loudspeaker!

The circuit of Figure 12.14(b) uses two complementary (*npn-pnp*) transistors in a *push-pull* configuration to avoid this problem. Note that this circuit requires a symmetrical power supply (the single-ended collector supply of Figure 12.14(a) is no longer adequate). When the input signal (assume a sinusoidal signal for simplicity) is in the positive half of the cycle, the top (*npn*) transistor enters the active region (once the input voltage exceeds the junction voltage) and amplifies the base current, while the bottom transistor is in the cutoff region. During the input signal negative half cycle, the bottom (*pnp*) transistor acts as a linear amplifier, while the top transistor is in cutoff. The advantage of this configuration is that if the input signal is zero, both transistors are off. Thus, there is no quiescent power consumption. Note that no resistors are required to bias this amplifier configuration; thus, there will be no I^2R power consumption either. The primary disadvantage of the push-pull amplifier is illustrated in Figure 12.14(c). The figure clearly shows that the output of the amplifier is zero until one of the transistors has entered the active region. This leads to the **crossover distortion** shown in the figure. The crossover distortion can be eliminated with the circuit of Figure 12.14(d), in which two diodes are included such that their forward bias voltages cause the two base emitter junctions to be always forward biased. Now the two BJTs are immediately in the active region whenever the appropriate half of the cycle is on.

Comments: The push-pull amplifier is the most common form of output stage in audio amplifier, and finds use in may other linear amplifier configurations.

LO3

EXAMPLE 12.3 Efficiency of Power Amplifiers

Problem

Example 12.2 illustrates the advantage of a push-pull amplifier output stage in reducing the quiescent power consumption of an amplifier. In this example we compare the efficiency (power delivered to the load for a given input power) of three different amplifier configurations: the common-emitter, emitter-follower, and push-pull amplifiers.

Solution

Known Quantities:

Find: The power dissipation $P_L = V_L I_L$ of each of the transistor amplifiers in Figures 12.15(a), (b), and (c). Calculate the efficiency η of each amplifier based on the following definition (ratio of load to input rms power):

$$\eta = \frac{\tilde{P}_L}{\tilde{P}_{CC}}$$

Schematics, Diagrams, Circuits, and Given Data: Figure 12.15.

Assumptions: In all amplifiers we assume a sinusoidal input, and we also assume that the amplifier is biased to permit maximum symmetrical load voltage swing, from $-V_{CC}/2$ to $+V_{CC}/2$ for the common-emitter and emitter-follower amplifiers, and $-V_{CC}$ to V_{CC} for the push-pull configuration. This last assumption is not very realistic, but it simplifies the calculations a great deal.

Analysis:

1. *Common-emitter amplifier.* Since we have assumed that the load voltage swing is from $-V_{CC}/2$ to $+V_{CC}/2$, we can write the following expression for the load voltage:

$$V_L = \frac{V_{CC}}{2} + \frac{V_{CC}}{2} \sin \omega t$$

and the current will be

$$I_L = \frac{V_{CC}}{2R_L} + \frac{V_{CC}}{2R_L} \sin \omega t$$

Now, since the input power is $P_{CC} = V_{CC} I_C$ and $I_C = I_L$, we can write

$$P_{CC} = V_{CC}\left(\frac{V_{CC}}{2R_L} + \frac{V_{CC}}{2R_L} \sin \omega t\right) = \frac{V_{CC}^2}{2R_L}(1 + \sin \omega t)$$

Conversely, the load power is

$$P_L = V_L I_L = \left(\frac{V_{CC}}{2} + \frac{V_{CC}}{2} \sin \omega t\right)\left(\frac{V_{CC}}{2R_L} + \frac{V_{CC}}{2R_L} \sin \omega t\right) = \frac{V_{CC}^2}{4R_L}(1 + \sin \omega t)^2$$

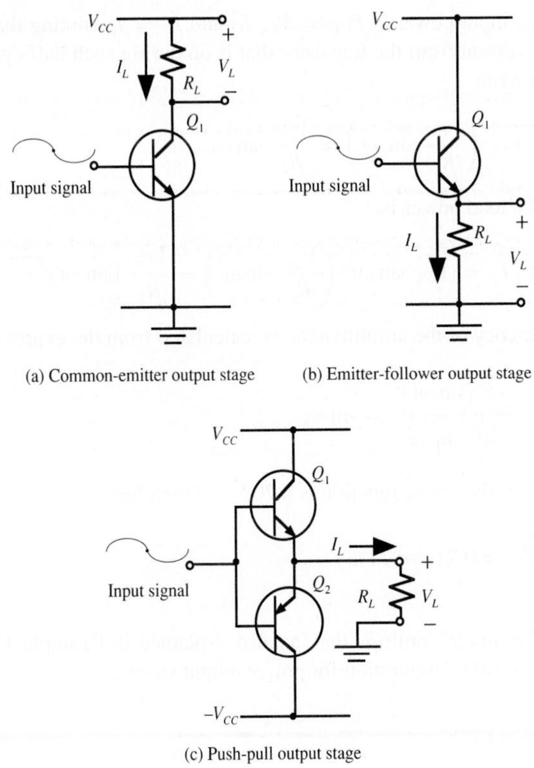

Figure 12.15 Three amplifier output stages

Thus, the efficiency of the amplifier can be calculated from the expression:

$$\frac{P_L}{P_{CC}} = \frac{\frac{V_{CC}^2}{4R_L}(1+\sin\omega t)^2}{\frac{V_{CC}^2}{2R_L}(1+\sin\omega t)} = \frac{1}{2}(1+\sin\omega t)$$

The rms value of the function $(1 + \sin\omega t)$ can be computed to be equal to $\sqrt{1.5} = 1.2247$, and therefore

$$\eta = \frac{\tilde{P}_L}{\tilde{P}_{CC}} \frac{1.2247}{2} \approx 0.61, \text{ or } 61\%.$$

2. *Emitter-follower amplifier.* Since we have assumed that the amplifier is biased to produce a symmetrical swing, there is very little difference between common-emitter and emitter-follower configurations. In the emitter-follower, the load current is the emitter current (rather than the collector current). A similar derivation can be followed, as shown in part (a), to arrive at the same result.
3. *Push-pull amplifier.* Here, we assume that the load voltage swing is from $-V_{CC}$ to $+V_{CC}$, for maximum amplification. Then, we can write the following expression for the load voltage:

$$V_L = V_{CC}\sin\omega t$$

and the current will be

$$I_L = \frac{V_{CC}}{R_L}\sin\omega t$$

Now, since the input power is $P_{CC} = V_{CC} I_C$ and $I_C = I_L$ (noting that the load current is the emitter current from the transistor that is on during each half cycle—see Example 12.2), we can write

$$P_{CC} = V_{CC}\left(\frac{V_{CC}}{R_L}\sin\omega t\right) = \frac{V_{CC}^2}{R_L}\sin\omega t$$

Conversely, the load power is

$$P_L = V_L I_L = (V_{CC}\sin\omega t)\left(\frac{V_{CC}}{R_L}\sin\omega t\right) = \frac{V_{CC}^2}{R_L}(\sin\omega t)^2$$

Thus, the efficiency of the amplifier can be calculated from the expression:

$$\frac{P_L}{P_{CC}} = \frac{\frac{V_{CC}^2}{R_L}(\sin\omega t)^2}{\frac{V_{CC}^2}{R_L}\sin\omega t} = \sin\omega t$$

The rms value of the above function is 0.707, and therefore

$$\eta = \frac{\tilde{P}_L}{\tilde{P}_{CC}} \approx 0.71, \text{ or } 71\%.$$

Comments: This example confirms that, as also explained in Example 12.2, the push-pull amplifier is a very useful configuration for power output stages.

BJT Switching Characteristics

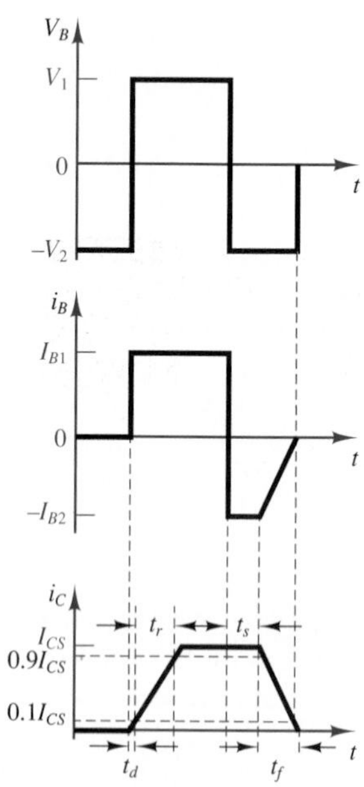

Figure 12.16 BJT switching waveforms

In addition to their application in power amplifiers, power BJTs can serve as controlled power switches, taking advantage of the switching characteristic described in Chapter 10 (see Figure 10.30). In addition to the properties already discussed, it is important to understand the phenomena that limit the switching speed of bipolar devices. The parasitic capacitances C_{CB} and C_{BE} that exist at the *CB* and *BE* junctions have the effect of imposing a charging time constant; since the transistor is also characterized by an internal resistance, you see that it is impossible for the transistor to switch from the cutoff to the saturation region instantaneously, because the inherent *RC* circuits physically present inside the transistor must first be charged. Figure 12.16 illustrates the behavior of the base and collector currents in response to a step change in base voltage. If a step voltage up to amplitude V_1 is applied to the base of the transistor and a base current begins to flow, the collector current will not begin to flow until after a delay, because the base capacitance needs to charge up before the *BE* junction voltage reaches V_γ; this **delay time** t_d is an important parameter. After the *BE* junction finally becomes forward-biased, the collector current will rise to the final value in a finite time, called the **rise time** t_r. An analogous process (though the physics are different) takes place when the base voltage is reversed to drive the BJT into cutoff. Now the excess charge that had been accumulated in the base must be discharged before the *BE* junction can be reverse-biased. This discharge takes place over a **storage time** t_s. To accelerate this process, the base voltage is usually driven to negative values ($-V_2$), so that the negative base current can accelerate the discharge of the charge stored in the base. Finally, the reverse-biased *BE* junction capacitance must now be charged to the negative base voltage value before the switching transient is complete; this process takes place during the **fall time** t_f. In the figure, I_{CS} represents the collector saturation current. Thus, the turn-on time of the BJT is given by

$$t_{\text{on}} = t_d + t_r \qquad \textbf{(12.3)}$$

and the turnoff time by

$$t_{\text{off}} = t_s + t_f \qquad \textbf{(12.4)}$$

Power MOSFETs

MOSFETs can also be used as power switches, as BJTs can. The preferred mode of operation of a power MOSFET when operated as a switch is in the ohmic region, where substantial drain current can flow for relatively low drain voltages. Thus, a MOSFET switch is driven from cutoff to the ohmic state by the gate voltage. In an enhancement MOSFET, positive gate voltages are required to turn the transistor on; in depletion MOSFETs, either positive or negative voltages can be used.

To understand the switching behavior of MOSFETs, recall once again the parasitic capacitances that exist between pairs of terminals: C_{GS}, C_{GD}, and C_{DS}. As a consequence of these capacitances, the transistor experiences a **turn-on delay** $t_{d(\text{on})}$ corresponding to the time required to charge the equivalent input capacitance to the threshold voltage V_T. As shown in Figure 12.17, the rise time t_r is defined as the time it takes to charge the gate from the threshold voltage to the gate voltage required to have the MOSFET in the ohmic state V_{GSP}. The **turnoff delay time** $t_{d(\text{off})}$ is the time required for the input capacitance to discharge, so that the gate voltage can drop and v_{DS} can begin to rise. As v_{GS} continues to decrease, we define the **fall time** t_f as the time required for v_{GS} to drop below the threshold voltage and turn the transistor off.

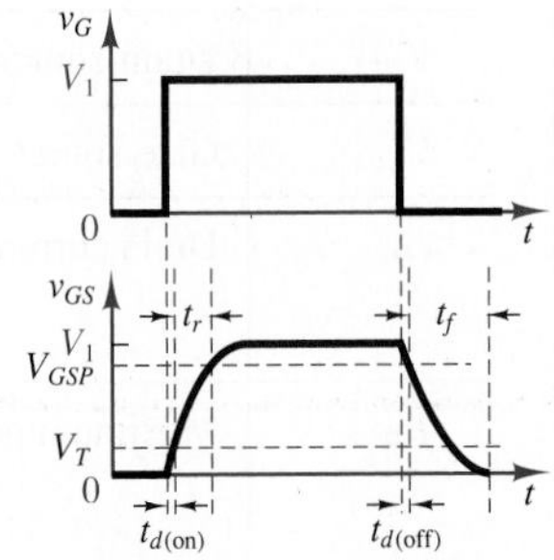

Figure 12.17 MOSFET switching waveforms

FOCUS ON METHODOLOGY

USING DEVICE DATA SHEETS

One of the most important design tools available to engineers is the **device data sheet.** In this box we illustrate the use of a device data sheet for the NDS8410 power MOSFET transistor. Excerpts from the data sheet are shown below, with some words of explanation.

FIND IT ON THE WEB

NDS8410
Single N-Channel Enhancement Mode Field-Effect Transistor

General Description

These *N*-channel enhancement mode power field-effect transistors are produced using Fairchild's proprietary, high-cell-density, DMOS technology. This very high-density process is especially tailored to minimize on-state resistance and provide superior switching performance. These devices are particularly suited for low-voltage applications such as notebook computer power management and other battery-powered circuits where fast switching, low in-line power loss, and resistance to transients are needed.

(Continued)

(*Concluded*)

Features

- 10 A, 30 V. $R_{DS(\text{ON})} = 0.015\ \Omega$ @ $V_{GS} = 10$ V.
  $R_{DS(\text{ON})} = 0.020\ \Omega$ @ $V_{GS} = 4.5$ V.
- High-density cell design for extremely low $R_{DS(\text{ON})}$.
- High power and current handling capability in a widely used surface-mount package.

ABSOLUTE MAXIMUM RATINGS

This table summarizes the limitations of the device. For example, one can find the maximum allowable gate-source and drain-source voltages and the **power rating.**

Absolute Maximum Ratings $T_A = 25°\text{C}$ unless otherwise noted

Symbol	Parameter	NDS8410	Units
V_{DSS}	Drain-source voltage	30	V
V_{GSS}	Gate-source voltage	20	V
I_D	Drain current—continuous	±10	A
	—pulsed	±50	
P_D	Maximum power dissipation	2.5	W
		1.2	
		1	
T_J, T_{STG}	Operating and storage temperature range	−55 to 150	°C

ELECTRICAL CHARACTERISTICS

The table summarizing electrical characteristics is divided into various sections, including "on" characteristics, "off" characteristics, dynamic characteristics, and switching characteristics. We focus on the last of these, and make reference to Figure 12.17. Note how all of the relevant parameters shown in this figure are listed in the data sheet.

Electrical Characteristics $T_A = 25°\text{C}$ unless otherwise noted

Symbol	Parameter	Conditions	Min.	Typ.	Max.	Units
Off Characteristics						
BV_{DSS}	Drain-source breakdown voltage	$V_{GS} = 0$ V, $I_D = 250\ \mu$A	30			V
I_{DSS}	Zero gate voltage drain current	$V_{DS} = 24$ V, $V_{GS} = 0$ V			1	μA
		$T_j = 55°$C			10	μA
I_{GSSF}	Gate-body leakage, forward	$V_{GS} = 20$ V, $V_{DS} = 0$ V			100	nA
I_{GSSR}	Gate-body leakage, reverse	$V_{GS} = -20$ V, $V_{DS} = 0$ V			−100	nA

(*Continued*)

On Characteristics						
$V_{GS(ON)}$	Gate-threshold voltage	$V_{DS} = V_{GS}$, $I_D = 250\ \mu A$	1	1.5		V
$R_{DS(ON)}$	Static drain-source on-resistance	$V_{GS} = 10$ V, $I_D = 10$ A		0.013	0.015	Ω
		$V_{GS} = 4.5$ V, $I_D = 9$ A		0.018	0.02	
$I_{D(ON)}$	On-state drain current	$V_{GS} = 10$ V, $V_{DS} = 5$ V	20			A
g_{FS}	Forward transconductance	$V_{DS} = 10$ V, $I_D = 10$ A		22		S
Electrical Characteristics $T_A = 25°C$ unless otherwise noted						
Symbol	**Parameter**	**Conditions**	**Min.**	**Typ.**	**Max.**	**Units**
Dynamic Characteristics						
C_{ISS}	Input capacitance	$V_{DS} = 15$ V, $V_{GS} = 0$ V,		1,350		pF
C_{DSS}	Output capacitance	$f = 1.0$ MHz		800		pF
C_{ISS}	Reverse transfer capacitance			300		pF
Switching Characteristics						
$t_{D(on)}$	Turn-on delay time	$V_{DD} = 10$ V, $I_D = 1$ A,		14	30	ns
t_T	Turn-on rise time	$V_{GEN} = 10$ V, $R_{GEN} = 6\ \Omega$		20	25	ns
$t_{D(off)}$	Turnoff delay time			56	100	ns
t_F	Turnoff fall time			31	80	ns
Q_g	Total gate charge	$V_{DS} = 15$ V,		46	60	nC
Q_{gs}	Gate-source charge	$I_D = 10$ A, $V_{GS} = 10$ V		5.6		nC
Q_{gd}	Gate-drain charge			14		nC

Insulated-Gate Bipolar Transistors

The insulated-gate bipolar transistor, or IGBT, is a hybrid device, combining features of both field-effect and bipolar devices. The circuit symbol of the IGBT is shown in Figure 12.1; a simplified equivalent circuit is shown in Figure 12.18. The IGBT is a voltage-controlled device, like a MOSFET, but its performance is closer to that of a BJT. The switching and conduction losses of the IGBT are lower than those of a MOSFET, and the switching speed is greater than that of a BJT (but somewhat lower than that of a MOSFET); the convenience of a MOSFET-like gate drive is an advantage over BJTs.

IGBTs can be rated up to 400 A and 1,200 V, and they can have switching frequencies as high as 20 kHz. These devices in recent years have found increasing application in medium-power applications, such as AC and DC motor drives.

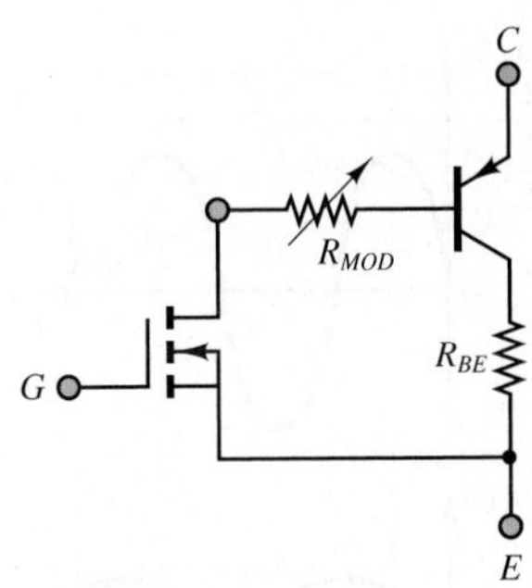

Figure 12.18 IGBT simplified equivalent circuit

12.5 RECTIFIERS AND CONTROLLED RECTIFIERS (AC-DC CONVERTERS)

As explained in Chapter 9, one of the most immediate applications of the semiconductor diode is rectification of AC voltages and currents, to convert AC waveforms to DC. Rectification can be achieved both with conventional diodes and with controlled diodes, such as thyristors. A simple diode rectifier can provide only a fixed DC voltage level; however, variable DC supplies can be easily obtained with the aid of thyristors. The aim of this section is to illustrate the basic features of diode rectifiers and to introduce thyristor-based controlled rectifiers.

The basic diode half-wave rectifier and also full-wave and bridge rectifiers were discussed in Section 9.3. In addition to the considerations noted in Chapter 9, one often has to take into account the nature of the load seen by such DC supplies.

In practice, loads are not always resistive, as will be seen in Chapters 18 through 20, where circuit models for electromechanical actuators and electric motors are introduced. A very common occurrence consists of a DC voltage supply providing current to a *DC motor*. For the purpose of the present discussion, it will suffice to state that a DC motor presents an inductive impedance to the voltage supply and requires a constant current from the supply to operate at a constant speed. The circuit of Figure 12.19 illustrates, as an example, a simple half-wave rectifier connected to an *RL* load.

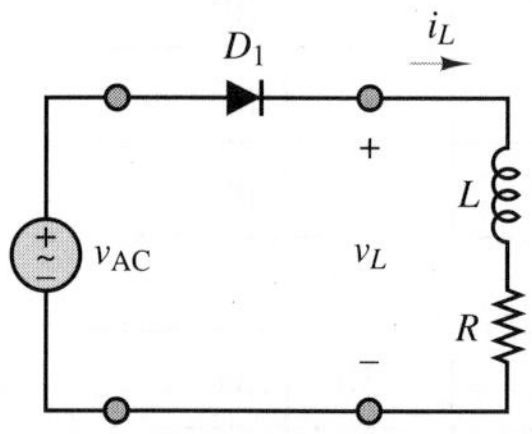

Simple half-wave rectifier

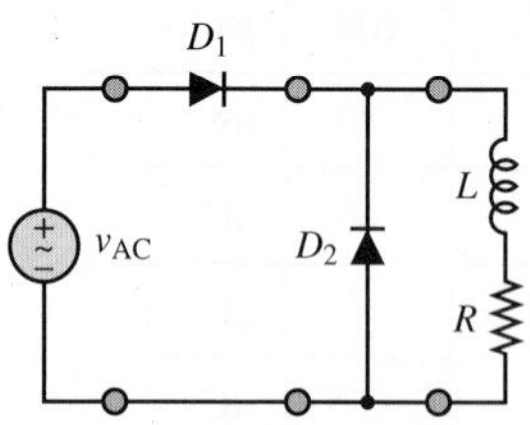

Same arrangement with freewheeling diode

Figure 12.19 Rectifier connected to an inductive load

The circuit on top in Figure 12.19, assuming an ideal diode, would present a serious problem during the negative half-cycle of the source voltage, since the requirement for continuity of current in the inductor (recall the discussion on continuity of inductor currents and capacitor voltages in Chapter 5) would be violated with D_1 off. Whenever the current flow through the inductor is interrupted (during the negative half-cycles of v_{AC}), the inductor attempts to build a **flyback voltage** proportional to di_L/dt. Since the rectifier does not provide any current during the negative half-cycle of the source voltage, the instantaneous inductor voltage could be very large and could lead to serious damage to either the motor or the rectifier.

The circuit shown on the bottom in Figure 12.19 contains a **freewheeling diode** D_2. This circuit is also called a **snubber.** The role of D_2 is to provide continuity of current when D_1 is in the off state. Diode D_2 is off during the positive half-cycle but turns on when D_1 ceases to conduct, because of the flyback voltage $L\, di_L/dt$. Rather than build up a large voltage, the inductor now has a path for current to flow, through D_2, when D_1 is off. Thus, the energy stored by the inductor during the positive half-cycle of v_{AC} is utilized to preserve a continuous current through the inductor during the off period. Figure 12.20 depicts the load current for the circuit including the diode. Note that D_2 allows the energy storage properties of the inductor to be utilized to smooth the pulselike supply current and to produce a nearly constant load current.

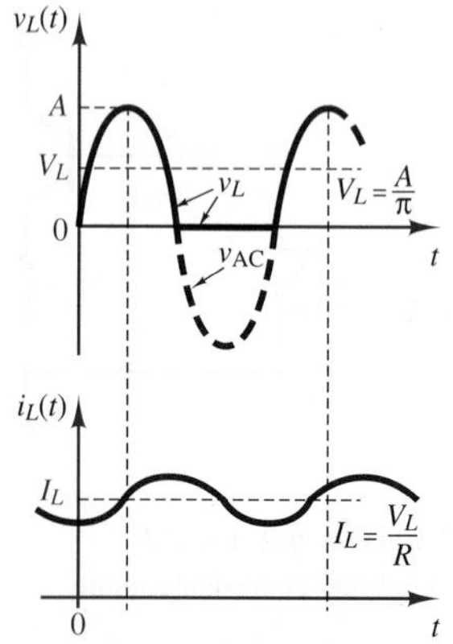

Figure 12.20 Operation of a freewheeling diode

Analyzing the circuit on the bottom of Figure 12.19, with

$$v_{AC}(t) = A \sin \omega t \tag{12.5}$$

(and assuming that both D_1 and D_2 are ideal), we conclude that the DC component of the load voltage V_L must appear across the load resistor R (no steady-state DC voltage can appear across the inductor, since $v_L = L\, di_L/dt$). Thus, the approximate DC flowing through the load is given by

$$I_L = \frac{A}{\pi R} \tag{12.6}$$

since the average output voltage of a half-wave rectifier is A/π V for an AC source of peak amplitude A (see Chapter 9). The AC component of the load current (or "ripple" current) is not as simple to compute, since it is due to the AC component of v_L, which is not a pure sinusoid. The exact analysis would require the use of a Fourier series expansion. For the purposes of this discussion, it is not unreasonable to assume that most of the energy is at a frequency equal to that of the AC source:

$$i_L(t) \approx I_L + I_{AC}\cos(\omega t + \theta) \qquad \textbf{(12.7)}$$

where I_L is the average load current, I_{AC} is the peak value of the ripple current, and θ is its phase. An acceptable approximation from which the amplitude of I_{AC} may be computed is

$$v_L(t) \approx \frac{A}{2\pi} + \frac{A}{2\pi}\sin\omega t \qquad \textbf{(12.8)}$$

Figure 12.21 graphically illustrates the extent of this approximation.

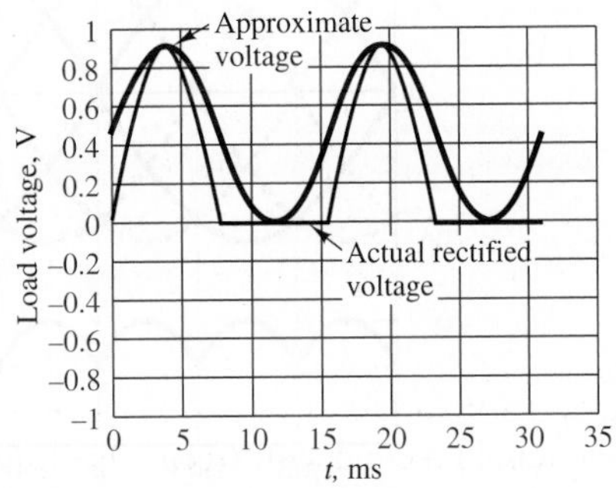

Figure 12.21 Approximation of ripple voltage for a half-wave rectifier

A common alternative to the half-wave rectifier is the full-wave rectifier, which was discussed in Chapter 9.

CHECK YOUR UNDERSTANDING

Using the approximation given in equation 12.8, find the DC and AC loads for the circuit of Figure 12.19 if $R = 10\ \Omega$, $L = 0.3$ H, $A = 170$ V, and $\omega = 377$ rad/s.

Answer: $I_L = 5.4$ A, $I_{AC} = 0.75$ A, $\alpha = 84.95°$

Three-Phase Rectifiers

It is important to realize that the same type of circuit that can be used for single-phase rectifiers can also be employed to design multiphase rectifiers. Recall the analysis of three-phase AC power systems in Section 7.4. In many high-power applications, three-phase voltages need to be rectified to give rise to a single DC supply; such rectification can be achieved by means of an extension of the bridge rectifier. Consider the balanced three-phase circuit shown in Figure 12.22. The three-phase wye-connected source is connected to a resistive load by means of a three-phase transformer, with a delta-connected primary and a wye-connected secondary. The circuit could also operate without the transformer. The three secondary currents i_a, i_b, and i_c flow through pairs of diodes D_1 to D_6 in a manner very similar to the single-phase bridge rectifier described in Figure 9.42. The diodes will conduct in pairs depending on the relative line voltages, according to the following sequence: D_1-D_2, D_2-D_3, D_3-D_4, D_4-D_5, D_5-D_6, and D_6-D_1. Recall from the analysis of Section 7.4, equation 7.54, that the line-to-line voltage is $\sqrt{3}$ times the phase voltage in a three-phase wye-connected source. The instantaneous source voltages and the related diode conduction periods, as well as the load voltage, are shown in Figure 12.23.

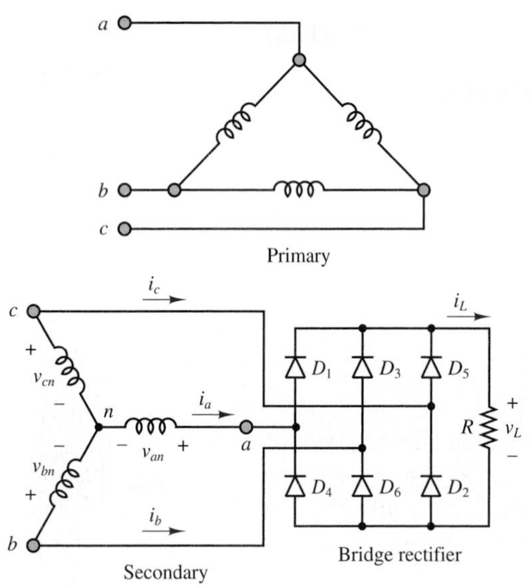

Figure 12.22 Three-phase diode bridge rectifier

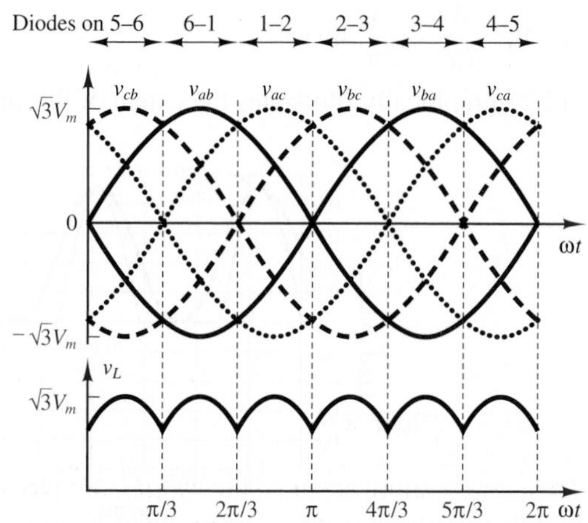

Figure 12.23 Waveforms and conduction times of three-phase bridge rectifier

It can be shown that the average output voltage is given by

$$V_L = \frac{2}{2\pi/6}\int_0^{\pi/6} \sqrt{3}V_m \cos\omega t \, d(\omega t) = \frac{3\sqrt{3}}{\pi} V_m = 1.654 V_m \tag{12.9}$$

where V_m is the peak phase voltage. The rms output voltage can be calculated to be

$$V_{rms} = \sqrt{\frac{2}{2\pi/6}\int_0^{\pi/6} \sqrt{3}V_m^2 \cos^2\omega t \, d(\omega t)} = \frac{3}{2} + \frac{9\sqrt{3}}{4\pi} \tag{12.10}$$

$$V_{rms} = 1.6554 V_m$$

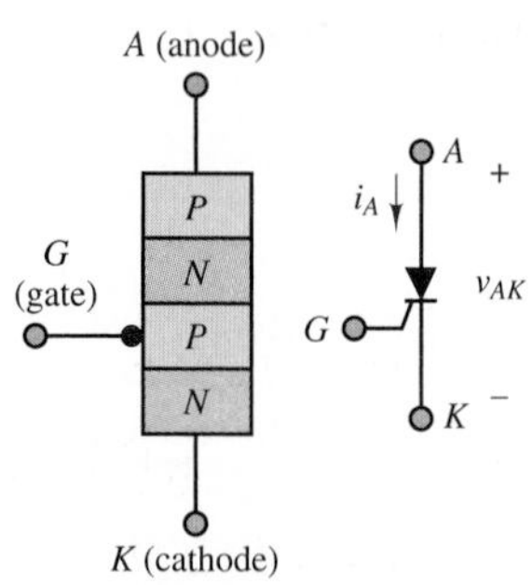

Figure 12.24 Thyristor structure and circuit symbol

Thyristors and Controlled Rectifiers

In a number of applications, it is useful to be able to externally control the amount of current flowing from an AC source to the load. A family of power semiconductor devices called **controlled rectifiers** allows for control of the rectifier state by means of a third input, called the **gate.** Figure 12.24 depicts the appearance of a **thyristor,** or **silicon controlled rectifier (SCR),** illustrating how the physical structure of this device consists of four layers, alternating p-type and n-type material. Note that the circuit symbol for the thyristor suggests that this device acts as a diode, with provision for an additional external control signal.

The operation of the thyristor can be explained in an intuitive fashion as follows. When the voltage v_{AK} is negative (i.e., providing reverse bias), the thyristor acts just as a conventional *pn* junction in the off state. When v_{AK} is forward-biased *and* a small amount of current is injected into the gate, the thyristor conducts forward current. The thyristor then continues to conduct (even in the absence of gate current), provided that v_{AK} remains positive. Figure 12.25 depicts the i–v curve for the thyristor. Note that the thyristor has two stable states, determined by the bias v_{AK} and by the gate current. In summary, the thyristor acts as a diode with a control gate that determines the time when conduction begins.

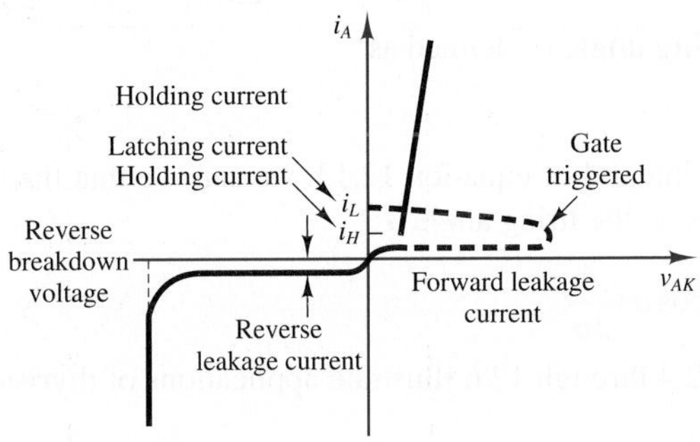

Figure 12.25 Thyristor i–v characteristic

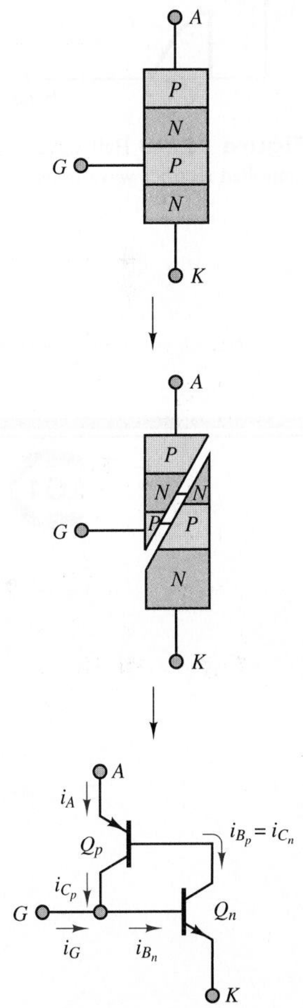

Figure 12.26 Thyristor two-transistor model

A somewhat more accurate description of thyristor operation may be provided if we realize that the four-layer *pnpn* device can be modeled as a *pnp* transistor connected to an *npn* transistor. Figure 12.26 clearly shows that, physically, this is a realistic representation. Note that the anode current i_A is equal to the emitter current of the *pnp* transistor (labeled Q_p) and the base current of Q_p is equal to the collector current of the *npn* transistor Q_n. Likewise, the base current of Q_n is the sum of the gate current and the collector current of Q_p. The behavior of this transistor model is explained as follows. Suppose, initially, i_G and i_{B_n} are both zero. Then it follows that Q_n is in cutoff, and therefore $i_{C_n} = 0$. But if $i_{C_n} = 0$, then the base current going into Q_p is also zero and Q_p is also in cutoff, and $i_{C_p} = 0$, consistent with our initial assumption. Thus, this is a stable state, in the sense that unless an external condition perturbs the thyristor, it will remain off.

Now, suppose a small pulse of current is injected at the gate. Then $i_{B_n} > 0$, and Q_n starts to conduct, provided, of course, that $v_{AK} > 0$. At this point i_{C_n}, and therefore i_{B_p}, must be greater than zero, so that Q_p conducts. It is important to note that once the gate current has turned Q_n on, Q_p also conducts, so that $i_{C_P} > 0$. Thus, even though i_G may cease, once this "on" state is reached, $i_{C_p} = i_{B_n}$ continues to drive Q_n so that the on state is also self-sustaining. The only condition that will cause the thyristor to revert to the off state is the condition in which v_{AK} becomes negative; in this case, both transistors return to the cutoff state.

In a typical controlled rectifier application, the device is used as a half-wave rectifier that conducts only after a trigger pulse is applied to the gate. Without concerning ourselves with how the trigger pulse is generated, we can analyze the general waveforms for the circuit of Figure 12.27 as follows. Let the voltage v_{trigger} be applied to the gate of the thyristor at $t = \tau$. The voltage v_{trigger} can be a short pulse, provided by a suitable trigger-timing circuit (Chapter 15 will discuss timing and switching circuits). At $t = \tau$, the thyristor begins to conduct, and it continues to do so until the AC source enters its negative cycle. Figure 12.28 depicts the relevant waveforms.

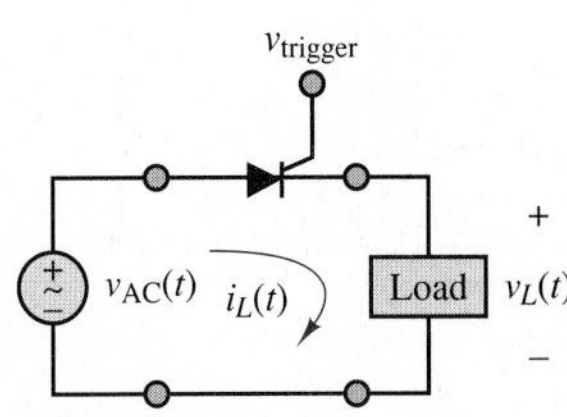

Figure 12.27 Controlled rectifier circuit

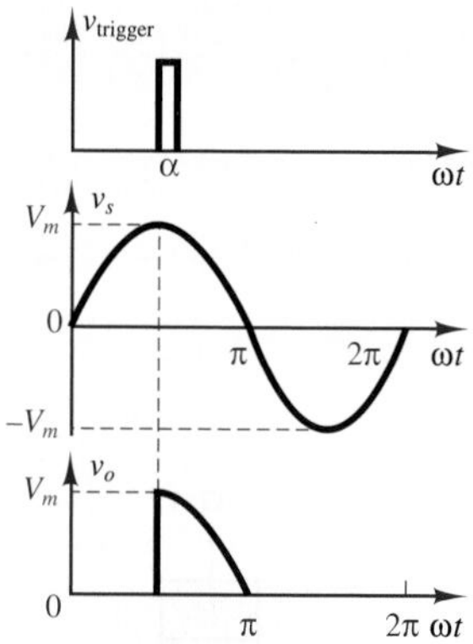

Figure 12.28 Half-wave controlled rectifier waveforms

Note how the DC load voltage is controlled by the firing time τ, according to

$$\langle v_L \rangle = V_L = \frac{1}{T}\int_{\tau}^{T/2} v_{AC}(t)\,dt \tag{12.11}$$

where T is the period of $v_{AC}(t)$. Now, if we let

$$v_{AC}(t) = A\sin\omega t \tag{12.12}$$

we can express the average (DC) value of the load voltage

$$V_L = \frac{1}{T}\int_{\tau}^{T/2} A\sin\omega t\,dt = (1+\cos\omega t)\frac{A}{2\pi} \tag{12.13}$$

in terms of the **firing angle** α, defined as

$$\alpha = \omega\tau \tag{12.14}$$

By evaluating the integral of equation 12.13, we can see that the (DC) load voltage amplitude depends on the firing angle α:

$$V_L = (1+\cos\alpha)\frac{A}{2\pi} \tag{12.15}$$

Examples 12.4 through 12.6 illustrate applications of thyristor circuits.

LO4

EXAMPLE 12.4 Thyristor-Based Variable Voltage Supply

Problem

Analyze the thyristor-based variable voltage supply shown in Figure 12.29. Determine (1) the rms load voltage as a function of the firing angle and (2) the power supplied to the resistive load at zero firing angle and at firing angles equal to $\pi/2$ and π.

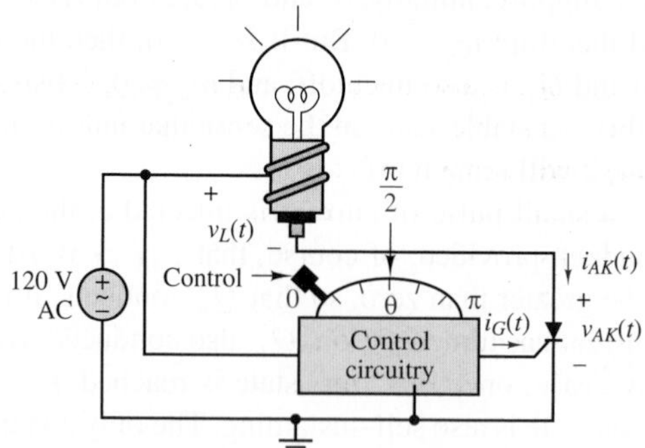

Figure 12.29

Solution

Known Quantities: Load resistance.

Find: $\tilde{V}_L$, $P_L|_{\alpha=0}$, $P_L|_{\alpha=\pi/2}$, $P_L|_{\alpha=\pi}$.

Schematics, Diagrams, Circuits, and Given Data: $V_{AK\,(on)} = 0$ V; $R_L = 240\ \Omega$. The pulsed gate current $i_G(t)$ is timed as shown in Figure 12.30.

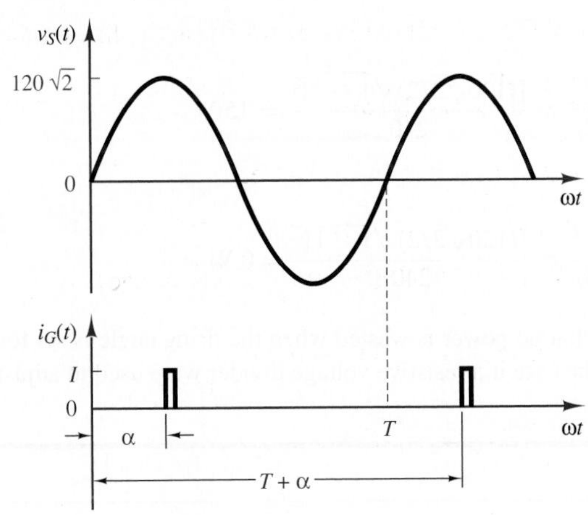

Figure 12.30

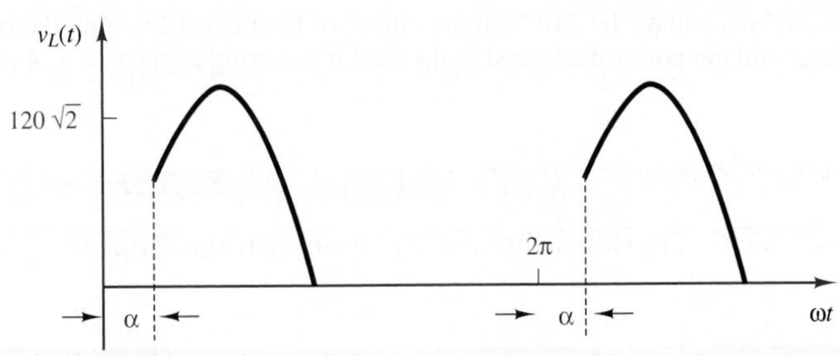

Figure 12.31

Assumptions: The thyristor acts as an ideal diode when on ($V_{AK} > 0$).

Analysis:

1. *Load voltage calculation.* As explained in the preceding section, the load voltage will have the appearance shown in Figure 12.31. The rms value of the load voltage as a function of the firing angle α is therefore computed as follows:

$$\tilde{V}_L(\alpha) = \sqrt{\frac{(120\sqrt{2})^2}{2\pi}\int_\alpha^\pi \sin^2 \omega t'\, d(\omega t')}$$

$$= \frac{120\sqrt{2}}{2}\sqrt{\frac{1}{\pi}\int_\alpha^\pi [1 - \cos(2\omega t')]\, d(\omega t')}$$

$$= \frac{120\sqrt{2}}{2}\sqrt{1 - \frac{\alpha}{\pi} + \frac{1}{2\pi}\sin(2\alpha)}$$

2. *Load power calculation.* We can now compute the load power for each of the three values of α:

$$P_L = \frac{\tilde{V}^2}{R_L}$$

For $\alpha = 0$:

$$P_L = \frac{\tilde{V}^2}{R_L} = \frac{(120\sqrt{2}/2)^2}{240} = 30 \text{ W}$$

for $\alpha = \pi/2$:

$$P_L = \frac{\tilde{V}^2}{R_L} = \frac{[(120\sqrt{2}/2)\sqrt{1-\frac{1}{2}}]^2}{240} = 15 \text{ W}$$

for $\alpha = \pi$:

$$P_L = \frac{\tilde{V}^2}{R_L} = \frac{[(120\sqrt{2}/2)\sqrt{1-1}]^2}{240} = 0 \text{ W}$$

Comments: Note that no power is wasted when the firing angle is set for zero load voltage. This would not be the case if a resistive voltage divider were used to adjust the load voltage.

CHECK YOUR UNDERSTANDING

Calculate the rms load voltage in the circuit of Figure 12.28 for $A = 100$ V and $\alpha = \pi/3$ rad.
Let the input AC rms voltage be 240 V in the circuit of Example 12.6. Find the rms value of the load voltage and the power dissipated in the load if the firing angle $\alpha = \pi/4$ rad.

Answers: $V_{L\,\text{rms}} = 23.87$ V; $V_{L\,\text{rms}} = 120$ V, $P_L = 60$ W

EXAMPLE 12.5 Automotive Battery Charger

Problem

FIND IT ON THE WEB

Qualitatively explain the operation of the automotive battery charger shown in Figure 12.32.

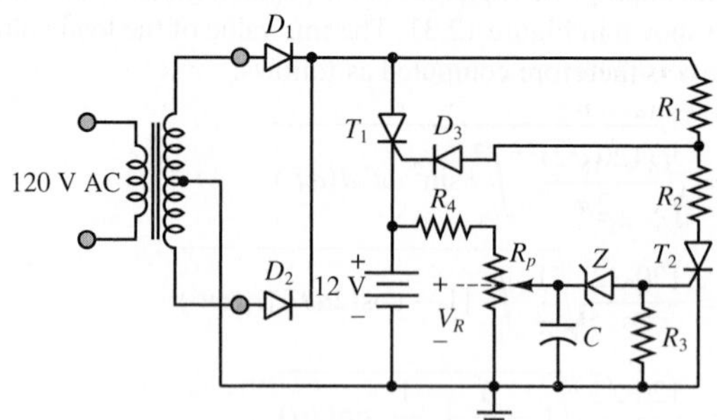

Figure 12.32 Automotive battery charger

Solution

Analysis: The charging circuit is connected to a standard 110-V single-phase supply. Diodes D_1 and D_2 form a full-wave rectifier (see Figure 9.39); resistors R_1 and R_2 and thyristor T_2 form a variable voltage divider.

Assume that thyristor T_2 is not in the conducting state and that the anode voltage of D_3 is such that D_3 conducts. Then T_1 will be fired near the beginning of the positive half-cycle of the AC source voltage, and its period of conduction will be long, providing a substantial current to the battery (resistors R_4 and R_p are sufficiently large that most of the current flowing through T_1 will go to the battery).

The potentiometer R_p is set so that when the battery voltage is low, the voltage V_R is not sufficient to turn on the Zener diode Z. Thus, Z is effectively an open circuit, and T_2 remains off (recall that we had initially assumed T_2 to be off—this confirms the correctness of the assumption). As the battery charges to a progressively higher value, Z will eventually conduct; when Z conducts, a gate current is injected into T_2, which is then turned on.

When T_2 conducts, the voltage across the R_2-T_2 series connection becomes significantly lower, because T_2 is now nearly a short circuit. Resistors R_1 and R_2 are selected so that when T_2 conducts, D_3 becomes reverse-biased. Once this condition occurs, T_1 is turned off and charging stops. You see that the circuit has built-in overcharging protection.

The reader will find simpler battery charging circuits in Examples 10.5 and 11.6.

EXAMPLE 12.6 Thyristor Circuit

Problem

Determine the value of R in the circuit of Figure 12.33 such that the average load current through the thyristor is 1 A.

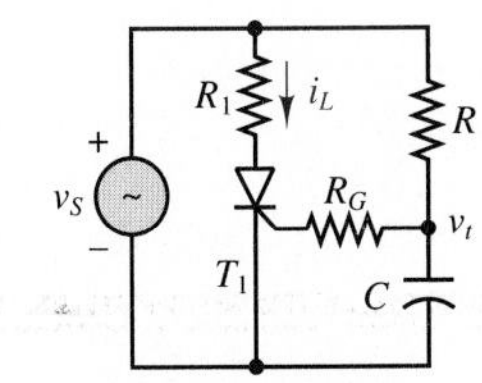

Figure 12.33

Solution

Known Quantities: Resistances and source voltage.

Find: Resistor R such that $\langle i_L \rangle = 1$ A.

Schematics, Diagrams, Circuits, and Given Data: $v_S = 200$ V rms, 250 Hz; $V_{AK\ (\text{on})} = 0$ V; $R_1 = 75\ \Omega$; $R_G = 1\ \text{k}\Omega$; $C = 1\ \mu$F.

Assumptions: The thyristor acts as an ideal diode when on ($V_{AK} > 0$).

Analysis: Figure 12.34 depicts the relative timing of the source voltage $v_S(t)$, thyristor current $i_L(t)$, and triggering voltage $v_t(t)$. The expression for the source voltage is

$$v_S(t) = \sqrt{2} \times 200 \sin(2\pi \times 250t)$$

The load current through the 74-Ω resistor is

$$i_L(t) = \begin{cases} \dfrac{\sqrt{2} \times 200}{75} \sin(2\pi \times 250t) & \alpha \le \omega t \le \pi \\ 0 & \pi \le \omega t \le 2\pi \end{cases}$$

and the triggering voltage is

$$v_t(t) = V_t \sin(2\pi \times 250t - \alpha)$$

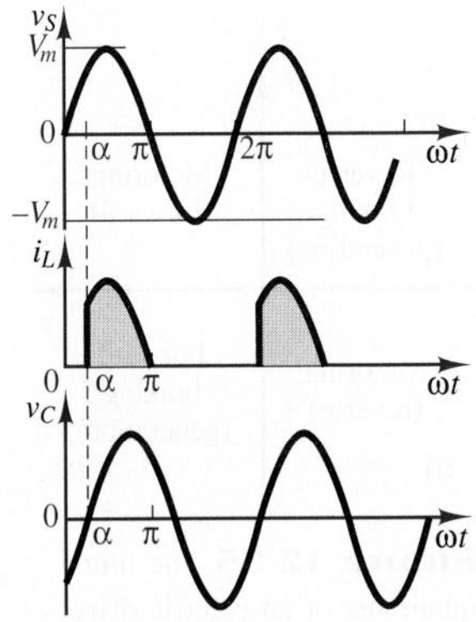

Figure 12.34

The triggering voltage will go positive at the desired firing angle α, thus injecting a current into the gate of the thyristor, turning it on. Thus, the requirement that the average load current be equal to 1 A is equivalent to requiring that

$$\langle i_L(t) \rangle = \frac{1}{\pi} \int_{\alpha}^{\pi} \frac{\sqrt{2} \times 200}{75} \sin(2\pi\omega t')\, d(\omega t') = 1 \text{ A}$$

Performing the integration, we determine that the requirement is

$$\frac{\sqrt{2} \times 200}{2\pi \times 75}(1 + \cos\alpha) = 1$$

Solving for α, we find $\alpha = 48.23°$.

Now, to determine the value of R, we observe that the AC source voltage appears across the RC circuit; thus, v_t can be computed from an impedance voltage divider by using phasor methods:

$$\mathbf{V}_t(j\omega) = \frac{1/j\omega C}{R + 1/j\omega C} V_S(j\omega)$$

$$= \frac{V_S}{\sqrt{1 + \omega^2 R^2 C^2}} \angle[-\arctan(\omega RC)]$$

We then observe that the phase of $\mathbf{V}_t(j\omega)$ is the firing angle α, and we can therefore determine α by setting

$$-\arctan(\omega RC) = \alpha = 48.23°$$

$$R = \frac{\tan(\alpha)}{\omega C} = \frac{\tan(48.23)}{2\pi \times 250 \times 10^{-6}} = 713\ \Omega$$

CHECK YOUR UNDERSTANDING

Compute the rms value of the load current in the circuit of Figure 12.33.

Answer: $\frac{V_m}{2}\sqrt{\frac{1}{2\pi}(2\pi - 2\alpha + \sin 2\alpha)}$

12.6 ELECTRIC MOTOR DRIVES

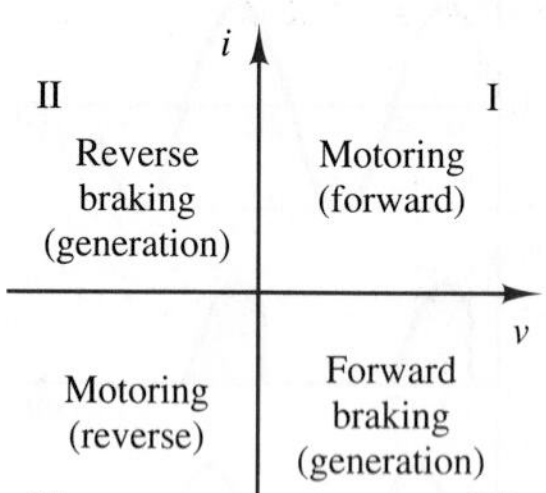

Figure 12.35 The four quadrants of an electric drive

The advent of high-power semiconductor devices has made it possible to design effective and relatively low-cost electronic supplies that take full advantage of the capabilities of the devices introduced in this chapter. Electronic power supplies for **DC and AC motors** have become one of the major fields of application of power electronic devices. The last section of this chapter is devoted to an introduction to two families of power supplies, or **electric drives: choppers,** or **DC-DC converters;** and **inverters,** or **DC-AC converters.** These circuits find widespread use in the control of AC and DC motors in a variety of applications and power ranges.

Before we delve into the discussion of the electronic supplies, it will be helpful to introduce the concept of quadrants of operation of a drive. Depending on the direction of current flow, and on the polarity of the voltage, an electronic drive can operate in one of four possible modes, as indicated in Figure 12.35.

DC-DC Converters

As the name suggests, a DC-DC converter is capable of converting a fixed DC supply to a variable DC supply. This feature is particularly useful in the control of the speed of a DC motor (described in greater detail in Chapter 19). In a DC motor, shown schematically in Figure 12.36, the developed torque T_m is proportional to the current supplied to the motor **armature** I_a, while the **electromotive force (emf)** E_a, which is the voltage developed across the armature, is proportional to the speed of rotation of the motor ω_m. A DC motor is an electromechanical energy conversion system; that is, it converts electrical to mechanical energy (or vice versa if it is used as a generator). If we recall that the product of torque and speed is equal to power in the mechanical domain, and that current times voltage is equal to power in the electrical domain, we conclude that in the ideal case of complete energy conversion, we have

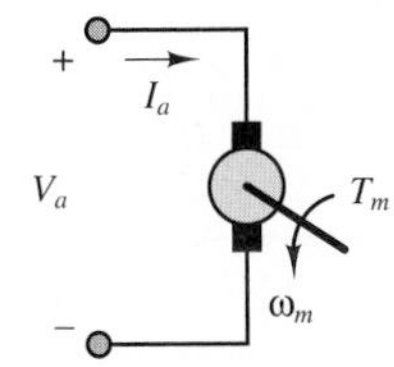

Figure 12.36 DC motor

$$E_a \times I_a = T_m \times \omega_m \tag{12.16}$$

Naturally, such ideal energy conversion cannot take place; however, we can see that there is a correspondence between the four electrical quadrants of Figure 12.35 and the mechanical power output of the motor: namely, if the voltage and current are both positive or both negative, the electrical power will be positive, and so will the mechanical power. This corresponds to the **forward** (i and v both positive) and **reverse** (i and v both negative) **motoring** operation. Forward motoring corresponds to quadrant I and reverse motoring to quadrant III in Figure 12.35. If the voltage and current are of opposite polarity (quadrants II and IV), electrical energy is flowing back to the electric drive; in mechanical terms this corresponds to a braking condition. Operation in the fourth quadrant can lead to **regenerative braking,** so called because power is regenerated by making current flow back to the source. This mode could be useful, for example, to recharge a battery supply, because the braking energy can be regenerated by returning it to the electric supply.

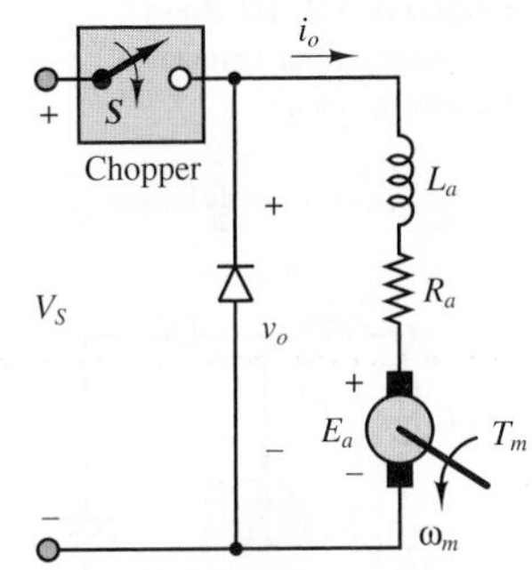

Figure 12.37 Buck converter (step-down chopper)

A simple circuit that can accomplish the task of providing a variable DC supply from a fixed DC source is the **buck converter (step-down chopper),** shown in Figure 12.37. The circuit consists of a switch, denoted by the symbol S, and a snubber diode, such as the one described in Section 12.5. The switch can be any of the power switches described in this chapter, for example, a power BJT or MOSFET, or a thyristor; see, for example, the BJT switch of Figure 12.4. The circuit to the right of the diode is a model of a DC motor, including the inductance and resistance of the armature windings, and the effect of the back-emf E_a. When the switch is turned on (say, at $t = 0$), the supply V_S is connected to the load and $v_o = V_S$. The load current i_o is determined by the motor parameters. When the switch is turned off, the load current continues to flow through the snubber diode, but the output voltage is now $v_o = 0$. At time T, the switch is turned on again, and the cycle repeats.

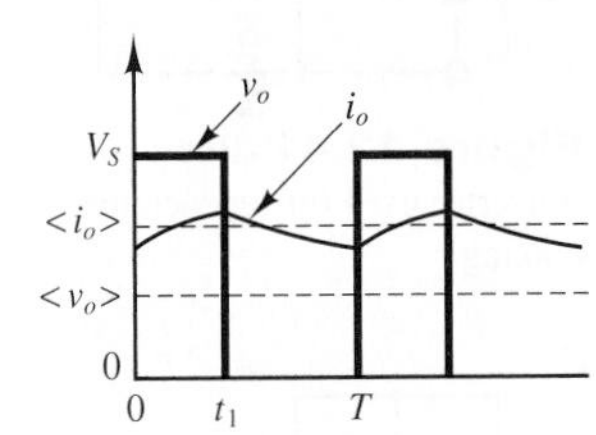

Figure 12.38 Step-down chopper waveforms

Figure 12.38 depicts the v_o and i_o waveforms. The average value of the output voltage $\langle v_o \rangle$ is given by

$$\langle v_o \rangle = \frac{t_1}{T} V_S = \delta V_S \tag{12.17}$$

where δ is the **duty cycle** of the chopper. The step-down chopper has a useful range

$$0 \leq \langle v_o \rangle \leq V_S \tag{12.18}$$

It is also possible to increase the range of a DC-DC converter to above the supply voltage by making use of the energy storage properties of an inductor; the resulting circuit is shown in Figure 12.39. When the chopper switch S is on, the supply current flows through the inductor and the closed switch, storing energy in the inductor; the output voltage v_o is zero, since the switch is a short circuit. When the switch is open, the supply current will flow through the load via the diode; but the inductor voltage is negative during the transient following the opening of the switch and therefore

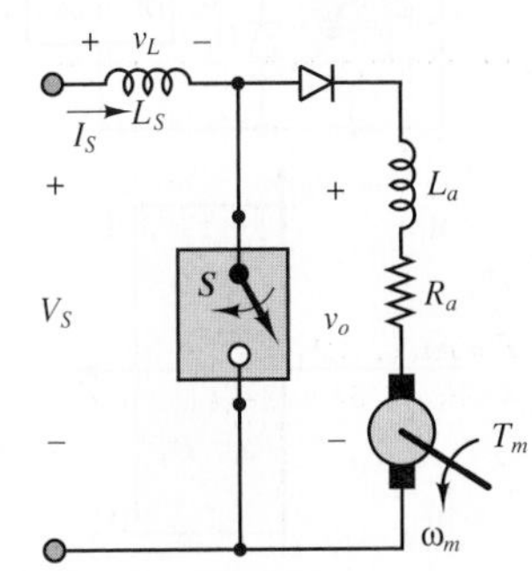

Figure 12.39 Boost converter (step-up chopper)

adds to the source voltage: the energy stored in the inductor while the switch was closed is now released and transferred to the load. This stored energy makes it possible for the output voltage to be higher than the supply voltage for a finite period of time.

To maintain a constant average load current, the current increase between 0 and t_1 must equal the current decrease from t_1 to T. Therefore,

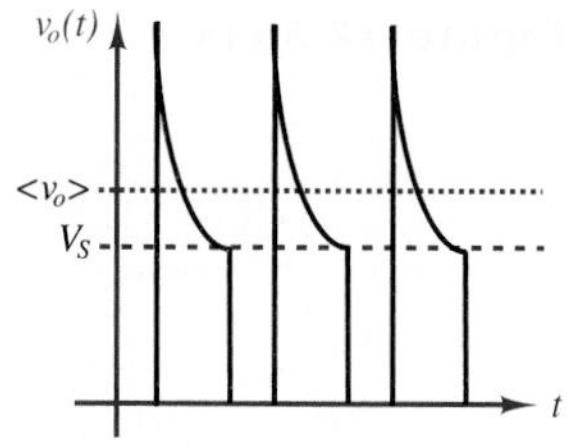

Figure 12.40 Boost converter output voltage waveform (ideal)

$$\frac{1}{L}\int_0^{t_1} V_S\,dt = \frac{1}{L}\int_{t_1}^{T} (\langle v_o\rangle - V_S)\,dt \tag{12.19}$$

from which we can calculate

$$V_S t_1 = (\langle v_o\rangle - V_S)(T - t_1) \tag{12.20}$$

This results in an average output voltage given by

$$\langle v_o\rangle = \frac{T}{T - t_1}V_S = \frac{1}{1 - t_1/T}V_S = \frac{1}{1-\delta}V_s \geq V_S \tag{12.21}$$

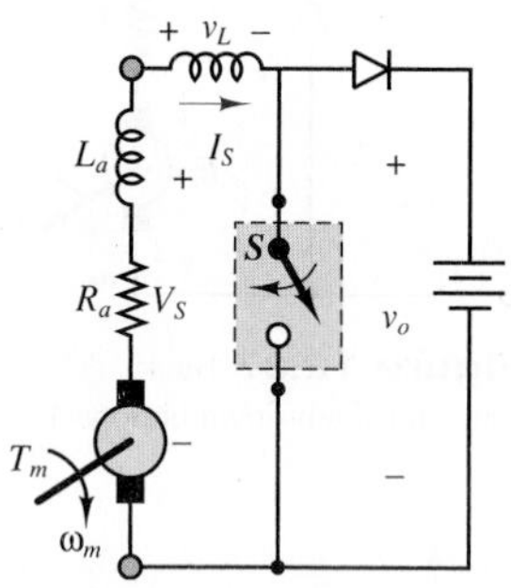

Figure 12.41 (Boost converter used for regenerative braking

Since the duty cycle δ is always less than 1, the theoretical range of the supply is

$$V_S \leq \langle v_o\rangle < \infty \tag{12.22}$$

The waveforms for the boost converter are shown in Figure 12.40.

A **boost converter (step-up chopper)** can also be used to provide regenerative braking: if the "supply" voltage is the motor armature voltage and the output voltage is the fixed DC supply (battery) voltage, then power can be made to flow from the motor to the DC supply (i.e., recharging the battery). This configuration is shown in Figure 12.41.

Finally, the operation of the step-down and step-up choppers can be combined into a **two-quadrant chopper,** shown in Figure 12.42. The circuit shown schematically in Figure 12.42 can provide both regenerative braking and motoring operation in a DC motor. When switch S_2 is open and switch S_1 serves as a chopper, the circuit operates as a step-down chopper, precisely as was described earlier in this section (convince yourself of this by redrawing the circuit with S_2 and D_2 replaced by open circuits). Thus, the drive and motor operate in the first quadrant (motoring operation). The output voltage v_o will switch between V_S and zero, as shown in Figure 12.38, and the load current will flow in the direction indicated by the arrow in Figure 12.42; diode D_1 freewheels whenever S_1 is open. Since both output voltage and current are positive, the system operates in the first quadrant.

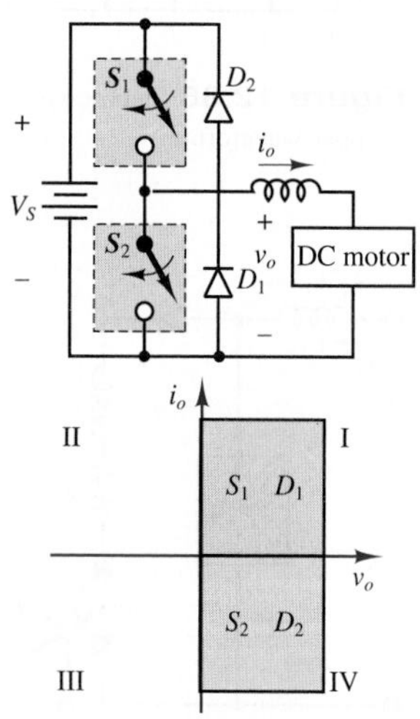

Figure 12.42 Two-quadrant DC-DC converter

When switch S_1 is open and switch S_2 serves as a chopper, the circuit resembles a step-up chopper. The source is the motor emf E_a, and the load is the battery; this is the situation depicted in Figure 12.41. The current will now be negative, since the sum of the motor emf and the voltage across the inductor (corresponding to the energy stored during the on cycle of S_2) is greater than the battery voltage. Thus, the drive operates in the fourth quadrant.

Example 12.7 illustrates the operation of a DC-DC converter as a DC motor power supply.

EXAMPLE 12.7 Two-Quadrant Chopper

LO5

Problem

1. Determine the turn-on time of the chopper of Figure 12.42 in the motoring mode if $n = 500$ r/min and $i_o = 90$ A. Also determine the power absorbed by the motor armature winding, the power absorbed by the motor, and the power delivered by the source.
2. Determine the turn-on time of the chopper in the regenerative mode if $n = 380$ r/min and $i_o = -90$ A. Also determine the power absorbed by the motor armature winding, the power absorbed by the motor, and the power delivered by the source.

Solution

Known Quantities: Supply voltage; motor parameters; chopping frequency armature resistance and inductance.

Find: For each of the two cases: t_1, P_a, P_m, P_S.

Schematics, Diagrams, Circuits, and Given Data:

1. $V_S = 120$ V; $E_a = 0.1n$; $R_a = 0.2\ \Omega$; $1/T =$ chopping frequency $= 300$ Hz.
2. $V_S = 120$ V; $E_a = 0.1n$; $R_a = 0.2\ \Omega$; $L_S \to \infty$; $1/T =$ chopping frequency $= 300$ Hz.

Assumptions: The switches in the chopper of Figure 12.42 act as ideal switches. Assume that the motor inductance is sufficiently small to be neglected in the calculations (i.e., assume a short circuit).

Analysis:

1. *Analysis of motoring operation.* To analyze motoring operation of the chopper, we refer to Figure 12.37 and apply KVL to the motor side:

$$\langle v_o \rangle = R_a I_a + E_a = R_a \langle i_o \rangle + 0.1n = 0.2 \times 90 + 0.1 \times 500 = 68 \text{ V}$$

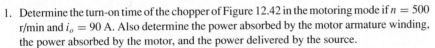

From equation 12.17 we can then compute the duty cycle of the chopper δ:

$$\delta = \frac{t_1}{T} = \frac{\langle v_o \rangle}{V_S} = \frac{68}{120} = 0.567$$

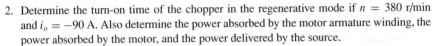

Since the chopping frequency is 300 Hz, we can compute t_1:

$$t_1 = \frac{T}{\delta} = \frac{1}{300 \times 0.567} = 1.89 \text{ ms}$$

The power absorbed by the armature is

$$P_a = R_a I_a^2 = R_a \langle i_o \rangle^2 = 0.2 \times 90^2 = 1.62 \text{ kW}$$

The power absorbed by the motor is

$$P_m = E_a I_a = 0.1n \times \langle i_o \rangle = 0.1 \times 500 \times 90 = 4.5 \text{ kW}$$

The power delivered by the voltage supply is

$$P_S = \delta V_S \langle i_o \rangle = 0.567 \times 120 \times 90 = 6.12 \text{ kW}$$

2. *Analysis of regenerative operation.* To analyze the regenerative operation of the chopper, we refer to Figure 12.39 and apply KVL to the motor side, noting that now the current is flowing in the reverse direction:

$$\langle v_o \rangle = R_a I_a + E_a = R_a \langle i_o \rangle + E_a = -90 \times 0.2 + 0.1 \times 380 = 20 \text{ V}$$

We now turn to equation 12.22 and observe that in this equation the motor acts as the source, and the supply voltage as the load; thus

$$V_S = \frac{1}{1 - t_1/T}\langle v_o \rangle \qquad \text{or} \qquad 120 = \frac{1}{1 - 300t_1} 20$$

We can then compute $t_1 = 2.8$ ms from the above equation. The duty cycle for the step-up chopper is now

$$\delta = \frac{t_1}{T} = \frac{5}{6} = 0.833$$

The power absorbed by the armature is

$$P_a = R_a I_a^2 = R_a \langle i_o \rangle^2 = 0.2 \times (-90)^2 = 1.62 \text{ kW}$$

The power absorbed by the motor will now be negative, since current is flowing in the reverse direction; the motor is in fact generating power:

$$P_m = E_a I_a = 0.1n \times \langle i_o \rangle = 0.1 \times 380 \times (-90) = -3.42 \text{ kW}$$

The power delivered by the voltage supply is

$$P_S = \delta V_S \langle i_o \rangle = 0.567 \times 120 \times (-90) = -1.8 \text{ kW}$$

This power is negative because the supply is actually absorbing power, not delivering it.

Comments:

1. Note that the sum of the motor and armature power losses is equal to the power delivered by the source; this is to be expected, since we have assumed ideal (lossless) switches. In a practical chopper, the chopping circuit would actually absorb power; heat dissipation is therefore an important issue in the design of choppers. This chopper operates in quadrant I (see Figure 12.42).
2. Note that in the regenerative case the equivalent duty cycle is greater than 1. Note also that now the power absorbed by the motor is a negative quantity; that is, the motor delivers power to the rest of the circuit. However, the power absorbed by the armature resistance is still a positive quantity because the armature resistance dissipates power regardless of the direction of the current flow through it. Here V_S might, for example, represent a battery pack in an electric vehicle, which would be recharged at the rate of 1.8 kW. The source of energy capable of producing this power is the inertial energy stored in the vehicle: when the vehicle decelerates, this mechanical energy causes the electric motor to act as a generator (see Chapter 19), producing the 90-A current in the reverse direction. This chopper operates in quadrant IV (see Figure 12.42).

Inverters (DC-AC Converters)

As explained in Chapter 19, variable-speed drives for AC motors require a multiphase variable-frequency, variable-voltage supply. Such drives are called *DC-AC converters,* or *inverters*. Inverter circuits can be quite complex, so the objective of this section is to present a brief introduction to the subject, with the aim of illustrating the basic principles. A **voltage source inverter (VSI)** converts the output of a fixed DC supply (e.g., a battery) to a variable-frequency AC supply. Figure 12.43 depicts a **half-bridge VSI;** once again, the switches can be either bipolar or MOS transistors, or thyristors. The operation of this circuit is as follows. When switch S_1 is turned on, the output voltage is in the positive half-cycle, and $v_o = V_S/2$. To generate the negative half-cycle, switch S_2 is turned on, and $v_o = -V_S/2$. The switching sequence of S_1 and S_2 is shown in Figure 12.44. It is important that each switch be turned off before the other is turned on; otherwise, the DC supply will be short-circuited. Since the load is always going to be inductive in the case of a motor drive, it is important to observe that the load current i_o will lag the voltage waveform, as shown in Figure 12.44. As shown in this figure, there will be some portions of the cycle in which the voltage is positive but the current is negative. The function of diodes D_1 and D_2 is precisely to conduct the load current whenever it is of direction opposite to the polarity of the voltage. Without these diodes, there would be no load current in this case. Figure 12.44 also shows which element is conducting in each portion of the cycle.

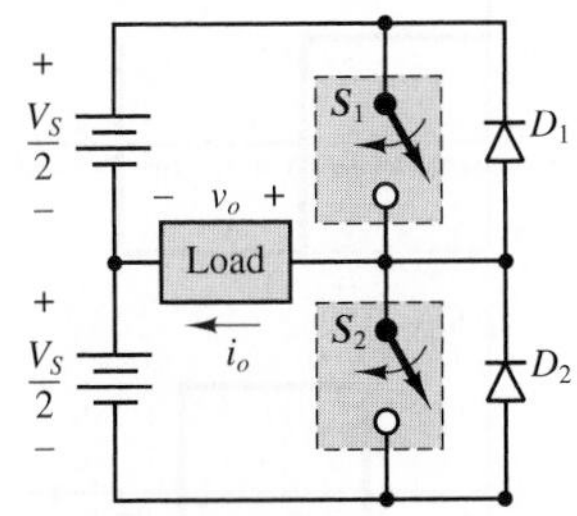

Figure 12.43 Half-bridge voltage source inverter

A full-bridge version of the VSI can also be designed as shown in Figure 12.45; the associated output voltage waveform is shown in Figure 12.46. The operation of this circuit is analogous to that of the half-bridge VSI; switches S_1 and S_2 are fired during the first half-cycle, and switches S_3 and S_4 during the second half. Note that the full-bridge configuration allows the output voltage to swing from V_S to $-V_S$. The diodes provide a path for the load current whenever the load voltage and current are of opposite polarity.

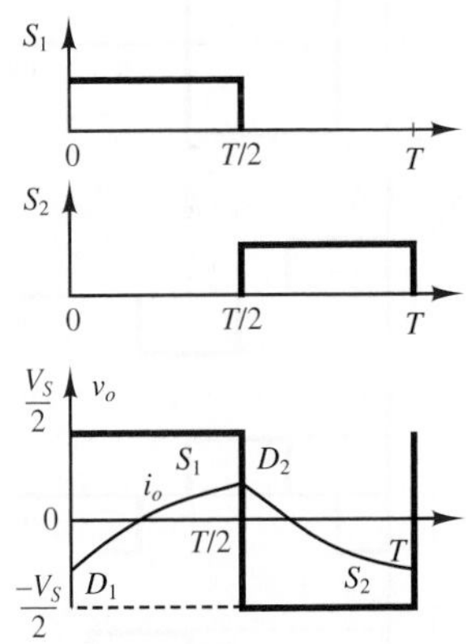

Figure 12.44 Half-bridge voltage source inverter waveforms

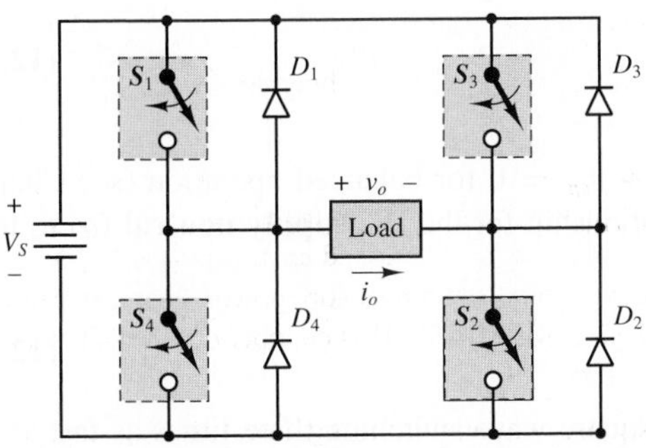

Figure 12.45 Full-bridge voltage source inverter

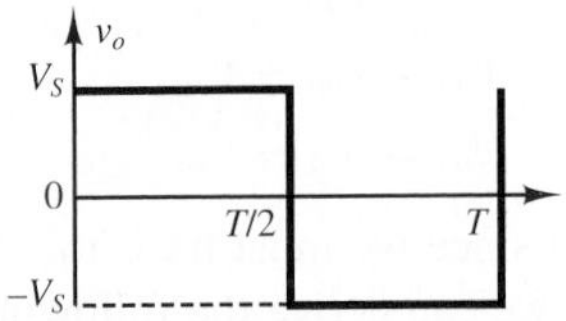

Figure 12.46 Half-bridge voltage source inverter output waveform

A three-phase version of the VSI is shown in Figure 12.47. Once again, the operation is analogous to that of the VSI circuits just presented. The related waveforms are shown in Figure 12.48. The top three waveforms depict the **pole voltages,** which

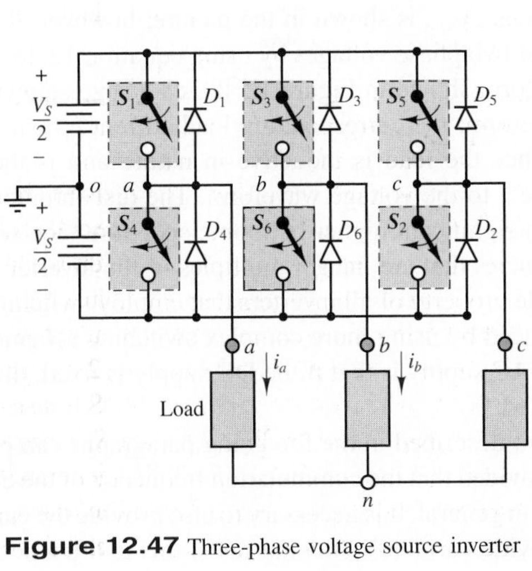

Figure 12.47 Three-phase voltage source inverter

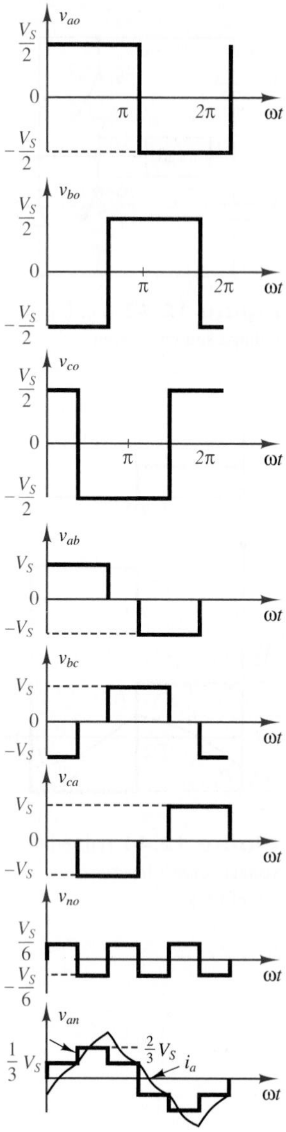

Figure 12.48 Three-phase voltage source inverter waveforms

are referenced to the DC supply neutral point o. The pole voltages are obtained by firing switches S_1 through S_6 at appropriate times. For example, if S_1 is fired at $\omega t = 0$, then pole a is connected to the positive side of the DC supply and $v_{ao} = V_S/2$; if S_4 is subsequently turned on at $\omega t = \pi$, then pole a is connected to the negative side of the DC supply and $v_{ao} = -V_S/2$. The other pairs of switches are then fired in an analogous sequence, shifted by 120 electrical degrees with respect to each other, to obtain the waveforms shown in the top three graphs of Figure 12.48. The **line voltages** are obtained from the pole voltages by using the relations

$$\begin{aligned} v_{ab} &= v_{ao} - v_{bo} \\ v_{bc} &= v_{co} - v_{co} \\ v_{ca} &= v_{co} - v_{ao} \end{aligned} \tag{12.23}$$

and are shown in the second set of three diagrams in Figure 12.48. These are also phase-shifted by 120°. Now, we can also express the pole voltages in terms of the **load phase voltages** v_{an}, v_{bn}, and v_{cn}:

$$\begin{aligned} v_{ao} &= v_{an} - v_{no} \\ v_{bo} &= v_{bn} - v_{no} \\ v_{co} &= v_{cn} - v_{no} \end{aligned} \tag{12.24}$$

and since we must have $v_{an} + v_{bn} + v_{cn} = 0$ for balanced operation (see Chapter 7), we can derive the following relationship for the DC **supply neutral** (o) to **load neutral** (n) voltage:

$$v_{no} = \frac{v_{ao} + v_{bo} + v_{co}}{3} \tag{12.25}$$

This voltage is also shown to be a square wave switching three times as fast as the inverter output voltage. Finally, to obtain the phase voltages, we make use of the relations

$$\begin{aligned} v_{an} &= v_{ao} - v_{no} = \tfrac{2}{3}v_{ao} - \tfrac{1}{3}(v_{bo} + v_{co}) \\ v_{bn} &= v_{bo} - v_{no} = \tfrac{2}{3}v_{bo} - \tfrac{1}{3}(v_{ao} + v_{co}) \\ v_{cn} &= v_{co} - v_{no} = \tfrac{2}{3}v_{bo} - \tfrac{1}{3}(v_{ao} + v_{bo}) \end{aligned} \tag{12.26}$$

Only one phase voltage, v_{an}, is shown in the picture; however, it is straightforward to construct the other two phase voltages by using equation 12.26. Note that the load phase voltage waveform shown in Figure 12.48 is a coarse stepwise approximation of a sinusoidal waveform; the corresponding load current i_a is a filtered version of the load voltage, since the load is inductive in nature and is therefore somewhat smoothed with respect to the voltage waveform. The discontinuous nature of these waveforms creates a significant higher harmonic spectrum (see section 6.2, Fourier Analysis), at frequencies that are integer multiples of the inverter output frequency; this is an unavoidable property of all inverters that employ switching circuits, but the problem can be reduced by using more complex switching schemes. Another major shortcoming of this AC supply is that if the DC supply is fixed, the amplitude of the inverter output is fixed.

The VSI circuit described in the foregoing paragraphs can provide a variable-frequency supply provided that the commutation frequency of the electronic switches can be varied. Thus, in general, it is necessary to also provide the capability for timing circuits that can provide variable switching rates; this is often accomplished with a microcontroller (discussed in Chapter 14).

The limitations of the VSI of Figure 12.47 can be overcome with the use of more advanced switching schemes, such as *pulse-width modulation* (*PWM*) and *sinusoidal PWM*. The complexity of these schemes is beyond the scope of this book, and the interested reader is invited to explore a more advanced power electronics text to learn about advanced inverter circuits. We simply mention that it is possible to significantly reduce the harmonic content of the inverter waveforms and to provide variable-frequency, variable-amplitude, three-phase supplies for AC motors by means of power switching circuits under microprocessor control. These advances are finding a growing field of application in the electric vehicle arena. This subject is approached again in Chapter 19.

Conclusion

This chapter introduces field-effect transistors and, by way of a set of universal curves, describes the operation of these devices as amplifiers. Switching models of the transistors are also introduced, and applications to analog and digital gates are presented. Upon completing this chapter, you should have mastered the following learning objectives:

1. *Learn the classification of power electronic devices and circuits.* Power electronic devices can handle up to a few thousand volts and up to several hundred amperes and are used in many industrial applications. The various families of power electronics devices and systems are introduced in this section.
2. *Analyze the operation of practical voltage regulators.* Voltage regulators are a basic element of DC power supplies and are used to provide a stable voltage output. The principal element of a voltage regulator is the Zener diode, used as a voltage reference.
3. *Understand the principal limitations of transistor power amplifiers.* Transistors, both BJT and MOSFET, as well as IGBTs are commonly used as power amplifiers and switches. The principal limitations in these applications are the allowable power dissipation and the switching time.
4. *Analyze the operation of single- and three-phase controlled rectifier circuits.* Rectifiers, like voltage regulators, are essential elements in DC power supplies. Controlled rectifiers can provide a variable DC voltage, and they are therefore very useful in the design of power converters for DC motors, and in other industrial applications that require variable DC voltage supplies.
5. *Understand the operation of power converters used in electric motor control, and perform simplified analysis on DC-DC converters.* Modern control of electric machines requires the use of power converters that can provide variable-voltage, variable-current, and sometimes variable-frequency output. DC motors can be speed- and torque-controlled through the use of DC-DC converters, which can be implemented by controlled regulators or by choppers. AC motors require power converters that can provide variable-frequency AC; these are called inverters. All power converters are based on high-power switching devices described in this chapter.

HOMEWORK PROBLEMS

Section 12.3: Voltage Regulators

12.1 Repeat Example 12.1 for a 7-V Zener diode.

12.2 For the current regulator circuit shown in Figure P12.2, derive an expression for R_S in terms of V_Z and I.

12.3 For the shunt-type voltage regulator shown in Figure P12.3, find the expression for the output voltage V_{out}.

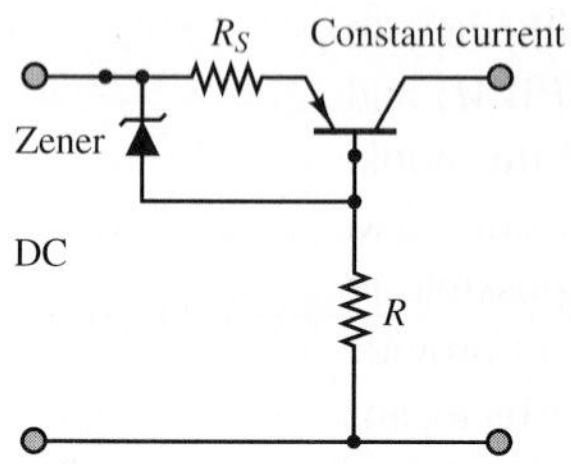

Figure P12.2

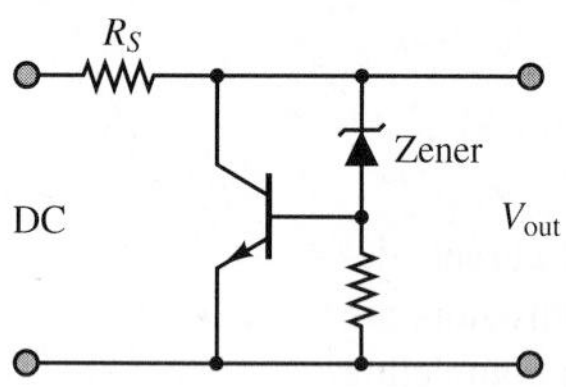

Figure P12.3

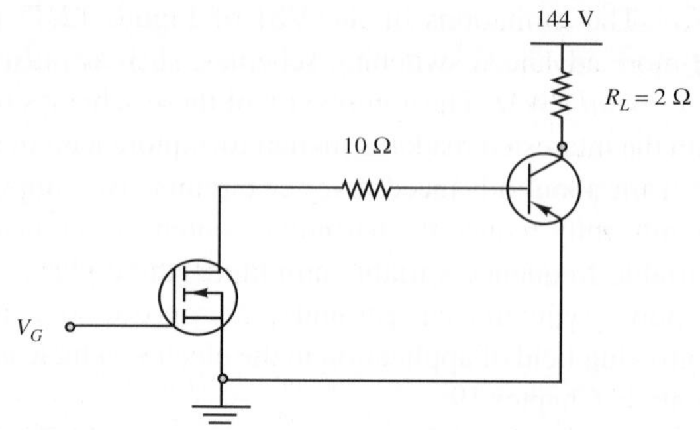

Figure P12.5

Section 12.4: Power Amplifiers

12.4 The circuit of Figure P12.4 is a very effective battery charger. Its operation is simple, and the TIP-33C *n-p-n* power transistor can sink A0 A amps if a big enough heat sink is used. Assuming that the transistor is in the linear operating region, determine the power delivered to the 1.2 V rechargeable battery in the circuit.

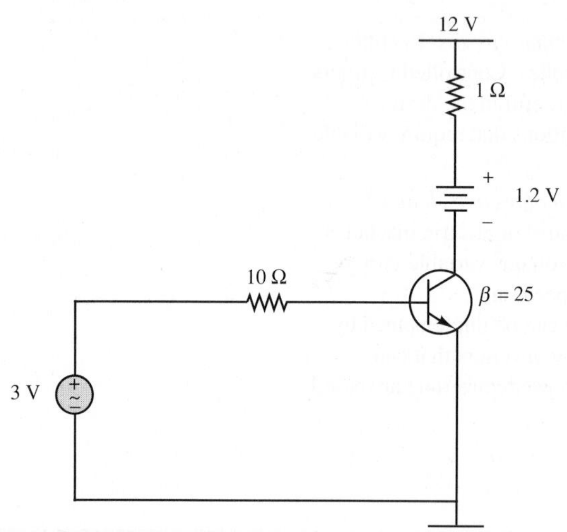

Figure P12.4

12.5 An IGBT can be modeled as shown in the circuit of Figure P12.5. With $V_T = 4$ and $K = 0.01$ A/V^2 for MOSFET, and $\beta = 200$ for the BJT, determine the current through R_L and the voltage across it. Let $V_G = 8$ V.

Section 12.5: Rectifiers and Controlled Rectifiers (AC-DC Converters)

12.6 For the circuit shown in Figure 12.19 in the text, if the *LR* load is replaced by a capacitor, draw the output waveform and label the values.

12.7 Draw $v_L(t)$ and label the values for the circuit in Figure 12.19 in the text if the diode forward resistance is 50 Ω, the forward bias voltage is 0.7 V, and the load consists of a resistor $R = 10\ \Omega$ and an inductor $L = 2$ H.

12.8 For the circuit shown in Figure P12.8, v_{AC} is a sinusoid with 10-V peak amplitude, $R = 2$ kΩ, and the forward-conducting voltage of D is 0.7 V.

a. Sketch the waveform of $v_L(t)$.

b. Find the average value of $v_L(t)$.

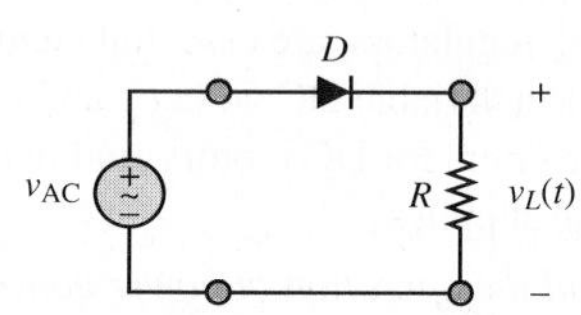

Figure P12.8

12.9 A vehicle battery charge circuit is shown in Figure P12.9. Describe the circuit, and draw the output waveform (L_1 and L_2 represent the inductances of the windings of the alternator).

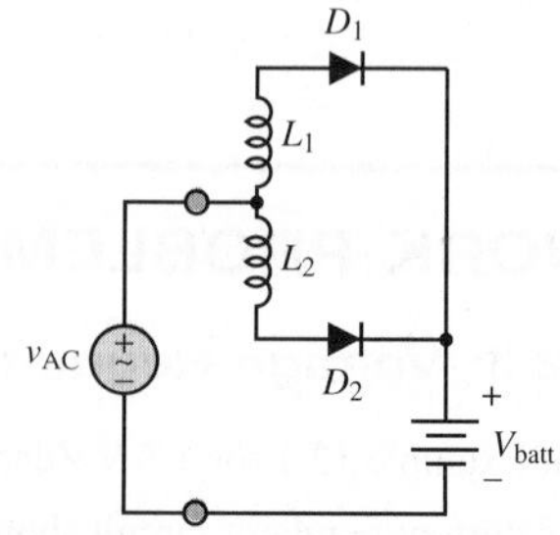

Figure P12.9

12.10 Repeat Example 12.4 for $\alpha = \pi/3$ and $\pi/6$.

12.11 The circuit shown in Figure P12.11 is a speed control system for a DC motor. Assume that the thyristors are fired at $\alpha = 60°$ and that the motor current is 20 A and is ripple-free. The supply is 110 V AC (rms).

a. Sketch the output voltage waveform v_o.

b. Compute the power absorbed by the motor.

c. Determine the volt-amperes generated by the supply.

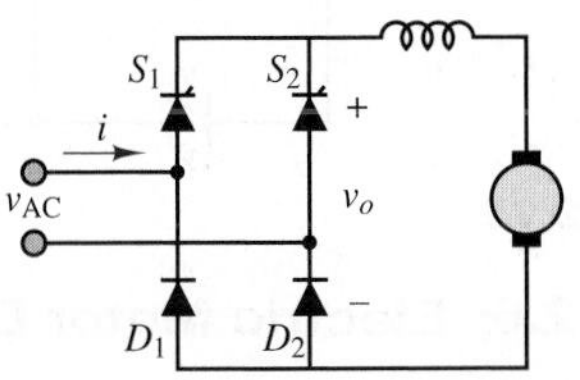

Figure P12.11

12.12 A full-wave, single-phase controlled rectifier is used to control the speed of a DC motor. The circuit is similar to that of Figure 12.2 in the text, except for replacing the resistive load with a DC motor. The motor operates at 110 V and absorbs 4 kW of power. The AC supply is 80 V, 60 Hz. Assume that the motor inductance is very large (i.e., the motor current is ripple-free) and that the motor constant is 0.055 V/(r/min). If the motor runs at 1,000 r/min at rated current

a. Determine the firing angle of the converter.

b. Determine the rms value of the supply current.

12.13 For the light dimmer circuit of Example 12.4, determine the load power at firing angles $\alpha = 0°, 30°, 60°, 90°, 120°, 150°$, and $180°$, and plot the load power as a function of α.

12.14 In the circuit shown in Figure P12.14, if

$$V_L = 10 \text{ V} \qquad V_r = 10\% = 1 \text{ V}$$

$$I_L = 650 \text{ mA} \qquad v_{\text{line}} = 170 \cos \omega t$$

$$\omega = 2{,}513 \text{ rad/s}$$

and if the diodes are fabricated from silicon, determine the conduction angle of the diodes.

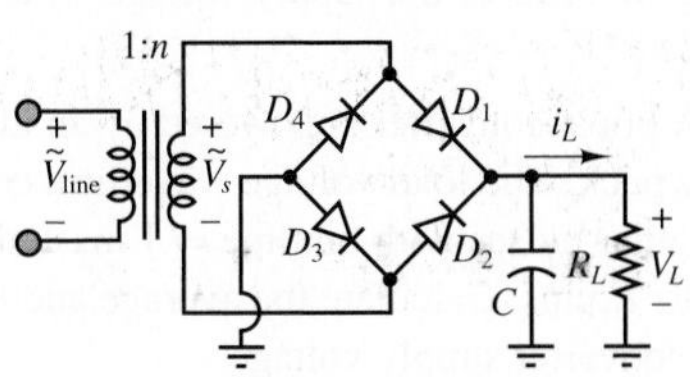

Figure P12.14

12.15 Assume that the conduction angle of the silicon diodes shown in the circuit of Figure P12.15 is

$$\phi = 23°$$

$$v_{s1}(t) = v_{s2}(t) = 8 \cos \omega t \qquad \text{V}$$

$$\omega = 377 \text{ rad/s} \qquad R_L = 20 \text{ k}\Omega$$

$$C = 0.5 \ \mu\text{F}$$

Determine the rms value of the ripple voltage.

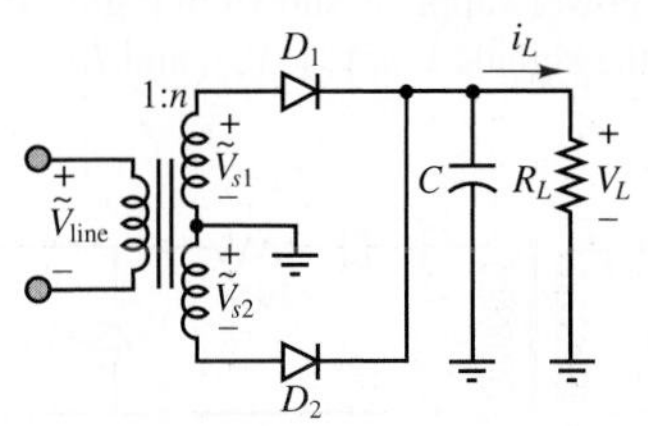

Figure P12.15

12.16 The diodes in the full-wave DC power supply shown in Figure P12.14 are silicon. If

$$I_L = 85 \text{ mA} \qquad V_L = 5.3 \text{ V}$$

$$V_r = 0.6 \text{ V} \qquad \omega = 377 \text{ rad/s}$$

$$v_{\text{line}} = 156 \cos \omega t \qquad \text{V}$$

$$C = 1{,}023 \ \mu\text{F}$$

$$\phi = \text{conduction angle} = 23.90°$$

determine the value of the average and peak current through each diode.

12.17 The diodes in the full-wave DC power supply shown in Figure P12.15 are silicon. If

$$I_L = 600 \text{ mA} \qquad V_L = 50 \text{ V}$$

$$V_r = 8\% = 4 \text{ V} \qquad v_{\text{line}} = 170 \cos \omega t \qquad \text{V}$$

determine the value of the conduction angle for the diodes and the average and peak current through the diodes. The load voltage waveform is shown in Figure P12.17.

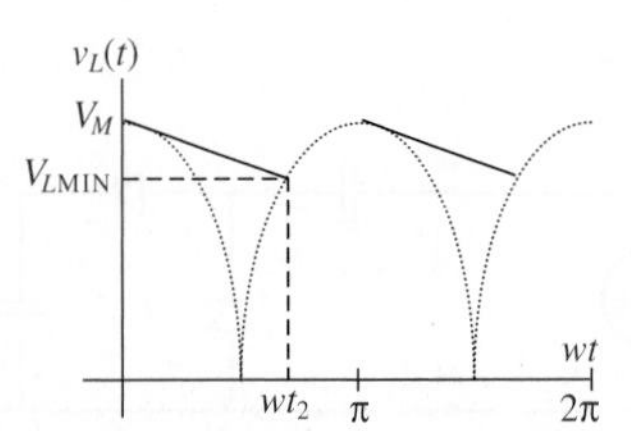

Figure P12.17

12.18 A power supply is shown in Figure P12.18. Sketch the signals V_{ab}, V_{cd}, V_{ef}, and V_{gh}.

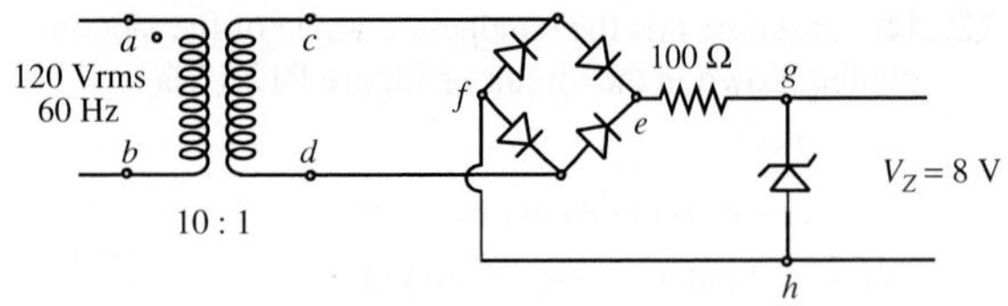

Figure P12.18

12.19 A power supply is shown in Figure P12.19. Sketch the signals V_{ab}, V_{cd}, V_{ef}, and I_Z.

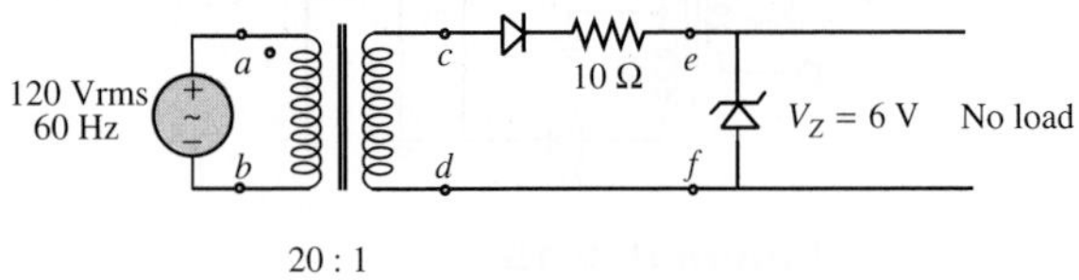

Figure P12.19

12.20 Figure P12.20 depicts a low-cost full-wave rectifier with a Zener diode voltage regulator. Sketch the voltages across terminals *a*-*b*, *c*-*d*, and *e*-*f*.

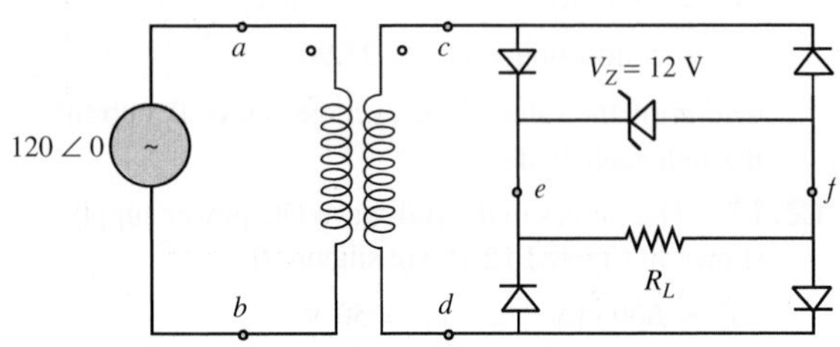

Figure P12.20

12.21 In the DC power supply shown in Figure P12.21, sketch the voltage across *a*-*b*, *c*-*a*, and *d*-*e*, assuming that *R* is so large as to make any ripple not noticeable.

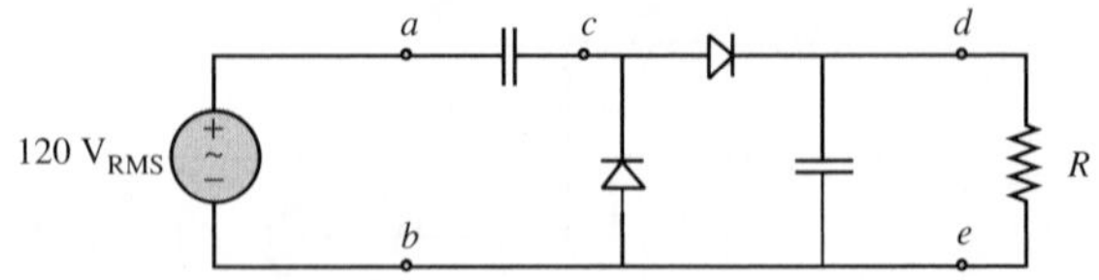

Figure P12.21

12.22 A DC power supply known as a *voltage doubler* is shown in Figure P12.22. It is assumed that the capacitors are large enough that the ripple is not significant in the output voltage. Sketch the signals V_{ab} and V_{cd}. Assume the input is at 60 Hz.

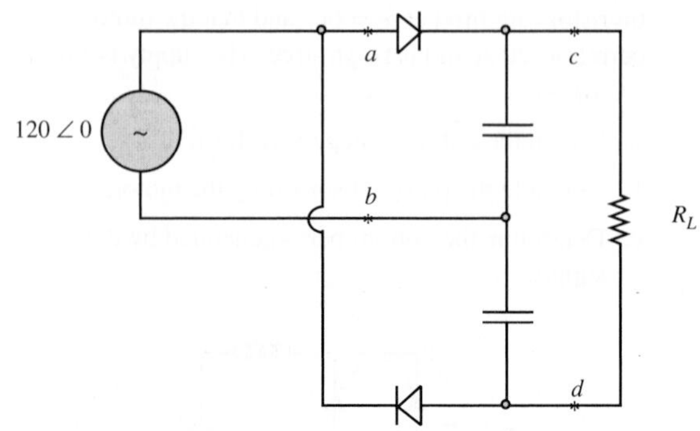

Figure P12.22

Section 12.6: Electric Motor Drives

12.23 The DC-DC converter of Figure 12.37 in the text, is used to control the speed of a DC motor. Let the supply voltage be 120 V and the armature resistance of the motor be 0.15 Ω. The motor back-emf constant is 0.05 V/(r/min), and the switching frequency is 250 Hz. Assume that the motor current is free of ripple and equal to 125 A at 120 r/min.

a. Determine the duty cycle of the converter, δ and the converter on time t_1.

b. Determine the power absorbed by the motor.

c. Determine the power generated by the supply.

12.24 The circuit of Figure 12.41 in the text, is used to provide regenerative braking in a traction motor. The motor constant is 0.3 V/(r/min), and the supply voltage is 600 V. The armature resistance is $R_a = 0.2\ \Omega$. The motor speed is 800 r/min and the motor current is 300 A.

a. Determine the duty cycle δ of the converter.

b. Determine the power fed back to the supply (battery).

12.25 For the two-quadrant converter of Figure 12.42 in the text, assume that thyristors S_1 and S_2 are turned on for time t_1 and off for time $T - t_1$ (T is the switching period). Derive an expression for the average output voltage in terms of the supply voltage V_S and the duty cycle δ.

12.26 A boost converter is powered by an ideal 100-V battery pack. The load voltage waveform consists of rectangular pulses with on time = 1 ms and period equal to 2.5 ms. Calculate the average and rms values of the converter supply voltage.

12.27 A buck converter connected to a 100-V battery pack supplies an RL load with $R = 0.5\ \Omega$ and $L =$

1 mH. The switch (a thyristor) is switched on for 1 ms, and the period of the switching waveform is 3 ms. Calculate the average value of the load voltage and the power supplied by the battery.

12.28 The converter of Problem 12.27 is used to supply a separately excited DC motor with $R_a = 0.2\ \Omega$ and $L_a = 1$ mH. At the lowest speed of operation, the back-emf E_a is equal to 10 V. Calculate the average value of the load current and voltage for this condition if the switching period is 3 ms and the duty cycle is $\frac{1}{3}$.

12.29 A separately excited DC motor with $R_a = 0.33\ \Omega$ and $L_a = 15$ mH is controlled by a DC-DC converter in the range of 0 to 2,000 r/min. The DC supply is 220 V. If the load torque is constant and requires an average armature current of 25 A, calculate the range of duty cycles required if the motor armature constant is $K_a\phi = 0.00167$ V-s/r.

12.30 A separately excited DC motor is rated at 10 kW, 240 V, 1,000 r/min, and is supplied by a single-phase controlled bridge rectifier. The power supply is sinusoidal and rated at 240 V, 60 Hz. The motor armature resistance is 0.42 Ω, and the motor constant is $K_a = 2$ V-s/rad. Calculate the speed, power factor, and efficiency for SCR firing angles α of 0° and 20° if the load torque is constant. Assume that additional inductance is present to ensure continuous conduction.

12.31 A separately excited DC motor is rated at 10 kW, 300 V, 1,000 r/min, and is supplied by a three-phase controlled bridge rectifier. The power supply is sinusoidal and rated at 220 V, 60 Hz. The motor armature resistance is 0.2 Ω, and the motor constant is $K_a = 1.38$ V-s/rad. The motor delivers rated power at $\alpha = 0°$. Calculate the speed, power factor, and efficiency for a firing angle $\alpha = 30°$ if the load torque is constant. Assume that additional inductance is present to ensure continuous conduction.

12.32 Sketch the current through the load in the switched-mode power supply of Figure P12.32.

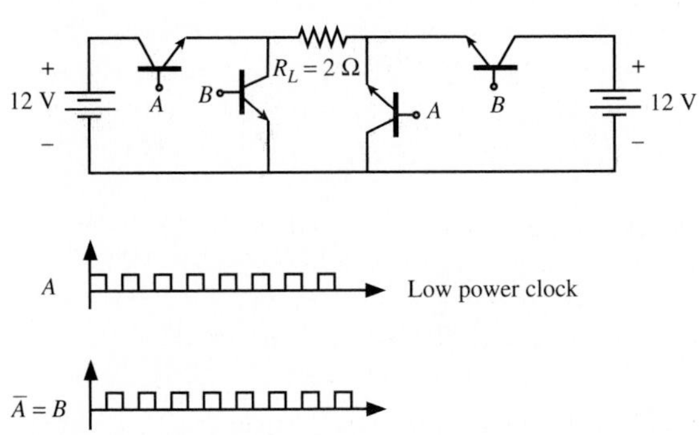

Figure P12.32

12.33 In the switched-mode power supply of Figure P12.33, sketch the load voltage signal, V_L.

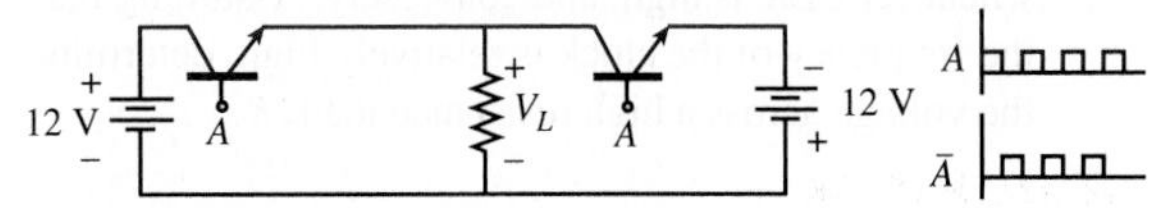

Figure P12.33

12.34 The switched mode power supply of Figure P12.34 will convert DC to three-phase AC. Sketch timing diagrams for the three low-power clock inputs to generate a balanced three-phase source. Also sketch the current in the neutral return wire. Assume the period of the cycle is normalized.

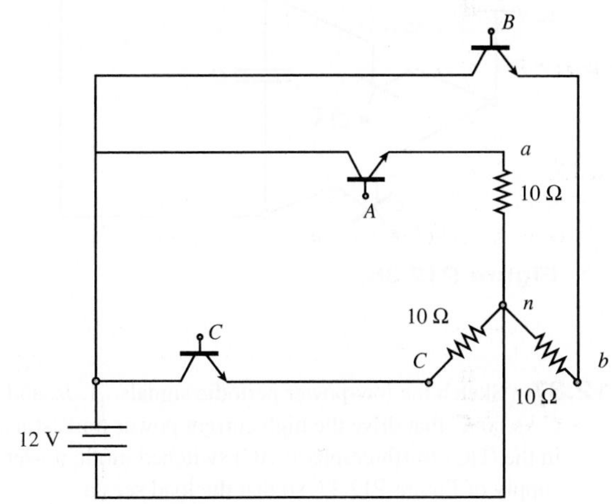

Figure P12.34

12.35 A DC-to-DC converter can also be thought of, and analyzed as, a DC transformer! Figure P12.35 shows a particular DC-to-DC converter configuration that converts the 1.2 V of a Ni-Cd battery cell to a desired 12 V DC supply. Using the usual transformer "reflecting theorems" (see Chapter 7), determine the power supplied by the 1.2 V source, and to the 10 Ω load.

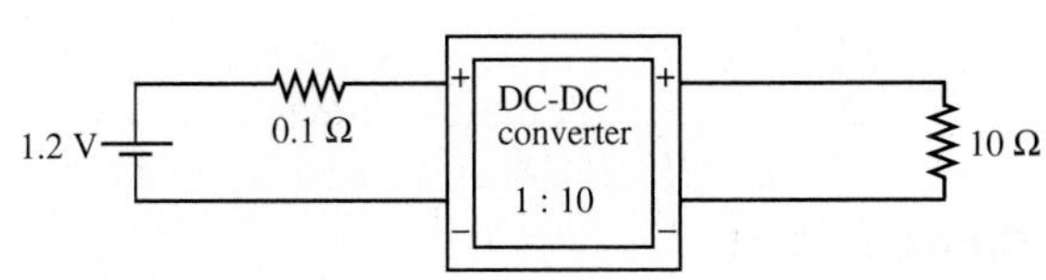

Figure P12.35

12.36 Shown in Figure P12.36 is a "charge pump" circuit for a switched-mode power supply (with all transistors acting in the switched-mode). A 555-timer chip drives the two inputs, which are the timer's clock (CLK), and its inverse, clock-not$(\overline{CLK})$. $\overline{CLK}$ is low whenever CLK is high, and conversely. Assuming that the frequency of the clock is relatively high, determine the voltage across a high resistance load, R.

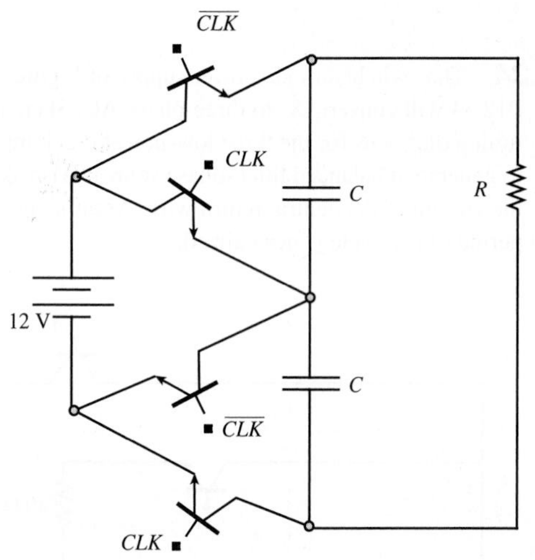

Figure P12.36

12.37 Sketch the low-power periodic signals, A, B, and C vs. *time,* that drive the high current power transistors in the (DC)-to-(three-phase-AC) switched mode power supply of Figure P12.37, so that the load sees a balanced three-phase square-wave source.

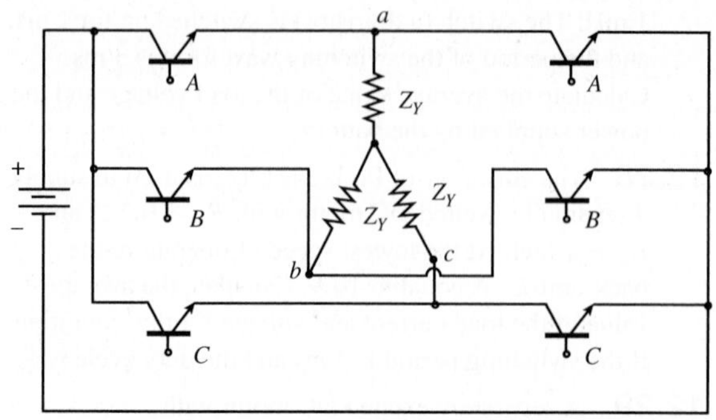

Figure P12.37

12.38 In the switched mode power supply of Figure P12.38, sketch the load voltage signal, V_L.

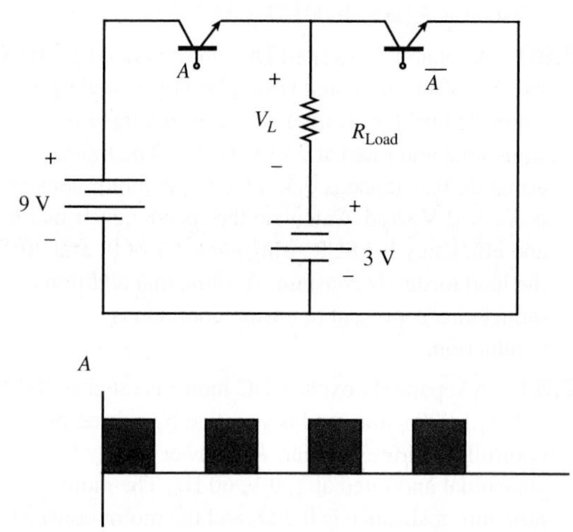

Figure P12.38

CHAPTER

13

DIGITAL LOGIC CIRCUITS

Digital computers have taken a prominent place in engineering and science over the last three decades, performing a number of essential functions such as numerical computations and data acquisition. It is not necessary to further stress the importance of these electronic systems in this book, since you are already familiar with personal computers and programming languages. The objective of the chapter is to discuss the essential features of digital logic circuits, which are at the heart of digital computers, by presenting an introduction to *combinational logic circuits*.

The chapter starts with a discussion of the binary number system and continues with an introduction to boolean algebra. The self-contained treatment of boolean algebra will enable you to design simple logic functions using the techniques of combinational logic, and several practical examples are provided to demonstrate that even simple combinations of logic gates can serve to implement useful circuits in engineering practice. In a later section, we introduce a number of logic modules which can be described by using simple logic gates but which provide more advanced functions. Among these, we discuss read-only memories, multiplexers, and decoders.

Throughout the chapter, simple examples are given to demonstrate the usefulness of digital logic circuits in various engineering applications.

This chapter provides the background needed to address the study of digital systems, which will be undertaken in Chapter 14. Upon completion of the chapter, you should be able to:

Learning Objectives

1. Understand the concepts of analog and digital signals and of quantization. *Section 13.1.*
2. Convert between decimal and binary number systems and use the hexadecimal system and BCD and Gray codes. *Section 13.2.*
3. Write truth tables, and realize logic functions from truth tables by using logic gates. *Section 13.3.*
4. Systematically design logic functions using Karnaugh maps. *Section 13.4.*
5. Study various combinational logic modules, including multiplexers, memory and decoder elements, and programmable logic arrays. *Section 13.5.*

13.1 ANALOG AND DIGITAL SIGNALS

One of the fundamental distinctions in the study of electronic circuits (and in the analysis of any signals derived from physical measurements) is that between analog and digital signals. As discussed in Chapter 12, an **analog signal** is an electric signal whose value varies in analogy with a physical quantity (e.g., temperature, force, or acceleration). For example, a voltage proportional to a measured variable pressure or to a vibration naturally varies in an analog fashion. Figure 13.1 depicts an analog function of time $f(t)$. We note immediately that for each value of time t, $f(t)$ can take one value among any of the values in a given range. For example, in the case of the output voltage of an op-amp, we expect the signal to take any value between $+V_{\text{sat}}$ and $-V_{\text{sat}}$, where V_{sat} is the supply-imposed saturation voltage.

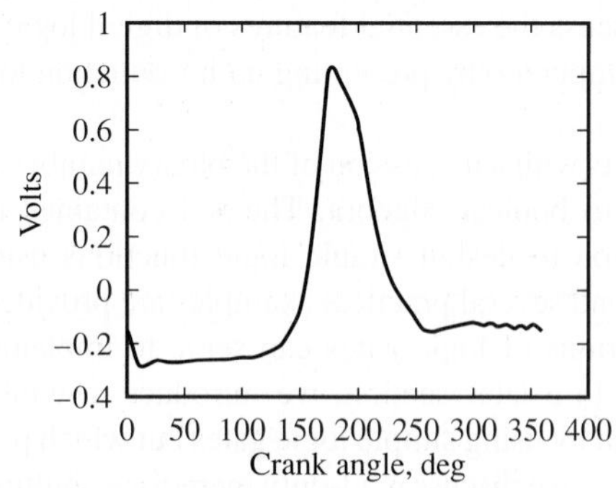

Figure 13.1 Voltage analog of internal combustion engine in-cylinder pressure

A **digital signal,** on the other hand, can take only a *finite number of values.* This is an extremely important distinction, as will be shown shortly. An example of a digital signal is a signal that allows display of a temperature measurement on a digital readout. Let us hypothesize that the digital readout is three digits long and can display numbers from 0 to 100, and let us assume that the temperature sensor is correctly calibrated to measure temperatures from 0 to 100°C. Further, the output of the sensor ranges from 0 to 5 V, where 0 V corresponds to 0°C and 5 V to 100°C. Therefore, the calibration constant of the sensor is

$$k_T = \frac{100°\text{C} - 0°\text{C}}{5\text{ V} - 0\text{ V}} = 20°\text{C/V}$$

Clearly, the output of the sensor is an analog signal; however, the display can show only a finite number of readouts (101, to be precise). Because the display itself can only take a value out of a discrete set of states—the integers from 0 to 100—we call it a digital display, indicating that the variable displayed is expressed in digital form.

Now, each temperature on the display corresponds to a *range of voltages:* each digit on the display represents one-hundredth of the 5-V range of the sensor, or 0.05 V = 50 mV. Thus, the display will read 0 if the sensor voltage is between 0 and 49 mV, 1 if it is between 50 and 99 mV, and so on. Figure 13.2 depicts the staircase function relationship between the analog voltage and the digital readout. This **quantization** of the sensor output voltage is in effect an approximation. If one wished to know the temperature with greater precision, a greater number of display digits could be employed.

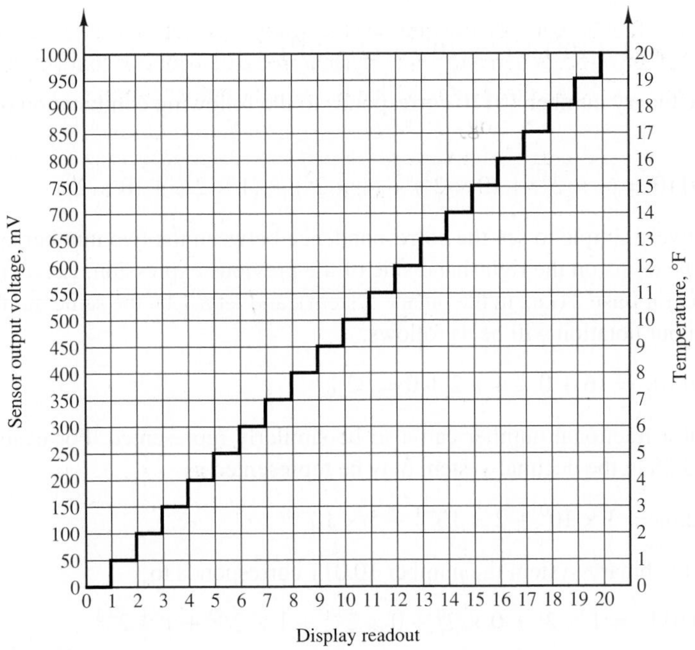

Figure 13.2 Digital representation of an analog signal

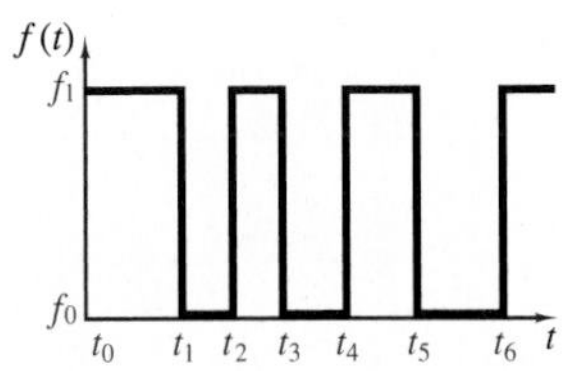

Figure 13.3 A binary signal

The most common digital signals are binary signals. A **binary signal** is a signal that can take only one of two discrete values and is therefore characterized by transitions between two states. Figure 13.3 displays a typical binary signal. In binary arithmetic (which we discuss in Section 13.2), the two discrete values f_1 and f_0 are represented, respectively, by the numbers 1 and 0. In binary voltage waveforms, these values are represented by two voltage levels. For example, in the TTL convention (see Chapter 10), these values are (nominally) 5 and 0 V, respectively; in CMOS circuits, these values can vary substantially. Other conventions are also used, including reversing the assignment, for example, by letting a 0-V level represent a logic 1 and a 5-V level represent a logic 0. Note that in a binary waveform, knowledge of the transition between one state and another (e.g., from f_0 to f_1 at $t = t_2$) is equivalent to knowledge of the state. Thus, digital logic circuits can operate by detecting transitions between voltage levels. The transitions are often called **edges** and can be positive (f_0 to f_1) or negative (f_1 to f_0). Virtually all the signals handled by a computer are binary. From here on, whenever we speak of digital signals, you may assume that the text is referring to signals of the binary type, unless otherwise indicated.

LO2

13.2 THE BINARY NUMBER SYSTEM

The binary number system is a natural choice for representing the behavior of circuits that operate in one of two states (on or off, 1 or 0, or the like). The diode and transistor gates and switches studied in Chapters 10 and 11 fall in this category. Table 13.1 shows the correspondence between decimal and binary number systems for integer decimal numbers up to 16.

Table 13.1 Conversion from decimal to binary

Decimal number n_{10}	Binary number n_2
0	0
1	1
2	10
3	11
4	100
5	101
6	110
7	111
8	1000
9	1001
10	1010
11	1011
12	1100
13	1101
14	1110
15	1111
16	10000

Binary numbers are based on powers of 2, whereas the decimal system is based on powers of 10. For example, the number 372 in the decimal system can be expressed as

$$372 = (3 \times 10^2) + (7 \times 10^1) + (2 \times 10^0)$$

while the binary number 10110 corresponds to the following combination of powers of 2:

$$10110 = (1 \times 2^4) + (0 \times 2^3) + (1 \times 2^2) + (1 \times 2^1) + (0 \times 2^0)$$

It is relatively simple to see the correspondence between the two number systems if we add the terms on the right-hand side of the previous expression. Let n_2 represent the number n **base 2** (i.e., in the binary system), and let n_{10} be the same number **base 10.** Then our notation will be as follows:

$$10110_2 = 16 + 0 + 4 + 2 + 0 = 22_{10}$$

Note that a fractional number can also be similarly represented. For example, the number 3.25 in the decimal system may be represented as

$$3.25_{10} = 3 \times 10^0 + 2 \times 10^{-1} + 5 \times 10^{-2}$$

while in the binary system the number 10.011 corresponds to

$$\begin{aligned}10.011_2 &= 1 \times 2^1 + 0 \times 2^0 + 0 \times 2^{-1} + 1 \times 2^{-2} + 1 \times 2^{-3} \\ &= 2 + 0 + 0 + \tfrac{1}{4} + \tfrac{1}{8} = 2.375_{10}\end{aligned}$$

Table 13.1 shows that it takes four binary digits, also called **bits,** to represent the decimal numbers up to 15. Usually, the rightmost bit is called the **least significant bit,** or **LSB,** and the leftmost bit is called the **most significant bit,** or **MSB.** Since binary numbers clearly require a larger number of digits than decimal numbers do, the digits are usually grouped in sets of 4, 8, or 16. Four bits are usually termed a **nibble,** 8 bits is called a **byte,** and 16 bits (or 2 bytes) is a **word.**

CHECK YOUR UNDERSTANDING

LO2

Convert the following decimal numbers to binary form.

a.	39	b.	59
c.	512	d.	0.4475
e.	$\frac{25}{32}$	f.	0.796875
g.	256.75	h.	129.5625
i.	4,096.90625		

Convert the following binary numbers to decimal.

a.	1101	b.	11011
c.	10111	d.	0.1011
e.	0.001101	f.	0.001101101
g.	111011.1011	h.	1011011.001101
i.	10110.0101011101		

Answers: (a) 100111, (b) 111011, (c) 1000000000, (d) 0.011100101001, (e) 0.11001, (f) 0.110011, (g) 100000000.11, (h) 10000001.1001, (i) 1000000000000.11101; (a) 13, (b) 27, (c) 23, (d) 0.6875, (e) 0.203125, (f) 0.212890625, (g) 59.6875, (h) 91.203125, (i) 22.340820312

Addition and Subtraction

LO2

The operations of addition and subtraction are based on the simple rules shown in Table 13.2. Note that, just as is done in the decimal system, a carry is generated whenever the sum of two digits exceeds the largest single-digit number in the given number system, which is 1 in the binary system. The carry is treated exactly as in the decimal system. A few examples of binary addition are shown in Figure 13.4, with their decimal counterparts.

Table 13.2 Rules for addition

$0 + 0 = 0$
$0 + 1 = 1$
$1 + 0 = 1$
$1 + 1 = 0$ (with a carry of 1)

Decimal	Binary	Decimal	Binary	Decimal	Binary
5	101	15	1111	3.25	11.01
+6	+110	+20	+10100	+5.75	+101.11
11	1011	35	100011	9.00	1001.00

(Note that in this example, $3.25 = 3\frac{1}{4}$ and $5.75 = 5\frac{3}{4}$.)

Figure 13.4 Examples of binary addition

LO2

The procedure for subtracting binary numbers is based on the rules of Table 13.3. A few examples of binary subtraction are given in Figure 13.5, with their decimal counterparts.

Table 13.3 Rules for subtraction

$0 - 0 = 0$
$1 - 0 = 1$
$1 - 1 = 0$
$0 - 1 = 1$ (with a borrow of 1)

Decimal	Binary	Decimal	Binary	Decimal	Binary
9	1001	16	10000	6.25	110.01
−5	−101	−3	−11	−4.50	−100.10
4	0100	13	01101	1.75	001.11

Figure 13.5 Examples of binary subtraction

CHECK YOUR UNDERSTANDING

Perform the following additions and subtractions. Express the answer in decimal form for (a) through (d) and in binary form for (e) through (h).

a.	$1001.1_2 + 1011.01_2$	b.	$100101_2 + 100101_2$
c.	$0.1011_2 + 0.1101_2$	d.	$1011.01_2 + 1001.11_2$
e.	$64_{10} - 32_{10}$	f.	$127_{10} - 63_{10}$
g.	$93.5_{10} - 42.75_{10}$	h.	$(84\frac{9}{32})_{10} - (48\frac{5}{16})_{10}$

Answer: (a) 20.75_{10}, (b) 74_{10}, (c) 1.5_{10}, (d) 21_{10}, (e) 100000_2, (f) 1000000_2, (g) 110010.11_2, (h) 100011.11111_2

Multiplication and Division

Whereas in the decimal system the multiplication table consists of $10^2 = 100$ entries, in the binary system we only have $2^2 = 4$ entries. Table 13.4 represents the complete multiplication table for the binary number system.

Division in the binary system is also based on rules analogous to those of the decimal system, with the two basic laws given in Table 13.5. Once again, we need be concerned with only two cases, and just as in the decimal system, division by zero is not contemplated.

Table 13.4 **Rules for multiplication**

$0 \times 0 = 0$
$0 \times 1 = 0$
$1 \times 0 = 0$
$1 \times 1 = 1$

Table 13.5 **Rules for division**

$0 \div 1 = 0$
$1 \div 1 = 1$

Conversion from Decimal to Binary

The conversion of a decimal number to its binary equivalent is performed by successive division of the decimal number by 2, checking for the remainder each time. Figure 13.6 illustrates this idea with an example. The result obtained in Figure 13.6 may be easily verified by performing the opposite conversion, from binary to decimal:

$$110001 = 2^5 + 2^4 + 2^0 = 32 + 16 + 1 = 49$$

Remainder
$49 \div 2 = 24 + 1$
$24 \div 2 = 12 + 0$
$12 \div 2 = 6 + 0$
$6 \div 2 = 3 + 0$
$3 \div 2 = 1 + 1$
$1 \div 2 = 0 + 1$
$49_{10} = 110001_2$

Figure 13.6 Example of conversion from decimal to binary

The same technique can be used for converting decimal fractional numbers to their binary form, provided that the whole number is separated from the fractional part and each is converted to binary form (separately), with the results added at the end. Figure 13.7 outlines this procedure by converting the number 37.53 to binary form. The procedure is outlined in two steps. First, the integer part is converted; then, to convert the fractional part, one simple technique consists of multiplying the decimal fraction by 2 in successive stages. If the result exceeds 1, a 1 is needed to the right of the binary fraction being formed (100101 . . . , in our example). Otherwise, a 0 is added. This procedure is continued until no fractional terms are left. In this case, the decimal part is 0.53_{10}, and Figure 13.7 illustrates the succession of calculations. Stopping the procedure outlined in Figure 13.7 after 11 digits results in the following approximation:

$$37.53_{10} = 100101.10000111101$$

Greater precision could be attained by continuing to add binary digits, at the expense of added complexity. Note that an infinite number of binary digits may be required to represent a decimal number *exactly.*

Remainder
$37 \div 2 = 18 + 1$
$18 \div 2 = 9 + 0$
$9 \div 2 = 4 + 1$
$4 \div 2 = 2 + 0$
$2 \div 2 = 1 + 0$
$1 \div 2 = 0 + 1$
$37_{10} = 100101_2$

$2 \times 0.53 = 1.06 \rightarrow 1$
$2 \times 0.06 = 0.12 \rightarrow 0$
$2 \times 0.12 = 0.24 \rightarrow 0$
$2 \times 0.24 = 0.48 \rightarrow 0$
$2 \times 0.48 = 0.96 \rightarrow 0$
$2 \times 0.96 = 1.92 \rightarrow 1$
$2 \times 0.92 = 1.84 \rightarrow 1$
$2 \times 0.84 = 1.68 \rightarrow 1$
$2 \times 0.68 = 1.36 \rightarrow 1$
$2 \times 0.36 = 0.72 \rightarrow 0$
$2 \times 0.72 = 1.44 \rightarrow 1$
$0.53_{10} = 0.10000111101_2$

Figure 13.7 Conversion from decimal to binary

Complements and Negative Numbers

To simplify the operation of subtraction in digital computers, **complements** are used almost exclusively. In practice, this corresponds to replacing the operation $X - Y$ with the operation $X + (-Y)$. This procedure results in considerable simplification, since the computer hardware need include only adding circuitry. Two types of complements are used with binary numbers: the **ones complement** and the **twos complement.**

The ones complement of an n-bit binary number is obtained by subtracting the number itself from $2^n - 1$. Two examples are as follows:

$$
\begin{aligned}
a &= 0101 \\
\text{Ones complement of } a &= (2^4 - 1) - a \\
&= (1111) - (0101) \\
&= 1010 \\
b &= 101101 \\
\text{Ones complement of } b &= (2^6 - 1) - b \\
&= (111111) - (101101) \\
&= 010010
\end{aligned}
$$

The twos complement of an n-bit binary number is obtained by subtracting the number itself from 2^n. Twos complements of the same numbers a and b used in the preceding illustration are computed as follows:

$$
\begin{aligned}
a &= 0101 \\
\text{Twos complement of } a &= 2^4 - a \\
&= (10000) - (0101) \\
&= 1011 \\
b &= 101101 \\
\text{Twos complement of } b &= 2^6 - b \\
&= (1000000) - (101101) \\
&= 010011
\end{aligned}
$$

A simple rule that may be used to obtain the twos complement directly from a binary number is the following: Starting at the least significant (rightmost) bit, copy each bit *until the first 1 has been copied,* and then replace each successive 1 by a 0 and each 0 by a 1. You may wish to try this rule on the two previous examples to verify that it is much easier to use than subtraction from 2^n.

Different conventions exist in the binary system to represent whether a number is negative or positive. One convention, called the **sign-magnitude convention,** makes use of a *sign bit,* usually positioned at the beginning of the number, for which a value of 1 represents a minus sign and a value of 0 represents a plus sign. Thus, an 8-bit binary number would consist of 1 sign bit followed by 7 *magnitude bits,* as shown in Figure 13.8(a). In a digital system that uses 8-bit signed integer words, we could

Sign bit b_7	b_6	b_5	b_4	b_3	b_2	b_1	b_0
	← Actual magnitude of binary number →						

(a)

Sign bit b_7	b_6	b_5	b_4	b_3	b_2	b_1	b_0
	← Actual magnitude of binary number (if $b_7 = 0$) → ← Ones complement of binary number (if $b_7 = 1$) →						

(b)

LO2

Sign bit b_7	b_6	b_5	b_4	b_3	b_2	b_1	b_0
	← Actual magnitude of binary number (if $b_7 = 0$) → ← Twos complement of binary number (if $b_7 = 1$) →						

(c)

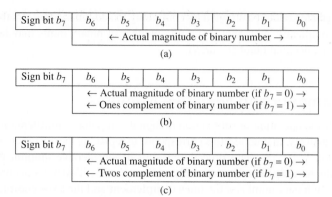

Figure 13.8 (a) An 8-bit sign-magnitude binary number; (b) an 8-bit ones complement binary number; (c) an 8-bit twos complement binary number

represent integer numbers (decimal) in the range

$$-(2^7 - 1) \le N \le +(2^7 - 1)$$

or

$$-127 \le N \le +127$$

A second convention uses the ones complement notation. In this convention, a sign bit is also used to indicate whether the number is positive (sign bit $= 0$) or negative (sign bit $= 1$). However, the magnitude of the binary number is represented by the true magnitude if the number is positive and by its *ones complement* if the number is negative. Figure 13.8(b) illustrates the convention. For example, the number 91_{10} would be represented by the 7-bit binary number 1011011_2 with a leading 0 (the sign bit): $\mathbf{0}1011011_2$. On the other hand, the number -91_{10} would be represented by the 7-bit ones complement binary number 0100100_2 with a leading 1 (the sign bit): $\mathbf{1}0100100_2$.

Most digital computers use the twos complement convention in performing integer arithmetic operations. The twos complement convention represents positive numbers by a sign bit of 0, followed by the true binary magnitude; negative numbers are represented by a sign bit of 1, *followed by the twos complement of the binary number,* as shown in Figure 13.8(c). The advantage of the twos complement convention is that the algebraic sum of twos complement binary numbers is carried out very simply by adding the two numbers *including the sign bit.* Example 13.1 illustrates twos complement addition.

CHECK YOUR UNDERSTANDING

How many possible numbers can be represented in a 12-bit word?

If we use an 8-bit word with a sign bit (7 magnitude bits plus 1 sign bit) to represent voltages -5 and $+5$ V, what is the smallest increment of voltage that can be represented?

Answers: 4,096; 39 mV

EXAMPLE 13.1 Twos Complement Operations

LO2

Problem

Perform the following subtractions, using twos complement arithmetic.

1. $X - Y = 1011100 - 1110010$
2. $X - Y = 10101111 - 01110011$

Solution

Analysis: The twos complement subtractions are performed by replacing the operation $X - Y$ with the operation $X + (-Y)$. Thus, we first find the twos complement of Y and add the result to X in each of the two cases:

$$\begin{aligned} X - Y &= 1011100 - 1110010 = 1011100 + (2^7 - 1110010) \\ &= 1011100 + 0001110 = 1101010 \end{aligned}$$

Next, we add the *sign bit* (in boldface type) in front of each number (1 in first case since the difference $X - Y$ is a negative number):

$$X - Y = \mathbf{1}1101010$$

Repeating for the second subtraction gives

$$\begin{aligned} X - Y &= 10101111 - 01110011 = 10101111 + (2^8 - 01110011) \\ &= 10101111 + 10001101 = 00111100 \\ &= \mathbf{0}00111100 \end{aligned}$$

where the first digit is a 0 because $X - Y$ is a positive number.

CHECK YOUR UNDERSTANDING

Find the twos complement of the following binary numbers.

a. 11101001 b. 10010111
c. 1011110

Answer: (a) 00010111, (b) 01101001, (c) 0100010

The Hexadecimal System

It should be apparent by now that representing numbers in base-2 and base-10 systems is purely a matter of convenience, given a specific application. Another base frequently used is the **hexadecimal system,** a direct derivation of the binary number system. In the hexadecimal (or hex) code, the bits in a binary number are subdivided into groups of 4. Since there are 16 possible combinations for a 4-bit number, the natural digits in the decimal system (0 through 9) are insufficient to represent a hex digit. To solve this

Table 13.6 **Hexadecimal code**

0	0000
1	0001
2	0010
3	0011
4	0100
5	0101
6	0110
7	0111
8	1000
9	1001
A	1010
B	1011
C	1100
D	1101
E	1110
F	1111

problem, the first six letters of the alphabet are used, as shown in Table 13.6. Thus, in hex code, an 8-bit word corresponds to just two digits; for example,

$$1010\ 0111_2 = \mathrm{A}7_{16}$$

$$0010\ 1001_2 = 29_{16}$$

Binary Codes

In this subsection, we describe two common binary codes that are often used for practical reasons. The first is a method of representing decimal numbers in digital logic circuits that is referred to as **binary-coded decimal,** or **BCD, representation.** In effect, the simplest BCD representation is just a sequence of 4-bit binary numbers that stops after the first 10 entries, as shown in Table 13.7. There are also other BCD codes, all reflecting the same principle: Each decimal digit is represented by a fixed-length binary word. One should realize that although this method is attractive because of its direct correspondence with the decimal system, it is not efficient. Consider, for example, the decimal number 68. Its binary representation by direct conversion is the 7-bit number 1000100. However, the corresponding BCD representation would require 8 bits:

$$68_{10} = 01101000_{\mathrm{BCD}}$$

Table 13.7 **BCD code**

0	0000
1	0001
2	0010
3	0011
4	0100
5	0101
6	0110
7	0111
8	1000
9	1001

Table 13.8 **Three-bit Gray code**

Binary	Gray
000	000
001	001
010	011
011	010
100	110
101	111
110	101
111	100

Another code that finds many applications is the **Gray code.** This is simply a reshuffling of the binary code with the property that any two consecutive numbers differ only by 1 bit. Table 13.8 illustrates the 3-bit Gray code. The Gray code can be very useful in practical applications, because in counting up or down according to this code, the binary representation of a number changes only 1 bit at a time. The following box illustrates an application of the Gray code to a practical engineering problem.

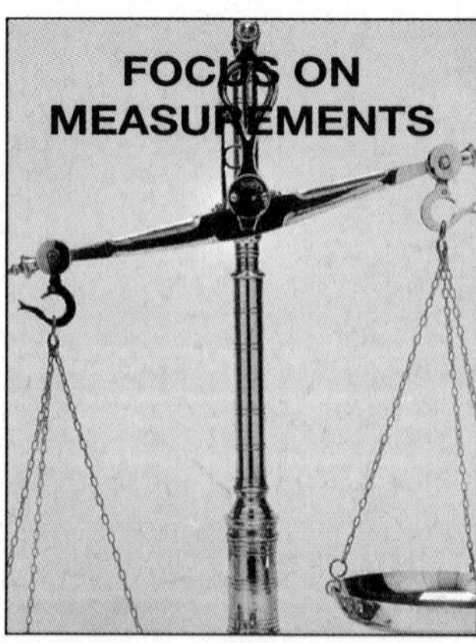

Digital Position Encoders

FIND IT ON THE WEB

Position encoders are devices that output a digital signal proportional to their (linear or angular) position. These devices are very useful in measuring instantaneous position in *motion control* applications. Motion control is a technique used when it is necessary to accurately control the motion of a moving object; examples are found in robotics, machine tools, and servomechanisms. For example, in positioning the arm of a robot to pick up an object, it is very important to know its exact position at all times. Since one is usually interested in both rotational and translational motion, two types of encoders are discussed in this example: *linear* and *angular* position encoders.

(Continued)

(Concluded)

An optical position encoder consists of an *encoder pad,* which is either a strip (for translational motion) or a disk (for rotational motion) with alternating black and white areas. These areas are arranged to reproduce some binary code, as shown in Figure 13.9, where both the conventional binary and Gray codes are depicted for a 4-bit linear encoder pad. A fixed array of photodiodes (see Chapter 9) senses the reflected light from each of the cells across a row of the encoder path; depending on the amount of light reflected, each photodiode circuit will output a voltage corresponding to a binary 1 or 0. Thus, a different 4-bit word is generated for each row of the encoder.

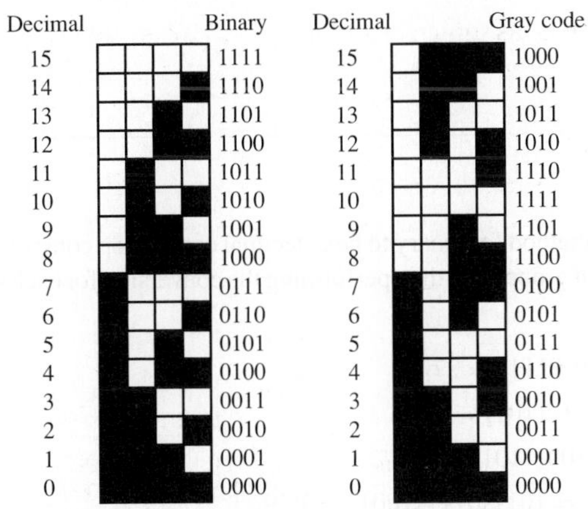

Figure 13.9 Binary and Gray code patterns for linear position encoders

Suppose the encoder pad is 100 mm in length. Then its resolution can be computed as follows. The pad will be divided into $2^4 = 16$ segments, and each segment corresponds to an increment of $100/16$ mm $= 6.25$ mm. If greater resolution were necessary, more bits could be employed: an 8-bit pad of the same length would attain a resolution of $100/256$ mm $= 0.39$ mm.

A similar construction can be employed for the 5-bit angular encoder of Figure 13.10. In this case, the angular resolution can be expressed in degrees of rotation, where $2^5 = 32$ sections correspond to 360°. Thus, the resolution is $360°/32 = 11.25°$. Once again, greater angular resolution could be obtained by employing a larger number of bits.

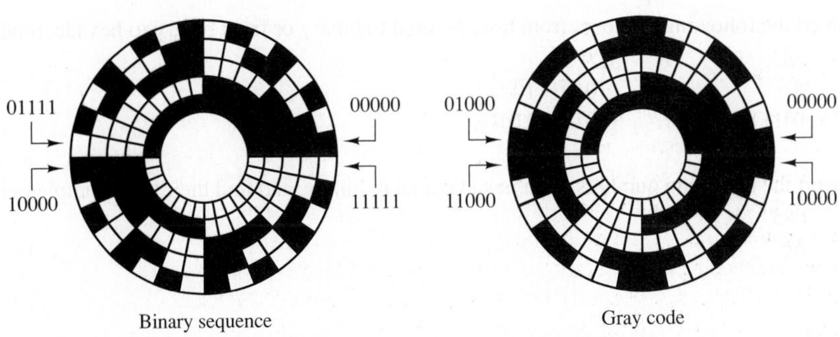

Figure 13.10 Binary and Gray code patterns for angular position encoders

EXAMPLE 13.2 Conversion from Binary to Hexadecimal

LO2

Problem

Convert the following binary numbers to hexadecimal form.

1. 100111
2. 1011101
3. 11001101
4. 101101111001
5. 100110110
6. 1101011011

Solution

Analysis: A simple method for binary to hexadecimal conversion consists of grouping each binary number into 4-bit groups and then performing the conversion for each 4-bit word following Table 13.6:

1. $100111_2 = 0010_2 0111_2 = 27_{16}$
2. $1011101_2 = 0101_2 1101_2 = 5D_{16}$
3. $11001101_2 = 1100_2 1101_2 = CD_{16}$
4. $101101111001_2 = 1011_2 0111_2 1001_2 = B79_{16}$
5. $100110110_2 = 0001_2 0011_2 0100_2 = 136_{16}$
6. $1101011011_2 = 0011_2 0101_2 1011_2 = 35B_{16}$

Comments: Note that we start grouping always from the right-hand side. The reverse process is equally easy: To convert from hexadecimal to binary, replace each hexadecimal number with the equivalent 4-bit binary word.

CHECK YOUR UNDERSTANDING

Convert the following numbers from hexadecimal to binary or from binary to hexadecimal.

a. F83 b. 3C9
c. A6 d. 110101110_2
e. 10111001_2 f. 11011101101_2

Convert the following numbers from hexadecimal to binary, and find their twos complements.

a. F43 b. 2B9
c. A6

Answers: (a) 111110000011, (b) 001111001001, (c) 10100110, (d) 1AE, (e) B9, (f) 6ED
(a) 0000 1011 1101, (b) 0001 0100 0111, (c) 0101 1010

13.3 BOOLEAN ALGEBRA

The mathematics associated with the binary number system (and with the more general field of logic) is called *boolean,* in honor of the English mathematician George Boole, who published a treatise in 1854 entitled *An Investigation of the Laws of Thought, on Which Are Founded the Mathematical Theories of Logic and Probabilities.* The development of a *logical algebra,* as Boole called it, is one of the results of his investigations. The variables in a boolean, or logic, expression can take only one of two values, usually represented by the numbers 0 and 1. These variables are sometimes referred to as true (1) and false (0). This convention is normally referred to as **positive logic.** There is also a **negative logic** convention in which the roles of logic 1 and logic 0 are reversed. In this book we employ only positive logic.

Analysis of **logic functions,** that is, functions of logical (boolean) variables, can be carried out in terms of truth tables. A truth table is a listing of all the possible values that each of the boolean variables can take, and of the corresponding value of the desired function. In the following paragraphs we define the basic logic functions upon which boolean algebra is founded, and we describe each in terms of a set of rules and a truth table; in addition, we introduce **logic gates.** Logic gates are physical devices (see Chapters 10 and 11) that can be used to implement logic functions.

Table 13.9 Rules for logical addition (OR)

$0 + 0 = 0$
$0 + 1 = 1$
$1 + 0 = 1$
$1 + 1 = 1$

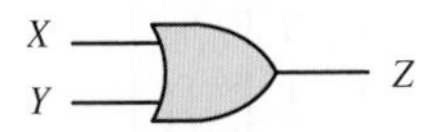

OR gate

X	Y	Z
0	0	0
0	1	1
1	0	1
1	1	1

Truth table

Figure 13.11 Logical addition and the OR gate

AND and OR Gates

The basis of **boolean algebra** lies in the operations of **logical addition,** or the **OR** operation; and **logical multiplication,** or the **AND** operation. Both of these find a correspondence in simple logic gates, as we shall presently illustrate. Logical addition, although represented by the symbol +, differs from conventional algebraic addition, as shown in the last rule listed in Table 13.9. Note that this rule also differs from the last rule of binary addition studied in Section 13.2. Logical addition can be represented by the logic gate called an **OR gate,** whose symbol and whose inputs and outputs are shown in Figure 13.11. The OR gate represents the following logical statement:

If either X or Y is true (1), then Z is true (1). Logical OR **(13.1)**

This rule is embodied in the electronic gates discussed in Chapters 10 and 11, in which a logic 1 corresponds, say, to a 5-V signal and a logic 0 to a 0-V signal.

Logical multiplication is denoted by the center dot $\cdot$ and is defined by the rules of Table 13.10. Figure 13.12 depicts the **AND gate,** which corresponds to this operation. The AND gate corresponds to the following logical statement:

If both X and Y are true (1), then Z is true (1). Logical AND **(13.2)**

One can easily envision logic gates (AND and OR) with an arbitrary number of inputs; three- and four-input gates are not uncommon.

The rules that define a logic function are often represented in tabular form by means of a **truth table.** Truth tables for the AND and OR gates are shown in Figures 13.11 and 13.12. A truth table is nothing more than a tabular summary of all possible outputs of a logic gate, given all possible input values. If the number of inputs is 3, the number of possible combinations grows from 4 to 8, but the basic idea is unchanged. Truth tables are very useful in defining logic functions. A typical logic design problem might specify requirements such as "the output Z shall be logic 1 only when the condition ($X = 1$ AND $Y = 1$) OR ($W = 1$) occurs, and shall be logic 0

Table 13.10 Rules for logical multiplication (AND)

$0 \cdot 0 = 0$
$0 \cdot 1 = 0$
$1 \cdot 0 = 0$
$1 \cdot 1 = 1$

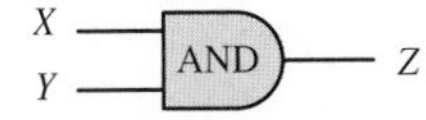

AND gate

X	Y	Z
0	0	0
0	1	0
1	0	0
1	1	1

Truth table

Figure 13.12 Logical multiplication and the AND gate

otherwise." The truth table for this particular logic function is shown in Figure 13.13 as an illustration. The design consists, then, of determining the combination of logic gates that exactly implements the required logic function. Truth tables can greatly simplify this procedure.

The AND and OR gates form the basis of all logic design in conjunction with the **NOT gate.** The NOT gate is essentially an inverter (which can be constructed by using bipolar or field-effect transistors, as discussed in Chapters 10 and 11, respectively), and it provides the complement of the logic variable connected to its input. The complement of a logic variable X is denoted by $\overline{X}$. The NOT gate has only one input, as shown in Figure 13.14.

To illustrate the use of the NOT gate, or inverter, we return to the design example of Figure 13.13, where we required that the output of a logic circuit be $Z = 1$ only if $X = 0$ AND $Y = 1$ OR if $W = 1$. We recognize that except for the requirement $X = 0$, this problem would be identical if we stated it as follows: "The output Z shall be logic 1 only when the condition ($\overline{X} = 1$ AND $Y = 1$) OR ($W = 1$) occurs, and shall be logic 0 otherwise." If we use an inverter to convert X to $\overline{X}$, we see that the required condition becomes ($\overline{X} = 1$ AND $Y = 1$) OR ($W = 1$). The formal solution to this elementary design exercise is illustrated in Figure 13.15.

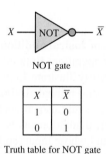

Logic gate realization of the statement "the output Z shall be logic 1 only when the condition ($X = 1$ AND $Y = 1$) OR ($W = 1$) occurs, and shall be logic 0 otherwise."

X	Y	W	Z
0	0	0	0
0	0	1	1
0	1	0	0
0	1	1	1
1	0	0	0
1	0	1	1
1	1	0	1
1	1	1	1

Truth table

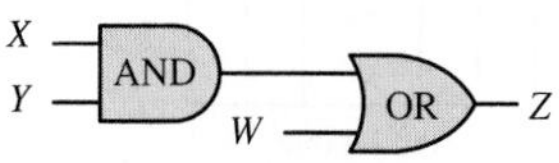

Solution using logic gates

Figure 13.13 Example of logic function implementation with logic gates

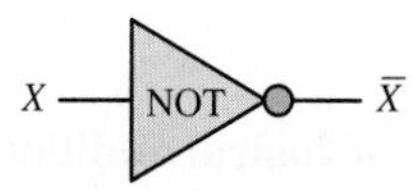

NOT gate

X	$\overline{X}$
1	0
0	1

Truth table for NOT gate

Figure 13.14 Complements and the NOT gate

X	$\overline{X}$	Y	W	Z
0	1	0	0	0
0	1	0	1	1
0	1	1	0	1
0	1	1	1	1
1	0	0	0	0
1	0	0	1	1
1	0	1	0	0
1	0	1	1	1

(Required logic function)

Truth table

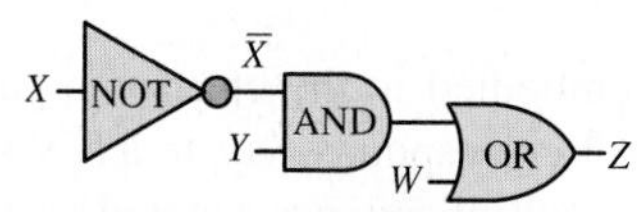

Solution using logic gates

Figure 13.15 Solution of a logic problem using logic gates

In the course of the discussion of logic gates, we make frequent use of truth tables to evaluate logic expressions. A set of basic rules will facilitate this task. Table 13.11 lists some of the rules of boolean algebra; each of these can be proved by using a truth table, as will be shown in examples and exercises. An example proof for rule 16 is given in Figure 13.16 in the form of a truth table. This technique can be employed to prove any of the laws of Table 13.11. From the simple truth table in Figure 13.16, which was obtained step by step, we can clearly see that indeed $X \cdot (X + Y) = X$. This methodology for proving the validity of logical equations is called **proof by perfect induction.** The 19 rules of Table 13.11 can be used to simplify logic expressions.

LO3

Table 13.11 **Rules of boolean algebra**

1.	$0 + X = X$	
2.	$1 + X = 1$	
3.	$X + X = X$	
4.	$X + \overline{X} = 1$	
5.	$0 \cdot X = 0$	
6.	$1 \cdot X = X$	
7.	$X \cdot X = X$	
8.	$X \cdot \overline{X} = 0$	
9.	$\overline{\overline{X}} = X$	
10.	$X + Y = Y + X$	Commutative law
11.	$X \cdot Y = Y \cdot X$	
12.	$X + (Y + Z) = (X + Y) + Z$	Associative law
13.	$X \cdot (Y \cdot Z) = (X \cdot Y) \cdot Z$	
14.	$X \cdot (Y + Z) = X \cdot Y + X \cdot Z$	Distributive law
15.	$X + X \cdot Z = X$	Absorption law
16.	$X \cdot (X + Y) = X$	
17.	$(X + Y) \cdot (X + Z) = X + Y \cdot Z$	
18.	$X + \overline{X} \cdot Y = X + Y$	
19.	$X \cdot Y + Y \cdot Z + \overline{X} \cdot Z = X \cdot Y + \overline{X} \cdot Z$	

X	Y	$X+Y$	$X \cdot (X+Y)$
0	0	0	0
0	1	1	0
1	0	1	1
1	1	1	1

Figure 13.16 Proof of rule 16 by perfect induction

To complete the introductory material on boolean algebra, a few paragraphs need to be devoted to two very important theorems, called **De Morgan's theorems.** These are stated here in the form of logic functions:

$$\overline{(X + Y)} = \overline{X} \cdot \overline{Y} \tag{13.3}$$

DeMorgan's Theorems

$$\overline{(X \cdot Y)} = \overline{X} + \overline{Y} \tag{13.4}$$

These two laws state a very important property of logic functions:

> Any logic function can be implemented by using only OR and NOT gates, or only AND and NOT gates.

De Morgan's laws can easily be visualized in terms of logic gates, as shown in Figure 13.17. The associated truth tables are proofs of these theorems.

The importance of De Morgan's laws lies in the statement of the **duality** that exists between AND and OR operations: Any function can be realized by just one of the two basic operations, plus the complement operation. This gives rise to two families of logic functions, **sums of products** and **product of sums,** as shown in Figure 13.18. Any logical expression can be reduced to one of these two forms. Although the two forms are equivalent, it may well be true that one of the two has a simpler implementation (fewer gates). Example 13.3 illustrates this point.

LO3

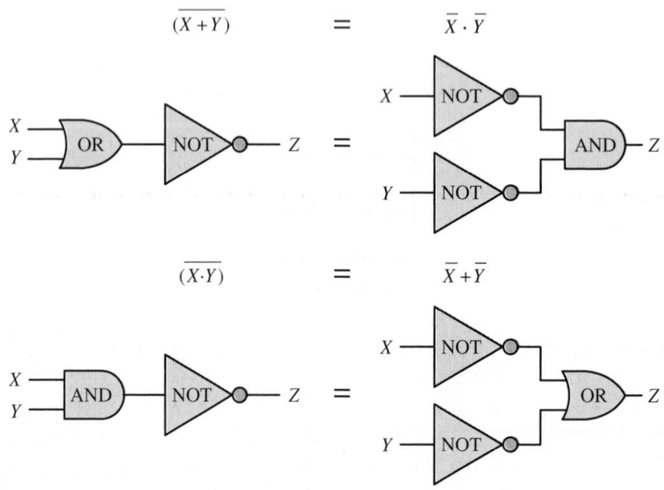

Truth table

X	Y	$Z = \overline{(X+Y)} = \overline{X} \cdot \overline{Y}$
0	0	1
0	1	0
1	0	0
1	1	0

Truth table

X	Y	$Z = \overline{(X+Y)} = \overline{X} + \overline{Y}$
0	0	1
0	1	1
1	0	1
1	1	0

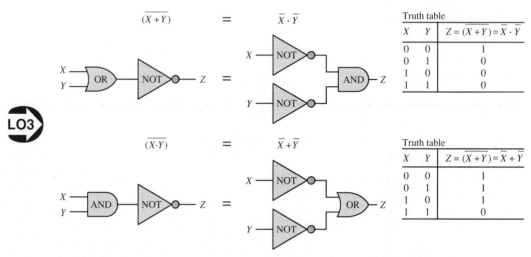

Figure 13.17 De Morgan's laws

LO3

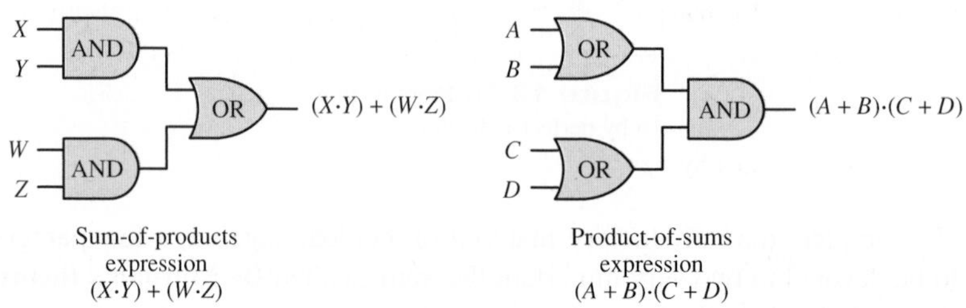

Figure 13.18 Sum-of-products and product-of-sums logic functions

LO3

EXAMPLE 13.3 Simplification of Logical Expression

Problem

Using the rules of Table 13.11, simplify the following function.

$$f(A, B, C, D) = \overline{A} \cdot \overline{B} \cdot D + \overline{A} \cdot B \cdot D + B \cdot C \cdot D + A \cdot C \cdot D$$

Solution

Find: Simplified expression for logical function of four variables.

Analysis:

$$
\begin{aligned}
f &= \overline{A}\cdot\overline{B}\cdot D + \overline{A}\cdot B\cdot D + B\cdot C\cdot D + A\cdot C\cdot D \\
&= \overline{A}\cdot D\cdot\left(\overline{B}+B\right) + B\cdot C\cdot D + A\cdot C\cdot D && \text{Rule 14} \\
&= \overline{A}\cdot D + B\cdot C\cdot D + A\cdot C\cdot D && \text{Rule 4} \\
&= \left(\overline{A} + A\cdot C\right)\cdot D + B\cdot C\cdot D && \text{Rule 14} \\
&= \left(\overline{A} + C\right)\cdot D + B\cdot C\cdot D && \text{Rule 18} \\
&= \overline{A}\cdot D + C\cdot D + B\cdot C\cdot D && \text{Rule 14} \\
&= \overline{A}\cdot D + C\cdot D\cdot(1+B) && \text{Rule 14} \\
&= \overline{A}\cdot D + C\cdot D = \left(\overline{A}+C\right)\cdot D && \text{Rules 2 and 6}
\end{aligned}
$$

FOCUS ON MEASUREMENTS

Fail-Safe Autopilot Logic

This example aims to illustrate the significance of De Morgan's laws and of the duality of the sum-of-products and product-of-sums forms. Suppose that a fail-safe autopilot system in a commercial aircraft requires that, prior to initiating a takeoff or landing maneuver, the following check be passed: two of three possible pilots must be available. The three possibilities are the pilot, the copilot, and the autopilot. Imagine further that there exist switches in the pilot and copilot seats that are turned on by the weight of the crew, and that a self-check circuit exists to verify the proper operation of the autopilot system. Let the variable X denote the pilot state (1 if the pilot is sitting at the controls), Y denote the same condition for the copilot, and Z denote the state of the autopilot, where $Z = 1$ indicates that the autopilot is functioning. Then since we wish two of these conditions to be active before the maneuver can be initiated, the logic function corresponding to "system ready" is

$$f = X\cdot Y + X\cdot Z + Y\cdot Z$$

This can also be verified by the truth table shown below.

Pilot	Copilot	Autopilot	System ready
0	0	0	0
0	0	1	0
0	1	0	0
0	1	1	1
1	0	0	0
1	0	1	1
1	1	0	1
1	1	1	1

The function f defined above is based on the notion of a *positive check;* that is, it indicates when the system is ready. Let us now apply De Morgan's laws to the function f, which is in sum-of-products form:

$$\overline{f} = g = \overline{X\cdot Y + X\cdot Z + Y\cdot Z} = (\overline{X}+\overline{Y})\cdot(\overline{X}+\overline{Z})\cdot(\overline{Y}+\overline{Z})$$

The function g, in product-of-sums form, conveys exactly the same information as the function f, but it performs a negative check; in other words, g verifies the *system not ready condition.* Clearly, whether one chooses to implement the function in one form or another is simply a matter of choice; the two forms give exactly the same information.

LO3

EXAMPLE 13.4 Realizing Logic Functions from Truth Tables

Problem

Realize the logic function described by the truth table below.

A	B	C	y
0	0	0	0
0	0	1	1
0	1	0	0
0	1	1	1
1	0	0	1
1	0	1	1
1	1	0	1
1	1	1	1

Solution

Known Quantities: Value of function $y(A, B, C)$ for each possible combination of logical variables A, B, C.

Find: Logical expression realizing the function y.

Analysis: To determine a logical expression for the function y, first we need to convert the truth table to a logical expression. We do so by expressing y as the sum of the products of the three variables for each combination that yields $y = 1$. If the value of a variable is 1, we use the uncomplemented variable. If it's 0, we use the complemented variable. For example, the second row (first instance of $y = 1$) would yield the term $\overline{A} \cdot \overline{B} \cdot C$. Thus,

$$\begin{aligned} y &= \overline{A} \cdot \overline{B} \cdot C + \overline{A} \cdot B \cdot C + A \cdot \overline{B} \cdot \overline{C} + A \cdot \overline{B} \cdot C + A \cdot B \cdot \overline{C} + A \cdot B \cdot C \\ &= \overline{A} \cdot C(\overline{B} + B) + A \cdot \overline{B} \cdot (\overline{C} + C) + A \cdot B \cdot (\overline{C} + C) \\ &= \overline{A} \cdot C + A \cdot \overline{B} + A \cdot B = \overline{A} \cdot C + A \cdot (\overline{B} + B) = \overline{A} \cdot C + A = A + C \end{aligned}$$

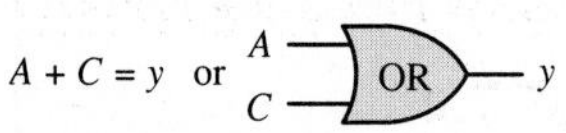

Figure 13.19

Thus, the function is a two-input OR gate, as shown in Figure 13.19.

Comments: The derivation above has made use of two rules from Table 13.11: rules 4 and 18. Could you have predicted that the variable B would not be used in the final realization?

CHECK YOUR UNDERSTANDING

Prepare a step-by-step truth table for the following logic expressions.

a. $\overline{(X + Y + Z)} + (X \cdot Y \cdot Z) \cdot \overline{X}$ b. $\overline{X} \cdot Y \cdot Z + Y \cdot (Z + W)$

c. $(X \cdot \overline{Y} + Z \cdot \overline{W}) \cdot (W \cdot X + \overline{Z} \cdot Y)$

(*Hint:* Your truth table must have 2^n entries, where n is the number of logic variables.)

Answer: a.

X	Y	Z	Result
0	0	0	1
0	0	1	0
0	1	0	0
0	1	1	0
1	0	0	1
1	0	1	1
1	1	0	1
1	1	1	0

b.

X	Y	Z	W	Result
0	0	0	0	0
0	0	0	1	0
0	0	1	0	0
0	0	1	1	0
0	1	0	0	0
0	1	0	1	1
0	1	1	0	1
0	1	1	1	1
1	0	0	0	0
1	0	0	1	0
1	0	1	0	0
1	0	1	1	0
1	1	0	0	0
1	1	0	1	1
1	1	1	0	1
1	1	1	1	1

c.

X	Y	Z	W	Result
0	0	0	0	0
0	0	0	1	0
0	0	1	0	0
0	0	1	1	0
0	1	0	0	0
0	1	0	1	0
0	1	1	0	0
0	1	1	1	0
1	0	0	0	0
1	0	0	1	1
1	0	1	0	0
1	0	1	1	1
1	1	0	0	0
1	1	0	1	1
1	1	1	0	0
1	1	1	1	1

EXAMPLE 13.5 De Morgan's Theorem and Product-of-Sums Expressions

LO3

Problem

Realize the logic function $y = A + B \cdot C$ in product-of-sums form. Implement the solution, using AND, OR, and NOT gates.

Solution

Known Quantities: Logical expression for the function $y(A, B, C)$.

Find: Physical realization using AND, OR, and NOT gates.

Analysis: We use the fact that $\overline{\overline{y}} = y$ and apply De Morgan's theorem as follows:

$$\overline{y} = \overline{A + (B \cdot C)} = \overline{A} \cdot \overline{(B \cdot C)} = \overline{A} \cdot \left(\overline{B} + \overline{C}\right)$$

$$\overline{\overline{y}} = y = \overline{\overline{A} \cdot \left(\overline{B} + \overline{C}\right)}$$

The preceding sum-of-products function is realized using complements of each variable (obtained using NOT gates) and is finally complemented as shown in Figure 13.20.

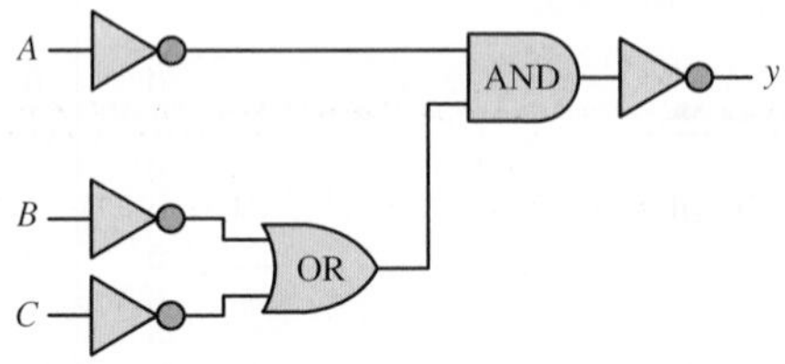

Figure 13.20

Comments: It should be evident that the original sum-of-products expression, which could be implemented with just one AND and one OR gate, has a much more efficient realization. In the next section we show a systematic approach to function minimization.

NAND and NOR Gates

In addition to the AND and OR gates we have just analyzed, the complementary forms of these gates, called NAND and NOR, are very commonly used in practice. In fact, NAND and NOR gates form the basis of most practical logic circuits. Figure 13.21 depicts these two gates and illustrates how they can be easily interpreted in terms of AND, OR, and NOT gates by virtue of De Morgan's laws. You can readily verify that the logic function implemented by the NAND and NOR gates corresponds, respectively, to AND and OR gates followed by an inverter. It is very important to note that, by De Morgan's laws, the NAND gate performs a *logical addition* on the *complements* of the inputs, while the NOR gate performs a *logical multiplication* on the *complements* of the inputs. Functionally, then, any logic function could be implemented with either NOR or NAND gates only.

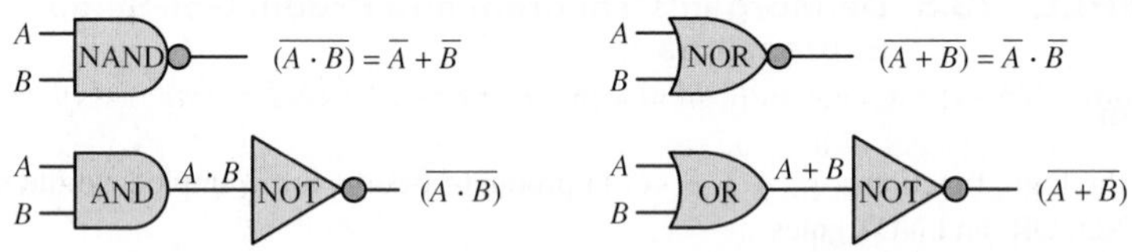

A	B	$\overline{A}$	$\overline{B}$	$\overline{(A \cdot B)}$
0	0	1	1	1
0	1	1	0	1
1	0	0	1	1
1	1	0	0	0

NAND gate

A	B	$\overline{A}$	$\overline{B}$	$\overline{(A + B)}$
0	0	1	1	1
0	1	1	0	0
1	0	0	1	0
1	1	0	0	0

NOR gate

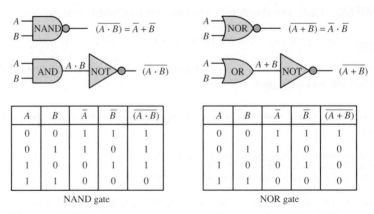

Figure 13.21 Equivalence of NAND and NOR gates with AND and OR gates

The next section shows how to systematically approach the design of logic functions. First, we provide a few examples to illustrate logic design with NAND and NOR gates.

EXAMPLE 13.6 Realizing the AND Function with NAND Gates

LO3

Problem

Use a truth table to show that the AND function can be realized using only NAND gates, and show the physical realization.

Solution

Known Quantities: AND and NAND truth tables.

Find: AND realization using NAND gates.

Assumptions: Consider two-input functions and gates.

Analysis: The truth table below summarizes the two functions:

A	**B**	**NAND** $\overline{A \cdot B}$	**AND** $A \cdot B$
0	0	1	0
0	1	1	0
1	0	1	0
1	1	0	1

A	$B\ (= A)$	$A \cdot B$	$\overline{(A \cdot B)}$
0	0	0	1
1	1	1	0

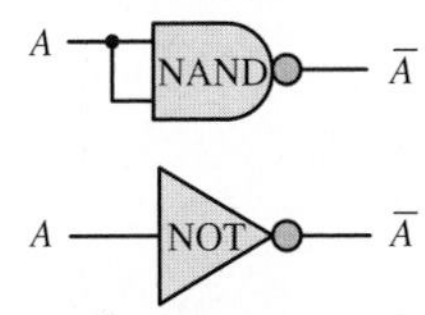

Figure 13.22 NAND gate as an inverter

Clearly, to realize the AND function, we need to simply invert the output of a NAND gate. This is easily accomplished if we observe that a NAND gate with its inputs tied together acts as an inverter; you can verify this in the above truth table by looking at the NAND output for the input combinations 0–0 and 1–1, or by referring to Figure 13.22. The final realization is shown in Figure 13.23.

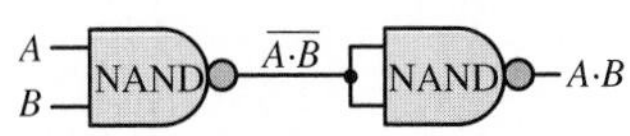

Figure 13.23

Comments: NAND gates naturally implement functions that contain complemented products. Gates that employ negative logic are a natural consequence of the inverting characteristics of transistor switches (refer to Section 10.5). Thus, one should expect that NAND (and NOR) gates are very commonly employed in practice.

CHECK YOUR UNDERSTANDING

Show that one can obtain an OR gate by using NAND gates only. (*Hint:* Use three NAND gates.)

EXAMPLE 13.7 Realizing the AND Function with NOR Gates

LO3

Problem

Show analytically that the AND function can be realized using only NOR gates, and determine the physical realization.

Solution

Known Quantities: AND and NOR functions.

Find: AND realization using NOR gates.

Assumptions: Consider two-input functions and gates.

Analysis: We can solve this problem using De Morgan's theorem. The output of an AND gate can be expressed as $f = A \cdot B$. Using De Morgan's theorem, we write

$$f = \overline{\overline{f}} = \overline{\overline{A \cdot B}} = \overline{\overline{A} + \overline{B}}$$

The above function is implemented very easily if we see that a NOR gate with its input tied together acts as a NOT gate (see Figure 13.24). Thus, the logic circuit of Figure 13.25 provides the desired answer.

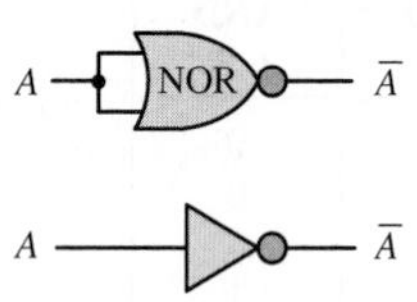

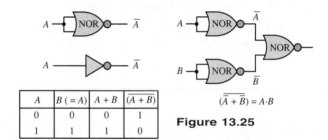

A	$B\,(=A)$	$A+B$	$\overline{(A+B)}$
0	0	0	1
1	1	1	0

Figure 13.24 NOR gate as an inverter

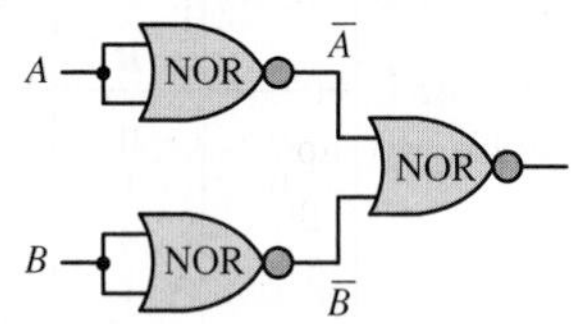

$\overline{(\overline{A} + \overline{B})} = A \cdot B$

Figure 13.25

Comments: NOR gates naturally implement functions that contain complemented sums. Gates that employ negative logic are a natural consequence of the inverting characteristics of transistor switches (refer to Section 10.4). Thus, one should expect that NOR (and NAND) gates are very commonly employed in practice.

CHECK YOUR UNDERSTANDING

Implement the three logic functions of the Check Your Understanding exercise accompanying Example 13.4, using the smallest number of AND, OR, and NOT gates only.

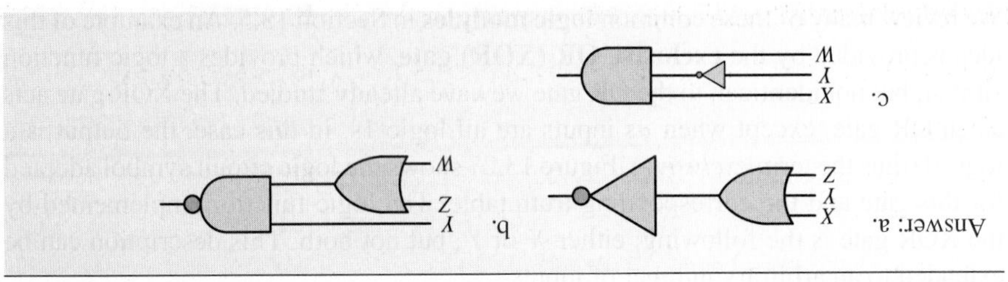

EXAMPLE 13.8 Realizing a Function with NAND and NOR Gates

LO3

Problem

Realize the following function, using only NAND and NOR gates:

$$y = \overline{\overline{(A \cdot B)} + C}$$

Solution

Known Quantities: Logical expression for y.

Find: Realization of y using only NAND and NOR gates.

Assumptions: Consider two-input functions and gates.

Analysis: On the basis of Examples 13.6 and 13.7, we see that we can realize the term $Z = \overline{(A \cdot B)}$ using a two-input NAND gate, and the term $\overline{Z + C}$ using a two-input NOR gate. The solution is shown in Figure 13.26.

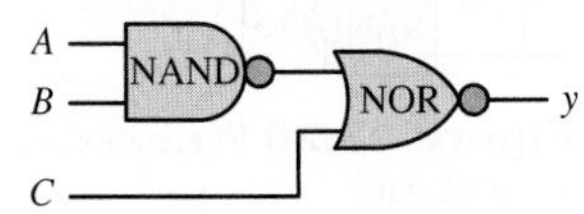

Figure 13.26

CHECK YOUR UNDERSTANDING

Implement the three logic functions of the Check Your Understanding exercise accompanying Example 13.4, using the least number of NAND and NOR gates only. (*Hint:* Use De Morgan's theorems and the fact that $\overline{\overline{f}} = f$.)

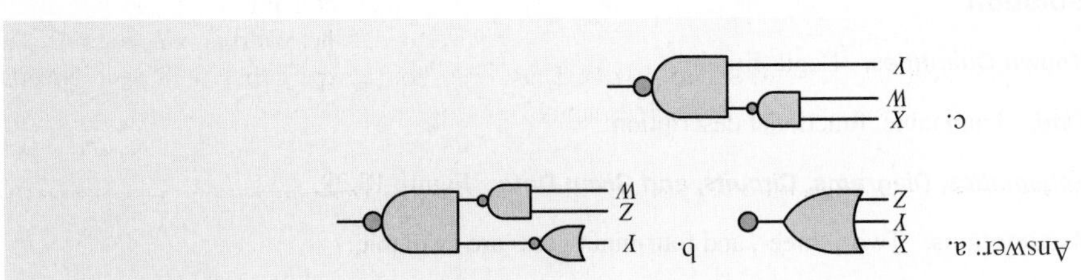

The XOR (Exclusive OR) Gate

It is rather common practice for a manufacturer of integrated circuits to provide common combinations of logic circuits in a single integrated-circuit (IC) package.

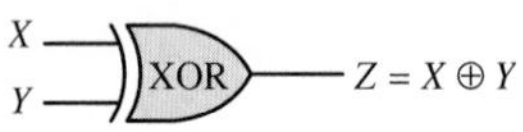

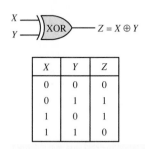

X	Y	Z
0	0	0
0	1	1
1	0	1
1	1	0

Truth table

Figure 13.27 XOR gate

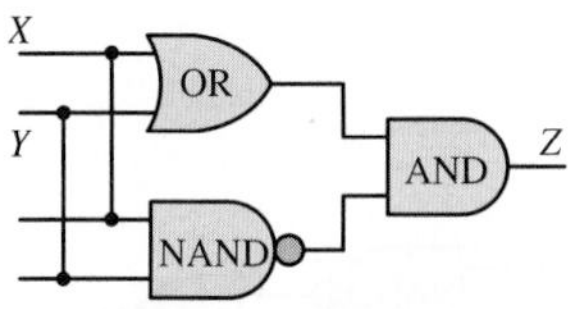

Figure 13.28 Realization of an XOR gate

We review many of these common **logic modules** in Section 13.5. An example of this idea is provided by the **exclusive OR (XOR) gate,** which provides a logic function similar, but not identical, to the OR gate we have already studied. The XOR gate acts as an OR gate, except when its inputs are all logic 1s; in this case, the output is a logic 0 (thus the term *exclusive*). Figure 13.27 shows the logic circuit symbol adopted for this gate and the corresponding truth table. The logic function implemented by the XOR gate is the following: either X or Y, but not both. This description can be extended to an arbitrary number of inputs.

The symbol adopted for the exclusive OR operation is $\oplus$, and so we write

$$Z = X \oplus Y$$

to denote this logic operation. The XOR gate can be obtained by a combination of the basic gates we are already familiar with. For example, if we observe that the XOR function corresponds to $Z = X \oplus Y = (X + Y) \cdot (\overline{X \cdot Y})$, we can realize the XOR gate by means of the circuit shown in Figure 13.28.

Common IC logic gate configurations, device numbers, and data sheets are included in the CD-ROM that accompanies this book. These devices are typically available in both of the two more common device families, TTL and CMOS. The devices listed in the CD-ROM are available in CMOS technology under the numbers SN74AHXX. The same logic gate ICs are also available as TTL devices.

CHECK YOUR UNDERSTANDING

Show that the XOR function can also be expressed as $Z = X \cdot \overline{Y} + Y \cdot \overline{X}$. Realize the corresponding function, using NOT, AND, and OR gates. [*Hint:* Use truth tables for the logic function Z (as defined in the exercise) and for the XOR function.]

EXAMPLE 13.9 Half Adder

Problem

Analyze the half adder circuit of Figure 13.29.

Solution

Known Quantities: Logic circuit.

Find: Truth table, functional description.

Schematics, Diagrams, Circuits, and Given Data: Figure 13.29.

Assumptions: Two-, three-, and four-input gates are available.

Table 13.2 **Rules for addition**

$0 + 0 = 0$
$0 + 1 = 1$
$1 + 0 = 1$
$1 + 1 = 0$ (with a carry of 1)

Analysis: The addition of two binary digits was illustrated in Section 13.2 (see Table 13.2, repeated below). It is important to observe that when both A and B are equal to 1, the sum requires two digits: the lower digit is a 0, and there also is a carry of 1. Thus, the circuit representing this operation must give an output consisting of two digits. Figure 13.29 shows a circuit called *half adder,* that performs binary addition providing two output bits: the sum, S, and the carry, C.

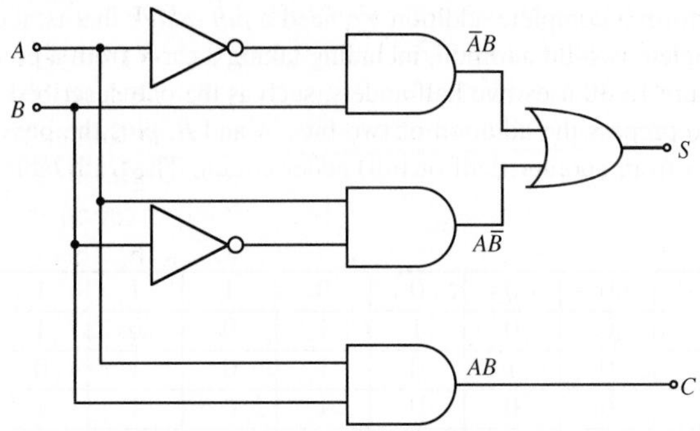

Figure 13.29 Logic circuit realization of a half adder

We can write the rule for addition in the form of a logic statement, as follows: *S is 1 if A is 0 and B is 1, or if A is 1 and B is 0; C is 1 if A and B are 1*. In terms of a logic function, we can express this statement with the following logical expressions:

$$S = \bar{A}B + A\bar{B} \text{ and } C = AB$$

The circuit of Figure 13.29 clearly implements this function using NOT, AND, and OR gates.

Comments: Various other implementations of the half adder are explored in the homework problems.

EXAMPLE 13.10 Full Adder

Problem

Analyze the full adder circuit of Figure 13.30.

Solution

Known Quantities: Logic circuit.

Find: Truth table, functional description.

Schematics, Diagrams, Circuits, and Given Data: Figure 13.30.

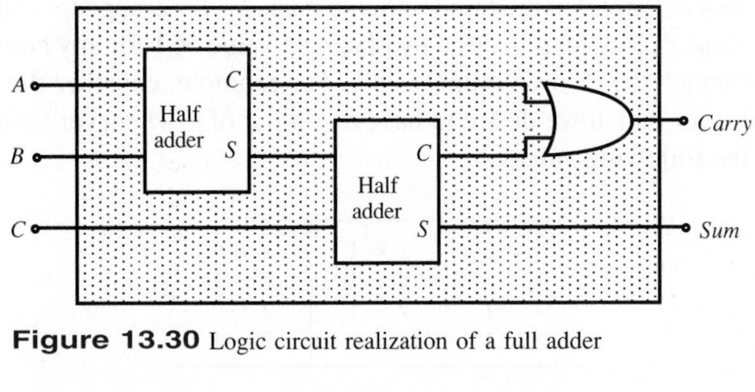

Figure 13.30 Logic circuit realization of a full adder

Analysis: To perform a complete addition we need a *full adder,* that is, a circuit capable of performing a complete two-bit addition, including taking a carry from a preceding operation. The circuit of Figure 13.30 uses two half adders, such as the one described in Example 13.9, and an OR gate to process the addition of two bits, A and B, plus the possible carry from a preceding addition from another (half or full) adder circuit. The truth table below illustrates this operation.

A	0	0	0	0	1	1	1	1
B	0	0	1	1	0	0	1	1
C	0	1	0	1	0	1	0	1
Sum	0	0	0	1	0	1	1	1
Carry	0	1	1	0	1	0	0	1

Truth table for full adder

Comments: To perform the addition of two four-bit "words," we would need a half adder for the first column (LSB), and a full adder for each additional column, that is, three full adders.

13.4 KARNAUGH MAPS AND LOGIC DESIGN

In examining the design of logic functions by means of logic gates, we have discovered that more than one solution is usually available for the implementation of a given logic expression. It should also be clear by now that some combinations of gates can implement a given function more efficiently than others. How can we be assured of having chosen the most efficient realization? Fortunately, there is a procedure that utilizes a map describing all possible combinations of the variables present in the logic function of interest. This map is called a **Karnaugh map,** after its inventor. Figure 13.31 depicts the appearance of Karnaugh maps for two-, three-, and four-variable expressions in two different forms. As can be seen, the row and column assignments for two or more variables are arranged so that all adjacent terms change by only 1 bit. For example, in the two-variable map, the columns next to column 01 are columns 00 and 11. Also note that each map consists of 2^N **cells,** where N is the number of logic variables.

Each cell in a Karnaugh map contains a **minterm,** that is, a product of the N variables that appear in our logic expression (in either uncomplemented or complemented form). For example, for the case of three variables ($N = 3$), there are $2^3 = 8$ such combinations, or minterms: $\overline{X} \cdot \overline{Y} \cdot \overline{Z}$, $\overline{X} \cdot \overline{Y} \cdot Z$, $\overline{X} \cdot Y \cdot \overline{Z}$, $\overline{X} \cdot Y \cdot Z$, $X \cdot \overline{Y} \cdot \overline{Z}$, $X \cdot \overline{Y} \cdot Z$, $X \cdot Y \cdot \overline{Z}$, and $X \cdot Y \cdot Z$. The content of each cell—that is, the minterm—is the product of the variables appearing at the corresponding vertical and horizontal coordinates. For example, in the three-variable map, $X \cdot Y \cdot \overline{Z}$ appears at the intersection of $X \cdot Y$ and $\overline{Z}$. The map is filled by placing a value of 1 for any combination of variables for which the desired output is a 1. For example, consider the function of three variables for which we desire to have an output of 1 whenever variables X, Y, and Z have the following values:

$X = 0$	$Y = 1$	$Z = 0$
$X = 0$	$Y = 1$	$Z = 1$
$X = 1$	$Y = 1$	$Z = 0$
$X = 1$	$Y = 1$	$Z = 1$

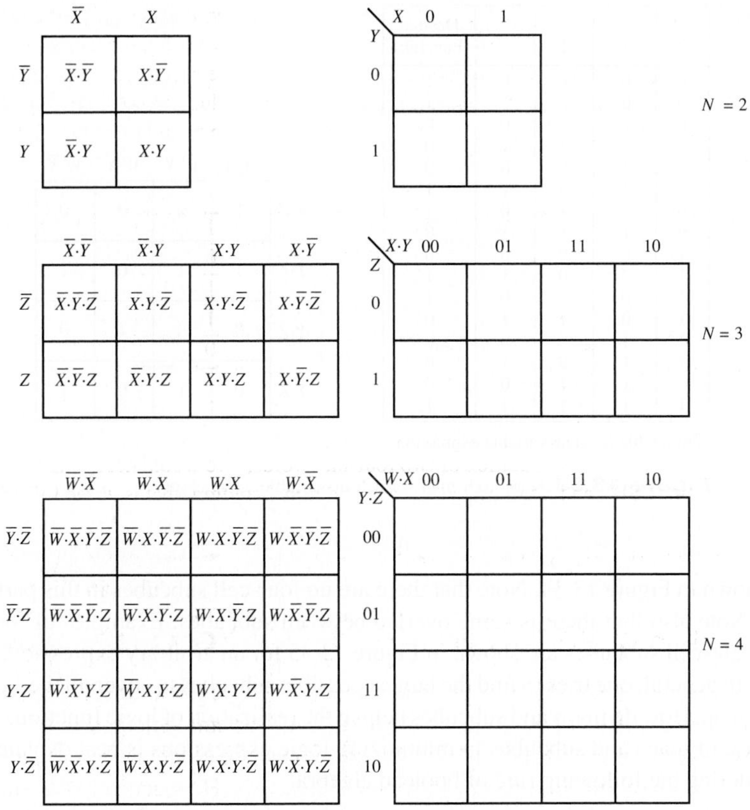

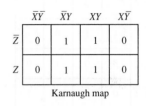

Figure 13.31 Two-, three-, and four-variable Karnaugh maps

	$\overline{X}\overline{Y}$	$\overline{X}Y$	XY	$X\overline{Y}$
$\overline{Z}$	0	1	1	0
Z	0	1	1	0

Karnaugh map

X	Y	Z	Desired Function
0	0	0	0
0	0	1	0
0	1	0	1
0	1	1	1
1	0	0	0
1	0	1	0
1	1	0	1
1	1	1	1

Truth table

Figure 13.32 Truth table and Karnaugh map representations of a logic function

The same truth table is shown in Figure 13.32 together with the corresponding Karnaugh map.

The Karnaugh map provides an immediate view of the values of the function in graphical form. Further, the arrangement of the cells in the Karnaugh map is such that any two adjacent cells contain minterms that vary in only one variable. This property, as will be verified shortly, is quite useful in the design of logic functions by means of logic gates, especially if we consider the map to be continuously wrapping around itself, as if the top and bottom, and right and left, edges were touching. For the three-variable map given in Figure 13.31, for example, the cell $X \cdot \overline{Y} \cdot \overline{Z}$ is adjacent to $\overline{X} \cdot \overline{Y} \cdot \overline{Z}$ if we "roll" the map so that the right edge touches the left. Note that these two cells differ only in the variable X, a property that we earlier claimed adjacent cells have.[1]

Shown in Figure 13.33 is a more complex, four-variable logic function, which will serve as an example in explaining how Karnaugh maps can be used directly to implement a logic function. First, we define a *subcube* as a set of 2^m adjacent cells with logical value 1, for $m = 1, 2, 3, \ldots, N$. Thus, a subcube can consist of 1, 2, 4, 8, 16, 32, ... cells. All possible subcubes for the four-variable map of Figure 13.31

[1]A useful rule to remember is that in a two-variable map, there are two minterms adjacent to any given minterm; in a three-variable map, three minterms are adjacent to any given minterm; in a four-variable map, the number is four, and so on.

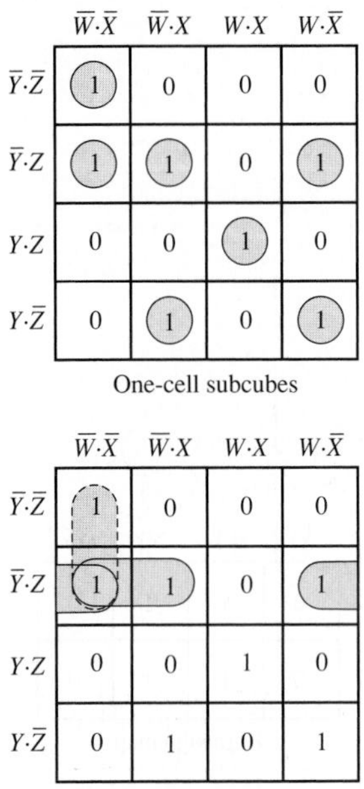

One-cell subcubes

Two-cell subcubes

Figure 13.34 One- and two-cell subcubes for the Karnaugh map of Figure 13.31

X	Y	Y	Z	Desired Function
0	0	0	0	1
0	0	0	1	1
0	0	1	0	0
0	0	1	1	0
0	1	0	0	0
0	1	0	1	1
0	1	1	0	1
0	1	1	1	0
1	0	0	0	0
1	0	0	1	1
1	0	1	0	1
1	0	1	1	0
1	1	0	0	0
1	1	0	1	0
1	1	1	0	0
1	1	1	1	1

Truth table for four-variable expression

	$\overline{W}\cdot\overline{X}$	$\overline{W}\cdot X$	$W\cdot X$	$W\cdot\overline{X}$
$\overline{Y}\cdot\overline{Z}$	1	0	0	0
$\overline{Y}\cdot Z$	1	1	0	1
$Y\cdot Z$	0	0	1	0
$Y\cdot\overline{Z}$	0	1	0	1

Figure 13.33 Karnaugh map for a four-variable expression

are shown in Figure 13.34. Note that there are no four-cell subcubes in this particular case. Note also that there is some overlap between subcubes. Examples of four-cell and eight-cell subcubes are shown in Figure 13.35 for an arbitrary expression.

In general, one tries to find the largest possible subcubes to cover all the 1 entries in the map. How do maps and subcubes help in the realization of logic functions, then? The use of maps and subcubes in minimizing logic expressions is best explained by considering the following rule of boolean algebra:

$$Y \cdot X + Y \cdot \overline{X} = Y$$

where the variable Y could represent a product of logic variables [e.g., we could similarly write $(Z \cdot W) \cdot X + (Z \cdot W) \cdot \overline{X} = Z \cdot W$ with $Y = Z \cdot W$]. This rule is easily proved by factoring Y

$$Y \cdot (X + \overline{X})$$

and observing that $X + \overline{X} = 1$ always. Then it should be clear that variable X need not appear in the expression at all.

Let us apply this rule to a more complex logic expression, to verify that it can also apply to this case. Consider the logic expression

$$\overline{W} \cdot X \cdot \overline{Y} \cdot Z + \overline{W} \cdot \overline{X} \cdot \overline{Y} \cdot Z + W \cdot \overline{X} \cdot \overline{Y} \cdot Z + W \cdot X \cdot \overline{Y} \cdot Z$$

and factor it as follows:

$$\overline{W} \cdot Z \cdot \overline{Y} \cdot (X + \overline{X}) + W \cdot \overline{Y} \cdot Z \cdot (\overline{X} + X) = \overline{W} \cdot Z \cdot \overline{Y} + W \cdot \overline{Y} \cdot Z$$
$$= \overline{Y} \cdot Z \cdot (\overline{W} + W) = \overline{Y} \cdot Z$$

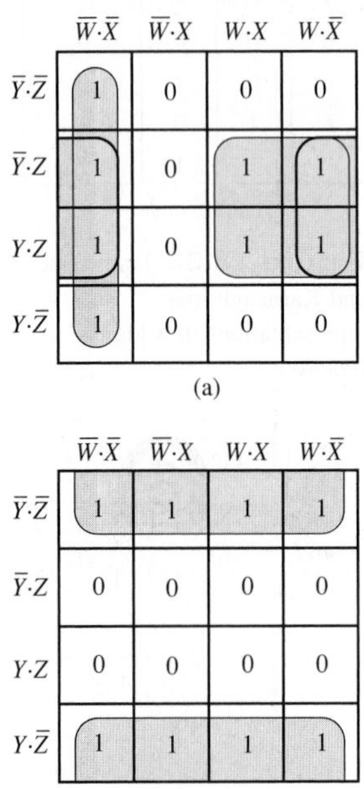

(a)

(b)

Figure 13.35 Four- and eight-cell subcubes for an arbitrary logic function

That is quite a simplification! If we consider, now, a map in which we place a 1 in the cells corresponding to the minterms $\overline{W} \cdot X \cdot \overline{Y} \cdot Z$, $\overline{W} \cdot \overline{X} \cdot \overline{Y} \cdot Z$, $W \cdot \overline{X} \cdot \overline{Y} \cdot Z$, and $W \cdot X \cdot \overline{Y} \cdot Z$, forming the previous expression, we obtain the Karnaugh map of Figure 13.36. It can easily be verified that the map of Figure 13.36 shows a single four-cell subcube corresponding to the term $\overline{Y} \cdot Z$.

We have not established formal rules yet, but it definitely appears that the map method for simplifying boolean expressions is a convenient tool. In effect, the map has performed the algebraic simplification automatically! We can see that in any subcube, one or more of the variables present will appear in both complemented *and* uncomplemented forms in all their combinations with the other variables. These variables can be eliminated. As an illustration, in the *eight-cell* subcube case of Figure 13.37, the full-blown expression would be

$$\overline{W}\cdot\overline{X}\cdot\overline{Y}\cdot\overline{Z}+\overline{W}\cdot X\cdot\overline{Y}\cdot\overline{Z}+W\cdot X\cdot\overline{Y}\cdot\overline{Z}+W\cdot\overline{X}\cdot\overline{Y}\cdot\overline{Z}$$
$$+\overline{W}\cdot\overline{X}\cdot Y\cdot\overline{Z}+\overline{W}\cdot X\cdot Y\cdot\overline{Z}+W\cdot X\cdot Y\cdot\overline{Z}+W\cdot\overline{X}\cdot Y\cdot\overline{Z}$$

	$\overline{W}\cdot\overline{X}$	$\overline{W}\cdot X$	$W\cdot X$	$W\cdot\overline{X}$
$\overline{Y}\cdot\overline{Z}$	0	0	0	0
$\overline{Y}\cdot Z$	1	1	1	1
$Y\cdot Z$	0	0	0	0
$Y\cdot\overline{Z}$	0	0	0	0

Figure 13.36 Karnaugh map for the function $\overline{W}\cdot X\cdot\overline{Y}\cdot Z+\overline{W}\cdot\overline{X}\cdot\overline{Y}\cdot Z+W\cdot\overline{X}\cdot\overline{Y}\cdot Z+W\cdot X\cdot\overline{Y}\cdot Z$

However, if we consider the eight-cell subcube, we note that the three variables X, W, and Z appear in both complemented and uncomplemented form in all their combinations with the other variables and thus can be removed from the expression. This reduces the seemingly unwieldy expression simply to $\overline{Y}$! In logic design terms, a simple inverter is sufficient to implement the expression.

The example just shown is a particularly simple one, but it illustrates how simple it can be to determine the minimal expression for a logic function. Clearly, the larger a subcube, the greater the simplification that will result. For subcubes that do not intersect, as in the previous example, the solution can be found easily and is unique.

	$\overline{W}\cdot\overline{X}$	$\overline{W}\cdot X$	$W\cdot X$	$W\cdot\overline{X}$
$\overline{Y}\cdot\overline{Z}$	1	1	1	1
$\overline{Y}\cdot Z$	1	1	1	1
$Y\cdot Z$	0	0	0	0
$Y\cdot\overline{Z}$	0	0	0	0

Figure 13.37

Sum-of-Products Realizations

Although not explicitly stated, the logic functions of the preceding section were all in sum-of-products form. As you know, it is also possible to realize logic functions in product-of-sums form. This section discusses the implementation of logic functions in sum-of-products form and gives a set of design rules. The next section will do the same for product-of-sums form logical expressions. The following rules are a useful aid in determining the minimal sum-of-products expression:

FOCUS ON METHODOLOGY

LO4

SUM-OF-PRODUCTS REALIZATIONS

1. Begin with isolated cells. These must be used as they are, since no simplification is possible.
2. Find all cells that are adjacent to only one other cell, forming two-cell subcubes.
3. Find cells that form four-cell subcubes, eight-cell subcubes, and so forth.
4. The minimal expression is formed by the collection of the *smallest number of maximal subcubes*.

Examples 13.11 through 13.15 illustrate the application of these principles to a variety of problems.

LO4

EXAMPLE 13.11 Logic Circuit Design Using Karnaugh Maps

Problem

Design a logic circuit that implements the truth table of Figure 13.38.

A	B	C	D	y
0	0	0	0	1
0	0	0	1	1
0	0	1	0	1
0	0	1	1	0
0	1	0	0	0
0	1	0	1	1
0	1	1	0	0
0	1	1	1	0
1	0	0	0	1
1	0	0	1	1
1	0	1	0	0
1	0	1	1	1
1	1	0	0	0
1	1	0	1	1
1	1	1	0	0
1	1	1	1	1

Figure 13.38

Solution

Known Quantities: Truth table for $y(A, B, C, D)$.

Find: Realization of y.

Assumptions: Two-, three-, and four-input gates are available.

Analysis: We use the Karnaugh map of Figure 13.39, which is shown with values of 1 and 0 already in place. We recognize four subcubes in the map; three are four-cell subcubes, and one is a two-cell subcube. The expressions for the subcubes are $\overline{A} \cdot \overline{B} \cdot \overline{D}$ for the two-cell subcube; $\overline{B} \cdot \overline{C}$ for the subcube that wraps around the map; $\overline{C} \cdot D$ for the 4-by-1 subcube; and $A \cdot D$ for the square subcube at the bottom of the map. Thus, the expression for y is

$$y = \overline{A} \cdot \overline{B} \cdot \overline{D} + \overline{B} \cdot \overline{C} + \overline{C}D + AD$$

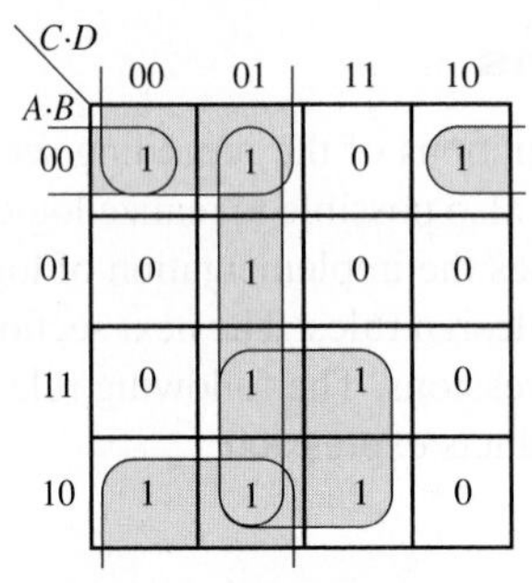

Figure 13.39 Karnaugh map for Example 13.11

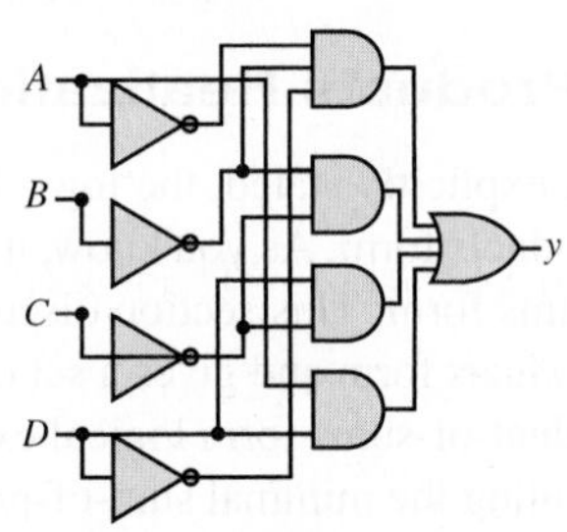

Figure 13.40 Logic circuit realization of Karnaugh map of Figure 13.39

The implementation of the above function with logic gates is shown in Figure 13.40.

Comments: The Karnaugh map covering of Figure 13.39 is a sum-of-products expression because we covered the map using the 1s.

CHECK YOUR UNDERSTANDING

Simplify the following expression, using a Karnaugh map.

$$\overline{W} \cdot \overline{X} \cdot \overline{Y} \cdot \overline{Z} + \overline{W} \cdot \overline{X} \cdot Y \cdot \overline{Z} + W \cdot X \cdot \overline{Y} \cdot \overline{Z} + W \cdot \overline{X} \cdot \overline{Y} \cdot \overline{Z} + W \cdot \overline{X} \cdot Y \cdot \overline{Z}$$
$$+ W \cdot X \cdot Y \cdot \overline{Z}$$

Simplify the following expression, using a Karnaugh map.

$$\overline{W}\cdot\overline{X}\cdot\overline{Y}\cdot\overline{Z}+\overline{W}\cdot\overline{X}\cdot Y\cdot\overline{Z}+W\cdot X\cdot\overline{Y}\cdot\overline{Z}+W\cdot\overline{X}\cdot\overline{Y}\cdot\overline{Z}$$
$$+W\cdot\overline{X}\cdot Y\cdot\overline{Z}+\overline{W}\cdot X\cdot\overline{Y}\cdot\overline{Z}$$

Answers: $W\cdot\overline{Z}+\overline{X}\cdot\overline{Z}$; $\overline{Y}\cdot\overline{Z}+\overline{X}\cdot\overline{Z}$

EXAMPLE 13.12 Deriving a Sum-of-Products Expression from a Logic Circuit

LO4

Problem

Derive the truth table and minimum sum-of-products expression for the circuit of Figure 13.41.

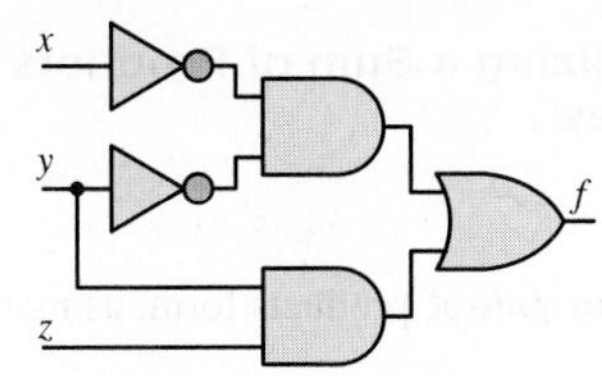

Figure 13.41

Solution

Known Quantities: Logic circuit representing $f(x, y, z)$.

Find: Expression for f and corresponding truth table.

Analysis: To determine the truth table, we write the expression corresponding to the logic circuit of Figure 13.41:

$$f=\overline{x}\cdot\overline{y}+y\cdot z$$

The truth table corresponding to this expression and the corresponding Karnaugh map with sum-of-products covering are shown in Figure 13.42.

x	y	z	f
0	0	0	1
0	0	1	1
0	1	0	0
0	1	1	1
1	0	0	0
1	0	1	0
1	1	0	0
1	1	1	1

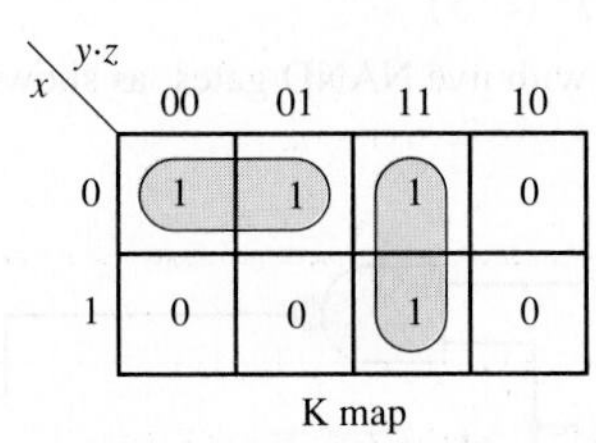

Figure 13.42

Comments: If we used 0s in covering the Karnaugh map for this example, the resulting expression would be a product of sums. You may verify that, in the case of this example, the complexity of the circuit would be unchanged. Note also that there exists a third subcube ($x = 0$, $yz = 01, 11$) that is not used because it does not help minimize the solution.

CHECK YOUR UNDERSTANDING

The function y of Example 13.11 can be obtained with fewer gates if we use gates with three or four inputs. Find the minimum number of gates needed to obtain this function.

Answer: Nine gates

LO4

EXAMPLE 13.13 Realizing a Sum of Products Using Only NAND Gates

Problem

Realize the following function in sum-of-products form, using only two-input NAND gates.

$$f = (\overline{x} + \overline{y}) \cdot (y + \overline{z})$$

Solution

Known quantities: $f(x, y, z)$.

Find: Logic circuit for f using only NAND gates.

Analysis: The first step is to convert the expression for f into an expression that can be easily implemented with NAND gates. We observe that direct application of De Morgan's theorem yields

$$\overline{x} + \overline{y} = \overline{x \cdot y}$$

$$y + \overline{z} = \overline{z \cdot \overline{y}}$$

Thus, we can write the function as

$$f = (\overline{x \cdot y}) \cdot \left(\overline{z \cdot \overline{y}}\right)$$

and implement it with five NAND gates, as shown in Figure 13.43.

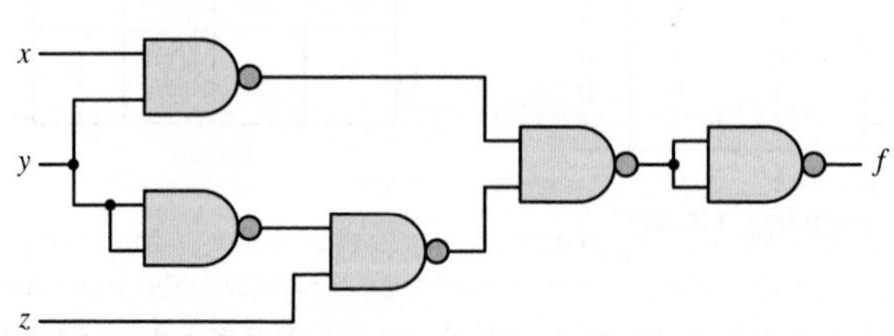

Figure 13.43

Comments: Note that we used two NAND gates as inverters—one to obtain $\overline{y}$, the other to invert the output of the fourth NAND gate, equal to $\overline{\overline{(x \cdot y)} \cdot \overline{(z \cdot \overline{y})}}$.

EXAMPLE 13.14 Simplifying Expressions by Using Karnaugh Maps

Problem

Simplify the following expression by using a Karnaugh map.

$$f = x \cdot y + \overline{x} \cdot z + y \cdot z$$

Solution

Known Quantities: $f(x, y, z)$.

Find: Minimal expression for f.

Analysis: We cover a three-term Karnaugh map to reflect the expression given above. The result is shown in Figure 13.44. It is clear that the Karnaugh map can be covered by using just two terms (subcubes): $f = x \cdot y + \overline{x} \cdot z$. Thus, the term $y \cdot z$ is redundant.

Comments: The Karnaugh map covering clearly shows that the term $y \cdot z$ corresponds to covering a third two-cell subcube vertically intersecting the two horizontal two-cell subcubes already shown. Clearly, the third subcube is redundant.

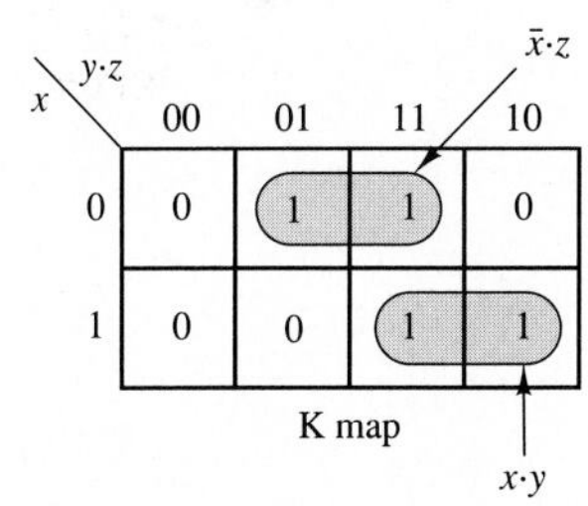

Figure 13.44

EXAMPLE 13.15 Simplifying a Logic Circuit by Using the Karnaugh Map

Problem

Derive the Karnaugh map corresponding to the circuit of Figure 13.45, and use the resulting map to simplify the expression.

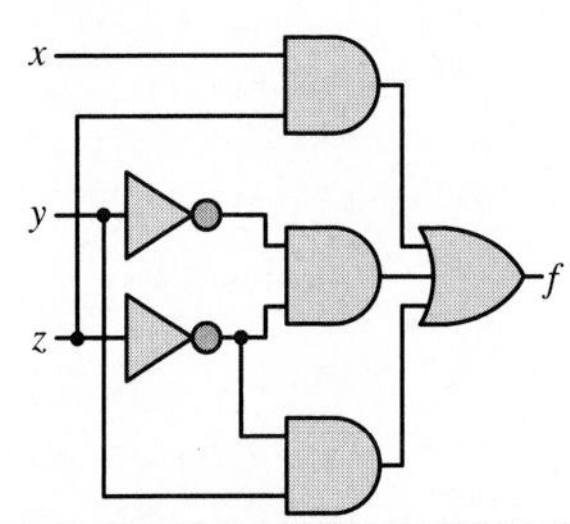

Figure 13.45

Solution

Known Quantities: Logic circuit.

Find: Simplified logic circuit.

Analysis: We first determine the expression $f(x, y, z)$ from the logic circuit:

$$f = (x \cdot z) + (\overline{y} \cdot \overline{z}) + (y \cdot \overline{z})$$

This expression leads to the Karnaugh map shown in Figure 13.46. Inspection of the Karnaugh map reveals that the map could have been covered more efficiently by using four-cell subcubes. The improved map covering, corresponding to the simpler function $f = x + \overline{z}$, and the resulting logic circuit are shown in Figure 13.47.

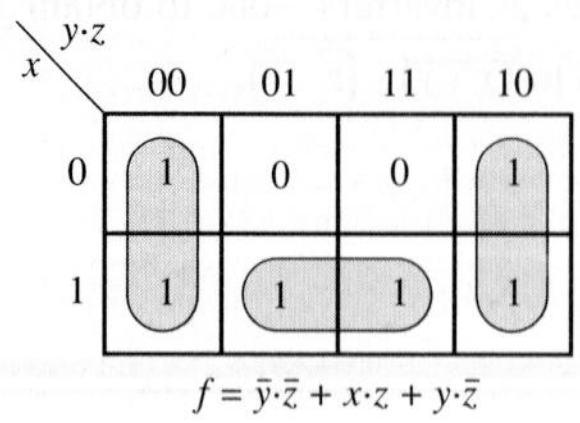

y·z
x
00
01
11
10
0
1
0
0
1
1
1
1
1
1
f = y̅·z̅ + x·z + y·z̅

Figure 13.46

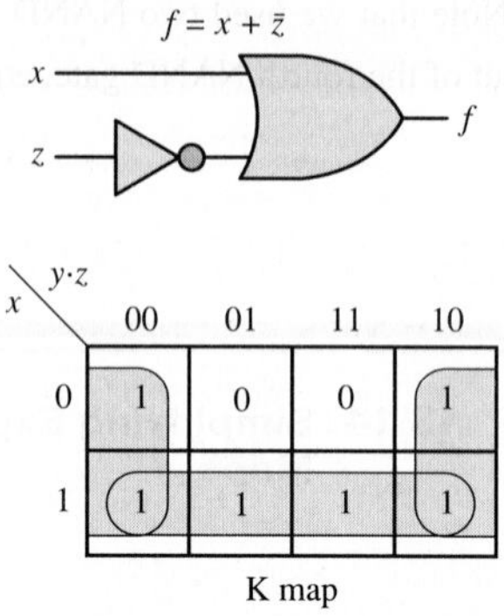

Figure 13.47

Comments: In general, one wishes to cover the largest possible subcubes in a Karnaugh map.

Product-of-Sums Realizations

Thus far, we have exclusively worked with sum-of-products expressions, that is, logic functions of the form $A \cdot B + C \cdot D$. We know, however, that De Morgan's laws state that there is an equivalent form that appears as a product of sums, for example, $(W + Y) \cdot (Y + Z)$. The two forms are completely equivalent logically, but one of the two forms may lead to a realization involving a smaller number of gates. When using Karnaugh maps, we may obtain the product-of-sums form very simply by following these rules:

LO4

FOCUS ON METHODOLOGY

PRODUCT-OF-SUMS REALIZATIONS

1. Solve for the 0s exactly as for the 1s in sum-of-products expressions.
2. Complement the resulting expression.

The same principles stated earlier apply in covering the map with subcubes and determining the minimal expression. Examples 13.16 and 13.17 illustrate how one form may result in a more efficient solution than the other.

LO4

EXAMPLE 13.16 Comparison of Sum-of-Products and Product-of-Sums Designs

Problem

Realize the function f described by the accompanying truth table, using both 0 and 1 coverings in the Karnaugh map.

x	y	z	f
0	0	0	0
0	0	1	1
0	1	0	1
0	1	1	1
1	0	0	1
1	0	1	1
1	1	0	0
1	1	1	0

Solution

Known Quantities: Truth table for logic function.

Find: Realization in both sum-of-products and product-of-sums forms.

Analysis:

1. *Product-of-sums expression.* Product-of-sums expressions use 0s to determine the logical expression from a Karnaugh map. Figure 13.48 depicts the Karnaugh map covering with 0s, leading to the expression

$$f = (x + y + z) \cdot (\overline{x} + \overline{y})$$

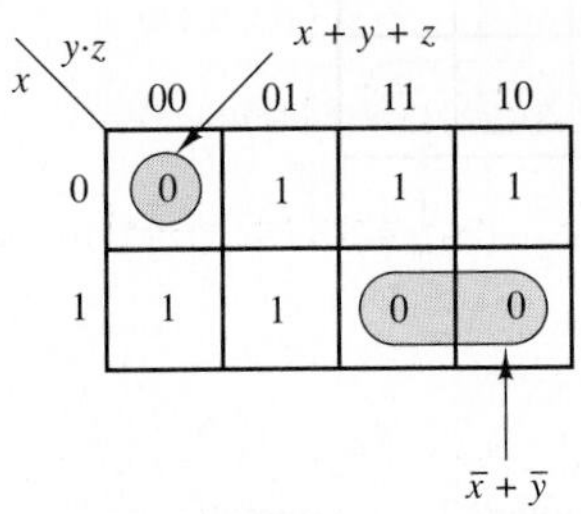

Figure 13.48

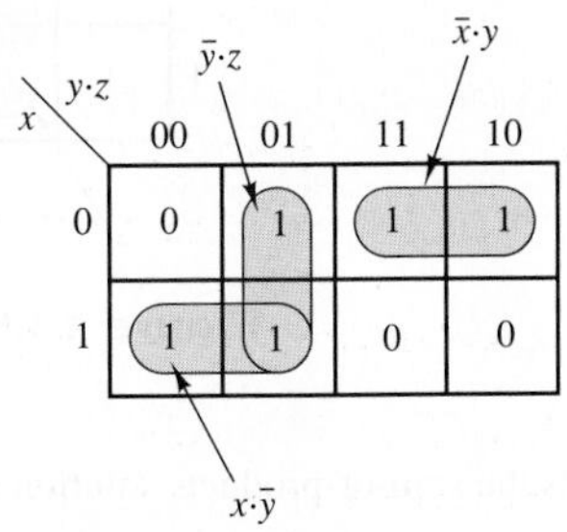

Figure 13.49

2. *Sum-of-products expression.* Sum-of-products expressions use 1s to determine the logical expression from a Karnaugh map. Figure 13.49 depicts the Karnaugh map covering with 1s, leading to the expression

$$f = (\overline{x} \cdot y) + (x \cdot \overline{y}) + (\overline{y} \cdot z)$$

Comments: The product-of-sums solution requires the use of five gates (two OR, two NOT, and one AND), while the sum-of-products solution will use six gates (one OR, two NOT, and three AND). Thus, solution 1 leads to the simpler design.

CHECK YOUR UNDERSTANDING

Verify that the product-of-sums expression for Example 13.16 can be realized with fewer gates.

EXAMPLE 13.17 Product-of-Sums Design

LO4

Problem

Realize the function f described by the accompanying truth table in minimal product-of-sums form. Draw the corresponding Karnaugh map.

x	y	z	f
0	0	0	1
0	0	1	0
0	1	0	1
0	1	1	0
1	0	0	1
1	0	1	0
1	1	0	0
1	1	1	0

Solution

Known Quantities: Truth table for logic function.

Find: Realization in minimal product-of-sums forms.

Analysis: We cover the Karnaugh map of Figure 13.50 using 0s, and we obtain the following function:

$$f = \bar{z} \cdot (\bar{x} + \bar{y})$$

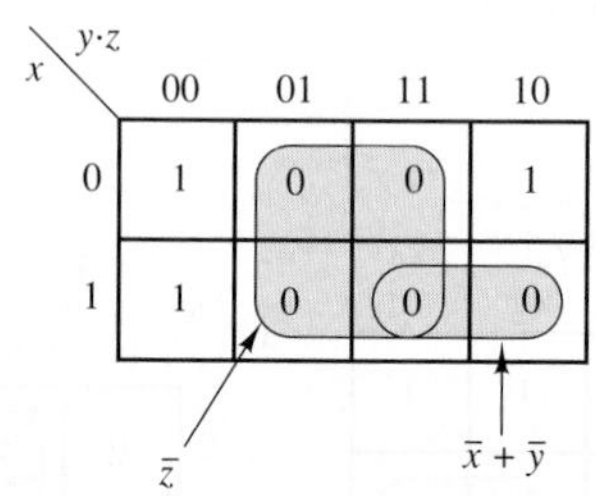

Figure 13.50

Comments: Is the sum-of-products solution simpler? Try it yourself.

CHECK YOUR UNDERSTANDING

Would a sum-of-products realization for Example 13.17 require fewer gates?

Answer: No

Safety Circuit for Operation of a Stamping Press

Problem:

In this example, the techniques illustrated in the preceding examples are applied to a practical situation. To operate a stamping press, an operator must press two buttons (b_1 and b_2) 1 m apart from each other and away from the press (this ensures that the operator's hands cannot be caught in the press). When the buttons are pressed, the logical variables b_1 and b_2 are equal to 1. Thus, we can define a new variable $A = b_1 \cdot b_2$; when $A = 1$, the operator's hands are safely away from the press. In addition to the safety requirement, however, other conditions must be satisfied before the operator can activate the press. The press is designed to operate on one of two workpieces, part I and part II, but not both. Thus, acceptable logic states for the press to be operated are "part I is in the press, but not part II" and "part II is in the press, but not part I." If we denote the presence of part I in the press by the logical variable $B = 1$ and the presence of part II by the

(Continued)

logical variable $C = 1$, we can then impose additional requirements on the operation of the press. For example, a robot used to place either part in the press could activate a pair of switches (corresponding to logical variables B and C) indicating which part, if any, is in the press. Finally, for the press to be operable, it must be "ready," meaning that it has to have completed any previous stamping operation. Let the logical variable $D = 1$ represent the ready condition. We have now represented the operation of the press in terms of four logical variables, summarized in the truth table of Table 13.12. Note that only two combinations of the logical variables will result in operation of the press: $ABCD = 1011$ and $ABCD = 1101$. You should verify that these two conditions correspond to the desired operation of the press. Using a Karnaugh map, realize the logic circuitry required to implement the truth table shown.

Table 13.12 Conditions for operation of stamping press

(A) $b_1 \cdot b_2$*	(B) Part I is in press	(C) Part II is in press	(D) Press is operable	Press operation 1 = pressing; 0 = not pressing
0	0	0	0	0
0	0	0	1	0
0	0	1	0	0
0	0	1	1	0
0	1	0	0	0
0	1	0	1	0
0	1	1	0	0
0	1	1	1	0
1	0	0	0	0
1	0	0	1	0
1	0	1	0	0
1	0	1	1	1
1	1	0	0	0
1	1	0	1	1
1	1	1	0	0
1	1	1	1	0

*Both buttons (b_1, b_2) must be pressed for this to be a 1.

Solution:

Table 13.12 can be converted to a Karnaugh map, as shown in Figure 13.51. Since there are many more 0s than 1s in the table, the use of 0s in covering the map will lead to greater simplification. This will result in a product-of-sums expression. The four subcubes shown in Figure 13.51 yield the equation

$$A \cdot D \cdot (C + B) \cdot (\overline{C} + \overline{B})$$

By De Morgan's law, this equation is equivalent to

$$A \cdot D \cdot (C + B) \cdot \overline{(C \cdot B)}$$

which can be realized by the circuit of Figure 13.52.

For the purpose of comparison, the corresponding sum-of-products circuit is shown in Figure 13.53. Note that this circuit employs a greater number of gates and will therefore lead to a more expensive design.

(*Continued*)

(*Concluded*)

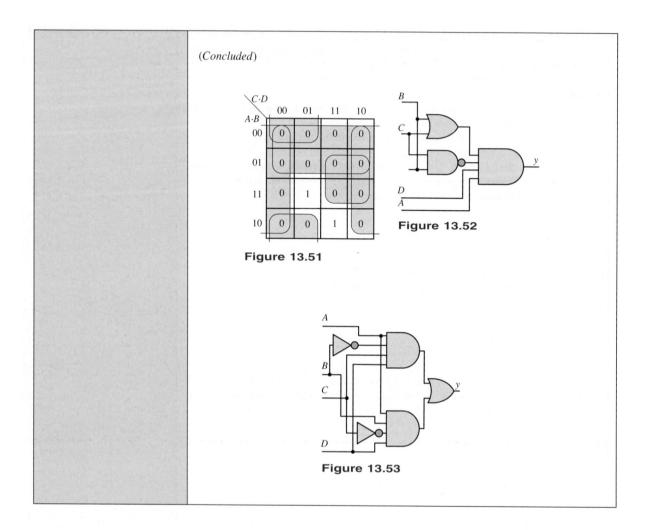

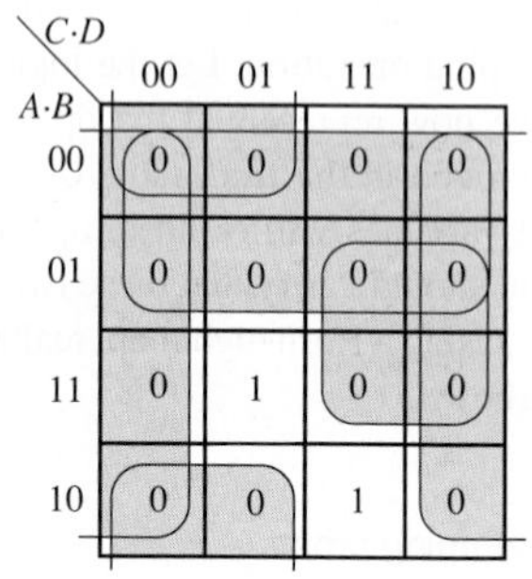

Figure 13.51

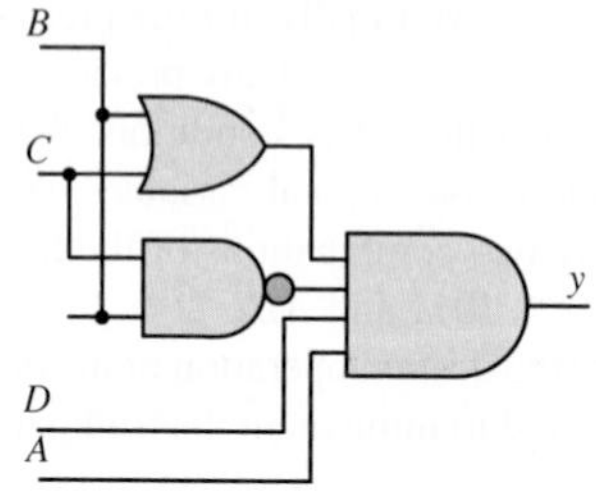

Figure 13.52

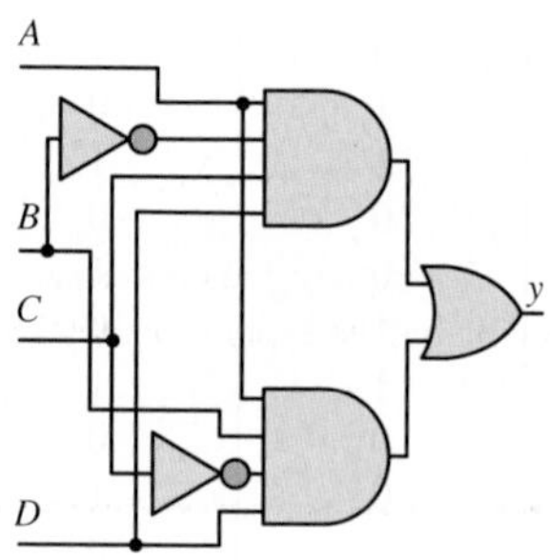

Figure 13.53

CHECK YOUR UNDERSTANDING

Prove that the circuit of Figure 13.53 can also be obtained from the sum of products.

Don't Care Conditions

Another simplification technique may be employed whenever the value of the logic function to be implemented can be either a 1 or a 0. This condition may result from the specification of the problem and is not uncommon. Whenever it does not matter whether a position in the map is filled by a 1 or a 0, we use a **don't care** entry, denoted by an ***x***. Then the don't care can be used as either a 1 or a 0, depending on which results in a greater simplification (i.e., helps in forming the smallest number of maximal subcubes). The following examples illustrate the use of don't care conditions.

EXAMPLE 13.18 Using Don't Care Conditions to Simplify Expressions—1

LO4

Problem

Use don't care entries to simplify the expression

$$f(A, B, C, D) = \overline{A} \cdot \overline{B} \cdot \overline{C} \cdot D + \overline{A} \cdot \overline{B} \cdot C \cdot \overline{D} + \overline{A} \cdot \overline{B} \cdot C \cdot D + \overline{A} \cdot B \cdot \overline{C} \cdot D + A \cdot \overline{B} \cdot C \cdot D + A \cdot B \cdot \overline{C} \cdot \overline{D}$$

Solution

Known Quantities: Logical expression; don't care conditions.

Find: Minimal realization.

Schematics, Diagrams, Circuits, and Given Data: Don't care conditions: $f(A, B, C, D) = \{0100, 0110, 1010, 1110\}$.

Analysis: We cover the Karnaugh map of Figure 13.54 using 1s, and also using x entries for each don't care condition. Treating all the x entries as 1s, we complete the covering with two four-cell subcubes and one two-cell subcube, to obtain the following simplified expression:

$$f(A, B, C, D) = B \cdot \overline{D} + \overline{B} \cdot C + \overline{A} \cdot \overline{C} \cdot D$$

Comments: Note that we could have also interpreted the don't care entries as 0s and tried to solve in product-of-sums form. Verify that the expression obtained above is indeed the minimal one.

Note that the x's never occur, and so they may be assigned a 1 or a 0, whichever will best simplify the expression.

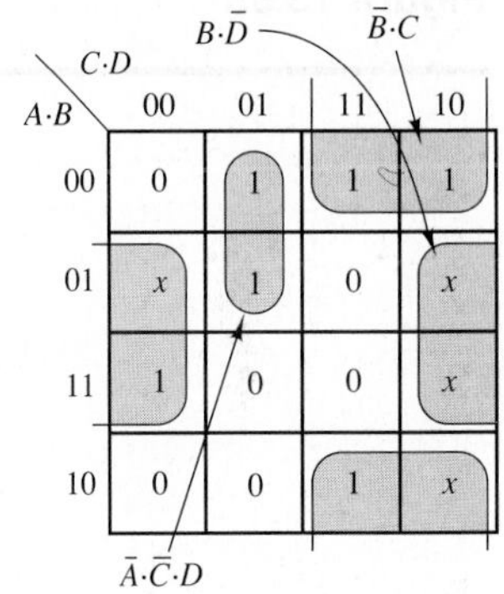

Figure 13.54

CHECK YOUR UNDERSTANDING

In Example 13.18, assign a value of 0 to the don't care terms and derive the corresponding minimal expression. Is the new function simpler than the one obtained in Example 13.18?

Answer: $f = A \cdot B \cdot \overline{C} \cdot \overline{D} + \overline{A} \cdot \overline{C} \cdot D + \overline{A} \cdot \overline{B} \cdot C \cdot D$

EXAMPLE 13.19 Using Don't Care Conditions to Simplify Expressions—2

LO4

Problem

Find a minimum product-of-sums realization for the expression $f(A, B, C)$.

Solution

Known Quantities: Logical expression, don't care conditions.

Find: Minimal realization.

Schematics, Diagrams, Circuits, and Given Data:

$$f(A,B,C) = \begin{cases} 1 & \text{for } \{A,B,C\} = \{000, 010, 011\} \\ x & \text{for } \{A,B,C\} = \{100, 101, 110\} \end{cases}$$

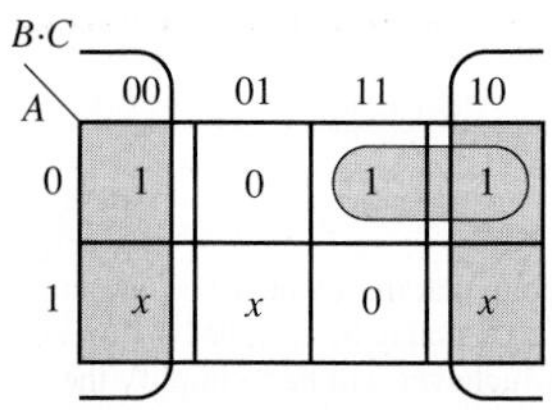

Figure 13.55

Analysis: We cover the Karnaugh map of Figure 13.55 using 1s, and also using x entries for each don't care condition. By appropriately selecting two of the three don't care entries to be equal to 1, we complete the covering with one four-cell subcube and one two-cell subcube, to obtain the following minimal expression:

$$f(A,B,C) = \overline{A} \cdot B + \overline{C}$$

Comments: Note that we have chosen to set one of the don't care entries equal to 0, since it would not lead to any further simplification.

CHECK YOUR UNDERSTANDING

In Example 13.19, assign a value of 0 to the don't care terms and derive the corresponding minimal expression. Is the new function simpler than the one obtained in Example 13.19?

In Example 13.19, assign a value of 1 to all don't care terms and derive the corresponding minimal expression. Is the new function simpler than the one obtained in Example 13.19?

Answers: $f = \overline{A} \cdot B + \overline{A} \cdot \overline{C}$; No; $f = \overline{A} \cdot B + A \cdot \overline{B} + \overline{C}$; No

LO4

EXAMPLE 13.20 Using Don't Care Conditions to Simplify Expressions—3

Problem

Find a minimum sum-of-products realization for the expression $f(A,B,C,D)$.

Solution

Known Quantities: Logical expression; don't care conditions.

Find: Minimal realization.

Schematics, Diagrams, Circuits, and Given Data

$$f(A,B,C,D) = \begin{cases} 1 & \text{for } \{A,B,C,D\} = \{0000, 0011, 0110, 1001\} \\ x & \text{for } \{A,B,C,D\} = \{1010, 1011, 1101, 1110, 1111\} \end{cases}$$

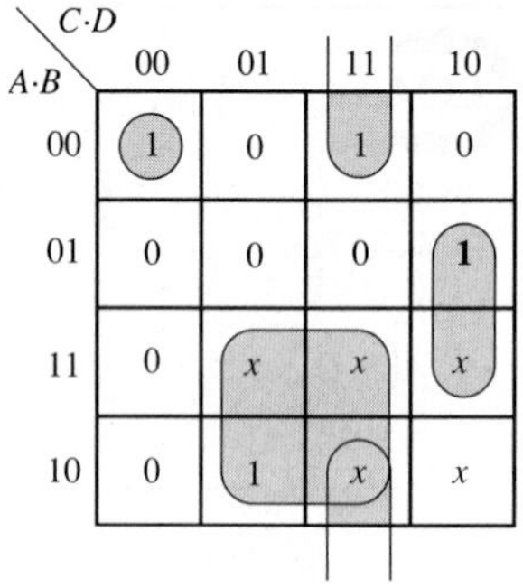

Figure 13.56

Analysis: We cover the Karnaugh map of Figure 13.56 using 1s, and using x entries for each don't care condition. By appropriately selecting three of the four don't care entries to be equal to 1, we complete the covering with one four-cell subcube, two two-cell subcubes, and one

one-cell subcube, to obtain the following expression:

$$f(A, B, C) = \overline{A} \cdot \overline{B} \cdot \overline{C} \cdot \overline{D} + B \cdot C \cdot \overline{D} + A \cdot D + \overline{B} \cdot C \cdot D$$

Comments: Would the product-of-sums realization be simpler? Verify.

CHECK YOUR UNDERSTANDING

In Example 13.20, assign a value of 0 to all don't care terms and derive the corresponding minimal expression. Is the new function simpler than the one obtained in Example 13.20?

In Example 13.20, assign a value of 1 to all don't care terms and derive the corresponding minimal expression. Is the new function simpler than the one obtained in Example 13.20?

Answers: $f = \overline{A} \cdot \overline{B} \cdot \overline{C} \cdot \overline{D} + \overline{A} \cdot \overline{B} \cdot C \cdot D + A \cdot \overline{B} \cdot \overline{C} \cdot D + \overline{A} \cdot B \cdot C \cdot \overline{D}$; No; $f = \overline{A} \cdot \overline{B} \cdot \overline{C} \cdot \overline{D} + B \cdot C \cdot \overline{D} + A \cdot D + \overline{B} \cdot C \cdot D + A \cdot C$; No

13.5 COMBINATIONAL LOGIC MODULES

The basic logic gates described in the previous section are used to implement more advanced functions and are often combined to form logic modules, which, thanks to modern technology, are available in compact integrated-circuit packages. In this section and the next, we discuss a few of the more common **combinational logic modules,** illustrating how these can be used to implement advanced logic functions.

Multiplexers

Multiplexers, or **data selectors,** are combinational logic circuits that permit the selection of one of many inputs. A typical multiplexer (MUX) has 2^n **data lines,** n **address** (or **data select**) **lines,** and one output. In addition, other control inputs (e.g., enables) may exist. Standard, commercially available MUXs allow for n up to 4; however, two or more MUXs can be combined if a greater range is needed. The MUX allows for one of 2^n inputs to be selected as the data output; the selection of which input is to appear at the output is made by way of the address lines. Figure 13.57 depicts the block diagram of a four-input MUX. The input data lines are labeled D_0, D_1, D_2, and D_3; the **data select,** or address, **lines** are labeled I_0 and I_1; and the output is available in both complemented and uncomplemented form and is thus labeled F, or $\overline{F}$. Finally, an **enable** input, labeled E, is also provided, as a means of enabling or disabling the MUX: if $E = 1$, the MUX is disabled; if $E = 0$, it is enabled. The negative logic (MUX off when $E = 1$ and on when $E = 0$) is represented by the small "bubble" at the enable input, which represents a complement operation (just as at the output of NAND and NOR gates). The enable input is useful whenever one is interested in a cascade of MUXs; this would be of interest if we needed to select a line from a large number, say, $2^8 = 256$. Then two four-input MUXs could be used to provide the data selection of 1 of 8.

The material described in previous sections is quite adequate to describe the internal workings of a multiplexer. Figure 13.58 shows the internal construction of a

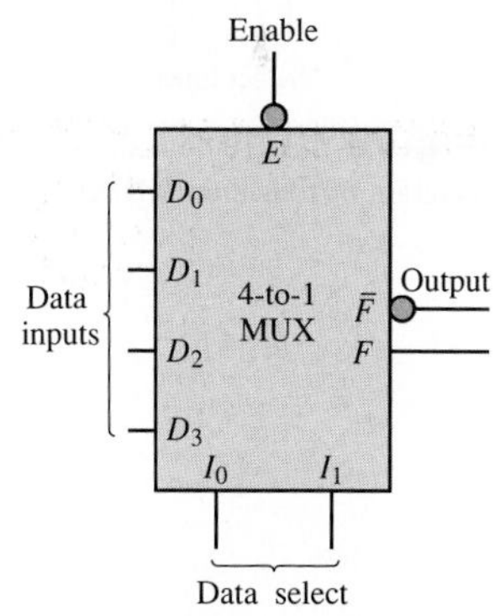

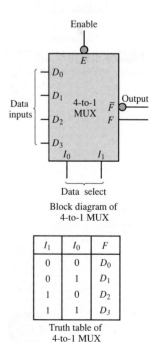

Block diagram of 4-to-1 MUX

I_1	I_0	F
0	0	D_0
0	1	D_1
1	0	D_2
1	1	D_3

Truth table of 4-to-1 MUX

Figure 13.57 4-to-1 MUX

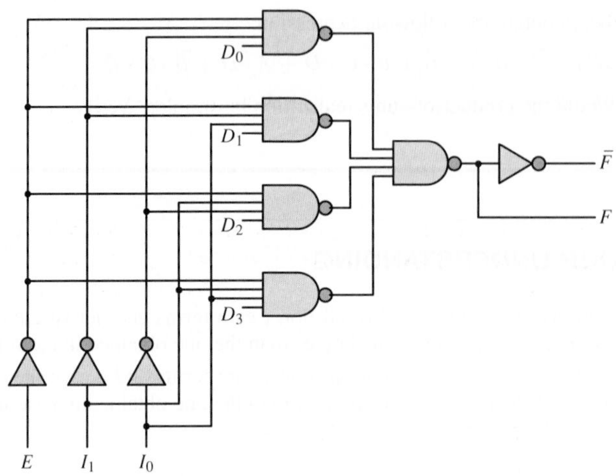

Figure 13.58 Internal structure of the 4-to-1 MUX

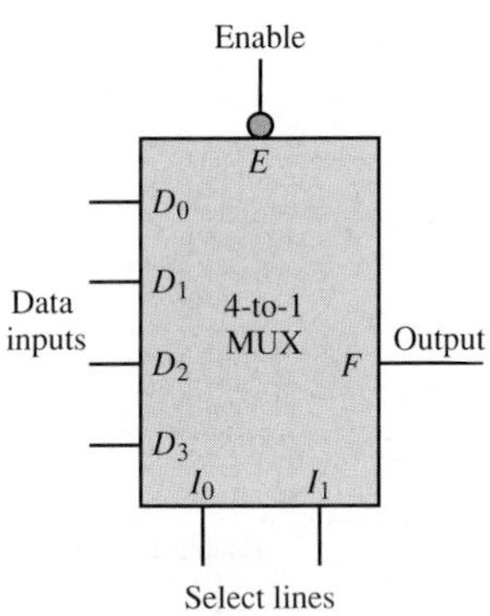

Figure 13.59 Functional diagram of four-input MUX

4-to-1 MUX using exclusively NAND gates (inverters are also used, but the reader will recall that a NAND gate can act as an inverter if properly connected).

In the design of digital systems (e.g., microprocessors), a single line is often required to carry two or more different digital signals. However, only one signal at a time can be placed on the line. A MUX will allow us to select, at different instants, the signal we wish to place on this single line. This property is shown here for a 4-to-1 MUX. Figure 13.59 depicts the functional diagram of a 4-to-1 MUX, showing four data lines, D_0 through D_3, and two select lines, I_0 and I_1.

The data selector function of a MUX is best understood in terms of Table 13.13. In this truth table, the x's represent don't care entries. As can be seen from the truth table, the output selects one of the data lines depending on the values of I_1 and I_0, assuming that I_0 is the least significant bit. As an example, $I_1 I_0 = 10$ selects D_2, which means that the output F will select the value of the data line D_2. Therefore $F = 1$ if $D_2 = 1$ and $F = 0$ if $D_2 = 0$.

Table 13.13

I_1	I_0	D_3	D_2	D_1	D_0	F
0	0	x	x	x	0	0
0	0	x	x	x	1	1
0	1	x	x	0	x	0
0	1	x	x	1	x	1
1	0	x	0	x	x	0
1	0	x	1	x	x	1
1	1	0	x	x	x	0
1	1	1	x	x	x	1

LO5

CHECK YOUR UNDERSTANDING

Which combination of the control lines will select the data line D_3 for a 4-to-1 MUX?

Show that an 8-to-1 MUX with eight data inputs (D_0 through D_7) and three control lines (I_0 through I_2) can be used as a data selector. Which combination of the control lines will select the data line D_5?

Which combination of the control lines will select the data line D_4 for an 8-to-1 MUX?

Answers: $I_1I_0 = 11$; For the first part, use the same method as in the preceding Check Your Understanding exercise, but for an 8-to-1 MUX. For the second part, $I_2I_1I_0 = 101$; $I_2I_1I_0 = 100$

Read-Only Memory (ROM)

Another common technique for implementing logic functions uses a **read-only memory,** or **ROM.** As the name implies, a ROM is a logic circuit that holds information in storage ("memory")—in the form of binary numbers—that cannot be altered but can be "read" by a logic circuit. A ROM is an array of memory cells, each of which can store either a 1 or a 0. The array consists of $2^m \times n$ cells, where n is the number of bits in each word stored in ROM. To access the information stored in ROM, m address lines are required. When an address is selected, in a fashion similar to the operation of the MUX, the binary word corresponding to the address selected appears at the output, which consists of n bits, that is, the same number of bits as the stored words. In some sense, a ROM can be thought of as a MUX that has an output consisting of a word instead of a single bit.

Figure 13.60 depicts the conceptual arrangement of a ROM with $n = 4$ and $m = 2$. The ROM table has been filled with arbitrary 4-bit words, just for the purpose of illustration. In Figure 13.60, if one were to select an enable input of 0 (i.e., on) and values for the address lines of $I_0 = 0$ and $I_1 = 1$, the output word would be $W_2 = 0110$, so that $b_0 = 0$, $b_1 = 1$, $b_2 = 1$, $b_3 = 0$. Depending on the content of the ROM and the number of address and output lines, one could implement an arbitrary logic function.

Unfortunately, the data stored in read-only memories must be entered during fabrication and cannot be altered later. A much more convenient type of read-only memory is the **erasable programmable read-only memory (EPROM),** the content of which can be easily programmed and stored and may be changed if needed. EPROMs find use in many practical applications, because of their flexibility in content and ease of programming. The following example illustrates the use of an EPROM to perform the linearization of a nonlinear function.

ROM address		ROM content (4-bit words)				
I_1	I_0	b_3	b_2	b_1	b_0	
0	0	0	1	1	0	W_0
0	1	1	0	0	1	W_1
1	0	0	1	1	0	W_2
1	1	1	1	1	1	W_3

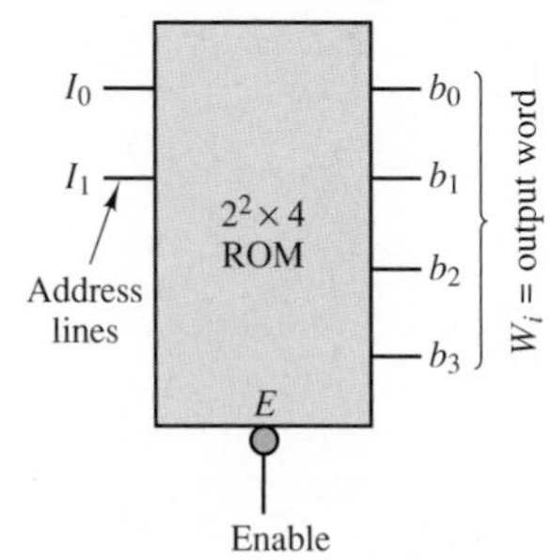

Figure 13.60 Read-only memory

CHECK YOUR UNDERSTANDING

How many address inputs do you need if the number of words in a memory array is 16?

Answer: Four

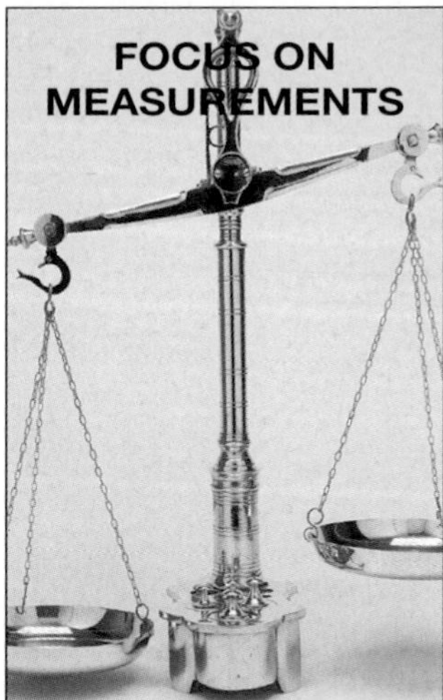

EPROM-Based Lookup Table for Automotive Fuel Injection System Control

One of the most common applications of EPROMs is the *arithmetic lookup table.* A lookup table is similar in concept to the familiar multiplication table and is used to store precomputed values of certain functions, eliminating the need for actually computing the function. A practical application of this concept is present in every automobile manufactured in the United States since the early 1980s, as part of the **exhaust emission control system.** In order for the catalytic converter to minimize the emissions of exhaust gases (especially hydrocarbons, oxides of nitrogen, and carbon monoxide), it is necessary to maintain the *air-to-fuel ratio* A/F as close as possible to the stoichiometric value, that is, 14.7 parts of air for each part of fuel. Most modern engines are equipped with fuel injection systems that are capable of delivering accurate amounts of fuel to each individual cylinder—thus, the task of maintaining an accurate A/F amounts to measuring the mass of air that is aspirated into each cylinder and computing the corresponding mass of fuel. Many automobiles are equipped with a *mass airflow sensor,* capable of measuring the mass of air drawn into each cylinder during each engine cycle. Let the output of the mass airflow sensor be denoted by the variable M_A, and let this variable represent the mass of air (in grams) actually entering a cylinder during a particular stroke. It is then desired to compute the mass of fuel M_F (also expressed in grams) required to achieve an A/F of 14.7. This computation is simply

$$M_F = \frac{M_A}{14.7}$$

Although the above computation is a simple division, its actual calculation in a low-cost digital computer (such as would be used on an automobile) is rather complicated. It would be much simpler to tabulate a number of values of M_A, to precompute the variable M_F, and then to store the result of this computation in an EPROM. If the EPROM address were made to correspond to the tabulated values of air mass, and the content at each address to the corresponding fuel mass (according to the precomputed values of the expression $M_F = M_A/14.7$), it would not be necessary to perform the division by 14.7. For each measurement of air mass into one cylinder, an EPROM address is specified and the corresponding content is read. The content at the specific address is the mass of fuel required by that particular cylinder.

In practice, the fuel mass needs to be converted to a time interval corresponding to the duration of time during which the fuel injector is open. This final conversion factor can also be accounted for in the table. Suppose, for example, that the fuel injector is capable of injecting K_F g/s of fuel; then the time duration T_F during which the injector should be open in order to inject M_F g of fuel into the cylinder is given by

$$T_F = \frac{M_F}{K_F} \quad \text{s}$$

Therefore, the complete expression to be precomputed and stored in the EPROM is

$$T_F = \frac{M_A}{14.7 \times K_F} \quad \text{s}$$

Figure 13.61 illustrates this process graphically.

(Continued)

(*Concluded*)

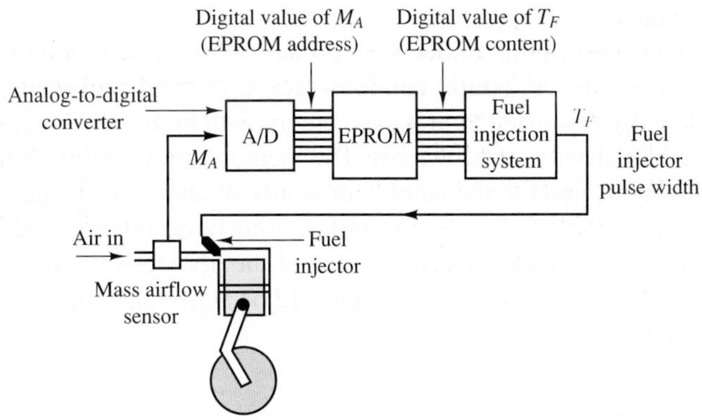

Figure 13.61 Use of EPROM lookup table in automotive fuel injection system

To provide a numerical illustration, consider a hypothetical engine capable of aspirating air in the range $0 < M_A < 0.51$ g and equipped with fuel injectors capable of injecting at the rate of 1.36 g/s. Thus, the relationship between T_F and M_A is

$$T_F = 50 \times M_A \text{ ms} = 0.05 M_A \quad \text{s}$$

If the digital value of M_A is expressed in decigrams (dg, or tenths of g), the lookup table of Figure 13.62 can be implemented, illustrating the conversion capabilities provided by the EPROM. Note that in order to represent the quantities of interest in an appropriate binary format compatible with the 8-bit EPROM, the units of air mass and of time have been scaled.

M_A (g) $\times 10^{-2}$	Address (digital value of M_A)	Content (digital value of T_F)	T_F (ms) $\times 10^{-1}$
0	00000000	00000000	0
1	00000001	00000101	5
2	00000010	00001010	10
3	00000011	00001111	15
4	00000100	00010100	20
5	00000101	00011001	25
⋮	⋮	⋮	⋮
51	00110011	11111111	255

Figure 13.62 Lookup table for automotive fuel injection application

Decoders and Read and Write Memory

Decoders, which are commonly used for applications such as address decoding or memory expansion, are combinational logic circuits as well. Our reason for introducing decoders is to show some of the internal organization of semiconductor memory devices. An important application of decoders in the organization of a memory system is discussed in Chapter 14.

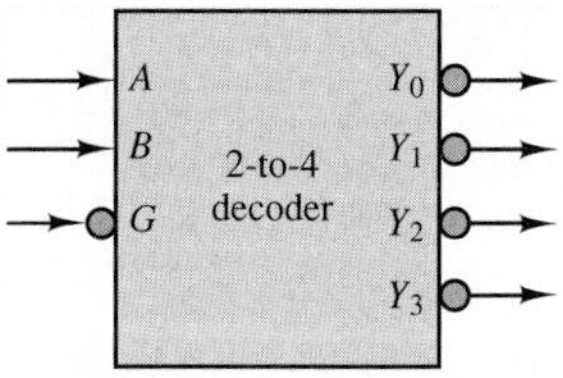

Inputs			Outputs			
Enable	Select					
$\overline{G}$	A	B	Y_0	Y_1	Y_2	Y_3
1	x	x	1	1	1	1
0	0	0	0	1	1	1
0	0	1	1	0	1	1
0	1	0	1	1	0	1
0	1	1	1	1	1	0

Figure 13.63 A 2-to-4 decoder

LO5

Figure 13.63 shows the truth table for a 2-to-4 decoder. The decoder has an enable input $\overline{G}$ and select inputs B and A. It also has four outputs, Y_0 through Y_3. When the enable input is logic 1, all decoder outputs are forced to logic 1 regardless of the select inputs.

This simple description of decoders permits a brief discussion of the internal organization of an **SRAM (static random-access,** or **read and write, memory).** SRAM is internally organized to provide memory with high speed (i.e., short access time), a large bit capacity, and low cost. The memory array in this memory device has a column length equal to the number of words W and a row length equal to the number of bits per word N. To select a word, an n-to-W decoder is needed. Since the address inputs to the decoder select only one of the decoder's outputs, the decoder selects one word in the memory array. Figure 13.64 shows the internal organization of a typical SRAM.

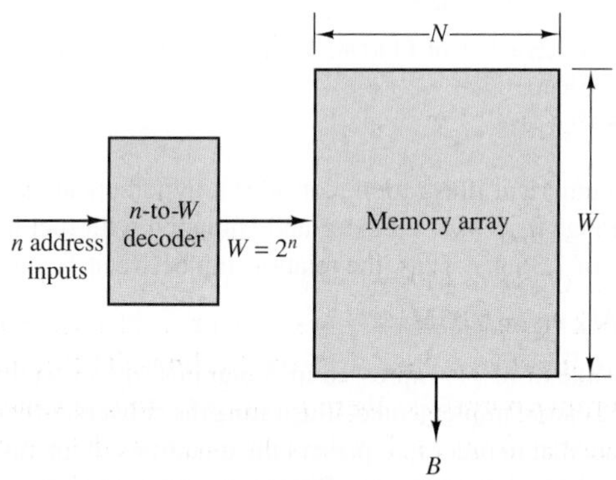

Figure 13.64 Internal organization of SRAM

Thus, to choose the desired word from the memory array, the proper address inputs are required. As an example, if the number of words in the memory array is 8, a 3-to-8 decoder is needed. Data sheets for 2-to-4 and 3-to-8 decoders from a CMOS family data book may be found on the book website.

Gate Arrays and Programmable Logic Devices

Digital logic design is performed today primarily using **programmable logic devices (PLDs).** These are arrays of gates having interconnections that can be programmed to perform a specific logical function. PLDs are large combinational logic modules consisting of arrays of AND and OR gates that can be programmed using special programming languages called **Hardware description languages (HDLs)**. Figure 13.65 shows the block diagram of one type of high-density PLD. We define three types of PLDs:

PROM (programmable read-only memory) offers high speed and low cost for relatively small designs.

PLA (programmable logic array) offers flexible features for more complex designs.

PAL/GAL (programmable array logic/generic array logic) offers good flexibility and is faster and less expensive than a PLA.

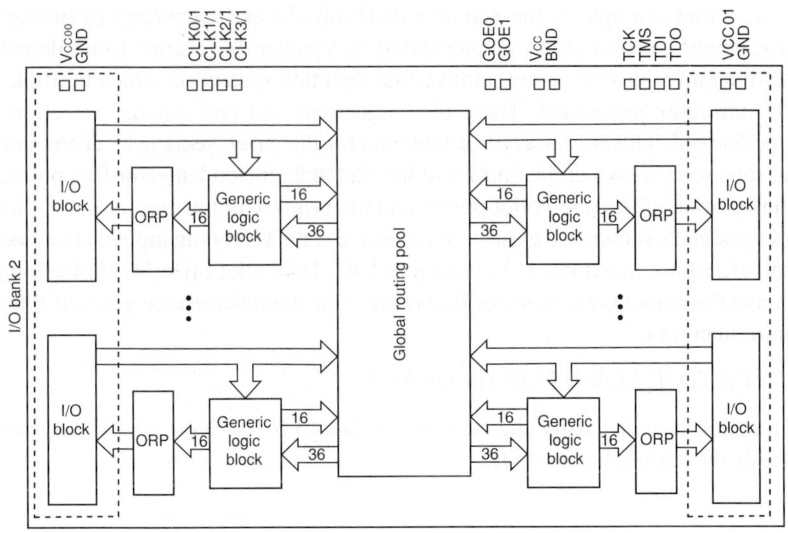

Figure 13.65 High-density PLD

To illustrate the concept of a logic design using a PLD, we employ a generic array logic (ispGAL16V8) to realize an output signal from three ANDed input signals. The functional block diagram of the GAL is shown in Figure 13.66(a). Notice that the device has eight input lines and eight output lines. The output lines also provide a clock input for timing purposes (there is more on this in Chapter 14). The sample code is shown in Figure 13.66(b). The code first defines the inputs and outputs; and it states the equation describing the function to be implemented, O14=I11&I12&I13, defining which output and inputs are to be used, and the functional relationship. Note that the symbol & represents the logical function AND.

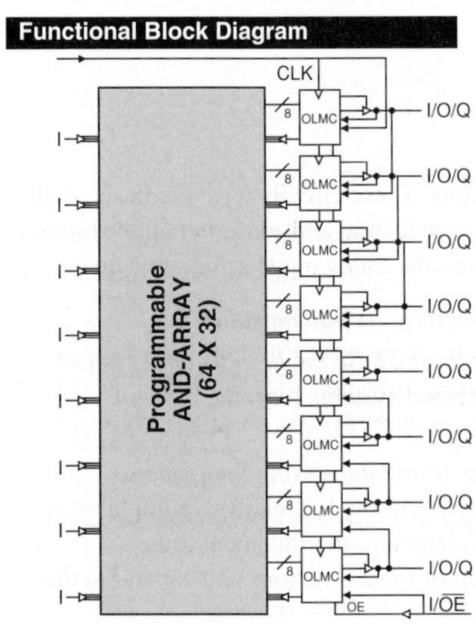

(a)

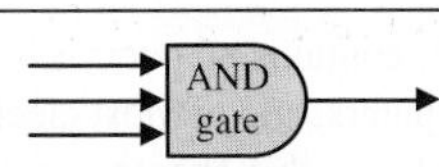


```
MODULE and_GATE_DEMO
    I11,I12,I13         PIN 11,12,13;
    O14                 PIN 14ISTYPE'COM,BUFFER';
Equations
    O14 = I11 & I12 & I13;
END
```

(b)

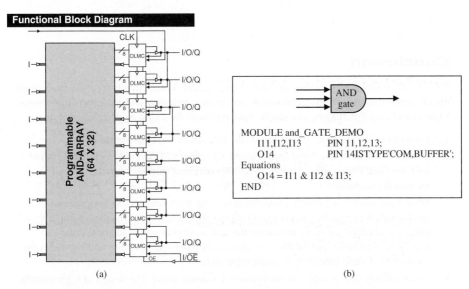

Figure 13.66 (a) The ispGAL16V8 connection diagram; (b) sample code for AND function (for ispGAL16V8).

A second example of the use of a PLD introduces the concept of timing diagrams, which are covered in greater detail in Chapter 14. Figure 13.67 depicts a timing diagram related to an automotive fuel injection system, in which multiple injections are to be performed. Three *pilot* injections and one *primary* injection are to be performed. The *master control* line enables the entire sequence. The resulting output sequence, shown at the bottom of the plot and labeled "injector fuel pulse," is the combination of the three pilot pulses and the primary pulse. Based on the timing plot of the signals shown in Figure 13.67(a), we use the following inputs: I11=master control, I12=pilot inject #1, I13=pilot inject #2, I14=pilot inject #3, I14=primary inject, and the output O14=injector fuel pulse. You should convince yourself that the required function is

I11 AND [I12 OR I13 OR I14 OR I14]

This function is realized by the code in Figure 13.67(b). Note that the symbol | represents the logical function OR.

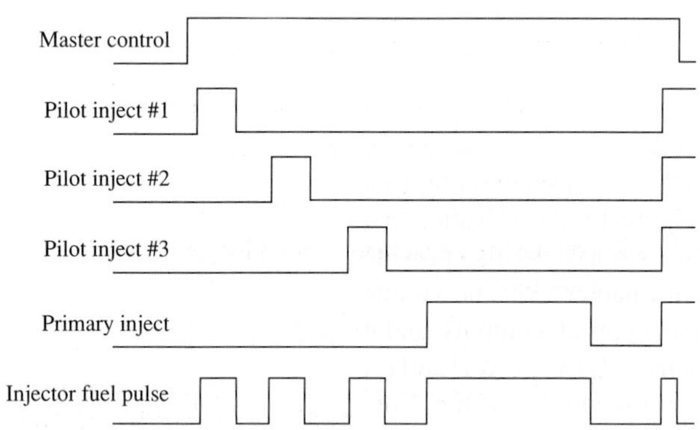

(a)

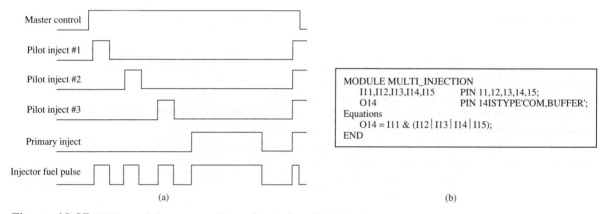

```
MODULE MULTI_INJECTION
    I11,I12,I13,I14,I15        PIN 11,12,13,14,15;
    O14                        PIN 14ISTYPE'COM,BUFFER';
Equations
    O14 = I11 & (I12 | I13 | I14 | I15);
END
```

(b)

Figure 13.67 (a) Injector timing sequence; (b) sample code for multiple-injection sequence

Conclusion

This chapter contains an overview of digital logic circuits. These circuits form the basis of all digital computers, and of most electronic devices used in industrial and consumer applications. Upon completing this chapter, you should have mastered the following learning objectives:

1. *Understand the concepts of analog and digital signals and of quantization.*
2. *Convert between decimal and binary number systems and use the hexadecimal system and BCD and Gray codes.* The binary and hexadecimal systems form the basis of numerical computing.
3. *Write truth tables, and realize logic functions from truth tables using logic gates.* Boolean algebra permits the analysis of digital circuits through a relatively simple set of rules. Digital logic gates are the means through which one can implement logic functions; truth tables permit the easy visualization of logic functions and can aid in the realization of these functions by using logic gates.
4. *Systematically design logic functions using Karnaugh maps.* The design of logic circuits can be systematically approached by using an extension of truth tables called the

Karnaugh map. Karnaugh maps facilitate the simplification of logic expressions and their realization through logic gates in either sum-of-products or product-of-sums form.

5. *Study various combinational logic modules, including multiplexers, memory and decoder elements, and programmable logic arrays.* Practical digital logic circuits rarely consist of individual logic gates; gates are usually integrated into combinational logic modules that include memory elements and gate arrays.

HOMEWORK PROBLEMS

Section 13.2: The Binary Number System

13.1 Convert the following base-10 numbers to hexadecimal and binary:

a. 401 b. 273 c. 15 d. 38 e. 56

13.2 Convert the following hexadecimal numbers to base-10 and binary:

a. A b. 66 c. 47 d. 21 e. 13

13.3 Convert the following base-10 numbers to binary:

a. 271.25 b. 53.375 c. 37.32 d. 54.27

13.4 Convert the following binary numbers to hexadecimal and base 10:

a. 1111 b. 1001101 c. 1100101 d. 1011100

e. 11101 f. 101000

13.5 Perform the following additions, all in the binary system:

a. $11001011 + 101111$

b. $10011001 + 1111011$

c. $11101001 + 10011011$

13.6 Perform the following subtractions, all in the binary system:

a. $10001011 - 1101111$

b. $10101001 - 111011$

c. $11000011 - 10111011$

13.7 Assuming that the most significant bit is the sign bit, find the decimal value of the following sign-magnitude form 8-bit binary numbers:

a. 11111000 b. 10011111 c. 01111001

13.8 Find the sign-magnitude form binary representation of the following decimal numbers:

a. 126 b. −126 c. 108 d. −98

13.9 Find the twos complement of the following binary numbers:

a. 1111 b. 1001101 c. 1011100 d. 11101

13.10 Assuming you have 10 fingers, including thumbs:

a. How high can you count on your fingers in a binary (base 2) number system?

b. How high can you count on your fingers in base 6, using one hand to count units and the other hand for the carries?

Section 13.3: Boolean Algebra

13.11 Use a truth table to prove that $B = AB + \overline{A}B$.

13.12 Use truth tables to prove that

$$BC + B\overline{C} + \overline{B}A = A + B.$$

13.13 Using the method of proof by perfect induction, show that

$$(X + Y) \cdot (\overline{X} + X \cdot Y) = Y$$

13.14 Using De Morgan's theorems and the rules of boolean algebra, simplify the following logic function:

$$F(X, Y, Z) = \overline{X} \cdot \overline{Y} \cdot \overline{Z} + \overline{X} \cdot Y \cdot Z + X \cdot (\overline{Y + Z})$$

13.15 Simplify the expression

$$f(A, B, C, D) = ABC + \overline{A}CD + \overline{B}CD.$$

13.16 Simplify the logic function $F(A, B, C) = \overline{A} \cdot B \cdot \overline{C} + \overline{A} \cdot B \cdot C + A \cdot B \cdot \overline{C} + A \cdot B \cdot C$, using boolean algebra.

13.17 Find the logic function defined by the truth table given in Figure P13.17.

A	B	C	F
0	0	0	0
0	0	1	1
0	1	0	0
0	1	1	1
1	0	0	1
1	0	1	1
1	1	0	1
1	1	1	1

Figure P13.17

13.18 Determine the boolean function describing the operation of the circuit shown in Figure P13.18.

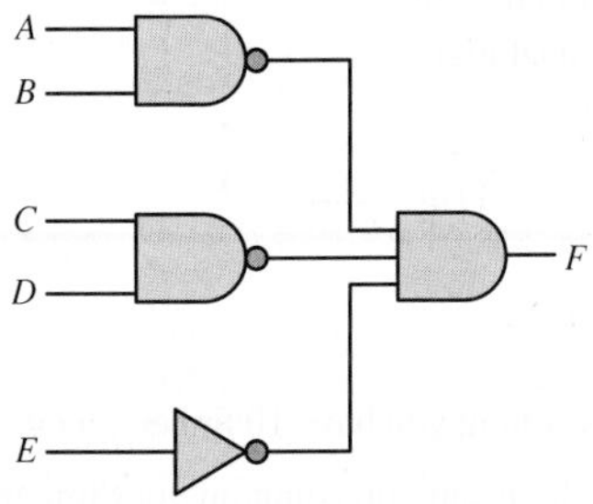

Figure P13.18

13.19 Use a truth table to show when the output of the circuit of Figure P13.19 is 1.

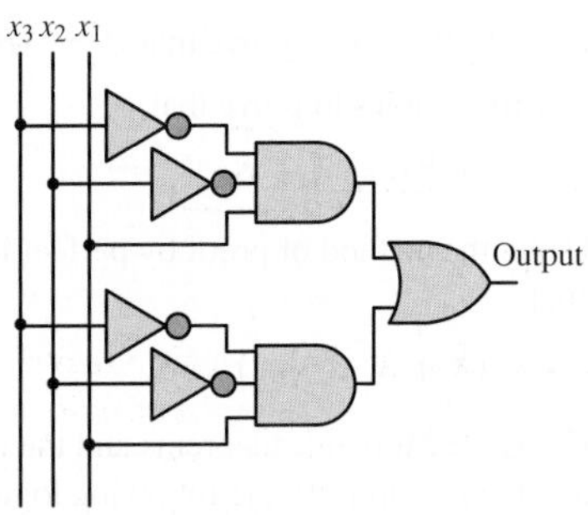

Figure P13.19

13.20 Baseball is a complicated game, and often the manager has a difficult time keeping track of all the rules of thumb that guide decisions. To assist your favorite baseball team, you have been asked to design a logic circuit that will flash a light when the manager should give the steal sign. The rules have been laid out for you by a baseball fan with limited knowledge of the game as follows: Give the steal sign if there is a runner on first base and

a. There are no other runners, the pitcher is right-handed, and the runner is fast or

b. There is one other runner on third base, and one of the runners is fast or

c. There is one other runner on second base, the pitcher is left-handed, and both runners are fast.

Under no circumstances should the steal sign be given if all three bases have runners. Design a logic circuit that implements these rules to indicate when the steal sign should be given.

13.21 A small county board is composed of three commissioners. Each commissioner votes on measures presented to the board by pressing a button indicating whether the commissioner votes for or against a measure. If two or more commissioners vote for a measure, it passes. Design a logic circuit that takes the three votes as inputs and lights either a green or a red light to indicate whether a measure passed.

13.22 A water purification plant uses one tank for chemical sterilization and a second, larger tank for settling and aeration. Each tank is equipped with two sensors that measure the height of water in each tank and the flow rate of water into each tank. When the height of water or the flow rate is too high, the sensors produce a logic high output. Design a logic circuit that sounds an alarm whenever the height of water in both tanks is too high and either of the flow rates is too high, or whenever both flow rates are too high and the height of water in either tank is also too high.

13.23 Many automobiles incorporate logic circuits to alert the driver to problems or potential problems. In one particular car, a buzzer is sounded whenever the ignition key is turned and either a door is open or a seat belt is not fastened. The buzzer also sounds when the key is not turned but the lights are on. In addition, the car will not start unless the key is in the ignition, the car is in park, and all doors are closed and seat belts fastened. Design a logic circuit that takes all the inputs listed and sounds the buzzer and starts the car when appropriate.

13.24 An on/off start-up signal governs the compressor motor of a large commercial air conditioning unit. In general, the start-up signal should be on whenever the output of a temperature sensor S exceeds a reference temperature. However, you are asked to limit the compressor start-ups to certain hours of the day and also enable service technicians to start up or shut down the compressor through a manual override. A time-of-day indicator D is available with on/off outputs, as is a manual override switch M. A separate timer T prohibits a compressor start-up within 10 min of a previous shutdown. Design a logic diagram that incorporates the state of all four devices (S, D, M, and T) and produces the correct on/off condition for the motor start-up.

13.25 NAND gates require one less transistor than AND gates. They are often used exclusively to construct logic circuits. One such logic circuit that uses three-input NAND gates is shown in Figure P13.25.

a. Determine the truth table for this circuit.

b. Give the logic equation that represents the circuit (you do not need to reduce it).

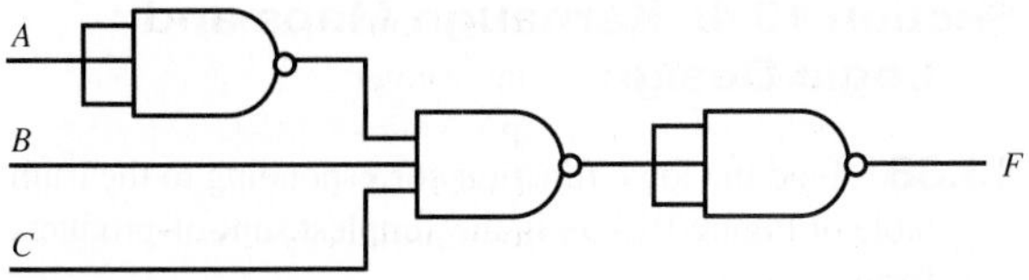

Figure P13.25

13.26 Draw a logic circuit that will accomplish the equation:

$$F = (A + \overline{B}) \cdot \overline{(C + \overline{A})} \cdot B.$$

13.27 The circuit shown in Figure P13.27 is called a half adder for two single bit inputs, giving a two-bit sum as outputs. Build a truth table and verify that it indeed acts as a summer.

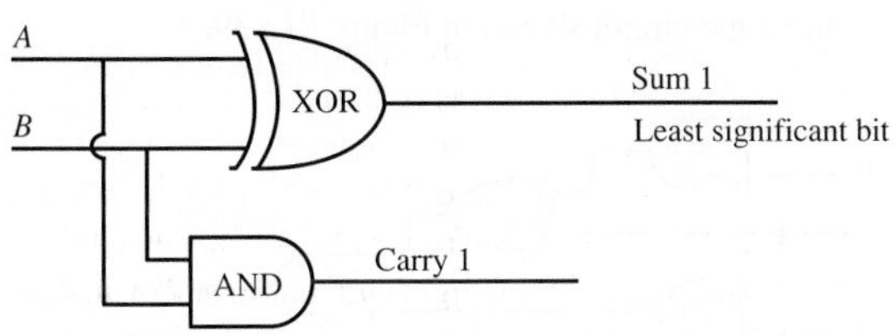

Figure P13.27

13.28 Draw a logic circuit that will accomplish the equation

$$F = \left[(A + C \cdot \overline{B}) + A \cdot \overline{B} \cdot \overline{C}\right] \cdot \overline{(B + C)}$$

13.29 Determine the truth table (F given A, B, C, & D) and the logical expression for the circuit of Figure P13.29.

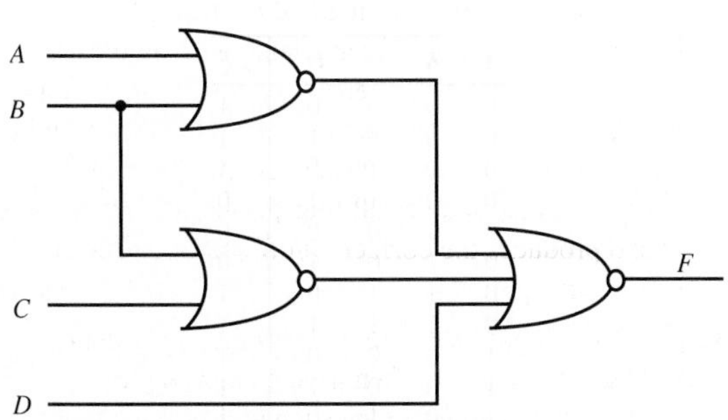

Figure P13.29

13.30 Determine the truth table (F given A, B, & C) and the logical expression for the circuit of Figure P13.30.

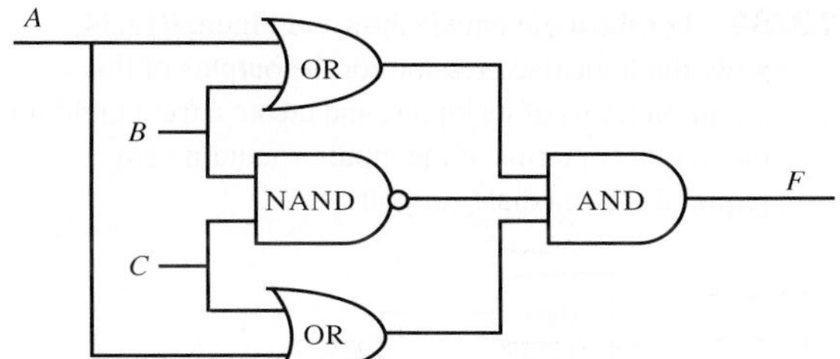

Figure P13.30

13.31 A "vote taker" logic circuit forces its output to agree with a majority of its inputs. Such a circuit is shown in Figure P13.31 for the three voters. Write the logic expression for the output of this circuit in terms of its inputs. Also create a truth table for the output in terms of the inputs.

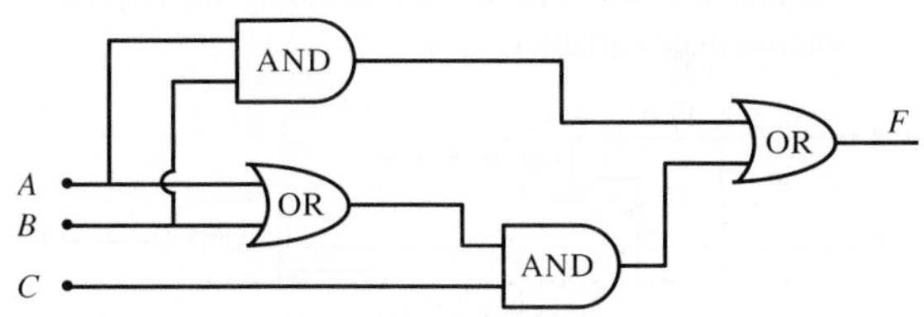

Figure P13.31

13.32 A "consensus indicator" logic circuit is shown in Figure P13.32. Write the logical expression for the output of this circuit in terms of its input. Also create a truth table for the output in terms of the inputs.

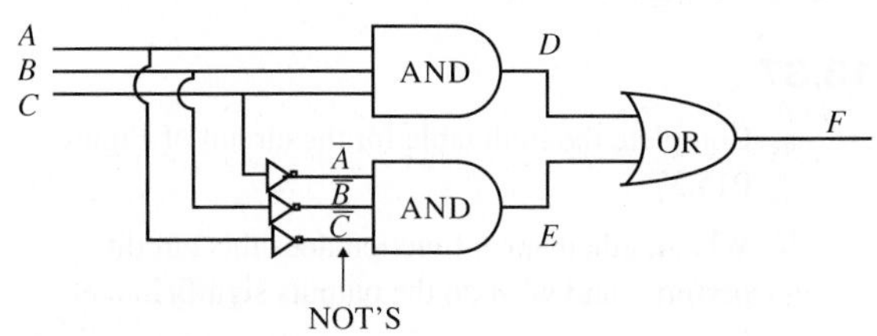

Figure P13.32

13.33 A half-adder circuit is shown in Figure P13.33. Write the logical expression for the outputs of this circuit in terms of its inputs. Also create a truth table for the outputs in terms of the inputs.

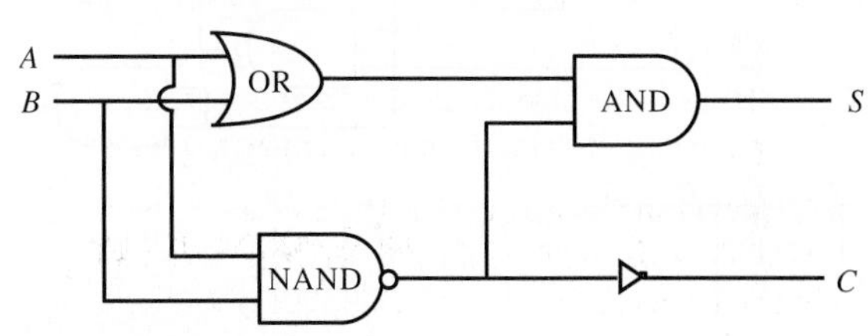

Figure P13.33

13.34 For the logic circuit shown in Figure P13.34, write the logical expression for the outputs of this circuit in terms of its inputs, and create a truth table for the outputs in terms of the inputs, including any required intermediate variables.

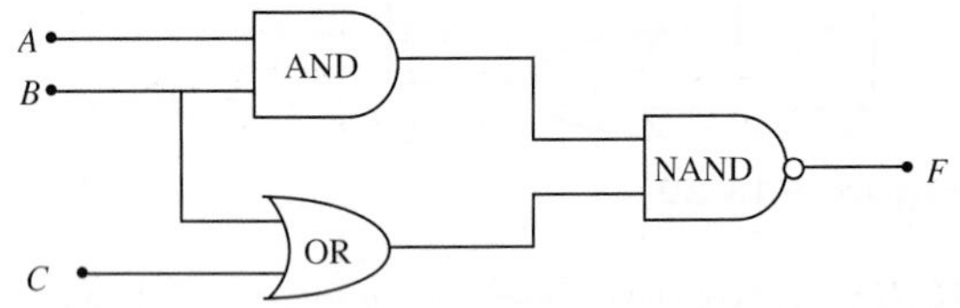

Figure P13.34

13.35 For the logic circuit in Figure P13.35, write the logical expression for the outputs of this circuit in terms of its inputs, and create a truth table for the outputs in terms of the inputs, including any required intermediate variables.

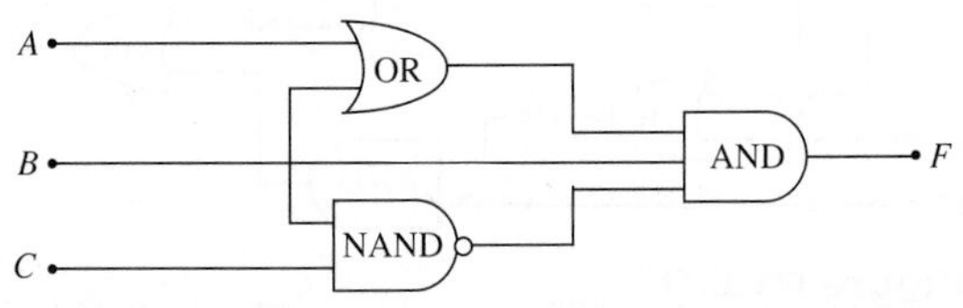

Figure P13.35

13.36 Determine the minimum expression for the following logic function, simplifying the expression:

$$f(A, B, C) = (A + B) \cdot A \cdot B + \overline{A} \cdot C + A \cdot \overline{B} \cdot C + \overline{B} \cdot \overline{C}$$

13.37

a. Complete the truth table for the circuit of Figure P13.37.

b. What mathematical function does this circuit perform, and what do the outputs signify?

c. How many standard 14-pin ICs would it take to construct this circuit?

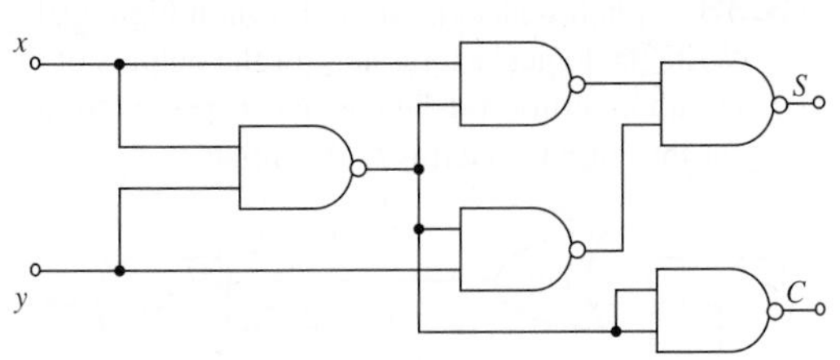

x	y	C	S
0	0		
0	1		
1	0		
1	1		

Figure P13.37

Section 13.4: Karnaugh Maps and Logic Design

13.38 Find the logic function corresponding to the truth table of Figure P13.38 in the simplest sum-of-products form.

A	B	C	F
0	0	0	1
0	0	1	0
0	1	0	0
0	1	1	0
1	0	0	1
1	0	1	0
1	1	0	1
1	1	1	1

Figure P13.38

13.39 Find the minimum expression for the output of the logic circuit shown in Figure P13.39.

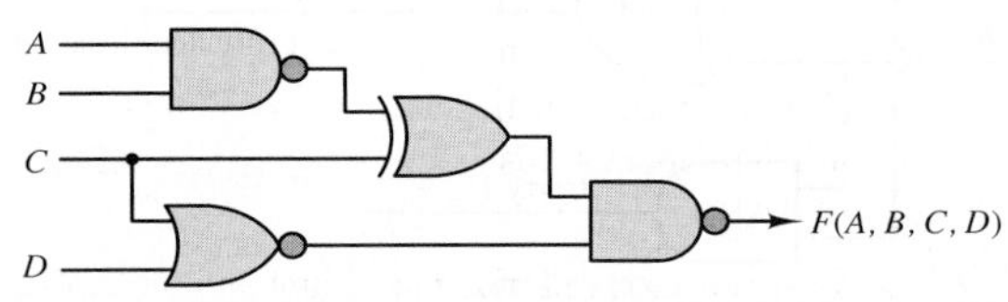

Figure P13.39

13.40 Use a Karnaugh map to minimize the function $f(A, B, C) = ABC + AB\overline{C} + A\overline{B}\,\overline{C}$.

13.41

a. Build the Karnaugh map for the logic function defined by the truth table of Figure P13.41.

b. What is the minimum expression for this function?

c. Realize F, using AND, OR, and NOT gates.

A	B	C	D	F
0	0	0	0	1
0	0	0	1	1
0	0	1	0	1
0	0	1	1	0
0	1	0	0	0
0	1	0	1	0
0	1	1	0	1
0	1	1	1	1
1	0	0	0	1
1	0	0	1	1
1	0	1	0	1
1	0	1	1	1
1	1	0	0	0
1	1	0	1	0
1	1	1	0	0
1	1	1	1	0

Figure P13.41

13.42 Fill in the Karnaugh map for the function defined by the truth table of Figure P13.42, and find the minimum expression for the function.

A	B	C	f(A,B,C)
0	0	0	0
0	0	1	1
0	1	0	1
0	1	1	0
1	0	0	1
1	0	1	0
1	1	0	0
1	1	1	1

Figure P13.42

13.43 A function F is defined such that it equals 1 when a 4-bit input code is equivalent to any of the decimal numbers 3, 6, 9, 12, or 15. Function F is 0 for input codes 0, 2, 8, and 10. Other input values cannot occur. Use a Karnaugh map to determine a minimal expression for this function. Design and sketch a circuit to implement this function, using only AND and NOT gates.

13.44 The function described in Figure P13.44 can be constructed using only two gates. Design the circuit.

Input			Output
A	B	C	F
0	0	0	0
0	0	1	0
0	1	0	0
0	1	1	1
1	0	0	0
1	0	1	0
1	1	0	1
1	1	1	x

Figure P13.44

13.45 Design a logic circuit which will produce the one's complement of an 8-bit signed binary number.

13.46 Construct the Karnaugh map for the logic function defined by the truth table of Figure P13.46, and find the minimum expression for the function.

13.47 Modify the circuit for Problem 13.45 so that it produces the twos complement of the 8-bit signed binary input.

13.48 Find the minimum output expression for the circuit of Figure P13.48.

A	B	C	D	F
0	0	0	0	1
0	0	0	1	0
0	0	1	0	1
0	0	1	1	0
0	1	0	0	0
0	1	0	1	0
0	1	1	0	0
0	1	1	1	1
1	0	0	0	1
1	0	0	1	0
1	0	1	0	1
1	0	1	1	0
1	1	0	0	1
1	1	0	1	1
1	1	1	0	1
1	1	1	1	0

Figure P13.46

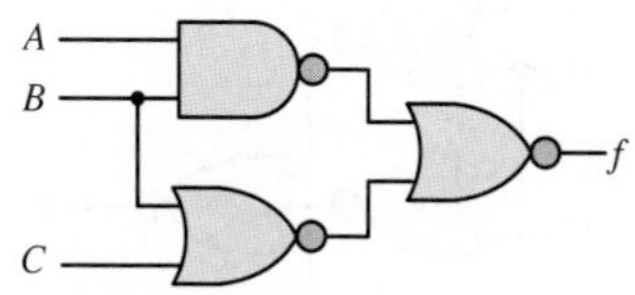

Figure P13.48

13.49 Design a combinational logic circuit which will add two 4-bit binary numbers.

13.50 Minimize the expression described in the truth table of Figure P13.50, and draw the circuit.

A	B	C	F
0	0	0	1
0	0	1	1
0	1	0	0
0	1	1	1
1	0	0	1
1	0	1	1
1	1	0	1
1	1	1	0

Figure P13.50

13.51 Find the minimum expression for the output of the logic circuit of Figure P13.51.

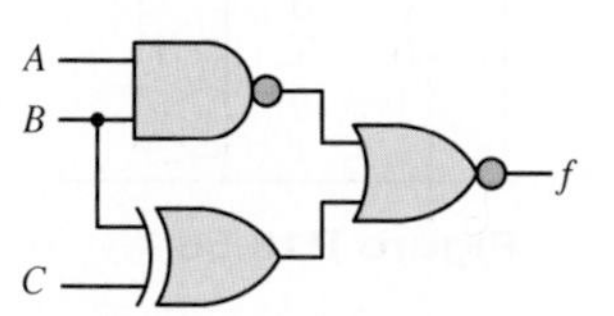

Figure P13.51

13.52 The objective of this problem is to design a combinational logic circuit which will aid in determination of the acceptability of emergency blood transfusions. It is known that human blood can be categorized into four types: A, B, AB, and O. Persons with type A blood can donate to both A and AB types and can receive blood from both A and O types. Persons with type B blood can donate to both B and AB and can receive from both B and O types. Persons with type AB blood can donate only to type AB, but can receive from any type. Persons with type O blood can donate to any type, but can receive only from type O. Make appropriate variable assignments, and design a circuit that will approve or disapprove any particular transfusion based on these conditions.

13.53 Find the minimum expression for the logic function at the output of the logic circuit of Figure P13.53.

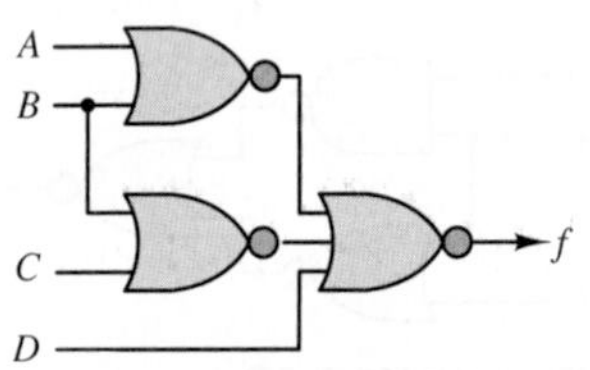

Figure P13.53

13.54 Design a combinational logic circuit which will accept a 4-bit binary number and if the number is even, divide it by 2_{10} and produce the binary result; if the number is odd, multiply it by 2_{10} and produce the binary result.

13.55

a. Fill in the Karnaugh map for the function defined in the truth table of Figure P13.55.

b. What is the minimum expression for the function?

c. Draw the circuit, using AND, OR, and NOT gates.

A	*B*	*C*	*f*(*A*,*B*,*C*)
0	0	0	1
0	0	1	1
0	1	0	0
0	1	1	1
1	0	0	1
1	0	1	1
1	1	0	1
1	1	1	0

Figure P13.55

13.56

a. Fill in the Karnaugh map for the logic function defined by the truth table of Figure P13.56.

b. What is the minimum expression for the function?

A	*B*	*C*	*D*	*F*
0	0	0	0	1
0	0	0	1	0
0	0	1	0	1
0	0	1	1	0
0	1	0	0	0
0	1	0	1	0
0	1	1	0	0
0	1	1	1	1
1	0	0	0	1
1	0	0	1	0
1	0	1	0	1
1	0	1	1	0
1	1	0	0	1
1	1	0	1	1
1	1	1	0	1
1	1	1	1	0

Figure P13.56

13.57

a. Fill in the Karnaugh map for the logic function defined by the truth table of Figure P13.57.

b. What is the minimum expression for the function?

c. Realize the function, using only NAND gates.

A	*B*	*C*	*D*	*F*
0	0	0	0	1
0	0	0	1	1
0	0	1	0	1
0	0	1	1	1
0	1	0	0	0
0	1	0	1	1
0	1	1	0	0
0	1	1	1	1
1	0	0	0	1
1	0	0	1	1
1	0	1	0	0
1	0	1	1	0
1	1	0	0	0
1	1	0	1	1
1	1	1	0	1
1	1	1	1	0

Figure P13.57

13.58 Design a circuit with a 4-bit input representing the binary number $A_3A_2A_1A_0$. The output should be 1 if the input value is divisible by 3. Assume that the circuit is to be used only for the digits 0 through 9 (thus, values for 10 to 15 can be don't care conditions).

a. Draw the Karnaugh map and truth table for the function.

b. Determine the minimum expression for the function.

c. Draw the circuit, using only AND, OR, and NOT gates.

13.59 Find the simplified sum-of-products representation of the function from the Karnaugh map shown in Figure P13.59. Note that x is the don't care term.

$C \cdot D$ \ $A \cdot B$	00	01	11	10
00	0	1	0	0
01	1	1	0	0
11	0	x	1	0
10	0	0	1	0

Figure P13.59

13.60 Can the circuit for Problem 13.54 be simplified if it is known that the input represents a BCD (binary-coded decimal) number, that is, it can never be greater than 10_{10}? If not, explain why not. Otherwise, design the simplified circuit.

13.61 Find the simplified sum-of-products representation of the function from the Karnaugh map shown in Figure P13.61.

$C \cdot D$ \ $A \cdot B$	00	01	11	10
00	0	1	x	0
01	0	1	x	0
11	0	1	0	1
10	x	x	1	0

Figure P13.61

13.62 One method of ensuring reliability in data transmission systems is to transmit a parity bit along with every nibble, byte, or word of binary data transmitted. The parity bit confirms whether an even or odd number of 1s were transmitted in the data. In even-parity systems, the parity bit is set to 1 when the number of 1s in the transmitted data is odd. Odd-parity systems set the parity bit to 1 when the number of 1s in the transmitted data is even. Assume that a parity bit is transmitted for every nibble of data. Design a logic circuit that checks the nibble of data and transmits the proper parity bit for both even- and odd-parity systems.

13.63 Assume that a parity bit is transmitted for every nibble of data. Design two logic circuits that check a nibble of data and its parity bit to determine if there may have been a data transmission error. Assume first an even-parity system, then an odd-parity system.

13.64 Design a logic circuit that takes a 4-bit Gray code input from an optical encoder and translates it into two 4-bit nibbles of BCD.

13.65 Design a logic circuit that takes a 4-bit Gray code input from an optical encoder and determines if the input value is a multiple of 3.

13.66 The 4221 code is a base 10–oriented code that assigns the weights 4221 to each of 4 bits in a nibble of data. Design a logic circuit that takes a BCD nibble as input and converts it to its 4221 equivalent. The logic circuit should also report an error in the BCD input if its value exceeds 1001.

13.67 The 4-bit digital output of each of two sensors along an assembly line conveyor belt is proportional to the number of parts that pass by on the conveyor belt in a 30-s period. Design a logic circuit that reports an error if the outputs of the two sensors differ by more than one part per 30-s period.

Section 13.5: Combinational Logic Modules

13.68 A function, F, is defined such that it equals 1 when a 4-bit input code is equivalent to any of the decimal numbers 3, 6, 9, 12, or 15. F is 0 for input codes 0, 2, 8, and 10. Other input values cannot occur. Use a Karnaugh map to determine a minimal expression for this function. Design and sketch a circuit to implement this function using only AND and NOT gates.

13.69

a. Fill in the Karnaugh map for the logic function defined by the truth table of Figure P13.69.

b. What is the minimum expression for the function?

c. Realize the function using a 1-of-8 multiplexer.

A	B	C	D	$f(A,B,C,D)$
0	0	0	0	1
0	0	0	1	0
0	0	1	0	1
0	0	1	1	1
0	1	0	0	0
0	1	0	1	1
0	1	1	0	0
0	1	1	1	0
1	0	0	0	0
1	0	0	1	1
1	0	1	0	0
1	0	1	1	0
1	1	0	0	1
1	1	0	1	0
1	1	1	0	1
1	1	1	1	1

Figure P13.69

13.70

a. Fill in the truth table for the multiplexer circuit shown in Figure P13.70.

b. What binary function is performed by these multiplexers?

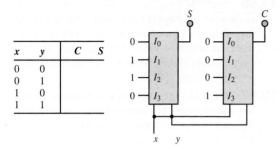

x	y	C	S
0	0		
0	1		
1	0		
1	1		

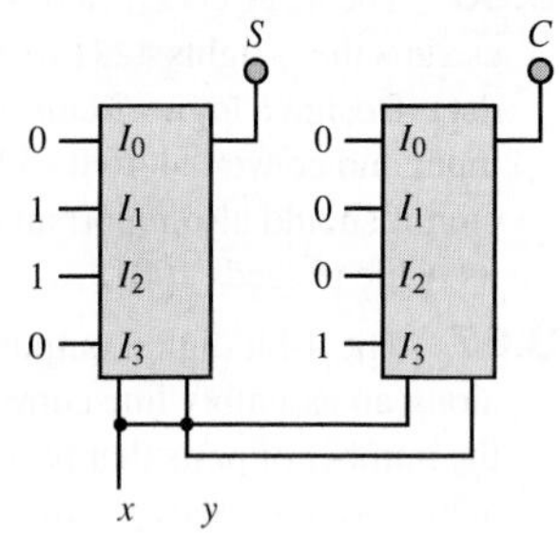

Figure P13.70

13.71 The circuit of Figure P13.71 can operate as a 4-to-16 decoder. Terminal EN denotes the enable input. Describe the operation of the 4-to-16 decoder. What is the role of logic variable A?

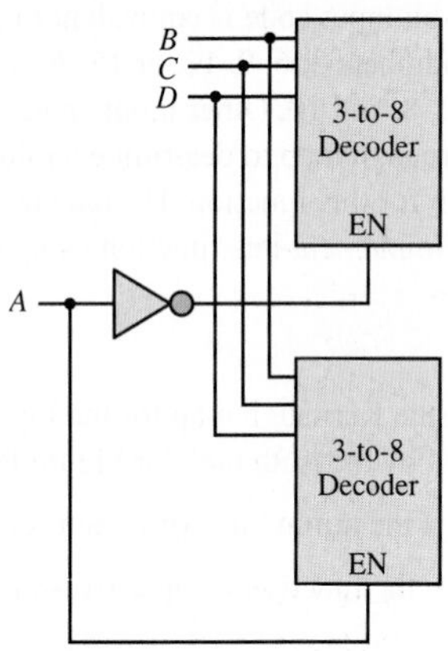

Figure P13.71

13.72 Show that the circuit given in Figure P13.72 converts 4-bit binary numbers to 4-bit Gray code.

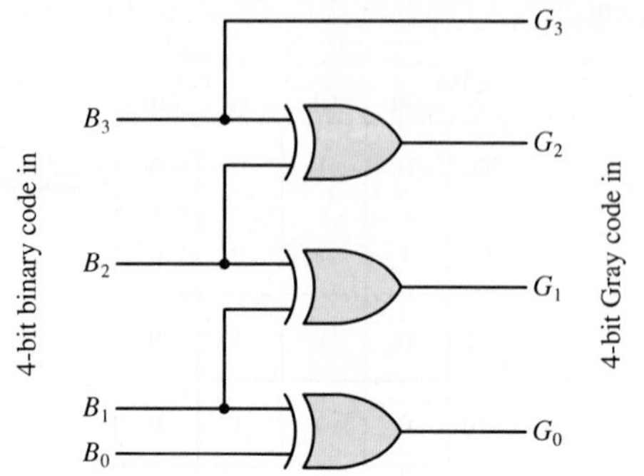

Figure P13.72

13.73 Suppose one of your classmates claims that the following boolean expressions represent the conversion from 4-bit Gray code to 4-bit binary numbers:

$$B_3 = G_3$$
$$B_2 = G_3 \oplus G_2$$
$$B_1 = G_3 \oplus G_2 \oplus G_1$$
$$B_0 = G_3 \oplus G_2 \oplus G_1 \oplus G_0$$

a. Show that your classmate's claim is correct.

b. Draw the circuit which implements the conversion.

13.74 Select the proper inputs for a four-input multiplexer to implement the function $f(A, B, C) = \overline{A}B\overline{C} + A\overline{BC} + AC$. Assume inputs I_0, I_1, I_2, and I_3 correspond to $\overline{AB}$, $\overline{A}B$, $A\overline{B}$, and AB, respectively, and that each input may be 0, 1, $\overline{C}$, or C.

13.75 Select the proper inputs for an 8-bit multiplexer to implement the function $f(A, B, C, D) = \sum(2, 5, 6, 8, 9, 10, 11, 13, 14)_{10}$. Assume the inputs I_0 through I_7 correspond to $\overline{ABC}$, $A\overline{BC}$, $\overline{A}B\overline{C}$, $AB\overline{C}$, $\overline{AB}C$, $A\overline{B}C$, $\overline{A}BC$, and ABC, respectively, and that each input may be 0, 1, $\overline{D}$, or D.

CHAPTER 14

DIGITAL SYSTEMS

The first half of Chapter 14 continues the analysis of digital circuits that was begun in Chapter 13 by focusing on sequential logic circuits, such as flip-flops, counters, and shift registers. The second half of the chapter is devoted to an overview of the basic functions of microcontrollers and microprocessors. During the last decade, microprocessors have become a standard tool in the analysis of engineering data, in the design of experiments, and in the control of plants and processes. No longer a specialized electronic device to be used only by appropriately trained computer engineers, today's microprocessor—perhaps more commonly represented by the ubiquitous *personal computer*—is a basic tool in the engineering profession. The common thread in its application in various engineering fields is its use in digital data acquisition instruments and digital controllers.

Modern microprocessors are relatively easy to program, have significant computing power and excellent memory storage capabilities, and can be readily interfaced with other instruments and electronic devices, such as transducers, printers, and other computers. The basic functions performed by the microprocessor in a typical digital data acquisition or control application are easily described: input signals (often analog, sometimes already in digital form) are acquired by the computer and processed by means of suitable software to produce the desired result (i.e., they undergo some kind of mathematical manipulation), which is then outputted to either a display or a storage device, or is used in controlling a process, a plant, or an experiment. The objective of this chapter is to describe these various processes, with the aim of giving the reader enough background information to understand the notation used in data books and instruction manuals.

Learning Objectives

1. Analyze the operation of sequential logic circuits. *Section 14.1.*
2. Understand the operation of digital counters. *Section 14.1.*
3. Design simple sequential circuits using state transition diagrams. *Section 14.2.*
4. Study the basic architecture of microprocessors and microcontrollers. *Sections 14.3, 14.4, 14.5.*

14.1 SEQUENTIAL LOGIC MODULES

The discussion of logic devices in Chapter 13 focuses on the general family of combinational logic devices. The feature that distinguishes combinational logic devices from the other major family—**sequential logic devices**—is that combinational logic circuits provide outputs that are based on a combination of present inputs only. On the other hand, sequential logic circuits depend on present and past input values. Because of this "memory" property, sequential circuits can store information; this capability opens a whole new area of application for digital logic circuits.

Latches and Flip-Flops

The basic information storage device in a digital circuit is called a **flip-flop.** There are many different varieties of flip-flops; however, all flip-flops share the following characteristics:

1. A flip-flop is a **bistable device;** that is, it can remain in one of two stable states (0 and 1) until appropriate conditions cause it to change state. Thus, a flip-flop can serve as a memory element.
2. A flip-flop has two outputs, one of which is the complement of the other.

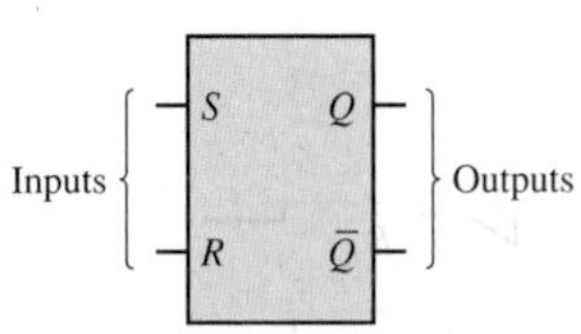

S	R	Q
0	0	Present state
0	1	Reset
1	0	Set
1	1	Disallowed

Figure 14.1 *RS* flip-flop symbol and truth table

RS Flip-Flop

It is customary to depict flip-flops by their block diagram and a name, such as Q or X, representing the output variable. Figure 14.1 represents the ***RS* flip-flop,** which has two inputs, denoted by S and R, and two outputs Q and $\overline{Q}$. The value at Q is called the *state* of the flip-flop. If $Q = 1$, we refer to the device as *being in the* 1 *state.* Thus, we need define only one of the two outputs of the flip-flop. The two inputs R and S are used to change the state of the flip-flop, according to the following rules:

1. When $R = S = 0$, the flip-flop remains in its present state (whether 1 or 0).
2. When $S = 1$ and $R = 0$, the flip-flop is *set* to the 1 state (thus, S, for **set**).
3. When $S = 0$ and $R = 1$, the flip-flop is *reset* to the 0 state (thus, R, for **reset**).
4. It is not permitted for both S and R to be equal to 1. (This would correspond to requiring the flip-flop to set and reset at the same time.)

LO1

The rules just described are easily remembered by noting that 1s on the S and R inputs correspond to the set and reset commands, respectively.

A convenient means of describing the series of transitions that occur as the signals sent to the flip-flop inputs change is the **timing diagram.** A timing diagram is a graph of the inputs and outputs of the *RS* flip-flop (or any other logic device) depicting the transitions that occur over time. In effect, one could also represent these transitions in tabular form; however, the timing diagram provides a convenient visual representation of the evolution of the state of the flip-flop. Figure 14.2 depicts a table of transitions for an *RS* flip-flop Q as well as the corresponding timing diagram.

S	R	Q
1	0	1
0	0	1
0	0	1
0	1	0
0	0	0
0	0	0
0	1	0
0	0	0
1	0	1
0	0	1

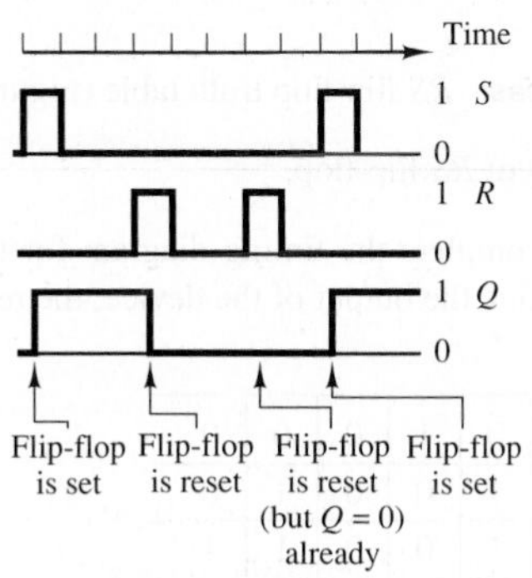

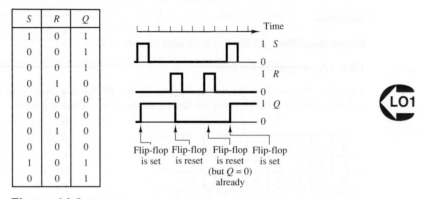

Figure 14.2 Timing diagram for the *RS* flip-flop

It is important to note that the *RS* flip-flop is **level-sensitive.** This means that the set and reset operations are completed only after the R and S inputs have reached the appropriate levels. Thus, in Figure 14.2 we show the transitions in the Q output as occurring with a small delay relative to the transitions in the R and S inputs.

It is instructive to illustrate how an *RS* flip-flop can be constructed using simple logic gates. For example, Figure 14.3 depicts a realization of such a circuit consisting of four gates: two inverters and two NAND gates (actually, the same result could be achieved with four NAND gates). Consider the case in which the circuit is in the initial state $Q = 0$ (and therefore $\overline{Q} = 1$). If the input $S = 1$ is applied, the top NOT gate will see inputs $\overline{Q} = 1$ and $\overline{S} = 0$, so that $Q = \overline{(\overline{S} \cdot \overline{Q})} = \overline{(0 \cdot 1)} = 1$—that is, the flip-flop is set. Note that when Q is set to 1, $\overline{Q}$ becomes 0. This, however, does not affect the state of the Q output, since replacing $\overline{Q}$ with 0 in the expression

$$Q = \overline{(\overline{S} \cdot \overline{Q})}$$

does not change the result:

$$Q = \overline{(0 \cdot 0)} = 1$$

Thus, the cross-coupled feedback from outputs Q and $\overline{Q}$ to the input of the NAND gates is such that the set condition sustains itself. It is straightforward to show (by symmetry) that a 1 input on the R line causes the device to reset (i.e., causes $Q = 0$) and that this condition is also self-sustaining.

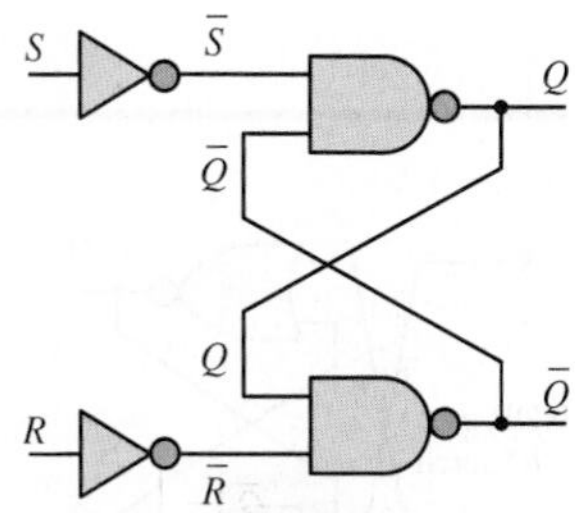

Figure 14.3 Logic gate implementation of the *RS* flip-flop

LO1

EXAMPLE 14.1 *RS* Flip-Flop Timing Diagram

Problem

Determine the output of an *RS* flip-flop for the series of inputs given in the table below.

R	0	0	0	1	0	0	0
S	1	0	1	0	0	1	0

Solution

Known Quantities: *RS* flip-flop truth table (Figure 14.1).

Find: Output *Q* of *RS* flip-flop.

Analysis: We complete the timing diagram for the *RS* flip-flop, following the rules stated earlier to determine the output of the device; the result is summarized below.

R	0	0	0	1	0	0	0
S	1	0	1	0	0	1	0
Q	1	1	1	0	0	1	1

A sketch of the waveforms, shown below, can also be generated to visualize the transitions.

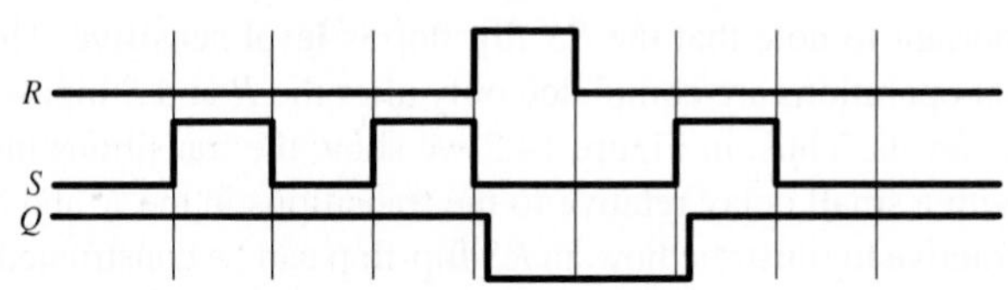

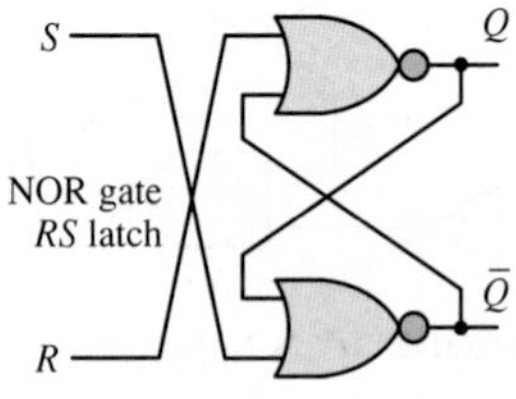

CHECK YOUR UNDERSTANDING

The circuit shown in the figure also serves as an *RS* flip-flop and requires only two NOR gates. Analyze the circuit to prove that it operates as an *RS* flip-flop. (*Hint:* Use a truth table with two variables, *S* and *R*.)

An extension of the *RS* flip-flop includes an additional enable input that is *gated* into each of the other two inputs. Figure 14.4 depicts an *RS* flip-flop consisting of two NOR gates. In addition, an enable input is connected through two AND gates to the *RS* flip-flop, so that an input to the *R* or *S* line will be effective only when the enable input is 1. Thus, any transitions will be controlled by the enable input, which acts as a synchronizing signal. The enable signal may consist of a **clock,** in which case the flip-flop is said to be **clocked** and its operation is said to be **synchronous.**

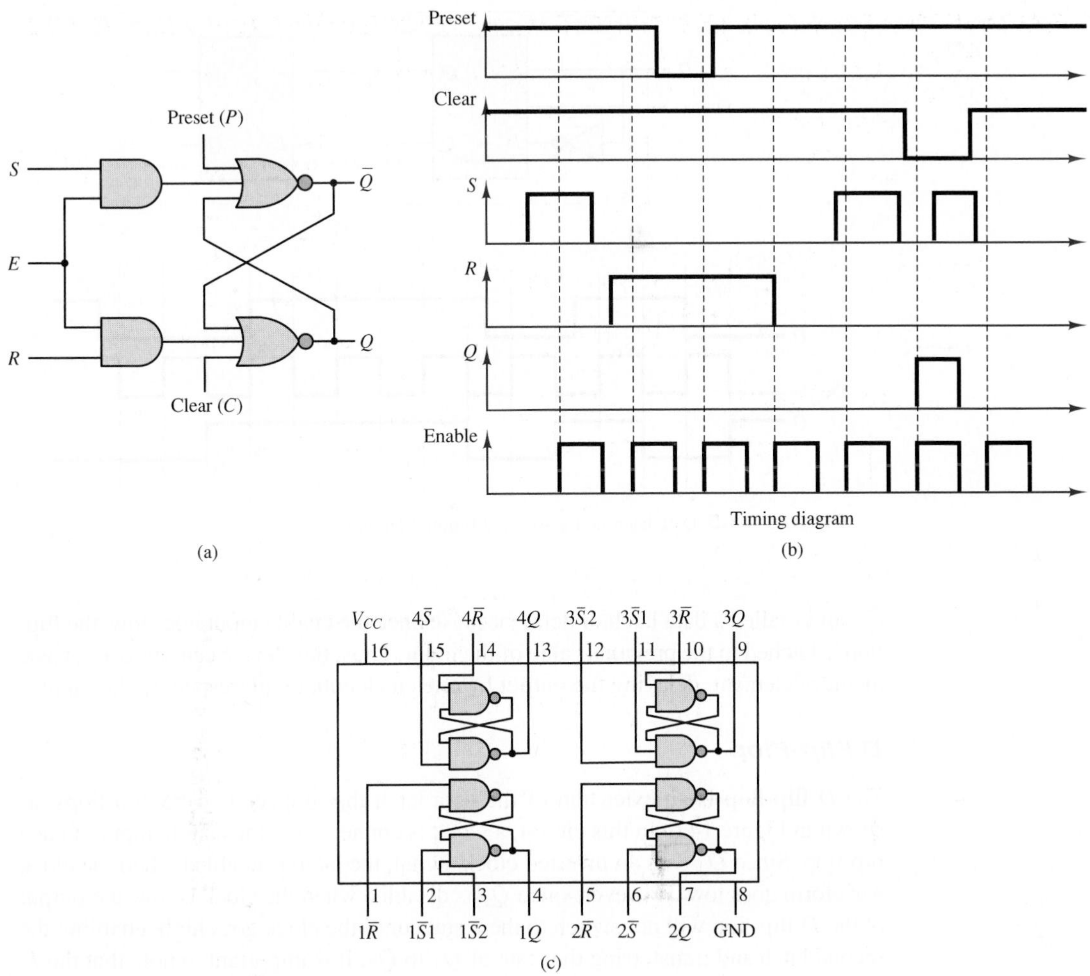

Figure 14.4 The *RS* flip-flop with enable, preset, and clear lines: (a) logic diagram, (b) timing diagram, (c) IC schematic

The same circuit of Figure 14.4 can be used to illustrate two additional features of flip-flops: the **preset** and **clear** functions, denoted by the inputs P and C, respectively. When P and C are 0, they do not affect the operation of the flip-flop. Setting $P = 1$ corresponds to setting $S = 1$ and therefore causes the flip-flop to go into the 1 state. Thus, the term *preset:* this function allows the user to preset the flip-flop to 1 at any time. When C is 1, the flip-flop is reset, or *cleared* (that is, Q is made equal to 0). Note that these direct inputs are, in general, asynchronous; therefore, they allow the user to preset or clear the flip-flop at any time. A set of timing waveforms illustrating the function of the enable, preset, and clear inputs is also shown in Figure 14.4. Note how transitions occur only when the enable input goes high (unless the preset or clear inputs are used to override the *RS* inputs).

Another extension of the *RS* flip-flop, called the **data latch,** or **delay element,** is shown in Figure 14.5. In this circuit, the R input is always equal to the inverted S input, so that whenever the enable input is high, the flip-flop is set. This device has the dual advantage of avoiding the potential conflict that might arise if both R and S were high and reducing the number of input connections by eliminating the reset input. This

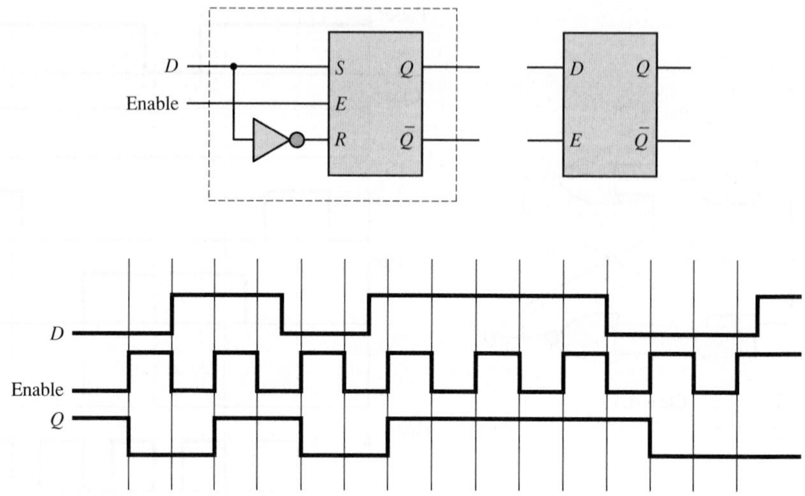

Figure 14.5 Data latch and associated timing diagram

circuit is called a data latch or delay because once the enable input goes low, the flip-flop is latched to the previous value of the input. Thus, this device can serve as a basic memory element, delaying the output by one clock count with respect to the input.

D Flip-Flop

The ***D* flip-flop** is an extension of the data latch that utilizes two *RS* flip-flops, as shown in Figure 14.6. In this circuit, a clock is connected to the enable input of each flip-flop. Since Q_1 sees an inverted clock signal, the latch is enabled when the clock waveform goes low. However, since Q_2 is disabled when the clock is low, the output of the *D* flip-flop will not switch to the 1 state until the clock goes high, enabling the second latch and transferring the state of Q_1 to Q_2. It is important to note that the *D* flip-flop changes state only on the positive edge of the clock waveform: Q_1 is set on the negative edge of the clock, and Q_2 (and therefore Q) is set on the positive edge of the clock, as shown in the timing diagram of Figure 14.6. This type of device is said to be **edge-triggered.** This feature is indicated by the "knife-edge" drawn next to the CLK input in the device symbol. The particular device described here is said to be positive edge–triggered, or **leading edge–triggered,** since the final output of the flip-flop is set on a positive-going clock transition.

On the basis of the rules stated in this section, the state of the *D* flip-flop can be described by the following truth table:

LO1

D	CLK	*Q*
0	↑	0
1	↑	1

where the symbol ↑ indicates the occurrence of a positive transition.

JK Flip-Flop

Another very common type of flip-flop is the ***JK* flip-flop,** shown in Figure 14.7 on p. 722. The *JK* flip-flop operates according to the following rules:

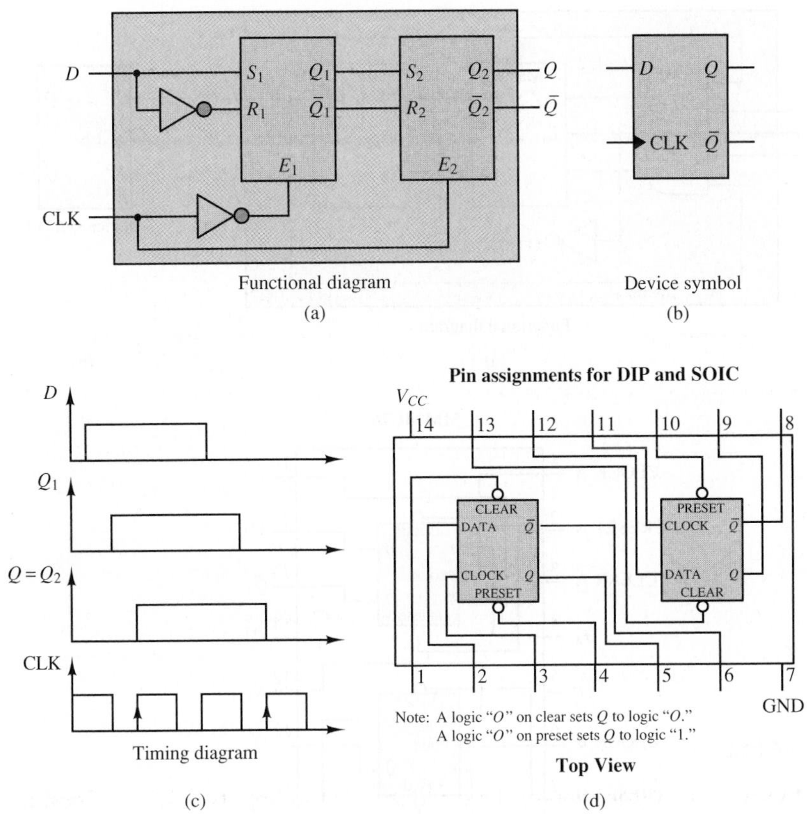

Figure 14.6 The *D* flip-flop: (a) functional diagram, (b) symbol, (c) timing waveforms, and (d) IC schematic

- When *J* and *K* are both low, no change occurs in the state of the flip-flop.
- When $J = 0$ and $K = 1$, the flip-flop is reset to 0.
- When $J = 1$ and $K = 0$, the flip-flop is set to 1.
- When both *J* and *K* are high, the flip-flop will toggle between states at every negative transition of the clock input, denoted from here on by the symbol $\downarrow$.

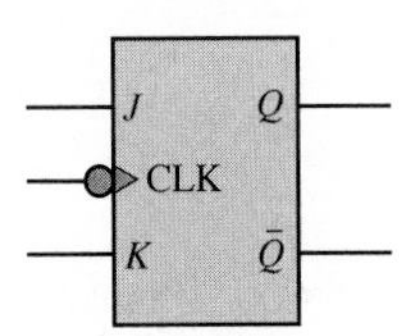

JK flip-flop

Note that, functionally, the operation of the *JK* flip-flop can also be explained in terms of two *RS* flip-flops. When the clock waveform goes high, the *master* flip-flop is enabled; the *slave* receives the state of the master upon a negative clock transition. The *bubble* at the clock input signifies that the device is negative or **trailing edge–triggered.** This behavior is similar to that of an *RS* flip-flop, except for the $J = 1$, $K = 1$ condition, which corresponds to a toggle mode rather than to a disallowed combination of inputs.

J_n	K_n	Q_{n+1}
0	0	Q_n
0	1	0 (reset)
1	0	1 (set)
1	1	$\bar{Q}_n$ (toggle)

Figure 14.8 Truth table for the *JK* flip-flop

Figure 14.8 depicts the truth table for the *JK* flip-flop. It is important to note that when both inputs are 0, the flip-flop remains in its previous state at the occurrence of a clock transition; when either input is high and the other is low, the *JK* flip-flop behaves as the *RS* flip-flop, whereas if both inputs are high, the output "toggles" between states every time the clock waveform undergoes a negative transition.

Data sheets for various types of flip-flops may be found in the accompanying CD-ROM.

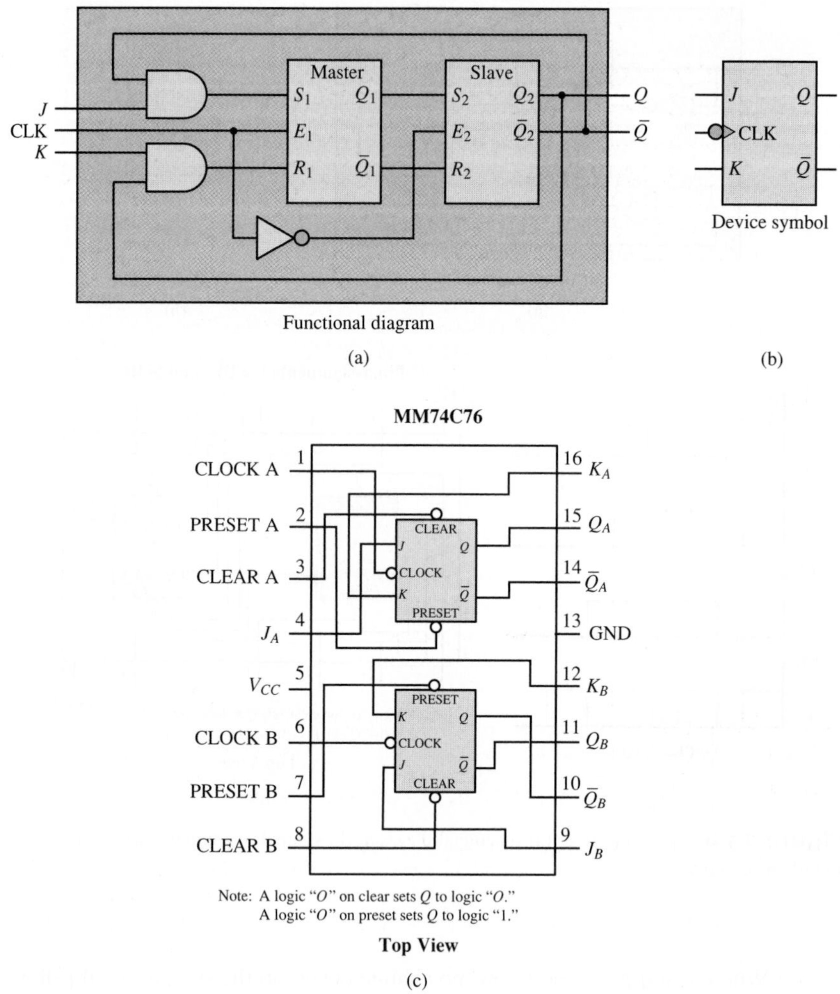

Figure 14.7 The *JK* flip-flop: (a) functional diagram, (b) device symbol, and (c) IC schematic

---

CHECK YOUR UNDERSTANDING

Derive the detailed truth table and draw a timing diagram for the *JK* flip-flop, using the model of Figure 14.7 with two flip-flops.

LO1

EXAMPLE 14.2 The *T* Flip-Flop

Problem

Determine the truth table and timing diagram of the ***T* flip-flop** of Figure 14.9. Note that the *T* flip-flop is a *JK* flip-flop with its inputs tied together.

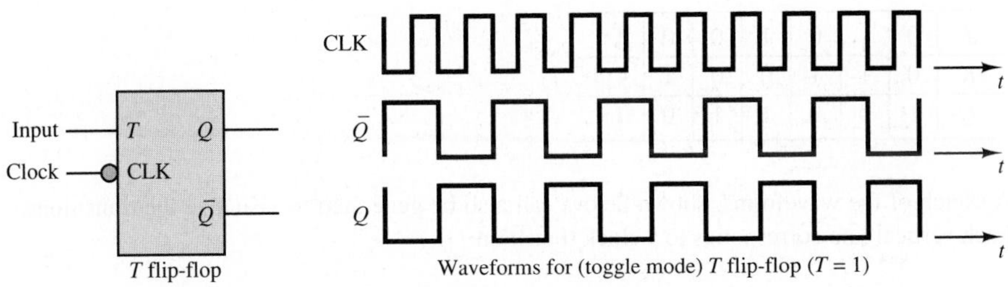

Figure 14.9 The T flip-flop symbol and timing waveforms

Solution

Known Quantities: JK flip-flop rules of operation (Figure 14.8).

Find: Truth table and timing diagram for T flip-flop.

Analysis: We recognize that the T flip-flop is a JK flip-flop with its inputs tied together. Thus, the flip-flop will need only a two-element truth table to describe its operation, corresponding to the top and bottom entries in the JK flip-flop truth table of Figure 14.8. The truth table is shown below. A timing diagram is also included in Figure 14.9.

T	CLK	Q_{k+1}
0	↓	Q_k
1	↓	$\overline{Q_k}$

Comments: The T flip-flop takes its name from the fact that it *toggles* between the high and low states. Note that the toggling frequency is one-half that of the clock. Thus the T flip-flop also acts as a *divide-by-2* counter. Counters are explored in greater detail in the next section.

EXAMPLE 14.3 The *JK* Flip-Flop Timing Diagram

LO1

Problem

Determine the output of a JK flip-flop for the series of inputs given in the table below. The initial state of the flip-flop is $Q_0 = 1$.

J	0	1	0	1	0	0	1
K	0	1	1	0	0	1	1

Solution

Known Quantities: JK flip-flop truth table (Figure 14.8).

Find: Output of RS flip-flop Q as a function of the input transitions.

Analysis: We complete the timing diagram for the JK flip-flop, following the rules of Figure 14.8; the result is summarized next.

J	0	1	0	1	0	0	1
K	0	1	1	0	0	1	1
Q	1	0	0	1	1	0	1

A sketch of the waveforms, shown below, can also be generated to visualize the transitions. Each vertical line corresponds to a clock transition.

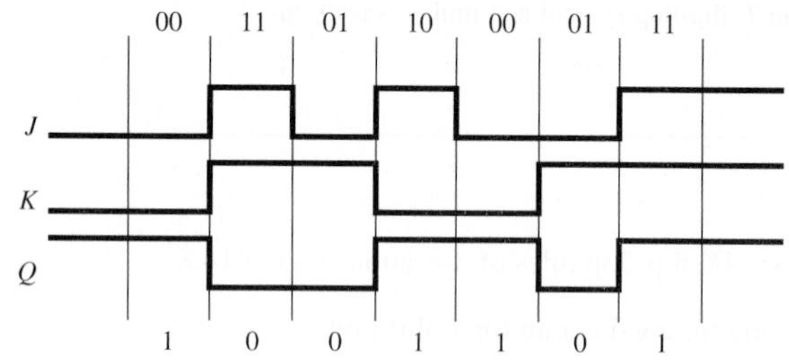

Comments: How would the timing diagram change if the initial state of the flip-flop were $Q_0 = 1$?

Digital Counters

One of the more immediate applications of flip-flops is in the design of **counters.** A counter is a sequential logic device that can take one of N possible states, stepping through these states in a sequential fashion. When the counter has reached its last state, it resets to 0 and is ready to start counting again. For example, a 3-bit **binary up counter** would have $2^3 = 8$ possible states, and might appear as shown in the functional block of Figure 14.10. The input clock waveform causes the counter to step through the eight states, making one transition for each clock pulse. We shall shortly see that a string of JK flip-flops can accomplish this task exactly. The device shown in Figure 14.10 also displays a reset input, which forces the counter output to equal 0: $b_2b_1b_0 = 000$.

Although binary counters are very useful in many applications, one is often interested in a **decade counter,** that is, a counter that counts from 0 to 9 and then resets. A 4-bit binary counter can easily be configured in principle to provide this function by means of simple logic that resets the counter when it has reached the count $1001_2 = 9_{10}$. As shown in Figure 14.11, if we connect bits b_3 and b_1 to a four-input AND gate, along with $\overline{b}_2$ and $\overline{b}_0$, the output of the AND gate can be used to reset the counter after a count of 10. Additional logic can provide a *carry* bit whenever a reset condition is reached, which could be passed along to another decade counter, enabling counts up to 99. Decade counters can be cascaded so as to represent decimal digits in succession.

Although the decade counter of Figure 14.11 is attractive because of its simplicity, this configuration would never be used in practice, because of the presence of **propagation delays.** These delays are caused by the finite response time of the individual transistors in each logic device and cannot be guaranteed to be identical for each gate and flip-flop. Thus, if the reset signal—which is presumed to be

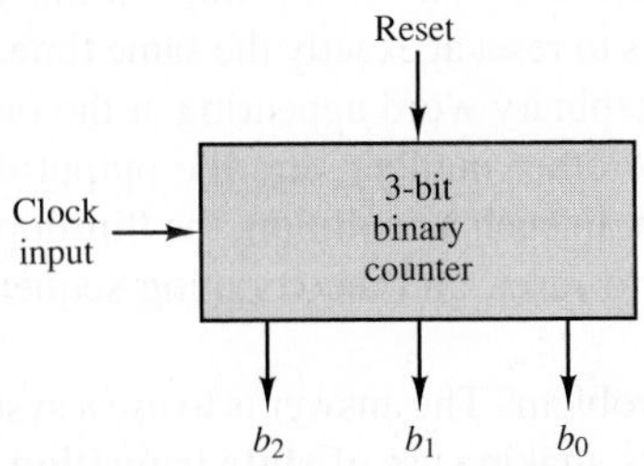

Functional representation of binary counter

Input pulses	State b_2	b_1	b_0
0	0	0	0
1	0	0	1
2	0	1	0
3	0	1	1
4	1	0	0
5	1	0	1
6	1	1	0
7	1	1	1

Timing table

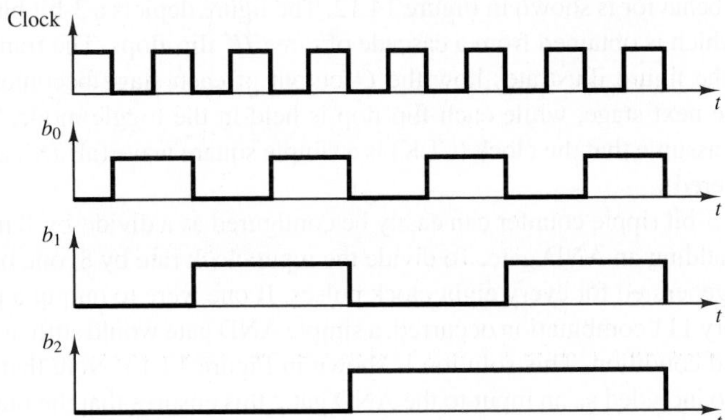

Timing diagram

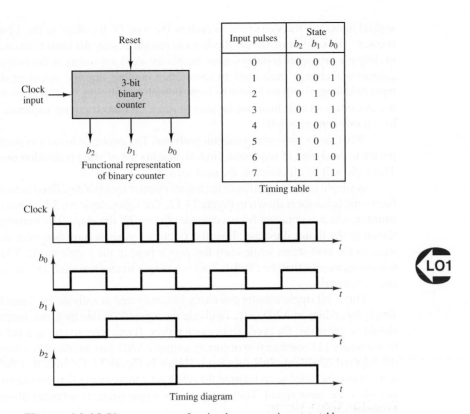

Figure 14.10 Binary up counter functional representation, state table, and timing waveforms

Input pulses	b_3	b_2	b_1	b_0
0	0	0	0	0
1	0	0	0	1
2	0	0	1	0
3	0	0	1	1
4	0	1	0	0
5	0	1	0	1
6	0	1	1	0
7	0	1	1	1
8	1	0	0	0
9	1	0	0	1
10	1	0	1	0

(a)

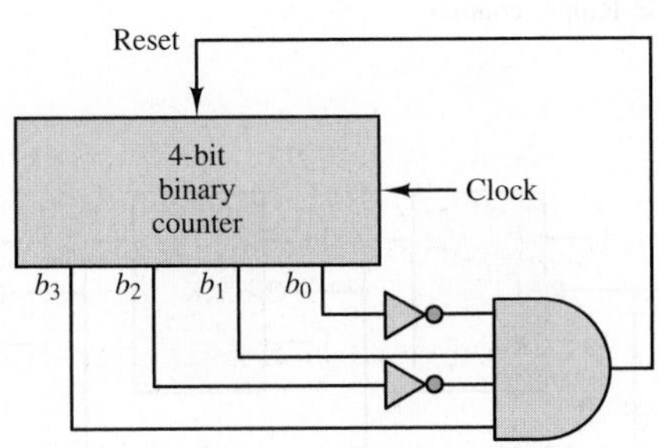

(b)

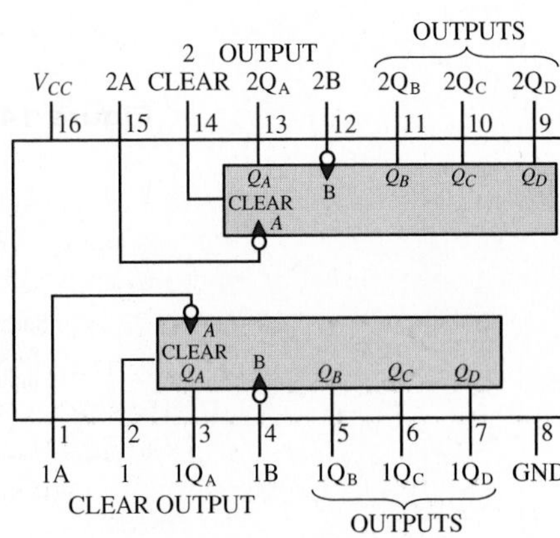

(c)

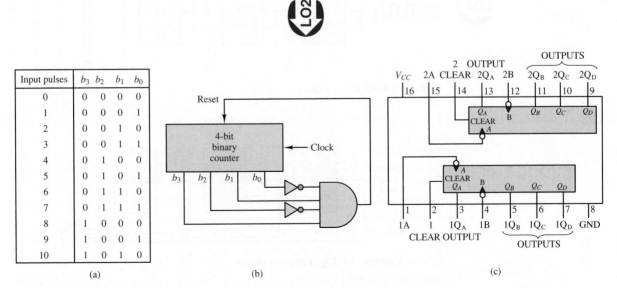

Figure 14.11 Decade counter: (a) counting sequence; (b) functional diagram; and (c) IC schematic

applied at exactly the same time to each of the four *JK* flip-flops in the 4-bit binary counter—does not cause the *JK* flip-flops to reset at exactly the same time on account of different propagation delays, then the binary word appearing at the output of the counter will change from 1001 to some other number, and the output of the four-input NAND gate will no longer be high. In such a condition, the flip-flops that have not already reset will then not be able to reset, and the counting sequence will be irreparably compromised.

What can be done to obviate this problem? The answer is to use a systematic approach to the design of sequential circuits, making use of **state transition diagrams.** This topic will be discussed in the next section.

A simple implementation of the binary counter we have described in terms of its functional behavior is shown in Figure 14.12. The figure depicts a 3-bit binary **ripple counter,** which is obtained from a cascade of three *JK* flip-flops. The transition table shown in the figure illustrates how the Q output of each stage becomes the clock input to the next stage, while each flip-flop is held in the toggle mode. The output transitions assume that the clock (CLK) is a simple square wave (all *JK*s are negative edge–triggered).

This 3-bit ripple counter can easily be configured as a divide-by-8 mechanism, simply by adding an AND gate. To divide the input clock rate by 8, one output pulse should be generated for every eight clock pulses. If one were to output a pulse every time a binary 111 combination occurred, a simple AND gate would suffice to generate the required condition. This solution is shown in Figure 14.13. Note that the square wave is also included as an input to the AND gate; this ensures that the output is only as wide as the input signal. This application of ripple counters is further illustrated in Example 14.4

LO2

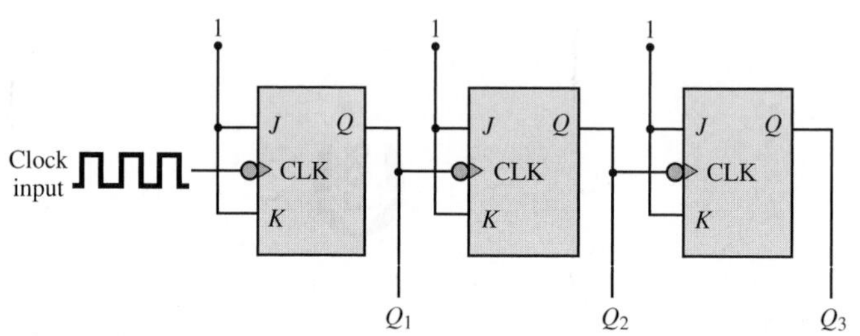

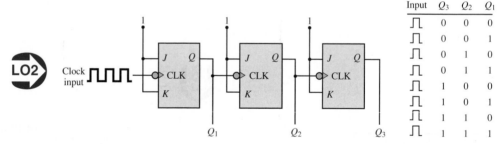

Input	Q_3	Q_2	Q_1
⎍	0	0	0
⎍	0	0	1
⎍	0	1	0
⎍	0	1	1
⎍	1	0	0
⎍	1	0	1
⎍	1	1	0
⎍	1	1	1

Figure 14.12 Ripple counter

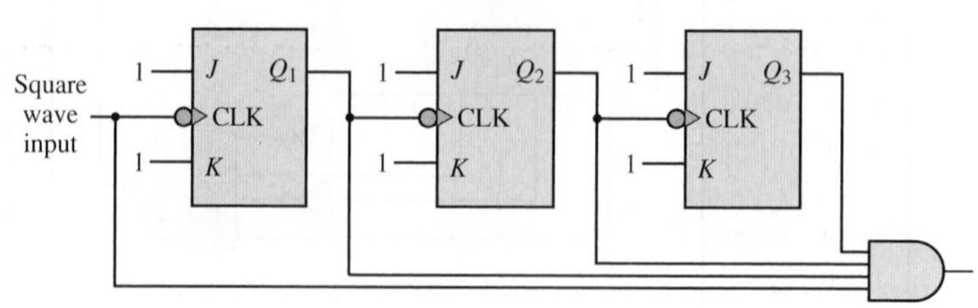

Figure 14.13 Divide-by-8 circuit

EXAMPLE 14.4 Divider Circuit

Problem

A binary ripple counter provides a means of dividing the fixed output rate of a clock by powers of 2. For example, the circuit of Figure 14.14 is a divide-by-2 or divide-by-4 counter. Draw the timing diagrams for the clock input, Q_0, and Q_1 to demonstrate these functions.

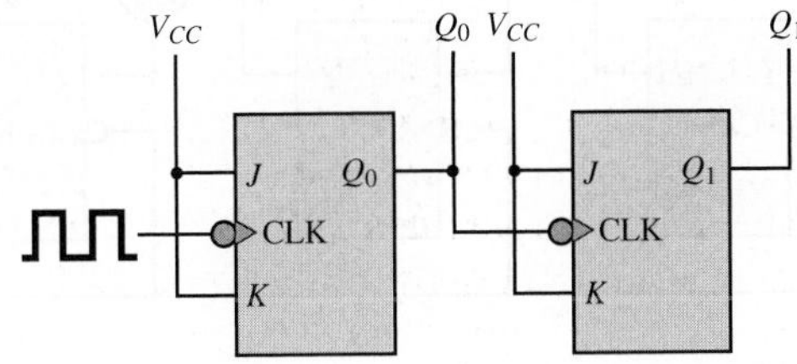

Figure 14.14

Solution

Known Quantities: *JK* flip-flop truth table (Figure 14.8).

Find: Output of each flip-flop *Q* as a function of the input clock transitions.

Assumptions: Assume positive edge–triggered devices. The DC supply voltage is V_{CC}.

Analysis: Following the timing diagram of Figure 14.15, we see that Q_0 switches at one-half the frequency of the clock input, and that Q_1 switches at one-half the frequency of Q_0, hence the timing diagram shown.

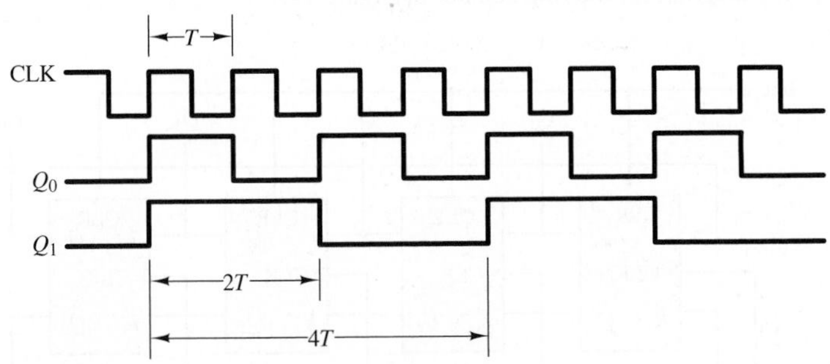

Figure 14.15 Divider circuit timing diagram

A slightly more complex version of the binary counter is the **synchronous counter,** in which the input clock drives all the flip-flops simultaneously. Figure 14.16 depicts a 3-bit synchronous counter. In this figure, we have chosen to represent each flip-flop as a *T* flip-flop. The clocks to all the flip-flops are incremented simultaneously. The reader should verify that Q_0 toggles to 1 first, and then Q_1 toggles to 1, and that the AND gate ensures that Q_2 will toggle only after Q_0 and Q_1 have both reached the 1 state ($Q_0 \cdot Q_1 = 1$).

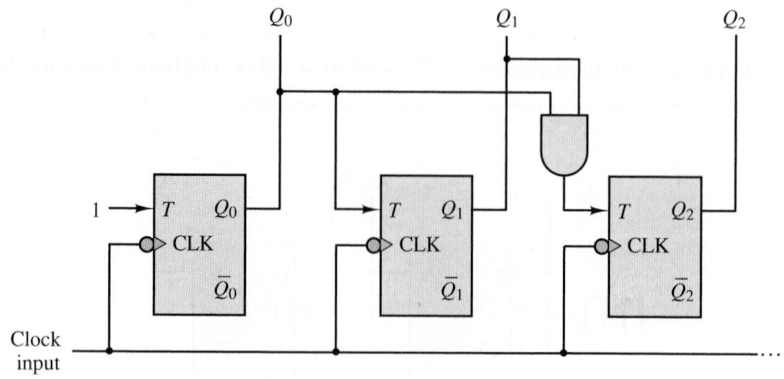

Figure 14.16 Three-bit synchronous counter

Other common counters are the **ring counter,** illustrated in Example 14.5, and the **up-down counter,** which has an additional select input that determines whether the counter counts up or down. Data sheets for various counters may be found in the accompanying CD-ROM.

LO2

EXAMPLE 14.5 Ring Counter

Problem

Draw the timing diagram for the ring counter of Figure 14.17.

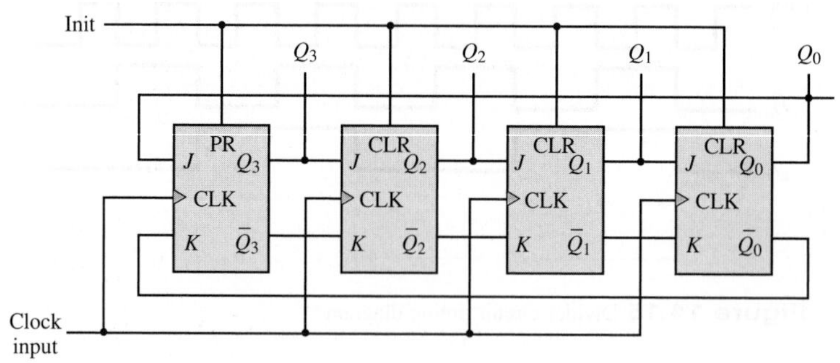

Figure 14.17 Ring counter

Solution

Known Quantities: *JK* flip-flop truth table (Figure 14.8).

Find: Output of each flip-flop Q as a function of the input clock transitions.

Assumptions: Prior to application of the clock input, the Init line sees a positive transition [this initializes the counter by setting the state of the first flip-flop to 1 through a PR (preset) input, and all other states to zero through a CLR (clear) input].

Analysis: With the initial state of $Q_3 = 0$, a clock transition will *set* $Q_3 = 1$. The clock also causes the other three flip-flops to see a *reset* input of 1, since $Q_3 = Q_2 = Q_1 = Q_0 = 0$ at the time of the first clock pulse. Thus, Q_2, Q_1, and Q_0 remain in the 0 state. At the second clock pulse, since Q_3 is now 1, the second flip-flop will see a *set* input of 1, and its output will become $Q_2 = 1$. Both Q_1 and Q_0 remain in the 0 state, and Q_3 is reset to 0. The pattern continues, causing the 1 state to ripple from left to right and back again. This rightward rotation gives the counter its name. The transition table is shown below.

CLK	Q_3	Q_2	Q_1	Q_0
↑	1	0	0	0
↑	0	1	0	0
↑	0	0	1	0
↑	0	0	0	1
↑	1	0	0	0
↑	0	1	0	0
↑	0	0	1	0

Comments: The shifting function implemented by the ring counter is used in the shift registers discussed in the following section.

Digital Measurement of Angular Position and Velocity

Another type of angular position encoder, besides the angular encoder discussed in Chapter 13 in the Focus on Measurements box "Position Encoders," is the slotted encoder shown in Figure 14.18. This encoder can be used in conjunction with a pair of counters and a high-frequency clock to determine the speed of rotation of the slotted wheel. As shown in Figure 14.19, a clock of known frequency is connected to a counter while another counter records the number of slot pulses detected by an optical slot detector as the wheel rotates. Dividing the counter values, one could obtain the speed of the rotating wheel in radians per second. For example, assume a clocking frequency of 1.2 kHz. If both counters are started at zero and at some instant the timer counter reads 2,850 and the encoder counter reads 3,050, then the speed of the rotating encoder is found to be

$$1{,}200 \frac{\text{cycles}}{\text{s}} \cdot \frac{2{,}850 \text{ slots}}{3{,}050 \text{ cycles}} = 1{,}121.3 \frac{\text{slots}}{\text{s}}$$

and

$$1{,}121.3 \text{ slots/s} \times 1^\circ \text{ per slot} \times \frac{2\pi}{360} \text{ rad/deg} = 19.6 \text{ rad/s}$$

(Continued)

If this encoder is connected to a rotating shaft, it is possible to measure the angular position and velocity of the shaft. Such shaft encoders are used in **measuring the speed of rotation of electric motors, machine tools, engines,** and other rotating machinery.

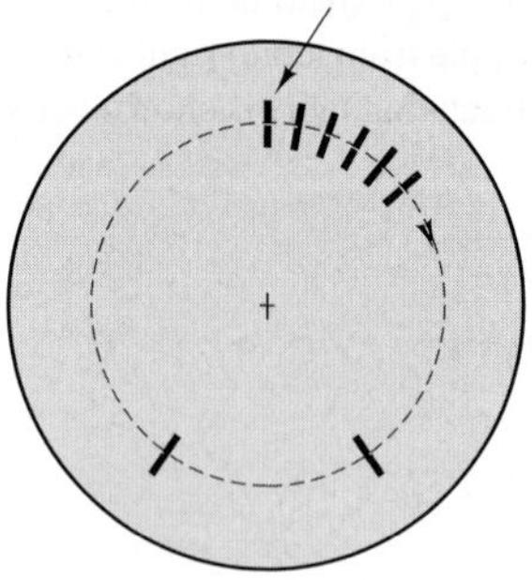

Figure 14.18

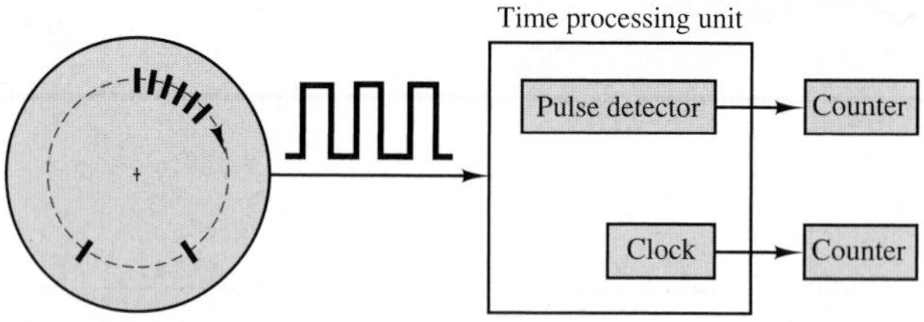

Figure 14.19 Calculating the speed of rotation of the slotted wheel

A typical application of the slotted encoder is to compute the ignition and injection timing in an automotive engine. In an automotive engine, information related to speed is obtained from the camshaft and the flywheel, which have known reference points. The reference points determine the timing for the ignition firing points and fuel injection pulses, and are identified by special slot patterns on the camshaft and crankshaft. Two methods are used to detect the special slots (reference points): *period measurement with additional transition detection (PMA)* and *period measurement with missing transition detection (PMM)*. In the PMA method, an additional slot (reference point) determines a known reference position on the crankshaft or camshaft. In the PMM method, the reference position is determined by the absence of a slot. Figure 14.20 illustrates a typical PMA pulse sequence, showing the presence of an additional pulse. The additional slot may be used to determine the timing for the ignition pulses relative to a known position of the crankshaft. Figure 14.21 depicts a typical PMM pulse sequence. Because the period of the pulses is known, the additional slot or the missing slot can be easily detected and used as a reference position. How would you implement these pulse sequences, using ring counters?

(*Continued*)

(Concluded)

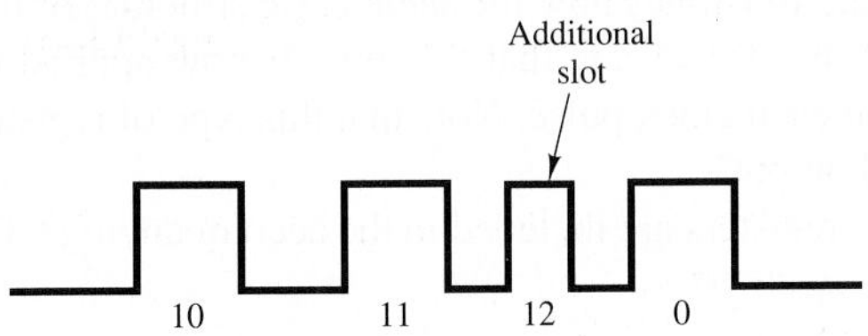

Figure 14.20 PMA pulse sequence

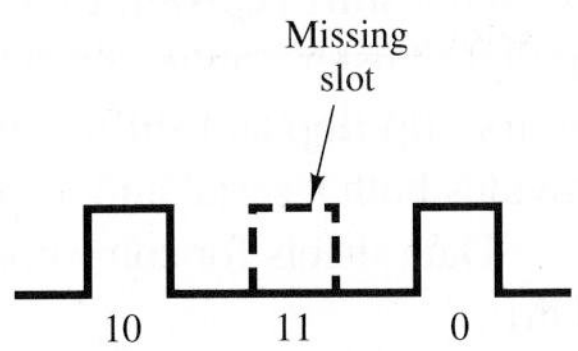

Figure 14.21 PMM pulse sequence

CHECK YOUR UNDERSTANDING

The speed of the rotating encoder of the Focus on Measurements box "Digital Measurement of Angular Position and Velocity" is found to be 9,425 rad/s. The encoder timer reads 10, and the clock counter reads 300. Assuming that both the timer counter and the encoder counter started at zero, find the clock frequency.

Answer: 4.5 kHz

Registers

A register consists of a cascade of flip-flops that can store binary data, 1 bit in each flip-flop. The simplest type of register is the parallel input–parallel output register shown in Figure 14.22. In this register, the *load* input pulse, which acts on all clocks simultaneously, causes the parallel inputs $b_0b_1b_2b_3$ to be transferred to the respective flip-flops. The D flip-flop employed in this register allows the transfer from b_n to Q_n to occur very directly. Thus, D flip-flops are very commonly used in this type of application. The binary word $b_3b_2b_1b_0$ is now "stored," each bit being represented by the state of a flip-flop. Until the load input is applied again and a new word appears at the parallel inputs, the register will preserve the stored word.

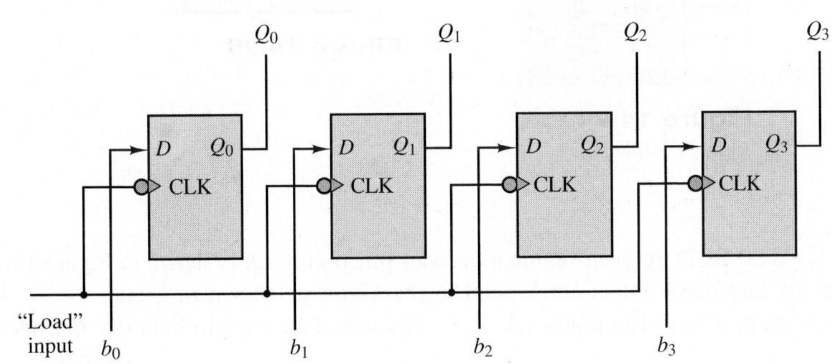

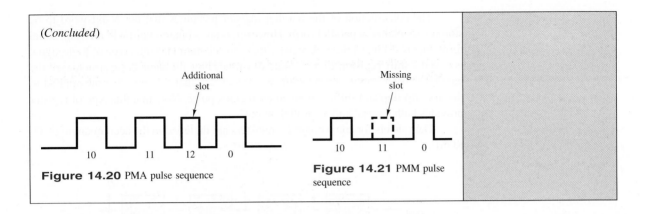

Figure 14.22 A 4-bit parallel register

The construction of the parallel register presumes that the N-bit word to be stored is available in parallel form. However, often a binary word will arrive in serial form, that is, 1 bit at a time. A register that can accommodate this type of logic signal is called a **shift register.** Figure 14.23 illustrates how the same basic structure of the parallel register applies to the shift register, except that the input is now applied to the first flip-flop and shifted along at each clock pulse. Note that this type of register provides both a serial and a parallel output.

Data sheets for some common registers are included in the accompanying CD-ROM.

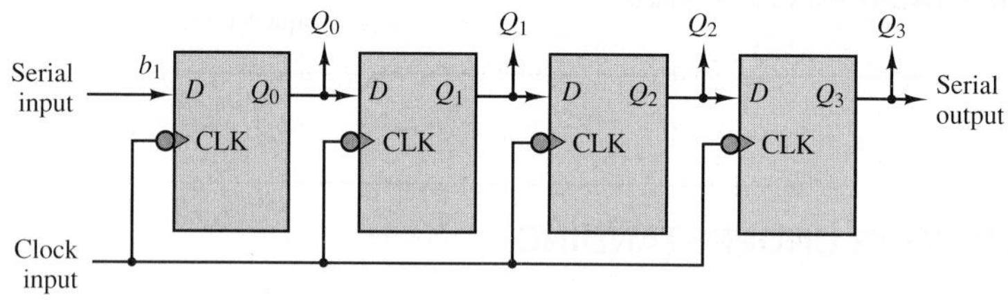

Figure 14.23 A 4-bit shift register

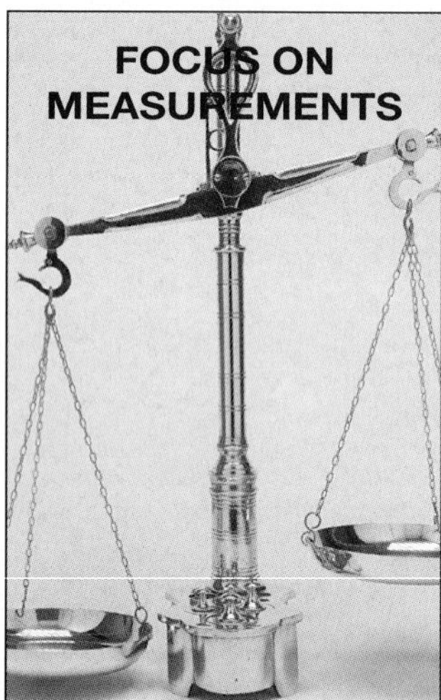

FOCUS ON MEASUREMENTS

Seven-Segment Display

A **seven-segment display** is a very convenient device for displaying digital data. The display is shown in Figure 14.24. Operation of a seven-segment display requires a decoder circuit to light the proper combinations of segments corresponding to the desired decimal digit.

This display, with the appropriate decoder driver, is capable of displaying values ranging from 0 to 9.

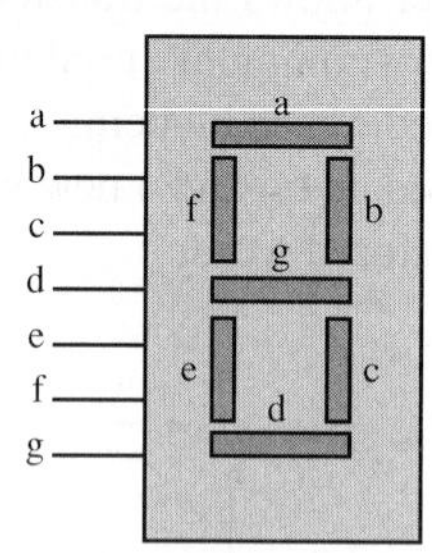

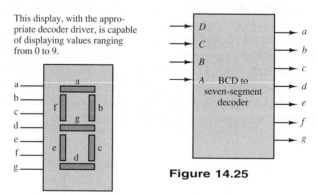

Figure 14.24 Seven-segment display

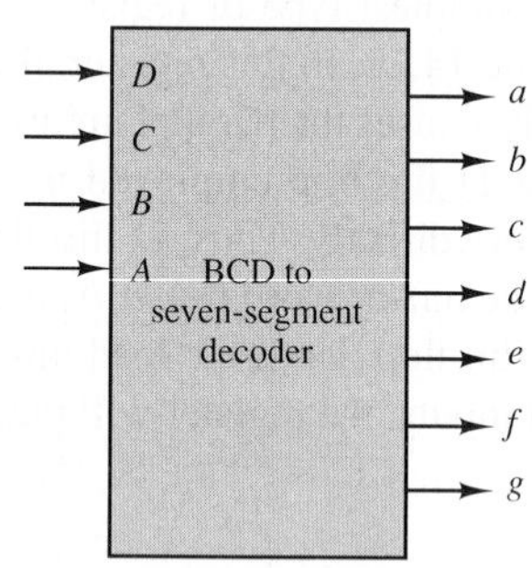

Figure 14.25

A typical BCD to seven-segment decoder function block is shown in Figure 14.25, where the lowercase letters correspond to the segments shown in Figure 14.24. The decoder features four data inputs (A, B, C, D), which are used to light the appropriate

(*Continued*)

(Concluded)

segment(s). The outputs of the decoder are connected to the seven-segment display. The decoder will light up the appropriate segments corresponding to the incoming value. A BCD to seven-segment decoder function is similar to the 2-to-4 decoder function described in Chapter 13 and shown in Figure 13.61.

FIND IT ON THE WEB

14.2 SEQUENTIAL LOGIC DESIGN

The design of sequential circuits, just like the design of combinational circuits, can be carried out by means of a systematic procedure. You will recall how the Karnaugh map, introduced in Chapter 13, allowed us to formalize the design procedures for an arbitrary combinational circuit. The equivalent of a Karnaugh map for a sequential circuit is the **state diagram,** with its associated **state transition table.** To illustrate these concepts, it is best to proceed with an example. Consider the 3-bit binary counter of Figure 14.26, which is made up of three T flip-flops. You can easily verify that the input equations for this counter are $T_1 = 1$, $T_2 = q_1$, and $T_3 = q_1 q_2$. Knowing the inputs, we can determine the three outputs from these relationships at any time. The outputs Q_1, Q_2, and Q_3 form the **state** of the machine. It is straightforward to show that as the clock goes through a series of cycles, the counter will go through the transitions shown in Table 14.1, where we indicate the current state by lowercase q and the next state by an uppercase Q. Note that the state diagram of Figure 14.26 provides information regarding the sequence of states assumed by the counter in graphical form. In a state diagram, each state is denoted by a circle called a **node,** and the transition from one state to another is indicated by a **directed edge,** that is, a line with a directional arrow. The analysis of sequential circuits consists of determining either their transition table or their state diagram.

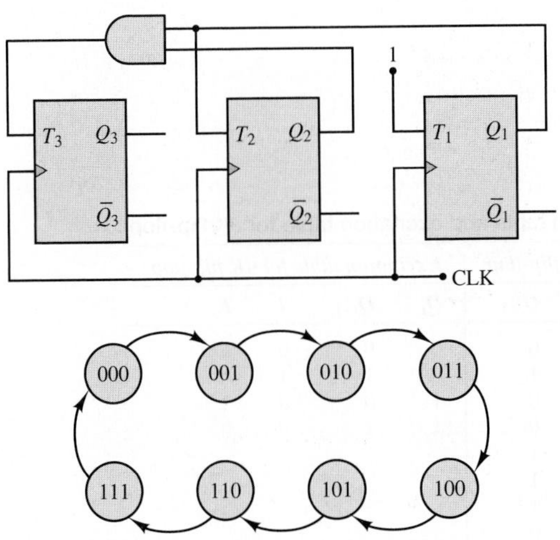

Figure 14.26 A 3-bit binary counter and state diagram

LO3

Table 14.1 State transition table for 3-bit binary counter

Current state			Input			Next state		
q_3	q_2	q_1	T_3	T_2	T_1	Q_3	Q_2	Q_1
0	0	0	0	0	1	0	0	1
0	0	1	0	1	1	0	1	0
0	1	0	0	0	1	0	1	1
0	1	1	1	1	1	1	0	0
1	0	0	0	0	1	1	0	1
1	0	1	0	1	1	1	1	0
1	1	0	0	0	1	1	1	1
1	1	1	1	1	1	0	0	0

The reverse of this analysis process is the design process. How can one systematically arrive at the design of a sequential circuit, such as a counter, by employing state transition tables and state diagrams? The design procedure will be explained in this section.

The initial specification for a logic circuit is usually in the form of either a transition table or a state diagram. The design will differ depending on the type of flip-flop used. Therefore one must first choose a flip-flop and define its behavior in the form of an excitation table. Truth tables and excitation tables for the *RS*, *D*, and *JK* flip-flops are given in Tables 14.2, 14.3, and 14.4, respectively.

Table 14.2 Truth table and excitation table for *RS* flip-flop

Truth table for RS flip-flop				*Excitation table for RS flip-flop*			
S	R	Q_t	Q_{t+1}	Q_t	Q_{t+1}	S	R
0	0	0	0	0	0	0	$d^\dagger$
0	0	1	1	0	1	1	0
0	1	0	0	1	0	0	1
0	1	1	0	1	1	d	0
1	0	0	1				
1	0	1	1				
1	1	x*	x				
1	1	x	x				

*An x indicates that this combination of inputs is not allowed.
†A *d* denotes a don't care entry.

Table 14.3 Truth table and excitation table for *D* flip-flop

Truth table for D flip-flop			*Excitation table for D flip-flop*		
D	Q_t	Q_{t+1}	Q_t	Q_{t+1}	D
0	0	0	0	0	0
0	1	0	0	1	1
1	0	1	1	0	0
1	1	1	1	1	1

LO3

LO3

LO3

Table 14.4 Truth table and excitation table for *JK* flip-flop

Truth table for JK flip-flop				*Excitation table for JK flip-flop*			
J	K	Q_t	Q_{t+1}	Q_t	Q_{t+1}	J	K
0	0	0	0	0	0	0	$d^\dagger$
0	0	1	1	0	1	1	d
0	1	0	0	1	0	d	1
0	1	1	0	1	1	d	0
1	0	0	1				
1	0	1	1				
1	1	0	1				
1	1	1	0				

†A *d* denotes a don't care entry.

The use of excitation tables will now be demonstrated through an example. Let us design a **modulo-4 binary up-down counter,** that is, a counter that can change state while counting up or down in the binary sequence from 0 to 3. For example, if the current state of the counter is 2, an input of 1 will cause the counter to change state "up" to 3, while an input of 0 will cause the counter to count "down" to 1. The state diagram for this counter is given in Figure 14.27. We choose two *RS* flip-flops for the implementation (the number of flip-flops must be sufficient to cover all the necessary states—two flip-flops are sufficient for a four-state machine) and begin constructing Table 14.5 by listing the possible inputs, denoted by the variable x, and their effect on the counter. Since the counter can have four states and there are two inputs, we must look at eight possible combinations. The first five columns of Table 14.5 describe the behavior of the counter for all possible inputs and present states; the behavior of the counter consists of determining the next state, denoted by Q_1Q_2, given the input x and the current state q_1q_2. Note that the first five columns of Table 14.5 contain exactly the same information that is given in the diagram of Figure 14.27. Now we can refer to the excitation table of the *RS* flip-flop to see what R and S inputs are required to obtain the desired counter function. For example, if $q_1 = 1$ and we wish to have $Q_1 = 0$, we must have $S_1 = 0$ and $R_1 = 1$ (we are resetting the first flip-flop). An entire state transition is handled by considering each flip-flop independently; for example, if we desire a transition from $q_1q_2 = 10$ to $Q_1Q_2 = 01$, we must have $S_1 = 0$ and $R_1 = 1$, as already stated, and $S_2 = 1$ and $R_2 = 0$. Repeating this analysis for each possible transition, we can then fill the next four columns of Table 14.5 with the values shown, where d represents a don't care condition.

LO3

Table 14.5 State transition table for modulo-4 binary up-down counter

Input x	Current state q_1	Current state q_2	Next state Q_1	Next state Q_2	S_1	R_1	S_2	R_2	Output y
0	0	0	1	1	1	0	1	0	1
0	0	1	0	0	0	*d*	0	1	0
0	1	0	0	1	0	1	1	0	1
0	1	1	1	0	*d*	0	0	1	0
1	0	0	0	1	0	*d*	1	0	1
1	0	1	1	0	1	0	0	1	0
1	1	0	1	1	*d*	0	1	0	1
1	1	1	0	0	0	1	0	1	0

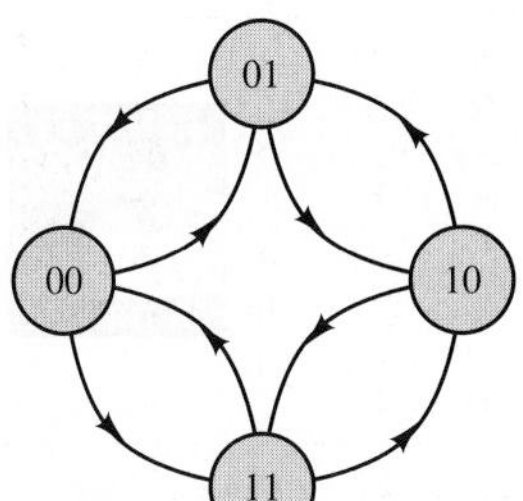

Figure 14.27 State diagram of a modulo-4 binary up-down counter

So far, we have been able to determine the desired inputs for each flip-flop based on the counter input and on the desired state transition. Now we need to design a logic circuit that will cause the flip-flop inputs to be as stated in Table 14.5 in response to the input x. This is a rather simple combinational logic problem, illustrated by the Karnaugh maps of Figure 14.28. From the Karnaugh maps we obtain the expressions

$$S_1 = \overline{x}\overline{q}_1\overline{q}_2 + x\overline{q}_1q_2 = (\overline{x}\overline{q}_2 + xq_2)\overline{q}_1$$
$$R_1 = \overline{x}q_1\overline{q}_2 + xq_1q_2 = (\overline{x}\overline{q}_2 + xq_2)q_1$$
$$S_2 = \overline{q}_2$$
$$R_2 = q_2$$

which allow us to complete the design, as shown in Figure 14.29.

S_1

x \ q_1q_2	00	01	11	10
0	1	0	d	0
1	0	1	0	d

R_1

x \ q_1q_2	00	01	11	10
0	0	d	0	1
1	d	0	1	0

S_2

x \ q_1q_2	00	01	11	10
0	1	0	0	1
1	1	0	0	1

R_2

x \ q_1q_2	00	01	11	10
0	0	1	1	0
1	0	1	1	0

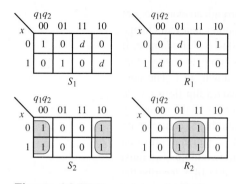

Figure 14.28 Karnaugh maps for flip-flop inputs in modulo-4 counter

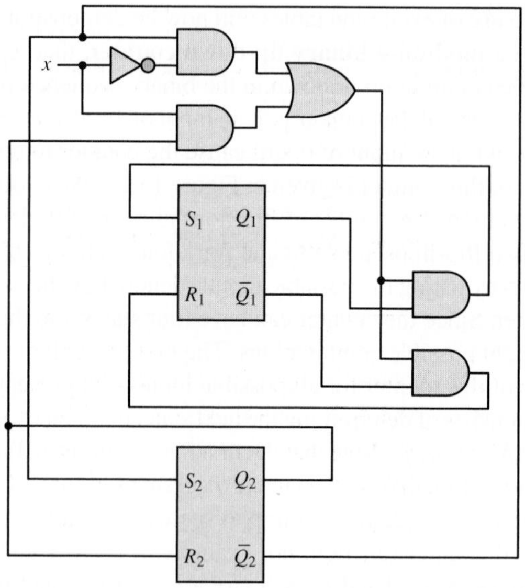

Figure 14.29 Implementation of modulo-4 counter

The procedure outlined in this section can be applied to more complex sequential circuits using the same basic steps. More advanced problems are explored in the homework problems.

Programmable Logic Controllers

FIND IT ON THE WEB

One of the most widely encountered applications of sequential logic designs and state machines are **programmable logic controllers,** or **PLCs.** PLCs are finite-state machines that are used in a variety of industrial applications to implement logic functions. For example, machining, packaging, material handling, and automated assembly are some of the example applications in which these systems are encountered. PLCs are specialized computers that are very effective at executing a series of complex logical decisions.

14.3 MICROPROCESSORS

To bring the broad range of applicability of microprocessors in engineering into perspective, it will be useful to stop for a moment to consider the possible application of microprocessor systems to different fields. The following list—by no means exhaustive—provides a few suggestions; it would be a useful exercise to imagine other likely applications in your own discipline.

Civil engineering	Measurement of stresses and vibration in structures
Chemical engineering	Process control
Industrial engineering	Control of manufacturing processes
Material and metallurgical engineering	Measurement of material properties

(Continued)

(Concluded)

Marine engineering	Instrumentation to determine ship location, ship propulsion control
Aerospace engineering	Instrumentation for flight control and navigation
Mechanical engineering	Mechanical measurements, robotics, control of machine tools
Nuclear engineering	Radiation measurement, reactor instrumentation
Biomedical engineering	Measurement of physiological functions (e.g., electrocardiography and electroencephalography), control of experiments

The massive presence of microprocessors in engineering laboratories and in plants and production facilities can be explained by considering the numerous advantages the computer can afford over more traditional instrumentation and control technologies. Consider, for example, the following points:

- A single microprocessor can perform computations and send signals from many different sensors measuring different parameters to many different display, storage, or control devices, under control of a single software program.
- The microprocessor is easily reprogrammed for any changes or adjustments to the measurement or control procedures, or in the computations.
- A permanent record of the activities performed by the microprocessor can be easily stored and retained.

It should be evident that microprocessors can perform repetitive tasks, or tasks that require great accuracy and repeatability, far better than could be expected of human operators and analog instruments. What, then, does constitute a **digital data acquisition and control system?** Figure 14.30 depicts the basic blocks that form such a system. In the figure, the user of the microprocessor system is shown to interact with the microprocessor by means of software, often called **application software.** Application software is a collection of programs written either in **high-level languages,** such as C, C^{++}, or Unix shell, or in **assembly language** (a programming language very close to the internal code used by the microprocessor). The particular

FIND IT ON THE WEB

LO4

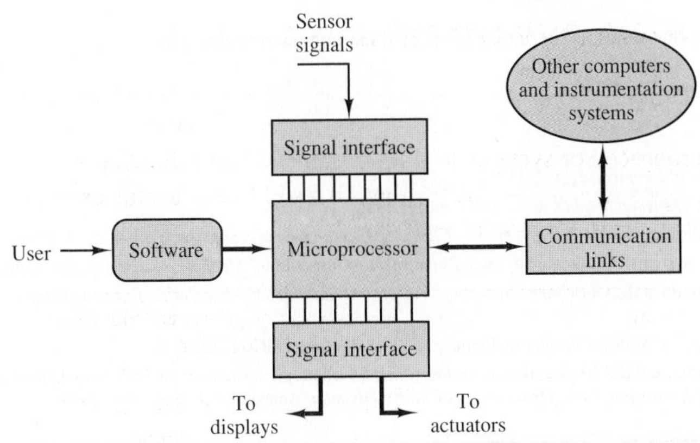

Figure 14.30 Structure of a digital data acquisition and control system

Mechatronics

Industry and the consumer market require engineering processes and products that are more reliable, more efficient, smaller, faster, and less expensive. The production and development of these devices require engineers who can understand and design systems from an integrated perspective. A discipline that shows particular promise in this arena is ***mechatronic design,*** based on the integration of mechanical engineering, electrical engineering, and computer science (Figure 14.31). Most major programs in the United States don't emphasize mechatronics as a primary curriculum component, but there is an industry-motivated push to change this situation.

FIND IT ON THE WEB

Mechatronics is an especially important and interesting domain for modern industry for a number of reasons. The automotive, aerospace, manufacturing, power systems, test and instrumentation, consumer, and industrial electronics industries make use of and contribute to mechatronics. Mechatronic design has surfaced as a new philosophy of design, based on the integration of existing disciplines, primarily mechanical and electrical, electronic, and software engineering.[1–7] Design elements from these traditional disciplines don't simply exist side by side, but are deeply integrated in the design process. Whether a given functionality should be achieved electronically, by software, or by elements from electrical or mechanical engineering domains requires mastery of analysis and synthesis techniques from the different areas. Being a successful mechatronics design engineer requires an in-depth understanding of many of, if not all, its constituent disciplines.

One of the distinguishing features of the mechatronic approach to the design of products and processes is the use of *embedded microcontrollers*. These microcontrollers replace many mechanical functions with electronic ones, resulting in much greater flexibility, ease of redesign or reprogramming, the ability to implement distributed control in complex systems, and the ability to conduct automated data collection and reporting. Mechatronic design represents the fusion of traditional mechanical, electrical, and software engineering design methods with sensors and instrumentation technology, electric drive and actuator technology, and embedded real-time microprocessor systems and real-time software. Mechatronic systems range from heavy industrial machinery, to vehicle propulsion systems, to precision electromechanical motion control devices.

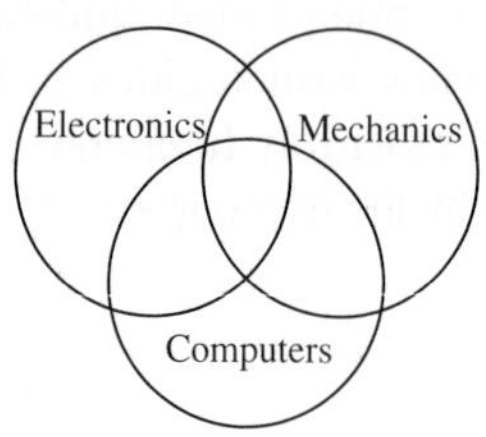

Figure 14.31 Mechatronics as the intersection of three engineering disciplines

[1] S. Ashley, "Getting Hold on Mechatronics," *Mechanical Engineering,* vol. 119, no. 5, 1997.

[2] D. Auslander, "What Is Mechatronics?" *IEEE/ASME Trans. on Mechatronics,* vol. 1, no. 1, 1996.

[3] F. Harashima, M. Tomizuka, and T. Fukuda, "Mechatronics—What Is It and How?" *IEEE/ASME Trans. on Mechatronics,* vol. 1, no. 2, 1996.

[4] G. Rizzoni and A. Keyhani, "Design of Mechatronic Systems: An Integrated, Inter-Departmental Curriculum," *Mechatronics,* vol. 5, no. 7, July 1995.

[5] G. Rizzoni, "Development of a Mechatronics Curriculum at the Ohio State University," *ASME International Mechanical Engineering Congress and Exposition,* San Francisco, 1996.

[6] D. Auslander and C. J. Kempf, *Mechatronics: Mechanical System Interfacing,* Prentice-Hall, Upper Saddle River, NJ, 1996.

[7] G. Rizzoni, A. Keyhani, G. Washington, G. Baumgartner, and B. Chandrasekaran, "Education in Mechatronic Systems at the Ohio State University," *ASME International Mechanical Engineering Congress and Exposition, Proc. Dyn. Sys. and Control Division,* Anaheim, CA, November 1998.

application software used may be commercially available or may be provided by the user; a combination of these two cases is the norm.

The signals that originate from real-world sensors—signals related to temperatures, vibration, or flow, for example—are interfaced to the microprocessor by means of specialized circuitry that converts analog signals to digital form and times the flow of information into the microprocessor using a clock reference, which may be internal to the microprocessor or externally provided. The heart of the signal interface unit is the *analog-to-digital converter,* or *ADC,* which will be discussed in some detail in Chapter 15. Not all sensor signals are analog, though. For example, the position of a switch or an on/off valve might be of interest; signals of this type are binary in nature, and the signal interface unit can route such signals directly to the microprocessor. Once the sensor data have been acquired and converted to digital form, the microprocessor can perform computations on the data and either display or store the results, or generate command outputs to actuators through another signal interface. Actuators are devices that can generate a physical output (e.g., force, heat, flow, pressure) from an electrical input. Some actuators can be controlled directly by means of a digital signal (e.g., an on/off valve), but some require an analog input voltage or current, which can be obtained from the digital signal generated by the microprocessor by means of a *digital-to-analog converter,* or *DAC*.

In addition to the program control exercised by the user, the microprocessor may also respond to inputs originating from other computer and instrumentation systems through appropriate **communication links,** which also permit communication in the reverse direction. Thus, a microprocessor system dedicated to a complex task may consist of several microprocessors tied over a **communication network.**

This chapter describes the basic architecture and operation of a special class of microprocessors, called microcontrollers, while Chapter 15 explores instrumentation-related issues.

14.4 COMPUTER SYSTEM ARCHITECTURE

Prior to delving into a description of how microprocessors interface with external devices (such as sensors and actuators) and communicate with the outside world, we will find it useful to discuss the general architecture of a microprocessor, in order to establish a precise nomenclature and paint a clear picture of the major functions required in the operation of a typical microprocessor.

The general structure of a computer system is shown in Figure 14.32. Note immediately that each of the blocks that are part of the computer system is connected with the **CPU bus,** which is the physical wire connection allowing each of the subsystems to communicate with the others. In effect, the CPU bus is simply a set of wires; note, however, that since only one set of signals can travel over the data bus at any one time, it is extremely important that the transmission of data between different parts of the computer (e.g., from the A/D unit to memory) be managed properly, to prevent interference with other functions (e.g., the display of unwanted data on a video terminal). As will be explained shortly, the task of managing the operation of the CPU bus resides within the **central processing unit,** or **CPU.** The CPU has the task of managing the flow of data and coordinating the different functions of the computer, in addition to performing the data processing—in effect, the CPU is the heart and brains of the computer. Some of the major functions of the CPU will be discussed in detail shortly.

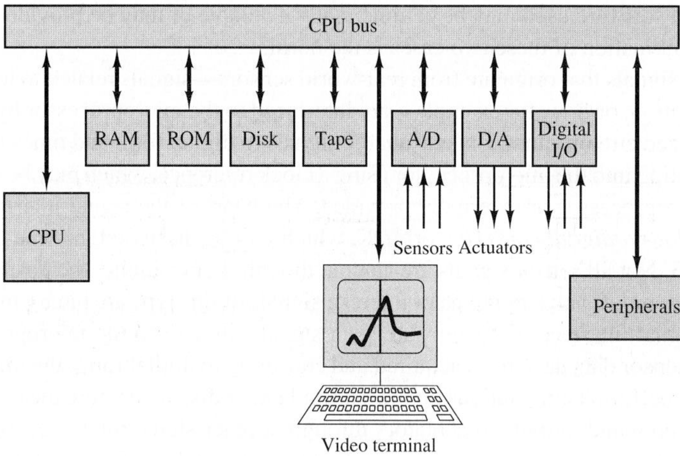

Figure 14.32 Computer architecture

One of the important features of a digital computer is its ability to store data. This is made possible by different types of **memory** elements: **read-only memory,** or **ROM; read and write memory (random-access memory),** or **RAM;** and **mass storage memory,** such as hard drive or floppy disk, tape, or optical drives. ROM is *nonvolatile memory:* it will retain its data whether the operating power is on or off. ROM memory contains software programs that are used frequently; one example is the *bootstrap program,* which is necessary to first start up the computer when power is turned on. RAM is memory that can be accessed very rapidly by the CPU; data can be either read from or written into RAM very rapidly. RAM is therefore used primarily during the execution of programs to store partial or permanent results, as well as to store all the software currently in use by the computer. The main difference between RAM and mass storage memory, such as a hard drive or a tape drive, therefore lies in the speed of access: RAM can be accessed in tens of nanoseconds, whereas a hard drive requires an access time on the order of microseconds, and tape drive, on the order of seconds. Another important distinction between RAM and mass memory is that the latter is far less expensive for an equivalent storage capability, the price typically being lower for longer access time.

A video terminal enables the user to enter programs and to display the data acquired by the computer. The video terminal is one of many **peripherals** that enable the computer to communicate information to the outside world. Among these peripherals are printers and devices that enable communication between computers, such as *modems* (a modem enables the computer to send and receive data over a telephone line). Finally, Figure 14.32 depicts real-time input/output (I/O) devices, such as analog-to-digital and digital-to-analog converters and digital I/O devices. These are the devices that allow a computer to read signals from external sensors, to output signals to actuators, and to exchange data with other computers.

14.5 MICROCONTROLLERS

A **microcontroller** is a special-purpose microprocessor-based system, designed to perform the functions illustrated in Figure 14.33. Microcontrollers have become an essential part of many engineering products, processes, and systems, and are often

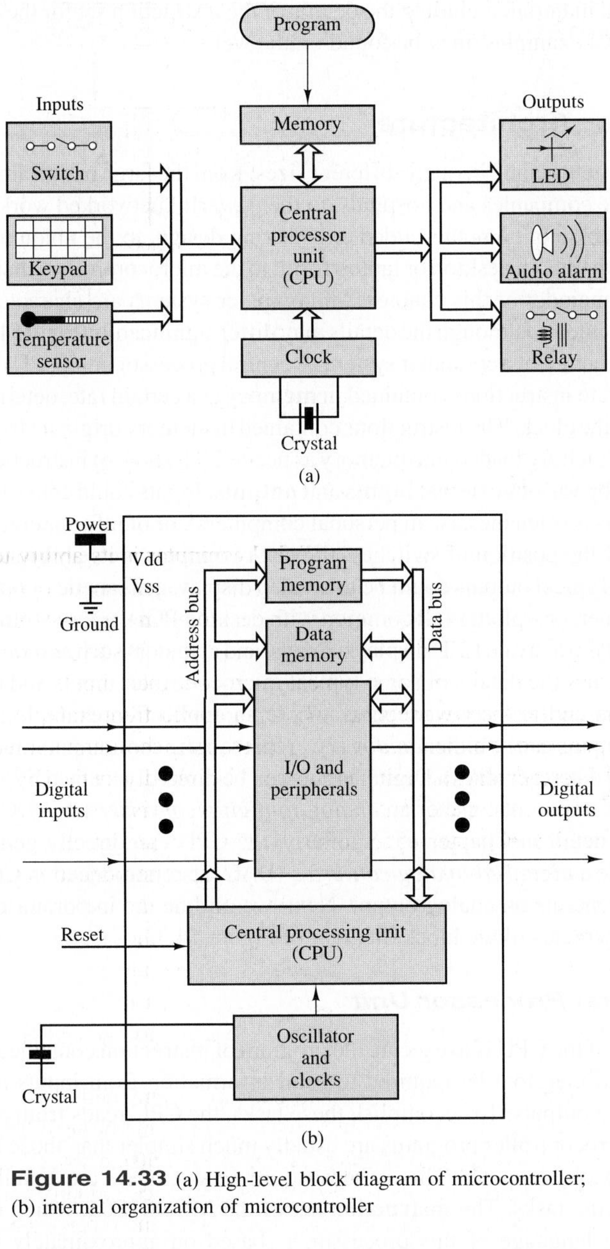

Figure 14.33 (a) High-level block diagram of microcontroller; (b) internal organization of microcontroller

LO4

deeply embedded in the inner workings of many products and systems we use daily (e.g., in automobile control systems and in many consumer products and appliances, such as autofocus cameras and washing machines). This section introduces the operation of a general-purpose microcontroller using as an example the architecture of the Motorola MC68HC05. Although much more powerful microcontrollers are available, the 8-bit, 2-MHz, HCMOS technology (where H stands for high speed) MC68HC05 contains all the essential elements that make microcontrollers so useful. The material presented in this section and in the next sections is intended to serve as an overview.

Supplemental material, including the details of the instruction set for the MC68HC05 and illustrative examples, may be found on the web.

Computer Architecture

Computer systems come in many different sizes, from the large mainframe computers that run entire companies and hospitals, to the powerful networked workstations that are the workhorse of computer-aided engineering design, to the ubiquitous personal computers, whether in desktop or laptop form, to the microcontrollers that are the subject of the remainder of this chapter. All computer systems are characterized by the same basic elements, although the details may differ significantly. Figure 14.32 depicts the overall structure of a computer system. A central processor unit (CPU) is timed by a **clock** to execute instructions contained in **memory** at a certain rate, determined by the frequency of the clock. The instructions contained in memory originate from computer **programs,** which are loaded into memory as needed. The flow of instruction execution is controlled by various external **inputs** and **outputs.** Inputs could consist of keyboard commands (as is often the case in personal computers), or of information provided by sensors, or of the position of switches (the last two inputs are very typical of microcontrollers). Typical outputs could be to a video display, a magnetic or optical storage device, a printer, or a plotter (all common with desktop PCs); microcontroller outputs are more likely to activate LED displays, relays, and actuators such as motors or valves.

We discuss the details of some typical microcontroller inputs and outputs later in this section and in the next section, where an application example is presented. Inputs and outputs can be either *analog* (i.e., representing continuous values) or *digital* (representing discrete values). Digital inputs can be directly accepted by a CPU, while analog inputs require the use of an *analog-to-digital converter* (or ADC). ADCs are described in detail in Chapter 15. Similarly, the CPU can directly generate digital outputs, while a *digital-to-analog converter* (DAC, also introduced in Chapter 15) is required to generate an analog output. Next, we outline the important properties of each of the elements of the block diagram of Figure 14.33.

The Central Processor Unit

The function of the CPU is to execute the program of instruction contained in memory. The CPU will therefore be required to **read** information from inputs and to **write** information to outputs. To accomplish these tasks, the CPU reads from and writes to memory. Microcontroller programs are usually much simpler than those that operate, say, in a desktop computer. This is because microcontrollers are usually dedicated to a few specific tasks. The **instruction set** of the MC68HC05, which is the native programming language of this processor, is based on approximately 60 different instructions. We shall see in a later section ("Operation of the Central Processing Unit") how these instructions are used to execute desired functions.

The Clock

The **clock** represents the heartbeat of the microcontroller. The clock function is typically implemented by a **crystal oscillator** that determines the basic clock cycle for the execution of each instruction step. Each step takes one clock cycle to complete. We are all familiar with the rating of processor speeds in megahertz. Typical microcontrollers are capable of speeds in the megahertz to tens of megahertz range.

Memory

The CPU needs to have access to different kinds of memory to execute programs. **Read-only memory (ROM)** is used for permanent programs and data that are necessary, for example, to boot and initialize the system. Information stored in ROM remains unchanged even when power to the computer is turned off. **Random access read/write memory (RAM)** is used to temporarily store data and instructions. For example, the program that is executed by the CPU and the intermediate results of the calculations are stored in RAM. Many microcontrollers also employ **erasable programmable read-only memory (EPROM)** or **electrically erasable programmable read-only memory (EEPROM);** these types of memory are used as ROM is, but can be reprogrammed relatively easily using special "EPROM burners." EPROM and EEPROM can be very useful if one wishes to make small but important changes to the functions executed by the microcontroller. For example, the microcontrollers used in automotive applications are usually the same from one model year to the next and across various vehicle platforms; but differences in control strategies and calibration data among vehicles, and changes and fixes required from one model year to the next are usually accommodated by means of EPROMs. Section 14.5 describes an automotive application.

Computer memory is arranged on the basis of **bits,** that is, a single digital variable with value of 0 or 1. Bits are grouped in **bytes,** consisting of 8 bits, and in **words,** consisting of 16 or 32 bits. While the size of a word can vary, 1 byte always consists of 8 bits. Small microcontrollers such as the MC68HC05 have access to a relatively limited amount of memory (for example, 64 Kbytes); more powerful microcontrollers may access as much as 1 Mbyte. Note that the memory capacity of a microcontroller is significantly smaller than that of the personal computers you are likely to use in your work.

Mass storage devices (magnetic and optical storage devices, such as hard drives and CD-ROMs) can also be used to increase a computer's access to data and information. Access to such external devices is much slower than access to ROM and RAM and is therefore usually not practical in embedded microcontrollers.

Computer Programs

A computer program is a listing of instructions to be executed by the CPU. The instructions are coded in a special **machine language** that consists of combinations of bytes. To assist the programmer, each CPU instruction is associated to a **mnemonic instruction code,** which associated a short word or abbreviated code to each instruction (e.g., the instruction ASL in the MC68HC05 stands for arithmetic shift left). More sophisticated software development systems allow the instructor to program in a higher-level language (often the C programming language); the high-level language program is then translated to machine code by a **compiler.** We shall devote one of the next sections to programming issues.

Number Systems and Number Codes in Digital Computers

Number Systems

It should be already clear that computers operate on the basis of the binary number system. The binary and hexadecimal number systems were introduced in Chapter 13.

The hexadecimal system is particularly well suited for use in computer codes, because it allows a much more compact notation than the binary code would. The hexadecimal code, as you will recall, permits expression of a 4-bit binary number as a single digit, using the numbers from 0 to 9 and the letters A to F. The range of possible combinations that can be expressed in a 16-bit word, for example, can be represented in decimal numbers as the range from 0_{10} to $2^{16} - 1 = 65{,}535_{10}$; as $0000\ 0000\ 0000\ 0000_2$ to $1111\ 1111\ 1111\ 1111_2$ in binary code; and as 0000_{16} to $FFFF_{16}$ in hexadecimal code.

It's very common to precede a hexadecimal number with the symbol $ to differentiate it from a decimal representation; for example, $32 would be interpreted as a hexadecimal word, and 32 as a decimal number.

Computer Codes

In addition to the codes described in Chapter 13 (binary, octal, hexadecimal, binary-coded decimal), a standard convention adopted by all computer manufacturers is the **ASCII,**[8] defined in Table 14.6. The ASCII defines the alphanumeric characters that are typically associated with text used in programming.

Instructions to the CPU are coded as **operation codes,** or **opcodes.** Each opcode instructs the CPU to perform a sequence of steps that correspond to an operation (e.g., an addition). Although all computers perform essentially the same basic tasks at the binary level, the manner in which these tasks are performed varies according to the computer manufacturer, and therefore opcodes vary from manufacture to manufacturer. The **instruction set** of a specific computer is the set of all basic operations that the computer can perform. For example, the MC68HC05 can execute 62 basic instructions, which are arranged into 210 unique opcodes. The difference between a basic instruction and an opcode is that the same basic instruction can be used in slightly different ways (in conjunction with other instructions) to perform a specific operation. Thus, opcodes are the basic building blocks of the programming language of a computer.

Mnemonics and Assemblers

To assist the programmer in remembering and identifying the function of opcodes, **mnemonics** are used. A mnemonic is an alphabetic abbreviation that corresponds to a specific opcode. Thus, the programmer writes a program using mnemonics, and the program is translated into **machine code** (consisting of opcodes and data) by a computer program called an **assembler.** A more detailed discussion of programming issues follows in a later section.

Memory Organization

Memory performs an essential function in microcontrollers. Different kinds of memory are used to store information of different types. The three basic types of memory are described in Section 13.5. ROM and EPROM are used to store the operating system and the programs used by the controller. RAM is used by the CPU to read and write instructions during the execution of a computer program.

Memory is usually organized in the form of a **memory map,** which is a graphical representation of the allocation of the memory used by a particular microcontroller. Example 14.6 describes the use of memory in a typical microcontroller.

---

[8] American Standard Code for Information Interchange.

Table 14.6 ASCII

Graphic or control	ASCII (hex)	Graphic or control	ASCII (hex)	Graphic or control	ASCII (hex)
NUL	00	+	2B	V	56
SOH	01	,	2C	W	57
STX	02	−	2D	X	58
ETX	03	.	2E	Y	59
EOT	04	/	2F	Z	5A
ENQ	05	0	30	[	5B
ACK	06	1	31	\	5C
BEL	07	2	32	]	5D
BS	08	3	33	↑	5E
HT	09	4	34	←	5F
LF	0A	5	35	`	60
VT	0B	6	36	a	61
FF	0C	7	37	b	62
CR	0D	8	38	c	63
SO	0E	9	39	d	64
SI	0F	:	3A	e	65
DLE	10	;	3B	f	66
DC1	11	<	3C	g	67
DC2	12	=	3D	h	68
DC3	13	>	3E	i	69
DC4	14	?	3F	j	6A
NAK	15	@	40	k	6B
SYN	16	A	41	l	6C
ETB	17	B	42	m	6D
CAN	18	C	43	n	6E
EM	19	D	44	o	6F
SUB	1A	E	45	p	70
ESC	1B	F	46	q	71
FS	1C	G	47	r	72
GS	1D	H	48	s	73
RS	1E	I	49	t	74
US	1F	J	4A	u	75
SP	20	K	4B	v	76
!	21	L	4C	w	77
”	22	M	4D	x	78
#	23	N	4E	y	79
$	24	O	4F	z	7A
%	25	P	50	{	7B
&	26	Q	51	|	7C
’	27	R	52	}	7D
(	28	S	53	~	7E
)	29	T	54	DEL	7F
*	2A	U	55		

EXAMPLE 14.6 Writing Data to and Reading Data from I/O Ports

LO4

Problem

1. Write specified data to an I/O address.
2. Assume that an input device is connected to address E6H. Write the code necessary to read a byte from this input device.

Solution

Known Quantities: Desired data, I/O address.

Find: Write the appropriate sequence of commands, using the MC68HC05 instruction set.

Schematics, Diagrams, Circuits, and Given Data: The data to be written to the I/O port are 36H (decimal number 36); the I/O address is A6H.

Assumptions: Data are written to the accumulator register first.

Analysis:

1. The command to write to an I/O port is *STA$⟨address⟩*. The command assumes that the data are in the accumulator register; thus we first must load the accumulator with the desired value:

```
LDA #$36     ; load accumulator with 36H
STA $00A6    ; write accumulator to I/O port A6H
```

2. The command to read from an I/O port is *LDA⟨address⟩*. The value read from the input port is stored in the accumulator. To store the byte into the accumulator register, we use the command

```
LDA $00E6    ; load accumulator with E6H
```

Operation of the Central Processing Unit

The MC68HC05 is organized as follows. Five **CPU registers** can be directly accessed by the CPU (i.e., without the need to access memory); the memory map defines the names and types of the memory locations that are accessible to the CPU in addition to the registers.

The **accumulator,** or **A register,** is used to hold the results of arithmetic operations performed by the CPU.

The **index,** or **X register,** is used to point to an address in memory where the CPU will read or write information. This register is used to perform a function called *indexed addressing*, which is described in greater detail in the MC68HC05 instruction set found in the accompanying CD-ROM.

The **program counter (PC) register** keeps track of the address of the next instruction to be executed by the CPU.

The **condition code register (CCR)** holds information that reflects the status of prior CPU operations. For example, *branch instructions* look at the CCR to make either/or decisions.

The **stack pointer (SP) register** contains return address information and the previous content of all CPU registers, so that if the CPU is *interrupted* or a subroutine is initiated (we shall visit this concept soon), the status of the program prior to the **interrupt** or prior to branching to the subroutine is retained. Once the CPU has finished *servicing* the interrupt or has completed the subroutine, it can resume its previous operations by loading the contents of the SP register.

Interrupts

Interrupts perform a very important function in microcontrollers by allowing the CPU to interrupt its normal flow of operations to respond to an external event. For example, an interrupt request may occur when an *analog-to-digital converter* (described in greater detail in Chapter 15) has completed the conversion of an analog signal to digital form, so that the digital value of a sensor reading may be made available to the CPU for further processing. The following Focus on Measurements box illustrates an automotive application of this concept.

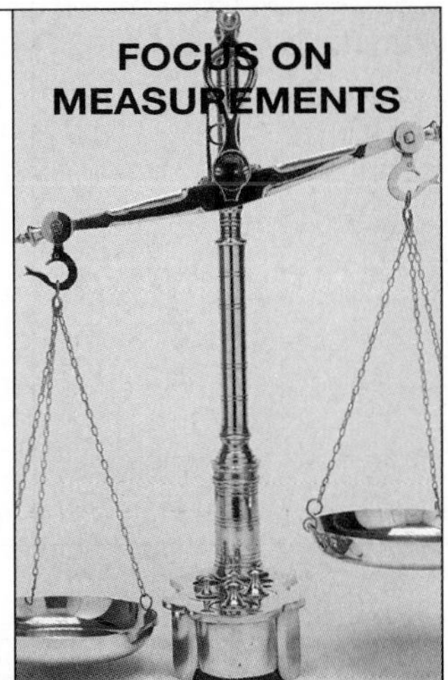

Reading Sensor Data by Using Interrupts

In modern automotive instrumentation, a microcontroller performs all the signal processing operations for several measurements. A block diagram for such instrumentation is given in Figure 14.34. Depending on the technology used, the sensors' outputs can be either digital or analog. If the sensor signals are analog, they must be converted to digital format by means of an analog-to-digital converter, as shown in Figure 14.35. The analog-to-digital conversion process requires an amount of time that depends on the individual ADC, as will be explained in Chapter 15. After the conversion is completed, the ADC then signals the computer by changing the logic state on a separate line that sets its *interrupt request* flip-flop. This flip-flop stores the ADC's interrupt request until it is acknowledged (see Figure 14.36).

When an interrupt occurs, the processor automatically jumps to a designated program location and executes the interrupt service subroutine. For the ADC, this would be a subroutine to read the conversion results and store them in some appropriate location, or to perform an operation on them. When the processor responds to the interrupt, the interrupt request flip-flop is cleared by a direct signal from the processor. To resume execution of the program at the proper point upon completion of the ADC service subroutine, the program counter content is automatically saved before control is transferred to the service subroutine. The service subroutine saves in a stack the content of any registers it uses, and restores the registers' content before returning.

The interrupt may occur at any point in a program's execution, independent of the internal clock; it is therefore referred to as an *asynchronous* event.

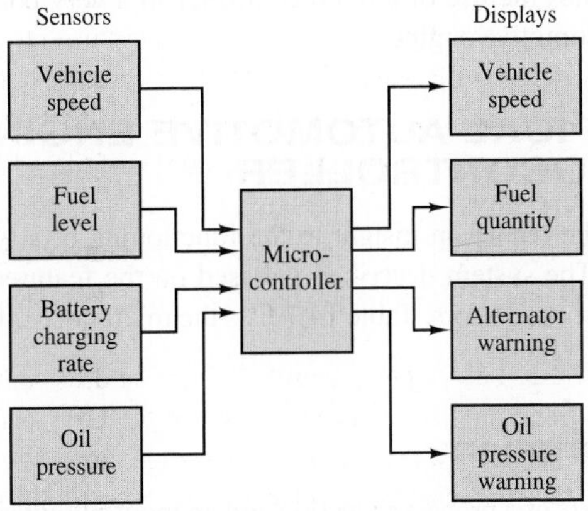

Figure 14.34 Automotive instrumentation

(Continued)

(Concluded)

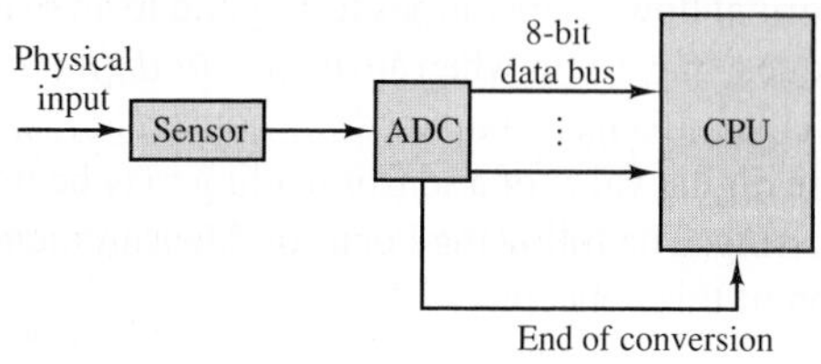

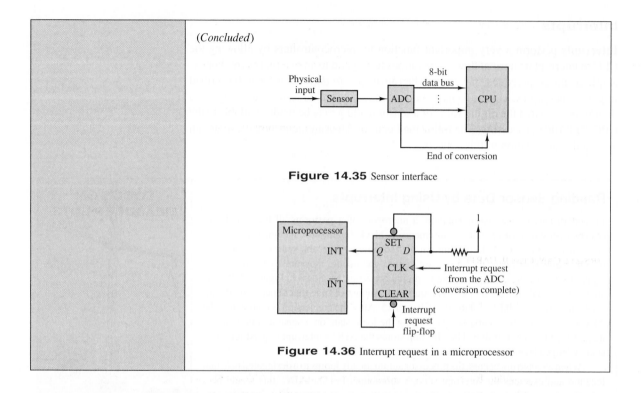

Figure 14.35 Sensor interface

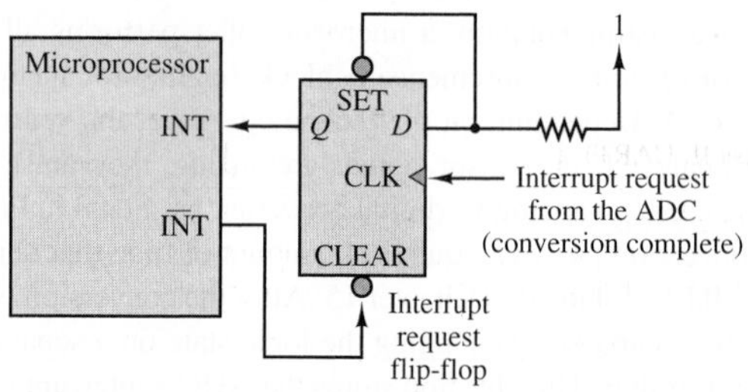

Figure 14.36 Interrupt request in a microprocessor

Instruction Set for the MC68HC05 Microcontroller

The complete instruction set for the MC68HC05 and MC68HC11 microcontrollers may be found on the web.

Programming and Application Development in a Microcontroller

Section 14.6 illustrates the use of a microcontroller in a very common application: the control of an automotive engine.

14.6 A TYPICAL AUTOMOTIVE ENGINE MICROCONTROLLER

FIND IT ON THE WEB

This section gives the reader an insight to the functioning of a typical **automotive engine controller.** The system described is based on the features of commercially available 32-bit microprocessors. Table 14.7 lists the microcontroller's major characteristics.

General Description

The controller consists of a processor section and an input/output section, mounted in an enclosure. These two sections are usually combined onto one printed-circuit board for cost considerations. A generic controller may be programmed with different software for a wide range of production engine applications, or specific hardware/software

Table 14.7 Characteristics of automotive microcontroller

Processor section	I/O section inputs	I/O section outputs
Microcoded timing channels (TPU—time processor unit)	Variable reluctance sensor interface*	Discrete low-side drivers
Discrete I/O channels	Hall sensor interfaces (cam, crank)	PWM low-side drivers
PWM channels	Analog input	Low-side-driven fuel injectors‡
8-bit ADC channels	Exhaust gas oxygen sensors	Low-side-driven coil drivers
10-bit ADC channels	Discrete pull-ups to ignition	High-side drivers
Boot memory (flash)	Discrete pull-downs to ground	High- and low-side current-controlled outputs
RAM	PWM/frequency inputs	Stepper motor driver
Serial communication (RS-232, CAN, Class II, UART)	Power and ground	H-bridge driver

*See Chapter 16 "Focus on Measurements: Magnetic Reluctance Position Sensor" for an explanation of the operation of this sensor.
‡See Chapter 12 for an example of a fuel injector driver.

designs may be provided for each vehicle application. Through embedded software, the controller is able to communicate with a personal computer (PC) for software debugging and development. Serial bus–based instrumentation is also available to assist in the development process, before an application is released for vehicle production. A block diagram of a typical system configuration is given in Figure 14.37.

Processor Section

The processor comprises a 32-bit microprocessor, memory, analog and digital I/O, timing channels, serial communications, and all required supporting logic.

Microprocessor

Typical of many systems, the controller printed-circuit board is designed to accept a plug-in replacement microprocessor, in anticipation of updates for higher performance and/or additional features. Such modifications may require a new crystal, different-value resistors, capacitors, etc. Typical crystal frequencies for automotive applications would be around 5 MHz, for a clock frequency of about 20 MHz.

A popular family of processors consists of a central processing unit and three integrated modules. Each of these modules can function on its own once it is initialized by the central processing unit.

Central Processing Unit

The processor is equipped with a background mode, which can be used for program download or system diagnostics. No target software is required for operation of this system, provided proper hardware and software for a personal computer (desktop or laptop) are present.

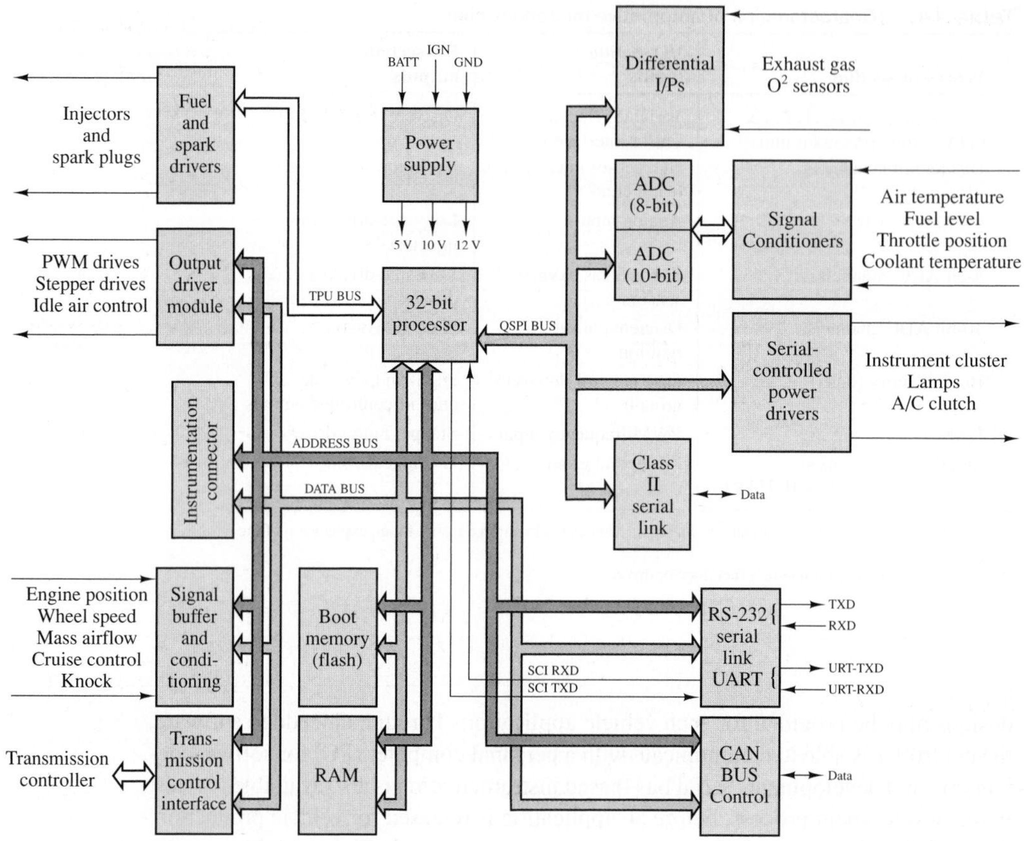

Figure 14.37 System block diagram

Timer Processor Unit (TPU)

The TPU has 16 channels of microcode-controlled pulse I/O. It comes with a predefined set of channel control programs called *primitives*. The TPU can be reprogrammed by using the on-chip RAM in the emulation mode. Programming for the TPU can be done by the end user.

System Integration Module (SIM)

The SIM performs various functions, some of which include

- External bus interface
- Interrupt detection
- Chip select activation
- Programmable timer interrupt
- Clock generation
- System protection
- System testing

The system is set up through the SIM setup registers.

Queued Serial Module (QSM)

The QSM contains a serial communications interface (SCI) and a queued serial port interface (QSPI) module. The queuing feature of the QSPI bus, and its associated chip selects, enhances the overall performance of the processors.

Memory

Boot Memory

Single-chip processors contain their own flash boot memory. Alternatively, expanded mode processors support external boot memory, which may be a variety of technologies, including EPROM, EEPROM, or NOVRAM (nonvolatile RAM). Because they are connected to the boot chip select on the processor, they must contain the boot program to set up the processor control registers. The remaining memory in the parts may be used for any of a number of purposes (program, parameter storage, data logging, etc.).

RAM

Up to 1 Mbyte of RAM is supported by a typical processor. This should provide sufficient RAM to facilitate data logging and parameter storage.

Memory Map

Processors control peripheral chips via programmable chip selects, which allow the user to change the memory map through control registers. The memory map is therefore not fixed until initialization has taken place. On power-up, the boot chip select is set to be active in the address range \$000000 to \$100000. After power-up, the chip selects can be set to create any desired memory map within the limits set by the hardware design.

Chip Select

Examples of such chip selects are CSBOOT and CS0 to CS10. CSBOOT is always a chip select since it must be used before the registers are configured. CS0 to CS10 may be used for a variety of functions, as they are programmable.

The hardware configuration of the board uses some of these chip selects for specific purposes. Table 14.8 shows a typical use of the chip selects.

Analog-to-Digital Converters

The system includes 10-bit and 8-bit ADCs. Both of these converters communicate with the processor via the QSPI bus. There is a low-pass *RC* filter on each of the ADC channels.

Communications

Several different types of serial communications are available on microprocessor chips. Some examples are SCI, QSPI, Class II, CAN, and RS-232.[9] All interfaces are made available to the user at the interface connector.

---

[9]See Chapter 15 for an introduction to the RS-232 communication protocol.

Table 14.8 Chip select allocation

CSBOOT	Boot memory select. This chip select is combined with other signals in a PAL to produce chip selects for the boot memories.
CS0	RAM select.
CS1	Enable signal buffering and conditioning circuit.
CS2	ADC enable.
CS3	CAN serial interface IC.
CS4	Output driver module enable.
CS5	Transmission controller enable.
CS6	Memory control.
CS7	Spare.
CS8	Spare.
CS9	Spare.
CS910	Spare.

The SCI has both TXD (digital transmit) and RXD (digital receive) lines. The SCI lines are available at the interface connector directly and are routed to a level shifter circuit to provide the appropriate signal levels.

Class II is a single-wire serial bus for communication between microprocessor-controlled modules. The bus allows any module to communicate with any other module on the bus. A data link controller IC is used as an interface between the processor and the bus.

CAN is a European-standard, high-speed communications link using a two-wire interface. The controller communicates with the CAN controller over the parallel bus. Serial lines are found on the interface connectors.

Interrupts

Many microprocessors have several interrupt input lines which, when set, will interrupt the software process being executed in favor of a new process. Some of these interrupt request lines may be software-maskable, such that their effect is ignored, while others may be nonmaskable, indicating that the processor will always service a request seen on these inputs. Each interrupt line will be prioritized to ensure an orderly process in the event that the processor receives more than one interrupt request. The nonmaskable interrupts will be ranked with the highest priority. Unused interrupt lines may be left disconnected, while a software fault routine may be serviced in the event the microprocessor experiences an input on one of these unused inputs.

Input/Output Section

For safety reasons, all outputs are disabled until such time as the processor passes its self-test routine and the system is able to establish proper control of all outputs. This is accomplished by use of the "computer not operating properly" (CNOP) signal, which is used by all significant devices within the controller. A custom circuit generates the CNOP signal.

A brief description of the features of the I/O section follows.

Inputs

Discrete

Discrete inputs may be "pulled up" to 12 V or "pulled down" to ground potential to ensure a recognizable default condition or to enable fault detection of failed inputs. All discrete inputs normally have some hardware filtering to provide noise immunity.

Analog

Several 8-bit and 10-bit ADC inputs provide a means of feeding the processor digital signals from analog sensors. For maximum accuracy the ADCs should be powered and referenced to the same power supply that will be providing the voltage reference for the associated sensors that it handles. A 10-bit ADC provides a higher signal resolution (1024 bits) than the 8-bit ADC (256 bits). The ADCs communicate with the processor via the QSPI bus.

PWM Frequency

Pulse-width modulation frequency inputs are read by the TPU preprocessor which relieves the main processor of the burden of handling large amounts of time-dependent data (e.g., spark and fuel information). Microcode, embedded within the preprocessor, defines the way in which these types of signals are processed.

Knock

Knock is a damaging, audible phenomenon that results from preignition of the fuel-air mixture in the combustion chamber. Piezoelectric sensors,[10] mounted on the engine, interface with a custom integrated circuit to detect this damaging condition so that the processor can adjust spark timing to eliminate the knock.

Outputs

Discrete

All discrete outputs have self-shutdown and diagnostic capability.

PWM

Outputs that are capable of being pulse-width–modulated can generally be used as a discrete output. These outputs also have self-shutdown and diagnostic capability. Alternatively, diagnostics can be implemented via a sense resistor fed back to an amplifier circuit and then on to an ADC.

Output Driver Module

The output driver module is an application-specific integrated circuit (ASIC) designed for discrete I/O processing. It has programmable discrete I/O lines, programmable PWM lines, drives for external *pnp* switches, and set logic circuitry. It has a time-out

[10]See Chapter 12, Focus on Measurements box "Charge Amplifiers" for an introduction to piezoelectric sensors.

line which is used as an active low "computer not operating properly" (CNOP) signal to other parts in the system and as a turnoff delay (TOD) for the power supply.

Fuel Injectors

The fuel and spark driver IC controls the fuel injectors, under command of software, via the TPU bus. Diagnostic feedback capability is available by latching the feedback lines with a parallel to serial shift register and then shifting out the data via SPI control.

Spark Coil

Ignition options include two different drive options. The first is driving external ignition coils. The second is high-side driving an external ignition (IGN) module.

Insulated-gate bipolar transistor (IGBT) coil drivers are available in the I/O section. These provide the ability to drive ignition coils directly. These IGBTs also have both analog and discrete feedback capabilities. A sense resistor from the IGBT is buffered and sent to two different circuits.

Exhaust Gas Recirculation (EGR) Valve Drivers

A traditional discrete low-side drive or high-side drive is available for the linear EGR valve interface.

Current-Controlled Circuit

A current-controlled circuit is provided to drive a force motor for direct application to a transmission control. The force motor circuit controls both the high and low sides of the load. The force motor circuit is controlled by a PWM signal. Sampling the current and converting that to an analog value provide the feedback. An ADC can read this analog value.

Stepper Motor

A stepper motor[11] driver is available for applications using a stepper-controlled idle air valve.

Brushless Motor

A brushless motor[12] driver (BMD) has been provided for applications using this type of actuator. It is configurable for different drive modes.

Power Supply

The power supply uses battery (BATT) and ignition (IGN) to provide different supply lines used by the controller circuitry. One example would be an independent, close-tolerance, tracking voltage reference, for all sensors that are being read by the ADC. The supply will operate reliably down to a battery voltage of 4 to 5 V. The CNOP signal provides a simple means of allowing the controller to perform housekeeping tasks prior to shutdown after IGN goes low.

---

[11] See Section 19.2 for an introduction to stepper motors.

[12] See Section 19.1 for an introduction to brushless DC motors.

Ground Structure

Care must be taken to avoid ground loops[13] within the controller that could result in interference with signal levels. Typical methods employed for this include separation of signal grounds, power grounds, and radio-frequency grounds. All three grounds have their own dedicated circuit board layer.

14.7 Conclusion

This chapter presents an overview of digital logic circuits. These circuits form the basis of all digital computers and of most electronic devices used in industrial and consumer applications. Upon completing this chapter, you should have mastered the following learning objectives:

1. *Analyze the operation of sequential logic circuits.* Sequential logic circuits are digital logic circuits with memory capabilities; their operation is described by state transition tables and state diagrams. Counters and registers are the two principal classes of sequential circuits.
2. *Understand the operation of digital counters.* Counters are a very important class of digital circuits and are based on sequential logic elements.
3. *Design simple sequential circuits using state transition diagrams.* Sequential circuits can be designed using formal design procedures employing state diagrams. These methods are equivalent to the Karnaugh map logic design for combinational circuits.
4. *Study the basic architecture of microprocessors and microcontrollers.* Digital systems play a prominent role in today's engineering products and processes. The microprocessor in particular has become an integral part of instrumentation and control systems. Microprocessors can be programmed to execute many different tasks. A special type of microprocessor, the microcontroller, is equipped with special input/output capabilities and finds widespread application in a number of instrumentation and control functions.

HOMEWORK PROBLEMS

Section 14.1: Sequential Logic Modules

14.1 The input to the circuit of Figure P14.1 is a square wave having a period of 2 s, maximum value of 5 V, and minimum value of 0 V. Assume all flip-flops are initially in the RESET state.

a. Explain what the circuit does.

b. Sketch the timing diagram, including the input and all four outputs.

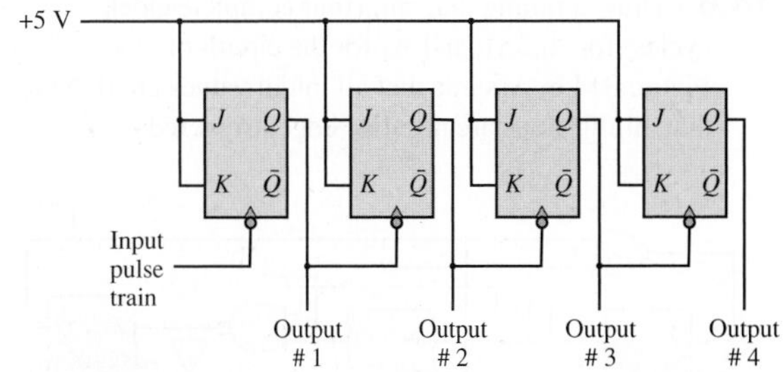

Figure P14.1

[13]See Section 15.2 for an introduction to grounding and noise issues in circuit design.

14.2 A binary pulse counter can be constructed by interconnecting T-type flip-flops in an appropriate manner. Assume it is desired to construct a counter which can count up to 100_{10}.

a. How many flip-flops would be required?

b. Sketch the circuit needed to implement this counter.

14.3 Explain what the circuit of Figure P14.3 does and how it works. (*Hint:* This circuit is called a 2-bit synchronous binary up-down counter.)

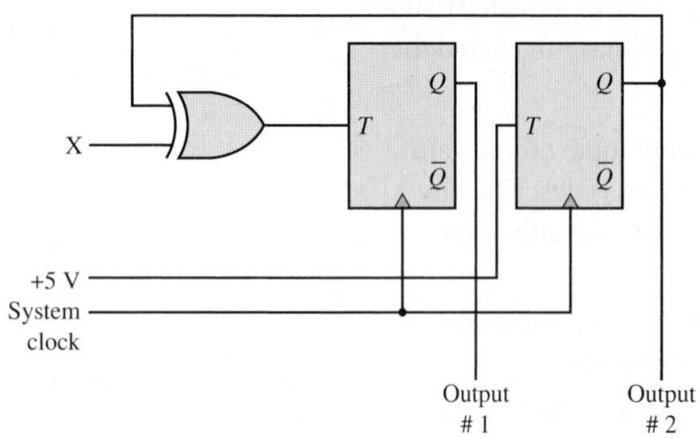

Figure P14.3

14.4 Suppose a circuit is constructed from three D-type flip-flops, with

$$D_0 = Q_2 \qquad D_1 = Q_2 \oplus Q_0 \qquad D_2 = Q_1$$

a. Draw the circuit diagram.

b. Assume the circuit starts with all flip-flops SET. Sketch a timing diagram which shows the outputs of all three flip-flops.

14.5 Suppose that you want to use a D flip-flop for a laboratory experiment. However, you have only T flip-flops. Assuming that you have all the logic gates available, make a D flip-flop using a T flip-flop and some logic gate(s).

14.6 Draw a timing diagram (four complete clock cycles) for A_0, A_1, and A_2 for the circuit of Figure P14.6. Assume that all initial values are 0. Note that all flip-flops are negative edge–triggered.

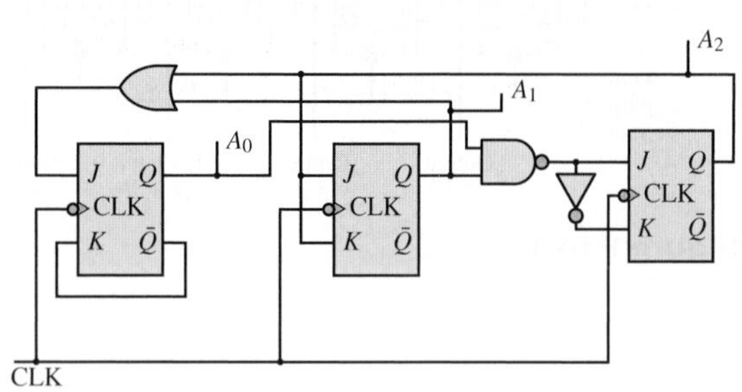

Figure P14.6

14.7 Assume that the slotted encoder shown in Figure P14.7 has a length of 1 m and a total of 1,000 slots (i.e., there is one slot per millimeter). If a counter is incremented by 1 each time a slot goes past a sensor, design a digital counting system that determines the speed of the moving encoder (in meters per second).

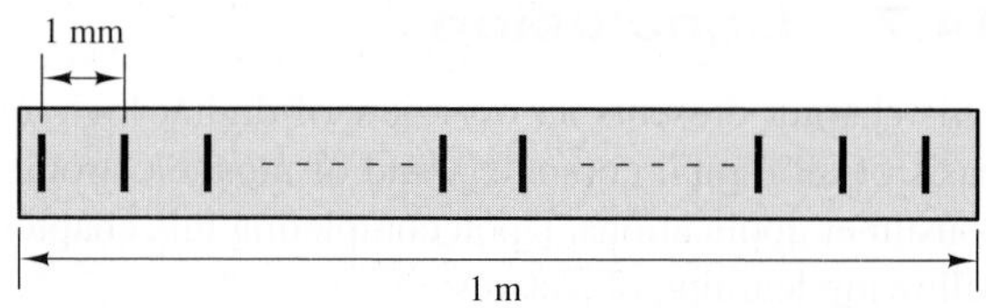

Figure P14.7

14.8 Find the output Q for the circuit of Figure P14.8.

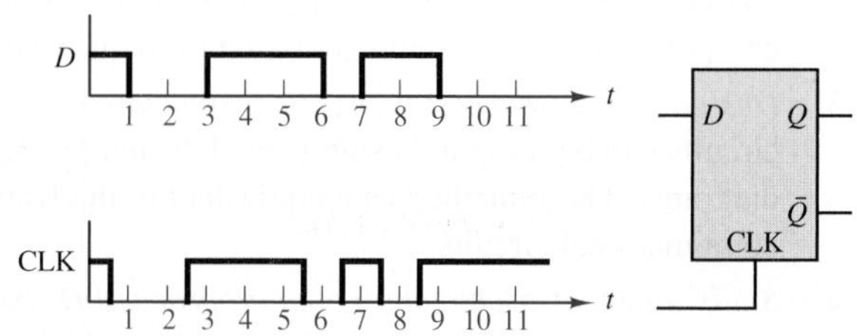

Figure P14.8

14.9 Describe how the ripple counter works. Why is it so named? What disadvantages can you think of for this counter?

14.10 Write the truth table for an RS flip-flop with enable (E), preset (P), and clear (C) lines.

14.11 A JK flip-flop is wired as shown in Figure P14.11 with a given input signal. Assuming that Q is at logic 0 initially and the trailing-edge triggering is effective, sketch the output Q.

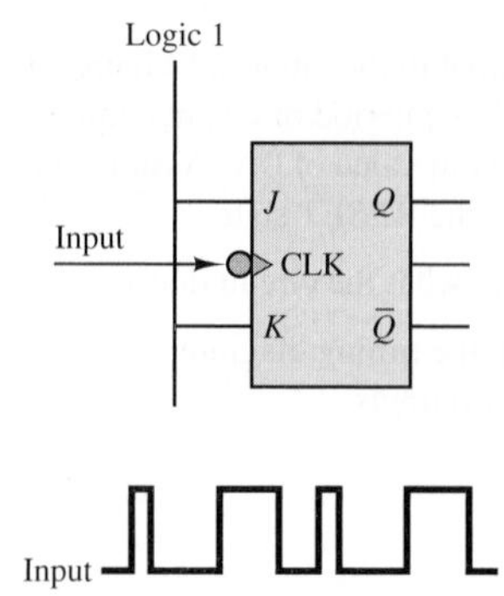

Figure P14.11

14.12 With reference to the JK flip-flop of Problem 14.11, assume that the output at the Q terminal is made to serve as the input to a second JK flip-flop wired exactly as the first. Sketch the Q output of the second flip-flop.

14.13 Assume that there is a flip-flop with the characteristic given in Figure P14.13, where A and B are the inputs to the flip-flop and Q is the next state output. Using necessary logic gates, make a T flip-flop from this flip-flop.

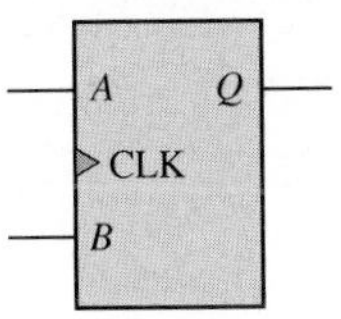

A	B	Q
0	0	$\bar{q}$
0	1	q
1	0	q
1	1	0

Figure P14.13

Section 14.3: Microprocessors

14.14 Draw a block diagram that shows the bus structure of a microprocessor-based system and the subsystems connected by the bus connects. For each subsystem, indicate the nature of the components that make up the system—that is, ROM or RAM, for example.

14.15 Explain the purpose of the ALU.

14.16 Name the internal registers of a microprocessor, and explain their functions.

14.17 Name the three different systems buses, and explain their functions.

14.18 Suppose a microprocessor has n registers.

a. How many control lines do you need to connect each register to all other registors?

b. How many control lines do you need if a bus is used?

14.19 Explain the function of the status register (flag register), and give an example.

14.20 What is the distinction between volatile and nonvolatile memory?

Section 14.4: Computer System Architecture; Section 14.5: Microcontrollers

14.21 A typical PC has 32 Mbytes of standard memory.

a. How many words is this?

b. How many nibbles is this?

c. How many bits is this?

14.22 Suppose a microprocessor has n registers.

a. How many control lines do you need to connect each register to all other registers?

b. How many control lines do you need if a bus is used?

14.23 Suppose it is desired to implement a 4-Kbyte 16-bit memory.

a. How many bits are required for the memory address register?

b. How many bits are required for the memory data register?

14.24 What is the distinction between volatile and nonvolatile memory?

14.25 Suppose a particular magnetic tape can be formatted with eight tracks per centimeter of tape width. The recording density is 200 bits/cm, and the transport mechanism moves the tape past the read heads at a velocity of 25 cm/s. How many bytes per second can be read from a 2-cm-wide tape?

14.26 Draw a block diagram of a circuit that will interface two interrupts, INT0 and INT1, to the INT input of a CPU so that INT1 has the higher priority and INT0 has the lower. In other words, a signal on INT1 is to be able to interrupt the CPU even when the CPU is currently handling an interrupt generated by INT0, but not vice versa.

CHAPTER

15

ELECTRONIC INSTRUMENTATION AND MEASUREMENTS

This chapter introduces measurement and instrumentation systems and summarizes important concepts by building on the foundation provided in earlier chapters. The development of the chapter follows a logical thread, starting from the physical sensors and proceeding through wiring and grounding to signal conditioning and analog-to-digital conversion, and finally to digital data transmission.

Section 15.1 presents an overview of sensors commonly used in engineering measurements. Some sensing devices have already been covered in earlier chapters, and others will be discussed in later chapters; the main emphasis in this chapter will be on classifying physical sensors, and on providing additional details on some sensors not presented elsewhere in this book—most notably, temperature transducers. Section 15.2 describes the common signal connections and proper wiring and grounding techniques, with emphasis on noise sources and techniques for reducing undesired interference. Section 15.3 provides an essential introduction to digital signal conditioning, namely, a discussion of instrumentation amplifiers and active filters. Sections

15.4 through 15.6 introduce analog-to-digital conversion, other integrated circuits used in instrumentation systems, and digital data transmission, respectively.

Learning Objectives

1. Review the major classes of sensors. *Section 15.1.*
2. Learn how to properly ground circuits, and methods for noise shielding and reduction. *Section 15.2.*
3. Design signal conditioning amplifiers and filters. *Section 15.3.*
4. Understand A/D and D/A conversion, and select the specifications of the appropriate conversion system for a given application. *Section 15.4.*
5. Analyze and design simple comparator and timing circuits using integrated circuits. Review other common instrumentation integrated circuits. *Sections 15.5 and 15.6.*

15.1 MEASUREMENT SYSTEMS AND TRANSDUCERS

Measurement Systems

In virtually every engineering application there is a need for measuring some physical quantities, such as forces, stresses, temperatures, pressures, flows, or displacements. These measurements are performed by physical devices called **sensors** or **transducers,** which are capable of converting a physical quantity to a more readily manipulated electrical quantity. Most sensors, therefore, convert the change of a physical quantity (e.g., humidity, temperature) to a corresponding (usually proportional) change in an electrical quantity (e.g., voltage or current). Often the direct output of the sensor requires additional manipulation before the electrical output is available in a useful form. For example, the change in resistance resulting from a change in the surface stresses of a material—the quantity measured by the resistance strain gauges described in Chapter 2[1]—must be first converted to a change in voltage through a suitable circuit (the Wheatstone bridge) and then amplified from the millivolt to the volt level. The manipulations needed to produce the desired end result are referred to as *signal conditioning*. The wiring of the sensor to the signal conditioning circuitry requires significant attention to *grounding* and *shielding* procedures, to ensure that the resulting signal is as free from noise and interference as possible. Very often, the conditioned sensor signal is then converted to *digital* form and recorded in a computer for additional manipulation, or is displayed in some form. The apparatus used in manipulating a sensor output to produce a result that can be suitably displayed or stored is called a **measurement system.** Figure 15.1 depicts a typical computer-based measurement system in block diagram form.

[1] See the Focus on Measurements box, "The Wheatstone Bridge and Force Measurements" on pp. 55–56.

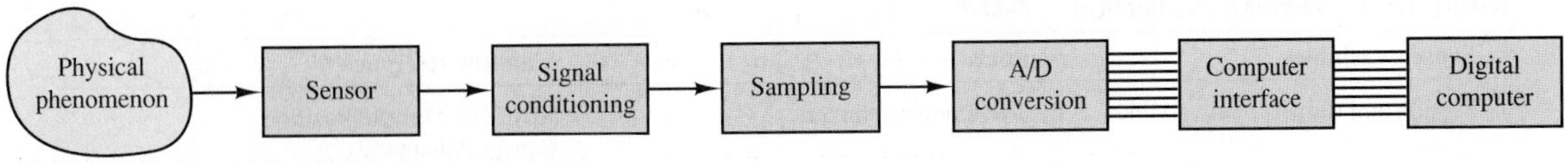

Figure 15.1 Measurement system

Sensor Classification

There is no standard and universally accepted classification of sensors. Depending on one's viewpoint, sensors may be grouped according to their physical characteristics (e.g., electronic sensors, resistive sensors) or by the physical variable or quantity measured by the sensor (e.g., temperature, flow rate). Other classifications are also possible. Table 15.1 presents a partial classification of sensors grouped according to the variable sensed; we do not claim that the table is complete, but we can safely state that most of the engineering measurements of interest to the reader are likely to fall in the categories listed in Table 15.1. Also included in the table are references to the Focus on Measurements boxes describing some of these sensors in other chapters of this book.

A sensor is usually accompanied by a set of specifications that indicate its overall effectiveness in measuring the desired physical variable. The following definitions will help the reader understand sensor data sheets:

Accuracy: conformity of the measurement to the true value, usually in percent of full-scale reading

Error: difference between measurement and true value, usually in percent of full-scale reading

Precision: number of significant figures of the measurement

Resolution: smallest measurable increment

Span: linear operating range

Range: the range of measurable values

Linearity: conformity to an ideal linear calibration curve, usually in percent of reading or of full-scale reading (whichever is greater)

LO1

Motion and Dimensional Measurements

FIND IT ON THE WEB

The **measurement of motion and dimension** is perhaps the most commonly encountered engineering measurement. Measurements of interest include absolute position, relative position (displacement), velocity, acceleration, and jerk (the derivative of acceleration). These can be either translational or rotational measurements; usually, the same principle can be applied to obtain both kinds of measurements. These measurements are often based on changes in elementary properties, such as changes in the resistance of an element (e.g., strain gauges, potentiometers), in an electric field (e.g., capacitive sensors), or in a magnetic field (e.g., inductive, variable-reluctance, or eddy current sensors). Other mechanisms may be based on special materials (e.g., piezoelectric crystals) or on optical signals and imaging systems. Table 15.1 lists several examples of dimensional and motion measurement that can be found in this book.

LO1

Table 15.1 Sensor classification

Sensed variables	Sensors	Chapter reference
Motion and dimensional variables	Resistive potentiometers	Resistive Throttle Position Sensor (Chapter 2)
	Strain gauges	Resistance Strain Gauges, The Wheatstone Bridge; and Force Measurements (Chapter 2)
	Differential transformers (LVDTs)	Linear Variable Differential Transformer (LVDT) (Chapter 16)
	Variable-reluctance sensors	Magnetic Reluctance Position Sensor (Chapter 16)
	Capacitive sensors	Capacitive Displacement Transducer and Microphone (Chapter 4); Peak Detector Circuit for Capacitive Displacement Transducer (Chapter 9)
	Piezoelectric sensors	Piezoelectric Sensor and Charge Amplifiers (Chapter 8)
	Electro-optical sensors	Digital Position Encoders; Digital Measurement of Angular Position and Velocity (Chapter 13)
	Moving-coil transducers	Seismic Transducer (Chapter 16)
	Seismic sensors	Seismic Transducer (Chapter 6)
Force, torque, and pressure	Strain gauges	Resistance Strain Gauges, The Wheatstone Bridge; and Force Measurements (Chapter 2)
	Piezoelectric sensors	Piezoelectric Sensor and Charge Amplifiers (Chapter 8)
	Capacitive sensors	Capacitive Displacement Transducer and Microphone (Chapter 4); Peak Detector Circuit for Capacitive Displacement Transducer (Chapter 9)
Flow	Pitot tube	
	Hot-wire anemometer	Hot-Wire Anemometer (Chapter 15)
	Differential pressure sensors	Differential Pressure Sensor (Chapter 15)
	Turbine meters	Turbine Meters (Chapter 15)
	Vortex shedding meters	
	Ultrasonic sensors	
	Electromagnetic sensors	
	Imaging systems	
Temperature	Thermocouples	Thermocouples (Chapter 15)
	Resistance thermometers (RTDs)	Resistance Thermometers (RTDs) (Chapter 15)
	Semiconductor thermometers	Diode Thermometer (Chapter 9)
	Radiation detectors	
Liquid level	Motion transducers	
	Force transducers	
	Differential pressure measurement devices	
Humidity	Semiconductor sensors	
Chemical composition	Gas analysis equipment	
	Solid-state gas sensors	

Force, Torque, and Pressure Measurements

FIND IT ON THE WEB

Another very common class of **measurements** is that **of pressure and force,** and the related **measurement of torque.** Perhaps the single most common family of force and pressure transducers comprises those based on strain gauges (e.g., load cells, diaphragm pressure transducers). Also very common are piezoelectric transducers.

Capacitive transducers again find application in the measurement of pressure. Table 15.1 indicates where the reader can find examples of these measurements in this book.

Flow Measurements

In many engineering applications it is desirable to sense the flow rate of a fluid, whether compressible (gas) or incompressible (liquid). The **measurement of fluid flow rate** is a complex subject; in this section we simply summarize the concepts underlying some of the most common measurement techniques. Shown in Figure 15.2 are three different types of flow rate sensors. The sensor in Figure 15.2(a) is based on **differential-pressure measurement** and on a **calibrated orifice:** the relationship between pressure across the orifice $p_1 - p_2$ and flow rate through the orifice q is predetermined through the calibration; therefore, measuring the differential pressure is equivalent to measuring flow rate.

FIND IT ON THE WEB

The sensor in Figure 15.2(b) is called a **hot-wire anemometer,** because it is based on a heated wire that is cooled by the flow of a gas. The resistance of the wire changes with temperature, and a Wheatstone bridge circuit converts this change in resistance to a change in voltage while the current is kept constant. Also commonly used are **hot-film anemometers,** where a heated film is used in place of the more delicate wire. A very common application of the latter type of sensor is in automotive engines, where control of the air-to-fuel ratio depends on measurement of the engine intake mass airflow rate.

Figure 15.2(c) depicts a **turbine flowmeter** in which the fluid flow causes a turbine to rotate; the velocity of rotation of the turbine (which can be measured by a noncontact sensor, e.g., a magnetic pickup)[2] is related to the flow velocity.

Besides the techniques discussed in this chapter, many other techniques exist for measuring fluid flow, some of significant complexity.

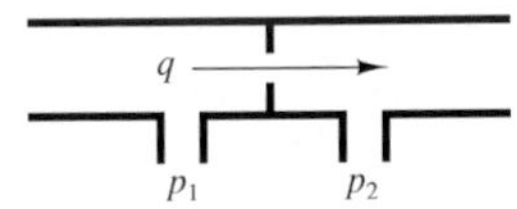

Differential-pressure flowmeter: A calibrated orifice and a pair of pressure transducers permit the measurement of flow rate.
(a)

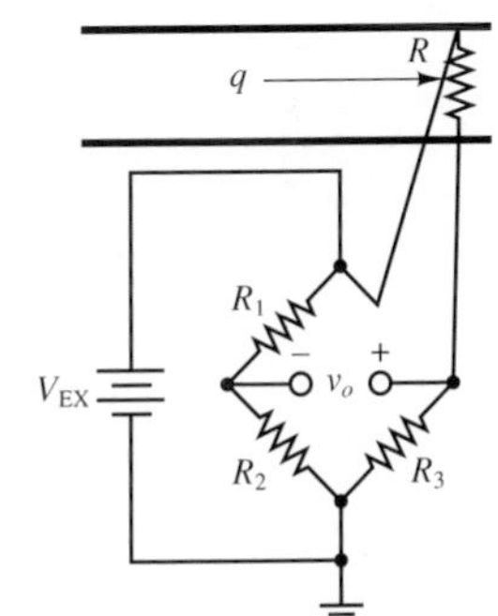

Hot-wire anemometer: A heated wire is cooled by the gas flow. The resistance of the wire changes with temperature.
(b)

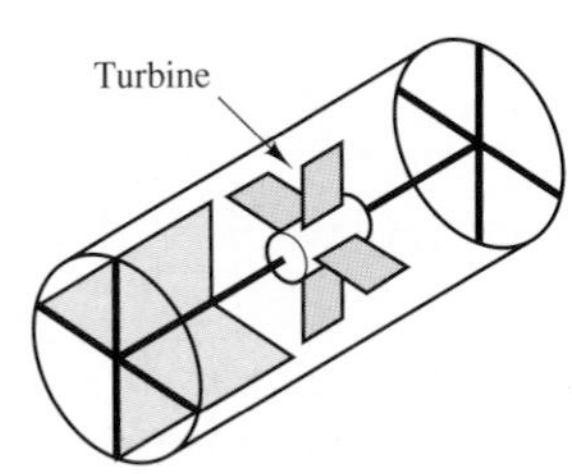

Turbine flowmeter: Fluid flow induces rotation of the turbine; measurement of turbine velocity provides an indication of flow rate.
(c)

Figure 15.2 Devices for the measurement of flow

Temperature Measurements

One of the most frequently measured physical quantities is temperature. The need to measure temperature arises in just about every field of engineering. This subsection is devoted to summarizing two common **temperature sensors**—the **thermocouple** and the **resistance temperature detector (RTD)**—and their related signal conditioning needs.

FIND IT ON THE WEB

Thermocouples

A thermocouple is formed by the junction of two dissimilar metals. This junction results in an open-circuit **thermoelectric voltage** due to the **Seebeck effect,** named after Thomas Seebeck, who discovered the phenomenon in 1821. Various types of thermocouples exist; they are usually classified according to the data of Table 15.2. The Seebeck coefficient shown in the table is specified at a given temperature because the output voltage of a thermocouple v has a nonlinear dependence on temperature. This dependence is typically expressed in terms of a polynomial of the following

[2]See the Focus on Measurements box "Magnetic Reluctance Position Sensors" in Chapter 16.

Table 15.2 Thermocouple data

Type	Elements +/–	Seebeck coefficient (μV/°C)	Range (°C)	Range (mV)
E	Chromel/constantan	58.70 at 0°C	−270 to 1000	−9.835 to 76.358
J	Iron/constantan	50.37 at 0°C	−210 to 1200	−8.096 to 69.536
K	Chromel/alumel	39.48 at 0°C	−270 to 1372	−6.548 to 54.874
R	Pt(10%)—Rh/Pt	10.19 at 600°C	−50 to 1768	−0.236 to 18.698
T	Copper/constantan	38.74 at 0°C	−270 to 400	−6.258 to 20.869
S	Pt(13%)—Rh/Pt	11.35 at 600°C	−50 to 1768	−0.226 to 21.108

form:

$$T = a_0 + a_1 v + a_2 v^2 + a_3 v^3 + \cdots + a_n v^n \tag{15.1}$$

For example, the coefficients of the J thermocouple in the range of −100 to +1000°C are as follows:

$$a_0 = -0.048868252 \qquad a_1 = 19{,}873.14503 \qquad a_2 = -128{,}614.5353$$
$$a_3 = 11{,}569{,}199.78 \qquad a_4 = -264{,}917{,}531.4 \qquad a_5 = 2{,}018{,}441{,}314$$

The use of a thermocouple requires special connections, because the junction of the thermocouple wires with other leads (such as voltmeter leads) creates additional thermoelectric junctions that in effect act as additional thermocouples. For example, in the J thermocouple circuit of Figure 15.3, junction J_1 is exposed to the temperature to be measured, but junctions J_2 and J_3 also generate a thermoelectric voltage, which is dependent on the temperature at these junctions, that is, the temperature at the voltmeter connections. One would therefore have to know the voltages at these junctions as well, in order to determine the actual thermoelectric voltage at J_1. To obviate this problem, a reference junction at known temperature can be employed; a traditional approach involves the use of a **cold junction,** so called because it often consists of an ice bath, one of the easiest means of obtaining a known reference temperature. Figure 15.4 depicts a thermocouple measurement using an ice bath. The voltage measured in Figure 15.4 is dependent on the temperature difference $T_1 - T_{ref}$, where $T_{ref} = 0$°C. The connections to the voltmeter are made at an *isothermal block,* kept at a constant temperature; note that the same metal is used in both of the connections to the isothermal block. Thus (still assuming a J thermocouple), there is no difference

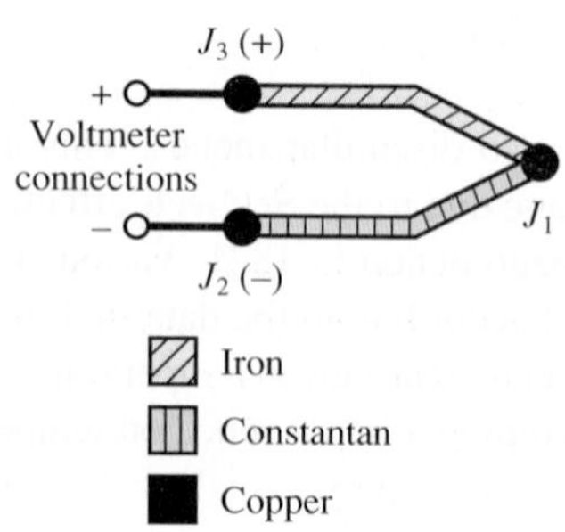

Figure 15.3 J thermocouple circuit

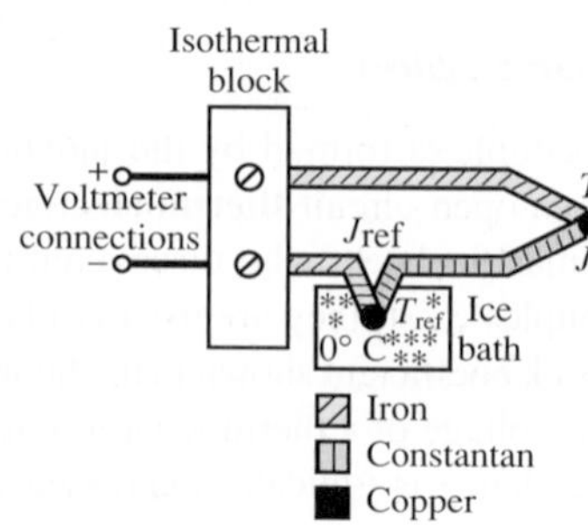

Figure 15.4 Cold junction–compensated thermocouple circuit

between the thermoelectric voltages at the two copper-iron junctions; these will add to zero at the voltmeter. The voltmeter will therefore read a voltage proportional to $T_1 - T_{\text{ref}}$.

An ice bath is not always a practical solution. Other cold junction temperature compensation techniques employ an additional temperature sensor to determine the actual temperature of junctions J_2 and J_3 of Figure 15.3.

Resistance Temperature Detectors

A resistance temperature detector (RTD) is a variable-resistance device whose resistance is a function of temperature. RTDs can be made with both positive and negative temperature coefficients and offer greater accuracy and stability than thermocouples. **Thermistors** are part of the RTD family. A characteristic of all RTDs is that they are *passive* devices; that is, they do not provide a useful output unless excited by an external source. The change in resistance in an RTD is usually converted to a change in voltage by forcing a current to flow through the device. An indirect result of this method is a **self-heating error,** caused by the i^2R heating of the device. Self-heating of an RTD is usually denoted by the amount of power that will raise the RTD temperature by 1°C. Reducing the excitation current can clearly help reduce self-heating, but it also reduces the output voltage.

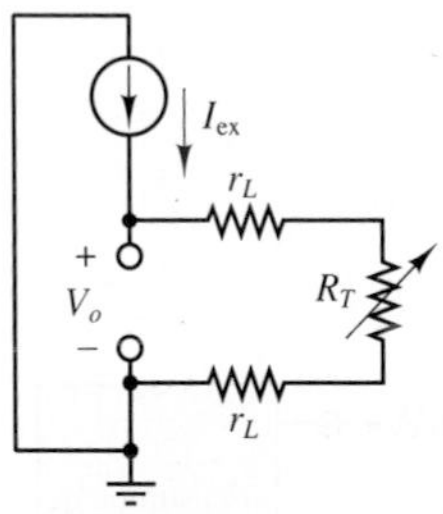

Figure 15.5 Effect of connection leads on RTD measurement

The RTD resistance has a fairly linear dependence on temperature; a common definition of the **temperature coefficient** of an RTD is related to the change in resistance from 0 to 100°C. Let R_0 be the resistance of the device at 0°C and R_{100} the resistance at 100°C. Then the temperature coefficient α is defined to be

$$\alpha = \frac{R_{100} - R_0}{100 - 0} \frac{\Omega}{°\text{C}} \tag{15.2}$$

A more accurate representation of RTD temperature dependence can be obtained by using a nonlinear (cubic) equation and published tables of coefficients. As an example, a platinum RTD could be described either by the temperature coefficient $\alpha = 0.003911$ or by the equation

$$\begin{aligned} R_T &= R_0(1 + AT - BT^2 - CT^3) \\ &= R_0(1 + 3.6962 \times 10^{-3}T - 5.8495 \times 10^{-7}T^2 \\ &\qquad - 4.2325 \times 10^{-12}T^3) \end{aligned} \tag{15.3}$$

where the coefficient C is equal to zero for temperatures above 0°C.

Because RTDs have fairly low resistance, they are sensitive to error introduced by the added resistance of the lead wires connected to them; Figure 15.5 depicts the effect of the lead resistances r_L on the RTD measurement. Note that the measured voltage includes the resistance of the RTD as well as the resistance of the leads. If the leads used are long (greater than 3 m is a good rule of thumb), then the measurement will have to be adjusted for this error. Two possible solutions to the lead problems are the *four-wire* RTD measurement circuit and the *three-wire* Wheatstone bridge circuit, shown in Figure 15.6(a) and (b), respectively. In the circuit of Figure 15.6(a), the resistances of the lead wires from the excitation r_{L1} and r_{L4} may be arbitrarily large, since the measurement is affected by the resistances of only the output lead wires r_{L2} and r_{L3}. The circuit of Figure 15.6(b) takes advantage of the properties of the Wheatstone bridge to cancel out the unwanted effect of the lead wires while still producing an output dependent on the change in temperature.

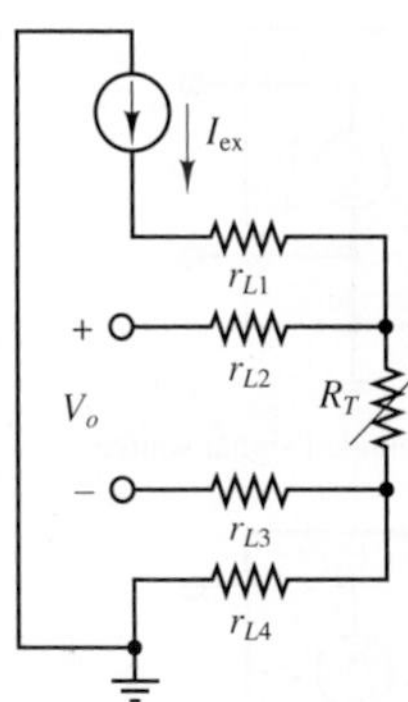

(a)

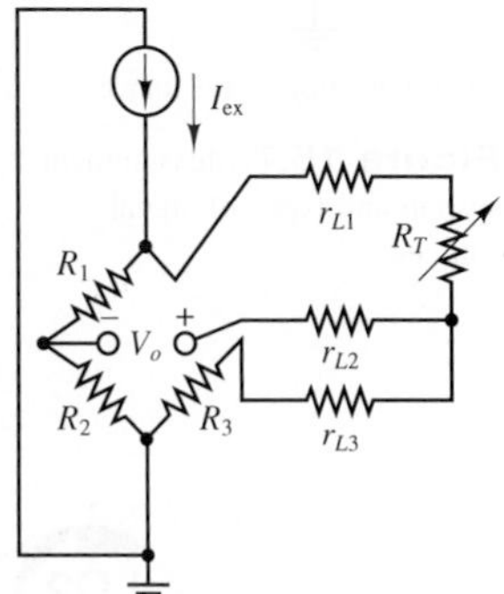

(b)

Figure 15.6 (a) Four-wire RTD circuit and (b) three-wire Wheatstone bridge RTD circuit

15.2 WIRING, GROUNDING, AND NOISE

The importance of proper circuit connections cannot be overemphasized. Unfortunately, this is a subject that is rarely taught in introductory electrical engineering courses. This section summarizes some important considerations regarding signal source connections, various types of input configurations, noise sources and coupling mechanisms, and means of minimizing the influence of noise on a measurement.

Signal Sources and Measurement System Configurations

Before proper connection and wiring techniques can be presented, we must examine the difference between **grounded** and **floating signal sources.** Every sensor can be thought of as some kind of signal source; a general representation of the connection of a sensor to a measurement system is shown in Figure 15.7(a). The sensor is modeled as an ideal voltage source in series with a source resistance. Although this representation does not necessarily apply to all sensors, it will be adequate for the purposes of this section. Figure 15.7(b) and (c) shows two types of signal sources: grounded and floating. A grounded signal source is one in which a ground reference is established, for example, by connecting the *signal low* lead to a case or housing. A floating signal source is one in which neither signal lead is connected to ground; since ground potential is arbitrary, the signal source voltage levels (*signal low* and *signal high*) are at an unknown potential relative to the case ground. Thus, the signal is said to be *floating*. Whether a sensor can be characterized as a grounded or a floating signal source ultimately depends on the connection of the sensor to its case, but the choice of connection may depend on the nature of the source. For example, the thermocouple described in Section 15.1 is *intrinsically* a floating signal source, since the signal of interest is a difference between two voltages. The same thermocouple *could* become a grounded signal source if one of its two leads were directly connected to ground, but this is usually not a desirable arrangement for this particular sensor.

In analogy with a signal source, a measurement system can be either **ground-referenced** or **differential.** In a ground-referenced system, the signal low connection is tied to the instrument case ground; in a differential system, neither of the two signal connections is tied to ground. Thus, a differential measurement system is well suited to measuring the difference between two signal levels (such as the output of an ungrounded thermocouple).

One of the potential dangers in dealing with grounded signal sources is the introduction of **ground loops.** A ground loop is an undesired current path caused by the connection of two reference voltages to each other. This is illustrated in Figure 15.8, where a grounded signal source is shown connected to a ground-referenced

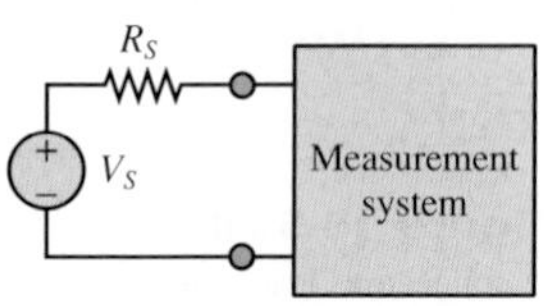

(a) Ideal signal source connected to measurement system

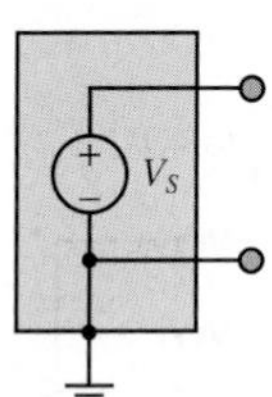

(b) Grounded signal source

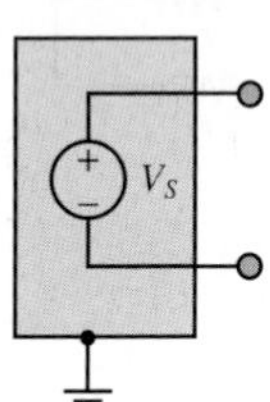

(c) Floating signal source

Figure 15.7 Measurement system and types of signal sources

LO2

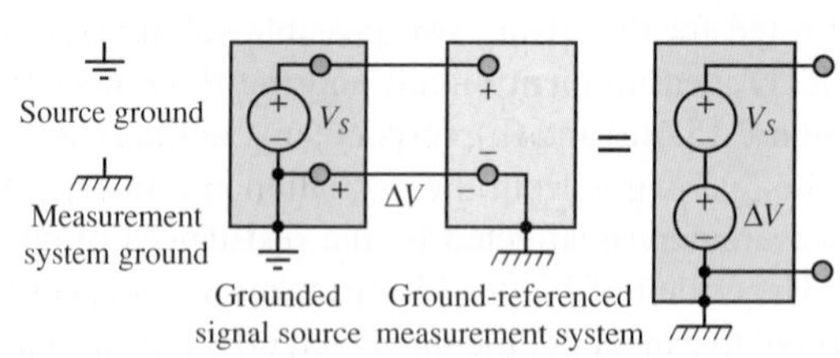

Figure 15.8 Ground loop in ground-referenced measurement system

measurement system. Notice that we have purposely denoted the signal source ground and the measurement system ground by two distinct symbols, to emphasize that these are not necessarily at the same potential—as also indicated by the voltage difference ΔV. Now, one might be tempted to tie the two grounds to each other, but this would only result in a current flowing from one ground to the other, through the small (but nonzero) resistance of the wire connecting the two. The net effect of this ground loop is that the voltage measured by the instrument would include the unknown ground voltage difference ΔV, as shown in Figure 15.8. Since this latter voltage is unpredictable, you can see that ground loops can cause substantial errors in measuring systems. In addition, ground loops are the primary cause of conducted noise, as explained later in this section.

A differential measurement system is often a way to avoid ground loop problems, because the signal source and measurement system grounds are not connected to each other, and especially because the signal low input of the measuring instrument is not connected to either instrument case ground. The connection of a grounded signal source and a differential measurement system is depicted in Figure 15.9.

If the signal source connected to the differential measurement system is floating, as shown in Figure 15.10, it is often a recommended procedure to reference the signal to the instrument ground by means of two identical resistors that can provide a return path to ground for any currents present at the instrument. An example of such input currents would be the input bias currents inevitably present at the input of an operational or instrumentation amplifier.

The simple concepts illustrated in the preceding paragraphs and figures can assist the user and designer of instrumentation systems in making the best possible wiring connections for a given measurement.

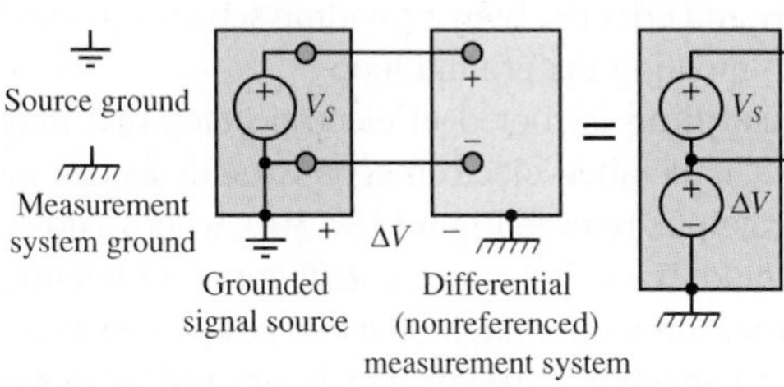

Figure 15.9 Differential (nonreferenced) measurement system

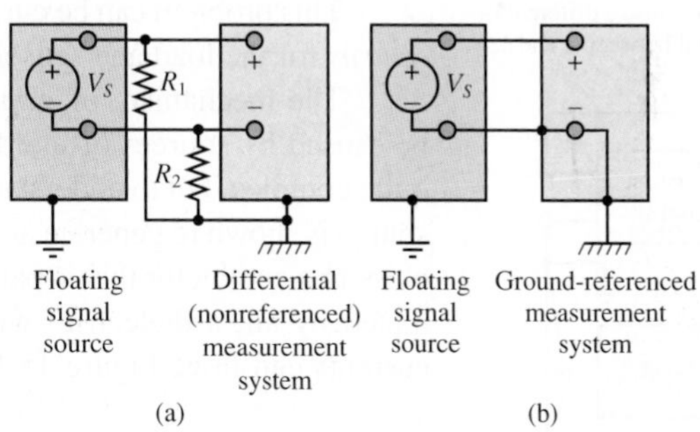

Figure 15.10 Measuring signals from a floating source: (a) differential input; (b) single-ended input

Noise Sources and Coupling Mechanisms

Noise—meaning any undesirable signal interfering with a measurement—is an unavoidable element of all measurements. Figure 15.11 depicts a block diagram of the three essential stages of a noisy measurement: a **noise source,** a **noise coupling mechanism,** and a sensor or associated signal conditioning circuit. Noise sources are always present and are often impossible to eliminate completely; typical sources of noise in practical measurements are the electromagnetic fields caused by fluorescent light fixtures, video monitors, power supplies, switching circuits, and high-voltage (or current) circuits. Many other sources exist, of course, but often the simple sources in our everyday environment are the most difficult to defeat.

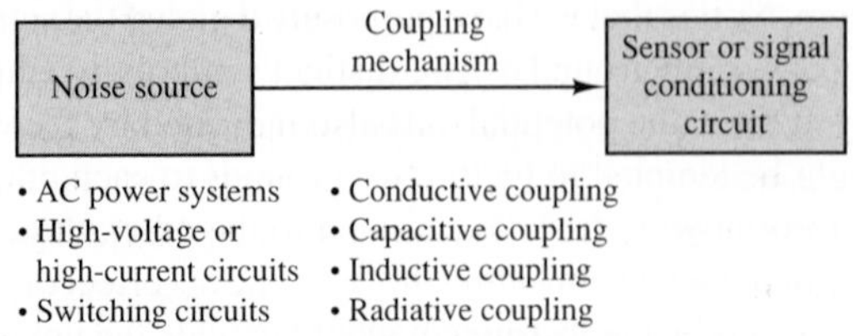

Figure 15.11 Noise sources and coupling mechanisms

Figure 15.11 also indicates that various coupling mechanisms can exist between a noise source and an instrument. Noise coupling can be conductive; that is, noise currents may actually be conducted from the noise source to the instrument by physical wires. Noise can also be coupled capacitively, inductively, and radiatively.

Figure 15.12 illustrates how interference can be **conductively coupled** by way of a ground loop. In the figure, a power supply is connected to both a load and a sensor. We shall assume that the load may be switched on and off, and that it carries substantial currents. The top circuit contains a ground loop: the current i from the supply divides between the load and sensor; since the wire resistance is nonzero, a large current flowing through the load may cause the ground potential at point a to differ from the potential at point b. In this case, the measured sensor output is no longer v_o, but it is now equal to $v_o + v_{ba}$, where v_{ba} is the potential difference from point b to point a. Now, if the load is switched on and off and its current is therefore subject to large, abrupt changes, these changes will be manifested in the voltage v_{ba} and will appear as noise on the sensor output.

This problem can be cured simply and effectively by providing separate *ground returns* for the load and sensor, thus eliminating the ground loop.

The mechanism of **capacitive coupling** is rooted in electric fields that may be caused by sources of interference. The detailed electromagnetic analysis can be quite complex, but to understand the principle, refer to Figure 15.13(a), where a noise source is shown to generate an electric field. If a noise source conductor is sufficiently close to a conductor that is part of the measurement system, the two conductors (separated by air, a dielectric) will form a capacitor, through which any time-varying currents can flow. Figure 15.13(b) depicts an equivalent circuit in which the noise

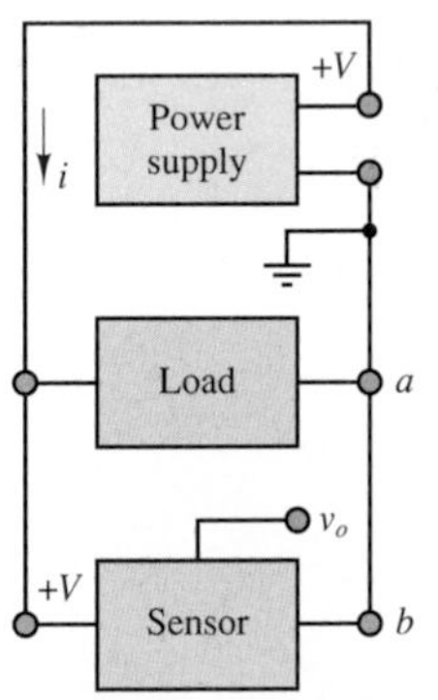

The ground loop created by the load circuit can cause a different ground potential between *a* and *b*.

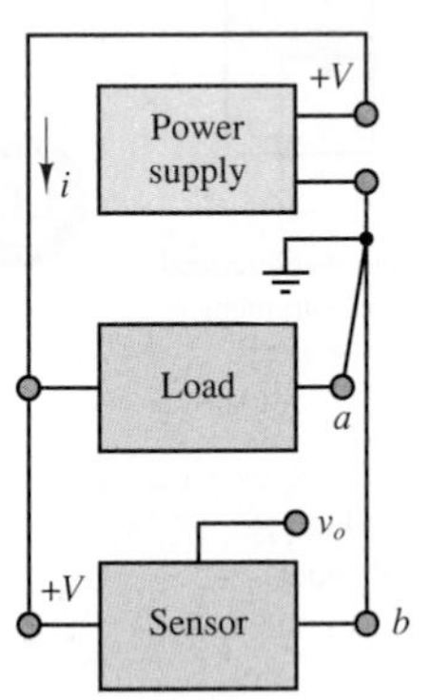

Separate ground returns for the load and the sensor circuit eliminate the ground loop.

Figure 15.12 Conductive coupling: ground loop and separate ground returns

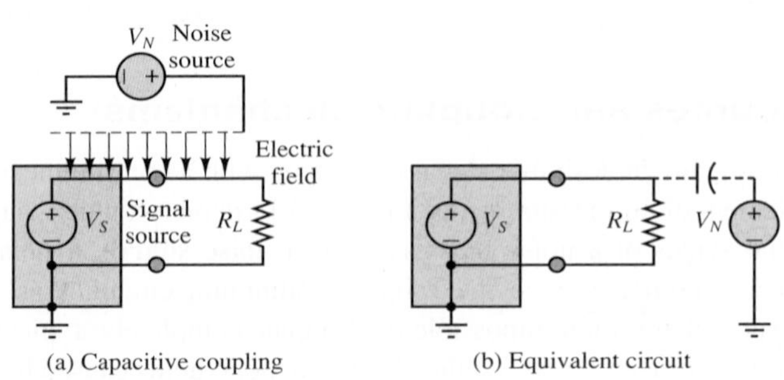

Figure 15.13 Capacitive coupling and equivalent-circuit representation

voltage V_N couples to the measurement circuit through an imaginary capacitor, representing the actual capacitance of the noise path.

The dual of capacitive coupling is **inductive coupling.** This form of noise coupling is due to the magnetic field generated by current flowing through a conductor. If the current is large, the magnetic fields can be significant, and the **mutual inductance** (see Chapters 5 and 16) between the noise source and the measurement circuit causes the noise to couple to the measurement circuit. Thus, inductive coupling, as shown in Figure 15.14, results when undesired (unplanned) magnetic coupling ties the noise source to the measurement circuit.

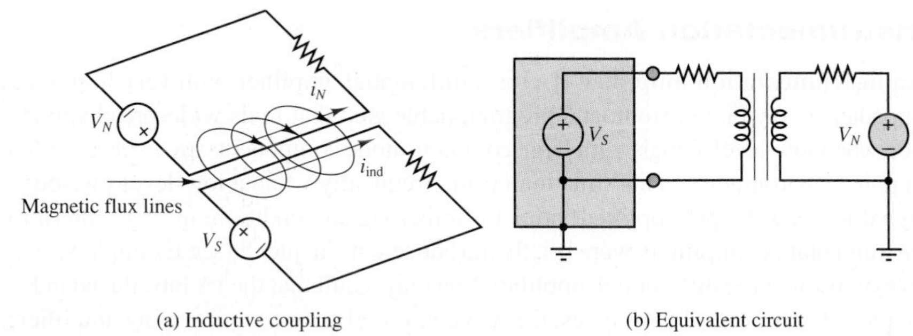

Figure 15.14 Inductive coupling and equivalent-circuit representation

Noise Reduction

Various techniques exist for minimizing the effect of undesired interference, in addition to proper wiring and grounding procedures. The two most common methods are **shielding** and the use of **twisted-pair wire.** A shielded cable is shown in Figure 15.15. The shield is made of a copper braid or of foil and is usually grounded at the source end *but not at the instrument end,* because this would result in a ground loop. The shield can protect the signal from a significant amount of electromagnetic interference, especially at lower frequencies. Shielded cables with various numbers of conductors are available commercially. However, shielding cannot prevent inductive coupling. The simplest method for minimizing inductive coupling is the use of twisted-pair wire; the reason for using twisted-pair wire is that untwisted wire can offer large loops that can couple a substantial amount of electromagnetic radiation (see Section 16.1). Twisting drastically reduces the loop area, and with it the interference. Twisted pair is available commercially.

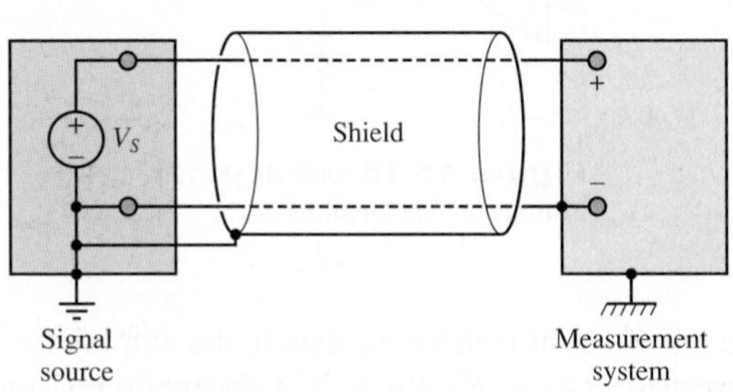

Figure 15.15 Shielding

15.3 SIGNAL CONDITIONING

A properly wired, grounded, and shielded sensor connection is a necessary first stage of any well-designed measurement system. The next stage consists of any **signal conditioning** that may be required to manipulate the sensor output into a form appropriate for the intended use. Very often, the sensor output is meant to be fed into a digital computer, as illustrated in Figure 15.1. In this case, it is important to condition the signal so that it is compatible with the process of data acquisition. Two of the most important signal conditioning functions are *amplification* and *filtering*. Both are discussed in this section.

Instrumentation Amplifiers

An **instrumentation amplifier (IA)** is a differential amplifier with very high input impedance, low bias current, and programmable gain that finds widespread application when low-level signals with large common-mode components are to be amplified in noisy environments. This situation occurs frequently when a low-level transducer signal needs to be preamplified, prior to further signal conditioning (e.g., filtering). Instrumentation amplifiers were briefly introduced in Chapter 8 (see Example 8.3), as an extension of the differential amplifier. You may recall that the IA introduced in Example 8.3 consisted of two stages, the first composed of two noninverting amplifiers, the second of a differential amplifier. Although the design in Chapter 8 is useful and is sometimes employed in practice, it suffers from a few drawbacks, most notably the requirement for very precisely matched resistors and source impedances to obtain the maximum possible cancellation of the common-mode signal. If the resistors are not matched exactly, the common-mode rejection ratio of the amplifier is significantly reduced, as the following will demonstrate.

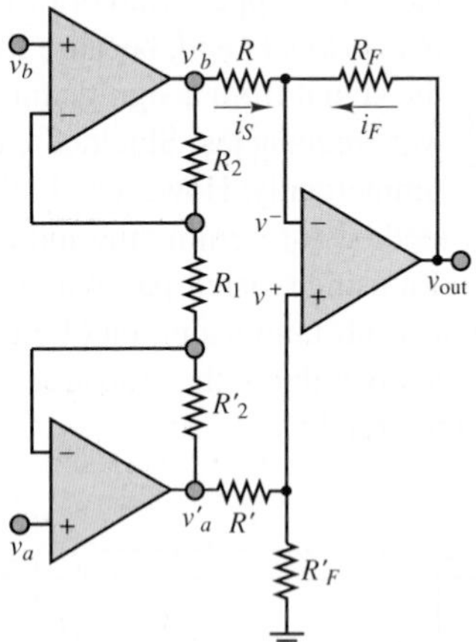

Figure 15.16 Discrete op-amp instrumentation amplifier

To illustrate the effect of resistor mismatch, the amplifier of Figure 15.16 has properly matched resistors ($R_2 = R'_2, R_F = R'_F$), except for resistors R and R', which differ by an amount ΔR such that $R' = R + \Delta R$. Let us compute the closed-loop gain for the amplifier. As shown in Example 8.3, the input-stage noninverting amplifiers have a closed-loop gain given by

$$A = \frac{v_b'}{v_b} = \frac{v_a'}{v_a} = 1 + \frac{2R_2}{R_1} \tag{15.4}$$

To compute the output voltage, observe that the voltage at the noninverting terminal is

$$v^+ = \frac{R_F}{R_F + R + \Delta R} v_a' \tag{15.5}$$

and since the inverting-terminal voltage is $v^- = v^+$, the feedback current is given by

$$i_F = \frac{v_{\text{out}} - v^-}{R_F} = \frac{v_{\text{out}} - [R_F/(R_F + R + \Delta R)]v_a'}{R_F} \tag{15.6}$$

and the source current is

$$i_S = \frac{v_b' - v^-}{R} = \frac{v_b' - [R_F/(R_F + R + \Delta R)]v_a'}{R} \tag{15.7}$$

Applying KCL at the inverting node (under the usual assumption that the input current going into the op-amp is negligible), we set $i_F = -i_S$ and obtain the expression

$$\frac{v_{\text{out}}}{R_F} = \frac{v_a'}{R_F + R + \Delta R} - \frac{v_b'}{R} + \frac{R_F}{R}\frac{v_a'}{R_F + R + \Delta R}$$
$$= \left(1 + \frac{R_F}{R}\right)\frac{v_a'}{R_F + R + \Delta R} - \frac{v_b'}{R}$$

so that the output voltage may be computed to be

$$v_{\text{out}} = R_F\left(\frac{R + R_F}{R}\right)\left(\frac{v_a'}{R_F + R + \Delta R}\right) - \frac{R_F}{R}v_b' \tag{15.8}$$

Note that if the term ΔR in the denominator were zero, the same result would be obtained as in Example 8.3: $v_{\text{out}} = (R_F/R)(v_a' - v_b')$; however, because of the resistor mismatch, there is a corresponding mismatch between the gains for the two differential signal components. Further—and more important—if the original signals v_a and v_b contained both differential-mode and common-mode components

$$v_a = v_{a,\text{dif}} + v_{\text{com}} \qquad v_b = v_{b,\text{dif}} + v_{\text{com}} \tag{15.9}$$

such that

$$v_a' = A(v_{a,\text{dif}} + v_{\text{com}}) \qquad v_b' = A(v_{b,\text{dif}} + v_{\text{com}}) \tag{15.10}$$

then the common-mode components would not cancel out in the output of the amplifier, because of the gain mismatch, and the output of the amplifier would be given by

$$v_{\text{out}} = R_F\left(\frac{R + R_F}{R}\right)\left[\frac{A(v_{a,\text{dif}} + v_{\text{com}})}{R_F + R + \Delta R}\right] - \frac{R_F}{R}A(v_{b,\text{dif}} + v_{\text{com}}) \tag{15.11}$$

resulting in the output voltage of

$$v_{\text{out}} = v_{\text{out, dif}} + v_{\text{out, com}} \tag{15.12}$$

with

$$v_{\text{out, dif}} = R_F\left(\frac{R + R_F}{R}\right)\left(\frac{Av_{a,\text{dif}}}{R_F + R + \Delta R}\right) - \frac{R_F}{R}Av_{b,\text{dif}} \tag{15.13}$$

and

$$v_{\text{out, com}} = R_F\left(\frac{R+R_F}{R}\right)\left(\frac{Av_{\text{com}}}{R_F+R+\Delta R}\right) - \frac{R_F}{R}Av_{\text{com}}$$
$$= \frac{R_F}{R}\left(\frac{R+R_F}{R_F+R+\Delta R} - 1\right)Av_{\text{com}} \tag{15.14}$$

The common-mode rejection ratio (CMRR; see Section 8.6) is given in units of decibels by

$$\text{CMRR}_{\text{dB}} = \left|\frac{A_{\text{dif}}}{A_{\text{com}}}\right| = 20\log\left|\frac{A_{\text{dif}}}{v_{\text{out, com}}/v_{\text{com}}}\right|$$
$$= 20\log\left|\frac{A_{\text{dif}}}{\frac{R_F}{R}\left(\frac{R+R_F}{R_F+R+\Delta R} - 1\right)A}\right| \tag{15.15}$$

where A_{dif} is the *differential gain* (which is usually assumed equal to the nominal design value). Since the common-mode gain $v_{\text{out, com}}/v_{\text{com}}$ should ideally be zero, the theoretical CMRR for the instrumentation amplifier with perfectly matched resistors is infinite. In fact, even a small mismatch in the resistors used would dramatically reduce the CMRR, as the Check Your Understanding exercises at the end of this subsection illustrate. Even with resistors having 1 percent tolerance, the maximum CMRR that could be attained for typical values of resistors and an overall gain of 1,000 would be only 60 dB. In many practical applications, a requirement for a CMRR of 100 or 120 dB is not uncommon, and these would demand resistors of 0.01 percent tolerance. It should be evident, then, that the "discrete" design of the IA, employing three op-amps and discrete resistors, will not be adequate for the more demanding instrumentation applications.

LO3

EXAMPLE 15.1 Common-Mode Gain and Rejection Ratio

Problem

Compute the common-mode gain and common-mode rejection ratio for the amplifier of Figure 15.16.

Solution

Known Quantities: Amplifier nominal closed-loop gain; resistance values; resistor tolerance.

Schematics, Diagrams, Circuits, and Given Data: $A = 10$; $R_F = 10\text{ k}\Omega$; $R = 1\text{ k}\Omega$; $\Delta R = 20\ \Omega$.

Find: $v_{\text{out, com}}/v_{\text{com}}$, CMRR_{dB}

Analysis: The common-mode gain is equal to the ratio of the common-mode output signal to the common-mode input; from equation 15.14, we can write

$$\frac{v_{\text{out, com}}}{v_{\text{com}}} = \frac{R_F}{R}\left(\frac{R+R_F}{R+R_F+\Delta R} - 1\right)A = 10\left(\frac{11}{11.02} - 1\right)10 = -0.1815$$

The CMRR (in units of decibels) can be computed from equation 15.15, where

$$A_{\text{dif}} = A \times \frac{R_F}{R} = 100$$

and therefore,

$$\text{CMRR} = \left|\frac{A_{\text{dif}}}{A_{\text{com}}}\right|_{\text{dB}} = 20\log\left|\frac{A_{\text{dif}}}{v_{\text{out, com}}/v_{\text{com}}}\right| = 20\log\left|\frac{A_{\text{dif}}}{\frac{R_F}{R}\left(\frac{R+R_F}{R+R_F+\Delta R}-1\right)A}\right|$$

$$= 20\log\left|\frac{100}{\frac{10}{1}\left(\frac{11}{11.02}-1\right)10}\right| = 54.82\text{ dB}$$

Comments: Note that, in general, it is difficult to determine exactly the level of resistor mismatch ΔR in an instrumentation amplifier.

CHECK YOUR UNDERSTANDING

Use the definition of the common-mode rejection ratio given in equation 15.16 to compute the CMRR (in decibels) of the amplifier of Example 15.1 if $R_F/R = 100$ and $A = 10$, and if $\Delta R = 5$ percent of R. Assume $R = 1\text{ k}\Omega$, $R_F = 100\text{ k}\Omega$.

Repeat for a 1 percent variation in R.

Repeat for a 0.01 percent variation in R.

Calculate the mismatch in gains for differential components for the 5 percent resistance mismatch.

Answers: 66 dB; 80 dB; 120 dB; −6.1 dB

The general expression for the CMRR of the instrumentation amplifier of Figure 15.16, without assuming any of the resistors are matched, except for R_2 and R_2', is

$$\text{CMRR} = \left|\frac{A_{\text{dif}}}{A_{\text{com}}}\right| = \left|\frac{(R_F/R)(1+2R_2/R_1)}{\frac{R_F}{R}\left[\frac{R_F'}{R_F}\left(\frac{R_F+R}{R_F'+R'}\right)-1\right]}\right| \tag{15.16}$$

and it can easily be shown that the CMRR is infinite if the resistors are perfectly matched.

Example 15.1 illustrated some of the problems encountered in the design of instrumentation amplifiers using discrete components. Many of these problems can be dealt with very effectively if the entire instrumentation amplifier is designed into a single *monolithic integrated circuit,* where the resistors can be carefully matched by appropriate fabrication techniques and many other problems can also be avoided. The functional structure of an IC instrumentation amplifier is depicted in Figure 15.17. Specifications for a common IC instrumentation amplifier (and a more accurate circuit

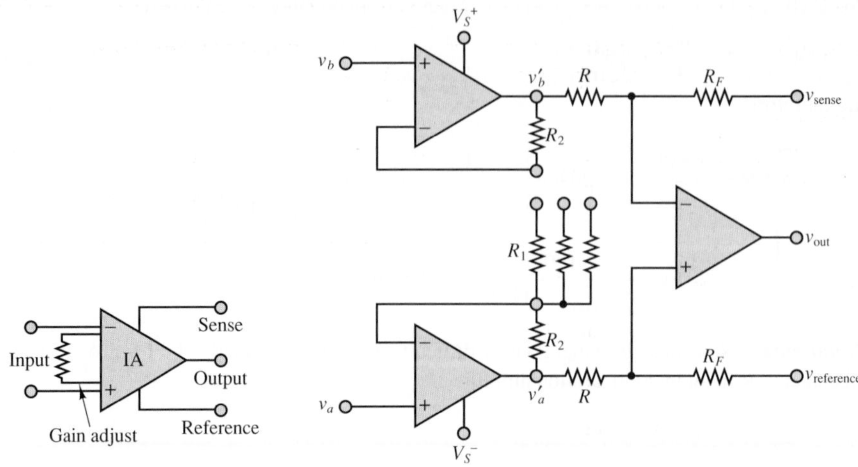

Figure 15.17 IC instrumentation amplifier

description) are shown in Figure 15.18. Among the features worth mentioning here are the programmable gains, which the user can set by suitably connecting one or more of the resistors labeled R_1 to the appropriate connection. Note that the user may also choose to connect additional resistors to control the amplifier gain, without adversely affecting the amplifier's performance, since R_1 requires no matching. In addition to the pin connection that permits programmable gains, two additional pins are provided, called **sense** and **reference.** These additional connections are provided to the user for the purpose of referencing the output voltage to a signal other than ground, by means of the reference terminal, or of further amplifying the output current (e.g., with a transistor stage), by connecting the sense terminal to the output of the current amplifier.

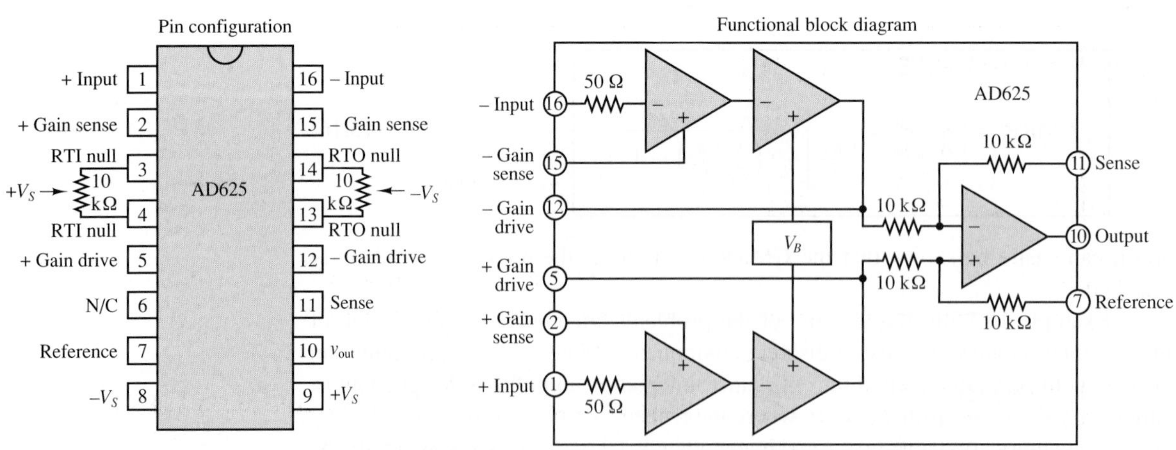

Figure 15.18 AD625 instrumentation amplifier

EXAMPLE 15.2 Instrumentation Amplifier Gain Configuration Using Internal Resistors

LO3

Problem

Determine the possible input-stage gains that can be configured using the choice of resistor values given for the instrumentation amplifier of Figure 15.17.

Solution

Known Quantities: IA resistor values.

Find: A, for different resistor combinations.

Schematics, Diagrams, Circuits, and Given Data: $R_F = R = 10\ \text{k}\Omega$; $R_2 = 20\ \text{k}\Omega$; $R_1 = 80.2, 201, 404\ \Omega$.

Analysis: Recall that the gain of the input stage (for each of the differential inputs) can be calculated according to equation 15.4:

$$A = 1 + \frac{2R_2}{R_1}$$

Thus, by connecting each of the three resistors, we can obtain gains

$$A_1 = 1 + \frac{40{,}000}{80.2} = 500 \qquad A_2 = 1 + \frac{40{,}000}{201} = 200 \qquad A_3 = 1 + \frac{40{,}000}{404} = 100$$

It is also possible to obtain additional input-stage gains by connecting resistors in parallel:

$$80.2 \| 201 = 57.3\ \Omega\ (A_4 \approx 700) \qquad 80.2 \| 404 = 66.9\ \Omega\ (A_5 \approx 600)$$

$$404 \| 201 = 134.2\ \Omega\ (A_6 \approx 300)$$

Comments: The use of resistors supplied with the IA package is designed to reduce the uncertainty introduced by the use of external resistors, since the value of the internally supplied resistors can be controlled more precisely.

CHECK YOUR UNDERSTANDING

Calculate the mismatch in gains for the differential components for the 1 percent resistance mismatch of Check Your Understanding on p. 773.

What value of resistance R_1 would permit a gain of 1,000 for the IA of Example 15.2?

Answers: −20.1 dB; 40 Ω

Active Filters

The need to filter sensor signals that may be corrupted by noise or other interfering or undesired inputs has already been approached in two earlier chapters. In Chapter 6, simple passive filters made of resistors, capacitors, and inductors were analyzed. It was shown that three types of filter frequency response characteristics can be achieved with these simple circuits: low-pass, high-pass, and bandpass. In Chapter 8, the concept of active filters was introduced, to suggest that it may be desirable to exploit the properties of operational amplifiers to simplify filter design, to more easily match source and load impedances, and to eliminate the need for inductors. The aim of this section is to discuss more advanced active filter designs, which find widespread application in instrumentation circuits.

Figure 15.19 depicts the general characteristics of a low-pass active filter, indicating that within the passband of the filter, a certain deviation from the nominal filter gain A is accepted, as indicated by the minimum and maximum passband gains $A + \varepsilon$ and $A - \varepsilon$. The width of the passband is indicated by the cutoff frequency ω_C. On the other hand, the stopband, starting at the frequency ω_S, does not allow a gain greater than A_{min}. Different types of filter designs achieve different types of frequency responses, which are typically characterized by having a particularly flat passband frequency response **(Butterworth filters)** or by a very rapid transition between passband and stopband (**Chebyshev filters,** and **Cauer,** or **elliptical, filters**), or by some other characteristic, such as a linear phase response **(Bessel filters).** Achieving each of these properties usually involves tradeoffs; for example, a very flat passband response will usually result in a relatively slow transition from passband to stopband.

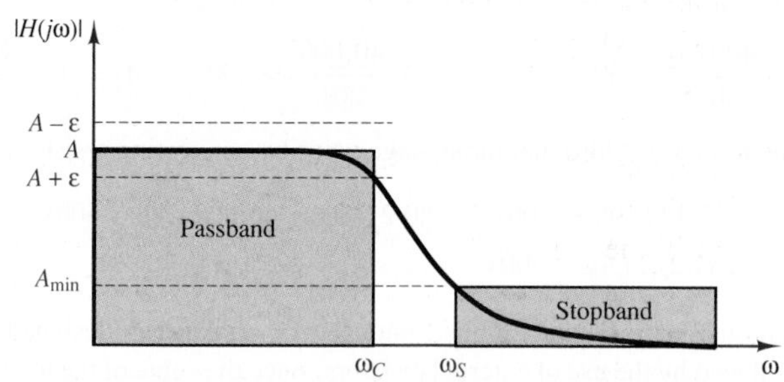

Figure 15.19 Prototype low-pass filter response

In addition to selecting a filter from a certain family, it is possible to select the *order* of the filter; this is equal to the order of the differential equation that describes the input-output relationship of a given filter. In general, the higher the order, the faster the transition from passband to stopband (at the cost of greater phase shifts and amplitude distortion, however). Although the frequency response of Figure 15.19 pertains to a low-pass filter, similar definitions also apply to the other types of filters.

Butterworth filters are characterized by a *maximally flat* passband frequency response characteristic; their response is defined by a magnitude-squared function of frequency

$$|H(j\omega)|^2 = \frac{H_0^2}{1 + \varepsilon^2 \omega^{2n}} \tag{15.17}$$

where $\varepsilon = 1$ for maximally flat response and n is the order of the filter. Figure 15.20 depicts the frequency response (normalized to $\omega_C = 1$) of first-, second-, third-, and fourth-order Butterworth low-pass filters. The **Butterworth polynomials,** given in Table 15.3 in factored form, permit the design of the filter by specifying the denominator as a polynomial in s. For $s = j\omega$, one obtains the frequency response of the filter. Examples 15.4 and 15.5 illustrate filter design procedures that make use of these tables.

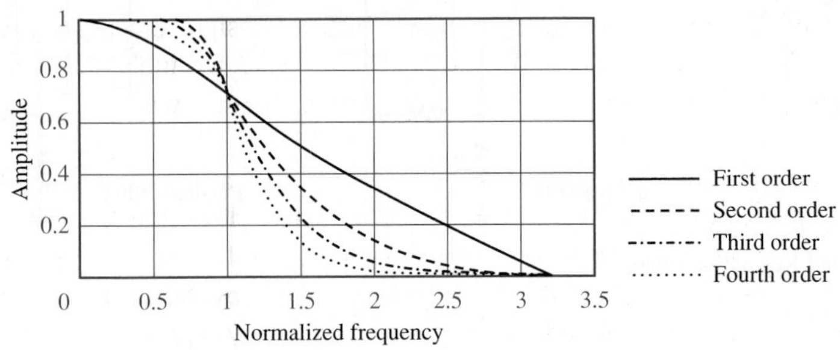

Figure 15.20 Butterworth low-pass filter frequency response

Table 15.3 Butterworth polynomials in quadratic form

Order n	Quadratic factors
1	$s+1$
2	$s^2+\sqrt{2}s+1$
3	$(s+1)(s^2+s+1)$
4	$(s^2+0.7654s+1)(s^2+1.8478s+1)$
5	$(s+1)(s^2+0.6180s+1)(s^2+1.6180s+1)$

Figure 15.21 depicts the normalized frequency response of first- to fourth-order low-pass Chebyshev filters ($n = 1$ to 4), for $\varepsilon = 1.06$. Note that a certain amount of ripple is allowed in the passband; the amplitude of the ripple is defined by the parameter ε and is constant throughout the passband. Thus, these filters are also called **equiripple filters.** Cauer, or elliptical, filters are similar to Chebyshev filters, except for being characterized by equiripple both in the passband and in the stopband. Design tables exist to select the appropriate order of Butterworth, Chebyshev, or Cauer filter for a specific application.

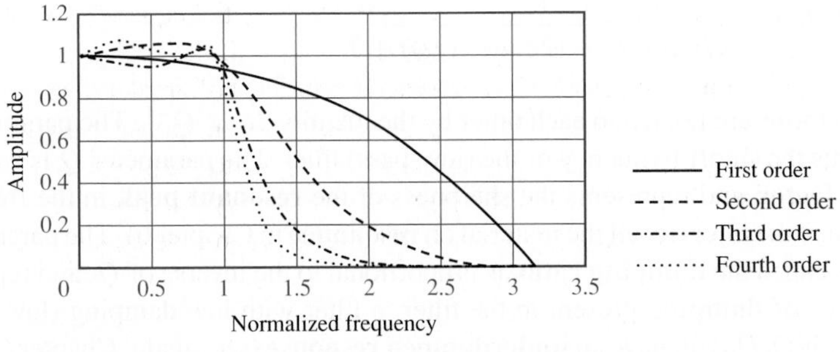

Figure 15.21 Chebyshev low-pass filter frequency response

Three common configurations of second-order active filters, which can be used to implement **second-order** (or **quadratic**) **filter sections** using a single op-amp, are shown in Figure 15.22. These filters are called **constant-*K*,** or **Sallen and Key, filters** (after the names of the inventors). The analysis of these active filters, although

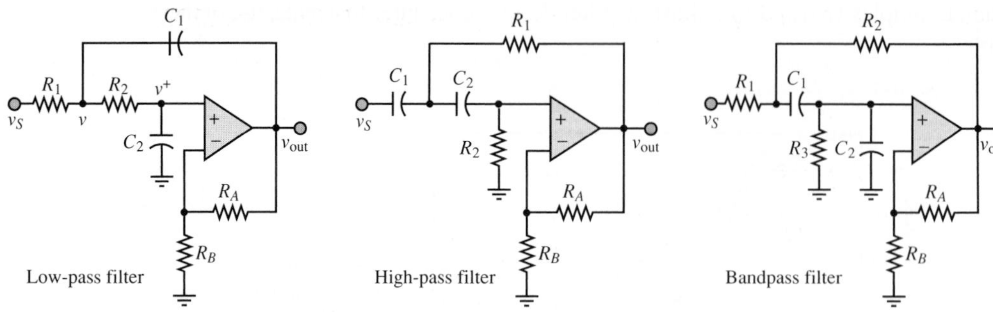

Figure 15.22 Sallen and Key active filters

somewhat more involved than that of the active filters presented in Chapter 14, is based on the basic properties of the ideal operational amplifier discussed earlier. Consider, for example, the low-pass filter of Figure 15.22. The first unusual aspect of the filter is the presence of both negative and **positive feedback;** that is, feedback connections are provided to both the inverting and the noninverting terminals of the op-amp. The analysis method consists of finding expressions for the input terminal voltages of the op-amp v^+ and v^- and using these expressions to derive the input-output relationship for the filter. This analysis is left as a homework problem. The frequency response of the low-pass filter is given by

$$H(j\omega) = \frac{K(1/R_1R_2C_1C_2)}{(j\omega)^2 + \left[\frac{1}{R_1C_1} + \frac{1}{R_2C_1} + \frac{1}{R_2C_2}(K-1)\right]j\omega + \frac{1}{R_1R_2C_1C_2}} \quad \textbf{(15.18)}$$

The above frequency response can be expressed in one of two more general forms:

$$H(j\omega) = \frac{K\omega_C^2}{(j\omega)^2 + (\omega_C/Q)(j\omega) + \omega_C^2}$$

and **(15.19)**

$$H(j\omega) = \frac{K}{(j\omega)^2/\omega_C^2 + (2\zeta/\omega_C)(j\omega) + 1}$$

The two forms are related to each other by the identity $2\zeta = Q^{-1}$. The parameter ω_C represents the cutoff frequency of the (low-pass) filter. The parameter Q is called the **quality factor,** and represents the sharpness of the **resonant peak** in the frequency response of the filter (recall the material on **resonance** in Chapter 6). The parameter ζ, which is called the **damping ratio,** is proportional to the inverse of Q, and represents the degree of **damping** present in the filter: a filter with low damping (low ζ), and therefore high Q, will have an **underdamped response** (see, again, Chapter 6), while a low-Q filter will be highly damped.

The relationships between the three parameters of the second-order filter (ω_C, ζ, and K) and the resistors and capacitors are defined below for the low-pass Sallen and Key filter. A very desirable property of the Sallen and Key, or **constant-*K*, filter** is the fact that the low-frequency gain of the filter is independent of the cutoff frequency and is determined simply by the ratio of resistors R_A and R_B. The other four components define the cutoff frequency and damping ratio (or Q), as shown in equations 15.20.

$$
\begin{aligned}
K &= 1 + \frac{R_A}{R_B} \\
\omega_C &= \frac{1}{\sqrt{R_1 R_2 C_1 C_2}} \\
\frac{1}{Q} &= 2\zeta = \sqrt{\frac{R_2 C_2}{R_1 C_1}} + \sqrt{\frac{R_1 C_2}{R_2 C_1}} + (K-1)\sqrt{\frac{R_1 C_1}{R_2 C_2}}
\end{aligned}
\tag{15.20}
$$

The quadratic filter sections of Figure 15.22 can be used to implement filters of arbitrary order and of different characteristics. For example, a fourth-order Butterworth filter can be realized by connecting two second-order Sallen and Key quadratic sections in cascade, and by observing that the component values of each section can be specified given the desired gain, cutoff frequency, and damping ratio (or quality factor). Examples 15.3, 15.4, and 15.5 illustrate these procedures.

Data sheets for integrated-circuit filters may be found on the book website.

FIND IT ON THE WEB

EXAMPLE 15.3 Determining the Order of a Butterworth Filter

LO3

Problem

Determine the required order of a filter, given the filter specifications.

Solution

Known Quantities: Filter gain at cutoff frequencies (passband and stopband).

Find: Order n of filter.

Schematics, Diagrams, Circuits, and Given Data: Passband gain: −3 dB at $\omega_C = 1$ rad/s; stopband gain: −40 dB at $\omega_C = 4\omega_C$.

Assumptions: Use a Butterworth filter response. Assume that the low-frequency gain $H_0 = 1$.

Analysis: Using the magnitude-squared response for the Butterworth filter (equation 15.17),

$$|H(j\omega)|^2 = \frac{H_0^2}{1 + \varepsilon^2 \omega^{2n}}$$

With $\varepsilon = 1$, we obtain the following expression at the passband cutoff frequency ω_C:

$$|H(j\omega = j\omega_C)| = \frac{H_0}{\sqrt{1 + \omega_C^{2n}}} = \frac{H_0}{\sqrt{1 + 1^{2n}}} = \frac{H_0}{\sqrt{2}}$$

This is already the desired value for the passband gain (3 dB below the low-frequency gain), since

$$20 \log_{10} \frac{1}{\sqrt{2}} = -3 \text{ dB}$$

Note that the first requirement is automatically satisfied because of the nature of the Butterworth filter response.

The requirement for the stopband gain imposes that the gain at frequencies at or above ω_S be less than -40 dB:

$$20\log_{10}|H(j\omega = j\omega_S)| = 20\log_{10}\frac{H_0}{\sqrt{1+\omega_S^{2n}}} = 20\log_{10}\frac{H_0}{\sqrt{1+4^{2n}}} \le -40$$

Thus,

$$20\log_{10} H_0 - 20\log_{10}\sqrt{1+4^{2n}} \le -40$$

or

$$\log_{10}(1+4^{2n}) \ge 4$$
$$1+4^{2n} \ge 10^4$$
$$2n\log_{10} 4 \ge \log(10^4 - 1)$$

Solving the above inequality, we obtain $n \ge 3.32$. Since n must be an integer, we choose $n = 4$. Note that for $n = 4$, the actual gain at the stopband frequency can be calculated to be

$$|H(j\omega = j\omega_S)| = \frac{H_0}{\sqrt{1+\omega_C^{2n}}} = \frac{1}{\sqrt{1+4^{2\times 4}}} = -48.16 \text{ dB}$$

which is lower than the minimum desired gain of -40 dB, thus satisfying the specification.

Comments: Note that the -3-dB gain at the passband cutoff frequency is always satisfied in a Butterworth filter, since $\varepsilon = 1$.

CHECK YOUR UNDERSTANDING

Determine the order of the filter required to satisfy the requirements of Example 15.3 if the stopband frequency is moved to $\omega_S = 2\omega_C$. What is the actual attenuation of the filter at the stopband frequency ω_S?

Answer: $n = 7$; 42.1 dB

LO3

EXAMPLE 15.4 Design of Sallen and Key Filter

Problem

Determine the cutoff frequency, DC gain, and quality factor for the Sallen and Key filter of Figure 15.22.

Solution

Known Quantities: Filter resistor and capacitor values.

Find: K; ω_C; Q.

Schematics, Diagrams, Circuits, and Given Data: All resistors are 500 Ω, all capacitors are 2 μF.

Assumptions: None.

Analysis: Using the definitions given in equation 15.20, we compute

$$K = 1 + \frac{R_A}{R_B} = 1 + \frac{500}{500} = 2$$

$$\omega_C = \frac{1}{\sqrt{R_1 R_2 C_1 C_2}} = \frac{1}{\sqrt{(500)^2 (2 \times 10^{-6})^2}} = 1{,}000 \text{ rad/s}$$

$$\frac{1}{Q} = 2\zeta = \sqrt{\frac{R_2 C_2}{R_1 C_1}} + \sqrt{\frac{R_1 C_2}{R_2 C_1}} + (1 - K)\sqrt{\frac{R_1 C_1}{R_2 C_2}} = 1$$

Comments: What type of response does the filter analyzed above have? We can compare the filter response to that of a quadratic Butterworth filter (or other filter family) by determining the Q of the filter. Once the gain and cutoff frequency have been defined, Q is the parameter that distinguishes, say, a Butterworth from a Chebyshev filter. The Butterworth polynomial of order 2 is given in Table 15.3 as $s^2 + \sqrt{2}s + 1$. If we compare this expression to the denominator of equation 15.19, we obtain

$$H(s) = \frac{K\omega_C^2}{s^2 + (\omega_C/Q)s + \omega_C^2} = \frac{1}{s^2 + \sqrt{2}s + 1}$$

Since the expressions for the quadratic polynomials of Table 15.3 are normalized to unity gain and cutoff frequency, we know that $K = 1$ and $\omega_C = 1$, and therefore we can solve for the value of Q in a Butterworth filter by setting

$$\frac{1}{Q} = \sqrt{2} \quad \text{or} \quad Q = \frac{1}{\sqrt{2}} = 0.707$$

Thus, every second-order Butterworth filter is characterized by a Q of 0.707; this corresponds to a damping ratio $\zeta = 0.5Q^{-1} = 0.707$, that is, to a lightly underdamped response. Example 15.5 considers the characteristics of a fourth-order Butterworth filter.

CHECK YOUR UNDERSTANDING

Design a quadratic filter section with $Q = 1$ and a cutoff frequency of 10 rad/s. Note that there can be many solutions, depending on your design.

Answer: $R_1 = R_2 = 1\text{ k}\Omega$; $C_1 = C_2 = 100\ \mu\text{F}$; $K = 2$

EXAMPLE 15.5 Design of Fourth-Order Butterworth Filter

LO3

Problem

Design a fourth-order low-pass Butterworth filter using two quadratic Sallen and Key sections.

Solution

Known Quantities: Filter response; desired gain and cutoff frequency.

Find: Component values $R_1, R_2, C_1, C_2, R_A, R_B$ for each filter section.

Schematics, Diagrams, Circuits, and Given Data: Gain = 100; cutoff frequency = 400 rad/s.

Assumptions: Use low-pass Sallen and Key filter prototype. In the sinusoidal steady state, $s \to j\omega$.

Analysis: Table 15.3 suggests that a Butterworth fourth-order filter is composed of the product of two quadratic responses. Our first objective is to determine the Q of each of these two quadratic responses, so that we can design each of the two Sallen and Key quadratic sections. Comparing the standard quadratic low-pass filter response to the first of the two Butterworth polynomials for a normalized filter with $K = 1$ and $\omega_C = 1$, we have

$$H(s) = \frac{K\omega_C^2}{s^2 + (\omega_C/Q)s + \omega_C^2} = \frac{1}{s^2 + 0.7654s + 1}$$

and for a normalized filter with $K = 1$ and $\omega_C = 1$, we can solve for the value of Q_1, the Q of the first section:

$$\frac{1}{Q_1} = 0.7654 \qquad \text{or} \qquad Q_1 = \frac{1}{0.7654} = 1.3065$$

Repeating the procedure for the second section, we obtain

$$\frac{1}{Q_2} = 1.8478 \qquad \text{or} \qquad Q_2 = \frac{1}{1.8478} = 0.5412$$

Having determined these values, we can now proceed to design two separate quadratic sections with the values of Q computed above, and each with gain $K = 10$ (so that the product of the two sections yields a low-frequency gain of 100, as specified), and cutoff frequency $\omega_C = 400$ rad/s. The responses for the two sections are

$$H(s) = \frac{K\omega_C^2}{s^2 + (\omega_C/Q_1)s + \omega_C^2} = \frac{1.6 \times 10^6}{s^2 + 306.16s + 1.6 \times 10^5}$$
$$= \frac{10}{6.25 \times 10^{-6}s^2 + 1.914 \times 10^{-3}s + 1}$$

and

$$H(s) = \frac{K\omega_C^2}{s^2 + (\omega_C/Q_2)s + \omega_C^2} = \frac{1.6 \times 10^6}{s^2 + 739.12s + 1.6 \times 10^5}$$
$$= \frac{10}{6.25 \times 10^{-6}s^2 + 4.62 \times 10^{-3}s + 1}$$

One of the important features of the Sallen and Key filter prototype is that we can choose the values for the resistors that set the circuit gain independently of the values of the resistors that set the cutoff frequency (the converse is not true). Thus, we can separately select $K = 10$ for both stages by requiring that $1 + R_A/R_B = 10$, for example, $R_A = 100\ \text{k}\Omega$, $R_B = 11.1\ \text{k}\Omega$.

Next, we compute the component values, using equations 15.20. Since we have only two equations and four unknowns, two values will have to be selected arbitrarily. We can write

$$\omega_C = \sqrt{\frac{1}{R_1 R_2 C_1 C_2}} \quad \Rightarrow \quad \omega_C \sqrt{C_1 C_2} = \frac{1}{\sqrt{R_1 R_2}} \quad \Rightarrow \quad \sqrt{R_1 R_2} = \frac{1}{\omega_C \sqrt{C_1 C_2}}$$

and

$$\frac{1}{Q} = \sqrt{\frac{R_2}{R_1}} \cdot \sqrt{\frac{C_2}{C_1}} + \sqrt{\frac{R_1}{R_2}} \cdot \sqrt{\frac{C_2}{C_1}} + (1-K)\sqrt{\frac{R_1}{R_2}} \cdot \sqrt{\frac{C_1}{C_2}}$$

Rearrange the last equation as

$$\frac{1}{Q} = \sqrt{\frac{R_2}{R_1}} \cdot \sqrt{\frac{C_2}{C_1}} + \sqrt{\frac{R_1}{R_2}} \cdot \left[\sqrt{\frac{C_2}{C_1}} + (1-K)\sqrt{\frac{C_1}{C_2}}\right]$$

and operate a change of variables to obtain

$$x = \sqrt{\frac{R_2}{R_1}} \qquad c = \left[\sqrt{\frac{C_2}{C_1}} + (1-K)\sqrt{\frac{C_1}{C_2}}\right] \qquad a = \sqrt{\frac{C_2}{C_1}} \qquad b = \frac{1}{Q}$$

$$b = ax + \frac{c}{x}$$

or

$$ax^2 - bx + c = 0$$

There is always a positive root in the preceding equation, as a and b are both positive. One can easily show that there is another positive root if

$$\sqrt{\frac{1}{4Q^2} + K - 1} < \sqrt{\frac{C_2}{C_1}} < \sqrt{K-1}$$

and in that case there are two solutions for R_1, R_2; also, there are real solutions only if $C_2 < (K-1)C_1$.

Solving the preceding equations gives

$$x = \sqrt{\frac{R_2}{R_1}} = \frac{b + \sqrt{b^2 - 4ac}}{2a}$$

or

$$x = \sqrt{\frac{R_2}{R_1}} = \frac{\frac{1}{Q} + \sqrt{\frac{1}{Q^2} - 4\frac{C_2}{C_1} + 4(K-1)}}{2\sqrt{\frac{C_2}{C_1}}}$$

Now, we have the new system

$$\sqrt{\frac{R_2}{R_1}} = \frac{\frac{1}{Q} + \sqrt{\frac{1}{Q^2} - 4\frac{C_2}{C_1} + 4(K-1)}}{2\sqrt{\frac{C_2}{C_1}}}$$

$$\sqrt{R_1 R_2} = \frac{1}{w_C \sqrt{C_1 C_2}}$$

by substitution, as follows:

$$\sqrt{R_2} = \frac{\dfrac{1}{Q} + \sqrt{\dfrac{1}{Q^2} - 4\dfrac{C_2}{C_1} + 4(K-1)}}{2\sqrt{\dfrac{C_2}{C_1}}}\sqrt{R_1}$$

$$R_1 \cdot \frac{\dfrac{1}{Q} + \sqrt{\dfrac{1}{Q^2} - 4\dfrac{C_2}{C_1} + 4(K-1)}}{2\sqrt{\dfrac{C_2}{C_1}}} = \frac{1}{\omega_C\sqrt{C_1C_2}}$$

That is,

$$R_1 = \frac{2\sqrt{\dfrac{C_2}{C_1}}}{\omega_C\sqrt{C_1C_2}\left[\dfrac{1}{Q} + \sqrt{\dfrac{1}{Q^2} - 4\dfrac{C_2}{C_1} + 4(K-1)}\right]}$$

$$R_2 = \frac{\dfrac{1}{Q} + \sqrt{\dfrac{1}{Q^2} - 4\dfrac{C_2}{C_1} + 4(K-1)}}{2\sqrt{\dfrac{C_2}{C_1}} \cdot \omega_C\sqrt{C_1C_2}}$$

If we assume 0.1-μF values for both C_1 and C_2 in each section, we can compute the value of the resistances required to complete the design:

First section:

$R_1 = 7{,}723\ \Omega$ (nearest standard 5% resistor value: 8.2 kΩ)
$R_2 = 80{,}923\ \Omega$ (nearest standard 5% resistor value: 82 kΩ)

Second section:

$R_1 = 6{,}411\ \Omega$ (nearest standard 5% resistor value: 6.8 kΩ)
$R_2 = 97{,}484\ \Omega$ (nearest standard 5% resistor value: 100 kΩ)

The designer may choose to employ high-precision resistors or adjustable resistors, if desired.

Comments: We have chosen to fix the values of the capacitors and compute the required values of the resistors because of the greater availability of resistor sizes.

15.4 ANALOG-TO-DIGITAL AND DIGITAL-TO-ANALOG CONVERSION

To take advantage of the capabilities of a microprocessor, it is necessary to suitably interface signals to and from external devices with the microprocessor. Depending on the nature of the signal, either an analog or a digital interface circuit will be required. The advantages in memory storage, programming flexibility, and computational power afforded by today's digital computers are such that the instrumentation designer often chooses to convert an analog signal to an equivalent digital representation, to exploit the capabilities of a microprocessor in processing the signal. In many cases, the data converted from analog to digital form remain in digital form for ease of storage or for further processing. In some instances it is necessary to convert the data back to analog form. The latter condition arises frequently in the context of control system design, where an analog measurement is converted to digital form and processed by a digital computer to generate a control action (e.g., raising or lowering the temperature of a process, or exerting a force or a torque); in such cases, the output of the digital computer is converted back to analog form, so that a continuous signal becomes available to the actuators. Figure 15.23 illustrates the general appearance of a digital measuring instrument and of a digital controller acting on a plant or process.

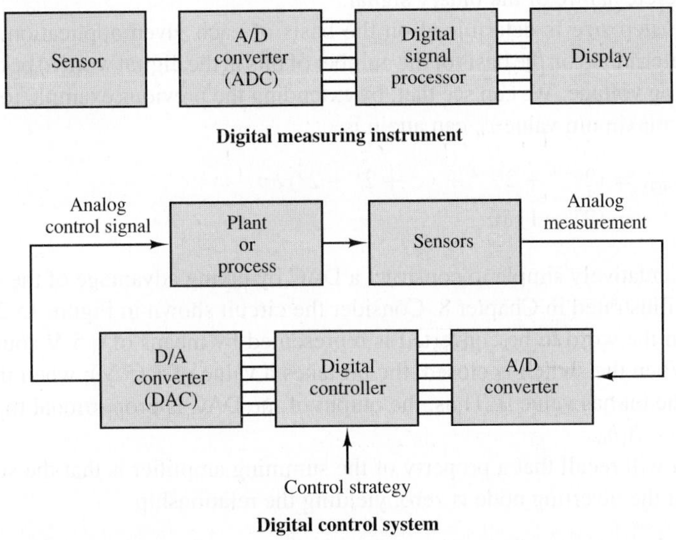

Figure 15.23 Block diagrams of a digital measuring instrument and a digital control system

The objective of this section is to describe how the digital-to-analog converter (DAC) and analog-to-digital converter (ADC) blocks of Figure 15.23 function. After illustrating discrete circuits that can implement simple analog-to-digital and digital-to-analog converters, we shall emphasize the use of ICs specially made for these tasks. Nowadays, it is uncommon (and impractical) to design such circuits using discrete components: the performance and ease of use of IC packages make them the preferred choice in virtually all applications.

Digital-to-Analog Converters

We discuss digital-to-analog conversion first because it is a necessary part of analog-to-digital conversion in some conversion schemes. A **digital-to-analog converter (DAC)** will convert a binary word to an analog output voltage (or current). The binary word is represented in terms of 1s and 0s, where typically (but not necessarily) 1s correspond to a 5-V level and 0s to a 0-V signal. As an example, consider a 4-bit binary word representing a positive (or unsigned) integer number

$$B = (b_3 b_2 b_1 b_0)_2 = (b_3 \cdot 2^3 + b_2 \cdot 2^2 + b_1 \cdot 2^1 + b_0 \cdot 2^0)_{10} \tag{15.21}$$

The analog voltage corresponding to the digital word B would be

$$v_a = (8b_3 + 4b_2 + 2b_1 + b_0)\,\delta v \tag{15.22}$$

where δv is the smallest *step size* by which v_a can increment. This least step size will occur whenever the least significant bit (LSB) b_0 changes from 0 to 1 and is the smallest increment the digital number can make. We shall also shortly see that the analog voltage obtained by the D/A conversion process has a "staircase" appearance because of the discrete nature of the binary signal.

The step size is determined on the basis of each given application, and it is usually determined on the basis of the number of bits in the digital word to be converted to an analog voltage. We can see that, by extending the previous example for an n-bit word, the maximum value v_a can attain is

$$\begin{aligned} v_{a\,\max} &= (2^{n-1} + 2^{n-2} + \cdots + 2^1 + 2^0)\,\delta v \\ &= (2^n - 1)\,\delta v \end{aligned} \tag{15.23}$$

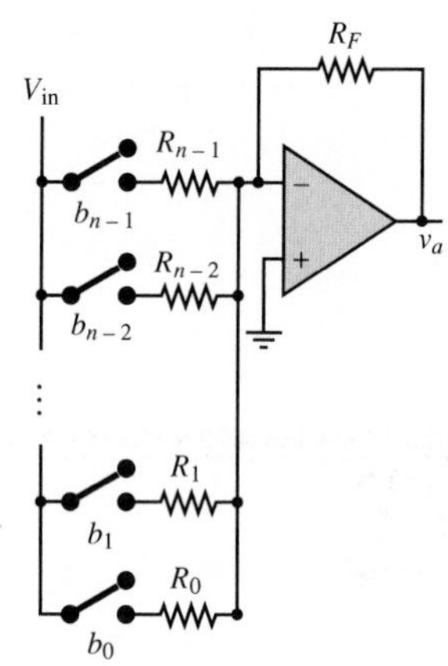

Figure 15.24 An n-bit digital-to-analog converter

It is relatively simple to construct a DAC by taking advantage of the summing amplifier illustrated in Chapter 8. Consider the circuit shown in Figure 15.24, where each bit in the word to be converted is represented by means of a 5-V source and a switch. When the switch is closed, the bit takes a value of 1 (5 V); when the switch is open, the bit has value 0. Thus, the output of the DAC is proportional to the word $b_{n-1}b_{n-2}\cdots b_1 b_0$.

You will recall that a property of the summing amplifier is that the sum of the currents at the inverting node is zero, yielding the relationship

$$v_a = -\left(\frac{R_F}{R_i} \cdot b_i \cdot V_{in}\right) \qquad i = 0, 1, \ldots, n-1 \tag{15.24}$$

where R_i is the resistor associated with each bit and b_i is the decimal value of the ith bit (that is, $b_0 = 2^0$, $b_1 = 2^1$, and so on). It is easy to verify that if we select

$$R_i = \frac{R_0}{2^i} \tag{15.25}$$

we can obtain weighted gains for each bit so that

$$v_a = -\frac{R_F}{R_0}(2^{n-1}b_{n-1} + \cdots + 2^1 b_1 + 2^0 b_0)V_{in} \tag{15.26}$$

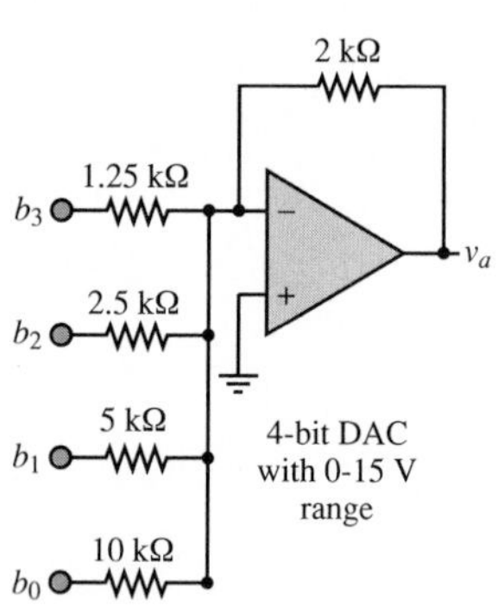

Figure 15.25 A 4-bit DAC

and so that the analog output voltage is proportional to the decimal representation of the binary word. As an illustration, consider the case of a 4-bit word; a reasonable choice for R_0 might be $R_0 = 10$ kΩ, yielding a resistor network consisting of 10-, 5-, 2.5-, and 1.25-kΩ resistors, as shown in Figure 15.25. The largest decimal value of a 4-bit word is $2^4 - 1 = 15$, and so it is reasonable to divide this range into steps of 1 V (that is, $\delta v = 1$ V). Thus, the full-scale value of v_a is 15 V

$$0 \le v_a \le 15 \text{ V}$$

and we select R_F according to the following expression:

$$R_F = \frac{\delta v\, R_0}{V_{\text{in}}} = \frac{1 \cdot 10^4}{5} = 2\ \text{k}\Omega$$

The corresponding 4-bit DAC is shown in Figure 15.25.

The DAC transfer characteristic is such that the analog output voltage v_a has a steplike appearance, because of the discrete nature of the binary signal. The coarseness of the "staircase" can be adjusted by selecting the number of bits in the binary representation.

The practical design of a DAC is generally not carried out in terms of discrete components, because of problems such as the accuracy required of the resistor value. Many of the problems associated with this approach can be solved by designing the complete DAC circuit in integrated-circuit form. The specifications stated by the IC manufacturer include the **resolution,** that is, the minimum nonzero voltage; the **full-scale accuracy;** the **output range;** the **output settling time;** the **power supply requirements;** and the **power dissipation.** The following examples illustrate the use of integrated-circuit DACs.

CHECK YOUR UNDERSTANDING

If the maximum analog voltage $V_{a,\max}$ of a 12-bit digital-to-analog converter (DAC) is 15 V, find the smallest step size δv by which v_a can increment.

Answer: 3.66 mV

EXAMPLE 15.6 DAC Resolution

LO4

Problem

Determine the smallest step size, or *resolution,* of an 8-bit DAC.

Solution

Known Quantities: Maximum analog voltage.

Find: Resolution δv.

Schematics, Diagrams, Circuits, and Given Data: $v_{a,\max} = 12$ V.

Analysis: Using equation 15.23, we compute

$$\delta v = \frac{v_{a,\max} - v_{a,\min}}{2^8 - 1} = \frac{12 - 0}{2^8 - 1} = 47.1\ \text{mV}$$

Comments: Note that the resolution is dependent not only on the number of bits, but also on the analog voltage range (12 V in this case).

CHECK YOUR UNDERSTANDING

Repeat Example 15.6 for the case of an 8-bit word with $R_0 = 10\ \text{k}\Omega$ and the same range of v_a. Find the values of δv and R_F. Assume that ideal resistor values are available.

For Figure 15.25, find V_{max} if $V_{\text{in}} = 4.5$ V.

For Figure 15.25, find the resolution if $V_{\text{in}} = 3.8$ V.

Answers: $\delta v = 47.1$ mV; $R_F = 94.2\ \Omega$; -13.5 V; 0.76 V

LO4 EXAMPLE 15.7 Determining the Required Number of Bits in a DAC

Problem

Find an expression for the required number of bits in a DAC, using the definitions of *range* and *resolution*.

Solution

Known Quantities: Range and resolution of DAC. Voltage level corresponding to logic 1.

Find: Number of DAC bits required.

Schematics, Diagrams, Circuits, and Given Data:

Range: the analog voltage range of the DAC $= v_{a,\max} - v_{a,\min}$

Resolution: the minimum step size δv

V_{in} = voltage level corresponding to logic 1

0 V = voltage level corresponding to logic 0

Analysis: The maximum analog voltage output of the DAC is obtained when all bits are set to 1. Using equation 15.26, we can determine $v_{a,\max}$:

$$v_{a,\max} = V_{\text{in}} \frac{R_F}{R_0}(2^n - 1)$$

The minimum analog voltage output is realized when all bits are set to logic 0. In this case, since the voltage level associated with a logic 0 is 0 V, $v_{a,\min} = 0$. Thus, the range of this DAC is $v_{a,\max} - v_{a,\min} = v_{a,\max}$.

The resolution was defined in Example 15.6 as

$$\delta v = \frac{v_{a,\max} - v_{a,\min}}{2^n - 1}$$

Knowing both range and resolution, we can solve for the number of bits n as follows:

$$n = \frac{\log\left[(v_{a,\max} - v_{a,\min})/\delta v + 1\right]}{\log 2} = \frac{\log\,(\text{range}/\text{resolution} + 1)}{\log 2}$$

Since n must be an integer, the result of the above expression will be rounded up to the nearest integer. For example, if we require a 10-V range DAC with a resolution of 10 mV, we can compute the required number of bits to be

$$n = \frac{\log\left(10/10^{-2}+1\right)}{\log 2} = 9.97 \rightarrow 10 \text{ bits}$$

Comments: The result of this example is of direct use in the design of practical DAC circuits.

CHECK YOUR UNDERSTANDING

Find the minimum number of bits required in a DAC if the range of the DAC is from 0.5 to 15 V and the resolution of the DAC is 20 mV.

Answer: 10 bits

EXAMPLE 15.8 Using DAC Device Data Sheets

LO4

Problem

Using the data sheets for the AD7524 (see course website), answer the following questions:

1. What is the best (smallest) resolution attainable for a range of 10 V?
2. What is the maximum allowable conversion frequency of this DAC?

Solution

Known Quantities: Desired range of DAC.

Find: Resolution and maximum conversion frequency.

Schematics, Diagrams, Circuits, and Given Data: Range = 10 V. DAC specifications found in device data sheet.

Assumptions: The DAC is operated at full-scale range.

Analysis:

1. From the data sheet we determine that the AD7524 is an 8-bit converter. Thus, the best resolution that can be obtained is

$$\delta v = \frac{v_{a,\max} - v_{a,\min}}{2^n - 1} = \frac{10}{2^8 - 1} = 39.2 \text{ mV}$$

2. The maximum frequency of the DAC depends on the *settling time*. This is defined as the time required for the output to settle to within one-half of the least significant bit of its final value. Only one conversion can be performed during the settling time. The settling time is dependent on the voltage range, and for the 10-V range indicated in this problem it is equal to $T_S = 1\ \mu s$. The corresponding maximum *sampling frequency* is $F_S = 1/T_S = 1$ MHz.

Comments: The significance of the *sampling frequency* is discussed in the next subsection in connection with the *Nyquist sampling criterion*.

Analog-to-Digital Converters

The device that makes conversion of analog signals to digital form is the **analog-to-digital converter (ADC),** and, just like the DAC, it is also available as a single IC package. This section will illustrate the essential features of four types of ADCs: the tracking ADC, which utilizes a DAC to perform the conversion; the integrating ADC; the flash ADC; and the successive-approximation ADC. In addition to discussing analog-to-digital conversion, we shall introduce the *sample-and-hold amplifier.*

LO4

Quantization

The process of converting an analog voltage (or current) to digital form requires that the analog signal be quantized and encoded in binary form. The process of **quantization** consists of subdividing the range of the signal into a finite number of intervals; usually, one employs $2^n - 1$ intervals, where n is the number of bits available for the corresponding binary word. Following this quantization, a binary word is assigned to each interval (i.e., to each range of voltages or currents); the binary word is then the digital representation of any voltage (current) that falls within that interval. You will note that the smaller the interval, the more accurate the digital representation is. However, some error is necessarily always present in the conversion process; this error is usually referred to as **quantization error.** Let v_a represent the analog voltage and v_d its quantized counterpart, as shown in Figure 15.26 for an analog voltage in the range of 0 to 16 V. In the figure, the analog voltage v_a takes on a value of $v_d = 0$ whenever it is in the range of 0 to 1 V; for $1 \le v_a < 2$, the corresponding value is $v_d = 1$; for $2 \le v_a < 3$, $v_d = 2$; and so on, until for $15 \le v_a < 16$, we have $v_d = 15$. You see that if we now represent the quantized voltage v_d by its binary counterpart, as shown in the table of Figure 15.26, each 1-V analog interval corresponds to a unique binary word. In this example, a 4-bit word is sufficient to represent the analog voltage, although the representation is not very accurate. As the number of bits increases, the quantized voltage is closer and closer to the original analog signal; however, the number of bits required to represent the quantized value increases. The issue of *ADC resolution* is discussed later in this section.

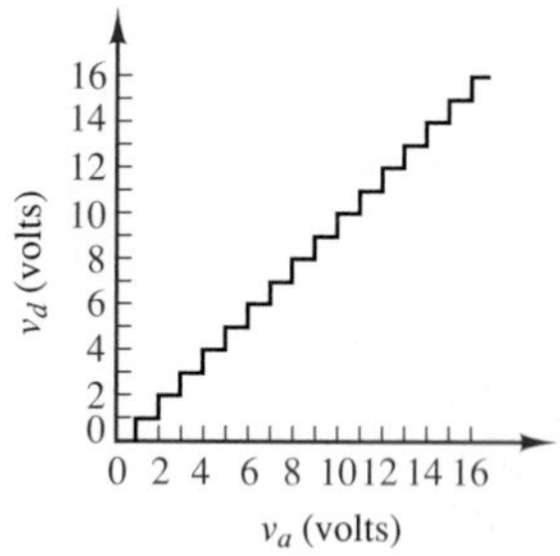

Quantized voltage | Binary representation

v_d	b_3	b_2	b_1	b_0
0	0	0	0	0
1	0	0	0	1
2	0	0	1	0
3	0	0	1	1
4	0	1	0	0
⋮		⋮		
14	1	1	1	0
15	1	1	1	1

Figure 15.26 A digital voltage representation of an analog voltage

Tracking ADC

Although not the most efficient in all applications, the **tracking ADC** is an easy starting point to illustrate the operation of an ADC, in that it is based on the DAC presented in the previous section. The tracking ADC, shown in Figure 15.27, compares the analog input signal with the output of a DAC; the comparator output determines whether the DAC output is larger or smaller than the analog input to be converted to binary form. If the DAC output is smaller, then the comparator output will cause an up-down counter (see Chapter 14) to count up until it reaches a level close to the analog signal; if the DAC output is larger than the analog signal, then the counter is forced to count down. Note that the rate at which the up-down counter is incremented is determined by the external clock, and that the binary counter output corresponds to the binary representation of the analog signal. A feature of the tracking ADC is that it follows ("tracks") the analog signal by changing 1 bit at a time.

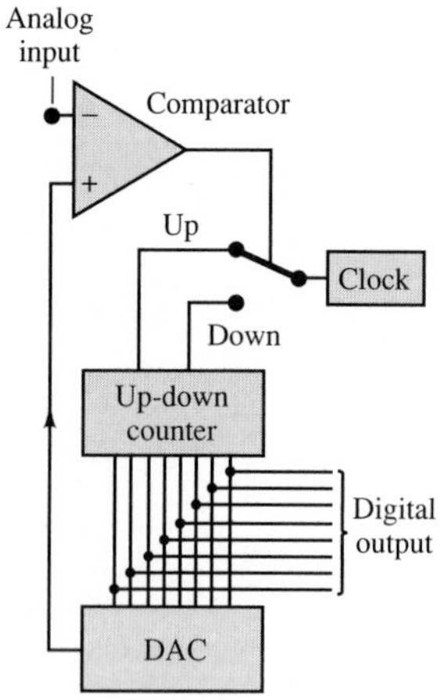

Figure 15.27 Tracking ADC

Integrating ADC

The **integrating ADC** operates by charging and discharging a capacitor, according to the following principle: If one can ensure that the capacitor charges (discharges)

linearly, then the time it will take for the capacitor to discharge is linearly related to the amplitude of the voltage that has charged the capacitor. In practice, to limit the time it takes to perform a conversion, the capacitor is not required to charge fully. Rather, a clock is used to allow the input (analog) voltage to charge the capacitor for a short time, determined by a fixed number of clock pulses. Then the capacitor is allowed to discharge through a known circuit, and the corresponding clock count is incremented until the capacitor is fully discharged. The latter condition is verified by a comparator, as shown in Figure 15.28. The clock count accumulated during the discharge time is proportional to the analog voltage.

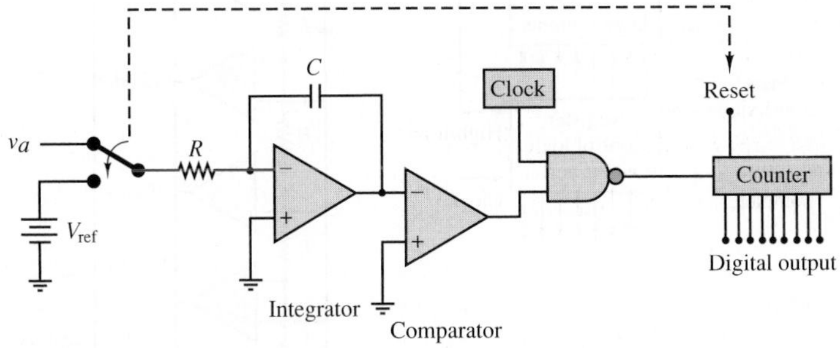

Figure 15.28 Integrating ADC

LO4

In Figure 15.28, the switch causes the counter to reset when it is connected to the reference voltage V_{ref}. The reference voltage is used to provide a known, linear discharge characteristic through the capacitor (see the material on the op-amp integrator in Chapter 8). When the comparator detects that the output of the integrator is equal to zero, it switches state and disables the NAND gate, thus stopping the count. The binary counter output is now the digital counterpart of the voltage v_a.

Other common types of ADC are the **successive-approximation ADC** and the **flash ADC.**

Successive-Approximation ADC

Successive-approximation ADCs are the most commonly used. A block diagram of the successive-approximation ADC is shown in Figure 15.29(a). This type of ADC uses a single comparator, and its performance depends strongly on the accuracy of the DAC used in the circuit. The analog output of a high-speed DAC is compared against the analog input signal. The digital result of the comparison, that is, the output of the comparator [C in Figure 15.29(a)] is used to control the contents of a digital buffer that both drives the DAC and provides the digital output word. The digital word corresponding to the output of the ADC is obtained by using n bit-by-bit comparisons, where n is the length of the binary word.

Flash ADC

The **flash ADC** is fully parallel and is used for high-speed conversion. A resistive divider network of 2^n resistors divides the known voltage range into that many equal increments. A network of $2^n - 1$ comparators then compares the unknown voltage with that array of test voltages. All comparators with inputs exceeding the unknown are *on*; all others are *off*. This comparator code can be converted to conventional

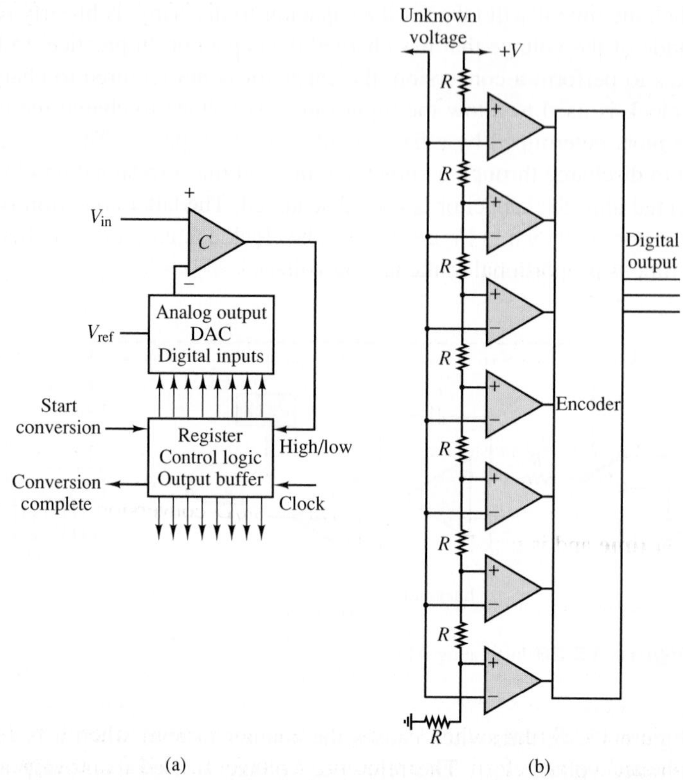

Figure 15.29 (a) Block diagram of 8-bit successive-approximation ADC; (b) A 3-bit flash ADC

LO4

binary by a digital priority encoder circuit. For example, assume that the 3-bit flash ADC of Figure 15.29(b), is set up with $V_{ref} = 8$ V. An input of 6.2 V is provided. If we number the comparators from the top of Figure 15.29(b), the state of each of the seven comparators is as given in Table 15.4.

Table 15.4 State of comparators in a 3-bit flash ADC

Comparator	Input on + line	Input on − line	Output
1	7 V	6.2 V	H
2	6 V	6.2 V	L
3	5 V	6.2 V	L
4	4 V	6.2 V	L
5	3 V	6.2 V	L
6	2 V	6.2 V	L
7	1 V	6.2 V	L

LO4

EXAMPLE 15.9 Flash ADC

Problem

How many comparators are needed in a 4-bit flash ADC?

Solution

Known Quantities: ADC resolution.

Find: Number of comparators required.

Analysis: The number of comparators needed is $2^n - 1 = 15$.

Comments: The flash ADC has the advantage of high speed because it can simultaneously determine the value of each bit, thanks to the parallel comparators. However, because of the large number of comparators, flash ADCs tend to be expensive.

In the preceding discussion, we explored a few different techniques for converting an analog voltage to its digital counterpart; these methods—and any others—require a certain amount of time to perform the A/D conversion. This is the ADC **conversion time** and is usually quoted as one of the main specifications of an ADC device. A natural question at this point would be: If the analog voltage changes during the analog-to-digital conversion and the conversion process itself takes a finite time, how fast can the analog input signal change while still allowing the ADC to provide a meaningful digital representation of the analog input? To resolve the uncertainty generated by the finite ADC conversion time of any practical converter, it is necessary to use a sample-and-hold amplifier. The objective of such an amplifier is to "freeze" the value of the analog waveform for a time sufficient for the ADC to complete its task.

A typical sample-and-hold amplifier is shown in Figure 15.30. It operates as follows. A MOSFET analog switch (see Chapter 11) is used to "sample" the analog waveform. Recall that when a voltage pulse is provided to the sample input of the MOSFET switch (the gate), the MOSFET enters the ohmic region and in effect becomes nearly a short circuit for the duration of the sampling pulse. While the MOSFET conducts, the analog voltage v_a charges the "hold" capacitor C at a fast rate through the small "on" resistance of the MOSFET. The duration of the sampling pulse is sufficient to charge C to the voltage v_a. Because the MOSFET is virtually a short circuit for the duration of the sampling pulse, the charging (RC) time constant is very small, and the capacitor charges very quickly. When the sampling pulse is over, the MOSFET returns to its nonconducting state, and the capacitor holds the sampled voltage without discharging, thanks to the extremely high input impedance of the voltage-follower (buffer) stage. Thus, v_{SH} is the sampled-and-held value of v_a at any given sampling time.

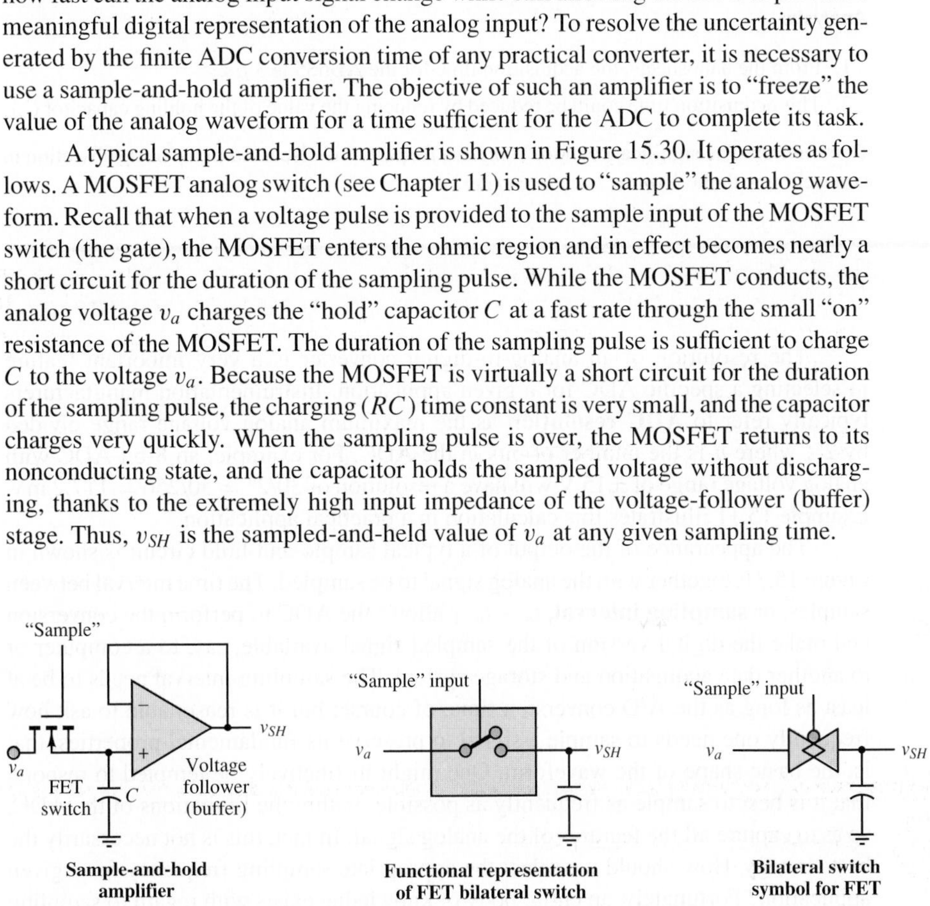

Figure 15.30 Description of the sample-and-hold process

LO4

EXAMPLE 15.10 Sample-and-Hold Amplifier

Problem

Using the data sheets for the AD585 sample-and-hold amplifier (found on the web), answer the following questions:

1. What is the acquisition time of the AD585?
2. How could the acquisition time be reduced?

Solution

Known Quantities: AD585 device data sheets.

Find: Acquisition time.

Schematics, Diagrams, Circuits, and Given Data: DAC specifications are found in the device data sheet. *Definition:* The *acquisition time T* is the time required for the output of the sample-and-hold amplifier to reach its final value, within a specified error bound, after the amplifier has switched from the *sample mode* to the *hold mode*. The time T includes the switch delay time, the slewing interval, and the amplifier settling time.

Analysis:

1. From the data sheets, the acquisition time for the AD585 is 3 μs.
2. This acquisition time could be reduced by reducing the value of the holding capacitor C_H.

Comments: The significance of the *sampling frequency* is discussed in the next subsection in connection with the *Nyquist sampling criterion*.

The resolution of an analog-to-digital converter is a very important feature in selecting a specific ADC for a given application. Instrumentation manufacturers typically refer to **ADC resolution** as the maximum analog voltage range divided by 2^n, where n is the number of bits in the ADC. For example, an 8-bit ADC with analog voltage range of ± 15 V will have a resolution of $30/2^8 = 30/256 = 117.2$ mV. Example 15.11 illustrates this calculation in a practical application.

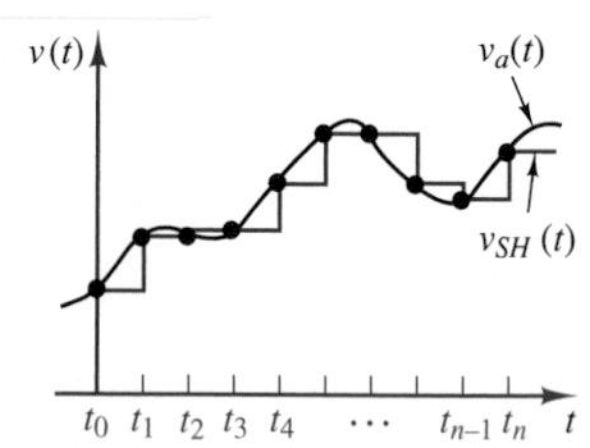

Figure 15.31 Sampled data

The appearance of the output of a typical sample-and-hold circuit is shown in Figure 15.31, together with the analog signal to be sampled. The time interval between samples, or **sampling interval,** $t_n - t_{n-1}$ allows the ADC to perform the conversion and make the digital version of the sampled signal available, say, to a computer or to another data acquisition and storage system. The sampling interval needs to be at least as long as the A/D conversion time, of course; but it is reasonable to ask how frequently one needs to sample a signal to preserve its fundamental properties, that is, the basic shape of the waveform. One might instinctively be tempted to respond that it is best to sample as frequently as possible, within the limitations of the ADC, so as to capture all the features of the analog signal. In fact, this is not necessarily the best strategy. How should we select the appropriate sampling frequency for a given application? Fortunately, an entire body of knowledge exists with regard to sampling theory, which enables the practicing engineer to select the best sampling rate for any given application. Given the scope of this chapter, we have chosen not to delve into the details of sampling theory but, rather, to provide the student with a statement of the fundamental result: the **Nyquist sampling criterion.**

The Nyquist criterion states that to prevent aliasing[3] when sampling a signal, *the sample rate should be selected to be at least twice the highest-frequency component present in the signal.*

LO4

Thus, if we were sampling an audio signal (say, music), we would have to sample at a frequency of at least 40 kHz (twice the highest audible frequency, 20 kHz). In practice, it is advisable to select sampling frequencies substantially greater than the Nyquist rate; a good rule of thumb is 5 to 10 times greater. Example 15.11 illustrates how the designer might take the Nyquist criterion into account in designing a practical A/D conversion circuit.

EXAMPLE 15.11 Performance Analysis of an Integrated-Circuit ADC

LO4

Problem

Using the data sheets for the AD574 (found on the web), answer the following questions:

1. What is the accuracy (in volts) of the AD574?
2. What is the highest-frequency signal that can be converted by this ADC without violating the Nyquist criterion?

Solution

Known Quantities: ADC supply voltage; input voltage range.

Find: ADC accuracy; maximum signal frequency for undistorted A/D conversion.

Schematics, Diagrams, Circuits, and Given Data: $V_{CC} = 15$ V; $0 \le V_{\text{in}} \le 15$ V. ADC specifications are found in device data sheet.

Analysis:

1. From the data sheet we determine that the AD574 is a 12-bit converter. The accuracy is limited by the LSB. For a range of 0 to 15 V, we can calculate the magnitude of the LSB to be

$$\frac{V_{\text{in, max}} - V_{\text{in, min}}}{2^n} = \frac{15}{2^{12}} \times (\pm 1 \text{ bit}) = \pm 3.66 \text{ mV}$$

2. The data sheet states that the maximum guaranteed conversion time of the ADC is 35 μs; therefore the highest conversion frequency for this ADC is

$$f_{\max} = \frac{1}{35 \times 10^{-6}} = 28.57 \text{ kHz}$$

[3]*Aliasing* is a form of signal distortion that occurs when an analog signal is sampled at an insufficient rate.

Since the Nyquist criterion states that the maximum signal frequency that can be sampled without aliasing distortion is one-half of the sampling frequency, we conclude that the maximum signal frequency that can be acquired by this ADC is approximately 14 kHz.

Comments: In practice, it is a good idea to *oversample* by a certain amount. A reasonable rule of thumb is to oversample by a factor of 2 to 5. Suppose we chose to oversample by a factor of 2; then we would not expect to have signal content above 7 kHz.

One way to ensure that the signal being sampled is limited to a 7-kHz bandwidth is to prefilter the signal with a low-pass filter having a cutoff frequency at or below 7 kHz. The active filters discussed in an earlier section of this chapter are often used for this purpose.

CHECK YOUR UNDERSTANDING

In Example 15.11, if the maximum conversion time available to you were 50 μs, what would be the highest-frequency signal you could expect to sample on the basis of the Nyquist criterion?

Answer: $f_{max} = 10$ kHz

Data Acquisition Systems

The structure of a data acquisition system, shown in Figure 15.32, can now be analyzed, at least qualitatively, since we have explored most of the basic building blocks. A typical data acquisition system often employs an *analog multiplexer,* to process several different input signals. A bank of bilateral analog MOSFET switches, such as the one we described together with the sample-and-hold amplifier, provides a simple and effective means of selecting which of the input signals should be sampled and converted to digital form. Control logic, employing standard gates and counters, is used to select the desired *channel* (input signal) and to trigger the sampling circuit and the ADC. When the conversion is completed, the ADC sends an appropriate

LO4

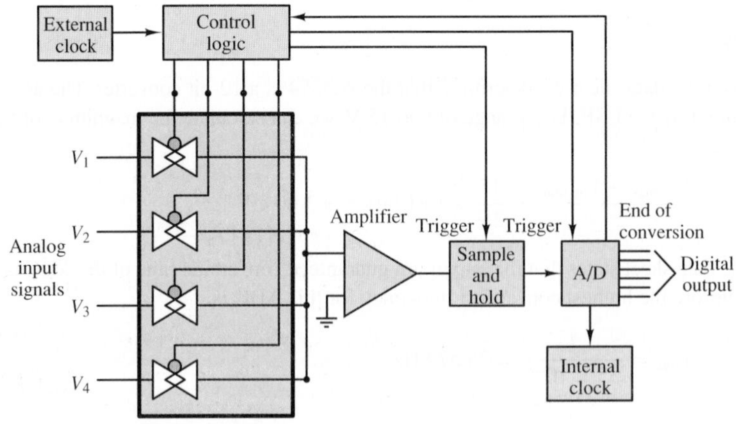

Figure 15.32 Data acquisition system

end-of-conversion signal to the control logic, thereby enabling the next channel to be sampled.

In the block diagram of Figure 15.32, four analog inputs are shown; if these were to be sampled at regular intervals, the sequence of events would appear as depicted in Figure 15.33. We notice, from a qualitative analysis of the figure, that the effective sampling rate for each channel is one-fourth the actual external clock rate; thus, it is important to ensure that the sampling rate for each individual channel satisfies the Nyquist criterion. Further, although each sample is held for four consecutive cycles of the external clock, we must notice that the ADC can use only one cycle of the external clock to complete the conversion, since its services will be required by the next channel during the next clock cycle. Thus, the internal clock that times the ADC must be sufficiently fast to allow for a complete conversion of any sample within the design range. These and several other issues are discussed in the next Focus on Measurements box.

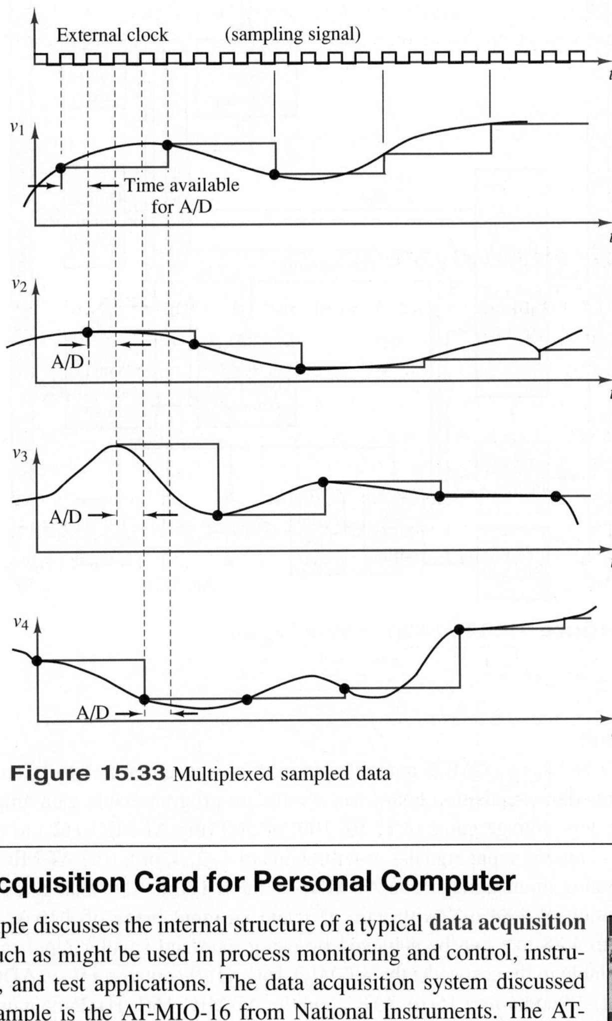

Figure 15.33 Multiplexed sampled data

Data Acquisition Card for Personal Computer

This example discusses the internal structure of a typical **data acquisition system,** such as might be used in process monitoring and control, instrumentation, and test applications. The data acquisition system discussed in this example is the AT-MIO-16 from National Instruments. The AT-MIO-16 is a high-performance, multifunction analog, digital, and timing

FIND IT ON THE WEB

(Continued)

input/output (I/O) board for the IBM PC/AT and compatibles. It contains a 12-bit ADC with up to 16 analog inputs, two 12-bit DACs with voltage outputs, eight lines of transistor-transistor-logic (TTL)–compatible digital I/O, and three 16-bit counter/timer channels for timing I/O. If additional analog inputs are required, the AMUX-64T analog multiplexer can be used. By cascading up to four AMUX-64Ts, 256 single-ended or 128 differential inputs can be obtained. The AT-MIO-16 also uses the RTSI bus (real-time system interface bus) to synchronize multiboard analog, digital, and counter/timer operations by communicating system-level timing signals between boards.

Figure 15.34 is a block diagram of the AT-MIO-16 circuitry. Its major functions are described next.

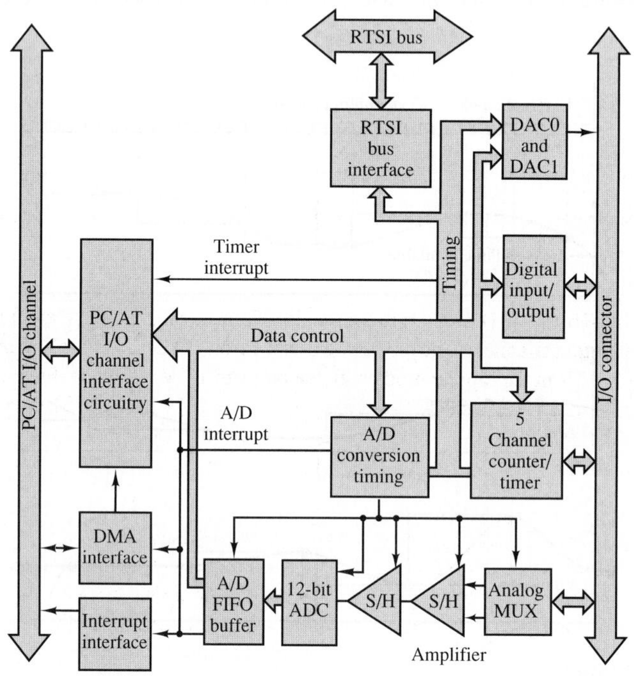

Figure 15.34 AT-MIO-16 block diagram

Analog Input:

The AT-MIO-16 has two CMOS analog input multiplexers connected to 16 analog input channels. This data acquisition board has a software-programmable gain amplifier that can be used with voltage gains of 1, 10, 100, or 500 (the AT-MIO-16L) to accommodate low-level analog input signals; or with gains of 1, 2, 4, or 8 (the AT-MIO-16H) for high-level analog input signals. The AT-MIO-16 has a 12-bit ADC that gives an analog signal resolution of 4.88 mV with gain of 1 and an input range of ± 10 V. Finer resolutions up to 4.88 mV can be achieved by using gain and smaller input ranges. The board is available in three speeds: the AT-MIO-16(L/H)-9 contains a 9-μs ADC; the AT-MIO-16(L/H)-15 contains a 15-μs ADC; and the AT-MIO-16(L/H)-25 contains a 25-μs ADC. These conversions of the board have the following data acquisition sample rates on a

(Continued)

single analog input channel:

	Sampling rate	
Model number	**Typical case**	**Worst case**
AT-MIO-16(L/H)-9	100 ksamples/s	91 ksamples/s
AT-MIO-16(L/H)-15	71 ksamples/s	59 ksamples/s
AT-MIO-16(L/H)-25	45 ksamples/s	37 ksamples/s

The timing of multiple A/D conversion is controlled either by the onboard counter/timer or by external timing signals. The onboard sample rate clock and sample counter control the onboard A/D timing. The AT-MIO-16 can generate both interrupts and DMA (direct memory access) requests on the PC/AT I/O channel. The interrupt can be generated when

1. An A/D conversion is available to be read from the A/D buffer.
2. The sample counter reaches its terminal count.
3. An error occurs.
4. One of the onboard timer clocks generates a pulse.

On the other hand, DMA requests can be generated whenever an A/D measurement is available from the A/D buffer.

Analog Output:
The AT-MIO-16 has two double-buffered multiplying 12-bit DACs that are connected to two analog output channels. The resolution of the 12-bit DACs is 2.44 mV in the unipolar mode or 4.88 mV in the bipolar mode with the onboard 10-V reference. Finer resolutions can be achieved by using smaller voltages on the external reference. The analog output channels have an accuracy of ±0.5 LSB and a differential linearity of ±1 LSB. Voltage offset and gain error can be trimmed to zero.

Digital I/O:
The AT-MIO-16 has eight digital I/O lines that are divided into two 4-bit ports. The digital input circuitry has an 8-bit register that continuously reads the eight digital I/O lines, thus making read-back capability possible for the digital output ports, as well as reading incoming signals. The digital I/O lines are TTL-compatible.

Counter/Timer:
The AT-MIO-16 uses the AM9513A counter/timer for time-related functions. The AM9513A contains five independent 16-bit counter/timers. A 1-MHz clock is the time baseline. Two of the AM9513A counter/timers are for multiple A/D conversion timing. The three remaining counters can be used for special data acquisition timing, such as expanding to a 32-bit sample counter or generating interrupts at user-programmable time intervals.

RTSI Bus Interface:
The AT-MIO-16 is interfaced to the RTSI bus. You can send or receive the external analog input control signal; the waveform generation timing signals; the output of counters 1, 2, and 5; the gate of counter 1; and the source of counter 5. You can send to the RTSI bus the frequency output of the AM9513A.

PC/AT I/O Channel Interface:
The PC/AT I/O channel interface circuitry includes address latches, address-decoding circuitry, data buffers, and interface timing and control signals.

(*Continued*)

(Concluded)

I/O Connector:

The I/O connector is a 50-pin male ribbon cable connector.

Software Support:

The AT-MIO-16 also has software packages that control data acquisition functions on the PC-based data acquisition boards.

15.5 COMPARATOR AND TIMING CIRCUITS

Timing and comparator circuits find frequent application in instrumentation systems. The aim of this section is to introduce the foundations that will permit the student to understand the operation of op-amp comparators and multivibrators, and of an integrated circuit timer.

The Op-Amp Comparator

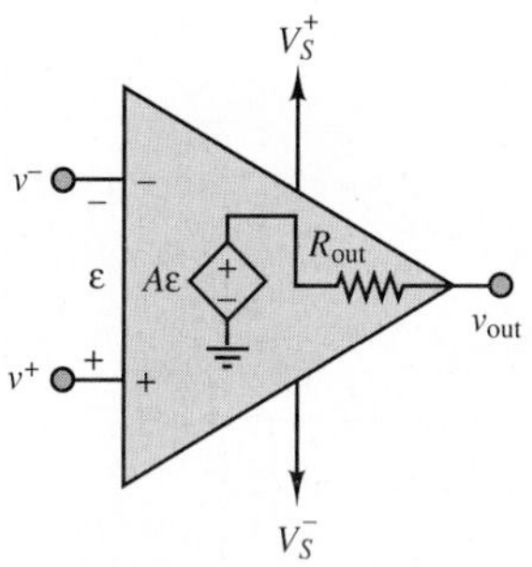

Figure 15.35 Op-amp in open-loop mode

The prototype of op-amp switching circuits is the op-amp comparator of Figure 15.35. This circuit, you will note, *does not employ feedback.* As a consequence of this,

$$v_{\text{out}} = A_{V(\text{OL})}(v^+ - v^-) \quad \textbf{(15.27)}$$

Because of the large gain that characterizes the open-loop (OL) performance of the op-amp ($A_{V(\text{OL})} > 10^5$), any small difference between input voltages ε will cause large outputs. In particular, for ε on the order of a few tens of microvolts, the op-amp will go into saturation at either extreme, according to the voltage supply values and the polarity of the voltage difference (recall the discussion of the op-amp voltage supply limitations in Section 8.6). For example, if ε were a 1-mV potential difference, the op-amp output would ideally be equal to 100 V, for an open-loop gain $A_{V(\text{OL})} = 10^5$ (and in practice the op-amp would saturate at the voltage supply limits). Clearly, any difference between input voltages will cause the output to saturate toward either supply voltage, depending on the polarity of ε.

One can take advantage of this property to generate switching waveforms. Consider, for example, the circuit of Figure 15.36, in which a sinusoidal voltage source $v_{\text{in}}(t)$ of peak amplitude V is connected to the noninverting input. In this circuit, in which the inverting terminal has been connected to ground, the differential input voltage is given by

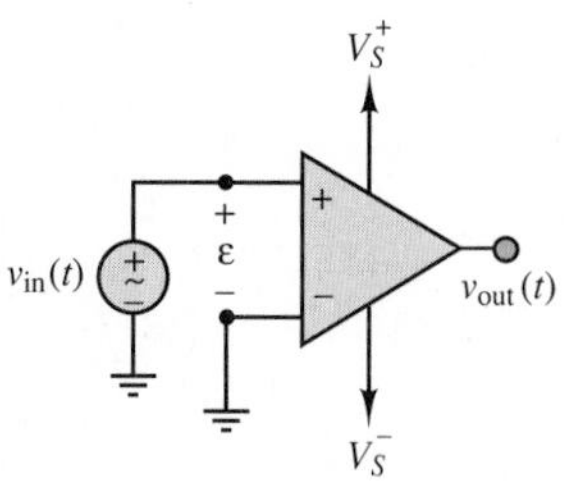

Figure 15.36 Noninverting op-amp comparator

$$\varepsilon = V \cos \omega t \quad \textbf{(15.28)}$$

and will be positive during the positive half-cycle of the sinusoid and negative during the negative half-cycle. Thus, the output will saturate toward V_S^+ or V_S^-, depending on the polarity of ε: the circuit is, in effect, *comparing* $v_{\text{in}}(t)$ and ground, producing a positive v_{out} when $v_{\text{in}}(t)$ is positive and a negative v_{out} when $v_{\text{in}}(t)$ is negative, independent of the amplitude of $v_{\text{in}}(t)$ (provided, of course, that the peak amplitude of the sinusoidal input is at least 1 mV or so). The circuit just described is therefore called a **comparator,** and in effect it performs a binary decision, determining whether $v_{\text{in}}(t) > 0$ or $v_{\text{in}}(t) < 0$. The comparator is perhaps the simplest form of an analog-to-digital converter, that is, a circuit that converts a continuous waveform to discrete values. The comparator output consists of only two discrete levels: greater than and less than a reference voltage.

The input and output waveforms of the comparator are shown in Figure 15.37, where it is assumed that $V = 1$ V and that the saturation voltage corresponding to the ±15-V supplies is approximately ±13.5 V. This circuit is termed a **noninverting comparator,** because a positive voltage differential ε gives rise to a positive output voltage. It should be evident that it is also possible to construct an inverting comparator by connecting the noninverting terminal to ground and connecting the input to the inverting terminal. Figure 15.38 depicts the waveforms for the **inverting comparator.** The analysis of any comparator circuit is greatly simplified if we observe that the output voltage is determined by the voltage difference present at the input terminals of the op-amp, according to the following relationship:

$$\begin{aligned} \varepsilon > 0 \quad &\Rightarrow \quad v_{out} = V^{+}_{sat} \\ \varepsilon < 0 \quad &\Rightarrow \quad v_{out} = V^{-}_{sat} \end{aligned} \qquad \text{Operation of op-amp comparator} \qquad (15.29)$$

LO5

where V_{sat} is the saturation voltage for the op-amp (somewhat lower than the supply voltage, as discussed in Chapter 8). Typical values of supply voltages for practical op-amps are ±5 to ±24 V.

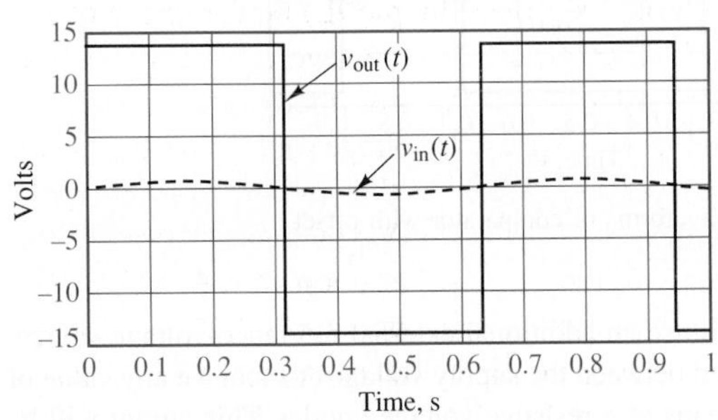

Figure 15.37 Input and output of noninverting comparator

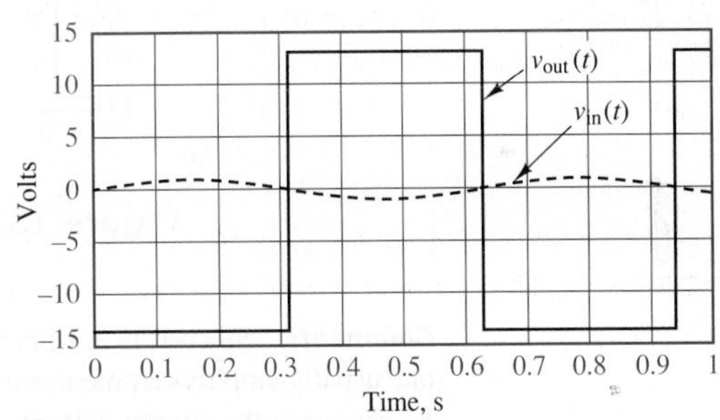

Figure 15.38 Input and output of inverting comparator

A simple modification of the comparator circuit just described consists of connecting a fixed reference voltage to one of the input terminals; the effect of the reference voltage is to raise or lower the voltage level at which the comparator will switch from one extreme to the other. Example 15.12 describes one such circuit.

EXAMPLE 15.12 Comparator with Offset

LO5

Problem

Sketch the input and output waveforms of the comparator with offset shown in Figure 15.39.

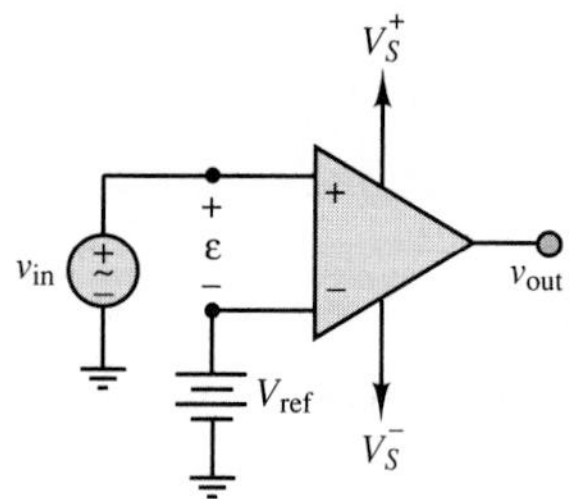

Figure 15.39 Comparator with offset

Solution

Known Quantities: Input voltage, voltage offset.

Find: Output voltage $v_{out}(t)$.

Schematics, Diagrams, Circuits, and Given Data: $v_{in}(t) = \sin \omega t$; $V_{ref} = 0.6$ V.

Analysis: We first compute the differential voltage across the inputs of the op-amp:

$$\varepsilon = v_{in} - V_{ref}$$

Then, using equation 15.29, we determine the switching conditions for the comparator:

$$v_{in} > V_{ref} \quad \Rightarrow \quad v_{out} = V_{sat}^+$$
$$v_{in} < V_{ref} \quad \Rightarrow \quad v_{out} = V_{sat}^-$$

Thus, the comparator will switch whenever the sinusoidal voltage rises above or falls below the reference voltage. Figure 15.40 depicts the appearance of the comparator output voltage. Note that comparator output waveform is no longer a symmetric square wave.

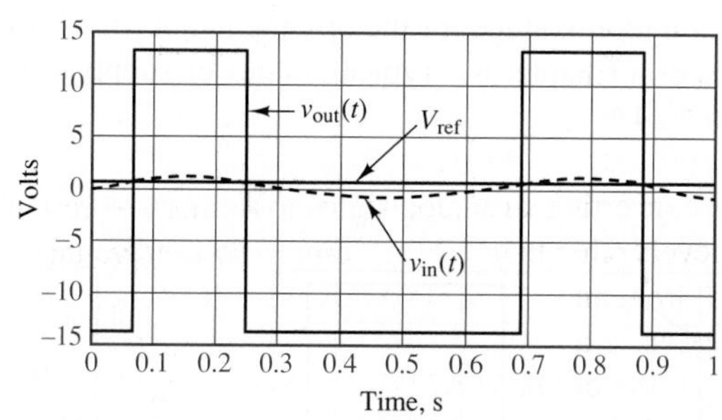

Figure 15.40 Waveforms of comparator with offset

Comments: Since it is not practical to use an additional external reference voltage source, one usually employs a potentiometer tied between the supply voltages to achieve any value of V_{ref} between the supply voltages by means of a resistive voltage divider. This circuit will be explored later in this chapter.

CHECK YOUR UNDERSTANDING

For the comparator circuit of Figure 15.39, sketch the waveforms $v_{out}(t)$ and $v_S(t)$ if $v_S(t) = 0.1 \cos \omega t$ and $V_{ref} = 50$ mV. Assume that $|V_S| = 15$ V.

Another useful interpretation of the op-amp comparator can be obtained by considering its **input-output transfer characteristic.** Figure 15.41 displays a plot of v_{out} versus v_{in} for a noninverting zero-reference (no offset) comparator. This circuit is often called a **zero-crossing comparator,** because the output voltage goes through a transition (V_{sat} to $-V_{sat}$, or vice versa) whenever the input voltage crosses the horizontal axis. You should be able to verify that Figure 15.42 displays the transfer characteristic for a comparator of the inverting type with a nonzero reference voltage.

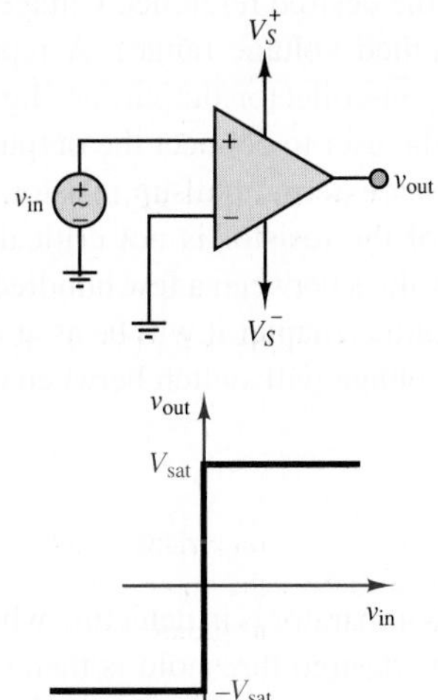

Figure 15.41 Transfer characteristic of zero-crossing comparator

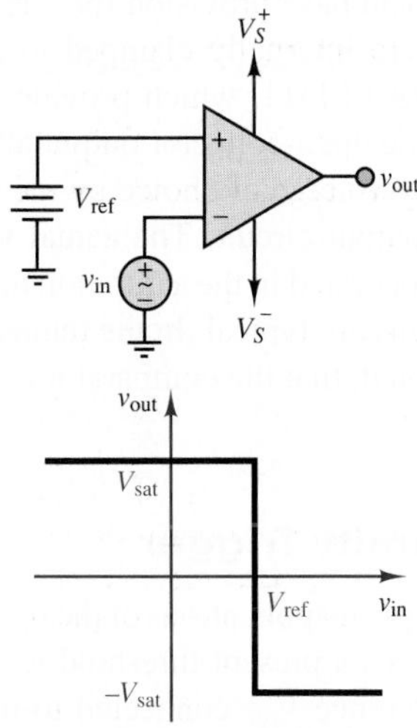

Figure 15.42 Transfer characteristic of inverting comparator with offset

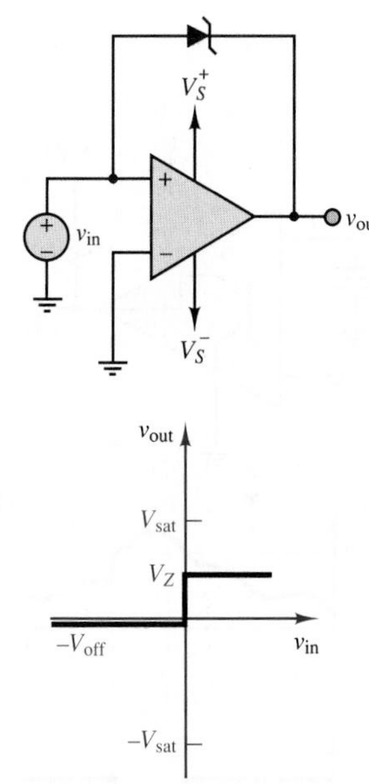

Figure 15.43 Level-clamped comparator

Very often, in converting an analog signal to a binary representation, one would like to use voltage levels other than $\pm V_{sat}$. Commonly used voltage levels in this type of switching circuit are 0 and 5 V. This modified voltage transfer characteristic can be obtained by connecting a Zener diode between the output of the op-amp and the noninverting input, in the configuration sometimes called a **level** or **Zener clamp.** The circuit shown in Figure 15.43 is based on the fact that a reversed-biased Zener diode will hold a constant voltage V_Z, as was shown in Chapter 9. When the diode is forward-biased, on the other hand, the output voltage becomes the negative of the offset voltage V_γ. An additional advantage of the level clamp is that it reduces the switching time. Input and output waveforms for a Zener-clamped comparator are shown in Figure 15.44, for the case of a sinusoidal $v_{in}(t)$ of peak amplitude 1 V and Zener voltage equal to 5 V.

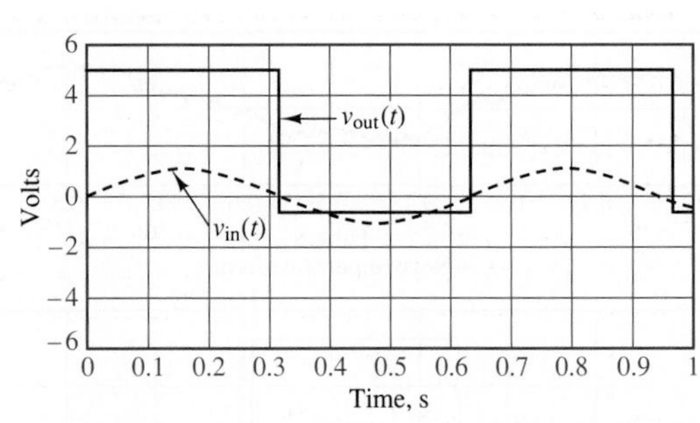

Figure 15.44 Zener-clamped comparator waveforms

Although the Zener-clamped circuit illustrates a specific issue of interest in the design of comparator circuits, namely, the need to establish desired reference output

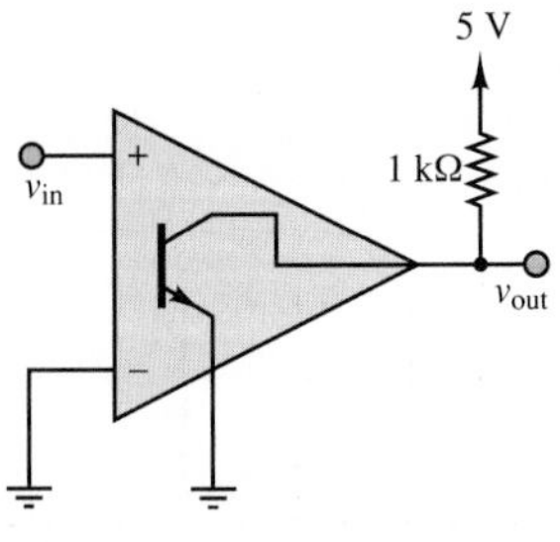

Figure 15.45 Open-collector comparator output with representative external supply connection

voltages other than the supply saturation voltages, this type of circuit is rarely employed in practice. Special-purpose integrated-circuit packages are available that are designed specifically to serve as comparators. These can typically accept relatively large inputs and have provision for selecting the desired reference voltage levels (or, sometimes, are internally clamped to a specified voltage range). A representative product is the LM311, which provides an open-collector output, as shown in Figure 15.45. The open-collector output allows the user to connect the output transistor to any supply voltage of choice by means of an external pull-up resistor, thus completing the output circuit. The actual value of the resistor is not critical, since the transistor is operated in the saturation mode; values between a few hundred and a few thousand ohms are typical. In the remainder of the chapter it will be assumed, unless otherwise noted, that the comparator output voltage will switch between 0 and 5 V.

The Schmitt Trigger

One of the typical applications of the op-amp comparator is in detecting when an input voltage exceeds a present threshold level. The desired threshold is then represented by a DC reference V_{ref} connected to the noninverting input, and the input voltage source is connected to the inverting input, as in Figure 15.42. Under ideal conditions, for noise-free signals, and with an infinite slew rate for the op-amp, the operation of such a circuit would be as depicted in Figure 15.46. In practice, the presence of noise and the finite slew rate of practical op-amps will require special attention.

Two improvements of this circuit will be discussed in this section: how to improve the switching speed of the comparator, and how to design a circuit that can operate correctly even in the presence of noisy signals. If the input to the comparator is changing slowly, the comparator will not switch instantaneously, since its open-loop gain is not infinite and, more important, its slew rate further limits the switching speed. In fact, commercially available comparators have slew rates that are typically much lower than those of conventional op-amps. In this case, the comparator output would not switch very quickly at all. Further, in the presence of noisy inputs, a conventional comparator is inadequate, because the input signal could cross the reference voltage level repeatedly and cause multiple triggering. Figure 15.47 depicts the latter occurrence.

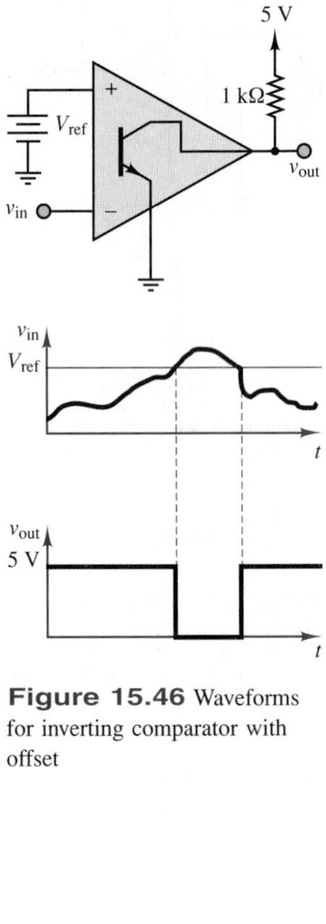

Figure 15.46 Waveforms for inverting comparator with offset

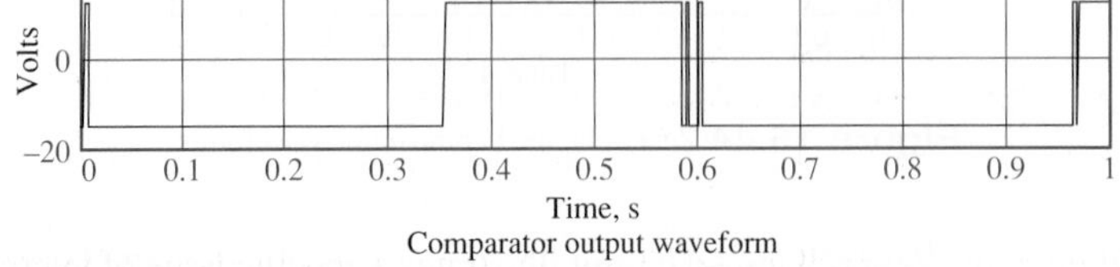

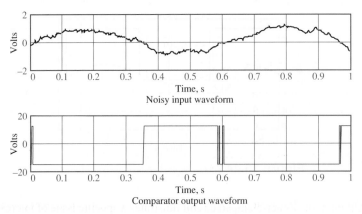

Figure 15.47 Comparator response to noisy inputs

One very effective way of improving the performance of the comparator is by introducing positive feedback. As will be explained shortly, positive feedback can increase the switching speed of the comparator and provide noise immunity at the same time. Figure 15.48 depicts a comparator circuit in which the output has been tied back to the *noninverting* input (thus the terminology *positive* feedback) by means of a resistive voltage divider. The effect of this positive-feedback connection is to provide a reference voltage at the noninverting input equal to a fraction of the comparator output voltage; since the comparator output is equal to either the positive or the negative saturation voltage $\pm V_{sat}$, the reference voltage at the noninverting input can be either positive or negative.

Consider, first, the case when the comparator output is $v_{out} = +V_{sat}$. It follows that

$$v^+ = \frac{R_2}{R_2 + R_1} V_{sat} \qquad \textbf{(15.30)}$$

and therefore the differential input voltage is

$$\varepsilon = v^+ - v^- = \frac{R_2}{R_2 + R_1} V_{sat} - v_{in} \qquad \textbf{(15.31)}$$

For the comparator to switch from the positive to the negative saturation state, the differential voltage ε must then become negative; that is, the condition for the comparator to switch state becomes

$$v_{in} > \frac{R_2}{R_2 + R_1} V_{sat} \qquad \textbf{(15.32)}$$

Since $[R_2/(R_2 + R_1)]V_{sat}$ is a positive voltage, the comparator will not switch when the input voltage crosses the zero level; but it will switch when the input voltage exceeds some positive voltage, which can be determined by appropriate choice of R_1 and R_2.

Consider, now, the case when the comparator output is $v_{out} = -V_{sat}$. Then

$$v^+ = -\frac{R_2}{R_2 + R_1} V_{sat} \qquad \textbf{(15.33)}$$

and therefore

$$\varepsilon = v^+ - v^- = -\frac{R_2}{R_2 + R_1} V_{sat} - v_{in} \qquad \textbf{(15.34)}$$

For the comparator to switch from the negative to the positive saturation state, the differential voltage ε must then become positive; the condition for the comparator to switch state is now

$$v_{in} < -\frac{R_2}{R_2 + R_1} V_{sat} \qquad \textbf{(15.35)}$$

Thus, the comparator will not switch when the input voltage crosses the zero level (from the negative direction), but it will switch when the input voltage becomes more negative than a threshold voltage, determined by R_1 and R_2. Figure 15.48 depicts the effect of the different thresholds on the voltage transfer characteristic, showing the switching action by means of arrows.

The circuit just described finds frequent application and is called a **Schmitt trigger.**

If it is desired to switch about a voltage other than zero, a reference voltage can also be connected to the noninverting terminal, as shown in Figure 15.49. Now the

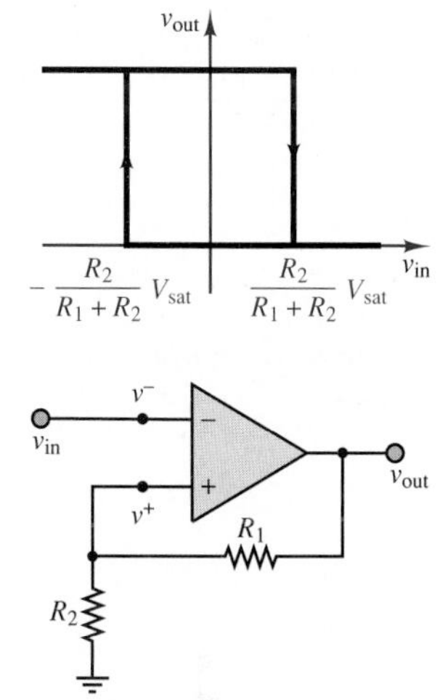

Figure 15.48 Transfer characteristic of the Schmitt trigger

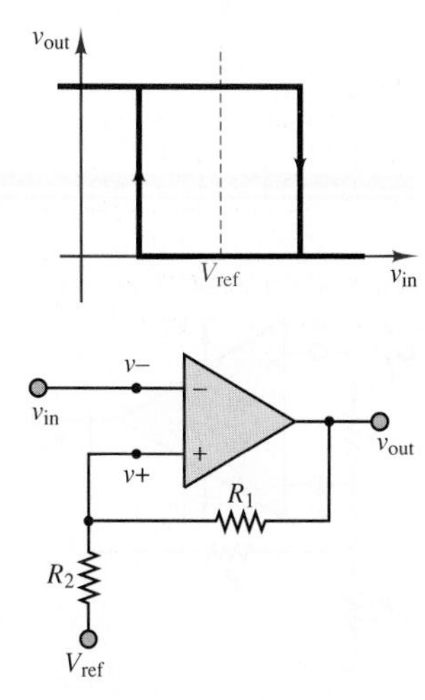

Figure 15.49 Schmitt trigger (general circuit)

expression for the noninverting terminal voltage is

$$v^+ = \frac{R_2}{R_2 + R_1} v_{out} + V_{ref} \frac{R_1}{R_2 + R_1} \quad \textbf{(15.36)}$$

and the switching levels for the Schmitt trigger are

$$v_{in} > \frac{R_2}{R_2 + R_1} V_{sat} + V_{ref} \frac{R_1}{R_2 + R_1} \quad \textbf{(15.37)}$$

for the positive-going transition and

$$v_{in} < -\frac{R_2}{R_2 + R_1} V_{sat} + V_{ref} \frac{R_1}{R_2 + R_1} \quad \textbf{(15.38)}$$

for the negative-going transition. In effect, the Schmitt trigger provides a noise rejection range equal to $\pm[R_2/(R_2 + R_1)]V_{sat}$ within which the comparator cannot switch. Thus, if the noise amplitude is contained within this range, the Schmitt trigger will prevent multiple triggering. Figure 15.50 depicts the response of a Schmitt trigger with appropriate switching thresholds to a noisy waveform. Example 15.13 provides a numerical illustration of this process.

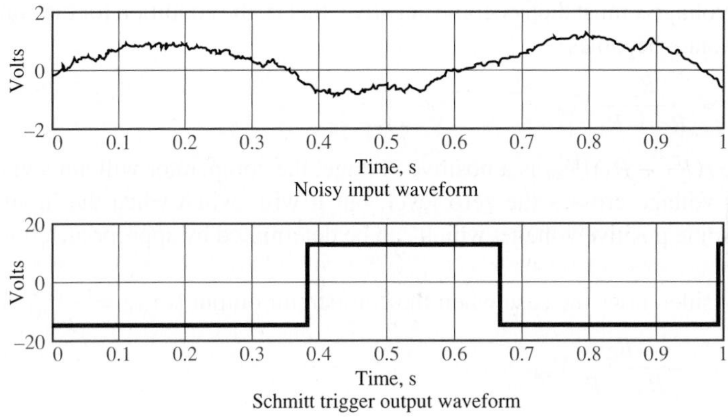

Figure 15.50 Schmitt trigger response to noisy waveforms

EXAMPLE 15.13 Analysis and Design of Schmitt Trigger

LO5

Problem

Find the required resistor values for the Schmitt trigger circuit shown in Figure 15.51.

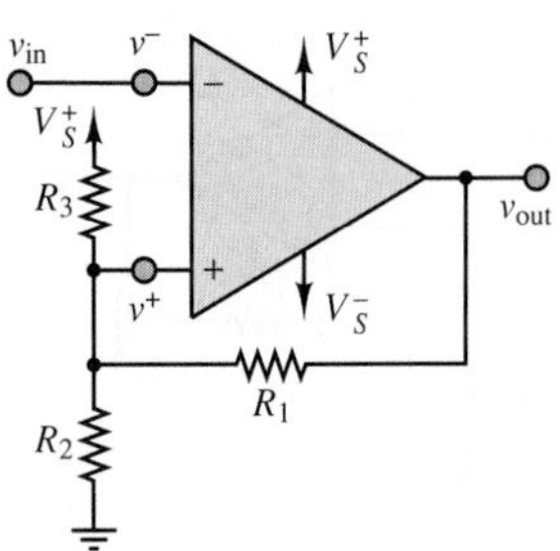

Figure 15.51 Schmitt trigger

Solution

Known Quantities: Supply voltages and supply saturation voltages; reference voltage (offset); noise amplitude.

Find: R_1, R_2, R_3.

Schematics, Diagrams, Circuits, and Given Data: $|V_S| = 18$ V; $|V_{sat}| = 16.5$ V; $V_{ref} = 2$ V.

Assumptions: $|v_{noise}| = 100$ mV.

Analysis: We first observe that the offset voltage has been obtained by tying two resistors, forming a voltage divider, between the positive supply voltage and ground. This procedure avoids requiring a separate reference voltage source. From the circuit of Figure 15.51 we can calculate the noninverting voltage to be

$$v^+ = \frac{R_2}{R_1 + R_2} v_{out} + \frac{R_2}{R_2 + R_3} V_S^+$$

Since the required noise protection level (the width of the transfer characteristic, symmetrically placed about V_{ref} in Figure 15.49) is $\Delta V = \pm 100$ mV, we can compute R_1 and R_2 from

$$\frac{\Delta v}{2} = \frac{R_2}{R_1 + R_2} V_{sat} = \frac{R_2}{R_1 + R_2} \times 16.5 = 0.1 \text{ V}$$

or

$$\frac{R_2}{R_1 + R_2} = \frac{0.1}{16.5}$$

The top half of Figure 15.52 depicts the ± 100-mV noise protection band around the reference voltage. If we select a large value for one of the resistors, say, $R_1 = 100$ kΩ, we can calculate $R_2 \approx 610$ Ω.

To determine R_3, we note that

$$V_{ref} = \frac{R_2}{R_2 + R_3} V_S^+$$

or

$$2 = \frac{610}{610 + R_3} \times 18$$

and calculate $R_3 = 4.88$ kΩ. The design is complete. The transfer characteristic of the comparator and the associated waveforms are shown in Figure 15.52.

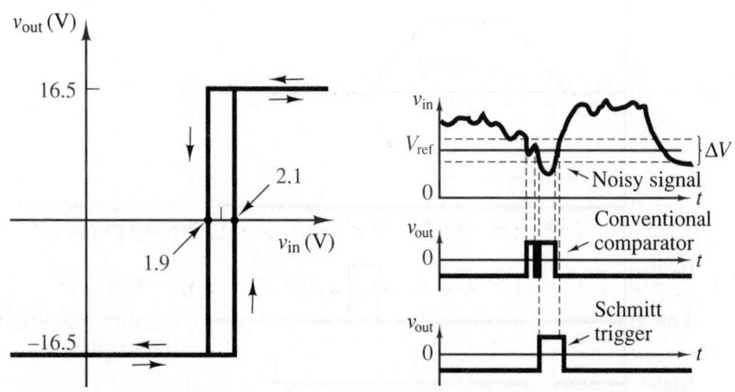

Figure 15.52 Schmitt trigger waveforms and transfer characteristics

Comments: In Figure 15.52 the Schmitt trigger output is compared to that of a comparator without noise protection. Note that a common comparator would be triggered twice in the presence of noise.

CHECK YOUR UNDERSTANDING

Derive the expressions for the switching thresholds of the Schmitt trigger of Figure 15.48.

Multivibrators

Timing circuits

A number of instrumentation applications require the implementation of timing functions. One important timing function is the generation of a fixed frequency clock waveform, that is, the generation of a periodic waveform, usually a square wave or pulse train, with a known period. Another is the one-shot, or monostable multivibrator function, in which a pulse of known duration and amplitude is generated as explained in this section.

Monostable multivibrators are usually employed in IC package form. An IC one-shot can generate voltage pulses when triggered by a **rising** or a **falling edge,** that is, by a transition in either direction in the input voltage. Thus, a one-shot IC offers the flexibility of external selection of the type of transition that will cause a pulse to be generated: a rising edge (from low voltage to high, typically 0 V to some threshold level) or a falling edge (high-to-low transition). Various input connections are usually provided for selecting the preferred triggering mode, and the time constant is usually set by selection of an external *RC* circuit. The output pulse that may be generated by the one-shot can also occur as a positive or a negative transition. Figure 15.53 shows the response of a one-shot to a triggering signal for the four conditions that may be attained with a typical one-shot.

LO5

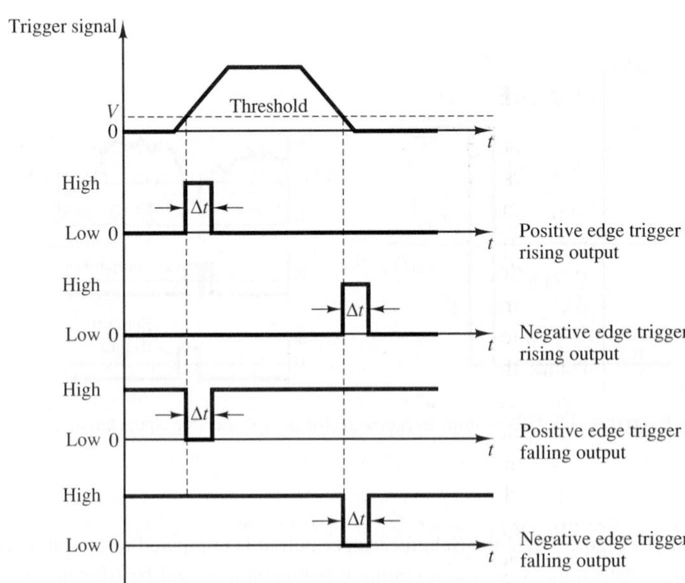

Figure 15.53 IC monostable multivibrator waveforms

A typical IC one-shot circuit based on the 74123 (see the data sheet found on the web) is displayed in Figure 15.54. The 74123 is a **dual one-shot,** meaning that the package contains two monostable multivibrators, which can be used independently. The outputs of the one-shot are indicated by the symbols Q_1, $\overline{Q}_1$, Q_2, and $\overline{Q}_2$, where the overbar indicates the complement of the output. For example, if Q_1 corresponds to a positive-going output pulse, $\overline{Q}_1$ indicates a negative-going output pulse of equal duration.

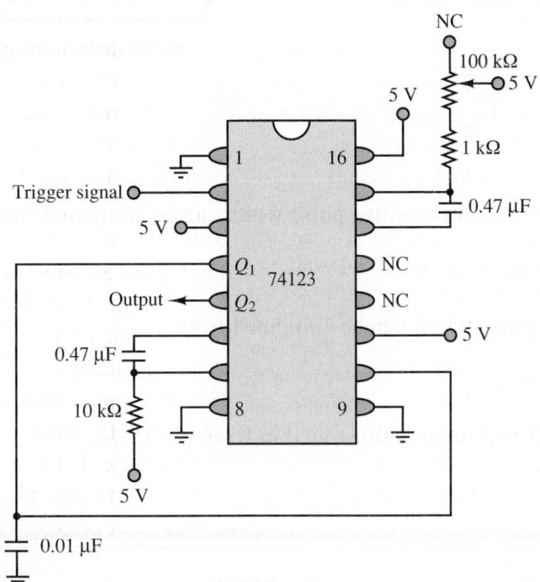

Figure 15.54 Dual one-shot circuit

Timer ICs: The NE555

Timing circuits fall—for our purposes—into one of two classes: pulse generators and clock waveform generators. Chapters 13 and 14 delve into a more detailed analysis of digital timing circuits, a family to which the circuits of the previous sections belong. This section will now introduce a multipurpose integrated circuit that can perform both the monostable and clock functions. The main advantage of the integrated-circuit implementation of these circuits (as opposed to the discrete op-amp version previously discussed) lies in the greater accuracy and repeatability one can obtain with ICs, their ease of application, and the flexibility provided in the integrated-circuit packages. The NE555 is a timer circuit capable of producing accurate time delays (pulses) or oscillation. In the time-delay, or monostable, mode, the time delay or pulse duration is controlled by an external RC network. In the astable, or clock generator, mode, the frequency is controlled by two external resistors and one capacitor. Figure 15.55 depicts typical circuits for monostable and astable operation of the NE555. Note that the threshold level and the trigger level can also be externally controlled. For the monostable circuit, the pulse width can be computed from

$$T = 1.1R_1C \qquad \textbf{(15.39)}$$

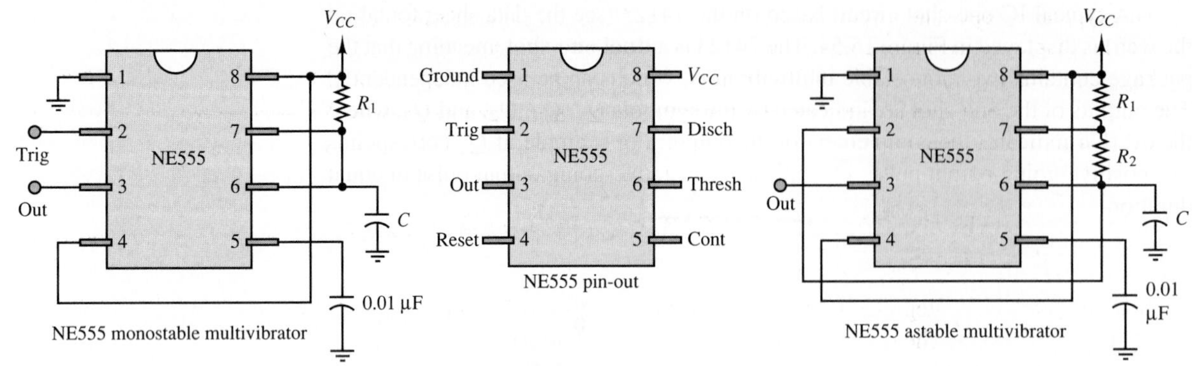

Figure 15.55 NE555 timer

For the astable circuit, the positive pulse width can be computed from

$$T_+ = 0.69(R_1 + R_2)C \quad \textbf{(15.40)}$$

and the negative pulse width can be computed from

$$T_- = 0.69R_2C \quad \textbf{(15.41)}$$

The use of the NE555 timer is illustrated in Example 15.14.

LO5

EXAMPLE 15.14 Analysis of the 555 Timer

Problem

Calculate the component values required to obtain a 0.421-ms pulse using the 555 timer monostable configuration of Figure 15.55.

Solution

Known Quantities: Desired pulse duration T.

Find: Values of R_1 and C.

Schematics, Diagrams, Circuits, and Given Data: $T = 0.421$ ms.

Assumptions: Assume a value for C.

Analysis: Using equation 15.39,

$$T = 1.1R_1C$$

And assuming $C = 1\ \mu$F, we calculate

$$0.421 \times 10^{-3} = 1.1R_1 \times 10^{-6}$$

or

$$R_1 = 382.73\ \Omega$$

Comments: Any reasonable combination of R_1 and C values can yield the desired design value of T. Thus, the component selection shown in this example is not unique.

15.6 OTHER INSTRUMENTATION INTEGRATED CIRCUITS

The advent of low-cost integrated electronics and microprocessors has revolutionized instrumentation design. The transition from the use of analog, discrete circuits (e.g., discrete transistor amplifiers) for signal conditioning to (often digital) integrated circuits and to microprocessor-based instrumentation systems has taken place over a period of two decades, and is now nearly complete, with the exception of very specialized applications (e.g., very high-frequency or low-noise circuits). The aim of this section is to present a summary of some of the signal conditioning and processing functions that are readily available in low-cost integrated-circuit form. The list is by no means exhaustive, and the reader is referred to the websites of the numerous integrated-circuit manufacturers for more detailed information.

The nonelectrical engineer interested in the design of special-purpose instrumentation circuits can benefit from the wealth of information contained in the application notes available from IC manufacturers (often directly downloadable from the web).

In the preceding sections of this chapter we have already explored some IC instrumentation elements, namely, instrumentation amplifiers, op-amp active filters, digital-to-analog and analog-to-digital converters, sample-and-hold amplifiers, voltage comparators, and timing ICs. Further, Chapter 8 delves into the basic operation of operational amplifiers, and Chapters 13 and 14 contain information on digital logic circuits and microprocessors and microcontrollers. In this section we briefly survey a number of common instrumentation applications not yet mentioned in this book, and we provide references for some applications that have already been discussed.

Amplifiers

A number of special-purpose IC amplifiers are available to perform a variety of functions. Although most of these amplifiers could be realized by using op-amps, these specially designed packages can save much design effort and provide better performance. The following list indicates some of the products available from one manufacturer (Analog Devices):

- Instrumentation amplifiers
- Logarithmic amplifiers
- RF amplifiers
- Sample-and-hold, track-and-hold amplifiers
- Variable-gain amplifiers
- (Audio) microphone preamplifiers
- (Audio) power amplifiers
- (Audio) voltage-controlled amplifiers

DACs and ADCs

Digital-to-analog and analog-to-digital converters are also available in a variety of packages intended for general use or tailored to special applications:

General-purpose ADCs ($\leq$ 1 Msample/s)
High-speed ADCs (> 1 Msample/s)
DACs
(Audio) Nyquist DAC
(Audio) sigma-delta DAC
(Audio) stereo ADC
Digital radio ADCs

Frequency-to-Voltage, Voltage-to-Frequency Converters and Phase-Locked Loops

The need for converting changes in instantaneous frequency to changes in an analog voltage (i.e., frequency demodulation) arises frequently in instrumentation applications. For example, *optical position encoders* (Chapter 13) and *magnetic position sensors* (Chapter 16) represent instantaneous velocity information as a frequency-modulated (FM) signal that can be demodulated by a frequency-to-voltage converter. Similarly, it is sometimes useful to encode analog voltage signals in FM form using a voltage-to-frequency converter.

Phase-locked loops (PLLs) can be used for FM demodulation as well as for a number of other related functions, such as tone decoding.

Other Sensor and Signal Conditioning Circuits

Sensor and signal conditioning IC technology often permits the integration of sensing and signal conditioning on the same integrated circuit. In addition, specialized signal conditioning modules enable the design of rather complex instrumentation systems with relatively few IC building blocks. The book website includes device data sheets and application notes for a number of commercially available IC products, including **rms-to-DC converters,** accelerometers and other **integrated sensors for industrial and automotive applications,** and **signal conditioning subsystems.** The following example, reprinted with permission of Analog Devices, consists of an application note discussing the use of integrated acceleration and tilt sensors in a car antitheft alarm. This is only an example of the wealth of electronic information appended to this book.

Using the ADXL202 Accelerometer as a Multifunction Sensor (Tilt, Vibration, and Shock) in Car Alarms

by Harvey Weinberg and Christophe Lemaire, Analog Devices

By using an intelligent algorithm, the ADXL202 ($\pm 2g$ dual-axis accelerometer) can serve as a low-cost, multifunction sensor for vehicle security systems, capable of acting

(Continued)

simultaneously as a shock/vibration detector as well as a tilt sensor (to detect towing or jacking up of the car). The accelerometer's output is passed through two parallel filters: a bandpass filter to extract shock/vibration information and a low-pass filter to extract tilt information. This application note describes the basics of such an implementation.

Introduction:

The ADXL202 is a low-cost, low-power, complete dual-axis accelerometer with a measurement range of $\pm 2\ g$. The ADXL202 outputs analog and digital signals proportional to acceleration in each of the sensitive axes (see Figure 15.56).

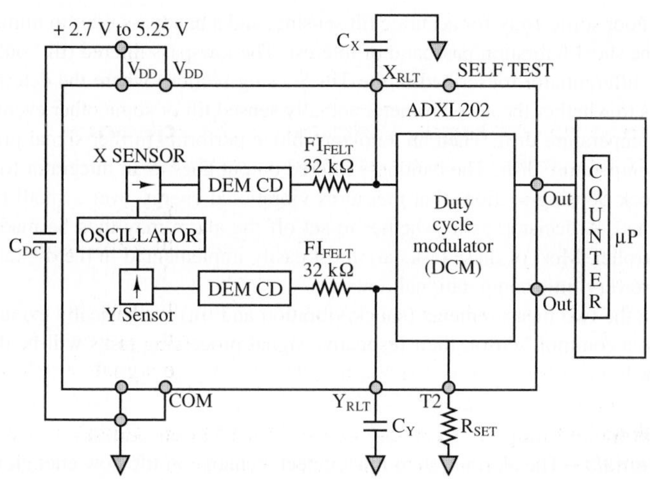

Figure 15.56 ADXL202 block diagram

Currently automotive security systems use shock/vibration sensors to detect collision or forced intrusion into the car. Typically, these sensors are based on magneto-inductive sensing. Sensors of this type generally have adequate sensitivity, but fall short in other areas. Often a fair amount of signal conditioning and trimming is required between the shock sensor and microcontroller due to variations in magnetic material and Hall effect sensor sensitivity, and their frequency response is fairly unpredictable due to inconsistency in mounting. In addition, such sensors have no response to gravity-induced acceleration, so they are incapable of sensing inclination (a static acceleration). Tilt sensing is the most direct way of detecting if a vehicle is being jacked up, about to be towed, or being loaded onto a flatbed truck. These are some of the most common methods of car theft today.

The ADXL202 is a true accelerometer, easily capable of shock/vibration sensing with virtually no external signal conditioning circuitry. Since the ADXL202 is sensitive to static (gravitational) acceleration, tilt sensing is also possible. Tilt sensing requires a very low noise floor which usually necessitates restricting the bandwidth of the accelerometer, while shock/vibration sensing requires wide bandwidth. These conflicting requirements may be met by using clever design techniques.

Principle of Operation:

The ADXL202 is set up to acquire acceleration from 0 to 200 Hz (the maximum frequency of interest). Figure 15.57 shows a block diagram of the system. The accelerometer's output is fed into two filters; a low-pass filter with a corner frequency at 12.5 Hz used to lower

(Continued)

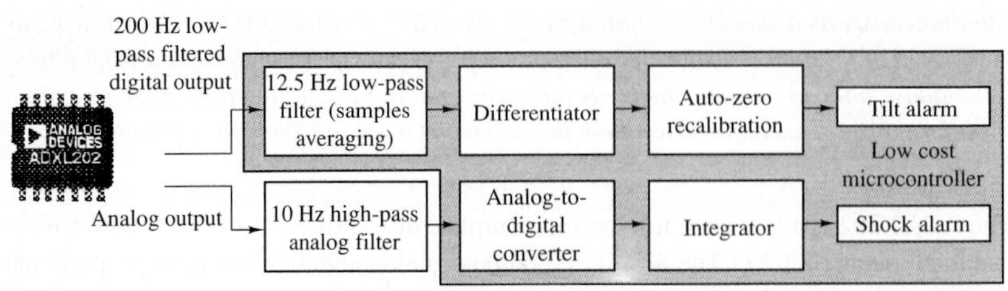

Figure 15.57 Shock and tilt sensing using the ADXL202

the noise floor sufficiently for accurate tilt sensing, and a bandpass filter to minimize the noise in the shock/vibration passband of interest. The low-pass filtered (tilt) output then goes to a differentiator (described in the Tilt Sensing section) where the determination is made as to whether the accelerometer actually sensed tilt or some other event such as noise or temperature drift. Then an auto-zero block performs further signal processing to reject temperature drift. The bandpass filtered output goes to an integrator (described in the Shock Sensing section) that measures vibrational energy over a small period of time (40 ms). A decision as to whether to set off the alarm may then be made by the microcontroller. Most of these tasks are most easily implemented in the digital domain and require very little computational power.

Since the two measurements (shock/vibration and tilt) are basically exclusive and only share a common sensor, their respective signal processing tasks will be described separately.

Tilt Sensing:

Fundamentals—The alarm system must detect a change in tilt slow enough to be the result of the vehicle being towed or jacked up, but must be immune to temperature changes and movement due to passing vehicles or wind. Note that the ADXL202 is most sensitive to tilt when its sensitive axes are perpendicular to the force of gravity, that is, parallel to the earth's surface. Figure 15.58 shows that the change in projection of a 1 g gravity-induced acceleration vector on the axis of sensitivity of the accelerometer will be more significant if the axis is tilted 10° from the horizontal than if it is tilted by the same amount from the vertical.

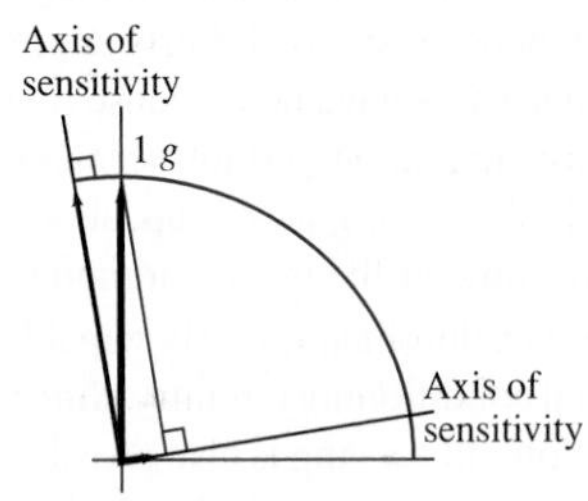

Figure 15.58 Tilt sensitivity

However, the car may not always be level when the alarm is activated, and while the 0 g offset can be recalibrated for any initial inclination, effectively the farther from the horizontal the axes of sensitivity are, the less sensitive the system will be to tilt (see ADXL202 data sheet). In most cases, this should not be of great concern, since the

(Continued)

sensitivity only declines by about 2.5 mg per degree of tilt when inclination goes from 0° (horizontal) to 30° of tilt. Nevertheless, installation guidelines should recommend that the tilt-sensing module containing the accelerometer be mounted such that the axes of sensitivity be as level as possible.

***Implementation*—** In general, we are interested in knowing if the inclination of the car has changed more than $\pm 5°$ from its inclination when initially parked. When the car is turned off, a measurement of the car's inclination is made. If the car's inclination is changed by more than $\pm 5°$, an alarm is triggered. Alternatively, the rate of change of tilt may be evaluated and if its absolute value is above 0.2° per second for several seconds, the alarm may be triggered.

Each technique has certain advantages. The former algorithm is better at false-alarm rejection due to jostling of the car, while the rate-of-change algorithm may be set up to react more quickly. Algorithms using a combination of both techniques may be used as well. It is left to the readers to decide which technique is best for their applications. While all the concepts presented here are valid for both algorithms, for consistency this application note will describe the former (absolute inclination) algorithm.

For the purpose of the following discussion, we will assume a less than perfect tilt sensitivity for the accelerometer of 15 mg per degree of tilt, or 75 mg for 5°. The ADXL202 will be set up to have a bandwidth of 200 Hz so that vibration may be detected. A 200-Hz bandwidth will result in a noise floor of:

$$\text{Noise} = 500\ \mu g\sqrt{\text{Hz}} \times (\sqrt{200 \times 1.5})\ \text{rms}$$
$$= 8.5\ \text{m}g\ \text{rms}$$

or 34 mg peak-to-peak noise (using a peak-to-peak to rms ratio of 4:1), well within our 75 mg requirement. For reliability purposes, we would like to have a noise floor about 10 times lower than this, or around 8 mg. Since towing a car takes at least a few seconds, we are free to narrow the bandwidth to lower the noise floor. An analog or digital low-pass filter may be used, but since low-pass filtering in the digital domain is very simple, it is the preferred method. By taking the average of 16 samples we reduce the effective bandwidth to 12.5 Hz (200 Hz/16 samples). The resulting noise performance is approximately 8.7 mg peak to peak, close enough to our target.

Lowering the noise floor even further, by taking up to 128 samples, for example, would result in about 3 mg peak-to-peak noise, which would allow us to easily detect the 15 mg of static acceleration resulting from a change in tilt of less than a degree.

The typical 0 g drift due to temperature for the ADXL202 is 2 mg/°C. Since our trigger point for a tilt alarm could be as low as 15 mg, it is conceivable that temperature drift alone would cause a false alarm (a car parked overnight could easily experience more than 7.5°C in ambient temperature change). Therefore we will include a differentiator to reject temperature drift.

In the event of the car being jacked up or lifted for towing, we would expect the rate of change in tilt to be faster than 5° or 75 mg/min (or 1.25 mg/s). Each time the acceleration is measured, it is compared to the previous reading. If the change is less than 1.25 mg/s, we know that the change in accelerometer output is due to temperature drift. We can now add an auto-zero block that adjusts our 0 g reference (that is the static acceleration sensed when the car was initially parked) to compensate for 0 g drift due to temperature.

Shock Sensing:

Generally for automotive shock/vibration sensing we are interested in signals between 10 and 200 Hz. Since the response of the ADXL202 extends from DC to 5 kHz, a band-pass filter will have to be added to remove out-of-band signals. This bandpass filter is most

(*Continued*)

(Concluded)

easily implemented in the analog domain (Figure 15.59 shows a simple 10-Hz high-pass filter). When coupled with the 200-Hz low-pass filter (from Xfilt and Yfilt on the ADXL202), a 10- to 200-Hz bandpass filter is realized.

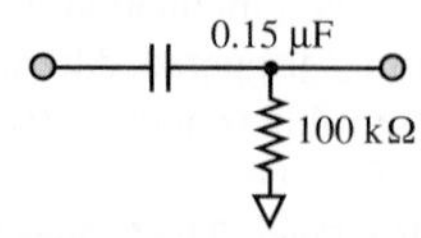

Figure 15.59 10-Hz high-pass filter

While analog bandpass filtering is very simple and requires no software overhead from the microcontroller, it does necessitate having an analog-to-digital converter. Today, even low-cost microcontrollers can commonly be found, integrating an AD converter on board. Bandpass filtering in the digital domain may be more effective, but may require a more powerful processor than one normally finds in automobile security systems. There are several methods for implementing bandpass filters in the digital domain. Specific recommendations will not be given here since processor selection will influence what method will be most efficient.

Whether a digital or analog bandpass filter is used, the Nyquist criteria for signal sampling must be satisfied. That is, we must sample at least twice the maximum frequency of interest. Sampling at 400 Hz (for our 200-Hz passband) gives us one sample every 2.5 ms. Our very simple software integrator will take the sum of the absolute value of 16 samples and evaluate if there is sufficient energy in that 40-ms period of time to warrant setting off the alarm (i.e., is the sum of 16 samples greater than some set point?). It is assumed that no events will be missed in 40 ms.

Design Tradeoffs:

The ADXL202 has digital (pulse-width-modulated) as well as analog (312 mV/g) outputs. In theory, either output may be used. Using the PWM interface for tilt sensing is recommended for two reasons:

1. We are interested in very small acceleration signals (on the order of 3 mg). This would correspond to approximately 0.94 mV. It is probably not resolvable by the onboard ADC of any microcontroller likely to be used in this application. The resolution of the pulse width modulator of the ADXL202 is around 14 bits and is sufficient for resolution of 3 mg acceleration signals.
2. All signal processing will be done in the digital domain.

An analog interface for the shock/vibration sensor is recommended since, as previously mentioned, bandpass filtering in the digital domain may be beyond the capability of many microcontrollers. In addition, using the PWM interface to acquire 200-Hz bandwidth requires that the PWM frequency be at least 4 kHz. The 10-bit resolution implies that the microcontroller must have a timer resolution of approximately 250 ns. Once again, this is probably beyond the capability of most microcontrollers.

Conclusion

- Measurements and instrumentation are among the most important areas of electrical engineering because virtually all engineering disciplines require the ability to perform measurements of some kind.
- A measurement system consists of three essential elements: a sensor, signal conditioning circuits, and recording or display devices. The last are often based on digital computers.
- Sensors are devices that convert a change in a physical variable to a corresponding change in an electrical variable, typically a voltage. A broad range of sensors exist to measure virtually all physical phenomena. Proper wiring, grounding, and shielding techniques are required to minimize undesired interference and noise.
- Often, sensor outputs need to be conditioned before further processing can take place. The most common signal conditioning circuits are instrumentation amplifiers and active filters.
- If the conditioned sensor signals are to be recorded in digital form by a computer, it is necessary to perform an analog-to-digital conversion process; timing and comparator circuits are also often used in this context.

HOMEWORK PROBLEMS

Section 15.1 Measurement Systems and Transducers

15.1 Most motorcycles have engine speed tachometers, as well as speedometers, as part of their instrumentation. What differences, if any, are there between the two in terms of transducers?

15.2 Explain the differences between the engineering specifications you would write for a transducer to measure the frequency of an audible sound wave and a transducer to measure the frequency of a visible light wave.

15.3 A measurement of interest in the summer is the temperature-humidity index, consisting of the sum of the temperature and the relative humidity percentage. How would you measure this? Sketch a simple schematic diagram.

15.4 Consider a capacitive displacement transducer as shown in Figure P15.4. Its capacitance is determined by the equation

$$C = \frac{0.255A}{d} \quad \text{F}$$

where A = cross-sectional area of the transducer plate (in^2) and d = air-gap length (in). Determine the change in voltage Δv_0 when the air gap changes from 0.01 to 0.015 in.

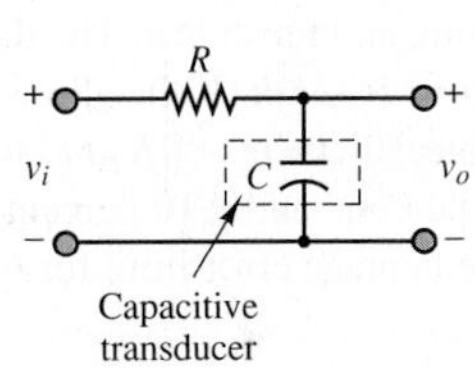

Figure P15.4

15.5 The circuit of Figure P15.5 may be used for operation of a photodiode. The voltage V_D is a reverse-bias voltage large enough to make the diode current i_D proportional to the incident light intensity H. Under this condition, $i_D/H = 0.5\ \mu\text{A-m}^2/\text{W}$.

a. Show that the output voltage V_{out} varies linearly with H.

b. If $H = 1{,}500\ \text{W/m}^2$, $V_D = 7.5$ V, and an output voltage of 1 V is desired, determine an appropriate value for R_L.

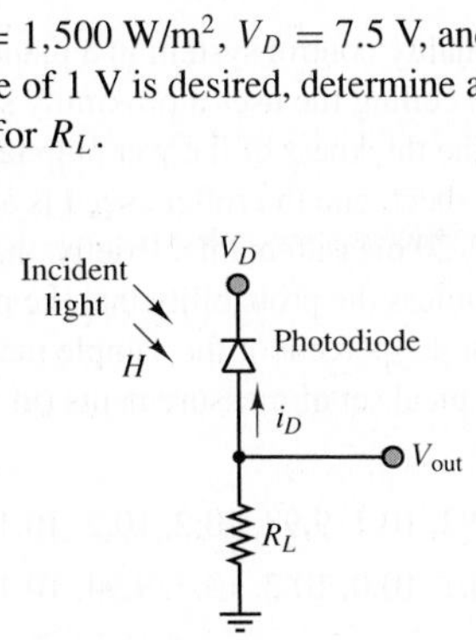

Figure P15.5

15.6 A material constant G is equal to 0.055 V-m/N for quartz in compressive stress and 0.22 V-m/N for polyvinylidene fluoride in axial stress.

a. A force sensor uses a piezoelectric quartz crystal as the sensing element. The quartz element is 0.25 in thick and has a rectangular cross section of 0.09 in^2. The sensing element is compressed, and the output voltage is measured across the thickness. What is the output of the sensor in volts per newton?

b. A polyvinylidene fluoride film is used as a piezoelectric load sensor. The film is 30 μm thick, 1.5 cm wide, and 2.5 cm in the axial direction. It is stretched by the load in the axial direction, and the output voltage is measured across the thickness. What is the output of the sensor in volts per newton?

15.7 Let b be the damping constant of the mounting structure of a machine as pictured in Figure P15.7. It must be determined experimentally. First, the spring constant K is determined by measuring the resultant displacement under a static load. The mass m is directly measured. Finally, the damping ratio ζ is measured using an impact test. The damping constant is given by $b = 2\zeta\sqrt{Km}$. If the allowable levels of error in the measurements of K, m, and ζ are ± 5 percent, ± 2 percent, and ± 10 percent respectively, estimate a percentage error limit for b.

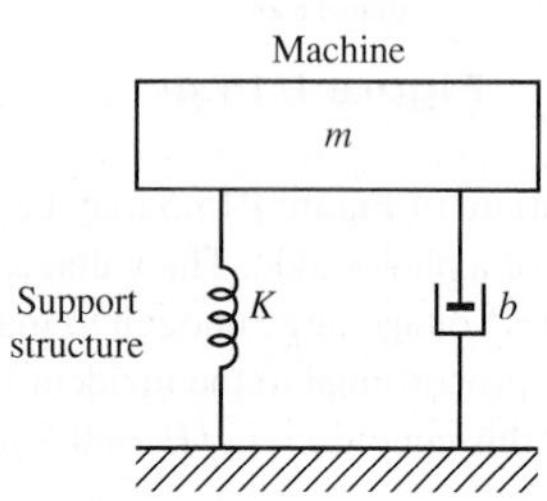

Figure P15.7

15.8 The quality control system in a plant that makes acoustical ceiling tile uses a proximity sensor to measure the thickness of the wet pulp layer every 2 ft along the sheet, and the roller speed is adjusted based on the last 20 measurements. Briefly, the speed is adjusted unless the probability that the mean thickness lies within ± 2 percent of the sample mean exceeds 0.99. A typical set of measurements (in millimeters) is as follows:

8.2, 9.8, 9.92, 10.1, 9.98, 10.2, 10.2, 10.16, 10.0, 9.94,
9.9, 9.8, 10.1, 10.0, 10.2, 10.3, 9.94, 10.14, 10.22, 9.8

Would the speed of the rollers be adjusted based on these measurements?

15.9 Discuss and contrast the following terms:

a. Measurement accuracy.
b. Instrument accuracy.
c. Measurement error.
d. Precision.

15.10 Four sets of measurements were taken on the same response variable of a process using four different sensors. The true value of the response was known to be constant. The four sets of data are shown in Figure P15.10. Rank these data sets (and hence the sensors) with respect to

a. Precision.
b. Accuracy.

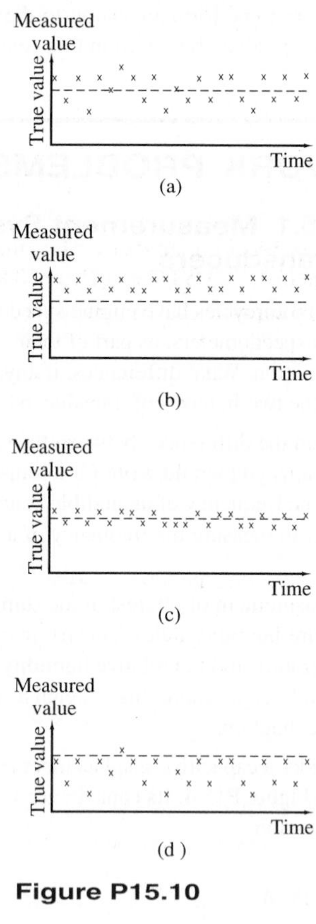

Figure P15.10

Section 15.3: Signal Conditioning

15.11 For the instrumentation amplifier of Figure P15.11, find the gain of the input stage if $R_1 = 1$ kΩ and $R_2 = 5$ kΩ.

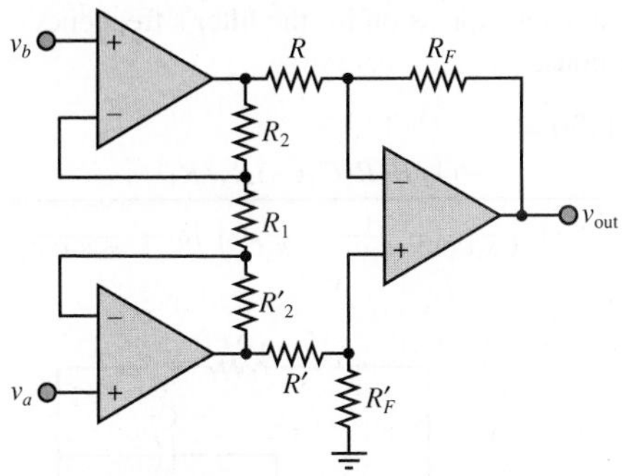

Figure P15.11

15.12 Consider again the instrumentation amplifier of Figure P15.11. Let $R_1 = 1$ kΩ. What value of R_2 should be used to make the gain of the input stage equal 50?

15.13 Again consider the instrumentation amplifier of Figure P15.11. Let $R_2 = 10$ kΩ. What value of R_1 will yield an input-stage gain of 16?

15.14 For the IA of Figure P15.11, find the gain of the input stage if $R_1 = 1$ kΩ and $R_2 = 10$ kΩ.

15.15 For the IA of Figure P15.11, find the gain of the input stage if $R_1 = 1.5$ kΩ and $R_2 = 80$ kΩ.

15.16 Find the differential gain for the IA of Figure P15.11 in the text if $R_2 = 5$ kΩ, $R_1 = R' = R = 1$ kΩ, and $R_F = 10$ kΩ.

15.17 Suppose, for the circuit of Figure P15.11, that $R_F = 200$ kΩ, $R = 1$ kΩ, and $\Delta R = 2$ percent of R. Calculate the common-mode rejection ratio of the instrumentation amplifier. Express your result in decibels.

15.18 Given the instrumentation amplifier of Figure P15.11, with the component values of Problem 15.17, calculate the mismatch in gains for the differential components. Express your result in decibels.

15.19 Given $R_F = 10$ kΩ and $R_1 = 2$ kΩ for the IA of Figure 15.11, find R and R_2 so that a differential gain of 900 can be achieved.

15.20 Replace the cutoff frequency specification of Example 15.3 with $\omega_C = 10$ rad/s, and determine the order of the filter required to achieve 40-dB attenuation at $\omega_S = 24$ rad/s.

15.21 The circuit of Figure P15.21 represents a low-pass filter with gain.

a. Derive the relationship between output amplitude and input amplitude.

b. Derive the relationship between output phase angle and input phase angle.

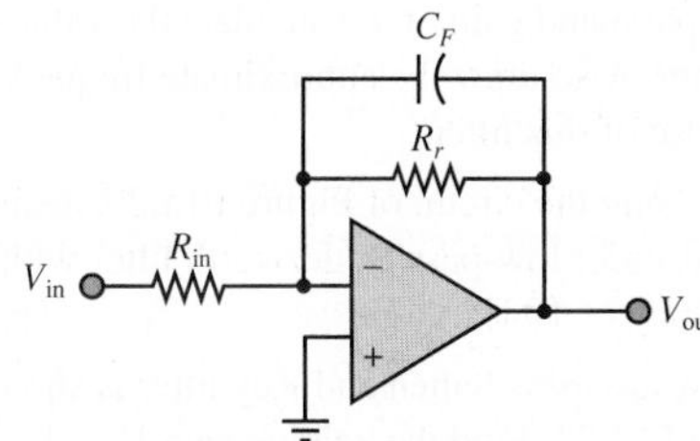

Figure P15.21

15.22 Consider again the circuit of Figure P15.21. Let $R_{in} = 20$ kΩ, $R_F = 100$ kΩ, and $C_F = 100$ pF. Determine an expression for $v_{out}(t)$ if $v_{in}(t) = 2\sin(2{,}000\pi t)$ V.

15.23 Derive the frequency response of the low-pass filter of Figure 15.22 in the text.

15.24 Derive the frequency response of the high-pass filter of Figure 15.22 in the text.

15.25 Derive the frequency response of the bandpass filter of Figure 15.22 in the text.

15.26 Consider again the circuit of Figure P15.21. Let $C_F = 100$ pF. Determine appropriate values for R_{in} and R_F if it is desired to construct a filter having a cutoff frequency of 20 kHz and a gain magnitude of 5.

15.27 Design a second-order Butterworth high-pass filter with a 10-kHz cutoff frequency, a DC gain of 10, $Q = 5$, and $V_S = \pm 15$ V.

15.28 Design a second-order Butterworth high-pass filter with a 25-kHz cutoff frequency, a DC gain of 15, $Q = 10$, and $V_S = \pm 15$ V.

15.29 The circuit shown in Figure P15.29 is claimed to exhibit a second-order Butterworth low-pass voltage gain characteristic. Derive the characteristic and verify the claim.

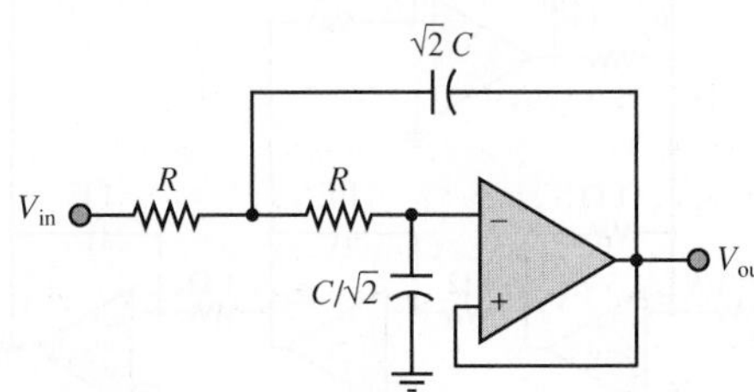

Figure P15.29

15.30 Design a second-order Butterworth low-pass filter with a 15-kHz cutoff frequency, a DC gain of 15, $Q = 5$, and $V_S = \pm 15$ V.

15.31 Design a bandpass filter with a low cutoff frequency of 200 Hz, a high cutoff frequency of 1 kHz, and a passband gain of 4. Calculate the value of Q for the filter. Also draw the approximate frequency response of this filter.

15.32 Using the circuit of Figure P15.29, design a second-order low-pass Butterworth filter with a cutoff frequency of 10 Hz.

15.33 A low-pass Sallen and Key filter is shown in Figure P15.33. Find the voltage gain V_{out}/V_{in} as a function of frequency and generate its Bode magnitude plot. Show and observe that the cutoff frequency is $1/2\pi RC$ and that the low-frequency gain is R_4/R_3.

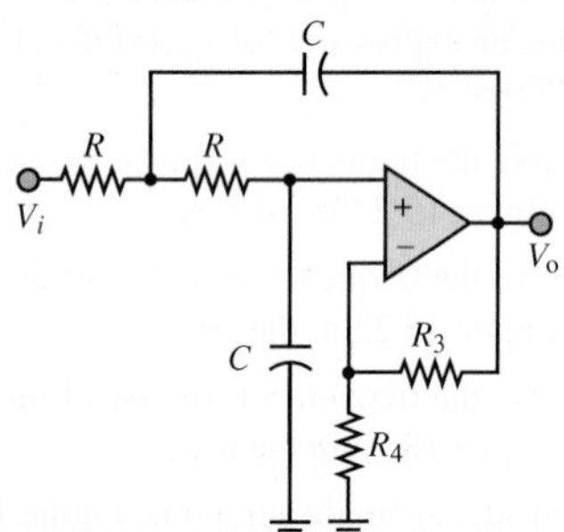

Figure P15.33

15.34 The circuit shown in Figure P15.34 exhibits low-pass, high-pass, and bandpass voltage gain characteristics, depending on whether the output is taken at node 1, node 2, or node 3. Find the transfer functions relating each of these outputs to V_{in}, and determine which is which.

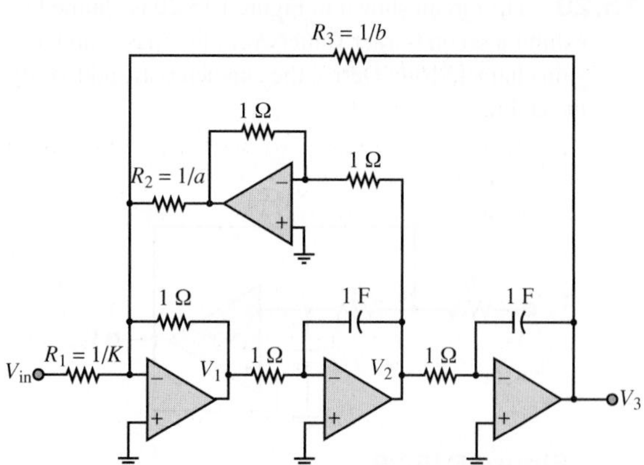

Figure P15.34

15.35 The filter shown in Figure P15.35 is called an *infinite-gain multiple-feedback filter.* Derive the following expression for the filter's frequency response:

$$H(j\omega) = \frac{-(1/R_3R_2C_1C_2)R_3/R_1}{(j\omega)^2 + \left(\frac{1}{R_1C_1} + \frac{1}{R_2C_1} + \frac{1}{R_3C_1}\right) j\omega + \frac{1}{R_3R_2C_1C_2}}$$

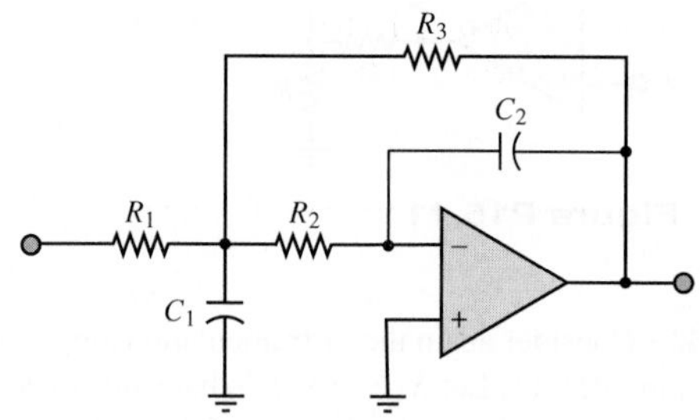

Figure P15.35

15.36 The filter shown in Figure P15.36 is a Sallen and Key bandpass filter circuit, where K is the DC gain of the filter. Derive the following expression for the filter's frequency response:

$$H(j\omega) = \frac{j\omega K/R_1C_1}{(j\omega)^2 + j\omega\left(\frac{1}{R_1C_1} + \frac{1}{R_3C_2} + \frac{1}{R_3C_1} + \frac{1-K}{R_2C_1}\right) + \frac{R_1+R_2}{R_1R_2R_3C_1C_2}}$$

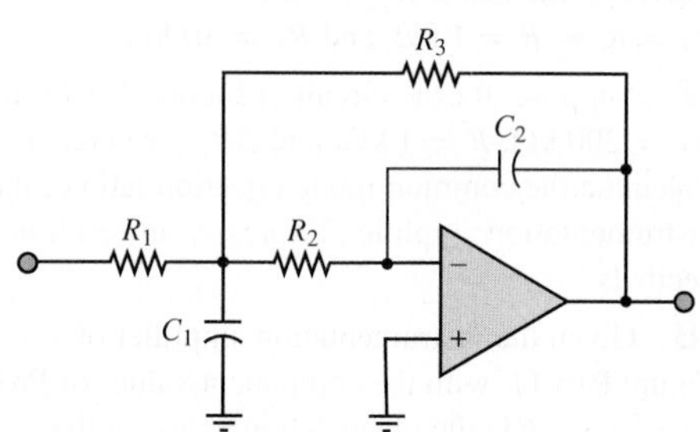

Figure P15.36

15.37 Show that the expression for Q in the filter of Problem 15.35 is given by

$$\frac{1}{Q} = \sqrt{R_2R_3\frac{C_2}{C_1}}\left(\frac{1}{R_1} + \frac{1}{R_2} + \frac{1}{R_3}\right)$$

Section 15.4: Analog-to-Digital and Digital-to-Analog Conversion

15.38 List two advantages of digital signal processing over analog signal processing.

15.39 Discuss the role of a multiplexer in a data acquisition system.

15.40 Discuss the purpose of using sample-and-hold circuits in data acquisition systems.

15.41 The circuit shown in Figure P15.41 represents a sample-and-hold circuit, such as might be used in a successive-approximation ADC. Assume that the JFET is turned *on* when V_G is high and *off* when V_G is low. Explain the operation of the circuit.

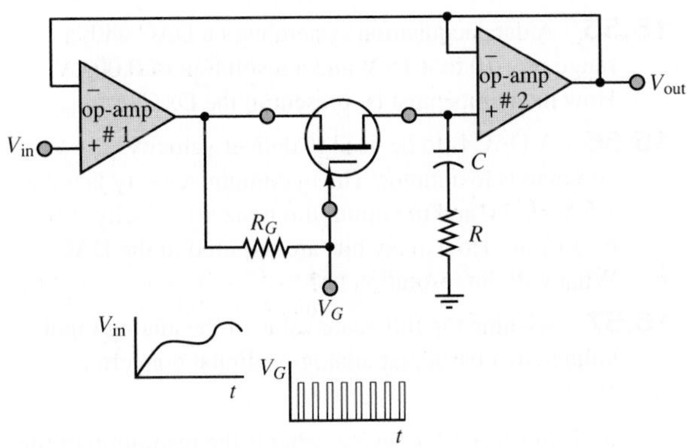

Figure P15.41

15.42 For the circuit shown in Figure P15.41, let V_{in} be a 1-kHz sinusoidal signal with 0° phase angle, 0-V DC offset, and 20-V peak-to-peak amplitude. Let V_G be a rectangular pulse train, with pulse width 10 μs and period 100 μs with the leading edge of the first pulse at $t = 0$.

a. Sketch V_{out} if the RC circuit has a time constant equal to 20 μs.

b. Sketch V_{out} if the RC circuit has a time constant equal to 1 ms.

15.43 The unsigned decimal number 12_{10} is inputted to a 4-bit DAC. Given that $R_F = R_0/15$, logic 0 corresponds to 0 V, and logic 1 corresponds to 4.5 V,

a. What is the output of the DAC?

b. What is the maximum voltage that can be outputted from the DAC?

c. What is the resolution over the range 0 to 4.5 V?

d. Find the number of bits required in the DAC if an improved resolution of 20 mV is desired.

15.44 The unsigned decimal number 215_{10} is inputted to an 8-bit DAC. Given that $R_F = R_0/255$, logic 0 corresponds to 0 V, and logic 1 corresponds to 10 V,

a. What is the output of the DAC?

b. What is the maximum voltage that can be outputted from the DAC?

c. What is the resolution over the range 0 to 10 V?

d. Find the number of bits required in the DAC if an improved resolution of 3 mV is desired.

15.45 The circuit shown in Figure P15.45 represents a simple 4-bit digital-to-analog converter. Each switch is controlled by the corresponding bit of the digital number—if the bit is 1, the switch is up; if the bit is 0, the switch is down. Let the digital number be represented by $b_3b_2b_1b_0$. Determine an expression relating v_o to the binary input bits.

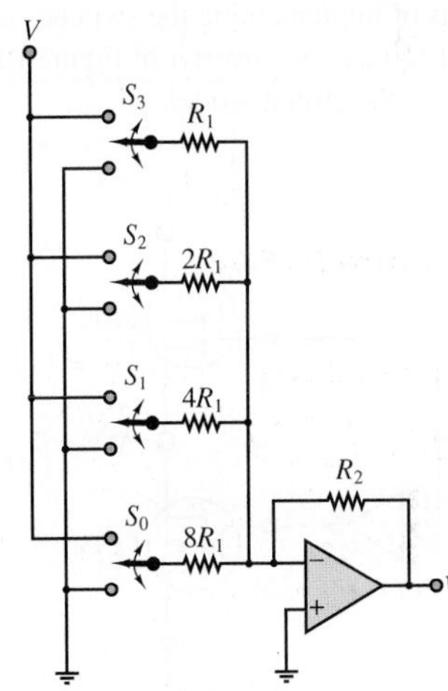

Figure P15.45

15.46 The unsigned decimal number 98_{10} is inputted to an 8-bit DAC. Given that $R_F = R_0/255$, logic 0 corresponds to 0 V, and logic 1 corresponds to 4.5 V,

a. What is the output of the DAC?

b. What is the maximum voltage that can be outputted from the DAC?

c. What is the resolution over the range 0 to 4.5 V?

d. Find the number of bits required in the DAC if an improved resolution of 0.5 mV is desired.

15.47 For the DAC circuit shown in Figure P15.47 (using an ideal op-amp), what value of R_F will give an output range of $-10 \le V_0 \le 0$ V? Assume that logic 0 = 0 V and logic 1 = 5 V.

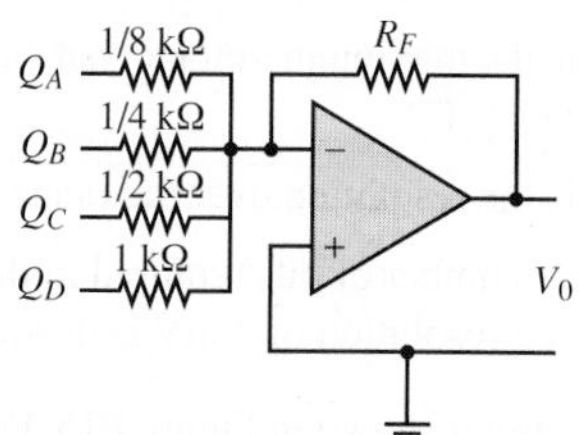

Figure P15.47

15.48 Explain how to redesign the circuit of Figure P15.45 so that the overall circuit is a "noninverting" device.

15.49 The circuit of Figure P15.49 has been suggested as a means of implementing the switches needed for the digital-to-analog converter of Figure P15.45. Explain how the circuit works.

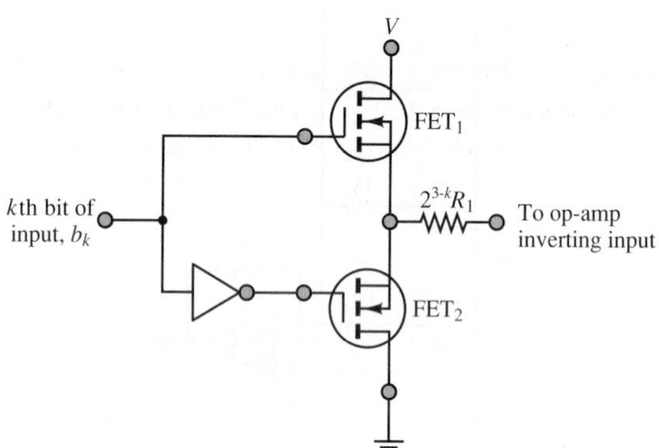

Figure P15.49

15.50 The unsigned decimal number 345_{10} is inputted to a 12-bit DAC. Given that $R_F = R_0/4{,}095$, logic 1 corresponds to 10 V, and logic 0 corresponds to 0 V,

a. What is the output of the DAC?

b. What is the maximum voltage that can be outputted from the DAC?

c. What is the resolution over the range 0 to 10 V?

d. Find the number of bits required in the DAC if an improved resolution of 0.5 mV is desired.

15.51 For the DAC circuit shown in Figure P15.47 (using an ideal op-amp), what value of R_F will give an output range of $-15 \le V_0 \le 0$ V?

15.52 Using the model of Figure P15.45, design a 4-bit digital-to-analog converter whose output is given by

$$v_o = -\frac{1}{10}(8b_3 + 4b_2 + 2b_1 + b_0)V$$

15.53 A data acquisition system uses a DAC with a range of ±15 V and a resolution of 0.01 V. How many bits must be present in the DAC?

15.54 A data acquisition system uses a DAC with a range of ±10 V and a resolution of 0.04 V. How many bits must be present in the DAC?

15.55 A data acquisition system uses a DAC with a range of −10 to +15 V and a resolution of 0.004 V. How many bits must be present in the DAC?

15.56 A DAC is to be used to deliver velocity commands to a motor. The maximum velocity is to be 2,500 r/min, and the minimum nonzero velocity is to be 1 r/min. How many bits are required in the DAC? What will the resolution be?

15.57 Assume the full-scale value of the analog input voltage to a particular analog-to-digital converter is 10 V.

a. If this is a 3-bit device, what is the resolution of the output?

b. If this is an 8-bit device, what is its resolution?

c. Make a general comment about the relationship between the number of bits and the resolution of an ADC.

15.58 The voltage range of feedback signal from a process is −5 to +15 V, and a resolution of 0.05 percent of the voltage range is required. How many bits are required for the DAC?

15.59 Eight channels of analog information are being used by a computer to close eight control loops. Assume that all analog signals have identical frequency content and are multiplexed into a single ADC. The ADC requires 100 μs per conversion. The closed-loop software requires 500 μs of computation and output time for four of the loops, and for the other four it requires 250 μs. What is the maximum frequency content that the analog signal can have according to the Nyquist criterion?

15.60 A rotary potentiometer is to be used as a remote rotational displacement sensor. The maximum displacement to be measured is 180°, and the potentiometer is rated for 10 V and 270° of rotation.

a. What voltage increment must be resolved by an ADC to resolve an angular displacement of 0.5°?

How many bits would be required in the ADC for full-range detection?

b. The ADC requires a 10-V input voltage for full-scale binary output. If an amplifier is placed between the potentiometer and the ADC, what amplifier gain should be used to take advantage of the full range of the ADC?

15.61 Suppose it is desired to digitize a 250-kHz analog signal to 10 bits using a successive-approximation ADC. Estimate the maximum permissible conversion time for the ADC.

15.62 A torque sensor has been mounted on a farm tractor engine. The voltage produced by the torque sensor is to be sampled by an ADC. The rotational speed of the crankshaft is 800 r/min. Because of speed fluctuation caused by the reciprocating action of the engine, frequency content is present in the torque signal at twice the shaft rotation frequency. What is the minimum sampling period that can be used to ensure that the Nyquist criterion is satisfied?

15.63 The output voltage of an aircraft altimeter is to be sampled using an ADC. The sensor outputs 0 V at 0-m altitude and outputs 10 V at 10,000-m altitude. If the allowable error in sensing ($\pm\frac{1}{2}$ LSB) is 10 m, find the minimum number of bits required for the ADC.

15.64 Consider a circuit that generates interrupts at fixed time intervals. Such a device is called a *real-time clock* and is used in control applications to establish the sample period as T seconds for control algorithms. Show how this can be done with a square wave (clock) that has a period equal to the desired time interval between interrupts.

15.65 Find the minimum number of bits required to digitize an analog signal with a resolution of

a. 5 percent

b. 2 percent

c. 1 percent

Section 15.5 Comparator and Timing Circuits

15.66 A useful application that exploits the open-loop characteristics of op-amps is known as a comparator. One particularly simple example, known as a window comparator, is shown in Figure P15.66(a) and (b). Show that $V_{out} = 0$ whenever $V_{low} < V_{in} < V_{high}$ and that $V_{out} = +V$ otherwise.

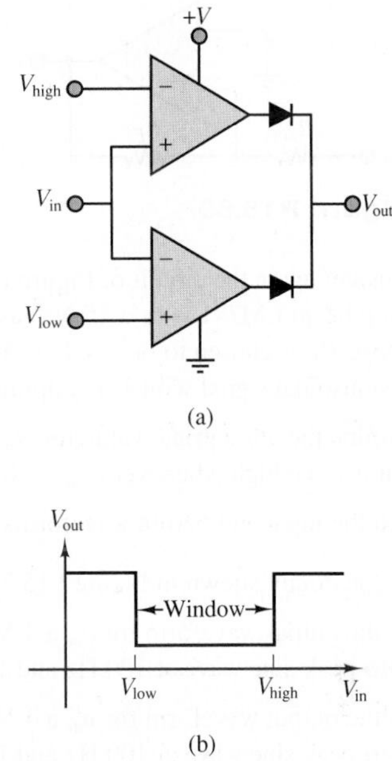

Figure P15.66

15.67 Design a Schmitt trigger to operate in the presence of noise with peak amplitude = ±150 mV. The circuit is to switch around the reference value −1 V. Assume an op-amp with ±10-V supplies ($V_{sat} = 8.5$ V).

15.68 In the circuit of Figure P15.68, $R_1 = 100\ \Omega$, $R_2 = 56\ \text{k}\Omega$, $R_i = R_1 \| R_2$, and v_{in} is a 1-V peak-to-peak sine wave. Assuming that the supply voltages are ±15 V, determine the threshold voltages (positive and negative v^+) and draw the output waveform.

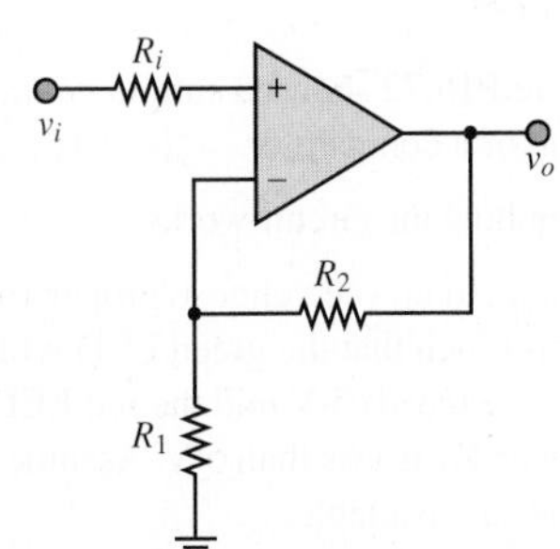

Figure P15.68

15.69 The circuit in Figure P15.69 shows how a Schmitt trigger might be constructed with an op-amp. Explain the operation of this circuit.

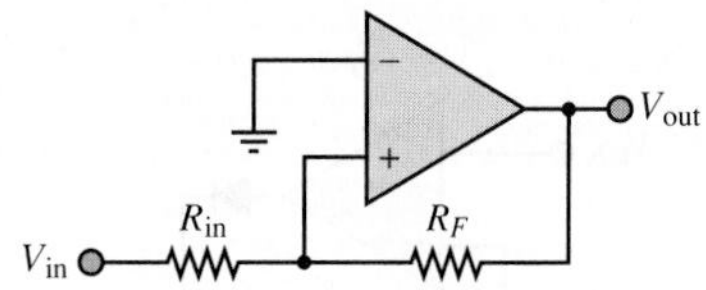

Figure P15.69

15.70 Consider again the circuit of Figure P15.69. Let the op-amp be an LM741 with ±15-V bias supplies, and suppose R_F is chosen to be 104 kΩ. Assume V_{in} is a 1-kHz sinusoidal signal with 1-V amplitude.

a. Determine the appropriate value for R_{in} if the output is to be high whenever $|V_{in}| \geq 0.25$ V.

b. Sketch the input and output waveforms.

15.71 For the circuit shown in Figure P15.71,

a. Draw the output waveform for v_{in} a 4-V peak-to-peak sine wave at 100 Hz and $V_{ref} = 2$ V.

b. Draw the output waveform for v_{in} a 4-V peak-to-peak sine wave at 100 Hz and $V_{ref} = -2$ V.

Note that the diodes placed at the input ensure that the differential voltage does not exceed the diode offset voltage.

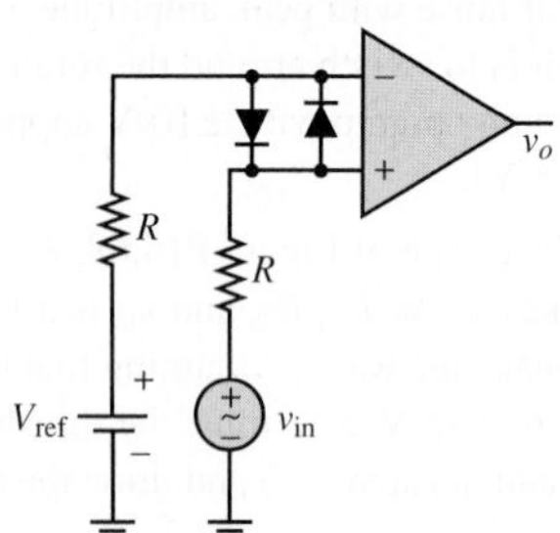

Figure P15.71

15.72 Figure P15.72 shows a simple *go–no go* detector application of a comparator.

a. Explain how the circuit works.

b. Design a circuit (i.e., choose proper values for the resistors) such that the green LED will turn on when V_{in} exceeds 5 V and the red LED will be on whenever V_{in} is less than 5 V. Assume only 15-V supplies are available.

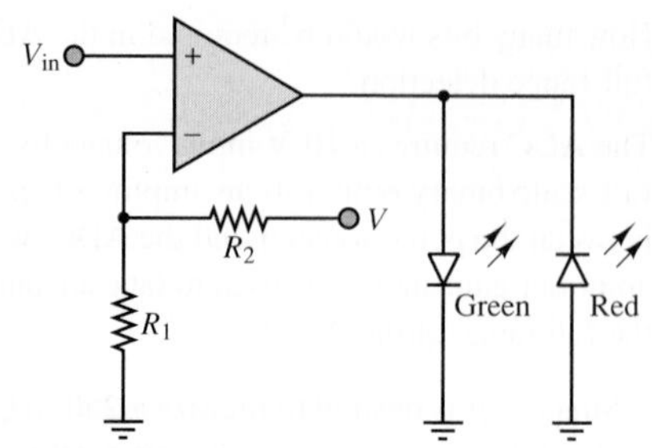

Figure P15.72

15.73 For the circuit of Figure P15.73, v_{in} is a 100-mV-peak sine wave at 5 kHz, $R = 10$ kΩ, and D_1 and D_2 are 6.2-V Zener diodes. Draw the output voltage waveform.

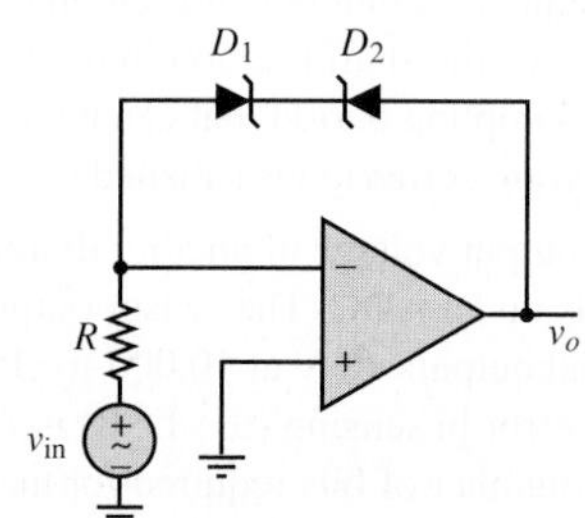

Figure P15.73

15.74 Show that the period of oscillation of an op-amp astable multivibrator is given by the expression

$$T = 2R_1 C \log_e \left(\frac{2R_2}{R_3} + 1 \right)$$

15.75 Use the data sheets for the 74123 monostable multivibrator to analyze the connection shown in Figure 15.54 in the text. Draw a timing diagram indicating the approximate duration of each pulse, assuming that the trigger signal consists of a positive-going transition.

15.76 In the NE 555 timer circuit of Figure 15.55 in the text, which represents a one-shot multivibrator circuit, $R_1 = 10$ kΩ and the output pulse width is $T = 10$ ms. Determine the value of C.

PART III

COMMUNICATION SYSTEMS

CHAPTER

16

ANALOG COMMUNICATION SYSTEMS

This chapter introduces the foundations of electrical communication systems, emphasizing basic analog communications ideas. An overview of digital communications concepts is provided in the last section.

The subject of electrical communications is one that touches everyone's life: telephones, TV, and radio have been a part of our lives for many decades. Today, new means of communications are becoming as pervasive as the traditional ones. Computer networks, satellite weather systems, and personal communication systems (pagers, cellular phones, etc.) are becoming essential parts of our everyday lives. The aim of this chapter is to present the basic mathematics of spectrum analysis, which are the foundations of all communication systems, and the basic operation of amplitude- and frequency-modulation systems. The explanation of these concepts is supplemented by the use of computer-aided tools. In addition, the chapter also includes an overview of different types of commonly used communication systems.

Learning Objectives

1. Be familiar with the most common types of communication systems in block diagram form. *Sections 16.1 and 16.5.*
2. Be capable of performing spectral analysis of simple signals using analytical and computer-aided tools. *Section 16.2.*
3. Understand the principles of AM modulation and demodulation, and perform basic calculations and numerical computations on AM signals. *Section 16.3.*
4. Understand the principles of FM modulation and demodulation, and perform basic calculations and numerical computations on FM signals. *Section 16.4.*

16.1 INTRODUCTION TO COMMUNICATION SYSTEMS

The modern era of communications began with **the telegraph and the Morse code** and rapidly moved toward radio and television. Table 16.1 summarizes some of the major dates in the history of communication systems.

Table 16.1 **A brief history of communications**

Date	Event
1838	Samuel F. B. Morse demonstrates telegraph
1876	Alexander Graham Bell patents the telephone
1897	Marconi patents a complete wireless telegraph system
1906	Lee DeForest invents the triode amplifier
1915	Bell System completes a transcontinental telephone line
1918	B.H. Armstrong perfects the superheterodyne receiver
1937	Alec Reeves conceives pulse code modulation
1938	Television broadcasting begins
WW II	Radar and microwave systems are developed
1948	The transistor is invented; Claude Shannon publishes Mathematical Theory of Communications
1956	First transoceanic telephone cable
1960	First communications satellite, Telstar I, is launched
1962–66	High-speed digital communications
1965	Mariner IV transmits pictures from Mars to Earth
1970	Color TV
1970	Commercial relay satellite telecommunications
1975	Intercontinental computer communication networks

Information, Modulation, and Carriers

The purpose of communication systems is to communicate *information;* the four most common sources of information are *speech* (or *sound*), *video,* and *data*. Regardless of the source, the information that is transmitted and received in a communication system consists of a signal, which has the information encoded in some appropriate fashion. Figure 16.1 depicts the general layout of a communication system: an

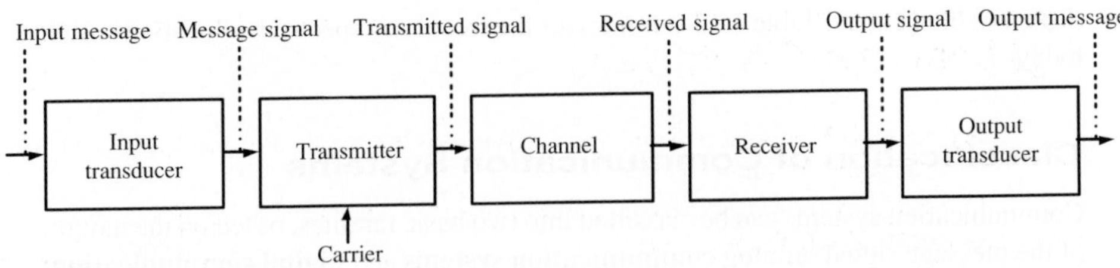

Figure 16.1 Block diagram of a communication system

input transducer (e.g., a microphone) converts the **input message** into a **message signal** (e.g., a time varying voltage) that is transmitted over a **channel,** and converted by a **receiver** into an **output signal.** An **output transducer** (e.g., a loudspeaker) converts the received signal into an **output message** (e.g., sound). The transmitter performs a very important function on communication signals by encoding the signals in some fashion making use of a **carrier signal.** The information is contained in a so-called **modulating signal** that *modulates* a carrier signal. For example, in FM radio the modulating signal consists of speech and music, and the carrier is a sinusoidal wave of predetermined frequency, much higher than the modulating signal frequency. Table 16.2 summarizes the **frequency band allocation** and typical applications in each frequency band. More detailed information can be found online.

There are two principal reasons for the use of a very broad *spectrum* of carrier frequencies. The first is that allowing for a broad spectrum permits many simultaneous users to broadcast information at different frequencies without interference among different transmissions; the second is that depending on the frequency of the carrier, the electromagnetic waves that are transmitted have different propagation characteristics. Thus, different carrier frequencies are better suited for propagating over long

FIND IT ON THE WEB

Table 16.2 **Frequency bands**

Frequency Band	Name	Medium	Applications
3–30 kHz	Very low frequency (VLF)	Wire pairs	Long-range navigation, sonar
30–300 kHz	Low frequency (LF)	Wire pairs	Navigational aids, radio beacons
300–3,000 kHz	Medium frequency (MF)	Coaxial cable	Maritime radio, direction finding, Coast Guard, commercial AM radio
3–30 MHz	High frequency (HF)	Coaxial cable	Search and rescue, aircraft communications with ships, telegraph, telephone and facsimile
30–300 MHz	Very high frequency (VHF)	Coaxial cable	VHF television channels, FM radio, private aircraft, air traffic control, taxi cabs, police
0.3–3 GHz	Ultra high frequency (UHF)	Coaxial cable, waveguide	UHF television channels, surveillance radar, satellite communications
3–30 GHz	Super high frequency (SHF)	Waveguide	Satellite communications, airborne radar, approach radar, weather radar, land mobile
30–300 GHz	Extremely high frequency (EHF)	Waveguide	Railroad service, radar landing systems, experimental
>300 GHz	Optical frequencies	Optical fiber	Wideband data, experimental

distances than others. Table 16.2 summarizes the frequency spectrum allocations used today.

Classification of Communication Systems

Communication systems can be classified into two basic families, based on the nature of the message signal: **analog communication systems** and **digital communication systems.** In this chapter we shall primarily focus on analog communications, although it should be noted that digital communications are taking an increasingly prominent role even in the most common applications.[1] Digital communications are covered in Chapter 17. Another classification may be made based on the type of transmission: *light wave* vs. *radio frequency,* or RF transmission, as is explained in the next section. A third classification is that of *carrier* vs. *direct baseband* transmission system. This latter classification is based on whether the signal of interest is directly transmitted (e.g., as in the case of the telegraph), or whether the signal *modulates* a *carrier wave,* as in the case of AM and FM radio transmission.

Communication Channels

The modulated transmitted signal can reach the receiver in a number of ways. In some cases, communication systems are hard wired. Examples of this configuration are local area computer networks, local telephone systems, and local cable TV networks. Depending on the frequency range, the transmitted signal can be carried by **twisted wire pair, coaxial cable, waveguides,** or **optical fiber.** However, in most communications systems, after the signal has been carried over a wire or cable, it is eventually broadcast over air by an antenna, to be received by a similar antenna elsewhere. Figure 16.2 depicts some typical communication system components.

The range of transmission can be significant—consider that signals can be received from the far reaches of the solar system via **radio astronomy.** The most common means of transmission of communication signals is via the broadcast of **radio frequency waves** over the air. To understand the different types of wave propagation, we need to briefly explain the geometry of the earth's atmosphere. With reference to Figure 16.3, the atmosphere is composed of layers, of which the **troposphere** and the **ionosphere** are the most important for radio wave transmission. The troposphere (up to about 20 km above sea level) is where the earth's air is contained; air density, temperature, and humidity decrease with increasing altitude. The propagation of radio waves in air depends on various properties of the medium. The speed of propagation of electromagnetic waves and the refractive index of the medium (causing the deflection of the wave) increase with altitude. As a consequence, radio waves tend to bend back toward the earth as they propagate through the troposphere.

The ionosphere is so called because of the ionization of the small amounts of air present at these altitudes (50 to 600 km). Electromagnetic waves reaching the ionosphere may propagate through it with some losses (attenuation), or may be

[1] An example of this phenomenon is the changeover from analog to digital systems in cellular telephony. Both systems coexist at the present time, but it is reasonable to forecast that in a few years all *personal communication systems* will be digital.

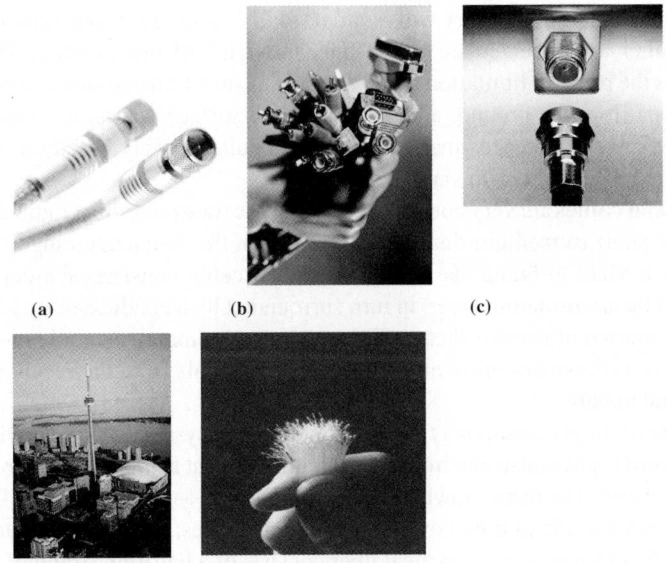

Figure 16.2 Communication system components; clockwise from top left: (a) coaxial cables; (b) RF cabling components; (c) detail of coaxial cable; (d) monopole antenna; (e) optical fiber bundle. Photos: (a) © Vol. 127/Getty; (b) © EP041/Eyewire/Getty; (c) © PhotoDisc Red/Getty; (d) © Sylvain Grandadam/Getty; (e) © Vol. 1 PhotoLink/Getty

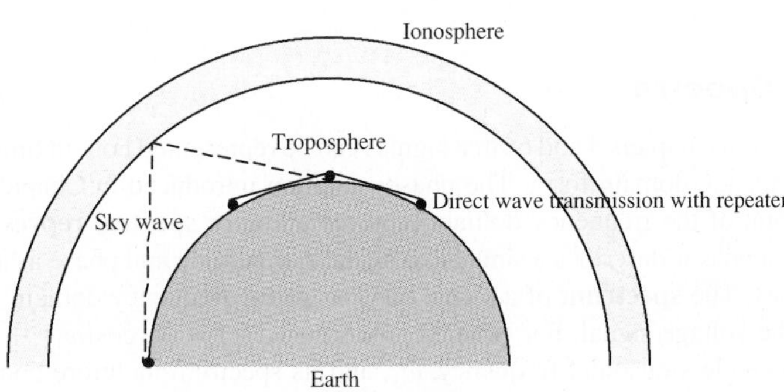

Figure 16.3 Propagation of radio frequency waves

reflected down to earth, depending on the frequency of the transmissions. In general, frequencies above 30 MHz will propagate through the ionosphere, and are therefore suitable for **space communications**.

To achieve long range communications over the earth, use is made of **sky waves.** These are waves that are reflected by the ionosphere and permit reaching points beyond the horizon. The frequencies used for these waves are below 30 MHz to permit reflection from the ionosphere. Short-wave radio makes use of the sky wave. **Tropospheric waves** can also propagate beyond the horizon, but instead of being reflected, as in the case of sky waves, they bend around the earth because of diffusion (scattering). **Direct waves** are used in *line-of-sight* transmission, where the transmitter and receiver are in the line of "sight" of one another. The earth's curvature is the primary limitation to the distance of such transmissions; however, due to reflections from the ground, and to ground and surface waves, this transmission can achieve greater distances than one would calculate simply based on the earth's curvature and the height of the antennas.

Coaxial cables are very commonly used for the transmission of radio-frequency waves over short to medium distances, typically in the frequency range between a fraction of a MHz to hundreds of MHz. Coaxial cable consists of a copper core, surrounded by an insulating layer, in turn surrounded by a conductive (ground) layer and by an external protective sheath. Today, the most common example of the use of coaxial cable is the distribution of cable television signals from the receiving station to individual homes.

An increasingly common type of communication system is based on **light wave transmission.** Light is also electromagnetic radiation, but at much higher frequencies than radio waves. The main drawback in the use of light as a carrier is that it needs to be enclosed in a guide to travel over significant distances; **optical fibers** are used to achieve such transmission. An optical fiber consists of a hair-thin strand of glass, the *core,* surrounded by a protective layer, the *cladding*. Snell's law of optics ensures that if light enters the fiber at a sufficiently low angle of incidence, the transmission benefits from *total internal reflection,* confining the light signal to the core with minimal losses. High-speed computer communications networks are increasingly making use of optical fibers.

16.2 SPECTRAL ANALYSIS

Signal Spectra

You know from Chapters 4 and 6[2] that signals can be represented both in time-domain and in frequency-domain form. The phasor notation introduced in Chapter 4 is the starting point of the frequency domain representation, or **spectral representation,** of signals: a phasor describes a sinusoidal signal's amplitude and phase as a function of frequency. The **spectrum** of a signal consists of the frequency domain representation of the voltage signal. For example, the signal $x(t) = A_1 \cos(\omega_1 t + \phi_1)$ only contains a single sinusoidal frequency, ω_1, and its spectrum therefore consists of a pair of **spectral lines** at the frequency $\pm\omega_1$. Figures 16.4(a) and (b)-(c) depict the representation of a sinusoidal signal in the *time domain* and in the *frequency domain.* Note that to completely represent the frequency domain signal one needs to consider both *magnitude* and *phase,* as was discussed in Chapter 4. Note also that the spectrum of the signal exists at both positive and negative frequencies;[3] this is a mathematical consequence of the definition of the Fourier transform, as will be shown soon.

[2]See Sections 4.4 and 6.2.

[3]Although negative frequencies have no physical significance, the mathematical form of the Fourier series and transform requires that we consider the spectrum of the signal at both positive and negative frequencies.

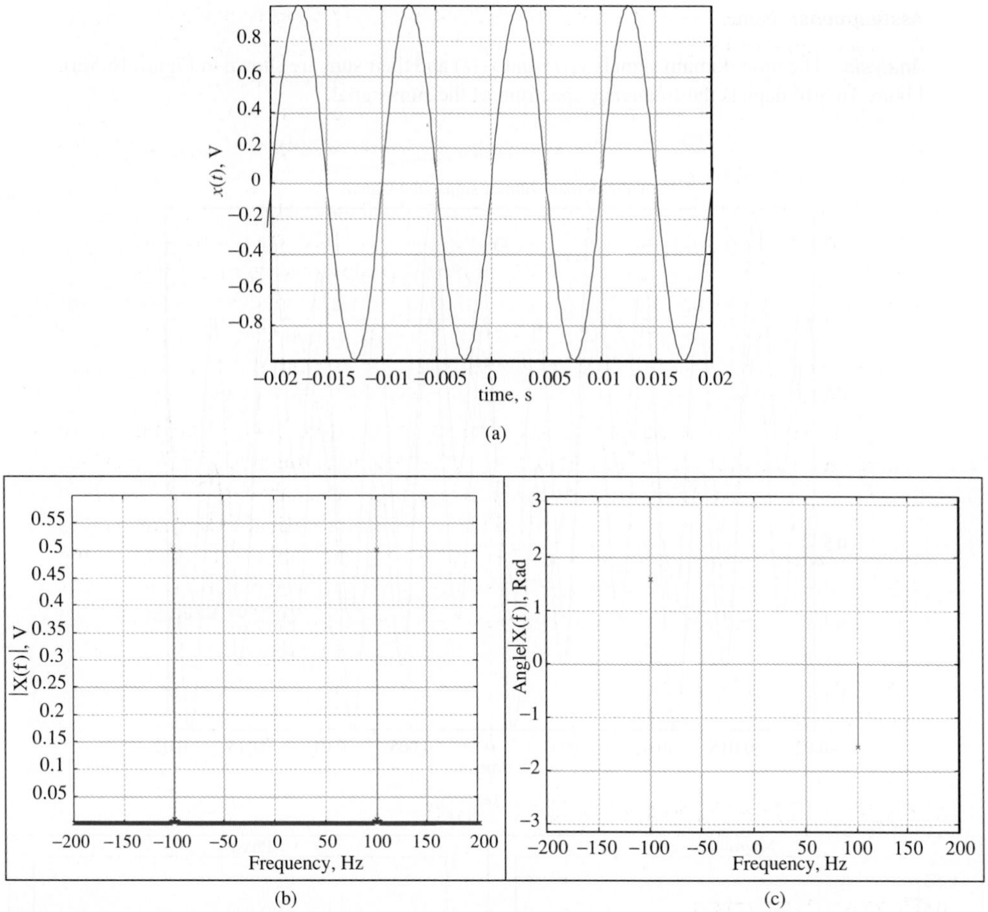

Figure 16.4 Time-domain (a) and frequency-domain [(b) magnitude, (c) phase] representation of a sinusoidal voltage (amplitude: 1 V-peak; phase: 0 rad)

EXAMPLE 16.1 Sinusoidal Signal Spectrum

Problem

Generate the spectrum of a signal consisting of the addition of two unity-amplitude sine waves for different frequencies and phases. Plot the time-domain sum and the spectrum of the sum.

Solution

Known Quantities: Sine wave amplitude, frequency, and phase.

Find: Plot the time-domain sum of the signals and the frequency domain spectrum.

Schematics, Diagrams, Circuits, and Given Data: $\omega_1 = 300$ rad/s; $\omega_2 = 500$ rad/s; $\phi_1 = 0$ rad; $\phi_2 = \pi/4$ rad/s.

Assumptions: None.

Analysis: The time-domain signals $x_1(t)$ and $x_2(t)$ and their sum are shown in Figure 16.5(a). Figure 16.5(b) depicts the frequency spectrum of the sum signal.

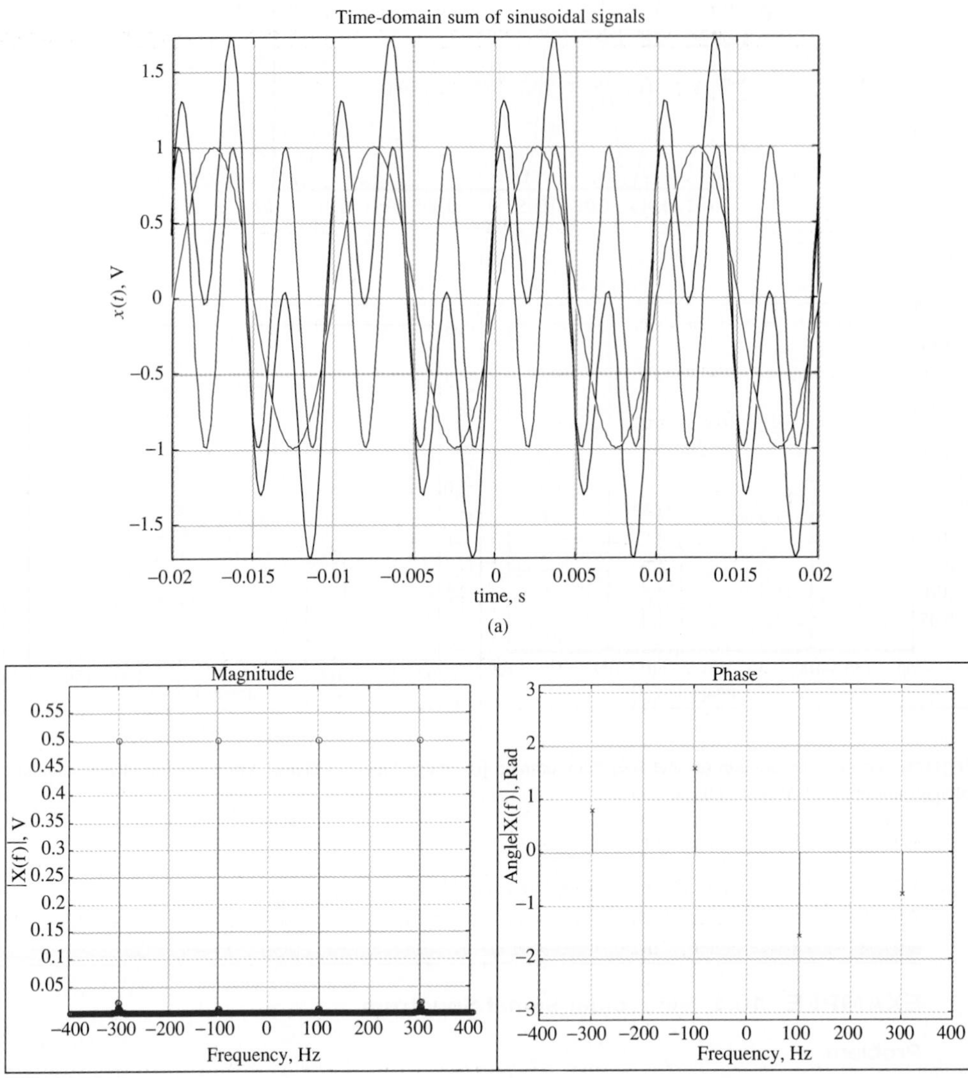

Figure 16.5 Time-domain (a) and frequency-domain (b) representation of the sum of two sinusoidal voltages

Comments: Note that the signal amplitude in the time domain is divided between two spectral lines at each signal component frequency, and at the corresponding negative frequency; thus, signal power is preserved. The phase angle (at the positive frequencies) is shown to be $-\pi/2$ for the signal $x_1(t)$ because the signal is a sine wave (in Chapter 4 we defined the cosine as the reference function, with zero phase angle); thus, $x_2(t)$ has phase angle $-\pi/2 + \pi/4 = -\pi/4$.

Periodic Signals: Fourier Series

Periodic signals (see Chapter 4 for an introduction to signal waveforms) can be represented by the infinite summation of sinusoidal signals, as explained in Section 6.2. Example 16.2 provides a review of the basic concepts.

EXAMPLE 16.2 Fourier Series of Pulse Train

Problem

Compute the complete Fourier series of the periodic pulse train shown in Figure 16.6.

Solution

Known Quantities: Amplitude, period, and functional form of the signal.

Find: b_n and c_n coefficients as a function of n.

Schematics, Diagrams, Circuits, and Given Data: $\delta = \frac{\tau}{T} = 0.2$.

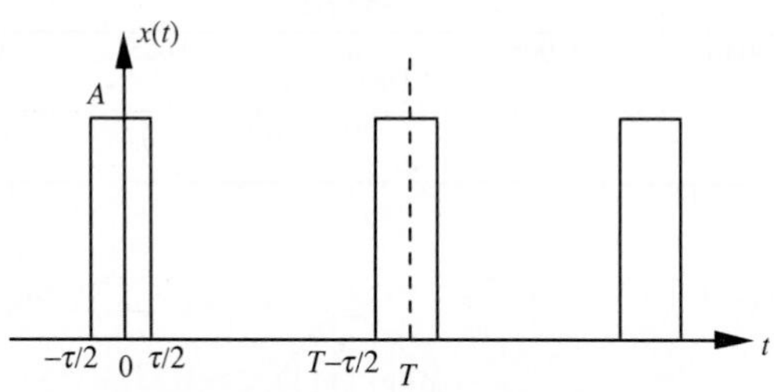

Figure 16.6 Periodic pulse train

Assumptions: The function repeats periodically.

Analysis: The function of Figure 16.6 is neither even nor odd. Since there is nothing to be gained by using the quadrature (odd-even) representation, we shall use the complex form. The expression for $x(t)$ is very simple in this case:

$$x(t) = A \quad \text{for} \quad -\frac{\tau}{2} < t \leq \frac{\tau}{2}$$
$$x(t) = 0 \quad \text{for} \quad \frac{\tau}{2} < t \leq T - \frac{\tau}{2}$$

To determine the complex coefficient of the Fourier series, we evaluate the integral below (see Chapter 6 for a definition of Fourier coefficients):

$$\gamma_n = \frac{1}{T}\int_{-\frac{T}{2}}^{\frac{T}{2}} x(t)e^{-jn\frac{2\pi}{T}t}\,dt = \frac{1}{T}\int_{-\frac{\tau}{2}}^{\frac{\tau}{2}} Ae^{-jn\frac{2\pi}{T}t}\,dt$$

$$\gamma_n = \frac{A}{T}\frac{1}{\left(-jn\frac{2\pi}{T}\right)}\left[e^{-jn\frac{2\pi}{T}t}\right]_{-\frac{\tau}{2}}^{\frac{\tau}{2}} = \frac{A}{\pi n}\left[\frac{e^{-jn\frac{\pi\tau}{T}} - e^{jn\frac{\pi\tau}{T}}}{-2j}\right] = \frac{A}{\pi n}\left[\frac{e^{jn\pi\delta} - e^{-jn\pi\delta}}{2j}\right]$$

$$\gamma_n = \frac{A}{\pi n}\sin(n\pi\delta) \qquad n = 0, \pm1, \pm2, \cdots$$

We may simplify the notation by using the *sinc function*, defined as:

$$\text{sinc}(x) = \frac{\sin(\pi x)}{\pi x}$$

Thus we may rewrite the coefficients as:

$$\gamma_n = \frac{A}{\pi n}\sin(n\pi\delta) = \delta\text{sinc}(n\delta)$$

Figure 16.7(a) depicts the pulse train corresponding to the numerical values given above, and Figure 16.7(b) its Fourier series coefficients up to $n = 1{,}000$. The *envelope* of the discrete-frequency coefficients is the sinc function defined above.

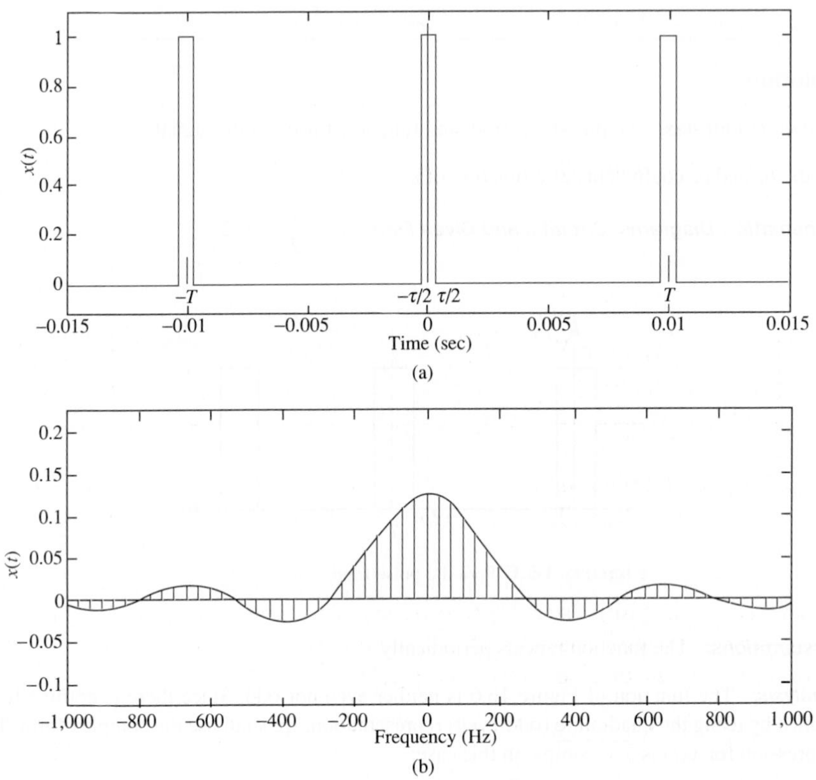

Figure 16.7 Time-domain representation and Fourier series spectrum of periodic pulse train with $\delta = 0.2$

Comments: Note that in the complex form of the Fourier series, the coefficients range from negative infinity to positive infinity.

Non-Periodic Signals: Fourier Transform

Practical communication signals have both periodic and non-periodic components. Typically, the carrier waveform is periodic (usually a sine wave), while the modulating signal, consisting of speech, music, video, or data, is non-periodic. The analysis of non-periodic signals uses a mathematical tool different from (but related to) the Fourier series: the **Fourier transform.** The Fourier transform, also named after the French mathematician Jean-Baptiste Joseph Fourier, is an integral transform, so called because it is mathematically represented by an integral, and because it performs a *transformation* between two domains: the **time domain** and the **frequency,** or **spectral, domain.**

The Fourier transform of a function $x(t)$ is the function $X(\omega)$ defined by the integral

$$X(\omega) = \int_{-\infty}^{\infty} x(t)e^{-j\omega t}dt \qquad \text{Fourier transform} \tag{16.1}$$

Conversely, if the function $X(\omega)$ is known, the **inverse Fourier transform** is defined by:

$$x(t) = \frac{1}{2\pi}\int_{-\infty}^{\infty} X(\omega)e^{j\omega t}d\omega \qquad \text{Inverse Fourier transform} \tag{16.2}$$

The pair $x(t)$ and $X(\omega)$ represent a Fourier transform pair, a relationship usually denoted by $x(t) \leftrightarrow X(\omega)$. Table 16.3 provides a useful summary of common Fourier transform pairs.

Before proceeding with the Fourier transform, it will be useful to define a function called the **unit impulse** or **delta function.** This function plays a very important role in Fourier analysis. The unit impulse function, $\delta(t)$, can be defined as the derivative of the unit step function (see Appendix B, equation B.13).

$$\delta(t) = \frac{du(t)}{dt} \quad \text{or} \quad u(t) = \int_{-\infty}^{t} \delta(t')dt' \qquad \text{Delta or unit impulse function} \tag{16.3}$$

The unit impulse function has three important properties: (1) it has area equal to one; (2) it has infinite amplitude; and (3) it has zero duration, that is, its occurrence is concentrated at one instant in time. Clearly, such a signal is a mathematical abstraction, since it is impossible to physically generate a signal that has zero duration and infinite amplitude. Figure 16.8 shows how one can think of the delta function as the limit of a sequence of rectangular pulses that are increasingly narrow and tall, such that the product of height ($1/\varepsilon$) and width (ε) is always equal to 1: the delta function can be thought of as the limit of this sequence as ε approaches zero.

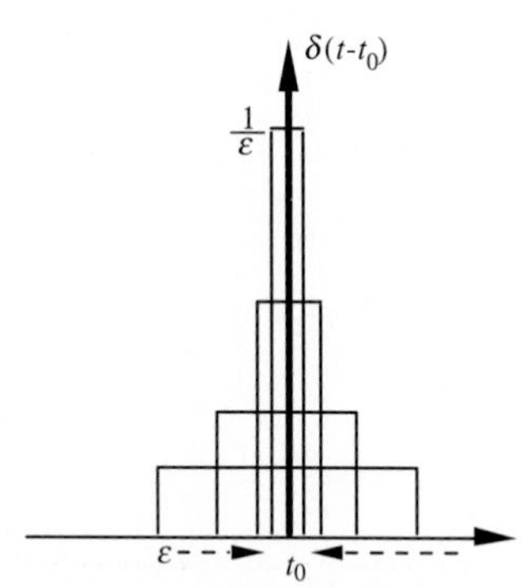

Figure 16.8 Delta function as the limit of a sequence of rectangular pulses of unit area

Table 16.3 Properties of Fourier transforms

Property	Signal	Fourier transform
Time shift	$x(t-t_0)$	$e^{-j\omega t_0}X(\omega)$
Frequency shift	$e^{j\omega t_0}x(t)$	$X(\omega-\omega_0)$
Complex conjugate	$x^*(t)$	$X^*(-\omega)$
Time reflection	$x(-t)$	$X(-\omega)$
Scaling	$x(at)$	$\frac{1}{a}X\left(\frac{\omega}{a}\right)$
Convolution	$x(t)^*y(t)$	$X(\omega)\cdot Y(\omega)$
Multiplication	$x(t)\cdot y(t)$	$\frac{1}{2\pi}X(\omega)*Y(\omega)$
Differentiation	$\frac{d}{dt}x(t)$	$j\omega X(\omega)$
Integration	$\int_{-\infty}^{t} x(t)dt$	$\frac{1}{j\omega}X(\omega)+\pi X(0)\delta(\omega)$

The delta function has one further property that is of interest in signal analysis:

$$\int_{-\infty}^{\infty} x(t)\delta(t-t_0)\,dt = x(t_0) \tag{16.4}$$

that is, the delta function "samples" the function $x(t)$ at the time of the occurrence of the impulse.

A number of properties of the Fourier transform can assist in the computation of Fourier transforms. The most important properties are summarized in Table 16.3. Some Fourier transform pairs are listed in Table 16.4.

Table 16.4 Fourier transform pairs

	x(t)	$X(\omega)$
1	$\delta(t)$	1
2	1	$2\pi\delta(\omega)$
3	$u(t)$ (unit step)	$2\pi\delta(\omega)+\frac{1}{j\omega}$
4	$tu(t)$ (unit ramp)	$2\pi\frac{d\delta(\omega)}{d\omega}-\frac{1}{\omega^2}$
5	$e^{-\alpha t}u(t)\quad \alpha>0$	$\frac{1}{\alpha+j\omega}$
6	$te^{-\alpha t}u(t)\quad \alpha>0$	$\frac{1}{(\alpha+j\omega)^2}$
7	$e^{j\omega_0 t}$	$2\pi\delta(\omega-\omega_0)$
8	$\cos(\omega_0 t)$	$\pi[\delta(\omega-\omega_0)+\delta(\omega+\omega_0)]$
9	$\sin(\omega_0 t)$	$-j\pi[\delta(\omega-\omega_0)-\delta(\omega+\omega_0)]$
10	$\cos(\omega_0 t)\,u(t)$	$\frac{\pi}{2}[\delta(\omega-\omega_0)+\delta(\omega+\omega_0)]+\frac{j\omega}{\omega_0^2-\omega^2}$
11	$\sin(\omega_0 t)\,u(t)$	$-j\frac{\pi}{2}[\delta(\omega-\omega_0)-\delta(\omega+\omega_0)]+\frac{\omega_0}{\omega_0^2-\omega^2}$

The importance of the Fourier transform is that it allows us to view signals in the frequency or spectral domain. As you shall see shortly, the spectral representation of signals is much more convenient and effective in representing communication signals, among other reasons because it permits defining important concepts such as *bandwidth* and *spectrum allocation*. You already know that sinusoidal signals are represented by single frequencies in the spectrum, from the Fourier series discussion in the preceding subsection. In the following three examples, we compute the Fourier transform of a well-known signal, a sinusoid; of a signal we have not yet analyzed, a single rectangular pulse; and of a signal that occurs very frequently in communication systems, a sine wave burst, or RF pulse. These examples will help you develop a feel for the spectral content of different types of signals.

EXAMPLE 16.3 Fourier Transform of Sine Wave

Problem

Compute the Fourier transform of an arbitrary sinusoidal signal.

Solution

Known Quantities: Functional form of the signal $x(t)$.

Find: $X(\omega)$.

Schematics, Diagrams, Circuits, and Given Data: Figure 16.9.

Assumptions: None

Analysis: The signal shown in Figure 16.9, is defined as

$$x(t) = \sin(\omega_0 t)$$

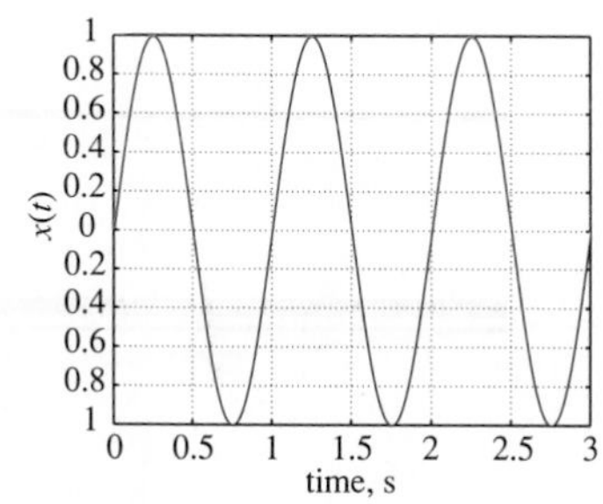

Figure 16.9 Sinusoidal signal of frequency ω_0.

The Fourier transform for the signal is calculated using the complex representation

$$\gamma_n = \frac{1}{T}\int_{-\frac{T}{2}}^{\frac{T}{2}} x(t)\, e^{-jn\frac{2\pi}{T}t} dt$$

$$= \frac{1}{T}\int_{-\frac{T}{2}}^{\frac{T}{2}} \sin(\omega_0 t)\, e^{-jn\frac{2\pi}{T}t} dt$$

$$= \frac{1}{T2j}\int_{-\frac{T}{2}}^{\frac{T}{2}} \left[e^{j\frac{2\pi}{T}t} - e^{-j\frac{2\pi}{T}t}\right] e^{-jn\frac{2\pi}{T}t} dt$$

$$\gamma_n = \frac{1}{2j}\frac{1}{T}\int_{-\frac{T}{2}}^{\frac{T}{2}} \left[e^{j\frac{2\pi}{T}(1-n)t} - e^{j\frac{2\pi}{T}(-n-1)t}\right] dt$$

Using the relation from Table 16.3 for the defined Fourier transform pairs, the above integral can be evaluated as:

$$\gamma_n = \frac{1}{2j}\left[2\pi\,\delta\,(\omega - \omega_0) - 2\pi\,\delta\,(\omega + \omega_0)\right]$$

$$\gamma_n = \frac{\pi}{j}\left[\delta\,(\omega - \omega_0) - \delta(\omega + \omega_0)\right]$$

Thus, the Fourier transform of an arbitrary sinusoid consists of a pair of delta functions in the frequency domain, as shown in Figure 16.10.

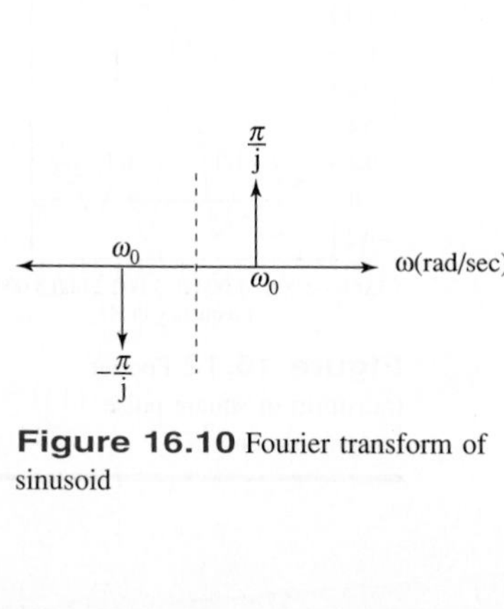

Figure 16.10 Fourier transform of sinusoid

Comments: We can extend the result of this example to an arbitrary periodic signal since we know that any periodic signal can be represented by the sum of an infinite number of sinusoidal functions. The Fourier transform $X(\omega)$ of a periodic signal $x(t)$ is a train of impulses occurring at the harmonically related frequencies and for which the area of the impulse at the nth harmonic frequency $n\omega_0$ is 2π times the nth Fourier series coefficient a_n. The Fourier series coefficients for this sinusoid signal are

$$a_1 = \frac{1}{2j}, \quad a_{-1} = -\frac{1}{2j}, \quad a_k = 0, \quad |k| \neq 1$$

Hence, the Fourier transform coefficients are given by

$$a_1 = \frac{\pi}{j}, \quad a_{-1} = -\frac{\pi}{j}, \quad a_k = 0 \quad |k| \neq 1$$

EXAMPLE 16.4 Fourier Transform of Rectangular Pulse Signal

Problem

Compute the Fourier transform of a square pulse signal.

Solution

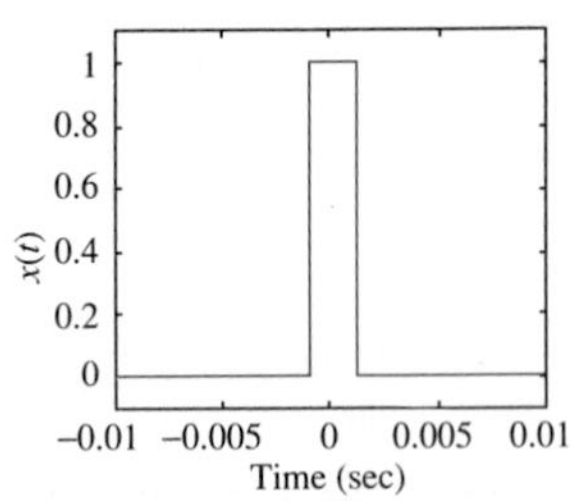

Figure 16.11 Rectangular pulse

Known Quantities: Functional form of the signal $x(t)$.

Find: $X(\omega)$.

Schematics, Diagrams, Circuits, and Given Data: Figure 16.11.

Assumptions: None.

Analysis: Consider the rectangular pulse $x(t)$ of duration T and unity amplitude shown in Figure 16.11. We define this pulse mathematically as follows:

$$x(t) = \begin{cases} 1 & \text{for} \quad |t| \leq \frac{T}{2} \\ 0 & \text{for} \quad |t| > \frac{T}{2} \end{cases}$$

The Fourier transform of $x(t)$ is

$$X(\omega) = \int_{-\frac{T}{2}}^{\frac{T}{2}} e^{-j\omega t} dt = \left[\frac{e^{-j\omega t}}{-j\omega}\right]_{-\frac{T}{2}}^{\frac{T}{2}} = \frac{2}{\omega}\left[\frac{e^{j\omega\frac{T}{2}} - e^{-j\omega\frac{T}{2}}}{2j}\right] = \frac{2}{\omega}\sin(\omega\frac{T}{2})$$

$$X(\omega) = T\,\text{sinc}\left(\omega\frac{T}{2}\right) \qquad \text{where,} \ \text{sinc}(x) = \frac{\sin(x)}{x}$$

$$X(f) = T\,\text{sinc}(\pi f T)$$

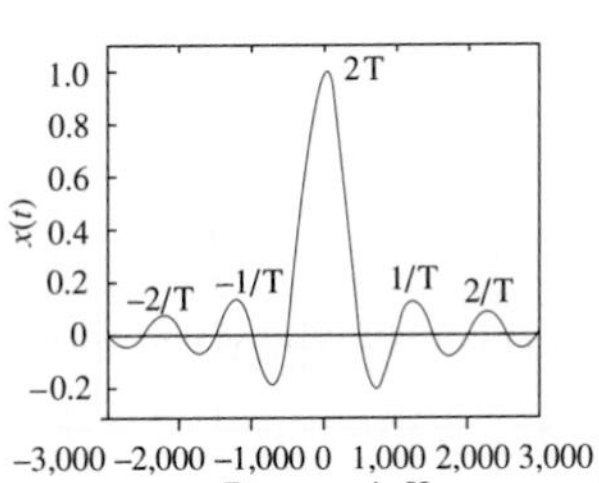

Figure 16.12 Fourier transform of square pulse (magnitude only)

A plot of the spectrum is shown in Figure 16.12. The figure illustrates the characteristics of the sinc function, with zero crossings at integer multiples of π/T rad/s, and peak amplitude of $2T$.

Comments: Single and repetitive square bursts occur commonly in communication systems. The analysis completed in this example will be useful in the following sections.

EXAMPLE 16.5 Fourier Transform of Sine Burst (RF Pulse)

Problem

Compute the Fourier transform of the sine wave burst shown in Figure 16.13.

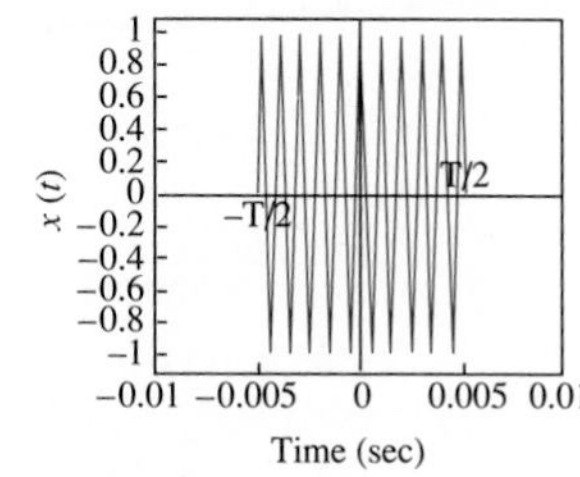

Figure 16.13 Radio-frequency (RF) burst

Solution

Known Quantities: Functional form of the signal $x(t)$.

Find: $X(\omega)$.

Schematics, Diagrams, Circuits, and Given Data: Figure 16.13.

Assumptions: None.

Analysis: The pulse signal $x(t)$ shown in Figure 16.13 consists of a sinusoidal wave of unit amplitude and frequency f_c, for a duration $t = -T/2$ to $t = T/2$. The signal $x(t)$ can be defined mathematically as follows:

$$x(t) = \begin{cases} A\cos(2\pi f_c t) & \text{for} \quad |t| \le \frac{T}{2} \\ 0 & \text{for} \quad |t| > \frac{T}{2} \end{cases}$$

The Fourier transform of $x(t)$ is

$$X(\omega) = \int_{-\frac{T}{2}}^{\frac{T}{2}} x(t)\, e^{-j\omega t}\, dt = \int_{-\frac{T}{2}}^{\frac{T}{2}} \frac{A}{2}\left(e^{j\omega_c t} + e^{-j\omega_c t}\right) e^{-j\omega t}\, dt$$

$$= \frac{A}{2}\left[\frac{e^{-j(\omega-\omega_c)t}}{-j(\omega-\omega_c)} + \frac{e^{-j(\omega+\omega_c)t}}{-j(\omega+\omega_c)}\right]_{-\frac{T}{2}}^{\frac{T}{2}}$$

$$= A\left[-\frac{e^{-j(\omega-\omega_c)\frac{T}{2}}}{2j(\omega-\omega_c)} - \frac{e^{-j(\omega+\omega_c)\frac{T}{2}}}{2j(\omega+\omega_c)} + \frac{e^{j(\omega-\omega_c)\frac{T}{2}}}{2j(\omega-\omega_c)} + \frac{e^{j(\omega+\omega_c)\frac{T}{2}}}{2j(\omega+\omega_c)}\right]$$

$$= A\left[\frac{\sin\pi\,(f+f_c)\,T}{2\pi\,(f+f_c)} + \frac{\sin\pi\,(f-f_c)\,T}{2\pi\,(f-f_c)}\right]$$

$$X(\omega) = \frac{AT}{2}\,[\mathrm{sinc}(f+f_c)T + \mathrm{sinc}(f-f_c)T]$$

Note that we could have used the frequency shifting property of the Fourier transform to obtain the above result without carrying out the integration explicitly. The magnitude spectrum of the RF pulse is shown in Figure 16.14. It clearly illustrates the *frequency-shifting property* of the Fourier transform.

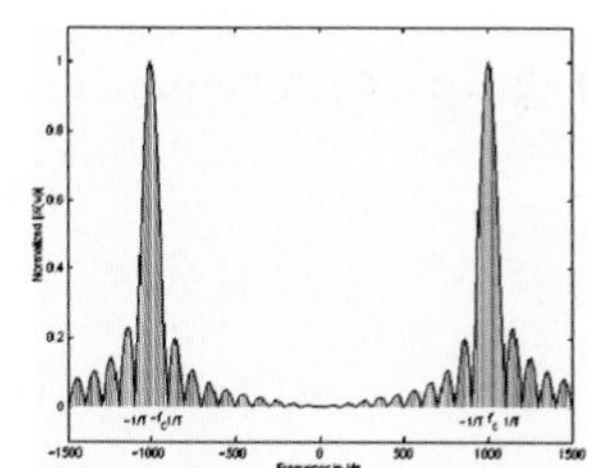

Figure 16.14 Magnitude spectrum of RF burst

Comments: This signal is specifically referred to as a *RF pulse* when the frequency f_c of the sinusoid wave falls in the radio frequency range.

Bandwidth

The **bandwidth** of a signal is the range of frequencies comprising the spectrum of the signal. Bandwidth is a very important concept in communication systems, as the allocation of radio frequency spectrum for different communication systems permits the transmission of signal within a certain specified bandwidth. For example,

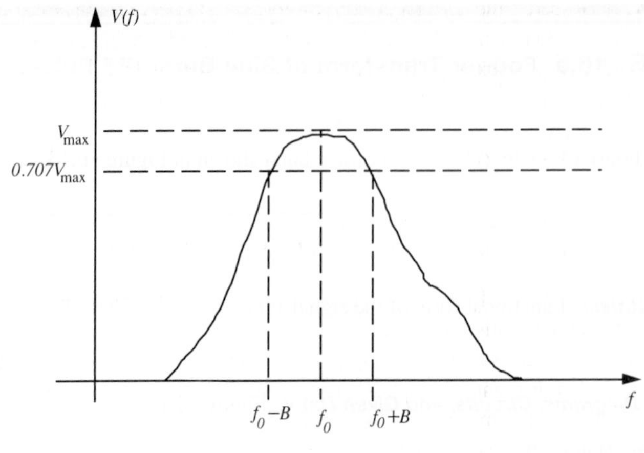

Figure 16.15 Definition of 3-dB (half-power) bandwidth

standard FM radio allows a bandwidth of 200 kHz for each radio station. The most common definition of bandwidth is that of **3-dB bandwidth,** also called **half-power bandwidth.** The 3-dB bandwidth of a signal is defined as the frequency range between points where the signal level is 3 dB below its maximum pass-band value. This informal definition is illustrated in Figure 16.15, where an arbitrary voltage signal is shown to have a spectrum $V(f)$, with *center frequency* f_0 and 3-dB bandwidth $2B$.

You will recall from the definitions given in Chapter 6 that the 3-dB point in a frequency plot is the frequency where the amplitude has dropped to a value equal to $1/\sqrt{2}$, or 0.707, times the maximum value. Since signal power is proportional to the square of the voltage, the 3-dB bandwidth is also called the half-power bandwidth. Thus, half of the signal power is contained in the frequency band $f_0 - B$ to $f_0 + B$; we call $2B$ the bandwidth of the signal. Please observe that this informal definition assumes that the signal spectrum has a band-pass shape. This is usually the case for most, if not all, communication signals.

How much bandwidth does a signal require? This depends on two facts: (1) the bandwidth of the signal itself, and (2) the type of modulation. We shall revisit the concept of bandwidth when we explore amplitude- and frequency-modulation systems.

EXAMPLE 16.6 Bandwidth of Commercial AM (or TV, or FM) Signals

Problem

Analyze the bandwidth of the signal from a commercial AM station and determine how many stations can be assigned frequencies over the frequency band assigned to commercial AM.

Solution

Schematics, Diagrams, Circuits, and Given Data: see http://www.ntia.doc.gov/osmhome/allochrt.pdf

Analysis: As can be seen in the diagram displayed in the above website, the AM band frequency allocation goes from 535 to 1,605 kHz. Each channel is allocated a bandwidth of approximately 10 kHz (we shall see in the next section what this calculation includes). Thus, the total number of stations that can operate in the same region is approximately equal to:

$$N = \frac{1605 - 535}{10} = 107$$

Each AM station can operate with a total bandwidth of 10 kHz. As we shall see in the next section, this actually corresponds to a signal bandwidth of only 5 kHz.

Comments: You are probably aware of the fact that the FCC licenses many more than 107 AM stations in the U.S.A. This is possible because AM broadcast has a limited range, and two stations can be assigned the same frequency if they are located sufficiently far apart. You may also have noticed that at night it is possible to receive AM radio signals from much greater distances (for example, in Ohio one can tune in stations from as far as New York City and New Orleans late at night). This is a consequence of the change in ionization density in the ionosphere during the night, permitting reflection of radio waves over a longer range. The FCC regulates not only the frequency allocated, but also the power allocated to a given station; a station may be required to switch to a lower-power transmitter during certain times of the day.

CHECK YOUR UNDERSTANDING

a. Compute all the coefficients of the Fourier series expansion for the signal $x(t) = 1.5\cos(100t)$.

b. Sketch the Fourier transform of a square pulse with unity amplitude and with a duration of 10 μs.

c. The spectrum of a signal can be described by the function $X(\omega) = \frac{\alpha\omega}{\omega^2+\alpha^2}$. Let $\alpha = 10^3$, and calculate the 3-dB bandwidth of the signal. What is the center frequency of the signal?

16.3 AMPLITUDE MODULATION AND DEMODULATION

The concept of **amplitude modulation** (AM) was introduced in Chapter 4 (*Focus on measurements: capacitive displacement transducer*), where it was shown that the signal produced by a capacitive microphone (displacement transducer) inserted in a Wheatstone bridge circuit, modulated the amplitude of a sinusoidal excitation signal. In that example, the pressure changes sensed by the microphone constituted the *modulation,* while the sinusoidal excitation provided a *carrier*. In Chapter 8 (*Focus on measurements: peak detector for capacitive displacement transducer*), it was shown that a diode circuit was capable of demodulating the AM signal, and of recovering the desired information (pressure changes corresponding to acoustic waves, that is, sound). In this section, the same basic principles introduced in the above mentioned examples will be discussed more formally, as they apply to AM communication systems.

The most common manifestation of amplitude modulation in communication systems is commercial AM radio, or *standard AM*. The *Federal Communications*

Commission (FCC), a body that regulates the usage and allocation of the radio frequency spectrum in the U.S. has assigned the frequency band between 540 and 1,600 kHz to commercial AM radio transmission. Each station can occupy a bandwidth of 10 kHz centered around its carrier. As we shall see, this corresponds to an effective signal bandwidth of 5 kHz—sufficient for good reproduction of speech, and acceptable reproduction of music.

Basic Principle of AM

AM signals are generated by modulating the amplitude of a carrier signal. Let the carrier signal be a sinusoid at frequency ω_c:

$$c(t) = A_c \cos(\omega_c t) \qquad \textbf{Carrier signal} \tag{16.5}$$

and, for illustration purposes, let the modulation also be a single tone (sinusoid), at a frequency $\omega_m \ll \omega_c$:

$$m(t) = A_m \cos(\omega_m t) \qquad \textbf{Modulating signal} \tag{16.6}$$

With these definitions, we can define the basic AM signal as follows:

$$s(t) = [A_c + m(t)] \cos(\omega_c t) \tag{16.7}$$

or

$$s(t) = A_c \left[1 + \frac{A_m}{A_c} \cos(\omega_m t)\right] \cos(\omega_c t) \qquad \textbf{AM signal} \tag{16.8}$$

The **modulation index**, μ, is defined to be the ratio of the modulation to carrier signal amplitudes:

$$\mu = \frac{A_m}{A_c} \qquad \textbf{Modulation index} \tag{16.9}$$

and for proper amplitude modulation should be less than 1. If equation 16.8 is expanded, we see that an AM signal is composed of two terms: a sinusoidal carrier wave, plus a wave that is the product of two sinusoidal terms. Using trigonometric identities, we can write the following expression:

$$\begin{aligned} s(t) &= A_c \cos(\omega_c t) + \mu A_c \cos(\omega_c t) \cos(\omega_m t) \\ &= A_c \cos(\omega_c t) + \mu \frac{A_c}{2} \cos(\omega_c - \omega_m)t + \mu \frac{A_c}{2} \cos(\omega_c + \omega_m)t \end{aligned} \tag{16.10}$$

Equation 16.10 shows that the AM signal is really composed of three sinusoidal waveforms: a carrier wave; a **lower sideband** signal, at frequency $(\omega_c - \omega_m)$, containing the modulating signal; and an **upper sideband** signal, at frequency $(\omega_c + \omega_m)$, also containing information (the modulation). Example 16.7 illustrates some important properties of an AM signal with pure sinusoidal modulation.

EXAMPLE 16.7 Single-Tone Amplitude Modulation

Problem

Analyze the spectrum of a single-tone modulation signal based on the WOSU AM 820 radio station. Use both analytical and computational tools.

Solution

Known Quantities: Carrier frequency; modulation index.

Find: Derive expressions for and plot time- and frequency-domain waveforms of the AM signal.

Schematics, Diagrams, Circuits, and Given Data: $f_c = 0.82$ MHz; $\mu = 0.5$.

Assumptions: Assume unity amplitude for the carrier, $A_c = 1$, and a modulating frequency $f_m = 10$ kHz.

Analysis: Define the modulating signal $m(t)$ and the carrier signal $c(t)$ as

$$m(t) = A_m \cos(2\pi f_m t)$$
$$c(t) = A_c \cos(2\pi f_c t)$$

These waveforms are plotted in Figures 16.16(a) and (b), respectively. The spectra of these signals are plotted in Figures 16.16(d) and (e).

The AM wave $s(t)$ is given by

$$s(t) = A_c [1 + \mu \cos(2\pi f_m t)] \cos(2\pi f_c t)$$

and is plotted in Figure 16.16(c). Using the Fourier transform pairs given in Table 16.3 (Pair 8), the Fourier transform of $s(t)$ can be expressed as the sum of three delta functions, centered at

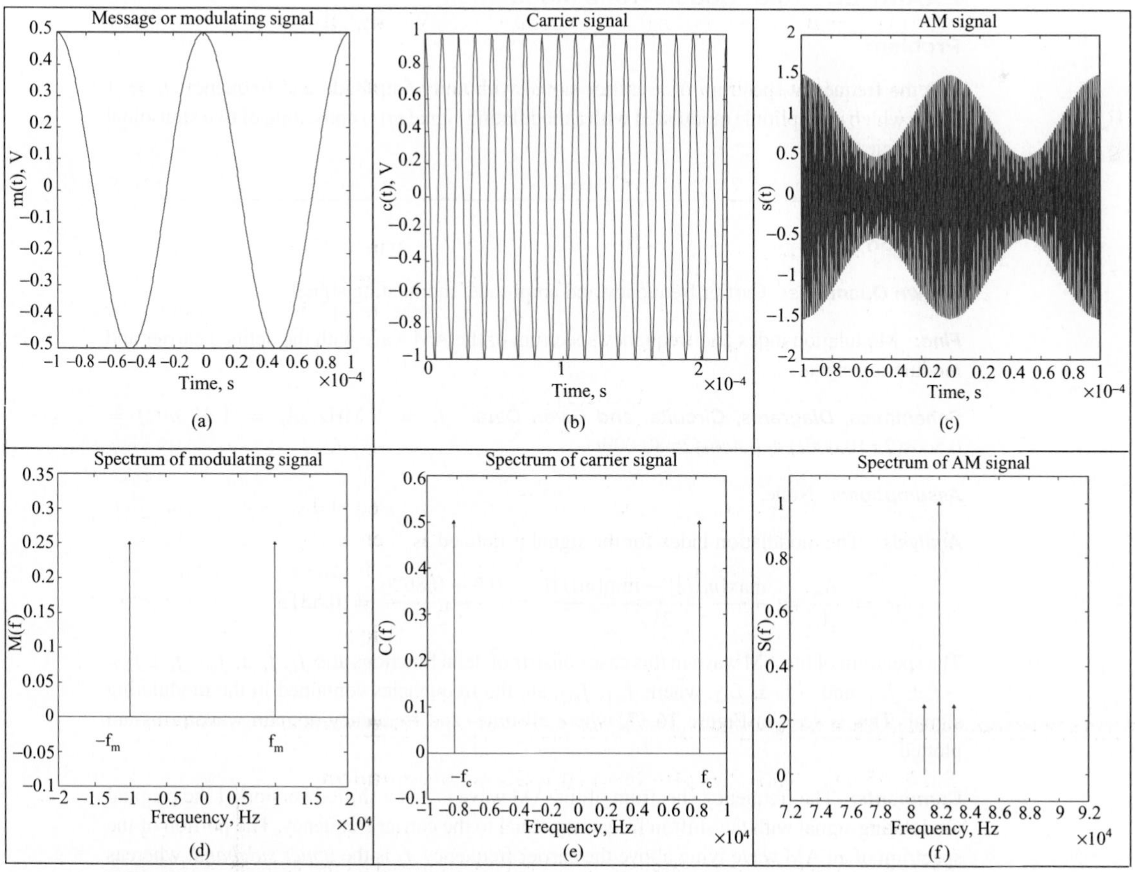

Figure 16.16 Time- and frequency-domain waveforms of single-tone AM signal

the carrier and at the sum (upper sideband) and difference (lower sideband) frequencies. Note also that the spectrum is repeated for negative frequencies, as explained earlier.

$$S(f) = \frac{A_c}{2}[\delta(f - f_c) + \delta(f + f_c)] + \mu\frac{A_c}{4}[\delta(f - f_c - f_m) + \delta(f + f_c + f_m)]$$
$$+\mu\frac{A_c}{4}[\delta(f - f_c + f_m) + \delta(f + f_c - f_m)]$$

Thus, the spectrum of an AM wave for the special case of sinusoidal modulation, consists of delta function at $\pm f_c$, $f_c \pm f_m$, and $-f_c \pm f_m$, as shown in Fig. 16.16(f).

Comments: If you like, you may experiment with the value of the modulation index and see its effect on the AM wave. It is recommended that the modulation index, μ, be nearly equal to 1, but not greater. If the modulation index is greater than 1 for any t, the carrier wave becomes overmodulated, resulting in carrier phase reversals whenever the function $[1 + \mu m(t)]$ crosses zero.

The single-tone modulation example is very valuable to understand the basic properties of an AM signal. In the next two examples we progress to the double-tone modulation, and then to a general, non-periodic modulating signal to explore the waveforms and spectrum of more realistic AM signals.

EXAMPLE 16.8 Double-Tone Modulation

Problem

Plot the frequency spectrum of a carrier signal with unity amplitude and frequency $f_c = 1$ MHz, which is amplitude modulated with a modulating signal $m(t)$ consisting of two sinusoidal frequencies.

Solution

Known Quantities: Carrier frequency and amplitude; modulating signal.

Find: Modulation index and frequency spectrum of the AM wave with the defined carrier and modulating signal.

Schematics, Diagrams, Circuits, and Given Data: $f_c = 1$ MHz; $A_c = 1$ V; $m(t) = 0.5\cos(2\pi\,10{,}000t) + 0.4\cos(2\pi\,80{,}000t)$.

Assumptions: None.

Analysis: The modulation index for the signal is defined as

$$\mu = \frac{A_m}{A_c} = \frac{\max[m(t)] - \min[m(t)]}{2V_c} = \frac{0.9 + 0.8626}{2} = 0.8813$$

The spectrum of the AM wave in this case consists of delta functions at $\pm f_c$, $f_c \pm f_{m1}$, $f_c \pm f_{m2}$, $-f_c \pm f_{m1}$, and $-f_c \pm f_{m2}$, where f_{m1}, f_{m2}, are the frequencies contained in the modulating signal. This is seen in Figure 16.17, where all time- and frequency-domain waveforms are plotted.

Comments: The frequency spectrum of the AM wave is just a shifted version of the original modulating signal with the shift in frequency equal to the carrier frequency. The portion of the spectrum of an AM wave lying above the carrier frequency f_c is the *upper sideband*, whereas the symmetric portion below f_c is called the *lower sideband*.

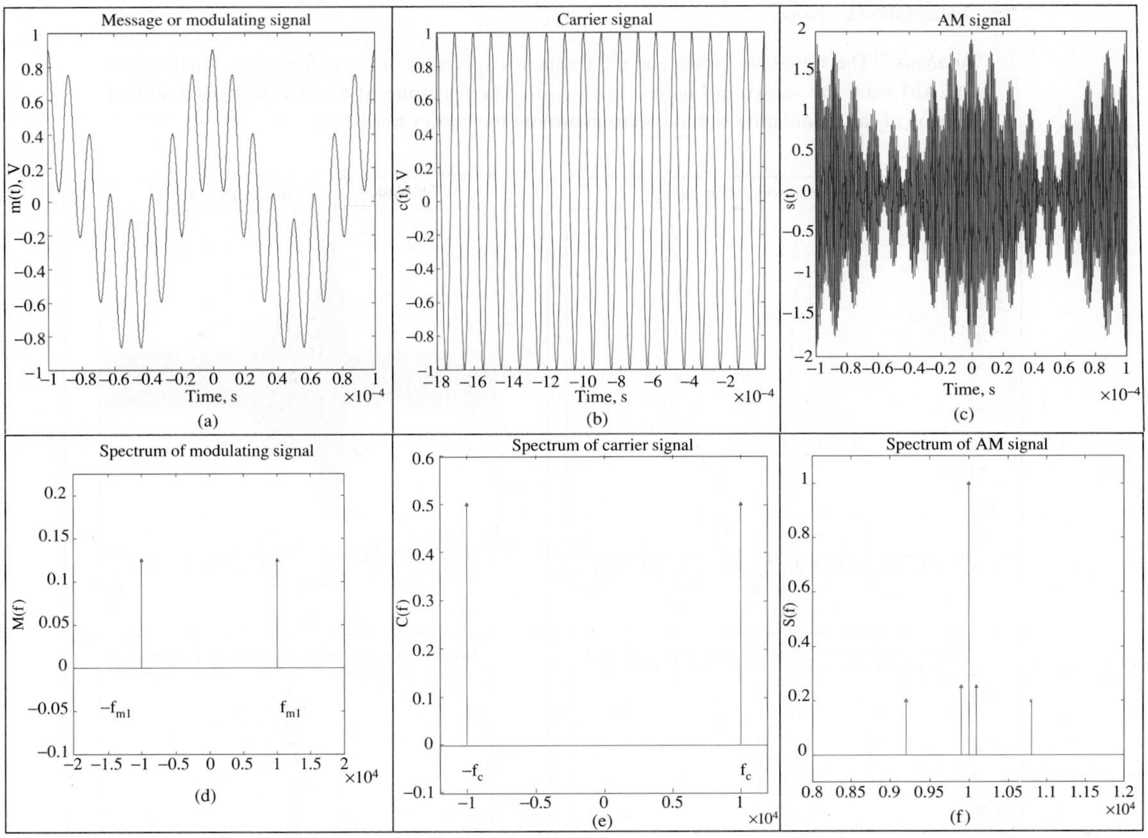

Figure 16.17 Time- and frequency-domain waveforms of double-tone AM signal

EXAMPLE 16.9 Non-Periodic Amplitude Modulation

Problem

Plot the frequency spectrum of a carrier signal with unity amplitude and frequency $f_c = 0.1$ MHz, which is amplitude modulated with a non-periodic modulating signal $m(t)$ having a defined shape.

Solution

Known Quantities: Carrier frequency, and amplitude; modulating wave $m(t)$ defined for a certain interval of time.

Find: Frequency spectrum of the AM wave.

Schematics, Diagrams, Circuits, and Given Data:

$$m(t) = \begin{cases} \sin(2\pi f_1 t) + \sin(2\pi f_2 t) + \sin(2\pi f_3 t) + \sin(2\pi f_4 t) + u(t) & \text{for} \quad t \le T \\ 0 \quad \text{otherwise} \end{cases}$$

$$u(t) = 1 \quad \text{for} \quad t \le T$$

$$c(t) = A_c \cos(2\pi f_c t)$$

$$f_c = 0.1\text{MHz};\ A_c = 1;\ f_1 = 1\text{kHz},\ f_2 = 2\text{kHz},\ f_3 = 3\text{kHz},\ f_4 = 4\text{kHz}.$$

Assumptions: None.

Analysis: The signal waveform and the frequency spectrum of the modulating signal and of the AM wave are shown in Figures 16.18(a–d). The spectrum of the AM wave is a shifted version of the modulating signal spectrum around the carrier frequency.

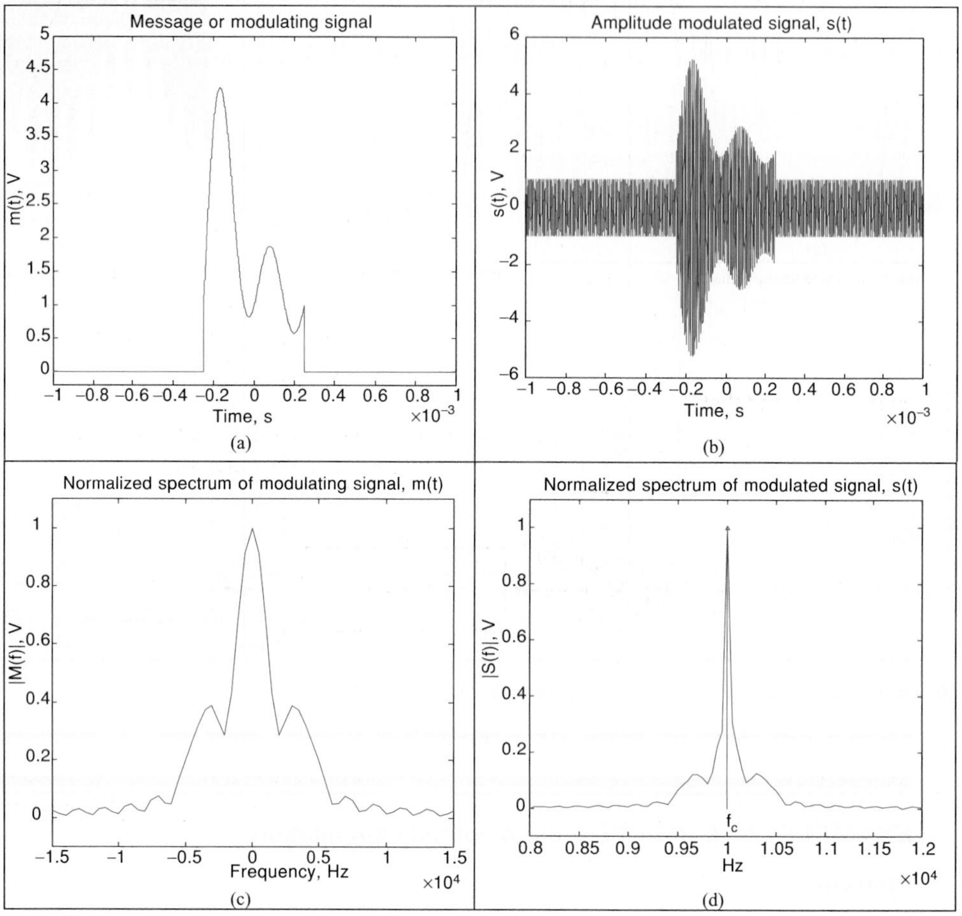

Figure 16.18 Time- and frequency-domain waveforms of non-periodic AM signal

Comments: If B_{max} is the bandwidth of the modulating signal, (the highest frequency in the modulating signal), the bandwidth of the AM wave is defined as *twice the highest frequency in the modulating signal,* i.e., $B = 2B_{max}$.

AM Demodulation; Integrated Circuit Receivers

Demodulation is the process of recovering the modulating signal from a received modulated signal. With reference to Figure 16.1, one can think of the transmitter in an AM signal as the device that imposes the modulation on a carrier, while the receiver extracts the modulating signal from a received AM signal. To understand the basic principle of modulation and demodulation, we observe that amplitude modulation consists in effect of *multiplying* the carrier signal times the modulating signal. This process is often called *mixing,* and a **mixer** is the device that implements this function, that is, multiplication.

Consider the AM signal of Equation (16.8), and multiply it by a second signal at the same frequency as the carrier signal:

$$s(t) \cdot c(t) = \left\{ A_c \left[1 + \frac{A_m}{A_c} \cos(\omega_m t) \right] \cos(\omega_c t) \right\} \cdot \cos(\omega_c t) \qquad \textbf{(16.11)}$$

The resulting squared term, $\cos^2(\omega_{ct})$ can be expanded to yield:

$$\begin{aligned} s(t) \cdot c(t) &= A_c \left[1 + \frac{A_m}{A_c} \cos(\omega_m t) \right] \cos^2(\omega_c t) \\ &= \frac{A_c}{2} \left[1 + \frac{A_m}{A_c} \cos(\omega_m t) \right] [1 + \cos(2\omega_c t)] \\ &= \frac{A_c}{2} + m(t) + \left[\frac{A_c}{2} + m(t) \right] \cos(2\omega_c t) \end{aligned} \qquad \textbf{(16.12)}$$

You see that the result of this mixing operation consists of two terms: a constant plus the modulation signal—what we desire to recover—and an amplitude modulated term at a frequency equal to twice the carrier frequency. Note that the modulation signal is back to *baseband,* that is, low frequencies (for example, 0–5 kHz for speech and music), and that it is therefore easy to recover the modulating signal by low-pass filtering the output of the mixer. Figure 16.19 depicts a block diagram of a conceptual AM demodulator as well as the spectra of the AM signal before and after mixing.

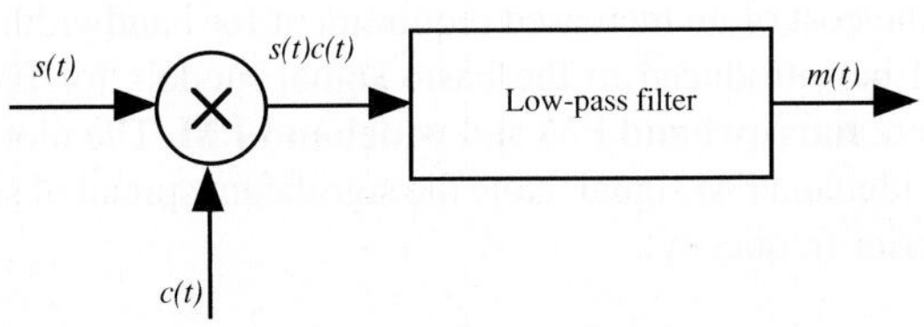

(a)

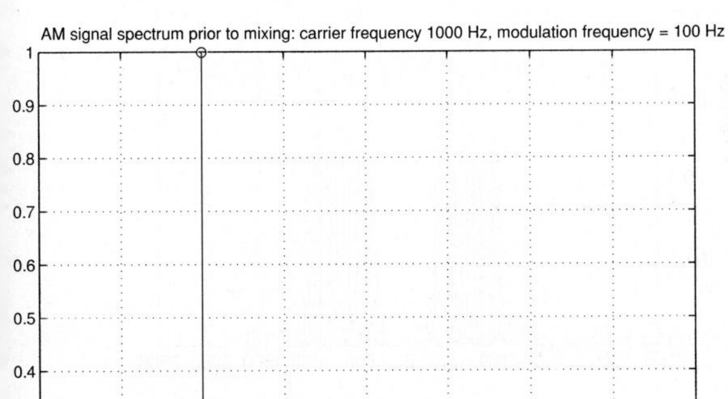

(b)

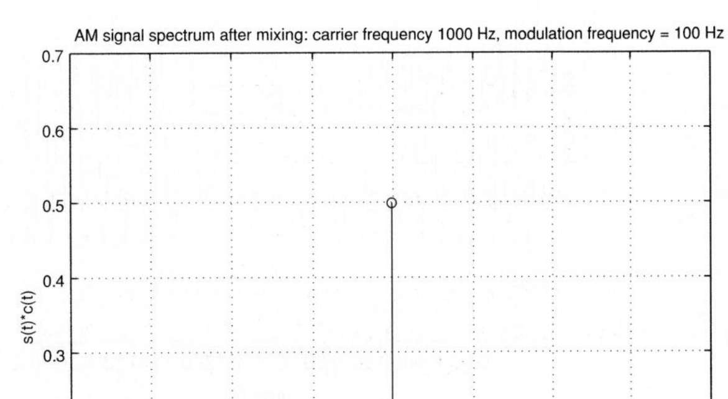

(c)

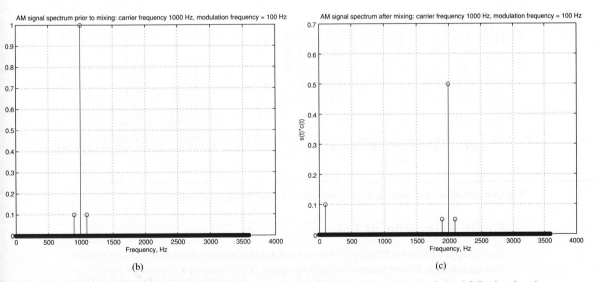

igure 16.19 (a): block diagram of AM demodulator; (b): spectrum of AM signal; (c) spectrum of signal following the mixer nultiplier) – note that the modulation signal (at 100 Hz) can now be recovered via low-pass filtering.

FIND IT ON THE WEB

The process of demodulating AM signals is carried out today by means of **integrated circuit receivers.** Additional information may be found on the Web.

CHECK YOUR UNDERSTANDING

a. Use the Matlab files that accompany Examples 16.8 and 16.9 to plot only the positive spectrum of the single- and double-tone AM signals. Determine the bandwidth of the AM signal in each case.

b. Determine the bandwidth of the modulating signal in Figure 16.18(c); what is the bandwidth of the AM signal? Is this consistent with commercial AM practice? Would the FCC allow a commercial station to broadcast such a signal?

16.4 FREQUENCY MODULATION AND DEMODULATION

You are certainly familiar with the term **frequency modulation** because of the great diffusion of FM radio. As its name implies, frequency modulation consists of encoding the information contained in a modulating signal in the frequency of a carrier signal. Figure 16.20 depicts a sinusoidally-modulated FM waveform and its corresponding magnitude spectrum. FM transmission permits significant improvements over AM, but at the cost of an increased requirement for bandwidth. In the next subsections you will be introduced to the basic signal models for FM. Two different cases are discussed: **narrowband FM** and **wideband FM.** The plots of Figure 16.20 correspond to a wideband FM signal. Note the significant spread of signal frequencies relative to the carrier frequency!

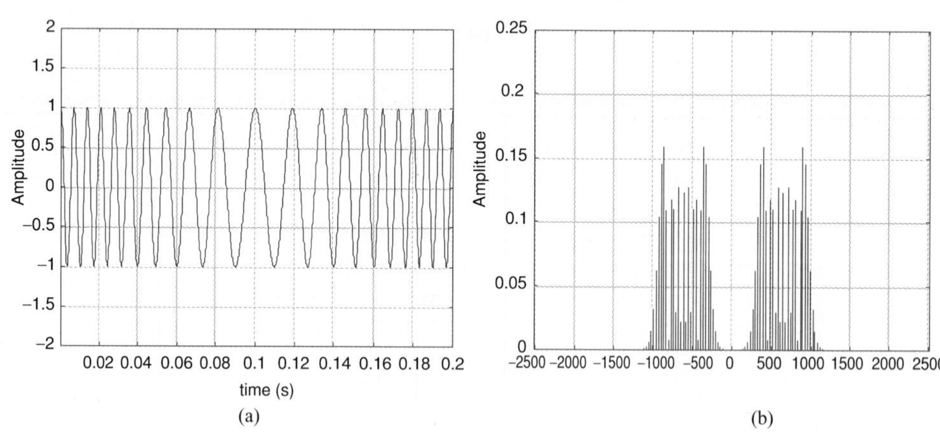

Figure 16.20 (a) FM signal time waveform; (b) FM signal magnitude spectrum

Basic Principle of FM

The basic principle underlying FM is that the **instantaneous frequency** of the carrier is modulated by the information-carrying signal. If we assume a sinusoidal carrier, as is usually the case, say $c(t) = \cos(2\pi f_c t)$, the modulation will cause the frequency f_c to be a function of time. In the signal of Figure 16.20, the carrier frequency varies sinusoidally as well. Before proceeding with the analysis of FM signals, it will be useful to examine the relationship between the instantaneous phase and frequency of a sinusoidal signal. We first define the relationship between the instantaneous phase and frequency of a sine wave as follows:

$$\theta_i = 2\pi \int_0^t f_i(t)dt; \quad f_i(t) = \frac{d\theta_i(t)}{dt} \tag{16.13}$$

To illustrate this definition, consider the case of a simple sinusoidal signal, $v(t) = A\cos(\omega t)$; in this signal we recognize that the phase angle (the argument of the cosine function) is $\theta_i(t) = \omega t$, and therefore the instantaneous signal frequency is given by $f_i(t) = \frac{d\theta_i(t)}{dt} = \omega$. This result should not surprise you: all sinusoidal signals must have a phase angle that increases linearly with time, so that their instantaneous frequency is constant. In the case of an FM signal, we might have a phase angle that varies sinusoidally with time, thus causing the instantaneous frequency to also vary in a sinusoidal fashion; this is the simplest case of an FM signal, and is treated next.

Single Tone Modulation

Consider the case of single tone modulation, where the modulating signal is:

$$m(t) = A_m \cos(\omega_m t) = A_m \cos(2\pi f_m t) \tag{16.14}$$

The instantaneous frequency of the FM signal varies linearly with the modulation; that is, the carrier frequency increases and decreases with the modulating signal, as shown in equation 16.15:

$$f(t) = f_c + k_f A_m \cos(2\pi f_m t) = f_c + \Delta f \cos(2\pi f_m t) \tag{16.15}$$

In the above expression we have implicitly defined the **frequency deviation,** Δf:

$$\Delta f = k_f A_m \qquad \text{frequency deviation} \tag{16.16}$$

The frequency deviation is a very important characteristic of an FM signal; Δf represents the maximum instantaneous deviation of the FM signal frequency from the carrier frequency. Δf is dependent on the *amplitude* of the modulating signal, and is *independent of the modulating signal frequency*. The constant k_f depends on the technique used for generating the modulated signal.

The instantaneous phase of the FM signal is calculated using equation 16.13:

$$\theta_i = 2\pi \int_0^t f_i(t)dt = 2\pi f_c t + \frac{\Delta f}{f_m} \sin(2\pi f_m t) \tag{16.17}$$

Note that equation 16.17 introduces another important constant: the ratio of the frequency deviation to the modulating frequency is called the **modulation index,** β:

$$\beta = \frac{\Delta f}{f_m} \tag{16.18}$$

The parameter β represents the maximum instantaneous deviation of the phase angle of the FM signal from the angle of the carrier. Now we can write the instantaneous angle of the FM signal as follows.

$$\theta_i = 2\pi f_c t + \beta \sin(2\pi f_m t) \tag{16.19}$$

Finally, the sinusoidally modulated FM signal is defined in equation 16.20:

$$s_{FM}(t) = A_c \cos[2\pi f_c t + \beta \sin(2\pi f_m t)] \tag{16.20}$$

It is now possible to make a formal distinction between the two cases mentioned earlier, *narrowband FM* and *wideband FM,* on the basis of the modulation index.

Narrowband FM

Narrowband FM corresponds to FM signals with a small modulation index (i.e., $\beta \ll 1$). We can use a trigonometric identity to expand equation 16.20:

$$s_{FM}(t) = A_c \cos(2\pi f_c t)\cos[\beta \sin(2\pi f_m t)] - A_c \sin(2\pi f_c t)\sin[\beta \sin(2\pi f_m t)] \tag{16.21}$$

and if $\beta \ll 1$ we can use the small-angle approximations $\cos[\beta \sin(2\pi f_m t)] \approx 1$ and $\sin[\beta \sin(2\pi f_m t)] \approx \beta \sin(2\pi f_m t)$ to write

$$\begin{aligned} s_{FM}(t) &\approx A_c \cos(2\pi f_c t) - \beta A_c \sin(2\pi f_c t)\sin(2\pi f_m t) \\ &\approx A_c \cos(2\pi f_c t) - \frac{1}{2}\beta A_c\{\cos[2\pi(f_c + f_m)t] - \cos[2\pi(f_c - f_m)t]\} \end{aligned} \tag{16.22}$$

Note the similarity between the expression for narrowband FM and that we derived earlier for the AM signal, repeated below for convenience:

$$s_{AM}(t) = A_c \cos(\omega_c t) + \mu\frac{A_c}{2}\cos(\omega_c - \omega_m)t + \mu\frac{A_c}{2}\cos(\omega_c + \omega_m)t \tag{16.10}$$

A reasonable rule of thumb is that the approximation given in equation 16.22 holds for $\beta < 0.3$. Figure 16.21(a)–(d) depicts the spectra of FM signals with various values

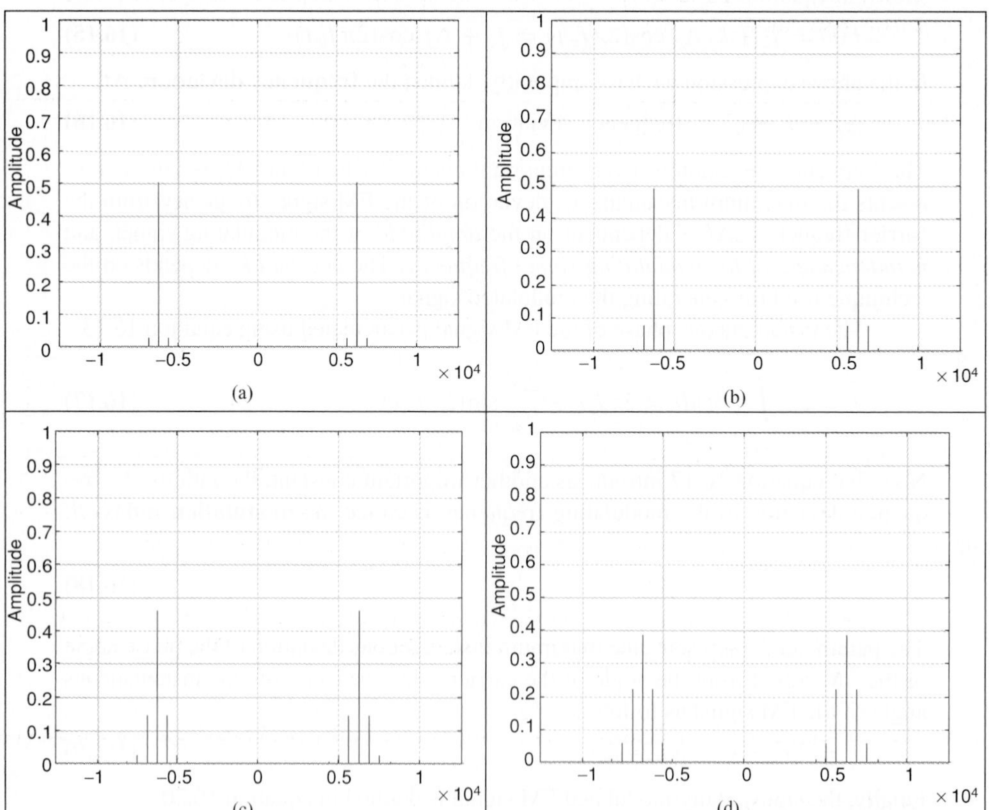

Figure 16.21 Bandwidth increases with modulation index in FM signals: (a) $\beta = 0.1$; (b) $\beta = 0.3$; (c) $\beta = 0.6$; (d) $\beta = 1$

of β. Note how the bandwidth increases with the value of the modulation index. Only in (a) and (b) is the signal narrowband FM.

EXAMPLE 16.10 Narrowband FM

Problem

Compare the spectrum of a narrowband FM waveform with that of an AM waveform with the same modulating and carrier frequencies.

Solution

Known Quantities: Carrier frequency, modulation frequency, modulation index.

Find: Plot the frequency domain waveforms of the narrowband FM and AM signals.

Schematics, Diagrams, Circuits, and Given Data: $f_c = 1{,}000$ Hz; $f_m = 100$ Hz; $A_c = 1$ V; $A_m = 0.2$ V; $\mu = 0.2$; $\beta = 0.2$.

Assumptions: Assume sinusoidal modulation and unity amplitude for the carrier, $A_c = 1$.

Analysis: Figure 16.22 depicts two signals: the first signal is an FM waveform with $\beta = 0.2$; the second is an AM signal with $\mu = 0.2$. The resulting spectra are plotted in Figures 16.22(a) and (b), respectively.

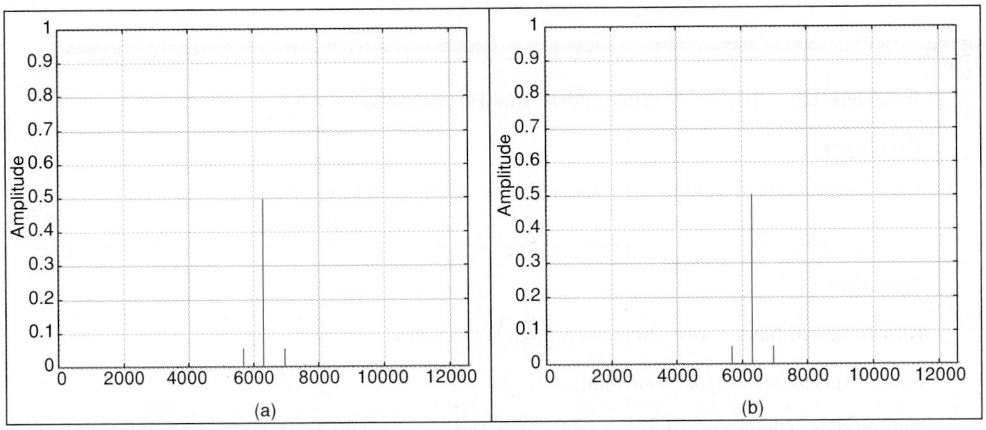

Figure 16.22 Comparison of narrowband FM and AM signal spectra: (a) FM, $\beta = 0.2$; (b) AM, $\mu = 0.2$

Comments: Note how the two amplitude spectra are virtually identical. Equations 16.22 and 16.10 predict this result. The only difference between the two spectra is in the phase angle of the signals.

Wideband FM

The mathematical representation of wideband FM signals is far more complex than the approximation of equation 16.22. The nonlinearity of the wideband FM signal is described using Bessel functions. This analysis is beyond the scope of this chapter, and the interested reader is referred to any one of a number of excellent textbooks in electrical communications.

Transmission Bandwidth of FM Signals

The transmission bandwidth of a frequency-modulated signal is theoretically infinite; however, practical approximations are possible. In the case of narrowband FM, we have already seen that we have the same transmissions bandwidth as in an AM signal: $B = 2f_m$. For large values of the modulation index β, the bandwidth of the FM signal can be experimentally observed to be close to the total frequency excursion $\pm\Delta f$ or $2\Delta f$. These observations lead to the well-known Carson's rule, relating the approximate transmission bandwidth to the frequency deviation and to the modulation index:

$$B = 2\Delta f + 2f_m = 2\Delta f\left(1 + \frac{1}{\beta}\right) \qquad \text{Carson's rule} \tag{16.23}$$

Carson's rule suggests that as β becomes larger, the bandwidth approaches $2\Delta f$, while as β decreases, the bandwidth becomes closer to $2f_m$.

EXAMPLE 16.11 Commercial FM Broadcast

Problem

Use Carson's rule to analyze the bandwidth of a commercial FM station.

Solution

Known Quantities: Carrier frequency, modulation index.

Find: Approximate signal bandwidth.

Schematics, Diagrams, Circuits, and Given Data: $f_c = 90.5$ MHz; $A_c = 1$; $A_m = 1$; $f_m = 10$ kHz; $k_f = 6{,}000$; $\beta = 0.2$.

Assumptions: Assume sinusoidal modulation.

Analysis: For sinusoidal modulation, the frequency deviation is $\Delta f = k_f A_m = 6$ kHz. Then, Carson's rule predicts a bandwidth

$$B = 2\Delta f + 2f_m = 2\Delta f\left(1 + \frac{1}{\beta}\right) = 1.2 \cdot 10^4\left(1 + \frac{1}{0.2}\right) = 72 \text{ kHz}$$

EXAMPLE 16.12

Problem

Given an FM message signal, find:

a. The bandwidth of the message signal B_m.
b. The bandwidth of the modulated carrier signal B_c.
c. The band of frequencies occupied by the signal B.

Solution

Known Quantities: Modulation frequency, carrier frequency, modulation constant.

Find: B_m, B_c.

Schematics, Diagrams, Circuits, and Given Data: $v_m = 5\cos(200\pi t)$; $f_c = 100 f_m$; $k_f = \frac{1000}{2\pi}$ Hz / V.

Assumptions: Assume sinusoidal modulation.

Analysis:

a. The message signal is $v_m = 5\cos(200\pi t)$. Hence,

$$f_m = 100 \text{ Hz}$$
$$A_m = 5$$
$$f_c = 100 f_m = 10 \text{ kHz}$$

The bandwidth of the message signal is

$$B_m = 2 f_m$$
$$B_m = 200 \text{ Hz}$$

b. The maximum frequency deviation is given by

$$\Delta f = k_f A_m$$
$$\Delta f = \frac{5000}{2\pi} \text{ Hz} \approx 795 \text{ Hz}$$

Thus, the bandwidth of the modulated carrier signal is approximately given by

$$B_c = 2(\Delta f + f_m)$$
$$B_c = 2(795 + 100) = 1790 \text{ Hz}$$

c. The carrier frequency is $f_c = 100$ kHz. The frequency band is centered about the carrier frequency. Therefore, the band of frequencies occupied spans from

$$\left(f_c - \frac{B_c}{2}\right) \quad \text{to} \quad \left(f_c + \frac{B_c}{2}\right).$$

Hence, the frequency band is

$$\left(10000 - \frac{1790}{2}\right) \text{ to } \left(10000 + \frac{1790}{2}\right),$$

which equals a band from 9.9105 kHz to 10.895 kHz.

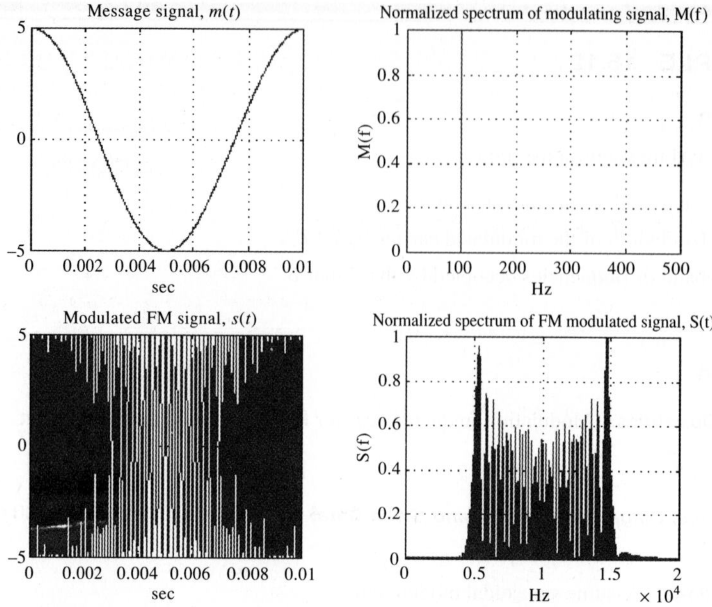

Figure 16.23 Time-domain and spectral plots for Example 16.12

Figure 16.23 depicts the modulating signal and the FM spectrum of the FM signal examined in this example.

EXAMPLE 16.13

Problem

In the United States the assigned band for FM commercial broadcast is from 88.0 MHz to 108.0 MHz with 100 possible channels. Find the bandwidth for each channel.

Solution

Known Quantities: Band for FM commercial broadcast, number of channels.

Find: The bandwidth of each channel.

Schematics, Diagrams, Circuits, and Given Data: See problem statement.

Assumptions: None.

Analysis: The bandwidth for each channel is defined as:

$$B = \frac{\text{Total band}}{\text{Number of channels}} = \frac{108.0 - 88}{100} = 200 \text{ kHz}$$

Comments: Each commercial FM radio station has a bandwidth allocation of 200 kHz.

FM Demodulation

Demodulation of an FM signal is accomplished by performing a **frequency-to-voltage conversion,** that is, by converting the frequency modulation into a voltage signal. This is the reverse of the modulation process, and can be realized in a number of ways. We describe two basic approaches in the following subsections.

Frequency-to-Voltage Conversion

If a pulse of fixed amplitude A and fixed duration τ is generated at each zero crossing of the sensor waveform, it can be readily shown that a voltage proportional to the instantaneous signal frequency may be obtained. Figure 16.24 depicts the functional form of a frequency-to-voltage converter.

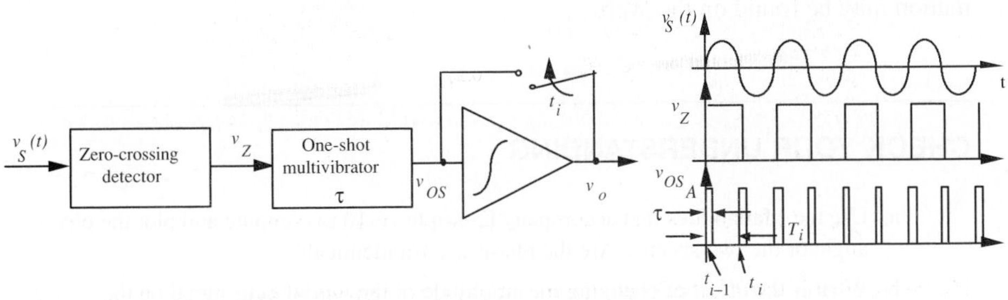

Figure 16.24 Frequency-to-voltage conversion

Ideally frequency to voltage (F-V) conversion could be obtained by computing the following integral:

$$\int_{t_{i-1}}^{t_i} v_{os}(t)dt = \frac{At}{\Delta T_i} = 2A\tau f_i \tag{16.24}$$

yielding a voltage proportional to the frequency of $v_S(t)$ during the *ith* cycle of the carrier waveform. In practice it is quite difficult to reset the integrator of Figure 16.23 at each zero crossing, so practical F-V converters employ a low-pass filter in place of the ideal integrator.

Phase-Locked Demodulation

Another method for implementing F-V conversion utilizes a **phase locked loop** (PLL) as a frequency-to-voltage converter, or FM demodulator. The PLL can act as an FM demodulator once it is phase-locked to an input signal waveform. When the PLL is in lock, any change in the input signal frequency generates an error voltage at the output of the phase detector, which can be either analog (a mixer) or digital, consisting of a pair of zero crossing detectors.

The output voltage of the PLL $v_o(t)$ is the voltage which is required to maintain a voltage-controlled oscillator (VCO) running at the same frequency as the input signal, and changes in the input signal frequency are matched by changes in $v_o(t)$. In this sense, the PLL acts as an F-V converter, with the input-output characteristic shown in Figure 16.25. Note that the PLL can offer only a finite *lock range*.

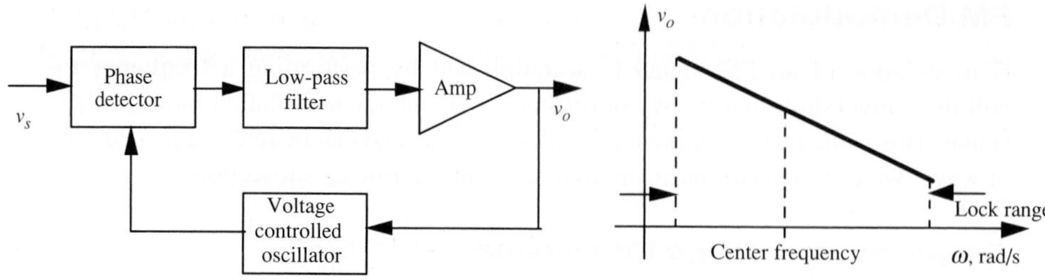

Figure 16.25 Phase-locked loop F-V conversion

Integrated Circuit Receivers

FM demodulation is performed using **integrated circuit receivers;** additional information may be found on the Web.

CHECK YOUR UNDERSTANDING

a. Use the Matlab files that accompany Example 16.10 to compute and plot the phase angle of the two spectra. Are the phase spectra identical?

b. What is the effect of changing the amplitude of the modulating signal on the bandwidth of the FM modulated signal?

c. Investigate the effect of changing β on the FM modulated signal bandwidth and analyze the spectrum of the FM signal of Example 16.10 with $\beta = 0.5$.

d. Find the carrier frequency for Channel 11 used by WCBE Radio, Columbus, OH, per the FCC regulations for the commercial broadcast in the US (use data in Example 16.13).

16.5 EXAMPLES OF COMMUNICATION SYSTEMS

The objective of this chapter is to summarize some important applications of modern communication systems. The overview given in this chapter is certainly not exhaustive, but will give you a good summary of the technology underlying some of today's communication systems and of their capabilities.

Global Positioning System (GPS)

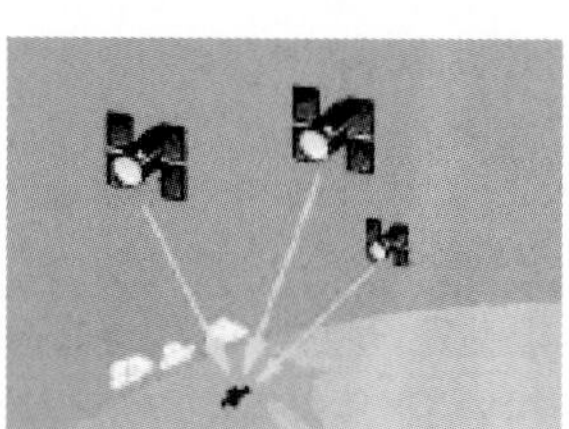

Figure 16.26 GPS principle (*Courtesy: Trimble Navigation*)

The **Global Positioning System (GPS)**, illustrated in Figure 16.26, is rapidly supplanting older navigation technologies based on radio communications used by the aircraft and marine industries, such as Loran-C. GPS is based on 24 satellites linked to ground stations, and effectively replaces—with much greater accuracy—the century-old system of navigation based on star position. You can think of the satellites as "man-made stars." Differential GPS is capable of position measurements with accuracy of a few centimeters. In recent years, GPS receivers have been miniaturized, and today amateur sailors, private pilots, and other private users have the ability to purchase hand-held GPS units. Among the uses of GPS we list navigation systems for cars, boats, planes, and guiding agricultural machining for "precision agriculture." The operation of GPS is explained in five basic steps, as shown in Table 16.5.

Table 16.5 **Basic operation of GPS**

1.	GPS operates based on triangulation of signals received from satellites.
2.	Triangulation is performed by the GPS receiver by measuring distance knowing the travel time of radio signals.
3.	Travel time is computed with the aid of accurate timing references.
4.	The exact position of the satellites in space is used to calculate distance.
5.	Delays experienced by the signals in traveling from the satellite to the ground stations are corrected.

Triangulation

The technique of triangulation is based on distance (range) measurements from the receiver location to the satellite. Four satellites are needed to determine exact position of any receiver location on earth.

Measuring Distance

To measure distance, GPS computes the time required to receive each satellite signal. The receiver and satellite both generate a synchronized signal; comparison of the signal received from the satellite with that in the receiver is used to calculate time of travel. Since the speed of travel of electromagnetic waves is known, distance can be calculated.

Timing Accuracy

Satellites carry atomic clocks on board to provide accurate timing. The clock in a low-cost GPS receiver need not be as accurate, as an additional range (distance) measurement can be used to remove the timing error in the receiver.

Satellite Positions

The position of the satellites is essential in calculating distance. The orbits of GPS satellites are known, and any deviations are measured by the Department of Defense; any errors are transmitted to the satellites, which in turn transmit this error information to the receivers along with their timing signals.

Correcting Errors

Signals traveling through the ionosphere experience additional delays, which turn into transmission errors. There are many methods for correcting errors; the technique called **differential GPS** can eliminate almost all errors. Additional information on these topics may be found on the Web.

FIND IT ON THE WEB

Radar

The acronym **RADAR** stands for RAdio Direction And Ranging. Radar systems operate by radiating electromagnetic waves (typically at microwave frequencies) and by detecting the echo returned from reflecting objects (targets). Radar technology

Figure 16.27 Radar antenna (top) and displays (*Courtesy: ProNav*)

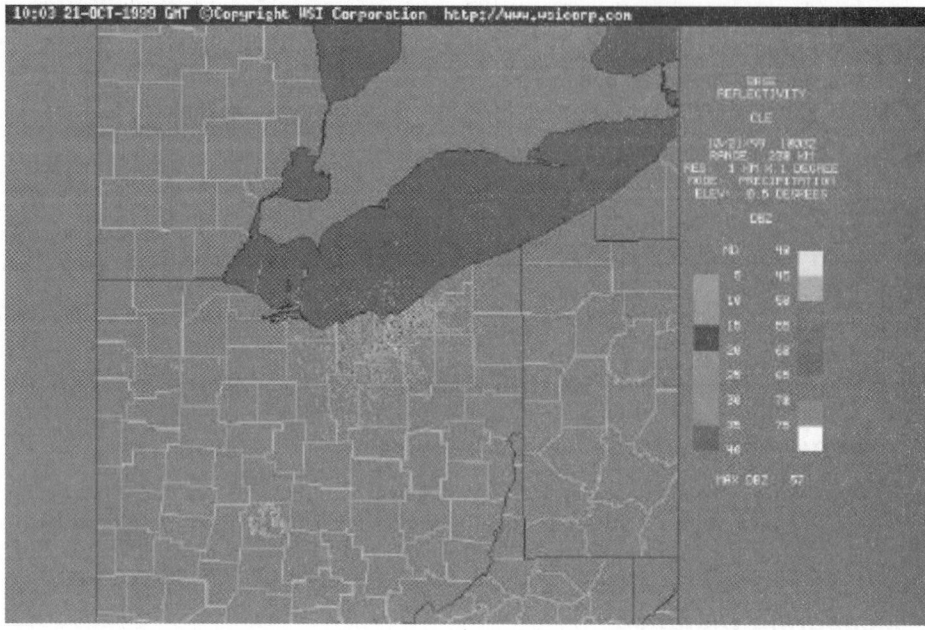

Figure 16.28 Weather radar (*Courtesy: WSI Corporation http://www.wsicorp.com*)

was developed during World War II and played a significant part in the success of the Allied Forces. While military applications are obvious, today Radar finds widespread civilian application in air traffic control and in tracking weather conditions, as well as in marine navigation (see Figure 16.27 for some examples of radar technology used in the latter application, and Figure 16.28 for an example of weather radar). In addition to detecting the position of a stationary target, Radar is capable of determining the trajectory of a moving target, thus predicting its future location. This function is, for example, very useful in weather radar where one wishes to predict weather conditions. The principle on which radar operation is based is that of the **doppler shift,** a concept with which you are probably already intuitively familiar (think of the sound of a train whistle as the train moves by a stationary observer). The doppler shift permits distinguishing a moving target from a stationary one, thus allowing the radar system to discern the echo of a stationary target from that of a moving target. Additional information may be found on the Web.

FIND IT ON THE WEB

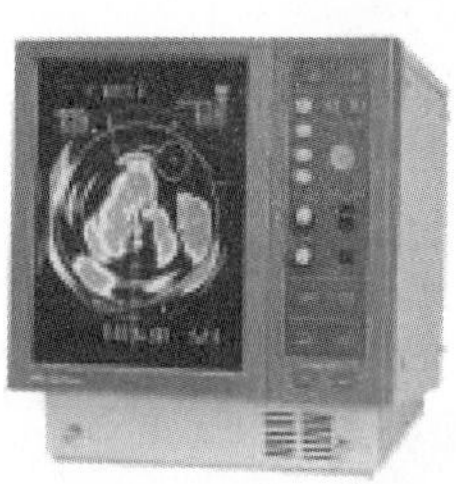

Figure 16.29 Sonar displays for marine navigation (*Courtesy: ProNav*)

Sonar

The term **SONAR** stands for SOund NAvigation and Ranging. Sonar is conceptually similar to radar, in that it uses information about the reflection and transmission of waves to determine the behavior of targets or the properties of the environment. The principal difference between Sonar and Radar is that the former is based on the reflection and propagation of acoustic (sound) waves, while the latter is based on radio frequency electromagnetic waves.

The applications of Sonar are numerous, ranging from inexpensive depth finders and imaging systems used to aid the navigation of small and large sea vessels (see, for example, Figure 16.29), to underwater navigation, to ocean thermal mapping. You will find a number of interesting resources related to Sonar on the Web.

FIND IT ON THE WEB

HOMEWORK PROBLEMS

Section 16.2: Fourier Series and Transform

16.1 Compute the Fourier series coefficients for a periodic square wave of unit amplitude, time period τ, and duty cycle η as shown in Figure P16.1 and defined mathematically as:

$$x(t) = \begin{cases} 1\, for|t| \leq \eta\tau \\ -1\, for|t| < \tau \end{cases}$$

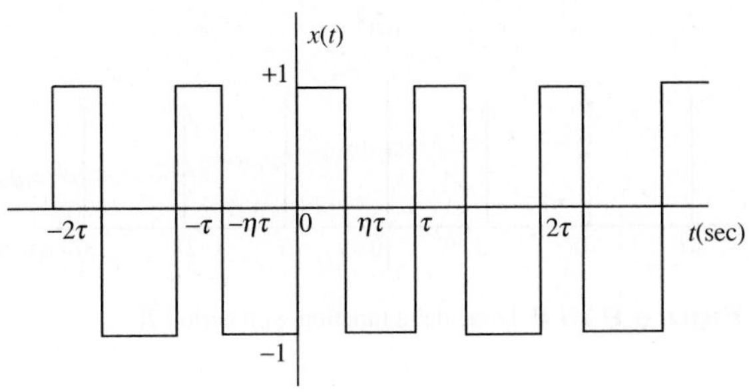

Figure P16.1 Square wave signal of period τ and duty cycle η

Use Matlab to plot the frequency spectrum of this signal with time period, $\tau = \frac{1}{300}$ sec and duty cycle

a. $\eta = 50\%$

b. $\eta = 30\%$

16.2 A full wave rectified sinusoidal signal of natural frequency ω_0 rad/sec is shown in Figure P16.2.

a. Find the Fourier series coefficients for the full wave rectified sinusoid.

b. Generate the frequency spectrum for a full wave rectified sinusoid of natural frequency $\omega_0 = 200\pi$ rad/sec.

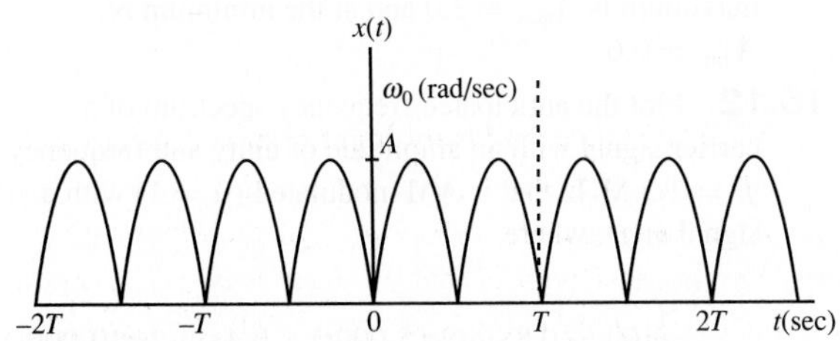

Figure P16.2 Full wave rectified sine wave of natural frequency ω_0 rad/sec

16.3 A full wave rectified cosine signal of natural frequency ω_0 rad/sec is shown in Figure P16.3.

a. Find the Fourier series coefficients for the full wave rectified cosine.

b. Generate the frequency spectrum for a full wave rectified cosine with natural frequency $\omega_0 = 150\pi$ rad/sec.

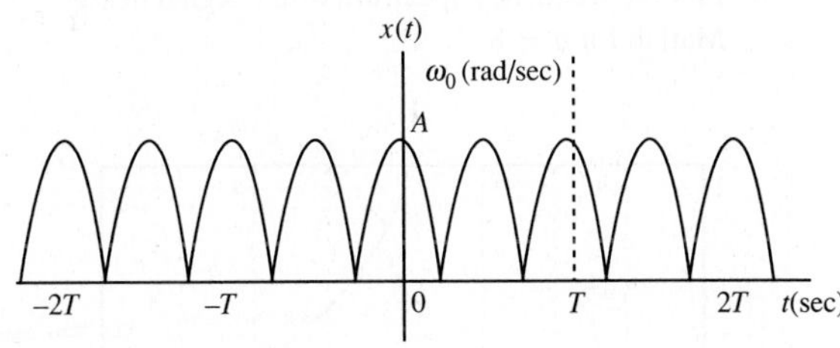

Figure P16.3 Full wave rectified cosine wave of natural frequency ω_0 rad/sec

16.4 Compute the Fourier series coefficients for the cosine-burst signal shown in Figure P16.4.

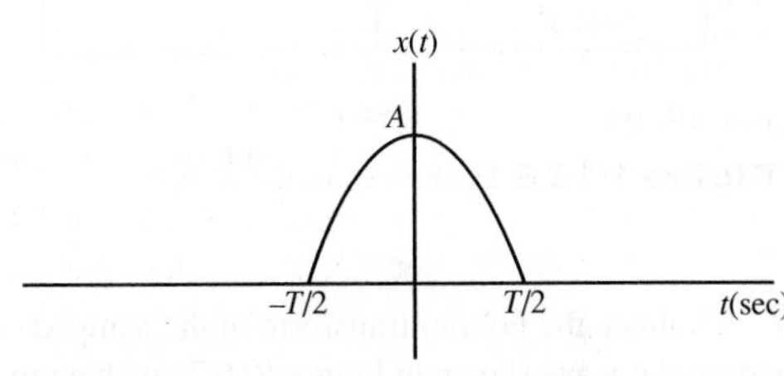

Figure P16.4 Cosine-burst signal

16.5 The triangular pulse signal shown in Figure P16.5 is mathematically defined as:

$$x(t) = A\left[1 - \frac{|t|}{T}\right](u(t+T) - u(t-T))$$

a. Find the Fourier transform of the triangular pulse.

b. Plot the frequency spectrum of a triangular pulse of period $T = 0.01$ sec and amplitude $A = 0.5$.

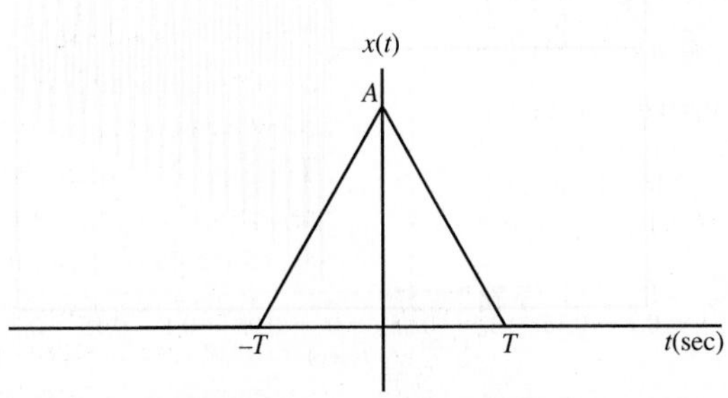

Figure P16.5 Triangular pulse $x(t)$ of duration $2T$

16.6 A double exponential signal shown in Figure P16.6 is defined mathematically by:

$$x(t) = \begin{cases} e^{-at}, & \text{for } t > 0 \\ 0, & \text{for } t = 0 \\ -e^{at}, & \text{for } t < 0 \end{cases}$$

a. Compute the Fourier transform of the signal. (*Hint:* Can use the linearity property of Fourier transform.)

b. Plot the frequency spectrum of the signal using Matlab for $a = 8$.

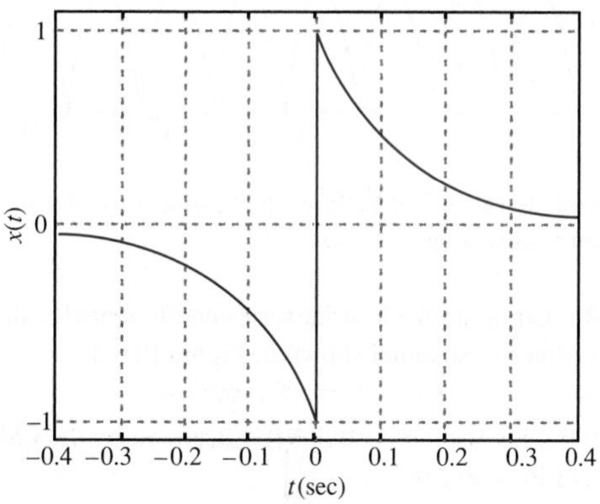

Figure P16.6 Double exponetial signal

16.7 Evaluate the Fourier transform of the damped sinusoidal wave shown in Figure P16.7 and having the functional form:

$$x(t) = \exp(-at)\cos(2\pi f_c t)u(t)$$

where $u(t)$ is the unit step function.

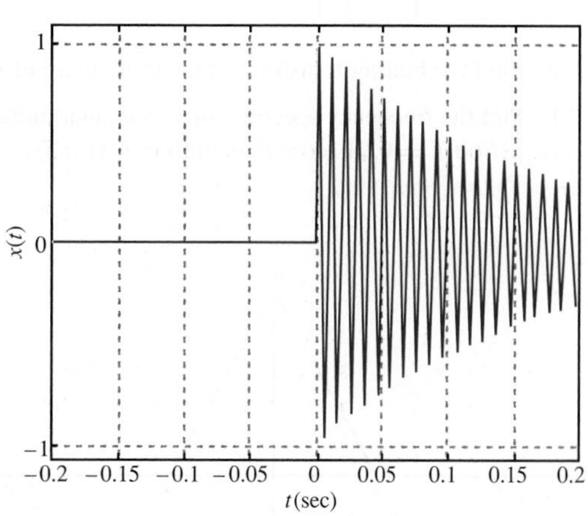

Figure P16.7 Damped sinusoidal signal

16.8 An ideal sampling function consists of an infinite sequence of uniformly spaced delta functions and is mathematically defined as:

$$\delta_{T_0}(t) = \sum_{m=-\infty}^{\infty} \delta(t - mT_0)$$

a. Compute the Fourier transform of ideal sampling function shown in Figure P16.8.

b. Also, use Matlab to generate the time domain signal and its amplitude spectrum for $T_o = 0.01$ sec.

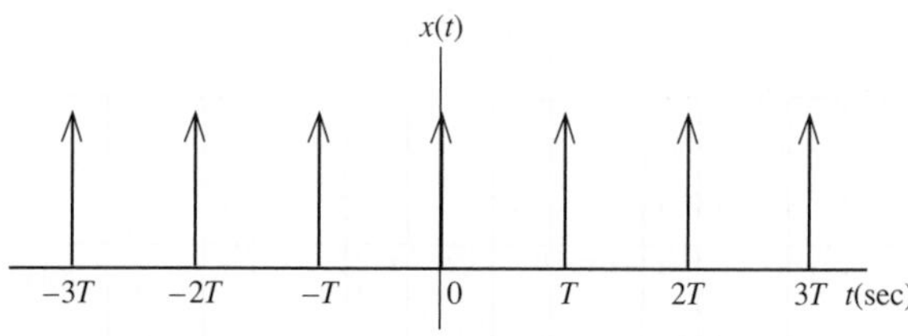

Figure P16.8 Dirac delta function with period T_0

16.9 Download the utterance signal utter.au and use the Matlab command **auread('utter.au')** to load it to the workspace. Use the FFT tools in Matlab to identify the frequency components of the signal.

16.10 A **bat echolocation chirp** signal is provided to you on the book website. Do a frequency analysis of the signal and explain what you observe. (Courtesy Digital Signal Processing Group, Rice University)

Section 16.3: Amplitude Modulation and Demodulation

16.11 Find the modulation index μ for an AM modulated signal having a carrier of amplitude $A_c = 1.0$, and the amplitude of the carrier at the maximum is $A_{\max} = 3.0$ and at the minimum is $A_{\min} = 0.6$.

16.12 Plot the anticipated frequency spectrum of a carrier signal with an amplitude of unity and frequency $f_c = 1.3$ MHz that is AM modulated ($\mu = 1$) with a signal $m(t)$, where

$$m(t) = 0.8\sin(2\pi 5{,}000t) + 0.4\sin(2\pi 10{,}000t) + 0.2\sin(2\pi 20{,}000t)$$

16.13 Plot the anticipated time domain response of a carrier signal with an amplitude of unity and frequency

$f_c = 10$ MHz that is AM modulated ($\mu = 1$) with a signal $m(t)$, where

$$m(t) = A\left[1 - \frac{|t|}{T}\right] \quad T = 0.01 \text{ sec}$$

Hint: The message signal is a triangular wave of time period T.

16.14 An AM radio station uses a carrier signal of unity amplitude and frequency $f_c = 1.6$ MHz. The message signal is a voice signal having certain frequency components and defined as:

$$\begin{aligned} m(t) &= 0.4\sin(2\pi 340t) + 0.35\sin(2\pi 960t) \\ &\quad +0.3\sin(2\pi 1345t) + 0.2\sin(2\pi 2230t) \\ &\quad +0.1\sin(2\pi 2890t) \end{aligned}$$

Plot the time domain and the frequency domain AM modulated signal of modulation index $\mu = 1$.

16.15 A non-periodic message signal $m(t)$ is amplitude modulated ($\mu = 1$) by a carrier signal of unity amplitude and frequency $f_c = 0.5$ MHz. Plot the time and frequency domain signal.

$$\begin{aligned} m(t) &= 0.4\sin(2\pi 340t) + 0.35\sin(2\pi 960t) \\ &\quad +0.3\sin(2\pi 1345t) + 0.2\sin(2\pi 2230t) \\ &\quad +0.1\sin(2\pi 2890t) + u(t) \end{aligned}$$

$$u(t) = \begin{cases} 1 & \text{for} \quad t \leq 0.01 \\ 0 & \text{otherwise} \end{cases}$$

16.16 Consider a modulating wave $m(t)$that consists of a single frequency component and defined as:

$$m(t) = A_m \cos(2\pi f_m t)$$

where A_m is the amplitude of the modulating wave and f_m is the frequency. The sinusoidal carrier wave has amplitude A_c and frequency f_c. The signal is amplitude modulated to produce the signal $s(t)$. Find the average power delivered to a 1-ohm resistor by $s(t)$.

16.17 The carrier frequency of the W-OSU channel is 0.82 MHz. If the upper sideband of the AM modulated signal has frequency components of amplitude 0.4 at 0.825 MHz, 0.2 at 0.83 MHz, and 0.25 at 0.84 MHz,

a. Find the modulating signal equation.

b. Plot the spectrum of the modulating signal.

c. Plot the spectrum of the AM signal including the lower sideband.

16.18 The AM commercial radio band in the U.S. is authorized to operate from 525 kHz to 1.7 MHz. A carrier frequency is assigned to each station, and regulations require them to be separated by 10 kHz. Find:

a. The number of channels that can be accommodated in the given frequency range.

b. The maximum modulating frequency that can be transmitted without overlap.

16.19 The speech signal utter.au is to be AM modulated for transmission on an AM commercial radio band in the U.S. Plot the frequency spectrum of the AM signal. Use any channel according to its separations and a modulation index $\mu = 1$ in your design.

Section 16.4: Frequency Modulation and Demodulation

16.20 The message signal given by $m(t) = 5\cos(750\pi t)$ is frequency modulated by a carrier frequency 105 times the message frequency and the modulation constant is $k_f = 1{,}005$. Find the bandwidth of the message signal.

16.21 If the message signal given by $m(t) = 2\cos(360\pi t)$ is frequency modulated by a carrier frequency 100 times the message frequency and the modulation constant $k_f = 1{,}000$, find the bandwidth of the modulated carrier signal.

16.22 Find the band of frequencies occupied by the FM signal of Problem 16.21.

16.23 A message signal $m(t)$ is FM modulated by a carrier of unity amplitude and frequency $f_c = 10.0$ MHz, with modulating constant $k_f = 1{,}000$. Plot the time and frequency domain FM modulated signal if $m(t) = 0.8\sin(2\pi 5{,}000t)$.

16.24 A packet of information is sent on an FM channel of frequency $f_c = 15.0$ MHz that uses a modulating constant $k_f = 6{,}000$. Plot the frequency spectrum of the FM modulated signal.

$$\begin{aligned} m(t) &= 0.4\sin(2\pi 340t) + 0.35\sin(2\pi 960t) + 0.3\sin(2\pi 1345t) \\ &\quad +0.2\sin(2\pi 2230t) + 0.1\sin(2\pi 2890t) + u(t) \end{aligned}$$

$$u(t) = \begin{cases} 1 & \text{for} \quad t \leq 0.001 \\ 0 & \text{otherwise} \end{cases}$$

16.25 WOSU-FM uses a carrier frequency of 90.5 MHz and modulating constant $k_f = 66{,}000$. The speech signal utter.au is transmitted on this channel. Plot the frequency spectrum of the FM modulated speech signal.

16.26 Consider Example 16.13. If Channel 2 is allocated for country music and the message signal may be considered to be $m(t) = 10\cos(2\pi 10^3 t)$, find:

a. The carrier frequency.

b. The value of k_f.

CHAPTER 17

DIGITAL COMMUNICATIONS

This chapter expands upon the analog communications concepts of Chapter 16 and introduces digital communications, or the idea of using digital processing techniques to simultaneously maximize information throughput while minimizing the probability of error, with the added constraint of limited transmitting power.

The field of digital communications was ushered in with the invention of the telegraph by Samuel Morse. After overhearing two of his colleagues speaking about electromagnets, Morse was struck with the idea of communicating information by using an electromagnetic clacker to indicate the flow of current in a long wire loop, where the flow of current was controlled by a switch at the other end (see Figure 17.1). Because such a machine can only communicate using the presence ("high") or absence ("low") of electric current, any message passed through the machine needed to be coded using these two states. To solve this problem, Morse invented a simple digital coding technique consisting of variable length high and low states. High states that were shorter in duration were called *dots,* those that were longer in duration were called *dashes,* and combinations of the two were used to indicate each letter of the alphabet. Short low states were used to indicate breaks between letters, while long low states were used to indicate breaks between words. This technique, later to

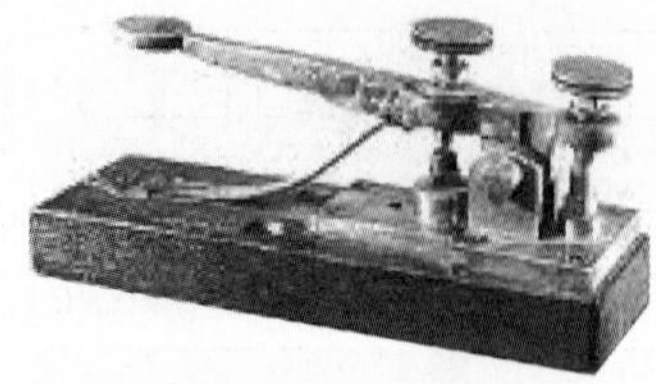

Figure 17.1 An early telegraph key (*Courtesy of the National Museum of American History, from the U.S. Patent office.*)

become the well-known Morse code, was first publicly demonstrated by transmitting the phrase "What hath God wrought," from the Supreme Court room at the Capitol to the railway depot in Baltimore on May 24, 1844.

Digital communication improves upon analog communication in many ways. In general, digital signals can be designed to be much more resistant to the effects of noise. Digital communication can have enhanced privacy through the use of encryption. And digital communication systems are usually less expensive to implement. Most importantly, digital communication allows the engineer to optimize a system, given a limited amount of transmitter power, for maximum information throughput while having predictable control over the integrity of the information at the receiver. On the downside, digital communication systems are generally more complex to design and require more bandwidth than their analog counterparts.

17.1 A TYPICAL DIGITAL COMMUNICATIONS SYSTEM

Although there are a great variety of digital communication systems, many of them can be broadly represented by the system of Figure 17.2. The input to the system is some *information source,* which may supply information in an analog (continuous) or digital (discrete) form. In the analog case, an *analog-to-digital converter* is used to convert the information into digital format. This is followed by a *source coder* that removes unnecessary redundancy and reduces the information signal to its most basic form. A *channel coder* then re-introduces a measured amount of redundancy that is carefully designed to improve the robustness of the information when exposed to noise. Next, the *baseband modulator* converts the digital information into an analog waveform that is shifted up to the desired carrier frequency by the *upconverter*. Once at the carrier frequency, the information is carried over the *channel,* which represents the physical medium over which we are communicating. On the receiver side, the counterparts to the transmitter components serve to recover the original information accurately.

17.2 INTRODUCTORY PROBABILITY

It is impossible to discuss digital communications without encountering probability theory, so in this section we present a very brief introduction to the topic. We include only enough material to allow a basic understanding of the digital communication concepts to follow. For a more thorough treatment of probability theory, the reader should examine a full text on the topic.

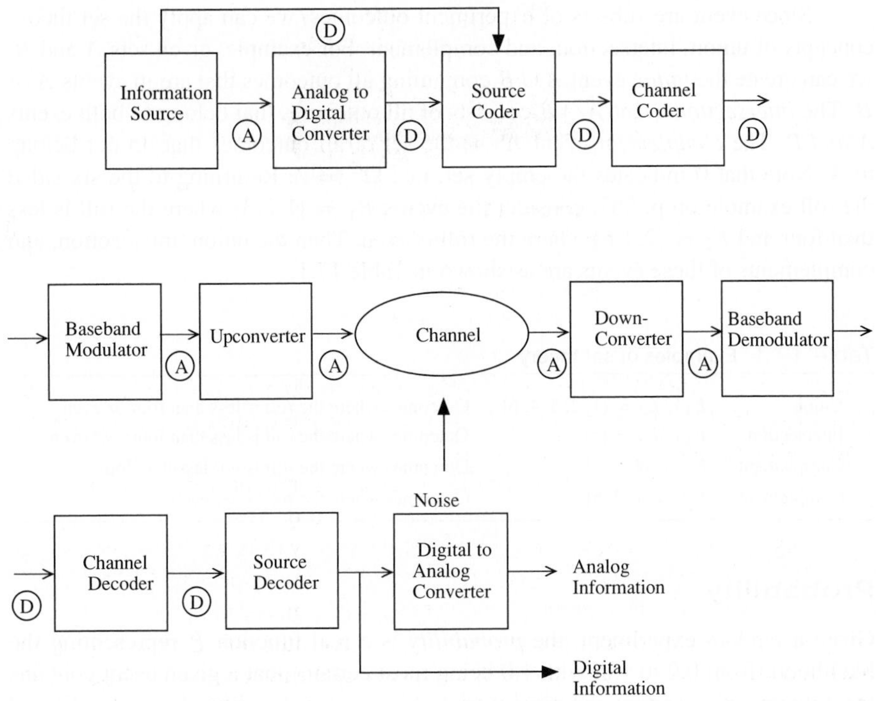

Figure 17.2 A typical communications system

Random Experiments

A *random experiment* is any experiment where the outcome cannot be exactly predicted and may differ from trial to trial. Flipping a coin, tossing a die, or spinning a roulette wheel are all examples of random experiments. A random experiment is described by the *sample space* Ω of all possible *outcomes* the experiment may produce. For example, flipping a coin once has the sample space

$$\Omega = \{\text{head, tail}\},$$

while the experiment consisting of flipping a coin twice has sample space

$$\Omega = \{\{\text{tail,tail}\}, \{\text{head, tail}\}, \{\text{tail,head}\}, \{\text{head,head}\}\}.$$

Rolling a six-sided die has the sample space

$$\Omega = \{1, 2, 3, 4, 5, 6\}.$$

The *outcome* obtained during a particular trial is denoted by ω.

An *event* E is a subset of Ω. For example, the event of getting at least one tail in two flips is

$$E = \{\{\text{tail,tail}\}, \{\text{head, tail}\}, \{\text{tail,head}\}\}.$$

The event of rolling a number less than four on the die above would be

$$E = \{1, 2, 3\}.$$

Since event are subsets of experiment outcomes, we can apply the set theory concepts of union, intersection, and complement. For example, given sets A and B, we can create the *union* event $A \cup B$ containing all outcomes that are in events A *or* B. The *intersection* event $A \cap B$ consists of all outcomes that belong to both events A *and* B. The *complement* event A^C is the set of all outcomes that do not belong to A. Note that $\emptyset$ indicates the empty set, i.e., $\Omega^C = \emptyset$. Returning to the six-sided die roll example on p. 867, consider the events $E_1 = \{1,2,3\}$ where the roll is less than four and $E_2 = \{2,4,6\}$ where the roll is even. Then the union, intersection, and complements of these events are as shown in Table 17.1.

Table 17.1 **Examples of set theory**

Union	$E_1 \cup E_2 = \{1, 2, 3, 4, 6\}$	Outcomes where the roll is less than four *or* even
Intersection	$E_1 \cap E_2 = \{2\}$	Outcomes where the roll is less than four *and* even
Complement	$E_1^C = \{4, 5, 6\}$	Outcomes where the roll is *not* less than four
Complement	$E_2^C = \{1, 3, 5\}$	Outcomes where the roll is *not* even

Probability

Given a random experiment, the *probability* is a real function P representing the likelihood (from 0.0 to 1.0, with 1.0 being most certain) that a given event contains the outcome of a random experiment trial. As an example, given the events E_1 and E_2 above, we have the probabilities shown in Table 17.2.

Table 17.2 **Examples of probabilities**

Event	Probability
$E_1 = \{1, 2, 3\}$	$P(E_1) = 1/2$
$E_2 = \{2, 4, 6\}$	$P(E_2) = 1/2$
$E_1 \cup E_2 = \{1, 2, 3, 4, 6\}$	$P(E_1 \cup E_2) = 5/6$
$E_1 \cap E_2 = \{2\}$	$P(E_1 \cap E_2) = 1/6$
$E_1^C = \{4, 5, 6\}$	$P(E_1^C) = 1/2$
$E_2^C = \{1, 3, 5\}$	$P(E_2^C) = 1/2$

Probabilities adhere to the following properties:

1. $P(E) \geq 0$
2. $P(\Omega) = 1$
3. $P(A \cup B) = P(A) + P(B)$ if $A \cap B = \emptyset$
4. $P(\emptyset) = 0$
5. $P(E) = 1 - P(E^C)$
6. $P(A \cup B) = P(A) + P(B) - P(A \cap B)$

Conditional Probability

The conditional probability $P(E_1|E_2)$ is the probability of an outcome ω being in E_1 if we already know that ω is in E_2. For example, if we define the events E_1 and E_2 as above, then $P(E_1|E_2)$ asks the question "What is the probability that the outcome is less than four if we already know that the outcome is even?" In this context, you can think of E_2 as a filter of sorts, narrowing down the possible outcomes from which E_1

might be selected. In this case, if we know that the outcome of the die roll is even, then we can intuitively see that there is a one-in-three chance that is it is less than four. This can be represented mathematically by the relation

$$P(E_1|E_2) = \frac{P(E_1 \cap E_2)}{P(E_2)}$$

In the example above, the intersection of the two sets is

$$P(E_1 \cap E_2) = \{2\}$$

and therefore

$$P(E_1|E_2) = \frac{1/6}{1/2} = \frac{1}{3}$$

as we found above.

Random Variables

A *random variable* is a one-to-one mapping from a random experiment's sample space Ω onto the real number line. Considering the example of the coin toss given above; for example, we might map heads to $+1$ while mapping tails to -1. A graphical depiction of this is given in Figure 17.3. We can express this mapping mathematically by writing

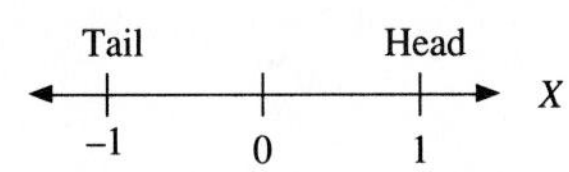

Figure 17.3 Defining a random variable for the coin toss

$$X(\omega) = \begin{cases} +1 & \omega = \text{heads} \\ -1 & \omega = \text{tails} \end{cases},$$

which is referred to as a *discrete random variable* because each outcome is mapped to a discrete number on the real axis.

Once a random variable has been defined for an experiment, we can express its *cumulative distribution function* (CDF) as

$$F_X(x) = P(X(\omega) \le x) \tag{17.1}$$

which represents the probability that the value of a random variable for a given trial will be less than or equal to the argument of the CDF. For example, the CDF of the coin toss example above is given by

$$F_X(x) = \begin{cases} 0 & x < -1 \\ 1/2 & -1 \le x < 1 \\ 1 & x \ge 1 \end{cases}$$

and illustrated in Figure 17.4. From this figure, we see that there is no chance of the random value being less than -1, which is obvious because the minimum possible value of X is -1. There is a 50 percent chance ($F_X = 05$) that X will be less than 1, which is its value when the outcome is tails. Finally, there is a 100 percent chance ($F_X = 1.0$) that the value of X will be less than or equal to $+1$, since there is no possible outcome for which the random value is more than $+1$.

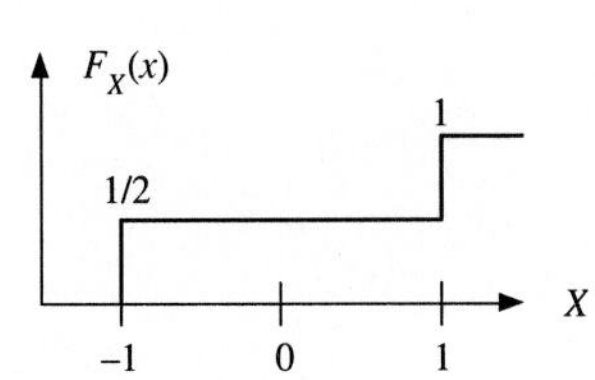

Figure 17.4 CDF of a coin toss random variable

From the CDF, one can define the *probability mass function* (PMF) given by

$$p_i = P(X = x_i)$$

The PMF tells us the probability that a discrete random variable is equal to a particular value. For the example above, we can write the PMF as

$$p_1 = P(X = -1) = 1/2$$
$$p_2 = P(X = +1) = 1/2$$

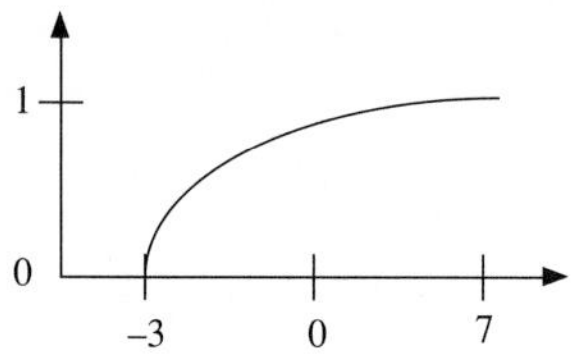

Figure 17.5 CDF of continuous random variable

Similarly, a *continuous random variable* is one that is mapped to some continuous range of the real number line, such as if we were to measure the amount of rainfall within a particular area on any given day. Such a random experiment might have a CDF that looks like Figure 17.5.

For continuous random variables, it is normal to define the *probability density function* (PDF) as

$$f_X(x) = \frac{d}{dx} F_X(x) \tag{17.2}$$

One commonly encountered PDF is that for a *Gaussian* random variable, given by

$$f_X(x) = \frac{1}{\sigma\sqrt{2\pi}} e^{\frac{-(x-m)^2}{2\sigma^2}} \tag{17.3}$$

where m is the mean, or average value, and σ^2 is the *variance,* or spreading factor. The Gaussian is encountered so often that it is frequently written as $N(m, \sigma^2)$.

The *standard normal* Φ is a special case of the Gaussian where $m = 0$ and $\sigma^2 = 1$. Plots of a few Gaussian distributions are given in Figure 17.6.

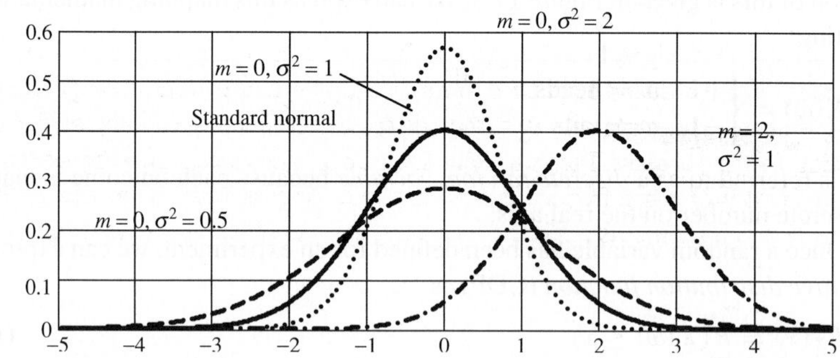

Figure 17.6 PDFs of Gaussian distributions

For any random variable, we can use the PDF to find the probability that X falls between two values for a given outcome by integrating the area under the PDF for that interval. When X is the standard normal, for example, the probability that X is greater than some value a is given by the integral

$$P(\Phi > a) = \int_a^\infty f_\Phi(x)dx \tag{17.4}$$

The value of the integral above can be found from widely available tabulated values of the *tail function* (also called the *Q-function*) defined by

$$Q(a) = \frac{1}{\sqrt{2\pi}} \int_a^\infty e^{-\frac{x^2}{2}} dx \tag{17.5}$$

Although the tail function is tabulated for the standard normal, it can be applied for any Gaussian through a simple variable transformation given by

$$P(X > a) = Q\left(\frac{a-m}{\sigma}\right)$$

where X is a Gaussian random variable.

17.3 PULSE CODE MODULATION

Pulse code modulation (PCM) is simply the process of converting an analog signal (continuous-time and continuous-amplitude) into a digital signal (discrete-time and discrete-amplitude) by applying sampling and quantization as previously described in section 15.4 and illustrated in Figure 17.7.

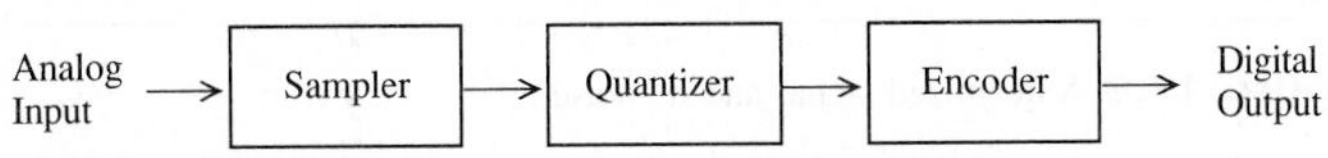

Figure 17.7 Block diagram of a pulse code modulator

For a well-designed digital system, it is important to know the extent of any errors that are introduced into the system as a byproduct of the PCM process. Assuming that the sampling rate is sufficient for the frequency content of the signal according to the Nyquist Theorem, any such error introduced into the system as a result of PCM will come from quantization alone. As you recall from section 15.4, a quantizer's resolution Δ is set by the number of bits used to represent its output. This resolution is

$$\Delta = \frac{v_{\max} - v_{\min}}{2^b} \tag{17.6}$$

where $v_{\max}$ and $v_{\min}$ define the dynamic range of the quantizer and b is the number of bits. The *quantization noise* is then defined as the difference between the sampled signal and the quantized version. To find a relationship between the number of bits and the quantization noise, consider the sampled continuous-amplitude signal and its quantized version shown in Figure 17.8. From this figure, we can see that the maximum possible error due to quantization noise is $\pm\Delta/2$, meaning that the maximum average quantization noise power $P_{qn}^{\max}$ is $(\Delta/2)^2$ or

$$P_{qn}^{\max} = \frac{1}{2^{2b}}\frac{(v_{\max} - v_{\min})^2}{4}$$

Assuming some signal power P_s, the signal to quantization noise ratio (SQNR) is therefore proportional to the number of bits and is given by

$$SQNR = \frac{P_s}{P_{qn}} \propto 2^{2b} \text{ or } \boxed{SQNR_{dB} \propto 10\log 2^{2b} = 6b} \tag{17.7}$$

From the above equation we can discover a very important rule of thumb: each additional bit of resolution we add to a quantizer will increase the SQNR of its output by about 6 dB. Thus, we can expect the output of a 10-bit quantizer to have a SQNR of about 60 dB.

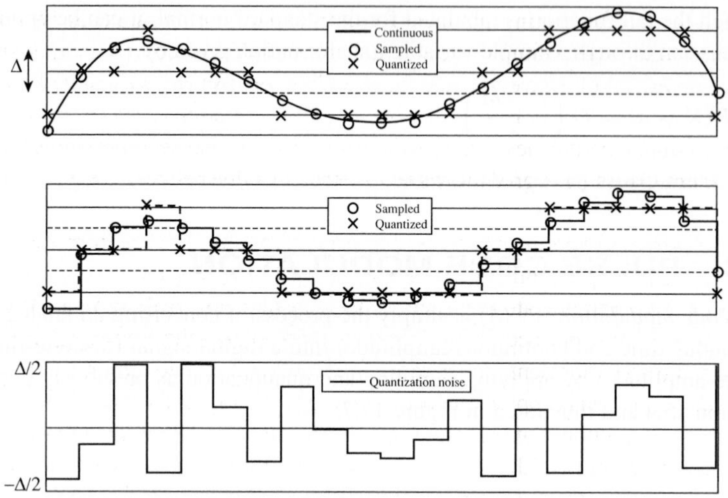

Figure 17.8 A quantized signal and its noise

17.4 SOURCE CODING

Source coding is the process of removing the redundancy of a signal and reducing it to its most basic form, retaining only the basic information that is necessary to communicate the message.

Redundancy is any message content carried by a signal beyond the fundamental information being communicated. A source of information contains redundency if its messages can be represented with complete fidelity in a more terse form. Redundancy may or may not serve a purpose. For example, when announcing over a public address system that a parked vehicle has its lights on, it is redundant, but useful, to announce the license plate number several times.

Human speech is one of the best examples of redundancy in nature—have you ever listened to a scratchy AM weather radio broadcast during a thunderstorm? Even if you haven't, you can imagine the quality of the audio you might hear as the electrons contained within the thunderstorm contribute noise to the signal as it travels from the radio station to our receiver. Yet even with the extra noise, the details of the weather report can usually still be understood by the listener. Even though a tiny fraction of the original signal survives the trip from the speaker's mouth to our ears, we are still able to decode the information contained within it. Signal redundancy, therefore, often serves to increase the probability that a message is understood by the receiver.

Unfortunately, the price paid for redundancy is efficiency. Consider the same radio example as above, except with a strong signal and no thunderstorm. Although the audio has much more clarity, we receive the same weather report details as before. Therefore, we can conclude that in this case redundancy is unnecessary, and with the absence of noise in the communications channel the message could be communicated just as accurately without it.

Compressibility

Source coding can also be thought of as *data compression* because it results in a message that can be represented using fewer bits. *Lossless* compression codes the signal so that it can be reconstructed exactly by the receiver, while *lossy* compression allows some of the message to become unrecoverable in order to gain additional compression by representing the message with fewer bits. The field of *information theory* allows us to determine a signal's fundamental information content, or *entropy,* which is a measure of its average uncertainty. From the entropy we can determine the signal's *compressibility,* or how much extraneous information can be removed before some of the fundamental message contained within the signal becomes unrecoverable. Thus, the entropy of the signal allows us to establish a line between lossy and lossless compression.

Consider an experiment where the result X can be chosen from the set $\{a_1, a_2, \ldots, a_N\}$ and the probability of that result being a_n is $p_n = P(X = a_n)$. The amount of information communicated by each possible result is inversely proportional to the probability of that result occurring. For example, knowing the stock market is going to crash tomorrow represents much more information than knowing it's not going to crash. This same idea is applicable to any experiment, not just the stock market, so we conclude that the amount of information contained by a particular result is dependent only on the probability of that result occurring and not the result itself. This measure of *self-information* is notated by I and defined as (see Figure 17.9)

$$I(X = p_n) = \log_2\left(\frac{1}{p_n}\right). \tag{17.8}$$

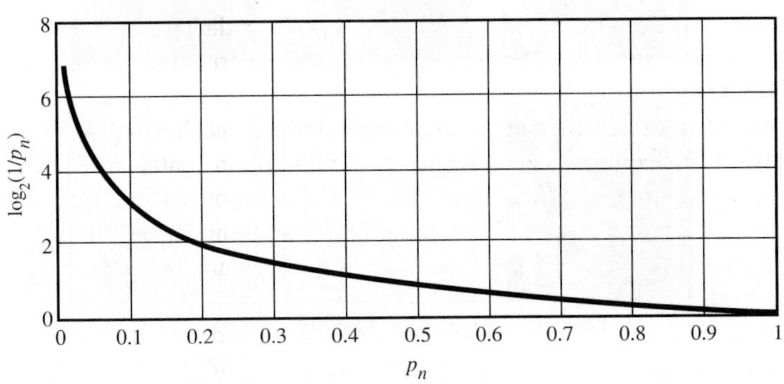

Figure 17.9 Self-information vs. probability of occurrence

The total entropy $H(X)$ represented by the experiment is then the weighted average of the self-information of each result and the probability of that result occurring, as given by

$$H(X) = \sum_{n=1}^{N} p_n \log_2\left(\frac{1}{p_n}\right) \tag{17.9}$$

A good example of lossy vs. lossless compression can be found by examining the storage of digital images. The two most common image storage formats in use today are the *Graphic Interchange Format* (GIF) created by CompuServe (now America Online) in 1987, and the format named after the *Joint Photographic Experts Group* (JPEG) created by that organization in 1986. While both formats serve to compress digital images, they go about it in much different ways. The GIF standard uses an implementation of the Lempel-Ziv-Welch (LZW) algorithm which is lossless and therefore guarantees that the original image can be recovered exactly. The

JPEG standard uses a discrete cosine transformation (DCT) to convert the image to the spectral domain where high-frequency coefficients below a threshold set by the user are filtered out. Once these coefficients are filtered out they cannot be recovered, and therefore the technique is lossy. Determining which compression technique to use usually depends on the application. For photographic images, JPEG usually results in a smaller file with little noticeable loss. However, for images of technical diagrams involving thin lines, JPEG usually results in larger files and may introduce unacceptable artifacts (see Figure 17.10).

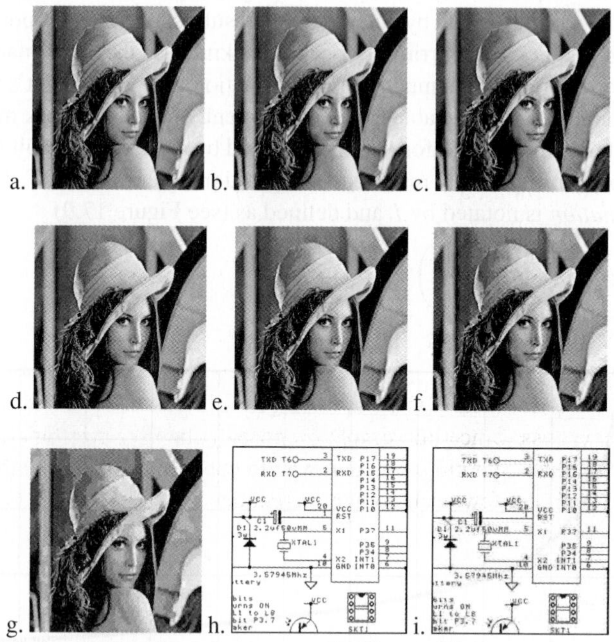

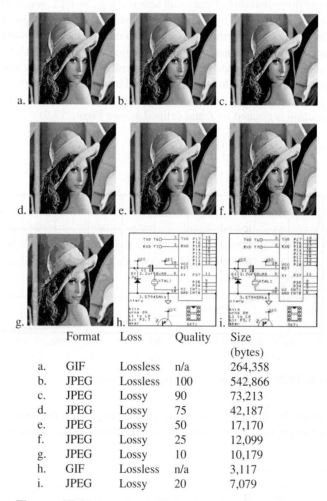

	Format	Loss	Quality	Size (bytes)
a.	GIF	Lossless	n/a	264,358
b.	JPEG	Lossless	100	542,866
c.	JPEG	Lossy	90	73,213
d.	JPEG	Lossy	75	42,187
e.	JPEG	Lossy	50	17,170
f.	JPEG	Lossy	25	12,099
g.	JPEG	Lossy	10	10,179
h.	GIF	Lossless	n/a	3,117
i.	JPEG	Lossy	20	7,079

Figure 17.10 Examples of GIF and JPEG file sizes vs. quality (Reproduced by Special Permission of *Playboy* magazine. Copyright ©1972 by Playboy.)

Source Coding Techniques

Once the entropy of a signal is known, a code must be devised so that individual events can be represented by sequences of bits, or *codewords*, for transmission. Consider the event sequence

$$X = \{a_1, a_2, a_1, a_4, a_3, a_1, a_1, a_2\}$$

At first glance, one might try coding each event with a unique binary codeword of fixed length, as shown in Table 17.3.

Table 17.3 Simple fixed-width coding scheme

Event	Probability	Codeword
a_1	$p_1 = 1/2$	00
a_2	$p_2 = 1/4$	01
a_3	$p_3 = 1/8$	10
a_4	$p_4 = 1/8$	11

The assembled bit sequence representing the message X would then be

$$\{00, 01, 00, 11, 10, 00, 00, 01\} = 0001001110000001$$

with two bits per event for a total of 16 bits to represent the entire sequence message. By computing the entropy of the above signal, we can determine the efficiency of the simple coding scheme above. By using equation 17.6 and Table 17.3, we find the entropy to be

$$H(X) = \frac{1}{2}\log_2 2 + \frac{1}{4}\log_2 4 + \frac{1}{8}\log_2 8 + \frac{1}{8}\log_2 8 = 1.75$$

meaning that the theoretical minimum bound is 1.75 bits per event to represent the message without loss. Since our simple proposed code uses two bits per pixel, we know that there must be a more efficient way to code the message. In the following sections we examine a few such coding techniques that are more efficient than our simple method above.

Huffman Source Coding

Huffman source coding is a lossless technique that represents frequently occurring events with short codewords and rarely occurring events with longer ones. In practical application, Huffman coding is generally used as a secondary coder following some other coding technique. For example, ZIP (a popular computer compression format), JPEG, and MP3 use Huffman encoding as a secondary step.

For an example of Huffman coding, consider the sequence above with the code shown in Table 17.4.

Table 17.4 Huffman coding scheme

Event	Probability	Codeword
a_1	$p_1 = 1/2$	1
a_2	$p_2 = 1/4$	01
a_3	$p_3 = 1/8$	001
a_4	$p_4 = 1/8$	000

The scheme defined in Table 17.4 would lead to a bit pattern of

$$\{1,01,1,000,001,1,1,01\} = 10110000011101$$

with a total of 14 bits or 1.75 bits per event, which is a savings of 12.5 percent over the fixed-width code above and meets the theoretical limit given by the entropy computed above. Because Huffman codewords are not fixed length, however, care must be taken when choosing the sequences so that they will be *uniquely decodable*. In order to be uniquely decodable, codewords should be constructed in such a way that the decoder can distinguish between consecutive events in a bit sequence. In the example p. 873, the decoder knows that an event is ending when it either detects a 1 in the bit sequence or when three 0's are received in a row (indicating a a_4).

The procedure for creating a Huffman code is as follows:

1. Gather statistical data about the frequency of occurrence of each potential event.
2. Combine the two least probable events (assign them bits 0 and 1) into a new single event that occurs when either of the two original events occur. The probability of this new event occurring is then the sum of the probabilities of the original events.
3. If only two events remain, go to step 4. Otherwise go back to step 2.
4. Assign 0 and 1 as bits representing the remaining two events. These steps are illustrated graphically in Figure 17.11.

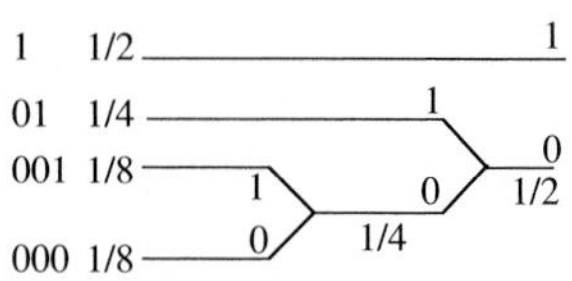

Figure 17.11 Creating a Huffman code

EXAMPLE 17.1 Construction of a Huffman Code

Problem

Considering the events and statistics given in Table 17.5, first compute the entropy. Then create an efficient Huffman code and find its average code length.

Table 17.5 **Huffman coding scheme**

Event	Probability
a_1	$p_1 = 0.6$
a_2	$p_2 = 0.3$
a_3	$p_3 = 0.1$

Solution

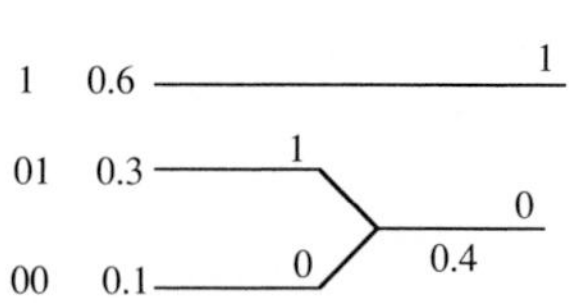

Figure 17.12 Creating a Huffman code for Example 17.1

The entropy of this experiment is given by

$$H(X) = 0.6 \log_2 \frac{1}{0.6} + 0.3 \log_2 \frac{1}{0.3} + 0.1 \log_2 \frac{1}{0.1} = 1.3$$

The Huffman code is created as shown in Figure 17.12. The average codeword length is then computed from Table 17.6 as

$$E(L) = 0.6 \cdot 1 + 0.3 \cdot 2 + 0.1 \cdot 2 = 1.4,$$

Table 17.6 **Example Huffman code**

Event	Probability	Codeword
a_1	$p_1 = 0.6$	1
a_2	$p_2 = 0.3$	01
a_3	$p_3 = 0.1$	00

which is close to the theoretical minimum of 1.3 bits per event given by the entropy computation p. 876.

Lempel-Ziv-Welch Source Coding

While Huffman coding gives very good results that are close to the theoretical limit given by the entropy, it has two major disadvantages. First, a detailed knowledge of the statistical probability of each event occurring must be known beforehand. With many common information sources such as photographs, music and speech recordings, and streaming video, these statistics cannot be predicted. Secondly, Huffman coding is efficient only for codewords that represent a single event. If events are known to occur frequently in patterns (say $\{a_1, a_1, a_2\}$), then building a Huffman code library for all such patterns quickly becomes unmanageable regardless of how efficient such a code might be.

The Lempel-Ziv-Welch (LZW) code solves these two problems by building a running *dictionary* of event patterns rather than the events themselves. This dictionary is built by examining the event sequence to identify unique recurring patterns and adding them to the dictionary as they are encountered. Then, the pattern is replaced with a fixed-width codeword indicating the matching entry is the dictionary.

To keep the size of the dictionary small, new event pattern entries are represented as the concatenation of an existing dictionary entry with the "new" event. Note that unlike Huffman encoding, the LZW approach requires the dictionary to be transmitted along with the message.

As an example, consider the event sequence given by

$$X = \{a_1, a_2, a_1, a_1, a_2, a_3, a_1, a_2, a_1, a_1, a_4\}$$

To create a LZW code for this sequence, we add unique patterns to the dictionary as they are encountered. The resulting grouped sequence and dictionary are given by

$$X = \{\{a_1\}, \{a_2\}, \{a_1, a_1\}, \{a_2, a_3\}, \{a_1, a_2\}, \{a_1, a_1, a_4\}\}$$

By substituting codes from Table 17.7, the encoded bit sequence is found to be

$$X = \{000, 001, 010, 011, 100, 101\} = 000001010011100101$$

and uses a total of 18 bits or 1.64 bits per event.

Table 17.7 **Example Lempel-Ziv-Welch coding scheme**

Entry	Existing entry	New event	(Effective sequence)
000	n/a	a_1	a_1
001	n/a	a_2	a_2
010	000	a_1	a_1, a_1
011	001	a_3	a_2, a_3
100	000	a_2	a_1, a_2
101	010	a_4	a_1, a_1, a_4

LZW coding is particularly useful in applications where very long or repetitive event patterns exist, such as images of technical drawings where there may be many consecutive pixels of the same color. It is typically less useful in applications where event patterns are largely random or occur infrequently, such as with photographs.

EXAMPLE 17.2 Construction of a Lempel-Ziv-Welch Code

Problem

For the event sequence

$$X = \{a_3, a_3, a_3, a_2, a_3, a_3, a_1, a_1, a_2, a_3\}$$

create a LZW dictionary and encode the sequence to bits.

Solution

First, the sequence is broken into unique patterns as they occur. This results in the grouped sequence and dictionary given by

$$X = \{\{a_3\}, \{a_3, a_3\}, \{a_2\}, \{a_3, a_3, a_1\}, \{a_1\}, \{a_2, a_3\}\}$$

From Table 17.8, the encoded bit sequence is then

$$\{000, 001, 010, 011, 100, 101\} = 000001010011100101$$

Note that the resulting bit sequence is identical to the sequence found for Example 17.1 in the text. Only the dictionary has changed.

Table 17.8 **Example Lempel-Ziv-Welch coding scheme**

Entry	Existing entry	New event	(Effective sequence)
000		a_3	a_3
001	000	a_3	a_3, a_3
010		a_2	a_2
011	001	a_1	a_3, a_3, a_1
100		a_1	a_1
101	010	a_3	a_2, a_3

Parametric Source Coding

Both of the coding methods described above are lossless and rely on the logical layout of the data to achieve compression rates near the entropy limit. For many types of sources these codes function very close to the theoretical limit determined by the entropy of the source. For sources such as photographs and music that have little logical structure, however, these codes usually result in poor performance.

Fortunately, it is frequently possible to achieve very efficient compression rates for such sources by leveraging knowledge about the ultimate application of the data. For example, fine details in a photograph that are imperceptible to the naked eye can be removed, as with JPEG, without significant detriment. Codes using such biometric techniques belong to a class known as *parametric source codes*. Parametric source codes generally achieve their impressive compression ratios by transforming a sequence to be encoded into some spectral domain and removing components that are unlikely to be missed. Because of this property, such codes are almost always lossy, and unlike the Huffman and LZW codes, they must be custom tailored for the intended receiver.

One of the most famous parametric codes of recent times was created in 1987 by Professor Dieter Seitzer of the University of Erlangen in association with the *Motion*

Picture Experts Group for compression of audio, technically referred to as "Layer III." This standard, commonly known as MP3, can reduce the size of music files by a factor of 10 or more by filtering out certain components of sound that are inaudible to the ear in much the same way that JPEG removes photographic details that are sharper than the eye can perceive.

Because of the enormous complexity of parametric encoding, an example is beyond the scope of this book.

17.5 DIGITAL BASEBAND MODULATION

We live in a world that is physically analog, so information must be in an analog form before being transmitted across the physical medium. However, this is not to be confused with analog communication systems where the information to be transmitted and the waveform that is transmitted are both analog (e.g., an analog voice signal). In digital communication systems, the information to be sent from the transmitter to the receiver is digital, but analog while traveling over the physical medium. Baseband digital modulation converts a binary signal into an analog signal that can be transmitted over the physical communication channel.

Digital *baseband modulation* transmits bits of information across a communication channel by mapping those bits to particular analog message waveforms to indicate the states of those bits. The analog waveforms used could represent voltage on a coaxial cable, electric-field potential in a wireless link, acoustic pressure waves in an underwater link, or light pulse amplitude in a fiber-optic line. By stringing together a number of these waveforms, one after the other, an entire sequence of bits can be transmitted. The receiver, using *baseband demodulation,* then decodes the message by matching the received waveforms to an internal library of previously agreed-upon waveforms in order to recover the original bits.

Pulse Position Modulation

One set of waveforms that is commonly used for digital baseband modulation is shown in Figure 17.13, where a 0 bit is represented by $m_0(t)$ and a 1 is represented by $m_1(t)$. Because the state of the bit is represented by the position of the pulse within the overall duty cycle T, these are known as *pulse position modulated* (PPM) waveforms.

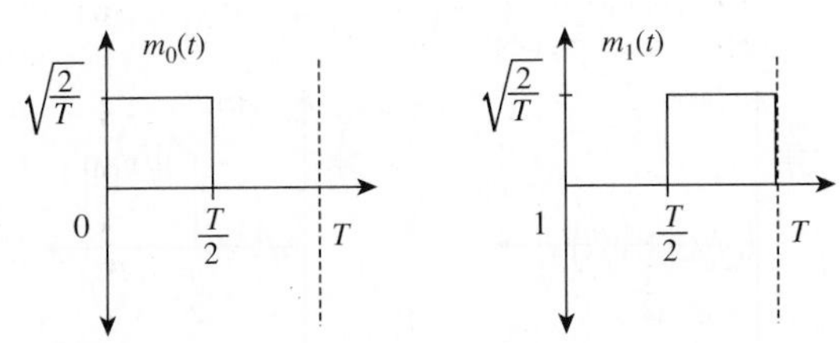

Figure 17.13 Binary pulse position modulated (PPM) message waveforms with duty cycle T

To send information in chunks of M simultaneous bits, the set with $K = 2^M$ message waveforms can be used, each with duration of T seconds. For example, to send two bits at a time we can use four message waveforms to convey all the possible bit states, as illustrated in Figure 17.14.

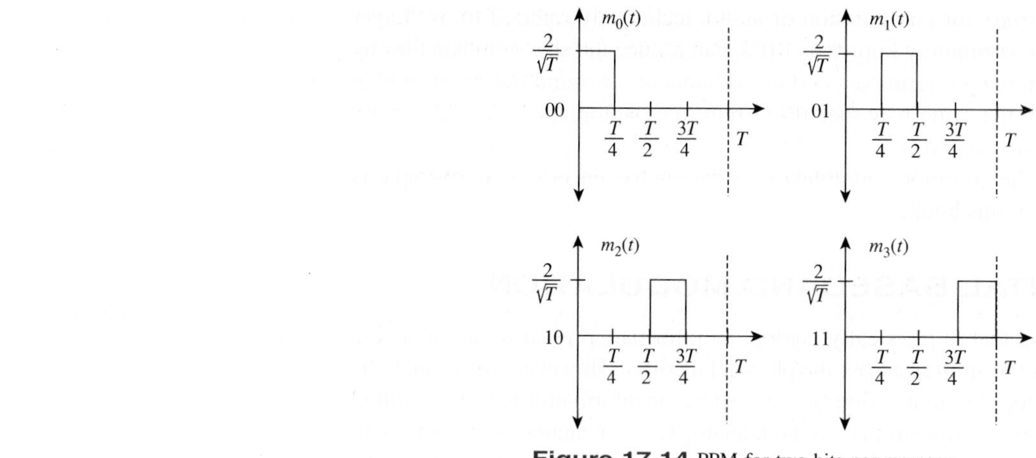

Figure 17.14 PPM for two bits per message waveform

The transmission rate of a digital system depends on the duty cycle T as well as the number of bits that are being transmitted simultaneously. The *baud rate* is $1/T$, or the number of message waveform transmissions per second. The *bit rate* is M/T, or M times the baud rate since M bits are sent per transmission. For example, the V.90 telephone modem standard specifies a baud rate of 8,000 message waveform transmissions per second and a bit rate of 56,000 bits per second. Thus, each message waveform conveys 7 bits by transmitting one out of a set of 128 message waveforms.

At the receiver, noise added to the signal while traveling through the channel corrupts the message waveform, possibly making a decision difficult. Mathematically, the received signal $r(t)$ is written

$$r(t) = m_i(t) + w(t)$$

where $m_i(t)$ is one of K message waveforms and $w(t)$ is an additive noise signal. For example, Figure 17.15a shows a noise signal $w(t)$ and Figure 17.15b shows the message waveform corrupted by the noise $w(t)$, where the message waveform is chosen from the binary PPM waveform set shown in Figure 17.13. By examining the received signal in Figure 17.15b, we see that it more closely resembles $m_1(t)$ than $m_0(t)$ in Figure 17.13. Thus, we conclude that a 1 was sent.

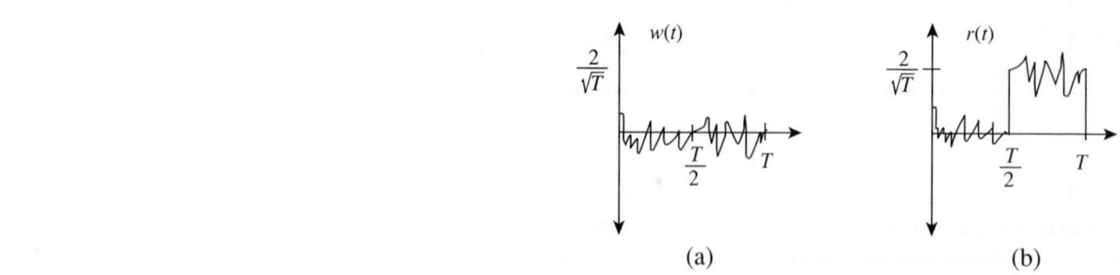

Figure 17.15 (a) Gaussian noise; (b) signal corrupted by noise

In using our intuition to determine which waveform was transmitted, we eventually come to the conclusion that the received signal $r(t)$ was somehow "closer" to one message waveform than the other. By measuring this "distance" between waveforms, receiver automatically decides which message waveform was probably transmitted. One approach is to calculate the *root-mean squared* (RMS) *distance* by adding up the squared absolute point-wise differences between the waveforms. The RMS distance d between $r(t)$ and a given waveform $m_i(t)$ is given by

$$d\,[r(t), m_i(t)] = \sqrt{\int_0^T |r(t) - m_i(t)|^2\,dt} \tag{17.11}$$

Once the distances between the received signal and each of the candidate waveforms are calculated, the receiver chooses the candidate with the minimum RMS distance from the received waveform. Detection theory tells us that if the message waveforms are equally likely to be sent, then this *minimum distance* (MD) detection algorithm is optimal in the sense of minimizing the probability of choosing the wrong waveform. Mathematically, the MD detector is written

$$\hat{\imath} = \arg \min_{i\in\{0,1,\ldots,K-1\}} \left[\int_0^T |r(t) - m_i(t)|^2\,dt \right] \tag{17.12}$$

where we have removed the square-root since it doesn't affect finding the minimum. Use of the *arg* operator returns the index of the minimum value and not the minimum value itself. A block diagram of the MD detector is shown in Figure 17.16.

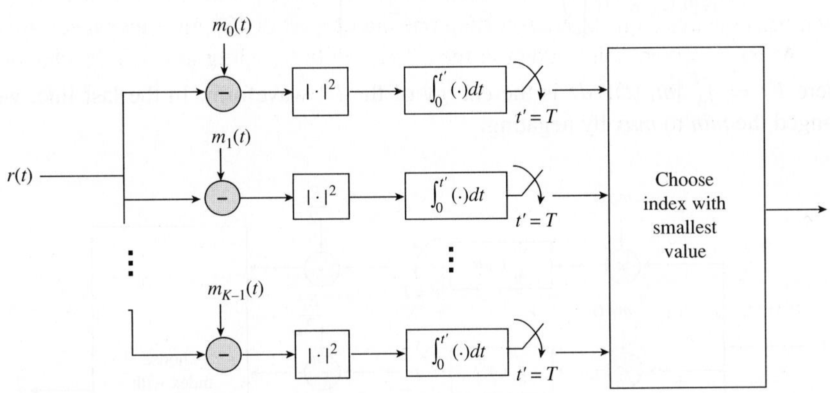

Figure 17.16 Minimum distance detector

In the absence of noise, the MD detector will find a distance of zero between the matching $m_j(t)$ and the received signal, and a non-zero distance between the received signal and all other message waveforms. As a result, the message waveforms should be designed to be as far apart from each other as possible so that noise disturbance cannot easily make the received signal closer to a message waveform that was not sent. In the ideal case there would be infinite distance between message waveforms in order to make the error probability go to zero. Unfortunately, the transmit power is constrained by the *average energy* E of the message waveform set, calculated as

$$E = \frac{1}{K} \sum_{i=0}^{K-1} \int_0^T |m_i(t)|^2\,dt \tag{17.13}$$

If two message waveform sets have the same average energy and one has a greater minimum distance between message waveforms, then in general it will have lower error probability. This is explored further in the chapter problems.

The MD detector can be changed to an equivalent form, called the *correlation receiver,* shown in Figure 17.17. In some cases, the correlation receiver can be easier to implement than the MD detector. The correlation receiver multiplies the samples of the received signal with each message waveform, integrates the result, samples at $t' = T$ seconds, subtracts the signal energy, and then chooses the index of the largest value. The multiply-integrate operation is called *correlation*. The correlation receiver can be derived from the MD detector through the following algebraic manipulation:

$$\begin{aligned}\hat{i} &= \arg \min_{i\in\{0,1,\ldots,K-1\}} \left[\int_0^T |r(t) - m_i(t)|^2\, dt\right] \\ &= \arg \min_{i\in\{0,1,\ldots,K-1\}} \left[\int_0^T |r(t)|^2 - 2r(t)\, m_i(t) + |m_i(t)|^2\, dt\right] \\ &= \arg \min_{i\in\{0,1,\ldots,K-1\}} \left[-2\int_0^T r(t)\, m_i(t) - \frac{|r(t)|^2}{2} - \frac{|m_i(t)|^2}{2} dt\right] \\ &= \arg \max_{i\in\{0,1,\ldots,K-1\}} \left[\int_0^T r(t)\, m_i(t) - \frac{E_i}{2} dt\right]\end{aligned} \tag{17.14}$$

where $E_i = \int_0^T |m_i(t)|^2\, dt$ is the *energy* of the ith waveform. In the last line, we changed the *min* to *max* by negating.

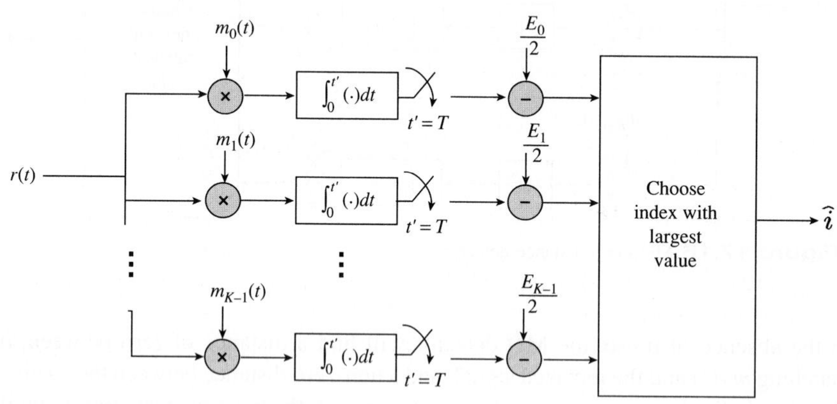

Figure 17.17 Correlation detector

Pulse-Amplitude Modulation

In *pulse-amplitude modulation* (PAM), the set of K message waveforms are built by specifying amplitudes A_i of a prototype pulse $m(t)$. Thus, the ith waveform is

$$m_i(t) = A_i m(t).$$

If we assume that the prototype pulse $m(t)$ has unit energy, i.e.,

$$\int_0^T |m(t)|^2\, dt = 1$$

then each message waveform has energy $E_i = A_i^2$. The correlation detector shown above can now be simplified to

$$\hat{i} = \arg \max_{i \in \{0,1,\ldots,K-1\}} \left[A_i \underbrace{\int_0^T r(t)m(t)dt}_{y} - \frac{A_i^2}{2} \right]$$

$$= \arg \max_{i \in \{0,1,\ldots,K-1\}} \left[A_i y - \frac{A_i^2}{2} \right]$$

$$= \arg \min_{i \in \{0,1,\ldots,K-1\}} \frac{1}{2} \left[A_i^2 - 2yA_i \right]$$

where we changed the *max* to a *min* by negating. We complete the square and ignore terms common to all indices to write the decision as

$$\hat{i} = \arg \min_{i \in \{0,1,\ldots,K-1\}} (A_i - y)^2 .$$

From this equation we see that when using PAM the decision is made simply by finding the amplitude that has minimum distance to the correlation output y. This result greatly simplifies the receiver structure, as seen in Figure 17.18.

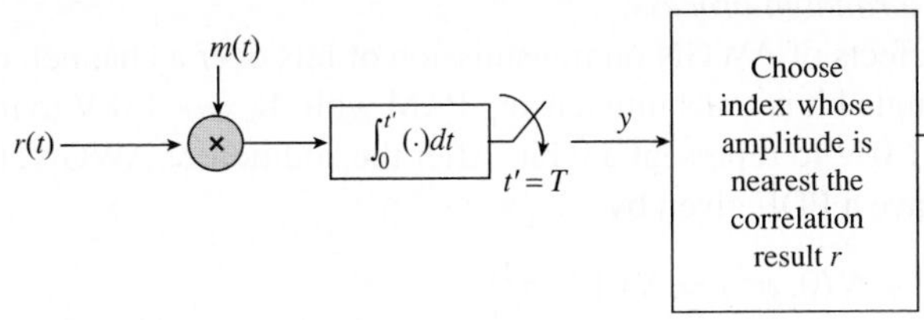

Figure 17.18 PAM detector

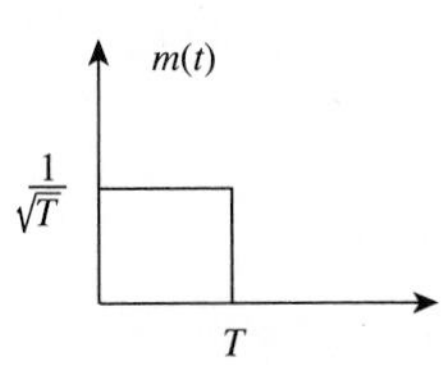

Figure 17.19 Unit energy prototype pulse for PAM signal set

For a moment, consider a PAM signal set with a unit energy square prototype pulse shown in Figure 17.19.

For binary PAM, we choose amplitudes $A_0 = +1$ and $A_1 = -1$. The resulting message waveforms are shown in Figure 17.20.

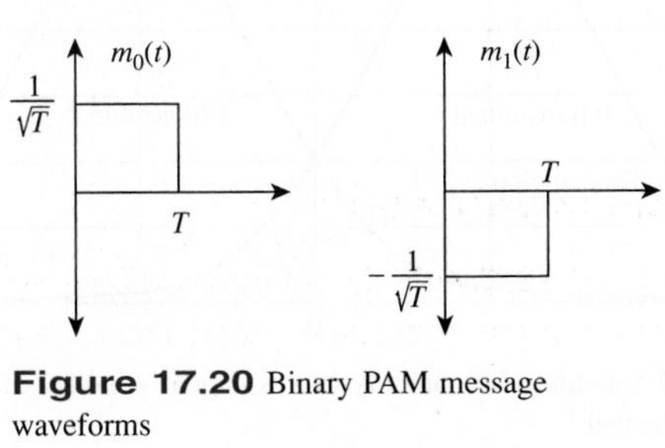

Figure 17.20 Binary PAM message waveforms

In this case, the correlation output y will be

$$y = \pm 1 + z$$

where

$$z = \int_0^T m(t)\, w(t)\, dt \tag{17.15}$$

is the noise. So far in our discussion we have avoided discussing the specific nature of the noise signal z. As it turns out, noise sources such as thermal noise, atmospheric interference, and channel distortion can be modeled as *white noise,* or noise with a constant power at all frequencies. Specifically, the power spectral density for such noise is given as

$$S_n(f) = \frac{kT}{2} \tag{17.16}$$

where k is *Boltzmann's constant* ($1.38 \cdot 10^{-23}\, J/K$) and T is the *noise temperature* in Kelvin. This model is actually non-physical because a constant power spectral density at all frequencies would mean that the total power of the noise signal is infinite. As we saw in Chapter 16, however, engineers are usually only interested in a very narrow band of the spectrum at a time. As a result, as long as the white noise model is applied within that narrow band we can avoid breaking any physical laws. The ratio of signal power to noise power is known as the *signal to noise ratio* (SNR).

Additive white Gaussian noise (AWGN) is also a signal with constant power density at all frequencies, but with the additional property that for any particular sample n_0 its value is a random variable Z with Gaussian PDF. Because the value of the signal z for a given sample n_0 is described by a random variable, the noise signal itself is known as a *random process.*

To see the effects of AWGN on transmission of bits over a channel, consider a system where a single bit is transmitted using PAM with $A_0 = -1.0$ V to represent a 1 bit and $A_1 = -1.0$ V to represent a 0 bit. After the addition of AWGN, the output y for bit n_0 will have a PDF given by

$$y[n_0] = \pm 1 + N(0, \sigma^2) = N(\pm 1, \sigma^2) \tag{17.17}$$

and resemble Figure 17.21, where the output for bit n_0 can take on any value according to the probability prescribed by the Gaussian distribution.

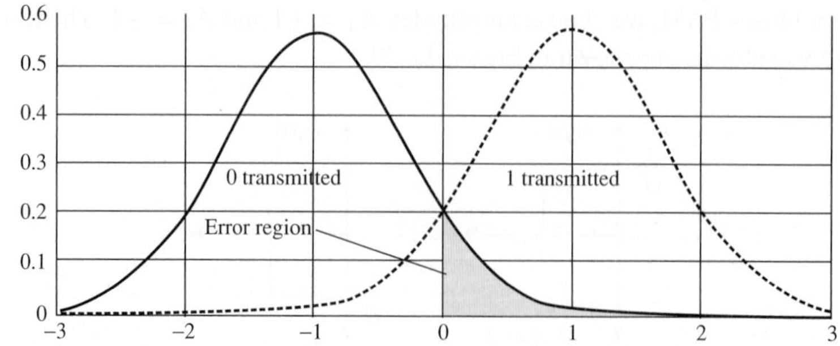

Figure 17.21 Likelihood probability curves (signal + noise) of bits transmitted over an AWGN channel

Now, consider a decoder that is designed to recover the intended value of the transmitted bit. Such a decoder $d[n]$ might use a simple comparator to decide whether the output bit is a 1 or 0, such as

$$d[n] = \begin{cases} 0 & y[n] \geq 0 \\ 1 & y[n] < 0 \end{cases}$$

The decoder interprets an output less than 0.0 V to indicate a 0 bit and more than 0.0 V to indicate a 1 bit. Unfortunately, the output of this decoder will be in error whenever the noise signal causes the output to cross over the zero threshold, as identified by the shaded region in the figure for a transmitted 0 bit (−1.0 V). The probability of this occurring is the area of the shaded region, as given by

$$\begin{aligned} P(d = 1|x = -1) &= P(y > 0|x = -1) \\ &= \frac{1}{\sqrt{2\pi}}\frac{1}{\sigma}\int_0^{\infty} e^{-\frac{(x+1)^2}{2\sigma^2}}\,dx \\ &= Q\left(\frac{1}{\sigma}\right) \end{aligned}$$

Through symmetry, the same probability can be found for $P(d = 0|x = 1)$.

Binary Symmetric Channel Model

The *binary symmetric* channel model simplifies the AWGN model by encapsulating the error mechanism into a single parameter, $P(y|x)$, or the probability that event y was received given that x was transmitted. Mathematically, the binary symmetric channel is described by specifying

$$\begin{aligned} a &= P(0|0) = P(1|1) \\ b &= P(1|0) = P(0|1) \end{aligned}$$

where a is the probability that the output is equal to the input, meaning that the bit was transmitted successfully, and b is the *crossover probability,* meaning that the bit was "flipped" during transmission (see Figure 17.22). This simpler model is often used to model computer storage or transmission systems.

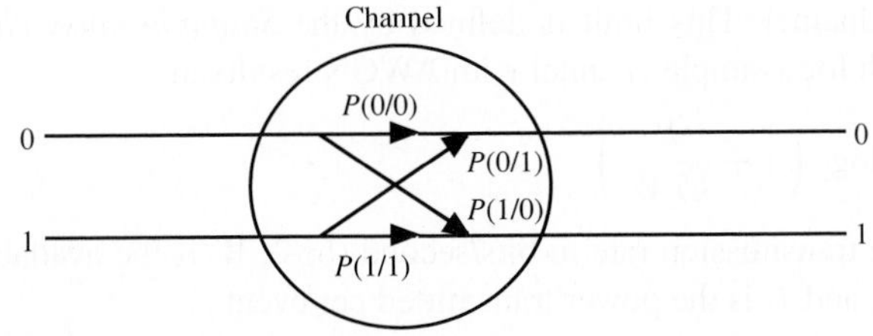

Figure 17.22 Binary symmetric channel

17.6 CHANNEL CODING

In a previous section, we showed how source coding is used to represent a sequence of events using the minimum possible number of bits. We also showed how the entropy equation set a limit on lossless compression and could be used to compute the efficiency of the coder. As we will see in the following section, a similar theorem exists to show that for a given transmitting power and bandwidth, there is a transmission rate at which one can accomplish effectively error-free transmission of data over a noisy channel. This transmission rate is called the *channel capacity*. Just as source coding is the mechanism that allows sources to approach the theoretical limit of compressibility, *channel coding* is the mechanism used to approach this theoretical error-free transmission rate by adding sufficient redundancy to improve the likelihood that the message can pass accurately through the communications channel over which it is transmitted or stored.

As an example, consider storage of music on a compact disc (CD). From the listener's perspective such discs can store about 74 minutes of audio, or the equivalent of 650 megabytes of data, stored in the form of microscopic pits in a layer of fragile metal foil encased in a piece of plastic. To reproduce the music, the CD player must provide the means to accurately recover the CD's data even in the presence of normal wear and tear which may include contamination of its surface with dust, oil, or scratches. To provide this functionality, an additional 200 MB of redundant data is stored on the CD to allow the player to perform automatic error detection and correction as the data bits are read from the disc. For CDs used for computer data storage, the percentage of redundant data is even higher.

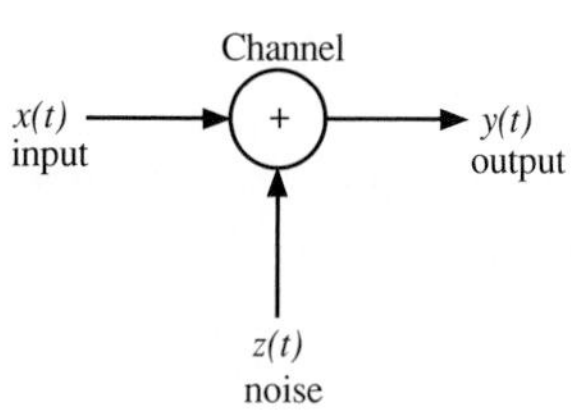

Figure 17.23 Modeling the additive communications channel

The Channel Model and Additive White Gaussian Noise

Before discussing channel coding schemes, we must characterize the nature of the communication channel over which communication will take place. Although there are many complicated methods for modeling communication channels, the most common (and simplest) is the *additive channel*. Using the system diagram of Figure 17.23, we can characterize the output $y[n]$ of an additive channel as the linear superposition of the message signal $x[n]$ with some noise signal $z[n]$, expressed mathematically as

$$y[n] = x[n] + z[n]$$

Shannon Noisy Channel Capacity

Just as the entropy of the message defined the absolute limit on lossless compression, a similar limit exists for the maximum error-free transmission rate that can be achieved over a given channel. This limit is defined by the *Shannon noisy channel coding theorem,* which for a simple channel with AWGN results in

$$C = W \log_2 \left(1 + \frac{P}{kTW}\right) \quad \textbf{(17.18)}$$

where C is the transmission rate in bits/second (bps), W is the available bandwidth of the channel, and P is the power transmitted per event.

EXAMPLE 17.3 Capacity of a AWGN channel

Problem

For a channel with an available bandwidth of 10 kHz, find the maximum bit rate at which an event sequence can be transmitted if AWGN is present with a noise temperature of 20 C and transmit power is 1.0 μW/event.

Solution

From the AWGN channel capacity theorem on p. 886, the channel capacity for the parameters given is

$$C = 10^4 \log\left[1 + \frac{10^{-6}}{(1.38 \cdot 10^{-23}) \cdot (274 + 20) \cdot 10^4}\right] \approx 345 \text{ kbps}$$

Linear Block Channel Coding

Linear block coding is the simplest of all the channel coding techniques. As with source coding, linear block coding relies on a dictionary of codes that will be substituted for the input presented to the coder. However, while source coders use dictionaries that reduce the total number of bits, channel codes use carefully chosen codewords that actually increase the number of bits and therefore add redundancy. As previously mentioned, adding redundancy does not come for free—the additional bits require either more time to transmit or increased bandwidth to accommodate the higher bit rate, not to mention increased system complexity.

In general, (n, k) linear block codes map 2^k input sequences of k bits into n bit output sequences. Consider the following very simple (2, 1) code as described by the dictionary in Table 17.9.

Table 17.9 An even parity error detecting channel code

Input	Codeword
0	00
1	11

This code converts each input bit into two output bits, essentially doubling the number of bits that must be transmitted. This simple code improves the reliability of the code by providing *error-detection* capability. In this case, the extra bit in the pair functions as a *parity bit,* meaning that the received code should always contain an even number of 1's, or *even parity*. If it doesn't, then the decoder knows that one of the bits was distorted in the channel. This simple code doesn't protect against two errors occurring in a single codeword (meaning both bits were flipped in the channel), and it also doesn't provide any capability to correct errors. However, it does give the receiver a rudimentary ability to detect when errors occur.

To examine a code that provides *error-correction* capability, consider the (5, 2) code of Table 17.10.

Table 17.10 An error correcting channel code

Input	Codeword
00	00000
01	01110
10	10011
11	11101

This code converts each pair of input bits into five output bits. From the receiver's perspective it is easy to decode a received bit pattern if it exactly matches one in the dictionary. But what happens when an error in the channel causes the received pattern not to match any of the available codewords? In this case we begin by computing the *Hamming distance,* or the number of nonmatching bits between the received pattern and each of the available codewords. Then, the codeword with the smallest Hamming distance is taken as the received output. For example, assuming that the decoder received the pattern 01111, the Hamming distances are computed as shown in Table 17.11.

Table 17.11 **Hamming distances for received bit pattern of 01111**

Input	Codeword	Hamming distance
00	00000	4
01	01110	1
10	10011	3
11	11101	2

Taking the smallest Hamming distance, the decoder would choose 01110 as the received codeword and therefore 01 as the decoded bit sequence.

17.7 ADVANCED TOPICS

One of the technologies that has rapidly expanded the field of digital communication is the cellular telephone. In this case the channel is the atmosphere and all cellular traffic within a geographic area must therefore share the same allocated frequency band, or *multiplex,* often at the same time. In the U.S. 1.9 GHz is one such band. The difficulty becomes how to multiplex without causing interference between two phones, or at least how a phone might understand the signal despite the interference.

The answer to these problems is channel sharing, used by wired and wireless systems alike to support multiple simultaneous users. There are several techniques by which the channel can be shared among users to limit interference. Two of the most common ones are frequency division multiple access and time division multiple access.

Frequency Division Multiple Access

As the name suggests, *frequency division multiple access* (FDMA) transmits multiple signals by separating them in frequency, assuming that the channel bandwidth is much larger than the actual bandwidth required for transmitting the signal. In other words, $B_C \gg B_U$ where the bandwidth of the channel is B_C and the bandwidth of each user is B_U. Under these conditions, we can transmit approximately (B_C/B_U) users on the channel simultaneously. As an example, consider a 3 KHz voice signal on a 100 KHz channel. Then we can theoretically transmit $(100/3) \approx 33$ users on this channel.

In practice, a frequency gap called a *guard band* is introduced between each pair of consecutive channels to ensure there is no interference between the two. Each of the signals is individually modulated and upconverted to the frequency of transmission. If the start of the channel's frequency band is f_C, then in our example, the first user would transmit at $f_C + 3$ KHz with a guard band from $f_C + 3$ KHz to $f_C + 4$ KHz.

The second user would transmit between $f_C + 4$ KHz and $f_C + 7$ KHz and so on. In this way 25 channels could be actually transmitted on the channel. Figure 17.24 illustrates the frequency layout of a typical FDMA system.

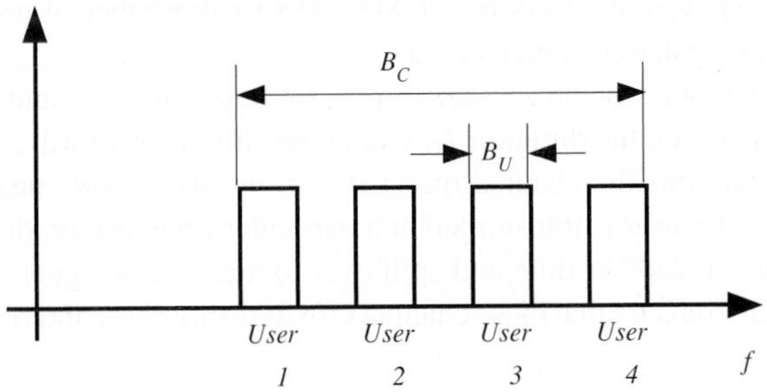

Figure 17.24 Frequency division multiple access

Examples of systems using FDMA are analog radio, analog television and cable service, and until recently the long distance telephone system. The low cost and high quality of available FDMA equipment, especially that intended for television signals, make it a reasonable choice for many purposes. The downside is that the required bandwidth increases as more and more users are added to the system. Since the channel often has a fixed bandwidth, it is not possible to add users limitlessly.

Time Division Multiple Access

Another way to achieve the sharing of the medium is through *time division multiple access* (TDMA). With this approach multiple users are isolated from each other by transmitting at different points in time. Assuming that all users' signals require the same bandwidth, say B_U Hz, we know from the Nyquist theorem that each of them requires the same sampling frequency $f_S = 2B_U$ and therefore two samples from the same user are $T_S = 1/f_S$ seconds apart. If we make the duration between the samples of two users $T_U = (T_S/N)$ then we can accommodate the samples of all N users in one *time slot* of T_S seconds by sampling each user's signal T_U seconds apart and interleaving the samples of all users. After interleaving, this multiplexed signal is modulated and upconverted to the final transmission frequency. Figure 17.25 illustrates the timing for a typical TDMA system.

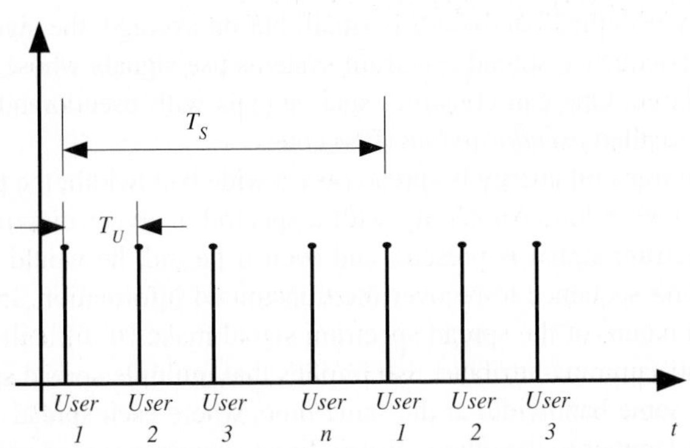

Figure 17.25 Time division multiple access

The Global System for Mobile Communications (GSM) cellular standard uses TDMA, with individual channels that use 200 KHz of bandwidth for one-way or *simplex* (i.e., mobile to base station or base station to mobile) communication. In all, there are 124 pairs of such simplex channels, where each simplex channel can be used to support 8 separate users by TDMA. The total number of users that can be supported by this system is therefore 992.

TDMA is more efficient, easier to operate, less complex, and less expensive than FDMA. However, the duration T_U cannot be chosen arbitrarily. This is because under some conditions the channel may suffer from *dispersion,* meaning that even though a signal of a user is transmitted at a particular time instant, the channel itself may cause it to "spread" in time and spill over to next user's signal. So T_U must be carefully chosen to ensure that these channel effects do not cause interference between users.

Spread Spectrum and Code Division Multiple Access

Spread spectrum had its beginnings during World War II when beautiful Hollywood movie star Hedi Lamarr and talented pianist George Antheil patented a covert system for guiding torpedoes. Their idea was to "hop" the torpedo control signal between many different frequencies to make it difficult for an enemy to jam. The torpedo receiver would be synchronized to the transmitter and know exactly when the next frequency hop would occur. Unfortunately for Lamarr and Antheil, the U.S. Navy was not impressed by their invention, and they never saw a dime for their efforts. Today, their basic spread spectrum idea is used in millions of mobile phones across the world.

A wireless communications system whose transmission occupies a bandwidth W much greater than the information rate R of the data source is called a *spread spectrum* system. For example, a voice channel in a cellular IS-95 system has an information rate of $R = 9.6$ kbps, but occupies a bandwidth of $W = 1.2288$ MHz, a bandwidth expansion of $W/R = 128$. The ratio W/R is often called the *processing gain* of the system. Third-generation cellular standards use spread-spectrum waveforms that occupy bandwidths of up to 5 MHz and have variable processing gains to enable transmission of low and high data rates.

There are several ways to spread the bandwidth of the data source. Two of the most popular include *frequency hopped* spread spectrum and *direct sequence* spread spectrum. Frequency hopping is straightforward: a narrowband transmit signal hops in a pseudorandom sequence from frequency to frequency over a wide bandwidth. The instantaneously occupied bandwidth is small, but on average, the signal bandwidth is large. Direct-sequence spread spectrum systems use signals whose instantaneous bandwidth is large. One can construct such signals with pseudorandom spreading code sequences called *pseudorandom* (PN) codes.

Since the transmit energy is spread over a wide bandwidth, the power per unit of bandwidth is very low. An enemy with a spectral analyzer may not even know the spread spectrum signal is present, and even if he did, he would have to know the PN spreading sequence to recover the transmitted information. In addition, the pseudorandom nature of the spread spectrum signal makes it difficult for an enemy to jam. The anti-jamming attribute also implies that multiple spread spectrum users can access the same bandwidth at the same time, where each spread spectrum user is considered a jammer to the others. When direct-sequence spread spectrum is used for simultaneous access to common bandwidth, this is called *code-division multiple access* (CDMA).

Ultrawideband

Ultrawideband (UWB) digital communication is a very recent development using waveforms that distribute their power across a very wide bandwidth. Because UWB is a form of extreme spread spectrum, dozens of devices can operate in the same location at the same time. In addition, the transmitted power per Hertz is so low (below the noise floor, in many cases) that data from the devices is extremely difficult for a third party to detect or intercept.

There are currently two design philosophies proposed for UWB communication system standardization. The first is *impulse* radio, using waveforms with subnanosecond duty cycle and occupying several Gigahertz of bandwidth. Impulse radio is mainly intended for longer-range low bit rate communication systems, though efforts are underway to increase transmission rates. The second proposal is to transmit a *multiband* waveform where each carrier occupies greater than 500 MHz. This approach is intended for shorter-range higher-rate communication systems. Both proposals have their benefits and drawbacks, but it is hoped that a consensus between them can be reached quickly.

Potential applications for UWB radio communication range from military covert communications to commercial indoor wireless links. UWB devices can operate with low-power for low-cost short-range communications, and may some day wirelessly connect hand-held digital devices with personal computers at rates of up to 1 Gigabit per second. Another application for UWB technology is wirelessly streaming multimedia content between components of a home entertainment system. In addition to being used for communications purposes, UWB signals have already been used for applications such as ground-penetrating radar, location sensing, and through-wall imaging.

Because UWB signals occupy several Gigahertz of bandwidth, it is inevitable that they will overlap with incumbent narrowband systems such as cellular telephony and global positioning service (GPS). Studies show that the interference is minimal because UWB transmission power per Megahertz is so low. Even so, GPS is so critical to many applications that as an extra precaution the FCC has mandated that UWB devices may not emit energy in the GPS band. A large push for commercial applications came after the Federal Communications Commission (FCC) amended its Part 15 rules in February 2002 to allow unlicensed UWB spectral emissions.

17.8 DATA TRANSMISSION IN DIGITAL INSTRUMENTS

One of the necessary aspects of data acquisition and control systems is the ability to transmit and receive data. Often, a microcomputer-based data acquisition system is interfaced to other digital devices, such as digital instruments or other microprocessors. In these cases it is necessary to transfer data directly in digital form. In fact, it is usually preferable to transmit data that are already in digital form, rather than analog voltages or currents. Among the chief reasons for the choice of digital over analog is that digital data are less sensitive to noise and interference than analog signals: In receiving a binary signal transmitted over a data line, the only decision to be made is whether the value of a bit is 0 or 1. Compared with the difficulty in obtaining a precise measurement of an analog voltage or current, either of which could be corrupted by noise or interference in a variety of ways, the probability of making an error in discerning between binary 0s and 1s is very small. Further, as will be shown shortly, digital data are often coded in such a way that many transmission errors may be detected and corrected. Finally, storage and processing of digital data are much more readily accomplished than would be the case with analog signals. This section explores a few of the methods that are commonly employed in transmitting digital data; both parallel and serial interfaces are considered.

Digital signals in a microprocessor are carried by a bus, consisting of a set of parallel wires, each carrying 1 bit of information. In addition to the signal-carrying wires, there are control lines that determine under what conditions transmission may occur. A typical computer data bus consists of eight parallel wires and therefore enables the transmission of 1 byte; digital data are encoded in binary according to one of a few standard codes, such as the BCD code described in Chapter 13 or the ASCII code introduced in Chapter 14 (see Table 14.6). This bus configuration is usually associated with **parallel transmission,** whereby all the bits are transmitted simultaneously, along with some control bits.

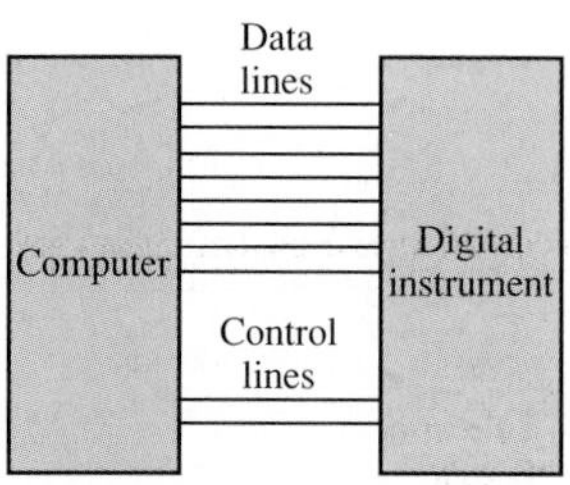

Figure 17.26 Parallel data transmission

Figure 17.26 depicts the general appearance of a parallel connection. Parallel data transmission can take place in one of two modes: **synchronous** or **asynchronous.** In synchronous transmission, a timing clock pulse is transmitted along with the data over a control line. The arrival of the clock pulse indicates that valid data have also arrived. While parallel synchronous transmission can be very fast, it requires the added complexity of a synchronizing clock and is typically employed only for internal computer data transmission. Further, this type of communication can take place only over short distances (approximately 4 m). Asynchronous data transmission, on the other hand, does not take place at a fixed clock rate, but requires a **handshake protocol** between sending and receiving ends. The handshake protocol consists of the transmission of *data ready* and *acknowledge* signals over two separate control wires. Whenever the sending device is ready to transmit data, it sends a pulse over the *data ready* line. When this signal reaches the receiver, and if the receiver is ready to receive the data, an *acknowledge* pulse is sent back, indicating that the transmission may occur; at this point, the parallel data are transmitted.

Perhaps the most common parallel interface is based on the **IEEE 488 standard,** leading to the IEEE 488 bus, also referred to as **GPIB** (for **general-purpose instrument bus**). Other increasingly common standard interfaces are the **Universal Serial Bus** (**USB**) and **Controller Area Network** (**CAN**). These are described in Chapter 15.

The IEEE 488 Bus

The IEEE 488 bus, shown in Figure 17.27, is an 8-bit parallel asynchronous interface that has found common application in digital instrumentation applications. The physical bus consists of 16 lines, of which 8 are used to carry the data, 3 for the handshaking protocol, and the rest to control the data flow. The bus permits connection of up to 15 instruments and data rates of up to 1 Mbyte/s. There is a limitation, however, in the maximum total length of the bus cable, which is 20 m. The signals transmitted are TTL-compatible and employ negative logic (see Chapter 13), whereby a logic 0 corresponds to a TTL high state (>2 V) and a logic 1 to a TTL low state (<0.8 V). Often, the 8-bit word transmitted over an IEEE 488 bus is coded in ASCII format (see Table 14.6), as illustrated in Example 17.4.

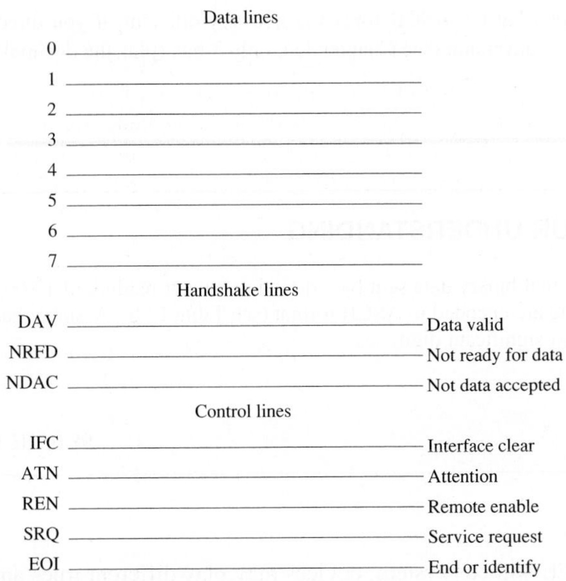

Figure 17.27 IEEE 488 bus

EXAMPLE 17.4 ASCII to Binary Data Conversion over IEEE 488 Bus

LO6

Problem

Determine the actual binary data sent by a digital voltmeter over an IEEE 488 bus.

Solution

Known Quantities: Digital voltmeter reading V.

Find: Binary data sequence.

Schematics, Diagrams, Circuits, and Given Data: $V = 3.405$ V. ASCII conversion table (Table 14.6).

Assumptions: Data are encoded in ASCII format. Sequence is sent from most to least significant digit.

Analysis: Using Table 14.6, we can tabulate the conversion as follows:

Control character	ASCII (hex)
3	33
.	2E
4	34
0	30
5	35

The actual binary data sent can therefore be determined by converting the hexadecimal ASCII sequence into binary data (see Chapter 13):

$$33\ 2E\ 34\ 30\ 35 \leftrightarrow 110011\ 101110\ 110100\ 110000\ 110101$$

Comments: Note that the ASCII format is not very efficient; if you directly performed a base-10 to binary conversion (see Chapter 13), only 8 bits (plus the decimal point) would be required.

CHECK YOUR UNDERSTANDING

Determine the actual binary data sent by a digital voltmeter reading of 17.06 V over an IEEE 488 bus if the data are encoded in ASCII format (see Table 14.6). Assume that the sequence is from most to least significant digit.

Answer: 31 35 2E 30 36

In an IEEE 488 bus system, devices may play different roles and are typically classified as *controllers,* which manage the data flow; *talkers* (e.g., a digital voltmeter), which can only send data; *listeners* (e.g., a printer), which can only receive data; and *talkers/listeners* (e.g., a digital oscilloscope), which can receive as well as transmit data. The simplest system configuration might consist of just a talker and a listener. If more than two devices are present on the bus, a controller is necessary to determine when and how data transmission is to occur on the bus. For example, one of the key rules implemented by the controller is that only one talker can transmit at a time; it is possible, however, for several listeners to be active on the bus simultaneously. If the data rates of the different listeners are different, the talker will have to transmit at the slowest rate, so that all the listeners are assured of receiving the data correctly.

The set of rules by which the controller determines the order in which talking and listening are to take place is determined by a **protocol.** One aspect of the protocol is the handshake procedure, which enables the transmission of data. Since different devices (with different data rate capabilities) may be listening to the same talker, the handshake protocol must take into account these different capabilities. Let us discuss a typical handshake sequence that leads to transmission of data on an IEEE 488 bus. The three handshake lines used in the IEEE 488 have important characteristics that give the interface system wide flexibility, allowing interconnection of multiple devices that may operate at different speeds. The slowest active device controls the rate of data transfer, and more than one device can accept data simultaneously. The timing diagram of Figure 17.28 and the list below illustrate the sequence in which the handshake and data transfer are performed.

1. All active listeners use the *not ready for data* (NRFD) line to indicate their state of readiness to accept a new piece of information. Nonreadiness to accept data is indicated if the NRFD line is held at 0 V. If even one active listener is not ready, the NRFD line of the entire bus is kept at 0 V and the active talker will not transmit the next byte. When all active listeners are ready and they have released the NRFD line, it goes high.
2. The designated talker drives all eight data input/output lines, causing valid data to be placed on them.

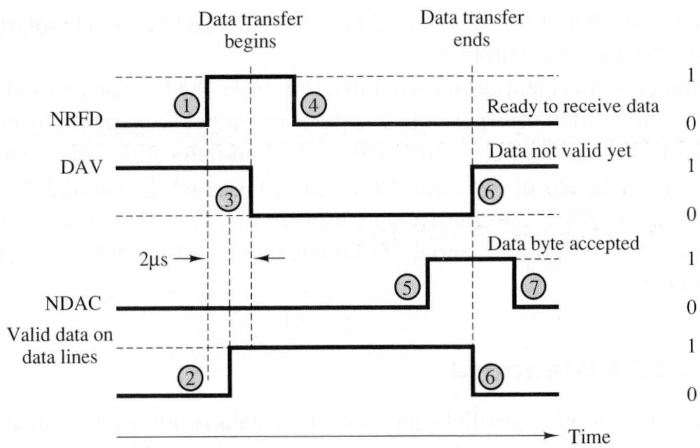

Figure 17.28 IEEE 488 data transmission protocol

LO6

3. Two microseconds after putting valid data on the data lines, the active talker pulls the *data valid* (DAV) line to 0 V and thereby signals the active listeners to read the information on the data bus. The 2-μs interval is required to allow the data put on the data lines to reach (settle to) valid logic levels.
4. After the DAV is asserted, the listeners respond by pulling the NRFD line back down to zero. This prevents any additional data transfers from being initiated. The listeners also begin accepting the data byte at their own rates.
5. When each listener has accepted the data, it releases the *not data accepted* (NDAC) line. Only when the last active listener has released its hold on the NDAC line will that line go to its high-voltage-level state.
6. (a) When the active talker sees that NDAC has come up to its high state, it stops driving the data line. (b) At the same time, the talker releases the DAV line, ending the data transfer. The talker may now put the next byte on the data bus.
7. The listeners pull down the NDAC line back to 0 V and put the byte "away."

Each of the instruments present on the data bus is distinguished by its own address, which is known to the controller; thus, the controller determines who the active talkers and listeners are on the bus by *addressing* them. To implement this and other functions, the controller uses the five control lines. Of these, ATN (attention) is used as a switch to indicate whether the controller is addressing or instructing the devices on the bus, or whether data transmission is taking place: when ATN is logic 1, the data lines contain either control information or addresses; with ATN = 1, only the controller is enabled to talk. When ATN = 0, only the devices that have been addressed can use the data lines. The IFC (interface clear) line is used to initialize the bus, or to clear it and reset it to a known condition in case of incorrect transmission. The REN (remote enable) line enables a remote instrument to be controlled by the bus; thus, any function that might normally be performed manually on the instrument (e.g., selecting a range or mode of operation) is now controlled by the bus via the data lines. The SRQ (service request) line is used by instruments on the bus whenever the instrument is ready to send or receive data; however, it is the controller that decides when to service the request. Finally, the EOI (end or identify) line can be used in two modes: When it is used by a talker, it signifies the end of a message; when it is used by the controller, it serves as a *polling* line, that is, a line used to interrogate the instrument about its data output.

Although it was mentioned earlier that the IEEE 488 bus can be used only over distances of up to 20 m, it is possible to extend its range of operation by connecting remote IEEE 488 bus systems over telephone communication lines. This can be accomplished by means of *bus extenders,* or by converting the parallel data to serial form (typically, in RS-232 format) and by transmitting the serial data over the phone lines by means of a modem. Serial communications and the RS-232 standard are discussed next.

The RS-232 Standard

The primary reason why parallel transmission of data is not used exclusively is the limited distance range over which it is possible to transmit data on a parallel bus. Although there are techniques which permit extending the range for parallel transmission, these are complex and costly. Therefore, **serial transmission** is frequently used, whenever data are to be transmitted over a significant distance. Since serial data travel along one single path and are transmitted 1 bit at a time, the cabling costs for long distances are relatively low; further, the transmitting and receiving units are also limited to processing just one signal and are also much simpler and less expensive. Two modes of operation exist for serial transmission: **simplex,** which corresponds to transmission in one direction only; and **duplex,** which permits transmission in either direction. Simplex transmission requires only one receiver and one transmitter, at each end of the link; on the other hand, duplex transmission can occur in one of two manners: **half-duplex** and **full-duplex.** In the former, although transmission can take place in both directions, it cannot occur simultaneously in both directions; in the latter case, both ends can simultaneously transmit and receive. Full-duplex transmission is usually implemented by means of four wires.

The data rate of a serial transmission line is measured in bits per second, since the data are transmitted 1 bit at a time. The unit of 1 bit/s is 1 **baud** (abbreviated as 1 Bd); thus, reference is often made to the baud rate of a serial transmission. The baud rate can be translated to a parallel transmission rate in words per second if the structure of the word is known; for example, if a word consists of 10 bits (start and stop bits plus an 8-bit data word) and the transmission takes place at 1,200 Bd, 120 words are being transmitted every second. Typical data rates for serial transmission are standardized; the most common rates (familiar to the users of personal computer modem connections) are 300, 600, 1,200, and 2,400 Bd. Baud rates can be as low as 50 Bd or as high as 19,200 Bd.

Like parallel transmission, serial transmission can also occur either synchronously or asynchronously. In the serial case, it is also true that asynchronous transmission is less costly but not as fast. A handshake protocol is also required for asynchronous serial transmission, as explained in the following. The most popular data-coding scheme for serial transmission is, once again, ASCII, consisting of a 7-bit word plus a **parity bit,** for a total of 8 bits per character. The role of the parity bit is to permit error detection in the event of erroneous reception (or transmission) of a bit. To see this, let us discuss the sequence of handshake events for asynchronous serial transmission and the use of parity bits to correct for errors. In serial asynchronous systems, handshaking is performed by using start and stop bits at the beginning and end of each character that is transmitted. The beginning of the transmission of a serial asynchronous word is announced by the *start* bit, which is always a 0 state bit. For the next five to eight successive bit times (depending on the code and the number of bits that specifies the word length in that code), the line is switched to the 1 and 0 states required to represent the character being sent. Following the last bit of the data and the parity bit (which will be explained next), there is 1 bit or more in the 1 state, indicating "idle." The time period associated with this transmission is called the *stop* bit interval.

If noise pulses affect the transmission line, it is possible that a bit in the transmission could be misread. Thus, following the 5 to 8 transmitted data bits, there is a parity bit that is used for error detection. Here is how the parity bit works. If the transmitter keeps track of the number of 1s in the word being sent, it can send a parity bit, a 1 or a 0, to ensure that the total number of 1s sent is always even (even parity) or odd (odd parity). Similarly, the receiver can keep track of the 1s received to see whether there was an error with the transmission. If an error is detected, retransmission of the word can be requested.

Serial data transmission occurs most frequently according to the **RS-232 standard.** The RS-232 standard is based on the transmission of voltage pulses at a preselected baud rate; the voltage pulses are in the range −3 to −15 V for a logic 0 and in the range +3 to +15 V for a logic 1. It is important to note that this amounts to a negative logic convention and that the signals are *not* TTL-compatible. The distance over which such transmission can take place is up to approximately 17 m (50 ft). The RS-232 standard was designed to make the transmission of digital data compatible with existing telephone lines; since phone lines were originally designed to carry analog voice signals, it became necessary to establish some standard procedures to make digital transmission possible over them. The resulting standard describes the mechanical and electrical characteristics of the interface between *data terminal equipment* (DTE) and *data communication equipment* (DCE). DTE consists of computers, terminals, digital instruments, and related peripherals; DCE includes all those devices that are used to encode digital data in a format that permits their transmission over telephone lines. Thus, the standard specifies how data should be presented by the DTE to the DCE so that digital data can be transmitted over voice lines.

A typical example of DCE is the **modem.** A modem converts digital data to audio signals that are suitable for transmission over a telephone line and is also capable of performing the reverse function, by converting the audio signals back to digital form. The term *modem* stands for *mo*dulate-*dem*odulate, because a modem modulates a sinusoidal carrier using digital pulses (for transmission) and demodulates the modulated sinusoidal signal to recover the digital pulses (at reception). Three methods are commonly used for converting digital pulses to an audio signal: **amplitude-shift keying, frequency-shift keying,** and **phase-shift keying,** depending on whether the amplitude, phase, or frequency of the sinusoid is modulated by the digital pulses. Figure 17.29 depicts the essential block of a data transmission system based on the RS-232 standard as well as examples of digital data encoded for transmission over a voice line.

In addition to the function just described, however, the RS-232 standard provides a very useful set of specifications for the direct transmission of digital data between computers and instruments. In other words, communication between digital terminal instruments may occur directly in digital form (i.e., without digital communication devices encoding the digital data in a form compatible with analog

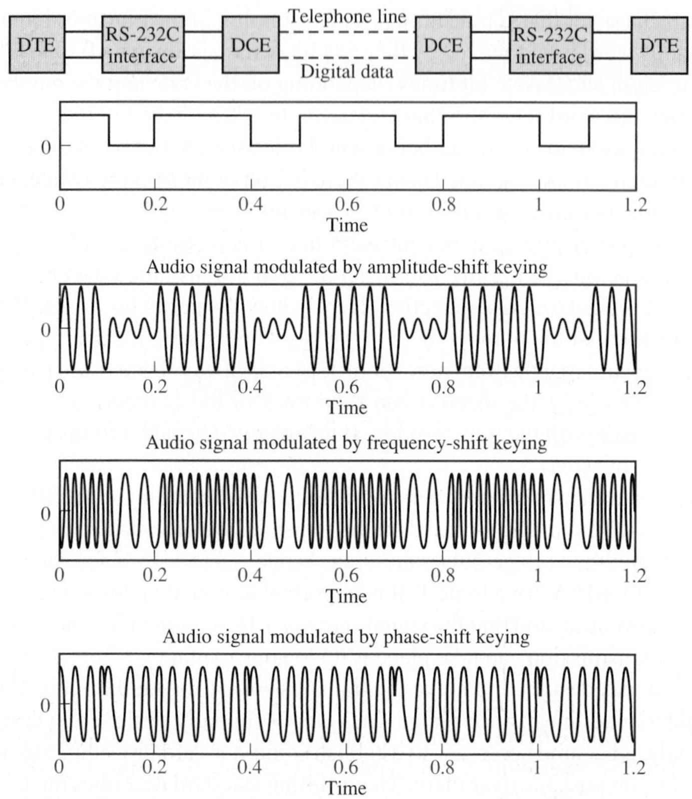

Figure 17.29 Digital data encoded for analog transmission

voice lines). Thus, this standard is also frequently used for direct digital communication.

LO6

The RS-232 standard is summarized as follows:

- Data signals are encoded according to a negative logic convention using voltage levels of −3 to −15 V for logic 1 and +3 to +15 V for logic 0.
- Control signals use a positive logic convention (opposite to that of data signals).
- The maximum shunt capacitance of the load cannot exceed 2,500 pF; this, in effect, limits the maximum length of the cables used in the connection.
- The load resistance must be between 300 Ω and 3 kΩ.
- Three wires are used for data transmission. One wire each is used for receiving and transmitting data; the third wire is a signal return line (signal ground). In addition, there are 22 wires that can be used for a variety of control purposes between the DTE and DCE.
- The male part of the connector is assigned to the DTE and the female part to the DCE. Figure 17.30 labels each of the wires in the 25-pin connector. Since each side of the connector has a *receive* and a *transmit* line, it has been decided by convention that the DCE transmits on the transmit line and receives on the receive line, while the DTE receives on the transmit line and transmits on the receive line.

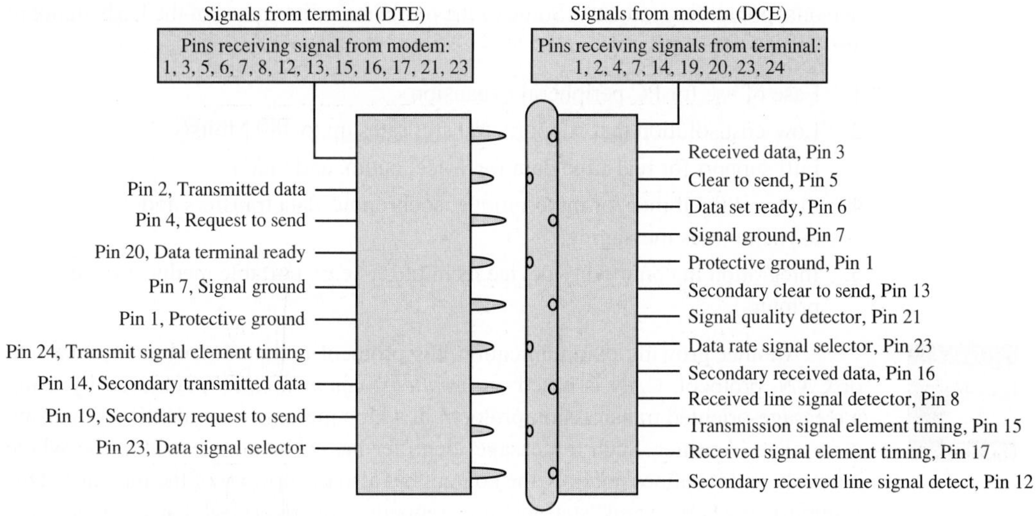

Figure 17.30 RS-232 connections

- The baud rate is limited by the length of the cable; for a 17-m length, any rate from 50 Bd to 19.2 kBd is allowed. If a longer cable connection is desired, the maximum baud rate will decrease according to the length of the cable, and **line drivers** can be used to amplify the signals, which are transmitted over twisted-pair wires. Line drivers are simply signal amplifiers that are used directly on the digital signal, prior to encoding. For example, the signal generated by a DTE device (say, a computer) may be transmitted over a distance of up to 3,300 m (at a rate of 600 Bd) prior to being encoded by the DCE.
- The serial data can be encoded according to any code, although the ASCII is by far the most popular.

Other Communication Network Standards

In addition to the popular RS-232 and IEEE bus standards, we should mention other communication standards that are commonplace or are rapidly becoming so. One is **Ethernet,** which operates at 10 Mbits/s and is based on IEEE Standard 802.3. This is commonly used in office networks. Higher-speed networks include **fiber-distributed data interface** (FDDI), which specifies an optical fiber ring with a data rate of 100 Mbits/s, and **asynchronous transfer mode** (ATM), a packet-oriented transfer mode moving data in fixed packets called *cells.* ATM does not operate at fixed speed. A typical speed is 155 Mbits/s, but there are implementations running as fast as 2 Gbits/s.

FIND IT ON THE WEB

Another bus architecture that is finding increasingly common use is the **Universal Serial Bus (USB).** The USB standard has been motivated by the increasing need to have a flexible protocol for personal computer peripheral interface. The goal of the USB standard is *to enable peripheral devices from different vendors to interoperate in an open architecture.* Thus, the USB technical specification describes the bus attributes, protocol definition, types of transactions, bus management, and programming interface that are consistent with the requirements of today's personal computing industry. Some of the more immediate goals of the USB standard are

1. Ease of use for PC peripheral expansion.
2. Low-cost solution that supports transfer rates up to 480 Mbits/s.
3. Full support for real-time data for voice, audio, and video.
4. Protocol flexibility for mixed-mode isochronous data transfers and asynchronous messaging.
5. Integration in commodity device technology (e.g., portable media storage, printers).

Another growing communication bus protocol is the **control area network,** or **CAN,** protocol. CAN is based on *broadcast communication,* achieved by using a message-oriented transmission protocol in which messages are identified by using a message identifier. Such a message identifier has to be unique within the whole network, and it defines not only the content but also the priority of the message. This is important when several "stations" are competing for bus access. The CAN protocol supports two message frame formats, the only essential difference being in the length of the identifier. The *CAN standard frame,* also known as **CAN 2.0 A,** supports a length of 11 bits for the identifier; and the *CAN extended frame,* also known as **CAN 2.0 B,** supports a length of 29 bits for the identifier. The structure of a CAN message is shown in Figure 17.31.

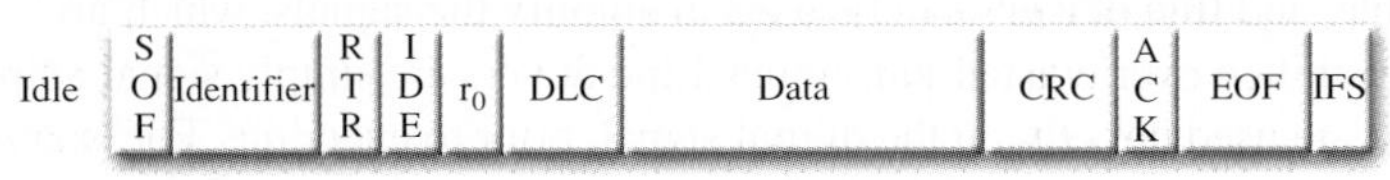

Figure 17.31 Structure of CAN message

It is very easy to add stations to an existing CAN network without making any hardware or software modifications to the existing stations as long as the new stations are purely receivers. Thus allows the concept of modular electronics and also permits multiple reception and the synchronization of distributed processes: data needed as information by several stations can be transmitted via the network, in such a way that it is unnecessary for each station to have to know who is the producer of the data. Thus makes it easy to service and upgrade networks, as data transmission is not based on the availability of specific types of stations.

CAN was originally developed for passenger car applications. CAN networks used in engine management connect several electronic control units (ECUs). In addition, some passenger cars are equipped with CAN-based multiplex systems connecting body electronic ECUs. These multiplex networks link door and roof control units as well as lighting control units and seat control units. In some passenger cars, there is implemented a CAN-based diagnostic interface. The different CAN-based in-vehicle networks are connected via gateways. In many system designs, the gateway functionality is implemented in the dashboard. In the future, the dashboard itself may use a local CAN network to connect the different display and control units. The following microcontrollers, commonly used in automotive applications, support CAN interfaces: Infineon C16XC series, Motorola 683xx series and MPC5xx series, and Hitachi SH-2. Due to its widespread use in automobiles, the popularity of the CAN bus has grown significantly in recent years, and CAN has become a de facto standard for the automotive industry.

HOMEWORK PROBLEMS

Section 17.2: Introductory Probability

17.1 The Q-function is often used in the communications literature to express the probability of error. Unfortunately, MATLAB does not have a built-in Q-function; however, the built-in *complementary error function* erfc(x) can be used to obtain the Q-function:

$$\text{erfc}(x) = \frac{2}{\sqrt{\pi}} \int_{x}^{\infty} e^{-t^2} dt$$

Obtain an expression for $Q(x)$ in terms of erfc(x).

Section 17.3: Pulse Code Modulation

17.2 Consider that you have been assigned the task of quantizing for transmission a voice signal whose bandwidth is 3.4 kHz. You are constrained by a channel which can support a maximum bit rate of 35,000 bits/sec. What is the maximum number of bits that you can employ for quantization? Also find the corresponding sampling rate of the signal.

Section 17.4: Source Coding

17.3 A remote color sensor observes a color manufacturing process and transmits observations to a control center. The observed colors occur with the following probabilities:

P (Red) = 3/4
P (Green) = 1/8
P (Blue) = 1/16
P (Yellow) = 1/16

A computer at the sensor encodes the color observations according to the following code:

Red → 00
Green → 01
Blue → 10
Yellow → 11

These codewords are then sent to the control center (leftmost bit first).

a. What is the average length of the code?

b. The sequence (00000001000000110000000000000100) was received at the control center with the leftmost bit received first. What is the observed sequence of colors?

c. Notice that the observed sequence from part b is highly redundant. To reduce this redundancy, devise a code that has an average length of less than two. What is the average length of your code?

d. Encode the following sequence of color observations (leftmost observed first) with your code from part c: {Red, Red, Red, Green, Red, Red, Red, Yellow, Red, Red, Red, Red, Red, Red, Green, Red}. Do you see an improvement?

17.4 In music CD recording, each of the two stereo signals is sampled with a 16-bit ADC at 44.1 kHz.

a. What is the output signal to quantization noise ratio for a sinusoidal input signal?

b. The music bit stream is appended with error correction bits, clock extraction bits, and display and control bits. Assume that these bits correspond to an overhead of 50 percent. What is the playback bit rate of a CD?

c. Let's assume a CD records an hour's worth of music. Determine the number of bits recorded on a CD.

d. Now let's compare data storage to the storage of music on a CD. Consider a history textbook which contains 1,000 pages, 50 lines per page, 15 words per line, 6 letters per word and 7 bits per word on average. Determine the number of bits required to digitally store this textbook. Now estimate the number of such textbooks that can be stored on a CD.

17.5 Suppose you have a sinusoidal signal with a maximum amplitude of 2 that has to be quantized such that the signal to quantization noise ratio is at least 10 dB. What is the minimum number of bits required to perform this quantization?

17.6 High-definition TV's today use a format in which there are 1920 × 1080 pixels on the TV screen. Each of these pixels has 16 different brightness levels, and pictures are repeated at the rate of 30 frames per second. Assuming that all brightness levels are equally likely to occur, calculate the average rate of information conveyed by the TV. (*Hint*: If the rate at which source X emits symbols is r symbols/sec, the information rate of the source is given by $R = rH(X)$ bits/sec. You can think of the TV screen as a source and all the brightness levels of the pixels as the set of possible outcomes of the source.)

Section 17.5: Digital Baseband Modulation

17.7 Suppose a cable modem communications standard specifies a baud rate of one million message waveform transmissions per second. Further, suppose there are 256 waveforms in the message waveform set. What is the bit rate?

17.8 Referring to the waveforms of Figure P17.8

a. Compute the average energy of each of the three message waveform sets. Note that for all sets the duty cycle is $T = 1$.

b. Compute the average RMS distance between the waveforms in each set.

c. Which waveforms would you choose for binary signaling? Why?

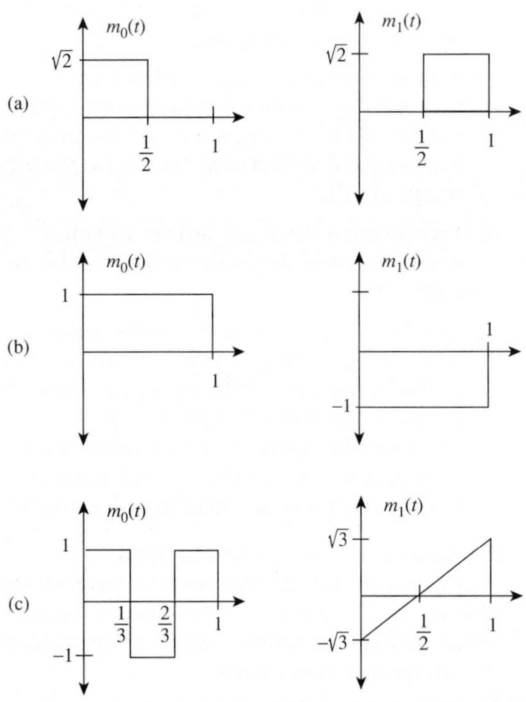

Figure P17.8 Three sets of binary message waveforms

Section 17.6: Channel Coding

17.9 It can be shown that the theoretical probability of error for a digital communication system is given by

$$P_e = Q\left(\sqrt{\frac{2E_b}{kT}}\right)$$

where $Q(x)$ is the Q-function, E_b is the energy per bit, and T is the noise temperature of the additive white Gaussian noise. We would like to verify this formula by performing the following experiment:

a. Generate a vector of equi-probable random bits $\{0, 1\}$.

b. Map each bit to a transmitted symbol in the following manner:

$$0 \longrightarrow \sqrt{E_b}$$
$$1 \longrightarrow -\sqrt{E_b}$$

where E_b is the energy of the symbol.

c. Add Gaussian noise with zero mean and variance $kT/2$ to the transmitted symbol.

d. Perform detection by checking the sign of the received signal. If the sign is greater than zero, then decide that a 0 was sent; otherwise, decide that a 1 was sent.

e. Count the number of errors made.

f. Repeat until you have counted at least 100 errors. (The more errors you count, the closer will be your estimate of the BER to the theoretical results).

The estimated probability of error is

$$P_e \approx \frac{\text{number of errors counted}}{\text{number of symbols generated}}$$

Do steps a–f for signal-to-noise ratios (SNRs) of $E_b/kT = 0, 2, \ldots, 10$ dB. Once you are done collecting the estimated probabilities of error, plot your results on a log scale versus SNR in dB, i.e., $10\log_{10}(E_b/N_o)$. Show each data point as an x-mark. Also plot the theoretical probability of error P_e for comparison. Check your results with Figure P17.9.

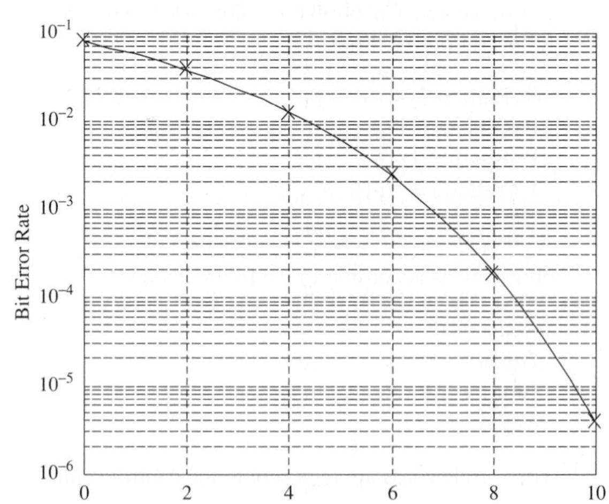

Figure P17.9 Theoretical bit error rate

17.10 You have been given the task of designing a point-to-point wireless communications link between two buildings spaced 2 km apart. Tests show that the signal-to-noise ratio P/kT at the receiver is 20 dB.

a. What is the minimum amount of bandwidth needed to support a 1 Mbps link?

b. Suppose you have a bandwidth limitation of 5 MHz. Is the SNR sufficient to support a data rate of 5 Mbps?

17.11 Your task is to transmit a voice signal of bandwidth 3.4 kHz over an AWGN channel. Assume the signal is sampled at 1.5 times the Nyquist rate and each sample is quantized into one of 256 equally likely levels.

a. What is the rate at which this voice source is generating information?

b. Can you transmit the output of this source without error over an AWGN channel with a bandwidth of 10 kHz and a signal to noise ratio of 20 dB? (*Hint*: Signal-to-noise ratio is P/kT.)

c. What is the minimum signal-to-noise ratio required for error-free transmission of this information over the channel?

d. Determine the minimum AWGN channel bandwidth required for error-free transmission of the output of this source if the signal-to-noise ratio is 20 dB.

17.12 A binary symmetric channel (BSC) has two inputs $x_1 = 0$ and $x_2 = 1$ and two outputs $y_1 = 0$ and $y_2 = 1$. The channel is symmetric because the probability of receiving a 1 if a 0 is transmitted is the same as the probability of receiving a 0 if a 1 is sent. This common transition probability is denoted as b. Consider a simple repetition coding scheme over the BSC in which each bit is repeated n times where $n = 2m + 1$, is (an odd integer). For decoding the received bits, a majority rule is employed. In other words, if in a block of n received bits the number of zeros is greater than the number of ones, the decoder decides that a 0 was transmitted. Otherwise, the decoder decides that a 1 was transmitted. Therefore an error occurs whenever $m + 1$ or more bits out of n bits are received incorrectly.

a. For this coding-decoding scheme, derive an expression for the probability of bit error, P_e.

b. Calculate P_e when $b = 0.05$ and $n = 3, 5, 7$.

17.13 Figure P17.13 compares the *bit error rate* (BER) performance of BPSK transmission over an AWGN channel with and without error control coding.

a. At a BER of 10^{-5} what is the SNR required to transmit bits over the channel without coding?

b. What is the SNR required to obtain BER 10^{-5} over the channel with coding employed?

c. The *coding gain* is defined as the difference between the SNRs required to obtain a certain BER using an uncoded and a coded system. The coding gain represents the reduction in signal power when using coding to achieve the same level of performance as that of a scheme without coding. What is the coding gain of this coding scheme at a BER 10^{-5}?

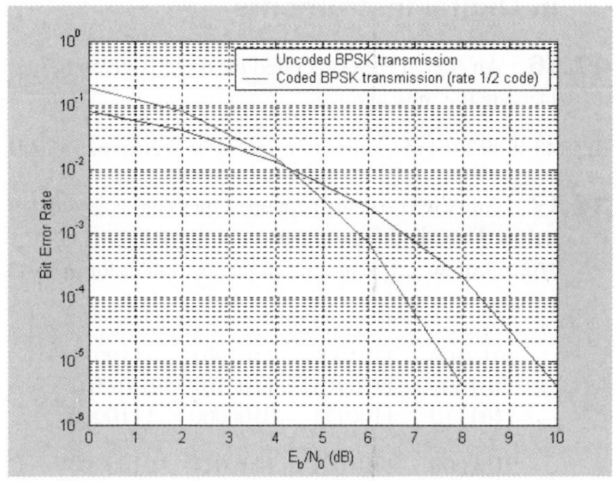

Figure P17.13

17.14 Your task is to transmit data from one computer to another through an AWGN channel. For this purpose you employ a bipolar binary signal which is a +1 V or a −1 V pulse during an interval [0, T] depending on whether the information bit is a 0 or a 1. The power spectral density of the noise added to the signal in the channel is $kT/2 = 10^{-3}$ W/Hz. Since you are transmitting important messages over this system, you require that you obtain at most 100 errors in every 125 kbyte file. Determine the maximum rate at which you can transmit bits over this channel. (*Hint*: The average probability of error of this system or BER is (number of bits in error)/(total number of bits sent). Compute the energy per bit, E_b. Then the average probability of error is given by $P_e = Q(\sqrt{(2E_b/kT)})$.

Section 17.7: Multiuser Channel Access

17.15 In this problem, we study a T1 carrier system used in digital telephony. Each voice signal is usually filtered using a low-pass filter of cut-off frequency 3.4 kHz and is then sampled at 8 kHz. This system multiplexes samples from 24 voice signals in a single sampling period. Each voice signal is coded with 8-bit PCM, and each frame (one sampling period) consists of 24 samples plus a single bit added for synchronization.

a. Calculate the duration of each bit. Note that this system multiplexes 24 samples each with 8 bits plus a single bit onto the time duration between two consecutive samples $T_s = 1/f_s$.

b. Calculate the resulting transmission rate.

Section 17.8: Data Transmission in Digital Instruments

17.16 An ASCII (hex) encoded message is given below. Decode the message.

41 53 43 49 49 20 64 65 63 6F 64 69 6E 67 20 69 73 20 65 61 73 79 21

17.17 An ASCII (binary) encoded message is given below. Decode the messsage.

Hint: Follow a line-by-line sequence, *not* column-by-column.

1010100	1101000	1101001	1110011
1101001	1101101	1100101	0101101
1101110	1100111	0100000	1110000
0100000	1101001	1110011	0100000
1100011	1101111	1101110	1110011
1110010	1101111	1100010	1101100
1100001	0100000	1110100	
1110101	1101101	1101001	
1100101	1101101	0101110	

17.18 Express the following decimal numbers in ASCII form:

a. 12

b. 345.2

c. 43.5

17.19 Express the following words in ASCII form:

a. Digital

b. Computer

c. Ascii

d. ASCII

17.20 Explain why data transmission over long distances is usually done via a serial scheme rather than parallel.

17.21 A certain automated data-logging instrument has 16K words of on-board memory. The device samples the variable of interest once every 5 min. How often must data be downloaded and the memory cleared in order to avoid losing any data?

17.22 Explain why three wires are required for the handshaking technique employed by IEEE 488 bus systems.

17.23 A CD-ROM can hold 650 Mbytes of information. Suppose the CD-ROMs are packed 50 per box. The manufacturer ships 100 boxes via commercial airliner from Los Angeles to New York. The distance between the two cities is 2,500 mi by air, and the airliner flies at a speed of 400 mi/h. What is the data transmission rate between the two cities in bits per second?

PART IV

ELECTROMECHANICS

CHAPTER

18

PRINCIPLES OF ELECTROMECHANICS

The objective of this chapter is to introduce the fundamental notions of electromechanical energy conversion, leading to an understanding of the operation of various electromechanical transducers. The chapter also serves as an introduction to the material on electric machines to be presented in Chapters 19 and 20. The foundations for the material introduced in this chapter will be found in the circuit analysis chapters (1 through 7). In addition, the material on power electronics (Chapter 12) is also relevant, especially with reference to Chapters 19 and 20.

The subject of electromechanical energy conversion is one that should be of particular interest to the *non–electrical* engineer, because it forms one of the important points of contact between electrical engineering and other engineering disciplines. Electromechanical transducers are commonly used in the design of industrial and aerospace control systems and in biomedical applications, and they form the basis of many common appliances. In the course of our exploration of electromechanics, we shall illustrate the operation of practical devices, such as loudspeakers, relays, solenoids, sensors for the measurement of position and velocity, and other devices of practical interest.

Learning Objectives

1. Review the basic principles of electricity and magnetism. *Section 18.1.*
2. Use the concepts of reluctance and magnetic circuit equivalents to compute magnetic flux and currents in simple magnetic structures. *Section 18.2.*
3. Understand the properties of magnetic materials and their effects on magnetic circuit models. *Section 18.3.*
4. Use magnetic circuit models to analyze transformers. *Section 18.4.*
5. Model and analyze force generation in electromagnetomechanical systems. Analyze moving-iron transducers (electromagnets, solenoids, relays) and moving-coil transducers (electrodynamic shakers, loudspeakers, and seismic transducers). *Section 18.5.*

18.1 ELECTRICITY AND MAGNETISM

The notion that the phenomena of electricity and magnetism are interconnected was first proposed in the early 1800s by H. C. Oersted, a Danish physicist. Oersted showed that an electric current produces magnetic effects (more specifically, a magnetic field). Soon after, the French scientist André Marie Ampère expressed this relationship by means of a precise formulation known as *Ampère's law*. A few years later, the English scientist Faraday illustrated how the converse of Ampère's law also holds true, that is, that a magnetic field can generate an electric field; in short, *Faraday's law* states that a changing magnetic field gives rise to a voltage. We shall undertake a more careful examination of both Ampère's and Faraday's laws in the course of this chapter.

As will be explained in the next few sections, the magnetic field forms a necessary connection between electrical and mechanical energy. Ampère's and Faraday's laws will formally illustrate the relationship between electric and magnetic fields, but it should already be evident from your own individual experience that the magnetic field can also convert magnetic energy to mechanical energy (e.g., by lifting a piece of iron with a magnet). In effect, the devices we commonly refer to as *electromechanical* should more properly be referred to as electro*magneto*mechanical, since they almost invariably operate through a conversion from electrical to mechanical energy (or vice versa) by means of a magnetic field. Chapters 18 through 20 are concerned with the use of electricity and magnetic materials for the purpose of converting electrical to mechanical energy, and back.

The Magnetic Field and Faraday's Law

The quantities used to quantify the strength of a magnetic field are the **magnetic flux** ϕ, in units of **webers** (Wb); and the **magnetic flux density B,** in units of webers per square meter (Wb/m^2), or **teslas** (T). The latter quantity and the associated **magnetic field intensity H** (in units of amperes per meter, or A/m) are vectors.[1] Thus, the density of the magnetic flux and its intensity are in general described in vector form, in terms of the components present in each spatial direction (e.g., on the x, y, and

[1] We will use the boldface symbols **B** and **H** to denote the vector forms of B and H; the standard typeface will represent the scalar flux density or field intensity in a given direction.

z axes). In discussing magnetic flux density and field intensity in this chapter and Chapter 19, we shall almost always assume that the field is a *scalar field*, that is, that it lies in a single spatial direction. This will simplify many explanations.

It is customary to represent the magnetic field by means of the familiar *lines of force* (a concept also due to Faraday); we visualize the strength of a magnetic field by observing the density of these lines in space. You probably know from a previous course in physics that such lines are closed in a magnetic field, that is, that they form continuous loops exiting at a magnetic north pole (by definition) and entering at a magnetic south pole. The relative strengths of the magnetic fields generated by two magnets could be depicted as shown in Figure 18.1.

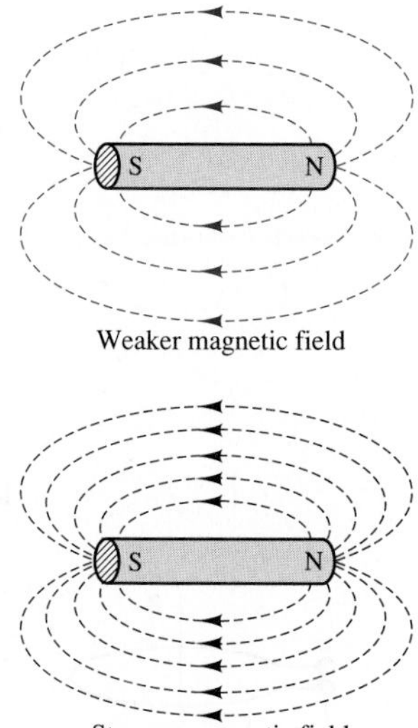

Figure 18.1 Lines of force in a magnetic field

Magnetic fields are generated by electric charge in motion, and their effect is measured by the force they exert on a moving charge. As you may recall from previous physics courses, the vector force **f** exerted on a charge of q moving at velocity **u** in the presence of a magnetic field with flux density **B** is given by

$$\mathbf{f} = q\mathbf{u} \times \mathbf{B} \tag{18.1}$$

where the symbol × denotes the (vector) cross product. If the charge is moving at a velocity **u** in a direction that makes an angle θ with the magnetic field, then the magnitude of the force is given by

$$f = quB \sin\theta \tag{18.2}$$

and the direction of this force is at right angles with the plane formed by the vectors **B** and **u.** This relationship is depicted in Figure 18.2.

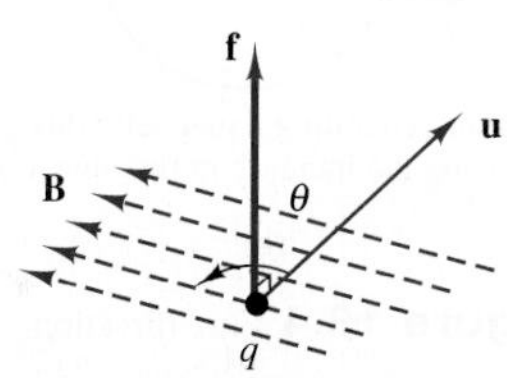

Figure 18.2 Charge moving in a constant magnetic field

The magnetic flux ϕ is then defined as the integral of the flux density over some surface area. For the simplified (but often useful) case of magnetic flux lines perpendicular to a cross-sectional area A, we can see that the flux is given by the integral

$$\phi = \int_A B\,dA \tag{18.3}$$

in webers, where the subscript A indicates that the integral is evaluated over surface A. Furthermore, if the flux were to be uniform over the cross-sectional area A (a simplification that will be useful), the preceding integral could be approximated by the following expression:

$$\boxed{\phi = B \cdot A} \tag{18.4}$$

LO1

Figure 18.3 illustrates this idea, by showing hypothetical magnetic flux lines traversing a surface, delimited in the figure by a thin conducting wire.

Faraday's law states that if the imaginary surface A were bounded by a conductor—for example, the thin wire of Figure 18.3—then a *changing* magnetic field would induce a voltage, and therefore a current, in the conductor. More precisely, Faraday's law states that a time-varying flux causes an induced **electromotive force,** or **emf,** e as follows:

$$e = -\frac{d\phi}{dt} \tag{18.5}$$

A little discussion is necessary at this point to explain the meaning of the minus sign in equation 18.5. Consider the one-turn coil of Figure 18.4, which forms

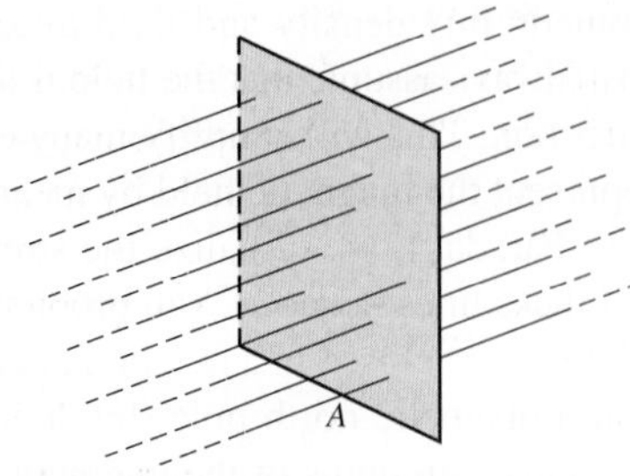

Figure 18.3 Magnetic flux lines crossing a surface

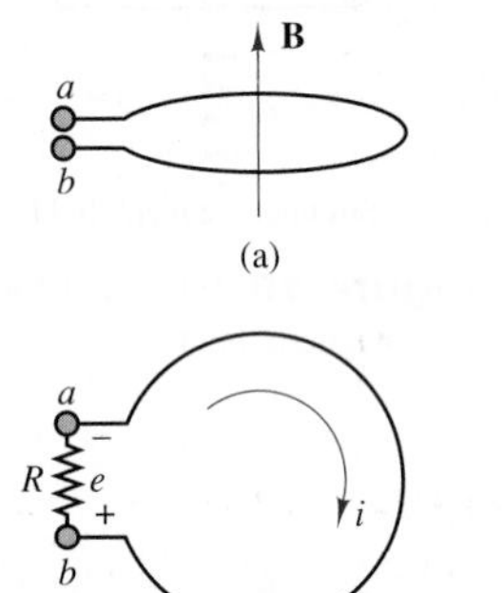

Current generating a magnetic flux opposing the increase in flux due to **B**

(b)

Figure 18.4 Flux direction

a circular cross-sectional area, in the presence of a magnetic field with flux density **B** oriented in a direction perpendicular to the plane of the coil. If the magnetic field, and therefore the flux within the coil, is constant, no voltage will exist across terminals a and b; if, however, the flux were increasing and terminals a and b were connected—for example, by means of a resistor, as indicated in Figure 18.4(b)—current would flow in the coil in such a way that *the magnetic flux generated by the current would oppose the increasing flux*. Thus, the flux induced by such a current would be in the direction opposite to that of the original flux density vector **B.** This principle is known as **Lenz's law.** The reaction flux would then point downward in Figure 18.4(a), or into the page in Figure 18.4(b). Now, by virtue of the **right-hand rule,** this reaction flux would induce a current flowing clockwise in Figure 18.4(b), that is, a current that flows out of terminal b and into terminal a. The resulting voltage across the hypothetical resistor R would then be negative. If, on the other hand, the original flux were decreasing, current would be induced in the coil so as to reestablish the initial flux; but this would mean that the current would have to generate a flux in the upward direction in Figure 18.4(a) [or out of the page in Figure 18.4(b)]. Thus, the resulting voltage would change sign.

The polarity of the induced voltage can usually be determined from physical considerations; therefore the minus sign in equation 18.5 is usually left out. We will use this convention throughout the chapter.

In practical applications, the size of the voltages induced by the changing magnetic field can be significantly increased if the conducting wire is coiled many times around, so as to multiply the area crossed by the magnetic flux lines many times over. For an N-turn coil with cross-sectional area A, for example, we have the emf

$$e = N \frac{d\phi}{dt} \quad \textbf{(18.6)}$$

LO1

CHECK YOUR UNDERSTANDING

A coil having 100 turns is immersed in a magnetic field that is varying uniformly from 80 to 30 mWb in 2 s. Find the induced voltage in the coil.

Answer: $e = -2.5$ V

Figure 18.5 shows an N-turn coil *linking* a certain amount of magnetic flux; you can see that if N is very large and the coil is tightly wound (as is usually the case in the construction of practical devices), it is not unreasonable to presume that each turn of the coil links the same flux. It is convenient, in practice, to define the **flux linkage** λ as

$$\lambda = N\phi \quad \textbf{(18.7)}$$

so that

LO1

$$e = \frac{d\lambda}{dt} \quad \textbf{(18.8)}$$

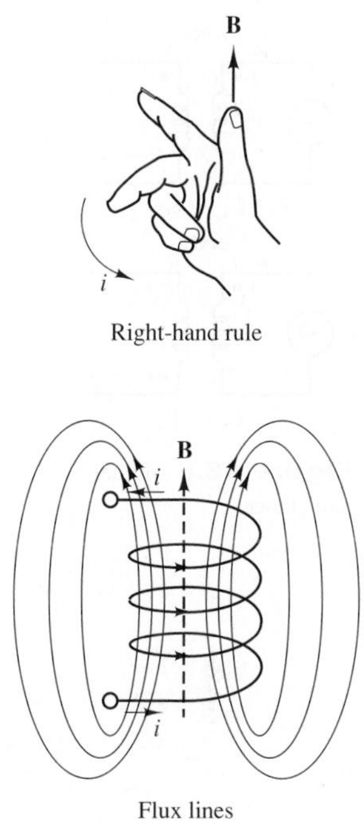

Figure 18.5 Concept of flux linkage

Note that equation 18.8, relating the derivative of the flux linkage to the induced emf, is analogous to the equation describing current as the derivative of charge:

$$i = \frac{dq}{dt} \quad \textbf{(18.9)}$$

In other words, flux linkage can be viewed as the dual of charge in a circuit analysis sense, provided that we are aware of the simplifying assumptions just stated in the preceding paragraphs, namely, a uniform magnetic field perpendicular to the area delimited by a tightly wound coil. These assumptions are not at all unreasonable when applied to the inductor coils commonly employed in electric circuits.

What, then, are the physical mechanisms that can cause magnetic flux to change, and therefore to induce an electromotive force? Two such mechanisms are possible. The first consists of physically moving a permanent magnet in the vicinity of a coil, for example, so as to create a time-varying flux. The second requires that we first produce a magnetic field by means of an electric current (how this can be accomplished is discussed later in this section) and then vary the current, thus varying the associated magnetic field. The latter method is more practical in many circumstances, since it does not require the use of permanent magnets and allows variation of field strength by varying the applied current; however, the former method is conceptually simpler to visualize. The voltages induced by a moving magnetic field are called **motion voltages;** those generated by a time-varying magnetic field are termed **transformer voltages.** We shall be interested in both in this chapter, for different applications.

In the analysis of linear circuits in Chapter 4, we implicitly assumed that the relationship between flux linkage and current was a linear one

$$\lambda = Li \quad \textbf{(18.10)}$$

so that the effect of a time-varying current was to induce a transformer voltage across an inductor coil, according to the expression

$$v = L\frac{di}{dt} \quad \textbf{(18.11)}$$

This is, in fact, the defining equation for the ideal **self-inductance** L. In addition to self-inductance, however, it is important to consider the **magnetic coupling** that can occur between neighboring circuits. Self-inductance measures the voltage induced in a circuit by the magnetic field generated by a current flowing in the same circuit. It is also possible that a second circuit in the vicinity of the first may experience an induced voltage as a consequence of the magnetic field generated in the first circuit. As we shall see in Section 18.4, this principle underlies the operation of all transformers.

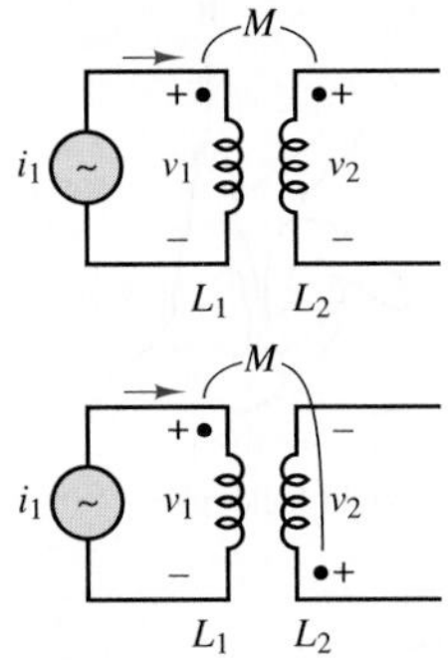

Figure 18.6 Mutual inductance

Self- and Mutual Inductance

Figure 18.6 depicts a pair of coils, one of which, L_1, is excited by a current i_1 and therefore develops a magnetic field and a resulting induced voltage v_1. The second coil, L_2, is not energized by a current, but links some of the flux generated by current i_1 around L_1 because of its close proximity to the first coil. The magnetic coupling between the coils established by virtue of their proximity is described by a quantity called **mutual inductance** and defined by the symbol M. The mutual inductance is defined by the equation

$$v_2 = M \frac{di_1}{dt} \tag{18.12}$$

The dots shown in the two drawings indicate the polarity of the coupling between the coils. If the dots are at the same end of the coils, the voltage induced in coil 2 by a current in coil 1 has the same polarity as the voltage induced by the same current in coil 1; otherwise, the voltages are in opposition, as shown in the lower part of Figure 18.6. Thus, the presence of such dots indicates that magnetic coupling is present between two coils. It should also be pointed out that if a current (and therefore a magnetic field) were present in the second coil, an additional voltage would be induced across coil 1. The voltage induced across a coil is, in general, equal to the sum of the voltages induced by self-inductance and mutual inductance.

In practical electromagnetic circuits, the self-inductance of a circuit is not necessarily constant; in particular, the inductance parameter L is not constant, in general, but depends on the strength of the magnetic field intensity, so that it will not be possible to use such a simple relationship as $v = L\, di/dt$, with L constant. If we revisit the definition of the transformer voltage

$$e = N \frac{d\phi}{dt} \tag{18.13}$$

we see that in an inductor coil, the inductance is given by

$$L = \frac{N\phi}{i} = \frac{\lambda}{i} \tag{18.14}$$

This expression implies that the relationship between current and flux in a magnetic structure is linear (the inductance being the slope of the line). In fact, the properties of ferromagnetic materials are such that the flux–current relationship is nonlinear, as we shall see in Section 18.3, so that the simple linear inductance parameter used in electric circuit analysis is not adequate to represent the behavior of the magnetic circuits of this chapter. In any practical situation, the relationship between the flux linkage λ and the current is nonlinear, and might be described by a curve similar to that shown in Figure 18.7. Whenever the i-λ curve is not a straight line, it is more convenient to analyze the magnetic system in terms of energy calculations, since the corresponding circuit equation would be nonlinear.

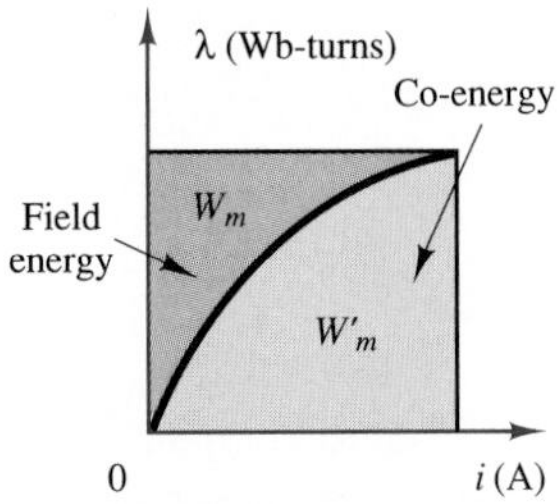

Figure 18.7 Relationship between flux linkage, current, energy, and co-energy

In a magnetic system, the energy stored in the magnetic field is equal to the integral of the instantaneous power, which is the product of voltage and current, just as in a conventional electric circuit:

$$W_m = \int ei\, dt' \tag{18.15}$$

However, in this case, the voltage corresponds to the induced emf, according to Faraday's law,

$$e = \frac{d\lambda}{dt} = N \frac{d\phi}{dt} \tag{18.16}$$

and is therefore related to the rate of change of the magnetic flux. The energy stored in the magnetic field could therefore be expressed in terms of the current by the integral

$$W_m = \int ei\, dt' = \int \frac{d\lambda}{dt} i\, dt' = \int i\, d\lambda' \qquad \textbf{(18.17)}$$

It should be straightforward to recognize that this energy is equal to the area above the λ-i curve of Figure 18.7. From the same figure, it is also possible to define a fictitious (but sometimes useful) quantity called **co-energy,** equal to the area under the curve and identified by the symbol W'_m. From the figure, it is also possible to see that the co-energy can be expressed in terms of the stored energy by means of the following relationship:

$$W'_m = i\lambda - W_m \qquad \textbf{(18.18)}$$

Example 18.1 illustrates the calculation of energy, co-energy, and induced voltage, using the concepts developed in these paragraphs.

The calculation of the energy stored in the magnetic field around a magnetic structure will be particularly useful later in the chapter, when the discussion turns to practical electromechanical transducers and it will be necessary to actually compute the forces generated in magnetic structures.

EXAMPLE 18.1 Energy and Co-Energy Calculation for an Inductor

LO1

Problem

Compute the energy, co-energy, and incremental linear inductance for an iron-core inductor with a given λ-i relationship. Also compute the voltage across the terminals, given the current through the coil.

Solution

Known Quantities: λ-i relationship; nominal value of λ; coil resistance; coil current.

Find: W_m; W'_m; L_Δ; v.

Schematics, Diagrams, Circuits, and Given Data: $i = (\lambda + 0.5\lambda^2)$ A; $\lambda_0 = 0.5$ V-s; $R = 1\ \Omega$; $i(t) = 0.625 + 0.01 \sin(400t)$.

Assumptions: Assume that the magnetic equation can be linearized, and use the linear model in all circuit calculations.

Analysis:

1. *Calculation of energy and co-energy.* From equation 18.17, we calculate the energy as follows.

$$W_m = \int_0^\lambda i(\lambda')\, d\lambda' = \frac{\lambda^2}{2} + \frac{\lambda^3}{6}$$

The above expression is valid in general; in our case, the inductor is operating at a nominal flux linkage $\lambda_0 = 0.5$ V-s, and we can therefore evaluate the energy to be

$$W_m(\lambda = \lambda_0) = \left(\frac{\lambda^2}{2} + \frac{\lambda^3}{6}\right)\Bigg|_{\lambda=0.5} = 0.1458 \text{ J}$$

Thus, after equation 18.18, the co-energy is given by

$$W'_m = i\lambda - W_m$$

where

$$i = \lambda + 0.5\lambda^2 = 0.625 \text{ A}$$

and

$$W'_m = i\lambda - W_m = (0.625)(0.5) - (0.1458) = 0.1667 \text{ J}$$

2. *Calculation of incremental inductance.* If we know the nominal value of flux linkage (i.e., the operating point), we can calculate a linear inductance L_Δ, valid around values of λ close to the operating point λ_0. This incremental inductance is defined by the expression

$$L_\Delta = \left(\frac{di}{d\lambda}\right)^{-1}\Bigg|_{\lambda=\lambda_0}$$

and can be computed to be

$$L_\Delta = \left(\frac{di}{d\lambda}\right)^{-1}\Bigg|_{\lambda=\lambda_0} = (1+\lambda)^{-1}\Big|_{\lambda=\lambda_0} = \frac{1}{1+\lambda}\Bigg|_{\lambda=0.5} = 0.667 \text{ H}$$

The above expressions can be used to analyze the circuit behavior of the inductor when the flux linkage is around 0.5 V-s, or, equivalently, when the current through the inductor is around 0.625 A.

3. *Circuit analysis using linearized model of inductor.* We can use the incremental linear inductance calculated above to compute the voltage across the inductor in the presence of a current $i(t) = 0.625 + 0.01\sin(400t)$. Using the basic circuit definition of an inductor with series resistance R, the voltage across the inductor is given by

$$\begin{aligned} v = iR + L_\Delta\frac{di}{dt} &= [0.625 + 0.01\sin(400t)] \times 1 + 0.667 \times 4\cos(400t) \\ &= 0.625 + 0.01\sin(400t) + 2.668\cos(400t) \\ &= 0.625 + 2.668\sin(400t + 89.8^\circ) \qquad \text{V} \end{aligned}$$

Comments: The linear approximation in this case is not a bad one: the small sinusoidal current is oscillating around a much larger average current. In this type of situation, it is reasonable to assume that the inductor behaves linearly. This example explains why the linear inductor model introduced in Chapter 4 is an acceptable approximation in most circuit analysis problems.

CHECK YOUR UNDERSTANDING

The relation between the flux linkages and the current for a magnetic material is given by $\lambda = 6i/(2i+1)$ Wb-turns. Determine the energy stored in the magnetic field for $\lambda = 2$ Wb-turns.

Answer: $W_m = 0.648$ J

Linear Variable Differential Transformer (LVDT)

FOCUS ON MEASUREMENTS

FIND IT ON THE WEB

The **linear variable differential transformer** (LVDT) is a displacement transducer based on the mutual inductance concept just discussed. Figure 18.8 shows a simplified representation of an LVDT, which consists of a primary coil, subject to AC excitation (v_{ex}), and of a pair of identical secondary coils, which are connected so as to result in the output voltage

$$v_{out} = v_1 - v_2$$

The ferromagnetic core between the primary and secondary coils can be displaced in proportion to some external motion x and determines the magnetic coupling between primary and secondary coils. Intuitively, as the core is displaced upward, greater coupling will occur between the primary coil and the top secondary coil, thus inducing a greater voltage in the top secondary coil. Hence, $v_{out} > 0$ for positive displacements. The converse is true for negative displacements. More formally, if the primary coil has resistance R_p and self-inductance L_p, we can write

$$iR_p + L_p \frac{di}{dt} = v_{ex}$$

and the voltages induced in the secondary coils are given by

$$v_1 = M_1 \frac{di}{dt}$$

$$v_2 = M_2 \frac{di}{dt}$$

so that

$$v_{out} = (M_1 - M_2) \frac{di}{dt}$$

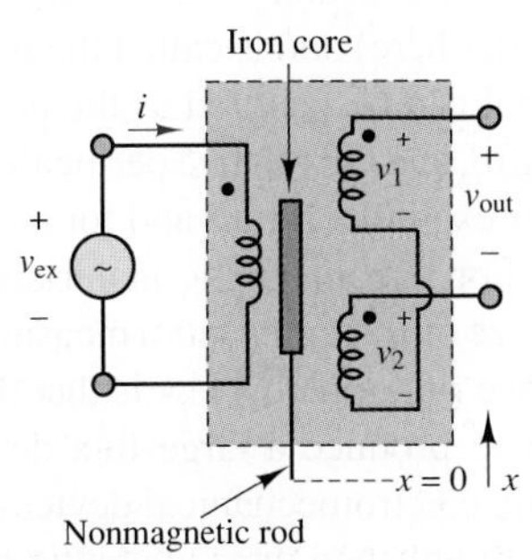

Figure 18.8 Linear variable differential transformer

where M_1 and M_2 are the mutual inductances between the primary and the respective secondary coils. It should be apparent that each of the mutual inductances is dependent on the position of the iron core. For example, with the core at the *null position*, $M_1 = M_2$ and $v_{out} = 0$. The LVDT is typically designed so that $M_1 - M_2$ is linearly related to the displacement of the core x.

(Continued)

(Concluded)

Because the excitation is by necessity an AC signal (why?), the output voltage is actually given by the difference of two sinusoidal voltages at the same frequency and is therefore itself a sinusoid, whose amplitude and phase depend on the displacement x. Thus, v_{out} is an *amplitude-modulated* (AM) signal, similar to the one discussed in the Focus on Measurements box "Capacitive Displacement Transducer" in Chapter 4. To recover a signal proportional to the actual displacement, it is therefore necessary to use a demodulator circuit, such as the one discussed in the Focus on Measurements box "Peak Detector for Capacitive Displacement Transducer" in Chapter 9.

Ampère's Law

As explained in the previous section, Faraday's law is one of two fundamental laws relating electricity to magnetism. The second relationship, which forms a counterpart to Faraday's law, is **Ampère's law.** Qualitatively, Ampère's law states that the magnetic field intensity **H** in the vicinity of a conductor is related to the current carried by the conductor; thus Ampère's law establishes a dual relationship with Faraday's law.

In the previous section, we described the magnetic field in terms of its flux density **B** and flux ϕ. To explain Ampère's law and the behavior of magnetic materials, we need to define a relationship between the magnetic field intensity **H** and the flux density **B.** These quantities are related by

$$\mathbf{B} = \mu\mathbf{H} = \mu_r\mu_0\mathbf{H} \qquad \text{Wb/m}^2 \text{ or T} \tag{18.19}$$

where the parameter μ is a scalar constant for a particular physical medium (at least, for the applications we consider here) and is called the **permeability** of the medium. The permeability of a material can be factored as the product of the permeability of free space $\mu_0 = 4\pi \times 10^{-7}$ H/m, and the relative permeability μ_r, which varies greatly according to the medium. For example, for air and for most electrical conductors and insulators, μ_r is equal to 1. For ferromagnetic materials, μ_r can take values in the hundreds or thousands. The size of μ_r represents a measure of the magnetic properties of the material. A consequence of Ampère's law is that the larger the value of μ, the smaller the current required to produce a large flux density in an electromagnetic structure. Consequently, many electromechanical devices make use of ferromagnetic materials, called iron cores, to enhance their magnetic properties. Table 18.1 gives approximate values of μ_r for some common materials.

Table 18.1 Relative permeabilities for common materials

Material	μ_r
Air	1
Permalloy	100,000
Cast steel	1,000
Sheet steel	4,000
Iron	5,195

Conversely, the reason for introducing the magnetic field intensity is that it is independent of the properties of the materials employed in the construction of magnetic circuits. Thus, a given magnetic field intensity **H** will give rise to different flux densities in different materials. It will therefore be useful to define *sources* of magnetic energy in terms of the magnetic field intensity, so that different magnetic structures and materials can then be evaluated or compared for a given source. In analogy with electromotive force, this "source" will be termed the **magnetomotive force (mmf).** As stated earlier, both the magnetic flux density and the field intensity are vector quantities; however, for ease of analysis, scalar fields will be chosen by appropriately selecting the orientation of the fields, wherever possible.

Ampère's law states that the integral of the vector magnetic field intensity **H** around a closed path is equal to the total current linked by the closed path i:

$$\oint \mathbf{H} \cdot d\mathbf{l} = \sum i \qquad \textbf{(18.20)}$$

where $d\mathbf{l}$ is an increment in the direction of the closed path. If the path is in the same direction as the direction of the magnetic field, we can use scalar quantities to state that

$$\int H\, dl = \sum i \qquad \textbf{(18.21)}$$

Figure 18.9 illustrates the case of a wire carrying a current i and of a circular path of radius r surrounding the wire. In this simple case, you can see that the magnetic field intensity **H** is determined by the familiar right-hand rule. This rule states that if the direction of current i points in the direction of the thumb of one's right hand, the resulting magnetic field encircles the conductor in the direction in which the other four fingers would encircle it. Thus, in the case of Figure 18.9, the closed-path integral becomes equal to $H \cdot 2\pi r$, since the path and the magnetic field are in the same direction, and therefore the magnitude of the magnetic field intensity is given by

$$H = \frac{i}{2\pi r} \qquad \textbf{(18.22)}$$

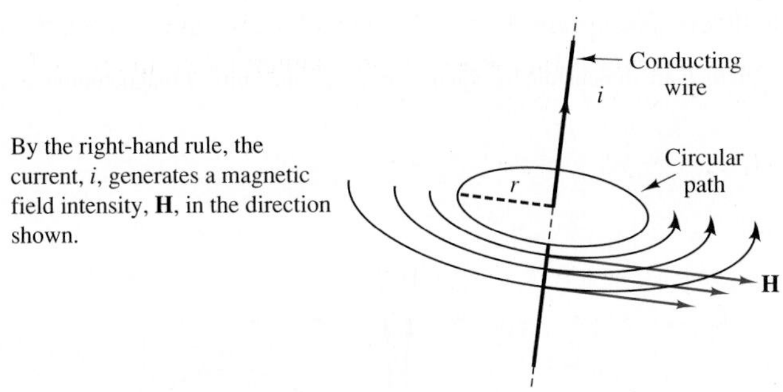
Conducting wire
i
By the right-hand rule, the current, i, generates a magnetic field intensity, H, in the direction shown.
Circular path
r
H

Figure 18.9 Illustration of Ampère's law

CHECK YOUR UNDERSTANDING

LO1

The magnitude of **H** at a radius of 0.5 m from a long linear conductor is 1 A-m^{-1}. Find the current in the wire.

Answer: $I = \pi$ A

Now, the magnetic field intensity is unaffected by the material surrounding the conductor, but the flux density depends on the material properties, since $B = \mu H$. Thus, the density of flux lines around the conductor would be far greater in the presence of a magnetic material than if the conductor were surrounded by air. The field generated by a single conducting wire is not very strong; however, if we arrange the wire into a tightly wound coil with many turns, we can greatly increase the strength of the magnetic field. For such a coil, with N turns, one can verify visually that the lines of force associated with the magnetic field link all the turns of the conducting coil, so that we have effectively increased the current linked by the flux lines N-fold. The product $N \cdot i$ is a useful quantity in electromagnetic circuits and is called the **magnetomotive force,**[2] $\mathcal{F}$ (often abbreviated mmf), in analogy with the electromotive force defined earlier:

LO1

$$\mathcal{F} = Ni \quad \text{A-turns} \qquad \text{Magnetomotive force} \tag{18.23}$$

Figure 18.10 illustrates the magnetic flux lines in the vicinity of a coil. The magnetic field generated by the coil can be made to generate a much greater flux density if the coil encloses a magnetic material. The most common ferromagnetic materials are steel and iron; in addition to these, many alloys and oxides of iron—as well as nickel—and some artificial ceramic materials called **ferrites** exhibit magnetic properties. Winding a coil around a ferromagnetic material accomplishes two useful tasks at once: It forces the magnetic flux to be concentrated within the coil and—if the shape of the magnetic material is appropriate—completely confines the flux within the magnetic material, thus forcing the closed path for the flux lines to be almost entirely enclosed within the ferromagnetic material. Typical arrangements are the iron-core

[2]Note that although they are dimensionally equal to amperes, the units of magnetomotive force are ampere-turns.

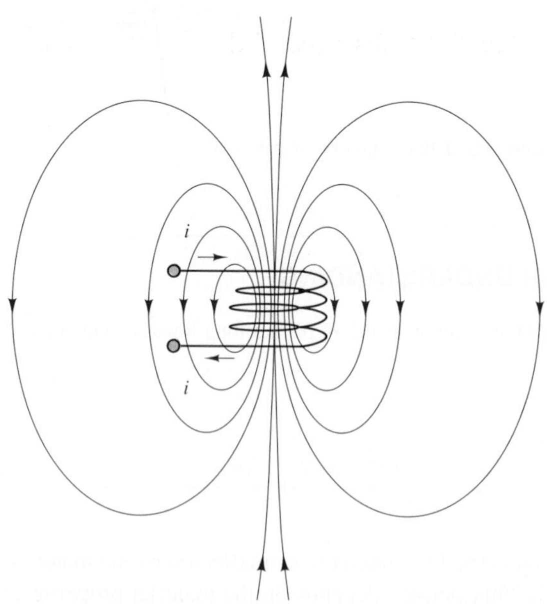

Figure 18.10 Magnetic flux lines in the vicinity of a current-carrying coil

inductor and the toroid of Figure 18.11. The flux densities for these inductors are given by

$$B = \frac{\mu N i}{l} \qquad \text{Flux density for tightly wound circular coil} \qquad \textbf{(18.24)}$$

$$B = \frac{\mu N i}{2\pi r_2} \qquad \text{Flux density for toroidal coil} \qquad \textbf{(18.25)}$$

In equation 18.24, l represents the length of the coil wire; Figure 18.11 defines the parameter r_2 in equation 18.25.

Intuitively, the presence of a high-permeability material near a source of magnetic flux causes the flux to preferentially concentrate in the high-μ material, rather than in air, much as a conducting path concentrates the current produced by an electric field in an electric circuit. In the course of this chapter, we shall continue to develop this analogy between electric circuits and magnetic circuits. Figure 18.12 depicts an example of a simple electromagnetic structure which, as we shall see shortly, forms the basis of the practical transformer.

Table 18.2 summarizes the variables introduced thus far in the discussion of electricity and magnetism.

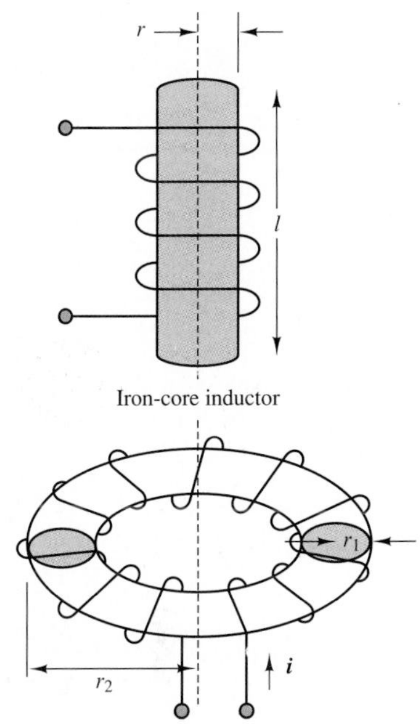

Figure 18.11 Practical inductors

Table 18.2 Magnetic variables and units

Variable	Symbol	Units
Current	I	A
Magnetic flux density	B	$\text{Wb/m}^2 = \text{T}$
Magnetic flux	ϕ	Wb
Magnetic field intensity	H	A/m
Electromotive force	e	V
Magnetomotive force	$\mathcal{F}$	A-turns
Flux linkage	λ	Wb-turns

LO1

18.2 MAGNETIC CIRCUITS

It is possible to analyze the operation of electromagnetic devices such as the one depicted in Figure 18.12 by means of magnetic equivalent circuits, similar in many respects to the equivalent electric circuits of earlier chapters. Before we can present this technique, however, we need to make a few simplifying approximations. The first of these approximations assumes that there exists a **mean path** for the magnetic flux, and that the corresponding mean flux density is approximately constant over the cross-sectional area of the magnetic structure. Using equation 18.4, we see that a coil wound around a core with cross-sectional area A will have flux density

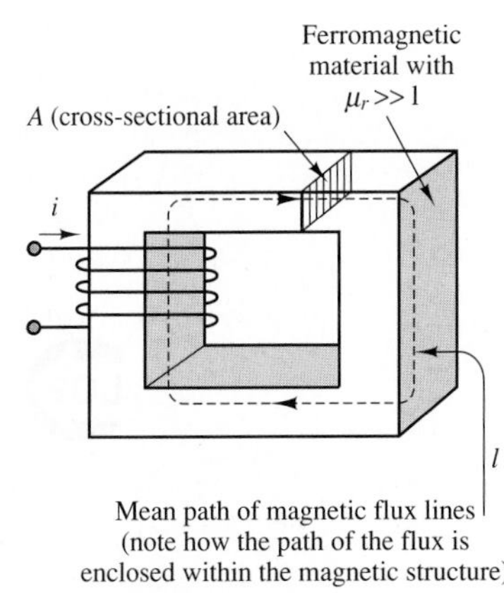

Figure 18.12 A simple electromagnetic structure

$$B = \frac{\phi}{A} \qquad \textbf{(18.26)}$$

where A is assumed to be perpendicular to the direction of the flux lines. Figure 18.12 illustrates such a mean path and the cross-sectional area A. Knowing the flux density, we obtain the field intensity:

$$H = \frac{B}{\mu} = \frac{\phi}{A\mu} \qquad \textbf{(18.27)}$$

But then, knowing the field intensity, we can relate the mmf of the coil $\mathcal{F}$ to the product of the magnetic field intensity H and the length of the magnetic (mean) path

l; we can use equations 18.24 and 18.19 to derive

$$\mathcal{F} = N \cdot i = H \cdot l \qquad \textbf{(18.28)}$$

In summary, the mmf is equal to the magnetic flux times the length of the magnetic path, divided by the permeability of the material times the cross-sectional area:

$$\mathcal{F} = \phi \frac{l}{\mu A} \qquad \textbf{(18.29)}$$

A review of this formula reveals that the magnetomotive force $\mathcal{F}$ may be viewed as being analogous to the voltage source in a series electric circuit, and that the flux ϕ is then equivalent to the electric current in a series circuit and the term $l/\mu A$ to the *magnetic resistance* of one leg of the magnetic circuit. You will note that the term $l/\mu A$ is very similar to the term describing the resistance of a cylindrical conductor of length l and cross-sectional area A, where the permeability μ is analogous to the conductivity σ. The term $l/\mu A$ occurs frequently enough to be assigned the name of **reluctance** and the symbol $\mathcal{R}$. It is also important to recognize the *relationship between the reluctance of a magnetic structure and its inductance*. This can be derived easily starting from equation 18.14:

LO2

$$L = \frac{\lambda}{i} = \frac{N\phi}{i} = \frac{N}{i}\frac{Ni}{\mathcal{R}} = \frac{N^2}{\mathcal{R}} \quad \text{H} \qquad \textbf{(18.30)}$$

In summary, when an N-turn coil carrying a current i is wound around a magnetic core such as the one indicated in Figure 18.12, the mmf $\mathcal{F}$ generated by the coil produces a flux ϕ that is *mostly* concentrated within the core and is assumed to be uniform across the cross section. Within this simplified picture, then, the analysis of a magnetic circuit is analogous to that of resistive electric circuits. This analogy is illustrated in Table 18.3 and in the examples in this section.

LO2

Table 18.3 **Analogy between electric and magnetic circuits**

Electrical quantity	Magnetic quantity
Electrical field intensity E, V/m	Magnetic field intensity H, A-turns/m
Voltage v, V	Magnetomotive force $\mathcal{F}$, A-turns
Current i, A	Magnetic flux ϕ, Wb
Current density J, A/m^2	Magnetic flux density B, Wb/m^2
Resistance R, Ω	Reluctance $\mathcal{R} = l/\mu A$, A-turns/Wb
Conductivity σ, 1/Ω-m	Permeability μ, Wb/A-m

The usefulness of the magnetic circuit analogy can be emphasized by analyzing a magnetic core similar to that of Figure 18.12, but with a slightly modified geometry. Figure 18.13 depicts the magnetic structure and its equivalent-circuit analogy. In the figure, we see that the mmf $\mathcal{F} = Ni$ excites the magnetic circuit, which is composed of four legs: two of mean path length l_1 and cross-sectional area $A_1 = d_1 w$, and the other two of mean length l_2 and cross-sectional area $A_2 = d_2 w$. Thus, the reluctance

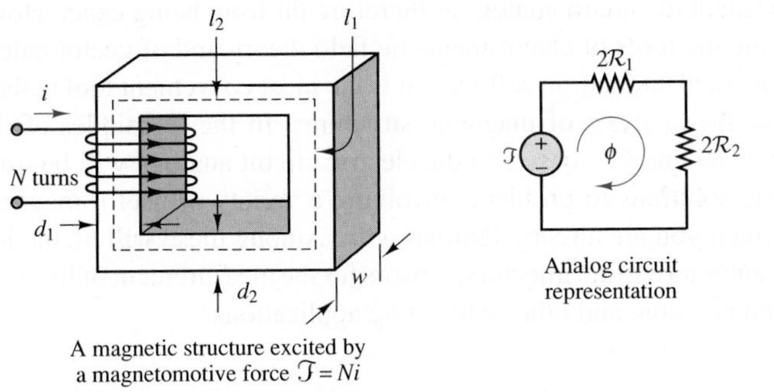

Figure 18.13 Analogy between magnetic and electric circuits

encountered by the flux in its path around the magnetic core is given by the quantity $\mathcal{R}_{\text{series}}$, with

$$\mathcal{R}_{\text{series}} = 2\mathcal{R}_1 + 2\mathcal{R}_2$$

and

$$\mathcal{R}_1 = \frac{l_1}{\mu A_1} \qquad \mathcal{R}_2 = \frac{l_2}{\mu A_2}$$

It is important at this stage to review the assumptions and simplifications made in analyzing the magnetic structure of Figure 18.13:

1. All the magnetic flux is linked by all the turns of the coil.
2. The flux is confined exclusively within the magnetic core.
3. The density of the flux is uniform across the cross-sectional area of the core.

LO2

You can probably see intuitively that the first of these assumptions might not hold true near the ends of the coil, but that it might be more reasonable if the coil is tightly wound. The second assumption is equivalent to stating that the relative permeability of the core is infinitely higher than that of air (presuming that this is the medium surrounding the core); if this were the case, the flux would indeed be confined within the core. It is worthwhile to note that we make a similar assumption when we treat wires in electric circuits as perfect conductors: The conductivity of copper is substantially greater than that of free space, by a factor of approximately 10^{15}. In the case of magnetic materials, however, even for the best alloys, we have a relative permeability only on the order of 10^3 to 10^4. Thus, an approximation that is quite appropriate for electric circuits is not nearly as good in the case of magnetic circuits. Some of the flux in a structure such as those of Figures 18.12 and 18.13 would thus not be confined within the core (this is usually referred to as **leakage flux**). Finally, the assumption that the flux is uniform across the core cannot hold for a finite-permeability medium, but it is very helpful in giving an approximate *mean* behavior of the magnetic circuit.

The magnetic circuit analogy is therefore far from being exact. However, short of employing the tools of electromagnetic field theory and of vector calculus, or advanced numerical simulation software, it is the most convenient tool at the engineer's disposal for the analysis of magnetic structures. In the remainder of this chapter, the approximate analysis based on the electric circuit analogy will be used to obtain approximate solutions to problems involving a variety of useful magnetic circuits, many of which you are already familiar with. Among these will be the loudspeaker, solenoids, automotive fuel injectors, sensors for the measurement of linear and angular velocity and position, and other interesting applications.

LO2

EXAMPLE 18.2 Analysis of Magnetic Structure and Equivalent Magnetic Circuit

Problem

Calculate the flux, flux density, and field intensity on the magnetic structure of Figure 18.14.

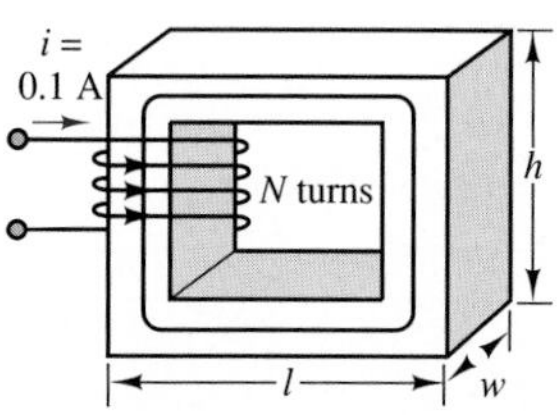

Figure 18.14

Solution

Known Quantities: Relative permeability; number of coil turns; coil current; structure geometry.

Find: ϕ; B; H.

Schematics, Diagrams, Circuits, and Given Data: $\mu_r = 1{,}000$; $N = 500$ turns; $i = 0.1$ A. The cross-sectional area is $A = w^2 = (0.01)^2 = 0.0001\ \text{m}^2$. The magnetic circuit geometry is defined in Figures 18.14 and 18.15.

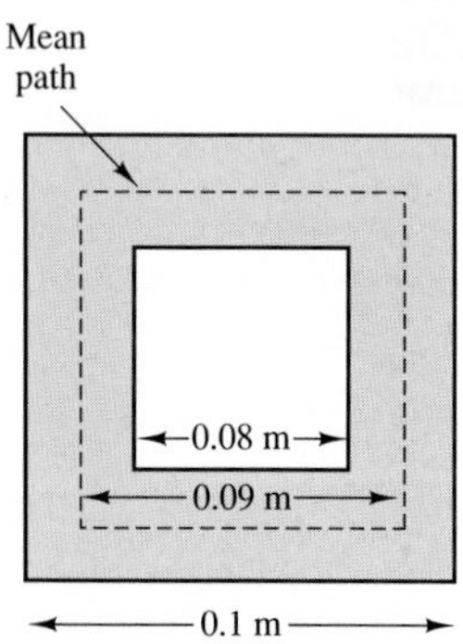

Figure 18.15

Assumptions: All magnetic flux is linked by the coil; the flux is confined to the magnetic core; the flux density is uniform.

Analysis:

1. *Calculation of magnetomotive force.* From equation 18.28, we calculate the magnetomotive force:

$$\mathcal{F} = \text{mmf} = Ni = (500\ \text{turns})(0.1\ \text{A}) = 50\ \text{A-turns}$$

2. *Calculation of mean path.* Next, we estimate the mean path of the magnetic flux. On the basis of the assumptions, we can calculate a mean path that runs through the geometric center of the magnetic structure, as shown in Figure 18.15. The path length is

$$l_c = 4 \times 0.09\ \text{m} = 0.36\ \text{m}$$

3. *Calculation of reluctance.* Knowing the magnetic path length and cross-sectional area, we can calculate the reluctance of the circuit:

$$\mathcal{R} = \frac{l_c}{\mu A} = \frac{l_c}{\mu_r \mu_0 A} = \frac{0.36}{1{,}000 \times 4\pi \times 10^{-7} \times 0.0001}$$
$$= 2.865 \times 10^6\ \text{A-turns/Wb}$$

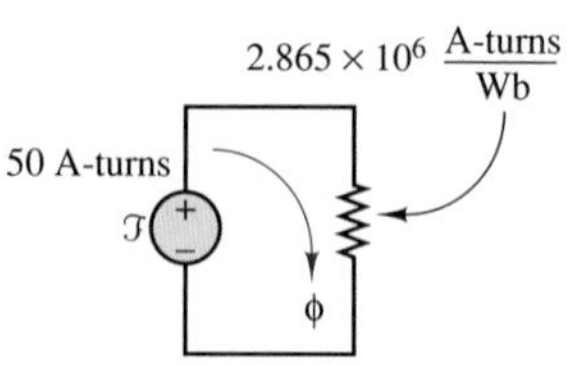

Figure 18.16

The corresponding equivalent magnetic circuit is shown in Figure 18.16.

4. *Calculation of magnetic flux, flux density, and field intensity.* On the basis of the assumptions, we can now calculate the magnetic flux

$$\phi = \frac{\mathcal{F}}{\mathcal{R}} = \frac{50 \text{ A-turns}}{2.865 \times 10^6 \text{ A-turns/Wb}} = 1.75 \times 10^{-5} \text{ Wb}$$

the flux density

$$B = \frac{\phi}{\text{A}} = \frac{\phi}{w^2} = \frac{1.75 \times 10^{-5} \text{ Wb}}{0.0001 \text{ m}^2} = 0.175 \text{ Wb/m}^2$$

and the magnetic field intensity

$$H = \frac{B}{\mu} = \frac{B}{\mu_r \mu_0} = \frac{0.175 \text{ Wb/m}^2}{1{,}000 \times 4\pi \times 10^{-7} \text{ H/m}} = 139 \text{ A-turns/m}$$

Comments: This example has illustrated all the basic calculations that pertain to magnetic structures. Remember that the assumptions stated in this example (and earlier in the chapter) simplify the problem and make its approximate numerical solution possible in a few simple steps. In reality, flux leakage, fringing, and uneven distribution of flux across the structure would require the solution of three-dimensional equations using finite-element methods. These methods are not discussed in this book, but are necessary for practical engineering designs.

The usefulness of these approximate methods is that you can, for example, quickly calculate the approximate magnitude of the current required to generate a given magnetic flux or flux density. You shall soon see how these calculations can be used to determine electromagnetic energy and magnetic forces in practical structures.

The methodology described in this example is summarized in the following Focus on Methodology box.

CHECK YOUR UNDERSTANDING

Determine the equivalent reluctance of the structure of Figure 18.17 as seen by the "source" if μ_r for the structure is 1,000, $l = 5$ cm, and all the legs are 1 cm on a side.

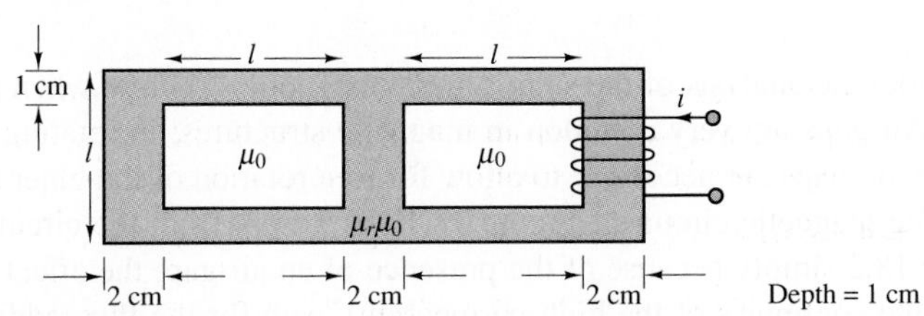

Figure 18.17

Answer: Assuming a mean path 1 cm from the edges of the structure, $\mathcal{R}_{eq} = 1.6 \times 10^6$ A-turns/Wb

FOCUS ON METHODOLOGY

MAGNETIC STRUCTURES AND EQUIVALENT MAGNETIC CIRCUITS

Direct Problem

Given—The structure geometry and the coil parameters.

Calculate—The magnetic flux in the structure.

1. Compute the mmf.
2. Determine the length and cross section of the magnetic path for each continuous *leg* or section of the path.
3. Calculate the equivalent reluctance of the *leg*.
4. Generate the equivalent magnetic circuit diagram, and calculate the total equivalent reluctance.
5. Calculate the flux, flux density, and magnetic field intensity, as needed.

Inverse Problem

Given—The desired flux or flux density and structure geometry.

Calculate—The necessary coil current and number of turns.

1. Calculate the total equivalent reluctance of the structure from the desired flux.
2. Generate the equivalent magnetic circuit diagram.
3. Determine the mmf required to establish the required flux.
4. Choose the coil current and number of turns required to establish the desired mmf.

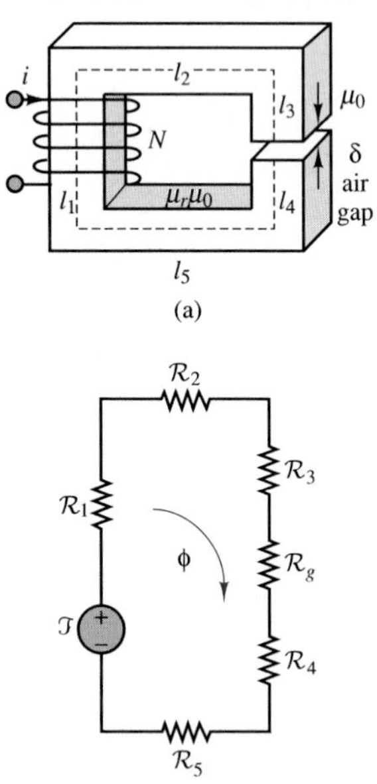

Figure 18.18 (a) Magnetic circuit with air gap; (b) equivalent representation of magnetic circuit with an air gap

Consider the analysis of the same simple magnetic structure when an **air gap** is present. Air gaps are very common in magnetic structures; in rotating machines, for example, air gaps are necessary to allow for free rotation of the inner core of the machine. The magnetic circuit of Figure 18.18(a) differs from the circuit analyzed in Example 18.2 simply because of the presence of an air gap; the effect of the gap is to break the continuity of the high-permeability path for the flux, adding a high-reluctance component to the equivalent circuit. The situation is analogous to adding a very large series resistance to a series electric circuit. It should be evident from Figure 18.18(a) that the basic concept of reluctance still applies, although now two different permeabilities must be taken into account.

The equivalent circuit for the structure of Figure 18.18(a) may be drawn as shown in Figure 18.18(b), where $\mathcal{R}_n$ is the reluctance of path l_n, for $n = 1, 2, \ldots, 5$, and $\mathcal{R}_g$ is the reluctance of the air gap. The reluctances can be expressed as follows, if we assume that the magnetic structure has a uniform cross-sectional area A:

$$\mathcal{R}_1 = \frac{l_1}{\mu_r \mu_0 A} \qquad \mathcal{R}_2 = \frac{l_2}{\mu_r \mu_0 A} \qquad \mathcal{R}_3 = \frac{l_3}{\mu_r \mu_0 A}$$
$$\mathcal{R}_4 = \frac{l_4}{\mu_r \mu_0 A} \qquad \mathcal{R}_5 = \frac{l_5}{\mu_r \mu_0 A} \qquad \mathcal{R}_g = \frac{\delta}{\mu_0 A_g} \tag{18.31}$$

Note that in computing $\mathcal{R}_g$, the length of the gap is given by δ and the permeability is given by μ_0, as expected, but A_g is different from the cross-sectional area A of the structure. This is so because the flux lines exhibit a phenomenon known as **fringing** as they cross an air gap. The flux lines actually *bow out* of the gap defined by the cross section A, not being contained by the high-permeability material any longer. Thus, it is customary to define an area A_g that is greater than A, to account for this phenomenon. Example 18.3 describes in greater detail the procedure for finding A_g and also discusses the phenomenon of fringing.

EXAMPLE 18.3 Magnetic Structure with Air Gaps

LO2

Problem

Compute the equivalent reluctance of the magnetic circuit of Figure 18.19 and the flux density established in the bottom bar of the structure.

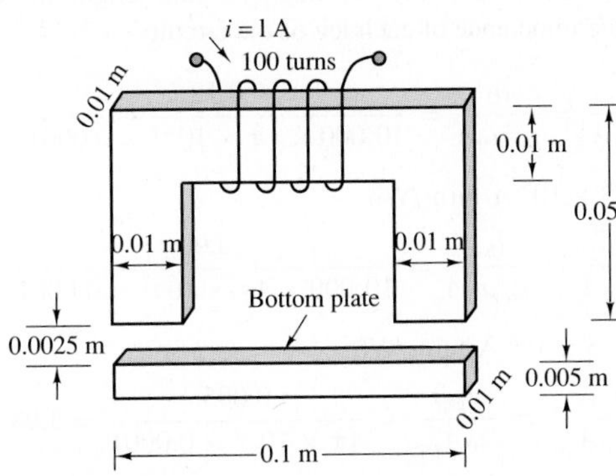

Figure 18.19 Electromagnetic structure with air gaps

Solution

Known Quantities: Relative permeability; number of coil turns; coil current; structure geometry.

Find: $\mathcal{R}_{\text{eq}}$; B_{bar}.

Schematics, Diagrams, Circuits, and Given Data: $\mu_r = 10{,}000$; $N = 100$ turns; $i = 1$ A.

Assumptions: All magnetic flux is linked by the coil; the flux is confined to the magnetic core; the flux density is uniform.

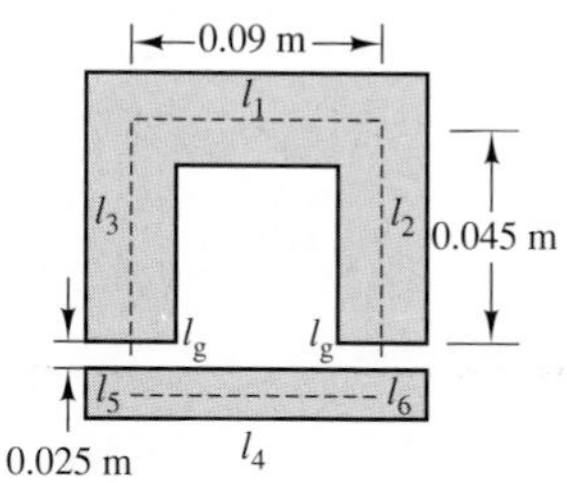

Figure 18.20

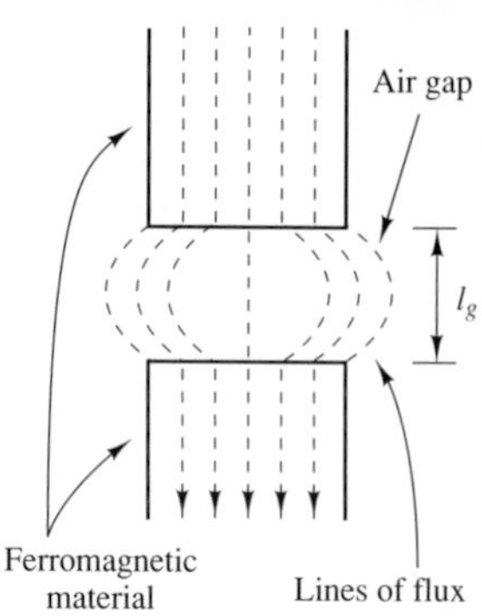

Figure 18.21 Fringing effects in air gap

Analysis:

1. *Calculation of magnetomotive force.* From equation 18.28, we calculate the magnetomotive force:

$$\mathcal{F} = \text{mmf} = Ni = (100 \text{ turns})(1 \text{ A}) = 100 \text{ A-turns}$$

2. *Calculation of mean path.* Figure 18.20 depicts the geometry. The path length is

$$l_c = l_1 + l_2 + l_3 + l_4 + l_5 + l_6 + l_g + l_g$$

However, the path must be broken into three legs: the upside-down U-shaped element, the air gaps, and the bar. We cannot treat these three parts as one because the relative permeability of the magnetic material is very different from that of the air gap. Thus, we define the following three paths, neglecting the very small (half bar thickness) lengths l_5 and l_6:

$$l_U = l_1 + l_2 + l_3 \qquad l_{\text{bar}} = l_4 + l_5 + l_6 \approx l_4 \qquad l_{\text{gap}} = l_g + l_g$$

where

$$l_U = 0.18 \text{ m} \qquad l_{\text{bar}} = 0.09 \text{ m} \qquad l_{\text{gap}} = 0.005 \text{ m}$$

Next, we compute the cross-sectional area. For the magnetic structure, we calculate the square cross section to be $A = w^2 = (0.01)^2 = 0.0001 \text{ m}^2$. For the air gap, we will make an empirical adjustment to account for the phenomenon of *fringing,* that is, to account for the tendency of the magnetic flux lines to bow out of the magnetic path, as illustrated in Figure 18.21. A rule of thumb used to account for fringing is to add the length of the gap to the actual cross-sectional area. Thus

$$A_{\text{gap}} = (0.01 \text{ m} + l_g)^2 = (0.0125)^2 = 0.15625 \times 10^{-3} \text{ m}^2$$

3. *Calculation of reluctance.* Knowing the magnetic path length and cross-sectional area, we can calculate the reluctance of each leg of the circuit:

$$\mathcal{R}_U = \frac{l_U}{\mu_U A} = \frac{l_U}{\mu_r \mu_0 A} = \frac{0.18}{10{,}000 \times 4\pi \times 10^{-7} \times 0.0001}$$
$$= 1.43 \times 10^5 \text{ A-turns/Wb}$$

$$\mathcal{R}_{\text{bar}} = \frac{l_{\text{bar}}}{\mu_{\text{bar}} A} = \frac{l_{\text{bar}}}{\mu_r \mu_0 A} = \frac{0.09}{10{,}000 \times 4\pi \times 10^{-7} \times 0.0001}$$
$$= 0.715 \times 10^5 \text{ A-turns/Wb}$$

$$\mathcal{R}_{\text{gap}} = \frac{l_{\text{gap}}}{\mu_{\text{gap}} A_{\text{gap}}} = \frac{l_{\text{gap}}}{\mu_0 A_{\text{gap}}} = \frac{0.005}{4\pi \times 10^{-7} \times 0.0001} = 3.98 \times 10^7 \text{ A-turns/Wb}$$

Note that the reluctance of the air gap is dominant with respect to that of the magnetic structure, in spite of the small dimension of the gap. This is so because the relative permeability of the air gap is much smaller than that of the magnetic material.

The equivalent reluctance of the structure is

$$\mathcal{R}_{\text{eq}} = \mathcal{R}_U + \mathcal{R}_{\text{bar}} + \mathcal{R}_{\text{gap}} = 1.43 \times 10^5 + 0.715 \times 10^5 + 2.55 \times 10^7$$
$$= 4 \times 10^7 \text{ A-turns/Wb}$$

Thus,

$$\mathcal{R}_{\text{eq}} \approx \mathcal{R}_{\text{gap}}$$

Since the gap reluctance is two orders of magnitude greater than the reluctance of the magnetic structure, it is reasonable to neglect the magnetic structure reluctance and work only with the gap reluctance in calculating the magnetic flux.

4. *Calculation of magnetic flux and flux density in the bar.* From the result of the preceding subsection, we calculate the flux

$$\phi = \frac{\mathcal{F}}{\mathcal{R}_{\text{eq}}} \approx \frac{\mathcal{F}}{\mathcal{R}_{\text{gap}}} = \frac{100 \text{ A-turns}}{2.55 \times 10^7 \text{ A-turns/Wb}} = 2.51 \times 10^{-6} \text{ Wb}$$

and the flux density in the bar

$$B_{\text{bar}} = \frac{\phi}{A} = \frac{3.92 \times 10^{-6} \text{ Wb}}{0.0001 \text{ m}^2} = 25.1 \times 10^{-3} \text{ Wb/m}^2$$

Comments: It is very common to neglect the reluctance of the magnetic material sections in these approximate calculations. We shall make this assumption very frequently in the remainder of the chapter.

CHECK YOUR UNDERSTANDING

Find the equivalent reluctance of the magnetic circuit shown in Figure 18.22 if μ_r of the structure is infinite, $\delta = 2$ mm, and the physical cross section of the core is 1 cm^2. Do not neglect fringing.

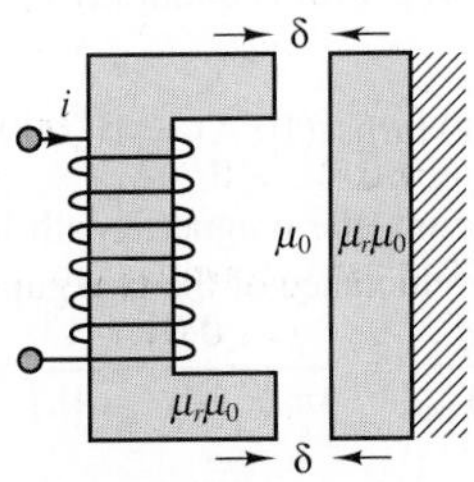

Figure 18.22

Answer: $\mathcal{R}_{\text{eq}} = 22 \times 10^6$ A-turns/Wb

EXAMPLE 18.4 Magnetic Structure of Electric Motor

LO2

Problem

Figure 18.23 depicts the configuration of an electric motor. The electric motor consists of a *stator* and a *rotor*. Compute the air gap flux and flux density.

Solution

Known Quantities: Relative permeability; number of coil turns; coil current; structure geometry.

Find: ϕ_{gap}; B_{gap}.

Schematics, Diagrams, Circuits, and Given Data: $\mu_r \to \infty$; $N = 1{,}000$ turns; $i = 10A$; $l_{\text{gap}} = 0.01$ m; $A_{\text{gap}} = 0.1$ m^2. The magnetic circuit geometry is defined in Figure 18.23.

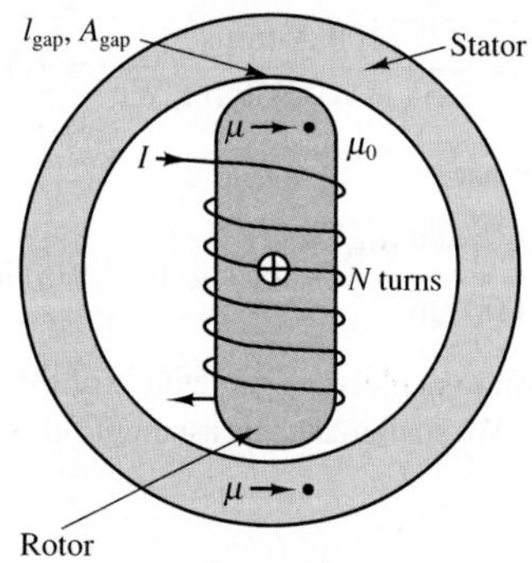

Figure 18.23 Cross-sectional view of synchronous motor

Assumptions: All magnetic flux is linked by the coil; the flux is confined to the magnetic core; the flux density is uniform. The reluctance of the magnetic structure is negligible.

Analysis:

1. *Calculation of magnetomotive force.* From equation 18.28, we calculate the magnetomotive force:

$$\mathcal{F} = \text{mmf} = Ni = (1{,}000 \text{ turns})(10 \text{ A}) = 10{,}000 \text{ A-turns}$$

2. *Calculation of reluctance.* Knowing the magnetic path length and cross-sectional area, we can calculate the equivalent reluctance of the two gaps:

$$\mathcal{R}_{gap} = \frac{l_{gap}}{\mu_{gap} A_{gap}} = \frac{l_{gap}}{\mu_0 A_{gap}} = \frac{0.01}{4\pi \times 10^{-7} \times 0.1} = 7.96 \times 10^4 \text{ A-turns/Wb}$$

$$\mathcal{R}_{eq} = 2\mathcal{R}_{gap} = 1.59 \times 10^5 \text{ A-turns/Wb}$$

3. *Calculation of magnetic flux and flux density.* From the results of steps 1 and 2, we calculate the flux

$$\phi = \frac{\mathcal{F}}{\mathcal{R}_{eq}} = \frac{10{,}000 \text{ A-turns}}{1.59 \times 10^5 \text{ A-turns/Wb}} = 0.0628 \text{ Wb}$$

and the flux density

$$B_{bar} = \frac{\phi}{A} = \frac{0.0628 \text{ Wb}}{0.1 \text{ m}^2} = 0.628 \text{ Wb/m}^2$$

Comments: Note that the flux and flux density in this structure are significantly larger than those in Example 18.3 because of the larger mmf and larger gap area of this magnetic structure.
The subject of electric motors will be formally approached in Chapter 19.

LO2

EXAMPLE 18.5 Equivalent Circuit of Magnetic Structure with Multiple Air Gaps

Problem

Figure 18.24 depicts the configuration of a magnetic structure with two air gaps. Determine the equivalent circuit of the structure.

Solution

Known Quantities: Structure geometry.

Find: Equivalent-circuit diagram.

Assumptions: All magnetic flux is linked by the coil; the flux is confined to the magnetic core; the flux density is uniform. The reluctance of the magnetic structure is negligible.

Analysis:

1. *Calculation of magnetomotive force.*

$$\mathcal{F} = \text{mmf} = Ni$$

2. *Calculation of reluctance.* Knowing the magnetic path length and cross-sectional area, we can calculate the equivalent reluctance of the two gaps:

$$\mathcal{R}_{\text{gap}-1} = \frac{l_{\text{gap}-1}}{\mu_{\text{gap}-1} A_{\text{gap}-1}} = \frac{l_{\text{gap}-1}}{\mu_0 A_{\text{gap}-1}}$$

$$\mathcal{R}_{\text{gap}-1} = \frac{l_{\text{gap}-2}}{\mu_{\text{gap}-2} A_{\text{gap}-2}} = \frac{l_{\text{gap}-2}}{\mu_0 A_{\text{gap}-2}}$$

3. *Calculation of magnetic flux and flux density.* Note that the flux must now divide between the two legs, and that a different air-gap flux will exist in each leg. Thus

$$\phi_1 = \frac{Ni}{\mathcal{R}_{\text{gap}-1}} = \frac{Ni\mu_0 A_{\text{gap}-1}}{l_{\text{gap}-1}}$$

$$\phi_2 = \frac{Ni}{\mathcal{R}_{\text{gap}-2}} = \frac{Ni\mu_0 A_{\text{gap}-2}}{l_{\text{gap}-2}}$$

and the total flux generated by the coil is $\phi = \phi_1 + \phi_2$.

The equivalent circuit is shown in the bottom half of Figure 18.24.

Comments: Note that the two legs of the structure act as resistors in a parallel circuit.

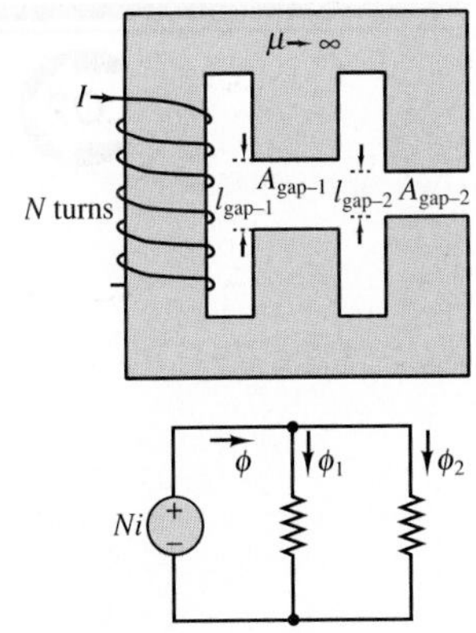

Figure 18.24 Magnetic structure with two air gaps

CHECK YOUR UNDERSTANDING

Find the equivalent magnetic circuit of the structure of Figure 18.25 if μ_r is infinite. Give expressions for each of the circuit values if the physical cross-sectional area of each of the legs is given by

$$A = l \times w$$

Do not neglect fringing.

Answer: $\mathcal{R}_\delta = \mathcal{R}_1 = \mathcal{R}_2 = \mathcal{R}_3 = \delta/\mu_0(l+\delta)(w+\delta)$; $\mathcal{F}_1 = Ni_1$; $\mathcal{F}_2 = Ni_2$

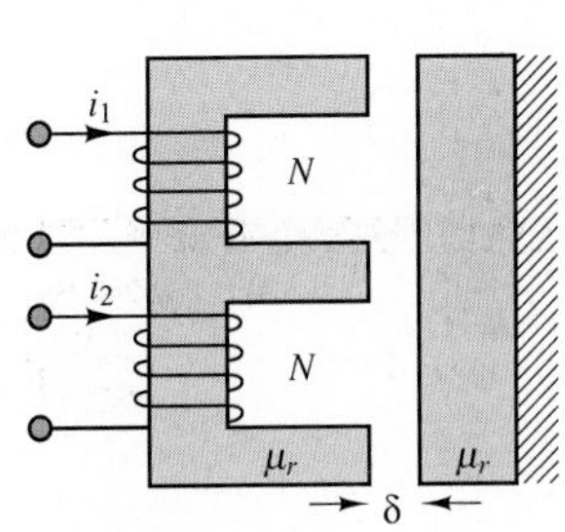

Figure 18.25

LO2

EXAMPLE 18.6 Inductance, Stored Energy, and Induced Voltage

Problem

1. Determine the inductance and the magnetic stored energy for the structure of Figure 18.18(a). The structure is identical to that of Example 18.2 except for the air gap.
2. Assume that the flux density in the air gap varies sinusoidally as $B(t) = B_0 \sin(\omega t)$. Determine the induced voltage across the coil e.

Solution

Known Quantities: Relative permeability; number of coil turns; coil current; structure geometry; flux density in air gap.

Find: L; W_m; e.

Schematics, Diagrams, Circuits, and Given Data: $\mu_r \to \infty$; $N = 500$ turns; $i = 0.1$ A. The magnetic circuit geometry is defined in Figures 18.14 and 18.15. The air gap has $l_g = 0.002$ m. $B_0 = 0.6$ Wb/m^2.

Assumptions: All magnetic flux is linked by the coil; the flux is confined to the magnetic core; the flux density is uniform. The reluctance of the magnetic structure is negligible.

Analysis:

1. To calculate the inductance of this magnetic structure, we use equation 18.30:

$$L = \frac{N^2}{\mathcal{R}}$$

Thus, we need to first calculate the reluctance. Assuming that the reluctance of the structure is negligible, we have

$$\mathcal{R}_{\text{gap}} = \frac{l_{\text{gap}}}{\mu_{\text{gap}} A_{\text{gap}}} = \frac{l_{\text{gap}}}{\mu_0 A_{\text{gap}}} = \frac{0.002}{4\pi \times 10^{-7} \times 0.0001} = 1.59 \times 10^7 \text{ A-turns/Wb}$$

and

$$L = \frac{N^2}{\mathcal{R}} = \frac{500^2}{1.59 \times 10^7} = 0.157 \text{ H}$$

Finally, we can calculate the stored magnetic energy as follows:

$$W_m = \tfrac{1}{2} L i^2 = \tfrac{1}{2} \times (0.157 \text{ H}) \times (0.1 \text{ A})^2 = 0.785 \times 10^{-3} \text{ J}$$

2. To calculate the induced voltage due to a time-varying magnetic flux, we use equation 18.16:

$$e = \frac{d\lambda}{dt} = N\frac{d\phi}{dt} = NA\frac{dB}{dt} = NAB_0\omega\cos(\omega t)$$

$$= 500 \times 0.0001 \times 0.6 \times 377\cos(377t) = 11.31\cos(377t) \qquad \text{V}$$

Comments: The voltage induced across a coil in an electromagnetic transducer is a very important quantity called the *back electromotive force*, or back emf. We shall make use of this quantity in Section 18.5.

Magnetic Reluctance Position Sensor

A simple magnetic structure, very similar to those examined in the previous examples, finds very common application in the **variable-reluctance position sensor,** which, in turn, finds widespread application in a variety of configurations for the measurement of linear and angular velocity. Figure 18.26 depicts one particular configuration that is used in many applications. In this structure, a permanent magnet with a coil of wire wound around it forms the sensor; a steel disk (typically connected to a rotating shaft) has a number of tabs that pass between the pole pieces of the sensor. The area of the tab is assumed equal to the area of the cross section of the pole pieces and is equal to a^2. The reason for the name *variable-reluctance sensor* is that the reluctance of the magnetic structure is variable, depending on whether a ferromagnetic tab lies between the pole pieces of the magnet.

FIND IT ON THE WEB

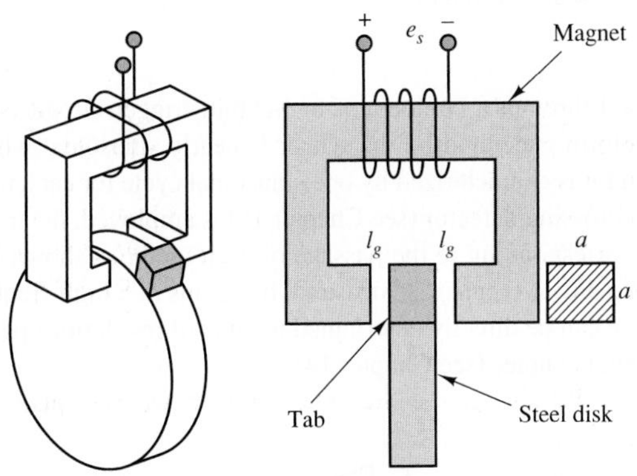

Figure 18.26 Variable-reluctance position sensor

The principle of operation of the sensor is that an electromotive force e_S is induced across the coil by the change in magnetic flux caused by the passage of the tab between the pole pieces when the disk is in motion. As the tab enters the volume between the pole pieces, the flux will increase, because of the lower reluctance of the configuration, until it reaches a maximum when the tab is centered between the poles of the magnet. Figure 18.27 depicts the approximate shape of the resulting voltage, which, according to Faraday's law, is given by

$$e_S = -\frac{d\phi}{dt}$$

The rate of change of flux is dictated by the geometry of the tab and of the pole pieces and by the speed of rotation of the disk. It is important to note that, since the flux is changing only if the disk is rotating, this sensor cannot detect the static position of the disk.

One common application of this concept is in the measurement of the speed of rotation of rotating machines, including electric motors and internal combustion engines. In these applications, use is made of a *60-tooth wheel*, which permits the conversion of the speed rotation directly to units of revolutions per minute. The output of a variable-reluctance position sensor magnetically coupled to a rotating disk equipped with 60 tabs

(*Continued*)

(Concluded)

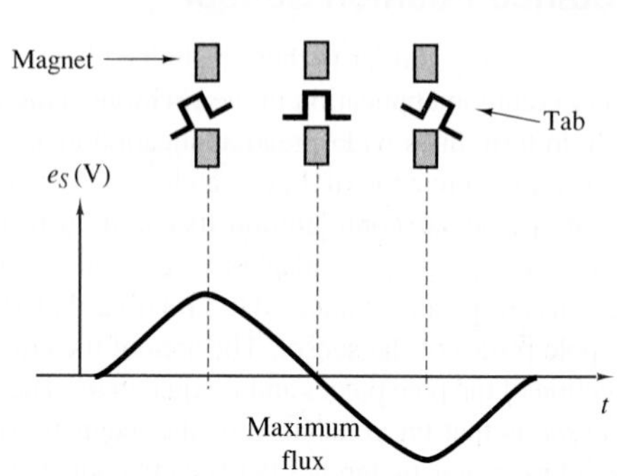

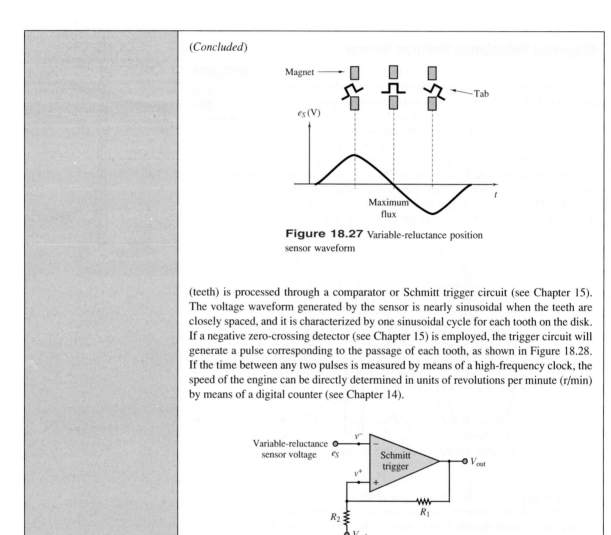

Figure 18.27 Variable-reluctance position sensor waveform

(teeth) is processed through a comparator or Schmitt trigger circuit (see Chapter 15). The voltage waveform generated by the sensor is nearly sinusoidal when the teeth are closely spaced, and it is characterized by one sinusoidal cycle for each tooth on the disk. If a negative zero-crossing detector (see Chapter 15) is employed, the trigger circuit will generate a pulse corresponding to the passage of each tooth, as shown in Figure 18.28. If the time between any two pulses is measured by means of a high-frequency clock, the speed of the engine can be directly determined in units of revolutions per minute (r/min) by means of a digital counter (see Chapter 14).

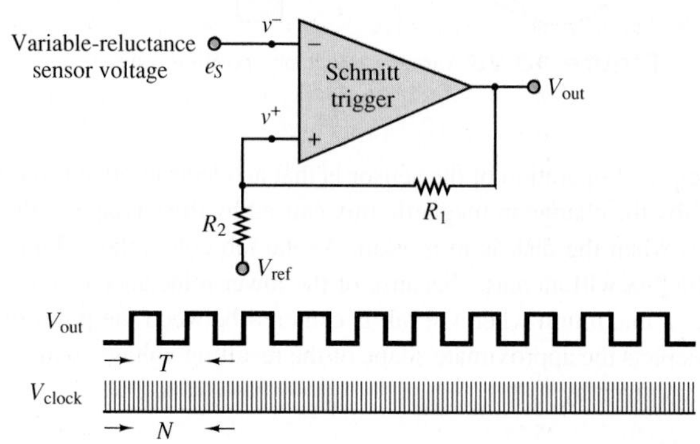

Figure 18.28 Signal processing for a 60-tooth wheel rpm sensor

Voltage Calculation in Magnetic Reluctance Position Sensor

Problem:

This example illustrates the calculation of the voltage induced in a magnetic reluctance sensor by a rotating toothed wheel. In particular, we will find an approximate expression for the reluctance and the induced voltage for the position sensor shown in Figure 18.29,

(Continued)

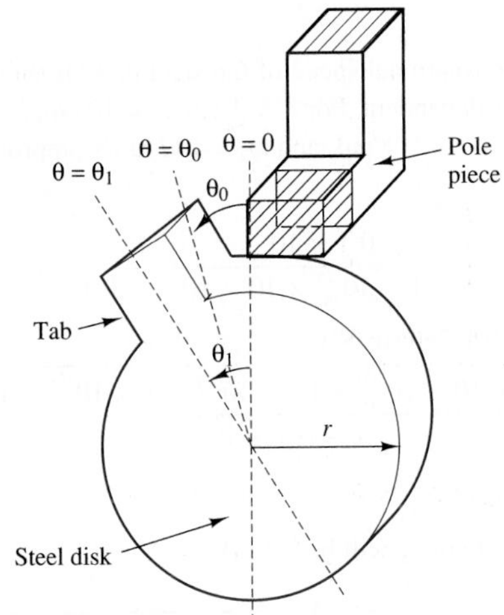

Figure 18.29 Reluctance sensor for measurement of angular position

and we will show that the induced voltage is speed-dependent. It will be assumed that the reluctance of the core and fringing at the air gaps are both negligible.

Solution:

From the geometry shown in the preceding Focus on Measurements box, the equivalent reluctance of the magnetic structure is twice that of one gap, since the permeability of the tab and the magnetic structure are assumed infinite (i.e., they have negligible reluctance). When the tab and the poles are aligned, the angle θ is zero, as shown in Figure 18.29, and the area of the air gap is maximum. For angles greater than $2\theta_0$, the magnetic length of the air gaps is so large that the magnetic field may reasonably be taken as zero.

To model the reluctance of the gaps, we assume the following simplified expression, where the area of overlap of the tab with the magnetic poles is assumed proportional to the angular displacement:

$$\mathcal{R} = \frac{2l_g}{\mu_0 A} = \frac{2l_g}{\mu_0 ar(\theta_1 - \theta)} \qquad \text{for } 0 < \theta < \theta_1$$

Naturally, this is an approximation; however, the approximation captures the essential idea of this transducer, namely, that the reluctance will decrease with increasing overlap area until it reaches a minimum, and then the reluctance will increase as the overlap area decreases. For $\theta = \theta_1$, that is, with the tab outside the magnetic pole pieces, we have $\mathcal{R}_{\text{max}} \to \infty$. For $\theta = 0$, that is, with the tab perfectly aligned with the pole pieces, we have $\mathcal{R}_{\text{min}} = 2l_g/\mu_0 ar\theta_1$. The flux ϕ may therefore be computed as follows:

$$\phi = \frac{Ni}{\mathcal{R}} = \frac{Ni\mu_0 ar(\theta_1 - \theta)}{2l_g}$$

The induced voltage e_S is found by

$$e_S = -\frac{d\phi}{dt} = -\frac{d\phi}{d\theta}\frac{d\theta}{dt} = -\frac{Ni\mu_0 ar}{2l_g}\omega$$

(Continued)

(Concluded)

where $\omega = d\theta/dt$ is the rotational speed of the steel disk. It should be evident that the induced voltage is speed-dependent. For $a = 1$ cm, $r = 10$ cm, $l_g = 0.1$ cm, $N = 1{,}000$ turns, $i = 10$ mA, $\theta_1 = 6° \approx 0.1$ rad, and $\omega = 400$ rad/s (approximately 3,800 r/min), we have

$$\mathcal{R}_{\max} = \frac{2 \times 0.1 \times 10^{-2}}{4\pi \times 10^{-7} \times 1 \times 10^{-2} \times 10 \times 10^{-2} \times 0.1}$$
$$= 1.59 \times 10^7 \text{ A-turns/Wb}$$
$$e_{S\ \text{peak}} = \frac{1{,}000 \times 10 \times 10^{-3} \times 4\pi \times 10^{-7} \times 1 \times 10^{-2} \times 10^{-1}}{2 \times 0.1 \times 10^{-2}} \times 400$$
$$= 2.5 \text{ mV}$$

That is, the peak amplitude of e_S will be 2.5 mV.

18.3 MAGNETIC MATERIALS AND *B-H* CURVES

In the analysis of magnetic circuits presented in the previous sections, the relative permeability μ_r was treated as a constant. In fact, the relationship between the magnetic flux density **B** and the associated field intensity **H**

$$\mathbf{B} = \mu \mathbf{H} \tag{18.32}$$

is characterized by the fact that the relative permeability of magnetic materials is not a constant, but is a function of the magnetic field intensity. In effect, all magnetic materials exhibit a phenomenon called **saturation,** whereby the flux density increases in proportion to the field intensity until it cannot do so any longer. Figure 18.30 illustrates the general behavior of all magnetic materials. You will note that since the B-H curve shown in the figure is nonlinear, the value of μ (which is the slope of the curve) depends on the intensity of the magnetic field.

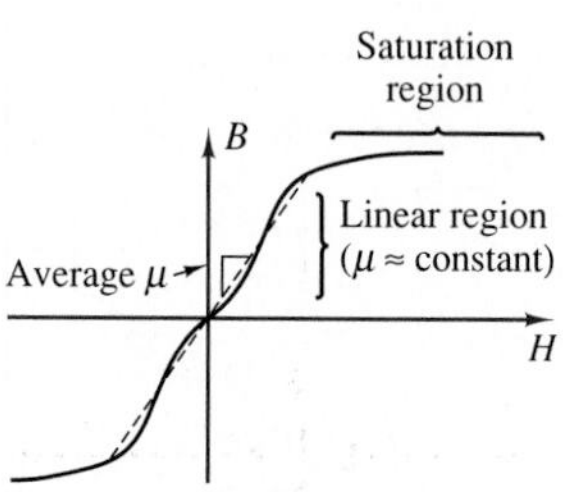

Figure 18.30 Permeability and magnetic saturation effects

To understand the reasons for the saturation of a magnetic material, we need to briefly review the mechanism of magnetization. The basic idea behind magnetic materials is that the spin of electrons constitutes motion of charge, and therefore leads to magnetic effects, as explained in the introductory section of this chapter. In most materials, the electron spins cancel out, on the whole, and no net effect remains. In ferromagnetic materials, on the other hand, atoms can align so that the electron spins cause a net magnetic effect. In such materials, there exist small regions with strong magnetic properties, called **magnetic domains,** the effects of which are neutralized in unmagnetized material by other, similar regions that are oriented differently, in a random pattern. When the material is magnetized, the magnetic domains tend to align with one another, to a degree that is determined by the intensity of the applied magnetic field.

In effect, the large number of miniature magnets within the material is *polarized* by the external magnetic field. As the field increases, more and more domains become aligned. When all the domains have become aligned, any further increase in magnetic field intensity does not yield an increase in flux density beyond the increase that would be caused in a nonmagnetic material. Thus, the relative permeability μ_r approaches 1 in the saturation region. It should be apparent that an exact value of μ_r

cannot be determined; the value of μ_r used in the earlier examples is to be interpreted as an average permeability, for intermediate values of flux density. As a point of reference, commercial magnetic steels saturate at flux densities around a few teslas. Figure 18.33, shown later in this section, will provide some actual B-H curves for common ferromagnetic materials.

The phenomenon of saturation carries some interesting implications with regard to the operation of magnetic circuits: The results of the previous section would seem to imply that an increase in the mmf (i.e., an increase in the current driving the coil) would lead to a proportional increase in the magnetic flux. This is true in the *linear region* of Figure 18.30; however, as the material reaches saturation, further increases in the driving current (or, equivalently, in the mmf) do not yield further increases in the magnetic flux.

There are two more features that cause magnetic materials to further deviate from the ideal model of the linear B-H relationship: **eddy currents** and **hysteresis.** The first phenomenon consists of currents that are caused by any time-varying flux in the core material. As you know, a time-varying flux will induce a voltage, and therefore a current. When this happens inside the magnetic core, the induced voltage will cause *eddy* currents (the terminology should be self-explanatory) in the core, which depend on the resistivity of the core. Figure 18.31 illustrates the phenomenon of eddy currents. The effect of these currents is to dissipate energy in the form of heat. Eddy currents are reduced by selecting high-resistivity core materials, or by *laminating* the core, introducing tiny, discontinuous air gaps between core layers (see Figure 18.31). Lamination of the core reduces eddy currents greatly without affecting the magnetic properties of the core.

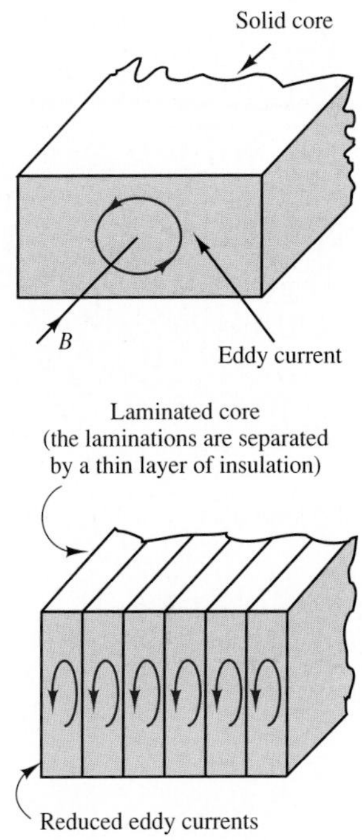

Figure 18.31 Eddy currents in magnetic structures

It is beyond the scope of this chapter to quantify the losses caused by induced eddy currents, but it will be important in Chapters 19 and 20 to be aware of this source of energy loss.

Hysteresis is another loss mechanism in magnetic materials; it displays a rather complex behavior, related to the magnetization properties of a material. The curve of Figure 18.32 reveals that the B-H curve for a magnetic material during magnetization (as H is increased) is displaced with respect to the curve that is measured when the material is demagnetized. To understand the hysteresis process, consider a core that has been energized for some time, with a field intensity of H_1 A-turns/m. As the current required to sustain the mmf corresponding to H_1 is decreased, we follow the hysteresis curve from the point α to the point β. When the mmf is exactly zero, the material displays the **remanent** (or **residual**) **magnetization** B_r. To bring the flux density to zero, we must further decrease the mmf (i.e., produce a negative current) until the field intensity reaches the value $-H_0$ (point γ on the curve). As the mmf is made more negative, the curve eventually reaches the point α'. If the excitation current to the coil is now increased, the magnetization curve will follow the path $\alpha' = \beta' = \gamma' = \alpha$, eventually returning to the original point in the B-H plane, but via a different path.

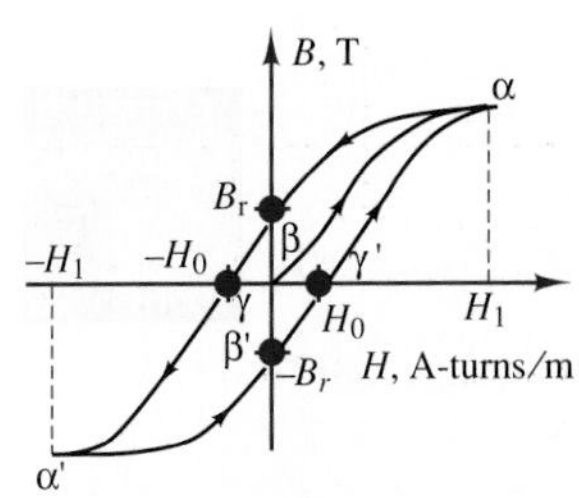

Figure 18.32 Hysteresis in magnetization curves

The result of this process, by which an *excess magnetomotive force* is required to magnetize or demagnetize the material, is a net energy loss. It is difficult to evaluate this loss exactly; however, it can be shown that it is related to the area between the curves of Figure 18.32. There are experimental techniques that enable the approximate measurement of these losses.

Figure 18.33(a) through (c) depicts magnetization curves for three very common ferromagnetic materials: cast iron, cast steel, and sheet steel. These curves will be useful in solving some of the homework problems.

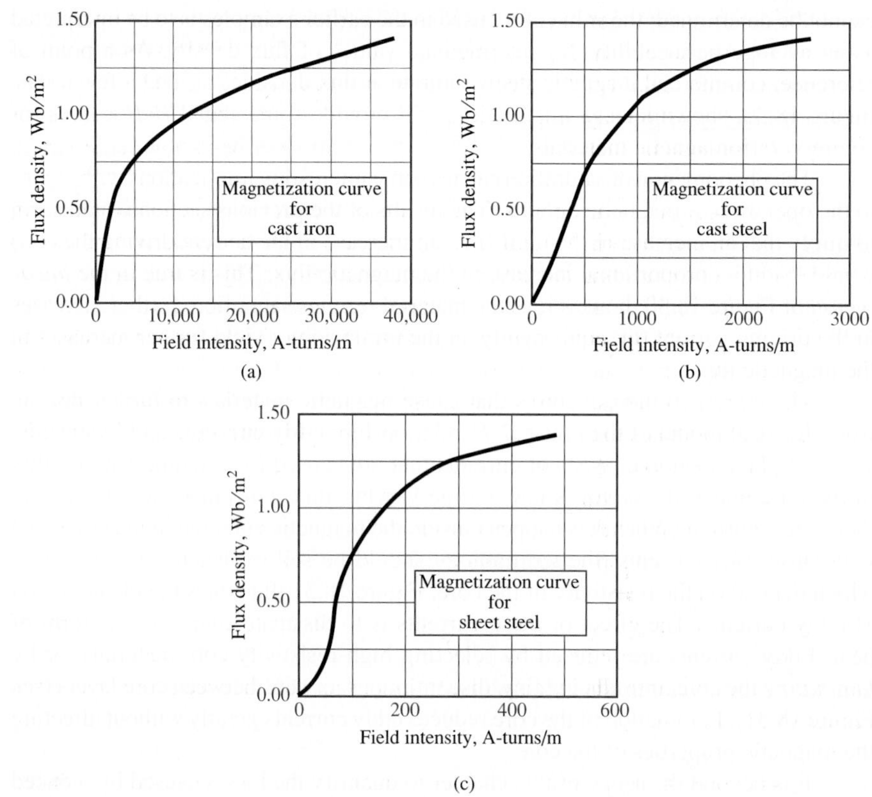

Figure 18.33 (a) Magnetization curve for cast iron; (b) magnetization curve for cast steel; (c) magnetization curve for sheet steel

18.4 TRANSFORMERS

FIND IT ON THE WEB

One of the more common magnetic structures in everyday applications is the **transformer.** The ideal transformer was introduced in Chapter 7 as a device that can step an AC voltage up or down by a fixed ratio, with a corresponding decrease or increase in current. The structure of a simple magnetic transformer is shown in Figure 18.34, which illustrates that a transformer is very similar to the magnetic circuits described earlier in this chapter. Coil L_1 represents the input side of the transformer, while coil L_2 is the output coil; both coils are wound around the same magnetic structure, which we show here to be similar to the "square doughnut" of the earlier examples.

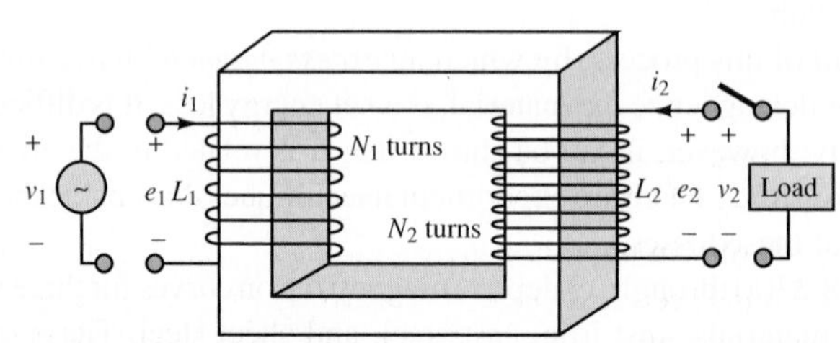

Figure 18.34 Structure of a transformer

The ideal transformer operates on the basis of the same set of assumptions we made in earlier sections: The flux is confined to the core, the flux links all turns of both coils, and the permeability of the core is infinite. The last assumption is equivalent to stating that an arbitrarily small mmf is sufficient to establish a flux in the core. In addition, we assume that the ideal transformer coils offer negligible resistance to current flow.

The operation of a transformer requires a time-varying current; if a time-varying voltage is applied to the primary side of the transformer, a corresponding current will flow in L_1; this current acts as an mmf and causes a (time-varying) flux in the structure. But the existence of a changing flux will induce an emf across the secondary coil! Without the need for a direct electrical connection, the transformer can couple a source voltage at the primary to the load; the coupling occurs by means of the magnetic field acting on both coils. Thus, a transformer operates by converting electric energy to magnetic, and then back to electric. The following derivation illustrates this viewpoint in the ideal case (no loss of energy) and compares the result with the definition of the ideal transformer in Chapter 7.

If a time-varying voltage source is connected to the input side, then by virtue of Faraday's law, a corresponding time-varying flux $d\phi/dt$ is established in coil L_1:

$$e_1 = N_1 \frac{d\phi}{dt} = v_1 \tag{18.33}$$

But since the flux thus produced also links coil L_2, an emf is induced across the output coil as well:

$$e_2 = N_2 \frac{d\phi}{dt} = v_2 \tag{18.34}$$

This induced emf can be measured as the voltage v_2 at the output terminals, and one can readily see that the ratio of the open-circuit output voltage to input-terminal voltage is

$$\frac{v_2}{v_1} = \frac{N_2}{N_1} = N \tag{18.35}$$

If a load current i_2 is now required by the connection of a load to the output circuit (by closing the switch in the figure), the corresponding mmf is $\mathcal{F}_2 = N_2 i_2$. This mmf, generated by the load current i_2, would cause the flux in the core to change; however, this is not possible, since a change in ϕ would cause a corresponding change in the voltage induced across the input coil. But this voltage is determined (fixed) by the source v_1 (and is therefore $d\phi/dt$), so that the input coil is forced to generate a **counter-mmf** to oppose the mmf of the output coil; this is accomplished as the input coil draws a current i_1 from the source v_1 such that

$$i_1 N_1 = i_2 N_2 \tag{18.36}$$

or

$$\frac{i_2}{i_1} = \frac{N_1}{N_2} = \alpha = \frac{1}{N} \tag{18.37}$$

where α is the ratio of primary to secondary turns (the transformer ratio) and N_1 and N_2 are the primary and secondary turns, respectively. If there were any net difference between the input and output mmf, the flux balance required by the input voltage source would not be satisfied. Thus, the two magnetomotive forces must be equal.

As you can easily verify, these results are the same as in Chapter 7; in particular, the ideal transformer does not dissipate any power, since

$$v_1 i_1 = v_2 i_2 \tag{18.38}$$

Note the distinction we have made between the induced voltages (emf's) e and the terminal voltages v. In general, these are not the same.

The results obtained for the ideal case do not completely represent the physical nature of transformers. A number of loss mechanisms need to be included in a practical transformer model, to account for the effects of leakage flux, for various magnetic core losses (e.g., hysteresis), and for the unavoidable resistance of the wires that form the coils.

Commercial transformer ratings are usually given on the **nameplate,** which indicates the normal operating conditions. The nameplate includes the following parameters:

- Primary-to-secondary voltage ratio
- Design frequency of operation
- (Apparent) rated output power

For example, a typical nameplate might read 480:240 V, 60 Hz, 2 kVA. The voltage ratio can be used to determine the turns ratio, while the rated output power represents the continuous power level that can be sustained without overheating. It is important that this power be rated as the apparent power in kilovoltamperes, rather than real power in kilowatts, since a load with low power factor would still draw current and therefore operate near rated power. Another important performance characteristic of a transformer is its **power efficiency,** defined by

$$\text{Power efficiency } \eta = \frac{\text{Output power}}{\text{Input power}} \tag{18.39}$$

Examples 18.7 and 18.8 illustrate the use of the nameplate ratings and the calculation of efficiency in a practical transformer, in addition to demonstrating the application of the circuit models.

LO4

EXAMPLE 18.7 Transformer Nameplate

Problem

Determine the turns ratio and the rated currents of a transformer from nameplate data.

Solution

Known Quantities: Nameplate data.

Find: $\alpha = N_1/N_2$; I_1; I_2.

Schematics, Diagrams, Circuits, and Given Data: Nameplate data: 120 V/480 V; 48 kVA; 60 Hz.

Assumptions: Assume an ideal transformer.

Analysis: The first element in the nameplate data is a pair of voltages, indicating the primary and secondary voltages for which the transformer is rated. The ratio α is found as follows:

$$\alpha = \frac{N_1}{N_2} = \frac{480}{120} = 4$$

To find the primary and secondary currents, we use the kilovoltampere rating (apparent power) of the transformer:

$$I_1 = \frac{|S|}{V_1} = \frac{48 \text{ kVA}}{480 \text{ V}} = 100 \text{ A} \qquad I_2 = \frac{|S|}{V_2} = \frac{48 \text{ kVA}}{120 \text{ V}} = 400 \text{ A}$$

Comments: In computing the rated currents, we have assumed that no losses take place in the transformer; in fact, there will be losses due to coil resistance and magnetic core effects. These losses result in heating of the transformer and limit its rated performance.

CHECK YOUR UNDERSTANDING

The high-voltage side of a transformer has 500 turns, and the low-voltage side has 100 turns. When the transformer is connected as a step-down transformer, the load current is 12 A. Calculate (a) the turns ratio α and (b) the primary current. (c) Calculate the turns ratio if the transformer is used as a step-up transformer.

The output of a transformer under certain conditions is 12 kW. The copper losses are 189 W, and the core losses are 52 W. Calculate the efficiency of this transformer.

Answers: (a) $\alpha = 5$; (b) $I_1 = I_2/\alpha = 2.4$ A; (c) $\alpha = 0.2$; $\eta = 98$ percent

EXAMPLE 18.8 Impedance Transformer

LO4

Problem

Find the equivalent load impedance seen by the voltage source (i.e., reflected from secondary to primary) for the transformer of Figure 18.35.

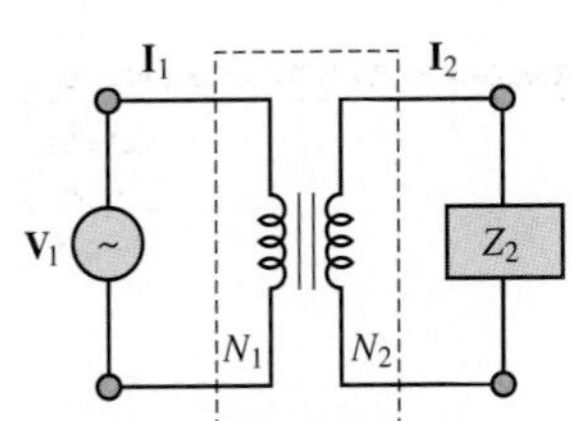

Figure 18.35 Ideal transformer

Solution

Known Quantities: Transformer turns ratio α.

Find: Reflected impedance Z_2'.

Assumptions: Assume an ideal transformer.

Analysis: By definition, the load impedance is equal to the ratio of secondary phasor voltage and current:

$$Z_2 = \frac{\mathbf{V}_2}{\mathbf{I}_2}$$

To find the reflected impedance, we can express the above ratio in terms of primary voltage and current:

$$Z_2 = \frac{\mathbf{V}_2}{\mathbf{I}_2} = \frac{\mathbf{V}_1/\alpha}{\alpha \mathbf{I}_1} = \frac{1}{\alpha^2}\frac{\mathbf{V}_1}{\mathbf{I}_1}$$

where the ratio $\mathbf{V}_1/\mathbf{I}_1$ is the impedance seen by the source at the primary coil, that is, the *reflected load impedance* seen by the primary (source) side of the circuit. Thus, we can write the load impedance Z_2 in terms of the primary circuit voltage and current; we call this the *reflected impedance* Z'_2:

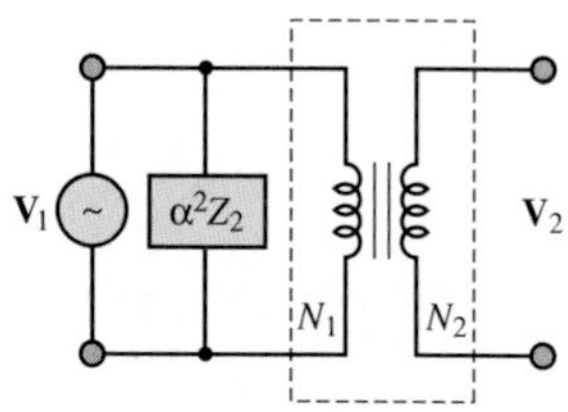

Figure 18.36

$$Z_2 = \frac{1}{\alpha^2}\frac{\mathbf{V}_1}{\mathbf{I}_1} = \frac{1}{\alpha^2}Z_1 = \frac{1}{\alpha^2}Z'_2$$

Thus, $Z'_2 = \alpha^2 Z^2$. Figure 18.36 depicts the equivalent circuit with the load impedance reflected back to the primary.

Comments: The equivalent reflected circuit calculations are convenient because all circuit elements can be referred to a single set of variables (i.e., only primary or secondary voltages and currents).

CHECK YOUR UNDERSTANDING

The output impedance of a servo amplifier is 250 Ω. The servomotor that the amplifier must drive has an impedance of 2.5 Ω. Calculate the turns ratio of the transformer required to match these impedances.

Answer: $\alpha = 10$

18.5 ELECTROMECHANICAL ENERGY CONVERSION

From the material developed thus far, it should be apparent that electromagneto-mechanical devices are capable of converting mechanical forces and displacements to electromagnetic energy, and that the converse is also possible. The objective of this section is to formalize the basic principles of energy conversion in electromagnetomechanical systems, and to illustrate its usefulness and potential for application by presenting several examples of **energy transducers.** A transducer is a device that can convert electrical to mechanical energy (in this case, it is often called an **actuator**), or vice versa (in which case it is called a **sensor**).

Several physical mechanisms permit conversion of electrical to mechanical energy and back, the principal phenomena being the **piezoelectric effect,**[3] consisting of the generation of a change in electric field in the presence of strain in certain crystals (e.g., quartz), and **electrostriction** and **magnetostriction,** in which changes in the dimension of certain materials lead to a change in their electrical (or magnetic) properties. Although these effects lead to many interesting applications, this chapter is concerned only with transducers in which electric energy is converted to mechanical energy through the coupling of a magnetic field. It is important to note that all rotating machines (motors and generators) fit the basic definition of electromechanical transducers we have just given.

[3]See the Focus on Measurements box "Charge Amplifiers" in Chapter 8.

Forces in Magnetic Structures

Mechanical forces can be converted to electric signals, and vice versa, by means of the coupling provided by energy stored in the magnetic field. In this subsection, we discuss the computation of mechanical forces and of the corresponding electromagnetic quantities of interest; these calculations are of great practical importance in the design and application of electromechanical actuators. For example, a problem of interest is the computation of the current required to generate a given force in an electromechanical structure. This is the kind of application that is likely to be encountered by the engineer in the selection of an electromechanical device for a given task.

As already seen in this chapter, an electromechanical system includes an electrical system and a mechanical system, in addition to means through which the two can interact. The principal focus of this chapter has been the coupling that occurs through an electromagnetic field common to both the electrical system and the mechanical system; to understand electromechanical energy conversion, it will be important to understand the various energy storage and loss mechanisms in the electromagnetic field. Figure 18.37 illustrates the coupling between the electrical and mechanical systems. In the mechanical system, energy loss can occur because of the heat developed as a consequence of *friction*, while in the electrical system, analogous losses are incurred because of *resistance*. Loss mechanisms are also present in the magnetic coupling medium, since *eddy current losses* and *hysteresis losses* are unavoidable in ferromagnetic materials. Either system can supply energy, and either system can store energy. Thus, the figure depicts the flow of energy from the electrical to the mechanical system, accounting for these various losses. The same flow could be reversed if mechanical energy were converted to electrical form.

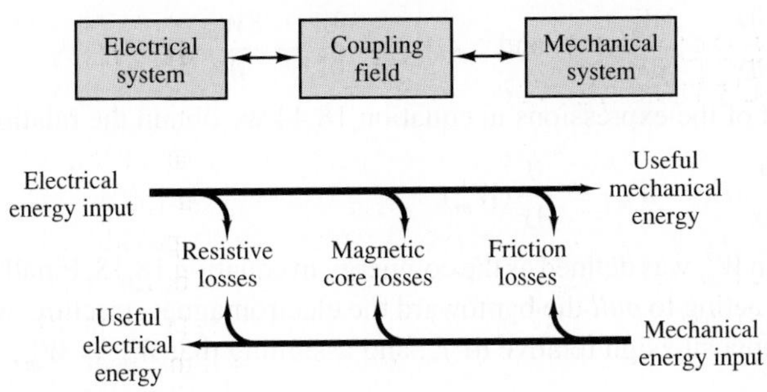

Figure 18.37

Moving-Iron Transducers

One important class of electromagnetomechanical transducers is that of **moving-iron transducers.** The aim of this section is to derive an expression for the magnetic forces generated by such transducers and to illustrate the application of these calculations to simple, yet common devices such as electromagnets, solenoids, and relays. The simplest example of a moving-iron transducer is the **electromagnet** of Figure 18.38, in which the U-shaped element is fixed and the bar is movable. In the following paragraphs, we shall derive a relationship between the current applied to the coil, the displacement of the movable bar, and the magnetic force acting in the air gap.

FIND IT ON THE WEB

The principle that will be applied throughout the section is that in order for a mass to be displaced, some work needs to be done; this work corresponds to a change in the energy stored in the electromagnetic field, which causes the mass to be

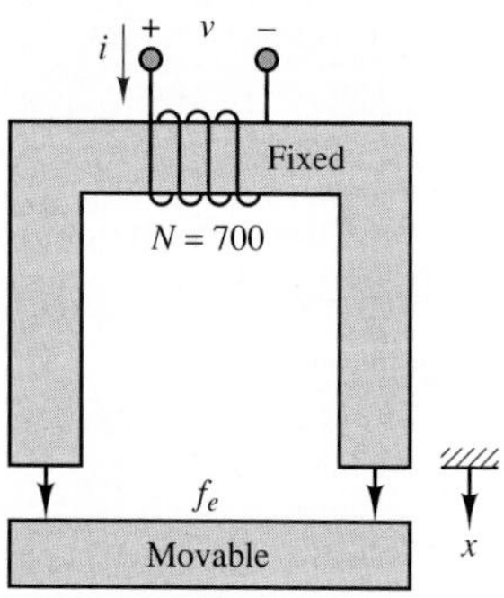

Figure 18.38

displaced. With reference to Figure 18.38, let f_e represent the magnetic force acting on the bar and x the displacement of the bar, in the direction shown. Then the net work into the electromagnetic field W_m is equal to the sum of the work done by the electric circuit plus the work done by the mechanical system. Therefore, for an incremental amount of work, we can write

$$dW_m = ei\,dt - f_e\,dx \tag{18.40}$$

where e is the electromotive force across the coil and the minus sign is due to the sign convention indicated in Figure 18.38. Recalling that the emf e is equal to the derivative of the flux linkage (equation 18.16), we can further expand equation 16.40 to obtain

$$dW_m = ei\,dt - f_e\,dx = i\frac{d\lambda}{dt}dt - f_e\,dx = i\,d\lambda - f_e\,dx \tag{18.41}$$

or

$$f_e\,dx = i\,d\lambda - dW_m \tag{18.42}$$

Now we must observe that the flux in the magnetic structure of Figure 18.38 depends on two variables, which are in effect independent: the current flowing through the coil and the displacement of the bar. Each of these variables can cause the magnetic flux to change. Similarly, the energy stored in the electromagnetic field is also dependent on both current and displacement. Thus we can rewrite equation 18.42 as follows:

$$f_e\,dx = i\left(\frac{\partial\lambda}{\partial i}di + \frac{\partial\lambda}{\partial x}dx\right) - \left(\frac{\partial W_m}{\partial i}di + \frac{\partial W_m}{\partial x}dx\right) \tag{18.43}$$

Since i and x are independent variables, we can write

$$f_e = i\frac{\partial\lambda}{\partial x} - \frac{\partial W_m}{\partial x} \qquad \text{and} \qquad 0 = i\frac{\partial\lambda}{\partial i} - \frac{\partial W_m}{\partial i} \tag{18.44}$$

From the first of the expressions in equation 18.44 we obtain the relationship

$$f_e = \frac{\partial}{\partial x}(i\lambda - W_m) = \frac{\partial}{\partial x}(W'_m) \tag{18.45}$$

where the term W'_m was defined as the co-energy in equation 18.18. Finally, we observe that the force acting to *pull* the bar toward the electromagnet structure, which we will call f, is of opposite sign relative to f_e, and assuming that $W_m = W'_m$, we can write

$$f = -f_e = -\frac{\partial}{\partial x}(W'_m) = -\frac{\partial W_m}{\partial x} \tag{18.46}$$

Equation 18.46 includes a very important assumption: that the energy is equal to the co-energy. If you refer to Figure 18.8, you will realize that in general this is not true. Energy and co-energy are equal only if the λ-i relationship is linear. Thus, the useful result of equation 18.46, stating that the magnetic force acting on the moving iron is proportional to the rate of change of stored energy with displacement, applies only for *linear magnetic structures*.

Thus, to determine the forces present in a magnetic structure, it will be necessary to compute the energy stored in the magnetic field. To simplify the analysis, it will be assumed hereafter that the structures analyzed are magnetically linear. This is, of course, only an approximation, in that it neglects a number of practical aspects of electromechanical systems (e.g., the nonlinear λ-i curves described earlier, and the core losses typical of magnetic materials), but it permits relatively simple analysis of many useful magnetic structures. Thus, although the analysis method presented in this section is only approximate, it will serve the purpose of providing a feeling for the direction and the magnitude of the forces and currents present in electromechanical devices. On the basis of a linear approximation, it can be shown that the stored energy in a magnetic structure is given by

$$W_m = \frac{\phi \mathcal{F}}{2} \tag{18.47}$$

and since the flux and the mmf are related by the expression

$$\phi = \frac{Ni}{\mathcal{R}} = \frac{\mathcal{F}}{\mathcal{R}} \tag{18.48}$$

the stored energy can be related to the reluctance of the structure according to

$$W_m = \frac{\phi^2 \mathcal{R}(x)}{2} \tag{18.49}$$

where the reluctance has been explicitly shown to be a function of displacement, as is the case in a moving-iron transducer. Finally, then, we shall use the following approximate expression to compute the magnetic force acting on the moving iron:

$$f = -\frac{dW_m}{dx} = -\frac{\phi^2}{2}\frac{d\mathcal{R}(x)}{dx} \qquad \text{magnetic force} \tag{18.50}$$

LO5

The examples 18.9, 18.10, and 18.12 illustrate the application of this approximate technique for the computation of forces and currents (the two problems of practical engineering interest to the user of such electromechanical systems) in some common devices. The Focus on Methodology box outlines the solution techniques for these classes of problems.

FOCUS ON METHODOLOGY

LO5

ANALYSIS OF MOVING-IRON ELECTROMECHANICAL TRANSDUCERS

a. Calculation of current required to generate a given force

1. Derive an expression for the reluctance of the structure as a function of air gap displacement: $\mathcal{R}(x)$.
2. Express the magnetic flux in the structure as a function of the mmf (i.e., of the current I) and of the reluctance $\mathcal{R}(x)$:

$$\phi = \frac{\mathcal{F}(i)}{\mathcal{R}(x)}$$

3. Compute an expression for the force, using the known expressions for the flux and for the reluctance:

$$|f| = \frac{\phi^2}{2}\frac{d\mathcal{R}(x)}{dx}$$

(Continued)

FOCUS ON METHODOLOGY

(Concluded)

4. Solve the expression in step 3 for the unknown current i.

b. Calculation of force generated by a given transducer geometry and mmf

Repeat steps 1 through 3 above, substituting the known current to solve for the force f.

LO5

EXAMPLE 18.9 An Electromagnet

Problem

An electromagnet is used to collect and support a solid piece of steel, as shown in Figure 18.38. Calculate the *starting current* required to lift the load and the *holding current* required to keep the load in place once it has been lifted and is attached to the magnet.

Solution

Known Quantities: Geometry, magnetic permeability, number of coil turns, mass, acceleration of gravity, initial position of steel bar.

Find: Current required to lift the bar; current required to hold the bar in place.

Schematics, Diagrams, Circuits, and Given Data:

$N = 500$

$\mu_0 = 4\pi \times 10^{-7}$

$\mu_r = 10^4$ (equal for electromagnet and load)

Initial distance (air gap) = 0.5 m

Magnetic path length of electromagnet = $l_1 = 0.60$ m

Magnetic path length of movable load = $l_2 = 0.30$ m

Gap cross sectional area = 3×10^{-4} m^2

m = mass of load = 5 kg

$g = 9.8$ m/s^2

Assumptions: None.

Analysis: To compute the current we need to derive an expression for the force in the air gap. We use the equation

$$f_{\text{mech}} = \frac{\phi^2}{2} \frac{\partial \mathcal{R}(x)}{\partial x}$$

and calculate the reluctance, flux and force as follows:

$$\mathcal{R}(x) = \mathcal{R}_{Fe} + \mathcal{R}_{\text{gap}}$$

$$\mathcal{R}(x) = \frac{2x}{\mu_0 A} + \frac{l_1 + l_2}{\mu_0 \mu_r A}$$

$$\phi = \frac{\mathcal{F}}{\mathcal{R}(x)} = \frac{Ni}{\left(\frac{2x}{\mu_0 A} + \frac{l_1+l_2}{\mu_0 \mu_r A}\right)}$$

$$\frac{\partial \mathcal{R}(x)}{\partial x} = \frac{2}{\mu_0 A} \Rightarrow f_{\text{mag}} = \frac{\phi^2}{2}\frac{\partial \mathcal{R}(x)}{\partial x} = \frac{(Ni)^2}{\left(\frac{2x}{\mu_0 A} + \frac{l_1+l_2}{\mu_0 \mu_r A}\right)^2}\frac{1}{\mu_0 A}$$

With this expression we can now calculate the current required to overcome the gravitational force when the load is 0.5 m away. The force we must overcome is $mg = 98$ N.

$$f_{\text{mag}} = \frac{(Ni)^2}{\left(\frac{2x}{\mu_0 A} + \frac{l_1+l_2}{\mu_0 \mu_r A}\right)^2}\frac{1}{\mu_0 A} = f_{\text{gravity}}$$

$$i^2 = f_{\text{gravity}} \frac{\frac{\mu_0 A}{2}\left(\frac{2x}{\mu_0 A} + \frac{l_1+l_2}{\mu_0 \mu_r A}\right)^2}{N^2} = 6.5 \times 10^4 \text{ A}^2 \quad i = 255 \text{ A}$$

Finally, we calculate the holding current by letting $x = 0$:

$$f_{\text{mag}} = \frac{(Ni)^2}{\left(\frac{l_1+l_2}{\mu_0 \mu_r A}\right)^2}\frac{1}{\mu_0 A} = f_{\text{gravity}}$$

$$i^2 = f_{\text{gravity}} \frac{\frac{\mu_0 A}{2}\left(\frac{l_1+l_2}{\mu_0 \mu_r A}\right)^2}{N^2} = 2.1056 \times 10^{-3} \text{ A}^2$$

$$i = 0.0459 \text{ A}$$

Comments: Note how much smaller the holding current is than the lifting current.

One of the more common practical applications of the concepts discussed in this section is the **solenoid.** Solenoids find application in a variety of electrically controlled valves. The action of a solenoid valve is such that when it is energized, the plunger moves in such a direction as to permit the flow of a fluid through a conduit, as shown schematically in Figure 18.39.

FIND IT ON THE WEB

Examples 18.10 and 18.11 illustrate the calculations involved in the determination of forces and currents in a solenoid.

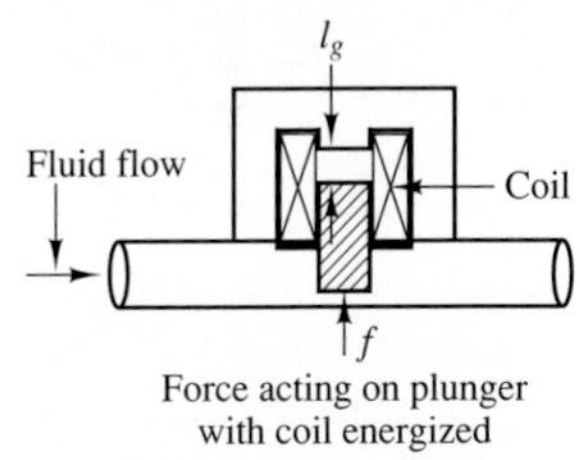

Figure 18.39 Application of the solenoid as a valve

EXAMPLE 18.10 A Solenoid

LO5

Problem

Figure 18.40 depicts a simplified representation of a solenoid. The restoring force for the plunger is provided by a spring.

1. Derive a general expression for the force exerted on the plunger as a function of the plunger position x.
2. Determine the mmf required to pull the plunger to its end position ($x = a$).

Solution

Known Quantities: Geometry of magnetic structure; spring constant.

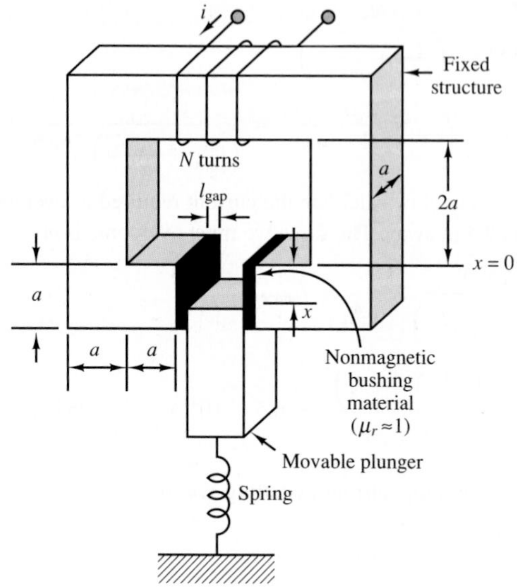

Figure 18.40 A solenoid

Find: f; mmf.

Schematics, Diagrams, Circuits, and Given Data: $a = 0.01$ m; $l_{gap} = 0.001$ m; $k = 10$ N/m.

Assumptions: Assume that the reluctance of the iron is negligible; neglect fringing. At $x = 0$ the plunger is in the gap by an infinitesimal displacement ε.

Analysis:

1. *Force on the plunger.* To compute a general expression for the magnetic force exerted on the plunger, we need to derive an expression for the force in the air gap. Using equation 18.50, we see that we need to compute the reluctance of the structure and the magnetic flux to derive an expression for the force.

   Since we are neglecting the iron reluctance, we can write the expression for the reluctance as follows. Note that the area of the gap is variable, depending on the position of the plunger, as shown in Figure 18.41.

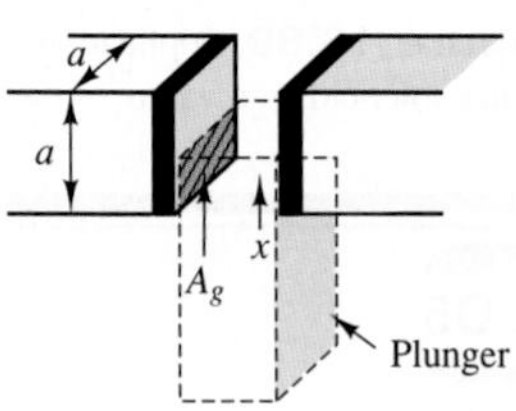

Figure 18.41

$$\mathcal{R}_{gap}(x) = 2 \times \frac{l_{gap}}{\mu_0 A_{gap}} = \frac{2l_{gap}}{\mu_0 a x}$$

The derivative of the reluctance with respect to the displacement of the plunger can then be computed to be

$$\frac{d\mathcal{R}_{gap}(x)}{dx} = \frac{-2l_{gap}}{\mu_0 a x^2}$$

Knowing the reluctance, we can calculate the magnetic flux in the structure as a function of the coil current:

$$\phi = \frac{Ni}{\mathcal{R}(x)} = \frac{Ni\mu_0 a x}{2l_{gap}}$$

The force in the air gap is given by

$$f_{gap} = \frac{\phi^2}{2}\frac{d\mathcal{R}(x)}{dx} = \frac{(Ni\mu_0 a x)^2}{8l_{gap}^2}\frac{-2l_{gap}}{\mu_0 a x^2} = -\frac{\mu_0 a (Ni)^2}{4l_{gap}}$$

Thus, the force in the gap is proportional to the square of the current and does not vary with plunger displacement.

2. *Calculation of magnetomotive force*. To determine the required magnetomotive force, we observe that the magnetic force must overcome the mechanical (restoring) force generated by the spring. Thus, $f_{gap} = kx = ka$. For the stated values, $f_{gap} = (10 \text{ N/m}) \times (0.01 \text{ m}) = 0.1$ N, and

$$Ni = \sqrt{\frac{4l_{gap} f_{gap}}{\mu_0 a}} = \sqrt{\frac{4 \times 0.001 \times 0.1}{4\pi \times 10^{-7} \times 0.01}} = 56.4 \text{ A-turns}$$

The required mmf can be most effectively realized by keeping the current value relatively low and using a large number of turns.

Comments: The same mmf can be realized with an infinite number of combinations of current and number of turns; however, there are tradeoffs involved. If the current is very large (and the number of turns small), the required wire diameter will be very large. Conversely, a small current will require a small wire diameter and a large number of turns. A homework problem explores this tradeoff.

CHECK YOUR UNDERSTANDING

A solenoid is used to exert force on a spring. Estimate the position of the plunger if the number of turns in the solenoid winding is 1,000 and the current going into the winding is 40 mA. Use the same values as in Example 18.10 for all other variables.

Answer: $x = 0.5$ cm

EXAMPLE 18.11 Transient Response of a Solenoid

LO5

Problem

Analyze the current response of the solenoid of Example 18.10 to a step change in excitation voltage. Plot the force and current as a function of time.

Solution

Known Quantities: Coil inductance and resistance; applied current.

Find: Current and force response as a function of time.

Schematics, Diagrams, Circuits, and Given Data: See Example 18.10. $N = 1{,}000$ turns. $V = 12$ V. $R_{coil} = 5\ \Omega$.

Assumptions: The inductance of the solenoid is approximately constant and is equal to the midrange value (plunger displacement equal to $a/2$).

Analysis: From Example 18.10, we have an expression for the reluctance of the solenoid:

$$\mathcal{R}_{gap}(x) = \frac{2l_{gap}}{\mu_0 ax}$$

Using equation 18.30 and assuming $x = a/2$, we calculate the inductance of the structure:

$$L \approx \frac{N^2}{\mathcal{R}_{\text{gap}}|_{x=a/2}} = \frac{N^2 \mu_0 a^2}{4 l_{\text{gap}}} = \frac{10^6 \times 4\pi \times 10^{-7} \times 10^{-4}}{4 \times 10^{-3}} = 31.4 \text{ mH}$$

The equivalent solenoid circuit is shown in Figure 18.42. When the switch is closed, the solenoid current rises exponentially with time constant $\tau = L/R = 6.3$ ms. As shown in Chapter 5, the response is of the form

$$i(t) = \frac{V}{R}(1 - e^{-t/\tau}) = \frac{V}{R}(1 - e^{-Rt/L}) = \frac{12}{5}(1 - e^{-t/6.3\times 10^{-3}}) \quad \text{A}$$

To determine how the magnetic force responds during the turn-on transient, we return to the expression for the force derived in Example 18.10:

$$f_{\text{gap}}(t) = \frac{\mu_0 a (Ni)^2}{4 l_{\text{gap}}} = \frac{4\pi \times 10^{-7} \times 10^{-2} \times 10^6}{4 \times 10^{-3}} i^2(t) = \pi i^2(t)$$

$$= \pi \left[\frac{12}{5}(1 - e^{-t/6.3\times 10^{-3}}) \right]^2$$

The two curves are plotted in Figure 18.42(b).

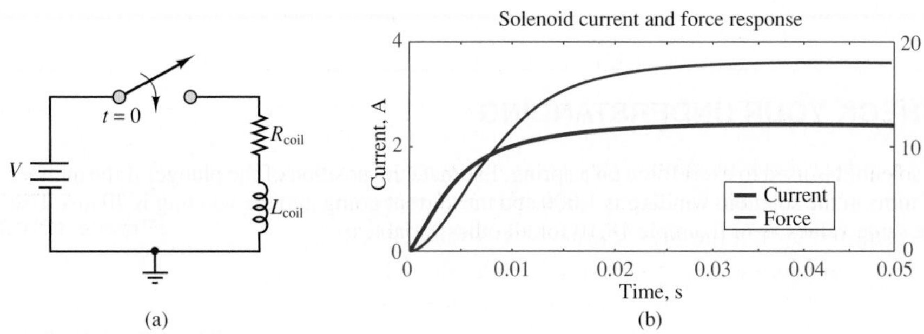

Figure 18.42 Solenoid equivalent electric circuit and step response

Comments: The assumption that the inductance is approximately constant is not quite accurate. The reluctance (and therefore the inductance) of the structure will change as the plunger moves into position. However, allowing for the inductance to be a function of plunger displacement causes the problem to become nonlinear, and requires numerical solution of the differential equation (i.e., the transient response results of Chapter 5 no longer apply). This issue is explored in the homework problems.

Practical Facts About Solenoids

Solenoids can be used to produce linear or rotary motion, in either the *push* or the *pull* mode. The most common solenoid types are listed here:

1. *Single-action linear* (push or pull). Linear stroke motion, with a restoring force (e.g., from a spring), to return the solenoid to the neutral position.
2. *Double-acting linear.* Two solenoids back to back can act in either direction. Restoring force is provided by another mechanism (e.g., a spring).
3. *Mechanical latching solenoid* (bistable). An internal latching mechanism holds the solenoid in place against the load.

(*Continued*)

(Concluded)

4. *Keep solenoid.* Fitted with a permanent magnet so that no power is needed to hold the load in the pulled-in position. Plunger is released by applying a current pulse of opposite polarity to that required to pull in the plunger.
5. *Rotary solenoid.* Constructed to permit rotary travel. Typical range is 25 to 95°. Return action via mechanical means (e.g., a spring).
6. *Reversing rotary solenoid.* Rotary motion is from one end to the other; when the solenoid is energized again, it reverses direction.

Solenoid power ratings are dependent primarily on the current required by the coil, and on the coil resistance. The I^2R is the primary power sink, and solenoids are therefore limited by the heat they can dissipate. Solenoids can operate in continuous or pulsed mode. The power rating depends on the mode of operation, and can be increased by adding *hold-in resistors* to the circuit to reduce the *holding current* required for continuous operation. The hold resistor is switched into the circuit once the *pull-in* current required to pull the plunger has been applied and the plunger has moved into place. The holding current can be significantly smaller than the pull-in current.

A common method to reduce the solenoid holding current employs a normally closed (NC) switch in parallel with a hold-in resistor. In Figure 18.43, when the pushbutton (PB) closes the circuit, full voltage is applied to the solenoid coil, bypassing the resistor through the NC switch, connecting the resistor in series with the coil. The resistor will now limit the current to the value required to hold the solenoid in position. Note the diode "snubber" circuit to shunt the reverse current when the solenoid is deenergized (see Figure 12.19 and related text for an explanation of this circuit).

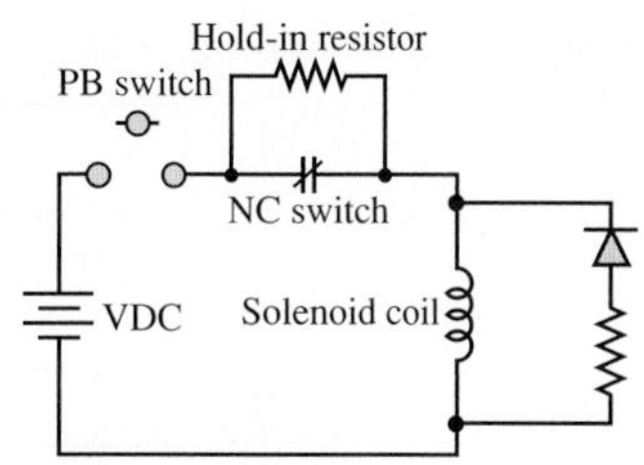

Figure 18.43

Another electromechanical device that finds common application in industrial practice is the **relay.** The relay is essentially an electromechanical switch that permits the opening and closing of electrical contacts by means of an electromagnetic structure similar to those discussed earlier in this section.

A relay such as would be used to start a high-voltage single-phase motor is shown in Figure 18.44. The magnetic structure has dimensions equal to 1 cm on all sides, and the transverse dimension is 8 cm. The relay works as follows. When the pushbutton is pressed, an electric current flows through the coil and generates a field in the magnetic structure. The resulting force draws the movable part toward the fixed part, causing an electrical contact to be made. The advantage of the relay is that a relatively low-level current can be used to control the opening and closing of a circuit that can carry large currents. In this particular example, the relay is energized by a 120-V AC contact, establishing a connection in a 240-V AC circuit. Such relay circuits are commonly employed to remotely switch large industrial loads.

Circuit symbols for relays are shown in Figure 18.45. An example of the calculations that would typically be required in determining the mechanical and electrical characteristics of a simple relay are given in Example 18.12.

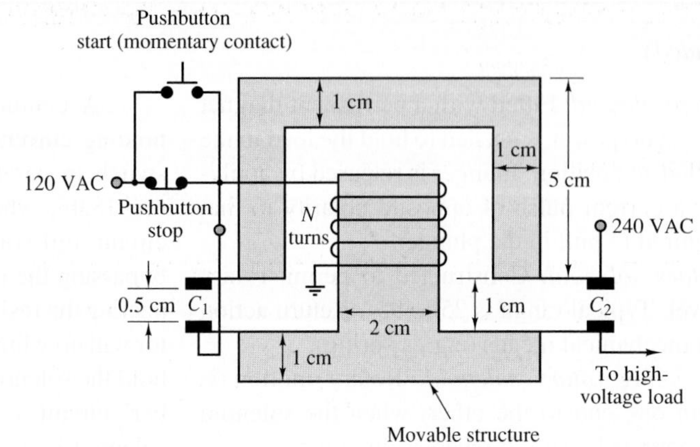

Figure 18.44 A relay

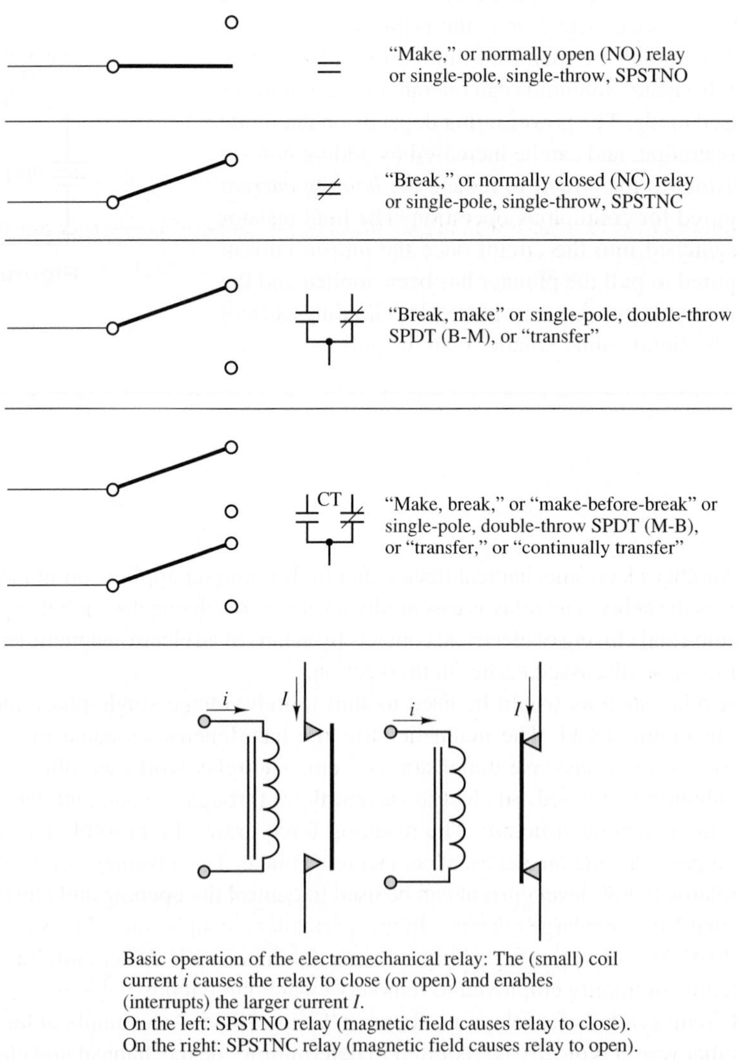

Figure 18.45 Circuit symbols and basic operation of relays

EXAMPLE 18.12 A Relay

LO5

Problem

Figure 18.46 depicts a simplified representation of a relay. Determine the current required for the relay to make contact (i.e., pull in the ferromagnetic plate) from a distance x.

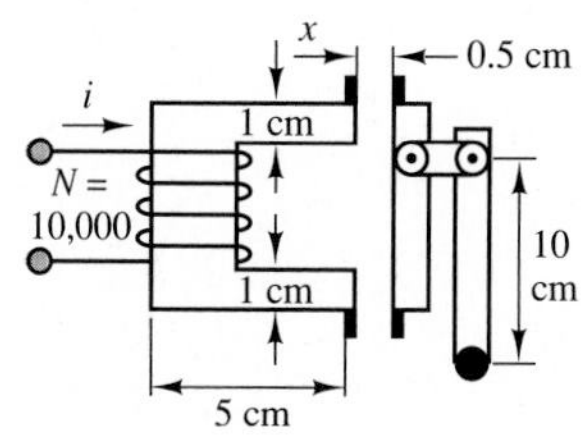

Figure 18.46

Solution

Known Quantities: Relay geometry; restoring force to be overcome; distance between bar and relay contacts; number of coil turns.

Find: i.

Schematics, Diagrams, Circuits, and Given Data: $A_{gap} = (0.01 \text{ m})^2$; $x = 0.05$ m; $f_{restore} = 5$ N; $N = 10{,}000$.

Assumptions: Assume that the reluctance of the iron is negligible; neglect fringing.

Analysis:

$$\mathcal{R}_{gap}(x) = \frac{2x}{\mu_0 A_{gap}}$$

The derivative of the reluctance with respect to the displacement of the plunger can then be computed as

$$\frac{d\mathcal{R}_{gap}(x)}{dx} = \frac{2}{\mu_0 A_{gap}}$$

Knowing the reluctance, we can calculate the magnetic flux in the structure as a function of the coil current:

$$\phi = \frac{Ni}{\mathcal{R}(x)} = \frac{Ni\mu_0 A_{gap}}{2x}$$

and the force in the air gap is given by

$$f_{gap} = \frac{\phi^2}{2}\frac{d\mathcal{R}(x)}{dx} = \frac{(Ni\mu_0 A_{gap})^2}{8x^2}\frac{2}{\mu_0 A_{gap}} = \frac{\mu_0 A_{gap}(Ni)^2}{4x^2}$$

The magnetic force must overcome a mechanical holding force of 5 N; thus,

$$f_{gap} = \frac{\mu_0 A_{gap}(Ni)^2}{4x^2} = f_{restore} = 5 \text{ N}$$

or

$$i = \frac{1}{N}\sqrt{\frac{4x^2 f_{restore}}{\mu_0 A_{gap}}} = \frac{1}{10{,}000}\sqrt{\frac{4(0.05)^2 5}{4\pi \times 10^{-7} \times 0.0001}} = \pm 2 \text{ A}$$

Comments: The current required to close the relay is much larger than that required to hold the relay closed, because the reluctance of the structure is much smaller once the gap is reduced to zero.

Moving-Coil Transducers

Another important class of electromagnetomechanical transducers is that of **moving-coil transducers.** This class of transducers includes a number of common devices, such as microphones, loudspeakers, and all electric motors and generators. The aim of this section is to explain the relationship between a fixed magnetic field, the emf across the moving coil, and the forces and motions of the moving element of the transducer.

The basic principle of operation of electromechanical transducers was presented in Section 18.1, where we stated that a magnetic field exerts a force on a charge moving through it. The equation describing this effect is

$$\mathbf{f} = q\mathbf{u} \times \mathbf{B} \qquad \textbf{(18.51)}$$

which is a vector equation, as explained earlier. To correctly interpret equation 18.51, we must recall the right-hand rule and apply it to the transducer, illustrated in Figure 18.47, depicting a structure consisting of a sliding bar which makes contact with a fixed conducting frame. Although this structure does not represent a practical actuator, it will be a useful aid in explaining the operation of moving-coil transducers such as motors and generators. In Figure 18.47, and in all similar figures in this section, a small cross represents the "tail" of an arrow pointing into the page, while a dot represents an arrow pointing out of the page; this convention will be useful in visualizing three-dimensional pictures.

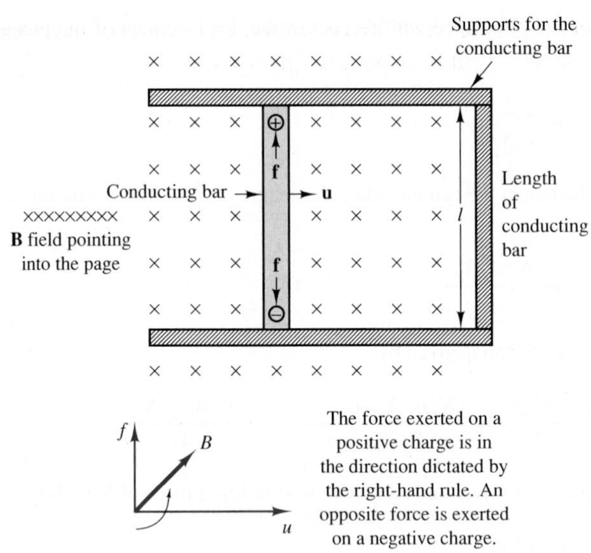

Figure 18.47 A simple electromechanical motion transducer

LO5

CHECK YOUR UNDERSTANDING

In the circuit in Figure 18.47, the conducting bar is moving with a velocity of 6 m/s. The flux density is 0.5 Wb/m^2, and $l = 1.0$ m. Find the magnitude of the resulting induced voltage.

Answer: 3 V

Motor Action

A moving-coil transducer can act as a motor when an externally supplied current flowing through the electrically conducting part of the transducer is converted to a force that can cause the moving part of the transducer to be displaced. Such a current would flow, for example, if the support of Figure 18.47 were made of conducting material, so that the conductor and the right-hand side of the support "rail" were to form a loop (in effect, a 1-turn coil). To understand the effects of this current flow in the conductor, one must consider the fact that a charge moving at a velocity u' (along the conductor and perpendicular to the velocity of the conducting bar, as shown in Figure 18.48) corresponds to a current $i = dq/dt$ along the length l of the conductor. This fact can be explained by considering the current i along a differential element dl and writing

$$i\,dl = \frac{dq}{dt} \cdot u'\,dt \qquad \textbf{(18.52)}$$

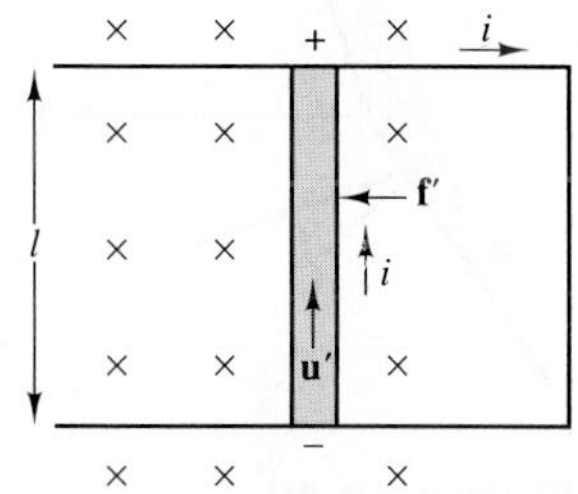

Figure 18.48

since the differential element dl would be traversed by the current in time dt at a velocity u'. Thus we can write

$$i\,dl = dq\,u' \qquad \textbf{(18.53)}$$

or

$$il = qu' \qquad \textbf{(18.54)}$$

for the geometry of Figure 18.48. From Section 18.1, the force developed by a charge moving in a magnetic field is, in general, given by

$$\mathbf{f} = q\mathbf{u} \times \mathbf{B} \qquad \textbf{(18.55)}$$

For the term $q\mathbf{u}'$ we can substitute $i\mathbf{l}$, to obtain

$$\mathbf{f}' = i\mathbf{l} \times \mathbf{B} \qquad \textbf{(18.56)}$$

Using the right-hand rule, we determine that the force $\mathbf{f}'$ generated by the current i is in the direction that would push the conducting bar to the left. The magnitude of this force is $f' = Bli$ if the magnetic field and the direction of the current are perpendicular. If they are not, then we must consider the angle γ formed by $\mathbf{B}$ and $\mathbf{l}$; in the more general case,

$$f' = Bli \sin\gamma = Bli \text{ if } \gamma = 90° \qquad Bli \text{ law} \qquad \textbf{(18.57)}$$

LO5

The phenomenon we have just described is sometimes referred to as the ***Bli* law.**

Generator Action

The other mode of operation of a moving-coil transducer occurs when an external force causes the coil (i.e., the moving bar, in Figure 18.47) to be displaced. This external force is converted to an emf across the coil, as will be explained in the following paragraphs.

Since positive and negative charges are forced in opposite directions in the transducer of Figure 18.47, a potential difference will appear across the conducting bar; this potential difference is the electromotive force, or emf. The emf must be equal to the force exerted by the magnetic field. In short, the electric force per unit charge (or electric field) e/l must equal the magnetic force per unit charge $f/q = Bu$. Thus, the relationship

$$e = Blu \qquad Blu \text{ law} \tag{18.58}$$

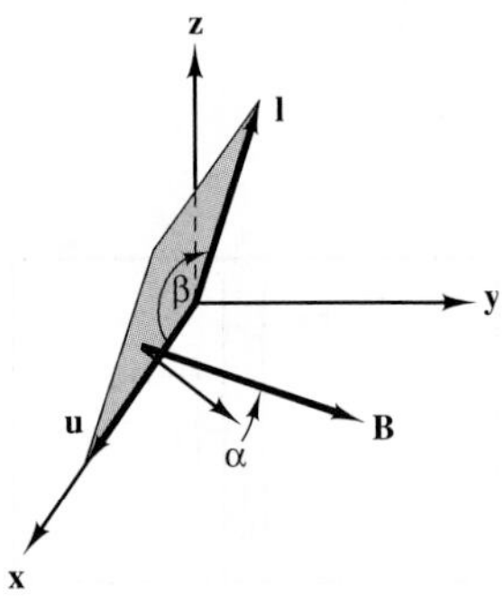

Figure 18.49

holds whenever **B, l,** and **u** are mutually perpendicular, as in Figure 18.49. If equation 18.58 is analyzed in greater depth, it can be seen that the product lu (length times velocity) is the area crossed per unit time by the conductor. If one visualizes the conductor as "cutting" the flux lines into the base in Figure 18.48, it can be concluded that the electromotive force is equal to the *rate at which the conductor "cuts" the magnetic lines of flux*. It will be useful for you to carefully absorb this notion of conductors cutting lines of flux, since this will greatly simplify the understanding of the material in this section and in Chapter 19.

In general, **B, l,** and **u** are not necessarily perpendicular. In this case one needs to consider the angles formed by the magnetic field with the normal to the plane containing **l** and **u,** and the angle between **l** and **u.** The former is angle α of Figure 18.49; the latter is angle β in the same figure. It should be apparent that the optimum values of α and β are 0° and 90°, respectively. Thus, most practical devices are constructed with these values of α and β. Unless otherwise noted, it will be tacitly assumed that this is the case. The ***Bli* law** just illustrated explains how a moving conductor in a magnetic field can generate an electromotive force.

To summarize the electromechanical energy conversion that takes place in the simple device of Figure 18.47, we must note now that the presence of a current in the loop formed by the conductor and the rail requires that the conductor move to the right at a velocity u (Blu law), thus cutting the lines of flux and generating the emf that gives rise to current i. On the other hand, the same current causes a force f' to be exerted on the conductor (Bli law) in the direction opposite to the movement of the conductor. Thus, it is necessary that an *externally applied force* f_{ext} exist to cause the conductor to move to the right with a velocity u. The external force must overcome the force f'. This is the basis of electromechanical energy conversion.

An additional observation we must make at this point is that the current i flowing around a closed loop generates a magnetic field, as explained in Section 18.1. Since this additional field is generated by a 1-turn coil in our illustration, it is reasonable to assume that it is negligible with respect to the field already present (perhaps established by a permanent magnet). Finally, we must consider that this coil links a certain amount of flux, which changes as the conductor moves from left to right. The area crossed by the moving conductor in time dt is

$$dA = lu\,dt \tag{18.59}$$

so that if the flux density B is uniform, the rate of change of the flux linked by the 1-turn coil is

$$\frac{d\phi}{dt} = B\frac{dA}{dt} = Blu \tag{18.60}$$

In other words, the *rate of change* of the flux linked by the conducting loop is equal to the emf generated in the conductor. You should realize that this statement simply confirms Faraday's law.

It was briefly mentioned that the Blu and Bli laws indicate that, thanks to the coupling action of the magnetic field, a conversion of mechanical to electrical energy—or the converse—is possible. The simple structures of Figures 18.47 and 18.48 can, again, serve as an illustration of this energy conversion process, although we have not yet indicated how these idealized structures can be converted to a practical device. In this section we begin to introduce some physical considerations. Before we proceed any further, we should try to compute the power—electric and mechanical—that is generated (or is required) by our ideal transducer. The electric power is given by

$$P_E = ei = Blui \qquad \text{W} \tag{18.61}$$

while the mechanical power required, say, to move the conductor from left to right is given by the product of force and velocity:

$$P_M - f_{\text{ext}}u = Bliu \qquad \text{W} \tag{18.62}$$

The principle of conservation of energy thus states that in this ideal (lossless) transducer we can convert a given amount of electric energy to mechanical energy, or vice versa. Once again we can utilize the same structure of Figure 18.47 to illustrate this reversible action. If the closed path containing the moving conductor is now formed from a closed circuit containing a resistance R and a battery V_B, as shown in Figure 18.50, the externally applied force f_{ext} generates a positive current i into the battery provided that the emf is greater than V_B. When $e = Blu > V_B$, the ideal transducer acts as a *generator*. For any given set of values of B, l, R, and V_B, there will exist a velocity u for which the current i is positive. If the velocity is lower than this value—that is, if $e = Blu < V_B$—then the current i is negative, and the conductor is forced to move to the right. In this case the battery acts as a source of energy and the transducer acts as a *motor* (i.e., electric energy drives the mechanical motion).

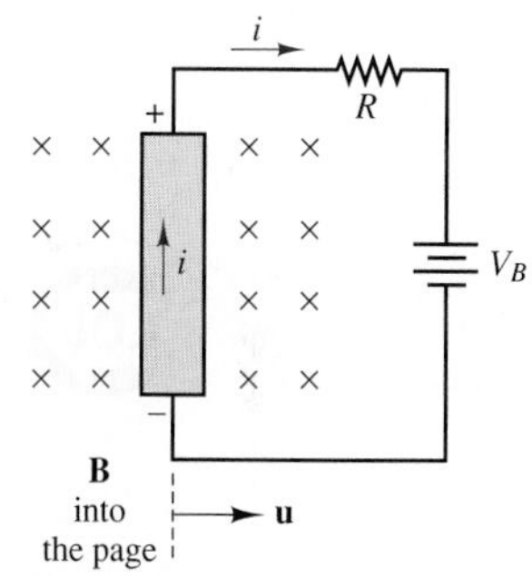

Figure 18.50 Motor and generator action in an ideal transducer

In practical transducers, we must be concerned with the inertia, friction, and elastic forces that are invariably present on the mechanical side of the transducer. Similarly, on the electrical side we must account for the inductance of the circuit, its resistance, and possibly some capacitance. Consider the structure of Figure 18.51. In the figure, the conducting bar has been placed on a surface with a coefficient of sliding friction b; it has a mass m and is attached to a fixed structure by means of a spring with spring constant k. The equivalent circuit representing the coil inductance and resistance is also shown.

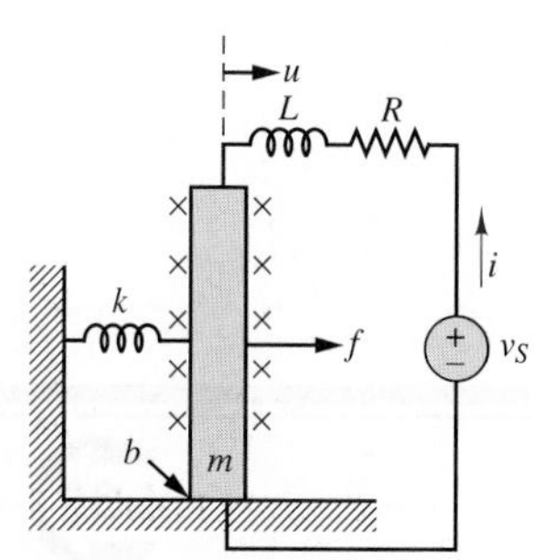

Figure 18.51 A more realistic representation of the transducer of Figure 16.50

If we recognize that $u = dx/dt$ in the figure, we can write the equation of motion for the conductor as

$$m\frac{du}{dt} + bu + \frac{1}{k}\int u\,dt = f = Bli \tag{18.63}$$

where the Bli term represents the driving input that causes the mass to move. The driving input in this case is provided by the electric energy source v_S; thus the transducer acts as a motor, and f is the electromechanical force acting on the mass of the conductor. On the electrical side, the circuit equation is

$$v_S - L\frac{di}{dt} - Ri = e = Blu \tag{18.64}$$

Equations 18.63 and 18.64 could then be solved by knowing the excitation voltage v_S and the physical parameters of the mechanical and electric circuits. For example, if the excitation voltage were sinusoidal, with

$$v_S(t) = V_S \cos \omega t$$

and the field density were constant

$$B = B_0$$

then we could postulate sinusoidal solutions for the transducer velocity u and current i:

$$u = U \cos(\omega t + \theta_u) \qquad i = I \cos(\omega t + \theta_i) \tag{18.65}$$

and use phasor notation to solve for the unknowns $(U, I, \theta_u, \theta_i)$.

The results obtained in the present section apply directly to transducers that are based on translational (linear) motion. These basic principles of electromechanical energy conversion and the analysis methods developed in the section will be applied to practical transducers in a few examples. A Focus on Methodology box outlines the analysis procedure for moving-coil transducers.

The methods introduced in this section will later be applied in Chapters 19 and 20 to analyze rotating transducers, that is, electric motors and generators.

LO5

FOCUS ON METHODOLOGY

ANALYSIS OF MOVING-COIL ELECTROMECHANICAL TRANSDUCERS

1. Apply KVL to write the differential equation for the electrical subsystem, including the back emf ($e = Blu$) term.
2. Apply Newton's second law to write the differential equation for the mechanical subsystem, including the magnetic force $f = Bli$ term.
3. Use a Laplace transform on the two coupled differential equations to formulate a system of linear algebraic equations, and solve for the desired mechanical and electrical variables.

LO5

EXAMPLE 18.13 A Loudspeaker

Problem

A loudspeaker, shown in Figure 18.52, uses a permanent magnet and a moving coil to produce the vibrational motion that generates the pressure waves we perceive as sound. Vibration of the loudspeaker is caused by changes in the input current to a coil; the coil is, in turn, coupled to a magnetic structure that can produce time-varying forces on the speaker diaphragm. A simplified model for the mechanics of the speaker is also shown in Figure 18.52. The force exerted on the coil is also exerted on the mass of the speaker diaphragm, as shown in Figure 18.53, which depicts a free-body diagram of the forces acting on the loudspeaker diaphragm.

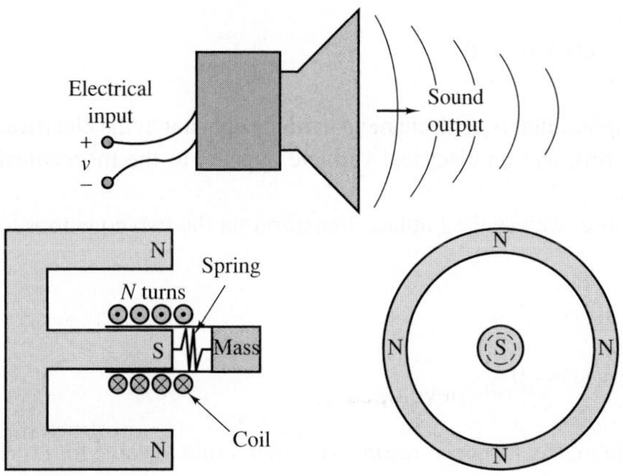

Figure 18.52 Loudspeaker

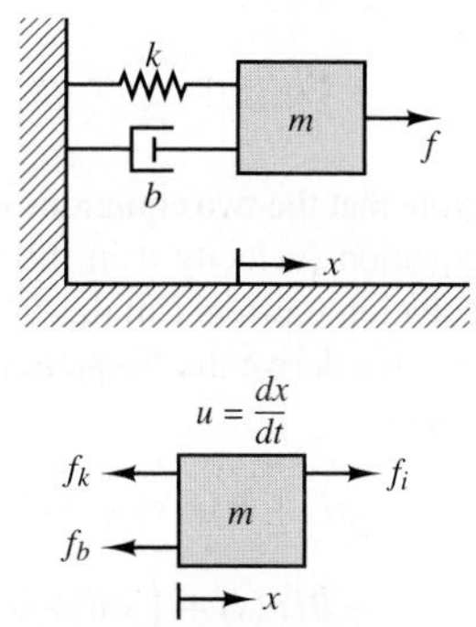

Figure 18.53 Forces acting on loudspeaker diaphragm

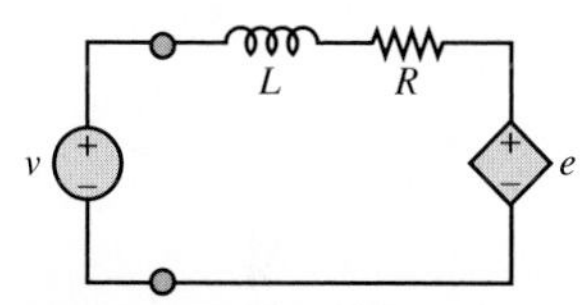

Figure 18.54 Model of transducer electrical side

The force exerted on the mass f_i is the magnetic force due to current flow in the coil. The electric circuit that describes the coil is shown in Figure 18.54, where L represents the inductance of the coil, R represents the resistance of the windings, and e is the emf induced by the coil moving through the magnetic field.

Determine the frequency response $U(j\omega)/V(j\omega)$ of the speaker.

Solution

Known Quantities: Circuit and mechanical parameters; magnetic flux density; number of coil turns; coil radius.

Find: Frequency response of loudspeaker $U(j\omega)/V(j\omega)$.

Schematics, Diagrams, Circuits, and Given Data: Coil radius = 0.05 m; $L = 10$ mH; $R = 8\ \Omega$; $m = 0.01$ kg; $b = 22.75$ N-s^2/m; $k = 5 \times 10^4$ N/m; $N = 47$; $B = 1$ T.

Analysis: To determine the frequency response of the loudspeaker, we write the differential equations that describe the electrical and mechanical subsystems. We apply KVL to the electric circuit, using the circuit model of Figure 18.54, in which we have represented the Blu term (motional voltage) in the form of a *back electromotive force e*:

$$v - L\frac{di}{dt} - Ri - e = 0$$

or

$$L\frac{di}{dt} + Ri + Blu = v$$

Next, we apply Newton's second law to the mechanical system, consisting of a lumped mass representing the mass of the moving diaphragm m; an elastic (spring) term, which represents the elasticity of the diaphragm k; and a damping coefficient b, representing the frictional losses and aerodynamic damping affecting the moving diaphragm.

$$m\frac{du}{dt} = f_i - f_d - f_k = f_i - bu - kx$$

where $f_i = Bli$ and therefore

$$-Bli + m\frac{du}{dt} + bu + k\int_{-\infty}^{t} u(t')\,dt' = 0$$

Note that the two equations are *coupled*; that is, a mechanical variable appears in the electrical equation (velocity u in the Blu term), and an electrical variable appears in the mechanical equation (current i in the Bli term).

To derive the frequency response, we use the Laplace-transform on the two equations to obtain

$$(sL + R)I(s) + BlU(s) = V(s)$$
$$-BlI(s) + \left(sm + b + \frac{k}{s}\right)U(s) = 0$$

We can write the above equations in matrix form and resort to Cramer's rule to solve for $U(s)$ as a function of $V(s)$:

$$\begin{bmatrix} sL + R & Bl \\ -Bl & sm + b + \frac{k}{s} \end{bmatrix} \begin{bmatrix} I(s) \\ U(s) \end{bmatrix} = \begin{bmatrix} V(s) \\ 0 \end{bmatrix}$$

with solution

$$U(s) = \frac{\det\begin{bmatrix} sL + R & V(s) \\ -Bl & 0 \end{bmatrix}}{\det\begin{bmatrix} sL + R & Bl \\ -Bl & sm + b + \frac{k}{s} \end{bmatrix}}$$

or

$$\frac{U(s)}{V(s)} = \frac{Bl}{(sL + R)(sm + b + k/s) + (Bl)^2}$$
$$= \frac{Bls}{(Lm)s^3 + (Rm + Lb)s^2 + [Rb + kL + (Bl)^2]s + kR}$$

To determine the frequency response of the loudspeaker, we let $s \to j\omega$ in the above expression:

$$\frac{U(j\omega)}{V(j\omega)} = \frac{jBl\omega}{kR - (Rm + Lb)\omega^2 + j\{[(Rb + kL + (Bl)^2]\omega - (Lm)\omega^3\}}$$

where $l = 2\pi Nr$, and substitute the appropriate numerical parameters:

$$\frac{U(j\omega)}{V(j\omega)} = \frac{j14.8\omega}{4 \times 10^5 - (0.08 + 0.2275)\omega^2 + j[(182 + 500 + 218)\omega - (10^{-4})\omega^3]}$$
$$= \frac{j14.8\omega}{4 \times 10^5 - 0.3075\omega^2 + j[(900)\omega - (10^{-4})\omega^3]}$$

The resulting frequency response is plotted in Figure 18.55.

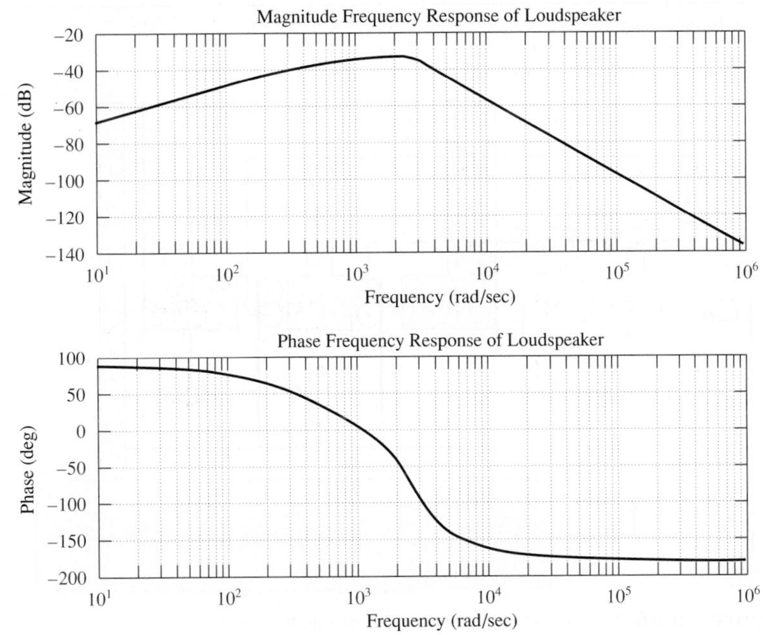

Figure 18.55 Frequency response of loudspeaker

CHECK YOUR UNDERSTANDING

In Example 18.13, we examined the frequency response of a loudspeaker. However, over time, permanent magnets may become demagnetized. Find the frequency response of the same loudspeaker if the permanent magnet has lost its strength to a point where $B = 0.95$ T.

Answer: $U(j\omega)/V(j\omega) = 0.056(j\omega/15{,}950)/(1 + j\omega/15{,}950)(1 + j\omega/31{,}347)$

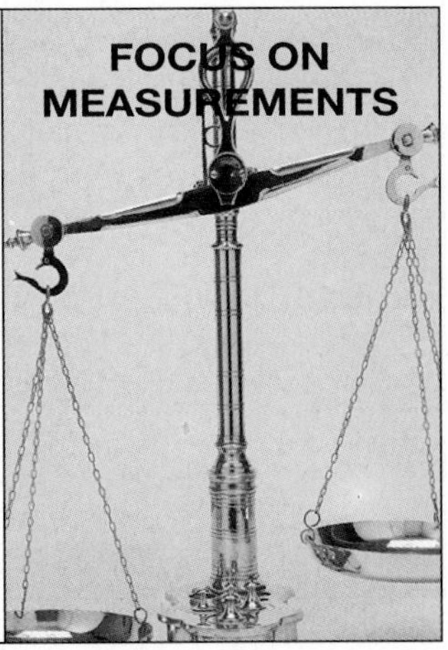

Seismic Transducer

Problem:

FIND IT ON THE WEB

The device shown in Figure 18.56 is called a **seismic transducer** and can be used to measure the displacement, velocity, or acceleration of a body. The permanent magnet of mass m is supported on the case by a spring k, and there is some viscous damping b between the magnet and the case; the coil is fixed to the case. You may assume that the coil has length l and resistance and inductance R_{coil} and L_{coil}, respectively; the magnet exerts a magnetic field B. Find the transfer function between the output voltage v_{out} and the velocity of the body dx_c/dt. Note that $x(t)$ is not equal to zero when the system is at rest. We shall ignore this offset displacement.

(Continued)

(Concluded)

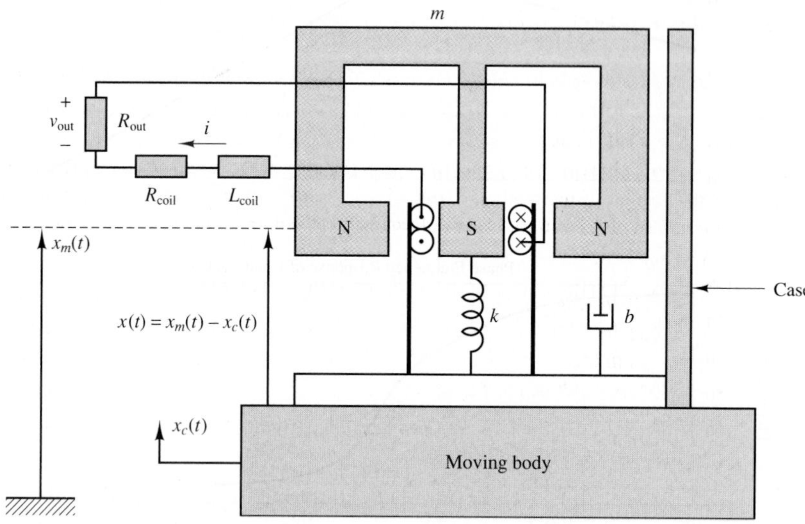

Figure 18.56 An electromagnetomechanical seismic transducer

Solution:

First we apply KVL around the electric circuit to write the differential equation describing the electrical systems:

$$L\frac{di}{dt} + (R_{coil} + R_{out})i + Bl\frac{dx}{dt} = 0$$

Also note that $v_{out} = -R_{out}i$. Next, we observe that the displacement of the magnet, x_m, is equal to the sum of the case displacement, x_m, and the relative displacement between the magnet and the case, $x(t)$: $x_m = x + x_c$. Apply Newton's Second law to the mass of the magnet, m, we obtain

$$m\frac{d^2x_m}{dt^2} = -k(x_m - x_c) - b\left(\frac{dx_m}{dt} - \frac{dx_c}{dt}\right) - Bli$$

Substituting the relation $x_m = x + x_c$, we obtain

$$m\left(\frac{d^2x}{dt^2} + \frac{d^2x_c}{dt^2}\right) + kx + b\frac{dx}{dt} = -Bli$$

From this expression we can now derive the transfer function between the displacement of the case, $X_c(s)$, and the output voltage, $V_{out}(s)$. Let $R = R_{coil} + R_{out}$. Then

$$(Ls + R)I(s) + BlsX(s) = 0$$
$$-BlI(s) + (ms^2 + bs + k)X(s) = -ms^2X_c(s)$$
$$I(s) = \frac{Blms^3X_c(s)}{mLs^3 + (bL + mR)s^2 + (kL + Rb + B^2l^2)s + kR}$$

Now, let the velocity of the case be $U_c(s) = sX_c(s)$; since $V_{out}(s) = -R_{out}I(s)$, the transfer function from case velocity to output voltage becomes

$$\frac{V_{out}(s)}{U_c(s)} = -\frac{BlmR_{out}s^2}{mLs^3 + (bL + mR)s^2 + (kL + Rb + B^2l^2)s + kR}$$

Conclusion

This chapter introduces electromechanical systems. Electromechanical devices include a variety of sensors and transducers that find common engineering application in many fields. All electromechanical devices use the coupling between mechanical and electrical systems provided by a magnetic field. This magnetic coupling makes it possible to convert energy from electric to mechanical form, and back. Devices that convert electric to mechanical energy include all forms of electromagnetomechanical actuators, such as electromagnets, solenoids, relays, electrodynamic shakers, linear motors, and loudspeakers. Conversion from mechanical to electric energy results in generators, and various sensors that can detect mechanical displacement, velocity, or acceleration. Upon completing this chapter, you should have mastered the following learning objectives:

1. *Review the basic principles of electricity and magnetism.* The basic laws that govern electromagnetomechanical energy conversion are Faraday's law, stating that a changing magnetic field can induce a voltage, and Ampère's law, stating that a current flowing through a conductor generates a magnetic field.
2. *Use the concepts of reluctance and magnetic circuit equivalents to compute magnetic flux and currents in simple magnetic structures.* The two fundamental variables in the analysis of magnetic structures are the magnetomotive force and the magnetic flux; if some simplifying approximations are made, these quantities are linearly related through the reluctance parameter, in much the same way as voltage and current are related through resistance according to Ohm's law. This simplified analysis permits approximate calculation of forces and currents in electromagnetomechanical structures.
3. *Understand the properties of magnetic materials and their effects on magnetic circuit models.* Magnetic materials are characterized by a number of nonideal properties, which must be considered in a detailed analysis of any electromechanical transducer. The most important phenomena are saturation, eddy currents, and hysteresis.
4. *Use magnetic circuit models to analyze transformers.* One of the most common magnetic structures in use in electric power systems is the transformer. The methods developed in the earlier sections provide all the tools needed to perform an analysis of these important devices.
5. *Model and analyze force generation in electromagnetomechanical systems. Analyze moving-iron transducers (electromagnets, solenoids, relays) and moving-coil transducers (electrodynamic shakers, loudspeakers, and seismic transducers).* Electromagnetomechanical transducers can be broadly divided into two categories: moving-iron transducers, which include all electromagnets, solenoids, and relays; and moving-coil transducers, which include loudspeakers, electrodynamic shakers, and all electric motors. Section 18.5 develops analysis and design methods for these devices.

HOMEWORK PROBLEMS

Section 18.1: Electricity and Magnetism

18.1 For the electromagnet of Figure P18.1:

a. Find the flux density in the core.

b. Sketch the magnetic flux lines and indicate their direction.

c. Indicate the north and south poles of the magnet.

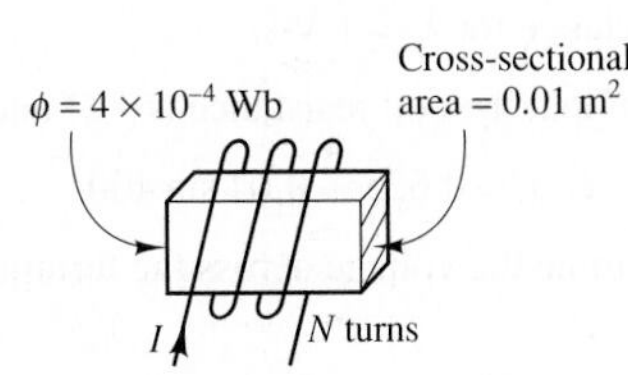

Figure P18.1

18.2 A single loop of wire carrying current I_2 is placed near the end of a solenoid having N turns and carrying current I_1, as shown in Figure P18.2. The solenoid is fastened to a horizontal surface, but the single coil is free to move. With the currents directed as shown, is there a resultant force on the single coil? If so, in what direction? Why?

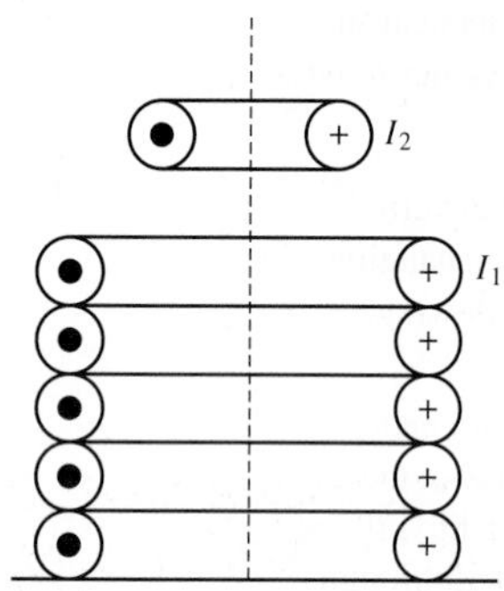

Figure P18.2

18.3 A practical LVDT is typically connected to a resistive load. Derive the LVDT equations in the presence of a resistive load R_L connected across the output terminals, using the results of the Focus on Measurements box "Linear Variable Differential Transformer." Let R_S, L_S be the secondary coil parameters.

18.4 On the basis of the equations of the Focus on Measurements box "Linear Variable Differential Transformer," and of the results of Problem 18.3, derive the frequency response of the LVDT, and determine the range of frequencies for which the device will have maximum sensitivity for a given excitation. *Hint:* Compute $dv_{\text{out}}/dv_{\text{ex}}$, and set the derivative equal to zero to determine the maximum sensitivity.

18.5 An iron-core inductor has the following characteristic:

$$i = \frac{\lambda}{0.5 + \lambda}$$

a. Determine the energy, co-energy, and incremental inductance for $\lambda = 1$ V-s.

b. Given that the coil resistance is 1 Ω and that

$$i(t) = 0.625 + 0.01 \sin 400t \qquad \text{A}$$

determine the voltage across the terminals on the inductor.

18.6 Repeat Problem 18.5 if

$$i = \frac{\lambda^2}{0.5 + \lambda^2}$$

18.7 An iron-core inductor has the characteristic shown in Figure P18.7:

a. Determine the energy and the incremental inductance for $i = 1.0$ A.

b. Given that the coil resistance is 2 Ω and that $i(t) = 0.5 \sin 2\pi t$, determine the voltage across the terminals of the inductor.

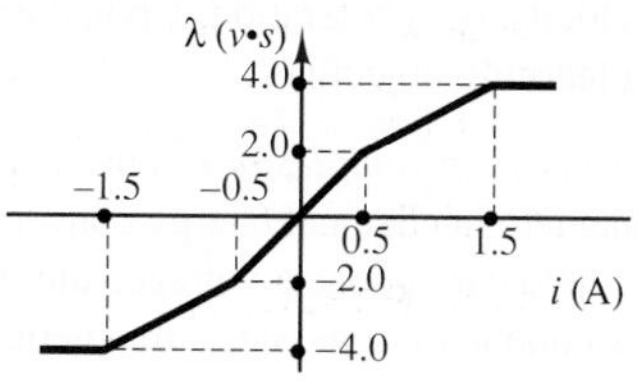

Figure P18.7

18.8 Determine the reluctance of the structure of Figure 18.12 in the text if the cross-sectional area is $A = 0.1\ \text{m}^2$ and $\mu_r = 2{,}000$. Assume that each leg is 0.1 m in length and that the mean magnetic path runs through the exact center of the structure.

Section 18.2: Magnetic Circuits

18.9

a. Find the reluctance of a magnetic circuit if a magnetic flux $\phi = 4.2 \times 10^{-4}$ Wb is established by an impressed mmf of 400 A-turns.

b. Find the magnetizing force H in SI units if the magnetic circuit is 6 in long.

18.10 For the circuit shown in Figure P18.10:

a. Determine the reluctance values and show the magnetic circuit, assuming that $\mu = 3{,}000\mu_0$.

b. Determine the inductance of the device.

c. The inductance of the device can be modified by cutting an air gap in the magnetic structure. If a gap of 0.1 mm is cut in the arm of length l_3, what is the new value of inductance?

d. As the gap is increased in size (length), what is the limiting value of inductance? Neglect leakage flux and fringing effects.

18.11 The magnetic circuit shown in Figure P18.11 has two parallel paths. Find the flux and flux density in each leg of the magnetic circuit. Neglect fringing at the air gaps and any leakage fields. $N = 1{,}000$ turns, $i = 0.2$ A, $l_{g1} = 0.02$ cm, and $l_{g2} = 0.04$ cm. Assume the reluctance of the magnetic core to be negligible.

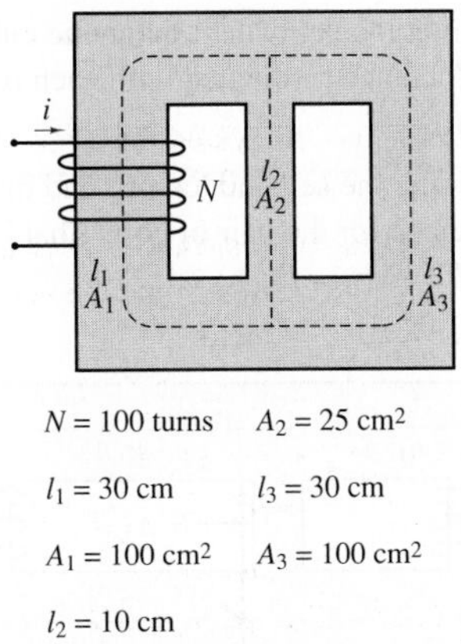

Figure P18.10

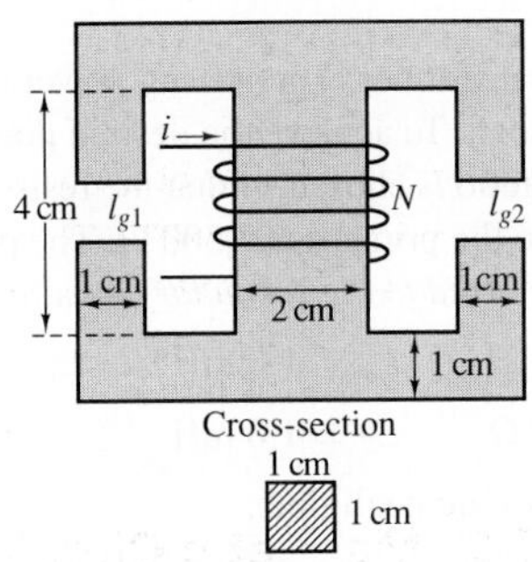

Figure P18.11

18.12 Find the current necessary to establish a flux of $\phi = 3 \times 10^{-4}$ Wb in the series magnetic circuit of Figure P18.12. Here $l_{\text{iron}} = l_{\text{steel}} = 0.3$ m, area (throughout) $= 5 \times 10^{-4}$ m², and $N = 100$ turns. Assume $\mu_r = 5{,}195$ for cast iron and $\mu_r = 1{,}000$ for cast steel.

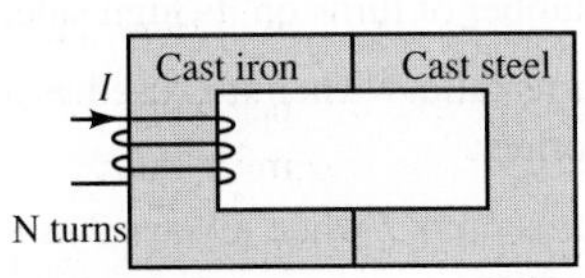

Figure P18.12

18.13 Find the magnetic flux ϕ established in the series magnetic circuit of Figure P18.13.

18.14

a. Find the current I required to establish a flux $\phi = 2.4 \times 10^{-4}$ Wb in the magnetic circuit of Figure P18.14. Here area(throughout) $=$ 2×10^{-4} m², $l_{ab} = l_{ef} = 0.05$ m, $l_{af} = l_{be} = 0.02$ m, $l_{bc} = l_{dc}$, and the material is sheet steel.

b. Compare the mmf drop across the air gap to that across the rest of the magnetic circuit. Discuss your results, using the value of μ for each material.

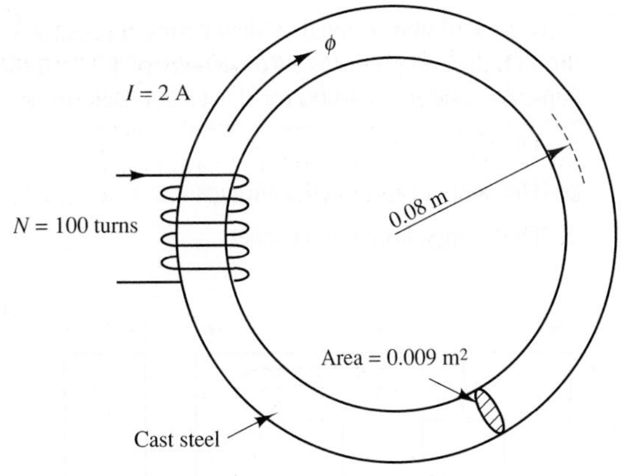

Figure P18.13

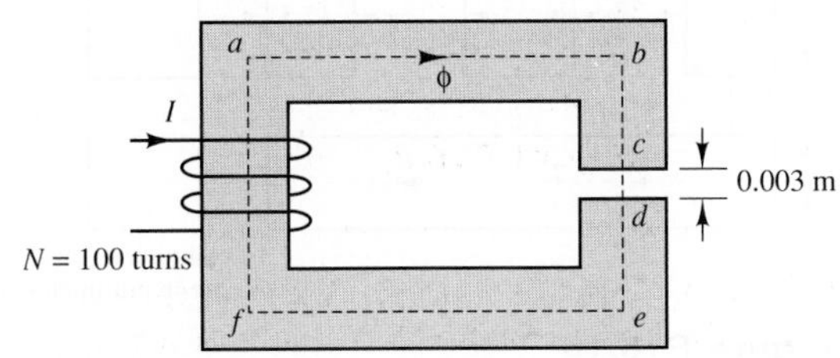

Figure P18.14

18.15 For the series-parallel magnetic circuit of Figure P18.15, find the value of I required to establish a flux in the gap of $\phi = 2 \times 10^{-4}$ Wb. Here, $l_{ab} = l_{bg} = l_{gh} = l_{ha} = 0.2$ m, $l_{bc} = l_{fg} = 0.1$ m, $l_{cd} = l_{ef} = 0.099$ m, and the material is sheet steel.

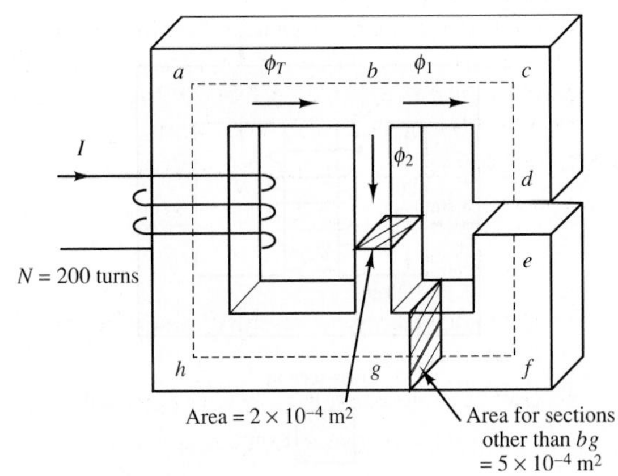

Figure P18.15

18.16 Refer to the actuator of Figure P18.16. The entire device is made of sheet steel. The coil has 2,000 turns. The armature is stationary so that the length of the air gaps, $g = 10$ mm, is fixed. A direct current passing through the coil produces a flux density of 1.2 T in the gaps. Assume $\mu_r = 4{,}000$ for sheet steel. Determine

a. The coil current.

b. The energy stored in the air gaps.

c. The energy stored in the steel.

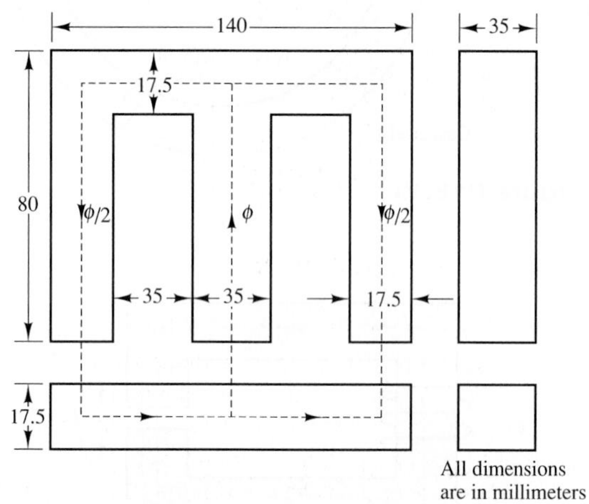

Figure P18.16

18.17 A core is shown in Figure P18.17, with $\mu_r = 2{,}000$ and $N = 100$. Find

a. The current needed to produce a flux density of 0.4 Wb/m^2 in the center leg.

b. The current needed to produce a flux density of 0.8 Wb/m^2 in the center leg.

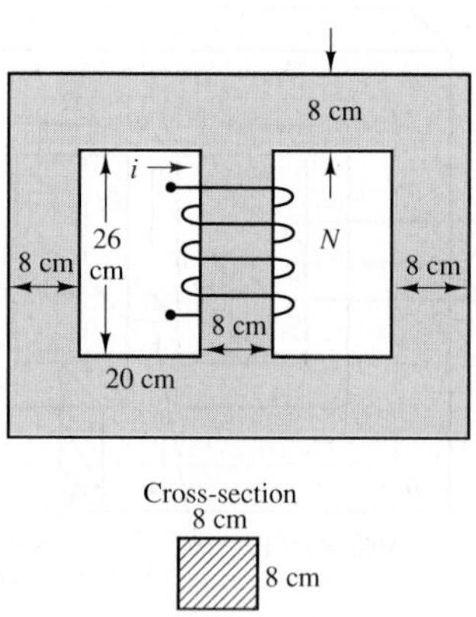

Figure P18.17

Section 18.4: Transformers

18.18 For the transformer shown in Figure P18.18, $N = 1{,}000$ turns, $l_1 = 16$ cm, $A_1 = 4$ cm^2, $l_2 = 22$ cm, $A_2 = 4$ cm^2, $l_3 = 5$ cm, and $A_3 = 2$ cm^2. The relative permeability of the material is $\mu_r = 1{,}500$.

a. Construct the equivalent magnetic circuit, and find the reluctance associated with each part of the circuit.

b. Determine the self-inductance and mutual inductance for the pair of coils (that is, L_{11}, L_{22}, and $M = L_{12} = L_{21}$).

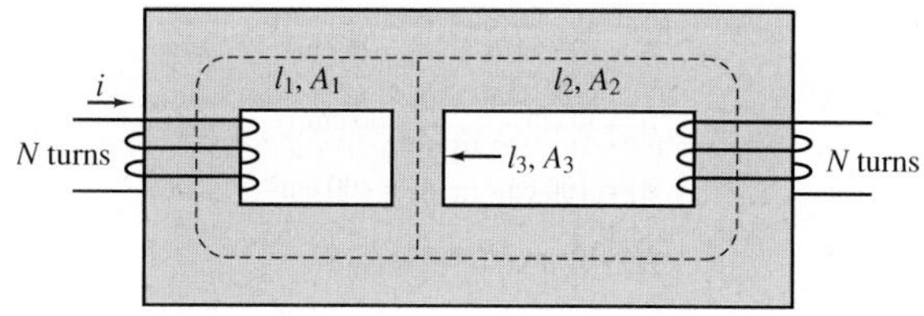

Figure P18.18

18.19 A transformer is delivering power to a 300-Ω resistive load. To achieve the desired power transfer, the turns ratio is chosen so that the resistive load referred to the primary is 7,500 Ω. The parameter values, *referred to the secondary winding,* are:

$r_1 = 20\ \Omega$ $\quad L_1 = 1.0$ mH $\quad L_m = 25$ mH

$r_2 = 20\ \Omega$ $\quad L_2 = 1.0$ mH

Core losses are negligible.

a. Determine the turns ratio.

b. Determine the input voltage, current, and power and the efficiency when this transformer is delivering 12 W to the 300-Ω load at a frequency $f = 10{,}000/2\pi$ Hz.

18.20 A 220/20-V transformer has 50 turns on its low-voltage side. Calculate

a. The number of turns on its high side.

b. The turns ratio α when it is used as a step-down transformer.

c. The turns ratio α when it is used as a step-up transformer.

18.21 The high-voltage side of a transformer has 750 turns, and the low-voltage side has 50 turns. When the high side is connected to a rated voltage of 120 V, 60 Hz, a rated load of 40 A is connected to the low side. Calculate

a. The turns ratio.

b. The secondary voltage (assuming no internal transformer impedance voltage drops).

c. The resistance of the load.

18.22 A transformer is to be used to match an 8-Ω loudspeaker to a 500-Ω audio line. What is the turns ratio of the transformer, and what are the voltages at the primary and secondary terminals when 10 W of audio power is delivered to the speaker? Assume that the speaker is a resistive load and that the transformer is ideal.

18.23 The high-voltage side of a step-down transformer has 800 turns, and the low-voltage side has 100 turns. A voltage of 240 V AC is applied to the high side, and the load impedance is 3 Ω (low side). Find

a. The secondary voltage and current.

b. The primary current.

c. The primary input impedance from the ratio of primary voltage to current.

d. The primary input impedance.

18.24 Calculate the transformer ratio of the transformer in Problem 18.23 when it is used as a step-up transformer.

18.25 A 2,300/240-V, 60-Hz, 4.6-kVA transformer is designed to have an induced emf of 2.5 V/turn. Assuming an ideal transformer, find

a. The numbers of high-side turns N_h and low-side turns N_l.

b. The rated current of the high-voltage side I_h.

c. The transformer ratio when the device is used as a step-up transformer.

Section 18.5: Electromechanical Transducers; (a) Moving-Iron Transducers

18.26 Calculate the current required to lift the load for the electromagnet of Example 18.9. Calculate the holding current required to keep the load in place once it has been lifted and is attached to the magnet. Assume: $N = 700$; $\mu_0 = 4\pi \times 10^{-7}$; $\mu_r = 10^4$ (equal for electromagnet and load); initial distance (air gap) $= 0.5$ m; magnetic path length of electromagnet $= l_1 = 0.80$ m; magnetic path length of movable load $= l_2 = 0.40$ m; gap cross sectional area $= 5 \times 10^{-4}$ m^2; m = mass of load $= 10$ kg; $g = 9.8$ m/s^2.

18.27 For the electromagnet of Example 18.9:

a. Calculate the current required to keep the bar in place. (*Hint:* The air gap becomes zero, and the iron reluctance cannot be neglected.) Assume $\mu_r = 1{,}000$, $L = 1$ m.

b. If the bar is initially 0.1 m away from the electromagnet, what initial current would be required to lift the magnet?

18.28 The electromagnet of Figure P18.28 has reluctance given by $\mathcal{R}(x) = 7 \times 10^8(0.002 + x)\ \text{H}^{-1}$, where x is the length of the variable gap in meters. The coil has 980 turns and 30-Ω resistance. For an applied voltage of 120 V DC, find

a. The energy stored in the magnetic field for $x = 0.005$ m.

b. The magnetic force for $x = 0.005$ m.

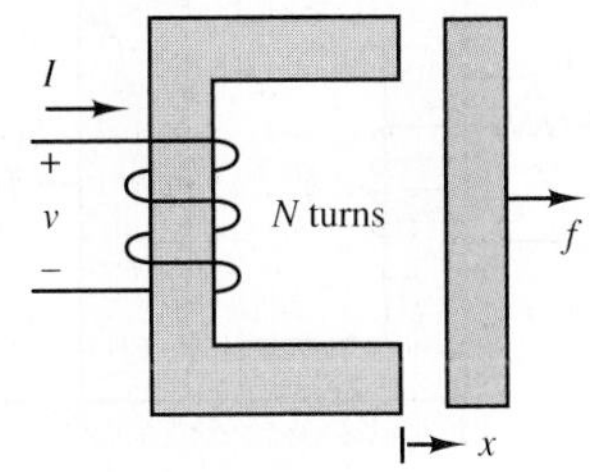

Figure P18.28

18.29 With reference to Example 18.10, determine the best combination of current magnitude and wire diameter to reduce the volume of the solenoid coil to a minimum. Will this minimum volume result in the lowest possible resistance? How does the power dissipation of the coil change with the wire gauge and current value? To solve this problem, you will need to find a table of wire gauge diameter, resistance, and current ratings. Table 2.2 in this book contains some information. The solution can only be found numerically.

18.30 Derive the same result obtained in Example 18.10, using equation 18.46 and the definition of inductance given in equation 18.30. You will first compute the inductance of the magnetic circuit as a function of the reluctance, then compute the stored magnetic energy, and finally write the expression for the magnetic force given in equation 18.46.

18.31 Derive the same result obtained in Example 18.11, using equation 18.46 and the definition of inductance given in equation 18.30. You will first compute the inductance of the magnetic circuit as a function of the reluctance, then compute the stored magnetic energy, and finally write the expression for the magnetic force given in equation 18.46.

18.32 With reference to Example 18.11, generate a simulation program (e.g., using Simulink™) that accounts for the fact that the solenoid inductance is not constant, but is a function of plunger position. Compare graphically the current and force step responses of the constant-L simplified solenoid model to the step responses obtained in Example 18.11. Assume $\mu_r = 1{,}000$.

18.33 With reference to Example 18.12, calculate the required holding current to keep the relay closed. The mass of the moving element is $m = 0.05$ kg. Neglect damping. The initial position is $x = \epsilon = 0.001$ m.

18.34 The relay circuit shown in Figure P18.34 has the following parameters: $A_{\text{gap}} = 0.001\ \text{m}^2$; $N = 500$ turns; $L = 0.02$ m; $\mu = \mu_0 = 4\pi \times 10^{-7}$ (neglect the iron reluctance); $k = 1{,}000$ N/m; $R = 18\ \Omega$. What is the minimum DC supply voltage v for which the relay will make contact when the electrical switch is closed?

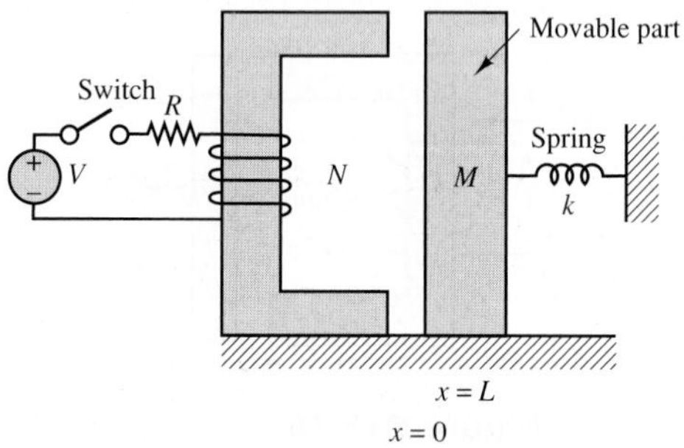

Figure P18.34

18.35 The magnetic circuit shown in Figure P18.35 is a very simplified representation of devices used as *surface roughness sensors*. The stylus is in contact with the surface and causes the plunger to move along with the surface. Assume that the flux ϕ in the gap is given by the expression $\phi = \beta/\mathcal{R}(x)$, where β is a known constant and $\mathcal{R}(x)$ is the reluctance of the gap. The emf e is measured to determine the surface profile. Derive an expression for the displacement x as a function of the various parameters of the magnetic circuit and of the measured emf. (Assume a frictionless contact between the moving plunger and the magnetic structure and that the plunger is restrained to vertical motion only. The cross-sectional area of the plunger is A.)

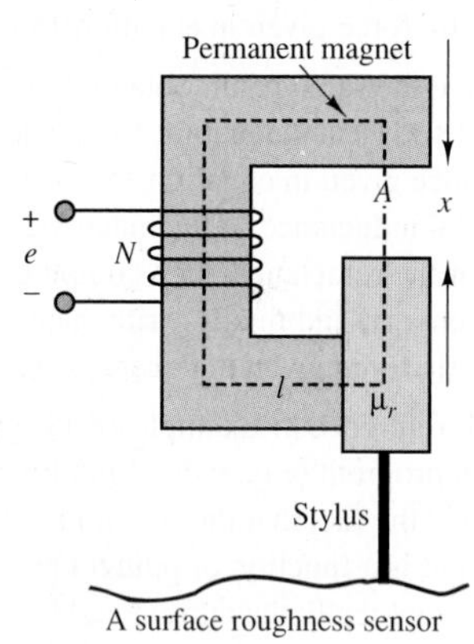

Figure P18.35 A surface roughness sensor

18.36 A cylindrical solenoid is shown in Figure P18.36. The plunger may move freely along its axis. The air gap between the shell and the plunger is uniform and equal to 1 mm, and the diameter d is 25 mm. If the exciting coil carries a current of 7.5 A, find the force acting on the plunger when $x = 2$ mm. Assume $N = 200$ turns, and neglect the reluctance of the steel shell. Assume l_g is negligible.

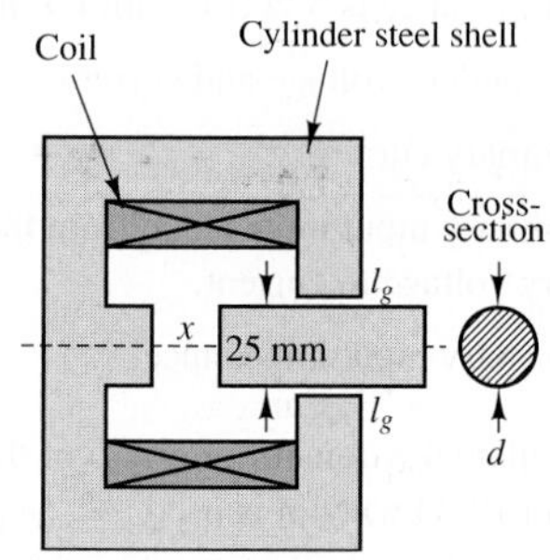

Figure P18.36

18.37 The double-excited electromechanical system shown in Figure P18.37 moves horizontally. Assume that resistance, magnetic leakage, and fringing are negligible; the permeability of the core is very large; and the cross section of the structure is $w \times w$. Find

a. The reluctance of the magnetic circuit.

b. The magnetic energy stored in the air gap.

c. The force on the movable part as a function of its position.

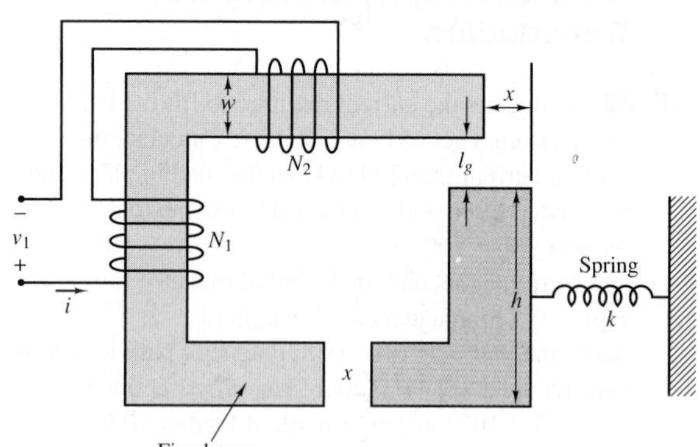

Figure P18.37

18.38 Determine the force F between the faces of the poles (stationary coil and plunger) of the solenoid pictured in Figure P18.38 when it is energized. When energized, the plunger is drawn into the coil and comes to rest with only a negligible air gap separating the two. The flux density in the cast steel pathway is 1.1 T. The diameter of the plunger is 10 mm. Assume that the reluctance of the steel is negligible.

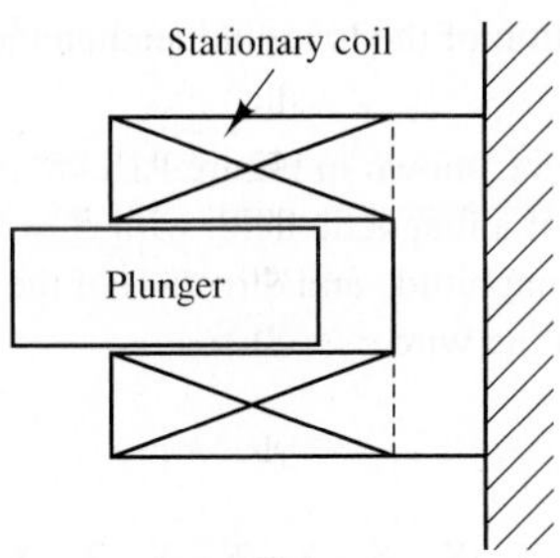

Figure P18.38

18.39 An electromagnet is used to support a solid piece of steel, as shown in Example 18.9. A force of 10,000 N is required to support the weight. The cross-sectional area of the magnetic core (the fixed part) is 0.01 m^2. The coil has 1,000 turns. Determine the minimum current that can keep the weight from falling for $x = 1.0$ mm. Assume negligible reluctance for the steel parts and negligible fringing in the air gaps.

18.40 The armature, frame, and core of a 12-V DC control relay are made of sheet steel. The average length of the magnetic circuit is 12 cm when the relay is energized, and the average cross section of the magnetic circuit is 0.60 cm^2. The coil is wound with 250 turns and carries 50 mA. Determine

a. The flux density $\mathcal{B}$ in the magnetic circuit of the relay when the coil is energized.

b. The force $\mathcal{F}$ exerted on the armature to close it when the coil is energized.

18.41 A relay is shown in Figure P18.41. Find the differential equations describing the system.

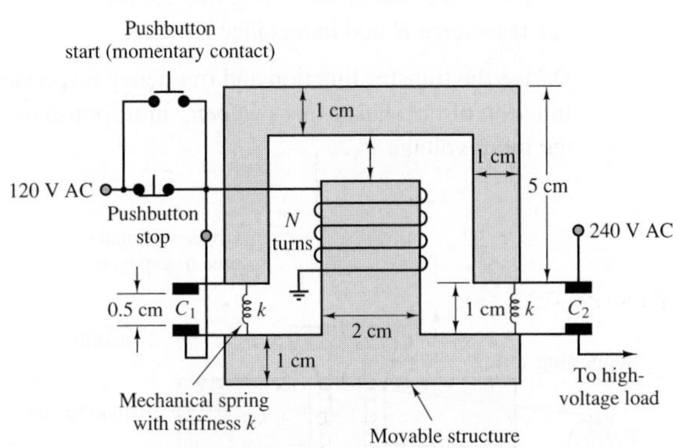

Figure P18.41

18.42 A solenoid having a cross section of 5 cm^2 is shown in Figure P18.42.

a. Calculate the force exerted on the plunger when the distance x is 2 cm and the current in the coil (where $N = 100$ turns) is 5 A. Assume that the fringing and leakage effects are negligible. The relative permeabilities of the magnetic material and the nonmagnetic sleeve are 2,000 and 1.

b. Develop a set of differential equations governing the behavior of the solenoid.

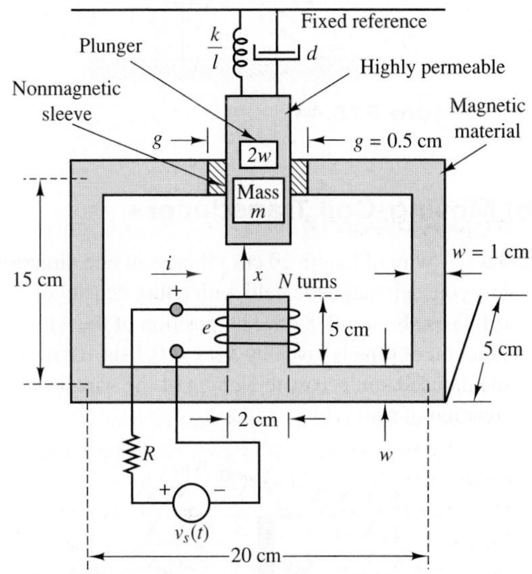

Figure P18.42

18.43 Derive the differential equations (electrical and mechanical) for the relay shown in Figure P18.43. Do not assume that the inductance is fixed; it is a function of x. You may assume that the iron reluctance is negligible.

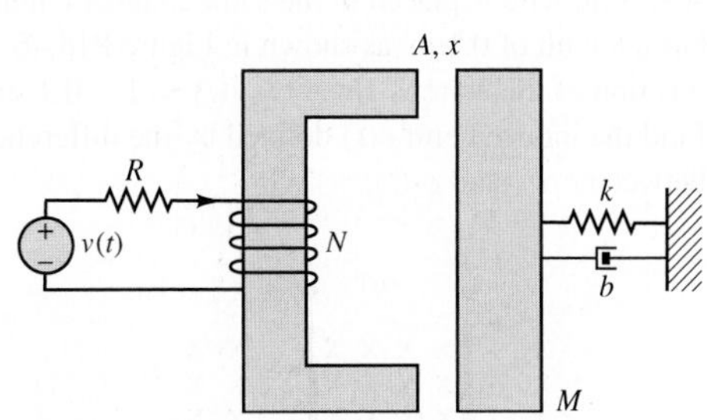

Figure P18.43

18.44 Derive the complete set of differential equations describing the relay shown in Figure P18.44.

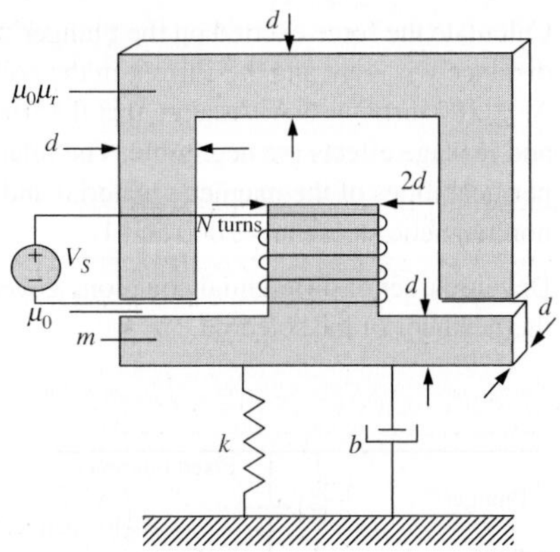

Figure P18.44

(b) Moving-Coil Transducers

18.45 A wire of length 20 cm vibrates in one direction in a constant magnetic field with a flux density of 0.1 T; see Figure P18.45. The position of the wire as a function of time is given by $x(t) = 0.1 \sin 10t$ m. Find the induced emf across the length of the wire as a function of time.

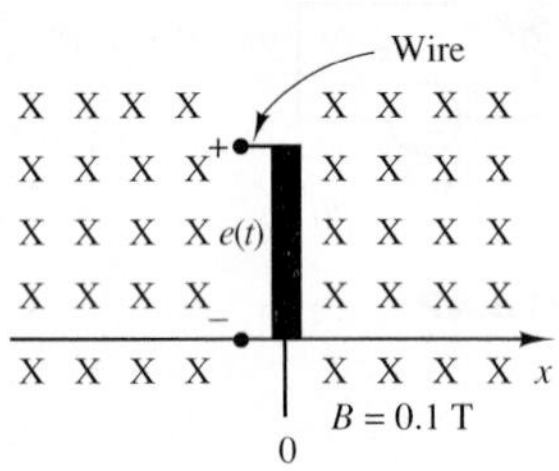

Figure P18.45

18.46 The wire of Problem 18.45 induces a time-varying emf of

$$e_1(t) = 0.02 \cos 10t$$

A second wire is placed in the same magnetic field but has a length of 0.1 m, as shown in Figure P18.46. The position of this wire is given by $x(t) = 1 - 0.1 \sin 10t$. Find the induced emf $e(t)$ defined by the difference between $e_1(t)$ and $e_2(t)$.

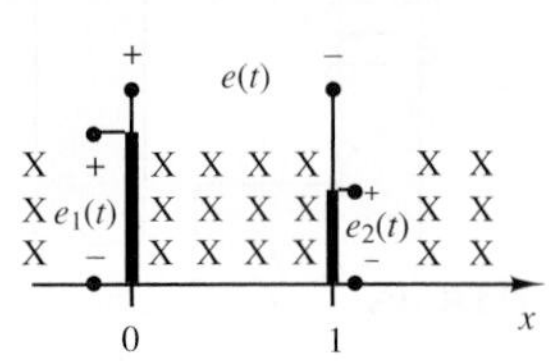

Figure P18.46

18.47 A conducting bar shown in Figure 18.48 in the text is carrying 4 A of current in the presence of a magnetic field; $B = 0.3$ Wb/m^2. Find the magnitude and direction of the force induced on the conducting bar.

18.48 A wire, shown in Figure P18.48, is moving in the presence of a magnetic field, with $B = 0.4$ Wb/m^2. Find the magnitude and direction of the induced voltage in the wire.

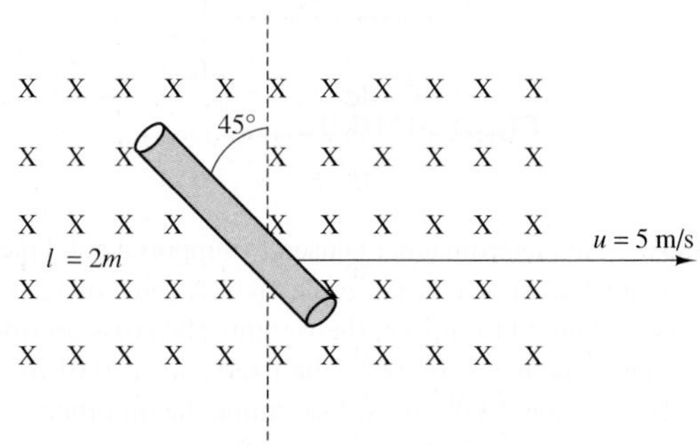

Figure P18.48

18.49 The electrodynamic shaker shown in Figure P18.49 is commonly used as a vibration tester. A constant current is used to generate a magnetic field in which the armature coil of length l is immersed. The shaker platform with mass m is mounted in the fixed structure by way of a spring with stiffness k. The platform is rigidly attached to the armature coil, which slides on the fixed structure thanks to frictionless bearings.

a. Neglecting iron reluctance, determine the reluctance of the fixed structure, and hence compute the strength of the magnetic flux density B in which the armature coil is immersed.

b. Knowing B, determine the dynamic equations of motion of the shaker, assuming that the moving coil has resistance R and inductance L.

c. Derive the transfer function and frequency response function of the shaker mass *velocity* in response to the input voltage V_S.

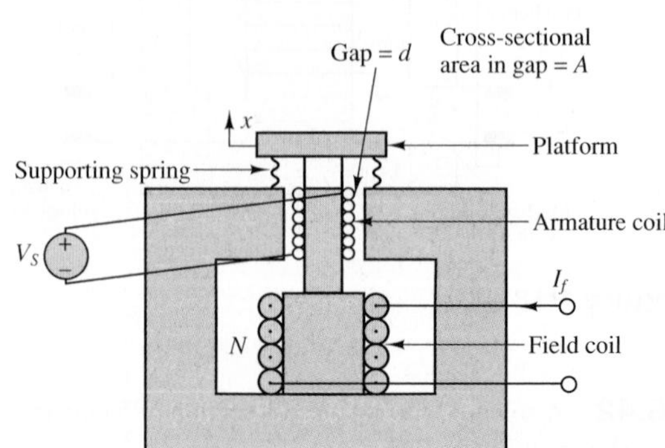

Figure P18.49 Electrodynamic shaker

18.50 The electrodynamic shaker of Figure P18.49 is used to perform vibration testing of an electrical connector. The connector is placed on the test platform (with mass m), and it may be assumed to have negligible mass when compared to the platform. The test consists of shaking the connector at the frequency $\omega = 2\pi \times 100$ rad/s.

Given the parameter values $B = 1{,}000$ Wb/m^2, $l = 5$ m, $k = 1{,}000$ N/m, $m = 1$ kg, $b = 5$ N-s/m, $L = 0.8$ H, and $R = 0.5\ \Omega$, determine the peak amplitude of the sinusoidal voltage V_S required to generate an acceleration of $5g$ (49 m/s^2) under the stated test conditions.

18.51 Derive and sketch the frequency response of the loudspeaker of Example 18.13 for (1) $k = 50{,}000$ N/m and (2) $k = 5 \times 10^6$ N/m. Describe qualitatively how the loudspeaker frequency response changes as the spring stiffness k increases and decreases. What will the frequency response be in the limit as k approaches zero? What kind of speaker would this condition correspond to?

18.52 The loudspeaker of Example 18.13 has a midrange frequency response. Modify the mechanical parameters of the loudspeaker (mass, damping, and spring rate), so as to obtain a loudspeaker with a bass response centered on 400 Hz. Demonstrate that your design accomplishes the intended task, using frequency response plots. *Note: This is an open-ended design problem.*

18.53 The electrodynamic shaker shown in Figure P18.53 is used to perform vibration testing of an electronic circuit. The circuit is placed on a test table with mass m, and is assumed to have negligible mass when compared to the table. The test consists of shaking the circuit at the frequency $\omega = 2\pi(100)$ rad/s.

a. Write the dynamic equations for the shaker. Clearly indicate system input(s) and output(s).

b. Find the frequency response function of the table acceleration in response to the applied voltage.

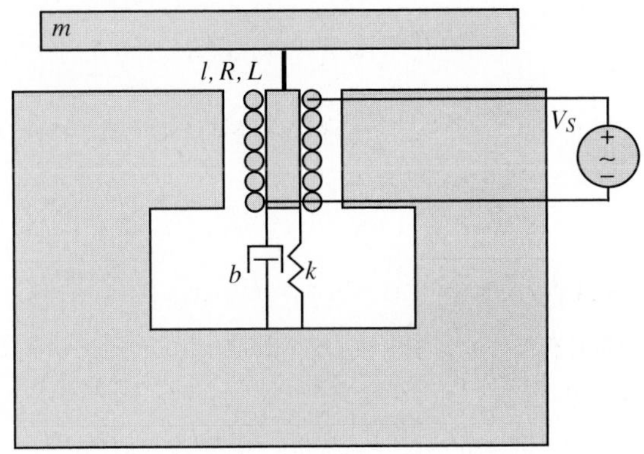

Figure P18.53

c. Given the following parameter values:

$B = 200$ Wb/m^2; $l = 5$ m; $k = 100$ N/m; $m = 0.2$ kg;
$b = 5$ N − s/m; $L = 8$ mH; $R = 0.5\ \Omega$;

Determine the peak amplitude of the sinusoidal voltage V_S required to generate an acceleration of 5g (49 m/s^2) under the stated test conditions.

CHAPTER

19

INTRODUCTION TO ELECTRIC MACHINES

The objective of this chapter is to introduce the basic operation of rotating electric machines. The operation of the three major classes of electric machines—DC, synchronous, and induction—first is described as intuitively as possible, building on the material presented in Chapter 18. The second part of the chapter is devoted to a discussion of the applications and selection criteria for the different classes of machines.

The emphasis of this chapter is on explaining the properties of each type of machine, with its advantages and disadvantages with regard to other types; and on classifying these machines in terms of their performance characteristics and preferred field of application.

Learning Objectives

1. Understand the basic principles of operation of rotating electric machines, their classification, and basic efficiency and performance characteristics. *Section 19.1*.
2. Understand the operation and basic configurations of separately excited, permanent-magnet, shunt and series DC machines. *Section 19.2*.
3. Analyze DC generators at steady state. *Section 19.3*.
4. Analyze DC motors under steady-state and dynamic operation. *Section 19.4*.
5. Understand the operation and basic configuration of AC machines, including the synchronous motor and generator, and the induction machine. *Sections 19.5 through 19.8*.

19.1 ROTATING ELECTRIC MACHINES

The range of sizes and power ratings and the different physical features of rotating machines are such that the task of explaining the operation of rotating machines in a single chapter may appear formidable at first. Some features of rotating machines, however, are common to all such devices. This introductory section is aimed at explaining the common properties of all rotating electric machines. We begin our discussion with reference to Figure 19.1, in which a hypothetical rotating machine is depicted in a cross-sectional view. In the figure, a box with a cross inscribed in it indicates current flowing into the page, while a dot represents current out of the plane of the page.

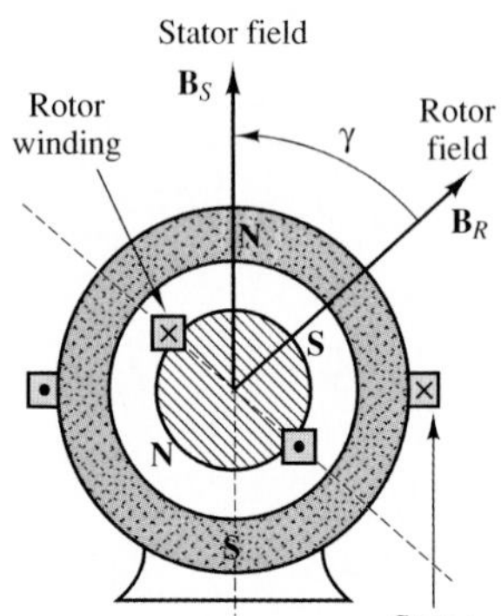

Figure 19.1 A rotating electric machine

In Figure 19.1, we identify a **stator,** of cylindrical shape, and a **rotor,** which, as the name indicates, rotates inside the stator, separated from the latter by means of an air gap. The rotor and stator each consist of a magnetic core, some electrical insulation, and the windings necessary to establish a magnetic flux (unless this is created by a permanent magnet). The rotor is mounted on a bearing-supported shaft, which can be connected to *mechanical loads* (if the machine is a motor) or to a *prime mover* (if the machine is a generator) by means of belts, pulleys, chains, or other mechanical couplings. The windings carry the electric currents that generate the magnetic fields and flow to the electrical loads, and also provide the closed loops in which voltages will be induced (by virtue of Faraday's law, as discussed in Chapter 18).

Basic Classification of Electric Machines

An immediate distinction can be made between different types of windings characterized by the nature of the current they carry. If the current serves the sole purpose of providing a magnetic field and is independent of the load, it is called a *magnetizing,* or excitation, current, and the winding is termed a **field winding.** Field currents are nearly always direct current (DC) and are of relatively low power, since their only purpose is to magnetize the core (recall the important role of high-permeability cores in generating large magnetic fluxes from relatively small currents). On the other hand, if the winding carries only the load current, it is called an **armature.** In DC and alternating-current (AC) synchronous machines, separate windings exist to carry field and armature currents. In the induction motor, the magnetizing and load currents flow in the same winding, called the *input winding,* or *primary;* the output winding is then called the *secondary*. As we shall see, this terminology, which is reminiscent of transformers, is particularly appropriate for induction motors, which bear a significant analogy to the operation of the transformers studied in Chapters 7 and 18. Table 19.1 characterizes the principal machines in terms of their field and armature configuration.

LO1

Table 19.1 Configurations of the three types of electric machines

Machine type	Winding	Winding type	Location	Current
DC	Input and output	Armature	Rotor	AC (winding) DC (at brushes)
	Magnetizing	Field	Stator	DC
Synchronous	Input and output	Armature	Stator	AC
	Magnetizing	Field	Rotor	DC
Induction	Input	Primary	Stator	AC
	Output	Secondary	Rotor	AC

It is also useful to classify electric machines in terms of their energy conversion characteristics. A machine acts as a **generator** if it converts mechanical energy from a prime mover, say, an internal combustion engine, to electrical form. Examples of generators are the large machines used in power generating plants, or the common automotive alternator. A machine is classified as a **motor** if it converts electrical energy to mechanical form. The latter class of machines is probably of more direct interest to you, because of its widespread application in engineering practice. Electric motors are used to provide forces and torques to generate motion in countless industrial applications. Machine tools, robots, punches, presses, mills, and propulsion systems for electric vehicles are but a few examples of the application of electric machines in engineering.

Note that in Figure 19.1 we have explicitly shown the direction of two magnetic fields: that of the rotor $\mathbf{B}_R$ and that of the stator $\mathbf{B}_S$. Although these fields are generated by different means in different machines (e.g., permanent magnets, alternating currents, direct currents), the presence of these fields is what causes a rotating machine to turn and enables the generation of electric power. In particular, we see that in Figure 19.1 the north pole of the rotor field will seek to align itself with the south pole of the stator field. It is this magnetic attraction force that permits the generation of torque in an electric motor; conversely, a generator exploits the laws of electromagnetic induction to convert a changing magnetic field to an electric current.

To simplify the discussion in later sections, we now introduce some basic concepts that apply to all rotating electric machines. Referring to Figure 19.2, we note that for all machines the force on a wire is given by the expression

$$\mathbf{f} = i_w \mathbf{l} \times \mathbf{B} \qquad \textbf{(19.1)}$$

where i_w is the current in the wire, $\mathbf{l}$ is a vector along the direction of the wire, and $\times$ denotes the cross product of two vectors. Then the torque for a multiturn coil becomes

$$T = KBi_w \sin\alpha \qquad \textbf{(19.2)}$$

where

B = magnetic flux density caused by stator field

K = constant depending on coil geometry

α = angle between $\mathbf{B}$ and normal to plane of coil

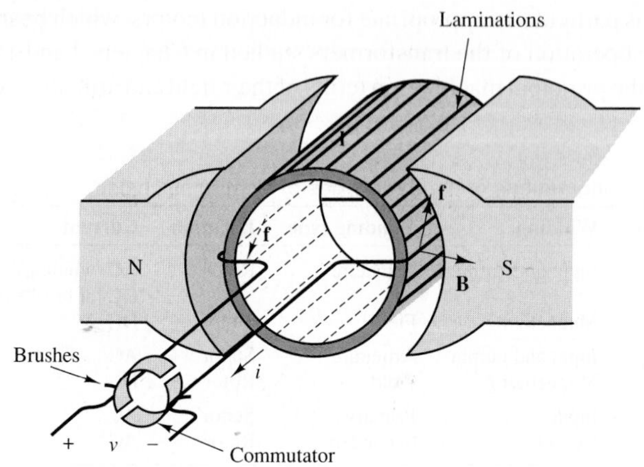

Figure 19.2 Stator and rotor fields and the force acting on a rotating machine

In the hypothetical machine of Figure 19.2, there are two magnetic fields: one generated within the stator, the other within the rotor windings. Either (but not both) of these fields could be generated by a current or by a permanent magnet. Thus, we could replace the permanent-magnet stator of Figure 19.2 with a suitably arranged winding to generate a stator field in the same direction. If the stator were made of a toroidal coil of radius R (see Chapter 18), then the magnetic field of the stator would generate a flux density B, where

$$B = \mu H = \mu \frac{Ni}{2\pi R} \tag{19.3}$$

and where N is the number of turns and i is the coil current. The direction of the torque is always the direction determined by the rotor and stator fields as they seek to align to each other (i.e., counterclockwise in the diagram of Figure 19.1).

It is important to note that Figure 19.2 is merely a general indication of the major features and characteristics of rotating machines. A variety of configurations exist, depending on whether each of the fields is generated by a current in a coil or by a permanent magnet and whether the load and magnetizing currents are direct or alternating. The type of excitation (AC or DC) provided to the windings permits a first classification of electric machines (see Table 19.1). According to this classification, one can define the following types of machines:

LO1

- *Direct-current machines:* DC in both stator and rotor
- *Synchronous machines:* AC in one winding, DC in the other
- *Induction machines:* AC in both

In most industrial applications, the induction machine is the preferred choice, because of the simplicity of its construction. However, the analysis of the performance of an induction machine is rather complex. On the other hand, DC machines are quite complex in their construction but can be analyzed relatively simply with the analytical tools we have already acquired. Therefore, the progression of this chapter is as follows. We start with a section that discusses the physical construction of DC machines, both motors and generators. Then we continue with a discussion of synchronous machines, in which one of the currents is now alternating, since these can easily be understood as an extension of DC machines. Finally, we consider the case where both rotor and stator currents are alternating, and we analyze the induction machine.

Performance Characteristics of Electric Machines

As already stated earlier in this chapter, electric machines are **energy conversion devices,** and we are therefore interested in their energy conversion **efficiency.** Typical applications of electric machines as motors or generators must take into consideration the energy losses associated with these devices. Figure 19.3(a) and (b) represents the various loss mechanisms you must consider in analyzing the efficiency of an electric machine for the case of direct-current machines. It is important for you to keep in mind this conceptual flow of energy when analyzing electric machines. The sources of loss in a rotating machine can be separated into three fundamental groups: electrical (I^2R) losses, core losses, and mechanical losses.

Usually I^2R losses are computed on the basis of the DC resistance of the windings at 75°C; in practice, these losses vary with operating conditions. The difference

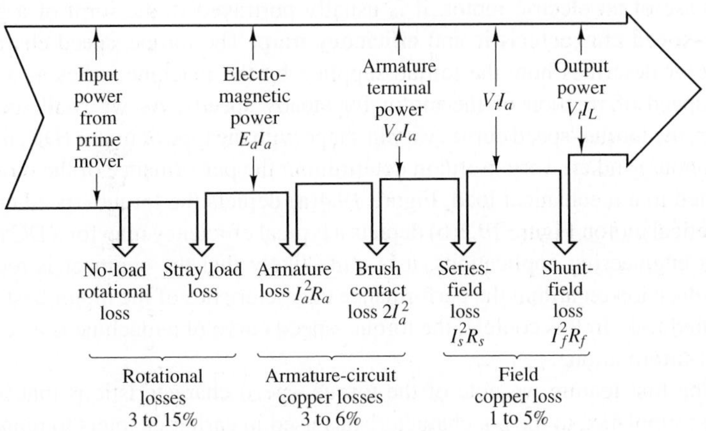

Figure 19.3a Generator losses, direct current

LO1

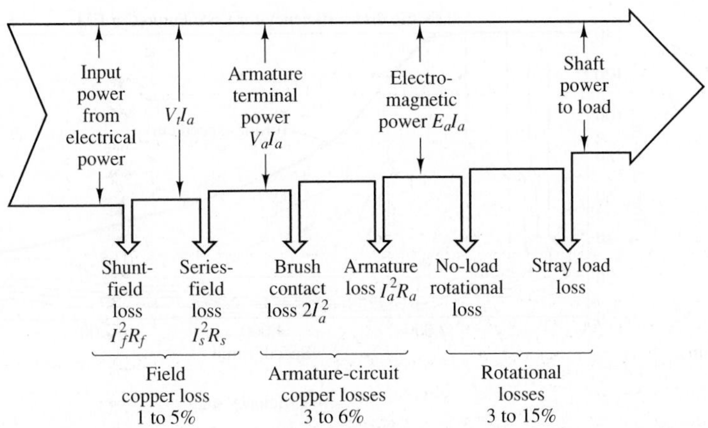

Figure 19.3b Motor losses, direct current

LO1

between the nominal and actual I^2R loss is usually lumped under the category of *stray-load loss*. In direct-current machines, it is also necessary to account for the *brush contact loss* associated with slip rings and commutators.

Mechanical losses are due to *friction* (mostly in the bearings) and *windage*, that is, the air drag force that opposes the motion of the rotor. In addition, if external devices (e.g., blowers) are required to circulate air through the machine for cooling purposes, the energy expended by these devices is included in the mechanical losses.

Open-circuit core losses consist of *hysteresis* and *eddy current* losses, with only the excitation winding energized (see Chapter 18 for a discussion of hysteresis and eddy currents). Often these losses are summed with friction and windage losses to give rise to the *no-load rotational loss*. The latter quantity is useful if one simply wishes to compute efficiency. Since open-circuit core losses do not account for the changes in flux density caused by the presence of load currents, an additional magnetic loss is incurred that is not accounted for in this term. *Stray-load losses* are used to lump the effects of nonideal current distribution in the windings and of the additional core losses just mentioned. Stray-load losses are difficult to determine exactly and are often assumed to be equal to 1.0 percent of the output power for DC machines; these losses can be determined by experiment in synchronous and induction machines.

The performance of an electric machine can be quantified in a number of ways. In the case of an electric motor, it is usually portrayed in the form of a graphical **torque–speed characteristic** and **efficiency map.** The torque–speed characteristic of a motor describes how the torque supplied by the machine varies as a function of the speed of rotation of the motor for steady speeds. As we shall see in later sections, the torque–speed curves vary in shape with the type of motor (DC, induction, synchronous) and are very useful in determining the performance of the motor when connected to a mechanical load. Figure 19.4(a) depicts the torque–speed curve of a hypothetical motor. Figure 19.4(b) depicts a typical efficiency map for a DC machine. In most engineering applications, it is quite likely that the engineer is required to make a decision regarding the performance characteristics of the motor best suited to a specified task. In this context, the torque–speed curve of a machine is a very useful piece of information.

The first feature we note of the torque–speed characteristic is that it bears a strong resemblance to the i-v characteristics used in earlier chapters to represent the

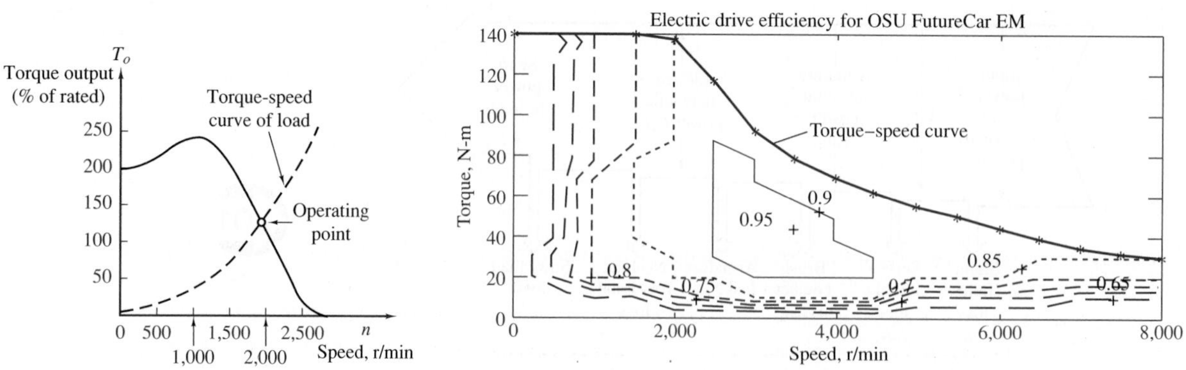

(a) Torque-speed curve (b) Efficiency map

Figure 19.4 Torque–speed and efficiency curves for an electric motor behavior of electrical sources. It should be clear that, according to this torque–speed curve, the motor is not an ideal source of torque (if it were, the curve would appear as a horizontal line across the speed range). One can readily see, for example, that the hypothetical motor represented by the curves of Figure 19.4(a) would produce maximum torque in the range of speeds between approximately 800 and 1,400 r/min. What determines the actual speed of the motor (and therefore its output torque and power) is the torque–speed characteristic of the load connected to it, much as a resistive load determines the current drawn from a voltage source. In the figure, we display the torque–speed curve of a load, represented by the dashed line; the operating point of the motor-load pair is determined by the intersection of the two curves.

Another important observation pertains to the fact that the motor of Figure 19.4(a) produces a nonzero torque at zero speed. This fact implies that as soon as electric power is connected to the motor, the latter is capable of supplying a certain amount of torque; this zero-speed torque is called the **starting torque.** If the load the motor is connected to requires less than the starting torque the motor can provide, then the motor can accelerate the load, until the motor speed and torque settle to a stable value, at the operating point. The motor-load pair of Figure 19.4(a) would behave in the manner just described. However, there may well be circumstances in which a motor might not be able to provide a sufficient starting torque to overcome the static load torque that opposes its motion. Thus, we see that a torque–speed characteristic can offer valuable insight into the operation of a motor. As we discuss each type of machine in greater detail, we shall devote some time to the discussion of its torque–speed curve.

The efficiency of an electric machine is also an important design and performance characteristic. The **1995 Department of Energy Energy Policy Act,** also known as EPACT, has required electric motor manufacturers to guarantee a minimum efficiency. The efficiency of an electric motor is usually described using a contour plot of the efficiency value (a number between 0 and 1) in the torque–speed plane. This representation permits a determination of the motor efficiency as a function of its performance and operating conditions. Figure 19.4(b) depicts the efficiency map of an electric drive used in a hybrid-electric vehicle—a 20-kW permanent-magnet AC (or brushless DC) machine. We shall discuss this type of machine in Chapter 20. Note that the peak efficiency can be as high as 0.95 (95 percent), but that the efficiency decreases significantly away from the optimum point (around 3,500 r/min and 45 N-m), to values as low as 0.65.

FIND IT ON THE WEB

The most common means of conveying information regarding electric machines is the *nameplate*. Typical information conveyed by the nameplate includes

1. Type of device (e.g., DC motor, alternator)
2. Manufacturer
3. Rated voltage and frequency
4. Rated current and voltamperes
5. Rated speed and horsepower

The **rated voltage** is the terminal voltage for which the machine was designed, and which will provide the desired magnetic flux. Operation at higher voltages will increase magnetic core losses, because of excessive core saturation. The **rated current** and **rated voltamperes** are an indication of the typical current and power levels at the terminal that will not cause undue overheating due to copper losses (I^2R losses) in the windings. These ratings are not absolutely precise, but they give an indication of the range of excitations for which the motor will perform without overheating.

Peak power operation in a motor may exceed rated torque, power, or currents by a substantial factor (up to as much as 6 or 7 times the rated value); however, continuous operation of the motor above the rated performance will cause the machine to overheat and eventually to sustain damage. Thus, it is important to consider both peak and continuous power requirements when selecting a motor for a specific application. An analogous discussion is valid for the speed rating: While an electric machine may operate above rated speed for limited periods of time, the large centrifugal forces generated at high rotational speeds will eventually cause undesirable mechanical stresses, especially in the rotor windings, leading eventually even to self-destruction.

Another important feature of electric machines is the **regulation** of the machine speed or voltage, depending on whether it is used as a motor or as a generator, respectively. Regulation is the ability to maintain speed or voltage constant in the face of load variations. The ability to closely regulate speed in a motor or voltage in a generator is an important feature of electric machines; regulation is often improved by means of feedback control mechanisms, some of which are briefly introduced in this chapter. We take the following definitions as being adequate for the intended purpose of this chapter:

$$\text{Speed regulation} = \frac{\text{Speed at no load} - \text{Speed at rated load}}{\text{Speed at rated load}} \qquad \textbf{(19.4)}$$

$$\text{Voltage regulation} = \frac{\text{Voltage at no load} - \text{Voltage at rated load}}{\text{Voltage at rated load}} \qquad \textbf{(19.5)}$$

Please note that the rated value is usually taken to be the nameplate value, and that the meaning of *load* changes depending on whether the machine is a motor, in which case the load is mechanical, or a generator, in which case the load is electrical.

LO1

EXAMPLE 19.1 Regulation

Problem

Find the percentage of speed regulation of a shunt DC motor.

Solution

Known Quantities: No-load speed; speed at rated load.

Find: Percentage speed regulation, denoted by SR%.

Schematics, Diagrams, Circuits, and Given Data:

$$n_{\text{nl}} = \text{no-load speed} = 1{,}800 \text{ r/min}$$

$$n_{\text{rl}} = \text{rated load speed} = 1{,}760 \text{ r/min}$$

Analysis:

$$\text{SR\%} = \frac{n_{\text{nl}} - n_{\text{rl}}}{n_{\text{rl}}} \times 100 = \frac{1{,}800 - 1{,}760}{1{,}800} \times 100 = 2.27\%$$

Comments: Speed regulation is an intrinsic property of a motor; however, external speed controls can be used to regulate the speed of a motor to any (physically achievable) desired value. Some motor control concepts are discussed later in this chapter.

CHECK YOUR UNDERSTANDING

The percentage of speed regulation of a motor is 10 percent. If the full-load speed is 50π rad/s, find (a) the no-load speed in radians per second, and (b) the no-load speed in revolutions per minute, (c) the percentage of voltage regulation for a 250-V generator is 10 percent. Find the no-load voltage of the generator.

Answer: (a) $\omega = 55\pi$ rad/s; (b) $n = 1{,}650$ r/min; (c) $V_{\text{no-load}} = 275$ V

EXAMPLE 19.2 Nameplate Data

LO1

Problem

Discuss the nameplate data, shown below, of a typical induction motor.

Solution

Known Quantities: Nameplate data.

Find: Motor characteristics.

Schematics, Diagrams, Circuits, and Given Data: The nameplate appears below.

MODEL	19308 J-X		
TYPE	CJ4B	FRAME	324TS
VOLTS	230/460	°C AMB.	40
		INS. CL.	B
FRT. BRG	210SF	EXT. BRG	312SF
SERV FACT	1.0	OPER INSTR	C-517
PHASE | 3	Hz | 60	CODE | G	WDGS | 1
H.P.	40		
R.P.M.	3,565		
AMPS	106/53		
NEMA NOM.	EFF		
NOM. P.F.			
DUTY	CONT.	NEMA DESIGN	B

Analysis: The nameplate of a typical induction motor is shown in the preceding table. The model number (sometimes abbreviated as MOD) uniquely identifies the motor to the manufacturer. It may be a style number, a model number, an identification number, or an instruction sheet reference number.

The term *frame* (sometimes abbreviated as FR) refers principally to the physical size of the machine, as well as to certain construction features.

Ambient temperature (abbreviated as AMB, or MAX. AMB) refers to the maximum ambient temperature in which the motor is capable of operating. Operation of the motor in a higher ambient temperature may result in shortened motor life and reduced torque.

Insulation class (abbreviated as INS. CL.) refers to the type of insulation used in the motor. Most often used are class A (105°C) and class B (130°C).

The duty (DUTY), or time rating, denotes the length of time the motor is expected to be able to carry the rated load under usual service conditions. "CONT." means that the machine can be operated continuously.

The "CODE" letter sets the limits of starting kilovoltamperes per horsepower for the machine. There are 19 levels, denoted by the letters A through V, excluding I, O, and Q.

Service factor (abbreviated as SERV FACT) is a term defined by NEMA (the National Electrical Manufacturers Association) as follows: "The service factor of a general-purpose alternating-current motor is a multiplier which, when applied to the rated horsepower, indicates a permissible horsepower loading which may be carried under the conditions specified for the service factor."

The voltage figure given on the nameplate refers to the voltage of the supply circuit to which the motor should be connected. Sometimes two voltages are given, for example, 230/460. In this case, the machine is intended for use on either a 230-V or a 460-V circuit. Special instructions will be provided for connecting the motor for each of the voltages.

The term "BRG" indicates the nature of the bearings supporting the motor shaft.

CHECK YOUR UNDERSTANDING

The nameplate of a three-phase induction motor indicates the following values:

$$\text{H.P.} = 10 \qquad \text{Volt} = 220\ \text{V}$$

$$\text{R.P.M.} = 1{,}750 \qquad \text{Service factor} = 1.15$$

$$\text{Temperature rise} = 60°\text{C} \qquad \text{Amp} = 30\ \text{A}$$

Find the rated torque, rated voltamperes, and maximum continuous output power.

Answer: $T_{\text{rated}} = 40.7$ N-m; rated VA $= 11{,}431$ VA; $P_{\text{max}} = 11.5$ hp.

LO1

EXAMPLE 19.3 Torque–Speed Curves

Problem

Discuss the significance of the torque–speed curve of an electric motor.

Solution

A variable-torque variable-speed motor has a torque output that varies directly with speed; hence, the horsepower output varies directly with the speed. Motors with this characteristic are commonly used with fans, blowers, and centrifugal pumps. Figure 19.5 shows typical torque–speed curves for this type of motor. Superimposed on the motor torque–speed curve is the torque–speed curve for a typical fan where the input power to the fan varies as the cube of the fan speed. Point *A* is the desired operating point, which could be determined graphically by plotting the load line and the motor torque–speed curve on the same graph, as illustrated in Figure 19.5.

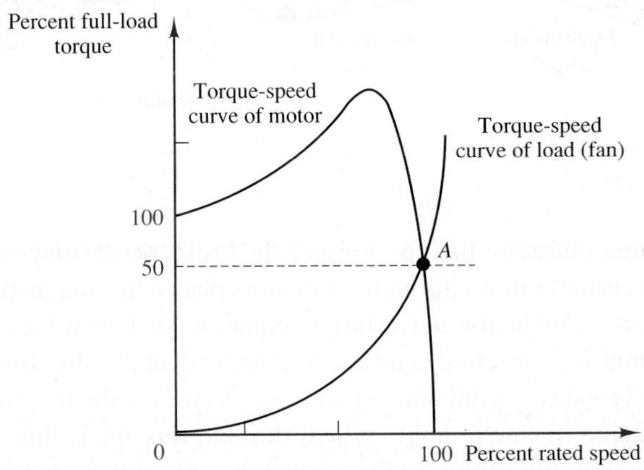

Figure 19.5 Torque-speed curves of electric motor and load

CHECK YOUR UNDERSTANDING

A motor having the characteristics shown in Figure 19.4(a) is to drive a load; the load has a linear torque–speed curve and requires 150 percent of rated torque at 1,500 r/min. Find the operating point for this motor-load pair.

Answer: 170 percent of rated torque; 1,700 r/min.

Basic Operation of All Rotating Machines

We have already seen in Chapter 18 how the magnetic field in electromechanical devices provides a form of coupling between electrical and mechanical systems. Intuitively, one can identify two aspects of this coupling, both of which play a role in the operation of electric machines:

1. Magnetic attraction and repulsion forces generate mechanical torque.
2. The magnetic field can induce a voltage in the machine windings (coils) by virtue of Faraday's law.

Thus, we may think of the operation of an electric machine in terms of either a motor or a generator, depending on whether the input power is electrical and mechanical power is produced (motor action), or the input power is mechanical and the output power is electrical (generator action). Figure 19.6 illustrates the two cases graphically.

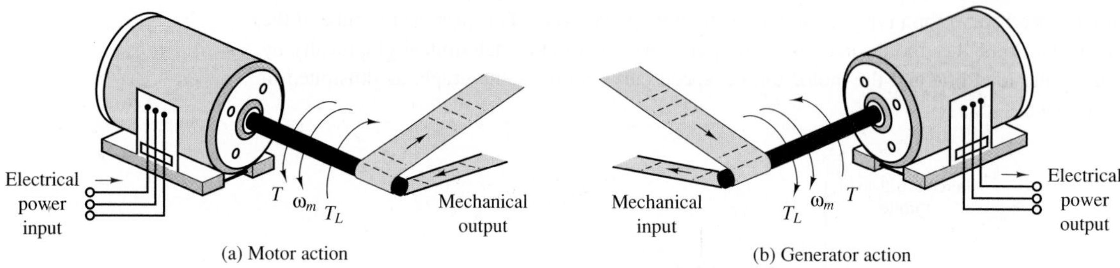

Figure 19.6 Generator and motor action in an electric machine

The coupling magnetic field performs a dual role, which may be explained as follows. When a current i flows through conductors placed in a magnetic field, a force is produced on each conductor, according to equation 19.1. If these conductors are attached to a cylindrical structure, a torque is generated; and if the structure is free to rotate, then it will rotate at an angular velocity ω_m. As the conductors rotate, however, they move through a magnetic field and cut through flux lines, thus generating an electromotive force in opposition to the excitation. This emf is also called *counter-emf,* as it opposes the source of the current i. If, on the other hand, the rotating element of the machine is driven by a prime mover (e.g., an internal combustion engine), then an emf is generated across the coil that is rotating in the magnetic field (the armature). If a load is connected to the armature, a current i will flow to the load, and this current flow will in turn cause a reaction torque on the armature that opposes the torque imposed by the prime mover.

You see, then, that for energy conversion to take place, two elements are required:

1. A coupling field **B**; usually generated in the field winding.
2. An armature winding that supports the load current i and the emf e.

Magnetic Poles in Electric Machines

Before discussing the actual construction of a rotating machine, we should spend a few paragraphs to illustrate the significance of **magnetic poles** in an electric machine. In an electric machine, torque is developed as a consequence of magnetic forces of attraction and repulsion between magnetic poles on the stator and on the rotor; these poles produce a torque that accelerates the rotor and a reaction torque on the stator. Naturally, we would like a construction such that the torque generated as a consequence of the magnetic forces is continuous and in a constant direction. This can be accomplished if the number of rotor poles is equal to the number of stator poles. It is also important to observe that the number of poles must be even, since there have to be equal numbers of north and south poles.

The motion and associated electromagnetic torque of an electric machine are the result of two magnetic fields that are trying to align with each other so that the south pole of one field attracts the north pole of the other. Figure 19.7 illustrates this action by analogy with two permanent magnets, one of which is allowed to rotate about its center of mass.

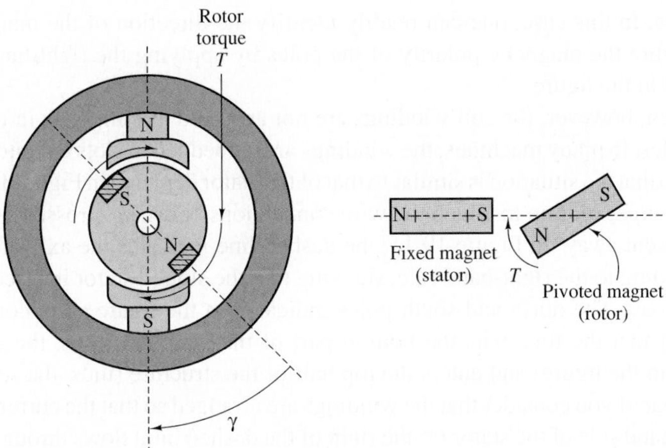

Figure 19.7 Alignment action of poles

Figure 19.8 depicts a two-pole machine in which the stator poles are constructed in such a way as to project closer to the rotor than to the stator structure. This type of construction is rather common, and poles constructed in this fashion are called **salient poles.** Note that the rotor could also be constructed to have salient poles.

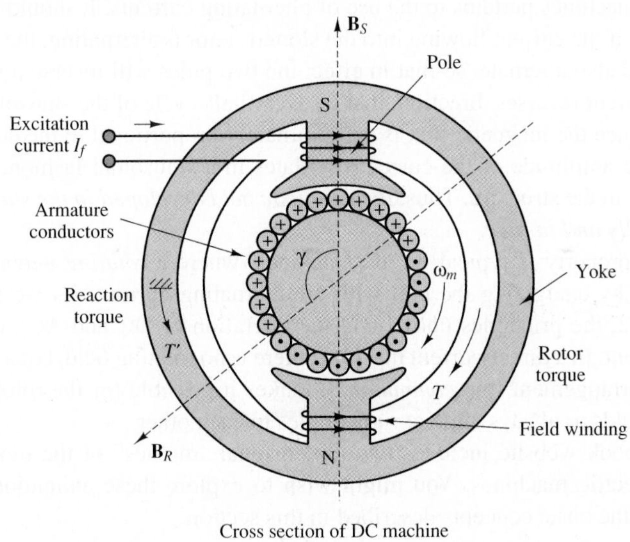

Figure 19.8 A two-pole machine with salient stator poles

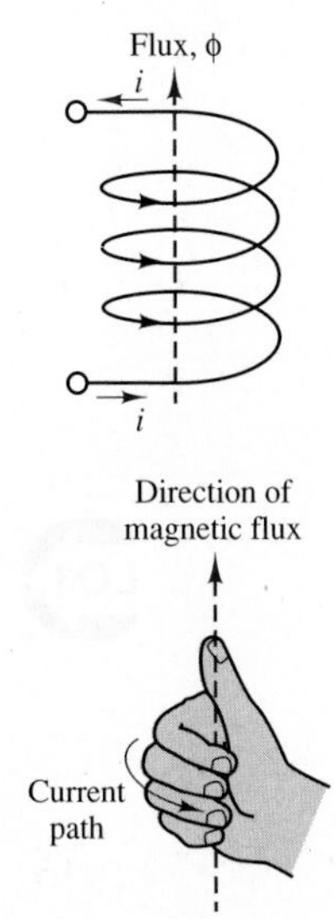

Figure 19.9 Right-hand rule

To understand magnetic polarity, we need to consider the direction of the magnetic field in a coil carrying current. Figure 19.9 shows how the *right-hand rule* can be employed to determine the direction of the magnetic flux. If one were to grasp the coil with the right hand, with the fingers curling in the direction of current flow, then the thumb would be pointing in the direction of the magnetic flux. Magnetic flux by convention is viewed as entering the south pole and exiting from the north pole. Thus, to determine whether a magnetic pole is north or south, we must consider the direction of the flux. Figure 19.10 shows a cross section of a coil wound around a pair of salient rotor poles. In this case, one can readily identify the direction of the magnetic flux and therefore the magnetic polarity of the poles by applying the right-hand rule, as illustrated in the figure.

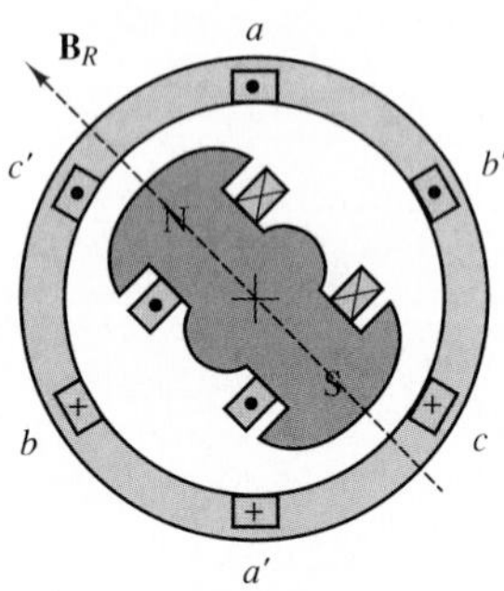

Figure 19.10 Magnetic field in a salient rotor winding

Often, however, the coil windings are not arranged as simply as in the case of salient poles. In many machines, the windings are embedded in slots cut into the stator or rotor, so that the situation is similar to that of the stator depicted in Figure 19.11. This figure is a cross section in which the wire connections between "crosses" and "dots" have been cut away. In Figure 19.11, the dashed line indicates the axis of the stator flux according to the right-hand rule, showing that the slotted stator in effect behaves as a pole pair. The north and south poles indicated in the figure are a consequence of the fact that the flux exits the bottom part of the structure (thus, the north pole indicated in the figure) and enters the top half of the structure (thus, the south pole). In particular, if you consider that the windings are arranged so that the current entering the right-hand side of the stator (to the right of the dashed line) flows through the back end of the stator and then flows outward from the left-hand side of the stator slots (left of the dashed line), you can visualize the windings in the slots as behaving in a manner similar to the coils of Figure 19.10, where the flux axis of Figure 19.11 corresponds to the flux axis of each of the coils of Figure 19.10. The actual circuit that permits current flow is completed by the front and back ends of the stator, where the wires are connected according to the pattern a-a', b-b', c-c', as depicted in the figure.

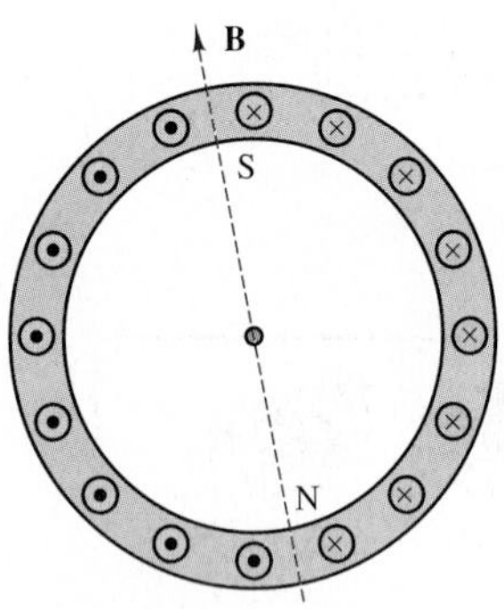

Figure 19.11 Magnetic field of stator

Another important consideration that facilitates understanding of the operation of electric machines pertains to the use of alternating currents. It should be apparent by now that if the current flowing into the slotted stator is alternating, the direction of the flux will also alternate, so that in effect the two poles will reverse polarity every time the current reverses direction, that is, every half-cycle of the sinusoidal current. Further—since the magnetic flux is approximately proportional to the current in the coil—as the amplitude of the current oscillates in a sinusoidal fashion, so will the flux density in the structure. Thus, *the magnetic field developed in the stator changes both spatially and in time*.

This property is typical of AC machines, where a *rotating magnetic field* is established by energizing the coil with an alternating current. As we shall see in Section 19.2, the principles underlying the operation of DC and AC machines are quite different: In a direct-current machine, there is no rotating field, but a mechanical switching arrangement (the *commutator*) makes it possible for the rotor and stator magnetic fields to always align at right angles to each other.

LO1

The book website includes two-dimensional "movies" of the most common types of electric machines. You might wish to explore these animations to better understand the basic concepts described in this section.

19.2 DIRECT-CURRENT MACHINES

As explained in the introductory section, direct-current machines are easier to analyze than their AC counterparts, although their actual construction is made rather complex by the need to have a commutator, which reverses the direction of currents and fluxes to produce a net torque. The objective of this section is to describe the major construction features and the operation of direct-current machines, as well as to develop simple circuit models that are useful in analyzing the performance of this class of machines.

Physical Structure of DC Machines

A representative DC machine was depicted in Figure 19.8, with the magnetic poles clearly identified, for both the stator and the rotor. Figure 19.12 is a photograph of the same type of machine. Note the salient pole construction of the stator and the slotted rotor. As previously stated, the torque developed by the machine is a consequence of the magnetic forces between stator and rotor poles. This torque is maximum when the angle γ between the rotor and stator poles is 90°. Also, as you can see from the figure, in a DC machine the armature is usually on the rotor, and the field winding is on the stator.

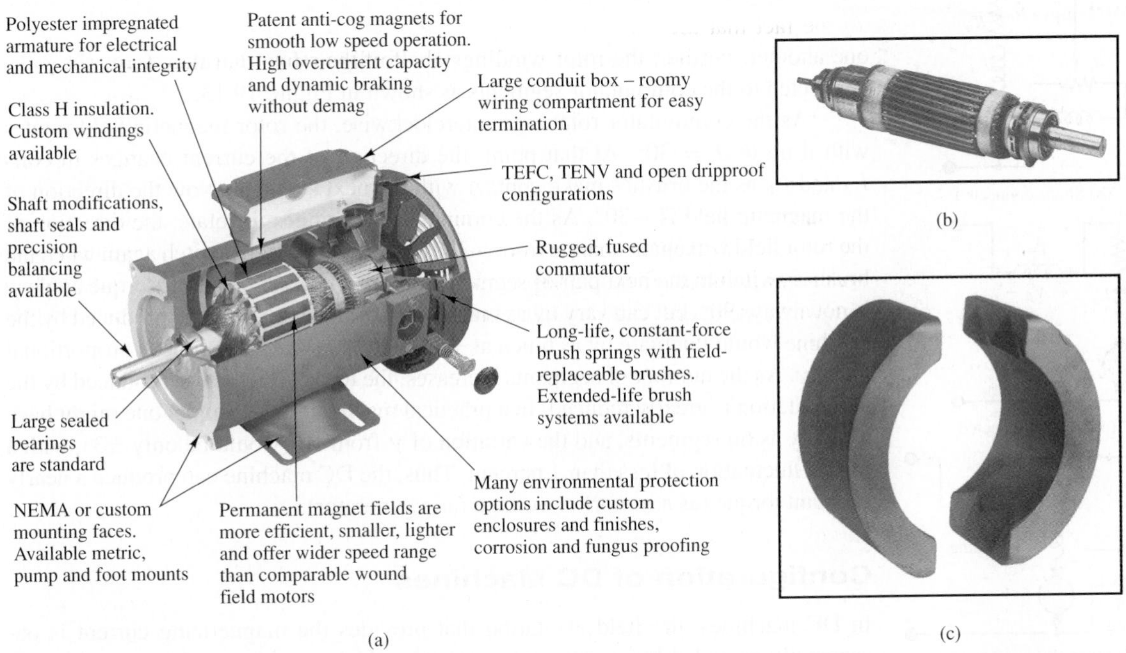

Polyester impregnated armature for electrical and mechanical integrity
Patent anti-cog magnets for smooth low speed operation. High overcurrent capacity and dynamic braking without demag
Large conduit box – roomy wiring compartment for easy termination
Class H insulation. Custom windings available
TEFC, TENV and open dripproof configurations
Shaft modifications, shaft seals and precision balancing available
Rugged, fused commutator
Long-life, constant-force brush springs with field-replaceable brushes. Extended-life brush systems available
Large sealed bearings are standard
Many environmental protection options include custom enclosures and finishes, corrosion and fungus proofing
NEMA or custom mounting faces. Available metric, pump and foot mounts
Permanent magnet fields are more efficient, smaller, lighter and offer wider speed range than comparable wound field motors
(a)
(b)
(c)

Figure 19.12 (a) DC machine; (b) rotor; (c) permanent-magnet stator *(Photos copyright © 2005, Rockwell Automation. All rights reserved. Used with permission.)*

To keep this torque angle constant as the rotor spins on its shaft, a mechanical switch, called a **commutator,** is configured so the current distribution in the rotor winding remains constant, and therefore the rotor poles are consistently at 90° with respect to the fixed stator poles. In a DC machine, the magnetizing current is DC, so that there is no spatial alternation of the stator poles due to time-varying currents. To understand the operation of the commutator, consider the simplified diagram of Figure 19.13. In the figure, the brushes are fixed, and the rotor revolves at an angular velocity ω_m; the instantaneous position of the rotor is given by the expression $\theta = \omega_m t - \gamma$.

The commutator is fixed to the rotor and is made up in this example of six segments that are made of electrically conducting material but are insulated from

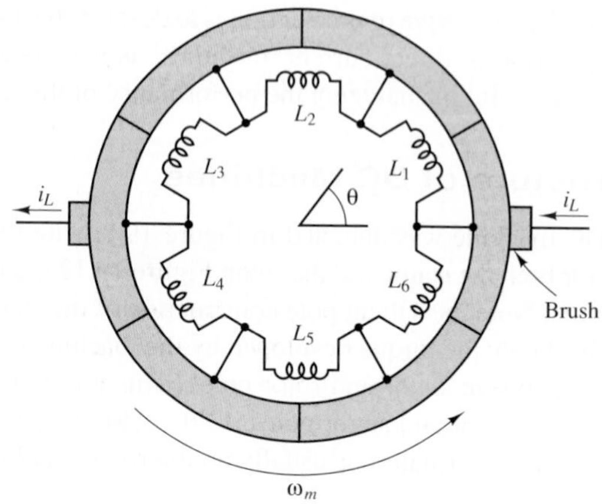

Figure 19.13 Rotor winding and commutator

one another. Further, the rotor windings are configured so that they form six coils, connected to the commutator segments as shown in Figure 19.13.

As the commutator rotates counterclockwise, the rotor magnetic field rotates with it up to $\theta = 30°$. At that point, the direction of the current changes in coils L_3 and L_6 as the brushes make contact with the next segment. Now the direction of the magnetic field is $-30°$. As the commutator continues to rotate, the direction of the rotor field will again change from $-30°$ to $+30°$, and it will switch again when the brushes switch to the next pair of segments. In this machine, then, the torque angle γ is not always $90°$, but can vary by as much as $\pm 30°$; the actual torque produced by the machine would fluctuate by as much as ± 14 percent, since the torque is proportional to $\sin \gamma$. As the number of segments increases, the torque fluctuation produced by the commutation is greatly reduced. In a practical machine, for example, one might have as many as 60 segments, and the variation of γ from $90°$ would be only $\pm 3°$, with a torque fluctuation of less than 1 percent. Thus, the DC machine can produce a nearly constant torque (as a motor) or voltage (as a generator).

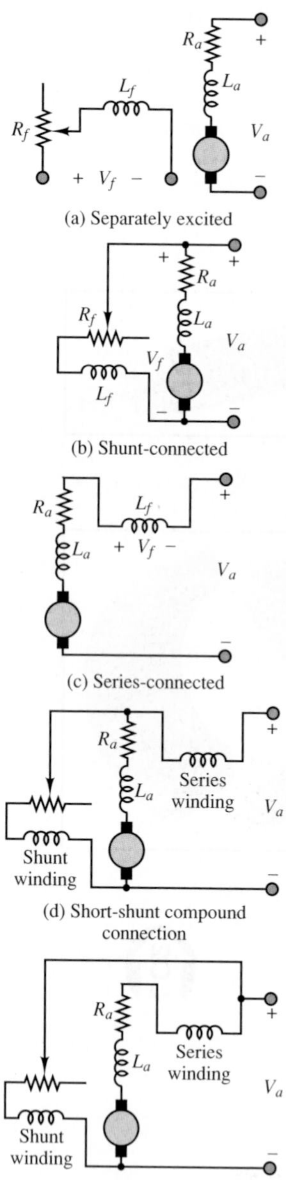

(a) Separately excited

(b) Shunt-connected

(c) Series-connected

(d) Short-shunt compound connection

(e) Long-shunt compound connection

Figure 19.14

Configuration of DC Machines

In DC machines, the field excitation that provides the magnetizing current is occasionally provided by an external source, in which case the machine is said to be **separately excited** [Figure 19.14(a)]. More often, the field excitation is derived from the armature voltage, and the machine is said to be **self-excited.** The latter configuration does not require the use of a separate source for the field excitation and is therefore frequently preferred. If a machine is in the separately excited configuration, an additional source V_f is required. In the self-excited case, one method used to provide the field excitation is to connect the field in parallel with the armature; since the field winding typically has significantly higher resistance than the armature circuit (remember that it is the armature that carries the load current), this will not draw excessive current from the armature. Further, a series resistor can be added to the field circuit to provide the means for adjusting the field current independent of the armature voltage. This configuration is called a **shunt-connected** machine and is depicted in Figure 19.14(b). Another method for self-exciting a DC machine consists of connecting the field in series with the armature, leading to the **series-connected** machine, depicted in Figure 19.14(c); in this case, the field winding will support the entire armature current, and thus the field coil must have low resistance (and therefore relatively few turns). This configuration is rarely used for generators, since the generated voltage and the load voltage must always differ by the voltage drop across the field coil, which varies with the load current. Thus, a series generator would have poor (large) regulation. However, series-connected motors are commonly used in certain applications, as will be discussed in a later section.

The third type of DC machine is the **compound-connected** machine, which consists of a combination of the shunt and series configurations. Figure 19.14(d) and (e) shows the two types of connections, called the **short shunt** and the **long shunt,** respectively. Each of these configurations may be connected so that the series part of the field adds to the shunt part (**cumulative compounding**) or so that it subtracts (**differential compounding**).

DC Machine Models

As stated earlier, it is relatively easy to develop a simple model of a DC machine, which is well suited to performance analysis, without the need to resort to the details of the construction of the machine itself. This section illustrates the development of such models in two steps. First, steady-state models relating field and armature currents and voltages to speed and torque are introduced; second, the differential equations describing the dynamic behavior of DC machines are derived.

When a field excitation is established, a magnetic flux ϕ is generated by the field current I_f. From equation 19.2, we know that the torque acting on the rotor is proportional to the product of the magnetic field and the current in the load-carrying wire; the latter current is the armature current I_a (i_w in equation 18.2). Assuming that, by virtue of the commutator, the torque angle γ is kept very close to 90°, and therefore $\sin\gamma = 1$, we obtain the following expression for the torque (in units of newton-meters) in a DC machine:

$$T = k_T \phi I_a \qquad \text{for } \gamma = 90^\circ \qquad \text{DC machine torque} \tag{19.6}$$

LO2

You may recall that this is simply a consequence of the Bli law of Chapter 18. The mechanical power generated (or absorbed) is equal to the product of the machine torque and the mechanical speed of rotation ω_m rad/s, and is therefore given by

$$P_m = \omega_m T = \omega_m k_T \phi I_a \tag{19.7}$$

Recall now that the rotation of the armature conductors in the field generated by the field excitation causes a **back emf** E_b in a direction that opposes the rotation of the armature. According to the Blu law (see Chapter 18), then, this back emf is given by

$$E_b = k_a \phi \omega_m \qquad \text{DC machine back emf} \tag{19.8}$$

LO2

where k_a is called the **armature constant** and is related to the geometry and magnetic properties of the structure. The voltage E_b represents a countervoltage (opposing the DC excitation) in the case of a motor and the generated voltage in the case of a generator. Thus, the electric power dissipated (or generated) by the machine is given by the product of the back emf and the armature current:

$$P_e = E_b I_a \tag{19.9}$$

The constants k_T and k_a in equations 19.6 and 19.8 are related to geometry factors, such as the dimension of the rotor and the number of turns in the armature winding; and to properties of materials, such as the permeability of the magnetic materials. Note that in the ideal energy conversion case $P_m = P_e$, and therefore $k_a = k_T$. We shall in general assume such ideal conversion of electrical to mechanical energy (or vice versa) and will therefore treat the two constants as being identical: $k_a = k_T$. The constant k_a is given by

$$k_a = \frac{pN}{2\pi M} \tag{19.10}$$

where

p = number of magnetic poles

N = number of conductors per coil

M = number of parallel paths in armature winding

An important observation concerning the units of angular speed must be made at this point. The equality (under the no-loss assumption) between the constants k_a and k_T in equations 19.6 and 19.8 results from the choice of consistent units, namely, volts and amperes for the electrical quantities and newton-meters and radians per second for the mechanical quantities. You should be aware that it is fairly common practice to refer to the speed of rotation of an electric machine in units of revolutions per minute (r/min).[1] In this book, we shall uniformly use the symbol n to denote angular speed in revolutions per minute; the following relationship should be committed to memory:

$$n \text{ (r/min)} = \frac{60}{2\pi}\omega_m \quad \text{rad/s} \tag{19.11}$$

If the speed is expressed in revolutions per minute, the armature constant changes as follows:

$$E_b = k'_a \phi n \tag{19.12}$$

where

$$k'_a = \frac{pN}{60M} \tag{19.13}$$

Having introduced the basic equations relating torque, speed, voltages, and currents in electric machines, we may now consider the interaction of these quantities in a DC machine at steady state, that is, operating at constant speed and field excitation. Figure 19.15 depicts the electric circuit model of a separately excited DC machine, illustrating both motor and generator action. It is very important to note the reference direction of armature current flow, and of the developed torque, in order to make a distinction between the two modes of operation. The field excitation is shown as a voltage V_f generating the field current I_f that flows through a variable resistor R_f and through the field coil L_f. The variable resistor permits adjustment of the field excitation. The armature circuit, on the other hand, consists of a voltage source representing the back emf E_b, the armature resistance R_a, and the armature voltage V_a. This model is appropriate both for motor and for generator action. When $V_a < E_b$, the

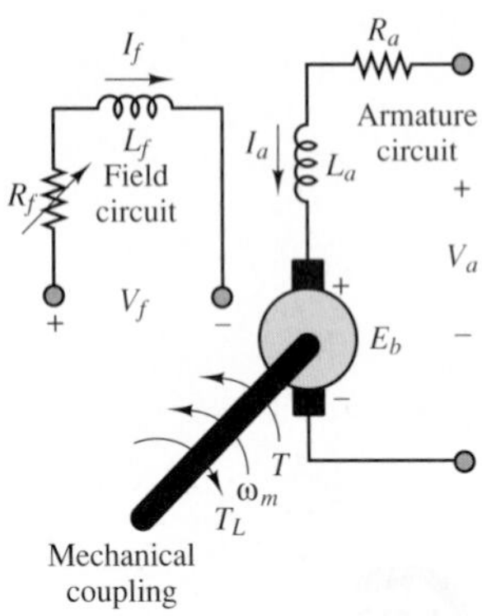

(a) Motor reference direction

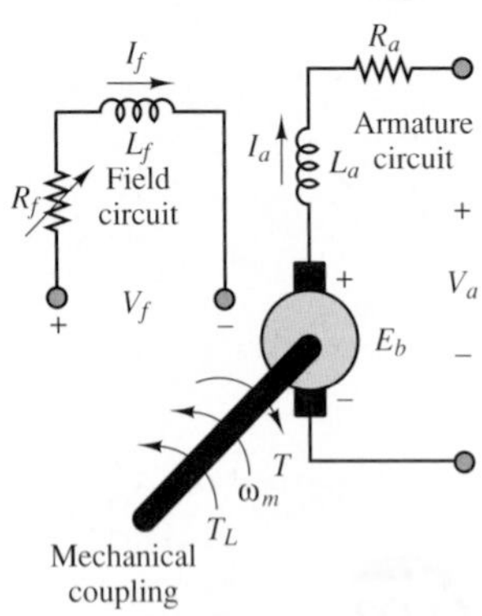

(b) Generator reference direction

Figure 19.15 Electric circuit model of a separately excited DC machine

[1] Note that the abbreviation *rpm*, although certainly familiar to the reader, is not a standard unit, and its use should be discouraged.

machine acts as a generator (I_a flows out of the machine). When $V_a > E_b$, the machine acts as a motor (I_a flows into the machine). Thus, according to the circuit model of Figure 19.15, the operation of a DC machine at steady state (i.e., with the inductors in the circuit replaced by short circuits) is described by the following equations:

$$-I_f + \frac{V_f}{R_f} = 0 \quad \text{and} \quad V_a - R_a I_a - E_b = 0 \quad \text{(motor action)}$$
$$-I_f + \frac{V_f}{R_f} = 0 \quad \text{and} \quad V_a + R_a I_a - E_b = 0 \quad \text{(generator action)} \qquad \textbf{(19.14)}$$

Equations 19.14 together with equations 19.6 and 19.8 may be used to determine the steady-state operating condition of a DC machine.

The circuit model of Figure 19.15 permits the derivation of a simple set of differential equations that describe the *dynamic* analysis of a DC machine. The dynamic equations describing the behavior of a separately excited DC machine are as follows:

$$V_a(t) - I_a(t)R_a - L_a \frac{dI_a(t)}{dt} - E_b(t) = 0 \quad \text{(armature circuit)} \qquad \textbf{(19.15}\boldsymbol{a}\textbf{)}$$

$$V_f(t) - I_f(t)R_f - L_f \frac{dI_f(t)}{dt} = 0 \quad \text{(field circuit)} \qquad \textbf{(19.15}\boldsymbol{b}\textbf{)}$$

These equations can be related to the operation of the machine in the presence of a load. If we assume that the motor is rigidly connected to an inertial load with moment of inertia J and that the friction losses in the load are represented by a viscous friction coefficient b, then the torque developed by the machine (in the motor mode of operation) can be written as

$$T(t) = T_L + b\omega_m(t) + J \frac{d\omega_m(t)}{dt} \qquad \textbf{(19.16)}$$

where T_L is the load torque. Typically T_L is either constant or some function of speed ω_m in a motor. In the case of a generator, the load torque is replaced by the torque supplied by a prime mover, and the machine torque $T(t)$ opposes the motion of the prime mover, as shown in Figure 19.15. Since the machine torque is related to the armature and field currents by equation 19.6, equations 19.16 and 19.17 are coupled to each other; this coupling may be expressed as follows:

$$T(t) = k_a \phi I_a(t) \qquad \textbf{(19.17)}$$

or

$$k_a \phi I_a(t) = T_L + b\omega_m(t) + J \frac{d\omega_m(t)}{dt} \qquad \textbf{(19.18)}$$

The dynamic equations described in this section apply to any DC machine. In the case of a *separately excited* machine, a further simplification is possible, since the flux is established by virtue of a separate field excitation, and therefore

$$\phi = \frac{N_f}{\mathcal{R}} I_f = k_f I_f \qquad \textbf{(19.19)}$$

where N_f is the number of turns in the field coil, $\mathcal{R}$ is the reluctance of the structure, and I_f is the field current.

19.3 DIRECT-CURRENT GENERATORS

To analyze the performance of a DC generator, it would be useful to obtain an open-circuit characteristic capable of predicting the voltage generated when the machine is driven at a constant speed ω_m by a prime mover. The common arrangement is to drive the machine at rated speed by means of a prime mover (or an electric motor). Then, with no load connected to the armature terminals, the armature voltage is recorded as the field current is increased from zero to some value sufficient to produce an armature voltage greater than the rated voltage. Since the load terminals are open-circuited, $I_a = 0$ and $E_b = V_a$; and since $k_a\phi = E_b/\omega_m$, the magnetization curve makes it possible to determine the value of $k_a\phi$ corresponding to a given field current I_f for the rated speed.

Figure 19.16 depicts a typical magnetization curve. Note that the armature voltage is nonzero even when no field current is present. This phenomenon is due to the *residual magnetization* of the iron core. The dashed lines in Figure 19.16 are called **field resistance curves** and are a plot of the voltage that appears across the field winding plus rheostat (variable resistor; see Figure 19.15) versus the field current, for various values of field winding plus rheostat resistance. Thus, the slope of the line is equal to the total field circuit resistance R_f.

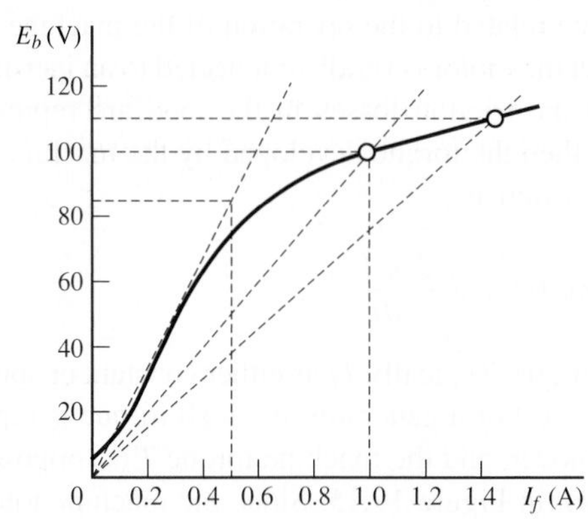

Figure 19.16 DC machine magnetization curve

The operation of a DC generator may be readily understood with reference to the magnetization curve of Figure 19.16. As soon as the armature is connected across the shunt circuit consisting of the field winding and the rheostat, a current will flow through the winding, and this will in turn act to increase the emf across the armature. This **buildup** process continues until the two curves meet, that is, until the current flowing through the field winding is exactly that required to induce the emf. By changing the rheostat setting, the operating point at the intersection of the two curves can be displaced, as shown in Figure 19.16, and the generator can therefore be made to supply different voltages. Examples 19.4 and 19.5 illustrate the operation of the separately excited DC generator.

EXAMPLE 19.4 Separately Excited DC Generator

LO3

Problem

A separately excited DC generator is characterized by the magnetization curve of Figure 19.16.

1. If the prime mover is driving the generator at 800 r/min, what is the no-load terminal voltage V_a?
2. If a 1-Ω load is connected to the generator, what is the generated voltage?

Solution

Known Quantities: Generator magnetization curve and ratings.

Find: Terminal voltage with no load and 1-Ω load.

Schematics, Diagrams, Circuits, and Given Data: Generator ratings: 100 V, 100 A, 1,000 r/min. Circuit parameters: $R_a = 0.14\ \Omega$; $V_f = 100$ V; $R_f = 100\ \Omega$.

Analysis:

1. The field current in the machine is

$$I_f = \frac{V_f}{R_f} = \frac{100\text{ V}}{100\ \Omega} = 1\text{ A}$$

From the magnetization curve, it can be seen that this field current will produce 100 V at a speed of 1,000 r/min. Since this generator is actually running at 800 r/min, the induced emf may be found by assuming a linear relationship between speed and emf. This approximation is reasonable, provided that the departure from the nominal operating condition is small. Let n_0 and E_{b0} be the nominal speed and emf, respectively (that is, 1,000 r/min and 100 V); then

$$\frac{E_b}{E_{b0}} = \frac{n}{n_0}$$

and therefore

$$E_b = \frac{n}{n_0}E_{b0} = \frac{800\text{ r/min}}{1{,}000\text{ r/min}} \times 100\text{ V} = 80\text{ V}$$

The open-circuit (output) terminal voltage of the generator is equal to the emf from the circuit model of Figure 19.15; therefore,

$$V_a = E_b = 80\text{ V}$$

2. When a load resistance is connected to the circuit (the practical situation), the terminal (or load) voltage is no longer equal to E_b, since there will be a voltage drop across the armature winding resistance. The armature (or load) current may be determined from

$$I_a = I_L = \frac{E_b}{R_a + R_L} = \frac{80\text{ V}}{(0.14 + 1)\ \Omega} = 70.2\text{ A}$$

where $R_L = 1\ \Omega$ is the load resistance. The terminal (load) voltage is therefore given by

$$V_L = I_L R_L = 70.2 \times 1 = 70.2\text{ V}$$

CHECK YOUR UNDERSTANDING

A 24-coil, two-pole DC generator has 16 turns per coil in its armature winding. The field excitation is 0.05 Wb per pole, and the armature angular velocity is 180 rad/s. Find the machine constant and the total induced voltage.

Answer: $k_a = 5.1$; $E_b = 45.9$ V

LO3

EXAMPLE 19.5 Separately Excited DC Generator

Problem

Determine the following quantities for a separately excited DC:

1. Induced voltage
2. Machine constant
3. Torque developed at rated conditions

Solution

Known Quantities: Generator ratings and machine parameters.

Find: E_b, k_a, T.

Schematics, Diagrams, Circuits, and Given Data: Generator ratings: 1,000 kW, 2,000 V, 3,600 r/min. Circuit parameters: $R_a = 0.1\ \Omega$; flux per pole $\phi = 0.5$ Wb.

Analysis:

1. The armature current may be found by observing that the rated power is equal to the product of the terminal (load) voltage and the armature (load) current; thus,

$$I_a = \frac{P_{\text{rated}}}{V_L} = \frac{1{,}000 \times 10^3}{2{,}000} = 500 \text{ A}$$

The generated voltage is equal to the sum of the terminal voltage and the voltage drop across the armature resistance (see Figure 19.14):

$$E_b = V_a + I_a R_a = 2{,}000 + 500 \times 0.1 = 2{,}050 \text{ V}$$

2. The speed of rotation of the machine in units of radians per second is

$$\omega_m = \frac{2\pi n}{60} = \frac{2\pi \times 3{,}600 \text{ r/min}}{60 \text{ s/min}} = 377 \text{ rad/s}$$

Thus, the machine constant is found to be

$$k_a = \frac{E_b}{\phi \omega_m} = \frac{2{,}050 \text{ V}}{0.5 \text{ Wb} \times 377 \text{ rad/s}} = 10.876 \frac{\text{V-s}}{\text{Wb-rad}}$$

3. The torque developed is found from equation 19.6:

$$T = k_a \phi I_a = 10.875 \text{ V-s/Wb-rad} \times 0.5 \text{ Wb} \times 500 \text{ A} = 2{,}718.9 \text{ N-m}$$

Comments: In many practical cases, it is not actually necessary to know the armature constant and the flux separately, but it is sufficient to know the value of the product $k_a\phi$. For example, suppose that the armature resistance of a DC machine is known and that, given a known field excitation, the armature current, load voltage, and speed of the machine can be measured. Then the product $k_a\phi$ may be determined from equation 19.8, as follows:

$$k_a\phi = \frac{E_b}{\omega_m} = \frac{V_L + I_a(R_a + R_S)}{\omega_m}$$

where V_L, I_a, and ω_m are measured quantities for given operating conditions.

CHECK YOUR UNDERSTANDING

A 1,000-kW, 1,000-V, 2,400 r/min separately excited DC generator has an armature circuit resistance of 0.04 Ω. The flux per pole is 0.4 Wb. Find (a) the induced voltage, (b) the machine constant, and (c) the torque developed at the rated conditions.

A 100-kW, 250-V shunt generator has a field circuit resistance of 50 Ω and an armature circuit resistance of 0.05 Ω. Find (a) the full-load line current flowing to the load, (b) the field current, (c) the armature current, and (d) the full-load generator voltage.

Answers: (a) $E_b = 1{,}040$ V; (b) $k_a = 10.34$; (c) $T = 4{,}138$ N-m
(a) 400 A; (b) 5 A; (c) 405 A; (d) 270.25 V

Since the compound-connected generator contains both a shunt and a series field winding, it is the most general configuration and the most useful for developing a circuit model that is as general as possible. In the following discussion, we consider the so-called short-shunt, compound-connected generator, in which the flux produced by the series winding adds to that of the shunt winding. Figure 19.17 depicts the equivalent circuit for the compound generator; circuit models for the shunt generator and for the rarely used series generator can be obtained by removing the shunt or series field winding element, respectively. In the circuit of Figure 19.17, the generator armature has been replaced by a voltage source corresponding to the induced emf and a series resistance R_a, corresponding to the resistance of the armature windings. The

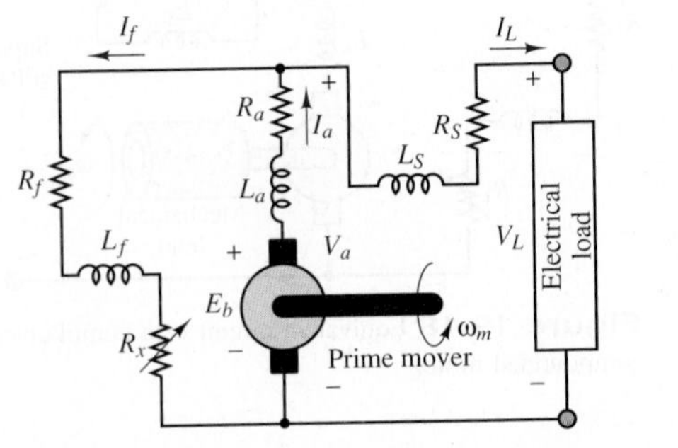

Figure 19.17 Compound generator circuit model

equations describing the DC generator at steady state (i.e., with the inductors acting as short circuits) are as follows:

DC Generator Steady-State Equations

LO3

$$E_b = k_a \phi \omega_m \qquad \text{V} \tag{19.20}$$

$$T = \frac{P}{\omega_m} = \frac{E_b I_a}{\omega_m} = k_a \phi I_a \qquad \text{N-m} \tag{19.21}$$

$$V_L = E_b - I_a R_a - I_S R_S \tag{19.22}$$

$$I_a = I_S + I_f \tag{19.23}$$

Note that in the circuit of Figure 19.17, the load and armature voltages are not equal, in general, because of the presence of a series field winding, represented by the resistor R_S and by the inductor L_S, where the subscript S stands for "series." The expression for the armature emf is dependent on the air gap flux ϕ to which the series and shunt windings in the compound generator both contribute, according to the expression

$$\phi = \phi_{\text{sh}} \pm \phi_S = \phi_{\text{sh}} \pm k_S I_a \tag{19.24}$$

19.4 DIRECT-CURRENT MOTORS

DC motors are widely used in applications requiring accurate speed control, for example, in servo systems. Having developed a circuit model and analysis methods for the DC generator, we can extend these results to DC motors, since these are in effect DC generators with the roles of input and output reversed. Once again, we analyze the motor by means of both its magnetization curve and a circuit model. It will be useful to begin our discussion by referring to the schematic diagram of a cumulatively compounded motor, as shown in Figure 19.18. The choice of the compound-connected motor is the most convenient, since its model can be used to represent either a series or a shunt motor with minor modifications.

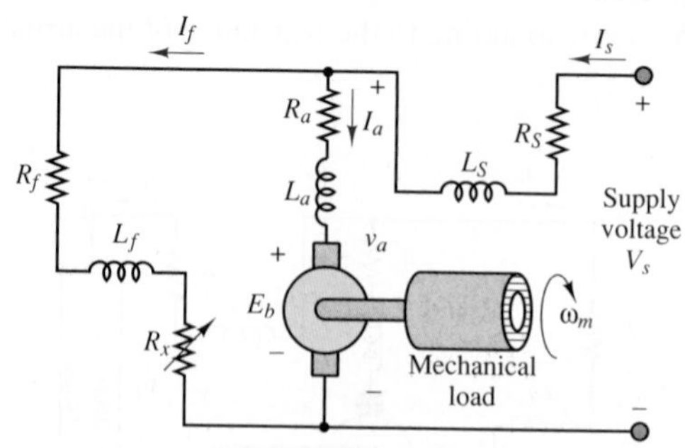

Figure 19.18 Equivalent circuit of a cumulatively compounded motor

The equations that govern the behavior of the DC motor follow and are similar to those used for the generator. Note that the only differences between these equations and those that describe the DC generator appear in the last two equations in the group, where the source voltage is equal to the *sum* of the emf and the voltage drop across the series field resistance and armature resistance, and where the source current now equals the *sum* of the field shunt and armature series currents.

DC Motor Steady-State Equations

$$E_b = k_a\phi\omega_m \quad \textbf{(19.25)}$$

$$T = k_a\phi I_a \quad \textbf{(19.26)}$$

$$V_s = E_b + I_a R_a + I_s R_S \quad \textbf{(19.27)}$$

$$I_s = I_f + I_a \quad \textbf{(19.28)}$$

LO4

Note that in these equations we have replaced the symbols V_L and I_L, used in the generator circuit model to represent the generator load current and voltage, with the symbols V_s and I_s, indicating the presence of an external source.

Speed–Torque and Dynamic Characteristics of DC Motors

The Shunt Motor

In a shunt motor (similar to the configuration of Figure 19.18, but with the series field short-circuited), the armature current is found by dividing the net voltage across the armature circuit (source voltage minus back emf) by the armature resistance:

$$I_a = \frac{V_s - k_a\phi\omega_m}{R_a} \quad \textbf{(19.29)}$$

An expression for the armature current may also be obtained from equation 16.26, as follows:

$$I_a = \frac{T}{k_a\phi} \quad \textbf{(19.30)}$$

It is then possible to relate the torque requirements to the speed of the motor by substituting equation 19.29 in equation 19.30:

$$\frac{T}{k_a\phi} = \frac{V_s - k_a\phi\omega_m}{R_a} \quad \textbf{(19.31)}$$

Equation 19.31 describes the steady-state torque–speed characteristic of the shunt motor. To understand this performance equation, we observe that if V_s, k_a, ϕ, and R_a are fixed in equation 19.31 (the flux is essentially constant in the shunt motor for a fixed V_s), then the speed of the motor is directly related to the armature current. Now consider the case where the load applied to the motor is suddenly increased, causing the speed of the motor to drop. As the speed decreases, the armature current increases, according to equation 19.29. The excess armature current causes the motor to develop additional torque, according to equation 19.30, until a new equilibrium is reached between the higher armature current and developed torque and the lower speed of rotation. The equilibrium point is dictated by the balance of mechanical and electrical power, in accordance with the relation

$$E_b I_a = T\omega_m \tag{19.32}$$

Thus, the shunt DC motor will adjust to variations in load by changing its speed to preserve this power balance. The torque–speed curves for the shunt motor may be obtained by rewriting the equation relating the speed to the armature current:

LO4

$$\omega_m = \frac{V_s - I_a R_a}{k_a \phi} = \frac{V_s}{k_a \phi} - \frac{R_a T}{(k_a \phi)^2} \qquad T\text{-}\omega \text{ curve for shunt motor} \tag{19.33}$$

To interpret equation 19.33, one can start by considering the motor operating at rated speed and torque. As the load torque is reduced, the armature current will also decrease, causing the speed to increase in accordance with equation 19.33. The increase in speed depends on the extent of the voltage drop across the armature resistance $I_a R_a$. The change in speed will be on the same order of magnitude as this drop; it typically takes values around 10 percent. This corresponds to a relatively good speed regulation, which is an attractive feature of the shunt DC motor (recall the discussion of regulation in Section 19.1). Normalized torque and speed versus power curves for the shunt motor are shown in Figure 19.19. Note that, over a reasonably broad range of powers, up to rated value, the curve is relatively flat, indicating that the DC shunt motor acts as a reasonably constant-speed motor.

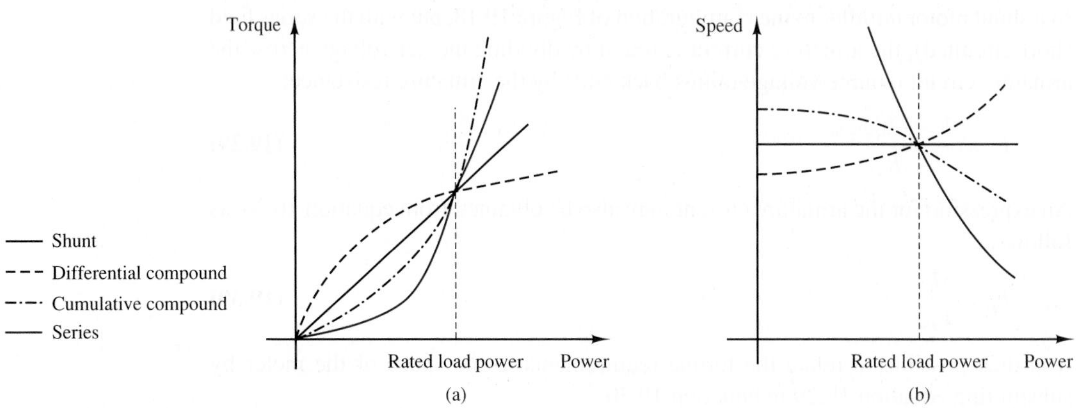

Figure 19.19 DC motor operating characteristics

The dynamic behavior of the shunt motor is described by equations 19.15 through 19.18, with the additional relation

$$I_a(t) = I_s(t) - I_f(t) \tag{19.34}$$

Compound Motors

It is interesting to compare the performance of the shunt motor with that of the compound-connected motor; the comparison is easily made if we recall that a series field resistance appears in series with the armature resistance and that the flux is due to the contributions of both series and shunt fields. Thus, the speed equation becomes

$$\omega_m = \frac{V_s - I_a(R_a + R_S)}{k_a(\phi_{\text{sh}} \pm \phi_S)} \tag{19.35}$$

where

A plus in the denominator is for a cumulatively compounded motor.

A minus in the denominator is for a differentially compounded motor.

ϕ_{sh} is the flux set up by the shunt field winding, assuming that it is constant.

ϕ_S is the flux set up by the series field winding $\phi_S = k_S I_a$.

For the cumulatively compound motor, two effects are apparent: The flux is increased by the presence of a series component ϕ_S, and the voltage drop due to I_a in the numerator is increased by an amount proportional to the resistance of the series field winding R_S. As a consequence, when the load to the motor is reduced, the numerator increases more dramatically than in the case of the shunt motor, because of the corresponding decrease in armature current, while at the same time the series flux decreases. Each of these effects causes the speed to increase; therefore, it stands to reason that the speed regulation of the compound-connected motor is poorer than that of the shunt motor. Normalized torque and speed versus power curves for the compound motor (both differential and cumulative connections) are shown in Figure 19.19.

The differential equation describing the behavior of a compound motor differs from that for the shunt motor in having additional terms due to the series field component:

$$\begin{aligned} V_s &= E_b(t) + I_a(t)R_a + L_a\frac{dI_a(t)}{dt} + I_s(t)R_S + L_S\frac{dI_s(t)}{dt} \\ &= V_a(t) + I_s(t)R_S + L_S\frac{dI_s(t)}{dt} \end{aligned} \tag{19.36}$$

The differential equation for the field circuit can be written as

$$V_a = I_f(t)(R_f + R_x) + L_f\frac{dI_f(t)}{dt} \tag{19.37}$$

We also have the following basic relations:

$$I_a(t) = I_s(t) - I_f(t) \tag{19.38}$$

and

$$E_b(t) = k_a I_a(t)\omega_m(t) \qquad \text{and} \qquad T(t) = k_a\phi I_a(t) \tag{19.39}$$

Series Motors

The series motor [see Figure 19.14(c)] behaves somewhat differently from the shunt and compound motors because the flux is established solely by virtue of the series current flowing through the armature. It is relatively simple to derive an expression for the emf and torque equations for the series motor if we approximate the relationship between flux and armature current by assuming that the motor operates in the linear region of its magnetization curve. Then we can write

$$\phi = k_S I_a \tag{19.40}$$

and the emf and torque equations become, respectively,

$$E_b = k_a \omega_m \phi = k_a \omega_m k_S I_a \quad \textbf{(19.41)}$$

$$T = k_a \phi I_a = k_a k_S I_a^2 \quad \textbf{(19.42)}$$

The circuit equation for the series motor becomes

$$V_s = E_b + I_a(R_a + R_S) = (k_a \omega_m k_S + R_T) I_a \quad \textbf{(19.43)}$$

where R_a is the armature resistance, R_S is the series field winding resistance, and R_T is the total series resistance. From equation 19.43, we can solve for I_a and substitute in the torque expression (equation 19.42) to obtain the following torque–speed relationship:

LO4

$$T = k_a k_S \frac{V_s^2}{(k_a \omega_m k_S + R_T)^2} \qquad T\text{-}\omega \text{ curve for series DC motor} \quad \textbf{(19.44)}$$

which indicates the inverse squared relationship between torque and speed in the series motor. This expression describes a behavior that can, under certain conditions, become unstable. Since the speed increases when the load torque is reduced, one can readily see that if one were to disconnect the load altogether, the speed would tend to increase to dangerous values. To prevent excessive speeds, series motors are always mechanically coupled to the load. This feature is not necessarily a drawback, though, because series motors can develop very high torque at low speeds and therefore can serve very well for traction-type loads (e.g., conveyor belts or vehicle propulsion systems). Torque and speed versus power curves for the series motor are also shown in Figure 19.19.

The differential equation for the armature circuit of the motor can be given as

$$\begin{aligned} V_s &= I_a(t)(R_a + R_S) + L_a \frac{dI_a(t)}{dt} + L_S \frac{dI_a(t)}{dt} + E_b \\ &= I_a(t)(R_a + R_S) + L_a \frac{dI_a(t)}{dt} + L_S \frac{dI_a(t)}{dt} + k_a k_S I_a \omega_m \end{aligned} \quad \textbf{(19.45)}$$

Permanent-Magnet DC Motors

Permanent-magnet (PM) DC motors have become increasingly common in applications requiring relatively low torques and efficient use of space. The construction of PM direct-current motors differs from that of the motors considered thus far in that the magnetic field of the stator is produced by suitably located poles made of magnetic materials. Thus, the basic principle of operation, including the idea of commutation, is unchanged with respect to the wound-stator DC motor. What changes is that there is no need to provide a field excitation, whether separately or by means of the self-excitation techniques discussed in the preceding sections. Therefore, the PM motor is intrinsically simpler than its wound-stator counterpart.

The equations that describe the operation of the PM motor follow. The torque produced is related to the armature current by a torque constant k_{PM}, which is determined by the geometry of the motor:

$$T = k_{T,\mathrm{PM}} I_a \quad \textbf{(19.46)}$$

As in the conventional DC motor, the rotation of the rotor produces the usual counter- or back emf E_b, which is linearly related to speed by a voltage constant $k_{a,\text{PM}}$:

$$E_b = k_{a,\text{PM}}\omega_m \tag{19.47}$$

The equivalent circuit of the PM motor is particularly simple, since we need not model the effects of a field winding. Figure 19.20 shows the circuit model and the torque–speed curve of a PM motor.

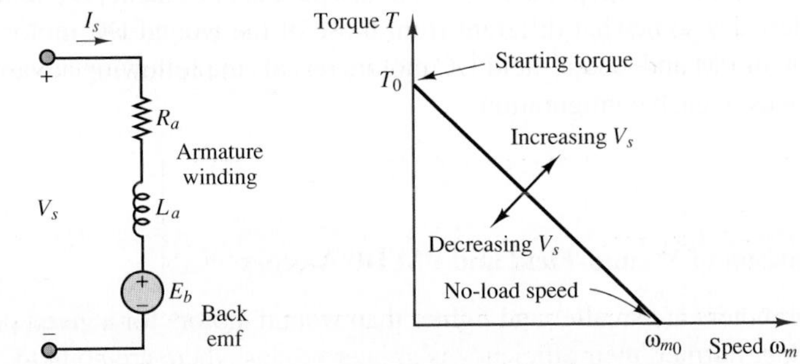

Circuit model for PM motor

Torque-speed curve for PM motor

Figure 19.20 Circuit model and torque–speed curve of PM motor

We can use the circuit model of Figure 19.20 to predict the torque–speed curve shown in the same figure as follows. From the circuit model, for a constant speed (and therefore constant current), we may consider the inductor a short circuit and write the equation

$$\begin{aligned} V_s &= I_a R_a + E_b = I_a R_a + k_{a,\text{PM}}\omega_m \\ &= \frac{T}{k_{T,\text{PM}}} R_a + k_{a,\text{PM}}\omega_m \end{aligned} \tag{19.48}$$

thus obtaining the equations relating speed and torque:

$$\omega_m = \frac{V_s}{k_{a,\text{PM}}} - \frac{T R_a}{k_{a,\text{PM}} k_{T,\text{PM}}} \qquad T\text{-}\omega \text{ curve for PM DC motor} \tag{19.49}$$

and

$$T = \frac{V_s}{R_a} k_{T,\text{PM}} - \frac{\omega_m}{R_a} k_{a,\text{PM}} k_{T,\text{PM}} \tag{19.50}$$

From these equations, one can extract the stall torque T_0, that is, the zero-speed torque

$$T_0 = \frac{V_s}{R_a} k_{T,\text{PM}} \tag{19.51}$$

and the no-load speed ω_{m0}:

$$\omega_{m0} = \frac{V_s}{k_{a,\text{PM}}} \tag{19.52}$$

Under dynamic conditions, assuming an inertia plus viscous friction load, the torque produced by the motor can be expressed as

$$T = k_{T,\text{PM}} I_a(t) = T_{\text{load}}(t) + d\omega_m(t) + J\frac{d\omega_m(t)}{dt} \tag{19.53}$$

The differential equation for the armature circuit of the motor is therefore given by

$$
\begin{aligned}
V_s &= I_a(t)R_a + L_a \frac{dI_a(t)}{dt} + E_b \\
&= I_a(t)R_a + L_a \frac{dI_a(t)}{dt} + k_{a,\mathrm{PM}}\omega_m(t)
\end{aligned}
\tag{19.54}
$$

The fact that the airgap flux is constant in a permanent-magnet DC motor makes its characteristics somewhat different from those of the wound DC motor. A direct comparison of PM and wound-field DC motors reveals the following advantages and disadvantages of each configuration.

LO4

Comparison of Wound-Field and PM DC Motors

1. PM motors are smaller and lighter than wound motors for a given power rating. Further, their efficiency is greater because there are no field winding losses.
2. An additional advantage of PM motors is their essentially linear speed–torque characteristic, which makes analysis (and control) much easier. Reversal of rotation is also accomplished easily, by reversing the polarity of the source.
3. A major disadvantage of PM motors is that they can become demagnetized by exposure to excessive magnetic fields, application of excessive voltage, or operation at excessively high or low temperatures.
4. A less obvious drawback of PM motors is that their performance is subject to greater variability from motor to motor than is the case for wound motors, because of variations in the magnetic materials.

FIND IT ON THE WEB

In summary, four basic types of **DC motors** are commonly used. Their principal operating characteristics are summarized as follows, and their general torque and speed versus power characteristics are depicted in Figure 19.19, assuming motors with identical voltage, power, and speed ratings.

Shunt wound motor: Field is connected in parallel with the armature. With constant armature voltage and field excitation, the motor has good speed regulation (flat speed–torque characteristic).

Compound wound motor: Field winding has both series and shunt components. This motor offers better starting torque than the shunt motor, but worse speed regulation.

Series-wound motor: The field winding is in series with the armature. The motor has very high starting torque and poor speed regulation. It is useful for low-speed, high-torque applications.

Permanent-magnet motor: Field windings are replaced by permanent magnets. The motor has adequate starting torque, with speed regulation somewhat worse than that of the compound wound motor.

EXAMPLE 19.6 DC Shunt Motor Analysis

LO4

Problem

Find the speed and torque generated by a four-pole DC shunt motor.

Solution

Known Quantities: Motor ratings; circuit and magnetic parameters.

Find: ω_m, T.

Schematics, Diagrams, Circuits, and Given Data:
Motor ratings: 3 hp, 240 V, 120 r/min.
Circuit and magnetic parameters: $I_S = 30$ A; $I_f = 1.4$ A; $R_a = 0.6\ \Omega$; $\phi = 20$ mWb; $N = 1{,}000$; $M = 4$ (see equation 19.10).

Analysis: We convert the power to SI units:

$$P_{\text{RATED}} = 3 \text{ hp} \times 746 \frac{\text{W}}{\text{hp}} = 2{,}238 \text{ W}$$

Next we compute the armature current as the difference between source and field current (equation 19.34):

$$I_a = I_s - I_f = 30 - 1.4 = 28.6 \text{ A}$$

The no-load armature voltage E_b is given by

$$E_b = V_s - I_a R_a = 240 - 28.6 \times 0.6 = 222.84 \text{ V}$$

and equation 19.10 can be used to determine the armature constant:

$$k_a = \frac{pN}{2\pi M} = \frac{4 \times 1{,}000}{2\pi \times 4} = 159.15 \frac{\text{V-s}}{\text{Wb-rad}}$$

Knowing the motor constant, we can calculate the speed, after equation 19.25:

$$\omega_m = \frac{E_a}{k_a \phi} = \frac{222.84 \text{ V}}{(159.15 \text{ V-s/Wb-rad})(0.02 \text{ Wb})} = 70 \frac{\text{rad}}{\text{s}}$$

Finally, the torque developed by the motor can be found as the ratio of the power to the angular velocity:

$$T = \frac{P}{\omega_m} = \frac{2{,}238 \text{ W}}{70 \text{ rad/s}} = 32 \text{ N-m}$$

CHECK YOUR UNDERSTANDING

A 200-V DC shunt motor draws 10 A at 1,800 r/min. The armature circuit resistance is 0.15 Ω, and the field winding resistance is 350 Ω. What is the torque developed by the motor?

Answer: $T = \frac{P}{\omega_m} = 9.93$ N-m

LO4

EXAMPLE 19.7 DC Shunt Motor Analysis

Problem

Determine the following quantities for the DC shunt motor, connected as shown in the circuit of Figure 19.21:

1. Field current required for full-load operation.
2. No-load speed.
3. Plot of the speed torque curve of the machine in the range from no-load torque to rated torque.
4. Power output at rated torque.

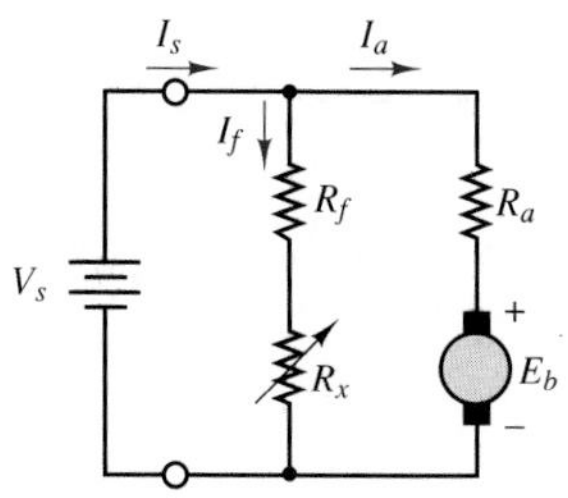

Figure 19.21 Shunt motor configuration

Solution

Known Quantities: Magnetization curve, rated current, rated speed, circuit parameters.

Find: I_f; $n_{\text{no-load}}$; T-n curve, P_{rated}.

Schematics, Diagrams, Circuits, and Given Data:
Figure 19.22 (magnetization curve)
Motor ratings: 8 A, 120 r/min
Circuit parameters: $R_a = 0.2\ \Omega$; $V_s = 7.2$ V; N = number of coil turns in winding = 200

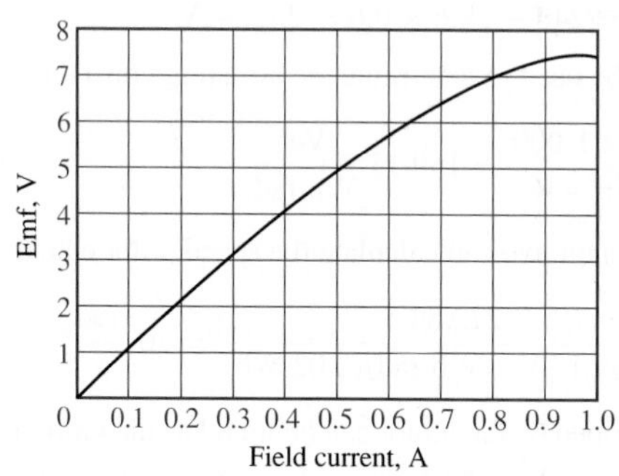

Figure 19.22 Magnetization curve for a small DC motor

Analysis:

1. To find the field current, we must find the generated emf since R_f is not known. Writing KVL around the armature circuit, we obtain

$$V_s = E_b + I_a R_a$$
$$E_b = V_s - I_a R_a = 7.2 - 8(0.2) = 5.6 \text{ V}$$

Having found the back emf, we can find the field current from the magnetization curve. At $E_b = 5.6$ V, we find that the field current and field resistance are

$$I_f = 0.6 \text{ A} \qquad \text{and} \qquad R_f = \frac{7.2}{0.6} = 12\ \Omega$$

2. To obtain the no-load speed, we use the equations

$$E_b = k_a\phi\frac{2\pi n}{60} \qquad T = k_a\phi I_a$$

leading to

$$V_s = I_aR_a + E_b = I_aR_a + k_a\phi\frac{2\pi}{60}n$$

or

$$n = \frac{V_s - I_aR_a}{k_a\phi(2\pi/60)}$$

At no load, and assuming no mechanical losses, the torque is zero, and we see that the current I_a must also be zero in the torque equation ($T = k_a\phi I_a$). Thus, the motor speed at no load is given by

$$n_{\text{no-load}} = \frac{V_s}{k_a\phi(2\pi/60)}$$

We can obtain an expression for $k_a\phi$, knowing that, at full load,

$$E_b = 5.6 \text{ V} = k_a\phi\frac{2\pi n}{60}$$

so that, for constant field excitation,

$$k_a\phi = E_b\left(\frac{60}{2\pi n}\right) = 5.6\left[\frac{60}{2\pi(120)}\right] = 0.44563\,\frac{\text{V-s}}{\text{rad}}$$

Finally, we may solve for the no-load speed.

$$n_{\text{no-load}} = \frac{V_s}{k_a\phi(2\pi/60)} = \frac{7.2}{(0.44563)(2\pi/60)}$$
$$= 154.3 \text{ r/min}$$

3. The torque at rated speed and load may be found as follows:

$$T_{\text{rated load}} = k_a\phi I_a = (0.44563)(8) = 3.565 \text{ N-m}$$

Now we have the two points necessary to construct the torque–speed curve for this motor, which is shown in Figure 19.23.

4. The power is related to the torque by the frequency of the shaft:

$$P_{\text{rated}} = T\omega_m = (3.565)\left(\frac{120}{60}\right)(2\pi) = 44.8 \text{ W}$$

or, equivalently,

$$P = 44.8 \text{ W} \times \frac{1}{746}\frac{\text{hp}}{\text{W}} = 0.06 \text{ hp}$$

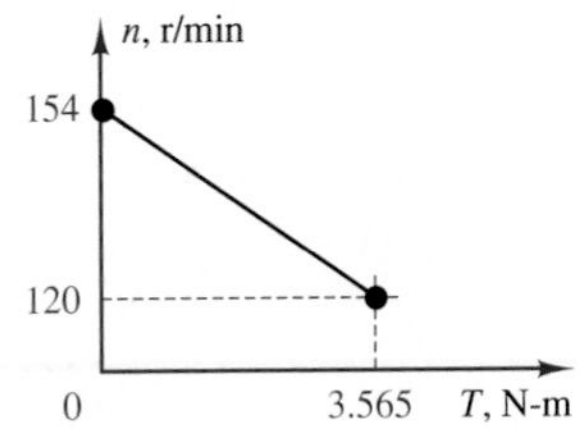

Figure 19.23

EXAMPLE 19.8 DC Series Motor Analysis

LO4

Problem

Determine the torque developed by a DC series motor when the current supplied to the motor is 60 A.

Solution

Known Quantities: Motor ratings; operating conditions.

Find: T_{60}, torque delivered at 60-A series current.

Schematics, Diagrams, Circuits, and Given Data:
Motor ratings: 10 hp, 115 V, full-load speed = 1,800 r/min
Operating conditions: motor draws 40 A

Assumptions: The motor operates in the linear region of the magnetization curve.

Analysis: Within the linear region of operation, the flux per pole is directly proportional to the current in the field winding. That is,

$$\phi = k_S I_a$$

The full-load speed is

$$n = 1{,}800 \text{ r/min}$$

or

$$\omega_m = \frac{2\pi n}{60} = 60\pi \qquad \text{rad/s}$$

Rated output power is

$$P_{\text{rated}} = 10 \text{ hp} \times 746 \text{ W/hp} = 7{,}460 \text{ W}$$

and full-load torque is

$$T_{40\text{ A}} = \frac{P_{\text{rated}}}{\omega_m} = \frac{7{,}460}{60\pi} = 39.58 \text{ N-m}$$

Thus, the machine constant may be computed from the torque equation for the series motor:

$$T = k_a k_S I_a^2 = K I_a^2$$

At full load,

$$K = k_a k_S = \frac{39.58 \text{ N-m}}{40^2 \text{ A}^2} = 0.0247 \frac{\text{N-m}}{\text{A}^2}$$

and we can compute the torque developed for a 60-A supply current to be

$$T_{60\text{ A}} = K I_a^2 = 0.0247 \times 60^2 = 88.92 \text{ N-m}$$

CHECK YOUR UNDERSTANDING

A series motor draws a current of 25 A and develops a torque of 100 N-m. Find (a) the torque when the current rises to 30 A if the field is unsaturated and (b) the torque when the current rises to 30 A and the increase in current produces a 10 percent increase in flux.

Answer: (a) 144 N-m; (b) 132 N-m

EXAMPLE 19.9 Dynamic Response of PM DC Motor

LO4

Problem

Develop a set of differential equations and a transfer function describing the dynamic response of the motor angular velocity of a PM DC motor connected to a mechanical load.

Solution

Known Quantities: PM DC motor circuit model; mechanical load model.

Find: Differential equations and transfer functions of electromechanical system.

Analysis: The dynamic response of the electromechanical system can be determined by applying KVL to the electric circuit (Figure 19.20) and Newton's second law to the mechanical system. These equations will be coupled to one another, as you shall see, because of the nature of the motor back emf and torque equations.

Applying KVL and equation 19.47 to the electric circuit, we obtain

$$V_L(t) - R_a I_a(t) - L_a \frac{dI_a(t)}{dt} - E_b(t) = 0$$

or

$$L_a \frac{dI_a(t)}{dt} + R_a I_a(t) + K_{a,\text{PM}}\omega_m(t) = V_L(t)$$

Applying Newton's second law and equation 19.46 to the load inertia, we obtain

$$J\frac{d\omega(t)}{dt} = T(t) - T_{\text{load}}(t) - b\omega$$

or

$$-K_{T,\text{PM}} I_a(t) + J\frac{d\omega(t)}{dt} + b\omega(t) = T_{\text{load}}(t)$$

These two differential equations are coupled because the first depends on ω_m and the second on I_a. Thus, they need to be solved simultaneously.

To derive the transfer function, we use the Laplace transform on the two equations to obtain

$$(sL_a + R_a)I_a(s) + K_{a,\text{PM}}\Omega(s) = V_L(s)$$
$$-K_{T,\text{PM}} I_a(s) + (sJ + b)\Omega(s) = T_{\text{load}}(s)$$

We can write the above equations in matrix form and resort to Cramer's rule to solve for $\Omega_m(s)$ as a function of $V_L(s)$ and $T_{\text{load}}(s)$.

$$\begin{bmatrix} sL_a + R_a & K_{a,\text{PM}} \\ -K_{T,\text{PM}} & sJ + b \end{bmatrix} \begin{bmatrix} I_a(s) \\ \Omega_m(s) \end{bmatrix} = \begin{bmatrix} V_L(s) \\ T_{\text{load}}(s) \end{bmatrix}$$

with solution

$$\Omega_m(s) = \frac{\det\begin{bmatrix} sL_a + R_a & V_L(s) \\ K_{T,\text{PM}} & T_{\text{load}}(s) \end{bmatrix}}{\det\begin{bmatrix} sL_a + R_a & K_{a,\text{PM}} \\ -K_{T,\text{PM}} & sJ + b \end{bmatrix}}$$

or

$$\Omega_m(s) = \frac{sL_a + R_a}{(sL_a + R_a)(sJ + b) + K_{a,\text{PM}}K_{T,\text{PM}}} T_{\text{load}}(s)$$

$$+ \frac{K_{T,\text{PM}}}{(sL_a + R_a)(sJ + b) + K_{a,\text{PM}}K_{T,\text{PM}}} V_L(s)$$

Comments: Note that the dynamic response of the motor angular velocity depends on both the input voltage and the load torque. This problem is explored further in the homework problems.

DC Drives and DC Motor Speed Control

FIND IT ON THE WEB

The advances made in power semiconductors have made it possible to realize low-cost **speed control systems for DC motors.** The basic operation of *controlled rectifier* and *chopper* drives for DC motors was described in Chapter 12. In this section we describe some of the considerations that are behind the choice of a specific drive type, and some of the loads that are likely to be encountered.

Constant-torque loads are quite common and are characterized by a need for constant torque over the entire speed range. This need is usually due to friction; the load will demand increasing horsepower at higher speeds, since power is the product of speed and torque. Thus, the power required will increase linearly with speed. This type of loading is characteristic of conveyors, extruders, and surface winders.

Another type of load is one that requires *constant horsepower* over the speed range of the motor. Since torque is inversely proportional to speed with constant horsepower, this type of load will require higher torque at low speeds. Examples of constant-horsepower loads are machine tool spindles (e.g., lathes). This type of application requires very high starting torques.

Variable-torque loads are also common. In this case, the load torque is related to the speed in some fashion, either linearly or geometrically. For some loads, for example, torque is proportional to the speed (and thus horsepower is proportional to speed squared); examples of loads of this type are positive displacement pumps. More common than the linear relationship is the squared-speed dependence of inertial loads such as centrifugal pumps, some fans, and all loads in which a flywheel is used for energy storage.

To select the appropriate motor and adjustable speed drive for a given application, we need to examine how each method for speed adjustment operates on a DC motor. Armature voltage control serves to smoothly adjust speed from 0 to 100 percent of the nameplate rated value (i.e., base speed), provided that the field excitation is also equal to the rated value. Within this range, it is possible to fully control motor speed for a constant-torque load, thus providing a linear increase in horsepower, as shown in Figure 19.24. Field weakening allows for increases in speed of up to several times the base speed; however, field control changes the characteristics of the DC motor from constant torque to constant horsepower, and therefore the torque output drops with speed, as shown in Figure 19.24. Operation above base speed requires special provision for field control, in addition to the circuitry required for armature voltage control, and is therefore more complex and costly.

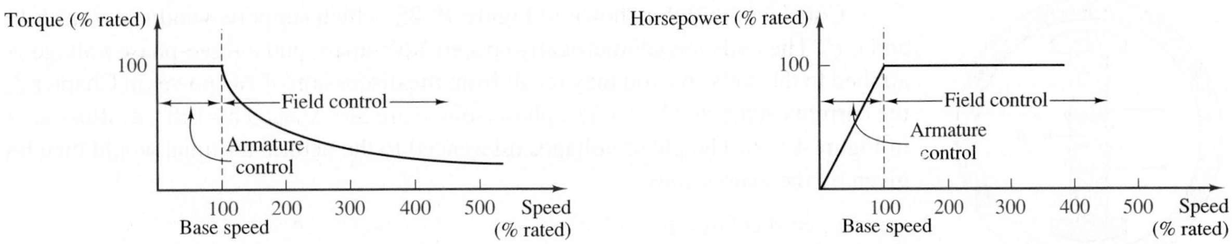

Figure 19.24 Speed control in DC motors

CHECK YOUR UNDERSTANDING

Describe the cause-and-effect behavior of the speed control method of changing armature voltage for a shunt DC motor.

Answer: Increasing the armature voltage leads to an increase in armature current. Consequently, the motor torque increases until it exceeds the load torque, causing the speed to increase as well. The corresponding increase in back emf, however, causes the armature current to drop and the motor torque to decrease until a balance condition is reached between motor and load torque and the motor runs at constant speed.

19.5 AC MACHINES

From the previous sections, it should be apparent that it is possible to obtain a wide range of performance characteristics from DC machines, as both motors and generators. A logical question at this point should be, Would it not be more convenient in some cases to take advantage of the single- or multiphase AC power that is available virtually everywhere than to expend energy and use additional hardware to rectify and regulate the DC supplies required by direct-current motors? The answer to this very obvious question is certainly a resounding yes. In fact, the AC induction motor is the workhorse of many industrial applications, and synchronous generators are used almost exclusively for the generation of electric power worldwide. Thus, it is appropriate to devote a significant portion of this chapter to the study of AC machines and of induction motors in particular. The objective of this section is to explain the basic operation of both synchronous and induction machines and to outline their performance characteristics. In doing so, we also point out the relative advantages and disadvantages of these machines in comparison with direct-current machines. The motor "movies" available on the book website may help you visualize the operation of AC machines.

Rotating Magnetic Fields

As mentioned in Section 19.1, the fundamental principle of operation of AC machines is the generation of a rotating magnetic field, which causes the rotor to turn at a speed that depends on the speed of rotation of the magnetic field. We now explain how a rotating magnetic field can be generated in the stator and air gap of an AC machine by means of alternating currents.

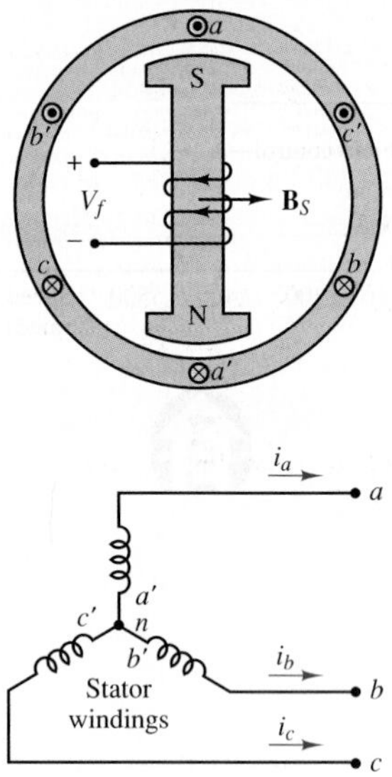

Figure 19.25 Two-pole three-phase stator

Consider the stator shown in Figure 19.25, which supports windings a-a', b-b' and c-c'. The coils are geometrically spaced 120° apart, and a three-phase voltage is applied to the coils. As you may recall from the discussion of AC power in Chapter 7, the currents generated by a three-phase source are also spaced by 120°, as illustrated in Figure 19.26. The phase voltages referenced to the neutral terminal would then be given by the expressions

$$v_a = A\cos(\omega_e t)$$

$$v_b = A\cos\left(\omega_e t - \frac{2\pi}{3}\right)$$

$$v_c = A\cos\left(\omega_e t + \frac{2\pi}{3}\right)$$

where ω_e is the frequency of the AC supply, or line frequency. The coils in each winding are arranged in such a way that the flux distribution generated by any one winding is approximately sinusoidal. Such a flux distribution may be obtained by appropriately arranging groups of coils for each winding over the stator surface. Since the coils are spaced 120° apart, the flux distribution resulting from the sum of the contributions of the three windings is the sum of the fluxes due to the separate windings, as shown in Figure 19.27. Thus, the flux in a three-phase machine rotates in space according to the vector diagram of Figure 19.28, and the flux is constant in amplitude. A stationary observer on the machine's stator would see a sinusoidally varying flux distribution, as shown in Figure 19.27.

Since the resultant flux of Figure 19.27 is generated by the currents of Figure 19.26, the speed of rotation of the flux must be related to the frequency of the sinusoidal phase currents. In the case of the stator of Figure 19.25, the number of magnetic poles

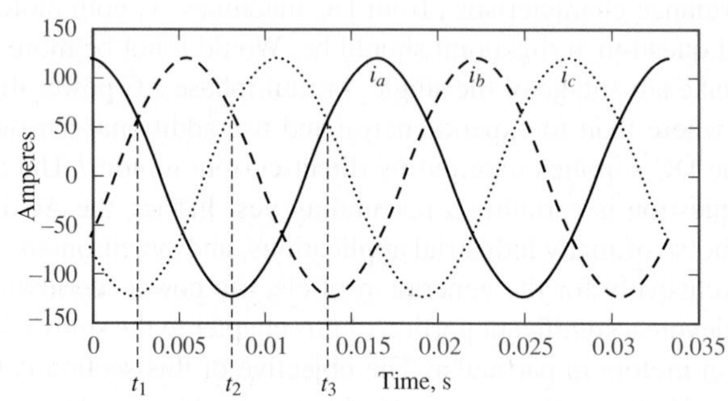

Figure 19.26 Three-phase stator winding currents

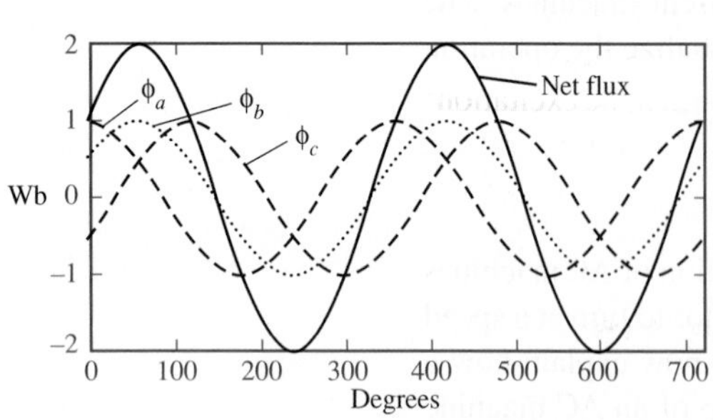

Figure 19.27 Flux distribution in a three-phase stator winding as a function of angle of rotation

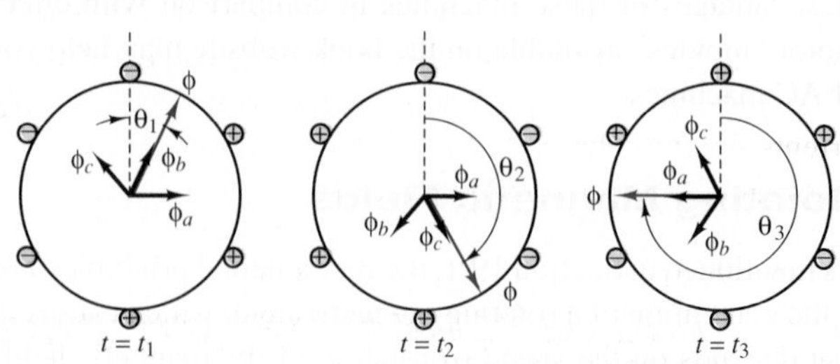

Figure 19.28 Rotating flux in a three-phase machine

resulting from the winding configuration is 2; however, it is also possible to configure the windings so that they have more poles. For example, Figure 19.29 depicts a simplified view of a four-pole stator.

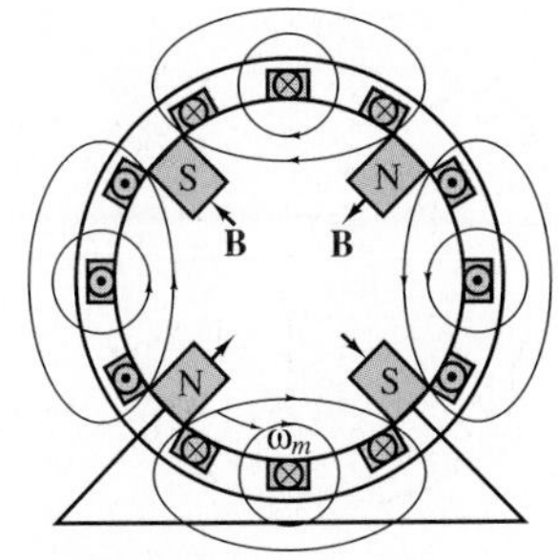

Figure 19.29 Four-pole stator

In general, the speed of the rotating magnetic field is determined by the frequency of the excitation current f and by the number of poles present in the stator p according to

$$n_s = \frac{120f}{p} \text{ r/min} \qquad \text{Synchronous speed}$$

or **(19.55)**

$$\omega_s = \frac{2\pi n_S}{60} = \frac{2\pi \times 2f}{p} \qquad \text{Synchronous speed}$$

LO5

where n_s (or ω_s) is usually called the **synchronous speed.**

Now, the structure of the windings in the preceding discussion is the same whether the AC machine is a motor or a generator; the distinction between the two depends on the direction of power flow. In a generator, the electromagnetic torque is a reaction torque that opposes rotation of the machine; this is the torque against which the prime mover does work. In a motor, on the other hand, the rotational (motional) voltage generated in the armature opposes the applied voltage; this voltage is the counter- (or back) emf. Thus, the description of the rotating magnetic field given thus far applies to both motor and generator action in AC machines.

As described a few paragraphs earlier, the stator magnetic field rotates in an AC machine, and therefore the rotor cannot "catch up" with the stator field and is in constant pursuit of it. The speed of rotation of the rotor will therefore depend on the number of magnetic poles present in the stator and in the rotor. The magnitude of the torque produced in the machine is a function of the angle γ between the stator and rotor magnetic fields; precise expressions for this torque depend on how the magnetic fields are generated and will be given separately for the two cases of synchronous and induction machines. What is common to all rotating machines is that the number of stator and rotor poles must be identical if any torque is to be generated. Further, the number of poles must be even, since for each north pole there must be a corresponding south pole.

One important desired feature in an electric machine is an ability to generate a constant electromagnetic torque. With a constant-torque machine, one can avoid torque pulsations that could lead to undesired mechanical vibration in the motor itself and in other mechanical components attached to the motor (e.g., mechanical loads, such as spindles or belt drives). A constant torque may not always be achieved, although it will be shown that it is possible to accomplish this goal when the excitation currents are multiphase. A general rule of thumb, in this respect, is that it is desirable, insofar as possible, to produce a constant flux per pole.

19.6 THE ALTERNATOR (SYNCHRONOUS GENERATOR)

One of the most common AC machines is the **synchronous generator,** or **alternator.** In this machine, the field winding is on the rotor, and the connection is made by means of brushes, in an arrangement similar to that of the DC machines studied earlier. The rotor field is obtained by means of a direct current provided to the rotor winding, or by permanent magnets. The rotor is then connected to a mechanical source of power and rotates at a speed that we will consider constant to simplify the analysis.

Figure 19.30 depicts a two-pole three-phase synchronous machine. Figure 19.31 depicts a four-pole three-phase alternator, in which the rotor poles are generated by means of a wound salient pole configuration and the stator poles are the result of windings embedded in the stator according to the simplified arrangement shown in the figure, where each of the pairs a/a', b/b', and so on contributes to the generation of the magnetic poles, as follows. The group a/a', b/b', c/c' produces a sinusoidally distributed flux (see Figure 17.27) corresponding to one of the pole pairs, while the group $-a/-a'$, $-b/-b'$, $-c/-c'$ contributes the other pole pair. The connections of the coils making up the windings are also shown in Figure 19.31. Note that the coils form a wye connection (see Chapter 7). The resulting flux distribution is such that the flux completes two sinusoidal cycles around the circumference of the air gap. Note also that each arm of the three-phase wye connection has been divided into two coils, wound in different locations, according to the schematic stator diagram of

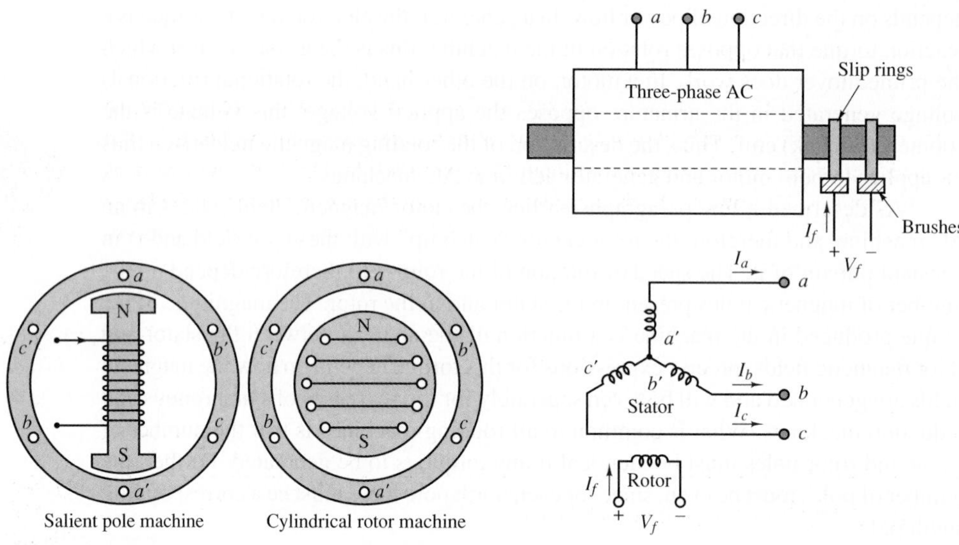

Figure 19.30 Two-pole synchronous machine

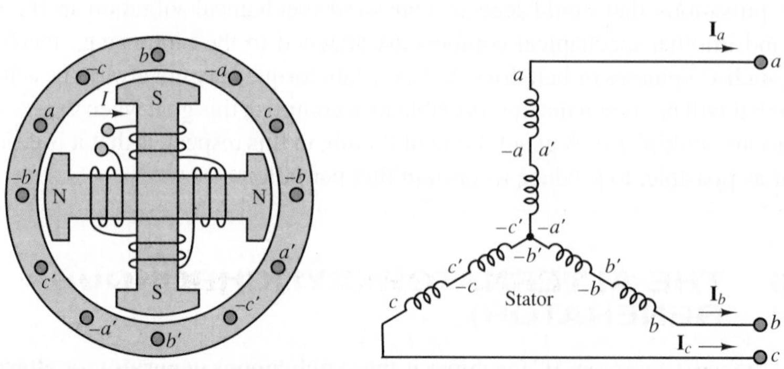

Figure 19.31 Four-pole three-phase alternator

Figure 19.31. One could then envision analogous configurations with greater numbers of poles, obtained in the same fashion, that is, by dividing each arm of a wye connection into more windings.

The arrangement shown in Figure 19.31 requires that a further distinction be made between mechanical degrees θ_m and electrical degrees θ_e. In the four-pole alternator, the flux will see two complete cycles during one rotation of the rotor, and therefore the voltage that is generated in the coils will also oscillate at twice the frequency of rotation. In general, the electrical degrees (or radians) are related to the mechanical degrees by the expression

$$\theta_e = \frac{p}{2}\theta_m \tag{19.56}$$

where p is the number of poles. In effect, the voltage across a coil of the machine goes through one cycle every time a pair of poles moves past the coil. Thus, the frequency of the voltage generated by a synchronous generator is

$$f = \frac{p}{2}\frac{n}{60} \text{ Hz} \tag{19.57}$$

where n is the mechanical speed in revolutions per minute. Alternatively, if the speed is expressed in radians per second, we have

$$\omega_e = \frac{p}{2}\omega_m \tag{19.58}$$

where ω_m is the mechanical speed of rotation in radians per second. The number of poles employed in a synchronous generator is then determined by two factors: the frequency desired of the generated voltage (for example, 60 Hz, if the generator is used to produce AC power) and the speed of rotation of the prime mover. In the latter respect, there is a significant difference, for example, between the speed of rotation of a steam turbine generator and that of a hydroelectric generator, the former being much greater.

A common application of the alternator is seen in automotive battery-charging systems, in which, however, the generated AC voltage is rectified to provide the DC required for charging the battery. Figure 19.32 depicts an automotive alternator.

Figure 19.32 Automotive alternator (*Courtesy: Delphi Automotive Systems*)

LO5

CHECK YOUR UNDERSTANDING

A synchronous generator has a multipolar construction that permits changing its synchronous speed. If only two poles are energized, at 50 Hz, the speed is 3,000 r/min. If the number of poles is progressively increased to 4, 6, 8, 10, and 12, find the synchronous speed for each configuration. Draw the complete equivalent circuit of a synchronous generator and its phasor diagram.

Answer: 1,500, 1,000, 750, 600, and 500 r/min

19.7 THE SYNCHRONOUS MOTOR

Synchronous motors are virtually identical to synchronous generators with regard to their construction, except for an additional winding for helping start the motor and minimizing motor speed over- and undershoots. The principle of operation is, of course, the opposite: An AC excitation provided to the armature generates a magnetic field in the air gap between stator and rotor, resulting in a mechanical torque. To generate the rotor magnetic field, some direct current must be provided to the field windings; this is often accomplished by means of an **exciter,** which consists of a small DC generator propelled by the motor itself, and therefore mechanically connected to it. It was mentioned earlier that to obtain a constant torque in an electric motor, it is necessary to keep the rotor and stator magnetic fields constant relative to each other. This means that the electromagnetically rotating field in the stator and the mechanically rotating rotor field should be aligned at all times. The only condition for which this is possible occurs if both fields are rotating at the synchronous speed $n_s = 120f/p$. Thus, synchronous motors are by their very nature constant-speed motors.

For a non–salient pole (cylindrical rotor) synchronous machine, the torque can be written in terms of the stator alternating current $i_S(t)$ and the rotor direct current, I_f:

LO5

$$T = k i_S(t) I_f \sin(\gamma) \qquad \text{Synchronous motor torque} \tag{19.59}$$

where γ is the angle between the stator and rotor fields (see Figure 19.7). Let the angular speed of rotation be

$$\omega_m = \frac{d\theta_m}{dt} \quad \text{rad/s} \tag{19.60}$$

where $\omega_m = 2\pi n/60$, and let ω_e be the electrical frequency of $i_S(t)$, where $i_S(t) = \sqrt{2}I_S \sin(\omega_e t)$. Then the torque may be expressed as

$$T = k\sqrt{2}I_S \sin(\omega_e t)\, I_f \sin(\gamma) \tag{19.61}$$

where k is a machine constant, I_S is the rms value of the stator current, and I_f is the rotor direct current. Now, the rotor angle γ can be expressed as a function of time by

$$\gamma = \gamma_0 + \omega_m t \tag{19.62}$$

where γ_0 is the angular position of the rotor at $t = 0$; the torque expression then becomes

$$\begin{aligned} T &= k\sqrt{2}I_S I_f \sin(\omega_e t)\sin(\omega_m t + \gamma_0) \\ &= k\frac{\sqrt{2}}{2} I_S I_f \cos[(\omega_m - \omega_e)t - \gamma_0] - \cos[(\omega_m + \omega_e)t + \gamma_0] \end{aligned} \tag{19.63}$$

It is a straightforward matter to show that the average value of this torque, denoted by $\langle T \rangle$, is different from zero only if $\omega_m = \pm\omega_e$, that is, only if the motor is turning at the synchronous speed. The resulting average torque is then given by

$$\langle T \rangle = k\sqrt{2}I_S I_f \cos(\gamma_0) \tag{19.64}$$

Note that equation 19.63 corresponds to the sum of an average torque plus a fluctuating component at twice the original electrical (or mechanical) frequency. The fluctuating component results because, in the foregoing derivation, a single-phase current was assumed. The use of multiphase currents reduces the torque fluctuation to zero and permits the generation of a constant torque.

A per-phase circuit model describing the synchronous motor is shown in Figure 19.33, where the rotor circuit is represented by a field winding equivalent resistance and inductance, R_f and L_f, respectively, and the stator circuit is represented

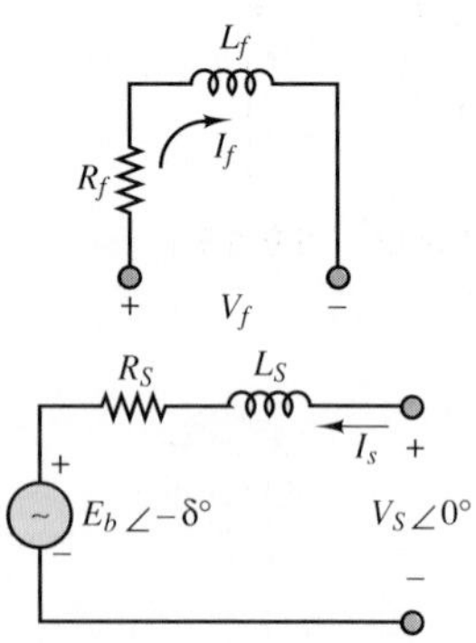

Figure 19.33 Per-phase circuit model

by equivalent stator winding inductance and resistance, L_S and R_S, respectively, and by the induced emf E_b. From the exact equivalent circuit as given in Figure 19.33, we have

$$V_S = E_b + I_S(R_S + jX_S) \quad \textbf{(19.65)}$$

where X_S is known as the *synchronous reactance* and includes magnetizing reactance.

The motor power is

$$P_{\text{out}} = \omega_S T = |V_S||I_S|\cos(\theta) \quad \textbf{(19.66)}$$

for each phase, where T is the developed torque and θ is the angle between the stator voltage and current, V_S and I_S.

When the phase winding resistance R_S is neglected, the circuit model of a synchronous machine can be redrawn as shown in Figure 19.34. The input power (per phase) is equal to the output power in this circuit, since no power is dissipated in the circuit; that is,

$$P_\phi = P_{\text{in}} = P_{\text{out}} = |\mathbf{V}_S||\mathbf{I}_S|\cos(\theta) \quad \textbf{(19.67)}$$

Also by inspection of Figure 19.34, we have

$$d = |\mathbf{E}_b|\sin(\delta) = |\mathbf{I}_S|X_S\cos(\theta) \quad \textbf{(19.68)}$$

Then

$$|\mathbf{E}_b||\mathbf{V}_S|\sin(\delta) = |\mathbf{V}_S||\mathbf{I}_S|X_S\cos(\theta) = X_S P_\phi \quad \textbf{(19.69)}$$

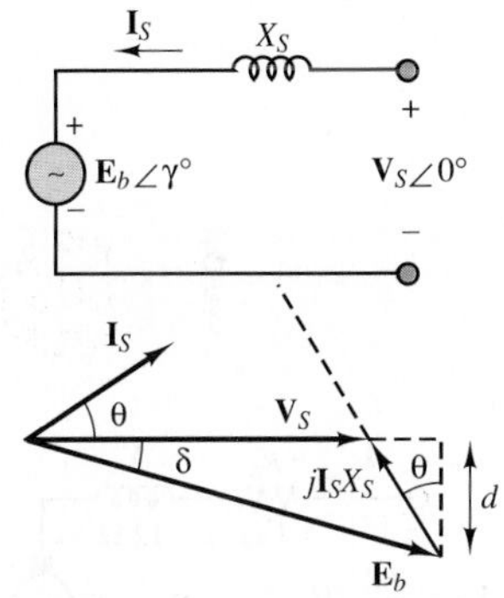

Figure 19.34

The total power of a three-phase synchronous machine is then given by

$$P = 3\frac{|\mathbf{V}_S||\mathbf{E}_b|}{X_S}\sin(\delta) \quad \textbf{(19.70)}$$

Because of the dependence of the power upon the angle δ, this angle has come to be called the **power angle.** If δ is zero, the synchronous machine cannot develop useful power. The developed power has its maximum value at δ equal to 90°. If we assume that $|\mathbf{E}_b|$ and $|\mathbf{V}_S|$ are constant, we can draw the curve shown in Figure 19.35, relating the power and power angle in a synchronous machine.

A synchronous generator is usually operated at a power angle varying from 15° to 25°. For synchronous motors and small loads, δ is close to 0°, and the motor torque is just sufficient to overcome its own windage and friction losses; as the load increases, the rotor field falls further out of phase with the stator field (although the two are still rotating at the same speed), until δ reaches a maximum at 90°. If the load torque exceeds the maximum torque, which is produced for $\delta = 90°$, the motor is forced to slow down below synchronous speed. This condition is undesirable, and provisions are usually made to shut down the motor automatically whenever synchronism is lost. The maximum torque is called the **pull-out torque** and is an important measure of the performance of the synchronous motor.

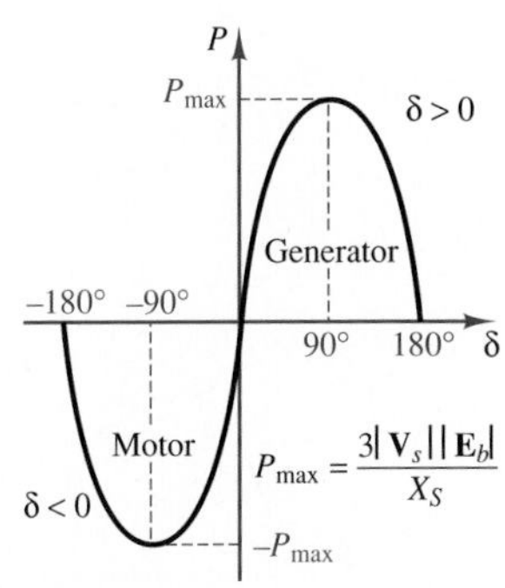

Figure 19.35 Power versus power angle for a synchronous machine

Accounting for each of the phases, the total torque is given by

$$T = \frac{m}{\omega_s}|\mathbf{V}_S||\mathbf{I}_S|\cos(\theta) \quad \textbf{(19.71)}$$

where m is the number of phases. From Figure 19.34, we have $E_b\sin(\delta) = X_S I_S\cos(\theta)$. Therefore, for a three-phase machine, the developed torque is

$$T = \frac{P}{\omega_s} = \frac{3}{\omega_s}\frac{|\mathbf{V}_S||\mathbf{E}_b|}{X_S}\sin(\delta) \qquad \text{N-m} \quad \textbf{(19.72)}$$

Typically, analysis of multiphase motors is performed on a per-phase basis, as illustrated in Examples 19.10 and 19.11.

LO5

EXAMPLE 19.10 Synchronous Motor Analysis

Problem

Find the kilovoltampere rating, the induced voltage, and the power angle of the rotor for a fully loaded synchronous motor.

Solution

Known Quantities: Motor ratings; motor synchronous impedance.

Find: S; $\mathbf{E}_b$; δ.

Schematics, Diagrams, Circuits, and Given Data: Motor ratings: 460 V; three-phase; power factor = 0.707 lagging; full-load stator current: 12.5 A. $Z_S = 1 + j12\ \Omega$.

Assumptions: Use per-phase analysis.

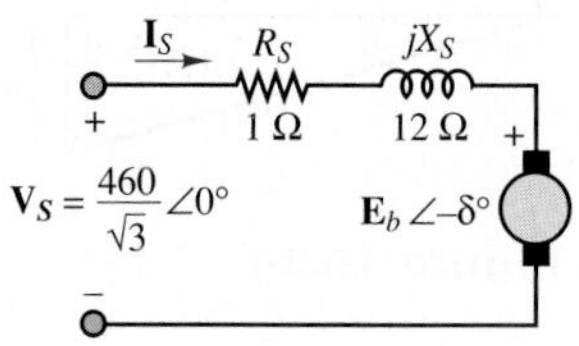

Figure 19.36

Analysis: The circuit model for the motor is shown in Figure 19.36. The per-phase current in the wye-connected stator winding is

$$I_S = |\mathbf{I}_S| = 12.5 \text{ A}$$

The per-phase voltage is

$$V_S = |\mathbf{V}_S| = \frac{460 \text{ V}}{\sqrt{3}} = 265.58 \text{ V}$$

The kilovoltampere rating of the motor is expressed in terms of the apparent power S (see Chapter 7):

$$S = 3V_S I_S = 3 \times 265.58 \text{ V} \times 12.5 \text{ A} = 9{,}959 \text{ W}$$

From the equivalent circuit, we have

$$\mathbf{E}_b = \mathbf{V}_S - \mathbf{I}_S(R_S + jX_S)$$
$$= 265.58 - (12.5\angle -45° \text{ A}) \times (1 + j12\ \Omega) = 179.31\angle -32.83° \text{ V}$$

The induced line voltage is defined to be

$$V_{\text{line}} = \sqrt{3}E_b = \sqrt{3} \times 179.31 \text{ V} = 310.57 \text{ V}$$

From the expression for $\mathbf{E}_b$, we can find the power angle:

$$\delta = -32.83°$$

Comments: The minus sign indicates that the machine is in the motor mode.

LO5

EXAMPLE 19.11 Synchronous Motor Analysis

Problem

Find the stator current, the line current, and the induced voltage for a synchronous motor.

Solution

Known Quantities: Motor ratings; motor synchronous impedance.

Find: $\mathbf{I}_S$; $\mathbf{I}_{\text{line}}$; $\mathbf{E}_b$.

Schematics, Diagrams, Circuits, and Given Data: Motor ratings: 208 V; three-phase; 45 kVA; 60 Hz; power factor = 0.8 leading; $Z_S = 0 + j2.5\ \Omega$. Friction and windage losses: 1.5 kW; core losses: 1.0 kW; load power: 15 hp.

Assumptions: Use per-phase analysis.

Analysis: The output power of the motor is 15 hp; that is,

$$P_{\text{out}} = 15 \text{ hp} \times 0.746 \text{ kW/hp} = 11.19 \text{ kW}$$

The electric power supplied to the machine is

$$\begin{aligned} P_{\text{in}} &= P_{\text{out}} + P_{\text{mech}} + P_{\text{core loss}} + P_{\text{elec loss}} \\ &= 11.19 \text{ kW} + 1.5 \text{ kW} + 1.0 \text{ kW} + 0 \text{ kW} = 13.69 \text{ kW} \end{aligned}$$

As discussed in Chapter 7, the resulting line current is

$$I_{\text{line}} = \frac{P_{\text{in}}}{\sqrt{3} V \cos\theta} = \frac{13{,}690 \text{ W}}{\sqrt{3} \times 208 \text{ V} \times 0.8} = 47.5 \text{ A}$$

Because of the delta connection, the armature current is

$$\mathbf{I}_S = \frac{1}{\sqrt{3}} \mathbf{I}_{\text{line}} = 27.4\angle 36.87^\circ \text{ A}$$

The emf may be found from the equivalent circuit and KVL:

$$\begin{aligned} \mathbf{E}_b &= \mathbf{V}_S - jX_S\mathbf{I}_S \\ &= 208\angle 0^\circ - (j2.5\ \Omega)(27.4\angle 36.87^\circ \text{ A}) = 255\angle -12.4^\circ \text{ V} \end{aligned}$$

The power angle is

$$\delta = -12.4^\circ$$

CHECK YOUR UNDERSTANDING

Find an expression for the maximum pull-out torque of the synchronous motor.

Answer: $T_{\text{max}} = \dfrac{3V_S E_b}{\omega_m X_S}$

Synchronous motors are not very commonly used in practice, for various reasons, among which are that they are essentially required to operate at constant speed (unless a variable-frequency AC supply is available) and that they are not self-starting. Further, separate AC and DC supplies are required. It will be seen shortly that the induction motor overcomes most of these drawbacks.

19.8 THE INDUCTION MOTOR

The induction motor is the most widely used electric machine, because of its relative simplicity of construction. The stator winding of an induction machine is similar to that of a synchronous machine; thus, the description of the three-phase winding of Figure 19.25 also applies to induction machines. The primary advantage of the induction machine, which is almost exclusively used as a motor (its performance as a generator is not very good), is that no separate excitation is required for the rotor. The rotor typically consists of one of two arrangements: a **squirrel cage** or a **wound rotor.** The former contains conducting bars short-circuited at the end and embedded within it; the latter consists of a multiphase winding similar to that used for the stator, but electrically short-circuited.

In either case, the induction motor operates by virtue of currents induced from the stator field in the rotor. In this respect, its operation is similar to that of a transformer, in that currents in the stator (which acts as a primary coil) induce currents in the rotor (acting as a secondary coil). In most induction motors, no external electrical connection is required for the rotor, thus permitting a simple, rugged construction without the need for slip rings or brushes. Unlike the synchronous motor, the induction motor operates not at synchronous speed, but at a somewhat lower speed, which is dependent on the load. Figure 19.37 illustrates the appearance of a squirrel

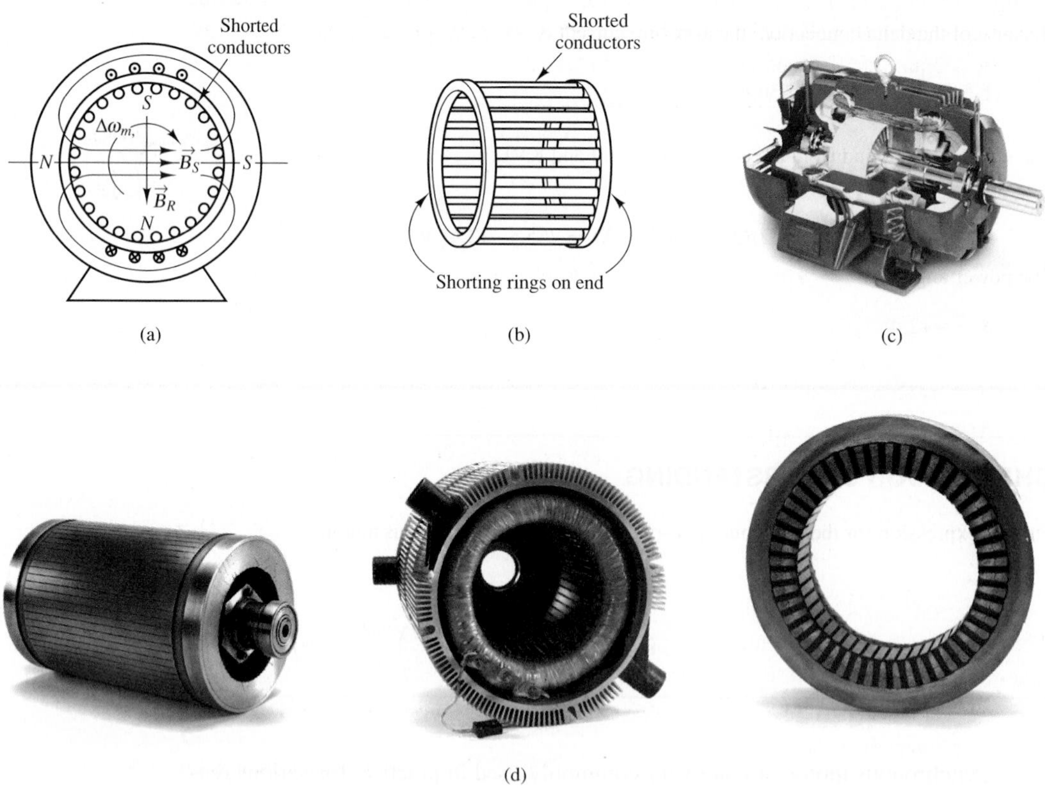

Figure 19.37 (a) Squirrel cage induction motor; (b) conductors in rotor; (c) photograph of squirrel cage induction motor; (d) views of Smokin' Buckeye motor: rotor, stator, and cross section of stator (*Photos Courtesy: David H. Koether Photography*)

cage induction motor. The following discussion focuses mainly on this very common configuration.

By now you are acquainted with the notion of a rotating stator magnetic field. Imagine now that a squirrel cage rotor is inserted in a stator in which such a rotating magnetic field is present. The stator field will induce voltages in the cage conductors, and if the stator field is generated by a three-phase source, the resulting rotor currents—which circulate in the bars of the squirrel cage, with the conducting path completed by the shorting rings at the end of the cage—are also three-phase and are determined by the magnitude of the induced voltages and by the impedance of the rotor. Since the rotor currents are induced by the stator field, the number of poles and the speed of rotation of the induced magnetic field are the same as those of the stator field, *if the rotor is at rest*. Thus, when a stator field is initially applied, the rotor field is synchronous with it, and the fields are stationary with respect to one another. Thus, according to the earlier discussion, a *starting torque* is generated.

If the starting torque is sufficient to cause the rotor to start spinning, the rotor will accelerate up to its operating speed. However, an induction motor can never reach synchronous speed; if it did, the rotor would appear to be stationary with respect to the rotating stator field, since it would be rotating at the same speed. But in the absence of relative motion between the stator and rotor fields, no voltage would be induced in the rotor. Thus, an induction motor is limited to speeds somewhere below the synchronous speed n_s. Let the speed of rotation of the rotor be n; then the rotor is losing ground with respect to the rotation of the stator field at a speed $n_s - n$. In effect, this is equivalent to backward motion of the rotor at the **slip speed,** defined by $n_s - n$. The **slip** s is usually defined as a fraction of n_s

$$s = \frac{n_s - n}{n_s} \qquad \text{Slip in induction machine} \tag{19.73}$$

LO5

which leads to the following expression for the rotor speed:

$$n = n_s(1 - s) \tag{19.74}$$

The slip s is a function of the load, and the amount of slip in a given motor is dependent on its construction and rotor type (squirrel cage or wound rotor). Since there is a relative motion between the stator and rotor fields, voltages will be induced in the rotor at a frequency called the **slip frequency,** related to the relative speed of the two fields. This gives rise to an interesting phenomenon: The rotor field travels relative to the rotor at the slip speed sn_s, but the rotor is mechanically traveling at the speed $(1 - s)n_s$, so that the net effect is that the rotor field travels at the speed

$$sn_s + (1 - s)n_s = n_s \tag{19.75}$$

that is, at synchronous speed. The fact that the rotor field rotates at synchronous speed—although the rotor itself does not—is extremely important, because it means that the stator and rotor fields will continue to be stationary with respect to each other, and therefore a net torque can be produced.

As in the case of DC and synchronous motors, important characteristics of induction motors are the starting torque, the maximum torque, and the torque–speed curve. These will be discussed shortly, after some analysis of the induction motor is performed.

LO5

EXAMPLE 19.12 Induction Motor Analysis

Problem

Find the full-load rotor slip and frequency of the induced voltage at rated speed in a four-pole induction motor.

Solution

Known Quantities: Motor ratings.

Find: s; f_R.

Schematics, Diagrams, Circuits, and Given Data: Motor ratings: 230 V; 60 Hz; full-load speed: 1,725 r/min.

Analysis: The synchronous speed of the motor is

$$n_s = \frac{120f}{p} = \frac{60f}{p/2} = \frac{60 \text{ s/min} \times 60 \text{ r/s}}{4/2} = 1{,}800 \text{ r/min}$$

The slip is

$$s = \frac{n_s - n}{n_s} = \frac{1{,}800 \text{ r/min} - 1{,}725 \text{ r/min}}{1{,}800 \text{ r/min}} = 0.0417$$

The rotor frequency f_R is

$$f_R = sf = 0.0417 \times 60 \text{ Hz} = 2.5 \text{ Hz}$$

CHECK YOUR UNDERSTANDING

A three-phase induction motor has six poles. (a) If the line frequency is 60 Hz, calculate the speed of the magnetic field in revolutions per minute. (b) Repeat the calculation if the frequency is changed to 50 Hz.

Answer: (a) $n = 1{,}200$ r/min; (b) $n = 1{,}000$ r/min

The induction motor can be described by means of an equivalent circuit, which is essentially that of a rotating transformer. (See Chapter 18 for a circuit model of the transformer.) Figure 19.38 depicts such a circuit model, where

R_S = stator resistance per phase, R_R = rotor resistance per phase

X_S = stator reactance per phase, X_R = rotor reactance per phase

X_m = magnetizing (mutual) reactance

R_C = equivalent core-loss resistance

E_S = per-phase induced voltage in stator windings

E_R = per-phase induced voltage in rotor windings

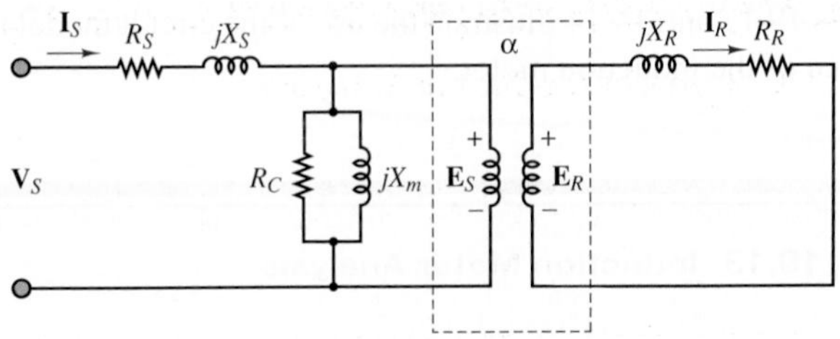

Figure 19.38 Circuit model for induction machine

The primary internal stator voltage $\mathbf{E}_S$ is coupled to the secondary rotor voltage $\mathbf{E}_R$ by an ideal transformer with an effective turns ratio of α. For the rotor circuit, the induced voltage at any slip will be

$$\mathbf{E}_R = s\mathbf{E}_{R0} \tag{19.76}$$

where $\mathbf{E}_{R0}$ is the induced rotor voltage at the condition in which the rotor is stationary. Also, $X_R = \omega_R L_R = 2\pi f_R L_R = 2\pi s f L_R = sX_{R0}$, where $X_{R0} = 2\pi f L_R$ is the reactance when the rotor is stationary. The rotor current is given by

$$\mathbf{I}_R = \frac{\mathbf{E}_R}{R_R + jX_R} = \frac{s\mathbf{E}_{R0}}{R_R + jsX_{R0}} = \frac{\mathbf{E}_{R0}}{R_R/s + jX_{R0}} \tag{19.77}$$

The resulting rotor equivalent circuit is shown in Figure 19.39.

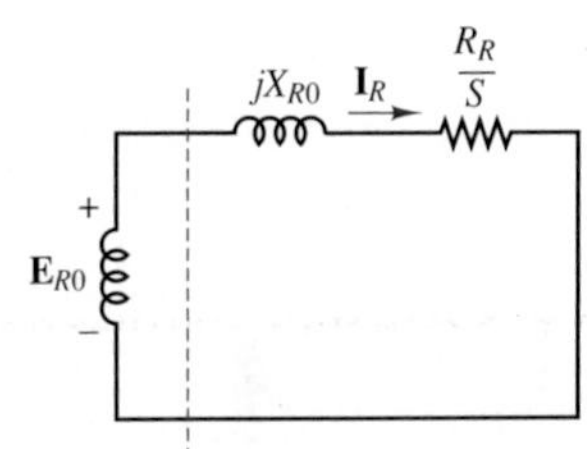

Figure 19.39 Rotor circuit

The voltages, currents, and impedances on the secondary (rotor) side can be reflected to the primary (stator) by means of the effective turns ratio. When this transformation is effected, the transformed rotor voltage is given by

$$\mathbf{E}_2 = \mathbf{E}'_R = \alpha\mathbf{E}_{R0} \tag{19.78}$$

The transformed (reflected) rotor current is

$$\mathbf{I}_2 = \frac{\mathbf{I}_R}{\alpha} \tag{19.79}$$

The transformed rotor resistance can be defined as

$$R_2 = \alpha^2 R_R \tag{19.80}$$

and the transformed rotor reactance can be defined by

$$X_2 = \alpha^2 X_{R0} \tag{19.81}$$

The final per-phase equivalent circuit of the induction motor is shown in Figure 19.40.

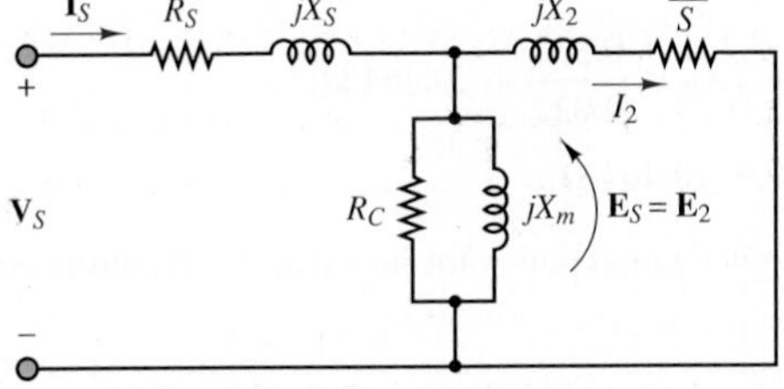

Figure 19.40 Equivalent circuit of an induction machine

Examples 19.13 and 19.14 illustrate the use of the circuit model in determining the performance of the induction motor.

LO5

EXAMPLE 19.13 Induction Motor Analysis

Problem

Determine the following quantities for an induction motor, using the circuit model of Figures 19.38 to 19.40.

1. Speed
2. Stator current
3. Power factor
4. Output torque

Solution

Known Quantities: Motor ratings; circuit parameters.

Find: n; ω_m; $\mathbf{I}_S$; power factor (pf); T.

Schematics, Diagrams, Circuits, and Given Data: Motor ratings: 460 V; 60 Hz; four poles; $s = 0.022$; $P_{\text{out}} = 14$ hp; $R_S = 0.641\ \Omega$; $R_2 = 0.332\ \Omega$; $X_S = 1.106\ \Omega$; $X_2 = 0.464\ \Omega$; $X_m = 26.3\ \Omega$

Assumptions: Use per-phase analysis. Neglect core losses ($R_C = 0$).

Analysis:

1. The per-phase equivalent circuit is shown in Figure 19.40. The synchronous speed is found to be

$$n_s = \frac{120f}{p} = \frac{60\text{ s/min} \times 60\text{ r/s}}{4/2} = 1{,}800\text{ r/min}$$

or

$$\omega_s = 1{,}800\frac{\text{r}}{\text{min}} \times \frac{2\pi\text{ rad}}{60\text{ s/min}} = 188.5\text{ rad/s}$$

The rotor mechanical speed is

$$n = (1 - s)n_s = 1{,}760\text{ r/min}$$

or

$$\omega_m = (1 - s)\omega_s = 184.4\text{ rad/s}$$

2. The reflected rotor impedance is found from the parameters of the per-phase circuit to be

$$Z_2 = \frac{R_2}{s} + jX_2 = \frac{0.332}{0.022} + j0.464\ \Omega$$
$$= 15.09 + j0.464\ \Omega$$

The combined magnetization plus rotor impedance is therefore equal to

$$Z = \frac{1}{1/jX_m + 1/Z_2} = \frac{1}{-j0.038 + 0.0662\angle -1.76^\circ} = 12.94\angle 31.1^\circ\ \Omega$$

and the total impedance is

$$Z_{\text{total}} = Z_S + Z = 0.641 + j1.106 + 11.08 + j6.68$$
$$= 11.72 + j7.79 = 14.07\angle 33.6^\circ\ \Omega$$

Finally, the stator current is given by

$$\mathbf{I}_S = \frac{\mathbf{V}_S}{Z_{\text{total}}} = \frac{460/\sqrt{3}\angle 0^\circ \text{ V}}{14.07\angle 33.6^\circ\ \Omega} = 18.88\angle -33.6^\circ \text{ A}$$

3. The power factor is

$$\text{pf} = \cos 33.6^\circ = 0.883 \text{ lagging}$$

4. The output power P_{out} is

$$P_{\text{out}} = 14 \text{ hp} \times 746 \text{ W/hp} = 10.444 \text{ kW}$$

and the output torque is

$$T = \frac{P_{\text{out}}}{\omega_m} = \frac{10{,}444 \text{ W}}{184.4 \text{ rad/s}} = 56.64 \text{ N-m}$$

CHECK YOUR UNDERSTANDING

A four-pole induction motor operating at a frequency of 60 Hz has a full-load slip of 4 percent. Find the frequency of the voltage induced in the rotor (a) at the instant of starting and (b) at full load.

Answer: (a) $f_R = 60$ Hz; (b) $f_R = 2.4$ Hz

EXAMPLE 19.14 Induction Motor Analysis

LO5

Problem

Determine the following quantities for a three-phase induction motor, using the circuit model of Figures 19.39 to 19.41.

1. Stator current.
2. Power factor.
3. Full-load electromagnetic torque.

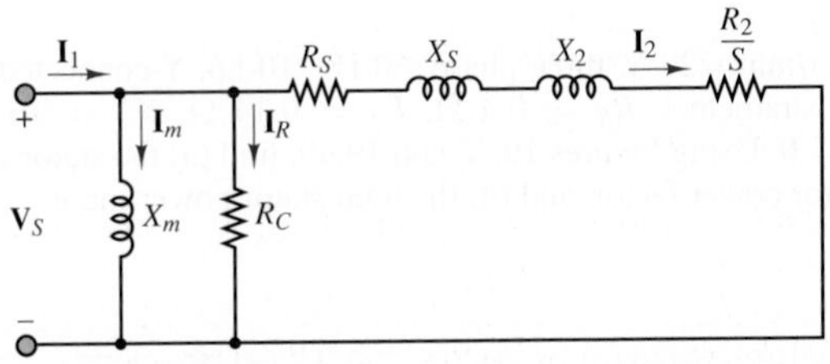

Figure 19.41

Solution

Known Quantities: Motor ratings; circuit parameters.

Find: $\mathbf{I}_S$; pf; T.

Schematics, Diagrams, Circuits, and Given Data: Motor ratings: 500 V; three-phase; 50 Hz; $p = 8$; $s = 0.05$; $P = 14$ hp.
Circuit parameters: $R_S = 0.13\ \Omega$; $R'_R = 0.32\ \Omega$; $X_S = 0.6\ \Omega$; $X'_R = 1.48\ \Omega$;
$Y_m = G_C + jB_m$ = magnetic branch admittance describing core loss and mutual inductance $= 0.004 - j0.05\ \Omega^{-1}$; stator/rotor turns ratio $= 1 : \alpha = 1 : 1.57$.

Assumptions: Use per-phase analysis. Neglect mechanical losses.

Analysis: The approximate equivalent circuit of the three-phase induction motor on a per-phase basis is shown in Figure 19.41. The parameters of the model are calculated as follows:

$$R_2 = R'_R \times \left(\frac{1}{\alpha}\right)^2 = 0.32 \times \left(\frac{1}{1.57}\right)^2 = 0.13\ \Omega$$

$$X_2 = X'_R \times \left(\frac{1}{\alpha}\right)^2 = 1.48 \times \left(\frac{1}{1.57}\right)^2 = 0.6\ \Omega$$

$$Z = R_S + \frac{R_2}{s} + j(X_S + X_2)$$

$$= 0.13 + \frac{0.13}{0.05} + j(0.6 + 0.6) = 2.73 + j1.2\ \Omega$$

Using the approximate circuit, we have

$$\mathbf{I}_2 = \frac{\mathbf{V}_S}{Z} = \frac{(500/\sqrt{3})\angle 0^\circ\ \text{V}}{2.73 + j1.2\ \Omega} = 88.8 - j39\ \text{A}$$

$$\mathbf{I}_R = \mathbf{V}_S G_S = 288.7\ \text{V} \times 0.004\ \Omega^{-1} = 1.15\ \text{A}$$

$$\mathbf{I}_m = -j\mathbf{V}_S B_m = 288.7\ \text{V} \times (-j0.05)\Omega = -j14.4\ \text{A}$$

$$\mathbf{I}_1 = \mathbf{I}_2 + \mathbf{I}_R + \mathbf{I}_m = 89.95 - j53.4\ \text{A}$$

$$\text{Input power factor} = \frac{\text{Re}[\mathbf{I}_1]}{|\mathbf{I}_1|} = \frac{89.95}{104.6} = 0.86\ \text{lagging}$$

$$\text{Torque} = \frac{3P}{\omega_S} = \frac{3I_2^2 R_2/s}{4\pi f/p} = 935\ \text{N-m}$$

CHECK YOUR UNDERSTANDING

A four-pole, 1,746 r/min, 220-V, three-phase, 60-Hz, 10-hp, Y-connected induction machine has the following parameters: $R_S = 0.4\ \Omega$, $R_2 = 0.14\ \Omega$, $X_m = 16\ \Omega$, $X_S = 0.35\ \Omega$, $X_2 = 0.35\ \Omega$, $R_C = 0$. Using Figures 19.38 and 19.39, find (a) the stator current, (b) the rotor current, (c) the motor power factor, and (d) the total stator power input.

Answer: (a) $25.92\angle -22.45^\circ$ A; (b) $24.39\angle -6.51^\circ$ A; (c) 0.9243; (d) 8,476 W

Performance of Induction Motors

The performance of induction motors can be described by torque–speed curves similar to those already used for DC motors. Figure 19.42 depicts an induction motor torque–speed curve, with five torque ratings marked a through e. Point a is the *starting torque,* also called **breakaway torque,** and is the torque available with the rotor "locked," that is, in a stationary position. At this condition, the frequency of the voltage induced in the rotor is highest, since it is equal to the frequency of rotation of the stator field; consequently, the inductive reactance of the rotor is greatest. As the rotor accelerates, the torque drops off, reaching a maximum value called the **pull-up torque** (point b); this typically occurs somewhere between 25 and 40 percent of synchronous speed. As the rotor speed continues to increase, the rotor reactance decreases further (since the frequency of the induced voltage is determined by the relative speed of rotation of the rotor with respect to the stator field). The torque becomes a maximum when the rotor inductive reactance is equal to the rotor resistance; maximum torque is also called **breakdown torque** (point c). Beyond this point, the torque drops off, until it is zero at synchronous speed, as discussed earlier. Also marked on the curve are the *150 percent torque* (point d) and the *rated torque* (point e).

A general formula for the computation of the induction motor steady-state torque–speed characteristic is

$$T = \frac{1}{\omega_e} \frac{mV_S^2 R_R/s}{(R_S + R_R/s)^2 + (X_S + X_R)^2} \qquad T\text{-}\omega \text{ equation for induction machine} \tag{19.82}$$

LO5

where m is the number of phases.

FIND IT ON THE WEB

Different construction arrangements permit the design of **induction motors** with different torque–speed curves, thus permitting the user to select the motor that best suits a given application. Figure 19.43 depicts the four basic classifications—classes A, B, C, and D—as defined by NEMA. The determining features in the classification are the locked-rotor torque and current, the breakdown torque, the pull-up torque, and the percentage of slip. Class A motors have a higher breakdown torque than class B motors, and a slip of 5 percent or less. Motors in this class are often designed for a specific application. Class B motors are general-purpose motors; this

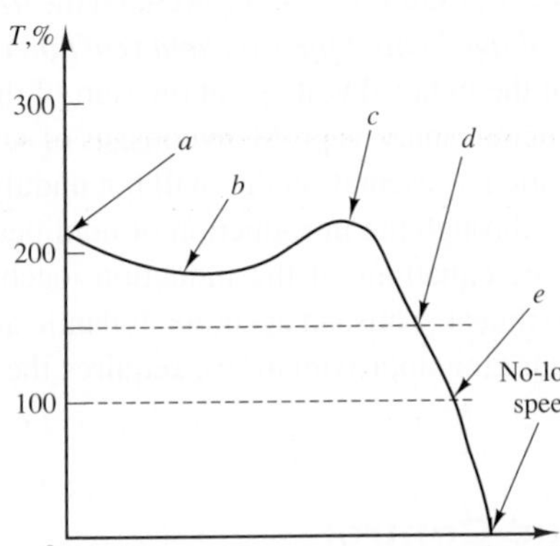

Figure 19.42 Performance curve for induction motor

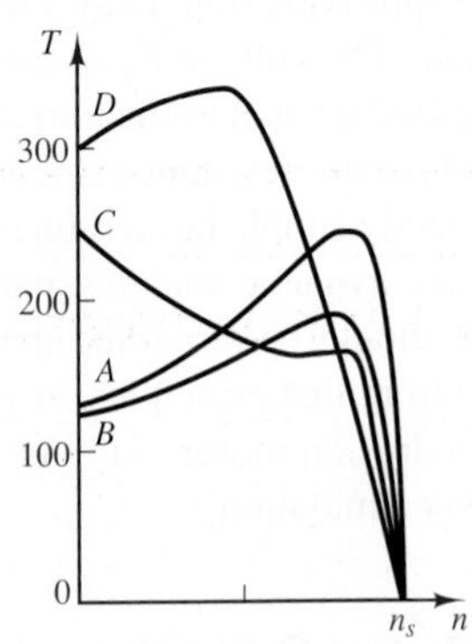

Figure 19.43 Induction motor classification

is the most commonly used type of induction motor, with typical values of slip of 3 to 5 percent. Class C motors have a high starting torque for a given starting current, and a low slip. These motors are typically used in applications demanding high starting torque but having relatively normal running loads, once the running speed has been reached. Class D motors are characterized by high starting torque, high slip, low starting current, and low full-load speed. A typical value of slip is around 13 percent.

Factors that should be considered in the selection of an AC motor for a given application are the *speed range,* both minimum and maximum, and the speed variation. For example, it is important to determine whether constant speed is required; what variation might be allowed, either in speed or in torque; or whether variable-speed operation is required, in which case a variable-speed drive will be needed. The torque requirements are obviously important as well. The starting and running torque should be considered; they depend on the type of load. Starting torque can vary from a small percentage of full-load torque to several times full-load torque. Furthermore, the excess torque available at start-up determines the *acceleration characteristics* of the motor. Similarly, *deceleration characteristics* should be considered, to determine whether external braking might be required.

Another factor to be considered is the *duty cycle* of the motor. The duty cycle, which depends on the nature of the application, is an important consideration when the motor is used in repetitive, noncontinuous operation, such as is encountered in some types of machine tools. If the motor operates at zero or reduced load for periods of time, the duty cycle—that is, the percentage of the time the motor is loaded—is an important selection criterion. Last, but by no means least, are the *heating properties* of a motor. Motor temperature is determined by internal losses and by ventilation; motors operating at a reduced speed may not generate sufficient cooling, and forced ventilation may be required.

Thus far, we have not considered the dynamic characteristics of induction motors. Among the integral-horsepower induction motors (i.e., motors with horsepower rating greater than 1), the most common dynamic problems are associated with starting and stopping and with the ability of the motor to continue operation during supply system transient disturbances. Dynamic analysis methods for induction motors depend to a considerable extent on the nature and complexity of the problem and the associated precision requirements. When the electric transients in the motor are to be included as well as the motion transients, and especially when the motor is an important element in a large network, the simple transient equivalent circuit of Figure 19.44 provides a good starting approximation. In the circuit model of Figure 19.44, X'_S is called the *transient reactance*. The voltage E'_S is called the *voltage behind the transient reactance* and is assumed to be equal to the initial value of the induced voltage, at the start of the transient. The stator resistance is R_S. The dynamic analysis problem consists of selecting a sufficiently simple but reasonably realistic representation that will not unduly complicate the dynamic analysis, particularly through the introduction of nonlinearities.

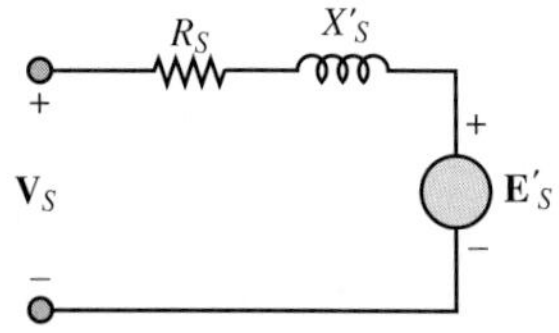

Figure 19.44 Simplified induction motor dynamic model

It should be remarked that the basic equations of the induction machine, as derived from first principles, are quite nonlinear. Thus, an accurate dynamic analysis of the induction motor, without any linearizing approximations, requires the use of computer simulation.

AC Motor Speed and Torque Control

As explained in an earlier section, AC machines are constrained to fixed-speed or near fixed-speed operation when supplied by a constant-frequency source. Several simple methods exist to provide limited **speed control in AC induction machines;** more complex methods, involving the use of advanced power electronics circuits, can be used if the intended application requires wide-bandwidth control of motor speed or torque. In this subsection we provide a general overview of available solutions.

FIND IT ON THE WEB

Pole Number Control

The (conceptually) easiest method to implement speed control in an induction machine is by *varying the number of poles.* Equation 19.55 explains the dependence of synchronous speed in an AC machine on the supply frequency and on the number of poles. For machines operated at 60 Hz, the following speeds can be achieved by varying the number of magnetic poles in the stator winding:

Number of poles	2	4	6	8	12
n, r/min	3,600	1,800	1,200	800	600

Motor stators can be wound so that the number of pole pairs in the stators can be varied by switching between possible winding connections. Such switching requires that care be taken in timing it to avoid damage to the machine.

Slip Control

Since the rotor speed is inherently dependent on the slip, *slip control* is a valid means of achieving some speed variation in an induction machine. Since motor torque falls with the square of the voltage (see equation 19.82), it is possible to change the slip by changing the motor torque through a reduction in motor voltage. This procedure allows for speed control over the range of speeds that allow for stable motor operation. With reference to Figure 19.42, this is possible only above point c, that is, above the *breakdown torque*.

Rotor Control

For motors with wound rotors, it is possible to connect the rotor slip rings to resistors; adding resistance to the rotor increases the losses in the rotor and therefore causes the rotor speed to decrease. This method is also limited to operation above the *breakdown torque,* although it should be noted that the shape of the motor torque–speed characteristic changes when the rotor resistance is changed.

Frequency Regulation

The last two methods cause additional losses to be introduced in the machine. If a variable-frequency supply is used, motor speed can be controlled without any additional losses. As seen in equation 19.55, the motor speed is directly dependent on the supply frequency, as the supply frequency determines the speed of the rotating magnetic field. However, to maintain the same motor torque characteristics over a range of speeds, the motor voltage must change with frequency, to maintain a constant torque. Thus, generally, the volts/hertz ratio should be held constant. This condition is difficult to achieve at start-up and at very low frequencies, in which cases the voltage must be raised above the constant volts/hertz ratio that will be appropriate at higher frequency.

Adjustable-Frequency Drives

FIND IT ON THE WEB

The advances made in the last two decades in power electronics and microcontrollers (see Chapters 13 and 14) have made AC machines employing **adjustable-frequency drives** well suited to many common engineering applications that until recently required the use of the more easily speed-controlled DC drives. An adjustable-frequency drive consists of four major subsystems, as shown in Figure 19.45.

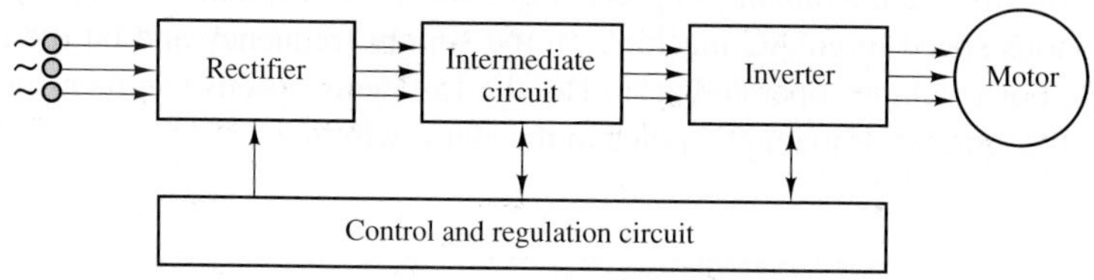

Figure 19.45 General configuration of adjustable-frequency drive

The diagram of Figure 19.45 assumes that a three-phase AC supply is available; the three-phase AC voltage is rectified using a controlled or uncontrolled **rectifier** (see Chapter 9 for a description of uncontrolled rectifiers and Chapter 12 for a description of controlled rectifiers). An **intermediate circuit** is sometimes necessary to further condition the rectified voltage and current. An **inverter** is then used to convert the fixed DC voltage to a variable frequency and variable-amplitude AC voltage. This is accomplished via **pulse-amplitude modulation** (PAM) or increasingly via **pulse-width modulation** (PWM) techniques. Figure 19.46 illustrates how approximately sinusoidal currents and voltages of variable frequency can be obtained by suitable shaping of a train of pulses. It is important to understand that the technology used to generate such wave shapes is based on the simple power-switching concepts underlying the voltage-source inverter (VSI) drive described in Chapter 12. DC-AC inverters come in many different configurations; the interested reader will find additional information and resources in the accompanying CD-ROM.

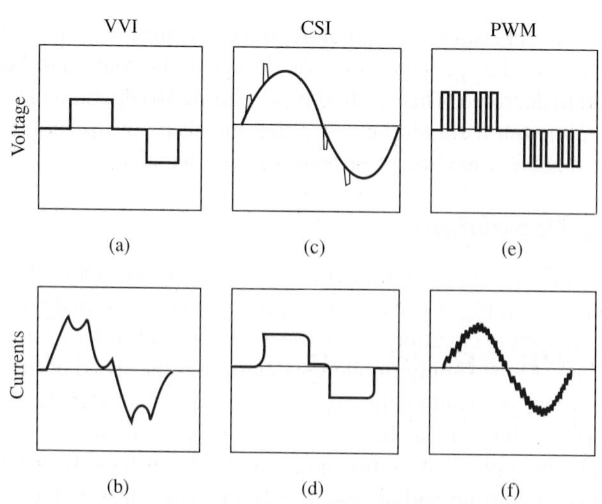

Figure 19.46 Typical adjustable-frequency controller voltage and current waveforms (*Courtesy: Rockwell Automation, Reliance Electric*)

Conclusion

This chapter introduces the most common classes of rotating electric machines. These machines, which can range in power from the milliwatt to the megawatt range, find common application in virtually every field of engineering, from consumer products to heavy-duty industrial applications. The principles introduced in this chapter can give you a solid basis from which to build upon.

Upon completing this chapter, you should have mastered the following learning objectives:

1. *Understand the basic principles of operation of rotating electric machines, their classification, and basic efficiency and performance characteristics.* Electric machines are defined in terms of their mechanical characteristics (torque–speed curves, inertia, friction and windage losses) and their electrical characteristics (current and voltage requirements). Losses and efficiency are an important part of the operation of electric machines, and it should be recognized that machines will suffer from electrical, mechanical, and magnetic core losses. All machines are based on the principle of establishing a magnetic field in the stationary part of the machine (stator) and a magnetic field in the moving part of the machine (rotor); electric machines can then be classified according to how the stator and rotor fields are established.
2. *Understand the operation and basic configurations of separately excited, permanent-magnet, shunt and series DC machines.* Direct-current machines, operated from a DC supply, are among the most common electric machines. The rotor (armature) circuit is connected to an external DC supply via a commutator. The stator electric field can be established by an external circuit (separately excited machines), by a permanent magnet (PM machines), or by the same supply used for the armature (self-excited machines).
3. *Analyze DC generators at steady state.* DC generators can be used to supply a variable direct current and voltage when propelled by a prime mover (engine, or other thermal or hydraulic machine).
4. *Analyze DC motors under steady-state and dynamic operation.* DC motors are commonly used in a variety of variable-speed applications (e.g., electric vehicles, servos) which require speed control; thus, their dynamics are also of interest.
5. *Understand the operation and basic configuration of AC machines, including the synchronous motor and generator and the induction machine.* AC machines require an alternating-current supply. The two principal classes of AC machines are the synchronous and induction types. Synchronous machines rotate at a predetermined speed, which is equal to the speed of a rotating magnetic field present in the stator, called the *synchronous speed.* Induction machines also operate based on a rotating magnetic field in the stator; however, the speed of the rotor is dependent on the operating conditions of the machine and is always less than the synchronous speed. Variable-speed AC machines require more sophisticated electric power supplies that can provide variable voltage/current and variable frequency. As the cost of power electronics is steadily decreasing, variable-speed AC drives are becoming increasingly common.

HOMEWORK PROBLEMS

Section 19.1: Rotating Electric Machines

19.1 The power rating of a motor can be modified to account for different ambient temperature, according to the following table:

Ambient temperature	30°C	35°C	40°C
Variation of rated power	+8%	+5%	0

Ambient temperature	45°C	50°C	55°C
Variation of rated power	−5%	−12.5%	−25%

A motor with $P_e = 10$ kW is rated up to 85°C. Find the actual power for each of the following conditions:

a. Ambient temperature is 50°C.

b. Ambient temperature is 30°C.

19.2 The speed–torque characteristic of an induction motor has been empirically determined as follows:

Speed, r/min	1,470	1,440	1,410	1,300	1,100
Torque, N-m	3	6	9	13	15

Speed, r/min	900	750	350	0
Torque, N-m	13	11	7	5

The motor will drive a load requiring a starting torque of 4 N-m and increase linearly with speed to 8 N-m at 1,500 r/min.

a. Find the steady-state operating point of the motor.

b. Equation 19.82 predicts that the motor speed can be regulated in the face of changes in load torque by adjusting the stator voltage. Find the change in voltage required to maintain the speed at the operating point of part a if the load torque increases to 10 N-m.

Section 19.2: Direct-Current Machines

19.3 Calculate the force exerted by each conductor, 6 in. long, on the armature of a DC motor when it carries a current of 90 A and lies in a field the density of which is 5.2×10^{-4} Wb/in^2.

19.4 In a DC machine, the air gap flux density is 4 Wb/m^2. The area of the pole face is 2 cm × 4 cm. Find the flux per pole in the machine.

Section 19.3: Direct-Current Generators

19.5 A 120-V, 10-A shunt generator has an armature resistance of 0.6 Ω. The shunt field current is 2 A. Determine the voltage regulation of the generator.

19.6 A 20-kW, 230-V separately excited generator has an armature resistance of 0.2 Ω and a load current of 100 A. Find

a. The generated voltage when the terminal voltage is 230 V.

b. The output power.

19.7 A 10-kW, 120-V DC series generator has an armature resistance of 0.1 Ω and a series field resistance of 0.05 Ω. Assuming that it is delivering rated current at rated speed, find (a) the armature current and (b) the generated voltage.

19.8 The armature resistance of a 30-kW, 440-V shunt generator is 0.1 Ω. Its shunt field resistance is 200 Ω. Find

a. The power developed at rated load.

b. The load, field, and armature currents.

c. The electric power loss.

19.9 A four-pole, 450-kW, 4.6-kV shunt generator has armature and field resistances of 2 and 333 Ω. The generator is operating at the rated speed of 3,600 r/min. Find the no-load voltage of the generator and terminal voltage at half load.

19.10 A 30-kW, 240-V generator is running at half load at 1,800 r/min with an efficiency of 85 percent. Find the total losses and input power.

19.11 A self-excited DC shunt generator is delivering 20 A to a 100-V line when it is driven at 200 rad/s. The magnetization characteristic is shown in Figure P19.11. It is known that $R_a = 1.0\ \Omega$ and $R_f = 100\ \Omega$. When the generator is disconnected from the line, the drive motor speeds up to 220 rad/s. What is the terminal voltage?

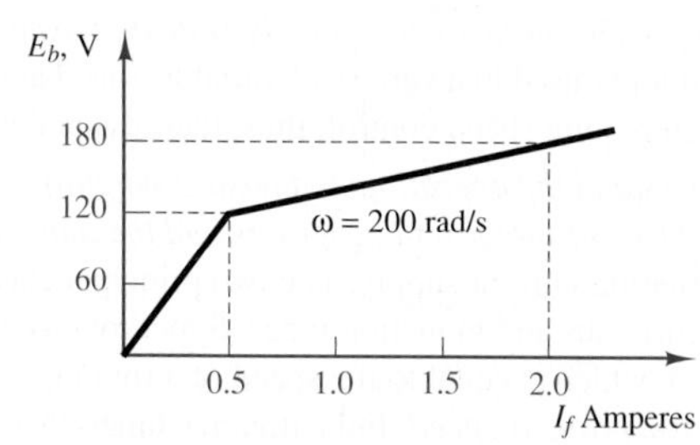

Figure P19.11

19.12 A high-pressure supply and a hydraulic motor are used as a prime mover to generate electricity through a DC generator. The system diagram is sketched in Figure P19.12. Assume that an ideal pressure source, P_S, is available, and that a hydraulic motor is connected to it through a linear "fluid resistor," used to regulate the average flow rate. An accumulator is inserted just upstream of the hydraulic motor to smooth pressure pulsations. The combined inertia of the hydraulic motor and of the DC generator is represented by the parameter, J. The DC generator is of the permanent magnet type, and has armature constants $k_a = k_T$. The permanent magnet flux is ϕ. Assume a resistive load for the generator R_L.

a. Derive the system differential equations.

b. Compute the transfer function of the system from supply pressure, P_S to load voltage, V_L.

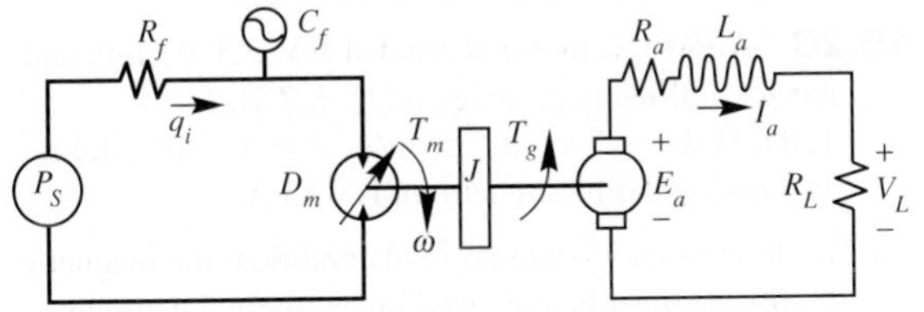

Figure P19.12

Section 19.4: Direct-Current Motors

19.13 A 220-V shunt motor has an armature resistance of 0.32 Ω and a field resistance of 110 Ω. At no load the armature current is 6 A and the speed is 1,800 r/min. Assume that the flux does not vary with load, and calculate

a. The speed of the motor when the line current is 62 A (assume a 2-V brush drop).

b. The speed regulation of the motor.

19.14 A 50-hp, 550-V shunt motor has an armature resistance, including brushes, of 0.36 Ω. When operating at rated load and speed, the armature takes 75 A. What resistance should be inserted in the armature circuit to obtain a 20 percent speed reduction when the motor is developing 70 percent of rated torque? Assume that there is no flux change.

19.15 A shunt DC motor has a shunt field resistance of 400 Ω and an armature resistance of 0.2 Ω. The motor nameplate rating values are 440 V, 1,200 r/min, 100 hp, and full-load efficiency of 90 percent. Find

a. The motor line current.

b. The field and armature currents.

c. The counter-emf at rated speed.

d. The output torque.

19.16 A 240-V series motor has an armature resistance of 0.42 Ω and a series-field resistance of 0.18 Ω. If the speed is 500 r/min when the current is 36 A, what will be the motor speed when the load reduces the line current to 21 A? (Assume a 3-V brush drop and that the flux is proportional to the current.)

19.17 A 220-V DC shunt motor has an armature resistance of 0.2 Ω and a rated armature current of 50 A. Find

a. The voltage generated in the armature.

b. The power developed.

19.18 A 550-V series motor takes 112 A and operates at 820 r/min when the load is 75 hp. If the effective armature-circuit resistance is 0.15 Ω, calculate the horsepower output of the motor when the current drops to 84 A, assuming that the flux is reduced by 15 percent.

19.19 A 200-V DC shunt motor has the following parameters:

$$R_a = 0.1\ \Omega \qquad R_f = 100\ \Omega$$

When running at 1,100 r/min with no load connected to the shaft, the motor draws 4 A from the line. Find E and the rotational losses at 1,100 r/min (assuming that the stray-load losses can be neglected).

19.20 A 230-V DC shunt motor has the following parameters:

$$R_a = 0.5\ \Omega \qquad R_f = 75\ \Omega$$
$$P_{\text{rot}} = 500\ \text{W} \qquad \text{at } 1{,}120\ \text{r/min}$$

When loaded, the motor draws 46 A from the line. Find

a. The speed, P_{dev}, and T_{sh}.

b. If $L_f = 25$ H, $L_a = 0.008$ H, and the terminal voltage has a 115-V change, find $i_a(t)$ and $\omega_m(t)$.

19.21 A 200-V DC shunt motor with an armature resistance of 0.1 Ω and a field resistance of 100 Ω draws a line current of 5 A when running with no load at 955 r/min. Determine the motor speed, the motor efficiency, the total losses (i.e., rotational and I^2R losses), and the load torque T_{sh} that will result when the motor draws 40 A from the line. Assume rotational power losses are proportional to the square of shaft speed.

19.22 A 50-hp, 230-V shunt motor has a field resistance of 17.7 Ω and operates at full load when the line current is 181 A at 1,350 r/min. To increase the speed of the motor to 1,600 r/min, a resistance of 5.3 Ω is "cut in" via the field rheostat; the line current then increases to 190 A. Calculate

a. The power loss in the field and its percentage of the total power input for the 1,350 r/min speed.

b. The power losses in the field and the field rheostat for the 1,600 r/min speed.

c. The percent losses in the field and in the field rheostat at 1,600 r/min.

19.23 A 10-hp, 230-V shunt-wound motor has a rated speed of 1,000 r/min and full-load efficiency of 86 percent. Armature circuit resistance is 0.26 Ω; field-circuit resistance is 225 Ω. If this motor is operating under rated load and the field flux is very quickly reduced to 50 percent of its normal value, what will be the effect upon counter-emf, armature current, and torque? What effect will this change have upon the operation of the motor, and what will be its speed when stable operating conditions have been regained?

19.24 The machine of Example 19.7 is being used in a series connection. That is, the field coil is connected in series with the armature, as shown in Figure P19.24. The machine is to be operated under the same conditions as in Example 19.7, that is, $n = 120$ r/min and $I_a = 8$ A. In the operating region, $\phi = kI_f$ and $k = 200$. The armature resistance is 0.2 Ω, and the resistance of the field winding is negligible.

a. Find the number of field winding turns necessary for full-load operation.

b. Find the torque output for the following speeds:

1. $n' = 2n$
2. $n' = 3n$
3. $n' = n/2$
4. $n' = n/4$

c. Plot the speed–torque characteristic for the conditions of part b.

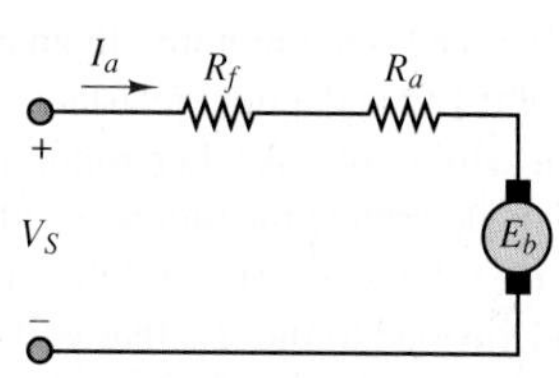

Figure P19.24

19.25 With reference to Example 19.9, assume that the load torque applied to the PM DC motor is zero. Determine the speed response of the motor speed to a step change in input voltage. Derive expressions for the natural frequency and damping ratio of the second-order system. What determines whether the system is over- or underdamped?

19.26 A motor with polar moment of inertia J develops torque according to the relationship $T = a\omega + b$. The motor drives a load defined by the torque–speed relationship $T_L = c\omega^2 + d$. If the four coefficients are all positive constants, determine the equilibrium speeds of the motor-load pair, and whether these speeds are stable.

19.27 Assume that a motor has known friction and windage losses described by the equation $T_{FW} = b\omega$. Sketch the T-ω characteristic of the motor if the load torque T_L is constant, and the T_L-ω characteristic if the motor torque is constant. Assume that T_{FW} at full speed is equal to 30 percent of the load torque.

19.28 A PM DC motor is rated at 6 V, 3,350 r/min and has the following parameters: $r_a = 7\ \Omega$, $L_a = 120$ mH, $k_T = 7 \times 10^{-3}$ N-m/A, $J = 1 \times 10^{-6}$ kg-m^2. The no-load armature current is 0.15 A.

a. In the steady-state no-load condition, the magnetic torque must be balanced by an internal damping torque; find the damping coefficient b. Now sketch a model of the motor, write the dynamic equations, and determine the transfer function from armature voltage to motor speed. What is the approximate 3-dB bandwidth of the motor?

b. Now let the motor be connected to a pump with inertia $J_L = 1 \times 10^{-4}$ kg-m^2, damping coefficient $b_L = 5 \times 10^{-3}$ N-m-s, and load torque $T_L = 3.5 \times 10^{-3}$ N-m. Sketch the model describing the motor-load configuration, and write the dynamic equations for this system; determine the new transfer function from armature voltage to motor speed. What is the approximate 3-dB bandwidth of the motor/pump system?

19.29 A PM DC motor with torque constant k_{PM} is used to power a hydraulic pump; the pump is a positive displacement type and generates a flow proportional to the pump velocity: $q_p = k_p\omega$. The fluid travels through a conduit of negligible resistance; an accumulator is included to smooth out the pulsations of the pump. A hydraulic load (modeled by a fluid resistance R) is connected between the pipe and a reservoir (assumed at zero pressure). Sketch the motor-pump circuit. Derive the dynamic equations for the system, and determine the transfer function between motor voltage and the pressure across the load.

19.30 A shunt motor in Figure P19.30 is characterized by a field coefficient $k_f = 0.12$ V-s/A-rad, such that the back emf is given by the expression $E_b = k_f I_f \omega$ and the motor torque by the expression $T = k_f I_f I_a$. The motor drives an inertia/viscous friction load with parameters $J = 0.8$ kg-m^2 and $b = 0.6$ N-m-s/rad. The field equation may be approximated by $V_S = R_f I_f$. The armature resistance is $R_a = 0.75\ \Omega$, and the field resistance is $R_f = 60\ \Omega$. The system is perturbed around the nominal operating point $V_{S0} = 150$ V, $\omega_0 = 200$ rad/s, $I_{a0} = 186.67$ A, respectively.

a. Derive the dynamic system equations in *symbolic form*.

b. Linearize the equations you obtained in part a.

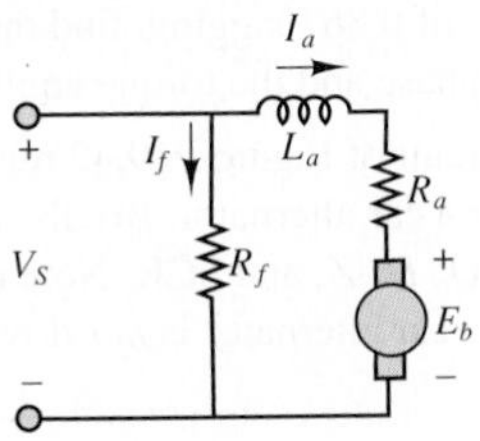

Figure P19.30

19.31 A PM DC motor is rigidly coupled to a fan; the fan load torque is described by the expression $T_L = 5 + 0.05\omega + 0.001\omega^2$, where torque is in Newton-meters and speed in radians per second. The motor has $k_a\phi = k_T\phi = 2.42$; $R_a = 0.2\ \Omega$, and the inductance is negligible. If the motor voltage is 50 V, what will be the speed of rotation of the motor and fan?

19.32 A separately excited DC motor has the following parameters:

$R_a = 0.1\ \Omega$	$R_f = 100\ \Omega$	$L_a = 0.2$ H
$L_f = 0.02$ H	$K_a = 0.8$	$K_f = 0.9$

The motor load is an inertia with $J = 0.5$ kg-m^2 and $b = 2$ N-m-rad/s. No external load torque is applied.

a. Sketch a diagram of the system and derive the (three) differential equations.

b. Sketch a simulation block diagram of the system (you should have three integrators).

c. Code the diagram, using Simulink.

d. Run the following simulations:
Armature control. Assume a constant field with $V_f = 100$ V; now simulate the response of the system when the armature voltage changes in step fashion from 50 to 75 V. Save and plot the current and angular speed responses.
Field control. Assume a constant armature voltage with $V_a = 100$ V; now simulate the response of the system when the field voltage changes in step fashion from 75 to 50 V. This procedure is called *field weakening*. Save and plot the current and angular speed responses.

19.33 Determine the transfer functions from *input voltage* to *angular velocity* and from *load torque* to *angular velocity* for a PM DC motor rigidly connected to an inertial load. Assume resistance and inductance parameters R_a, L_a let the armature constant be k_a. Assume ideal energy conversion, so that $k_a = k_T$. The motor has inertia J_m and damping coefficient b_m, and it is rigidly connected to an inertial load with inertia J and damping coefficient b. The load torque T_L acts on the load inertia to oppose the magnetic torque.

19.34 Assume that the coupling between the motor and the inertial load of Problem 19.33 is flexible (e.g., a long shaft). This can be modeled by adding a torsional spring between the motor inertia and the load inertia. Now we can no longer lump together the two inertias and damping coefficients as if they were one; we need to write separate equations for the two inertias. In total, there will be three equations in this system the motor electrical equation, the motor mechanical equation (J_m and B_m), and the load mechanical equation (J and B).

a. Sketch a diagram of the system.

b. Use free-body diagrams to write each of the two mechanical equations. Set up the equations in matrix form.

c. Compute the transfer function from input voltage to load inertia speed, using the method of determinants.

19.35 A wound DC motor is connected in both a shunt and a series configuration. Assume generic resistance and inductance parameters R_a, R_f, L_a, L_f; let the field magnetization constant be k_f and the armature constant be k_a. Assume ideal energy conversion, so that $k_a = k_T$. The motor has inertia J_m and damping coefficient b_m, and it is rigidly connected to an inertial load with inertia J and damping coefficient b.

a. Sketch a system-level diagram of the two configurations that illustrates both the mechanical and electrical systems.

b. Write an expression for the torque–speed curve of the motor in each configuration.

c. Write the differential equations of the motor-load system in each configuration.

d. Determine whether the differential equations of each system are linear; if one (or both) is (are) nonlinear, could they be made linear with some simple assumption? Explain clearly under what conditions this would be the case.

19.36 Derive the differential equations describing the electrical and mechanical dynamics of a shunt-connected DC motor, shown in Figure P19.36; and draw a simulation block diagram of the system. The motor parameters are k_a, k_T = armature and torque reluctance constant and k_f = field flux constant.

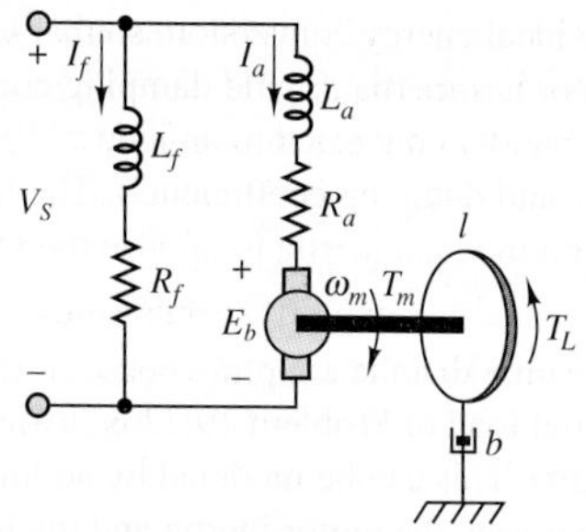

Figure P19.36

19.37 Derive the differential equations describing the electrical and mechanical dynamics of a series-connected DC motor, shown in Figure P19.37, and draw a simulation block diagram of the system. The motor parameters are k_a, k_T = armature and torque reluctance constant and k_f = field flux constant.

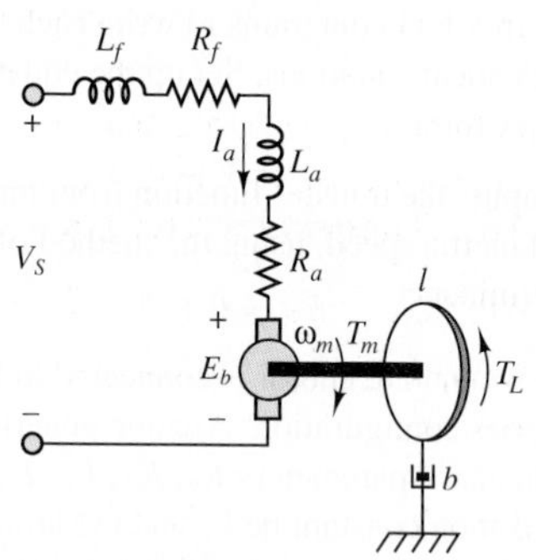

Figure P19.37

19.38 Develop a Simulink simulator for the shunt-connected DC motor of Problem 19.36. Assume the following parameter values: $L_a = 0.15$ H; $L_f = 0.05$ H; $R_a = 1.8\ \Omega$; $R_f = 0.2\,\Omega$; $k_a = 0.8$ V-s/rad; $k_T = 20$ N-m/A; $k_f = 0.20$ Wb/A; $b = 0.1$ N-m-s/rad; $J = 1$ kg-m^2.

19.39 Develop a Simulink simulator for the series-connected DC motor of Problem 19.37. Assume the following parameter values: $L = L_a + L_f = 0.2$ H; $R = R_a + R_f = 2\ \Omega$; $k_a = 0.8$ V-s/rad; $k_T = 20$ N-m/A; $k_f = 0.20$ Wb/A; $b = 0.1$ N-m-s/rad; $J = 1$ kg-m^2.

Section 19.6: The Alternator (Synchronous Generator)

19.40 An automotive alternator is rated 500 VA and 20 V. It delivers its rated voltamperes at a power factor of 0.85. The resistance per phase is 0.05 Ω, and the field takes 2 A at 12 V. If the friction and windage loss is 25 W and the core loss is 30 W, calculate the percent efficiency under rated conditions.

19.41 It has been determined by test that the synchronous reactance X_s and armature resistance r_a of a 2,300-V, 500-VA, three-phase synchronous generator are 8.0 and 0.1 Ω, respectively. If the machine is operating at rated load and voltage at a power factor of 0.867 lagging, find the generated voltage per phase and the torque angle.

19.42 The circuit of Figure P19.42 represents a voltage regulator for a car alternator. Briefly, explain the function of Q, D, Z, and SCR. Note that unlike other alternators, a car alternator is *not* driven at constant speed.

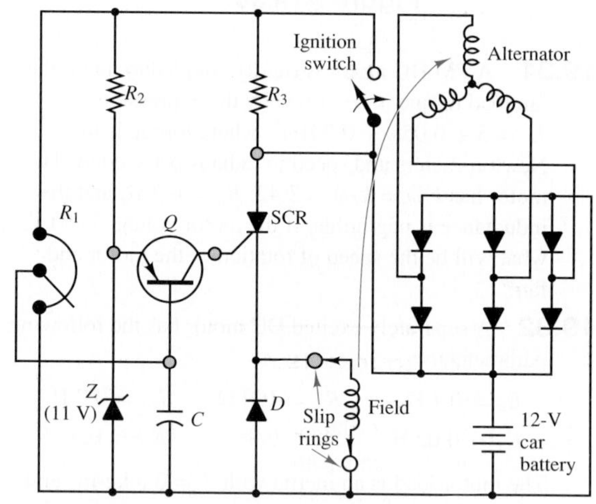

Figure P19.42

Section 19.7: The Synchronous Motor

19.43 A non–salient pole, Y-connected, three-phase, two-pole synchronous machine has a synchronous reactance of 7 Ω and negligible resistance and rotational losses. One point on the open-circuit characteristic is given by $V_o = 400$ V (phase voltage) for a field current of 3.32 A. The machine is to be operated as a motor, with a terminal voltage of 400 V (phase voltage). The armature current is 50 A, with power factor 0.85, leading. Determine E_b, field current, torque developed, and power angle δ.

19.44 A factory load of 900 kW at 0.6 power factor lagging is to be increased by the addition of a synchronous motor that takes 450 kW. At what power factor must this motor operate, and what must be its kilovoltampere input if the overall power factor is to be 0.9 lagging?

19.45 A non–salient pole, Y-connected, three-phase, two-pole synchronous generator is connected to a 400-V (line to line), 60-Hz, three-phase line. The stator impedance is $0.5 + j1.6\ \Omega$ (per phase). The generator is delivering rated current (36 A) at unity power factor to the line. Determine the power angle for this load and the value of E_b for this condition. Sketch the phasor diagram, showing $\mathbf{E}_b$, $\mathbf{I}_S$, and $\mathbf{V}_S$.

19.46 A non–salient pole, three-phase, two-pole synchronous motor is connected in parallel with a three-phase, Y-connected load so that the per-phase equivalent circuit is as shown in Figure P19.46. The parallel combination is connected to a 220-V (line to line), 60-Hz, three-phase line. The load current $\mathbf{I}_L$ is 25 A at a power factor of 0.866 inductive. The motor has $X_S = 2\ \Omega$ and is operating with $I_f = 1$ A and $T = 50$ N-m at a power angle of $-30°$. (Neglect all losses for the motor.) Find $\mathbf{I}_S$, P_{in} (to the motor), the overall power factor (i.e., angle between $\mathbf{I}_1$ and $\mathbf{V}_S$), and the total power drawn from the line.

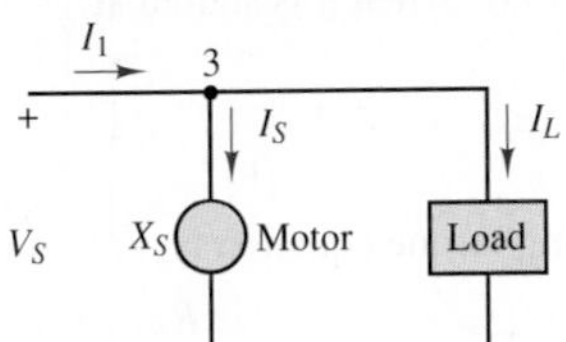

Figure P19.46

19.47 A four-pole, three-phase, Y-connected, non–salient pole synchronous motor has a synchronous reactance of 10 Ω. This motor is connected to a $230\sqrt{3}$ V (line to line), 60-Hz, three-phase line and is driving a load such that $T_{\text{shaft}} = 30$ N-m. The line current is 15 A, leading the phase voltage. Assuming that all losses can be neglected, determine the power angle δ and E for this condition. If the load is removed, what is the line current, and is it leading or lagging the voltage?

19.48 A 10-hp, 230-V, 60 Hz, three-phase, Y-connected synchronous motor delivers full load at a power factor of 0.8 leading. The synchronous reactance is 6 Ω, the rotational loss is 230 W, and the field loss is 50 W. Find

a. The armature current.

b. The motor efficiency.

c. The power angle.

Neglect the stator winding resistance.

19.49 A 2,000-hp, unity power factor, three-phase, Y-connected, 2,300-V, 30-pole, 60-Hz synchronous motor has a synchronous reactance of 1.95 Ω per phase. Neglect all losses. Find the maximum power and torque.

19.50 A 1,200-V, three-phase, Y-connected synchronous motor takes 110 kW (exclusive of field winding loss) when operated under a certain load at 1,200 r/min. The back emf of the motor is 2,000 V. The synchronous reactance is 10 Ω per phase, with negligible winding resistance. Find the line current and the torque developed by the motor.

19.51 The per-phase impedance of a 600-V, three-phase, Y-connected synchronous motor is $5 + j50\ \Omega$. The motor takes 24 kW at a leading power factor of 0.707. Determine the induced voltage and the power angle of the motor.

Section 19.8: The Induction Motor

19.52 A 74.6-kW, three-phase, 440-V (line to line), four-pole, 60-Hz induction motor has the following (per-phase) parameters referred to the stator circuit:

$R_S = 0.06\ \Omega$ $\quad X_S = 0.3\ \Omega$ $\quad X_m = 5\ \Omega$

$R_R = 0.08\ \Omega$ $\quad X_R = 0.3\ \Omega$

The no-load power input is 3,240 W at a current of 45 A. Determine the line current, input power, developed torque, shaft torque, and efficiency at $s = 0.02$.

19.53 A 60-Hz, four-pole, Y-connected induction motor is connected to a 400-V (line to line), three-phase, 60-Hz line. The equivalent circuit parameters are

$R_S = 0.2\ \Omega$ $\quad R_R = 0.1\ \Omega$

$X_S = 0.5\ \Omega$ $\quad X_R = 0.2\ \Omega$

$X_m = 20\ \Omega$

When the machine is running at 1,755 r/min, the total rotational and stray-load losses are 800 W. Determine the slip, input current, total input power, mechanical power developed, shaft torque, and efficiency.

19.54 A three-phase, 60-Hz induction motor has eight poles and operates with a slip of 0.05 for a certain load. Determine

a. The speed of the rotor with respect to the stator.

b. The speed of the rotor with respect to the stator magnetic field.

c. The speed of the rotor magnetic field with respect to the rotor.

d. The speed of the rotor magnetic field with respect to the stator magnetic field.

19.55 A three-phase, two-pole, 400-V (per phase), 60-Hz induction motor develops 37 kW (total) of mechanical power P_m at a certain speed. The rotational loss at this speed is 800 W (total). (Stray-load loss is negligible.)

a. If the total power transferred to the rotor is 40 kW, determine the slip and the output torque.

b. If the total power into the motor P_{in} is 45 kW and R_S is 0.5 Ω, find I_S and the power factor.

19.56 The nameplate speed of a 25-Hz induction motor is 720 r/min. If the speed at no load is 745 r/min, calculate

a. The slip.

b. The percent regulation.

19.57 The nameplate of a squirrel cage four-pole induction motor has the following information: 25 hp, 220 V, three-phase, 60 Hz, 830 r/min, 64-A line current. If the motor draws 20,800 W when operating at full load, calculate

a. Slip.

b. Percent regulation if the no-load speed is 895 r/min.

c. Power factor.

d. Torque.

e. Efficiency.

19.58 A 60-Hz, four-pole, Y-connected induction motor is connected to a 200-V (line to line), three-phase, 60-Hz line. The equivalent circuit parameters are

$R_S = 0.48\ \Omega$ Rotational loss torque = 3.5 N-m

$X_S = 0.8\ \Omega$ $R_R = 0.42\ \Omega$ (referred to stator)

$X_m = 30\ \Omega$ $X_R = 0.8\ \Omega$ (referred to stator)

The motor is operating at slip $s = 0.04$. Determine the input current, input power, mechanical power, and shaft torque (assuming that stray-load losses are negligible).

19.59

a. A three-phase, 220-V, 60-Hz induction motor runs at 1,140 r/min. Determine the number of poles (for minimum slip), the slip, and the frequency of the rotor currents.

b. To reduce the starting current, a three-phase squirrel cage induction motor is started by reducing the line voltage to $V_s/2$. By what factor are the starting torque and the starting current reduced?

19.60 A six-pole induction motor for vehicle traction has a 50-kW input electric power rating and is 85 percent efficient. If the supply is 220 V at 60 Hz, compute the motor speed and torque at a slip of 0.04.

19.61 An AC induction machine has six poles and is designed for 60-Hz, 240-V (rms) operation. When the machine operates with 10 percent slip, it produces 60 N-m of torque.

a. The machine is now used in conjunction with a friction load which opposes a torque of 50 N-m. Determine the speed and slip of the machine when used with the above-mentioned load.

b. If the machine has an efficiency of 92 percent, what minimum rms current is required for operation with the load of part a?

Hint: you may assume that the speed–torque curve is approximately linear in the region of interest.

19.62 A blocked-rotor test was performed on a 5-hp, 220-V, four-pole, 60-Hz, three-phase induction motor. The following data were obtained: $V = 48$ V, $I = 18$ A, $P = 610$ W. Calculate

a. The equivalent stator resistance per phase R_S.

b. The equivalent rotor resistance per phase R_R.

c. The equivalent blocked-rotor reactance per phase X_R.

19.63 Calculate the starting torque of the motor of Problem 19.62 when it is started at

a. 220 V

b. 110 V

The starting torque equation is

$$T = \frac{m}{\omega_e} \cdot V_S^2 \cdot \frac{R_R}{(R_R + R_S)^2 + (X_R + X_S)^2}$$

19.64 A four-pole, three-phase induction motor drives a turbine load. At a certain operating point the machine has 4 percent slip and 87 percent efficiency. The motor drives a turbine with torque–speed characteristic given by $T_L = 20 + 0.006\omega^2$. Determine the torque at the motor-turbine shaft and the total power delivered to the turbine. What is the total power consumed by the motor?

19.65 A four-pole, three-phase induction motor rotates at 1,700 r/min when the load is 100 N-m. The motor is 88 percent efficient.

a. Determine the slip at this operating condition.

b. For a constant-power, 10-kW load, determine the operating speed of the machine.

c. Sketch the motor and load torque–speed curves on the same graph. Show numerical values.

d. What is the total power consumed by the motor?

19.66 Find the speed of the rotating field of a six-pole, three-phase motor connected to (a) a 60-Hz line and (b) a 50-Hz line, in revolutions per minute and radians per second.

19.67 A six-pole, three-phase, 440-V, 60-Hz induction motor has the following model impedances:

$R_S = 0.8\ \Omega$ $X_S = 0.7\ \Omega$

$R_R = 0.3\ \Omega$ $X_R = 0.7\ \Omega$

$X_m = 35\ \Omega$

Calculate the input current and power factor of the motor for a speed of 1,200 r/min.

19.68 An eight-pole, three-phase, 220-V, 60-Hz induction motor has the following model impedances:

$R_S = 0.78\ \Omega \quad X_S = 0.56\ \Omega \quad X_m = 32\ \Omega$

$R_R = 0.28\ \Omega \quad X_R = 0.84\ \Omega$

Find the input current and power factor of this motor for $s = 0.02$.

19.69 A nameplate is given in Example 19.2. Find the rated torque, rated voltamperes, and maximum continuous output power for this motor.

19.70 A three-phase induction motor, at rated voltage and frequency, has a starting torque of 140 percent and a maximum torque of 210 percent of full-load torque. Neglect stator resistance and rotational losses and assume constant rotor resistance. Determine

a. The slip at full load.

b. The slip at maximum torque.

c. The rotor current at starting as a percentage of full-load rotor current.

19.71 A 60 Hz, four-pole, three-phase induction motor delivers 35 kW of mechanical (output) power. At a certain operating point the machine has 4 percent slip and 87 percent efficiency. Determine the torque delivered to the load and the total electrical (input) power consumed by the motor.

19.72 A four-pole, three-phase induction motor rotates at 16,800 rev/min when the load is 140 N-m. The motor is 85 percent efficient.

a. Determine the slip at this operating condition.

b. For a constant-power, 20-kW load, determine the operating speed of the machine.

c. Sketch the motor and load torque-speed curves for the load of part b. on the same graph. Show numerical values.

19.73 An AC induction machine has six poles and is designed for 60-Hz, 240-V (rms) operation. When the machine operates with 10 percent slip, it produces 60 N-m of torque.

a. The machine is now used in conjunction with an 800-W constant power load. Determine the speed and slip of the machine when used with the above-mentioned load.

b. If the machine has an efficiency of 89 percent, what minimum rms current is required for operation with the load of part a?

[*Hint:* You may assume that the speed torque curve is approximately linear in the region of interest]

CHAPTER

20

SPECIAL-PURPOSE ELECTRIC MACHINES

The objective of this chapter is to introduce the operating principles and performance characteristics of a number of special-purpose electric machines that find widespread engineering application in a variety of fields, ranging from robotics to vehicle propulsion, aerospace, and automotive control. In Chapters 18 and 19, you were introduced to the operating principles of the major classes of electric machines: DC machines, synchronous machines, and induction motors. The machines discussed in this chapter operate according to the essential principles described earlier, but are also characterized by unique features that set them apart from the machines described in Chapter 19. The first of these special-purpose machines is the brushless DC motor. Next, we discuss stepping motors, illustrating a very natural match between electromechanical devices and digital logic. The switched reluctance motor is presented next. A discussion of universal motors and single-phase induction motors follows, with a brief description of the types of electronic drives used to supply power to these machines. The discussion of the electronic drives ties the electromechanics material with the subject of power electronics introduced in Chapter 12.

Section 20.5 of this chapter covers design and performance specifications related to the application of electric machines.

The machines introduced in this chapter are used in many applications requiring fractional horsepower, or the ability to accurately control position, velocity, or torque.

Learning Objectives

1. Understand the basic principles of operation of brushless DC motors, and the trade-offs between these and brush-type DC motors. *Section 20.1.*
2. Understand the operation and basic configurations of step motors as well as step sequences for the different classes of step motors. *Section 20.2.*
3. Understand the operating principles of switched reluctance machines. *Section 20.3.*
4. Classify and analyze single-phase AC motors, including the universal motor and various types of single-phase induction motors, using simple circuit models. *Section 20.4.*
5. Outline the selection process for an electric machine, given an application; perform calculations related to load inertia, acceleration, efficiency, and thermal characteristics. *Section 20.5.*

20.1 BRUSHLESS DC MOTORS

FIND IT ON THE WEB

In spite of its name, the **brushless DC motor** is actually not a DC motor, but (typically) a permanent-magnet synchronous machine; the name is actually due not to the construction of the machine, but to the fact that its operating characteristics resemble those of a shunt DC motor with constant field current. This characteristic can be obtained by providing the motor with a power supply whose electrical frequency is always identical to the mechanical frequency of rotation of the rotor. To generate a source of variable frequency, use is made of DC-to-AC converters (inverters), consisting of banks of transistors that are switched on and off at a frequency corresponding to the rotor speed; thus, the inverter converts a DC source to an AC source of variable frequency. As far as the user is concerned, then, the source of excitation of a brushless DC motor is DC, although the current that actually flows through the motor windings is AC. (The operation of the inverters will be explained shortly.) In effect, the brushless DC motor is a synchronous motor in which the torque angle δ is kept constant by an appropriate excitation current.

Brushless DC motors also require measurement of the position of the rotor to determine its speed of rotation, and to generate a supply current at the same frequency. This function is accomplished by means of a position-sensing arrangement that usually consists either of a magnetic Hall-effect position sensor, which senses the passage of each pole in the rotor, or of an optical encoder similar to the encoders discussed in Chapter 13.

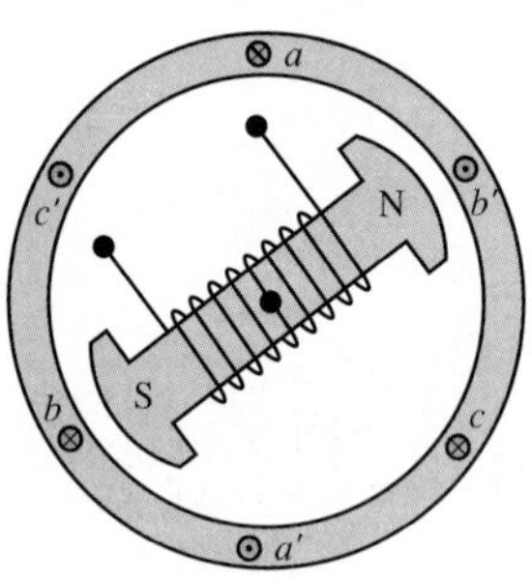

Figure 20.1a Two-pole brushless DC motor with three-phase stator winding

Figure 20.1(a) depicts the appearance of a brushless DC motor. Note how the multiphase winding is similar to that of the synchronous motor of Chapter 19. Figure 20.1(b) depicts the construction of a typical brushless DC servomotor. The brushless motor consists of a stator with a multiphase winding, usually three-phase; a permanent-magnet rotor; and a rotor position sensor. It is interesting to observe that since the commutation is performed electronically by switching the current to the motor—rather than by brushes, as in DC motors—the brushless motor can be produced in many different configurations, including, for example, very flat ("pancake") motors. Figure 20.1 shows the classical configuration of inside rotor, outside stator.

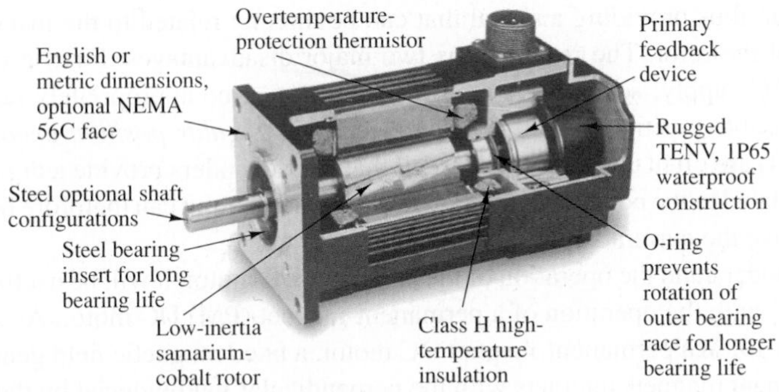

Figure 20.1b A typical brushless DC servomotor (*Courtesy: Pacific Scientific Motor Products Division.*)

For simple machines, it is also possible to resort to an outside rotor, with greater ease of magnet attachment and inherently smoother rotation, but with inferior thermal characteristics, since a stator encased within the rotor structure cannot be cooled efficiently.

In conventional DC motors, the supply voltage is limited by brush wear and sparking that can occur at the commutator, often resulting in the need for transformers to step down the supply voltage. In brushless DC motors, on the other hand, such a concern does not arise, because the commutation is performed electronically without the need for brushes. Further, since, in general, the armature (load-carrying winding) is on the stator and thus the losses are concentrated in the stator, liquid cooling (if required) is feasible and does not involve excessive complexity. You will recall that in a conventional DC motor the armature is on the rotor, and therefore auxiliary liquid cooling is very difficult to implement.

Another important advantage of brushless DC motors is that by sealing the stator, submersible units can be built. In addition to these operational advantages, note that these motors are also characterized by easier construction: The construction of the stator in a brushless DC motor is similar to that in traditional induction motors and is therefore suitable for automated production. The windings may also be fitted with temperature sensors, providing the possibility of additional thermal protection.

The permanent-magnet rotor is typically made either of rare-earth magnets (Sm-Co) or of ceramic magnets (ferrites). Rare-earth magnets have outstanding magnetic properties, but they are expensive and in limited supply, and therefore the more commonly employed materials are ceramic magnets. Rare-earth magnet motors can be a cost-effective solution—since they allow much greater fluxes to be generated by a given supply current—in applications where high speed, high efficiency, and small size are important. Brushless DC motors can be rated up to 250 kW at 50,000 r/min. The rotor position sensor must be designed for operation inside the motor, and must withstand the backlash, vibrations, and temperature range typical of motor operation.

Brushless DC motors do require a position-sensing device, though, to permit proper switching of the supply current. Recall that the brushless DC motor replaces the cumbersome mechanical commutation arrangement with electronic switching of the supply current. The most commonly used position-sensing devices are *position encoders* and **resolvers.** The resolver, shown in Figure 20.2, is a rotating machine that is mechanically coupled to the rotor of the brushless motor and consists of two stator and two rotor windings; the stator windings are excited by an AC signal, and the resulting rotor voltages are proportional to the sine and cosine of the angle of rotation

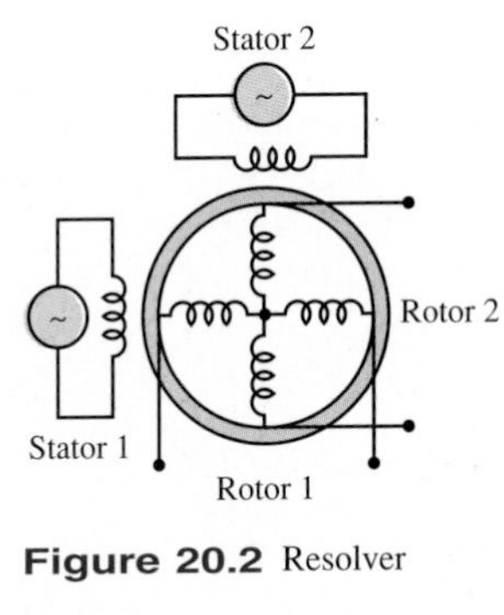

Figure 20.2 Resolver

of the rotor, thus providing a signal that can be directly related to the instantaneous position of the rotor. The resolver has two major disadvantages: First, it requires a separate AC supply; second, the resolver output must be appropriately decoded to obtain a usable position signal. For these reasons, *angular position encoders* (see Chapter 13) are often used. You will recall that such encoders provide a digital signal directly related to the position of a rotating shaft. Their output can therefore be directly used to drive the current supply for a brushless motor.

To understand the operation of the brushless DC motor, it will be useful to make an analogy with the operation of a permanent-magnet (PM) DC motor. As discussed in Chapter 19, in a permanent-magnet DC motor, a fixed magnetic field generated by the permanent magnets interacts with the perpendicular field induced by the currents in the rotor windings, thus creating a mechanical torque. As the rotor turns in response to this torque, however, the angle between the stator and rotor fields is reduced, so that the torque would be nullified within a rotation of 90 electrical degrees. To sustain the torque acting on the rotor, permanent-magnet DC motors incorporate a commutator, fixed to the rotor shaft. The commutator switches the supply current to the stator so as to maintain a constant angle $\delta = 90^\circ$ between interacting fields. Because the current is continually switched between windings as the rotor turns, the current in each stator winding is actually alternating, at a frequency proportional to the number of motor magnetic poles and the speed.

The basic principle of operation of the brushless DC motor is essentially the same, with the important difference that the supply current switching takes place electronically, instead of mechanically. Figure 20.3 depicts a transistor switching circuit capable of switching a DC supply so as to provide the appropriate currents to a three-phase rotor winding. The electronic switching device consists of a rotor position sensor, fixed on the motor shaft, and an electronic switching module that can supply each stator winding. Diagrams of the phase-to-phase back emf's and the switching sequence of the inverter are shown in Figure 20.4. The back emf waveforms shown in Figure 20.4 are called *trapezoidal*; the total back emf of the inverter is obtained by piecewise addition of the motor phase voltages and is a constant voltage, proportional

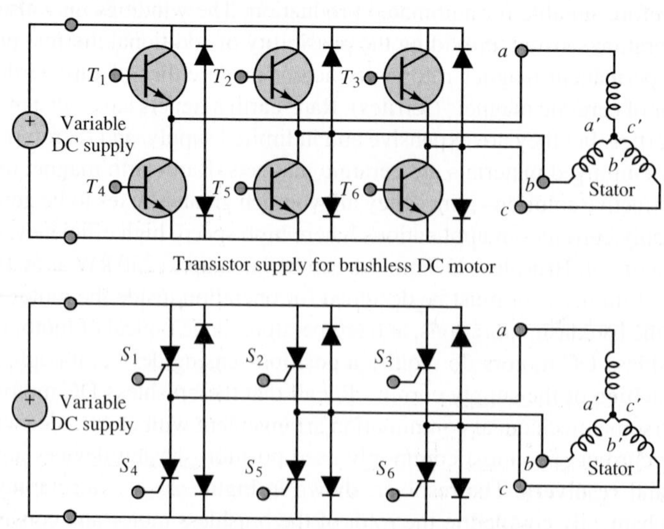

Figure 20.3 Transistor and SCR drives for a brushless DC motor

to motor speed. You should visually verify that the addition of the three phase voltages of Figure 20.4 leads to a constant voltage. The brushless DC motor is therefore similar to a standard permanent-magnet DC motor, and it can be described by the following simplified equations:

$$V = k_a \omega_m + R_w I \qquad (20.1)$$

$$T = k_T I \qquad (20.2)$$

BLDC motor equations

LO1

where

$$k_a = k_T$$

and where

V = motor voltage
k_a = armature constant
ω_m = mechanical speed
R_w = winding resistance
T = motor torque
k_T = torque constant
I = motor (armature) current

The speed and torque of a brushless DC motor can therefore be controlled with any variable-speed DC supply, such as one of the supplies briefly discussed in Chapter 12. Further, since the brushless motor has intrinsically higher torque and lower inertia than its DC counterpart, its response speed is superior to that obtained from traditional DC motors. Figure 20.5 depicts the (a) torque-speed and (b) efficiency curves of a commercially produced brushless DC motor.

One important difference between the conventional DC motor and the brushless motor, however, is due to the coarseness of the electronic switching compared with the mechanical switching of the brush-type DC motor (recall the discussion of torque

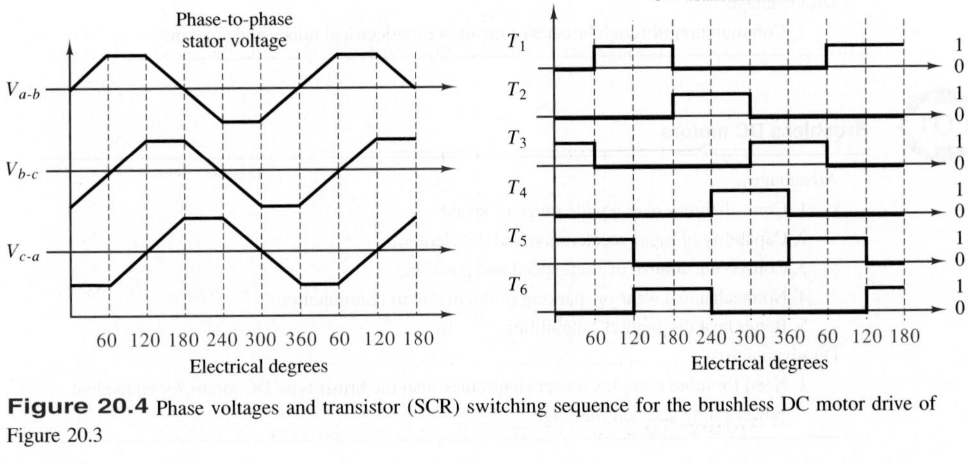

LO1

Figure 20.4 Phase voltages and transistor (SCR) switching sequence for the brushless DC motor drive of Figure 20.3

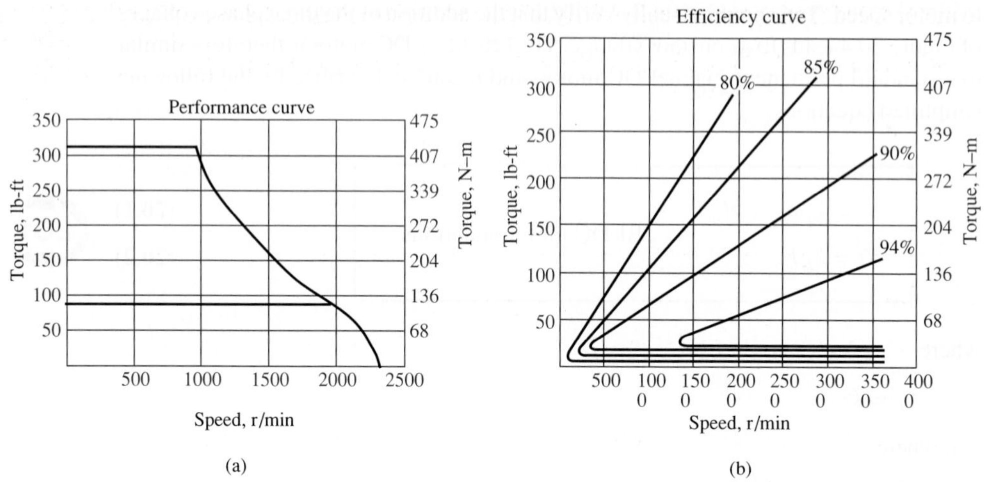

Figure 20.5 Performance and efficiency characteristics of brushless DC motor *(Courtesy Pacific Scientific)*

ripple due to the commutation effect in DC motors in Chapter 19). In practice, one cannot obtain the exact trapezoidal emf of Figure 20.4 by means of the transistor switching circuit of Figure 20.3, and a voltage ripple results as a consequence, leading to a torque ripple in the motor. Additional phase windings on the stator could solve the problem, at the expense of further complexity in the drive electronics, since the switching sequence would be more complex. Thus, brushless motors suffer from an inherent tradeoff between torque ripple and drive complexity.

Among other applications, brushless DC motors find use in the design of servo loops in control systems, for example, in computer disk drives, and in propulsion systems for electric vehicles. The comparisons between the conventional DC motor and the brushless DC motor are summarized in the following table:

LO1

Conventional DC motors

Advantages

1. Controllability over a wide range of speeds.
2. Capability of rapid acceleration and deceleration.
3. Convenient control of shaft speed and position by servo amplifiers.

Disadvantage

1. Commutation (through brushes) causing wear, electrical noise, and sparking.

LO1

Brushless DC motors

Advantages

1. Controllability over a wide range of speeds.
2. Capability of rapid acceleration and deceleration.
3. Convenient control of shaft speed and position.
4. No mechanical wear or sparking problem due to commutation.
5. Better heat dissipation capabilities.

Disadvantage

1. Need for more complex power electronics than the brush-type DC motor for equivalent power rating and control range.

EXAMPLE 20.1 Sinusoidal Torque Generation in Brushless DC Motors

LO1

Problem

Show that the use of sinusoidal currents in a brushless DC motor can result in a ripple-free torque.

Solution

Known Quantities: Coil (phase) currents.

Find: Total output torque T.

Schematics, Diagrams, Circuits, and Given Data: $I_{m1} = I_m \sin\theta$; $I_{m2} = I_m \cos\theta$.

Assumptions: The field coil is wound in a two-phase circuit; each winding is sinusoidally spaced. Sinusoidal currents can be generated by suitable power electronics circuits.

Analysis: Using equation 20.2, we determine that the torques generated by the currents in each of the two coils of the two-phase stator are

$$T_1 = k_T I_{m1} \sin\theta$$

$$T_2 = k_T I_{m2} \cos\theta$$

The sinusoidal form of the torques is due to the sinusoidal distribution of the stator windings in each phase, which are spaced 90° out of phase with one another so as to produce sine-cosine components, and is shown in Figure 20.6.

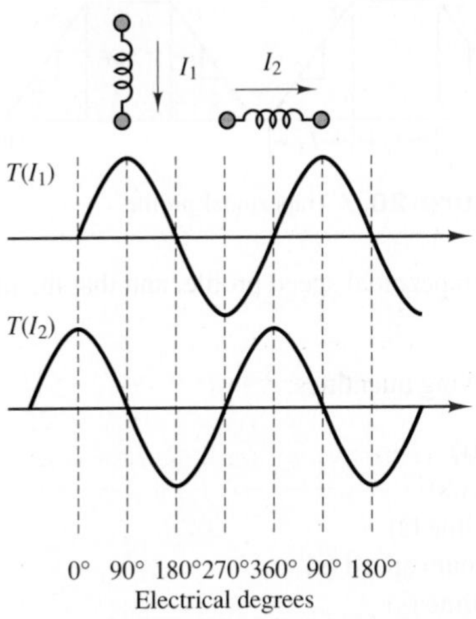

Figure 20.6 Sinusoidal torque generation circuit and current waveforms for a brushless DC motor

The net torque produced by the motor is the sum of the two phase torques:

$$T = T_1 + T_2 = k_T I_{m1} \sin\theta + k_T I_{m2} \cos\theta = k_T[(I_m \sin\theta)\sin\theta + (I_m \cos\theta)\cos\theta]$$

$$= k_T I_m(\sin^2\theta + \cos^2\theta) = k_T I_m$$

Thus, the torque generated by the motor is constant, or ripple-free.

Comments: Note that this scheme requires two features: sinusoidally spaced two-phase windings and sinusoidal phase currents. It is also very important that both the windings and the currents be exactly 90° out of phase.

LO1

EXAMPLE 20.2 Selecting a Trapezoidal Speed Profile to Match a Desired Motion Profile

Problem

Determine the trapezoidal speed profile required to move a load 0.5 m in 5 s. Analyze the motion of the motor.

Solution

Known Quantities: Desired load motion profile.

Find: Required trapezoidal speed profile.

Schematics, Diagrams, Circuits, and Given Data: The motor covers 0.5 m in 100 r. Trapezoidal profile characteristics as shown in Figure 20.7.

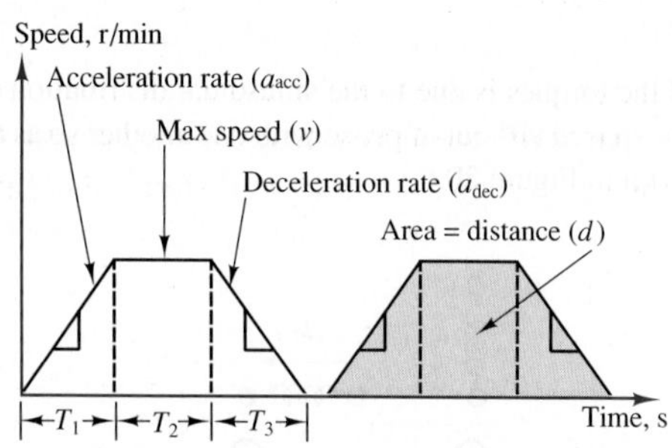

Figure 20.7 Trapezoidal profile

Assumptions: Assume a trapezoidal speed profile, and that the motor will accelerate for 1 s and decelerate for 1 s.

Analysis: Define the following quantities:

d = motor travel (r)
v = motor speed (r/s)
T_1 = acceleration time (s)
T_2 = time at maximum speed (s)
T_3 = deceleration time (s)
a = acceleration or deceleration rate (r/s^2)

From the above definitions, we can calculate the maximum rotational velocity of the motor as follows. For constant acceleration, the expressions for the motor displacement and velocity are

$$d = \tfrac{1}{2}at^2 \qquad \text{and} \qquad v = d' = at$$

From the above expressions, we can relate the maximum velocity to the acceleration and deceleration rates:

$$a_{\text{acc}} = \frac{v}{T_1} \qquad a_{\text{dec}} = \frac{v}{T_3}$$

Now we can write an expression for the total motor travel (100 r):

$$d = \frac{1}{2}a_{\text{acc}}T_1^2 + vT_2 + \frac{1}{2}a_{\text{dec}}T_3^2 = \frac{1}{2}\frac{v}{T_1}T_1^2 + vT_2 + \frac{1}{2}\frac{v}{T_3}T_3^2$$
$$= \frac{1}{2}vT_1 + vT_2 + \frac{1}{2}vT_3 = v\left(\frac{1}{2}T_1 + T_2 + \frac{1}{2}T_3\right)$$

Note that the above expression is quite general and could be used also for asymmetric profiles. Using the given numbers, we calculate the maximum velocity to be

$$v = \frac{d}{\frac{1}{2}T_1 + T_2 + \frac{1}{2}T_3} = \frac{100 \text{ r}}{(0.5 + 3 + 0.5) \text{ s}} = 25 \text{ r/s}$$

which corresponds to $25 \times 60 = 1{,}500$ r/min.

Comments: The results derived in this example are very useful—trapezoidal speed profiles are very common in servomotors.

20.2 STEPPING MOTORS

Stepping, or **stepper, motors** are motors that convert digital information to mechanical motion. The principles of operation of stepping motors have been known since the 1920s; however, their application has seen a dramatic rise with the increased use of digital computers. Stepping motors, as the name suggests, rotate in distinct steps, and their position can be controlled by means of logic signals. Typical applications of stepping motors are in line printers, positioning of heads in magnetic disk drives, and any other situation where continuous or stepwise displacements are required.

Stepping motors can generally be classified in one of three categories: variable-reluctance, permanent-magnet, and hybrid types. It will soon be shown that the principles of operation of each of these devices bear a definite resemblance to those of devices already encountered in this book. Stepping motors have a number of special features that make them particularly useful in practical applications. Perhaps the most important feature of a stepping motor is that the angle of rotation of the motor is directly proportional to the number of input pulses; further, the angle error per step is very small and does not accumulate. Stepping motors are also capable of rapid responses—starting, stopping, and reversing commands—and can be driven directly by digital signals. Another important feature is a self-holding capability that makes it possible for the rotor to be held in the stopped position without the use of brakes. Finally, a wide range of rotating speeds—proportional to the frequency of the pulse signal—may be attained in these motors.

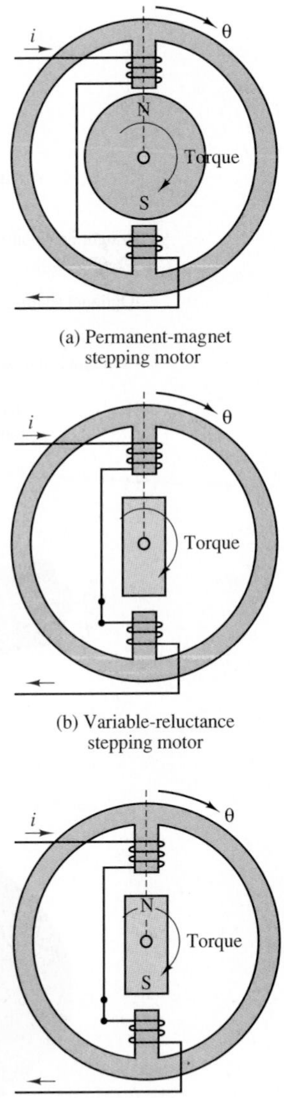

(a) Permanent-magnet stepping motor

(b) Variable-reluctance stepping motor

(c) Hybrid stepping motor

Figure 20.8 Stepping motor configurations

Figure 20.8 depicts the general appearance of three types of stepping motors. The **permanent-magnet rotor stepping motor,** seen in Figure 20.8(a), permits a nonzero holding torque when the motor is not energized. Depending on the construction of the motor, it is typically possible to obtain step angles of 7.5, 11.25, 15, 18, 45, or 90°. The angle of rotation is determined by the number of stator poles, as is illustrated in Example 20.3. The **variable-reluctance stepping motor,** seen in Figure 20.8(b), has an iron multipole rotor and a laminated wound stator, and it rotates when the teeth on the rotor are attracted to the electromagnetically energized stator teeth. The rotor inertia of a variable-reluctance stepping motor is low, and the response is very quick, but the allowable load inertia is small. When the windings are not energized, the static torque of this type of motor is zero. Generally, the step angle of the variable-reluctance stepping motor is 15° (see examples 20.4 and 20.5).

The **hybrid stepping motor,** seen in Figure 20.8(c), is characterized by multitoothed stator and rotor, the rotor having an axially magnetized concentric magnet around its shaft. It can be seen that this configuration is a mixture of the variable-reluctance (VR) and permanent-magnet types. This type of motor generally has high accuracy and high torque and can be configured to provide a step angle as small as 1.8°. Figure 20.9(a) through (e) depicts the construction of a VR step motor.

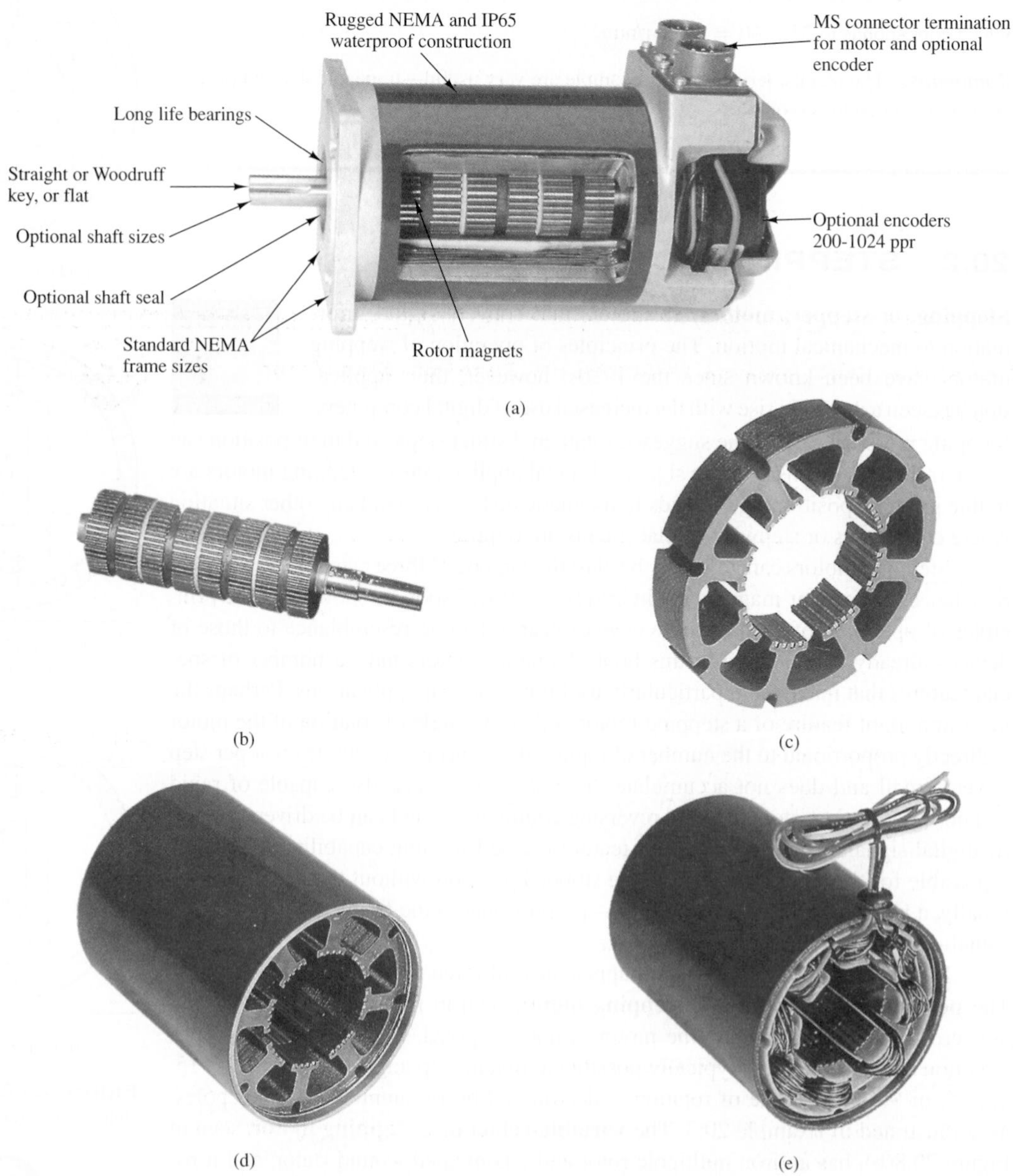

Figure 20.9 VR stepper motor (a) Complete motor assembly; (b) PM rotor; (c) stator cross section; (d) fully assembled stator; (e) stator with windings. (*Courtesy: Pacific Scientific Motor Products Division.*)

For any of these configurations, the principle of operation is essentially the same: When the coils are energized, magnetic poles are generated in the stator, and the rotor will align in accordance with the direction of the magnetic field developed in the stator. By reversing the phase of the currents in the coils, or by energizing only some of the coils (this is possible in motors with more than two stator poles), the alignment of the stator magnetic field can take one of a discrete number of positions; if the currents in the coils are pulsed in the appropriate sequence, the rotor will advance in a step-by-step fashion. Thus, this type of motor can be very useful whenever precise incremental motion must be attained. As mentioned earlier, typical applications are in printer wheels, computer disk drives, and plotters. Other applications are found in the control of the position of valves (e.g., control of the throttle valve in an engine, or of a hydraulic valve in a fluid power system) and in drug-dispensing apparatus for clinical applications.

Examples 20.3, 20.4, 20.5, illustrate the operation of a four-pole, two-phase permanent-magnet stepping motor and of a similar motor of the variable-reluctance type. The operation of these motors is representative of all stepping motors.

EXAMPLE 20.3 Analysis of Two-Phase, Four-Pole Step Motor

LO2

Problem

Determine the full-step single-phase, full-step two-phase, and half-step current excitation sequences for the PM step motor of Figure 20.10.

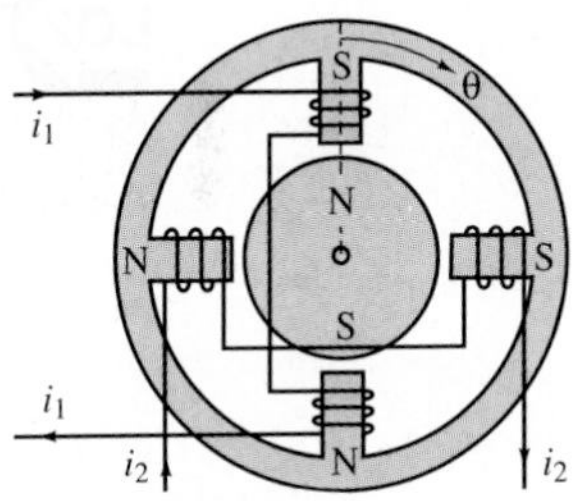

Figure 20.10 Two-phase four-pole PM stepper motor

Solution

Known Quantities: Phase currents.

Find: Full-step sequence for the motor.

Assumptions: The motor currents at the start of the sequence are $i_1 > 0$ and $i_2 = 0$.

Analysis: With the initial currents assumed (phase 1 energized), the motor will be at rest if the rotor is in the position shown in Figure 20.10. A single-phase sequence consists of turning on each of the two coils in sequence, reversing the polarity of the currents every other time. Then the PM rotor will align with the stator poles according to the polarity of the magnetic field generated by each coil's pole pair. For example, if coil 1 is turned off and coil 2 is turned on with a positive current polarity, the rotor will rotate clockwise by 90°. Table 20.1 depicts the (bipolar) sequence of coil currents and the corresponding motor position.

If both coils are activated, it is possible to cause the rotor to align between stator poles, also in increments of 90°, but shifted in phase by 45° with respect to the single-phase stepping sequence. Table 20.2 illustrates this stepping sequence.

Finally, if one combines the two sequences (easily accomplished, since the current commands for the two sequences are distinct), it is possible to obtain increments of 45°. Table 20.3 depicts the half-step sequence. Any finer resolution would require an increase in the number of windings and teeth in the stator.

Comments: The simplicity of the electronic controls required by this type of machine is one of the very attractive features of step motors.

Table 20.1 Full-step, single-phase sequence

i_1	i_2	θ
+	0	0°
0	+	90°
−	0	180°
0	−	270°
+	0	0°

Table 20.2 Full-step, two-phase sequence

i_1	i_2	θ
+	+	45°
−	+	135°
−	−	225°
+	−	315°
+	+	45°

Table 20.3 Half-step sequence

i_1	i_2	θ
+	0	0°
+	+	45°
0	+	90°
−	+	135°
−	0	180°
−	−	225°
0	−	270°
+	−	315°
+	0	0°

CHECK YOUR UNDERSTANDING

Determine the smallest increment in angular position that can be achieved with a PM stepper motor with six stator teeth and three-phase current excitation.

Answer: $\Delta\theta = 20°$

LO2

EXAMPLE 20.4 Analysis of Variable-Reluctance Step Motor

Problem

Determine the current excitation sequences required to achieve 45° steps in the VR step motor of Figure 20.11.

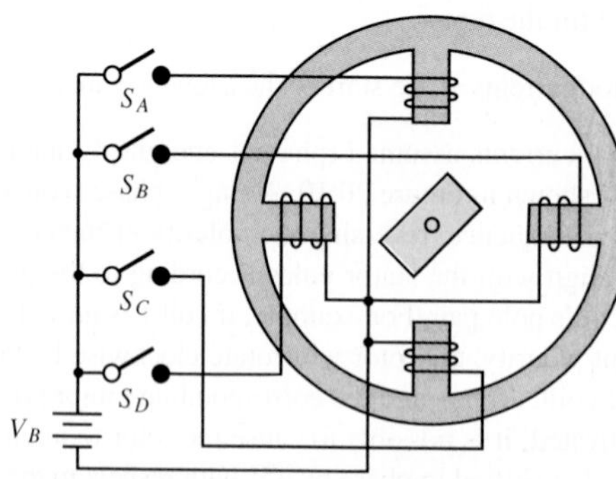

Figure 20.11 Two-phase, four-pole VR stepping motor

Solution

Known Quantities: Phase currents.

Find: Current excitation sequence for 45° steps.

Assumptions: The motor currents at the start of the sequence are $i_1 > 0$ and $i_2 = 0$.

Analysis: The operation of the variable-reluctance step motor (with a salient pole rotor) is simpler than that of the PM type, because the rotor is not magnetically polarized, and therefore it is not necessary to have bipolar currents to achieve the desired rotor motion. The stator of Figure 20.10 is excited by direct currents supplied by a single (unipolar) voltage supply. The switches shown in the figure could be controlled by a logic circuit similar to the ones described in Chapters 13 and 14. Note that four separate coils are used.

Figure 20.12 depicts how the first three steps of the sequence could be achieved. These are summarized in Table 20.4.

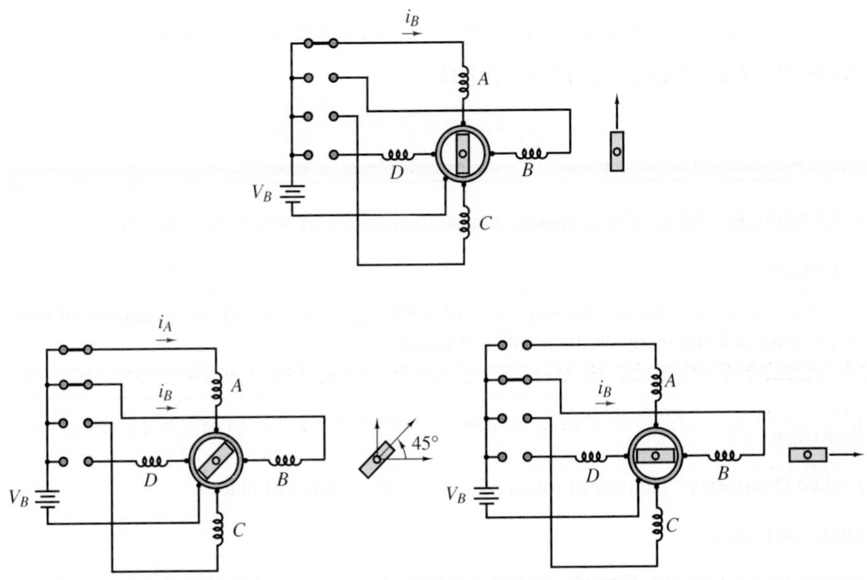

Figure 20.12 Two-phase, four-pole VR motor positioning sequence

Table 20.4 Current excitation sequence for VR step motor

S_A	S_B	S_C	S_D	Rotor position
1	0	0	0	0°
1	1	0	0	45°
0	1	0	0	90°
0	1	1	0	135°
0	0	1	0	180°
0	0	1	1	225°
0	0	0	1	270°
1	0	0	1	315°
1	0	0	0	360°

Comments: Note that the circuit required to drive this circuit is even simpler than the one required by the PM step motor.

CHECK YOUR UNDERSTANDING

Express the stepping sequence of the variable-reluctance stepping motor of Example 20.4 as a four-digit binary sequence.

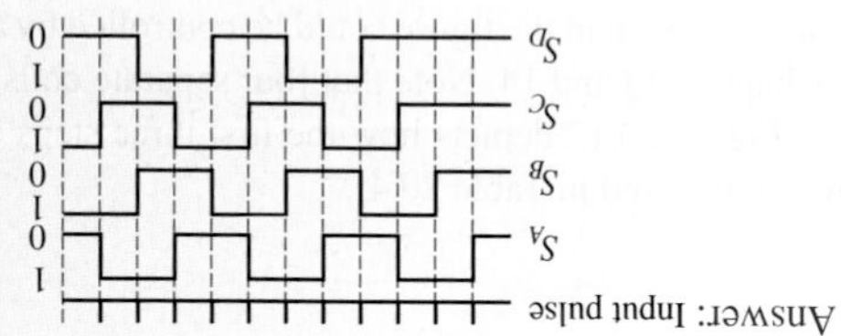

EXAMPLE 20.5 Step Angle Determination of VR Step Motor

LO2

Problem

Determine an expression for the step angle of a VR step motor based on the number of teeth on the rotor and stator and on the number of phases.

Solution

Known Quantities: Number of rotor and stator teeth; number of phases.

Find: Step angle.

Schematics, Diagrams, Circuits, and Given Data: t = number of teeth = 4; m = number of phases = 3.

Analysis: The number of steps in a revolution N is given by the product of the number of teeth and the number of phases (e.g., in Example 20.4 it is equal to 2 teeth × 4 phases = 8 steps). Thus, $N = tm$.

The step angle increment, or *resolution*, is equal to $\Delta\theta = 360°/N$. For the motor described in this example,

$$\Delta\theta = \frac{360°}{N} = \frac{360°}{tm} = \frac{360°}{12 \times 3} = 10°$$

EXAMPLE 20.6 Torque Equation of Step Motor

LO2

Problem

Calculate the torque generated by a step motor.

Solution

Known Quantities: t = number of teeth per phase; L = axial length of rotor; g = rotor-to-stator radial air gap; r = rotor radius; $\mathcal{F}$ = mmf developed across the two air gaps (in series) through which a line of flux must pass in one phase. Expression for the motor torque.

Find: Torque developed by the motor.

Schematics, Diagrams, Circuits, and Given Data: $t = 16$ (48 steps, three-phase excitation); $L = 6.35 \times 10^{-3}$ m; $g = 6.35 \times 10^{-5}$ m; $r = 1.29 \times 10^{-2}$ m; $\mathcal{F} = 720$ A-turns.

$$T = 0.314 \times 10^{-6} \frac{tL(r + g/2)\mathcal{F}^2}{g} \quad \text{N-m}$$

Analysis: Using the expression given above gives

$$\begin{aligned} T &= 0.314 \times 10^{-6} \frac{tL(r + g/2)\mathcal{F}^2}{g} \\ &= 0.314 \times 10^{-6} \frac{16 \times 6.35 \times 10^{-3}(1.29 \times 10^{-2} + 3.175 \times 10^{-5})(720^2)}{6.35 \times 10^{-5}} \\ &= 3.37 \text{ N-m} \end{aligned}$$

CHECK YOUR UNDERSTANDING

Express the torque in Example 20.6 in units of pound-inches.

Answer: 29.875 lb-in

From the preceding examples, you should now have a feeling for the operation of variable-reluctance and PM stepping motors. The hybrid configuration is characterized by multitooth rotors that are made of magnetic materials, thus providing a variable-reluctance geometry in conjunction with a permanent-magnet rotor.

An ideal torque–speed characteristic for a stepper motor is shown in Figure 20.13. Two distinct modes of operation are marked on the curve: the **locked-step mode** and the **slewing mode.** In the first mode, the rotor comes to rest (or at least decelerates) between steps; this is the mode commonly used to achieve a given rotor position. In the locked-step mode, the rotor can be started, stopped, and reversed. The slewing mode, on the other hand, does not allow stopping or reversal of the rotor, although the rotor still advances in synchronism with the stepping sequence, as described in the preceding examples. This second mode could be used, for example, in rewinding or fast-forwarding a tape drive.

The power supply, or driver, required by a stepping motor is shown in block diagram form in Figure 20.14; it includes a DC power supply, to provide the required current to drive the motor, in addition to logic and switching circuits to provide the appropriate inputs at the right time. One of the important considerations in driving

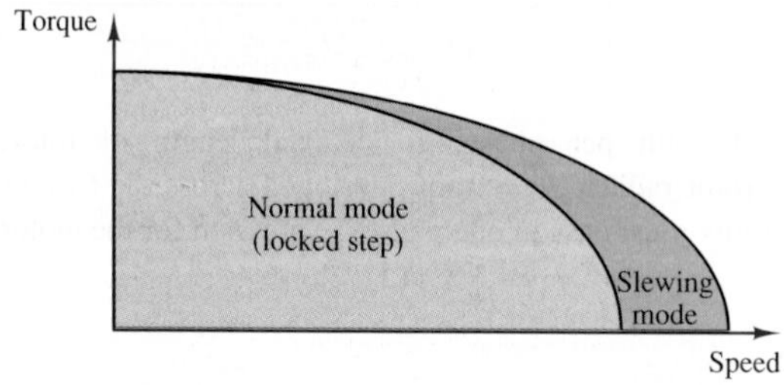

Figure 20.13 Ideal torque–speed characteristic of a stepping motor

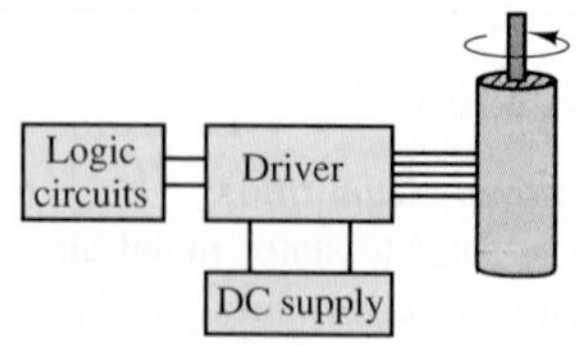

Figure 20.14 Power supply for stepping motor

a stepping motor is the excitation mode, which can be one-phase or two-phase. The driver is the circuit that arranges, distributes, and amplifies pulse trains from the logic circuit determining the stepping sequence; the driver excites each winding of the stepping motor at specified times. In the **one-phase excitation mode,** current is supplied to one phase at a time, with the advantages of low power consumption and good step-angle accuracy. Input signal pulses and the change in the condition of each phase excitation are shown in Figure 20.15. In the **two-phase excitation mode,** current is simultaneously provided to two phases. Input signal pulses and the change in the condition of each phase excitation are also shown in Figure 20.16.

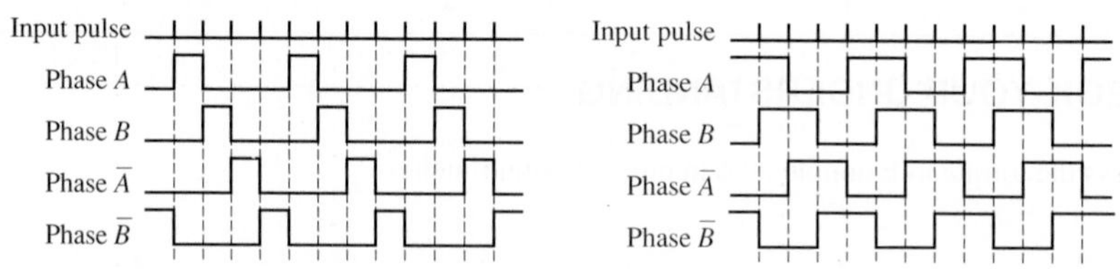

Input signal pulses and the change in phase excitation of the one-phase excitation mode

Input signal pulses and the change in phase excitation of the two-phase excitation mode

Figure 20.15 One- and two-phase excitation waveforms for stepper motors

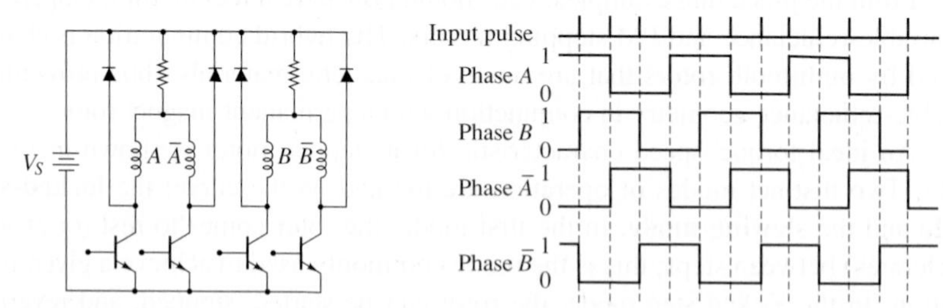

Simplified unipolar drive circuit diagram

Input signal pulses and the change in phase excitation of unipolar drive

Figure 20.16 Unipolar drive for stepper motor

CHECK YOUR UNDERSTANDING

Derive the excitation waveforms corresponding to the direction of rotation opposite to that caused by the stepping sequence shown in Figure 20.14.

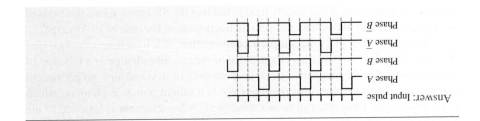

In addition to the classification of the excitation by phase, stepping motor drives are classified according to whether the drive supplies are unipolar or bipolar, that is, whether they can supply current in one or both directions. Unipolar excitation is clearly simpler, although in the case of the two-phase excitation mode, only one-half of the motor windings are used, with an obvious decrease in performance. Figure 20.17 shows a circuit diagram of the unipolar drive and the sequence of phase excitation.

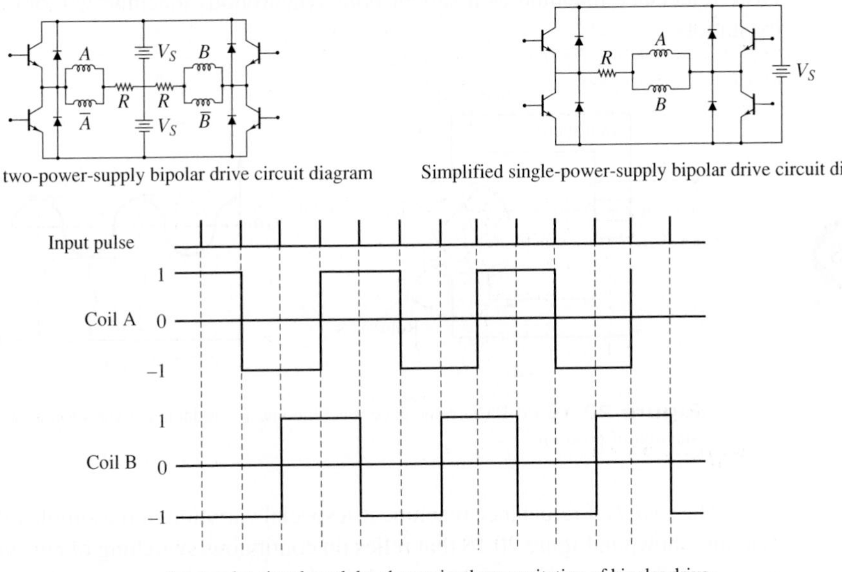

Simplified two-power-supply bipolar drive circuit diagram

Simplified single-power-supply bipolar drive circuit diagram

Input pulse signals and the change in phase excitation of bipolar drive

Figure 20.17 Bipolar drive for stepper motors

When a bipolar drive is used, motor windings are used effectively, because of the bidirectional exciting current; when operated in this mode, a stepping motor can generate a large output torque at low speed compared with the unipolar drive. Figure 20.17 shows two versions of the bipolar drive. The first requires two power supplies, one for each polarity, while the second requires only one power supply but needs four switching transistors per phase to reverse the polarity.

20.3 SWITCHED RELUCTANCE MOTORS

FIND IT ON THE WEB

The **switched reluctance (SR) machine** is the simplest electric machine that permits variable-speed operation. Today, this machine finds increasingly common application in variable-speed drives for industrial applications and in traction drives for automotive propulsion. It is a widely held belief that the SR motor forms the basis of an ideal electric and hybrid-electric vehicle traction drive because of its low cost.

Figure 20.18 depicts the simplest configuration of a reluctance machine and illustrates how the reluctance and inductance of the machine change as a function of position. Note that the magnetic circuit consists only of iron and air—no permanent magnets are required! Note also that the rotor is a salient pole iron element, which is the lowest-cost rotor that can be manufactured. When a current is supplied to the coil, the rotor will experience a torque seeking to align it with the magnetic poles of the stator; when $\theta = 0$, the torque is zero and the rotor will no longer move, having reached its minimum reluctance position. Note that minimum reluctance corresponds to minimum stored energy in the system. Thus, the torque in the motor is developed because of the change in reluctance with rotor position. This principle makes the reluctance machine different from all other (AC or DC) machines discussed so far. Note also that this machine is one of a few machines, along with the induction motor and VR step motor, to be *singly excited,* that is, to have a single source of magnetic field (whether generated by a coil or by a permanent magnet). One can think of the basic reluctance machine as a salient pole synchronous machine without any field excitation.

LO3

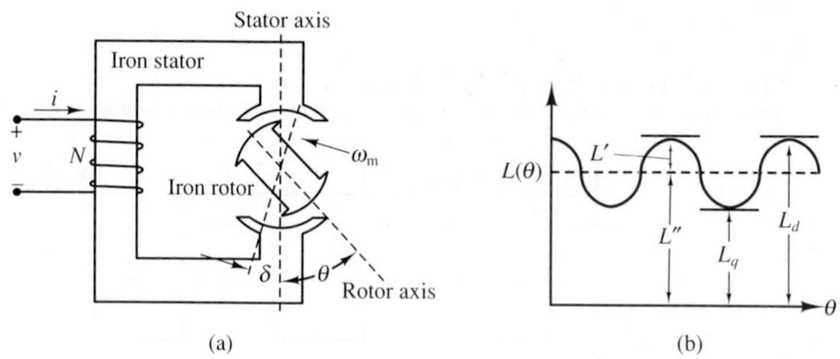

Figure 20.18 (a) Basic reluctance machine and (b) inductance variation as a function of position

The *switched* reluctance machine is a special variation of the simple reluctance machine shown in Figure 20.18 that relies on continuous switching of currents in the stator to guarantee motion of the rotor. It is also a true reluctance machine in that it has *salient poles* both in the rotor and in the stator. The configuration of a typical SR machine is shown in Figure 20.19. Note that the configuration of the SR machine is very similar to that of a VR step motor, discussed in Section 20.2. The primary difference between the two is that the SR machine is designed for continuous and not stepped (discrete) motion. The advent of low-cost power semiconductors, especially GTOs, IGBTs and power MOSFETs (see Chapter 12) has made it possible to reliably control SR machines. With reference to Figure 20.19, you can see that the stator of an SR machine is wound through slots, with simple solenoid-type windings, and is similar to that of an induction or synchronous AC machine. This stator can be excited by any multiphase source, such as the three-phase sources described in Chapter 19. The SR machine is excited by discrete current pulses that must be timed with respect to the position of the rotor poles with respect to the stator poles, thus requiring **position feedback.** The speed of the rotor is determined by the switching frequency of the stator coil currents.

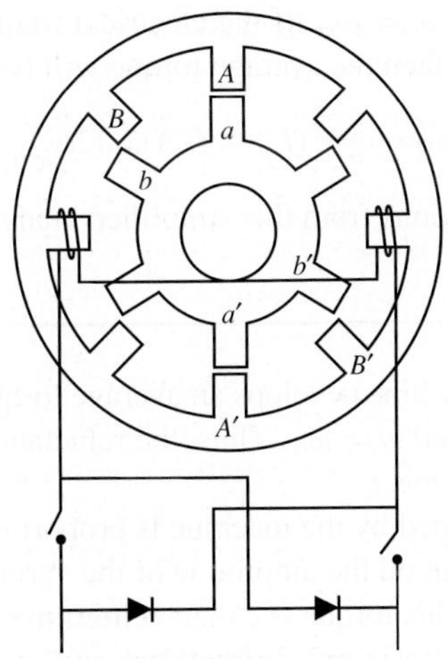

Figure 20.19 Configuration of switched reluctance machine

Operating Principles of SR Machine

Torque production in an SR machine depends on the variation in stored magnetic energy as a function of position. Consider the simple reluctance machine of Figure 20.18, and assume that the variation in winding inductance with rotor position is sinusoidal. The inductance will vary at twice the excitation frequency, because of the number of poles:

$$L(\theta) = L'' + L' \cos 2\theta \tag{20.3}$$

We shall determine the torque generated by the machine, given the excitation current

$$i(t) = I_m \sin(\omega t) \tag{20.4}$$

The magnetic stored energy (see Chapter 18) is given by

$$W_m = \tfrac{1}{2} L(\theta) i^2(t) \tag{20.5}$$

and the flux linkage is

$$\lambda(\theta) = L(\theta) i(t) \tag{20.6}$$

From Chapter 16, we know that the torque can be written as follows:

$$T_m = -\frac{\partial W_m}{\partial \theta} + i\frac{\partial \lambda}{\partial \theta} = -\frac{1}{2} i^2 \frac{\partial L}{\partial \theta} + i^2 \frac{\partial L}{\partial \theta} = \frac{1}{2} i^2 \frac{\partial L}{\partial \theta} \tag{20.7}$$

Given the known sinusoidal current and inductance variations, we can write the torque expression as

$$T_m = -I_m^2 L' \sin(2\theta) \sin^2(\omega t) \tag{20.8}$$

It can be shown, with the use of trigonometric identities, that if the rotor rotates at angular velocity ω_m, such that $\theta = \omega_m t - \theta_0$ (with θ_0 equal to the rotor position at $t = 0$), the torque of the SR machine will be nonzero only if the frequency of the sinusoidal stator current is $\omega = \omega_m$. If the electrical frequency is synchronous with the mechanical frequency, then the average torque will be given by

$$\langle T_m \rangle = \tfrac{1}{4} I_m^2 L' \sin(2\theta_0) = \tfrac{1}{8} I_m^2 (L_d - L_q) \sin(2\theta_0) \qquad \textbf{(20.9)}$$

We can draw some conclusions from this simplified analysis of the SR machine:

1. The reluctance machine develops an average torque only at one particular (synchronous) speed $\omega = \omega_m$. Thus, the reluctance machine is a synchronous machine.
2. The torque developed by the machine is proportional to $L_d - L_q$ and is therefore dependent on the amplitude of the variation in inductance (or reluctance); thus, this torque is called **reluctance torque.** The values L_d and L_q are called **direct axis inductance** and **quadrature axis inductance,** respectively.
3. The torque varies with angle θ_0, which is therefore equivalent to the *torque angle* δ defined in Chapter 19 for synchronous machines. The maximum torque occurs at $\theta_0 = \pi/4$ and is called the **pull-out torque.**

The above equations have been derived for a continuous reluctance machine; a switched reluctance machine has discontinuous currents, and will therefore have non-sinusoidal reluctance (inductance) variations and a discontinuous torque. Figure 20.20 depicts the typical appearance of the L and T_m curves for an SR machine. It can be shown that the magnetic torque generated by an SR machine may be expressed in the form

$$\langle T_m \rangle = \frac{1}{4\pi}(KmP)\tilde{I}^2(L_{\max} - L_{\min}) \qquad \text{SR machine torque} \qquad \textbf{(20.10)}$$

where $\tilde{I}$ is the root-mean-square (rms) current, P is the number of pulses per revolution, m is the number of phases, $L_{\max}$ and $L_{\min}$ are the maximum and minimum inductances seen by the exciting coils, and K is a physical constant. Note that the rms value in equation 20.10 includes also the higher harmonics.

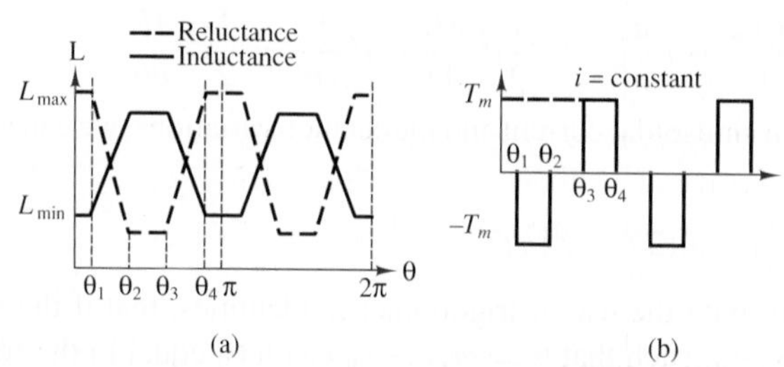

Figure 20.20 Inductance and torque variation in switched reluctance machine

20.4 SINGLE-PHASE AC MOTORS

In Chapter 19, two types of AC machines were discussed: synchronous and induction. In the discussion of these devices, especially in their motor applications, three-phase excitation was assumed; however, in many practical applications—and especially in small household appliances and small industrial motors—three-phase sources are not readily available, and it would be desirable to use single-phase excitation. Unfortunately, single-phase power does not lend itself to the generation of a rotating magnetic field: single-phase currents in the winding of an AC machine lead to a magnetic field that pulsates in amplitude but does not rotate in space. Thus, it would not be possible to use the AC machines described in Chapter 19 if only single-phase power were available. This section discusses the construction and the operating and performance characteristics of single-phase AC motors. The discussion focuses mainly on the **universal motor** and **single-phase induction motors.**

Fractional-horsepower (as opposed to **integral-horsepower) motors** represent by far the major share of the industrial production of electric motors. Many fractional-horsepower motors are designed for single-phase use, since single-phase AC power is readily available practically anywhere. Many applications are related to household appliances: refrigerator compressors, air conditioners, fans, electric tools, washer and dryer motors, and others. For the rest of this chapter, we shall examine qualitatively the principle of operation of single-phase motors and look at a few applications. The variety of designs for practical single-phase motors is such that it would not be possible to present the detailed principles of operation for all common types. However, it is hoped that the introduction provided in this chapter will help you in decoding the manufacturer's specifications for a given motor, and in making a preliminary selection for a given application.

Fractional-Horsepower Motors

A *small motor,* as defined by the American Standards Association (ASA) and the National Electrical Manufacturers Association (NEMA; see Chapter 19), is a "motor built in a frame smaller than that having a continuous rating of 1 hp, open type, at 1700 to 1800 r/min." Small motors are generally considered *fractional-horsepower motors*. However, since the determination is based on frame size and on a given speed range, the classification of a motor is not always obvious. Let us give two examples.

1. Consider a $\frac{3}{4}$-hp, 1,200 r/min motor. This motor is not considered a fractional-horsepower motor, because of its frame size. If the same frame size were used for an 1,800 r/min motor, it would produce a rating of more than 1 hp. Thus, it is considered an integral-horsepower motor of

$$0.75 \text{ hp} \times \frac{1{,}800}{1{,}200} = 1.125 \text{ hp}$$

In other words, since the motor is capable of integral-horsepower performance at speeds of 1,700 to 1,800 r/min, it is classified as an integral-horsepower motor.

2. Consider now a 1.25-hp, 3,600 r/min motor. This motor is classified as a fractional-horsepower motor, in spite of the fact that its power output is actually greater than 1 hp. If the same motor were used at a speed of 1,800 r/min, it would produce a rating of less than 1 hp:

$$1.25 \times \frac{1{,}800}{3{,}600} = 0.625 \text{ hp}$$

Thus, we see once again that some attention must be paid to the speed of operation of the motor in determining its classification. The term *fractional horsepower* relates more to the physical size of the machine than to the actual power output rating.

The Universal Motor

If it were possible to operate a DC motor from a single-phase AC supply, a wide range of simple applications would become readily available. Recall that the direction of the torque produced by a DC machine is determined by the direction of current flow in the armature conductors and by the polarity of the field; torque is developed in a DC machine because the commutator arrangement permits the field and armature currents to remain in phase, thus producing torque in a constant direction. A similar result can be obtained by using an AC supply, and by connecting the armature and field windings in series, as shown in Figure 20.21. A series DC motor connected in this configuration can therefore operate on a single-phase AC supply, and it is referred to as a **universal motor.** An additional consideration is that, because of the AC excitation, it is necessary to reduce AC core losses by laminating the stator; thus, the universal motor differs from the series DC motor discussed in Chapter 19 in its construction features. Typical torque–speed curves for AC and DC operation of a universal motor are shown in Figure 20.22. As shown in Figure 20.21, the load current is sinusoidal and therefore reverses direction each half-cycle; however, the torque generated by the motor is always in the same direction, resulting in a pulsating torque, with nonzero average value.

As in the case of a DC series motor, the best method for controlling the speed of a universal motor is to change its (rms) input voltage. The higher the rms input voltage, the greater the resulting speed of the motor. Approximate torque–speed characteristics of a universal motor as a function of voltage are shown in Figure 20.23.

LO4

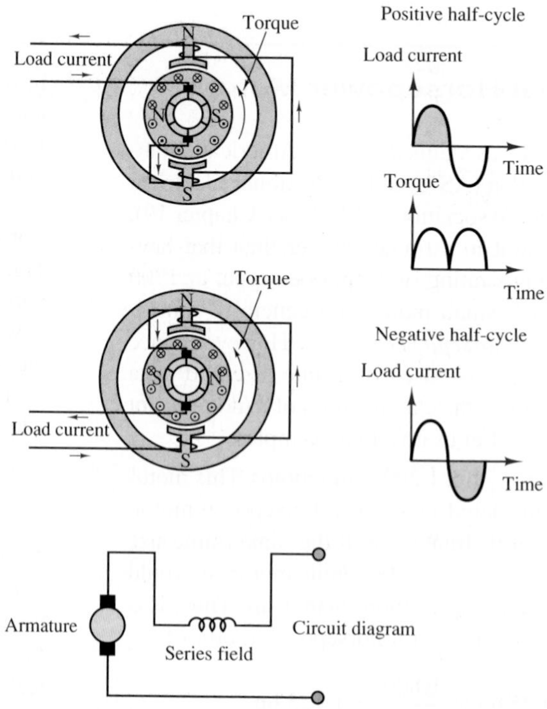

Figure 20.21 Operation and circuit diagram of a universal motor

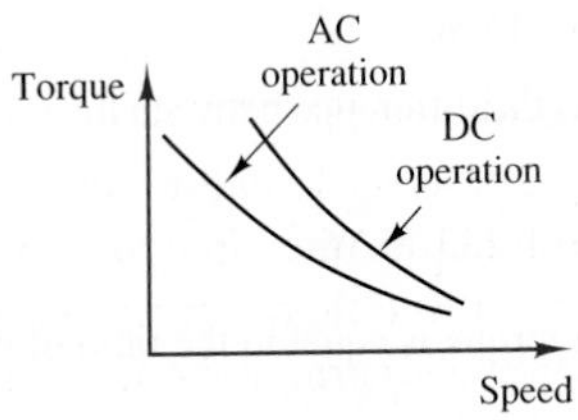

Figure 20.22 Torque–speed curve of a universal motor

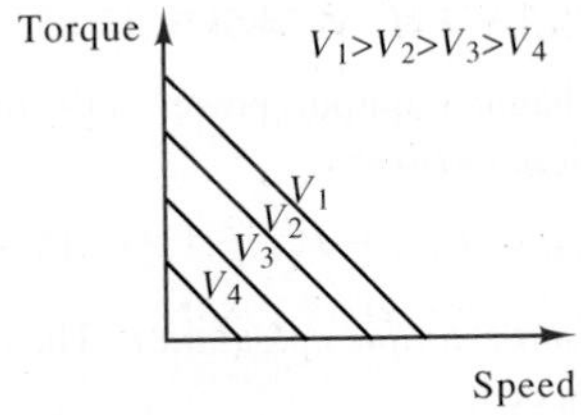

Figure 20.23 Torque–speed characteristics of a universal motor

EXAMPLE 20.7 Analysis of Universal Motor

Problem

Find the following quantities for a universal motor:

1. Back emf
2. Power output
3. Shaft torque
4. Motor efficiency

Solution

Known Quantities: Motor operating data and circuit parameters.

Find: $\mathbf{E}_b$; P_{out}; T_{out}; η (efficiency).

Schematics, Diagrams, Circuits, and Given Data: Motor operating data: 120 V; 60 Hz; two poles; 800 r/min; 17.85 A (full load); pf = 0.912 (lagging). Circuit parameters: $R_f = 0.65\ \Omega$; $X_f = 1.2\ \Omega$; $R_a = 1.36\ \Omega$; $X_{fa} = 1.6\ \Omega$.

Assumptions: Use the circuit model for the series motor described in Chapter 19. The rotational losses amount to 80 W.

Analysis: The circuit model for the series machine is shown in Figure 20.24. We shall use this model with the understanding that all currents and voltages are now phasors.

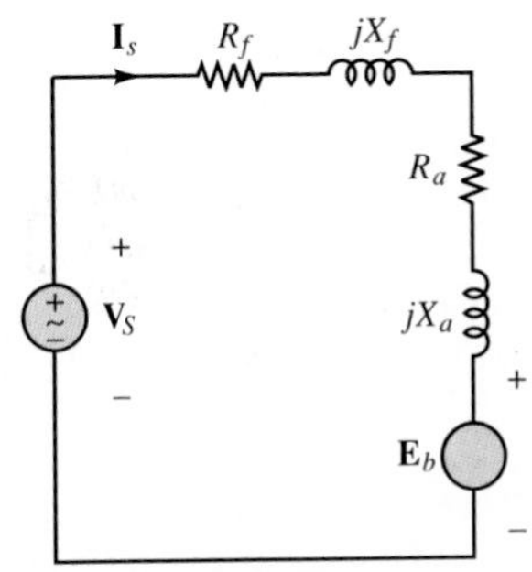

Figure 20.24 Equivalent circuit of a universal motor

1. *Back emf computation.* To determine the back emf, we need to calculate the voltage across the armature coil and subtract it from the supply voltage:

$$\begin{aligned}\mathbf{E}_b &= \mathbf{V}_S - \mathbf{I}_S(R_f + jX_f + R_a + jX_a)\\ &= \mathbf{V}_S - (I_S\angle\theta)(R_f + jX_f + R_a + jX_a)\end{aligned}$$

The impedance angle is the only unknown quantity, and it may be found from the power factor:

$$\text{pf} = \cos(\theta) = 0.912 \text{ (lagging)} \qquad \theta = \arccos(0.912) = -24.22^\circ$$

Thus,

$$\begin{aligned}\mathbf{E}_b &= \mathbf{V}_S - (I_S\angle\theta)(R_f + jX_f + R_a + jX_a)\\ &= 120 - (17.85\angle 22.24^\circ)(0.65 + j1.2 + 1.36 + j1.6)\\ &= 73.56\angle -24.8^\circ \text{ V}\end{aligned}$$

2. *Output power calculation.* The total power developed by the motor is equal to the product of the back emf and the series current:

$$P_{\text{total}} = E_b I_S = 73.56 \times 17.85 = 1{,}313.15 \text{ W}$$

The mechanical (output) power of the motor is the difference between the total power and the rotational losses:

$$P_{\text{out}} = P_{\text{total}} - P_{\text{rot}} = 1{,}313.15 - 80 = 1{,}233.15 \text{ W}$$

3. *Output (shaft) torque calculation.* The output torque is equal to the ratio of output power to shaft speed:

$$T_{\text{out}} = \frac{P_{\text{out}}}{\omega} = \frac{1{,}233.15 \text{ W}}{(2\pi \times 800/60) \text{ rad/s}} = 14.72 \text{ N-m}$$

4. *Efficiency calculation.* The efficiency of the motor is defined as the ratio of output power to input power:

$$\eta = \frac{P_{\text{out}}}{P_{\text{in}}} = \frac{P_{\text{out}}}{V_S I_S \cos\theta} = \frac{1{,}233.15}{1{,}953.5} = 63.12\%$$

Comments: Note that the analysis of this machine is very similar to that of the series DC motor, except for the use of phasors. It is very important to notice that in calculating the input power, one has to consider the power factor of the motor to obtain the *real power*.

LO4

EXAMPLE 20.8 Universal Motor Torque Expression

Problem

Compute an expression for the average torque generated by a universal motor, based on the circuit diagram of Figure 20.24.

Solution

Known Quantities: Circuit model of motor.

Find: Expression for average torque T_{av}.

Assumptions: The motor operates in the linear region of the magnetization curve.

Analysis: With reference to Chapter 19, we know that the flux produced in a series motor by the series current $i_S(t)$ is $\phi = k_S i_S(t)$. The instantaneous torque produced by the machine is given by

$$T(t) = k_T \phi(t) i_S(t)$$

If the source waveform has period $\tau = 2\pi/\omega$, we can calculate the average power by integrating the instantaneous torque over one period:

$$T_{\text{av}}(t) = \frac{\omega}{2\pi}\int_0^{2\pi/\omega} k_T k_S i_S^2(t')\,dt' = \frac{\omega}{2\pi}\int_0^{2\pi/\omega} k_T k_S I_S^2(\sin^2 \omega t')\,dt' = \tfrac{1}{2} k_T k_S I_S^2$$

where I_S is the rms value of the (series) armature current.

Comments: The series motor can produce a nonzero average torque when excited by an alternating current because of the quadratic nature of the instantaneous torque. A permanent-magnet DC machine, which has a linear torque–current relationship, would generate zero average torque if driven from an AC supply.

Single-Phase Induction Motors

A typical single-phase induction motor bears close resemblance to the polyphase squirrel cage induction motor discussed in Chapter 19, the major difference being in the configuration of the stator winding. A simplified schematic diagram of such a motor, with a single winding, is shown in Figure 20.25; the winding is typically distributed around the stator so as to produce an approximately sinusoidal mmf.

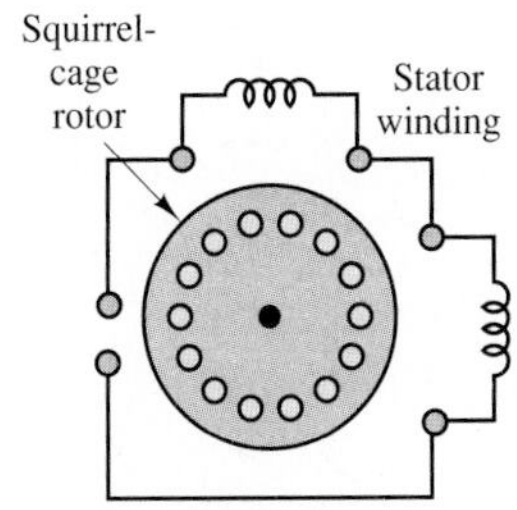

Figure 20.25 Single-phase induction motor

Assume that the mmf for a practical motor can be generated so as to approximate the following function:

$$\mathcal{F} = F_{\max}\cos(\omega t)\cos(\theta_m)$$

This function can be written as the sum of two components, as follows:

$$\mathcal{F}^{+} = \tfrac{1}{2}F_{\max}\cos(\theta_m - \omega t)$$
$$\mathcal{F}^{-} = \tfrac{1}{2}F_{\max}\cos(\theta_m + \omega t)$$

These two components may be interpreted as representing two mmf waves traveling in opposite directions around the stator. Each mmf produces torque according to the induction principles described in Chapter 19; however, the two components are equal and opposite, and no net torque results if the rotor is at rest. The resulting mmf is pulsating (i.e., changing in amplitude), but not rotating in space, as it would be in a polyphase stator. If the rotor is made to turn in either direction, however, the two mmf's will not be equal any longer, because the motion of the rotor will induce an additional mmf, which will add to one of the two mmf's and subtract from the other. Thus, a net torque will be established, causing the motor to continue its rotation in the same direction in which it was started. In particular, if the rotor is started in the forward direction, the forward mmf $\mathcal{F}^{+}$ will be greater than the backward mmf, and the motor will continue to rotate in the forward direction.

Figure 20.26 depicts an equivalent circuit for the single-phase induction motor *with stationary rotor,* where the parameters in the circuit are defined as follows:

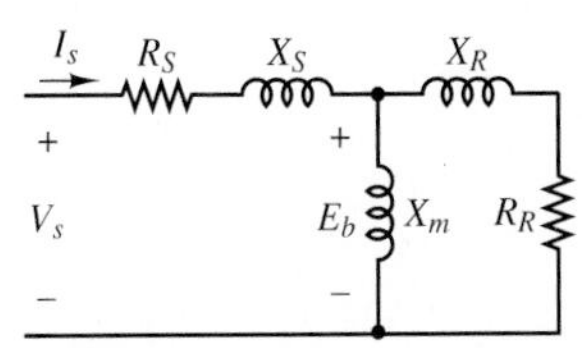

Figure 20.26 Circuit model for single-phase induction motor with rotor at standstill

V_s = supply voltage
R_S = resistance of stator winding
X_S = leakage reactance of stator winding
X_m = magnetizing reactance of stator winding
X_R = leakage reactance of rotor referred to stator at standstill
R_R = leakage resistance of rotor referred to stator at standstill
E_b = voltage induced in stator winding by (stationary) pulsating flux in air gap

Figure 20.27 depicts the equivalent circuit for the same motor with the rotor rotating with slip s. Note that the circuit is asymmetric, because of the different air gap flux forward and backward components E_f and E_b, respectively. The factors of 0.5 come from the resolution of the pulsating stator mmf into forward and backward components. Note further that the reflected rotor impedance is asymmetric because of the presence of the slip parameter in the expression for the reflected rotor resistance. Further, the circuit model also confirms that the forward induced voltage E_f must be greater than the backward voltage E_b since the slip is always less than 1.

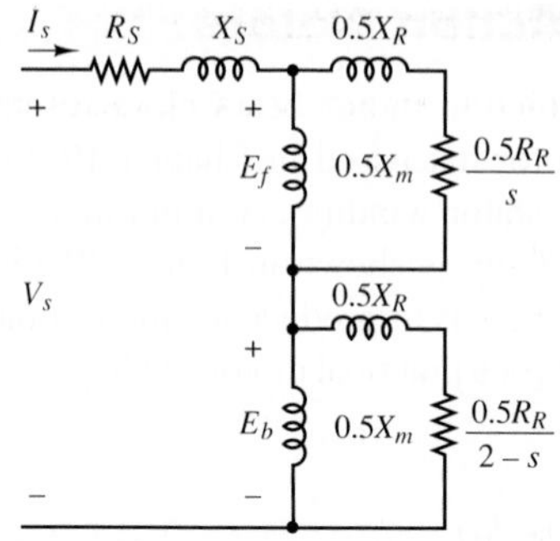

Figure 20.27 Circuit model for single-phase induction motor with rotor in motion

It can be shown that the torque components in the forward and backward directions are given by

$$T_f = \frac{P_f}{\omega_s} \tag{20.11}$$

and

$$T_b = \frac{P_b}{\omega_s} \tag{20.12}$$

where ω_s is the synchronous speed and

$$P_f = I_s^2 R_f \tag{20.13}$$

Here R_f is the resistive component of the forward field impedance; also

$$P_b = I_s^2 R_b \tag{20.14}$$

where R_b is the resistive component of the backward field impedance. Since the torque produced by the backward field is in the opposite direction to that produced by the forward field, the net torque will consist of the difference between the two:

$$T = T_f - T_b = \frac{I_s^2(R_f - R_b)}{\omega_s} \tag{20.15a}$$

The mechanical power developed by the motor is

$$\begin{aligned} P_{\text{mech}} &= T\omega_m = T\omega_s(1-s) = (P_f - P_b)(1-s) \\ &= I_s^2(R_f - R_b)(1-s) \end{aligned} \tag{20.15b}$$

CHECK YOUR UNDERSTANDING

(a) What is the zero-speed torque for a single-phase induction motor? (b) If a starting torque is applied to the machine, what final speed will the machine reach?

Answers: (a) Zero, without a start winding; (b) $n = (1 - s)n_s$, where $n_s = 120f/p$ and s = slip (determined by shaft load).

EXAMPLE 20.9 Slip in a Single-Phase Induction Motor

LO4

Problem

Find the slip of the field in the forward and backward directions for a single-phase induction machine.

Solution

Known Quantities: Motor operating characteristics.

Find: Forward slip s_f; backward slip s_b.

Schematics, Diagrams, Circuits, and Given Data: Motor operating characteristics: 115 V; 60 Hz; four poles; 1710 r/min.

Analysis: We first determine the synchronous speed of the motor:

$$n_s = \frac{120f}{p} = \frac{120 \times 60}{4} = 1{,}800 \text{ r/min}$$

The slip in the forward direction (direction of rotation of the motor) can now be computed:

$$s_f = \frac{n_s - n}{n_s} = \frac{1{,}800 - 1{,}710}{1{,}800} = 0.05 = 5\%$$

The slip in the backward direction can be computed as follows, with reference to Figure 20.26:

$$s_b = 2 - s_f = 2 - 0.05 = 1.95$$

EXAMPLE 20.10 Analysis of Single-Phase Induction Motor

LO4

Problem

Find the input current and generated torque for a single-phase induction motor.

Solution

Known Quantities: Motor operating characteristics and circuit parameters.

Find: Motor input (stator) current I_S; motor torque T.

Schematics, Diagrams, Circuits, and Given Data: Motor operating data: $\frac{1}{4}$ hp; 110 V; 60 Hz; four poles. Circuit parameters: $R_S = 1.5\ \Omega$; $X_S = 2\ \Omega$; $R_R = 3\ \Omega$; $X_R = 2\ \Omega$; $X_m = 50\ \Omega$; $s = 0.05$.

Assumptions: The motor is operated at rated voltage and frequency.

Analysis: With reference to the equivalent circuit of Figure 20.26, you can easily show that the impedance seen by the back emf E_b is much smaller than that seen by the forward emf E_f. This corresponds to stating that the backward component of the magnetizing impedance (which is in parallel with the backward component of the rotor impedance) is much larger than the backward component of the rotor impedance, and can therefore be neglected. This approximation is generally true for values of slip less than 0.15 and corresponds to stating that

$$0.5Z_b = 0.5\frac{jX_m\,[R_R/(2-s) + jX_R]}{R_R/(2-s) + j(X_m + X_R)}$$

$$\approx 0.5\left(\frac{R_R}{2-s} + jX_R\right)$$

The approximate circuit based on this simplification is shown in Figure 20.28.

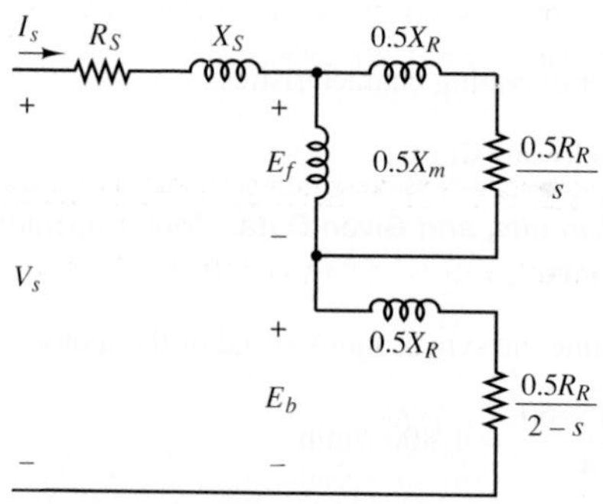

Figure 20.28 Approximate circuit model for single-phase induction motor

Using the approximate circuit, we find that the impedance seen by the backward emf is given by

$$0.5Z_b = 0.5\left(\frac{R_R}{2-s} + jX_R\right) = 0.5(1.538 + j2) = 0.5(R_b + jX_b)$$

The impedance seen by the forward emf is, on the other hand, given by the exact expression:

$$0.5Z_f = 0.5\frac{jX_m\,(R_R/s + jX_R)}{R_R/s + j(X_m + X_R)}$$

$$= 0.5\frac{j50(60 + j2)}{60 + j(50 + 2)} = 11.9 + j14.69 = 0.5R_f + j0.5X_f$$

If we let $Z_S = R_S + jX_S = 1.5 + j2\ \Omega$, we can write an expression for the total impedance of the motor as follows:

$$Z = Z_S + 0.5Z_f + 0.5Z_b = 14.169 + j17.69 = 22.66\angle 51.3^\circ\ \Omega$$

Knowing the total series impedance, we can calculate the stator current:

$$\mathbf{I}_S = \frac{\mathbf{V}_S}{Z} = \frac{110\ \text{V}}{22.66\angle 51.3^\circ\ \Omega} = 4.85\angle -51.3^\circ\ \text{A}$$

We can now calculate the power absorbed by the motor by separately computing the real power absorbed in the forward and backward fields:

$$P_f = I_S^2 \times 0.5R_f = (4.85)^2 \times 11.9 = 279.9\ \text{W}$$

$$P_b = I_S^2 \times 0.5R_b = (4.85)^2 \times 0.769 = 18.1\ \text{W}$$

The net power is the difference between the two components; thus, $P = P_f - P_b = 261.8$ W, and the torque developed by the motor is equal to the ratio of the power to the motor speed.

The synchronous speed can be computed to be

$$\omega_s = \frac{4\pi f}{p} = 188.5 \text{ rad/s}$$

and if we assume negligible rotational losses, we have

$$T = \frac{P}{\omega} = \frac{P}{(1-s)\times\omega_s} = \frac{261.8 \text{ W}}{0.95 \times 188.5 \text{ rad/s}} = 1.46 \text{ N-m}$$

Comments: Note that the power factor of the motor is pf $= \cos(-51.3°) = 0.625$. Such low power factors are typical of single-phase motors.

EXAMPLE 20.11 Analysis of Single-Phase Induction Motor

LO4

Problem

Find the following quantities for the single-phase machine of Example 20.10:

1. Output torque
2. Output power
3. Efficiency

Solution

Known Quantities: Motor operating characteristics.

Find: Motor torque T; output power P_{out}; efficiency η.

Schematics, Diagrams, Circuits, and Given Data: Motor operating data: $\frac{1}{4}$ hp; 110 V; 60 Hz; four poles; $s = 0.05$.

Assumptions: The motor is operated at rated voltage and frequency. The combined rotational and core losses are $P_{rot} + P_{core} = 30$ W.

Analysis:

1. *Output power calculation.* The motor generated power is the difference between the forward and backward components, as explained in Example 20.10. Thus,

   $$P = P_f - P_b = 261.8 \text{ W}$$

   The motor power is the difference between the generated power and the losses:

   $$P_{out} = P - P_{loss} = 261.8 - 30 = 231.8 \text{ W}$$

2. *Shaft torque calculation.* The shaft speed is

   $$\omega = (1-s)\times\omega_s = (1-s)\times\frac{4\pi f}{p} = 179 \text{ rad/s}$$

   and the torque is

   $$T = \frac{P_{out}}{\omega} = \frac{231.8 \text{ W}}{179 \text{ rad/s}} = 1.29 \text{ N-m}$$

3. *Efficiency calculation.* To calculate the overall efficiency of the motor, we must account for three loss mechanisms: mechanical losses, core losses, and electrical losses. The first

two are given as a lumped number; the electrical losses can be computed by calculating the I^2R losses in the stator and forward and backward circuits:

$$P_{S\,\text{loss}} = I_S^2 R_S = (4.85)^2 \times 1.5 = 35.3 \text{ W}$$

$$P_{R_f\,\text{loss}} = sP_f = 0.05 \times 279.9 = 14 \text{ W}$$

$$P_{R_b\,\text{loss}} = (1 - s)P_b = 1.95 \times 18.1 = 35.3 \text{ W}$$

$$P_{\text{elec}} = P_{S\,\text{loss}} + P_{R_f\,\text{loss}} + P_{R_b\,\text{loss}} = 114.6 \text{ W}$$

The efficiency can be calculated according to the following expression:

$$\eta = 1 - \frac{\sum \text{losses}}{P_{\text{out}} + \sum \text{losses}} = 1 - \frac{P_{\text{rot}} + P_{\text{core}} + P_{\text{elec}}}{P_{\text{out}} + P_{\text{rot}} + P_{\text{core}} + P_{\text{elec}}}$$

$$= 1 - \frac{(30 + 114.16) \text{ W}}{(2131.8 + 30 + 114.16) \text{ W}} = 0.617 = 61.7\%$$

Comments: Note that the overall efficiency of this machine is fairly low. Multiphase AC machines can achieve significantly higher efficiencies.

The equations and circuit models in the preceding examples suggest that a single-phase induction motor is capable of sustaining a torque, and of reaching its operating speed, once it is started by external means. However, because the magnetic field in a single-phase winding is stationary, a single-phase motor is not self-starting. The speed–torque characteristic of a typical single-phase induction motor shown in Figure 20.29 clearly shows that the starting torque for this motor is zero. The curve also shows that the motor can operate in either direction, depending on the direction of the initial starting torque, which must be provided by separate means.

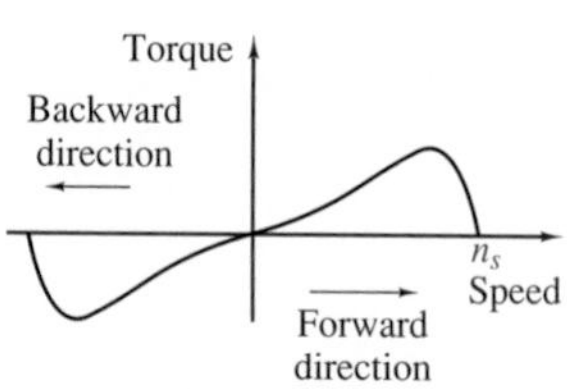

Figure 20.29 Torque–speed curve of a single-phase induction motor

Classification of Single-Phase Induction Motors

Thus far, we have not mentioned how the initial starting torque can be provided to a single-phase motor. In practice, single-phase motors are classified by their starting and running characteristics, and several methods exist to provide nonzero starting torque. The aim of this section is to classify single-phase motors by describing their configuration on the basis of the method of starting. For each class of motor, a torque–speed characteristic is also described.

Split-Phase Motors

Split-phase motors are constructed with two separate stator windings, called **main** and **auxiliary windings;** the axes of the two windings are actually at 90° with respect to each other, as shown in Figure 20.30. The auxiliary winding current is designed to be out of phase with the main winding current, as a result of the different reactances of the two windings. Different winding reactances can be attained by having a different ratio of resistance to inductance, for example, by increasing the resistance of the auxiliary winding. In particular, the auxiliary winding current $\mathbf{I}_{\text{aux}}$ leads the main winding current $\mathbf{I}_{\text{main}}$. The net effect is that the motor sees a two-phase (unbalanced) current that results in a rotating magnetic field, as in any polyphase stator arrangement. Thus, the motor has a nonzero starting torque, as shown in Figure 20.31. Once the motor has started, a centrifugal switch is used to disconnect the auxiliary winding,

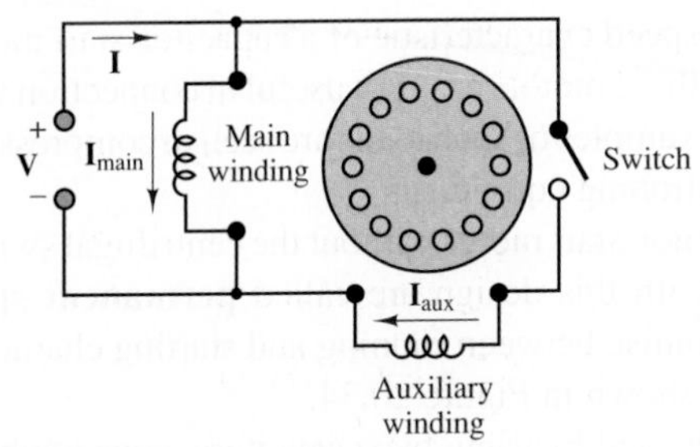

Figure 20.30 Split-phase motor

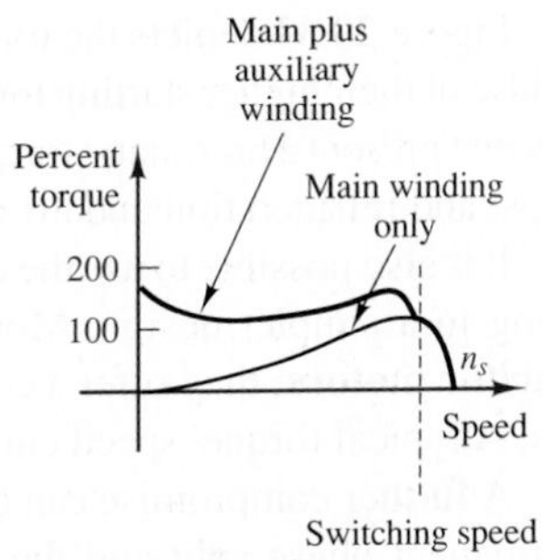

Figure 20.31 Torque–speed curve of split-phase motor

since a single winding is sufficient to sustain the motion of the rotor. The switching action permits the use of relatively high-resistance windings, since these are not used during normal operation, and therefore one need not be concerned with the losses associated with a higher-resistance winding. Figure 20.31 also depicts the combined effect of the two modes of operation of the split-phase motor.

Split-phase motors have appropriate characteristics (at very low cost) for fans, blowers, centrifugal pumps, and other applications in the range of $\frac{1}{20}$ to $\frac{1}{2}$ hp.

Capacitor-Type Motors

Another method for obtaining a phase difference between two currents that will give rise to a rotating magnetic field is by the addition of a capacitor. Motors that use this arrangement are termed **capacitor-type motors.** These motors make different use of capacitors to provide starting or running capabilities, or a combination of the two. The **capacitor-start motor** is essentially identical to the split-phase motor, except for the addition of a capacitor in series with the auxiliary winding, as shown in Figure 20.32. The addition of the capacitor changes the reactance of the auxiliary circuit in such a way as to cause the auxiliary current to lead the main current. The advantage of using the capacitor as a means for achieving a phase split is that greater starting torque may be obtained than with the split-phase arrangement. A centrifugal switching arrangement is used to disconnect the auxiliary winding above a certain speed, in the neighborhood of 75 percent of synchronous speed.

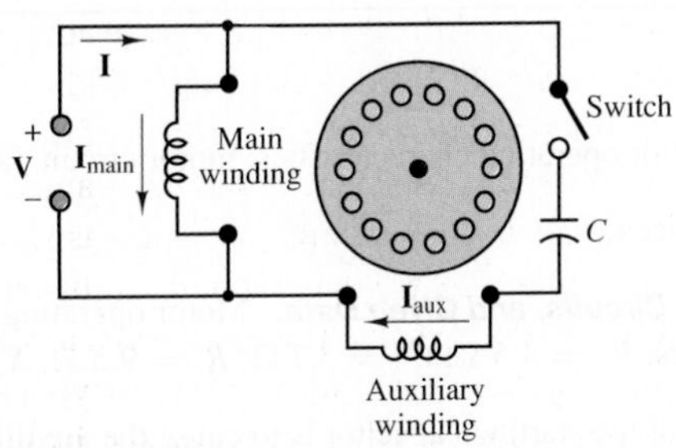

Figure 20.32 Capacitor-start motor

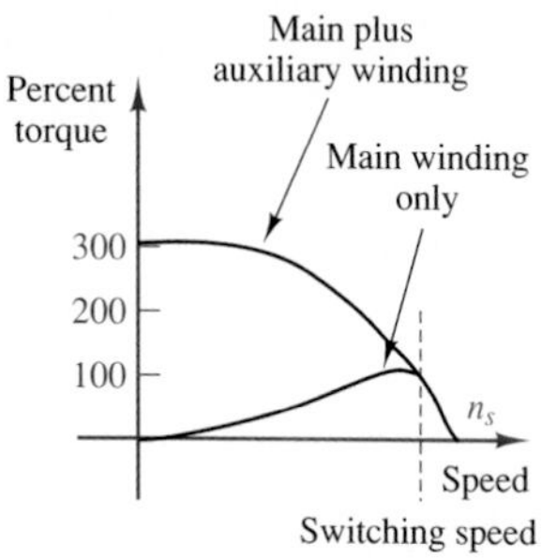

Figure 20.33 Torque–speed curve for a capacitor-start motor

Figure 20.33 depicts the torque–speed characteristic of a capacitor-start motor. Because of their higher starting torque, these motors are very useful in connection with loads that present a high static torque. Examples of such loads are seen in compressors, pumps, and refrigeration and air-conditioning equipment.

It is also possible to use the capacitor-start motor without the centrifugal switch, leading to a simpler design. Motors with this design are called **permanent split-capacitor motors;** they offer a compromise between running and starting characteristics. A typical torque–speed curve is shown in Figure 20.34.

A further compromise can be achieved by using two capacitors—one to obtain a permanent phase split and the resulting improvement in running characteristics, the other to improve the starting torque. A small capacitance is sufficient to improve the running performance, while a much larger capacitor provides the temporary improvement in starting torque. A motor with this design is called a **capacitor-start capacitor-run motor;** its schematic diagram is shown in Figure 20.35. Its torque–speed characteristic is similar to that of a capacitor-start motor.

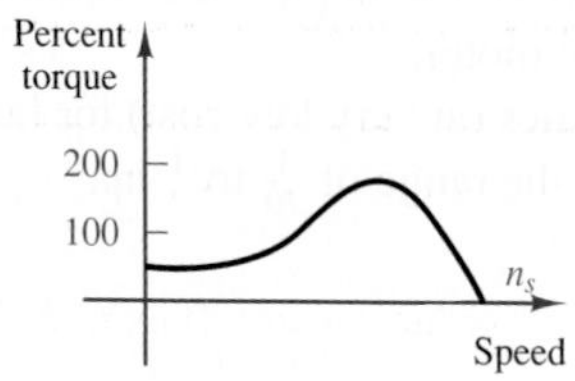

Figure 20.34 Torque–speed curve for a permanent split-capacitor motor

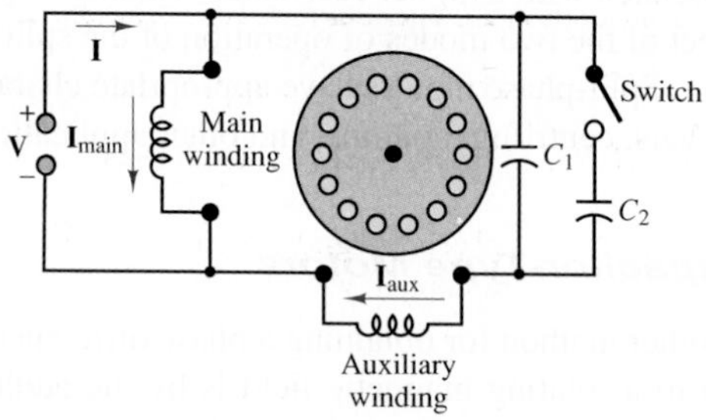

Figure 20.35 Capacitor-start capacitor-run motor

LO4

EXAMPLE 20.12 Analysis of Capacitor-Start Motor

Problem

With reference to Figure 20.32, find the required starting capacitance.

Solution

Known Quantities: Motor operating characteristics; motor circuit parameters.

Find: Starting capacitance C.

Schematics, Diagrams, Circuits, and Given Data: Motor operating data: $\frac{1}{3}$ hp; 120 V; 60 Hz. Circuit parameters: $R_m = 4.5\ \Omega$; $C_m = 3.7\ \Omega$; $R_a = 9.5\ \Omega$; $X_a = 3.5\ \Omega$.

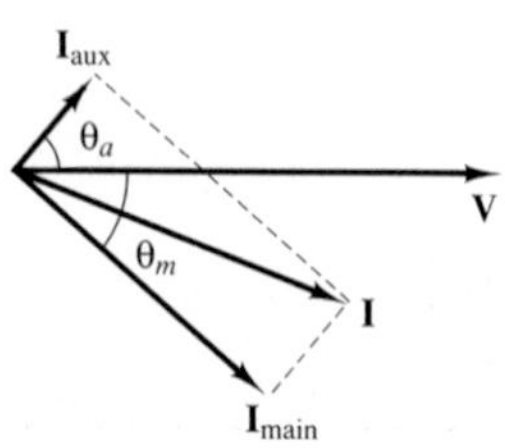

Figure 20.36 Starting phasor diagram for capacitor-start motor

Analysis: The purpose of the starting capacitor is to cause the auxiliary winding current $\mathbf{I}_{aux}$ at standstill to lead the main winding current $\mathbf{I}_{main}$ by 90°. The 90° phase lead will provide the maximum starting torque. Figure 20.36 shows the phasor diagram for these two currents and the voltage. The impedance angle of the main winding is

$$\theta_m = \arctan\left(\frac{X_m}{R_m}\right) = \arctan\left(\frac{3.7\ \Omega}{4.5\ \Omega}\right) = 39.4^\circ$$

Knowing that the desired phase shift between the main and auxiliary impedance angles is -90° (see Figure 20.36), we compute the impedance angle of the auxiliary winding:

$$\theta_a = 39.4^\circ - 90^\circ = -50.6^\circ$$

The minus sign indicates that $\mathbf{I}_{\text{aux}}$ leads the terminal voltage. The required capacitance can now be calculated from the relationship

$$\arctan\left(\frac{X_a - X_C}{R_a}\right) = -50.6^\circ$$

$$X_C = -R_a \times \tan(-50.6^\circ) + X_a = -9.5 \times (-1.21) + 3.5 = 15.07\ \Omega$$

and we can compute the desired capacitance as

$$C = \frac{1}{\omega X_C} = \frac{1}{377 \times 15.07} = 176 \times 10^{-6}\ \text{F} = 176\ \mu\text{F}$$

CHECK YOUR UNDERSTANDING

Draw the starting phasor diagram relating $\mathbf{V}$, $\mathbf{I}$, $\mathbf{I}_{\text{main}}$, and $\mathbf{I}_{\text{aux}}$ for the circuit of Figure 20.36, and sketch the time-domain waveforms.

Answer:

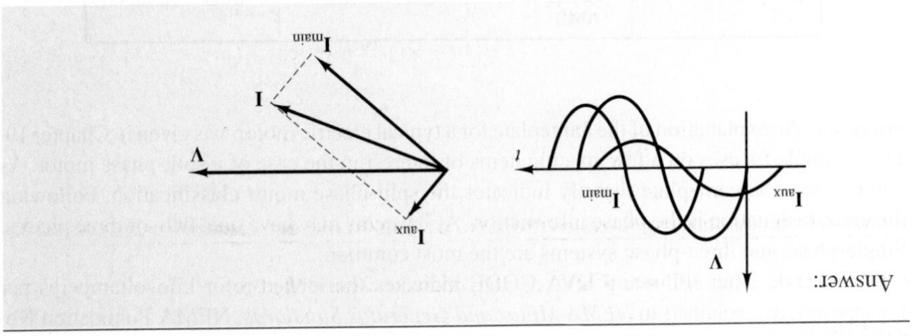

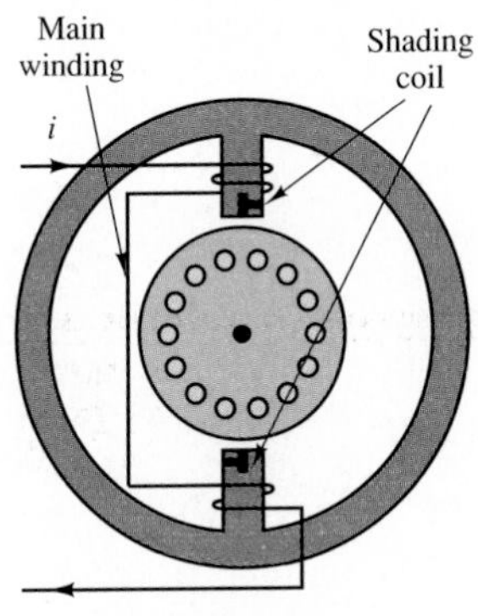

Figure 20.37 Shaded-pole motor

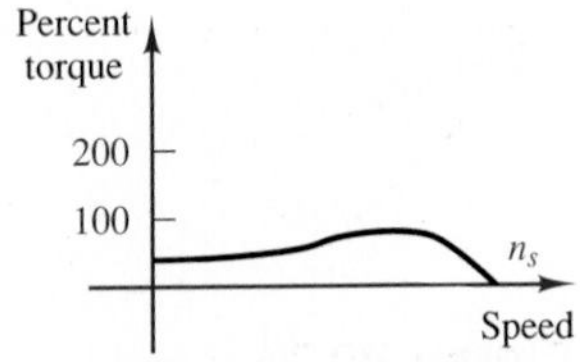

Figure 20.38 Torque–speed curve of a shaded-pole motor

Shaded-Pole Motors

The last type of single-phase induction motor discussed in this chapter is the **shaded-pole motor.** This type of motor operates on a different principle from the motors discussed thus far. The stator of a shaded-pole motor has a salient pole construction, as shown in Figure 20.37, that includes a shading coil consisting of a copper band wound around part of each pole. The flux in the shaded portion of the pole lags behind the flux in the unshaded part, achieving an effect similar to a rotation of the flux in the direction of the shaded part of the pole. This flux rotation in effect produces a rotating field that enables the motor to have a starting torque. This construction technique is rather inexpensive and is used in motors up to about $\frac{1}{20}$ hp.

A typical torque–speed characteristic for a shaded-pole motor is given in Figure 20.38.

LO4

EXAMPLE 20.13 Split-Phase Motor Nameplate Analysis

Problem

The table below depicts a split-phase motor nameplate. Determine the following quantities, using nameplate data:

1. Rated slip
2. Synchronous speed
3. Rated torque

Solution

Known Quantities: Nameplate data.

Find: s; ω_S; T.

Schematics, Diagrams, Circuits, and Given Data:

Thermal protected		Split-phase fan and blower motor	
MOD 4k800	HP $\frac{1}{3}$	RPM 1,725	KVA CODE N
V 115		A. 5.5	
FR 48Y	Hz 60	PH 1	DUTY CONT.
INS.CL. B	MAX. AMB 40 C	S. F. 1.35	BRG SLEEVE

Analysis: An explanation of the nameplate for a typical electric motor was given in Chapter 19. This example focuses on a few specific items of interest in the case of a split-phase motor. As you can see, the nameplate directly indicates the split-phase motor classification. Following the hertz designation is the phase information. AC systems may have one, two, or three phases. Single-phase and three-phase systems are the most common.

The code letter following KVA CODE indicates the locked-rotor kilovoltamperes per horsepower, as explained in *NEMA Motor and Generator Standards,* NEMA Publication No. MG1-10.37. The symbol "N" means that this motor has a maximum locked-rotor kilovoltamperes per horsepower of 12.5. Since the motor is rated at $\frac{1}{3}$ hp, the maximum locked-rotor kilovoltampere value is $12.5/3 = 4.167$. The maximum locked-rotor ampere value at 115 V will be 4.167 kVA/115 V = 36.23 A.

A large percentage of fractional-horsepower motors are now provided with built-in thermal protection. The use of such protection will also be indicated in the motor nameplate, for example, here by "Thermal Protected."

Bearing is abbreviated as "BRG." Fractional-horsepower motors normally use one of two types of bearings: sleeve or ball.

A variety of additional information may appear on the nameplate. This may include instructions for connecting the motor to a source of supply, reversing the direction of rotation, lubricating the motor, or operating it safely.

For the machine in this example, the synchronous speed is

$$n_s = 1{,}800 \text{ r/min}$$

The slip at rated speed is

$$s = \frac{n_s - n}{n_s} = \frac{1{,}800 - 1{,}725}{1{,}800} = 0.042$$

The power is

$$P = \tfrac{1}{3}\ \text{hp} = \tfrac{1}{3} \times 746\ \text{W/hp} = 248.7\ \text{W}$$

The rated torque is

$$T = \frac{K \times P}{n}$$

where the constant K is given by

$$K = \begin{cases} 0.97376 & T \text{ in meter-kilograms} \\ 9.549 & T \text{ in Newton-meters} \end{cases}$$

$$T = 9.549 \times \frac{248.7}{1{,}725} = 1.377\ \text{N-m}$$

Summary of Single-Phase Motor Characteristics

Four basic classes of single-phase motors are commonly used:

1. Single-phase induction motors are used for the larger home and small business tasks, such as furnace oil burner pumps, or hot water or hot air circulators. Refrigerator compressors, lathes, and bench-mounted circular saws are also powered with induction motors.
2. Shaded-pole motors are used in the smaller sizes for quiet, low-cost applications. The size range is from $\frac{1}{30}$ hp (24.9 W) to $\frac{1}{2}$ hp (373 W), particularly for fans and similar drives in which the starting torque is low.
3. Universal motors will operate on any household AC frequency or on DC without modification or adjustment. They can develop very high speed while loaded and very high power for their size. Vacuum cleaners, sewing machines, kitchen food mixers, portable electric drills, portable circular saws, and home motion-picture projectors are examples of applications of universal motors.
4. The capacitor-type motor finds its widest field of application at low speeds (below 900 r/min) and in ratings from $\frac{3}{4}$ hp (0.5595 kW) to 3 hp (2.238 kW) at all speeds, especially in fan drives.

20.5 MOTOR SELECTION AND APPLICATION

The objectives of this section are to outline the selection process of a motor for application to an electrical drive and to summarize the characteristics of the most common drive motors, with emphasis on fractional-horsepower applications. An electric motor should satisfy a set of precise requirements to be considered for a specific application. These include

1. Starting characteristics (torque and current).
2. Acceleration characteristics (dependent on the load).
3. Efficiency at rated load.
4. Overload capability.
5. Electrical and thermal safety.
6. Cost.

These requirements suggest that the specific details of the application should be clear in the designer's mind. For example, the nature of the load, the available electrical supplies, and the ambient conditions should be carefully investigated before the motor selection process is initiated. Once the application environment is known, it is usually possible to narrow the selection of a drive motor to a few choices. In this section we provide some insight into the motor selection process.

Motor Performance Calculations

FIND IT ON THE WEB

To better understand the **motor selection process,** it is important to review some of the basic ideas underlying the motion of the motor and of the load. For rotational systems, the relationships of Table 20.5 hold. Figure 20.39 summarizes the various types of load profiles that are likely to be encountered in practical applications. These include constant-torque loads; viscous friction-type loads, where torque is proportional to speed; loads in which the torque is proportional to a power of speed (e.g., fans, pumps); and constant-power loads, where torque is inversely proportional to speed.

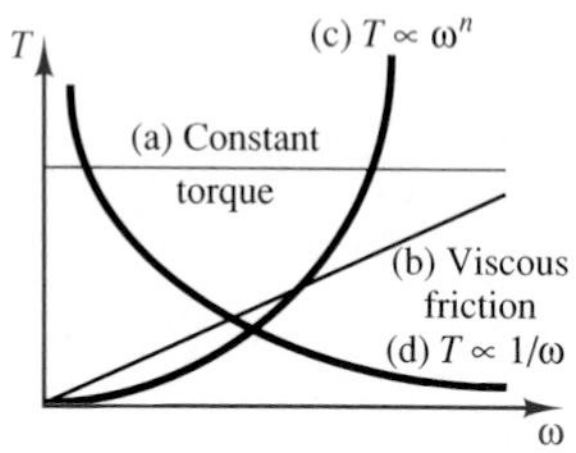

Figure 20.39 Typical load torque–speed curves

Table 20.5 **Equations of motion and definitions of variables**

LO5

Equations of motion	Definition of terms
$\omega_2 = \omega_1 + \alpha t$	ω_1 = initial velocity (rad/s)
$\theta = \omega_1 t + \frac{1}{2}\alpha t^2$	ω_2 = final velocity (rad/s)
$\omega_2^2 = \omega_1^2 + 2\alpha\theta$	α = acceleration (rad/s^2)
$P = T\omega$	θ = angular displacement (rad)
$T = Fr = J\alpha$	n = speed of rotation (r/min)
$W = T\theta$	P = output power (W)
	W = work (J)
$\omega = \dfrac{2\pi n}{60} = 0.105n$	F = force (N)
	T = torque (N-m)
$J = mk^2$	J = polar moment of inertia (kg-m^2)
	r = radius of arm (m)
	k = radius of gyration (m)
	m = mass (kg)

Reflected Load Inertia Calculations

To calculate the motor requirements, one must compute the required torque referenced to the motor output shaft. Since gearing systems are often employed, the inertias of all rotating components must be referred to the motor shaft. Using the terminology of Table 20.5, we then conclude that the reflected load torque at the motor shaft T_r is related to the load torque T_L by the relationship

$$T_r = \frac{\omega_L}{\omega_r} T_L \qquad \textbf{(20.16)}$$

where $\omega_r = \omega_m$ is the motor shaft speed, and the ratio of load speed to motor speed is equal to the gear ratio:

$$\frac{\omega_L}{\omega_r} = \frac{n_L}{n_r} \qquad \textbf{(20.17)}$$

If we equate the kinetic energy on the motor side to that on the load side, we can also derive an expression for the reflected load inertia:

$$J_L \omega_L^2 = J_r \omega_r^2 \qquad \textbf{(20.18)}$$

or

$$J_r = J_L \left(\frac{n_L}{n_r}\right)^2 \quad \text{Reflected inertia} \qquad \textbf{(20.19)}$$

LO5

Thus, the reflected inertia seen by the motor at the shaft is equal to the load inertia times the square of the gear ratio. Note that this is a mechanical impedance transformation similar to that used in the case of transformers. For all practical purposes, one can think of a gearing system as a mechanical transformer that, in the ideal case, conserves power. Under this ideal gearing assumption, it can be shown that the acceleration of the load is given by

$$\alpha = \frac{T_{m\ \text{peak}}}{(\omega_r/\omega_L)\left[J_m + \dfrac{J_L}{(\omega_r/\omega_L)^2}\right]} \qquad \textbf{(20.20)}$$

where the numerator on the right-hand side is the peak torque the motor can produce and J_m is the motor inertia. If one wished to determine what gear ratio were required to obtain maximum acceleration of the load, equation 20.20 would be differentiated and set equal to zero, to obtain

$$\frac{\omega_r}{\omega_L} = \sqrt{\frac{J_L}{J_m}} \qquad \textbf{(20.21)}$$

This expression implies that the load inertia should be made equivalent to the motor inertia by appropriate gearing, in order to obtain the best acceleration. Substituting equation 20.21 in 20.20, we can show that the maximum acceleration is given by

$$\alpha_{\max} = \frac{1}{2}\frac{T_{m\ \text{peak}}\,\omega_L}{J_m \omega_r} \quad \text{Maximum acceleration} \qquad \textbf{(20.22)}$$

LO5

Equations 20.16 through 20.22 are very useful in sizing a motor and in determining whether any gearing will be necessary to achieve the desired performance.

Torque Definitions

Although the definition of various torques was introduced in Chapter 19, it will be useful to briefly review these definitions in light of the preceding subsection. In sizing a motor, it is important to ensure that the motor is capable of overcoming the static friction at start-up, to accelerate to the desired speed in an acceptable fashion, and to handle any overloads that may occur. The following definitions will help in the analysis:

LO5

1. **Locked-rotor,** or **static, torque:** The minimum torque the motor will develop at rest for all angular positions under rated conditions.
2. **Breakdown torque:** The maximum torque a motor will develop under rated conditions without an abrupt drop in speed.
3. **Full-load torque:** The torque necessary to produce rated power output at full-load speed.
4. **Acceleration torque:** At any specified speed, acceleration torque, that is, the torque available for acceleration, is $T_{acc} = T_m - T_L - T_F$, where T_m is the motor torque, T_L is the load torque, and T_F is the frictional load torque.

Clearly, in order for the motor to accelerate to full-speed operation, the motor torque at standstill must exceed the total static-load torque. When the motor torque–speed curve intersects the load torque–speed curve, then a balanced operating condition has been reached.

Acceleration Calculations

The equation that defines the acceleration characteristics of a motor-load pair is

$$T_m - T_L = J\frac{d\omega}{dt} \qquad \textbf{(20.23)}$$

where T_L is the total load torque. From this equation we can calculate the time required to accelerate the load from a speed ω_1 to a speed ω_2 as follows:

$$t = J\int_{\omega_1}^{\omega_2} \frac{d\omega}{T_m - T_L} \quad \text{s} \qquad \textbf{(20.24)}$$

or, in units of revolutions per minute,

LO5

$$t = \frac{2\pi J_T}{60T}(n_1 - n_2) \quad \text{s} \quad \text{Acceleration} \qquad \textbf{(20.25)}$$

where T is the net torque (motor torque minus load torque) and J_T is the total system inertia (motor inertia plus reflected load inertia).

Efficiency Calculations

The efficiency of a motor is the ratio of the mechanical power output to the electrical power input, that is, the effectiveness of the electromechanical energy conversion.

We have already discussed the sources of loss in Chapter 19 and classified them as electrical losses, magnetic losses, and mechanical losses; refer to Section 19.1 for these definitions. The efficiency of a motor η is defined by

$$\eta = 1 - \frac{P_{\text{loss}}}{P_{\text{input}}} = 1 - \frac{P_{\text{loss}}}{VI} \qquad \text{Efficiency} \tag{20.26}$$

LO5

Thermal Calculations

The calculation of the temperature rise and thermal dissipation in a motor can be quite complex and depends very much on the motor construction. For the purpose of illustration, we briefly discuss only the thermal characteristics of a DC motor and perform some thermal calculations for this type of machine.

Thermal dissipation is one of the most important limiting factors in the operation of DC machines. We assume that most power losses take place in the armature (a reasonable assumption, since most of the electric power flows through the armature circuit), and we use the thermal-electrical system analogy where the thermal difference $\Delta\theta^\circ$ is given by

$$\Delta\theta^\circ = I_a^2 R_a \times R_T \tag{20.27}$$

and where I_a is the armature current, R_a the armature resistance, and R_T the thermal resistance of the rotor. The thermal time constant of the motor is then defined to be the time (in seconds) taken by the armature to reach 63 percent of the temperature rise corresponding to a given constant power dissipation. Now, the maximum continuous torque the motor can develop is related to the power dissipation, because the motor torque is proportional to the armature current:

$$T_{\max} = K_T I_{a\max} = K_T \sqrt{\frac{P_{\text{diss}}}{R_{\max}}} = K_T \sqrt{\frac{\Delta\theta^\circ}{R_T R_{\max}}} \tag{20.28}$$

where P_{diss} is the dissipated power and $R_{\max}$ the rotor resistance at the maximum temperature, R_T is the rotor thermal resistance at ambient temperature, and $\Delta\theta$ is the temperature rise. The temperature rise of copper can be determined from the known resistance of the wound rotor by computing the maximum temperature as follows:

$$\theta^\circ_{\max} = \frac{R_{\max}}{R_T}(\theta^\circ_{\text{ambient}} + 235) - 235 \qquad \text{Temperature rise} \tag{20.29}$$

LO5

and by computing $\Delta\theta^\circ = \theta^\circ_{\max} - \theta^\circ_{\text{ambient}}$, it is possible to use equation 20.28 to determine the maximum acceptable torque.

Conversely, to ensure that a given motor can operate within its thermal limits, one can calculate an average rms current requirement I_{rms}, consisting of the acceleration, deceleration, and running current, and use it to compute the temperature as follows:

$$\Delta\theta^\circ = I_{\text{rms}}^2 R_a R_T \tag{20.30}$$

Motor Selection

The range of electric motor applications is so broad that it is difficult to establish precise rules for motor selection. The differences between applications such as vehicle traction, robot motion, micromotors, disk drives, manufacturing machines, and pump systems, for example, are so many that it is virtually impossible to specify what the best motor would be, unless the application and its environment were clearly specified. The aim of this subsection is simply to outline a procedure that can help in narrowing down the choice of a suitable drive motor to a few most likely candidates.

The first step in selecting a motor is the analysis of the requirements imposed by the application; these can be divided into three groups: (1) motor requirements, (2) load requirements, and (3) control requirements. Table 20.6 summarizes the important considerations for each of these.

Table 20.6 **Motor selection requirements**

LO5

Motor requirements	Load requirements	Control requirements
Operating speed	Determine worst-case operating conditions	Available power (AC, DC)
Life span and maintenance	Dynamic acceleration, full-load and overload conditions	Motor operating voltage and current
Torque characteristics	Starting conditions	Open- or closed-loop
Mechanical aspects (size, weight, noise level, environment)	Transients	Forward/reverse operation
Applicable standards (e.g., radio-frequency interference)	Need for gearing, selection of optimum gear ratio	Motoring and/or braking
Overload characteristics	Frictional characteristics	Torque, position, or speed control
Thermal dissipation characteristics		Accuracy of speed or position control
		Controller complexity and cost

On the basis of the requirements listed in Table 20.6, one can undertake the task of selecting a motor for a specific application.

Motion Requirements

The first step in the drive selection process is to understand the application-driven specifications, that is, issues such as the type of motion, duty cycle, required acceleration and gearing system, and type of control that may be required (position, velocity, torque).

Motor Sizing

The second step in the drive selection process concerns the sizing of the motor itself. This is done first by calculating the maximum speed; next, the reflected inertia of the load and drive components is calculated, as discussed earlier in this section. From the inertia calculations, the peak torque required by the application can be calculated. The maximum speed and torque requirements thus obtained will narrow the field significantly. Next, one should determine the appropriate constants for each of the candidate motors; these include, in general, inertias, resistances (electrical and thermal), and torque and back emf constants. With these constants it becomes possible to determine that the motor can operate within its thermal specifications by calculating the **temperature rise** of the machine in operation. This, of course, can be a greater limitation during certain portions of the motion cycle, for example, during a hard acceleration.

Defining the Power Requirements

This step involves calculating peak voltage and current, to determine the supply requirements.

Choosing a Transmission

Although we have been assuming that the mechanical drive system was known beforehand, so that the reflected inertia and peak torque could be calculated, there are many issues that need to be investigated in establishing the drive system, for example, the effect that elastic couplings might have in creating mechanical resonances; noise and vibration characteristics; and backlash due to gearing system imperfections.

Summary

It should be apparent from this brief discussion that the process of selecting an electromechanical drive is quite complex, and it requires a good understanding of many aspects of engineering, including heat transfer, kinematics, dynamics, electronics, systems, and, of course, electromechanics. We hope that this brief introduction will provide the motivation to pursue further studies in this subject area.

Conclusion

This chapter introduces a number of special-purpose electric machines that find widespread application in industry. The operating principles and analysis methods used in this chapter build directly on the foundations of Chapters 18 and 19.

Upon completing this chapter on electric machines, you should have mastered the following learning objectives:

1. *Understand the basic principles of operation of brushless DC motors and the tradeoffs between these and brush-type DC motors.* The brushless DC motor is a PM synchronous motor in which the mechanical commutation of conventional DC motors is replaced by electronic commutation. Brushless DC motors can be made quite compact, and they find application in vehicle propulsion and motion control.
2. *Understand the operation and basic configurations of step motors, and understand step sequences for the different classes of step motors.* Stepping motors—of the variable-reluctance, PM, or hybrid type—permit fine angular displacement control by moving in fixed, discrete steps. Typical applications are seen in robotics and control systems.
3. *Understand the operating principles of switched reluctance machines.* Switched reluctance machines are gaining more widespread acceptance because of their simplicity, since they require no permanent magnets and have very simple stator windings and rotor construction. Possible applications of switched reluctance machines include low-cost industrial applications and vehicular propulsion.
4. *Classify and analyze single-phase AC motors, including the universal motor and various types of single-phase induction motors, using simple circuit models.* The universal motor is very similar in construction to a DC motor, but can operate on AC supplies; its speed can be controlled by electronic circuits of modest complexity. Thus, the universal motor finds common application in both variable- and fixed-speed appliances, such as power drills and vacuum cleaners, respectively. Induction motors can operate on a single-phase

AC supply as a means is provided to establish a starting torque. Various techniques are commonly employed, such as split-phase, capacitor-start, and shaded-pole construction. The different types are characterized by differing torque–speed characteristics that make the single-phase induction motor a very versatile device. This is probably the most commonly employed electric machine.

5. *Outline the selection process for an electric machine, given an application; perform calculations related to load inertia, acceleration, efficiency, and thermal characteristics.* The selection of the appropriate motor for a given application should take into consideration many factors, including cost and packaging, the nature of the load, the performance specifications, and thermal considerations.

HOMEWORK PROBLEMS

Section 20.1: Brushless DC Motors

20.1 It is found that $\lambda_m = 0.1$ V-s for a permanent-magnet, six-pole, two-phase synchronous machine. Calculate the amplitude (peak value) of the open-circuit phase voltage measured when the rotor is turned at 60 r/sec.

20.2 A four-pole, two-phase brushless dc motor is driven by a mechanical source at $n = 3{,}600$ r/min. The open-circuit voltage across one of the phases is 50 V rms.

a. Calculate λ.

b. The mechanical source is removed, and the following voltages are applied: $V_a = \sqrt{2}\,25\cos\theta$ and $V_b = \sqrt{2}\,25\sin\theta$, where $\theta = \omega_e t$. Calculate the no-load rotor speed ω in radians per second.

20.3 With reference to Example 20.2, we wish to shorten the trapezoidal speed profile cycle time by accelerating the motor to a maximum speed of 1,800 r/min. If we still allow 1 s for acceleration and deceleration, how long will the cycle time be?

20.4 With reference to the triangular speed profile of Figure P20.4, determine the speed profile required to move a load 0.5 m (or 100 r) in 3 s.

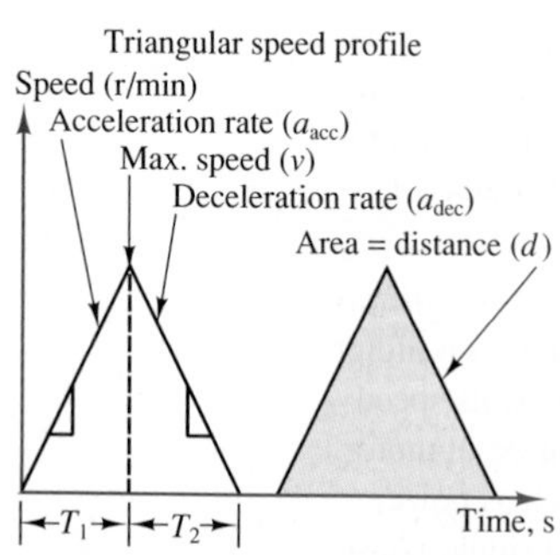

Figure P20.4

Section 20.2 Stepping Motors

20.5 With reference to Example 20.4, design a logic circuit that uses the logic design principles of Chapters 13 and 14 to achieve the step sequence given in Table 20.4.

Hint: Use a counter and logic gates.

20.6 A permanent-magnet stepper motor has six poles and a bipolar supply (i.e., the current into each coil pair can be either positive or negative). Figure 20.10 depicts a four-pole stepper motor as an example; the motor described in this problem has two additional poles. The spacing between the poles is uniform. Determine the size of the smallest achievable step in degrees.

20.7 Derive the dynamic equation for a stepping motor coupled to a load. The motor moment of inertia is J_m, the load moment of inertia is J_L, the viscous damping coefficient is D, and motor friction torque is T_f.

20.8 Sketch the rotor-stator configuration of a hybrid stepper motor capable of 18° steps.

Hint: The rotor will have five teeth.

20.9 Use a binary counter and logic gates to implement the stepping motor binary sequence of Check Your Understanding following Example 20.4.

20.10 A two-phase permanent-magnet stepper motor has 50 rotor teeth. When the rotor is driven by an external mechanical source at $\omega = 100$ rad/s, the measured open-circuit phase voltage is 25 V peak to peak. Calculate λ. If $i_a = 1$ A and $i_b = 0$, express the developed torque. Assume the winding resistance is 0.1 Ω.

20.11 The schematic diagram of a four-phase, two-pole permanent-magnet stepper motor is shown in Figure P20.11. The phase coils are excited in sequence by means of a logic circuit. Find

a. The logic schedule for full stepping of this motor.

b. The displacement angle of the full step.

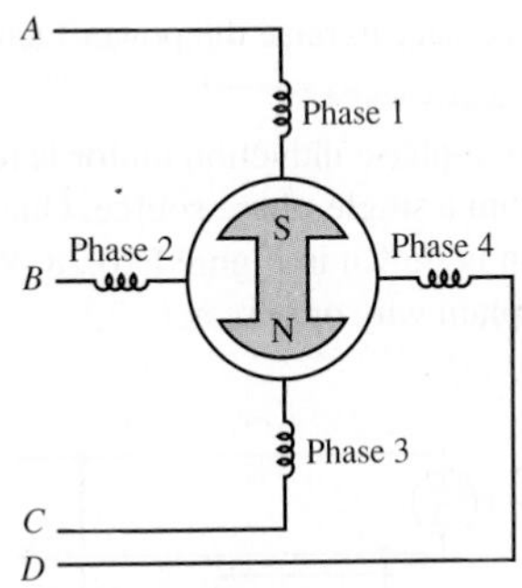

Figure P20.11

20.12 A permanent-magnet stepper motor is designed to provide a full-step angle of 15°. Find the number of stator and rotor poles.

20.13 A bridge driver scheme for a two-phase stepping motor is shown in Figure P20.13. Find the excitation sequences of the bridge operation (fill in the blanks of the table).

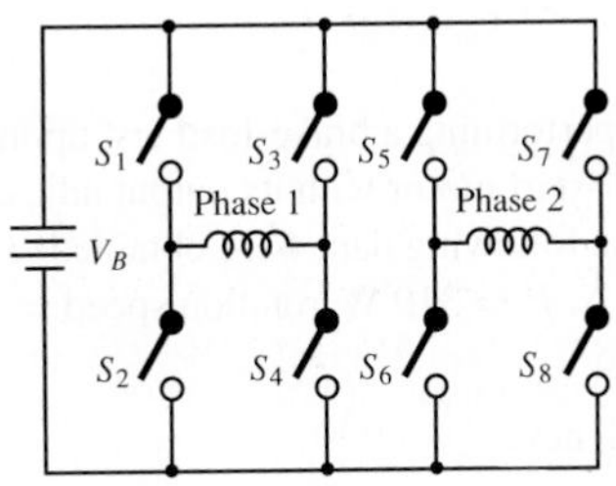

Clock state	Reset	1	2	3	4	5	6	7	8
S_1									
S_2									
S_3									
S_4									
S_5									
S_6									
S_7									
S_8									

Figure P20.13

20.14 A permanent-magnet stepper motor with a 15° step angle is used to directly drive a 0.100-in lead screw. Determine

a. The resolution of the stepper motor in steps per revolution.

b. The distance the lead screw travels (in inches) for each 15° step of the stepper motor.

c. The number of full 15° steps required to move the lead screw and the stepper motor shaft through 17.5 r.

d. The shaft speed (r/min) when the stepping frequency is 220 pulses per second.

Section 20.4: Single-Phase AC Motors

20.15 Determine whether the following motors are integral- or fractional-horsepower motors:

a. $\frac{3}{4}$ hp, 900 r/min

b. $1\frac{1}{2}$ hp, 3,600 r/min

c. $\frac{3}{4}$ hp, 1,800 r/min

d. $1\frac{1}{2}$ hp, 6,000 r/min

20.16 The spatial fluctuation of the stator mmf $\mathcal{F}_1$ is expressed as

$$\mathcal{F}_1 = F_{1(\text{peak})} \cos\theta$$

where θ is the electrical angle measured from the stator coil axis and $F_{1(\text{peak})}$ is the instantaneous value of the mmf wave at the coil axis and is proportional to the instantaneous stator current. If the stator current is a cosine function of time, the instantaneous value of the spatial peak of the pulsating mmf wave is

$$F_{1(\text{peak})} = F_{1(\text{max})} \cos\omega t$$

where $F_{1(\text{max})}$ is the peak value corresponding to maximum instantaneous current. Derive the expression for $\mathcal{F}_1$, and verify that for a single-phase winding, both forward and backward components are present.

20.17 A 200-V, 60-Hz, 10-hp, single-phase induction motor operates at an efficiency of 0.86 and a power factor of 0.9. What capacitor should be placed in parallel with the motor so that the feeder supplying the motor will operate at unity power factor?

20.18 A 230-V, 50-Hz, two-pole, single-phase induction motor is designed to run at 3 percent slip, Find the slip in the opposite direction of rotation. What is the speed of the motor in the normal direction of rotation?

20.19 Determine the amount of time (in seconds) it will take for a stepper motor with a 15° step angle, operating in one-phase excitation mode, to rotate through 28 rad when the pulse rate is 180 pps. *Note:* $t = \theta/\omega$.

20.20 A $\frac{1}{4}$-hp, 110-V, 60-Hz, four-pole capacitor-start motor has the following parameters:

$$R_S = 2.02\ \Omega \qquad X_S = 2.8\ \Omega$$
$$R_R = 4.12\ \Omega \qquad X_R = 2.12\ \Omega$$
$$X_m = 66.8\ \Omega \qquad s = 0.05$$

Find

a. The stator current.

b. The mechanical power.

c. The rotor speed.

20.21 A $\frac{1}{4}$-hp, four-pole, 110-V, 60-Hz, single-phase induction motor has the following data:

$R_S = 1.86\ \Omega$ $X_S = 2.56\ \Omega$

$R_R = 3.56\ \Omega$ $X_R = 2.56\ \Omega$

$X_m = 53.5\ \Omega$ $s = 0.05$

Find the mechanical power output.

20.22 A one-phase, 115-V, 60-Hz, four-pole induction motor has the following parameters:

$R_S = 0.5\ \Omega$ $X_S = 0.4\ \Omega$

$R_R = 0.25\ \Omega$ $X_R = 0.4\ \Omega$

$X_m = 35\ \Omega$

Find the input current and developed torque when the motor speed is 1,730 r/min.

20.23 The no-load test of a single-phase induction motor is made by running the motor without load at rated voltage and rated frequency. Derive the equivalent circuit of a single-phase induction motor for the no-load test.

Hint: The no-load slip is very small.

20.24 Derive the equivalent circuit of a single-phase induction motor for the locked-rotor test. Neglect the magnetizing current.

20.25 The design for a $\frac{1}{8}$-hp, two-pole, 115-V universal motor gives the effective resistances of the armature and series field as 4 and 6 Ω, respectively. The output torque is 0.17 N-m when the motor is drawing rated current of 1.5 A (rms) at a power factor of 0.88 at rated speed. Find

a. The full-load efficiency.

b. The rated speed.

c. The full-load copper losses.

d. The combined windage, friction, and iron losses.

e. The motor speed when the rms current is 0.5 A, neglecting phase differences and saturation.

20.26 A 240-V, 60-Hz, two-pole universal motor operates at a speed of 12,000 r/min on full load and draws a current of 6.5 A at 0.94 power factor lagging. The series field-winding impedance is $4.55 + j3.2\ \Omega$, and the armature circuit impedance is $6.15 + j9.4\ \Omega$. Find

a. The back emf of the motor.

b. The mechanical power developed by the motor.

c. The power output if the rotational loss is 65 W.

d. The efficiency of the motor.

20.27 A single-phase motor is drawing 20 A from a 400-V, 50-Hz supply. The power factor is 0.8 lagging. What value of capacitor connected across the circuit will be necessary to raise the power factor to unity?

20.28 A three-phase induction motor is required to operate from a single-phase source. One possible connection is shown in Figure P20.28. Will the motor work? Explain why or why not.

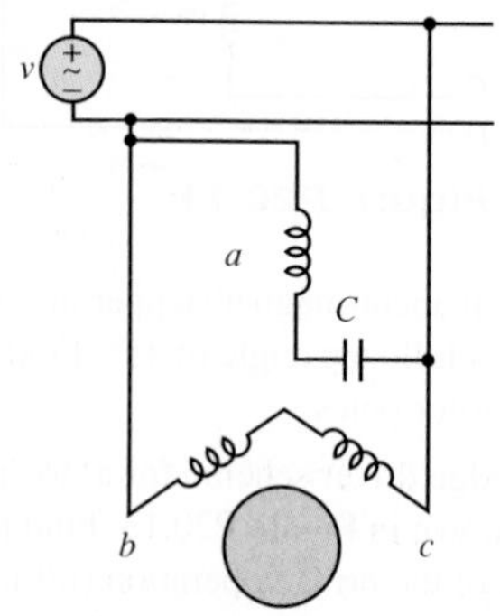

Figure P20.28

20.29 In performing a brake-load test upon a $\frac{1}{4}$-hp capacitor-start motor with its output adjusted to rated value, the following data were obtained: $E = 115$ V, $I = 3.8$ A, $P = 310$ W, rotation speed = 1,725 r/min. Calculate

a. Efficiency.

b. Power factor.

c. Torque in pound-inches.

Section 20.5: Motor Selection and Application

20.30 What type of motor would you select to perform the following tasks? Justify your selection.

a. Vacuum cleaner

b. Refrigerator

c. Air conditioner compressor

d. Air conditioner fan

e. Variable-speed sewing machine

f. Clock

g. Electric drill

h. Tape drive

i. X-Y plotter

20.31 A 5-hp, 1,150-r/min shunt motor has its speed controlled by means of a tapped field resistor, as shown in Figure P20.31. With the tap at position 3, determine the speed of the motor and the torque available at the maximum permissible load.

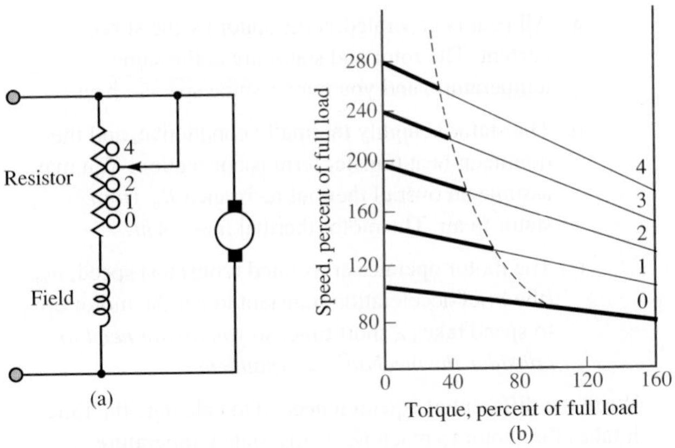

Figure P20.31

20.32 Which single-phase motor would you choose for the following applications?

a. Inexpensive analog electric clock.

b. Bathroom ventilator fan.

c. Escalator which must start under all load conditions.

d. Kitchen blender.

e. Table-model circular saw operating at about 3,500 r/min.

f. Handheld circular saw operating at 15,000 r/min.

g. Water pump.

20.33 The power required to drive a fan varies as the cube of the speed. If a motor driving a shaft-mounted fan is loaded to 100 percent of its horsepower rating on the top speed connection, find the horsepower output in percent of rating

a. At a speed reduction of 20 percent.

b. At a speed reduction of 30 percent.

c. At a speed reduction of 50 percent.

20.34 An industrial plant has a load of 800 kW at a power factor of 0.8 lagging. It is desired to purchase a synchronous motor of sufficient capacity to deliver a load of 200 kW and also serve to correct the overall plant power factor to 0.92. Assuming that the synchronous motor has an efficiency of 91 percent, determine its kilovoltampere input rating and the power factor at which it will operate.

20.35 An electric machine is controlled so that its torque–speed characteristics exhibit a constant-torque region and a constant-power region as shown in Figure P20.35. The average efficiency of the electric drive (combination of machine, plus power electronics, plus control electronics) is 87 percent. The machine torque is constant from 0 to 2,500 r/min, and is equal to 150 N-m. The constant-power region is valid from 2,500 to 6,000 r/min. The machine drives a constant-torque load requiring 75 N-m.

a. Determine the operating speed of the machine.

b. Determine the electric power needed to operate the machine.

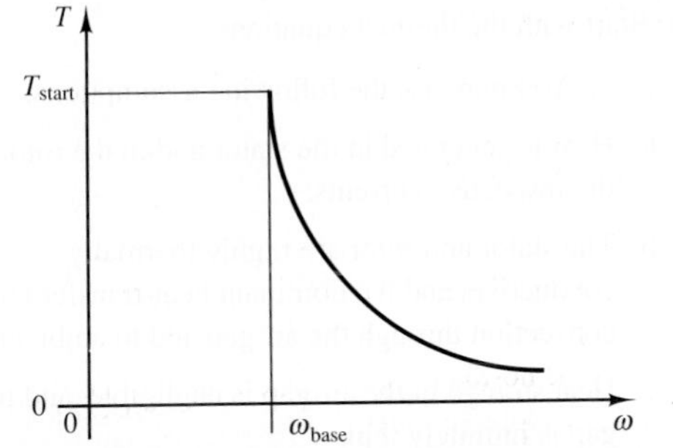

Figure P20.35

20.36 A PM synchronous (brushless DC) motor (wound stator, PM rotor, as shown in Figure P20.36) needs to be rated in terms of its thermal dissipation characteristics. To help in developing a rating, you are asked to write the dynamic equations linking the electrical dynamics to the thermal dynamics. Write the differential equations describing the electrothermomechanical dynamics of the system. You may make the following assumptions:

- All heat is generated in the stator by the stator current (i.e., the heat generated in the rotor is negligible). The rotor and stator are at the same temperature, and you may assume specific heat c.
- The stator is highly thermally conductive, and the dominant heat-transfer term is convection. You may assume an overall thermal resistance R_t from stator to air. The motor thermal mass is m.
- The motor generates torque according to the equation $T_m = kI_S$, and the back emf is equal to $E_b = k\omega$.

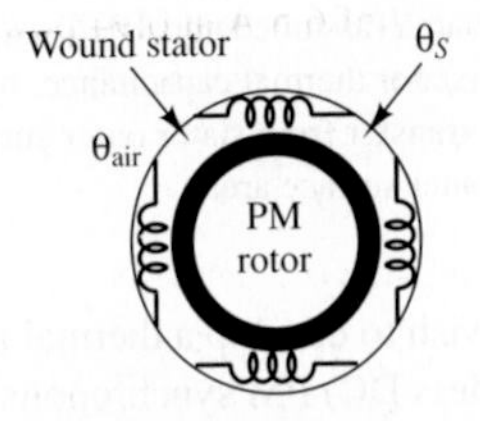

Figure P20.36 Thermoelectromechanical system. Electrical subsystem parameters: R_S, L_S, k (motor constant), $V_S(t)$, $I_S(t)$. Mechanical subsystem parameters: inertia and damping coefficient, J, b. Thermal subsystem parameters: thermal resistance, specific heat, mass, R_t, c, m.

20.37 A wound separately excited DC motor (wound stator and rotor, as shown in Figure P20.37) needs to be rated in terms of its thermal dissipation characteristics. To help in developing a rating, you are asked to write the dynamic equations linking the electrical dynamics to the thermal dynamics. Write the differential equations describing the electrothermomechanical dynamics of the system

Hint: Start with the thermal equations.

You may use the following assumptions:

a. Heat is generated in the stator and in the rotor by the respective currents.

b. The stator and rotor are highly thermally conductive, and the dominant heat-transfer term is convection through the air gap and to ambient.

c. Heat storage in the air gap is negligible, and the air gap is infinitely thin.

d. The motor generates torque according to the equation $T_m = k_T I_a$, and the back emf is equal to $E_b = k_a \omega$.

e. The stator and rotor each act as a lumped thermal mass.

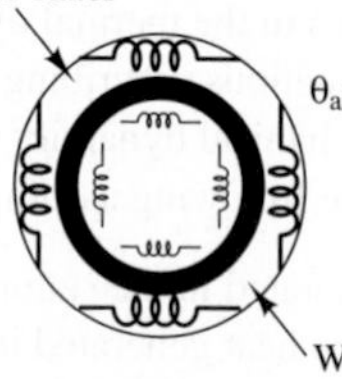

Figure P20.37 Thermoelectromechanical system. Electrical subsystem parameters: R_f, L_f, R_a, L_a (motor field and armature electrical parameters), k_f, k_a, k_T (motor field and armature constants), $V_S(t), V_f(t), I_a(t), I_f(t)$. Mechanical subsystem parameters: load inertia, damping coefficient, load torque, J, b, T_L. Thermal subsystem parameters: $C_{t-\text{rotor}}, h_{t-\text{rotor}}, A_{\text{rotor}}$ [rotor thermal capacitance, film coefficient of heat transfer from rotor surface to air and from air to stator inner surface, rotor and inner stator surface area (assumed equal)]. $C_{t-\text{stator}}, h_{t-\text{stator}}, A_{\text{stator}}$ (stator thermal capacitance, film coefficient of heat transfer from stator outer surface to air, stator outer surface area).

20.38 We wish to develop a thermal power rating for the (brushless DC) PM synchronous motor of Problem 20.36. To help in developing a rating, you are asked to write a simplified set of dynamic equations linking the electrical dynamics to the thermal dynamics. You may use the following assumptions:

a. All heat is generated in the stator by the stator current. The rotor and stator are at the same temperature, and you may assume specific heat c.

b. The stator is highly thermally conductive, and the dominant heat transfer term is convection. You may assume an overall thermal resistance R_m from stator to air. The motor thermal mass is m.

c. The motor operates at its rated (constant) speed, ω_m (the brief acceleration transient to get the motor up to speed takes a short time, so *you do not need to consider the mechanical dynamics*).

Write the differential equation needed to calculate the time it takes the motor to reach its steady-state temperature. Determine, symbolically, the time constant for the temperature rise.

20.39 An electric machine is controlled so that its torque-speed characteristics exhibit a constant-torque region and a constant-power region, as shown in the sketch of Figure P20.39. The average efficiency of the electric drive (combination of machine, plus power electronics, plus control electronics) is 87 percent. The machine torque is constant from 0 to 2,500 rpm, and is equal to 150 N-m. The constant power region is valid from 2,500 to 6,000 rpm. The machine drives a constant torque load requiring 75 N-m.

a. Determine the operating speed of the machine.

b. Determine the electric power needed to operate the machine.

Sketch the solution on the curve of Figure P20.39 with exact numerical values.

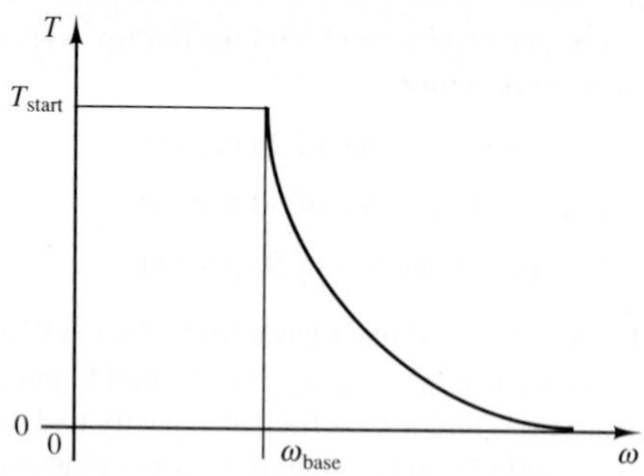

Figure P20.39

20.40 A PM DC motor is rigidly coupled to a fan; the fan load torque is described by the expression $T_L = 5 + 0.05\omega + 0.001\omega^2$ where torque is in N-m and speed in rad/s. The motor has $k_a\phi = k_T\phi = 2.42$. $R_a = 0.2\ \Omega$, and the inductance is negligible. If the motor voltage is 50 V, what will be the speed of rotation of the motor and fan?

APPENDIX A

LINEAR ALGEBRA AND COMPLEX NUMBERS

A.1 SOLVING SIMULTANEOUS LINEAR EQUATIONS, CRAMER'S RULE, AND MATRIX EQUATION

The solution of simultaneous equations, such as those that are often seen in circuit theory, may be obtained relatively easily by using Cramer's rule. This method applies to 2 × 2 or larger systems of equations. Cramer's rule requires the use of the concept of determinant. Linear, or matrix, algebra is valuable because it is systematic, general, and useful in solving complicated problems. A determinant is a scalar defined on a square array of numbers, or matrix, such as

$$\det(A) = |A| = \begin{vmatrix} a_{11} & a_{12} \\ a_{21} & a_{22} \end{vmatrix} \tag{A.1}$$

In this case the matrix is a 2 × 2 array with two rows and two columns, and its determinant is defined as

$$\det = a_{11}a_{22} - a_{12}a_{21} \tag{A.2}$$

A third-order, or 3 × 3, determinant such as

$$\det(A) = \begin{vmatrix} a_{11} & a_{12} & a_{13} \\ a_{21} & a_{22} & a_{23} \\ a_{31} & a_{32} & a_{33} \end{vmatrix} \tag{A.3}$$

is given by

$$\begin{aligned} \det &= a_{11}(a_{22}a_{33} - a_{23}a_{32}) - a_{12}(a_{21}a_{33} - a_{23}a_{31}) \\ &\quad + a_{13}(a_{21}a_{32} - a_{22}a_{31}) \end{aligned} \tag{A.4}$$

For higher-order determinants, you may refer to a linear algebra book. To illustrate Cramer's method, a set of two equations in general form will be solved here. A set of two linear simultaneous algebraic equations in two unknowns can be written in the form

$$\begin{aligned} a_{11}x_1 + a_{12}x_2 &= b_1 \\ a_{21}x_1 + a_{22}x_2 &= b_2 \end{aligned} \tag{A.5}$$

where x_1 and x_2 are the two unknowns. The coefficients a_{11}, a_{12}, a_{21}, and a_{22} are known quantities. The two quantities on the right-hand sides, b_1 and b_2, are also known (these are typically the source currents and voltages in a circuit problem). The set of equations can be arranged in matrix form, as shown in equation A.6.

$$\begin{bmatrix} a_{11} & a_{12} \\ a_{21} & a_{22} \end{bmatrix} \begin{bmatrix} x_1 \\ x_2 \end{bmatrix} = \begin{bmatrix} b_1 \\ b_2 \end{bmatrix} \tag{A.6}$$

In equation A.6, a coefficient matrix multiplied by a vector of unknown variables is equated to a right-hand-side vector. Cramer's rule can then be applied to find x_1 and x_2, using the following formulas:

$$x_1 = \frac{\begin{vmatrix} b_1 & a_{12} \\ b_2 & a_{22} \end{vmatrix}}{\begin{vmatrix} a_{11} & a_{12} \\ a_{21} & a_{22} \end{vmatrix}} \qquad x_2 = \frac{\begin{vmatrix} a_{11} & b_1 \\ a_{21} & b_2 \end{vmatrix}}{\begin{vmatrix} a_{11} & a_{12} \\ a_{21} & a_{22} \end{vmatrix}} \tag{A.7}$$

Thus, the solution is given by the ratio of two determinants: the denominator is the determinant of the matrix of coefficients, while the numerator is the determinant of the same matrix with the right-hand-side vector ($\begin{bmatrix} b_1 \\ b_2 \end{bmatrix}$ in this case) substituted in place of the column of the coefficient matrix corresponding to the desired variable (i.e., first column for x_1, second column for x_2, etc.). In a circuit analysis problem, the coefficient matrix is formed by the resistance (or conductance) values, the vector of unknowns is composed of the mesh currents (or node voltages), and the right-hand-side vector contains the source currents or voltages.

In practice, many calculations involve solving higher-order systems of linear equations. Therefore, a variety of computer software packages are often used to solve higher-order systems of linear equations.

CHECK YOUR UNDERSTANDING

A.1 Use Cramer's rule to solve the system

$$\begin{aligned} 5v_1 + 4v_2 &= 6 \\ 3v_1 + 2v_2 &= 4 \end{aligned}$$

A.2 Use Cramer's rule to solve the system

$$i_1 + 2i_2 + i_3 = 6$$
$$i_1 + i_2 - 2i_3 = 1$$
$$i_1 - i_2 + i_3 = 0$$

A.3 Convert the following system of linear equations into a matrix equation as shown in equation A.6, and find matrices A and b.

$$2i_1 - 2i_2 + 3i_3 = -10$$
$$-3i_1 + 3i_2 - 2i_3 + i_4 = -2$$
$$5i_1 - i_2 + 4i_3 - 4i_4 = 4$$
$$i_1 - 4i_2 + i_3 + 2i_4 = 0$$

Answers: $v_1 = 2$, $v_2 = -1$; $i_1 = 1$, $i_2 = 2$, $i_3 = 1$;

$$A = \begin{bmatrix} 2 & -2 & 3 & 0 \\ -3 & 3 & -2 & 1 \\ 5 & -1 & 4 & -4 \\ 1 & -4 & 1 & 2 \end{bmatrix}, \quad b = \begin{bmatrix} -10 \\ -2 \\ 4 \\ 0 \end{bmatrix}.$$

A.2 INTRODUCTION TO COMPLEX ALGEBRA

From your earliest training in arithmetic, you have dealt with real numbers such as 4, -2, $\frac{5}{9}$, π, e, etc., which may be used to measure distances in one direction or another from a fixed point. However, a number that satisfies the equation

$$x^2 + 9 = 0 \tag{A.8}$$

is not a real number. Imaginary numbers were introduced to solve equations such as equation A.8. Imaginary numbers add a new dimension to our number system. To deal with imaginary numbers, a new element, j, is added to the number system having the property

$$j^2 = -1$$

or

$$j = \sqrt{-1} \tag{A.9}$$

Thus, we have $j^3 = -j$, $j^4 = 1$, $j^5 = j$, etc. Using equation A.9, you can see that the solutions to equation A.8 are $\pm j3$. In mathematics, the symbol i is used for the imaginary unit, but this might be confused with current in electrical engineering. Therefore, the symbol j is used in this book.

A complex number (indicated in boldface notation) is an expression of the form

$$\mathbf{A} = a + jb \tag{A.10}$$

where a and b are real numbers. The complex number **A** has a real part a and an imaginary part b, which can be expressed as

$$a = \text{Re}\,\mathbf{A}$$
$$b = \text{Im}\,\mathbf{A} \tag{A.11}$$

It is important to note that a and b are both real numbers. The complex number $a + jb$ can be represented on a rectangular coordinate plane, called the *complex plane*, by interpreting it as a point (a, b). That is, the horizontal coordinate is a in real axis, and the vertical coordinate is b in imaginary axis, as shown in Figure A.1. The complex number $\mathbf{A} = a + jb$ can also be

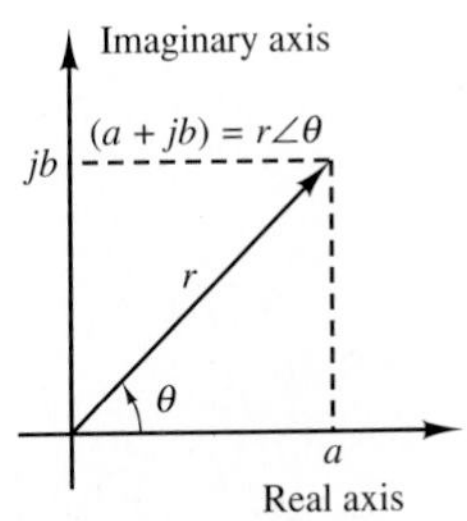

Figure A.1 Polar form representation of complex numbers

uniquely located in the complex plane by specifying the distance r along a straight line from the origin and the angle θ, which this line makes with the real axis, as shown in Figure A.1. From the right triangle of Figure A.1, we can see that

$$
\begin{aligned}
r &= \sqrt{a^2 + b^2} \\
\theta &= \tan^{-1}\left(\frac{b}{a}\right) \\
a &= r\cos\theta \\
b &= r\sin\theta
\end{aligned}
\tag{A.12}
$$

Then we can represent a complex number by the expression

$$
\mathbf{A} = re^{j\theta} = r\angle\theta \tag{A.13}
$$

which is called the polar form of the complex number. The number r is called the magnitude (or amplitude), and the number θ is called the angle (or argument). The two numbers are usually denoted by $r = |\mathbf{A}|$ and $\theta = \arg \mathbf{A} = \angle\mathbf{A}$.

Given a complex number $\mathbf{A} = a + jb$, the *complex conjugate* of $\mathbf{A}$, denoted by the symbol $\mathbf{A}^*$, is defined by the following equalities:

$$
\begin{aligned}
\text{Re}\,\mathbf{A}^* &= \text{Re}\,\mathbf{A} \\
\text{Im}\,\mathbf{A}^* &= -\text{Im}\,\mathbf{A}
\end{aligned}
\tag{A.14}
$$

That is, the sign of the imaginary part is reversed in the complex conjugate.

Finally, we should remark that two complex numbers are equal *if and only if* the real parts are equal and the imaginary parts are equal. This is equivalent to stating that two complex numbers are equal only if their magnitudes are equal and their arguments are equal.

The following examples and exercises should help clarify these explanations.

EXAMPLE A.1

Convert the complex number $\mathbf{A} = 3 + j4$ to its polar form.

Solution:

$$
r = \sqrt{3^2 + 4^2} = 5 \qquad \theta = \tan^{-1}\left(\frac{4}{3}\right) = 53.13^\circ
$$

$$
\mathbf{A} = 5\angle 53.13^\circ
$$

EXAMPLE A.2

Convert the number $\mathbf{A} = 4\angle(-60^\circ)$ to its complex form.

Solution:

$$
\begin{aligned}
a &= 4\cos(-60^\circ) = 4\cos(60^\circ) = 2 \\
b &= 4\sin(-60^\circ) = -4\sin(60^\circ) = -2\sqrt{3}
\end{aligned}
$$

Thus, $\mathbf{A} = 2 - j2\sqrt{3}$.

Addition and *subtraction* of complex numbers take place according to the following rules:

$$(a_1 + jb_1) + (a_2 + jb_2) = (a_1 + a_2) + j(b_1 + b_2)$$
$$(a_1 + jb_1) - (a_2 + jb_2) = (a_1 - a_2) + j(b_1 - b_2) \quad \textbf{(A.15)}$$

Multiplication of complex numbers in polar form follows the law of exponents. That is, the magnitude of the product is the product of the individual magnitudes, and the angle of the product is the sum of the individual angles, as shown below.

$$(\mathbf{A})(\mathbf{B}) = (Ae^{j\theta})(Be^{j\phi}) = ABe^{j(\theta+\phi)} = AB\angle(\theta + \phi) \quad \textbf{(A.16)}$$

If the numbers are given in rectangular form and the product is desired in rectangular form, it may be more convenient to perform the multiplication directly, using the rule that $j^2 = -1$, as illustrated in equation A.17.

$$\begin{aligned}(a_1 + jb_1)(a_2 + jb_2) &= a_1a_2 + ja_1b_2 + ja_2b_1 + j^2b_1b_2 \\ &= (a_1a_2 + j^2b_1b_2) + j(a_1b_2 + a_2b_1) \\ &= (a_1a_2 - b_1b_2) + j(a_1b_2 + a_2b_1)\end{aligned} \quad \textbf{(A.17)}$$

Division of complex numbers in polar form follows the law of exponents. That is, the magnitude of the quotient is the quotient of the magnitudes, and the angle of the quotient is the difference of the angles, as shown in equation A.18.

$$\frac{\mathbf{A}}{\mathbf{B}} = \frac{Ae^{j\theta}}{Be^{j\phi}} = \frac{A\angle\theta}{B\angle\phi} = \frac{\mathbf{A}}{\mathbf{B}}\angle(\theta - \phi) \quad \textbf{(A.18)}$$

Division in the rectangular form can be accomplished by multiplying the numerator and denominator by the complex conjugate of the denominator. Multiplying the denominator by its complex conjugate converts the denominator to a real number and simplifies division. This is shown in Example A.4. Powers and roots of a complex number in polar form follow the laws of exponents, as shown in equations A.19 and A.20.

$$\mathbf{A}^n = (Ae^{j\theta})^n = A^n e^{jn\theta} = A^n\angle n\theta \quad \textbf{(A.19)}$$

$$\begin{aligned}\mathbf{A}^{1/n} &= (Ae^{j\theta})^{1/n} = A^{1/n}e^{j1/n\theta} \\ &= \sqrt[n]{A}\angle\left(\frac{\theta + k2\pi}{n}\right) \qquad k = 0, \pm1, \pm2, \ldots\end{aligned} \quad \textbf{(A.20)}$$

EXAMPLE A.3

Perform the following operations, given that $\mathbf{A} = 2 + j3$ and $\mathbf{B} = 5 - j4$.

(a) $\mathbf{A} + \mathbf{B}$ (b) $\mathbf{A} - \mathbf{B}$ (c) $2\mathbf{A} + 3\mathbf{B}$

Solution:

$$\mathbf{A} + \mathbf{B} = (2 + 5) + j[3 + (-4)] = 7 - j$$
$$\mathbf{A} - \mathbf{B} = (2 - 5) + j[3 - (-4)] = -3 + j7$$

For part (c), $2\mathbf{A} = 4 + j6$ and $3\mathbf{B} = 15 - j12$. Thus, $2\mathbf{A} + 3\mathbf{B} = (4 + 15) + j[6 + (-12)] = 19 - j6$

EXAMPLE A.4

Perform the following operations in both rectangular and polar form, given that $\mathbf{A} = 3 + j3$ and $\mathbf{B} = 1 + j\sqrt{3}$.

(a) **AB** (b) $\mathbf{A} \div \mathbf{B}$

Solution:

(a) In rectangular form:

$$\begin{aligned}\mathbf{AB} &= (3 + j3)(1 + j\sqrt{3}) = 3 + j3\sqrt{3} + j3 + j^2 3\sqrt{3} \\ &= (3 + j^2 3\sqrt{3}) + j(3 + 3\sqrt{3}) \\ &= (3 - 3\sqrt{3}) + j(3 + 3\sqrt{3})\end{aligned}$$

To obtain the answer in polar form, we need to convert **A** and **B** to their polar forms:

$$\mathbf{A} = 3\sqrt{2}e^{j45^\circ} = 3\sqrt{2}\angle 45^\circ$$

$$\mathbf{B} = \sqrt{4}e^{j60^\circ} = 2\angle 60^\circ$$

Then

$$\mathbf{AB} = (3\sqrt{2}e^{j45^\circ})(\sqrt{4}e^{j60^\circ}) = 6\sqrt{2}\angle 105^\circ$$

(b) To find $\mathbf{A} \div \mathbf{B}$ in rectangular form, we can multiply **A** and **B** by $\mathbf{B}^*$.

$$\frac{\mathbf{A}}{\mathbf{B}} = \frac{3 + j3}{1 + j\sqrt{3}} \frac{1 - j\sqrt{3}}{1 - j\sqrt{3}}$$

Then

$$\frac{\mathbf{A}}{\mathbf{B}} = \frac{(3 + 3\sqrt{3}) + j(3 - 3\sqrt{3})}{4}$$

In polar form, the same operation may be performed as follows:

$$\frac{\mathbf{A}}{\mathbf{B}} = \frac{3\sqrt{2}\angle 45^\circ}{2\angle 60^\circ} = \frac{3\sqrt{2}}{2}\angle(45^\circ - 60^\circ) = \frac{3\sqrt{2}}{2}\angle(-15^\circ)$$

Euler's Identity

This formula extends the usual definition of the exponential function to allow for complex numbers as arguments. Euler's identity states that

$$e^{j\theta} = \cos\theta + j\sin\theta \tag{A.21}$$

All the standard trigonometry formulas in the complex plane are direct consequences of Euler's identity. The two important formulas are

$$\cos\theta = \frac{e^{j\theta} + e^{-j\theta}}{2} \qquad \sin\theta = \frac{e^{j\theta} - e^{-j\theta}}{2j} \tag{A.22}$$

EXAMPLE A.5

Using Euler's formula, show that

$$\cos\theta = \frac{e^{j\theta} + e^{-j\theta}}{2}$$

Solution:

Using Euler's formula gives

$$e^{j\theta} = \cos\theta + j\sin\theta$$

Extending the above formula, we can obtain

$$e^{-j\theta} = \cos(-\theta) + j\sin(-\theta) = \cos\theta - j\sin\theta$$

Thus,

$$\cos\theta = \frac{e^{j\theta} + e^{-j\theta}}{2}$$

CHECK YOUR UNDERSTANDING

A.4 In a certain AC circuit, $\boldsymbol{V} = \boldsymbol{ZI}$, where $\boldsymbol{Z} = 7.75\angle 90^\circ$ and $\boldsymbol{I} = 2\angle{-45^\circ}$. Find $\boldsymbol{V}$.

A.5 In a certain AC circuit, $\boldsymbol{V} = \boldsymbol{ZI}$, where $\boldsymbol{Z} = 5\angle 82^\circ$ and $\boldsymbol{V} = 30\angle 45^\circ$. Find $\boldsymbol{I}$.

A.6 Show that the polar form of $\boldsymbol{AB}$ in Example A.4 is equivalent to its rectangular form.

A.7 Show that the polar form of $\boldsymbol{A} \div \boldsymbol{B}$ in Example A.4 is equivalent to its rectangular form.

A.8 Using Euler's formula, show that $\sin\theta = (e^{j\theta} - e^{-j\theta})/2j$.

Answers: $\boldsymbol{V} = 15.5\angle 45^\circ$; $\boldsymbol{I} = 6\angle(-37^\circ)$

EXAMPLE A.6

Using Euler's formula, show that

$$\cos\theta = \frac{e^{j\theta} + e^{-j\theta}}{2}$$

Solution

Using Euler's formula gives

$$e^{j\theta} = \cos\theta + j\sin\theta$$

Extending the above formula, we can obtain

$$e^{-j\theta} = \cos(-\theta) + j\sin(-\theta) = \cos\theta - j\sin\theta$$

Thus

$$\cos\theta = \frac{e^{j\theta} + e^{-j\theta}}{2}$$

CHECK YOUR UNDERSTANDING

A.4 In a certain AC circuit, $V = ZI$, where $Z = 7.75\angle 68°$ and $I = 2\angle -15°$. Find V.

A.5 In a certain AC circuit, $V = ZI$, where $Z = 5\angle 82°$ and $V = 30\angle 45°$. Find I.

A.6 Show that the polar form of AB in Example A.4 is equivalent to its rectangular form.

A.7 Show that the polar form of $A \div B$ in Example A.4 is equivalent to its rectangular form.

A.8 Using Euler's formula, show that [illegible]

APPENDIX

B

THE LAPLACE TRANSFORM

B.1 COMPLEX FREQUENCY AND THE LAPLACE TRANSFORM

The transient analysis methods illustrated in Chapter 5 for first- and second-order circuits can become rather cumbersome when applied to higher-order circuits. Moreover, solving the differential equations directly does not reveal the strong connection that exists between the transient response and the frequency response of a circuit. The aim of this section is to introduce an alternate solution method based on the notions of complex frequency and of the **Laplace transform.** The concepts presented will demonstrate that the frequency response of linear circuits is but a special case of the general transient response of the circuit, when analyzed by means of Laplace methods. In addition, the use of the Laplace transform method allows the introduction of *systems* concepts, such as poles, zeros, and transfer functions, that cannot be otherwise recognized.

Complex Frequency

In Chapter 4, we considered circuits with sinusoidal excitations such as

$$v(t) = A\cos(\omega t + \phi) \quad \textbf{(B.1)}$$

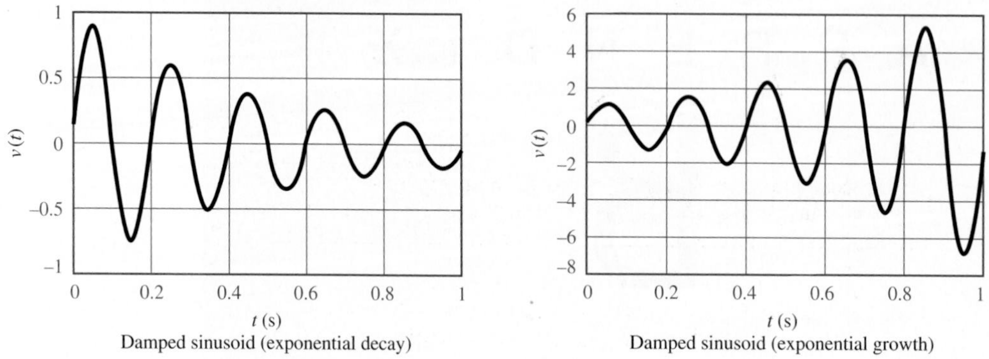

Figure B.1(a) Damped sinusoid: negative σ **Figure B.1(b)** Damped sinusoid: positive σ

which we also wrote in the equivalent phasor form

$$\mathbf{V}(j\omega) = Ae^{j\phi} = A\angle\phi \tag{B.2}$$

The two expressions just given are related by

$$v(t) = \text{Re}(\mathbf{V}e^{j\omega t}) \tag{B.3}$$

As was shown in Chapter 4, phasor notation is extremely useful in solving AC steady-state circuits, in which the voltages and currents are *steady-state sinusoids*. We now consider a different class of waveforms, useful in the transient analysis of circuits, namely, *damped sinusoids*. The most general form of a damped sinusoid is

$$v(t) = Ae^{\sigma t}\cos(\omega t + \phi) \tag{B.4}$$

As one can see, a damped sinusoid is a sinusoid multiplied by a real exponential $e^{\sigma t}$. The constant σ is real and is usually zero or negative in most practical circuits. Figure B.1(a) and (b) depicts the case of a damped sinusoid with negative σ and with positive σ, respectively. Note that the case of $\sigma = 0$ corresponds exactly to a sinusoidal waveform. The definition of phasor voltages and currents given in Chapter 4 can easily be extended to account for the case of damped sinusoidal waveforms by defining a new variable s, called the *complex frequency*:

$$s = \sigma + j\omega \tag{B.5}$$

Note that the special case of $\sigma = 0$ corresponds to $s = j\omega$, that is, the familiar steady-state sinusoidal (phasor) case. We shall now refer to the complex variable $\mathbf{V}(s)$ as the **complex frequency domain** representation of $v(t)$. It should be observed that from the viewpoint of circuit analysis, the use of the Laplace transform is analogous to phasor analysis; that is, substituting the variable s wherever $j\omega$ was used is the only step required to describe a circuit using the new notation.

CHECK YOUR UNDERSTANDING

B.1 Find the complex frequencies that are associated with

a. $5e^{-4t}$ b. $\cos 2\omega t$ c. $\sin(\omega t + 2\theta)$ d. $4e^{-2t}\sin(3t - 50°)$ e. $e^{-3t}(2 + \cos 4t)$

B.2 Find s and $\mathbf{V}(s)$ if $v(t)$ is given by

a. $5e^{-2t}$ b. $5e^{-2t}\cos(4t + 10°)$ c. $4\cos(2t - 20°)$

B.3 Find $v(t)$ if

a. $s = -2$, $\mathbf{V} = 2\angle 0°$ b. $s = j2$, $\mathbf{V} = 12\angle -30°$ c. $s = -4 + j3$, $\mathbf{V} = 6\angle 10°$

Answers: **B.1:** a. -4; b. $\mp j2\omega$; c. $\mp j\omega$; d. $-2 \mp j3$; e. -3 and $-3 \mp j4$. **B.2:** a. -2, $5\angle 0°$; b. $-2 + j4$, $5\angle 10°$; c. $j2$, $4\angle -20°$. **B.3:** a. $2e^{-2t}$; b. $12\cos(2t - 30°)$; c. $6e^{-4t}\cos(3t + 10°)$

All the concepts and rules used in AC network analysis (see Chapter 4), such as impedance, admittance, KVL, KCL, and Thévenin's and Norton's theorems, carry over to the damped sinusoid case exactly. In the complex frequency domain, the current $\mathbf{I}(s)$ and voltage $\mathbf{V}(s)$ are related by the expression

$$\mathbf{V}(s) = Z(s)\mathbf{I}(s) \tag{B.6}$$

where $Z(s)$ is the familiar impedance, with s replacing $j\omega$. We may obtain $Z(s)$ from $Z(j\omega)$ by simply replacing $j\omega$ by s. For a resistance R, the impedance is

$$Z_R(s) = R \tag{B.7}$$

For an inductance L, the impedance is

$$Z_L(s) = sL \tag{B.8}$$

For a capacitance C, it is

$$Z_C(s) = \frac{1}{sC} \tag{B.9}$$

Impedances in series or parallel are combined in exactly the same way as in the AC steady-state case, since we only replace $j\omega$ by s.

EXAMPLE B.1 Complex Frequency Notation

Problem:

Use complex impedance ideas to determine the response of a series *RL* circuit to a damped exponential voltage.

Solution:

Known Quantities: Source voltage, resistor, inductor values.

Find: The time-domain expression for the series current $i_L(t)$.

Schematics, Diagrams, Circuits, and Given Data: $v_s(t) = 10e^{-2t}\cos(5t)$ V; $R = 4\ \Omega$; $L = 2$ H.

Assumptions: None.

Analysis: The input voltage phasor can be represented by the expression

$$\mathbf{V}(s) = 10\angle 0 \text{ V}$$

The impedance seen by the voltage source is

$$Z(s) = R + sL = 4 + 2s$$

Thus, the series current is

$$\mathbf{I}(s) = \frac{\mathbf{V}(s)}{Z(s)} = \frac{10}{4+2s} = \frac{10}{4+2(-2+j5)} = \frac{10}{j10} = j1 = 1\angle\left(-\frac{\pi}{2}\right)$$

Finally, the time-domain expression for the current is

$$i_L(t) = e^{-2t}\cos(5t - \pi/2) \qquad \text{A}$$

Comments: The phasor analysis method illustrated here is completely analogous to the method introduced in Chapter 4, with the complex frequency $j\omega$ (steady-state sinusoidal frequency) related by s (damped sinusoidal frequency).

Just as frequency response functions $H(j\omega)$ were defined in this appendix, it is possible to define a **transfer function** $H(s)$. This can be a ratio of a voltage to a current, a ratio of a voltage to a voltage, a ratio of a current to a current, or a ratio of a current to a voltage. The transfer function $H(s)$ is a function of network elements and their interconnections. Using the transfer function and knowing the input (voltage or current) to a circuit, we can find an expression for the output either in the complex frequency domain or in the time domain. As an example, suppose $\mathbf{V}_i(s)$ and $\mathbf{V}_o(s)$ are the input and output voltages to a circuit, respectively, in complex frequency notation. Then

$$H(s) = \frac{\mathbf{V}_o(s)}{\mathbf{V}_i(s)} \qquad \textbf{(B.10)}$$

from which we can obtain the output in the complex frequency domain by computing

$$\mathbf{V}_o(s) = H(s)\mathbf{V}_i(s) \qquad \textbf{(B.11)}$$

If $\mathbf{V}_i(s)$ is a known damped sinusoid, we can then proceed to determine $v_o(t)$ by means of the method illustrated earlier in this section.

CHECK YOUR UNDERSTANDING

B.4 Given the transfer function $H(s) = 3(s+2)/(s^2+2s+3)$ and the input $\mathbf{V}_i(s) = 4\angle 0°$, find the forced response $v_o(t)$ if

a. $s = -1$ b. $s = -1 + j1$ c. $s = -2 + j1$

B.5 Given the transfer function $H(s) = 2(s+4)/(s^2+4s+5)$ and the input $\mathbf{V}_i(s) = 6\angle 30°$, find the forced response $v_o(t)$ if

a. $s = -4 + j1$ b. $s = -2 + j2$

Answers: **B.4:** a. $6e^{-t}$; b. $12\sqrt{2}e^{-t}\cos(t + 45°)$; c. $6e^{-2t}\cos(t + 135°)$.
B.5: a. $3e^{-4t}\cos(t + 165°)$; b. $8\sqrt{2}e^{-2t}\cos(2t - 105°)$

The Laplace Transform

The Laplace transform, named after the French mathematician and astronomer Pierre Simon de Laplace, is defined by

$$\mathcal{L}[f(t)] = F(s) = \int_0^\infty f(t)e^{-st}\,dt \quad \textbf{(B.12)}$$

The function $F(s)$ is the Laplace transform of $f(t)$ and is a function of the complex frequency $s = \sigma + j\omega$, considered earlier in this section. Note that the function $f(t)$ is defined only for $t \geq 0$. This definition of the Laplace transform applies to what is known as the **one-sided** or **unilateral Laplace transform,** since $f(t)$ is evaluated only for positive t. To conveniently express arbitrary functions only for positive time, we introduce a special function called the **unit step function** $u(t)$, defined by the expression

$$u(t) = \begin{cases} 0 & t < 0 \\ 1 & t > 0 \end{cases} \quad \textbf{(B.13)}$$

EXAMPLE B.2 Computing a Laplace Transform

Problem:

Find the Laplace transform of $f(t) = e^{-at}u(t)$.

Solution:

Known Quantities: Function to be Laplace-transformed.

Find: $F(s) = \mathcal{L}[f(t)]$.

Schematics, Diagrams, Circuits, and Given Data: $f(t) = e^{-at}u(t)$.

Assumptions: None.

Analysis: From equation B.12,

$$F(s) = \int_0^\infty e^{-at}e^{-st}\,dt = \int_0^\infty e^{-(s+a)t}\,dt = \frac{1}{s+a}e^{-(s+a)t}\Big|_0^\infty = \frac{1}{s+a}$$

Comments: Table B.1 contains a list of common Laplace transform pairs.

EXAMPLE B.3 Computing a Laplace Transform

Problem:

Find the Laplace transform of $f(t) = \cos(\omega t)\,u(t)$.

Solution:

Known Quantities: Function to be Laplace-transformed.

Find: $F(s) = \mathcal{L}[f(t)]$.

Schematics, Diagrams, Circuits, and Given Data: $f(t) = \cos(\omega t)\,u(t)$.

Assumptions: None.

Analysis: Using equation B.12 and applying Euler's identity to $\cos(\omega t)$ give:

$$F(s) = \int_0^{\infty} \frac{1}{2}(e^{j\omega t} + e^{-j\omega t})e^{-st}\,dt = \frac{1}{2}\int_0^{\infty} (e^{(-s+j\omega)t} + e^{(-s-j\omega)t})\,dt$$

$$= \frac{1}{-s+j\omega}e^{-(s+j\omega)t}\Bigg|_0^{\infty} + \frac{1}{-s-j\omega}e^{-(s-j\omega)t}\Bigg|_0^{\infty}$$

$$= \frac{1}{-s+j\omega} + \frac{1}{-s-j\omega} = \frac{s}{s^2+\omega^2}$$

Comments: Table B.1 contains a list of common Laplace transform pairs.

CHECK YOUR UNDERSTANDING

B.6 Find the Laplace transform of the following functions:

a. $u(t)$ b. $\sin(\omega t)\,u(t)$ c. $tu(t)$

B.7 Find the Laplace transform of the following functions:

a. $e^{-at}\sin\omega t\;u(t)$ b. $e^{-at}\cos\omega t\;u(t)$

Answers: **B.6:** a. $\frac{1}{s}$; b. $\frac{\omega}{s^2+\omega^2}$; c. $\frac{1}{s^2}$; **B.7:** a. $\frac{\omega}{(s+a)^2+\omega^2}$; b. $\frac{s+a}{(s+a)^2+\omega^2}$

Table B.1 Laplace transform pairs

$f(t)$	$F(s)$
$\delta(t)$ (unit impulse)	1
$u(t)$ (unit step)	$\frac{1}{s}$
$e^{-at}u(t)$	$\frac{1}{s+a}$
$\sin\omega t\;u(t)$	$\frac{\omega}{s^2+\omega^2}$
$\cos\omega t\;u(t)$	$\frac{s}{s^2+\omega^2}$
$e^{-at}\sin\omega t\;u(t)$	$\frac{\omega}{(s+a)^2+\omega^2}$
$e^{-at}\cos\omega t\;u(t)$	$\frac{s+a}{(s+a)^2+\omega^2}$
$tu(t)$	$\frac{1}{s^2}$

From what has been said so far about the Laplace transform, it is obvious that we may compile a lengthy table of functions and their Laplace transforms by repeated application of equation B.12 for various functions of time $f(t)$. Then we could obtain a wide variety of inverse transforms by matching entries in the table. Table B.1 lists some of the more common **Laplace transform pairs.** The computation of the **inverse Laplace transform** is in general rather complex if one wishes to consider arbitrary functions of s. In many practical cases, however, it is possible to use combinations of known transform pairs to obtain the desired result.

EXAMPLE B.4 Computing an Inverse Laplace Transform

Problem:

Find the inverse Laplace transform of

$$F(s) = \frac{2}{s+3} + \frac{4}{s^2+4} + \frac{4}{s}$$

Solution:

Known Quantities: Function to be inverse Laplace–transformed.

Find: $f(t) = \mathcal{L}^{-1}[F(s)]$.

Schematics, Diagrams, Circuits, and Given Data:

$$F(s) = \frac{2}{s+3} + \frac{4}{s^2+4} + \frac{4}{s} = F_1(s) + F_2(s) + F_3(s)$$

Assumptions: None.

Analysis: Using Table B.1, we can individually inverse-transform each of the elements of $F(s)$:

$$f_1(t) = 2\mathcal{L}^{-1}\left(\frac{1}{s+3}\right) = 2e^{-3t}u(t)$$

$$f_2(t) = 2\mathcal{L}^{-1}\left(\frac{2}{s^2+2^2}\right) = 2\sin(2t)\,u(t)$$

$$f_3(t) = 4\mathcal{L}^{-1}\left(\frac{1}{s}\right) = 4u(t)$$

Thus

$$f(t) = f_1(t) + f_2(t) + f_3(t) = \left(2e^{-3t} + 2\sin 2t + 4\right)u(t).$$

EXAMPLE B.5 Computing an Inverse Laplace Transform

Problem:

Find the inverse Laplace transform of

$$F(s) = \frac{2s+5}{s^2+5s+6}$$

Solution

Known Quantities: Function to be inverse Laplace–transformed.

Find: $f(t) = \mathcal{L}^{-1}[F(s)]$.

Assumptions: None.

Analysis: A direct entry for the function cannot be found in Table B.1. In such cases, one must compute a *partial fraction expansion* of the function $F(s)$ and then individually transform each term in the expansion. A partial fraction expansion is the inverse operation of obtaining a common denominator and is illustrated below.

$$F(s) = \frac{2s+5}{s^2+5s+6} = \frac{A}{s+2} + \frac{B}{s+3}$$

To obtain the constants A and B, we multiply the above expression by each of the denominator terms:

$$(s+2)F(s) = A + \frac{(s+2)B}{s+3}$$

$$(s+3)F(s) = \frac{(s+3)A}{s+2} + B$$

From the above two expressions, we can compute A and B as follows:

$$A = (s+2)F(s)|_{s=-2} = \left.\frac{2s+5}{s+3}\right|_{s=-2} = 1$$

$$B = (s+3)F(s)|_{s=-3} = \left.\frac{2s+5}{s+2}\right|_{s=-3} = 1$$

Finally,

$$F(s) = \frac{2s+5}{s^2+5s+6} = \frac{1}{s+2} + \frac{1}{s+3}$$

and using Table B.1, we compute

$$f(t) = \left(e^{-2t} + e^{-3t}\right)u(t)$$

CHECK YOUR UNDERSTANDING

B.8 Find the inverse Laplace transform of each of the following functions:

a. $F(s) = \dfrac{1}{s^2+5s+6}$ b. $F(s) = \dfrac{s-1}{s(s+2)}$

c. $F(s) = \dfrac{3s}{(s^2+1)(s^2+4)}$ d. $F(s) = \dfrac{1}{(s+2)(s+1)^2}$

Answer: a. $f(t) = (e^{-2t} - e^{-3t})u(t)$; b. $f(t) = (\frac{3}{2}e^{-2t} - \frac{1}{2})u(t)$; c. $f(t) = (\cos t - \cos 2t)u(t)$; d. $f(t) = (e^{-2t} + te^{-t} - e^{-t})u(t)$

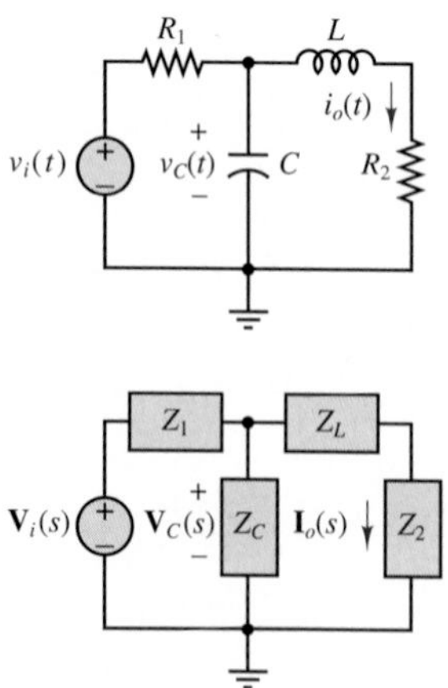

Figure B.2 A circuit and its Laplace transform domain equivalent

Transfer Functions, Poles, and Zeros

It should be clear that the Laplace transform can be quite a convenient tool for analyzing the transient response of a circuit. The Laplace variable s is an extension of the steady-state frequency response variable $j\omega$ already encountered in this appendix. Thus, it is possible to describe the input-output behavior of a circuit by using Laplace transform ideas in the same way in which we used frequency response ideas earlier. Now we can define voltages and currents in the complex frequency domain as $\mathbf{V}(s)$ and $\mathbf{I}(s)$, and we denote impedances by the notation $Z(s)$, where s replaces the familiar $j\omega$. We define an extension of the frequency response of a circuit, called the *transfer function,* as the ratio of any input variable to any output variable, i.e.,

$$H_1(s) = \frac{\mathbf{V}_o(s)}{\mathbf{V}_i(s)} \quad \text{or} \quad H_2(s) = \frac{\mathbf{I}_o(s)}{\mathbf{V}_i(s)} \quad \text{etc.} \qquad \textbf{(B.14)}$$

As an example, consider the circuit of Figure B.2. We can analyze it by using a method analogous to phasor analysis by defining impedances

$$Z_1 = R_1 \qquad Z_C = \frac{1}{sC} \qquad Z_L = sL \qquad Z_2 = R_2 \qquad \textbf{(B.15)}$$

Then we can use mesh analysis methods to determine that

$$\mathbf{I}_o(s) = \mathbf{V}_i(s)\frac{Z_C}{(Z_L + Z_2)Z_C + (Z_L + Z_2)Z_1 + Z_1 Z_C} \quad \textbf{(B.16)}$$

or, upon simplifying and substituting the relationships of equation B.15,

$$H_2(s) = \frac{\mathbf{I}_o(s)}{\mathbf{V}_i(s)} = \frac{1}{R_1 LCs^2 + (R_1 R_2 C + L)s + R_1 + R_2} \quad \textbf{(B.17)}$$

If we were interested in the relationship between the input voltages and, say, the capacitor voltage, we could similarly calculate

$$H_1(s) = \frac{\mathbf{V}_C(s)}{\mathbf{V}_i(s)} = \frac{sL + R_2}{R_1 LCs^2 + (R_1 R_2 C + L)s + R_1 + R_2} \quad \textbf{(B.18)}$$

Note that a transfer function consists of a *ratio of polynomials;* this ratio can also be expressed in factored form, leading to the discovery of additional important properties of the circuit. Let us, for the sake of simplicity, choose numerical values for the components of the circuit of Figure B.2. For example, let $R_1 = 0.5\ \Omega$, $C = \frac{1}{4}$ F, $L = 0.5$ H, and $R_2 = 2\ \Omega$. Then we can substitute these values into equation B.18 to obtain

$$H_1(s) = \frac{0.5s + 2}{0.0625s^2 + 0.375s + 2.5} = 8\left(\frac{s + 4}{s^2 + 6s + 40}\right) \quad \textbf{(B.19)}$$

Equation B.19 can be factored into products of first-order terms as follows:

$$H_1(s) = 8\left[\frac{s + 4}{(s - 3.0000 + j5.5678)(s - 3.0000 - j5.5678)}\right] \quad \textbf{(B.20)}$$

where it is apparent that the response of the circuit has very special characteristics for three values of s: $s = -4$; $s = +3.0000 + j5.5678$; and $s = +3.0000 - j5.5678$. In the first case, at the complex frequency $s = -4$, the numerator of the transfer function becomes zero, and the response of the circuit is zero, regardless of how large the input voltage is. We call this particular value of s a **zero** of the transfer function. In the latter two cases, for $s = +3.0000 \pm j5.5678$, the response of the circuit becomes infinite, and we refer to these values of s as **poles** of the transfer function.

It is customary to represent the response of electric circuits in terms of poles and zeros, since knowledge of the location of these poles and zeros is equivalent to knowing the transfer function and provides complete information regarding the response of the circuit. Further, if the poles and zeros of the transfer function of a circuit are plotted in the complex plane, it is possible to visualize the response of the circuit very effectively. Figure B.3 depicts the pole–zero plot of the circuit of Figure B.2; in plots of this type it is customary to denote zeros by a small circle and poles by an "×."

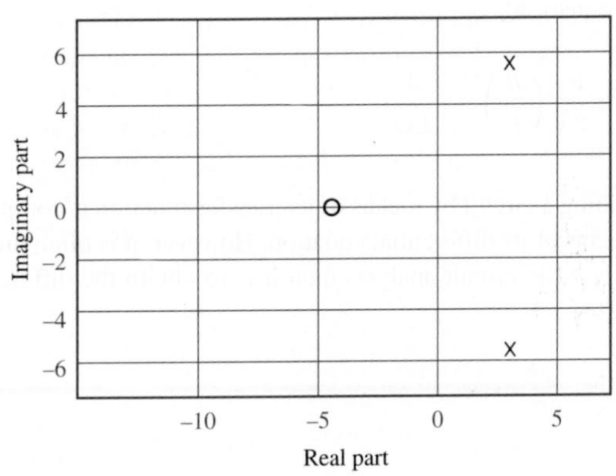

Figure B.3 Zero–pole plot for the circuit of Figure B.2

The poles of a transfer function have a special significance, in that they are equal to the roots of the natural response of the system. They are also called the **natural frequencies** of the circuit. Example B.6 illustrates this point.

EXAMPLE B.6 Poles of a Second-Order Circuit

Problem:

Determine the poles of a parallel *RLC* circuit. Express the homogeneous equation using i_L as the independent variable.

Solution:

Known Quantities: Values of resistor, inductor, and capacitor.

Find: Poles of the circuit.

Assumptions: None.

Analysis: The differential equation describing the natural response of the parallel *RLC* circuit is

$$\frac{d^2 i}{dt^2} + \frac{R}{L}\frac{di}{dt} + \frac{1}{LC} i = 0$$

with characteristic equation given by

$$s^2 + \frac{R}{L}s + \frac{1}{LC} = 0$$

Now, let us determine the transfer function of the circuit, say, $\mathbf{V}_L(s)/\mathbf{V}_S(s)$. Applying the voltage divider rule, we can write

$$\frac{\mathbf{V}_L(s)}{\mathbf{V}_S(s)} = \frac{sL}{1/sC + R + sL}$$

$$= \frac{s^2}{s^2 + (R/L)s + 1/LC}$$

The denominator of this function, which determines the poles of the circuit, is identical to the characteristic equation of the circuit: The poles of the transfer function are identical to the roots of the characteristic equation!

$$s_{1,2} = -\frac{R}{2L} \pm \frac{1}{2}\sqrt{\left(\frac{R}{L}\right)^2 - \frac{4}{LC}}$$

Comments: Describing a circuit by means of its transfer function is completely equivalent to representing it by means of its differential equation. However, it is often much easier to derive a transfer function by basic circuit analysis than it is to obtain the differential equation of a circuit.

APPENDIX C

FUNDAMENTALS OF ENGINEERING (FE) EXAMINATION

C.1 INTRODUCTION

The *Fundamentals of Engineering* (FE) examination[1] is one of four steps to be completed toward registering as a Professional Engineer (PE). Each of the 50 states in the United States has laws that regulate the practice of engineering; these laws are designed to ensure that registered professional engineers have demonstrated sufficient competence and experience. The same exam is administered at designated times throughout the country, but each state's Board of Registration administers the exam and supplies information and registration forms. **Additional information is available on the NCEES website.**

FIND IT ON THE WEB

Four steps are required to become a Professional Engineer:

1. *Education.* Usually this requirement is satisfied by completing a B.S. degree in engineering from an accredited college or university.
2. *Fundamentals of Engineering examination.* One must pass an 8-h examination described in Section C.2.

[1]This exam used to be called *Engineer in Training*.

3. *Experience.* Following successful completion of the Fundamentals of Engineering examination, 2 to 4 years of engineering experience are required.
4. *Principles and practices of engineering examination.* One must pass a second 8-h examination, also known as the Professional Engineer (PE) examination, which requires in-depth knowledge of one particular branch of engineering.

This appendix provides a review of the background material in electrical engineering required in the Electrical Engineering part of the FE exam. This exam is prepared by the National Council of Examiners for Engineering and Surveying[2] (NCEES).

C.2 EXAM FORMAT AND CONTENT

FIND IT ON THE WEB

The FE exam is a two-part national examination, administered by the **National Council of Examiners for Engineers and Surveyors** (NCEES) and given twice a year (in April and October). The exam is divided into two four-hour sessions, consisting of 120 questions in the four-hour morning session, and 60 questions in the four-hour afternoon session. The morning session covers general background in twelve different areas, one of which is *Electricity and Magnetism.* The afternoon session requires the examinee to choose among seven modules – Chemical, Civil, Electrical, Environmental, Industrial, Mechanical and Other/General engineering.

C.3 THE ELECTRICITY AND MAGNETISM SECTION OF THE MORNING EXAM

The *Electricity and Magnetism* part of the morning session consists of approximately 9% of the morning session, and covers the following topics:

A. Charge, energy, current, voltage, power
B. Work done in moving a charge in an electric field (relationship between voltage and work)
C. Force between charges
D. Current and voltage laws (Kirchhoff, Ohm)
E. Equivalent circuits (series, parallel)
F. Capacitance and inductance
G. Reactance and impedance, susceptance and admittance
H. AC circuits
I. Basic complex algebra

The remainder of Appendix C contains a review of the electrical circuits portion of the FE examination, including references to the relevant material in the book. In addition, Appendix C also contains a collection of sample problems – some including a full explanation of the solution, some with answers supplied separately.

C.4 REVIEW FOR THE ELECTRICITY AND MAGNETISM SECTION OF THE MORNING EXAM

A. Charge, energy, current, voltage and power.

B. Work done in moving a charge in an electric field (relationship between voltage and work)

[2]P.O. Box 1686 (1826 Seneca Road), Clemson, SC 29633-1686.

C. Force between charges

The basic definitions of these quantities and relevant examples can be found in Chapter 2 and in any undergraduate physics textbook. The following example problems will further assist you in reviewing the material.

CHECK YOUR UNDERSTANDING

A.1 Determine the total charge entering a circuit element between $t = 1$ s and $t = 2$ s if the current passing through the element is $i = 5t$.

A.2 A light bulb sees a 3-A current for 15 s. The light bulb generates 3 kJ of energy in the form of light and heat. What is the voltage drop across the light bulb?

A.3 How much energy does a 75-W electric bulb consume in six hours?

B.1 Find the voltage drop v_{ab} required to move a charge q from point a to point b if $q = -6$ C and it takes 30 J of energy to move the charge.

C.1 Two 2-C charges are separated by a dielectric with thickness of 4 mm, and with dielectric constant $\varepsilon = 10^{-12}$ F/m. What is the force exerted by each charge on the other?

C.2 The magnitude of the force on a particle of charge q placed in the empty space between two infinite parallel plates with a spacing d and a potential difference V is proportional to:

(a) qV/d^2 (b) qV/d (c) qV^2/d (d) q^2V/d (e) q^2V^2/d

Answers: **A.1:** $q = \int_{t=1}^{t=2} i\,dt = \int_{t=1}^{t=2} 5t\,dt = \left(\frac{5t^2}{2}\right)\Big|_{t=1}^{t=2} = 7.5$ C

A.2: The total charge is $\Delta q = i\,\Delta t = 3 \times 15 = 45$ C. The voltage drop is

$$v = \frac{\Delta w}{\Delta q} = \frac{3 \times 10^3}{45} = 66.67 \text{ V}$$

A.3: The energy used
is$w = pt = 75\,[W] \times 6\,[h] = 75\,[W] \times 6 \times 3600\,[s]\,450 = 1.62$ MJ

B.1 The voltage drop is $v_{ab} = \frac{w}{q} = \frac{30}{-6} = -5$ V

C.1: $F = \frac{q_1 q_2}{4\pi\varepsilon r^2} = \frac{(2 \times 10^{-3}) \times (2 \times 10^{-3})}{4\pi \times 10^{-12} \times (4 \times 10^{-3})^2} = 2 \times 10^{10}$ N

C.2: Answer is (a), since this is the only term that has a distance squared term in the denominator.

D. Current and Voltage Laws (Kirchhoff, Ohm)

The material related to this section is covered in Chapter 2, sections 2, 3, 4 and 6. Examples 2.3, 2.4, 2.6, 2.7, 2.8, 2.9, 2.10, and the related *Check Your Understanding exercises* will help you review the necesssary material.

E. Equivalent circuits

The analysis of DC circuits forms the foundation of electrical engineering. Chapters 2 and 3 cover this material with a wealth of examples. The following exercises illustrate the general type of questions that might be encountered in the FE exam.

CHECK YOUR UNDERSTANDING

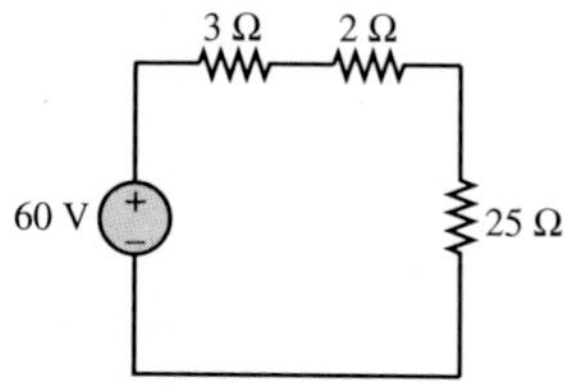

Figure E.1

E.1 Assuming the connecting wires and the battery have negligible resistance, the voltage across the 25-Ω resistance in Figure E.1 is

a. 25 V b. 60 V c. 50 V d. 15 V e. 12.5 V

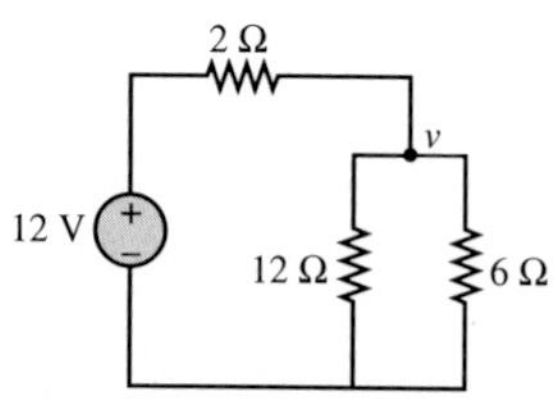

Figure E.2

E.2 Assuming the connecting wires and the battery have negligible resistance, the voltage across the 6-Ω resistor in Figure E.2 is

a. 6 V b. 3.5 V c. 12 V d. 8 V e. 3 V

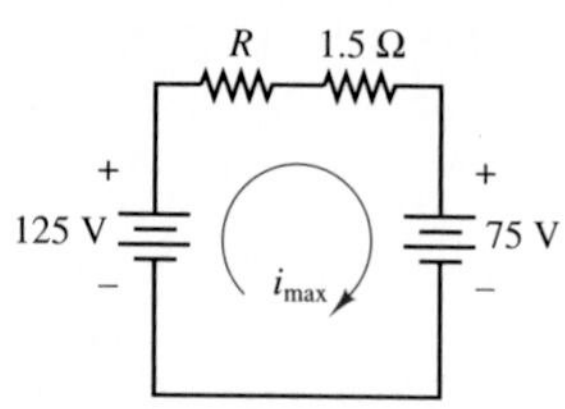

Figure E.3

E.3 A 125-V battery charger is used to charge a 75-V battery with internal resistance of 1.5 Ω. If the charging current is not to exceed 5 A, the minimum resistance in series with the charger must be

a. 10 Ω b. 5 Ω c. 38.5 Ω d. 41.5 Ω e. 8.5 Ω

Answers: **E.1:** This problem calls for application of the voltage divider rule, discussed in Section 2.6. Applying the voltage divider rule to the circuit of Figure E.2, we have

$$v_{25\,\Omega} = 60\left(\frac{25}{3+2+25}\right) = 50\text{ V}$$

Thus, the answer is c.

E.2: This problem can be solved most readily by applying nodal analysis (Section 3.1), since one of the node voltages is already known. Applying KCL at the node v, we obtain

$$\frac{12-v}{2} = \frac{v}{6} + \frac{v}{12}$$

This equation can be solved to show that $v = 8$ V. Note that it is also possible to solve this problem by mesh analysis (Section 3.2). You are encouraged to try this method as well.

E.3: The circuit of Figure E.3 describes the charging arrangement. Applying KVL to the circuit of Figure E.3, we obtain

$$i_{max}R + 1.5i_{max} - 125 + 75 = 0$$

and using $i = i_{max} = 5$ A, we can find R from the following equation:

$$5R + 7.5 - 125 + 75 = 0$$

$$R = 8.5\ \Omega$$

Thus, e is the correct answer.

F. Capacitance and Inductance

The material on capacitance and inductance pertains to two basic areas: energy storage in these elements and transient response of the circuits containing these elements. The exercises below deal with the former part (covered in Chapter 4); the latter is covered under the heading "Transients" in this appendix and in Chapter 5 of the book.

CHECK YOUR UNDERSTANDING

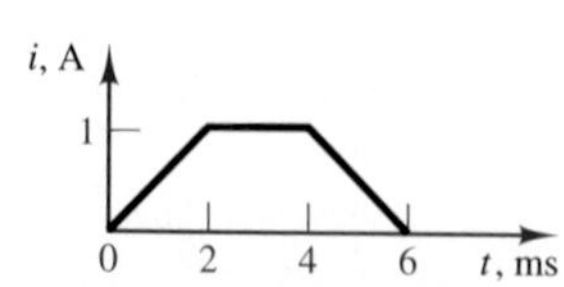

Figure F.1

F.1 A coil with inductance of 1 H and negligible resistance carries the current shown in Figure F.1. The maximum energy stored in the inductor is

a. 2 J b. 0.5 J c. 0.25 J d. 1 J e. 0.2 J

F.2 The maximum voltage that will appear across the coil is

a. 5 V b. 100 V c. 250 V d. 500 V e. 5,000 V

Answers: **F.1:** The energy stored in an inductor is $W = \frac{1}{2}Li^2$ (see Section 4.1). Since the maximum current is 1 A, the maximum energy will be $W_{max} = \frac{1}{2}Li^2_{max} = \frac{1}{2}$ J. Thus, b is the correct answer.

F.2: Since the voltage across an inductor is given by $v = L(di/dt)$, we need to find the maximum (positive) value of di/dt. This will occur anywhere between $t = 0$ and $t = 2$ ms. The corresponding slope is

$$\left.\frac{di}{dt}\right|_{max} = \frac{1}{2 \times 10^{-3}} = 500$$

Therefore $v_{max} = 1 \times 500 = 500$ V, and the correct answer is d.

G. Reactance and impedance, susceptance and admittance

The material related to this section is covered in Chapter 4, section 4. Examples 4.12, 4.13. 4.14. 4.15, and the related *Check Your Understanding* exercises will help you review the necessary material.

H. AC circuits

The material related to basic AC circuits is covered in Chapter 4, sections 2 and 4. Examples 4.8, 4.9, 4.16, 4.17. 4.18. 4.19, 4.20, 4.21 and the related *Check Your Understanding* exercises will help you review the necessary material. In addition, material on AC power may be found in Chapter 7, sections 1 and 2. Examples 7.1 through 7.11 and the accompanying *Check Your Understanding* exercises will provide additional review material. The rest of this section offers a number of sample FE Exam problems.

CHECK YOUR UNDERSTANDING

H.1 A voltage sine wave of peak value 100 V is in phase with a current sine wave of peak value 4 A. When the phase angle is 60° later than a time at which the voltage and the current are both zero, the instantaneous power is most nearly

a. 300 W b. 200 W c. 400 W d. 150 W e. 100 W

H.2 A sinusoidal voltage whose amplitude is $20\sqrt{2}$ V is applied to a 5-Ω resistor. The root-mean-square value of the current is

a. 5.66 A b. 4 A c. 7.07 A d. 8 A e. 10 A

H.3 The magnitude of the steady-state root-mean-square voltage across the capacitor in the circuit of Figure H.1 is

a. 30 V b. 15 V c. 10 V d. 45 V e. 60 V

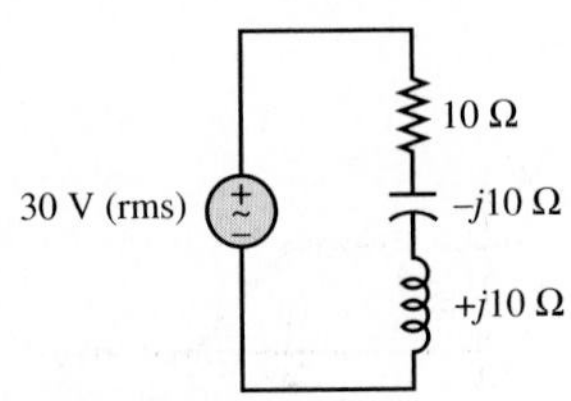

Figure H.1

Answers: **H.1:** As discussed in Section 7.1, the instantaneous AC power $p(t)$ is

$$p(t) = \frac{VI}{2}\cos\theta + \frac{VI}{2}\cos(2\omega t + \theta_V + \theta_I)$$

In this problem, when the phase angle is 60° later than a "zero crossing," we have $\theta_V = \theta_I = 0$, $\theta = \theta_V - \theta_I = 0$, $2\omega t = 120°$. Thus, we can compute the power at this instant as

$$p = \frac{100 \times 4}{2} + \frac{100 \times 4}{2}\cos(120°) = 300 \text{ W}$$

The correct answer is a.

H.2: From Section 4.2, we know that

$$V_{\text{rms}} = \frac{V}{\sqrt{2}} = \frac{20\sqrt{2}}{\sqrt{2}} = 20 \text{ V}$$

Thus, $I_{\text{rms}} = 20/5 = 4$ A. Therefore, b is the correct answer.

H.3: This problem requires the use of impedances (Section 4.4). Using the voltage divider rule for impedances, we write the voltage across the capacitor as

$$\mathbf{V} = 30\angle 0° \times \frac{-j10}{10 - j10 + j10}$$
$$= 30\angle 0° \times (-j1) = 30\angle 0° \times 1\angle -90° = 30\angle 90°$$

Thus, the rms amplitude of the voltage across the capacitor is 30 V, and a is the correct answer. Note the importance of the phase angle in this kind of problem.

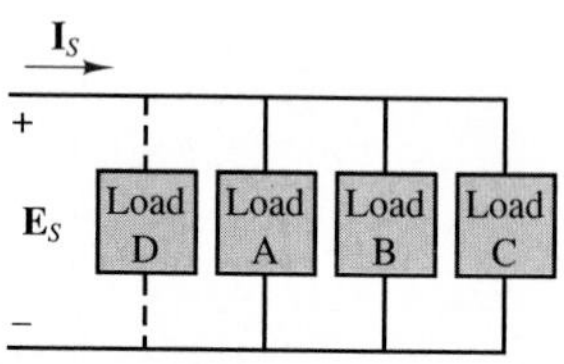

Figure H.2

The next set of questions (Exercises H.4 to H.8) pertain to single-phase AC power calculations and refer to the single-phase electrical network shown in Figure H.2. In this figure, $\mathbf{E}_S = 480\angle 0°$ V; $\mathbf{I}_S = 100\angle -15°$ A; $\omega = 120\pi$ rad/s. Further, load A is a bank of single-phase induction machines. The bank has an efficiency η of 80 percent, a power factor of 0.70 lagging, and a load of 20 hp. Load B is a bank of overexcited single-phase synchronous machines. The machines draw 15 kVA, and the load current leads the line voltage by 30°. Load C is a lighting (resistive) load and absorbs 10 kW. Load D is a proposed single-phase capacitor that will correct the source power factor to unity. This material is covered in Sections 7.1 and 7.2.

CHECK YOUR UNDERSTANDING

H.4 The root-mean-square magnitude of load A current, denoted by I_A, is most nearly

a. 44.4 A b. 31.08 A c. 60 A d. 38.85 A e. 55.5 A

H.5 The phase angle of $\mathbf{I}_A$ with respect to the line voltage $\mathbf{E}_S$ is most nearly

a. 36.87° b. 60° c. 45.6° d. 30° e. 48°

H.6 The power absorbed by synchronous machines is most nearly

a. 20,000 W b. 7,500 W c. 13,000 W d. 12,990 W e. 15,000 W

H.7 The power factor of the system before load D is installed is most nearly

a. 0.70 lagging b. 0.866 leading c. 0.866 lagging
d. 0.966 leading e. 0.966 lagging

H.8 The capacitance of the capacitor that will give a unity power factor of the system is most nearly

a. 219 μF b. 187 μF c. 132.7 μF d. 240 μF e. 132.7 pF

Answers: **H.4:** The output power P_o of the single-phase induction motor is $P_o = 20 \times 746 = 14{,}920$ W. The input electric power P_{in} is

$$P_{in} = \frac{P_o}{\eta} = \frac{14{,}920}{0.80} = 18{,}650 \text{ W}$$

P_{in} can be expressed as

$$P_{in} = E_S I_A \cos\theta_A$$

Therefore, the rms magnitude of the current $\mathbf{I}_A$ is found as

$$I_A = \frac{P_{in}}{E_S \cos\theta_A} = \frac{18{,}650}{480 \times 0.70} = 55.5015 \approx 55.5 \text{ A}$$

Thus, the correct answer is e.
H.5: The phase angle between $\mathbf{I}_A$ and $\mathbf{E}_S$ is

$$\theta = \cos^{-1} 0.70 = 45.57^\circ \approx 45.6^\circ$$

The correct answer is c.
H.6: The apparent power S is known to be 15 kVA, and θ is 30°. From the power triangle, we have

$$P = S\cos\theta$$

Therefore, the power drawn by the bank of synchronous motors is

$$P = 15{,}000 \times \cos 30^\circ = 12{,}990.38 \approx 12.99 \text{ kW}$$

The answer is d.
H.7: From the expression for the current $\mathbf{I}_S$, we have

$$\text{pf} = \cos\theta = \cos[0^\circ - (-15^\circ)] = \cos 15^\circ = 0.966 \text{ lagging}$$

The correct answer is e.
H.8: The reactive power Q_A in load A is

$$Q_A = P_A \times \tan\theta_A$$
$$\theta_A = \cos^{-1} 0.70 = 45.57^\circ$$

Therefore,

$$Q_A = 18{,}650 \times \tan 45.57^\circ = 19{,}025 \text{ VAR}$$

The total reactive power Q_B in load B is

$$Q_B = S \times \sin\theta_B = 15{,}000 \times \sin(-30^\circ) = -7{,}500 \text{ VAR}$$

The total reactive power Q is

$$Q = Q_A + Q_B = 19{,}025 - 7{,}500 = 11{,}525 \text{ VAR}$$

To cancel this reactive power, we set

$$Q_C = -Q = -11{,}525 \text{ VAR}$$

and

$$Q_C = -\frac{E_S^2}{X_C} \quad \text{and} \quad X_C = -\frac{1}{\omega C}$$

Therefore, the capacitance required to obtain a power factor of unity is

$$C = -\frac{Q_C}{\omega E_S^2} = \frac{11{,}525}{120\pi \times 480^2} = 132.7\ \mu\text{F}$$

The correct answer is c.

I. Basic complex algebra

A review of complex algebra is contained in Appendix A, section A.2. The examples and exercises included in this section will provide the needed review material.

APPENDIX

D

ANSWERS TO SELECTED PROBLEMS

Chapter 1

1.1 *Bathroom*

- Ventilation fan
- Electric toothbrush
- Hair dryer
- Electric shaver
- Electric heater fan

Kitchen

- Microwave fan
- Microwave turntable
- Mixer
- Food processor
- Blender
- Coffee grinder
- Garbage disposal
- Ceiling fan
- Electric clock
- Exhaust fan
- Refrigerator compressor
- Dishwasher

Utility Room
- Clothes washer
- Dryer
- Air conditioner
- Furnace blower
- Pump

Family Room
- VCR drive
- Cassette tape drive
- Treadmill

Miscellaneous
- Lawn tools
- Power tools

Chapter 2

2.3 a. 360,000 C b. 224.7×10^{22}

2.5 a. 15.12 MJ b. 1.26 MC

2.26 a. 2.53 KW b. 218.6 MJ c. $ 3.64

2.27 7 V; 7 mW; 58.33%

2.33 a. 220.4 Ω b. 2.009%

2.35 $P_{B60} = 23.44$ W; $P_{B100} = 14.06$ W

2.51 $I = 6.09$ A

2.64 With terminals *c-d* open, $R_{eq} = 400\ \Omega$; with terminals *c-d* shorted, $R_{eq} = 390\ \Omega$; with terminals *a-b* open, $R_{eq} = 360\ \Omega$; with terminals *a-b* shorted, $R_{eq} = 351\ \Omega$

2.65 AWG no. 8 or larger wire must be used.

2.66 $R_{eq} = 4.288$ kΩ

2.69 $V_{AB} = -11.999991$ V

2.71 a. A series resistor to drop excess voltage is required. b. $R_{series} = 800\ \Omega$ c. 20–100 kPa.

2.79 $F = 19.97$ kN

Chapter 3

3.7 $i_1 = -0.5$ A; $i_2 = -0.5$ A

3.9 $i = 0.491$ A

3.25 $v = 1.09$ V

3.28 $i = 0.491$ A

3.31 $A_V = \dfrac{v_2}{v_1} = -0.1818$

3.33 $V_{R1} = -36.39$ V; $V_{R2} = -109.2$ V; $V_{R3} = 72.81$ V; $V_{R4} = 0$; $V_F = 151.4$ V

3.37
$$I_{R1} = I_1 = \frac{\begin{vmatrix} V_{S1} + V_{S2} & -(R_2 + R_{W2}) \\ -V_{S2} + V_{S3} & R_2 + R_{W2} + R_3 + R_{W3} \end{vmatrix}}{\begin{vmatrix} R_1 + R_{W1} + R_2 + R_{W2} & -(R_2 + R_{W2}) \\ -(R_2 + R_{W2}) & R_2 + R_{W2} + R_3 + R_{W3} \end{vmatrix}}$$

3.40 $I_{R1-2} = 60.02$ mA

3.44 $I_{R1} = 54.25$ A

3.70 $\Delta V_o = -16.33$ V

3.74 a. $R_L = R_{eq} = 600\ \Omega$ b. $P_{RL} = 510.4$ mW c. $\eta = 50\%$

3.77 a. $I = 52.2$ mA; $V = 4.57$ V b. $R_{inc} = 43.8\ \Omega$; c. $I = 73$ mA; $V = 5.40$ V; $R_{inc} = 37\ \Omega$

3.80 $I_D = 12.5$ mA

Chapter 4

4.1 $v_L(t) = 377 \sin\left(377t - \dfrac{5\pi}{6}\right)$ V

4.4 For $-\infty < t < 0$, $w_L(t) = 0$ J; For $0 \le t < 10$ s, $w_L(t) = t^2$ J
For $10\text{ s} \le t < \infty$, $w_L(t) = 100$ J

4.7 For $-\infty < t < 0$, $w_C(t) = 0$ J; For $0 \le t < 10$ s, $w_C(t) = 0.05t^2$ J
For $10\text{ s} \le t < \infty$, $w_C(t) = 5$ J

4.12 For $0 < t \le 5$ ms, $i_C = -480$ mA; For 5 ms $< t < 10$ ms, $i_C = 0$; For $t > 10$ ms, $i_C = 0$

4.14 $i_L(t = 15\ \mu\text{s}) = -13.67$ mA

4.20 $C = 0.5\ \mu$F

4.29 $x_{\text{rms}} = 2.87$

4.35 $i_{\text{rms}} = 1.15$ A

4.48 $Z_{\text{eq}} = 7.05\angle 0.9393$

4.51 $i_3(t) = 20\cos(377t - 11°)$ A

4.53 $i_s(t) = 22.22\cos(\omega t + 0°)$ mA

4.59 $i_R(t) = 157\cos(200\pi t + 99.04°)$ mA

4.61 $Z = 1\ \Omega$

4.75 a. $X_4 = \dfrac{R_1 R_2}{1/\omega C_3 - \omega L_3}$ b. $C = 448\ \mu$F c. $\omega = \dfrac{1}{\sqrt{L_3 C_3}} = 1473\ \dfrac{\text{rad}}{\text{s}}$

4.77 $v_T(t) = 297\cos(100t + 21.8°)$ V

4.81 a. $-v_S + i_1 R_S + v_c(0) + \dfrac{1}{C}\displaystyle\int_0^t (i_1 - i_2)\,dt + (i_1 - i_2)R_1 = 0$

$(i_2 - i_1)R_1 - v_c(0) + \dfrac{1}{C}\displaystyle\int_0^t (i_2 - i_1)\,dt + L\dfrac{di_2}{dt} + i_2 R_2 = 0$

b. $-\mathbf{V}_S + \mathbf{I}_1 R_S + (\mathbf{I}_1 - \mathbf{I}_2)Z_C + (\mathbf{I}_1 - \mathbf{I}_2)R_1 = 0$

$(\mathbf{I}_2 - \mathbf{I}_1)R_1 + (\mathbf{I}_2 - \mathbf{I}_1)Z_C + \mathbf{I}_2 Z_L + \mathbf{I}_2 R_2 = 0$

4.83 $\mathbf{V}_{bc} = 779.5\angle 5.621°$ V

4.85 $L_1 = 98.29$ mH; $C_2 = 112.2\ \mu$F; $L_3 = 77.37$ mH; $R_3 = 29.11\ \Omega$

Chapter 5

5.21 The actual steady-state current through the inductor is larger than the current specified. Therefore, the circuit is not in a steady-state condition just before the switch is opened.

5.23 $i_C(0^-) = 0$; $i_C(0^+) = 17.65$ mA

5.25 $V_{R3}(0^+) = -4.080$ V

5.27 $i_C(0^+) = 3.234$ mA

5.29 $\tau = 5.105\ \mu$s

5.31 $\tau = 0.3750\ \mu$s

5.33 $i_{R3}(t) = 0.058 - 0.28e^{-t/69.7\times10^{-9}}$ A

5.35 $R_1 = 0.5170\ \Omega$; $L = 22.11$ H

5.37 $i_C(0^-) = 0$; $i_C(0^+) = 30$ A

5.39 $\tau = 3.433$ h

5.41 a. $V_C(0^+) = 0$ V b. $\tau = 48$ s c. $V_C(t) = \begin{cases} 8(1 - e^{-t/48}) & t \ge 0 \\ 0 & t < 0 \end{cases}$

d. $V_C(0) = 0$ V; $V_C(\tau) = 5.06$ V; $V_C(2\tau) = 6.9$ V; $V_C(5\tau) = 7.95$ V; $V_C(10\tau) = 8.0$ V

5.43 a. $V_C(0^+) = 11.67$ V

b. $V_C(t) = \begin{cases} V(0^+)e^{-t/\tau} = 11.67e^{-t/56} & 0 \le t < 3 \\ 11.67 - 0.61e^{-(t-3)/23.3} & t \ge 3 \end{cases}$

5.61 $i_L(\infty) = 31.03$ mA; $i_{RS2}(\infty) = 31.03$ mA; $V_C(\infty) = 0$

5.63 $i_L(\infty) = 18.3$ mA; $V_{R1}(\infty) = 42.04$ V; $V_C(\infty) = 42.04$ V

5.65 $i_L(\infty) = 226\ \mu$A; $V_C(\infty) = 4.98$ V

5.67 $i_L(0^+) = 146.4$ mA; $V_C(0^+) = 0$; $V_L(0^+) = -19.94$ V; $i_C(0^+) = -28.49$ mA; $i_{RS2}(0^+) = -99.79$ mA

5.69 $i(t) = 2 + e^{-0.041t}\{-3.641\cos[0.220(t-5)] + 1.77\sin[0.220(t-5)]\}$ A for $t \geq 5$ s

5.73 $L = 1.6\ \mu$H; $R = 0.56\ \Omega$

Chapter 6

6.9 a. As $\omega \to \infty$, $H_v \to 0\angle(-90°)$ b. $H_v(j\omega) = \dfrac{R_2}{R_1+R_2}\dfrac{1}{1+j\omega R_1R_2C/(R_1+R_2)}$ c. $\omega_c = 2.5$ krad/s, $H_o = 4.527$ dB

As $\omega \to 0$, $H_v \to \dfrac{R_2}{R_1+R_2}\angle 0°$

6.10 a. As $\omega \to 0$, $V_o \to 0$ b. $V_o = 1.806$ V $\angle 173.2°$

As $\omega \to \infty$, $V_o = \dfrac{V_iR_2}{R_1+R_2}$

c. $V_o = 4.303$ V $\angle 114.4°$ d. $V_o = 4.605$ V $\angle 75.05°$

6.50 The peak gain occurs at the resonant frequency of 4.6 rad/s. Bandwidth $\approx$ 2 rad/s.

6.52 a. low-pass b. high-pass c. high-pass d. low-pass

6.56 a. $H_v(j\omega) = \dfrac{R_L}{R_s+R_L}\dfrac{1}{1-jR_sR_L(1-\omega^2LC)/[(R_s+R_L)\omega L]}$ b. $\omega_r = 1.409$ Mrad/s

c. $\omega_{c1} = 1.405$ Mrad/s, $\omega_{c2} = 1.412$ Mrad/s d. BW $= 7.143$ krad/s; $Q = 197.2$

6.59 $H_v(j\omega) = \dfrac{R_L}{R_s+R_L}\dfrac{1}{1+j\omega L/[(R_s+R_L)(1-\omega^2LC)]}$

6.62 a. $H_v(j\omega) = \dfrac{R_cR_L}{R_s(R_c+R_L)+R_cR_L}\dfrac{1+j\dfrac{1}{R_c}\left(\omega L - \dfrac{1}{\omega C}\right)}{1+j\dfrac{(R_s+R_L)(\omega L - 1/\omega C)}{R_c(R_s+R_L)+R_sR_L}}$ b. As $\omega \to \infty$: $H_v(j\omega) \to \dfrac{R_L}{R_s+R_L}$

As $\omega \to 0$: $H_v(j\omega) \to \dfrac{R_L}{R_s+R_L}$

c. $\omega_r = 50$ Mrad/s d. $\omega_{c2} = 49.875$ Mrad/s; $\omega_{c3} = 50.125$ Mrad/s

6.81 $Z(j\omega) = \dfrac{R_c}{1-\omega^2LC}\dfrac{1+j\omega L/R_c}{1+j\omega R_cC/(1-\omega^2LC)}$

The circuit is a parallel resonant circuit and should exhibit maxima of impedance and minima of impedance at low and high frequencies.

1. At the resonant frequency, the impedance is real; i.e., the reactive part is zero. Set $f_1(\omega_r) = f_2(\omega_r)$ and solve for ω_r.
2. The magnitude of the impedance at the resonant frequency is Z_o evaluated at the resonant frequency: $Z_o = R_c/(1-\omega_r^2LC)$
3. The three cutoff (3-dB) frequencies are evaluated by making the functions of frequency equal to $+1$ or -1. $f_1(\omega_c) = 1 \Rightarrow \omega_{c3}$. $f_2(\omega_c) = \pm 1 \Rightarrow \omega_{c1}$ and ω_{c2}.
4. The magnitude of the impedance when the frequency is low can be determined in two ways. First, the circuit can be modeled at low frequencies by replacing the inductor with a short circuit and the capacitor with an open circuit. Under these conditions the impedance is equal to that of the resistor. Or the limit of the impedance as the frequency approaches zero can be determined. $Z_o = \lim_{\omega\to 0} Z(j\omega) = R_c$.

Chapter 7

7.39 $S_1 = 17.88 + j10.32$ kVA $= P_{av1} + jQ_1$
$S_2 = 9.56 + j1.17$ kVA $= P_{av2} + jQ_2$
$S_3 = 34.68 - j46.24$ kVA $= P_{av3} + jQ_3$

7.47 $N = \frac{1}{15}$

7.50 $R_c = 1.8\ \text{k}\Omega$; $L_c = 0.68$ H; $R_w = 0.9476\ \Omega$; $L_w = 0.348$ mH

7.51 $R_c = 106.38\ \text{k}\Omega$; $L_c = 17.46$ H; $R_w = 0.54\ \Omega$; $L_w = 0$

7.60 a. $\tilde{\mathbf{V}}_{RW} = 207.8\angle(-30°)$ V b. Same answer c. No difference
$\tilde{\mathbf{V}}_{WB} = 207.8\angle 90°$ V
$\tilde{\mathbf{V}}_{BR} = 207.8\angle(-150°)$ V

7.65 $\mathbf{I}_1 = 293\angle(-41.8°)$ A

7.66 $\mathbf{I}_R = 1.003\angle(-27.6°)$ A

7.67 $\mathbf{I}_1 = 98.1\angle(-30°)$ A; $\mathbf{I}_2 = 42.1\angle(-60°)$ A; $\mathbf{I}_3 = 26.8\angle(-180°)$ A

7.72 a. $P = 3541.3$ W b. $P_m = 2832.23$ W c. pf $= 0.64$ d. The company will face a 25% penalty.

Chapter 8

8.1 $G = 173.4$ dB

8.25 $A_v = -27.5$; $|A_v|_{\max} = 33.6$; $|A_v|_{\min} = 22.5$

8.26 $R_S = 10\ \text{k}\Omega$; $v_o(t) = 2 \times 10^{-3} \sin(\omega t)$ V

8.39 a. Use a voltage follower stage, followed by a differential amplifier stage. The differential amplifier must have voltage gain of 10.

$$A_{vy} = -\frac{R_2}{R_1} = -10;\ \text{Choose: } R_2 = 100\ \text{k}\Omega \qquad R_1 = 10\ \text{k}\Omega$$

$$A_{vr} = \frac{R_4(R_2 + R_1)}{R_1(R_4 + R_3)} = 10;\ \text{Choose: } R_3 = 10\ \text{k}\Omega \qquad R_4 = 100\ \text{k}\Omega$$

8.43 $R_3 \approx 102\ \text{k}\Omega$; $R_4 \approx 5\ \text{M}\Omega$

8.52 a. The gain (i.e., the magnitude of the transfer function) and output voltage increase continuously with frequency, at least until the output voltage tries to exceed the DC supply voltages and clipping occurs.
b. There is no specific cutoff frequency.
c. The filter behaves as a high-pass filter, in general. It in fact is a differentiator.

8.63 a. Applying KCL at the inverting terminal, $\dfrac{v_{\text{out}}}{v_{\text{in}}} = -\dfrac{1 + j\omega R_2 C}{j\omega R_1 C}$
b. Gain $= 0.043$ dB c. Gain $= 0.007$ dB; phase $= 177.71°$ d. $\omega > 196.5$ rad/s

8.68 $v_{\text{out}}(t) = -RC\dfrac{dV_S(t)}{dt}$

8.70 a. $t = 104$ ms b. $t = 120$ ms

8.71 a. $v_{\text{out}}(t) = -2000 + 1.9839 \sin(2{,}000\pi t + 90.57°)$ V b. $v_{\text{out}}(t) = -20 + 1.41 \sin(2{,}000\pi t + 135°)$ V
c. In order to have an ideal integrator, it is desirable to have $\tau \gg T$.

8.80 From the frequency response plot, one can approximate the minimum useful frequency for distortionless response to be (nominally) 1 Hz for the 10-MΩ case, 10 Hz for the 1-MΩ case, and 100 Hz for the 0.1-MΩ case. This amplifier cannot amplify direct current.

8.81 $\text{CMRR}_{\min} = 74$ dB

8.83 a. $A_{CL} = -99.899$ b. $A_{CL} = -990$ c. $A_{CL} = -9091$ d. As $A_{OL} \to \infty$, $A_{CL} = -10{,}000$

8.89 a. $v_{S\text{-}C} = 3.15\ \text{V} + 1.5\ \text{mV} \cos \omega t$; $v_{S\text{-}D} = -0.7\ \text{V} + 17\ \text{mV} \cos \omega t$
b. $A_{vc} = -3$; $A_{vd} = -11.5$
c. $v_{O\text{-}C} = -9.450 - 4.5 \times 10^{-3} \cos \omega t$ V; $v_{O\text{-}D} = 8.050 - 195.5 \times 10^{-3} \cos \omega t$ V
d. CMRR $= -11.67$ dB

Chapter 9

9.2 a. N type
b. Majority = conduction band free electrons = negative carriers. Minority = valence band holes = positive carriers.
$n_{no} \approx 4.910^{18}\ \text{m}^{-3}$, $p_{no} = 4.5910^{13}\ \text{m}^{-3}$

9.19 a. $R = 860\ \Omega$ b. $E_{min} = 1.56$ V

9.21 Reverse, forward, reverse, forward, forward

9.23 a. D_2 and D_4 are forward-biased; D_1 and D_3 are reverse-biased. $v_{out} = -5 + 0.7 = -4.3$ V
b. D_1 and D_2 are reverse-biased; D_3 is forward-biased. $v_{out} = -10 + 0.7 = -9.3$ V
c. D_1 is reverse-biased; D_2 is reverse-biased. $v_{out} = -10$ V

9.28 $V_{DQ} = V_{D\text{-on}} = 0.7$ V; $I_{DQ} = 1.0$ mA

9.32 $V_{S1\,min} = 3.7$ V

9.33 4.6 V

9.35 c. $v_{peak} = 157.2$ V e. $V_{in_{rms}} = 111.2$ V

9.39 a. -33.3 V
b. The actual peak reverse voltage (33.3 V) is greater than rated (30 V). Therefore, the diodes are not suitable for the specifications given.

9.41 $n = 0.31$; $C = 1{,}093\ \mu$F

9.44 a. -49.3 V
b. The diodes are not suitable because the rated and actual peak reverse voltages are too close.

9.47 $n = 0.04487$; $C = 1.023\ \mu$F

9.50 $R_{L_{min}} = 812.9\ \Omega$

9.52 $V_Z = 5$ V; $r_Z = 25\ \Omega$

9.54 $I_{Z\ max} = 32.6$ mA

9.58 $I_{L\ max} = 18.29$ mA; $I_{L\ min} = 1.07$ mA

9.65 For Figure P9.58(a):
a. At $t = t_1^-$, $I_{SW} = 0$; $I_S = I_B = 0.31$ A
b. At $t = t_1^+$, $I_S = 13$ A; $I_B = -0.96$ A; $I_{SW} = I_S - I_B = 13.96$ A
c. The battery voltage will drop quickly because of the small resistance in the circuit.

For Figure P9.58(b):
a. At $t = t_1^-$, we have $I_{SW} = 0$; $I_S = I_B = 0.25$ A
b. At $t = t_1^+$, $I_S = I_{SW} = 13$ A; $I_B = 0$
c. The battery will not be drained, because of the large reverse resistance of the diode.

Chapter 10

10.1 a. The transistor is in the active region. b. The transistor is in the cutoff region.
c. The transistor is in the saturation region. d. The transistor is in the active region.

10.3 The transistor is in the active region.

10.5 $I_E = -520\ \mu$A; $V_{CB} = 17.8$ V

10.6 The transistor is in the active region.

10.9 a. 170, 165, 143 b. $V_{CE} \approx 6.29$ V

10.11 $\beta \approx 150$

10.32 $V_{CEQ} = 10.55$ V. The transistor is in the active region.

10.34 $V_{CEQ} = 4.9$ V. The transistor is in the active region.

10.41

v_1	v_2	Q_1	Q_2	Q_3	v_{o1}	v_{o2}
0	0	Off	Off	On	0	5 V
0	5 V	Off	On	Off	5 V	0
5 V	0	On	Off	Off	5 V	0
5 V	5 V	On	On	Off	5 V	0

10.44

v_1	v_2	Q_1	Q_2	Q_3	v_{o1}	v_{o2}
0	0	Off	Off	On	0	5 V
0	5 V	Off	On	On	0	5 V
5 V	0	On	Off	On	0	5 V
5 V	5 V	On	On	Off	5 V	0

10.47 The answer is not unique. Choosing $\beta_2 = 10 \Rightarrow \beta_1 = 2.27$.

10.49 The circuit performs the function of a two-input NAND gate.

Chapter 11

11.1 a. Saturation b. Saturation c. Triode d. Saturation

11.3 a. Triode b. Triode or saturation c. Saturation

11.6 $i_D = 1.44$ mA

11.8 $i_D = 0.25$ mA

11.33

v_{in}	Q_1	Q_2	v_{out}
Low	Resistive	Open	High
High	Open	Resistive	Low

11.35

v_1	v_2	Q_1	Q_2	v_{out}
0	0	Off	Off	High
0	High	Off	On	High
High	0	On	Off	High
High	High	On	On	Low

Chapter 12

12.2 $R_S = \dfrac{V_Z + 0.6}{I}$

12.8 b. $\langle v_L \rangle = 2.841$ V

12.11 $P_m = 2.1$ kW; $P_S = 2{,}180$ W

12.14 $\phi = 23.66°$

12.16 Approximating the current by a triangular waveform, $I_L = \dfrac{1}{2\pi}\left(\dfrac{\phi i_{D-pk}}{2} + \dfrac{\phi i_{D-pk}}{2}\right)$; $i_{D-pk} = 1.280$ A

12.23 a. $\delta = 0.2063$; $t_1 = 825\ \mu$s b. $P_m = 750$ W c. $P_S = 3.094$ kW

12.25 $\langle v_o \rangle = \dfrac{t_1}{T} V_S = \dfrac{t_1}{t \cdot t_1} V_S = \dfrac{1}{t} V_S = \delta \cdot V_S$

12.27 $\langle V_L \rangle = 33.33$ V; $P_L = (\tilde{I}_L)^2 R$. The rms current can be computed given the duty cycle and the shape of the waveform.

Chapter 13

13.1 a. 191_{16}, 110010001_2 b. 111_{16}, 100010001_2 c. F_{16}, 1111_2 d. 26_{16}, 100110_2 e. 38_{16}, 111000_2

13.2 a 10_{10}, 1010_2 b. 102_{10}, 1100110_2 c. 71_{10}, 1000111_2 d. 33_{10}, 100001_2 e. 19_{10}, 10011_2

13.3 a. 100001111.01_2 b. 110101.011_2 c. 100101.01010_2 d. 110110.010001_2

13.4 a. F_{16}, 15_{10} b. $4D_{16}$, 77_{10} c. 65_{16}, 101_{10} d. $5C_{16}$, 92_{10} e. $1D_{16}$, 29_{10} f. 28_{16}, 40_{10}

13.5 a. 11111010 b. 100010100 c. 110000100

13.6 a. 11100 b. 1101110 c. 1000

13.12

A	B	C	BC	$B\overline{C}$	$\overline{B}A$	$BC + B\overline{C} + \overline{B}A$	$A + B$
1	1	1	1	0	0	1	1
1	1	0	0	1	0	1	1
1	0	1	0	0	1	1	1
1	0	0	0	0	1	1	1
0	1	1	1	0	0	1	1
0	1	0	0	1	0	1	1
0	0	1	0	0	0	0	0
0	0	0	0	0	0	0	0

13.15 $f(A, B, C, D) = ABC + CD(\overline{A} + \overline{B})$

13.18 $F = \overline{AB + CD + E}$

13.21 $f(A, B, C) = AB + BC + AC$

13.39 $F = (\overline{C} \cdot \overline{D})[A \cdot B \cdot C + (\overline{A} + \overline{B}) \cdot \overline{C}]$

13.42 $f(A, B, C) = \overline{A} \cdot \overline{B} \cdot C + \overline{A} \cdot B \cdot \overline{C} + A \cdot \overline{B} \cdot \overline{C} + A \cdot B \cdot C$

13.46 $F = \overline{B} \cdot \overline{D} + A \cdot \overline{D} + A \cdot B \cdot \overline{C} + \overline{A} \cdot B \cdot C \cdot D$

13.50 $F = \overline{B} + A \cdot \overline{C} + \overline{A} \cdot C$

13.53 $F = B \cdot \overline{D} + A \cdot C \cdot \overline{D}$

13.56 $F = \overline{B} \cdot \overline{D} + A \cdot \overline{D} + A \cdot B \cdot \overline{C} + \overline{A} \cdot B \cdot C \cdot D$

13.59 $F = \overline{A} \cdot \overline{C} \cdot D + \overline{A} \cdot B \cdot \overline{C} + A \cdot B \cdot C$

13.61 $F = A \cdot \overline{B} \cdot C \cdot D + B\overline{D} + \overline{A}B$

13.70 a.

x	y	C	S
0	0	0	0
0	1	0	1
1	0	0	1
1	1	1	0

b. Binary addition—S is the sum, and C is the carry.

13.72

Binary input $B_3B_2B_1B_0$	G_3	G_2 $B_2 \oplus B_3$	G_1 $B_1 \oplus B_2$	G_0 $B_0 \oplus B_1$
0000	0	0	0	0
0001	0	0	0	1
0010	0	0	1	1
0011	0	0	1	0
0100	0	1	1	0
0101	0	1	1	1
0110	0	1	0	1
0111	0	1	0	0
1000	1	1	0	0
1001	1	1	0	1
1010	1	1	1	1
1011	1	1	1	0
1100	1	0	1	0
1101	1	0	1	1
1110	1	0	0	1
1111	1	0	0	0

13.74

BC \ A	0	1
00	0	1
01	0	1
11	0	1
10	1	0

Chapter 14

14.1 The device is called a modulo-16 ripple counter. It can count clock pulses from 0 to $2^4 - 1$. The outputs divide the frequency by 2^1, 2^2, 2^3, and 2^4, respectively. Therefore, you can use this circuit as a divide-by-N counter, where N is 2, 4, 8, and 16.

14.3 The basic operation of the circuit is to count up when $X = 0$, and to count down when $X = 1$.

Clock	X	Output 2	T_1	Output 1
↑	0	0	0	No change
↑	1	1	0	No change
↑	0	1	1	Toggle
↑	1	0	1	Toggle

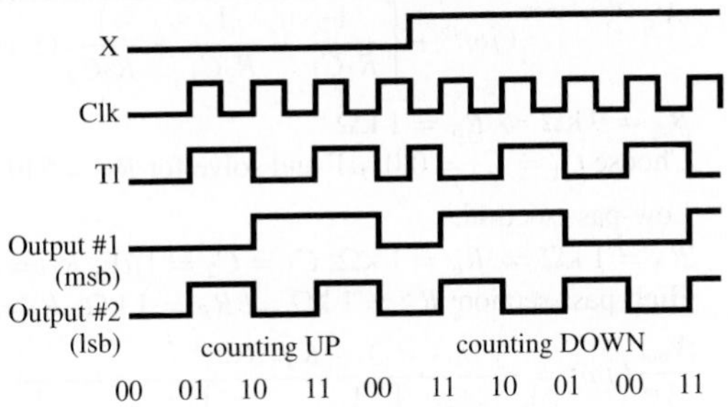

14.6

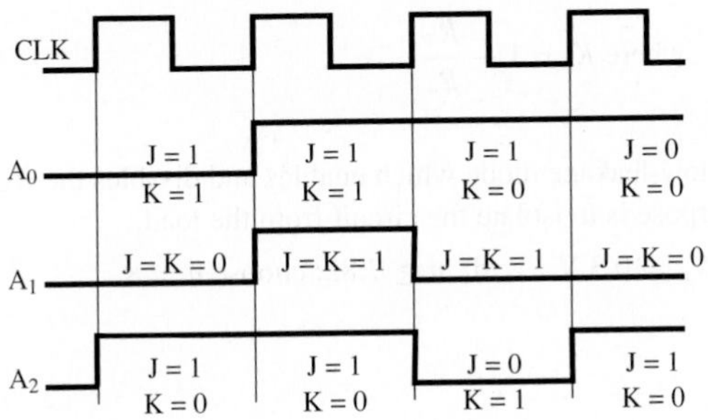

14.8

14.11

J_n	K_n	Q_{n+1}
1	1	$\overline{Q_n}$ (toggle)

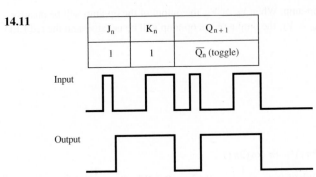

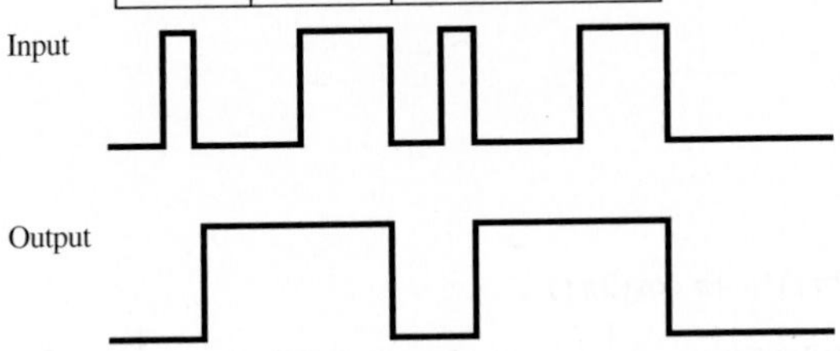

14.22 a. $n(n-1)$ b. $2n$

14.25 10,000 bytes/s

Chapter 15

15.4 $\Delta v_o = 1.5$ V

15.5 a. $V_{out} = 0.5 \times 10^{-6}\, HR_L$, hence varies linearly with H b. $R_L = 1333\ \Omega$

15.8 Measurement 1 exceeds the standard deviation σ probability $< 0.99 \Rightarrow$ roller speed will be adjusted.

15.10 (b) and (c) are precise, (a) and (d) are not.
(a) and (c) are accurate, (b) and (d) are not.

15.11 $A = 11$

15.17 $A_{dif} = 110$

15.19 The answer is not unique. We choose $R = 1\ \text{k}\Omega \Rightarrow R_2 = 89\ \text{k}\Omega$

15.23 $$\frac{V_{out}}{V_s}(j\omega) = \frac{K(1/R_1R_2C_1C_2)}{(j\omega)^2 + \left[\frac{1}{R_1C_1} + \frac{1}{R_2C_1} + \frac{1}{R_2C_2}(1-K)\right](j\omega) + \frac{1}{R_1R_2C_1C_2}} \quad \text{where } K = 1 + \frac{R_A}{R_B}$$

15.27 $R_A = 9\ \text{k}\Omega \Rightarrow R_B = 1\ \text{k}\Omega$
Choose $C_1 = C_2 = 0.01\mu\text{F}$ and solve for $R_1 = 540\ \Omega$ and $R_2 = 4.7\ \text{k}\Omega$. This answer is not unique.

15.31 Low-pass section:
$R_A = 1\ \text{k}\Omega \Rightarrow R_B = 1\ \text{k}\Omega$; $C_1 = C_2 = 1\mu\text{F}$; $R_1 = R_2 = 160\ \Omega$
High-pass section: $R_A = 1\ \text{k}\Omega \Rightarrow R_B = 1\ \text{k}\Omega$; $R_1 = R_2$ and $C_1 = C_2 = 1\mu\text{F}$; $R_1 = R_2 = 800\ \Omega$

15.33 $$\frac{V_{out}}{V_s}(j\omega) = \frac{K/(RC)^2}{(j\omega)^2 + \left[\frac{1}{RC}(3-K)\right](j\omega) + \frac{1}{(RC)^2}} \quad \text{where } K = 1 + \frac{R_3}{R_4}$$

15.41 Op-amp 1 is an input buffer. The JFET behaves as a low-leakage diode which enables and disables the RC holding circuit, and op-amp 2 is a voltage follower whose purpose is to isolate the circuit from the load.

15.43 a. $V_a = -3.6$ V b. $(V_a)_{max} = -4.5$ V c. $\delta V_a = 0.3$ V d. $n \geq 7.82$; choose $n = 8$

15.45 $v_O = -\dfrac{R_2V}{8R_1}(8b_3 + 4b_2 + 2b_1 + b_0)$

15.47 $R_f = 133.3\ \Omega$

15.53 Choose $n = 12$.

15.56 Choose $n = 12$. Resolution = 0.61 r/min

15.59 $f_{max} = \dfrac{f_s}{2} = 104.15$ Hz

15.63 9 bits

15.67 $R_1 = 100\ \text{k}\Omega$; $R_2 = 1.8\ \text{k}\Omega$; $R_3 = 16.2\ \text{k}\Omega$. The answer is not unique.

15.72 a. V_2 is the voltage at the inverting input of the op-amp. When $V_{in} > V_2$, the output of the op-amp will be positive and the green LED will turn on (go). When $V_{in} < V_2$, the output of the op-amp will be negative and the red LED will turn on (no go).
b. Choose $R_1 = 10\ \text{k}\Omega$ and $R_2 = 20\ \text{k}\Omega$

Chapter 18

18.3 $$\left[2(L - M_S)s + 1 + 2\frac{R_S}{R_L}\right]V_{out} = (M_1 - M_2)sI_L$$

18.7 a. $W_m = 1.25$ J; $L_\Delta = 2H$ b. $V_L(t) = \sin(2\pi t) + 4\pi\cos(2\pi t)$

18.11 $\phi_1 = \dfrac{Ni}{\Re_{g1}} = 1.26 \times 10^{-4}$ Wb; $B_1 = \dfrac{\phi_1}{A} = 1.26\ \text{Wb/m}^2$; $\phi_2 = \dfrac{1}{2}\phi_1 = 0.63 \times 10^{-4}$ Wb; $B_2 = \dfrac{1}{2}B_1 = 0.63\ \text{Wb/m}^2$

18.15 $i = 7.59$ A

18.17 a. $i = 1.257$ A b. Since the current is directly proportional to B, the current will have to be doubled.

18.19 a. $N = 5$ b. $\eta = 80.1\%$

18.21 a. $\alpha = 15$ b. $V_2 = 8$ V c. $R_L = 0.2\ \Omega$

18.25 a. $N_h = 920$ turns; $N_l = 96$ turns b. $I_h = 2$ A c. $\alpha = 0.1044$

18.28 a. $W_m = 1.568$ J b. $f = -224$ N

18.36 $f = 173.5$ N

18.37 a. $\Re = \dfrac{x}{\mu_0 w^2} + \dfrac{l_g}{\mu_0 w(w-x)}$ b. $W_m = \dfrac{(N_1+N_2)^2 i^2}{2\Re}$ c. $f = \dfrac{i^2}{2}\dfrac{(N_1+N_2)\left(\dfrac{x}{\mu_0 w^2} + \dfrac{l_g}{\mu_0 w(w-x)^2}\right)}{\left(\dfrac{x}{\mu_0 w^2} + \dfrac{l_g}{\mu_0 w(w-x)}\right)^2}$

18.41 $v = iR + \dfrac{N^2\mu_0 A}{2x}\dfrac{di}{dt}$; $m\dfrac{d^2x}{dt^2} + kx = \dfrac{N^2 i^2 \mu A}{4x^2}$

18.47 $f = 1.2l$ N force will be to the left if current flows upward.

18.51 For $k = 50{,}000$ N/m, the transfer function is $\dfrac{U}{V} = \dfrac{1.478\times 10^5 s}{s^3 + 3075s^2 + 9\times 10^6 s + 4\times 10^9}$

This response would correspond to a midrange speaker. For $k = 5\times 10^6$ N/m, the transfer function is
$\dfrac{U}{V} = \dfrac{1.478\times 10^5 s}{s^3 + 3075s^2 + 5.04\times 10^8 s + 4\times 10^{11}}$

This response would correspond to a treble range speaker. For $k = 0$, the transfer function is
$\dfrac{U}{V} = \dfrac{1.478\times 10^5}{s^2 + 3075s + 4\times 10^6}$

In this case, the speaker acts as a "woofer," emphasizing the low-frequency range.

Chapter 19

19.1 a. $P_e' = 8.75$ kW b. $P_e' = 10.8$ kW

19.4 $\phi = 3.2$ mWb

19.6 a. 250 V b. 23 kW

19.9 No-load: $V_L = E_b - i_a R_a = 4{,}820.4$ V; Half-load: $V_L = 4{,}810.7$ V

19.11 $V = 193$ V

19.14 $R_{\text{add}} = 2.15\ \Omega$

19.18 56.7 hp

19.22 a. $P_f = 2{,}988.7$ W; $\dfrac{P_f}{P_m} = 0.072 = 7.2\%$
b. $P_f = 1{,}770$ W; $P_R = 530$ W
c. $\%P_f = 4.05\%$; $\%P_R = 1.21\%$

19.25 $\omega_n = \sqrt{\dfrac{R_a b + K_{aPM}K_{TPM}}{JL_a}}$; $\zeta = \dfrac{1}{2}\dfrac{JR_a + bL_a}{JL_a}\sqrt{\dfrac{JL_a}{R_a b + K_{aPM}K_{TPM}}}$

From these expressions, we can see that both natural frequency and damping ratio are affected by each of the parameters of the system, and that one cannot predict the nature of the damping without knowing numerical values of the parameters.

19.30 a. $L_a\dfrac{dI_a(t)}{dt} + R_a I_a(t) + k_f\dfrac{V_S(t)}{R_f}\omega_m(t) = V_S(t)$

$-k_f\dfrac{V_S(t)}{R_f}I_a(t) + J\dfrac{d\omega_m(t)}{dt} + b\omega_m(t) = T_L(t)$

b. $L_a\dfrac{d\delta I_a(t)}{dt} + R_a\delta I_a(t) + \dfrac{k_f}{R_f}\bar{V}_S\delta\omega_m(t) = \delta V_S(t) - \dfrac{k_f}{R_f}\bar{\omega}_m\delta V_S(t)$

$-\dfrac{k_f}{R_f}\bar{V}_S\delta I_a(t) + J\dfrac{d\delta\omega_m(t)}{dt} + b\delta\omega_m(t) = \bar{I}_a\delta V_S(t) + \delta T_L(t)$

19.33 $$\left.\frac{\Omega_m(s)}{T_L(s)}\right|_{V_a(s)=0} = \frac{sL_a + R_a}{(sL_a + R_a)[s(J_m + J) + (b_m + b)] + k_a^2}$$

$$\left.\frac{\Omega_m(s)}{V_a(s)}\right|_{T_L(s)=0} = \frac{k_a}{(sL_a + R_a)[s(J_m + J) + (b_m + b)] + k_a^2}$$

19.44 $\text{pf}_m = 0.636$ leading; $S_m = 708$ kVA

19.47 $I = 13.495\angle 90°$ A. The current is leading the voltage.

19.50 $T = 875.1$ N-m

19.53 $s = 0.025$; $I_S = 54.6\angle(-19.98°)$ A; $P_{\text{in}} = 35.6$ kW
$P_m = (1 - s)P_t = 32.97$ kW; $T_{sh} = 175$ N-m; $\eta = 0.904$

19.56 a. slip $= 0.04 = 4\%$ b. reg $= 0.035 = 3.5\%$

19.60 $n = 1{,}152$ r/min; $T_{\text{out}} = 266.7$ N-m

19.67 $I_S = 7.11\angle(-88.7°)$ A; pf $= 0.0224$ lagging

19.70 a. $s_{FL} = 0.097$ b. $s_{MT} = 0.382$ c. 379%

Chapter 20

20.2 a. $\lambda = 0.094 \cos \omega t$ b. $\omega_m = 60\pi$ rad/s

20.3 The total trapezoidal profile is shortened by $\frac{2}{3}$ s.

20.6 30°

20.10 $\lambda = 0.00785$ V-s; $T = \dfrac{V - R_w I}{\omega_m}$ N-m

20.14 a. Steps/revolution = 24 b. $d = 0.0042$ in c. 420 steps d. 550 r/min

20.18 Slip in the opposite direction of rotation = 0.97; Motor speed = 2,910 r/min = 304.7 rad/s.

20.22 $I_S = 29.9\angle(-20.52°)$ A; $T_{\text{dev}} = 14.68$ N-m

20.26 a. $E_b = 156.05\angle(-19.95°)$ V b. $P_{\text{dev}} = 1{,}014.3$ W c. $P_{\text{out}} = 949.3$ W d. $\eta = 64.7\%$

20.29 a. $\eta = 60.2\%$ b. pf $= 0.709$ lagging c. $T = 9.13$ lb-in

20.30 The universal motor speed is easily controlled and thus it would be used for variable speed, that is, (e) and (g). The vacuum cleaner motors are often universal motors. This motor could also be used for the fan motors. A single-phase induction motor is used for (b) and (c). The clock should use a single-phase synchronous motor. The tape drive would be a single-phase synchronous motor also. An X-Y plotter uses a stepper motor.

20.33 a. 0.512 hp = 51.2% of rated b. 0.343 hp = 34.3% of rated c. 0.125 hp = 12.5% of rated

Index

T